A TEXTBOOK OF
POWER SYSTEM ENGINEERING

By the Same Author :

- **A Textbook of Electrical Machines**
- **Direct Current Machines**
- **Alternating Current Machines**
- **Utilisation of Electrical Power (Including "*Electrical Drives*" and "*Electrical Traction*")**
- **Electrical Engineering Materials**
- **Basic Electrical and Electronics Engineering**
- **Power Plant Engineering**

A TEXTBOOK OF

POWER SYSTEM ENGINEERING

[For Engineering Students of B.E./B.Tech.; A.M.I.E. Section B (India); Diploma and Competitive Examinations]

S.I. UNITS

By

Er. R.K. RAJPUT

M.E. (Hons.), *Gold Medallist*; Grad. (*Mech. Engg. & Elect. Engg.*);

M.I.E. (India); M.S.E.S.I.; M.I.S.T.E.; C.E. (India)

Recipient of :

"Best Teacher (Academic) Award"

"Distinguished Author Award"

"Jawahar Lal Nehru Memorial Gold Medal"

for an outstanding research paper

(Institution of Engineers–India)

Principal *(Formerly)*:

- *Thapar Polytechnic College;*
- *Punjab College of Information Technology,*

PATIALA (Punjab)

LAXMI PUBLICATIONS (P) LTD

(An ISO 9001:2008 Company)

BENGALURU • CHENNAI • COCHIN • GUWAHATI • HYDERABAD

JALANDHAR • KOLKATA • LUCKNOW • MUMBAI • RANCHI • NEW DELHI

BOSTON (USA) • ACCRA (GHANA) • NAIROBI (KENYA)

A TEXTBOOK OF POWER SYSTEM ENGINEERING

Printed and bound in India
Typeset at Goswami Associates, Delhi
First Edition: 2006; *Second Edition*: 2015 Reprint : 2016
ISBN 978-81-318-0879-5

Branches

©	Bengaluru	080-26 75 69 30	
©	Chennai	044-24 34 47 26,	24 35 95 07
©	Cochin	0484-237 70 04,	405 13 03
©	Guwahati	0361-254 36 69,	251 38 81
©	Hyderabad	040-27 55 53 83,	27 55 53 93
©	Jalandhar	0181-222 12 72	
©	Kolkata	033-22 27 43 84	
©	Lucknow	0522-220 99 16	
©	Mumbai	022-24 91 54 15,	24 92 78 69
©	Ranchi	0651-220 44 64	

PUBLISHED IN INDIA BY

Laxmi Publications (P) Ltd.
(An ISO 9001:2008 Company)
113, GOLDEN HOUSE, DARYAGANJ,
NEW DELHI - 110002, INDIA
Telephone : 91-11-4353 2500, 4353 2501
Fax : 91-11-2325 2572, 4353 2528
www.laxmipublications.com info@laxmipublications.com

C—10733/015/08
Printed at: Ajit Printing Press, Delhi

TO ALMIGHTY

Contents

Chapter *Pages*

PART I : GENERATION

1. INTRODUCTION **3—15**

1.1. Energy ... 3
1.2. Power ... 4
1.3. Sources of Energy ... 4
1.3.1. Fuels ... 4
1.3.2. Energy stored in water ... 10
1.3.3. Nuclear energy ... 10
1.3.4. Wind power ... 10
1.3.5. Solar energy ... 11
1.3.6. Tidal power ... 12
1.3.7. Geothermal energy ... 13
1.3.8. Thermo-electric power ... 14
1.4. Power Systems' Structure ... 14
Highlights ... 15
Theoretical Questions ... 15

2. STEAM POWER PLANT **16—69**

2.1. Introduction ... 16
2.2. Classification of Steam Power Plants ... 17
2.3. Layout of a Modern Steam Power Plant ... 17
2.4. Essential Requirements of Steam Power Station Design ... 19
2.5. Selection of Site for Steam Power Station ... 19
2.6. Capacity of Steam Power Plant ... 21
2.7. Choice of Steam Conditions ... 22
2.8. Fuel Handling ... 23
2.8.1. Introduction ... 23
2.8.2. Requirements of good coal handling plant ... 24
2.8.3. Coal handling systems ... 24
2.8.4. Coal handling ... 25
2.8.5. Layout of a fuel handling equipment ... 26
2.9. Combustion Equipment for Steam Boilers ... 27
2.9.1. General aspects ... 27
2.9.2. Combustion equipment for solid fuels—selection considerations ... 28
2.9.3. Burning of coal ... 28
2.9.4. Burners ... 38
2.10. Fluidised Bed Combustion (FBC) ... 40

Chapter *Pages*

2.11. Ash Handling ... 42
2.11.1. Ash handling equipment ... 42
2.11.2. Ash handling systems ... 43
2.12. Dust Collection ... 43
2.12.1. Introduction ... 43
2.12.2. Removal of smoke ... 44
2.12.3. Removal of dust and dust collectors ... 44
2.12.4. Uses of ash and dust ... 44
2.12.5. General layout of ash handling and dust collection system ... 45
2.13. Boilers ... 45
2.13.1. Introduction ... 45
2.13.2. Classification of boilers ... 46
2.13.3. Comparison between 'fire-tube and water tube' boilers ... 47
2.13.4. High pressure boilers ... 47
2.14. Steam Turbines ... 49
2.14.1. Introduction ... 49
2.14.2. Classification of steam turbines ... 49
2.14.3. Advantages of steam turbine over the steam engines ... 51
2.14.4. Description of common types of turbines ... 51
2.14.5. Methods of reducing wheel or rotor speed ... 53
2.14.6. Difference between impulse and reaction turbines ... 57
2.14.7. Steam turbine governing and control ... 57
2.15. Steam Condensers ... 57
2.15.1. Introduction ... 57
2.15.2. Vacuum ... 58
2.15.3. Organs of a steam condensing plant ... 58
2.15.4. Classification of condensers ... 58
2.15.5. Jet condensers ... 59
2.15.6. Surface condensers ... 59
2.15.7. Comparison between jet and surface condensers ... 59
2.16. Feed Water Treatment ... 59
2.16.1. Classification of impurities in water ... 60
2.16.2. Troubles caused by the impurities in water ... 60
2.16.3. Methods of feed water treatment ... 60
2.17. Advantages and Disadvantages of Steam Power Plants ... 61
2.18. Miscellaneous ... 61
2.18.1. Useful life of steam power plant components ... 61
2.18.2. Steam power plant pumps ... 62
2.18.3. Cost of steam power plant ... 62
2.18.4. Comparison of various types of power plants ... 62
2.18.5. Thermal power stations in India ... 64
Highlights ... 67
Theoretical Questions ... 68

3. DIESEL ENGINE POWER PLANT **70—80**

3.1. Introduction ... 70
3.2 Advantages and Disadvantages of Diesel Power Plants ... 70

Chapter			Pages
3.3.	Applications of Diesel Power Plant	...	71
3.4	Site Selection	...	71
3.5.	Heat Engines	...	71
3.6.	Classification of I.C. Engines	...	72
3.7.	Comparison between a Petrol Engine and a Diesel Engine	...	73
3.8.	Essential Components of a Diesel Power Plant	...	73
	3.8.1. Engine	...	74
	3.8.2. Air intake system	...	74
	3.8.3. Exhaust system	...	75
	3.8.4. Fuel system	...	75
3.9.	Operation of a Diesel Power Plant	...	77
3.10.	Types of Diesel Engines Used for Diesel Power Plants	...	78
3.11.	Layout of a Diesel Engine Power Plant	...	78
	Highlights	...	79
	Theoretical Questions	...	80
4.	**GAS TURBINE POWER PLANTS**		**81—103**
4.1.	Gas Turbines—General Aspects	...	81
4.2.	Applications of Gas Turbine Plants	...	82
4.3.	Advantages and Disadvantages of Gas Turbine Power Plants Over Diesel and Thermal Power Plants	...	82
4.4.	Site Selection	...	83
4.5.	The Simple Gas Turbine Plant	...	83
4.6.	Energy Cycle for a Simple-Cycle Gas Turbine	...	84
4.7.	Performance Terms	...	85
4.8.	Classification of Gas Turbine Power Plants	...	85
4.9.	Classification of Gas Turbines	...	86
4.10.	Combination Gas Turbine Cycles	...	86
	4.10.1. Combined turbine and steam power plants	...	86
	4.10.2. Combined gas turbine and diesel power plants	...	90
	4.10.3. Advantages of combined cycle	...	91
4.11.	Operation of a Gas Turbine	...	91
4.12.	Gas Turbine Power Plant Layout	...	92
4.13.	Components of a Gas Turbine Power Plant	...	93
4.14.	Various Arrangements of Gas Turbine Power Plants	...	97
4.15.	Effect of Thermodynamic Variables on Air Rate	...	100
	Highlights	...	102
	Theoretical Questions	...	102
5.	**HYDRO-ELECTRIC POWER PLANT**		**104—157**
5.1.	Introduction	...	104
5.2.	Application of Hydro-electric Plants	...	105
5.3.	Advantages and Disadvantages of Hydro-electric Plants	...	105
5.4.	Selection of Site for a Hydro-electric Plant	...	106
5.5.	Essential Features/Elements of Hydro-electric Power Plant	...	107

Chapter			Pages
5.6.	Classification of Hydro-electric Power Plants	...	108
	5.6.1. High head power plants	...	109
	5.6.2. Medium head power plants	...	110
	5.6.3. Low head power plants	...	110
	5.6.4. Base load plants	...	111
	5.6.5. Peak load plants	...	111
	5.6.6. Run-of-river plants without pondage	...	111
	5.6.7. Run-of-river plants with pondage	...	111
	5.6.8. Storage type plants	...	111
	5.6.9. Pumped storage plants	...	112
	5.6.10. Mini and microhydel plants	...	113
5.7.	Hydraulic Turbines	...	113
	5.7.1. Classification of hydraulic turbines	...	113
	5.7.2. Description of various types of turbines	...	115
	5.7.3. Specific speed of a turbine	...	121
	5.7.4. Efficiencies of a turbine	...	121
	5.7.5. Cavitation	...	122
	5.7.6. Performance of hydraulic turbines	...	123
	5.7.7. Governing of hydraulic turbines	...	126
	5.7.8. Selection of turbines	...	129
5.8.	Plant Layout	...	130
5.9.	Hydro-plant Auxiliaries	...	130
5.10.	Cost of Hydro-plant	...	131
5.11.	Average Life of Hydro-plant Components	...	131
5.12.	Hydro-plant Controls	...	131
5.13.	Electrical and Mechanical Equipment in a Hydro-plant	...	132
5.14.	Combined Hydro and Steam Power Plants	...	133
5.15.	Comparison of Hydro-power Station with Thermal Power Stations	...	134
5.16.	Underground Hydro-plants	...	135
5.17.	Automatic and Remote Control of Hydro-station	...	136
5.18.	Safety Measures in Hydro-electric Power Plants	...	137
5.19.	Preventive Maintenance of Hydro-plant	...	137
5.20.	Calculation of Available Hydro-power	...	138
5.21.	Cost of Hydro-power	...	138
5.22.	Hydrology	...	139
	5.22.1. Introduction	...	139
	5.22.2. The hydrologic cycle	...	139
	5.22.3. Measurement of run-off	...	140
	5.22.4. Hydrograph	...	141
	5.22.5. Flow duration curve	...	144
	5.22.6. Mass curve	...	145
5.23.	Hydro-power Development in India	...	146
	Highlights	...	155
	Theoretical Questions	...	156
	Unsolved Examples	...	157

Chapter *Pages*

6. NUCLEAR POWER PLANT **158—197**

6.1. General Aspects of Nuclear Engineering ... 158
6.1.1. Nuclear structure ... 158
6.1.2. Atomic mass unit ... 160
6.1.3. Isotopes ... 160
6.1.4. Radioactivity ... 161
6.1.5. Nuclear radiation ... 161
6.1.6. Binding energy ... 163
6.1.7. Radioactive decay ... 164
6.1.8. Nuclear reactions ... 166
6.1.9. Nuclear cross-sections ... 167
6.1.10. Fertile materials ... 168
6.1.11. Fission of nuclear fuel ... 169
6.1.12. Nuclear fusion ... 172
6.1.13. Comparison of fission and fusion processes ... 172
6.2. Nuclear Power Systems ... 173
6.3. Nuclear Reactors ... 175
6.3.1. Introduction ... 175
6.3.2. Classification of nuclear reactors ... 175
6.3.3. Essential components of a nuclear reactor ... 177
6.3.4. Power of a nuclear reactor ... 181
6.4. Main Components of a Nuclear Power Plant ... 182
6.5. Description of Reactors ... 182
6.5.1. Pressurised water reactor (PWR) ... 182
6.5.2. Boiling water reactor (BWR) ... 184
6.5.3. CANDU (Canadian-Deuterium-Uranium) reactor ... 184
6.5.4. Gas-cooled reactor ... 186
6.5.5. Liquid metal cooled reactors ... 188
6.5.6. Breeder reactor ... 189
6.6. Selection of Materials for Reactor Components ... 190
6.7. Advantages and Disadvantages of Nuclear Power Plants ... 193
6.8. Nuclear-plant Site Selection ... 193
6.9. Nuclear Power Plants in India ... 194
Highlights ... 194
Theoretical Questions ... 196

7. NON-CONVENTIONAL POWER GENERATION AND DIRECT ENERGY CONVERSION **198—239**

7.1. Introduction ... 198
7.2. Wind Power Plants ... 199
7.2.1. Introduction ... 199
7.2.2. Wind availability and measurement ... 199
7.2.3. Types of wind mills ... 200
7.2.4. Wind-electric generating power plant ... 202
7.2.5. Types of wind machines ... 203
7.2.6. Wind-powered battery chargers ... 204

Chapter			Pages
	7.2.7.	Wind electricity in small independent grids ...	205
	7.2.8.	Wind electricity economics ...	205
	7.2.9.	Problems in operating large wind power generators ...	206
7.3.		Tidal Power Plants ...	206
	7.3.1.	Introduction ...	206
	7.3.2.	Components of tidal power plants ...	206
	7.3.3.	Classification and operation of tidal power plants ...	207
	7.3.4.	Advantages and limitations of tidal power generation ...	209
7.4.		Solar Power Plants ...	209
	7.4.1.	Solar energy-general terms and introduction ...	209
	7.4.2.	Collectors in various ranges and applications ...	210
	7.4.3.	Flat plate collectors ...	211
	7.4.4.	Focusing (or concentrating) collectors ...	213
	7.4.5.	Solar pond technology ...	214
	7.4.6.	Low temperature thermal power generation ...	216
	7.4.7.	Medium temperature systems using focusing collectors ...	217
	7.4.8.	High temperature systems-solar farm and solar tower plant ...	218
7.5.		Geothermal Power Plants ...	219
	7.5.1.	Geothermal energy ...	219
	7.5.2.	Geothermal sources ...	220
	7.5.3.	Geothermal power estimates ...	222
	7.5.4.	Environmental problems ...	222
	7.5.5.	Applications of geothermal energy ...	222
	7.5.6.	Advantages and disadvantages of geothermal energy over other energy forms ...	223
7.6.		Direct Energy Conversion Systems ...	223
	7.6.1.	Thermoelectric conversion system ...	223
	7.6.2.	Thermionic conversion system ...	226
	7.6.3.	Photovoltaic power system ...	228
	7.6.4.	Magnetohydrodynamic (MHD) system ...	231
	7.6.5.	Electrostatic mechanical generators ...	234
	7.6.6.	Electro gas-dynamic generators (EGD) ...	234
	7.6.7.	Fuel cells ...	235
	7.6.8.	Nuclear batteries ...	237
		Highlights ...	239
		Theoretical Questions ...	239

8. COMBINED OPERATION OF DIFFERENT POWER PLANTS **240—257**

8.1.	General Aspects ...	240
8.2.	Advantages of Combined Operation of Plants ...	240
8.3.	Load Division between Power Stations ...	241
8.4.	Hydro-electric (storage type) Plant in Combination with Steam Plant ...	244
8.5.	Run-of-river Plant in Combination with Steam Plant ...	245
8.6.	Pump Storage Plant in Combination with Steam or Nuclear Power Plant ...	245
8.7.	Co-ordination of Hydro-electric and Gas Turbine Stations ...	246

Chapter			Pages
8.8.	Co-ordination of Different Types of Power Plants	...	246
	Theoretical Questions	...	255
	Unsolved Examples	...	256

9. ECONOMICS OF POWER GENERATION **258—339**

9.1.	Introduction	...	258
9.2.	Terms and Definitions	...	258
9.3.	Principles of Power Plant Design	...	261
9.4.	Location of Power Plant	...	262
9.5.	Layout of Power Plant Building	...	262
9.6.	Cost Analysis	...	263
9.7.	Selection of Type of Generation	...	267
9.8.	Selection of Power Plant Equipment	...	267
	9.8.1. Selection of boilers	...	267
	9.8.2. Selection of prime-movers	...	268
	9.8.3. Selection of size and number of generating units	...	268
9.9.	Economics in Plant Selection	...	270
9.10.	Factors Affecting Economics of Generation and Distribution of Power	...	271
9.11.	How to Reduce Power Generation Cost ?	...	272
9.12.	Power Plant—Useful Life	...	272
9.13.	Economics of Hydro-electric Power Plants	...	273
9.14.	Economics of Combined Hydro and Steam Power Plants	...	274
9.15.	Performance and Operating Characteristics of Power Plants	...	274
9.16.	Economic Load Sharing	...	276
9.17.	Tariff for Electrical Energy	...	307
	9.17.1. Introduction	...	307
	9.17.2. Objectives and requirements of tariff	...	307
	9.17.3. General tariff form	...	307
	Highlights	...	329
	Theoretical Questions	...	330
	Unsolved Examples	...	330

10. INTERCONNECTED SYSTEMS **340—380**

10.1.	Introduction	...	340
10.2.	Advantages and Limitations of Interconnected System	...	340
10.3.	Parallel Operation of Alternators	...	341
	10.3.1. Requirements for paralleling	...	341
	10.3.2. Alternator synchronising procedure	...	341
	10.3.3. Synchronising current, power and torque	...	344
	10.3.4. Effect of reactance	...	349
	10.3.5. Effect of increasing the excitation of one of the alternators	...	349
	10.3.6. Effect of increasing the driving torque of one of the alternators	...	350
	10.3.7. Effect of changes in speed of one of the alternators	...	350
	10.3.8. Effect of unequal voltages	...	350
	10.3.9. Distribution of load	...	351
	10.3.10. Load sharing between two alternators	...	351

Chapter *Pages*

10.3.11. Hunting of alternators ... 352
10.3.12. Synchronous machines on infinite bus-bars ... 353
10.4. Interconnected Stations ... 369
10.4.1. General aspects ... 369
10.4.2. Load sharing ... 369
10.4.3. Power limit of interconnectors ... 371
10.4.4. Voltage, frequency and load control ... 376
10.4.5. Load despatching ... 378
Highlights ... 378
Theoretical Questions ... 379
Unsolved Examples ... 379

PART II : TRANSMISSION AND DISTRIBUTION

11. ELECTRIC POWER SYSTEMS **383—417**

11.1. Introduction ... 383
11.2. Electric Supply System ... 383
11.3. Comparison between D.C. and A.C. Systems of Transmission and Distribution ... 385
11.4. Advantages and Limitations of High Transmission Voltage ... 388
11.5. Systems of Power Transmission ... 390
11.6. Comparison of Volume of Conductor Material ... 390
11.6.1. Overhead system ... 391
11.6.2. Underground system ... 396
11.7. Choice of Transmission Voltage ... 408
11.8. Conductor Size and Kelvin's Law ... 409
Highlights ... 415
Theoretical Questions ... 416
Unsolved Examples ... 416

12. MECHANICAL DESIGN OF OVERHEAD LINES **418—469**

12.1. Introduction ... 418
12.2. Main Component of Overhead Lines ... 418
12.3. Conductors ... 418
12.4. Line Supports ... 423
12.5. Overhead Line Insulators ... 426
12.5.1. Introduction ... 426
12.5.2. Insulating materials ... 426
12.5.3. Types of insulators ... 427
12.5.4. Causes of failure of insulators ... 429
12.5.5. Testing of insulators ... 429
12.5.6. Potential distribution over a string of suspension insulators ... 431
12.6. Sag in Overhead Lines ... 446
12.7. Stringing Chart ... 457
12.8. Vibration and Vibration Dampers ... 457

Chapter *Pages*

12.9. Corona ... 458
12.9.1. General aspects ... 458
12.9.2. Theory of corona formation ... 459
12.9.3. Factors affecting corona ... 459
12.9.4. Advantages and disadvantages of corona ... 459
12.9.5. Corona loss and its control ... 460
12.9.6. Important terms relating to corona ... 460
Highlights ... 465
Theoretical Questions ... 466
Unsolved Examples ... 467

13. CONSTANTS OF OVERHEAD TRANSMISSION LINES 470—517

13.1. Introduction ... 470
13.2. Inductance of a Transmission Line ... 471
13.2.1. Concept of inductance ... 471
13.2.2. Flux linkages ... 471
13.2.3. Inductance of 2-wire (1-phase) transmission line ... 474
13.2.4. Inductance of a 3-phase overhead transmission line ... 475
13.2.5. Self-GMD and mutual-GMD ... 478
13.3. Resistance of a Transmission Line ... 494
13.4. Skin Effect and Proximity Effect ... 495
13.5. Capacitance of Transmission Lines ... 496
13.5.1. Concept of capacitance ... 496
13.5.2. Electric potential ... 496
13.5.3. Capacitance of a single-phase two-wire overhead transmission line ... 498
13.5.4. Capacitance of a 3-phase overhead transmission line ... 499
13.5.5. Capacitance of a double-circuit 3-phase overhead line ... 503
13.5.6. Effect of earth on the capacitance of a transmission line ... 504
Highlights ... 513
Theoretical Questions ... 515
Unsolved Examples ... 515

14. PERFORMANCE OF TRANSMISSION LINES 518—589

14.1. Introduction ... 518
14.2. Classification of Transmission Lines ... 518
14.3. Regulation and Efficiency of a Transmission Line ... 519
14.4. Short Transmission Lines ... 519
14.4.1 Single-phase lines ... 519
14.4.2. Three-phase lines ... 521
14.4.3. Effect of load power factor on efficiency and regulation of a transmission line ... 522
14.5. Medium Transmission Lines ... 528
14.5.1. End condenser method ... 528
14.5.2. Nominal-T method ... 531
14.5.3. Nominal-π method ... 536

Chapter *Pages*

14.6. Generalised Circuit Constants ... 541
14.7. Long Transmission Lines ... 550
14.7.1. General aspects ... 550
14.7.2. Analysis of long transmission line-rigorous method ... 551
14.7.3. Evaluation of A, B, C, D constants ... 553
14.7.4. Surge impedance and surge impedance loading of transmission lines ... 554
14.7.5. Transmission line with series impedance at receiving end ... 558
14.7.6. Transmission line with series impedance at the sending end ... 559
14.7.7. Transmission line with series impedance at both ends ... 560
14.7.8. Transmission line with transformers at both ends ... 562
14.7.9. Equivalent T-network of a long transmission line ... 564
14.7.10. Equivalent π-network of a long transmission line ... 565
14.7.11. Constant voltage transmission ... 566
14.7.12. Power circle diagram ... 567
14.7.13. Synchronous phase modifiers ... 574
14.8. Ferranti Effect ... 585
14.9. Charging Current and Losses in an Open-circuited Line ... 586
Highlights ... 586
Theoretical Questions ... 587
Unsolved Examples ... 588

15. INTERFERENCE OF POWER LINES WITH NEIGHBOURING COMMUNICATION CIRCUITS **590—596**

15.1. Introduction ... 590
15.2. Electromagnetic Effect ... 590
15.3. Electrostatic Effect ... 592
15.4. Transposition ... 595
Highlights ... 596
Theoretical Questions ... 596
Unsolved Examples ... 596

16. UNDERGROUND CABLES **597—641**

16.1. Introduction ... 597
16.2. General Construction of Cable ... 597
16.3. Insulating Materials for Cables ... 598
16.4. Classification of Cables ... 599
16.4.1. Low voltage (or L.T.) cables ... 600
16.4.2. High voltage (H.T.) cables upto 11 kV ... 601
16.4.3. Super tension/screened cables ... 601
16.4.4. Extra high tension (E.H.T.) cables ... 603
16.5. Laying of Underground Cables ... 605
16.6. Insulation Resistance of a Single-core Cable ... 607
16.7. Capacitance of a Single-core Cable ... 609
16.8. The Most Economical Conductor Size in a Cable ... 612
16.9. Grading of Cables ... 613

16.10. Capacitance of 3-core Cables ... 622
16.11. Heating of Cables ... 624
16.12. Thermal Resistance of Cables ... 630
16.13. Permissible Current Loading ... 631
16.14. Types of Cable Faults and their Location ... 632
16.14.1. Types of cable faults ... 632
16.14.2. Location of cable fault ... 632
16.15. Causes of Failure of Underground Cables ... 636
16.16. Operating Problems with Cables ... 637
16.17. Direct Current (D.C.) Cables ... 637
16.18. Cryogenic Cables ... 637
16.19. Comparison between Overhead Lines and Underground Cables ... 638
Highlights ... 638
Theoretical Questions ... 639
Unsolved Examples ... 640

17. POWER SYSTEM STABILITY **642—669**

17.1. Introduction ... 642
17.2. Classification of Stability Conditions ... 643
17.3. Steady State Stability ... 644
17.3.1. Maximum steady state power ... 644
17.3.2. Power-angle diagram ... 646
17.3.3. General aspects of steady state stability ... 647
17.4. Transient Stability ... 651
17.4.1. General aspects ... 651
17.4.2. The swing equation ... 652
17.4.3. Equal area criterion ... 656
17.4.4. Critical clearing angle ... 662
17.4.5. Transient stability in a multi-machine system ... 663
17.4.6. Transient stability improvement ... 663
Highlights ... 667
Theoretical Questions ... 668
Unsolved Examples ... 668

18. CONTROL OF REACTIVE POWER, VOLTAGE AND LOAD FREQUENCY IN TRANSMISSION SYSTEMS **670—677**

18.1. Generation and Absorption of Reactive Power ... 670
18.2. Effect of Reactive Power on Voltage and Voltage Regulation ... 671
18.3. Relations between δQ, δP and V at a Node ... 672
18.4. Reactive Compensation in Power System ... 673
18.5. Types of Compensators ... 673
18.6. Methods of Voltage Control in Transmission System ... 674
18.7. Relations between Voltage and Reactive Power, Frequency and Active Power ... 675

Chapter *Pages*

18.8. Control of Load Frequency ... 676
Highlights ... 677
Theoretical Questions ... 677

19. POWER DISTRIBUTION SYSTEMS **678—759**

19.1. Distribution Systems–General Aspects ... 678
19.1.1. Classification of distribution systems ... 679
19.1.2. Requirements of a distribution system ... 679
19.1.3. Overhead versus underground system ... 680
19.2. D.C. Distribution ... 680
19.2.1. Types of distributors ... 681
19.2.2. Types of loading ... 683
19.2.3. D.C. distributor fed at one end ... 683
19.2.4. D.C. distributor fed at both ends ... 689
19.2.5. Ring distributor ... 709
19.2.6. Three-wire D.C. system ... 715
19.3. A.C. Distribution ... 728
19.3.1. Introduction ... 728
19.3.2. A.C. distribution calculations ... 732
19.3.3. A.C. distributors with concentrated loads ... 733
19.3.4. Three-phase unbalanced loads ... 742
Highlights ... 755
Theoretical Questions ... 756
Unsolved Examples ... 756

20. EXTRA HIGH VOLTAGE TRANSMISSION **760—765**

20.1. Introduction ... 760
20.2. Extra High Voltage A.C. Transmission ... 761
20.2.1. Advantages of E.H.V. transmission ... 761
20.2.2. Limitations/Problems involved in E.H.V. transmission ... 761
20.3. High Voltage Direct Current (H.V.D.C) Transmission ... 763
20.3.1. Classification of H.V.D.C. transmission systems ... 763
20.3.2. Advantages, limitations and applications of H.V.D.C. transmission systems ... 764
Highlights ... 765
Theoretical Questions ... 765

PART III : SWITCHGEAR AND PROTECTION

21. SYMMETRICAL COMPONENTS AND FAULT CALCULATIONS **769—864**

21.1. Introduction ... 769
21.2. Types of Faults ... 770
21.3. Symmetrical Fault Calculations ... 770
21.3.1. Percentage reactance ... 771
21.3.2. Reactances of synchronous machines ... 773

21.3.3. Reactor control of short-circuit currents ... 774
21.3.4. Steps for symmetrical fault calculations ... 776
21.4. Unsymmetrical Fault Calculations ... 798
21.4.1. Unsymmetrical faults ... 798
21.4.2. Symmetrical components ... 798
21.4.3. Sequence impedances ... 828
21.4.4. Analysis of unsymmetrical faults ... 832
21.4.5. Sequence networks ... 849
21.4.6. Open conductor faults ... 860
Highlights ... 861
Theoretical Questions ... 861
Unsolved Examples ... 862

22. SWITCHES, FUSES AND CIRCUIT BREAKERS **865—911**

22.1. Switchgear—General Aspects ... 865
22.1.1. Definition of switchgear ... 865
22.1.2. Functions of switchgear ... 865
22.1.3. Essential features of switchgear ... 866
22.1.4. Principles of layout of switchgear ... 866
22.1.5. Types of switchgear ... 866
22.1.6. Short-circuits in electrical installations and limiting methods ... 867
22.2. Switches ... 868
22.3. Fuses ... 869
22.3.1. Definition of fuse ... 869
22.3.2. Advantages and disadvantages of a fuse ... 869
22.3.3. Desirable characteristics of fuse element ... 870
22.3.4. Fuse element materials ... 870
22.3.5. Important terms relating to fuses ... 870
22.3.6. Types of fuses ... 872
22.3.7. Fuse selection ... 875
22.3.8. Difference between sub-circuit fuses, main circuit fuses, meter board fuses and pole fuses in an electrical installation ... 875
22.3.9. Difference between fuse and circuit breaker ... 876
22.4. Circuit Breakers ... 877
22.4.1. Introduction ... 877
22.4.2. Function of a circuit breaker ... 877
22.4.3. Principle of circuit breaker ... 877
22.4.4. Arc phenomenon ... 878
22.4.5. Methods of arc extinction ... 879
22.4.6. Arc, restriking and recovery voltages ... 881
22.4.7. Current chopping ... 883
22.4.8. Resistance switching ... 884
22.4.9. Classification of circuit breakers ... 886
22.4.10. Oil circuit breakers ... 887
22.4.11. Water type circuit breaker ... 891
22.4.12. Air blast circuit breakers ... 891
22.4.13. Sulphur hexafluoride (SF_6) circuit breakers ... 893

22.4.14. Vacuum circuit breakers ... 895
22.4.15. Low voltage circuit breakers ... 896
22.4.16. Ratings of circuit breakers ... 897
22.4.17. Testing of circuit breakers ... 899
22.4.18. Autoreclosing ... 901
22.4.19. Maintenance of circuit breakers ... 901
22.4.20. Switchgear components ... 902
22.4.21. Bus-bar arrangements ... 905
22.5. Control Room–Power Station ... 906
Highlights ... 908
Theoretical Questions ... 909
Unsolved Examples ... 911

23. PROTECTIVE RELAYS **912—946**

23.1. Protective System—General Aspects ... 912
23.2. Protective Relays—Introduction ... 913
23.3. Classification of Relays ... 916
23.4. Electromagnetic Attraction Relays ... 918
23.4.1. Attracted armature type relay ... 918
23.4.2. Solenoid type relay ... 919
23.4.3. Balanced beam type relay ... 919
23.5. Thermal Relay ... 920
23.6. Buchholz Relay ... 920
23.7. Electromagnetic Induction Relays ... 921
23.8. Induction Type Non-directional Over-current Relay ... 924
23.9. Induction Type Directional Power Relay ... 926
23.10. Induction Type Directional Overcurrent Relay ... 927
23.11. Differential Relays ... 931
23.11.1. Current differential relay ... 931
23.11.2. Voltage balance differential relay ... 933
23.12. Translay Relay ... 934
23.13. Distance or Impedance Relays ... 935
23.13.1. Definite-distance type impedance relay ... 936
23.13.2. Time-distance impedance relay ... 939
23.14. Reactance Type Distance Relay ... 940
23.15. Mho Type Distance Relay ... 941
23.16. Universal Torque Equation ... 942
23.17. Static Relays ... 943
23.17.1. Introduction ... 943
23.17.2. Merits and demerits of static relays ... 943
23.17.3. Applications of static relays ... 944
23.17.4. Electronic circuits commonly used in static relays ... 944
23.17.5. Analog circuits used in static relay ... 944
23.17.6. Different static relays ... 944
Highlights ... 945
Theoretical Questions ... 946

Chapter *Pages*

24. PROTECTION OF ALTERNATORS AND TRANSFORMERS **947—968**

24.1. Protection of Alternators/Synchronous Generators ... 947
24.1.1. Types of faults in alternators ... 947
24.1.2. Protection against stator faults by percentage differential relays (Merz-Price protection system) ... 949
24.1.3. Balanced earth-fault protection ... 950
24.1.4. Stator inter-turn protection ... 951
24.1.5. Unbalanced loading of the alternator ... 957
24.1.6. Overload (or overcurrent) protection ... 957
24.1.7. Failure of prime-mover ... 957
24.1.8. Overvoltage protection ... 958
24.1.9. Restricted earth fault protection of alternator ... 958
24.1.10. Stand-by earth fault protection ... 959
24.1.11. Rotor fault protection ... 959
24.2. Protection of Transformers ... 960
24.2.1. Buchholz protection ... 960
24.2.2. Core-balance leakage protection ... 961
24.2.3. Combined leakage and overload protection ... 961
24.2.4. Differential protection for transformers ... 962
24.2.5. Interlocked overcurrent protection in the transformer protection scheme ... 966
Highlights ... 967
Theoretical Questions ... 967
Unsolved Examples ... 968

25. PROTECTION OF BUS-BARS AND LINES/FEEDERS **969—983**

25.1. Introduction ... 969
25.2. Bus-bar Protection ... 969
25.2.1. Differential protection ... 970
25.2.2. Fault bus protection ... 970
25.3. Protection of Lines/Feeders ... 971
25.3.1. Requirements of line protection ... 971
25.3.2. Protection schemes for line/feeder protection ... 971
Highlights ... 982
Theoretical Questions ... 982

26. PROTECTION AGAINST OVERVOLTAGES AND INSULATION CO-ORDINATION **984—1000**

26.1. Introduction ... 984
26.2. Causes of Overvoltages ... 984
26.3. Internal Causes of Overvoltages ... 985
26.4. External Causes of Overvoltages–Lightning ... 986
26.4.1. Process of lightning discharge ... 987
26.4.2. Types of lightning strokes ... 988
26.4.3. Protection against overvoltages and lightning ... 989
26.4.4. Types of lightning arresters ... 991

Chapter *Pages*

26.4.5. Surge absorber ... 996
26.4.6. Rating of lightning arrester ... 997
26.4.7. Tests on lightning arresters ... 998
26.5. Insulation Co-ordination ... 998
Highlights ... 999
Theoretical Questions ... 1000

27. NEUTRAL EARTHING (GROUNDING) 1001—1015

27.1. Introduction ... 1001
27.2. Objects of Earthing ... 1001
27.3. Advantages of Earthing ... 1002
27.4. Necessity of Neutral Earthing ... 1002
27.5. Methods of Neutral Earthing ... 1004
27.6. Earthed Transformer ... 1009
27.7. Specifications Required for Earthing as per I.S.I. ... 1009
27.8. Sizes of Earth Wire and Earth Plate for Domestic and Motor Installations ... 1009
27.9. Measurement of Earth Resistance by Earth Tester ... 1010
27.10. Indian Electricity Rules ... 1011
Highlights ... 1014
Theoretical Questions ... 1014
Unsolved Examples ... 1015

28. SUB-STATIONS 1016—1026

28.1. Introduction ... 1016
28.2. Functions of Sub-stations ... 1016
28.3. Classification of Sub-stations ... 1016
28.4. Indoor Sub-stations ... 1017
28.5. Outdoor Sub-stations ... 1017
28.6. Transformer Sub-stations ... 1019
28.7. Symbols for Circuit Elements/Equipment in Sub-stations ... 1020
28.8. Equipment for Sub-stations and Switchgear Installations ... 1020
28.9. Key Diagram of 11 kV/400 V Indoor Sub-station ... 1023
28.10. Earthing ... 1023
Highlights ... 1026
Theoretical Questions ... 1026

29. UNIVERSITIES' QUESTIONS (LATEST) WITH "ANSWERS AND SOLUTIONS" 1027—1090

"COMPETITIVE EXAMINATIONS' OBJECTIVE TYPE QUESTIONS" WITH "ANSWERS" 1091—1144

INDEX 1145—1149

Preface to The Second Edition

I am pleased to present the "***Second Edition***" of this book on "**Power System Engineering**". The warm reception which the previous edition has enjoyed all over India is a matter of satisfaction to me.

Besides revising the whole book thoroughly, a new "*Chapter 29*" namely **Universities' Questions (Latest) with "Answers and Solutions"** has been added to make the book still more a useful unit.

Any suggestions for the improvement of the book will be warmly received and incorporated in the next edition.

Er.R.K.Rajput
(Author)

Preface to The First Edition

This treatise on **"Power System Engineering"** contains comprehensive treatment of the subject matter in a simple, lucid and direct language. It envelopes a large number of solved problems properly graded, including typical worked examples from examination point of view.

The book comprises *three parts* : ***Part-I*** deals with **"Generation"**, ***Part-II*** with **"Transmission and Distribution"** while ***Part-III*** includes **"Switchgear and Protection"**. The book contains 28 chapters ; all the chapters are effectively saturated with much needed text supported by simple and self explanatory figures and an ample number of solved examples. At the end of each chapter, ***Highlights, Theoretical Questions*** and ***Unsolved Examples*** have been added; besides this an **"Objective Type Questions Bank"** (including *Indian Universities* and *Competitive Examinations Questions*) has also been included to make the book a comprehensive and a complete unit in all respects.

The author's thanks are due to his wife Ramesh Rajput for extending all cooperation during preparation of the manuscript and proofreading of the book.

In the end the author wishes to express his gratitude to Sh. R.K. Gupta, Chairman and Sh. Saurabh Gupta, Managing Director, Laxmi Publications Pvt. Ltd., New Delhi for taking a lot of pains in bringing out the book with good presentation, in a short span of time.

Although every care has been taken to make the book free of errors both in text as well as in solved examples, yet the author shall feel obliged if any errors present are brought to his notice. Constructive criticism of the book will be warmly received.

Er.R.K.Rajput
(Author)

PART-I : GENERATION

Chapters :

1. Introductory
2. Steam Power Plant
3. Diesel Engine Power Plant
4. Gas Turbine Power Plants
5. Hydro-electric Power Plant
6. Nuclear Power Plant
7. Non-Conventional Power Generation and Direct Energy Conversion
8. Combined Operation of Different Power Plants
9. Economics of Power Generation
10. Interconnected Systems

1

Introductory

1.1. Energy 1.2. Power 1.3. Sources of energy—Fuels—solid fuels—liquid fuels—gaseous fuels—calorific or heating values of fuels—Energy stored in water—Nuclear energy—Wind power—Solar energy—Tidal power—Geothermal energy—Thermo-electric power. 1.4. Power systems' structure. Highlights—Theoretical Questions.

1.1. ENERGY

- **Energy** probably was the *original stuff or creation.* It appears in many forms, but has one thing in common— *energy is possessed of the ability to produce a dynamic, vital effect.* It shows itself by *excited animated state* assumed by material which receives energy.
- Energy exists in various forms, *e.g., mechanical, thermal, electrical* etc. One form of energy can be converted into other by the use of suitable arrangements.
- **Electric energy** is an essential gradient for the industrial and all-around development of any country. It is preferred one to the following *advantages* :

  (*i*) Can be generated centrally in bulk.

  (*ii*) Can be easily and economically transported from one place to another over long distances.

  (*iii*) Losses in transport are minimum.

  (*iv*) Can be easily sub-divided.

  (*v*) Can be adapted easily and efficiently to domestic and mechanical work.
- **Electric energy** is obtained, conventionally, by *conversion from fossil fuels (coal, oil, natural gas), the nuclear and hydro sources.* Heat energy released by burning fossil fuels or by fusion of nuclear material is converted to electricity by first converting heat energy to the mechanical form through a thermocycle and then converting mechanical energy through generators to the electrical form. Thermo-cycle is basically a low efficiency process—highest efficiencies for modern large size plants range upto 40%, while smaller plants may have considerably lower efficiencies. The *earth has fixed non-replenishable resources of fossil fuels and nuclear materials. Hydro-energy*, though replenishable, is also *limited in terms of power.*
- In view of the ever increasing per capita energy consumption and exponentially rising population, the earth's non-replenishable fuel resources are not likely to last for a long time. Thus a coordinated world-wide action plan is, therefore, necessary to ensure that energy supply to humanity at large is assured for a long time and at low economic cost. The following factors needs to considered and actions to be taken accordingly : (*i*) *Energy consumption curtailment* ; (*ii*) *To initiate concerted efforts to develop alternative sources of energy including unconventional sources like solar, tidal, geothermal energy etc.* ; (*iii*) Recycling of *nuclear wastes* ; (*iv*) *Development and application of antipollution technologies.*

1.2. POWER

- Any physical unit of energy when divided by a unit of time automatically becomes a *unit of power*. However, it is in connection with the mechanical and electrical forms of energy that the term "*power*" is generally used. The rate of production or consumption of heat energy and, to a certain extent, of radiation energy is not ordinarily thought of as power. *Power is primarily associated with mechanical work and electrical energy*. Therefore, **power** can be defined as the *rate of flow of energy* and can state that a *power plant* is a ***unit built*** *for production and delivery of a flow of mechanical and electrical energy*.
- In common usage, a machine or assemblage of equipment that produces and delivers a flow of mechanical or electrical energy is a **power plant.** Hence, an internal combustion engine is a power plant, a water wheel is a power plant, etc. However, what we generally mean by the term is that assemblage of equipment, permanently located on some chosen site which receives raw energy in the form of a substance capable of being operated on in such a way as to produce electrical energy for delivery from the power plant.

1.3. SOURCES OF ENERGY

The various sources of energy are :

1. Fuels	Solids—Coal, coke anthracite etc. Liquids—Petroleum and its derivatives Gases—Natural gas, blast furnace gas etc.	Conventional
2. Energy stored in water 3. Nuclear energy 4. Wind power 5. Solar energy 6. Tidal power 7. Geothermal energy 8. Thermoelectric power.		Non-conventional

1.3.1. Fuels

Fuels may be chemical or nuclear. Here we shall consider chemical fuels only.

A ***chemical fuel*** *is a substance which releases heat energy on combustion.* The principal combustible elements of each fuel are *carbon* and *hydrogen*. Though *sulphur* is a combustible element too but its presence in the fuel is considered to be *undesirable*.

Classification of fuels :

Fuels can be *classified* according to whether :

1. They *occur in nature* called *primary fuels* or are *prepared* called *secondary fuels*.
2. They are *in solid, liquid* or *gaseous state*.

The detailed classification of fuels can be given in a summary form as below :

Type of fuel	*Natural (Primary)*	*Prepared (Secondary)*
Solid	Wood Peat Lignite coal	Coke Charcoal Briquettes.

Liquid	Petroleum	Gasoline Kerosene Fuel oil Alcohol Benzol Shale oil.
Gaseous	Natural gas	Petroleum gas Producer gas Coal gas Coke-oven gas Blast furnace gas Carburetted gas Sewer gas.

1.3.1.1. Solid fuels

Coal. Its main constituents are carbon, hydrogen, oxygen, nitrogen, sulphur, moisture and ash. Coal passes through different *stages* during its formation from vegetation. These stages are enumerated and discussed below :

Plant debris-Peat-Lignite-Brown coal-Sub-bituminous coal-Bituminous coal-Semi-bituminous coal-Semi-anthracite coal-Anthracite coal-graphite.

Peat. It is the first stage in the formation of coal from wood. It contains huge amount of moisture and therefore it is dried for about 1 to 2 months before it is put to use. It is used as a domestic fuel in Europe and for power generation in Russia. In India it does not come in the categories of good fuels.

Lignites and brown coals. These are intermediate stages between peat and coal. They have a woody or often a clay like appearance associated with high moisture, high ash and low heat contents. Lignites are usually amorphous in character and impose transport difficulties as they break easily. They burn with a smoky flame. Some of this type are suitable for local use only.

Bituminous coal. It burns with long yellow and smoky flames and has high percentages of volatile matter. The average calorific value of bituminous coal is about 31350 kJ/kg. It may be of two types, namely *caking* or *non-caking*.

Semi-bituminous coal. It is softer than the anthracite. It burns with a very small amount of smoke. It contains 15 to 20 per cent volatile matter and has a tendency to break into small sizes during storage or transportation.

Semi-anthracite. It has less fixed carbon and less lustre as compared to true anthracite and gives out longer and more luminous flames when burnt.

Anthracite. It is very hard coal and has a shining black lustre. It ignites slowly unless the furnace temperature is high. It is non-caking and has high percentage of fixed carbon. It burns either with very short blue flames or without flames. The calorific value of this fuel is high to the tune of 35500 kJ/kg and as such is *very suitable for steam generation*.

Wood charcoal. It is obtained by destructive distillation of wood. During the process the volatile matter and water are expelled. The physical properties of the residue (charcoal) however, depends upon the rate of heating and temperature.

Coke. It consists of carbon, mineral matter with about 2% sulphur and small quantities of hydrogen, nitrogen and phosphorus. It is solid residue left after the destructive distillation of certain kinds of coals. It is smokeless and clear fuel and can be produced by several processes. It is mainly used in blast furnace to produce heat and at the same time to reduce the iron ore.

Briquettes. These are prepared from fine coal or coke by compressing the material under high pressure.

Analysis of coal :

The following two types of analysis is done on the coal :

1. Proximate analysis.

2. Ultimate analysis.

1. **Proximate analysis.** In this analysis, *individual elements are not determined ; only the percentage of moisture, volatile matters, fixed carbon and ash are determined.*

Example. Moisture = 4.5%, volatile matter = 5.5%, fixed carbon = 20.5%.

This type of analysis is easily done and is for *commercial purposes* only.

2. **Ultimate analysis.** In the ultimate analysis, the percentage of various elements are determined.

Example. Carbon = 90%, hydrogen = 2%, oxygen = 4%, nitrogen = 1%, sulphur = 15% and ash = 1.5%.

This type of analysis is useful for *combustion calculations.*

Properties of coal :

Important properties of coal are given below :

1. Energy content or heating value.
2. Sulphur content.
3. Burning characteristics.
4. Grindability.
5. Weatherability.
6. Ash softening temperature.

A *good coal should have* :

(*i*) Low ash content and high calorific value.

(*ii*) Small percentage of sulphur (less than 1%).

(*iii*) Good burning characteristics (*i.e.* should burn freely without agitation) so that combustion will be complete.

(*iv*) High grindability index (in case of ball mill grinding).

(*v*) High weatherability.

Ranking of coal :

ASME and ASTM have accepted a specification based on the fixed carbon and heating value of the mineral matter free analysis.

- **Higher ranking** is done on the basis of fixed carbon percentage (*dry basis*).
- **Lower ranking** is done on the heating value on the *moist basis.*

Example. 62% C and a calorific value of 5000 kcal/kg is ranked as (62—500) rank.

Rank is an inherent property of the fuel depending upon its relative progression in the classification process.

Grading of coal :

Grading is done on the following basis :

(*i*) Size

(*ii*) Heating value

(*iii*) Ash content

(*iv*) Ash softening temperature

(*v*) Sulphur content.

Example. A grade written as 5—10 cm, 5000-A8-F24-S1.6 indicate the coal as having :

—a size of 5–10 cm,

—heating value of 5000 kcal/kg,

—8 to 10% ash,

—ash softening temperature of 2400—2590°F, and

—a sulphur content of 1.4 to 1.6%.

A rank and grade of a coal gives a complete report of the material. Thus the following are the rank and grade of the coal described above :

(62—500), 5—10 cm, 500-A8-F24-S1.6.

1.3.1.2. Liquid fuels

The chief source of liquid fuels is petroleum which is obtained from wells under the earth's crust. These fuels have proved *more advantageous in comparison to sold fuels* in the following respects.

Advantages :

1. Require less space for storage.
2. Higher calorific value.
3. Easy control of consumption.
4. Staff economy.
5. Absence of danger from spontaneous combustion.
6. Easy handling and transportation.
7. Cleanliness.
8. No ash problem.
9. Non-deterioration of the oil in storage.

Petroleum. There are different opinions regarding the origin of petroleum. However, now it is accepted that petroleum has originated probably from organic matter like fish and plant life etc., by bacterial action or by their distillation under pressure and heat. It consists of a mixture of gases, liquids and solid hydrocarbons with small amounts of nitrogen and sulphur compounds. In India the main sources of petroleum are Assam and Gujarat.

Heavy fuel oil or crude oil is imported and then refined at different refineries. The refining of crude oil supplies the most important product called *petrol*. Petrol can also be made by polymerization of refinery gases.

Other liquid fuels are kerosene, fuels oils, colloidal fuels and alcohol.

The following table gives *composition* of some common liquid fuels used in terms of the elements in *weight percentage.*

Fuel	*Carbon*	*Hydrogen*	*Sulphur*	*Ash*
Petrol	85.5	14.4	0.1	—
Benzene	91.7	8.0	0.3	—
Kerosene	86.3	13.6	0.1	—
Diesel oil	86.3	12.8	0.9	—
Light fuel oil	86.2	12.4	1.4	—
Heavy fuel oil	88.3	9.5	1.2	1.0

Important properties of liquid fuels :

(1) Specific gravity
(2) Flash point
(3) Fire point
(4) Volatility
(5) Pour point
(6) Viscosity
(7) Carbon residue
(8) Octane number
(9) Cetane number
(10) Corrosive property
(11) Ash content
(12) Gum content
(13) Heating value
(14) Sulphur content.

The requisite properties vary from device to device which uses the fuel to generate power. For example, *higher the octane number, higher can be the compression ratio and the thermal efficiency will be higher.* Similarly, *the cetane number of a diesel oil should be as high as possible.*

In general the liquid fuels should have :

(*i*) Low ash content
(*ii*) High heating value
(*iii*) Low gum content
(*iv*) Less corrosive tendency
(*v*) Low sulphur content
(*vi*) Low pour point.

Viscosity and other properties vary from purpose to purpose to which the fuel is employed.

1.3.1.3. Gaseous fuels

Natural gas. The main constituents of natural gas are methane (CH_4) and ethane (C_2H_6). It has calorific value nearly 21000 kJ/m^3. Natural gas is used alternately or simultaneously with oil for internal combustion engines.

Coal gas. This gas mainly consists of hydrogen, carbon monoxide and hydrocarbons. It is prepared by carbonisation of coal. It finds its use in boilers and sometimes used for commercial purposes.

Coke-oven gas. It is obtained during the production of coke by heating the bituminous coal. The volatile content of coal is driven off by heating and major portion of this gas is utilised in heating the ovens. This gas must be thoroughly filtered before using in gas engines.

Blast furnace gas. It is obtained from smelting operation in which air is forced through layers of coke and iron ore, the example being that of pig iron manufacture where this gas is produced as by product and contains about 20% carbon monoxide (CO). After filtering it may be blended with richer gas or used in gas engines directly. The heating value of this gas is very low.

Producer gas. It results from the partial oxidation of coal, coke or peat when they are burnt with an insufficient quantity of air. It is produced in specially designed retorts. It has low heating value and in general is suitable for large installations. It is also used in steel industry for firing open hearth furnaces.

Water or Illuminating gas. It is produced by blowing steam into white hot coke or coal. The decomposition of steam takes place liberating free hydrogen and oxygen in the steam combines with carbon to form carbon monoxide according to the reaction :

$$C + H_2O \longrightarrow CO + H_2$$

The gas composition varies as the hydrogen content if the coal is used.

Sewer gas. It is obtained from sewage disposal vats in which fermentation and decay occur. It consists of mainly marsh gas (CH_4) and is collected at large disposal plants. It works as a fuel for gas engines which is turn drive the plant pumps and agitators.

Gaseous fuels are becoming popular because of following *advantages* they possess :

Advantages :

1. Better control of combustion.

2. Much less excess air is needed for complete combustion.
3. Economy in fuel and more efficiency of furnace operation.
4. Easy maintenance of oxidizing or reducing atmosphere.
5. Cleanliness.
6. No problem of storage if the supply is available from public supply line.
7. The distribution of gaseous fuels even over a wide area is easy through the pipe lines and as such handling of the fuel is altogether eliminated. Gaseous fuels give economy of heat and produce higher temperatures as they can be preheated in regenerative furnaces and thus heat from hot flue gases can be recovered.

Important properties of gaseous fuels :
1. Heating value or calorific value.
2. Viscosity.
3. Specific gravity.
4. Density.
5. Diffusibility.

Typical composition of some gaseous fuels is given below :

Fuel	H_2	CO	CH_4	C_2H_4	C_2H_6	C_4H_8	O_2	CO_2	N_2
Natural gas	—	1	93	—	3	—	—	—	3
Coal gas	53.6	9.0	25	—	—	3	0.4	3	6
Blast furnace gas	2	27	—	—	—	—	—	11	—

1.3.1.4. Calorific or heating values of fuels

The **calorific value** *of the fuel is defined as the energy liberated by the complete oxidation of a unit mass or volume of a fuel.* It is expressed in kJ/kg for solid and liquid fuels and kJ/m³ for gases.

Fuels which contain hydrogen have two calorific values, the *higher* and the *lower*. The *'lower calorific value'* is the heat liberated per kg of fuel after deducting the heat necessary to vaporise the steam, formed from hydrogen. The *'higher or gross calorific value' of the fuel is one indicated by a constant-volume calorimeter in which the steam is condensed and the heat of vapour is recovered.*

The lower or net calorific value is obtained by subtracting latent heat of water vapour from gross calorific value. In other words, the relation between Lower Calorific Value (L.C.V.) and Higher Calorific Value (H.C.V.) can be expressed in the following way :

$$\text{L.C.V.} = (\text{H.C.V.} - 2465\, m_w) \qquad \text{...(1.1)}$$

where m_w is the mass of water vapour produced by combustion of 1 kg of fuel and 2465 kJ/kg is the latent heat corresponding to standard temperature (saturation) of 15°C.

In MKS units:

$$\text{L. C. V.} = (\text{H. C. V.} - 588.76\, m_w)$$

where m_w is the mass of water vapour produced by combustion of 1 kg of fuel and 588.76 is the latent heat value in kcal as read from steam tables for 1 kg of water vapour

Dulong's formula (Solid/liquid fules). Dulong suggested a formula for the calculation of the calorific value of the solid or liquid fuels from their chemical composition which is as given below :

Gross calorific value,

$$\text{H.C.V.} = \frac{1}{100}\left[33800\,\text{C} + 144000\left(\text{H} - \frac{\text{O}}{8}\right) + 9270\,\text{S}\right] \text{ kJ/kg} \qquad \text{...(1.2)}$$

$$\left[\begin{array}{l}\textbf{In MKS units:}\\ \text{H.C.V.} = \dfrac{1}{100}\left[8080\,\text{C} + 34500\left(\text{H} - \dfrac{\text{O}}{8}\right) + 2240\,\text{S kcal/kg}\right]\end{array}\right]$$

where C, H, O and S are carbon, hydrogen, oxygen and sulphur in percentages respectively in 100 kg of fuel. In the above formula, the oxygen is assumed to be in combination with hydrogen and only extra surplus hydrogen supplies the necessary heat.

1.3.2. Energy Stored in Water

- The energy contained in flowing streams of water is a form of mechanical energy. It may exist as the kinetic energy of a moving stream or as potential energy of water at some elevation with respect to a lower datum level, an example of which would be the water held behind a dam. *Hydraulic plants* are slowly increasing in number, although the number of new plants of this type built is quite small compared with those which exploit heat energy. As a usual thing, the most desirable hydroelectric sites are the first to be utilized, consequently, as more hydroelectric plants are built, the owners must pay increasingly higher development costs.
- From the stand point of capitalistic economics, it is often hard to justify the development of hydroelectric power in comparison with steam power, but from the stand point of the conservation of a fixed natural resource, namely, its mineral fuels, it is obvious that every effort should be made to harness the water power of the country, since if unharnessed it goes to waste, whereas fuel, if unmined, remains intact and undiminished in value in the ground.
- Water power is quite cheap where water is available in abundance. Although *capital cost of hydroelectric power plants is higher as compared to other types of power plants yet their operating costs are quite low.*

1.3.3. Nuclear Energy

- One of the outstanding facts about nuclear power is the *large amount of energy that can be released from a small mass of active material.* Complete fission of one kg of uranium contains the energy equivalent of 4500 tonnes of coal or 2000 tonnes of oil. The *nuclear power is not only available in abundance but it is cheaper than the power generated by conventional sources.*

  The following factors go in *favour of nuclear energy* :

  (*i*) Practically independent of geographical factors.

  (*ii*) No combustion products.

  (*iii*) Clean source of power which does not contribute to air pollution.

  (*iv*) Fuel transportation networks and large storage facilities not required.
- The economic advantage of nuclear power can be realised only if one can ensure a guaranteed base load of about 75%. The number of electro-chemical processes (fertiliser plants), desalination of water and use of electricity for pumping water from tube wells assure a constant base load. Therefore, such type of power requirements must be developed before the adoption of nuclear power in the country.

1.3.4. Wind Power

- The man has been served by the power from winds for many centuries but the total amount of energy generated in this manner is *small*. The expense of installation and variability of operation have tended to *limit* the use of the windmill to intermittent services where *its variable output has no serious disadvantage.* The principal services of this nature are the *pumping of water into storage tanks and the charging of storage batteries.*

- Windmill power equipment may be *classified* as follows :
  1. *The multi-bladed turbine wheel.* This is the foremost type in use and its efficiency is about 10 per cent of the kinetic energy of the wind passing through it.
  2. *The high-speed propeller type.*
  3. *The rotor.*
- The propeller and rotor types are *suitable for the generation of electrical energy,* as both of them possess the ability to start in very low winds. The *Propeller type is more likely to be used in small units* such as the driving of small battery charging generators, whereas the *rotor,* which is rarely seen, is more practical *for large installations,* even of several hundred kilowatts' capacity.
- In india, the wind velocity along coastline has a range 10-16 kmph and a survey of wind power has revealed that wind power is capable of exploitation for pumping water from deep wells or for generating small amounts of electric energy.

Modern windmills are capable of working on velocities as low as 3-7 kmph while *maximum efficiency is attained at 10-12 kmph.*

- A normal working life of 20 to 25 years is estimated for windmills.
- The great advantage of this source of energy is that *no operator is needed and no maintenance and repairs are necessary for long intervals.*

Characteristics of wind power/energy. Some characteristics of wind energy are given below :

1. No fuel provision and transport are required in wind energy systems.
2. It is a renewable source of energy.
3. Wind power systems are non-polluting.
4. Wind power systems, upto a few kW, are less costly, but on a large scale, costs can be competitive with conventional electricity. Lower costs can be achieved by mass production.

Problems associated with wind energy :

1. Wind energy systems are *noisy in operation.*
2. *Large areas* are needed to install *wind farms* for electrical power generators.
3. Wind energy available is *dilute and fluctuating in nature.* Because of dilute form, conversion machines have to be necessarily large.
4. Wind energy *needs storage means* because of its irregularity.

1.3.5. Solar Energy

- A lot of work to utilise solar energy for generation of steam has been done in some countries, and it is likely that this could be developed on commercial scale.
- A serious fault of this source of energy is, of course, that it is *effective only during the day,* so that if a continuous output is needed, some large reservoir of energy, such as a storage battery or a heat accumulator tank, must be drawn upon at night. Also, the output is handicapped if there is cloudy weather. Nevertheless, there are some locations in the world where strong solar radiation is received very regularly, and where the sources of mineral fuel are either scanty or entirely lacking. Such locations offer more interest to the solar power plant builder than the more favoured regions of earth.
- For developing solar energy two ways have been explored *viz.,* the *glass lens and the reflector.* These devices concentrate the solar rays to a focal point which is characterised by a high degree of heat which can be utilised to boil water and generate steam. The reflector is the better of the two methods due to the convenience with which it can be

manufactured in different shapes and sizes. If an arrangement is provided to turn the reflector with the sun, so that the rays can constantly concentrate at the focal point, a continuous supply of heat is made available during the hours of the day. However, a great deal of practical research is still necessary before the solar energy can be *commercially exploited at a cheaper rate.*

- Conditions for utilisation of solar energy, in India, are favourable since for nearly six months of the year sunshine is uninterrupted during the day, while in the other six months cloudy weather and rain provide conditions suitable for water power. Thus, a *coordination of solar energy with water power can provide a workable plan for most places in India.*

1.3.6. Tidal Power

- The rise and fall of tides offers a means for storing water at the rise and discharging the same at fall. Of course the head of water available under such cases is very low but with increased catchment area considerable amounts of power can be generated at a negligible cost.

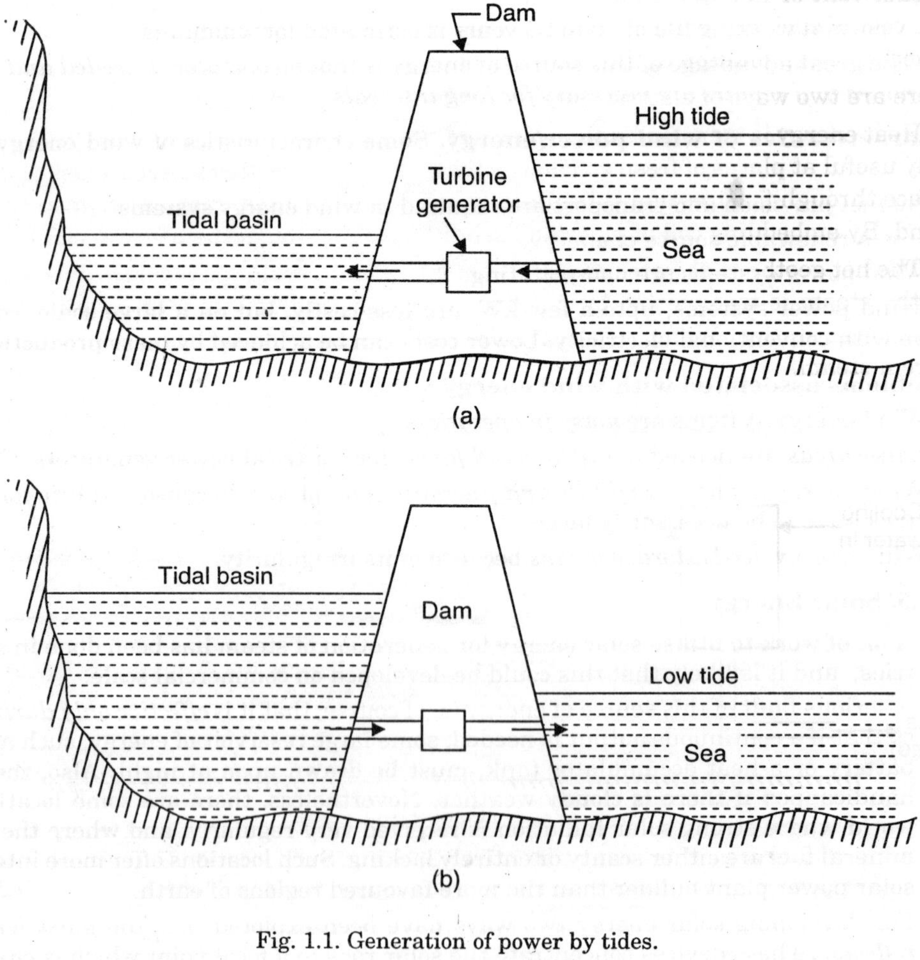

Fig. 1.1. Generation of power by tides.

- The use of tides for electric power generation is practical in a few favourably situated sites where the geography of an inlet of bay favours the construction of a large scale hydroelectric plant. To harness the tides, a dam would be built across the mouth of the bay in which large gates and low head hydraulic turbines would be installed. At the time of high tide the gates are opened and after storing water in the tidal basin the gates are closed. After the tide has recided, there is a working hydraulic head between the basin water and open sea/ocean and the water is allowed to flow back to the sea through water turbines installed in the dam. With this type of arrangement, the generation of electric power is *not continuous*. However by using reversible water turbine the turbine can be run continuously as shown in Fig. 1.1.

1.3.7. Geothermal Energy

In many places on the earth natural steam escapes from surface vents. Such natural steam wells suggest the possibility of tapping terrestrial heat (or geothermal energy) in this form and using it for the development of power. Unfortunately, the locations where the steam-producing sub-strata seem to be fairly close to the surface are far removed from centres of civilization where the power could be usefully employed. Nevertheless, there are probably many places where, although no natural steam vent or hot springs are showing, deep drillings might tap a source of underground steam. The cost of such explorations and the great likelihood of an unsuccessful conclusion are not very conductive to exploitation of this source of energy.

There are two ways of electric power production from geothermal energy :

(*i*) Heat energy is transferred to a working fluid which operates the power cycle. This may be particularly useful at places of fresh volcanic activity where the molten interior mass of earth vents to the surface through fissures and substantially high temperatures, such as between 450 to 550°C can be found. By embedding coil of pipes and sending water through them steam can be raised.

(*ii*) The hot geothermal water and/or steam is used to operate the turbines directly. From the well-head the steam is transmitted by pipelines upto 1 m in diameter over distances upto about 3 km to the power station. Water separators are usually employed to separate moisture and solid particles from steam.

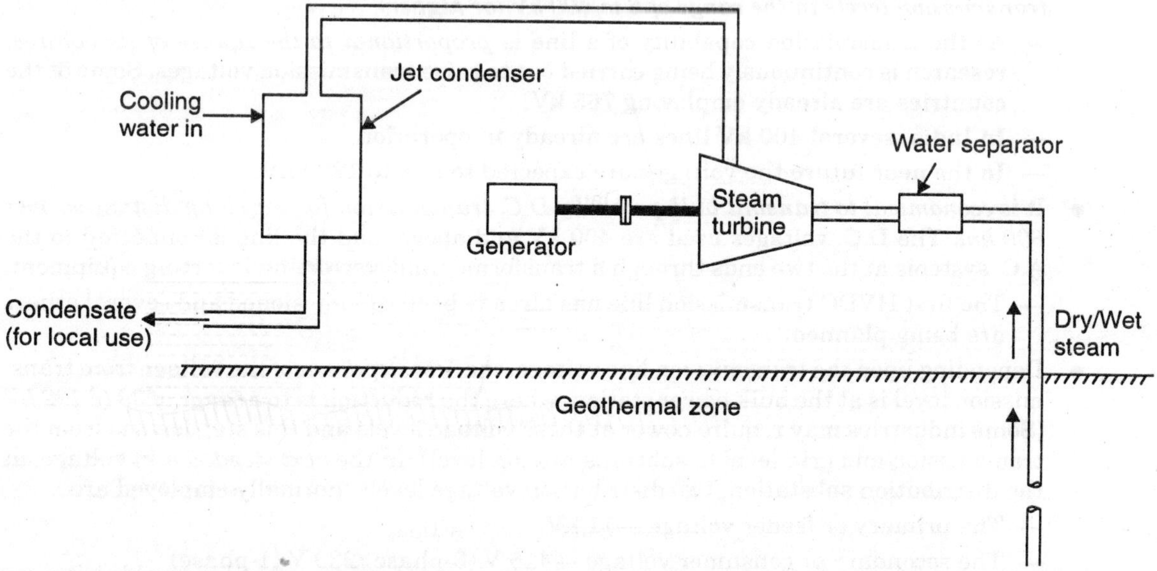

Fig. 1.2. Geothermal power plant.

Presently, only steam coming out of the ground is used to generate electricity, the hot water is discarded, because it contains as much as 30% dissolved salts and minerals, and these cause serious rust damage to the turbine. The water, however, contains more than $\frac{1}{3}$rd of the available thermal energy.

1.3.8. Thermo-electric Power

According to *Seebeck effect*, when the two ends of a loop of two dissimilar metals are held at different temperatures, an electromotive force is developed and the current flows in loop. This method, by selection of suitable materials, can also be used for power generation. This method involves *low initial cost* and *negligible operating cost.*

1.4. POWER SYSTEMS' STRUCTURE

- The main components of an electric power system are :
  (*i*) *Generating stations.*
  (*ii*) *Transmission systems.*
  (*iii*) *Distributon systems.*
  — Generating stations and a distribution system are connected through transmission lines, which also connect one power system (grid, area) to another.
  — A distribution system connects all the loads in a particular area to the transmission lines.
- Individual power systems, for economical and technological reasons, are organised in the form of electrically connected areas or ***regional grids*** (also called **power pools**). Each area or regional grid operates technically and economically independently but these are eventually interconnected (*Interconnection* has the *economic advantage of reducing the reserve generation capacity in each area*) to form a *national grid* (which may even form an *international grid*) *so that each area is contractually tied to other areas in respect to certain generation and scheduling features.*
- *Electric power* is generated at a *voltage of 11 to 25 kV which then is stepped up to the transmission levels in the range of 6 to 400 kV (or higher).*
  — As the transmission capability of a line is *proportional to the square of its voltage,* research is continuously being carried out to raise transmission voltages. Some of the countries are already employing 765 kV.
  — In India, several 400 kV lines are already in operation.
  — In the near future the voltages are expected to rise to 1200 kV.
- It is *economical to transmit bulk power by D.C. transmission for very long distances, over 600 km.* The D.C. voltages used are 400 kV and above, and the line is connected to the A.C. systems at the two ends through a transformer and converting/inverting equipment.
  — The first HVDC transmission line has already been commissioned and several others are being planned.
- Depending upon the transmission line voltage, the *first stepdown* of the voltage from transmission level is at the bulk power station, where the reduction is to a *range of 33 to 132 kV* (Some industries may require power at these voltage levels and this stepdown is from the transmission and grid level to subtransmission level). In the *next stepdown* in voltage, at the distribution substation, two distribution voltage levels, normally employed are :
  — The primary or feeder voltage —11 kV.
  — The secondary or consumer voltage —415 V (3-phase)/230 V (1-phase)

  In the chapters to follow, different types of generating stations will be discussed.

HIGHLIGHTS

1. **Energy appears in many forms, but has one thing in common—energy is possessed of the ability to produce a dynamic, vital effect.**

2. **Sources of energy are :**

(*i*) Fuels (*ii*) Energy stored in water (*iii*) Nuclear energy
(*iv*) Wind power (*v*) Solar energy (*vi*) Tidal power
(*vii*) Geothermal energy (*viii*) Thermoelectric power.

3. **A chemical fuel is a substance which releases heat energy on combustion. The principal element of each fuel are carbon and hydrogen. Though sulphur is a combustible element too but its presence in the fuel is considered to be undesirable.**

4. **The principal types of power plants are :**

(*i*) Steam plants using coal, oil or nuclear fission (*ii*) Internal combustion engine plants
(*iii*) Gas turbine plants (*iv*) Hydro-electric plants.

5. **The main components of an electric power system are :**

(*i*) Generating stations ; (*ii*) Transmission systems ; (*iii*) Distribution systems.

THEORETICAL QUESTIONS

1. **Enumerate sources of energy.**
2. **What is a chemical fuel ? How does it differ from a nuclear fuel ?**
3. **How are chemical fuels classified ?**
4. **Explain briefly the following solid fuels :**
Lignites and brown coals, bituminous coal, and coke.
5. **List the advantages of liquid fuels.**
6. **Describe the following gaseous fuels :**
Coal gas, Coke-oven gas, Blast furnace gas and Producer gas.
7. **What are the advantages of gaseous fuels ?**
8. **State the factors which go in favour of nuclear energy.**
9. **Write short notes on :**
Tidal power, wind power and thermoelectric power.
10. **Name the principal types of power plants.**
11. **Why do we feel the necessity of using excess air for burning fuels ?**
12. **How can the following be calculated ?**
(*i*) Weight of carbon in flue gases.
(*ii*) Weight of flue gas per kg of fuel burnt.

2

Steam Power Plant

2.1. Introduction. 2.2. Classification of steam power plants. 2.3. Layout of a modern steam power plant. 2.4. Essential requirements of steam power station design. 2.5. Selection of site for steam power station. 2.6. Capacity of steam power plant. 2.7. Choice of steam conditions. 2.8. **Fuel handling** : Introduction—requirements of good coal handling plant—coal handling systems—coal handling—layout of a fuel handling equipment. 2.9. **Combustion equipment for steam boilers** : General aspects—combustion equipment for solid fuels—selection considerations—burning of coal—stoker firing, pulverised fuel firing—pulverised fuel burners—oil burners—gas burners. 2.10. **Fluidised Bed Combustion (FBC).** 2.11. **Ash handling** : Ash handling equipment—ash handling systems. 2.12. **Dust collection** : Introduction, removal of smoke, removal of dust and dust collectors— uses of ash and dust—general layout of ash handling and dust collection system. 2.13. **Boilers** : Introduction—classification of boilers—comparison between 'fire tube and water tube' boilers—high pressure boilers. 2.14. **Steam turbines** : Introduction—classification of steam turbines—advantages of steam turbine over the steam engines—description of common types of turbines—methods of reducing wheel or rotor speed— difference between impulse and reaction turbines—steam turbine governing and control. 2.15. **Steam condensers** : Introduction—vacuum—organs of a steam condensing plant—classification of condensers—jet condensers—surface condensers—comparison between jet and surface condensers. 2.16. **Feed water treatment** : Classification of impurities in water—troubles caused by the impurities in water—methods of feed water treatment. 2.17. **Advantages and disadvantages of steam power plants.** 2.18. **Miscellaneous** : Useful life of steam power plant components—steam power plant pumps—cost of steam power plant—comparison of various types of power plants—thermal power stations in India. Highlights—Theoretical Questions.

2.1. INTRODUCTION

A **steam power plant** *converts the chemical energy of the fossil fuels (coal, oil, gas) into mechanical / electrical energy.* This is achieved by raising the steam in the boilers, expanding it through the turbines and coupling the turbines to the generators which convert mechanical energy to electrical energy as shown in Fig. 2.1.

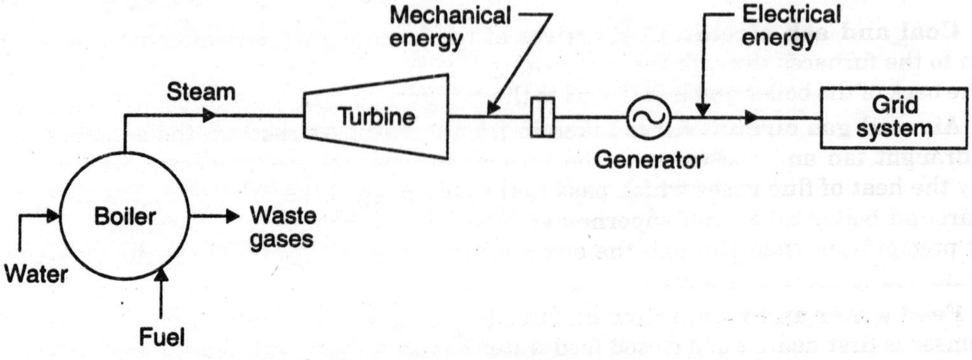

Fig. 2.1. Production of electric energy by steam power plant.

The following two purposes can be served by a steam power plant :

1. To produce electric power.

2. To produce steam for industrial purposes besides producing electric power. The steam may be used for varying purposes in the industries such as *textiles, food manufacture, paper mills, sugar mills* and *refineries* etc.

2.2. CLASSIFICATION OF STEAM POWER PLANTS

The steam power plants may be *classified* as follows :

1. Central stations.

2. Industrial power stations or captive power stations.

1. **Central stations.** The electrical energy available from these stations is meant for general sale to the customers who wish to purchase it. Generally, these stations are *condensing type* where the exhaust steam is discharged into a condenser instead of into the atmosphere. In the condenser the pressure is maintained below the atmospheric pressure and the exhaust steam is condensed.

2. **Industrial power stations or captive power stations.** This type of power station is run by a manufacturing company for its own use and its output is not available for general sale. Normally these plants are *non-condensing* because a large quantity of steam (low pressure) is required for different manufacturing operations.

In the *condensing steam power plants* the following *advantages* accrue :

(*i*) The amount of energy extracted per kg of steam is *increased* (a given size of the engine or turbine develops more power).

(*ii*) The steam which has been condensed into water in the condenser, can be recirculated to the boilers with the help of pumps.

In non-condensing steam power plants a continuous supply of fresh feed water is required which becomes a problem at places where there is a shortage of pure water.

2.3. LAYOUT OF A MODERN STEAM POWER PLANT

Refer Fig. 2.2. The layout of a modern steam power plant comprises of the following *four circuits* :

1. Coal and ash circuit.
2. Air and gas circuit.
3. Feed water and steam flow circuit.
4. Cooling water circuit.

The brief descripation of these circuits is given below :

1. **Coal and ash circuit.** Coal arrives at the storage yard and after necessary handling, passes on to the furnaces through the *fuel feeding device.* Ash resulting from combustion of coal collects at the back of the boiler and is removed to the ash storage yard through *ash handling equipment.*

2. **Air and gas circuit.** Air is taken in from atmosphere through the action of a forced or induced draught fan and passes on to the furnace through the *air preheater*, where it has been heated by the heat of flue gases which pass to the chimney *via* the preheater. The flue gases after passing around boiler tubes and superheater tubes in the furnace pass through a *dust* catching device or precipitator, then through the economiser, and finally through the air preheater before being exhausted to the atmosphere.

3. **Feed water and steam flow circuit.** In the water and steam circuit condensate leaving the condenser is first heated in a closed feed water heater through extracted steam from the lowest pressure extraction point of the turbine. It then passes through the *deaerator* and a few more water heaters before going into the boiler through *economiser.*

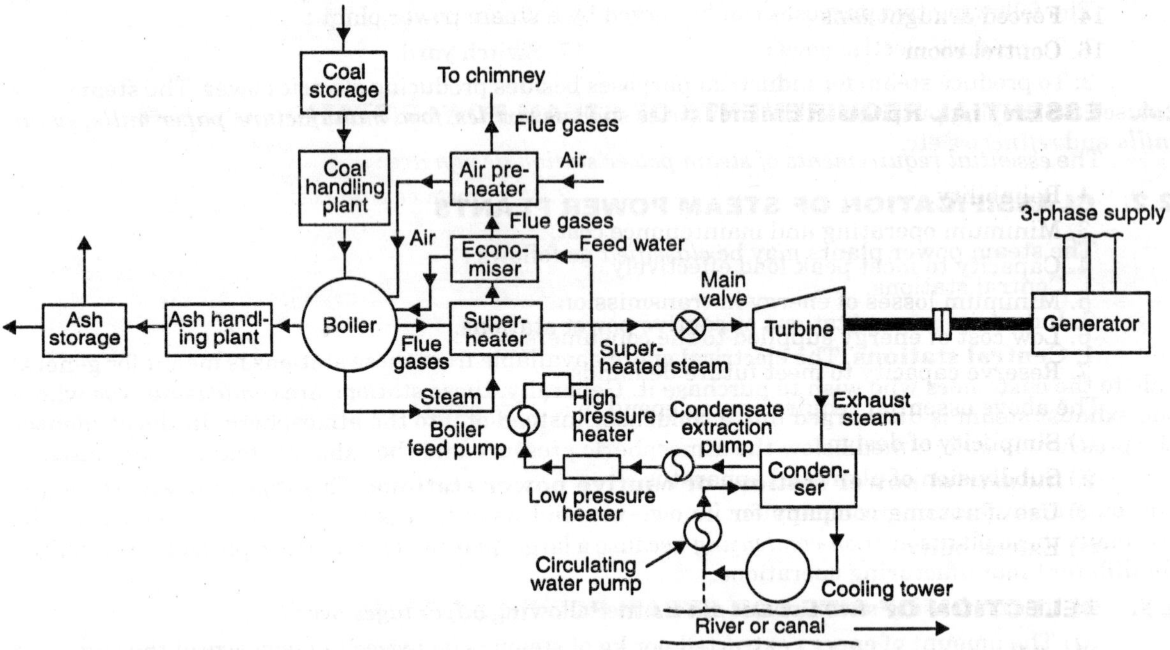

Fig. 2.2. Layout of a steam power plant.

In the boiler drum and tubes, water circulates due to the difference between the density of water in the lower temperature and the higher temperature sections of the boiler. Wet steam from the drum is further heated up in the superheater before being supplied to the primemover. After expanding in high pressure turbine steam is taken to the reheat boiler and brought to its original dryness or superheat before being passed on to the low pressure turbine. From there it is exhausted through the condenser into the hot well. The condensate is heated in the feed heaters using the steam trapped (bled steam) from different points of turbine.

A part of steam and water is lost while passing through different components and this is compensated by supplying additional feed water. This feed water should be purified before hand, to avoid the scaling of the tubes of the boiler.

4. Cooling water circuit. The cooling water supply to the condenser helps in maintaining a low pressure in it. The water may be taken from a natural source such as river, lake or sea or the same water may be cooled and circulated over again. In the later case the cooling arrangement is made through spray pond or cooling tower.

Components of a Modern Steam Power Plant :

A modern steam power plant consists of the following *components* :

1. Boiler

(*i*) Superheater | **(*ii*) Reheater**

(*iii*) Economiser | **(*iv*) Air-heater**

2. Steam turbine
3. Generator
4. Condenser
5. Cooling towers
6. Circulating water pump
7. Boiler feed pump
8. Wagon tippler
9. Crusher house
10. Coal mill
11. Induced draught fans
12. Ash precipitators
13. Boiler chimney

14. Forced draught fans
15. Water treatment plant
16. Control room
17. Switch yard.

2.4. ESSENTIAL REQUIREMENTS OF STEAM POWER STATION DESIGN

The *essential requirements of steam power station design are* :

1. Reliability.
2. Minimum capital cost.
3. Minimum operating and maintenance cost.
4. Capacity to meet peak load effectively.
5. Minimum losses of energy in transmission.
6. Low cost of energy supplied to the consumers.
7. Reserve capacity to meet future demands.

The above essential requirements depend to a large extent on the following :

(*i*) Simplicity of design.
(*ii*) Subdivision of plant and apparatus.
(*iii*) Use of automatic equipment.
(*iv*) Extensibility.

2.5. SELECTION OF SITE FOR STEAM POWER STATION

The following points should be taken into consideration while *selecting the site* for a steam power station :

1. Availability of raw material.
2. Nature of land.
3. Cost of land.
4. Availability of water.
5. Transport facilities.
6. Ash disposal facilities.
7. Availability of labour.
8. Size of the plant.
9. Load centre.
10. Public problems.
11. Future extensions.

1. **Availability of raw material.** Modern steam power stations using coal or oil as fuel require huge quantity of it per annum. A thermal power plant of 400 MW capacity requires 5000 to 6000 tonnes of coal per day. Therefore, it is necessary to locate the plant as far as possible near the coalfields in order to save the transportation charges. Besides transportation charges, a plant located away from the coalfields, cannot always depend on the coal deliveries in time (*i*) as there may be failure of transportation system, (*ii*) there may be strike etc. at the mines. For these reasons a considerable amount of coal must be stored at the power stations, this results in :

(*i*) increased investment ;
(*ii*) increased space required at the site of the plant for the storage ;
(*iii*) losses in storage ; and
(*iv*) additional staff requirement.

If it is not possible to locate the plant near the coalfields then the plant should be located as near the railway station as possible. Even if this is not possible then at least arrangement should be made for railway siding to the power plant so that the coal wagons can be shunted from the station to the site of power plant. This applies to plants using oil as fuel, as well.

2. **Nature of land.** The type of the land to be selected should have *good bearing capacity* as it has to withstand the *dead load* of the plant and the *forces transmitted to the foundation* due to the machine operations. The minimum bearing capacity of the land should be 1 MN/m^2.

3. **Cost of land.** Considerable area is required for the power stations. The cost of the land for that purpose should be reasonable. The large plants in the heart of big cities and near the load centre are not economical as the cost of land is very high.

4. **Availability of water.** Steam power stations use water as the working fluid which is repeatedly evaporated and condensed. Theoretically there should be no loss of water, but in fact some make up water is required. Besides this, considerable amount of water is required for condensers. A large quantity of water is also required for disposing the ash if hydraulic system is used. It is, therefore, necessary to locate the power plant near the water source which will be able to supply the required quantity of water throughout the year.

5. **Transport facilities.** Availability of proper transport facilities is another important consideration in locating the thermal power station. It is always necessary to have a railway line available near the power station for bringing in heavy machinery for installation and for bringing the fuel.

6. **Ash disposal facilities.** The ash handling problem is more serious than coal handling because it comes out in hot condition and it is highly corrosive. Its effect on atmospheric pollution are more serious as the human health is concerned. Therefore, there must be sufficient space to dispose of large quantity of ash.

In a power station of 400 MW capacity 10 hectares area is required per year if the ash is dumped to a height of 6.5 metres.

7. **Availability of labour.** During construction of plant enough labour is required. The labour should be available at the proposed site at cheap rate.

8. **Size of the plant.** The expenses involved in electric transmission from a *small plant* are *relatively severe*, owing to the impracticability of using high voltages, so that the electric transmission feature alone becomes dominant in the location of the plant. It case of *large plants*, the costs of transporting enormous quantities of coal and water are considerably high ; therefore, the plant must be located near the pit head provided the required water quantity must be available as near as possible. The large plants should be located close to the railroad offering adequate services. The economic significance of the large plant with small one is much greater than the mere ratio of size.

9. **Load centre.** A power station must be located *near the loads* to which it is supplying power. However, a plant cannot be located near all loads. As such C.G. of the loads is determined with reference to two arbitrarily chosen axes, this C. G. is known as the *load centre* (see Fig. 2.3).

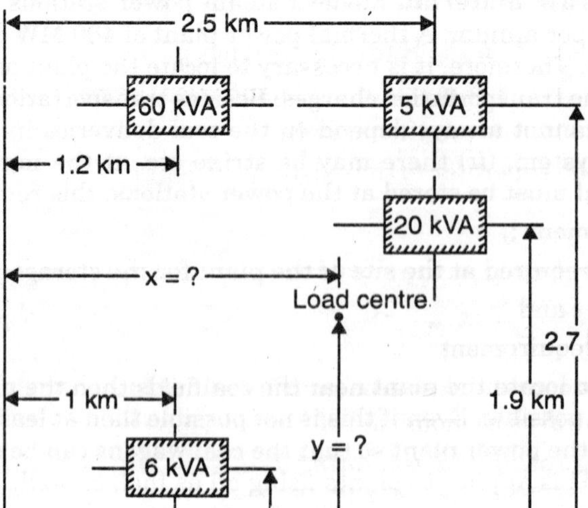

Fig. 2.3. Load centre.

The power plant should be located, as far as possible near the load centre in order to *minimize the cost of transmission lines and also the losses occuring in them.*

10. **Public problems.** In order to avoid the nuisance from smoke, fly ash and heat discharged from the power plant, it should be located far away from the towns.

11. **Future extensions.** The choice of the site should allow for economical extensions consistent with the estimated growth of load.

12. Consent of Town Planning Department must be sought in case urban area is selected for the purpose.

2.6. CAPACITY OF STEAM POWER PLANT

The power plant capacity can be determined by studying *the load duration curve and the anticipated future demands. The minimum capacity of the plant must be equal to at least the peak load.*

(*i*) In case of "*Small loads*", it may be economical to install two units of equipment, each being capable of supplying the maximum demand independently. In the event of failure of one unit or during maintenance etc., at least one unit can be used to maintain uninterrupted supply of energy.

(*ii*) In case of "*medium power plants*", usually the number of units is more than two with the total installed capacity equal to the *maximum demand plus the capacity of two large units.*

(*iii*) "*Large power plants*" are generally *conservately rated.* In the case of steam turbines, there is an overload capacity of 10 to 15% of the rated capacity. With a number of units, peak load can be easily adjusted by overloading some units temporarily.

The power plant load can be reduced by dropping the supply voltage. Thus a 5 percent reduction in supply voltage results in similar reduction in the load. An electric supply undertaking has to maintain the voltage within 10 percent of the declared pressure as per Indian Electricity Act ; so *during peak hours the voltage can be reduced within the allowable limits in order to meet the demand without use of additional units.* By using this technique saving of capital cost is materialised.

When a new unit is to be added to the existing power station, its size is decided on the following considerations :

1. Effect of additional unit on the thermal efficiency of the plant.
2. Expected rate of increase of maximum demand over the next few years.
3. The room available for the additional unit.
4. The suitability of the generator to the existing system regarding temperature, pressure etc.

Rating of Units :

Normally the output of units is classified under the following heads :

(*i*) Economical rating.

(*ii*) Maximum continuous rating.

A generator need not operate most economically at full load. For the most economical operation, the present trend is towards economical running at 75-85 percent of full load.

Maximum continuous rating of a generating unit is the maximum load at which it can be run continuously for several hours. It is normally 10-15 percent less than the maximum capacity of the unit.

2.7. CHOICE OF STEAM CONDITIONS

The choice of steam conditions depends upon the following *factors* :

1. Price of coal.
2. Capital cost of the plant.
3. Time available for erection.
4. Thermal efficiency obtainable.
5. The station 'load factor'.

The present trend is towards adoption of high pressures and high temperatures. The effect of increased pressure and temperature on the efficiency and cost of plant is illustrated with the help of Figs. 2.4 and 2.5. It is evident from the curves that :

(*i*) *With the increase in pressure the efficiency obeys the 'law of diminishing returns'.*

(*ii*) With the increase in temperature the efficiency obeys the '*straight line law*' indicating the desirability of adopting the highest possibe temperature. The strength of material available limits the adoption of high temperatures. Beyond 500°C there is a very rapid change in the physical properties of the material and the problem becomes complicated. *With the increase in pressures the degree of superheat should be decreased in order to keep the total temperature within limits.*

For entirely new stations, present practice favours the use of steam pressures around 60 bar, but there is a profitable field for higher pressures of the order of 100 bar, when the problem is that of *increasing thermal efficiency of existing medium pressure units.*

It may be noted that *consumption of steam per kilo-watt hour decreases with the increased pressure.*

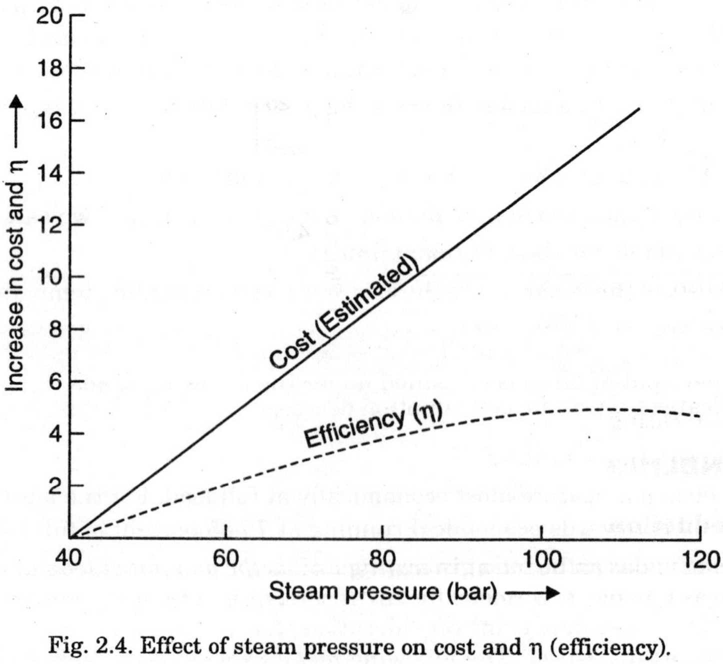

Fig. 2.4. Effect of steam pressure on cost and η (efficiency).

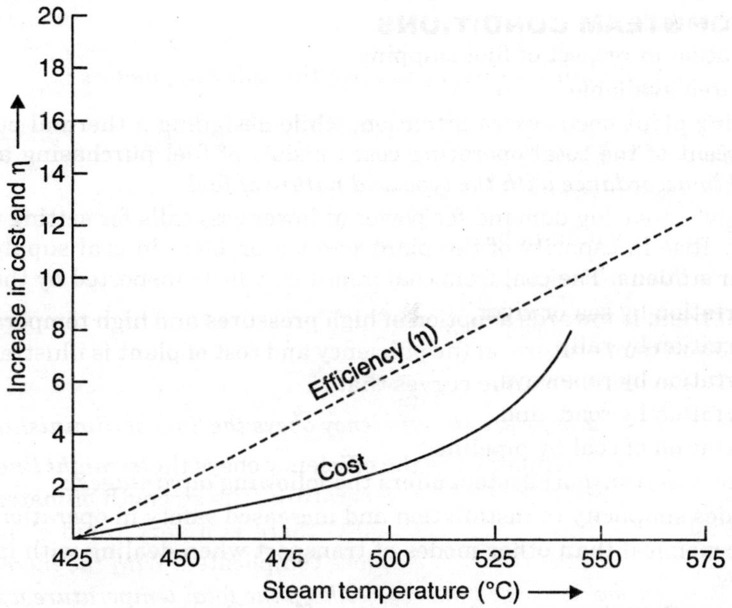

Fig. 2.5. Effect of steam temperature on cost and η (efficiency).

Fig. 2.6 shows that *as the average output of plant units increases the percentage total capital cost decreases.*

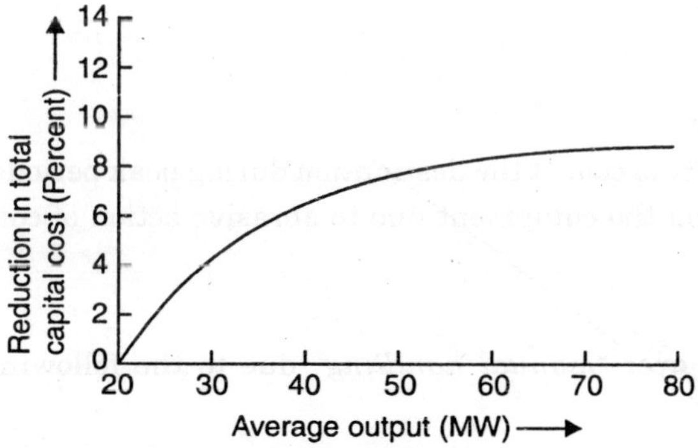

Fig. 2.6

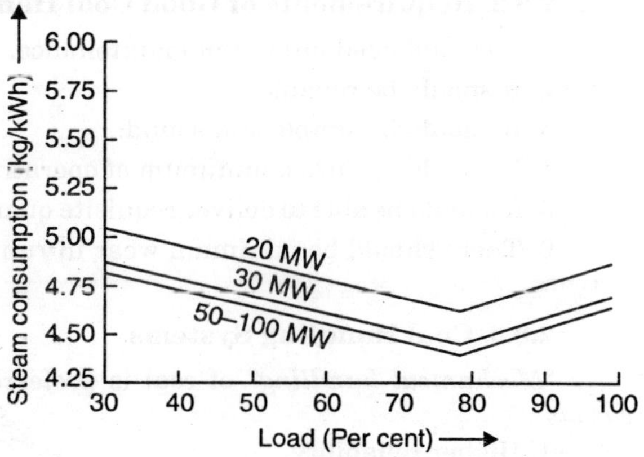

Fig. 2.7

Fig. 2.7 indicates that steam consumption decreases with increase in the capacity of the unit.

2.8. FUEL HANDLING

2.8.1. Introduction

Three types of fuels can be burnt in any type of steam generating plant : 1. Solid fuel such as *coal*, 2. Liquid fuel as oil, and 3. Gaseous fuel as gas. Supply of these fuels to the power plants from various sources is one of the important considerations for a power plant engineer. The handling of these fuels is an important aspect. The following *factors* should be considered in *selecting the fuel handling system* :

1. Plant fuel rate.
2. Plant location in respect of fuel shipping.
3. Storage area available.

Fuel handling plant needs extra attention, while designing a thermal power station, as almost 50 to 60 percent of the total operating cost consists of fuel purchasing and handling. *Fuel system is designed in accordance with the type and nature of fuel.*

Continuously increasing demand for power at lower cost calls for setting up of higher capacity power stations. Rise in capacity of the plant poses a problem in coal supply system from coal mines to the power stations. The coal from coal mines may be transported by the following *means* :

1. Transportation by sea or river,
2. Transportation by rail,
3. Transportation by ropeways,
4. Transportation by road, and
5. Transportation of coal by pipeline.

The pipeline coal transport system offers the following *advantages* :

1. It provides simplicity in installation and increased safety in operation.
2. More economical than other modes of transport when dealing with large volume of coal over long distances.
3. This system is continuous as it remains unaffected by the vagaries of climate and weather.
4. High degree of reliability.
5. Loss of coal during transport due to theft and pilferage is totally eliminated.
6. Manpower requirement is low.

2.8.2. Requirements of Good Coal Handling Plant

1. It should need minimum maintenance.
2. It should be reliable.
3. It should be simple and sound.
4. It should require a minimum of operatives.
5. It should be able to deliver requisite quantity of coal at the destination during peak periods.
6. There should be minimum wear in running the equipment due to abrasive action of coal particles.

2.8.3. Coal Handling Systems

"*Mechanical handling*" of coal is preferred over "*manual handling*" due to the following *reasons* :

1. Higher reliability.
2. Less labour required.
3. Economical for medium and large capacity plants.
4. Operation is easy and smooth.
5. Can be easily started and can be economically adjusted according to the need.
6. With reduced labour, management and control of the plant becomes easy and smooth.
7. Minimum labour is put to unhealthy condition.
8. Losses in transport are minimised.

Disadvantages :

1. Needs continuous maintenance and repair.
2. Capital cost of the plant is increased.

3. In mechanical handling some power generated is usually consumed, resulting in less net power available for supply to consumers.

2.8.4. Coal Handling

Refer to Fig. 2.8.

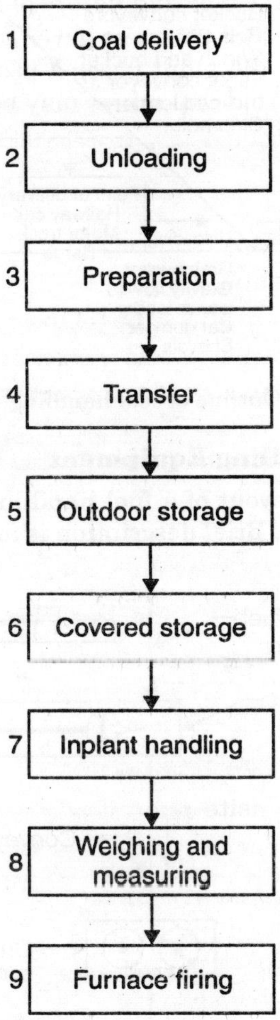

Fig. 2.8. Various stages in coal handling.

The following *stages / steps* are involved in handling the coal :

1. Coal delivery
2. Unloading
3. Preparation
4. Transfer
5. Storage of coal
6. In-plant handling
7. Weighing and measuring
8. Furnace firing.

Fig. 2.9 shows the outline of coal handling equipment.

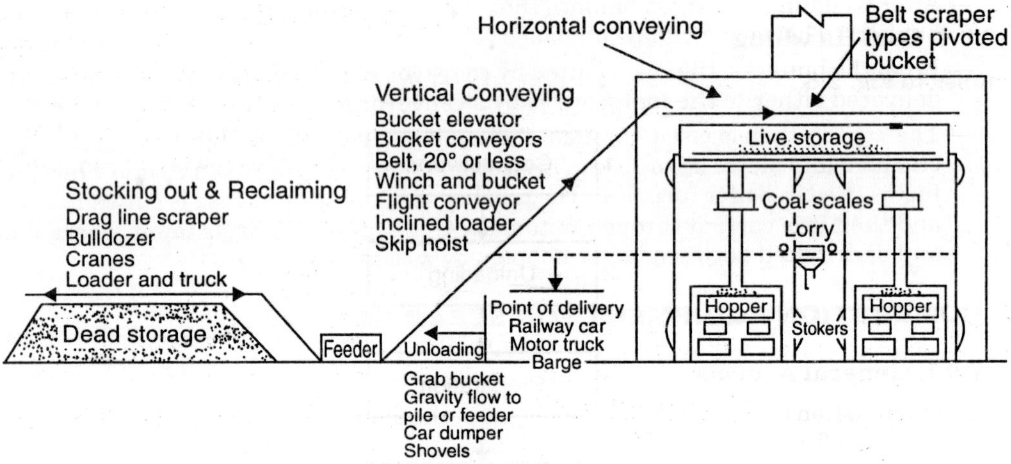

Fig. 2.9. Outline of coal handling equipment.

2.8.5. Layout of a Fuel Handling Equipment

Fig. 2.10 shows a schematic layout of a fuel handling equipment of a modern steam power plant where coal (a solid fuel) is used. Brief description is as follows :

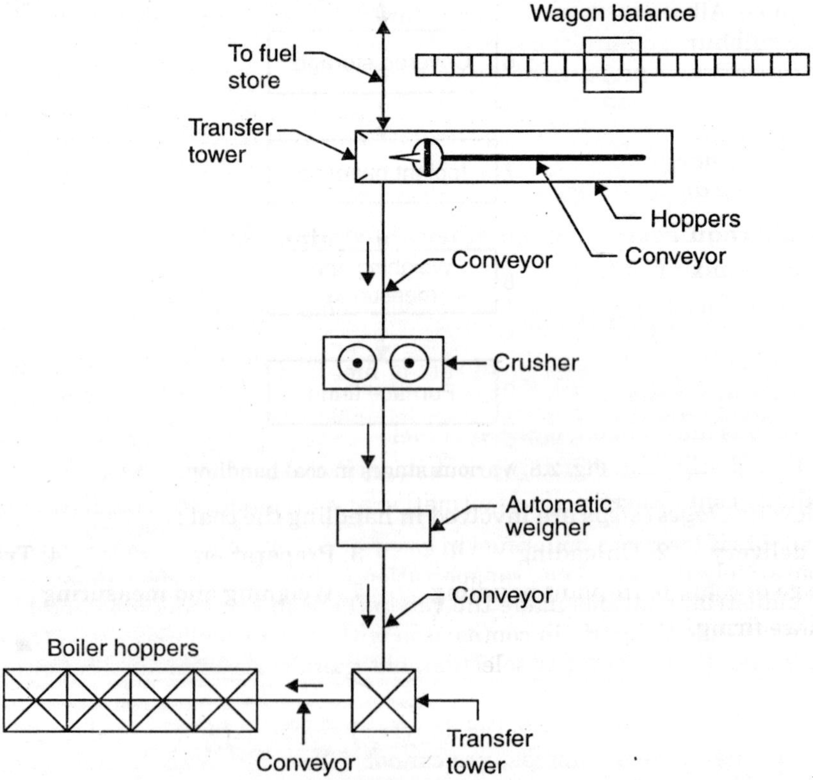

Fig. 2.10. Layout of a fuel handling equipment.

— Coal is supplied to the power plant in railway wagons.

— After weighing on wagon balance the coal is then unloaded into underground hoppers or bunkers. The wagon can be unloaded either manually or through rotary wagon tipplers.

— From the bunkers, the coal is lifted by conveyor to the transfer tower from where it can be delivered either to the fuel store or by a conveyor to a crusher.

— The coal is then passed through the magnetic separators and screens and crushed in crushers into pieces 25 to 30 mm in size for stoker firing and 10 to 20 mm when pulverished fuel is fired in boiler furnaces. The crushed coal in the later case is *milled to a fine powder* and then it is carried through automatic weigher to a transfer tower where fuel is lifted and distributed between boiler hoppers by a conveyor.

2.9. COMBUSTION EQUIPMENT FOR STEAM BOILERS

2.9.1. General Aspects

The combustion equipment is a component of the steam generator. Since the source of heat is the combustion of a fuel, a working unit must have, whatever, equipment is necessary to receive the fuel and air, proportioned to each other and to the boiler steam demand, mix, ignite, and perform any other special combustion duties, such as distillation of volatile from coal prior to ignition.

- *Fluid fuels are handled by burners ; solid lump fuels* by *stokers*.
- In boiler plants hand firing on grates is *practically unheard* of nowdays in new plants, although there are many small industrial plants still in service with hand firing.
- The fuels are mainly bituminous coal, fuel oil and natural gas mentioned in order of importance. All are composed of hydrocarbons, and coal has, as well, much fixed carbon and little sulphur. To burn these fuels to the desired end products, CO_2 and H_2O, requires (*i*) *air in sufficient proportions*, (*ii*) *a good mixing of the fuel and air*, (*iii*) *a turbulence* or *relative motion between fuel and air*. The combustion equipment *must fulfill these requirements and, in addition, be capable of close regulation of rate of firing the fuel, for boilers which ordinarily operate on variable load*. Coal-firing equipment *must also have a means for holding and discharging the ash residue*.

The basic requirements of combustion equipment :

1. Thorough mixing of fuel and air.

2. Optimum fuel-air ratios leading to most complete combustion possible maintained over full load range.

3. Ready and accurate response of rate of fuel feed to load demand (usually as reflected in boiler steam pressure).

4. Continuous and reliable ignition of fuel.

5. Practical distillation of volatile components of coal.

6. Adequate control over point of formation and accumulation of ash, when coal is the fuel.

Natural gas is used as a boiler fuel in gas well regions where fuel is relatively cheap and coal sources comparatively distant. The transportation of natural gas over land to supply cities with domestic and industrial heat has made the gas in the well more valuable and the gas-fired steam generator more difficult to justify in comparison with coal, or fuel cost alone. Cleanliness and convenience in use are other criteria of selection, but more decisive in small plants in central power stations.

Transportation costs add less to the delivery price of oil than gas ; also fuel oil may be stored in tanks at a reasonable cost, whereas, gas cannot. Hence although fuel oil is usually more costly than coal per kg of steam generated, many operators select fuel oil burners rather than stokers because of the simplicity and cleanliness of storing and transporting the fuel from storage to burner.

Depending on the type of combustion equipment, boilers may be *classified* as follows :

1. Solid fuels fired :

(*a*) Hand fired

(*b*) Stoker fired

(*i*) Overfeed stokers

(*ii*) Underfeed strokers.

(*c*) Pulverised fuel fired

(*i*) Unit system

(*ii*) Central system

(*iii*) Combination of (*i*) and (*ii*).

2. Liquid fuel fired :

(*a*) Injection system

(*b*) Evaporation system

(*c*) Combination of (*a*) and (*b*).

3. Gaseous fuel fired :

(*a*) Atmospheric pressure system

(*b*) High pressure system.

2.9.2. Combustion Equipment for Solid Fuels—Selection Considerations

While selecting combustion equipment for solid fuels the following *considerations* should be taken into account :

1. Initial cost of the equipment.
2. Sufficient combustion space and its ability to withstand high flame temperature.
3. Area of the grate (over which fuel burns)
4. Operating cost
5. Minimum smoke nuisance.
6. Flexibility of operation.
7. Arrangements for thorough mixing of air with fuel for efficient combustion.

2.9.3. Burning of Coal

The two most commonly used methods for the burning of coal are :

1. Stroker firing
2. Pulverised fuel firing.

The selection of one of the above methods depends upon the following *factors* :

(*i*) Characteristics of the coal available.

(*ii*) Capacity of the boiler unit.

(*iii*) Load fluctuations.

(*iv*) Station load factor.

(*v*) Reliability and efficiency of the various types of combustion equipment available.

2.9.3.1. Stoker Firing

A "**stoker**" *is a power operated fuel feeding mechanism and grate.*

Advantages of stoker firing :

1. A cheaper grade of fuel can be used.

2. A higher efficiency attained.
3. A greater flexibility of operations assured.
4. Less smoke produced.
5. Generally less building space is necessary.
6. Can be used for small or large boiler units.
7. Very reliable, maintenance charges are reasonably low.
8. Practically immune from explosions.
9. Reduction in auxiliary plant.
10. Capital investment as compared to pulverised fuel system is less.
11. Some reserve is gained by the large amount of coal stored on the grate in the event of coal handling plant failure.

Disadvantages :

1. Construction is complicated.
2. In case of very large units the initial cost may be rather higher than with pulverised fuel.
3. There is always a certain amount of loss of coal in the form of riddling through the grates.
4. Sudden variations in the steam demand cannot be met to the same degree.
5. Troubles due to slagging and clinkering of combustion chamber walls are experienced.
6. Banking and standby losses are always present.
7. Structural arrangements are not so simple and surrounding floors have to be designed for heavy loadings.
8. There is excessive wear of moving parts due to abrasive action of coal.

Classification of stoker firing :

Automatic stokers are *classified* as follows :

1. Overfeed stokers.
2. Underfeed stokers.

In case of overfeed stokers, the coal is fed into the grate *above* the point of air admission and in case of underfeed stokers, the coal is admitted into the furnace *below* the point of air admission.

1. Overfeed stokers :

Principle of operation. Refer to Fig. 2.11. The principle of an overfeed stoker is discussed below :

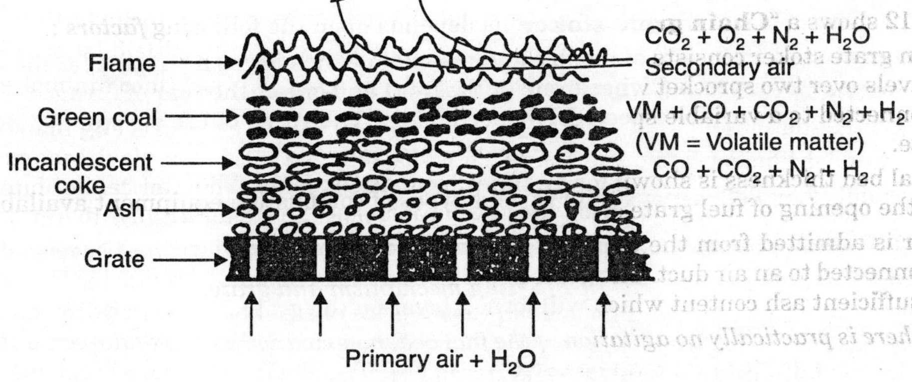

Fig. 2.11. Principle of overfeed stoker.

The fuel bed section receives fresh coal on top surface. The ignition plane lies between green coal and incandescent coke.

The air (with its water vapour content from atmosphere) enters the bottom of the grate under pressure. In flowing through the grate opening the air is heated while it cools the grate. The warm air then passes through a layer of hot ashes and picks up the heat energy.

The region immediately above the ashes contains a mixture of incandescent coke and ash, coke content increasing in upward direction. As the air comes in contact with incandescent coke, the oxygen reacts with carbon to form carbondioxide. Water vapour entering with the air reacts with coke to form CO_2, CO and free H_2. Upon further travel through the incandescent region some of the CO_2 reacts with coke to form CO. Hence no free O_2 will be present in the gases leaving the incandescent region.

Fresh fuel undergoing distillation of its volatile matter forms the top-most layer of the fuel bed. Heat for distillation and eventually ignition comes from the following *four sources* :

(*i*) By conduction from the incandescent coke below.

(*ii*) From high temperature gases diffusing through the surface of the bed.

(*iii*) By radiation from flames and hot gases in the furnace.

(*iv*) From the hot furnace walls.

The ignition zone lies directly below the raw fuel undergoing distillation.

To burn the combustible gases, additional secondary air must be fed into the furnace to supply the needed oxygen. The secondary air must be injected at considerable speed to create turbulence and to penetrate to all parts of the area above the fuel bed. The combustible gases then completely burn in the furnace.

Fuel, coke and ash in the fuel bed move in direction opposite to that of air and gases. Raw fuel continually drops on the surface of the bed. The rising air feed cools the ash until it finally rests in a plane immediately adjacent to the grate.

Types of overfeed stokers

The "overfeed stokers" are used for large capacity boiler installation where the coal is burnt with pulverisation.

These stokers are mainly of following two types :

(*i*) Travelling grate stoker

(*a*) Chain grate type (*b*) Bar grate type

(*ii*) Spreader stoker.

(i) Travelling grate stoker :

These stokers may be *chain grate type or bar grate type*. These two differ only in the details of grate construction.

Fig. 2.12 shows a "**Chain grate stoker**".

A chain grate stoker consists of flexible endless chain which forms a support for the fuel bed. The chain travels over two sprocket wheels one at the front and one at the rear of furnace. The front sprocket is connected to a variable speed drive mechanism. The speed of the stroker is 15 cm to 50 cm per minute.

The coal bed thickness is shown for all times by an index plate. This can be regulated either by adjusting the opening of fuel grate or by the speed control of the stoker driving motor.

The air is admitted from the underside of the grate which is divided into several compartments each connected to an air duct. The grate should be saved from being overheated. For this, coal should have sufficient ash content which will form a layer on the grate.

Since there is practically no agitation of the fuel bed, non-coking coals are best suited for chain grate stokers.

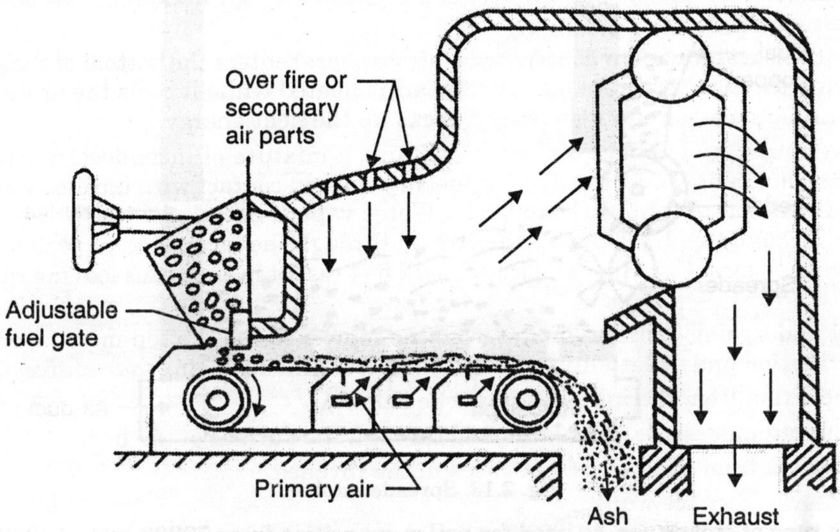

Fig. 2.12. Chain grate stoker.

The rate of burning with this stoker is 200 to 300 kg per m^2 per hour when forced draught is used.

Advantages of chain grate stoker :

1. Simple in construction.
2. Initial cost low.
3. Maintenance charges low.
4. Self-cleaning stoker.
5. Gives high release rates per unit volume of the furnace.
6. Heat release rates can be controlled just by controlling the speed of chain.

Disadvantages :

1. Preheated air temperatures are limited to 180°C maximum.
2. The clinker troubles are very common.
3. There is always some loss of coal in the form of fine particles through riddlings.
4. Ignition arches are required (to suit specific furnace conditions).
5. This cannot be used for high capacity boilers (200 tonnes/hr or more).

(ii) Spreader stoker. Refer to Fig. 2.13.

- In this type of stoker the coal is not fed into furnace by means of grate. The function of the grate is only to support a bed of ash and move it out of the furnace.
- From the coal hopper, coal is fed into the path of a rotor by means of a conveyer, and is thrown into the furnace by the rotor and is *burnt in suspension*. The *air for combustion* is supplied through the holes in the grate.
- The *secondary air* (or overfire air) to *create turbulence and supply oxygen for thorough combustion of coal* is supplied through *nozzles* located directly above the ignition arch.
- Unburnt coal and ash are deposited on the grate which can be moved periodically to remove ash out of the furnace.
- Spreader stokers *can burn any type of coal.*

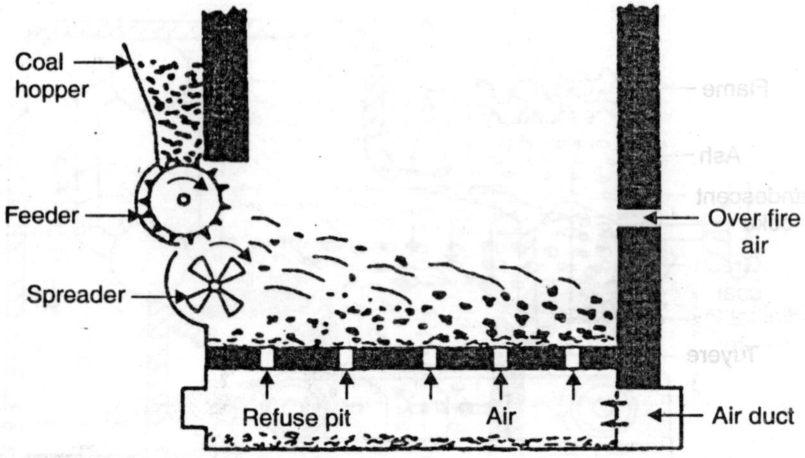

Fig. 2.13. Spreader stoker.

- This type of stoker can be used for boiler capacities from 70000 kg to 140000 kg of steam per hour. The heat release rate of 10×10^6 k cal/m^2-hr is possible with *stationary grate* and of 20×10^6 k cal/m^2-hr is possible with *travelling grate.*

Advantages :

1. A wide variety of coal can be burnt.
2. This stoker is simple to operate, easy to light up and bring into commission.
3. The use of high temperature preheated air is possible.
4. Operation cost is considerably low.
5. The clinking difficulties are reduced even with coals which have high clinkering tendencies.
6. Volatile matter is easily burnt.
7. Fire arches etc. are generally not required with this type of stokers.
8. As the depth of coal bed on the grate is usually limited to 10 to 15 cm only, fluctuating loads can be easily met with.

Disadvantages :

1. It is difficult to operate spreader with varying sizes of coal with varying moisture content.
2. Fly-ash is much more.
3. No remedy for clinker troubles.
4. There is a possibility of some fuel loss in the cinders up the stack because of the thin fuel bed and suspension burning.

2. Underfeed feeders :

Principle of operation. Refer to Fig. 2.14 (*a*).

- *The underfeed principle is suitable for burning the semi-bituminous and bituminous coals.*
- Air entering through the holes in the grate comes in contact with the raw coal (green coal). Then it passes through the incandescent coke where reactions similar to overfeed system take place. The gases produced then pass through a layer of ash. The secondary air is supplied to burn the combustible gases.

The underfeed stokers fall into two main groups, the *single retort* and *multi retort stokers.*

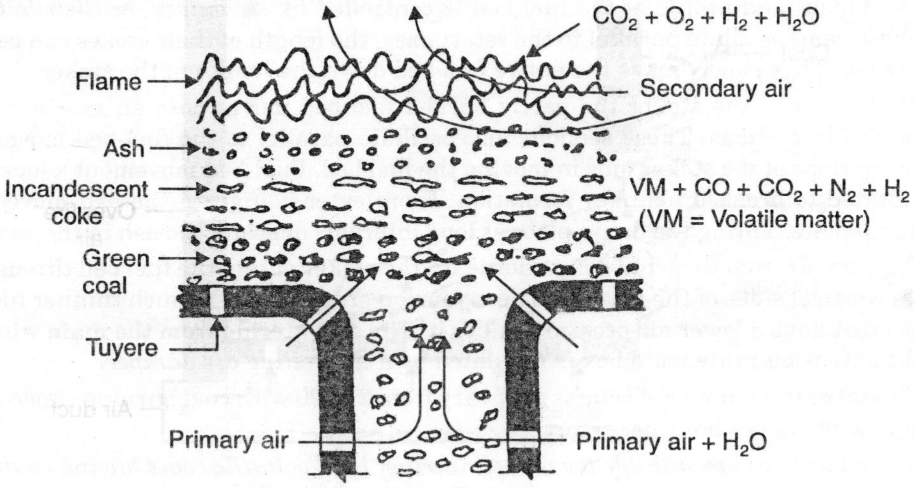

Fig. 2.14. (*a*) Principle of underfeed feeders.

Multi-retort underfeed stokers :
Refer to Fig. 2.14 (*b*).

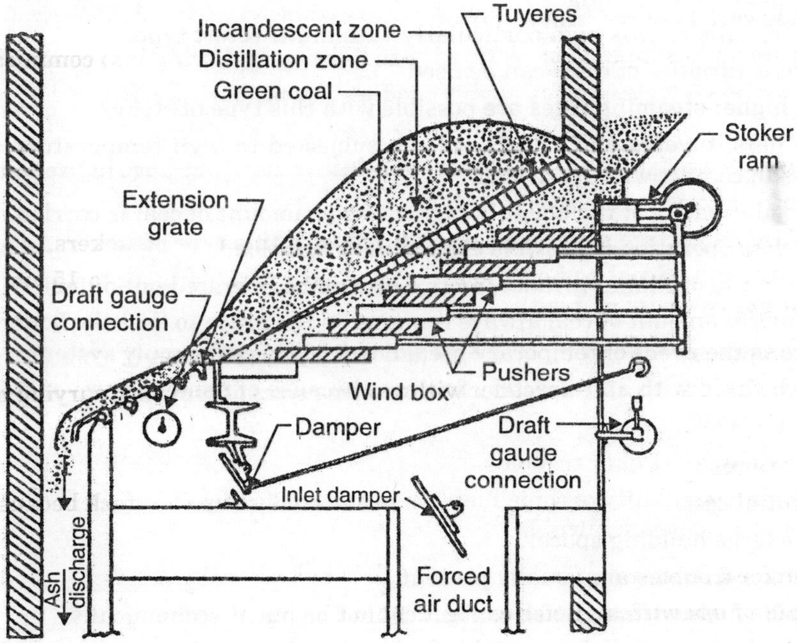

Fig. 2.14. (*b*) Multi-retort underfeed stokers.

- The stoker consists of a series of sloping parallel troughs formed by tuyere stacks. These troughs are called *retorts*. Under the coal hopper at the head end of the retorts, *feeding rams* reciprocate back and forth. With the ram in the outer position coal from the hopper falls into space vacated by the ram. On the inward stroke the ram forces the coal into the retort.

- The height and profile of the fuel bed is controlled by *secondary*, or *distributing rams*. These rams oscillate parallel to the retort axes, the length of their stokes can be varied as needed. They slowly move the entire fuel bed down the length of the stoker.
- At the rear of the stoker the partly burned fuel bed moves onto an *extension grate* arranged in sections. These sections also oscillate parallel to the fuel-bed movement. The sharp slope of the stoker aids in moving the fuel bed. Fuel-bed movement keeps it slightly agitated to break up clinker formation. From extension grate the ash moves onto ash dump plate. Tilting the dump plate at long intervals deposits the ash in the ashpit below.
- *Primary air* from the wind box underneath the stoker enters the fuel bed through holes in the vertical sides of the tuyeres. The extension grate carries a much thinner fuel bed and so must have a lower air pressure under it. The air entering from the main wind box into the extension-grate wind box is regulated by a *controlling air damper*.

In this stoker the number of retorts may vary from 2 to 20 with coal burning capacity ranging from 300 kg to 2000 kg per hour per retort.

Underfeed stokers are *suitable for non-clinkering, high voltatile coals having caking properties and low ash contents*.

Advantages :

1. High thermal efficiency (as compared to chain grate stokers).
2. Combustion rate is considerably higher.
3. The grate is self cleaning.
4. Part load efficiency is high particularly with multi-retort type.
5. Different varieties of coals can be used.
6. Much higher steaming rates are possible with this type of stoker.
7. Grate bars, tuyeres and retorts are not subjected to high temperature as they remain always in contact with fresh coal.
8. Overload capacity of the boiler is high as large amount of coal is carried on the grate.
9. Smokeless operation is possible even at very light load.
10. With the use of clinker grinder, more heat can be liberated out of fuel.
11. Substantial amount of coal always remains on the grate so that the boiler may remain in service in the event of temporary breakdown of the coal supply system.
12. It can be used with all refractory furnaces because of non-exposure of stoker mechanism to the furnace.

Disadvantages :

1. High initial cost.
2. Require large building space.
3. The clinker troubles are usually present.
4. Low grade fuels with high ash content cannot be burnt economically.

2.9.3.2. Pulverised fuel firing

In pulverised fuel firing system the *coal is reduced to a fine powder with the help of grinding mill and then projected into the combustion chamber with the help of hot air current*. The amount of air required (known as *secondary air*) to complete the combustion is supplied separately to the combustion chamber. The resulting turbulence in the combustion chamber helps for uniform mixing of fuel and air and thorough combustion. *The amount of air which is used to carry the coal and to dry it before entering into the combustion chamber is known as* **Primary air** *and the amount of air which is supplied separately for completing the combustion is known as* **Secondary air.**

The *efficiency* of the pulverised fuel firing system mostly *depends upon the size of the powder.* The fineness of the coal should be such as 70% of it would pass through a 200 mesh sieve and 90% through 50 mesh sieve.

Fig. 2.15 shows elements of pulverised coal system.

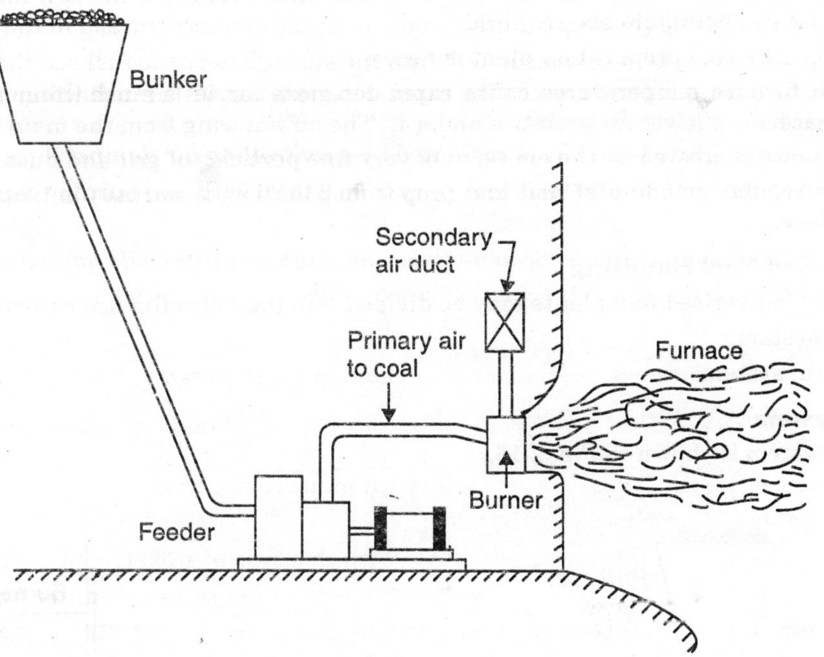

Fig. 2.15. Elements of pulverised coal system.

Advantages :

1. Any grade of coal can be used since coal is powdered before use.
2. The rate of feed of the fuel can be regulated properly resulting in fuel economy.
3. Since there is almost complete combustion of the fuel there is increased rate of evaporation and higher boiler efficiency.
4. Greater capacity to meet peak loads.
5. The system is practically free from sagging and clinkering troubles.
6. No standby losses due to banked fires.
7. Practically no ash handling problems.
8. No moving part in the furnace is subjected to high temperatures.
9. This system works successfully with or in combination with gas and oil.
10. Much smaller quantity of air is required as compared to that of stoker firing.
11. Practically free from clinker troubles.
12. The external heating surfaces are free from corrosion.
13. It is possible to use highly preheated secondary air (350°C) which helps for rapid flame propagation.
14. The furnace volume required is considerably less.

Disadvantages :

1. High capital cost.
2. Lot of fly-ash in the exhaust, which makes the removing of fine dust uneconomical.
3. The possibility of explosion is more as coal burns like gas.
4. The maintenance of furnace brickwork is costly.
5. Special equipment is needed to start this system.
6. The skilled operators are required.
7. A separate coal preparation plant is necessary.
8. High furnace temperatures cause rapid deterioration of the refractory surfaces of the furnace.
9. Nuisance is created by the emission of very fine particles of grit and dust.
10. Fine regular grinding of fuel and proper distribution to burners is usually difficult to achieve.

Pulverised Fuel Handling

Basically, pulverised fuel plants may be divided into the following *two systems* :

1. Unit system.
2. Central system.

Unit system :

A unit system is shown in Fig. 2.16.

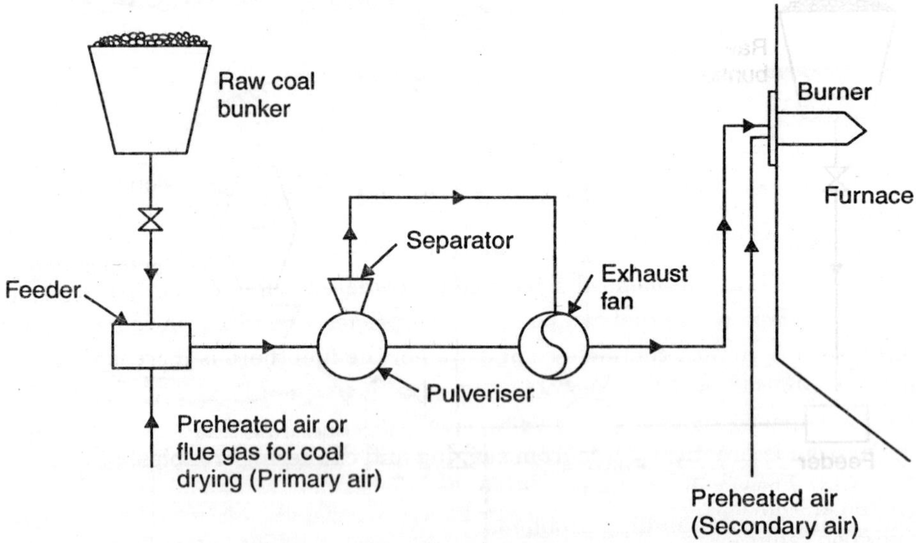

Fig. 2.16. Unit system.

Most pulverised coal plants are now being installed with unit pulveriser.

The *unit system* is so called from the fact that each burner or burner group and the pulveriser constitute a unit. Crushed coal is fed to the pulverising mill at a variable rate governed by the combustion requirements of the boiler and furnace. Primary air is admitted to the mill and becomes the transport air which carries the coal through the short delivery pipe to the burner. This air may be preheated if mill drying is desirable.

Advantages :

1. The layout is simple and permits easy operation.
2. It is cheaper than central system.
3. Less spaces are required.
4. It allows direct control of combustion from the pulveriser.
5. Maintenance charges are less.
6. There is no complex transportation system.
7. In a replacement of stokers, the old conveyor and bunker equipment may be used.
8. Coal which would require drying in order to function satisfactorily in the central system may usually be employed without drying in the unit system.

Disadvantages :

1. Firing aisle is obstructed with pulverising equipment, unless the latter is relegated to a basement.

2. The mills operate at variable load, a condition not especially conducive to best results.

3. With load factors in common practice, total mill capacity must be higher than for the central system.

4. Flexibility is less than central system.

Central system :

This system is illustrated in Fig. 2.17.

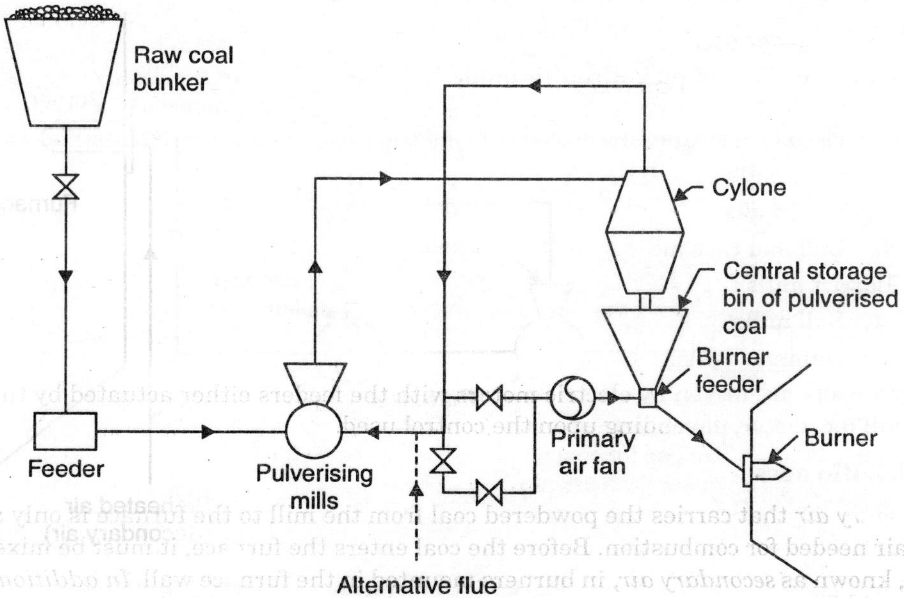

Fig. 2.17. Central system.

A central pulverising system employs a limited number of large capacity pulverisers at a central point to prepare coal for all the burners. Driers, if required, are conveniently installed at this point. From the pulverisers the coal is transported to a central storage bin where it is deposited and its transporting air vented from the bin through a *"cyclone"*. This bin may contain from 12 to 24 hours supply of pulverised coal. From the bin the coal is metered to the burners by motor-driven feeders of varied design. Primary air, added at the feeders, floats the coal to the burners.

Advantages :

1. Offers good control of coal fineness.
2. The pulverising mill may work at constant load because of the storage capacity between it and the burners.
3. The boiler aisels are unobstructed.
4. More latitude in the arrangement and number of burners is allowed to the designers.
5. The large storage is protection against interruption of fuel supply to the burners.
6. Less labour is required.
7. Power consumption per tonne of coal handled is low.
8. Burners can be operated independent of the operation of coal preparation plant.
9. Fans handle only air, as such, there is no problem of excessive wear as in case of unit system, where air and coal both are handled by the fan.

Disadvantages :

1. Driers are usually necessary.
2. Fire hazard of quantities of stored pulverised coal.
3. Central preparation may require a separate building.
4. Additional cost and complexity of coal transportation system.
5. Power consumption of auxiliaries is high.

Pulveriser. Coal is pulverised in order to increase its surface exposure, thus promoting rapid combustion without using large quantities of excess air. A pulveriser is the most important part of a pulverised coal system. Pulverisers (sometimes called *mills*) are classified as follows :

1. Attrition mills :
   (*i*) Bowl mills
   (*ii*) Ball and race mills.
2. Impact mills :
   (*i*) Ball mills
   (*ii*) Hammer mills.

Pulverisers are driven by electric motors with the feeders either actuated by the main drive or by a small d.c. motor, depending upon the control used.

2.9.4. Burners

Primary air that carries the powdered coal from the mill to the furnace is only about 20% of the total air needed for combustion. Before the coal enters the furnace, it must be mixed with additional air, known as *secondary air,* in burners mounted in the furnace wall. *In addition to the prime function of mixing, burners must also maintain stable ignition of fuel-air mix* and *control the flame shape and travel in the furnace.* Ignition depends on the rate of flame propagation. To prevent flash back into the burner, the coal-air mixture must move away from the burner at a rate equal to flame-front travel. *Too much secondary air can cool the mixture and prevent its heating to ignition temperature.*

The *requirements of a burner* can be summarised as follows :

(*i*) The coal and air should be so handled that there is stability of ignition.

(*ii*) The combustion is complete.

(*iii*) In the flame the heat is uniformly developed avoiding any superheat spots.

(*iv*) Adequate protection against overheating, internal fires and excessive abrasive wear.

2.9.4.1. Pulverised fuel burners

Pulverised fuel burners may be *classified* as follows :

1. Long flame burners.
2. Turbulent burners.
3. Tangential burners.
4. Cyclone burners.

2.9.4.2. Oil burners

Principle of oil firing. *The functions of an oil burner are to mix the fuel and air in proper proportion and to prepare the fuel for combustion.* Fig. 2.18 shows the principle of oil firing.

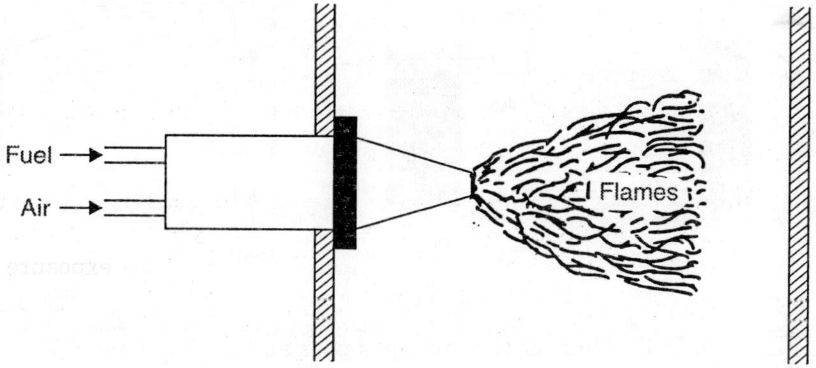

Fig. 2.18. Principle of oil firing.

Classification of oil burners. The oil burners may be *classified* as :

1. *Vapourising oil burners :*
   (*a*) Atmospheric pressure atomising burner
   (*b*) Rotating cup burner
   (*c*) Recirculation burner
   (*d*) Wick type burner.
2. *Atomising fuel burners :*
   (*a*) Mechanical or oil pressure atomising burner
   (*b*) Steam or high pressure air atomising burner
   (*c*) Low pressure air atomising burner.

2.9.4.3. Gas burners

Gas burning claims the following *advantages* :

(*i*) It is much simpler as the fuel is ready for combustion and requires no preparation.

(*ii*) Furnace temperature can be easily controlled.

(*iii*) A long slow burning flame with uniform and gradual heat liberation can be produced.

(*iv*) Cleanliness.

(*v*) High chimney is not required.

(*vi*) No ash removal is required.

For generation of steam, natural gas is invariably used in the following cases :

(*i*) Gas producing areas.

(*ii*) Areas served by gas transmission lines.

(*iii*) Where coal is costlier.

Typical gas burners used are shown in Fig. 2.19 to 2.20.

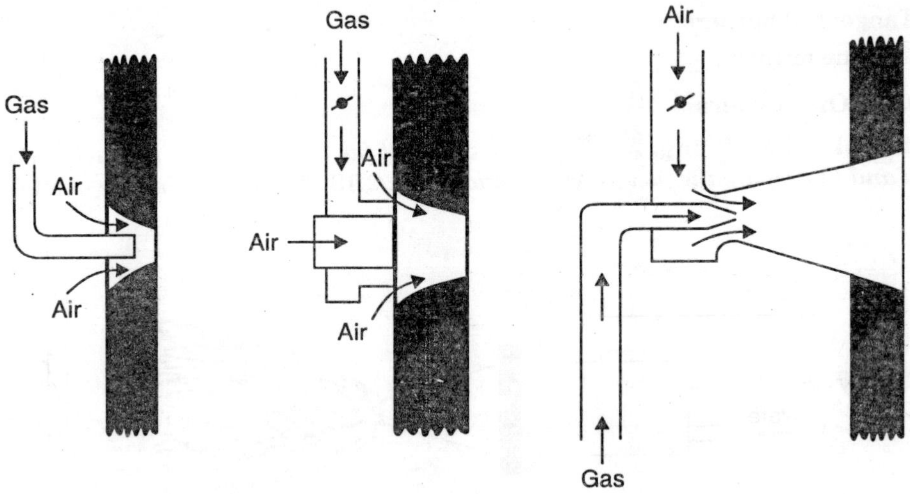

Fig. 2.19 Fig. 2.20 Fig. 2.21

Refer to Fig. 2.19. In this burner the mixing is poor and a fairly long flame results.

Refer to Fig. 2.20. This is a *ring type burner* in which a short flame is obtained.

Refer to Fig. 2.21. This arrangement is used when both gas and air are under pressure.

In order to prevent the flame from turning back the velocity of the gas should be more than the "*rate of flame propagation*".

2.10. FLUIDISED BED COMBUSTION (FBC)

A ***fluidised bed*** *may be defined as the bed of solid particles behaving as a fluid.* The *principle of FBC-system* is given below :

When a gas is passed through a packed bed of finely divided solid particles, it experiences a pressure drop across the bed. At low gas velocities, this pressure drop is small and does not disturb the particles. But if the gas velocity is increased further, a stage is reached, *when particles are suspended in the gas stream and the packed bed becomes a 'fluidised bed'*. With further increase in gas velocity, the bed becomes turbulent and rapid mixing of particles occurs. In general, the behavior of this mixture of solid particles and gas is like a fluid. Burning of a fuel in such a state is known as a *fluidised bed combustion.*

Fig. 2.22 shows the arrangement of the FBC system.

On the distributor plate are fed the fuel and inert material *dolomite* and from its bottom air is supplied. The high velocity of air keeps the solid feed material in suspending condition during burning. The generated heat is rapidly transferred to the water passing through the tubes immersed in the bed and generated steam is taken out. During the burning sulphur dioxide formed is absorbed by the dolomite and prevents its escape with the exhaust gases. The molten slag is tapped from the top surface of the bed.

The primary object of using the *inert material* is to control the bed temperature, it accounts for 90% of the bed volume. It is very necessary that the selection of an inert material should be done judiciously as it remains with the fuel in continuous motion and at high temperature to the tune of 800°C. Moreover, the inert material should not disintegrate coal, the parent material of the bed.

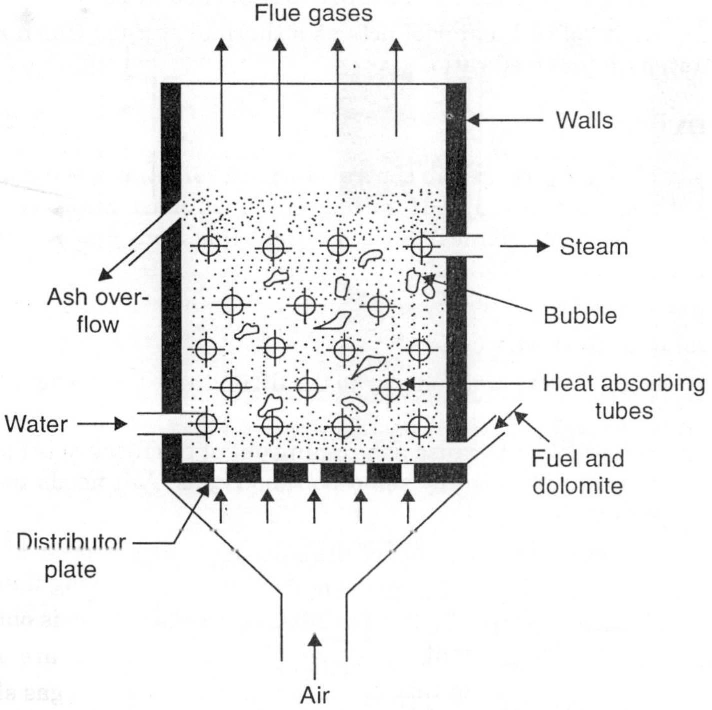

Fig. 2.22. Basic FBC system.

The cost economic shows that a saving of about 10% in operating cost and 15% capital cost could be achieved for a unit rating of 120 MW and it may be still higher for bigger units.

Advantages :

1. As a result of better heat transfer, the unit size and hence the *capital costs are reduced.*
2. It can respond rapidly to changes in load demand (since thermal equilibrium between air and coal particles in the bed is quickly established).
3. Low combustion temperatures (800 to 950°C) inhibits the formation of nitrogen oxides like nitric oxide and nitrogen dioxide.
4. Since combustion temperatures are low the fouling and corrosion of tubes is reduced considerably.
5. As it is not necessary to grind the coal very fine as is done in pulverised fuel firing, therefore, the cost of coal crushing is reduced.
6. Pollution is controlled and combustion of high-sulphur coal is possible.

7. FBC system can use solid, liquid or gaseous fuel or mix as well as domestic and industrial waste. Any variety of coal can be used successfully.
8. Combustion temperature can be controlled accurately.
9. The system can be readily designed for operation at raised combustion pressure, owing to the simplicity of arrangement, small size of the plant and reduced likelihood of corrosion or erosion of gas turbine blades.
10. The combustion in conventional system becomes unstable when the ash exceeds 48% but even 70% ash containing coal can be efficiently burned in FBC.
11. The large quantity of bed material acts as a thermal storage which reduces the effect of any fluctuation in fuel feed ratio.

2.11. ASH HANDLING

A huge quantity of ash is produced in central stations, sometimes being as much as 10 to 20% of the total quantity of coal burnt in a day. Hundreds of tonnes of ash may have to be handled every day in large power stations and mechanical devices become indispensable. A station using low grade fuel has to deal with large quantities of ash.

Handling of ash includes :

(*i*) Its removal from the furnace.

(*ii*) Loading on the conveyers and delivery to the fill or dump from where it can be disposed off by sale or otherwise.

Handling of ash is a problem because ash coming out of the furnace is too hot, it is dusty and irritating to handle and is accompanied by some poisonous gas. Ash needs to be *quenched* before handling due to following *reasons* :

(*i*) Quenching reduces corrosion action of the ash.

(*ii*) It reduces the dust accompanying the ash.

(*iii*) It reduces temperature of the ash.

(*iv*) Ash forms clinkers by fusing in large lumps and by quenching clinkers will disintegrate.

2.11.1. Ash Handling Equipment

A *good ash handling plant* should have the following *characteristics* :

1. It should have enough capacity to cope with the volume of ash that may be produced in a station.
2. It should be able to handle large clinkers, boiler refuse, soot etc., with little personal attention of the workmen.
3. It should be able to handle hot and wet ash effectively and with good speed.
4. It should be possible to minimise the corrosive or abrasive action of ashes and dust nuisance should not exist.
5. The plant should not cost much.
6. The operation charges should be minimum possible.
7. The operation of the plant should be noiseless as much as possible.
8. The plant should be able to operate effectively under all variable load conditions.
9. In case of addition of units, it should need minimum changes in original layout of plant.
10. The plant should have high rate of handling.

The *commonly used equipment for ash handling* in large and medium size plants may comprise of :

(*i*) Bucket elevator

(*ii*) Bucket conveyor

(*iii*) Belt conveyor

(*iv*) Pneumatic conveyor

(*v*) Hydraulic sluicing equipment

(*vi*) Trollies or rail cars etc.

Fig. 2.23 shows the outline of ash disposal equipment.

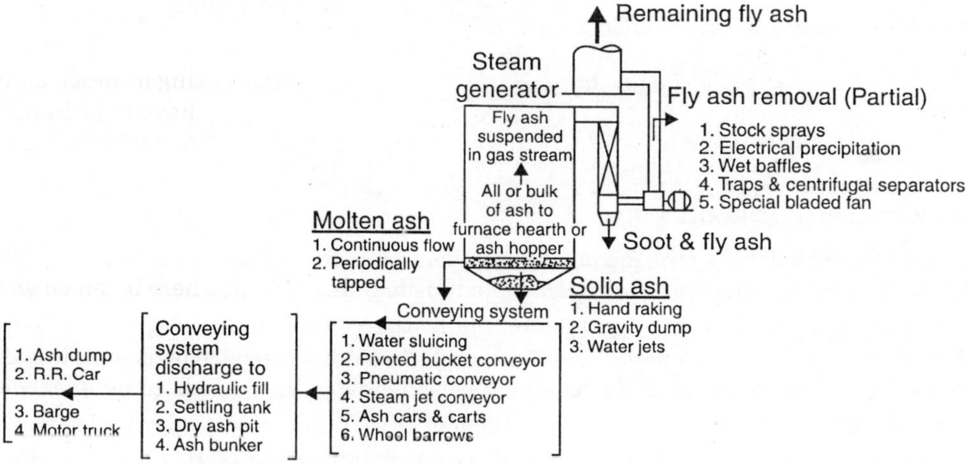

Fig. 2.23. Outline of ash disposal equipment.

2.11.2. Ash Handling Systems

The modern ash-handling systems are mainly *classified* into *four* groups :

1. Mechanical handling system
2. Hydraulic system
3. Pneumatic system
4. Steam jet system.

2.12. DUST COLLECTION

2.12.1. Introduction

The products of combustion of coal-fed fires contain particles of solid matter floating in suspension. This may be smoke or dust. If *smoke*, the indication is that combustion conditions are faulty, and the proper remedy is in the design and management of the furnace. If *dust*, the particles are mainly fine ash particles called "*Fly-ash*" intermixed with some quantity of carbon-ash material called "*cinder*". *Pulverised coal and spreader stoker firing units are the principle types causing difficulty from this source.* Other stokers may produce minor quantities of dust but generally not enough to demand special gas cleaning equipment. The two mentioned are troublesome because coal is burned in suspension—in a turbulent furnace atmosphere and every opportunity is offered for the gas to pick up the smaller particles and sweep them along with it.

The size of the dust particles is measured in *microns*. The micron is one millionth of a metre. As an indication of the scale of this measure, the diameter of a human hair is approximately 80 microns. Typical classification of particles by name is given in Fig. 2.24, but the limits shown are, for the most part, arbitrary. A critical characteristic of dust is its "*Settling Velocity*" in still air. *This is proportional to the product of the square of micron size and mass density.*

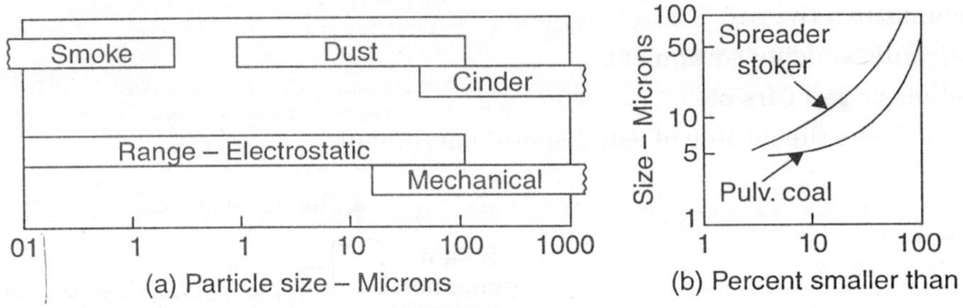

Fig. 2.24. Typical particle sizes : (*a*) Flue gas particles and ranges of collecting equipment. (*b*) Typical distribution of particle size in products of combustion.

2.12.2. Removal of Smoke

Smoke is produced due to the incomplete combustion of fuels. Smoke particles are less than 1 micron in size. The smoke disposal to the atmosphere is not desirable due to the following *reasons* :

(*i*) Smoke is produced due to incomplete combustion of coal. This will create a big economic loss due loss of heating value of coal.

(*ii*) A smoky atmosphere is unhealthy.

(*iii*) Smoke corrodes the metals, darkens the paints and gives lower standards of cleanliness.

In order to check the nuisance of smoke the coal should be completely burnt in the furnace. *The presence of dense smoke indicates poor furnace conditions and a loss in efficiency and capacity of a boiler plant.*

2.12.3. Removal of Dust and Dust Collectors

The removal of dust and cinders from flue gas can usually be effected to the required degree by commercial dust collectors.

The dust collectors may be *classified* as follows :

1. Mechanical dust collectors :

(*i*) Wet type (Scrubbers)

(*a*) Spray type

(*b*) Packed type

(*c*) Impingement type

(*ii*) Dry type

(*a*) Gravitational separators

(*b*) Cyclone separators

2. Electrical dust collectors :

(*i*) Rod type

(*ii*) Plate type.

2.12.4. Uses of Ash and Dust

The *uses of ash and dust* are listed below :

1. Ash is widely used in the production of cement.

2. Ash is used in the production of concrete. 20 percent fly-ash and 30 percent bottom ash are presently used constructively in U.S.A.

3. Because of their better alkali values, they are used for treating acidic soils. It has been found that if ash is used in limited quantity in soil, it increases the yield of corn, turnip etc.

4. From the ash, the metals such as Al, Fe, Si and titanium can be recovered.

2.12.5. General Layout of Ash Handling and Dust Collection System

Fig. 2.25 shows the general layout of ash handling and dust collection system which is self explanatory.

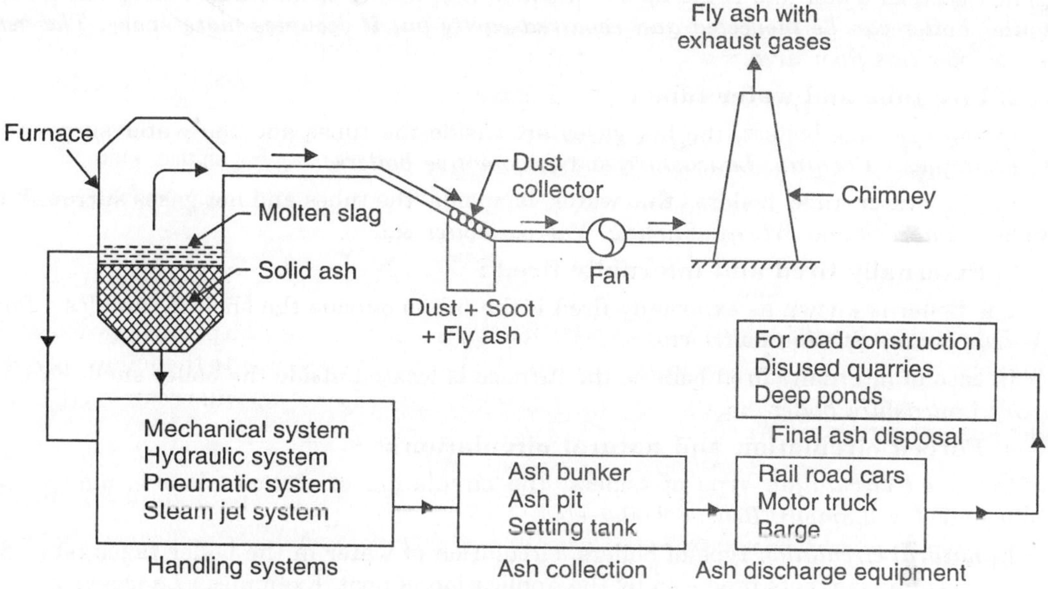

Fig. 2.25. General layout of ash handling and dust collection system.

2.13. BOILERS

2.13.1. Introduction

In simple a ***boiler*** *may be defined as a closed vessel in which steam is produced from water by combustion of fuel.*

According to American Society of Mechanical Engineers (A.S.M.E.) a *'steam generating unit'* is defined as :

"A combination of apparatus for producing, furnishing or recovering heat together with the apparatus for transferring the heat so made available to the fluid being heated and vapourised".

The steam generated is employed for the following *purposes* :

(*i*) For generating power in steam engines or steam turbines.

(*ii*) In the textile industries for sizing and bleaching etc., and many other industries like sugar mills ; chemical industries.

(*iii*) For heating the buildings in cold weather and for producing hot water for hot water supply.

The *primary requirements* of steam generators or boilers are :

(*i*) The water must be contained safely.

(*ii*) The steam must be safely delivered in desired condition (as regards its pressure, temperature, quality and required rate).

2.13.2. Classification of Boilers

The boilers may be *classified* as follows :

1. Horizontal, vertical or inclined :

If the axis of the boiler is horizontal, the boiler is called as horizontal, if the axis is vertical, it is called vertical boiler and if the axis is inclined it is known as inclined boiler. *The parts of a horizontal boiler can be inspected and repaired easily but it occupies more space. The vertical boiler occupies less floor area.*

2. Fire tube and water tube :

In the fire tube boilers, the hot gases are inside the tubes and the water surrounds the tubes. Examples : *Cochran, Lancashire* and *Locomotive boilers*.

In the water tube boilers, the water is inside the tubes and hot gases surround them. Examples : *Babcock* and *Wilcox, Stirling, Yarrow boiler etc.*

3. Externally fired and internally fired :

The boiler is known as externally fired if the fire is outside the shell. Examples : *Babcock and Wilcox boiler, Stirling boiler etc.*

In case of internally fired boilers, the furnace is located inside the boiler shell. Examples : *Cochran, Lancashire boiler etc.*

4. Forced circulation and natural circulation :

In *forced circulation* type of boilers, the circulation of water is done *by a forced pump. Examples* : *Velox, Lamont, Benson boiler etc.*

In *natural circulation* type of boilers, circulation of water in the boiler takes place due to *natural convention* currents produced by the application of heat. Examples : *Lancashire, Babcock* and *Wilcox boiler* etc.

5. High pressure and low pressure boilers :

The boilers which produce steam at *pressures of 80 bar and above* are called *high* pressure boilers. *Examples* : *Babcock* and *Wilcox, Velox, Lamont, Benson boilers.*

The boilers which produce steam at pressure *below 80 bar* are called low pressure boilers. *Examples* : *Cochran, Cornish, Lancashire and Locomotive boilers.*

6. Stationary and portable :

Primarily, the boilers are classified as either stationary (*land*) or mobile (*marine* and *locomotive*).

- Stationary boilers are used for power plant-steam, for central station utility power plants, for plant process steam etc.
- Mobile boilers or portable boilers include locomotive type, and other small units for temporary use at sites (just as in small coalfield pits).

7. Single tube and multi-tube boilers :

The fire tube boilers are classified as single-tube and multi-tube boilers, depending upon whether the fire tube is one or more than one. The examples of the former type are cornish, simple vertical boiler and rest of the boilers are multi-tube boilers.

2.13.3. Comparison between 'Fire-tube and Water tube' Boilers

S. No.	Aspects	Fire-tube boilers	Water-tube boilers
1.	*Position of water and hot gases*	Hot gases inside the tubes and water outside the tubes.	Water inside the tubes and hot gases outside the tubes.
2.	*Mode of firing*	Generally internally fired.	Externally fired.
3.	*Operating pressure*	Operating pressure limited to 16 bar.	Can work under as high pressure as 100 bar.
4.	*Rate of steam production*	Lower	Higher.
5.	*Suitability*	Not suitable for large power plants.	Suitable for large power plants.
6.	*Risk on bursting*	Involves lesser risk on explosion due to lower pressure.	Involves more risk on bursting due to high pressure.
7.	*Floor area*	For a given power it occupies more floor area.	For a given power it occupies less floor-area.
8.	*Construction*	Difficult	Simple
9.	*Transportation*	Difficult	Simple
10.	*Shell diameter*	Large for same power	Small for same power
11.	*Chances of explosion*	Less	More
12.	Treatment of water	Not so necessary	More necessary
13.	*Accessibility of various parts*	Various parts not so easily accessible for cleaning, repair and inspection.	Various parts are more accessible.
14.	*Requirement of skill*	Require less skill for efficient and economic working.	Require more skill and careful attention.

2.13.4. High Pressure Boilers

In applications where steam is needed at pressure, 30 bar, and individual boilers are required to raise less than about 30000 kg of steam per hour, *shell boilers are considerably cheaper than the water-tube boilers.* Above these limits, shell boilers (generally factory built) are difficult to transport if not impossible. There are no such limits to water-tube boilers. These can be site erected from easily transportable parts, and moreover the pressure parts are of smaller diameter and therefore can be thinner. The geometry can be varied to suit a wide range of situations and furnace is not limited to cylindrical form. Therefore, *water tube boilers are generally preferred for high pressure and high output whereas shell boilers for low pressure and low output.*

The modern high pressure boilers employed for power generation are for steam capacities 30 to 650 tonnes/h and above with a pressure up to 160 bar and maximum steam temperature of about 540°C.

2.13.4.1. Unique features of high pressure boilers

Following are the *unique features* of high pressure boilers :

1. Method of water circulation
2. Type of tubing
3. Improved method of heating.

1. *Method of water circulation.* The circulation of water through the boiler may be *natural circulation* due to density difference or *forced circulation.* In all modern high pressure boiler plants, the water circulation is maintained with the help of pump which forces the water through the boiler plant. The use of natural circulation is limited to sub-critical boilers due to its limitations.

2. *Type of tubing.* In most of the high pressure boilers, the water circulated through the tubes and their external surfaces are exposed to the flue gases. In water-tube boilers, if the flow takes place through one continuous tube, the large pressure drop takes place due to friction. This is *considerably reduced by arranging the flow to pass through parallel system of tubing.* In most of the cases, several sets of the tubings are used. This type of arrangement helps to *reduce the pressure loss,* and better control over the quality of the steam.

3. *Improved method of heating.* The following improved methods of heating may be used to increase the heat transfer :

(*i*) The saving of heat by *evaporation of water* above critical pressure of the steam.

(*ii*) The heating of water can be made by *mixing the superheated steam.* The mixing phenomenon *gives highest heat transfer co-efficient.*

(*iii*) The overall heat transfer coefficient can be increased by *increasing the water velocity inside the tube and increasing the gas velocity above sonic velocity.*

2.13.4.2. Advantages of high pressure boilers

The following are the *advantages* of high pressure boilers :

1. In high pressure boilers pumps are used to maintain forced circulation of water through the tubes of the boiler. This *ensures positive circulation of water and increases evaporative capacity of the boiler and less number of steam drums will be required.*
2. The *heat of combustion is utilised more efficiently* by the use of small diameter tubes in large number and in multiple circuits.
3. *Pressurised combustion* is used which *increases rate of firing of fuel* thus increasing the rate of heat release.
4. Due to compactness less floor space is required.
5. The tendency of scale formation is eliminated due to high velocity of water through the tubes.
6. All the parts are uniformly heated, therefore, the danger of overheating is reduced and thermal stress problem is simplified.
7. The differential expansion is reduced due to uniform temperature and this reduces the possibility of gas and air leakages.
8. The components can be arranged horizontally as high head required for natural circulation is eliminated using forced circulation. There is a greater flexibility in the components arrangement.
9. The steam can be raised quickly to meet the variable load requirements without the use of complicated control devices.
10. The *efficiency of plant is increased upto 40 to 42 percent* by using high pressure and high temperature steam.
11. A very rapid start from cold is possible if an external supply of power is available. Hence, the boiler can be used for carrying peak loads or standby purposes with hydraulic station.
12. Use of high pressure and high temperature steam is economical.

High pressure boilers are enumerates below :

1. Lamont boiler
2. Loeffler boiler
3. Benson boiler
4. Velox boiler
5. Supercritical boilers
6. Supercharged boiler

2.14. STEAM TURBINES

2.14.1. Introduction

The **steam turbine** *is a prime-mover in which the potential energy of the steam is transformed into kinetic energy, and latter in its turn is transformed into the mechanical energy of rotation of the turbine shaft.* The turbine shaft, directly or with the help of a reduction gearing, is connected with the driven mechanism. Depending on the type of the driven mechanism a steam turbine may be *utilised in most diverse fields of industry, for power generation* and *for transport.* Transformation of the potential energy of steam into the mechanical energy of rotation of the shaft is brought about by different means.

2.14.2. Classification of Steam Turbines

There are several ways in which the steam turbines may be classified. The most important and common division being with respect to the *action of the steam*, as :

(*a*) Impulse.

(*b*) Reaction.

(*c*) Combination of impulse and reaction.

Other classifications are :

1. **According to the number of pressure stages :**

(*i*) *Single-stage turbines* with one or more velocity stages usually of small-power capacities ; these turbines are mostly used for driving centrifugal compressors, blowers and other similar machinery.

(*ii*) *Multi-stage impulse and reaction turbines* ; they are made in a wide range of power capacities varying from small to large.

2. **According to the direction of steam flow :**

(*i*) *Axial turbines* in which steam flows in a direction parallel to the axis of the turbine.

(*ii*) *Radial turbines* in which steam flows in a direction perpendicular to the axis of the turbine ; one or more low-pressure stages in such turbines are made axial.

3. **According to the number of cylinders :**

(*i*) Single-cylinder turbines.

(*ii*) Double-cylinder turbines.

(*iii*) Three-cylinder turbines.

(*iv*) Four-cylinder turbines.

Multi-cylinder turbines which have their rotors mounted on one and the same shaft and coupled to a single generator are known as *single shaft turbines* ; turbines with separate rotor shafts for each cylinder placed parallel to each other are known as ***multiaxial turbines.***

4. **According to the method of governing :**

(i) *Turbines with throttle governing* in which fresh steam enters through one or more (depending on the power developed) simultaneously operated throttle valves.

(ii) *Turbines with nozzle governing* in which fresh steam enters through two or more consecutively opening regulators.

(iii) *Turbines with bypass governing* in which steam turbines besides being fed to the first stage is also directly fed to one, two or even three intermediate stages of the turbine.

5. **According to heat drop process :**

(i) *Condensing turbines with generators* ; in these turbines steam at a pressure less than atmospheric is directed to a condenser ; besides, steam is also extracted from intermediate stages for feed water heating, the number of such extractions usually being from 2-3 to as much 8-9. The latent heat of exhaust steam during the process of condensation is completely lost in these turbines.

(ii) *Condensing turbines with one or two intermediate stage extractions* at specific pressures for industrial and heating purposes.

(iii) *Back pressure turbines*, the exhaust steam from which is utilised for industrial or heating purposes ; to this type of turbines can also be added (in a relative sense) turbines with deteriorated vacuum, the exhaust steam of which may be used for heating and process purposes.

(iv) *Topping turbines* ; these turbines are also of the back pressure type with the difference that the exhaust steam from these turbines is further utilised in medium and low-pressure condensing turbines. These turbines, in general, operate at high initial conditions of steam pressure and temperature, and are mostly used during extension of power station capacities, with a view to obtain better efficiencies.

(v) *Back pressure turbines with steam extraction from intermediate stages at specific pressure* ; turbines of this type are meant for supplying the consumer with steam of various pressures and temperature conditions.

(vi) *Low pressure turbines* in which the exhaust steam from reciprocating steam engines, power hammers, etc., is utilised for power generation purposes.

(vii) *Mixed pressure turbines* with two or three pressure stages, with supply of exhaust steam to its intermediate stages.

6. **According to steam conditions at inlet to turbine :**

(i) *Low pressure turbines*, using steam at a pressure of 1.2 to 2 ata.

(ii) *Medium pressure turbines*, using steam at pressure of up to 40 ata.

(iii) *High pressure turbines*, utilising steam at pressures above 40 ata.

(iv) *Turbines of very high pressures*, utilising steam at pressures of 170 ata and higher and temperatures of 550°C and higher.

(v) *Turbines of supercritical pressures*, using steam at pressures of 225 ata and above.

7. **According to their usage in industry :**

(i) *Stationary turbines with constant speed of rotation* primarily used for driving alternators.

(*ii*) *Stationary steam turbines* with variable speed meant for driving turbo-blowers, air circulators, pumps, etc.

(*iii*) *Non-stationary* turbines with variable speed ; turbines of this type are usually employed in steamers, ships and railway locomotives.

2.14.3. Advantages of Steam Turbine over the Steam Engines

The following are the principal *advantages* of steam turbine over steam engines :

1. The thermal efficiency of a steam turbine is much higher than that of a steam engine.
2. The power generation in a steam turbine is at a uniform rate, therefore necessity to use a flywheel (as in the case of steam engine) is not felt.
3. Much higher speeds and greater range of speed is possible than in case of a steam engine.
4. In large thermal stations where we need higher outputs, the steam turbines prove very suitable as these can be made in big sizes.
5. With the absence of reciprocating parts (as in steam engine) the balancing problem is minimised.
6. No internal lubrication is required as there are no rubbing parts in the steam turbine.
7. In a steam turbine there is no loss due to initial condensation of steam.
8. It can utilise high vacuum very advantageously.
9. Considerable overloads can be carried at the expense of slight reduction in overall efficiency.

2.14.4. Description of Common Types of Turbines

The common types of steam turbines are :

1. Simple Impulse turbine.
2. Reaction turbine.

The main difference between these turbines lies *in the way in which the steam is expanded while it moves through them. In the former type steam expands in the nozzles and its pressure does not alter as it moves over the blades while in the latter type the steam expands continuously as it passes over the blades and thus there is gradual fall in the pressure during expansion.*

1. Simple impulse turbine :

Fig. 2.26 shows a simple impulse turbine diagrammatically. The top portion of the figure exhibits a longitudinal section through the upper half of the turbine, the middle portion shows one set of nozzles which is followed by a ring of moving blades, while lower part of the diagram indicates approximately changes in pressure and velocity during the flow of steam through the turbine. This turbine is called *'simple'* impulse turbine since the expansion of the steam takes place in one set of the nozzles.

As the steam flows through the nozzle its pressure falls from steam chest pressure to condenser pressure (or atmospheric pressure if the turbine is non-condensing). Due to this relatively higher ratio of expansion of steam in the nozzles the steam leaves the nozzle with a very high velocity. From Fig. 2.26, it is evident that the velocity of the steam leaving the moving blades is a large portion of the maximum velocity of the steam when leaving the nozzle. The loss of energy due to this higher exit velocity is commonly the ***"carry over loss"*** or ***"leaving loss"***.

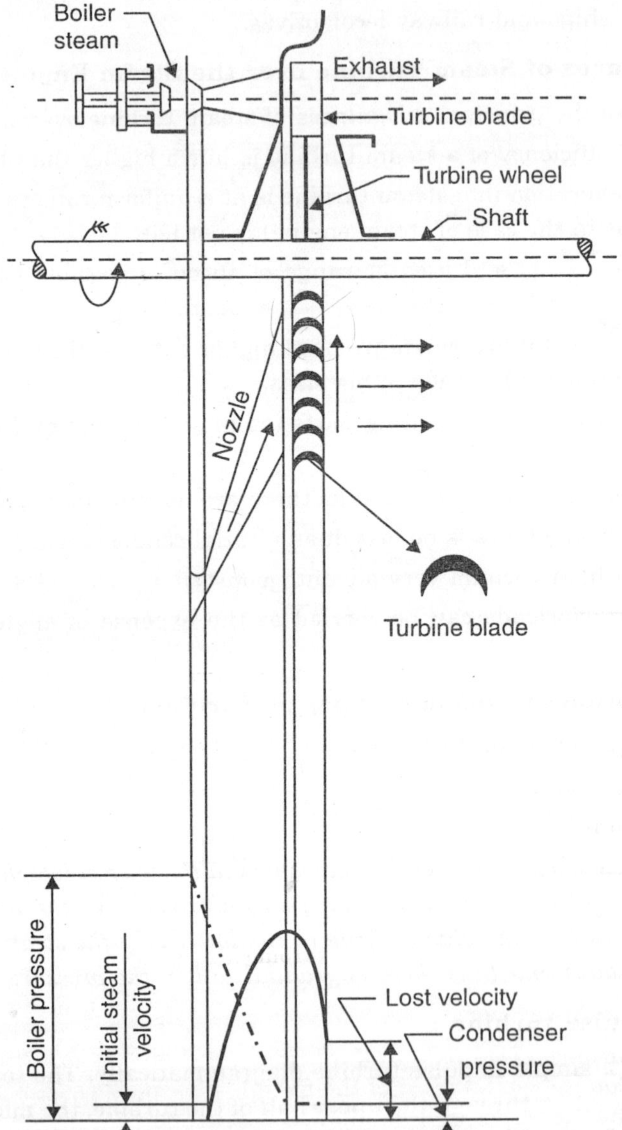

Fig. 2.26. Simple impulse turbine.

The principal example of this turbine is the well known **De laval turbine** and in this turbine the 'exit velocity' or 'leaving velocity' or 'lost velocity' may amount to 3.3 per cent of the nozzle outlet velocity. Also since all the kinetic energy is to be absorbed by one ring of the moving blades only, the velocity of wheel is too high (varying from 25,000 to 30,000 r.p.m.). This wheel or rotor speed, however, can be reduced by different methods (discussed in the following article).

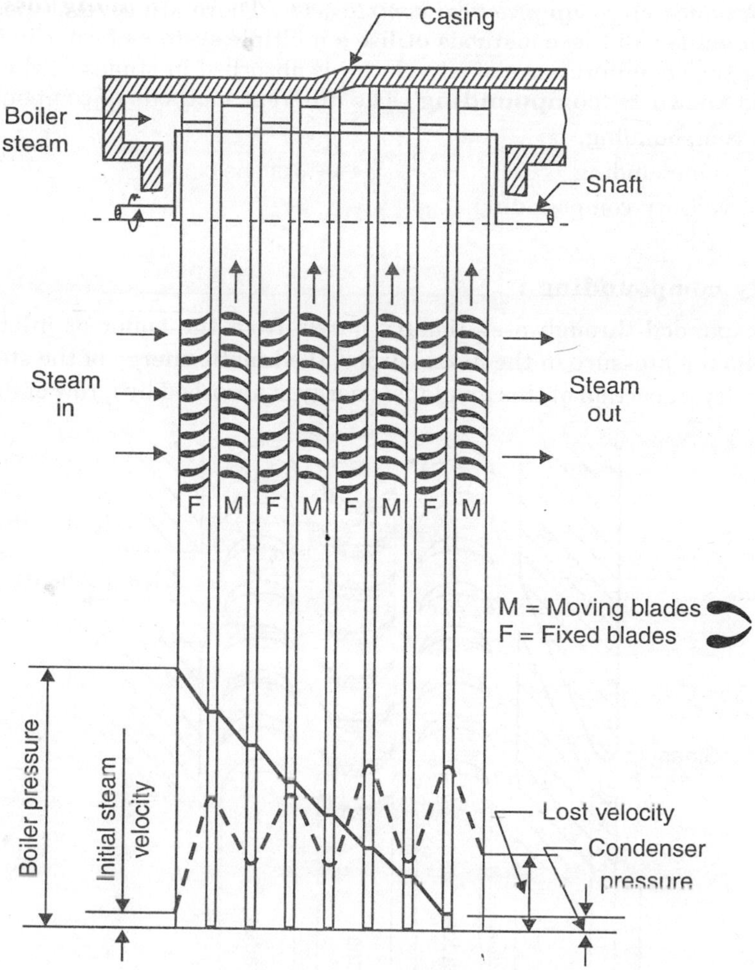

Fig. 2.27. Reaction turbine (three stage).

2. Reaction turbine :

In this type of turbine, *there is a gradual pressure drop and takes place continuously over the fixed and moving blades*. The function of the fixed blades is (the same as the nozzle) that they alter the direction of the steam as well as allow it to expand to a larger velocity. As the steam passes over the moving blades its kinetic energy (obtained due to fall in pressure) is absorbed by them. Fig. 2.27 shows a *multi-stage* reaction turbine. The changes in pressure and velocity are also shown therein.

As the volume of steam increases at lower pressures therefore, the diameter of the turbine must increase after each group of blade rings. It may be noted that in this turbine since the pressure drop per stage is small, therefore, the number of stages required is much higher than an impulse turbine of the same capacity.

2.14.5. Methods of Reducing wheel or Rotor Speed

As already discussed under the heading 'simple impulse turbine' that if the steam is expanded from the boiler pressure to condenser pressure in one stage the speed of the rotor becomes *tremendously high which crops up practical complicacies.* There are several methods of reducing this speed to lower value ; all these methods utilise a multiple system of rotor in series, keyed on a common shaft and the steam pressure of jet velocity is absorbed in stages as the steam flows over the blades. This is known as **'compounding'**. The different methods of compounding are :

1. Velocity compounding.
2. Pressure compounding.
3. Pressure velocity compounding.
4. Reaction turbine.

1. Velocity compounding :

Steam is expanded through a stationary nozzle from the boiler or inlet pressure to condenser pressure. So the pressure in the nozzle drops, the kinetic energy of the steam increases due to increase in velocity. A portion of this available energy is absorbed by a row of moving blades. The

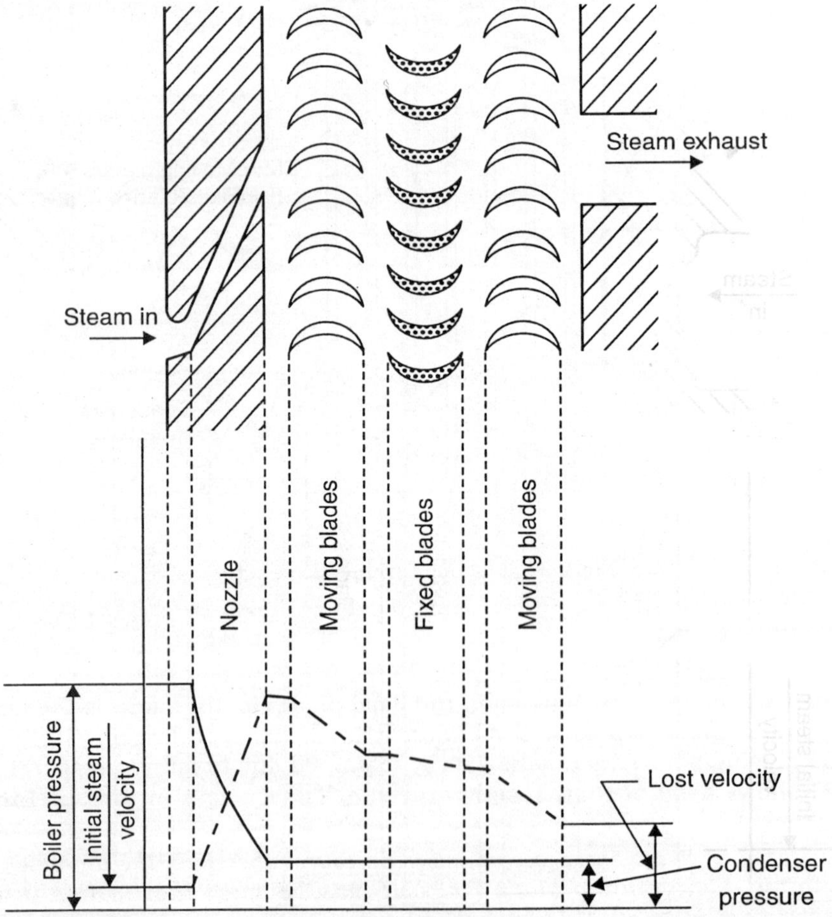

Fig. 2.28. Velocity compounding.

steam (whose velocity has decreased while moving over the moving blades) then flows through the second row of blades which are fixed. The function of these fixed blades is to redirect the steam flow without altering its velocity to the following next row moving blades where again work is done on them and steam leaves the turbine with a low velocity. Fig. 2.28 shows a cut away section of such a stage and changes in pressure and velocity as the steam passes through the nozzle, fixed and moving blades.

Though this method has the advantage that the initial cost is low due to lesser number of stages yet its efficiency is low.

2. Pressure compounding :

Fig. 2.29 shows rings of fixed nozzles incorporated between the rings of moving blades. The steam at boiler pressure enters the first set of nozzles and expands partially. The kinetic energy of the steam thus obtained is absorbed by the moving blades (stage 1). The steam then expands partially in the second set of nozzles where its pressure again falls and the velocity increases ; the

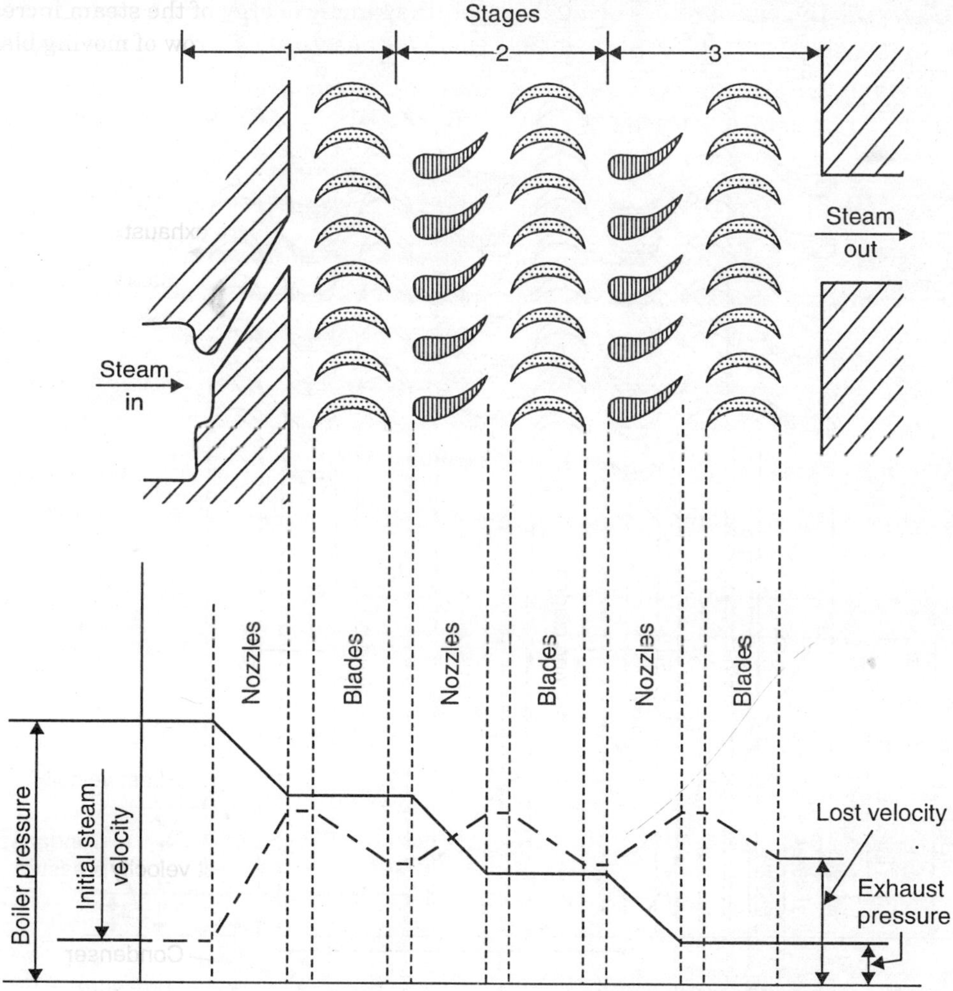

Fig. 2.29. Pressure compounding.

kinetic energy so obtained is absorbed by the second ring of moving blades (stage 2). This is repeated in stage 3 and steam finally leaves the turbine at low velocity and pressure. The number of stages (or pressure reductions) depends on the number of rows of nozzles through which the steam must pass.

This method of compounding is used in *Rateau and Zoelly turbine. This is most efficient turbine since the speed ratio remains constant but it is expensive owing to a large number of stages.*

3. Pressure velocity compounding :

This method of compounding is the combination of two previously discussed method. The total drop in steam pressure is divided into stages and the velocity obtained in each stage is also compounded. The rings of nozzles are fixed at the beginning of each stage and pressure remains constant during each stage. The changes in pressure and velocity are shown in Fig. 2.30.

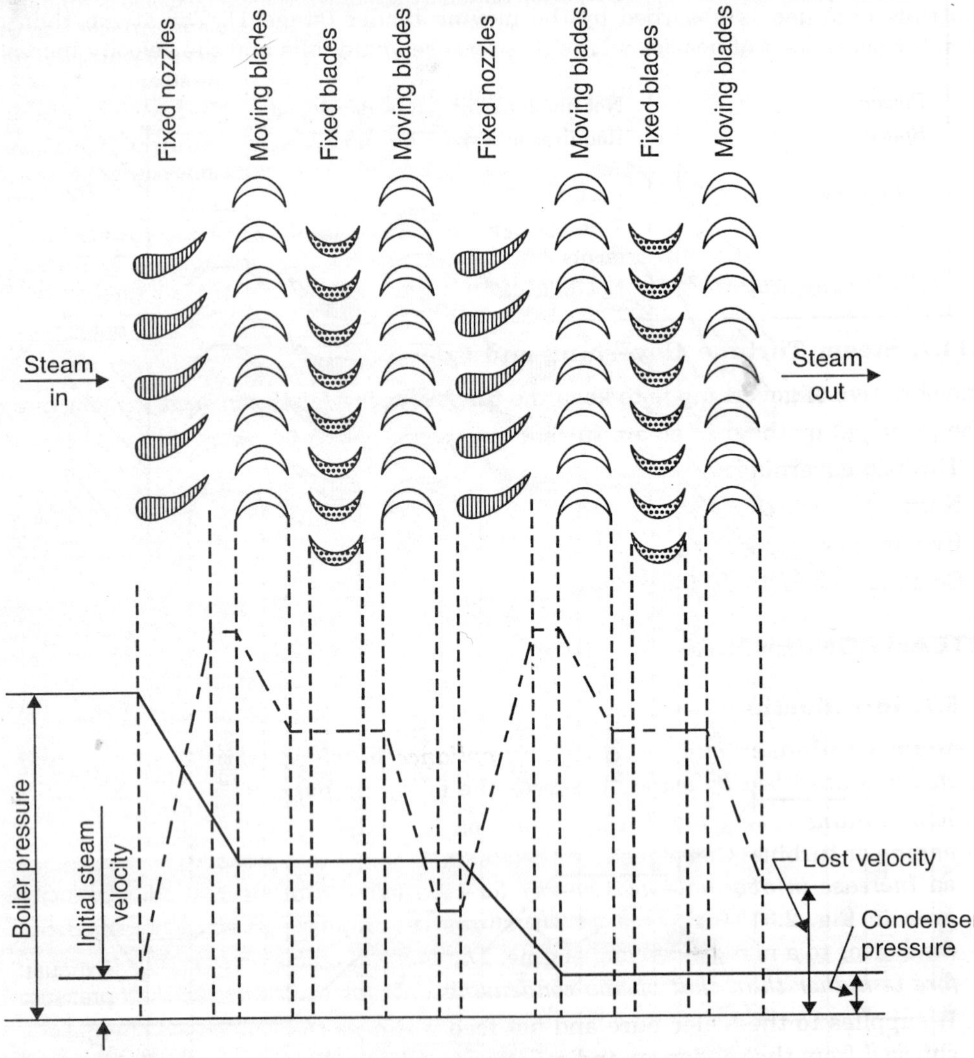

Fig. 2.30. Pressure velocity compounding.

This method of compounding is used in *Curits and Moore turbine.*

4. Reaction turbine :

It has been discussed in Article 2.14.4.

2.14.6. Difference Between Impulse and Reaction Turbines

S. No.	Aspects	Impulse turbine	Reaction turbine
1.	*Pressure drop*	Only in nozzles and not in moving blades.	In fixed blades (nozzles) as well as in moving blades.
2.	*Area of blade channels*	Constant.	Varying (converging type).
3.	*Blades*	Profile type.	Aerofoil type.
4.	*Admission of steam*	Not all round or complete.	All round or complete.
5.	*Nozzles / fixed blades*	Diaphram contains the nozzle.	Fixed blades similar to moving blades attached to the casing serve as nozzles and guide the steam.
6.	*Power*	Not much power can be developed.	Much power can be developed.
7.	*Space*	Requires less space for same power.	Requires more space for same power.
8.	*Efficiency*	Low.	High.
9.	*Suitability*	Suitable for small power requirements.	Suitable for medium and higher power requirements.
10.	*Blade manufacture*	Not difficult.	Difficult.

2.14.7. Steam Turbine Governing and Control

The objective of governing is to keep the turbine speed fairly constant irrespective of load. The principal methods of steam turbine governing are as follows :

1. Throttle governing
2. Nozzle governing
3. By-pass governing
4. Combination of 1 and 2 and 1 and 3.

2.15. STEAM CONDENSERS

2.15.1. Introduction

A **steam condenser** *is a device or an appliance in which steam condenses and heat released by steam is absorbed by water.* It serves the following *purposes :*

1. It *maintains a very low back pressure* on the exhaust side of the piston of the steam engine or turbine. Consequently, the steam expands to a greater extent which results in an *increase in available heat energy* for converting into mechanical work. The shaded area in Fig. 2.31. (*i.e.*, area 44′5′5) shows the increase in work obtained by fitting a condenser to a non-condensing engine. *The thermal efficiency of a condensing unit therefore is higher than that of non-condensing unit for the same available steam.*
2. It supplies to the boiler pure and hot feed water as the condensed steam which is discharged from the condenser and collected in a hot well, can be used as feed water for the boiler.

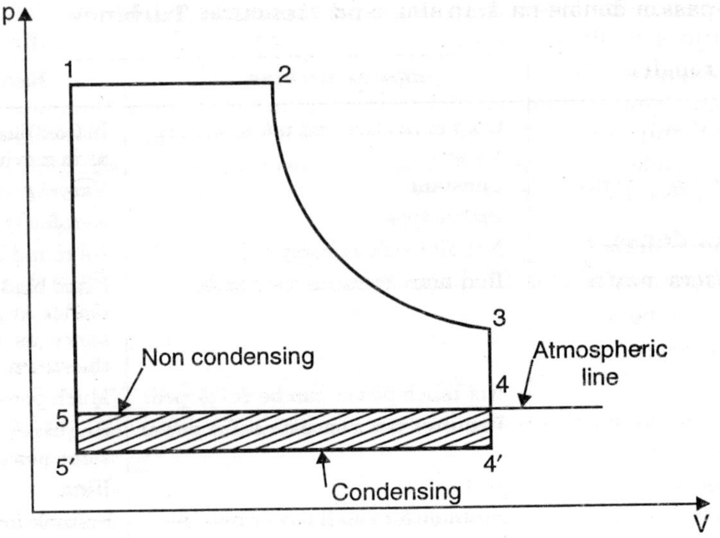

Fig. 2.31

2.15.2. Vacuum

Vacuum *is sub-atmospheric pressure.* It is measured as the pressure depression below atmospheric. The condensation of steam in a closed vessel produces a partial vacuum by reason of the great reduction in the volume of the low pressure steam or vapour. The back pressure in steam engine or steam turbine can be lowered from 0.013 to 0.2 bar abs. or even less. Since the steam engines are intermittent flow machines and as such cannot take the advantage of a very low vacuum, therefore, for most steam engines the exhaust pressure is about 0.2 to 0.28 bar abs. On the other hand, in steam turbines, which are continuous flow machines, the back pressure may be about 0.025 bar abs.

2.15.3. Organs of a Steam Condensing Plant

A steam condensing plant mainly consists of the following *organs/elements* :

1. Condenser (to condense the steam).
2. Supply of cooling (or injection) water.
3. Wet air pump (to remove the condensed steam, the air and uncondensed water vapour and gases from the condenser ; separate pumps may be used to deal with air and condensate).
4. Hot well (where the condensate can be discharged and from which the boiler feed water is taken).
5. Arrangement for recooling the cooling water in case surface condenser is employed.

2.15.4. Classification of Condensers

Mainly, condensers are of two types : (1) Jet condensers, (2) Surface condensers.

In ***jet condensers,*** *the exhaust steam and water come in direct contact with each other* and temperature of the condensate is the same as that of cooling water leaving the condenser. The cooling water is usually sprayed into the exhaust steam to cause rapid condensation.

In ***surface condensers,*** *the exhaust steam and water do not come into direct contact.* The steam passes over the outer surface of tubes through which a supply of cooling water is maintained.

There may be single-pass or double-pass. In single-pass condensers, the water flows in one direction only through all the tubes, while in two-pass condenser the water flows in one direction through the tubes and returns through the remainder.

A jet condenser is simpler and cheaper than a surface condenser. It should be installed when the cooling water is cheaply and easily made suitable for boiler feed or when a cheap source of boiler and feed water is available. *A surface condenser is most commonly used because the condensate obtained is not thrown as a waste but returned to the boiler.*

2.15.5. Jet Condensers

These condensers may be classified as :

(*a*) Parallel-flow type

(*b*) Counter-flow type

(*c*) Ejector type.

Parallel flow and counter flow condensers are further sub-divided into two types : (*i*) Low level type (*ii*) High level type.

In *parallel-flow type* of condenser, both the exhaust steam and cooling water find their entry at the top of the condenser and then flow downwards and condensate and water are finally collected at the bottom.

In *counter-flow type,* the steam and cooling water enter the condenser from opposite directions. Generally, the exhaust steam travels in upward direction and meet the cooling water which flows downwards.

2.15.6. Surface Condensers

Most condensers are generally classified on the direction of flow of condensate, the arrangement of the tubing and the position of the condensate extraction pump. The following is the main classification of surface condensers :

(*i*) Down flow type

(*ii*) Central flow type

(*iii*) Inverted flow type

(*iv*) Regenerative type

(*v*) Evaporative type.

2.15.7. Comparison Between Jet and Surface Condensers

S. No.	Jet Condenser	Surface Condenser
1.	Low manufacturing cost.	High manufacturing cost.
2.	Lower up keep.	Higher upkeep.
3.	Requires small floor space.	Requires large floor space.
4.	The condensate cannot be used as feed water in the boilers unless the cooling water is free from impurities.	Condensate can be reused as feed water as it does not mix with the cooling water.
5.	More auxiliary power required.	Less auxiliary power needed.

2.16. FEED WATER TREATMENT

For steam power plants water is one of the most important raw materials. In most of the cases, water used for steam power plants contains impurities which must be treated before use. All natural waters—even rain, snow, hail, treated municipal supplies contain impurities in one form or the other.

2.16.1. Classification of Impurities in Water

The impurities in water may be classified as follows :

1. Visible impurities :

(*i*) *Microbiological growth.* Presence of micro-organisms is always undesirable as they may produce *clogging troubles*.

(*ii*) *Turbidity and sediments. Turbidity* is the suspended insoluble matter whereas *sediments* are the coarse particles which settle down in stationary water, both are objectionable.

2. Dissolved gases :

(*i*) Carbondioxide
(*ii*) Oxygen
(*iii*) Nitrogen
(*iv*) Methane
(*v*) Hydrogen sulphide.

3. Minerals and salts :

(*i*) Iron and manganese
(*ii*) Sodium and potassium salts
(*iii*) Flourides
(*iv*) Silica.

4. Mineral acids. Their presence in water is always undesirable as it may result in the chemical reaction with the boiler material.

5. Hardness. The salts of calcium and magnesium as bicarbonates, chlorides, sulphates etc. are mainly responsible for the formation of a *very hard surface which resists heat transfer and clogs the passages in pipes*. Presence of these salts is known as *hardness*.

2.16.2. Troubles Caused by the Impurities in Water

The impurities in water may cause one or more of the following troubles :

1. Scale formation
2. Corrosion
3. Carry over
4. Embrittlement.

2.16.3. Methods of Feed Water Treatment

The different treatments adopted to remove the various impurities are enumerated below :

1. Mechanical treatment :

(*i*) Sedimentation
(*ii*) Coagulation
(*iii*) Filteration
(*iv*) Interior painting.

2. Thermal treatment :

(*i*) Deaeration
(*ii*) Distillation by evaporators.

3. Chemical treatment :

(*i*) Cold lime-soda softening process
(*ii*) Hot lime-soda softening process
(*iii*) Lime-phosphate softening process
(*iv*) Ion exchange process which may be sodium zeolite process or hydrogen zeolite process.

4. Demineralisation.

5. Blow down :

(*i*) Hot lime-soda and hot zeolite process

(*ii*) Adding acid to control alkalinity and *vice-versa*.

2.17. ADVANTAGES AND DISADVANTAGES OF STEAM POWER PLANTS

Advantages of Steam power plants :

1. They can respond to rapidly changing loads without difficulty.
2. A portion of the steam generated can be used as a process steam in different industries.
3. Can be located very conveniently near the load centre.
4. As these plants can be set up near the industry transmission costs are reduced.
5. Steam engines and turbines can work under 25 per cent of overload continuously.
6. Fuel used is cheaper.
7. Less space is required in comparison with that for hydro-electric plants.
8. Cheaper in production cost in comparison with that of diesel power stations.
9. Cheaper in initial cost in comparison with that of diesel power stations.

Disadvantages :

1. Maintenance and operating costs are high.
2. The cost of plant increases with increase in temperature and pressure.
3. Long time required for erection and putting into action.
4. A large quantity of water is required.
5. Great difficulty experienced in coal handling.
6. The plant efficiency decreases rapidly below about 75 per cent load.
7. Presence of troubles due to smoke and heat in the plant.

2.18. MISCELLANEOUS

2.18.1. Useful Life of Steam Power Plant Components

Approximate useful life of some of the components of a steam power plant is given below :

Components	Life (years)
1. Water tube boiler	20
2. Coal and ash machinery	10 to 20
3. Steam turbines	15 to 20
4. Steam condensers	20
5. Turbo-generators	10 to 20
6. Feed water heater	30
7. Pumps	15 to 20
8. Transformers	15 to 20
9. Motors	20
10. Air compressors	20 to 25
11. Buildings	50

2.18.2. Steam Power Plant Pumps

The pumps used in a steam power plant are *classified* as follows :

1. *Reciprocating* :
   (*i*) Direct acting (*ii*) Power.
2. *Rotary* :
   (*i*) Vane (*ii*) Screw
   (*iii*) Gear (*iv*) Lobe.
3. *Centrifugal* :
   (*i*) Volute (*ii*) Diffuser
   (*iii*) Axial flow (*iv*) Mixed flow.

The above mentioned pumps are used for the following *services* :

(*i*) Boiler feed (*ii*) Circulating water
(*iii*) Evaporator feed (*iv*) Condensate
(*v*) Well water (*vi*) Ash sluicing
(*vii*) Fuel oil.

2.18.3. Cost of Steam Power Plant

A typical subdivision of investment cost of steam power station is as follows :

Components	*Investment cost*
1. Building etc.	25%
2. Boiler Plant	18%
3. Turbo-generators and condensers	25%
4. Fuel handling	6%
5. Switch yard, switching and wiring	16%
6. Piping	5%
7. Miscellaneous	5%

2.18.4. Comparison of Various types of Power Plants

S. No.	Particulars	Steam power plants	Hydro-electric power plants	Diesel power plants	Nuclear power plants
1.	*Site*	Located near load centres but other factors water supply, land cost, transportation facilities are to be kept in view. Location of steam power plant is somewhat flexible as compared with that of hydro-plants.	The site should be such that sufficient area, huge quantity of water at a sufficient head, cheap and rocky land and transportation facilities should be available. These plants are usually at a far off distance from the load centres.	Can be installed anywhere.	Located near the load centre.

2.	*Initial cost*	*Low* as compared to hydroelectric and atomic power plants.	*High* as compared to steam power plant on account of large amount of excavation work and heavy cost of transportation of the plant and machinery.	Initial cost is the *minimum*.	The capital cost is *very high* as compared to all other types of power plants on account of heavy cost of nuclear reactors and heavy cost of erection as highly specialised and expert engineers are required for its erection work.
3.	*Fuel transportation cost*	*High,* especially when the power plants are away from the coal mines and has no railway siding.	*Nil* (as no fuel is required).	Higher than that of atomic power plant but lower than that for steam power plants.	Very low (since quantity of fuel required is very small).
4.	*Operating cost*	Very high as compared to hydro-plants and atomic power plants but low as compared to diesel power plant.	Practically nil (since no fuel is required for the operation of the plant).	Very high as compared to all other types of power plants.	Very low as compared to all other types of power plants except that of hydro-electric plants.
5.	*Maintenance cost*	Higher as compared with that of hydro-plants and diesel power plants (because large operating staff and more skilled engineers are required).	Maintenance cost is comparatively low (because few skilled engineers and small operating staff is required).	Maintenance cost is comparatively lower (because lesser operating and supervising staff is required).	Maintenance cost is comparatively higher (because skilled and well trained staff is required for its operation and maintenance).
6.	*Limit of source of power*	Source of fuel *i.e., coal* reserve all over the world is considered to be fixed and therefore coal mines are being exhausted and time may come when all of these might get exhausted.	Source of power *i.e., water* in case of hydro-plants is not dependable because it depends upon the rainfall which is at the whim of nature.	Source of fuel *i.e., diesel* is not available in plenty.	The source of power is unlimited since large deposits of fissionable materials all over the world are available.
7.	*Transmission and distribution cost.*	Low (because short transmission lines are required).	*Very high* (because long transmission lines are required).	Transmission cost is *nil* and distribution cost is also *very small*.	*Very low*.

8.	*Reliability*	Less reliable.	More reliable.	Less reliable.	Reliable.
9.	*Simplicity and cleanliness*	Atmosphere is polluted by fumes and residues of pulverised fuels.	Most simple and clean.	Simple and clean than steam plants and atomic power plants.	The handling of atomic power plant is quite complicated as radiation hazards are involved in it. Special outfits are designed and explored by the operating staff.
10.	*Field of application*	Most economical if sited near coal mines and by the side of river or canal.	Most economical where water source is available at a sufficient head.	These plants are installed to supply power is emergency.	Where neither water nor coal as a source of power is available, as in Rajasthan these plants are more feasible and adopted.
11.	*Standby losses*	Maximum as the boiler remains in operation even when the turbine is not working.	No standby losses.	Less standby losses.	Less standby losses.

2.18.5. Thermal Power Stations in India

Some of the thermal power stations installed in the country or under the process of installation are as follows :

S. No.	State	Name of power station	Capacity (MW)
1.	Andhra Pradesh	Kothagndam	240
2.	Assam	Gauhati	40
3.	Bihar	(*i*) Barauni	140
		(*ii*) Bokaro	225
		(*iii*) Patratu	400
4.	Delhi	(*i*) Rajghat and I.P. thermal power station	350
		(*ii*) Badarpur	300
5.	Gujrat	(*i*) Dhuvaran	530
		(*ii*) Ukai	240
6.	Haryana	(*i*) Faridabad	200
		(*ii*) Panipat	220
7.	Madhya Pradesh	(*i*) Kobra	420
		(*ii*) Satpura	300
		(*iii*) Amarkantak	180
8.	Maharashtra	(*i*) Nagpur (Koradi)	480
		(*ii*) Nasik	280
		(*iii*) paras	90

9.	Orissa	Talcher	254
10.	Tamil Nadu	(*i*) Neyveli	600
		(*ii*) Eunore	450
11.	Uttar Pradesh	(*i*) Hardnaganj	540
		(*ii*) Obra	550
12.	West Bengal	(*i*) Santaldih	480
		(*ii*) Chandrapur	420
		(*iii*) Bandel	330
		(*iv*) Durgapur	290

WORKED EXAMPLES

Example 2.1. (*a*) *Discuss the necessity of superheated steam, pulverised coal and preheated air in thermal power plant.*

(*b*) *Explain why thermal plants are not suitable for supplying fluctuating loads.*

(*c*) *A steam power station of 100 MW capacity uses coal of calorific value of 26800 kJ/kg. The thermal efficiency of the station is 30% and electrical generation efficiency is 92%. Determine the coal required per hour when the plant is working at full load.* **(M.U.)**

Solution. (*a*) (*i*) ***Necessity of superheated steam.*** The temperature and pressure plays an important part in increasing the thermal efficiency of the boiler. Also too much condensation in the last stages of the turbine is avoided. Thus transfer rates are considerably increased. The pressure level becomes more stable, the problem of erosion and corrosion are minimised.Thus to get all the advantages superheating of steam is used these days.

(*ii*) ***Necessity of pulverised coal.*** The conventional fuel firing method were found to be unable to take the fluctuating loads on the plant due to limited capacity of combustion. Further these conventional methods using stoker firing were found unsuitable for large capacity plants, containing high percentage of ash, due to difficulties experienced in removing large quantities of ash in the combustion process. To overcome those disadvantages necessity of pulverised coal was felt in which coal is dried and reduced to fine powder and after mixing with air is burnt in the furnace. The pulverised coal has the *advantages of high rate of combustion, high efficiency of boiler and the plant to easily meet with the fluctuating load demands.*

(*iii*) ***Necessity of preheated air.*** Since the entire heat of the flue gases cannot be extracted through the economizer, air preheaters necessity was felt to remove some of the heat in these gases, on an average an increase of 20°C in the air temperature results in an increase in the boiler efficiency by 1%. The use of air preheater is more economical with pulverised fuel boilers because the temperature of the gases going out is sufficiently large and high air temperature is possible to achieve.

(*b*) Thermal plants are not suitable for supplying fluctuating loads because any change in load demand requires corresponding change in the output energy. In the thermal plants input energy is produced by burning of coal and hence there is always a large time lapse between the change in energy at output and input, which is not desirable and hence such power stations are used only as base load stations, supplying constant power.

(*c*) *Given :* Capacity of the power station = 100 MW ; C.V. = 26800 kJ/kg ; $\eta_{th.} = 30\%$; $\eta_G = 92\%$.

$$\text{Input power to generator} = \frac{100 \times 10^3}{\eta_G} = \frac{100 \times 10^3}{0.92} = 108.7 \times 10^3 \text{ kW}$$

Input coal power required $= \frac{108.7 \times 10^3}{\eta_{th.}} = \frac{108.7 \times 10^3}{0.3} = 362.3 \times 10^3$ kW

$\therefore$ Coal required in kg/s $= \frac{362.3 \times 10^3}{26800} = 13.52$ kg/s

$\therefore$ Coal required per hour $= 13.52 \times 3600 = 48672$ kg or **48.672 tonnes. (Ans.)**

Example 2.2. *(a) What are the advantages and disadvantages of pulverising the fuel used in thermal power stations ? Name the various the methods of arresting ash from the flue gases ?*

(b) Why are cooling towers used in thermal power stations ?

(c) Why are deflectors used in a Pelton wheel in a hydropower station ?

(d) Why are pump storage employed in inter-connected power system and how is the site selected for such a scheme ? **(Banglore University)**

Solution. (*a*) **Advantages :** *The advantages of pulverising the fuel are :*

(*i*) Pulverising exposes more surface area for combustion and hence allows faster combustion.

(*ii*) Wide variety and low grade coal can be burnt more easily.

(*iii*) Rate of combustion can be controlled easily and immediately and hence gives fast response to load changes.

(*iv*) The system is perfectly free from clinker and slagging trouble.

(*v*) The system works successfully with gas and oil.

(*vi*) It is possible to use highly pre-treated secondary air which helps for rapid flame propagation.

(*vii*) The pulverising system can be repaired without cooling the unit, as the pulverising equipment is located outside the furnace.

(*viii*) Large amount of heat release is possible.

(*ix*) The banking losses are low.

(*x*) The boiler can be started from cold very rapidly and efficiently.

(*xi*) There are no moving parts in the furnace.

(*xii*) Practically no ash handling problems.

Disadvantages :

(*i*) The capital cost is high.

(*ii*) The system produces fly-ash which requires special and costly fly-ash removal equipments.

(*iii*) The possibilities of the explosion are more as coal burns like a gas.

(*iv*) The storage of powdered coal requires special attention.

(*v*) The building space required is large.

(*vi*) The skilled operators are required.

(*vii*) Special starting-up equipments are required.

(*viii*) The removal of liquid slag creates additional problems.

- *The different methods of arresting ash from the flue gases are :*

1. The setting chamber.
2. Impingement separators.
3. Cyclone separators.
4. Fabric filters.
5. Electrostatic precipitators.

(*b*) The cooling water system is one of the most important systems of power plant and its availability predominantly decides the plant site. As the cooling water takes the latent heat of steam in the condenser, the temperature of the water increases. The hot water coming out of the condenser can not be used again in a closed system without pre-cooling. This is because the hot water coming out if used again will not be able to absorb the heat, and the condenser vacuum cannot be maintained. Therefore it is absolutely necessary to pre-cool the water coming out of condenser before using again. When enough quantity of water is not available for cooling requirements from river, then closed type system is adopted. In closed system, the hot water coming out of condenser is cooled either spraying in the pond or passing through the cooling tower.

(*c*) *To avoid the water hammer effects due to rapid closing of nozzle during sudden fall of load, a deflection is introduced in the system.* The function of the deflector is to deflect the jet from the turbine runner when the load on turbine suddenly decreases. In this case, the quantity of water passing through the nozzle remains same but the part of water coming out from the nozzle is deflected and is not allowed to strike the bucket. The deflected water goes waste into the tail race level. The deflector thus comes into operation with the decrease in load and prevents the rise in speed of the turbine. The needle valve then closes the opening of the nozzle gradually and avoids the undue rise of pressure. *The combined operation of nozzle and deflector reduces the danger of heavy water hammer and minimises the wastage of water.*

(*d*) The large generating units of modern steam power stations and nuclear power stations operate most economically when they run day and night at constant load. The major problem with these power plants is the peak load. *Covering of short peak load demands with steam power station is economically undesirable and is not adaptable due to slow pick-up.* Pump storage plants for peak load operations in interconnected system are more suitable *where the quantity of water available for power generation is insufficient but natural site for high dam construction is most suitable.*

The pump storage power plant essentially consists of a high water pond and a tail water pond. The site selected for such a plant should therefore take into consideration of natural pond facilities in hilly area. *Water from the tail pond is lifted into the high head pond during off load period and the stored water at high level is used for generation of electrical power during peak load period.*

HIGHLIGHTS

1. A steam power plant converts the chemical energy of the fossil fuels (coal, oil, gas) into mechanical/electrical energy.
2. The layout of a modern steam power plant comprises of the following four circuits :
   (*i*) Coal and ash circuit (*ii*) Air and gas circuit
   (*iii*) Feed water and steam flow circuit (*iv*) Cooling water circuit.
3. The choice of steam conditions depends upon the following :
   (*i*) Price of a the coal (*ii*) Capital cost of the plant
   (*iii*) Time available for erection (*iv*) Thermal efficiency obtainable
   (*v*) The station 'load factor'.
4. Automatic stokers are classified as follows :
   (*i*) Overfeed stokers (*ii*) Underfeed stokers.
5. Pulverised fuel handling systems are classified as :
   (*i*) Unit system (*ii*) Central system.

6. Pulverised fuel burners are classified as follows :
   (*i*) Long flame burners (*ii*) Turbulent burners
   (*iii*) Tangential burners (*iv*) Cyclone burners.
7. Oil burners are classified as follows :
   1. Vapourising oil burners :
      (*a*) Atmospheric pressure atomising burner (*b*) Rotating cup burner
      (*c*) Recirculating burner (*d*) Wick type burner.
   2. Atomising fuel burners :
      (*a*) Mechanical or oil pressure atomising burner
      (*b*) Steam or high pressure atomising burner
      (*c*) Low pressure air atomising burner.
8. Fluidised bed may be defined as the bed of solid particles behaving as a fluid.
9. Ash handling systems may be classified as follows :
   (*i*) Mechanical handling system (*ii*) Hydraulic system
   (*iii*) Pneumatic system (*iv*) Steam jet system.
10. The dust collectors may be classified as follows :
   1. Mechanical just collectors
      (*i*) Wet type (scrubbers)
         (*a*) Spray type (*b*) Packed type
         (*c*) Impingement type.
      (*ii*) Dry type
         (*a*) Gravitational separators (*b*) Cyclone separators.
   2. Electrical dust collectors
      (*i*) Rod type (*ii*) Plate type.
11. The 'collection efficiency' of a dust collector is the amount of dust removed per unit weight of dust.
12. pH value of water is the logarithm of the reciprocal of hydrogen ion concentration. It is number from 0 to 14 with 7 indicating neutral number.
13. The small pressure difference which causes a flow of gas to take place is termed as a *draught*.
14. The most important classification of steam turbines is as follows :
   (*i*) Impulse turbines (*ii*) Reaction turbines
   (*iii*) Combination of impulse and reaction turbines.

THEORETICAL QUESTIONS

1. How are steam power plants classified ?
2. Give the layout of a modern steam power plant and explain it briefly.
3. What are the essential requirements of steam power station design ?
4. What factors should be taken into consideration while selecting the site for steam power plant ?
5. How can the capacity of a steam power plant be determined ?
6. On factors does the choice of steam conditions depend ?
7. Enumerate the means by which the coal from coal mines can be transported.
8. What are the requirements of good coal handling plant ?
9. Enumerate and explain the steps involved in handling of the coal.
10. Explain with the help of a neat diagram the arrangement of the Fluidised Bed Combustion (FBC) system.
11. State the characteristics of a good ash handling plant.
12. Enumerate and explain various modern ash-handling systems.
13. How are dust collectors classified ?
14. Explain with the help of a diagram the working of a 'cyclone separator'.

15. **How do you define the 'collection efficiency' of a dust separator ?**
16. **What are the uses of ash and dust ?**
17. **Give the general layout of ash handling and dust collection system.**
18. **Discuss various methods of compounding steam turbines ?**
19. **Describe briefly the various methods of 'steam turbine governing'.**
20. **Define a steam condenser and state its functions.**
21. **Explain the reasons for inefficiency in surface condensers.**
22. **Explain the effects of air leakage in a condenser.**
23. **List the advantages and disadvantages of steam power plants.**
24. **Give comparison between steam, hydro-electric, diesel and nuclear power plants.**

3

Diesel Engine Power Plant

3.1. Introduction. 3.2. Advantages and disadvantages of diesel power plants. 3.3. Applications of diesel power plant. 3.4. Site selection. 3.5. Heat engines. 3.6. Classification of I.C. engines. 3.7. Comparison between a petrol engine and a diesel engine. 3.8. Essential components of a diesel power plant : engine, air intake system, exhaust system, fuel system. 3.9. Operation of a diesel power plant. 3.10. Types of diesel engines used for diesel power plants. 3.11. Layout of a diesel engine power plant. Highlights—Theoretical Questions.

3.1. INTRODUCTION

- Diesel engine power plants are installed *where supply of coal and water is not available in sufficient quantity* or *where power is to be generated in small quantity* or *where standby sets are required for continuity of supply such as in hospitals, telephone exchanges, radio stations and cinemas*. These plants in the range of 2 to 50 MW capacity are used as *central stations* for supply authorities and works and they are universally adopted to supplement hydro-electric or thermal stations where standby generating plants are essential for starting from cold and under emergency conditions.
- In several countries, the demand for diesel power plants is increased for electric power generation because of difficulties experienced in construction of new hydraulic plants and enlargement of old hydro-plants. A long term planning is required for the development of thermo and hydro-plants which cannot keep the pace with many times the increased demand by the people and industries.
- The diesel units used for electric generation are *more reliable* and *long-lived piece of equipment* compared with other types of plants.

3.2. ADVANTAGES AND DISADVANTAGES OF DIESEL POWER PLANTS

The *advantages and disadvantages* of diesel power plants are listed below :

Advantages :

1. Design and installation are very simple.
2. Can respond to varying loads without any difficulty.
3. The standby losses are less.
4. Occupy less space.
5. Can be started and put on load quickly.
6. Require less quantity of water for cooling purposes.
7. Overall capital cost is lesser than that for steam plants.
8. Require less operating and supervising staff as compared to that for steam plants.
9. The efficiency of such plants at part loads does not fall so much as that of a steam plant.
10. The cost of building and civil engineering works is low.
11. Can burn fairly wide range of fuels.

12. These plants can be located very near to the load centres, many times in the heart of the town.
13. No problem of ash handling.
14. The lubrication system is more economical as compared with that of a steam power plant.
15. The diesel power plants are *more efficient than steam power plants* in the range of 150 MW capacity.

Disadvantages :

1. High operating cost.
2. High maintenance and lubrication cost.
3. Diesel units capacity is limited. These cannot be constructed in large size.
4. In a diesel power plant noise is a serious problem.
5. Diesel plants cannot supply overloads continuously whereas steam power plant can work under 25% overload continuously.
6. The diesel power plants are not economical where fuel has to be imported.
7. The life of a diesel power plant is quite small (2 to 5 years or less) as compared to that of a steam power plant (25 to 30 years).

3.3. APPLICATIONS OF DIESEL POWER PLANT

The diesel power plants find wide application in the following fields :

1. Peak load plant
2. Mobile plants
3. Standby units
4. Emergency plant
5. Nursery station
6. Starting stations
7. Central stations—where capacity required is small (5 to 10 MW)
8. Industrial concerns where power requirement is small say of the order of 500 kW, diesel power plants become more economical due to their higher overall efficiency.

3.4. SITE SELECTION

The following factors should be considered while *selecting the site for a diesel power plant* :

1. Foundation sub-soil condition. The conditions of sub-soil should be such that a foundation at a reasonable depth should be capable of providing a strong support to the engine.

2. Access to the site. The site should be so selected that it is accessible through rail and road.

3. Distance from the load centre. The location of the plant should be near the load centre. This reduces the cost of transmission lines and maintenance cost. The power loss is also minimised.

4. Availability of water. Sufficient quantity of water should be available at the site selected.

5. Fuel transportation. The site selected should be near to the source of fuel supply so that transportation charges are low.

3.5. HEAT ENGINES

Any type of engine or machine which derives heat energy from the combustion of fuel or any other sources and converts this energy into mechanical work is termed as a **heat engine.**

Heat engines may be classified into two main classes as follows :

1. External Combustion Engines.
2. Internal Combustion Engines.

1. External combustion engines (E.C. engines). In this case, *combustion of fuel takes place outside the cylinder* as in case of *steam engines* where the heat of combustion is employed to generate steam which is used to move a piston in a cylinder. Other examples of external combustion engines are *hot air engines, steam turbine* and *closed cycle gas turbine*. These engines are generally used for driving locomotives, ships, generation of electric power etc.

2. Internal combustion engines (I.C. engines). In this case, *combustion of the fuel with oxygen of the air occurs within the cylinder of the engine*. The internal combustion engines group includes engines employing mixtures of combustible gases and air, known as *gas engines*, those using lighter liquid fuel or spirit known as *petrol engines* and those using heavier liquid fuels, known as *oil compression ignition* or *diesel engines*.

3.6. CLASSIFICATION OF I.C. ENGINES

Internal combustion engines may be *classified* as given below :

1. **According to cycle of operation :**

(*i*) Two stroke cycle engines

(*ii*) Four stroke cycle engines.

2. **According to cycle of combustion :**

(*i*) Otto cycle engine (combustion at constant volume)

(*ii*) Diesel cycle engine (combustion at constant pressure)

(*iii*) Dual-combustion or Semi-Diesel cycle engine (combustion partly at constant volume and partly at constant pressure).

3. **According to arrangement of cylinder :**

(*i*) Horizontal engine (*ii*) Vertical engine

(*iii*) V-type engine (*iv*) Radial engine etc.

4. **According to their uses :**

(*i*) Stationary engine (*ii*) Portable engine

(*iii*) Marine engine (*iv*) Automobile engine

(*v*) Aero engine etc.

5. **According to the fuel employed and the method of fuel supply to the engine cylinder :**

(*i*) Oil engine (*ii*) Petrol engine

(*iii*) Gas engine (*iv*) Kerosene engine etc.

(*v*) Carburettor, hot bulb, solid injection and air injection engine.

6. **According to the speed of the engine :**

(*i*) Low speed engine (*ii*) Medium speed engine

(*iii*) High speed engine.

7. **According to method of ignition :**

(*i*) Spark ignition S.I. engine (*ii*) Compression ignition C.I. engine.

8. **According to method of cooling the cylinder :**

(*i*) Air-cooled engine (*ii*) Water-cooled engine.

9. **According to method of governing :**

(*i*) Hit and miss governed engine (*ii*) Quality governed engine

(*iii*) Quantity governed engine.

10. **According to valve arrangement :**

(*i*) Over head valve engine (*ii*) L-head type engine

(*iii*) T-head type engine (*iv*) F-head type engine.

11. **According to number of cylinders :**

(*i*) Single cylinder engine (*ii*) Multi-cylinder engine.

3.7. COMPARISON BETWEEN A PETROL ENGINE AND A DIESEL ENGINE

	Petrol engine	Diesel engine
1.	Air petrol mixture is sucked in the engine cylinder during suction stroke.	Only air is sucked during suction stroke.
2.	Spark plug is used.	Employs an injector.
3.	Power is produced by spark ignition.	Power is produced by compression ignition.
4.	Thermal efficiency up to 25%.	Thermal efficiency up to 40%.
5.	Occupies less space.	Occupies more space.
6.	More running cost.	Less running cost.
7.	Light in weight.	Heavy in weight.
8.	Fuel (Petrol) costlier.	Fuel (Diesel) cheaper.
9.	Petrol being volatile is dangerous.	Diesel is not dangerous as it is non-volatile.
10.	Pre-ignition possible.	Pre-ignition not possible.
11.	Works on Otto cycle.	Works on diesel cycle.
12.	Less dependable.	More dependable.
13.	Used in *cars* and *motor cycles*.	Used in heavy duty vehicles like *trucks*, *buses* and *heavy machinery*.

3.8. ESSENTIAL COMPONENTS OF A DIESEL POWER PLANT

Refer to Fig. 3.1. The essential components of a diesel power plant are listed and discussed as follows :

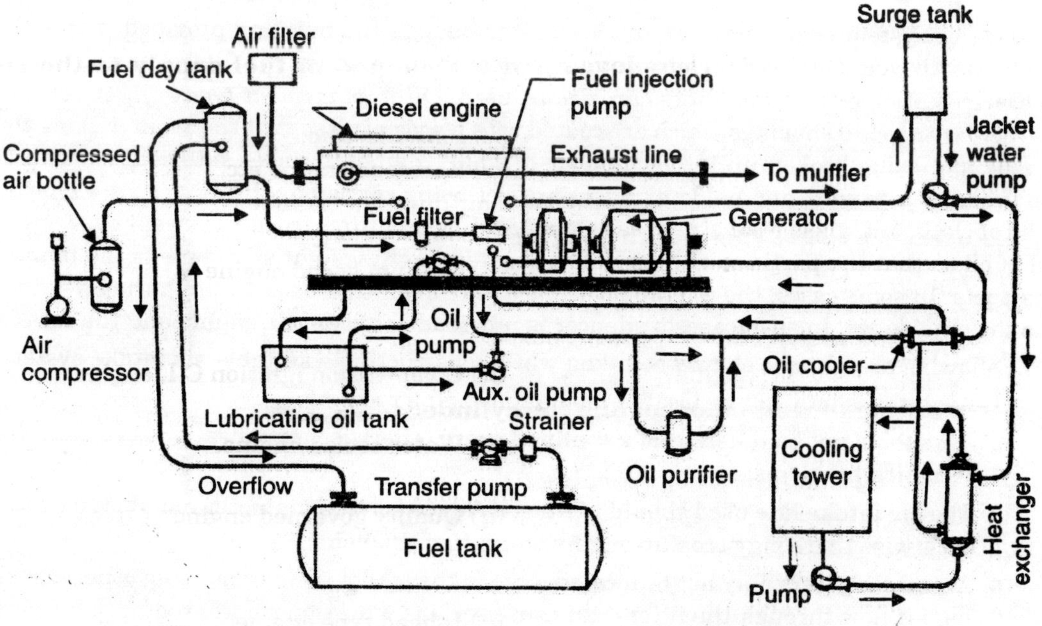

Fig. 3.1. Schematic arrangement of a diesel power plant.

1. Engine
2. Air intake system
3. Exhaust system
4. Fuel system
5. Cooling system
6. Lubrication system
7. Engine starting system
8. Governing system.

3.8.1. Engine

This is the main component of the plant which develops the required power. It is generally directly coupled to the generator as shown in Fig. 3.1.

3.8.2. Air Intake System

The air intake system conveys fresh air through pipes or ducts to : (*i*) Air intake manifold of four stroke engines (*ii*) The scavenging pump inlet of a two stroke engine and (*iii*) The supercharger inlet of a supercharged engine.

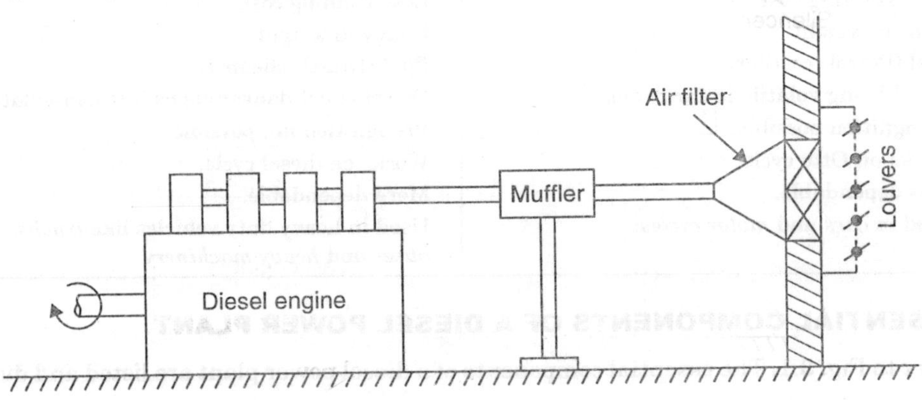

Fig. 3.2. Air intake system.

The air system begins with an intake located outside the building provided with a *filter* to catch dirt which would otherwise cause excessive wear in the engine. The filters may be of *dry or oil bath*. Electrostatic precipitator filters can also be used. *Oil impingement type of filter* consists of a frame filled with metal shavings which are coated with a special oil so that the air in passing through the frame and being broken up into a number of small filaments comes into contact with the oil whose property is to sieze and hold any dust particles being carried by the air. The *dry type* of filter is made of cloth, felt, glass wool etc. In case of *oil bath type of filter* the air is swept over or through a pool of oil so that the particles of dust become coated. Light weight steel pipe is the material for intake ducts. In some cases, the engine noise may be transmitted back through the air intake system to the outside air. In such cases a silencer is provided between the engine and the intake.

Following *precautions* should be taken while constructing a suitable air intake system :

1. Air intakes may not be located inside the engine room.
2. Air should not be taken from a confined space otherwise air pulsations can cause serious vibration problems.
3. The air-intake line used should neither have too small a diameter nor should be too long, otherwise there may crop up engine starvation problem.
4. Air intake filters may not be located close to the roof of the engine room otherwise pulsating air flow through the filters can cause serious vibrations of the roof.
5. Air intake filters should not be located in an inaccessible location.

3.8.3. Exhaust System

Refer to Fig. 3.3. The purpose of the exhaust system is to discharge the engine exhaust to the atmosphere outside the building. The exhaust manifold connects the engine cylinder exhausts outlets to the exhaust pipe which is provided with a muffler to reduce pressure in the exhaust line and eliminate most of the noise which may result if gases are discharged directly into the atmosphere.

The exhaust pipe leading out of the building should be short in length with minimum number of bends and should have one or two flexible tubing sections which take up the effects of expansion, and isolate the system from the engine vibration. Every engine should be provided with its independent exhaust system.

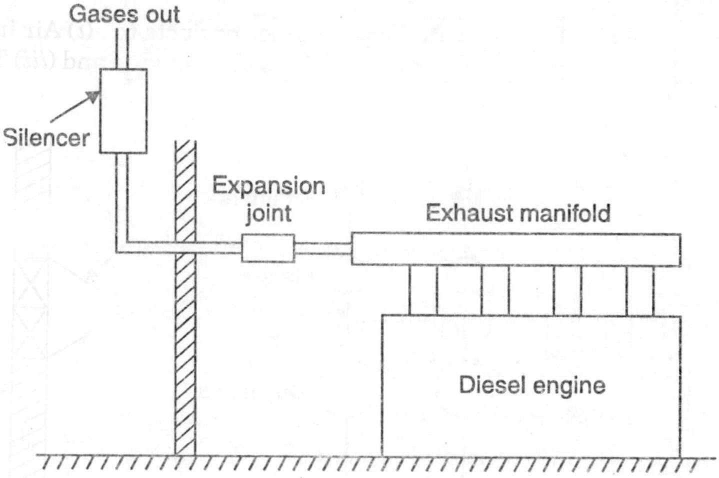

Fig. 3.3. Exhaust system.

The waste heat utilisation in a *diesel-steam* station may be done by providing waste-heat boilers in which most of the heat of exhaust gases from the engine is utilised to raise low pressure steam. Such application is common on *marine plants*. On the *stationary power plant* the heat of exhaust may be utilised to heat water in gas-to-water heat exchangers consisting of a water coil placed in exhaust muffler and using the water in the plant suitably. If air heating is required, the exhaust pipe from the engine is surrounded by the cold air jacket, and transfers the heat of exhaust gases to the air.

3.8.4. Fuel System

Refer to Fig. 3.4.

The fuel oil may be delivered at the plant site by trucks, railroad tank cars or barge and tankers. From tank car or truck the delivery is through the unloading facility to main storage tanks and then by transfer pumps to small service storage tanks known as *engine day tanks*. Large storage capacity allows purchasing fuel when prices are low. The main flow is made workable and practical by arranging the piping equipment with the necessary heaters, by passes, shut-offs, drain lines, relief valves, strainers and filters, flow meters and temperature indicators. The actual flow plans depend on type of fuel, engine equipment, size of the plant etc. The tanks should contain manholes for internal access and repair, fill lines to receive oil, vent lines to discharge vapours, overflow return lines for controlling oil flow and a suction line to withdraw oil. *Coils heated by hot water or steam reduce oil viscosity to lower pumping power needs.*

The minimum storage capacity of at least a month's requirement of oil should be kept in bulk, but where advantage of seasonal fluctuations in cost of oil is to be availed, it may be necessary to provide storage for a few month's requirements. *Day tanks* supply the daily fuel need of engines and may contain a minimum of about 8 hours of oil requirement of the engines. These tanks are usually placed high so that oil may flow to engines under gravity.

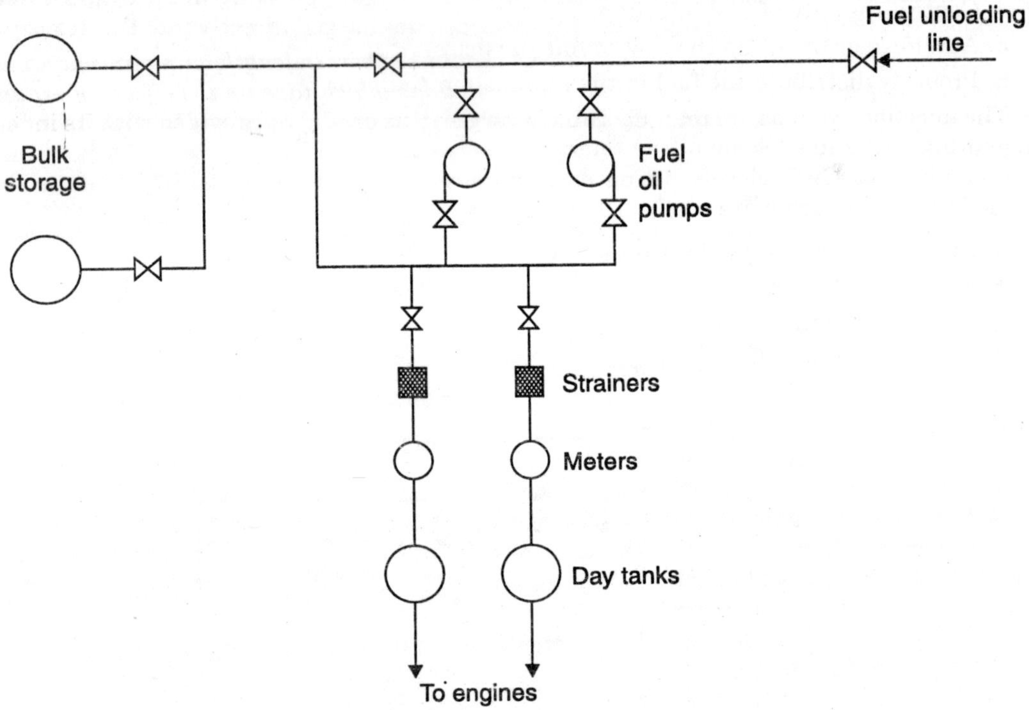

Fig. 3.4. System of fuel storage for a diesel power plant.

For satisfactory operation of a fuel oil supply system the following points should be taken care of :

1. There should be provisions for cleanliness and for changing over of lines during emergencies.
2. In all suction lines the pipe joints should be made tight.
3. Before being covered, all oil lines should be put under air pressure and the joints tested with soap solution. Small air leaks into the line can be the source of exasperating operating difficulties and are hard to remedy once the plant is in operation.
4. The piping between filter and the engine should be thoroughly oil flushed before being first placed in service.
5. Considerable importance should be given for cleanliness in handling bulk fuel oil. Dirt particles will ruin the fine lap of injection pumps or plug the injection nozzle orifices. So *high-grade filters* are of paramount importance to the diesel oil supply system.

3.8.4.1. Fuel injection system

The mechanical heart of the Diesel engine is the *fuel injection system*. The engine can perform no better than its fuel injection system. A very small quantity of fuel must be measured out, injected, atomised, and mixed with combustion air. The mixing problem becomes more difficult—the larger the cylinder and faster the rotative speed. Fortunately the high-speed engines are the small-bore automotive types ; however, special combustion arrangements such as precombustion chambers, air cells, etc., are necessary to secure good mixing. *Engines driving electrical generators have lower speeds and simple combustion chambers.*

3.8.4.2. Functions of a fuel injection system

1. Filter the fuel.
2. Meter or measure the correct-quantity of fuel to be injected.
3. Time the fuel injection.
4. Control the rate of fuel injection.
5. Automise or break up the fuel to fine particles.
6. Properly distribute the fuel in the combustion chamber.

The injection systems are manufactured with great accuracy, especially the parts that actually meter and inject the fuel. Some of the tolerances between the moving parts are very small of the order of *one micron.* Such closely fitting parts require special attention during manufacture and hence the injection systems are *costly.*

3.8.4.3. Types of fuel injection systems

The following fuel injection systems are commonly used in diesel power station :

1. Common-rail injection system.
2. Individual pump injection system.
3. Distributor.

Atomisation of fuel oil has been secured by (*i*) *air blast* and (*ii*) *pressure spray.* Early diesel engines used air fuel injection at about 70 bar. This is sufficient not only to inject the oil, but also to atomise it for a rapid and thorough combustion. The expense of providing an air compressor and tank lead to the development of "*solid*" injection, using a liquid pressure of between 100 and 200 bar which is sufficiently high to atomise the oil it forces through spray nozzles. Great advances have been made in the field of solid injection of the fuel through research and progress in fuel pump, spray nozzles, and combustion chamber design.

3.9. OPERATION OF A DIESEL POWER PLANT

When diesel alternator sets are put in parallel, "*hunting*" or "*phase swinging*" may be produced *due to resonance* unless due care is taken in the design and manufacture of the sets. This condition occurs *due to resonance between the periodic disturbing forces of the engine and natural frequency of the system.* The engine forces result from uneven turning moment on the engine crank which are corrected by the flywheel effect. "*Hunting" results from the tendency of each set trying to pull the other into synchronism* and is characterised by flickering of lights.

To ensure *most economical operation of diesel engines* of different sizes when working together and sharing load it is necessary that they should carry the same percentage of their full load capacity at all times as the fuel consumption would be lowest in this condition. For best operation performance the manufacturer's recommendations should be strictly followed.

In order to get good performance of a diesel power plant the following points should be taken care of :

1. It is necessary to maintain the *cooling temperature* within the prescribed range and use of very cold water should be avoided. The cooling water should be free from suspended impurities and suitably treated to be scale and corrosion free. If the ambient temperature approaches freezing point, the cooling water should be drained out of the engine when it is kept idle.

2. During operation the *lubrication system* should work effectively and requisite pressure and temperature maintained. The engine oil should be of the correct specifications and should be in a fit condition to lubricate the different parts. A watch may be kept on the consumption of lubricating oil as this gives an indication of the true internal condition of the engine.

3. *The engine should be periodically run even when not required to be used and should not be allowed to stand idle for more than 7 days.*

4. *Air filter, oil filters and fuel filters* should be periodically serviced or replaced as recommended by the manufacturers or if found in an unsatisfactory condition upon inspection.

5. Periodical checking of engine compression and firing pressures and also exhaust temperatures should be made.

- The engine exhaust usually provides a good indication of satisfactory performance of the engine. *A black smoke in the exhaust is a sign of inadequate combustion or engine overloading.*
- *The loss of compression resulting from wearing out of moving parts lowers the compression ratio causing inadequate combustion.* These defects can be checked by taking *indicator diagrams* of the engine after reasonable intervals.

3.10. TYPES OF DIESEL ENGINES USED FOR DIESEL POWER PLANTS

The diesel engines may be four-stroke or two stroke cycle engines. *The two-stroke cycle engines are favoured for diesel power plants.*

Efforts are being made to use "*dual fuel engines*" in diesel power plants for better economy and proper use of available gaseous fuels in the country. The gas may be a waste product as in the case of sewage treatment installations or oil fuels where the economic advantage is self evident. With the wider availability of natural gas, the dual fuel engines may become an attractive means of utilising gas as fuel at off-peak tariffs for the electric power generation.

Working of Dual Fuel Engines :

The various strokes of a dual fuel engine are as follows :

1. Suction stroke. During this stroke *air and gas are drawn* in the engine cylinder.

2. Compression stroke. During this stroke the pressure of the mixture drawn is increased. Near the end of this stroke the *'pilot oil' is injected into the engine cylinder. The compression heat first ignites the pilot oil* and then gas mixture.

3. Working/power stroke. During this stroke the gases (at high temperature) expand and thus power is obtained.

4. Exhaust stroke. The exhaust gases are released to the atmosphere during the stroke.

3.11. LAYOUT OF A DIESEL ENGINE POWER PLANT

Fig. 3.5 shows the layout of a diesel engine power plant.

The most common arrangement for diesel engines is with parallel centre lines, with some room left for extension in future. The repairs and usual maintenance works connected with such engines necessitate sufficient space around the units and consideration should be given to the need for dismantling and removal of large components of the engine generator set. The air intakes and filters as well as the exhaust mufflers are located outside the building or may be separated from the main engine room by a partition wall. The latter arrangement is not vibration free. Adequate space for oil storage and repair shop as well as for office should be provided close to the main engine room. Bulk storage of oil may be outdoor. The engine room should be well ventilated.

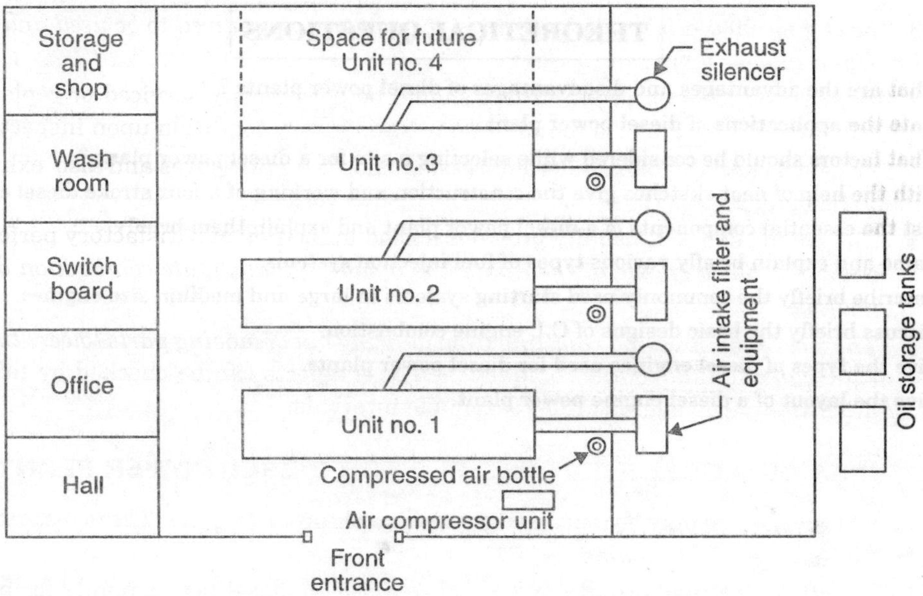

Fig. 3.5. Layout of a diesel engine power plant.

HIGHLIGHTS

1. Any type of engine or machine which derives heat energy from the combustion of fuel or any other source and converts this energy into mechanical work is termed as a *heat engine*.
2. Essential components of a diesel power plant are :
   (*i*) Engine (*ii*) Air intake system
   (*iii*) Exhaust system (*iv*) Fuel system
   (*v*) Cooling system (*vi*) Lubrication system
   (*vii*) Engine starting system (*viii*) Governing system.
3. Commonly used fuel injection system in a diesel power station :
   (*i*) Common-rail injection system
   (*ii*) Individual pump injection system
   (*iii*) Distribution system.
4. In liquid cooling following methods are used for circulating the water around the cylinder and cylinder head :
   (*i*) Thermo-system cooling (*ii*) Forced or pump cooling
   (*iii*) Cooling with thermostatic regulator (*iv*) Pressurised cooling
   (*v*) Evaporative cooling.
5. Various lubrication systems use for I.C. engines are :
   (*i*) Wet sump lubrication system
   (*ii*) Dry sump lubrication system
   (*iii*) Mist lubrication system.
6. The following three are the commonly used starting systems in large and medium size engines :
   (*i*) Starting by an auxiliary engine
   (*ii*) Use of electric motors or self starters
   (*iii*) Compressed air system.
7. The purpose of supercharging is to raise the volumetric efficiency above that value which can be obtained by normal aspiration.

THEORETICAL QUESTIONS

1. **What are the advantages and disadvantages of diesel power plants ?**
2. **State the applications of diesel power plant.**
3. **What factors should be considered while selecting a site for a diesel power plant ?**
4. **With the help of neat sketches give the construction and working of a four stroke diesel cycle engine.**
5. **List the essential components of a diesel power plant and explain them briefly.**
6. **Name and explain briefly various types of fuel injection systems.**
7. **Describe briefly the commonly used starting systems in large and medium size engines.**
8. **Discuss briefly the basic designs of C.I. engine combustion.**
9. **Give the types of diesel engines used for diesel power plants.**
10. **Give the layout of a diesel engine power plant.**

4

Gas Turbine Power Plants

4.1. Gas turbines—General aspects. 4.2. Applications of gas turbine plants. 4.3. Advantages and disadvantages of gas turbine power plants over diesel and thermal power plants. 4.4. Site selection. 4.5. The simple gas turbine plant. 4.6. Energy cycle for a simple-cycle gas turbine. 4.7. Performance terms. 4.8. Classification of gas turbine power plants. 4.9. Classification of gas turbines. 4.10. Combination gas turbine cycles—Combined turbine and steam power plants—Combined gas turbine and diesel power plants—Advantages of combined cycle. 4.11. Operation of a gas turbine. 4.12. Gas turbine power plant layout. 4.13. Components of a gas turbine power plant. 4.14. Various arrangements of gas turbine power plants. 4.15. Effect of thermodynamic variables on air rate. Highlights—Theoretical Questions.

4.1. GAS TURBINES—GENERAL ASPECTS

Probably a windmill was the first turbine to produce useful work, wherein there is no pre-compression and no combustion. The characteristic features of a gas turbine as we think of the name today include a *compression process* and a *heat-addition* (or combustion) process. The gas turbine represents perhaps the *most satisfactory way of producing very large quantities of power in a self-contained and compact unit*. The gas turbine may have an ample future use in conjunction with the oil engine. For smaller gas turbine units, the inefficiencies in compression and expansion processes become greater and to improve the thermal efficiency it is necessary to use a heat exchanger. In order that a small gas turbine may compete for economy with the small oil engine or petrol engine it is necessary that a compact effective heat exchanger be used in the gas turbine cycle. The thermal efficiency of the gas turbine alone is still quite modest 20 to 30% compared with that of a modern steam plant 38 to 40%. It is possible to construct *combined plants* whose efficiencies are of the order of 45% or more. Higher efficiencies might be attained in future.

The following are the major fields of application of gas turbines :

1. Aviation
2. Power generation
3. Oil and gas industry
4. Marine propulsion.

The efficiency of a gas turbine is not the criteria for the choice of this plant. A gas turbine is used in *aviation* and marine fields because *it is self contained, light weight not requiring cooling water and generally fit into the overall shape of the structure.* It is selected for '*power generation*' because of its *simplicity, lack of cooling water, needs quick installation and quick starting*. It is used in *oil and gas industry* because of *cheaper supply* of *fuel and low installation cost*.

The gas turbines have the following ***"limitations"*** :

1. They are not self starting.
2. Low efficiencies at part loads.

3. Non-reversibility.
4. Higher rotor speeds.
5. Low overall plant efficiency.

In the last two decides, rapid progress has been observed in the *development and improvement of the gas turbine plants for electric power production*. The major progress has been observed in the following *directions* :

(*i*) Increase in unit capacities of gas turbine units.
(*ii*) Increase in their efficiency.
(*iii*) Drop in capital cost.

4.2. APPLICATIONS OF GAS TURBINE PLANTS

Gas turbine plants for the purpose of power plant engineering find the following *applications* :

1. To drive generators and supply peak loads in steam, diesel or hydroplants.
2. To work as combination plants with conventional steam boilers.
3. To supply mechanical drive for auxiliaries.

- These plants are well suited for *peak load service* since the fuel costs are somewhat higher and initial cost low. Moreover, *peak load operation permits use of water injection which increases turbine work by about 40% with an increase in heat rate of about 20%.* The short duration of increase in heat rate does not prove of any much harm.
- The combination arrangement of gas turbines with conventional boilers may be supercharging or for heat recovery from exhaust gases. In the *supercharging system* air is supplied to the boiler under pressure by a compressor mounted on the common shaft with turbine and gases formed as result of combination after coming out of the boiler pass through the gas turbine before passing through the economiser and the chimney.
- The application of the gas turbine to drive the auxiliaries is not strictly included under direct electric power generation by the turbines and would not be discussed.

4.3. ADVANTAGES AND DISADVANTAGES OF GAS TURBINE POWER PLANTS OVER DIESEL AND THERMAL POWER PLANTS

A. Advantages Over Diesel Plants :

1. The work developed per kg of air is large compared with diesel plant.
2. Less vibrations due to perfect balancing.
3. Less space requirements.
4. Capital cost considerably less.
5. Higher mechanical efficiency.
6. The running speed of the turbine (40,000 to 100,000 r.p.m.) is considerably large compared to diesel engine (1000 to 2000 r.p.m.).
7. Lower installation and maintenance costs.
8. The torque characteristics of turbine plants are far better than diesel plants.
9. The ignition and lubrication systems are simpler.
10. The specific fuel consumption does not increase with time in gas turbine plant as rapidly as in diesel plants.
11. Poor quality fuels can be used.

Disadvantages :

1. Poor part load efficiency.
2. Special metals and alloys are required for different components of the plants.
3. Special cooling methods are required for cooling the turbine blades.
4. Short life.

B. Advantages Over Steam Power Plant :

1. No ash handling problem.
2. Low capital cost.
3. The gas turbine plants can be installed at selected load centre as space requirement is considerably less where steam plant cannot be accommodated.
4. Fewer auxiliaries required/used.
5. Gas turbines can be built relatively quicker. They require much less space and civil engineering works and water supply.
6. The gas turbine plant as peak load plant is more preferable as it can be brought on load quickly and surely.
7. The components and circuits of a gas turbine plant can be arranged to give the most economic results in any given circumstances which is not possible in case of steam power plants.
8. For the same pressure and initial temperature conditions the ratio of exhaust to inlet volume would be only 3.95 in case of gas turbine plant as against 250 for steam plant.
9. Above 550°C, the thermal efficiency of the gas turbine plant increases three times as fast the steam cycle efficiency for a given top temperature increase.
10. The site of the steam power plant is dictated by the availability of large cooling water whereas an open cycle gas turbine plant can be located near the load centre as no cooling water is required. The cooling water required for closed cycle gas turbine is hardly 10% of the steam power plant.
11. The gas turbine plants can work quite economically for short running hours.
12. Storage of fuel is much smaller and handling is easy.

4.4. SITE SELECTION

While selectinig the site for a gas turbine plant. The following *points* should be given due consideration :

1. The plant should be located near the load centre to avoid transmission costs and losses.
2. The site should be away from business centres due to noisy operations.
3. Cheap and good quality fuel should be easily available.
4. Availability of labour.
5. Availability of means of transportation.
6. The land should be available at a cheap price.
7. The bearing capacity of the land should be high.

4.5. THE SIMPLE GAS TURBINE PLANT

A gas turbine plant may be defined as one *"in which the principal primemover is of the turbine type and the working medium is a permanent gas"*.

Refer Fig. 4.1. A simple gas turbine plant consists of the following :

1. *Turbine.*
2. *A compressor* mounted on the same shaft or coupled to the turbine.
3. *The combustor.*
4. *Auxiliaries* such as starting device, auxiliary lubrication pump, fuel system, oil system and the duct system etc.

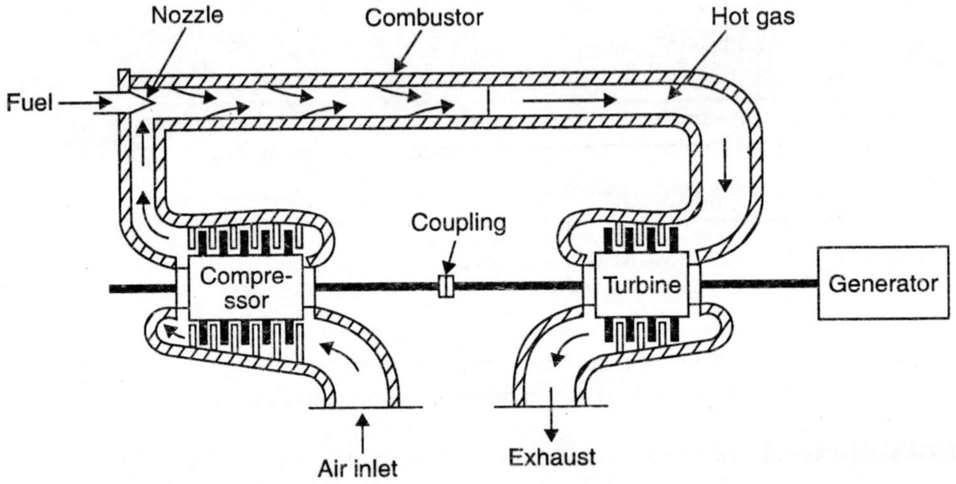

Fig. 4.1. Arrangement of a simple gas turbine plant.

A modified plant may have in addition to above an *intercooler, a regenerator, a reheater* etc.

The working fluid is compressed in a compressor which is generally rotary, multistage type. Heat energy is added to the compressed fluid in the chamber. This high energy fluid, at high temperature and presssure, then expands in the turbine unit thereby generating power. Part of the power generated is consumed in driving the generating compressor and accessories and the rest is utilised in electrical energy. The gas turbines work on open cycle, semi-closed cycle or closed cycle. In order to improve efficiency, compression and expansion of working fluid is carried out in multistages.

4.6. ENERGY CYCLE FOR A SIMPLE-CYCLE GAS TURBINE

Fig. 4.2 shows an energy-flow diagram for a simple-cycle gas turbine, the description of which is as follows :

- The air brings in minute amount of energy (measured above 0°C).
- Compressor adds considerable amount of energy.
- Fuel carries major input to cycle.
- Sum of fuel and compressed-air energy leaves combustor to enter turbine.
- In turbine smallest part of entering energy goes to useful output, largest part leaves in exhaust.

Shaft energy to drive compressor is about twice as much as the useful shaft output.

Actually the shaft energy keeps circulating in the cycle as long as the turbine runs. The important comparison is *the size of the output with the fuel input.* For the simple-cycle gas turbine the output may run about 20 per cent of the fuel input for pressure and temperature conditions at turbine inlet. This means 80% of the fuel energy is wasted. While the 20% thermal efficiency is not too bad, it can be improved by including *additional heat recovery apparatus.*

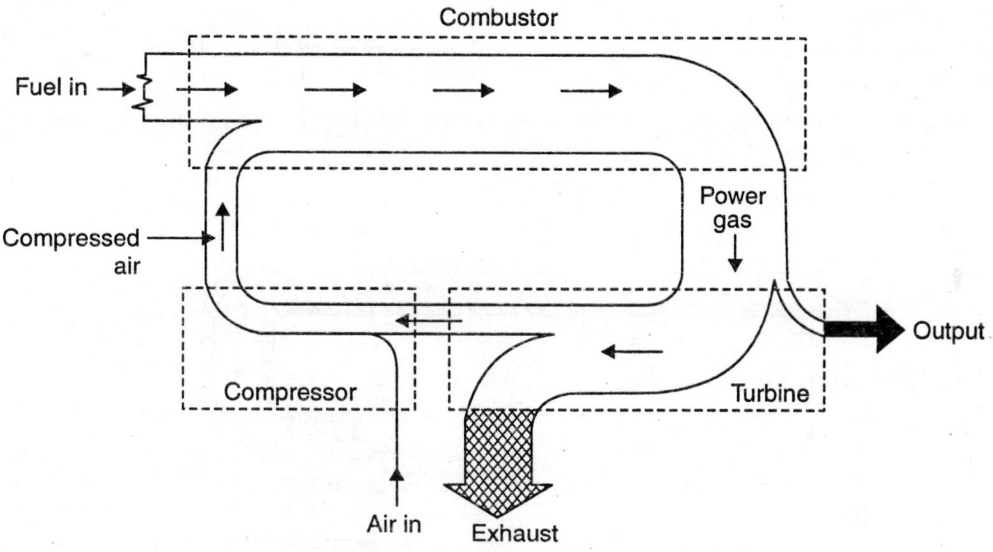

Fig. 4.2. Energy flow diagram for gas-turbine unit.

4.7. PERFORMANCE TERMS

Some of the important terms used to measure performance of a gas turbine are defined as follows :

1. Pressure ratio. It is the ratio of cycle's highest to its lowest pressure, usually highest-pressure-compressor discharges to the lowest-pressure-compressor inlet pressures.

2. Work ratio. It is the ratio of network output to the total work developed in the turbine or turbines.

3. Air ratio. Kg of air entering the compressor inlet per unit of cycle net output, for example, kg/kWh.

4. Compression efficiency. It is the ratio of work needed for ideal air compression through a given pressure range to work actually used by the compressor.

5. Engine efficiency. It is the ratio of the work actually developed by the turbine expanding hot power gas through a given pressure range to that would be yielded for ideal expansion conditions.

6. Machine efficiency. It is the collective term meaning both engine efficiency and compressor efficiency of turbine and compressor, respectively.

7. Combustion efficiency. It is the ratio of heat actually released by 1 kg of fuel to heat that would be released by complete perfect combustion.

8. Thermal efficiency. It is the percentage of total energy input appearing as net work output of the cycle.

4.8. CLASSIFICATION OF GAS TURBINE POWER PLANTS

The gas turbine power plants may be *classified* according to the following criteria :

1. By application :

(*i*) *In aircraft*

(*a*) Jet propulsion (*b*) Prop-jets

(*ii*) *Stationary*

(*a*) Peak load unit
(*b*) Standby unit
(*c*) End of transmission line unit
(*d*) Base load unit
(*e*) Industrial unit.

(*iii*) Locomotive
(*iv*) Marine
(*v*) Transport.

2. By cycle :

(*i*) Open cycle
(*ii*) Closed cycle
(*iii*) Semi-closed cycle.

3. According to arrangement :

(*i*) Simple
(*ii*) Single shaft
(*iii*) Multi-shaft
(*iv*) Intercooled
(*v*) Reheat
(*vi*) Regenerative
(*vii*) Combination.

4. According to combustion :

(*i*) Continuous combustion
(*ii*) Intermittent combustion.

5. By fuel :

(*i*) Solid fuel
(*ii*) Liquid fuel
(*iii*) Gaseous fuel.

4.9. CLASSIFICATION OF GAS TURBINES

The gas turbines are mainly divided into two groups :

1. **Constant pressure combustion gas turbine**

(*a*) Open cycle constant pressure gas turbine

(*b*) Closed cycle constant pressure gas turbine.

2. **Constant volume combustion gas turbine**

In almost *all the fields open cycle gas turbine plants are used. Closed cycle plants were introduced at one stage because of their ability to burn cheap fuel.* In between their progress remained slow because of availability of cheap oil and natural gas. Because of rising oil prices, now again, the attention is being paid to closed cycle plants.

4.10. COMBINATION GAS TURBINE CYCLES

4.10.1. Combined Turbine and Steam Power Plants

The characteristics of the gas turbine plants render these plants very well suited for use in combination with steam or hydro-plants. These plants can be quickly started for emergency or peak load service. The combination *'gas-turbine-steam cycles' aim at utilising the heat of exhaust gases from the gas turbine and thus, improve the overall plant efficiency.*

Three popular designs of combination cycle comprise of :

1. Heating feed water with exhaust gases.
2. Employing the gases from a supercharged boiler to expand in the gas turbine.
3. Employing the gases as combustion air in the steam boiler.

1. Heating feed water with exhaust gases :

Refer Fig. 4.3. By employing exhaust gases to heat the feed water it is possible to utilise the entire steam supply to the turbines to expand through entire range of expansion, and thus result in increasing the output of work, since bleeding would not be required.

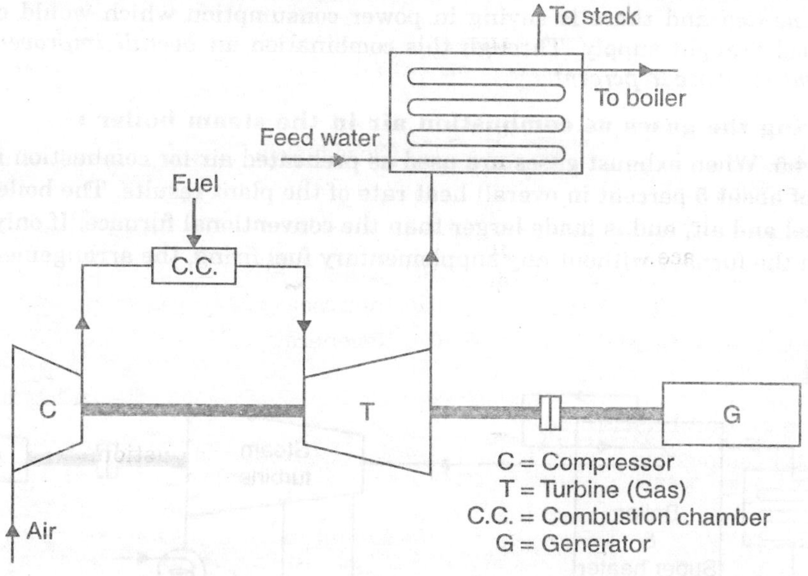

Fig. 4.3. Use of exhaust gases to heat feed water of steam cycle.

2. Employing the gases from a supercharged boiler to expand in the gas turbine :

Refer Fig. 4.4. The supercharged boiler is good application of the combined gas turbine-steam cycle. In this arrangement of the boiler furnace works under a pressure of about 5 bar and

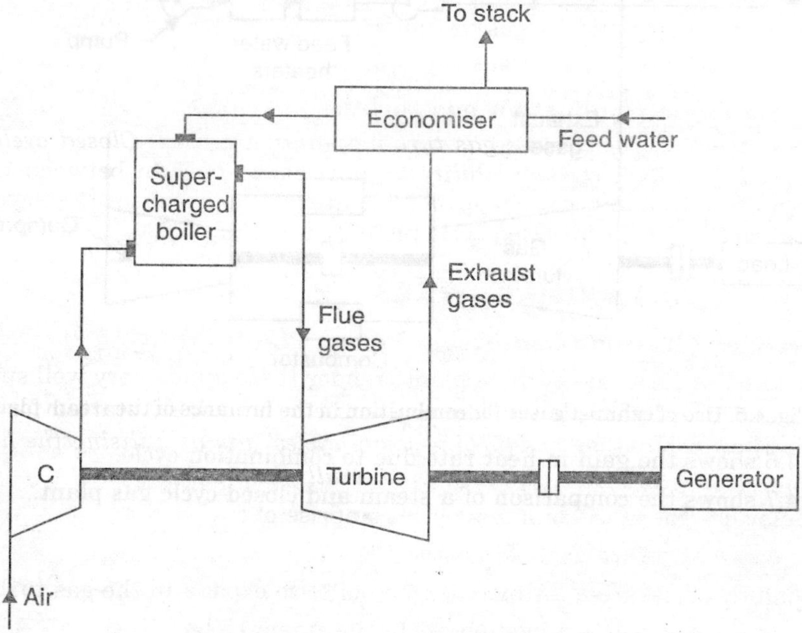

Fig. 4.4. Arrangement of supercharged boiler.

the gases are expanded in the gas turbine, its exhaust being used to heat feed water before being discharged through the stack. The *heat transfer rate in this boiler is very high as compared to that in a conventional boiler due to higher pressure of gases ; and a smaller size of steam generator is needed* for the same steam raising capacity as of the conventional plant. Further more, since the gases in the furnace are already under pressure, *no induced draught or forced draught fans are needed* and there is saving in power consumption which would otherwise be spent in mechanical draught supply. Through this combination an *overall improvement in heat rate is to the extent of above 7 percent.*

3. Employing the gases as combustion air in the steam boiler :

Refer Fig. 4.5. When exhaust gases are used as preheated air for combustion in the boiler, an improvement of about 5 percent in overall heat rate of the plant results. The boiler is fed with supplementary fuel and air, and is made larger than the conventional furnace. If only the turbine exhaust is used in the furnace without any supplementary fuel firing, the arrangement becomes a waste heat boiler.

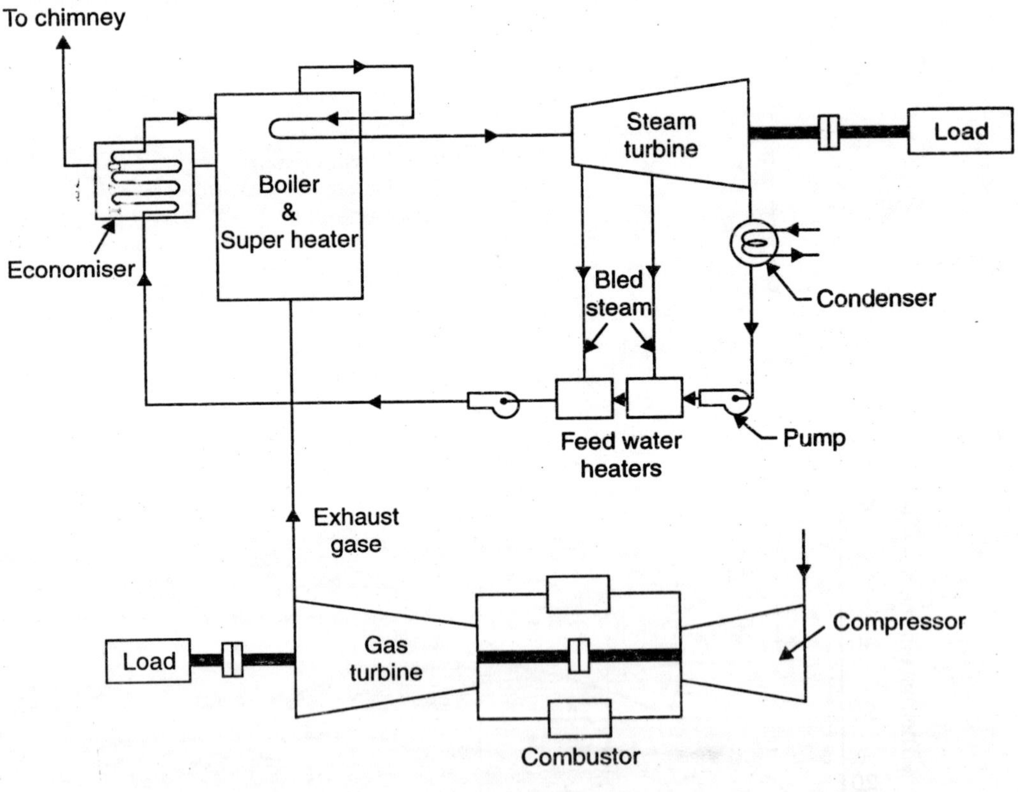

Fig. 4.5. Use of exhaust gases for combustion in the furnance of the steam plant.

- Fig. 4.6 shows the gain in heat rate due to combination cycle.
- Fig. 4.7 shows the comparison of a steam and closed cycle gas plant.

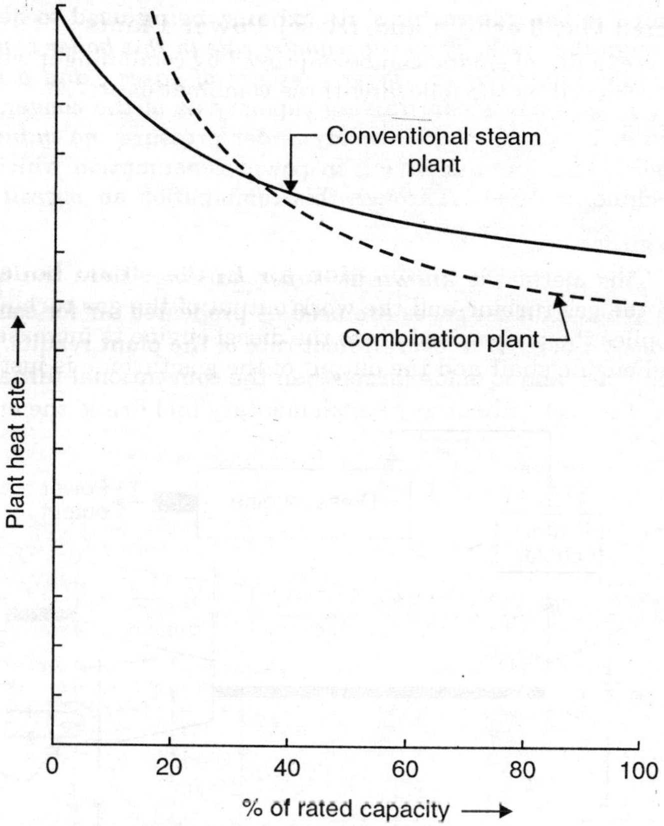

Fig. 4.6

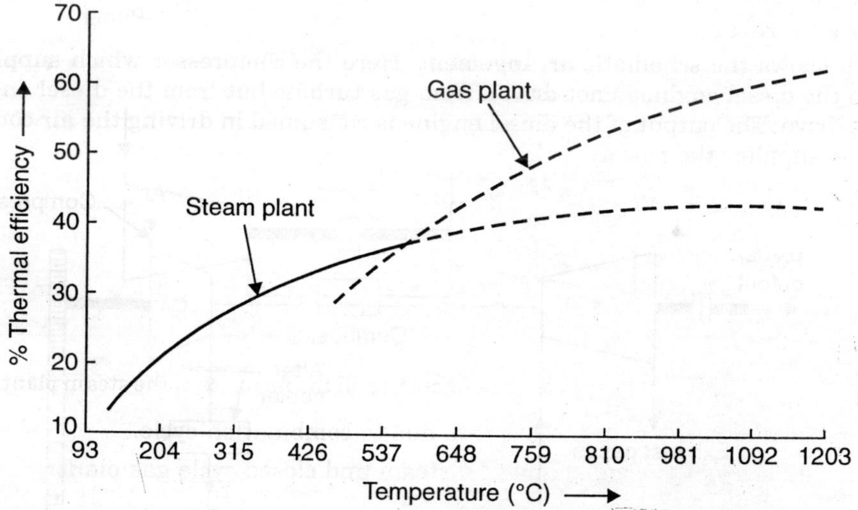

Fig. 4.7. Comparison between steam and closed cycle gas turbine plant.

4.10.2. Combined Gas Turbine and Diesel Power Plants

The performance of a diesel engine can be improved by combining it with an exhaust driven gas turbine. It can be achieved by the following three combinations :

1. Turbo-charging.
2. Gas-generator.
3. Compound engine.

1. Turbo-charging :

Refer Fig. 4.8. This method is known as *supercharging*. Here the exhaust of the diesel engine is expanded in the gas turbine and the work output of the gas turbine is utilised to run a compressor which supplies the pressurised air to the diesel engine to increase its output. The load is coupled to the diesel engine shaft and the output of the gas turbine is just sufficient to run the compressor.

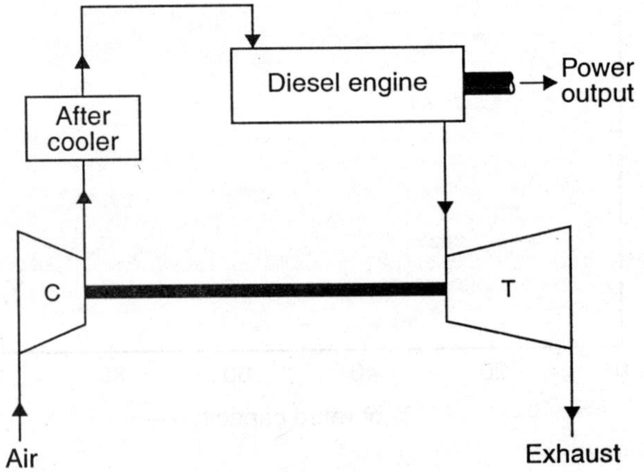

Fig. 4.8. Turbo-charging.

2. Gas-generator :

Fig. 4.9 shows the schematic arrangement. Here the compressor which supplies the compressed air to the diesel engine is not driven from gas turbine but from the diesel engine through some suitable drive. The output of the diesel engine is consumed in driving the air compressor and the gas turbine supplies the power.

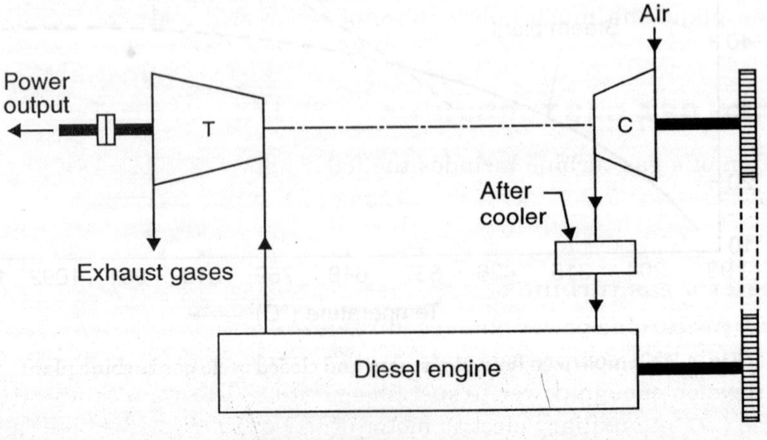

Fig. 4.9. Gas-generator.

3. Compound engine :

Refer Fig. 4.10. In this arrangement the air compressor is driven from both diesel engine and gas turbine through a suitable gearing and the power output is taken from the diesel engine shaft.

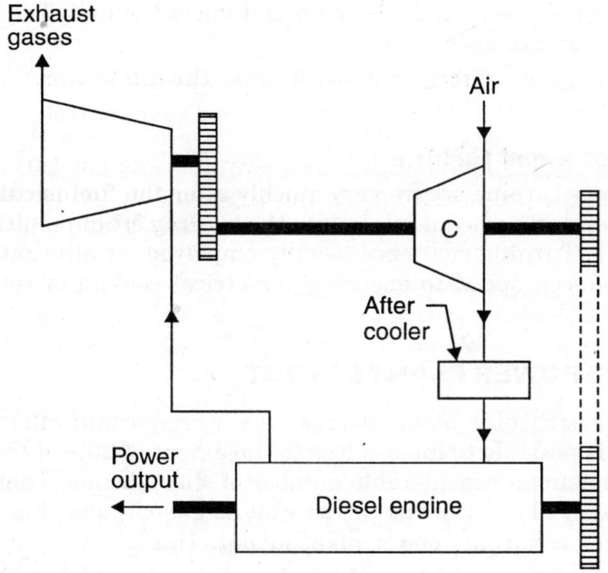

Fig. 4.10. Compound engine.

4.10.3. Advantages of Combined Cycle

The *advantages of combined cycle* are :

1. Improved efficiency.
2. More suitable for rapid start and shut down.
3. Less cooling water requirement.
4. It gives high ratio of power output to occupied ground space.
5. The combined system offers self-sustaining feature.
6. The capital cost of combined plant with supplementary firing is slightly higher than a simple gas turbine plant and much below those of a classical steam plant of the same power capacity.

4.11. OPERATION OF A GAS TURBINE

The operation of a gas turbine includes the following :

1. Starting
2. Shut down

1. Starting of a gas turbine :

Starting of a gas turbine power plant requires an *auxiliary power source*, till the plants own compressor inducts air and compresses it to a pressure such that expansion from reasonable temperature will develop enough power to sustain operation. The *starter* may be (*i*) an I.C. engine (*ii*) a steam turbine (*iii*) an auxiliary electric motor or (*iv*) another gas turbine. It must be coupled to the turbo-compressor shaft with a disengaging or over-running clutch. A main generator or its direct connected exciter may be pressed into temporary service as a motor.

Starting procedure :

1. Run the unit and induct air.

2. Actuate the combustion ignition system and inject the fuel. The fuel flow is controlled to obtain the necessary warm up.

3. Adjust the speed and voltage and synchronise the alternator.

4. Build up the load on the alternator by governor gear control.

2. Shut down of a gas turbine :

Shut down of a gas turbine occurs very quickly after the fuel is cut off from the combustor. A rapid shut down is desirable since it minimizes the *chilling effect* resulting from passing cold air through the hot turbine. Turning gears are usually employed, or alternatively the starting device may be operated at reduced speed to ensure symmetrical cooling of rotor. This avoids thermal '*kinking*' of the shaft.

4.12. GAS TURBINE POWER PLANT LAYOUT

The layout of a gas turbine plant exercises a very important effect on the overall performance of the plant, as it is possible to incur a loss to the extent of 20% of the power developed in the interconnecting ducts having a considerable number of sharp bends. Therefore ample care should be taken in the design and layout of the air as well as gas circuits. Fig. 4.11 shows a schematic simplified diagram of a gas turbine power plant layout. The following points are *worth noting* :

- In this case the main building is the turbine house in which major portion of the plant as well as auxiliaries are installed. Usually it is similar to the steam plant house in many respects.
- The storage tanks containing fuel oil are arranged outside but adjoining the turbine house. In some installations even heat exchangers are also out of the doors.

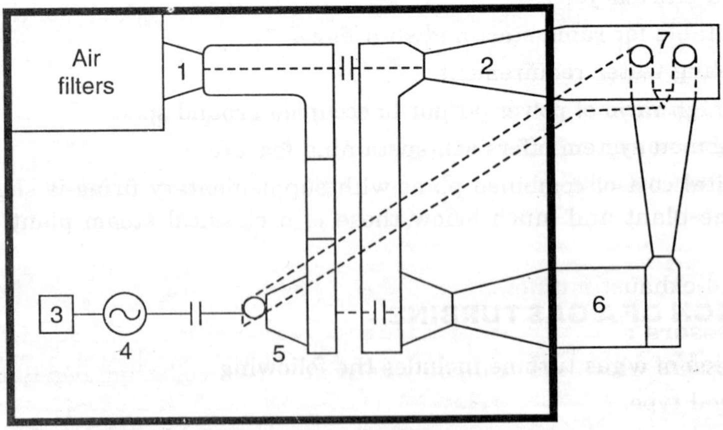

1 = L.P. compressor   2 = L.P. turbine
3 = Starting motor   4 = Alternator
5 = H.P. compressor   6 = Combustion chamber
7 = Heat exchangers.

Fig. 4.11. Gas turbine power plant layout.

- Whereas the major portion of the total space is occupied by the intercoolers, combustion chambers, heat exchangers, waste heat boilers and interconnecting duct work, the rotating parts of the plant form a very small part of the total volume of the plant.

4.13. COMPONENTS OF A GAS TURBINE POWER PLANT

The main components of a gas turbine power plant are enumerated and discussed as follows :

1. Gas turbines
2. Compressors
3. Combustor
4. Intercoolers and regenerators.

1. Gas turbines :

A turbine basically employs vanes or blades mounted on a shaft and enclosed in a casing. The flow of fluid through the turbine in *most design is axial and tangential to the rotor* at a nearly constant or increasing radius. The basic requirements of the turbines are : (*i*) *Light weight* (*ii*) *High efficiency* (*iii*) *Reliability in operation and* (*iv*) *Long working life*. Large work output can be obtained per stage with high blade speeds when the blades are designed to sustain higher stresses. *More stages of the turbine are always preferred in gas turbine power plant because it helps to reduce the stresses in the blades and increases the overall life of the turbine.*

It is essential to cool the gas turbine blades for long life as these are continuously subjected to high temperature gases. The blades can be cooled by different methods, the common method being the *air-cooling*. The air is passed through the holes provided through the blade.

The following **accessories** are fitted to the turbine :

(*i*) *Tachometer*. It shows the speed of the machine and also actuates the fuel regulator in case the speed shoots above or falls below the regulated speed, so that the fuel regulator admits less or more fuel into the combustor and varies the turbine power according to the demand. The tachometer is driven through a gear box.

(*ii*) *An overspeed governor*. The governor backs off fuel feed if exhaust temperature from the turbine exceeds the safe limit, thermal switches at the turbine exhaust acting on fuel control to maintain present maximum temperature.

(*iii*) *Lubricating oil pump*. It supplies oil to the bearings under pressure.

(*iv*) Starting motor or engine

(*v*) Starting set-up gear

(*vi*) Oil coolers

(*vii*) Filters

(*viii*) Inlet and exhaust mufflers.

2. Compressors :

The compressors which are commonly used are of the following two types :

1. Centrifugal type.
2. Axial flow type.

- The *'centrifugal compressor'* consists of an *impeller* and a *diffuser*. The *impeller imparts the high kinetic energy to the air* and *diffuser converts the kinetic energy into the pressure energy*. The pressure ratio of 2 to 3 is possible with single stage compressor and it can be increased upto 20 with 3-stage compressor. The compressors may have single or double inlet. The *single inlet* compressors are designed to handle the air in the range of 15 to 300 m^3/min. and *double inlets* are preferred above 300 m^3/min capacity. Fig. 4.12 shows a single stage centrifugal compressor.

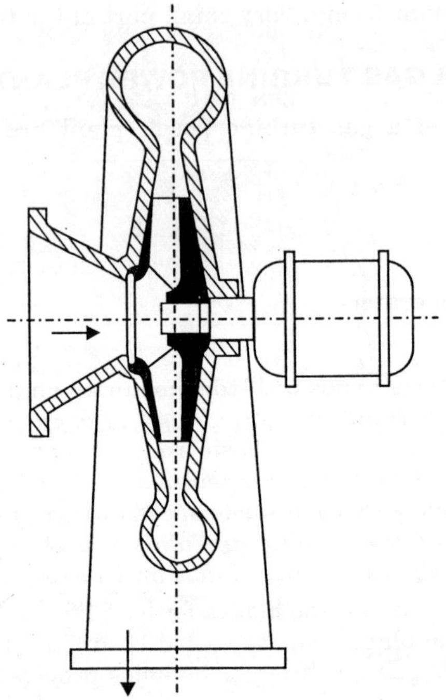

Fig. 4.12. Single stage centrifugal compressor.

In the centrifugal compressor :

(*i*) *Capacity* varies *directly as the speed ratio.*

(*ii*) *Total pressure* varies as the *square of speed ratio.*

(*iii*) *Power input* varies as the *cube of the speed ratio.*

(*iv*) Pressure is developed *independent of load*, but the *volume handled depends on the load.*

In a *multistage centrifugal compressor*, two or more impellers operating in series on a single shaft are provided in a single casing. The effect of multi-staging is to increase the delivery pressure of air as air compressed in one stage of machine is fed into the next stage for further compression and pressure is multiplied in each stage. The *efficiency of multistage compressor is lower than a single stage due to the losses.*

The centrifugal compressor is *superior to the axial flow machine in that a high pressure ratio can be obtained in a short rugged single stage machine, though at the cost of lower efficiency and increased frontal area.*

- The *'axial flow compressors'* are commonly *used in gas turbine installations.* An axial flow compressor consists of a series of rotor and stator stages with decreasing diameters along the flow of air. The blades are fixed on the rotors and rotors are fixed on the shaft. The stator blades are fixed on the stator casing. The stator blades guide the air flow to the next rotor stage coming from the previous rotor stage. The air flows along the axis of the rotor. The kinetic energy is given to the air as it passes through the rotor and part of it is converted into pressure. A multi-stage single flow axial compressor is shown in Fig. 4.13.

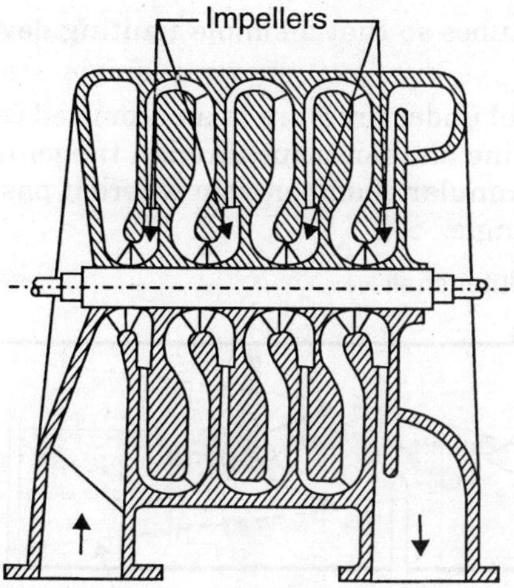

Fig. 4.13. Multi-stage single flow axial compressor.

An axial compressor is capable of *delivering constant volumes of air over varying discharge pressures*. These machines are well suited for large capacities at moderate pressures. If the impeller of a centrifugal compressor is designed to give an axial component of velocity at the exit, the design becomes a *mixed flow type*.

3. Combustor :

The primary function of the *combustor is to provide for the chemical reaction of the fuel and air being supplied by the compressor*. It must fulfil the following *conditions* :

(*i*) Combustion must take place at high efficiency because of the effect of the combustion efficiency on the thermal efficiency of the gas turbine cycle.

(*ii*) The pressure losses must be low.

(*iii*) Ignition must be reliable and accomplished with ease over a wide range of atmospheric condition especially in aircraft installation.

(*iv*) Thorough mixing of fuel and air.

(*v*) Carbon deposits must not be formed under any conditions.

The physical process of combustion may be divided into four important steps :

1. Formation of reactive mixture
2. Ignition
3. Flame propagation
4. Cooling of combustion products with air. Atomisation should be done for perfect burning.

Fig. 4.14 shows an arrangement of a typical combustor design which employs an *outer cylindrical shell* with a *conical inner sleeve* which is provided with ports or slots along the length. At the cone apex is fitted a *nozzle* through which fuel is sprayed in a conical pattern into the sleeve and near this is an *igniting device* or spark plug. A fuel line conveys the fuel to the nozzle. A few air ports provided close to the situation of the nozzle supply the combustion air directly to the fuel and are fitted with vanes to produce a whirling motion of oil and thereby create turbulence. The rest of the air admitted ahead of combustion zone serves to cool the combustor and outlet gases. The combustor is best located between the compressor outlet and turbine inlet and takes the shape of a cylinder. Alternatively, the 'can' arrangement may be used in which the flow is divided to pass through a number of smaller cylindrical chambers. In this latter design the adjacent chambers may be interconnected through small tubes so that a simple igniting device fitted in one of the chambers serves all the chambers.

The *nozzle* sprays the fuel under pressure in an atomised conical spray. The fuel is delivered to the nozzle through the fuel line and flows out through tangential slots in the nozzle, thus being given a whirling motion in an annular chamber from where it passes out through a small orifice in the conical pattern of desired angle.

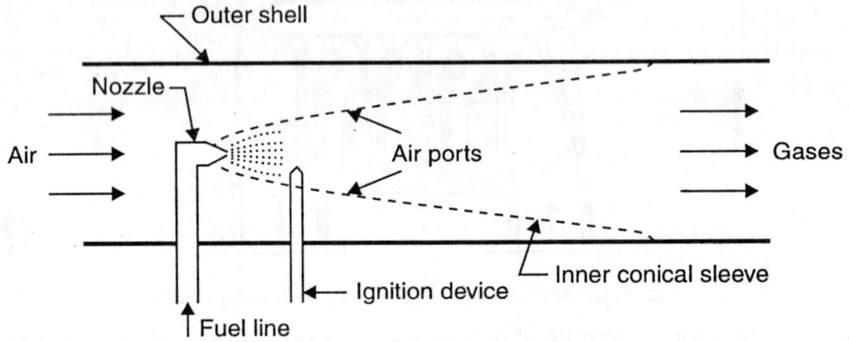

Fig. 4.14. Arrangement of a combustor.

4. Intercoolers and regenerators :

Intercoolers. In a gas turbine plant the *intercooler* is generally used when the pressure ratio used is sufficiently large and the compression is completed with two or more stages. The cooling of compressed air is generally done with the use of cooling water. A *cross-flow* type intercooler (Fig. 4.15) is generally preferred for effective heat transfer.

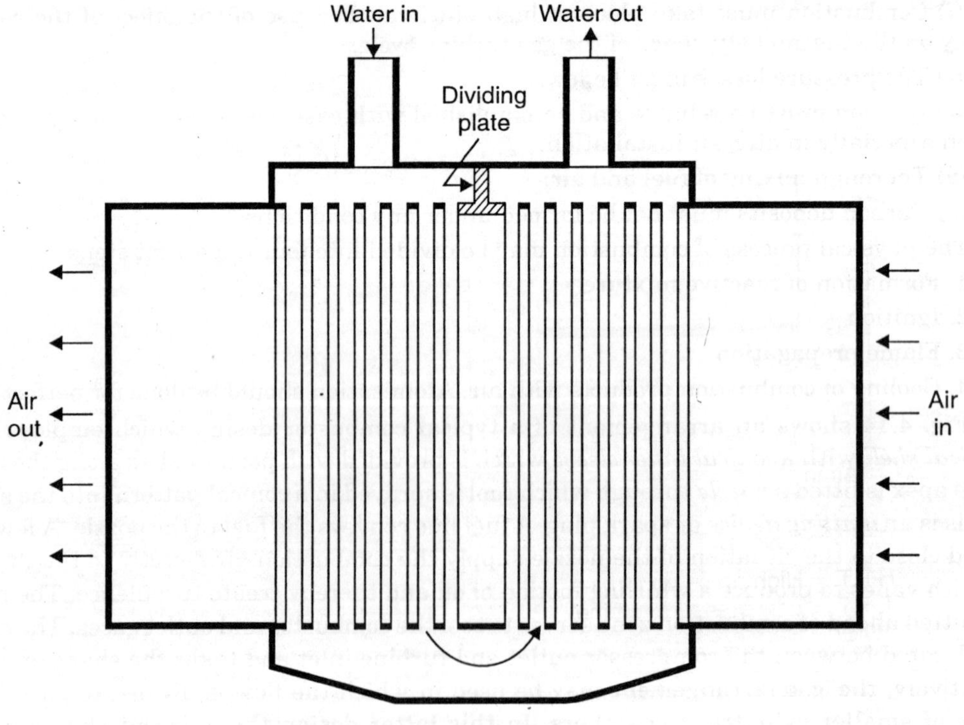

Fig. 4.15. Intercooler.

Regenerators. Refer to Fig. 4.16. In the regenerator heat transfer takes place between the exhaust gases and cool air. It is usually made in shell and tube construction with gas flowing inside the tubes and air outside the tubes, the two fluids being made to flow in opposite directions. Since the gas is bound to carry dust and deposit the same on the heat transfer surface, the internal flow through the tubes is convenient as the tube inside can be easily cleaned with brushes whereas it is very difficult to clean the outside surface of tubes. The effect of counterflow is the highest average temperature difference between the heating and heated medium with consequent high heat transfer between the two fluids. A number of baffles in the air put in the shell make the air to flow in contact with maximum heat transfer. However, *the pressure drop in both air and gas during the flow should be minimum possible*.

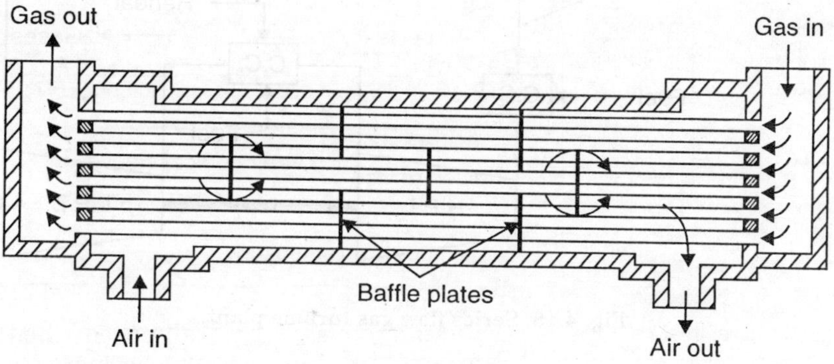

Fig. 4.16. Regenerator.

4.14. VARIOUS ARRANGEMENTS OF GAS TURBINE POWER PLANTS

The various arrangements of gas turbine plants are shown in Figs. 4.17 to 4.22.

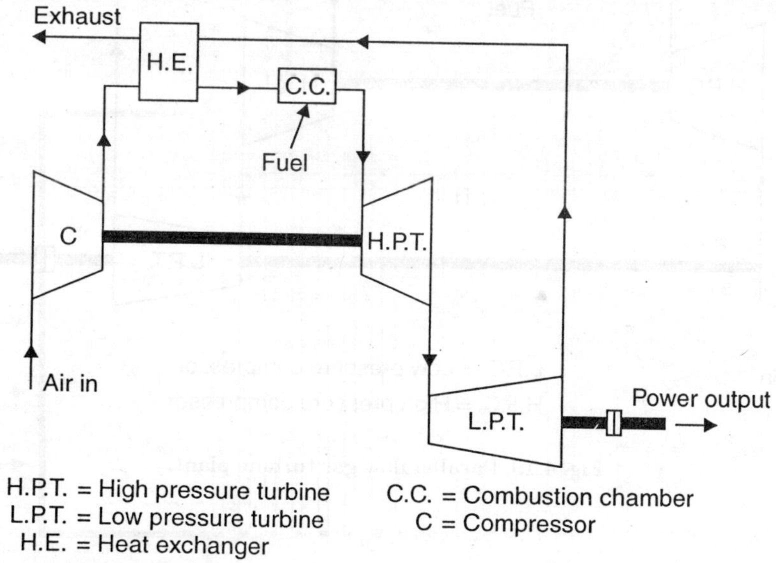

Fig. 4.17. Open cycle gas turbine with separate power turbine.

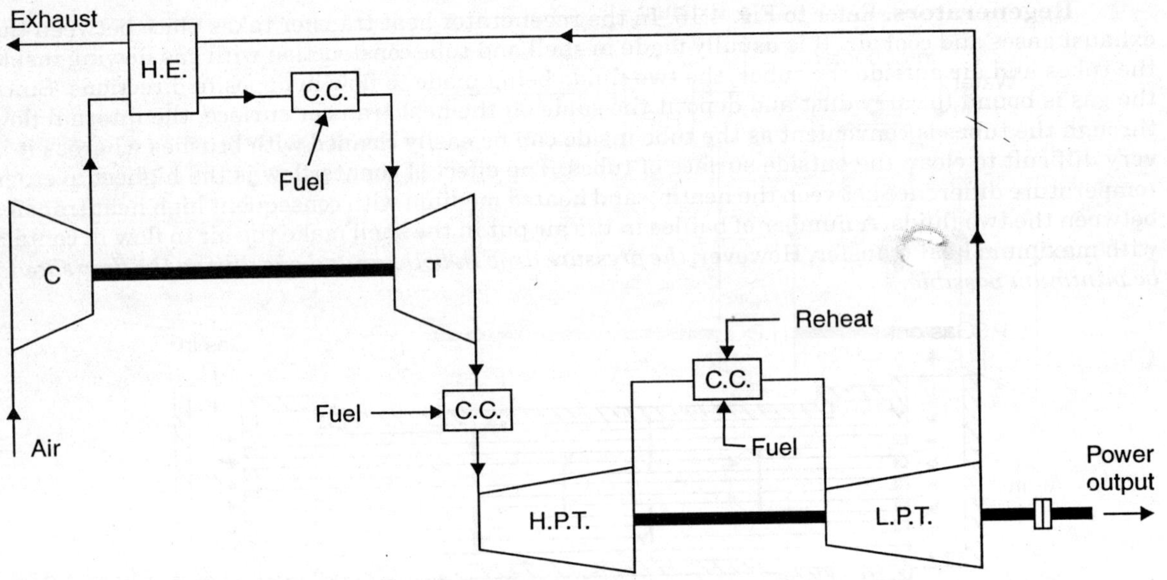

Fig. 4.18. Series flow gas turbine plant.

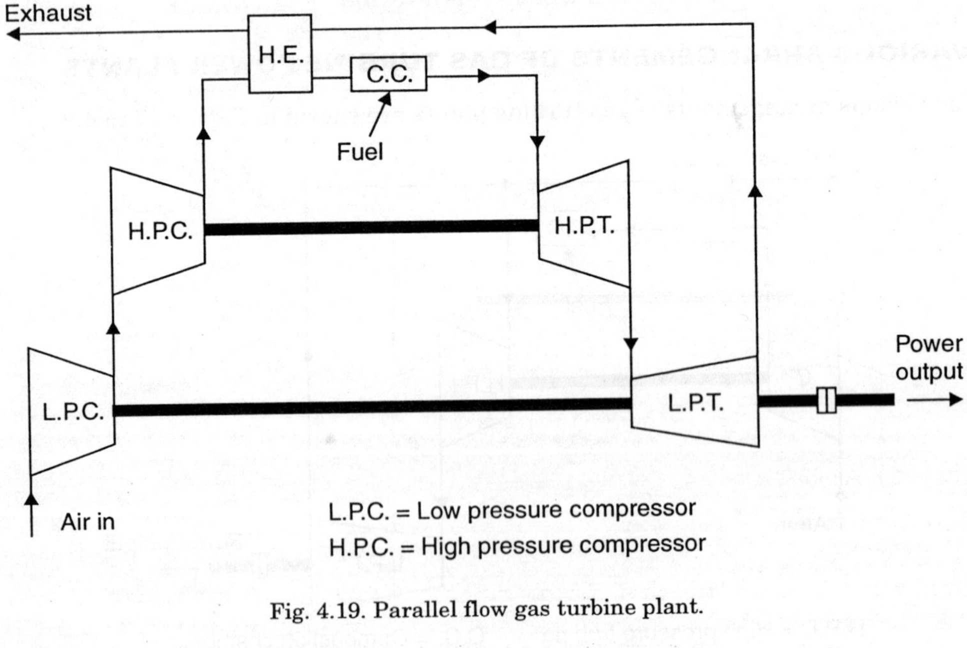

Fig. 4.19. Parallel flow gas turbine plant.

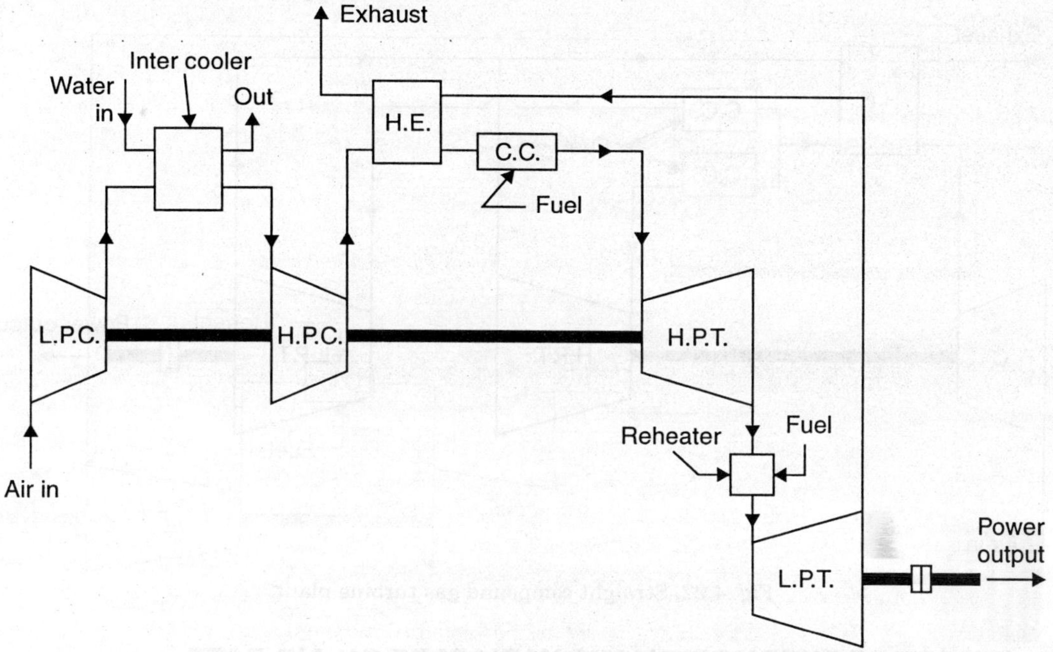

Fig. 4.20. Series flow plant with intercooled compression.

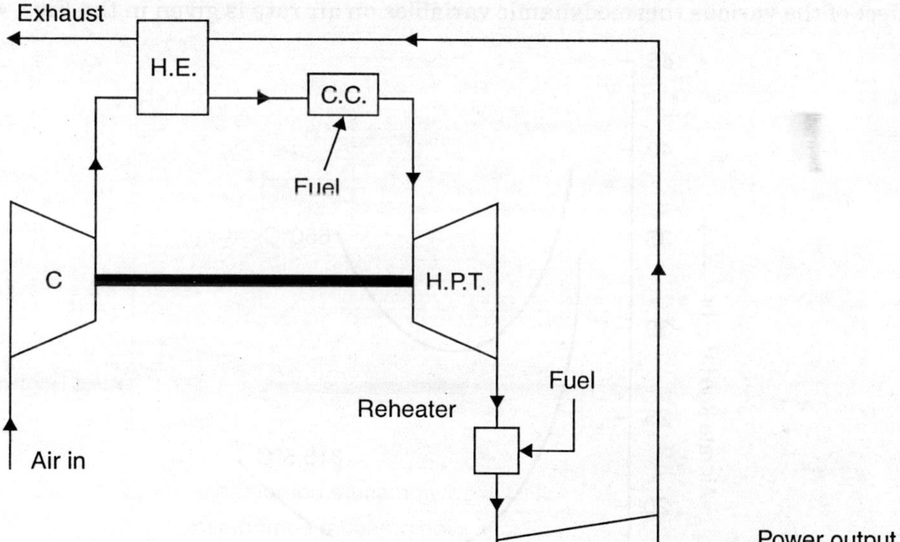

Fig. 4.21. Series flow plant with reheat between power turbine stages.

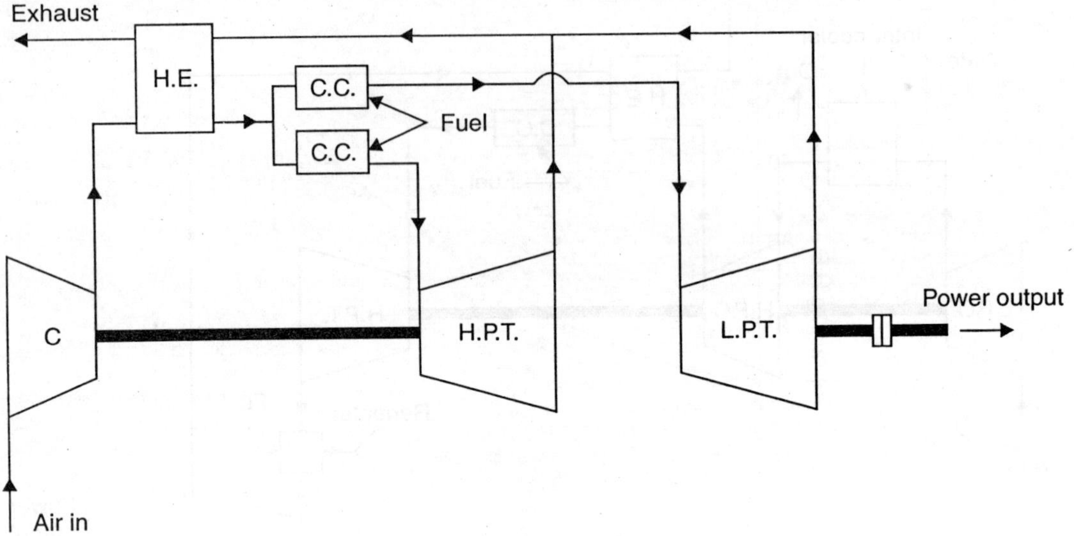

Fig. 4.22. Straight compound gas turbine plant.

4.15. EFFECT OF THERMODYNAMIC VARIABLES ON AIR RATE

The gas turbine plant size depends upon air rate. *The lower the air rate, the smaller the plant size*. The effect of the various thermodynamic variables on air rate is given in the Figs. 4.23 to 4.26.

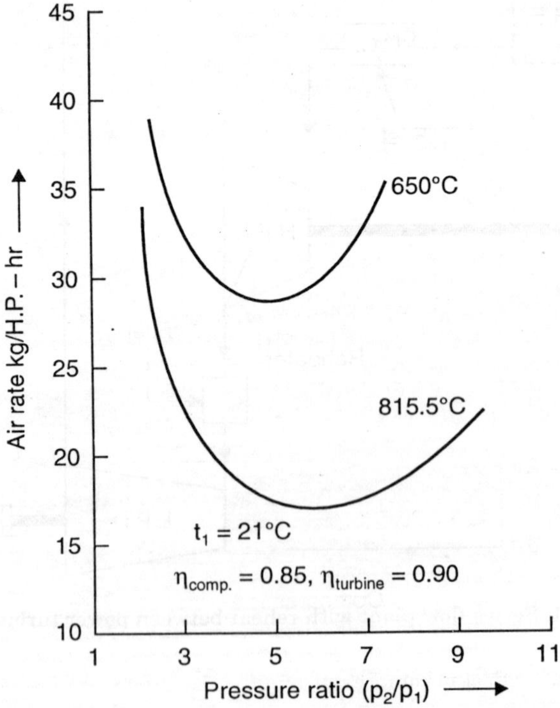

Fig. 4.23. Effect of turbine inlet temperature and pressure ratio on air rate.

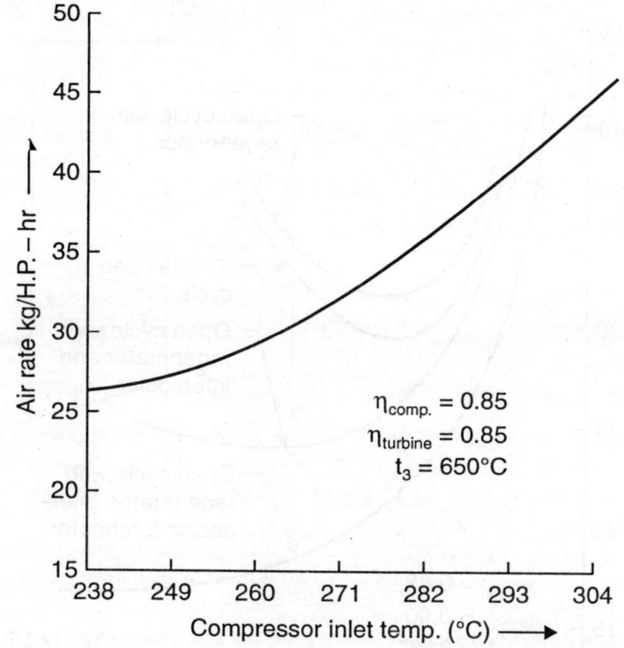

Fig. 4.24. Effect of the compressor inlet temperature on air rate.

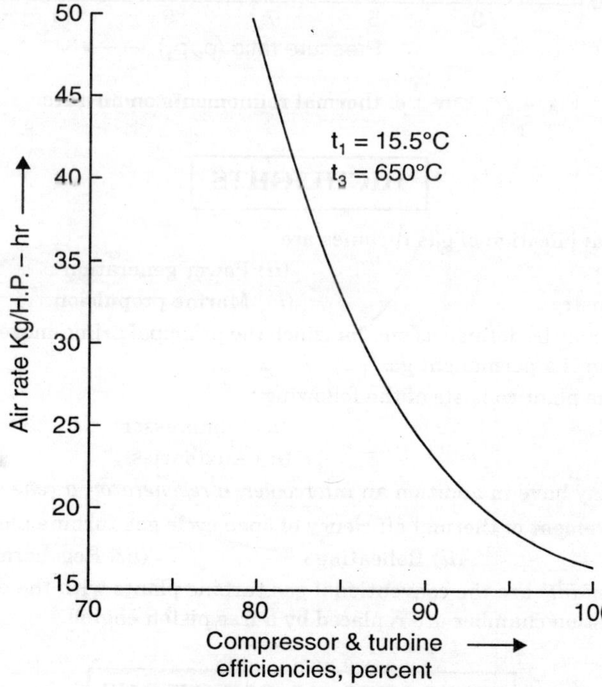

Fig. 4.25. Effect of compressor and turbine efficiencies on air rate.

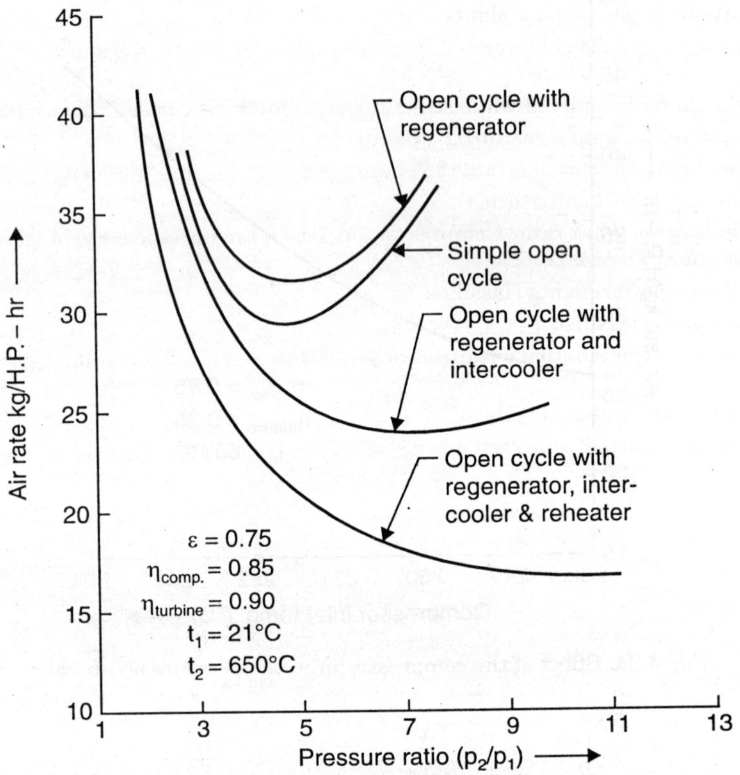

Fig. 4.26. Effect of thermal refinements on air rate.

HIGHLIGHTS

1. The major fields of application of gas turbines are :
   (*i*) Aviation (*ii*) Power generation
   (*iii*) Oil and gas industry (*iv*) Marine propulsion.
2. A gas turbine plant may be defined as one "in which the principal prime-mover is of the turbine type and the working medium is a permanent gas".
3. A simple gas turbine plant consists of the following :
   (*i*) Turbine (*ii*) Compressor
   (*iii*) Combustor (*iv*) Auxiliaries.
   A modified plant may have in addition an *intercooler, a regenerator, a reheater etc.*
4. Methods for improvement of thermal efficiency of open cycle gas turbine plant are :
   (*i*) Intercooling (*ii*) Reheating (*iii*) Regeneration.
5. *Free-piston engine plants* are the conventional gas turbine plants with the difference that the air compressor and combustion chamber are replaced by a free piston engine.

THEORETICAL QUESTIONS

1. What are the major fields of application of gas turbine ?
2. State the limitations of gas turbines.

3. List the applications of gas turbine plants.
4. State the advantages and disadvantages of gas turbine power plants over diesel and thermal power plants.
5. What factors should be considered while selecting a site for a gas turbine power plant ?
6. Give the description of a simple gas turbine plant.
7. Explain with the help of a neat diagram the energy cycle for a simple-cycle gas turbine.
8. Define the following performance terms :
   Air ratio, Pressure ratio, Work ratio, Compressor efficiency, Engine efficiency, Machine efficiency, Combustion efficiency and Thermal efficiency.
9. How are gas turbine power plants classified ?
10. How are gas turbines classified ?
11. What do you mean by "combination gas turbine cycles". Explain briefly combined gas turbine and steam power plants.
12. List the advantages of 'combined cycle'.
13. How is a gas turbine 'started' and 'shut down' ?
14. Explain with a neat sketch the layout of a gas turbine power plant.
15. Enumerate and explain briefly the components of a gas turbine power plant.

5

Hydro-Electric Power Plant

5.1. Introduction. 5.2. Application of hydro-electric plants. 5.3. Advantages and disadvantages of hydro-electric plants. 5.4. Selection of site for a hydro-electric plant. 5.5. Essential features/elements of hydro-electric power plant—Catchment area—Reservoir—Dam—Spillways—Conduits—Surge tanks—Primemovers—Draft tubes—Power house and equipment. 5.6. Classification of hydro-electric power plants—High head power plants—Medium head power plants—Low head power plants—Base load plants—Peak load plants—Run-of-river plants without pondage—Run-of-river plant with pondage—Storage type plants—Pumped storage plants—Mini and microhydel plants. 5.7. Hydraulic turbines—Classification of hydraulic turbines— Description of various types of turbines—Specific speed of a turbine—Efficiencies of a turbine—Cavitation—Performance of hydraulic turbines—Governing of hydraulic turbines—Selection of turbines. 5.8. Plant layout. 5.9. Hydro-plant auxiliaries. 5.10. Cost of hydro-plant. 5.11. Average life of hydro-plant components. 5.12. Hydro-plant controls. 5.13. Electrical and mechanical equipment in a hydro-plant. 5.14. Combined hydro and steam power plants. 5.15. Comparison of hydro-power stations with thermal power stations. 5.16. Underground hydro-plants. 5.17. Automatic and remote control of hydro-station. 5.18. Safety measures in hydro-electric power plants. 5.19. Preventive maintenance of hydro-plant. 5.20. Calculation of available hydro-power. 5.21. Cost of hydro-power. 5.22. Hydrology—Introduction—The hydrologic cycle—Measurement of run-off—Hydrograph—Flow duration curve—Mass curve. 5.23. Hydro-power development in India. Highlights—Theoretical Questions—Unsolved Examples.

5.1. INTRODUCTION

- In hydro-electric plants energy of water is utilised to move the turbines which in turn run the electric generators. The energy of water utilised for power generation may be kinetic or potential. The *kinetic energy* of water is its energy in motion and is a function of mass and velocity, while the *potential energy* is a function of the difference in level/head of water between two points. In either case continuous availability of a water is a basic necessity ; to ensure this, water collected in natural lakes and reservoirs at high altitudes may be utilised or water may be artificially stored by constructing dams across flowing streams. The ideal site is one in which a good system of natural lakes with substantial catchment area, exists at a high altitude. *Rainfall is the primary source of water* and depends upon such factors as temperature, humidity, cloudiness, wind etc. The usefulness of rainfall for power purposes further depends upon several complex factors which include its intensity, time distribution, topography of land etc. However it has been observed that only a small part of the rainfall can actually be utilised for power generation. A significant part is accounted for by *direct evaporation,* while another similar quantity *seeps* into the soil and forms the underground storage. Some water is also absorbed by vegetation. Thus, only a part of water falling as rain actually flows over the ground surface as direct run off and forms the streams which can be utilised for hydro-schemes.

- First hydro-electric station was probably started in America in 1882 and thereafter development took place very rapidly. In India the first major hydro-electric development of 4.5 MW capacity named as Sivasamudram Scheme in Mysore was commissioned in 1902. In 1914 a hydro-power plant named Khopoli project of 50 MW capacity was commissioned in Maharashtra. The hydro-power capacity, upto 1947, was nearly 500 MW.
- Hydro (water) power is a conventional renewable source of energy which is clean, free from pollution and generally has a good environmental effect. However the following factors are *major obstacles in the utilisation of hydro-power resources* :

(*i*) Large investments (*ii*) Long gestation period

(*iii*) Increased cost of power transmission.

Next to thermal power, hydro-power is important in regard to power generation. The hydro-electric power plants provide 30 per cent of the total power of the world. The total hydro-potential of the world is about 5000 GW. In some countries (like Norway) almost total power generation is hydrobased.

5.2. APPLICATION OF HYDRO-ELECTRIC PLANTS

- Earlier hydro-electric plants have been used as exclusive source of power, but the trend is towards use of hydropower in an *interconnected system with thermal stations.* As a self-contained and independent power source, a hydro-plant is most effective with adequate storage capacity otherwise the maximum load capacity of the station has to be based on minimum flow of stream and there is a great wastage of water over the dam for greater part of the year. This increases the per unit cost of installation. By interconnecting hydro-power with steam, a great deal of *saving in cost can be effected due to* :

(*i*) reduction in necessary reserve capacity,

(*ii*) diversity in construction programmes,

(*iii*) higher utilisation factors on hydroplants, and

(*iv*) higher capacity factors on efficient steam plants.

- *In an interconnected system the base load is supplied by hydropower when the maximum flow demand is less than the stream flow while steam supplies the peak. When stream flow is lower than the maximum demand the hydroplant supplies the peak load and steam plant the base load.*

5.3. ADVANTAGES AND DISADVANTAGES OF HYDRO-ELECTRIC PLANTS

Advantages of hydro-electric plant :

1. No fuel charges.
2. An hydro-electric plant is highly reliable.
3. Maintenance and operation charges are very low.
4. Running cost of the plant is low.
5. The plant has no stand by losses.
6. The plant efficiency does not change with age.
7. It takes a few minutes to run and synchronise the plant.

8. Less supervising staff is required.
9. No fuel transportation problem.
10. No ash problem and atmosphere is not polluted since no smoke is produced in the plant.
11. In addition to power generation these plants are also used for flood control and irrigation purposes.
12. Such a plant has comparatively a long life (100-125 years as against 20-45 years of a thermal plant).
13. The number of operations required is considerably small compared with thermal power plants.
14. The machines used in hydro-electric plants are more robust and generally run at low speeds at 300 to 400 r.p.m. where the machines used in thermal plants run at a speed 3000 to 4000 r.p.m. Therefore, there are no specialised mechanical problems or special alloys required for construction.
15. The cost of land is not a major problem since the hydro-electric stations are situated away from the developed areas.

Disadvantages :

1. The initial cost of the plant is very high.
2. It takes considerable long time for the erection of such plants.
3. Such plants are usually located in hilly areas far away from the load centre and as such they require long transmission lines to deliver power, subsequently the cost of transmission lines and losses in them will be more.
4. Power generation by the hydro-electric plant is only dependent on the quantity of water available which in turn depends on the natural phenomenon of rain. So if the rainfall is in time and proper and the required amount of water can be collected, the plant will function satisfactorily otherwise not.

5.4. SELECTION OF SITE FOR A HYDRO-ELECTRIC PLANT

The following *factors* should be considered while selecting the site for a hydro-electric plant :

1. Availability of water
2. Water storage
3. Water head
4. Accessibility of the site
5. Distance from load centre
6. Type of the land of site.

1. Availability of water :

The most important aspect of hydro-electric plant is the availability of water at the site since all other designs are based on it. Therefore the run-off data at the proposed site must be available before hand. It may not be possible to have run-off data at the proposed site but data concerning the rainfall over the large catchment area is always available. Estimate should be made about the average quantity of water available throughout the year and also about maximum and minimum quantity of water available during the year. These details are necessary to :

(*i*) decide the capacity of the hydro-electric plant,

(*ii*) setting up of peak load plant such as steam, diesel or gas turbine plant and to,

(*iii*) provide adequate spillways or gate relief during the flood period.

2. Water storage :

Since there is a wide variation in rainfall during the year, therefore, it is always necessary to store the water for continuous generation of power. The storage capacity can be calculated with the help of mass curve. Maximum storage should justify the expenditure on the project.

The *two types of storages in use* are :

(*i*) The storage is so constructed that it can make water available for power generation of one year only. In this case storage becomes full in the beginning of the year and becomes empty at the end of each year.

(*ii*) The storage is so constructed that water is available in sufficient quantity even during the worst dry periods.

3. Water head :

In order to generate a requisite quantity of power it is necessary that a large quantity of water at a *sufficient head* should be available. An increase in effective head, for a given output, reduces the quantity of water required to be supplied to the turbines.

4. Accessibility of the site :

The site where hydro-electric plant is to be constructed should be easily accessible. This is important if the electric power generated is to be utilised at or near the plant site. The site selected should have transportation facilities of rail and road.

5. Distance from the load centre :

It is of paramount importance that the power plant should be set up *near the load centre* ; this will *reduce the cost of erection and maintenance of transmission line.*

6. Type of the land of the site :

The land to be selected for the site should be cheap and rocky. The ideal site will be one where the dam will have largest catchment area to store water at high head and will be economical in construction.

The necessary requirements of the foundation rocks for a masonry dam are as follows :

(*i*) The rock should be strong enough to withstand the stresses transmitted from the dam structure as well as the thrust of the water when the reservoir is full.

(*ii*) The rock in the foundation of the dam should be reasonably impervious.

(*iii*) The rock should remain stable under all conditions.

5.5. ESSENTIAL FEATURES/ELEMENTS OF HYDRO-ELECTRIC POWER PLANT

The following are the *essential elements of hydro-electric power plant* :

1. Catchment area
2. Reservoir
3. Dam
4. Spillways
5. Conduits
6. Surge tanks
7. Primemovers
8. Draft tubes
9. Powerhouse and equipment.

Fig. 5.1 shows the flow sheet of hydro-electric power plant.

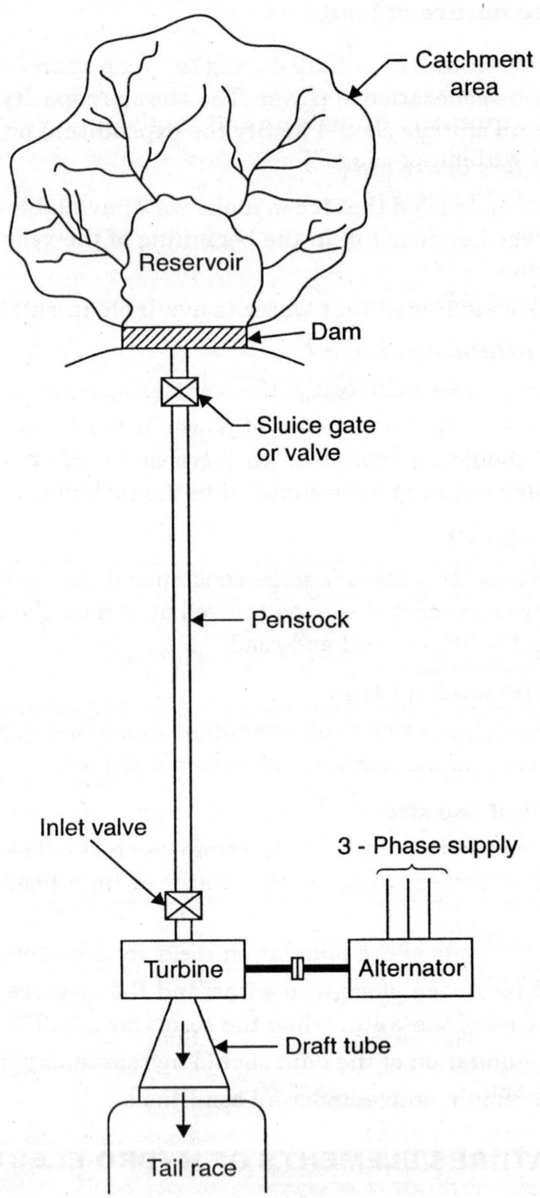

Fig. 5.1. Flow sheet of hydro-electric power plant.

5.6. CLASSIFICATION OF HYDRO-ELECTRIC POWER PLANTS

Hydro-electric power stations may be *classified* as follows :

A. According to availability of head :

1. High head power plants
2. Medium head power plants
3. Low head power plants.

B. According to the nature of load :

1. Base load plants
2. Peak load plants.

C. According to the quantity of water available :

1. Run-of-river plant without pondage
2. Run-of-river plant with pondage
3. Storage type plants
4. Pump storage plants
5. Mini and micro-hydel plants

A. *According to availability of head :*

The following figures give a rough idea of the heads under which the various types of plants work :

(*i*) High head power plants	100 m and above
(*ii*) Medium head power plants	30 to 100 m
(*iii*) Low head power plants	25 to 80 m.

Note. It may be noted that figures given above overlap each other. Therefore it is difficult to classify the plants directly on the basis of head alone. The basis, therefore, technically adopted is the *specific speed* of the turbine used for a particular plant.

5.6.1. High Head Power Plants

These types of plants work under heads ranging from 100 to 2000 metres. Water is usually stored up in lakes on high mountains during the rainy season or during the season when the snow melts. The rate of flow should be such that water can last throughout the year.

Fig. 5.2 shows high head power plant layout. Surplus water discharged by the spillway cannot endanger the stability of the main dam by erosion because they are separated. The tunnel through the mountain has a surge chamber excavated near the exit. Flow is controlled by head gates at the tunnel intake, butterfly valves at the top of the penstocks, and gate valves at the turbines. This type of site might also be suitable for an underground station.

The *Pelton wheel* is the common primemover used in high *head power plants*.

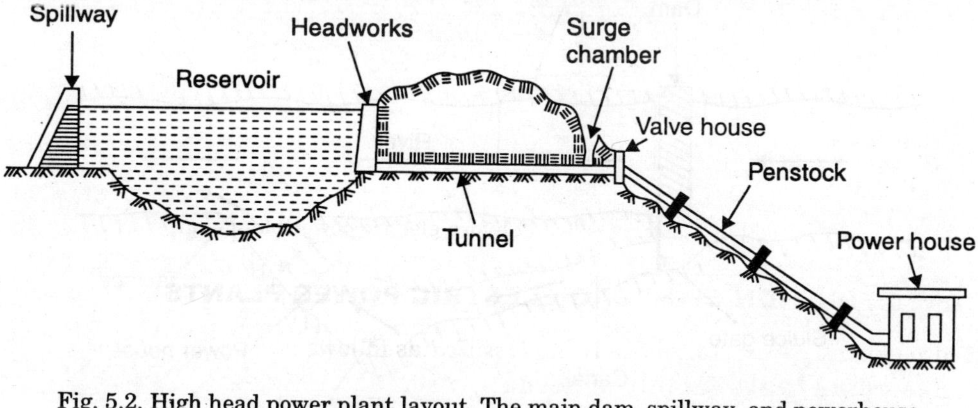

Fig. 5.2. High head power plant layout. The main dam, spillway, and powerhouse stand at widely separated locations. Water flows from the reservoir through a tunnel and penstocks to the turbines.

5.6.2. Medium Head Power Plants

Refer Fig. 5.3. When the operating head of water lies between 30 to 100 metres, the power plant is known as medium head power plant. This type of plant commonly uses *Francis turbines*. The forebay provided at the beginning of the penstock serves as water reservoir. In such plants, the water is generally carried in open canals from main reservoir to the forebay and then to the power-house through the penstock. The forebay itself works as a surge tank in this plant.

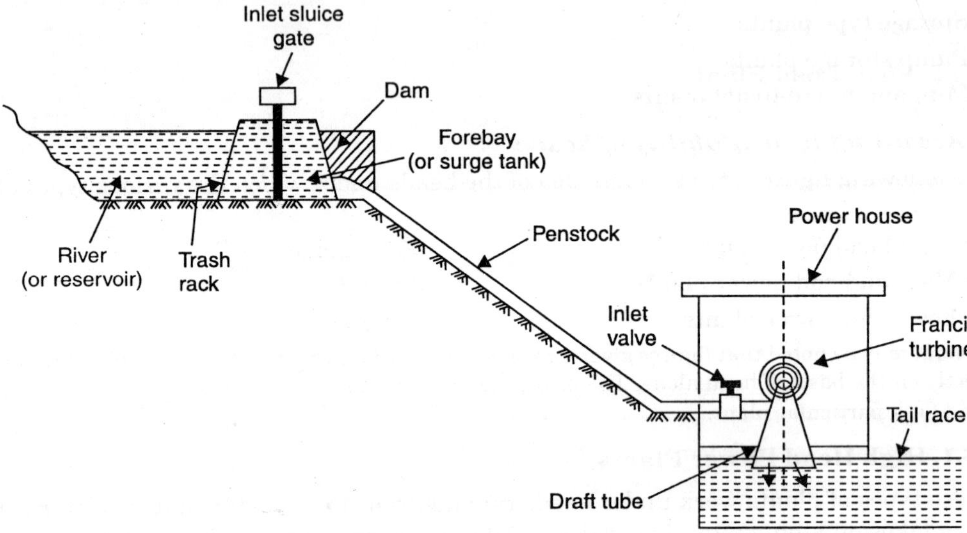

Fig. 5.3. Medium head power plant layout.

5.6.3. Low Head Power Plants

Refer Fig. 5.4. These plants usually consist of a dam across a river. A sideway stream diverges from the river at the dam. Over this stream the power house is constructed. Later this channel joins the river further downstream. This type of plant uses vertical shaft Francis turbine or Kaplan turbine.

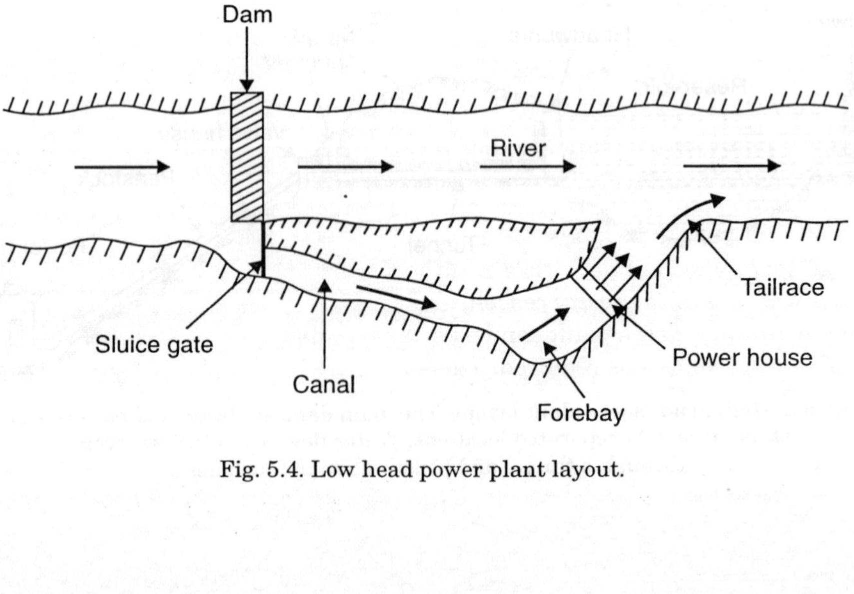

Fig. 5.4. Low head power plant layout.

B. *According to the nature of load :*

5.6.4. Base Load Plants

The plants which cater to the base load of the system are called **base load plants.** These plants are required to *supply a constant power when connected to the grid*. Thus they *run without stop* and are often remote-controlled with which *least staff* is *required for such plants*. Run-of-river plants without pondage may sometimes work as baseload plant, but the firm capacity in such cases, will be much less.

5.6.5. Peak Load Plants

The plants which can supply the power during peak loads are known as **peak load plants.** Some of such plants supply the power during average load but also supply peak load as and when it is there ; whereas other peak load plants are required to work during peak load hours only. The run-of-river plants may be made for the peak load by providing pondage.

C. *According to the quantity of water available :*

5.6.6. Run-of-river Plants without Pondage

A run-of-river plant without pondage, as the name indicates, does not store water and uses the water as it comes. There is *no control on flow of water* so that during high floods or low loads water is wasted while during low run-off the plant capacity is considerably reduced. *Due to non-uniformity of supply and lack of assistance from a firm capacity the utility of these plants is much less* than those of other types. The head on which these plants work *varies considerably*. Such a plant can be made a great deal more useful by providing sufficient storage at the plant to take care of the hourly fluctuations in load. This lends some *firm capacity* to the plant. During good flow conditions these plants may cater to base load of the system, when flow reduces they may supply the peak demands. *Head water elevation for plant fluctuates with the flow conditions*. These plants without storage may sometimes be made to supply the base load, but the firm capacity depends on the minimum flow of river. The run-of-river plant may be made for load service with pondage, though storage is usually seasonal.

5.6.7. Run-of-river Plants with Pondage

Pondage usually refers to the collection of water behind a dam at the plant and increases the stream capacity for a short period, say a week. *Storage* means collection of water in up stream reservoirs and this increases the capacity of the stream over an extended period of several months. Storage plants may work satisfactorily as base load and peak load plants.

This type of plant, as compared to that without pondage, is *more reliable* and its generating capacity *is less dependent on the flow rates of water available.*

5.6.8. Storage Type Plants

A storage type plant is one with a reservoir of sufficiently large size to permit carry-over storage from the wet reason to the dry reason, and thus to supply firm flow substantially more than the minimum natural flow. This plant can be used as base load plant as well as peak load plant as water is available with control as required. The majority of hydro-electric plants are of this type.

5.6.9. Pumped Storage Plants

Refer to Fig. 5.5.

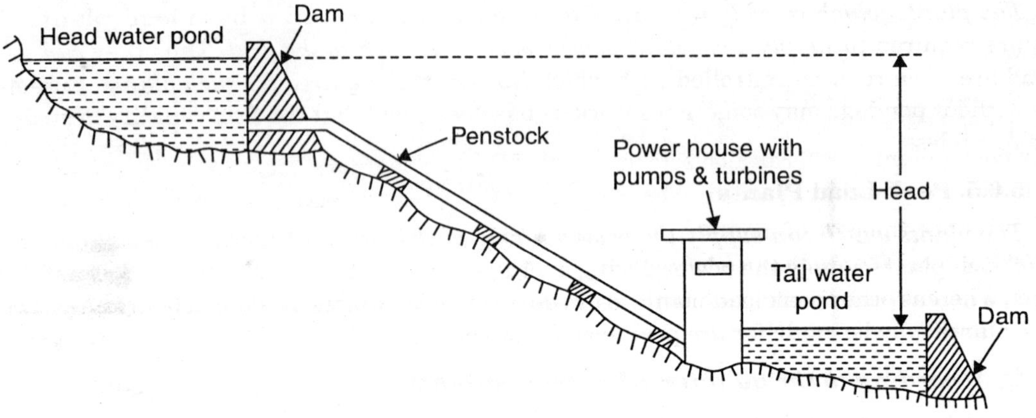

Fig. 5.5. Pumped storage plant.

Pumped storage plants are employed at the places where the quantity of water available for power generation is *inadequate*. Here the water passing through the turbines is stored in *'tail race pond'*. During low load periods this water is pumped back to the head reservoir using the extra energy available. This water can be again used for generating power during peak load periods. Pumping of water may be done seasonally or daily depending upon the conditions of the site and the nature of the load on the plant.

Such plants are *usually interconnected* with steam or diesel engine plants so that off peak capacity of interconnecting stations is used in pumping water and the same is used during peak load periods. Of course, the energy available from the quantity of water pumped by the plant is *less* than the energy input during pumped operation. Again while using pumped water the *power available is reduced* on account of losses occuring in primemovers.

Advantages. The pump storage plants entail the following *advantages* :

1. There is substantial increase in peak load capacity of the plant at comparatively low capital cost.
2. Due to load comparable to rated load on the plant, the operating efficiency of the plant is high.
3. There is an improvement in the load factor of the plant.
4. The energy available during peak load periods is higher than that of during off peak periods so that inspite of losses incurred in pumping there is *over-all gain*.
5. Load on the hydro-electric plant remains uniform.
6. The hydro-electric plant becomes partly independent of the stream flow conditions.

Under pump storage projects almost 70 percent power used in pumping the water can be recovered. In this field the use of *"Reversible Turbine Pump"* units is also worth noting. These units can be used as turbine while generating power and as pump while pumping water to storage. The generator in this case works as motor during reverse operation. The efficiency in such case is high and almost the same in both the operations. With the use of reversible turbine pump sets, additional capital investment on pump and its motor can be saved and the scheme can be worked more economically.

5.6.10. Mini and Microhydel Plants

In order to meet with the present energy crisis partly, a solution is to develop *mini* (5 m to 20 m head) and *micro* (less than 5 m head) hydel potential in our country. The low head hydro-potential is scattered in this country and estimated potential from such sites could be as much as 20,000 MW.

By proper planning and implementation, it is possible to commission a small hydro-generating set up of 5 MW within a period of one and half year against the period of a decade or two for large capacity power plants. Several such sets upto 1000 kW each have been already installed in Himachal Pradesh, U.P., Arunachal Pradesh, West Bengal and Bhutan.

To reduce the cost of micro-hydel stations than that of the cost of conventional installation the following *considerations are kept in view* :

1. The civil engineering work needs to be kept to a *minimum* and designed to fit in with already existing structures *e.g.*, irrigation, channels, locks, small dams etc.
2. The machines need to be manufactured in a small range of sizes of simplified design, allowing the use of unified tools and aimed at reducing the cost of manufacture.
3. These installations must be automatically controlled to reduce attending personnel.
4. The equipment must be simple and robust, with easy accessibility to essential parts for maintenance.
5. The units must be light and adequately subassembled for ease of handling and transport and to keep down erection and dismantling costs.

Micro-hydel plants (micro-stations) *make use of standardised bulb sets with unit output ranging from 100 to 1000 kW working under heads between 1.5 to 10 metres*.

5.7. HYDRAULIC TURBINES

A hydraulic turbine converts the potential energy of water into mechanical energy which in turn is utilised to run an electric generator to get electric energy.

5.7.1. Classification of Hydraulic Turbines

The hydraulic turbines are *classified* as follows :

(*i*) According to the head and quantity of water available.
(*ii*) According to the name of the originator.
(*iii*) According to the action of water on the moving blades.
(*iv*) According to the direction of flow of water in the runner.
(*v*) According to the disposition of the turbine shaft.
(*vi*) According to the specific speed N_s.

1. According to the head and quantity of water available :

(*i*) *Impulse turbine*—requires *high head* and *small quantity of flow*.
(*ii*) *Reaction turbine*—requires *low head* and *high rate of flow*.

Actually there are two types of reaction turbines, one for medium head and medium flow and the other for low head and large flow.

2. According to the name of the originator :

(*i*) *Pelton turbine*—named after Lester Allen Pelton of California (USA). It is an impulse type of turbine and is used for *high head* and *low discharge*.

(*ii*) *Francis turbine*—named after James Bichens Francis. It is a reaction type of turbine for *medium high to medium low heads* and *medium small to medium large quantities of water.*

(*iii*) *Kaplan turbine*—named after Dr. Victor Kaplan. It is a *reaction type of turbine for low heads and large quantities of flow.*

3. According to action of water on the moving blades :

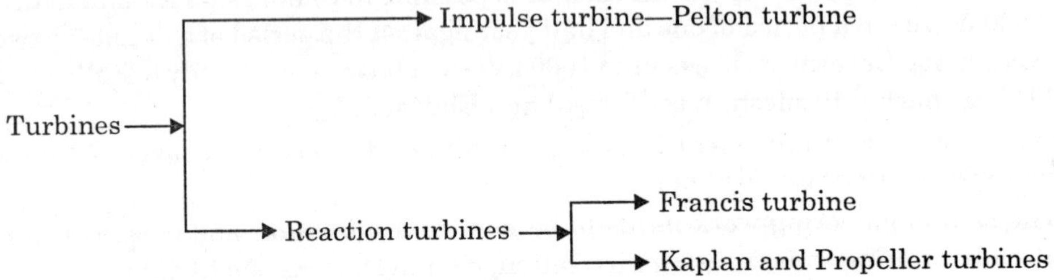

4. According to direction of flow of water in the runner :

(*i*) Tangential flow turbine (Pelton turbine)

(*ii*) Radial flow turbine (no more used)

(*iii*) Axial flow turbine (Kaplan turbine)

(*iv*) Mixed (radial and axial) flow turbine (Francis turbine)

In *tangential flow* turbine of Pelton type the water strikes the runner tangential to the path of rotation.

In *axial flow* turbine water flows parallel to the axis of the turbine shaft. Kaplan turbine is an axial flow turbine. In Kaplan turbine the runner blades are *adjustable and can be rotated* about pivots fixed to the boss of the runner. If the runner blades of the axial flow turbines are *fixed*, these are called "*Propeller turbines*".

In *mixed flow* turbines the water enters the blades radially and comes out axially, parallel to the turbine shaft. Modern Francis turbines have mixed flow runners.

5. According to the disposition of the turbine shaft :

Turbine shaft may be either vertical or horizontal. In modern turbine practice, Pelton turbines usually have horizontal shafts whereas the rest, especially the large units, have vertical shafts.

6. According to specific speed :

The *specific speed* of a turbine is defined as the speed of a geometrically similar turbine that would develop *one brake horsepower* under the *head of one metre.* All geometrically similar turbines (irrespective of their sizes) will have the same specific speed when operating under the same conditions of head and flow.

Specific speed $$N_s = \frac{N\sqrt{P_t}}{H^{5/4}}$$

where, N = The normal working speed in r.p.m.,

P_t = Power output of the turbine, and

H = The net or effective head in metres.

The specific speeds for the various types of runners are given below :

Type of turbine	*Type of runner*	*Specific speed* (N_s)
Pelton	Slow	10 to 20
	Normal	20 to 28
	Fast	28 to 35
Francis	Slow	60 to 120
	Normal	120 to 180
	Fast	180 to 300
Kaplan	—	300 to 1000

Turbines with low specific speeds work under a high head and low discharge condition, while high specific speed turbines work under low head and high discharge conditions.

5.7.2. Description of Various Types of Turbines

5.7.2.1. Impulse turbines

Pelton wheel, among the various impulse turbines that have been designed and utilized, is by far the important. The Pelton wheel or Pelton turbine is a *tangential flow impulse turbine*. It consists of a rotor, at the periphery of which are mounted equally spaced double-hemispherical or double-ellipsoidal buckets. Water is transferred from a high head source through penstock pipes. A branch pipe from each penstock pipe ends in a nozzle, through which the water flows out as a high speed jet. A needle or spear moving inside the nozzle controls the water flow through the nozzle and at the same time, provides a smooth flow with negligible energy loss. All the available *potential energy is thus converted into kinetic energy* before the jet strikes the buckets. The pressure all over the wheel is *constant and equal to atmosphere, so that energy transfer occurs due to purely impulse action.*

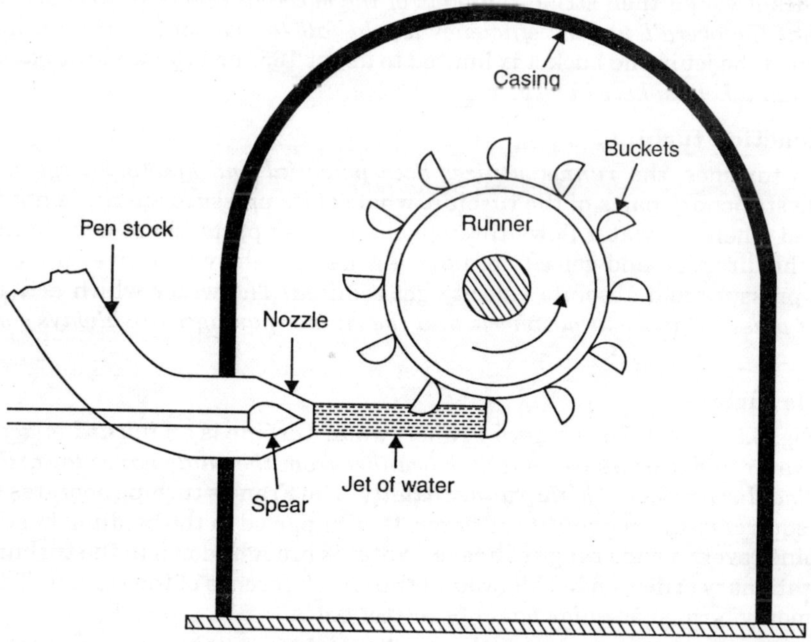

Fig. 5.6. Pelton wheel.

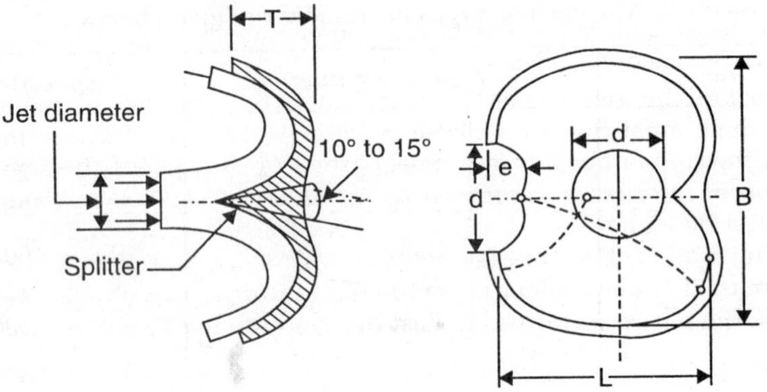

$$\frac{L}{d} = 2.5 - 2.8,\ \frac{D}{d} = 14 - 16,\ \frac{T}{d} = 0.95, \text{ Notch (width)} = 1.1d + 5 \text{ mm}$$

Fig. 5.7. The bucket dimensions.

Fig. 5.6 shows a schematic diagram of a Pelton wheel, while Fig. 5.7 shows two views of its bucket.

The jet emerging from the nozzle hits the splitter symmetrically and is equally distributed into the two halves of hemispherical bucket as shown. *The bucket centre-line cannot be made exactly like a mathematical cusp, partly because of manufacturing difficulties and partly because the jet striking the cusp invariably carries particles of sand and other abrasive materials which tend to wear it down*. The inlet angle of the jet is therefore between 1° and 3°, but it is always assumed to be zero in all calculations. Theoretically, if the bucket were exactly hemispherical, it would deflect the jet through 180°. Then, the relative velocity of the jet leaving the bucket, C_{r2}, would be opposite in direction to the relative velocity of the entering jet C_{r1}. This cannot be achieved in practice since the jet leaving the bucket would then strike the *back of the succeeding bucket to cause splashing and interference so that the overall turbine efficiency would fall to low values*. Hence, in practice, the angular deflection of the jet in the bucket is limited to about 165° or 170°, and the *bucket is therefore slightly smaller than a hemisphere in size*.

5.7.2.2. Reaction turbines

In reaction turbines, the *runner utilizes both potential and kinetic energies*. As the water flows through the stationary parts of the turbine, whole of its pressure energy is not transferred to kinetic energy and when the water flows through the moving parts, there is a change both in the pressure and in the direction and velocity of flow of water. As the water gives up its energy to the runner, both its pressure and absolute velocity get reduced. The water which acts on the runner blades is under *a pressure above atmospheric and the runner passages are always completely filled with water*.

A. Francis turbines

Refer to Fig. 5.8 (*a*), (*b*). The modern Francis water turbine is an *inward mixed flow reaction turbine i.e.,* the water under pressure, *enters the runner from the guide vanes towards the centre in radial direction and discharges out of the runner axially*. The Francis turbine operates under *medium heads* and also requires *medium quantity* of water. It is employed in the medium head power plants. This type of turbine covers a wide range of heads. Water is brought down to the turbine and directed to a number of stationary orifices fixed all around the circumference of the runner. These stationary orifices are commonly termed as *guide vanes or wicket gates*.

The head acting on the turbine is partly transformed into kinetic energy and the rest remains as pressure head. There is a difference of pressure between the guide vanes and the runner which is called the *reaction pressure* and is responsible for the motion of the runner. That is why a Francis turbine is also known as *reaction turbine*.

In Francis turbine the pressure at the inlet is more than that at the outlet. This means that the water in the turbine must flow in a closed conduit. Unlike the Pelton type, where the water strikes only a few of the runner buckets at a time, in the Francis turbine the *runner is always full of water. The movement of the runner is affected by the change of both the potential and the kinetic energies of water*. After doing its work the water is discharged to the tail race through a closed tube of gradually enlarging section. This tube is known as *draft tube*. It does not allow water to fall freely to tail race level as in the Pelton turbine. The free end of the draft tube is submerged deep in the tail water making, thus, the entire water passage, right from the head race upto the tail race, totally enclosed.

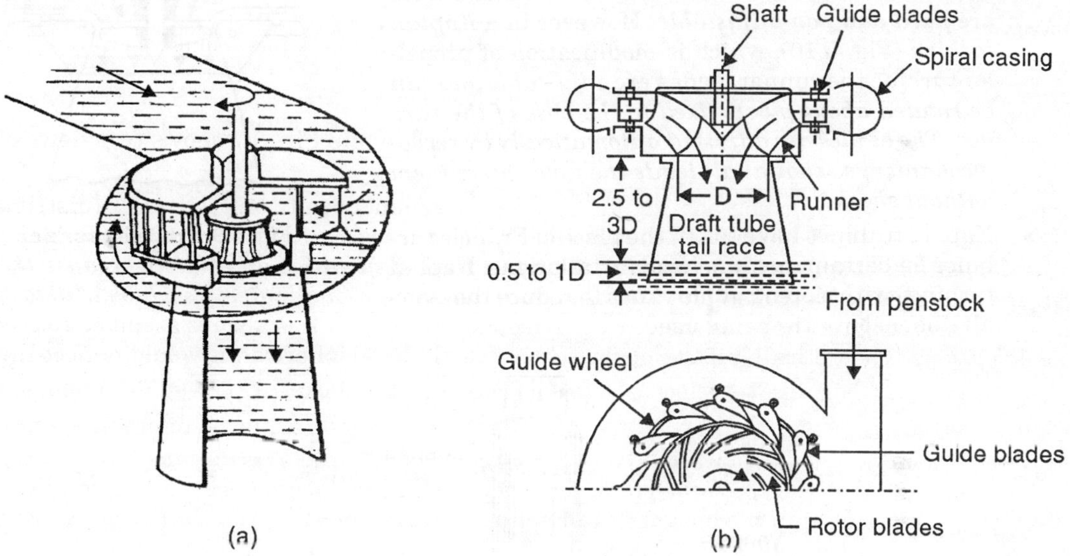

Fig. 5.8. Schematic diagram of a Francis turbine.

Draft tube :

The draft tube serves the following two purposes :

1. *It allows the turbine to be set above tail-water level, without loss of head, to facilitate inspection and maintenance.*

2. *It regains, by diffuser action, the major portion of the kinetic energy delivered to it from the runner.*

At rated load the velocity at the upstream end of the tube for modern units ranges from 7 to 9 m/s, representing from 2.7 to 4.8 m head. As the specific speed (it is the speed of a geometrically similar turbine running under a unit head and producing unit power) is increased and the head reduced, it becomes increasingly important to have an efficient draft tube. Good practice limits the velocity at the discharge end of the tube to 1.5 to 2.1 m/s, representing less than 0.3 m velocity head loss.

The following two types of *draft tubes* are commonly used :

(*i*) The straight conical or concentric tube.

(*ii*) The elbow type.

Properly designed, the two types are about equally efficient, over 85%.

B. Propeller and Kaplan turbines

- The need to utilize low heads where large volumes of water are available makes it essential to provide a large flow area and to run the machine at very low speeds. The propeller turbine is a reaction turbine used for heads between 4 m and 80 m, and has a specific speed ranging from 300 to 1000. It is purely *axial-flow* device providing the largest possible flow area that will utilize a large volume of water and still obtain flow velocities which are not too large.
- The propeller turbine (Fig. 5.9) consists of an axial-flow runner with four to six or at most ten blades of air-foil shape. The spiral casing and guide blades are similar to those in Francis turbines. In the **propeller turbine** as in Francis turbine the runner blades are *fixed* and *non-adjustable*. However in a *Kaplan turbine* (Fig. 5.10), which is modification of propeller turbine the runner blades are *adjustable and can be rotated about pivots fixed to the boss of the runner. The blades are adjusted automatically by servo-mechanism so that at all loads the flow enters them without shock.*

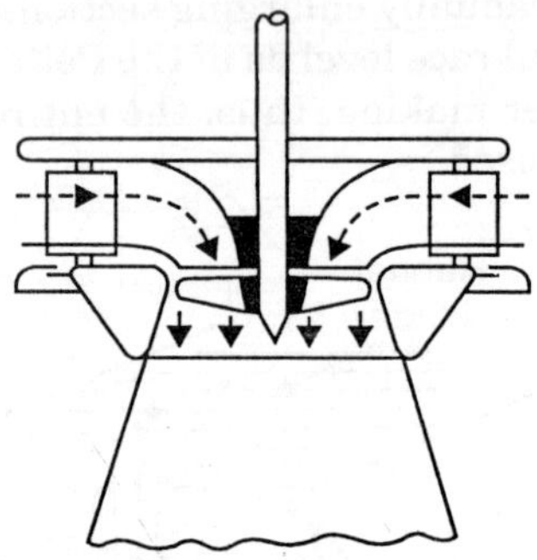

Fig. 5.9. Propeller turbine.

- Kaplan turbines have taken the place of Francis turbines for certain medium head installations. Kaplan turbines with sloping guide vanes to reduce the overall dimensions are being used.

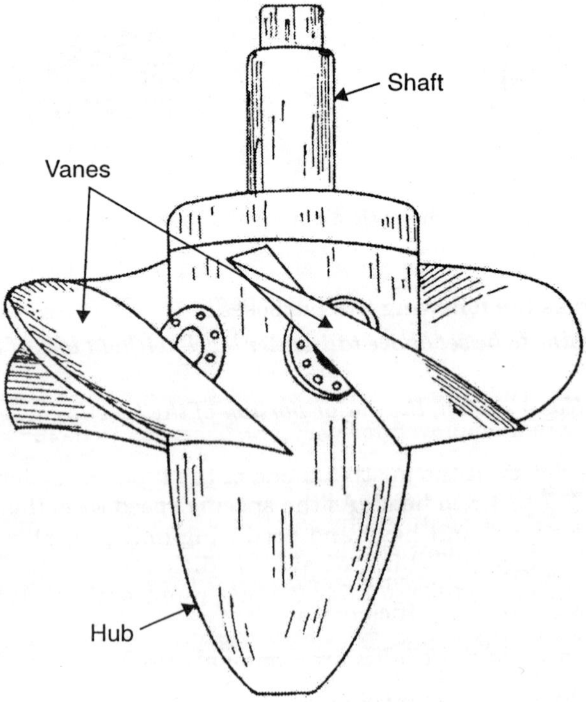

(a)

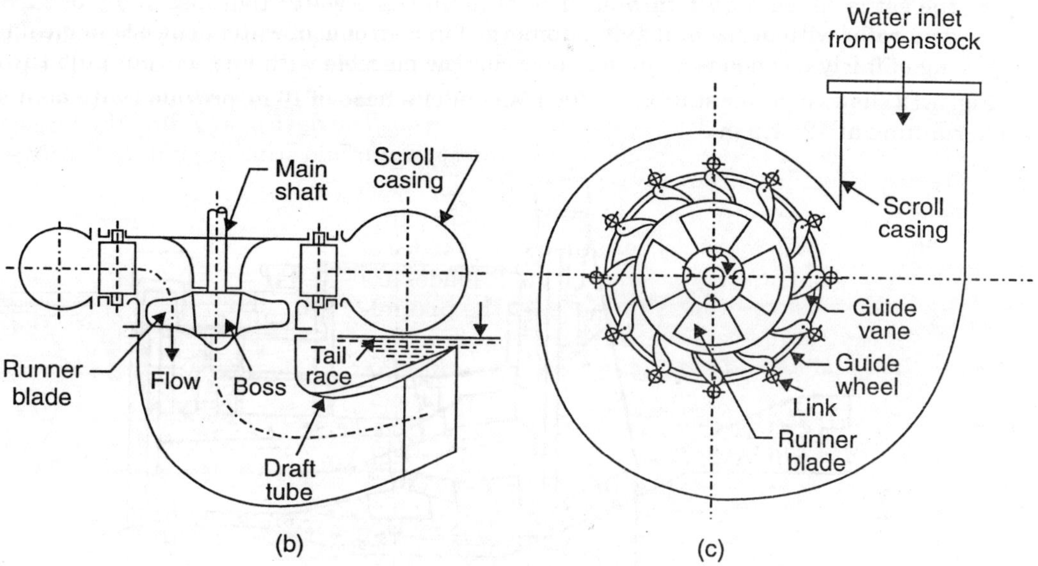

Fig. 5.10. Schematic diagram of a Kaplan turbine.

Fig. 5.11 shows the comparison of efficiencies of propeller (fixed blades) and Kaplan turbines.

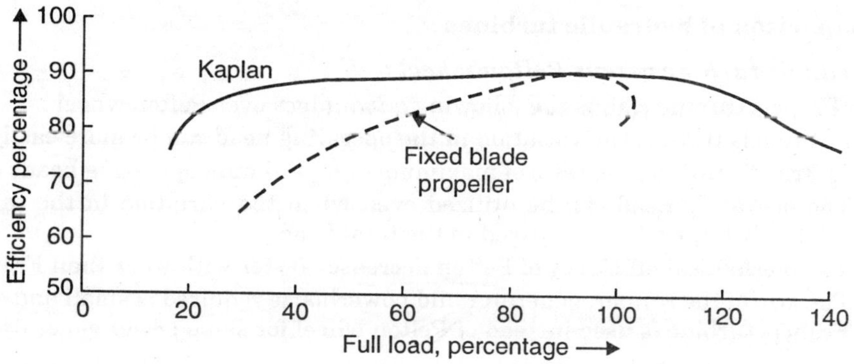

Fig. 5.11. Comparison of efficiencies of propeller (fixed blades) and Kaplan turbines.

C. Tubular (or Bulb) turbines

- Kaplan turbine when employed for very low head has to be installed below the tail race level, thus requiring a *deep excavation.* Further for Kaplan turbine installation there are a number of bends at inlet casing and the draft tube of elbow type through which the water flows describing '*Z*' path giving rise to *continuous losses at the bends.* Whenever the turbine is repaired or dismantled, the generator has to be removed first. The cost of turbine and that of civil engineering works using conventional Kaplan turbine with deep excavation is very high. The efficiency of such plants working under low head is less due to excessive losses at the bends. Therefore, efforts have been made to *reduce the overall cost* and *improve the efficiency of the power plant* keeping these two things in view.
- In 1937 Arno Fischer developed in Germany a modified axial flow turbine which is known as *tubular turbine.* The turbo-generator set using tubular turbine has an outer casing having the shape of a bulb. Such a set is now termed as *bulb set* and the turbine used for

the set is called a *bulb turbine.* The bulb unit is a water tight assembly of turbine and generator with horizontal axis, submerged in a stream of water. The economical harnessing of fairly low heads on major rivers is now possible with high-output bulb turbines.

Fig. 5.12 shows a power station (87300 kW) under a head of 10 m, provided with six 14550 kW bulb sets running at 125 r.p.m.

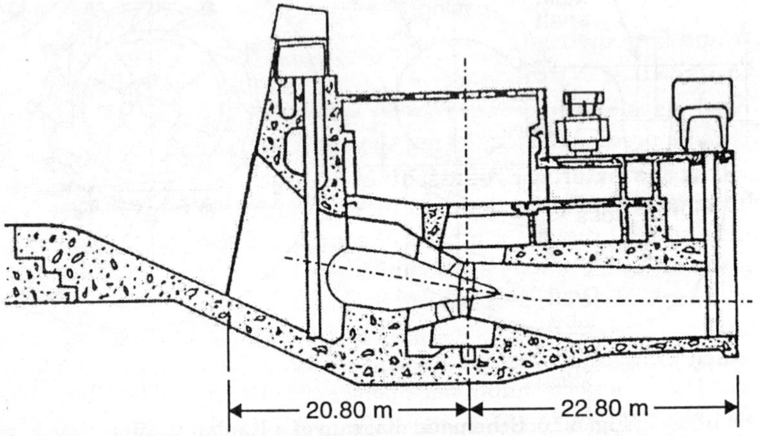

Fig. 5.12. Power station—using bulb turbines.

Comparison of hydraulic turbines :

I. *Francis turbine versus Pelton wheel :*

The Francis turbine claims the following *advantages* over Pelton wheel :

1. In Francis turbine the variation in the operating head can be more easily controlled.
2. In Francis turbine the ratio of maximum and minimum operating heads can be even two.
3. The operating head can be utilized even when the variation in the tail water level is relatively large when compared to the total head.
4. The mechanical efficiency of Pelton decreases faster with wear than Francis.
5. The size of the runner, generator and powerhouse required is small and economical if the Francis turbine is used instead of Pelton wheel for same power generation.

Drawbacks of Francis turbine :

As compared with Pelton wheel, the Francis turbine has the following *drawbacks* :

1. Water which is not clean can cause very rapid wear in high head Francis turbine.
2. The overhaul and inspection is much more difficult comparatively.
3. Cavitation is an ever-present danger.
4. The water hammer effect is more troublesome with Francis turbine.
5. If Francis turbine is run below 50% head for a long period it will not only lose its efficiency but also the cavitation danger will become more serious.

II. *Kaplan verses Francis turbine :*

Kaplan turbine claims the following *advantages* over Francis turbine :

1. For the same power developed Kaplan turbine is more compact in construction and smaller in size.
2. Part-load efficiency is considerably high.
3. Low frictional losses (because of small number of blades used).

5.7.3. Specific Speed of a Turbine

The **specific speed** of a turbine is defined *as the speed of a turbine which is identical in shape, geometrical dimensions, blade angles, gate opening etc., with the actual turbine but of such a size that it will develop unit horse power when working under unit head.*

$$\text{Specific speed } N_s = \frac{N\sqrt{P}}{H^{5/4}}$$

(where P is in kW and H in metres)

[N_s (S.I. Units) = 0.86 N_s (metric)]

Specific speed plays an important role for *selecting the type of the turbine.* Also the *performance of a turbine can be predicted* by knowing the specific speed of the turbine.

To compare the characteristics of machines of different types, it is necessary to know a characteristic of an imaginary machine identical in shape. The imaginary turbine is called a *specific turbine.* The specific speed provides a means of comparing the speed of all types of hydraulic turbines on the basis of head and horse power capacity.

If a runner of *high specific speed* is used for a given head horse power output, the *overall cost* of installation is *lower*. The selection of *too* high specific speed reaction runner would reduce the size of the runner to such an extent that the discharge velocity of water into the throat of draft tube would be excessive. This is objectionable because a *vacuum* may be created in the extreme case.

The runner of *too high* specific speed with available head *increases the cost of turbine* on account of high mechanical strength required. The runner of *too low* specific speed with low available head *increases the cost of generator* due to the low turbine speed.

An increase in specific speed of turbine is accompanied by lower maximum efficiency and greater depth of excavation of the draft tube. In choosing a high specific speed turbine, an increase in cost of excavation of foundation and draft tube should be considered in addition to the efficiency. The weighted efficiency over the operating range of turbine is more important in the selection of a turbine instead of maximum efficiency.

Table 5.1 gives the specific speeds for various turbines.

Table 5.1. Specific speeds

Type of turbine		*Specific speed (N_s)* *M.K.S. Units*	*S.I. Units*
Impulse (Pelton)	Slow	4—10	3.5—8.5
	Normal	10—25	8.5—21.5
	Fast	25—60	21.5—51.5
Radial and mixed flow (Francis and Deriaz)	Slow	60—150	51.5—130
	Normal	150—250	130—215
	Fast	250—400	215—345
Axial flow (Kaplan)	Slow	300—450	255—385
	Normal	450—700	385—600
	Fast	700—1000	600—860

5.7.4. Efficiencies of a Turbine

The important efficiencies of a turbine are as under :

1. Hydraulic efficiency, η_h

2. Mechanical efficiency, η_m
3. Volumetric efficiency, η_v
4. Overall efficiency, η_o.

1. Hydraulic efficiency, (η_h). It is defined as the *ratio of power developed by the runner of a turbine to the power supplied by the water at the inlet of the turbine.*

$$\eta_h = \frac{\text{Power developed by runner}}{\text{Power developed at inlet}} \qquad ...(5.1)$$

2. Mechanical efficiency (η_m). It is defined as the *ratio of power available at the shaft of the turbine* (known as S.H.P. or B.H.P.) *to the power developed by the runner.*

$$\eta_m = \frac{\text{Power available at the shaft of the turbine}}{\text{Power developed by the runner}} \qquad ...(5.2)$$

3. Volumetric efficiency (η_v). *The ratio of the volume of the water actually striking the runner to the volume of water supplied to the turbine is called volumetric efficiency.*

$$\eta_v = \frac{\text{Volume of water actually striking the runner}}{\text{Volume of water supplied to the turbine}} \qquad ...(5.3)$$

4. Overall efficiency (η_0). It is defined as the *ratio of power available at the shaft of the turbine to the power supplied by the water at the inlet of the turbine.*

$$\eta_0 = \frac{\text{Power available at the shaft of the turbine}}{\text{Power supplied at the inlet of the turbine}} = \frac{\text{Shaft power}}{\text{Water power}} = \frac{P}{wQH}$$

If η_g is the efficiency of a generator, then power output of a hydro-unit (turbine + hydro generators)

$$= (wQH) \times \eta_o \times \eta_g.$$

5.7.5. Cavitation

It is known that when velocity of flow increases, the pressure falls. In liquids, the pressure cannot fall below vapour pressure which depends upon the temperature and height above mean sea level of the site. *In any turbine part if the pressure drops below the evaporation pressure, the liquid boils and a large number of small bubbles of vapour are formed.* These bubbles mainly formed *on account of low pressure* are carried by the stream to higher pressure zones where the vapours condense and the *bubbles suddenly collapse*, as the vapours are condensed to liquid again. This results in the *formation of a cavity* and the surrounding liquid rushes to fill it. The streams of liquid coming from all directions collide at the centre of cavity giving rise to a *very high local pressure* whose magnitude may be as high as 7000 atmospheres. Formation of cavity and high pressure are repeated many thousand times a second. This causes *pitting on the metallic surface of runner blades or draft tube*. The material then *fails by fatigue*, and perhaps by corrosion. During this process some parts like runner blades may be torn away completely.

'Cavitation' may thus be defined as the "*phenomenon which manifests itself in the pitting of the metallic surfaces of turbine parts because of formation of cavities*".

Methods to avoid cavitation :

The following methods may be used to *avoid cavitation* :

1. Runner/turbine may be *kept under water*. But it is not advisable as the inspection and repair of the turbine is difficult. The other method to avoid cavitation zone without keeping the runner under water is *to use the runner of low specific speed*.
2. The *cavitation free runner* may be designed to fulfill the given conditions with extensive research.

3. It is possible to reduce the cavitation effect by *selecting materials* which can *resist better the cavitation effect*. The cast steel is better than cast iron and stainless steel or alloy steel is still better than cast steel.
4. The cavitation effect can be reduced by *polishing* the surfaces. That is why the cast steel runners and blades are coated with stainless steel.
5. The 'cavitation' may be avoided by *selecting a runner of proper specific speed for given head.*

5.7.6. Performance of Hydraulic Turbines

In dimensional analysis we may arrive at the following relations :

Head co-efficient or parameter, $\dfrac{gH}{N^2D^2}$

Capacity or discharge parameter, $\dfrac{Q}{ND^3}$ or $\dfrac{Q}{D^2\sqrt{H}}$

Power co-efficient, $\dfrac{P}{\rho N^3 D^5}$ or $\dfrac{P}{\rho D^2 H^{3/2}}$

Efficiency of operation, η

The laws of performance are stated as follows from the fact that each of these parameters should remain constant for the same machine or a family of similar machines :

		For the same machine	*For a similar machine*
Discharge Q	$\propto$	N	D^3
Head H	$\propto$	N^2	D^2
Power P	$\propto$	N^3	D^5

The specific speed $N_s = \dfrac{N\sqrt{P/\rho}}{(gH)^{5/4}}$

Which is the type number of fluid machines that must also remain the same for a family of similar machines operating at the optimum efficiency.

The terms '*unit speed*', '*unit power*' and '*unit discharge*' are frequently used to express the operational features of hydraulic turbines. The unit quantities are theoretical features for a head of 1 metre for the same turbine. Sometimes, the unit quantities are referred to a head of 1 metre for a turbine of 1 metre diameter.

	For unit head	*For unit head and unit diameter*
Unit speed N_u	$\dfrac{N}{\sqrt{H}}$	—
Unit power P_u	$\dfrac{P}{H^{3/2}}$	$\dfrac{P}{D^2 H^{3/2}}$
Unit discharge Q_u	$\dfrac{Q}{\sqrt{H}}$	$\dfrac{Q}{D^2\sqrt{H}}$

These definitions follow from the definitions of the head, discharge and power coefficients respectively.

The characteristic curves for hydraulic reaction turbines are usually plotted in terms of unit quantities *vs* '*unit*' speed. A set of performance curves are shown plotted in Fig 5.13 (*a*). A comparative set of curves for different types of turbines is shown in Fig. 5.13 (*b*). It can be seen that the *Deriaz and Kaplan turbines have the highest efficiency in the entire load range*. This is due to the fact that the runner blades of these two types of turbines are *adjustable during operation*. Consequently the flow is efficient and well-guided by the runner blades at all flow conditions unlike the other turbines where the rotor-vane adjustability is not provided. The guide vanes of each of reaction turbines are adjusted for varying the discharge.

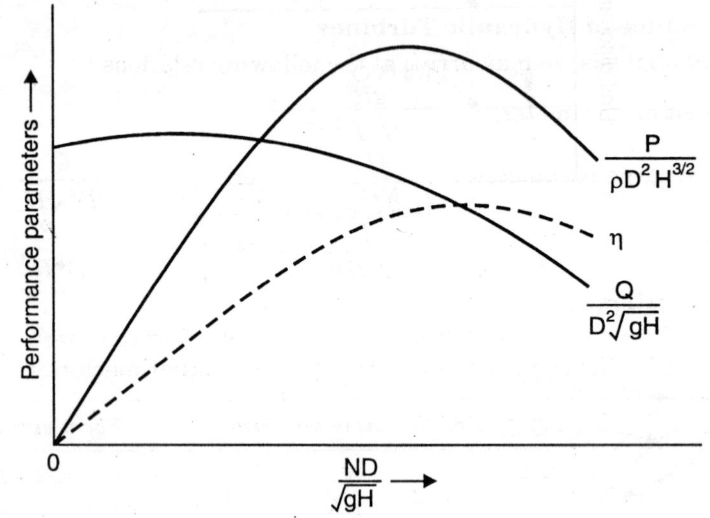

(a) Dimensionless turbine curves

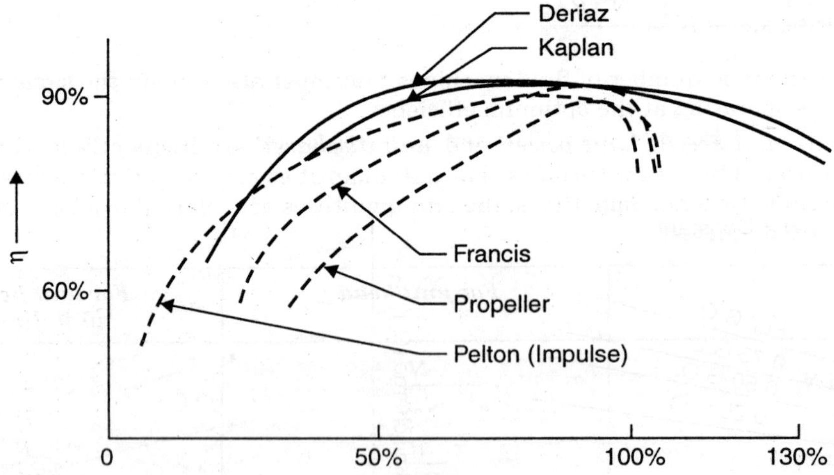

(b) Comparison of turbines

Fig. 5.13. Performance of hydraulic turbines.

Fig. 5.14 and Fig. 5.15 shows the *main characteristic curves* for a Pelton wheel and reaction turbine respectively.

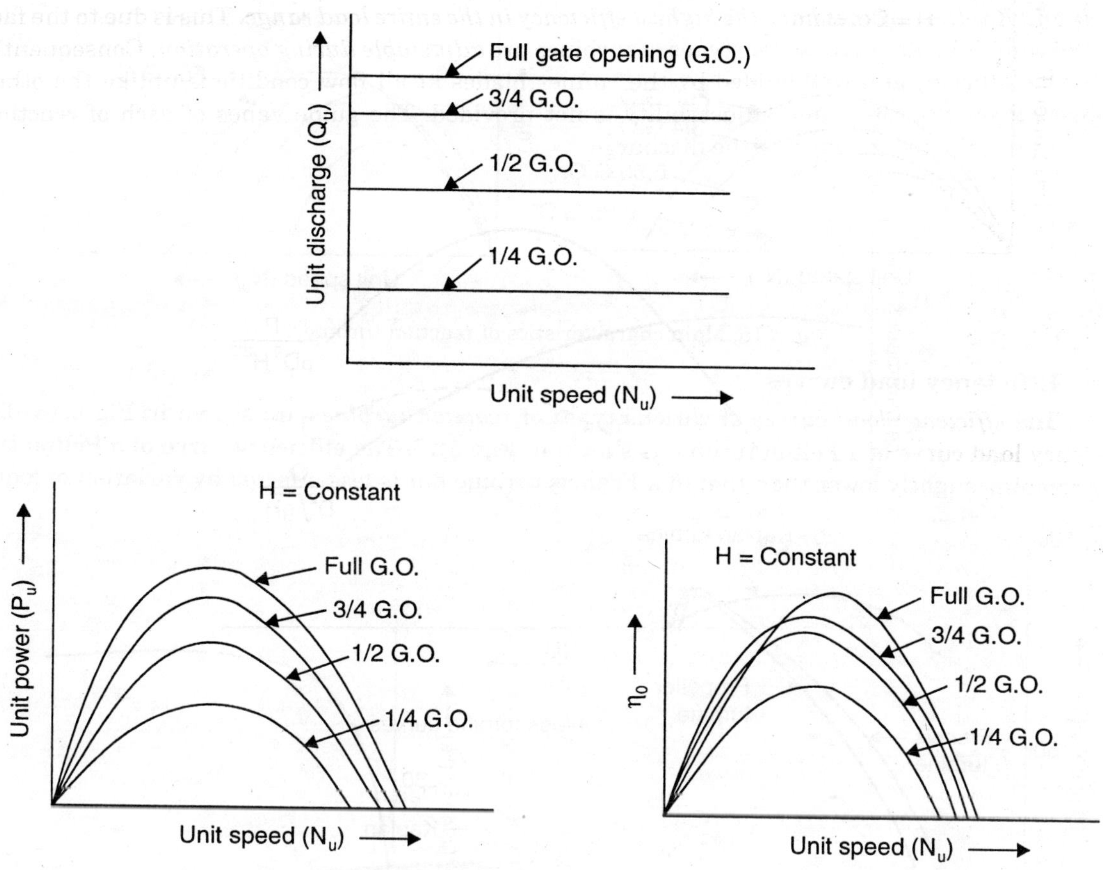

Fig. 5.14. Main characteristic curves for a Pelton wheel.

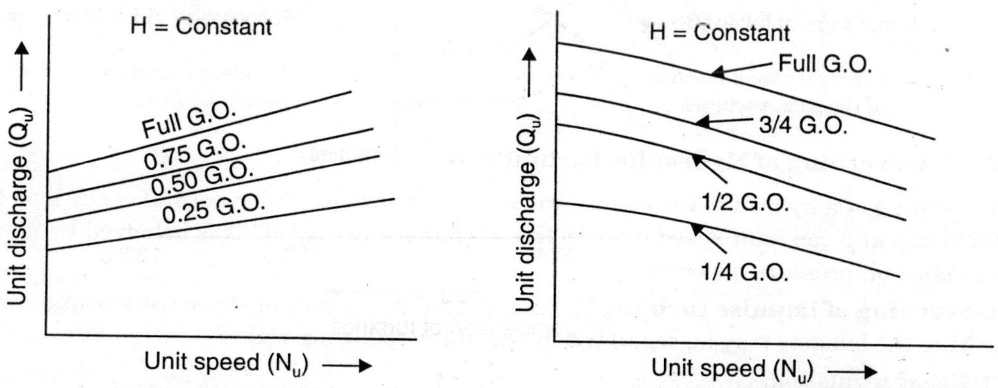

(a) For Kaplan turbine

(b) For Francis turbine

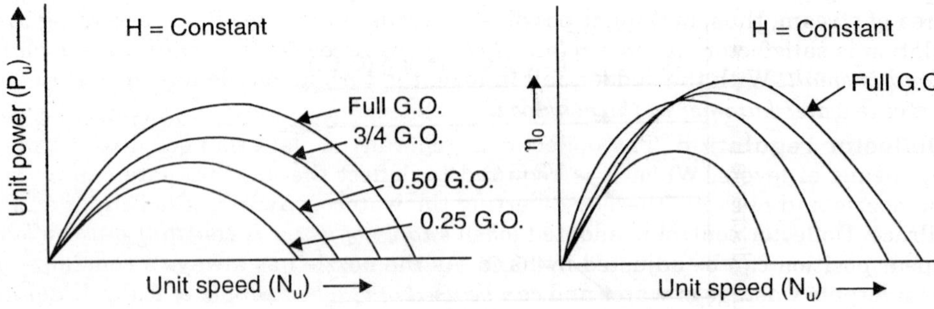

Fig. 5.15. Main characteristics of reaction turbines.

Efficiency load curves

The *efficiency load curves* of various types of *reaction turbines* are shown in Fig 5.16. The efficiency load curve of a Pelton turbine is shown in Fig. 5.17. The efficiency curve of a Pelton turbine remains slightly lower than that of a Francis turbine but is less affected by variation of load.

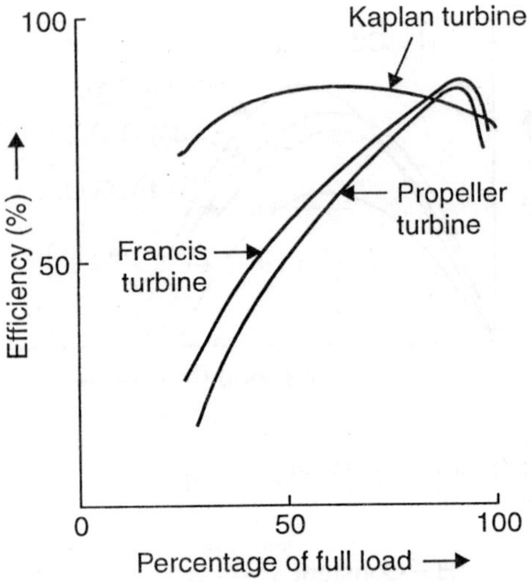

Fig. 5.16. Efficiency-load curves of reaction turbines.

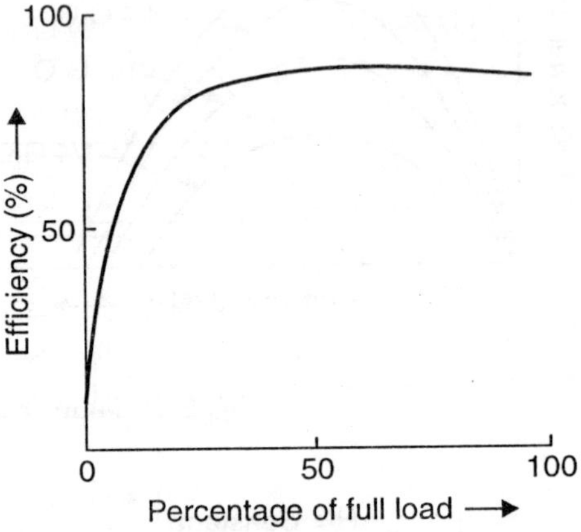

Fig. 5.17. Efficiency-load curve of a Pelton turbine.

5.7.7. Governing of Hydraulic Turbines

Governing of a hydraulic turbine means *speed regulation*. Under normal conditions the turbine should run at a constant speed irrespective of changes in load. This is achieved by means of a governor called oil pressure governor.

Governing of impulse turbine. The quantity of water rejected from the turbine nozzle and from striking the buckets may be regulated in one of the following ways :

1. Spear regulation.
2. Deflector regulation.
3. Combined spear and deflector regulation.

The spear and deflector in all cases are operated by the servomotor mechanism.

1. Spear regulation. To and fro movement of the spear inside the nozzle alters the cross-sectional area of stream, thus, making it possible to regulate the rate of flow according to the load. Spear regulation is satisfactory *when a relatively large penstock feeds a small turbine and the fluctuation of load is small.* With the sudden fall in load, the turbine nozzle has to be closed suddenly which way create *water hammer in the penstock.*

2. Deflector regulation. The deflector is generally a plate connected to the oil pressure governor by means of levers. When it is required to deflect the jet, the plate can be brought in between the nozzle and buckets, thereby diverting the water away from the runner and directing into the tailrace. Deflector control is adopted *when supply of water is constant* but the *load fluctuates.* The spear position can be adjusted by hand. As the nozzle has always a constant opening, it involves considerable wastage of water and *can be used only when supply of water is abundant.*

3. Combined spear and deflector regulation. Since both the above methods have some disadvantages, the modern turbines are provided with double regulation which is the combined spear and deflector control. Double regulation means regulation of speed and pressure. *The speed is regulated by spear and the pressure is regulated by deflector arrangement.*

Fig. 5.18 shows an arrangement for governing of Pelton turbine when the turbine is running at the *normal speed.*

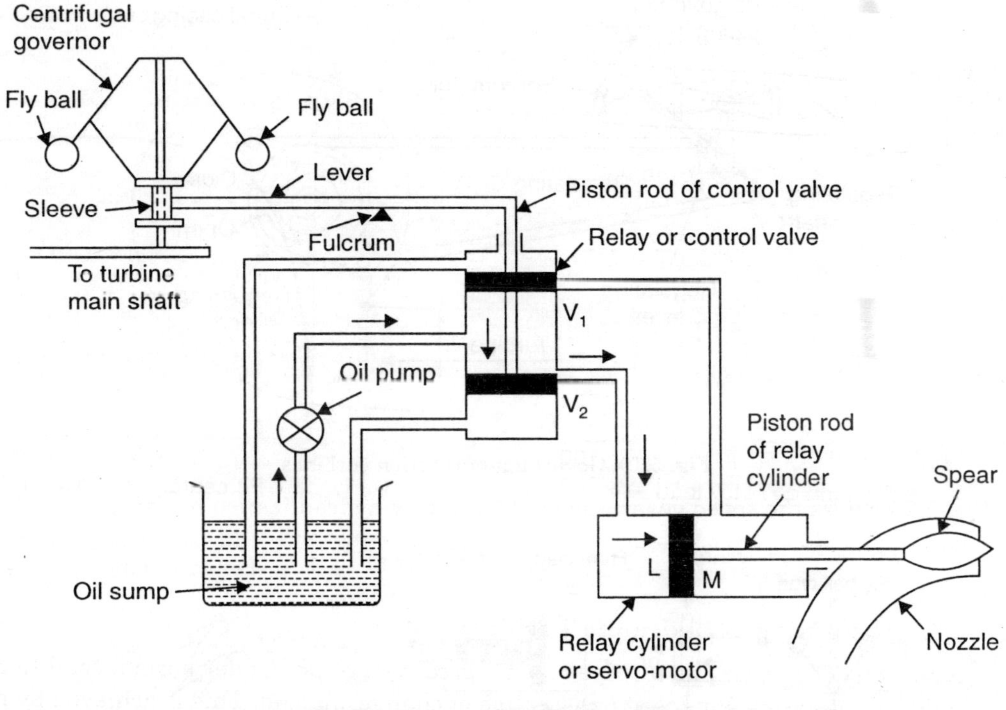

Fig. 5.18. Governing of Pelton turbine.

When the load on the generator decreases, the speed of the generator *increases* and consequently the speed of the turbine and hence centrifugal governor increases beyond the normal speed. Due to increased speed the fly-ball of the centrifugal governor move outwards/upwards (due to increased centrifugal force) causing upward movement of the sleeve. As the sleeve moves up, the lever (a horizontal lever, supported over a fulcrum, connects the sleeve and the piston rod of control valve) turns about the fulcrum and the piston rod of the control valve moves downward. Subsequently the V_1 closes and valve V_2 opens as shown in Fig. 5.18. The oil, pumped from the oil sump to the control valve or relay valve, under pressure will flow through the valve V_2 to the servomotor (or relay cylinder) and will exert force on the face L of the piston of the relay cylinder. The piston along with piston rod and spear will move towards *right*. This will decrease the area of flow of water at the outlet of the nozzle and as a consequence of this the flow rate to the turbine is reduced and the speed of the turbine falls. After the speed of the turbine becomes normal the fly balls, sleeve, lever etc. will come to normal position.

On the other hand, when the load on the generator increases, the speed of the generator and hence that of the turbine and the centrifugal governor decreases due to which its (governor) balls move downward, the sleeve moves down and piston rod of the control valve moves in the upward direction. Subsequently valve V_1 opens and valve V_2 closes. The oil under pressure will move through valve V_1 and exert a force on face M of the piston. This will make the piston move towards left thereby increasing the area of flow of water at the outlet of the nozzle and hence increase the rate of flow of water to the turbine. As a result, the speed of the turbine will increase till it becomes normal.

Governing of reaction turbines

Refer Fig. 5.19. The guide blades of a reaction turbine are pivoted and connected by levers and links to the regulating ring. To the regulating ring are attached two long regulating rods connected

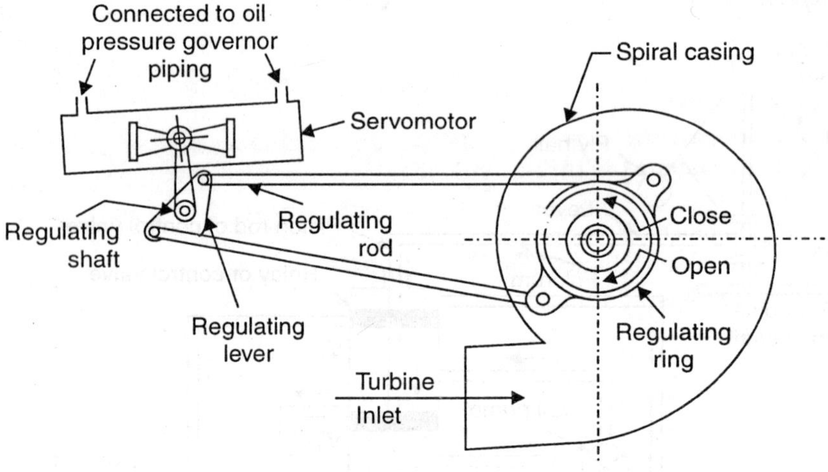

Fig. 5.19. Governing of reaction turbines.

Fig. 5.20 shows the speed governing systems for by hydraulic turbines schematically.

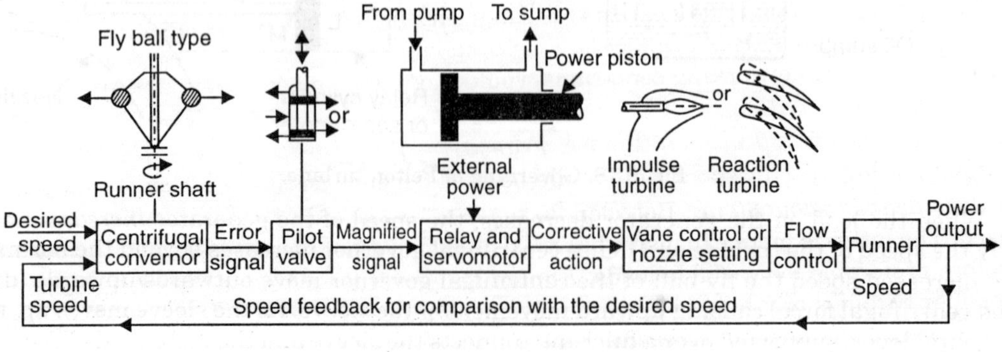

Fig. 5.20. Speed governing systems for hydraulic turbines.

to the regulating lever. The regulating lever is keyed to a regulating shaft which is turned by a servomotor piston of oil pressure governor. The penstock which feeds the turbine inlet, has a relief valve better known as *'Pressure Regulator'*. When the guide vanes have to be suddenly closed, the relief valve opens and diverts the water direct to the tailrace. Its function is therefore, similar to that of deflector in Pelton turbines. Thus the double regulation, which is the simultaneous operation, of two elements is accomplished by moving the guide vanes and relief valve in Francis turbines, by the governor.

5.7.8. Selection of Turbines

The following points should be considered while selecting the right type of hydraulic turbine :

1. Specific speed. High specific speed is essential where head is low and output is large, because otherwise the rotational speed will be low which means cost of turbo-generator and power house will be high. On the other hand there is practically no need of choosing a high value of specific speed for high head installations, because even with low specific speed, high rotational speed can be attained with medium capacity plants.

2. Rotational speed. Rotational speed depends on specific speed. Also the rotational speed of an electrical generator with which the turbine is to be directly coupled, depends on the frequency and number of pair of poles. The *value of specific speed adopted should be such that it will give the synchronous speed of the generator.*

3. Efficiency. The turbine selected should be such that it gives the *highest overall efficiency for various operating conditions.*

4. Part load operation. In general the efficiency at part-loads and overloads is less than normal. For the sake of economy the turbine should always run with maximum possible efficiency to get more revenue.

When the turbine has to run at part or overload conditions Deriaz turbine is employed. Similarly, for low heads, Kaplan turbine will be useful for such purposes in place of propeller turbine.

5. Cavitation. The installation of water turbines of reaction type over the tailrace level is effected by *cavitation*. The critical value of cavitation factor must be obtained to see that the turbine works in safe zone. Such a value of cavitation factor also effects the design of turbine, especially of Kaplan, propeller and bulb types.

6. Disposition of turbine shaft. Experience has shown that the vertical shaft arrangement is better for large-sized reaction turbines, therefore it is almost universally adopted. In case of large size impulse turbine, horizontal shaft arrangement is mostly employed.

7. Head. (*i*) *Very high heads (350 m and above).* For heads greater than 350 m, Pelton turbine is generally employed and there is practically no choice except in very special cases.

(*ii*) *High heads (150 m to 350 m).* In this range either Pelton or Francis turbine may by employed. *For higher specific speeds Francis turbine is more compact and economical than the Pelton turbine* which for the same working conditions would have to be much bigger and rather cumbersome.

(*iii*) *Medium heads (60 m to 150 m).* A *Francis turbine* is usually employed in this range. Whether a high or low specific speed unit would be used depends on the selection of speed.

(*iv*) *Low heads (below 60 m).* Between 30 to 60 *m* heads both Francis and Kaplan turbines may be used. The latter is more expensive but yields a higher efficiency at part loads and overloads. It is therefore preferable for variable loads. *Kaplan turbine is generally employed for heads under 30 m.* Propeller turbines are, however, commonly used for heads upto 15 m. *They are adopted only when there is practically no load variation.*

(*v*) *Very low heads.* For very low heads *bulb turbines* are employed these days. Although Kaplan turbines can also be used for heads from 2 m to 15 m but they are *not economical.*

5.8. PLANT LAYOUT

The plant layout is considerably affected by the nature of project.

In *low head plants*, the width of the dam is usually small and *suitable space* for fixing the turbines and generators is *limited*. On the other hand, in high head schemes the site for the power plants can be selected more conveniently to give desired layouts.

The *size and type of units* proposed to be installed also affect the layout of the power station.

For *reaction machines with vertical single runner*, the arrangement with the machines in a *line parallel to the length of the turbine house is usually preferred* ; the spacing between two units being controlled by the width of the flume or scroll case at entrance to the runner or the width of the draft tube at its month, or sometimes by the overall diameter of the alternator. *Horizontal machines of the reaction type* are best located at *right angles to the length of the turbine house* while for impulse machines with horizontal setting the wheel shaft is placed parallel to the longitudinal axis of the turbine house.

Unlike the reaction machines, the spacing of the impulse machines is fixed by the machine dimensions and necessary clearances rather than by the penstock flume or tailrace, as the amount of water discharged is comparatively small. For impulse machines with vertical setting the arrangement with centres on a line parallel to axis of turbine house is suggested. The use of such machines is rare.

5.9. HYDRO-PLANT AUXILIARIES

The following auxiliaries are essential for *starting the generating unit* :

(*i*) Exciter
(*ii*) Governor oil system
(*iii*) Lubricating oil pump.

The station can start usually independent of external power, but stand-by auxiliaries are needed for emergency service. Exciter for the main generator may be direct connected, motor-generator type engine driven or auxiliary water turbine driven. *Direct connected exciters make the unit independent of auxiliary power and also have higher efficiency than the other types*. Motor generator exciters depend upon electric power for starting, whereas an engine driven exciter is an independent unit. *The efficiency of water turbine driven exciter is very low and this is not suited for any other use except during emergencies*. The exciter is an essential piece of equipment and it is important to select a drive that is reliable and at the same time reasonably low in first cost and operation. The pressurised oil system for the governor is driven from the turbine main shaft and sufficient pressure is usually maintained in the oil tanks during the shut down, so that the turbine can be started. A stand-by motor driven pump is also installed for emergency service. The lubricating oil pumps may be driven from the turbine shaft and would work as soon as the unit starts. Sometimes a stand-by lubricating pump is also installed.

During starting of hydroplants the following auxiliaries are *not directly* needed :

(*i*) Coolant pumps
(*ii*) Air compressors
(*iii*) Drainage pumps
(*iv*) Fans and cooling oil pumps
(*v*) Cranes
(*vi*) Gate hoists
(*vii*) Valves
(*viii*) Battery charging units etc.

These auxiliaries are generally electrically driven. Water may be used to cool the bearings of the turbine and generator and transformers and is circulated through water pumps. *Air compressors* maintain a supply of air under pressure for operation of generator brakes and other uses in the power station. Unwatering of turbine pits may be required during repairs or check up for which *water pumps* are necessary. *Fans* are required for ventilation of the turbine and switch gear room or for cooling transformers. *Oil pumps* handle transformer oil through the cleaning and cooling system.

Cranes are required during machinery repairs and installations to lift heavy parts or place them in position. The head works *control gates* with *hoists* which may be manually or electrically operated. *Storage batteries* are needed to supply low voltage D.C. power for switch gear control. During failure of the main generators lighting and power for the station may also be obtained from the batteries.

Most of the station auxiliaries are electrically driven due to obvious advantages of using such drivers.

5.10. COST OF HYDRO-PLANT

A typical subdivision of investment under various items (*e.g.*, as dam, tailrace, reservoir, turbines, generators, land, building and foundations, switching and wiring, switchyard etc.) is as follows :

1. Reservoir, dam, intake, tailrace	35%
2. Turbines and generators	20%
3. Land, building and foundations	30%
4. Switching and wiring	5%
5. Switchyard	5%
6. Miscellaneous	5%

5.11. AVERAGE LIFE OF HYDRO-PLANT COMPONENTS

The average life (approximate) of various components of hydro-electric power plant is given below :

Components	*Average life (years)*
1. Reservoirs	70—80
2. Dams	
(*i*) Earthen, concrete or masonry	150
(*ii*) Loose rock	60
3. Water ways	
(*i*) Canals, tunnels	50—100
(*ii*) Penstocks	
(*a*) Steel	40—50
(*b*) Concrete	25—50
4. Power house and Equipment	
(*i*) Building	35—50
(*ii*) Generators	25
(*iii*) Transformers	30
(*iv*) Turbines (hydraulic)	5
(*v*) Pumps	20—25

5.12. HYDRO-PLANT CONTROLS

The various controls which are provided in an hydro-electric power plant are enumerated and discussed below :

1. Hydraulic controls.
2. Machine controls—Starting and stopping.
3. Machine controls—Loading and frequency.

4. Voltage control of generator and system.
5. Machine protection.

1. **Hydraulic controls.** In a hydro-plant the following *hydraulic controls are provided* :

(*i*) Storage level indicators—primary and secondary.
(*ii*) Flood control.
(*iii*) River flow control.
(*iv*) Intake gate control.

2. **Machine controls—Starting and stopping.** The control of water flowing to the turbine is exercised by providing gates and valves in the supplying conduit and at the turbine inlet. The quantity of water flowing to the turbine is regulated according to the load on the generator by the use of a governor system. While starting the turbines the casing should be filled gradually and to limit the rate of water flow by-pass valves are provided.

3. **Machine controls—Loading and frequency.** The load on the machine is controlled as follows :

(*i*) By adjusting the governor speed control.
(*ii*) By controlling system frequency.

4. **Voltage control of generator and system.** The voltage regulators are employed to ensure that electric power is supplied at proper voltage.

5. **Machine protection.** In a hydro-plant provision of protective devices should be made to guard against breakdown of turbo-generator and auxiliary services, like transformers, switchgears, overhead lines etc. Protection measures are also required to guard against incorrect operation and failure of control system.

Automatic controls are efficient, safe and reliable.

The control room should be designed for convenience of operation and the equipment should be so arranged/spaced that it is easily accessible.

5.13. ELECTRICAL AND MECHANICAL EQUIPMENT IN A HYDRO-PLANT

In an hydro-electric power plant the electrical and mechanical equipment comprise of the following :

A. *Electrical equipment :*

In electrical equipment the following elements items are included :

(*i*) Generators
(*ii*) Exciters
(*iii*) Voltage regulators
(*iv*) Transformers
(*v*) Switch gear
(*vi*) Control room equipment.

Generators. The generators employed in a hydro-plant are usually 3-phase synchronous machines and have either a vertical shaft arrangement or horizontal shaft arrangement ; but *vertical shaft arrangement is preferred.*

The generator cooling may be achieved by air circulation through the stator ducts. Cooling by water-cooled heat exchangers is common.

The power output of a 3-phase alternator is given by :

$$P = \sqrt{3}\ VI \cos \phi \times 10^{-6} \text{ MW}$$

where, P = Power output,
V = Voltage (in volts),

I = Current (in amperes), and

$\cos \phi$ = Power factor (varies from 0.9 to 0.95).

Transformers. These may be of single phase or three phase type. They are oil filled for insulation and cooling purposes. The generated voltage is stepped up by means of step up transformers. Transformers may be of power or distribution type.

Switchgear. A switchgear consists of *switches, isolators, surge arrestors and circuit breakers*. Its main function is to *make and break the circuits.*

For generated voltage it is preferred to locate switchgear indoors whereas outdoor location is used for transmission voltage.

When switchgears and transformers are located outside they should be provided adequate lightning protection.

Control room equipment. It performs the following *functions* : (*i*) Machine starting and stopping (*ii*) Generator and system voltage control (*iii*) Machine loading control (*iv*) Frequency control (*v*) Hydraulic control (*vi*) Machine running control.

B. Mechanical equipment :

Mechanical equipment include the following :

(*i*) Compressors and air ducts.

(*ii*) Shaft, coupling, bearings etc.

(*iii*) Braking equipment for the generator.

(*iv*) The oil circuits and pumps.

(*v*) Cranes and other lifting equipment.

(*vi*) Ventilation and cooling systems.

(*vii*) Equipment for water supply and drainage.

(*viii*) Equipment for power house lighting.

5.14. COMBINED HYDRO AND STEAM POWER PLANTS

An electrical power system should fulfil the following objectives :

1. To ensure an *adequate and reliable electric power supply* at all loads and at all times.
2. The source of energy should be such as to give the *minimum overall cost of the system as a whole.*

The above objectives (unless a country/region is rich either in abundant supply of cheap fuel or ample water power resources which can be developed at suitable site) can be best realised by a *judicious combination of both hydro and thermal power.* Hydropower represents a renewable source of energy which enjoys many intrinsic advantages as compared to thermal power. Although the cost of construction of hydro power plant is nearly same as that of a coal based steam power plant in terms of investment for MW, but hydro-power plant uses water for power generation which is available in abundance in nature.

It is known as that *hydro-plant can meet the demands of load variations more rapidly and easily.* Thus, *when the rate of flow of water is low, the steam plant can work at constant load producing a better efficiency and the hydro-plant will work most effectively as peak load plant and its output can be varied to meet the load fluctuations.* The steam and hydro-plants reverse their functions (*steam plant providing the peak load and the hydro-plant providing base load*) when high rate of water flow is available. But even under this condition, the *steam plant output will remain constant* and the hydro-plant output *will be varied to meet the load fluctuations.*

5.15. COMPARISON OF HYDRO-POWER STATION WITH THERMAL POWER STATIONS

The comparison between hydro-power stations and thermal power stations is given below :

S. No.	Aspects	Hydro-power station	Thermal power station
1.	*Raw material consumption*	Nil. Water power is inexhaustible and perpetual and is continuously replenished by the direct agency of sun.	Huge quantity of coal consumed, thereby exhausting "fuel reserves".
2.	*Cost of energy*	Cheaper.	Costlier.
3.	*Cost of energy generation*	Immune to inflation.	Very much influenced by the increase in the cost of fuel.
4.	*Life of plant*	Long useful life.	Not so long comparatively. The component parts deteriorate and become obsolete at a faster rate.
5.	*Pollution*	Free from problems of pollution.	Causes pollution and subsequently create health hazards.
6.	*Design, construction and reliability*	Simple in design, robust in construction and reliable in operation.	Comparatively more complicated in design, less robust in construction and less reliable in operation.
7.	*Running below a certain minimum load factor.*	Can be run.	Cannot be run.
8.	*Reserve capacity and variation in power demands.*	Particularly suited to provide reserve capacity as well as meeting the exacting needs of daily variation in power demands.	Comparatively not suited for the mentioned requirements.
9.	*Employment potential*	More. Affords a relatively high employment potential and better utilization of the available local talent and resources.	Less
10.	*Man power required ;* *Labour problem*	Small Less	Large More
11.	*Foreign exchange requirements for equipment*	Less	More
12.	*Construction time required*	Almost same as thermal power station.	Almost same as hydro-power station.
13.	*Overall capital expenditure requirements*	Low	High

5.16. UNDERGROUND HYDRO-PLANTS

The conventional hydro-plant is usually located overground at the foot of a dam or a hill slope on the banks of a river. The first of these plants was set up in U.S.A as far back as 1898. Since then several underground stations have been set up in Germany, France, Sweden, Switzerland and other countries.

Consideration supporting the construction of underground hydro-plants :

1. A suitable site for a conventional surface station/plant and good slope for penstock *not available.*
2. There may be danger of falling rocks and snow avalanches particularly in narrow valleys.
3. Availability of underground sound rock and avoidance of a long pressure tunnel and facility for a convenient tailrace outlet.
4. Possibility of elimination of surge tank required for surface station due to long pressure channel.
5. The rugged topographical features and the difficulties in finding a suitable short and steep slope for pipelines make it more economical to install the water conduit, the machine, transformer hall and tailrace system underground.

Types of layouts for underground hydro-plant :

The following are two distinctly different types of layouts for underground hydro-power stations :

(*i*) The head development (*ii*) The tail development.

In the *head development* the powerhouse is located at the intake and provides a long unlined tailrace tunnel. The necessity of surge shafts and penstocks is eliminated and quicker opening and closing of the turbine is possible. The same crew can operate the power house and headworks which gives an operational advantage.

In the *tail development* the power house is placed at the surge end of the low pressure tunnel. This requires a short tailrace. Access tunnels and shafts are necessary in both types. Between these two extreme layouts there is a wide range of intermediate possibilities for plant location.

Before taking a decision in favour of the underground station, *thorough geological investigations of the site are essential. Rock should be properly tested to devise economical tunneling method which may considerably affect the cost of construction of the power station.*

Advantages of the underground Power house/Station :

1. Better rock bearing properties.
2. Foundation difficulties are easier to overcome.
3. In planning the design and location of the principal works surface topography does not interfere at all.
4. There is greater freedom in positioning stations.
5. Maintenance cost of power station including painting, heating etc. are practically eliminated.
6. These stations are safer from defence point of view and are immune to bombardment. They are also protected against such natural hazards as rain, wind, loose rocks etc.
7. There is saving in cost of land.
8. Land acquisition problem is minimum.
9. There is greater uniformity in climatic conditions during all seasons.
10. In steep and rugged countryside, the site selection for an outdoor plant may cause difficulties due to the risk of avalanches and land slips. At some places there might be added hazard due to forest fire.

11. Uninterrupted construction work in adverse weather.
12. Shorter conduit lengths and therefore smaller surge tanks.
13. A tailrace tunnel may be easier to design and construct compared to the headrace pressure tunnel.
14. The useful life of the structures excavated in the rock is considerably longer than that of concrete and reinforced concrete structures.
15. It is possible to improve the governing of the turbines with the construction of underground power station.
16. The construction period is reduced mainly due to the possibility of full-scale construction work in winter.
17. At places, where there is extreme difference in climate of winter and summer, a uniform temperature can be maintained inside the plant with adequate control of ventilation regardless of outside weather.

Disadvantages :

1. The cost of construction of power station and accessories is more.
2. Increased cost of lighting, special ventilation and air-conditioning.
3. The tailrace tunnels and additional surge chambers are costly.
4. Additional cost of underground location of transformers etc.

5.17. AUTOMATIC AND REMOTE CONTROL OF HYDRO-STATION

Some automatic controlling devices such as *relays, governors, voltage regulators* etc. are employed in every hydro-station, big or small. However, a few operations may still be needed to start or stop the units and distribute load on two or more units in operation at a time. For stations where full shift operation is not required, the services of the operators are not fully utilized. A saving in working cost of plant can be effected if the functions of the operators are derived from an automatic control system.

An automatic system is more safe, efficient and reliable and usually works through the governor and the voltage regulator. Due to these reasons automatic hydrostations have been developed. Automatic plants may be divided into following three classes :

1. Fully automatic plant.
2. Partly automatic plant.
3. Remote controlled plant.

1. Fully automatic plant. This type of plant may be controlled by :

(*i*) A time switch (*ii*) A float switch (*iii*) A load sensitive device.

The *time switch* may start and stop the power station at preset clock timings.

The *float switch* works with the change in level of water in the reservoir, so that if the water level rises the generator output is increased and *vice versa*. Thus the station shares the load in interconnected system according to availability of water in the reservoir.

The *load sensitive device* is actuated by the demand for power in the area served by the plant, so that increase in power demand would increase the output of the generators automatically. In case of any trouble arising anywhere in the system, the plant is automatically shut-down till such time as the source of trouble is removed.

2. Partly automatic plant. In partly automatic system the units are manually started and synchronised, but in case of a fault shutdown automatically. The partly automatic system is cheaper than fully automatic control system.

3. Remote control plant. When the control on a power system is exercised from a distance, usually from another generating station or control centre, the automatic control system becomes a remote controlled one. In such a system the operator at the control point transmits a signal to the controlled station and this actuates the automatic system at the station to perform the desired effect such as starting and stopping of the units and load distribution among the sets in operation. The automatic remote control system also receives from the controlled stations all necessary information such as water level in the reservoir, share of load supplied by each unit etc.

Small hydro-plants which are not justified to be developed for manual control, can usually be justified by the application of remote automatic control.

5.18. SAFETY MEASURES IN HYDRO-ELECTRIC POWER PLANTS

Following safety measures need to be taken for the safe operation of an hydro-electric power plant :

1. Surge tanks
2. Screens
3. Sand traps
4. Jet dispersers
5. Pressure regulator.

Surge tanks. A surge tank is used to *prevent sudden increase of pressure* in the supply line or the penstock. It is placed as near as possible to the turbine. The tank may be open at the top or closed. In case it is open at the top, it must not be lower than the level of the water in the reservoir.

Screens. These are provided to prevent logs, fishes, ice blocks and other obstructive elements from entering the pipelines and turbines.

Sand traps. Sand traps are provided to prevent the sand flowing with water in pipes since sand blast action of solid matter in the water causes rapid wear of nozzles, spears, blades etc. of the turbine.

Jet dispersers. The discharged water at the bottom of the high dams possesses large amount of energy which is likely to cause scouring of the channel below the dam and consequent damage to the dam foundation unless some means are provided to dissipate it. The possible remedies for this are either *to discharge water into a cushion pool* or *to provide a jet dispersers* at the end of outlet pipe so that the end of the outlet pipe is such that the jet is broken up into a conical shower of drops and their energy is absorbed by air.

Pressure regulator. It is usually operated by a governor of the turbine. It is provided on the pipeline near the turbine inlet so that when the turbine gates are suddenly closed, pressure surges so produced are kept within the safe limits of the pipeline. The water discharged from the regulator is passed on to tailrace through a separate pipeline.

5.19. PREVENTIVE MAINTENANCE OF HYDRO-PLANT

The purpose of preventive maintenance is to minimise breakdown and excessive depreciation resulting from neglect. In a hydro-plant (using reaction turbines) monthly, quarterly, half yearly and yearly inspection and maintenance are carried out on the following parts :

Inspection/Maintenance	***Parts***
Monthly	*Turbine cover parts* (*e.g.*, leakage unit, drainage holes, servomotor connections, turbine shaft and cover, oil pump etc.)
	Operating ring of turbines
	Guide vane mechanism

Quarterly	*Servomotor*
	Governor oil system
	Ejector cabinet
	Feedback system
Half yearly	*Governor mechanism*
	Gauges
	Grease pumps for guide vanes and guide bearings
	Grease pipes connected to grease pumps
Yearly	*Turbine auxiliaries* (*e.g.*, oil pressure tank, turbine guide bearing, turbine instruments)
	Scroll casing runner with guide vanes
	Emergency slide valve
	Pit liner
	Draft tube
	Runner blades checked for cavitational effects, cracks and wearing out.

5.20. CALCULATION OF AVAILABLE HYDRO-POWER

The theoretical power available from falling water can be calculated using the following formula :

$$P_{th.} = \frac{wQH}{1000} \text{ kW} \quad \text{...(5.4)}$$

where, $P_{th.}$ = Theoretical output in *kW*,

w = Weight density of water in N/m^3,

Q = Flow through turbine (or quantity of water available for hydro-power generation) in m^3/s, and

H = Head available in *metres*.

The actual useful or *effective output* depends upon the efficiency of the various parts of the installation.

If, η_1 = Efficiency of pipelines, intake etc., and

η_2 = Efficiency of hydraulic turbine,

Then, *overall efficiency* $\eta_0 = \eta_1 \times \eta_2$.

Since the turbine and the generator are directly coupled on common shaft the hydro-electrical power available will be given by the equation :

$$P_{actual} = P_{th.} \times \eta_0 \quad \text{...(5.5)}$$

or,

$$P_{actual} = \frac{wQH}{1000} \times \eta_0 \text{ kW} \quad \text{...(5.6)}$$

5.21. COST OF HYDRO-POWER

The initial cost of any hydro-plant is very high but the power produced by it is the cheapest. The following costs are included in development of a hydro-plant :

1. Cost of land and riparian rights.
2. Cost of railways and highways required for the construction work.

3. Cost of construction.
4. Cost of engineering supervision of the project.
5. Cost of building etc.
6. Cost of equipment.
7. Cost of equipment used for power transmission.

5.22. HYDROLOGY

5.22.1. Introduction

Hydrology *may be defined as the science which deals with the depletion and replenishment of water resources.* It deals with the surface water as well as the ground water. It is also concerned with the *transportation of water* from one place to another, and from one form to another. It helps us in determining the occurence and availability of water.

Hydrology aims at answering the following major questions :

(*i*) How is the water going to precipitate ?

(*ii*) How is water going to behave ?

(*iii*) What will happen to water after precipitation ?

The basic knowledge of this science is a must for every civil engineer, particularly the one who is engaged in the design, planning or construction of irrigation structures, bridges and highway culverts, or flood control works, etc.

5.22.2. The Hydrologic Cycle

Most of the earth's water sources, such as rivers, lakes, oceans and underground sources, etc. get their supplies from rains, while the rain water itself is the evaporation from these sources. Water is lost to the atmosphere as vapour from earth, which is then precipitated back in the form of rain, snow, hail, dew, sleet or frost, etc. This evaporation and precipitation continues for ever, and thereby, a balance is maintained between the two. This process is known as *Hydrologic Cycle*. It can be represented graphically as shown in Fig. 5.21. Hydrologic equation is expressed as follows :

$$P = R + E \quad ...(5.7)$$

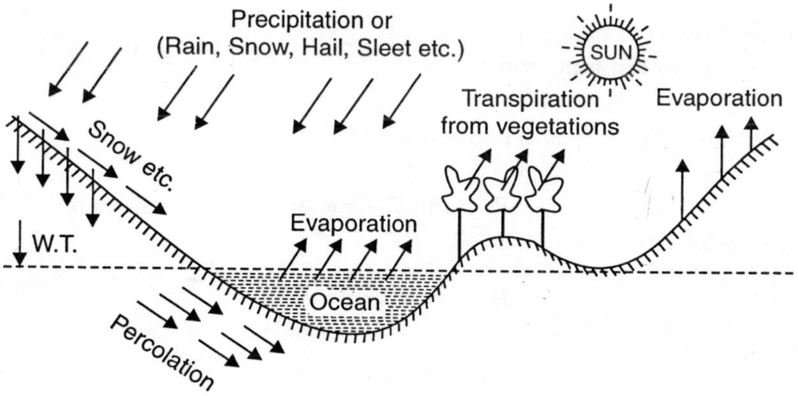

Fig. 5.21. Hydrologic cycle.

where, P = Precipitation,

R = Run-off, and

E = Evaporation.

Precipitation. It includes all the water that falls from atmosphere to earth surface. Precipitation is of two types : (*i*) Liquid precipitation (rain fall). (*ii*) Solid precipitation (snow, hail).

Run-off and surface run-off. Run-off and surface run-off are two different terms and should not be confused. *Run-off* includes all the water flowing in the stream channel at any given section. While the *surface run-off* includes only the water that reaches the stream channel without first percolating down to the water table.

Run-off can, therefore, also be named as Discharge or Stream flow. Rainfall duration, its intensity and a real distribution influence the rate and volume of run-off.

Evaporation. Transfer of water from liquid to vapour state is called *evaporation.*

Transpiration. The process by which water is released to the atmosphere by the plants is called *transpiration.*

5.22.3. Measurement of Run-off

Run-off can be measured daily, monthly, seasonal or yearly. It can be measured by the following *methods* :

1. From rainfall records.
2. Empirical formulae.
3. Run-off curves and tables.
4. Discharge observation method.

1. From rainfall records. In this method consistent rainfall record for a sufficiently long period is taken and then average depth of rainfall over the catchment is determined. Then considering all the factors which affect run-off process, a coefficient is arrived at for that catchment. Now a simple equation can be used to find out the run-off over the catchment.

$$\text{Run-off} = \text{Rainfall} \times \text{coefficient} \quad \text{...(5.8)}$$

2. Empirical formulae. In this method an attempt is made to derive a direct relationship between the rainfall and subsequent run-off. For this purpose some constants are established which give fairly accurate result for a specified region. Some important formulae are given below :

(*a*) Khosla's formula :

$$R = P - 4.811\,T$$

where, R = Annual run-off in mm,
P = Annual rainfall in mm, and
T = Mean temperature in °C.

(*b*) Inglis formulae for hilly and plain areas of Maharashtra :

For *Ghat region*

$$R = 0.88\,P - 304.8$$

For *plain region*

$$R = \frac{(P - 177.8) \times P}{2540}$$

(*c*) *Lacey's formula* :

$$R = \frac{P}{1 + \dfrac{3084\,F}{PS}}$$

where, R = Monsoon run-off in mm,
P = Monsoon rainfall in mm,
S = Catchment area factor, and
F = Monsoon duration factor.

Values of S for various types of catchment are given below :

Type of catchment	***Value of S***
Flat, cultivated and black cotton soils	0.25
Flat, partly cultivated, various soils	0.6
Average catchment	1.00
Hills and places with little cultivation	1.70
Very hilly and steep, with hardly any cultivation	3.45

Values of F for various durations of monsoon are given below :

Class of monsoon	***Value of F***
Very short	0.50
Standard length	1.00
Very long	1.50

3. Run-off curves and tables. Each region has its own catchment area and rainfall characteristics. Thus formulae given above and coefficients derived there in cannot be applied universally. However, for the same region the characteristics mostly remain unchanged. Based on this fact the run-off coefficients are derived once for all. Then a graph is plotted in which one axis represents rainfall and the other run-off. The curves obtained are called *run-off curves*. Alternatively a *table* can be prepared to give the run-off for a certain value of rainfall for a particular region.

4. Discharge observation method. By actual measurement of discharge at an outlet of a drainage basin run-off over a catchment can be computed. The complication in this method is that the discharge of the stream at the outlet comprises surface run-off as well as sub-surface flow. To find out the sub-surface run-off it is essential to separate the sub-surface flow from the total flow. The separation can be done on an approximate basis but with correct analysis.

Factors affecting the run-off

The following factors affect run-off :

1. *Rainfall pattern.*
2. Character of catchment area.
3. Topography.
4. Shape and size of the catchment area.
5. Vegetation.
6. Geology of the area.
7. Weather conditions.

5.22.4. Hydrograph

Hydrograph *is defined* as *a graph showing discharge (run-off) of flowing water with respect to time for a specified time.* Discharge graphs are known as flood or run-off graphs. Each hydrograph has a reference to a particular river site. The time period for discharge hydrograph may be hour, day, week or month.

Hydrograph of stream of river will depend on the characteristics of the catchment and precipitation over the catchment. Hydrograph will access the flood flow of rivers hence it is essential that anticipated hydrograph could be drawn for river for a given storm.

Hydrograph indicates the power available from the stream at different times of day, week or year.

Typical hydrographs are shown in Figs. 5.22 and 5.23.

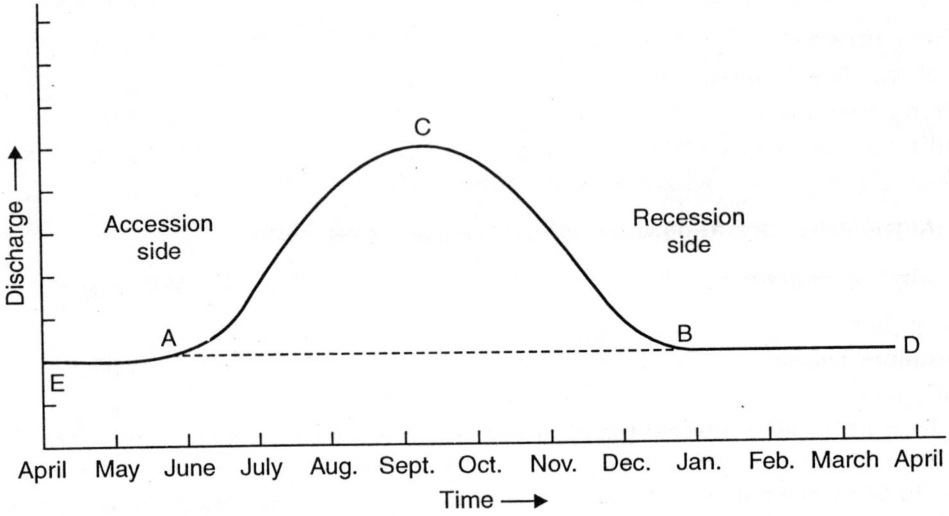

Fig. 5.22. Typical hydrograph.

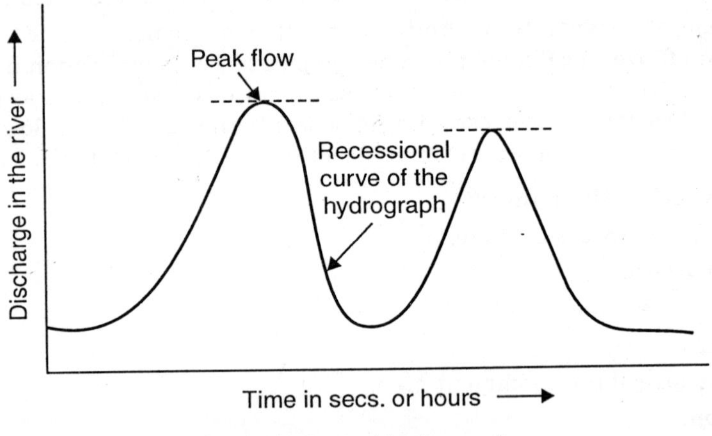

Fig. 5.23. Typical hydrograph.

The Unit hydrograph :

The peak flow represents a momentary value. Therefore the peak flow alone does give sufficient information about the run-off. It is necessary to understand the full hydrograph of flow. Introduction of unit hydrograph theory in 1932 made it possible to predict a run-off hydrograph corresponding to an observed or hypothetical storm. *The basic concept of unit hydrograph is that the hydrographs of run-off from two identical storms would be the same.* In practice identical storms occur very rarely. The rainfall generally varies in duration, amount and areal distribution. This makes it necessary to construct a typical hydrograph for a basin which could be used as a unit of measurement of run-off.

A **unit hydrograph** *may be defined as a hydrograph which represents unit run-off resulted from an intense rainfall of unit duration and specific areal distribution.*

The *following steps* are used for the *construction of unit hydrograph :*

1. Choose an isolated intense rainfall of unit duration *from past records.*
2. Plot the discharge hydrograph for outlet from the rainfall records.
3. Deduct the base flow from stream discharge hydrograph to get hydrograph of surface run-off.
4. Find out the volume of surface run-off and convert this volume into cm of run-off over the catchment area.
5. Measure the ordinates of surface run-off hydrograph.
6. Divide these ordinates by obtained run-off in cm to get ordinates of unit hydrograph.

Thus for any catchment unit hydrograph can be prepared once. Then whenever peak flow is to be found out, *multiply the maximum ordinate of unit hydrograph by the run-off value expressed in cm.* Similarly to obtain run-off hydrograph of the storm of same unit duration multiply the ordinates of the unit hydrograph by the run-off value expressed in cm. If the storm is of longer duration calculate the run-off in each unit duration of the storm. Then super-impose the run-off hydrographs in the same order giving a lag of unit period between each of them. Finally draw a summation hydrograph by adding all the overlapping ordinates. Generally the computations are done in a tabular form before the hydrograph is plotted.

Fig. 5.24 is self explanatory and shows how a run-off hydrograph is constructed from a unit hydrograph.

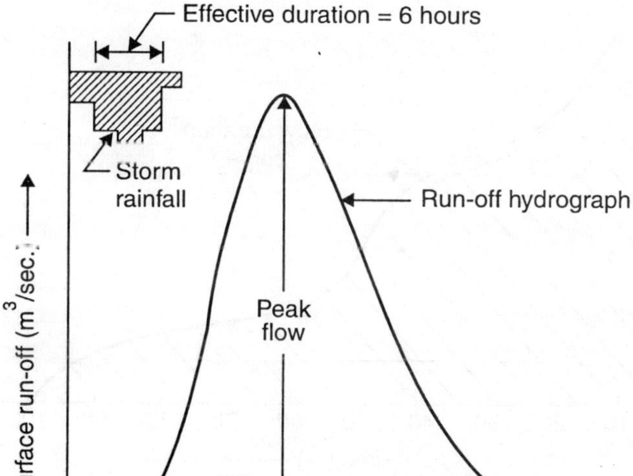

Fig. 5.24

Limitations to the use of unit hydrographs :

1. Its use is limited to areas about 5000 sq. kilometres since similar rainfall distribution over a large area from storm to storm is rarely possible.

2. The odd-shaped basins (particularly long and narrow) have very uneven rainfall distribution, therefore, unit hydrograph method is not adopted to such basins.
3. In mountain areas, the areal distribution is very uneven, even then unit hydrograph method is used because the distribution pattern remains same from storm to storm.

5.22.5. Flow Duration Curve

Refer to Fig. 5.25. Flow duration curve is another useful form *to represent the run-off data for the given time. This curve is plotted between flow available during a period versus the fraction of time.* If the magnitude on the ordinate is the potential power contained in the stream flow, then the curve is known as "*power duration curve*". This curve is a very useful tool in the analysis for the development of water power.

The flow duration curve is drawn with the help of a hydrograph from the available run-off data and, here it is necessary to find out the length of time duration which certain flows are available. This information either from run-off data or from hydrograph is tabulated. Now the flow duration curve taking 100 percent time on X-axis and run-off on Y-axis can be drawn.

The area under the flow duration curve (Fig. 5.25) gives the total quantity of run-off during that period as the flow duration curve is representation of graph with its flows arranged in order of descending magnitude.

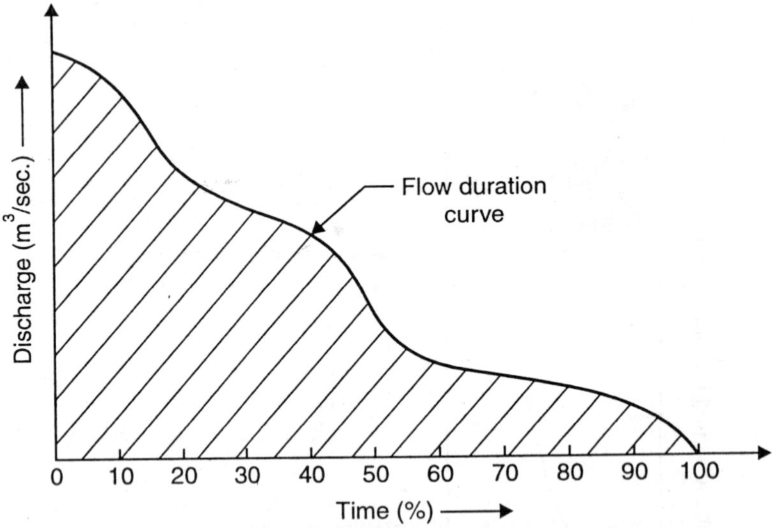

Fig. 5.25. Flow duration curve.

If the head of discharge is known, the possible power developed from water in kW can be determined from the following equation :

$$\text{Power (kW)} = \frac{wQH}{1000} \times \eta_0$$

where, Q = Discharge, m^3/sec.,

H = Head available, m,

w = Weight density of water, N/m^3, and

η_0 = Overall efficiency.

Thus the *flow duration curve can be converted to a power duration with some other scale on the same graph.*

Flow duration curves are most useful in the following cases :

(*i*) *For preliminary studies* (*ii*) *For comparison between streams.*

Uses of flow duration curve :

1. A flow duration curve allows the evaluation of low level flows.

2. It is highly useful in the planning and design of water resources projects. In particular, for hydropower studies, the flow duration curve serves to determine the potential for *firm power* generation. In the case of a run-of-the-river plant, with no storage facilities, the firm power is usually computed on the basis of flow available 90 to 97 percent of the time. The *firm power* is also known as the *primary power*. *Secondary power* is the power generated at the plant utilising water other than that used for the generation of firm power.

3. If a sediment rating curve is available for the given stream, the flow duration curve can be converted into cumulative sediment transport curve by multiplying each flow rate by its rate of sediment transport. The area under this curve represents the total amount of sediment transported.

4. The flow duration curve also finds use in the design of drainage systems and in flood control studies.

5. A flow duration curve plotted on a log-log paper provides a qualitative description of the run-off variability in the stream. If the curve is having steep slope throughout, it indicates a stream with highly variable discharge. This is typical of the conditions where the flow is mainly from surface run-off. A flat slope indicates small variability which is a characteristic of the streams receiving both surface run-off and ground water run-off. A flat portion at the lower end of the curve indicates substantial contribution from ground water run-off, while the flat portion at the upper end of the curve is characteristic of streams with large flood plain storage, such as lakes and swaps, or where the high flows are mainly derived from snowmelt.

6. The shape of the flow duration curve may change with the length of record. This aspect of the flow duration curve can be utilised for extrapolation of short records.

Shortcomings/Defects of flow duration curve

1. It does not present the flows in natural source of occurrence.

2. It is also not possible to tell from flow duration curve whether the lowest flows occurred in consecutive periods or were scattered throughout the considered period.

5.22.6. Mass Curve

A **'mass curve'** *is the graph of the cumulative values of water quantity (run-off) against time.* A mass curve is an *integral curve of the hydrograph* which expresses the area under the hydrograph from one time to another.

It is a convenient device *to determine storage requirement* that is needed to produce a certain dependable flow from fluctuating discharge of a river by a reservoir.

Mass curve can also be used to solve the reserve problem of determining the maximum demand rate that can be maintained by a given storage volume. However, it is a trial and error procedure.

The mass curve *will always have a positive shape* but of a greater or less degree depending upon the variations in the quantity of inflow water available. The negative inclination of mass curve would show that the amount of water flowing in the reservoir was less than the loss due to evaporation and seepage.

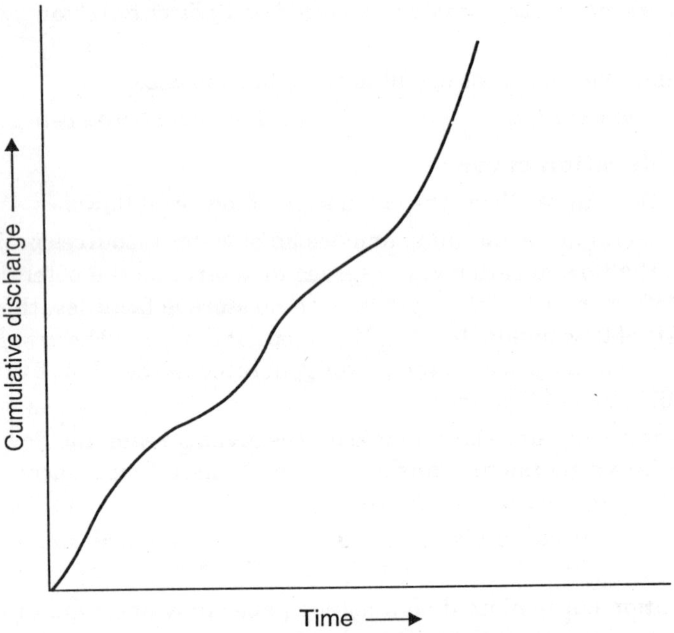

Fig. 5.26. Mass curve.

5.23. HYDRO-POWER DEVELOPMENT IN INDIA

Hydro-power is a renewable source of energy which entails many intrinsic advantages. In India the scope of water power development is tremendous. The first hydropower station in India dates back to year 1897 when a small power station of 200 kW capacity was constructed at Darjeeling. Since then many big and small hydropower stations have been installed in the country. Total hydro potential in our country is estimated to be equivalent to about 75000 MW at 60 percent load factor of which only about 20 percent has been exploited so far.

Important hydro plants in India

State/Name of power plant	*Installed capacity (MW)*
Andhra Pradesh	
Machkand (stage I and II)	114
Upper silern	120
Lower silern	600
Srisailam	770
Nagarjun sagar pumped storage	100
Assam	
Umiam	54
Gujarat	
Ukai	300
Himachal Pradesh	
Baira suil	200
Jammu and Kashmir	
Salal	270

Karnataka	
Tungabhadra	72
Sharavati	890
Kailindi	396
Kerala	
Parambikulam-Aliyar	185
Sabarigiri	300
Idikki (Stage I)	390
Maharashtra	
Koyna (stages I, II and III)	860
Manipur	
Loktak	70
Orissa	
Hirakud (stages I and II)	270
Balimela	480
Punjab	
Bhakra Nangal	1084
Beas-Sutlej link	780
Rajasthan	
Chambal	287
Uttar Pradesh	
Rihand	300
Yamuna (stages I and II)	424
Tamil Nadu	
Kundah (stages I, II and III)	425
Kodiar	100

Although the present utilization of hydropower in over country is relatively small with the present tempo of development and need for power resources it would not be long before the available potential is fully harnessed. Hydro-field provides immense scope for sophisticated study requiring application of modern mathematical and operational research techniques with the help of computers.

Example 5.1. *The following data relate to a proposed hydro-electric station :*

Available head = 28 m ; Catchment area = 420 sq. km ; rainfall = 140 cm/year ; percentage of total rainfall utilized = 68% ; Penstock efficiency = 94% ; turbine efficiency = 80% ; generator efficiency = 84% and load factor = 44%.

(i) Calculate the power developed.

(ii) Suggest suitable machines and specify the same.

Solution. Head available, $H = 28$ m

Catchment area, $A = 420$ sq. km $(= 420 \times 10^6 \text{ m}^2)$

Rainfall $= 140$ cm/year $(= 1.4 \text{ m})$

Rainfall utilized, $h = 68\%$ of the total rainfall

$= (0.68 \times 1.4)$ m per year

Penstock efficiency, $\eta_p = 94\%$

Turbine efficiency, $\eta_t = 80\%$

Generator efficiency, $\eta_g = 84\%$

Load factor $= 44\%$.

(*i*) **Power developed, P :**

Quantity of water available *per year*

$$= A \times h = (420 \times 10^6) \times (0.68 \times 1.4)$$
$$= 399.84 \times 10^6 \text{ m}^3$$

Hence the quantity of water available *per second,*

$$Q = \frac{399.84 \times 10^6}{(365 \times 24) \times 3600} = 12.6 \text{ m}^3$$

$\therefore$ $P = \eta_0 \times wQH$ (where, η_0 = Overall efficiency $= \eta_p \times \eta_t \times \eta_g$)

or $P = \eta_p \times \eta_t \times \eta_g \times wQH$

$$= 0.94 \times 0.8 \times 0.84 \times 9.81 \times 12.6 \times 28 = 2186 \text{ kW}$$

Hence *average output of generating units* = **2186 kW. (Ans.)**

(*ii*) **Machines to be used :**

Total ratings of generators $= \dfrac{2186}{0.44} = 4968$ kW

Providing *two machines* of equal rating,

Capacity of each unit $= \dfrac{4968}{2 \times 0.84} = 2957$ each.

As the available head is low, **Kaplan turbines** (propeller type) are suggested, each having a *generating capacity of* **2957 kW. (Ans.)**

Example 5.2. *The following data is available for a hydro-power plant :*

Available head = 140 m ; catchment area = 2000 sq. km ; annual average rainfall = 145 cm ; turbine efficiency = 85% ; generator efficiency = 90% ; percolation and evaporation losses = 16%.

Determine the following :

(i) Power developed.

(ii) Suggest type of turbine to be used if runner speed is to be kept below 240 r.p.m.

Solution. Head available, $H = 140$ m

Catchment area, $A = 200$ sq. km ($= 200 \times 10^6$ m^2)

Annual average rainfall, $h = 145$ cm ($= 1.45$ m)

Turbine efficiency, $\eta_t = 85\%$

Generator efficiency, $\eta_g = 90\%$

Percolation and evaporation losses, $z = 16\% = 0.16$

(*i*) **Power developed, P :**

Quantity of water available for power generation *per year*

$$= A \times h \times (1 - z)$$
$$= 200 \times 10^6 \times 1.45 \times (1 - 0.16) = 2.436 \times 10^8 \text{ m}^3\text{/year}$$

Hence, quantity of water available per second,

$$Q = \frac{2.436 \times 10^8}{(365 \times 24) \times 3600} = 7.72 \text{ m}^3\text{/s}$$

$\therefore$
$$P = \eta_0 \times wQH$$
$$= \eta_t \times \eta_g \times wQH$$

$= 0.85 \times 0.9 \times 9.81 \times 7.72 \times 140$

$= \mathbf{8111\ kW\ or\ 8.111\ MW.}$ **(Ans.)**

(*ii*) **Type of turbine to be used :**

Specific speed, $N_s = \dfrac{N\sqrt{P}}{H^{5/4}} = \dfrac{240\sqrt{8111}}{(140)^{5/4}} = 44.28$ r.p.m.

Single Pelton turbine with 4 jets can be used. Further, since head available is large and discharge is low, Pelton turbine will work satisfactorily.

Example 5.3. *(a) Name the common types of prime movers that are employed in a hydroelectric power plant and discuss the factors that govern their choice.*

(b) A hydroelectric station is to be designed to operate at a mean head of 205 m and supplied from a reservoir lake having a catchment area of 1000 km² with average annual rain fall of 125 cm of which 80% is available for power generation. The expected load factor at the plant is 75%. Allowing a head loss of 5 m and assuming efficiency of the turbine and generator to be respectively 90% and 95%, calculate suitable MW rating of the station. Comment also on the type of turbine to be installed.

(AMII Summer 2001)

Solution. (*a*) The following three different types of hydraulic turbines are used as prime movers in hydroelectric power plants :

(*i*) **Kaplan.** For heads lower than 30 m and for variable load operation. Fixed vane propellers are used for heads less than 70 m and for fairly constant load operations.

(*ii*) **Francis turbine.** Used for heads from 70 m to 500 m. It is a reaction type of turbine and mostly used inward flow. The efficiency of Francis turbine at full load is high. It is designed for higher speed.

(*iii*) **Pelton wheel.** Used for heads more than 500 m. Pelton wheel is invariably mounted with a horizontal shaft and has low number of revolutions per minute. The limiting power output for a pelton wheel is of the order of 65,000 H.P.

(*b*) *Given :* Gross head = 205 m ; Catchment area = 1000 km^2 ; Average annual rainfall = 125 cm = 0.125 m ; Load factor = 75% or 0.75 ; Loss of head = 5 m ; $\eta_T = 90\%$; $\eta_G = 95\%$.

Water available during the year = $(1000 \times 10^6) \times 1.25 \times 0.8 = 10^9$ m^3.

∴ Quantity of water available per sec.

$$= \frac{10^9}{8760 \times 3600} = 31.7 \text{ m}^3\text{/s}$$

Available head, $H = 205 - 5 = 200$ m

Now, *Average power produced,* $P = wQH \times \eta_T \times \eta_G$ kW

$= 9.81 \times 31.7 \times 200 \times 0.9 \times 0.95 = 53177$ kN

$\simeq 53.2$ MW (where, $w = 9.81$ kN/m^3)

With a load factor of 75%, *overall capacity of the station for which it should be designed*

$$= \frac{53.2}{0.75} = \mathbf{70.9\ MW.} \quad \textbf{(Ans.)}$$

The type of turbine for a head of 200 m recommended is **Francis turbine. (Ans.)**

Example 5.4. *(a) Discuss the function of 'surge tank' in a hydroelectric power plant.*

(b) What are pumped storage plants and where are they used ?

(c) What are the differences between a 'run-off-river plant' and a 'run-off-river plant with pondage' ?

(d) The daily load curve data for a certain area is as follows :

12 MN–5 AM	:	*100 MW*
5 AM–8 AM	:	*120 MW*
8 AM–12 Noon	:	*250 MW*
12 Noon–1 PM	:	*100 MW*
1 PM–5 PM	:	*250 MW*
5 PM–9 PM	:	*350 MW*
9 PM–12 MN	:	*150 MW*

It is proposed to install a run-off-river plant with pondage and a steam plant for supplying the above load. The run-off data indicates that a flow of 50 m^3/sec is available. The head is 90 m. Hydro-plant efficiency is 90% and transmission losses 5%. Determine the capacity of the hydro-plant and the steam plant. The steam plant is located near the load centre. **(Banglore University)**

Solution. (*a*) **Surge tank** in *a hydroelectric power plant is an open tank which is often used with the pressure conduit of considerable length.* The main purpose of providing surge tank is to *reduce the distance between the free water surface and turbine, thereby reducing the water hammer effect on penstock and also protect upstream tunnel from high pressure rises.* It also *serves as a supply tank to the turbine when the water to the pipe is accelerating during increased load conditions and as a storage tank when the water is decelerating during reduced load conditions.*

(*b*) The **pump storage power plant** essentially consists of a head water pond and a tail water pond. During the off-peak periods of an interconnected power plant system consisting of steam and pump storage, the water from the tail water pond is pumped with the help of a pump using extra energy available from thermal power plant during off-peak hours. With surplus available energy during off-peak period, energy is stored in the form of hydraulic potential energy by lifting the water from lower level to upper level. The same stored hydraulic energy is used during peak load period by supplying the water from the upper basin to the water turbine through the penstocks.

These plants are used wherever *base load stations are thermal stations and water is also available for hydro-generation.*

(*c*) **Run of river plant without pondage** *does not store the water and uses the water as it comes.* This plant has no control over the river flow. Therefore water is wasted during low load and high flood conditions. During dry seasons, the capacity of the plant goes down due to the low flow rates of the water. The utility of these plants is very less compared with other plants due to non-uniformity of supply and lack of assurance for continuous constant supply.

Run of river plant with pondage is *provided with a pond* in which water is stored during off-peak periods and the stored water is utilized during peak hours of the same day. *Pondage increases the stream capacity for a short period.* This type of plant *can be used as base load or peak load plant.*

(*d*) *Given :* $Q = 50\ m^3/s$; $H = 90$ m ; Hydro-plant efficiency = 91% ; Transmission losses = 5%.

Capacity of the hydro-plant and steam plant :

Total electrical energy required in 24 hours

$$= 5 \times 100 + 3 \times 120 + 4 \times 250 + 1 \times 100 + 4 \times 250 + 4 \times 350 + 3 \times 150$$

$$= 500 + 360 + 1000 + 100 + 1000 + 1400 + 450 = 4810 \text{ MWh.}$$

The average output of hydro-plant,

$$P = wQH \times \eta \times 10^{-3} \text{ MW} = 9.81 \times 50 \times 90 \times 0.91 \times 10^{-3} = \mathbf{40\ MW.} \quad \textbf{(Ans.)}$$

Since the transmission losses are given as 5%, therefore, total generation required

$$= 4810 \times 1.05 = 5050 \text{ MWh.}$$

Hydro-plant generation capacity = 40 × 24 = 960 MWh.

∴ The steam plant generating capacity required

$$= 5050 - 960 = 4090 \text{ MWh}$$

Assuming power required is for 24 hours,

$$\text{Steam plant-rating} = \frac{4090}{24} = \textbf{170 MW.} \quad \textbf{(Ans.)}$$

Example 5.5. *What is a pumped storage plant ? Explain the advantage of a pumped storage plant for short peak load operation. Compare its economics with an old steam plant.*

Solution. Pump storage plant *works as a peak load station.* In this station water is pumped into a high level reservoir at off-peak periods and is utilized to drive turbines and generate power when the peak of the system occurs.

The plant consists of a synchronous motor or an induction motor coupled on one side to a multistage centrifugal pump and on the other side to a turbine. During off-peak periods the water drives the pump and replenishes the reservoir. When the system peak occurs the motor runs as an alternator or an induction generator driven by turbine. The plant can change over from one function to another *i.e.* pumping to generating and vice versa very rapidly and the whole sequence of operation is automatically controlled. An induction generator always supplies the load at leading p.f. hence compensating the system p.f. The basic advantage of pumped storage may be justified because it permits available energy to be stored for later use when it is more valuable.

When an old steam plant is used for peak load operations as an alternative to a pumped storage plant, it is a costly proposition and more over it requires large starting time and shut down time. Hence old steam plants are not at all convenient for peak load demand. Further these systems can not be used in reversible mode as is possible with pumped storage plants.

HYDROLOGY

Example 5.6. *At a particular site the mean monthly discharge is as follows :*

Month	*Discharge, m³/s*	*Month*	*Discharge, m³/s*
January	*100*	*July*	*1000*
February	*225*	*August*	*1200*
March	*300*	*September*	*900*
April	*600*	*October*	*600*
May	*750*	*November*	*400*
June	*800*	*December*	*200*

Draw the following :

(i) Hydrograph

(ii) Flow duration curve.

Solution. (*i*) The **hydrograph** is plotted between discharge (m^3/sec) and time (months) as shown in Fig. 5.27.

(*ii*) **Flow duration curve :**

In order to draw *flow duration curve* it is essential to find the *length* of *time* during which certain flows are available, *e.g.,* 100 m^3/s is available for all 12 months, flow of 200 m^3/s for 11 months, 225 m^3/s for 10 months and so on. This information is indicated in the table below :

Discharge, m³/s	***Length of time, months***	***%age time***
100 (and more)	12	100
200 (and more)	11	91.7
225 (and more)	10	83.3

300 (and more)	9	75.0
400 (and more)	8	66.7
600 (and more)	7	58.3
750 (and more)	5	41.7
800 (and more)	4	33.3
900 (and more)	3	25.0
1000 (and more)	2	16.7
1200 (and more)	1	8.3

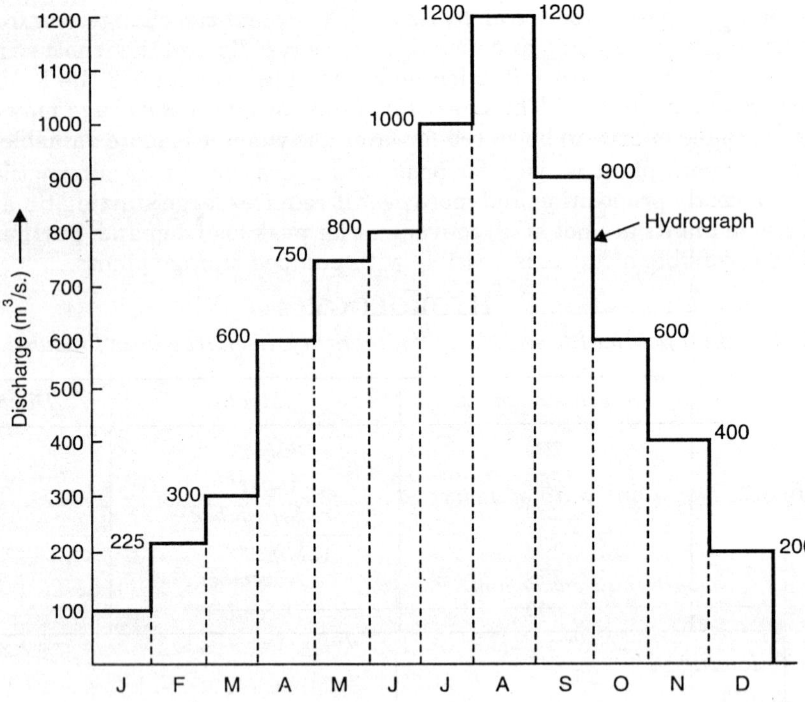

Fig. 5.27. Hydrograph.

The flow duration curve is then plotted as shown in Fig. 5.28.

Note. When selecting a suitable site for a hydropower plant the flow data for a number of years is collected and hydrographs and flow duration curves and the various periods are determined.

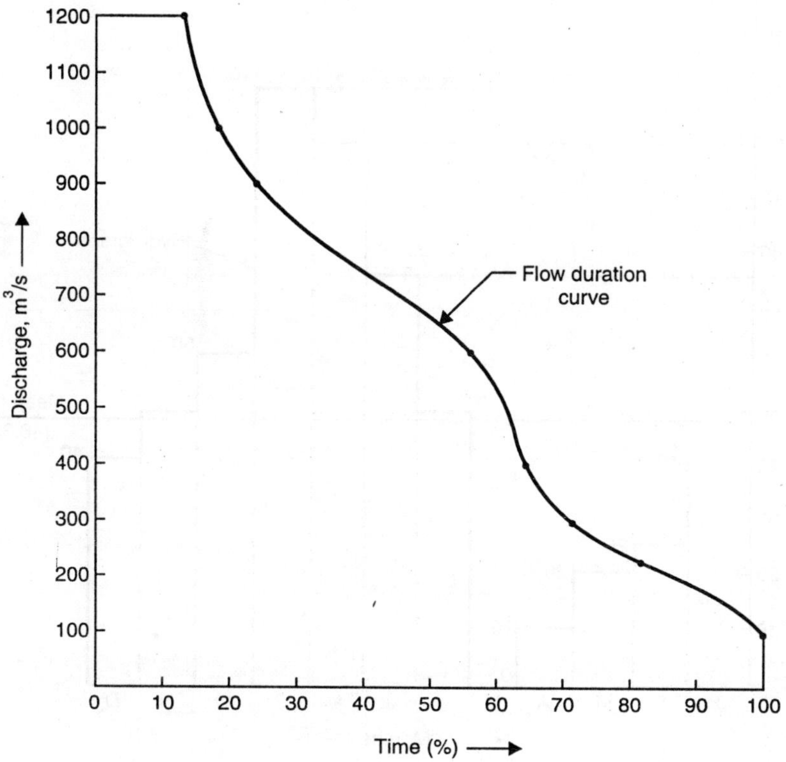

Fig. 5.28. Flow duration curve.

Example 5.7. *The runoff data of a river at a particular site is tabulated below :*

Month	*Mean discharge per month (millions of cum)*	*Month*	*Mean discharge per month (millions of cum)*
January	*40*	*July*	*75*
February	*25*	*August*	*100*
March	*20*	*September*	*110*
April	*10*	*October*	*60*
May	*0*	*November*	*50*
June	*50*	*December*	*40*

(*i*) *Draw a hydrograph and find the mean flow,*

(*ii*) *Also draw the flow duration curve,*

(*iii*) *Find the power in MW available at mean flow if the head available is 80 m and overall efficiency of generation is 85%.*

Take each month of 30 days.

Solution. (*i*) **Hydrograph :**

The *hydrograph* for the given data is drawn as shown in Fig. 5.29.

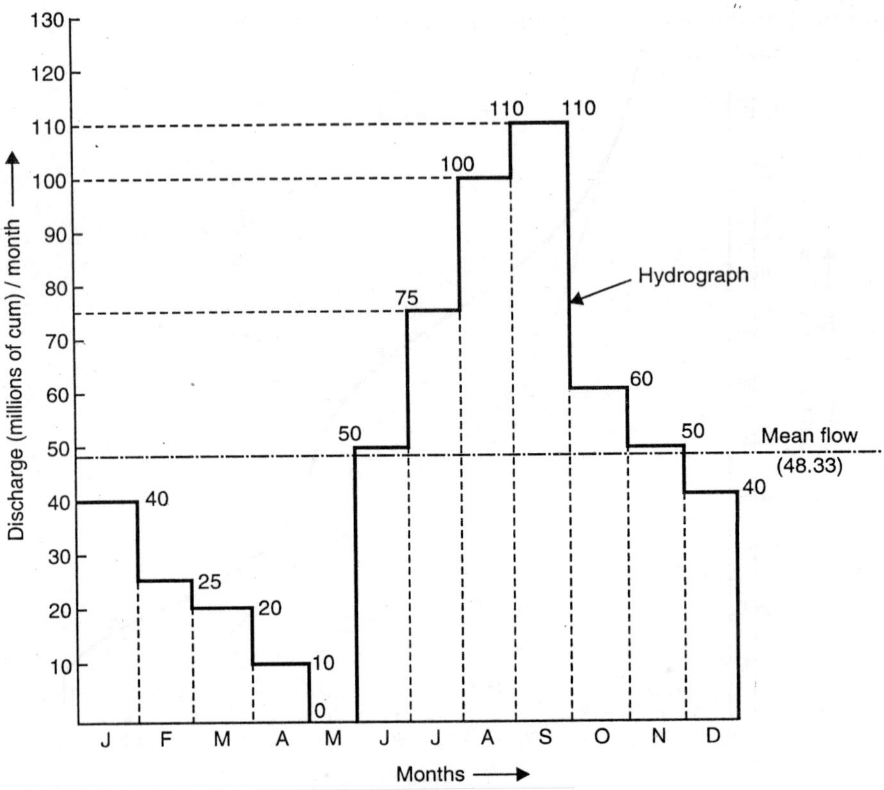

Fig. 5.29. Hydrograph.

The mean discharge for the given data

$$= \frac{40 + 25 + 20 + 10 + 0 + 50 + 75 + 100 + 110 + 60 + 50 + 40}{12}$$

$$= \frac{580}{12} = 48.33 \text{ millions of m}^3\text{/month.}$$

(*ii*) **Flow duration curve :**

To obtain the *flow duration curve* it is necessary to find the *lengths of time* during which certain flows are available. This information is tabulated, using the hydrograph, in the table below :

Discharge per month (millions of m³)	***Total number of months during which flow is available***	***Percentage time***
0	12	100
10	11	91.7
20	10	83.3
25	9	75
40	8	66.7
50	6	50
60	4	33.3
75	3	25.0
100	2	16.7
110	1	8.3

The flow duration curve can be drawn using the data tabulated, as shown in Fig. 5.30.

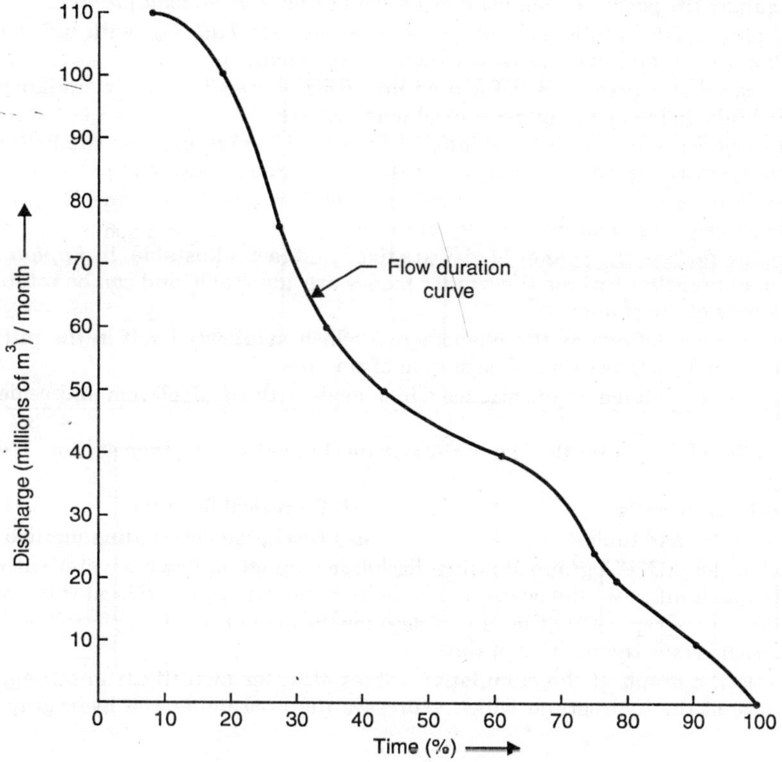

Fig. 5.30. Flow duration curve.

(*iii*) **Average MW energy available :**

$$= \eta_0 \, wQH \times \frac{1}{1000} \text{ MW} \qquad \text{(where } w = 9.81 \text{ kN/m}^3\text{)}$$

$$\left[\text{where } Q \text{ (discharge in m}^3\text{/s)} = \frac{48.33 \times 10^6}{30 \times 24 \times 3600}\right]$$

$$= 0.85 \times \frac{9.81 \times 48.33 \times 10^6 \times 80}{(30 \times 24 \times 3600)} \times \frac{1}{1000} \text{ MW}$$

$= \textbf{12.4 MW.}$ **(Ans.)**

HIGHLIGHTS

1. A *dam* is a barrier to confine or raise water for storage or diversion to create a hydraulic head.
2. A *canal* is an open waterway excavated in natural ground. A *flume* is an open channel excavated on the surface or supported above ground on a trestle. A *tunnel* is a closed channel excavated through a natural obstruction such as a ridge of higher land between the dam and the powerhouse.
3. A *surge tank* is a small reservoir or tank in which the water level rises or falls to reduce the pressure swings so that they are not transmitted in full to a closed circuit.
4. A *draft tube* serves the following two purposes :
   (*i*) It allows the turbine to be set above tail-water level, without loss of head, to facilitate inspection and maintenance.
   (*ii*) It regains, by diffuser action, the major portion of the kinetic energy delivered to it from the runner.

5. The plants which cater to the base load of the system are called *'base load plants'* whereas the plants which can supply the power during peak loads are known as *peak load plants.*
6. Microhydel plants (microstations) make use of standardized *bulb sets* with unit output ranging from 100 to 1000 kW working under heads between 1.5 to 10 metres.
7. The *specific speed* of a turbine is defined as the speed of a geometrically similar turbine that would develop one brake horse power under a head of one metre.
8. The *Pelton* turbine is a tangential flow impulse turbine. The pressure over the Pelton wheel is constant and equal to atmosphere, so that energy transfer occurs due to purely impulse action.
9. The modern *Francis* water turbine is an inward mixed flow reaction turbine. It operates under medium heads and also requires medium quantity of water.
10. In the *propeller turbine* the runner blades are fixed and non-adjustable. In *Kaplan turbine*, which is a modification of propeller turbine the runner blades are adjustable and can be rotated about the pivots fixed to the boss of the runner.
11. *'Cavitation'* may be defined as the phenomenon which manifests itself in the pitting of the metallic surfaces of turbine parts because of formation of cavities.
12. *'Hydrology'* may be defined as the science which deals with the depletion and replenishment of water resources.
13. Run-off includes all the water flowing in the stream channel at any given section. It can be measured by the following methods :
    (*i*) From rainfall records (*ii*) Empirical formulae
    (*iii*) Run-off curves and tables (*iv*) Discharge observation method.
14. *Hydrograph* is defined as a graph showing discharge (run-off) of flowing water with respect to time for a specified time. It indicates the power available from the stream at different times of day, week or year.
15. *Flow duration curve* represents the run-off data for the given time. It is plotted between flow available during a period *versus* the fraction of time.
16. *Mass curve* is the graph of the cumulative values of water quantity (run-off) against time. It is an integral curve of the hydrograph which expresses the area under the hydrograph from one time to another.

THEORETICAL QUESTIONS

1. Give the application of hydro-electric plants.
2. Enumerate advantages and disadvantages of hydro-plants.
3. Enumerate and explain briefly the factors which should be considered while selecting the site for hydro-electric plant.
4. Enumerate essential elements of hydro-electric power plant.
5. What is a catchment area ?
6. What is a reservoir ?
7. What is a dam ? What are its various types ?
8. Explain briefly any two of the following dams :
    (*i*) Rockfill dams (*ii*) Buttress dams
    (*iii*) Timber dams.
9. What is a spillway ? Explain any two types of spillways.
10. What is the difference between canal, flume and tunnel ?
11. What is a surge tank ?
12. Explain with a neat diagram any one of the following surge tanks :
    (*i*) Inclined surge tank (*ii*) Restricted orifice surge tank
    (*iii*) Differential surge tank.
13. What are the functions of a draft tube ?
14. How are hydro-electric power plants classified ?
15. Explain an high head power plant giving its layout clearly.
16. Explain with a neat sketch a pumped storage plant.

17. What is the function of a hydraulic turbine ? How are the turbines classified ?
18. Explain the working of a 'Pelton turbine' with the help of a neat diagram.
19. With the help of a schematic diagram explain the working of the modern Francis turbine.
20. What is a Kaplan turbine ? How does it differ from a propeller turbine ?
21. What are tubular or bulb turbines ?
22. What do you mean by 'specific speed' of a turbine ?
23. What is cavitation ? How can it be avoided/checked ?
24. Describe briefly the methods of governing an impulse turbine ?
25. What points should be considered while selecting a right type of turbine ?
26. Enumerate the various controls which are provided in an hydro-electric power plant.
27. Explain the advantages of combined operation of hydro-electric station and thermal station.
28. Compare hydro and thermal power plants.
29. List the advantages and disadvantages of underground power house/station.
30. What safety measures need to be taken for the safe operation of an hydro-electric plant ?
31. What do you mean by 'preventive maintenance' of hydro-plant ?
32. Define hydrology.
33. Draw and explain the hydrologic cycle.
34. Define run-off. How is it measured ?
35. List the factors which affect run-off.
36. What is a hydrograph ?
37. What is a unit hydrograph ? What are the limitations to the use of unit hydrographs ?
38. What is a flow duration curve ?
39. What is a mass curve ?
40. Write a short note on hydropower development in India.

UNSOLVED EXAMPLES

1. The following data is available for a hydro-power plant :
Available head = 130 m ; catchment area = 2200 sq. km. ; annual average rainfall = 150 cm ; turbine efficiency = 86% ; generator efficiency = 91% , percolation and evaporation losses = 18%.
Determine power developed in MW taking load factor as unity. **[Ans.** 8.546 MW]
2. From the investigation of a hydrosite the following data is available :
Available head = 50 m ; catchment area = 50 sq. km. ; rainfall = 150 cm per year ; 70% of rainfall can be utilized ; turbine efficiency = 80% ; generator efficiency = 91% ; penstock efficiency = 75% ; load factor = 60%.
Determine the suitable capacity of a turbo-generator.
[Ans. 750 kW (maximum rating), Francis or Kaplan turbine]

6

Nuclear Power Plant

6.1. General aspects of nuclear engineering—Atomic structure—Atomic mass unit—Isotopes—Radioactivity—Nuclear radiation—Binding energy—Radioactive decay—Nuclear reactions—Nuclear cross-sections—Fertile materials—Fission of nuclear fuel—Nuclear fusion—Comparison of fission and fusion processes. 6.2. Nuclear power systems. 6.3. Nuclear reactors—Introduction—Classification of nuclear reactors—Essential components of a nuclear reactor—Power of a nuclear reactor—6.4. Main components of a nuclear power plant. 6.5. Description of reactors—Pressurised water reactor (PWR)—Boiling water reactor (BWR)—CANDU (Canadian-Deuterium-Uranium) reactor—Gas cooled reactor—Liquid metal cooled reactors—Breeder reactor. 6.6. Selection of materials for reactor components. 6.7. Advantages and disadvantages of nuclear power plants. 6.8. Nuclear-plant site selection. 6.9. Nuclear power plants in India. Highlights—Theoretical Questions.

6.1. GENERAL ASPECTS OF NUCLEAR ENGINEERING

6.1.1. Atomic Structure

- An element is defined as a substance *which cannot be decomposed into other substances.* The *smallest particle* of an element which takes part in chemical reaction is known as an '*atom*'. The word atom is derived from Greek word 'Atom' which means indivisible and for a long time the atom was considered as such. *Dalton's atomic theory* states that (*i*) all the atoms of one element are precisely alike, have the same mass but differs from the atoms of other elements (*ii*) the chemical combination consists of the union of a small fixed number of atoms of one element with a small fixed number of other elements.
- Various atomic models proposed by scientists over the last few decades are : 1. Thompson's plum puddling model, 2. Rutherford's nuclear model, 3. Bohr's model, 4. Sommerfeld's model, 5. Vector model, 6. Wave-mechanical model.
- The complex structure of atom can be classified into *electrons* and *nucleus*. The nucleus consists of *protons* and *neutrons* both being referred as *nucleons*. *Protons* are *positively* charged and *neutrons* are *neutral*, thus making complete nucleus as positively charged.
- The *electrons* carry *negative* charge and circulate about the nucleus. As the positive charge on proton particle is equal to the negative charge on electron particle, and the *number of electrons is equal to the number of protons,* atom is a neutral element. Any addition of the number of electrons to the neutral atom will make it negatively charged. Similarly any subtraction of the electrons will make it positively charged. Such an atom is known as *ion* and the process of charging the atom is termed an *ionisation*.
- The nuclear power engineering is specially connected with *variation of nucleons in nucleus.* Protons and neutrons are the particles having the mass of about 1837 times and 1839 times the mass of an electron.

- The modern atomic theory tells that the atom has a diameter of about 10^{-7} mm. In a neutral atom the electrons are bound to the nucleus by the electrostatic forces, which follows the Coloumb's law of forces, *i.e.*, like charges repel and unlike charges attract each other. The function of electrostatic force is similar to the gravitational force.
- The atomic spectrum study has revealed that every electron in an atom is in one group of specific states of motion which is corresponding to its total energy. In an atom the electrons are spinning around the nucleus in orbits. These orbits are called *shells*, which represent the energy levels for the electrons. *All the electrons having very nearly the same total energy are said to be in the same shell.* The shells have been named as *K*, *L*, *M*, *N* etc. Each shell consists of the specific maximum number of electrons. The *K* shell (inner shell) contains **2** electrons, *L* shell has **8** electrons, *M* shell is limited to **18** and the *N* shell possesses **32** electrons. In fact, the number of electrons in any orbit is equal to $2n^2$ where *n* is the serial number of the orbit taking first orbit nearest to the nucleus, *with the exception that the outermost orbit cannot have more than eight electrons* . In a given atom all orbits may not be complete. It is obvious from the study that *amplitude difference in energy between two shells is much more than the difference in between energy levels in one shell.* In a *shell less than the specified number of electrons may exist but not a large number.* The inner shell is filled up first and then the other successive shells are completed.
- The *chemical properties of the atom varies with composition of number of electrons in various shells and the state of energies within the shells determine the electrical characteristics of the atom.* For example, *Hydrogen* (H_2) consists of one electron in the *first shell*, *Helium* (He) has two electrons in the first shell, *Lithium* (Li) has two electrons in first shell and one is second shell, *Carbon* (C) consists of two electrons in first and four in second shell.

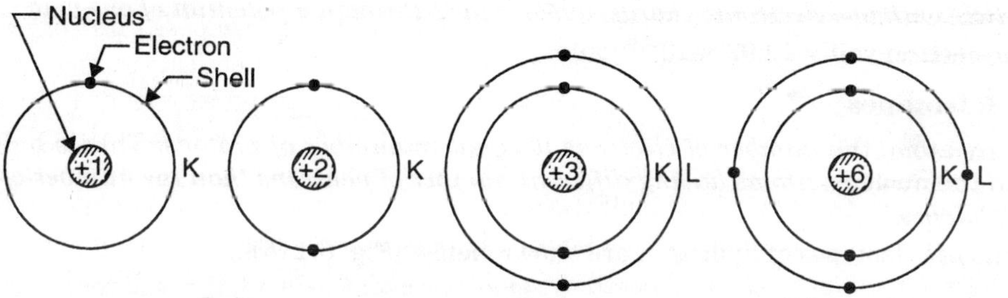

Fig. 6.1 (*a*). Atomic structure of H_2, He, Li and C.

- The electrons lying in the outermost shell are termed *valence electrons.* If the outermost shell is completely filled, the atom is stable and will not take any electron to fill up the gap. However, the incomplete outer shell will try to *snatch* the required number of electrons from the adjacent atom in a matter. The binding force between the electron and nucleus is the electrostatic force of attraction. *To emit one electron energy required is more than the electrostatic force of attraction. When the energy is supplied, the electron jumps from one discrete energy level to another permissible level. The process starts from outer shell.* The electron possesses the energy in two forms, *i.e.*, kinetic energy due to its motion and potential energy due to its position with respect to the nucleus. It is obvious that *electrons cannot exist in between the permissible orbits.*
- The charge of nucleus is represented by the *number of protons* present. This number is known as *atomic number* and designated by the letter *Z*. It also shows the position of atom in the periodic table. Hydrogen has only one number but natural uranium has ninety two. The atoms having *higher atomic number* have been developed artificially ranging from

93 to 102. These are einsteinium (Z = 99), Ferinium (Z = 100), and mendelevium (Z = 101). Platonium (Z = 94) is an important element to the nuclear power field.

The *mass number* (A) is the sum of total number of protons and neutrons in a nucleus. The number of electrons is represented by the letter N, *i.e.*, $N = (A - Z)$.

6.1.2. Atomic Mass Unit

The *mass of the atom is expressed in terms of the mass of the electron.* The unit of mass has been considered as $\frac{1}{16}$ th of the mass of neutral oxygen atom which contains 8 protons and 8 neutrons. The atomic mass unit (a.m.u.) is equal to $\frac{1}{16}$ th the mass of oxygen neutral atom.

One a.m.u. = 1.66×10^{-24} g

$$\text{Mass of proton} = 1837 \text{ me} = \frac{1837 \times 9.1 \times 10^{-28}}{1.66 \times 10^{-24}} = 1.00758 \text{ a.m.u.}$$

$$\text{Mass of neutron} = 1839 \text{ me} = \frac{1839 \times 9.1 \times 10^{-28}}{1.66 \times 10^{-24}} = 1.00893 \text{ a.m.u.}$$

It has been concluded that the density of matter in a nucleus is enormous. It has been investigated that the *radius of nucleus is equal* to $1.57 \times 10^{-3} \times 3\sqrt{A}$, where A is the number of nucleons in nucleus.

The density of uranium by calculations comes to 1.65×10^{14} g/cm^3. It has been found by calculations that *natural substance has density millions of times lower than that of nuclear matter.*

Electron volt. The energy is expressed in electron volt unit. An electron volt is equal to *work done in moving an electron by a potential difference of one volt.* Or *it is the amount of energy acquired by any particle with one electronic charge, when it falls through a potential of one volt.*

One electron volt = 1.602×10^{-19} joule.

6.1.3. Isotopes

In any atom, the *number of electrons is equal to number of protons.* This is independent of neutrons in the nucleus. *Atoms having different number of neutrons than the number of protons are known as 'Isotopes.'*

Example. Isotopes of hydrogen are shown below [Fig. 6.1 (*b*)].

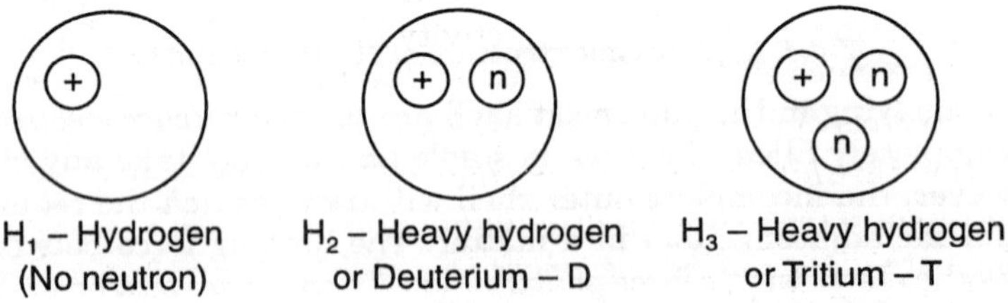

Fig. 6.1 (*b*).

These isotopes have the *same chemical properties* and *have the same atomic number* and *occupy the same place in the periodic table.* But the *nuclear properties* of each of the isotopes are different *because of the different number of neutrons in the nucleus.*

The isotopes of oxygen vary from O_{14} to O_{19}. The change of number of neutrons in nucleus affect the mass of atom.

Example. Weight of heavy hydrogen is twice the weight of simple hydrogen. This means a volume of H_2O weighs less than the same volume of D_2O.

The isotopes can be represented with mass number (*A*) as subscript and atomic number (*Z*) as subscript like $_ZH^A$.

Example. Hydrogen isotopes are represented as $_1H^2$ (Deuterium), $_1H^3$ (Tritium) and Uranium isotopes as U^{234}, U^{235}, U^{238}.

— The isotopes are *not stable* and *disintegrate* at a certain rate. The isotope which disintegrates at a fixed rate is called *Radioactive isotope* or *Radio isotope*.

The instability of the nucleus can be either by the separation of parent nucleus into 2 or more nuclei or by the rearrangement of nucleons in the matter so that there is an emission of particles or energy in the form of rays or by rearrangement of electrons. During this transformation there is emission of particles at a very high velocity. This is known as *radiation*.

We must note that for any specific isotope, the rate of radiation from a unit mass and also the energy distribution are fixed and cannot be changed by any method. Thus, for any isotope, the quantity of radiation per unit time can be determined easily.

6.1.4. Radioactivity

Radioactivity was originally discovered by Becquerel in 1896. This phenomenon is *confined almost entirely to the heaviest element* from 83 to 106 in the periodic table.

The phenomenon of spontaneous emission of powerful radiations exhibited by heavy elements is called **radioactivity.** Radioactivity is essentially a *nuclear phenomenon* and is a *drastic process* because the element changes its kind. It is *spontaneous* and an *irreversible self-disintegrating activity* because the element breaks itself up for good. Those elements which exhibit this activity are called *radioactive elements*. Examples are : *Uranium, polonium, radium, radon, ionium, thorium, actinium* and *mesothorium.*

The radioactive radiations emitted by the radioactive elements are found to consist of the following :

(*i*) Alpha (α) rays or α-particles

(*ii*) β rays or β-particles

(*iii*) γ-rays or photons.

The radioactivity may be *natural* or *artificial.*

Natural radioactivity. It is that *which is exhibited by elements as found in Nature.* It is always found in heavier elements in the periodic table.

Artificial or induced radioactivity. The modern techniques of artifical transmutation of elements have made it possible to produce radioactivity in many other elements much lighter than those that occur in Nature. Such type of radioactivity is known as *artificial or induced radioactivity*.

The general *properties* of radioactive radiations are :

1. These radiations are *highly penetrating, they affect photographic plates, ionise gases, cause scintillations on fluorescent screen, develop heat and produce chemical changes.*

2. As radiations are given out, *new elements are formed* in an irreversible process—the new elements themselves being usually radioactive.

3. The emission of radiations is *spontaneous* and is not affected by external agents.

4. The emission is *not instantaneous* but is prolonged *i.e.*, it is extented over a period of time otherwise it would not have been discovered at all.

5. Except for radioactivity, there is nothing abnormal about the radioactivity elements as regards their physical and chemical properties.

6.1.5. Nuclear Radiation

In nuclear power technology there are only *five* types of radiation of interest, but there many mechanisms by which these five are produced in reactor systems. The five types of radiation (with electrical charges indicated as + or – for positive or negative, respectively, and mass in atomic mass units) are :

1. **Gamma rays** (or photons) : electromagnetic radiation.
2. **Neutrons :** uncharged particles, mass approximately 1.
3. **Photons :** + 1 charged particles, mass approximately 1.
4. **Alpha particles :** helium nuclei, charge + 2, mass 4.
5. **Beta particles :** electrons (charge – 1), positrons (charge + 1), mass very small.

1. Gamma rays :

Prompt-fission gamma rays. These are produced as a result of the fissioning of a U^{235} (or other fissile-material) nucleus. The gamma rays are emitted within a fraction of a microsecond after fission takes place and are considered to be coincident with the fission process. Prompt-fission gamma rays carry off about 7 MeV/fission, with individual photon energies ranging from less than 0.5 MeV to greater than 1.5 MeV.

Fission-product-decay gamma rays. These are emitted from the fragments resulting from the fission process and their decay products. These radioactive fission products have half lives (the time it takes for one-half the atoms originally present to decay) from a fraction of a second to million of years. In most cases, they emit soft (low-energy) gamma rays and beta particles with energies lower than 1 MeV.

Capture gamma rays. These are emitted by nucleus of an atom instantaneously upon the capture of a neutron. The energy of these gamma rays is generally higher than those released by fission or decay. Many elements yield capture gamma rays in the 6 to 8 MeV range.

Activation-decay gamma rays. These are often emitted from the nucleus after a neutron-capture process, if the new nucleus formed is unstable. Most decays of this type are accomplished by electron emission accompanied by one or more gamma-rays photons. Each unstable (radioactive) isotope has a specific half-life and mode of decay which is an intrinsic property.

Inelastic-scattering gamma rays. These are emitted from a nucleus that has been excited to level above its ground state by interaction with an energetic neutron. These are emitted within an extremely short time after the interaction takes place, and the total energy carried off by these photons is less than or equal to the kinetic energy of the incident neutron.

2. Neutrons :

Prompt-fission neutrons. These are produced as a result of the fissioning of a fissile material and, as in the case of prompt gamma rays, are considered to be emitted coincidentally with the fission process.

Delayed neutrons. These are emitted from several of the fission products with apparent half-lives of upto about 2 min. Although half-lives are usually ascribed to the production of delayed neutrons, they are actually emitted within less than a microsecond after the formation of a highly excited nucleus. The half-life actually describes the decay of a fission fragment to the highly excited fission product.

Photoneutrons. These are produced when a photon with energy greater than the binding energy of a neutron (the energy required to bind a neutron to the nucleus) interacts with a nucleus and ejects a neutron. This process is generally not important, since most isotopes have a high threshold for the reaction and a low probability of occurence above the threshold. Two isotopes used in some reactor systems which have low thresholds and a fairly significant probability of photoneutron interaction are hydrogen 2 (deuterium) and beryllium 9.

Activation neutrons. Neutron decay of an activated material occurs occasionally in cases other than delayed neutron decay of fission products. The only case which is of some interest in nuclear-reactor technology is the neutron decay of nitrogen 17, formed by fast-neutron irridiation of oxygen 17.

Reaction neutrons. These are the neutrons ejected from a nucleus by interaction with one of several particles. There are known cases of neutron emission resulting from a nucleus interacting with a neutron, proton, or alpha particle. Important use is made of this process in producing neutron sources for reactor start-up.

3. Protons :

Protons are produced in a few radioactive-decay processes and more frequently by neutron-proton reactions in which an incident neutron causes a proton to be emitted from the nucleus.

4. Alpha particles :

Alpha particles are produced by the decay of several fission products and a few activated materials as well as by a few neutron-alpha reactions in which an incident neutron interacts with and causes an alpha particle to be emitted from the nucleus.

5. Beta particles :

Beta particles (electrons and positrons) are produced by several mechanisms, such as radioactive decay and pair production (in which a photon of high energy is converted into an electron-positron pair).

Effects of nuclear radiation on matter :

Nuclear radiation, when it interacts with any material, deposits energy in the material and can have various effects. In chemical components the chemical form will be changed, in solids the crystalline structure may be altered, in any case heat will be generated. For charged particles and gamma rays the mechanisms of energy transfer is ionisation of the material traversed by the radiation. Ionization is the production of electrically charged particles by stripping orbital electrons from the electrically neutral atoms. In the case of neutrons, the primary energy-transfer process is a kinetic-energy exchange caused by collision of neutrons with nuclei of the matter traversed.

6.1.6. Binding Energy

The nucleus of an atom is formed when the nucleons come closer to each other and this distance between the two nucleons is of the order of nearly 10^{-12} mm. At the *moment of combination there is a release of energy* and is known as *'binding energy'*. Further if it is required to separate out or to disintegrate two nucleons the equivalent amount of binding energy is to be supplied from the external source to overcome the force of attraction. The binding energy can also be defined as the *energy required to overcome the binding forces of nucleus.*

When two nuclear particles are combined to form a nucleus, it is observed that, there is a difference in the mass of the resultant nucleus and the sum of the masses of two parent nuclear particles. This decrement of mass is known as *'mass defect'*. The *amount of mass defect is directly proportional to the amount of energy released.*

The nuclear binding energy per nucleon increases with the increase of the number of nucleons in the nucleus. *Example* : The binding energy per nucleon for H^2 is 1.109 MeV and for He^4 it is $28.2 \div 4 \approx 7.05$ MeV. A curve representing the variation of nuclear binding energy per nucleon with the mass number is shown in Fig. 6.2. Here the average value of binding energy per nucleon has been considered. The curve indicates that the average binding energy per nucleon increases as the mass number increases initially with the peak value of about 8.8 MeV at nearly 60 mass number. The elements falling in this region are nickel and iron. As the mass number increases still further, the binding energy curve falls gradually to 7.6 MeV for U^{238}. For U^{235} from the Fig. 6.2 the binding energy per nucleon is 7.7 MeV. At the point where mass number is 117, binding energy per nucleon is nearly 8.6 MeV.

U^{235} nucleus is splitted into two approximately equal nuclei. The formation of two nuclei will release the energy of about 0.9 MeV per nucleon. There is a release of energy as the mass number decreases within the range of 60 to 250 mass number. This release of energy is corresponding to the increase of mass defect. In fission process, the U^{235} nuclei is splitted to two other nuclei and energy is liberated.

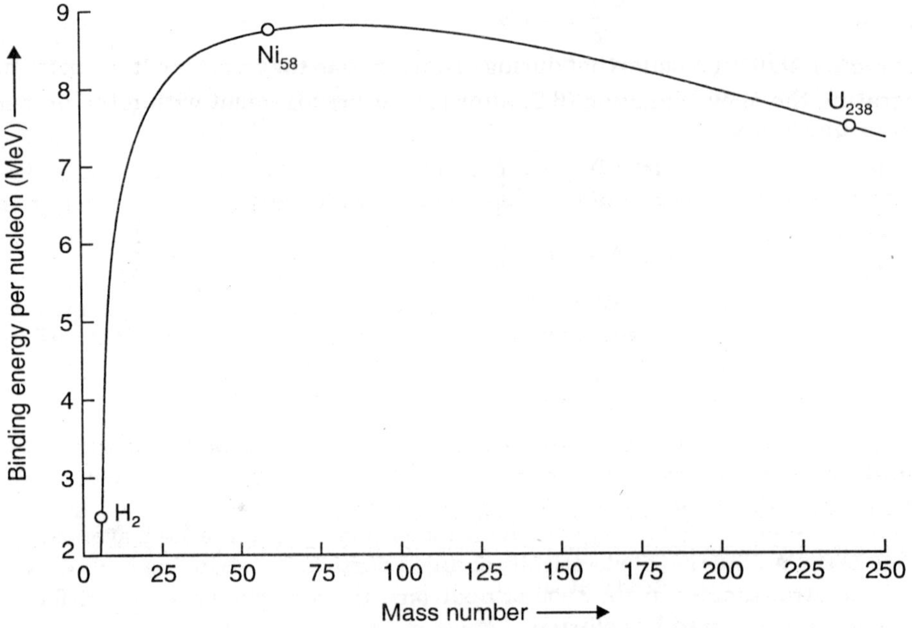

Fig. 6.2. Variation of average binding energy per nucleon with mass number.

It is evident from the above discussion that the nuclear transformations of other nucleus is also possible such as U^{235}, U^{233} and Pu^{239} (these are the important fuels used in the production of nuclear power).

An atom with even number of protons of mass number is more stable because of the pairing of protons and neutrons. This type of atom also possesses *higher binding energy per nucleon* and is represented as *even type of atom.* In nuclear physics, the first even or odd represents the even or odd number of protons respectively and second one represents the even or odd mass number. This is obvious from the practical data that the U^{235} is fissionable with slow neutrons (neutrons having less energy) but U^{238} is fissionable only when the neutrons are having energy more than 1 MeV.

6.1.7. Radioactive Decay

It has been observed that the emission of the particles in the form of alpha, beta or gamma radiations is not an instantaneous process. For various elements the decay time is different, which follows a certain *law*. Obviously the process is independent of the physical and chemical properties of the given isotope at a particular temperature and pressure.

The *law* states that the *small amount of disintegration of the isotope in a small period is directly proportional to the total number of radioactive nuclei and proportionality constant.*

If, N = Number of radioactive nuclei present at any time t,

N_0 = Initial number of such nuclei,

λ = Proportionality constant (also known as disintegration constant or the radioactive decay constant of the material),

Then the above law can be stated in the form of equation as follows :

$$\Delta N = -\lambda N \Delta t \qquad ...(6.1)$$

or,
$$\frac{dN}{dt} = -\lambda N \qquad ...(6.2)$$

The *negative* sign represents that during disintegration the number of the nuclei is *decreasing*.

Integrating the above equation (6.2) after proper arrangement within the proper limits, we get

$$\int_{N_0}^{N} \frac{dN}{N} = -\lambda \int_0^t dt \qquad ...(6.3)$$

or,
$$\log_e N - \log_e N_0 = -\lambda t, \qquad \text{or,} \qquad \log_e \frac{N}{N_0} = -\lambda t$$

or,
$$\frac{N}{N_0} = e^{-\lambda t}, \qquad \text{or,} \qquad N = N_0\, e^{-\lambda t} \qquad ...(6.4)$$

$$\frac{dN}{dt} = -\lambda N = -\lambda N_0\, e^{-\lambda t} \qquad ...(6.5)$$

The eqn. (6.5) represents that the *decay scheme follows the exponential law.*

Activity :

The intensity of emitted radiation is termed **activity.**

This is directly dependent on the rate of disintegration of the element.

If, A = Activity at time t,
A_1 = Initial activity,
k = Detection coefficient,

Then,
$$A = k\left(-\frac{dN}{dt}\right) = k\lambda N = k\lambda N_0\, e^{-\lambda t} = A_1\, e^{-\lambda t} \qquad ...(6.6)$$

Half-life :

Half-life *represents the rate of decay of the radioactive isotopes.* The half-life is the *time required for half of the parent nuclei to decay or to disintegrate.*

Putting $N = \frac{N_0}{2}$ and $t = t_{1/2}$ in eqn. (6.5), we get

$$\frac{N_0}{2} = N_0\, e^{-\lambda t_{1/2}} \qquad \therefore \qquad e^{-\lambda t_{1/2}} = 1/2$$

$$\therefore \qquad \lambda t_{1/2} = \log_e 2 = 0.693 \qquad \therefore \qquad t_{1/2} = \frac{0.693}{\lambda} \qquad ...(6.7)$$

Here $t_{1/2}$ is the half-life of radioactive nuclei. After passing every half-life the number of nuclei is reduced to half and so is the activity. This process is repeated for the several half lives till the activity becomes negligible. The variation of half-life is from fraction of seconds to million of years.

Half-life of some of the metals is given below :

Metal	*Half-life*
Po-214	170 μ sec
I-137	25 sec
Carbon-14	5100 years
Th-232	1.4×10^{10} years
Uranium-238	4.525×10^{9} years

Average (mean) life :

This indicates the average of total time for which the radioactive nuclei has disintegrated for several half lives. Hence this is *greater than half-life.* This is obtained by taking the sum of the decay time of the radioactive nuclei and then it is divided by the initial number of nuclei.

If T is the time of average life, then

$$T = \frac{-\int_0^{\infty} t\,dN}{N_0} = \frac{\lambda N_0 \int_0^{\infty} t\,e^{-\lambda t}\,dt}{N_0} \quad \text{...(6.8)}$$

Now integrating by parts, we get

$$T = \left[-t\,e^{-\lambda t} - \frac{e^{-\lambda t}}{\lambda} \right]_0^{\infty} = \frac{1}{\lambda} \quad \text{...(6.9)}$$

But, $$t_{1/2} = \frac{0.693}{\lambda}$$

From the above eqns. it is clear that mean life is 1.445 times greater than half-life.

Note. Number of disintegrations per second is the unit of radioactivity and is termed *curie*, as this phenomenon was first discovered by Curie.

6.1.8. Nuclear Reactions

During a nuclear reaction, the *change in the mass of the particle represents the release or an absorption of energy.* If the total mass of the particle after the reaction is reduced, the process releases the energy, consequently, the increase in the mass of the resultant particle, will cause the absorption of energy.

The equations of nuclear reactions are connected with the resettlement of protons and neutrons within the atom. The equations are much similar to chemical reactions. The energy variation is also of the order of MeV. In simple term the equation shows the *balance of neutron and proton.*

A nuclear reaction is written as follows :

(*i*) The bombarded nuclei or the target nuclei is written first from left hand side.

(*ii*) In the middle within brackets, first is the incident particle and second one the ejected.

(*iii*) On the right hand side, the resultant nucleus is placed.

A *neutron* is written as : ${}_0n^1$ because it has unit mass and it does not have any charge.

An *electron* is written as : ${}_{-1}e^0$ because its mass is negligible as compared to proton or neutron and its charge is equal but opposite to the charge of proton.

Some of the *examples of reactions* are given below :

(*i*) When ${}_{11}Na^{23}$ is bombarded with protons possessing high energy, it is converted to ${}_{12}Mg^{23}$

$${}_{11}Na^{23} + {}_1H^1 \longrightarrow {}_{12}Mg^{23} + {}_0n^1 + q \quad \text{...(6.10)}$$

(where q = release or absorption of energy in the reaction)

(*ii*) When ${}_{13}Al^{27}$ is bombarded with high energy protons it is transformed to ${}_{14}Si^{27}$.

$${}_{13}Al^{27} + {}_1H^1 \longrightarrow {}_{14}Si^{27} + {}_0n^1 + q \quad \text{...(6.11)}$$

(*iii*) When ${}_{13}Al^{27}$ is bombarded with deutrons, Al^{28} and proton may be produced.

$${}_{13}Al^{27} + {}_1H^2 \longrightarrow {}_{13}Al^{28} + {}_1H^1 \quad \text{...(6.12)}$$

The eqns. (6.10), (6.11) and (6.12) may be written in the equation form as given below :

$$Na^{23}(p, n)Mg^{23} \quad \text{...(6.13)}$$

$${}_{13}Al^{27}(p, n)Si^{27} \quad \text{...(6.14)}$$

$$Al^{27}(d, p)Al^{28} \quad \text{...(6.15)}$$

It is evident from the above mentioned reactions that the *nuclear reaction is followed by capturing a particle, resulting in a compound excited nucleus, which undergoes further transformation in a short period of time.*

The *transformation may adopt the following five main different paths :*

1. Elastic scattering. The neutron interacts with the nucleus and after transformation the compound nucleus emits a particle which is identical to the captured one. There is also no change in the resultant nucleus. The *total internal energy* of the bombarded nucleus and the restriking particle *will not change* at all. The process is known as *elastic scattering*. Elastic scattering is also termed as *elastic collision*. When the neutron strikes the nucleus, it imparts the part of initial kinetic energy and momentum to the nucleus which causes the displacement of the nucleus in the crystal lattice by a significant distance and can *change the structural properties of the material.*

In elastic scattering process the *kinetic energy of neutron is reduced and is beneficial to slow down the neutron in reactor*. In this transformation, *there is neither release nor absorption of energy but as a result of collision, redistribution of kinetic energy takes place.*

Example of elastic scattering. When a neutron strikes a light nucleus (*e.g.,* hydrogen nucleus), the velocity of the neutron is very much reduced and the energy is transferred to the proton. Here most of the energy is transferred because both the particles are having nearly the *same masses*. It has been observed that in such a single collision, the loss of energy of the proton is nearly 70 to 75 percent. In case the neutron impacts with the heavy nucleus, the energy loss in single collision is *less*. With carbon nucleus this loss amounts to nearly 12 to 17 percent of the initial value. The reaction is written as $C^{12}(n, n)C^{12}$.

2. Inelastic scattering. The composition of the incident particle and ejected particle remains unchanged. When the particle interacts with the nuclei it loses its kinetic energy and the target nucleus is excited. The energy is released in the form of gamma emission. This transformation is known as *inelastic scattering* or collision. The process is limited to the condition that the neutron should have minimum energy sufficient to excite the target nucleus. The reaction is completed with the absorption or release of energy. The neutron energy loss is of the order of 10 to 20 percent of the initial value.

When a fast moving neutron hits the U^{238} nucleus, the nucleus is excited and there is an emission of gamma quantum [U^{238} $(n, \eta\gamma)$ U^{238}].

3. Capture. In this process the incident particle may be captured or absorbed by the nucleus and may raise the mass number by unity. The nucleus is excited and the energy is emitted in the form of gamma quantum. The *artificial radioactive materials are produced by this process*. In a reactor, Co-60 isotope is produced by bombarding the natural Co-59 with neutrons. The reaction has both the possibilities of producing the stable and unstable nucleus and may result in (n, γ) or (p, γ) reactions. This transformation may take place with elastic scattering. When a neutron interacts with light hydrogen, it forms heavy hydrogen, deuterium. The mass of deuterium is less than its components. This mass defect is corresponding to the release of gamma quantum.

4. In this reaction, the *impinging particle is trapped in the nucleus but the ejected particle is a different one. The composition of the resultant nucleus is also different from the parent nucleus.*

5. Fission. *When the nucleus is excited too much, it splits into two mostly equal masses*. This particular reaction is suited only to the *heavy nucleus* such as U^{233}, U^{235}, Pu^{239} etc. The transformation is known as *fission*. The produced two nuclei are lighter nuclei ; they have *more binding energies per nucleon and hence this reaction always releases the energy* (Fig. 6.2).

6.1.9. Nuclear Cross-sections

Cross-sections (or attenuation coefficients) *are measures of the probability that a given reaction will take place between a nucleus or nuclei and incident radiation.*

Cross-sections are called either *microscopic* or *macroscopic*, depending on whether the reference is to a single nucleus or to the nuclei contained in a unit volume of material.

Microscopic cross-section :

It is a measure of the probability that a given reaction will take place between a single nucleus and an incident particle. Microscopic cross-section is usually denoted by the symbol σ and is expressed in terms of the *effective area that a single nucleus presents for the specified reaction.* Since these cross sections are usually quite small, in the range of 10^{-22} to 10^{-26} cm²/nucleus it is general practice to express them in terms of a unit called the *barn*, which is 10^{-24} cm²/nucleus.

Macroscopic cross-sections :

These are the products of microscopic cross-sections and the atomic density in nuclei per cubic centimeter and are equivalent to the total cross-section, for a specific reaction of all the nuclei in 1 cm³ of material. Macroscopic cross-sections are denoted by the symbol Σ for *neutrons* and μ for *gamma rays* and have the units cm^{-1}.

Gamma ray cross-sections :

Although there are a large number of interaction processes that take place between gamma rays and matter, the most commonly used are the energy-absorption cross-section (used to determine gamma heating and dose rates) and the total attenuation cross-section (used to determine material gamma-ray attenuation and for shielding design).

Neutron cross-sections :

Neutrons undergo a large number of different interaction processes with matter, and, unlike gamma rays, many of these individual interactions must be evaluated. Neutron cross-sections of general use are :

(*i*) Fission
(*ii*) Gamma-ray production
(*iii*) Activation
(*iv*) Elastic scattering
(*v*) Inelastic scattering
(*vi*) Reaction particle production
(*vii*) Total absorption
(*viii*) Total attenuation.

Both neutron and gamma-ray cross-sections are *energy-dependent properties.* Plots of gamma-ray cross-section *vs* photon energy for all materials are, over the energy range of interest, *smooth curves*, whereas for neutron cross sections the curves of many materials *show gross variations from a smooth curve.* The variations in neutron cross-sections show up as peaks and valleys on the cross-section plot ; these peaks are called *resonances.* When a material has a large number of resonance peaks over a portion of the energy range, this portion of the cross-section plot is called a *resonance region.* The *resonance region can have a significant effect on reactor design*, since the material U^{238} which is present in most fuels has a relatively wide resonance region which can cause extensive neutron absorption during the slowing down of neutrons to thermal energy.

The known cross-sections for materials potentially useful in reactor systems are used as primary criteria in materials selection. For example, high-neutron-absorption cross-section materials would not normally be used as materials of construction in the vicinity of a reactor core to prevent competition for the neutrons required to sustain the fission process ; and high activation cross-section materials would not be chosen, if they can be avoided, in a region exposed to a high neutron flux during operation, if that region is to be accessible after reactor shut-down.

6.1.10. Fertile Materials

It has been found that some materials are not fissionable by themselves but they *can be converted to the fissionable materials*, these are known as *fertile materials.*

Pu^{239} and U^{233} are not found in nature but U^{238} and Th^{232} can produce them by nuclear reactions. When U^{238} is bombarded with slow neutrons it produces ${}_{92}U^{239}$ with half-life of 23.5 days which is unstable and undergoes two beta disintegration. The resultant Pu^{239} has half-life of 2.44×10^4 yrs and is a good alpha emitter.

$$_{92}U^{238} + {_0}n^1 \longrightarrow {_{92}}U^{239} + \gamma \qquad ...(6.16)$$

$$_{92}U^{239} \xrightarrow{23.5 \text{ min.}} - {_{-1}}e^0 + {_{91}}Np^{231} \qquad ...(6.17)$$

$$_{93}Np^{239} \xrightarrow{2.3 \text{ days}} {_{-1}}e^0 + {_{94}}Pu^{239} \qquad ...(6.18)$$

During conversion the above noted reactions will take place. The other isotopes of neptunium such as 2.1 day Np^{238} and plutonium can also be produced by the bombardment of heavy particles accelerated by the cyclotron.

The nuclear transformations to convert $_{90}Th^{232}$ to U^{233} are given below :

$$_{90}Th^{232} + {_0}n^1 \longrightarrow {_{90}}Th^{233} + \gamma \qquad ...(6.19)$$

$$_{90}Th^{233} \xrightarrow{23.3 \text{ min.}} {_{91}}Pa^{233} + {_{-1}}e^0 \qquad ...(6.20)$$

$$_{91}Pu^{233} \xrightarrow{27.4 \text{ days}} {_{92}}U^{233} + {_{-1}}e^0 \qquad ...(6.21)$$

U^{235} is the source of neutrons required to derive Pu^{239} and U^{233} from Th^{232} and U^{238} respectively. This process of conversion is performed in the *breeder reactors.*

Other fissionable materials. Th^{227}, Pa^{232}, U^{231}, Np^{238} and Pu^{241} are the other nuclides which are having high cross-sections for neutron thermal fission. Pu^{241} is the important nuclide which is used in plutonium fueled power reactors.

6.1.11. Fission of Nuclear Fuel

Fission *is the process that occurs when a neutron collides with the nucleus of certain of the heavy atoms, causing the original nucleus to split into two or more unequal fragments which carry off most of the energy of fission as kinetic energy*. This process is *accompanied by the emission of neutron and gamma rays.*

Fig. 6.3 is a representation of the fission of uranium 235. The *energy released as a result of fission is the basis for nuclear-power generation.* The release of about *2.5 neutrons/fission* makes it possible to produce *sustained fissioning.*

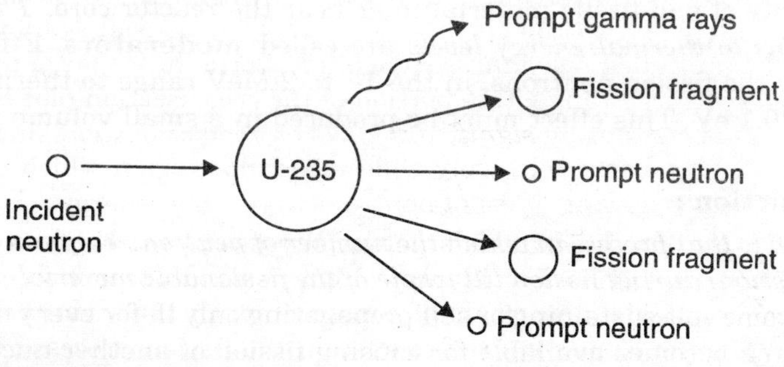

Fig. 6.3. Fission of uranium 235. Incident neutron, upon colliding with U^{235} nucleus, causes fission to take place, resulting in the production of fission fragments, prompt neutrons and prompt gamma rays.

The *fission fragments* that result from the fission process are *radioactive* and decay by emission of beta particles, gamma rays and to a lesser extent alpha particles and neutrons. The neutrons that are emitted after fission, by decay of some of the fission fragments, are called *delayed neutrons.* These are of the utmost importance, since *they permit the fission chain reaction to be easily controlled.*

The total detectable *energy released* owing to the fission of a single nucleus of uranium 235 is 193 MeV, distributed as shown below :

Distribution of fission energy

	MeV
Instantaneous energy release :	
Kinetic energy of fission fragments	168
Prompt-gamma-ray energy	7
Kinetic energy of prompt neutrons	5
Instantaneous total	180
Delayed energy release :	
Beta particle decay of fission products	7
Gamma-ray decay of fission products	6
Delayed total	13

As is shown above, the neutron emitted as a result of fission of a uranium 235 nucleus carry off 5 MeV of kinetic energy. Since on average there are about 2.5 neutrons emitted/U^{235} fission, the average neutron energy is 2 MeV. Actually fission neutrons are emitted with an energy speed of from nearly zero energy to approximately 16 MeV, the bulk of the them being in the 1- to 2-MeV energy region.

Note. Although not strictly a result of the fission process, there is an additional 5 to 8 MeV emitted per fission as a result of the capture of neutrons not used in the fission chain reaction. About 1 MeV of this total is emitted over a period of time owing to decay of activation products, and the remainder is emitted immediately upon neutron capture.

Most of the reactors in existence today or planned for the near future are called *thermal reactors*, since they depend on neutrons which are in or near thermal equilibrium with their surroundings to cause the bulk of fissions. *These reactors make use of the fact that the probability for fission is highest at low energy by slowing down the neutrons emitted as a result of fissioning to enhance fission captures in the fuel.* Loss of neutrons to non-fission-capture processes is lessened by minimising the quantity of non-fissile material in or near the reactor core. *The materials used to decelerate fast neutrons to thermal energy levels* are called **moderators.** Effective and efficient moderators must slow the fission neutrons, in the 1- to 2-MeV range to thermal energy at about 0.025 eV to less than 0.1 eV. This effect must be produced in a small volume and with very little absorption.

The Chain reaction :

A chain reaction is that process in which the number of neutrons keeps on multiplying rapidly (in geometrical progression) during fission till whole of the fissionable material is disintegrated. The chain reaction will become self-sustaining or self propagating only if, for every neutron absorbed, at least one fission neutron becomes available for causing fission of another nucleus. This condition can be conveniently expressed in the form of *multiplication factor or reproduction factor* of the system which may be definded as

$$K = \frac{\text{No. of neutrons in any particular generation}}{\text{No. of neutrons in the preceding generation}}$$

If $K > 1$, chain reaction will continue and if $K < 1$, **chain reaction cannot be maintained.**

Fig. 6.4 shows schematically a chain reaction which when set off ultimately leads to a rapidly growing avalanche having the characteristic of an explosion. The rate of growth of the chain process is shown in Fig. 6.5.

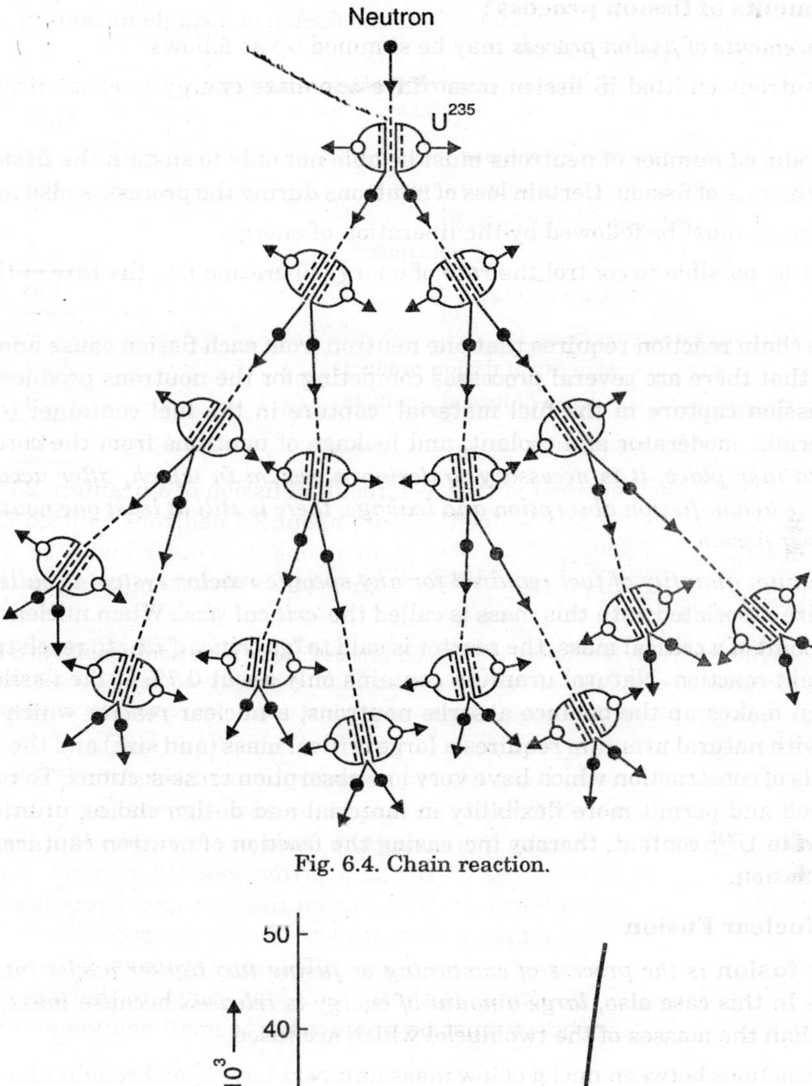

Fig. 6.4. Chain reaction.

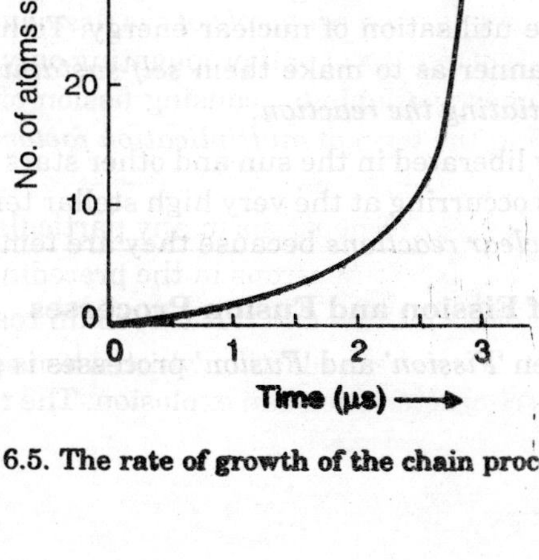

Fig. 6.5. The rate of growth of the chain process.

Requirements of fission process :

The *requirements of fission process* may be summed up as follows :

1. The neutrons emitted in fission must have adequate energy to cause fission of another nuclei.

2. The produced number of neutrons must be able not only to sustain the fission process but also to increase the rate of fission. Certain loss of neutrons during the process is also to be accounted.

3. The process must be followed by the liberation of energy.

4. It must be possible to control the rate of energy liberation *i.e.*, the rate of fission by some means.

Since the chain reaction requires that one neutron from each fission cause another fission, it is worth noting that there are several processes competing for the neutrons produced. These processes are non-fission capture in the fuel material, capture in the fuel container (cladding), core structural materials, moderator and coolant, and leakage of neutrons from the core. *To permit a chain reaction to take place, it is necessary to design a system in which, after accounting for all neutron losses due to non-fission absorption and leakage, there is still at least one neutron remaining to produce another fission.*

The minimum quantity of fuel required for any specific reactor system is called the **critical mass** and the size associated with this mass is called the *'critical size'*. When nuclear fuel is assembled just to the point of a critical mass, the reactor is said to *"go critical" i.e.*, to reach the point of just sustaining a chain reaction. Natural uranium contains only about 0.7% of the fissile isotope U^{235}. Since U^{238} which makes up the balance absorbs neutrons, a nuclear reactor which will sustain a chain reaction with natural uranium requires a large critical mass (and size) and the use of moderator and materials of construction which have very low absorption cross-sections. To reduce the critical mass required and permit more flexibility in material and design choice, uranium fuel is frequently *enriched* in U^{235} content, thereby increasing the fraction of neutron captures that occur in U^{235} and cause fission.

6.1.12. Nuclear Fusion

Nuclear fusion *is the process of combining or fusing two lighter nuclei into a stable and heavier nuclide.* In this case also, *large amount of energy is released* because mass of the product nucleus is *less* than the masses of the two nuclei which are fused.

Several reactions between nuclei of low mass numbers have been brought about by accelerating one or the other nucleus in a suitable manner. These are often fusion processes accompanied by release of energy. However, reactions involving artificially-accelerated particles cannot be regarded as of much significance for the utilisation of nuclear energy. To have practical value, fusion reactions must occur in such a manner as to make them *self-sustaining, i.e., more energy must be released than is consumed in initiating the reaction.*

It is thought the energy liberated in the sun and other stars of the main sequence type is due to the nuclear fusion reactions occurring at the very high stellar temperature of 30 million °K. Such processes are called *thermonuclear reactions* because they are temperature-dependent.

6.1.13. Comparison of Fission and Fusion Processes

The comparison between *'Fission'* and *'Fusion'* processes is given below :

Fission	*Fusion*
1. When heavy unstable nucleon is bombarded with neutrons, the nucleus *splits* into fragments of equal mass and energy is released.	1. Some light elements *fuse together* with the release of energy.
2. About *one thousandth* of the mass is *converted into energy*.	2. It is possible to have *four thousandths* of mass *converted into energy*.
3. Nuclear reaction *residual problem is great*.	3. Residual problem is *much less*.
4. Amount of radioactive material in a fission reactor is *high*.	4. A possible advantage is that the total amount of radioactive material in a working fusion reactor is likely to be *very much less* than that in a fission reactor.
5. Because of higher radioactive material, *health hazards is high in case of accidents*.	5. Because of lesser radioactive material, *health hazards is much less*.
6. It is *possible to construct self-sustained fission reactors* and have positive energy release.	6. It is *extremely difficult to construct controlled fusion reactors*.
7. *Manageable temperatures* are obtained.	7. Needs *unmanageable temperatures* like 30 million degrees for fusion process to occur.
8. Raw fissionable material is *not available in plenty*.	8. Reserves of deuterium, the fusion element, *is available in great quantity*.

6.2. NUCLEAR POWER SYSTEMS

A nuclear-fueled power-producing system consists essentially of the following :

(*i*) *A controlled fission heat source.*

(*ii*) *A coolant system to remove and transfer the heat produced.*

(*iii*) *Equipment to convert the thermal energy contained in the hot coolant to electric power.*

Regardless of the type of fission heat source used, the basic mechanism is fission of nuclear fuel to produce thermal energy. This thermal energy is removed from the heat source (reactor core) by contacting the fuel with a coolant which can be used directly as the working fluid in the power-conversion cycle or indirectly to heat another fluid to be used as the working fluid.

In some cases an intermediate heat-transfer loop is inserted between the reactor coolant and the working fluid, to increase isolation of the radioactive reactor coolant from the conventional power-producing equipment. The working fluid is then used to drive a turbo-generator set to produce electrical power.

Though several other methods are feasible for direct conversion of fission energy to electric power (*i.e.*, thermoelectric, thermionic, photoelectric, etc.), these methods are not at present suitable for the production of large quantities of power.

Schematic representations of nuclear power systems using the direct, indirect, and indirect with intermediate heat-transfer approaches are shown in Fig. 6.6.

Nuclear power systems differ in a number of respects from fossil-fuel systems. Some of the more important *considerations* that differentiate nuclear-fueled plants from fossil-fueled plants are listed below :

1. Nuclear fuel is charged to a power plant infrequently and has a relatively long life, usually measured in months or years, as compared with the continuous fuel-feed requirements for fossil-fueled plants.

2. Burned nuclear fuel is radioactive ; it requires remote handling and special processing and disposal.

3. Major portions of a nuclear plant are radioactive during and after operation, requiring special precautions for maintenance of much of the plant.

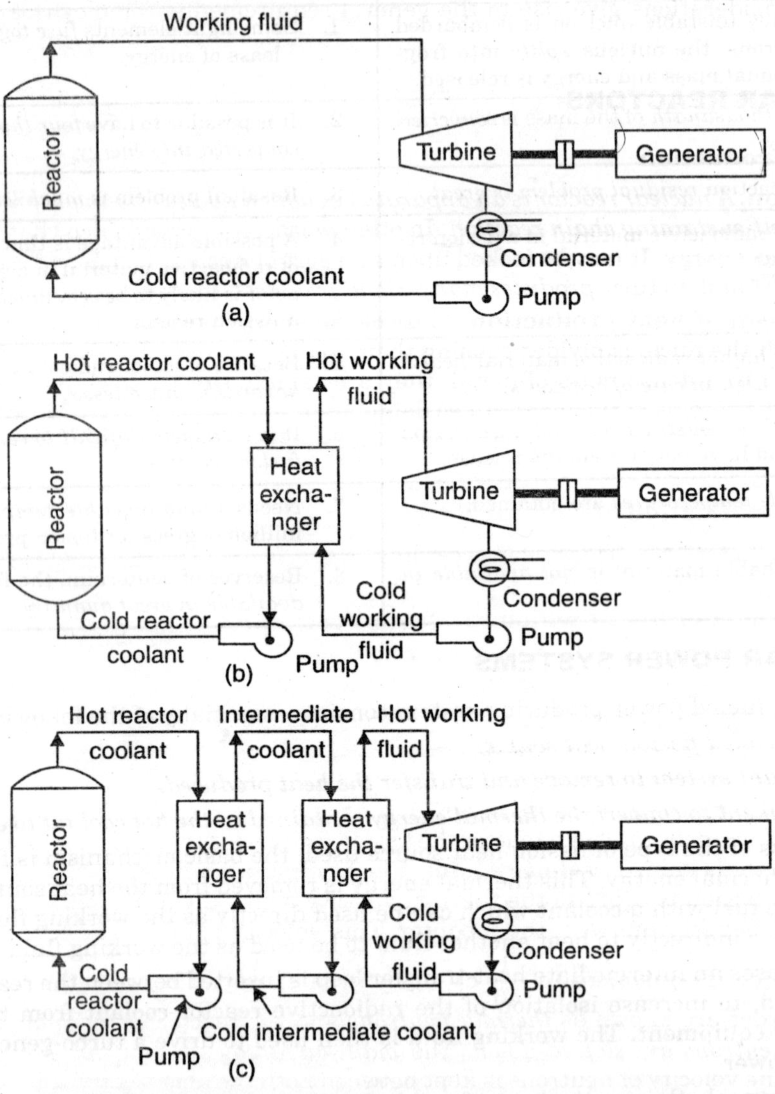

Fig. 6.6. Schematics of nuclear power systems. (*a*) Direct cycle, reactor coolant used as the working fluid ; (*b*) indirect cycle, reactor coolant transfers heat to separate working fluid ; (*c*) indirect cycle with intermediate loop, reactor coolant transfers heat through intermediate heat-transfer loop to working fluid.

4. Special system designs are required to prevent radioactivity release during normal operation or due to accidents.

5. Control and instrumentation requirements are strongly influenced by safety requirements and are related to reactor stability, load-following requirements, and the capability of a reactor to increase power output with no additional fuel input.

6. Nuclear fuel is highly processed material generally used in a precise fabricated form, as opposed to fossil fuels, which are essentially raw materials used with only minimal rough processing.

7. The use of nuclear fuel does not require combustion air, thus obviating thermal stack losses and related problems.

These considerations give rise to the general requirements, complexities, and problems of nuclear systems.

6.3. NUCLEAR REACTORS

6.3.1. Introduction

Definition. *A nuclear reactor is an apparatus in which nuclear fission is produced in the form of a controlled self-sustaining chain reaction.* In other words, it is a controlled chain-reacting system supplying nuclear energy. It may be looked upon as a sort of nuclear furnace which burns fuels like U^{235}, U^{233} or Pu^{239} and, in turn, *produces many useful products like heat, neutrons and radioisotopes.*

Mechanism of heat production. Most of the energy is imparted to the two fission fragments into which the nucleus divides causing them to move at high speed. However, because they have taken birth in a dense mass of metal, they are rapidly slowed down and brought to rest by colliding with other atoms of the metal. In so doing, their energy is converted into heat in much the same way as energy given up by a slowing motor can be converted into heat in the brake lining. In this way, the mass of uranium metal gets heated up.

6.3.2. Classification of Nuclear Reactors

Nuclear reactors are *"classified"* according to the *chain reacting system, use, coolants, fuel material etc.*

1. Neutron energies at which the fission occurs :

(*i*) Fast fission is caused by high energy neutrons	*Fast reactors*
(*ii*) Intermediate or epithermal	*Intermediate reactors*
(*iii*) Low energy *i.e.*, thermal	*Slow reactors*

On the basis of the energy of the neutrons to cause fission the reactors have been divided into three groups—*fast, intermediate and thermal*. (*a*) In ***fast reactors*** the high velocity neutrons produced by fission are *utilised directly* to cause fission of the fuel in the reactor. The velocity of the neutrons is *not reduced deliberately*. (*b*) If in a reactor the fission process is maintained due to the slow neutrons capture, the reactor is known as ***slow reactor***. The minimum velocity to which neutrons *are slowed down* before the fission is equal to the thermal velocity which the slow neutrons may acquire in a state of thermal equilibrium with the medium. This velocity is of the order of 2150 m/s at room temperature which is equivalent to $\frac{1}{40}$ eV or neutron energy. The neutrons associated with the energy of this order are known as *thermal neutrons* and the reactor as *thermal reactors*. With the *moderator the neutrons are slowed down*. The *main advantage is that the probability of reaction increases*. (*c*) If the velocity of neutrons is kept between both the above noted limits, the reactors are termed as *'intermediate reactors'*.

2. Fuel-moderator assembly :

(*i*) Homogeneous reactors (*ii*) Heterogeneous reactors.

In *'homogeneous reactor' the fuel and moderator are mixed to form a homogeneous material, i.e.*, uranium fuel salt forms a homogeneous solution in water which is a moderator or fine particles of uranium and carbon gives a mechanical mixture.

In *'heterogeneous reactor'* the fuel is used in the form of rods, plates or wires etc. and the moderator surrounds the each fuel element in the reactor core.

3. Fuel state :

(*i*) Solid (*ii*) Liquid (*iii*) Gas

The nuclear fuel is available in three states—solid, liquid and gas. In reactors the fuel is mostly used in solid state or in the form of solution dissolved in water. The *liquid metal reactors are in practical use.*

4. Fuel material :

(*i*) Natural uranium with U^{235} contents (occurs in nature)

(*ii*) Enriched uranium with more than 0.71 of U^{235}

(*iii*) Pu^{239}, Pu^{241} or Pu^{239} (man made)

(*iv*) U^{233} (man made)

Considering the necessary requirement of fission process and its availability economically the fuels used in reactors are *uranium, plutonium* and *thorium.* U^{235} is easily available in natural uranium (*i.e.*, 0.7%) and its content increases upto 90% in enriched uranium.

5. Moderator :

(*i*) Water (H_2O) (*ii*) Heavy water (D_2O)

(*iii*) Graphite (*iv*) Beryllium or beryllium oxide

(*v*) Hydrocarbons or hydrides.

A moderator's function is to *absorbs the part of the kinetic energy of the neutrons.* The neutrons collide directly with the moderator and thus slowed down. No ideal moderator is available in nature or has been produced artificially. The light weight nuclei materials are not suited at all as a moderator because they do not possess the property of absorption of neutrons.

Light water, heavy water and *graphite* are the *most common moderators* used in reactors.

6. Principal product :

(*i*) Research features to produce neutrons — *Research reactors*

(*ii*) Power reactor to produce heat — *Power reactors*

(*iii*) Breeder reactors to produce fissionable materials — *Breeder reactors*

(*iv*) Production reactors to produce isotopes — *Production reactors*

Research reactors. These are designed to produce the high neutron flux for research work and these neutrons are used to determine the neutron properties of interaction with the nuclei and the effect of bombardment of neutrons on the materials. The reactors are operated at high neutron flux and low power level otherwise the cooling will be a problem. The unit is cooled constantly during operation. The by-products are heat and fission products which are removed during operation.

Power reactors. In these reactors the energy is produced in heat form which is carried away to the heat exchanger by circulating the coolant through the reactor and heat exchanger. In the heat exchanger the coolant converts the water into steam to run the turbine. The by-products are fission products, neutrons and other radiation particles.

These reactors are useful to produce *huge amount of power* and are *widely used in power plant stations.* In such reactors *consumption is very low.*

Breeder reactors. A breeder reactor *converts fertile materials into fissionable materials* such as U^{238} and Th^{232} to Pu^{239} and U^{233} respectively *besides the power production*. It is worth noting that the amount of fissionable material produced is *more* than its consumption of fissionable material. By-products are the same as those of power reactors.

Production reactors. The output of such reactors is radioactive materials which are used as sources of radiation and tracers in research in all areas of science. By-products are the same as those of power reactors.

7. Coolant :

(*i*) Air, carbon or helium-cooled reactors (*ii*) Water or other liquid-cooled reactors

(*iii*) Liquid metal-cooled reactors.

In *gas-cooled reactors* the amount of gas required to extract the heat is too much and therefore these reactors are expensive. Gases have poor heat carrying capacity. CO_2 and He have been used in the early reactors. *Mostly water is used as a coolant.*

Liquid metal cooled reactors are also suitable as the metal is having high boiling point and low steam pressure. These are the power reactors.

8. Construction of core

(*i*) Cubical (*ii*) Cylindrical (*iii*) Octagonal
(*iv*) Spherical (*v*) Slab (*vi*) Annulus.

The proper shape to the core is given on the practical consideration and can have cubical, cylindrical or ring type construction.

6.3.3. Essential Components of a Nuclear Reactor

The essential components of a nuclear reactor are as follows :

1. Reactor core
2. Reflector
3. Control mechanism
4. Moderator
5. Coolants
6. Measuring instruments
7. Shielding.

1. Reactor core :

The reactor core is that part of a nuclear power plant where fission chain reaction is made to occur and where *fission energy is liberated in the form of heat for operating power conversion equipment.* The *core* of the reactor consists of an assemblage of fuel elements, control rods, coolant and moderator. Reactor cores generally have a shape approximating to a right circular cylinder with diameters ranging from 0.5 m to 15 m. The pressure vessels which houses the reactor core is also considered a part of the core (Fig. 6.7). The fuel elements are made of plates or rods of uranium metal. These plates or rods are usually clad in a thin sheath of stainless steel, zirconium

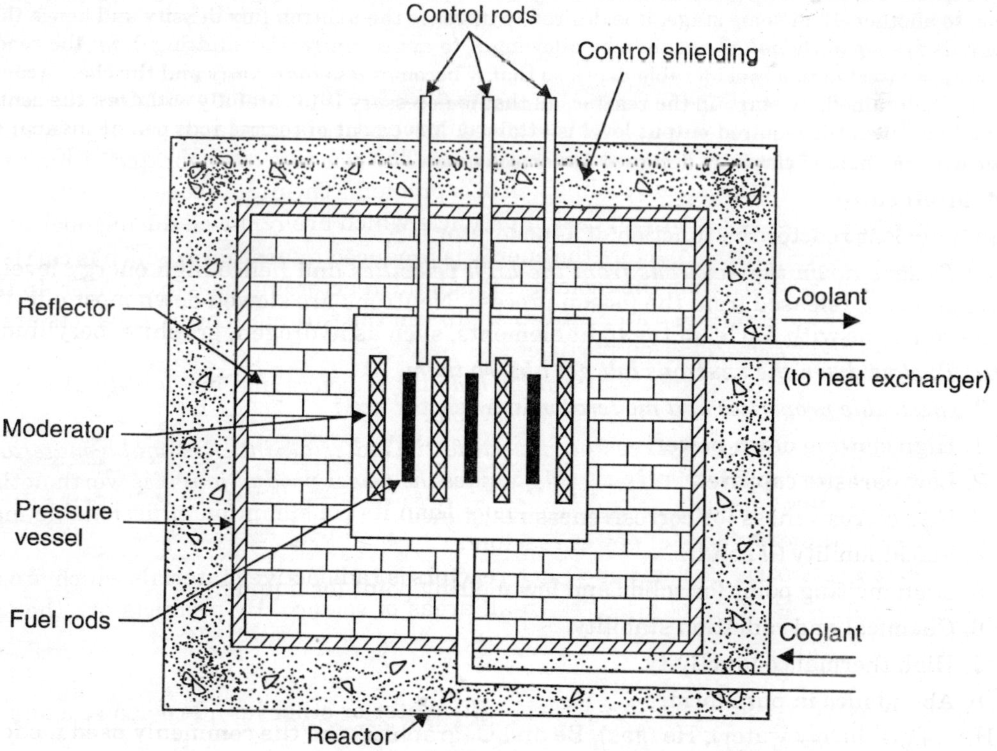

Fig. 6.7. Nuclear reactor.

or aluminium to provide corrosion resistance, retention of radioactivity and in some cases, structural support. Enough space is provided between individual plates or rods to allow free passage of the coolant.

2. Reflector :

A reflector is usually placed round the core to reflect back some of the neutrons that leak out from the surface of the core. It is generally made of the same material as the moderator.

3. Control mechanism :

It is an essential part of a reactor and serves the following *purposes* :

(*i*) For starting the reactor *i.e.*, to bring the reactor up to its *normal operating level.*

(*ii*) For maintaining at *that level i.e.*, keep power production at a steady state.

(*iii*) For shutting the reactor down under normal or emergency conditions.

The control system is also necessary to prevent the chain reaction from becoming violent and consequently damaging the reactor. The effective multiplication factor of the reactor is always kept *greater than unity* in order that the number of neutrons keeps on increasing in successive generations. As the number of neutrons and hence the neutron flux density increases, the temperature also increases. Unless the growth is checked at some point, the reactor is likely to be damaged as a result of too rapid liberation of energy.

Note. The *control system* works on the simple *principle of absorbing the excess neutrons* with the help of control rods either made of boron steel or cadmium strips. Both these materials have very large cross-section for thermal neutrons *i.e.*, they are very good absorbers of slow neutrons and also have the advantage of not becoming radioactive due to neutron capture. By pushing these rods deeper into the central core, any amount of excess neutrons can be absorbed. Once the reactor has reached pre-determined power level, these control rods serve to keep the value of $K = 1$ so that there is no further increase in the number of neutrons from one generation to another. If, at some stage, it is desired to increase the neutron flux density and hence the power level, the rods are *partially pulled out* thereby allowing K to *exceed unity*. For shutting down the reactor, the control rods are inserted to a considerable depth so that K becomes *less than unity* and the chain-reaction can no longer be maintained. To start up the reactor, all that is necessary is to carefully withdraw the control rods and then adjust them till required output level is attained. Movement of control rods can be manual or made automatic with the help of carefully designed *servomechanism*.

4. Moderator :

In a nuclear reactor the function of a moderator is :

(*i*) *To slow down the neutrons from the high velocities* and hence high energy level, which they have on being released from the fission process. Neutrons are slowed down most effectively in scattering collisions with nuclei of the light elements, such as hydrogen, graphite, beryllium etc.

(*ii*) *To slow down the neutrons but not absorb them.*

The *desirable properties of a moderator* in a reactor are :

1. High slowing down power.
2. Low parasite capture.
3. Non-corrosiveness (or corrosiveness resistance).
4. Machinability (if solid).
5. High melting point for solids and low melting point for liquids.
6. Chemical and radiation stability.
7. High thermal conductivity.
8. Abundance in pure form.

H_2O, D_2O (heavy water), He (gas), Be and C (graphite) are the commonly used moderators.

As a moderator D_2O is *the best material available* (moderating ratio of D_2O is 12000 as compared to 72 for H_2O and 170 for carbon) because (*i*) it has excellent neutron slowing properties (*ii*) it has very small cross-section for neutron capture (*iii*) it can be used as a coolant as well. Its *disadvantages* are : (*i*) it has low boiling point so that it *necessitates pressurisation*, and (*ii*) it is *very expensive*. But, the advantages of D_2O as moderator or moderator coolant outweigh its high cost.

5. Coolants :

The function of a coolant is to remove the intense heat produced in the reactor and to bring out for being utilised.

The desirable characteristics for a reactor coolant are :

1. Low parasite capture.
2. Low melting point.
3. High boiling point.
4. Chemical and radiation stability.
5. Low viscosity.
6. Non-toxicity.
7. Non-corrosiveness.
8. Minimum induced activity (short half lives, low energy emission).
9. High specific heat (reduces pumping power and thermal stresses).
10. High density (reduces pumping power and physical plant size).

Commonly used coolants : Santiwax R (organic, Hg, He, CO_2)

The most widely-used gaseous coolant is CO_2 particularly in large-power reactors. It is (*i*) cheap, (*ii*) does not attack metals at reasonable temperatures, and (*iii*) has small cross-section for neutron capture.

Some possible reactor cooling systems :

Some possible reactor cooling systems are illustrated schematically in Figs. 6.8, 6.9 and 6.10.

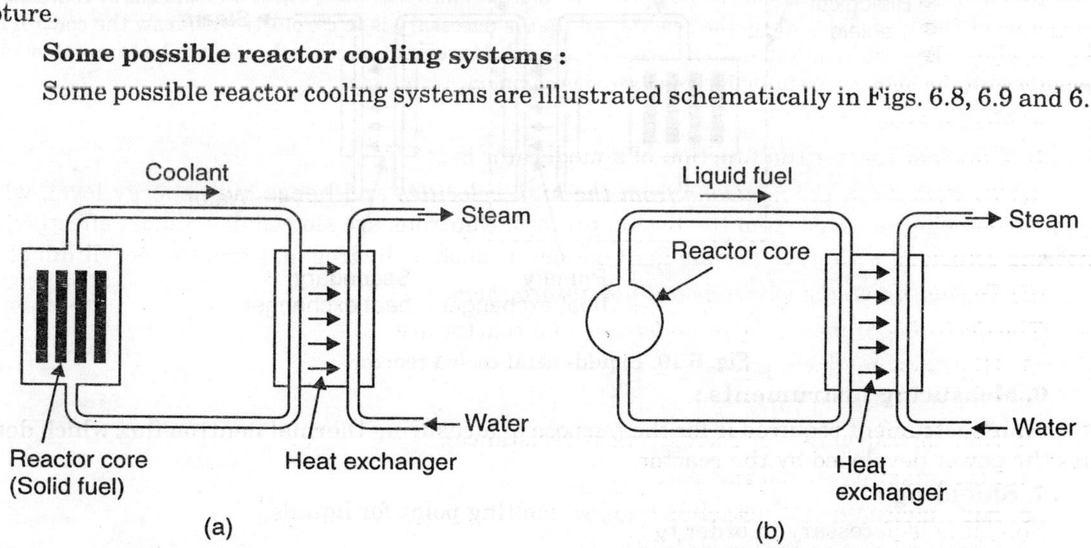

Fig. 6.8. Cooling systems : (*a*) Indirect cooling ; (*b*) Direct cooling.

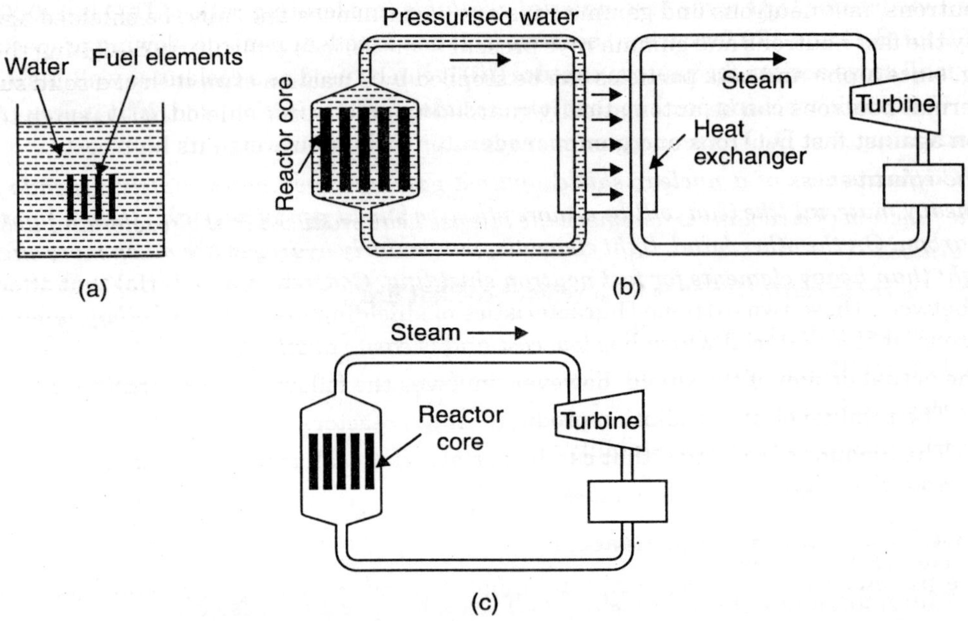

Fig. 6.9. Water-cooled reactors : (*a*) Swimming-pool reactor ; (*b*) Pressurised-water reactor ; (*c*) Boiling water reactor.

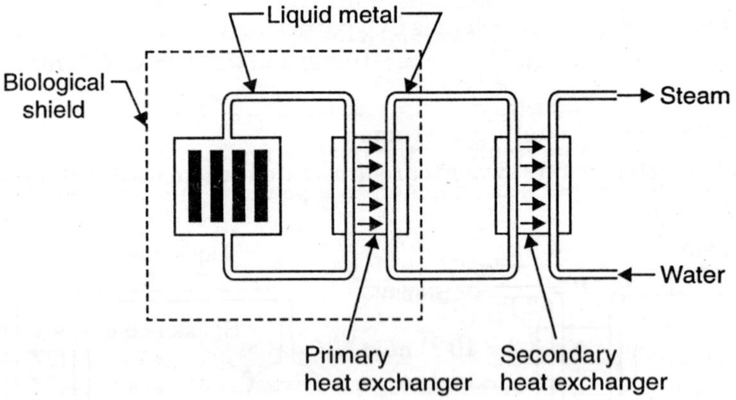

Fig. 6.10. Liquid-metal-cooled reactor.

6. Measuring instruments :

Main instrument required is for the purpose of measuring thermal neutron flux which determines the power developed by the reactor.

7. Shielding :

Shielding is necessary in order to :

(*i*) protect the walls of the reactor vessel from radiation damage, and also to

(*ii*) protect operating personnel from exposure to radiation.

The first known as *thermal shield* is provided through the *steel lining*, while the other called *external* or *biological* shield is generally made of *thick concrete surrounding the reactor installation*.

Among the nuclear radiations produced in a reactor the alpha and beta particles, thermal (slow) neutrons, fast neutrons and gamma rays are harmful ones and must be shielded against. Of these only the fast neutrons and gamma rays present some serious difficulty in designing the reactor shielding, since alpha and beta particles can be stopped by a fraction of an inch of a solid substance, while thermal neutrons can be automatically guarded against with a shield thick enough to provide protection against fast neutrons and gamma rays.

The effectiveness of a nuclear shield against gamma rays approximately depends upon its mass. A heavy material like lead will be a more effective shield per unit weight, than a light element such as carbon. On the other hand, light elements, particularly hydrogen are much more effective per unit weight than heavy elements for fast neutron shielding. Concrete is a material that offers a compromise between these two extreme characteristics of shielding material for *both gamma rays and fast neutrons*. It is a material which has *low cost* and is *easily available*.

The actual design of the shield, however, involves the following *considerations :*

(*i*) The total amount of radiation produced in the reactor.

(*ii*) The amount of radiation that can be permitted to leak through the shield.

(*iii*) The shielding properties of material.

6.3.4. Power of a Nuclear Reactor

The fission rate of a reactor *i.e.*, total number of nuclei undergoing fission per second in a reactor is

$$= nC\sigma NV = \phi_{nu}\ \sigma NV$$

where,

n = Average neutron density *i.e.*, number per m^3,

C = Average speed in m/s,

$\phi_{nu} = nC$ = Average neutron flux,

N = Number of fissile nuclei /m^3,

σ = Fission cross-section in m^2, and

V = Volume of the nuclear fuel.

Since 3.1×10^{10} fission per second generate a power of one watt, the power P of a nuclear reactor is given by,

$$P = \frac{nC\sigma NV}{3.1 \times 10^{10}} \text{ watt}$$

$$= 3.2 \times 10^{-11}\ nC\sigma\ NV \text{ watt}$$

$$= 3.2 \times 10^{-11}\ \phi_{nu}\ \sigma\ NV \text{ watt}$$

Now, NV = Total number of fissile nuclei in the reactor fuel

$$= m \times 6.02 \times 10^{26}/235$$

where, m is the mass of the U^{235} fuel. It is known that fission cross-section σ of U^{235} for thermal neutrons is 582 barns = 582×10^{-28} m^2.

$$\therefore \quad P = \frac{3.2 \times 10^{-11} \times \phi_{nu} \times 582 \times 10^{-28} \times m \times 6.02 \times 10^{26}}{235}$$

$$= 4.77 \times 10^{-12}\ m\ \phi_{nu} \text{ watt}$$

$$\simeq 4.8 \times 10^{-12}\ mnC \text{ watt.}$$

6.4. MAIN COMPONENTS OF A NUCLEAR POWER PLANT

Fig. 6.11 shows schematically a nuclear power plant.

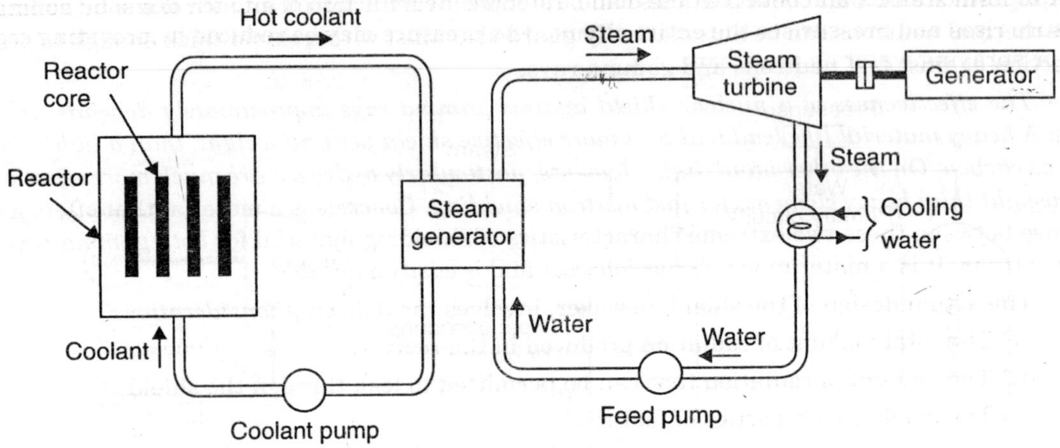

Fig. 6.11. Nuclear power plant.

The *main components of a nuclear power plant* are :

1. Nuclear reactor
2. Heat exchanger (steam generator)
3. Steam turbine
4. Condenser
5. Electric generator.

In a nuclear power plant the reactor performs the same function as that of the furnace of steam power plant (*i.e.*, produces heat). The heat liberated in the reactor as a result of the nuclear fission of the fuel is taken up by the coolant circulating through the reactor core. Hot coolant leaves the reactor at the top and then flows through the tubes of steam generator and passes on its heat to the feed water. The steam so produced expands in the steam turbine, producing work and thereafter is condensed in the condenser. The steam turbine in turn runs an electric generator thereby producing electrical energy. In order to maintain the flow of coolant, condensate and feed water pumps are provided as shown in Fig. 6.11.

6.5. DESCRIPTION OF REACTORS

6.5.1. Pressurised Water Reactor (PWR)

A pressurised water reactor, in its simplest form, is a light water-cooled and moderated thermal reactor having an unusual core design, using both natural and highly enriched fuel. The principal parts of the reactor are :

1. Pressure vessel
2. Reactor thermal shield
3. Fuel elements
4. Control rods
5. Reactor containment
6. Reactor pressuriser.

The components of the secondary system of pressurised water plant are similar to those in a normal steam station.

Refer to Fig. 6.12. In PWR, there are two circuits of water, one *primary circuit* which passes through the fuel core and is *radioactive*. This primary circuit then produces steam in a *secondary circuit* which consists of heat exchanger or the boiler and the turbine. As such the steam in the turbine is *not radioactive* and need not be shielded. The *pressure* in the primary circuit should be *high* so that the boiling of water takes place at high pressure. A *pressuring tank* keeps the water at about 100 kgf/cm^2 *so that it will not boil*. Electric heating coils in the pressuriser boil some of the water to form steam that collects in the dome. As more steam is forced into the dome by boiling, its pressure rises and pressurises the entire circuit. The pressure may be reduced by providing cooling coils or spraying water on the steam.

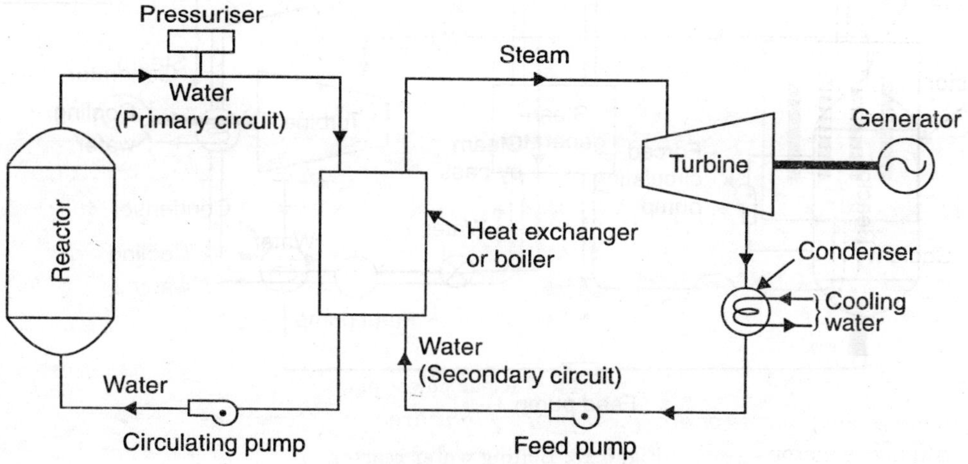

Fig. 6.12. Pressurised water reactor.

Water acts both as coolant as well as moderator. Either heavy water or the light water may be used for the above purpose.

A pressurised water reactor *can produce only saturated steam*. By providing a separate furnace, the steam formed from the reactor could be superheated.

Advantages of PWR :

1. Water used in reactor (as coolant, moderator and reflector) is cheap and easily available.
2. The reactor is compact and power density is high.
3. Fission products remain contained in the reactor and are not circulated.
4. A small number of control rods is required.
5. There is a complete freedom to inspect and maintain the turbine, feed heaters and condenser during operation.
6. This reactor allows to reduce the fuel cost extracting more energy per unit weight of fuel as it is ideally suited to the utilisation of fuel designed for higher burn-ups.

Disadvantages :

1. Capital cost is high as high primary circuit requires strong pressure vessel.
2. In the secondary circuit the thermodynamic efficiency of this plant is quite low.
3. Fuel suffers radiation damage and, therefore its reprocessing is difficult.
4. Severe corrosion problems.
5. It is imperative to shut down the reactor for fuel charging which requires a couple of month's time.
6. Low volume ratio of moderator to fuel makes fuel element design and insertion of control rods difficult.
7. Fuel element fabrication is expensive.

6.5.2. Boiling Water Reactor (BWR)

In a boiling water reactor *enriched fuel* is used. As compared to PWR, the arrangement of BWR plant is simple. The plant can be safely operated using natural convection within the core or forced circulation as shown in the Fig. 6.13. For the safe operation of the reactor the pressure in the forced circulation must be maintained constant irrespective of the load. In case of *part load operation of the turbine some steam is by-passed.*

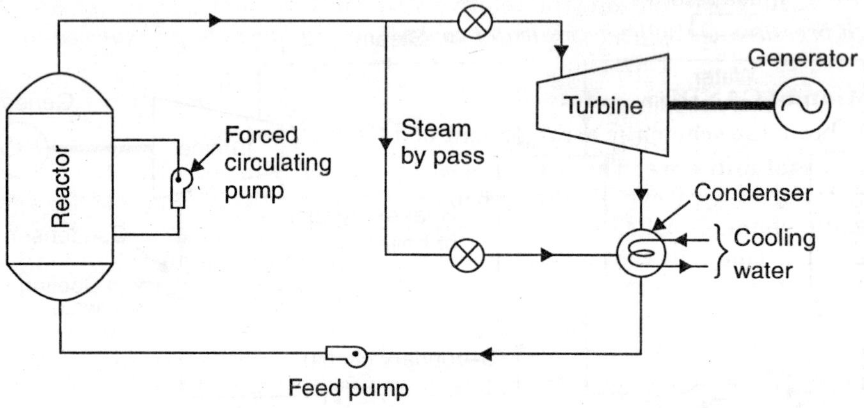

Fig. 6.13. Boiling water reactor.

Advantages of BWR :

1. Heat exchanger circuit is eliminated and consequently there is gain in thermal efficiency and gain in cost.
2. There is use of a lower pressure vessel for the reactor which further reduces cost and simplifies containment problems.
3. The metal temperature remains low for given output conditions.
4. The cycle for BWR is more efficient than PWR for given containment pressure, the outlet temperature of steam is appreciably higher in BWR.
5. The pressure inside the pressure vessel is not high so a thicker vessel is not required.

Disadvantages :

1. Possibility of radioactive contamination in the turbine mechanism, should there be any failure of fuel elements.
2. More elaborate safety precautions needed which are costly.
3. Wastage of steam resulting in lowering of thermal efficiency on part load operation.
4. Boiling limits power density ; only 3 to 5% by mass can be converted to steam per pass through the boiler.
5. The possibility of "burn out" of fuel is more in this reactor than PWR as boiling of water on the surface of the fuel is allowed.

6.5.3. CANDU (Canadian-Deuterium-Uranium) Reactor

CANDU is a thermal nuclear power reactor in which *heavy water* (99.8% deuterium oxide D_2O) *is the moderator and coolant, as well as the neutron reflector.* This reactor was developed in Canada and is being extensively used in this company. A few CANDU reactors are operating or under construction in some other countries as well.

In this type of reactor the *natural uranium* (0.7% U^{235}) *is used as fuel and heavy water as moderator.* These reactors are more economical to those countries which do not produce enriched uranium, as the *enrichment of uranium is very costly.*

CANDU (heavy water) reactor, differs basically from light-water reactors (LWRS) in that in the latter the *same water serves as both moderator and coolant*, whereas in the *CANDU reactor the moderator and coolant are kept separate.* Consequently unlike the pressure vessel of a LWR, the CANDU reactor vessel, which contains the relatively cool heavy water moderator, *does not have to withstand a high pressure. Only the heavy water coolant circuit has to be pressurised to inhibit boiling in the reactor core.*

Description of CANDU reactor

Fig. 6.14 shows the schematic arrangement of a CANDU reactor.

Reactor vessel and core. The reactor vessel is a steel cylinder with a horizontal axis ; the length and diameter of a typical cylinder being 6 m and 8 m respectively. The vessel is penetrated by some 380 horizontal channels called *pressure tubes* because they are designed to withstand a high internal pressure. The channels contain the fuel elements and the pressurised coolant flows along the channels and around the fuel elements to remove the heat generated by fission. Coolant flows in the opposite directions in adjacent channels.

The high pressure (10 MPa) and high temperature (370°C) coolant leaving the reactor core enters the steam generator. About 5% of fission heat is generated by fast neutrons escaping into the moderator, and this is removed by circulation through a separate heat exchanger.

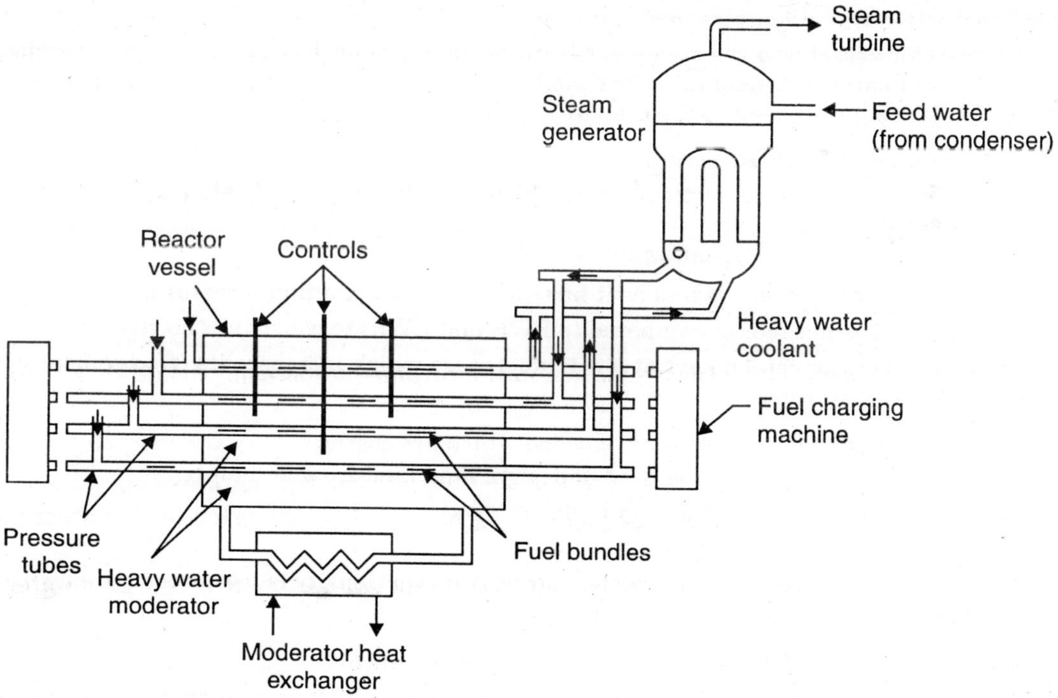

Fig. 6.14. CANDU reactor.

Fuel. In a CANDU reactor the fuel is *normal* (*i.e.*, unenriched) *uranium oxide* as small cylinder pellets. The pellets are packed in a corrosion resistance zirconium alloy tube, nearly 0.5 long and 1.3 cm diameter, to form a fuel rod. The relatively short rods are combined in bundles of 37 rods, and

12 bundles are placed end to end in each pressure tube. The total mass of fuel in the core is about 97,000 kg. The CANDU reactor is *unusual in that refueling is conducted while the reactor is operating.*

Control and protection system :

There are the various types of vertical control system incorporated in the CANDU reactor :

— A number of strong neutron absorber *rods of cadmium* which are used mainly for *reactor shut-down* and *start-up*.

— In addition to above there are other *less strongly*, absorbing rods to *control power variations* during reactor operation and *to produce an approximately uniform heat* (power) distribution throughout the core.

In an emergency situation, the shut-down rods would immediately drop into the core, followed, if necessary by the injection of a gadolinium nitrate solution into the moderator.

Steam system. Steam system is discussed below :

— The respective ends of the pressure tubes are all connected into inlet and outlet headers.

— The high temperature coolant leaving the reactor passes out the outlet header to a steam generator of the conventional inverted U-tube and is then pumped back into the reactor by way of the inlet header.

— Steam is generated at a temperature of about 265°C.

There are two coolant outlet (and two inlet) headers, one at each end of the reactor vessel, corresponding to the opposite directions of coolant flow through the core. Each inlet (and outlet) header is connected to a separate steam generator and pump loop. A single pressurizer (of the type used in pressurised water reactors) maintains an essentially constant coolant system pressure.

The reactor vessel and the steam generator system are enclosed by a *concrete containment structure*. A water spray in the containment would condense the steam and reduce the pressure that would result from a large break in the coolant circuit.

Advantages of CANDU reactor :

1. Heavy water is used as moderator, which has higher multiplication factor and low fuel consumption.
2. Enriched fuel is not required.
3. The cost of the vessel is less as it has not to withstand a high pressure.
4. Less time is needed (as compared to PWR and BWR) to construct the reactor.
5. The moderator can be kept at low temperature which increases its effectiveness in slowing down neutrons.

Disadvantages :

1. It requires a very high standard of design, manufacture and maintenance.
2. The cost of heavy water is very high.
3. There are leakage problems.
4. The size of the reactor is extremely large as power density is low as compared with PWR and BWR.

6.5.4. Gas-Cooled Reactor

In such a type of reactor, the coolant used can be air, hydrogen, helium or carbondioxide. Generally inert gases are used such as helium and carbondioxide. The moderator used is *graphite*. The problem of corrosion is reduced much in such reactors. This type of reactor is more safe specially in case of accidents and the failure of circulating pumps. The thickness of gas cooled reactor shield is much reduced as compared to the other types of reactors.

There are two principal types of gas-cooled reactors developed for centre station service and these are :

(*i*) The gas cooled, graphite moderator reactor (GCGM)

(*ii*) The high temperature gas cooled reactor (HTGC).

Both types are graphite moderated. The former (GCGM) uses natural uranium fuel while the latter (HTGC) employs highly enriched uranium carbide mixed with thorium carbide and clad with graphite.

The *coolant pressure* and *temperature* in *GCGM* are about 7 bar 336°C respectively, for *HTGC*, there figures are 15 to 30 bar and 700°C to 800°C.

Arrangement of high temperature, gas cooled reactor is shown in Fig. 6.15.

Advantages of gas-cooled reactor :

1. The processing of the fuel is simpler.
2. No corrosion problem.
3. As a result of low parasitic absorption it gives better neutron economy.
4. Graphite remains stable under irradiation at high temperatures.
5. The use of carbondioxide as coolant completely eliminates the possibility of explosion in the reactor which is always present in water-cooled plants.
6. The uranium carbide and graphite are able to resist high temperatures, and, therefore, the problem of limiting the fuel element temperature is not as serious as in other reactors.

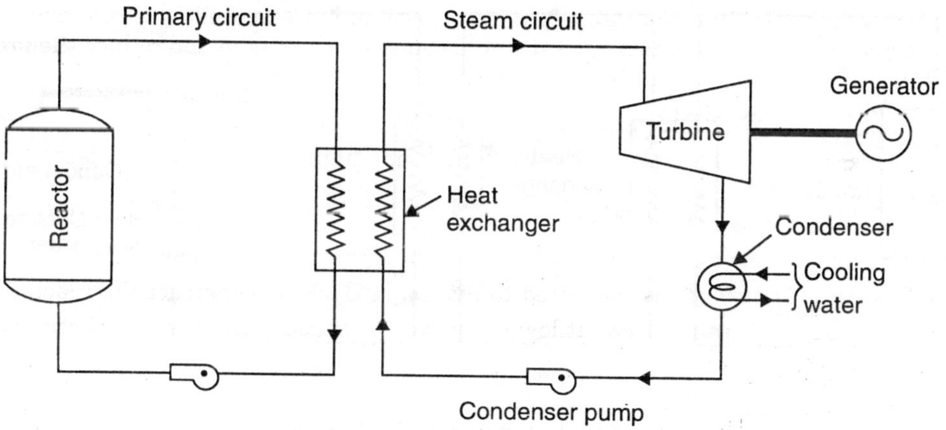

Fig. 6.15. Gas-cooled reactor.

Disadvantages :

1. Fuel loading is more elaborate and costly.
2. Power density is very low (due to low heat transfer coefficient), therefore large vessel is required.
3. Since the critical mass is high therefore large amount of fuel loading is initially required.
4. If helium is used instead of carbondioxide, the leakage of gas is a major problem.
5. More power is required for coolant circulation (as compared with water-cooled reactors).
6. The control is more complicated due to low negative coefficient as helium does not absorb neutrons.

6.5.5. Liquid Metal Cooled Reactors

Sodium-graphite reactor (SGR) is one of the typical *liquid metal reactors.* In this reactor *sodium works as a coolant* and *graphite works as moderator.*

Sodium boils at 880°C under atmospheric pressure and freezes at 95°C. Hence *sodium* is *first melted* by electric heating system and be pressurised to about 7 bar. The liquid sodium is then circulated by the circulation pump. The reactor will have *two coolant circuits or loops* :

(*i*) The *primary circuit* has *liquid sodium* which circulates through the fuel core and gets heated by the fissioning of the fuel. This liquid sodium gets cooled in the intermediate heat exchanger and goes back to the reactor vessel.

(*ii*) The *secondary circuit* has an *alloy* of *sodium and potassium in liquid form.* This coolant takes heat from the intermediate heat exchanger which gets heat from liquid sodium of primary circuit. The liquid sodium-potassium then passes through a boiler which is once through type having tubes only. The steam generated from this boiler will be superheated. Feed water from the condenser enters the boiler, the heated sodium-potassium passing through the tubes gives it heat to the water thus converting it into steam. The sodium-potassium liquid in the second circuit is then pumped back to the intermediate heat exchanger thus making it a closed circuit.

The reactor vessel, primary loop and the intermediate heat exchanger are to be shielded for radio-activity. The liquid metal be handled under the cover of an inert gas, such as helium, to prevent contact with air while charging or draining the primary or secondary circuit/loop.

The arrangement of a sodium-graphite reactor (SGR) is shown in Fig. 6.16.

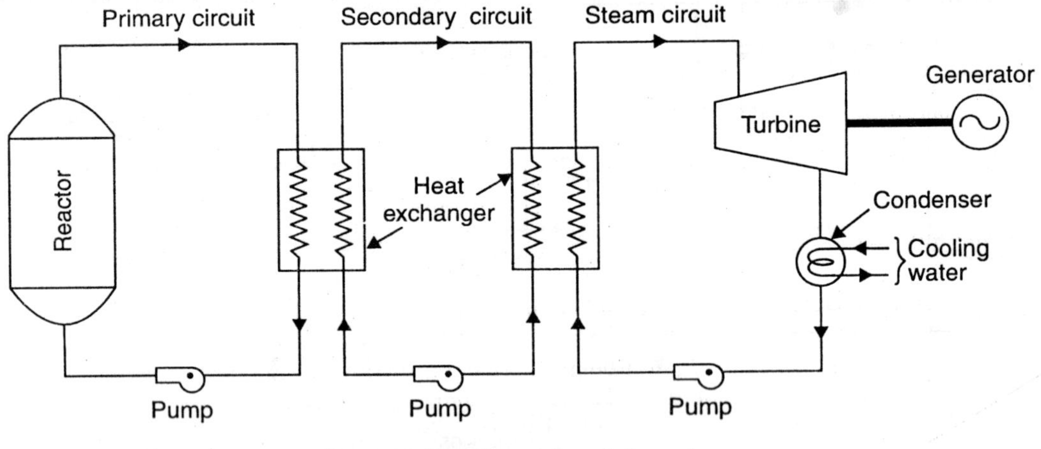

Fig. 6.16. Liquid metal-cooled reactor.

Advantages of SGR :

1. The sodium as a coolant need not be pressurised.
2. High thermal efficiency at low cost.
3. The low cost graphite moderator can be used as it can retain its mechanical strength and purity at high temperatures.
4. Excellent heat removal.
5. High conversion ratio.
6. Superheating of steam is possible.
7. The size of the reactor is comparatively small.
8. The neutron absorption cross-section of sodium is low and, therefore, it is best suited to thermal reactor with slightly enriched fuel.

Disadvantages :

1. Sodium reacts violently with water and actively with air.
2. Thermal stresses are a problem.
3. Intermediate system is necessary to separate active sodium from water.
4. Heat exchanger must be leak proof.
5. It is necessary to shield the primary and secondary cooling systems with concrete blocks as sodium becomes highly radioactive.
6. The leak of sodium is very dangerous as compared with other coolants.

6.5.6. Breeder Reactor

In its simplest form a *fast breeder reactor* is a small vessel in which necessary amount of *enriched plutonium* is kept *without using moderator.* A fissible material, which absorbs neutrons, surrounds the vessel. The reactor core is cooled by liquid metal. Necessary *neutron shielding* is provided by the use of *light water, oil* or *graphite.* Additional shielding is also provided for gamma rays. (It is worth noting that when U^{235} is fissioned, it produces heat and additional neutrons. If some U^{238} is kept in the same reactor, part of the additional neutrons available, after reaction with U^{235}, convert U^{238} into fissible plutonium).

Fig. 6.17 shows a schematic diagram of a breeder reactor.

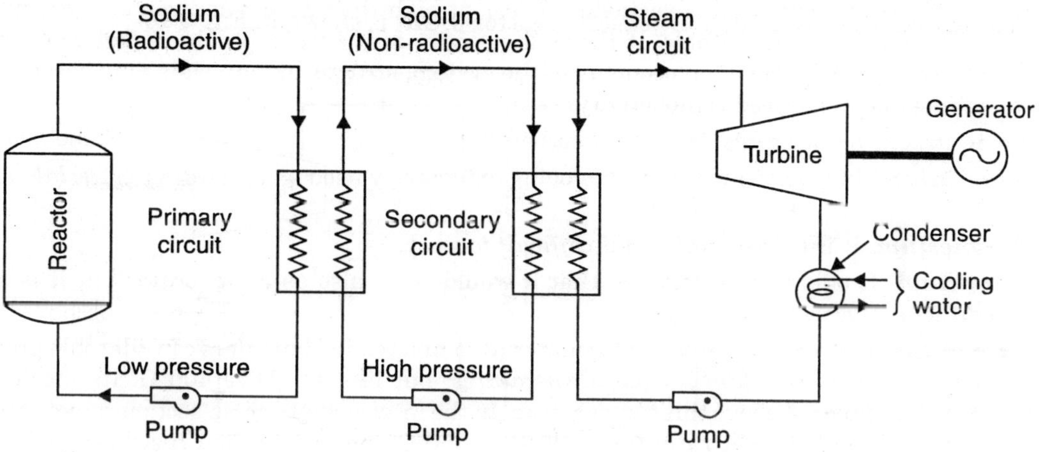

Fig. 6.17. Breeder reactor.

Advantages of a breeder reactor :

1. The moderator is not required.
2. High breeding is possible.
3. Small core is sufficient (since it gives high power density than any other reactor).
4. The parasite absorption of fuel is achievable.
5. High burn-up of fuel is achievable.
6. Absorption of neutrons is low.

Disadvantages :

1. Requires highly enriched (15 percent) fuel.
2. It is necessary to provide safety against *melt-down.*
3. Neutron flux is high at the centre of the core.

4. The specific power of the reactor is low.
5. There is a major problem of handling sodium as it becomes hot and radioactive.

6.6. SELECTION OF MATERIALS FOR REACTOR COMPONENTS

Some of the factors governing the selection of materials for the various reactor components are considered below :

1. Structural material :

The structural units implied are :

(*i*) The cladding material for fuel elements.

(*ii*) Fuel element assemblies.

(*iii*) Containers for coolants, ducts, etc., in the core.

(*iv*) Any general structure in the reaction zone.

Obviously the *purely mechanical properties must be adequate at the normal working temperature*, and the *material must be capable of the necessary fabrication and working*.

In case where liquids are involved, corrosion or erosion problems may be present. These may be especially troublesome in aqueous homogeneous reactors or in reactors employing liquid metals.

From the standpoint of the nuclear physics the *basic requirement is low neutron absorption cross-section.* Since absorption cross-sections are less at high than at low neutron energy levels, a material which is entirely unsuitable at thermal energies may be acceptable in fast reactors. In addition to the effect on the neutron economy of the system, structural materials may become dangerous to handle due to induced radioactivity.

Some *structural materials* are discussed below :

Aluminium. In a very pure state it is being extensively used as a *cladding material for fuel elements.*

Magnesium. It is *more costly* and is *difficult to work.*

Beryllium. It offers advantages in that it would *serve* also as a *moderator* but it is again *costly* and *difficult to work.*

Zirconium. It is a comparatively new material commercially, but offers considerable promise. It is *extremely corrosion resistant at low temperatures,* but less so above 400°C. Its mechanical properties are also impaired above this temperature. In the intermediate cross-section range *titanium* has a very good strength/weight ratio from 100°C to 450°C and is corrosion resistant to aqueous solutions of high temperatures.

Alloys. *Silicon steels* may be used in some fast reactors, but considerable care is needed in their heat treatment.

A number of nickel alloys are of interest and offer considerable resistance to attack by fused salts and alkali hydroxide.

Silicon, of absorption cross-section 0.1 barn and tin, absorption cross-section 0.6 barn, might prove useful for alloying with other material.

2. Reactor coolants :

Gaseous coolants. These coolants have much to recommend them from the standpoint of *general radiation* and *thermal stability* and *ease of handling.* Gases, however, from the standpoint of heat transfer and at high temperatures some of them, for example *oxygen* and *hydrogen,* may attack other materials present in the core. *Air* has been used in some research reactors operating on open cycle with discharge to the atmosphere at a high level to minimise the effect of Ar. *Helium* is attractive but is expensive. *Gases involve high pumping cost.*

Water. Water has *better thermal properties* than gases and the pumping power is roughly $\frac{1}{10}$ th that for gases operated at ten atmospheres pressure. Its moderating properties may be usefully utilised when used as a coolant. It has a *fairly large absorption cross section,* and *undergoes decomposition by radiation.* It is *subjected to induced radioactivity*, and it has a *corrosive action on metals,* and a *low boiling point. Degasification may be necessary.*

For efficient power production high pressure operation is necessary. This necessitates the use of a pressure vessel to enclose the reactor. Boiling water reactors (BWR) in which the coolant serves also as a moderator show promise.

Heavy water has a lower absorption cross-section than natural water and its use leads to economy in fissile material. If it were cheaper it would be attractive for use in boiling water type reactor.

Liquid metals are of special interest in relation to reactors operating at high thermal flux. Their main disadvantage is in *difficulty of handling,* and in their *corrosive properties.* Their *heat transfer properties are better than those of water*, but their volumetric heat capacity is not as good, so that the pumping power may be greater or less than that of water. ***Sodium*** is the most favoured at present, for *if free from oxygen it does not attack stainless steel, nickel and nickel alloys, beryllium or graphite at temperatures below 600°C.*

3. Moderators and reflectors :

Possible materials for use as moderator and reflectors are :

(*i*) Ordinary and heavy water (*ii*) Beryllium
(*iii*) Beryllia (*iv*) Graphite.

Ordinary water has *excellent slowing down properties for neutrons,* but unfortunately, neutron capture is also high. This means that enriched uranium fuel is required in ordinary water moderated reactors, but the small migration length still permits a reactor of relatively small size. Some of *disadvantages* of water are (*i*) *attendant corrosion problems,* (*ii*) *its relatively low boiling point,* and (*iii*) decomposition by nuclear radiations, resulting in the liberation of oxygen and hydrogen which may require to be recombined in ancillary plant. The low boiling point requires the use of a pressure vessel when high temperatures are involved.

Heavy water is an *excellent moderator* and *has a high moderating ratio compared with the other materials.*

Beryllium. The nuclear properties of beryllium are *eminently suitable, but it is expensive, brittle and difficult to fabricate and is corroded by water.*

Graphite. It has been used *most extensively as a reactor moderator* despite the fact that its moderating properties are not as good as heavy water or beryllium. However, (*i*) it is *reasonably cheap,* even when the necessary high degree of purity is achieved, (*ii*) it has *good mechanical properties* and *thermal stability*, and (*iii*) is a *good conductor of heat.*

Its chief *disadvantages* are the possibility of reaction with air at high temperatures and its relatively low mechanical strength.

4. Fuel :

The proportion of the fissile material in the fuel is of considerable importance in determining the *critical size* of the reactor. This is because the ratio of fissile to non-fissile material in the fuel determines the neutron economy at the source. The following table gives the average number of neutrons liberated per neutron absorbed in the fuel.

Neutron type	**U^{235}**	**Pu^{239}**	**Natural uranium**
Thermal	2.11	1.95	1.32
Fast	2	2	1

5. Shielding and radiation protection :

Protection of personnel is achieved partly by the use of remote control and, in some cases, the provision of a pressure vessel to contain the fission products which might result from an accident. However, shielding of the reactor itself is invariably required.

A biological shield must slow down the fast neutrons leaving the core of the reactor, must capture the slowed down neutrons and must absorb all gamma and similar radiation produced. It must be borne in mind that neutron capture in the shield itself may give rise to further gamma radiation. A combination of light (moderating or slowing down) and heavy elements is the best. The latter reduces the energy of very fast neutrons by inelastic scattering. The incorporation of a good neutron absorber (*e.g.,* boron 10) is beneficial. *Cadmium* is not particularly suitable since it emits high energy photons, with energies of the order 7.5 MeV. The *heavy elements* may be *metallic iron* or *lead*, or *iron oxide* or *barytes.* The *light element* is usually *hydrogen in the form of water,* often combined in concrete, but in research reactors in particular, possibly as water. A *homogeneous arrangement of heavy and light elements* is best, for example, the *use of a laminated construction or the use of iron concrete.* The latter consists of iron mixed in barytes concrete, or alternatively limonite (iron ore) is used partially to replace barytes in the mix.

Thermal shield. Every absorption in a shield is roughly exponential. Thus 90% of the radiation is absorbed in the first 10% of the thickness. This results in considerable liberation of heat. For this reason that part of the shield nearest to the core is usually of *iron* in thickness 5 to 10 cm which may be ***air-cooled.*** *This part of the shield is referred to as the* ***'thermal shield'.***

Summary of Materials for Nuclear Power Reactors

Structural :

(*i*) Aluminium
(*ii*) Stainless steel
(*iii*) Nickle alloys
(*iv*) Zirconium
(*v*) Magnesium.

Fuel :

(*i*) Uranium
(*ii*) Uranium ceramics
(*iii*) Thorium
(*iv*) Thorium oxide.

Coolant :

(*i*) Water
(*ii*) Liquid metals
(*iii*) Sodium, potassium
(*iv*) Mercury
(*v*) Lead bismuth
(*vi*) Gases
(*vii*) Helium
(*viii*) Nitrogen.
(*ix*) Carbondioxide

Control :

(*i*) Boron steel
(*ii*) Cadmium
(*iii*) Samarium oxide
(*iv*) Gadolinium oxide.

Mederator reflector :

(*i*) Water
(*ii*) Heavy water
(*iii*) Beryllium
(*iv*) Beryllium oxide
(*v*) Graphite
(*vi*) Metal hydrides.

Shielding :

(*i*) Water
(*ii*) Cement and concrete
(*iii*) Iron
(*iv*) Lead
(*v*) Tantalum
(*vi*) Bismuth
(*vii*) Boron.

6.7. ADVANTAGES AND DISADVANTAGES OF NUCLEAR POWER PLANTS

Some of the *major advantages* of nuclear power plants are :

1. A nuclear power plant *needs less space* as compared to other conventional power plant of equal size.
2. Nuclear power plants are well suited to meet large power demands. They give better performance at high load factors (80 to 90%).
3. Since the fuel consumption is very small as compared to conventional type of power plants, therefore, there is *saving in cost of the fuel transportation.*
4. The nuclear power plants, besides producing large amount of power, *produce valuable fissible material* which is produced when the fuel is renewed.
5. The operation of a nuclear power plant is *more reliable.*
6. Nuclear power plants are not affected by adverse weather conditions.
7. Bigger capacity of a nuclear power plant is an additional advantage.
8. The expenditure on metal structures piping, storage mechanisms is much lower for a nuclear power plant than a coal burning power plant.

Disadvantages/Limitations

1. The *capital cost* of a nuclear power station is always *high.*
2. The *danger of radioactivity* always persists in the nuclear stations (inspite of utmost precautions and care).
3. These plants *cannot be operated at varying load efficiently.*
4. The *maintenance cost is always high* (due to lack of standardisation and high salaries of the trained personnel in this field of specialisation).
5. The disposal of fission products is a big problem.
6. Working conditions in nuclear power station are always detrimental to the health of the workers.

Comparative Ratings of 10 MW Nuclear and Coal-burning Power Plant

S. No.	*Ratings*	*Nuclear plant*	*Coal-burning plant*
1.	Weight of machines and mechanisms, *tons*	700	2700
2.	Weight of metal structures, *tons*	900	1250
3.	Weight of pipes and fittings, *tons*	200	300
4.	Weight of masonary/graphite assembly, *tons*	500	1500
5.	Weight of fuel storage mechanism, *tons*	—	2500
6.	Weight of rolling stock, *tons*	—	300
7.	Volume of plain and reinforced concrete work, *cum*	9000	4000
8.	Capacity of buildings (without turbine room and electrical facilities), *cum*	50000	75000
9.	Area of construction site, *hectares*	5	15
10.	Internal power consumption, *kW*	5000	8000

6.8. NUCLEAR-PLANT SITE SELECTION

Nuclear power plants *must meet all the economic and technical and most of the legal criteria* that apply to the siting of conventional fossil-fuel-fired power plants. *In addition, the importance of site characteristics in the assessment of public safety results in greater concern in siting nuclear plants than with any other type of industrial facility*. Of particular concern are the population distribution with respect to the site and the natural factors which could affect the transport of radioactive material to the public, under normal operating conditions and in the highly unlikely event of an accident which could release radioactive material to the environment.

The various factors to be considered while selecting the site for a nuclear power plant are enumerated below :

1. Proximity to load centre
2. Population distribution
3. Land use
4. Meteorology
5. Geology
6. Seismology
7. Hydrology.

6.9. NUCLEAR POWER PLANTS IN INDIA

The various nuclear power plants situated in India are as follows :

1. Tarapur power plant
2. Rana Partap Sagar power plant
3. Kalpakkam power plant
4. Narora power plant
5. Kakrapar power plant.

The particulars of these plants are given in tabular form below :

Particulars	*Power plants*				
	Tarapur	*Rana Partap Sagar*	*Kalpakkam*	*Narora*	*Kakrapar*
1. Location	Maharashtra (65 miles north Mumbai)	Near Kota in Rajasthan	Near Chennai in Tamil Nadu state	U.P	Surat district, Gujrat
2. Capacity	380 MW	400 MW (2 × 200 MW)	470 MW (2 × 235 MW)	2 × 235 MW (under construction) 40 bar at 250°C	4 × 235 MW (proposed)
3. Steam pressure and temperature	35 bar at 240°C	40 bar at 250°C	—	—	—

HIGHLIGHTS

1. Those pairs of atoms which have the same atomic number and hence similar chemical properties but different atomic mass number are called *isotopes*.
2. Those atoms which have the same mass number but different atomic numbers are called *isobars*. Obviously, these atoms belong to different chemical elements.
3. Those pairs of atoms (nuclides) which have the same atomic number and atomic mass number but have different *radioactive properties* are called *isomers* and their existence is referred to as nuclear isomerism.
4. Those atoms whose nuclei have the same number of neutrons are called *isotones*.
5. The phenomenon of spontaneous emission of powerful radiations exhibited by heavy elements is called *radioactivity*. The radioactivity may be natural or artificial.

6. The five types of nuclear radiations are :
(*i*) Gamma rays (or photons) : electromagnetic radiation.
(*ii*) Neutrons : uncharged particles, mass approximately 1.
(*iii*) Protons : + 1 charged particles, mass approximately 1.
(*iv*) Alpha particles : helium nuclei, charge + 2, mass 4.
(*v*) Beta particles : electrons (charge – 1), positrons (charge + 1), mass very small.

7. *Half life* represents the rate of decay of the radioactive isotopes. The half life is the time required for half of the parent nuclei to decay or to disintegrate.

8. *Nuclear cross sections* (or attenuation co-efficients) are measures of the probability that a given reaction will take place between a nucleus or nuclei and incident *radiation.*

9. It has been found that some materials are not fissionable by themselves but they can be converted to the fissionable materials, these are known as *fertile materials.*

10. *Fission* is the process that occurs when a neutron collides with the nucleus of certain of the heavy atoms, causing the original nucleus to split into two or more unequal fragments which carry-off most of the energy of fission as kinetic energy. This process is accompanied by the emission of neutrons and gamma rays.

11. A *chain reaction* is that process in which the number of neutrons keeps on multiplying rapidly (in geometrical progression) during fission till whole the fissionable material is disintegrated. The multiplication or reproduction factor (K) is given by :

$$K = \frac{\text{No. of neutrons in any particular generation}}{\text{No. of neutrons in the preceding generation}}$$

If $K > 1$, chain reaction will continue and if $K < 1$, chain reaction cannot be maintained.

12. *Nuclear fusion* is the process of combining or fusing two lighter nuclei into a stable and heavier nuclide. In this case large amount of energy is released because mass of the product nucleus is less than the masses of the two nuclei which are fused.

13. A *nuclear reactor* is an apparatus in which nuclear fission is produced is the form of a controlled self-sustaining chain reaction.

14. Essential components of a nuclear reactor are :
(*i*) Reactor core (*ii*) Reflector
(*iii*) Control mechanism (*iv*) Moderator
(*v*) Coolants (*vi*) Measuring instruments
(*vii*) Shielding.

15. The main components of a nuclear power plant are :
(*i*) Nuclear reactor (*ii*) Heat exchanger (steam generator)
(*iii*) Steam turbine (*iv*) Condenser
(*v*) Electric generator.

16. Some important reactors are :
(*i*) Pressurised water reactor (PWR) (*ii*) Boiling water reactor (BWR)
(*iii*) Gas-cooled reactor (*iv*) Liquid metal-cooled reactor
(*v*) Breeder reactor.

17. Following factors should be considered while selecting the site for a nuclear power plant :
(*i*) Proximity to load centre (*ii*) Population distribution
(*iii*) Land use (*iv*) Meteorology
(*v*) Geology (*vi*) Seismology
(*vii*) Hydrology.

18. Typically, all costs of nuclear power plants are broken down into the following categories :
(*i*) Capital costs (total) (*ii*) Fuel costs (per year)
(*iii*) Other operating and maintenance cost (per year).

THEORETICAL QUESTIONS

1. Explain the following terms :
   (*i*) Atomic model (*ii*) Atomic mass unit
   (*iii*) Isotopes (*iv*) Isobars
   (*v*) Isomers (*vi*) Isotones.
2. What do you mean by the term 'Radioactivity' ?
3. What is the difference between 'Artificial radioactivity' and 'Natural radioactivity' ?
4. Name five types of radiation of interest, in nuclear power technology.
5. Explain briefly the following :
   (*i*) Prompt-fission gamma rays (*ii*) Fission-product-decay gamma rays
   (*iii*) Capture gamma rays (*iv*) Activation gamma rays
   (*v*) Inelastic scattering gamma rays.
6. Explain briefly the following types of neutrons :
   (*i*) Prompt-fission neutrons (*ii*) Delayed neutrons
   (*iii*) Photoneutrons (*iv*) Activation neutrons
   (*v*) Reaction neutrons.
7. What do you mean by 'Binding Energy' ? What are the total binding energy and binding energy per nucleon for the ${}_{6}C^{12}$ nucleus ?
8. Explain briefly the following terms relating radioactive decay :
   (*i*) Activity (*ii*) Half life
   (*iii*) Average (mean) life.
9. What do you mean by the following :
   (*i*) Elastic scattering (*ii*) Inelastic scattering
   (*iii*) Capture (*iv*) Fission.
10. Write a short note on 'Fertile materials'.
11. What do you mean by 'Fission of nuclear fuel' ?
12. What is a chain reaction ?
13. What are the requirement of fission process ?
14. How are the following defined ?
    (*i*) Critical mass (*ii*) Critical size.
15. What is 'nuclear fusion' ? How does it differ from '*nuclear fission*' ?
16. What is a nuclear reactor ?
17. How are nuclear reactors classified ?
18. Enumerate and explain essential components of a nuclear reactor.
19. Explain with help of neat diagram the construction and working of a nuclear power plant.
20. What is a moderator ? Name common moderators and discuss their advantages and limitations.
21. Give the functions and materials for the following :
    (*i*) Reflector (*ii*) Control rods
    (*iii*) Biological shield.
22. Describe with the help of a neat sketch the construction working of a Pressurised Water Reactor (PWR). What are its advantages and disadvantages ?
23. What is 'Boiling Water Reactor' (BWR) ? How does it differ from 'Pressurised Water Reactor' (PWR) ?
24. Give the construction and working of a 'Gas-cooled reactor'. What are its advantages and disadvantages ?

25. What is a 'Liquid metal-cooled reactor' ? Explain briefly a typical liquid metal reactor.
26. Describe a breeder reactor. What are its advantages and disadvantages ?
27. What factors must be considered while selecting materials for the various reactor components ?
28. List the advantages and disadvantages/limitations of nuclear power plants.
29. Discuss the various factors which must be considered while selecting a site for a nuclear power plant.
30. Give the application of nuclear power plants.
31. What do you mean by 'Economics of nuclear power plants' ?
32. List down some safety measures for nuclear power plants.
33. What is the future of nuclear power ?

7

Non-Conventional Power Generation and Direct Energy Conversion

7.1. Introduction. 7.2. Wind power plants—Introduction—Wind availability and measurement—Types of wind mills—Wind-electric generating power plant—Types of wind machines—Wind-powered battery chargers—Wind electricity in small independent grids—Wind electricity economics—Problems in operating large wind power generators. 7.3. Tidal power plants—Introduction—Components of tidal power plants—Classification and operation of tidal power plants—Advantages and limitations of tidal power generation. 7.4. Solar power plants—Solar energy—General terms and introduction—Collectors in various ranges and applications—Flat plate collectors—Focusing (or concentrating) collectors—Solar pond technology—Low temperature thermal power generation—Medium temperature systems using focusing collectors—High temperature systems—solar farm and solar tower plant. 7.5. Geothermal power plants—Geothermal energy—Geothermal sources—Geothermal power estimate—Environmental problems—Applications of geothermal energy—Advantages and disadvantages of geothermal energy over other energy forms. 7.6. Direct energy conversion systems—Thermoelectric conversion system—Thermionic conversion system—Photovoltaic power system—Magneto-hydrodynamics (MHD) system—Electrostatic mechanical generators—Electro gas-dynamic generators (EGD)—Fuel cells—Nuclear Batteries. Highlights—Theoretical Questions.

7.1. INTRODUCTION

A plenty of energy is needed to sustain industrial growth and agricultural production. The existing sources of energy such as coal, oil, uranium etc. may not be adequate to meet the ever increasing energy demands. These conventional sources of energy are also depleting and may be exhausted at the end of the century or beginning of the next century. Consequently sincere and untiring efforts shall have to be made by the scientists and engineers in exploring the possibilities of harnessing energy from several non-conventional energy sources. The various non-conventional energy sources are as follows :

(*i*) Solar energy
(*ii*) Wind energy
(*iii*) Energy from biomass and biogas
(*iv*) Ocean thermal energy conversion
(*v*) Tidal energy
(*vi*) Geothermal energy
(*vii*) Hydrogen energy
(*viii*) Fuel cells
(*ix*) Magneto-hydrodynamics generator
(*x*) Thermionic converter
(*xi*) Thermo-electric power.

Advantages of non-conventional energy sources :

The *leading advantages* of non-conventional energy sources are :

1. They do not pollute the atmosphere.
2. They are available in large quantities.
3. They are well suited for decentralised use.

According to energy experts the non-conventional energy sources can be used with advantage for *power generation* as well as other applications in a large number of locations and situations in our country.

7.2. WIND POWER PLANTS

7.2.1. Introduction

The electrical energy can be generated by wind energy. The wind energy, which is an indirect source of energy, can be used to run a wind will which in turn drives a generator to produce electricity. Although wind mills have been used for more than a dozen centuries for grinding grain and pumping water, interest in large scale power generation has developed over the past 50 years. A largest wind generator built in the past was 800 kW unit operated in France from 1958-60. The flexible 3 blades propeller was about 35 m in diameter and produced the rated power in a 60 km/hour wind with a rotation speed of 47 r.p.m. The maximum power developed was 12 MW. In India the interest in the wind mills was shown in the last fifties and early sixties. Apart from importing a few from outside, new designs were also developed, but they did not sustain. It is only in last 10—15 years that development work is going on in many institutions. An important reason for this lack of interest in wind energy must be that wind, in India is relatively low and vary appreciable with seasons. These low and seasonal winds imply a high cost of exploitation of wind energy. In our country high wind speeds are however available in coastal areas of Sourashtra, Western Rajsthan and some parts of central India. In these areas there could be a possibility of using medium and large sized wind mills for generation of electricity.

Characteristics of wind energy :

1. Wind-power systems do not pollute the atmosphere.
2. Fuel provision and transport are not required in wind-power systems.
3. Wind energy is a renewable source of energy.
4. Wind energy when produced on small scale is cheaper, but competitive with conventional power generating systems when produced on a large scale.

Wind energy entails following *shortcomings/problems* :

1. It is fluctuating in nature.
2. Due to its irregularity it needs storage devices.
3. Wind power generating systems produce ample *noise*.

7.2.2. Wind Availability and Measurement

Wind energy can only be economical in areas of good wind availability. Wind energy differs with region and season and also, possibly to an even greater degree with local terrain and vegetation. Although wind speeds generally increase with height, varying speeds are found over different kinds of terrain. Observations of wind speed are carried out at meteorological stations, airports and lighthouses and are recorded regularly with ten minute mean values being taken every three hours at a height of 10 m. But airports, sometimes are in valleys and many wind speed meters are situated low and combinations of various, other factors mean that reading can be misleading. It is difficult, therefore, to determine the real wind speed of a certain place without actual in-situ measurements.

The World Meteorological Organization (WMO) has accepted four methods of wind recording :

(*i*) Human observation and log book.

(*ii*) Mechanical cup-counter anemometers.

(*iii*) Data logger.

(*iv*) Continuous record of velocity and direction.

1. **Human observation and log book.** This involves using the Beaufort Scale of wind strengths which defines visible "symptoms" attributable to different wind speeds. The method is *cheap* and *easily implemented* but is *often unreliable*. The best that can be said of such records is that they are better than nothing.

2. **Mechanical cup-counter anemometers.** The majority of meteorological stations use mechanical cup-counter anemometers. By taking the readings twice or three times a day, it is possible to estimate the mean wind speed. This is a *low cost method*, but is *only relatively reliable*. The instrument has to be in good working order, it has to be correctly sited and should be reliably read at least daily.

3. **Data logger.** The equipment summarizes velocity frequency and direction. It is *more expensive and prone to technical failures but gives accurate data*. The method is tailored to the production of readily interpretable data of relevance to wind energy assessment. It does not keep a time series record but *presents the data in processed form*.

4. **Continuous record of velocity and direction.** This is how data is recorded at major airports of permanently manned meteorological stations. The *equipment is expensive and technically complex*, but it retains a detailed times-series record (second-by-second) of wind direction and wind speed. Results are given in copious quantities of data which require lengthy and expensive analysis.

7.2.3. Types of Wind Mills

The various types of wind mills which are practically useful are shown in Fig. 7.1.

1. **Multiple blade type.** It is the most widely used wind mill. It has 15 to 20 blades made from metal sheets. The *sail type* has three blades made by stitching out triangular pieces of convass cloth. Both these types run at low speeds of 60 to 80 r.p.m.

2. **Savonius type.** This type of windmill has hollow circular cylinder sliced in half and the halves are mounted on vertical shaft with a gap in between. Torque is produced by the pressure difference between the two sides of the half facing the wind. This is quite efficient but needs a large surface area.

3. **Darrieus type.** This wind mill needs much less surface area. It is shaped like an egg beater and has two or three blades shaped like aerofoils.

It may be noted that :

— Both the Savonius and Darrieus types are mounted on a *vertical axis* and hence they *can run independently of the direction of wind.*

— The horizontal axis mills *have to face the direction of the wind* in order to generate power.

Performance of wind mills :

The performance of a wind mill is defined as *'Co-efficient of performance'* (K_p).

$$K_p = \frac{\text{Power delivered by the rotor}}{\text{Maximum power available in the wind}}$$

or,

$$K_p = \frac{P}{P_{max}} = \frac{P}{\frac{1}{2}\rho A U_w^{\,3}}$$

where, ρ = Density of air,

A = Swept area, and

U_w = Velocity of wind.

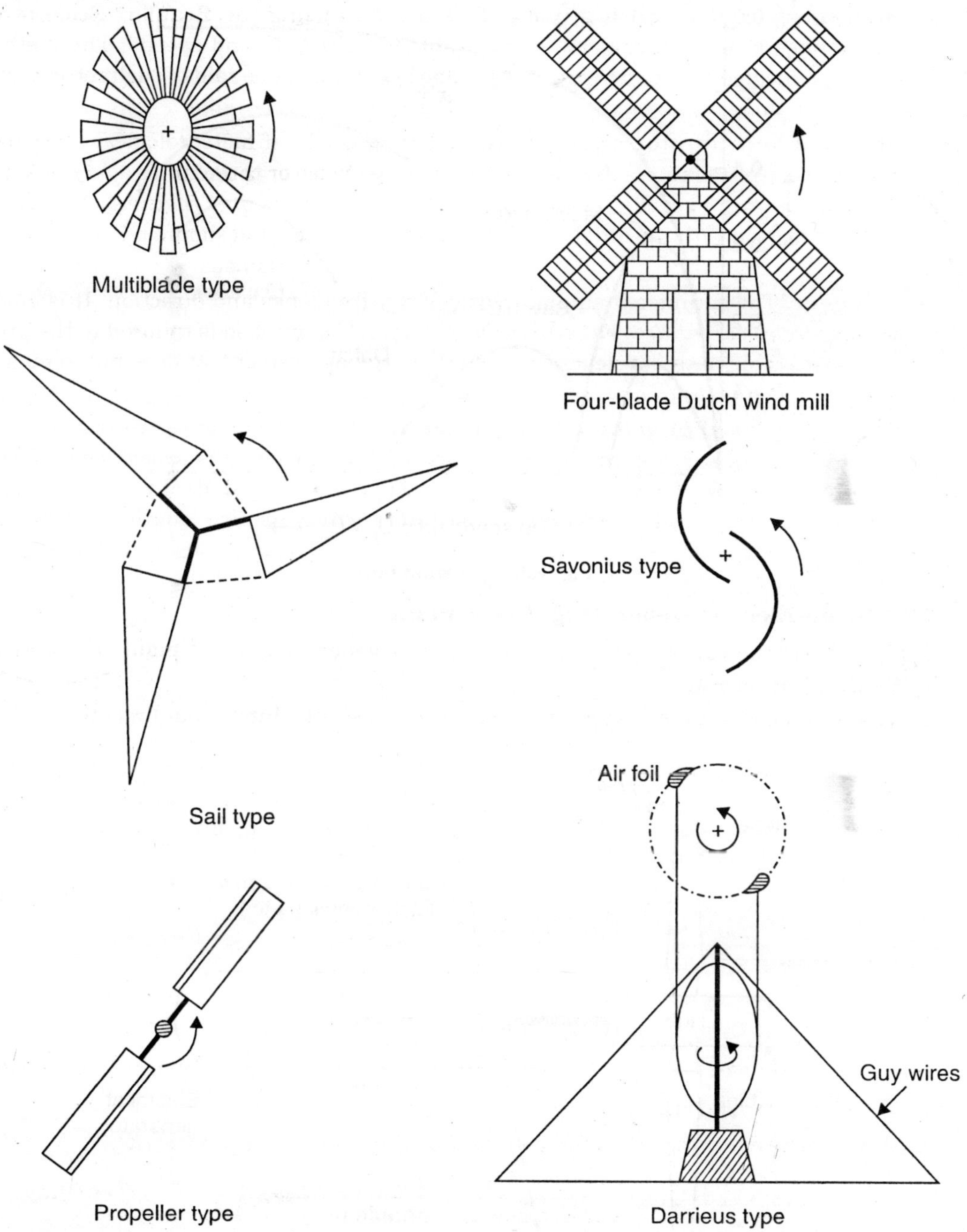

Fig. 7.1. Types of wind mills.

Fig. 7.2 shows a plot between K_p and tip speed ratio U_{bt}/U_w where U_{bt} = Speed of blade tip.

It can be seen that K_p is the *lowest of Savonius* and *Dutch types* whereas the *propeller types have the highest value*.

In the designing of wind mills, it is upper most to keep the power to weight ratio at the lowest possible level.

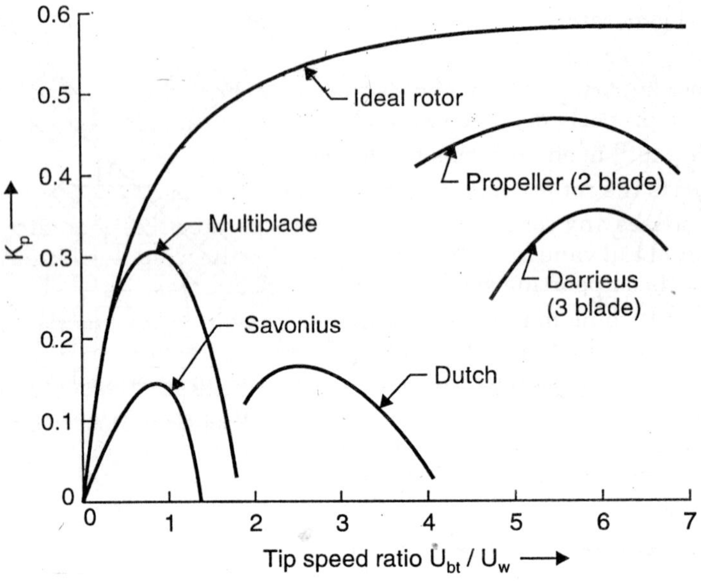

Fig. 7.2. K_p of wind mills.

7.2.4. Wind-Electric Generating Power Plant

Fig. 7.3 shows the various parts of a wind-electric generating power plant. These are :

1. Wind turbine or rotor.
2. Wind mill head—it houses speed increaser, drive shaft, clutch, coupling etc.

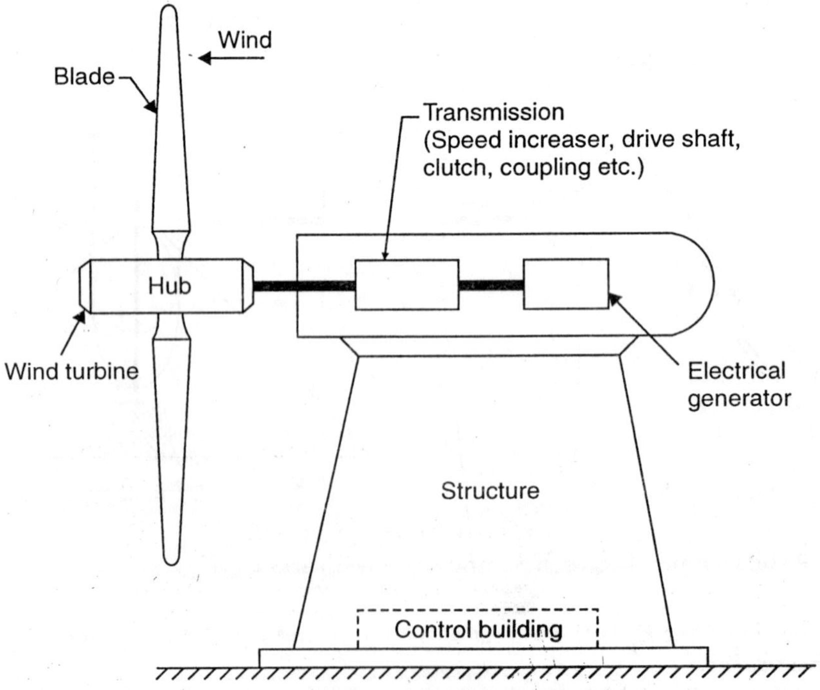

Fig. 7.3. Wind-electric generating power plant.

3. Electrical generator.
4. Supporting structure.

— The most important component is the **rotor.** For an effective utilisation, all components should be properly designed and matched with the rest of the components.

— The wind mill head performs the following *functions* :

(*i*) It supports the rotor housing and the rotor bearings.

(*ii*) It also houses any control mechanism incorporated like changing the pitch of the blades for safety devices and tail vane to orient the rotor to face the wind, the latter is facilitated by mounting it on the top of the supporting structure on suitable bearings.

— The wind turbine may be located either *unwind* or *downwind* of the power. In the unwind location the wind encounters the turbine before reaching the tower. *Downwind rotors are generally preferred especially for the large aerogenerators.*

— The **supporting structure** is designed to withstand the wind load during gusts. Its type and height is related to cost and transmission system incorporated. Horizontal axis wind turbines are mounted on towers so as to be above the level of turbulence and other ground related effects.

7.2.5. Types of Wind Machines

Wind machines (aerogenerators) are generally *classified* as follows :

1. Horizontal axis wind machines.
2. Vertical axis wind machines.

Horizontal axis wind machines. Fig 7.4 shows a schematic arrangement of a horizontal axis machine. Although the common wind turbine with a horizontal axis is simple in principle,

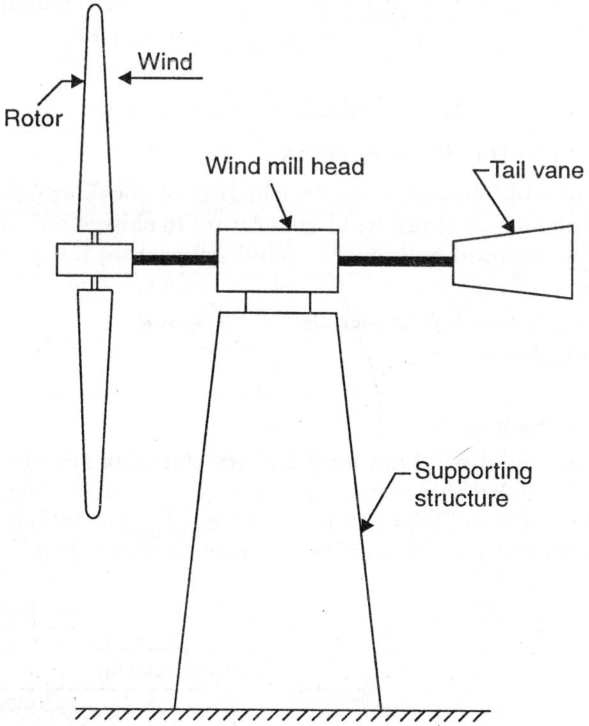

Fig. 7.4. Horizontal axis wind machine.

yet the design of a complete system, especially a large one that would produce electric power economically, is complex. It is of paramount importance that the components like rotor, transmission, generator and tower should not only be as efficient as possible but they must also function effectively in combination.

Vertical axis wind machines. Fig. 7.5 shows vertical axis type wind machine. One of the main advantages of vertical axis rotors is that they do not have to be turned into the windstream as the wind direction changes. Because their operation is independent of wind direction, vertical axis machines are called *panemones*.

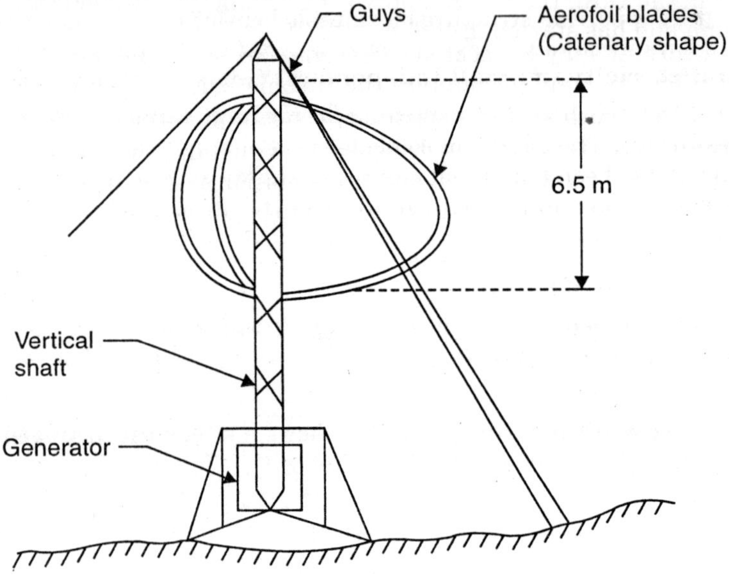

Fig. 7.5. Vertical axis wind machine.

7.2.6. Wind-Powered Battery Chargers

One application of wind energy systems which is of considerable potential importance (to developing countries) is the use of small wind generators to charge batteries for powering lighting, radio communication and hospital equipment. Wind generators have been in use in Europe and North America since the 1920s, although their use declined considerably.

A battery charging system has to include the following :

(*i*) A wind powered generator.

(*ii*) A converter.

(*iii*) A container for the batteries.

Fig. 7.6 shows a set-up of wind powered battery charging system. It is worthnoting that 12 volt batteries, which are rechargeable using wind generators, can be used to power fluorescent tube lighting which is six times more efficient than tungsten filament lamps. Such lighting opens up a number of important development opportunities in areas which normally have no lighting.

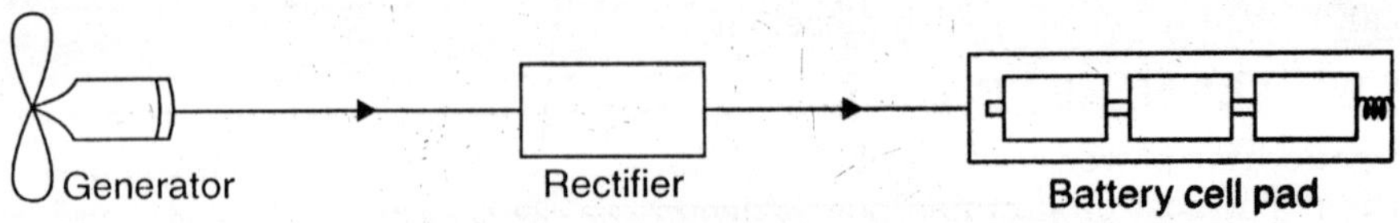

Fig. 7.6

For small wind generators the total system efficiency is made up as follows :

Wind regime matching efficiency	60% (approx.)
Rotor efficiency	35% (approx.)
Generator and wiring efficiency	70% (approx.)
Battery charge/discharge efficiency	70% (approx.)

Cumulatively, a total energy capture efficiency of about 10% is generally obtained from small wind generators utilized for battery charging.

Battery charging wind generators are produced in several countries, notably Australia, France, Sweden, Switzerland, the U.K., the U.S.A. and West Germany. In developing countries production is underway in China and has started in India.

7.2.7. Wind Electricity in Small Independent Grids

Refer to Fig. 7.7. In such systems electricity consumption fluctuates constantly as does the availability of wind energy. The degree of coincidence of supply and demand can be calculated by statistical means and it has been found that electricity supply with an acceptable degrees of reliability cannot be based solely on wind energy. If an extensive grid does not exist, electricity storage (batteries) or a back-up system (diesel) is required. Loads for remote systems of upto 6 kWh/day equivalent to an average power consumption of 250 W with a duty cycle of 24 hours, can be provided with battery storage.

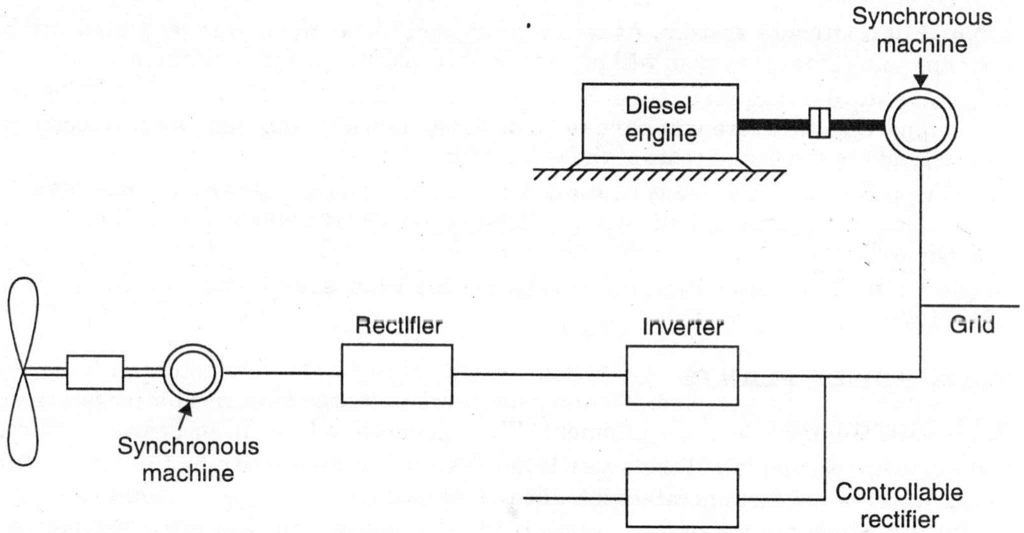

Fig. 7.7. Principle of combined wind/diesel power generation.

If a diesel and wind generator are used in conjunction with a grid, the *diesel generator should only be used when wind energy is absent*. Problems can occur, however, when the diesel generator is called on to change its output frequently as wind energy availability fluctuates. Besides decreasing the oil saving, diesel generation on this basis leads to more frequent overhauls of the generator. Both factors will increase costs. Several methods of overcoming these problems have been tried but there is not yet an established solution. Some development work has still to be done before wind generators can be run in parallel with diesel on a routine basis.

7.2.8. Wind Electricity Economics

Wind generator power costs are heavily linked to the characteristics of a wind resource in a specific location. The cost of supplied power declines as wind speeds increase, and the power supplied increases in proportion to the *cube of the wind speed.*

Matching available energy and load requirements is also important in wind energy economics. The correct size of wind generator must be chosen together with some kind of storage or co-generation with an engine or a grid to obtain the best economy. The ideal application is a task that can utilize a variable power supply, *e.g.,* ice making or water purification.

Regarding the *economics*, the choice of interest rate obviously has a major effect on the overall energy cost. With low interest rates, capital intensive power sources such as solar and wind are favoured. Other factors bearing a strong influence on the economics of wind electricity are the standard of maintenance and service facilities and the cost of alternative energy supplies in the particular area.

7.2.9. Problems in Operating Large Wind Power Generators

The operation of large wind power generators entails the following *problems* :

1. **Location of site.** The most important factor is locating a site big enough which has a reasonable average high wind velocity.

Sourastra and Coastal Regions in India are promising areas.

2. **Constant angular velocity.** A constant angular velocity is a must for generating A.C. (alternating current) power and this means very sensitive governing.

3. **Variation in wind velocity.** The wind velocity varies with time and varies in direction and also varies from the bottom to top of a large rotor (some rotors are as long as 50 meters). This causes fatigue in blades.

4. **Need of a storage system.** At zero velocity conditions, the power generated will be zero and this means some storage system will have to be incorporated along with the wind mill.

5. **Strong supporting structure.** Since the wind mill generator will have to be located at a height, the supporting structure will have to be designed to withstand high wind velocity and impacts. This will add to the initial costs of the wind mill.

6. **Occupation of large areas of land.** Large areas of land will become unavailable due to wind mill gardens (places where many wind mills are located). The whole area will have to be protected to avoid accidents.

Inspite of all these difficulties, interest to develop wind mills is there since this is a clean source of energy.

7.3. TIDAL POWER PLANTS

7.3.1. Introduction

The periodic rise and fall of the water level of sea which are carried by the action of the sun and moon on water of the earth is called the '**tide**'. Tidal energy can furnish a significant portion of all such energies which are renewable in nature. The large scale up and down movement of sea water represents an unlimited source of energy. If some part of this vast energy can be converted into electrical energy, it would be an important source of hydropower.

The *main feature of the tidal cycle is the difference in water surface elevations at the high tide and at the low tide.* If this differential head could be utilized in operating a hydraulic turbine, the tidal energy could be converted into electrical energy by means of an attached generator.

7.3.2. Components of Tidal Power Plants

The following are the *components* of a tidal power plant :

1. The dam or dyke (low wall) to form the pool or basin.
2. Sluice ways from the basins to the sea and *vice versa.*
3. The power house.

Dam or dyke. The function of dam or dyke is to form a barrier between the sea and the basin or between one basin and the other in case of multiple basins.

Sluice ways. These are used to fill the basin during the high tide or empty the basin during the low tide, as per operational requirement. These devices are controlled through gates.

Power house. A power house houses turbines, electric generators and other auxiliary equipment. As far as possible the power house and sluice ways should be in *alignment* with the dam or dyke.

7.3.3. Classification and Operation of Tidal Power Plants

Tidal power plants are *classified* as follows :

1. **Single basin arrangement**

(*i*) Single ebb-cycle system (*ii*) Single tide-cycle system

(*iii*) Double cycle system.

2. **Double basin arrangement**

In a *single basin arrangement* power can be generated only *intermittently*. In this arrangement only one basin interacts with the sea. The two are separated by a dam or dyke and the flow between them is through sluice ways located conveniently along the dam. The rise and fall of tidal water levels provide the potential head.

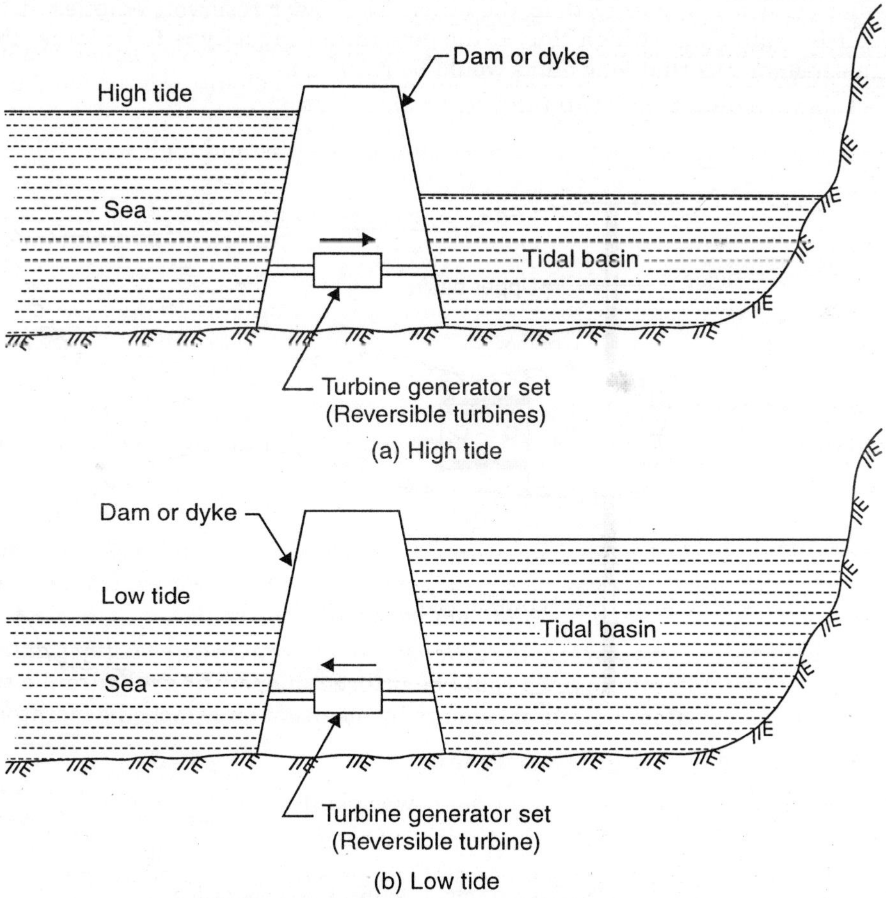

Fig. 7.8. General arrangement of tidal power plant.

Fig. 7.8 shows a general arrangement of single basin tidal power plant (double cycle system). Such plants generally use *reversible water turbines* so that power is generated on low tide as well high tide. The operation of the plant is as follows :

When the incoming tide sea level and tidal-basin level are equal, the turbine conduit is closed. When the sea level rises, and about half way to high tide the turbine valves are opened and the sea water flows into the basin through the turbine runner generating power. This also raises the level of water in the basin. The turbine continues to generate power until the tide passes through its high point and begins to drop. The water head then quickly diminishes till it is not enough to supply the no-load losses. By pass valve then quickly opens to let water into the basin to gain maximum water level. When sea and basin water level are again equal, the valves are closed as well as the turbine conduit. The basin level then stays constant while the tide continues to go out. After sufficient head has developed, the turbine valves are again opened and water now flows from basin to the sea, thereby generating power. The plant continues to generate power till the tide reaches its lowest level.

A *single basin plant cannot generate power continuously, though it might do so by using a pumped storage plant if the load it supplies fluctuates considerably*. A *double basin scheme* can provide power *continuously* or on demand, which is a *great advantage*. The drawback is that the civil works become more extensive. In the simplest *double-basin scheme* there must be a dam between each basin and the sea and also a dam between the basins, containing the power house. One basin is maintained always at a lower level than the other. The lower reservoir empties at low tide, the upper reservoir is replinshed at high tide. If the generating capacity is to be large, the reservoirs must be large which means that long dams would be required.

Fig. 7.9 shows a *tidal power plant-double basin operation.*

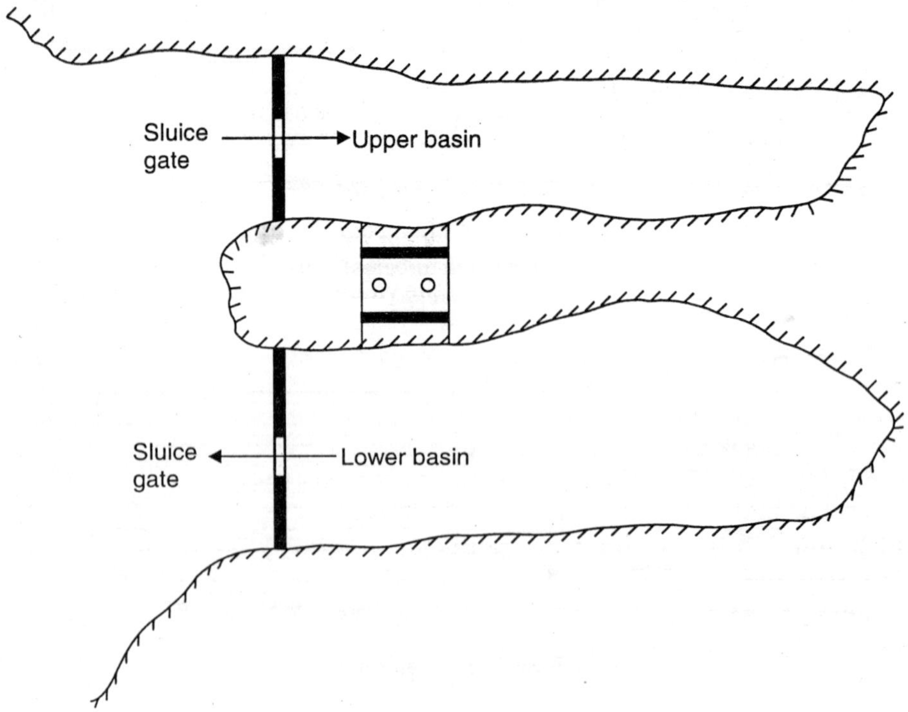

Fig. 7.9. Tidal power plant-double basin operation.

7.3.4. Advantages and Limitations of Tidal Power Generation

Advantages :

1. Tidal power is completely independent of the precipitation (rain) and its uncertainty, besides being inexhaustible.
2. Large area of valuable land is not required.
3. When a tidal power plant works in combination with thermal or hydro-electric system peak power demand can be effectively met with.
4. Tidal power generation is free from pollution.

Limitations :

1. Due to variation in tidal range the output is not uniform.
2. Since the turbines have to work on a wide range of head variation (due to variable tidal range) the plant efficiency is affected.
3. There is a fear of machinery being corroded due to corrosive sea water.
4. It is difficult to carry out construction in sea.
5. As compared to other sources of energy, the tidal power plant is costly.
6. Sedimentation and silteration of basins are the problems associated with tidal power plants.
7. The power transmission cost is high because the tidal power plants are located away from load centres.

— The first commercial tidal power station in the World was constructed in France in 1965 across the mouth of La Rance Estuary. It has a high capacity of 240 MW. The average tidal range at La Rance is 8.4 m and the dam built across the estuary encloses an area of 22 km^2.

7.4. SOLAR POWER PLANTS

7.4.1. Solar Energy—General Terms and Introduction

General Terms :

Solar constant. Solar constant is the energy from the sun, per unit time, received on a unit area of surface perpendicular to the radiation, in space, at the earth's mean distance from the sun. According to Thekaekara and Drummond (1971) the value of the solar constant is 1353 W/m^2 (1.940 Cal/cm^2 min, or 4871 kJ/m^2 hr.).

Beam radiation. The solar radiation received from the sun without change of direction is called beam radiation.

Diffuse radiation. It is the solar radiation received from the sun after its direction has been changed by reflection and scattering by the atmosphere.

Air mass. It is the path length of radiation through the atmosphere, considering the vertical path at sea level as unity.

Zenith angle. It is the angle between the beam from the sun and the vertical.

Solar altitude. It is the angle between the beam from the sun and horizontal *i.e.*, (90 – zenith angle).

Solar or Short-wave radiation. It is the radiation originating from the sun, at a source temperature of about 6000°K and in the wavelength range of 0.3 to 3.0 μm.

Long-wave radiation. Radiation originating from sources at temperatures near ordinary ambient temperatures and thus substantially all at wavelength greater than 3 μm.

Declination. It is the angular position of the sun at solar noon with respect to the plane of the equator (north positive).

Solar Energy—Introduction

The surface of the earth receives from the sun about 10^{14} kW of solar energy which is approximately five order of magnitude greater than currently being consumed from all resources. It is evident that sun will last for 10^{11} years. Even though the sun light is filtered by the atmosphere one square metre of the land exposed to direct sun light receives the energy equivalent of about 1 H.P or 1 kW. However, this vast amount of solar energy reaching earth is not easily convertible and certainly is not "free".

There are two obvious obstacles to harnessing solar energy. *Firstly* it is not constantly available on earth. Thus some form of storage is needed to sustain a solar power system through the night and during periods when local weather conditions obscure the sun. *Second* the solar energy is diffused. Although the total amount of energy is enormous, the collection and conservation of solar energy into useful forms must be carried out over a large area which entails a large capital investment for the conversion apparatus.

Solar energy, therefore, most likely will be developed not because it is cheaper than alternative energy sources but because these alternative sources sooner or later (*i*) will be exhausted, (*ii*) will become increasingly more expensive, (*iii*) will continue to political and economical control by the nations possessing them and (*iv*) will produce undesirable yet incompletely understood environmental consequences, especially on large scale that will be required to meet projected demands even with controlled growth.

Solar energy has some good advantages in comparison to the other sources of power. Solar radiation does not contaminate environment or endanger ecological balance. It avoids major problems like exploration, extraction and transportation.

7.4.2. Collectors in Various Ranges and Applications

Following list gives the thermal applications of solar energy and possible temperature ranges :

1. **Long temperature**
   (t = 100°C)
   (*i*) Water heating
   (*ii*) Space heating
   (*iii*) Space cooling
   (*iv*) Drying
   } ... Flat plate

2. **Medium temperature**
   (100 to 200°C)
   (*i*) Vapour engines and turbines
   (*ii*) Process heating
   (*iii*) Refrigeration
   (*iv*) Cooking
   } ... Cylindrical parabola

3. **High temperature**
   (> 200°C)
   (*i*) Steam engines and turbines
   (*ii*) Stirling engine
   (*iii*) Thermo-electric generator
   } ... Parabolloid mirror arrays

The above classification of low, medium and high temperature ranges is some what arbitrary. Heating water for domestic applications, space heating and cooling and drying of agricultural products (and industrial products) is generally at temperature below 100°C, achieved using flat plate collectors with one or two glass plate covers. Refrigeration for preservation of food products, heating for certain industrial processes, and operation of engines and turbines using low boiling organic vapours is possible at some what higher temperature of 100 to 200°C and may be achieved using focusing collectors with cylindrical-parabola reflectors requiring only one directional diurnal tracking. Conventional steam engines and turbines, stirling hot air engines, and thermoelectric generators require the solar collectors to operate at high temperatures. Solar collectors operating at temperature above 200°C generally consist of parabolloid reflector as an array of mirrors reflecting to a central target, and requiring two directional diurnal tracking.

The concentrators or focusing type collectors can give high temperatures than flat plate collectors, but they entail the following shortcomings/limitations.

1. Non-availability and high cost of materials required. These materials must be easily shapeable, yet have a long life ; they must be light weight and capable of retaining their brightness in tropical weather. Anodised aluminium and stainless steel are two such materials but they are expensive and not readily available in sufficient quantities.

2. They require direct light and are not operative when the sun is even partly covered with clouds.

3. They need tracking systems and reflecting surfaces undergo deterioration with the passage of time.

4. These devices are also subject to similar vibration and movement problems as radar antenna dishes.

Comparison between Flat plate and Focusing collectors :

1. The absorber area of a concentrator system is smaller than that of a flat-plate system of the same solar energy collection area and the *insolation intensity is therefore greater.*

2. Because the area from which heat is lost to the surroundings per unit of the solar energy collecting area is less than that for a flat plate collector and because the insolation on the absorber is more concentrated, the working fluid can attain higher temperatures in a concentrating system than in a flat-plate collector of the same solar energy collecting surface.

3. Owing to the small area of absorber per unit of solar energy collecting area, selective surface treatment and/or vacuum insulation to reduce heat losses and *improve collector efficiency are economically feasible.*

4. Since higher temperatures can be achieved, the focusing collector *can be used for power generation.*

5. *Little or no anti-freeze is required* to protect the absorber in a concentrator system whereas the entire solar energy collection surface requires anti-freeze protection in a flat-plate collector.

6. Out of the beam and diffuse solar radiation components, only beam component is collected in case of focusing collectors because diffuse component cannot be reflected and is thus lost.

7. *Costly orienting systems* have to be used to track the sun.

8. Non-uniform flux on the absorber whereas flux in flat-plate collectors is uniform.

7.4.3. Flat Plate Collectors

7.4.3.1. Description

Fig. 7.10 shows a Flat plate collector which consists of four essential *components :*

1. **An absorber plate.** It intercepts and absorbs solar radiation. This plate is usually metallic (Copper, aluminium or steel), although plastics have been used in some low temperature applications

In most cases it is coated with a material to enhance the absorption of solar radiation. The coating may also be tailored to minimise the amount of infrared radiation emitted.

A heat transport fluid (usually air or water) is used to extract the energy collected and passes over, under or through passages which form an integral part of the plate.

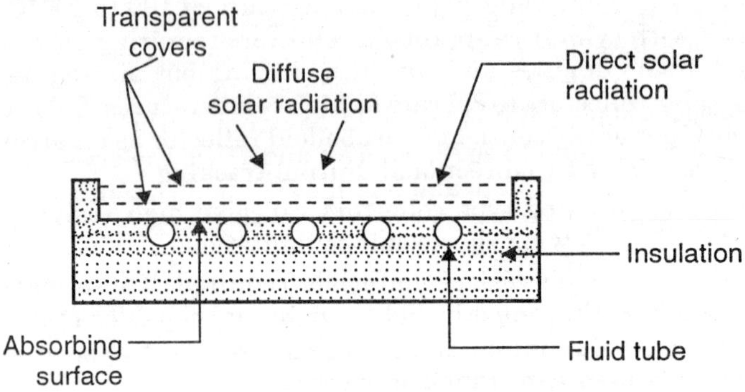

Fig. 7.10. Flat plate solar collector.

2. **Transparent covers.** These are one or more sheets of solar radiation transmitting materials and are placed above the absorber plate. They allow solar energy to reach the absorber plate while reducing convection, conduction and re-radiation heat losses.

3. **Insulation beneath the absorber plate.** It minimises and protects the absorbing surface from heat losses.

4. **Box like structure.** It contains the above components and keeps them in position.

Various types of flat plate collectors have been designed and studied. These include tube in plate, corrugated type, spiral wound type etc. Other criteria is single exposure, double exposure or exposure and reflector type. The collector utilizes sheets of any of the highly conducting material *viz.*, copper, aluminium, or galvanized iron. The sheets are painted dead black for increasing the absorbtivity. The sheets are provided with one or more glass or plastic covers with air gap in between to reduce the heat transfer losses. The sides which are not exposed to solar radiation are well insulated. The whole assembly is fixed in air tight wooden box which is mounted on simple device to give the desired angle of inclination. The dimensions of collectors should be such as to make their handling easy. The collector will absorb the sun energy (direct as well as diffused) and transfer it to the fluid (air, water or oil) flowing within the collector. Basically, a flat plate collector is *effective* most of time, *reliable* for good many years and also *inexpensive.*

Use of flat mirrors in the flat plate collectors improves the output, permitting higher temperatures of operation. Side mirrors are used either at north and south edges or at east and west edges of the collector or a combination of both. The mirrors may be of reversible or non-reversible type.

7.4.3.2. Analysis

Consider an object exposed to sun radiations of intensity P, per unit area at the surface of the body. These radiations, will partly be absorbed by the body, while the remaining will be partly transmitted and rest reflected. If we take the incident radiations equal to unity, then, the absorbed, reflected, and transmitted parts of energy will add up to unity. There parts are called *absorption* co-efficient, *reflection* co-efficient and *transmission* co-efficient and represented by the symbols α, ρ and τ respectively.

Using the above symbols we can write,

$$\alpha + \rho + \tau = 1 \quad ...(i)$$

The absorbed part of the solar radiations, which is equal to α is responsible for increasing the temperature of the body. However, the body also loses energy by conduction, convection and radiation. The equilibrium temperature of the body will be that at which the heat loses from the body are equal to the absorbed radiations. For analysis purposes, if we represent the body by a flat plate and assume that the convection and conduction loses are negligible to begin with, then at equilibrium temperature, the absorbed solar radiations should be equal to the radiation losses from the flat plate. The radiation losses are equal to $\varepsilon\sigma T^4$, where ε and T are the emission co-efficient and absolute temperature respectively of flat plate and σ is the Boltzman's constant.

Therefore, at equilibrium,

$$\alpha P = \varepsilon\sigma T^4 \quad ...(ii)$$

or,

$$\frac{\alpha P}{\varepsilon} = \sigma T^4 \quad ...(iii)$$

From equation (*iii*), it is evident that comparatively higher equilibrium temperature will be obtained where the quantity $\frac{\alpha}{\varepsilon}$ *i.e.*, the ratio of absorption co-efficient to emission co-efficient of the flat plate is more. However, this has been demonstrated by an equation obtained under idealised condition. In the realistic conditions too, its nature will remain the same, but it will get modified by other influencing factors.

The collectors for which ratio is equal to unity are called **'Neutral collectors'** and those for which the ratio is greater than unity are called **'Selective collectors'.**

The amount of energy collected, however, does not depend on $\frac{\alpha}{\varepsilon}$ ratio. It primarily depends on higher value of α. So to obtain higher energy collection, one should use such flat plate where absorption co-efficient is as high as possible.

A flat plate painted black is placed on a well insulated base. If it is exposed to solar radiations where $P = 800$ W/m^2, a typical summer value for a tropical region, we obtain from equation (*iii*) the equilibrium temperature as 70°C. Inspite of the simplifications here, it is a fair estimate of the temperature reached by a black plate left for a time in the tropical sun.

This method can be refined by including the convection losses and the energy gain as a result of absorption of diffused radiations by the flat plate.

If P' is the intensity of the diffused radiations and α' the absorption co-efficient, then equation (*ii*) becomes

$$\alpha P + \alpha' P' = h_c(T - T_a) + \varepsilon\sigma T^4 \quad ...(iv)$$

This is valid, where the base is insulated, hence conduction losses are neglected. Here T_a is the atmospheric temperature and h_c is the convection heat transfer co-efficient.

7.4.4. Focusing (or Concentrating) Collectors

The main types of focusing or concentrating collectors are as follows :

1. Parabolic trough collector
2. Mirror strip collector
3. Fresnel collector
4. Flat plate collector with adjustable mirrors
5. Compound parabolic concentrator.

Fig. 7.11 (*a*) shows the principle of the parabolic trough collector which is often used in focusing collectors. Solar radiation coming from the particular direction is collected over the area of

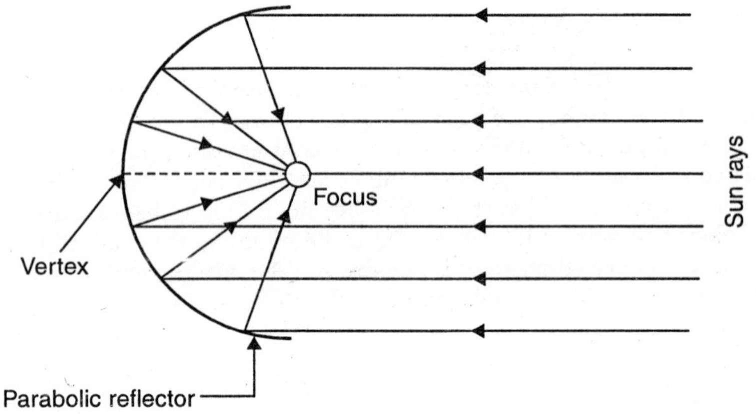

Fig. 7.11. (*a*) Cross-section of parabolic trough collector.

reflecting surface and is concentrated at the focus of the parabola, if the reflector is in the form of a trough with parabolic cross-section, the solar radiation is focused along a line. Mostly *cylindrical parabolic concentrators* are used in which absorber is placed along focus axis [Fig. 7.11 (*b*)].

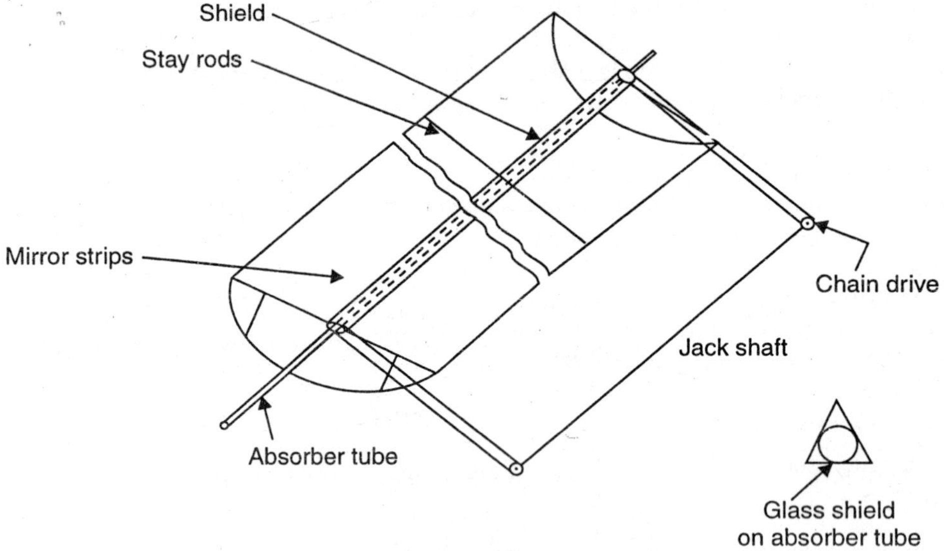

Fig. 7.11. (*b*) Cylindrical parabolic system.

7.4.5. Solar Pond Technology

Refer Fig. 7.12.

The vertical configuration of salt-gradient solar pond normally consists of following *three zones* :

1. Adjacent the surface there is a *homogeneous convective zone* that serves as a buffer zone between environmental fluctuations at the surface and conductive heat transport from the layer below. This is the *Upper Convective Zone* (UCZ).

2. At the bottom of the pond there is another convective zone, the Lower Convective Zone or LCZ. This is the layer with the *highest salt concentration* and where the high temperatures are built up.

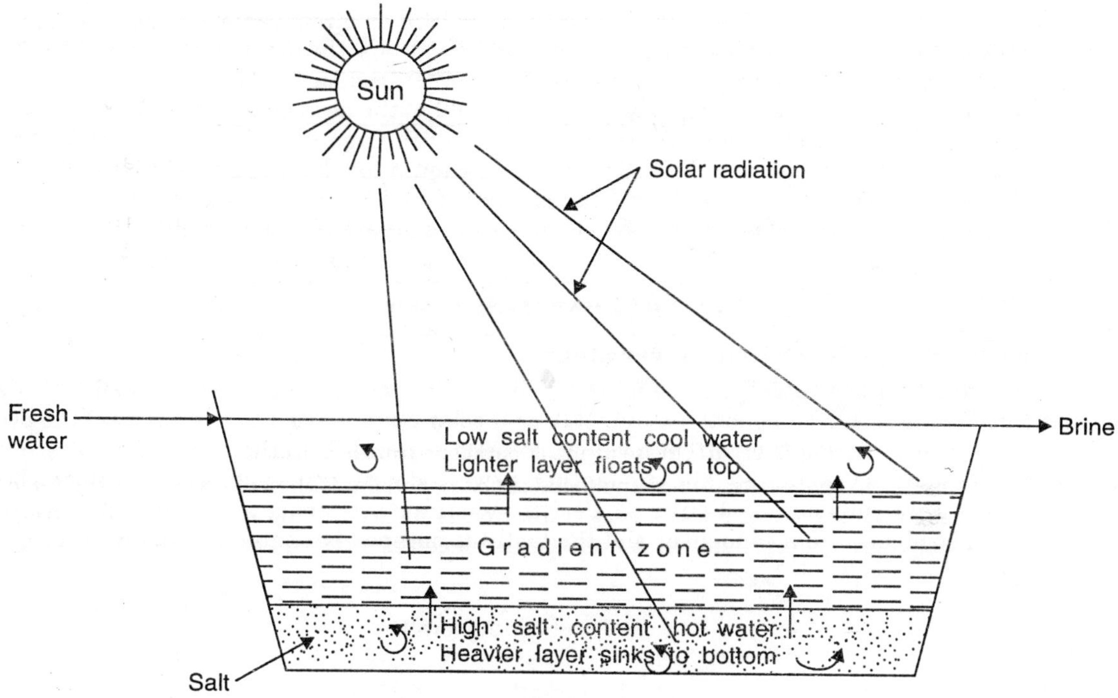

Fig. 7.12. Principle of solar pond.

3. For given salinites and temperatures in the upper and lower convective zones, there exists a stable *intermediate gradient zone*. This zone keeps the two convective zones apart and gives the solar pond its unique thermal performance. This intermediate zone provides excellent insulation for the storage layer, while simultaneously transmitting the solar radiation. To maintain a solar pond in this non-equilibrium stationary state, it is necessary to replace the amount of salt that is transported by molecular diffusion from the LCZ to the UCZ. This means that salt must be added to the LCZ, and fresh water to the UCZ whilst-brine is removed. The brine can be recycled, divided into water and salt (by solar distillation) and returned to the pond.

The major heat loss occurs from the surface of the solar pond. This heat loss can be prevented by spreading a plastic grid over the pond's surface to prevent disturbance by the wind. Disturbed water tends to lose heat transfer faster than when calm.

Due to the excessively high salt concentration of the LCZ, a *plastic liner* or impermeable soil must be used to *prevent infiltration* into the nearby ground water or soil. The liner is a factor that increases the *cost* of a solar pond. A site where the soil is naturally impermeable, such as the base of a natural pond or lake, or can be made impermeable by compaction or other means, will allow considerably lower power costs.

The optical transmission properties and related collection efficiency vary greatly and depend on the following :

(*i*) Salt concentration.

(*ii*) The quantity of suspended dust or other particles.

(*iii*) Surface impurities like leaves or debris, biological material like bacteria and algae.

(*iv*) The type of salt.

It becomes obvious that much higher efficiencies and storage can be achieved *through the utilization of refined or pure salt* whenever possible, as this *maximizes optical transmission*.

The solar pond is an effective collector of diffuse, as well as direct radiation, and will gather useful heat even on cloudy or overcast days. Under ideal conditions, the pond's absorption efficiency can reach 50% of incoming solar radiation, although actual efficiencies average about 20% due to heat losses. Once the lower layer of the pond reaches over 60°C the heat generated can be drawn off through a heat exchanger and used to drive a low temperature Organic Rankine Cycle (ORC) turbine. This harnesses the pressure differentials created when a low boiling point organic fluid (or gas) is boiled by heat from the pond *via* a heat exchanger and cooled by a condenser to drive a turbine to generate electricity. The conversion efficiency of an organic Rankine cycle turbine driving an electric generator is 5—8% (which mean 1—3% from insolation to electricity output).

7.4.6. Low Temperature Thermal Power Generation

7.4.6.1. Solar pond electric power plant

A low temperature thermal electric power production scheme using solar pond is shown schematically in Fig. 7.13. The energy obtained from a solar pond is used to drive a Rankine cycle heat engine. Hot water from the bottom level of the pond is pumped to the evaporator where the organic working fluid is vapourized. The vapour then flows under high pressure to the turbine where it expands and work thus obtained runs an electric generator producing electricity. The exhaust vapour is then condensed in a condenser and the liquid is pumped back to the evaporator and the cycle is repeated.

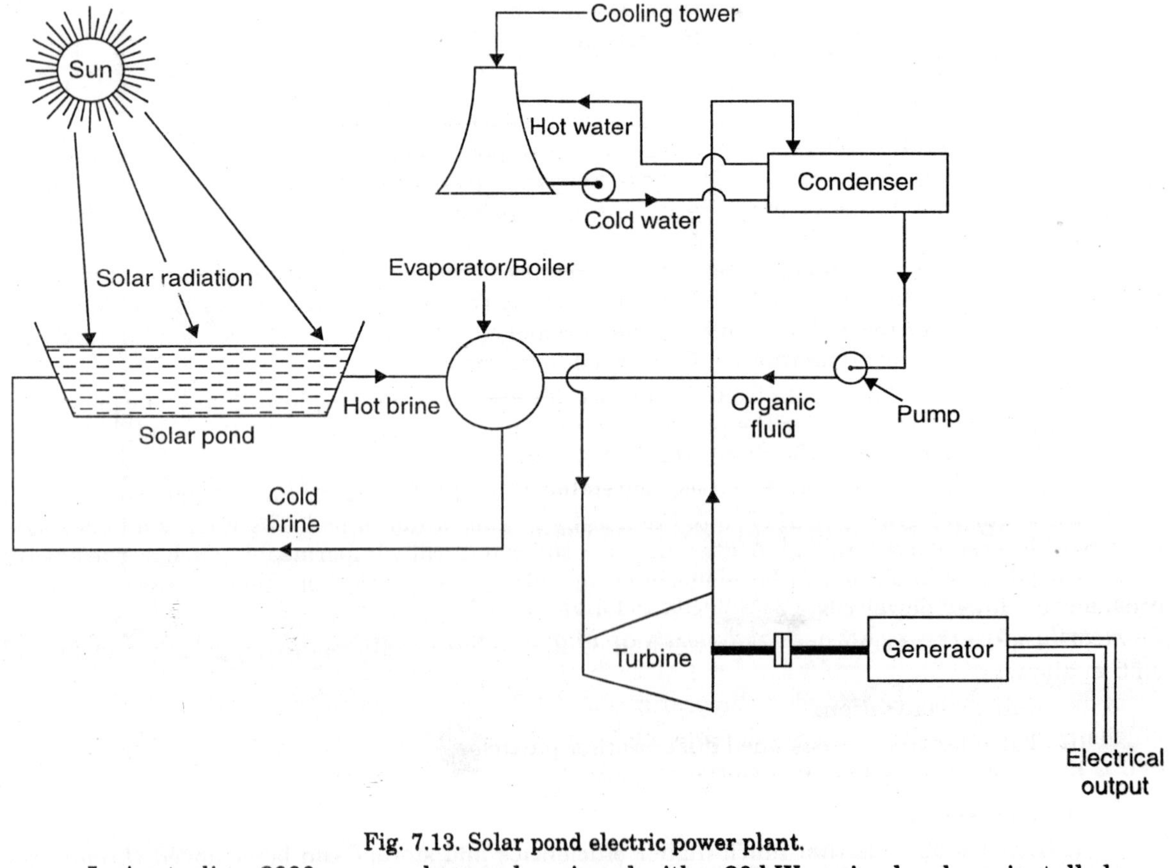

Fig. 7.13. Solar pond electric power plant.

In Australia a 2000 sq. m. solar pond equipped with a 20 kW engine has been installed.

7.4.6.2. Low temperature solar power plant

Fig. 7.14 shows a schematic diagram of a low temperature solar power plant. In this system an array of flat plate collectors is used to heat water to about 70°C and then this heat is used to boil butane in a heat exchanger. The high pressure butane vapour thus obtained runs a butane turbine which in term operates a hydraulic pump. The pump pumps the water from well which is used for irrigation purposes. The exhaust butane vapour (from butane turbine) is condensed with the help of water which is pumped by the pump and the condensate is returned to the heat exchanger (or boiler).

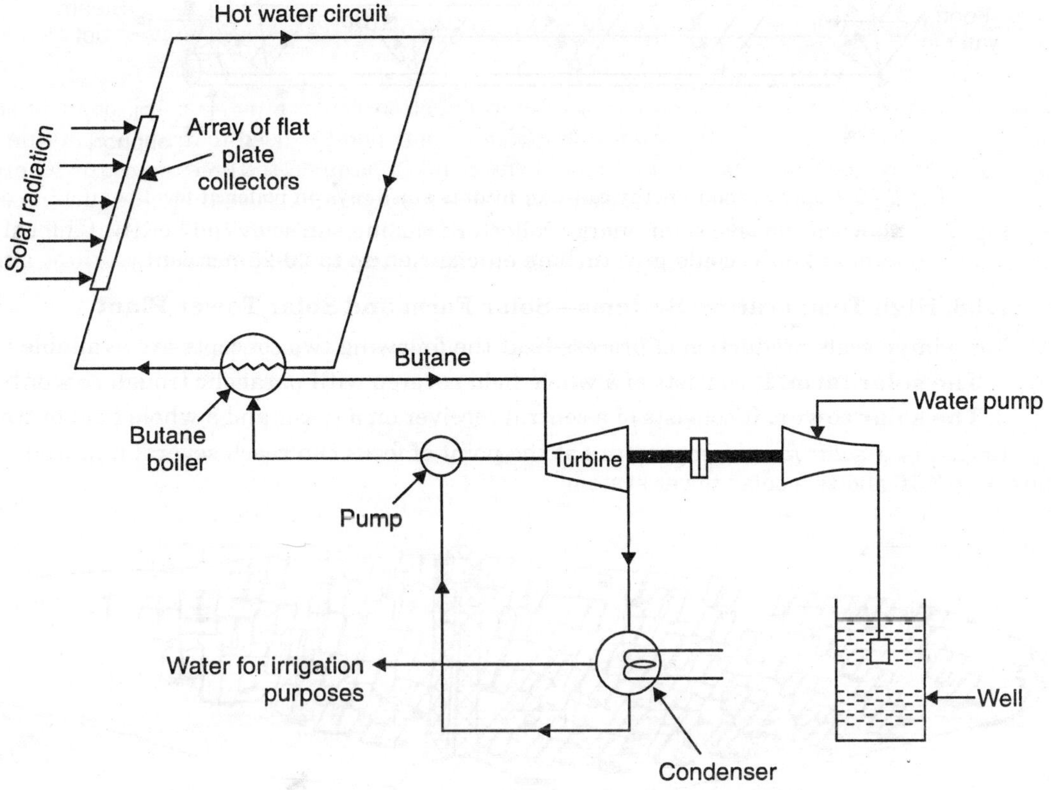

Fig. 7.14. Low temperature solar power plant.

7.4.7. Medium Temperature Systems Using Focusing Collectors

A circular or rectangular parabolic mirror can collect the radiation and focus it on to a small area, a mechanism for moving the collector to follow the sun being necessary. Such devices are used for *metallurgical research* where *high purity and high temperatures are essential*, an example being a 55 m diameter collector giving about 1 MW (th) at Mont Louis in Pyrenees. Smaller units having 20 m diameter reflector can give temperatures of 300°C over an area of about 50 m^2. The collector efficiency is about 50%. On a small scale, units about 1 m diameter giving temperatures of about 300°C have been used for cooking purposes.

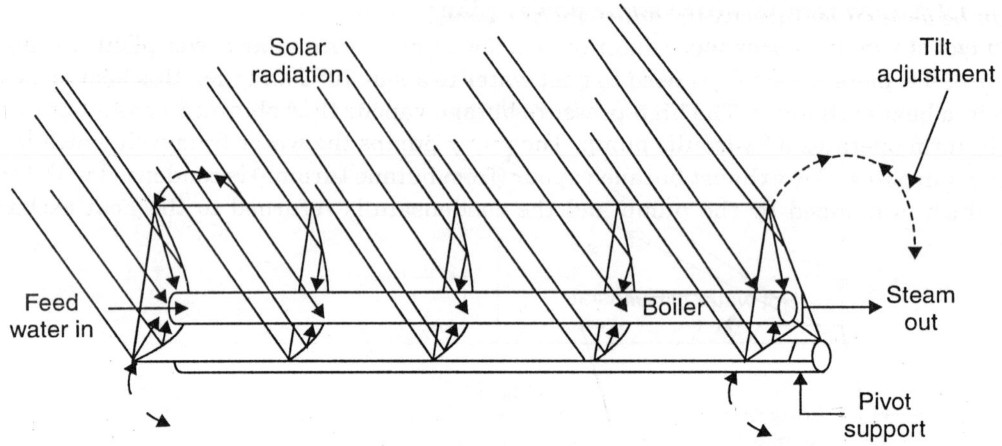

Fig. 7.15. Concave solar energy collector focuses sun's rays on boiler at focal point.

Fig. 7.15 shows a concave solar energy collector focusing sun's rays on boiler at a focal point. Generation of steam at 250°C could give turbine efficiencies up to 20-25 per cent.

7.4.8. High Temperature Systems—Solar Farm and Solar Tower Plant

For a large scale production of process-heat the following two concepts are available :

1. **The solar farm.** It consists of a whole field covered with parabolic trough concentrators.

2. **The solar tower.** It consists of a central receiver on a tower and a whole field of tracking.

In case of a '*solar farm*' temperature at the point of focus can reach several hundred degrees celsius. Fig. 7.16 shows a solar tower system.

Fig. 7.16. Solar tower system.

In case of central receiver "*solar tower*" concentrators, temperature can reach thousands of degrees celsius, since a field of reflectors (heliostats) are arranged separately on sun-tracking frames to reflect the sun on to a boiler mounted on a central tower (Figs. 7.17, 7.18).

With both systems ('solar farm' and 'solar tower'), *a heat transfer fluid or gas is passed through the point or line of insolation concentration to collect the heat and transfer it to the point of use. Such heat can be used either directly in industrial or commercial processes or indirectly in electricity production via. steam and a turbine.*

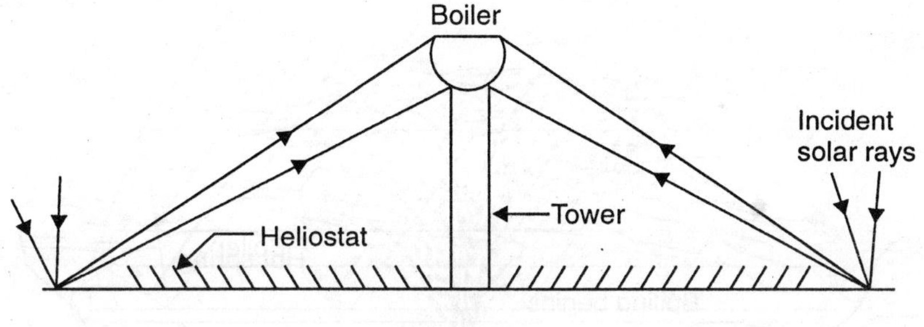

Fig. 7.17

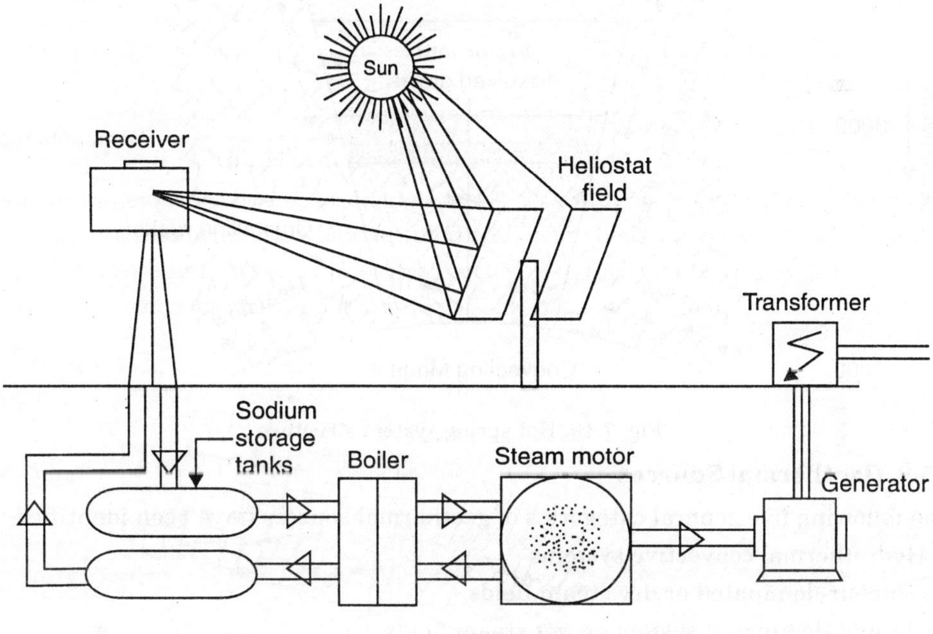

Fig. 7.18. Diagram of solar tower power plant.

The solar technologies such as the above two systems that produce very hot water or steam are currently still under development and, in general, these technologies are not cost competitive with conventional power sources such as oil or gas.

7.5. GEOTHERMAL POWER PLANTS

7.5.1. Geothermal Energy

Geothermal energy is the heat from high pressure steam coming from within the earth. It is a renewable source of energy derived from the rainwater in the earth heated to over 180°C by subterranean hot rocks. Geothermal sources show up hot water springs with temperature of steam upto 150°C ; these are widely distributed over the globe.

Fig. 7.19 shows a schematic diagram depicting how hot springs are produced through hot magma (molten mass), the fractured crystalline rocks, the permeable rocks and percolating ground water.

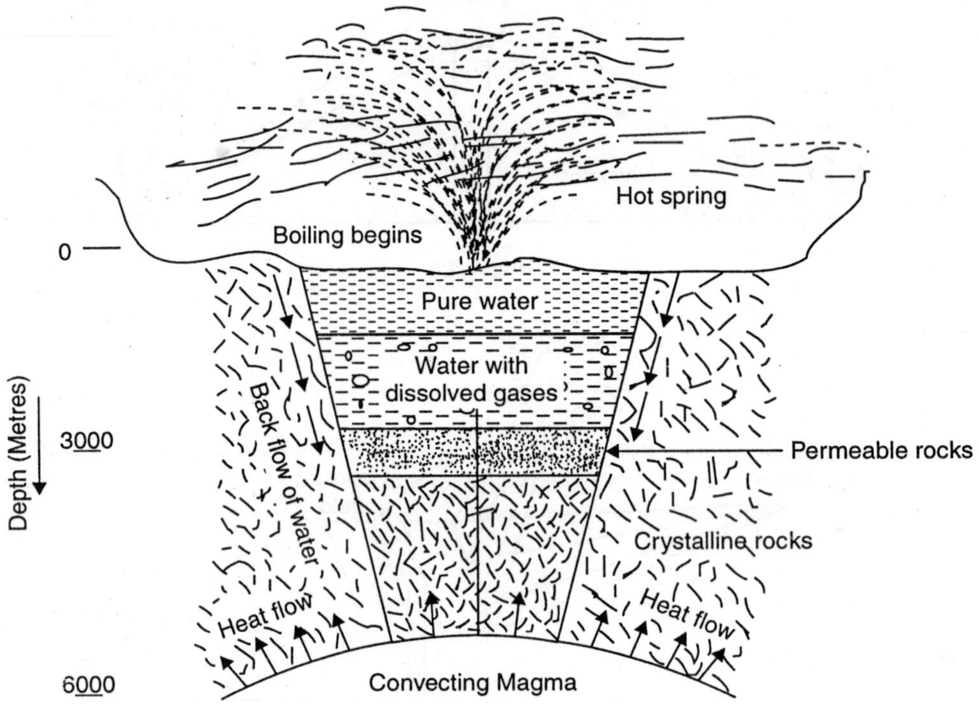

Fig. 7.19. Hot spring system structure.

7.5.2. Geothermal Sources

The following five general categories of geothermal sources have been identified :

1. Hydrothermal convective systems
   (*i*) Vapour-dominated or dry steam fields.
   (*ii*) Liquid-dominated system or wet steam fields.
   (*iii*) Hot-water fields.
2. Geopressure resources.
3. Petro-thermal or hot dry rocks (HDR).
4. Magma resources.
5. Volcanoes.

The hydro-thermal convective systems are best resources for geothermal energy exploitation at present. Hot dry rock is also being considered.

Fig. 7.20 shows a dry-steam open system used in Larderello (Italy) and Greyser (U.S.A.).

Fig. 7.21 shows a flash steam open type system used in Cerro Prietol Mexico, Otake (Japan).

Fig. 7.22 shows a hot water closed (Binary) system under development in U.S.S.R. and U.S.A.

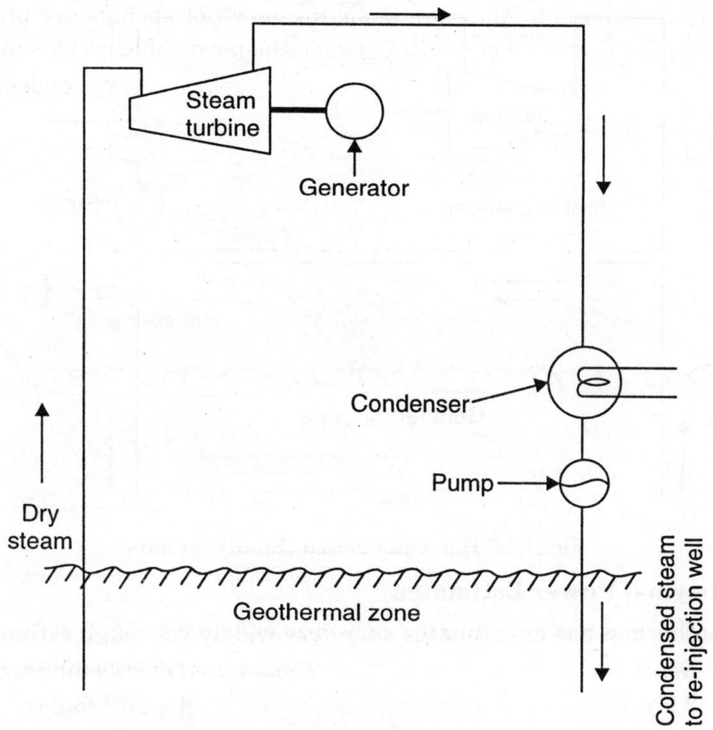

Fig. 7.20. Dry-steam open system.

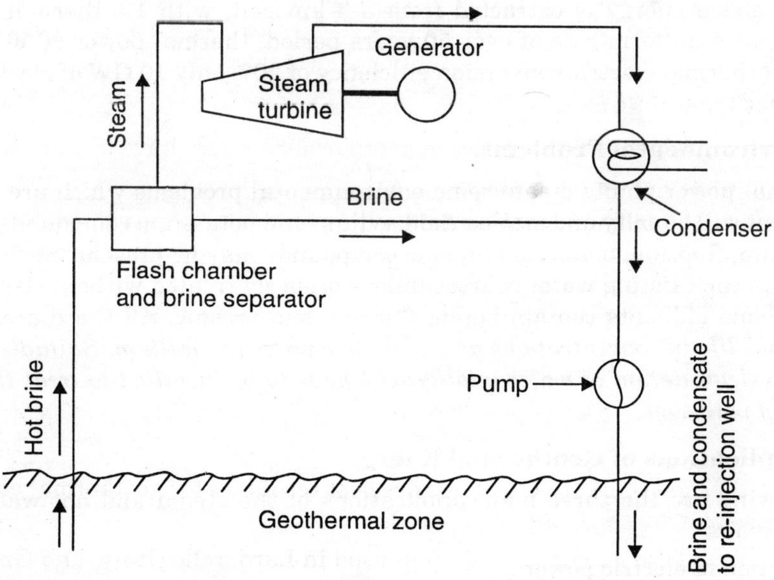

Fig. 7.21. Flash steam open type system.

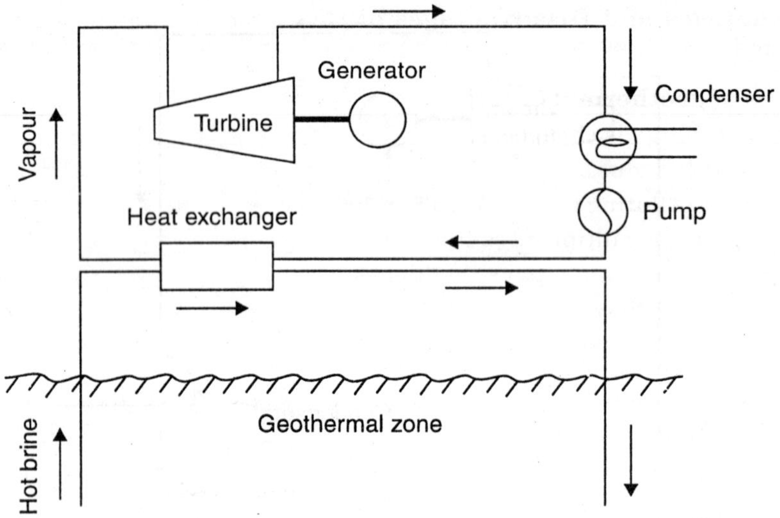

Fig. 7.22. Hot water closed (Binary) system.

7.5.3. Geothermal Power Estimates

Although geothermal power estimates vary very widely yet rough estimate is given below :

Depth	*Total stored energy* (approximately)
3 km	8×10^{21} joules
10 km	4×10^{22} joules

— The energy stored in hot springs = 10 per cent of the above quantities.

— If the above energy is extracted from a 3 km belt, with 1% thermal energy recovery factor, at a uniform rate of over 50 years period, thermal power of 50 GW is obtained. With a thermal electric conversion efficiency of 20% only 10 GW of electric power will be obtained.

7.5.4. Environmental Problems

Geothermal power plants create some environmental problems which are peculiar to them alone. The effluent will be salty and may contain sodium and potassium compounds. Additionally, in some cases lithium, fluorine, boron and arsenic compounds may be present. Such effluents cannot be discharged into the existing water courses unless properly treated without risking severe pollution problems. Some effluents contain boron, fluorine and arsenic. All these are *very harmful to plants and animal life in concentrations as low as two parts per million. Suitable waste treatment plants to prevent degradation of water quality will have to be installed to treat these new and increased sources of pollution.*

7.5.5. Applications of Geothermal Energy

The following are the three main applications of the steam and hot water from the wet geothermal reservoirs :

1. Generation of electric power.
2. Space heating for buildings.
3. Industrial process heat.

The major benefit of geothermal energy is its *varied application and versatility*.

7.5.6. Advantages and Disadvantages of Geothermal Energy over Other Energy Forms

Advantages of geothermal energy :

1. Geothermal energy is cheaper.
2. It is versatile in its use.
3. It is the least polluting as compared to other conventional energy sources.
4. It is amenable for multiple uses from a single resource.
5. Geothermal power plants have the highest annual load factors of 85 per cent to 90 per cent compared to 45 per cent to 50 per cent for fossil fuel plants.
6. It delivers greater amount of net energy from its system as compared to other alternative or conventional systems.
7. Geothermal energy from the earth's interior is almost as inexhaustible as solar or wind energy, so long as its sources are actively sought and economically tapped.

Disadvantages :

1. Low overall power production efficiency (about 15% as compared to 35% to 40% for fossil fuel plants).
2. Drilling operation is noisy.
3. Large areas are needed for exploitation of geo-thermal energy.
4. The withdrawal of large amounts of steam or water from a hydro-thermal reservoir may result in surface subsidence or settlement.

7.6. DIRECT ENERGY CONVERSION SYSTEMS

The energy conversion devices that have been in use for a long time are those that accept energy as heat and produce mechanical work, which is transformed into electric power distribution at large. *Direct energy conversion devices convert naturally available energy into electricity without an intermediate conversion into mechanical energy*. (The energy source may be thermal, solar or chemical). Until now, their use has been confined to small scale, special purpose applications, since the voltage output available with them is rather small and no inexpensive device that is reliable like a turbine or alternator has been built.

Under this topic the following systems will be discussed :

1. Thermoelectric conversion system
2. Thermionic conversion system
3. Photovoltaic power system
4. Magnetohydrodynamic system
5. Electrostatic mechanical generators
6. Electro Gas-Dynamic Generators (EGD)
7. Fuel cells
8. Nuclear batteries.

7.6.1. Thermoelectric Conversion System

The quest for a reliable, silent, energy converter with no moving parts that transforms heat to electrical power has led engineers to reconsider a set of phenomena called the *Thermoelectric effects*. These effects, known for over a hundred years, have permitted the development of small, self contained electrical power sources.

Seebeck (thermoelectric) effect. The German Scientist Seebeck (in 1822) discovered that *if two dissimilar materials* are joined to form a loop and the two junctions maintained at different temperatures, an e.m.f. will be set up around the loop. The magnitude of e.m.f. will be $E = \alpha \Delta T$ where ΔT is the temperature difference between the two junctions and α is the *Seebeck co-efficient*. This effect has long been used in *thermocouples to measure temperatures.*

This phenomenon offers one method of producing electrical energy directly from the heat of combustion, but its *thermal efficiency* is very low, of the order of 1 to 3 per cent. In any heat engine, the efficiency of thermoelectric generator depends upon the temperature of hot and cold junctions.

Thermoelectric Power Generator

Fig. 7.23 shows a schematic diagram of a thermoelectric power generator. The thermocouple materials A and B are joined at the hot end, but the other ends are kept cold ; an electric voltage or electromotive force is then generated between the cold ends. A D.C. (Direct Current) will flow in a circuit or load connected between these ends. The flow of current will continue as long as the heat is supplied to the hot junction and removed from the cold ends. For a given thermocouple, the voltage and electric power output are increased by increasing the temperature difference between the hot and cold ends.

In a practical thermoelectric converter, several thermocouples are convected in series to *increases both voltage and power* as shown in Fig. 7.24. If the output voltage is insufficient to operate a particular device or equipment, it can be increased, with little loss of power, by an inverter transformers combination. The direct current generated by the thermocouples is first changed into alternating current of essentially the same average by means of an inverter. The alternating current and voltage in then increased to the desired value with the help of a transformer. The high voltage alternating current can be reconverted into direct current if required, by the use of a rectifier.

The source of heat for a thermoelectric generator may be a small oil or gas burner, a radioisotope or direct solar radiation.

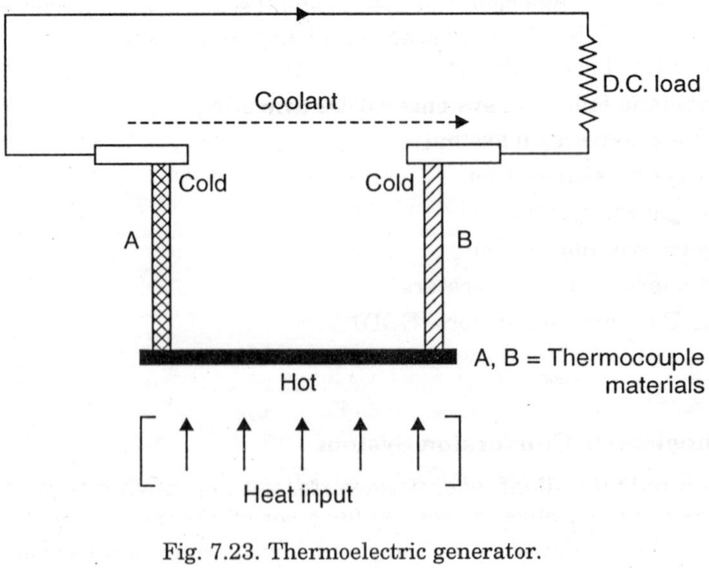

Fig. 7.23. Thermoelectric generator.

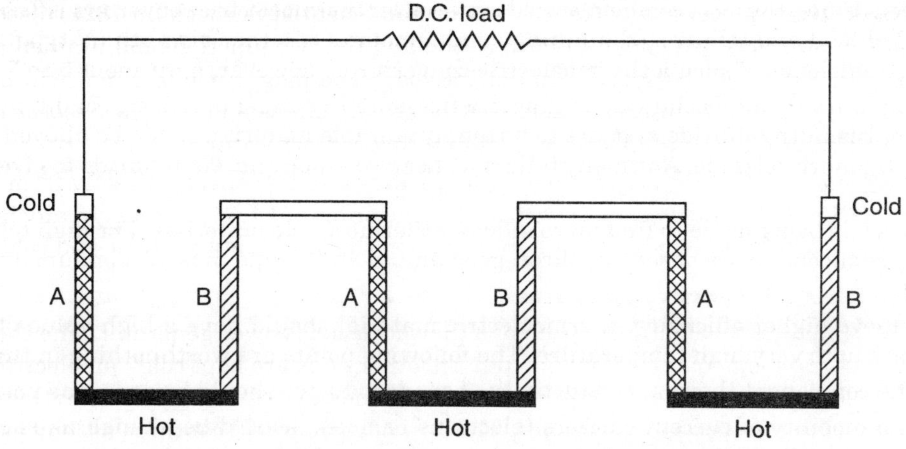

Fig. 7.24. Thermocouples in series (to increase voltage).

A typical couple operating with hot and cold junction temperatures of 600°C and 200°C could be designed to give about 0.1 V and 2 A *i.e.*, about 0.5 W, so that a 1 kW device could require about 5000 couples in series.

Taking into account mechanical characteristics, stability under operating conditions and ease of fabrication, Bismuth telluride appears to be most suitable material. It can be alloyed with such materials as Bismuth selenide, Antimony telluride, Lead selenide and Tin telluride to give improved properties.

Research is being carried out on the possibility of using thermoelectric devices within the core of a nuclear reactor. The hot junction would be located on the fuel element and the cold junctions in contact with the coolants.

Note. A thermoelectric converter is a form of *heat engine* which takes up heat at an *upper temperature* (hot junction) converts it partly into electrical energy and discharges the remaining part at a *lower temperature* (cold junction). The efficiency of a thermocouple, as in the case with other heat engines, increases by increasing the upper temperature and decreasing the lower temperature. Since the lower temperature is usually that of environment the efficiency of a thermocouple, practically, depends upon the *hot junction temperature.*

Thermoelectric materials and their selection :

The following materials find use in the making of thermoelectric elements :

Material	*Formula*	*Figure of merit, Z (°K⁻¹)*
Lead telluride	PbTe	1.5×10^{-3}
Bismuth telluride (doped with Sb or Se)	$Bi_2 Te_3$	4×10^{-3}
Germanium telluride (with bismuth)	Ge Te	1.5×10^{-3}
Cesium sulphide	Ce S	1.0×10^{-3}
Zinc antimonide (doped with silver)	Zn Sb	1.5×10^{-3}

where Z is an *index used in rating thermoelectric converters*. It depends on the properties of thermoelectric materials used. A high value of Z is obtained by using materials of :

(*i*) *Large Seebeck co-efficient* (*ii*) *Small thermal conductivity*

(*iii*) *Small electrical resistivity.*

In recent times, the most commonly used material for thermoelectric converters is *Lead telluride* [a compound of lead and tellurim., containing small amounts of either bismuth (N-type) or sodium (P-type)]. The efficiency of such a thermoelectric converter is, however, only about 5 to 7 per cent.

Taking into account mechanical characteristics, stability under operating conditions and ease of fabrication, Bismuth telluride appears to be amply suitable material. It can be alloyed with such materials as Bismuth selenide, Antimony telluride, Lead selenide and Tin telluride to give improved properties.

Research is being made to find more efficient thermocouple materials. For high temperature applications, semiconductors based on silicon-germanium and compounds of selenium appear to be promising.

To achieve higher efficiency, thermoelectric material should have a high value of Z and be able to operate upto very high temperature. The following points are worthnothing in this regard :

1. The component thermal conductivity of semiconductor should be as low as possible.
2. The mobility of current carriers (electrons or holes, should be as high as is compatible with condition 1).
3. One of the arms should consist of a purely hole type and the other of a purely electronic type semiconductor.
4. In the low temperature zone the impurity concentration should be lower than in the higher temperature zone.
5. So that the thermoelectric material may not crack under the effect of stresses it should possess the following properties :
   (*i*) It should be able to resist chemical influences such as oxidation etc.
   (*ii*) It should have good mechanical strength.
   (*iii*) It should be amply elastic.

7.6.2. Thermionic Conversion System

Introduction :

A thermionic converter can be analyzed from at least three different points of view :

1. In terms of *thermodynamics*, it may be viewed as a heat-engine that uses an electron gas as a working substance.

2. In terms of *electronics*, it may be viewed as a *diode* that transforms heat to electricity by the law of thermionic emission.

3. In terms of *thermoelectricity*, it may be viewed as a thermocouple in which an evacuated space or a plasma has been substituted for one of the conductors.

Regardless of the point of view adopted in analysis, a *thermionic converter works because of the phenomenon of 'thermionic emission'. Thermionic emission implies emission of electrons from the metal when it is heated.*

Work function (ϕ) :

It is defined as *the energy required to extract an electron from the metal*. It is measured in *electron volts*. The value of work function varies with the *nature of the metal and its surface condition.*

A thermionic converter, in principle, consists of two metals or electrodes with different work functions sealed into an evacuated vessel. The electrode with a large work function is maintained at a higher temperature than one with the smaller work function.

Thermionic generators. Refer Fig. 7.25.

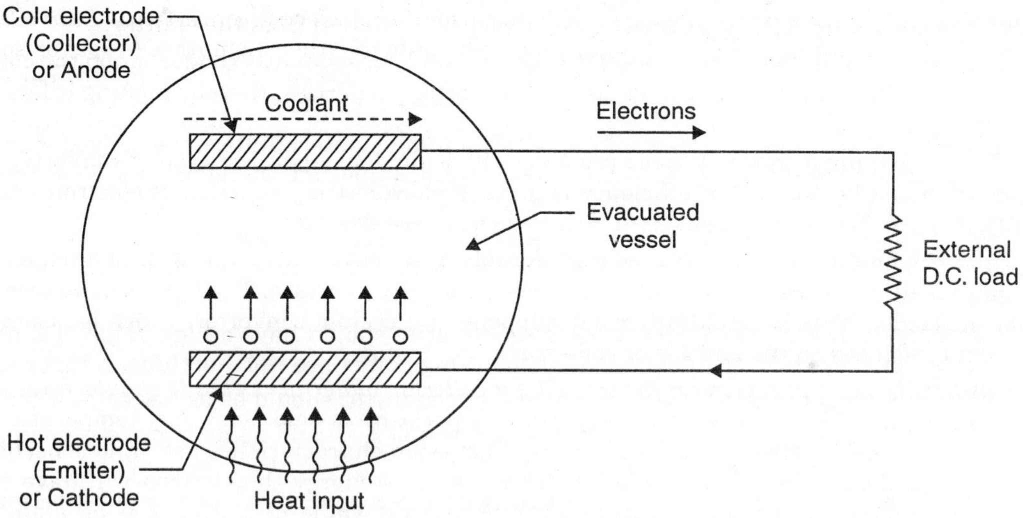

Fig. 7.25. Thermionic generator.

A thermionic converter/generator comprises a *heated cathode* (electron *emitter*) and an *anode* (electron *collector*) separated by a *vacuum*, the electrical output circuit being connected between the two as shown in Fig. 7.25. The heat which is supplied to the cathode raises the energy of its electrons to such a level that it enables them to escape from the surface and flow to the anode. *At the anode the energy of electrons appears partially as heat, removed by cooling and partially as electrical energy delivered to the circuit.* Although the distance between anode and cathode is only about *one millimetre*, the negative space charge with such an arrangement hinders the passage of the electrons and *must be reduced*, this can be achieved by introducing positive ions into the interelectrode space, cesium vapour being valuable source of such ions.

In order to materialise a substantial electron emission rate (per unit area of emitter), and hence a significant current output as well as a high efficiency, the emitter temperature in a thermionic converter containing cesium should be *at least 1000°C*, the efficiency is then 10 per cent. Efficiency as high as 40 per cent can be obtained by operating at still higher temperatures. Although temperature has little effect on the voltage generated, the increase in current (per unit emitter area) associated with a temperature increase results in increase in power. Electric power (P) is the product of voltage (E) and current (I) *i.e.*, $P = EI$.

Anode materials should have a *low work function e.g., barium and strontium oxides* while that of the *cathode* should be *considerably higher, tungsten impregnated with a barium compound* being a suitable material. Even with these materials temperatures upto 2000°C will be required to secure for the generator itself, efficiencies of 30-35 per cent. Electrical outputs of about 6 W/cm^2 of anode surface are envisaged with about 13 W/cm^2 removed by coolant.

A thermionic generator, in principal can make use of any fuel (may be fossil fuel, a nuclear fuel or solar energy) subject to the condition that sufficiently high temperatures are obtainable. The thermionic conversion can be utilized in several different situations—remote locations on the earth and in space.

Thermionic converter materials. The problem of developing materials suitable for use in thermionic converters ranks next to the space charge control problem in the development of efficient thermionic generators. Following *properties are desirable in materials suitable for converters :*

Emitter. A good emitter will :

(*i*) have high-electron emission capability coupled with a low rate of deterioration.

(*ii*) have *low emissivity*, to reduce heat transfer by radiation from the emitter.

(*iii*) be such that in the event some of it vaporizes and subsequently condenses on the collector it will not poison the collector (that is, change the collector properties, thereby making it less effective).

The relative importance of these properties is dependent upon the type of converter being designed. It should be noted that efficiency is a much slower rising function of electron emission capability if space charge is present than if there is no space charge.

The work function may be reduced considerably by an absorbed single layer of foreign atoms. This comes about by the establishment of a dipole layer at the surface. The layer can be formed by atoms or molecules. This is essentially what happens in a cesium converter, which is designed so that cesium condenses on the emitter or collector.

Collector. The *main criteria for choosing a collector material is that it should have as low a work function as possible.* Because the collector temperature is held below any temperature that will cause significant electron emission, its actual emission characteristics are of no consequence. The lower the collector work function (ϕ_c), however the less energy the electron will have to give up as it enters the collector surface. In practice the lowest value of ϕ_c that can be maintained stably is about 1.5 eV. For applications in which it is desirable to maintain the collector at elevated temperatures (greater than 900°K) such as space applications, an optimum value of ϕ_c may be determined. Molybdenum has been widely used as a collector ; it is frequently assumed to have a work function of 1.7 eV.

7.6.3. Photovoltaic Power System

Photovoltaic generators—Historical background. Edmond Becquerel in 1839 noted that a voltage was developed when light was directed onto one of the electrodes in an electrolytic solution. The effect was first observed in a solid in 1877 by W.G. Adams and R.E. Day, who conducted experiments with selenium. Other early workers with solids included Schottky, Lange and Grandahl, who did pioneering work in producing photovoltaic cells with selenium and cuprous oxide. This work led to the development of photoelectric exposure meters. 1954 researchers turned to the problem of utilizing the photovoltaic effect as a source of power. In that year several groups including the workers at Bell Telephone Laboratories achieved conversion efficiencies of about 6 per cent by means of junctions of *P*-type and *N*-type semiconductors. These early junctions, commonly called *P-N* junctions, were made of cadmium sulphide and silicon. Later workers in the area have achieved efficiencies more than 20 per cent by using improved silicon *P-N* junctions.

Photovoltaic cell. Solar energy can be directly converted to electrical energy by means of *photovoltaic effect* which is defined as the *generation of an electromotive force as a result of the absorption of ionizing radiation.* Energy conversion devices which are employed to convert sunlight into electricity by the use of the photovoltaic effect are called *solar cells.* A single converter cell is called a solar cell or a *photovoltaic cell.* To increase the electrical power output a number of such cells are combined and the combination is called a *solar array* (or solar module).

In a photovoltaic cell sensitive element is a *semiconductor* (not metal) which generates voltage in proportion to the light or any radiant energy incident on it. The most commonly used photovoltaic cells are barrier layer type like iron-selenium cells or Cu—CuO_2 cells.

Fig. 10.26 shows a typical widely used photo-voltaic cell—"*Selenium cell*". It consists of a metal electrode on which a layer of selenium is deposited ; on the top of this a barrier layer is formed which is coated with a very thin layer of gold. The latter serves as a translucent electrode through which light can impinge on the layer below. Under the influence of this light, a negative charge will build up on the gold electrode and a positive charge on the bottom electrode.

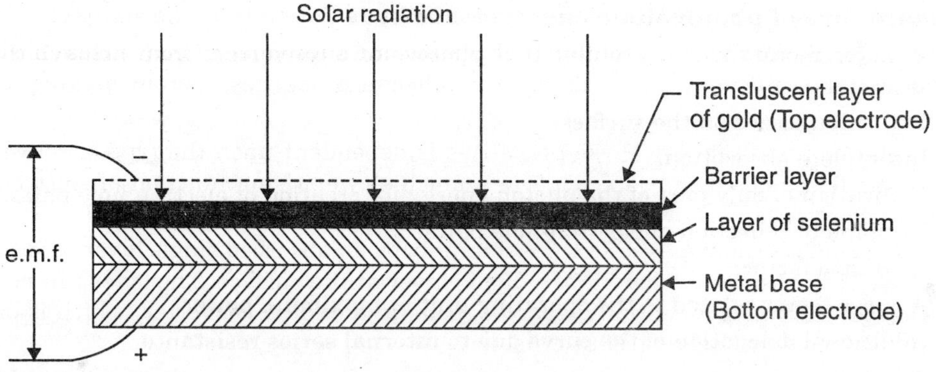

Fig. 7.26. Photovoltaic cell.

Photovoltaic cells are widely used in the following fields :

(*i*) Automatic control systems.

(*ii*) Television circuits.

(*iii*) Sound motion picture and reproducing equipment.

Basic photovoltaic system for power generation :

Fig 7.27 shows a basic photovoltaic system integrated with the utility grid. With the help of this system the generated electrical power can be delivered to the local load.

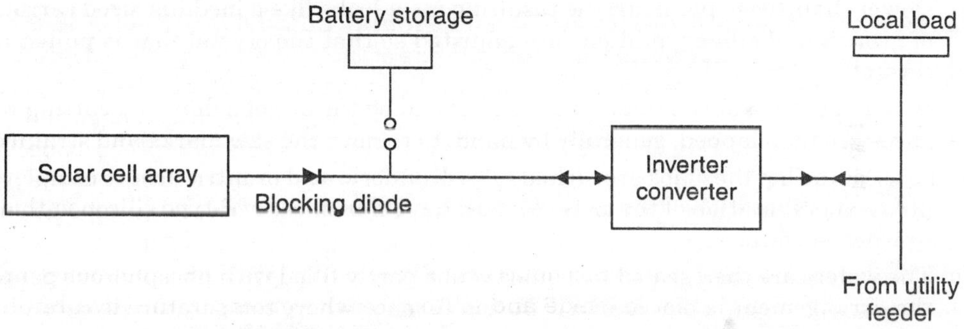

Fig. 7.27. Basic photovoltaic system integrated with power grid.

This system consists of the following :

1. Solar array
2. Blocking diode
3. Battery storage
4. Inverter converter
5. Switches and circuit breakers.

— The *solar array* (large or small) converts the insolation to useful D.C. electrical power.

— The *blocking diode* confines the electrical power generated by the solar array to flow towards the battery or grid only. In the absence of blocking diode the battery would discharge back (through the solar array) during the period when there is no insolation.

— *Battery storage* stores the electrical power generated through solar array.

— *Inverter/converter* (usually solid state) converts the battery bus voltage to A.C. of frequency and phase to match that needed to integrate with the utility grid. Thus it is typically a D.C., A.C. inverter.

— *Switches and circuit breakers* permit isolating parts of the system, as the battery.

Limitations of photovoltaic energy converters

The major factors which prohibit real photovoltaic converters from achieving the higher efficiencies are :

1. Reflection losses on the surface.
2. Incomplete absorption.
3. Utilization of only part of the photon energy for creation of electron hole pairs.
4. Incomplete collection of electron-hole pairs.
5. A voltage factor.
6. A curve factor related to the operating unit at maximum power.
7. Additional delegation of the curve due to internal series resistance.

Fabrication of Cells :

A. Silicon cells :

Silicon cells are most widely used. Next to oxygen, silicon is the most abundant element on earth, the pure silicon used in cell manufacture is extracted from sand which is mostly silicon dioxide (SiO_2). The silicon required for solar cell use, because of its high purity, is expensive.

The fabrication of silicon cells include the following steps :

(*i*) The pure silicon is placed in an induction furnace where boron is added to melt. This turns the crystal resulting from the melt into *P*-type material.

(*ii*) A small seed of single crystal silicon is dipped into the melt and withdrawn at a rate slower than 10 cm per hour, the resulting inset looks like a medium sized carrot. The rate of growth and other conditions are adjusted so that the crystal that is pulled is a single crystal.

(*iii*) Wafers are then sliced from the grown crystal by the use of a diamond cutting wheel. The slices are then lapped, generally by hand, to remove the saw marks and strained regions.

(*iv*) After a fine lap the slabs are etched in hydrofluoric acid or nitric acid to complete the first phase of preparation of the cells. We now have thin slices of *P*-type silicon with a carefully finished surface.

(*v*) The wafers are then sealed in a quartz tube partly filled with phosphorous pentoxide and the arrangement is placed in a diffusion furnace where temperature is carefully controlled ; this process causes the phosphorous to diffuse into the *P*-type silicon to a depth of about 10^{-4} cm to 10^{-5} cm.

(*vi*) The cells are then etched in a concentrated acid to remove unwanted coatings that formed during manufacture. Wax or Teflon masking tape is used to protect the surfaces not to be etched.

B. Thin film solar cells :

These cells have the following *advantages :*

(*i*) The material cost is low.

(*ii*) The manufacturing cost is low (possibly avoiding the need for single crystal growth).

(*iii*) High power-to-weight ratios.

(*iv*) Low array costs, because the number of connections needed will be greatly reduced.

The example of this type of cell is cadmium sulphide (CdS) cells. CdS cells having areas of 50 cm^2 have been made by evaporating the semiconductor on to a flexible substrate such as kapton, a metallized plastic substrate. A barrier layer of copper sulphide is then deposited on top of the CdS.

Power to weight ratios of 200 watts/kg are claimed for such cells. These cells have low efficiency and instability.

Advantages and disadvantages of Photovoltaic solar energy conversion :

Advantages :

(*i*) There are no moving parts.

(*ii*) Solar cells are easy to operate and need little maintenance.

(*iii*) They have longer life.

(*iv*) They are highly reliable.

(*v*) They do not create pollution problem.

(*vi*) Their energy source is unlimited.

(*vii*) They can be fabricated easily.

(*viii*) They have high power to weight ratio.

(*ix*) They can be used with or without sun tracking, making possible a wide range of application possibilities.

(*x*) They have ability to function unattended for long periods as evident in space programme.

Disadvantages :

(*i*) The cost of a solar cell is quite high.

(*ii*) The output of a solar cell is not constant, it varies with the time of day and weather.

(*iii*) Amount of power generated is small.

7.6.4. Magnetohydrodynamics (MHD) System

Introduction. Magnetohydrodynamics (MHD), as the name implies, is concerned with the flow of a conducting fluid in the presence of magnetic and electric field. The fluid may be gas at elevated temperature or liquid like sodium or potassium.

MHD generator is a device *which converts heat energy of a fuel directly into electrical energy without a conventional electric generator*. MHD converter system is a heat engine whose efficiency, like all heat engine, is increased by supplying the heat at the highest practical temperature and rejecting it at the lowest practical temperature. MHD generation looks the most promising of the direct conversion techniques for the large scale production of electric power.

Principle of MHD power generation :

Faraday's law of electromagnetic induction states that when a conductor and a magnetic field move in respect to each other, an electric voltage is induced in the conductor. The conductor need not be a solid—it may be a *gas or liquid*. The magnetohydrodynamics (MHD) generator uses this principle by forcing a *high-pressure, high temperature combustion gas* through a strong magnetic field.

Fig. 7.28 shows the comparison between a *turbogenerator* and the *MHD generator*.

MHD systems :

The broad classification of the MHD systems is as follows :

1. Open cycle systems

2. Closed cycle systems

(*i*) Seeded inert gas systems

(*ii*) Liquid metal systems.

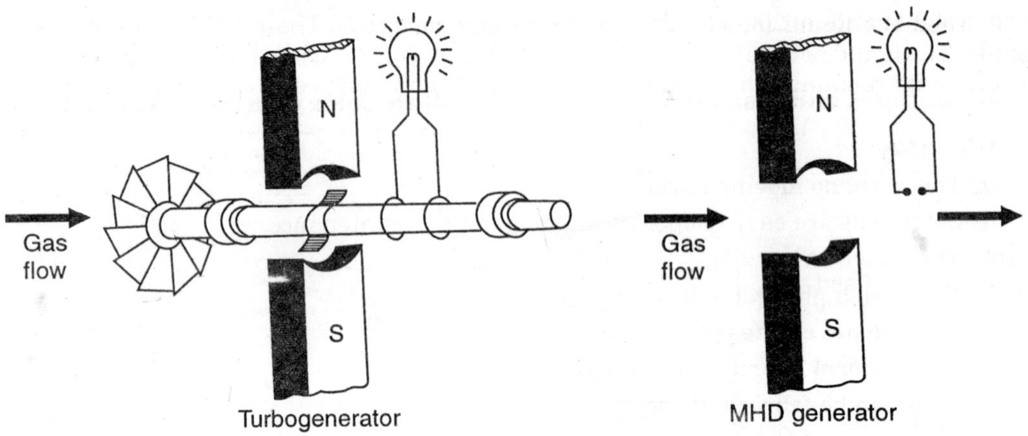

Fig. 7.28. Comparison between the conventional turbogenerator and the MHD generator.

Open cycle MHD system :

Fig. 7.29 shows an open cycle MHD system. Here the fuel (such an oil, coal, natural gas) is burnt in the combustion chamber, air required for combustion is supplied from *air preheater*. The hot gases produced by the combustion chamber are then seeded with a small amount of an ionized

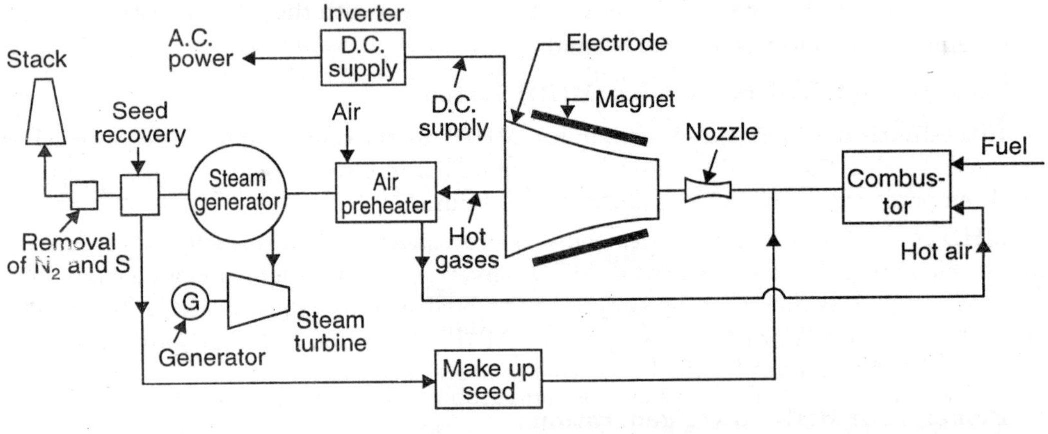

Fig. 7.29. Open cycle MHD system.

alkali metal (cesium or potassium) to increase the electrical conductivity of the gas. The ionization of potassium (generally potassium carbonate is used as seed material) takes place due to gases produced at temperature of about 2300-2700°C by combustion. The hot pressurised working fluid so produced leaves the combustion chamber and passes through a convergent divergent nozzle. The gases coming out the nozzle at high velocity then enter the MHD generator. The expansion of the hot gases take place in the generator surrounded by powerful magnents. The MHD generator produces direct current. By using an *inverter* this direct current can be converted into alternating current.

Closed cycle MHD system :

A liquid metal closed cycle system is shown in Fig. 7.30. A liquid metal (potassium) is used as working fluid in this system. The liquid potassium after being heated in the breeder reactor is passed through the nozzle where its velocity is increased. The vapour formed due to nozzle action are separated in the separator and condensed and then pumped back to the reactor as shown in

Fig. 7.30. Then the liquid metal with high velocity is passed through MHD generator to produce D.C. power. The liquid potassium coming out of MHD generator is passed through the heat exchanger (boiler) to use its remaining heat to run a turbine and then pumped back to the reactor.

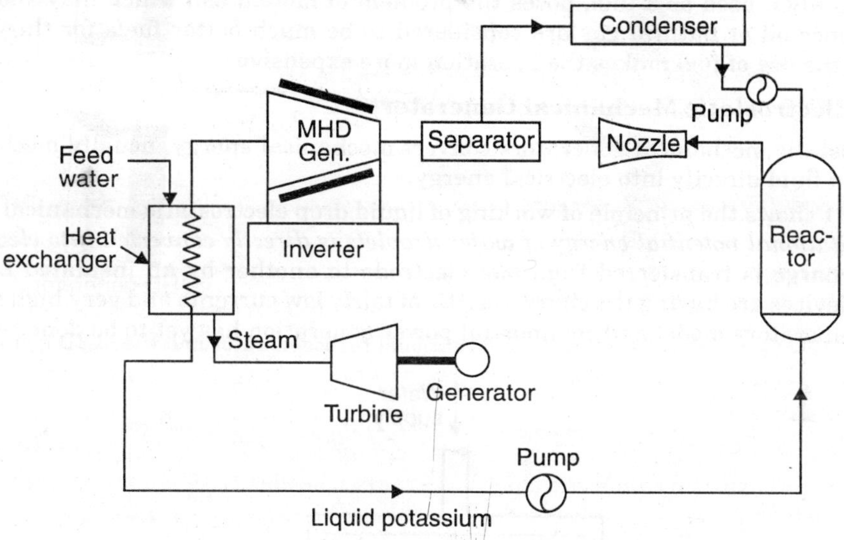

Fig. 7.30. Closed cycle (liquid metal) system.

This system entails many constructional and operational difficulties.

Advantages of MHD systems :

1. More reliable since there are no moving parts.
2. In MHD system the efficiency can be about 50% (still higher expected) as compared to less than 40% for most efficient steam plants.
3. Power produced is free of pollution.
4. As soon as it is started it can reach the full power level.
5. The size of plant is considerably smaller than conventional fossil fuel plants.
6. Less overall operational cost.
7. The capital cost of MHD plants is comparable to those of conventional steam plants.
8. Better utilization of fuel.
9. Suitable for peak power generation and emergency service.

Drawbacks of MHD system :

1. MHD systems suffer from the reverse flow (short circuits) of electrons through the conducting fluids around the ends of the magnetic field. This loss can be reduced by (*i*) increasing aspect ratio (*L*/*d*) of the generator, (*ii*) by permitting the magnetic field poles to extend beyond the end of electrodes and (*iii*) by using insulated vanes in the fluid ducts and at the inlet and outlet of the generator.

2. There will be high friction losses and heat transfer losses. The friction loss may be as high as 12% of the input.

3. The MHD system operates at very high temperatures to obtain high electrical conductivity. But the electrodes must be relatively at low temperatures and hence the gas in the vicinity of the electrodes is cooler. This increases the resistivity of the gas near the electrodes and hence there will

be a very large voltage drop across the gas film. By adding the seed material, the resistivity can be reduced.

4. The MHD system needs very large magnets and this is a major expense.

5. Coal, when used as a fuel, poses the problem of molten ash which may short circuit the electrodes. Hence oil or natural gas are considered to be much better fuels for this system. This restriction on the use of fuel makes the operation more expensive.

7.6.5. Electrostatic Mechanical Generators

Electrostatic mechanical generators convert mechanical energy, usually mechanical potential energy of a fluid directly into electrical energy.

Fig. 7.31 shows the principle of working of liquid drop electrostatic mechanical generator. In this, *the gravitational potential energy of water droplets is directly converted into electrical energy.* The electric charge is transferred from one electrode to another by an insulated belt. All these electrostatic devices are having the characteristic of fairly low currents and very high voltages. This is yet only a laboratory model and commercial power generation has yet to be done.

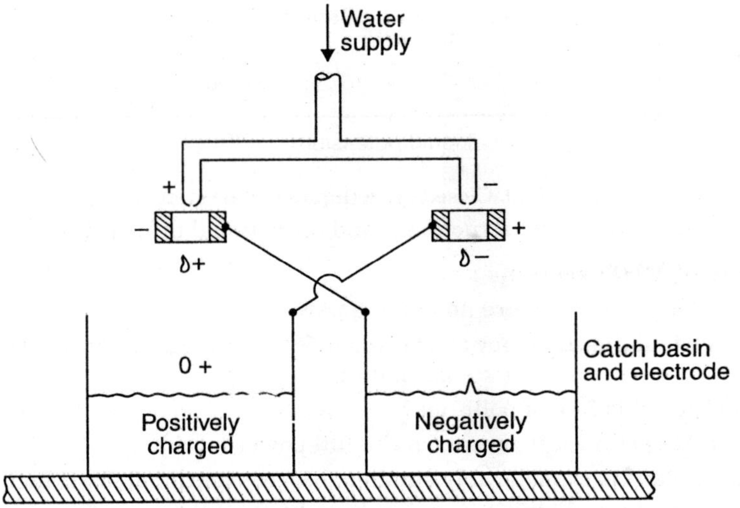

Fig. 7.31. Liquid drop electrostatic mechanical generator.

7.6.6. Electro Gas-dynamic Generators (EGD)

The EGD generator uses the potential energy of a high pressure gas to carry electrons from a low potential electrode to a high potential electrode, thereby doing work against an electric field. A schematic diagram of EGD is shown in Fig. 7.32.

Carona electrode at the entrance of the duct generates electrons. This ionised gas particles are carried down the duct with the neutral atoms and the ionized particles are neutralised by the collector electrode, at the end of the insulated duct. The working fluid in these systems are commonly, either combustion gases produced by burning of fuel at high pressures or it is a pressurised reactor gas coolant. The maximum power output from EGD is about 10 to 30 W per channel. Hence, several thousand channels are connected in series and parallel. The voltage produced is very high, of the order of 1,00,000 V to 2,00,000 V. Thus, it needs very good high voltage insulators. (Beryllium Oxide, Beo, is generally used).

EGD can produce a high efficiency equal to MHD-steam combination.

Advantages of EGD over MHD systems :

1. EGD systems operate at relatively low temperatures.
2. No need for injection and recovery of seed material.
3. It is self contained since it does not need a steam generator.
4. Energy can be extracted till the gases reach almost the stack temperature.
5. Does not need large quantities of condenser cooling water.

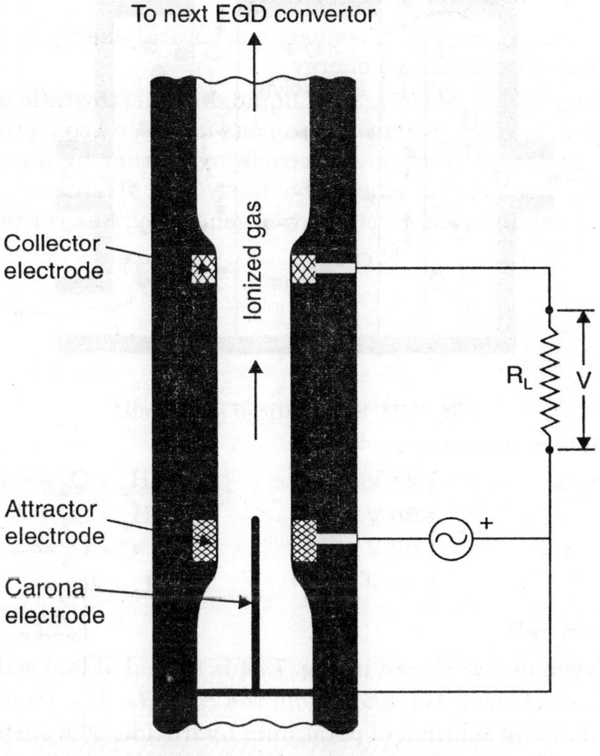

Fig. 7.32. Gas duct in an EGD convertor.

EGD and MHD hold the prospects of offering the best solutions for high efficiency, large capacity systems for the production of electricity.

7.6.7. Fuel Cells

A **fuel cell** *is an electrochemical device in which the chemical energy of a conventional fuel is converted directly and efficiently into low voltage, direct-current electrical energy*. One of the chief advantages of such a device is that because the conversion, atleast in theory, can be carried out isothermally, the Carnot limitation on efficiency does not apply. A fuel cell is often described as a primary battery in which the fuel and oxidizer are stored external to the battery and fed to it as needed.

Fig. 7.33 shows a schematic diagram of a fuel cell. *The fuel gas diffuses through the anode and is oxidized, thus releasing electrons to the external circuit ; the oxidizer diffuses through the cathode and is reduced by the electrons that have come from the anode by way of the external circuit.*

The fuel cell is a device that keeps the fuel molecules from mixing with the oxidizer molecules, permitting, however, the transfer of electrons by a metallic path that may contain a load.

Of the available fuels, *hydrogen* has so far given the most promising results, although cells consuming coal, oil or natural gas would be economically much more useful for large scale applications.

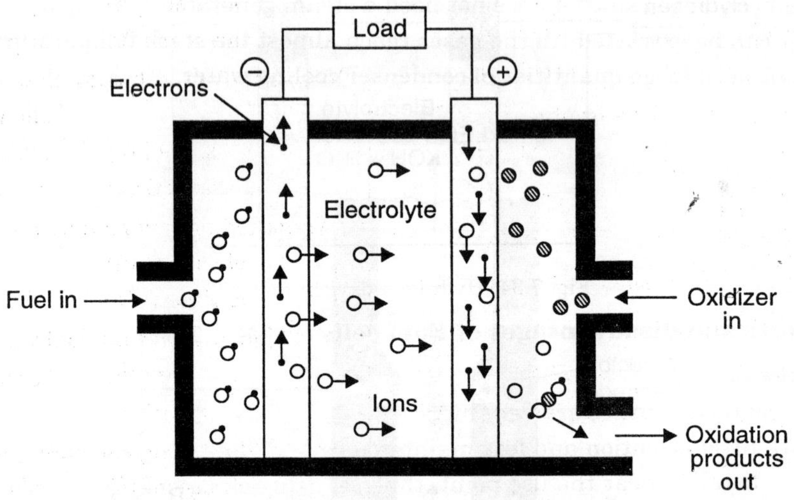

Fig. 7.33. Schematic of a fuel cell.

Some of the possible reactions are :

Hydrogen/oxygen	1.23 V	$2H_2 + O_2 \longrightarrow 2H_2O$
Hydrazine	1.56 V	$N_2H_4 + O_2 \longrightarrow 2H_2O + N_2$
Carbon (coal)	1.02 V	$C + O_2 \longrightarrow CO_2$
Methane	1.05 V	$CH_4 + 2O_2 \longrightarrow CO_2 + 2H_2O$

Hydrogen-oxygen cell :

The hydrogen-oxygen device shown in Fig. 7.34 is typical of fuel cells. It has three chambers separated by two porous electrodes, the *anode* and the *cathode.* The middle chamber between the electrodes is filled with a strong solution of potassium hydroxide. The surfaces of the electrodes are chemically treated to repel the electrolyte, so that there is minimum leakage of potassium hydroxide into the outer chambers. The gases diffuse through the electrodes, undergoing reactions as shown below :

$$4KOH \longrightarrow 4K^+ + 4(OH)^-$$

Anode : $2H_2 + 4(OH)^- \longrightarrow 4H_2O + 4e^-$

Cathode : $O_2 + 2H_2O + 4e^- \longrightarrow 4(OH)^-$

Cell reaction $2H_2 + O_2 \longrightarrow 2H_2O$

The water formed is drawn off from the side. The electrolyte provides the $(OH)^-$ ions needed for the reaction, and remains unchanged at the end, since these ions are regenerated. The electrons liberated at the anode find their way to the cathode through the external circuit. *This transfer is equivalent to the flow of a current from the cathode to the anode.*

Such cells when properly designed and operated, have an open circuit voltage of about 1.1 volt. Unfortunately, their life is limited since the water formed continuously dilutes the electrolyte. Fuel efficiencies as high as 60%-70% may be obtained.

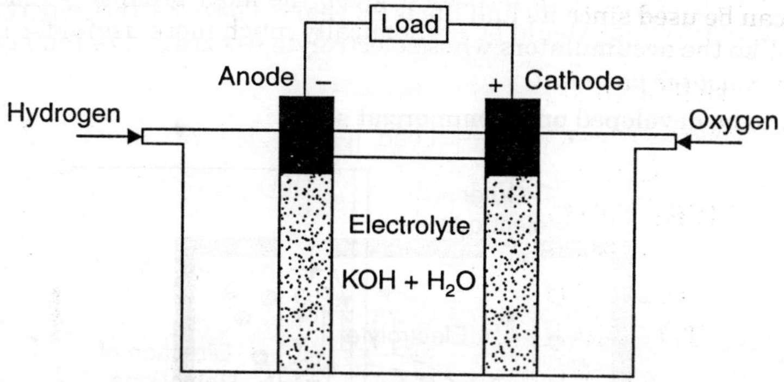

Fig. 7.34. Hydrogen-oxygen fuel cell.

Advantages and disadvantages of Fuel cells :

Advantages :

1. Conversion efficiencies are very high.
2. Require little attention and less maintenance.
3. Can be installed near the use point, thus reducing electrical transmission requirements and accompanying losses.
4. Fuel cell does not make any noise.
5. A little time is needed to go into operation.
6. Space requirement considerably less in comparison to conventional power plants.

Disadvantages :

1. High initial cost.
2. Low service life.

Applications of Fuel cells :

The applications of fuel cell relate to :

1. Domestic use.
2. Automotive vehicles.
3. Central power stations.
4. Special applications.

Note. The human body functions essentially like a fuel cell. Living things take in food (fuel) and oxygen to produce both thermal energy and work output. They are not heat engines.

7.6.8. Nuclear Batteries

A nuclear battery works on the principle that *beta emitter can produce the electrical energy.*

Nuclear batteries *are of the following two kinds* : (*i*) *High voltage* atomic battery and (*ii*) *Low voltage* atomic battery.

7.6.8.1. High voltage atomic battery

Fig. 7.34 shows the schematic diagram of a high voltage atomic battery.

It consists of an inner spherical electrode on the surface of which is deposited a powerful beta emitting (*i.e.,* fast electrons) substance. This is surrounded by another spherical condenser and the inner surface of this condenser becomes negatively charged. This acts as the outside electrode and is properly insulated at the opening. The inner and outer electrodes become the – ve and + ve terminals of the battery.

Sr^{90} isotope can be used since its half life is 28 years. These batteries perform independent of the temperature unlike the accumulators whose electrolyte freezes at low temperatures. These also supply a *very steady constant potential.*

This has yet to be developed on a commercial scale.

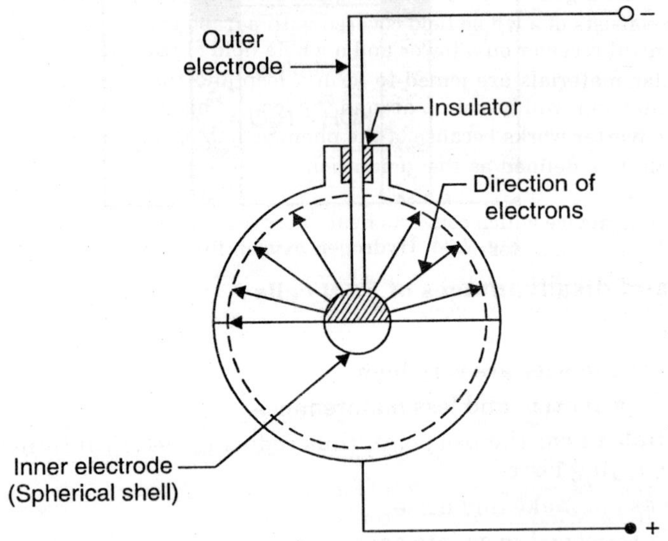

Fig. 7.35. High voltage atomic battery.

7.7.8.2. Low voltage atomic battery

Fig. 7.36 shows a low voltage battery.

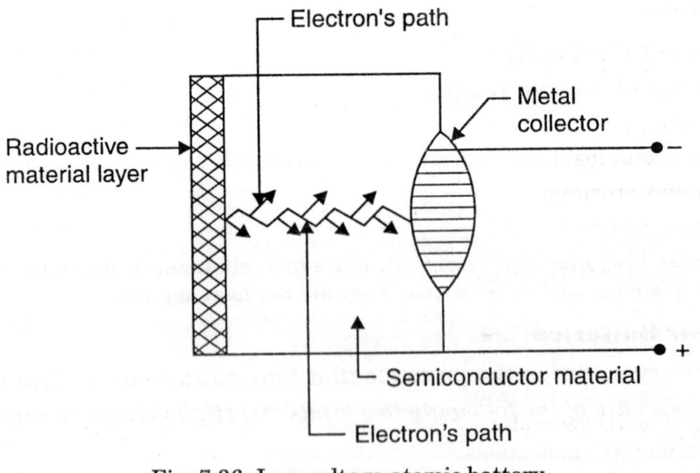

Fig. 7.36. Low voltage atomic battery

In this Sr^{90} (beta emitter) is deposited on the surface of a semiconductor (germanium or silicon) at one end and the other end is having a metallic collector. The semiconductor has the characteristic of undirectional current flow. The fast electrons pass through the semiconductor and strike the metallic disc at the other end and the two ends of the semiconductor become the – ve and + ve terminals of the battery. The power produced by this battery is a few microwatts at a potential difference of 1/10 volt. This power is sufficient to *feed a small radio.*

HIGHLIGHTS

1. The periodic rise and fall of the water level of sea which are carried by the action of sun and moon on water of the earth is called the *tide*.
2. In a single basin arrangement power can be generated only intermittently.
3. The solar farm consists of a whole field covered with parabolic trough concentrators and a 'solar tower' consists of a central receiver on a tower and a whole field of tracking.
4. "If two dissimilar materials are joined to form a loop and the two junctions maintained at different temperatures, an e.m.f. will be set up around the loop". This is called *Seebeck effect.*
5. A thermionic converter works because of the phenomenon of 'thermionic emission'.
6. 'Photovoltaic effect' is defined as the generation of an electromotive force as a result of absorption of ionizing radiation.
7. 'MHD generator' is device which converts heat energy of a fuel directly into electrical energy without a conventional electric generator.
8. A 'fuel cell' is an electrochemical device in which the chemical energy of a conventional fuel is converted directly and efficiently into low voltage, direct current electrical energy.

THEORETICAL QUESTIONS

1. List the various non-conventional energy sources.
2. Explain with a real diagram a wind electric generating power plant.
3. Write a shot-note no 'Wind Electricity Economics'.
4. Name the components of a tidal power plant.
5. Give the classification of tidal power plants.
6. Give the working of a single basin tidal power plant.
7. What are the advantages and limitations of tidal power generation ?
8. Explain with the help of a neat diagram a solar pond electric power plant.
9. Describe the working of a 'solar tower plant'.
10. Give the working of a geothermal plant.
11. What is 'Thermoelectric effect' ?
12. Write a short note on 'Thermoelectric generator'.
13. Define work function.
14. Explain with a neat diagram the working of a 'Thermoelectric generator'.
15. What is a photovoltaic cell ?
16. How are silicon cells fabricated ?
17. Write down the advantages and disadvantage of 'Photo-voltaic solar energy conversion'.
18. Describe open cycle MHD system.
19. What is a fuel cell ?
20. Describe a hydrogen-oxygen cell.
21. State the advantages and disadvantages of fuel cells.
22. Write short notes on the following :
    (*a*) Working of MHD generator.
    (*b*) Fuel cells and its applications.
23. Describe with the help a neat sketch, the working of a solar power plant. What are its salient features ?

8

Combined Operation of Different Power Plants

8.1. General aspects. 8.2. Advantages of combined operation of plants. 8.3. Load division between power stations. 8.4. Hydro-electric (storage type) plant in combination with steam plant. 8.5. Run-of-river plant in combination with steam plant. 8.6. Pump storage plant in combination with steam or nuclear power plant. 8.7. Co-ordination of hydro-electric and gas turbine stations. 8.8. Co-ordination of different types of power plants. Worked Examples—Theoretical Questions—Unsolved Examples.

8.1. GENERAL ASPECTS

The leading aim of the national economy is to make available maximum amount of generating capacity with the available funds and ensure power generation at the cheapest rate possible. Since a large investment is required in power supply industry, therefore, once generating facilities are created it is desirable to utilise them in optimum manner. It is also of paramount importance that most economic generating scheme should be selected to supply power at lowest cost before huge amount of money is invested. When the generating facilities are established it is a wiser step to think of having integrated operation of neighbouring power systems so that maximum energy generation takes place from power stations like *thermal* and *nuclear* and maximum energy and capacity are utilised from the *hydro-electric* power stations. This is possible only if we have close combined operation of different power systems which if operated individually cannot be utilized to the maximum advantage. This leads to conclusion that if maximum benefit is to be yielded then power systems of different states should be interconnected. It is beyond doubt that the rapid pace of interconnection between the power systems can greatly improve the *continuity, security* and *integrity of power supply* provided it is associated with sound mechanism for monitoring and control.

8.2. ADVANTAGES OF COMBINED OPERATION OF PLANTS

If several power stations (such as hydro, thermal, nuclear etc.) work together to meet the demand of the consumers then the system is known as **Interconnected system.** Such a combined system claims the following ***advantages*** over a single power plant/station :

1. Greater reliability of supply to the consumers.
2. When one of the stations fails to operate the consumers can be fed from the other station, thus avoiding complete shut down.
3. The overall cost of energy per unit of an interconnected system is less.
4. There is a more effective use of transmission line facilities at higher voltage.
5. Less capital investment required.
6. Less expenses on supervision, operation and maintenance.

7. The interconnection of different power plants reduces the amount of generating capacity required to be installed as compared to that which would be required without interconnection.
8. In an interconnected system the spinning reserve required is reduced.

8.3. LOAD DIVISION BETWEEN POWER STATIONS

Under the situation when the load curve has a very high peak value, it is usually supplied for two or more power stations/services by interconnection. In that case, total load as shown on load duration curve may be divided into following two parts :

(*i*) The base load (*ii*) The peak load.

Base load** is supplied by one power station and the other power station takes care of the **peak load. In such cases the *load is economically apportioned to various systems in order to achieve the overall economy.*

In such cases it is not very necessary to interconnect the two systems of the *same type*. For example, if the base load is supplied by the steam power station, then it is not very necessary that the peak load may also be supplied by the steam power station. A hydro-electric power station can very well be adopted for supplying the peak load. Similarly, a hydro-electric station can be used for supplying the base load and in that case the peak load can be supplied by steam power station or a diesel engine station or any other suitable unit. However the selection of the power stations for supplying the base load or peak load is made on the basis of the requirements and ability of the various power stations/services to meet those requirements.

Requirements of a plant supplying the '*Base load*' :

1. Minimum operation cost.
2. Continuous supply of the load.
3. Capital cost of the plant should be minimum.
4. Requirement of plant maintenance should be minimum.
5. Plant should be such that it can be easily located near the load centre.
6. The number of operators required should be minimum.
7. The spare parts etc. should be readily available.

Taking into view the above requirements, let us now consider various types of plants for their suitability to meet the base load.

Hydro-electric stations. (*i*) In these plants the *operating cost is minimum* as practically *no fuel* is required for the purpose of power generation, and as such there is no problem of procurement of the fuel.

(*ii*) *Maintenance cost is lower* than that of other plants.

(*iii*) *Initial cost of the plant is very high and sometimes prohibitive.*

(*iv*) These plants cannot be necessarily located near the load centre as the same can be located at the site suitable for it.

(*v*) In this case there is more or less dependence on availability of water, which in turn depends on the natural phenomenon of rain.

Steam power stations. (*i*) The capital investment in this case as compared to hydro-electric stations is less but with the modern trend of using higher pressures for the purpose, the cost of such stations has increased considerably. But this increased cost has resulted in lower operating costs so much so that even it may compete with that of hydro-electric power stations.

(*ii*) These plants can be easily located near the load centre, as such the cost of transmission lines and the losses occurring can be minimised which results in economical operation.

(*iii*) Maintenance requirement is slightly higher.

Diesel power stations. *Due to limited generation capacity of diesel power station it is not much suitable as a base load plant.*

Nuclear power station. (*i*) Initial cost is high although operating cost is comparable with that of steam power station.

(*ii*) Due to limited availability of the fuel these plants are not much favoured.

(*iii*) Whenever these plants are constructed they are invariably used as *base load* as otherwise also such plants suit to constant load conditions as economy in operation is such plants can be achieved only, when these are used as base load plants.

Requirements of a plant supplying the *'Peak load'* :

1. Low operating cost.
2. Minimum capital cost.
3. The plant should be capable of being started from cold conditions within minimum time.
4. In case of emergency the plant should have the capacity to withstand the peak load for sometime.
5. It must have quick response to the change in load.

In view of the above requirements let us consider various plants for their *suitability to meet the peak load.*

Hydro-electric stations. A hydro-electric station *can* be considered for this purpose :

(*i*) It can be easily started from cold conditions as no warming up period is required.

(*ii*) High initial cost is the major disadvantage.

(*iii*) However, in this case, there is no requirement of fuel, so whenever the output of hydro-electric station is to be utilised, it can be used as a *peak station also.*

(*iv*) Particularly in case where the quantity of water available for the purpose of power generation is limited, such stations are used for meeting the peak load in case of a interconnected system.

Nuclear power stations. They do not find their use as peak load plants.

Diesel power stations. (*i*) These plants *can be used for meeting peak load* as they can be easily started from cold conditions and the initial cost in this case is also not high.

(*ii*) In this case the additional advantage is that the number of auxiliaries required is also limited as a result of which the maintenance work required is also less.

(*iii*) These plants have got good overload capacity (minimum being 10 per cent as per Indian Standard Specifications).

(*iv*) Thermal efficiency of diesel engines being high, the operating cost is also less as compared to that of a steam power station of equivalent capacity.

Analysis of Load Sharing between *Base Load* and *Peak Load* Stations. The suitability of a plant not only depends on the above mentioned factor but also on local conditions for a particular application. However, if the combination of two plants/services is to be used, in that case the next problem is division of load between the plants. It is not desirable to transfer all the loads to one plant and also depending upon the operating characteristics of the various plants the load between the plants should be so decided that overall economy is achieved. This load division can be easily manipulated by viewing the load duration curve of the plants as follows :

Fig. 8.1 shows a load duration curve.

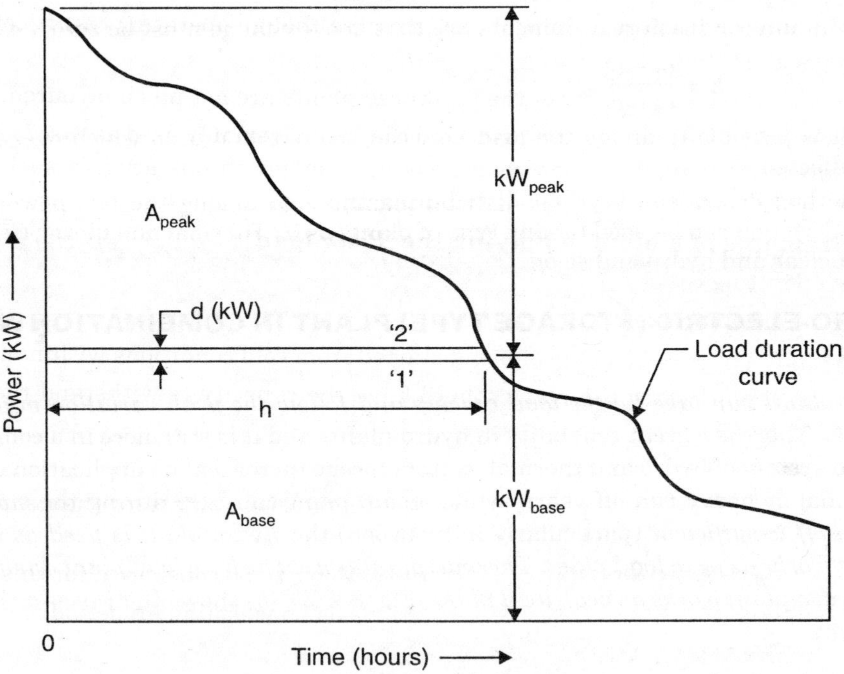

Fig. 8.1. Load duration curve of the plants.

Let,
A_{peak} = Area of curve for *peak load* plant,
A_{base} = Area of curve for *base load* plant,
kW_{peak} = Load for peak load plant,
kW_{base} = Load for base load plant,
C = Total operating cost of the combination, and
h = Hours per year.

Let the *base load* be supplied by a plant having the annual cost equation as

$$(\text{Rs.})_1 = a_1 + b_1 \text{ kW} + c_1 \text{ kWh} \quad ...(8.1)$$

For the plant supplying the *peak load* let the equation be

$$(\text{Rs.})_2 = a_2 + b_2 \text{ kW} + c_2 \text{ kWh} \quad ...(8.2)$$

Since the base load plant is operated most of the time, therefore, normally a plant having $c_1 < c_2$ is used for meeting the base load.

Let $b_1 > b_2$.

Let the load between the two plants (Fig. 8.1) be divided by arbitrary line drawn on the load duration curve represented by '1'. Under these conditions let kW_{base} be the kW for base load plant and let kW_{peak} be the load for peak load plant.

In this case the total operating cost of the combination is given as :

$$C_1 = a_1 + a_2 + b_1 \text{ kW}_{base} + b_2 \text{ kW}_{peak} + c_1 A_{base} + c_2 A_{peak} \quad ...(8.3)$$

Now, if the base power is extended by the amount of d (kW) to line '2', the total operating cost of the combination will modify as follows :

$$C_2 = a_1 + a_2 + b_1 (\text{kW}_{base} + d \text{ kW}) + b_2 (\text{kW}_{peak} - d \text{ kW}) + (A_{base} + d \text{ kW} \times h)\, c_1 + (A_{peak} - d \text{ kW} \times h)\, c_2 \quad ...(8.4)$$

The change in cost,

$$C_2 - C_1 = (b_1 - b_2)\, d\text{ kW} + (c_1 - c_2)\, d\text{ kW} \times h \qquad \text{...(8.5)}$$

The optimum condition requirements are that above change must be zero, *i.e.*,

$$h = \frac{b_1 - b_2}{c_2 - c_1} \qquad \text{...(8.6)}$$

Thus it is possible to divide the load between the plants due to which overall economy in operation is effected.

The method described above for distributing the load among the two power plants in an interconnected system can be used for any type of plants as (*i*) Thermal and diesel, (*ii*) Thermal and hydro, (*iii*) Nuclear and hydro and so on.

8.4. HYDRO-ELECTRIC (STORAGE TYPE) PLANT IN COMBINATION WITH STEAM PLANT

Hydro-plants can take up the load quickly and follow the peak variation much better than thermal plants. There is a great reliability in hydro-plants and it is still more in a combined system. In a combined system of hydro and thermal, water storage increases the application of more hydro-power in normal or heavy run-off years, while steam plant can help during the time of drought. *When the run-off is sufficient* (particularly in monsoon) the *hydro-plant is used as base load and thermal plant works as peak load plant. Thermal plant is used as base load plant during the drought period and hydro-plant works as peak load plant.* Fig. 8.2 (*a*), (*b*) shows their uses as base load or as peak load plant.

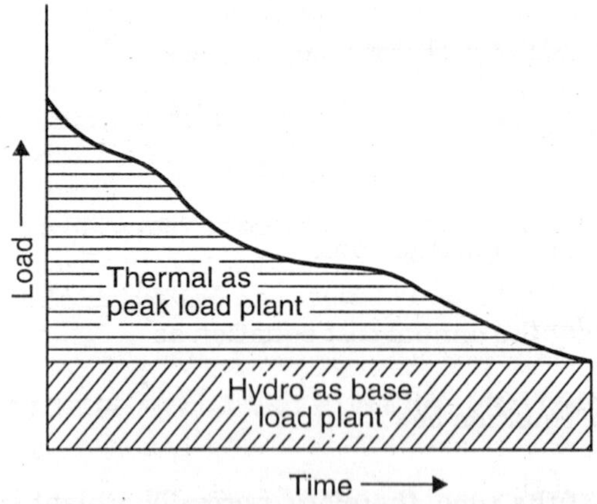

Fig. 8.2 (*a*). Hydro-plant used as base load plant during normal run-off in an interconnected system.

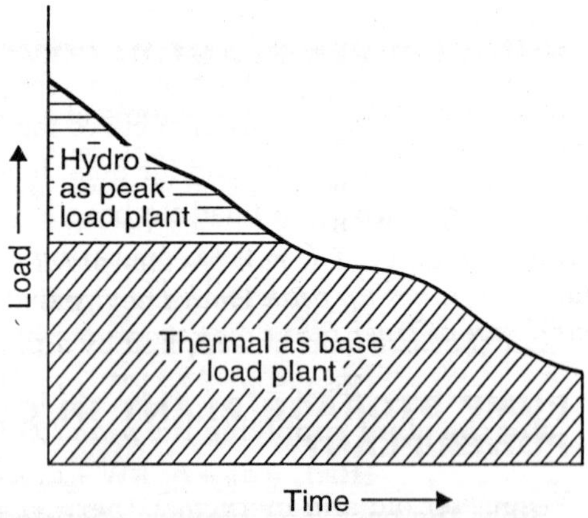

Fig. 8.2 (*b*). Hydro-plant used as peak load during drought period in an interconnected system.

Thermal plants can be used at any portion of the load duration curve but it is more expensive to use peak load station at low load factors.

Following cases will be discussed :

1. Predominant hydro.
2. Predominant thermal.
3. Hydro and thermal equally predominant.

Predominant Hydro :

Some of the hydro-plants, such as run-off river plants, are used as base load plants whereas some others are used as peak load plants. *When the hydro-plants carry the major demand throughout the year then the thermal plants are used in a combined system to improve the hydropower efficiency during the periods when there is low run-off.*

Predominant Thermal :

To develop hydro-plants to operate even at comparatively low annual load factors is always advantageous. This is due to the fact the cost of storage water forms a major portion of the capital investment which is independent of annual load factor and capital cost is less for low than for a high load factor. Thus, *in a predominantly thermal station,* it is *preferable to develop hydro-power at the lowest practical load factor.*

Hydro and Thermal Equally Predominant :

The economic balance between hydro and thermal power in an interconnected system at any time depends upon the *nature of load curve, run-off and its seasonal variation, cost of fuel, availability of condensing water* etc. There is an optimum ratio of hydro-power to total peak demand which gives minimum cost for power supply. This is particularly true for the areas where the cost of hydro-power development is high and fuel cost is low. In areas where fuel is cheap and cost of hydro-power development is not high, the economic power ratio lies between 0.25 to 0.4. In areas where fuel is costly and favourable hydro-power plant sites are abundant the ratio will be higher to the tune of 0.8-0.9.

The combined system of hydro and thermal plants is being adopted all the world over, and is particularly useful to developing countries like India where economy is desirable at every stage of development.

8.5. RUN-OF-RIVER PLANT IN COMBINATION WITH STEAM PLANT

Since during the year the supply of water is not regular in *run-of-river plants*, therefore,these plants cannot meet with variable load requirements. Further as the variation of run-off during the year does not match the variation of power demand during the year, therefore, it becomes necessary to combine such a hydro-plant with steam plant to supply the load according to requirement with maximum reliability. *The run-of-river plant can be used as base load plant during rainy season and thermal plant takes up peak load. During dry season ; the thermal stations can be used as base load plant and run-of-river plant may work as peak load plant.*

8.6. PUMP STORAGE PLANT IN COMBINATION WITH STEAM OR NUCLEAR POWER PLANT

Whenever old and inefficient thermal stations are available they are generally used to take up peak loads. If suitable plants are not available to take load it is desirable to develop pumped storage plant for the purpose. In an interconnected system a pumped storage plant is useful in supplying sudden peak loads of short duration (a few hours in the year). Such a plant (pumped storage) possesses the following advantages when used in interconnected system :

(*i*) Thermal plants are loaded more economically.

(*ii*) The wastage of off-peak energy of thermal plants is reduced.

(*iii*) A pumped storage plant stores the energy using off-peak energy of thermal plant and the same is supplied when demand arises.

A combined system of a pumped storage plant and nuclear power plant is proposed at Ramganga power station in Uttar Pradesh. A parallel development of nuclear power station

(400 MW capacity) near the hydro-site could provide cheap pumping energy to pump storage plant during off peak hours. This arrangement will enable the nuclear station to operate at high capacity factor and thereby make it competitive as compared to the conventional thermal station in that region where the coal prices are relatively high.

8.7. CO-ORDINATION OF HYDRO-ELECTRIC AND GAS TURBINE STATIONS

The working of gas turbine plants at peak load points is most economical under the following conditions :

(*i*) When the amount of energy supplied at peak load is small part of the total energy.

(*ii*) The load factor is less than 15%.

The following points are worth noting :

— The normal capacity of the gas turbine varies from 10 MW to 25 MW.

— The capital cost of a gas turbine plant is quite less (nearly 60%) compared with steam plant.

— The thermal efficiency of gas turbine plant (25 per cent) is less than that of a steam plant (32 per cent).

— The high working cost of a gas turbine is compensated by *lower fixed charges and lower operating and maintenance charges.*

When used as a peak load plant, the *gas turbine plant claims the following advantages over steam plant* :

1. A gas turbine plant occupies less space comparatively.
2. Heavy foundations are not required.
3. The construction and installation can be carried out in a smaller period.
4. The cooling water requirement is much less comparatively.
5. Less number of operators are required.
6. The response of gas turbine plant is quick.

8.8. CO-ORDINATION OF DIFFERENT TYPES OF POWER PLANTS

In a particular region if different types of power plants are available it becomes necessary to co-ordinate them and use them with maximum economy. The problem of co-ordination of different types of power plants (*e.g.,* hydro, thermal, nuclear, gas turbine and diesel plants) for best possible working and economy is very complicated as the *factors* to be considered for economical co-ordination are large in number ; some of these factors are listed below :

(*i*) Initial capital cost.

(*ii*) Fuel cost.

(*iii*) Operation and maintenance cost.

(*iv*) Availability of fuel.

(*v*) The economics of base load and peak load operations.

(*vi*) The working characteristics of the plants.

(*vii*) The transmission liability.

(*viii*) The cost of incremental power.

The best co-ordination, several times, depends on the *nature of load duration curve* and availability of fuels and resources in the country.

Fig. 8.3 shows an annual load curve showing loads allocated to different plants.

The local conditions may change the sequence depending upon the availability of fuels and resources.

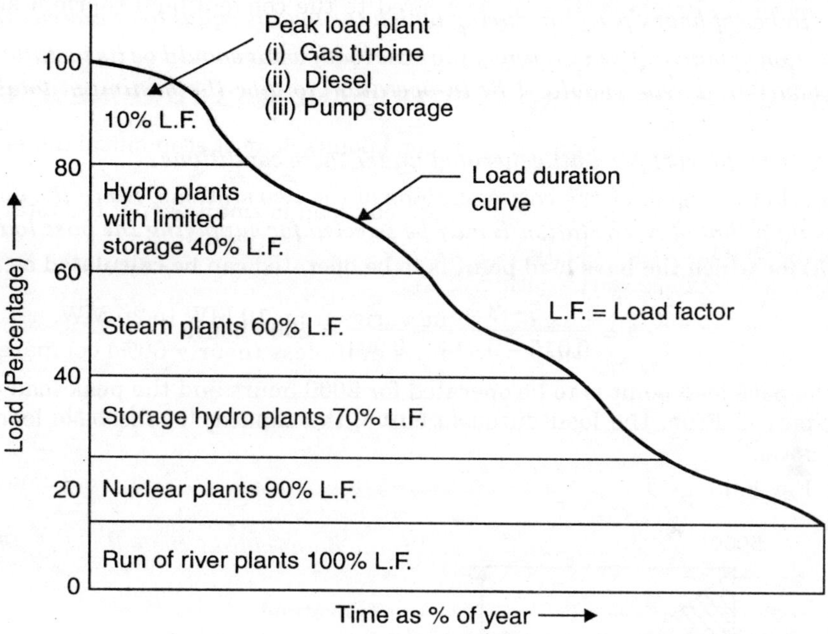

Fig. 8.3. Annual load duration curve showing loads allocated to different plants.

WORKED EXAMPLES

Example 8.1. *The estimated total annual operating costs and capital charges for two proposed power stations are given by the following expressions :*

Annual cost for station A = *Rs.* $(6,00,000 + 3.0\ kW + 0.015\ kWh)$

Annual cost for station B = *Rs.* $(7,50,000 + 5.0\ kW + 0.014\ kWh)$

where kW represents the capacity of the station and kWh represents the total annual output.

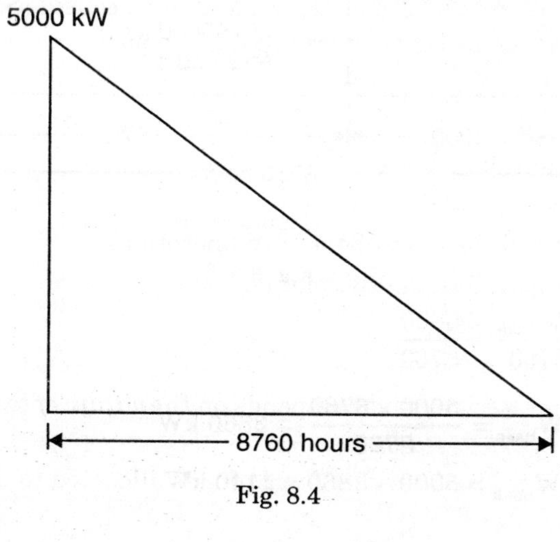

Fig. 8.4

The stations are to be used for supplying a load having a load duration curve as shown in Fig. 8.4. The ordinate of a point on this curve represents a certain load on the station and the abscissa represents the number of hours per year during which the load is equal to or exceeds this amount.

Which station should be used to supply the base load, what should be its installed capacity and for how many hours in a year should it be in operation to give the minimum total cost per unit generated ?

Calculate the total cost per unit generated under these conditions.

Solution. Let us consider the cost equations of the two stations. *Since the operating cost of station B is less than that of A, so station B may be selected for supplying the base load.*

Hours (h) for which the base load point is to be operated can be calculated as follows :

$$h = \frac{5-3}{0.015-0.014} = \frac{2}{0.001} = 2000 \text{ hours.}$$

Hence the base load point is to be operated for 2000 hours and the peak load point plant for the remaining period. From the load duration curve, the capacity of the base load plant can be computed as follows.

Refer to Fig. 8.5.

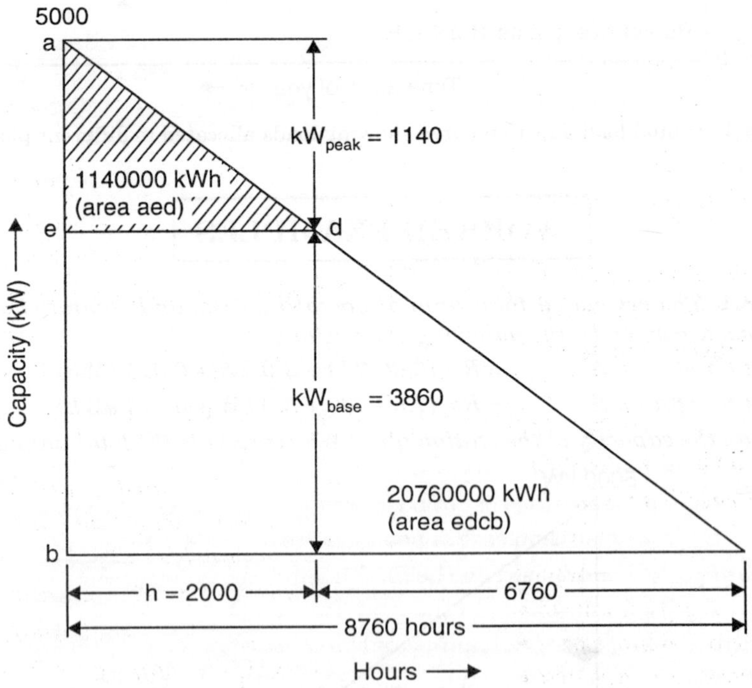

Fig. 8.5

$$\frac{kW_{base}}{6760} = \frac{5000}{8760}$$

$$\therefore \quad kW_{base} = \frac{5000 \times 6760}{8760} \simeq 3860 \text{ kW}$$

$$\therefore \quad kW_{peak} = 5000 - 3860 = 1140 \text{ kW}$$

Cost of generation

Total kWh supplied by peak load plant

$$= \frac{1140 \times 2000}{2} = 11,40,000 \text{ kWh}$$

Total kWh supplied by base load plant

$$= \frac{5000 \times 8760}{2} - 11,40,000 = 2,07,60,000 \text{ kWh}$$

Cost of generation per annum for *base load plant*

= Rs. (7,50,000 + 5 × 3,860 + 0.014 × 2,07,60,000)

= Rs. (7,50,000 + 19,300 + 2,90,640)

= Rs. 10,59,940.

Cost of generation per annum for *peak load plant*

= Rs. (6,00,000 + 3 × 1,140 + 0.015 × 11,40,000)

= Rs. (6,00,000 + 3,420 + 17,100)

= Rs. 6,20,520

Total cost = 10,59,940 + 6,20,520 = 16,80,460

$$\textit{Cost of generation per unit} = \frac{\text{Total cost}}{\text{Total kWh supplied}}$$

$$= \frac{16,80,460}{\left(\frac{5,000 \times 8,760}{2}\right)} = \frac{16,80,460}{2,19,00,000}$$

= Rs. 0.0767 or 7.67 P. (Ans.)

Example 8.2. *In a system the load duration curve is a straight line, the maximum and minimum loads being 120 MW and 24 MW respectively. The costs of the base load and peak load plants which supply the load, are given below :*

Base load plant : Rs. 240/kW-year + 6 p./kWh

Peak load plant : Rs. 60/kW-year + 12 p./kWh

Determine, for minimum overall cost of the following :

(*i*) *The load shared by peak load plant.*

(*ii*) *The annual load factors for both stations.*

Solution. (*i*) **Load shared by peak load plant :**

Let, C_{base} = Operating cost of base load plant,

C_{peak} = Operation cost of peak load plant,

P_{base} = Load (peak) on the base load plant, and

P_{peak} = Load (peak) on the peak load plant.

The operating costs of base load and peak load plants may be expressed as follows :

$$C_{base} = A_1 \text{ kW} + B_1 \text{ kWh}$$

$$C_{peak} = A_2 \text{ kW} + B_2 \text{ kWh}$$

From the given data :

$$A_1 = 240, \quad B_1 = 0.06$$

$$A_2 = 60, \quad B_2 = 0.12$$

The time, *h* hours, for which base load to be operated *for minimum overall cost* is given by :

$$h = \frac{A_1 - A_2}{B_2 - B_1} = \frac{240 - 60}{0.12 - 0.06} = 3000 \text{ hrs.}$$

Now, $P_{base} + P_{peak} = 120$

$\therefore$ $P_{peak} = 120 - P_{base}$

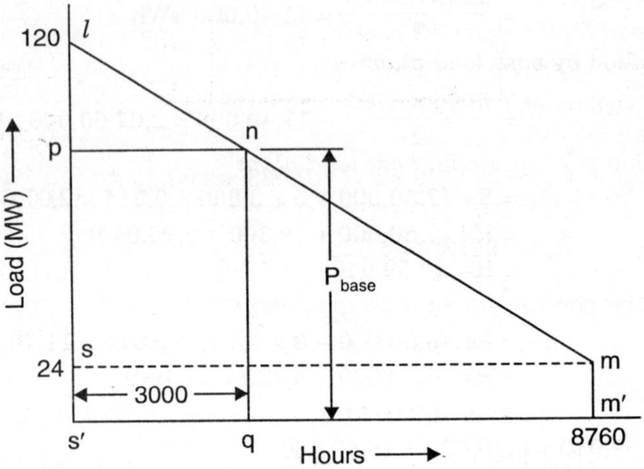

Fig. 8.6

Now, from Δs *lms* and *lnp*, we have

$$\frac{120 - P_{base}}{120 - 24} = \frac{3000}{8760}$$

$$P_{base} = 120 - (120 - 24) \times \frac{3000}{8760} = \mathbf{87\ MW.} \quad \textbf{(Ans.)}$$

$$\therefore \quad P_{peak} = 120 - 87 = \mathbf{33\ MW.} \quad \textbf{(Ans.)}$$

(*ii*) **Annual load factors (L.F.) for both the stations :**

$$\text{L.F. (base load plant)} = \frac{\text{Average load}}{\text{Peak load}} \times \frac{8760}{8760} = \frac{\text{Area '}pnmm's'p\text{'}}{P_{base} \times 8760}$$

$$= \frac{\text{Area '}pnmsp\text{' + area '}smm's's\text{'}}{P_{base} \times 8760}$$

$$= \frac{\frac{1}{2}(3000 + 8760) \times (87 - 24) + 24 \times 8760}{87 \times 8760}$$

$$= \frac{370440 + 210240}{762120} = 0.762 \text{ or } 76.2\%$$

i.e., Load factor for base load plant = **76.2%. (Ans.)**

$$\text{L.F. (peak load plant)} = \frac{\text{Area '}lnpl\text{'}}{P_{peak} \times 8760}$$

$$= \frac{\frac{1}{2} \times 3000 \times 33}{33 \times 8760} = 0.171 \text{ or } 17.1\%$$

i.e., Load factor for peak load plant = **17.1%. (Ans.)**

Example 8.3. *The two power stations X and Y supply to a system whose maximum loac 120 MW and minimum load is 12 MW during the year. The estimated costs of these stations are follows :*

$$C_X = Rs.\ (120 \times kW + 0.028 \times kWh)$$

$$C_Y = Rs. (115 \times kW + 0.032 \times kWh)$$

If the load varies as a straight line, find for minimum cost of generation :

(i) Installed capacity of each station.

(ii) The annual load factor, capacity factor and use factor of each machine.

(iii) The average cost of production per kWh for the entire system.

Assume reserve capacity of Y as 22%.

Solution. Given :

$$C_X = \text{Rs. } (120 \times \text{kW} + 0.028 \times \text{kWh})$$
$$C_Y = \text{Rs. } (115 \times \text{kW} + 0.032 \times \text{kWh})$$

i.e., $A_1 = 120,\ B_1 = 0.028$

and $A_2 = 115,\ B_2 = 0.032$

$$h = \frac{A_1 - A_2}{B_2 - B_1} = \frac{120 - 115}{0.032 - 0.028} = 1250 \text{ hours}$$

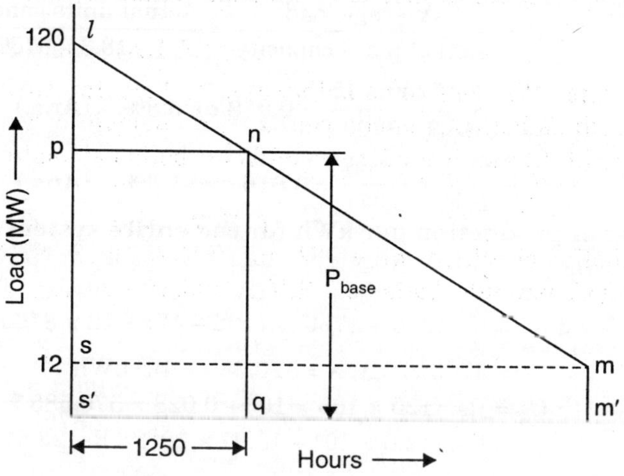

Fig. 8.7

From Δs *lms* and *lnp*, we have

$$\frac{120 - P_{base}}{120 - 12} = \frac{1250}{8760}$$

$$\therefore \quad P_{base} = 120 - (120 - 12) \times \frac{1250}{8760} \simeq 105 \text{ MW}$$

$$\therefore \quad P_{peak} = 120 - 105 = 15 \text{ MW}$$

(*i*) **Installed capacity of each station :**

Installed capacity of base load plant,

$$P_{base} = \mathbf{105\ MW.} \quad \textbf{(Ans.)}$$

Installed capacity of peak load plant = 15 × 1.22 = **18.3 MW. (Ans.)**

(*ii*) **The annual load, capacity and use factors for each station :**

For base load plant :

$$\textit{Load factor} \text{ (L.F.)} = \frac{\text{Actual units generated}}{P_{base} \times 8760}$$

$$= \frac{\frac{1}{2}(1250 + 8760) \times (105 - 12) + 112 \times 8760}{105 \times 8760}$$

$$= \frac{465465 + 105120}{919800} = \mathbf{0.62 \text{ or } 62\%.} \quad \textbf{(Ans.)}$$

As there is no reserve capacity of plant X,

$\therefore$ *Capacity factor (C.F.)* = L.F. = **62%.** **(Ans.)**

Use factor (U.F.) $= \frac{\text{C.F.}}{\text{L.F.}} = \frac{0.62}{0.62} = \mathbf{1.}$ **(Ans.)**

For peak load plant :

$$\text{L.F.} = \frac{\text{Actual units generated}}{P_{peak} \times 8760}$$

$$= \frac{\frac{1}{2} \times 1250 \times 15}{15 \times 8760} = \mathbf{0.071 \text{ or } 7.1\%.} \quad \textbf{(Ans.)}$$

$$\text{C.F.} = \frac{\text{Average load}}{\text{Actual plant capacity}} = \frac{\text{Actual units generated}}{18.3 \times 8760}$$

$$= \frac{\frac{1}{2} \times 1250 \times 15}{18.3 \times 8760} = \mathbf{0.058 \text{ or } 5.8\%.} \quad \textbf{(Ans.)}$$

$$\text{U.F.} = \frac{\text{C.F.}}{\text{L.F.}} = \frac{0.058}{0.071} = \mathbf{0.816 \text{ or } 81.6\%.} \quad \textbf{(Ans.)}$$

(*iii*) **Average cost of production per kWh for the entire system :**

For plant X (base) :

Total units generated $= \frac{1}{2}(1250 + 8760) \times (105 - 12) + 12 \times 8760$

$= 570585 \text{ MWh} = 570.585 \times 10^6 \text{ kWh}$

$\therefore$ $C_X = \text{Rs. } (120 \times 105 \times 10^3 + 0.028 \times 570.585 \times 10^6)$

$= \text{Rs. } (12.6 \times 10^6 + 15.97 \times 10^6) = \text{Rs. } 28.57 \times 10^6$

For plant Y (peak) :

Total units generated $= (\frac{1}{2} \times 1250 \times 15) = 9375 \text{ MWh}$

$= 9.375 \times 10^6 \text{ kWh}$

$\therefore$ $C_Y = \text{Rs. } (115 \times 15 \times 10^3 + 0.032 \times 9.375 \times 10^6)$

$= \text{Rs. } (1.725 \times 10^6 + 0.3 \times 10^6) = \text{Rs. } 2.025 \times 10^6$

Total units generated from both the plants

$= 570.585 \times 10^6 + 9.375 \times 10^6 \simeq 580 \times 10^6 \text{ kWh}$

Total generation cost,

$$C = C_X + C_Y = 28.57 \times 10^6 + 2.025 \times 10^6 = \text{Rs. } 30.595 \times 10^6$$

$\therefore$ *Average cost* $= \frac{\text{Total generation cost}}{\text{Total units generated}} = \frac{30.595 \times 10^6}{580 \times 10^6}$

= **Rs. 0.0527 or 5.3 paise/kWh.** **(Ans.)**

Example 8.4. *In an industry the maximum and minimum demands are 60 MW and 12 MW respectively and the variation is linear. The hydro-power plant can take the load of 72 MWh per day of the factory at the time of minimum regulated flow and the remaining is supplied by the thermal plant. It is proposed to pump the water from tailrace of the existing plant to reservoir during off-peak* *period of thermal plant and allow the thermal plant to run always of full load condition to economise the supply of power.*

If the overall efficiency of conversion of steam off-peak power to hydel potential power and then hydel power to electric power is 62%, calculate the capacity of steam and hydel plants considering the pumping of water during off-peak period of steam plants.

Solution. Fig. 8.8 shows the load duration curve.

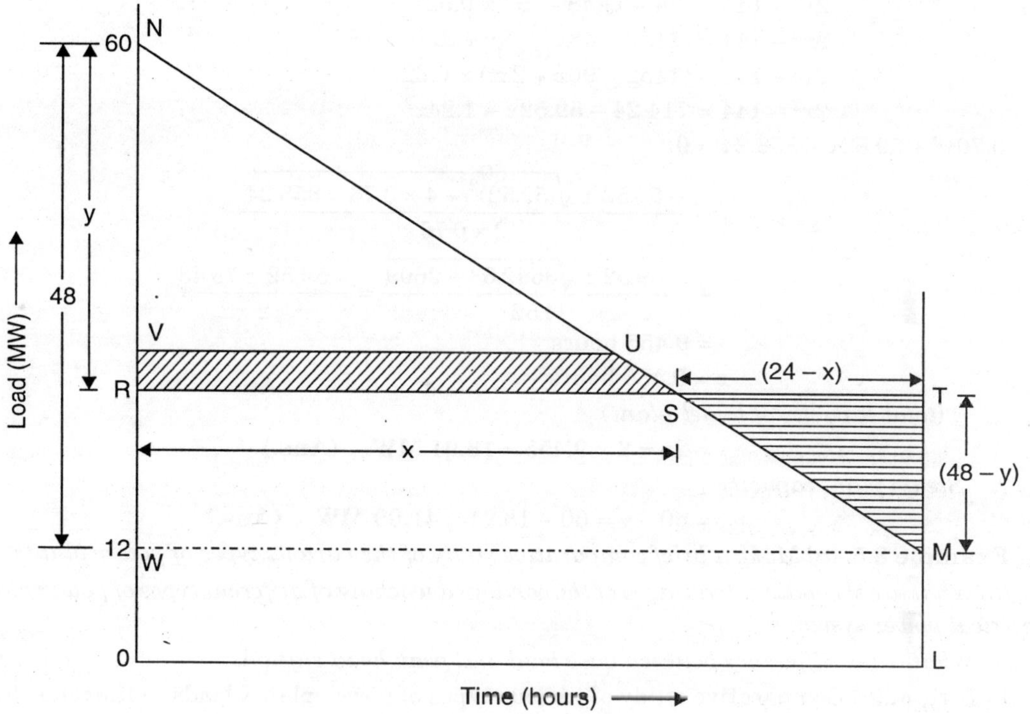

Fig. 8.8

Maximum supply = 60 MW

Let, y = The load taken by existing hydel plant and pump storage plant (in MW).

Then $(60 - y)$ = The power supplied by thermal plant (in MW).

The off-peak power of thermal plant represented by the area '*STMS*' can be used for pumping the water. The power supplied by the existing hydel plant will be given by the area '*NVUN*'.

Hydel potential in electrical form supplied by pump-storage plant must be equal to the area '*RSUVR*'.

∴ Area '*RSUVR*' = 0.62 × area '*STMS*' (data given)

∴ Area '*NSRN*' – area '*NUVN*' = 0.62 × area '*STMS*'.

But area '*NUVN*' represents the energy supplied by the existing hydel plant which is given as 72 MWh.

Let, x = Number of hours (out of 24 hours) for which the steam plant operates at full load condition.

Then, $(24 - x)$ = Number of hours for which the steam plant works under off-peak conditions.

$$\therefore \quad \frac{1}{2}xy - 72 = \frac{1}{2}(24 - x)(48 - y) \times 0.62$$

or, $$xy - 144 = (24 - x)(48 - y) \times 0.62 \qquad ...(i)$$

Now from Δs *NRS* and *NWM*, we can write

$$\frac{y}{48} = \frac{x}{24}$$

$$\therefore \quad y = 2x \qquad ...(ii)$$

Substituting the value of '*y*' in eqn. (*i*), we get

$$2x^2 - 144 = (24 - x)(48 - 2x) \times 0.62$$

or, $$2x^2 - 144 = (1152 - 48x - 48x + 2x^2) \times 0.62$$

or, $$2x^2 - 144 = (1152 - 96x + 2x^2) \times 0.62$$

$$2x^2 - 144 = 714.24 - 59.52x + 1.24x^2$$

or, $$0.76x^2 + 59.52x - 858.24 = 0$$

$$\therefore \quad x = \frac{-59.52 \pm \sqrt{(59.52)^2 + 4 \times 0.76 \times 858.24}}{2 \times 0.76}$$

$$= \frac{-59.52 \pm \sqrt{3542.63 + 2609}}{1.52} = \frac{-59.52 \pm 78.43}{2}$$

$$= 9.455 \text{ hours}$$

i.e., $$x = 9.455 \text{ hours}$$

∴ *y (total capacity of hydel-plant)*

$$= 2x = 2 \times 9.455 = \mathbf{18.91\ MW.} \quad \textbf{(Ans.)}$$

∴ *Steam plant capacity*

$$= 60 - y = 60 - 18.91 = \mathbf{41.09\ MW.} \quad \textbf{(Ans.)}$$

Example 8.5. *(a) Make a brief comparative study of the various types of power plants.*

(b) What are the main advantages of the combined working of different types of power plants in an electrical power system ?

(c) What is the difference between base load and peak head stations ? **(N.U.)**

Solution. (*a*) Comparative study of various types of power plants leads to the following *conclusions* :

1. The initial cost of a power plant using water as the source of power is more than for the plant where fuel is used as the source of power. But this cost is less than the cost of erecting a nuclear power plant.
2. Running cost is cheaper in the case of hydro electric power station than a steam generated power station, the cost of running a nuclear power station is least of all.
3. Of all the three sources the nuclear source for generating power is inexhaustible.
4. The clearest, simplest and easier way for producing power is from water. There is no danger of any hazards, which are common in nuclear sources. When coal is used there is nuisance of smoke, ashes etc.
5. The quantity of raw material used for generating a given amount of power is least in the case of nuclear power plants.

(*b*) Major advantages of the combined working of different types of power plants in an electrical power system are :

1. Increased reliability of supply.
2. Reduction in total installed capacity.
3. Economic operation.

Thus in case of interconnected power stations, if one of the station fails the supply can be restored from other generating stations. Total reserve capacity requirement is also reduced and depending upon the total power requirement the generators to be operated can be planned to have most economic generation.

(c) *If a plant runs continuously at constant load*, it is called as **base load plant**.

If the plant is run only for the duration to supply the peak load at the load centre it is called **peak load plant**. The plants which *can be run and shut off quickly are mostly used as peak load plants.*

- Commonly used *peak load plants* are *storage type hydro-plant, Diesel or gas turbine plants.*
- Commonly used *base load plants* are the *thermal and nuclear plants.*

Example 8.6. *Discuss the importance of load forecasting in power system. Briefly explain two methods of load forecasting.*

Solution. As power plant planning and construction requires a generation period of four to eight years or even longer for the present day super power stations, energy and load demand forecasting plays a crucial role in power system studies. If load is not estimated or forecasted for years to come it can not be made possible to meet the electrical demands, which will largely affect the industrial production and the comfort of living. It is therefore very essential to have a load forecast. Normally forecast is done for 20 years ahead.

Methods commonly adopted for load forecasting are :

1. Load survey.
2. Method of extrapolation.
3. Mathematical methods.
4. Mathematical methods using economic parameters.

Load survey method :

- In load survey method, loads are grouped under different categories such as residential or domestic, commercial, industrial etc.
- Detailed information category-wise and area-wise is collected to estimate the demand and its rate of growth, duration and time of occurrence of load and the energy requirements etc. This enables a forecast of electrical load and energy.
- This method normally covers a period of 8 to 10 years for planning.

Method of extrapolation :

Method of extrapolation estimates the yearly increase in energy consumption. These methods are essentially statistical. Per capita typical yearly increase in energy are estimated for different types of loads in forecasting.

THEORETICAL QUESTIONS

1. State the advantages of operating the power plants combinedly in electric power system.
2. Describe the working of hydro-electric plants having ample storage with steam power plants.
3. How would you make an economic analysis of the combined operation of the hydro and steam power plants ?
4. What are the advantages of pump storage plant as peak load plant in an interconnected system ?
5. State the advantages of gas turbine plant as peak load plant in an interconnected system.
6. Discuss the suitability of steam power plants to supply the load in case of an interconnected system.
7. List the factors which decide the distribution of plants for operation on different portions of the annual load duration curve of a power system.

UNSOLVED EXAMPLES

1. The two power stations I and II supply to a system whose maximum load is 120 MW and minimum load is 12 MW during the year. The estimated costs of these stations are as follows :

$$C_I = \text{Rs. } (125 \times \text{kW} + 0.0275 \times \text{kWh})$$
$$C_{II} = \text{Rs. } (120 \times \text{kW} + 0.03 \times \text{kWh})$$

If the load varies as a straight line, find for minimum cost of generation :

(*i*) Installed capacity of each station.
(*ii*) The annual load, capacity and capacity use factor of each station.
(*iii*) The average cost of production per kWh for entire system.

Assume reserve capacity of II as 20 per cent.

[**Ans.** (*i*) 80 MW, 24 MW (*ii*) 65%, 65%, U.F. = 1 ; 11.5%, 9.5%, U.F. = 82.5% (*iii*) 5.4 paise/kWh]

2. The annual load duration curve of a station varies uniformly from 64000 kW to zero. The load is supplied by two stations whose cost equations are given as :

$$C_1 = \text{Rs. } (84000 + 84 \text{ kW} + 0.0116 \text{ kWh})$$
$$C_2 = \text{Rs. } (50000 + 44 \text{ kW} + 0.02985 \text{ kWh})$$

Find the minimum cost of generation in paise/kWh for the system. [**Ans.** 3 paise/kWh (approx.)]

3. An annual load duration curve of a system of loads is a straight line with maximum of 12 MW at the beginning and 2 MW at the end of the year. Annual costs of base and peak load stations are given below :

$$C_1 = 8000 + \text{Rs. } 75/\text{kW} + 3 \text{ paise/kWh (base load)}$$
$$C_2 = 6000 + \text{Rs. } 55/\text{kW} + 4 \text{ paise/kWh (peak load)}$$

Determine the following :

(*i*) The duration of time when peak load station will work in order to obtain the minimum annual cost.
(*ii*) The lowest overall cost per kW (in paise). [**Ans.** (*i*) 2000 hours (*ii*) 4.66 paise/kWh]

4. In a system the load duration curve is a straight line with a maximum demand of 45 MW tapering to zero. The load is to be taken from two sources whose annual cost equations are : C_1 = Rs. $(100 \times 10^3 + 65.7 \times \text{kW} + 0.012 \times \text{kWh})$ and C_2 = Rs. $(60 \times 10^3 + 36.5 \times \text{kW} + 2.122 \times \text{kWh})$. Find the installed capacity and service hours for each station per year to give minimum cost per unit and cost per unit.

[**Ans.** 30 MW and 15 MW, 2920 hours and 58 hours, 2.72 paise/kWh]

5. In an industry the maximum and minimum demands are 50 MW and 10 MW respectively and variation takes place linearly. The hydro-power plant can take the load of 60 MWh per day of the factory at the time of minimum regulated flow and the remaining is supplied by the thermal plant. It is proposed to pump the water from tailrace of the existing plant to reservoir during off-peak period of thermal plant and allowing the thermal plant to run always at full load condition to economise the supply of power. If the overall efficiency of conversion of steam off-steam off-peak power to hydel potential power and then hydel power to electric power is 60 per cent, calculate the capacity of steam and hydel plants considering the pumping of water during off-peak period of steam plant.

6. (*a*) Explain load-duration curve. What are annual operating costs ? What are the factors that influence the depreciation of capital equipment ?

(*b*) The estimated total annual operating costs and capital charges for two stations are given by the following expressions :

Annual cost for station *A* :

$$\text{Rs. } (10^5 + 60 \text{ kW} + 0.01 \text{ kWh})$$

Annual cost for station *B* :

$$\text{Rs. } (6 \times 10^4 + 35 \text{ kW} + 0.02 \text{ kWh})$$

where kW represents the capacity of the station and kWh represents the total annual energy generated.

The stations are to be used to supply a common load having annual load duration curve approximated by a straight line, maximum and minimum loads being 50 MW and zero respectively. Find the following :

(*i*) Which station should be used to supply the peak load ?

(*ii*) What should be its installed capacity ?

(*iii*) For how many hours per year should it be in operation to give the minimum total cost per unit generated ?

Calculate also the total cost per unit generated under these conditions.

9

Economics of Power Generation

9.1. Introduction. 9.2. Terms and definitions. 9.3. Principles of power plant design. 9.4. Location of power plant. 9.5. Layout of power plant building. 9.6. Cost analysis. 9.7. Selection of type of generation. 9.8. Selection of power plant equipment : selection of boilers—selection of prime-movers—selection of size and number of generating units. 9.9. Economics in plant selection. 9.10. Factors affecting economics of generation and distribution of power. 9.11. How to reduce power generation cost ? 9.12. Power plant—useful life. 9.13. Economics of hydro-electric power plants. 9.14. Economics of combined hydro and steam power plants. 9.15. Performance and operating characteristics of power plants. 9.16. Economic load sharing—Worked Examples. 9.17. Tariff for electrical energy—Introduction—objectives and requirements of tariff—general tariff form. Highlights—Theoretical Questions—Unsolved Examples.

9.1. INTRODUCTION

In all fields of industry economics plays an important role. In power plant engineering economics of power system use certain well established techniques for choosing the most suitable system. The power plant design must be made on the basis of most economical condition and not on the most efficient condition as the profit is the main basis in the design of the plant and its effectiveness is measured financially. *The main purpose of design and operation of the plant is to bring the cost of energy produced to minimum.* Among many factors, the efficiency of the plant is one of the factors that determines the energy cost. In majority of cases, unfortunately, the most thermally efficient plant is not economic one.

9.2. TERMS AND DEFINITIONS

1. Connected load. The connected load on any system, or part of a system, *is the combined continuous rating of all the receiving apparatus on consumers' premises, which is connected to the system, or part of the system, under consideration.*

2. Demand. The demand of an installation or system *is the load that is drawn from the source of supply at the receiving terminals averaged over a suitable and specified interval of time.* Demand is expressed in kilowatts (kW), kilovolt-amperes (kVA), amperes (A), or other suitable units.

3. Maximum demand or Peak load. The maximum demand of an installation or system *is the greatest of all the demands that have occurred during a given period.* It is determined by measurement, according to specifications, over a prescribed interval of time.

4. Demand factor. The demand factor of any system, or part of a system, is *the ratio of maximum demand of the system, a part of the system, to the total connected load of the system, or of the part of the system, under consideration.* Expressing the definition mathematically,

$$\text{Demand factor} = \frac{\text{Maximum demand}}{\text{Connected load}} . \qquad ...(9.1)$$

5. Load factor. The load factor is *the ratio of the average power to the maximum demand.* In each case, the interval of maximum load and the period over which the average is taken should be definitely specified, such as a "half-hour monthly" load factor. The proper interval and period are usually *dependent upon local conditions and upon the purpose for which the load factor is to be used.* Expressing the definition mathematically,

$$\text{Load factor} = \frac{\text{Average load}}{\text{Maximum demand}} \quad \text{...(9.2)}$$

6. Diversity factor. The diversity factor of any system, or part of a system, *is the ratio of the maximum power demands of the subdivisions of the system, or part of a system, to the maximum demand of the whole system, or part of the system, under consideration, measured at the point of supply*. Expressing the definition mathematically,

$$\text{Diversity factor} = \frac{\text{Sum of individual maximum demands}}{\text{Maximum demand of entire group}} \quad \text{...(9.3)}$$

7. Utilization factor. The utilization factor is defined as *the ratio of the maximum generator demand to the generator capacity.*

8. Plant capacity factor. It is defined as the *ratio of actual energy produced in kilowatt hours (kWh) to the maximum possible energy that could have been produced during the same period.* Expressing the definition mathematically,

$$\text{Plant capacity factor} = \frac{E}{C \times t} \quad \text{...(9.4)}$$

where, E = Energy produced (kWh) in a given period,
C = Capacity of the plant in kW, and
t = Total number of hours in the given period.

9. Plant use factor. It is defined as *the ratio of energy produced in a given time to the maximum possible energy that could have been produced during the actual number of hours the plant was in operation.* Expressing the definition mathematically,

$$\text{Plant use factor} = \frac{E}{C \times t'} \quad \text{...(9.5)}$$

where, t' = Actual number of hours the plant has been in operation.

10. Types of loads.

(*i*) ***Residential load.*** This type of load includes domestic lights, power needed for domestic appliances such as radios, television, water heaters, refrigerators, electric cookers and small motors for pumping water.

(*ii*) ***Commercial load.*** It includes lighting for shops, advertisements and electrical appliances used in shops and restaurants etc.

(*iii*) ***Industrial load.*** It consists of load demand of various industries.

(*iv*) ***Municipal load.*** It consists of street lighting, power required for water supply and drainage purposes.

(*v*) ***Irrigation load.*** This type of load includes electrical power needed for pumps driven by electric motors to supply water to fields.

(*vi*) ***Traction load.*** It includes trams, cars, trolley, buses and railways.

11. Load curve. A load curve (or load graph) *is a graphic record showing the power demands for every instant during a certain time interval.* Such a record may cover 1 hour, in which case it would be an *hourly load graph* ; 24 hours, in which case it would be a *daily load graph* ; a month in which case it would be a *monthly load graph* ; or a year (8760 hours), in which case it would be a *yearly load graph*. The following points are worth noting :

Refer to Fig. 9.1.

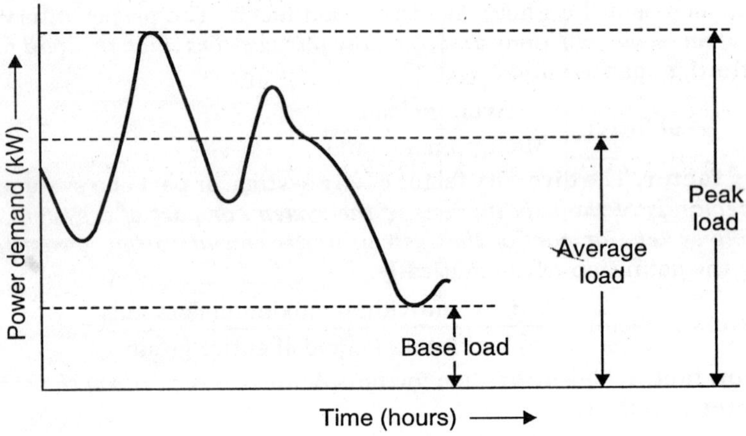

Fig. 9.1. Load curve.

(i) The area under the load curve represents the energy generated in the period considered.

(ii) The area under the curve divided by the total number of hours gives the average load on the power station.

(iii) The peak load indicated by the load curve/graph represents the maximum demand of the power station.

Significance of load curves :

- Load curves give full information about the incoming load and help to decide the installed capacity of the power station and to decide the economical sizes of various generating units.
- These curves also help to estimate the generating cost and to decide the operating schedule of the power station *i.e.,* the sequence in which different units should be run.

12. Load duration curve. A load duration curve represents re-arrangements of all the load elements of chronological load curve in order of descending magnitude. This curve is derived from the chronological load curve.

Fig. 9.2 shows a typical daily load curve for a power station. It may be observed that the maximum load on power station is 35 kW from 8 A.M to 2 P.M. This is plotted in Fig. 9.3. Similarly other loads of the load curve are plotted in *descending order* in the same figure. This is called *load duration curve* (Fig. 9.3).

The following points are worthnoting :

(i) The area under the load duration curve and the corresponding chronological load curve is equal and represents total energy delivered by the generating station.

(ii) Load duration curve gives a clear analysis of generating power economically. Proper selection of base load power plants and peak load power plants becomes easier.

13. Dump power. This term is used in hydroplants and it shows the *power in excess of the load requirements* and it is made available by *surplus water*.

14. Firm power. It is the power which *should always be available even under emergency conditions.*

15. Prime power. It is the power which may be mechanical, hydraulic or thermal that is *always available for conversion into electric power.*

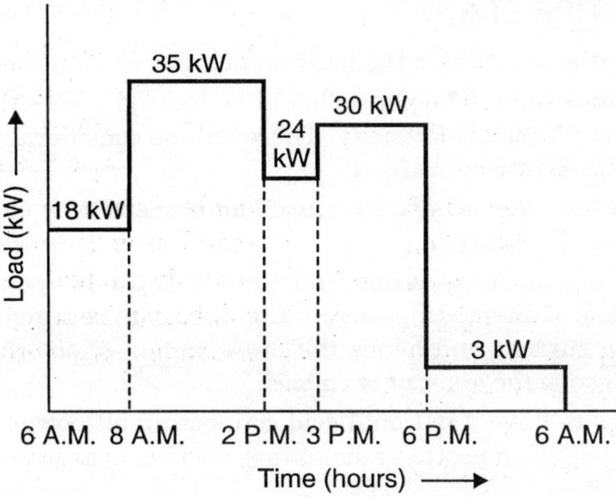

Fig. 9.2. Typical daily load curve.

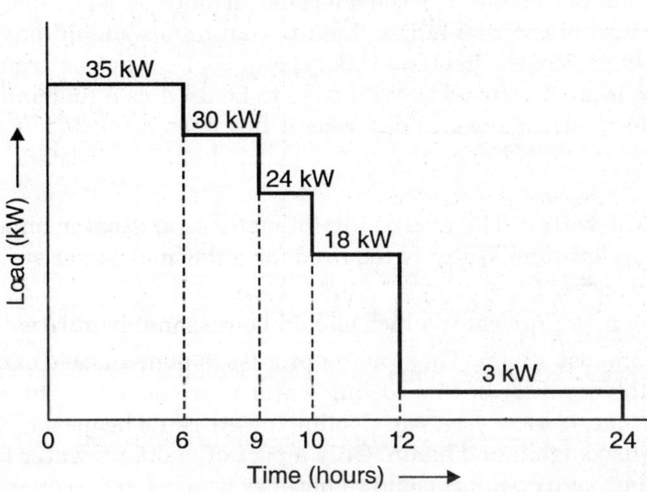

Fig. 9.3. Load duration curve.

16. Cold reserve. It is that *reserve generating capacity which is not in operation but can be made available for service.*

17. Hot reserve. It is that *reserve generating capacity which is in operation but not in service.*

18. Spinning reserve. It is that *reserve generating capacity which is connected to the bus and is ready to take the load.*

9.3. PRINCIPLES OF POWER PLANT DESIGN

The following *factors* should be considered while designing a power plant :

1. Simplicity of design.
2. Low capital cost.
3. Low cost of energy generated.
4. High efficiency.
5. Low maintenance cost.
6. Low operating cost.
7. Reliability of supplying power.
8. Reserve capacity to meet future power demand.

9.4. LOCATION OF POWER PLANT

Some of the considerations on which the location of a power plant depends are :

1. Centre of electrical load. The plant should be located where there are industries and other important consumption places of electricity. There will be considerable *advantage* in placing the power station *nearer to the centre of the load.*

— There will be *saving in the cost of copper* used for transmitting electricity as the distance of transmission line is reduced.

— The cross-section of the transmission line directly depends upon the maximum current to be carried. In case of alternating current the voltage to be transmitted can be increased thus reducing the current and hence the *cross-section of the transmission line can be reduced.* This will *save the amount of copper.*

— It is desirable now to have a national grid connecting all power stations. This provides for selecting a site which has other advantages such as nearer to fuel supply, condensing water available.

2. Nearness to the fuel source. The cost of transportation of fuel may be quite high if the distance of location of the power plant is considerable. It may be advisable to locate big thermal power plants at the mouth of the coal mines. Lignite coal mines should have centralised thermal power station located in the mines itself as this type of coal cannot be transported. Such type of power stations could be located near oil fields if oil is to be used as a fuel and near gas wells where natural gas is available in abundance. In any case it has been seen that *it is cheaper to transmit electricity than to transport fuel.* Hence the power plant should be located nearer the fuel supply source.

3. Availability of water. The availability of water is of greater importance than all other factors governing station location. Water is required for a thermal power station using turbines for the following two purposes :

(*i*) To supply the make-up water which should be reasonably pure water.

(*ii*) To cool the exhaust steam. This cooling process is done in case of diesel engines too. For *bigger power stations* the quantity of this cooling water is tremendous and requires *some natural source of water* such as *lake, river or even sea.* Cooling towers could be used economically as the same cooling water could be used again and again. Only a part of make-up water for cooling will then be required. For *small plants spray pounds* could sometimes be used. *It is economical to limit the rise in cooling-water temperature to a small value (between 6°C and 12°C), and to gain in cycle efficiency at the expense of increased cooling water pumping requirement.*

4. Type of soil available and land cost. While selecting a site for a power plant it is important to know about the character of the soil. If the soil is loose having low bearing power the pile foundations have to be used. Boring should be made at most of the projected site to have an idea of the character of the various strata as well as of the bearing power of the soil. *The best location is that for which costly and special foundation is not required.*

In case of power plants being situated near metropolitan load centres, the land there will be very costly as compared to the land at a distance from the city.

9.5. LAYOUT OF POWER PLANT BUILDING

The following points should be taken care of while deciding about power *plant building and its layout :*

1. The power plant structure should be simple and rugged with pleasing appearance.

2. Costly materials and ornamental work should be avoided.

3. The power plant interior should be clean, airy and attractive.

4. The exterior of the building should be impressive and attractive.

5. Generally the building should be *single storeyed*.

6. The layout of the power plant should first be made on paper, the necessary equipment well arranged and then design the covering structure. In all layout, allowances must be made for sufficient clearances and for walkways. Good clearance should be allowed around generators, boilers, heaters, condensers etc. Walkway clearances around hot objects and rapidly moving machinery should be wider than those just necessary to allow passage. Also the galleries in the neighbourhood of high tension bus bars should be sufficient as the space will permit.

7. Provision for future extension of the building should be made.

8. The height of the building should be sufficient so that overhead cranes could operate well and the overhauling of the turbines etc. is no problem. Sufficient room should be provided to lift the massive parts of the machines.

9. Each wall should receive a symmetrical treatment in window openings etc.

10. The principal materials used for building the power plant building are brick, stone, hollow tiles, concrete and steel.

11. In case of a *steam power plant*, there are distinct parts of the building *viz.*, *boiler room, turbine room* and *electrical bays*. Head room required in the boiler room should be greater than in the others. Ventilation in boiler room presents greater difficulty because of heat liberated from the boiler surfaces. The turbine room is actually the show room of the plant. Mezzanine flooring should be used in the power plant. The chimney height should be sufficient so as to release the flue gases sufficiently high so that the atmosphere is not polluted and the nearby buildings are not affected.

12. The foundation of a power plant is one of the most important considerations. For this the bearing capacity of the sub-soil, selection of a working factor of safety and proportioning the wall footings to economical construction should be well thought of and tested. The pile foundations may have to be used where the soils have low bearing values.

13. In any power plant *machine foundation* plays an important part. The machine foundation should be able to distribute the weight of the machine, bed plate and its own weight over a safe subsoil area. It must also provide sufficient mass *to absorb machine vibrations*.

14. Sufficient room for storage of fuel should be provided indoor as well as outdoor so as to ensure against any prolonged breakdown.

9.6. COST ANALYSIS

The cost of a power system depends upon whether :

(*i*) an entirely new power system has to be set up, or

(*ii*) an existing system has to be replaced, or

(*iii*) an extension has to be provided to the existing system. The cost interalia includes :

1. Capital Cost or Fixed Cost. It includes the following :

(*i*) Initial cost (*ii*) Interest

(*iii*) Depreciation cost (*iv*) Taxes

(*v*) Insurance.

2. Operational Cost. It includes the following :

(*i*) Fuel cost (*ii*) Operating labour cost

(*iii*) Maintenance cost (*iv*) Supplies

(*v*) Supervision (*vi*) Operating taxes.

The above mentioned costs are discussed as follows :

(*a*) **Initial Cost :**

Some of the several factors on which cost of a generating station or a power plant depends are :

(*i*) Location of the plant.
(*ii*) Time of construction.
(*iii*) Size of units.
(*iv*) Number of main generating units.
(*v*) The type of structure to be used.

The *initial cost* of a power station includes the following :

1. Land cost.
2. Building cost.
3. Equipment cost.
4. Installation cost.
5. Overhead charges which will include the transportation cost, stores and storekeeping charges, interest during construction etc.

— To *reduce the cost of building*, it is desirable to eliminate the superstructure over the boiler house and as far as possible on turbine house also.

— The *cost on equipment can be reduced* by adopting unit system where one boiler is used for one turbogenerator. Also by simplifying the piping system and elimination of duplicate system such as steam headers and boiler feed headers. The cost can be further reduced by eliminating duplicate or stand-by auxiliaries.

— When the power plant is not situated in the proximity to the load served, the cost of a primary distribution system will be a part of the initial investment.

(*b*) **Interest :**

All enterprises need investment of money and this money may be obtained as loan, through bonds and shares or from owners of personal funds. *Interest is the difference between money borrowed and money returned*. It may be charged at a simple rate expressed as % per annum or may be compounded, in which case the interest is reinvested and adds to the principal, thereby earning more interest in subsequent years. Even if the owner invests his own capital the charge of interest is necessary to cover the income that he would have derived from it through an alternative investment or fixed deposit with a bank. *Amortization* in the periodic repayment of the principal as a uniform annual expense.

(*c*) **Depreciation :**

Depreciation accounts for the deterioration of the equipment and decrease in its value due to corrosion, weathering and wear and tear with use. It also covers the decrease in value of equipment due to obsolescence. With rapid improvements in design and construction of plants, obsolescence factor is of enormous importance. Availability of better models with lesser overall cost of generation makes it imperative to replace the old equipment earlier than its useful life is spent. The actual life span of the plant has, therefore, to be taken as shorter than what would be normally expected out of it.

The following methods are used to calculate the depreciation cost :

(*i*) Straight line method.
(*ii*) Percentage method.
(*iii*) Sinking fund method.
(*iv*) Unit method.

(*i*) **Straight line method.** It is the *simplest* and *commonly used method*. The life of the equipment or the enterprise is first assessed as also the residual or salvage value of the same after the estimated life span. This salvage value is *deducted* from the initial capital cost and the balance is *divided by the life as assessed in years*. Thus, the annual value of decrease in cost of equipment is found and is set aside as depreciation annually from the income. *Thus, the rate of depreciation is uniform throughout the life of the equipment*. By the time the equipment has lived out its useful life, an amount equivalent to its net cost is accumulated which can be utilised for replacement of the plant.

(*ii*) **Percentage method.** In this method the deterioration in value of equipment from year to year is taken into account and the amount of depreciation calculated upon actual residual value for each year. It thus, reduces for successive years.

(*iii*) **Sinking fund method.** This method is based on the *conception that the annual uniform deduction from income for depreciation will accumulate to the capital value of the plant at the end of life of the plant or equipment.* In this method, the amount set aside per year consists of annual instalments and the interest earned on all the instalments.

Let, A = Amount set aside at the end of each year for n years,

n = Life of plant in years,

S = Salvage value at the end of plant life,

i = Annual rate of compound interest on the invested capital, and

P = Initial investment to install the plant.

Then, amount set aside at the end of first year = A

Amount at the end of second year

$$= A + \text{interest on } A = A + Ai = A(1+i)$$

Amount at the end of third year

$$= A(1+i) + \text{interest on } A(1+i)$$

$$= A(1+i) + A(1+i)i$$

$$= A(1+i)^2$$

∴ Amount at the end of nth year = $A(1+i)^{n-1}$

Total amount accumulated in n years (say x)

= Sum of the amounts accumulated in n years

i.e.,

$$x = A + A(1+i) + A(1+i)^2 + \ldots\ldots + A(1+i)^{n-1}$$

$$= A\,[1 + (1+i) + (1+i)^2 + \ldots\ldots + (1+i)^{n-1}] \quad \ldots(i)$$

Multiplying the above equation by $(1 + i)$, we get

$$x(1+i) = A\,[(1+i) + (1+i)^2 + (1+i)^3 + \ldots\ldots + (1+i)^n] \quad \ldots(ii)$$

Subtracting equation (*i*) from (*ii*), we get

$$x.i = [(1+i)^n - 1]\,A$$

$$\therefore \quad x = \left[\frac{(1+i)^n - 1}{i}\right] A$$

where, $x = (P - S)$

$$\therefore \quad P - S = \left[\frac{(1+i)^n - 1}{i}\right] A \quad \ldots(9.6)$$

or,

$$A = \left[\frac{i}{(1+i)^n - 1}\right] (P - S) \quad \ldots(9.7)$$

(*iv*) **Unit method.** In this method *some factor is taken as a standard one and depreciation is measured by that standard.* In place of years an equipment will last, the number of hours that an equipment will last is calculated. This total number of hours is then divided by the capital value of the equipment. This constant is then multiplied by the number of actual working hours each year to get the value of depreciation for that year. In place of number of hours, the number of units of production is taken as the measuring standard.

(*d*) **Operational cost :**

The elements that make up the operating expenditure of a power plant include the following costs :

(*i*) Cost of fuels. (*ii*) Labour cost.
(*iii*) Cost of maintenance and repairs. (*iv*) Cost of stores (other than fuel).
(*v*) Supervision. (*vi*) Taxes.

(*i*) ***Cost of fuels.*** In a thermal station fuel is the heaviest item of operating cost. The selection of the fuel and the maximum economy in its use are, therefore, very important considerations in thermal plant design. It is desirable to achieve the highest thermal efficiency for the plant so that fuel charges are reduced. *The cost of fuel includes not only its price at the site of purchase but its transportation and handling costs also*. In the hydroplants the absence of fuel factor in cost is responsible for lowering the operating cost. *Plant heat rate can be improved by the use of better quality of fuel or by employing better thermodynamic conditions in the plant design*.

The cost of fuel varies with the following :

- Unit price of the fuel.
- Amount of energy produced.
- Efficiency of the plant.

(*ii*) ***Labour cost.*** For plant operation labour cost is another item of operating cost. Maximum labour is needed in a thermal power plant using coal as a fuel. A hydraulic power plant or a diesel power plant of equal capacity require a lesser number of persons. In case of automatic power station the cost of labour is reduced to a great extent. However labour cost cannot be completely eliminated even with fully automatic station as they will still require some manpower for periodic inspection etc.

(*iii*) ***Cost of maintenance and repairs.*** In order to avoid plant breakdowns *maintenance* is necessary. *Maintenance* includes *periodic cleaning, greasing, adjustments and overhauling of equipment*. The material used for maintenance is also charged under this head. Sometimes an arbitrary percentage is assumed as maintenance cost. A good plan of maintenance would keep the sets in dependable condition and avoid the necessity of too many stand-by plants.

Repairs are necessitated when the plant breaks down or stops due to faults developing in the mechanisms. The repairs may be minor, major or periodic overhauls and are charged to the depreciation fund of the equipment. This item of cost is higher for thermal plants than for hydro-plants due to complex nature of principal equipment and auxiliaries in the former.

(*iv*) ***Cost of stores*** (other than fuel). The items of consumable stores other than fuel include such articles as lubricating oil and greases, cotton waste, small tools, chemicals, paints and such other things. The incidence of this cost is also higher in thermal stations than in hydro-electric power stations.

(*v*) ***Supervision.*** In this head the salary of supervising staff is included. A good supervision is reflected in lesser breakdowns and extended plant life. The supervising staff includes the station superintendent, chief engineer, chemist, engineers, supervisors, stores incharges, purchase officer and other establishment. Again, thermal stations, particularly coal fed, have a greater incidence of this cost than the hydro-electric power stations.

Taxes. The taxes under operating head includes the following :

(*i*) Income tax
(*ii*) Sales tax
(*iii*) Social security and employee's security etc.

9.7. SELECTION OF TYPE OF GENERATION

While choosing the type of generation the following points should be taken into consideration :

1. The type of fuel available or availability of suitable sites for water power generation.
2. Fuel transportation cost.
3. Land required.
4. Foundation cost.
5. The availability of cooling water.
6. The type of load to be taken by the power plant.
7. Reliability in operation.
8. Plant life.
9. Cost of transmitting the energy.

9.8. SELECTION OF POWER PLANT EQUIPMENT

Selection of some important power plant equipment is discussed below :

9.8.1. Selection of Boilers

It is now well known fact that only water tube boilers (fire tube boilers not suitable) should be used for all central power stations. While selecting a boiler the following points should be taken care of :

1. Type of fuel to be burnt.
2. Type of load.
3. Cost of fuel.
4. Desirability of heat-reclaiming equipment.
5. Availability of space for boiler installation.

- The design and efficiency of the boiler is considerably influenced by the *type of fuel* used in a boiler. *A high efficiency can be obtained with coal firing as compared to oil or gas firing.* This is due to increased hydrogen loss in gaseous fuels.
- *The location of the plant* will also decide the type of fuel to be used. If the plant is nearer the coal fields, coal will be cheaper. Power plants near to the oil fields and gas wells will naturally use these fuels.
- *Coal firing* will also *influence furnace design and hence the cost of boiler.* In case of low ranking fuel such as lignites etc., pulverised firing is used. Very low fusing temperatures of coal require water-cooled walls and in some cases the slag tap furnaces. The yearly minimum operating cost has to be considered which may include production cost and fixed charges. In case of anthracite coal or metallurgical coke etc. the wear on pulverising machinery is relatively much higher than that of bituminous coal.

The cost of boilers vary with the following :

(*i*) Type of boiler used. (*ii*) Operating pressure.
(*iii*) Operating temperature. (*iv*) Type of firing.
(*v*) Efficiency desired.

- '*Heat-reclaiming equipment*' such as *economisers* and *air preheaters* should be provided with boilers. With the addition of economisers and air preheaters the efficiency of the boiler increases from 75% to 90% and above.

- The *'increased pressure' affects the cost of boiler drum, boiler tubes, headers, economisers and other accessories.* Similarly high temperatures increase the cost of superheaters as higher pressure and higher temperatures require special alloy steels. *High pressures require forced circulation also. This also increases the cost but this forced circulation also increases the efficiency of the boiler.*
- The *method of firing* has also an influence on the percentage of total time for which the boiler will be available and should be considered when planning boiler capacity. Stoker firing is in general *slightly less costly than pulverised fuel firing. The pulverised fuel firing increases the efficiency.*
- *Economisers* improve the boiler efficiency by 4 to 10%. The *air preheater* further improves the boiler efficiency from 6 to 8%.
- The exhaust gases should not be cooled *below 150°C.* Below this temperature the condensation of moisture may take place and when mixed with SO_2 this moisture produces a dilute solution of H_2SO_4 which is finitely detrimental to the equipment.
- While selecting the proper economiser size as well as the size of the air preheaters *fixed* as well as *operating* charges should be considered. The fixed charges include the cost of heat recovery equipment, flue work, ducts and also increased fan cost and building cost.

9.8.2. Selection of Prime-movers

For proper selection of prime-mover it is of paramount importance to construct the following curves :

(*i*) A typical daily load curve.

(*ii*) A peak-load curve.

(*iii*) A probable future-load curve.

The prime-movers to be used for generating electricity could be *diesel engines, steam engines, steam turbines, gas turbines, water turbines* etc.

- While selecting a prime-mover the *initial cost of a unit erected* has to be taken into consideration. The *efficiency of this unit at various loads* is also to be taken into consideration. *As the capacity of the unit increases there is a corresponding reduction in floor space per kW.*
- The selection of the prime-mover depends also on the type of use whether it is used for *industrial purpose* or *for central power stations.*

Prime-movers used for industrial purpose should be *non-condensing so that steam after exhausting could be used for processing.*

In case of *central power stations condensing steam turbines should be used. Diesel engines* have an advantage of higher efficiency and low cost. It also requires less labour and the initial investment is also less. But the cost of coal is less as compared to diesel oil. Also the capacity of diesel engines is limited and hence for *bigger power stations they are unsuitable.* The diesel engines are used as *standby plant* in all the central power stations whether thermal or hydro.

- In places where water is in abundance and a certain head is available hydro-power plants/stations are installed. In rivers where there is a natural fall, the same could be used for driving a *water turbine* in a hydro power plant. The maintenance of hydro-power plant is the cheapest.

9.8.3. Selection of Size and Number of Generating Units

There can be no hard and fast rules, but however *looking at the load curve of the station one can guess for the total generating capacity, size and number of the units. Minimum generating capacity of a plant must be more than the predicted maximum demand.* Obviously, the minimum number of generators can be one but this will not be a wise suggestion. As the load on a power station is never constant, owing to variable demands at the different times of the day, the generator will have to run continuously at variable loads, which will be much less than the rated capacity of the generator for most of the times, *without any provision for the maintenance*. So a power station which is expected to be reliable in service, must have at *least two generators, irrespective of the total capacity of the plant.*

The following points are worth noting :

(*i*) The most appropriate way of deciding the size, and number of generating sets in a station is *to select the number of sets in such a way so as to fit in the load curve as closely as possible, so that the plant capacity may be used efficiently*.

(*ii*) Extra spare capacity is not desired as it increases the capital expenditure.

(*iii*) The main aim should be to have units of different capacities which will suitably fit in the load curve so that most of the generators when in use can be operated at nearly full load.

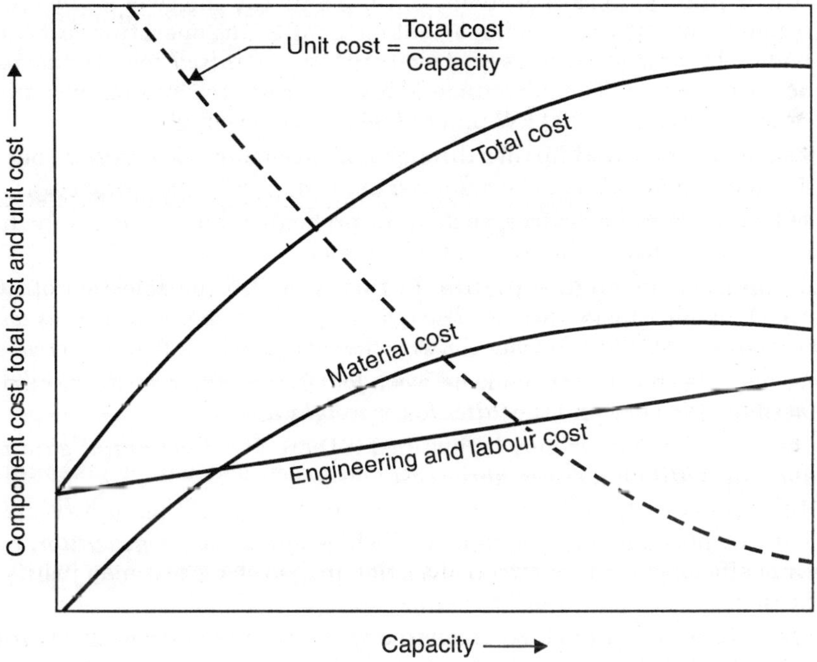

Fig. 9.4. Variation of costs of power plant *versus* its capacity.

The equipment prices are usually compared on the basis of price per unit of capacity, usually termed 'unit price'. *The unit price decreases as the capacity of the machine increases*. This is the main reason for adopting a large size generating unit in power plants. Fig. 9.4 shows the general trend and trend of the major cost components in building a given type of machine. The following points are worth noting :

- The labour and engineering cost curve *increases slightly* with the capacity of the unit.
- The material cost curve *decreases in slope* with an increase in capacity of plant.
- The total cost curve follows the pattern of material cost curve. The total cost curve shows the *positive intercept at zero capacity which represents the cost of just maintaining an organisation of men and plant ready to produce.*

- The dotted curve shows the *reduction in unit price with an increase in capacity* and this is the major argument for installing large units. *The large units are always preferred for the loads with higher load factor (0.8 to 1).*

9.9. ECONOMICS IN PLANT SELECTION

After selection of type of drive (such as steam, gas diesel or water power) which depends on availability of cheap fuels or water resources, further selection of the design and size of the equipment is primarily based upon economic consideration and a *plant that gives the lowest unit cost of production is usually chosen.* In case of all types of equipment the working *efficiency is generally higher with larger sizes of plants and with high load factor operation. Also, the capital cost per unit installation reduces as the plant is increased in size.* However, *a bigger size of plant would require greater investment, and possibilities of lower than optimum load factor usually increase with larger size of the plant.*

Steam power plants. In case of steam power plants the choice of *steam conditions* such as throttle pressure and temperature, is an important factor affecting operating costs and is, therefore, very carefully made. *As throttle pressure and temperature are raised the capital cost increases but the cycle efficiency is increased.* The advantage of higher pressures and temperatures is generally not apparent below capacity of 10,000 kW unless fuel cost is very high.

Heat rates may be improved further through *reheating* and *regeneration*, but again the capital cost of additional equipment has to be *balanced against gain in operating cost.*

The use of heat reclaiming devices, such as air preheaters and economisers, has to be considered from the point of economy in the consumption of fuel.

Internal combustion engine plants. In this case also the selection of I.C. engines also depends on thermodynamic considerations. *The efficiency of the engine improves with compression ratio but high pressures necessitate heavier construction of equipment which increases cost.*

The choice may also have to be made between *four-stroke* and *two-stroke* engines, the *former having higher thermal efficiency* and the *latter lower weight* and *cost.*

The cost of the *supercharger* may be justified if there *is a substantial gain in engine power which may balance the additional supercharge cost.*

Gas turbine power plant. The cost of the gas turbine power plant increases as the simple plant is modified by inclusion of other equipment such as *intercooler, regenerator, reheater*, etc. but the gain in thermal efficiency and thereby a reduction in operating cost may justify this additional expense in first cost.

Hydro-electric power plant. As compared with thermal stations an hydro-electric power plant has little operating cost and if sufficient water is available to cater to peak loads and special conditions for application of these plants justify, *power can be produced at a small cost.*

The *capital cost* per unit installed is *higher if the quantity of water is small.* Also, the unit cost of conveying water to the power house is greater if the quantity of water is small. The cost of storage per unit is also lower if the quantity of water stored is large.

An existing plant capacity may be *increased by storing additional water* through increasing the height of dam or by diverting water from other streams into the head reservoir. However, again it would be an economic study whether this additional cost of civil works would guarantee sufficient returns.

Some hydro-power plants may be made *automatic* or *remote controlled* to reduce the operating cost further, but the cost of automation has to be *balanced against the saving effected in the unit cost of generation.*

Interconnected hydro-steam system. In such a system where *peak loads are taken up by steam units*, the capacity of water turbine may be kept somewhat higher than the water flow capacity at peak loads, and lesser than or equal to maximum flow of river. This would make it possible for the water turbine to generate adequate energy at low cost during sufficient water flow.

Some of the principal characteristics of hydro-electric, steam and diesel power plants are listed below :

S. No.	*Characteristics*	*Hydro-plant*	*Steam plant*	*Diesel plant*
1.	*Planning and construction*	Difficult and takes long time	Easier than hydro-plant	Easiest
2.	*Civil works cost*	Highest	Lower than hydro-plant	Lowest
3.	*Running and maintenance cost* (as a fraction of total generation cost)	$\frac{1}{10}$	$\frac{1}{7}$	$\frac{1}{6}$
4.	*Overall generation cost*	Lowest	Lower than for diesel plants	Highest
5.	*Reliability*	Good	Good	Excellent

Advantages of interconnection :

Major advantages of interconnecting various power stations are :

1. Increased reliability of supply.
2. Reduction in total installed capacity.
3. Economic operation.
4. Operating savings.
5. Low capital and maintenance costs.
6. Peak loads of combined system can be carried at a *much lower cost* than what is possible with small individual system.

9.10. FACTORS AFFECTING ECONOMICS OF GENERATION AND DISTRIBUTION OF POWER

The economics of power plant operation is greatly influenced by :

(*i*) Load factor (*ii*) Demand factor

(*iii*) Utilisation factor.

Load factor. In a *hydro-electric power station* with water available and a fixed staff for maximum output, the cost per unit generated at 100% *load factor* would be *half* the cost per unit at 50% load factor. In a *steam power station* the difference would not be so pronounced since fuel cost constitutes the major item in operating costs and does not vary in the same proportion as load factor. The cost at 100% load factor in case of this station may, therefore, be about 2/3rd of the cost at 50% load factor. For a *diesel station* the cost per unit generated at 100% load factor may be about at 3/4th of the same cost at 50% load factor. From the above discussion it follows that :

(*i*) *Hydro-electric power station should be run at its maximum load continuously on all units.*

(*ii*) *Steam power station* should be run in such a way that *all its running units are economically loaded.*

(*iii*) *Diesel power station* should be worked for *fluctuating loads* or *as a stand by.*

Demand factor and utilisation factor. *A highly efficient station, if worked at low utilisation factor, may produce power at high unit cost.*

The time of maximum demand occurring in a system is also important. In an interconnected system, *a study of the curves of all stations is necessary to plan most economical operations.*

The endeavour should be to load the most efficient and cheapest power producing stations to the greatest extent possible. Such stations, called "*base load stations*" carry full load over 24 hours *i.e.*, for three shifts of 8 hours.

- The stations in the *medium range of efficiency* are operated only during the two shifts of 8 hours during 16 hours of average load.
- The older or *less efficient stations* are used *as peak or standby stations* only, and are operated *rarely* or for *short periods of time.*

Presently there is a tendency to use units of large capacities to reduce space costs and to handle larger loads. However, the *maximum economical benefit of large sets occurs only when these are run continuously at near full load. Running of large sets for long periods at lower than maximum continuous rating increases cost of unit generated.*

9.11. HOW TO REDUCE POWER GENERATION COST ?

The cost of power generation can be *reduced by* :

1. Using a plant of simple design that *does not need highly skilled personnel.*
2. Selecting equipment of *longer life and proper capacities.*
3. Carrying out *proper maintenance* of power plant equipment to avoid plant breakdowns.
4. Running the power stations at *high load factors.*
5. Increasing the efficiency of the power plant.
6. Keeping proper supervision, which ensures *less breakdowns* and extended plant life.

9.12. POWER PLANT—USEFUL LIFE

The useful life of a power plant is that *after which repairs become so frequent and extensive that it is found economical to replace the power plant by a new one.* Useful life of some of the power plants is given below :

Plant	*Useful life*
1. Conventional thermal power plant	20–25 years
2. Nuclear power plant	15–20 years
3. Diesel power plant	About 15 years.

The useful life of some of the equipment of a *steam power plant* is given below :

Equipment	*Useful life (years)*
1. Boilers	
(*i*) Fire tube	10–20
(*ii*) Water tube	20
2. Steam turbine	5–20
3. Steam turbo-generators	10–20
4. Condensers	20
5. Pumps	15–20
6. Coal and ash machinery	10–20
7. Feed water heaters	20–30
8. Stacks	10–30
9. Stokers	10–20

10. Transformers	15–20
11. Motors	20
12. Electric meters and instruments	10–15
13. Transmission lines	10–20

9.13. ECONOMICS OF HYDRO-ELECTRIC POWER PLANTS

The cost analysis of an hydro-electric power plant is different from those of the other plants in the respect that the fixed *coal is the major item of the total cost and the operating cost is relatively much smaller* whereas in the steam and other plants, the *operating cost is a large part of the total cost.*

In a hydro-plant the total annual cost can be divided into two following categories :

1. Fixed costs :

(*i*) Interest on capital (*ii*) Amortization of the capital

2. Running costs :

(*i*) Maintenance and repairs. (*ii*) Operating costs including salaries and wages.

(*iii*) Rates and taxes. (*iv*) Stores, oil and other supplies.

(*v*) Management expenses including insurance.

- The *fixed charges* of a hydro-plant are about 60 to 70% of the total cost of power and *these do not depend upon the station output.*
- The *running charges depend upon the station output, but not so much as in the thermal power plants.*

The following items go to form the *total capital outlay* or the *investment on a hydro-plant* :

(*i*) Preliminary surveys and investigations of the topography and geology of the proposed site of the plant.

(*ii*) Purchase of land (needed for adequate storage or pondage) and water rights.

(*iii*) Compensation to oustees.

(*iv*) Cost of preparation of detailed designs and specifications.

(*v*) Cost of testing the materials of construction.

(*vi*) Cost of carrying out experimental work and model tests or designs for hydraulic structures.

(*vii*) The actual cost of construction.

(*viii*) Cost relating to the purchase and installation of the equipment.

(*ix*) Interest on capital during construction.

(*x*) Working capital during the period of load development.

(*xi*) Cost in respect of new roads, railway lines, residential houses and even new towns which may have to be constructed.

A typical cost analysis of a hydro-plant is as follows :

S. No.	*Components*	*Cost*
1.	Reservoir, dam and water ways	55%
2.	Land	15%
3.	Structures	10%
4.	Power plant and equipment	20%

Besides this there is an another important item called *transmission liability* which refers to the *transmission charges for conveying the electricity from the plant site to the load centre.*

- The total cost of construction of hydro-plant is *invariably higher than that of a thermal plant of equal capacity*. Therefore the annual charges for interest and depreciation are comparatively higher.
- In case of a *hydro-plant*, the *smaller the quantity of water stored higher is the cost per kW.*

9.14. ECONOMICS OF COMBINED HYDRO AND STEAM POWER PLANTS

It has been established that if a region/country is neither rich in fuel reserves nor in hydro resources then a combined operation of hydro and steam power plants gives the best results in regard to generation of electricity at *the economical cost.* The following advantages accrue from combined plants :

1. Flexibility of operation.
2. Security of supply.
3. Improved utilisation of hydro-power.
4. Spare plant.

Practically all large power systems in the world have hydro-steam interconnected. *Hydro-plant may function as a capacity plant i.e., to supply system peak with minimum flow conditions*, or it may work as an *energy plant* to replace the costly steam generated electricity by *low cost hydro power.*

In such a system (interconnected) *there should be a daily or seasonal load allocation plan set prehand so that the use of the two power systems is made to the best advantage. A certain amount of forecasting of load and capacity of the system as well as flexibility are necessary for optimum results and experience is a big factor in good co-ordinated action.*

A knowledge of system plant loading schedules for minimum production cost is as important as that of reservoir levels, pond storage and stages of low flow and heavy loads.

9.15. PERFORMANCE AND OPERATING CHARACTERISTICS OF POWER PLANTS

The *performance* of generating power plants is compared by their *average efficiency* over a period of time. The average efficiency of a power plant is the *ratio of useful energy output to the total energy input during the period considered.* This *measure of performance varies with uncontrolled conditions viz.* (*i*) cooling water temperature (*ii*) quality of fuel and (*iii*) shape of load curve. Thus, unless all plant performances are corrected to the same controlled conditions it is not a satisfactory standard of comparison.

The performance of a plant can be precisely represented by the *input-output curve* from the tests conducted on individual power plant. The input-output curve is *graphical representation* between the net energy output (L) and input (I). The input is generally expressed in millions of kcal/h or kJ/h and load output is expressed as megawatts (MW). The input to hydro-plant is measured in cusecs or m^3/s of water.

In general input-output may be represented as follows :

$$I = a + bL + cL^2 + dL^3 \qquad \text{...(9.8)}$$

where, I = Input,

L = Output, and

a, b, c and d = Constants

Input-output curve. Fig. 9.5 (a) shows an *input-output curve*. In order to keep the apparatus functioning at zero load, a certain input (I_0) is required *to meet frictional and heat losses.*

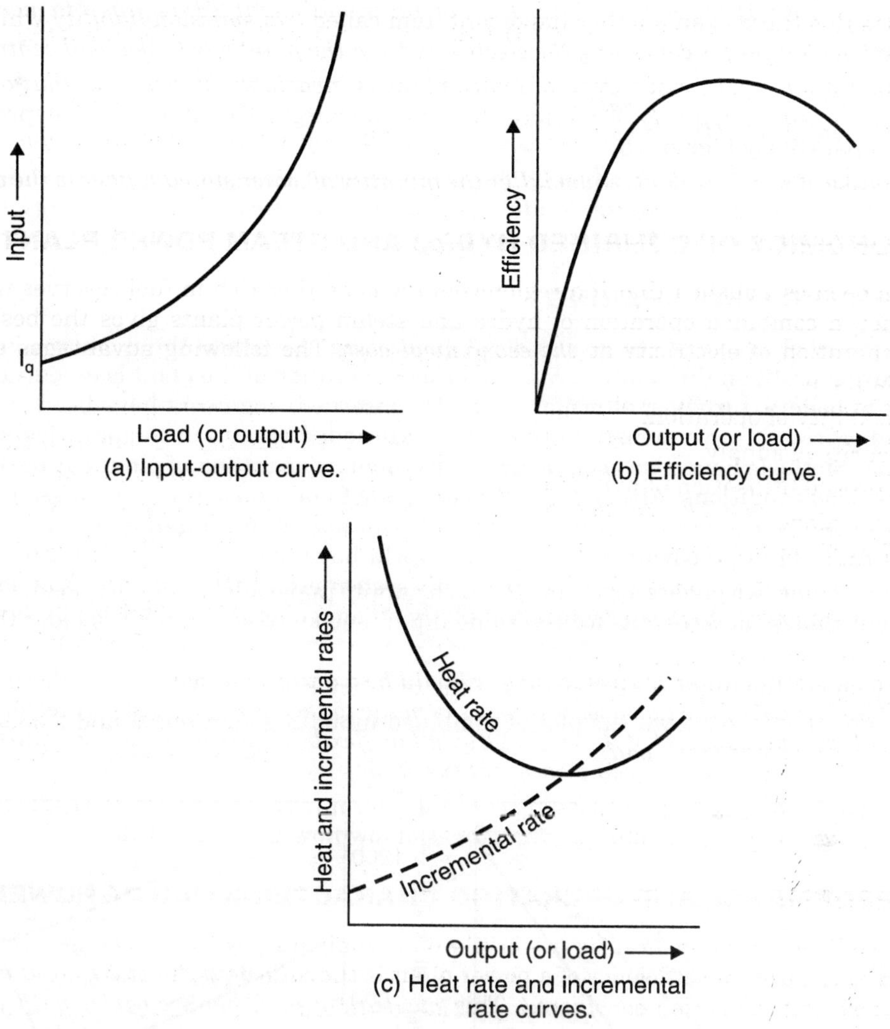

(a) Input-output curve.

(b) Efficiency curve.

(c) Heat rate and incremental rate curves.

Fig. 9.5.

Efficiency curve. The efficiency of the power plant is defined *as the ratio of output to input.*

$\therefore$ Efficiency, $$\eta = \frac{L}{I} = \frac{L}{a + bL + cL^2 + dL^3}$$

By using the above formula the efficiency *for any given load* can be calculated.

An *efficiency curve* is shown in Fig. 9.5 (*b*).

Heat rate and incremental rate curves. These curves can be derived from basic input-output curve.

Heat rate (HR) is defined as the *ratio of input to output.*

i.e., $$\textit{Heat rate (HR)} = \frac{I}{L} = \frac{a + bL + cL^2 + dL^3}{L} = \frac{a}{L} + b + cL + dL^2$$

Heat rate curve is obtained by plotting values of heat rate against corresponding values of output. Fig. 9.5 (*c*) shows a heat rate curve.

Incremental rate (IR) is defined as the *ratio of additional input (dI) required to increase additional output (dL).*

i.e., Incremental rate $(IR) = \frac{dI}{dL}$.

Incremental rate curve is obtained by plotting values of *IR* against corresponding values of output. Such a curve is shown in Fig. 9.5 (*c*). *This curve expresses additional energy required to produce an added unit of output at the given load.*

9.16. ECONOMIC LOAD SHARING

The primary objective of the design of all generating stations is the *economy*. For a power system to return a profit on the capital invested, proper operation of the plant is essential. As far as the efficiency of boilers, turbines, alternators etc. is concerned, engineers have been successful in increasing the efficiency continuously so that each unit added results in comparatively more efficient operation. Methods have also been devised for economic operation of plants at part loads and under variable load conditions. Attempts have been made to minimise the transmission losses too. Now the *only aspect that remains is the economic distribution of the output of a plant between the generators, or units within the plant.*

Let us consider two generators 1 and 2 which supply in parallel a common load. Generator 1 is more efficient than generator 2 as for the same input, output of generator 1 is more than that of generator 2.

Fig. 9.6 shows the input-output curves of the two generators/units.

In the Fig. 9.6 (*b*) is shown the plot of combined input of generators 1 and 2 *versus* load on generator 1, for a *constant total load.*

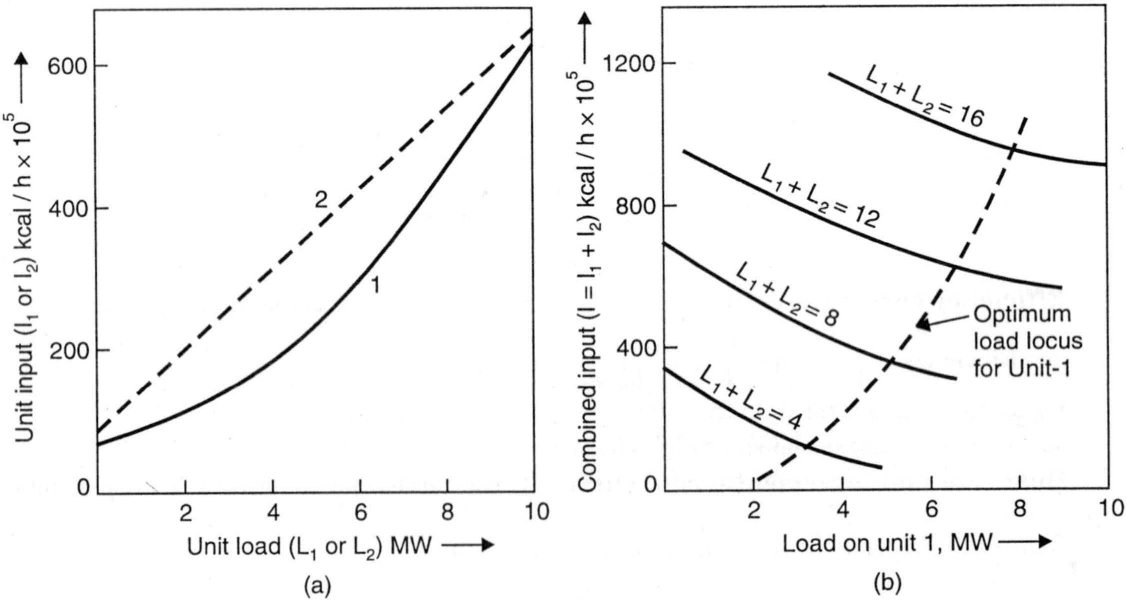

Fig. 9.6.

Although generator 1 requires less input for a given output it is not essential that unit 1 should be loaded first and then generator 2. For economical loading the combined input of units 1 and 2 should be plotted against load on unit 1 for a constant total load. Let a total load of 4 MW be supplied by generators 1 and 2.

$$L = L_1 + L_2$$

where, L = combined output,

L_1 = output of generator 1, and

L_2 = output of generator 2.

Let the generator 2 supply total load of 4 MW and generator 1 supply zero load. Now corresponding to zero load on generator 1 and 4 MW on generator 2 the values of input to generator 1 (I_1) and input to generator 2 (I_2) can be determined respectively from Fig. 9.6 (*a*) and thus value of $(I_1 + I_2)$ can be plotted against zero load on generator 1. Again let 2 MW be supplied by generator 1 and 2 MW be supplied by generator 2 (so that total load remains 4 MW) then values of I_1 and I_2 can be determined corresponding to 2 MW load on each generator and value of $(I_1 + I_2)$ can be plotted against 2 MW load on generator 1 as shown in Fig. 9.6 (*b*). In this way curve for a total load of 4 MW can be plotted corresponding to different output of generator 1.

Similarly curve for total load of 8 MW etc. can be plotted. In these curves *there is at least one point where combined input is minimum for a given total load. Corresponding to this point of minimum, the load of generator 1 can be found.* Then the load on generator 2 will be difference of total load and load on generator 1. *This load saving will be the most economical.*

This method is difficult to apply in practice as such because generally stations have got more than two generators / units and in that case the application of above principles becomes a cumbersome process.

Considering any combined constant input in the Fig. 9.6 (*b*), at the point of minimum input

$$\frac{dI}{dI_1} = 0 \qquad \text{...(i)}$$

where, $I = I_1 + I_2$ = Input of generator 1 + input of generator 2

= Combined input to generators 1 and 2

$L = L_1 + L_2$ = Output of generator 1 + output of generator 2

= Combined output of generators 1 and 2

Then, $$\frac{dI}{dI_1} = \frac{dI_1}{dL_1} + \frac{dI_2}{dL_1} = 0 \qquad \left(\because \frac{dI}{dI_1} = 0\right)$$

As I is constant.

$$\therefore \quad \frac{dI_1}{dL_1} = -\frac{dI_2}{dL_1} \qquad \text{...(ii)}$$

But, $$\frac{dI_2}{dL_1} = \frac{dI_2}{dL_2} \times \frac{dL_2}{dL_1} \qquad \text{...(iii)}$$

Also, $$L_2 = L - L_1$$

$$\therefore \quad \frac{dL_2}{dL_1} = \frac{dL}{dL_1} - \frac{dL_1}{dL_1}$$

Since L is constant.

$$\therefore \quad \frac{dL}{dL_1} = 0$$

Hence, $$\frac{dL_2}{dL_1} = -\frac{dL_1}{dL_1} = -1 \qquad \text{...(iv)}$$

Substituting in (*iii*), we get

$$\frac{dI_2}{dL_1} = -\frac{dI_2}{dL_2} \quad ...(v)$$

∴ From (*ii*) and (*v*),

$$\frac{dI_1}{dL_1} = \frac{dI_2}{dL_2} \quad ...(vi)$$

Thus, *for minimum combined input to carry a given combined output, the slopes of the input-output curve for each unit must be equal.*

If there are *n* units, supplying a constant load, then the required condition for the minimum input or maximum system efficiency is

$$\frac{dI_1}{dL_1} = \frac{dI_2}{dL_2} = \frac{dI_3}{dL_3} = ... = \frac{dI_n}{dL_n} \quad ...(vii)$$

Condition for maximum efficiency :

Refer Fig. 9.7. *The load at which efficiency will be maximum, the heat rate will be minimum at that load as efficiency is inverse of heat.*

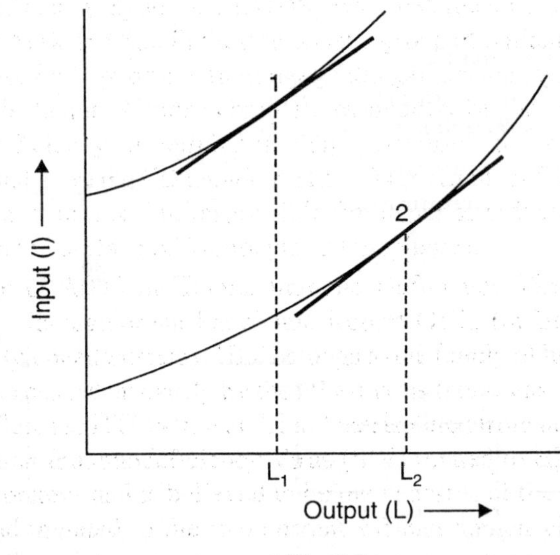

Fig. 9.7.

$$\text{Efficiency} = \frac{L}{I}$$

$$\text{Heat rate } (HR) = \frac{I}{L}$$

∴ For *minimum value of heat rate,*

$$\frac{d}{dL}(HR) = 0 \quad \text{or} \quad \frac{d}{dL}\left(\frac{I}{L}\right)$$

$$\therefore \quad \frac{LdI - IdL}{L^2} = 0 \quad \text{or} \quad LdI = IdL \quad \text{or} \quad \frac{I}{L} = \frac{dI}{dL}$$

This shows that efficiency will be maximum at a load where heat rate is equal to incremental heat rate.

WORKED EXAMPLES

Example 9.1. *The maximum demand of a power station is 96000 kW and daily load curve is described as follows :*

Time (hours)	*0—6*	*6—8*	*8—12*	*12—14*	*14—18*	*18—22*	*22—24*
Load (MW)	*48*	*60*	*72*	*60*	*84*	*96*	*48*

(*i*) *Determine the load factor of power station.*

(*ii*) *What is the load factor of standby equipment rated at 30 MW that takes up all load in excess of 72 MW ? Also calculate its use factor.*

Solution. Load curve is shown in Fig. 9.8.

Energy generated = Area under the load curve

$= 48 \times 6 + 60 \times 2 + 72 \times 4 + 60 \times 2 + 84 \times 4 + 96 \times 4 + 48 \times 2$

$= 1632 \text{ MWh} = 1632 \times 10^3 \text{ kWh}.$

(*i*) **Load factor :**

Average load $= \dfrac{1632 \times 10^3}{24} = 68000$ kW

Maximum demand = 96000 kW (given)

$\therefore$ Load factor $= \dfrac{\text{Average load}}{\text{Maximum demand}} = \dfrac{68000}{96000} =$ **0.71. (Ans.)**

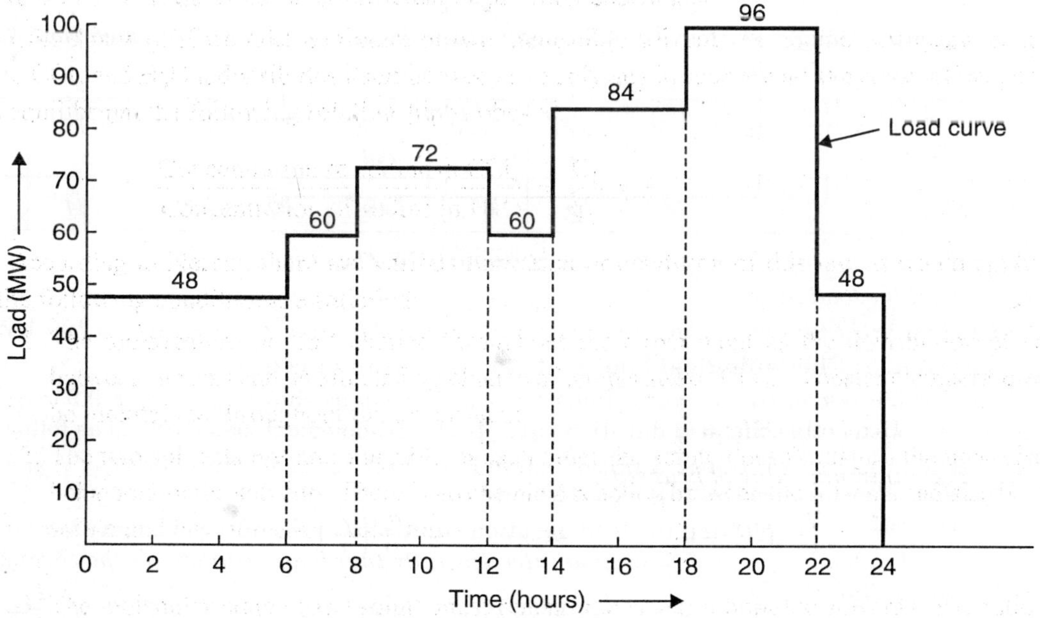

Fig. 9.8.

(*ii*) **Land factor of standby equipment :**

The standby equipment supplies

$$84 - 72 = 12 \text{ MW for 4 hours } (14 - 18)$$

$$96 - 72 = 24 \text{ MW for 4 hours } (18 - 22)$$

∴ Energy generated by standby equipment

$$= (12 \times 4 + 24 \times 4) \times 10^3 = 144 \times 10^3 \text{ kWh}$$

Time for which standby equipment *remains in operation* (from the load curve)

$$= 4 + 4 = 8 \text{ hours}$$

Average $= \dfrac{144 \times 10^3}{8} = 18 \times 10^3$ kW

Load factor $= \dfrac{18 \times 10^3}{24 \times 10^3} = \mathbf{0.75.}$ **(Ans.)**

Use factor $= \dfrac{E}{C \times t'}$

where, E = Energy generated,

C = Capacity of the standby equipment and

t' = Actual number of hours the plant has been in operation.

∴ **Use factor** $= \dfrac{144 \times 10^3}{30 \times 10^3 \times 8} = \mathbf{0.6.}$ **(Ans.)**

☞**Example 9.2.** *An electrical system experiences linear changes in load such that its daily load curve is defined as follows :*

Time	*12 PM*	*2 AM*	*6 AM*	*8 AM*	*12 AM*	*12.30 PM*	*1 PM*	*5 PM*	*6 PM*	*12 PM*
Load (MW)	*24*	*12*	*12*	*60*	*60*	*48*	*60*	*60*	*84*	*24*

(*i*) *Plot the chronological and load duration curve for the system.*

(*ii*) *Find the load factor.*

(*iii*) *What is the utilisation factor of the plant serving this load if its capacity is 120 MW.*

Solution. (*i*) **Chronological load and load duration curves :**

Chronological load and load duration curves are drawn as shown in the Fig. 9.9 (*a*), (*b*). The procedure for constructing the load duration curve from chronological load curve is as follows :

- The abscissa of the load duration curve is laid off *equal to the number of hours in the chronological curve,* in this case 24 hours.
- The criterion of plotting the load duration curve *makes the abscissa at any load ordinate equal to the length of the abscissa intercepted by that load ordinate on the chronological curve.* Thus :

(*i*) At the maximum demand or peak load, the intercept is one point which will be plotted at 0 hour.

(*ii*) At 70 MW load the intercept is x_1 hours and is plotted as x_1 hour on the load duration curve.

(*iii*) At 55 MW load the intercept is a total of $(x_2 + x_3)$ and is plotted accordingly.

(*iv*) At minimum load of 12 MW the intercept covers the entire period of 24 hours.

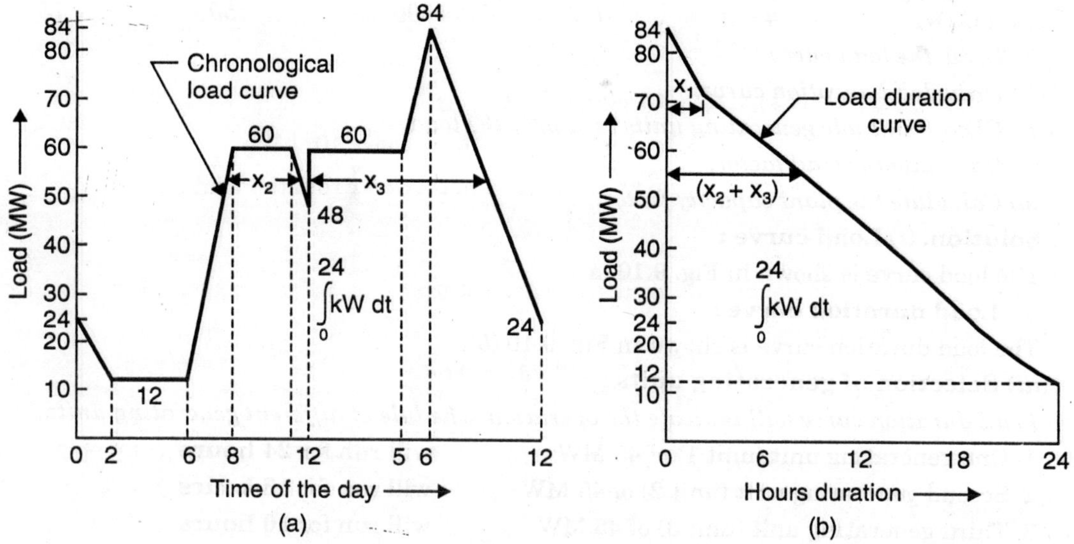

Fig. 9.9. (*a*) Chronological load curve, (*b*) Derived load duration curve.

Following points may be noted :

1. *Any point on the load duration curve is a measure of number of hours in a given period during which the given load and higher loads have prevailed.*
2. *If the chronological curve indicated a constant demand during the entire day, it would be of rectangular shape and load duration curve would be an exact duplicate.*

(*ii*) **Load factor :**

From the load duration curve, the average load can be estimated.

Average load for the period

$$= \frac{\text{Total energy in load curve for period}}{\text{Total number of hours in period}}$$

$$= \frac{\left(\frac{24+12}{2}\right)\times 2 + 12\times 4 + \left(\frac{12+60}{2}\right)\times 2 + 60\times 4 + \left(\frac{60+48}{2}\right)\times\frac{1}{2} + \left(\frac{48+60}{2}\right)\times\frac{1}{2} + 60\times 4 + \left(\frac{60+84}{2}\right)\times 1 + \left(\frac{84+24}{2}\right)\times 6}{24}$$

$$= \frac{36 + 48 + 72 + 240 + 27 + 27 + 240 + 72 + 324}{24}$$

$$= \frac{1086}{24} = 45.2 \text{ MW}$$

$\therefore$ **Load factor** $= \frac{45.2}{84} =$ **0.45 or 45%. (Ans.)**

(*iii*) **Utilisation factor :**

Utilisation factor $= \frac{\text{Maximum load}}{\text{Rated capacity of the plant}} = \frac{84}{120} =$ **0.70 or 70%. (Ans.)**

☞**Example 9.3.** *A power station has to supply load as follows :*

Time (hours) :	*0—6*	*6—12*	*12—14*	*14—18*	*18—24*
Load (MW) :	*45*	*135*	*90*	*150*	*75*

(*i*) *Draw the load curve.*

(*ii*) *Draw load duration curve.*

(*iii*) *Choose suitable generating units to supply the load.*

(*iv*) *Calculate the load factor.*

(*v*) *Calculate the plant capacity factor.*

Solution. (*i*) **Load curve :**

The load curve is shown in Fig. 9.10 (*a*).

(*ii*) **Load duration curve :**

The load duration curve is shown in Fig. 9.10 (*b*).

(*iii*) **Selection of generating units :**

Load duration curve will indicate the operation schedule of different generating units.

1. One generating unit (unit 1) of 45 MW will run for **24 hours**
2. Second generating unit (unit 2) of 45 MW will run for **18 hours**
3. Third generating unit (unit 3) of 45 MW will run for **10 hours**
4. Fourth generating unit (unit 4) of 15 MW will run for **4 hours**

One additional unit (unit 5) should be kept as standby. Its capacity should be equal to the capacity of biggest set, *i.e.*, 45 MW.

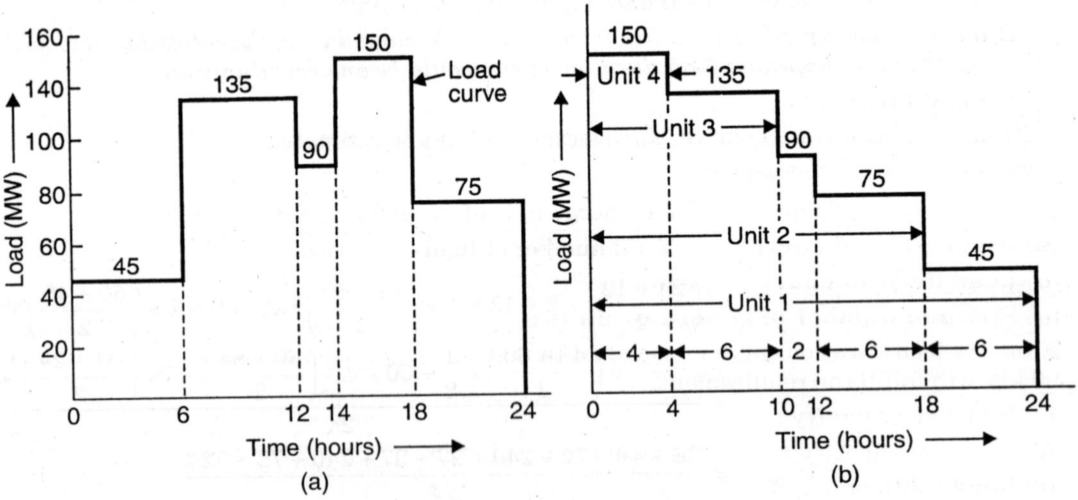

Fig. 9.10. (*a*) Load curve, (*b*) Load duration curve.

(*iv*) **Load factor :**

Energy generated $= 45 \times 6 + 135 \times 6 + 90 \times 2 + 150 \times 4 + 75 \times 6$
$= 270 + 810 + 180 + 600 + 450 = 2310$ MWh

Average load $= \dfrac{2310 \times 10^3}{24}$ kW = 96250 kW

Maximum demand $= 150 \times 10^3 = 150000$ kW

$\therefore$ **Load factor** $= \dfrac{\text{Average load}}{\text{Maximum demand}}$

$= \dfrac{96250}{150000} = \mathbf{0.64.}$ **(Ans.)**

(*v*) **Plant capacity factor :**

Plant capacity factor $= \dfrac{E}{C \times t}$

where, E = Energy generated (kWh),

C = Capacity of the plant (kW),

$= 45 \times 4 + 1 \times 15 = 195$ MW $= 195 \times 10^3$ kW, and

t = Number of hours in the given period = 24 hours.

$\therefore$ *Plant capacity factor* $= \dfrac{2310 \times 10^3}{195 \times 10^3 \times 24} = \mathbf{0.49.}$ **(Ans.)**

Example 9.4. *A generating station has a maximum demand of 5000 kW, and the daily load on the station is as follows :*

Load (MW)	*1000*	*1750*	*4000*	*1500*
Time	*11 PM to 6 AM*	*6 AM to 8 AM*	*8 AM to 12.00 Noon*	*12 PM to 1 PM*
Load (MW)	*3750*	*4250*	*5000*	*2250*
Time (hours)	*1 PM to 5 PM*	*5 PM to 7 PM*	*7 PM to 9 PM*	*9 PM to 11 PM*

(*i*) *Draw the load curve.*

(*ii*) *Draw the load duration curve.*

(*iii*) *Select the size and number of generator units.*

(*iv*) *What reserve plant would be necessary ?*

(*v*) *Load factor.*

(*vi*) *Plant capacity factor.*

Solution. (*i*) **Load curve** is shown in the Fig. 9.16 (*a*).

(*ii*) **Load duration curve** is shown in Fig. 9.16 (*b*).

(*iii*) **Size and number of generator units :**

From the load duration curve it is evident that generating sets of capacity 1000 kW, 1500 kW and 2500 kW will fulfill the requirement.

(*iv*) **Reserve capacity :**

Also, reserve capacity = largest size of the unit in the station = **2500 kW.** **(Ans.)**

(*v*) **Load factor :**

Area under the load curve gives the energy generated during 24 hours

$= 1000 \times 7 + 1750 \times 2 + 4000 \times 4 + 1500 \times 1 + 3750 \times 4 + 4250 \times 2 + 5000 \times 2 + 2250 \times 2$

$= 7000 + 3500 + 16000 + 1500 + 15000 + 8500 + 10000 + 4500$

$= 66000$ kWh

or, Average load $= \dfrac{66000}{24} = 2750$ kW

$\therefore$ *Load factor* $= \dfrac{\text{Average load}}{\text{Maximum demand}} = \dfrac{2750}{5000} = \mathbf{0.5.}$ **(Ans.)**

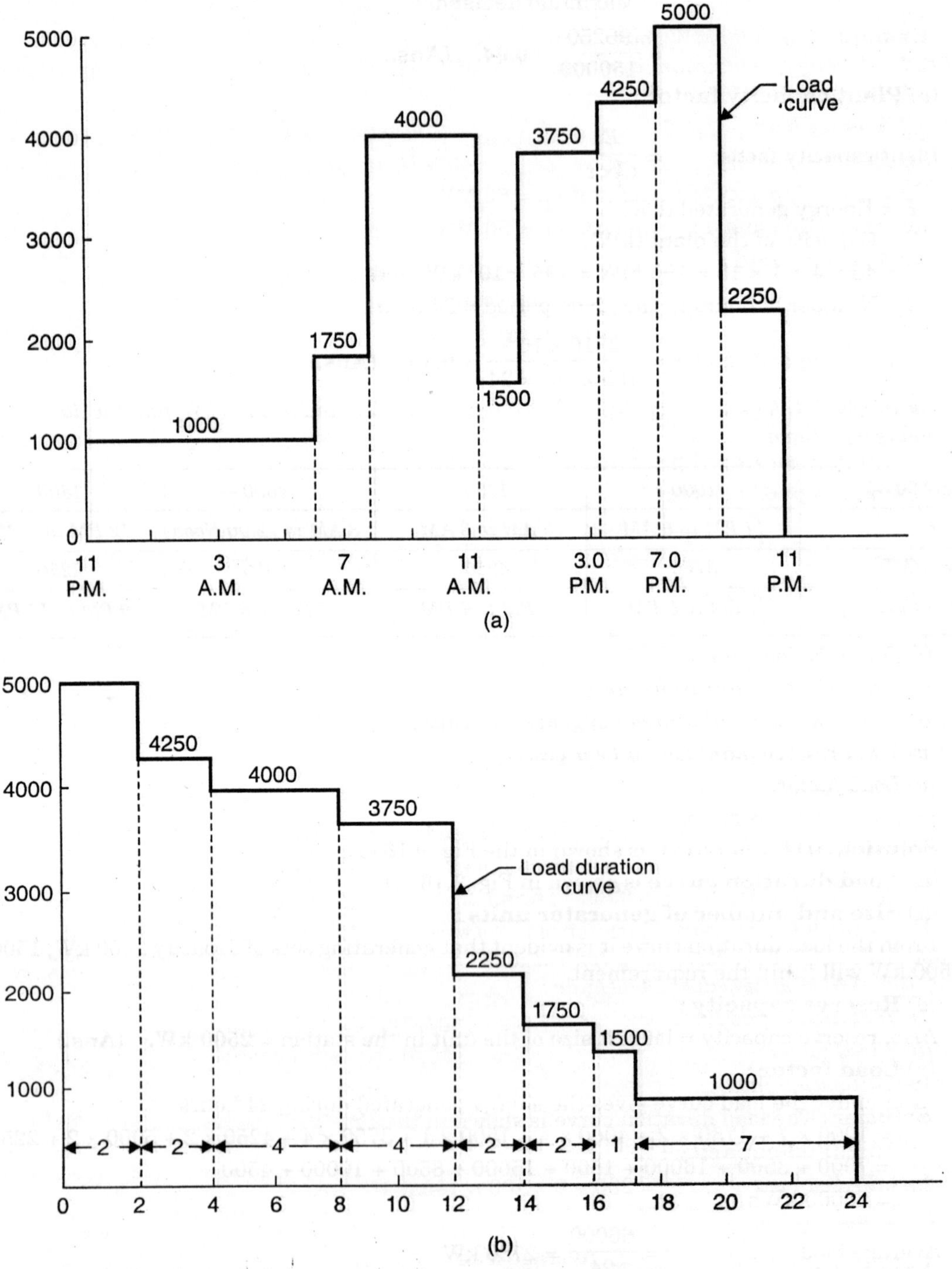

Fig. 9.11. (*a*) Load curve, (*b*) Load duration curve.

(*vi*) **Plant capacity factor :**

Plant capacity factor $= \dfrac{E}{C \times t} = \dfrac{66000}{(2500 + 1500 + 1000 + 2500) \times 24}$ **= 0.367. (Ans.)**

Example 9.5. *A 60 MW power station has an annual peak load of 50 MW. The power station supplies loads having maximum demands of 20 MW, 17 MW, 10 MW and 9 MW. The annual load factor is 0.45. Find :*

(*i*) *Average load.* (*ii*) *Energy supplied per year.*

(*iii*) *Diversity factor.* (*iv*) *Demand factor.*

Solution. Capacity of power station = 60 MW

Maximum demand on power station = 50 MW

(*i*) **Average load :**

Load factor $= \dfrac{\text{Average load}}{\text{Maximum demand}}$

i.e., $0.45 = \dfrac{\text{Average load}}{50}$

∴ *Average load* = 50 × 0.45 = **22.5 MW. (Ans.)**

(*ii*) **Energy supplied per year :**

Energy supplied per year

= Average load × number of hours in one year

= $(22.5 \times 10^3) \times 365 \times 24$ = **197.1×10^6 kWh. (Ans.)**

(*iii*) **Diversity factor :**

Diversity factor $= \dfrac{\text{Sum of individuals maximum demands}}{\text{Simultaneous maximum demand}}$

$= \dfrac{20 + 17 + 10 + 9}{50} = \dfrac{56}{50} = 1.12$

Hence, *diversity factor* = **1.12. (Ans.)**

(*iv*) **Demand factor :**

Demand factor $= \dfrac{\text{Maximum demand}}{\text{Connected load}} = \dfrac{50}{20 + 17 + 10 + 9} = \dfrac{50}{56} = 0.89$

Hence, *demand factor* = **0.89. (Ans.)**

Example 9.6. *The yearly duration curve of a certain plant can be considered as a straight line from 300 MW to 80 MW. Power is supplied with one generating unit of 200 MW capacity and two units of 100 MW capacity each. Determine :*

(*i*) *Installed capacity.* (*ii*) *Load factor.*

(*iii*) *Plant factor.* (*iv*) *Maximum demand.*

(*v*) *Utilization factor.*

Solution. The load duration curve is shown in the Fig. 9.12.

(*i*) **Installed capacity :**

Installed capacity = 200 + 2 × 100 = **400 MW. (Ans.)**

(*ii*) **Load factor :**

Load factor $= \dfrac{\text{Average load}}{\text{Maximum demand}}$

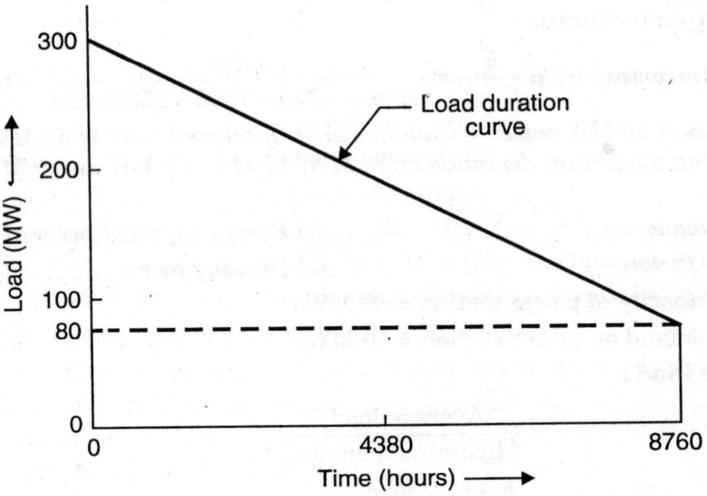

Fig. 9.12. Load duration curve.

Average load $= \dfrac{\text{Total energy in load curve for the period}}{\text{Total number of hours in the period}}$

$= \dfrac{80 \times 8760 + \frac{1}{2} \times (300 - 80) \times 8760}{8760}$

$= \dfrac{8760}{8760}\left[80 + \dfrac{1}{2}(300 - 80)\right] = 190 \text{ MW}$

∴ *Load factor* $= \dfrac{190}{300} = \mathbf{0.633.}$ **(Ans.)**

(*iii*) **Plant factor :**

Plant factor $= \dfrac{\text{Average load}}{\text{Capacity of the plant}} = \dfrac{190}{400}$

$= \mathbf{0.475.}$ **(Ans.)**

(*iv*) **Maximum demand = 300 MW. (Ans.)**

(*v*) **Utilization factor :**

Utilization factor $= \dfrac{\text{Maximum load}}{\text{Rated capacity of the plant}}$

$= \dfrac{300}{400} = \mathbf{0.75.}$ **(Ans.)**

☞**Example 9.7.** *A generating station has a maximum demand of 30 MW, a load factor of 0.6, a plant capacity of 0.48, and a plant use factor of 0.82. Find :*

(i) The daily energy produced.

(ii) The reserve capacity of the plant.

(iii) The maximum energy that could be produced if the plant were running all the time.

(iv) The maximum energy that could be produced daily, if the plant when running according to operating schedule were fully loaded.

Solution. Maximum demand of the power station = 30 MW

Load factor = 0.6

Plant capacity = 0.48

Plant use factor = 0.82

(*i*) **The daily energy produced :**

$$\text{Load factor} = \frac{\text{Average demand}}{\text{Maximum demand}}$$

$$0.6 = \frac{\text{Average demand}}{30}$$

$\therefore$ Average demand = 30 × 0.6 = 18 MW or 18000 kW

Daily energy product = Average demand × number of hours

= 18000 × 24 = **4.32 × 10⁵ kWh.** **(Ans.)**

(*ii*) **Reserve capacity of the plant :**

$$\text{Plant capacity factor} = \frac{\text{Average demand}}{\text{Installed capacity}}$$

$$0.48 = \frac{18000}{\text{Installed capacity}}$$

$$\therefore \quad \text{Installed capacity} = \frac{18000}{0.48} = 37500 \text{ kW}$$

$\therefore$ *Reserve capacity of the plant*

= Installed capacity – maximum demand

= 37500 – 30000 = **7500 kW.** **(Ans.)**

(*iii*) **Maximum daily energy produced when running all the time**

= **4.32 × 10⁵ kWh.** **(Ans.)**

(*iv*) **Maximum energy that could be produced daily :**

Maximum energy that could be produced, operating as per operating schedule

$$= \frac{\text{Actual energy produced}}{\text{Plant use factor}}$$

$$= \frac{4.32 \times 10^5}{0.82} = \mathbf{5.268 \times 10^5 \text{ kWh.}} \quad \textbf{(Ans.)}$$

Example 9.8. *A power station has the following loads :*

1. Residential lighting load :

Maximum demand	*= 1200 kW*
Load factor	*= 0.21*
Diversity between consumers	*= 1.32*

2. Commercial load :

Maximum demand	*= 2400 kW*
Load factor	*= 0.32*
Diversity between consumers	*= 1.2*

3. Industrial load :

Maximum demand	*= 6000 kW*
Load factor	*= 0.82*
Diversity between consumers	*= 1.22*

Overall diversity factor may be taken as 1.42.

Determine the following :

(*i*) *Maximum demand on system.*

(*ii*) *Daily energy consumption (total).*

(*iii*) *Overall load factor.*

(*iv*) *Connected load (total) assuming that demand factor for each load is unity.*

Solution. (*i*) **Maximum demand on system :**

$$\text{Group diversity factor} = \frac{\text{Sum of individual maximum demands}}{\text{Actual maximum demand of the group}}$$

$$\therefore \quad 1.42 = \frac{1200 + 2400 + 6000}{\text{Maximum demand on system}}$$

i.e., Maximum demand on system $= \dfrac{1200 + 2400 + 6000}{1.42}$

$=$ **6760.5 kW. (Ans.)**

(*ii*) **Daily energy consumption :**

$$\text{Load factor} = \frac{\text{Average demand}}{\text{Maximum demand}}$$

or, Average demand = Maximum demand × load factor

$= 1200 \times 0.21 + 2400 \times 0.32 + 6000 \times 0.82$

$= 5940$ kW

$\therefore$ *Daily energy consumption* $= 5940 \times 24 =$ **142560 kWh. (Ans.)**

(*iii*) **Overall load factor :**

$$\text{Overall load factor} = \frac{\text{Average demand}}{\text{Maximum demand}}$$

$= \dfrac{5940}{6760.5} =$ **0.878. (Ans.)**

(*iv*) **Connected load :**

Maximum demand $= 1200 \times 1.32 + 2400 \times 1.2 + 6000 \times 1.22 = 11784$ kW

Connected load $= \dfrac{\text{Maximum demand}}{\text{Demand factor}} = \dfrac{11784}{1}$

$=$ **11784 kW. (Ans.)**

Example 9.9. *The following data relate to a steam power plant :*

Maximum demand	= *30000 kW*
Load factor	= *0.42*
Coal consumption	= *1.1 kg/kWh*
Boiler efficiency	= *84%*
Turbine efficiency	= *88%*
Price of coal	= *Rs. 70 per tonne.*

Determine the following :

(*i*) *Thermal efficiency of the plant.*

(*ii*) *Coal bill of the plant for one year.*

Solution. (*i*) **Thermal efficiency of the plant :**

Thermal efficiency of the plant = Boiler efficiency × turbine efficiency

$= 0.84 \times 0.88 =$ **0.7392 or 73.92%. (Ans.)**

(*ii*) **Coal bill :**

Average demand on station = Maximum demand × load factor

= 30000 × 0.42 = 12600 kW

∴ Energy generated per year = 12600 × (365 × 24) kWh

∴ Coal consumption = 12600 × (365 × 24) × 1.1 kg per year

∴ *Coal bill* $= \frac{12600 \times (365 \times 24) \times 1.1 \times 70}{1000}$

= **Rs. 8498952. (Ans.)**

☞**Example 9.10.** *A power station is to supply for regions of load whose peak loads are 10 MW, 5 MW, 8 MW and 7 MW. The diversity factor of the load at the station is 1.5 and the average annual load factor is 0.6. Calculate :*

(i) Maximum demand on the station.

(ii) Annual energy supplied from the station.

Suggest the installed capacity and the number of units taking all aspects into account.

Solution. (*i*) **Maximum demand :**

Maximum demand on power station = $\frac{\text{Sum of individual maximum demands}}{\text{Diversity factor}}$

$= \frac{10 + 5 + 8 + 7}{1.5}$

= **20 MW or 20000 kW. (Ans.)**

(*ii*) **Annual energy supplied :**

Average load = Maximum demand × load factor

= 20000 × 0.6 = 12000 kW

Annual energy supplied from the station

= Average load × (365 × 24)

= 12000 × (365 × 24) = **105.12 × 10^6 kWh. (Ans.)**

Installed capacity and number of units :

Considering 50% increase in maximum demand on the power station in next five years, the *installed capacity* should be **30000 kW or 30 MW. (Ans.)**

Select, four similar units each of 7.5 MW capacity because minimum number of spare parts will be required to be stored, at the same time *three units can supply present maximum demand and fourth unit can be taken out for routine maintenance or during breakdown without any disruption in supply*.

Example 9.11. *The peak load on a 50 MW power station is 39 MW. It supplies power through four transformers whose connected loads are 17, 12, 9 and 10 MW. The maximum demands on these transformers are 15, 10, 8 and 9 MW respectively. If the annual load factor is 50% and the plant is operating for 65% of the period in a year, find out the following :*

(i) Average load on the station *(ii) Energy supplied per year*

(iii) Demand factor *(iv) Diversity factor*

(v) Power station use factor.

Solution. Power station rated capacity = 50 MW or 50000 kW

Maximum demand on the power station = 39 MW or 39000 kW

Sum of connected load = 17 + 12 + 9 + 10 = 48 MW or 48000 kW

Sum of maximum demands on the transformers = 15 + 10 + 8 + 9 = 42 MW or 42000 kW

Annual load factor = 50% or 0.5

Plant operating period = 0.65 × (365 × 24) = 5964 hours.

(*i*) **Average load on the station :**

Average load on the station = Maximum demand × load factor

$= 39000 \times 0.5 =$ **19500 kW. (Ans.)**

(*ii*) **Energy supplied per year :**

Energy supplied per year = Average load × (365 × 24)

$= 19500 \times 8760$

$=$ **170.82 × 10⁶ kWh. (Ans.)**

(*iii*) **Demand factor :**

$$\textit{Demand factor} = \frac{\text{Maximum demand}}{\text{Sum of connected load}}$$

$$= \frac{39000}{48000} = \mathbf{0.8125.}\ \textbf{(Ans.)}$$

(*iv*) **Diversity factor :**

$$\textit{Diversity factor} = \frac{\text{Sum of maximum demands}}{\text{Maximum demand}}$$

$$= \frac{42000}{39000} = \mathbf{1.077.}\ \textbf{(Ans.)}$$

(*v*) **Power station use factor :**

$$\textit{Use factor} = \frac{\text{Energy generated per year}}{\text{Rated capacity} \times \text{number of operating hours}}$$

$$= \frac{170.82 \times 10^6}{50000 \times 5694} = \mathbf{0.6\ or\ 60\%.}\ \textbf{(Ans.)}$$

Example 9.12. *A base load power station and standby power station share a common load as follows :*

Base load station annual output	$= 180 \times 10^6$ *kWh*
Base load station capacity	*= 42 MW*
Maximum demand on base load station	*= 36 MW*
Standby station capacity	*= 22 MW*
Standby station annual output	$= 17 \times 10^6$ *kWh*
Maximum demand (peak load) on standby station	*= 18 MW*

Determine the following for both power stations :

(i) Load factor. *(ii) Capacity (or plant) factor.*

Solution. Base load station :

$$\text{Average load} = \frac{180 \times 10^6}{365 \times 24} = 20548 \text{ kW}$$

$$(i)\ \textbf{Load factor} = \frac{\text{Average load}}{\text{Maximum demand}} = \frac{20548}{36 \times 10^3} = \mathbf{0.57.}\ \textbf{(Ans.)}$$

$$(ii)\ \textbf{Capacity factor} = \frac{\text{Energy generated}}{\text{Capacity of plant} \times (24 \times 365)}$$

$$= \frac{180 \times 10^6}{42 \times 1000 \times 24 \times 365} = \mathbf{0.489.}\ \textbf{(Ans.)}$$

Standby power station :

Annual average load $= \dfrac{17 \times 10^6}{365 \times 24} = 1940.6$ kW

(*i*) **Load factor** $= \dfrac{\text{Average load}}{\text{Maximum demand}} = \dfrac{1940.6}{18 \times 1000} = \mathbf{0.1078.}$ **(Ans.)**

(*ii*) **Capacity factor** $= \dfrac{\text{Energy generated}}{\text{Capacity} \times (24 \times 365)}$

$$= \frac{17 \times 10^6}{22 \times 1000 \times 24 \times 365} = \mathbf{0.088.} \quad \textbf{(Ans.)}$$

☞**Example 9.13.** *A base load station having a capacity of 18 MW and a standby station having a capacity of 20 MW share a common load. Find (i) annual load factor, (ii) use factor and (iii) capacity factor of the two power stations from the following data :*

Annual standby station output $= 7.35 \times 10^6$ *kWh*

Annual base load station output $= 101.35 \times 10^6$ *kWh*

Peak load on the standby station = *12 MW*

Hours of use of standby station during the year = *2190 hours.*

Solution. Standby station :

Capacity of standby station = 20 MW or 20000 kW

Maximum demand on standby station = 12 MW or 12000 kW

Annual standby station output $= 7.35 \times 10^6$ kWh

Hours of use of standby station during the year = 2190 hours

Annual average load of standby station $= \dfrac{\text{Output in kWh}}{365 \times 24} = \dfrac{7.35 \times 10^6}{365 \times 24} = 839$ kW

(*i*) **Annual load factor :**

Annual load factor $= \dfrac{\text{Annual average load}}{\text{Maximum demand}} = \dfrac{839}{12000} = \mathbf{0.07}$ **or 7%. (Ans.)**

(*ii*) **Use factor :**

$$\textit{Use factor} = \frac{\text{Total kWh generated}}{\text{Rated capacity of station} \times \text{number of operating hours}}$$

$$= \frac{7.35 \times 10^6}{20000 \times 2190} = \mathbf{0.1678} \quad \textbf{or 16.78\%.} \quad \textbf{(Ans.)}$$

(*iii*) **Capacity factor :**

$$\textit{Capacity factor} = \frac{\text{Average load}}{\text{Rated capacity}} = \frac{839}{20000}$$

$$= \mathbf{0.0419} \quad \textbf{or} \quad \textbf{4.19\%.} \quad \textbf{(Ans.)}$$

Base load station :

Capacity of base load station = 18 MW or 18000 kW

Assume maximum demand on base load station equal to its rated capacity *i.e.*, 18 MW.

Annual base load station output $= 101.35 \times 10^6$ kWh

$$\text{Annual average load of base load station} = \frac{\text{Output in kWh}}{365 \times 24}$$

$$= \frac{101.35 \times 10^6}{365 \times 24} = 11570 \text{ kW}$$

(*i*) **Annual load factor :**

$$\textit{Annual load factor} = \frac{\text{Annual average load}}{\text{Maximum demand}}$$

$$= \frac{11570}{18000} = \mathbf{0.643} \quad \textbf{or} \quad \mathbf{64.3\%.} \quad \textbf{(Ans.)}$$

(*ii*) **Use factor :**

$$\textit{Use factor} = \frac{\text{Total kWh generated}}{\text{Rated capacity} \times \text{number of operating hours}}$$

$$= \frac{101.35 \times 10^6}{18000 \times (365 \times 24)} = \mathbf{0.643} \quad \textbf{or} \quad \mathbf{64.3\%.} \quad \textbf{(Ans.)}$$

(*iii*) **Capacity factor :**

$$\textit{Capacity factor} = \frac{\text{Average load}}{\text{Rated capacity}} = \frac{11570}{18000}$$

$$= \mathbf{0.643} \quad \textbf{or} \quad \mathbf{64.3\%.} \quad \textbf{(Ans.)}$$

COST ANALYSIS

Example 9.14. *Determine the annual cost of a feed water softner from the following data .*

Cost	*= Rs. 96000*
Salvage value	*= 5%*
Life	*= 10 years*
Annual repair and maintenance cost	*= Rs. 3000*
Annual cost of chemicals	*= Rs. 6000*
Labour cost per month	*= Rs. 360*
Interest on sinking fund	*= 5%.*

Solution. Capital cost, P = Rs. 96000.

Salvage value, $S = \frac{5}{100} \times 96000 = \text{Rs. } 4800$

Rate of interest on sinking fund, i = 5% or 0.05

Life, n = 10 years

∴ Annual sinking fund payment

$$= (P - S)\left[\frac{i}{(1+i)^n - 1}\right]$$

$$= (9600 - 4800)\left[\frac{0.05}{(1+0.05)^{10} - 1}\right] = \text{Rs. } 7250.8$$

Total cost per year :

Annual sinking fund	= Rs. 7250.8
Annual repair and maintenance cost	= Rs. 3000

Annual cost of chemicals = Rs. 6000

Annual labour cost = (360 × 12) = Rs. 4320

∴ *Total cost per year* = 7250.8 + 3000 + 6000 + 4320

= **Rs. 20570.8. (Ans.)**

Example 9.15. *The output of a generating station is 500 × 10⁶ kWh per year and average load factor is 0.7. If the annual fixed charges are Rs. 50 per kW of installed plant and annual running charges are 5 per kWh, what is the cost per kWh of energy at the bus bar.*

Solution. Output energy per annum = 500×10^6 kWh

Average load $= \dfrac{\text{Annual average load}}{365 \times 24} = \dfrac{500 \times 10^6}{365 \times 24} = 57077$ kW

Maximum demand $= \dfrac{\text{Average load}}{\text{Load factor}} = \dfrac{57077}{0.7}$

= 81538 kW

Assuming installed capacity equal to maximum demand, we have :

Fixed charges = 50 × 81538 = Rs. 4076900

Running charges $= \text{Rs. } \dfrac{5}{100} \times 500 \times 10^6 = \text{Rs. } 25000000$

Total annual charges = Rs. 25000000 + Rs. 4076900

= Rs. 29076900

Cost of energy at bus-bar $= \dfrac{\text{Total annual charges}}{\text{Output energy per annum}}$

$= \dfrac{29076900}{500 \times 10^6}$

= **Rs. 0.058 or 5.8 p/kWh. (Ans.)**

Example 9.16. *From the following data calculate the cost of generation per unit delivered from the power plant :*

Installed capacity of the power plant = *200 MW*

Annual load factor = *0.4*

Capital cost of power plant = *Rs. 280 lacs*

Annual cost of fuel, oil, salaries, taxation = *Rs. 60 lacs.*

Interest and depreciation = *13%.*

Solution. Installed capacity of the power plant = 200 MW or 200×10^3 kW

Assuming maximum demand equal to installed capacity

Maximum demand = 200×10^3 kW

Annual load factor = 0.4

Total units generated per annum = Maximum demand × load factor × (365 × 24)

$= 200 \times 10^3 \times 0.4 \times (365 \times 24) = 700.8 \times 10^6$ kWh

Capital cost of the power plant = Rs. 280×10^5

Annual interest and depreciation = Rs. $280 \times 10^6 \times \dfrac{13}{100}$ = Rs. 3.64×10^6

Annual cost of fuel, oil, salaries, taxation etc. = Rs. 60×10^5 or 6×10^6

Total annual cost = Rs. 3.64×10^6 + Rs. 6×10^6 = Rs. 9.64×10^6

Generating cost $= \dfrac{\text{Total annual cost}}{\text{Total units generated per annum}}$

$= \dfrac{9.64 \times 10^6}{700.8 \times 10^6}$

= **Rs. 0.0137 or 1.37 p/kWh. (Ans.)**

☞**Example 9.17.** *The following data relate to a 10 MW power station :*

Cost of plant	*= Rs. 1200 per kW*
Interest, insurances and taxes	*= 5% per annum*
Depreciation	*= 5%*
Cost of primary distribution	*= Rs. 500000*
Interest, insurances, taxes and depreciation	*= 5%*
Cost of coal including transportation	*= Rs. 4.4 per kN*
Operating cost	*= Rs. 500000*
Plant maintenance cost :	
(i) Fixed	*= Rs. 20000 per annum*
(ii) Variable	*= Rs. 30000 per annum*
Installed plant capacity	*= 10000 kW*
Maximum demand	*= 9000 kW*
Annual load factor	*= 0.6*
Consumption of coal	*= 255000 kN*

Determine the following :

(i) Cost of power generation per kW per year.

(ii) Cost per kWh generated.

(iii) Total cost of generation per kWh.

Transmission or primary distribution chargeable to generation.

Solution. Installed capacity of plant = 10 MW or 10000 kW

Total cost of plant = Rs. 10000×1200 = Rs. 12×10^6

Annual interest, insurances and taxes = Rs. $0.05 \times 12 \times 10^6$ = Rs. 0.06×10^6
= Rs. 600000

Annual depreciation = Rs. $0.05 \times 12 \times 10^6$ = Rs. 0.6×10^6
= Rs. 600000

Annual interest, insurance, taxes and depreciation on primary distribution
= Rs. 0.05×500000 = Rs. 25000

Annual plant maintenance cost (fixed) = Rs. 20000

Total fixed cost = Rs. (600000 + 600000 + 25000 + 20000)
= Rs. 1245000

Annual operating cost = Rs. 500000

Annual plant maintenance cost (variable) = Rs. 30000

Annual cost of coal = Rs. 4.4×255000 = Rs. 1122000

Total annual running cost = Rs. (500000 + 300000 + 1122000)
= Rs. 1652000

Maximum demand = 9000 kW

Annual load factor = 0.6

Average load = 9000 × 0.6 = 5400 kW

Annual energy generated = 5400 × 365 × 24 = 47.3 × 10^6 kWh

(*i*) **Cost of power generation per kW per year :**

$$\text{Annual cost per kW of maximum demand} = \frac{\text{Fixed cost per annum}}{\text{Maximum demand}}$$

$$= \frac{1245000}{9000} = \textbf{Rs. 138.33.} \quad \textbf{(Ans.)}$$

(*ii*) **Cost per kWh generated :**

$$\text{Annual cost/kWh} = \frac{\text{Annual running cost}}{\text{Annual energy generated}}$$

$$= \frac{1652000}{47.3 \times 10^6} = \textbf{Rs. 0.035 or 3.5 p.} \quad \textbf{(Ans.)}$$

(*iii*) **Total cost per kWh :**

$$\text{Total cost per kWh} = \frac{\text{Total annual cost}}{\text{Annual energy generated}}$$

$$= \frac{1245000 + 1652000}{47.3 \times 10^6}$$

$$= \textbf{Rs. 0.06 or 6 p.} \quad \textbf{(Ans.)}$$

Example 9.18. *Annual 200 MW steam power station is estimated to cost Rs. 350 millions. Other costs are as follows :*

Insurance and taxes = *Rs. 5 lacs per annum*

Fuel and lubricants = *Rs. 70 lacs per annum*

Transport and storage = *Rs. 10 lacs per annum*

Salaries and wages = *Rs. 12 lacs per annum*

Miscellaneous = *Rs. 2 lacs per annum*

Reckoning interest and depreciation at 15% per annum of capital cost, determine :

The cost of energy generated per unit if the power station works at an average load factor of 0.6. What would it be, if the load factor be increased to 75% with a consequent increase in fuel costs by 10%, other costs remaining the same ?

Solution. Capacity of steam power station = 200 MW or 200000 kW

Cost of power station = Rs. 350 millions = Rs. 350 × 10^6

Annual cost of insurance and taxes = Rs. 5 lacs = Rs. 0.5 × 10^6

Annual cost of fuel and lubricants = Rs 70 lacs = Rs. 7 × 10^6

Annual cost of transport and storage = Rs. 10 lacs = Rs. 1 × 10^6

Annual cost of salaries and wages = Rs. 12 lacs = Rs 1.2 × 10^6

Annual miscellaneous cost = Rs. 2 lacs = Rs. 0.2 × 10^6

$$\text{Annual interest and depreciation} = \text{Rs. } 350 \times 10^6 \times \frac{15}{100} = \text{Rs. } 52.5 \times 10^6$$

Total annual cost $= \text{Rs. } (0.5 + 7 + 1 + 1.2 + 0.2 + 52.5) \times 10^6 = \text{Rs. } 62.4 \times 10^6$

Total energy generated per annum

$= \text{Maximum demand} \times \text{average load factor} \times (365 \times 24)$

$= 200000 \times 0.6 \times (365 \times 24) = 1051.2 \times 10^6 \text{ kWh}$

Cost of generation $= \dfrac{\text{Total annual cost}}{\text{Total units generated}}$

$= \dfrac{62.4 \times 10^6}{1051.2 \times 10^6}$ = **Rs. 0.0594 per kWh or 5.94 p. per kWh. (Ans.)**

If the load factor is improved to 75% :

Total energy generated per annum

$= 200000 \times 0.75 \times (365 \times 24) = 1314 \times 10^6 \text{ kWh}$

Annual operating cost $= \text{Rs. } (0.5 + 7 \times 1.1 \times 10^6 + 1 + 1.2 + 0.2 + 52.5) \times 10^6$

$= \text{Rs. } 63.1 \times 10^6$

Cost of generation $= \dfrac{\text{Total annual cost}}{\text{Total units generated}}$

$= \dfrac{63.1 \times 10^6}{1314 \times 10^6}$ = **Rs. 0.048 per kWh or 4.8 p. per kWh. (Ans.)**

Example 9.19. *A steam station has two 110 MW units. Following cost data are given :*

Particulars	*Units A*	*Units B*
Capital cost	*Rs. 2400 per kW*	*Rs. 3000 per kW*
Fixed charge rate	*10%*	*10%*
Capital factor	*0.55*	*0.60*
Fuel consumption	*1 kg/kWh*	*0.9 kg/kWh*
Fuel cost	*Rs. 96 per 1000 kg*	*Rs. 96 per 1000 kg*
Annual cost of operation, labour, maintenance and supplies	*20% of annual fuel cost*	*15% of annual fuel cost*
Utilisation factor	*1*	*1*

Calculate the following :

(*i*) *Annual plant cost and generation cost of unit A.*

(*ii*) *Annual plant cost and generation cost of unit B.*

(*iii*) *Overall generation cost of the station.*

Solution. (*i*) **Annual plant cost and generation cost of unit A :**

Annual fixed cost of unit A

$= \dfrac{10}{100} \times 2400 \times (100 \times 1000)$

$= \text{Rs. } 26.4 \times 10^6$

Annual energy output $= \text{Maximum demand} \times \text{capacity factor} \times \text{no. of hours}$

$= (100 \times 1000) \times 0.55 \times (356 \times 24)$

$= 52.998 \times 10^7 \text{ kWh}$

Annual fuel consumption $= 1 \times 52.998 \times 10^7$

$= 52.998 \times 10^7 \text{ kg}$

Fuel cost $= \frac{96}{1000} \times 52.998 \times 10^7 = \text{Rs. } 50.87 \times 10^7$

Annual cost of operating labour, maintenance and supplies

$= 20\%$ of annual cost

$= \frac{20}{100} \times 50.87 \times 10^6 = \text{Rs. } 10.174 \times 10^6$

The annual operating cost of unit A

= Annual fuel cost + annual cost of operation, labour and maintenance

$= \text{Rs. } (50.87 \times 10^6 + 10.174 \times 10^6) = \text{Rs. } 61.044 \times 10^6$

Annual plant cost of unit A

= Annual fixed cost + annual operating cost

$= \text{Rs. } (26.4 \times 10^6 + 61.044 \times 10^6) =$ **Rs. 87.444 × 10⁶. (Ans.)**

Generation cost of unit A

$= \frac{\text{Annual plant cost}}{\text{Annual energy output}}$

$= \frac{87.444 \times 10^6}{52.998 \times 10^7} =$ **Rs. 0.165 or 16.5 p/kWh. (Ans.)**

(*ii*) **Annual plant cost and generation cost of unit B :**

Annual fixed cost of unit B

$= \text{Rs. } \frac{10}{100} \times 3000 \times 110 \times 1000 = \text{Rs. } 33 \times 10^6$

Expected annual energy output

$= (110 \times 1000) \times (365 \times 24) \times 0.6 = 57.816 \times 10^7$ kWh

Annual fuel consumption

$= 0.9 \times 57.816 \times 10^7 = 52.0344 \times 10^7$ kg

Fuel cost $= \frac{96}{100} \times 52.0344 \times 10^7 = \text{Rs. } 49.95 \times 10^6$

Annual cost of maintenance, repair etc.

$= \text{Rs. } \frac{15}{100} \times 49.95 \times 10^6 = \text{Rs. } 7.4925 \times 10^6$

Annual operating cost

= Fuel cost + maintenance cost

$= \text{Rs. } (49.95 \times 10^6 + 7.4925 \times 10^6) = \text{Rs. } 57.4425 \times 10^6$

Annual plant cost of unit B

= Fixed cost + operating cost

$= \text{Rs. } 33 \times 10^6 + 57.4425 \times 10^6$

= **Rs. 90.4425 × 10⁶. (Ans.)**

Generation cost of unit B

$= \frac{\text{Annual plant cost}}{\text{Annual energy output}}$

$= \frac{90.4425 \times 10^6}{57.816 \times 10^7}$

= **Rs. 0.1564 or 15.64 p/kWh. (Ans.)**

(*iii*) **Overall generation cost of the station**

$$= \frac{\text{Sum of annual plant cost of both units}}{\text{Sum of energy supplied}}$$

$$= \frac{87.444 \times 10^6 + 90.4425 \times 10^6}{52.998 \times 10^7 + 57.816 \times 10^7}$$

= **Rs. 0.16 or 16 p/kWh. (Ans.)**

☞**Example 9.20.** *The annual costs of operating a 15000 kW thermal power station are as follows :*

Cost of plant	= *Rs. 1080 per kW*
Interest, insurance, taxes on plant	= *5 per cent*
Depreciation	= *5 per cent*
Cost of primary distribution system	= *Rs. 600000*
Interest, insurance, taxes and depreciation on primary distribution system	= *5 per cent*
Cost of secondary distribution system	= *Rs. 1080000*
Interest, taxes, insurance and depreciation on secondary distribution system	= *5 per cent*
Maintenance of secondary distribution system	= *Rs. 216000*
Plant maintenance cost	
(i) Fixed cost	= *Rs. 36000*
(ii) Variable cost	= *Rs. 48000*
Operating costs	= *Rs. 720000*
Cost of coal	= *Rs. 7.2 per kN*
Consumption of coal	= *300000 kN*
Dividend to stock holders	= *Rs. 1200000*
Energy loss in transmission	= *10 per cent*
Maximum demand	= *14000 kW*
Diversity factor	= *1.5*
Load factor	= *0.7*

Determine : (i) Charge per kW per year

(ii) Rate per kWh.

Solution. Maximum demand = 14000 kW

Load factor $= 0.7 = \frac{\text{Average load}}{\text{Maximum demand}}$

∴ Average load = 0.7 × 14000 = 9800 kW

∴ Energy generated per year = 9800 × (365 × 24)

= 85.8 × 10^6 kWh

Cost of plant = Capacity of plant × cost per kW

= 15000 × 1080 = Rs. 16.2 × 10^6

Interest, insurances, taxes on plant

$$= \frac{5}{100} \times 16.2 \times 10^6 = \text{Rs. } 810000$$

Plant depreciation $= \frac{5}{100} \times 16.2 \times 10^6 = \text{Rs. } 810000$

Cost of primary distribution system = Rs. 600000

Interest, insurance, taxes, depreciation on primary distribution system

$$= \frac{5}{100} \times 600000 = \text{Rs. } 30000$$

Cost of secondary distribution system = Rs. 1080000

$$= \frac{5}{100} \times 1080000 = \text{Rs. } 54000$$

Cost of coal $= 7.2 \times 300000 = \text{Rs. } 2160000.$

(*i*) **Charge per kW per year :**

Fixed costs :

Interest, taxes and insurance on plant	= Rs. 810000
Plant depreciation	= Rs. 810000
Interest, taxes, insurance and depreciation on primary distribution system	= Rs. 30000
Secondary distribution system	= Rs. 54000
Fixed part of plant maintenance	= Rs. 36000
Dividend of stock-holder	= Rs. 1200000
∴ Total fixed cost	= Rs. 2940000

Sum of maximum demand of consumers

$$= \text{Maximum demand} \times \text{diversity factor}$$

$$= 14000 \times 1.5 = 21000 \text{ kW}$$

Charge per kW per year $= \frac{2940000}{21000}$

= **Rs. 140 per kW. (Ans.)**

(*ii*) **Rate per kWh :**

Variable costs :

Cost of coal = Rs. 2160000

Plant maintenance = Rs. 48000

Operating costs	= Rs. 720000
Maintenance of secondary distribution system	= Rs. 216000
Total variable cost	= Rs. 3144000
Energy loss in transmission	= 10 percent

∴ Net energy transmitted

$$= 0.9 \times 85.8 \times 10^6 = 77.22 \times 10^6 \text{ kWh}$$

∴ **Rate per kWh** $= \frac{3144000}{77.22 \times 10^6}$ = **Rs. 0.0407 or 4.07 p/kWh. (Ans.)**

Example 9.21. *It is proposed to supply a load with a maximum demand of 100 MW and a load factor of 0.4. Choice is to be made from nuclear, hydro and steam power plants. Calculate the overall cost per kWh in each scheme.*

Cost	*Nuclear power plant*	*Hydro-power plant*	*Steam power plant*
Capital per kW installed	*Rs. 6000*	*Rs. 4320*	*Rs. 2160*
Interest	*10%*	*10%*	*12%*
Depriciation	*10%*	*8%*	*12%*
Operating cost per kWh	*12 paise*	*6 paise*	*18 paise*
Transmission and distribution cost/kWh	*0.24 paise*	*0.96 paise*	*0.24 paise*

Solution. (*i*) **Nuclear power plant :**

Capital cost $= 6000 \times (100 \times 10^3) = \text{Rs. } 60 \times 10^7$

Interest $= \dfrac{10}{100} \times 60 \times 10^7 = \text{Rs. } 6 \times 10^7$

Depreciation $= \text{Rs. } 6 \times 10^7$

Annual fixed cost (interest + depreciation) = Rs. 12×10^7

Energy generated per year

$= \text{Average load} \times (365 \times 24)$

$= \text{Load factor} \times \text{maximum demand} \times (365 \times 24)$

$= 0.4 \times (100 \times 10^3) \times (365 \times 24)$

$= 350.4 \times 10^6 \text{ kWh}$

Running cost per kWh = Operating cost per kWh + transmission and distribution cost per kWh

$= 12 + 0.24 = 12.24 \text{ p}$

∴ **Overall cost per kWh** = Running cost/kWh + fixed cost/kWh

$= 12.24 + \dfrac{12 \times 10^7}{350.4 \times 10^6} \times 100 =$ **46.48 p. (Ans.)**

(*ii*) **Hydro-electric power plant :**

Capital cost $= 4320 \times 100 \times 10^3 = \text{Rs. } 43.2 \times 10^7$

Interest $= \dfrac{10}{100} \times 43.2 \times 10^7 = \text{Rs. } 43.2 \times 10^6$

Depreciation $= \dfrac{8}{100} \times 43.2 \times 10^7 = 34.56 \times 10^6$

Annual fixed cost = Interest + depreciation

$= \text{Rs. } (43.2 + 34.56) \times 10^6 = \text{Rs. } 77.76 \times 10^6$

Running cost per kWh = (Operation cost + transmission cost)

$= (6 + 0.96) \text{ p} = 6.96 \text{ p}$

Overall cost per kWh = Running cost/kWh + annual fixed cost/kWh

$= 6.96 + \dfrac{77.76 \times 10^6}{350.4 \times 10^6} \times 100$

$= 6.96 + 22.2 =$ **29.16 p. (Ans.)**

(*iii*) **Steam power plant :**

Capital cost $= \text{Rs. } 2160 \times 100 \times 10^3 = \text{Rs. } 21.6 \times 10^7$

Interest $= \text{Rs. } \dfrac{12}{100} \times 21.6 \times 10^7 = \text{Rs. } 25.92 \times 10^6$

Depreciation $= \text{Rs. } \frac{12}{100} \times 21.6 \times 10^7 = \text{Rs. } 25.92 \times 10^6$

Annual fixed cost = Interest + depreciation

$= \text{Rs. } 2 \times 25.92 \times 10^6 = \text{Rs. } 51.84 \times 10^6$

Running cost/kWh $= 18 + 0.24 = 18.24$ p

Overall cost/kWh = Running cost/kWh + fixed cost/kWh

$$= 18.24 + \frac{51.84 \times 10^6}{350.4 \times 10^6} \times 100$$

$= 18.24 + 14.79 =$ **33.03 p. (Ans.)**

From the above calculations it is concluded that overall cost/kWh is *minimum in case of hydropower plant*.

☞**Example 9.22.** *A power plant of 180 MW installed capacity has the following data :*

Capital cost = *Rs. 2160/kW installed*

Interest and depreciation = *12 per cent*

Annual load factor = *0.6*

Annual capacity factor = *0.5*

Annual running charges = *Rs.* 36×10^6

Energy consumed by power auxiliaries = *6 per cent*

Calculate : (i) Reserve capacity

(ii) Generation capacity.

Solution. Load factor $= \frac{\text{Average load}}{\text{Maximum demand}}$

and, Capacity factor $= \frac{\text{Average load}}{\text{Rated capacity}}$

$$\therefore \quad \frac{\text{Load factor}}{\text{Capacity factor}} = \frac{\text{Rated capacity}}{\text{Maximum demand}}$$

$$\frac{0.6}{0.5} = \frac{180}{\text{Maximum demand}}$$

$\therefore$ Maximum demand $= \frac{0.5 \times 180}{0.6} = 150$ MW.

(*i*) **Reserved capacity :**

Reserved capacity = Installed/rated capacity – maximum demand

$= 180 - 150 =$ **30 MW. (Ans.)**

(*ii*) **Generation cost :**

Average load = Load factor × maximum demand

$= 0.6 \times 150 = 90$ MW

Energy generated per annum

$= 90 \times 10^3 \times (365 \times 24) = 788.4 \times 10^6$ kWh

Energy consumed by auxiliaries

$$= \frac{6}{100} \times 788.4 \times 10^6 = 47.3 \times 10^6 \text{ kWh}$$

Net energy available $= 788.4 \times 10^6 - 47.3 \times 10^6 = 741.1 \times 10^6$ kWh

Fixed cost of generation = Interest + depreciation

$$= \text{Rs. } \frac{12}{100} \times 2160 \times 180 \times 10^3 = \text{Rs. } 46.65 \times 10^6$$

Total annual cost = Running cost + fixed cost

$$= 36 \times 10^6 + 46.65 \times 10^6 = \text{Rs. } 82.65 \times 10^6$$

∴ *Generation cost* $= \dfrac{82.65 \times 10^6}{741.1 \times 10^6} \times 100 =$ **11.1 p. (Ans.)**

Example 9.23. *A power station has the installed capacity of 180 MW. Calculate the cost of generation, other data pertaining to power station are given below :*

Capital cost = *Rs.* 300×10^6

Rate of interest and depreciation = *18 per cent*

Annual cost of fuel oil, salaries and taxation = *Rs.* 36×10^6

Load factor = *0.4*

Also calculate the saving in cost per kWh if the annual load factor is raised to 0.5.

Solution. Assuming maximum demand equal to the capacity of the power plant,

$$\text{Load factor} = \frac{\text{Average load}}{\text{Maximum demand}}$$

$$0.4 = \frac{\text{Average load}}{180}$$

∴ Average load $= 0.4 \times 180 = 72$ MW

Energy generated per annum $= 72 \times 10^3 \times (365 \times 24) = 630.72 \times 10^6$ kWh

Fixed cost = Interest and depreciation on capital cost

$$= \frac{18}{100} \times 300 \times 10^6 = \text{Rs. } 54 \times 10^6$$

Running (operating cost) = Cost of fuel oil, salaries and taxation

$$= \text{Rs. } 36 \times 10^6$$

Total annual cost = Fixed cost + operating cost

$$= \text{Rs. } (54 + 36) \times 10^6$$

$$= \text{Rs. } 90 \times 10^6$$

∴ *Cost per kWh* $= \dfrac{90 \times 10^6}{630.72 \times 10^6} \times 100 =$ **14.27 p. per kWh. (Ans.)**

When the load factor is raised to 0.5 :

Average load = Load factor × maximum demand

$$= 0.5 \times 180 = 90 \text{ MW}$$

Energy produced per annum $= 90 \times 10^3 \times 365 \times 24$

$$= 788.4 \times 10^6 \text{ kWh}$$

Total annual cost will not change.

∴ Cost per kWh $= \dfrac{90 \times 10^6}{788.4 \times 10^6} \times 100$

$= 11.41$ p. per kWh

∴ *Saving in cost per kWh* $= 14.27 - 11.41$

= **2.86 p. (Ans.)**

Example 9.24. *A 60 MW generating station has the following data :*

Capital cost $= \text{Rs. } 18 \times 10^6$

Annual taxation $= \text{Rs. } 0.48 \times 10^6$

Annual salaries and wages $= \text{Rs. } 1.44 \times 10^6$

Cost of coal = *Rs. 72 per tonne*

Calorific value (C.V.) of coal = *23000 kJ/kg*

Rate of interest and depreciation = *12 per cent*

Plant heat rate = *138000 kJ/kWh at 100% capacity*

Calculate the generating cost/kWh at 100% capacity factor.

Solution. Maximum demand = 60 MW (= peak load)

Fixed cost = Interest and depreciation on capital cost

$$= \text{Rs. } \frac{12}{100} \times 18 \times 10^6 = \text{Rs. } 2.16 \times 10^6$$

Running cost = Annual salaries, wages and taxation

$$= \text{Rs. } (1.44 + 0.48)\, 10^6$$

$$= \text{Rs. } 1.92 \times 10^6$$

At 100% capacity factor :

Rated/Installed capacity = 60 MW

$$\text{Capacity factor} = \frac{\text{Average load}}{\text{Rated capacity}} = 1$$

or, Average load = Rated capacity = 60 MW

Average energy produced per annum

$$= 60 \times 10^3 \times (365 \times 24) \text{ kWh}$$

$$= 525.6 \times 10^6 \text{ kWh}$$

Total plant heat rate $= 138000 \times 525.6 \times 10^6$ kJ

$$\text{Weight of coal required per annum} = \frac{138000 \times 525.6 \times 10^6}{23000 \times 1000} \text{ tonnes}$$

$$= 3.15 \times 10^6 \text{ tonnes}$$

Cost of fuel $= \text{Rs. } 72 \times 3.15 \times 10^6$

$$= \text{Rs. } 226.8 \times 10^6$$

Total annual cost $= \text{Rs. } (2.16 + 1.92 + 226.8) \times 10^6$

$$= \text{Rs. } 230.88 \times 10^6$$

$$\text{Generation cost} = \frac{230.88 \times 10^6}{525.6 \times 10^6} \times 100$$

= **43.93 p.** **(Ans.)**

ECONOMIC LOAD SHARING

Example 9.25. *An input-output curve of a 10 MW station is expressed as follows :*

$$I = 4 \times 10^6 (10 + 8L + 0.4L^2)$$

where I is in kJ/hour and L is in megawatts.

(i) Without plotting any curve find the load at which the maximum efficiency occurs.

(ii) Find the increase in input required to increase station output from 3 to 5 MW by means of the input-output curve and also by incremental rate curve.

Solution. (*i*) **Load at which maximum efficiency occurs :**

$$I = 4 \times 10^6 (10 + 8L + 0.4L^2)$$

or
$$\frac{I}{L} = 4 \times 10^6 \left(\frac{10}{L} + 8 + 0.4L\right)$$

Efficiency
$$= \frac{\text{Output}}{\text{Input}} = \frac{L}{I}$$

∴ Efficiency,
$$\eta = \frac{1}{4 \times 10^6 \left(\frac{10}{L} + 8 + 0.4L\right)} \quad ...(i)$$

Now the *efficiency will be maximum* when $\left(\frac{10}{L} + 8 + 0.4L\right)$ is *minimum*

i.e.,
$$\frac{d}{dL}\left(\frac{10}{L} + 8 + 0.4L\right) = 0$$

∴
$$-\frac{10}{L^2} + 0.4 = 0 \quad \text{or} \quad L^2 = \frac{10}{0.4} = 25 \quad \text{or} \quad L = 5 \text{ MW}$$

Hence, *the load at which the maximum efficiency occurs* = **5 MW. (Ans.)**

(*ii*) **Increase in input :**

(*a*) ***By input output curve :***

When load, L = 3 MW

Input,
$$I_3 = 4 \times 10^6 (10 + 8 \times 3 + 0.4 \times 3^2)$$
$$= 150.4 \times 10^6 \text{ kJ/h}$$

When load, L = 5 MW

Input,
$$I_5 = 4 \times 10^6 (10 + 8 \times 5 + 0.4 \times 5^2)$$
$$= 240 \times 10^6 \text{ kJ/h}$$

Increase in input required
$$= I_5 - I_3$$
$$= (240 - 150.4) \times 10^6$$
$$= \mathbf{89.6 \times 10^6 \text{ kJ/h.}} \quad \textbf{(Ans.)}$$

(*b*) ***By incremental rate curve :***

When load varies from 3 to 5 MW, the incremental rate may be considered to be straight line and the average height of area under the curve between 3 MW and 5 MW would be

$$= \frac{3 + 5}{2} = 4 \text{ MW}$$

$$I = 4 \times 10^6 (10 + 8L + 0.4L^2)$$

Increment rate,
$$IR = \frac{dI}{dL} = 4 \times 10^6 (8 + 0.8L)$$

∴
$$IR = 4 \times 10^6 (8 + 0.8 \times 4) \text{ when load} = 4 \text{ MW}$$
$$= 4 \times 10^6 (8 + 3.2) = 4 \times 10^6 \times 11.2$$

Hence total increase in input $= 4 \times 10^6 \times 11.2 (5 - 3) = \mathbf{89.6 \times 10^6}$ **kJ/h. (Ans.)**

This shows that increase in input required to increase the required output in both cases (*a*) and (*b*) is same. This *indicates that the incremental rate curve can be taken as straight line for small increase in output.*

Example 9.26. *The input-output curve of a 50 MW power station is given by :*

$$I = 4 \times 10^6 (8 + 8L + 0.4L^2) \text{ kJ/hour}$$

where I is the input in kJ/hour and L is load in MW.

(i) Determine the heat input per day to the power station if it works for 20 hours at full load and remaining period at no load.

(ii) Also find the saving per kWh of energy produced if the plant works at full load for all 24 hours generating the same amount of energy.

Solution. (*i*) **Heat input per day :**

Total energy generated by the plant during 24 hours

$$= 20 \times 50 + 4 \times 0 = 1000 \text{ MWh}$$

Input to the plant when the plant is running at full load

$$I_{50} = 4 \times 10^6 (8 + 8 \times 50 + 0.4 \times 50^2) \times 20$$

$$= 4 \times 1408 \times 20 \text{ kJ during 20 hours when the plant was running at } \textit{full load.}$$

Input at no load, $I_0 = 4 \times 10^6 \times 8 \times 4$

$$= 128 \times 10^6 \text{ kJ during 4 hours when the plant was running at } \textit{no load.}$$

Total input to the plant during 24 hours

$$= I_{50} + I_0 = 4 \times 10^6 \times 1408 \times 20 + 128 \times 10^6$$

$$= 10^6 (5632 \times 20 + 128) = \mathbf{112768 \times 10^6 \text{ kJ/day.}} \quad \textbf{(Ans.)}$$

(*ii*) **Saving per kWh :**

Average heat supplied per kWh generated

$$= \frac{112768 \times 10^6}{1000 \times 10^3} = 112768 \text{ kJ/kWh}$$

If the same energy is generated within 24 hours, the average load is given by :

Average load $= \dfrac{1000}{24} = 41.67$ MW

Heat supplied during 24 hours in this case

$$I_{50} = 4 \times 10^6 (8 + 8 \times 50 + 0.4 \times 41.67^2) \times 24$$

$$= 4 \times 10^6 (8 + 400 + 694.5) \times 24$$

$$= 4 \times 10^6 \times 1102.5 \times 24 \text{ kJ/day} = 105840 \times 10^6 \text{ kJ/day}$$

Net saving per day

$$= 112768 \times 10^6 - 105840 \times 10^6 = 6928 \times 10^6 \text{ kJ/day}$$

$\therefore$ *Saving per kWh* $= \dfrac{6928 \times 10^6}{1000 \times 10^3} = \mathbf{6928 \text{ kJ/kWh.}}$ **(Ans.)**

Example 9.27. *The incremental fuel costs for two generating units 1 and 2 of a power plant are given by the following equations :*

$$\frac{dF_1}{dP_1} = 0.07 P_1 + 24$$

$$\frac{dF_2}{dP_2} = 0.075 P_2 + 22$$

where F is fuel cost in rupees per hour and P is power output in MW. Determine :

(i) The economic loading of the two units when the total load supplied by the power plants is 180 MW.

(ii) The loss in fuel cost per hour if the load is equally shared by both units.

Solution. (*i*) **Economic loading of two units :**

$$P_1 + P_2 = 180 \quad \text{...(Given)} \quad \text{...(i)}$$

The condition required for economic loading is given by :

$$\frac{dF_1}{dP_1} = \frac{dF_2}{dP_2}$$

$\therefore$ $$0.07\,P_1 + 24 = 0.075\,P_2 + 22 \quad ...(ii)$$

Substituting the value of P_2 (= $180 - P_1$) from (*i*) in (*ii*), we get

$$0.07\,P_1 + 24 = 0.075\,(180 - P_1) + 22$$

or, $$0.07\,P_1 + 24 = 13.5 - 0.075\,P_1 + 22$$

or, $$0.145\,P_1 = 11.5$$

or, $$P_1 = \frac{11.5}{0.145} = \mathbf{79.3\ MW.} \quad \textbf{(Ans.)}$$

and, $$P_2 = 180 - 79.3 = \mathbf{100.7\ MW.} \quad \textbf{(Ans.)}$$

(*ii*) **Loss in fuel cost :**

If the load is equally shared by both the units $\left(\text{supplying } \frac{180}{2} = 90 \text{ MW each}\right)$, then the increase in cost of fuel for *unit 1* is

$$= \int_{79.3}^{90} (0.07\,P_1 + 24)\,dP_1 = \left[\frac{0.07\,P_1^2}{2} + 24P_1\right]_{79.3}^{90}$$

$$= 0.035\,(90^2 - 79.3^2) + 24\,(90 - 79.3)$$

$$= 63.4 + 256.8 = \text{Rs. } 320.2\text{/hour}$$

Increase in cost of fuel for *unit 2*

$$= \int_{100.7}^{90} (0.075\,P_2 + 22)\,dP_2 = \left[\frac{0.075P_2^2}{2} + 22P_2\right]_{100.7}^{90}$$

$$= \frac{0.075}{2}(90^2 - 100.7^2) + 22\,(90 - 100.7)$$

$$= -76.5 - 235.4 = \text{Rs. } -311.9\text{/hour}$$

This indicates that the cost of fuel for unit 2 **decreases.**

Net increase in cost (or loss in fuel cost) due to departure from economic distribution of load

$$= 320.2 - 311.9 = \textbf{Rs. 8.3/hour.} \quad \textbf{(Ans.)}$$

Example 9.28. *Two steam turbines each of 30 MW capacity take a load 45 MW. The steam consumption rates in kg per hour for both turbines are given by the following equations :*

$$S_1 = 2400 + 12L_1 - 0.00012\,L_1^2$$

$$S_2 = 1200 + 8.4L_2 - 0.00006\,L_2^2$$

L represents the load in kW and S represents the steam consumption per hour. Find the most economical loading when the load taken by both units is 45 MW.

Solution. $$L_1 + L_2 = 45 \text{ MW} = 45000 \text{ kW} \quad ...(i)$$

For the most economical loading, the required condition is

$$\frac{dS_1}{dL_1} = \frac{dS_2}{dL_2}$$

$\therefore$ $$12 - 2 \times 0.00012\,L_1 = 8.4 - 2 \times 0.00006\,L_2$$

$$12 - 0.00024\,L_1 = 8.4 - 0.00012\,L_2 \quad ...(ii)$$

Substituting the value of L_2 (= $45000 - L_1$) from (*i*) in (*ii*), we get

$$12 - 0.00024\,L_1 = 8.4 - 0.00012\,(45000 - L_1)$$

$$12 - 0.00024\,L_1 = 8.4 - 5.4 + 0.00012\,L_1$$
$$0.00036\,L_1 = 9$$
$$\therefore \quad L_1 = \frac{9}{0.00036} = \mathbf{25000\ kW \quad or \quad 25\ MW.} \quad \textbf{(Ans.)}$$
and, $$L_2 = 45000 - 25000 = \mathbf{20000\ kW \quad or \quad 20\ MW.} \quad \textbf{(Ans.)}$$

9.17. TARIFF FOR ELECTRICAL ENERGY

9.17.1. Introduction

The cost of generation of electrical energy consists of *fixed cost and running cost.* Since the electricity generated is to be supplied to the consumers, the total cost of generation has to be recovered from the consumers. *Tariffs or energy rates are the different methods of charging the consumers for the consumption of electricity*. It is desirable to charge the consumer according to the maximum demand (kW) and the energy consumed (kWh). *The tariff chosen should recover the fixed cost, operating cost and profit etc. incurred in generating the electrical energy.*

9.17.2. Objectives and Requirements of Tariff

Objectives of tariff :

1. Recovery of cost of capital investment in generating equipment, transmission and distribution system.
2. Recovery of the cost of operation, supplies and maintenance of the equipment.
3. Recovery of the cost of material, equipment, billing and collection cost as well as for miscellaneous services.
4. A net return on the total capital investment must be ensured.

Requirements of tariff :

1. It should be easier to understand.
2. It should provide low rates for high consumption.
3. It should be uniform over large population.
4. It should encourage the consumers having high load factors.
5. It should take into account maximum demand charges and energy charges.
6. It should provide incentive for using power during off-peak hours.
7. It should provide less charges for power connection than lighting.
8. It should have a provision of penalty for low power factors.
9. It should have a provision for higher demand charges for high loads demanded at system peaks.
10. It should apportion equitably the cost of service to the different categories of consumers.

9.17.3. General Tariff Form

A large number of tariffs have been proposed from time to time and are in use. They are all derived from the following general equation :

$$z = a.x + b.y + c$$

where, z = Total amount of bill for the period considered,
x = Maximum demand in kW,
y = Energy consumed in kWh during the period considered,
a = Rate per kW of maximum demand,
b = Energy rate per kWh, and

c = Constant amount charged to the consumer during each billing period. This charge is independent of demand or total energy because a consumer that remains connected to the line incurs expenses even if he does not use energy.

Various types of tariffs :

The various types of tariffs are :

1. Flat demand rate.
2. Straight meter rate.
3. Block meter rate.
4. Hopkinson demand rate (Two-part tariff).
5. Doherty rate (Three-part tariff).
6. Wright demand rate.

1. Flat demand rate :

The flat demand rate is expressed as follows :

$$z = ax \quad ...(9.9)$$

i.e., the bill *depends only on the maximum demand irrespective of the amount of energy consumed*. It is based on the customer's installation of energy consuming devices which is generally denoted by so many kW per month or per year. It is probably one of the early systems of charging energy rates. It was based upon the total number of lamps installed and a fixed number of hours of use per year. Hence the rate could be expressed as a price per lamp or unit of installed capacity.

Now-a-days the use of *this tariff is restricted to signal system, street lighting etc.*, where the number of hours are fixed and energy consumption can be easily predicted. *Its use is very common to supplies to irrigation tubewells*, since the number of hours for which the tubewell feeders are switched on are fixed. The charge is made according to horse power of the motor installed.

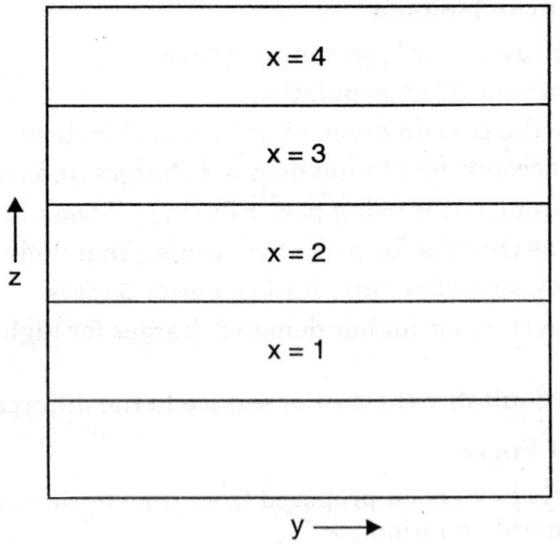

Fig. 9.13. Flat demand rate.

In this form of tariff the unit energy cost decreases progressively with an increased energy usage since the total cost remains constant. The variation in total cost and unit cost are given in Fig. 9.13.

By the use of this form of tariff the cost of metering equipment and meter reading is eliminated.

2. Straight meter rate :

The straight meter rate can be expressed in the form :

$$z = b \cdot y \qquad ...(9.10)$$

This is the simplest form of tariff. Here the charge per unit is constant. The charges depend on the energy used. This tariff is *sometimes used for residential and commercial consumer*. The variation of bill according to the variation of energy consumed is shown in Fig. 9.14.

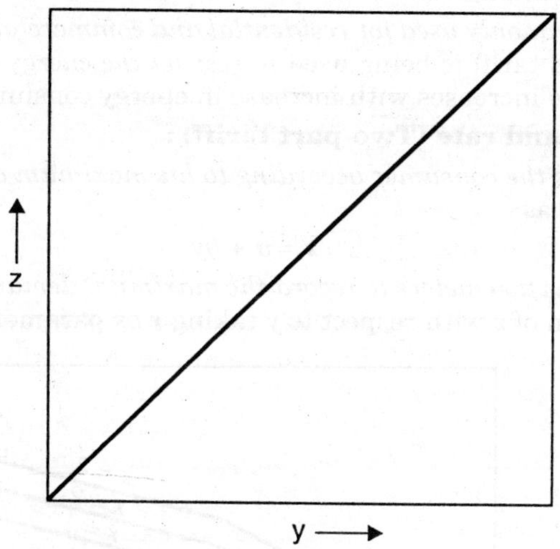

Fig. 9.14. Straight meter rate.

Advantage. *Simplicity.*

Disadvantages. 1. The consumer using no energy will not pay any amount although he has incurred some expenses to the power station.

2. This method does not encourage the use of electricity unless the tariff is very low.

3. Block motor rate :

In order to remove the inconsistency of straight meter rate, the block meter rate charges the consumers on a sliding scale. The term '*block*' indicates that a certain specified price per unit is

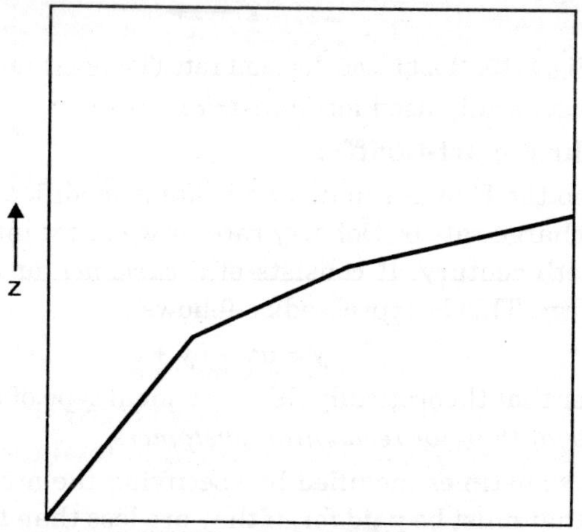

Fig. 9.15. Block meter rate.

charged for all or any part of such units. The reduced prices per unit are charged for all or any part of succeeding block of units, each such reduced price per unit applying only to a particular block or portion thereof.

The variation of bill according to this method is shown in Fig. 9.15.

The block meter rate accomplishes the same purpose of decreasing unit energy charges with increasing consumption as the step meter rate without its defect. Its *main defect is that it lacks a measure of the customer's demand.*

This tariff is very commonly used for residential and commercial customers. In many states of India, a *reverse form* of this tariff is being used *to restrict the energy consumption.* In this reverse form the unit energy charge increases with increase in energy consumption.

4. Hopkinson demand rate (Two-part tariff) :

This method *charges the consumer according to his maximum demand and energy consumption.* This can be expressed as

$$z = a + by \qquad ...(9.11)$$

This method requires *two meters to record the maximum demand and energy consumption of the consumer.* The variation of z with respect to y taking x as parameter is shown in Fig. 9.16.

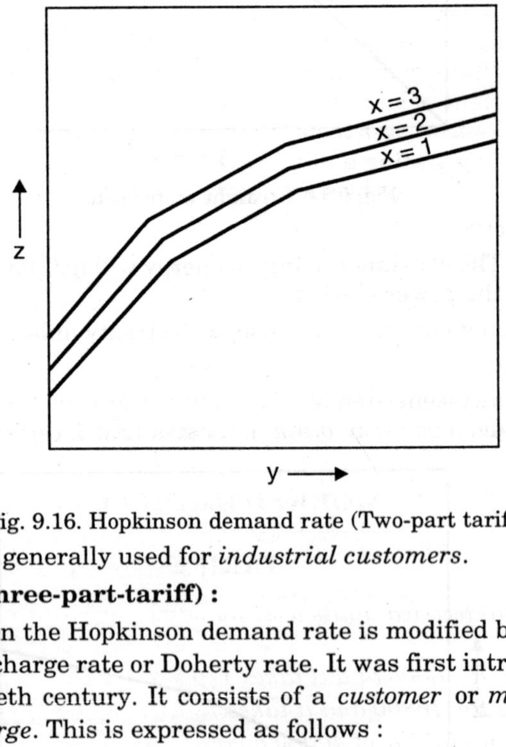

Fig. 9.16. Hopkinson demand rate (Two-part tariff).

This form of tariff is generally used for *industrial customers.*

5. Doherty rate (Three-part-tariff) :

Refer Fig. 9.17. When the Hopkinson demand rate is modified by the addition of a customer charge, it becomes a three charge rate or Doherty rate. It was first introduced by Henry L. Doherty at the beginning of twentieth century. It consists of a *customer* or *meter charge,* plus *a demand charge* plus *any energy charge.* This is expressed as follows :

$$y = ax + by + c$$

Many people consider that theoretically this is an ideal type of rate. As it requires *two meters, it is better suited for industrial than for residential customers.*

The Doherty rate is sometimes modified by specifying the minimum demand and the minimum energy consumption that must be paid for, if they are less than the minimum values specified. In this manner the customer charge is incorporated with the demand and energy component.

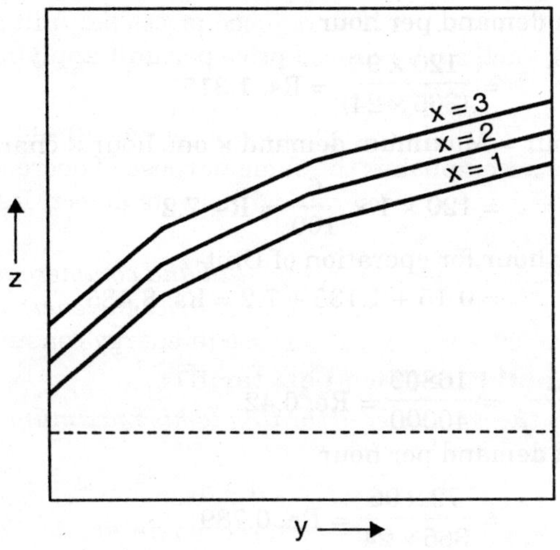

Fig. 9.17. Doherty rate (Three-part tariff).

6. Wright demand rate :

This tariff was introduced by Arthur Wright (of England) in 1896. This rate intensifies the inducement by lowering both the demand and energy charge for a reduction in maximum demand or in other words an improvement in load factor. *This rate is usually specified for industrial consumers who have some measure of control over their maximum demands.*

The rate is modified by stating a minimum charge which must be paid if the energy for the billing period falls below the amount by such charge. For allowing fair returns some adjustment in the rate forms are provided. Some of them are :

(*i*) Higher demand charges in summer.

(*ii*) Fuel price adjustment to provide a rate change when fuel prices deviate from the standard.

(*iii*) Wage adjustment.

(*iv*) Tax adjustment.

(*v*) Power factor adjustment.

(*vi*) Discount to be given to the customers for prompt payment of bills.

WORKED EXAMPLES

TARIFF

Example 9.29. *Two electrical units used for same purpose are compared for their economical working :*

(i) Cost of Unit-1 is Rs. 6000 and it takes 120 kW.

(ii) Cost of Unit-2 is Rs. 16800 and it takes 72 kW.

Each of them has a useful life of 40000 hours.

Which unit will prove economical if the energy is charged at Rs. 96 per kW of maximum demand per year and 6 p. per kWh ?

Assume both units run at full load.

Solution. (*i*) **Unit-1 :**

Capital cost per hour $= \dfrac{6000}{40000} =$ Re. 0.15

Maximum demand = 120 kW

Charge for maximum demand per hour

$$= \frac{120 \times 96}{(365 \times 24)} = \text{Rs. } 1.315$$

Energy charge per hour = Maximum demand × one hour × charge per kWh

$$= 120 \times 1 \times \frac{6}{100} = \text{Rs. } 7.2$$

∴ Total charges per hour for operation of Unit-1

$$= 0.15 + 1.135 + 7.2 = \text{Rs. } 8.485$$

(*ii*) **Unit-2 :**

Capital cost per hour $= \frac{16800}{40000} = \text{Re. } 0.42.$

Charge for maximum demand per hour

$$= \frac{72 \times 96}{365 \times 24} = \text{Re. } 0.789$$

Energy charge per hour $= 72 \times 1 \times \frac{6}{100} = \text{Rs. } 4.32$

Total charges per hour for the operation of Unit-2

$$= 0.42 + 0.789 + 4.32 = \text{Rs. } 5.529$$

The charges of operation for the Unit-2 per hour are less than the charges of operation for the Unit-1, therefore *Unit-2 is more economical* in this case. **(Ans.)**

Example 9.30. *The monthly electricity consumption of a residence can be approximated as under :*

Light load	*: 6 tube lights 40 watts each working for 4 hours daily*
Fan load	*: 6 fans 100 watts each working for 6 hours daily*
Refrigerator load	*: 2 kWh daily*
Miscellaneous load	*: 2 kW for 2 hours daily*

Find the monthly bill at the following tariff :

First 20 units	*Rs. 0.50/kWh*
Next 30 units	*Rs. 0.40/kWh*
Remaining units	*Rs. 0.30/kWh*
Constant charge	*Rs. 2.50 per month*

Discount for prompt payment = 5 per cent.

Solution. Total energy consumption in 30 days

$$= (6 \times 40 \times 4 \times 30 + 6 \times 100 \times 6 \times 30) \times \frac{1}{1000} + 2 \times 30 + 2 \times 2 \times 30$$

$$= (28800 + 108000) \times \frac{1}{1000} + 60 + 120 = 316.8 \text{ kWh per month}$$

The monthly bill $= \text{Rs. } [(20 \times 0.5 + 30 \times 0.4 + 266.8 \times 0.3) + 2.5]$

$$= \text{Rs. } [(10 + 12 + 80.04) + 2.5] = \text{Rs. } 104.54$$

$$\left[\because \text{ Remaining units per month} = 316.8 - 20 - 30 = 266.8\right]$$

Net monthly bill if the payment is made promptly

$$= 104.54 \times 0.9 = \textbf{Rs. 94.08.} \quad \textbf{(Ans.)}$$

Example 9.31. *An industrial undertaking has a connected load of 220 kW. The maximum demand is 180 kW. On an average each machine works for 60% time. Find the yearly expenditure on electricity if the tariff is :*

Rs. 1200 + Rs. 120 per kW of maximum demand per year + Re. 0.15 per kWh.

Solution. Energy consumption in one year

$$= 180 \times 0.6 \times (365 \times 24) = 946080 \text{ kWh}$$

Total electricity bill $= \text{Rs.}\ (1200 + 120 \times 180 + 0.15 \times 946080) =$ **Rs. 164712. (Ans.)**

☞**Example 9.32.** *A Hopkinson demand rate is quoted as follows :*

Demand rates :

First 1 kW of maximum demand	*= Rs. 6/kW/month*
Next 4 kW of maximum demand	*= Rs. 5/kW/month*
Excess 5 kW of maximum demand	*= Rs. 4/kW/month*

Energy rates :

First 50 kWh	*= 7 paise/kWh*
Next 50 kWh	*= 5 paise/kWh*
Next 200 kWh	*= 4 paise/kWh*
Next 400 kWh	*= 3 paise/kWh*
Excess over 700 kWh	*= 2 paise/kWh.*

Determine : (i) The monthly bill for a total consumption of 2000 kWh and a maximum demand of 15 kW. Also find out the unit energy cost.

(ii) Lowest possible bill for a month and a corresponding unit energy cost.

Solution. (*i*) **Monthly bill and energy cost :**

Demand charges per month $= \text{Rs.}\ (1 \times 6 + 4 \times 5 + 10 \times 4) = \text{Rs. } 66$

Energy charge $= \text{Rs.}\ [50 \times 7 + 50 \times 5 + 200 \times 4 + 400 \times 3 + 1300 \times 2] \times \dfrac{1}{100}$

$$= \text{Rs.}\ (350 + 250 + 800 + 1200 + 2600) \times \frac{1}{100} = \text{Rs. } 52$$

∴ *Monthly bill* $= 66 + 52 =$ **Rs. 118. (Ans.)**

Average unit energy cost $= \dfrac{118}{2000} \times 100 =$ **5.9 paise/kWh. (Ans.)**

(*ii*) **Lowest possible bill :**

The lowest possible bill will occur when average load

= Maximum load or at 100% load factor

∴ Maximum load $= \text{Average load} = \dfrac{2000}{30 \times 24} = 2.77 \text{ kW}$

∴ Demand charges $= \text{Rs.}\ (6 + 1.77 \times 5) = \text{Rs. } 14.85$

Energy charges will be same $= \text{Rs. } 52$

∴ *Minimum monthly bill* $= 14.85 + 52 =$ **Rs. 66.85. (Ans.)**

Unity energy cost for this condition

$= \dfrac{66.85}{2000} \times 100 =$ **3.34 paise/kWh. (Ans.)**

TARIFF AND COST ANALYSIS

☞**Example 9.33.** *A new factory requires a maximum demand of 700 kW and load factor of 25%. The following two suppliers are available :*

(i) Public supply tariff is Rs. 48 per kW of maximum demand plus 2.4 p. per kWh.

Capital cost	*= Rs. 84000*
Interest and depreciation	*= 10 per cent*

(ii) Private oil engine generating station :

Capital cost	*= Rs. 300000*
Fuel consumption	*= 3 N/kWh*
Cost of fuel	*= Rs. 8.4 per kN*
Wages	*= 0.48 p/kWh*
Maintenance cost	*= 0.36 p/kWh*
Interest and depreciation	*= 15 per cent.*

Find which supply will be more economical ?

Solution. Load factor $= \dfrac{\text{Average load}}{\text{Maximum demand}}$

∴ Average load = Load factor × maximum demand = 0.25 × 700 = 175 kW

Energy consumed per year = 175 × (365 × 24) = 1.533×10^6 kWh.

(*i*) **Public supply :**

Maximum demand charges per year

= 48 × 700 = Rs. 33600

Energy charge per year $= \dfrac{2.4}{100} \times 1.533 \times 10^6 =$ Rs. 36792

Interest and depreciation $= \dfrac{10}{100} \times 84000 =$ Rs. 8400

Total cost = Rs. (33600 + 36792 + 8400) = Rs. 78792

∴ *Energy cost per kWh* $= \dfrac{78792}{1.533 \times 10^6} \times 100 =$ **5.14 p.**

(*ii*) **Private oil engine generating station :**

Fuel consumption $= \dfrac{3 \times 1.533 \times 10^6}{1000} = 4599$ kN

Cost of fuel = 4599 × 8.4 = Rs. 38631

Cost of wages and maintenance $= \left(\dfrac{0.48 + 0.36}{100}\right) \times 1.533 \times 10^6 =$ Rs. 12877

Interest and depreciation $= \dfrac{15}{100} \times 300000 =$ Rs. 45000

Total cost = Rs. (38631 + 12877 + 45000) = Rs. 96508

Energy cost per kWh $= \dfrac{96508}{1.533 \times 10^6} \times 100 =$ **6.29 p.**

As the energy cost per kWh for oil engine is less than the public supply, the oil engine generation is more preferable. **(Ans.)**

☞**Example 9.34.** *A load having a maximum demand of 100 MW and a load factor of 0.4 may be supplied by one of the following schemes :*

(i) A steam station capable of supplying the whole load.

(ii) A steam station in conjunction with pump storage plant which is capable of supplying 130 × 10⁶ kWh energy per year with a maximum output of 40 MW.

Find out the cost of energy per unit in each of the two cases mentioned above.

Use the following data :

Capital cost of steam station	*= Rs. 2000/kW of installed capacity*
Capital cost of pump storage plant	*= Rs. 1300/kW of installed capacity*
Operating cost of steam plant	*= 6 p./kWh*
Operating cost of pump storage plant	*= 0.5 p./kWh*

Interest and depreciation together on the capital invested should be taken as 12 per cent. Assume that no spare capacity is required.

Solution. (*i*) **Steam station :**

Capital cost $= 100 \times 10^3 \times 2000 = \text{Rs. } 200 \times 10^6$

Interest and depreciation $= \dfrac{12}{100} \times 200 \times 10^6 = \text{Rs. } 24 \times 10^6$

Average load = Load factor × maximum demand

$= 0.4 \times 100 \times 10^3 = 40000$ kW

Energy supplied per year = Average load × (365 × 24)

$= 40000 \times 365 \times 24 = 350.4 \times 10^6$ kWh

∴ Interest and depreciation charges *per unit of energy*

$$= \frac{24 \times 10^6}{350.4 \times 10^6} \times 100 = 6.85 \text{ p/kWh}$$

∴ *Total cost per unit* $= 6 + 6.85 =$ **12.85 p/kWh. (Ans.)**

(*ii*) **Steam station in conjunction with pump-storage plant :**

The load supplied by the steam plant $= 100 - 40 = 60$ MW

∴ Capital cost of steam plant $= 60 \times 1000 \times 2000 = \text{Rs. } 120 \times 10^6$

Capital cost of pump storage plant $= 40 \times 1000 \times 1300 = \text{Rs. } 52 \times 10^6$

∴ Total capital cost of combined station $= 120 \times 10^6 + 52 \times 10^6 = \text{Rs. } 172 \times 10^6$

Interest and depreciation charges on capital investment

$$= \frac{12}{100} \times 172 \times 10^6 = \text{Rs. } 20.64 \times 10^6$$

∴ Operating cost of pump storage plant $= \dfrac{0.5}{100} \times 130 \times 10^6 = \text{Rs. } 0.65 \times 10^6$

The energy units supplied by steam station

= Total units required – energy units supplied by pump storage plant

$= 350.4 \times 10^6 - 130 \times 10^6 = 220.4 \times 10^6$ kWh

Operating cost of the steam station

$$= \frac{6}{100} \times 220.4 \times 10^6 = \text{Rs. } 13.22 \times 10^6$$

Total cost per year = Rs. $(20.64 \times 10^6 + 0.65 \times 10^6 + 13.22 \times 10^6) = \text{Rs. } 34.51 \times 10^6$

Total cost per unit $= \dfrac{34.51 \times 10^6}{350.4 \times 10^6} \times 100 =$ **9.85 p/kWh. (Ans.)**

Note : If the above example is repeated with a load factor of 0.7 it will be observed from the results that the *cost of generation becomes less with higher load factor irrespective of the type of the plant.*

Example 9.35. *The following data relate to a 2000 kW diesel power station :*

The peak load on the plant	= 1500 kW
Load factor	= 0.4
Capital cost per kW installed	= Rs. 1200
Annual costs	= 15 per cent of capital
Annual operating costs	= Rs. 50000
Annual maintenance costs :	
(i) Fixed	= Rs. 9000
(ii) Variable	= Rs. 18000
Cost of fuel	= Rs. 0.45 per kg
Cost of lubricating oil	= Rs. 1.3 per kg
C.V. of fuel	= 41800 kJ/kg
Consumption of fuel	= 0.45 kg/kWh
Consumption of lubricating oil	= 0.002 kg/kWh

Determine the following :

(i) The annual energy generated.

(ii) The cost of generation per kWh.

Solution. Capital cost of the plant $= 2000 \times 1200 =$ Rs. 2.4×10^6 per year

Interest on capital $= \dfrac{15}{100} \times 2.4 \times 10^6 =$ Re. 0.36×10^6 per year.

(*i*) **Annual energy generated** = Load factor × maximum demand × (365 × 24)

$= 0.4 \times 1500 \times 365 \times 24 =$ **5.256×10^6 kWh. (Ans.)**

(*ii*) **Cost of generation :**

Fuel consumption $= 0.45 \times 5.256 \times 10^6 = 2.365 \times 10^6$ kg per year

Cost of fuel = Rs. $0.45 \times 2.365 \times 10^6$ = Rs. 1.064×10^6 per year

Lubricant consumption $= 0.002 \times 5.256 \times 10^6 = 10512$ kg per year

Cost of lubricating oil $= 1.3 \times 10512$ = Rs. 13665 per year

Total fixed cost = Interest + maintenance (fixed)

$= 0.36 \times 10^6 + 9000$ = Rs. 369000 per year

Total running or variable costs

= Fuel cost + lubricant cost + maintenance (running) + annual operating costs

$= 1.064 \times 10^6 + 13665 + 18000 + 50000$ = Rs. 1145665 per year

Total cost = Fixed cost + running cost = 369000 + 1145665 = Rs. 1514665

Cost of generation $= \dfrac{1514665}{5.256 \times 10^6} \times 100 =$ **28.8 paise/kWh. (Ans.)**

☞**Example 9.36.** *The annual costs of operating a 15 MW thermal plant are given below :*

Capital cost of plant	= Rs. 1500/kW
Interest, insurance and depreciation	= 10 per cent of plant cost

Capital cost of primary and secondary distribution	$= Rs.\ 20 \times 10^6$
Interest, insurance and depreciation on the capital cost of primary and secondary distribution	= 5% the capital cost
Plant maintenance cost	$= Rs.\ 100 \times 10^3$ *per year*
Maintenance cost of primary and secondary equipment	$= Rs.\ 2.2 \times 10^5$ *per year*
Salaries and wages	$= Rs.\ 6.5 \times 10^5$ *per year*
Consumption of coal	$= 40 \times 10^4$ *kN per year*
Cost of coal	= *Rs. 9 per kN*
Dividend to stockholders	$= Rs.\ 1.5 \times 10^6$ *per year*
Energy loss in transmission	= *10 per cent*
Diversity factor	= *1.5*
Load factor	= *0.75*
Maximum demand	= *12 MW*

(*i*) *Devise a two-part tariff.*

(*ii*) *Find the average cost per kWh.*

Solution. (*i*) **Two-part tariff :**

Load factor $= \dfrac{\text{Average load}}{\text{Maximum demand}}$

∴ Average load = Load factor × maximum demand

$= 0.75 \times 12 \times 10^3 = 9000$ kW

Energy generated per year $= 9000 \times (365 \times 24) = 78.84 \times 10^6$ kWh

Cost of the plant $= 15 \times 10^3 \times 1500 = \text{Rs. } 22.5 \times 10^6$

Interest, insurance and depreciation charges of the plant

$$= \frac{10}{100} \times 22.5 \times 10^6 = \text{Rs. } 2.25 \times 10^6$$

Interest, insurance and depreciation charges of primary and secondary equipments

$$= \frac{5}{100} \times 20 \times 10^6 = \text{Rs. } 1.0 \times 10^6$$

Total fixed cost = Insurance, interest and depreciation costs + dividend to stock-holders

$= \text{Rs. } (2.25 \times 10^6 + 1.5 \times 10^6) = \text{Rs. } 3.75 \times 10^6$

Sum of individual maximum demand

= Maximum demand × diversity factor

$= 12 \times 10^3 \times 1.5 = 18000$ kW

∴ *Fixed charges per kW* $= \dfrac{3.75 \times 10^6}{18000}$ = **Rs. 208.3.**

Total variable charges = All maintenance costs + salaries and wages + fuel cost

$= (100 \times 10^3 + 2.2 \times 10^5) + 6.5 \times 10^5 + 40 \times 10^4 \times 9$

$= (1 \times 10^5 + 2.2 \times 10^5) + 6.5 \times 10^5 + 36 \times 10^5$

$= \text{Rs. } 45.7 \times 10^5$ or Rs. 4.57×10^6

Energy transmitted = Energy generated × transmission efficiency

$$= 78.84 \times 10^6 \times \left(\frac{100 - \text{energy loss in transmission}}{100}\right)$$

$$= 78.84 \times 10^6 \times \frac{90}{100} = 70.956 \times 10^6 \text{ kWh}$$

∴ *Charges for energy consumption*

$$= \frac{4.57 \times 10^6}{70.956 \times 10^6} \times 100 = \mathbf{6.44\ paise/kWh.}$$

∴ *Two-part tariff* = **Rs. 208.3/kW + 6.44 paise/kWh. (Ans.)**

(*ii*) **Average cost per kWh :**

Total charges = Fixed charges + variable charges

$= 3.75 \times 10^6 + 4.57 \times 10^6 =$ Rs. 8.32×10^6

Average cost of supply $= \frac{8.32 \times 10^6}{70.956 \times 10^6} \times 100 =$ **11.72 paise/kWh. (Ans.)**

Example 9.37. *A 10 MW thermal power plant has the following data :*

Peak load = *8 MW*

Plant annual load factor = *0.72*

Cost of the plant = *Rs. 800/kW installed capacity*

Interest, insurance and depreciation = *10 per cent of the capital cost*

Cost of transmission and distribution system = *Rs. 350 × 10³*

Interest, depreciation on distribution system = *5 per cent*

Operating cost = *Rs. 350 × 10³ per year*

Cost of coal = *Rs. 6 per kN*

Plant maintenance cost = *Rs. 30000/year (fixed)*

= *Rs. 40000/year (running)*

Coal used = *250000 kN/year*

Assume transmission and distribution costs are to be charged to generation

(i) Devise a two-part tariff.

(ii) Average cost of generation in paise/kWh.

Solution. (*i*) **Two-part tariff :**

S. No.	*Items*	*Fixed cost per year (in Rs.)*	*Running cost per year (in Rs.)*
1.	Interest, depreciation etc. of the plant	$\frac{10}{100} \times 10000 \times 800$ = Rs. 800×10^3	—
2.	Interest, depreciation etc. of the transmission and distribution	$\frac{5}{100} \times 350 \times 10^3$ = 17.5×10^3	—
3.	Annual cost of coal	—	250000×6 = 1500×10^3
4.	Operating cost	—	= 350×10^3
5.	Plant maintenance cost	= 30×10^3	= 40×10^3
	Total cost	847.5×10^3	1890×10^3

∴ Grand total cost = Fixed cost + running cost

$= 847.5 \times 10^3 + 1890 \times 10^3 =$ Rs. 2737.5×10^3

Energy generated/year = Average load × (365 × 24)

= (Peak load × load factor) × (365 × 24)

= $(8 \times 10^3 \times 0.72) \times (365 \times 24) = 50.46 \times 10^6$ kWh

$\therefore$ *Two-part tariff* $= \dfrac{\text{Fixed cost}}{\text{Maximum load}} + \dfrac{\text{Running cost}}{\text{Energy generated}}$

$$= \frac{847.5 \times 10^3}{8 \times 10^3} + \frac{1890 \times 10^3}{50.46 \times 10^6} \times 100$$

= **Rs. 105.9/kW + paise 3.74/kWh. (Ans.)**

(*ii*) **Average cost generation in paise/kWh :**

Average generation cost $= \dfrac{\text{Grand total cost}}{\text{Energy generated}}$

$$= \frac{2737.5 \times 10^3}{50.46 \times 10^6} \times 100 = \mathbf{5.42\ paise/kWh.}\ \ \textbf{(Ans.)}$$

Example 9.38. *Determine the load factor at which the cost of supplying a unit of electricity is same in Diesel station as in a steam station if the respective annual fixed and running charges are given below :*

Diesel : *Rs.* $(40/kW + 0.06/kWh)$

Steam : *Rs.* $(160/kW + 0.015/kWh)$.

Solution. Let, P = Maximum load in kW, and

x = Load factor (same for both the stations).

Then, Average load = $P \times x$

Cost of diesel station,

$$C_{\text{diesel}} = 40\,P + 0.06 \times P \times x \times (365 \times 24)$$

Cost of steam station,

$$C_{\text{steam}} = 160\,P + 0.015 \times P \times x \times (365 \times 24)$$

As given in the problem,

Unit energy cost (diesel station) = Unit energy cost (steam station)

$$\therefore \quad \frac{40\,P + 0.06\,Px \times (365 \times 24)}{Px \times (365 \times 24)} = \frac{160\,P + 0.015\,Px \times (365 \times 24)}{Px \times (365 \times 24)}$$

$$\therefore \quad 40\,P + 0.06\,Px \times 8760 = 160\,P + 0.015\,Px \times 8760$$

or
$$40\,P + 525.6\,Px = 160\,P + 131.4\,Px$$

or
$$120\,P = 394.2\,Px \quad \text{or} \quad x = \frac{120}{394.2} = 0.3$$

i.e., Load factor = **0.3. (Ans.)**

Example 9.39. *A motor of 25 H.P. connected to a condensate pump has been burnt beyond economical repairs. Two alternatives have been proposed to replace it by :*

	Cost	*η at full load*	*η at half load*
Motor A :	*Rs. 5000*	*90%*	*85%*
Motor B :	*Rs. 3500*	*86%*	*80%*

The life of each motor is 20 years and its salvage value is 12 per cent of the initial cost. The rate of interest is 5 percent annually. The motor operates at full load for 30% of time and at half load for the remaining period. The annual maintenance cost of motor A is Rs. 400 and that of motor B is Rs. 200. The energy rate is 12 paise/kWh.

Which motor will be economical ?

Solution. Motor A :

Salvage value $= \frac{12}{100} \times 5000 = \text{Rs. } 600$

Depreciation $= \frac{5000 - 600}{20} = \text{Rs. } 220\text{/year}$

Interest $= \frac{5}{100} \times 5000 = \text{Rs. } 250\text{/year}$

Maintenance = Rs. 400

Energy given to motor $= \frac{\text{Load on motor} \times \text{time in hours}}{\text{Efficiency of the motor}}$

$\therefore$ Energy cost

$$= \left[\left(25 \times 0.7355 \times (365 \times 24) \times \frac{30}{100} \times \frac{1}{0.9}\right) + \left((25 \times 0.7355) \times \frac{1}{2} \times (365 \times 24) \times \frac{70}{100} \times \frac{1}{0.85}\right)\right] \times \frac{12}{100}$$

$$= (53691.5 + 66324.8) \times \frac{12}{100} = \text{Rs. } 14402\text{/year}$$

$\therefore$ *Total cost of motor A* = 220 + 250 + 400 + 14402 = **Rs. 15272/year.**

Motor B :

Salvage value $= \frac{12}{100} \times 3500 = \text{Rs. } 420$

Depreciation $= \frac{3500 - 420}{20} = \text{Rs. } 154$

Interest $= \frac{5}{100} \times 3500 = \text{Rs. } 175$

Maintenance = Rs. 200

Energy cost

$$= \left[\left(25 \times 0.7355 \times (365 \times 24) \times \frac{30}{100} \times \frac{1}{0.86}\right) + \left(25 \times 0.7355 \times \frac{1}{2} \times (365 \times 24) \times \frac{70}{100} \times \frac{1}{0.8}\right)\right] \times \frac{12}{100}$$

$$= (56188.8 + 70470) \times \frac{12}{100} = \text{Rs. } 15199$$

Total cost of motor B = 154 + 175 + 200 + 15199 = **Rs. 15728/year.**

Hence ***Motor A is economical*** *since its annual cost is less than motor B.* **(Ans.)**

☞**Example 9.40.** *The following proposals are under consideration for an industry which has a maximum demand of 45 MW and a load factor of 0.45 :*

(i) A steam plant having an initial cost of Rs. 1200/kW and maintenance cost of 2.4 paise/kWh. The coal of C.V. of 2550 kJ/N is used. The overall efficiency of the plant is 24 per cent.

(ii) An hydro-plant having a capital cost of Rs. 3600/kW and a running cost of 0.6 paise/kWh.

Assuming interest and depreciation rate of 10 per cent for steam plant and 8 per cent for hydro-plant, determine the price of coal above which steam station is uneconomical.

Solution. Energy required per year

$= \text{Peak load} \times \text{load factor} \times (365 \times 24)$

$= 45 \times 10^3 \times 0.45 \times (365 \times 24) = 177.39 \times 10^6$ kWh/year

(*i*) **Steam plant :**

Interest and depreciation $= \dfrac{10}{100} \times (45 \times 10^3) \times 1200 = \text{Rs. } 5.4 \times 10^6$

Maintenance cost $= \dfrac{2.4}{100} \times 177.39 \times 10^6 = \text{Rs. } 4.257 \times 10^6$

Let, W_{coal} = Weight of coal in kN used/year, and

x = Cost of coal in Rs. per kN.

$\therefore \quad W_{coal} \times 10^3 \times \text{C.V.} \times \eta_{overall} = 177.39 \times 10^6 \times 3.6 \times 10^3$ $\quad (\because \; 1 \text{ kWh} = 3.6 \times 10^3 \text{ kJ})$

or, $\quad W_{coal} = \dfrac{177.39 \times 10^6 \times (3.6 \times 10^3)}{10^3 \times 2550 \times 0.24} = 1.043 \times 10^6$ kN/year

Now, total cost of steam plant = Interest + maintenance cost + fuel cost

$= 5.4 \times 10^6 + 4.257 \times 10^6 + 1.043 \times 10^6 \times x$ ...(1)

(*ii*) **Hydel plant :**

Interest and depreciation $= 45 \times 10^3 \times 3600 \times \dfrac{8}{100} = \text{Rs. } 12.96 \times 10^6$

Running cost $= \dfrac{0.6}{100} \times 177.39 \times 10^6 = \text{Rs. } 1.064 \times 10^6$

Total cost of hydel plant $= 12.96 \times 10^6 + 1.064 \times 10^6 = \text{Rs. } 14.024 \times 10^6$ ...(2)

The steam and hydel station will be equally economical if the total cost/year remains same.

$\therefore$ Equating the values of (1) and (2), we get

$$5.4 \times 10^6 + 4.257 \times 10^6 + 1.043 \times 10^6 \times x = 14.024 \times 10^6$$

or, $\quad 5.4 + 4.257 + 1.043\,x = 14.024$ (Dividing both sides by 10^6)

$\therefore \quad x = \dfrac{14.024 - 5.4 - 4.257}{1.043} = \text{Rs. } 4.19$ kN

Hence, *price coal above which steam station is uneconomical* = **Rs. 4.19 per kN. (Ans.)**

Example 9.41. *An industrial consumer has a choice between low and high voltage supply available at the following rates :*

High voltage : *Rs. 50/kW per year + paise 4/kWh*

Low voltage : *Rs. 55/kW per year + paise 5/kWh*

In order to have high voltage supply, consumer has to install his own transformer which costs Rs. 110/kW. The losses in the transformer are 4 per cent of full load. Determine the number of working hours per week above which the high voltage supply will be economical.

Assume : Interest and depreciation 12 per cent of capital, working weeks per year 50 and load of consumer as 1.5 MW.

Solution. Consumer load = 1.5 MW = 1500 kW

Required rating of transformer $= \dfrac{1500}{(1 - 0.4)} = 1562$ kW

Cost of the transformer to the consumer

$= 1562 \times 110 = \text{Rs. } 171820$

Annual interest and depreciation $= \dfrac{12}{100} \times 171820 = \text{Rs. } 20618$

Let the number of hours for which power is required by the consumer = x hours/week

∴ Number of hours for which power is used during the year = $50x$ hours

(*i*) Number of units consumed from low voltage side if the load is connected to low voltage

$$= 1500 \times 50x = 75000x \text{ kWh/year}$$

(*ii*) Number of units consumed from high voltage side if the load is connected to high voltage

$$= 1562 \times 50x = 78100x \text{ kWh/year}$$

Total cost from low voltage supply in rupees

$$= 1500 \times 55 + 75000x \times \frac{5}{100} = 82500 + 3750x \quad ...(1)$$

Total cost from high voltage supply in rupees

$$= 1562 \times 50 + 78100x \times \frac{4}{100} + 20618$$

$$= 98718 + 3124x \quad ...(2)$$

It both the systems cost the same to the consumer, then equating (1) and (2), we get

$$82500 + 3750x = 98718 + 3124x$$

i.e.,

$$x = \frac{98718 - 82500}{(3750 - 3124)} = \frac{16218}{626} = 25.9 \text{ hours.}$$

Hence the number of working hours above which the high voltage supply will be economical

= 25.9 hours. (Ans.)

ADDITIONAL TYPICAL EXAMPLES

Example 9.42. *A load curve of a factory follows a parabola and it works for 8 hours a day from 10 A.M. to 6 P.M. The maximum and minimum loads of the factory are $\sqrt{3}$ MW and 1 MW. The capacity of the diesel power plant supplying the power to the factory is 2 MW. Determine the following :*

(i) Load factor and capacity factor of the plant supplying power to the factory.

(ii) Energy consumption of the factory per month assuming it works for 26 days per month and 8 hours per day.

(iii) Electrical charges to be paid by the factory if the charges are Rs. 60 / kW for maximum load during a day and Rs. 2.75 / kWh.

The time at 6 A.M. may be taken as zero. **(M.U.)**

Solution. *Given* : Working hours per day = 8 (10 A.M. to 6 P.M.) ; Maximum load = $\sqrt{3}$ MW ; Minimum load = 1 MW ; Capacity of diesel power plant = 2 MW ; Tariff : Rs. 60/kW (maximum load) ; Rs. 2.75/kWh.

The load curve is shown in Fig. 9.18.

The load curve is given by :

$y^2 = ax$, where x and y represent hours and MW respectively.

The boundary conditions are :

At $x = 0, y = 0$; At $x = 4, y = 1$

$$\therefore \quad 1 = a \times 4 \quad \text{or} \quad a = \frac{1}{4}$$

$$\therefore \quad y^2 = \frac{x}{4} \quad \text{or} \quad y = \frac{\sqrt{x}}{2} \quad \text{(load curve)}$$

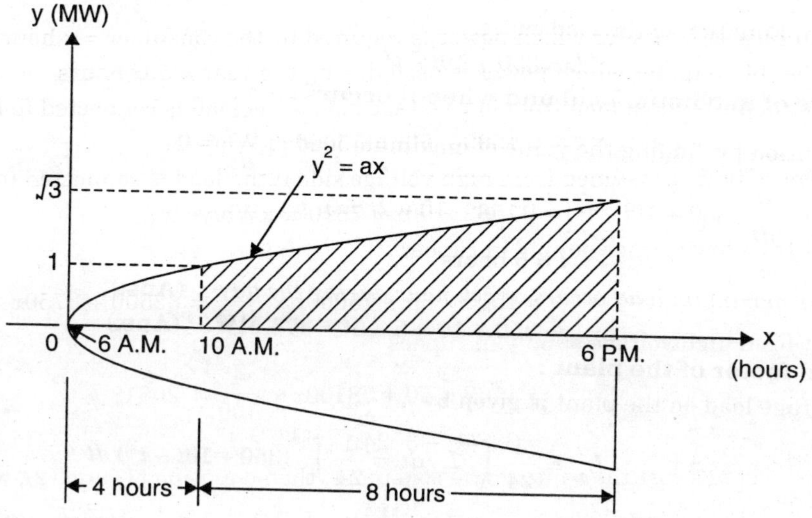

Fig. 9.18.

The above load curve also fulfills the another condition which is :

At $x = 12, \quad y = \sqrt{3}$

$$\therefore \qquad (\sqrt{3})^2 = \frac{12}{4} \quad \therefore \quad 3 = 3$$

The average load of the factory on the diesel power plant is given by :

$$L_{av} = \frac{1}{8}\int_4^{12} y \cdot dx = \frac{1}{8}\int_4^{12} \frac{\sqrt{x}}{2}\, dx$$

$$= \frac{1}{16}\left[\frac{2}{3}(x)^{1.5}\right]_4^{12} = \frac{1}{24}\,[(12)^{1.5} - (4)^{1.5}] = 1.4 \text{ MW}$$

(*i*) **Load factor and capacity factor :**

Load factor $= \frac{L_{av}}{L_{max}} = \frac{1.4}{\sqrt{3}} = 0.808.$ **(Ans.)**

Capacity factor $= \frac{L_{max}}{\text{Plant capacity}} = \frac{\sqrt{3}}{2} = 0.866.$ **(Ans.)**

(*ii*) **Energy consumption per month :**

Energy consumption per month,

$$E = (L_{av} \times 8) \times 26 = (1.4 \times 8 \times 1000) \times 26 = 291200 \text{ kWh}$$

(*iii*) **Electrical charges to be paid by the factory :**

Electrical charges to be paid by the factory

$$= L_{max} \times 60 + E \times 2.75$$

$$= \sqrt{3} \times 60 + 291200 \times 2.75 = \textbf{Rs. 800904.} \quad \textbf{(Ans.)}$$

Example 9.43. *The daily load curve for a power plant is given by the following equation :*

$$L = 350 + 10t - t^2$$

where, t is time in hours from 0 to 24 hours and L is in MW. Calculate :

(*i*) *Value of maximum load and when it occurs, and*

(*ii*) *Load factor of the plant.*

Draw load curve and load duration curve. **(P.U. May, 2001)**

Solution. Equation of the load curve,

$$L = 350 + 10t - t^2 \qquad ...(Given)$$

(*i*) **Value of maximum load and when it occurs :**

The condition for finding the value of maximum load is $\frac{dL}{dt} = 0$

$$\therefore \quad \frac{d}{dt}(350 + 10t - t^2) = 0 \quad \text{or} \quad 10 - 2t = 0$$

$$\therefore \quad t = 5 \text{ hours}$$

Thus, the *maximum load occurs at 5th hour during the day.* **(Ans.)**

$$\therefore \quad L_{max} = 350 + 10 \times 5 - 5^2 = \mathbf{375\ MW.} \quad \textbf{(Ans.)}$$

(*ii*) **Load factor of the plant :**

The average load on the plant is given by

$$L_{av} = \frac{1}{24}\int_0^{24} L\,.\,dt = \frac{1}{24}\int_0^{24}(350 + 10t - t^2)\,dt$$

$$= \frac{1}{24}\left[350t + 10 \times \frac{t^2}{2} - \frac{t^3}{3}\right]_0^{24} = \frac{1}{24}\left[350 \times 24 + 10 \times \frac{(24)^2}{2} - \frac{(24)^3}{3}\right] = 278 \text{ MW}$$

$$\therefore \quad \text{Load factor} = \frac{L_{av}}{L_{max}} = \frac{278}{375} = \mathbf{0.7413.} \quad \textbf{(Ans.)}$$

Load curve and load duration curve :

- The *'load curve is' the representation of load with respect to time.*
- The *'load duration curve' is the representation of load with respect to time in descending order.*

In order to draw these curves, we need to calculate the values of L when $t = 0, 1, 2, 3, \ldots .\ 24$ hours and these values are tabulated below :

(Eqn. of the load curve $L = 350 + 10t - t^2$), At $t \simeq 0$, $L = 350$ MW

t (h)	*L (MW)*	*t (h)*	*L (MW)*
1	359	13	311
2	366	14	294
3	371	15	275
4	374	16	254
5	375	17	231
6	374	18	206
7	371	19	179
8	366	20	150
9	359	21	119
10	350	22	86
11	339	23	51
12	326	24	14

From this data the load curve and load duration curve are drawn as shown in Fig. 9.19 (*a*) and (*b*) respectively.

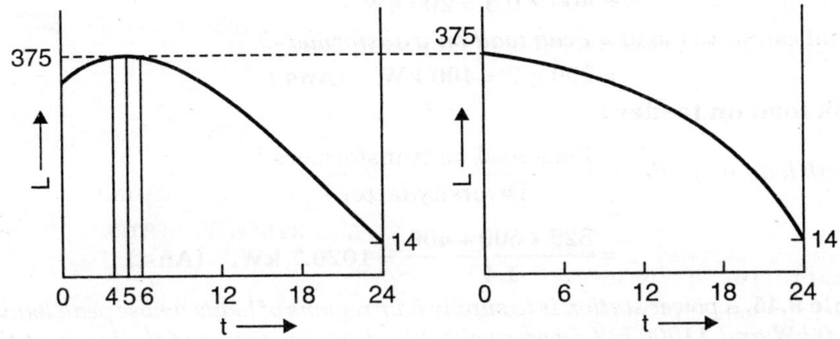

(a) Load curve. (b) Load duration curve.

Fig. 9.19

Example 9.44. *A feeder supplies 3 distribution transformers which feed the following connected loads :*

Transformer–1	:	*Motor loads 300 kW, demand factor 0.6 ; commercial loads 100 kW, demand factor 0.5*
Transformer–2	:	*Residential loads 500 kW, demand factor 0.4*
Transformer–3	:	*Residential loads 400 kW, demand factor 0.5*

The diversity factors for the loads on the three transformers may be taken as 2.3, 2.5 and 2.0. The diversity factors between transformers may be taken as 1.4.

Determine :

(i) Peak load on each transformer, and *(ii) Peak load on feeder.*

Solution. (*i*) **Peak load on each transformer :**

We know that, $\text{Demand factor} = \dfrac{\text{Maximum demand}}{\text{Connected load}}$

∴ *For transformer–1 :*

Maximum demand of motors = 300 × 0.6 = 180 kW

Maximum commercial loads = 100 × 0.5 = 50 kW

$$\text{Diversity factor} = \frac{\text{Total connected load}}{\text{Maximum demand on transformer}}$$

$$2.3 = \frac{\text{Total connected load}}{(180 + 50)}$$

∴ Total connected load = *Peak load on transformer–1*

$= 2.3 \times (180 + 50) =$ **529 kW. (Ans.)**

For transformer–2 :

Maximum demand of residential loads = 500 × 0.4 = 200 kW

∴ Total connected load = *Peak load on transformer–2*

$= 2.5 \times 200 =$ **500 kW. (Ans.)**

For transformer–3 :

Maximum demand of residential loads

$$= 400 \times 0.5 = 200 \text{ kW}$$

∴ Total connected load = *Peak load on transformer–3*

$$= 200 \times 2 = \mathbf{400\ kW.} \quad \textbf{(Ans.)}$$

(*ii*) **Peak load on feeder :**

$$\textit{Peak load on feeder} = \frac{\text{Peak load on transformers}}{\text{Diversity factor}}$$

$$= \frac{529 + 500 + 400}{1.4} = \mathbf{1020.7\ kW.} \quad \textbf{(Ans.)}$$

Example 9.45. *A power station is to supply four regions of loads whose peak loads are 15000 kW, 10000 kW, 9000 kW and 11000 kW respectively. The diversity factor of the loads at the station is 1.5 and the average annual load factor is 50%. Calculate the maximum demand on the station and the annual energy supplied from the station. Suggest the installed capacity and the number of units taking all aspects into account.* **(Madras University)**

Solution. *Given :* Peak loads = 15000 kW, 10000 kW, 9000 kW, 11000 kW

Diversity factor of the loads = 1.5

Average annual load factor = 50%

Maximum demand on power station

$$= \frac{\text{Sum of individual maximum demand}}{\text{Diversity factor}}$$

$$= \frac{15000 + 10000 + 9000 + 11000}{1.5} = \mathbf{30000\ kW.} \quad \textbf{(Ans.)}$$

$$\text{Load factor} = 0.5 = \frac{\text{No. of units generated in a year}}{\text{Maximum demand} \times 8760}$$

∴ No. of units generated in a year

$$= 30000 \times 0.5 \times 8760 = \mathbf{131.4 \times 10^6\ kWh.} \quad \textbf{(Ans.)}$$

Consider that there is an increase of maximum demand by 50% in next 5 years.

Thus maximum demand = 30000 × 1.5 = 45000 kW or 45 MW.

Hence we can select three similar units of 15 MW each, because it will necessiate the minimum number of spare parts to be stored and at the same time two units can supply the present maximum demand and the *third unit can be taken out for routine maintenance or during breakdown without any disruption of supply.*

Example 9.46. *(a) Define and explain the significance of the following terms with illustrations :*

(i) Load estimate ; *(ii) Demand factor ;*
(iii) Load factor ; *(iv) Diversity factor ;*
(v) Plant factor ; *(vi) Base load.*

(b) Make two points of consideration for favouring the choice of each of the following alternatives of unit sizes for a thermal power plant with a maximum demand of 1000 MW :

(i) One unit of 1000 MW ; *(ii) Two unit of 500 MW ;*
(iii) One unit of 500 MW and three units of 200 MW.

(c) Describe briefly the principle of operation of load despatch organisation co-ordinating different types of power plants in a power system.

Solution. (*a*) (*i*) **Load estimate.** Load estimates are the total load requirements by different categories of users such as :

— Private ;
— Public ;
— Cinema ;
— Shops and hostels ;
— Hotels and hospitals ;
— Railways ;
— Water works ;
— Industrial ;
— Street lights etc.

After estimating the load of different categories load sheets are prepared from which load curves are drawn and then a summation curve of all the loads is taken to estimate the total load of an area. Finally future demands are also considered to decide about the rating of the system.

(*ii*) **Demand factor.** Is defined as the ratio of average load to the maximum connected load.

i.e., $$\text{Demand factor} = \frac{\text{Average load}}{\text{Maximum connected load}}$$

The practical use of this factor is in deciding the capacity of the generator units to be asked for generating stations to meet the actual demand.

(*iii*) **Load factor.** The area under the load curve gives the number of units generated and the total area of the rectangle in which the load curve is contained the total number of units which could have been generated had the maximum demand remained at the highest ordinate of the load curve throughout this period. The ratio of the number of units actually generated in a given period to the number of units which could have been generated with the same maximum demand is called the load factor for the station. Alternately

$$\text{Load factor} = \frac{\text{Average demand}}{\text{Maximum demand}}$$

$$= \frac{\text{Energy generated in a given period}}{\text{(Max. demand) (Hours of operation in the given period)}}$$

If the period is considered is a day the load factor is the daily load factor, if period is a month it is monthly load factor and similarly for a year the yearly load factor.

(*iv*) **Diversity factor.** When considering the installed capacity of a generating station, it is seldom necessary to choose the rating of plant to meet all the load at the sametime for all the load connected to the station does not come simultaneously. A term commonly used in this connection is diversity factor being defined as

$$\text{Diversity factor} = \frac{\text{Sum of consumers' max. demands}}{\text{Max. load on the station}}$$

The effect of diversity factor is to reduce the simultaneous maximum demand on the station for the same individual demands.

(*v*) **Plant factor.** The plant factor is similar to that of load factor; but is differs from it in the following respect :

The load factor refers to the total load on the station and total capacity of the station, but the plant factor relates only to particular plant. The plant factor is the ratio of average demand to the rated capacity, i.e.,

$$\text{Plant factor} = \frac{\text{Station output in kW}}{\Sigma[(kW_1)h_1 + (kW_2)h_2 + (kW_3)h_3]}$$

where kW_1, kW_2, kW_3 etc. are the kilowatt rating of each generator and h_1, h_2 etc. are the number of hours for which they have been worked. The *plant factor indicates how best the plant capacity has been utilized.*

(*vi*) **Base load.** *It is that value of load which a station or system has to supply for all the time of operation and is the minimum load which will exist at all time.* The value of this can be obtained by plotting a load duration curve and dividing it into number of section. The section having a load factor of about 90% decides the base load of the station.

(*b*) (*i*) **One unit of 1000 MW.** Load factor will be high and capital and running cost will be less.

(*ii*) **Two units of 500 MW.** The spares to be stored will be less and working time of each unit can be well regulated.

(*iii*) **One unit of 500 MW and three units of 200 MW.** Reserve capacity is available and spares and maintenance cost will be less.

(*c*) When there are number of different types of power plants in a power system connected in parallel, it is necessary to coordinate the different stations for best possible economic operation. The *basic principles on which decisions are taken by the organisation are :*

1. Availability of fuel and other resources in the country.
2. Plant having low cost of operation should be used as base load plant.
3. Plants requiring low capital costs should be used as peak load plants.
4. Plants requiring quick starting and shutting down should be used to meet the peak-loads.

Based on these considerations :

- Nuclear plants are used as base load stations.
- Hydro-electric stations with ample storage and run of river plants are also used as *base load stations.*
- Hydro-electric stations with limited storage are used to meet the peak loads.
- Steam stations can be used both as peak and base load stations.
- Pumped storage plants, diesel and gas turbines are also used as *peak load plants only.*

Example 9.47. *(a) Mention the methods used to control reactive power.*

(b) Discuss the role of load factor on the cost of electric energy. **(N.U.)**

Solution. (*a*) Methods to control reactive power :

1. Use of static capacitor.
2. Use of synchronous capacitors or phase modifiers.
3. Use of phase advancers.

By use of any of these devices reactive power is controlled by injection of reactive power, which improves the power factor of the system. Normally such devices are installed at the receiving end or at the generating stations.

(*b*) Higher value of load factor (L.F.) means more number of units generated per year, and the maximum load factor will be 1. It will be at the instant when the maximum demand remains throughout the year at all times. The cost of labour per kilowatt of energy generated will decrease as the number of units generated increases because the labour cost remains same whether the alternator is worked at 20%, 50% or 100% load factor. Thus *with higher value of load factor the overall cost of electrical energy reduces.*

HIGHLIGHTS

1. Types of loads : (*i*) Residential load (*ii*) Commercial load (*iii*) Industrial load (*iv*) Municipal load (*v*) Irrigation load and (*vi*) Traction load.
2. A *load curve* is a graphic record showing the power demands for every instant during a certain time interval. The area under the load curve represents the energy generated in the period considered.
3. The cost of a power system includes the following :
   A. *Capital cost or fixed cost :*
   (*i*) Initial cost (*ii*) Interest
   (*iii*) Depreciation cost (*iv*) Taxes
   (*v*) Insurance
   B. *Operational cost :*
   (*i*) Fuel cost (*ii*) Operating labour cost
   (*iii*) Maintenance cost (*iv*) Supplies
   (*v*) Supervision (*vi*) Operating taxes.
4. The following methods are used to calculate the depreciation cost :
   (*i*) Straight line method. (*ii*) Percentage method.
   (*iii*) Sinking fund method. (*iv*) Unit method.
5. The economics of power plant is greatly influenced by :
   (*i*) Load factor (*ii*) Demand factor
   (*iii*) Utilisation factor.
6. The performance of a plant can be precisely represented by the *input-output* curve from the tests conducted on individual power plant. In general input-output may be represented as follows :
   $$I = a + bL + cL^2 + dL^3$$
   where, I = Input (in millions of kcal/h or kJ/h in case of thermal plants and m^3/s of water in case of hydro-plants),
   L = Output (in MW or kW), and
   a, b, c, d = Constants.
7. For minimum combined input to carry a given combined output, the slopes of the input-output curves for each unit must be equal. If there are n units supplying a constant load, then the required condition for the minimum input or maximum system efficiency is
   $$\frac{dI_1}{dL_1} = \frac{dI_2}{dL_2} = \frac{dI_3}{dL_3} = \ldots\ldots \frac{dI_n}{dL_n}.$$
8. The general tariff form is given by :
   $$z = a.x + b.y + c$$
   where, z = Total amount of bill for the period considered,
   x = Maximum demand in kW,
   y = Energy consumed in kWh during the period considered,
   a = Rate per kW of maximum demand,
   b = Energy rate per kWh, and
   c = Constant amount charged to the consumer during each billing period. This charge is independent of demand or total energy because a consumer that remains connected to the line incurs expenses even if he does not use energy.
9. Various types of *tariff* are :
   (*i*) Flat demand rate. (*ii*) Straight meter rate.
   (*iii*) Block meter rate. (*iv*) Hopkinson demand rate (Two-part tariff).
   (*v*) Doherty rate (Three-part tariff). (*vi*) Wright demand rate.

THEORETICAL QUESTIONS

1. Define the following terms :
   (*i*) Connected load (*ii*) Demand
   (*iii*) Demand factor (*iv*) Load factor
   (*v*) Diversity factor (*vi*) Utilisation factor.
2. Explain briefly the following :
   (*i*) Load curve ; (*ii*) Load duration curve.
3. What is the significance of load curves ?
4. Enumerate various types of loads.
5. List the factors which should be considered while designing a power plant.
6. What are the considerations on which the location of a power plant depends ?
7. List the points which should be taken care of while deciding about power plant building and its layout.
8. List the various costs which go to form the total cost of a power system.
9. Explain briefly the following :
   (*i*) Capital or fixed cost (*ii*) Operational cost.
10. What do you mean by depreciation ?
11. Enumerate and explain briefly various methods used to calculate the depreciation cost.
12. Name the elements that make up the operating expenditure of a power plant.
13. What points should be considered while choosing the type of generation ?
14. Discuss the economic loading of combined steam and hydro-plants.
15. How can the power generation cost be reduced ?
16. What do you understand by the term tariff ?
17. What are the objectives and requirements of tariff ?
18. Enumerate various types of tariff and explain any two of them.
19. Explain briefly the following tariff :
    (*i*) Straight meter rate (*ii*) Block meter rate
    (*iii*) Doherty rate (three-part tariff).

UNSOLVED EXAMPLES

1. The maximum demand of a power station is 80000 kW and the load curve is defined as follows :

Time (hours)	:	0—6	6—8	8—12	12—14	14—18	18—22	22—24
Load (MW)	:	40	50	60	50	70	80	40

   (*i*) Determine the load factor of power station.
   (*ii*) What is the load factor of standby equipment rated at 25 MW that takes up all load in excess of 60 MW ? Also calculate its use factor. **[Ans.** (*i*) 0.71 (*ii*) 0.75, 0.6]

2. The following load is to be supplied by a power station :

Load (MW)	:	30	90	60	100	50
Time (hours)	:	0—6	6—12	12—14	14—18	18—24

   (*i*) Draw the load curve.
   (*ii*) Draw the load duration curve.
   (*iii*) Choose suitable generating units to supply the load.
   (*iv*) Calculate the load factor.
   (*v*) Calculate plant capacity factor.
   [Ans. (*iii*) 30 MW (4 units including standby unit), 10 MW (one unit) (*iv*) 0.64 (*v*) 0.49]

3. The yearly duration curve of a certain plant can be considered as a straight line from 150 MW to 40 MW. Power is supplied with one generating unit of 100 MW capacity and two units of 50 MW capacity each. Determine :
(*i*) Installed capacity (*ii*) Load factor
(*iii*) Plant factor (*iv*) Maximum demand
(*v*) Utilization factor. **[Ans.** (*i*) 200 MW (*ii*) 0.633 (*iii*) 0.475 (*iv*) 150 MW (*v*) 0.75]

4. A generating station has a maximum demand of 20 MW, a load factor of 0.6, a plant capacity of 0.48 and a plant use factor of 0.80. Find :
(*i*) The daily energy produced.
(*ii*) The reserve capacity of the plant.
(*iii*) The maximum energy that could be produced if the plant were running all the time.
(*iv*) The maximum energy that could be produced daily, if the plant when running, according to operating schedule, were fully loaded.
[Ans. (*i*) 2.88×10^5 kWh (*ii*) 5000 kW (*iii*) 2.88×10^5 kWh (*iv*) 3.60×10^5 kWh]

5. A proposed power station has to supply load as follows :

Time (hours)	:	01—08	08—12	12—17	17—20	20—23	23—01
Load (MW)	:	10	20	25	18	35	20

After drawing the load curve, find out the load factor. Also choose suitable generating units to supply this load, maintaining reliability of supply. Prepare operation schedule for the machine and calculate plant use factor. **[Ans.** 0.56, 0.92]

6. A generating station supplies the following loads :
15 MW ; 12 MW ; 8 MW and 0.5 MW. The station has a maximum demand of 20 MW and the annual load factor is 0.5. Find :
(*i*) Number of units supplied annually. (*ii*) Diversity factor.
[Ans. (*i*) 876×10^5 kWh (*ii*) 1.775]

7. A base load power station and standby power station share a common load as follows :
Base load station annual output = 150×10^6 kWh ; Base load station capacity = 35 MW ; Maximum demand on base load station = 30 MW ; Standby station capacity = 18 MW ; Standby station annual output = 140×10^6 kWh ; Maximum demand (peak load) on standby station = 15 MW. Determine the following for both power stations :
(*i*) Load factor (*ii*) Capacity factor (plant factor).

[Ans. Base load station : (*i*) 0.57 (*ii*) 0.49
Standby power station : (*i*) 0.107 (*ii*) 0.09]

8. A power system has the following load particulars :

	Maximum demand	***Load factor***	***Diversity between consumers***
1. Residential load :	1000 kW	0.2	1.3
2. Commercial load :	2000 kW	0.3	1.1
3. Industrial load :	5000 kW	0.8	1.2

Overall diversity factor may be taken as 1.4.
Determine the following :
(*i*) Maximum demand on system.
(*ii*) Daily energy consumption (total).
(*iii*) Overall load factor.
(*iv*) Connected load (total) assuming that demand factor for each load is unity.

9. The following data is available for a steam power station :
Maximum demand = 25000 kW ; Load factor = 0.4 ; Coal consumption = 0.86 kg/kWh ; Boiler efficiency = 85% ; Turbine efficiency = 90% ; Price of coal = Rs. 55 per tonne.
Determine the following :
(*i*) Thermal efficiency of the station.
(*ii*) Coal bill of the plant for one year. **[Ans.** (*i*) 76.5% (*ii*) Rs. 4143480]

10. The daily load curve of a power plant is given by the table below :

Time :	12	2	4	6	8	10	12	2	4	6	8	10	12
Load (MW) :	2	2.5	3	4	6	6.5	6.5	5	6	8	9	5	2

(*i*) Find the daily load factor.

(*ii*) All loads in excess of 400 kW are carried out by unit No. 2 rated at 600 kW. Find its use factor.

[**Ans.** (*i*) 0.814 (*ii*) 0.417]

11. The annual peak load on a 30 MW power station is 25 MW. The power station supplies load having maximum demands of 10 MW, 8.5 MW, 5 MW and 4.5 MW. The annual load factor is 0.45. Find :

(*i*) Average load (*ii*) Energy supplied per year

(*iii* Diversity factor (*iv*) Demand factor.

[**Ans.** (*i*) 11.25 MW (*ii*) 98.55×10^6 kWh (*iii*) 1.12 (*iv*) 0.9]

12. A generating station supplies the following loads :

15 MW, 12 MW, 8.5 MW, 6 MW and 0.45 MW. The station has a maximum demand of 22 MW. The annual load factor of the station is 0.48. Calculate :

(*i*) The number of units supplied annually

(*ii*) The diversity factor. (*iii*) The demand factor.

[**Ans.** (*i*) 92.5×10^6 kWh (*ii*) 1.907 (*iii*) 0.525]

13. A power station has a maximum demand of 15 MW, a load factor of 0.7, a plant capacity factor of 0.525 and a plant use factor of 0.85. Find :

(*i*) The daily energy produced.

(*ii*) The reserve capacity of the plant.

(*iii*) The maximum energy that could be produced daily if the plant operating schedule is fully loaded when in operation. [**Ans.** (*i*) 252000 kWh (*ii*) 5000 kW (*iii*) 296470 kWh]

14. Determine the annual cost of a feed water softner from the following data :

Cost = Rs. 80000 ; Salvage value = 5%, Life = 10 years ; Annual repair and maintenance cost = Rs. 2500 ; Annual cost of chemicals = Rs. 5000 ; Labour cost per month = Rs. 300 ; Interest on sinking fund = 5%.

[**Ans.** Rs. 17,140]

15. Estimate the generating cost per kWh delivered from a generating station from the following data :

Plant capacity	= 50 MW
Annual load factor	= 0.4
Capital cost	= Rs. 1.2 crores
Annual cost of wages, taxation etc.	= Rs. 4 lacs
Cost of fuel, lubrication, maintenance etc.	= 1.0 paise per kWh generated.

Interest 5% per annum, depreciation 5% per annum of initial value. [**Ans.** 1.91 paise/kWh delivered]

16. A 100 MW, steam power station is estimated to cost Rs. 20 crores. The operating expenses are estimated as follows :

Cost of fuel and oil	= Rs. 140 lacs per annum
Transportation and storage	= Rs. 20 lacs per annum
Salaries and wages	= Rs. 20 lacs per annum
Miscellaneous	= Rs. 20 lacs per annum

Reckoning interest and depreciation at 10% of the capital cost, calculate the cost of generation per unit, if the average load factor of the power station is 0.6.

What economics could be affected if the load factor was improved to 0.8, the operating expenses increasing by only 10% thereby. [**Ans.** 6 p/kWh, 21% reduction in cost of generation]

17. A steam station has two 110 MW units. Following cost data are given :

Particulars	***Unit A***	***Unit B***
Capital cost	Rs. 2000 per kW	Rs. 2500 per kW
Fixed charge rate	10 per cent	10 per cent
Capacity factor	0.55	0.6
Fuel consumption	1 kg/kWh	0.9 kg/kWh

Fuel cost	Rs. 80 per 1000 kg	Rs. 80 per 1000 kg
Annual cost of operating, labour, maintenance and supplies	20 per cent of annual cost	15 per cent of annual cost
Utilisation factor	1	1

Calculate the following :

(*i*) Annual plant cost and generation cost of unit A.

(*ii*) Annual plant cost and generation cost of unit B.

(*iii*) Overall generation cost of the station.

[**Ans.** (*i*) Rs. 7,28,78,080 ; 13.75 p/kWh (*ii*) Rs. 75.371648×10^6 ; 13.036 p/kWh (*iii*) 13.378 p/kWh]

18. The annual costs of operating a 15,000 kW thermal power station are as follows :

Cost of plant = Rs. 900 per kW

Interest, insurance, taxes on plant = 5 per cent

Depreciation = 5 per cent

Cost of primary distribution system = Rs. 500000

Interest, insurance, taxes and depreciation on primary distribution system = 5 per cent

Cost of secondary distribution system = Rs. 900000

Interest, taxes, insurance and depreciation on secondary distribution system = 5 per cent

Maintenance of secondary distribution system = Rs. 180000

Plant maintenance cost

(*i*) Fixed cost = Rs. 30000

(*ii*) Variable cost = Rs. 40000

Operating costs = Rs. 600000

Cost of coal = Rs. 60 per tonne

Consumption of coal = 30000 tonnes

Dividend to stock-holders = Rs. 1000000

Energy loss in transmission = 10 per cent

Maximum demand = 14000 kW

Diversity factor = 1.5

Load factor = 0.7

Determine : (*i*) Charge per kW per year (*ii*) Rate per kWh. [**Ans.** (*i*) Rs. 116.6 (*ii*) 3.4 p/kWh]

19. It is proposed to supply a load with a maximum demand of 100 MW and a load factor of 0.4. Choice is to be made from steam, hydro and nuclear power plants. Calculate the overall cost per kWh in case of each machine :

Cost	*Steam power plant*	*Hydro power plant*	*Nuclear power plant*
Capital cost per kW installed	Rs. 1800	Rs. 3600	Rs. 5000
Interest	12%	10%	10%
Depreciation	12%	8%	10%
Operating cost per kWh	15 paise	5 paise	10 paise
Transmission and distribution cost/kWh	0.2 paise	0.8 paise	0.2 paise

[**Ans.** 27.52 p ; 23.57 p ; 38.73 p]

20. A power plant of 150 MW installed capacity has the following data :

Capital cost = Rs. 1800/kW installed ; Interest and depreciation = 12 per cent ;

Annual load factor = 0.6 ; Annual capacity factor = 0.5 ; Annual running charges = Rs. 30×10^6 ; Energy consumed by the power plant auxiliaries = 6 per cent. Calculate :

(*i*) Reserve capacity (*ii*) Generating cost. [**Ans.** (*i*) 25 MW (*ii*) 10.10 paise]

21. Compare the annual cost of supplying a factory load having a maximum demand of 1 MW at a load factor of 50% by energy obtained from

(*a*) Nuclear power plant (*b*) Public supply

Nuclear power plant :

Capital cost	= Rs. 50000
Cost of fuel	= Rs. 600 per 1000 kg
Fuel consumption	= 30 g per kWh generated
Cost of maintenance etc.	= Re. 0.005 per kWh generated
Wages	= Rs. 20000 per annum
Interest and depreciation	= 10 per cent.

Public supply : Rs. 50 per kW + Re. 0.03 per kWh generated. [**Ans.** Rs. 170740 ; Rs. 181400]

22. A system with a maximum demand of 100000 kW and a load factor of 30% is to be supplied by either (*a*) steam station alone or (*b*) a steam station in conjunction with a water storage scheme, the latter supplying 100 million units with a maximum output of 40000 kW. The capital cost of steam and storage stations are Rs. 600 per kW and Rs. 1200 per kW respectively. The corresponding operating costs are 15 paise and 3 paise per kWh respectively. The interest on capital cost is 15% per annum. Calculate the overall generating cost per kWh and state which of the two projects will be economical.

[**Ans.** 18.425 p/kWh, 15.23 p/kWh]

23. A power station has the installed capacity of 120 MW. Calculate the cost of generation, other data pertaining to power station are given below :

Capital cost	= Rs. 200×10^6
Rate of interest and depreciation	= 18 per cent
Annual cost of fuel oil, salaries and taxation	= Rs. 24×10^6
Load factor	= 0.4

Also calculate the saving in cost per kWh if the annual load factor is raised to 0.5.

[**Ans.** 14.25 paise ; 2.84 paise]

24. A 50 MW generating station has the following data :

Capital cost = Rs. 15×10^6 ; Annual taxation = Rs. 0.4×10^6 ; Annual salaries and wages = Rs. 1.2×10^6 ; Cost of coal = Rs. 65 per tonne ; Calorific value of coal = 5500 kcal/kg ; Rate of interest and depreciation = 12 per cent ; plant heat rate = 33000 kcal/kWh at 100% capacity factor. Calculate the generating cost/kWh at 100% capacity factor. [**Ans.** 39.77 p/kWh]

25. An input output curve of a 10 MW thermal station is given by an equation

$$I = 10^6(18 + 12L + 0.5L^2) \text{ kcal/hour}$$

where I is in kcal/hour and L is the load on power plant in MW.

Find : (*i*) The load at which the efficiency of the plant will be maximum.

(*ii*) The increase in output required to increase the station output from 5 MW to 7 MW by using the input-output equation and by incremental rate curve. [**Ans.** (*i*) 6 MW (*ii*) 36×10^6 kcal/hour]

26. The input-output curve of a 60 MW power station is given by :

$$I = 10^6(8 + 8L + 0.4L^2) \text{ kcal/hour}$$

where I is the input in kcal/hour and L is load in MW.

(*i*) Determine the heat input per day to the power station if it works for 20 hours at full load and remaining period at no load.

(*ii*) Also find the saving per kWh of energy produced if the plant works at full load for all 24 hours generating the same amount of energy. [**Ans.** (*i*) 38592×10^6 kcal/day (*ii*) 4000 kcal/kWh]

27. The incremental fuel costs for two generating units 1 and 2 of a power plant are given by the following equations :

$$\frac{dF_1}{dP_1} = 0.06P_1 + 11.4$$

$$\frac{dF_2}{dP_2} = 0.07P_2 + 10$$

where P is in megawatts and F is in rupees per hour.

(*i*) Find the economic loading of the two units when the total load to be supplied by the power station is 150 MW.

(*ii*) Find the loss in fuel costs per hour if the load is equally shared by the two units.

[**Ans.** (*i*) $P_1 = 70$ MW, $P_2 = 80$ MW (*ii*) Rs. 1.63 per hour]

28. The incremental fuel costs for two generating units 1 and 2 of a power plant are given by the following equations :

$$\frac{dF_1}{dP_1} = 0.065P_1 + 25$$

$$\frac{dF_2}{dP_2} = 0.08P_2 + 20$$

where, F is fuel cost in rupees per hour and P is power output in MW. Find :

(*i*) The economic loading of the two units when the total load supplied by the power plants is 160 MW.

(*ii*) The loss in fuel cost per hour if the load is equally shared by both units.

[**Ans.** $P_1 = 53.5$ MW, $P_2 = 106.5$ MW (*ii*) Rs. 35/hour]

29. Two steam turbines each of 20 MW capacity take a load of 30 MW. The steam consumption rates in kg per hour for both turbines are given by the following equations :

$$S_1 = 2000 + 10L_1 - 0.0001L_1^2$$
$$S_2 = 1000 + 7L_2 - 0.00005L_2^2$$

L represents the load in kW and S represents the steam consumption per hour.

Find the most economical loading when the load taken by both units is 30 MW.

[**Ans.** $L_1 = 20$ MW, $L_2 = 10$ MW]

30. Two electrical units used for the same purpose are compared for their economical working :

(*i*) Cost of Unit-1 is Rs. 5000 and it takes 100 kW.

(*ii*) Cost of Unit-2 is Rs. 14000 and it takes 60 kW.

Each of them has a useful life of 40000 hours. Which unit will prove economical if the energy is charged at Rs. 80 per kW of maximum demand per year and 5 p. per kWh ?

Assume both units run at full load.

[**Ans.** Unit-1 : Rs. 6.039 ; Unit-2 : Rs. 3.898, Unit-2 is more economical]

31. A new industry requires maximum demand of 800 kW at 30% load factor. The following two power supplies are available :

(*i*) *Public supply* charges Rs. 50/kW of maximum demand and 4 p. per kWh.

Capital cost	= Rs. 80000
Interest and depreciation	= 10 per cent.

(*ii*) *Private oil engine generating station :*

Capital cost	= Rs. 30000
Interest and depreciation	= 12 per cent
Maintenance and labour charges	= 1 p. per kWh energy generated
Fuel consumption	= 0.35 kg/kWh
Cost of fuel	= 8 paise/kg.

Find which supply will be more economical ?

[**Ans.** (*i*) 6.3 p/kWh (*ii*) 5.1 p/kWh ; oil engine generation is more preferable]

32. A load having a maximum demand of 80 MW and a load factor of 40% may be supplied by one of the following schemes :

(*i*) A steam station capable of supplying the whole load.

(*ii*) A steam station in conjunction with pump-storage plant which is capable of supplying 120×10^6 kWh energy per year with a maximum output of 30 MW.

Find out the cost of energy per unit in each of the two cases mentioned above. Use the following data : Capital cost of steam station = Rs. 1800/kW of installed capacity ; Capital cost of pump storage plant = Rs. 1200/kW of installed capacity ; Operating cost of steam plant = 0.5 p/kWh ; Operating cost of pump storage plant = 0.4 p./kWh.

Interest and depreciation together on capital invested should be taken as 12 per cent.
Assume that no spare capacity is required. **[Ans.** (*i*) 11.16 p./kWh (*ii*) 8.42 p./kWh]

33. The monthly electricity consumption of a residence can be approximated as under :
Light load : 4 tube lights 40 watts each working for 3 hours daily ; *Fan load* : 4 fans 100 watts each working for 5 hours daily ; *Refrigerator load* : 1 kWh daily ; *Miscellaneous load* : 1 kW for one hour daily.
Find the monthly bill at the following tariff :
First 15 units : Re. 0.50/kWh, Next 25 units : Re. 0.40 per kWh ; Remaining units : Re. 0.30 per kWh ; Constant charge : Rs. 2.50 per month. Discount for prompt payment = 5% . **[Ans.** Rs. 45.20]

34. An industrial undertaking has a connected load of 110 kW. The maximum demand is 90 kW. On an average each machine works for 60 per cent time. Find the yearly expenditure on electricity if the tariff is :
Rs. 1000 + Rs. 100 per kW of maximum demand per year + Re. 0.10 per kWh. **[Ans.** Rs. 67186]

35. A Hopkinson demand rate is quoted as follows :

Demand rates

First 1 kW of maximum demand	= Rs. 5/kW/month
Next 4 kW of maximum demand	= Rs. 4/kW/month
Excess over 5 kW of maximum demand	= Rs. 3/kW/month

Energy rates

First 50 kWh	= 6 paise/kWh
Next 50 kWh	= 4 paise/kWh
Next 200 kWh	= 3 paise/kWh
Next 400 kWh	= 2.5 paise/kWh
Excess over 700 kWh	= 2 paise/kWh

Determine : (*i*) The monthly bill for a total consumption of 1500 kWh and a maximum demand of 12 kW. Also find the unit energy cost.
(*ii*) Lowest possible bill for a month and corresponding unit energy cost.
[Ans. (*i*) Rs. 79, 5.26 paise/kWh (*ii*) Rs. 46.33, 3.09 paise/kWh]

36. Find the cost of generation per kWh from the following data :

Capacity of the plant	= 120 MW
Capital cost	= Rs. 1200 per kW installed
Interest and depreciation	= 10 per cent on capital
Fuel consumption	= 1.2 kg/kWh
Fuel cost	= Rs. 40 per tonne
Salaries, wages, repairs and maintenance	= Rs. 600000 per year
The maximum demand	= 80 MW
Load factor	= 40%.

[Ans. 10.18 paise/kWh]

37. The following data relate to a 2200 kW diesel power station :

The peak load on the plant	= 1600 kW
Load factor	= 45%
Capital cost per kW installed	= Rs. 1000
Annual costs	= 15 per cent of capital
Annual operating costs	= Rs. 60000
Annual maintenance cost :	
(*i*) Fixed	= Rs. 10000
(*ii*) Variable	= Rs. 20000
Cost of fuel	= Re. 0.4 per kg
Cost of lubricating oil	= Rs. 1.25 per kg
C.V. of fuel	= 10000 kcal/kg

Consumption of fuel = 0.5 kg/kWh
Consumption of lubricating oil = 0.0025 kg/kWh
Determine the following :
(*i*) The annual energy generated.
(*ii*) The cost of generation (per kWh). **[Ans.** (*i*) 6.3×10^6 kWh/year (*ii*) 27 paise/kWh]

38. The following data relate to a steam power station of 120 MW capacity which takes 100 MW peak demand at 80% load factor :
Annual cost towards interest and depreciation = Rs. 100/kW installed ; Operating costs = Rs. 1200×10^3/year ; Maintenance costs = Rs. 200×10^3/year (fixed) and = Rs. 400×10^3/year (variable) ; Miscellaneous costs = Rs. 100×10^3/year ; Cost of coal used = Rs. 32/ton ; C.V. of fuel used = 6400 kcal/kg ; Overall efficiency of the plant = 20 per cent ; Steam consumption in kg/kWh = (0.8 + 3.5 × load factor).
Determine the following :
(*i*) Coal cost per year.
(*ii*) Overall cost of generation (paise/kWh). **[Ans.** (*i*) Rs. 15×10^6/year (*ii*) 4.12 paise/kWh]

39. A power system requires a maximum load of 80 MW at 35% load factor. It can be supplied by any of the following schemes :
(*i*) A steam plant capable to supply whole load.
(*ii*) A steam plant with hydel plant where energy supplied by steam plant is 120×10^6 kWh/year with a maximum load of 50 MW.

Plant	*Capital cost*	*Operating cost*	*Transmission cost*
Steam plant	Rs. 600/kW installed	4.8 paise/kWh	Negligible
Hydro plant	Rs. 1400/kW installed	1.2 paise/kWh	0.3 paise/kWh

Assume interest and depreciation at 12 per cent of capital for steam plant and 10 per cent of capital for hydro plant. Calculate the overall cost per kWh.
(*iii*) If the whole load is supplied by a nuclear plant, determine annual cost. Take capital cost of Rs. 2500/kW and running cost of 2.5 paise/kWh. Assume interest and depreciation as 10 per cent per annum. **[Ans.** (*i*) 7.05 paise/kWh (*ii*) 5.98 paise/kWh (*iii*) 10.67 paise/kWh]

40. The following data relate to a power plant of 120 MW capacity :
Capital cost = Rs. 1500/kW ; Interest and depreciation = 10 per cent on capital ; Annual running charges = Rs. 20×10^6 ; Profit to be gained = 10 per cent of the capital ; The energy consumed by the power plant auxiliaries = 5 per cent of generated ; The annual load factor = 0.6 ; Annual capacity factor = 0.5.
Calculate the following :
(*i*) The reserve capacity.
(*ii*) The cost of generation per kWh. **[Ans.** (*i*) 20 MW (*ii*) 9.3 paise/kWh]

41. A small generating unit of 5,000 kW capacity supplies the following loads :
(*i*) Street-light load with maximum demand of 200 kW at 0.3 load factor.
(*ii*) Small industrial load with maximum demand of 1800 kW at a load factor of 0.5.
(*iii*) Domestic consumers with a maximum demand of 3000 kW at a load factor of 0.2.
Find the overall energy rate for each type of consumer using the following data :
Capital cost of the plant = Rs. 1800/kW of installed capacity
Total running cost = Rs. 620000 per year
Annual rate of depreciation and interest on capital cost is 10%.
[Ans. (*i*) 11.4 paise/kWh (*ii*) 8.65 paise/kWh (*iii*) 14.8 p/kWh]

42. The following data relate to a 15 MW thermal plant :
Capital cost of plant = Rs. 1200/kW ; Interest, insurance and depreciation = 10% of the plant ; Capital cost of primary and secondary distribution = Rs. 15×10^5 ; Interest, insurance and depreciation on the capital cost of primary and secondary distribution = 5% of capital cost ; Plant maintenance cost = Rs. 80×10^3 per year ; Maintenance cost of primary and secondary equipments = Rs. 2×10^5 per year ; Salaries and wages = Rs. 6×10^5 per year ; Consumption of coal = 40×10^3 tonnes per year ; Cost of coal

= Rs. 80 per tonne ; Dividend to stockholders = Rs. 12 × 10^5 per year ; Energy loss in transmission = 10% ; Diversity factor = 1.6 ; Load factor = 0.8 ; Maximum demand = 14 MW.

Devise a two-part tariff and find the average cost per kWh.

[**Ans.** Fixed charge per kW = 146, charge for energy consumption = 4.63 paise/kWh ; Average cost of supply = 8.1 paise/kWh]

43. The following data relate to a 12 MW capacity thermal plant :

Peak load	= 10 MW
Annual load factor	= 70%
Cost of the plant	= Rs. 700/kW installed capacity
Interest, insurance and depreciation	= 10 per cent of the capital cost
Cost of transmission and distribution system	= Rs. 300 × 10^3
Interest, depreciation on distribution system	= 5 per cent
Operating cost	= Rs. 300 × 10^3 per year
Cost of coal	= Rs. 50/ton
Plant maintenance cost	= Rs. 25,000 per year (fixed)
	= Rs. 35,000 per year (running)
Coal used	= 30,000 tons/year.

Assume transmission and distribution costs are to be charged to generation.

(*i*) Design a two-part tariff.

(*ii*) Determine overall cost of generation in paise/kWh.

[**Ans.** (*i*) Rs. 88/kW + 3 paise/kWh (*ii*) 4.43 paise/kWh]

44. Determine the load factor at which the cost of supplying a unit of electricity is same in Diesel station as in a Steam station if the respective annual fixed and running charges are as given below :

Diesel : Rs. (30/kW + 0.05/kWh)

Steam : Rs. (120/kW + 0.0125/kWh). [**Ans.** 0.275]

45. A 30 H.P. motor connected to a condensate pump has been burnt beyond economical repairs. Two alternatives have been proposed to replace it by :

	Cost	*η at full load*	*η at half load*
Motor A	Rs. 6000	90%	86%
Motor B	Rs. 4000	85%	82%

The life of each motor is 20 years and its salvage value is 10 per cent of the initial value. The rate of interest is 5 per cent annually. The motor operates at full load for 25 per cent of the time and at half load for the remaining period. The annual maintenance cost of motor *A* is Rs. 420 and that of motor *B* is Rs. 240. The energy rate is 10 paise/kWh.

Which motor will be economical ? [**Ans.** Motor A ; Total cost : Motor A = Rs. 14,777/year Motor B = Rs. 15135/year]

46. The following proposals are under consideration for an industry which has a maximum demand of 50 MW and a load factor of 0.4.

(*i*) A steam plant having an initial cost of Rs. 1000/kW and maintenance cost is 2 paise/kWh. The coal of C.V. of 6150 kcal/kg is used. The overall efficiency of the plant is 25 percent.

(*ii*) A hydro-plant having a capital cost of Rs. 3000/kW and running cost of 0.5 paise/kWh.

Assuming interest and depreciation rate of 12 per cent for steam plant and 9 per cent for hydro-plant, determine the price of coal above which steam station is uneconomical. [**Ans.** Rs. 49.7/ton]

47. An industrial consumer has a choice between low and high voltage supply available at the following rates :

High voltage : Rs. 45/kW per year + paise 3.5/kWh

Low voltage : Rs. 47/kW per year + paise 4/kWh

In order to have high voltage supply, consumer has to install his own transformer which costs Rs. 100/kW. The losses in the transformer are 3 per cent of full load. Determine the number of working hours per week above which the high voltage supply will be economical. Assume interest and depreciation 10 per cent of capital and working weeks per year 50. Assume the load of the consumer as 1 MW.

[**Ans.** 49 hours per week]

48. The expected annual cost of power system supplying the energy to 40000 consumers is tabulated below :

Fixed charges	= Rs. 2400×10^3
Energy charges	= Rs. 1716×10^3
Consumer charges	= Rs. 210×10^3
Profit	= Rs. 168×10^3
Maximum demand	= 5000 kW
Diversity factor	= 4
Energy supplied	= 17×10^6 kWh.

Devise a three-part tariff allowing 25% of the profit in fixed charges, 50% in energy charges and remaining 25% in customer charges.

[**Ans.** Rs. 122/kW per year + 11 paise/kWh + Rs. 6.3 per consumer per year]

10

Interconnected Systems

10.1. Introduction. 10.2. Advantages and limitations of interconnected system. 10.3. Parallel operation of alternators—requirements for paralleling—alternator synchronising procedure—synchronising current, power and torque—effect of reactance—effect of increasing the excitation of one of the alternators—effect of increasing the driving torque of one of the alternators—effect of changes in speed of one of the alternators—effect of unequal voltages—distribution of load—load sharing between two alternators—hunting of alternators—synchronous machines on infinite bus-bars. 10.4. Interconnected stations—general aspects—load sharing—power limit of interconnectors—voltage, frequency and load control—load despatching. Highlights—Theoretical Questions—Unsolved Examples.

10.1. INTRODUCTION

- The primary requirement of any power system is to give an uninterrupted continuous supply. If an isolated plant is used for such a purpose then a large amount of reserve capacity would be required, which would obviously increase the initial cost resulting in increase in the fixed cost due to interest and depreciation. Further, some units are required either to be lightly loaded or floating on the lines in order to meet the unpredicted sudden increase in demand. Moreover, since the efficiency of the lightly loaded units is low, therefore, running cost per unit increases. Also in some parts of the country there are large resources of generation but their requirements are low whereas in other parts requirement of power is more in comparison to the resources of power available.

  The above mentioned difficulties/problems faced by the power engineers are considerably reduced by *interconnecting different power stations in parallel.*
- *The connection of several generating stations in parallel is known as* **Interconnected grid system.**

10.2. ADVANTAGES AND LIMITATIONS OF INTERCONNECTED SYSTEM

Some of the important *advantages and limitations / demerits of interconnected system* are listed below :

Advantages :

1. Reliability of the supply is increased.
2. Reserve capacity required is reduced.
3. Reduction in total installed capacity.
4. The load factor, diversity factor and efficiency are improved.
5. The interconnected system makes it possible to use the older and less efficient plants to carry peak loads of short durations.
6. Economical operation of stations is ensured.
7. Low capital and maintenance costs.
8. Peak loads of combined system can be carried at a *much lower cost* than that what is possible with small individual system.

Limitations/Demerits :

1. The interconnection of stations *involves extra cost.*
2. The *problems of load and frequency control are more difficult* in large interconnected system with many power stations scattered over a wide area in comparison with a system having one or two generating stations.

10.3. PARALLEL OPERATION OF ALTERNATORS

- The parallel operation of power stations is essentially an extension of parallel operation of alternators, with the exception of resistances and reactances of the inter-connector.
- *The process of connecting an alternator in parallel with another alternator with common bus-bars (bus-bars to which a number of alternators are connected) is called* **synchronizing.**

10.3.1. Requirements for Paralleling

For proper synchronizing of alternators, the following *conditions* must be fulfilled :

1. The *voltages* must be the *same* at the *paralleling point* or *junction* even though not the same at the alternators.

2. The *phase sequence* for three phase (or any multiple phase) must be the *same* at the paralleling point.

3. The incoming machine must be *in phase* at the moment of paralleling. It will stay in phase under normal conditions after paralleling. It is important to recognize that *phase sequence* and *in phase* are *not the same thing.*

4. The line frequencies must be *identical* at the paralleling point. In the vast majority of cases, this means the same frequency at the generator because frequency changing is not economic. Mixed frequencies must be paralleled through some frequency conversion means for compatibility at the point of interconnection.

5. The primemovers must have relatively similar and *drooping speed-load characteristics.* This is to prevent a machine with a rising speed load characteristic from taking more and more of the load until it fails from overload.

Violation of these requirements for paralleling would result in circulating currents between the machines varying from uneconomic, to serious, to disastrous.

10.3.2. Alternator Synchronising Procedure

It may logically be assumed that one alternator is placed in parallel with one or more other alternators only when additional load requires it. Those alternators already carrying load are known as the *running machines,* while that which is to be placed in the system is known as the *incoming machine.* At the time of synchronizing, the following *conditions* must be met :

1. The effective voltage of the incoming alternator *must be exactly equal to that of the others, or of the bus-bars connecting them.*

2. The *phase rotation,* or sequence of the running and incoming alternators, *must be the same.*

3. The individual phase voltages which are to be connected to each other must be in *exact phase opposition.* This is the same as saying that D.C. generators must be connected + to + and – to –.

4. The frequencies should be the same, although it is more desirable that the frequencies at the *instant* of paralleling be almost, but not quite, identical.

Synchronizing lamps :

- To satisfy the first condition of paralleling (*i.e.*, effective voltages be the same) a voltmeter can be used as shown in (Fig. 10.1).
- Satisfaction of the other conditions of *phase sequence, voltage opposition,* and *frequency* may be *determined by the use of the incandescent lamps connected between the two machines* (Fig. 10.1). The auxiliary equipment is omitted to simplify the sketch.

At any instant it is seen that the voltage across the lamp is the sum of the individual phase voltages.

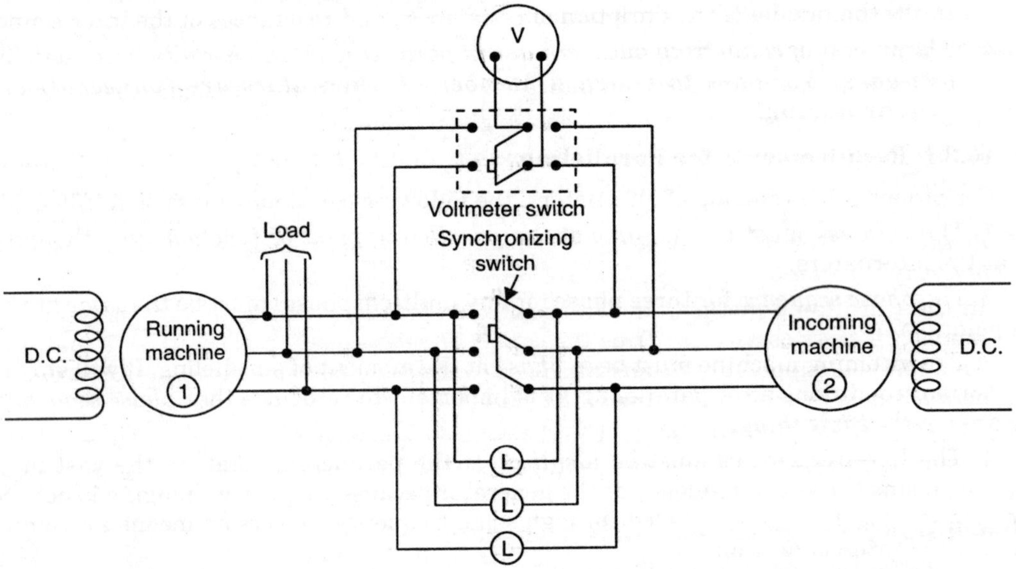

Fig. 10.1. Parallel operation of alternators : Synchronizing lamps and voltmeter.

The procedural steps for putting incoming alternator in parallel with the running machine are as follows :

Step 1. The primemover of the incoming machine is started, and the alternator is brought up to near its rated speed.

Step 2. By adjusting the field current, the terminal voltage of incoming machine is made the same as that of the running alternator.

The lamp in the circuit will now flicker at a rate equal to the difference in frequency of the two alternators. *If the phases are properly connected, all lamps will be bright and dark at the same time*. If this is not the case, then it means that the phase sequences are not correct, and it is merely necessary to interchange two of the line leads of the incoming machine.

Step 3. Further adjustment of the incoming primemover is now necessary, until the lamps flicker at a very low rate, *usually less than one dark period per second.*

Step 4. Final adjustment of the incoming voltage again made and the synchronizing switch is thrown in the middle of a dark period. The voltage across the lamps varies from zero to twice the phase voltage, and therefore, the lamps must be rated for this higher voltage. It is not convenient, however, to stock special lamps for this purpose, and two lamps of standard voltage ratings may be connected in series in each line.

The *lamp method* has the following *advantages* and *disadvantages :*

Advantages :

(*i*) Equipment is inexpensive.

(*ii*) Proper phase sequence is readily obtained.

Disadvantages :

(*i*) Lamps go dark at somewhat less than half their rated voltage, and so the synchronizing switch might be closed when there is a considerable phase difference between the machines, with a high circulating current resulting in possible damage to the machines. An experienced operator, however, can minimize this danger, since he can quickly learn to estimate the middle of the dark period.

(*ii*) The lamp filament can burn out.

(*iii*) The flicker of the lamps does not indicate which machine has the higher frequency.

There are *two other lamp methods* which are given below :

- If the lamps are connected across the phases, that is, P to Q', Q to R', R to P', they will again flicker in unison. This time, however, the proper synchronizing moment occurs when all three are in the middle of a bright period. This has the *advantage of avoiding synchronizing when the lamps may have burned out, but it is more difficult to estimate the middle of a bright period than the middle of a dark one.*
- In the third lamp method, known as the '*two-bright one-dark* method', only two of the lamps are cross-connected. Thus the connections are phase P to P', Q to R' and R to Q'. This method supposedly *avoids the disadvantages of both previous lamps methods, but actually may be more confusing to the eye.*

In case of high-voltage alternators, the lamps are connected through transformers to obtain the nominal voltages for which lamps are rated.

Synchronoscope synchronization. In large central station installations, an additional device called a *synchronoscope* (Fig. 10.2) is used. A synchronoscope has a rotating hand and a dial labelled with slow and fast direction arrows to show the incoming machine speed relation. In addition, an index point shows the actual in-phase position. During synchronization, as the incoming machine rotational speed approaches near synchronism, the speed of the synchronoscope hand drops enough to become visible. The *hand speed is proportional to the difference in speed.* The slow indication is accompanied with an arrow showing that *counter-clockwise* hand rotation means *below* synchronous speed. Similarly *clockwise* rotation means above speed. When the speeds are matched as that the hand speed is very slow, the hand is matched until it points to the index mark, where upon the paralleling switch may be closed.

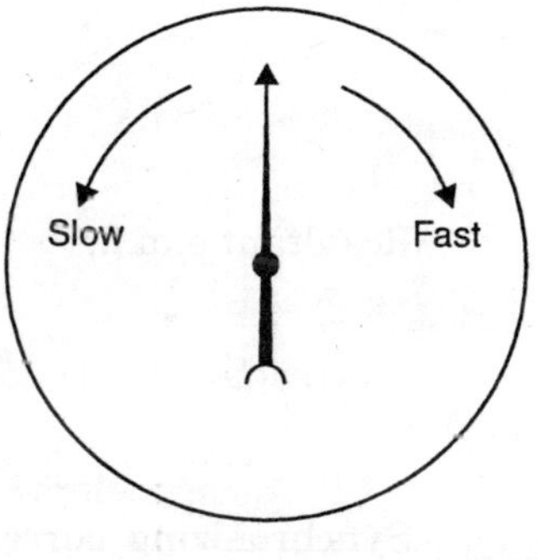

Fig. 10.2. Synchronoscope.

There are following three principal types of synchronoscopes :

(*i*) Polarized vane. (*ii*) Moving iron.

(*iii*) Crossed coil.

All are used similarly. However, *none* of the types will detect an *out-of-phase sequence condition.* They are used as a convenient and accurate means of routinely achieving synchronous speed and in-phase indications. The phase sequence problem is really an original test problem since, once sorted out, the phase sequence will remain correct until repair or other circuit changes are needed.

10.3.3. Synchronising Current, Power and Torque

Synchronising current :

- When in exact synchronism, the two alternators have equal inducted e.m.fs. and are in exact phase opposition as shown in Fig. 10.3.

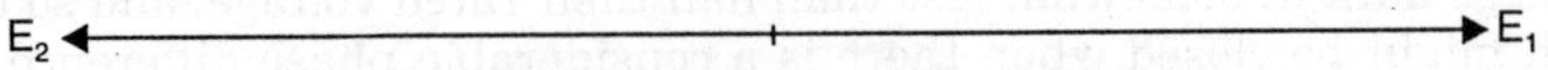

Fig. 10.3

- When the induced e.m.fs., of the two alternators are equal but *not in exact phase opposition,* as shown in Fig. 10.4, their resultant e.m.f. acts round the local circuit and causes flow of current called the '*synchronizing current*'.

Let, $(180° - \alpha)$ = Angular phase difference between the two induced e.m.fs. E_1 and E_2, and

E = Each e.m.f. (phase e.m.f.)

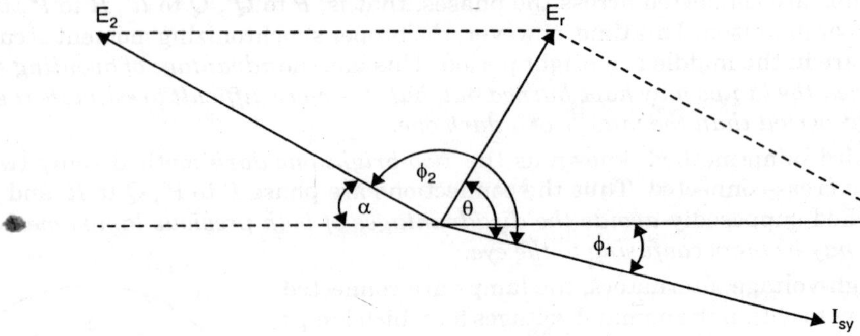

Fig. 10.4

Resultant e.m.f., $$E_r = 2E \cos\left(\frac{180° - \alpha}{2}\right) = 2E \cos\left(90° - \frac{\alpha}{2}\right) = 2E \sin\frac{\alpha}{2}$$

$$= 2E \times \frac{\alpha}{2} \qquad [\because \alpha \text{ is very small}]$$

$$= E\alpha$$

Synchronizing current, $I_{sy} = \dfrac{E_r}{Z} = \dfrac{E\alpha}{Z}$ lagging behind the resultant e.m.f. E_r by θ, where

$\theta = \tan^{-1}\dfrac{X_s}{R}$, where Z is the *combined* impedance per phase of the two alternators or of one alternator only if, it is connected to infinite bus bars. If resistance is very small as compared to synchronous reactance of the alternator, then

Synchronising current, $I_{sy} = \dfrac{E\alpha}{X_s}$ lagging behind E_r by 90°.

Synchronizing power :

Refer to Fig. 10.4.

Power supplied by machine No. 1 = $E_1 I_{sy} \cos\phi_1$

Power supplied by machine No. 2 = $E_2 I_{sy} \cos\phi_2$

Also, Power supplied by machine No. 1 = Power supplied by machine No. 2 + copper losses

or $$E_1 I_{sy} \cos\phi_1 = E_2 I_{sy} \cos\phi_2 + \text{copper losses}$$

The power supplied by machine No. 1 is called *synchronizing power* and is given by the expression :

$$P_{sy} = E_1 I_{sy} \cos\phi_1 = E_1 I_{sy}$$

$$= E \times \frac{E\alpha}{X_s} \qquad [\because \; E_1 = E \text{ and } \phi_1 \text{ is very small}]$$

$$= \frac{\alpha E^2}{X_s} \qquad ...(10.1)$$

Total synchronising power for 3-phases

$$= 3P_{sy} = \frac{3\alpha E^2}{X_s} \qquad ...(10.2)$$

Synchronizing torque :

If T_{sy} is the synchronizing torque in N-m, the total synchronizing power,

$$3P_{sy} = \frac{T_{sy} \times 2\pi N_s}{60}$$

or

$$T_{sy} = \frac{3P_{sy} \times 60}{2\pi N_s} \qquad ...(10.3)$$

where, N_s is the synchronous speed in r.p.m. $\left(= \frac{120f}{P}\right)$.

Synchronizing Power and Torque

Example 10.1. *A 3 MVA, 6-pole alternator runs at 1000 r.p.m. in parallel with other machines on 3.3 kV bus-bars. The synchronous reactance is 20%. Calculate the synchronizing power per one mechanical degree of displacement and the corresponding synchronizing torque.*

Solution. Phase voltage $E_{ph} = \frac{3.3 \times 1000}{\sqrt{3}} = 1905$ V

Speed of the alternator, $N_s = 1000$ r.p.m.

Synchronous reactance, $X_s = 20\%$

Rotor displacement, $\alpha = 1°$ (mech.)

Number of poles, $p = 6$

Synchronizing power ; synchronizing torque :

$\alpha = 1°$ (mech.) $= 1° \times$ number of pair of poles (degrees elect.)

$$= 1° \times \left(\frac{6}{2}\right) = 3° \text{ (electrical)} = 3 \times \frac{\pi}{180} = \frac{\pi}{60} \text{ radian}$$

Full-load current

$$= \frac{3 \times 10^6}{\sqrt{3} \times 3.3 \times 1000} = 524.9 \text{ A}$$

Synchronous impedance,

$$Z_s = \frac{\text{Phase voltage}}{\text{Full-load current}} = \frac{1905}{524.9} = 3.63 \ \Omega$$

Also, $X_s = 0.2Z_s = 0.2 \times 3.63 = 0.726 \ \Omega$

Synchronizing power, P_{sy} :

Using the relation :

$$3P_{sy} = \frac{3\alpha E_{ph}^2}{X_s} = \frac{3 \times \frac{\pi}{60} \times (1905)^2}{0.726 \times 1000} \text{ kW}$$

$$= \mathbf{785.19 \ kW} \text{ (for three phases).} \quad \textbf{(Ans.)}$$

Synchronizing torque, T_{sy} :

Using the relation :
$$T_{sy} = \frac{3P_{sy} \times 60}{2\pi N_s} = \frac{785.19 \times 1000 \times 60}{2\pi \times 1000} = \mathbf{7498\ N\text{-}m.} \quad \textbf{(Ans.)}$$

☞**Example 10.2.** *A 4 MVA, 10 kV, 1500 r.p.m., 50 Hz alternator runs in parallel with other machines. Its synchronous reactance is 25%. Find for (i) no-load, (ii) full-load, at p.f. 0.8 lagging, synchronizing power per unit mechanical angle of phase displacement, and calculate the synchronizing torque if the mechanical displacement is 0.7°.*

Solution. Phase voltage, $E_{ph} = \frac{10 \times 1000}{\sqrt{3}} = 5774$ V

Speed of the alternator, $N_s = 1500$ r.p.m.

Frequency, $f = 50$ Hz

Synchronous reactance, $X_s = 25\%$

$$\text{Full-load current} = \frac{4 \times 10^6}{\sqrt{3} \times 10000} = 231 \text{ A}$$

Synchronous impedance,
$$Z_s = \frac{E_{ph}}{\text{Full-load current}} = \frac{5774}{231} = 25\ \Omega$$

∴ $X_s = 0.25 Z_s = 0.25 \times 25 = 6.25\ \Omega$

Also $f = \frac{Np}{120}$

∴ $p = \frac{120f}{N} = \frac{120 \times 50}{1500} = 4$

∴ No. of pair of poles $= \frac{p}{2} = \frac{4}{2} = 2$

∴ $\alpha = 1° \times 2 = 2°$ (elect.) $= 2 \times \frac{\pi}{180} = \frac{\pi}{90}$ rad.

(*i*) **At No-load :**

Synchronizing power *(for 3-phases)*,
$$3P_{sy} = \frac{3\alpha E_{ph}^2}{X_s} = 3 \times \frac{\pi}{90} \times \frac{(5774)^2}{6.25 \times 1000} \text{ kW} = \mathbf{558.6\ kW.} \quad \textbf{(Ans.)}$$

∴ *Synchronizing torque,* $\mathbf{T_{sy}} = \frac{3P_{sy} \times 60}{2\pi N_s} = \frac{558.6 \times 1000 \times 60}{2 \times \pi \times 1500} = \mathbf{3556.2\ N\text{-}m.}$ **(Ans.)**

∴ $\mathbf{T_{sy}}$ **for 0.7°** = 3556.2 × 0.7 = **2489.3 N-m. (Ans.)**

(*ii*) **At full-load 0.8 lagging :**

Let us first find E_0.

Refer to Fig. 10.5.

- Draw $OL = 5774$ V, $\phi°$ (*i.e.* 36° 52′) ahead of current vector OI.
- LM represents reactance drop IX_s and is at right angles to OI.

Hence, OM is ahead of V by an angle α.

∴ $$E_0 = \sqrt{(V \cos \phi)^2 + (V \sin \phi + IX_s)^2}$$

Now, $\cos \phi = 0.8$ and $\sin \phi = 0.6$

$X_s = 6.25\ \Omega$

$$E_0 = \sqrt{(5774 \times 0.8)^2 + (5774 \times 0.6 + 231 \times 6.25)^2}$$
$$= \sqrt{(4619.2)^2 + (4908.15)^2} \simeq 6740 \text{ V}$$

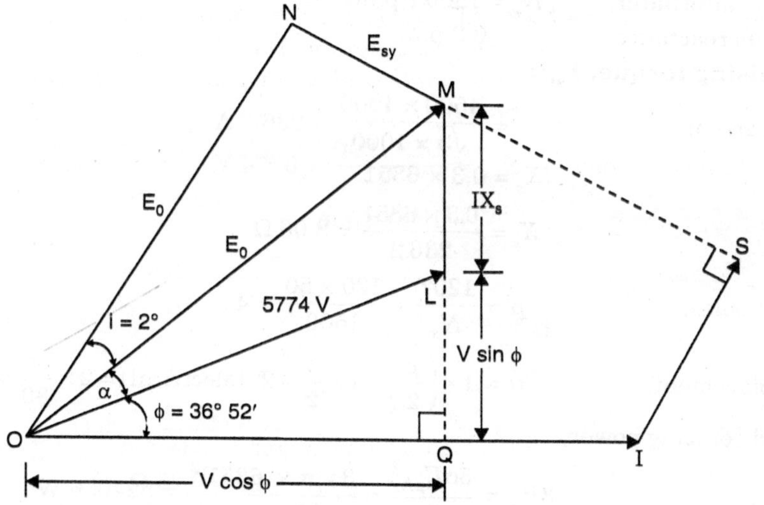

Fig. 10.5

In ΔOMQ, $\tan (\alpha + \phi) = \dfrac{MQ}{OQ} = \dfrac{IX_s + V \sin \phi}{V \cos \phi} = \dfrac{231 \times 6.25 + 5774 \times 0.6}{5774 \times 0.8} = 1.0625$

$\therefore$ $\alpha + \phi = 46.7° \text{ or } 46° \ 40'$

or $\alpha = 46° \ 40' - \phi = 46° \ 40' - 36° \ 52' = 9° \ 48'$

In Fig. 10.5 if E_0 is further shifted through an angle of 2° (elect.) then an additional e.m.f., E_{sy} = MN is created which produces a synchronizing current I_{sy}. This current (shown by vector IS) lags behind E_{sy} by 90° as shown.

Obviously, $E_{sy} = 2E_0 \sin \left(\dfrac{2}{2}\right) = 2 \times 6740 \sin 1° = 235.26 \text{ V}$

$$I_{sy} = \frac{E_{sy}}{X_s} = \frac{235.26}{6.25} = 37.64 \text{ A}$$

It can be shown from the diagram that E_{sy} leads E_0 by an angle $\left(90 + \dfrac{2}{2}\right)$ and I_{sy} is lagging E_{sy} by 90°, hence I_{sy} is leading V by an angle $(\alpha + 1) = 9° \ 48' + 1° = 10° \ 48'$.

$\therefore$ $P_{sy}/\text{phase} = VI_{sy} \cos 10° \ 48'$

$= 5774 \times 37.64 \times 0.9823 \text{ W} = 213.48 \text{ kW}$

Total synchronizing power $= 3 \times 213.48 =$ **640.44 kW. (Ans.)**

T$_{sy}$ */unit displacement* $= \dfrac{640.44 \times 1000 \times 60}{2\pi \times 1500} =$ **4077.2 N-m. (Ans.)**

T$_{sy}$ *for 0.7° displacement* $= 4077.2 \times 0.7 =$ **2854 N-m. (Ans.)**

Example 10.3. *A 4500 kVA, 50 Hz, 3-phase, synchronous generator having a synchronous reactance of 0.3 p.u. is running at 1500 r.p.m. and is excited to give 11000 V. If the rotor deviates slightly from its equilibrium position, what is the synchronizing torque in N-m per degree mechanical displacement.*

Solution. Induced e.m.f./phase $= \dfrac{11000}{\sqrt{3}} = 6351$ V (assuming machine star-connected)

Frequency, $f = 50$ Hz

Speed of the alternator, $N_s = 1500$ r.p.m.

Synchronous reactance $= 0.3$ p.u.

Synchronising torque, T_{sy} :

Full-load current $= \dfrac{4500 \times 1000}{\sqrt{3} \times 1000} = 236.2$ A

Now, $IX_s = 0.3 \times 6351$

$\therefore$ $X_s = \dfrac{0.3 \times 6351}{236.2} = 8.06\ \Omega$

Number of poles, $p = \dfrac{120f}{N_s} = \dfrac{120 \times 50}{1500} = 4$

Rotor displacement, $\alpha = 1 \times \left(\dfrac{p}{2}\right) = 1 \times \dfrac{4}{2} = 2°$ (electrical) $= 2 \times \dfrac{\pi}{80} = \dfrac{\pi}{90}$ rad

Total synchronizing power,

$$3P_{sy} = \frac{3\alpha E_{ph}^2}{X_s} = \frac{3 \times \pi \times (6351)^2}{90 \times 8.06} = 524056 \text{ W}$$

Synchronizing torque, $T_{sy} = \dfrac{3P_{sy} \times 60}{2\pi N_s} = \dfrac{524056 \times 60}{2 \times \pi \times 1500} \simeq$ **3336 N-m. (Ans.)**

Example 10.4. *A 2-pole, 50-Hz, 3-phase, turbo-alternator is excited to generate the bus-bar voltage of 11 kV on no-load. The machine is star-connected and the short-circuit current for this excitation is 1000 A.*

Calculate the synchronizing power per degree of mechanical displacement of the rotor and the corresponding synchronizing torque.

Solution. Number of poles, $p = 2$

Frequency, $f = 50$ Hz

E.m.f. per phase, $E_{ph} = \dfrac{11000}{\sqrt{3}} = 6351$ V

Short-circuit current, $I_{sc} = 1000$ A

Rotor displacement $= 1°$ mech.

$$= 1 \times \frac{p}{2} = 1 \times \frac{2}{2} = 1° \text{ (electrical)} = 1 \times \frac{\pi}{180} = \frac{\pi}{180} \text{ rad.}$$

Synchronous speed, $N_s = \dfrac{120f}{p} = \dfrac{120 \times 50}{2} = 3000$ r.p.m.

Synchronous reactance, $X_s = \dfrac{E_{phase}}{I_{sc}} = \dfrac{6351}{1000} = 6.35\ \Omega$

Total synchronizing power for three phases,

$$3P_{sy} = \frac{3\alpha E_{ph}^2}{X_s} = \frac{3 \times \pi \times (6351)^2}{180 \times 6.35} = \textbf{332590 W or 332.6 kW.} \quad \textbf{(Ans.)}$$

Synchronizing torque, T_{sy} $= \dfrac{3P_{sy} \times 60}{2\pi N_s} = \dfrac{332590 \times 60}{2 \times \pi \times 3000} =$ **1058.67 N-m. (Ans.)**

Example 10.5. *3 MVA, 6-pole alternator runs at 1000 r.p.m. on 3.3 kV bus-bars. The synchronous reactance is 25%. Calculate the synchronizing power and torque per mechanical degree of displacement when alternator is supplying full-load at 0.8 power factor lagging.*

(Madras University)

Solution. *Given* : Rating : 3 MVA ; $p = 6$; $N_s = 1000$ r.p.m. ;

Bus-bar voltage = 3.3 kV ; $X_s = 25\%$; $\cos \phi = 0.8$.

Synchronising power and torque :

Voltage per phase, $$V = \frac{3.3 \times 10^3}{\sqrt{3}} = 1905 \text{ V}$$

Full-load current, $$I = \frac{3 \times 10^6}{\sqrt{3} \times 3.3 \times 10^3} = 524.9 \text{ A}$$

Now, $$IX_s = \frac{25}{100} \times 1905 = 476.25 \text{ V}$$

$\therefore$ Synchronous reactance, $$X_s = \frac{476.25}{524.9} = 0.9074 \ \Omega$$

Power factor, $\cos \phi = 0.8$; $\phi = \cos^{-1}(0.8) = 36.87°$ and $\sin 36.87° = 0.6$.

Rotor displacement in electrical degrees,

$$\alpha = \text{Rotor displacement in mechanical degrees} \times \frac{1}{2}$$

$$= 1 \times \frac{6}{2} = 3° \quad \text{or} \quad 3 \times \frac{\pi}{180} \text{ rad.} = 0.05236 \text{ rad.}$$

Induced e.m.f., $$E = V + jIX_s$$
$$= 1905 + j524.9(0.8 - j0.6) \times 0.9074$$
$$= 1905 + j381 + 285.8 = 2190.8 + j381 = 2224\angle 9.87°$$

Total synchronising power (*for 3-phases*),

$$3P_{sy} = \frac{3\alpha EV}{X_s} = \frac{3 \times 0.05236 \times 2224 \times 1905}{0.9074} = \mathbf{733418 \text{ W.}} \quad \textbf{(Ans.)}$$

Synchronising torque, $$\mathbf{T}_{sy} = \frac{3P_{sy}}{(2\pi N_s/60)}$$
$$= \frac{3 \times 733418}{(2\pi \times 1000/60)} = \mathbf{21010.8 \text{ Nm.}} \quad \textbf{(Ans.)}$$

10.3.4. Effect of Reactance

It has been observed that *without reactance, the synchronous machines cannot develop motor and generator power, respectively to restore synchronism and consequently parallel operation is not possible.*

10.3.5. Effect of increasing the Excitation of one of the Alternators

- If the *excitation of an alternator* operating in parallel with other alternators is *increased above its normal value of excitation, its power factor changes in the lagging direction*, and *its current output increases with no appreciable change in its kW load.*

When the alternator is *under-excited* its power factor becomes *more leading* and its *current output increases with no change in kW output*. This increase in current in either case is *not applied to the load but circulates between the alternators connected to the system, thereby increasing their losses and reducing their useful capacity*. It is desirable in most cases, therefore, to *operate at the same factor (at or near the load power factor), keeping the circulating currents to the minimum.*

- The proper amount of excitation for alternators operating in parallel, in general, is the amount of excitation each alternator would need if it were carrying its load alone at the same frequency and voltage.
- The voltage of the system in which several alternators are operating in parallel may be raised or lowered by varying the excitation of all the machines up or down as needed ; this is accomplished automatically through the voltage regulators.

10.3.6. Effect of increasing the driving Torque of One of the Alternators

- When two alternators are running in parallel and if the driving torque of one of the alternators is increased, it is further loaded and other is relieved of its load. If the output of the alternator, whose driving torque has been increased, becomes more than the total load supplied, then the other alternator runs as a synchronous motor.
- Governor controls are adjusted to make kW load division between the alternators.

10.3.7. Effect of Changes in Speed of One of the Alternators

- When in exact synchronizm, the two alternators have equal induced e.m.fs. and are in exact phase opposition as shown in Fig. 10.6. *No circulating current flows round the local circuit.*
- When the speed of alternator 2 is *reduced*, E_2 falls back by a phase angle α electrical degrees as shown in Fig. 10.4, the magnitude of induced e.m.fs., E_1 and E_2 remaining the same. It will cause resultant e.m.f. E_r acting in the local circuit ; which will *cause a flow of synchronizing current I_{sy} in the local circuit. This synchronizing current sets up a synchronizing torque tending to retard the generating alternator 1* and accelerating the motoring alternator 2. Thus the *two alternators come again in synchronism.*

  Thus by increasing or reducing the speed of all the alternators simultaneously, the frequency of the system can be raised or lowered.

10.3.8. Effect of Unequal Voltages

Let there be two alternators having e.m.fs. E_1 and E_2 ($E_1 > E_2$) exactly in phase opposition relative to external load circuit (Fig. 10.6). The resultant e.m.f. (E_r) equal to E_1–E_2 acts in the local circuit and causes synchronising current I_{sy} around the local circuit. The current I_{sy} lags behind E_r or E_1 by 90° and leads E_2 by 90°. Thus current I_{sy} produces *demagnetising effect on machine No. 1, resulting thereby reduction in E_1* and *magnetising effect on machine No. 2, resulting thereby increase in E_2*. Thus difference of E_1 and E_2 is reduced and stable condition is established.

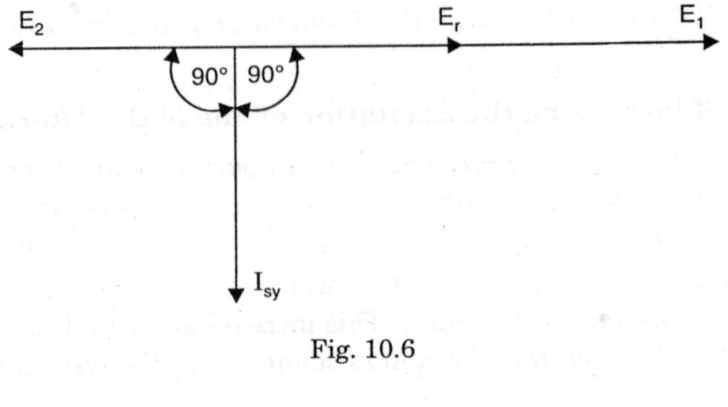

Fig. 10.6

10.3.9. Distribution of Load

Under this heading we shall discuss the *following* :

(*i*) ***Effect of change in driving torque.***

(*ii*) ***Effect of change in excitation.***

It will be shown that :

(*i*) Amount of load taken up by an alternator running in parallel with other machines is solely determined by its *driving torque i.e.,* by the *power input* to its primemover.

(*ii*) *Any change in its excitation merely changes its kVA output but not its kW output.* In other words, it *merely changes the power factor* at which load is delivered.

- Let us consider two single-phase alternators running in parallel with generated e.m.fs. E_1 and E_2 respectively. If they are in phase with external circuit, then they will be in direct opposition with respect to their local circuit.
- Let their excitation and power inputs to their primemovers are changed such that they deliver currents of I_1 and I_2 at respective power factors of cos ϕ_1 and cos ϕ_2. Then, total current supplied to the load = vector sum of I_1 and I_2.

Consider that the *power input to the primemover of the second machine is increased.* Then its e.m.f. vector swings ahead by a certain angle as shown in Fig. 10.7. Now, a resultant voltage E_r or E_{sy} is produced which, acting on the local circuit, sets flowing a current I_{sy} which lags almost by 90° behind E_r. The new armature current I_2' of second machine is vector sum of I_{sy} and I_2. It will be noted that *I_2' is greater than I_2 and is less lagging than it.* Hence, the second alternator takes up increased share of the load. This *synchronising current is taken up by the first machine.* Its *armature current also changes from I_1 to I_1'.* It may be noted that *I_1' is lesser and is at a greater angle from E_1.* Hence, this alternator is taking up a *lesser share of its load.*

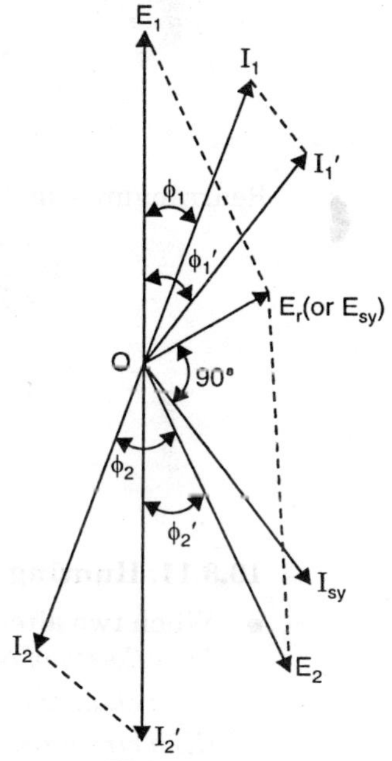

Fig. 10.7

Thus we find that by *increasing the input to its prime-mover, an alternator can be made to take a greater share of load, though at different power factor.*

The following points are worth noting :

- *If excitation is changed, merely power factor at which load is delivered is changed.*
- *The load taken up by an alternator directly depends upon its driving torque* (or angular advance of the rotor).
- By *keeping the input to the primemover of an alternator constant,* if excitation is changed, *then only kVA component of the output is changed and not kW.*

10.3.10. Load Sharing between Two Alternators

Let us consider two machines with identical speed characteristics running in parallel with a common terminal voltage V and load impedance Z.

$$I = I_1 + I_2 \quad \text{...}(i)$$

$$I = \frac{V}{Z} \quad \text{...}(ii)$$

From Fig. 10.8,

$$I_1 = \frac{E_1 - V}{Z_1} \quad \text{...}(iii)$$

$$I_2 = \frac{E_2 - V}{Z_2} \quad \text{...}(iv)$$

From eqns. (*i*), (*ii*), (*iii*) and (*iv*), we have

$$V = (I_1 + I_2)Z = \left(\frac{E_1 - V}{Z_1} + \frac{E_2 - V}{Z_2}\right)Z$$

$$= \left(\frac{E_1}{Z_1} + \frac{E_2}{Z_2}\right)Z - V\left(\frac{1}{Z_1} + \frac{1}{Z_2}\right)Z \qquad ...(v)$$

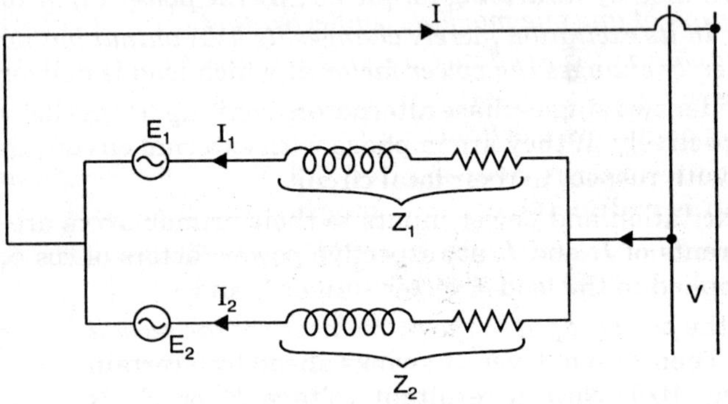

Fig. 10.8. Equivalent circuit for two alternators in parallel.

Rearranging eqn. (*v*), we get

$$V\left(\frac{1}{Z} + \frac{1}{Z_1} + \frac{1}{Z_2}\right) = \frac{E_1}{Z_1} + \frac{E_2}{Z_2}$$

or

$$V = \frac{\left(\frac{E_1}{Z_1} + \frac{E_2}{Z_2}\right)}{\left(\frac{1}{Z} + \frac{1}{Z_1} + \frac{1}{Z_2}\right)} \qquad ...(10.4)$$

10.3.11. Hunting of Alternators

- When two alternators are operating in parallel, any *instantaneous reduction in the angular velocity of one machine* causes :
  - (*i*) *a change in load division between them* ; and
  - (*ii*) *a circulating current*.

The *circulating current acts as additional load on one machine and lightens the load on the other.* This *retards the former* and permits the *latter to accelerate* until the two are once more in the proper relative phase positions where no circulating current flows if the excitation have been equal. The change to correct phase position cannot be accomplished without some *'overshooting'* on the part of the rotors, accompanied by a retardation, with a repetition of the entire cycle. That is, the alternators hunt, their actions being exactly equivalent to those of the synchronous motors under similar conditions. The period of the swing agrees with the natural oscillating period of the rotor as a torsional pendulum.

- The torque output will *pulsate* if the primemover of one of the alternators is a reciprocating engine. If this pulsation has a forced frequency within 20 per cent of that of the natural oscillating frequency of the alternator rotor, the oscillation following any load change will be *cumulative* and the machines may be thrown out of synchronism.

- It is customary to specify, for alternators to be operated in parallel, the allowable *torque-angle variation.*
- *Machines driven by internal combustion engines must have large flywheels or heavy damping windings to prevent excessive oscillation.*

Note. For economic loading of alternators/generator, phase refer to Art. 9.16.

10.3.12. Synchronous Machines on Infinite Bus-bars

The bus-bars whose frequency and the phase magnitude of potential differences are not affected by changes in condition of any one machine connected to it are called **infinite bus bars.**

A network having zero impedance and infinite rotational inertia is also termed as **Infinite bus-bars.**

The following points are *worth noting* :

1. When the alternator is **over-excited** (*i.e.* $E > V$) it *operates at lagging power factor* and when its **excitation is reduced** (*i.e.* $E < V$) *it operates at leading power factor.*

- *On infinite bus-bars no change in excitation can change the terminal voltage which is fixed by the network, the change in excitation merely affecting the power factor at which the particular machine will work without any change in power output.*

2. On increasing the torque of the prime-mover of the alternator connected to infinite bus-bars, the alternator connected to it will swing ahead in phase will respect to other alternators or alternator connected to the common bus-bars. It will cause a phase difference between V and E, the magnitude of E remaining unchanged. The resultant of these two voltages, E_r will cause a synchronous current I_{sy}. Thus the power delivered by the alternator, whose input power has been increased, will increase by an amount equal to $EI_{sy} \cos \theta$.

- Hence if the *torque of the prime-mover of one alternator is increased, it is further loaded and an equivalent load is removed from the other unit or units with which this machine was paralleled.*
- *Further, if the output of the alternator, whose prime-mover torque has been increased, becomes more than the total load being supplied then the other machine or machines will operate as synchronous motor or motors.*
- The only way in which the prime-mover *while operating at a fixed synchronous speed can deliver more or less power is to change its governor setting in such a manner as to increase or decrease the power given to the prime-mover at the fixed speed.*

Parallel Operation of Alternators

Example 10.6. *A number of alternators are working in parallel with their terminal voltage equal to the rated value. One of the machines, which has a synchronous reactance of 50% and a resistance of 1%, delivers a power output in kW equal to 70% of its rated kVA. If the e.m.f. of this unit equals 1.2 times the terminal voltage, find out the power factor at which the machine is operating.*

(I.A.S.)

Solution. *Given* : $\frac{IR_a}{V} = 1\% = 0.01$; $\frac{IX_s}{V} = 50\% = 0.5$; $E = 1.2$ V

Power factor, cos ϕ :

We know that, $$E = \sqrt{(V \cos \phi + IR_a)^2 + (V \sin \phi + IX_s)^2}$$

(where E and V are e.m.f. and terminal voltage per phase respectively)

Squaring both sides, we get

$$E^2 = (V \cos \phi + IR_a)^2 + (V \sin \phi + IX_s)^2$$

$$= V^2\left[\left(\cos\phi + \frac{IR_a}{V}\right)^2 + \left(\sin\phi + \frac{IX_s}{V}\right)^2\right]$$

Here, $I = 0.7\, I_{rated}$

$\therefore$ $(1.2V)^2 = V^2[(\cos\phi + 0.01 \times 0.7)^2 + (\sin\phi + 0.5 \times 0.7)^2]$

$$1.44 = [(\cos^2\phi + 0.007^2 + 2 \times 0.007 \times \cos\phi) + (\sin^2\phi + 0.35^2 + 2 \times 0.35 \sin\phi)]$$

$$= 0.014\cos\phi + 0.7\sin\phi + 1.12255$$

or, $0.014\cos\phi + 0.7\sin\phi = 0.31745$ ...(*i*)

We know that, $A\cos\phi + B\sin\phi = \sqrt{(A^2 + B^2)}\,\sin\left(\phi + \tan^{-1}\frac{A}{B}\right)$

$\therefore$ Eqn. (*i*) can be written as

$$\sqrt{[(0.014)^2 + (0.7)^2]}\,\sin\left\{\phi + \tan^{-1}\left(\frac{0.014}{0.7}\right)\right\} = 0.31745$$

$$0.70014\sin(\phi + 1.14577°) = 0.31745$$

or, $\sin(\phi + 1.14577°) = 0.4534$

or, $\phi + 1.14577° = \sin^{-1}(0.4534) = 26.962°$

or, $\phi = 25.82°$

$\therefore$ Power factor, $\cos\phi = \cos(25.82°) = \mathbf{0.9}$ **(lagging). (Ans.)**

☞ **Example 10.7.** *Two 1-phase alternators operate in parallel and supply a load impedance of (3 + j4) Ω. If the impedance of each machine is (0.2 + j2) and e.m.fs. are (200 + j0) and (220 + j0) volts respectively, determine for each machine :*

(i) Terminal voltage, *(ii) Power factor, and* *(iii) Output.*

Solution. Load impedance, $Z_L = (3 + j4)\ \Omega$

Impedance of each machine $= (0.2 + j2)\ \Omega$

E.m.f. of machine 1, $E_1 = (200 + j0)$

E.m.f. of machine 2, $E_2 = (220 + j0)$

Let, V = Terminal voltage,

I_1 = Current supplied by machine 1, and

I_2 = Current supplied by machine 2.

Now, $V = E_1 - I_1 Z_1$

or, $V = (200 - j0) - I_1(0.2 + j2)$ ...(*i*)

Similarly, $V = (E_2 - I_2 Z_2) = (220 + j0) - I_2(0.2 + j2)$ ...(*ii*)

Also, $V = IZ_L = (I_1 + I_2)Z_L = (I_1 + I_2)(3 + j4)$ ...(*iii*)

Adding, (*i*) and (*ii*), we get

$$2V = (420 + j0) - (I_1 + I_2)(0.2 + j2)$$

or, $V = (210 + j0) - (I_1 + I_2)(0.1 + j1)$ ...(*iv*)

Multiply (*iii*) by $(0.1 + j1)$ and (*iv*) by $(3 + j4)$, we get

$$(0.1 + j1)V = (I_1 + I_2)(3 + j4)(0.1 + j1) \quad ...(v)$$

$$(3 + j4)V = 210(3 + j4) - (I_1 + I_2)(3 + j4)(0.1 + j1) \quad ...(vi)$$

Adding (*v*) and (*vi*), we get

$$(3.1 + j5)V = 210(3 + j4) = 630 + j840$$

or,
$$V = \frac{630 + j840}{3.1 + j5} = \frac{630 + j840}{3.1 + j5} \times \frac{(3.1 - j5)}{(3.1 - j5)}$$
$$= \frac{630 \times 3.1 + j840 \times 3.1 - 630 \times j5 - j^2\,(840 \times 5)}{(3.1)^2 + (5)^2}$$
$$= \frac{1953 + j2604 - j3150 + 4200}{34.61} = \frac{6153 - j546}{34.61}$$
$$= 177.8 - j15.77 = \mathbf{178.5 \angle -5.07°.} \quad \textbf{(Ans.)}$$

Machine 1 :

(*i*) ∴ **Terminal voltage, V = 178.5 ∠– 5.07°. (Ans.)**

Current,
$$I_1 = \frac{E_1 - V}{Z_1}$$
$$= \frac{(200 + j0) - (177.8 - j15.77)}{(0.2 + j2)} = \frac{22.2 + j15.77}{(0.2 + j2)}$$
$$= \frac{22.2 + j15.77}{(0.2 + j2)} \times \frac{(0.2 - j2)}{(0.2 - j2)}$$
$$= \frac{22.2 \times 0.2 + j15.77 \times 0.2 - 22.2 \times j2 - j^2\,(2 \times 15.77)}{(0.2)^2 + (2)^2}$$
$$= \frac{4.44 + j3.154 - j44.4 + 31.54}{4.04} = \frac{35.98 - j41.25}{4.04}$$
$$= 8.9 - j10.27 = 13.58 \angle -49.08° \text{ A}$$

(*ii*) **Power factor** $= \cos \phi_1$

$= \cos [-49.08 - (-5.07)] = \mathbf{0.7192}$ **(lag). (Ans.)**

(*iii*) **Output** $= VI_1 \cos \phi_1 = 178.5 \times 13.58 \times 0.7192 = \mathbf{1743.4\ W.}$ **(Ans.)**

Machine 2 :

(*i*) **Terminal voltage, V = 178.5 ∠– 5.07°** (*same as machine 1*)

Current,
$$I_2 = \frac{E_2 - V}{Z_2} = \frac{(220 + j0) - (177.8 - j15.77)}{(0.2 + j2)}$$
$$= \frac{42.2 + j15.77}{(0.2 + j2)} = \frac{(42.2 + j15.77)(0.2 - j2)}{(0.2 + j2)(0.2 - j2)}$$
$$= \frac{42.2 \times 0.2 + j15.77 \times 0.2 - 42.2 \times j2 - j^2(15.77 \times 2)}{(0.2)^2 + (2)^2}$$
$$= \frac{8.44 + j3.154 - j84.4 + 31.54}{4.04} = \frac{39.98 - j81.246}{4.04}$$
$$= 9.89 - j20.1 = 22.4 \angle -63.8° \text{ A}$$

(*ii*) **Power factor** $= \cos \phi_2$

$= \cos [-63.8° - (-5.07°)] = \cos 58.73° = \mathbf{0.519}$ **(lag). (Ans.)**

(*iii*) **Output** $= VI_2 \cos \phi_2 = 178.5 \times 22.4 \times 0.519 = \mathbf{2075\ W.}$ **(Ans.)**

Example 10.8. *Two alternators working in parallel supply a lighting load of 3000 kW and a motor load aggregating to 5000 kW at a p.f. 0.72. One machine is loaded up to 5000 kW at 0.8 p.f. lagging. What is the load and power factor of the other machine ?*

Solution. Lighting load = 3000 kW at unity p.f.

Motor load = 5000 kW at 0.72 p.f. (lag)

Load of machine 1 = 5000 kW at 0.8 p.f. (lag)

Load and p.f. of the machine-2 :

Phase angle of light load, $\phi_l = \cos^{-1} 1.0 = 0°$

Phase angle of motor load, $\phi_m = \cos^{-1} 0.7 = 43.94°$

Reactive kVA of load = Load in kW × tan of phase angle

Reactive kVA of lighting load = 0

Reactive kVA of motor load = 5000 tan 43.94°

= 5000 × 0.964 = 4820 kVAR

Total reactive kVA = 0 + 4820 = 4820 kVAR

Total load = 3000 + 5000 = 8000 kW

Load supplied by machine-1 = 5000 kW

Reactive kVA supplied by machine, 1

= 5000 tan (cos^{-1} 0.8) = 3750 kVAR

Load supplied by machine-2 = 8000 – 5000 = **3000 kW. (Ans.)**

kVAR supplied by other machine

= 4820 – 3750 = 1070 kVAR

Phase angle of machine-2, $\phi_2 = \tan^{-1} \frac{\text{kVAR}}{\text{kW}} = \tan^{-1} \frac{1070}{3000} = 19.63°$

∴ **Power factor of machine-2** = cos ϕ_2 = cos 19.63° = **0.942. (Ans.)**

Example 10.9. *The following loads are supplied by two alternators running in parallel :*

(*i*) *1400 kW at p.f. 0.86 lagging,* (*ii*) *900 kW at p.f. 0.8 lagging.*

(*iii*) *800 kW at p.f. unity.* (*iv*) *500 kW at p.f. 0.8 leading.*

If the load on one machine is adjusted to 2100 kW at p.f. 0.92, find the load and p.f. of the other machine.

Solution. Phase angle of *first* load, $\phi_1 = \cos^{-1} 0.86 = 30° \, 41'$

Phase angle of *second* load, $\phi_2 = \cos^{-1} 0.8 = 36° \, 52'$

Phase angle of *third* load, $\phi_3 = \cos^{-1} = 0°$

Phase angle of *fourth* load, $\phi_4 = \cos^{-1} 0.8 = (-36° \, 52')$

Reactive kVA of each load, kVAR

= load in kW × tan. of phase angle

∴ Reactive kVA of *first* load, kVAR

= 1400 × tan 30° 41′ = 830.7

Reactive kVA of *second* load, kVAR_2

= 900 × tan 36° 52′ = 675

Reactive kVA of *third* load, kVAR_3

= 800 × tan 0° = 0

Reactive kVA of *fourth* load, kVAR_4

= 500 × tan (– 36° 52′) = – 375

Reactive kVA of total load, kVAR

= 830.7 + 675 + 0 + (– 375) = 1130.7

Total load = 1400 + 900 + 800 + 500 = 3600 kW.

Load supplied by machine-1 = 2100 kW

Reactive kVA supplied by machine

= 2100 tan ($\cos^{-1}$ 0.92) = 894.6 kVAR

Load supplied by machine-2

= 3600 – 2100 = **1500 kW. (Ans.)**

kVAR supplied by machine-2 = 1130.7 – 894.6 = 236.1 kVAR

∴ Phase angle of machine-2,

$$\phi = \tan^{-1}\left(\frac{\text{kVAR}}{\text{kW}}\right) = \tan^{-1}\left(\frac{236.1}{1500}\right) = 8.94°$$

∴ **Power factor of machine-2**

= cos φ = cos 8.94° = **0.9878. (Ans.)**

Example 10.10. *The following data pertain to two similar alternators–3-phase, 6600 V, 60-Hz, 1200 kVA, star-connected :*

Resistance/phase = *1.05 Ω*

Synchronous reactance/phase = *5 Ω*

Saturation curve :

Field current, A	*: 150*	*200*	*250*	*300*	*350*	*500*
Terminal voltage	*: 5600*	*6490*	*7000*	*7400*	*7750*	*8500*

When operating in parallel with a terminal voltage of 6600 V the first machine supplies 90 A at p.f. 0.8 lagging. If the load power factor is 0.707 lagging and total load is 1600 kW, determine the excitation of second alternator.

Solution. Terminal voltage of each machine = 6600 V

Phase angle of the load, $\phi = \cos^{-1}(0.707) = 45°$

Reactive power of the load = Load in kW × tan φ

= 1600 × tan 45° = 1600 kVAR

kW supplied by machine-1 $= \frac{\sqrt{3}\, VI_1 \cos\phi_1}{1000} = \frac{\sqrt{3} \times 6600 \times 90 \times 0.8}{1000} = 823$ kW

Reactive power supplied by machine-1

= Load in kW × tan ϕ_1

= 823 × tan ($\cos^{-1}$ 0.8) = 617.2 kVAR

Power supplied by machine-2 = 1600 – 823 = 777 kW

Reactive power supplied by machine-2

= 1600 – 617.2 = 982.8 kVAR

Since power supplied $= \frac{\sqrt{3}\, VI_2 \cos\phi_2}{1000}$

$$\therefore \quad I_2 \cos\phi_2 = \frac{\text{Power supplied (kW)} \times 1000}{\sqrt{3}V} = \frac{777 \times 1000}{\sqrt{3} \times 6600} = 67.97 \text{ A}$$

Since reactive power supplied $= \frac{\sqrt{3}\, VI_2 \sin\phi_2}{1000}$

$$\therefore \quad I_2 \sin\phi_2 = \frac{\text{Reactive power (kVAR)} \times 1000}{\sqrt{3}\, V} = \frac{982.8 \times 1000}{\sqrt{3} \times 6600} = 85.97 \text{ A.}$$

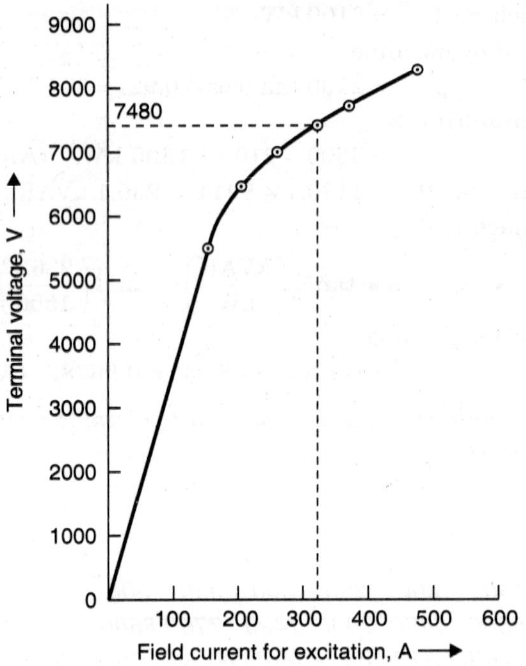

Fig. 10.9

Phase voltage $= \frac{6600}{\sqrt{3}} = 3810$ V

Now, $E_0 = [(V + I_2R\cos\phi_2 + I_2X\sin\phi_2)^2 + (I_2X\cos\phi_2 - I_2R\sin\phi_2)^2]^{1/2}$

$= [(3810 + 67.97 \times 1.05 + 85.97 \times 5)^2 + (67.97 \times 5 - 85.97 \times 1.05)^2]^{1/2}$

$= [(4311.2)^2 + (249.6)^2]^{1/2} = 4318.4$ V

E_0 between lines $= 4318.4 \times \sqrt{3} = 7480$ V

From saturation curve (Fig. 10.9) *excitation required to induce an e.m.f. of 7480 V* = **308 A. (Ans.)**

☞**Example 10.11.** *Two identical 3-phase alternators work in parallel and supply a total load of 1600 kW at 11000 V at a power factor of 0.92. Each machine supplies half the total power. The synchronous reactance of each is 50 Ω/phase and resistance is 2.5 Ω/phase. The field excitation of the first machine is adjusted so that armature current is 50 A lagging. Determine the armature current of the second alternator, the power factor at which each machine is working and generated voltage of the first alternator.*

Solution. Total load supplied = 1600 kW (at 11000 V and 0.92 p.f.)

Synchronous reactance of each machine/phase = 50 Ω

Resistance/phase = 2.5 Ω

Load current at 0.92 p.f. $= \frac{1600 \times 1000}{\sqrt{3} \times 11000 \times 0.92} = 91.3$ A

$\cos\phi = 0.92, \sin\phi = 0.392$

∴ Working component $= 91.3 \times \cos\phi = 91.3 \times 0.92 = 84$ A

Wattless component $= 91.3 \times \sin\phi = 91.3 \times 0.392 = 35.8$ A

Each alternator supplies half of each component when conditions are identical as shown in current diagram of Fig. 10.10.

Total current supplied per machine = $\frac{91.3}{2}$ = 45.65 A.

Since the steam supply of machine 1 is not changed hence the working components of both machines would remain the same at 42 A. But the *reactive/wattless components will be redivided due to change in excitation.*

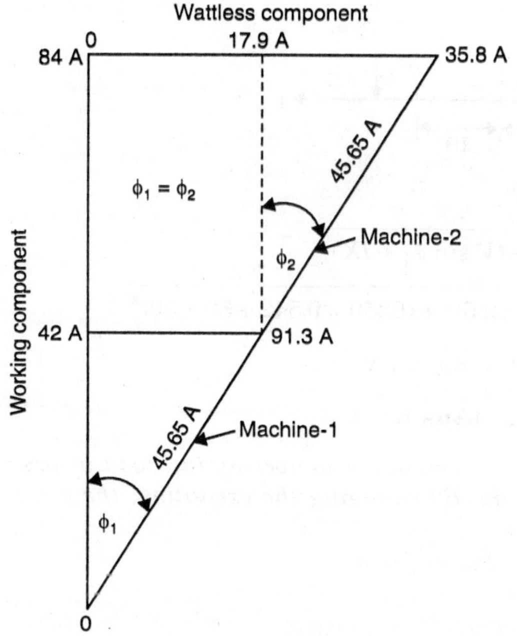

Fig. 10.10

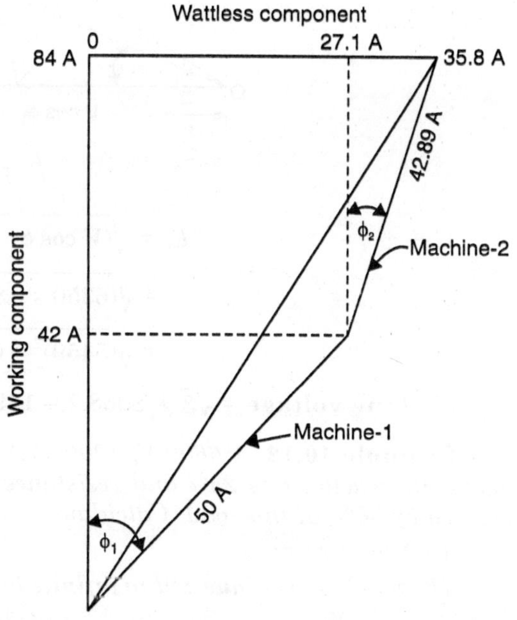

Fig.10.11

The total current taken by machine 1 is changed from 45.65 A to 50 A.

∴ Wattless component for machine 1

$$= \sqrt{50^2 - 42^2} = 27.1 \text{ A}$$

∴ Wattless current delivered by machine 2

$$= 35.8 - 27.1 = 8.7 \text{ A}$$

New current diagram is shown in Fig. 10.11.

(*i*) **New total current for machine-2,**

$$\mathbf{I_2} = \sqrt{(42)^2 + (8.7)^2} = \mathbf{42.89\ A.}\quad \textbf{(Ans.)}$$

(*ii*)
$$\cos \phi_1 = \frac{42}{50} = \mathbf{0.84.\ (Ans.)}$$

$$\cos \phi_2 = \frac{42}{42.89} = \mathbf{0.979.}\quad \textbf{(Ans.)}$$

(*iii*) **Generated voltage of machine-1 :**

$$\text{Terminal voltage/phase} = \frac{11000}{\sqrt{3}} = 6350 \text{ V}$$

$$\cos \phi_1 = 0.84, \sin \phi_1 = 0.542$$

Refer to Fig. 10.12.

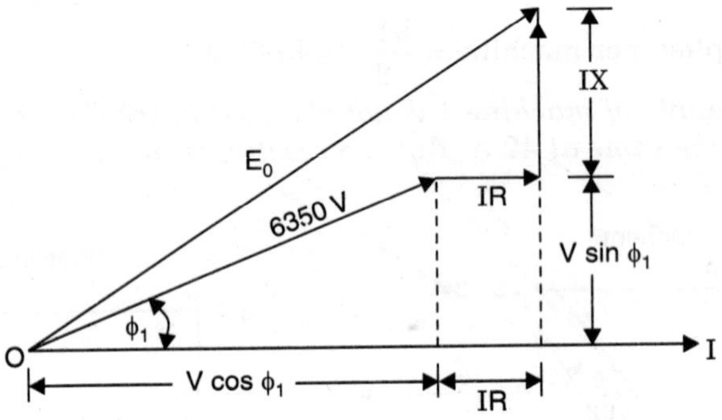

Fig. 10.12

$$E_0 = \sqrt{(V \cos \phi_1 + IR)^2 + (V \sin \phi_1 + IX)^2}$$

$$= \sqrt{(6350 \times 0.84 + 50 \times 2.5)^2 + (6350 \times 0.542 + 50 \times 50)^2}$$

$$= \sqrt{(5459)^2 + (5941.7)^2} = 8068.7 \text{ V}$$

∴ **Line voltage** = $\sqrt{3} \times 8068.7$ = **13975.4 V. (Ans.)**

Example 10.12. *A 6600 V, 1200 kVA, 3-phase alternator is delivering full-load at 0.8 p.f. lagging. Its reactance is 25% and resistance negligible. By changing the excitation, the e.m.f. is increased by 30%, at this load. Calculate :*

(i) New current *(ii) Power factor.*

The machine is connected to infinite bus-bars.

Solution. Voltage/phase $= \dfrac{6600}{\sqrt{3}} = 3810$ V

Power factor of load = 0.8 (lag)

Reactance = 25%

(*i*) **New current :**

Full-load current $= \dfrac{1200 \times 1000}{\sqrt{3} \times 6600} \simeq 105$ A

Reactance $= \dfrac{3810 \times 25}{105 \times 100} = 9.07\ \Omega$

In Fig. 10.13 current vector *OI* is taken along *X*-axis. *OV* represents bus-bar or terminal voltage and is hence constant.

$$I_R = \text{Active component of current } I$$
$$= 105 \times 0.8 = 84 \text{ A}$$
$$I_X = \text{Reactive component of current } I$$
$$= 105 \times 0.6 = 63 \text{ A}$$
$$VL_1 = I_X \,.\, X = 63 \times 9.07 = 571.4 \text{ V}$$
$$L_1M_1 = I_R \,.\, X = 84 \times 9.07 = 761.9 \text{ V}$$
$$E_0 = OM_1 = \sqrt{(V + I_X \,.\, X)^2 + (I_R \,.\, X)^2}$$
$$= \sqrt{(3810 + 571.4)^2 + (761.9)^2} = 4447 \text{ V}$$

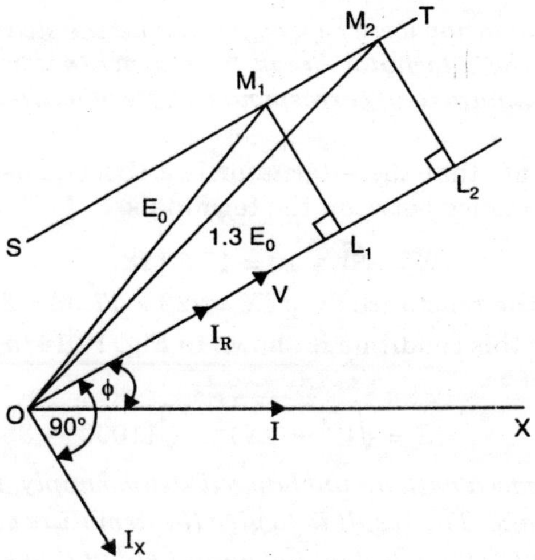

Fig. 10.13

When e.m.f. is increased by 30% e.m.f. becomes equal to

$$4447 \times 1.3 = 5781 \text{ V}$$

The locus of the extremity of E_0 lies on the line ST, which is parallel to OV. Since the kW is unchanged, I_R and hence I_RX will remain the same. It is the only I_X. X component which will be changed. Let OM_2 be the new value of E_0. Then $L_2M_2 = L_1M_1 = I_RX$ as shown before. But the I_X. X component will change. Let I' be the new line current having active component I_R (the same as before) and the new reactive component I'_X. Then

$$I'_X X = VL_2$$

From right angled triangle OL_2M_2,

$$OM_2^{\,2} = OL_2^{\,2} + L_2M_2^{\,2}$$

$$(5781)^2 = (3810 + VL_2)^2 + (761.9)^2$$

or, $$(3810 + VI_2)^2 = (5781)^2 - (761.9)^2$$

or, $$3810 + VL_2 = 5730$$

$\therefore$ $$VL_2 = 1920 \text{ V}$$

or, $$I_X'X = 1920 \quad \text{or} \quad I'_X = \frac{1920}{9.07} = 211.7 \text{ A.}$$

(*i*) ∴ **New line current,** $$I' = \sqrt{(84)^2 + (211.7)^2} = 227.7 \text{ A.}$$

(*ii*) **New power factor :**

Now angle of lag, $$\phi' = \tan^{-1}\frac{211.7}{84} = 68°\ 21'$$

∴ *Power factor,* cos ϕ' = cos 68° 21′ = **0.3689. (Ans.)**

[As a check, the new power $= \sqrt{3} \times 6600 \times 227.7 \times 0.3689 = 960$ kW
which is same as before $= 1200 \times 0.8 = 960$ kW]

☞**Example 10.13.** *A turbo-alternator having a reactance of 10 Ω has an armature current of 220 amperes at unity power factor when running on 11000 V constant frequency bus bars. If the steam admission be unchanged and the e.m.f. raised by 25 per cent determine graphically or otherwise the new value of machine current and power factor.*

If this higher value of excitation were kept constant and the steam supply gradually increased, at what power output would the alternator break from synchronism ? Find also the current and power factor to which this maximum load corresponds. State whether this power factor is lagging or leading.

Solution. Let us assume that the alternator is a three-phase one and that the reactance given is per phase. Hence reactance between the terminals,

$$X = \sqrt{3} \times 10 = 17.32 \ \Omega.$$

At unity power factor the reactance drop $IX = 220 \times 17.32 = 3810$ V.

The vector diagram for this condition is shown in Fig. 10.14 (*a*) and the e.m.f. of the alternator is

$$E = \sqrt{V^2 + (IX)^2} = \sqrt{11000^2 + 3810^2} = 11640 \text{ V}.$$

If *the excitation is increased with an unchanged steam supply, i.e. with the same power input, the power output will not change. The result is to give the armature current a lagging reactive component which exerts a demagnetising effect on the main field and neutralizes the increase in excitation.* The vector diagram is Fig. 10.14 (*b*), which shows that the locus of the e.m.f. vector is dotted line perpendicular to $I_R X$. I_R is the power component of the total current = 220 A as before and I_X is the reactive demagnetising component.

With the increased excitation the e.m.f. is

$$E_1 = 1.25\, E = 1.25 \times 11640 = 14550 \text{ V}$$

From Fig. 10.14 (*b*), we have

$$V + I_X X = \sqrt{E_1^2 - (I_R X)^2}$$

$$11000 + I_X X = \sqrt{14550^2 - 3810^2} = 14040 \text{ V}$$

$$\therefore \quad I_X X = 14040 - 11000 = 3040 \text{ V or } I_X = \frac{3040}{17.32} = 175.5 \text{ A}$$

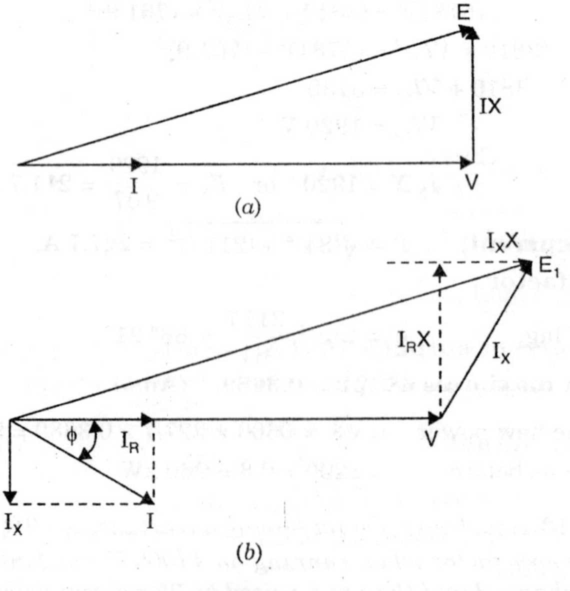

Fig. 10.14

New value of the machine current,

$$I = \sqrt{I_R^2 + I_X^2} = \sqrt{220^2 + 175.5^2} = \mathbf{281.4\ A.} \quad \textbf{(Ans.)}$$

New power factor, $\cos\phi = \frac{I_R}{I} = \frac{220}{281.4} = \mathbf{0.78\ lagging.}$ **(Ans.)**

When the excitation is kept constant and steam supply is increased the locus of the e.m.f. vector is a circle (Fig. 10.15). A series of dotted lines may be drawn as shown, each of which represents the locus of the e.m.f. vector for a certain constant power output at varying excitation. *Maximum power output conditions are reached when the circular e.m.f. locus is tangential to this family of loci, and if the steam supply is increased still further the alternator will break from synchronism.*

Therefore maximum output is being given when

$$I_R X = E_1 \quad \text{and} \quad I_X X = V$$

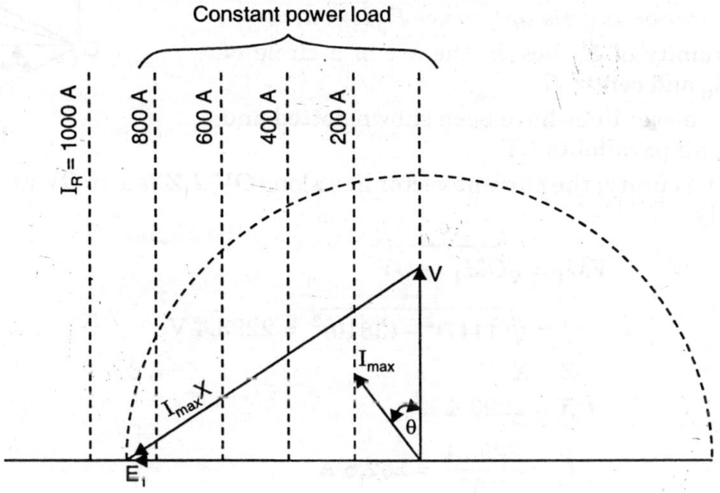

Fig. 10.15

Hence $\quad I_R = \frac{14550}{17.32} = 840$ A

$$I_X = \frac{11000}{17.32} = 635.1 \text{ A}$$

$$I_{max} = \sqrt{I_R^2 + I_X^2} = \sqrt{840^2 + (635.1)^2} = 1053 \text{ A}$$

Phase angle for maximum output, $\theta = \tan^{-1}\frac{I_X}{I_R} = \tan^{-1}\frac{635.1}{840} = 37.1°$

$\therefore$ **Maximum current output = 1053 A. (Ans.)**

Power factor at maximum output = $\cos\theta$ = **0.797 leading. (Ans.)**

Maximum power output $= \frac{\sqrt{3} \times 11000 \times 1053 \times 0.797}{1000}$ kW = **15989 kW. (Ans.)**

☞**Example 10.14.** *A 6600 V, 1200 kVA alternator has a reactance of 25% and is delivering full-load at 0.8 p.f. lagging. It is connected to constant frequency bus-bar. If the steam supply is gradually increased, calculate :*

(*i*) *At what output will the power factor become unity.*

(*ii*) *The maximum load which it can supply without dropping out of synchronism and the corresponding power factor.*

Solution. Phase voltage $= \frac{6600}{\sqrt{3}} = 3810$ V

Reactance $= 25\%$

In example 10.12, we have found that,

$I = 105$ A, $X = 9.07\ \Omega$

$E_0 = 4447$ V, $I_R = 84$ A, $I_X = 63$ A

and $IX = 105 \times 9.07 = 952.4$ V

By using this data, vector diagram of Fig. 10.16 can be constructed.

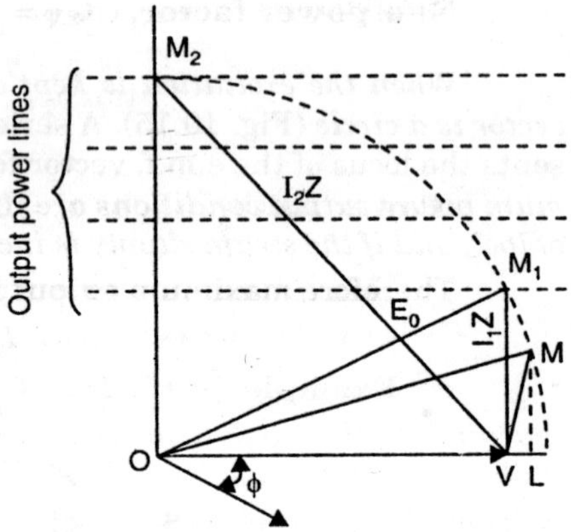

Fig. 10.16

- *Since excitation is constant, hence E_0 is constant.*
- The extremity of E_0 lies on the arc of a circle of radius E_0 and centre *O*.
- Constant power lines have been shown dotted and they are all parallel to *OV*.
- When p.f. is unity, the current vector lies along *OV*, I_1Z is $\perp$ to *OV* and cuts the arc at M_1. Obviously,

$$VM_1 = \sqrt{OM_1^2 - OV^2}$$

$$= \sqrt{(4447)^2 - (3810)^2} = 2293.4 \text{ V}$$

Now, $Z = X$

$\therefore$ $I_1X = 2293.4$

$\therefore$ $I_1 = \frac{2293.4}{9.07} = 252.8$ A

(*i*) $\therefore$ **Power output at unity p.f.**

$$= \frac{\sqrt{3} \times 6600 \times 252.8}{1000} = \mathbf{2889.9\ kW.}\quad \textbf{(Ans.)}$$

(*ii*) A vector *OM* moves *upwards* along the arc, *output power goes on increasing i.e.*, point *M* shifts on to a higher output power line. Maximum output power is reached when *OM* reaches the position OM_2, where it is vertical to *OV*. The output power line passing through M_2 represents the maximum output for that excitation. If *OM* is further rotated, then point M_2 shifts down to a lower point line *i.e.*, power is decreased. Hence $M_2V = I_2Z$ where I_2 is the new current corresponding to maximum output.

From triangle OM_2V, we have

$$M_2V = \sqrt{(OV)^2 + (OM_2)^2} = \sqrt{(3810)^2 + (4447)^2} = 5856 \text{ V}$$

$\therefore$ $I_2Z = 5856$

$$I_2 = \frac{5856}{9.07} = 645.6 \text{ A}$$

Let, I_{2R} = Power component of I_2

I_{2X} = Wattless component of I_2

Then, $I_{2R} X = OM_2 = 4447$ V

$\therefore \quad I_{2R} = \frac{4447}{9.07} = 490.3$ A

Similarly, $I_{2X} = \frac{3810}{9.07} = 420$ A

$$\tan \phi_2 = \frac{420}{490.3} = 0.8566$$

$\therefore \quad \phi_2 = 40.58°$

Power factor, $\cos \phi_2 =$ **0.759. (Ans.)**

$\therefore$ **Maximum power output** $= \frac{\sqrt{3} \times 6600 \times 645.6 \times 0.759}{1000} =$ **5601.57 kW. (Ans.)**

☞**Example 10.15.** *Two 3-phase synchronous mechanically coupled generators operate in parallel on the same load. Determine the kW output and power factor of each machine under the following conditions : Synchronous impedance of each generator = 0.2 + j2 Ω/phase ; Equivalent impedance of the load = 3 + j4 Ω/phase ; Induced e.m.f./phase = 2000 + j0 volts for machine-1 and 2200 + j100 for machine-2.*

Solution. Synchronous impedance of each generator

$$= 0.2 + j2 \ \Omega/\text{phase}$$

Equivalent impedance of load

$$= 3 + j4 \ \Omega/\text{phase}$$

Induced e.m.f./phase for machine-1,

$$E_1 = 2000 + j0 \text{ volts}$$

Induced e.m.f./phase for machine-2,

$$E_2 = 2200 + j100 \text{ volts}$$

kW output and p.f. of each machine :

Current of machine-1,

$$I_1 = \frac{E_1 - V}{0.2 + j2}$$

or, $E_1 - V = I_1(0.2 + j2)$

or, $2000 - V = I_1(0.2 + j2)$ ...(*i*)

Similarly, $E_2 - V = I_2(0.2 + j2)$

or, $(2200 + j100) - V = I_2 (0.2 + j2)$ ...(*ii*)

and, $V = (I_1 + I_2)(3 + j4)$ ...(*iii*)

Solving from above, we get

$$I_1 = 68.2 - j102.5$$

and, $I_2 = 127 - j196.4$

and, $I = I_1 + I_2 = 195.2 - j299$

Now, $V = IZ = (195.2 - j299)(3 + j4)$

$= [195.2 \times 3 - j299 \times 3 + 195.2 \times j4 - j^2(299 \times 4)]$

$= 585.6 - j897 + j780.8 + 1196 = 1781 - j116$

Output of machine-1 :

Output $= VI_1$ = Dot product of V and $I_1 = (1781 - j116)(68.2 - j102.5)$

$= 1781 \times 68.2 + 116 \times 102.5$

$= 133.35$ kW per phase

$= 3 \times 133.35 =$ **400 kW** (*for 3 phases*). **(Ans.)**

Now, $\tan^{-1} \dfrac{102.5}{68.2} = 56.36°$ or $56° 22'$

$\tan^{-1} \dfrac{116}{1781} = 3.72°$ or $3° 43'$

∴ **For machine-1, power factor**

= cos φ = cos (56° 22′ – 3°43′) = 0.6068. (Ans.)

For machine-2 :

Output = Dot product of V and I_2

$= (1781 - j116)(127 - j196.4) = 1781 \times 127 + 116 \times 196.4$

$= 248.97$ kW per phase $= 3 \times 248.97$

= **746.91 kW** (*for 3 phases*). **(Ans.)**

$\tan^{-1} \dfrac{196.4}{127} = 57.11°$ or $57° 6'$

Power factor, $\cos \phi = \cos (57° 6' - 3° 43') =$ **0.596. (Ans.)**

Example 10.16. *Two exactly similar turbo-alternators are rated at 25 MW each. They are running in parallel. The speed-load characteristics of the driving turbines are such that the frequency of alternator 1 drops uniformly from 50-Hz on no-load to 48-Hz on full-load, and that of alternator 2 from 50-Hz to 48.5-Hz. How will the two machines share a load of 30 MW ?*

Solution. Rating of each alternator = 25 MW

Change of frequency of alternator 1 = 50 Hz to 48 Hz

Change of frequency of alternator 2 = 50 Hz to 48.5 Hz

Total load to be shared = 30 MW

Load shared by each alternator :

Fig. 10.17 shows the speed-load characteristics for driving turbines of alternators 1 and 2.

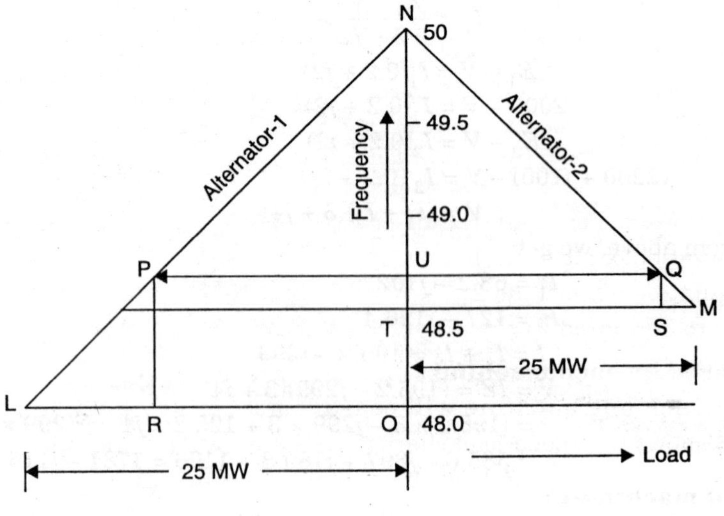

Fig. 10.17

- NL represents the speed-load characteristic for turbine driving alternator-1.
- NM represents the speed-load characteristic for turbine driving alternator-2.
  Since the machines are running in parallel, therefore their frequency *must* be same.
- Let PQ be drawn through U so that total load supplied is 30 MW (or 30000 kW).

Load supplied by alternator-1

$$= UP = OL - LR$$

$$= 25 - \frac{PR}{ON} \times OL$$

$$= 25 - \frac{OU}{50 - 48} \times 25$$

$$\left[\begin{array}{l} \text{From similar } \Delta\text{s } PLR \text{ and } NLO, \\ \dfrac{PR}{LR} = \dfrac{ON}{OL} \\ \therefore \quad LR = \dfrac{PR}{ON} \times OL \end{array}\right]$$

i.e., $$UP = 25 - 12.5\ OU \quad ...(i)$$

Load supplied by alternator-2

$$= UQ = TM - MS = 25 - \frac{QS}{NT} \times MT = 25 - \frac{OU - 0.5}{50 - 48.5} \times 25$$

i.e., $$UQ = 25 - \frac{25}{1.5}(OU - 0.5) \quad ...(ii)$$

Total load supplied $= UP + UQ = 30$

i.e., $$(25 - 12.5OU) + \left[25 - \frac{25}{1.5}(OU - 0.5)\right] = 30$$

$$25 - 12.5OU + 25 - 16.67OU + 8.33 = 30$$

$$29.17OU = 28.33$$

$$\therefore \quad OU = 0.9712$$

Load supplied/shared by alternator-1

$= 25 - 12.5 \times 0.9712 =$ **12.86 MW. (Ans.)**

Load supplied/shared by alternator-2

$= 25 - \frac{25}{1.5}(0.9712 - 0.5) =$ **17.14 MW. (Ans.)**

☞**Example 10.17.** *Two 800 kW alternators operate in parallel. The speed regulation of one set is 100% to 103% from full-load to no-load and that of the other is 100% to 104%. How will the two alternators share a load of 1200 kW and at what load will one machine cease to supply any portion of the load ?*

Solution. Rating of each alternator = 800 kW

Speed regulation of machine-1 = 100% to 103%

Speed regulation of machine-2 = 100% to 104%

Total load to be shared = 1200 kW

Load shared by each machine :

Load at which one machine will cease to supply any portion of the load :

Fig. 10.18 shows the speed-load characteristics for driving turbines of machines 1 and 2.

- NL represents the speed-load characteristic for turbine driving machine-1.
- NM represents the speed-load characteristic for turbine driving machine-2.
  Since the machines are running in parallel, therefore, their frequency or speed *must* be same.

- Let *PQ* be drawn through *U* so that total load supplied is 1200 kW.

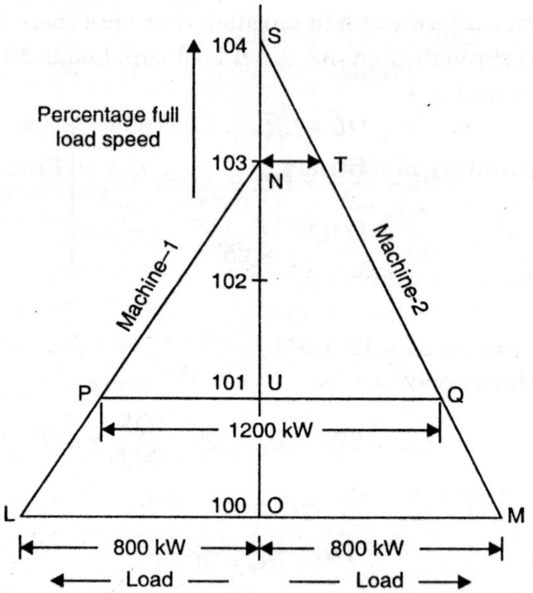

Fig. 10.18

Machine-1 :

Load supplied by machine, 1 = *UP*

From Δs *NPU* and *NLO*, we get

$$\frac{UP}{OL} = \frac{NU}{NO}$$

$$\therefore \quad UP = \frac{NU}{NO} \times OL = \frac{3 - OU}{103 - 100} \times 800 = \frac{800}{3}(3 - OU)$$

$$\therefore \quad UP = \frac{800}{3}(3 - OU) \quad \text{...(i)}$$

Machine-2 :

Load supplied by machine 2 = *UQ*

From Δs *SUQ* and *SOM*, we get

$$\frac{UQ}{OM} = \frac{SU}{SO}$$

$$\therefore \quad UQ = \frac{SU}{SO} \times OM = \frac{(4 - OU)}{104 - 100} \times 800$$

$$= \frac{800}{4}(4 - OU) = 200(4 - OU)$$

$$\therefore \quad UQ = 200(4 - OU) \quad \text{...(ii)}$$

Since *UP* + *UQ* = *PQ*, total load supplied

$$= 1200 \text{ kW}$$

$$\therefore \quad \frac{800}{3}(3 - OU) + 200(4 - OU) = 1200$$

$$2400 - 800OU + 2400 - 600U = 3600$$

$$1400OU = 1200$$

$$\therefore \quad OU = \frac{1200}{1400} = 0.857$$

Load shared/supplied by machine-1

$$= \frac{800}{3}(3 - 0.857) = \mathbf{571.4\ kW.} \quad \textbf{(Ans.)}$$

Load shared/supplied by machine-2

$$= 200(4 - 0.857) = \mathbf{628.6\ kW.} \quad \textbf{(Ans.)}$$

Machine-1 will cease to supply any load when the line PQ is shifted to point N. In this case only machine-2 will supply load equal to NT which is found as follows :

From Δs SNT and SOM, we get

$$\frac{NT}{OM} = \frac{SN}{SO}$$

$$\therefore \quad NT = \frac{SN}{SO} \times OM = \frac{104 - 103}{104 - 100} \times 800 = \frac{1}{4} \times 800 = \mathbf{200\ kW.} \quad \textbf{(Ans.)}$$

10.4. INTERCONNECTED STATIONS

10.4.1. General Aspects

In an interconnected system when it is required to transfer large loads between stations, it is essential to load the lines in such a fashion that they are not overloaded and the loads are shared almost equally.

Consider two generating stations S-1 and S-2 supplying a receiving station (RS) through Line-1 and Line-2, in addition to their local loads. The generating stations are also connected by an interconnector as shown in Fig. 10.19. So that the Line-1 and Line-2 deliver equal power and the system operates at the same terminal voltage, it is essential that the active (in-phase) components of the line currents I_1 and I_2 must be equal, the sum of active components of I_1 and I_2 corresponding to total power received at RS. In order that the total current in lines 1 and 2 be a minimum, the currents I_1 and I_2 should have not only equal active components but these currents must have same phase as well. This is achieved, in practice, by *installing regulating equipment (RE) at the sending end of each transmission line and in the interconnector, which take care of impedance drops in the transmission lines and interconnector.*

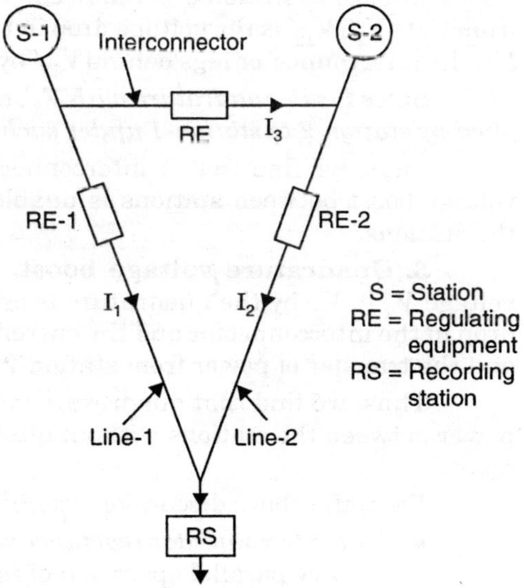

Fig. 10.19

10.4.2. Load Sharing

Whereas the parallel operation and load sharing between the two alternators is made possible only because of internal reactances of the alternates themselves, in case of interconnected stations it is the reactance of the interconnector which allows for angular displacement between two stations enabling power to flow from one machine to the other in consonance with the operating conditions. The incorporation of an impedance, however, causes unnecessary power loss.

1. Angular displacement between stations. The transfer of power by angular displacement is shown in Fig. 10.20. V_1 and V_2 are the respective voltages of stations 1 and 2, equal in magnitude but displaced from each other by an angle θ. The resultant of voltages V_1 and V_2, V_{12} acts along the interconnector and causes a current I to flow through it. The current I lags behind the voltage V_{12} by almost 90° since most of the interconnectors are designed to have reactance, the resistance being negligibly small. The current I, thus, is nearly in phase with voltage V_2 and the transfer of power takes place. No power transfer takes place if V_1 and V_2 are in phase. Thus power transfer from station-2 to station-1 will take place only if phasor V_2 is advanced by an angle θ (depending upon the magnitude of the current to be carried by the interconnector and its impedance).

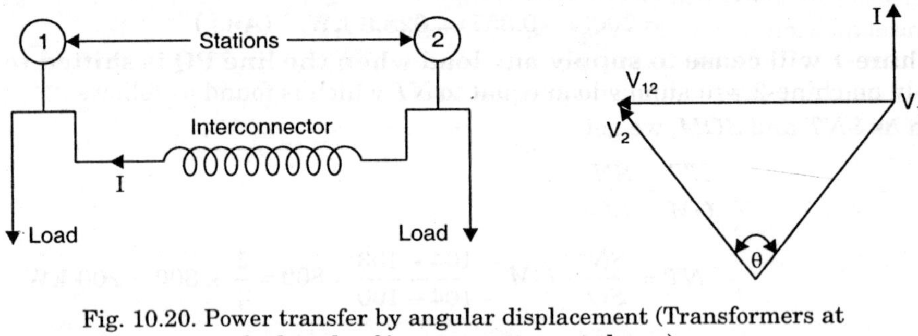

Fig. 10.20. Power transfer by angular displacement (Transformers at both ends of interconnector not shown).

2. In-phase voltage boost. Refer to Fig. 10.21. Let us assume that station-2 has surplus available capacity and so power flow in a direction from station-2 to station-1 is required. The voltage of station-2 is boosted from V_2 to V_2', V_{12}' is the voltage drop in the interconnector and the current I in the interconnector lags behind V_{12}' by almost 90° (in case it is reactive).

Since *I is in quadrature with V_2', therefore no real power can be supplied by station-2 to station-1 under such a condition.*

Thus, we find that if interconnector has only *reactance*, in phase voltage boost between stations is unable to transfer real power between the stations.

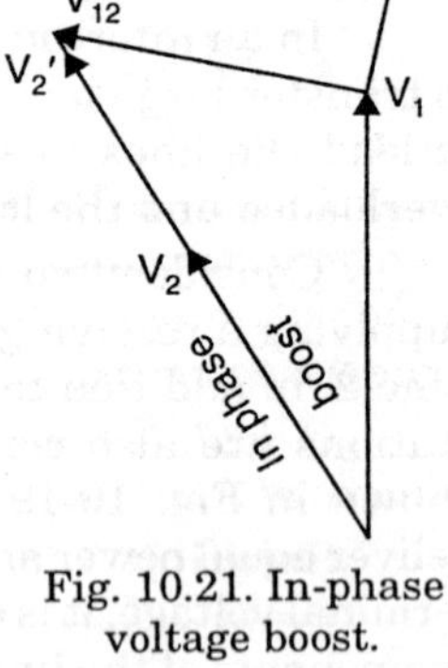

Fig. 10.21. In-phase voltage boost.

3. Quadrature voltage boost. Fig. 10.22 shows the increase of voltage V_2 to $V_{2'}$ by the quadrature boost $V_2V_{2'}$. Here $V_{2'}V_2$ is the voltage drop in the interconnector and the current I is very nearly in phase with V_2 and the transfer of power from station-2 of station-1 takes place.

Thus, we find that quadrature voltage enables the transfer of real power between the stations without alterations in rotor angular displacement.

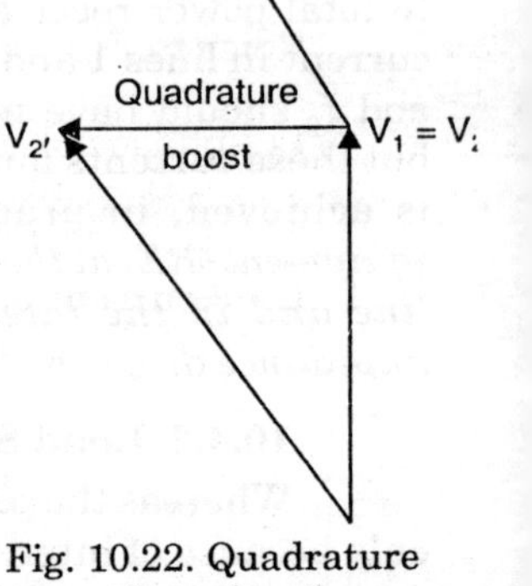

Fig. 10.22. Quadrature voltage boost.

From the above discussion, the *following conclusions may be made :*

- The *interconnector reactance* which is instrumental in the satisfactory parallel operation of interconnected stations should not be *too great* otherwise a large phase displacement will be needed for a comparatively small current to flow through the interconnector. Thus, when a slightly large amount of current is required to flow (for the transfer of power between stations), the station may fall out of step as a result of large phase displacement between them.

- In-phase voltage boost helps little in the transfer of real power as it materialises transfer of wattless currents.
- The quadrature voltage boost causes the flow of real power, between stations. This method *increases the inherent stability* and sudden load changes etc. can also be dealt with easily.

10.4.3. Power Limit of Interconnectors

Let us assume that the interconnected stations 1 and 2 are of large capacity (as compared to the loads to be transferred between them) and are coupled by an interconnector which has reactance only. The *maximum power will be transmitted by the interconnector when the voltage V_1 and V_2 displaced by 90°* (see Fig. 10.23). Let the station-2 transmit power to station-1. Under the conditions of maximum power in the interconnector, current I lags the voltage V_2 by 45°, and leads the voltage V_1 by 45°, the former station being at the sending end and the latter being at the receiving end.

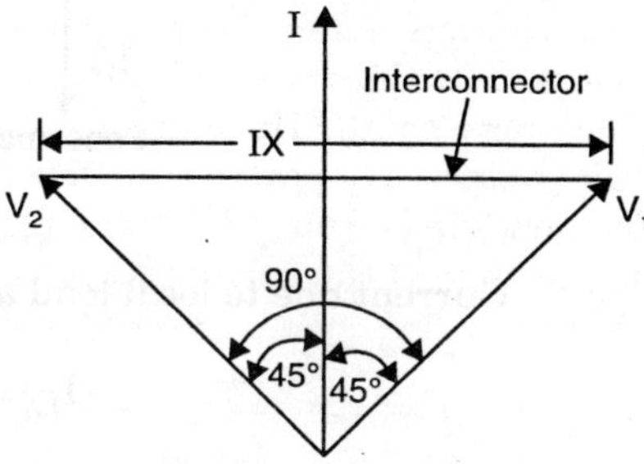

Fig. 10.23. Condition of maximum power transfer.

Synchronous capacity of interconnector

or maximum power transmitted = $VI \cos 45°$

$$= V \times \frac{\sqrt{2}\,V}{X} \times \frac{1}{\sqrt{2}} \qquad \left[\because\ I = \frac{V_{12}}{X} = \frac{\sqrt{V^2 + V^2}}{X} = \frac{\sqrt{2}\,V}{X}\right]$$

$$= \frac{V^2}{X} \text{ watts}$$

$$= \frac{V^2}{1000} \text{ kW}$$

- **Synchronous capacity** *of an interconnector is defined as the change of kW transmitted per radian change of angular displacement of the two voltages of the two stations.*

Example 10.18. *Two 3-phase generating stations A and B are linked through a 33 kV interconnector having a resistance of 0.84 Ω and a reactance of 4.2 Ω per phase. At station-A the load on the generators is 72 MW at a p.f. of 0.8 lagging and the local load taken by the consumers connected to the A bus-bars is 48 MW at a p.f. of 0.707 lagging.*

Calculate the kW received from station-A by station-B, its p.f. and phase difference between the voltages of A and B.

Solution. *Given :* Interconnector voltage = 33 kV

Interconnector impedance = $0.84 + j4.2$

Load on the generator at station-A = 72 MW at 0.8 p.f. lagging

Load connected to A bus-bars = 48 MW at 0.707 p.f. lagging.

Refer to Fig. 10.24.

Let phase voltage at $A \left(= \frac{33000}{\sqrt{3}} \simeq 19050 \text{ V} \right)$ be taken as reference phasor, so that

$$V_A = 19050\,(1 + j0) \text{ volts}$$

Load current on generators at station-A,

$$I_A = \frac{72 \times 10^6}{\sqrt{3} \times 33 \times 10^3 \times 0.8} = 1574.6 \text{ A}$$

$$= 1574.6\,(0.8 - j0.6) = (1259.7 - j944.8) \text{ A}$$

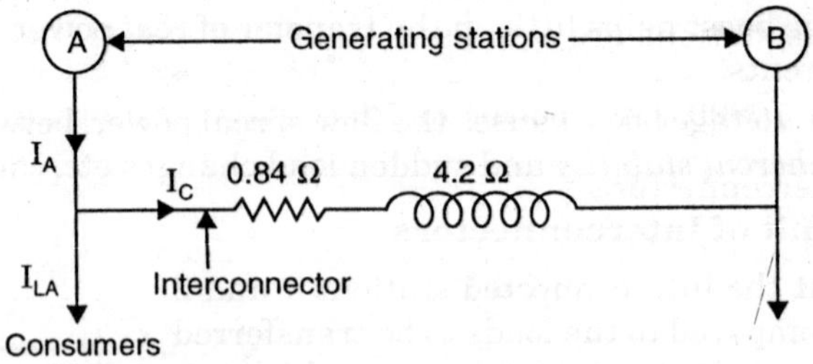

Fig. 10.24

Current due to local load at *A*,

$$I_{LA} = \frac{48 \times 10^6}{\sqrt{3} \times 33 \times 10^3 \times 0.707} = 1187.8 \text{ A}$$

$$= 1187.8\,(0.707 - j0.707) = (839.8 - j839.8)$$

Current transferred through the interconnector,

$$I_C = I_A - I_{LA}$$

$$= (1259.7 - j944.8) - (839.8 - j839.8)$$

$$= 419.9 - j105 = 432.8 \angle -14.04°$$

Voltage drop in the interconnector

$$= (419.9 - j105)(0.84 + j4.2)$$

$$= 352.72 + j1763.58 - j88.2 + 441 = (793.72 + j\,1675.38) \text{ V}$$

Voltage at generating station-*B*

$$V_B = 19050(1 + j0) - (793.72 + j1675.38)$$

$$= 18256.28 - j1675.38 = 18333 \angle -5.24°$$

Phase difference between V_B and I_C

$$= -14.04° - (-5.24°) = -8.8°$$

i.e. 8.8° with I_C *lagging* V_B

Power factor of the current received by station B

= cos 8.8° = **0.988 (lagging). (Ans.)**

Power received by station B

$$= 3V_B I_C \cos 8.8° = 3 \times 18333 \times 432.8 \times 0.988 \times 10^{-6} \text{ MW}$$

= **23.5 MW. (Ans.)**

Phase angle between voltages of station-A and station-B

= **5.24° with V_A lagging. (Ans.)**

Example 10.19. *Two generating stations A and B are linked by an interconnector cable and reactor having a combined reactance of 4.8 Ω per phase with negligible resistance. Station-A and station-B supply in their own areas loads of 12000 kW at a lagging p.f. of 0.8 and 9600 kW at a p.f. of 0.9 lagging, each at a bus-bar voltage of 11 kV, respectively. The station loads are equalised by the flow of power in the interconnector cable.*

Calculate the power factors of the stations.

Solution. *Given :* Interconnector reactance = 3.36 Ω

Load supplied by station-*A* in its area = 12000 kW, 0.8 p.f. lagging

Load supplied by station *B* in its area = 9600 kW, 0.9 p.f. lagging

Bus-bar voltage of each station = 11 kV

As the loads on two stations are equalised, therefore each station will supply half the total load *i.e.* $\dfrac{12000 + 9600}{2} = 10800$ kW ; power of 1200 kW will be transferred from station B to station A through the interconnector.

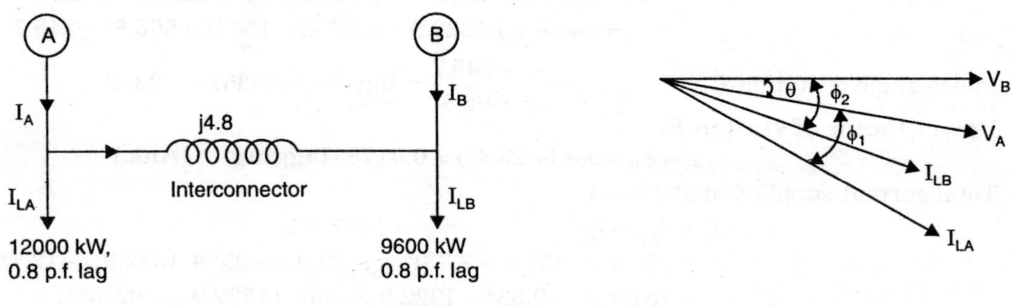

(*a*) (*b*)

Fig. 10.25

Let phase voltage of station $B = \left(= \dfrac{11000}{\sqrt{3}} = 6350 \text{ V} \right)$ be taken as reference vector.

Then $V_A = 6350(1 + j0)$ volts $= 6350 \angle 0°$ V.

The phase voltage of station-A will be equal to that of station-B in magnitude but displaced by angle θ.

$$\therefore \qquad V_A = 6350 \angle -\theta \ V$$

The load currents I_{LA} and I_{LB} will lag behind their respective booster voltages by angles $\phi_1(= \cos^{-1} 0.8 = 36.8°)$ and $\phi_2(= \cos^{-1} 0.9 = 25.8°)$, as shown in Fig. 10.25 (*b*).

$$I_{LA} = \frac{12000 \times 10^3}{\sqrt{3} \times 11000 \times 0.8} = 787.3 \angle -(36.8 + \theta)° \text{ A}$$

$$I_{LB} = \frac{9600 \times 10^3}{\sqrt{3} \times 11000 \times 0.9} = 559.8 \angle -25.8°$$

Current in the interconnector,

$$I_C = \frac{V_B - V_A}{4.8 \angle 90°} = \frac{6350 - 6350 \angle -\theta}{4.8 \angle 90°}$$
$$= 1322.9 \angle -90° - 1332.9 \angle -(\theta + 90)°$$

Total current supplied by station-B,

$$I_B = I_{LB} + I_C$$
$$= 559.8 \angle -25.8° + 1322.9 \angle -90° - 1322.9 \angle -(\theta + 90)°$$
$$= 559.8(0.9 - j0.436) - j1322.9 - 1322.9(-\sin\theta - j\cos\theta)$$
$$= 503.8 - j244.1 - j1322.9 + 1322.9 \sin\theta + j1322.9 \cos\theta$$
$$= (503.8 + 1322.9 \sin\theta) + j(1322.9 \cos\theta - 1567)$$

Real component of current $= 503.8 + 1322.9 \sin\theta$

Power generated at station B,

$$\sqrt{3} \times 11000\,(503.8 + 1322.9 \sin\theta) = 10800 \times 10^3$$

or,
$$503.8 + 1322.9 \sin\theta = \frac{10800 \times 10^3}{\sqrt{3} \times 11000} = 566.8 \text{ A}$$

or, $$\sin\theta = \frac{566.8 - 503.8}{1322.9} = 0.04762$$

$\therefore$ $$\theta = \sin^{-1}(0.04762) = 2.73°$$

Total current supplied at station-B,

$$I_B = (503.8 + 1322.9 \times 0.04762) + j(1322.9 \times \cos 2.73° - 1567)$$
$$= 566.8 + j(1322.9 \times 0.9988 - 1567) = 566.8 - j245.7$$

Phase angle of station-B, $\phi_B = \tan^{-1}\dfrac{-245.7}{566.8} = \tan^{-1}(-0.4335) = -23.4°$

Power factor of station-B,

$$\cos\phi_B = \cos(-23.4°) = \mathbf{0.9178\ (lagging).}\quad \textbf{(Ans.)}$$

Total current supplied at station-A,

$$I_A = I_{LA} - I_C$$
$$= 787.3 \angle -(36.8 + 2.73)° - 1322.9 \angle -90° + 1322.9 \angle -(2.73 + 90)°$$
$$= 787.3 \angle -39.53° - 1322.9 \angle -90° + 1322.9 \angle -92.73°$$
$$= 787.3(0.7713 - j0.6365) + j322.9 + 1322.9(-0.04763 - j0.9988)$$
$$= 607.2 - j501.1 + j1322.9 - 63.0 - j1321.3$$
$$= 544.2 - j499.5$$

Phase angle of power station-A, $\phi_A = \tan^{-1}\left(\dfrac{-499.5}{544.2}\right) = -42.5°$

Power factor of station-A, $\cos\phi_A = \cos(-42.5°) = \mathbf{0.737\ (lagging).}$ **(Ans.)**

Example 10.20. *Two power stations A and B operate in parallel and are interconnected by a short transmission line. The station capacities are 15 MW and 7.5 MW respectively and the generating sets have uniform speed regulation (from no-load to full-load) of 2 per cent and 4 per cent respectively. When the load of each station bus-bar is 9 MW, calculate :*

(i) Load on the interconnector, and

(ii) Output of each station. **(P.U.)**

Solution. *Given :* Capacity of station-A = 15 MW

Capacity of station-B = 7.5 MW

Speed regulation of generator set A = 2%

Speed regulation of generator set B = 4%

Load of each station busbar = 9 MW

(*i*) **Load on the interconnector :**

Let, x = Load on the interconnector.

Output of power station-A = $(9 + x)$ MW

Output of power station-B = $(9 - x)$ MW

%age drop in the generators of power station-A

$$= \left(\frac{9 + x}{15}\right) \times 2$$

%age drop in the generators of power station-B

$$= \left(\frac{9 - x}{7.5}\right) \times 4$$

The speed drop in the generators of both the power stations A and B should be same,

i.e., $$\left(\frac{9 + x}{15}\right) \times 2 = \left(\frac{9 - x}{7.5}\right) \times 4 \quad \text{or} \quad \mathbf{x = 5.4\ MW.}\ \textbf{(Ans.)}$$

(*ii*) **Output of each station :**

Output of power station-A = 9 + 5.4 = **14.4 MW.** **(Ans.)**

Output of power station-B = 9 – 5.4 = **3.6 MW.** **(Ans.)**

Example 10.21. *Two 10 kV generating stations 1 and 2 generating 20 MW each at unity p.f. are interconnected by a reactor with a voltage drop of 30 per unit for transfer rate of 12 MW. The loads connected to the stations 1 and 2 are 24 MW and 16 MW each, at unity p.f., respectively. Calculate the phase angle between the bus-bar voltages of the two generating stations.* **(N.U)**

Solution. Voltage of each of the generating stations 1 and 2 = 10 kV

Power generated by station-1 = 20 MW

Power generated by station-2 = 20 MW

Voltage drop in the interconnected (for transfer rate of 12 MW) = 30%

Load connected to station-1 = 24 MW

Load connected to station-2 = 16 MW

Under the condition shown in Fig. 10.26 (*a*) a power transfer of 4 MW is taking place from station-2 to station-1. Hence the percentage voltage drop for transfer of power of 4 MW at unity p.f.

$$= \frac{4}{12} \times 30 = 10\%$$

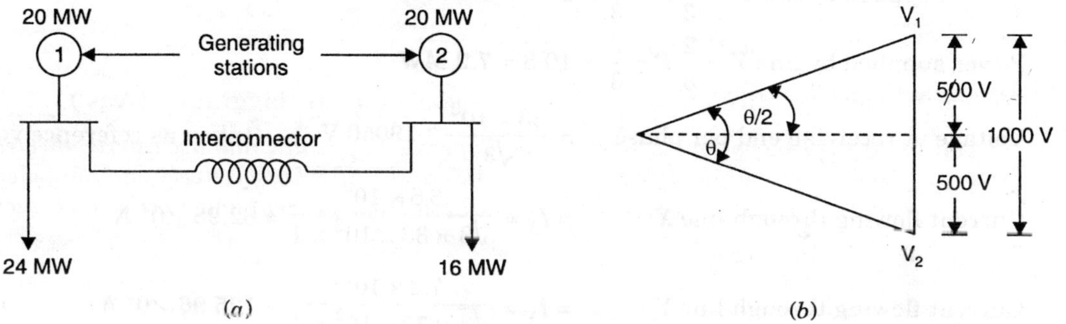

Fig. 10.26

$$\text{Voltage drop in the interconnector} = \frac{10}{100} \times 10000 = 1000 \text{ V}$$

Let the displacement between voltages V_1 and V_2 be θ. In Fig. 10.26 (*b*) $V_1 = V_2 = 10000$ V. Phasor V_2V_1 is the voltage drop in the interconnector.

From this figure, $\sin \frac{\theta}{2} = \frac{500}{10000} = 0.05$ or $\theta = 2(\sin^{-1} 0.05) = 5.73°$

i.e. V_1 *and* V_2 *are displaced by* **5.73°**, V_1 *leading* V_2. **(Ans.)**

Example 10.22. *From a substation, a load of 10.8 MW is supplied via two 3-φ transmission circuits X and Y having impedances per phase of (2.4 + j9.6) ohms and (4.8 + j4.8) ohms respectively. In-phase and quadrature boosters are used at the sending end of one line to deliver* $\frac{1}{3}$ *rd and* $\frac{2}{3}$ *rd of the total power at unity power factor through circuits X and Y respectively. Determine the necessary booster voltages.*

Solution. Total load to be supplied = 10.8 MW

Impedance of circuit $X = (2.4 + j9.6)\ \Omega$

Impedance of circuit $Y = (4.8 + j4.8)\ \Omega$

Booster voltages :

Refer to Fig. 10.27. *RE* is any regulating equipment in line *Y* which can provide in-phase and/or quadrature voltage boost required.

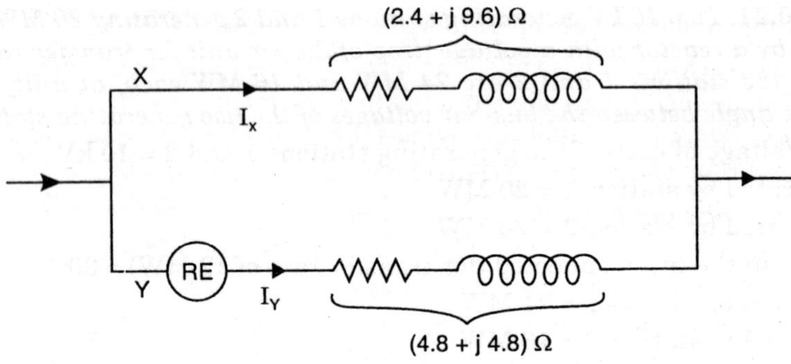

Fig. 10.27

Total load, $P = 10.8$ MW

Power supplied by line $X = \frac{1}{3} P = \frac{1}{3} \times 10.8 = 3.6$ MW

Power supplied by line $Y = \frac{2}{3} P = \frac{2}{3} \times 10.8 = 7.2$ MW

Voltage at receiving end per phase $= \frac{33 \times 10^3}{\sqrt{3}} \simeq 19050$ V (taken as reference vector)

Current flowing through line X $= I_X = \frac{3.6 \times 10^6}{\sqrt{3} \times 33 \times 10^3 \times 1} = 62.98 \angle 0°$ A

Current flowing through line Y $= I_Y = \frac{7.2 \times 10^6}{\sqrt{3} \times 33 \times 10^3 \times 1} = 125.96 \angle 0°$ A

Impedance drop in line X $= I_X Z_X = 62.98(2.4 + j9.3) = 151.15 + j604.61$

Impedance drop in line Y $= I_Y Z_Y = 125.96(4.8 + j4.8) = 604.61 + j604.61$

∴ *RE* (Regulating equipment) must provide [assuming that sending end voltage is kept as 19050 + (151.15 + *j*604. 61) = 19201.15 + *j*604.61 V/phase] :

(*i*) *In-phase voltage boost* = 604.61 – 151.15 = **453.46 V/phase.** **(Ans.)**

(*ii*) *Quadrature voltage boost* = 604.61 – 604.61 = **0 V/phase.** **(Ans.)**

10.4.4. Voltage, Frequency and Load Control

Voltage Control. The following methods may be used for voltage control :

1. *Injection of reactive power :*
   (*i*) Shunt capacitors and reactors.
   (*ii*) Series capacitors.
   (*iii*) Synchronous compensator.

This is the most fundamental method and is used only when transformers alone will not suffice.

2. *Tap changing transformers*. This is the most widely used method of voltage control.

3. *Booster transformers.* Boosters are oftenly used in distribution feeders when the cost of tap-changing transformers becomes excessive.

Frequency and Load Control. In an interconnected system where several power stations and generators run in parallel, it is imperative that the system frequency remains constant. In order to maintain frequency constant, both the governor action as well as supplementary controls are required for stable and economic operation. The regulation may be *manual* or *automatic*.

Manual Regulation : In case of a single generator supplying a group of isolated loads, the supply frequency can be easily controlled by adjusting the setting of the turbine governor. But if two or more generators operate in parallel serving a common load, the frequency control becomes more complex/difficult and then the following control methods are employed :

1. *Flat-frequency control.* Refer to Fig. 10.28. Two generators X and Y are operating in parallel and interconnected by a *tie line.* Here only one generator (X) is used for maintaining the frequency constant ; it absorbs all variations in the system loads while the other generator Y supplies constant load. This type of regulation/control is known as flat frequency regulation at station-X.

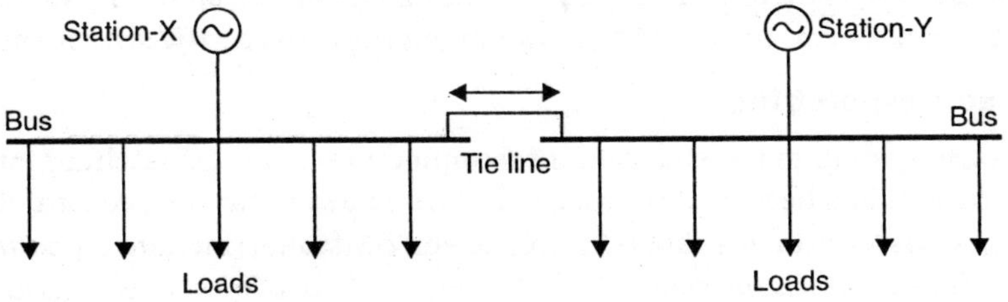

Fig. 10.28. Two generators in parallel.

- The **advantage** of this method is that the *new and more efficient generators can be made to carry the base load and less efficient generator(s) can be used for meeting the load variations.*
- This method has the main **disadvantage** that it *results in random variations in the line powers.*

2. *Parallel-frequency control.* In this method both generators/stations X and Y are regulated to maintain constant frequency when the load at X increases the governor characteristics and generation at both the stations are adjusted (by the operators) so that the system frequency is maintained constant.

3. *Flat tie-line control.* In this method of frequency control, station-X is used to control the frequency while the station-Y meets all load variations in its own area by varying its generation, thus power flow in the tie-line is kept constant irrespective of load demands.

- This method is used when a small system and a large system are interconnected through a tie line ; the large system maintains the system frequency constant while the small system keeps the tie-line power constant. This method is not suitable when two or more large system are connected.

4. *Tie-line bias control.* This is the most widely used method for large inter-connected stations. All power systems contribute towards regulation of frequency and tie-line power flow regardless of where from the frequency variation originates. The control equipment consists of *load frequency controller and tie-line load recorder-controller.*

Automatic regulation. Since in manual regulation a continuous watch on the frequency and loading of various generators is to be kept by the operators, it is not feasible to regulate manually the modern large interconnected systems. The *manual controls are sluggish and involve inherent human time lags.* On the other hand, "*automatic control*" is *more convenient and efficient.*

"*Automatic equipment*" besides providing all manual controls (discussed earlier), also permits *selective-frequency* and the *tie-line load bias* controls.

'*Automatic regulation*' envisages the following controls :

1. *Flat frequency control.* In this method the power input to the primemover is varied to correct the system frequency to the pre-determined value.

2. *Flat tie-line control.* Here the power input to primemover is varied so that the tie-line load is connected to a predetermined schedule. In this type of control another system is required to maintain frequency constant.

3. *Tie-line bias control.* It works well on power systems where a large number of generators and station are required to be kept under control. In case of frequency variation being more than pre-determined value, the control can be made to change from automatic to manual automatically with the sounding of an alarm, to enable the system operator to rectify/correct the fault.

10.4.5. Load Despatching

Due to electric system expansion and interconnection, load despatching (or centralised control of system operation) has become a necessity. In large power stations, central load despatcher is essential. The *functions* of a load despatcher or power controller include : *Loading of generating stations ; loading of units within the generating stations ; control of system voltage and reactive power flow ; control of system frequency and active power flow ; loading of various interconnectors and transfer of power from one system to another ; and maintaining continuity of service.*

Very large systems are sometimes divided into load districts each having its own load despatcher, often with a central load despatching agency for general supervision over the districts.

Centralised operation supervises *reliability of service, quality of service,* and *economy of production.*

HIGHLIGHTS

1. The connection of several generating stations in parallel is known as "Interconnected grid system".
2. The process of connecting an alternator in parallel with another alternator with common bus-bars is called "*synchronizing*".
3. The bus-bars whose frequency and the phase magnitude of potential differences are not affected by changes in condition of any one machine connected to it is called *infinite bus-bars.*
4. *Synchronous capacity* of an interconnector is defined as the change of kW transmitted per radian change of angular displacement of the two voltages of the two stations.
5. *Voltage control* may be exercised by the following methods :
   (*i*) Injection of reactive power.
   (*ii*) Tap changing transformers.
   (*iii*) Booster transformers.
   Frequency and load control may be carried by :
   (*i*) Manual regulation
   (*ii*) Automatic regulation.
6. In load despatching, centralised operation supervises reliability of service, quality of service, and economy of production.

THEORETICAL QUESTIONS

1. State the advantages and limitations of interconnected system of power stations.
2. What are the necessary conditions under which power is transmitted through an interconnector between power stations ?
3. (*a*) What is the necessity of parallel operation of alternators ?
   (*b*) State the requirements for paralleling alternators.
4. Describe briefly alternator synchronizing procedure.
5. Derive an expression for the synchronizing torque of an alternator operation with a given power angle.
6. Explain the necessity of interconnecting reactor for successful parallel operation of two generating stations. Obtain an expression for the power limit of the interconnector.
7. What are the necessary conditions under which power is transmitted through an interconnector between two power stations ?
8. Describe in outline the action of an automatic frequency control system of flat frequency control. Indicate the advantages and drawbacks.
9. Explain the term load despatch and discuss how economic generation scheduling is achieved in modern power system.
10. Write short notes on the following :
    (*i*) Interconnected system.
    (*ii*) Economic loading of alternators.
    (*iii*) Load sharing and power transfer between stations.
    (*iv*) Power limit of interconnectors.
    (*v*) Synchronous machines on infinite busbars.

UNSOLVED EXAMPLES

Parallel Operation of Alternators

1. Two alternators working in parallel supply a lighting load of 300 kW and a motor load aggregating to 5000 kW at a p.f. of 0.71. One machine is loaded up to 5000 kW at 0.8 p.f. lagging. What is the load and power factor of the other machine ? **[Ans.** 3000 kW ; 0.928]
2. Two alternators running in parallel supply the following loads :
   (*i*) 1500 kW at p.f. 0.9 lagging. (*ii*) 1000 kW at p.f. 0.8 lagging.
   (*iii*) 800 kW at p.f. unity. (*iv*) 600 kW at p.f. 0.8 leading.
   The load on one machine is adjusted to 2000 kW at p.f. 0.95 lagging. Find the load and p.f of the other machine. **[Ans.** 1900 kW ; 0.98]
3. Two similar alternators display the following data : 3-phase 6600 V, 60 Hz, 1000 kVA, resistance per phase of each of the star connected alternator = 1.02 Ω.
   Saturation Curve :

Field current :	150	200	250	300	350	500
Terminal voltage :	5000	6490	7000	7400	7750	8500

   Synchronous reactance per phase = 4Ω.
   When operating in parallel with a terminal voltage of 6600 V, the first alternator supplies 85 A at a p.f. of 0.85 lagging. If the load power factor is 0.78 lagging and total load is 1500 kW, determine the excitation of the second alternator. **[Ans.** 265 A]
4. A 6.6 kV, 1 MVA, 2-phase alternator is delivering full-load at 0.8 p.f. lagging. Its reactance is 20% and resistance negligible. By changing the excitation, the e.m.f. is increased by 25% at this load. Calculate the new current and the power factor.
   The machine is connected to infinite bus-bars. **[Ans.** 191 A ; 0.3665]
5. A 6.6 kV, 1 MVA alternator has a reactance of 20% and is delivering full-load at 0.8 p.f. lagging. It is connected to constant frequency bus-bar. If the steam supply is gradually increased, calculate :
   (*i*) At what output will the power factor become unity ?
   (*ii*) The maximum lead which it can supply without dropping out of synchronism, and the corresponding power factor. **[Ans.** (*i*) 2652 kW ; (*ii*) 5658 kW ; 0.749]

6. Two exactly similar turbo-alternators are rated 20 MW each. They are running in parallel. The speed-load characteristics of the driving turbines are such that the frequency of alternator 1 drops uniformly from 50 Hz on no-load to 48 Hz on full-load, and that of alternator 2 from 50 Hz to 48.5 Hz. How will the two machines share a load of 30 MW ? [**Ans.** 12.8572 MW ; 17.1428 MW]

7. Two 750 kW alternators operate in parallel. The speed regulation of one set is 100% to 103% from full-load to no-load and that of the other is 100% to 104%. How will the two alternators share a load of 1000 kW and at what load will one machine cease to supply any portion of the load ? [**Ans.** 464.3 kW ; 535.7 kW ; 187.5 kW]

8. Two 50 MVA, 3-phase alternators operate in parallel. The settings of the governors are such that the rise in speed from full-load to no-load is 2% in one machine and 3% in the other, the characteristics being straight lines in both cases. If each machine is fully loaded when the total load is 100 MW, what will be the load on each machine when the total load is reduced 60 MW ? [**Ans.** 26 MW, 34 MW]

9. Two 30 MVA, 3-phase alternators operate in parallel. The settings of governors are such that the rise in speed from full-load to no-load is one per cent in one machine and two per cent in the other, the characteristics being straight lines in both cases. If each machine is fully loaded when the total load is 60 MW, what will be no load on each machine when the total load is reduced to 40 MW ? [**Ans.** 16.7 MW (app.), 23.3 MW (app.)]

10. Two identical 2000 kVA alternators operate in parallel. The governor of first machine is such that the frequency drops uniformly for 50 Hz on no-load to 48 Hz on full-load. The corresponding uniform speed drop of the second machine is 50 to 47.5 Hz.

(*i*) How will the two machines share a load of 3000 kW ?

(*ii*) What is the maximum load at unity p.f. that can be delivered without overloading either machine ? [**Ans.** 1333 kW assuming u.p.f. ; 1667 kW assuming u.p.f. ; 3600 kW]

Interconnected Stations

11. Two 3-phase generating stations *A* and *B* are linked through a 33 kV interconnector having a resistance of 0.8 Ω and a reactance of 4.0 Ω per phase. At station-*A* the load on the generators is 80 MW at a p.f. of 0.8 lagging and the local load taken by the consumers connected to the *A* bus-bars is 50 MW at a p.f. of 0.707 lagging. Calculate the load in kW received from the station-*A* by station-*B*, its p.f. and phase difference between the voltages of *A* and *B*. [**Ans.** 29.26 MW, 0.977 (lag), 6°14′]

12. Two generating stations *A* and *B* are linked by an interconnector cable and reactor having a combined reactance of 4 Ω per phase with negligible resistance. Station-*A* and station-*B* supply in their own areas loads of 10000 kW at a lagging p.f. of 0.8 and 8000 kW at a p.f. of 0.9 lagging, respectively each at a bus-bar voltage of 11 kV. The station loads are equalised by the flow of power in the interconnector cable. Calculate the power factors of the station *A* and station *B*. [**Ans.** 0.748 (lag) ; 0.9178 (lag)]

13. Two power stations 1 and 2 operate in parallel and are interconnected by a short transmission line. The station capacities are 10 MW and 5 MW respectively and the generating sets have uniform speed regulation (from no-load to full-load) of 2 percent and 4 percent respectively. When the load of each station bus-bar is 6 MW, calculate :

(*i*) Load on the interconnector, and

(*ii*) Output of each station. [**Ans.** (*i*) 3.6 MW ; (*ii*) 9.6 MW , 2.4 MW)]

14. Two 11 kV generating stations 1 and 2 generating 25 MW each at unity p.f. are interconnected by a reactor with a voltage drop of 30 percent for transfer rate of 15 MW. The loads connected to the stations 1 and 2 are 30 MW and 20 MW each at unity p.f. respectively. Calculate the phase angle between the bus-bar voltages of the two generating stations. [**Ans.** 5°44′]

15. From a sub-station, a load of 9 MW is supplied via two 3-ϕ transmission circuits *X* and *Y* having impedances per phase of $(2 + j8)$ ohms and $(4 + j4)$ ohms respectively. In-phase and quadrature boosters are used at the sending end of one line to deliver $\frac{1}{3}$ rd and $\frac{2}{3}$ rd of the total power at unity power factor through circuits *X* and *Y* respectively. Determine the necessary booster voltages.

[**Ans.** 314.94 V/phase (in-phase) ; 0 V/phase (quadrature)]

PART-II : TRANSMISSION AND DISTRIBUTION

Chapters :

11. Electric Power Systems
12. Mechanical Design of Overhead Lines
13. Constants of Overhead Transmission Lines
14. Performance of Transmission Lines
15. Interference of Power Lines with Neighbouring Communication Circuits
16. Underground Cables
17. Power System Stability
18. Control of Reactive Power, Voltage and Load Frequency in Transmission Systems
19. Power Distribution Systems
20. Extra High Voltage Transmission

11

Electric Power Systems

11.1. Introduction. 11.2. Electric supply system. 11.3. Comparison between D.C. and A.C. systems of transmission and distribution. 11.4. Advantages and limitations of high transmission voltage. 11.5. Systems of power transmission. 11.6. Comparison of volume of conductor material—Overhead system—Underground system. 11.7. Choice of transmission voltage. 11.8. Conductor size and Kelvin's law. Highlights—Theoretical Questions—Unsolved Examples.

11.1. INTRODUCTION

Electrical energy is generated in large hydroelectric, thermal and nuclear power stations. These stations are mostly situated away from the load centres. Therefore an extensive power supply network is necessary between the generating plants and consumers' loads.

By *transmission and distribution of electrical power is meant its converyance from the central station where it is generated to the places where it is demanded by the consumers* (like pumping stations, residential and commercial buildings, mills, factories etc.)

- The maximum voltage in advanced countries is 33 kV while that in India is 11 kV.
- The amount of power that has to be transmitted through transmission lines is very large and if this power is transmitted at 11 kV (or 33 kV) *the line current and the power loss would be very large*. Therefore this *voltage is stepped up to a higher value by using step-up transformers* located in *sub-stations.*
- The transmission voltages in India are 400 kV, 220 kV and 132 kV.
- All *transmission and distribution systems are 3 phase systems.*
- The transmission lines and feeders are 3-phase 3-wire circuits.
- The distributors are 3-phase 4-wire circuits because a neutral wire is necessary to supply the single-phase loads of domestic and commercial consumers.

The transmission network is commonly known as **Grid.**

11.2. ELECTRIC SUPPLY SYSTEM

An electric supply system comprises the following *three principal components :*

1. Power station.
2. Transmission lines.
3. Distribution lines.

The electric supply system is broadly *classified* as follows :

(*i*) D.C. or A.C. system.

(*ii*) Overhead or underground system.

The overhead system is less expensive than the underground one. In our country this system is mostly adopted for transmission and distribution of power.

Fig. 11.1 shows a typical layout of a power system between geneation and use of electric power.

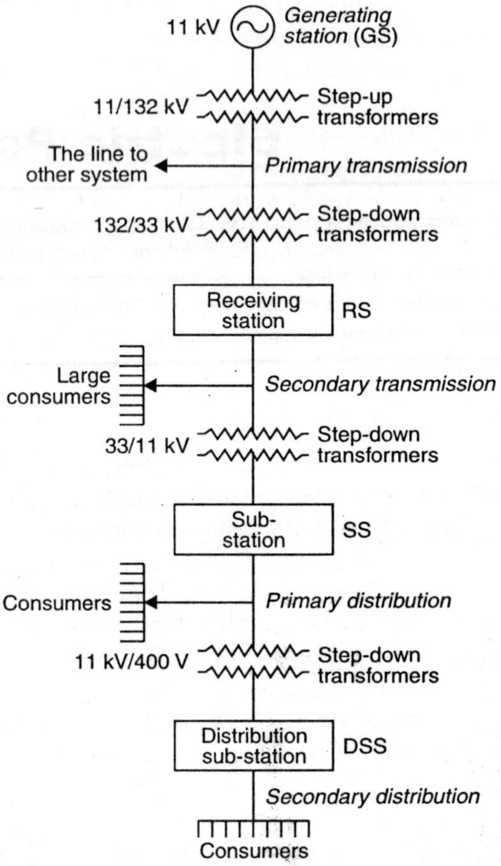

Fig. 11.1. A typical power supply network.

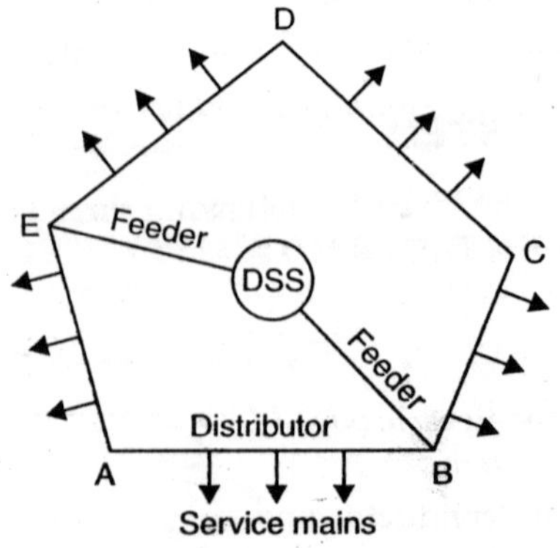

Fig. 11.2. Low voltage (L.V.) distribution system.

The various components are discussed below :

(*i*) *Generating station* (GS). In the generating station electric power is produced by 3-phase alternators operating in parallel. The normal generàtion voltage is 11 kV (it may be 6.6 kV or even 33 kV in some cases). This voltage is stepped upto 132 kV (or more) with the help of 3-phase transformers. Generally the primary transmission is carried at 66 kV, 132 kV, 220 kV or 400 kV.

Note. *Whereas the use of high voltage leads to several advantages including saving of conductor material and high transmission efficiency, on the other hand, introduces insulation problems and increases cost of switchgear and transformer equipment.*

(*ii*) *Primary transmission.* In this type of transmission, the electric power at 132 kV is supplied to other system by 3-phase, 3-wire overhead system.

(*iii*) *Secondary transmission.* The primary transmission line terminates at the receiving station (RS) where the voltage is reduced to 33 kV by step-down transformers. From this station, the electric power is transmitted at 33 kV to large consumers by 3-phase, 3-wire overhead system, and this forms the secondary transmission.

(*iv*) *Primary distribution.* The secondary transmission line terminates at sub-station (SS) where voltage is reduced from 33 kV to 11 kV. The 11 kV (3-phase, 3-wire) lines run along the important road sides of the city.

Note. The large/big consumers (having demand more than 50 kW) are normally supplied power at 11 kV which they handle it individually with their own sub-stations.

(*v*) *Secondary distribution.* The electric power available at 11 kV from the primary distribution line is delivered to distribution substations (DSS) (located near the localities of the consumers) which step down the voltage to 400 V, 3-phase, 4-wire secondary distribution. The single phase residental lighting load is connected between *any phase and neutral* (230 V) and 3-phase motor load is connected a across 3-phase lines (400 V) directly.

The seconadary distribution system consists of *feeders, distributors* and *service mains* (See Fig. 11.2).

- ***Feeders.*** *These are line conductors which connect the stations to the areas, to be fed by those stations. Normally no tappings are taken from feeders.* They are *designed* mainly from point of view of their *current carrying capacities.*
- ***Distributors.*** *These are the conductors from which several tappings for the supply to the consumers are taken.* They are designed from the point of view of the *voltage drops in them.*
- ***Service mains.*** *These are the conductors which connect the consumers terminals to the distributor.*

11.3. COMPARISON BETWEEN D.C. AND A.C. SYSTEMS OF TRANSMISSION AND DISTRIBUTION

D.C. System :

Advantages. Transmission of electric power by high voltage D.C. systems claim the following *advantages over high voltage A.C. system :*

1. There is greater power per conductor and simple line construction.
2. These systems are economical for long distance bulk power transmission by overhead lines.
3. Ground return is possible.

4. The voltage regulation problem is much less serious for D.C. since only IR drop is involved, IX drops is nil.
5. There is easy reversibility and controllability of power flow through a D.C. link.
6. No inductance, capacitance, phase displacement and surge problem in D.C. transmission.
7. There is no skin effect in D.C., *X*-section of line conductor is, therefore, fully utilized.
8. A D.C. line has less corona loss and reduced interference with communication circuits.
9. Underground cables can be used because of less potential stress and negligible dielectric loss.
10. No stability problems and synchronizing difficulties.
11. Since the potential stress on the insulation in case of D.C. system is $\frac{1}{\sqrt{2}}$ times that in A.C. system for same working voltage, therefore, less insulation is required in D.C. system.

Disadvantages. The high voltage D.C. systems entail the following *disadvantages* :

1. The systems are costly since installation of complicated converters and D.C. switch gear is expensive.
2. Converters require considerable reactive power.
3. Hormonics are generated which require filters.
4. Converters do not have overload capability.
5. The D.C. voltage cannot be stepped up for transmission of power at high voltages.
6. Electric power cannot be generated at high D.C. voltage due to commutator problems.

A.C. System :

These days electrical energy is almost exclusively generated, transmitted and distributed in the form of A.C.

Advantages :

1. The power can be generated at high voltages.
2. The sub-stations can be maintained easily and at a lesser cost.
3. The A.C. voltage can be stepped-up and stepped-down easily and efficiently with the use of transformers.

Disadvantages :

1. The construction of transmission lines is comparatively difficult.
2. The quantity of copper required is more.
3. In order to provide adequate insulation and to avoid corona loss in case of overhead lines, more spacing between the conductors is required.
4. The alternators need to be synchronized before they are put in parallel.
5. In A.C. system, the resistance of the line is increased due to skin effect.
6. Since an A.C. line has capacitance, therefore, there is a continuous power loss due to charging current even when the line is open.

- *The best method is to employ A.C. system for generation and distribution and D.C. system for transmission.*
- Now-a-days it has become possible to transmit electric power by D.C. system because of introduction of mercury are rectifiers and thyratrons, which can convert A.C. into D.C. and vice-versa directly and at a reasonable cost. Such devices can handle 30 MW at 400 V.

Fig. 11.3 shows a single line diagram of high voltage D.C. (H.V.D.C.) transmission.

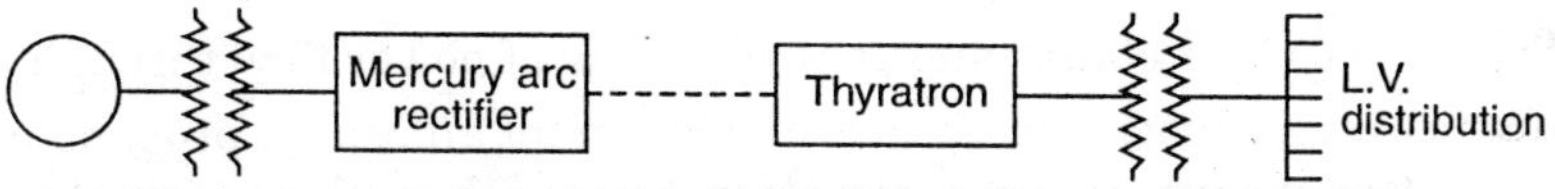

Fig. 11.3. Line diagram of typical H.V.D.C. transmission system.

— Generating station generates electric power which is A.C., stepped to high voltage by the step-up transformers.

— This A.C. power at high voltage is fed to the *mercury arc rectifiers which convert. A.C. into D.C.*

— The transmission of electric power is carried at high D.C. voltage. At the receiving end *D.C. power is converted to A.C. power using thyratrons.* The A.C. supply is then stepped down to low voltage for distribution by step-down transformers.

Classification of Transmission Lines :

For transmission of electrical power three-phase circuits are generally used becuase of economical reasons. Transmission lines may be *classified* as follows :

1. Single line
2. Parallel lines
3. Radial lines
4. Ring system
5. Network.

1. **Single line :**

- The simplest form is the single line, such as obtained from a power plant supplying its entire output to one load centre over a single-circuit line.
- Such a system has the *disadvantage* that in case of damage to the line the service is *interrupted.*
- Its use is more or less confined to small power systems and is therefore becoming more and more uncommon.

2. **Parallel lines :**

- Where continuity of service is necessary, it is best to use at least two circuits in parallel, placed either on the same supports or on separate supports.
- *Separate supports afford greater safety* against both lines being damaged at the same time, but the cost is much higher than when two circuits are placed on one support.
- In some cases, where very large quantities of power must be handled, more than two circuits may be run in parallel.

3. **Radial lines :**

- Invariably a power plant or substation supplies power to the neighbouring territory by means of radial lines.
- These radial lines may be either single circuit for the less important loads or double circuit for the more important loads.

4. **Ring system :**

- For *systems covering a large territory* the ring system of transmission is very important.
- With this system the main high-voltage power line makes a closed ring, taps being taken off at any advantageous points of the ring, thus supplying a large territory.

- In case of damage to any section of the ring, that section may be disconnected for repairs, and power will be supplied from both ends of the rings, thereby maintaining continuity of service.

5. **Network :**

- A network often constitutes several ring systems with sections of single, parallel, or radial lines.

11.4. ADVANTAGES AND LIMITATIONS OF HIGH TRANSMISSION VOLTAGE

Advantages. The following are the *advantages of using high transmission voltage* :

1. Lesser conductor material required.
2. Transmission efficiency is increased.
3. Percentage line drop decreases, this leads to better voltage regulation.

Limitations. The use of high transmission voltage entails the following *limitations* :

1. The cost of insulating the conductor increases.
2. The cost of transformers, switchgear and other terminal apparatus increases.

— The limit to the use of higher transmission voltage (which can be economically used in a particular case) *is reached when the saving in cost of conductor material due to higher voltage is offset by the increased cost of insulation, transformer, switchgear* etc. Thus, the choice of proper transmission voltage is essentially a question of economics.

- The facts mentioned under advantages of higher transmission voltage can be clearly shown as follows :

Let, P = Power transmitted,

V = Line voltage,

l = Length of the line,

ρ = Resistivity of conductor material,

a = X-sectional area of conductor,

$\cos \phi$ = Power factor of the load, and

J = Current density (I/a)

Load current, $$I = \frac{P}{\sqrt{3}\, V \cos \phi}$$

Resistance per condutor, $$R = \frac{\rho l}{a}$$

Total power loss, $P_{loss} = 3I^2R$

$$= 3 \times \left(\frac{P}{\sqrt{3}\, V \cos \phi} \right)^2 \times \frac{\rho l}{a} = \frac{P^2 \rho l}{V^2 \cos^2 \phi \times a}$$

$\therefore$ Cross sectional-area of the condutor, $$a = \frac{P^2 \rho l}{P_{loss} \times V^2 \cos^2 \phi}$$

(*i*) ***Volume of conductor material required :***

Total volume of conductor material required

$$= 3\, a\, l = 3 \left(\frac{P^2 \rho l}{P_{loss} \times V^2 \cos^2 \phi} \right) l = \frac{3P^2 \rho l^2}{P_{loss} \times V^2 \cos^2 \phi} \quad ...(1)$$

From expression (1) it is clear that for given values of P, l, ρ and P_{loss}, the *volume of conductor required is inversely proportional to the square of the transmission voltage and power factor.* In other words, with *the increase in transmission voltage the volume of conductor material required decreases.*

(*ii*) ***Transmission efficiency :***

Input power $= P$ + total losses

$$= P + \frac{P^2 \rho l}{V^2 \cos^2 \phi \times a} = P + \frac{P^2 \rho l J}{V^2 \cos^2 \phi \times I} \qquad (\because a = I/J)$$

$$= P + \frac{P^2 \rho l J}{V^2 \cos^2 \phi} \times \frac{\sqrt{3}\, V \cos \phi}{P} \qquad \left(\because I = \frac{P}{\sqrt{3} V \cos \phi}\right)$$

$$= P + \frac{\sqrt{3} P J \rho l}{V \cos \phi} = P\left(1 + \frac{\sqrt{3} J \rho l}{V \cos \phi}\right)$$

$$\therefore \text{ Transmission effeciency} = \frac{\text{Output power}}{\text{Input power}}$$

$$= \frac{P}{P\left(1 + \frac{\sqrt{3} J \rho l}{V \cos \phi}\right)} = \frac{1}{1 + \frac{\sqrt{3} J \rho l}{V \cos \phi}}$$

$$= \left(1 - \frac{\sqrt{3} J \rho l}{V \cos \phi}\right) \text{ approx.} \qquad \text{...(2)}$$

... By Bionomial theorem

From expression (2) is it is clear that for given values of J, ρ and l when the *transmission line voltage is increased, the transmission efficiency increases.*

(*iii*) ***Percentage line drop :***

Line drop $= IR = I \times \frac{\rho l}{a}$

$$= I \times \rho l \times \frac{J}{I} = \rho l J \qquad (\because a = I/J)$$

$$\therefore \text{ Percentage line drop } = \frac{J \rho l}{V} \times 100 \qquad \text{...(3)}$$

The expression (3) indicates that, J, ρ and l being constants, *with the increase in transmission voltage, the percentage line drop decreases.*

Note. (*i*) A considerable economy can be affected in conductor material if the transmission line is operated at high power factor.

(*ii*) With the increase in the transmission voltage :

— The insulation required between the conductors and the earthed tower increases due to which the *cost of line supports increases ;*

— More clearance is required between the conductors and ground which necessitates the use of *high towers* ;

— More distance is required between the conductors, therefore longer cross-arms need to be used.

11.5. SYSTEMS OF POWER TRANSMISSION

Although 3-phase, 3-wire A.C. system is universally employed for transmission and 3-phase, 4-wire A.C. system for distribution of electric power, yet under special circumstances other systems can also be used. The *various systems of power transmission are* :

1. **D.C. system :**
   (*i*) D.C. two-wire.
   (*ii*) D.C. two-wire with mid-point earthed.
   (*iii*) D.C. three-wire system.
2. **Single phase A.C. system :**
   (*i*) Single phase two-wire.
   (*ii*) Single phase two-wire with mid-point earthed.
   (*iii*) Single phase three-wire.
3. **Two-phase A.C. system :**
   (*i*) Two-phase four-wire.
   (*ii*) Two-phase three-wire.
4. **Three-phase A.C. system :**
   (*i*) Three-phase 3-wire.
   (*ii*) Three-phase 4-wire.

It is quite obvious that, out of such a large number of possible systems of transmission it is difficult to decide the best system without making comparison. For various systems of transmission, the basis of comparison is usually the cost of the conductor ; the cost depends upon the volume of the conductor which should be *minimum*. In order to compare the volume of copper required for different systems, the following two cases need to be considered :

(*i*) *Transmission by overhead system :*

In the overhead system of transmission the conductors are insulated from the cross-arms and supporting towers. Since the towers and cross-arms are earthed, therefore, the *maximum voltage between each conductor and earth forms the basis of comparison of volume of conductor.*

(*ii*) *Transmission by underground system :*

In underground system the maximum disruptive stress is between the two conductors (of the cable), so the maximum voltage between conductors forms the basis of comparison.

11.6. COMPARISON OF VOLUME OF CONDUCTOR MATERIAL

The comparison of volume of conductor material will be made in the following *two cases* :

I. Overhead system.

II. Underground system.

For calculating the volume of conductor material required, the following **assumptio** are made for both the above mentioned systems :

1. In all the cases, same power is transmitted (P watts).
2. The distance over which the power is to be transmitted is same (l metres).
3. In each case, the line loss is same (P_{loss} watts).
4. The maximum voltage between any conductor and earth (V_m volts) is same in each case.

11.6.1. Overhead System

I. D.C. system :

1. ***D.C. two-wire system with one conductor earthed :***

Fig. 11.4 shows a D.C. two-wire system, with one conductor earthed.

Maximum voltage between conductors = V_m *Volts*

Power to be transmitted = P

Load current, $I_1 = \frac{P}{V_m}$

Let R_1 is the resistance of each line conductor.

Then, $R_1 = \frac{\rho l}{a_1}$, where a_1 is the X-sectional area of the conductor.

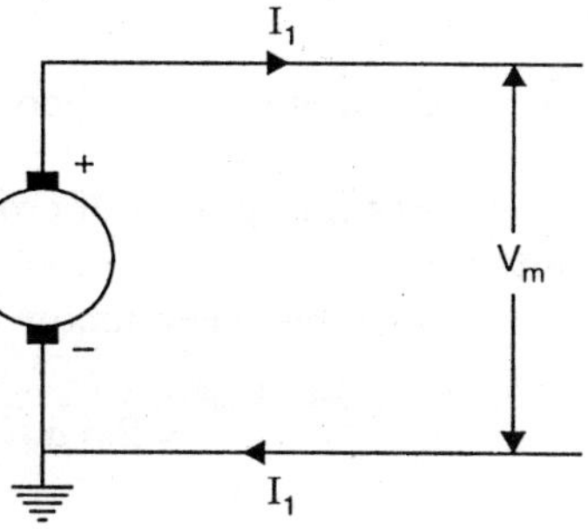

Fig. 11.4

Line losses, $$P_{loss} = 2I_1^2R_1 = 2\left(\frac{P}{V_m}\right)^2 \frac{\rho l}{a_1} \quad \text{or} \quad a_1 = \frac{2P^2\rho l}{P_{loss}V_m^2}$$

$\therefore$ Volume of conductor material required $= 2\,a_1 l = 2\left(\frac{2P^2\rho l}{P_{loss}V_m^2}\right)l = \frac{4P^2\rho l^2}{P_{loss}\,V_m^2} = \text{K (say)}$

Usually this system is made as a basis for comparison with other systems.

2. ***D.C. two-wire system with mid-point earthed :*** Refer to Fig. 11.5.

Maximum voltage between the conductors = $2V_m$ volts

Load current, $I_2 = \frac{P}{2V_m}$

Line losses, $P_{loss} = 2I_2^2R_2$

$$= 2\left(\frac{P}{2V_m}\right)^2 \times \frac{\rho l}{a_2} = \frac{P^2\rho l}{2a_2V_m^2} \qquad \left(\because R_2 = \frac{\rho l}{a_2}\right)$$

or, Area of cross-section, $a_2 = \frac{P^2\rho l}{2P_{loss}V_m^2}$

$\therefore$ Volume of conductor material required

$$2a_2l = 2\left(\frac{P^2\rho l}{2P_{loss}V_m^2}\right)l = \frac{P^2\rho l^2}{P_{loss}V_m^2} = \frac{K}{4} \qquad \left(\because K = \frac{4P^2\rho l^2}{P_{loss}V_m^2}\right)$$

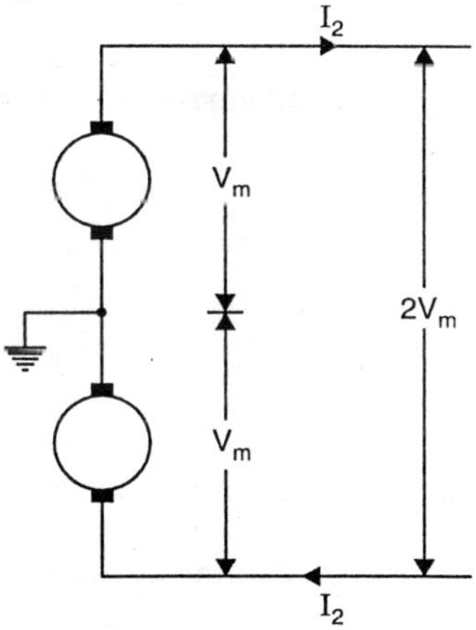

Fig. 11.5

Hence, the volume of conductor material required in this system is *one-fourth of that required in two-wire D.C. system with one conductor earthed.*

3. ***D.C. three-wire system :*** Refer to Fig. 11.6.

Maximum voltage between the outer and earth = V_m volts

Load current, $I_3 = \frac{P/2}{V_m} = \frac{P}{2V_m}$

Line losses, $P_{loss} = 2I_3^2R_3 = 2\left(\frac{P}{2V_m}\right)^2 \times \frac{\rho l}{a_3} = \frac{P^2\rho l}{2V_m^2 a_3}$

or, Area of cross-section, $a_3 = \frac{P^2\rho l}{2P_{loss}V_m^2}$

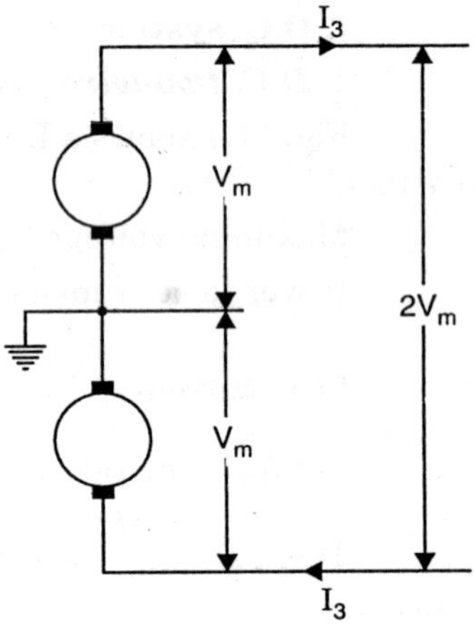

Fig. 11.6

Assuming area of cross-section of neutral wire as half of any outer,

Volume of conductor material required

$$= 2.5\, a_3 l_3 = 2.5\left(\frac{P^2\rho l}{2P_{loss}V_m^2}\right) l$$

$$= \frac{5}{4}\left(\frac{P^2\rho l^2}{P_{loss}V_m^2}\right) = \frac{5}{16}K \qquad \left(\because K = \frac{4P^2\rho l^2}{P_{loss}V_m^2}\right)$$

Hence, the volume of conductor material required in this system is $\frac{5}{16}$*th of that required in two-wire D.C. system with one conductor earthed.*

II. A.C. system :

4. ***A.C. single-phase two-wire system with one conductor earthed :*** Refer to Fig. 11.7.

R.M.S. value of voltage between conductor $= \frac{V_m}{\sqrt{2}}$ volts

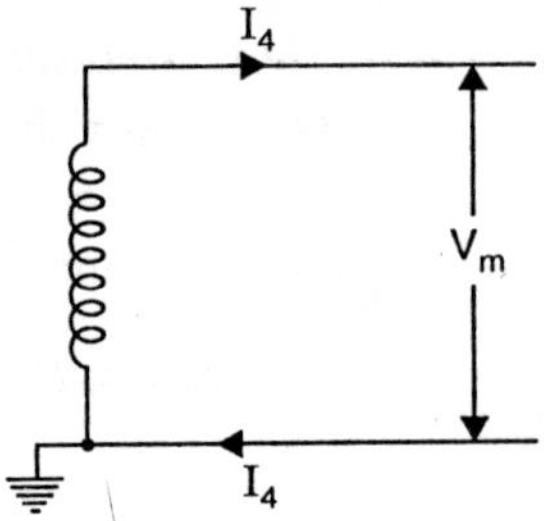

Fig. 11.7

Load current, $I_4 = \frac{P}{(V_m/\sqrt{2})\cos\phi} = \frac{\sqrt{2}P}{V_m\cos\phi}$

Line losses, $P_{loss} = 2I_4^2R_4 = 2\left(\frac{\sqrt{2}P}{V_m\cos\phi}\right)^2 \times \frac{\rho l}{a_4}$

$$= \frac{4P^2\rho l}{V_m^2\cos\phi \times a_4}$$

or, Area of cross-section, $a_4 = \frac{4P^2\rho l}{P_{loss}V_m^2\cos^2\phi}$

∴ Volume of conductor material required $= 2a_4 l$

$$= 2 \times \frac{4P^2\rho l}{P_{loss}V_m^2\cos^2\phi} \times l = \frac{2}{\cos^2\phi} \times K$$

Hence, volume of conductor material required in this system is $\frac{2}{\cos^2\phi}$ *times that of two-wire D.C. system with one conductor earthed.*

5. ***A.C. single-phase two-wire system with mid-point earthed :*** Refer to Fig. 11.8.

Maximum voltage between the two wires = 2 V_m

R.M.S. value of voltage between conductors $= \frac{2V_m}{\sqrt{2}} = \sqrt{2}\,V_m$ volts

Load current, $I_5 = \dfrac{P}{(\sqrt{2}V_m)\cos\phi}$

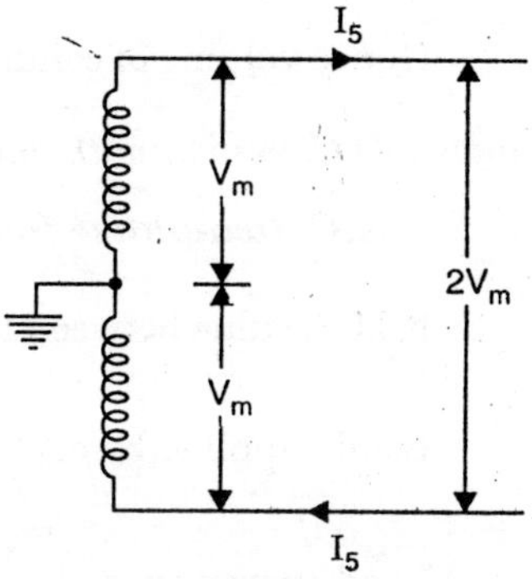

Fig. 11.8

Line losses, $P_{loss} = 2I_5^2R_5 = 2\left[\dfrac{P}{(\sqrt{2}V_m)\cos\phi}\right]^2 \times \dfrac{\rho l}{a_5}$

$$= \frac{P^2\rho l}{V_m^2 \cos^2\phi \times a_5}$$

or, Area of cross-section, $a_5 = \dfrac{P^2\rho l}{P_{loss}V_m^2\cos^2\phi}$

∴ Volume of conductor material required

$$= 2a_5 l = 2\left(\frac{P^2\rho l}{P_{loss}V_m^2\cos\phi}\right)\times l = \frac{2}{\cos^2\phi}\times\frac{P^2\rho l^2}{P_{loss}V_m^2}$$

$$= \frac{K}{2\cos^2\phi} \qquad \left(\because K = \frac{4P^2\rho l^2}{P_{loss}V_m^2}\right)$$

Hence, volume of conductor material required in this system is $\dfrac{1}{2\cos^2\phi}$ *times that of two-wire D.C. system with one conductor earthed.*

6. ***A.C. single-phase three-wire system :*** Refer to Fig. 11.9.

R.M.S. value of voltage between outer and earth = $\dfrac{V_m}{\sqrt{2}}$ volts

Load current, $I_6 = \dfrac{P/2}{(V_m/\sqrt{2})\cos\phi} = \dfrac{P}{\sqrt{2}V_m\cos\phi}$, assuming balanced load.

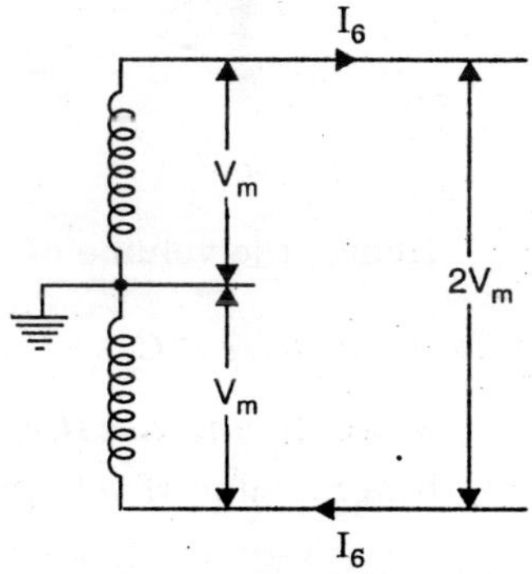

Fig. 11.9

Line losses, $P_{loss} = 2I_6^2R_6 = 2\times\left(\dfrac{P}{\sqrt{2}V_m\cos\phi}\right)^2\times\dfrac{\rho l}{a_6}$

$$= \frac{P^2\rho l}{V_m^2\cos^2\phi\times a_6}$$

or, Area of cross-section, $a_6 = \dfrac{P^2\rho l}{P_{loss}V_m^2\cos^2\phi}$

Assuming the area of cross-section of neutral wire to be half that of any of the outers,
Volume of conductor material required

$$= 2.5\,a_6 l = 2.5\left(\frac{P^2\rho l}{P_{loss}V_m^2\cos^2\phi}\right)\times l = \frac{2.5P^2\rho l^2}{P_{loss}V_m^2\cos^2\phi}$$

$$= \frac{2.5}{\cos^2\phi}\times\frac{P^2\rho l^2}{P_{loss}V_m^2} = \frac{5}{8\cos^2\phi}\times K \qquad \left(\because K = \frac{4P^2\rho l^2}{P_{loss}V_m^2}\right)$$

Hence volume of conductor material required in this system is $\frac{5}{8\cos^2\phi}$ *times that required in two-wire D.C. system with one conductor earthed.*

7. ***A.C. two-phase four-wire system :*** Refer to Fig. 11.10.

R.M.S. value between the outers $= \frac{2V_m}{\sqrt{2}} = \sqrt{2}V_m$

Load supplied by each phase $= \frac{P}{2}$

Load current per phase, $I_7 = \frac{P}{2\sqrt{2}V_m \cos\phi}$

Line losses, $P_{loss} = 4I_7^2R_7$

$$= 4 \times \left(\frac{P}{2\sqrt{2}V_m \cos\phi}\right)^2 \times \frac{\rho l}{a_7}$$

$$= \frac{P^2\rho l}{2V_m^2 \cos^2\phi \times a_7}$$

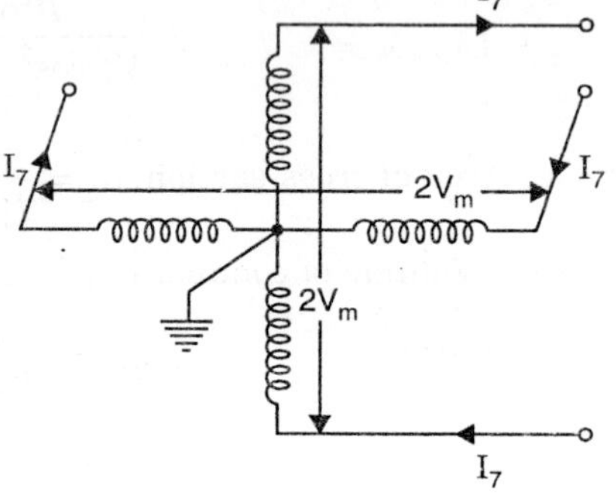

Fig. 11.10

$\therefore$ Area of cross-section, $a_7 = \frac{P^2\rho l}{2P_{loss}V_m^2 \cos^2\phi}$

$\therefore$ Volume of conductor material required

$$= 4a_7l = 4 \times \frac{P^2\rho l}{2P_{loss}V_m^2 \cos^2\phi} \times l$$

$$= \frac{1}{2\cos^2\phi} \times \frac{4P^2\rho l^2}{P_{loss}V_m^2} = \frac{1}{2\cos^2\phi} \times K$$

Hence, the volume of conductor material required for this system is $\frac{1}{2\cos^2\phi}$ *times that required in two-wire D.C. system with one conductor earthed.*

8. ***A.C. two-phase three-wire system :*** Refer to Fig. 11.11.

R.M.S. value of voltage between outer and neutral, considering balance load, $= \frac{V_m}{\sqrt{2}}$ volts

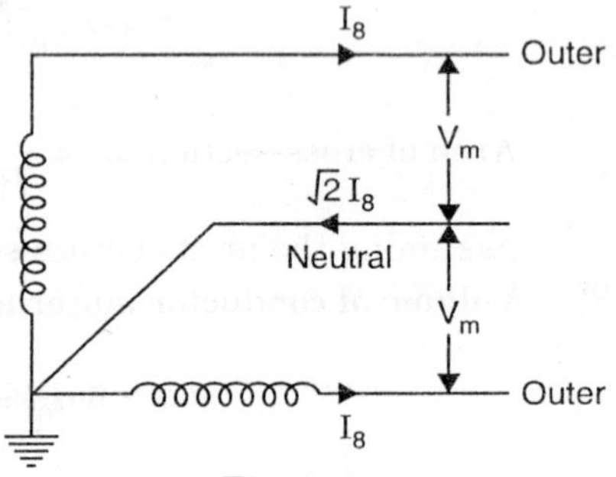

Fig. 11.11

Current in each of the outer conductors,

$$I_8 = \frac{P/2}{(V_m/\sqrt{2})\cos\phi} = \frac{P}{\sqrt{2}V_m \cos\phi}$$

Current in the neutral wire

= phasor sum of currents in outer wires

$= \sqrt{I_8^2 + I_8^2} = \sqrt{2}I_8$ (currents in the outers being 90° apart)

Assuming the *current density to be constant,* the area of cross-section of the neutral wire will be $\sqrt{2}$ times that of either of the outers (as the neutral wire carries $\sqrt{2}$ times the current in each of the outers, its cross-section must be increased in the same ratio to maintain the same current density).

$$\text{Rasistance of neutral wire} = \frac{R_8}{\sqrt{2}} = \frac{\rho l}{\sqrt{2}\, a_8}$$

$$\text{Line losses, } P_{loss} = 2I_8^2R_8 + (\sqrt{2}I_8)^2\frac{R_8}{\sqrt{2}} = I_8^2R_8\,(2+\sqrt{2})$$

$$= \left(\frac{P}{\sqrt{2}V_m\cos\phi}\right)^2 \times \frac{\rho l}{a_8}(2+\sqrt{2}) = \frac{P^2\rho l}{2V_m^2\cos^2\phi \times a_8}(2+\sqrt{2})$$

or, $$\text{Area of } X\text{-section, } a_8 = \frac{P^2\rho l}{2P_{loss}V_m^2\cos^2\phi}(2+\sqrt{2})$$

∴ Volume of conductor material required

$$= 2a_8l + \sqrt{2}a_8l = a_8l\,(2+\sqrt{2})$$

$$= \frac{P^2\rho l^2}{2P_{loss}V_m^2\cos^2\phi}(2+\sqrt{2})^2 = \frac{1.457}{\cos^2\phi}\left(\frac{4P^2\rho l^2}{P_{loss}V_m^2}\right) = \frac{1.457}{\cos^2\phi}K$$

Hence, volume of conductor material required in this system is $\frac{1.457}{\cos^2\phi}$ *times that of two-wire D.C. system with one conductor earthed.*

9. ***A.C. three-phase, three-wire system :*** Refer to Fig. 11.12.

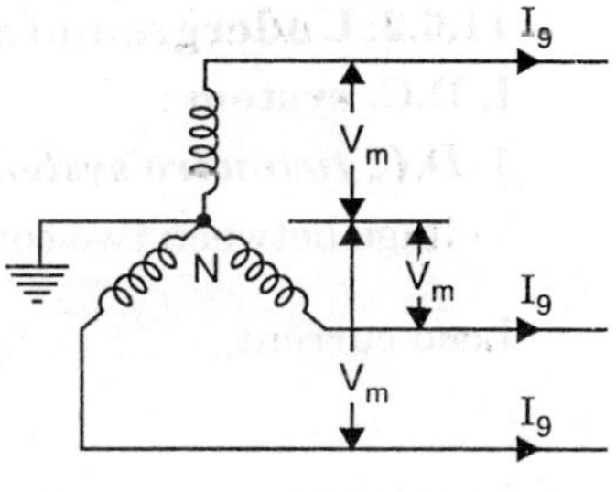

Fig. 11.12

$$\text{R.M.S. voltage per phase} = \frac{V_m}{\sqrt{2}}$$

$$\text{Power transmitted per phase} = \frac{P}{3}$$

$$\text{Load current per phase, } I_9 = \frac{(P/3)}{(V_m/\sqrt{2})\cos\phi} = \frac{\sqrt{2}P}{3V_m\cos\phi}$$

$$\text{Line losses, } P_{loss} = 3I_9^2R_9$$

$$= 3\left(\frac{\sqrt{2}P}{3V_m\cos\phi}\right)^2 \times \frac{\rho l}{a_9} = \frac{2P^2\rho l}{3V_m^2\cos^2\phi \times a_9}$$

or, $$\text{Area of cross-section, } a_9 = \frac{2P^2\rho l}{3P_{loss}V_m^2\cos^2\phi}$$

∴ Volume of conductor material required

$$= 3a_9l = 3\left(\frac{2P^2\rho l}{3P_{loss}V_m^2\cos^2\phi}\right)\times l = \frac{2P^2\rho l^2}{P_{loss}V_m^2\cos^2\phi}$$

$$= \frac{1}{2\cos^2\phi}K$$

Hence volume of conductor material required is $\frac{1}{2\cos^2\phi}$ *times that required for two-wire D.C. system with one conductor earthed.*

10. ***A.C. three-phase, four-wire system :*** Refer to Fig. 11.13.

Assuming balanced load, there will be no current in neutral wire and copper losses will be same as in three-phase, three-wire system,

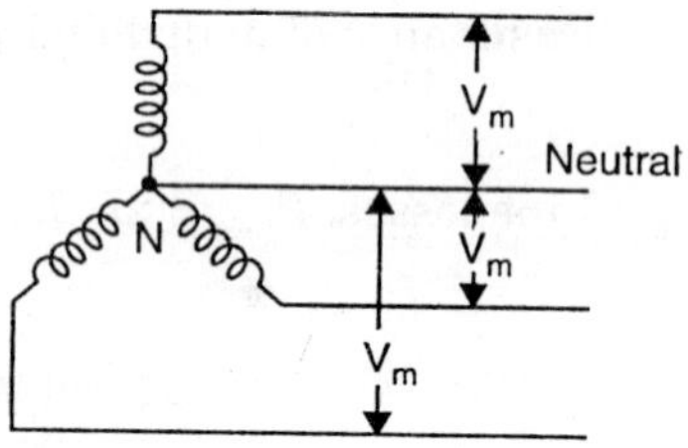

Fig. 11.13

i.e.,
$$P_{loss} = \frac{2P^2\rho l}{3V_m^2 \cos^2\phi \times a_{10}}$$

$\therefore$ Area of cross-section, $a_{10} = \frac{2P^2\rho l}{3P_{loss}V_m^2\cos^2\phi}$

As the area of cross-section of neutral wire is one-half that of any line conductor, therefore,
Voltage of conductor material required

$$= 3.5\, a_{10}l = 3.5 \times \left(\frac{2P^2\rho l}{3P_{loss}V_m^2\cos^2\phi}\right) \times l = \frac{7P^2\rho l^2}{3P_{loss}V_m^2\cos^2\phi}$$

$$= \frac{7}{12\cos^2\phi} \times K \qquad \left(\because K = \frac{4P^2\rho l^2}{P_{loss}V_m^2}\right)$$

Hence, volume of the conductor material required for this system is $\frac{7}{12\cos^2\phi}$ *times that required for two-wire D.C. system with one conductor earthed.*

11.6.2. Underground System

I. D.C. system :

1. ***D.C. two-wire system :*** Refer to Fig. 11.14.

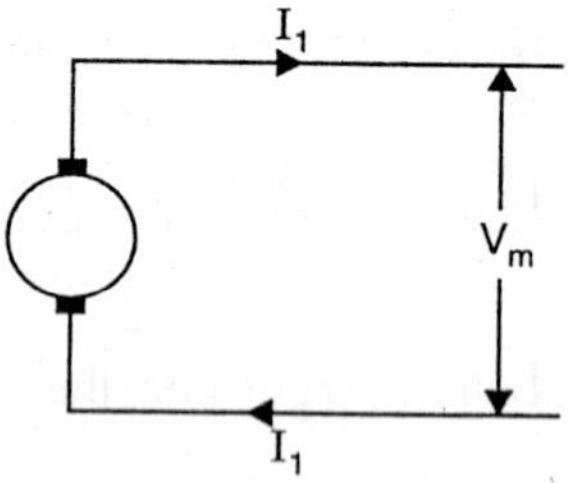

Fig. 11.14

Voltage between two conductors = V_m volts

Load current, $I_1 = \frac{P}{V_m}$

Line losses, $P_{loss} = 2\,I_1^2R_1 = 2\left(\frac{P}{V_m}\right)^2 \times \frac{\rho l}{a_1}$

$$= \frac{2P^2\rho l}{V_m^2 \times a_1}$$

or, Area of cross-section, $a_1 = \frac{2P^2\rho l}{P_{loss}V_m^2}$

$\therefore$ Volume of conductor material required

$$= 2a_1 l = 2\left(\frac{2P^2\rho l}{P_{loss}V_m^2}\right) \times l = \frac{4P^2\rho l^2}{P_{loss}V_m^2} = K \text{ (say)}$$

This volume will be taken as the basic quantity for comparison with other systems.

2. ***D.C. two-wire system with mid-point earthed :*** Refer to Fig. 11.15.

Load current $$I_2 = \frac{P}{V_m}$$

Lines losses, $$P_{loss} = 2I_2^2R_2 = 2\left(\frac{P}{V_m}\right)^2 \times \frac{\rho l}{a_2}.$$

$$= \frac{2P^2\rho l}{V_m^2 \times a_2}$$

or, Area of cross-section, $$a_2 = \frac{2P^2\rho l}{P_{loss}V_m^2}$$

∴ Volume of conductor material required

$$= 2a_2l = 2\left(\frac{2P^2\rho l}{P_{loss}V_m^2}\right) \times l = \frac{4P^2\rho l^2}{P_{loss}V_m^2} = K$$

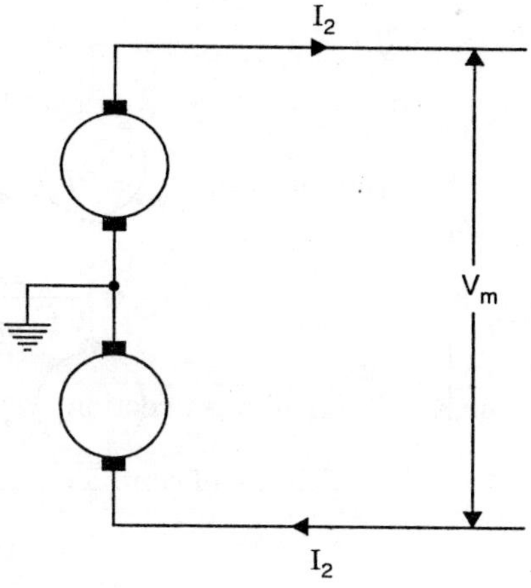

Fig. 11.15

Hence, the volume of conductor material required in this system is the *same as that for two-wire D.C. system.*

3. ***D.C. three-wire system :*** Refer to Fig. 11.16.

Assume load is balanced, (*i.e.*, current through the neutral is zero).

Maximum voltage between the outers = V_m volts

Load current, $$I_3 = \frac{P}{V_m}$$

Line losses, $$P_{loss} = 2I_3^2R_3 = 2\left(\frac{P}{V_m}\right)^2 \times \frac{\rho l}{a_3}$$

$$= \frac{2P^2\rho l}{V_m^2 a_3}$$

or, Area of cross-section, $$a_3 = \frac{2P^2\rho l}{P_{loss}V_m^2}$$

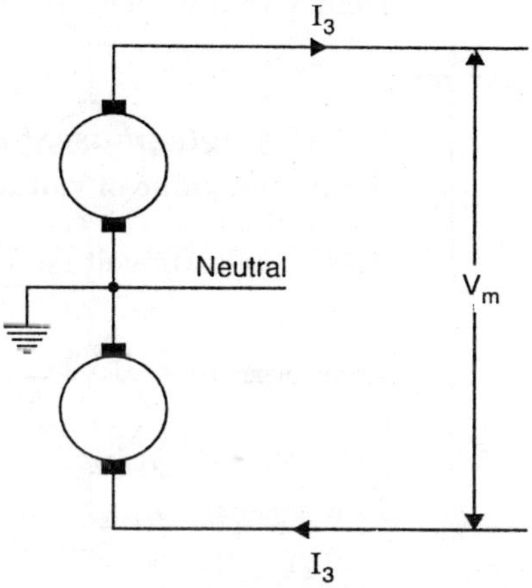

Fig. 11.16

Assuming cross-section of the neutral as half of that of either outer,

Volume of conductor material required

$$= 2.5\, a_3l = 2.5\left(\frac{2P^2\rho l}{P_{loss}V_m^2}\right) \times l = \frac{5P^2\rho l^2}{P_{loss}V_m^2}$$

$$= 1.25\, K \qquad \left(\because\ K = \frac{4P^2\rho l^2}{P_{loss}V_m^2}\right)$$

Hence, volume of conductor material required is *1.25 times that required for two-wire D.C. system.*

II. A.C. system :

4. ***A.C. single phase two-wire system :*** Refer to Fig. 11.17.

Maximum voltage between outer = V_m volts

R.M.S. value of voltage between outers $= \dfrac{V_m}{\sqrt{2}}$ volt

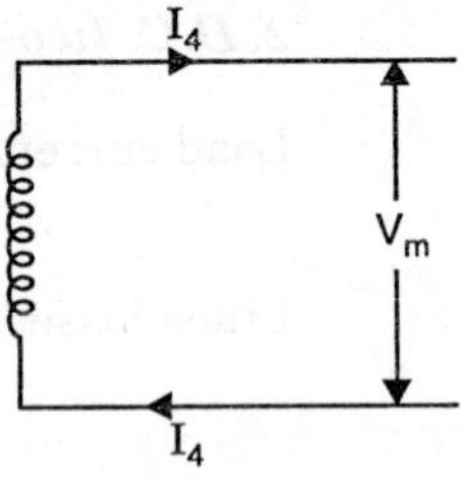

Fig. 11.17

Load current, $I_4 = \dfrac{P}{V_m / \sqrt{2} \cos\phi} = \dfrac{\sqrt{2}P}{V_m \cos\phi}$

Line losses, $P_{loss} = 2I_4{}^2R_4 = 2\left(\dfrac{\sqrt{2}P}{V_m \cos\phi}\right)^2 \times \dfrac{\rho l}{a_4}$

$$= \frac{4P^2\rho l}{V_m^2 \cos^2\phi \times a_4}$$

or, Area of cross-section, $a_4 = \dfrac{4P^2\rho l}{P_{loss} V_m^2 \cos^2\phi}$

∴ Volume of conductor material required

$$= 2a_4 l = 2\left(\frac{4P^2\rho l}{P_{loss} V_m \cos^2\phi}\right) \times l = \frac{2}{\cos^2\phi} \times \frac{4P^2\rho l^2}{P_{loss} V_m} = \frac{2}{\cos^2\phi} \times K$$

Hence, volume of conductor material required is $\dfrac{2}{cos^2\phi}$ *times that required for two-wire D.C. system.*

5. ***A.C. single-phase, two-wire system with mid-point earthed :*** Refer to Fig. 11.18

Maximum value of voltage between outers $= V_m$ volts

R.M.S. value of voltage between outers $= \dfrac{V_m}{\sqrt{2}}$ volts

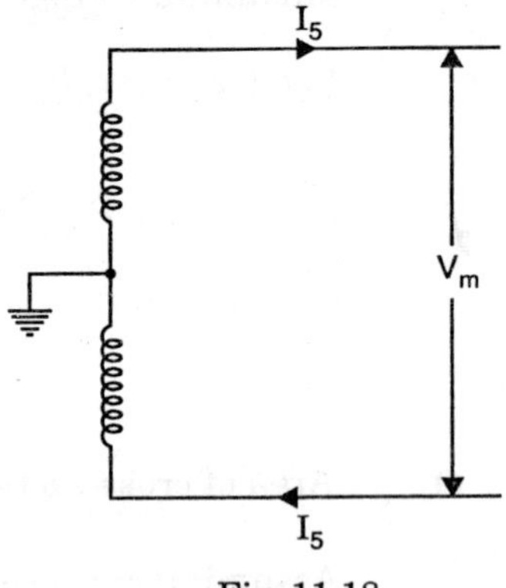

Fig. 11.18

Load current, $I_5 = \dfrac{P}{(V_m / \sqrt{2}) \cos\phi} = \dfrac{\sqrt{2}P}{V_m \cos\phi}$

Line losses, $P_{loss} = 2I_5{}^2R_5 = \left(\dfrac{\sqrt{2}P}{V_m \cos\phi}\right)^2 \times \dfrac{\rho l}{a_5}$

$$= \frac{4P^2\rho l}{V_m^2 \cos^2\phi \times a_5}$$

or, Area of cross-section, $a_5 = \dfrac{4P^2\rho l}{P_{loss} V_m^2 \cos^2\phi}$

∴ Volume of conductor material required

$$= 2a_5 l = 2 \times \left(\frac{4P^2\rho l}{P_{loss} V_m^2 \cos^2\phi}\right) \times l = \frac{2}{\cos^2\phi}\left(\frac{4P^2\rho l^2}{P_{loss} V_m^2}\right) = \frac{2}{\cos^2\phi} \times K$$

Hence, volume of conductor material required in this system is $\dfrac{2}{cos^2\phi}$ *times that required in two-wire D.C. system.*

6. ***A.C. single phase, three-wire system.*** Refer to Fig. 11.19.

Assuming load to be balanced, the system reduces to a single-phase, two-wire A.C. system except that a neutral wire is provided additionally. Assuing the cross-sectional area of the neutral wire to be half of either of the outers,

Volume of conductor material required

$= 2.5\, a_4 l$ ...(a_4 = area of cross-section of conductor for single-phase two-wire system)

$$= 2.5\left(\frac{4P^2\rho l}{P_{loss}V_m^2\cos^2\phi}\right)\times l$$

$$= \frac{10P^2\rho l^2}{P_{loss}V_m^2\cos^2\phi} = \frac{2.5}{\cos^2\phi}\times\frac{4P^2\rho l^2}{P_{loss}V_m^2}$$

$$= \frac{2.5}{\cos^2\phi}K$$

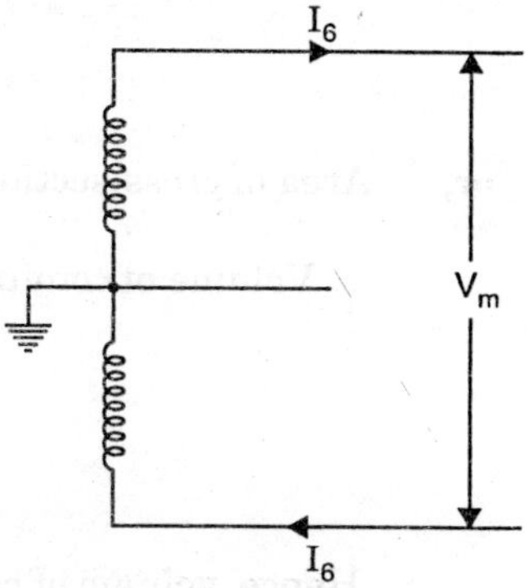

Fig. 11.19

Hence, volume of conductor material required in this system is $\frac{2.5}{\cos^2\phi}$ *times that required in two-wire D.C. system.*

7. ***A.C. two-phase, four-wire system.*** Refer to Fig. 11.20.

In this system each phase shares the half of the total load. This system is equivalent to two single-phase, two-wire A.C. systems. In this case cross-sectional area of each conductor is taken half of that of single-phase, two-wire A.C. system but four wires are required in place of two wires, so the same volume of conductor material is required *i.e.,* $\frac{2}{\cos^2\phi}$ *times that required is case of two-wire D.C. system.*

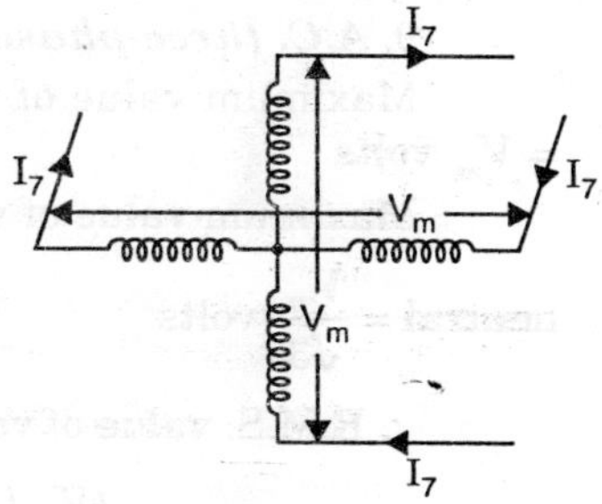

Fig. 11.20

8. ***A.C. two-phase, three-wire system.*** Refer to Fig. 11.21.

Maximum value of voltage between outers = V_m volts

Maximum voltage between either outer and neutral wire = $\frac{V_m}{\sqrt{2}}$

... (the voltages in two-phase windings are 90° out of phase)

R.M.S. voltage between the outer and neutral

$$= \frac{(V_m/\sqrt{2})}{\sqrt{2}} = \frac{V_m}{2}$$

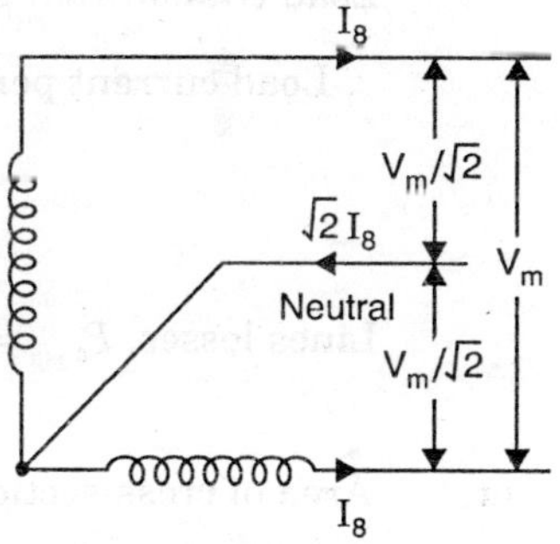

Fig. 11.21

Load current is each outer, $I_8 = \frac{P/2}{(V_m/2)\cos\phi} = \frac{P}{V_m\cos\phi}$

Current is neutral wire $= \sqrt{I_8^2 + I_8^2} = \sqrt{2}\, I_8$

Assuming the *current density to be constant,* the cross-sectional area of neutral wire will be $\sqrt{2}$ times that of either of the outers.

$\therefore$ Resistance of neutral wire, $\frac{R_8}{\sqrt{2}} = \frac{\rho l}{\sqrt{2}a_8}$

Line losses, $P_{loss} = 2I_8{}^2R_8 + (\sqrt{2}I_8)^2\,\frac{R_8}{\sqrt{2}} = I_8{}^2\,R_8(2+\sqrt{2})$

$$= \left(\frac{P}{V_m \cos\phi}\right)^2 \times \frac{\rho l}{a_8}(2+\sqrt{2}) = \frac{P^2 \rho l}{V_m^2 \cos^2\phi \times a_8}(2+\sqrt{2})$$

or, Area of cross-section, $a_8 = \dfrac{P^2 \rho l}{P_{loss} V_m^2 \cos^2\phi}(2+\sqrt{2})$

∴ Volume of conductor material required

$$= 2a_8 l + \sqrt{2}\, a_8 l = a_8\, l(2+\sqrt{2})$$

$$= \frac{P^2 \rho l^2}{P_{loss} V_m^2 \cos^2\phi}(2+\sqrt{2})^2 = \frac{2.914}{\cos^2\phi} \times K \qquad \left(\because\ K = \frac{4P^2\rho l^2}{P_{loss} V_m^2}\right)$$

Hence, volume of conductor material required in this system is $\dfrac{2.914}{\cos^2\phi}$ *times that required in two-wire D.C. system.*

9. ***A.C. three-phase, three-wire system.*** Refer to Fig. 11.22.

Maximum value of voltage between the conductors $= V_m$ volts

Maximum value of voltage between each phase and neutral $= \dfrac{V_m}{\sqrt{3}}$ volts

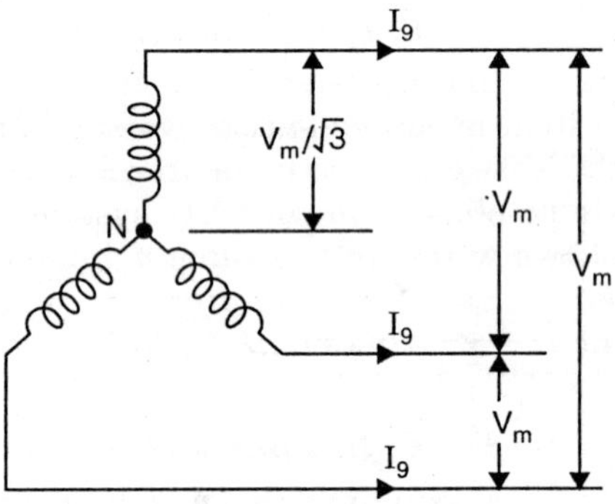

Fig. 11.22

∴ R.M.S. value of voltage per phase

$$= \frac{(V_m / \sqrt{3})}{\sqrt{2}} = \frac{V_m}{\sqrt{6}} \text{ volts}$$

Load transmitted per phase $= \dfrac{P}{3}$

∴ Load current per phase,

$$I_9 = \frac{(P/3)}{(V_m / \sqrt{6})\cos\phi} = \frac{\sqrt{6}P}{3V_m \cos\phi}$$

Lines losses, $P_{loss} = 3I_9^2 R_9 = 3 \times \left(\dfrac{\sqrt{6}P}{3V_m \cos\phi}\right)^2 \times \dfrac{\rho l}{a_9} = \dfrac{2P^2\rho l}{V_m^2 \cos^2\phi \times a_9}$

or, Area of cross-section, $a_9 = \dfrac{2P^2 \rho l}{P_{loss} V_m^2 \cos^2\phi}$

∴ Volume of conductor material required

$$= 3\, a_9 l = 3 \times \left(\frac{2P^2\rho l}{P_{loss} V_m^2 \cos^2\phi}\right) \times l = \frac{6P^2 \rho l^2}{P_{loss} V_m^2 \cos^2\phi}$$

$$= \frac{1.5}{\cos^2\phi} \times \frac{4P^2\rho l^2}{P_{loss} V_m^2} = \frac{1.5}{\cos^2\phi} \times K \qquad \left(\because\ K = \frac{4P^2\rho l^2}{P_{loss} V_m^2}\right)$$

Hence, volume of conductor material required in this system is $\dfrac{1.5}{\cos^2\phi}$ *times that required in two-wire D.C. system.*

10. ***D.C. three-phase, four-wire system.*** Refer to Fig. 11.23.

Assuming balanced load, this system reduces to three-phase three-wire A.C. system except that an additional wire known an neutral wire is provided. Further assuming the cross-sectional area of the neutral wire to be half that of the line conductor,

Volume of conductor material required = 3.5 $a_9 l$

$$= 3.5 \left(\frac{2P^2 \rho l}{P_{loss} V_m^2 \cos^2 \phi} \right) \times l$$

$$= \frac{7}{\cos^2 \phi} \times \frac{P^2 \rho l^2}{P_{loss} V_m^2} = \frac{1.75}{\cos^2 \phi} \times K \qquad \left(\because \; K = \frac{4P^2 \rho l^2}{P_{loss} V_m^2} \right)$$

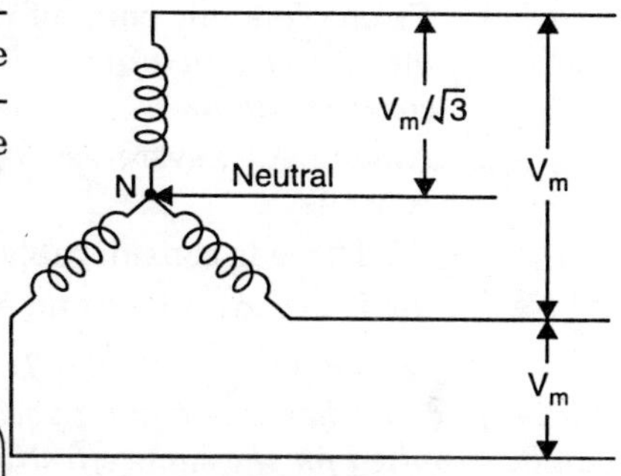

Fig. 11.23

Hence, volume of conductor material required in this system is $\frac{1.75}{\cos^2 \phi}$ *times that required in two-wire D.C. system.*

Summary of Results of Comparison of Various Systems of Transmission

The summary of results of comparison of various systems of transmission is given below :

System	***Volume of conductor material required***	
	Maximum voltage between conductor and earth	***Maximum voltage between any two conductors***
1. ***D.C. system***		
(*i*) Two-wire	1	1
(*ii*) Two-wire with mid-point earthed	0.25	1
(*iii*) Three-wire	0.3125	1.25
2. ***A.C. single-phase system***		
(*i*) Two-wire	$\frac{2}{\cos^2 \phi}$	$\frac{2}{\cos^2 \phi}$
(*ii*) Two-wire with mid-point earthed	$\frac{0.5}{\cos^2 \phi}$	$\frac{2}{\cos^2 \phi}$
(*iii*) Three-wire	$\frac{0.625}{\cos^2 \phi}$	$\frac{2.5}{\cos^2 \phi}$
3. ***A.C. two-phase system***		
(*i*) Two-phase four-wire	$\frac{0.5}{\cos^2 \phi}$	$\frac{2}{\cos^2 \phi}$
(*ii*) Two-phase three-wire	$\frac{1.457}{\cos^2 \phi}$	$\frac{2.914}{\cos^2 \phi}$
4. ***A.C. three-phase system***		
(*i*) Three-phase three-wire	$\frac{0.5}{\cos^2 \phi}$	$\frac{1.5}{\cos^2 \phi}$
(*ii*) Three-phase four-wire	$\frac{0.583}{\cos^2 \phi}$	$\frac{1.75}{\cos^2 \phi}$

From this comparison the following points *emerge* :

1. From economic point of view (saving in conductor material) *D.C. system is ideal for transmission, particularly when p.f. of the load which is usually less than unity is taken into consideration.*
2. Among A.C. systems, *3-phase, 3-wire system is most suitable* because of the following reasons :
   (*i*) There is considerable saving in conductor material.
   (*ii*) It is convenient and efficient.

Example 11.1. *A 3-wire D.C. system is converted to a 3-phase, 4-wire, A.C. system by the addition of another wire equal in section to one of the outers. Find the percentage additional load that can be supplied for the same effective voltage between outers and neutral at the consumer's terminals and the percentage loss.*

Assume that load is balanced and in the A.C. system the p.f. is 0.95.

Solution. Let, V = Voltage between outer and neutral in both the cases,
I_1 = Current in 3-wire D.C. system, and
I_2 = Current in 4-wire A.C. system.

3-wire D.C. system :

Power suplied, $P_1 = 2VI_1$

Power loss, $P_{loss} = 2\,I_1^2R$ ...(R = resistance of each conductor)

$$\%\text{age power loss} = \frac{2I_1^2R}{2VI_1} \times 100 = \frac{100I_1R}{V} \quad ...(i)$$

3-phase, 4-wire A.C. system :

Power supplied, $P_2 = 3VI_2 \cos\phi = 3VI_2 \times 0.95 = 2.85VI_2$

Power loss, $P_{loss} = 3I_2^2R$

$$\%\text{age power loss} = \frac{3I_2^2R}{2.85VI_2} \times 100 = \frac{100I_2R}{0.95V} \quad ...(ii)$$

(Since load is balance, therefore the current in the neutral will be zero)

As percentage loss is same (*given*), therefore,

$$\frac{100I_1R}{V} = \frac{100I_2R}{0.95V} \quad \text{or} \quad I_2 = 0.95\,I_1$$

∴ Power supplied in 3-phase, 4-wire system,

$$P_2 = 2.85\,V \times (0.95\,I_1) \simeq 2.71\,VI_1$$

Hence, percentage increase in load supplied

$$= \frac{P_2 - P_1}{P_1} \times 100 = \frac{2.71VI_1 - 2VI_1}{2VI_1} \times 100 = \mathbf{35.5\%}. \quad \textbf{(Ans.)}$$

Example 11.2. *A three-phase, four-wire A.C. system is employed for lighting. Compare the amount of copper required with that needed for a two-wire D.C. system with the same lamp voltage.*

Assume balanced load, same losses and neutral having half the cross-section of the respective outers. **(M.U.)**

Solution. ***2-wire D.C. system :***

Load current, $I_1 = \dfrac{P}{V}$ amp.

Line losses, $$P_{loss} = 2I_1^2R_1 = 2\left(\frac{P}{V}\right)^2 \times \frac{\rho l}{a_1} = \frac{2P^2\rho l}{V^2 a_1} \quad ...(i)$$

where, ρ, l and a_1 are the resistivity of copper, length of line and a_1 is the cross-sectional area of the conductor respectively,

or, Cross-sectional area, $$a_1 = \frac{2P^2\rho l}{P_{loss}V^2}$$

$$\therefore \text{ Volume of copper required} = 2a_1 l = 2\left(\frac{2P^2\rho l}{P_{loss}V^2}\right) \times l = \frac{2P^2\rho l^2}{P_{loss}V^2} \quad ...(ii)$$

Three-phase, four-wire A.C. system :

Power supplied by each phase $= \frac{P}{3}$ (Since load is balanced)

R.M.S. voltage between line and neutral = V volts

Power factor of lighting load, cos φ = 1 (assumed)

Load current, $$I_2 = \frac{P}{3V}$$

Line losses, $$P_{loss} = 3I_2^2R_2 = 3\left(\frac{P}{3V}\right)^2 \times \frac{\rho l}{a_2} = \frac{P^2\rho l}{3V^2 a_2}$$

or, cross-sectional area, $$a_2 = \frac{P^2\rho l}{3P_{loss}V^2}$$

$$\therefore \text{ Volume of copper required} = 3.5\, a_2 l = 3.5\left(\frac{P^2\rho l}{3P_{loss}V^2}\right) \times l$$

$$= \frac{3.5P^2\rho l^2}{3P_{loss}V^2} = \frac{3.5}{3 \times 4} \times \frac{4P^2\rho l^2}{P_{loss}V^2}$$

$$= 0.2917 \times \frac{4P^2\rho l^2}{P_{loss}V^2}$$ (Since neutral is one-half the cross-section of the outer)

Hence, in case of 3-phase, 4-wire A.C. system the volume of copper required will **0.2917 times** *that required in 2-wire D.C. system.* **(Ans.)**

Example 11.3. *A 60 km long transmission line supplies a load of 6 MVA at 0.8 p.f. lagging at 33 kV, with a transmission efficiency of 88 percent. Calculate the volume of aluminium conductor required for the line in the following cases :*

(*i*) *Single-phase two-wire system.*

(*ii*) *Three-phase, three-wire system.*

Take resistivity of aluminium conductor as $2.84 \times 10^{-8}\ \Omega m$.

Solution. *Given :* Length of the transmission line = 6 km ; Load supplied = 6 MVA, 0.8 p.f. lagging ; Voltage, V = 33 kV ; Transmission efficiency = 88 per cent.

(*i*) ***1-phase, 2-wire system :***

Power transmitted = MVA × cos φ = 6 × 0.8 = 4.8 MW = 4.8×10^6 W

Line losses, $$P_{loss} = \left(1 - \frac{88}{100}\right) \times \text{Power transmitted}$$

$$= (1 - 0.88) \times 4.8 \times 10^6 = 0.576 \times 10^6 \text{ W}$$

Line current, $I_1 = \dfrac{\text{Apparent power}}{V} = \dfrac{6 \times 10^6}{33 \times 10^3} = 181.8$ A

Let, a_1 = Cross-sectional area of the conductor, then,

Line losses, $P_{loss} = 2I_1^{\,2}R_1 = 2I_1^{\,2}\left(\dfrac{\rho l}{a_1}\right)$

or, cross-sectional area, $a_1 = \dfrac{2I_1^2\rho l}{P_{loss}} = \dfrac{2 \times (181.8)^2 \times (2.84 \times 10^{-8}) \times 60 \times 10^3}{0.576 \times 10^6}$

$= 1.955 \times 10^{-4}$ m^2

∴ *Volume of conductor required*

$= 2a_1l = 2 \times (1.955 \times 10^{-4}) \times 60 \times 10^3 =$ **23.46 m³. (Ans.)**

(*ii*) ***3-phase, 3-wire system :***

Line current, $I_2 = \dfrac{\text{Apparent power}}{\sqrt{3}V} = \dfrac{6 \times 10^6}{\sqrt{3} \times 33 \times 10^3} = 104.97$ A

Let, a_2 = Cross-sectional area of the conductor, then

Line losses, $P_{loss} = 3I_2^{\,2}R_2 = 3 \times I_2^{\,2} \times \left(\dfrac{\rho l}{a_2}\right)$

or, cross-sectional area, $a_2 = \dfrac{3I_2^2\rho l}{P_{loss}} = \dfrac{3 \times (104.97)^2 \times (2.84 \times 10^{-8}) \times 60 \times 10^3}{0.576 \times 10^6}$

$= 9.78 \times 10^{-5}$ m^2

∴ *Volume of conductor required* $= 3a_2l = 3 \times (9.78 \times 10^{-5}) \times 60 \times 10^3 =$ **17.60 m³. (Ans.)**

It may be noted that *volume of conductor and hence weight required is less in case of 3-phase 3-wire system.*

Example 11.4. *Determine the weight of copper required for a 3-phase transmission system supplying at 380 kV between lines a load of 80 MW at lagging p.f. of 0.8. The length of the line is 120 km. The efficiency of transmission is 90 percent. The neutral point is earthed. The resistance of 1 km long conductor having 1 cm² X-section is 0.046 Ω. The density of copper is 0.0981 N/cm³.*

Also calculate the weight of copper required for D.C. transmission of the same power with equal losses working at a voltage between the lines double that of the peak voltage to the earth of the 3-phase supply.

Solution. *Given :* Line voltage, V = 380 kV ; Load = 80 MW ; p.f. = 0.8 ; length of the line = 120 km ; Efficiency of transmission = 90% ; Resistance of the conductor having 1 km length and 1 cm² cross-section = 0.046 Ω ; Density of copper = 0.0981 N/ cm³.

Weight of copper required :

Case I : *A.C. transmission :*

Line current, $I = \dfrac{80 \times 10^6}{\sqrt{3} \times 380 \times 10^3 \times 0.8} = 151.9$ A

Power loss in the line $= \left(1 - \dfrac{90}{100}\right) \times 80 = 8$ MW $= 8 \times 10^6$ W

As I^2R loss in the conductors is 8×10^6 W,

or, I^2R loss per conductor $= \frac{1}{3} \times (8 \times 10^6)$

$$\therefore \text{Resistance per conductor} = \frac{8 \times 10^6}{3 \times (151.9)^2} = 115.6\ \Omega$$

or, Resistance per conductor per km $= \frac{115.6}{120} = 0.963\ \Omega$

$\therefore$ Conductor area $= \frac{0.046}{0.963} = 0.0478\ \text{cm}^2$

Volume of copper conductor per metre run $= 0.0478 \times 100 = 4.78\ \text{cm}^3$

$\therefore$ Weight of copper metre run $= 4.78 \times 0.0981$ N

Hence, weight of copper for 3 conductors of 120 km length

$= 3 \times (120 \times 1000) \times 4.78 \times 0.0981$

$= 168810$ N or **168.81 kN. (Ans.)**

Case II : *D.C. transmission :*

Let, V_m = Maximum A.C. volts to neutral, and

P = Power supplied in watts.

Then, D.C. voltage between lines $= 2\ V_m$ *(Given)*

$$\text{Losses (D.C.)} = 2I^2R_1 = 2\left(\frac{P}{2V_m}\right)^2 R_1 = \frac{P^2R_1}{2V_m^2}$$

$$\text{R.M.S. value of phase voltage} = \frac{V_m}{\sqrt{2}}$$

$$\text{R.M.S. value of line voltage} = \sqrt{3} \times \frac{V_m}{\sqrt{2}}$$

$$\therefore \text{Line current, } I_L = \frac{P}{\sqrt{3} \times \frac{\sqrt{3}}{\sqrt{2}} V_m \cos\phi} = \frac{P}{\frac{\sqrt{3}}{\sqrt{2}} V_m \cos\phi}$$

$$\text{Losses (A.C.)} = 3\ I_L^2R_2 = 3\left(\frac{P}{\frac{3}{\sqrt{2}} V_m \cos\phi}\right)^2 \times R_2 = \frac{2P^2R_2}{3V_m^2\cos^2\phi}$$

Since the losses are same, therefore,

$$\frac{P^2R_1}{2V_m^2} = \frac{2P^2R_2}{3V_m^2\cos^2\phi} \quad \text{or} \quad \frac{R_1}{R_2} = \frac{4}{3\cos^2\phi}$$

As resistance $\propto \frac{1}{X\text{-sectional area}}$,

$$\frac{\text{Area of conductor in D.C.}}{\text{Area of conductor in A.C.}} = \frac{3}{4}\cos^2\phi, \text{ and}$$

$$\frac{\text{Weight of copper in D.C.}}{\text{Weight of copper in A.C.}} = \frac{2}{3} \times \frac{3}{4}\cos^2\phi = \frac{\cos^2\phi}{2}$$

... Since in D.C. there are 2 wires and in A.C. there are 3 wires.

Hence, weight of conductors in D.C. = $\frac{0.8^2}{2} \times 168.81 =$ **54.02 kN. (Ans.)**

Example 11.5. *A double circuit three-phase, three-wire line is converted into a triple circuit two-pole two-wire D.C. line with same maximum voltage to earth as that of the A.C. line. Calculate the percentage increase in the capacity of the line if the transmission loss remains unchanged. Assume power factor of the A.C. load to be unity.* **(P.U.)**

Solution. As the existing double-circuit three-phase three-wire line containing 6 conductors is converted into triple-circuit two-pole two-wire D.C. line, it is *assumed that the line resistance will remain the same.*

(i) ***D.C. line-wire system :***

Fig. 11.24 shows a 2-wire D.C. system, for which I_1 is the line current, R is the resistance of conductor.

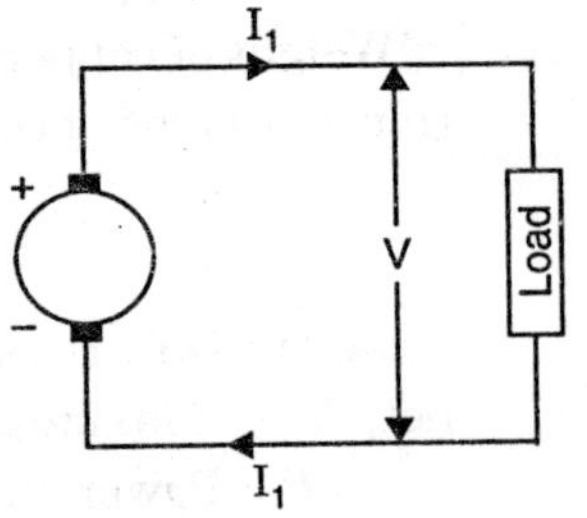

Fig. 11.24

Power transmitted, $P_1 = VI_1$

Line current, $I_1 = \frac{P_1}{V}$

Total I^2R loss for both the conductors $= 2I_1^2R = 2\left(\frac{P_1}{V}\right)^2 \times R$

$\therefore$ Total line losses for all the 6 conductors,

$$(P_{loss})_1 = 3 \times 2\left(\frac{P_1}{V}\right)^2 \times R = 6\left(\frac{P_1}{V}\right)^2 R$$

(ii) ***Three-phase, three-wire star-connected system :***

This system is shown in Fig. 11.25. Maximum value of voltage between the outer and the neutral = V volts

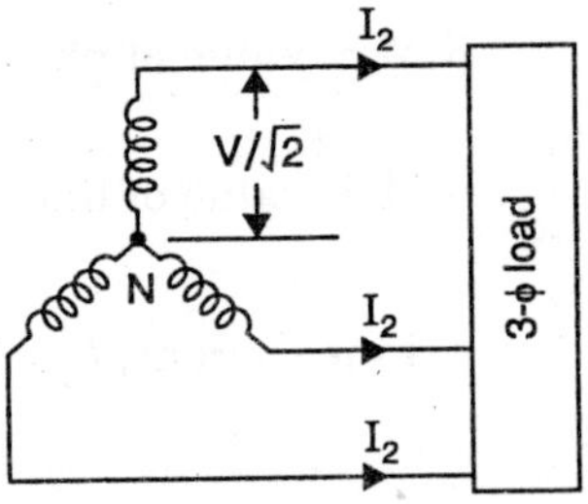

Fig. 11.25

$\therefore$ R.M.S. value of voltage per phase $= \frac{V}{\sqrt{2}}$ volts

If I_2 is the r.m.s. phase current, R the resistance of each conductor and $\cos\phi$, the power factor of load, then

$$P_2 = 3\,V_{ph}\,I_{ph}\cos\phi = 3 \times \frac{V}{\sqrt{2}} \times I_2 \times 1$$

or, $$I_2 = \frac{\sqrt{2}P_2}{3V}$$

Line losses $= 3I_2^2R = 3 \times \left(\frac{\sqrt{2}P_2}{3V}\right)^2 R$

Total line losses for 2 such 3-ϕ circuits, $(P_{loss})_2 = 2 \times 3\left(\frac{\sqrt{2}P_2}{3V}\right)^2 R = 6\left(\frac{\sqrt{2}P_2}{3V}\right)^2 R$

Since $(P_{loss})_1 = (P_{loss})_2$ ...(*Given*)

$\therefore$ $$6\left(\frac{P_1}{V}\right)^2 R = 6\left(\frac{\sqrt{2}P_2}{3V}\right)^2 R, \quad \text{or,} \quad \frac{P_1}{V} = \frac{\sqrt{2}P_2}{3V}$$

$\therefore$ $$\frac{P_2}{P_1} = \frac{3}{\sqrt{2}} = 2.12$$

Hence, *percentage increase in the capacity of 3-phase double circuit lines is by* **112 (%). (Ans.)**

Example 11.6. *An existing single-phase A.C. system comprising of two overhead conductors is to be converted into a three-phase, 3-wire system by providing an additional conductor of same size. Caculate the percentage of additional load that can be transmitted by the three-phase system if the operating line voltage and percentage line losses remain the same in both the systems.*

Solution. Let, I_1 = Single-phase current,
I_3 = Three-phase current,
R = Ressistance of each conductor, and
$\cos\phi$ = Power factor in both the systems.

(*i*) ***Single-phase system :***

Power transmitted, $P_1 = VI_1 \cos\phi$

Line loss $= 2\,I_1^2R$

$$\therefore \text{ \%age line loss} = \frac{2I_1^2 R}{VI_1 \cos\phi} \times 100$$

(*ii*) ***Three-phase system :***

Power transmitted, $P_3 = \sqrt{3}\,VI_3 \cos\phi$

Line loss $= 3\,I_3^2R$

$$\text{\%age line loss} = \frac{3I_3^2 R}{\sqrt{3}VI_3 \cos\phi} \times 100$$

Since the %age line loss remains same in both the systems (*given*), therefore,

$$\frac{2I_1^2 R}{VI_1 \cos\phi} \times 100 = \frac{3I_3^2 R}{\sqrt{3}VI_3 \cos\phi} \times 100 \quad \text{or} \quad I_3 = \frac{2}{\sqrt{3}} I_1$$

$$\therefore \qquad P_3 = \sqrt{3}\,VI_3 \cos\phi = \sqrt{3}\,V\left(\frac{2}{3}I_1\right)\cos\phi = 2VI_1\cos\phi$$

$\therefore$ %age additional load that can be transmitted

$$= \frac{P_3 - P_1}{P_1} = \frac{2VI_1\cos\phi - VI_1\cos\phi}{VI_1\cos\phi} \times 100 = 100$$

Hence, **100 per cent** *additional load can be transmitted by converting single-phase line to three-phase line.* **(Ans.)**

Example 11.7. *Show that for a given amount of power to be transmitted over a given distance, the volume of copper is inversely proportional to the square of the voltage.*

Solution. Consider a n-phase system of transmission as shown in Fig. 11.26.

Let, V = Voltage to neutral,
I = Current lagging V by angle, ϕ on each phase of the system, and
R = Resistance of each line conductor.

Then, power transmitted by each phase,

$$P = \frac{VI\cos\phi}{100} \text{ kW, and}$$

$$\text{Power loss per phase, } P_{loss} = \frac{I^2R}{1000} \text{ kW/phase}$$

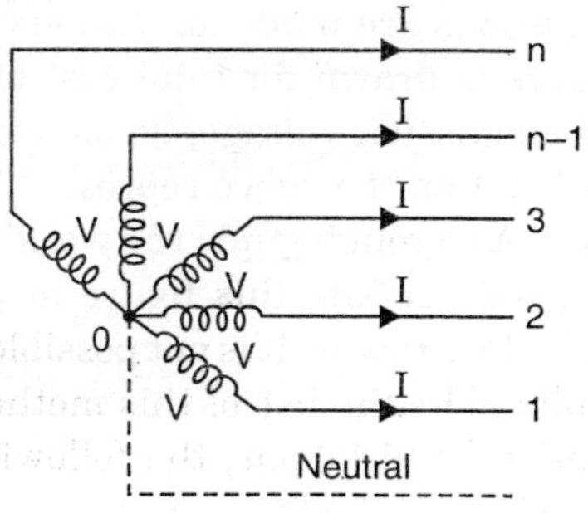

Fig. 11.26

$$= \left(\frac{1000P}{V \cos\phi}\right)^2 \times \frac{R}{1000} \quad \text{...(in terms of power transmitted)}$$

$$= \frac{1000P^2R}{V^2 \cos^2 \phi} \text{ kW/phase}$$

If ρ, l and a denote the resistivity of conductor, lenght of line and the cross-sectional area of the conductor respectively, then

$R = \frac{\rho l}{a}$, so we may write

$$P_{loss} = \frac{1000P^2\rho l}{V^2 \cos^2 \phi \times a}, \quad \text{or,} \quad a = \frac{1000P^2\rho l}{P_{loss} V^2 \cos^2 \phi}$$

$$\text{Volume of copper} = a\, l = \left(\frac{1000P^2\rho l}{P_{loss}}\right) \times \frac{1}{V^2 \cos^2 \phi},$$

i.e., for a fixed given amount of power to be transmitted over a given length with a fixed power loss, the figures wihin the bracket is constant ; and the volume of copper required is *inversely proportional to the square of the voltage.* ......**(Proved).**

11.7. CHOICE OF TRANSMISSION VOLTAGE

In case of any transmission line there is a specific limit fixed for the voltage to be employed, beyond which nothing is gained economically. The limit reaches when the *cost of conductor, insulator, supports, transformer, switchgear lighting arrestors and the erection cost is "minimum".*

The economical transmission voltage (assuming power to be transmitted, generation voltage and length of transmission are known) is found as follows :

We tentatively choose a standard voltage of transmission and work out the following costs :

(*i*) *Transformers,* at the generating and receiving ends.

(*ii*) *Switchgear* (consisting of isolators, bus-bars, circuit breakers, relays etc.).

(*iii*) *Lightning arrestors.*

(*iv*) *Insulators.*

(*v*) *Supports* (including the *X*-arms and other fittings).

(*vi*) *Conductor cost.*

The total sum of all the costs gives the total capital cost of transmission for the voltage considered. Similar calculations are made for various standard voltages. Then a curve is drawn for total cost of transmission against the transmission voltage, as shown in Fig. 11.27. The lowest point *A* on the curve represent *optimum voltage* to be chosen (As a rough guide the working voltage may be taken as 0.6 kV per km, this figure in practice varies from 0.4 to 0.9). Practically, it is not possible to determine the economic voltage by the use of this method. Hence, to avoid complications and labour, the following empirical formulae are used :

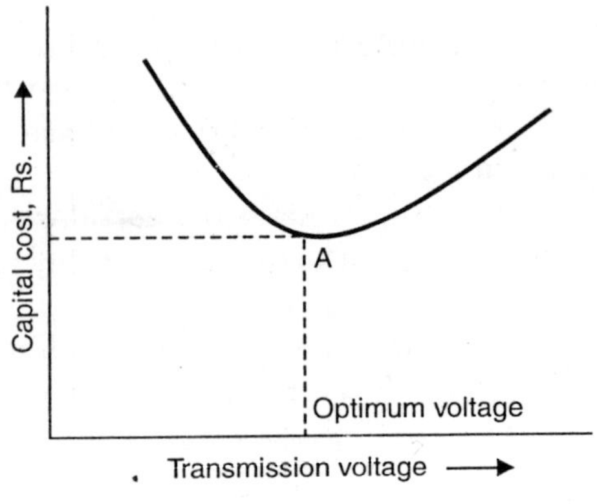

Fig. 11.27

(*i*) According to modern American practice (based on empirical formula) empirical line-to-line voltage (kV),

$$V = 5.5\sqrt{0.62l + \frac{3P}{100}}$$

where, l = Distance of transmission in km, and

P = Extimated maximum kW/phase to be delivered over a single circuit.

(*ii*) Voltage in kV (line-to-line), $V = 5.5\sqrt{0.62l + \frac{kVA}{150}}$

where, l = Distance of transmission, km, and

kVA = Total power.

11.8. CONDUCTOR SIZE AND KELVIN'S LAW

The determination of economical size of the conductor is of paramount importance since the cost of conductor material generally forms a considerable part of the total cost of a transmission line. The most economical conductor size is that for which the total *annual cost of transmission line is minimum.* The total annual cost consists of two parts *viz.*, annual charge on capital outlay and annual cost of energy wasted in the conductor.

(*i*) *Annual charge on capital outlay.* It is on account of annual interest and depreciation on capital cost of conductors, supports and insulators and their erection cost in case of *overhead systems.* For *underground system,* it is on account of annual interest and depreciation on cost of conductors, insulation and the cost of laying the cables.

For a particular voltage cost of insulation is practically constant and does not change with X-section of the conductor, but the *cost of conductor varies "directly"* as *the X-section of the conductor, irrespective of the system of transmission.* In case of *"overhead system"* the *cost of supports and their erection partly varies as the X-section of the conductors and partly constant.*

Thus, the total annual charge on an overhead transmission line can be expressed as :

Total annual charge = $P_1 + P_2a$ ...(*i*)

where, P_1 and P_2 are constants and a is the area of X-section of the conductor.

(*ii*) *Annual cost-of energy wasted in the conductor.* It is on account of energy lost in conductor due to its ohmic resistance, *i.e.*, I^2R losses, losses in insulating material and in metallic sheaths (for insulated cables). The *resistance of the conductor being inversely proportional to its X-sectional area, the energy loss due to ohmic* resistance may be represented as :

$$\text{Annual cost of energy wasted} = \frac{P_3}{a} \quad ...(ii)$$

where P_3 is constant.

$$\therefore \text{ Total annual cost, } C = (P_1 + P_2a) + \frac{P_3}{a} \quad \text{[Adding expression (i) and (ii)]}$$

The value of C will be minimum if differentiation with respect a is zero, *i.e.*, $\frac{dC}{da} = 0$

or, $$\frac{d}{da}\left(P_1 + P_2a + \frac{P_3}{a}\right) = 0$$

or, $$P_2 - \frac{P_3}{a^2} = 0$$

or, $$P_2a = \frac{P_3}{a} \quad \text{or,} \quad a = \sqrt{\frac{P_3}{P_2}}$$

In other words, the ***variable part of anual charge should be equal to the cost of annual losses due to energy wasted in the conductor for most economical working.***

The above statement is **Kelvin's law.** Although this law hold true theorectically, but in actual working its application involves complicacies/difficulties.

Fig. 11.28 shows the *graphical representation of Kelvin's law.* The lowest point *A* on the sum curve ($P_1 + P_2a + P_3/a$) which coresponds to the point of intersection of (P_2a) and $\frac{P_3}{a}$ curves gives the most economical conductor size.

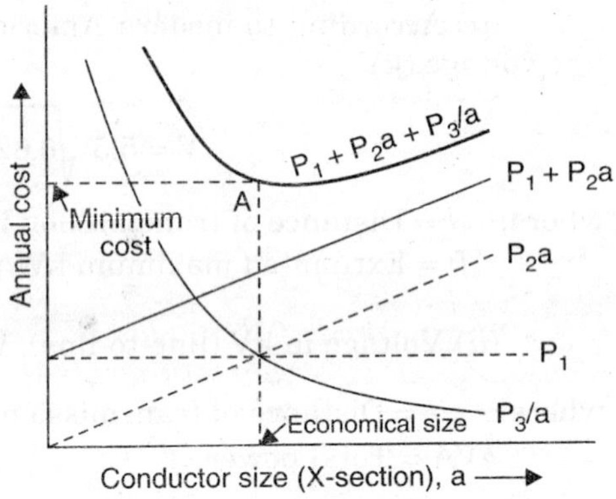

Fig. 11.28

Limitations of Kelvin's law :

Following are the *limitations* of Kelvin's law :

1. It is difficult to estimate the loss of energy in line without actual load curves, which are not available at the time of estimation.
2. The assumption that annual charge is of form ($P_1 + P_2a$) is strictly speaking not true. For example, in underground cables neither the cost of laying the cables nor the cost of cable dielectric and sheath vary in this manner.
3. The Kelvin's law does not take into account factor like mechanical strength, safe current density, corona loss etc
4. The conductor size determined may be of such a small *X*-section that it may cause too much voltage drops in the line.
5. Interest and depreciation on the capital outlay cannot be accurately determined.

Example 11.8. *A two-conductor cable one km long, is required to supply a constant load of 180 A throughout the year. The cost of cable is Rs. (120 a + 60) per metre , where 'a' is the area of X-section of the conductor in cm². The cost of enegy is 20 P per kWh and interest and depreciation charges amount to 10 percent. Specific resistivity of copper is 1.84 μΩ cm. Find the most economical X-section of the cable.*

Solution. *Given :* $l = 1$ km ; $I = 180$ A ; cost of cable per metre length = Rs ($120a + 60$) ; cost of energy = 20 *p*/kWh ; Interest and depreciation charges = 10% ; $\rho = 1.84\ \mu\Omega$ cm.

Most economical X-section of the cable, a (cm²) :

Resistance of each conductor, *R*

$$R = \frac{\rho l}{a} = \frac{1.84 \times 10^{-6} \times (1000 \times 100)}{a} = \frac{0.184}{a}\ \Omega$$

$$\text{Energy lost per annum} = \frac{2I^2R \times 8760}{1000}$$

$$= \frac{2 \times (180)^2 \times 0.184 \times 8760}{a \times 1000} = \frac{104447}{a}\ \text{kWh}$$

$$\text{Annual cost due to energy lost} = \text{Rs.}\ \frac{20}{100} \times \frac{104447}{a} = \text{Rs.}\ \frac{20889.4}{a}$$

Capital cost (variable) of the cable = Rs. $120a \times 1000 = 120000\ a$

Annual charges on account of interest and depreciation on variable cost of cable

$$= \text{Rs.}\ \frac{10}{100} \times 120000a = 12000\ a$$

According to Kelvin's law, for most economical X-section of the conductor,

Variable annual charge = Annual cost of energy lost

$$12000\,a = \frac{20889.4}{a}$$

or,
$$a = \sqrt{\frac{20889.4}{12000}} = \mathbf{1.319\ cm^2}. \quad \textbf{(Ans.)}$$

Example 11.9. *A 3-phase overhead transmission line supplies a constant load of 6 MW at 33 kV and 0.8 p.f. logging throughout the year. The cost of the line is Rs. (80000 a + 3500) per km, where 'a' is the X-section of each conductor in cm². The cost of energy is 25 P per kWh and interest and depreciation total 10 percent per annum. Assuming specific resistance of conductor material as 1.8 μΩ cm, find the most economical size of the conductor.*

Solution. *Given :* Load transmitted = 6 MW at 33 kV, 0.8 p.f. logging ;

Cost of the line = Rs (80000 a + 3500) per km, 'a' in cm² ;

Cost of energy = 25P/kWh ; Interest and depreciation = 10% per annum ;

ρ for the conductor material = 1.8 μΩ cm.

Most economical conductor size, a :

Resistance of each conductor, $R = \frac{\rho l}{a} = \frac{1.8 \times 10^{-6} \times (1000 \times 100)}{a} = \frac{0.18}{a}$

Line current,
$$I = \frac{P}{\sqrt{3} V \cos\phi} = \frac{6 \times 10^6}{\sqrt{3} \times (33 \times 10^3) \times 0.8} = 131.2 \text{ A}$$

Energy lost per annum
$$= \frac{3I^2R \times 8760}{1000} \text{ kWh}$$
$$= \frac{3 \times (131.2)^2 \times 0.18 \times 8760}{a \times 1000} = \frac{81426}{a} \text{ kWh}$$

Annual cost of energy lost $= \text{Rs. } \frac{25}{100} \times \frac{81426}{a} = \frac{20356.5}{a}$

Capital cost (variable) of the cable = Rs. 80000 a ...(*Given*)

Variable annual charge = 10% of capital cost (variable) of line

$$= \text{Rs } \frac{10}{100} \times 80000\,a = 8000\,a$$

According to Kelvin's law, for most economical conductor size,

Variable annual charge = Annual cost of energy lost

$$8000a = \frac{20356.5}{a}$$

or
$$a = \sqrt{\frac{20356.5}{8000}} \approx \mathbf{1.6\ cm^2}. \quad \textbf{(Ans.)}$$

Example 11.10. *The following data relate to a 2-wire feeder ;*

Current carried thougth out the year = *220 A*

The portion of the capital cost which is proportional to X-sectional area

= *Rs. 6 per kg of copper conductor*

Cost of energy = *6 P per kWh*

Interest and depreciation charges = *10% per annum*

Density of copper $= 8.93\ g/cm^3$
Specific resistance of copper $= 1.8\ \mu\Omega\ cm$
Find the most economical X-section of the conductor.

Solution. Let, a = Most economical X-sectional area of each conductor, cm^2
Consider 1 m length of the feeder.

$$\text{Resistance of each conductor, } R = \frac{\rho l}{a} = \frac{1.8 \times 10^{-6} \times 100}{a} = \frac{1.8 \times 10^{-4}}{a}\,\Omega$$

$$\text{Energy lost per annum} = \frac{2I^2R \times 8760}{1000}\ \text{kWh} = \frac{2 \times (220)^2 \times 1.8 \times 10^{-4} \times 8760}{a \times 1000}$$

$$= \frac{152.6}{a}\ \text{kWh}$$

$$\text{Annual cost of energy lost} = \text{Rs. } \frac{6}{100} \times \frac{152.6}{a} = \text{Rs. } \frac{9.156}{a}$$

Mass of 1 metre feeder $= 2 \times$ (Volume $\times$ density)

$$= \frac{2 \times (a \times 100) \times 8.93}{1000} = 1.786\ a\ \text{kg}$$

Capital cost (variable) $=$ Rs. $6 \times 1.786\ a =$ Rs. $10.72\ a$

$$\text{Variable annual charges} = \frac{10}{100} \times 10.72a = 1.072\ a$$

For *most economical conductor size,*

Variable annual charges = Annual cost of energy lost

$$1.072a = \frac{9.156}{a}$$

or,
$$a = \sqrt{\frac{9.156}{1.072}} = \mathbf{2.92\ cm^2.} \quad \textbf{(Ans.)}$$

Example 11.11. *A 2-core, 11 kV cable is to supply 1 MW at 0.8 p.f. lag for 3000 hours in a year. Capital cost of the cable is Rs. (20 + 400 a) per metre where 'a' is the X-sectional area of core in cm². Interest and depreciation total 10% and cost per unit of energy is 15 P. If the length of cable is 1 km, calculate the most economical X-section of the conductor. The specific resistance of copper is 1.75 μΩ cm.* **(N.U.)**

Solution. *Given* : Load to be supplied = 1 MW at 11 kV and 0.8 p.f. lag.
Capital cost of the cable = Rs. (20 + 400 a), a being the X-sectional area
Interest and depreciation rate = 10%
Cost per unit of energy = 15 P/kWh
Length of the cable = 1 km
Specific resistance of copper = 1.75 μΩ cm.

Most economical X-section of the conductor, a (cm^2) :

Cost of 1 km length of cable = Rs. $(20 + 400\ a) \times 1000 = (20000 + 400 \times 10^3\ a)$

$$\text{Resistance of 1 km length of cable} = \frac{1.75 \times 10^{-6} \times (1000 \times 100)}{a} = \frac{0.175}{a}\,\Omega$$

$$\text{Full load current, } I = \frac{P}{V \cos\phi} = \frac{1 \times 10^6}{11 \times 10^3 \times 0.8} = 113.6\ \text{A}$$

Power loss of the cable per annum $= \frac{2I^2R \times 3000}{1000} = \frac{2 \times (113.6)^2 \times 0.175 \times 3000}{a \times 1000}$

$= \frac{13550.2}{a}$ kWh

Annual cost of energy loss $= \frac{15}{100} \times \frac{13550.2}{a} = \frac{2032.5}{a}$

Capital cost (variable) of the cable $= 400 \times 10^3 a$ ...(*Given*)

Variable annual charges = 10% of capital cost. (variable) of line

$= \frac{10}{100} \times 400 \times 10^3 a = 40000a$

Using Kelvin's law,

Variable annual charges = Annual cost of energy lost

$$40000\, a = \frac{2032.5}{a}$$

or

$$a = \sqrt{\frac{2032.5}{40000}} = \mathbf{0.2254\ cm^2}. \quad \textbf{(Ans.)}$$

$$\left[\text{or, } d = \sqrt{\frac{0.2254 \times 4}{\pi}} = 0.536 \text{ cm}\right]$$

Example 11.12. *The following data refer to a 3-phase cable supplying power to an industry at 11 kV:*

Maximum demand	*= 1000 kW*
Total annual energy consumption	*= 5×10^6 kWh at 0.85 p.f.*
Capital cost of the cable	*= Rs. (80000 a + 20000) per km, 'a' being X-sectional area in cm^2/per core*
Energy cost	*= 5 P/kWh*
Rate of interest and depreciation	*= 10%*
Specific resistance of copper	*= 1.72×10^{-6} Ω per cm^3*
Voltage at which power is supplied	*= 11 kV*
Loss load factor	*= 0.25 load factor + 0.75 (load factor)^2*

Determine the most economical section of the conductor. **(Banglore University)**

Solution. Let, l = Length of the cable, km.

Annual load factor $= \frac{\text{Annual energy consumption}}{\text{Maximum demand} \times 8760}$

$= \frac{5 \times 10^6}{1000 \times 8760} = 0.57$

Loss load factor $= 0.25 \times 0.57 + 0.75\,(0.57)^2 = 0.386$

Current corresponding to maximum demand,

$$I = \frac{1000 \times 1000}{\sqrt{3} \times 11000 \times 0.85} = 61.7 \text{ A}$$

Capital cost (variable) of the cable = Rs. 80000 al

Annual cost on account of interest and depreciation on capital cost (variable) of the cable

$$= \text{Rs. } \frac{10}{100} \times 80000\, a\,l = 8000\, a\, l$$

Energy lost per annum $= \dfrac{3I^2R \times 8760}{1000} \times$ loss load factor

$$= \frac{3 \times (61.7)^2 \times (1.72 \times 10^{-6} \times l \times 1000 \times 100) \times 8760}{a \times 1000} \times 0.386$$

$$= \frac{6642.2\, l}{a}$$

Annual cost due to energy lost $= \text{Rs. } \dfrac{5}{100} \times \dfrac{6642.2\, l}{a} = \text{Rs. } \dfrac{332.11\, l}{a}$

For most economical section of the conductor,

Variable annual charge = Annual cost of energy lost

i.e.,

$$8000\, al = \frac{332.11\, l}{a}$$

$$a = \sqrt{\frac{332.11}{8000}} = \mathbf{0.204\ cm^2}. \quad \textbf{(Ans.)}$$

Example 11.13. *Assume the following daily load cycle for a 11 kV, 3-phase line :*

(i) 2.4 MW at unity p.f. for 6 hours,

(ii) 0.8 MW at 0.8 p.f. (lag) for 6 hours, and

(iii) 0.4 MW at unity p.f. for 12 hours.

If the resistance of the line conductor of 1 cm² X-sectional area is 0.19 ohm and energy cost is Re 1. per unit, calculate the annual energy lost. **(M.U.)**

Solution. Currents corresponding to loads of different values for given load cycle are :

$$I_1 = \frac{2.4 \times 10^6}{\sqrt{3} \times 11 \times 10^3 \times 1} = 126 \text{ A}$$

$$I_2 = \frac{0.8 \times 10^6}{\sqrt{3} \times 11 \times 10^3 \times 0.8} = 52.5 \text{ A}$$

$$I_3 = \frac{0.4 \times 10^6}{\sqrt{3} \times 11 \times 10^3 \times 1} = 21 \text{ A}$$

∴ Energy loss during the period of 24 hours in each line

$$= (126)^2 \times 0.19 \times 6 + (52.5)^2 \times 0.19 \times 6 + (21)^2 \times 0.19 \times 12$$

$$= 18099 + 3142 + 1005 = 22246 \text{ Wh} = 22.246 \text{ kWh per day}$$

∴ Total energy loss annually in all the three conductors

$$= 3 \times 22.246 \times 365 = 24359 \text{ kWh/year}$$

∴ *Annual energy cost* = Rs. 24359 × 1 = **Rs. 24359. (Ans.)**

Example 11.14. *Determine the most economical size of a 3-phase line which supplies the following loads at 10 kV :*

(i) 1,000 kW at 0.8 p.f. (lag) for 10 hours,

(ii) 500 kW at 0.9 p.f. (lag) for 8 hours, and

(iii) 100 kW at u. p. f. for 6 hours.

The above gives the daily load cycle. The cost per km of the completely erected line is Rs. (8000 a + 1500) where a is the area of X-section of each conductor in cm². The combined interest and depreciation is 10 percent per annum of capital cost. Cost of energy losses is 5 paise per kWh. Resistivity of conductor material = 1.72 × 10⁻⁶ Ω-cm. **(AMIE, Power Systems)**

Solution. Let, l = Length of the line, km

Capital cost of the line = Rs. $(8000\,a + 1500) \times l$, where '$a$' is in cm^2

Annual cost on account of interest and depreciation on capital (variable) cost

$$= \text{Rs. } \frac{10}{100} \times 8000\,a \times l = 800\,a\,l$$

Currents corresponding to loads of different values for given load cycle are :

$$I_1 = \frac{1000 \times 10^3}{\sqrt{3} \times 10 \times 10^3 \times 0.8} = 72.17 \text{ A}$$

$$I_2 = \frac{500 \times 10^3}{\sqrt{3} \times 10 \times 10^3 \times 0.9} = 32.07 \text{ A}$$

$$I_3 = \frac{100 \times 10^3}{\sqrt{3} \times 10 \times 10^3 \times 1} = 5.77 \text{ A}$$

$$\text{Resistance of line conductor, } R = \frac{\rho l}{a} = \frac{1.72 \times 10^{-6} \times 1000 \times 100 \times l}{a}\ \Omega = \frac{0.172\,l}{a}\ \Omega$$

$$\text{Energy lost per annum} = \frac{3I^2R \times 365}{1000} \text{ kWh}$$

$$= \frac{3 \times 0.172\,l}{a} \times \frac{365}{1000}\,[(72.17)^2 \times 10 + (32.07)^2 \times 8 + (5.77)^2 \times 6]$$

$$= \frac{11397}{a} \times l$$

$$\text{Cost of energy wasted per annum} = \text{Rs. } \frac{5}{100} \times \frac{11397}{a} \times l = \frac{567.85}{a} \times l$$

For most economical X-section of the conductor,

Variable annual charge = Annual cost of energy lost

$$800\,a\,l = \frac{569.85}{a} \times l$$

or

$$a = \sqrt{\frac{569.85}{800}} = \mathbf{0.844\ cm^2}. \quad \textbf{(Ans.)}$$

HIGHLIGHTS

1. By transmission and distribution of electric power is meant its conveyance from the central station where it is generated to the places where it is demanded by the consumers.
2. An electric supply system comprises the following three principal components :
   (*i*) Power station (*ii*) Transmission lines (*iii*) Distribution lines.
3. A.C. system should be used for generation and distribution and D.C. system for tranmission of the electric power.
4. The transmission lines may be classified as :
   (*i*) Single lines (*ii*) Parallel lines (*iii*) Radial lines
   (*iv*) Ring system (*v*) Network.
5. "The variable part of annual charge should be equal to the cost of annual losses due to energy wasted in the conductor for most economical working". This statement is *Kelvin's law*.

THEORETICAL QUESTIONS

1. What are the principal components of an electric sypply system ? Draw and explain briefly a typical power supply network.
2. Give the comparison between D.C. and A.C. systems of transmission and distribution.
3. How are transmission lines classified ?
4. List the advantages and limitations of high transmission voltage.
5. What are the various systems of power transmission ?
6. Write a short note on "Choice of transmission voltage".
7. Explain Kelvin's law for determination of conductor size. Give its practical limitations,.
8. Discuss the technology of transmitting large amounts of power over long distances by means of high voltage direct current (H.V.D.C.) What are the advantages and difficulties of H.V.D.C. transmission as compared to the usual 3-phase A.C. transmission.
9. Distinguish between a feeder, a distributor and service mains.
10. Compare the conductor material (volume) for 3-phase, 4-wire distribution with that of 2-wire D.C. distribution system. State assumptions made.
11. Discribe the various systems of power transmission and compare the following as regard the amount of copper used for the same distance, the same power transfer, the same maximum voltage to ground and the same power loss :
    (*i*) 3-phase, 3-wire A.C. (*ii*) 3-wire D.C.
    (*iii*) 1-phase, 2-wire A.C.
12. Show that for overhead system the ratios of the volume of conductor in D.C., single-phase A.C. (mid-point earthed), and three-phase A.C. are given by :
    $$V_1 : V_2 : V_3 :: 1 : \frac{2}{\cos^2 \phi} : \frac{2}{\cos^2 \phi}$$
    where, cos ϕ is the power factor of the load.
    Assume equal power transmitted over equal length with equal losses and maximum voltage to earth to be same in all cases.
13. Compare the volume of copper required for the distributor cable in a low voltage distribution network in *a* D.C. 3-wire system, with a 3-phase, 4-wire system. Assume the same consumer voltage, same percentage loss, unity power factor and balanced load. The neutrals are of half the *X*-section of corresponding outer.
14. Show that for the transmision of power over a given length :
    (*i*) Power loss is inversely proportional to the square of transmission voltage and also inversely proportinal to the square of power factor (in case of A.C.).
    (*ii*) Volume of conductor is inversely proportional to the square of the voltage and also inversely proportional to the square of the p.f. (in case of A.C. system).
    Assume that the line is working at constant efficiency.
15. State Kelvin's law and explain why in practice the law is usually not strictly observed.
16. State the economic considerations which influence the selection of the size of the conductor of a transmission line.

UNSOLVED EXAMPLES

1. A 3-wire D.C. system is converted to a 3-phase, 4-wire A.C. system by the addition of another wire equal in section to one of the outers. Find the percentage additional load that can be supplied for the same effective voltage between outers and neutral at the consumer's terminals and percentage loss. Assume that load is balanced and in the A.C. system p.f. is 0.9. [**Ans.** 21.5%]
2. A three-phase, four-wire A.C. system is employed for lighting. Compare the amount of copper required with that needed for a two-wire D.C. system with the same lamp voltage. Assume balanced loads, same losses and neutral having half the X-section of one of the respective outers. [**Ans.** 0.2917]

3. A 50 km long transmission line supplied a load of 5 MVA at 0.8 p.f. lagging at 33 kV, with a transmission efficiency of 90 percent. Calculate the volume of aluminium conductor required for the line in the following cases :
   (*i*) Single-phase two-wire system.
   (*ii*) Three-phase, three-wire system.
   Take resistivity of aluminium conductor as 2.85×10^{-8} Ωm. [**Ans.** (*i*) 16.35 m^3, (*ii*) 12.27 m^3]
4. Determine the weight of copper required for a 3-phase transmission system supplying at 380 kV between lines a load of 100 MW at lagging p.f. of 0.9. The length of line is 150 km. The efficiency of transmission is 92 per cent. The resistance of 1 km long conductor having 1 cm^2 *X*-section is 0.045 Ω. The density of copper 0.0981 N/cm^3.
   Also calculate the weight of copper required for D.C. transmission of the same power with equal losses working at a voltage between the lines double that of the peak voltage to the earth of the 3-phase supply. [**Ans.** 317.84 kN; 128.7 kN]
5. A two-conductor cable, one km long, is required to supply a constant load of 200A throughout the year. The cost of cable is Rs. (100 *a* + 40) per metre, where '*a*' is the area of cross-section of the conductor in cm^2. The cost of energy is 10P per kWh and interest and depreciation charges amount to 10 percent. Specific resistivity of copper is 1.85 μΩ cm. Find the most economical *X*-section of the cable.
   [**Ans.** 1.14 cm^2]
6. A 3-phase overhead transmission line supplies a constant load of 5 MW at 33 kV, and 0.8 p.f. lagging throughout the year. The cost of the line is Rs. (25000*a* + 2500) per km, where '*a*' is the cross-section of each conductor in cm^2. The cost of energy is 4P per kWh and interest and depreciation total 10 per cent per annum. Assuming specific resistance of conductor material as 1 μΩ cm, Find the most economical size of the conductor. [**Ans.** 0.71 cm^2]
7. The following data relate to a 2-wire feeder :
   Current carried throughout the year = 250 A
   The portion of the capital cost which is proportional to *X*-sectional area
   = Rs. 5 per kg of copper conductor.
   Cost of energy = 5ρ per kWh
   Interest and depreciation charges = 10 per cent per annum
   Density of copper = 8.93 g/cm^3
   Specific resistance of copper = 1.73 μΩ cm
   Find the most economical size of the conductor. [**Ans.** 3.25 cm^2]

12

Mechanical Design of Overhead Lines

12.1. Introduction. 12.2. Main component of overhead lines. 12.3. Conductors. 12.4. Line supports. 12.5. Overhead line insulators—Introduction—Insulating materials—Types of insulators—Causes of failure of insulators—Testing of insulators—Potential distribution over a string of suspension insulators. 12.6. Sag in overhead lines. 12.7. Stringing chart. 12.8. Vibration and vibration dampers. 12.9. Corona—General aspects—Theory of corona formation—Factors affecting corona—Advantages and disadvantages of corona—Corona loss and its control—Important terms relating to corona. Highlights—Theoretical Questions—Unsolved Examples.

12.1. INTRODUCTION

The transmission or distribution of electric power can be carried out by means of overhead lines or underground cables. The underground cables are rarely employed for power transmission because of the two *reasons* : (*i*) Power is generally transmitted over long distances to load centres ; (*ii*) The installation costs are very high. The use of underground system is limited to congested areas where safety and good appearance are the main consideration. Thus, as a rule, transmission of power over long distances is carried out by using overhead lines.

In this chapter we shall discuss topics relating to mechanical design of overhead lines *viz.*, line supports, conductor materials, overhead line insulators, sag etc.

12.2. MAIN COMPONENT OF OVERHEAD LINES

The main components of an overhead line are :

1. *Conductors.* They carry power from sending end station to receiving end station.

2. *Supports.* These may be poles or towers (depending upon the working voltage and the region where these are used) and keep the conductors at a suitable level above the ground.

3. *Cross-arms.* They provide support to the insulators.

4. *Insulators*. They provide insulation to high voltage wire with the metal structure and also provide support to the conductor. They also provide support to bus-bar conductors and other live high voltage equipment terminals.

5. *Miscellaneous items :*

(*i*) Phase plates

(*ii*) Lightning arrestors

(*iii*) Vee guards

(*iv*) Guard wires

(*v*) Anticlimbing wires

(*vi*) Fuse and isolating switches etc.

12.3. CONDUCTORS

The material of the conductor to be used for transmission and distribution of electric power should possess the following *characteristics* :

(*i*) High electrical conductivity (*i.e.*, low resistivity).

(*ii*) High tensile strength (to withstand mechanical stress).

(*iii*) Low specific gravity so that weight per unit volume is small.

(*iv*) Low cost (so that it can be used over long distances).

All these characteristics are not found in a single material, as such, while selecting a conductor material for a particular case, a *compromise is made between cost and electrical and mechanical properties.*

Materials used to transmission lines :

The leading and most important material used for transmission lines is *copper* since it has high conductivity and high tensile strength. *Aluminium* is used to a large extent especially with a *steel core for high voltage line*. The choice of materials depends upon : (*i*) cost of materials (*ii*) required electrical properties (*iii*) local conditions, and (*iv*) required mechanical strength.

Other materials employed for transmission lines are :

(*i*) Cadmium copper materials.

(*ii*) Copper weld materials.

(*iii*) Phospher bronze materials.

(*iv*) Galvanised steel materials.

(*v*) Galvanised iron.

(*vi*) Steel core copper.

(*vii*) Steel cored aluminium materials.

The most commonly used conductor materials for overhead lines are briefly discussed below :

1. Copper :

(*i*) *Hard drawn copper conductor is the best conductor due its high electrical conductivity and high tensile strength for all types of transmission.* Although hard drawing decreases the electrical conductivity slightly, yet it increases the tensile strength considerably.

(*ii*) It has a high current density so lesser X-sectional area of conductor is required (and so lesser area is subjected to wind loads).

(*iii*) It is quite homogeneous, durable and has a high scrap value.

Copper conductors having a *steel core* are employed for long span transmission lines, where a combination of high conductivity, small sag and minimum cross-section are desired. In such conductors, an insulating tape over the wire has to be provided in order to prevent the corrosion action of steel on copper.

Though copper is an ideal material for transmission and distribution of electric power, yet due to its higher cost and non-availability, it is rarely used for such purposes.

2. Aluminium :

(*i*) Aluminium, as compared to copper, is lighter in weight but has smaller conductivity and tensile strength.

(*ii*) The resistivity of aluminium being 1.6 times that of copper, for the same loss and length of conductor an aluminium conductor should have a X-sectional area of 60% greater than that of copper conductor (the diameter of aluminium conductor is about 1.26 times the diameter of copper conductor). The increased X-section of aluminium exposes a greater surface to wind pressure and, therefore, supporting towers must be designed for greater transverse strength. As a consequence of greater sag, the use of higher towers is often required.

(*iii*) Since the density of copper is about 3.3 times that of aluminium, weight of *aluminium conductor will equal to 48% of weight of copper*. Owing to this reason, the supporting structures for aluminium used not be made so strong as that of copper.

(*iv*) Aluminium conductor being light, is liable to swing, requires large cross-arms.

(*v*) The sag is greater in aluminium conductors (due to lower tensile strength and higher coefficient of linear expansion of aluminium).

(*vi*) Aluminium conductors are particularly suitable for operation in very high ambient temperature.

Usually aluminium conductors are used for transmission of heavy current, where conductor size is large and its cost forms a large proportion of the total cost of the complete installation.

Table 12.1 shows the comparison between copper and aluminium.

Table 12.1. Comparison of Copper and Aluminium

Nature of property	*Copper*	*Aluminium*
Density g/cm^3	8.94	2.7
Melting point, °C	1083	657
Thermal conductivity (watts per m^2 per °C)	3.50	2.00
Resistivity (ohm per m and mm^2)	0.017	0.0287
Resistance temperature co-efficient at 20°C	0.0039	0.0035

Table 12.2 shows the comparison of the properties of aluminium cable of a given size with those of copper cable of the same size.

Table 12.2. Comparison of the Properties of Aluminium Cable of a Given Size with Those of Copper Cable of the Same Size

Nature of property	*Aluminium cable (Hard drawn)*	*Copper cable (Hard drawn)*
Conductivity percent	61	97
Size for equal conductance	1.10	1.00
Strength for equal size	0.50	0.75
Weight for equal conductance	0.50	1.03
Volume resistivity	More	Less
Ductility	Good	Poor

3. Steel cored aluminium (A.C.S.R.) :

Owing to the low tensile strength of aluminium, aluminium conductors produce greater sag which prohibits their use for longer spans and make them unsuitable for long distance transmission. The tensile strength of aluminium conductor is increased with a core of galvanised steel wires. The composite conductor thus obtained is called as A.C.S.R. (*Aluminium conductor steel reinforced*). Fig. 12.1 shows steel cored conductor having one steel wire surrounded by 18 wires of aluminium. The *steel core takes a greater percentage of mechanical stresses while the aluminium carries bulk of current.*

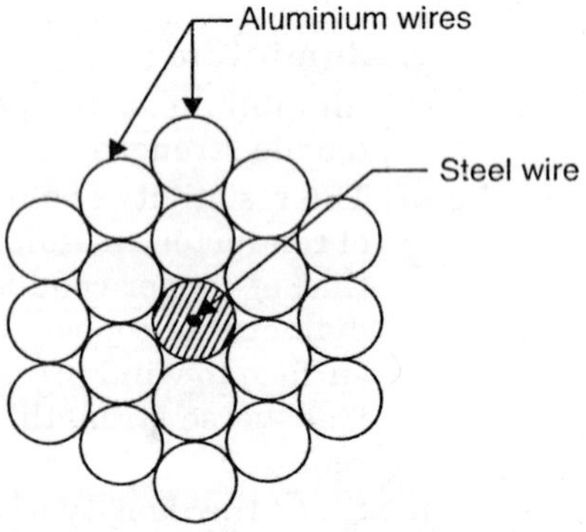

Fig. 12.1. A.C.S.R. conductor.

- A.C.S.R conductor being of high tensile strength and lighter in weight produces *small sag* and therefore *longer spans can be used.*
- A.C.S.R. conductor has a *largest diameter* than any other type of conductor of same resistance, so *corona losses are reduced but stronger supports are required for a given span.*

- A.C.S.R. conductor gets *deteriorated in service due to atmospheric corrosion* (as a result of bimetallic action of zinc and aluminium, electro-chemical in nature).
- Since the *skin effect is very predominant in* A.C.S.R. conductor, therefore, its *resistance (composite conductor) is taken equal to that of the aluminium covering alone. The reactance* of A.C.S.R. conductor is taken to be equal to that of a non-magnetic conductor having diameter equal to that outside of the A.C.S.R. conductor.

4. Galvanised steel :

- Galvanised steel conductors are used for extremely long spans. These are most suitable for rural areas where cheapness is the primary consideration.
- These conductors, due to poor conductivity and high resistance of steel, are *not suitable for transmitting large power over a long distance.*

Iron or steel wire use is *most advantageous for transmission of small power over a short distance*, where the size of copper conductor desirable from economical consideration comes out to be smaller than 8 S.W.G, which cannot be used due to poor mechanical strength.

5. Cadmium copper :

- In certain cases copper alloyed with cadmium is used. An addition of 1 or 2 percent of cadmium to copper increases the tensile strength by about 40 per cent and reduces the conductivity only by 17 per cent. Cadmium copper is costlier than copper.
- Cadmium copper conductor will be economical for a line with long spans and small cross-section.

Stranded conductors :

***Stranded conductors** are the conductors made of thin wires of small cross-section and bunched together*. They are flexible, not rigid and can be coiled very easily. They eliminate to a large extent the risk of breaking through insulation.

The following points are worth noting about *stranded conductors* :

(*i*) A stranded conductor is made by twisting the wires, called the *strands*, together to form layers.

(*ii*) The wires of each layer are laid in helical fashion round the preceding layer. The process is called '*stranding*.'

(*iii*) Generally *stranding is done in opposite directions for successive layers.* This means, if the wires of one layer are twisted in left hand direction, the next layer of wires will be twisted in the right hand direction, and so on.

(*iv*) A standard stranding consists of 6 wires around one wire, then 12 wires around the previous 6, then 18 wires around 12, then 24 wires around 18 and so on. The number of layers to be provided depends upon the number of wires to be provided. The centre wire is not counted as a layer. Instead of single, 3 or 4 stranded wires may also be put in the centre and over them layers may be formed. If 3 stranded wires are put in the centre, 9 wires will be in the first layer, then 15 wires in the second layer and so on. The increase in the number of wires in each successive layer is 6 in each of the above cases.

In the Table 12.3 are given the formulae which gives the number of wires for a particular layer, total number of wires in particular stranded conductor and the diameter of stranded conductor.

Table 12.3. Formulae for Stranded Conductor

1. *No. of wires in the centre*	1 wire	3 wires	4 wires
2. *No. of wires in the nth layer from the centre*	$6n$	$3 + 6n$	$4 + 6n$
3. *Total no. of wires in a stranded conductor having n layers*	$1 + 3n(1 + n)$	$3(1 + n)^2$	$(4 + 3n)(1 + n)$
4. *Diameter over the nth layer in cm, where, d = dia. of each wire in cm.*	$(1 + 2n)\,d$	$(2.155 + 2n)d$	$(2.414 + 2n)\,d$

The stranded conductors are expressed as follows : 7/2.24, 19/2.50, 37/2.06, and so on.

First number (*i.e.*, 7, 19, 37, etc.) indicates ...... Total no. of wires.

Second number (*i.e.*, 2.24, 2.50, 2.06, etc.) indicates... Diameter of each wire in mm. Let us consider *19/2.50 stranded conductor :*

No. of wires at the centre $= 1$

No. of wires in the first layer $= 6$

No. of wires in the second layer $= 12$

Total no. of layer, $n = 2$

Total no. of wires in the strand $= 1 + 3n(1 + n)$

$= 1 + 3 \times 2(1 + 2) = 1 + 18 = 19$

Overall diameter of the stranded conductor

$= (1 + 2n)d = (1 + 2 \times 2)\ 2.50$

$= 5 \times 2.50 = 12.50$ mm

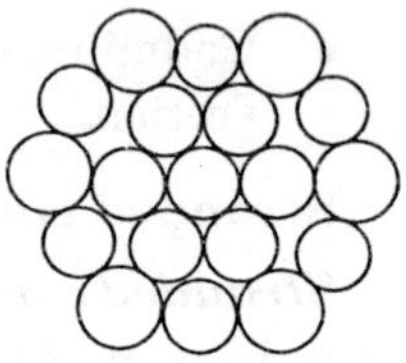

Fig. 12.2. 19/2.50 stranded conductor.

Fig. 12.2 shows the cross-section of 19/2.50 stranded conductor, Figs. 12.3 and 12.4 show *circular stranded conductors and compact circular stranded conductors respectively*. The former require more insulating and protective materials than those required by the latter. Overall dimensions of compacted conductors become less as is evident from the Fig. 12.4.

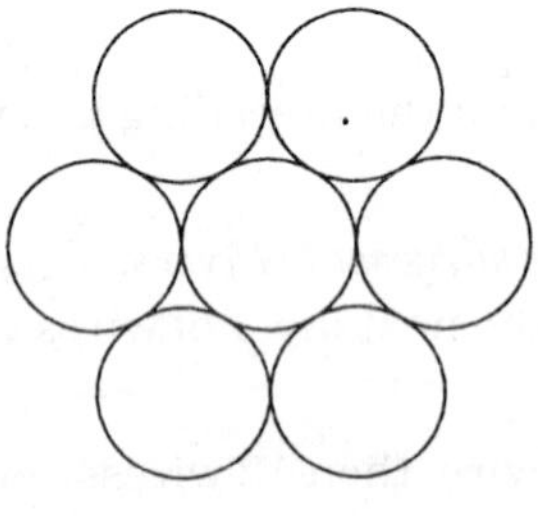

Fig. 12.3. Circular stranded conductor.

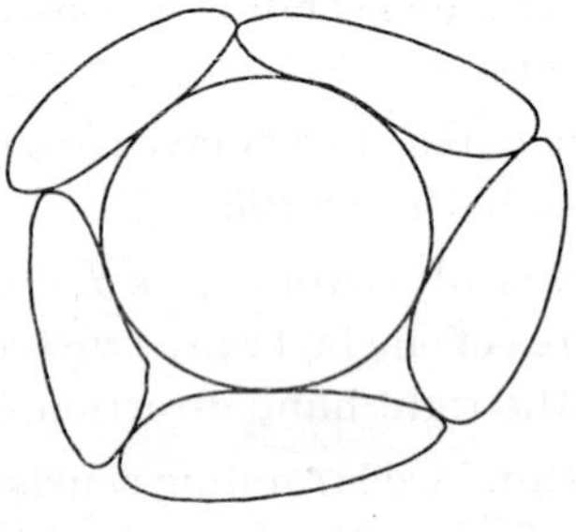

Fig. 12.4. Compact circular stranded conductor.

- *Circular stranded conductors are normally used for single-phase system.*
- *Compact circular stranded conductors are employed in the manufacture of cables.*

Example 12.1. *A 132 kV, 3-phase, 3-wire transmission line transmits over a distance of 144 km an electric power of 36 MW at a p.f. of 0.8 lagging, with an efficiency of 90%. Calculate the weight of material required for the following metallic conductors :*

(*i*) *Copper having resistivity of* 1.78×10^{-8} Ω-*m and specific gravity 8.9, and*

(*ii*) *Aluminium having resistivity of* 2.6×10^{-8} Ω-*m and specific gravity of 2.*

Solution. *Given* : Power to be transmitted, $P = 36$ MW

Power factor, $\cos \phi = 0.8$

Line voltage, $V_L = 13$ kV

Length of the transmission line, $l = 144$ km

Transmission efficiency $= 90\%$

Resistivity of copper, $\rho = 1.78 \times 10^{-8}$ Ω-m

Resistivity of aluminium, $\rho = 2.6 \times 10^{-8}$ Ω-m

Specific gravity of copper $= 8.9$

∴ Density of copper $= 8.9 \times 9.81 = 87.31$ kN/m³

Specific gravity of aluminium $= 2$

∴ Density of copper $= 2 \times 9.81 = 19.62$ kN/m³

W_{cu}, W_{al} :

Line current, $I_L = \dfrac{P}{\sqrt{3}\, V_L \cos \phi}$

Power loss, $P_{loss} = 3 I_L^2 R = 3 \times \left(\dfrac{P}{\sqrt{3}\, V_L \cos \phi} \right)^2 \times \dfrac{\rho l}{a} = \dfrac{P^2 \rho l}{V_L^2 \cos^2 \phi \times a}$

or,

$$a = \frac{P^2 \rho l}{P_{loss} V_L^2 \cos^2 \phi} = \frac{P^2 \rho l}{(1 - \eta) P V_L^2 \cos^2 \phi}$$

$$= \frac{P \rho l}{(1 - \eta) V_L^2 \cos^2 \phi} \qquad [\because P_{loss} = (1 - \eta) P]$$

Volume of conductor material required

$$= 3\, al = \frac{3 P \rho l^2}{(1 - \eta) V_L^2 \cos^2 \phi}$$

Weight of conductor material required

$$= \frac{3 P \rho l^2}{(1 - \eta) V_L^2 \cos^2 \phi} \times \text{Density of material}$$

∴ Weight of *copper* required,

$$W_{cu} = \frac{3 \times (36 \times 10^6) \times 1.78 \times 10^{-8} \times (144 \times 1000)^2}{(1 - 0.9) \times (132 \times 1000)^2 \times (0.8)^2} \times 87.31 = \mathbf{3121\ kN.} \quad \textbf{(Ans.)}$$

Weight of aluminium required,

$$W_{al} = \frac{3 \times 36 \times 10^6 \times 2.6 \times 10^{-8} \times (144 \times 1000)^2}{(1 - 0.9) \times (132 \times 1000)^2 \times (0.8)^2} \times 19.62 = \mathbf{1024.4\ kN.} \quad \textbf{(Ans.)}$$

12.4. LINE SUPPORTS

The line supports (Poles and towers) employed for transmission and distribution of electric power should have the following *characteristics* :

1. Cheaper in cost.
2. High mechanical strength.
3. Light in weight.
4. Longer life.

5. Easily accessible for erection of line conductors, painting etc.
6. Low maintenance cost.
7. Good looking.

The line supports may be wooden poles, steel poles, R.C.C. poles and steel towers.

1. Wooden poles :

- These poles (made of seasoned wood-sal or chir) are easily available, cheap, provide insulating properties and are widely used for distribution purposes in *rural areas* as an economical proposition.
- They are used for short spans, upto 60 metres.
- The portion of the poles below the ground level is impregnated with preservative compounds like creosite oil.
- Double pole structures of the '*A*' or '*H*' type [Fig. 12.5 (*d*)] are often used to obtain a high transverse strength than could be economically provided by means of single pole.

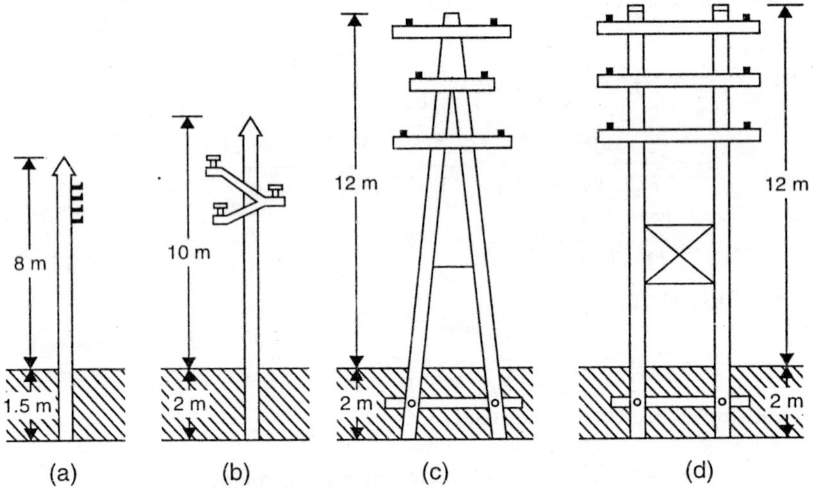

Fig. 12.5. Wooden poles : (*a*) 400 V, 4-phase, 4-wire distributor, (*b*) 11 kV, 3-phase feeder, (*c*) '*A*' pole, double circuit, 33 kV feeder, (*d*) '*H*' pole double circuit 33 kV feeder.

- A well impregnated wooden pole has a life between 25 and 30 years.
- These poles have the disadvantage that they need periodical inspection because they tend to rot and have a short life.

2. Steel poles :

- These poles possess great mechanical strength and thus permit that use of longer spans (50-80 metres), but they are costly.
- These poles are of three types :

  (*i*) Rail poles, (*ii*) Tubular poles, and

  (*iii*) Rolled steel joists.
- These are set in concrete muffs at the foundation in order to protect them from chemical action.
- Such poles are generally used for distribution purposes in cities.
- Their average life is more than 40 years.

3. R.C.C. poles :

- They possess greater mechanical strength and permit longer span than steel poles (80-200 metres).
- They give good look, require no maintenance and have good insulating properties.
- These poles are of the following two types :
— One type is of square X-section from bottom to top but sides of square decreasing from bottom to top.
— Other type has rectangular bottom and square top with rectangular holes in it to facilitate the climbing of poles and reduce the weight of poles.
- As these poles are very heavy, therefore their transportation cost is heavy and as such they are manufactured at site.
- They have a very long life.

Fig. 12.6 shows a R.C.C. pole.

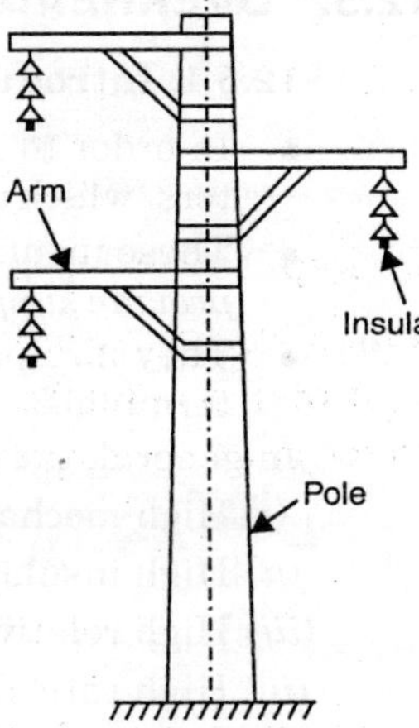

Fig. 12.6. R.C.C. pole.

4. Steel towers :

- Whereas wooden, steel and R.C.C. poles are used for distribution purposes at low voltage (say 11 kV), steel towers are invariably employed for long distance transmission at higher voltage.
- They have greater mechanical strength, longer life and can withstand most severe climatic conditions and permit the use of longer spans.
- Due to long spans, the risk of interrupted service owing to broken or punctured insulation is considerably reduced.
- Lightning troubles are also minimised as each tower acts as a lightning conductor.

Fig. 12.7 shows a single circuit power. However, at a moderate additional cost, double circuit power can provided as shown in Fig. 12.8. Steel towers are generally four legged, each leg anchored properly.

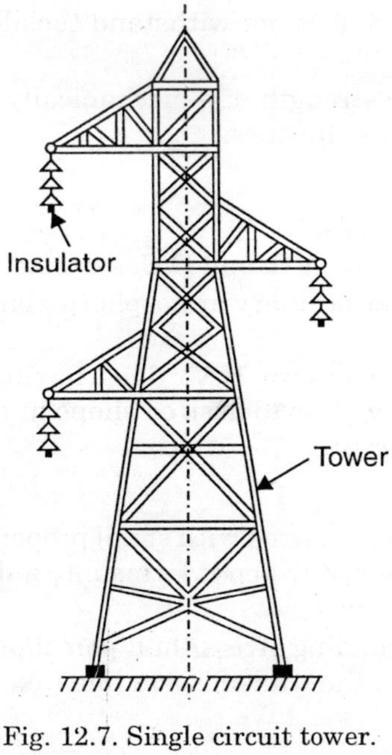

Fig. 12.7. Single circuit tower.

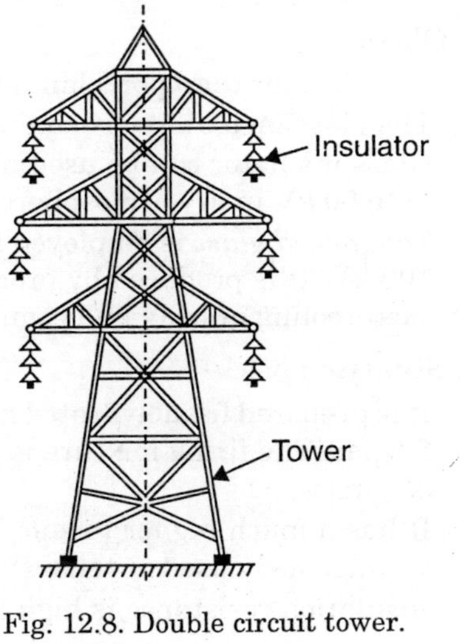

Fig. 12.8. Double circuit tower.

12.5. OVERHEAD LINE INSULATORS

12.5.1. Introduction

- In order to maintain safety and necessary clearance between live transmission conductors (which are bare and do not have any insulated coating over it) *insulators* are mounted.
- These insulators *provide insulation of high voltage wire with the metal structure and also provide support to the conductor.*
- They also provide support to bus-bar conductors and other live high voltage equipment terminals.

In general, the insulators should have the following *characteristics* :

(*i*) High mechanical strength (to withstand conductor load, wind load etc.).

(*ii*) High insulation resistance (to avoid leakage of currents to earth).

(*iii*) High relative permittivity (to give high dielectric strength).

(*iv*) High ratio of rupture strength to flash-over voltage.

(*v*) The insulating material used should not be porous, free from internal impurities and cracks and impervious to the fluids and gases in the atmosphere. It should also withstand temperature changes.

12.5.2. Insulating Materials

The following materials are commonly used for the manufacture of insulators :

1. Porcelain 2. Glass 3. Steatite 4. Synthetic resin.

1. Porcelain :

- It is the most commonly used material for insulators de on overhead lines. It is made from china clay, ball clay, quartz and felspar, finally powedered and mixed with water. The plasticity of mixture is regulated by clay. Quartz and felspar act as filler and flux respectively. The ingredients are processed in the mill, shaked and covered with glaze. *Glazing of insulator is very much necessary to keep it relatively free from dust and moisture.*
- It is mechanically stronger than glass.
- This material is usually weak in tension and does not withstand tensile strength more than 50 MN/m^2.
- The compressive strength and the dielectric strength of a mechanically sound insulator are about 7000 MN/m^2 and 60 kV per cm of its thickness.

2. Glass :

- It is cheaper than porcelain in the simple shapes.
- The glass insulators are mainly used for E.H.V., A.C. and D.C. systems.
- Glass insulator can be used upto 25 kV under ordinary atmospheric conditions and well upto 50 kV in dry atmosphere.
- *Toughened glass* is employed for insulation in extra H.V. lines having voltage above 100 kV. It is produced by processing ordinary glass to desired shape at about 100°C and then cooling quickly in a compressed air current.

3. Steatite :

- It is prepared from hydrated magnesium silicate mixed with small proportions of clay and felspar. The firing mixture is carried upto 1400°C when it forms into a dense homogeneous mass.
- It has a much higher tensile strength and bending stress than porcelain and can be advantageously used at tension towers or when the transmission line takes a sharp turn.
- Insulation resistance is high.

4. Synthetic resin :

- Synthetic insulators contain compounds of silicon, rubber, resin etc.
- These insulators have high strength and lower weight. Leakage current is higher and longevity is low.
- They are comparatively cheaper.
- These insulators are used in various indoor applications. They are extensively used as bushings.

12.5.3. Types of Insulators

Transmission line insulators may be *classified* as follows :

1. Pin type insulators
2. Suspension type insulators
3. Strain type insulators
4. Shackle insulators.

1. Pin type insulators. Refer to Fig. 12.9.

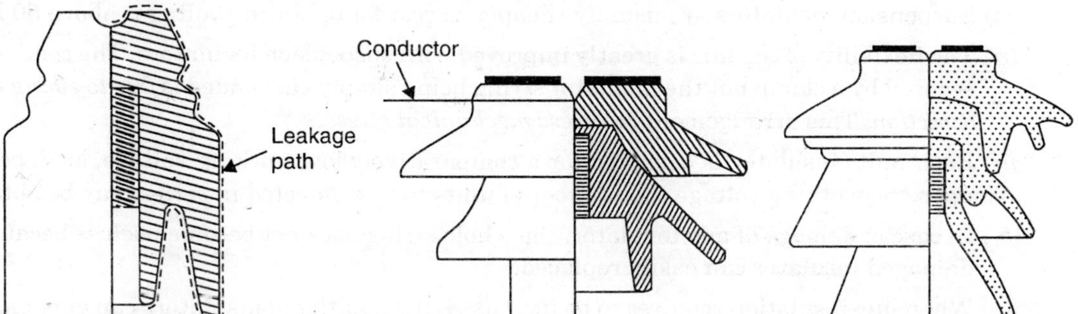

Fig. 12.9. Pin type insulator : (*i*) 11 kV , (*ii*) 22 kV ; (*iii*) 66 kV.

- As the name suggests, the pin type insulator is designed to be mounted on a pin which in turn is installed on the cross-arm of the pole. The insulator is screwed on the pin and electrical conductor is placed in the groove at the top of the insulator and is tied down with soft copper or soft aluminium binding wire according to material of the conductor.
- For low voltages, pin type insulators made of *glass* are generally used. Pin type insulators made of *porcelain* are designed for voltages upto about 90 kV but are seldom used on lines above 60 kV.

2. Suspension type insulators :

- Since the cost of pin type insulator increases rapidly as the working voltage is increased therefore, this type of insulator is not economical beyond 33 kV. For voltages higher than 33 kV, it is a usual practice to use suspension type insulators as shown in Fig. 12.10.
- The suspension insulator hangs from the cross arm, as opposed to the pin insulator which sits on the top of it. The line conductor is attached to its lower end. These insulators consists of *a number of porcelain discs connected in series by metal links in the form of a "string"*. The number of insulators comprising a string depends upon the following *factors* : (*i*) Working voltage, (*ii*) Type of transmission construction, (*iii*) Size of the insulator used, and (*iv*) Weather conditions.

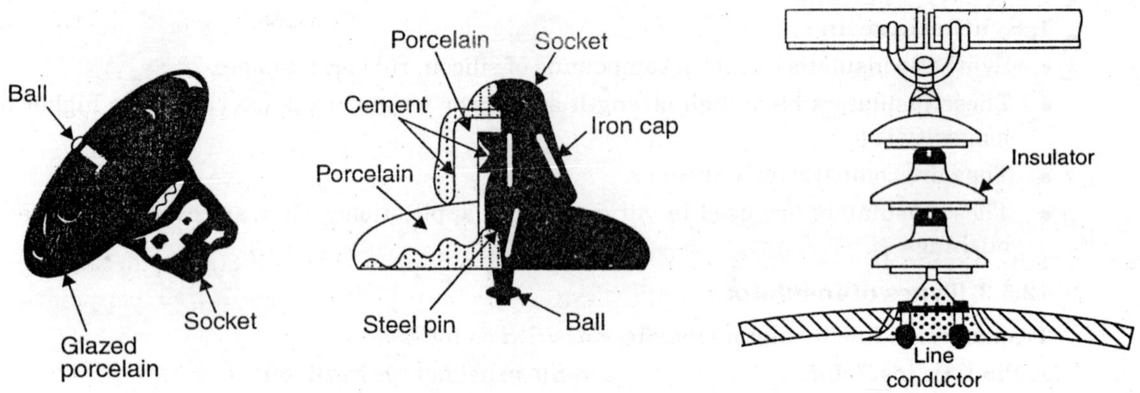

Fig. 12.10. Suspension type insulators.

The following are the *advantages* of suspension type insulators over pin type insulators :

(*i*) Suspension insulators are usually cheaper in cost for operating voltages above 50 kV.

(*ii*) The flexibility of the line is greatly improved with suspension insulators. The conductor is secured by a clamp but the insulator string being simply suspended is *free to swing in any direction.* This arrangement *reducess mechanical stresses*.

(*iii*) Each unit / insulator is designed for a comparatively low working voltage, so depending upon the working voltage the number of units to be connected in series may be obtained.

(*iv*) In case of damage of any insulator, the whole string does not become useless because the damaged insulator can easily replaced.

(*v*) When line insulation requires to be increased, the additional insulators can very easily be added to the string.

(*vi*) As the conductor is hung below the earthed cross arms the tower functions as a lightning rod, this arrangement provides partial protection from lightning.

The *disadvantage* (not a serious one) of suspension insulators is that large spacings between conductors are required than with pin type insulators due to large amplitude of the swing of the conductors.

Following three types of suspensions insulators are generally used :

(*i*) Hewlett or Interlinking type ;

(*ii*) Cemented cap type ;

(*iii*) Core-and-line type.

3. Strain insulators :

- *Strain insulators may be of pin or suspension type.* Upto about 30 kV pin type insulators are satisfactory, but for higher voltages the suspension type are generally used.
- *These insulators are used on dead-end towers at bends or corners of transmission lines, or when making very long spans.* Extra heavy suspension units are made for such service, but often standard units may be used. On ordinary straight line dead-end towers a single string is often sufficient but for severe service two more strings may be connected in parallel.

4. Shackle insulators :

- Fig 12.11 shows a shackle insulator mounted in a clamp. These insulators can be directly fixed to the pole with a bolt or to the cross arm. The conductor in the groove is fixed with a soft binding wire.
- Earlier, the shackle insulators were used as strain insulators. These days they are used for low voltage distribution lines.

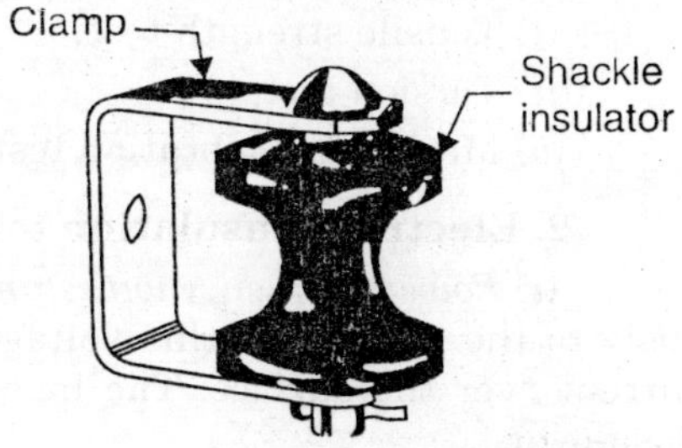

Fig. 12.11. Shackle insulator mounted in a clamp.

12.5.4. Causes of Failure of Insulators

The following are the *causes of failure of insulators* :

1. Mechanical stresses
2. Porosity of material
3. Cracking of insulator
4. Improper vitrification
5. Short circuits
6. Flash-over
7. Dust deposition.

12.5.5. Testing of Insulators

The insulators are tested by the following *tests* :

1. Mechanical tests
2. Electrical insulation tests
3. Environmental tests
4. Temporary cycle tests
5. Corona and radio interference tests.

The tests are *also categorised as follows* :

(*a*) ***Design tests.*** These tests are carried out to ascertain the electrical and mechanical performance of a particular type of insulators ; these tests include :

(*i*) Power-frequency dry flash-over test,

(*ii*) Power-frequency wet test,

(*iii*) Power-frequency flash-over test, and

(*iv*) Impulse frequency flash-over test.

(*b*) ***Sample tests.*** These test include :

(*i*) Puncture test,

(*ii*) Porosity test,

(*iii*) Electro-mechanical test,

(*iv*) Mechanical test, and

(*v*) Temperature cycle test.

(*c*) ***Routine tests.*** These tests are performed to weed out the defective insulators ; these test include :

(*i*) Proof-load test,

(*ii*) Corrosion test, and

(*iii*) High voltage test.

The commonly used tests are briefly discussed below :

1. Mechanical tests. These tests are conducted on the insulators to ascertain their capabilities to withstand the mechanical stresses. The following mechanical tests are performed on the insulators and their fittings :

(*i*) Tensile strength test, (*ii*) Compression tests,
(*iii*) Torsion test, (*iv*) Bending moment test, and
(*v*) Mechanical vibration test.

2. Electrical insulation tests :

(*i*) *Power frequency withstand test.* In this test normal frequency voltage is applied continuously to the insulator ; this voltage aligns the dust particles on the surface and causes leakage current over the surface. The insulators are provided with sufficient creepage distance to avoid flash-over.

During this test twice the rated voltage is applied for a period of one minute, there should be no flash-over or puncture (The test voltage at site is about 70 per cent of that at works)

This test is of the following *two types* :

- One-minute power frequency withstand test–*Dry*.
- One-minute power frequency withstand test–*Wet*.

(*ii*) *Impulse voltage withstand test :*

- In this test, a standard lightning impulse voltage wave, called 1.2/50 wave (1.2 μs wave front and 50 μs wave tail) is applied to the insulator. Here the clearance, voltage distribution between winds, stress concentration at sharp points on the insulators, smoothness of metal fittings, etc. are worth consideration.
- The standard switching impulse wave is 250/2500 μ s. Switching impulse tests are applicable for rated voltage above 220 kV.

3, 4. Environmental and temporary cycle tests :

- During these tests the electrical insulators are subjected to pollution and environmental stresses, alternate temperature cycles and sudden temperature changes.

The following tests are mandatory for important extra high voltage (E.H.V.) insulators :

(*i*) Pollution test (*ii*) Extremely low temperature test
(*iii*) Sudden temperature drop test (thermal shock tests).

- Insulators also undergo temperature cycle test ; the specimen is exposed to the specified temperature cycle.

5. Corona and radio interference test. In order to check corona discharge, the voltage pattern is analysed while designing the clamps, conductors, metal fittings and the strings of insulators. By the provision of suitable voltage grading rings and smooth surface, the possibility of corona discharge and radio interference are eliminated for certain voltage range.

Other Tests :

(*i*) ***Puncture test :***

- This test (destructive one) is *used to determine puncture voltage*.
- Here, the insulator is fully immersed in an insulating oil and is subjected to the voltage being gradually increased to a certain specified value till the puncture takes place. The voltage at which conduction starts in called the puncture voltage. A good insulator should not puncture under this test.

(*ii*) ***Porosity test.*** In this test (destructive one) the specimen insulator is broken into pieces which are immersed in 1 per cent solution of dye in alcohol under a pressure (15 N/mm^2) for a sufficient period (about 24 hours). The specimen pieces are removed from the testing pot and examined. The slightest degree of porosity is indicated by a deep penetration of the dye into the pieces.

(*iii*) ***Galvanising test.*** This test is performed to assess the galvanisation strength of the galvanised metal parts of the insulators.

12.5.6. Potential Distribution Over a String of Suspension Insulators

In case of overhead lines operating on higher voltages, a *string of insulators*, consisting of a number of disc. (insulators) connected in series, is used (See Fig. 12.12 (*a*)). The porcelain portion of each disc is in between the two metal links, therefore, each disc forms a capacitor C as shown in Fig. 12.12 (*b*). This capacitance (C) is known an *mutual capacitance* (*or self capacitance*). Had there been only the mutual capacitance, the charging current would have the same through all the discs and as a consequence the voltage across each disc in this case could have been $\frac{V}{3}$. But, since there is also a capacitance between metal fitting of each unit and tower or earth, known as *shunt capacitance* (C_1), the voltage across each disc will be different. *The disc near the line conductor will have the maximum voltage* (since the charging current through the string has the maximum value at disc nearest to the conductor), this voltage progressively *decreases* as the disc nearest the tower is approached. With the increase in insulator units the inequality of voltage distribution between individual units becomes more pronounced ; it also depends on the ratio of capacity of insulator to capacity to earth.

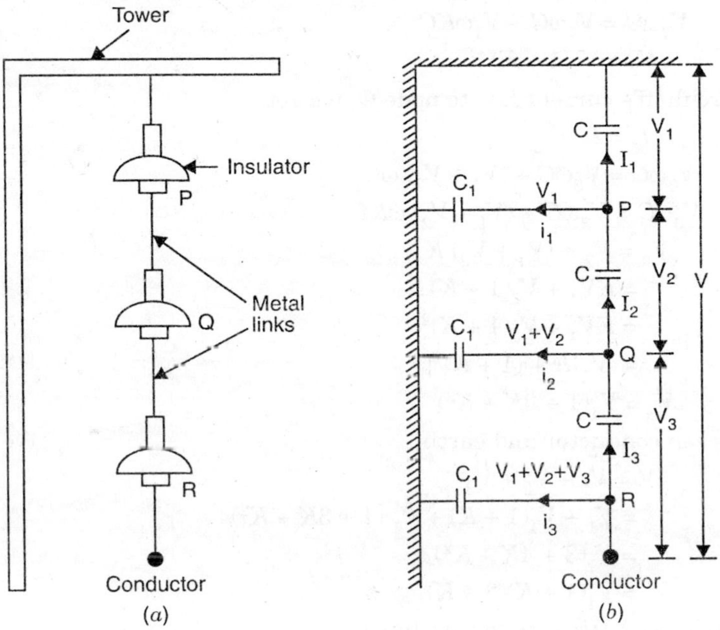

Fig. 12.12. Equivalent circuit of a string of suspension insulators.

The unequal potential distribution is undesirable and is usually expressed in terms of string efficiency.

The ratio of voltage across the whole string to the product of number of discs and the voltage across the disc nearest to the conductor is known as **string efficiency.**

i.e.,
$$\text{String efficiency} = \frac{\text{Voltage across the string}}{n \times \text{Voltage across disc nearest to the conductor}}$$

100 percent string efficiency means that voltage across each disc will be exactly the same ; this value is impossible to achieve, however, efforts may be made to approach this value.

Fig. 12.12 (*b*) shows the equivalent circuit of a string of suspension insulators (discs).

Let, C = Mutual capacitance of each disc,

C_1 = Shunt capacitance,

V_1 = Voltage across the first disc (nearest the cross-arm),

V_2 = Voltage across the second disc,

V_3 = Voltage across the third disc (nearest the line conductor), and

V = Voltage between the conductor and earth (*i.e.* tower).

Further, Let = $\frac{C_1}{C} = K$ or $C_1 = KC$

Applying Kirchhoff's current law (KCL) to node P, we get

$$I_2 = I_1 + i_1$$

$$V_2\omega C = V_1\omega C + V_1\omega C_1 \qquad \left(\because\ I = \frac{V}{1/\omega C} = V\omega C\right)$$

or, $$V_2\omega C = V_1\omega C + V_1\omega KC$$

or, $$V_2 = V_1(1 + K) \qquad \text{...}(i)$$

Applying Kirchhoff's current law to node Q, we get

$$I_3 = I_2 + i_2$$

or, $$V_3\omega C = V_2\omega C + (V_1 + V_2)\,\omega C_1$$

or, $$V_3\omega C = V_2\omega C + (V_1 + V_2)\,\omega KC$$

or, $$V_3 = V_2 + (V_1 + V_2)\,K$$

$$= KV_1 + V_2(1 + K)$$

$$= KV_1 + V_1(1 + K)^2 \qquad [\because\ V_2 = V_1(1 + K)]$$

$$= V_1[K + (1 + K)^2]$$

or, $$V_3 = V_1(1 + 3K + K^2) \qquad \text{...}(ii)$$

Voltage between conductor and earth,

$$V = V_1 + V_2 + V_3$$

$$= V_1 + V_1(1 + K) + V_1\,(1 + 3K + K^2)$$

$$= V_1(3 + 4K + K^2)$$

$$= V_1(1 + K)(3 + K) \qquad \text{...}(iii)$$

From expressions (*i*), (*ii*) and (*iii*), we have

$$\frac{V_1}{1} = \frac{V_2}{1+K} = \frac{V_3}{1+3K+K^2} = \frac{V}{(1+K)(3+K)} \qquad \text{...(12.1)}$$

$\therefore$ Voltage across first (top) disc, $$V_1 = \frac{V}{(1+K)(3+K)} \qquad \text{...[12.1 }(a)]$$

Voltage across second disc, $$V_2 = V_1(1 + K) \qquad \text{...[2.1 }(b)]$$

Voltage across third disc, $$V_3 = V_1(1 + 3K + K^2) \qquad \text{...[12.1 }(c)]$$

$\therefore$ *Percentage string efficiency* = $\dfrac{\text{Voltage across string}}{n \times \text{Voltage across the unit nearest to the conductor}}$

$$= \frac{V}{3 \times V_3} \times 100 \qquad \text{...(12.2)}$$

The approximate number of insulators (discs) used (in clear atmospheres) for various transmission voltages is as follows :

Line voltage (kV)	*No. of discs*
66	5
132	10
220	14
400	21
765	35

Note. When the insulators are *wet* the value of C (mutual capacitance) increases while that of C_1 (shunt capacitance) remains constant (except for the disc nearest the cross-arm) so that value of K decreases, more uniform potential distribution is obtained and there is *increase in string efficiency*.

Methods of improving string efficiency :

Following are the *methods of improving string efficiency* :

1. Long cross-arm method.
2. Capacitance grading method.
3. Static shielding.

1. *Long cross-arm method.* The *value of string efficiency increases with the decrease in the value of K.* The value of K can be decreased by reducing the shunt capacitance. This requires *long cross-arms* and hence taller supporting structures ; as such it is uneconomical to go beyond certain value of the length of cross-arm. It has been found that, practically it is *not possible* to obtain the value of K *less than 0.1.*

2. *Capacitance grading method.* Unequal distribution of voltage is due to the leakage current from the insulator pin to the tower structure. This makes the voltages unequal. To make the voltages equal the discs of the string are used of different capacities, such that the product of their capacitive reactance and the current flowing through the respective unit is same. This requires that the *unit nearest the cross-arm should have the minimum capacitance and as we go towards the power conductor the capacitance should increase.* By this grading complete equality of voltage can be obtained.

The *disadvantage* of this method is that a large number of different-sized insulators are required. Good results, however, can be obtained by using standard insulators for most of the string and larger units for that near to the line conductor. This method is *very inconvenient and not used in practice below 200 kV.*

3. *Static shielding.* In static shielding the idea is to *cancel exactly the pin to tower charging currents so that the same current flows through the units of identical capacities to produce equal voltage drop across each unit.* The arrangement is shown in Fig. 12.13.

- In this method a guard ring or grading ring is connected round to the power conductor such that this surrounds the bottom unit. Since identical units are being used, their mutual capacities are equal. Similarly ground capacities are equal. The design of the ring should be such that shunt capacitance currents i_1, i_2, etc. are equal to metal fitting line capacitance currents i_1', i_2' etc. ; consequently the same charging current I flows through each unit of string, leading to uniform potential distribution across the units.

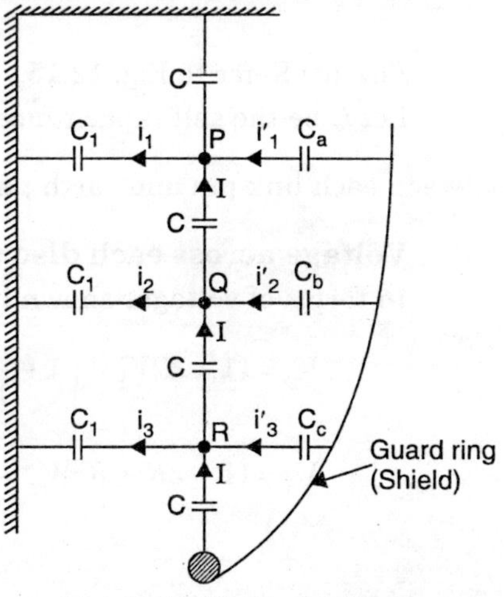

Fig. 12.13. Static shielding.

With this method also it is impossible to obtain in practice an equal distribution of voltage but it is possible to have considerable improvements.

Example 12.1. *An insulator string has three units each having a safe working voltage of 15 kV. The ratio of unit self capacitance to stray capacitance of earth is 10 : 1. Calculate :*

(*i*) *The maximum safe working voltage.*

(*ii*) *The string efficiency.* **(Panjab University, AMIE Winter, 2001)**

Solution. Refer to Fig. 12.14. *Given* : No. of units $n = 3$; safe working voltage of each unit = 15 kV ; Ratio of self capacitance to shunt capacitance,

$$\frac{C}{C_1} = 10 \quad \text{or} \quad K = \frac{C_1}{C} = 0.1$$

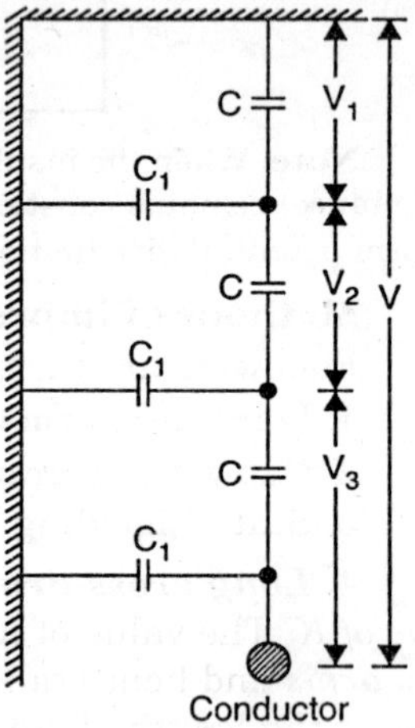

Fig. 12.14

(*i*) **The maximum safe working voltage, V :**

Voltage across the lowest (bottom) unit, V_3 = Safe working voltage of the unit = 15 kV.

Also, $$\frac{V_1}{1} = \frac{V_2}{1+K} = \frac{V_3}{1+3K+K^2} \quad \text{...[Eqn. (12.1)]}$$

$$\therefore \quad V_1 = \frac{V_3}{1+3K+K^2} = \frac{15}{1+3\times 0.1+(0.1)^2} = 11.45 \text{ kV}$$

$$V_2 = V_3 \times \left(\frac{1+K}{1+3K+K^2}\right) = 15 \times \left[\frac{1+0.1}{1+3\times 0.1+(0.1)^2}\right]$$

$$= 12.595 \text{ kV}$$

$\therefore$ Maximum and safe working voltage of the string,

$$V = V_1 + V_2 + V_3$$
$$= 11.45 + 12.595 + 15 = \mathbf{39\ kV.} \quad \textbf{(Ans.)}$$

(*ii*) **String efficiency :**

$$\% \text{ String efficiency} = \frac{V \times 100}{n \times V_3} = \frac{39 \times 100}{3 \times 15} = \mathbf{86.67\%.} \quad \textbf{(Ans.)}$$

Example 12.2. (*a*) *For a string insulator with four discs, the capacitance of the disc is 10 times the capacitance between the pin and earth. Calculate the voltage across each disc when used on a 66 kV line. Also calculate the string efficiency.*

(*b*) *List the basic tests that are carried out on insulators.*

(Madras University)

Sol. (*a*) Refer to Fig. 12.15.

Let C be the self capacitance of each disc and KC be the capacitance between each link pin and earth ; $K = \frac{1}{10}$ (given).

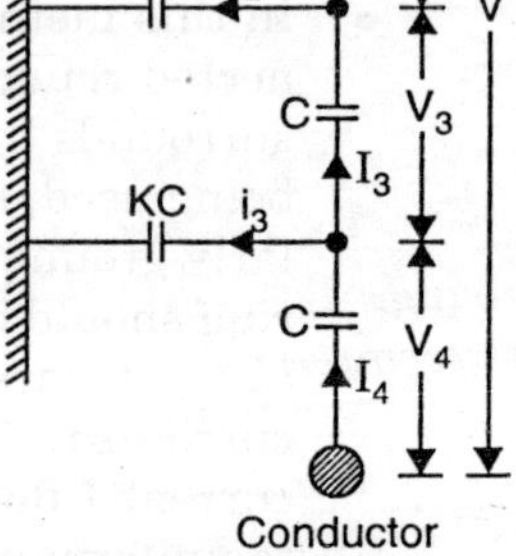

Fig. 12.15

Voltage across each disc :

In terms of voltages shown in the figure, we know that

$$V_2 = (1+K)V_1 = \left(1+\frac{1}{10}\right)V_1 = \frac{11}{10}V_1$$

$$V_3 = (1+3K+K^2)V_1 = \left[1+3\times\frac{1}{10}+\left(\frac{1}{10}\right)^2\right]V_1 = \frac{131}{100}V_1$$

$$V_4 = (1 + 6K + 5K^2 + K^3)\, V_1$$

$$= \left[1 + 6 \times \frac{1}{10} + 5 \times \left(\frac{1}{10}\right)^2 \times \left(\frac{1}{10}\right)^3\right] V_1$$

$$= \left(1 + \frac{6}{10} + \frac{5}{100} + \frac{1}{1000}\right) V_1 = \frac{1651}{1000} V_1$$

and,
$$V = V_1 + V_2 + V_3 + V_4$$

$$= V_1 + \frac{11}{10} V_1 + \frac{131}{100} V_1 + \frac{1651}{1000} V_1$$

$$= V_1\left(1 + \frac{11}{10} + \frac{131}{100} + \frac{1651}{1000}\right)$$

$$V = V_1\left(\frac{1000 + 1100 + 1310 + 1651}{1000}\right) = \frac{5061}{1000} V_1$$

$\therefore$
$$\mathbf{V_1} = V \times \frac{1000}{5061} = \frac{66}{\sqrt{3}} \times \frac{1000}{5061} = \mathbf{7.53\ kV.} \quad \textbf{(Ans.)}$$

$$\mathbf{V_2} = \frac{11}{10} V_1 = \frac{11}{10} \times 7.53 = \mathbf{8.28\ kV.} \quad \textbf{(Ans.)}$$

$$\mathbf{V_3} = \frac{131}{100} V_1 = \frac{131}{100} \times 7.53 = \mathbf{9.86\ kV.} \quad \textbf{(Ans.)}$$

$$\mathbf{V_4} = \frac{1651}{1000} V_1 = \frac{1651}{1000} \times 7.53 = \mathbf{12.43\ kV.} \quad \textbf{(Ans.)}$$

String efficiency :

$$\text{String efficiency} = \frac{V}{4 \times V_4} = \frac{(66/\sqrt{3})}{4 \times 12.43} = 0.7664 \quad \text{or} \quad \mathbf{76.64\ \%.} \quad \textbf{(Ans.)}$$

(*b*) The basic tests which are carried out on insulators are :

1. *Design tests* :
   - Dry flash over tests.
   - Wet flash over tests.
2. *Performance tests :*
   - Puncture test.
   - Mechanical test.
   - Porosity test.
3. *Routine tests :*
   - High voltage test.
   - Proof load test.
   - Corrosion test.

Example 12.3. *A 3-phase transmission line is supported by a 3-unit suspension insulator string. The voltage across the line unit is 20 kV and that across the adjacent unit is 15 kV. Determine :*

(*i*) *Ratio of ground to mutual capacitance.*

(*ii*) *System line voltage.*

(*iii*) *String efficiency.*

(Banglore University)

Solution. Refer to Fig. 12.16.

Given : No. of units, $n = 3$; $V_3 = 20$ kV, $V_2 = 15$ kV

(*i*) **Ratio of mutual to ground capacitance, $\mathbf{K = \frac{C_1}{C}}$:**

Applying *KCL* to node *P*, we get

$$I_2 = I_1 + i_1$$

$$V_2\omega C = V_1\omega C + V_1\omega C_1$$

$$= V_1\omega C + V_1\omega KC$$

$$= V_1\omega C\,(1 + K)$$

or, $$V_2 = V_1(1 + K) \quad \text{...(i)}$$

Applying *KCL* to node *Q*, we get

$$I_3 = I_2 + i_2$$

$$V_3\omega C = V_2\omega C + (V_1 + V_2)\,\omega C_1$$

$$= V_2\omega C + (V_1 + V_2)\,\omega KC$$

$$V_3\omega C = \omega C\,[V_2 + (V_1 + V_2)\,K]$$

or, $$V_3 = V_1(1 + K)\,[V_1 + \{V_1(1 + K)\}]\,K$$

$$= V_1(1 + K) + V_1K + V_1K(1 + K)$$

or, $$V_3 = V_1\,[1 + K + K + K\,(1 + K)] = V_1(1 + 3K + K^2) \quad \text{...(ii)}$$

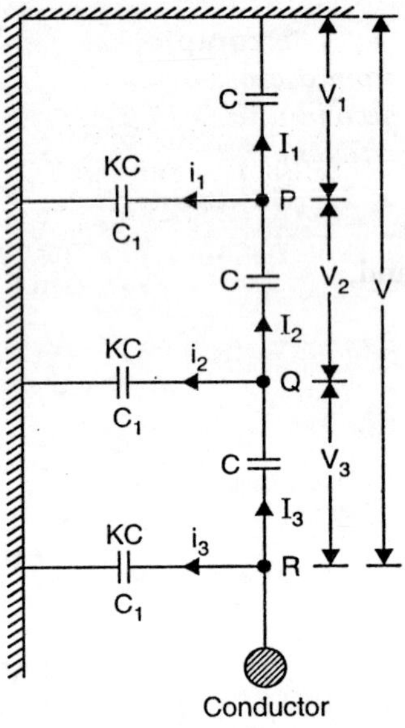

Fig. 12.16

Substituting the values for (*i*) and (*ii*), we have

$$15 = V_1(1 + K) \quad \text{...(iii)}$$

$$20 = V_1(1 + 3K + K^2) \quad \text{...(iv)}$$

Dividing (*iv*) by (*iii*), we have

$$\frac{1 + 3K + K^2}{1 + K} = \frac{20}{15} = \frac{4}{3}$$

or, $$3 + 9K + 3K^2 = 4 + 4K$$

or, $$3K^2 + 5K - 1 = 0$$

or, $$K = \frac{-5 \pm \sqrt{25 + 12}}{6} = \frac{-5 \pm 6.08}{6} = 0.18 \quad \text{[considering + ve value only]}$$

Hence $K =$ **0.18. (Ans.)**

(*ii*) **System line voltage, V :**

$$V = V_1 + V_2 + V_3$$

$$= \frac{V_2}{1 + K} + V_2 + V_3 = \frac{15}{1 + 0.18} + 15 + 20 = \mathbf{47.7\ kV.}\ \textbf{(Ans.)}$$

(*iii*) **String efficiency :**

$$\text{String efficiency} = \frac{\text{Voltage across the string}}{n \times \text{Voltage across the string adjacent to the line conductor}}$$

$$= \frac{47.7}{3 \times 20} = 0.795 \text{ or } \mathbf{79.5\%.}\ \textbf{(Ans.)}$$

Example 12.4. *Each conductor of a three-phase overhead transmission line is suspended from a cross-arm of a steel tower by a string of four suspension insulators. The voltage across the second unit is 15 kV and across the third 20 kV. Find the voltage between conductors and the string efficiency.*

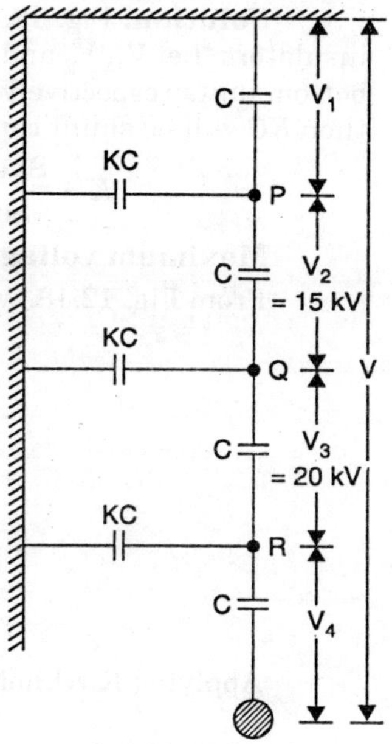

Fig. 12.17

Solution. *Given :* V_2 = 15 kV : V_3 = 20 kV.

We know that at any node X in case of string of insulators,

$$V_{x+1} = V_x + KV_x$$

i.e., $$V_{1+1} = V_1 + KV_1$$

$$V_2 = V_1(1 + K)$$

$$V_{2+1} = V_3 = V_2 + K(V_1 + V_2)$$

$$= V_2\,(1 + K) + KV_1$$

or, $$V_3 = V_1(1 + K)\,(1 + K) + KV_1$$

$$= V_1(1 + 2K + K^2) + KV_1$$

$$= V_1(1 + 3K + K^2)$$

and $$V_{3+1} = V_3 + K(V_1 + V_2 + V_3)$$

$$= V_3 + KV_1 + KV_2 + KV_3$$

$$= KV_1 + KV_2 + V_3(1 + K)$$

$$= KV_1 + V_1K(1 + K) + V_1(1 + K)\,(1 + 3K + K^2)$$

$$= KV_1 + V_1(K + K^2) + V_1(1 + 3K + K^2 + K + 3K^2 + K^3)$$

$$= V_1[K + (K + K^2) + (1 + 4K + 4K^2 + K^3)]$$

or, $$V_4 = V_1(1 + 6K + 5K^2 + K^3)$$

Also, $$\frac{V_3}{V_2} = \frac{V_1(1+3K+K^2)}{V_1\,(1+K)} = \frac{1+3K+K^2}{1+K} = \frac{20}{15} = \frac{4}{3}$$

$$3(1 + 3K + K^2) = 4(1 + K)$$

or, $$3 + 9K + 3K^2 = 4 + 4K$$

or, $$3K^2 + 5K - 1 = 0$$

or, $$K = \frac{-5 \pm \sqrt{25 + 4 \times 3 \times 1}}{6} = \frac{-5 \pm \sqrt{37}}{6} = \frac{-5 \pm 6.08}{6} = \frac{1.08}{6}$$

$$= 0.18 \text{ considering +ve only}$$

$\therefore$ $$V_1 = \frac{V_2}{1+K} = \frac{15}{1+0.18} = 12.71 \text{ kV}$$

$$V_4 = V_1(1 + 6K + 5K^2 + K^3)$$

$$= 12.71\,[1 + 6 \times 0.18 + 5 \times (0.18)^2 + (0.18)^3] = 28.57 \text{ kV}$$

$\therefore$ Voltage between line and earth (*i.e.* phase voltage)

$$= V_1 + V_2 + V_3 + V_4$$

$$= 12.71 + 15 + 20 + 28.57 = 76.28 \text{ kV}$$

Voltage between conductors (*i.e.* line voltage)

$$= \sqrt{3} \times 76.28 = \mathbf{132.12\ kV.} \quad \textbf{(Ans.)}$$

String efficiency $$= \frac{V}{4 \times V_4} = \frac{76.28}{4 \times 28.57} = 0.6675 \text{ or } \mathbf{66.75\%.} \quad \textbf{(Ans.)}$$

Example 12.5. *A string insulator has three units. The capacitance from each joint to tower is 12.5% of the capacitance of each unit. The voltage across each unit should not exceed 11 kV. Find the maximum voltage for the string application.* **(N.U.)**

Solution. Fig. 12.18 shows the equivalent circuit of string insulators. Let V_1, V_2 and V_3 be the voltages across top, middle and bottom units respectively. If C is the self-capacitance of each unit, then KC will be shunt capacitance.

$$K = \frac{\text{Shunt capacitance}}{\text{Self capacitance}} = \frac{12.5}{100} = 0.125$$

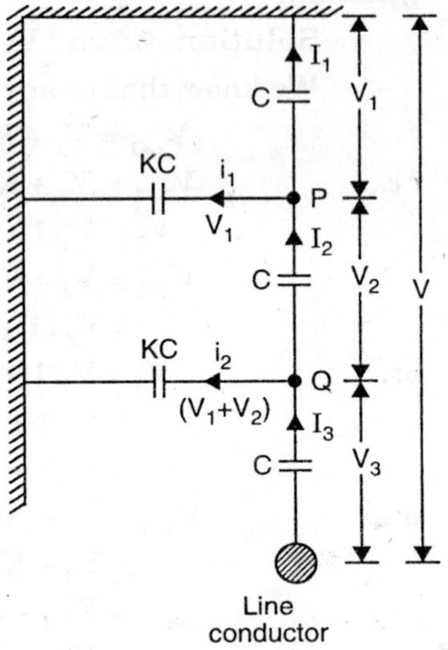

Fig. 12.18

Maximum voltage for the string, V :

From Fig. 12.18, we have

$$I_1 = \frac{V_1}{1/\omega C} = V_1 \omega C$$

$$i_1 = \frac{V_1}{1/\omega KC} = V_1 \omega KC$$

$$I_2 = \frac{(V_1 + V_2)}{1/\omega C} = (V_1 + V_2)\omega C$$

$$i_2 = \frac{(V_1 + V_2)}{1/\omega KC} = (V_1 + V_2)\omega KC$$

Applying Kirchhoff's current law (*KCL*) to node *P*, we get

$$I_2 = I_1 + i_1$$

$$V_2 \omega C = V_1 \omega C + V_1 \omega KC$$

or,
$$V_2 = V_1(1 + K)$$

Applying *KCL* to node *Q*, we get

$$I_3 = I_2 + i_2$$

$$V_3 \omega C = (V_1 + V_2)\omega C + (V_1 + V_2)\omega KC$$

or,
$$V_3 = (V_1 + V_2)(1 + K) = V_1(1 + K) + V_2(1 + K)$$

$$= V_1(1 + K) + V_1(1 + K^2)^2 = V_1[(1 + K) + (1 + 2K + K^2)]$$

$$= V_1(1 + 3K + K^2)$$

Also,
$$V = V_1 + V_2 + V_3$$

$$= V_1 + V_1(1 + K) + V_1(1 + 3K + K^2)$$

$$= V_1(1 + 1 + K + 1 + 3K + K^2) = V_1(3 + 4K + K^2)$$

$$= V_1(3 + 4 \times 0.125 + 0.125^2) = 3.515\ V_1$$

or,
$$V_1 = \frac{V}{3.515} = 0.284\ \text{V}$$

$\therefore$
$$V_3 = V_1(1 + 3K + K^2)$$

$$= 0.284\ \text{V}\ (1 + 3 \times 0.125 + 0.125^2) = 0.395\ \text{V}$$

But
$$V_3 = 11\ \text{kV} \quad (\textit{Given})$$

$\therefore$
$$11 = 0.395\ \text{V}$$

or,
$$V = \frac{11}{0.395} = \mathbf{27.85\ kV.} \quad \textbf{(Ans.)}$$

Example 12.6. *A string of six insulators (suspension type) is to be graded to obtain uniform distribution of voltage across the string. If the capacitance of the top unit is 10 times the capacitance to ground of each unit, determine the capacitance of the remaining units.* **(AMIE Winter, 2002)**

Solution. Let C be the capacitance to ground of each unit, then capacitance of the top unit is mC *i.e.* 10 C (where $m = 10$).

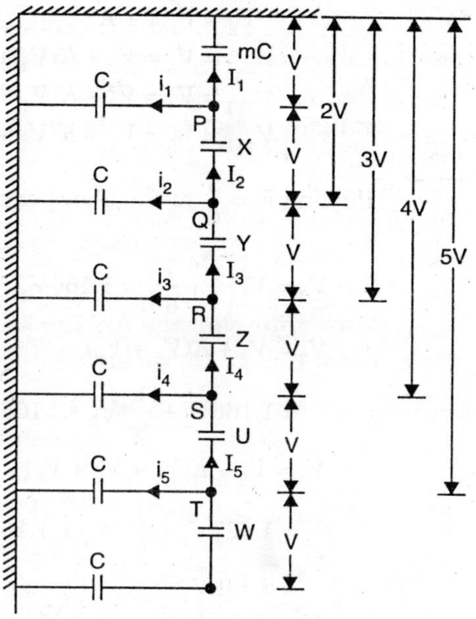

Fig. 12.19

At anode P : $I_2 = I_1 + i_1$

$$VX = VmC + VC$$

$$X = C(1 + m) = C + mC$$

At anode Q : $I_3 = I_2 + i_2$

$$VY = VX + 2VC$$

or, $$Y = X + 2C = C(1 + m) + 2C$$

$$= (1 + 2 + m)C = (1 + 2)\, C + mC$$

At anode R : $I_4 = I_3 + i_3$

$$VZ = VY + 3VC$$

or, $$Z = Y + 3C = (1 + 2 + 3 + m)C$$

$$= (1 + 2 + 3)C + mC$$

At anode S : $I_5 = I_4 + i_4$

$$VU = VZ + 4VC$$

or, $$U = Z + 4C = (1 + 2 + 3 + 4 + m)C$$

$$= (1 + 2 + 3 + 4)C + mC$$

and so on, in general, $C_n = [1 + 2 + 3 + \ldots\ldots + (n - 1)C] + mC$

It is given that $m = 10$

$\therefore$ $$X = C + 10C = \mathbf{11\ C};$$

$$Y = (1 + 2)C + 10 \times C = \mathbf{13\ C};$$

$$Z = (1 + 2 + 3)C + 10C = \mathbf{16\ C};$$

$$U = (1 + 2 + 3 + 4)C + 10C = \mathbf{20\ C};$$

$$W = (1 + 2 + 3 + 4 + 5)C + 10C = \mathbf{25\ C}. \quad \textbf{(Ans.)}$$

Example 12.7. *In a 5 insulator disc string capacitance between each unit and earth is 1/6 of the mutual capacitance.*

(i) Find the voltage distribution a cross each insulator in the string as %age of voltage of the conductor to earth.

(ii) Find string efficiency. How is this efficiency affected by rain ?

(iii) If the insulators in the string are designed each to withstand 35 kV maximum (peak), find the operating voltage of the line where 5 suspension insulators string can be used.

Solution. No. of discs = 5

Mutual capacitance of each unit = C

Capacitance between each unit and earth = C_1

$$K = \frac{C_1}{C} = \frac{C_1}{6C_1} = \frac{1}{6} \qquad (\because\ C = 6C_1 \ldots \textit{Given})$$

(*i*) **Voltage distribution across each insulator :**

We know that at any node X in case of a string of insulators,

$$V_{x+1} = V_x + KV_x$$

i.e., $$V_{1+1} = V_1 + KV_1$$

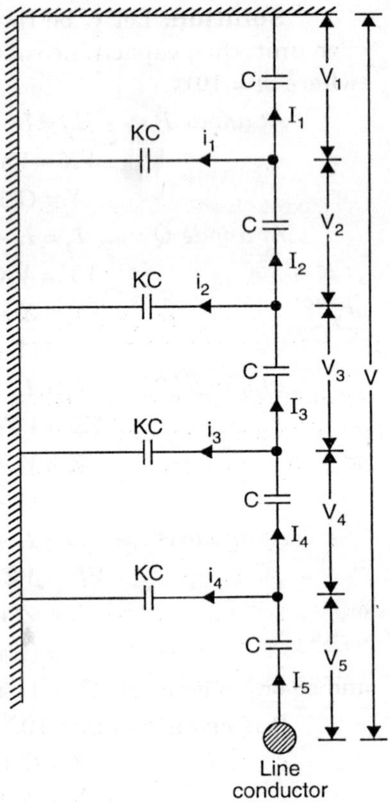

Fig. 12.20

or, $V_2 = V_1(1 + K)$ ...(*i*)

$V_{2+1} = V_3 = V_2 + K(V_1 + V_2)$ ...(*ii*)

$V_{3+1} = V_4 = V_3 + K(V_1 + V_2 + V_3)$ ...(*iii*)

$V_{4+1} = V_5 = V_4 + K(V_1 + V_2 + V_3 + V_4)$ ...(*iv*)

Putting $K = \frac{1}{6}$ in the above expression, we get

$$V_2 = V_1\left(1 + \frac{1}{6}\right) = 1.166V_1$$

$$V_3 = V_2 + K(V_1 + V_2)$$

$$= 1.166V_1 + \frac{1}{6}(V_1 + 1.166\,V_1) = 1.527V_1$$

$$V_4 = V_3 + K(V_1 + V_2 + V_3)$$

$$= 1.527V_1 + \frac{1}{6}(V_1 + 1.166V_1 + 1.527V_1)$$

$$= 2.142V_1$$

$$V_5 = V_4 + K(V_1 + V_2 + V_3 + V_4)$$

$$= 2.142V_1 + \frac{1}{6}(V_1 + 1.166V_1 + 1.527V_1 + 2.142V_1)$$

$$= 3.114V_1$$

Now, $V = V_1 + V_2 + V_3 + V_4 + V_5$

$= V_1 + 1.166V_1 + 1.527V_1 + 2.142V_1 + 3.114V_1$

$= 8.949V_1$

$\therefore$ $V_1 = \frac{V}{8.494}$ = **11.17% V ;**

$V_2 = 1.166 \times 11.17$ = **13.02% V ;**

$V_3 = 1.527 \times 11.17$ = **17.06% V ;**

$V_4 = 2.142 \times 11.17$ = **23.93% V ;**

$V_5 = 3.114 \times 11.17$ = **34.78% V.** **(Ans.)**

(*ii*) **String efficiency :**

String efficiency $= \frac{V}{5 \times V_5} = \frac{V}{5 \times 0.3478\,V} = 0.575$ or **57.5%.** **(Ans.)**

- When the insulators get wet in rainy weather, the capacitance between the unit (C) gets *increased.* This *decreases the value of K making the distribution of voltage between the units more uniform thus resulting in higher string efficiency.*

(*iii*) **Operating voltage of the line :**

Voltage across the bottom unit,

$V_5 = 0.3478$ V

= Safe working voltage of the unit

$= 35$ kV (maximum) $= \frac{35}{\sqrt{2}} = 24.75$ kV (r.m.s.)

Voltage between conductor and earth,

$$V = \frac{V_5}{0.3478} = \frac{24.75}{0.3478} = 71.16 \text{ kV}$$

Operating line voltage $= \sqrt{3}\, V = \sqrt{3} \times 71.16 \simeq$ **123 kV. (Ans.)**

Example 12.8. (*a*) *Each conductor at 66 kV, is suspended by a string of 5 similar insulators, the capacitance of each disc being 5 times the capacitance of earth. Determine the % string efficiency.*

(*b*) *If in* (*a*) *the distance between the insulators and tower structure is increased so as to make the capacitance of the disc 6 times the earth capacitance, will the string efficiency improve and how much ?*

Solution. (*a*) Refer to Fig. 12.21.

Number of units = 5

Mutual capacitance of each unit = C

Capacitance between each unit and earth = C_1

$$K = \frac{C_1}{C} = \frac{C_1}{5C_1} = \frac{1}{5} = 0.2 \qquad (\because \; C = 5C_1 \; ... \; \textit{Given})$$

We know that at any node X in case of string of insulators,

$$V_{x+1} = V_x + KV_x$$

i.e. $\quad V_{1+1} = V_1 + KV_1$

$$V_2 = V_1(1 + K) \qquad ...(i)$$

$$V_{2+1} = V_3 = V_2 + K(V_1 + V_2) \qquad ...(ii)$$

or, $\quad V_{3+1} = V_4 = V_3 + K(V_1 + V_2 + V_3) \qquad ...(iii)$

$$V_{4+1} = V_5 = V_4 + K(V_1 + V_2 + V_3 + V_4) \qquad ...(iv)$$

Putting $K = 0.2$ in the above expressions, we have

$$V_2 = V_1(1 + 0.2) = 1.2\, V_1$$

$$V_3 = V_2 + K(V_1 + V_2) = 1.2V_1 + 0.2(V_1 + 1.2V_1) = 1.64V_1$$

$$V_4 = V_3 + K(V_1 + V_2 + V_3)$$

$$= 1.64V_1 + 0.2(V_1 + 1.2V_1 + 1.64V_1) = 2.408V_1$$

$$V_5 = V_4 + K(V_1 + V_2 + V_3 + V_4)$$

$$= 2.408V_1 + 0.2(V_1 + 1.2V_1 + 1.64V_1 + 2.408\, V_1)$$

$$= 3.6576V_1$$

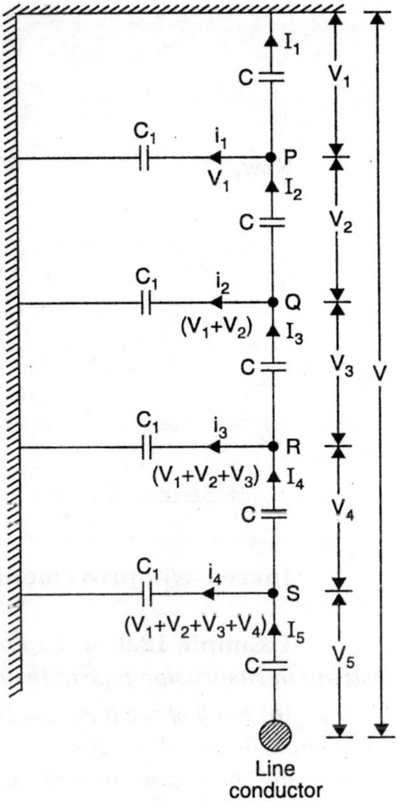

Fig. 12.21

Now, $\quad V = V_1 + V_2 + V_3 + V_4 + V_5 = 66$

$$= V_1(1 + 1.2 + 1.64 + 2.408 + 3.6576) = 9.9056V_1 = 66$$

$$\therefore \quad V_1 = \frac{66}{9.9056} = 6.663 \text{ kV}$$

$$V_2 = 1.2 \times 6.663 = 7.996 \text{ kV}$$

$$V_3 = 1.64 \times 6.663 = 10.927 \text{ kV}$$

$$V_4 = 2.408 \times 6.663 = 16.044 \text{ kV}$$

$$V_5 = 3.6576 \times 6.663 = 24.37 \text{ kV}$$

$\therefore$ **% String efficiency** $= \dfrac{V}{5 \times V_5} = \dfrac{66}{5 \times 24.37} \times 100 =$ **54.16%. (Ans.)**

(*b*) For $K = \frac{1}{6}$, we have

$$V_2 = V_1\left(1+\frac{1}{6}\right) = 1.166V_1$$

$$V_3 = V_2 + K(V_1 + V_2) = 1.166V_1 + \frac{1}{6}(V_1 + 1.166V_1) = 1.527V_1$$

$$V_4 = V_3 + K(V_1 + V_2 + V_3)$$

$$= 1.527V_1 + \frac{1}{6}(V_1 + 1.166V_1 + 1.527V_1) = 2.142V_1$$

$$V_5 = V_4 + K(V_1 + V_2 + V_3 + V_4)$$

$$= 2.142V_1 + \frac{1}{6}(V_1 + 1.166V_1 + 1.527V_1 + 2.142V_1) = 3.114V_1$$

Now, $$V = V_1 + V_2 + V_3 + V_4 + V_5$$

$$= V_1 + 1.166V_1 + 1.527V_1 + 2.142V_2 + 3.114V_1 = 8.949V_1$$

∴ $$V_1 = \frac{66}{8.949} = 7.375 \text{ kV}$$

$$V_2 = 1.166 \times 7.375 = 8.599 \text{ kV}$$

$$V_3 = 1.527 \times 7.375 = 11.262 \text{ kV}$$

$$V_4 = 2.142 \times 7.375 = 15.797 \text{ kV}$$

$$V_5 = 3.114 \times 7.375 = 22.966 \text{ kV}$$

∴ % String efficiency $$= \frac{V}{5 \times V_5} \times 100 = \frac{66}{5 \times 22.966} \times 100 = 57.48\%$$

Increase/improvement in efficiency $= \frac{57.48 - 54.16}{54.16} \times 100 =$ **6.13%. (Ans.)**

Example 12.9. (*a*) *Explain why the potential distribution is not, in general, uniform over the string in suspension type of insulators.*

(*b*) *Each of the three insulators forming a string has a self capacitance of C farads. The shunting capacitance of the connecting metal work of each insulator is 0.2 C to earth and 0.05 C to line. Calculate the voltage across each insulator as a percentage of the line voltage to earth and also the string efficiency.* **(Panjab University)**

Solution. (*a*) The potential distribution in a string of insulators depends upon the capacitance formed by the metal parts of the insulators and the tower structure. These capacitances could be made negligibly small and equal by increasing the distance between the insulators and the tower structure which require large lengths of cross-arms and bigger size of the towers and hence it becomes uneconomical. Therefore, in practice the insulators are not very far from the tower structure and hence these capacitances affect the voltage distribution across the string. Since the mutual capacitance of each disc is same and the current through the topmost unit is minimum, the voltage drop across that unit will be minimum. As we go down the unit towards the power conductor the current goes on increasing being the maximum in the lowest unit, the voltage drop is maximum there.

(*b*) Fig. 12.22 shows the arrangement of a string of three insulators *using guard ring*.

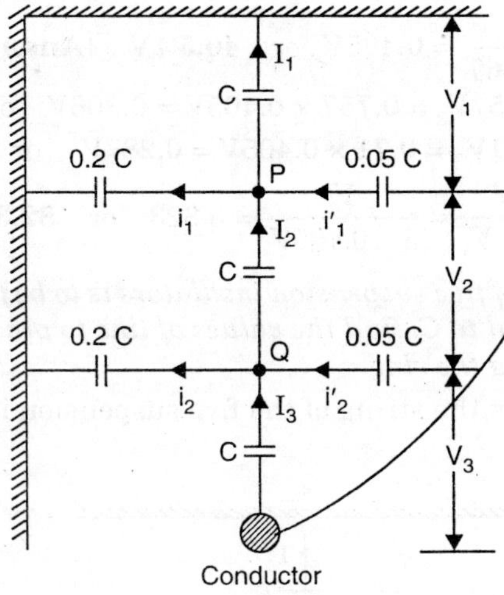

Fig. 12.22

Let V_1, V_2 and V_3 be the voltages across the discs as shown in the figure.

Applying *KCL* at node *P*, we get

$$I_2 + i'_1 = I_1 + i_1$$

or $$V_2\omega C + (V_2 + V_3)\omega \times 0.05C = V_1\omega C + 0.2V_1\omega C$$

or $$V_2 + (V_2 + V_3) \times 0.05 = V_1 + 0.2V_1 = 1.2V_1$$

or $$1.2V_1 - 1.05V_2 - 0.05V_3 = 0 \quad ...(i)$$

Applying *KCL* at node *Q*, we have

$$I_3 + i'_2 = I_2 + i_2$$

or $$V_3\omega C + 0.05\ V_3\omega C = V_2\omega C + (V_1 + V_2) \times 0.2C$$

or $$V_3 + 0.05V_3 = V_2 + (V_1 + V_2) \times 0.2$$

or $$0.2V_1 + 1.2V_2 - 1.05V_3 = 0 \quad ...(ii)$$

There are three unknowns and two equations only. Hence we divide both the equations by V_3 and rewrite them as

$$1.2x - 1.05y = 0.05 \quad \text{or} \quad 12x - 10.5y = 0.5 \quad ...(1)$$

and $$0.2x + 1.2y = 1.05 \quad \text{or} \quad 2x + 12y = 10.5 \quad ...(2)$$

where $x = \dfrac{V_1}{V_3}$, and $y = \dfrac{V_2}{V_3}$.

Multiplying eqn. (2) by 6 and subtracting from eqn. (1), we have

$$-82.5y = -62.5$$

$$\therefore \quad y = \frac{62.5}{82.5} = 0.757 = \frac{V_2}{V_3}$$

and $$2x + 12 \times 0.757 = 10.5$$

$$\therefore \quad x \simeq 0.71 = \frac{V_1}{V_3}$$

$\therefore$ $V_2 = 0.757V_3$, and $V_1 = 0.71V_3$

Now, $V = V_1 + V_2 + V_3 = 0.71V_3 + 0.757V_3 + V_3 = 2.467V_3$

$\therefore$ $$\mathbf{V_3} = \frac{V}{2.467} \doteq 0.405V \text{ or } \mathbf{40.5\%V} \text{ (Ans.)}$$

$$\mathbf{V_2} = 0.757V_3 = 0.757 \times 0.405V = 0.306V \text{ or } \mathbf{30.6\%V.} \text{ (Ans.)}$$

$$\mathbf{V_1} = 0.71V_3 = 0.71 \times 0.405V = 0.287V \text{ or } \mathbf{28.7\%V.} \text{ (Ans.)}$$

String efficiency $= \frac{V}{3 \times V_3} = \frac{V}{3 \times 0.405V} = 0.823$ or **82.3%. (Ans.)**

Example 12.10. *A string of five suspension insulators is to be fitted with a grading ring. If the pin to earth capacitances are equal to C, find the values of line to pin capacitances that would give a uniform voltage distribution along the ring.*

Solution. Fig. 12.23 shows the string of the five suspension insulators fitted with a grading ring.

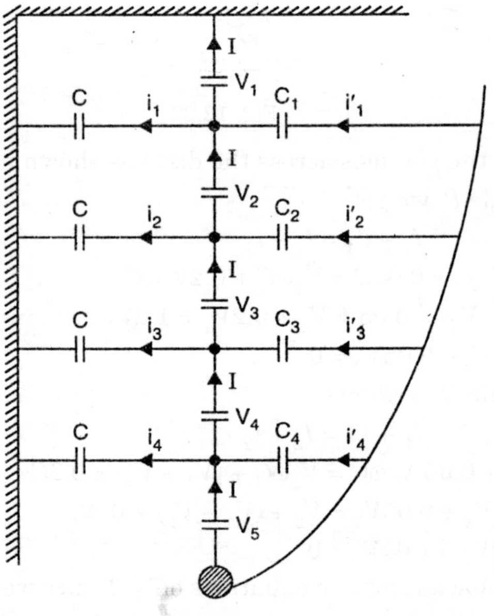

Fig. 12.23

For voltage distribution to be uniform,

$$i_1 = i_1' ;\quad i_2 = i_2'$$
$$i_3 = i_3' ;\quad i_4 = i_4'$$

Also $V_1 = V_2 = V_3 = V_4 = V_5$

Now, $$i_1 = V_1\omega C = \frac{V}{5}\omega C ;$$

$$i_1' = \frac{4V}{5}\omega C_1$$

Since, $i_1 = i_1'$

$$\frac{V}{5}\omega C = \frac{4V}{5}\omega C_1 \quad \text{or,} \quad \mathbf{C_1 = \frac{C}{4}}. \text{ (Ans.)}$$

Similarly, $i_2 = i_2'$ gives,

$$\frac{2V}{5}\omega C = \frac{3V}{5}\omega C_2 \quad \text{or} \quad \mathbf{C_2 = \frac{2}{3}C.} \quad \textbf{(Ans.)}$$

and, $i_3 = i_3'$ gives,

$$\frac{3V}{5}\omega C = \frac{2V}{5}\omega C_3 \quad \text{or} \quad \mathbf{C_3 = \frac{3}{2}C.} \quad \textbf{(Ans.)}$$

and, $i_4 = i_4'$ gives,

$$\frac{4V}{5}\omega C = \frac{V}{5}\omega C_4 \quad \text{or} \quad \mathbf{C_4 = 4C.} \quad \textbf{(Ans.)}$$

Example 12.11. *A string of 8 suspension insulators is to be fitted with a grading ring. If the pin-to-earth capacitances are equal to C find the value of line-to-pin capacitance that would give a uniform voltage distribution over the string.*

Solution. No. of suspension insulators, $n = 8$

Pin-to-earth capacitance = C

Capacitance of each disc = mC

Line-to-pin capacitances (A, B, C, D, E, F, G) :

Since voltage across each unit is same (V), current flowing in each unit, *i.e.* $I_1, I_2, I_2 \ldots I_6$ etc. will be *equal*.

At node P : $\quad I_1 + i_1 = I_2 + i_1'$

or, $\quad i_1 = i_1' \quad (\because I_1 = I_2)$

or, $\quad VC = (n-1)VA \quad \text{or} \quad A = \dfrac{C}{n-1}$

Similarly, at *anode Q* :

$$I_2 + i_2 = I_3 + i_2'$$

or, $\quad i_2 = i_2' \quad (\because I_2 = I_3)$

$$2VC = (n-2)VB \quad \text{or} \quad B = \frac{2C}{n-2}$$

Similarly, $\quad C = \dfrac{3C}{n-3}$; $D = \dfrac{4C}{n-4}$; $E = \dfrac{5C}{n-5}$;

$$F = \frac{6C}{n-6} \quad \text{and} \quad G = \frac{7C}{n-7}$$

In general, capacitance from the shield to the pth link from the top is given by

$$C_p = \frac{pC}{n-p}$$

Putting $n = 8$, we get

$$A = \frac{C}{7}\ ;\ B = \frac{C}{3}\ ;\ C = \frac{3}{5}C\ ;\ D = 1C\ ;$$

$$E = \frac{5}{3}C\ ;\ F = 3C\ ;\ G = \mathbf{7C.}\ \textbf{(Ans.)}$$

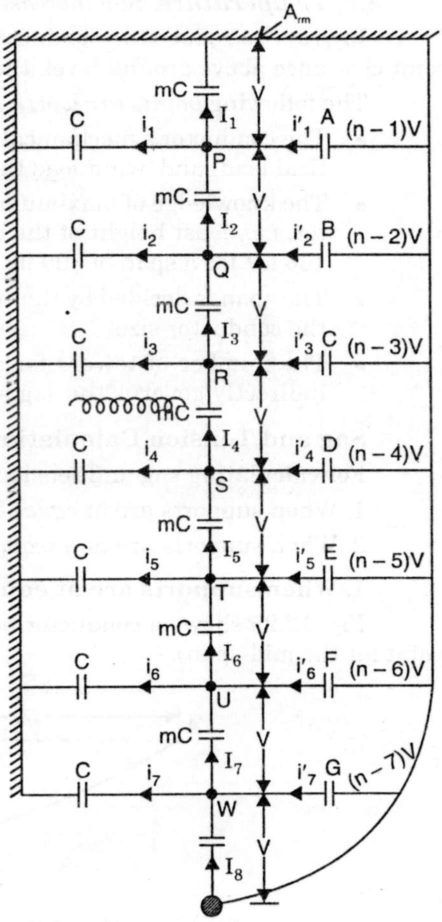

Fig. 12.24

12.6. SAG IN OVERHEAD LINES

The conductors, in overhead transmission lines, are supported at the towers (or poles for low voltage lines). The conductors are pulled and stringing effected. When the conductor is supported in this fashion it will *sag* or dip under its own weight and it takes the shape of catenary. *The distance between the adjacent supporting tower is called the "**span**".*

"The difference in level between the points of supports and the lowest point is known as **sag.**

In an overhead line, the following *factors affect the sag* :

(*i*) ***Weight of the conductor.*** The sag, is directly affected by this factor ; *heavier the conductor, greater will be the sag.*

(*ii*) **Span length.** Sag is *directly proportional to the square of the span length* other conditions remaining unchanged.

(*iii*) ***Working tensile strength.*** Other conditions remaining the same, the sag is *inversely proportional to the working tensile strength.*

Under varying weather conditions of ambient temperature, the conductor tension, which is maximum near the tower ends, should not exceed the permissible limit (= breaking strength of conductor/factor of safety of 2 to 2.5).

(*iv*) ***Temperature.*** Sag *increases with increase in temperature.*

In order to reduce the conductor material required and to avoid extra pole height for sufficient clearance above ground level, the sag should be kept to a *minimum.*

The following *points are worth noting* :

- The conductor's mechanical loading is due to its own weight (vertical), weight of ice (vertical load) and wind load (horizontal).
- The knowledge of maximum sag is essential in determine the ground clearance of conductor, *i.e.,* least height of the conductor above ground which is 7 m for voltages more than 166 kV for a span of 300 m.
- The span is decided by the economic factors with due consideration to the line voltage and the conductor size.
- The number of towers for a given line length are determined by the span which also indirectly governs the sag and height of the tower.

Sag and Tension Calculations :

For calculating sag and tension of a conductor the following *two cases* will be considered :

1. When supports are at *equal levels.*
2. When supports are at *unequal levels.*

1. **When supports are at equal levels :**

Fig. 12.25 shows a conductor between two equilevel supports *A* and *B* with *O* as the lowest point (at the mid-span).

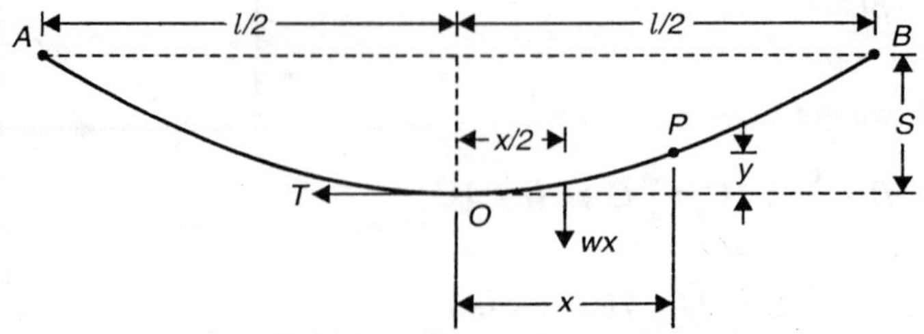

Fig. 12.25. Supports at equal level.

Let, l = Span length,

w = Weight per unit length of conductor, and

T = Tension in the conductor.

Consider a point P on the conductor. Its coordinates are x and y, taking O as the origin. If the curvature is considered so small that curved length is equal to its horizontal projection, *i.e.* $OP = x$, then the forces acting on the portion OP are :

(*i*) The weight $w.x$ of the conductor acting at a distance $\frac{x}{2}$ from O.

(*ii*) The tension T acting at O.

Taking moments of these forces about point P, we get

$$T \times y = w.x \times \frac{x}{2} \quad \text{or} \quad y = \frac{wx^2}{2T}$$

The maximum sag (dip) is represented by the value of y at either of the supports A and B.

At support A (or B), $x = l/2$ and $y = S$

$$\therefore \quad \text{Sag, } S = \frac{w\,(l/2)^2}{2T} = \frac{wl^2}{8T} \qquad ...(12.3)$$

2. **When supports are at unequal levels :**

Consider an overhead line AOB supported over the supports A and B, at unequal levels as shown in Fig. 12.26 ; O being the lowest point.

Let, l = Span length,

h = Difference in levels between two supports,

x_1 = Distance of support A (lower level) from O,

x_2 = Distance of support B (higher level) from O,

w = Weight per unit length of conductor, and

T = Tension in the conductor.

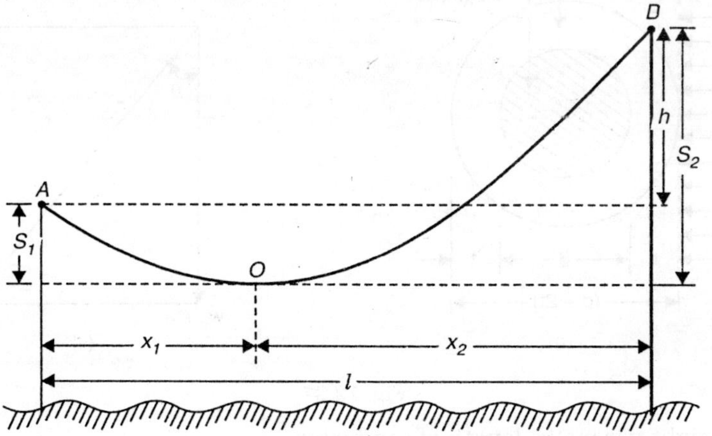

Fig. 12.26. Supports at unequal levels.

Now, $\text{Sag } S_1 = \frac{wx_1^2}{2T}$, and $S_2 = \frac{wx_2^2}{2T}$

Also, $x_1 + x_2 = l$ ...(*i*)

$$\therefore \quad S_2 - S_1 = \frac{wx_2^2}{2T} - \frac{wx_1^2}{2T} = \frac{w}{2T}(x_2^2 - x_1^2) = \frac{w}{2T}(x_2 + x_1)(x_2 - x_1)$$

or, $$S_2 - S_1 = \frac{wl}{2T}(x_2 - x_1) \qquad (\because \; x_1 + x_2 = l)$$

But, $S_2 - S_1 = h$

$$\therefore \quad h = \frac{wl}{2T}(x_2 - x_1)$$

or, $$(x_2 - x_1) = \frac{2Th}{wl} \qquad ...(ii)$$

Adding (*i*) and (*ii*), we get

$$2x_2 = l + \frac{2Th}{wl} \quad \text{or} \quad x_2 = \frac{l}{2} + \frac{Th}{wl}$$

and, $$x_1 = l - \left[\frac{l}{2} + \frac{Th}{wl}\right] = \frac{l}{2} - \frac{Th}{wl} \qquad ...[\text{From eqn. } (i)]$$

After finding x_1 and x_3 the values of S_1 and S_2 can be calculated.

Effect of wind and ice on sag :

The above formulae for sag hold good only for still air at normal temperature conditions and the conductor is acted upon by its weight alone. In actual practice the conductor may be coated with ice (at low temperature) and subjected to wind load simultaneously. Under such conditions the weight of the conductor, together with weight of ice acts vertically downwards while the wind acts in horizontal direction as shown in Fig. 12.27 (*i* and *ii*).

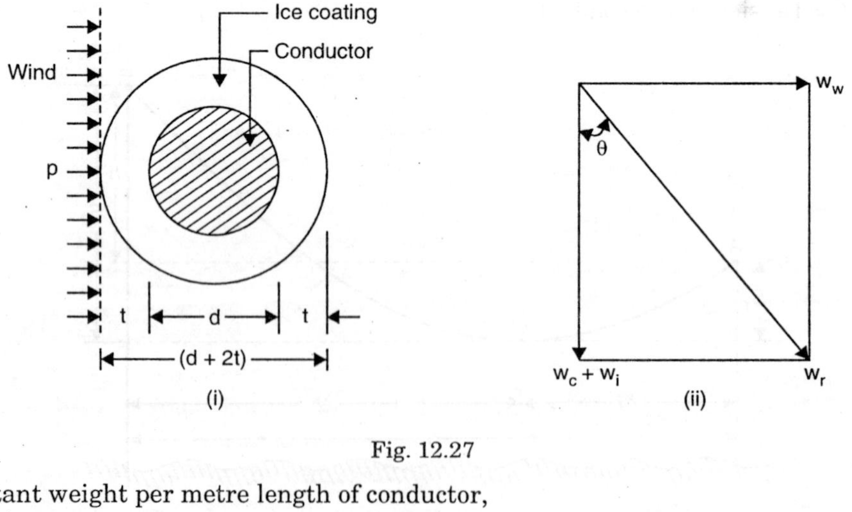

Fig. 12.27

Resultant weight per metre length of conductor,

$$w_r = \sqrt{(w_c + w_i)^2 + (w_w)^2} \qquad ...(12.4)$$

where, w_C = Weight of conductor per unit length

= Density of conductor material × volume per unit length,

w_i = Weight of ice per unit length

$= \text{Density of ice} \times \frac{\pi}{4} [(d + 2t)^2 - d^2] \times 1$

w_w = Wind force per unit length

= Wind pressure (p) per unit area × projected area per unit length

$= p \times [(d + 2t) \times 1]$, and

$$\tan \theta = \frac{w_w}{w_c + w_i} \quad ...(12.5)$$

The *slant sag* in the conductor is given by,

$$S = \frac{w_r l^2}{8T} \quad ...(12.6)$$

The *vertical sag* $= S \cos \theta$ ...(12.7)

Example 12.12. *A transmission line has a span of 240 metres. Find the weight of the conductor per metre length if the sag, ultimate tensile strength and factor of safety are 1.6 metres, 5200 kg and 2 respectively.*

Solution. *Given :* l = 240 m ; S = 1.6 m ; Ultimate tensile strength = 5200 kg ; F.O.S. = 2

Weight of the conductor per metre length, w :

Working tension, $T = \frac{5200}{2} = 2600$ kg

We know that, $S = \frac{wl^2}{8T}$ or $1.6 = \frac{w \times (240)^2}{8 \times 2600}$

$\therefore \quad w = \frac{1.6 \times 8 \times 2600}{(240)^2}$ = **0.578 kg/m. (Ans.)**

Example 12.13. (*a*) *Discuss the various methods employed for measuring and checking 'sag' during erection of a transmission line.*

(*b*) *An overhead line has a span of 180 metres between level supports. The conductor has a cross-sectional area of 1.22 cm² and weighs 1.12 kg/metre and has a breaking stress of 4150 kg/cm². Allowing a wind pressure of 110 kg/m², calculate the sag for a factor safety of 4.*

Solution. (*a*) Following are the methods usually employed for *checking the sag* :

(*i*) **Sight method.** In this case wooden strips are used as targets, the length of each of this target is equal to the minimum clearance needed from ground. These wooden strips are placed along the poles vertically at each end of the span so that the distance of the upper end of the targets and the conductors at the poles is equal to the required sag. The conductor is correctly sagged when its lower point in the span is in line with top of the targets when viewed from one end of the span to other.

(*ii*) **Stop watch or Return wave method.** In this method a sharp blow is given to the conductor. This will cause a wave to travel to the other pole, from where the wave is again reflected back. Thus wave will travel back and forth until it finally dies out by absorption of energy by the conductor. There exists a definite relation between the time in seconds taken by the wave to return to the first support and the amount of sag. The return wave can be felt by a man on the first pole by keeping his finger lightly on the conductor.

(*b*) *Given :* l = 180 m ; a = 1.22 cm² ; w_c = 1.12 kg/m ; Breaking stress = 4150 kg/cm² ;

p = 110 kg/m² ; F.O.S. = 4.

Slant and vertical sag :

Working tension $= \dfrac{4150 \times a}{\text{F.O.S.}} = \dfrac{4150 \times 1.22}{4} = 1265.75$ kg

Diameter of the conductor, $d = \sqrt{\dfrac{4 \times a}{\pi}} = \sqrt{\dfrac{4 \times 1.22}{\pi}} \simeq 1.25$ cm

Wind force/metre length, w_w = Pressure × projected area
$= 110 \times (1.25 \times 10^{-2} \times 1) = 1.375$ kg

Resultant weight/force per metre conductor length,

$$w_r = \sqrt{w_c^2 + w_w^2} = \sqrt{(1.12)^2 + (1.375)^2} = 1.773 \text{ kg}$$

∴ *Slant sag* $= \dfrac{w_r l^2}{8T} = \dfrac{1.773 \times (180)^2}{8 \times 1265.75} =$ **5.67 m. (Ans.)**

The slant sag makes an angle θ (Fig. 12.28) with the vertical where value of θ is given by :

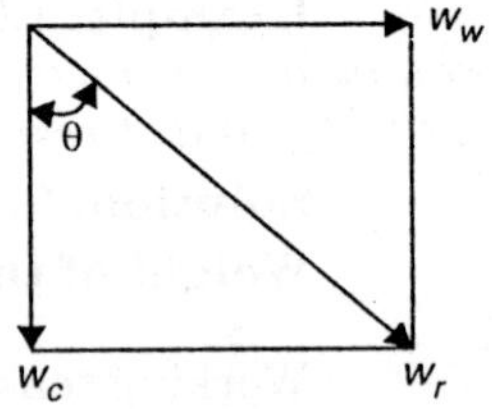

Fig. 12.28

$$\tan\theta = \frac{w_w}{w_c} \quad \text{or} \quad \theta = \tan^{-1}\left(\frac{w_w}{w_c}\right)$$

or, $$\theta = \tan^{-1}\left(\frac{1.375}{1.12}\right) = 50.8°$$

Vertical sag = $S \cos\theta = 5.67 \times \cos(50.8°) =$ **3.58 m. (Ans.)**

Example 12.14. *A transmission line has a span of 160 metres between the level supports. With the following data, calculate the height of the conductor above the ground level at which it should be supported if a minimum clearance of 7 m is to be left between the ground and the conductor :*

X-sectional area of the conductor	*= 2.2 cm² ;*
The density of the material	*= 8.9 g/cc ;*
The wind force	*= 1.6 kg/m run ;*
The ultimate strength	*= 4980 kg/cm² ;*
Factor of safety	*= 4.*

Solution. *Given :* $a = 2.2$ cm² ; sp. gr. = 8.9 g/cc ; wind force = 1.6 kg/m,
T (ultimate) = 4980 kg/cm² ; F.O.S. = 4 ; Minimum clearance = 7 m

Height of the conductor above the ground level :

Weight of the conductor per metre run, w_c = X-sectional area × 1 m length × density
$= 2.2 \text{ cm}^2 \times 100 \text{ cm} \times 8.9 \text{ g} \times 10^{-3} \text{ kg/cm}^3 \simeq 1.96$ kg

Working tension, $T = \dfrac{4980 \times 2.2}{4} = 2739$ kg

Resultant force per metre conductor length,

$$w_r = \sqrt{w_c^2 + w_w^2} = \sqrt{(1.96)^2 + (1.6)^2} = 2.53 \text{ kg}$$

Slant sag, $S = \dfrac{w_r l^2}{8T} = \dfrac{2.53 \times (160)^2}{8 \times 2739} = 2.95$ m

Vertical sag $= S\cos\theta = 2.95 \times \dfrac{w_c}{w_r} = 2.95 \times \dfrac{1.96}{2.53} = 2.28$ m $\quad \left(\because \cos\theta = \dfrac{w_c}{w_r}\right)$

∴ Height of the conductor above the ground level = 7 + 2.28 = **9.28 m. (Ans.)**

Example 12.15. *The following data relate to a transmission line :*

Span length = 200 m (between level supports) ; X-sectional area of the conductor = 2.9 cm² ; Vertical sag = 2.28 m ; Wind force = 1.4 kg/m run ; Breaking stress = 2480 kg/cm² ; Weight of the conductor = 1.12 kg/m run.

Calculate the factor of safety.

Solution. *Given :* l = 200 m ; a = 2.9 cm² ; Vertical sag = 2.28 m ; w_w = 1.4 kg/m run ; Breaking stress = 2480 kg/cm²; w_c = 1.12 kg/m run.

Factor of safety, F.O.S. :

Resultant force per metre conductor length,

$$w_r = \sqrt{w_c^2 + w_w^2} = \sqrt{(1.12)^2 + (1.4)^2} = 1.793 \text{ kg}$$

Working tension, $$T = \frac{\text{Breaking stress} \times a}{\text{F.O.S.}} = \frac{2480 \times 2.9}{\text{F.O.S.}} = \frac{7192}{\text{F.O.S.}}$$

Slant slag, $$S = \frac{\text{Vertical sag}}{\cos\theta} = \frac{2.28}{(w_c/w_r)} = \frac{2.28}{(1.12/1.793)} = 3.65 \text{ m}$$

Also, $$S = \frac{w_r l^2}{8T} \quad \text{or} \quad T = \frac{w_r l^2}{8S} \quad \text{or} \quad \frac{7192}{\text{F.O.S.}} = \frac{1.793 \times (200)^2}{8 \times 3.65}$$

$$\therefore \quad \text{F.O.S.} = \frac{7192 \times 8 \times 3.65}{1.793 \times (200)^2} = \mathbf{2.93.} \quad \textbf{(Ans.)}$$

Example 12.16. *The following data refer to a transmission line :*

Length of span = 250 m ; Effective diameter of the conductor = 1.88 cm ; Weight of the conductor = 0.87 kg/m run ; Ultimate strength = 8100 kg ; Radial thickness of ice coating on the conductor = 1.2 cm ; Wind pressure = 38 kg/m² of projected area ; F.O.S. = 2 ; Density of ice = 913 kg/m³. Calculate the sag.

Solution. *Given :* l = 250 m ; d = 1.88 cm ; w_c = 0.87 kg/m run ; Ultimate strength = 8100 kg ; t = 1.2 cm ; p = 38 kg/m² ; F.O.S. = 2 ; Density of ice = 913 kg/m³.

Slant and vertical slag :

Working tension, $$T = \frac{\text{Ultimate strength}}{\text{F.O.S.}} = \frac{8100}{2} = 4050 \text{ kg}$$

Weight of ice per metre length of conductor,

$$w_i = \text{Volume of conductor per metre length} \times \text{density of ice}$$
$$= \frac{\pi}{4}[(d + 2t)^2 - d^2] \times 1 \times 913$$
$$= \frac{\pi}{4}[(1.88 + 2 \times 1.2)^2 - (1.88)^2] \times 10^{-4} \times 1 \times 913 = 1.06 \text{ kg/m run}$$

Horizontal wind force per metre length,

$$w_w = \text{Wind pressure} \times \text{projected area of the ice covered conductor}$$
$$= 38 \times [(d + 2t) \times 10^{-2} \times 1] \qquad \text{[where } d \text{ and } t \text{ are in cm]}$$
$$= 38 \times [(1.88 + 2 \times 1.2) \times 10^{-2}] \times 1 = 1.626 \text{ kg/m}$$

Resultant force per metre conductor length,

$$w_r = \sqrt{(w_c + w_i)^2 + w_w^2}$$
$$= \sqrt{(0.87 + 1.06)^2 + (1.626)^2} = 2.524 \text{ kg/m}$$

Slant sag, $$S = \frac{w_r l^2}{8T} = \frac{2.524 \times (250)^2}{8 \times 4050} = \mathbf{4.869 \ m.} \quad \textbf{(Ans.)}$$

Vertical sag $= S \cos\theta = S \times \frac{(w_c + w_i)}{w_r} = 4.869 \times \frac{(0.87 + 1.06)}{2.524}$ = **3.72 m. (Ans.)**

Example 12.17. *The following data refer to a transmission line supported on level supports :*

Span length = *220 metres ;*

Hard drawn copper conductor : X-sectional area = *120* mm^2 *; 37/2.11 mm ;* $w_c = 1.2$ *kg/m.*

Ultimate tensile stress = *42.2 kg/*mm^2

Factor of safety = *4*

Wind pressure = *55 kg/*m^2

Thickness of ice coating = *12 mm*

Density of ice = *913 kg/*m^3

Find the vertical sag.

Solution. *Given :* $l = 220$ m ; $a = 120$ mm^2 ; 37/2.11 mm ; $w_c = 1.2$ kg/m

Ultimate tensile stress = 42.2 kg/mm^2 ; F.O.S. = 4 ; $p = 55$ kg/m^2

$t = 12$ mm, density of ice = 0.913 g/cc.

Vertical sag :

For the given conductor (73/2.11 mm), let us find the overall diameter of the conductor. In a stranded conductor number of strands = $1 + 3n(1 + n) = 3n^2 + 3n + 1$, where n is the number of layers (Refer to Table 12.3).

$$37 = 3n^2 + 3n + 1 \quad \text{or} \quad 3n^2 + 3n - 36 = 0$$

or
$$n = \frac{-3 \pm \sqrt{(3)^2 + 4 \times 3 \times 36}}{6} = \frac{-3 \pm 21}{6} = 3 \text{ considering +ve value only.}$$

The number of strands in different layers are : 1, 6, 12, 18 ; Total = 37

Overall diameter of the conductor,

$$d = (1 + 2n)d_s \qquad \text{(where } d_s = \text{diameter of each strand)}$$
$$= (1 + 2 \times 3) \times 2.11 = 14.77 \text{ mm}$$

Ultimate strength
$$= \text{Ultimate stress} \times X\text{-sectional area}$$
$$= 42.2 \times 120 = 5064 \text{ kg}$$

$\therefore$ Working tension, $T = \frac{5064}{4} = 1266$ kg

Weight of ice per metre length of conductor,

$$w_i = \text{Volume of conductor per metre length} \times \text{density of ice}$$
$$= \frac{\pi}{4}[(d + 2t)^2 - d^2] \times 1 \times \text{density of ice}$$
$$= \frac{\pi}{4}[(14.77 + 2 \times 12)^2 - (14.77)^2] \times 10^{-6} \times 1 \times 913 = 0.92 \text{ kg/m run}$$

Horizontal wind force per metre length,

$$w_w = \text{Wind pressure} \times \text{projected area of the ice covered conductor}$$
$$= p \times (d + 2t) \times 1$$
$$= 55 \times [(14.77 + 2 \times 12) \times 10^{-3}] \times 1 = 2.13 \text{ kg/m run}$$

Resultant force per metre conductor length,

$$w_r = \sqrt{(w_c + w_i)^2 + {w_w}^2} = \sqrt{(1.2 + 0.92)^2 + (2.13)^2} = 3.0 \text{ kg}$$

Slant or total sag, $S = \frac{w_r l^2}{8T} = \frac{3.0 \times (220)^2}{8 \times 1266} = 14.34$ m

Vertical sag $= S \cos \theta = S \times \left(\frac{w_c + w_i}{w_r} \right)$

$$= 14.34 \times \left(\frac{1.2 + 0.92}{3.0} \right) = \mathbf{10.13\ m.} \quad \textbf{(Ans.)}$$

Example 12.18. *A transmission line conductor crossing a river is supported from two towers at heights of 30 m and 80 m above the water level. The horizontal distance between the towers is 450 metres. If the tension in the conductor is 1500 kg and weight of the conductor is 1.4 kg/m length, find the minimum clearance of the conductor and water and clearance mid-way between the supports.*

Solution. Refer to Fig. 12.29.

Given : $l = 300$ m ; $T = 1500$ kg ; $w_c = 1.4$ kg/m

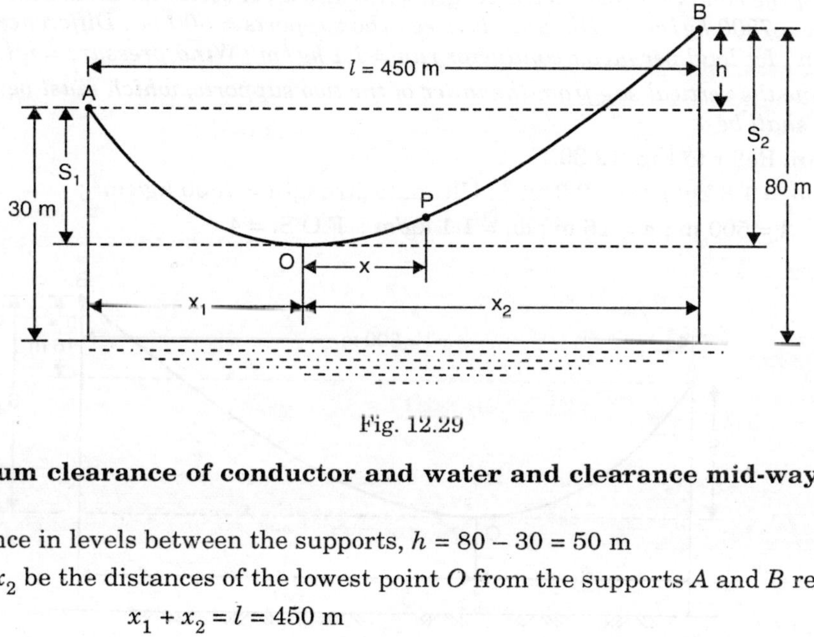

Fig. 12.29

Minimum clearance of conductor and water and clearance mid-way between the supports :

Difference in levels between the supports, $h = 80 - 30 = 50$ m

Let x_1, x_2 be the distances of the lowest point O from the supports A and B respectively.

Then $\quad x_1 + x_2 = l = 450$ m ...(*i*)

Now, $\quad S_1 = \frac{w_c x_1^2}{2T}, \quad S_2 = \frac{w_c x_2^2}{2T}$

$\therefore \quad h = S_2 - S_1 = \frac{w_c x_2^2}{2T} - \frac{w_c x_1^2}{2T}$

or $\quad 50 = \frac{w_c}{2T}(x_2^2 - x_1^2) = \frac{w_c}{2T}(x_2 + x_1)(x_2 - x_1)$

$\therefore \quad (x_2 - x_1) = \frac{50 \times 2 \times 1500}{1.4 \times 450} = 238$ m ...(*ii*)

Solving (*i*) and (*ii*), we get

$$x_1 = 106 \text{ m}, x_2 = 344 \text{ m}$$

Now, $$S_1 = \frac{w_c x_1^2}{2T} = \frac{1.4 \times (106)^2}{2 \times 1500} = 5.24 \text{ m}$$

∴ *Clearance of conductor (i.e. lowest point O) and water* = 30 – 5.24 = **24.76 m. (Ans.)**

Let the mid-point P be at a distance x from the lowest point O.

Then, $$x = \frac{450}{2} - x_1 = 225 - 106 = 119 \text{ m}$$

Sag at mid-point P, $$S_{mid} = \frac{w_c x^2}{2T} = \frac{1.4 \times 119^2}{2 \times 1500} \simeq 6.61 \text{ m}$$

∴ *Clearance of mid-point P from water level*

$$= 24.76 + 6.61 = 31.37 \text{ m.}$$ **(Ans.)**

Example 12.19. *The following data refer to an overhead transmission line having parabolic configuration :*

Weight of the conductor per metre length = 1.8 kg/m ; X-sectional area = 2.3 cm² ; Ultimate tensile strength = 7800 kg/cm² ; Distance between the supports = 500 m ; Difference of levels of the supports = 16 m ; Ice load per meter conductor run = 1.1 kg/m ; Wind pressure = nil.

Determine the vertical sag from the taller of the two supports, which must be allowed so that factor of safety shall be 4.

Solution. Refer to Fig. 12.30.

Given : w_c = 1.8 kg/m ; a = 2.3 cm² ; Ultimate strength = 7800 kg/cm² ;
l = 500 m ; h = 16 m ; w_i = 1.1 kg/m ; F.O.S. = 4

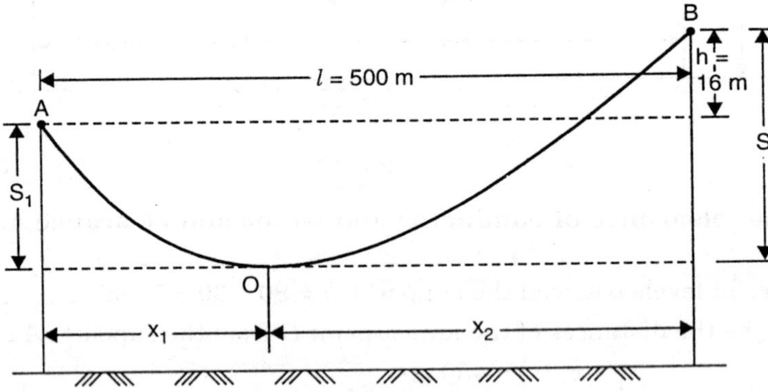

Fig. 12.30

Vertical sag from the taller of the two supports :

Working tension, $$T = \frac{7800 \times 2.3}{4} = 4485 \text{ kg}$$

Total weight of the conductor per metre length,

$$w_{tc} = w_c + w_i = 1.8 + 1.1 = 2.9 \text{ kg/m}$$

Now, let the lowest point O be the at distances x_1 and x_2 from the supports A and B respectively.

Then, $$x_1 + x_2 = l = 500 \text{ m} \quad \text{...}(i)$$

$$h = S_2 - S_1 = \frac{w_{tc} x_2^2}{2T} - \frac{w_{tc} x_1^2}{2T} = \frac{w_{tc}}{2T}(x_2^2 - x_1^2)$$

or, $$16 = \frac{w_{tc}}{2T}(x_2 + x_1)(x_2 - x_1) = \frac{2.9}{2 \times 4485} \times 500 \times (x_2 - x_1)$$

$$\therefore \quad (x_2 - x_1) = \frac{16 \times 2 \times 4485}{2.9 \times 500} \simeq 99 \text{ m} \quad ...(ii)$$

Solving (*i*) and (*ii*), we get

$$x_2 = 299.5 \text{ m and } x_1 = 200.5 \text{ m}$$

∴ Sag from the taller of the two supports,

$$S_2 = \frac{w_{tc}\, x_2^{\,2}}{2T} = \frac{2.9 \times (299.5)^2}{2 \times 4485} = 29 \text{ m.} \quad \textbf{(Ans.)}$$

Example 12.20. *An overhead transmission line at river crossing is supported from two towers at heights of 40 m and 80 m above the water level. The horizontal distance between the towers is 250 m. If the weight of the conductor is 1.1 kg/m and the allowable tension is 1100 kg, find the clearance between the conductor and water at a point mid-way between the towers.*

Sol. Refer to Fig. 12.31.

Given : l = 400 m ; h = 80 – 40 = 40 m ; w_c = 1.1 kg/m ; T = 1100 kg.

Clearance of mid-point above water level :

We know that, $x_1 + x_2 = 250$ m ...(*i*)

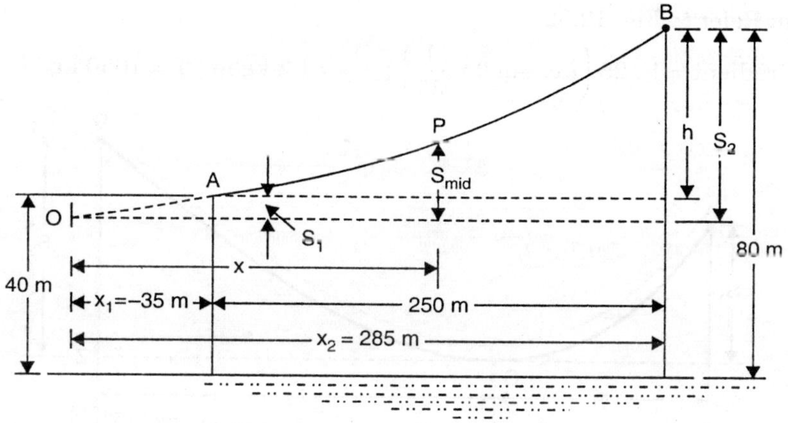

Fig. 12.31

$$h = S_2 - S_1 = \frac{w_c\, x_2^{\,2}}{2T} - \frac{w_c\, x_1^{\,2}}{2T}$$

where x_1 and x_2 are the distances of the lowest point O from the supports A and B.

$$40 = \frac{w_c}{2T}(x_2^{\,2} - x_1^{\,2}) = \frac{w_c}{2T}(x_2 + x_1)(x_2 - x_1)$$

$$= \frac{1.1}{2 \times 1100} \times 250 \times (x_2 - x_1)$$

or, $$(x_2 - x_1) = \frac{40 \times 2 \times 1100}{1.1 \times 250} = 320 \text{ m} \quad ...(ii)$$

Solving (*i*) and (*ii*), we get

$$x_2 = 285 \text{ m and } x_1 = -35 \text{ m}$$

Negative sign means that point *A* lies on the same side of *O* as *B* (Fig. 12.31).

Horizontal distance of mid-point *P* from the lowest point *O*,

$$x = x_1 + \frac{250}{2} = 35 + \frac{250}{2} = 160 \text{ m}$$

$\therefore$ Sag at point *P*, $S_{mid} = \frac{w_c x^2}{2T} = \frac{1.1 \times 160^2}{2 \times 1100} = 12.8 \text{ m}$

Now, $\quad$ sag $S_2 = \frac{w_c x_2^2}{2T} = \frac{1.1 \times (285)^2}{2 \times 1100} = 40.6 \text{ m}$

Height of point *B* above mid-point *P*

$$= S_2 - S_{mid} = 40.6 - 12.8 = 27.8 \text{ m}$$

$\therefore$ Clearance of mid-point *P* above water level

$$= 80 - 27.8 = \mathbf{52.2 \text{ m.}} \quad \textbf{(Ans.)}$$

Example 12.21. *A transmission line over a hillside, where the gradient is 1 : 25, is supported by two 25 metres high towers with a distance of 300 m between them. The lowest conductor is fixed 2 m below the top of each tower. If the conductor weighs 1.2 kg/m and allowable tension is 1650 kg find the clearance of the conductor from the ground.*

Solution. Refer to Fig. 12.32.

Given : Gradient = 1 : 25 $\left(i.e. \sin\theta = \frac{1}{25}\right)$; $w_c = 1.2$ kg/m ; T = 1650 kg

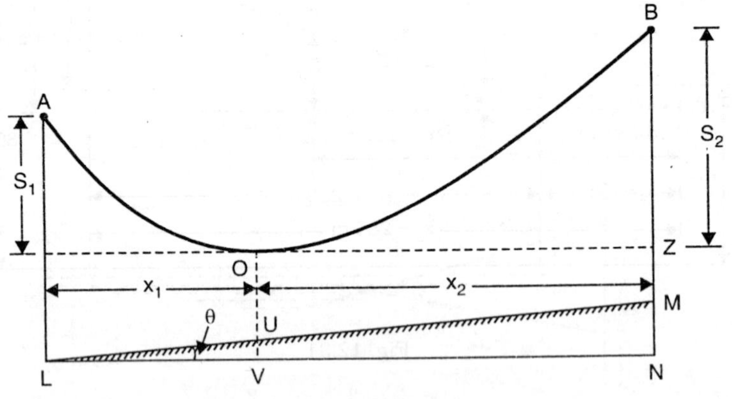

Fig. 12.32

Clearance of the conductor (lowest point O) from the ground :

Effective height of each tower (*AL* or *BM*) = 25 – 2 = 23 m

Vertical distance between the towers,

$$h = MN = LM \sin\theta = 300 \times \frac{1}{25} = 12 \text{ m}$$

Horizontal distance between towers is,

$$LN = \sqrt{LM^2 - MN^2} = \sqrt{(300)^2 - (12)^2} \simeq 300 \text{ m}$$

or, $\quad x_1 + x_2 = 300 \text{ m} \quad ...(i)$

Now, $$h = 12 = \frac{w_c x_2^2}{2T} - \frac{w_c x_1^2}{2T} = \frac{w_c}{2T}(x_2 + x_1)(x_2 - x_1)$$

or, $$(x_2 - x_1) = \frac{12 \times 2T}{w_c (x_2 + x_1)} = \frac{12 \times 2 \times 1650}{1.2 \times 300} = 110 \text{ m} \quad ...(ii)$$

Solving (*i*) and (*ii*), we get

$$x_2 = 205 \text{ m, and } x_1 = 95 \text{ m}$$

$$\text{Sag } S_2 = \frac{w_c x_2^2}{2T} = \frac{1.2 \times (205)^2}{2 \times 1650} = 15.28 \text{ m}$$

Now, $$BN = BM + MN = 23 + 12 = 35 \text{ m}$$

Clearance of the lowest point O from the ground,

$$OU = BN - S_2 - UV = BN - S_2 - x_1 \tan \theta \quad (\because \; UV = x_1 \tan \theta)$$

$$= 35 - 15.28 - 95 \times \frac{1}{25} \quad (\because \; \sin \theta = \tan \theta \text{ when } \theta \text{ is very small})$$

$$= \mathbf{15.92 \text{ m.}} \quad \textbf{(Ans.)}$$

12.7. STRINGING CHART

The conditions of the line at the time of erection are not same under which it is required to operate. In still air, the conductor is acted upon by its own weight ; if the temperature is high, the resulting sag gives a low tension. Probably this is the easiest condition. But sometimes, when the conductor gets coated with ice due to low temperature (which reduces the sag) and simultaneously it is subjected to wind pressure, the situation gets amply aggravated. Further, as per regulations, in view of the above severe conditions, suitable factors of safety have to be provided. In order to deal with the problem it is pertinent to prepare plots of tension and sag versus temperature for varying conditions. The *curves of tension and sag versus temperature are called* **"Stringing charts"**. These chart are quite useful while erecting the transmission line conductor for adjusting the sag and tension properly. Fig. 12.33 shows a stringing chart.

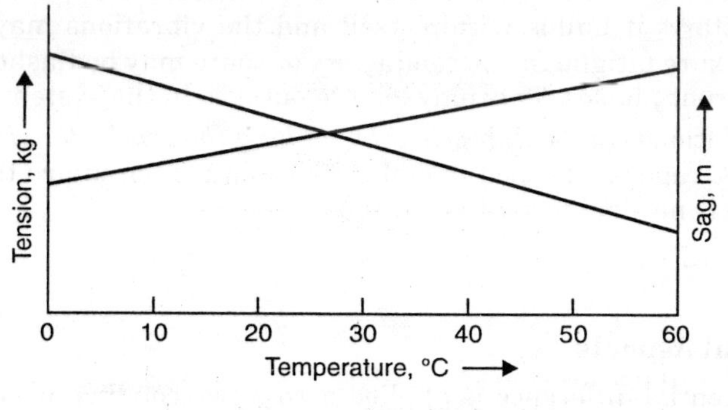

Fig. 12.33. Stringing chart.

12.8. VIBRATION AND VIBRATION DAMPERS

In overhead line conductors the vibrations may be of the following *three types* :

1. Simple swinging.
2. Aeoline vibration.
3. Galloping or dancing of conductors (vibration).

1. Simple swing :

Simple swing of conductors in the horizontal plane due to simple wind deflection is harmless provided the clearance is adequately large so that the conductors do not approach within sparking distance of each other.

2. Aeoline vibration :

- This type of vibration (forced type) is of low amplitude (2 cm to 5 cm, loop length of 1 m to 10 m) and high frequency (5 to 40 c/s).
- While these vibrations are not of themselves harmful, however, they may give rise to troubles at points where free vibration is restricted. The harmful effects of this vibration takes place at the clamps or supports where the conductor suffers fatigue and breaks eventually.
- Besides the design of clamp and accessories, the vibration problem has been solved by the use of the following *special devices* :
  (*i*) *Reinforcement or armour rods.*
  (*ii*) *Dampers.* Fig. 12.34 shows stock-bridge damper an extremely simple, nevertheless effective device for suppressing high frequency vibrations. It consists of two weights attached to a piece of stranded cable 30 cm to 50 cm long, which is clamped to the line conductor. The energy of vibration is absorbed by the stranded cable, and the vibration is damped out rapidly.

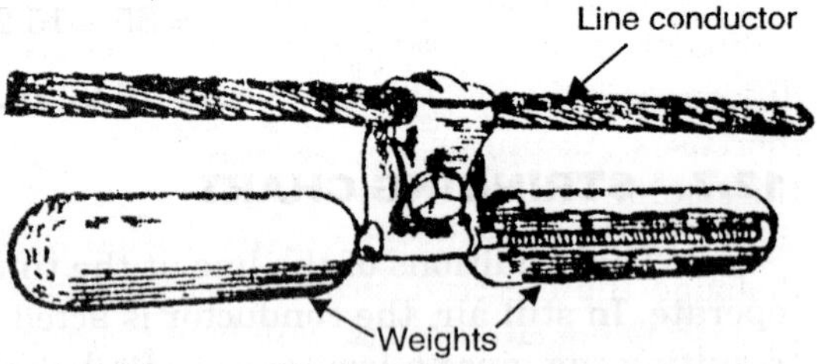

Fig. 12.34. Stock-bridge damper.

3. Galloping or dancing of conductors (vibration) :

- The galloping or dancing of conductors are the low frequency $\left(\frac{1}{4} \text{ to } 1\frac{1}{2} \text{ c/s}\right)$ and high amplitude (about 6 metres) vibrations (*self-excited type*) of the conductors.
- These vibrations are caused due to the wind when it blows past a non-circular conductor. Once this starts it builds within itself and the vibrations may become very large and result in flexure fatigue in the conductors or there may be flashover between the conductors of different phases. This may lead to outages in the system.
- These oscillations can be *dampened by making the conductor circular*. For stranded conductors PVC tape is wrapped to make the conductor circular ; this method is useful only when there is no sleet formation on the conductor.

12.9. CORONA

12.9.1. General Aspects

When A.C. potential difference is applied across two conductors, where spacing is large in comparison to diameter, then the electrostatic stresses are set up in the atmospheric air surrounding the conductors. If the voltage is low, there will be no change in the condition of air but when the potential difference is gradually increased, a stage reaches when a faint violent luminous glow (around the conductor) will appear together with a hissing noise. This phenomenon is called "*Visual corona*" ; it is accompanied by the production of ozone which is detected by its characteristic odour. Further increase of potential difference will increase the glow and noise intensity until the spark-over eventually occurs.

The *phenomenon of appearance of violet luminous glow and production of hissing noise and ozone gas in an overhead transmission line is known as* **corona.**

The *corona effects* include the following :

(*i*) A violet luminous glow appears around the conductor (the glow is maximum over rough and dirty surfaces of the conductor).

(*ii*) Hissing noise is produced. (*iii*) Ozone gas is produced.

(*iv*) Power loss occurs, it depends upon weather condition.

- In E.H.V. A.C. transmission lines the visible glow of corona is *mostly uniform about both the conductors.* Whereas in H.V.D.C. system transmission the glow is *uniform about the positive wire but spotty about the negative wire.*
- The A.C. corona, viewed through a stroboscope, has the same appearance as D.C. corona.

12.9.2. Theory of Corona Formation

The phenomenon of corona is the *ionization of air surrounding the power conductor.* Owing to radioactivity and cosmic rays, free electrons are normally present in free space. When the potential difference between the conductors is increased the gradient around the surface of the conductor increases. Assuming the spacing between the conductors to be large as compared to the diameter of the conductors, the free electrons will move with certain velocity depending upon the strength of the field. These electrons will have collision with the molecules of air, and in case the speed is large, will dislodge the electrons from these molecules and thereby increase the number of electrons. This process being cumulative, will lead to an avalanche of electrons. Eventually there will be ionization of air surrounding the conductor.

If the spacing of the conductors is less than the 15 times the diameter of conductor, the flash-over will take place between the conductors before corona phenomenon occurs ; in practice the spacing is much large as compared to diameter so the flash-over is almost impossible to occur.

The corona phenomenon is very much evident in power lines of 100 kV and above.

12.9.3. Factors Affecting Corona

Corona depends upon the following factors :

1. Atmosphere. Since the corona formation is due to ionization of air surrounding the conductors, therefore, it is affected by the state of atmosphere. When the weather is stormy, the number of ions / electrons is more than the normal, and as such the *corona may occur at much less voltage as compared with fair weather.*

2. Conductor size. The shape and conditions of the conductors affect corona formation. The unevenness of the surface decreases the value of breakdown voltage. Thus in case of stranded conductor which has irregular surface the corona formation is more pronounced than a solid conductor.

3. Spacing between conductors. Spacing between conductors affect the corona formation. If the spacing between conductors is made very large as compared to their diameters, there may not be any corona effect due to reduction of electrostatic stresses.

4. Line voltage. Corona is largely affected by line voltage. At low voltage there is no corona effect, but as the voltage increases the electrostatic stresses developed at the conductor surface make the air surrounding the conductor conducting, which leads to corona formation.

12.9.4. Advantages and Disadvantages of Corona

Following are the *advantages* and *disadvantages* of corona :

Advantages :

1. Owing to corona formation the air surrounding the conductor becomes condu ive, consequently the virtual diameter of the conductor increases resulting in reduced electrostatic

stress between the conductors. Thus flash-over possibility is reduced and there is an improvement in the system performance.

2. The effects of transients produced by surges are reduced.

Disadvantages :

1. There is a loss of power (more prominent under abnormal weather conditions) which affects transmission efficiency of the line.
2. Due to corona formation ozone gas is produced which chemically reacts with the conductor and causes corrosion.
3. Due to non-sinusoidal corona current there is a non-sinusoidal voltage drop which may cause some interference with neighbouring communication circuits due to electromagnetic and electrostatic induction.
4. Due to the distortion of wave form, the harmonics (predominantly 3rd harmonic) are introduced into the transmission line. These harmonics increase corona loss.

12.9.5. Corona Loss and its Control

The following *points* relating to corona loss are important :

1. Corona loss is a function of *frequency* ; the higher the frequency of the supply the higher are corona losses. Thus *D.C. corona loss is less as compared with A.C. corona.*

2. When transmission lines are *irregularly spaced*, the surface gradients of the conductors and hence the corona losses if any are *unequal.*

3. Corona loss (a function of air density correction factor) is *more on hilly areas than on plain areas.*

4. Corona loss is *increased during bad atmospheric conditions* (rains, snow, hailstorm etc.) due to reduced critical disruptive voltage.

5. *Larger the size of the conductor, lower is the corona loss.*

6. By bundling the conductors the self GMD (Geometric Mean Diameter) of the conductors is increased thereby, the critical disruptive voltage is *increased* and hence corona loss is *reduced.*

7. Because of *field uniformity in case of cylindrical conductor*, the corona loss is *less*, as compared to any other shape.

8. Flow of load current increases the temperature of conductor. Thus it *prevents* deposition of dew or snow on conductor's surface. This reduces corona loss.

Control of corona loss :

Corona loss can be *reduced* as follows :

(*i*) By using *large diameter conductors* (*ii*) By using *hollow conductors* ;

(*iii*) By using *bundled conductors.*

Note. A typical transmission line may have a fair weather loss of 1 kW per 3-phase mile and foul weather loss of 20 kW per 3-phase mile.

12.9.6. Important Terms Relating to Corona

1. Critical disruptive voltage :

Critical disruptive voltage *is defined as* the *voltage at which complete disruption of dielectric occurs*. This voltage corresponds to the gradient at the surface equal to breakdown strength of air.

Consider two parallel conductors, each of radius r cm and separated by distance D cm. If V is the phase-neutral potential, then potential gradient at the surface of the conductor is given by,

$$g = \frac{V}{r \ln\left(\frac{D}{r}\right)} \text{ volts/cm} \quad \text{...(12.8)}$$

The corona formation will take place if the value of g is made equal to the breakdown strength of air ; this dielectric strength is normally denoted by g_0 and is equal to 30 kV/cm peak (or 21.21 kV/cm r.m.s.) at N.T.P. *i.e.* 25°C and 76 cm of Hg.

At any other temperature and pressure,

$$g'_0 = g_0.\delta \quad ...(12.9)$$

where δ is the air density correction factor and is given by,

$$\delta = \frac{3.92b}{273 + t} \quad ...(12.10)$$

where, b = Barometric pressure, cm of Hg, and

t = Temperature, °C.

Therefore, critical disruptive voltage (V_c) is given by,

$$V_C = g_0 \delta r \, ln\left(\frac{D}{r}\right) \quad ...(12.11)$$

The above expression is based on the assumption that the conductor is solid and the surface is smooth. But in practice it is necessary to allow for the surface condition of the conductor, this is done by introducing irregularity factor m_0 so that

$$V_c = g_0 \, \delta m_0 r \ln\left(\frac{D}{r}\right) \; kV/\text{phase} \quad ...(12.12)$$

where, m_0 = 1.0 for polished wires,

= 0.98 to 0.93 for roughened or weathered wires,

= 0.87 to 0.83 for seven strand cable, and

= 0.90 approx. for large cables with more than seven strands.

2. Visual critical voltage :

Visual critical voltage *is the minimum phase-neutral voltage at which corona glow appears all along the line conductors.*

In case of parallel conductors it has been observed that corona glow does not appear at disruptive critical voltage (V_c) but it appears at a higher voltage, called the visual critical voltage (V_v). The value of visual critical voltage is given by the empirical formula :

$$V_v = g_0 \, \delta \, m_v \, r\left(1 + \frac{0.3}{\sqrt{\delta r}}\right) \ln\left(\frac{D}{r}\right) \text{kV/phase} \quad ...(12.13)$$

where m_v is *another irregularity factor* and has the following values :

According to Peek, m_v = 1.0 for polished wires

= 0.98 to 0.93 for rough conductor exposed to atmospheric severities, and

= 0.72 for local corona on stranded conductors.

3. Power loss due to corona :

The corona formation is accompanied by power loss which is dissipated in the form of heat, light, sound and chemical action. Power loss due to corona is given by the following empirical relation (based on experiments conducted by ***Peek***) :

$$P = (241 \times 10^{-5})\left(\frac{f + 25}{\delta}\right)\sqrt{\frac{r}{D}}\,(V_{ph} - V_c)^2 \text{ kW/km/phase} \quad ...(12.14)$$

where, f = Supply frequency, Hz

V_{ph} = Phase-neutral voltage (r.m.s.), and

V_c = Disruptive voltage (r.m.s.) per phase.

- The above relation is applicable to a fair weather condition. The approximate loss under foul weather condition is obtained by taking V_c as 0.8 times the fair weather value.
- The empirical formula [Eqn. (12.14)] gives correct result only if supply frequency lies between 25 to 120 Hz, the conductor radius is greater than 0.25 cm and the ratio $\frac{V_{ph}}{V_c} > 1.8$.

Peek's formula is useful only when the corona losses are predominant and is not applicable when the losses are low. When the ratio $\frac{V_{ph}}{V_c} < 1.8$ *Peterson's* formula holds good and is given by :

$$P = \frac{21 \times 10^{-6} f(V_{ph})^2}{\left(\log_{10} \frac{D}{r}\right)^2} \times F \text{ kW/km/phase}$$

where F is a factor which varies with the ratio of $\frac{V_{ph}}{V_c}$ as given below :

V_{ph}/V_c	0.6	0.8	1.0	1.2	1.4	1.6	1.8	2.0	2.2
F	0.012	0.018	0.05	0.08	0.3	1.0	3.0	6.0	8.0

Example 12.22. *In a 3-phase overhead line the conductors have each a radius of 1.2 cm and are spaced symmetrically 3 metres apart. If the dielectric strength of air is 30 kV per cm (peak), find the disruptive critical voltage.*

Solution. *Given :* $r = 1.2$ cm ; $D = 3$ m or 300 cm ; $g_0 = \frac{30}{\sqrt{2}}$ kV/cm = 21.21 kV/cm

Disruptive critical voltage to neutral, V_c

We know that, $$V_c = g_0 r \ln\left(\frac{D}{r}\right)$$ ...[Eqn. (12.8)]

$$= 21.21 \times 1.2 \ln\left(\frac{300}{1.2}\right) = 140.53 \text{ kV per phase}$$

$$= 140.53 \times \sqrt{3} = \mathbf{243.4\ kV}\ (\textit{line to line}).$$ **(Ans.)**

Example 12.23. *In a 3-phase overhead line the conductors are arranged in the form of an equilateral triangle. The diameter of each conductor is 30 mm. Assuming fair weather conditions, air density factor of 0.94, irregularity factor 0.93 and breakdown strength of air 30 kV per cm (peak), find the minimum spacing between conductors, if the disruptive critical voltage is not to exceed 220 kV between the lines.*

Solution. *Given :* Radius of each conductor, $r = \frac{30}{2} = 15$ mm = 1.5 cm

Air density factor, $\delta = 0.94$

Irregularity factor, $m_0 = 0.93$

Breakdown strength of air, $g_0 = \frac{30}{\sqrt{2}}$ kV/cm = 21.21 kV/cm

Disruptive critical voltage, $V_c = 220$ kV (line to line) $= \frac{200}{\sqrt{3}} = 127$ kV per phase

Minimum spacing between the conductors, D :

We know that, $V_c = g_0 \delta m_0 r \ln\left(\frac{D}{r}\right)$ kV/phase ...[Eqn. (12.12)]

$$127 = 21.21 \times 0.94 \times 0.93 \times 1.5 \ln\left(\frac{D}{1.5}\right)$$

or, $$\ln\left(\frac{D}{1.5}\right) = \frac{127}{21.21 \times 0.94 \times 0.93 \times 1.5} = 4.566$$

or, $$\frac{D}{1.5} = e^{4.566} = 96.16$$

$\therefore$ $D = 1.5 \times 96.16 = 144.24$ cm or **1.44 m. (Ans.)**

Example 12.24. *A 3-phase, 220 kV, 50 Hz transmission line consists of 12 mm diameter conductor spaced 2.1 metres apart in the form of an equilateral triangle. If the temperature is 38°C and atmospheric pressure is 76 cm, calculate the corona loss per km of the line. Assume the irregularity factor as 0.83.*

Solution. *Given :* $f = 50$ Hz ; $r = \frac{12}{2} = 6$ mm $= 0.6$ cm ; $D = 2.1$ m $= 210$ cm ;

$t = 38°$ C ; $b = 76$ cm, $m_0 = 0.83$

Corona loss per km of the line, P :

We know that, corona loss is given by,

$$P = (241 \times 10^{-5}) \left(\frac{f + 25}{\delta}\right) \sqrt{\frac{r}{D}} (V_{ph} - V_c)^2 \text{ kV/km/phase}$$ ...[Eqn. (12.14)]

Now, $$\delta = \frac{3.92\, b}{273 + t}$$ ...[Eqn. (12.10)]

$$= \frac{3.92 \times 76}{273 + 38} = 0.958$$

Assuming $g_0 = 21.21$ kV/cm (r.m.s.), we have critical disruptive voltage,

$$V_c = g_0 \delta m_0 r \ln\left(\frac{D}{r}\right) \text{ kV/phase}$$

$$= 21.21 \times 0.958 \times 0.83 \times 0.6 \times \ln\left(\frac{210}{0.6}\right) = 59.3 \text{ kV/phase}$$

Supply voltage per phase, $V_{ph} = \frac{220}{\sqrt{3}} = 127$ kV

Substituting the values, we get

$$P = (241 \times 10^{-5}) \left(\frac{50 + 25}{0.958}\right) \sqrt{\frac{0.6}{210}} (127 - 59.3)^2 = 46.22 \text{ kW/km/phase}$$

$\therefore$ Total corona loss per km for three phases

$= 3 \times 46.22 =$ **138.66 kW. (Ans.)**

Example 12.25. *A certain 3-phase equilateral transmission line has a total corona loss of 53 kW at 106 kV and a loss of 98 kW at 110.9 kV. What is the disruptive critical voltage ? What is the corona loss at 113 kV ?* **(Gujarat University)**

Solution. The power loss due to corona for three phases is given by :

$$P = 3 \times (241 \times 10^{-5}) \left(\frac{f + 25}{\delta}\right) \sqrt{\frac{r}{D}} \, (V_{ph} - V_c)^2 \text{ kW/km}$$

As f, δ, r and D are the same for two cases,

$$\therefore \quad P \propto (V_{ph} - V_c)^2$$

For *first case* : $P = 53$ kW and $V_{ph} = \dfrac{106}{\sqrt{3}} = 61.2$ kV

For *second case* : $P = 98$ kW and $V_{ph} = \dfrac{110.9}{\sqrt{3}} = 64$ kV

$$\therefore \quad 53 \propto (61.2 - V_c)^2 \qquad \text{...}(i)$$

and, $$98 \propto (64 - V_c)^2 \qquad \text{...}(ii)$$

Dividing (*ii*) by (*i*), we get

$$\frac{98}{53} = \left(\frac{64 - V_c}{61.2 - V_c}\right)^2$$

or, $$\frac{64 - V_c}{61.2 - V_c} = \sqrt{\frac{98}{53}} = 1.36$$

or, $$64 - V_c = 1.36(61.2 - V_c) = 83.23 - 1.36\, V_c$$

or, $$V_c = \left(\frac{83.23 - 64}{1.36 - 1}\right) = \mathbf{53.4 \text{ kV.}} \quad \textbf{(Ans.)}$$

Let P' kW be the power loss at 113 kV

Then $$P' \propto \left(\frac{113}{\sqrt{3}} - V_c\right)^2 \propto (65.2 - 53.4)^2 \qquad \text{...}(iii)$$

Dividing (*iii*) by (*i*), we get

$$\frac{P'}{53} = \frac{(65.2 - 53.4)^2}{(61.2 - 53.4)^2} = \left(\frac{11.8}{7.8}\right)^2 = 2.288$$

or, $$P' = 53 \times 2.288 = \mathbf{121.26 \text{ kW.}} \quad \textbf{(Ans.)}$$

Example 12.26. *A 132 kV, 50 Hz 3-phase transmission line 190 km long consists of three 1.2 cm diameter stranded copper conductors spaced in 2.8 metres delta arrangement. Assuming g_0 = 21.21 kV/cm, temperature = 30°C, barometric pressure = 75 cm, m_0 = 0.86, m_v for local corona = 0.74 and m_v for general corona = 0.84, find the corona characteristics of the transmission line.*

Solution. *Given :* $f = 50$ Hz ; $r = \dfrac{1.2}{2} = 0.6$ cm ; $D = 2.8$ m $= 280$ cm ;

$g_0 = 21.21$ kV/cm ; $t = 30$°C ; $b = 75$ cm ; $m_0 = 0.86$, $m_v = 0.74$, 0.84.

Disruptive critical voltage,

$$V_c = g_0\, \delta m_0\, r \ln\left(\frac{D}{r}\right) \text{ kV-to-neutral}$$

Here, air density correction factor, $\delta = \dfrac{3.92b}{273 + t} = \dfrac{3.92 \times 75}{273 + 30} = 0.97$

$\therefore \qquad V_c = 21.21 \times 0.97 \times 0.86 \times 0.6 \ln\left(\frac{280}{0.6}\right) = 65.2$ kV/phase (r.m.s.) ...(i)

For local corona ($m_0 = 0.74$) and for general corona ($m_v = 0.84$), we have the visual critical voltages (V_v) as follows :

We know that, $$V_v = g_0 \delta m_v r \left(1 + \frac{0.3}{\sqrt{\delta r}}\right) \ln\left(\frac{D}{r}\right) \text{ kV/phase}$$

For $m_v = 0.74$: $$V_v = 21.21 \times 0.97 \times 0.74 \times 0.6 \left(1 + \frac{0.3}{\sqrt{0.97 \times 0.6}}\right) \ln\left(\frac{280}{0.6}\right)$$

$$= 9.135(1 + 0.393) \times 6.146 = 78.2 \text{ kV/phase} \qquad ...(ii)$$

For $m_v = 0.84$, $$V_u = 21.21 \times 0.97 \times 0.84 \times 0.6 \left(1 + \frac{0.3}{\sqrt{0.97 \times 0.6}}\right) \ln\left(\frac{280}{0.6}\right)$$

$$= 88.8 \text{ kV/phase (r.m.s.)} \qquad ...(iii)$$

Voltage to which conductors are subjected,

$$V = \frac{132}{\sqrt{3}} = 76.2 \text{ kV/phase (r.m.s.)}$$

Hence from eqns. (*i*), (*ii*) and (*iii*) and (*iv*) it is obvious that corona will take place but it *will not be visual*.

Power loss due to corona :

(*i*) Using Peek's formula for fair weather conditions, we have

$$P = (241 \times 10^{-5}) \left(\frac{f + 25}{\delta}\right) \sqrt{\frac{r}{D}} (V_{ph} - V_c)^2 \text{ kW/km/phase}$$

$$= (241 \times 10^{-5}) \left(\frac{50 + 25}{0.97}\right) \sqrt{\frac{0.6}{280}} (76.2 - 65.2)^2 = 1.0437 \text{ kW/km/phase}$$

$\therefore$ Total power loss $= 3 \times 1.0437 \times 190 = 594.9$ kW

(*ii*) Using Peek's formula for stormy conditions, we have

$$P = (241 \times 10^{-5}) \left(\frac{50 + 25}{0.97}\right) \sqrt{\frac{0.6}{280}} (76.2 - 0.8 \times 65.2)^2$$

$$= 4.985 \text{ kW/km/phase}$$

$\therefore$ Total power loss $= 3 \times 4.985 \times 190 = 2841.45$ kW.

HIGHLIGHTS

1. The main components of overhead lines are
   (*i*) Conductors (*ii*) Supports
   (*iii*) Cross arms (*iv*) Insulators.
2. The most commonly used conductor materials for overhead lines are
   (*i*) Copper (*ii*) Aluminium
   (*iii*) Steel cored aluminium (A.C.S.R.) (*iv*) Galvanised steel
   (*v*) Cadmium copper.
3. *Stranded conductors* are be the conductors made of thin wires of small cross-section and bunched together.
4. The line supports may *wooden poles*, *steel poles*, *R.C.C. poles* and *steel towers*.

5. The insulating materials commonly used for the manufacture of insulators are :
   (*i*) Porcelain (*ii*) Glass
   (*iii*) Steatite (*iv*) Synthetic resin.
6. Transmission line insulators may be classified as :
   (*i*) Pin type insulators (*ii*) Suspension type insulators
   (*iii*) Strain type insulators (*iv*) Shackle insulators.
7. The insulators are tested by the following tests :
   (*i*) Mechanical tests (*ii*) Electrical insulation tests
   (*iii*) Environmental tests (*iv*) Temporary cycle tests
   (*v*) Corona and radio interference tests.
8. The ratio of voltage across the whole string to the product of number of discs and the voltage across the disc nearest to the conductor is known as "*string efficiency*".
9. Methods of improving string efficiency :
   (*i*) Long cross-arm method (*ii*) Capacitance grading method
   (*iii*) Static shielding.
10. The difference in level between the points of supports and the lowest point is known as *sag*.
11. The sag is affected by the following factors :
    Weight of the conductor ; Span length ; Working tensile strength ; Temperature.
12. The curves of tension and sag versus temperature are called *stringing charts*.
13. *Aeoline vibration* is of forced type while *galloping of conductors* vibration is of self-excited type.
14. The phenomenon of appearance of violet luminous glow and production of hissing noise and ozone gas in an overhead transmission line is known as "*corona*".
15. Corona is affected by the following factors :
    Atmosphere ; Conductor size ; Spacing between conductors ; Line voltage.
16. Corona loss can be controlled by using *large diameter conductors, hollow conductors* and *bundled conductors*.

THEORETICAL QUESTIONS

1. List the main components of overhead lines and explain them briefly.
2. Enumerate the materials used for transmission lines.
3. Give brief description of A.C.S.R. conductor.
4. What are stranded conductors ?
5. State the characteristics which the line supports (poles and towers) should possess.
6. Describe briefly any two of the following :
   (*i*) Wooden poles (*ii*) Steel poles
   (*iii*) R.C.C. poles (*iv*) Steel towers.
7. List the characteristics which the insulators should possess.
8. Discuss briefly the following materials which are commonly used for the manufacture of insulators :
   (*i*) Porcelain (*ii*) Glass
   (*iii*) Steatite (*iv*) Synthetic resin.
9. How are the transmission line insulators classified ?
10. Describe briefly any two of the following types of insulators :
    (*i*) Pin type insulators (*ii*) Suspension type insulators
    (*iii*) Strain type insulators (*iv*) Shackle insulators.
11. List the advantages of suspension type insulators over pin type insulators.
12. What are the causes of failure of insulators ?
13. Enumerate and briefly discuss the tests used for testing of insulators.
14. How can it be improved ? What do you mean by "string efficiency"?
15. What electrical and mechanical characteristics are required for a good insulator for use in high voltage transmission lines ?

16. Explain why suspension type of insulators are preferred for high voltage overhead lines. Sketch a sectional view of one unit of the suspension type of insulators and describe its construction.

17. Draw neat sketches of pin type and suspension type insulators and label their parts. State applications and advantages/disadvantages of suspension insulator over pin insulator.

18. Explain what is meant by a string efficiency of a suspension insulator consisting of a number of units. What causes the string efficiency to be less than 100 per cent ? Describe any one method of improving the same.

19. Explain the use of grading rings and arcing horns on suspension insulators.

20. Discuss various methods by which the voltage across the units can be equalised. Illustrate your answer by a string of 3 insulator units.

21. Why the voltage does not divide equally across the various units of a string insulator ?

22. What is static shielding ?

23. Write short notes on the following :

(*i*) Causes of failure of insulators.

(*ii*) Methods of equalising voltage drop across insulator strings.

(*iii*) Use of guard rings and arcing horns.

(*iv*) Shackle insulators.

Sag

24. What is sag ?

25. Why is it disadvantageous to have either too low or too high a sag ?

26. How do you calculate the sag of a conductor ?

27. What is the effect of temperature on sag ?

28. Discuss briefly the factors which affect the sag.

29. How is sag calculated in the following cases :

(*i*) When supports are at equal levels.

(*ii*) When supports are at unequal levels.

30. Discuss the effect of wind and ice on sag.

31. What is a stringing chart ? What is its utility ?

32. Write a short note on vibration and vibration dampers.

Corona

33. What is corona ?

34. Explain briefly the factors which affect corona.

35. What are the advantages and disadvantages of corona ?

36. What is corona loss ? How can it be reduced ?

37. What is meant by the disruptive critical voltage and visual critical voltage ? State the effects of conductor size, spacing and condition of the surface of conductors on these voltages ?

38. Explain the phenomenon of corona.

39. How can the corona loss be minimised in transmission lines.

UNSOLVED EXAMPLES

Insulators

1. A 3-phase overhead transmission line is being supported by three disc insulators. The potentials across top unit and middle unit are 9 kV and 1 kV respectively. Calculate :

(*i*) The ratio of capacitance between pin and earth to the self-capacitance of each unit.

(*ii*) The line voltage.

(*iii*) The string efficiency. [**Ans.** (*i*) 0.22 ; (*ii*) 61.35 kV ; (*iii*) 76%]

2. A three-phase overhead transmission line is being supported by three disc suspension type insulators, the potential across the first and second insulators are 10 kV and 14 kV respectively. Calculate :

(*i*) The line voltage.

(*ii*) The ratio of capacitance between pin and earth to the self-capacitance of each unit.

(*iii*) The string efficiency. [**Ans.** (*i*) 47.6 kV ; (*ii*) 0.4 ; (*iii*) 67.2%]

3. A string of suspension insulators has four units. If the capacitance between each unit and earth is $\frac{C}{6}$ farads and capacitance between adjacent units is C farads, calculate the string efficiency.

[**Ans.** 69.5%]

4. Each line of a three-phase system is suspended by a string of 3 similar insulators. If the voltage across the line unit is 17.5 kV, calculate the line to neutral voltage. Assume that the shunt capacitance between each insulator and earth is $\frac{1}{8}$ th of the capacitance of the insulator itself. Also find the string efficiency.

[**Ans.** 44.25 kV ; 84.28%]

5. A string of 6 insulator units has a self-capacitance equal to 10 times the pin-to-earth capacitance. Find :
(*i*) The voltage distribution across various units as a percentage of total voltage across the string.
(*ii*) The string efficiency.

[**Ans.** (*i*) 0.0980 V, 0.108 5V, 0.1305 V, 0.1640 V, 0.2140 V, 0.2850 V ; (*ii*) 58.5%]

6. The self-capacitance of each unit in a string of three suspension insulators is C. The shunting capacitance of the connecting metal work of each insulator to earth is 0.15 C while for line it is 0.1 C. Calculate :
(*i*) The voltage across each insulator as a percentage of the line voltage to earth.
(*ii*) The string efficiency. [**Ans.** (*i*) 32.6%, 30.7%, 36.4% (*ii*) 91.5%]

Sag and Tension calculations

7. A transmission line has a span of 250 metres. Find the weight of the conductor per metre length if the sag, ultimate tensile strength and factor of safety are 1.5 metres, 5758 kg and 2 respectively.

[**Ans.** 0.5528 kg/m]

8. An overhead line has a span of 200 metres between level supports. The conductor has a cross-sectional area of 1.29 cm^2 and weighs 1.17 kg/m and has a breaking stress of 4218 kg/cm^2. Allowing a wind pressure of 122 kg/m^2, calculate the sag for a factor of safety of 5.

[**Ans.** Slant sag, 8.96 m ; $\theta = 53.13°$; vertical sag = 5.37 m]

9. The following data relate to a transmission line :
Span length = 220 m (between level supports) ; X-sectional area of the conductor = 3.225 cm^2 ; Vertical sag = 2.35 m ; Wind force = 1.5 kg/m run ; Breaking stress = 2540 kg/cm^2 ; Weight of the conductor = 1.125 kg/m run.
Calculate the factor of safety. [**Ans.** 2.84]

10. A transmission line conductor consists of hard drawn copper 240 mm^2 in X-section and has a span of 160 metres, the supporting structures being level. The conductor has an ultimate strength of 42.2 kg/mm^2 and the desired factor of safety is 5. Find the vertical sag with a wind pressure of 1.35 kg/metre length and an ice coating of 1.25 cm. Take the density of hard drawn copper as 8.9 g/cc and weight of ice as 915 kg/m^3. [**Ans.** 4.22 m]

11. A transmission line conductor crossing a river is supported from two towers at heights of 30 m and 90 m above the water level. The horizontal distance between the towers is 500 metres. If the tension in the conductor is 1600 kg and weight of the conductor is 1.5 kg/m length, find the minimum clearance of the conductor and water and clearance mid-way between the supports. [**Ans.** 23 m ; 30.68 m]

12. An overhead transmission line at river crossing is supported from two towers at heights of 40 m and 90 m above water level. The horizontal distance between the towers is 400 m. If the weight of the conductor is 1 kg/m and the allowable tension is 2000 kg, find the clearance between the conductor and water at a point mid-way between the towers. [**Ans.** 55 m]

13. A transmission line over a hill side, where the gradient is 1 : 20, is supported by two 22 metres high towers with a distance of 300 m between them. The lowest conductor is fixed 2 m below the top of each tower. If the conductor weighs 1 kg/m and allowable tension is 1500 kg find the clearance of the conductor from the ground. [**Ans.** 14.38 m]

Corona

14. A 3-phase overhead line has conductors, each 20 mm in diameter, spaced equilaterally 1 metre apart. If the dielectric strength of air is 30 kV per cm (peak), find the disruptive critical voltage for the line. Assume air density factor and irregularity factor as 0.952 and 0.9 respectively. [**Ans.** 144.8 kV]

15. In a 3-phase overhead line the conductors are arranged in the form of an equilateral triangle. The diameter of each conductor is 30 mm. Assuming fair weather conditions, air density factor 0.95, irregularity factor 0.95 and breakdown strength of air 30 kV per cm (peak), find the minimum spacing between conductors if the disruptive critical voltage is not to exceed 230 kV between lines.

[**Ans.** 1.575 m]

16. A 3-phase, 220 kV, 50 Hz transmission line consists of 15 mm radius conductor spaced 2 m apart in the form of an equilateral triangle. If the temperature is 40°C and atmospheric pressure is 76 cm, calculate the corona loss per km of the line. Assume irregularity factor as 0.85. [**Ans.** 0.06 kW]

17. Find the disruptive critical and visual corona voltages of a grid-line operating at 132 kV. The following data is given :
Conductor diameter = 1.9 cm ; conductor spacing = 3.81 m ; temperature = 44°C ; barometric pressure = 73.7 cm ; conductor surface factor : Fine weather = 0.8, rough weather = 0.66.

[**Ans.** 155.4 kV/phase, 157.6 kV/phase]

13

Constants of Overhead Transmission Lines

13.1. Introduction. 13.2. Inductance of a transmission line—Concept of inductance—Flux linkages—Flux linkages due to a single carrying conductor—Flux linkages in parallel current-carrying conductors—Inductance of a 2-wire (1-phase) transmission line—Inductance of a 3-phase overhead transmission line—Self-GMD and mutual-GMD. 13.3. Resistance of a transmission line. 13.4. Skin effect and proximity effect. 13.5. Capacitance of transmission lines—Concept of capacitance—Electric potential—Potential at a charged single conductor—Potential at a charged conductor in a group of charged conductors—Capacitance of a single-phase two-wire overhead transmission line—Capacitance of a 3-phase overhead transmission line—Symmetrical spacing—Unsymmetrical spacing with transposed conductors—Unsymmetrical spaced line—Capacitance of a double-circuit 3-phase overhead line—Effect of earth on the capacitance of a transmission line. Highlights—Theoretical Questions—Unsolved Examples.

13.1. INTRODUCTION

The performance of a transmission line as an element of a power system is affected by the following *four parameters* :

(*i*) Inductance (L) (*ii*) Resistance (R)

(*iii*) Capacitance (C) (*iv*) Conductance (G).

- *Inductance* is by far the most dominant line parameter. It is the inductive reactance which limits the transmission capacity of a line.
- The *capacitance* together with *conductance* forms the shunt admittance of a transmission line. Shunt conductance, which is mostly due to leakage over the surface of insulators, is almost always neglected in power transmission lines.
- — When an alternating voltage is applied to the line, the line capacitance draws a leading sinusoidal current called the *charging current* which is drawn even when the line is open circuited at the far end. As the line capacitance is proportional to its length, the charging current is *negligible for lines less than 100 km long.* The capacitance for longer lines becomes increasingly important and has to be taken into account.
- Though the contribution of line *resistance* to series line impedance can be neglected in most cases, it is main source of line power loss.
- — R and G do not affect much the total equivalent impedance of the line and hence the transmission capacity. These are of course very much important when transmission efficiency and economy are to be evaluated as they completely determine the real transmission line losses.

Fig. 13.1 shows a complete representation of the parameters of a transmiss[illegible] line in distributed form.

- In a 2-wire transmission system *line parameters* are usually represented on l[illegible] length basis, whereas in case of 3-phase transmission systems these are represented on per conductor basis.

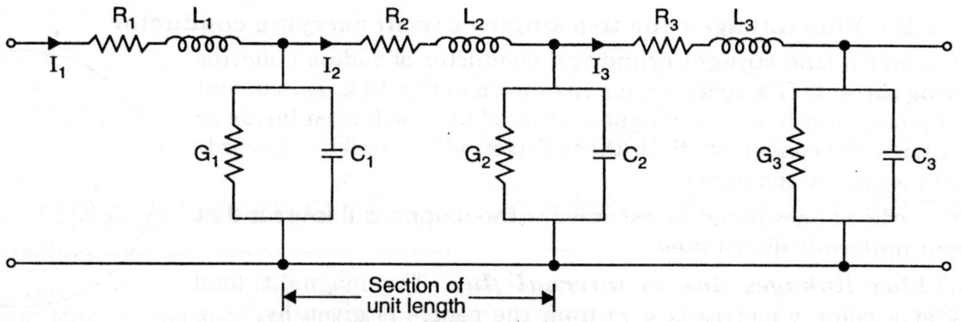

Fig. 13.1. Complete representation of line parameters of a transmission line in distributed form.

- ***Short or long line does not refer to the physical length of the line but it is based on comparison with the wavelength of the source.***

13.2. INDUCTANCE OF A TRANSMISSION LINE

13.2.1. Concept of Inductance

The **inductance** *of a circuit is defined as the flux linkages per unit current ;* it is expressed in henrys (H).

i.e.,
$$L = \frac{\psi}{I} \; H \qquad ...(13.1)$$

where, L = Inductance of the circuit,

ψ = Flux linkages (weber-turns), and

I = Current (amperes)

or,
$$\psi = LI \text{ Wb-T} \qquad ...[13.1\ (a)]$$

If the current is alternating, the above equation can be written as,

$$\lambda = LI$$

where, λ and I are the r.m.s. values of flux linkages and current respectively ; these are of course in phase.

The *mutual inductance* between two circuits is defined as the *flux linkages of one circuit per unit current in the other circuit.* Thus, if current I_2 produces λ_{12} flux linkages with circuit 1, the mutual inductance is

$$M_{12} = \frac{\lambda_{12}}{I_2} \; H \qquad ...(13.2)$$

- The two parallel conductors of a transmission line form a rectangular loop of one turn. The changing flux in the line links the loop and hence the line has inductance.
- The inductance is uniformly distributed along the length of the line ; for the convenience of analysis, it can be taken to be lumped.

13.2.2. Flux Linkages

It is known that a current carrying conductor is surrounded by concentric circles of magnetic lines. In A.C. system this field set up around the conductor changes and links with the same conductor as well as with other conductors. *Due to these flux linkages the line possesses inductance.*

There are *two* flux linkages : (*i*) Due to internal flux, and (*ii*) due to external flux.

13.2.2.1. Flux linkages due to a single current carrying conductor

Consider a long straight cylindrical conductor of radius r metres and carrying current of I amperes (r.m.s.) as shown in Fig. 13.2. The current I will set up magnetic field ; the magnetic lines of force will exist inside as well as outside the conductor. Both these fluxes will contribute towards the inductance of the conductor.

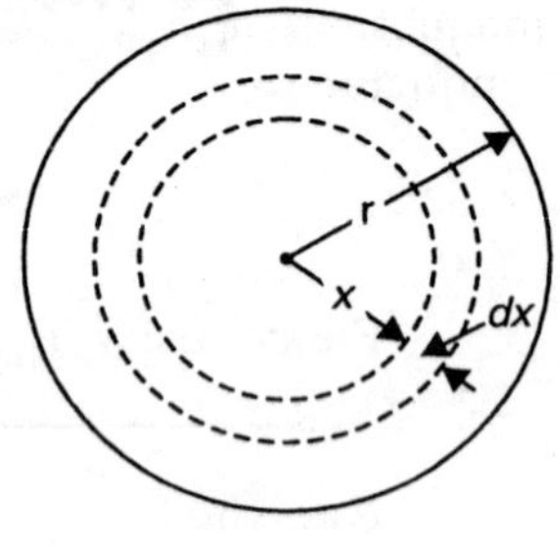

Fig. 13.2

In overhead lines it may be assumed without appreciable error that the current uniformly distributed.

(*i*) ***Flux linkages due to internal flux.*** The magnetic field intensity at a point x metres ($x < r$) from the centre is given by,

$$H_x = \frac{I_x}{2\pi x}$$

[As per Ampere's law, m.m.f. (ampere-turns) around any closed path equals the current enclosed by the path, *i.e.,* in this case current is I_x and m.m.f. is $H_x \times 2\pi x$, $\therefore \quad H_x \times 2\pi x = I_x$]

Assuming a uniform current density, we have

$$\frac{I_x}{\pi x^2} = \frac{I}{\pi r^2} \quad \text{or} \quad I_x = I \times \frac{\pi x^2}{\pi r^2} = I \times \frac{x^2}{r^2}$$

$$\therefore \qquad H_x = \frac{1}{2\pi x} \times I \times \frac{x^2}{r^2} = \frac{x}{2\pi r^2} I \text{ AT/m}$$

If μ is the permeability of the conductor, then flux density at the considered point is given by,

$$B_x = \mu H_x = \mu_0 \mu_r H_x \text{ Wb/m}^2 \qquad (\because \quad \mu = \mu_0 \mu_r)$$

$$= \frac{\mu_0 x I}{2\pi r^2} \qquad (\because \quad \mu_r = 1 \text{ for non-magnetic material})$$

Now, the flux through a cylindrical shell of radial thickness dx and axial length 1 m is given by,

$$d\phi = B_x \times 1 \times dx = \frac{\mu_0 x I}{2\pi r^2} dx \text{ webers}$$

This flux links with only the current lying within the circle of radius x *i.e.,* I_x or $\frac{x^2}{r^2} I$

$$\therefore \quad \text{Linkages of the shell,} \qquad d\lambda = \frac{x^2}{r^2} d\phi = \frac{x^2}{r^2} \times \frac{\mu_0 x I}{2\pi r^2} dx$$

$$= \frac{\mu_0 I x^3}{2\pi r^4} dx \text{ weber-turns (Wb-T)}$$

Total flux linkages from centre of the conductor and upto the surface of the conductor,

$$\lambda_{\text{int.}} = \int_0^r d\lambda = \frac{\mu_0 I}{2\pi r^4} \int_0^r x^3 dx = \frac{\mu_0 I}{2\pi r^4} \times \frac{r^4}{4}$$

or,
$$\lambda_{\text{int.}} = \frac{\mu_0 I}{8\pi} \text{ Wb-T/m} \qquad \text{...(13.3)}$$

From this expression, it is clear that the flux linkages due to internal flux is *independent of the conductor size.*

(*ii*) ***Flux linkages due to external flux.*** Refer to Fig. 13.3. Now let us consider external flux, which extends from surface of the conductor to infinity. The magnetic field intensity at any distance x from the centre of the conductor ($x > r$),

$$H_x = \frac{I}{2\pi x} \text{ AT/m}$$

Flux density, $B_x = \mu_0\mu_r H_x = \dfrac{\mu_0 I}{2\pi x}$ Wb/m²

($\because$ $\mu_r = 1$, medium being air)

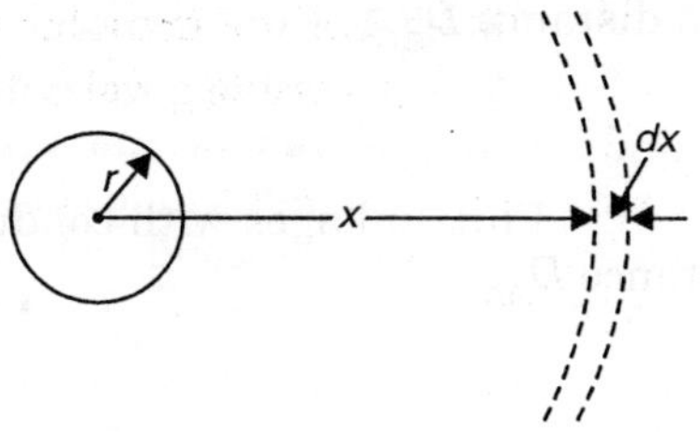

Fig. 13.3

Now the flux through a cylindrical shell of radial thickness dx and axial length 1 m is given by,

$$d\phi = B_x \times 1 \times dx = \frac{\mu_0 I}{2\pi x} dx \text{ webers}$$

Now the flux linkage per metre is equal to $d\phi$ since flux external to conductor links all the current in the conductor only once,

$$\therefore \quad d\lambda = \frac{\mu_0 I}{2\pi x} \text{ Wb-T}$$

Total flux linkages of the conductor from surface to infinity,

$$\lambda_{\text{ext.}} = \int_r^{\infty} \frac{\mu_0 I}{2\pi x} dx = \frac{\mu_0 I}{2\pi} \left| \ln x \right|_r^{\infty} \text{ Wb-T/m} \quad \text{...(13.4)}$$

Total flux linkages due to one conductor,

λ = Total internal flux linkages + total external flux linkages

or
$$\lambda = \lambda_{\text{int.}} + \lambda_{\text{ext.}} = \frac{\mu_0 I}{8\pi} + \int_r^{\infty} \frac{\mu_0 I}{2\pi x} dx$$

$$\lambda = \frac{\mu_0 I}{2\pi}\left[\frac{1}{4} + \int_r^{\infty} \frac{dx}{x}\right] \text{ Wb-T/m length} \quad \text{...(13.5)}$$

13.2.2.2. Flux linkages in parallel current-carrying conductors

Consider a group of parallel conductors A, B, C, D, E, ..., N carrying currents $I_A, I_B, I_C, I_D, I_E, ..., I_N$ respectively as shown in Fig. 13.4. Let us consider the flux linkages with any conductor say *A*. The flux linkages with conductor *A* will include flux linkages due to its own current and flux linkages due to mutual inductance effects of currents $I_B, I_C, I_D, I_E, ..., I_N$ respectively.

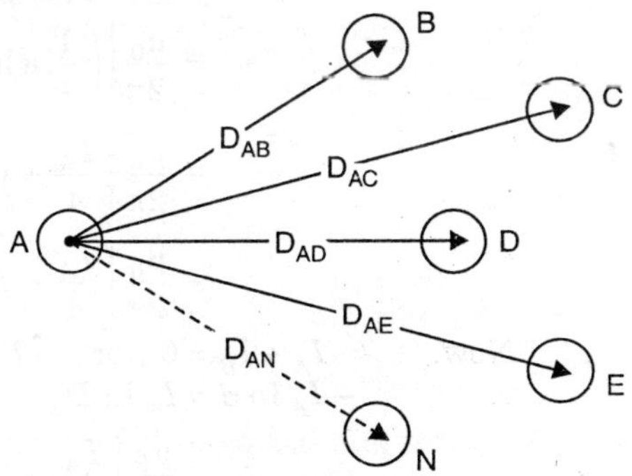

Fig. 13.4

Flux linkages with conductor *A* due to its own current

$$= \frac{\mu_0 I_A}{2\pi}\left[\frac{1}{4} + \int_r^{\infty} \frac{dx}{x}\right] \quad \text{...(}i\text{)}$$

Flux linkages with conductor *A* due to current I_B flowing in conductor *B* situated at distance d_1

$$= \frac{\mu_0 I_B}{2\pi} \int_{D_{AB}}^{\infty} \frac{dx}{x} \quad \text{...(}ii\text{)}$$

(Since only external flux due to I_B links with conductor *A* from D_{AB} to ∞)

Similarly flux linkages with conductor A due to current I_C flowing in conductor C situated at a distance D_{AC}

$$= \frac{\mu_0 I_C}{2\pi} \int_{D_{AC}}^{\infty} \frac{dx}{x} \qquad ...(iii)$$

Flux linkages with conductor A due to current I_N flowing in conductor N situated at a distance D_{AN}

$$= \frac{\mu_0 I_N}{2\pi} \int_{D_{AN}}^{\infty} \frac{dx}{x}$$

Total flux linkages with conductor A,

$$\lambda_{A(t)} = \frac{\mu_0 I_A}{2\pi}\left[\frac{1}{4} + \int_r^{\infty} \frac{dx}{x}\right] + \frac{\mu_0 I_B}{2\pi}\int_{D_{AB}}^{\infty}\frac{dx}{x} + \frac{\mu_0 I_C}{2\pi}\int_{D_{AC}}^{\infty}\frac{dx}{x} + ... + \frac{\mu_0 I_N}{2\pi}\int_{D_{AN}}^{\infty}\frac{dx}{x} \text{ Wb-T/m} \qquad ...(13.6)$$

13.2.3. Inductance of 2-wire (1-phase) Transmission Line

Consider a single phase overhead line consisting of two parallel conductors A and B, each of radius r metres and spaced D metres apart ($D >> r$) as shown in Fig. 13.5. Conductors A and B carry the same current (*i.e.* $I_A = I_B$) in magnitude but opposite in direction as one forms the return path for the other.

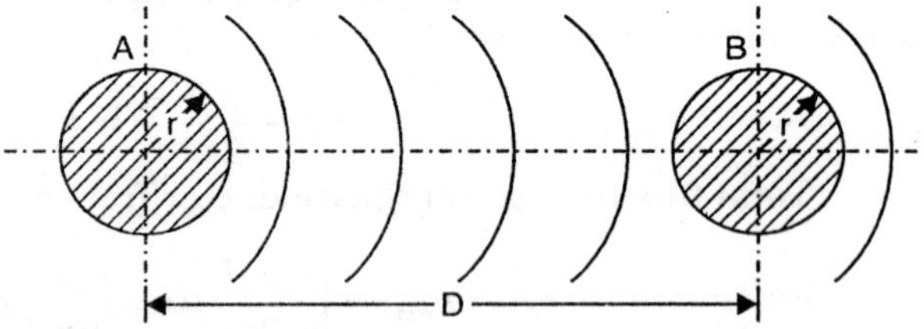

Fig. 13.5. A 1–phase two wire line.

$$\therefore \qquad I_A + I_B = 0$$

Total flux linkages with conductor A,

λ_A = Flux linkages with conductor A due it own current + Flux linkages with conductor A due to current I_B

$$= \frac{\mu_0 I_A}{2\pi}\left[\frac{1}{4} + \int_r^{\infty}\frac{dx}{x}\right] + \frac{\mu_0 I_B}{2\pi}\int_D^{\infty}\frac{dx}{x}$$

$$= \frac{\mu_0}{2\pi}\left[\left(\frac{1}{4} + \int_r^{\infty}\frac{dx}{x}\right) I_A + I_B \int_D^{\infty}\frac{dx}{x}\right]$$

$$= \frac{\mu_0}{2\pi}\left[\left(\frac{1}{4} + \ln\infty - \ln r\right) I_A + (\ln\infty - \ln D)\, I_B\right]$$

$$= \frac{\mu_0}{2\pi}\left[\frac{I_A}{4} + \ln\infty\,(I_A + I_B) - I_A \ln r - I_B \ln D\right]$$

$$= \frac{\mu_0}{2\pi}\left[\frac{I_A}{4} - I_A \ln r - I_B \ln D\right] \qquad [\because\ I_A + I_B = 0]$$

Now, $\qquad I_A + I_B = 0 \quad \text{or} \quad -I_B = I_A$

$$\therefore \qquad -I_B \ln d = I_A \ln D$$

$$\therefore \qquad \lambda_A = \frac{\mu_0}{2\pi}\left[\frac{I_A}{4} + I_A \ln D - I_A \ln r\right]$$

$$= \frac{\mu_0}{2\pi}\left[\frac{I_A}{4} + I_A \ln\left(\frac{D}{r}\right)\right]$$

$$= \frac{\mu_0 A}{2\pi}\left[\frac{1}{4} + \ln\frac{D}{r}\right] \text{ Wb-T/m}$$

Inductance of conductor A, $L_A = \frac{\lambda_{A(t)}}{I_A}$

$$= \frac{\mu_0}{2\pi}\left[\frac{1}{4} + \ln\frac{D}{r}\right] \text{ H/m}$$

$$= \frac{4\pi \times 10^{-7}}{2\pi}\left[\frac{1}{4} + \ln\frac{D}{r}\right] \text{ H/m}$$

$\therefore$ $$L_A \text{ (Inductance per conductor)} = 10^{-7}\left[\frac{1}{2} + 2\ln\frac{D}{r}\right] \text{ H/m} \quad ...(13.7)$$

$$\textit{Loop inductance} = 2L_A \text{ H/m} = 10^{-7}\left(1 + 4\ln\frac{D}{r}\right) \text{ H/m} \quad ...(13.8)$$

- The expression for the inductance of a conductor can be put in the following form :

$$L_A = 10^{-7}\left(\frac{1}{2} + 2\ln\frac{D}{r}\right) \text{ H/m}$$

$$= 2 \times 10^{-7}\left(\frac{1}{4} + \ln\frac{D}{r}\right) \text{ H/m} = 2 \times 10^{-7}\left[\ln(e^{1/4}) + \ln\frac{D}{r}\right]$$

$\therefore$ $$L_A = 2 \times 10^{-7} \ln\left(\frac{D}{re^{-1/4}}\right)$$

Putting $re^{-1/4} = r'$, then

$$L_A = 2 \times 10^{-7} \ln\frac{D}{r'} \text{ H/m} \quad ...(13.9)$$

The radius r' is that of a fictitious conductor assumed to have no internal flux linkages but with the same inductance as the actual conductor with radius r. The quantity $e^{-1/4} = 0.7788$, so that

$$r' = 0.7788r.$$

The term r' $(= re^{-1/4} = 0.7788r)$ is called the *geometric mean radius* (GMR) of the wire.

Note. $r' = 0.7788r$ is applicable to *only solid round conductor.*

13.2.4. Inductance of a 3-phase Overhead Transmission Line

Consider three conductors A, B and C (each of radius r metres) of a 3-phase transmission line carrying currents I_A, I_B and I_C respectively and having spacings between them D_{AB}, D_{BC}, D_{CA} respectively, as shown in Fig. 13.6. Assume the load is balanced *i.e.* $I_A + I_B + I_C = 0$.

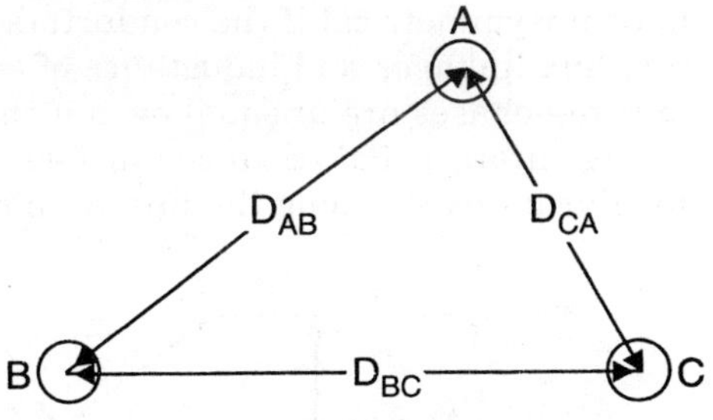

Fig. 13.6

Total flux linkages with conductor A,

λ_A = Flux linkages due to its own current
+ flux linkages due to current I_B
+ flux linkages due to current I_C

$$= \frac{\mu_0 I_A}{2\pi}\left(\frac{1}{4} + \int_r^{\infty}\frac{dx}{x}\right) + \frac{\mu_0 I_B}{2\pi}\int_{D_{AB}}^{\infty}\frac{dx}{x} + \frac{\mu_0 I_C}{2\pi}\int_{D_{CA}}^{\infty}\frac{dx}{x}$$

$$= \frac{\mu_0}{2\pi}\left[\left(\frac{1}{4} + \int_r^{\infty}\frac{dx}{x}\right)I_A + I_B\int_{D_{AB}}^{\infty}\frac{dx}{x} + I_C\int_{D_{CA}}^{\infty}\frac{dx}{x}\right]$$

$$= \frac{\mu_0}{2\pi}\left[\left(\frac{1}{4}\ \ln\infty - \ln r\right)I_A + I_B(\ln\infty - \ln D_{AB}) + I_C(\ln\infty - \ln D_{CA})\right].$$

$$= \frac{\mu_0}{2\pi}\left[\left(\frac{1}{4} - \ln r\right) I_A - I_B \ln D_{AB} - I_C \ln D_{CA} + \ln \infty \,(I_A + I_B + I_C)\right]$$

$$\therefore \quad \lambda_A = \frac{\mu_0}{2\pi}\left[\left(\frac{1}{4} - \ln r\right) I_A - I_B \ln D_{AB} - I_C \ln D_{CA}\right] \qquad (\because\ I_A + I_B + I_C = 0) \quad ...(13.10)$$

(*i*) **Symmetrical spacing.** If the three conductors *A*, *B* and *C* are placed symmetrically at the corners of an equilateral triangle of the side *D* then, $D_{AB} = D_{BC} = D_{CA} = D$. Under such a condition the total flux linkages with the conductor *A*,

$$\lambda_A = \frac{\mu_0}{2\pi}\left[\left(\frac{1}{4} - \ln r\right) I_A - I_B \ln D - I_C \ln D\right]$$

$$= \frac{\mu_0}{2\pi}\left[\left(\frac{1}{4} - \ln r\right) I_A - (I_B + I_C) \ln D\right]$$

$$= \frac{\mu_0}{2\pi}\left[\left(\frac{1}{4} - \ln r\right) I_A + I_A \ln D\right] \qquad (\because\ I_B + I_C = -I_A)$$

or

$$\lambda_A = \frac{\mu_0 I_A}{2\pi}\left[\frac{1}{4} + \ln \frac{D}{r}\right] \text{ Wb-T} \qquad ...(13.11)$$

Inductance of conductor *A*, $L_A = \frac{\lambda_A}{I_A}$ H/m $= \frac{\mu_0}{2\pi}\left[\frac{1}{4} + \ln \frac{D}{r}\right]$ H/m

$$= \frac{4\pi \times 10^{-7}}{2\pi}\left[\frac{1}{4} + \ln \frac{D}{r}\right] \text{ H/m}$$

or

$$L_A = 10^{-7}\left(0.5 + 2 \ln \frac{D}{r}\right) \text{ H/m} \qquad ...(13.12)$$

Derived on the same lines, the expressions for inductances of conductors *B* and *C* are same as that of conductor *A*.

(*ii*) **Unsymmetrical spacing.** The conductor spacing in a 3-phase transmission line is said to be unsymmetrical if the conductors are not equidistant from each other. Under such conditions the, flux linkages and inductance of each phase are not the same and subsequently voltage drops in three phases are unequal even if the currents in the conductors are balanced. In order to make voltage drops equal in all conductors, we generally interchange the positions of the conductors at regular intervals along the line so that *each conductor occupies the original position of every other*

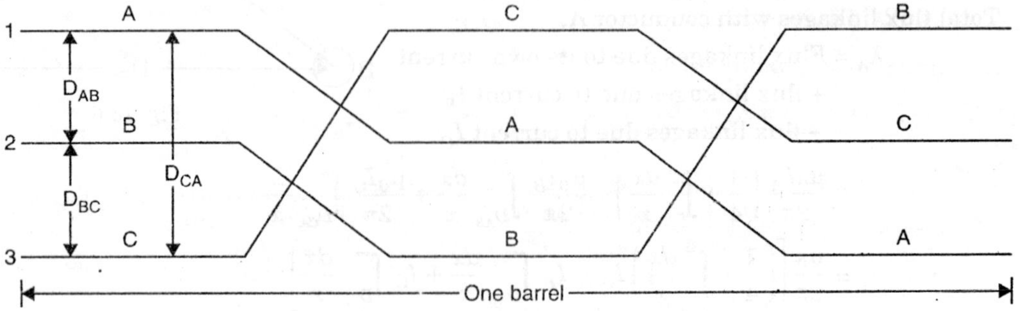

Fig. 13.7. Transposition of conductors.

conductor over an equal distance. Such an exchange of conductor positions is called **transposition.** A complete transposition cycle is shown in Fig. 13.7. The phase conductors are designated as A, B and C and positions occupied are numbered 1, 2 and 3. This arrangement causes *each conductor to have the same average inductance over the transposition cycle.* Let us assume the balanced conditions, *i.e.* $I_A = I_B = I_C = I$ (say) in magnitude.

Taking I_A as the reference vector, the currents are represented as :

$$\bar{I}_A = I(1 + j0)$$

$$\bar{I}_B = I(-0.5 - j0.866)$$

$$\bar{I}_C = I(-0.5 + j0.866)$$

As proved earlier, the total flux linkages per metre length of conductor A is,

$$\lambda_A = \frac{\mu_0}{2\pi}\left[\left(\frac{1}{4} - \ln r\right)\bar{I}_A - \bar{I}_B \ln D_{AB} - \bar{I}_C \ln D_{CA}\right]$$

Putting the values of $\bar{I}_A$, $\bar{I}_B$ and $\bar{I}_C$, we get

$$\lambda_A = \frac{\mu_0}{2\pi}\left[\left(\frac{1}{4} - \ln r\right) I - I(-0.5 - j0.866)\ln D_{AB} - I(-0.5 + j0.866)\ln D_{CA}\right]$$

$$= \frac{\mu_0}{2\pi}\left[\frac{1}{4} I - I \ln r + 0.5\, I \ln D_{AB} + j0.866\, I \ln D_{AB} + 0.5\, I \ln D_{CA} - j0.866\, I \ln D_{CA}\right]$$

$$= \frac{\mu_0}{2\pi}\left[\frac{1}{4} I - I \ln r + 0.5\, I (\ln D_{AB} + \ln D_{CA}) + j0.866\, I (\ln D_{AB} - \ln D_{CA})\right]$$

$$= \frac{\mu_0}{2\pi}\left[\frac{1}{4} I - I \ln r + 0.5\, I \{\ln (D_{AB} \times D_{CA})\} + j0.866\, I \left\{\ln\left(\frac{D_{AB}}{D_{CA}}\right)\right\}\right]$$

$$= \frac{\mu_0}{2\pi}\left[\frac{1}{4} I - I \ln r + I \ln \sqrt{D_{AB} D_{CA}} + j0.866\, I \ln\left(\frac{D_{AB}}{D_{CA}}\right)\right]$$

$$= \frac{\mu_0}{2\pi}\left[\frac{1}{4} I + I \ln\left(\frac{\sqrt{D_{AB} D_{CA}}}{r}\right) + j0.866 \ln\left(\frac{D_{AB}}{D_{CA}}\right)\right]$$

$$= \frac{\mu_0 I}{2\pi}\left[\frac{1}{4} + \ln\left(\frac{\sqrt{D_{AB} D_{CA}}}{r}\right) + j0.866 \ln\left(\frac{D_{AB}}{D_{CA}}\right)\right]$$

$\therefore$ Inductance of conductor A,

$$L_A = \frac{\lambda_A}{I_A} = \frac{\lambda_A}{I}$$

$$= \frac{\mu_0}{2\pi}\left[\frac{1}{4} + \ln\left(\frac{\sqrt{D_{AB} D_{CA}}}{r}\right) + j0.866 \ln\left(\frac{D_{AB}}{D_{CA}}\right)\right] \text{ H/m}$$

$$= \frac{4\pi \times 10^{-7}}{2\pi}\left[\frac{1}{4} + \ln\left(\frac{\sqrt{D_{AB} D_{CA}}}{r}\right) + j0.866 \ln\left(\frac{D_{AB}}{D_{CA}}\right)\right] \text{ H/m}$$

or, $$L_A = 10^{-7}\left[\frac{1}{2} + 2\ln\left(\frac{\sqrt{D_{AB}\,D_{CA}}}{r}\right) + j1.732\ln\left(\frac{D_{AB}}{D_{CA}}\right)\right] \text{ H/m} \quad ...(13.13)$$

[**Note.** *The imaginary part of the flux-linkage or inductance represents exchange of energy between phases.*]

Similarly, $$L_B = 10^{-7}\left[\frac{1}{2} + 2\ln\left(\frac{\sqrt{D_{BC}\,D_{AB}}}{r}\right) + j1.732\ln\left(\frac{D_{BC}}{D_{AB}}\right)\right] \text{ H/m, and} \quad ...(13.14)$$

$$L_C = 10^{-7}\left[\frac{1}{2} + 2\ln\left(\frac{\sqrt{D_{CA}\,D_{BC}}}{r}\right) + j1.732\ln\left(\frac{D_{CA}}{D_{BC}}\right)\right] \text{ H/m} \quad ...(13.15)$$

Inductance of *each line conductor*

$$= \frac{1}{3}(L_A + L_B + L_C)$$

$$= 10^{-7}\left[\frac{1}{2} + 2\ln\left(\frac{\sqrt[3]{D_{AB}\,D_{BC}\,D_{CA}}}{r}\right)\right] \text{ H/m} \quad ...(13.16)$$

......after substitution and simplification.

If the formula of inductance of an unsymmetrically spaced transposed line is composed with that of symmetrically placed line, we find that inductance of each line conductor in the two cases will be equal if $D = \sqrt[3]{D_{AB}\,D_{BC}\,D_{CA}}$. The distance D is known as *equivalent equilateral spacing* for unsymmetrically transposed line.

Note. Normally, modern power lines are *not* transposed. The transposition, however, *may be affected in the intermediate switching station.* It is to be noted that the difference in the inductances of the three phases is *negligibly small* due to asymmetrical spacing and the inductance of the untransposed line is taken equal to the average value of the inductance of one phase of the same line correctly transposed.

13.2.5. Self-GMD and Mutual-GMD

The inductance calculations relating to multiconductor arrangements can be amply simplified by the use of *self geometrical mean distance* (self-GMD) and *mutual geometrical mean distance* (mutual-GMD). Self-GMD and mutual-GMD are generally represented by the symbols D_s and D_m respectively, and are duscussed below :

(*i*) **Self-GMD (D_s).** It is also sometimes called *"Geometrical mean radius* (GMR)".

Inductance per conductor per metre length is given by,

$$L = 2 \times 10^{-7}\left(\frac{1}{4} + \ln\frac{D}{r}\right) \quad ...[\text{Eqn. (13.7)}]$$

or, $$L = \left(2 \times 10^{-7} \times \frac{1}{4}\right) + \left(2 \times 10^{-7}\ln\frac{D}{r}\right) \quad ...(i)$$

In the above expression the term $\left[2 \times 10^{-7} \times \left(\frac{1}{4}\right)\right]$ is the inductance due to *flux within the solid conductor.* The concept of self-GMD or GMR is introduced to eliminate this term for several purposes. It can be proved mathematically that for a solid round conductor of radius r, the self-GMD or GMR = 0.7788r. Using self-GMD the eqn. (*i*) reduces to

$$L = 2 \times 10^{-7} \ln \frac{D}{D_s}$$

where, D_s = Self-GMD or GMR = $0.7788r$.

Sometimes GMR is denoted by r'.

It may be observed that self-GMD of a conductor depends upon the *size and shape of the conductor and is independent of the spacing between the conductors.*

(*ii*) **Mutual-GMD (D_m).** The mutual-GMD is the geometrical mean of the distances from one conductor to the others, must be between the largest and smallest such distances. It simply represents the *equivalent geometrical spacing.*

- Assuming the spacing between the conductors to be large compared to the diameter of each conductor, the mutual-GMD between two conductors is equal to the distance between their centres, *i.e.* $D_m = D$ (spacing between the conductors)
- For a single circuit 3-phase transmission line, the mutual-GMD is equal to the equivalent equilaterial spacing, $D_m = (D_{AB} \times D_{BC} \times D_{CA})^{1/3}$.
- The concept of geometrical mean distance can be applied to 3-phase double circuit lines as discussed below :

The conductor arrangement of the double circuit is shown in Fig. 13.8. Let the radius of each conductor be r.

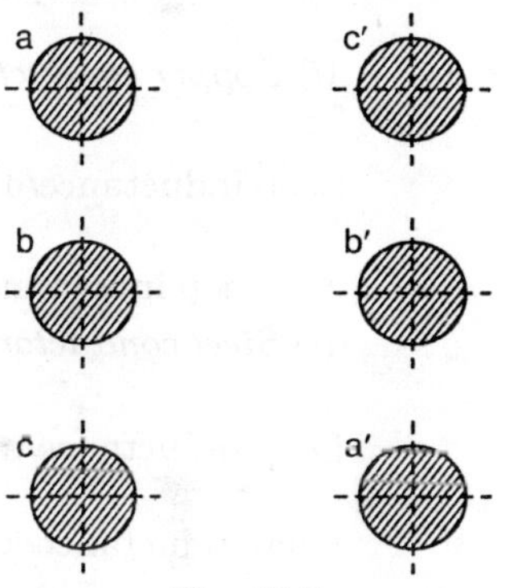

Fig. 13.8

Self-GMD of conductor = $0.7788r$.

Self-GMD of combination aa',

$$D_{s1} = (D_{aa} \times D_{aa'} \times D_{a'a'} \times D_{a'a})^{1/4}$$

(D_{aa} or $D_{a'a'}$ means self-GMD of the conductor, whereas $D_{aa'}$ means distance between a and a')

Self-GMD of combination bb',

$$D_{s2} = (D_{bb}\, D_{bb'}\, D_{b'b'}\, D_{b'b})^{1/4}$$

Self-GMD of combination cc',

$$D_{s3} = (D_{cc}\, D_{cc'}\, D_{c'c'}\, D_{c'c})^{1/4}$$

Equivalent self-GMD of one phase,

$$D_s = (D_{s1} \times D_{s2} \times D_{s3})^{1/3}$$

Since each conductor has the same radius, therefore the value of D_s is same for all the phases.

Mutual-GMD between phases A and B,

$$D_{AB} = (D_{ab} \times D_{ab'} \times D_{a'b} \times D_{a'b'})^{1/4}$$

Mutual-GMD between phases B and C,

$$D_{BC} = (D_{bc} \times D_{bc'} \times D_{b'c} \times D_{b'c'})^{1/4}$$

$$D_{CA} = (D_{ca} \times D_{ca'} \times D_{c'a} \times D_{c'a'})^{1/4}$$

Equivalent mutual-GMD, $D_m = (D_{AB} \times D_{BC} \times D_{CA})^{1/3}$

It may be noted that *mutual-GMD* is substantially independent of exact size, shape and orientation of the conductor and *depends only upon the "spacing".*

Inductance formulae in terms of GMD :

(*i*) *Single phase line :*

$$\text{Inductance/conductor/m} = 2 \times 10^{-7} \ln \frac{D_m}{D_s}$$

where, $D_m = D$ (spacing between the conductors), and $D_s = 0.7788r$.

(*ii*) *Single circuit 3-phase line :*

$$\text{Inductance/phase/m} = 2 \times 10^{-7} \ln \frac{D_m}{D_s}$$

where, $D_m = (D_{AB} \times D_{BC} \times D_{CA})^{1/3}$, and $D_s = 0.7788r$.

(*iii*) *Double circuit 3-phase line :*

$$\text{Inductance/phase/m} = 2 \times 10^{-7} \ln \frac{D_m}{D_s}$$

where, $D_m = (D_{AB} \times D_{BC} \times D_{CA})^{1/3}$ and $D_s = (D_{s1} \times D_{s2} \times D_{s3})^{1/3}$.

Example 13.1. *A single phase transmission line has two parallel conductors, each of 1.2 cm diameter, and 2.5 metres apart. Calculate the loop inductance per km length of the line if the material of the conductor is (i) copper ; (ii) steel with relative permeability of 200.*

Solution. *Given :* $r = \frac{1.2}{2} = 0.6$ cm ; $D = 2.5$ m $= 250$ cm ; μ_r (steel) $= 200$

$$\text{Loop inductance} = 10^{-7}\left(\mu_r + 4 \ln \frac{D}{r}\right)$$

(*i*) *Copper conductors :* $\mu_r = 1$

$$\text{Loop inductance/m} = 10^{-7}\left(1 + 4 \ln \frac{D}{r}\right) = 10^{-7}\left[1 + 4 \ln\left(\frac{250}{0.6}\right)\right] \text{H} = 25.13 \times 10^{-7} \text{ H}$$

$\therefore$ Loop inductance/km $= 25.13 \times 10^{-7} \times 1000 = 2.513 \times 10^{-3} =$ **2.513 mH. (Ans.)**

(*ii*) *Steel conductors :* $\mu_r = 200$

$$\text{Loop inductance/m} = 10^{-7}\left[200 + 4 \ln\left(\frac{250}{0.6}\right)\right] \text{H} = 224.13 \times 10^{-7} \text{ H}$$

Loop inductance/km $= 224.13 \times 10^{-7} \times 1000 = 22.413 \times 10^{-3}$ H $=$ **22.413 mH. (Ans.)**

Example 13.2. *In a 3-phase transmission line the conductors are placed at the corners of an equilateral triangle of each side 2.5 m. If the radius of each conductor is 0.8 cm find inductance per phase per km length of the line.*

Solution. Fig. 13.9 shows the positions of the conductors.

Here $D_{12} = D_{23} = D_{31} = D = 2.5$ m or 250 cm and radius $r = 0.8$ cm

$$\text{Inductance/phase/m} = 10^{-7}\left(0.5 + 2 \ln \frac{D}{r}\right) H$$

...[Eqn. (13.7)]

$$= 10^{-7}\left[0.5 + 2 \ln\left(\frac{250}{0.8}\right)\right] = 11.99 \times 10^{-7} \text{H}$$

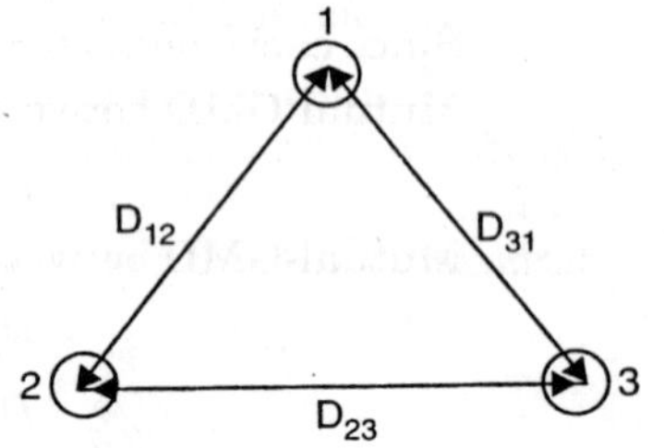

Fig. 13.9

Inductance/phase/km $= 11.99 \times 10^{-7} \times 1000$ H $=$ **1.199 mH. (Ans.)**

Example 13.3. *In a 3-phase transmission line the three conductors are placed at the corners of a triangle of sides 1.5 m, 3 m and 2.6 m respectively. If the diameter of each conductor is 1.4 cm and the conductors are regularly transposed, calculate the inductance/phase/km length of the line.*

Solution. Refer to Fig. 13.10.

Given : $D_{12} = 1.5$ m ; $D_{23} = 3$ m ; $D_{31} = 2.6$ m,

$$r = \frac{1.4}{2} = 0.7 \text{ cm} ;$$

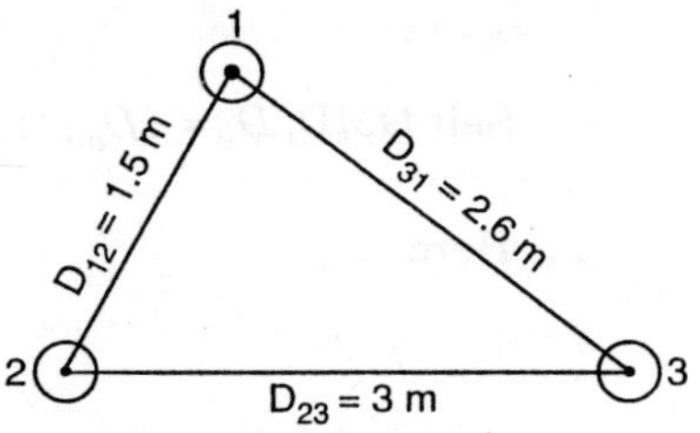

Fig. 13.10

Equivalent equilateral spacing,

$$D_{eq} = (D_{12} \times D_{23} \times D_{31})^{1/3}$$
$$= (1.5 \times 3 \times 2.6)^{1/3} = 2.27 \text{ m} = 227 \text{ cm}$$

$$\text{Inductance/phase/m} = 10^{-7}\left[0.5 + 2 \ln\left(\frac{D_{eq}}{r}\right)\right] \text{H}$$

$$= 10^{-7}\left[0.5 + 2 \ln\left(\frac{227}{0.7}\right)\right] \text{H} = 12.06 \times 10^{-7} \text{ H}$$

Inductance/phase/km $= 12.06 \times 10^{-7} \times 1000 =$ **1.206 mH. (Ans.)**

Example 13.4. *The conductors of 3-phase, 3-wire system are transposed and are arranged in a horizontal plane (Fig. 13.11), such that $D_{12} = D_{31} = 1.8$ m ; $D_{23} = 3.6$ m. If the diameter of each conductor is 2.2 cm, calculate the inductance / phase / km length of the line.*

Solution. Refer to Fig. 13.11.

Given : $D_{12} = D_{31} = 1.8$ m ; $D_{23} = 3.6$ m ; $r = \frac{2.2}{2} = 1.1$ cm

Equivalent equilateral spacing,

$$D_{eq} = (D_{12} \times D_{23} \times D_{31})^{1/3}$$
$$= (1.8 \times 3.6 \times 1.8)^{1/3} = 2.27 \text{ m} = 227 \text{ cm}$$

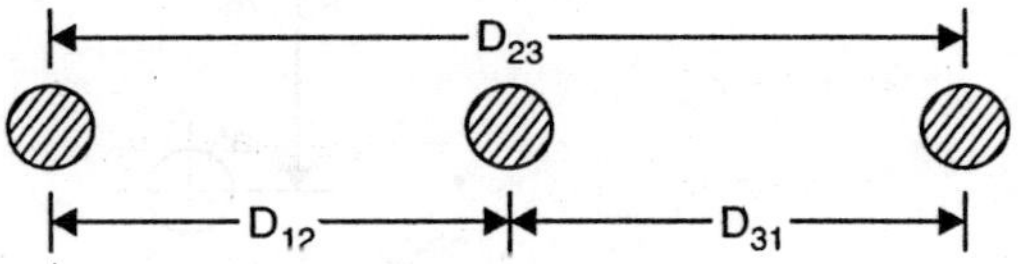

Fig. 13.11

Inductance/phase/m

$$= 10^{-7}\left[0.5 + 2 \ln\left(\frac{D_{eq}}{r}\right)\right] H$$

$$= 10^{-7}\left[0.5 + 2 \ln\left(\frac{227}{1.1}\right)\right] H = 11.16 \times 10^{-7} \text{ H}$$

Inductance/phase/km $= 11.16 \times 10^{-7} \times 1000\, H =$ **1.116 mH. (Ans.)**

Example 13.5. *In a single phase line (See Fig. 13.12), conductors a and a′ in parallel form one circuit while conductors b and b′ in parallel form the return path. Calculate the total inductance of the line per km assuming that the current is equally shared by the two parallel conductors. Conductor diameter is 1.8 cm.*

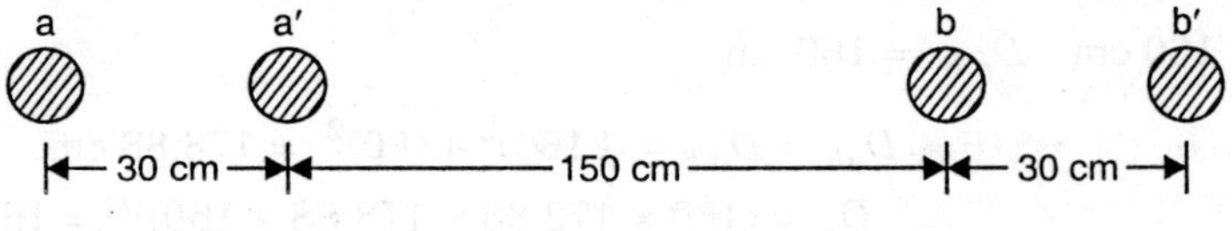

Fig. 13.12

Solution. Loop inductance/km, $L = 4 \times 10^{-7} \ln \frac{D_m}{D_s}$ H/m $= 4 \times 10^{-4} \ln\left(\frac{D_m}{D_s}\right)$ H/km

Mutual-GMD, $D_m = (D_{ab} \times D_{ab'} \times D_{a'b} \times D_{a'b'})^{1/4}$

$$= (180 \times 210 \times 150 \times 180)^{1/4} = 178.7 \text{ cm}$$

Self-GMD, $D_s = (D_{aa} \times D_{aa'} \times D_{a'a'} \times D_{a'a})^{1/4}$

Here $$D_{aa} = D_{a'a'} = 0.7788r = 0.7788 \times \frac{1.8}{2} = 0.7 \text{ cm}$$

$$D_{aa'} = D_{a'a} = 30 \text{ cm}$$

∴ $$D_s = (0.7 \times 30 \times 0.3 \times 30)^{1/4} = 4.58 \text{ cm}$$

∴ $$L = 4 \times 10^{-4} \times \ln\left(\frac{178.7}{4.58}\right) = 1.465 \times 10^{-3} \text{ H/km}$$

$$= \mathbf{1.465 \text{ mH/km.}} \quad \textbf{(Ans.)}$$

Example 13.6. *Fig. 13.13, shows the arrangement of a double-circuit single-phase line. Conductors a, a′ form one connection and conductors b, b′ form return connection. The distance between a and a′ is 80 cm and between a and b is 160 cm. If the diameter of each conductor is 2.2 cm determine the total inductance per km of the line.*

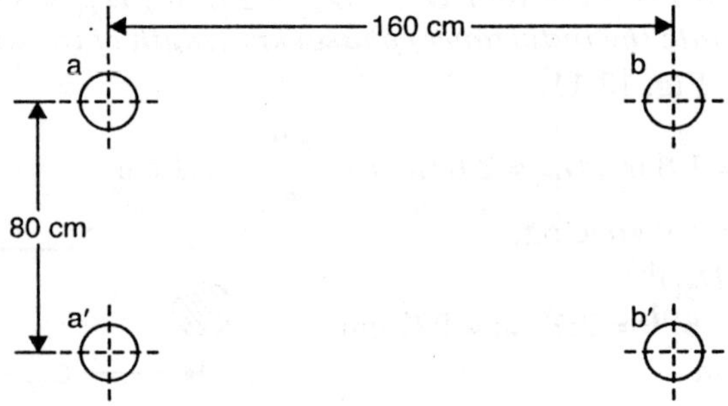

Fig. 13.13

Solution. The conductor radius, $r = \frac{2.2}{2} = 1.1$ cm

GMR of conductor $= 0.7788r = 0.7788 \times 1.1 = 0.8567$ cm

Self-GMD of aa' combination, $D_s = (D_{aa} \times D_{aa'} \times D_{a'a'} \times D_{a'a})^{1/4}$

$$= (0.8567 \times 80 \times 0.8567 \times 80)^{1/4} = 8.279 \text{ cm}$$

$(\because D_{aa} = D_{a'a'} = 0.8567 \text{ cm} ; D_{aa'} = D_{a'a} = 80 \text{ cm})$

Mutual-GMD between a and b,

$$D_m = (D_{ab} \times D_{ab'} \times D_{a'b} \times D_{a'b'})^{1/4}$$

Here $D_{ab} = 160$ cm ; $D_{a'b'} = 160$ cm

$$D_{ab'} = D_{a'b} = \sqrt{(160)^2 + (80)^2} = 178.88 \text{ cm}$$

∴ $$D_m = (160 \times 178.88 \times 178.88 \times 160)^{1/4} = 169.18 \text{ cm}$$

Loop inductance $= 4 \times 10^{-7} \ln \frac{D_m}{D_s}$ H/m

$$= 4 \times 10^{-7} \ln\left(\frac{169.18}{8.279}\right) = 12.07 \times 10^{-7} \text{ H/m}$$

∴ Loop inductance per km $= 12.07 \times 10^{-7} \times 10^3 = $ **1.207 mH/km.** **(Ans.)**

Example 13.7. *The three conductors of a 3-phase transmission line are arranged in a horizontal plan and are 4 metres apart. The diameter of each conductor is 2.5 cm. Determine the inductance per km of each conductor (line-to-neutral). Assume balanced load and R, Y, R phase sequence. Determine average inductance per phase for regularly transposed line.*

Solution. Radius of conductor, $r = \frac{2.5}{2} = 1.25$ cm

Refer to Fig. 13.14. $D_{AB} = 400$ cm ; $D_{BC} = 400$ cm ;
$D_{CA} = 800$ cm

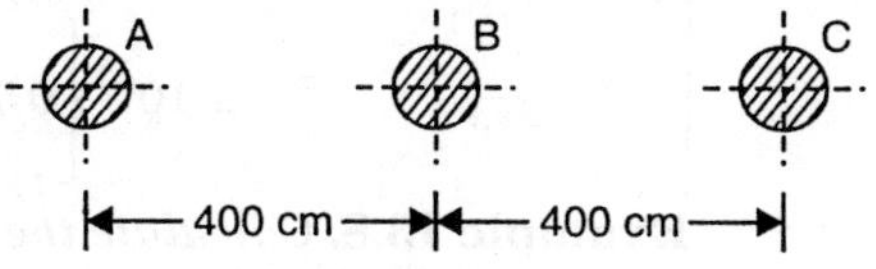

Fig. 13.14

We know that, inductance of conductor A,

$$\mathbf{L_A} = 10^{-7}\left[\frac{1}{2} + 2\ln\left(\frac{\sqrt{D_{AB}\,D_{CA}}}{r}\right) + j1.732\ln\left(\frac{D_{AB}}{D_{CA}}\right)\right] \text{ H/m}$$

...[Eqn. (13.13)]

$$= 10^{-7}\left[\frac{1}{2} + 2\ln\left(\frac{\sqrt{400\times 800}}{1.25}\right) + j1.732\ln\left(\frac{400}{800}\right)\right] \text{ H/m}$$

$$= 10^{-7}\,[0.5 + 12.23 + j1.732 \times (-\,0.693)] \text{ H/m}$$

$$= 10^{-7}\,(12.73 - j1.2) \text{ H/m}$$

$$= 10^{-7}\,(12.73 - j1.2) \times 1000$$

$$= \mathbf{(1.273 - j0.12)\ mH/km.}\quad \textbf{(Ans.)}$$

Similarly,

$$\mathbf{L_B} = 10^{-7}\left[\frac{1}{2} + 2\ln\left(\frac{\sqrt{D_{BC}\,D_{AB}}}{r}\right) + j1.732\ln\frac{D_{RC}}{D_{AR}}\right] \text{ H/m} \quad \text{...Eqn. (13.14)}$$

$$= 10^{-7}\left[0.5 + 2\ln\left(\frac{\sqrt{400\times 400}}{1.25}\right) + j1.732\ln\left(\frac{400}{400}\right)\right] \text{ H/m}$$

$$= 10^{-7}\,[0.5 + 11.54 + 0] = 12.04 \times 10^{-7} \text{ H/m}$$

$$= 12.04 \times 10^{-7} \times 1000 = \mathbf{1.204\ mH/km.}\quad \textbf{(Ans.)}$$

and,

$$\mathbf{L_C} = 10^{-7}\left[\frac{1}{2} + 2\ln\left(\frac{\sqrt{D_{CA}\,D_{BC}}}{r}\right) + j1.732\ln\left(\frac{D_{CA}}{D_{BC}}\right)\right] \text{ H/m}$$

...Eqn. (13.15)

$$= 10^{-7}\left[\frac{1}{2} + 2\ln\left(\frac{\sqrt{800\times 400}}{1.25}\right) + j1.732\ln\left(\frac{800}{400}\right)\right] \text{ H/m}$$

$$= 10^{-7}\,[0.5 + 12.23 + j1.2] = 10^{-7}\,(12.73 + j1.2) \text{ H/m}$$

$$= 10^{-7}\,(12.73 + j1.2) \times 1000 = \mathbf{(1.273 + j0.12)\ mH/km.}\quad \textbf{(Ans.)}$$

Average inductance per phase for regularly transposed line,

$$L_{avg.} = \frac{1}{3}(L_A + L_B + L_C)$$

$$= \frac{1}{3}[(1.273 - j0.12) + 1.204 + (1.273 + j0.12)] = \mathbf{1.25\ mH.}\quad \textbf{(Ans.)}$$

Alternatively, average inductance $L_{avg.} = 10^{-7}\left[\frac{1}{2} + 2\ln\left(\frac{\sqrt{D_{AB}D_{BC}D_{CA}}}{r}\right)\right]$ H/m

...Eqn. (13.16)

$$= 10^{-7}\left[0.5 + 2\ln\left(\frac{\sqrt[3]{400\times400\times800}}{1.25}\right)\right] = 1000\text{ H/km} = 1.25\text{ mH/km.}$$

Example 13.8. *Calculate the inductance per phase per km length of the system of conductors shown in Fig. 13.15. Self-GMD of one conductor is 0.8 cm. Assume line regularly transposed.*

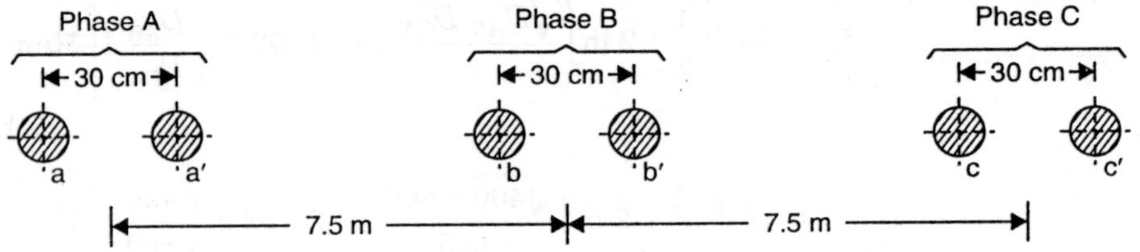

Fig. 13.15

Solution. Refer to Fig. 13.15.

Self-GMD of conductor a, $D_{aa} = 0.8$ cm

Self-GMD of each phase, $D_{s1} = D_{s2} = D_{s3}$ because in all the three positions of phase A distance between aa' is the same.

$$D_{s1}\,(= D_{s2} = D_{s3}) = (D_{aa}\ D_{aa'}\ D_{a'a'}\ D_{a'a})^{1/4}$$

$$= (0.8 \times 30 \times 0.8 \times 30)^{1/4} \simeq 4.9\text{ cm}$$

$\therefore$ $$D_s = (D_{s1} \times D_{s2} \times D_{s3})^{1/3} = (4.9 \times 4.9 \times 4.7)^{1/3} = 4.9\text{ cm}$$

Mutual-GMD, $D_m = (D_{AB} \times D_{BC} \times D_{CA})^{1/3}$

where, $$D_{AB} = (D_{ab} \times D_{ab'} \times D_{a'b} \times D_{a'b'})^{1/4}$$

$$= (7.5 \times 7.8 \times 7.2 \times 7.5)^{1/4} \simeq 7.5\text{ m}$$

$$D_{BC} = D_{AB} = 7.5\text{ m},\ D_{CA} = 15\text{ m}$$

$\therefore$ $$D_m = (7.5 \times 7.5 \times 15)^{1/3} = 9.45\text{ m}$$

Inductance per phase per km $= 2 \times 10^{-4} \ln\left(\frac{D_m}{D_s}\right)$ H/km

$$= 2 \times 10^{-4} \ln\left(\frac{9.45 \times 100}{4.9}\right)$$

$$= 1.05 \times 10^{-3}\text{ H} \quad \text{or} \quad \textbf{1.05 mH.} \quad \textbf{(Ans.)}$$

Example 13.9. *The arrangement of a symmetrical double-circuit, 3-phase line having six conductors is shown in Fig. 13.16. If the diameter of each conductor is 2.5 cm and the line is completely transposed, calculate the inductive reactance per phase in ohms per km.* **(Agra University)**

Solution. Radius of each conductor $= \frac{25}{2} = 1.25$ cm

GMR of conductor $= 0.7788 \times 1.25 = 0.9735$ cm

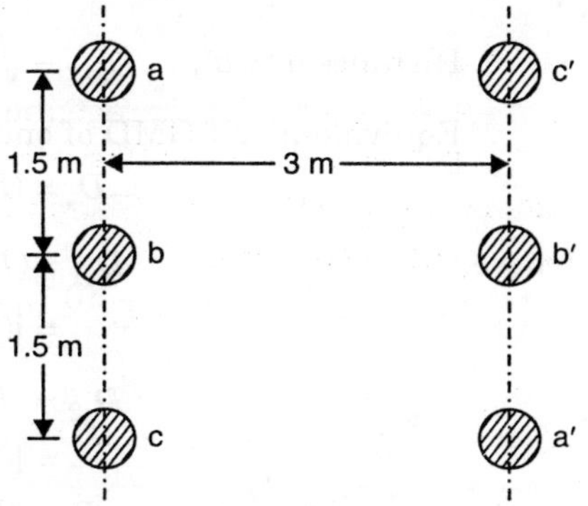

Fig. 13.16

Distance a to b', $D_{ab'} = \sqrt{(1.5)^2 + (3)^2} = 3.354$ m

Distance a to a', $D_{aa'} = \sqrt{(3)^2 + (3)^2} = 4.243$ m

Equivalent self-GMD of one phase,

$$D_s = (D_{s1} \times D_{s2} \times D_{s3})^{1/3}$$

where, D_{s1}, D_{s2} and D_{s3} represent the self-GMD in positions 1, 2 and 3 respectively. Also D_s is the same for all the phases.

Now, $$D_{s1} = (D_{aa} \times D_{aa'} \times D_{a'a'} \times D_{a'a})^{1/4}$$
$$= (0.9735 \times 10^{-2} \times 4.243 \times 0.9735 \times 10^{-2} \times 4.243)^{1/4}$$
$$= 0.2 \text{ m} = D_{s3}$$

$$D_{s2} = (D_{bb} \times D_{bb'} \times D_{b'b'} \times D_{b'b})^{1/4}$$
$$= (0.9735 \times 10^{-2} \times 3 \times 0.9735 \times 10^{-2} \times 3)^{1/4} = 0.171 \text{ m}$$

$\therefore$ $$D_s = (D_{s1} \times D_{s2} \times D_{s3})^{1/3} = (0.2 \times 0.171 \times 0.2)^{1/3} \simeq 0.19 \text{ m}$$

Equivalent mutual-GMD,

$$D_m = (D_{AB} \times D_{BC} \times D_{CA})^{1/3}$$

where, D_{AB}, D_{BC} and D_{CA} represent the mutual-GMD between phases A and B, B and C and C and A respectively.

Now, $$D_{AB} = (D_{ab} \times D_{ab'} \times D_{a'b} \times D_{a'b'})^{1/4}$$
$$= (1.5 \times 3.354 \times 3.354 \times 1.5)^{1/4} = 2.243 \text{ m} = D_{BC}$$

$$D_{CA} = (D_{ca} \times D_{ca'} \times D_{c'a} \times D_{c'a'})^{1/4}$$
$$= (3 \times 3 \times 3 \times 3)^{1/4} = 3 \text{ m}$$

$\therefore$ $$D_m = (2.243 \times 2.243 \times 3)^{1/3} = 2.47 \text{ m}$$

$\therefore$ Inductance per phase per metre length

$$= 2 \times 10^{-7} \ln \left(\frac{D_m}{D_s}\right) \text{ H/m} = 2 \times 10^{-7} \ln \left(\frac{2.47}{0.19}\right) = 5.13 \times 10^{-7} \text{ H/m}$$

Inductance/phase/km = $5.13 \times 10^{-7} \times 1000 = 0.513$ mH

$\therefore$ Inductive reactance per phase/km of the line, assuming supply frequency 50 Hz,

$$\mathbf{X_L} = 2\pi f L = 2\pi \times 50 \times (0.513 \times 10^{-3}) = \mathbf{0.161\ \Omega/km.} \quad \textbf{(Ans.)}$$

Example 13.10. *Find the inductance per phase per km of double-circuit 3-phase line shown in Fig. 13.17. The conductors are transposed and are of radius 0.8 cm each.*

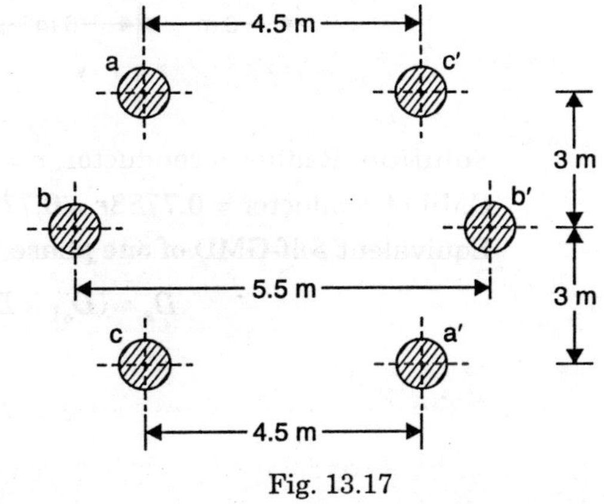

Fig. 13.17

Solution. Radius of conductor, $r = 0.8$ cm

GMR of conductor $= 0.7788 \times 0.8 = 0.623$ cm

Distance a to b, $D_{ab} = \sqrt{3^2 + (0.5)^2} = 3.04$ m

Distance a to b', $D_{ab'} = \sqrt{3^2 + 5^2} = 5.83$ m

Distance a to a', $D_{aa'} = \sqrt{6^2 + 4.5^2} = 7.5$ m

Equivalent self-GMD of one phase,

$$D_s = (D_{s1} \times D_{s2} \times D_{s3})^{1/3}$$

where,

$$D_{s1} = (D_{aa} \times D_{aa'} \times D_{a'a'} \times D_{a'a})^{1/4}$$

$$= [(0.623 \times 10^{-2}) \times 7.5 \times (0.623 \times 10^{-2}) \times 7.5]^{1/4} \simeq 0.216 \text{ m} = D_{s3}$$

$$D_{s2} = (D_{bb} \times D_{bb'} \times D_{b'b'} \times D_{b'b})^{1/4}$$

$$= [(0.623 \times 10^{-2}) \times 5.5 \times (0.623 \times 10^{-2}) \times 5.5]^{1/4} = 0.185 \text{ m}$$

$$\therefore \quad D_s = (0.216 \times 0.185 \times 0.216)^{1/3} = 0.205 \text{ m}$$

Equivalent mutual-GMD,

$$D_m = (D_{AB} \times D_{BC} \times D_{CA})^{1/3}$$

where,

$$D_{AB} = (D_{ab} \times D_{ab'} \times D_{a'b} \times D_{a'b'})^{1/4}$$

$$= (3.04 \times 5.83 \times 5.83 \times 3.04)^{1/4} = 4.33 \text{ m} = D_{BC}$$

$$D_{CA} = (D_{ca} \times D_{ca'} \times D_{c'a} \times D_{c'a'})^{1/4}$$

$$= (6 \times 4.5 \times 4.5 \times 6)^{1/4} = 5.196 \text{ m}$$

$$\therefore \quad D_m = (4.33 \times 4.33 \times 5.196)^{1/3} = 4.6 \text{ m}$$

$$\therefore \quad \text{Inductance/phase/m} = 2 \times 10^{-7} \ln \left(\frac{D_m}{D_s} \right) \text{ H/m}$$

$$= 2 \times 10^{-7} \ln \left(\frac{4.6}{0.205} \right) = 6.22 \times 10^{-7} \text{ H/m}$$

Inductance/phase/km = $6.22 \times 10^{-7} \times 1000$ = **0.622 mH/km. (Ans.)**

Example 13.11. *The horizontal arrangement of the conductors for 3-phase double-circuit line is shown in Fig. 13.18. If the radius of each conductor is 4.5 cm find the inductance per phase per metre of the line.*

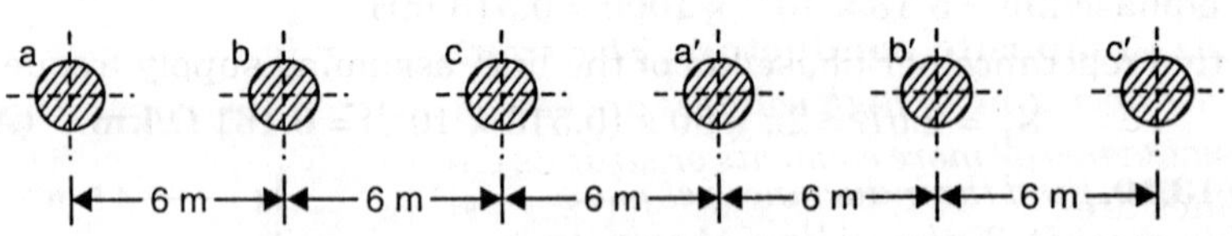

Fig. 13.18

Solution. Radius of conductor, $r = 4.5$ cm

GMR of conductor = $0.7788r = 0.7788 \times 4.5 = 3.5$ cm $= 0.035$ m

Equivalent self-GMD of one phase,

$$D_s = (D_{s1} \times D_{s2} \times D_{s3})^{1/3}$$

where, $D_{s1} = (D_{aa} \times D_{aa'} \times D_{a'a'} \times D_{a'a})^{1/4}$

$$= (0.035 \times 18 \times 0.035 \times 18)^{1/4} = 0.7937 \text{ m}$$

$$D_{s2} = (D_{bb} \times D_{bb'} \times D_{b'b'} \times D_{b'b})^{1/4}$$

$$= (0.035 \times 18 \times 0.035 \times 18)^{1/4} = 0.7937 \text{ m}$$

Similarly, $D_{s3} = (D_{cc} \times D_{cc'} \times D_{c'c'} \times D_{c'c})$

$$= (0.035 \times 18 \times 0.035 \times 18)^{1/4} = 0.7937 \text{ m}$$

$\therefore$ $D_s = (0.7937 \times 0.7937 \times 0.7937)^{1/3} = 0.7937$ m

Equivalent mutual-GMD

$$D_m = (D_{AB} \times D_{BC} \times D_{CA})^{1/3}$$

where, $D_{AB} = (D_{ab} \times D_{ab'} \times D_{a'b} \times D_{a'b'})^{1/4}$

$$= (6 \times 24 \times 12 \times 6)^{1/4} = 10.1 \text{ m}$$

$$D_{BC} = (D_{bc} \times D_{bc'} \times D_{b'c} \times D_{b'c'})^{1/4}$$

$$= (6 \times 24 \times 12 \times 6)^{1/4} = 10.1 \text{ m}$$

$$D_{CA} = (D_{ca} \times D_{ca'} \times D_{c'a} \times D_{c'a'})^{1/4}$$

$$= (12 \times 6 \times 30 \times 12)^{1/4} = 12.69 \text{ m}$$

$\therefore$ $D_m = (10.1 \times 10.1 \times 12.69)^{1/3} \simeq 10.898$ m

$$\text{Inductance/phase/m} = 2 \times 10^{-7} \times \ln\left(\frac{D_m}{D_s}\right)$$

$$= 2 \times 10^{-7} \times \ln\left(\frac{10.898}{0.7937}\right) = \mathbf{5.24 \times 10^{-7}\ H/m.} \quad \textbf{(Ans.)}$$

Example 13.12. (*a*) *What are composite conductors ?*

(*b*) *Find the inductive reactance of a three-phase bundled conductor line with 2 conductors per phase with spacing of 40 cm. Phase-to-phase separation is 7 m in horizontal configuration. All conductors are ACSR with dia. of 3.5 cm.*

Compare the above value with that of an 'equivalent' single-conductor line.

(Baroda University)

Solution. (*a*) **Composite conductors.** *The standard conductors which are normally used for transmission lines operating at high voltages are known as* "**composite conductors.** They are so called as they consist of *two or more elements or strands electrically in parallel*. The conductors used for transmission lines are :

(*i*) Stranded copper conductors.

(*ii*) Hollow copper conductors.

(*iii*) ACSR conductors.

(*iv*) Copper weld and copper weld-copper conductors.

Steel cored, reinforced aluminium conductors (ACSR) which combine the lightness, electrical conductivity and rustlessness of aluminium with high tensile strength of steel find wide application on every kind of system, low voltage distribution to the most important long distance transmission lines.

(b) Radius of each conductor, $r = \frac{3.5}{2} = 1.75$ cm

GMR of each conductor, $r' = 0.7788r = 0.7788 \times 1.75 = 1.363$ cm

$$D_{s1} = (D_{aa} \times D_{aa'} \times D_{a'a'} \times D_{a'a})^{1/4}$$
$$= (1.363 \times 40 \times 1.363 \times 40)^{1/4} = 7.38 \text{ cm} = 0.0738 \text{ m}$$

Similarly, $D_{s2} = D_{s3} = 0.1044$ m

$\therefore$ $D_s^{\,b} = (D_{s1} \times D_{s2} \times D_{s3})^{1/3} = 0.0738$ m

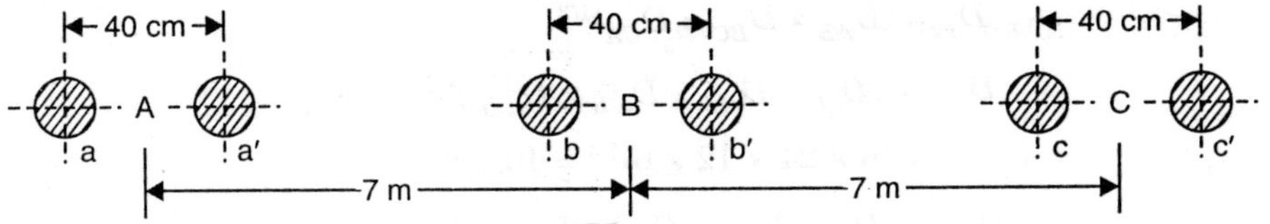

Fig. 13.19

GMD between conductors of one bundle and those of another

$$D_m^{\,b} = (D_{AB} \times D_{BC} \times D_{CA})^{1/3}$$

where ,
$$D_{AB} = (D_{ab} \times D_{ab'} \times D_{a'b} \times D_{a'b'})^{1/4}$$
$$= (7 \times 7.4 \times 6.6 \times 7)^{1/4} \simeq 7 \text{ m}$$
$$D_{AB} = D_{BC} = 7 \text{ m}$$
$$D_{CA} = (D_{ca} \times D_{ca'} \times D_{c'a'} \times D_{c'a})^{1/4}$$
$$= (14 \times 13.6 \times 14 \times 14.4)^{1/4} = 14 \text{ m}$$

$\therefore$ $D_m = (7 \times 7 \times 14)^{1/3} = 8.82$ m

The inductance of bundled conductor line,

$$L_b = 2 \times 10^{-7} \ln\left(\frac{D_m}{D_s}\right) = 2 \times 10^{-7}\left(\frac{8.82}{0.0738}\right) = 9.567 \times 10^{-7} \text{ H/m}$$

The inductive reactance of the bundled conductor line per phase,

$$X_{L_b} = 2\pi f L_b = 2\pi \times 50 \times 9.567 \times 10^{-7} = 3005 \times 10^{-7} \ \Omega/\text{m}$$
$$= 3005 \times 10^{-7} \times 1000 = 0.3005 \ \Omega/\text{km}$$

Equivalent single conductor line :

Let r_1 be the radius of conductor in the equivalent line with single conductor per phase. Then for equal areas in both systems,

$$n\pi r^2 = \pi r_1^{\,2}$$

where n is the number of sub-conductors in a bundles. Here $n = 2$

$\therefore$ $2 \times \pi r^2 = \pi r_1^{\,2}$ or $r_1 = r\sqrt{2}$

$$= 1.75 \times \sqrt{2} = 2.475 \text{ cm} = 0.02475 \text{ m}$$

GMR of each conductor, $D_{s1} = 0.7788 r_1 = 0.7788 \times 0.02475 = 0.01927$ m

GMD, $D_{m1} = (D_{AB} \times D_{BC} \times D_{CA})^{1/3} = (7 \times 7 \times 14)^{1/3} = 8.82$ m

The inductance of the line with one conductor per phase,

$$L_1 = 2 \times 10^{-7} \ln \left(\frac{D_{m1}}{D_{s1}}\right)$$

$$= 2 \times 10^{-7} \ln \left(\frac{8.82}{0.01927}\right) = 12.25 \times 10^{-7} \text{ H/m}$$

The inductive reactance of the line with one conductor per phase,

$$X_{L_1} = 2\pi f L_1 = 2\pi \times 50 \times 12.25 \times 10^{-7} \ \Omega/\text{m}$$

$$= 2\pi \times 50 \times 12.25 \times 10^{-7} \times 10^3 \ \Omega/\text{km} = 0.3848 \ \Omega/\text{km}$$

$\therefore$ *Reduction in inductive reactance due to bundling*

$$= X_{L_1} - X_{L_b} = 0.3848 - 0.3005 = \mathbf{0.0843 \ \Omega/km.} \quad \textbf{(Ans.)}$$

Example 13.13. (*a*) *What are bundled conductors ? What is the basic difference between a bundle conductor and a composite conductor ?*

(*b*) *In the Fig. 13.20 is shown the arrangement of conductors of a single-phase transmission line, wherein the forward circuit is composed of three solid wires 2.4 mm in radius and the return circuit of two-wires of radius 4.8 mm placed symmetrically with respect to the forward circuit. Determine the inductance of each side of the line and that of the complete line.*

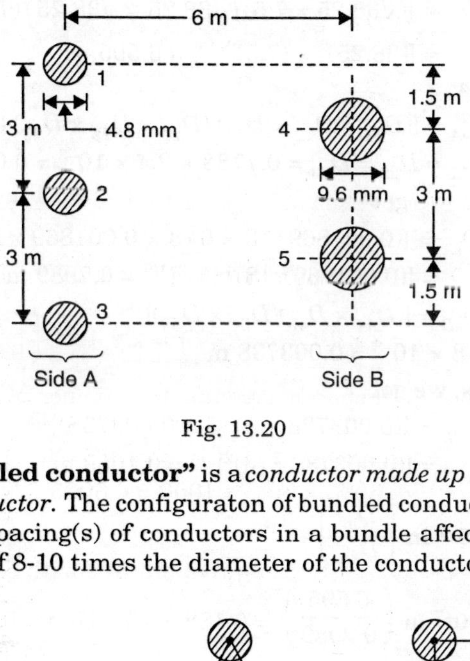

Fig. 13.20

Solution. (*a*) A **"bundled conductor"** is a *conductor made up of two or more sub-conductors and is used as one phase conductor.* The configuraton of bundled conductors is shown in Fig. 13.21. It has been observed that the spacing(s) of conductors in a bundle affects voltage gradient and optimum spacing is of the order of 8-10 times the diameter of the conductor, irrespective of the number of conductors in the bundle.

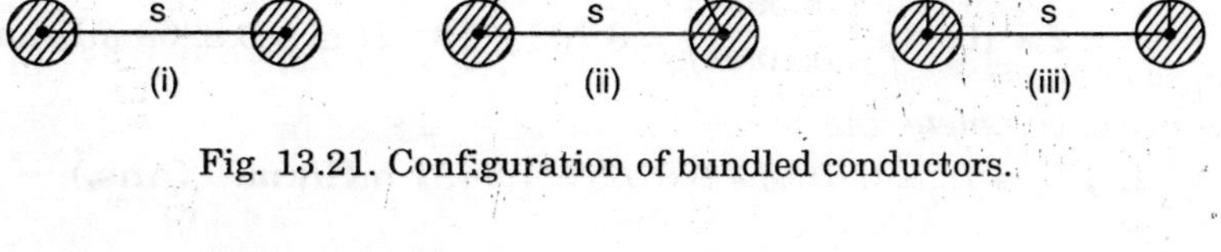

Fig. 13.21. Configuration of bundled conductors.

Further, more the number of conductors in a bundle, the more is the self-GMD. Because of increased GMD line inductance is reduced considerably which results in *increased transmission capacity of the line.*

The *use of bundle conductors entails the following "advantages :*

(*i*) Reduced surge impedance.

(*ii*) Reduced reactance.

(*iii*) Reduced voltage gradient.

(*iv*) Reduced corona loss.

(*v*) Reduced radio interference.

- The *basic difference between a "bundled conductor" and a "composite conductor"* is that the sub-conductors of a bundled conductor are separated from each other by a distance of almost 30 cm or more whereas the wires of a composite conductor touch each other.

(*b*) Refer to Fig. 13.20.

$$D_{14} = D_{24} = D_{25} = D_{35} = \sqrt{(6)^2 + (1.5)^2} = \sqrt{38.25} \text{ m}$$

$$D_{15} = D_{34} = \sqrt{(6)^2 + (4.5)^2} = 7.5 \text{ m}$$

Mutual-GMD between sides A and B,

$$D_m = [(D_{14} \times D_{15})(D_{24} \times D_{25})(D_{34} \times D_{35})]^{1/6}$$

$$= [(\sqrt{38.25} \times 7.5)(\sqrt{38.25} \times \sqrt{38.25})(7.5 \times \sqrt{38.25})]^{1/6}$$

$$= [(38.25)^2 \times (7.5)^2]^{1/6} = 6.595 \text{ m}$$

The self-GMD for side A,

$$D_{sA} = [(D_{11} \times D_{12} \times D_{13})(D_{21} \times D_{22} \times D_{23})(D_{31} \times D_{32} \times D_{33})]^{1/9}$$

Here, $D_{11} = D_{22} = D_{33} = 0.7788 \times 2.4 \times 10^{-3} = 0.001869 \text{ m}$

Substituting the values, we get

$$D_{sA} = [(0.001869 \times 3 \times 6)(3 \times 0.001869 \times 3)(6 \times 3 \times 0.001869)]^{1/9}$$

$$= [(0.001869)^3 (3)^4 (6)^2]^{1/9} = 0.2989 \text{ m}$$

Similarly, $D_{sB} = [(D_{44} \times D_{45})(D_{54} \times D_{55})]^{1/4}$

where, $D_{44} = D_{55} = 0.7788 \times 4.8 \times 10^{-3} = 0.003738 \text{ m}$

Substituting the values, we get

$$D_{sB} = [(0.003738 \times 3)(3 \times 0.003738)]^{1/4}$$

$$= [(0.003738)^2 (3)^2]^{1/4} = 0.1059 \text{ m}$$

Inductance, $\mathbf{L_A} = 2 \times 10^{-7} \ln\left(\frac{D_m}{D_{sA}}\right)$

$$= 2 \times 10^{-7} \ln\left(\frac{6.595}{0.2989}\right) = 6.188 \times 10^{-7} \text{ H/m} = 0.6188 \text{ mH/km}$$

Inductance, $\mathbf{L_B} = 2 \times 10^{-7} \ln\left(\frac{D_m}{D_{sB}}\right)$

$$= 2 \times 10^{-7} \ln\left(\frac{6.595}{0.1059}\right) = 8.263 \times 10^{-7} \text{ H/m} = 0.8263 \text{ mH/km}$$

∴ *Inductance of complete line*

$$L = L_A + L_B = 0.6188 + 0.8263 = \mathbf{1.4451 \text{ mH/km}}. \quad \textbf{(Ans.)}$$

Example 13.14. (*a*) *Discuss the importance of standard voltages and frequency in transmission and distribution.*

(*b*) *How the size of the conductor and spacing between conductors affect the line performance, give reasons?*

(*c*) *A three-phase overhead line is designed with an equilateral spacing of 3.5 m with a conductor diameter of 1.2 cm. If the line is constructed with horizontal spacing with suitably transposed conductors, find spacing between adjacent conductors which would give the same value of inductance as in the equilateral arrangement.* **(UPTU)**

Solution. (*a*) *Importance of standard voltages and frequency :*

(*i*) It provides better facilities for research and development.

(*ii*) The equipment can be manufactured economically with greater reliability.

(*iii*) The maximum possible use and interconnection can be made.

(*iv*) Use of equipments can be made interchangeably.

(*b*) For conductors of radius r and having distance of separation of D in between them, the value of inductance and capacitance between two parallel conductors is given by

$$L \propto \log_e \frac{D}{r}, \text{ and } C \propto \frac{1}{\log_e D/r}$$

- Hence with *increased value* of D *i.e.*, large value of distance of separation, *inductance increases and capacitance decreases.*
- *With large value of r i.e.*, with larger size of conductor, *inductance decreases and capacitance increases.*

Thus with changes in values of inductance and capacitance the performance of line will be accordingly affected.

(*c*) *Given :* Spacing, $D = 3.5$ m ; $r = \frac{1.2}{2} = 0.6$ cm $= 0.006$ m

For equilateral triangular arrangement, the inductance of each conductor is given by :

$$L = 10^{-7}\left[0.5 + 2\ln\left(\frac{D}{r}\right)\right] \text{ H/m} \qquad \text{...[Eqn. (13.12)]}$$

$$= 10^{-7}\left[0.5 + 2\ln\left(\frac{3.5}{0.006}\right)\right] = 13.24 \times 10^{-7} \text{ H/m}$$

If the line is constructed with horizontal spacing with suitably transposed conductors, then inductance of each conductor is given by

$$L = 10^{-7}\left[\frac{1}{2} + 2\ln\left(\frac{\sqrt[3]{D_{AB} \times D_{BC} \times D_{CA}}}{r}\right)\right] \text{ H/m} \qquad \text{...[Eqn. (13.16)]}$$

Let, distance be D metres, then

$$D_{AB} = D_{BC} = D \text{ and } D_{CA} = 2D$$

$$\therefore \qquad L = 10^{-7}\left[\frac{1}{2} + 2\ln\left(\frac{\sqrt[3]{D \times D \times 2D}}{r}\right)\right] \text{ H/m}$$

$$= 10^{-7}\left[\frac{1}{2} + 2\ln\frac{\sqrt[3]{2}\,D}{r}\right] = 10^{-7}\left[\frac{1}{2} + 2\ln\left(\frac{1.26D}{r}\right)\right] \text{ H/m}$$

For same value of inductance in both the arrangements, we have

$$10^{-7}\left[\frac{1}{2}+2\ln\left(\frac{1.26D}{r}\right)\right]=13.24\times 10^{-7}$$

or, $$\ln\left(\frac{1.26D}{r}\right)=\left(\frac{13.24-0.5}{2}\right)=6.37$$

or, $$\frac{1.26D}{r}=e^{6.37}=584$$

or, $$D=\frac{584\times r}{1.26}=\frac{584\times 0.006}{1.26}=\mathbf{2.78\ m.}\quad\textbf{(Ans.)}$$

Example 13.15. *Determine the inductance per phase per km of a double-circuit 3-phase line. The radius of each conductor is 2 cm and the conductor are placed on the circumference of an imaginary circle of radius 7 m forming a regular hexagon figure.*

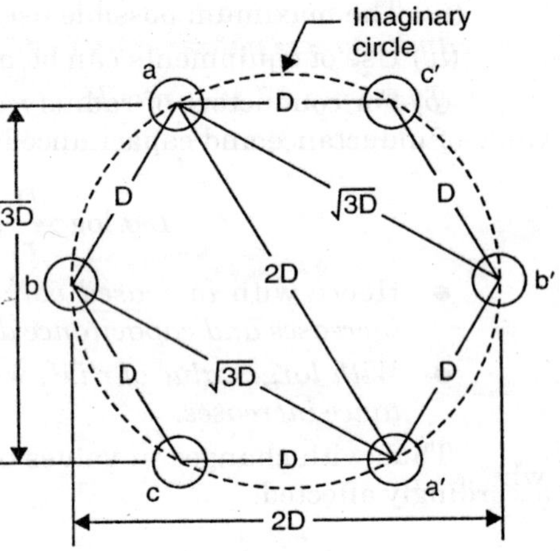

Fig. 13.22

Solution. Radius of each conductor, r = 2 cm

Self-GMD of conductor, $r = 0.7788 \times r = 0.7788 \times 2$
= 1.5576 cm = 0.015576 m

Radius of imaginary circle = D = 7 m

Self-GMD of combination aa',

$$D_{s1}=(D_{aa}\times D_{aa'}\times D_{a'a'}\times D_{a'a})^{1/4}$$
$$=(r'\times 2D\times r'\times 2D)^{1/4}=(2r'D)^{1/2}$$

Self-GMD of combination bb',

$$D_{s2}=(D_{bb}\times D_{bb'}\times D_{b'b'}\times D_{b'b})^{1/4}$$
$$=(r'\times 2D\times r'\times 2D)^{1/4}=(2r'D)^{1/2}$$

Self-GMD of combination cc',

$$D_{s3}=(D_{cc}\times D_{cc'}\times D_{c'c'}\times D_{c'c})^{1/4}$$
$$=(r'\times 2D\times r'\times 2D)^{1/4}=(2r'D)^{1/2}$$

Equivalent self-GMD of one phase,

$$D_s=(D_{s1}\times D_{s2}\times D_{s3})^{1/3}$$
$$=[(2r'D)^{1/2}\times(2r'D)^{1/2}\times(2r'D)^{1/2}]^{1/3}=(2r'D)^{1/2}$$

Since each conductor has the same radius, therefore the value of D_s is same for all the phases.

Mutual-GMD between phases A and B,

$$D_{AB}=(D_{ab}\times D_{ab'}\times D_{a'b}\times D_{a'b'})^{1/4}$$
$$=(D\times\sqrt{3}D\times\sqrt{3}D\times D)^{1/4}=D\times(3)^{1/4}$$

Similarly, $$D_{BC}=(D_{bc}\times D_{bc'}\times D_{b'c}\times D_{b'c'})$$
$$=(D\times\sqrt{3}D\times\sqrt{3}D\times D)^{1/4}=D\times(3)^{1/4}$$

and, $$D_{CA}=(D_{ca}\times D_{ca'}\times D_{c'a}\times D_{c'a'})$$
$$=(\sqrt{3}D\times D\times D\times\sqrt{3}D)^{1/4}=D\times(3)^{1/4}$$

Equivalent mutual-GMD,

$$D_m = (D_{AB} \times D_{BC} \times D_{CA})^{1/3}$$
$$= [\{D \times (3)^{1/4}\} \times \{D \times (3)^{1/4}\} \times \{D \times (3)^{1/4}\}]^{1/3} = D \times (3)^{1/4}$$

Inductance/phase/m $= 2 \times 10^{-7} \ln \left(\dfrac{D_m}{D_s}\right)$

$$= 2 \times 10^{-7} \ln \left[\frac{D \times (3)^{1/4}}{(2r'D)^{1/2}}\right] \text{ H/m}$$

$$= 2 \times 10^{-7} \ln \left[\frac{7 \times (3)^{1/4}}{(2 \times 0.015576 \times 7)^{1/2}}\right] = 5.964 \times 10^{-7} \text{ H/m}$$

Inductance / phase / km = 5.964 × 10^{-7} × 1000 H/m = **0.5964 mH/km. (Ans.)**

Example 13.16. *Calculate the GMR of a 6 / 3 mm Al, 1 / 3 mm steel ACSR conductor.*

Solution. Refer to Fig. 13.23.

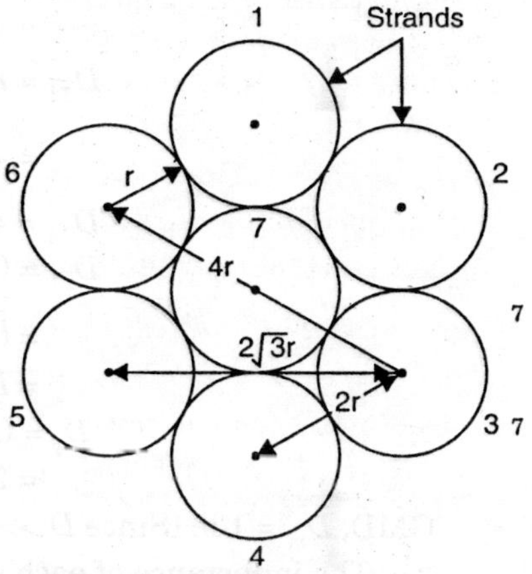

Fig. 13.23. Stranded conductor.

$$D_{11} \text{ (GMR)} = r' = 0.7788r = D_{77}$$
$$D_{12} = D_{16} = D_{17} = 2r$$
$$D_{13} = D_{15} = D_{14}{}^2 - D_{34}{}^2$$
$$= \sqrt{(4r)^2 - (2r)^2} = 2\sqrt{3}r$$
$$D_{14} = 4r$$

Self-GMD $= (D_{s1} \times D_{s2} \times D_{s3} \times D_{s4} \times D_{s5} \times D_{s6} \times D_{s7})^{1/}$

where, $D_{s1} = (D_{11} \times D_{12} \times D_{13} \times D_{14} \times D_{15} \times D_{16} \times D_{17})^{1/7}$

$$= (0.7788r \times 2r \times 2\sqrt{3}\,r \times 4r \times 2\sqrt{3}\,r \times 2r \times 2r)^{1/}$$
$$= (299.06)^{1/7}r$$
$$= D_{s2} = D_{s3} = D_{s4} = D_{s5} = D_{s6}$$
$$D_{s7} = (D_{77} \times D_{71} \times D_{72} \times D_{73} \times D_{74} \times D_{75} \times D_{76})^{1/7}$$
$$= (0.7788r \times 2r \times 2r \times 2r \times 2r \times 2r \times 2r)^{1/7}$$
$$= (49.84)^{1/7}r$$
$$D_s = (D_{s1} \times D_{s2} \times D_{s3} \times D_{s4} \times D_{s5} \times D_{s6} \times D_{s7})^{1/7}$$
$$= [(299.06)^6 \times 49.84]^{1/49}\, r = 2.177 \times r = 2.177 \times 3$$

= **6.531 mm. (Ans.)**

Example 13.17. *In the Fig. 13.24 is shown an ACSR conductor in which the outside diameter of the single larger of aluminium strand is 50.4 mm. The radius of each strand is 8.4 mm. Determine the 50 Hz reactance at 1 m spacing, neglect the effect of the central strand of steel and state reasons for the same.*

Solution. Refer to Fig. 13.24.

Radius of each strand, r = 8.4 mm = 0.0084 m

GMR, $r' = 0.7788r$

Diameter of steel strand $= 50.4 - 4 \times r$

$$= 50.4 - 4 \times 8.4 = 16.8 \text{ mm} = 0.0168 \text{ m}$$

The effect of central strand of steel can be neglected since the *conductivity of steel is much smaller than that of copper and the internal inductance of steel strands is μ-times that of aluminium strands ; as such the current carried by the steel strand can be assumed to be nil.*

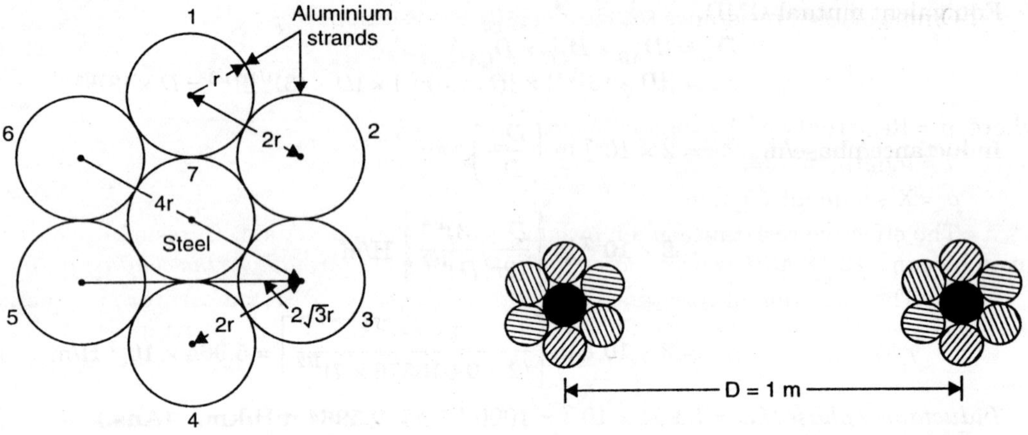

(a) Cross-section of ACSR conductor. (b) Line consisting of two ACSR conductors.

Fig. 13.24

$$D_{11} = r' = 0.7788r \;;\; D_{12} = D_{16} = 2r \;;\; D_{13} = D_{15}$$

$$= \sqrt{{D_{14}}^2 - {D_{34}}^2} = \sqrt{(4r)^2 - (2r)^2} = 2\sqrt{3}r$$

$$D_{14} = 4r$$

$$D_{s1} = (D_{11} \times D_{12} \times D_{13} \times D_{14} \times D_{15} \times D_{16})^{1/6}$$

$$= (0.7788r \times 2r \times 2\sqrt{3}r \times 4r \times 2\sqrt{3}r \times 2r)^{1/6} = 2.304r$$

$$= D_{s2} = D_{s3} = D_{s4} = D_{s5} = D_{s6}$$

$$\therefore \quad D_s = (D_{s1} \times D_{s2} \times D_{s3} \times D_{s4} + D_{s5} \times D_{s6})^{1/6} = [(2.304r)^6]^{1/6}$$

$$= 2.304r = 2.304 \times 0.0084 = 0.01935 \text{ m}$$

GMD, $D_m = 1$ m (Since $D >> r$)

$\therefore$ The inductance of each conductor,

$$L = 2 \times 10^{-7} \ln \left(\frac{D_m}{D_s}\right)$$

$$= 2 \times 10^{-7} \ln \left(\frac{1}{0.01935}\right) = 7.89 \times 10^{-7} \text{ H/m} \quad \text{or} \quad \mathbf{0.789\ mH/km}$$

Loop inductance $= 2L = 2 \times 0.789 = 1.578$ mH/km

or, Loop reactance $= 2\pi \times 50 \times 1.578 \times 10^{-3} =$ **0.495 Ω/km. (Ans.)**

13.3. RESISTANCE OF A TRANSMISSION LINE

The *resistance* of transmission line conductors is the *main source of line power loss* (although the contribution of line resistance of series line impedance can be neglected in most cases). Thus, while considering the transmission line economy, the presence of line resistance must be taken into consideration.

$$\text{The effective AC resistance, } R = \frac{\text{Average power loss in conductor (in watts)}}{I^2} \text{ ohms} \quad \text{...(13.17)}$$

where, I = R.m.s. current in the conductor, amperes

Ohmic or D.C. resistance (R_0) is given by the relation :

$$R_0 = \frac{\rho l}{a} \text{ ohms} \qquad ...(13.18)$$

where, ρ = Resistivity of the conductor, ohm-m

l = length, m, and

a = X-sectional area, m^2.

The effective resistance given by eqn. (13.17) is equal to the D.C. resistance of the conductor given by eqn. (13.18) *if the current distribution is uniform throughout the conductor.*

- The variation of resistance of metallic conductors with temperature is practically linear over the normal range of operation. Let R_1 and R_2 are the resistances of a conductor at t_1 °C and t_2 °C ($t_2 > t_1$) respectively and α_1 is the temperature coefficient at t_1 °C, then

$$R_2 = R_1 [1 + \alpha_1(t_2 - t_1)]$$

where, $\alpha_1 = \frac{\alpha_0}{1 + \alpha_0 t}$, α_0 being temperature coefficient at 0°C.

— In a 1-phase or 2-wire D.C. line, the total resistance (known as *loop resistance*) is equal to double the resistance of either conductor.

— In case of 3-phase transmission line, resistance per phase is the resistance of one conductor.

13.4. SKIN EFFECT AND PROXIMITY EFFECT

Skin Effect :

Direct currents distribute themselves uniformly over the cross-section of the conductor and therefore use the centre of conductor just as effectively as they use the periphery. *Alternating currents,* however, *owing to inductance effects within the conductor, crowd toward the outside of the conductor. This behaviour is known as* **'skin effect'**. It *raises the apparent resistance of the conductor,* because power loss varies as the current and the higher density of current flowing near the skin of the conductor produces more power loss than the lower density of current flowing through the centre of the conductor.

The *resistivity of the conducting material is unchanged by skin effect* ; only the total resistance of the conductor increases as the frequency of the current and the diameter of the conductor increase. The increase in resistance should not be neglected in very large conductors at commercial power frequencies and in most conductors at carrier and radio frequencies. *Since skin effect reduces the amount of current carried by the centre portion of the conductor, hollow conductors are sometimes employed to use them more effectively.* Aluminium cable steel inforced (ACSR) is used even more than hollow conductors. *This cable has steel core which strengthens the cable, and an outer covering of aluminium which, because of skin effect and high resistivity of the steel, carries most of the current.* These conductors also have the advantage of a *large surface area available for cooling.*

The skin effect is negligible when the supply frequency is low (< 50 Hz) and conductor diameter is small (< 1 cm).

Proximity Effect :

The *alternating magnetic flux in a conductor caused by the current flowing in a neighbouring conductor gives rise to circulating currents which cause an apparent increase in the resistance of a conductor. This phenomenon is called* **'proximity effect'**. Like skin effect, proximity effect also *results in non-uniformity of current in a conductor cross-section.*

For normal spacing of overhead lines, this effect is always of negligible order. However, for *underground cables* where conductors are located close to each other, proximity effect causes an *appreciable increase in effective conductor resistance.*

- Both skin and proximity effects depend upon the following *factors* :

(*i*) Size of the conductor.

(*ii*) Distance between the conductors.

(*iii*) Frequency of the current.

(*iv*) Permeability of the conductor material.

For a circular conductor of diameter d the increase of effective resistance is proportional to $\frac{d^2 f \mu_r}{\rho}$. Thus for small conductors and at power frequencies the effect is negligible but these effects may be quite important for *large conductors and high frequencies.*

13.5. CAPACITANCE OF TRANSMISSION LINES

13.5.1. Concept of Capacitance

Any two conductors separated by an insulating medium constitute a *condenser* or a *capacitor*. Since the two conductors of an overhead transmission line are separated by air, which acts as an insulation, there exists capacitance between them. This capacitance is uniformly distributed over the total length of the line and may be regarded as a uniform series of capacitors connected between the conductors, as shown in Fig. 13.25.

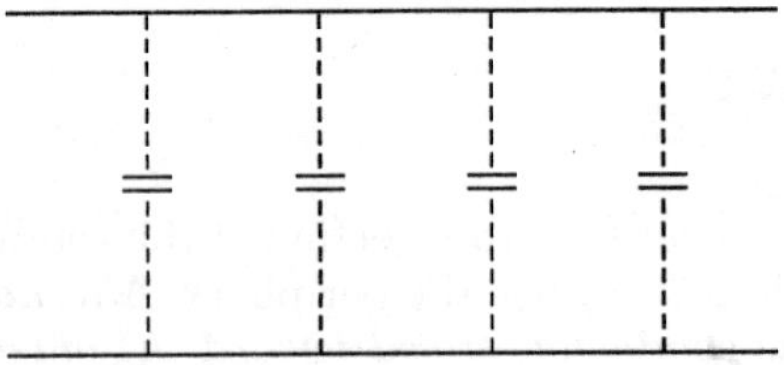

Fig. 13.25

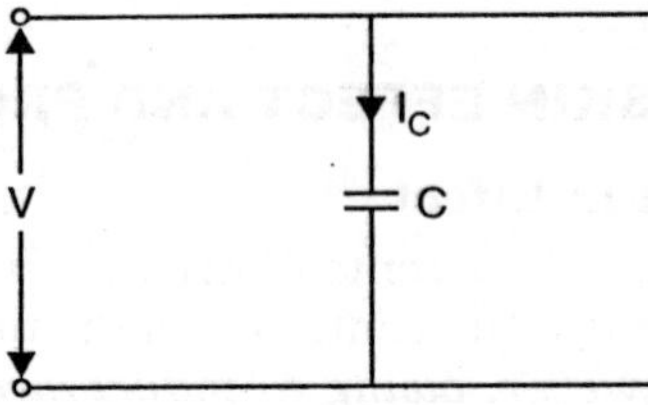

Fig. 13.26

When an alternating potential difference is impressed upon a transmission line (see Fig. 13.26), the line capacitance draws a leading sinusoidal current called the *charging current* (I_c) which is drawn even when the line is open circuited at the far end. The strength of charging current (I_c) depends upon *frequency* (f) of the A.C. supply, the *capacitance* (C) of the line and the *voltage* (V) of transmission and is given by :

$$I_c = 2\pi f C V \qquad ...(13.19)$$

- The line capacitance being proportional to its length, the charging current is negligible for lines less than 100 km long. For longer lines the capacitance becomes increasingly important and should be taken into consideration.
- When the capacitance of an overhead transmission line is high, the line draws more charging current, which compensates or cancels the lagging (reactive) component of the load current (normally the load is inductive one). This *results in* :

(*i*) Reduction of losses and so *increase of transmission efficiency.*

(*ii*) *Improved voltage regulation* (owing to reduction in voltage drop).

(*iii*) *Improved power factor.*

(*iv*) *Increased load capacity.*

13.5.2. Electric Potential

"Electric potential or potential" at *any point in an electric field* is *generally defined* as *the work done in moving a unit positive charge from infinity to that point.*

The concept of electric potential is quite significant for the determination of capacitance (defined as charge per unit potential *i.e.*, $C = Q/V$).

We shall consider the following *two cases* :

(*i*) Potential at a charged single conductor.

(*ii*) Potential at a charged conductor in a group of charged conductors.

13.5.2.1. Potential at a charged single conductor

Consider a long straight cylindrical conductor A of radius r metres and having a charge Q_A, coulomb per metre of its length. The expression for V_A is derived as follows :

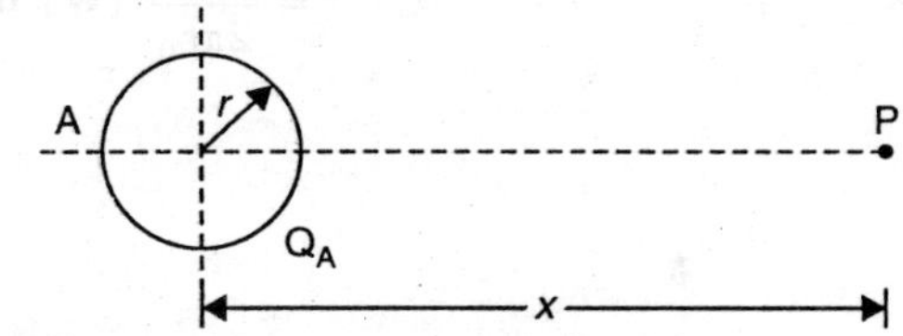

Fig. 13.27

The electric field intensity at a distance x from the centre of the conductor,

$$E = \frac{Q_A}{2\pi x \varepsilon_0 \varepsilon_r} \text{ volts/m}$$

or,
$$E = \frac{Q_A}{2\pi \varepsilon_0} \text{ volts/m, taking air as medium, } \varepsilon_r = 1$$

where, ε_0 = Permittivity of free space.

As x approaches infinity, the value of E approaches zero. Therefore, the potential difference between conductor A and infinity distant neutral plane (a plane where E and therefore potential is zero) will be equal to work done in bringing a unit positive charge against E from infinity to conductor surface is given as :

$$V_A = \int_r^\infty \frac{Q_A}{2\pi\varepsilon_0 x}\, dx = \frac{Q}{2\pi\varepsilon_0}\int_r^\infty \frac{dx}{x} \quad ...(13.20)$$

13.5.2.2. Potential at a charged conductor in a group of charged conductors

Consider a group of long straight conductors $A, B, C, D, E, \ldots\ldots N$ having charges $Q_A, Q_B, Q_C, Q_D, Q_E, \ldots\ldots Q_N$ per metre length respectively as shown in Fig. 13.28.

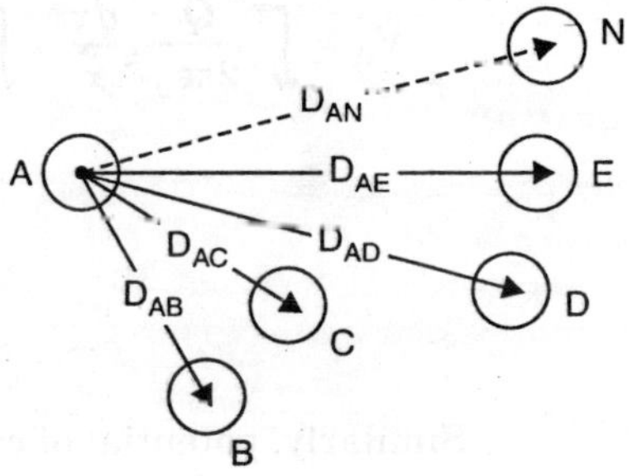

Fig. 13.28

Potential of conductor A due to its own charge

$$= \int_r^\infty \frac{Q_A}{2\pi\varepsilon_0} \cdot \frac{dx}{x}$$

Potential of conductor A due to charge Q_B

$$= \int_{D_{AB}}^\infty \frac{Q_B}{2\pi\varepsilon_0} \cdot \frac{dx}{x}$$

since the field due to charge Q_B extends from infinity upto distance D_{AB} from the conductor A.

Potential of conductor A due to charge Q_C

$$= \int_{D_{AC}}^\infty \frac{Q_C}{2\pi\varepsilon_0} \cdot \frac{dx}{x}$$

Potential of conductor A due to charge Q_N

$$= \int_{Q_{AN}}^\infty \frac{Q_N}{2\pi\varepsilon_0} \cdot \frac{dx}{x}$$

∴ Overall potential difference between conductor A and infinite neutral plane is,

$$V_A = \int_r^\infty \frac{Q_A}{2\pi\varepsilon_0} \cdot \frac{dx}{x} + \int_{D_{AB}}^\infty \frac{Q_B}{2\pi\varepsilon_0} \cdot \frac{dx}{x} + \int_{D_{AC}}^\infty \frac{Q_{AC}}{2\pi\varepsilon_0} \cdot \frac{dx}{x} + \ldots\ldots \int_{D_{AN}}^\infty \frac{Q_N}{2\pi\varepsilon_0} \cdot \frac{dx}{x}$$

$$= \frac{1}{2\pi\varepsilon_0} \; [Q_A(\ln \infty - \ln r) + Q_B (\ln \infty - \ln D_{AB}) + Q_C (\ln \infty - \ln D_{AC})$$

$$+ \ldots\ldots Q_N (\ln \infty - \ln D_{AN})]$$

$$= \frac{1}{2\pi\varepsilon_0} \left[Q_A \ln\left(\frac{1}{r}\right) + Q_B \ln\left(\frac{1}{D_{AB}}\right) + Q_C \ln\left(\frac{1}{D_{AC}}\right) + \ldots\ldots \right.$$

$$\left. \ldots\ldots Q_N \ln\left(\frac{1}{D_{AN}}\right) + \ln \infty \, (Q_A + Q_B + Q_C + \ldots\ldots Q_N) \right]$$

$$\left[\because \; -\ln (r) = \ln (r)^{-1} = \ln\left(\frac{1}{r}\right) \right]$$

Assuming balanced conditions *i.e.*, $Q_A + Q_B + Q_C + \ldots\ldots Q_N = 0$, we have

$$V_A = \frac{1}{2\pi\varepsilon_0} \left[Q_A \ln\left(\frac{1}{r}\right) + Q_B \ln\left(\frac{1}{D_{AB}}\right) + Q_C \ln\left(\frac{1}{D_{AC}}\right) + \ldots\ldots Q_N \ln\left(\frac{1}{D_{AN}}\right) \right] \qquad \ldots(13.21)$$

13.5.3. Capacitance of a Single-phase Two-wire Overhead Transmission Line

Consider a single-phase overhead transmission line with two parallel conductors A and B, each of radius r metres placed at a distance D metres in air ($D >> r$). Let Q_1 and Q_2 coulombs per metre length be their respective charges.

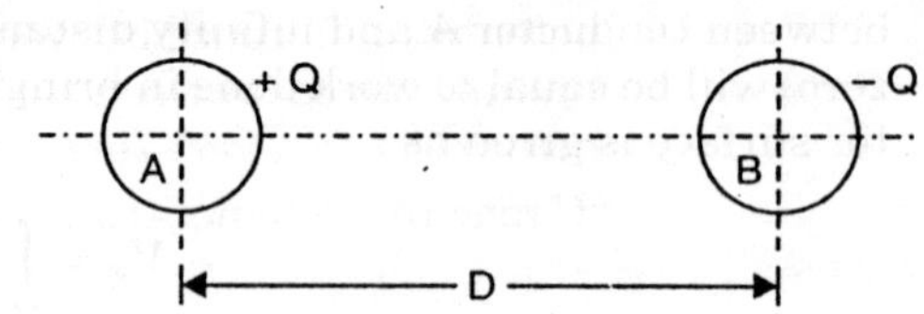

Fig. 13.29

Potential of conductor A,

$$V_A = \int_r^{\infty} \frac{Q}{2\pi\varepsilon_0} \cdot \frac{dx}{x} + \int_D^{\infty} \frac{(-Q)}{2\pi\varepsilon_0} \cdot \frac{dx}{x}$$

$$= \frac{Q}{2\pi\varepsilon_0} \; [(\ln \infty - \ln r) - (\ln \infty - \ln D)]$$

$$= \frac{Q}{2\pi\varepsilon_0} \left[\ln\left(\frac{\infty}{r}\right) - \ln\left(\frac{\infty}{D}\right) \right] = \frac{Q}{2\pi\varepsilon_0} \ln\left(\frac{D}{r}\right) \text{ volts}$$

Similarly, potential of conductors B,

$$V_B = \int_r^{\infty} \frac{-Q}{2\pi\varepsilon_0} \cdot \frac{dx}{x} + \int_D^{\infty} \frac{Q}{2\pi\varepsilon_0} \cdot \frac{dx}{x}$$

$$= \frac{-Q}{2\pi\varepsilon_0} \left[\ln\left(\frac{\infty}{r}\right) - \ln\left(\frac{D}{r}\right) \right] = \frac{-Q}{2\pi\varepsilon_0} \ln \frac{D}{r} \text{ volts}$$

Potential difference between conductor A and B,

$$V_{AB} = V_A - V_B = \frac{Q}{2\pi\varepsilon_0} \ln\left(\frac{D}{r}\right) - \left[\left(\frac{-Q}{2\pi\varepsilon_0}\right) \ln\left(\frac{D}{r}\right) \right]$$

or,

$$V_{AB} = \frac{Q}{\pi\varepsilon_0} \ln\left(\frac{D}{r}\right) \text{ volts} \qquad \ldots(13.22)$$

Capacitance, $$C_{AB} = \frac{Q}{V_{AB}} = \frac{Q}{\frac{Q}{\pi\varepsilon_0} \ln\left(\frac{D}{r}\right)} \text{ F/m}$$

$\therefore$ $$C_{AB} = \frac{\pi\varepsilon_0}{\ln\left(\frac{D}{r}\right)} \text{ F/m} \quad ...(13.23)$$

Capacitance to neutral :

- Fig. 13.30 shows the capacitance between the conductors of a two-wire line.

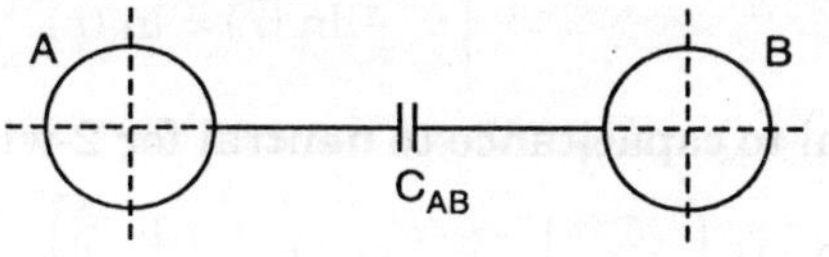

Fig. 13.30. Line-to-line capacitance.

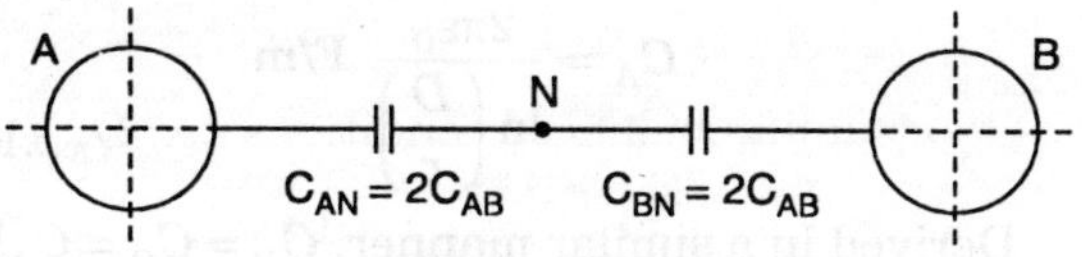

Fig. 13.31. Line-to-neutral capacitance.

- Refer to Fig. 13.31. Since potential of the mid-point between the conductors is zero, the potential difference between each conductor and the ground or neutral is half the potential difference between the conductors. Thus the *capacitance to ground* or *capacitance to neutral for two-wire line is twice the line-to-line capacitance.*

$\therefore$ Capacitance to neutral, $C_N = C_{AN} = C_{BN} = 2C_{AB}$

$\therefore$ $$C_N = \frac{2\pi\varepsilon_0}{\ln\left(\frac{D}{r}\right)} \text{ F/m} \quad ...(13.24)$$

— This equation *applies only to a solid round conductor.*

— One difference between the equation of capacitance and inductance is that the *radius in the equation for capacitance is the actual radius of the conductor and not GMR of the conductor as in the inductance.*

13.5.4. Capacitance of a 3-phase Overhead Transmission Line

13.5.4.1. Symmetrical spacing

Fig. 13.32 shows a 3-phase line with conductors *A*, *B* and *C*, each of radius *r* metres, and spaced at the equilateral triangle, each side being *D*. Let Q_A, Q_B and Q_C coulombs per metre length be their respective charges.

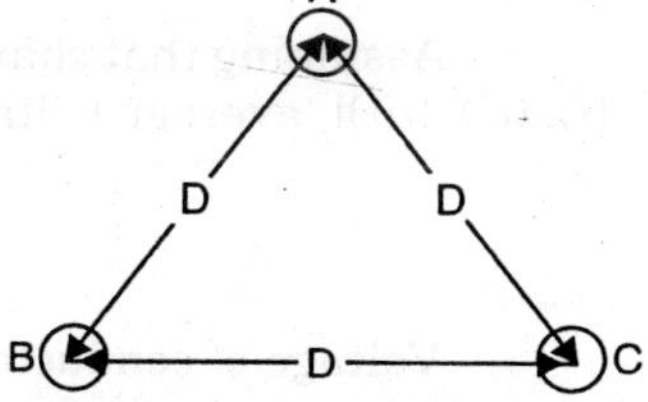

Fig. 13.32. Equilateral spacing of a 3-phase line.

Potential of conductor *A*,

$$V_A = \int_r^{\infty} \frac{Q_A}{2\pi\varepsilon_0} \cdot \frac{dx}{x} + \int_D^{\infty} \frac{Q_B}{2\pi\varepsilon_0} \cdot \frac{dx}{x} + \int_D^{\infty} \frac{Q_C}{2\pi\varepsilon_0} \cdot \frac{dx}{x}$$

$$= \frac{1}{2\pi\varepsilon_0} [Q_A \ln(\infty) - Q_A \ln(r) + Q_B \ln(\infty) - \ln(D) + Q_C \ln(\infty) - \ln(D)]$$

$$= \frac{1}{2\pi\varepsilon_0} \left[Q_A \ln\left(\frac{1}{r}\right) + Q_B \ln\left(\frac{1}{D}\right) + Q_C \ln\left(\frac{1}{D}\right) + (Q_A + Q_B + Q_C) \ln(\infty)\right]$$

Assuming balanced supply, we have, $Q_A + Q_B + Q_C = 0$

$$\therefore \quad Q_B + Q_C = - Q_A$$

$$\therefore \quad V_A = \frac{1}{2\pi\varepsilon_0}\left[Q_A \ln\left(\frac{1}{r}\right) - Q_A \ln\left(\frac{1}{D}\right)\right] = \frac{Q_A}{2\pi\varepsilon_0} \ln\left(\frac{D}{r}\right) \text{ volts}$$

$\therefore$ Capacitance of conductor w.r.t. neut 'al,

$$C_A = \frac{Q_A}{V_A} = \frac{Q_A}{\frac{Q_A}{2\pi\varepsilon_0} \ln\left(\frac{D}{r}\right)} \text{ F/m}$$

or

$$C_A = \frac{2\pi\varepsilon_0}{\ln\left(\frac{D}{r}\right)} \text{ F/m} \quad ...(13.25)$$

...Identical to capacitance to neutral for 2-wire line.

Derived in a similar manner, $C_B = C_C = C_A$.

13.5.4.2. Unsymmetrical spacing with transposed conductors

This type of line presents difficulty in the treatment on account of the fact that *capacitances* from each phase of neutral are *unequal. Transposition results in the same average capacitance to neutral to various phases over the entire length of the line.*

The arrangement is shown in Fig. 13.33.

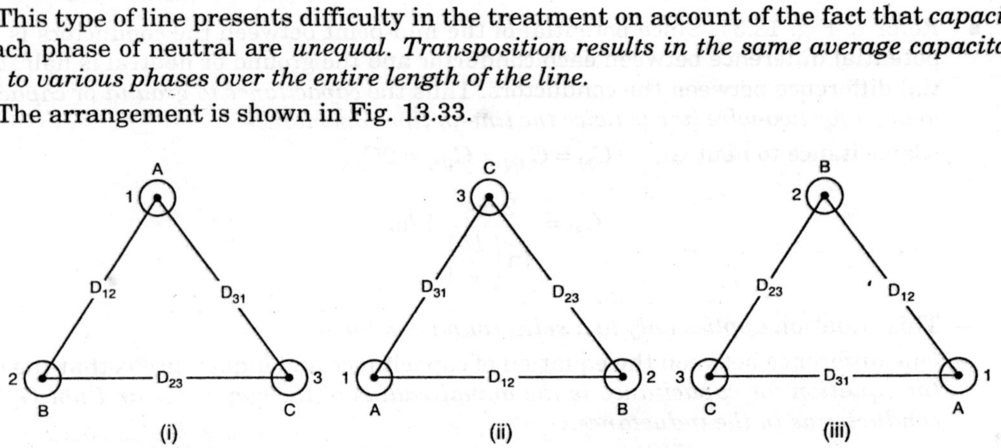

Fig. 13.33. Unsymmetrical but transposed spacing of three-phase conductors.

Assuming that charge per unit length is same in every part of the transposed cycle [Fig. 13.33 (*i*), (*ii*), (*iii*)], average voltage of conductor *A* will be

$$\frac{V_{A1} + V_{A2} + V_{A3}}{3}$$

Voltage of conductor *A* in position (*i*),

$$V_{A1} = \frac{1}{2\pi\varepsilon_0}\left[Q_{A1} \ln\left(\frac{1}{r}\right) + Q_{A2} \ln\left(\frac{1}{D_{12}}\right) + Q_{A3} \ln\left(\frac{1}{D_{31}}\right)\right]$$

$$V_{A2} = \frac{1}{2\pi\varepsilon_0}\left[Q_{A1} \ln\left(\frac{1}{r}\right) + Q_{A2} \ln\left(\frac{1}{D_{23}}\right) + Q_{A3} \ln\left(\frac{1}{D_{12}}\right)\right]$$

$$V_{A3} = \frac{1}{2\pi\varepsilon_0}\left[Q_{A1} \ln\left(\frac{1}{r}\right) + Q_{A2} \ln\left(\frac{1}{D_{31}}\right) + Q_{A3} \ln\left(\frac{1}{D_{23}}\right)\right]$$

$$\therefore \quad V_A = \frac{V_{A1} + V_{A2} + V_{A3}}{3}$$

$$= \frac{1}{6\pi\varepsilon_0}\left[Q_{A1} \ln\left(\frac{1}{r^3}\right) + Q_{A2} \ln\left(\frac{1}{D_{12}\, D_{23}\, D_{31}}\right) + Q_{A3} \ln\left(\frac{1}{D_{12}\, D_{23}\, D_{31}}\right)\right]$$

$$= \frac{1}{6\pi\varepsilon_0}\left[Q_{A1} \ln\left(\frac{1}{r^3}\right) + (Q_2 + Q_3) \ln\left(\frac{1}{D_{12}\, D_{23}\, D_{31}}\right)\right]$$

Substituting $Q_{A2} + Q_{A3} = -\, Q_{A1}$ in the above expression, we get

$$V_A = \frac{1}{6\pi\varepsilon_0}\left[Q_{A1} \ln\left(\frac{1}{r^3}\right) - Q_{A1} \ln\left(\frac{1}{D_{12}\, D_{23}\, D_{31}}\right)\right]$$

$$= \frac{Q_{A1}}{6\pi\varepsilon_0}\left[\ln\left(\frac{D_{12}\, D_{23}\, D_{31}}{r^3}\right)\right] = \frac{1}{3} \times \frac{Q_{A1}}{2\pi\varepsilon_0}\left[\ln\left(\frac{D_{12}D_{23}D_{31}}{r^3}\right)\right]$$

$$= \frac{Q_{A1}}{2\pi\varepsilon_0}\left[\ln\left(\frac{D_{12}D_{23}D_{31}}{r^3}\right)\right]^{1/3}$$

or, $$V_A = \frac{Q_{A1}}{2\pi\varepsilon_0} \ln\left[\frac{(D_{12}D_{23}D_{31})^{1/3}}{r}\right] \quad \text{...(13.26)}$$

$\therefore$ Capacitance from conductor to neutral,

$$C_A = \frac{Q_{A1}}{V_A} = \frac{Q_{A1}}{\dfrac{Q_{A1}}{2\pi\varepsilon_0} \ln\left[\dfrac{(D_{12}D_{23}D_{31})^{1/3}}{r}\right]}$$

or, $$C_A = \frac{2\pi\varepsilon_0}{\ln\left[\dfrac{(D_{12}D_{23}D_{31})^{1/3}}{r}\right]} \text{ F/m} \quad \text{...(13.27)}$$

13.5.4.3. Unsymmetrically spaced line

Let the charges on conductors A, B and C be Q_A, Q_B and Q_C coulombs per metre respectively.

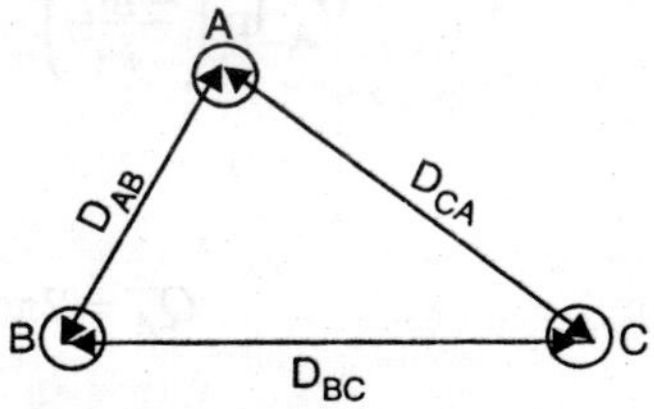

Fig. 13.34. Unsymmetrically spaced line.

Potential of conductor A,

$$V_A = \frac{Q_A}{2\pi\varepsilon_0}\int_r^{\infty}\frac{dx}{x} + \frac{Q_B}{2\pi\varepsilon_0}\int_{D_{AB}}^{\infty}\frac{dx}{x} + \frac{Q_C}{2\pi\varepsilon_0}\int_{D_{CA}}^{\infty}\frac{dx}{x}$$

$$= \frac{1}{2\pi\varepsilon_0}\,[\{Q_A \ln(\infty) - Q_A \ln(r)\} + \{Q_B \ln(\infty) - Q_B \ln(D_{AB})\} + \{Q_C \ln(\infty) - Q_C \ln(D_{CA})\}]$$

$$= \frac{1}{2\pi\varepsilon_0}\left[Q_A \ln\left(\frac{1}{r}\right) + Q_B \ln\left(\frac{1}{D_{AB}}\right) + Q_C \ln\left(\frac{1}{D_{CA}}\right) + (Q_A + Q_B + Q_C)\ln(\infty)\right]$$

or, $$V_A = \frac{1}{2\pi\varepsilon_0}\left[Q_A \ln\left(\frac{1}{r}\right) + Q_B \ln\left(\frac{1}{D_{AB}}\right) + Q_C \ln\left(\frac{1}{D_{CA}}\right)\right] \quad \text{...}(i)\ (\because\ Q_A + Q_B + Q_C = 0)$$

Similarly,
$$V_B = \frac{1}{2\pi\varepsilon_0}\left[Q_B \ln\left(\frac{1}{r}\right) + Q_C \ln\left(\frac{1}{D_{BC}}\right) + Q_A \ln\left(\frac{1}{D_{AB}}\right)\right] \quad ...(ii)$$

and,
$$V_C = \frac{1}{2\pi\varepsilon_0}\left[Q_C \ln\left(\frac{1}{r}\right) + Q_A \ln\left(\frac{1}{D_{CA}}\right) + Q_B \ln\left(\frac{1}{D_{BC}}\right)\right] \quad ...(iii)$$

Substituting $Q_C = -(Q_A + Q_B)$ in (*i*) and (*ii*), we get
$$V_A = \frac{1}{2\pi\varepsilon_0}\left[Q_A \ln\left(\frac{1}{r}\right) + Q_B \ln\left(\frac{1}{D_{AB}}\right) - (Q_A + Q_B)\ln\left(\frac{1}{D_{CA}}\right)\right]$$
$$= \frac{1}{2\pi\varepsilon_0}\left[Q_A \ln\left(\frac{D_{CA}}{r}\right) + Q_B \ln\left(\frac{D_{CA}}{D_{AB}}\right)\right] \quad ...(iv)$$

and,
$$V_B = \frac{1}{2\pi\varepsilon_0}\left[Q_B \ln\left(\frac{1}{r}\right) - (Q_A + Q_B)\ln\left(\frac{1}{D_{BC}}\right) + Q_A \ln\left(\frac{1}{D_{AB}}\right)\right]$$
$$= \frac{1}{2\pi\varepsilon_0}\left[Q_A \ln\left(\frac{D_{BC}}{D_{AB}}\right) + Q_B \ln\left(\frac{D_{BC}}{r}\right)\right] \quad ...(v)$$

Multiplying (*iv*) by $\ln\left(\frac{D_{BC}}{r}\right)$ and (*v*) by $\ln\left(\frac{D_{CA}}{D_{AB}}\right)$, we get
$$V_A \ln\left(\frac{D_{BC}}{r}\right) = \frac{1}{2\pi\varepsilon_0}\left[Q_A \ln\left(\frac{D_{CA}}{r}\right).\ln\left(\frac{D_{BC}}{r}\right) + Q_B \ln\left(\frac{D_{CA}}{D_{AB}}\right).\ln\left(\frac{D_{BC}}{r}\right)\right] \quad ...(vi)$$

and,
$$V_B \ln\left(\frac{D_{CA}}{D_{AB}}\right) = \frac{1}{2\pi\varepsilon_0}\left[Q_A \ln\left(\frac{D_{BC}}{D_{AB}}\right)\ln\left(\frac{D_{CA}}{D_{AB}}\right) + Q_B \ln\left(\frac{D_{BC}}{r}\right).\ln\left(\frac{D_{CA}}{D_{AB}}\right)\right] \quad ...(vii)$$

Subtracting (*vii*) from (*vi*), we get
$$V_A \ln\left(\frac{D_{BC}}{r}\right) - V_B \ln\left(\frac{D_{BC}}{D_{AB}}\right) = \frac{Q_A}{2\pi\varepsilon_0}\left[\ln\left(\frac{D_{CA}}{r}\right)\ln\left(\frac{D_{BC}}{r}\right) - \ln\left(\frac{D_{BC}}{D_{AB}}\right).\ln\left(\frac{D_{CA}}{D_{AB}}\right)\right]$$

or,
$$Q_A = 2\pi\varepsilon_0\left[\frac{V_A \ln\left(\frac{D_{BC}}{r}\right) - V_B \ln\left(\frac{D_{CA}}{D_{AB}}\right)}{\ln\left(\frac{D_{CA}}{r}\right)\ln\left(\frac{D_{BC}}{r}\right) - \ln\left(\frac{D_{BC}}{D_{AB}}\right)\ln\left(\frac{D_{CA}}{D_{AB}}\right)}\right] \quad ...(viii)$$

Capacitance of conductor *A*,
$$C_A = \frac{Q_A}{V_A} = 2\pi\varepsilon_0\left[\frac{\ln\left(\frac{D_{BC}}{r}\right) - \frac{V_B}{V_A}\ln\left(\frac{D_{CA}}{D_{AB}}\right)}{\ln\left(\frac{D_{CA}}{r}\right)\ln\left(\frac{D_{BC}}{r}\right) - \ln\left(\frac{D_{BC}}{D_{AB}}\right)\ln\left(\frac{D_{CA}}{D_{AB}}\right)}\right] \text{ F/m} \quad ...(13.28)$$

Similarly, capacitance of conductor B,

$$C_B = 2\pi\varepsilon_0 \left[\frac{\ln\left(\frac{D_{CA}}{r}\right) - \frac{V_C}{V_B}\ln\left(\frac{D_{AB}}{D_{BC}}\right)}{\ln\left(\frac{D_{AB}}{r}\right)\ln\left(\frac{D_{CA}}{r}\right) - \ln\left(\frac{D_{CA}}{D_{BC}}\right)\ln\left(\frac{D_{AB}}{D_{BC}}\right)} \right] \text{ F/m} \quad ...(13.29)$$

and, capacitance, $$C_C = 2\pi\varepsilon_0 \left[\frac{\ln\left(\frac{AB}{r}\right) - \frac{V_A}{V_C}\left(\frac{D_{BC}}{D_{CA}}\right)}{\ln\left(\frac{D_{BC}}{r}\right)\ln\left(\frac{D_{AB}}{r}\right) - \ln\left(\frac{D_{AB}}{D_{CA}}\right)\ln\left(\frac{D_{BC}}{D_{CA}}\right)} \right] \text{ F/m} \quad ...(13.30)$$

13.5.5. Capacitance of a Double-Circuit 3-phase Overhead Line

Consider a 3-phase double circuit connected in parallel—conductors A, B and C forming one circuit and conductors A', B' and C' forming an other circuit (conductors symmetrically placed). Let the charges over conductors A, B and C be Q_A, Q_B and Q_C coulombs per metre ; then charge over conductors A', B' and C' will also be Q_A, Q_B and Q_C coulombs per metre and $Q_A + Q_B + Q_C = 0$.

The capacitances are calculated when the conductors are transposed, as shown in Fig. 13.35.

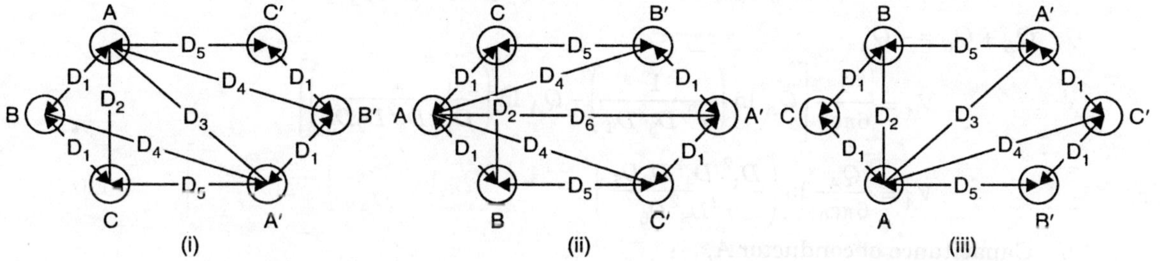

Fig. 13.35. Double circuit 3-phase line with transposed conductors.

Voltage of conductor A in position (i) :

$$V_{A1} = \frac{1}{2\pi\varepsilon_0}\left[Q_A\int_r^\infty \frac{dx}{x} + Q_B\int_{D_1}^\infty \frac{dx}{x} + Q_C\int_{D_2}^\infty \frac{dx}{x} + Q_A\int_{D_3}^\infty \frac{dx}{x} + Q_B\int_{D_4}^\infty \frac{dx}{x} + Q_C\int_{D_5}^\infty \frac{dx}{x}\right]$$

$$= \frac{1}{2\pi\varepsilon_0} [Q_A \ln(\infty) - Q_A \ln(r) + Q_B \ln(\infty) - Q_B \ln(D_1) + Q_C \ln(\infty) - Q_C \ln(D_2) +$$
$$Q_A \ln(\infty) - Q_A \ln(D_3) + Q_B \ln(\infty) - Q_B \ln(D_4) + Q_C \ln(D_5) - Q_C \ln(\infty)]$$

$$= \frac{1}{2\pi\varepsilon_0}\left[2(Q_A + Q_B + Q_C)\ln\infty + Q_A \ln\left(\frac{1}{rD_3}\right) + Q_B \ln\left(\frac{1}{D_1D_4}\right) + Q_C \ln\left(\frac{1}{D_2D_5}\right)\right]$$

or, $$V_{A1} = \frac{1}{2\pi\varepsilon_0}\left[Q_A \ln\left(\frac{1}{rD_3}\right) + Q_B \ln\left(\frac{1}{D_1D_4}\right) + Q_C \ln\left(\frac{1}{D_2D_5}\right)\right] \quad ...(13.31)$$

$(\because \; Q_A + Q_B + Q_C = 0)$

Similarly, voltage of conductor A in position (ii),

$$V_{A2} = \frac{1}{2\pi\varepsilon_0}\left[Q_A\int_r^\infty \frac{dx}{x} + Q_B\int_{D_1}^\infty \frac{dx}{x} + Q_C\int_{D_4}^\infty \frac{dx}{x} + Q_A\int_{D_6}^\infty \frac{dx}{x} + Q_B\int_{D_4}^\infty \frac{dx}{x} + Q_C\int_{D_1}^\infty \frac{dx}{x}\right]$$

or, $$V_{A2} = \frac{1}{2\pi\varepsilon_0}\left[Q_A \ln\left(\frac{1}{rD_6}\right) + Q_B \ln\left(\frac{1}{D_1D_4}\right) + Q_C \ln\left(\frac{1}{D_1D_4}\right)\right] \quad ...(13.32)$$

and, voltage of conductor A in position (*iii*),

$$V_{A3} = \frac{1}{2\pi\varepsilon_0}\left[Q_A \ln\left(\frac{1}{rD_3}\right) + Q_B \ln\left(\frac{1}{D_2D_5}\right) + Q_C \ln\left(\frac{1}{D_1D_4}\right)\right] \quad ...(13.33)$$

The average voltage of conductor A,

$$V_A = \frac{V_{A1} + V_{A2} + V_{A3}}{3}$$

$$= \frac{1}{3 \times 2\pi\,\varepsilon_0}\left[Q_A\left(\ln\frac{1}{rD_3} + \ln\frac{1}{rD_6} + \ln\frac{1}{rD_3}\right) + Q_B\left(\ln\frac{1}{D_1D_4} + \ln\frac{1}{D_1D_4} + \ln\frac{1}{D_2D_5}\right)\right.$$

$$\left. + Q_C\left(\ln\frac{1}{D_2D_5} + \ln\frac{1}{D_1D_4} + \ln\frac{1}{D_1D_4}\right)\right]$$

or, $$V_A = \frac{1}{6\pi\varepsilon_0}\left[Q_A \ln\left(\frac{1}{r^3\,D_3{}^2D_6}\right) + (Q_B + Q_C)\ln\left(\frac{1}{D_1{}^2D_4{}^2D_2D_5}\right)\right]$$

$\because \quad Q_B + Q_C = -Q_A$

$\therefore$ $$V_A = \frac{1}{6\pi\varepsilon_0}\left[Q_A \ln\left(\frac{1}{r^3\,D_3{}^2D_6}\right) - Q_A \ln\left(\frac{1}{D_1{}^2\,D_4{}^2D_2D_5}\right)\right]$$

or, $$V_A = \frac{Q_A}{6\pi\varepsilon_0}\ln\left(\frac{D_1{}^2\,D_4{}^2D_2D_5}{r^2D_3{}^2D_6}\right)$$

$\therefore$ Capacitance of conductor A,

$$C_A = \frac{Q_A}{V_A} = \frac{Q_A}{\dfrac{Q_A}{6\,\pi\varepsilon_0}\ln\left(\dfrac{D_1{}^2\,D_4{}^2D_2D_5}{r^2D_3{}^2D_6}\right)}$$

or, $$C_A = \frac{2\pi\varepsilon_0}{\ln\left(\dfrac{\sqrt[3]{D_1{}^2\,D_4{}^2D_2D_5/D_3{}^2D_6}}{r}\right)} = \frac{2\pi\varepsilon_0}{\ln\left(\dfrac{D_{eq}}{r}\right)} \text{ F/m} \quad ...(13.34)$$

where, $$D_{eq} = \sqrt[3]{\frac{D_1{}^2\,D_4{}^2D_2D_5}{D_3{}^2D_6}}$$

Derived in the similar manner, $C_B = C_C = C_A$.

13.5.6. Effect of Earth on the Capacitance of a Transmission Line

The presence of earth changes the electric field of the transmission line and consequently its capacitance is affected. Earth may be assumed to be a perfect conductor in the form of horizontal plane of infinite extent. Therefore the electric field of charged conductors is forced to conform to a presence of this equipotential surface. *The effect of earth on capacitance can be conveniently taken into account by **method of images.***

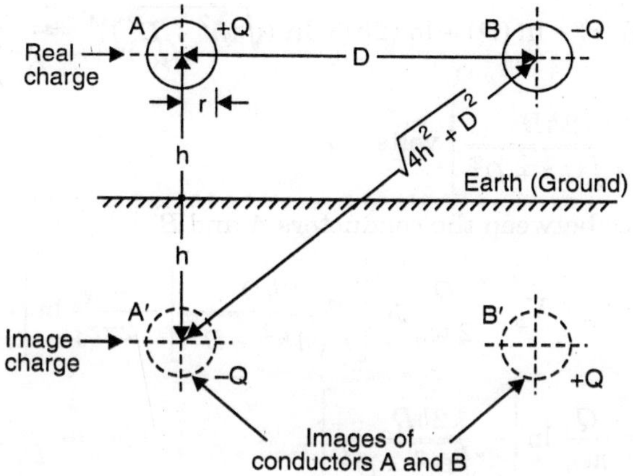

Fig. 13.36. Single-phase transmission line with images.

Let us consider the simple case of a single-phase transmission line, assuming conductors A' and B' as image conductors of conductors A and B (each of radius r) respectively, as shown in Fig. 13.36. Let $+ Q$ coulombs per metre length and $- Q$ coulombs per metre length be the charges on conductors A and B respectively and h be the height of the conductors above the ground.

The earth is considered to be at zero potential and it is possible only if there is an image conductor A' having a charge of $- Q$ coulombs per metre length at a depth of h metres below the earth. Similarly there is a conductor B' below the earth having a charge of $+ Q$ coulombs per metre length.

Potential of conductor A,

$$V_A = \int_r^{\infty} \frac{Q}{2\pi\varepsilon_0} \cdot \frac{dx}{x} + \int_D^{\infty} \frac{(-Q)}{2\pi\varepsilon_0} \cdot \frac{dx}{x} + \int_{2h}^{\infty} \frac{(-Q)}{2\pi\varepsilon_0} \cdot \frac{dx}{x} + \int^{\infty} \sqrt{4h^2 + D^2}\, \frac{Q}{2\pi\varepsilon_0} \cdot \frac{dx}{x}$$

$$= \frac{Q}{2\pi\varepsilon_0} [\{\ln(\infty) - \ln(r)\} - \{\ln(\infty) - \ln(D)\} - \{\ln(\infty) - \ln(2h)\} - \{\ln(\infty) - \ln(\sqrt{4h^2 + D^2})\}]$$

$$= \frac{Q}{2\pi\varepsilon_0} [\ln(\infty) - \ln(r) - \ln(\infty) + \ln(D) - \ln(\infty) + \ln(2h) + \ln(\infty) - \ln(\sqrt{4h^2 + D^2})]$$

$$= \frac{Q}{2\pi\varepsilon_0} \left[\ln\left(\frac{D}{r}\right) + \ln \frac{2h}{\sqrt{4h^2 + D^2}} \right]$$

or,

$$V_A = \frac{Q}{2\pi\varepsilon_0} \ln \left[\frac{2hD}{r\sqrt{4h^2 + D^2}} \right] \text{volts} \qquad ...(i)$$

Similarly, potential of conductor B,

$$V_B = \int_r^{\infty} \frac{(-Q)}{2\pi\varepsilon_0} \cdot \frac{dx}{x} + \int_D^{\infty} \frac{Q}{2\pi\varepsilon_0} \cdot \frac{dx}{x} + \int_{2h}^{\infty} \frac{Q}{2\pi\varepsilon_0} \cdot \frac{dx}{x} + \int_{\sqrt{4h^2 + D^2}}^{\infty} \frac{(-Q)}{2\pi\varepsilon_0} \cdot \frac{dx}{x}$$

$$= \frac{Q}{2\pi\varepsilon_0} [-\ln(\infty) + \ln(r) + \ln(\infty) - \ln(D) + \ln(\infty) - \ln(2h) - \ln(\infty) + \ln(\sqrt{4h^2 + D^2})]$$

$$= \frac{Q}{2\pi\varepsilon_0} [\ln(r) - \ln(D) - \ln(2h) + \ln(\sqrt{4h^2 + D^2})]$$

$$= \frac{-Q}{2\pi\varepsilon_0} [-\ln(r) + \ln(D) + \ln(2h) - \ln(\sqrt{4h^2 + D^2})]$$

or, $$V_B = \frac{(-Q)}{2\pi\varepsilon_0} \ln\left[\frac{2hD}{r\sqrt{4h^2 + D^2}}\right] \text{ volts} \quad ...(ii)$$

Potential difference between the conductors A and B,

$$V_{AB} = V_A - V_B = \frac{Q}{2\pi\varepsilon_0} \ln\left[\frac{2hD}{r\sqrt{4h^2 + D^2}}\right] - \frac{-Q}{2\pi\varepsilon_0} \ln\left[\frac{2hD}{r\sqrt{4h^2 + D^2}}\right]$$

or, $$V_{AB} = \frac{Q}{\pi\varepsilon_0} \ln\left[\frac{2hD}{r\sqrt{4h^2 + D^2}}\right]$$

$$= \frac{Q}{\pi\varepsilon_0} \ln\left[\frac{2hD}{r \times 2h\sqrt{1 + (D^2/4h^2)}}\right] = \frac{Q}{\pi\varepsilon_0} \ln\left[\frac{D}{r\sqrt{1 + (D^2/4h^2)}}\right] \text{ volts}$$

∴ Capacitance between conductors A and B,

$$C_{AB} = \frac{Q}{V_{AB}} = \frac{Q}{\frac{Q}{\pi\varepsilon_0} \ln\left[\frac{D}{r\sqrt{1 + (D^2/4h^2)}}\right]} = \frac{\pi\varepsilon_0}{\ln\left[\frac{D}{r\sqrt{1 + (D^2/4h^2)}}\right]} \text{ F/m line-to-line} \quad ...(13.35)$$

$$C_N = \frac{2\pi\varepsilon_0}{\ln\left[\frac{D}{r\sqrt{1 + (D^2/4h^2)}}\right]} \text{ F/m to neutral} \quad ...(13.35\ a)$$

From eqn. [13.35 (*a*)] it may be observed that the presence of earth modifies radius r to $r\sqrt{1 + (D^2/4h^2)}$. Since normally h is large compared to D, the effect of earth on line capacitance is *negligible*.

A similar treatment can be adopted for 3-phase transmission lines.

Example 13.18. *A single-phase overhead transmission line consists of two parallel straight conductors spaced 2.5 metres apart. The radius of each conductor is 1.2 cm. Calculate the capacitance of the line per km. Given that* $\varepsilon_0 = 8.854 \times 10^{-12}$ *F/m.*

Solution. *Given :* $r = 1.2$ cm $= 0.012$ m ; $D = 2.50$ cm ; $\varepsilon_0 = 8.854 \times 10^{-12}$ F/m

Capacitance of the line, $$\mathbf{C} = \frac{\pi\varepsilon_0}{\ln\left(\frac{D}{r}\right)} \text{ F/m} \quad ...[\text{Eqn. } (13.23)]$$

$$= \frac{\pi \times 8.854 \times 10^{-12}}{\ln\left(\frac{2.5}{0.012}\right)} = 5.21 \times 10^{-12} \text{ F/m}$$

$= 5.21 \times 10^{-9}$ F/km $=$ **0.00521μ F/m. (Ans.)**

Example 13.19. *A single-phase, 25 km long overhead line consists of two conductors 1.8 metres apart, diameters of each conductor being 6 mm. If the line voltage is 33 kV, 50 Hz determine the charging current of the open circuited line.*

Solution. *Given :* $l = 25$ km, $r = \frac{6}{2} = 3$ mm $= 0.003$ m ; $D = 1.8$ m ; $V = 33$ kV ; $f = 50$ Hz

Capacitance of the line, $C = \dfrac{\pi\varepsilon_0}{\ln\left(\dfrac{D}{r}\right)} \times l$

$$= \frac{\pi \times 8.854 \times 10^{-12}}{\ln\left(\dfrac{1.8}{0.003}\right)} \times (25 \times 10^3) \times 10^6 \text{ μF/km} = 0.1087 \text{ μF}$$

Charging current, $I_c = 2\pi fCV$

$= 2\pi \times 50 \times (0.1087 \times 10^{-6}) \times 33000 =$ **1.127 A. (Ans.)**

Example 13.20. *Calculate capacitance of a single phase transmission line 35 km long consisting of two parallel wires each 5 mm in diameter and 1.8 m apart. The height of each conductor above ground is 7.5 m.*

Solution. Refer to Fig. 13.37.

Given : $l = 35$ km ; $r = \frac{5}{2} = 2.5$ mm $= 0.0025$ m ; $D = 1.8$ m ; $h = 7.5$ m.

The capacitance of the line, taking into consideration the presence of ground, is given as :

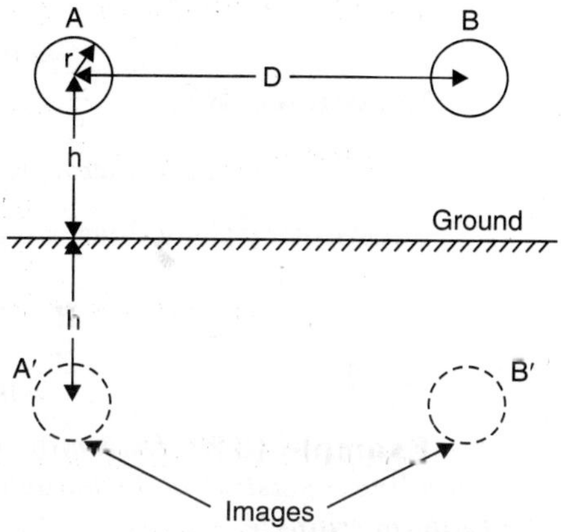

Fig. 13.37

$$C = \frac{\pi\varepsilon_0}{\ln\left[\dfrac{D}{r\sqrt{1 + D^2/4h^2}}\right]} \text{ F/m (line-to-line)}$$

...[Eqn. (13.35]

$$= \frac{\pi \times 8.854 \times 10^{-12}}{\ln\left[\dfrac{1.8}{0.0025\sqrt{1 + (1.8^2/4 \times 7.5^2)}}\right]}$$

$$= \frac{\pi \times 8.854 \times 10^{-12}}{6.572} = 4.23 \times 10^{-12} \text{ F/m}$$

Capacitance for 35 km $= 4.23 \times 10^{-12} \times (35 \times 1000)$ F

$=$ **0.148 μF. (Ans.)**

Example 13.21. *A 3-phase, 132 kV, 50 Hz overhead transmission line has steel cored aluminium conductors of equivalent copper area of 1.5 cm² and effective diameter of 39.2 mm spaced equilaterally 8 metres apart. Calculate the line constants per km length of the line. The resistivity of copper is 1.73 μΩ cm.*

Solution. *Given* : Supply voltage = 132 kV ; $f = 50$ Hz ; $a = 1.5$ cm^2 ;

$r = \frac{39.2}{2} = 19.6$ mm $= 0.0196$ m ; $D = 8$ m ; $l = 1$ km $= 1000$ m, $\rho = 1.73$ μΩ cm.

Line constants, L, C, R :

Inductance, L : For a 3-phase overhead transmission line spaced equilaterally, we know that,

$$L = 2 \times 10^{-7} \ln\left(\frac{D}{D_s}\right)$$

where, D_s = Self-GMD or GMR = 0.7788r

$$\mathbf{L} = 2 \times 10^{-7} \ln\left(\frac{8}{0.7788 \times 0.0196}\right) = 12.52 \times 10^{-7} \text{ H/m}$$

$$= \mathbf{1.252 \text{ mH/km.}} \quad \textbf{(Ans.)}$$

Capacitance C :

We know that, $$\mathbf{C_N} = \frac{2\pi\varepsilon_0}{\ln\left(\frac{D}{r}\right)} \text{ F/m} \qquad \text{...[Eqn.(13.24)]}$$

$$= \frac{2\pi \times 8.854 \times 10^{-12}}{\ln\left(\frac{8}{0.0196}\right)} = 9.254 \times 10^{-12} \text{ F/m}$$

$$= \mathbf{0.009254\ \mu F/km.} \quad \textbf{(Ans.)}$$

Resistance, R :

Resistance of a conductor, $R = \frac{\rho l}{a}$

Substituting the values, we get

$$\mathbf{R} = \frac{1.73 \times 10^{-6} \times (1000 \times 100)}{1.5}$$

$$= \mathbf{0.1153\ \Omega/km} \text{ (per conductor).} \quad \textbf{(Ans.)}$$

Example 13.22. *Determine the capacitance of a 80 km long 3-phase, 50 Hz overhead transmission line consisting of 3-conductors, each of 1.8 cm diameter and spaced 2.4 m at the corners of an equilateral triangle.* **(Banglore University)**

Solution. *Given* : $r = \frac{1.8}{2}$ = 0.9 cm = 0.009 m ; D = 2.4 m

Capacitance of the line, $$C = \frac{2\pi\varepsilon_0}{\ln\left(\frac{D}{r}\right)} \text{ F/m} \qquad \text{...[Eqn. (13.25)]}$$

$$= \frac{2\pi \times 8.854 \times 10^{-12}}{\ln\left(\frac{2.4}{0.009}\right)} \text{ F/m} = 9.96 \times 10^{-12} \text{ F/m}$$

Capacitance of 80 km long line

$$= 9.96 \times 10^{-12} \times (80 \times 1000) = \mathbf{0.7968\ \mu F.} \quad \textbf{(Ans.)}$$

Example 13.23. *A 3-phase, 50 Hz, 132 kV overhead line has conductors placed in a horizontal plane 4.56 metres apart. Conductor diameter is 22.4 mm. If the line length is 100 km, calculate the charging current per phase assuming complete transposition.* **(N.U.)**

Solution. *Given* : $f = 50$ Hz, $V = 132$ kV ; $D_{12} = D_{23} = 4.56$ m ;

$$D_{31} = 4.56 \times 2 = 9.12 \text{ m} = 4.56 \text{ m} ; r = \frac{22.4}{2} = 11.2 \text{ mm} = 0.0112 \text{ m}, l = 100 \text{ km}$$

Capacitance from conductor to neutral is

$$= \frac{2\pi\varepsilon_0}{\ln\left[\frac{(D_{12}D_{23}D_{31})^{1/3}}{r}\right]} \text{ F/m} \quad \text{...[Eqn. (13.26)]}$$

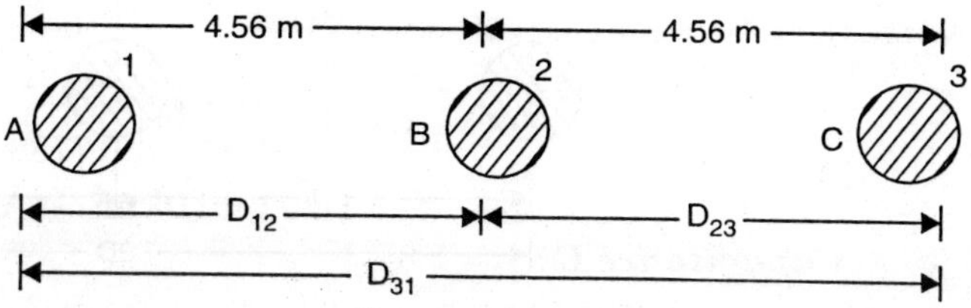

Fig. 13.38

$$= \frac{2\pi \times 8.854 \times 10^{-12}}{\ln\left[\frac{(4.56 \times 4.56 \times 9.12)^{1/3}}{0.0112}\right]}$$

$= 8.915 \times 10^{-12}$ F/m

∴ Line to neutral capacitance for 100 km length,

$$C = 8.915 \times 10^{-12} \times (100 \times 1000) = 0.8915 \times 10^{-6} \text{ F}$$

∴ *Charging current per phase,*

$$\mathbf{I_C} = 2\pi f C V_{ph}$$

$$= 2\pi \times 50 \times 0.8915 \times 10^{-6} \times \frac{(132 \times 10^3)}{\sqrt{3}} = \mathbf{21.34\ A.} \quad \textbf{(Ans.)}$$

Example 13.24. *A single 3-phase line operated at 50 Hz is arranged as shown in Fig. 13.39. The conductor diameter is 8 mm and the line is regularly transposed. Determine the inductance and capacitance per km.*

Solution. *Given* : $r = \frac{8}{2} = 4$ mm $= 0.004$ m ; $D_{12} = 1.5$ m ; $D_{23} = 3$ m ; $D_{31} = 2.6$ m.

Inductance of the line per km, L :

$$D_m = (D_{12} \times D_{23} \times D_{31})^{1/3}$$
$$= (1.5 \times 3 \times 2.6)^{1/3} = 2.27 \text{ m}$$
$$D_s = 0.7789\, r = 0.7788 \times 0.004$$
$$= 0.003115 \text{ m}$$
$$\mathbf{L} = 2 \times 10^{-7} \ln\left(\frac{D_m}{D_s}\right)$$
$$= 2 \times 10^{-7} \ln\left(\frac{2.27}{0.003115}\right)$$
$$= 13.18 \times 10^{-7} \text{ H/m}$$
$$= \mathbf{1.318\ mH/km.} \quad \textbf{(Ans.)}$$

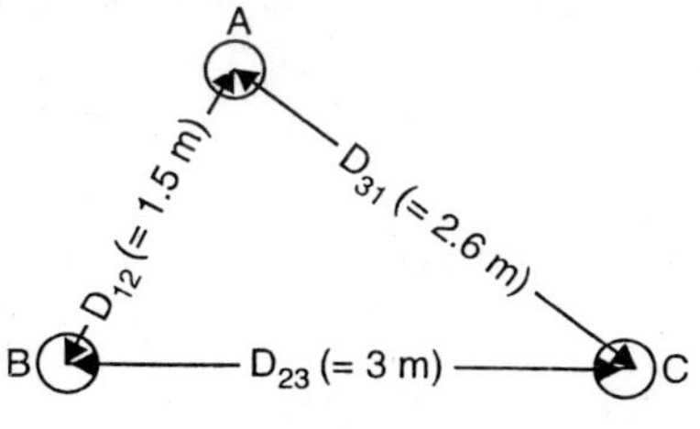

Fig. 13.39

Capacitance of the line per km C :

$$\mathbf{C} = \frac{2\pi\varepsilon_0}{\ln\left[\frac{(D_{12}D_{23}D_{31})^{1/3}}{r}\right]} \text{ F/m} \quad \text{...Eqn. (13.27)}$$

$$= \frac{2\pi \times 8.854 \times 10^{-12}}{\ln\left(\frac{2.27}{0.004}\right)} = 8.773 \times 10^{-12} \text{ F/m} = \mathbf{8.773 \times 10^{-9}\ F/km.} \quad \textbf{(Ans.)}$$

Example 13.25. *Derive an expression for the charge (complex) value per metre length of conductor of an untransposed three phase line shown in Fig. 13.40. The applied voltage is balanced three phase, 50 Hz. Take the voltage of phase-a as reference phasor. All conductors have the same radii. Also find the charging current of phase-a. Neglect the effect of ground.*

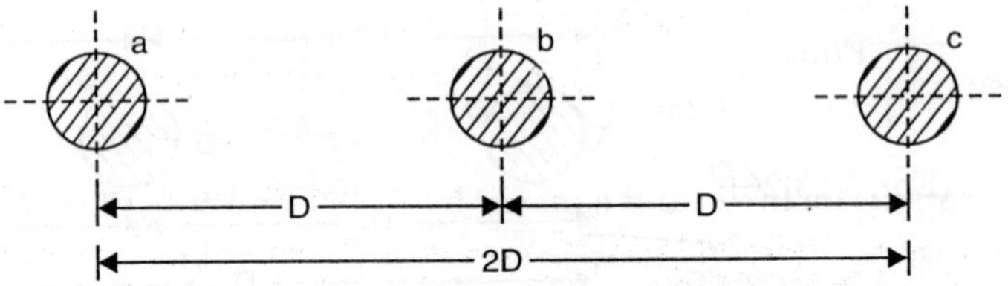

Fig. 13.40

Solution. Let Q_1, Q_2 and Q_3 coulombs per metre be charges on the respective conductors a, b and c (due to the applied balanced voltage on each conductor).

The potential of conductor a is equal to work done to bring a unit charge from infinity to the surface of the conductor, but as the unit charge is brought from infinity the work will have to be done against the electrostatic forces due to charges Q_2 and Q_3 on conductors b and c respectively.

∴ The work done to bring the unit charge from infinity to the surface of conductor a (at that instant the distances of the unit charge from b and c are D and $2D$)

$$= \int_r^{\infty} \frac{Q_1}{2\pi\varepsilon_0 x}\,dx + \int_D^{\infty} \frac{Q_2}{2\pi\varepsilon_0 x}\,dx + \int_{2D}^{\infty} \frac{Q_3}{2\pi\varepsilon_0 x}\,dx$$

∴ $$V_1 = \int_r^{\infty} \frac{Q_1}{2\pi\varepsilon_0 x}\,dx + \int_D^{\infty} \frac{Q_2}{2\pi\varepsilon_0 x}\,dx + \int_{2D}^{\infty} \frac{Q_3}{2\pi\varepsilon_0 x}\,dx$$

or, $$V_1 = \frac{1}{2\pi\varepsilon_0}\left[Q_1 \log_e \frac{\infty}{r} + Q_2 \log_e \frac{\infty}{D} + Q_3 \log_e \frac{\infty}{2D}\right]$$

$$= \frac{1}{2\pi\varepsilon_0}\left[-Q_1 \ln(r) - Q_2 \ln(D) - Q_3 \ln(2D) + \ln\infty\,(Q_1 + Q_2 + Q_3)\right]$$

Since, $Q_1 + Q_2 + Q_3 = 0$,

∴ $$V_1 = \frac{1}{2\pi\varepsilon_0}\left[Q_1 \ln\frac{1}{r} + Q_2 \ln\frac{1}{D} + Q_3 \ln\frac{1}{2D}\right]$$

Similarly, $$V_2 = \frac{1}{2\pi\varepsilon_0}\left[Q_1 \ln\frac{1}{D} + Q_2 \ln\frac{1}{r} + Q_3 \ln\frac{1}{2D}\right]$$

and, $$V_3 = \frac{1}{2\pi\varepsilon_0}\left[Q_1 \ln\frac{1}{2D} + Q_2 \ln\frac{1}{D} + Q_3 \ln\frac{1}{r}\right]$$

Putting $Q_3 = -(Q_1 + Q_2)$, we have

$$V_1 = \frac{1}{2\pi\varepsilon_0}\left[Q_1 \ln\frac{1}{r} + Q_2 \ln\frac{1}{D} + (Q_1 + Q_2)\ln\frac{1}{2D}\right]$$

$$= \frac{1}{2\pi\varepsilon_0}\left[Q_1 \ln\frac{1}{r} + Q_2 \ln\frac{1}{D} - Q_1 \ln\frac{1}{2D} - Q_2 \ln\frac{1}{2D}\right]$$

$$= \frac{1}{2\pi\varepsilon_0}\, Q_1\left[\ln\frac{1}{r} - \ln\frac{1}{2D}\right] + Q_2\left[\ln\frac{1}{D} - \ln\frac{1}{2D}\right]$$

$$= \frac{1}{2\pi\varepsilon_0}\left[Q_1 \ln \frac{2D}{r} + Q_2 \ln \frac{2D}{D}\right] \quad ...(i)$$

and,

$$V_2 = \frac{1}{2\pi\varepsilon_0}\left[Q_1 \ln \frac{1}{D} + Q_2 \ln \frac{1}{r} - (Q_1 + Q_2) \ln \frac{1}{D}\right]$$

$$= \frac{1}{2\pi\varepsilon_0}\left[Q_1 \ln \frac{D}{D} + Q_2 \ln \frac{D}{r}\right] \quad ...(ii)$$

Multiplying eqn. (*i*) by $\ln \frac{D}{r}$ and eqn. (*ii*) by $\ln \frac{2D}{D}$ and substituting, we get

$$2\pi\varepsilon_0 V_1 \ln \frac{D}{r} = \left[Q_1 \ln \frac{2D}{r} \ln \frac{D}{r} + Q_2 \ln \frac{2D}{D} \times \ln \frac{D}{r}\right] \quad ...(iii)$$

$$2\pi\varepsilon_0 V_2 \ln \frac{2D}{D} = \left[Q_1 \ln \frac{D}{D} \ln \frac{2D}{D} + Q_2 \ln \frac{D}{r} \times \ln \frac{2D}{D}\right] \quad ...(iv)$$

Substracting eqn. (*iv*) from eqn. (*iii*)

$$Q_1\left(\ln \frac{2D}{r} \times \ln \frac{D}{r} - \ln \frac{D}{D} \times \ln \frac{2D}{D}\right) = 2\pi\varepsilon_0\left(V_1 \ln \frac{D}{r} - V_2 \ln \frac{2D}{D}\right)$$

or,

$$Q_1 = 2\pi\,\varepsilon_0 \frac{\left(V_1 \ln \frac{D}{r} - V_2 \ln \frac{2D}{D}\right)}{\ln \frac{2D}{r} \ln \frac{D}{r} - \ln \frac{D}{D} \times \ln \frac{2D}{D}}$$

or,

$$Q_1 = 2\pi\,\varepsilon_0 \left[\frac{V_1 \ln \frac{D}{r} - 0.693\, V_2}{\ln \frac{2D}{r} \ln \frac{D}{r}}\right]$$

Thus capacitance $C_1 = \frac{Q_1}{V_1}$ can be found and then the charging current (I_C) will be given as :

$$I_C = jwCV = \mathbf{j2\pi fCV}. \quad \textbf{(Ans.)}$$

Example 13.26. *The six conductors of a double-circuit transmission line are arranged as shown in Fig. 13.41.*

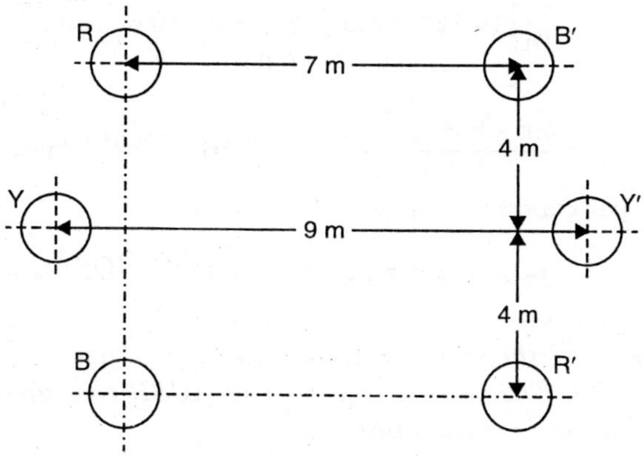

Fig. 13.41

The diameter cf each conductor is 25 mm. Find the charging current per km per conductor at 132 kV and 50 Hz assuming that line is regulatry transposed. Neglect the effect of earth.

(Agra University)

Solution. Radius of each conductor, $r = \frac{25}{2} = 12.5 \text{ mm} = 0.0125 \text{ m}$

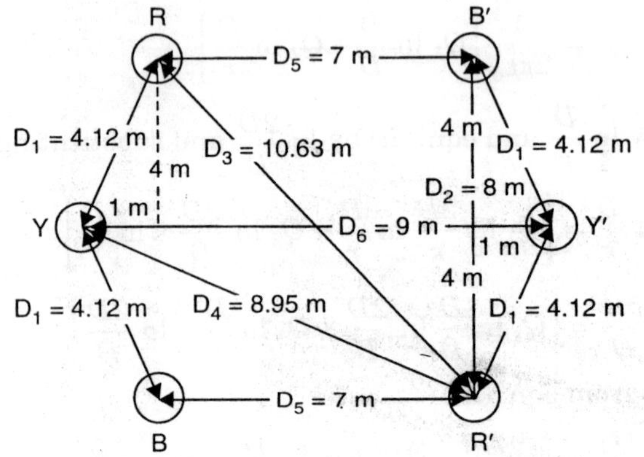

Fig. 13.42

Various distances between conductors (Fig. 3.42) are :

$$D_1 = RY\ (= YB = B'Y' = Y'R') = \sqrt{4^2 + 1^2} = 4.12 \text{ m}$$
$$D_2 = RB\ (= B'R') = 4 + 4 = 8 \text{ m}$$
$$D_3 = RR'\ (= BB') = \sqrt{(4+4)^2 + 7^2} = 10.\ 63 \text{ m}$$
$$D_4 = YR' = (YB') = \sqrt{(9-1)^2 + 4^2} = 8.95 \text{ m}$$

Capacitance of each conductor,

$$C = \frac{2\pi\varepsilon_0}{\ln\left(\dfrac{\sqrt[3]{D_1^{\,2} D_4^{\,2} D_2 D_5 / D_3^{\,2} D_6}}{r}\right)} \qquad \text{...[Eqn.(13.34)]}$$

$$= \frac{2\pi \times 8.854 \times 10^{-12}}{\ln\left[\dfrac{(4.12^2 \times 8.95^2 \times 8 \times 7/(10.6^2 \times 9))^{1/3}}{0.0125}\right]}$$

$$= \frac{2\pi \times 8.854 \times 10^{-12}}{5.82} = 9.56 \times 10^{-12} \text{ F/m} = 9.56 \times 10^{-9} \text{ F/km}$$

$\therefore$ Charging current per conductor per km, $\mathbf{I_c} = 2\pi f C V_{ph}$

$$= 2\pi \times 50 \times 9.56 \times 10^{-9} \times \left(\frac{132 \times 1000}{\sqrt{3}}\right) = \mathbf{0.229\ A.} \quad \textbf{(Ans.)}$$

Example 13.27. *Determine the capacitance per km per phase to neutral of a 3-phase line arranged as shown in Fig. 13.43. The outside diameter of ACSR conductor is 2.8 cm. The line is transposed. Take the effect of ground into account.*

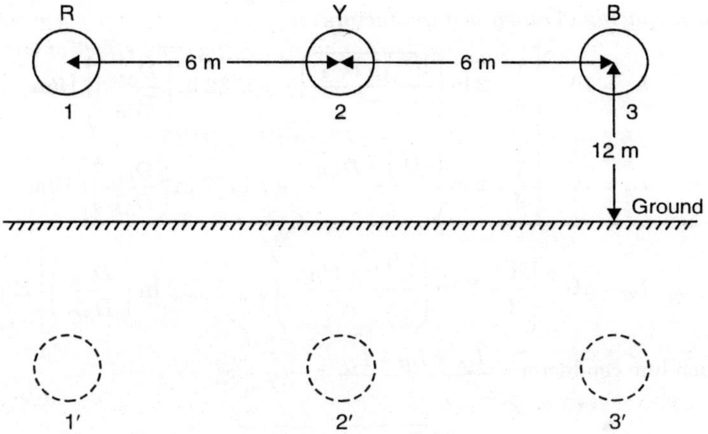

Fig. 13.43

Solution. Radius of conductor = $\frac{2.8}{2}$ = 1.4 cm = 0.014 m

The capacitance per phase to neutral can be shown to be equal to

$$C_N = \frac{2\pi\varepsilon_0}{\ln \frac{D_{eq}}{r} - \ln \sqrt[3]{\frac{h_{12}h_{23}h_{31}}{h_1h_2h_3}}} \text{ F/m}$$

where, $D_{eq} = (D_{12}D_{23}D_{31})^{1/3} = (6 \times 6 \times 12)^{1/3} = 7.56$ m

and, $h_{12} = h_{23} = \sqrt{(6)^2 + (24)^2} = 24.74$ m

$h_{31} = \sqrt{(12)^2 + (24)^2} = 26.83$ m

$h_1 = h_2 = h_3 = 24$ m

$$\therefore \quad C_N = \frac{2\pi \times 8.854 \times 10^{-12}}{\ln\left(\frac{7.56}{0.014}\right) - \ln\left(\frac{24.74 \times 24.74 \times 26.83}{24 \times 24 \times 24}\right)^{1/3}}$$

$= 8.92 \times 10^{-12}$ F/m = $\mathbf{8.92 \times 10^{-9}}$ **F/km. (Ans.)**

HIGHLIGHTS

1. Inductance is by far the most dominant line parameter.
2. *Important formulae for "inductance".*

Loop inductance of *1* – φ transmission line = $10^{-7}\left[1 + 4\ln\left(\frac{D}{r}\right)\right]$ H/m

Inductance of *3*-φ transmission line :

(*i*) *Symmetrical spacing :*

$$L_A = L_B = L_C = 10^{-7}\left[0.5 + 2\ln\left(\frac{D}{r}\right)\right]$$

(*ii*) *Unsymmetrical spacing* (Transposed conductors) :

$$L_A = 10^{-7}\left[\frac{1}{2} + 2\ln\left(\frac{\sqrt{D_{AB}D_{CA}}}{r}\right) + j\,1.732\ln\left(\frac{D_{AB}}{D_{CA}}\right)\right] \text{ H/m}$$

$$L_B = 10^{-7}\left[\frac{1}{2} + 2\ln\left(\frac{\sqrt{D_{BC}D_{AB}}}{r}\right) + j\,1.732\ln\left(\frac{D_{BC}}{D_{AB}}\right)\right] \text{ H/m}$$

$$L_C = 10^{-7}\left[\frac{1}{2} + 2\ln\left(\frac{\sqrt{D_{CA}D_{BC}}}{r}\right) + j\,1.732\ln\left(\frac{D_{CA}}{D_{BC}}\right)\right] \text{ H/m}$$

Inductance of each line conductor $= \dfrac{L_A + L_B + L_C}{3}$

$$= 10^{-7}\left[\frac{1}{2} + 2\ln\left(\frac{\sqrt[3]{D_{AB} \times D_{BC} \times D_{CA}}}{r}\right)\right] \text{ H/m}$$

3. Self-GMD or GMR = 0.7788 r.
4. Mutual-GMD represents the equivalent geometrical spacing.
5. The tendency of alternating current to concentrate near the surface of a conductor is known as *skin effect.*
6. The alternating magnetic flux in a conductor caused by the current flowing in a neighbouring conductor gives rise to circulating current which causes an apparent increase in the resistance of a conductor. This phenomenon is called *proximity effect.*
7. *Important formulate for "capacitance"* :

(*i*) *Capacitance of a single phase two-wire line* :

$$C = \frac{\pi\varepsilon_0}{\ln\left(\dfrac{D}{r}\right)} \text{ F/m}$$

$$C_N = \frac{2\pi\varepsilon_0}{\ln\left(\dfrac{D}{r}\right)} \text{ F/m}$$

(*ii*) *Capacitance of a 3-phase line* :

(*a*) *Symmetrical spacing* :

$$C_A = \frac{2\pi\varepsilon_0}{\ln\left(\dfrac{D}{r}\right)} \quad (= C_B = C_C)$$

(*b*) *Unsymmetrical spacing with transposed conductors* :

$$C_A = \frac{2\pi\varepsilon_0}{\ln\left[\dfrac{(D_{12}D_{23}D_{31})^{1/3}}{r}\right]} \quad (= C_B = C_C)$$

(*iii*) Effect of earth on the capacitance of single-phase transmission lines :

$$C_{AB} = \frac{\pi\varepsilon_0}{\ln\left[\dfrac{D}{r\sqrt{1+(D^2/4h^2)}}\right]} \text{ F/m line-to-line}$$

$$C_N = 2\,C_{AB} \text{ F/m to neutral}$$

where, D = Distance between the conductors, and

h = Height of the conductor above the ground.

THEORETICAL QUESTIONS

1. Name and explain briefly the four parameters on which the performance of a transmission line as an element of power system depends.
2. Show that the inductance per loop metre of a two-wire transmission line using solid round conductor is given by,

$$L = 4 \times 10^{-7} \ln \left(\frac{D}{r'}\right) \text{ H}$$

where, D is the distance between the conductors and r' is the GMR of the conductors.
3. Derive an expression for the loop inductance of a single-phase, two-wire (solid) system from the fundamentals.
4. Explain the meaning of self-GMD and mutual-GMD.
5. What do you understand by transposition of conductors ?
6. What are the factors which govern the inductance of a transmission line ?
7. Explain briefly the *'skin effect'* in a transmission line. On what factors does it depend ? What is its effect on the resistance of the line ?
8. Distinguish between A.C. and D.C. resistances of a conductor. Why the two differ ? Explain fully.
9. State the factors which govern the capacitance of a transmission line.
10. Derive from basic considerations an expression for capacitance and charging current per km length of a single-phase line made up of two solid round conductors of radius r metres and spaced at D metres. Neglect the effect of ground.
11. Derive from first principles the capacitance per km to neutral of a 3-phase overhead transmission line with unsymmetrical spacing of conductors assuming transposition.
12. Discuss how transposition helps in equalising the capacitances in an unsymmetrically spaced 3-phase overhead transmission line.
13. Show that when the effect of earth is considered, the capacitance of a two-wire transmission line is given by

$$C_{ab} = \frac{\pi \varepsilon}{\ln\left[\dfrac{D}{r\sqrt{1 + D^2/4h^2}}\right]} \text{ F/m line-to-line}$$

where, ε = Permittivity of the medium,
D = Spacing between conductors,
h = Height of conductor above ground, and
r = Radius of each conductor.

If the effect of earth is neglected, how would the capacitance be affected ?

UNSOLVED EXAMPLES

Inductance

1. In a 3-phase transmission line the three conductors are placed at the corners of a triangle of sides 2 m, 3.2 m and 2.5 m respectively. If the diameter of each conductor is 1.2 cm and the conductors are regularly transposed, calculate the inductance/phase/km length of the line. **[Ans.** 1.26 mH]
2. Find the inductance per phase per km of a 3-phase overhead transmission line using 2 cm diameter conductors when these are placed at the corners of an equilateral triangle of side 4 metres. **[Ans.** 1.25 mH]
3. A 3-phase 3-wire transmission system consisting of 2 cm diameter conductors, spaced 4 metres apart are arranged in a horizontal plane. Determine the inductance per km of each conductor (lint-to-neutral). Assume balanced load and R, Y, B phase sequence. Determine also the average inductance per phase for regularly transposed line.
[Ans. 1.3174 – j 0.12 mH/km ; 1.248 mH/km ; 1.3174 + j 0.12 mH/km ; L_{avg} = 1.295 mH/km]

4. Two conductors of a single-phase line, each of 2 cm diameter, are arranged in a vertical plane with one conductor mounted 1 m above the other. A second identical line is mounted at the same height as the first and spaced horizontally 0.25 m apart from it. The two upper and two lower conductors are connected in parallel. Determine the inductance per km of the resulting double-circuit line. [**Ans.** 0.7 mH]

5. Calculate the inductance per phase per km length of the system of conductors shown in Fig. 13.44. Self-GMD of one conductor is 0.9 cm. Assume line regularly transposed. [**Ans.** 1.023 mH]

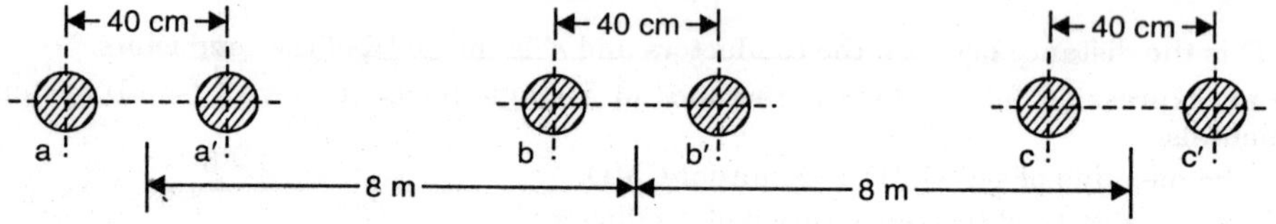

Fig. 13.44

6. The arrangement of a symmetrical double-circuit 3-phase line having six conductors is shown in Fig. 13.45. If the diameter of the conductor is 2.6 cm and the line is completely transposed, calculated the inductive reactance per phase in ohms per km. Assume supply frequency as 50 Hz. [**Ans.** 0.179 Ω/km]

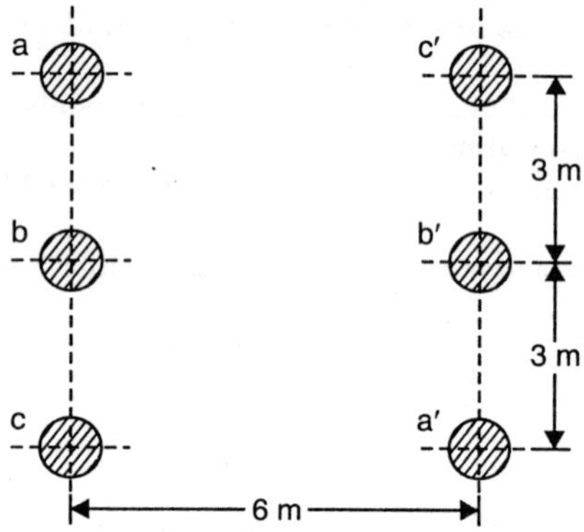

Fig. 13.45

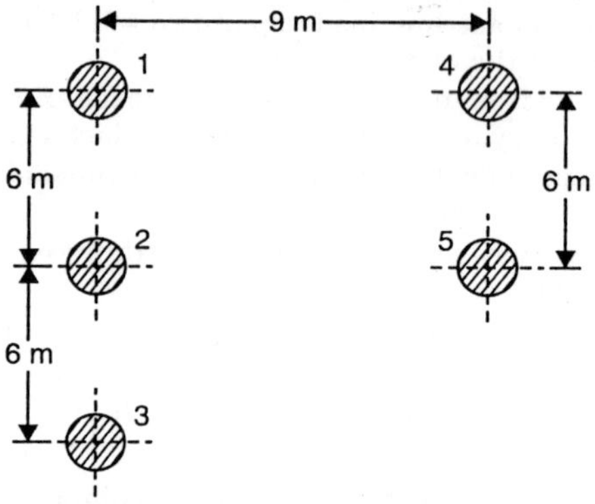

Fig. 13.46

7. Determine the inductance of a single-phase transmission line consisting of three conductors of 2.5 mm diameter radii in the 'go' conductor and two conductors of 5 mm radii in the 'return' conductor. The configuration of the line is as shown in the Fig. 13.46. [**Ans.** 1.42 mH/km]

Capacitance

8. Determine the capacitance of a single-phase line 20 km long consisting of two parallel wires each 5 mm in diameter and 1.5 m apart.

   Find the capacitance of the same line taking with account the effect of ground. The height of the conductors above the ground is 7 m. [**Ans.** 0.087 μF ; 0.088 μF]

9. Calculate the capacitance of a 90 km long, 3-phase, 50 Hz overhead transmission line consisting of 3-conductors, each of diameter 20 mm spaced 2.5 m at the corners of an equilateral triangle. [**Ans.** 0.9067 μF]

10. A 3-phase, 50 Hz, 132 kV overhead line has conductors placed in a horizontal plane 4 m apart. Conductor diameter is 20 mm. If the length of line is 100 km, calculate the charging current per phase assuming complete transposition. [**Ans.** 21.18 A]

11. The six conductors of a double-circuit three-phase line having an overall radius of 8.65 mm are arranged as shown in Fig. 13.47. Find the charging current per conductor per km at 110 kV, 50 Hz.

[Ans. 0.1805 A/km]

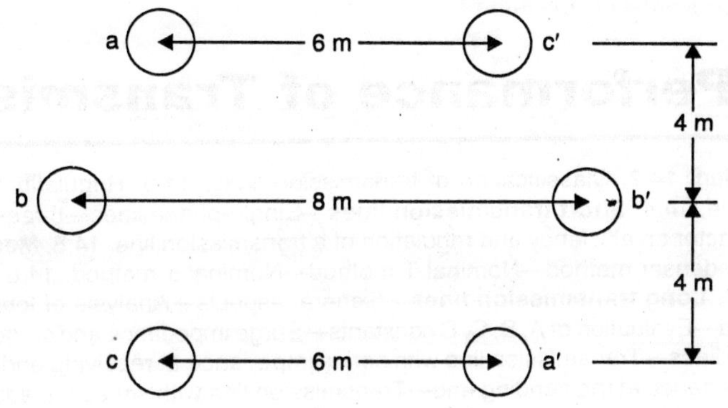

Fig. 13.47

14

Performance of Transmission Lines

14.1. Introduction. 14.2. Classification of transmission lines. 14.3. Regulation and efficiency of a transmission line. 14.4. **Short transmission lines**—Single-phase lines—three-phase lines—Effect of load power factor on efficiency and regulation of a transmission line. 14.5. **Medium transmission lines**—End condenser method—Nominal-T method—Nominal-π method. 14.6. Generalised circuit constants. 14.7. **Long transmission lines**—General aspects—Analysis of long transmission line-rigorous method—Evaluation of A, B, C, D constants—Surge impedance and surge impedance loading of transmission lines—Transmission line with series impedance at receiving end—Transmission line with series impedance at the sending end—Transmission line with series impedance at both ends—Transmission line with transformers at both ends—Equivalent T-network of a long transmission line—Equivalent π-network of a long transmission line—Constant voltage transmission—Power circle diagram—Synchronous phase modifiers. 14.8. Ferranti effect. 14.9. Charging current and losses in an open-circuited line. Highlights—Theoretical Questions—Unsolved Examples.

14.1. INTRODUCTION

- *By performance of transmission lines* is meant the *determination* of *efficiency and regulation of lines.*
- The voltage drop, power losses occurring in the line and transmission efficiency are largely influenced by the transmission line constants *i.e.*, R (resistance), L (inductance) and C (capacitance). The line *resistance* causes voltage drop (IR) and power loss (I^2R) in the line, the line *inductance* causes voltage drop in quadrature with the current flowing in the conductor ($2\pi fL \times I$) whereas the *capacitance* produces charging current ($2\pi fCV$), in quadrature with the voltage. The value of charging current is maximum at the sending end of the line and decreases continuously and its value becomes zero at the receiving end. It may be added that in case of overhead lines inductive reactance effects are relatively more important due to wide spacing of the conductors while in the case of underground cables capacitive reactance effects are more predominant in comparison to inductive reactance effects.

14.2. CLASSIFICATION OF TRANSMISSION LINES

The transmission lines are generally *classified* as follows :

1. Short transmission lines.
2. Medium transmission line.
3. Long transmission line.

1. Short transmission lines :

- Transmission lines having *length less than 50 km* and operating voltage *less that 20 kV are generally* considered as *"short transmission lines"*.
- Due to smaller distance and lower line voltage the capacitance effects are extremely small and can be neglected. Hence the performance of such lines depend upon resistance and

inductance only. Further, although resistance and inductance are distributed along the line, yet in short lines these constants are *assumed to be lumped at one phase.*

2. **Medium transmission lines :**

- The transmission lines having *length between 50 km and 160 km* and *line voltage between 21 kV and 100 kV* are considered as *"medium transmission lines".*
- Due to sufficient length and voltage of the line the capacitance effects need to be considered. Although capacitance is distributed uniformly over the entire length of the line, yet it is *assumed to be concentrated at one or more points.*

3. **Long transmission lines :**

- The transmission lines having *length more than 160 km* and *line voltage above 100 kV* are considered as *"long transmission lines".*
- Here the *line constants are considered uniformly distributed* (not lumped) *over the entire length of the line and rigorous methods are employed for solution.*

14.3. REGULATION AND EFFICIENCY OF A TRANSMISSION LINE

Regulation :

The ***"regulation of a line"*** is defined as the *change in receiving end voltage, expressed in percent of full load voltage, from no-load to full-load, keeping the sending end voltage and frequency constant.*

Mathematically, $$\% \text{ Regulation} = \frac{V_{R0} - V_R}{V_R} \times 100 \qquad ...(14.1)$$

where, V_{R0} = Receiving end voltage under *no-load* condition.
(= *sending end voltage in a short line*),
V_R = Receiving end voltage under *full-load* condition.

It may be noted here that V_{R0} and V_R are the magnitudes of voltages.

- Knowledge of regulation helps in maintaining the voltage at the load terminals *within presented limits* (± 5 percent of declared voltage) by using suitable *voltage control equipment.*

Efficiency :

"Efficiency of a transmission line" *is defind as the ratio of power delivered at the receiving end to the power sent from the sending end.*

Mathematically, %Transmission efficiency, $$\eta_T = \frac{V_R I_R \cos \phi_R}{V_S I_S \cos \phi_S} \times 100 \qquad ...(14.2)$$

where, V_R, I_R, $\cos \phi_R$ are the receiving end voltage, current and power factor while V_S, I_S, $\cos \phi_S$ are the corresponding values at the sending end.

14.4. SHORT TRANSMISSION LINES

14.4.1. Single-phase Lines

In short transmission lines, as earlier stated, the *capacitance is usually neglected* ; only the total line resistance and inductance are taken into account. The circuit diagram (resistance and inductance shown as lumped or concentrated instead of being distributed) of the line and the corresponding phasor diagram of a single-phase line are shown in Fig. 14.1 (*a*) and (*b*).

Let, I = Load current,
R = Loop resistance (*i.e.* resistance of both the conductors),

X = Loop reactance,
V_R = Receiving end voltage,
$\cos \phi_R$ = Receiving end power factor (lagging),
V_S = Sending end voltage, and
$\cos \phi_S$ = Sending end power factor.

Refer to Fig. 14.1. (*b*).

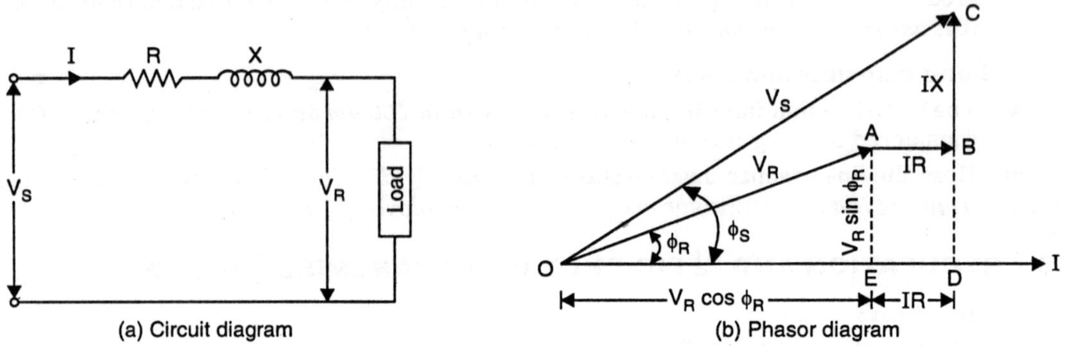

Fig. 14.1. Short transmission line.

$$V_S = OC = \sqrt{OD^2 + DC^2} = \sqrt{(OE + ED)^2 + (DB + BC)^2}$$

or,
$$V_S = \sqrt{(V_R \cos \phi + IR)^2 + (V_R \sin \phi_R + I_X)^2} \quad ...(14.3)$$

$\therefore$
$$\% \text{ Regulation} = \frac{V_S - V_R}{V_R} \times 100 \quad ...(14.4)$$

$$\text{Sending end power factor, } \cos \phi_S = \frac{OD}{OC} = \frac{V_S \cos \phi_R + IR}{V_S} \quad ...(14.5)$$

$$\% \text{ Efficiency} = \frac{\text{Power delivered}}{\text{Power sent out}} \times 100$$
$$= \frac{V_R I_R \cos \phi_R}{V_R I_R \cos \phi_R + I^2 R} \times 100 \quad ...(14.6)$$

An *approximate expression* for sending out voltage V_S can be obtained as follows :
Again, from the phasor diagram, we have

$$V_S \cos \phi_S = V_R \cos \phi_R + IR \quad ...(i)$$
$$V_S \sin \phi_S = V_R \sin \phi_R + IX \quad ...(ii)$$

Squaring and adding expressions (*i*) and (*ii*), we get

$$(V_S \cos \phi)^2 + (V_S \sin \phi_S)^2 = (V_R \cos \phi_R + IR)^2 + (V_R \sin \phi_R + IX)^2$$
$$V_S^2(\cos^2 \phi_S + \sin^2 \phi_S) = V_R^2 \cos^2 \phi_R + I^2R^2 + 2IR\, V_R \cos \phi_R + V_R^2 \sin^2 \phi_R + I^2X^2 + 2IX\, V_R \sin \phi_R$$
$$V_S^2 = V_R^2 + 2IR\, V_R \cos \phi_R + 2IX\, V_R \sin \phi_R + I^2(R^2 + X^2) \quad ...(iii)$$

or,
$$V_S = V_R \sqrt{1 + \frac{2IR \cos \phi_R}{V_R} + \frac{2IX \sin \phi_R}{V_R} + \frac{I^2}{V_R^2}(R^2 + X^2)}$$

In practice the *last term under the square root sign is generally negligible*, therefore,

$$V_S = V_R \left\{1 + \left(\frac{2IR}{V_R} \cos \phi_R + \frac{2I_X}{V_R} \sin \phi_R\right)\right\}^{1/2} \quad ...(iv)$$

The *terms within the simple brackets is small as compared to unity.* Using binomial expression and limiting only to second term, we have

$$V_S \simeq V_R + IR \cos \phi_R + IX \sin \phi_R \quad ...(14.7)$$

or,
$$V_S - V_R \simeq IR \cos \cos \phi_R + IX \sin \phi_R \quad ...[14.7(a)]$$

$$\therefore \quad \%\text{Regulation} = \frac{V_S - V_R}{V_R} \times 100 = \frac{IR \cos \phi_R + I_X \sin \phi_R}{V_R} \times 100 \quad ...(14.8)$$

$$= \frac{IR}{V_R} \cos \phi_R + \frac{IX}{V_R} \sin \phi_R \quad ...[14.8(a)]$$

$$\text{Regulation per unit} = \frac{IR}{V_R} \cos \phi_R + \frac{IX}{V_R} \sin \phi_R = v_R \cos \phi_R + v_X \sin \phi_R \quad ...(14.9)$$

where v_R and v_X are the per unit values of resistance and reactance of the line.

Alternatively, the approximate voltage drop $(V_S - V_R)$ can also be found as follows :

Refer to Fig. 14.2.

$$OC \simeq OF = OA + AF = OA + AD + DF$$

$$= OA + AD + BG$$

$$V_S \simeq V_R + IR \cos \phi_R + IX \sin \phi_R$$

$$\therefore \quad V_S - V_R \simeq IR \cos \phi_R + IX \sin \phi_R$$

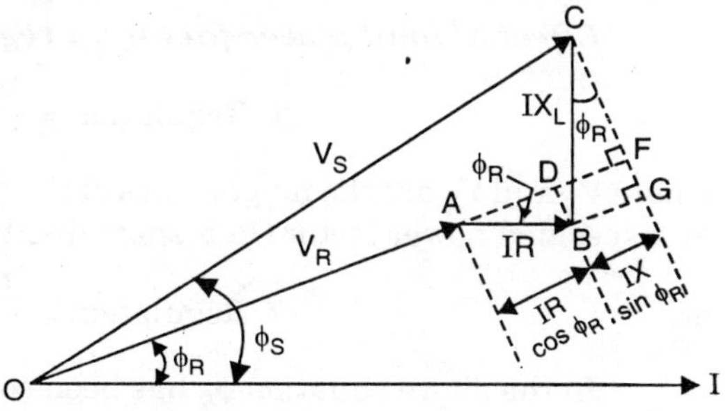

Fig. 14.2

It may be noted that approximate formula for $V_S (= V_R + IR \cos \phi_R + IX \sin \phi_R)$ gives *fairly correct results for lagging power factors; for leading power factors appreciable error is caused.* Hence *approximate expression for V_S should be used for lagging power factors only.*

14.4.2. Three-phase Lines

The electric power is transmitted by a 3-phase system, being economical one. This system may be regarded as consisting of three single-phase units, each wire transmitting one-third of the total power. Since similar conditions prevail in three phases, therefore, 3-phase system is generally analysed by considering one-phase only. Thus, expression for regulation, efficiency etc. derived for a single-phase line can also be applied to a 3-phase system. Fig. 14.3 shows a star-connected generator supplying a balanced star-connected load through a transmission line, each conductor having a resistance of $R\ \Omega$ and reactance $X\ \Omega$. Fig. 14.4 shows one phase separately.

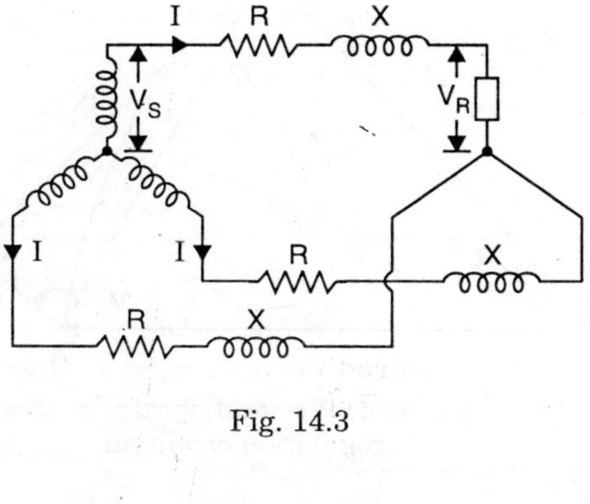

Fig. 14.3

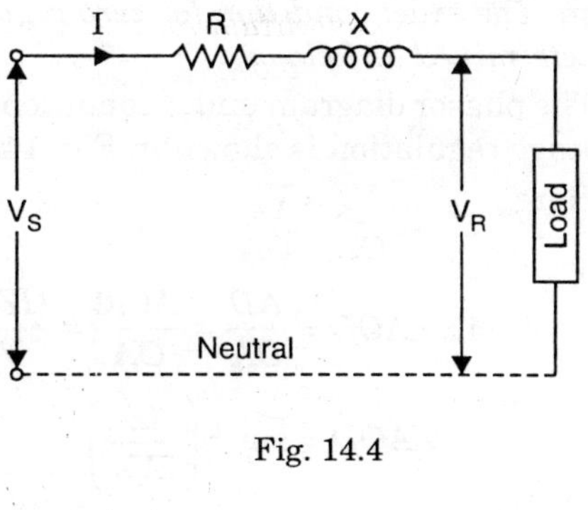

Fig. 14.4

14.4.3. Effect of Load Power Factor on Efficiency and Regulation of a Transmission Line

Effect of load power factor on efficiency of the line :

When, given power at a given voltage is delivered to the consumer, the line current is inversely proportional to the power factor of the load ($P = V_R I \cos \phi_R$ or $I = \dfrac{P}{V_R \cos \phi_R}$ for a single-phase line, or $I = \dfrac{P}{3V_R \cos \phi_R}$ for a 3-phase line). Thus, the current and hence power loss in the line goes on increasing as the power factor goes or decreasing. Hence the *"efficiency" of the line decreases with the fall in power factor* and vice versa.

Effect of load power factor on regulation of the short-line :

$$\% \text{ Regulation} = \frac{V_S - V_R}{V_R} \times 100$$

where V_S and V_R are the magnitudes of the no-load and full-load receiving end voltages respectively, at a specified power factor (In a short line V_s is equal to the no-load receiving end voltage).

or,
$$\% \text{ Regulation} = \frac{IR \cos \phi_R + IX \sin \phi_R}{V_R} \quad ...(i)$$

In the above equation ϕ_R has been considered + *ve* for *lagging load*. It will be – *ve* for *leading load*,

$$\% \text{ Regulation} = \frac{IR \cos \phi_R - IX \sin \phi_R}{V_R} \times 100 \quad ...(ii)$$

Voltage regulation becomes negative *i.e.* load voltage is more than no-load voltage when in equation (*ii*)

$$X \sin \phi_R > R \cos \phi_R \quad \text{or} \quad \tan \phi_R \text{ (leading)} > \frac{R}{X}$$

From eqn. (*ii*) it follows that for zero voltage regulation,

$$\tan \phi_R = \frac{R}{X} = \cot \theta$$

i.e.
$$\phi_R \text{ (leading)} = \frac{\pi}{2} - \theta \quad ...(iii)$$

where θ is the angle of the transmission line impedance. This is, however, an approximate condition. The *exact condition for zero regulation is determined as follows :*

The phasor diagram under conditions of zero voltage regulation is shown in Fig. 14.5.

Here,
$$V_S = V_R$$
$$OC = OA$$

$$\sin \angle AOD = \frac{AD}{OA} = \frac{AC/2}{OA} = \frac{(IZ/2)}{V_R}$$

$$\angle AOD = \sin^{-1}\left(\frac{IZ}{2V_R}\right)$$

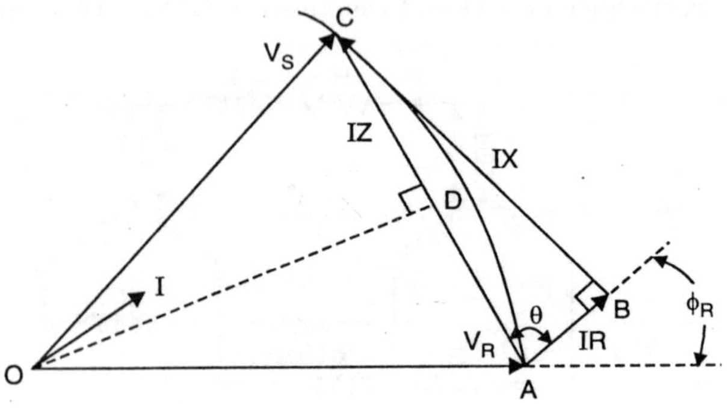

Fig. 14.5. Phasor diagram for zero regulation condition.

For zero voltage regulation, from the geometry of angles at A, we have

$$\phi_R \text{ (leading)} = \left[\frac{\pi}{2} - \theta + \sin^{-1}\left(\frac{IZ}{2V_R}\right)\right] \qquad \text{...(14.10)}$$

It may be observed from the above discussion that the voltage regulation of a line is amply dependent upon load power factor. The *voltage regulation of a transmission line improves/decreases as the power factor of the lagging load increased and it becames zero at a leading power factor given by eqn. 14.10.*

Example 14.1. *A single-phase overhead line is transmitting 1200 kW power to a factory at 11 kV and at 0.8 power factor lagging. The total resistance and loop reactance of the line are 3 Ω and 4.5 Ω respectively.*

Determine :

(i) The voltage at the sending end, *(ii) The sending end power factor,*

(iii) Percentage regulation, and *(iv) Transmission efficiency.*

Solution. *Given* : P = 1200 kW ; r = 3 Ω ; X = 4.5 Ω ; V_R = 11 kV = 16000 V ; cos ϕ_R = 0.8 (lag) the circuit diagram of the line and the corresponding phasor diagram are shown in Fig. 14.6 (*a*) and 14.6 (*b*).

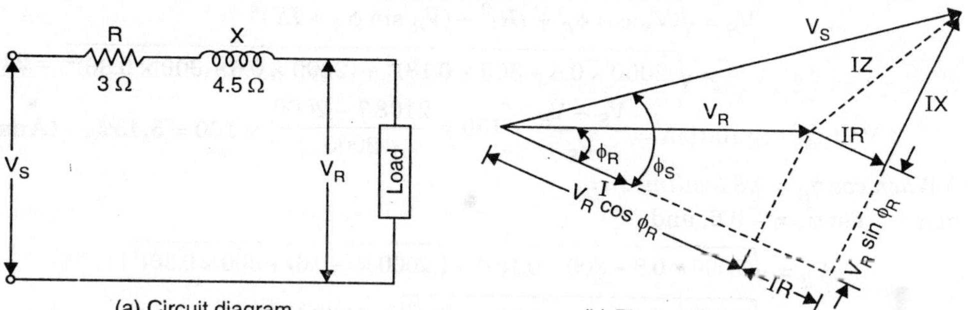

(a) Circuit diagram (b) Phasor diagram

Fig. 14.6

(*i*) **The voltage at the sending end, V_S :**

$$V_S = \sqrt{(V_R \cos\phi_R + IR)^2 + (V_R \sin\phi_R + IX)^2}$$

where, $$I = \frac{P}{V_R \cos\phi_R} = \frac{1200 \times 1000}{11000 \times 0.8} = 136.36 \text{ A}$$

and, $$\sin\phi_R = \sqrt{1 - \cos^2\phi_R} = \sqrt{1 - 0.8^2} = 0.6$$

$$\therefore \quad V_S = \sqrt{(11000 \times 0.8 + 136.36 \times 3)^2 + (11000 \times 0.6 + 136.36 \times 4.5)^2} = \mathbf{11698\ V.} \quad \textbf{(Ans.)}$$

(*ii*) **The sending end power factor, cos ϕ_s :**

$$\cos\phi_s = \frac{V_R \cos\phi_R + IR}{V_S} = \frac{11000 \times 0.8 + 136.36 \times 3}{11698} = \mathbf{0.7872\ lagging.} \quad \textbf{(Ans.)}$$

(*iii*) **Percentage regulation :**

$$\text{Percentage regulation} = \frac{V_S - V_R}{V_R} \times 100 = \frac{11698 - 11000}{11000} \times 100 = \mathbf{6.34\%.} \quad \textbf{(Ans.)}$$

($\because$ $V_S = V_R$ in a short line)

(*iv*) **Transmission efficiency, η_T :**

$$\eta_T = \frac{\text{Power delivered}}{\text{Power sent out}} = \frac{\text{Power delivered}}{\text{Power delivered + line losses}}$$

where, Line losses = $I^2R = (136.36)^2 \times 3 = 55782$ W or 55.782 kW

$$\therefore \qquad \eta_T = \frac{1200}{1200 + 55.782} = 0.9556 \text{ or } \mathbf{95.56\%.} \textbf{ (Ans.)}$$

Example 14.2. *A single-phase overhead transmission line is delivering 600 kVA load at 2 kV. Its resistance and reactance are 0.18 Ω and 0.36 Ω respectively. Determine the voltage regulation if the load power factor is (i) 0.8 lagging and (ii) 0.8 leading.*

Solution. *Given :* Power transmitted, P(kVA) = 600 ; V_R = 2kV = 2000 V ; R = 0.18 Ω, X = 0.36 Ω cos ϕ_R = 0.8 (lagging and leading).

Voltage regulation :

Load current, $$I = \frac{600 \times 1000}{2000} = 300 \text{ A}$$

(*i*) *When cos ϕ_R = 0.8 lagging :*

Then, sin ϕ_R = 0.6,

and, $$V_S = \sqrt{(V_R \cos\phi_R + IR)^2 + (V_R \sin\phi_R + IX)^2}$$

$$= \sqrt{(2000 \times 0.8 + 300 \times 0.18)^2 + (2000 \times 0.6 + 300 \times 0.36)^2} = 2108.7 \text{ V}$$

$$\therefore \quad \% \text{ Voltage regulation} = \frac{V_S - V_R}{V_S} \times 100 = \frac{2108.7 - 2000}{2000} \times 100 = \mathbf{5.43\%.} \quad \textbf{(Ans.)}$$

(*ii*) *When cos ϕ_R = 0.8 leading :*

Then, sin ϕ_R = – 0.6, and

$$V_S = \sqrt{(2000 \times 0.8 + 300 \times 0.18)^2 + [\{2000 \times (-0.6) + 300 \times 0.36\}^2]}$$

$$= \sqrt{(1600 + 54)^2 + (-1200 + 108)^2} \simeq 1982 \text{ V}$$

$$\therefore \quad \% \text{ Voltage regulation} = \frac{V_S - V_R}{V_R} \times 100 = \frac{1982 - 2000}{2000} \times 100 = \mathbf{-0.9\%.} \quad \textbf{(Ans.)}$$

Example 14.3. *An overhead 3-phase transmission line delivers 4000 kW at 11 kV at 0.8 power factor lagging. The resistance and reactance of earth conductor are 1.5 Ω and 4 Ω per phase respectively. Determine :*

(*i*) *The line sending end voltage.*

(*ii*) *Percentage regulation.*

(*iii*) *Transmission efficiency.* **(Banglore University)**

Solution. *Given* : P = 4000 kW ; V_{RL} = 11 kV = 11000 V ; cos ϕ_R = 0.8 lagging ; R = 1.5 Ω/phase ; X = 4Ω/phase.

Receiving end voltage/phase, $$V_R = \frac{11000}{\sqrt{3}} = 6350.8 \text{ V}$$

Line current, $$I = \frac{P}{3V_R \cos\phi_R} = \frac{4000 \times 1000}{3 \times 6350.8 \times 0.8} = 262.4 \text{ A}$$

(*i*) **The line sending end voltage V_{SL} :**

$$V_S \text{ (per phase)} = V_R + IR \cos\phi_R + IX \sin\phi_R \text{(approximately)}$$

$$= 6350.8 + 262.4 \times 1.5 \times 0.8 + 262.4 \times 4 \times 0.6 = 7295.4 \text{ V}$$

Line sending end voltage, $V_{SL} = \sqrt{3} \times 7295.4 = \mathbf{12636\ V.}$ **(Ans.)**

(*ii*) **Percentage regulation :**

$$\text{Percentage regulation} = \frac{V_S - V_R}{V_R} \times 100 = \frac{7295.4 - 6350.8}{6350.8} \times 100 = \mathbf{14.87\%.} \quad \textbf{(Ans.)}$$

(*iii*) **Transmission efficiency, η_T :**

$$\eta_T = \frac{\text{Power delivered}}{\text{Power sent out}} = \frac{\text{Power delivered}}{\text{Power delivered + line losses}}$$

where, Line losses = $3I^2R = 3 \times (262.4)^2 \times 1.5 = 309842$ W or 309.84 kW

$$\therefore \quad \eta_T = \frac{4000}{4000 + 309.84} = 0.928 = \mathbf{92.8\%.} \quad \textbf{(Ans.)}$$

Example 14.4. *A 3-phase, 50 Hz, 20 km long overhead line supplies 1100 kW at 11 kV, 0.8 power factor lagging. The line resistance and inductance are 0.04 Ω and 0.8 mH per phase per km. Determine :*

(*i*) *The line sending end voltage.*

(*ii*) *Percentage regulation.*

(*iii*) *Transmission efficiency of the line.*

Solution. *Given* : $f = 50$ Hz ; $l = 20$ km ; $P = 1100$ kW ; $V_{RL} = 11000$ V ; $\cos \phi_R = 0.8$; $R = 0.04$ Ω/km/phase ; $L = 0.8$ mH/km/phase.

Resistance of each conductor $R = 0.04 \times 20 = 0.8$ Ω

Reactance of each conductor, $X = 2\pi fL \times l = 2\pi \times 50 \times (0.8 \times 10^{-3}) \times 20 = 5.03$ Ω

$$\text{Phase voltage at receiving end } V_R = \frac{11000}{\sqrt{3}} = 6350.8\text{V}$$

$$\text{Line current, } I = \frac{P}{3V_R \cos \phi_R} = \frac{1100 \times 1000}{3 \times 6350.8 \times 0.8} = 72.17 \text{ A}$$

(*i*) **The line sending end voltage V_{SL} :**

Sending end voltage per phase,

$$V_S = V_R + IR \cos \phi_R + IX \sin \phi_R \text{ (approximately)}$$
$$= 6350.8 + 72.17 \times 0.8 \times 0.8 + 72.17 \times 5.03 \times 0.6 = 6614.8 \text{ V}$$

$$\therefore \quad V_{SL} = \sqrt{3}\, V_S = \sqrt{3} \times 6614.8 = \mathbf{11457.2 \ V.} \quad \textbf{(Ans.)}$$

(*ii*) **Percentage regulation :**

$$\text{Percentage regulation} = \frac{V_S - V_R}{V_R} \times 100 = \frac{6614.8 - 6350.8}{6350.8} \times 100 = \mathbf{4.16\%.} \quad \textbf{(Ans.)}$$

(*iii*) **Transmission efficiency, η_T :**

$$\eta_T = \frac{\text{Power delievered}}{\text{Power sent out}} = \frac{\text{Power delivered}}{\text{Power delivered – line losses}}$$

where, Line losses = $3I^2R = 3 \times (72.17)^2 \times 0.8$ W = 12.5 kW

$$\therefore \quad \eta_T = \frac{1100}{1100 + 12.5} = 0.988 \text{ or } \mathbf{98.8\%.} \quad \textbf{(Ans.)}$$

Example 14.5. *Calculate the distance over which a load of 16000 kW at 0.8 power factor lagging can be delivered by a 3-phase transmission line having conductors of steel cored aluminium each of resistance 1.2 Ω/km. The voltage at the receiving end is to be 132 kV and the loss in transmission is to be 8.5 percent of the load.* **(N.U.)**

Solution. *Given* : $P = 16000$ kW ; $\cos \phi_R = 0.8$; $R = 1.2$ Ω/km ; $V_{RL} = 132$ kV, Line losses = 8.5% of the load.

Length of the line, l :

Load current, $I = \dfrac{P}{\sqrt{3}\, V_{RL} \cos \phi_R} = \dfrac{16000}{\sqrt{3} \times 132 \times 0.8} = 87.48$ A

Line losses $= 3I^2R = 8.5\%$ of $16000 = 1360$ kW

or, $$R = \frac{1360 \times 1000}{3 \times (87.48)^2} = 59.24\ \Omega$$

$\therefore$ Length of line, $l = \dfrac{59.24}{1.2} =$ **49.4 km. (Ans.)**

Example 14.6. *A short 3-phase transmission line connected to a 33 kV, 50 Hz, generating station at the sending end is required to supply a load of 10 MW at 0.8 lagging power factor at 30 kV at the receiving end. If the minimum transmission efficiency is limited to 96 percent determine the per phase values of resistance and inductance of the line.* **(AMIE)**

Solution. *Given* : $V_{SL} = 33$ kV ; $f = 50$ Hz ; $P = 10$ MW $= 10000$ kW ; $\cos \phi_R = 0.8$ lagging $V_{RL} = 30$ kV ; $\eta_r = 96\%$.

R, L per phase :

Phase voltage at the receiving end, $V_R = \dfrac{30 \times 1000}{\sqrt{3}} = 17320$ V

Line current, $I = \dfrac{P}{3V_R \cos \phi_R} = \dfrac{10000 \times 1000}{3 \times 17320 \times 0.8} = 240.57$ A

Power sent out $= \dfrac{\text{Power delivered}}{\eta_T} = \dfrac{10000}{0.96} = 104.17$ kW

Line losses $= \dfrac{3I^2R}{1000} = 10417 - 10000 = 417$ kW

$\therefore$ $$R = \frac{417 \times 1000}{3 \times (240.57)^2} = \mathbf{2.4\Omega/phase.}\ \ \textbf{(Ans.)}$$

Phase voltage at the sending end, $V_S = \dfrac{V_{SL}}{\sqrt{3}} = \dfrac{33000}{\sqrt{3}} = 19052$ V

Also, $$V_S = V_R + IR \cos \phi_R + I X \sin \phi_R \text{ (app.)}$$

or, $$19052 = 17320 + 240.57 \times 2.4 \times 0.8 + 240.57 \times X \times 0.6 = 17781.9 + 144.3X$$

or, $$X = \frac{19052 - 17781.9}{144.3} = 8.8\ \Omega/\text{phase}$$

$\therefore$ $$L = \frac{8.8}{2\pi \times 50} = \mathbf{0.028\ H/phase.}\ \ \textbf{(Ans.)}$$

Example 14.7. *A 3-phase overhead line delivers 4200 kW at a power factor of 0.8 lagging to a load. If the sending end voltage is 33 kV, and resistance and reactance of each conductor are 4.5 Ω and 5.6 Ω respectively, determine :*

(i) The receiving end line voltage.

(ii) Line current.

(iii) Transmission efficiency.

Solution. *Given* : $P = 4200$ kW ; $\cos \phi_R = 0.8$ lagging ; $V_{SL} = 33$ kV ; $R = 4.5\ \Omega$; $X = 5.6\ \Omega$ per conductor.

(i) **Receiving end line voltage, V_{RL}.**

Sending end voltage/phase $V_S = \dfrac{33000}{\sqrt{3}} = 19052$ V

Let V_R be the phase voltage at the receiving end.

Line current, $I = \frac{P}{3V_R \cos \phi_R} = \frac{4200 \times 1000}{3\, V_R \times 0.8} = \frac{17.5 \times 10^5}{V_R}$, where V_R is in volts

Now, $V_S = V_R + IR \cos \phi_R + IX \sin \phi_R$ (appr.)

$$19052 = V_R + \frac{17.5 \times 10^5}{V_R} \times 4.5 \times 0.8 + \frac{17.5 \times 10^5}{V_R} \times 5.6 \times 0.6$$

or, $19052\, V_R = V_R^2 + 6300000 + 5880000$

or, $V_R^2 - 19052\, V_R + 12180000 = 0$

or, $V_R = \frac{19052 \pm \sqrt{(19052)^2 - 4 \times 1 \times 12180000}}{2} = \frac{19052 \pm 17727}{2}$

$= 18389$ V, considering +ve sign only

$\therefore \quad V_{RL} = \sqrt{3} \times 18389 = 31850$ V or **31.85 kV. (Ans.)**

(*ii*) **Line current, I :**

$$I = \frac{17.5 \times 10^5}{V_R} = \frac{17.5 \times 10^5}{18389} = \textbf{95.16 A.} \quad \textbf{(Ans.)}$$

(*iii*) **Transmission efficiency, η_T :**

Line losses $= 3I^2R = 3 \times (95.16)^2 \times 4.5 = 12248$ W or 122.248 kW

$$\therefore \quad \eta_T = \frac{\text{Power delivered}}{\text{Power delivered + Line losses}}$$

$$= \frac{4200}{4200 + 122.248} = 0.9717 \text{ or } \textbf{97.17\%.} \quad \textbf{(Ans.)}$$

Example 14.8. *A 3-phase load of 2200 kVA, 0.8 power factor lagging is supplied at 6.6 kV, 50 Hz by means of a 33 kV transmission line 25 km long and 33/6.6 kV step-down transformer. The resistance and reactance of each conductor are 0.35 Ω and 0.4 Ω per km respectively. The resistance and reactance of transformer primary are 6.4 Ω and 10.8 Ω, while those of secondary are 0.32 Ω and 0.6 Ω respectively. Determine the following when 6.6 kV is maintained at the receiving end :*

(*i*) *Sending end line voltage.*

(*ii*) *Sending end power factor.*

(*iii*) *Transmission efficiency.*

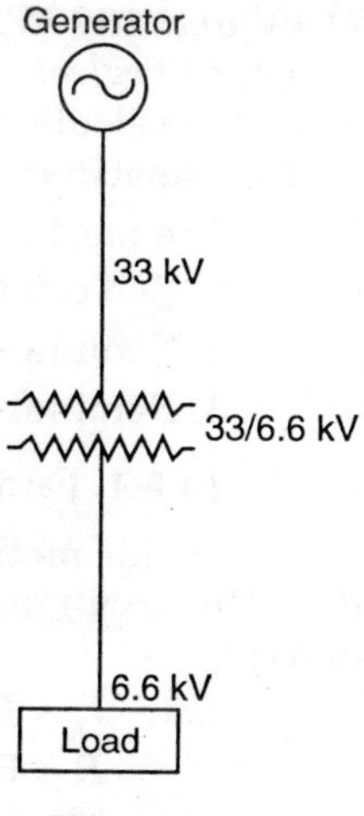

Fig. 14.7

Solution. *Given :* P(kVA) = 2200 ; $\cos \phi_R = 0.8$ lagging ;

Transformation ratio, $K = \frac{6.6}{33} = \frac{1}{5}$

Resistance of each conductor = 0.35 × 25 = 8.75 Ω

Reactance of each conductor = 0.4 × 25 = 10 Ω

Let us transfer the impedance of transformer secondary to H.V. side *i.e.* 33 kV side.

Equivalent resistance of transformer referred to 33 kV side

$= \text{Primary resistance} + 0.32 \times \frac{1}{K^2}$

$= 6.4 + 0.32 \times 25 = 14.4\ \Omega$

Equivalent reactance of transformer referred to 33 kV side

$$= \text{Primary reactance} + 0.6 \times \frac{1}{K^2}$$

$$= 10.8 + 0.6 \times 25 = 25.8\ \Omega$$

∴ Total resistance of the line and transformer, $R = 8.75 + 14.4 = 23.15\ \Omega$

Total reactance of the line and transformer, $X = 10 + 25.8 = 35.8\ \Omega$

(*i*) **Sending end line voltage, V_{SL} :**

Receiving end voltage/phase, $V_R = \frac{33 \times 1000}{\sqrt{3}} = 19052$ V

Line current, $I = \frac{2200 \times 1000}{\sqrt{3} \times 33000} = 38.49$ A

Also, sending end voltage/phase,

$$V_S = V_R + IR \cos \phi_R + IX \sin \phi_R \quad \text{(app.)}$$

$$= 19052 + 38.49 \times 23.15 \times 0.8 + 38.49 \times 35.8 \times 0.6$$

$$= 20592 \text{ V} \quad \text{or} \quad 20.592 \text{ kV}$$

∴ $V_{SL} = \sqrt{3} \times 20.592 = \mathbf{35.67\ kV.}$ **(Ans.)**

(*ii*) **Sending end power factor, $\cos \phi_s$:**

$$\cos \phi_s = \frac{V_R \cos \phi_R + IR}{V_S} = \frac{19052 \times 0.8 + 38.49 \times 23.15}{20592} = \mathbf{0.783\ lag.} \quad \textbf{(Ans.)}$$

(*iii*) **Transmission efficiency, η_T :**

Line losses $= 3I^2R = 3 \times (38.49)^2 \times 23.15 = 102889$ W or 102.889 kW

$$\eta_T = \frac{\text{Power delivered}}{\text{Power delivered + line losses}}$$

$$= \frac{2200 \times 0.8}{(2200 \times 0.8) + 102.889} = 0.9448 \quad \text{or} \quad \mathbf{94.48\%.} \quad \textbf{(Ans.)}$$

14.5. MEDIUM TRANSMISSION LINES

Medium transmission lines have *length between 50 km and 160 km and line voltage between 21 kV and 100 kV.* In such lines the *capacitance current of the system is appreciable and as such cannot be neglected.* The capacitance is distributed over the whole length of the line but in order to make the calculations simple, the line capacitance is *assumed to be lumped* or concentrated in the form of capacitors shunted across the line at one or more points.

The most commonly used methods known as "*localised capacitance methods*" are :

1. End condenser method.
2. Nominal-T method.
3. Nominal-π method.

14.5.1. End Condenser Method

In this method the capacitance of the transmission line is assumed to be lumped at the load end. The circuit diagram and the corresponding phasor diagram of the line are shown in Fig. 14.8 (*a*) and (*b*).

Let, I_R = Load current per phase,

R = Resistance per phase,

X = Inductive reactance per phase,

C = Capacitance per phase,

$\cos \phi_R$ = Receiving end power factor (lagging),
V_R = Receiving voltage per phase, and
V_S = Sending end voltage per phase.

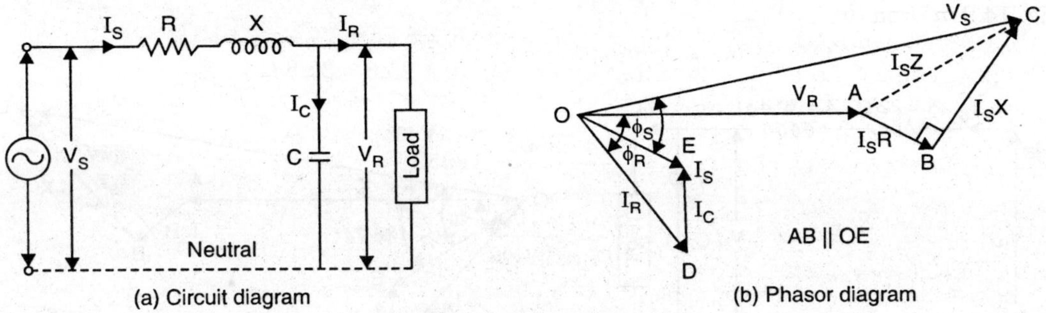

Fig. 14.8. End condenser method.

Taking the receiving end voltage $\overline{V}_R$ as the reference phasor/vector

$$\overline{V}_R = V_R(1 + j0)$$

Receiving end or load current, $\overline{I}_R = I_R(\cos \phi_R - j \sin \phi_R)$

Capacitive current, $\overline{I}_C = j\overline{V}_R\, \omega C = j\, 2\pi f C\, \overline{V}_R$

Sending end current, $\overline{I}_S = \overline{I}_R + \overline{I}_C = I_R(\cos \phi_R - j \sin \phi_R) + j2\pi f C \overline{V}_R$ ...(14.11)

Voltage drop per phase $= \overline{I}_S \overline{Z} = \overline{I}_S(R + jX)$

Sending end phase voltage, $\overline{V}_S = \overline{V}_R + \overline{I}_S \overline{Z} = V_R + \overline{I}_S(R + jX)$ ...(14.12)

Thus, the magnitude of sending end voltage V_s can be calculated.

$$\% \text{ Regulation} = \frac{V_S - V_R}{V_R} \times 100$$

$$\% \text{ Transmission efficiency, } \eta_T = \frac{\text{Power delivered/phase}}{\text{Power delivered phase + line losses/phase}} \times 100$$

$$= \frac{V_R I_R \cos \phi_R}{V_R I_R \cos \phi_R + I_S^2 R} \times 100 \quad \text{...(14.13)}$$

This method has the following *limitations / shortcomings* :

(*i*) It overestimates the effect of capacitance.

(*ii*) Since the distributed capacitance has been assumed to be lumped or concentrated there is considerable error (about 10 percent) in calculations.

Example 14.9. *A medium single-phase transmission line 110 km long has the following constants :*

Resistance / km = 0.2 Ω ; Reactance / km = 0.64 Ω ;

Susceptance / km = 15 × 10⁻⁶ ℧ ; Receiving end voltage = 66 kV.

Assuming that the total capacitance of the line is localised at the receiving end alone, determine :

(i) Sending end current, *(ii) Sending end voltage,*

(iii) Sending end power factor, *(iv) Percent regulation,*
(v) Line losses, and *(vi) Transmission efficiency.*
The line is delivering 16 MW at 0.8 power factor lagging.

Solution. The circuit diagram of the line and the corresponding phasor diagram are shown in Fig. 14.9 (*a*) and (*b*).

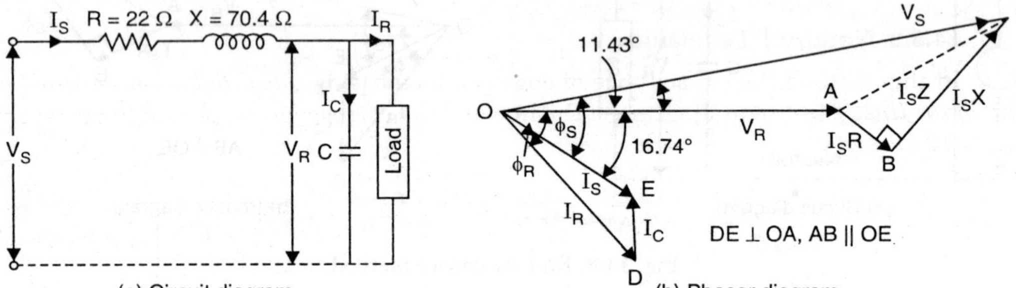

(a) Circuit diagram (b) Phasor diagram

Fig. 14.9. End condenser method.

Total resistance, $R = 0.2 \times 110 = 22\ \Omega$
Total reactance, $X = 0.64 \times 110 = 70.4\ \Omega$
Total susceptance, $Y = 15 \times 10^{-6} \times 110 = 16.5 \times 10^{-4}$ ℧
Receiving and voltage, $V_R = 66$ kV $= 66000$ V

$\therefore$ Load current, $I_R = \dfrac{16 \times 10^6}{66000 \times 0.8} = 303$ A

Taking receiving end voltage V_R as the rerference phasor, we have

$$\overline{V_R} = V_R(1 + j0)$$

$$\overline{I_R} = I_R(\cos \phi_R - j \sin \phi_R) = 303(0.8 - j\,0.6) = (242.4 - j\,181.8)$$

Charging (capacitive) current, $\overline{I_C} = jY \times V_R = j\,16.5 \times 10^{-4} \times 66000 = j\,108.9$ A

(*i*) **Sending end current, $\mathbf{I_S}$:**

$$\overline{I_S} = \overline{I_R} + \overline{I_C} = (242.4 - j\,181.8) + j\,108.9 = 242.4 - j\,72.9 = \mathbf{253.1 \angle -16.74°}. \quad \textbf{(Ans.)}$$

(*ii*) **Sending end voltage, $\overline{\mathbf{V_S}}$:**

$$\overline{V_S} = \overline{V_R} + \overline{I_S}\,\overline{Z} = \overline{V_R} + \overline{I_S}(R + jX) = (66000 + j0) + (242.4 - j72.9)(22 + j70.4)$$

$$= 66000 + (5332.8 + j17065 - j1603.8 + 5132.2)$$

$$= 76465 + j15461 = \mathbf{78012 \angle 11.43°}. \quad \textbf{(Ans.)}$$

(*iii*) **Sending end power factor, cos $\boldsymbol{\phi_s}$:**

$$\cos \phi_S = \cos\,[11.43 - (-16.74°)] = \cos 28.17° = \mathbf{0.8816\ lag.} \quad \textbf{(Ans.)}$$

(*iv*) **Percent regulation :**

$$\textit{Percent regulation} = \frac{V_S - V_R}{V_R} \times 100 = \frac{78012 - 66000}{66000} \times 100 = \mathbf{18.2\ \%}. \quad \textbf{(Ans.)}$$

(*v*) **Line Losses :**

$$\text{Line losses} = \frac{I_S{}^2 R}{1000}\ \text{kW} = \frac{(253.1)^2 \times 22}{1000} = \mathbf{1409.3\ kW.}\quad \textbf{(Ans.)}$$

(*vi*) **Transmission efficiency, η_T :**

$$\eta_T = \frac{\text{Power delivered}}{\text{Power delivered + line losses}} = \frac{16 \times 10^3}{16 \times 10^3 + 1409.3} = 0.919 \text{ or } \mathbf{91.9\ \%.}\quad \textbf{(Ans.)}$$

14.5.2. Nominal-T method

In this method the capacitance of each conductor is *assumed to be concentrated at the mid-point of the line*. The circuit diagram of the line and the corresponding phasor diagram are shown in Fig. 14.10 (*a*) and (*b*).

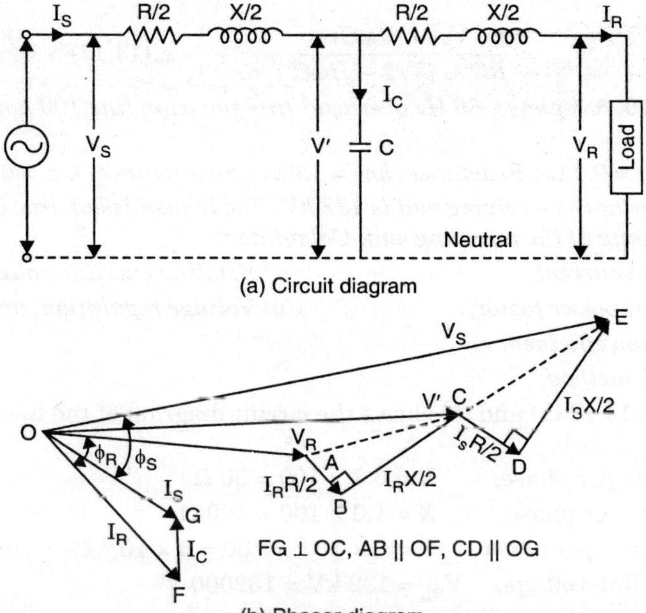

(a) Circuit diagram

(b) Phasor diagram

Fig. 14.10. Nominal T method.

Taking the receiving end voltage $\overline{V_R}$ as the reference phasor, we have,

$$\overline{V_R} = V_R\,(1 + j0), \text{ represented by phasor } OA.$$

Load current, $\quad \overline{I_R} = I_R\,(\cos \phi_R - j \sin \phi_R)$, represented by phasor OF

Voltage across the condenser C,

$$\overline{V'} = \overline{V_R} + \overline{I_R}\,\frac{\overline{Z}}{2} = \overline{V_R} + \overline{I_R}\left(\frac{R}{2} + j\,\frac{X}{2}\right)$$

$$= V_R(1 + j0) + I_R(\cos \phi_R - j \sin \phi_R)\left(\frac{R}{2} + j\,\frac{X}{2}\right),$$

represented by phasor OC ...(14.1[illegible])

Capacitance current,

$$\bar{I}_C = j\omega C\bar{V}' = j2\pi fC\bar{V}', \text{ represented by vector } FG(FG \perp OC) \quad ...(14.15)$$

Sending end current,

$$\bar{I}_S = \bar{I}_R + \bar{I}_C, \text{ represented by phasor } OG$$

Sending end voltage,

$$\bar{V}_S = \bar{V}' + \bar{I}_S \frac{\bar{Z}}{2} = \bar{V}' + \bar{I}_S \left(\frac{R}{2} + j\frac{X}{2}\right),$$

represented by phasor OE ...(14.16)

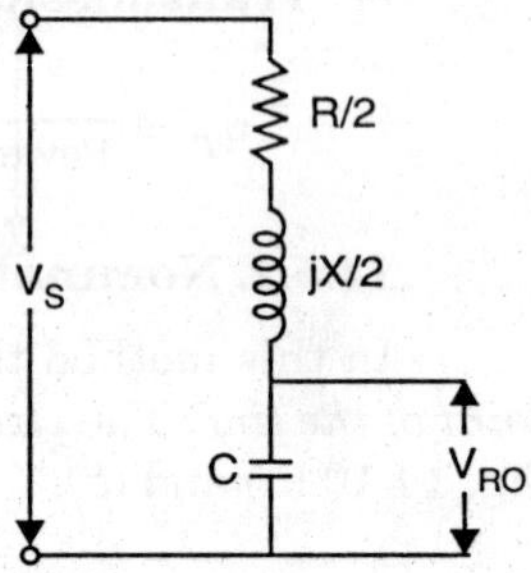

Fig. 14.11. Equivalent circuit under no-load.

In order to calculate *regulation* it is required to calculate V_{RO} the receiving end no-load voltage keeping V_S as calculated above fixed in magnitude. The nominal-T circuit for this reduces to the following (Refer to Fig. 14.11) :

$$V_{RO} = \frac{V_S\,(-j/\omega C)}{R/2 + jX/2 - j/\omega C} \quad ...(14.17)$$

Example 14.10. *A 3-phase, 50 Hz overhead transmission line 100 km long has the following constants :*

Resistance / km = 0.3 Ω ; Reactance / km = 1.0 Ω ; Susceptance / km = 6 × 10^{-6} ℧

The line voltage at the receiving end is 132 kV. The transmission line is delivering 50 MVA at 0.85 power factor lagging at the receiving end. Calculate :

(i) Sending end current, *(ii) Sending end line voltage,*
(iii) Sending end power factor, *(iv) Voltage regulation, and*
(v) Transmission efficinency.

Use nominal-T method.

Solution. Fig. 14.12 (*a*) and (*b*) shows the circuit diagram of the line and the corresponding phasor diagram.

Total resistance per phase, $R = 0.3 \times 100 = 30\ \Omega$

Total reactance per phase, $X = 1.0 \times 100 = 100\ \Omega$

Total susceptance per phase, $Y = 6 \times 10^{-6} \times 100 = 6 \times 10^{-4}$ ℧

Receiving end line voltage, $V_{RL} = 132\text{ kV} = 132000\text{ V}$

Receiving end phase voltage, $V_R = \dfrac{132000}{\sqrt{3}} = 76210\text{ V}$

Load current, $I_R = \dfrac{50 \times 10^6}{\sqrt{3} \times 132000} = 218.7\text{A}$

$$\cos\phi_R = 0.85,\ \sin\phi_R = \sqrt{1-\cos^2\phi_R} = \sqrt{1-(0.85)^2} = 0.527$$

Taking voltage at receiving end as the reference phasor, we have

$$\bar{V}_R = V_R(1 + j0) = 76210(1 + j0)V$$

Load current, $\bar{I}_R = I_R(\cos\phi_R - j\sin\phi_R)$

$$= 218.7(0.85 - j\,0.527 = (185.9 - j115.2)A$$

Line impedance, $\bar{Z} = R + jX = (30 + j100)\Omega$

$$\frac{\bar{Z}}{2} = (15 + j50)\ \Omega$$

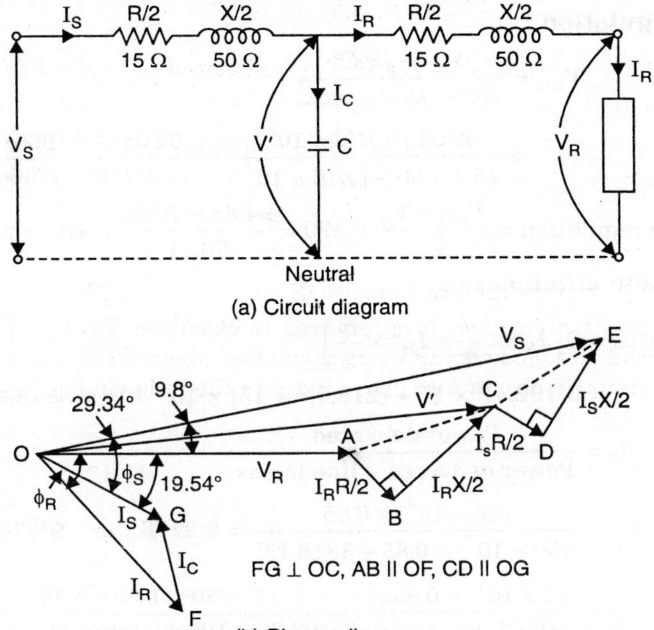

Fig. 14.12. Nominal-T method.

Phase voltage across the capacitor C, $\overline{V}' = \overline{V}_R + \overline{I}_R \dfrac{\overline{Z}}{2}$

$$= 76210 + (185.9 - j\,115.2)(15 + j\,50)$$
$$= 76210 + (2788.5 + j\,9295 - j1728 + 5760)$$
$$= 84758.5 + j\,7567$$

Charging current, $\overline{I}_C = jY\overline{V}' = j6 \times 10^{-4}(84758.5 + j\,7567)$

$$= j\,50.85 - 4.54$$

(*i*) **Sending ending current $\overline{I}_S$:**

$$\overline{I}_S = \overline{I}_R + \overline{I}_C = (185.9 - j\,115.2) + (j\,50.85 - 4.54)$$
$$= 181.36 - j\,64.35 = \mathbf{192.4 \angle -19.54°A.} \quad \textbf{(Ans.)}$$

(*ii*) **Sending ending line voltage, $\overline{V}_{SL}$:**

$$\overline{V}_S = \overline{V}' + \overline{I}_S \frac{\overline{Z}}{2}$$
$$= (84758.5 + j\,7567) + (181.36 - j64.35)(15 + j\,50)$$
$$= (84758.5 + j\,7567) + (2720.4 + j\,9068 - j965.25 + 3217.5)$$
$$= 90696.4 + j\,15669.8 = 92040 \angle 9.8°\text{V} \quad \text{or} \quad 92.04 \angle 9.8° \text{ kV}$$

$\therefore$ $\overline{V}_{SL} = \sqrt{3} \times 92.04 \angle 9.8° = 159.4 \times \angle 9.8°$ kV. **(Ans.)**

(*iii*) **Sending end power factor, $\cos \phi_S$:**

$$\cos \phi_S = \cos(9.8° - (-19.54)°) = \cos 29.34° = \mathbf{0.872.} \quad \textbf{(Ans.)}$$

(*iv*) **Voltage regulation :**

$$V_{RO} = \frac{V_S\,(-\,j/\omega C)}{R/2 + jX/2 - j/\omega C} \qquad ...(14.18)$$

$$= \frac{92.04\,[-\,j/(6 \times 10^4)]}{15 + j\,50 - [j/(6 \times 10^4)]} = \frac{92.04\,(-j\,1666)}{15 + j50 - j1666} = 94.88 \text{ kV}$$

$\therefore$ Percentage regulation $= \dfrac{V_{RO} - V_R}{V_R} \times 100 = \dfrac{94.88 - 76.21}{76.21} \times 100 = \mathbf{24.5\%.}$ **(Ans.)**

(*v*) **Transmission efficiency,** η_T **:**

$$\text{Line losses} = 3\left(I_S^{\,2} \times \frac{R}{2} + I_R^{\,2} \times \frac{R}{2}\right)$$

$$= 3[192.4)^2 \times 15 + (218.7)^2 \times 15] \times 10^{-3} \text{ kW} = 3818.13 \text{ kW}$$

$$\therefore \qquad \eta_T = \frac{\text{Power delivered}}{\text{Power delivered + line losses}}$$

$$= \frac{(50 \times 10^3) \times 0.85}{[(50 \times 10^3) \times 0.85 + 3818.13]} = 0.9176 \quad \text{or} \quad \mathbf{91.76\%.} \quad \textbf{(Ans.)}$$

$$\left[\text{Alternatively, } \eta_T = \frac{(50 \times 10^3) \times 0.85}{3V_S\, I_S \cos\phi_S} = \frac{(50 \times 10^3) \times 0.85}{3 \times 92.04 \times 192.4 \times \cos\,[9.8° - (-\,19.54°]} = 91.77\%\right]$$

Example 14.11. *A 220 kV, 3-phase, 150 km, 50 Hz transmission line delivers a load of 100 MW at 0.85 p.f. lagging. The line has per phase a total impedance of Z = (40 + j 125) ohms and a total shunt admittance Y = j 0.001 mho. Determine :*

(i) Sending end line voltage, *(ii) Sending end current,*
(iii) Sending end power factor, *(iv) Transmission efficiency,*
(v) Percentage regulation, and *(vi) Surge impedance loading of the line.*
Use nominal-T method. **(AMIE, 2000)**

Solution. Refer to Fig. 14.10. *Given* : V_{RL} = 220 kV ; f = 50 Hz ; l = 150 km,

Power delivered = 100 MW at 0.85 p.f. (lag) ; $\overline{Z}$ phase = (40 + j 125) Ω ; $\overline{Y}$/phase = j 0.001 ℧.

Receiving end phase voltage, $V_R = \dfrac{220}{\sqrt{3}} = 127$ kV

Load current, $I_R = \dfrac{100 \times 10^3}{\sqrt{3} \times 220 \times 0.85} = 308.74$ A

$$\cos\phi_R = 0.85,\ \sin\phi_R = \sqrt{1 - (0.85)^2} = 0.527$$

Taking voltage at the receiving end as the reference phasor, we have

$$\overline{V}_R = V_R(1 + j0) = 127(1 + j0)$$

Load current, $\overline{I}_R = I_R(\cos\phi_R - j\sin\phi_R) = 308.74(0.85 - j\,0.527) = (262.43 - j162.7)$ A

Phase voltage across the capacitor C, $\overline{V}' = \overline{V}_R + \overline{I}_R\,\dfrac{\overline{Z}}{2}$

$$= 127 + (262.43 - j\,162.7)\left(\frac{40 + j\,125}{2}\right) \times 10^{-3}$$

$$= 127 + 10^{-3}\,[(262.43 - j162.7)(20 + j62.5)] \text{ kV}$$

$$= 127 + 10^{-3}[5248.6 + j16402 - j3254 + 10168.8]$$
$$= 127 + 10^{-3}(15417 + j13148) = (142.42 + j13.15)\text{kV}$$

Charging current, $\bar{I}_C = jY\bar{V}' = j0.001(142.42 + j13.15) \times 10^3$ A

$$= (j142.42 - 13.15) \text{ A}$$

(*i*) **Sending end current, $\mathbf{I_S}$:**

$$\bar{I}_S = \bar{I}_R + \bar{I}_C = (262.43 - j162.7) + (j142.42 - 13.15)$$
$$= (249.28 - j20.28) = 250 \angle -4.65°$$

(*ii*) **Sending end line voltage, $\mathbf{V_{SL}}$:**

$$\bar{V}_S = \bar{V}' + \bar{I}_S \frac{\bar{Z}}{2}$$
$$= (142.42 + j13.15) + (249.28 - j20.28)\left(\frac{40 + j\,125}{2}\right)$$
$$= (142.42 + j13.15) + (249.28 - j20.28)(20 + j62.5) \times 10^{-3} \text{ kV}$$
$$= (142.42 + j13.15) + 10^{-3}(4985.6 + j15580 - j405.6 + 1267.5)$$
$$= (142.42 + j13.15) + 10^{-3}(6253 + j15174) = (148.67 + j28.32) \text{ kV}$$
$$= 151.34 \angle 10.78° \text{ kV}$$

$$\therefore \quad V_{SL} = \sqrt{3} \times 151.34 = \mathbf{262.13\ kV.} \quad \textbf{(Ans.)}$$

(*iii*) **Sending end power factor, cos ϕ_S :**

$$\cos \phi_S = \cos [10.78° - (-465°)] = \cos 15.43° = \mathbf{0.964.} \quad \textbf{(Ans.)}$$

(*iv*) **Transmission efficiency, η_T :**

$$\eta_T = \frac{\text{Receiving end power}}{\text{Sending end power}} = \frac{100 \times 10^3 \text{ (kW)}}{3V_S\, I_S \cos \phi_S \text{ (kW)}}$$
$$= \frac{100 \times 10^3}{3 \times 151.34 \times 250 \times 0.964} = 0.914 \quad \text{or} \quad \mathbf{91.4\%.} \quad \textbf{(Ans.)}$$

(*v*) **Percentage regulation :**

$$V_{RO} = \frac{V_S\,(-j/\omega C)}{R/2 + j\,X/2 - j/\omega C} = \frac{151.34\,(-j10^3)}{20 + j62.5 - j10^3} = \frac{151.34\,(-j10^3)}{20 - j\,937.5} = 161.39 \text{ kV}$$

$$\therefore \quad \text{Percentage regulation} = \frac{V_{RO} - V_R}{V_R} \times 100 = \frac{161.39 - 127}{127} \times 100 = \mathbf{27.08\%.} \quad \textbf{(Ans.)}$$

(*vi*) **Surge impedance of the line, $\mathbf{Z_{surge}}$:**

Surge impedance of a line is defined as the square root of Z/Y *i.e.* $Z_{\text{surge}} = \sqrt{\frac{Z}{Y}}$

where, Z = Series impedance (= $R + jX$), and

Y = Shunt admittance.

$$\therefore \quad Z_{\text{surge}} = \sqrt{\frac{40 + j125}{j0.001}} = \sqrt{\frac{(40 + j125) \times j0.001}{j0.001 \times j0.001}} = \sqrt{-10^6(j0.04 - 0.125)}$$
$$= 10^3\sqrt{(0.125 - j0.04)} = \mathbf{131.24\ \Omega.} \quad \textbf{(Ans.)}$$

14.5.3. Nominal-π Method

In this method the capacitance of each conductor is assumed to be divided into two halves, one half being shunted between conductor and neutral at the receiving end and other half at the sending end. The circuit diagram of the line and the corresponding phasor diagram are in shown in Fig. 14.13(*a*), (*b*).

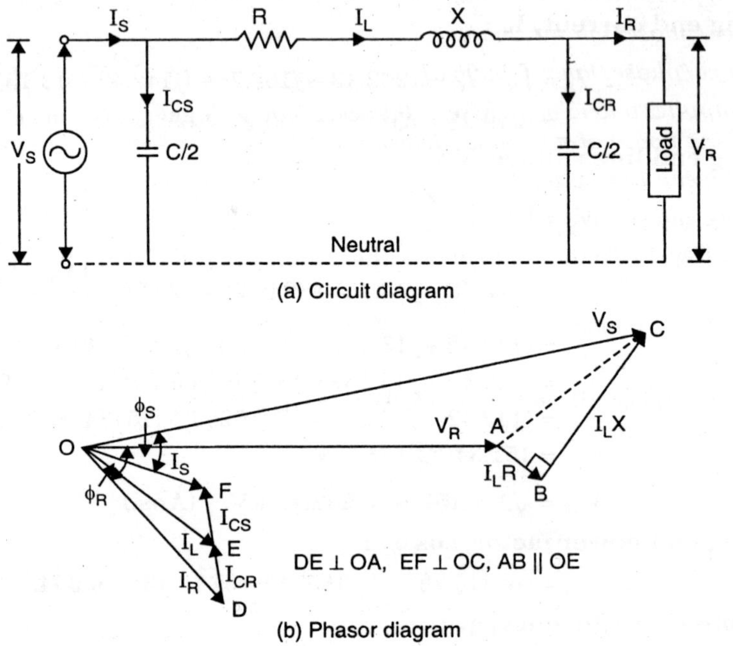

Fig. 14.13. Nominal π method.

Taking the receiving end voltage $\bar{V}_R$ as the reference phasor, we have

$$\bar{V}_R = V_R(1 + j0), \text{ represented by phasor } OA.$$

Receiving end current, $\bar{I}_R = I_R(\cos \phi_R - j \sin \phi_R)$, represented by phasor OD

Charging current at the receiving end $\bar{I}_{CR} = j\omega(C/2)\bar{V}_R$

$$= j\pi f C \bar{V}_R, \text{ represented by DE(DE} \perp \text{OA)}$$

Line current, $\bar{I}_L = \bar{I}_R + \bar{I}_{CR}$, represented by OE

Sending end voltage,

$$\bar{V}_S = \bar{V}_R + \bar{I}_L \bar{Z} = \bar{V}_R + \bar{I}_L\ (R + jX), \text{ represented by } OC. \quad ...(14.19)$$

Charging current at the sending ending,

$$\bar{I}_{CS} = j\omega\ (C/2)\ \bar{V}_S = j\pi f C \bar{V}_S, \text{ represented by } EF(EF \perp OC)$$

∴ Sending end current,

$$\bar{I}_S = I_L + \bar{I}_{CS}, \text{ represented by } OF \quad ...(14.20)$$

After calculating the sending end voltage, it is required to find out no-load receiving end voltage for regulation keeping sending end voltage constant in magnitude. The nominal – π-circuit for this reduces to Fig. 14.14.

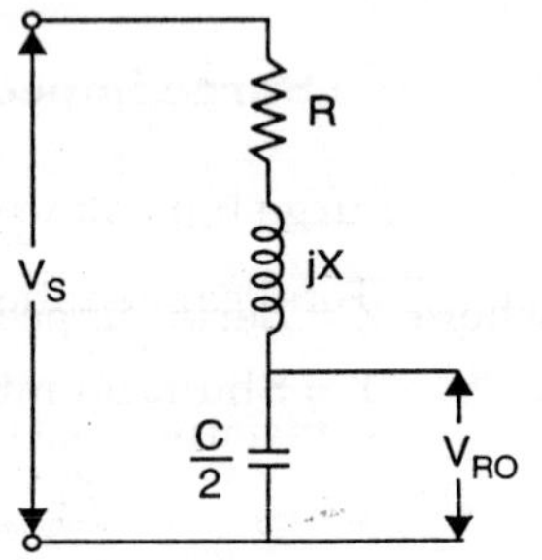

Fig. 14.14. Equivalent circuit under no-load.

$$V_{RO} = \frac{V_S\,[-\,j/(\omega C/2)]}{R + jX - j/(\omega C/2)} = \frac{V_S\,(-\,2j/\omega C)}{R + jX - 2j/\omega C} \qquad \text{...(14.21)}$$

Example 14.12. *A 3-phase, = 0.15 ohm 50 Hz, 150 km long overhead line has the following constants :*

Resistance/phase/km = 0.15 ohm

Inductance/phase/km = 1.2 mH

Capacitance/phase/km = 0.0096 μF

The line supplies a load of 30 MW at 0.9 power factor lagging at a line voltage of 132 kV at the receiving end. Using nominal-π method, determine :

(i) Sending end line voltage, *(ii) Sending end current, and*

(iii) Sending end power factor.

Solution. Fig. 14.15 (*a* and *b*) shows the circuit diagram of the line and the corresponding phasor diagram.

Total resistance/phase, $R = 0.15 \times 150 = 22.5\ \Omega$

Total inductance/phase $L = (1.2 \times 10^{-3}) \times 150 = 0.18$ H

Inductive reactance/phase, $X = 2\pi f L = 2\pi \times 50 \times 0.18 = 56.55\ \Omega$

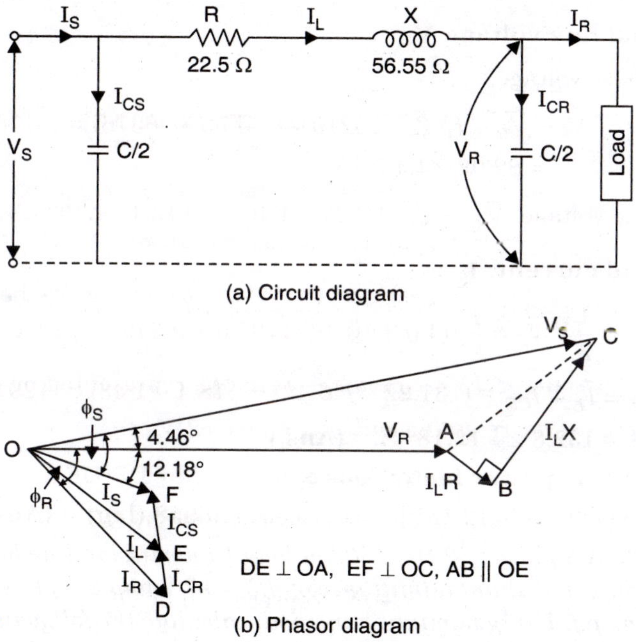

(a) Circuit diagram

(b) Phasor diagram

Fig. 14.15. Nominal-π method.

Total capacitance/phase, $C = (0.0096 \times 10^{-6}) \times 150 = 1.44 \times 10^{-6}$ F

Capacitive reactance/phase $X_C = \dfrac{1}{2\pi f C} = \dfrac{1}{2\pi \times 50 \times 1.44 \times 10^{-6}}$

$= 2210.5\ \Omega$

Total susceptance/phase, $\overline{Y} = j\dfrac{1}{Z} = j\dfrac{1}{X_C} = j\left(\dfrac{1}{2210.5}\right) = j4.5 \times 10^{-4}\ ℧$

Receiving end current, $I_R = \dfrac{30 \times 10^6}{\sqrt{3} \times (132 \times 1000) \times 0.9} = 145.8$ A

Phase voltage at the receiving end, $V_R = \dfrac{132 \times 1000}{\sqrt{3}} = 76210$ V (or 76.21 kV)

$$\cos \phi_R = 0.9, \sin \phi_R = 0.436$$

Taking phase voltage at receiving end as reference vector, we have

$$\overline{V_R} = V_R(1 + j0) = 76210(1 + j0) = (76210 + j0)$$

$$\overline{I_R} = I_R(\cos \phi_R - j \sin \phi_R) = 145.8(0.9 - j0.436) = (131.22 - j63.57) \text{ A}$$

$$\overline{I_{CR}} = \frac{1}{2}\overline{Y}\overline{V_R} = \frac{1}{2}(j4.5 \times 10^{-4}) \times (76210 + j0) = j17.15 \text{ A}$$

Line current, $\overline{I_L} = \overline{I_R} + \overline{I_{CR}} = (131.22 - j63.57) + j17.15 = (131.22 - j46.42)$ A

Impedance drop in line $= \overline{I_L}\overline{Z} = (131.22 - j46.42)(22.5 + 56.55)$

$= 2952.4 + j7420.5 - j1044.5 + 2625.1$

$= (5577.5 + j6376)$ V

(*i*) **Sending end line voltage $\overline{\mathbf{V}_{SL}}$:**

Sending end phase voltage,

$$\overline{V_S} = \overline{V_R} + \overline{I_L}\,\overline{Z} = 76210 + (5577.5 + j6376) = (81787.5 + j6376) \text{ V}$$

$$= 82.03 \angle 4.46° \text{ kV}$$

Sending end line voltage, $\overline{V_{SL}} = \sqrt{3}(82.03 \angle 4.46°) = \mathbf{142.1 \angle 4.46°}$ **kV. (Ans.)**

(*ii*) **Sending end current, $\overline{\mathbf{I}_S}$:**

$$\overline{I_{CS}} = \frac{1}{2}\overline{Y}\,\overline{V_S} = \frac{1}{2}j(4.5 \times 10^{-4})(81787.5 + j6376) = (j18.4 - 1.43) \text{ A}$$

$$\overline{I_S} = \overline{I_L} + \overline{I_{CS}} = (131.22 - j46.42) + (j18.4 - 1.43) = (129.8 - j28.02) \text{ A}$$

$= \mathbf{132.8 \angle -12.18°}$ **A. (Ans.)**

(*iii*) **Sending ending power factor, $\cos \phi_s$:**

$\cos [4.46° - (-12.18°)] = \cos 16.64° = \mathbf{0.958}$ **(lag). (Ans.)**

Example 14.13. *A 3-phase, 50 Hz, 150 km long transmission line has a series impedance of (15 + j75) ohms per phase and shunt admittance of 4.5 × 10⁻⁴ mho per phase. The load is 50 MW at 110 kV with 0.8 lagging p.f. Using nominal-π method, calculate the following :*

(*i*) *Sending end line voltage,* (*ii*) *Sending end current,*

(*iii*) *Sending end power factor,* (*iv*) *Percentage regulation, and*

(*v*) *Transmission efficiency.*

Solution. Refer to Fig. 14.13.

Resistance per phase, $R = 15\ \Omega$

Inductive reactance per phase, $X = 75\ \Omega$, } *i.e.*, $\overline{Z} = (15 + j75)\ \Omega$

Shunt admittance per phase, $\overline{Y} = j4.5 \times 10^{-4}$ ℧

Receiving end current, $I_R = \dfrac{50 \times 10^6}{\sqrt{3} \times (110 \times 1000) \times 0.8} = 328$ A

Phase voltage at the receiving end, $V_R = \dfrac{110}{\sqrt{3}} = 63.51$ kV

$\cos \phi_R = 0.8$, $\sin \phi_R = 0.6$

Taking phase voltage at the receiving end as the reference phasor, we have

$$V_R = 63.51(1 + j0)\text{kV}$$

$$\bar{I}_R = I_R(\cos \phi_R - j \sin \phi_R) = 328(0.8 - j0.6) = (262.4 - j196.8) \text{ A}$$

$$\bar{I}_{CR} = \frac{1}{2}\bar{Y}\,\bar{V}_R = \frac{1}{2} \times (j\,4.5 \times 10^{-4}) \times 63.51(1 + j0) \times 10^3 = j14.3 \text{ A}$$

Line current, $\bar{I}_L = \bar{I}_R + \bar{I}_{CR} = (262.4 - j196.8) + j14.3 = (262.4 - j182.5)$

$= 319.62 \angle -34.8°$ A

Impedance drop in line $= \bar{I}_L\,\bar{Z} = (262.4 - j182.5)(15 + j75)$

$= (3936 + j19680 - j2738 + 13688)\ \Omega$

$= (17624 + j16942)$ V

(*i*) **Sending end line voltage, $\bar{V}_{SL}$:**

Sending end phase voltage, $\bar{V}_S = \bar{V}_R + \bar{I}_L\bar{Z}$

$= 63.51 + (17624 + j16942) \times 10^{-3}$ kV

$= 63.51 + (17.62 + j\,16.94) = (81.13 + j\,16.94)$ kV

$= 82.88 \angle 11.8°$ A

$\therefore$ $\bar{V}_{SL} = \sqrt{3} \times 82.88 \angle 11.8° =$ **143.55 ∠11.8° kV. (Ans.)**

(*ii*) **Sending end current $\bar{I}_S$:**

$$\bar{I}_{CS} = \frac{1}{2}\bar{Y}\,\bar{V}_S = \frac{1}{2}j(4.5 \times 10^{-4})(81.13 + j16.94) \times 10^3$$

$= (j18.25 - 3.81)$ A

$\bar{I}_S = \bar{I}_L + I_{CS} = (262.4 - j182.5) + (j18.25 - 3.81)$

$= (258.59 - j164.25)$ A $=$ **306.3 ∠ – 32.4°. (Ans.)**

(*iii*) **Sending end power Factor, $\cos \phi_s$:**

$\cos[11.8° - (-32.4°)] = \cos 44.2° =$ **0.717 (lag). (Ans.)**

(*iv*) **Percentage regulation :**

$$V_{RO} = \frac{V_S(-2j/\omega C)}{R + jX - 2j/\omega C} \quad \text{...Eqn. (14.21)}$$

$$= \frac{82.88\,[-2j/(4.5 \times 10^{-4})]}{15 + j75 - [2j/(4.5 \times 10^{-4})]} = \frac{82.88(-j4444.4)}{15 + j75 - j4444.4}$$

$$= \frac{82.88(-j4444.4)}{15 - j4369.4} = 84.3 \text{ kV}$$

$\therefore$ Percentage regulation $= \dfrac{V_{RO} - V_R}{V_R} \times 100 = \dfrac{84.3 - 63.51}{63.51} \times 100 =$ **32.7%. (Ans.)**

(*v*) **Transmission efficiency, η_T :**

$$\eta_T = \frac{\text{Power delivered}}{\text{Power sent out}}$$

$$= \frac{50 \times 10^3}{3\, V_S I_S \cos \phi_S} = \frac{50 \times 10^3}{3 \times 82.88 \times 306.3 \times 0.717} = 0.9156\% \text{ or } \mathbf{91.56\%}. \quad \textbf{(Ans.)}$$

[*Alternatively*, $\eta_T = \dfrac{\text{Power delivered}}{\text{Power delivered + line losses}}$

$$= \frac{50 \times 10^3}{50 \times 10^3 + 3I_L^2 R} = \frac{50 \times 10^3}{50 \times 10^3 + 3 \times (319.62)^2 \times 15 \times 10^{-3}} = 0.9158 \text{ or } \mathbf{9158\%}. \quad \textbf{(Ans.)}$$]

Example 14.14. *Determine the sending end voltage and voltage regualtion of a 250 km 3-phase, 50 Hz, transmission line delivering 25 MVA at 0.8 lagging p.f. to a balanced load at 132 kV. The line conductors are spaced equilaterally 3 m apart. The conductor resistance is 0.11 ohm/km and its effective diameter is 16 mm. Neglect leakance*

Use the nominal-π method. **(M.U.)**

Solution. Length of the line = 250 km

Load delivered = 25 MVA at 0.8 p.f. lagging

Distance between line conductors, D = 3 m

Radius of each conductor, $r = \dfrac{16}{2}$ = 8 mm = 0.008 m

Conductor resistance = 0.11 Ω/km

∴ Total resistance/phase **R** = 0.11 × 250 = 27.5 Ω

Inductance of each conductor/km/phase,

$$L = 2 \times 10^{-7} \ln \left(\frac{D}{D_s} \right) \times 10^3 \text{ H/km}$$

where, $D_s = 0.7788r = 0.7788 \times 0.008$ m

$$\therefore \quad L = 2 \times 10^{-7} \ln \left(\frac{3}{0.7788 \times 0.008} \right) = 0.00124 \text{ H/m}$$

$$= 1.24 \text{ mH/km}$$

∴ Total inductive reactance/phase,

$$\mathbf{X} = 2\pi f L \times 250 = 2\pi \times 50 \times 0.00124 \times 250 = 97.4\ \Omega$$

Total impedance/phase, $\overline{Z} = R + jX = (27.5 + j\,97.4)\ \Omega$

Capacitance of each conductor/km/phase,

$$C = \frac{2\pi\varepsilon_0}{\ln\left(\dfrac{D}{r}\right)} \times 10^3 \text{ F/km}$$

$$= \frac{2\pi \times 8.854 \times 10^{-12}}{\ln\left(\dfrac{3}{0.008}\right)} \times 10^3 = 9.4 \times 10^{-9} \text{ F/km}$$

Total admittance/phase, $\overline{Y} = j\omega C \times 250 = j \times 2\pi f C \times 250$

$$= j \times 2\pi \times 50 \times 9.4 \times 10^{-9} \times 250 = j7.38 \times 10^{-4}\ ℧$$

Receiving end current, $I_R = \dfrac{25 \times 10^3}{\sqrt{3} \times 132} = 109.3$ A

Receiving end phase voltage, $V_R = \dfrac{132}{\sqrt{3}} = 76.2$ kV

$$\cos \phi_R = 0.8, \sin \phi_R = 0.6$$

Taking phase voltage at the receiving end as reference phase, we have (Refer to Fig. 14.13) :

$$\overline{V}_R = 76.2\,(1 + j0) V$$

$$\overline{I}_R = 109.3(0.8 - j0.6) = (87.44 - j\,65.58) \text{ A}$$

$$\overline{I}_{CR} = \frac{1}{2}\overline{Y}\,\overline{V}_R = \frac{1}{2} \times j7.38 \times 10^{-4} \times 76.2(1 + j0) \times 10^3 = j28.12 \text{ A}$$

Line current, $\overline{I}_L = \overline{I}_R + \overline{I}_{CR} = (87.44 - j65.58) + j28.12 = (87.44 - j37.46)$ A

Impedance drop in line $= \overline{I}_L\overline{Z} = (87.44 - j37.46)(27.5 + j97.4)$

$$= 2404.6 + j\,8516.7 - j1030.2 + 3648.6 = (6053.2 + j7486.5) \text{ V}$$

Sending end line voltage $\overline{V}_{SL}$:

Sending ending phase voltage, $\overline{V}_S = \overline{V}_R + \overline{I}_R\,\overline{Z} = 76.2 + (6053.2 + j7486.5) \times 10^{-3}$

$$= (82.25 + j7.49) \text{ kV}$$

$$= 82.6 \angle 5.2°$$

∴ **Sending end line voltage** $\overline{V}_{SL} = \sqrt{3} \times 82.6 \angle 5.2° =$ **143.1 ∠5.2° kV. (Ans.)**

Voltage regulation :

$$V_{RO} = \frac{V_S\,(-2j/\omega C)}{[R + jX - 2j/\omega C]} = \frac{82.6\,[-2j/(7.38 \times 10^{-4})}{27.5 + j97.4 - [-2j/(7.38 \times 10^{-4})]}$$

$$= \frac{82.6\,(-j2710)}{25 + j97.4 - j2710} = \frac{82.6\,(-j2710)}{25 - j2612.6} = 85.68 \text{ kV}$$

∴ Percentage regulation $= \dfrac{V_{RO} - V_R}{V_R} \times 100 = \dfrac{85.68 - 76.2}{76.2} \times 100 =$ **12.4%. (Ans.)**

14.6. GENERALISED CIRCUIT CONSTANTS

In case of any linear network having two input and two output terminals the input voltage and input current can be expressed in terms of output voltage and output current. A transmission can be represented as a network with two input terminals where the power enters the network, and two output terminals, where the power leaves, the network. This type of circuit is *passive* (since it does not contain any source of e.m.f.), *linear* (as its impedance is independent of current flowing and *bilateral* (as its impedance is independent of direction of flow of current).

The sending end input voltage per phase ($\overline{V}_S$) and the sending end input current ($\overline{I}_S$) of a 3-phase transmission line can be expressed as :

$$\overline{V}_S = A\overline{V}_R + B\overline{I}_R \qquad ...(14.22)$$

$$\overline{I}_S = C\overline{V}_R + D\overline{I}_R \qquad ...(14.23)$$

where, $\overline{V}_S$ = Sending end voltage per phase,

$\overline{I}_S$ = Sending end current per phase,

$\overline{V}_R$ = Receiving end voltage per phase, and

$\overline{I}_R$ = Receiving end current.

A, B, C, D are called the *Generalised Circuit Constants* (generally complex numbers) of the transmission line. The values of these constants depend upon the particular method of solution adopted. After determination of their values the performance of the line can be easily evaluated.

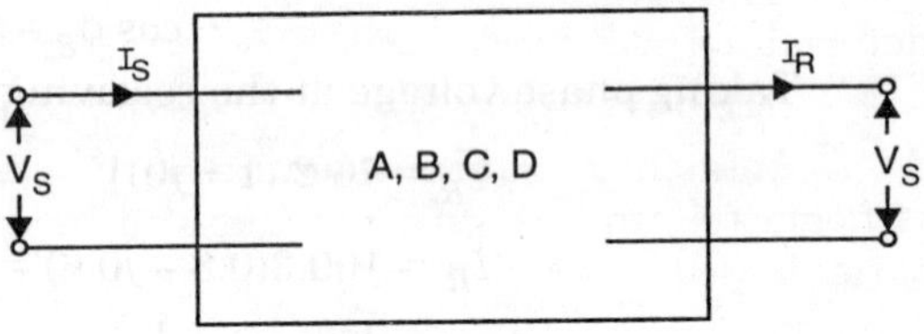

Fig. 14.16 (*a*)

The values of *A, B, C* and *D* can be determined as follows :

On open circuit :

Receiving end current, $\overline{I}_R = 0$

$$\therefore \quad A = \frac{\overline{V}_S}{\overline{V}_R} \quad ...(14.24)$$

and, $$C = \frac{\overline{I}_S}{\overline{V}_R} \quad ...(14.25)$$

On short circuit :

Receiving end voltage, $\overline{V}_R = 0$

$$\therefore \quad B = \frac{\overline{V}_S}{\overline{I}_R} \quad ...(14.26)$$

and, $$D = \frac{\overline{I}_S}{\overline{I}_R} \quad ...(14.27)$$

From the above expressions constants *A, B, C* and *D* can be defined as follows :

Constant A **:** It is defined as the *ratio of sending end voltage* $\overline{V}_S$ *to receiving end voltage* $\overline{V}_R$ *when the line is open circuited on receiving end.*

- This constant is *dimensionless* and usually a complex number.

Constant B **:** It is defined as the ratio of sending end voltage $\overline{V}_S$ to receiving end current $\overline{I}_R$ when line is short-circuited on receiving end.

- It is also a complex number and has the *unit of ohm.*

Constant C **:** It is defined as the *ratio of sending ending current* $\overline{I}_S$ *to receiving end voltage* $\overline{V}_R$ *when the line is open-circuted on the receiving end.*

- It is also a complex number and has the *unit of mho.*

Constant D **:** It is defined as the *ratio of sending end current* $\overline{I}_S$ *to the receiving end current* $\overline{I}_R$ *when the line is short-circuited.*

- It is *dimensionless* and usually a complex number.

For passive network : $AD - BC = 1$ ...(14.28)

For symmetrical network : $A = D$ ...(14.29)

Determination of Generalised Constants for Transmission Lines :

We know that,

$$\overline{V}_S = A\overline{V}_R + B\overline{I}_R \quad ...(i)$$

$$\overline{I}_S = C\overline{V}_R + D\overline{I}_R \quad ...(ii)$$

The values of generalised constants A, B, C and D for various methods of solution adopted are determined as follows :

1. Short lines (*Simple series impedance method*) :

In a short transmission lines, the effect of capacitance is neglected and as such the line can be considered to have series impedance Fig. 14.17 shows the circuit of a 3-phase transmission line on a single-phase basis.

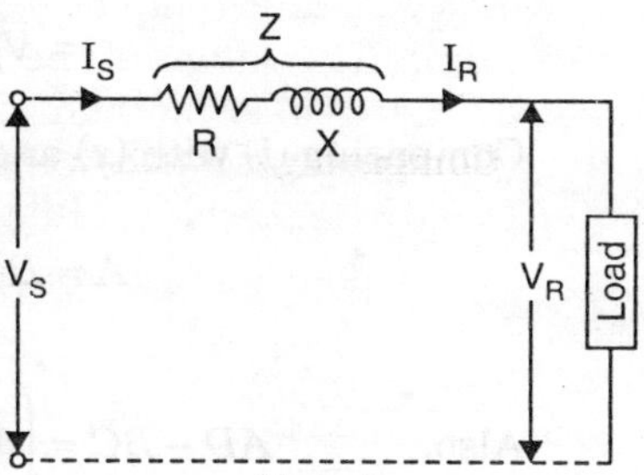

Fig. 14.17

Here, $\overline{I}_S = \overline{I}_R \quad ...(iii)$

and, $\overline{V}_S = \overline{V}_R + \overline{I}_R\overline{Z} \quad ...(iv)$

Comparing (*i*) with (*iv*) and (*ii*) with (*iii*), we have

$$A = 1 \; ; B = \overline{Z} \; ; C = 0 \; ; D = 1 \quad ...(14.30)$$

Incidentally, $A = D$, and $AD - BC = 1$.

2. Medium lines :

(*i*) ***End condenser method.*** In this method the capacitance of the line is assumed to be *lumped at the load end*, as shown in Fig. 14.18. The method overestimates the effect of capacitance.

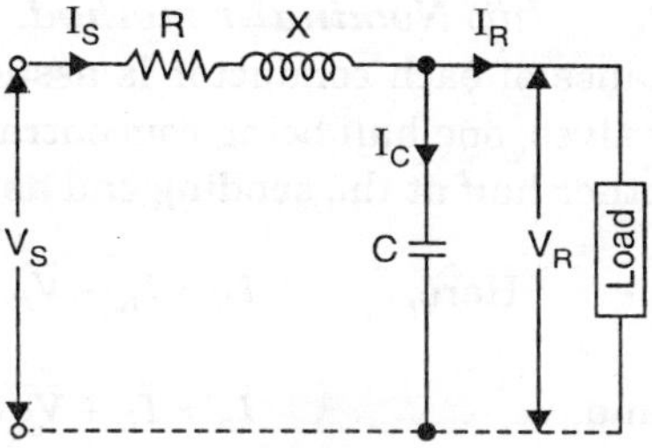

Fig. 14.18

Here we have, $\overline{I}_S = \overline{I}_C + \overline{I}_R = \overline{Y}\,\overline{V}_R + \overline{I}_R \quad ...(v)$

and,
$$\overline{V}_S = \overline{V}_R + (\overline{I}_C + \overline{I}_R)\,\overline{Z}$$
$$= \overline{V}_R + \overline{Y}\,\overline{V}_R\overline{Z} + \overline{I}_R\,\overline{Z}$$
$$= \overline{V}_R\,(1 + \overline{Y}\,\overline{Z}) + \overline{I}_R\,\overline{Z} \quad ...(vi)$$

Comparing (*i*) with (*vi*) and (*ii*) with (*v*), we have

$$A = 1 + \overline{Y}\,\overline{Z} \; ; B = \overline{Z} \; ; C = \overline{Y} \; ; D = 1 \quad ...(14.31)$$

Also, $AD - BC = (1 + \overline{Y}\,\overline{Z}) \times 1 - \overline{Z}\,\overline{Y} = 1.$

(*ii*) ***Nominal-T method.*** In this method the capacitance of each conductor is assumed to be concentrated at the mid point of the line, as shown in Fig. 14.19.

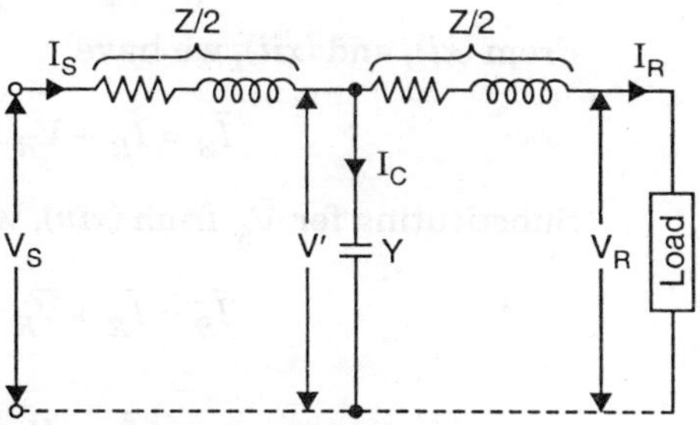

Fig. 14.19

Here, $\overline{V}_S = \overline{V}' + \overline{I}_S \dfrac{\overline{Z}}{2} \quad ...(vii)$

and, $\overline{V}' = \overline{V}_R + \overline{I}_R \dfrac{\overline{Z}}{2} \quad ...(viii)$

and,
$$\overline{I}_S = \overline{I}_R + \overline{I}_C = \overline{I}_R + \overline{V}'\overline{Y}$$
$$= \overline{I}_R + \left(\overline{V}_R + \overline{I}_R \frac{\overline{Z}}{2}\right)\overline{Y}$$
$$= \overline{V}_R\overline{Y} + \overline{I}_R + \left(1 + \frac{\overline{Y}\,\overline{Z}}{2}\right) \quad ...(ix)$$

Substituting the value of $\bar{V}'$ in eqn. (*vii*), we get

$$\bar{V}_S = \bar{V}_R + \bar{I}_R \frac{\bar{Z}}{2} + \bar{I}_S \frac{\bar{Z}}{2}$$

$$= \bar{V}_R + \bar{I}_R \frac{\bar{Z}}{2} + \left[\bar{V}_R \bar{Y} + \bar{I}_R \left(1 + \frac{\bar{Y}\bar{Z}}{2}\right)\right]\frac{\bar{Z}}{2}$$

$$= \bar{V}_R \left(1 + \frac{\bar{Y}\bar{Z}}{2}\right) + \bar{I}_R \left(\bar{Z} + \frac{\bar{Y}\bar{Z}^2}{4}\right) \quad ...(x)$$

Comparing (*i*) with (*x*) and (*ii*) with (*ix*), we get

$$A = D = \left(1 + \frac{\bar{Y}\bar{Z}}{2}\right); B = \bar{Z}\left(1 + \frac{\bar{Y}\bar{Z}}{4}\right); C = \bar{Y}; \quad ...(14.32)$$

Also, $$AD - BC = \left(1 + \frac{\bar{Y}\bar{Z}}{2}\right)^2 - \bar{Z}\left(1 + \frac{\bar{Y}\bar{Z}}{4}\right)\bar{Y}$$

$$= 1 + \frac{\bar{Y}^2\bar{Z}^2}{4} - \bar{Y}\bar{Z} - \bar{Z}\bar{Y} - \frac{\bar{Y}^2\bar{Z}^2}{4} = 1.$$

(*iii*) ***Nominal-π method.*** In this method, the capacitance of each conductor is assumed to be divided into two halves, one half being concentrated at the load end and the other half at the sending end as shown in Fig. 14.20.

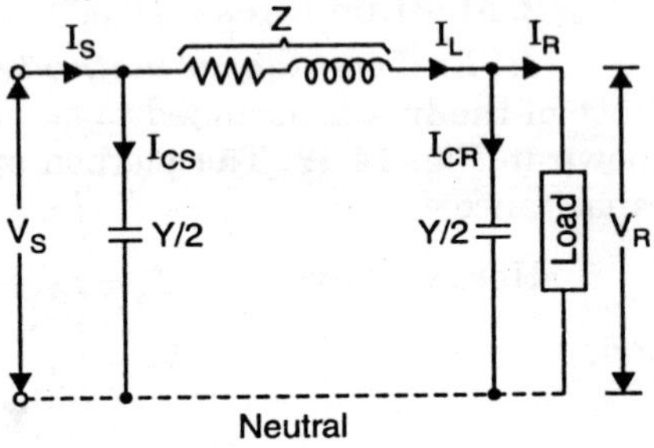

Fig. 14.20

Here, $$\bar{I}_L = \bar{I}_R + \bar{V}_R \times \frac{\bar{Y}}{2} \quad ...(xi)$$

and, $$\bar{I}_S = \bar{I}_L + \bar{V}_S \times \frac{\bar{Y}}{2} \quad ...(xii)$$

and, $$\bar{V}_S = \bar{V}_R + \bar{I}_L \bar{Z} \quad ...(xiii)$$

$$= \bar{V}_R + \left(\bar{I}_R + \bar{V}_R \times \frac{\bar{Y}}{2}\right)\bar{Z} \quad \text{... substituting for } \bar{I}_L$$

or, $$\bar{V}_S = \bar{V}_R\left(1 + \frac{1}{2}\bar{Y}\bar{Z}\right) + \bar{I}_R\bar{Z} \quad ...(xiv)$$

From (*xi*), and (*xii*), we have

$$\bar{I}_S = \bar{I}_R + \bar{V}_R \times \frac{\bar{Y}}{2} + \bar{V}_S \times \frac{\bar{Y}}{2}$$

Substituting for $\bar{V}_S$ from (*xiv*), we get

$$\bar{I}_S = \bar{I}_R + \bar{V}_R \times \frac{\bar{Y}}{2} + \left[\bar{V}_R\left(1 + \frac{1}{2}\bar{Y}\bar{Z}\right) + \bar{I}_R\bar{Z}\right] \times \frac{\bar{Y}}{2}$$

$$= \bar{I}_R + \bar{V}_R \times \frac{\bar{Y}}{2} + \bar{V}_R\left(1 + \frac{1}{2}\bar{Y}\bar{Z}\right) \times \frac{\bar{Y}}{2} + \bar{I}_R\bar{Z} \times \frac{\bar{Y}}{2}$$

$$= \bar{V}_R \times \frac{\bar{Y}}{2}\left(1 + 1 + \frac{1}{2}\bar{Y}\bar{Z}\right) + \bar{I}_R\left(1 + \frac{1}{2}\bar{Y}\bar{Z}\right)$$

$$= \bar{V}_R\bar{Y}\left(1 + \frac{1}{4}\bar{Y}\bar{Z}\right) + \bar{I}_R\left(1 + \frac{1}{2}\bar{Y}\bar{Z}\right) \quad ...(xv)$$

Comparing (*i*) with (*xiv*) and (*ii*) with (*xv*), we get

$$A = D = 1 + \frac{1}{2}\overline{Y}\,\overline{Z} \;;\; B = \overline{Z} \;;\; C = \overline{Y}\left(1 + \frac{1}{4}\overline{Y}\,\overline{Z}\right) \qquad ...(14.33)$$

Also, $$AD - BC = \left(1 + \frac{1}{2}\overline{Y}\,\overline{Z}\right)^2 - \overline{Z}\,\overline{Y}\left(1 + \frac{1}{4}\overline{Y}\,\overline{Z}\right)$$

$$= 1 + \frac{1}{4}\overline{Y}^2\,\overline{Z}^2 + \overline{Y}\,\overline{Z} - \overline{Y}\,\overline{Z} - \frac{1}{4}\overline{Y}^2\,\overline{Z}^2 = 1.$$

Example 14.15. *Using nominal -T method find the A, B, C, D parameters of a 3-phase 80 km, 50 Hz transmission line with series inductance of 0.15 + j0.78 ohm per km and a shunt admittance of 5 × 10⁻⁶ mho per km.*

Solution. *Given* : Total series impedance, $\overline{Z} = (0.15 + j0.78) \times 80 = (12 + j62.4)\ \Omega$

Total shunt admittance, $\overline{Y} = j5 \times 10^{-6} \times 80 = j(4 \times 10^{-4})$ ℧

For nominal-*T* method, various constants have the following values :

$$\mathbf{A = D} = 1 + \frac{\overline{Y}\,\overline{Z}}{2} = 1 + \frac{j\,(4 \times 10^{-4})\,(12 + j62.4)}{2}$$

$$= 1 + \frac{j\,0.0048 - 0.02496}{2} = 1 + j\,0.0024 - 0.0125 = \mathbf{0.9875 + j\,0.0024.} \quad \textbf{(Ans.)}$$

$$\mathbf{B} = \overline{Z}\left(1 + \frac{\overline{Y}\,\overline{Z}}{4}\right) = (12 + j\,62.4)\left[1 + \frac{j\,(4 \times 10^{-4})\,(12 + j\,62.4)}{4}\right]$$

$$= (12 + j\,62.4)\left(1 + \frac{j\,0.0048 - 0.02496}{4}\right)$$

$$= (12 + j\,62.4)(1 + j\,0.0012 - 0.00624) = (12 + j\,62.4)(0.9938 + j\,0.0012)$$

$$= 11.9256 + j\,0.0144 + j\,62.013 - 0.0749 = \mathbf{(11.851 + j\,62.027)\ \Omega.} \quad \textbf{(Ans.)}$$

$$C = \overline{Y} = j\,(4 \times 10^{-4}) = \boldsymbol{j\,0.0004}\ ℧. \quad \textbf{(Ans.)}$$

Example 14.16. *A 3-phase overhead transmission live delivers a load of 80 MW at 0.8 p.f. lagging and 220 kV between the lines. Its total series impedance per phase and total shunt admittance per phase are 200 ∠80° ohms and 0.0013 ∠90° mho per phase respectively. Using nominal-T method determine the following :*

(*i*) *A, B, C, D constants of the line,* (*ii*) *Sending end voltage,*

(*iii*) *Sending end current,* (*iv*) *Sending end power factor, and*

(*v*) *Transmission efficiency of the line.* **(Agra University)**

Solution. *Given* : Load delivered = 80 MW at 0.8 p.f. lagging

Recieving end phase voltage, $V_R = \frac{220}{\sqrt{3}} = 127$ kV

Total series impedance/phase, $\overline{Z} = 200\ \angle 80°\ \Omega$

Total shunt admittance/phase, $\overline{Y} = 0.0013\ \angle 90°$ ℧

(*i*) **A, B, C and D constants of the line :**

$$\text{Receiving end current} = \frac{80 \times 10^6}{\sqrt{3} \times 220 \times 1000 \times 0.8} = 262.43\ \text{A}$$

$$\cos\phi_R = 0.8, \quad \phi_R = \cos^{-1}(0.8) = 36.87°\ \text{(lag)}$$

Taking receiving end phase voltage as reference phasor, we have

$$\bar{V}_R = 127(1 + j0) = 127 \angle 0° \text{ kV}$$

and, $$\bar{I}_R = 262.43 \angle -36.87°\text{A}$$

For nominal-T method various constants have the following values :

$$\mathbf{A = D} = 1 + \frac{\bar{Y}\bar{Z}}{2} = 1 + \frac{0.0013\angle 90° \times 200 \angle 80°}{2} = 1 + 0.13\angle 170°$$

$$= 1 + 0.13(\cos 170° + j \sin 170°) = 1 - 0.128 + j\,0.02257$$

$$= 0.872 + j\,0.02257 = \mathbf{0.872 \angle 1.5°}. \quad \textbf{(Ans.)}$$

$$\mathbf{B} = \bar{Z}\left(1 + \frac{\bar{Y}\bar{Z}}{4}\right) = 200 \angle 80° \left(1 + \frac{0.0013\angle 90° \times 200 \angle 80°}{4}\right)$$

$$= 200\angle 80° \,(1 + 0.065\angle 170°) = 200\angle 80°(1 - 0.064 + j\,0.0113)$$

$$= 200\angle 80° \,(0.936 + j\,0.0113) = 200\angle 80° \times 0.936\angle 0.69°$$

$$= \mathbf{187.2\angle 80.69°\ \Omega}. \quad \textbf{(Ans.)}$$

$$\mathbf{C} = \bar{Y} = \mathbf{0.0013\angle 90°\ ℧}. \quad \textbf{(Ans.)}$$

(*ii*) **Sending end voltage :**

Sending ending phase voltage,

$$\bar{V}_S = A\bar{V}_R + B\bar{I}_R \qquad \text{...[Eqn. (14.22)]}$$

$$= 0.872 \angle 1.5° \times 127 \angle 0° + (187.2 \angle 80.69° \times 262.43 \angle -36.87° \times 10^{-3}) \text{ kV}$$

$$= 110.74 \angle 1.5° + 49.13 \angle 43.82°$$

$$= (110.7 + j\,2.9) + (35.45 + j\,34.02) = (146.15 + j\,36.92) \text{ kV}$$

$$= 150.74 \angle 14.18° \text{ kV}$$

∴ Sending end line voltage, $\bar{V}_{SL} = \sqrt{3} \times 150.73 \angle 14.18° = \mathbf{261 \angle 14.18°\ kV}$. **(Ans.)**

(*iii*) **Sending end current :**

$$\bar{I}_S = C\bar{V}_R + D\bar{I}_R \qquad \text{...[Eqn. (14.23)]}$$

$$= 0.0013 \angle 90° \times 127 \angle 0° \times 10^3 + (0.872 \angle 1.5° \times 262.43 \angle -36.87°).$$

$$= 165.1 \angle 90° + 228.84 \angle -35.37°$$

$$= (0 + j\,165.1) + (186.6 - j\,132.5) = 186.6 + j\,32.6 = \mathbf{189.4 \angle 9.91°}. \quad \textbf{(Ans.)}$$

(*iv*) **Sending end power factor, cos ϕ_s :**

$$\cos \phi_s = \cos(14.18° - 9.91°) = \cos 4.27° = \mathbf{0.9972\ (lag)}. \quad \textbf{(Ans.)}$$

(*v*) **Transmission efficiency, η_T :**

$$\eta_T = \frac{\text{Power delivered}}{\text{Power sent out}} = \frac{80 \times 10^3}{\sqrt{3}\, V_{SL}\, I_S \cos \phi_S}$$

$$= \frac{80 \times 10^3}{\sqrt{3} \times 261 \times 189.4 \times 0.9972} = 0.937 \quad \text{or} \quad \mathbf{93.7\%}. \quad \textbf{(Ans.)}$$

Example 14.17. *A 50 Hz, 3-phase 100 km long transmission line has a total resistance of 35 Ω, series reactance of 140 Ω and shunt admittance (line to neutral) 930 × 10^{-6} mho. It delivers 40 MW at 220 kV at 0.9 p.f. lagging. Using nominal-π method determine the following :*

(*i*) *A, B, C, D constant,* (*ii*) *Sending end voltage,*

(*iii*) *Sending end current,* (*iv*) *sending end power factor,*

(*v*) *Voltage regulation, and* (*vi*) *Transmission efficiency.*

Solution. *Given* : Series impedance, $\overline{Z} = 35 + j\,140 = 144.31 \angle 75.96°\ \Omega$

Shunt admittance, $\overline{Y} = j\,930 \times 10^{-6} = 9.3 \times 10^{-4}$ ℧

Power at the receiving end = 40 MW

Receiving end phaser voltage, $V_R = \dfrac{220}{\sqrt{3}} = 127$ kV

Power factor, $\cos \phi_R = 0.9$ or $\phi_R = -25.84°$

(*i*) **A, B, C, D constants :**

Receiving end current $I_R = \dfrac{40 \times 10^3}{\sqrt{3} \times 220 \times 0.9} = 116.6$ A

Taking receiving end phase voltage as the reference phasor, we have

$$\overline{V}_R = 127(1 + j\,0) = 127\angle 0° \text{ kV}$$

$$\overline{I}_R = 116.6\angle -25.84°$$

A, B, C, D constants for nominal-π method are

$$A = D = \left(1 + \frac{1}{2}\overline{Y}\,\overline{Z}\right) = 1 + \frac{1}{2}(9.3 \times 10^{-4} \angle 90° \times 144.31\angle 75.96°)$$

$$= 1 + 0.067 \angle 165.96° = 1 + (-0.065 + j\,0.01625)$$

$$= 0.935 + j\,0.01625 = \mathbf{0.935\angle 0.99°}. \quad \textbf{(Ans.)}$$

$$B = \overline{Z} = \mathbf{144.31 \angle 75.96°\ \Omega}. \quad \textbf{(Ans.)}$$

$$C = \overline{Y}\left(1 + \frac{1}{4}\overline{Y}\,\overline{Z}\right) = 9.3 \times 10^{-4} \angle 90° \left[1 + \frac{1}{4}(9.3 \times 10^{-4} \angle 90° \times 144.31 \angle 75.96°)\right]$$

$$= 9.3 \times 10^{-4}\angle 90°\ [1 + 0.0335\angle 165.96°]$$

$$= 9.3 \times 10^{-4}\angle 90°\ [1 + (-0.0325 + j\,0.00813)]$$

$$= 9.3 \times 10^{-4} \angle 90°\ (0.9675 + j0.00813)$$

$$= 9.3 \times 10^{-4} \angle 90° \times 0.9675 \angle 0.48° = \mathbf{900 \times 10^{-6} \angle 90.48°}\ ℧. \quad \textbf{(Ans.)}$$

(*ii*) **Sending end voltage :**

Sending end phase voltage,

$$\overline{V}_S = A\overline{V}_R + B\overline{I}_R$$

$$= (0.935 \angle 0.99° \times 127 \angle 0°) + (144.31 \angle 75.96° \times 116.6 \angle -25.84° \times 10^{-3}) \text{ kV}$$

$$= 118.74 \angle 0.99° + 16.83 \angle 50.12°$$

$$= (118.72 + j2.05) + (10.79 + j12.92)$$

$$= 129.51 + j14.97 = 130.37 \angle 6.6° \text{ kV}$$

Sending end line voltage, $\overline{V}_{SL} = \sqrt{3} \times 130.37 \angle 6.6° = \mathbf{225.8 \angle 6.6°kV}$. **(Ans.)**

(*iii*) **Sending end current :**

$$\overline{I}_S = C\overline{V}_R + D\overline{I}_R$$

$$= 900 \times 10^{-6} \angle 90.48° \times 127 \angle 0° \times 10^3 + 0.935 \angle 0.99° \times 116.6 \angle -25.84°$$

$$= 114.3 \angle 90.48° + 109.02 \angle -24.85°$$

$$= (-0.96 + j114.29) + (98.93 - j45.82)$$

$$= 97.97 + j68.47 = \mathbf{119.53\angle 35°}. \quad \textbf{(Ans.)}$$

(*iv*) **Sending end power factor, cos ϕ_S :**

$$\cos \phi_S = \cos (35° - 6.6°) = \cos 28.4° = \mathbf{0.8796.} \quad \textbf{(Ans.)}$$

(*v*) **Voltage regulation :**

"Regulation" is defined as the *change in voltage at the receiving end when full-load is thrown off.*

Now, $\bar{V}_S = A\bar{V}_R + B\bar{I}_R$

At no load, $\bar{I}_R = 0$

∴ $\bar{V}_S = A\bar{V}_{RO}$, where V_{RO} is the voltage at the receiving end at no-load.

or, $\bar{V}_{RO} = \dfrac{\bar{V}_S}{A}$

or, $V_{RO} = \dfrac{V_S}{A}$ (in magnitude)

$$\therefore \quad \text{Percentage regulation} = \frac{V_{RO} - V_R}{V_R} \times 100 = \frac{(V_S/A) - V_R}{V_R} \times 100$$

$$= \frac{(130.37/0.935) - 127}{127} \times 100 = \mathbf{9.79\%.} \quad \textbf{(Ans.)}$$

(*vi*) **Transmission efficiency η_T :**

$$\eta_T = \frac{\text{Power delivered}}{\text{Power sent out}} = \frac{40 \times 10^3\,(\text{kW})}{\sqrt{3} V_{SL} I_S \cos \phi_S\,(\text{kW})}$$

$$= \frac{40 \times 10^3}{\sqrt{3} \times 225.8 \times 119.53 \times 0.8796} = 0.9728 \quad \text{or} \quad \mathbf{97.28\%.} \quad \textbf{(Ans.)}$$

Example 14.18. *A 220 kV, 50 Hz, 3-phase transmission line delivers a load of 75 MW at 0.8 p.t. lagging at the receiving end. The generalised constants of the transmission line are :*

$$A = D = 0.9 \angle 0.6° \,;\, B = 153.2 \angle 84.6° \,;\, C = 0.0012 \angle 90°$$

Determine the following :

(i) Sending end line voltage, *(ii) Sending end current,*

(iii) Sending end power factor, *(iv) Percentage regulation, and*

(v) Charging current, using nominal-T method. **(Panjab University)**

Solution. *Given :* $\bar{V}_R = \dfrac{220}{\sqrt{3}} \angle 0° = 127 \angle 0°$ kV ; Load delivered = 75 MW, $\cos \phi_R = 0.8$ *i.e.*, $\phi_R = \cos^{-1}(0.8) = 36.87°$; $A = D = 0.9 \angle 0.6°$; $B = 153.2 \angle 84.6°$, $C = 0.0012 \angle 90°$.

(*i*) **Sending end line voltage, $\bar{V}_{SL}$:**

Receiving end current, $I_R = \dfrac{75 \times 10^3}{\sqrt{3} \times 220 \times 0.8} = 246\ A$

Taking receiving end phase voltage as the reference phasor, we have

$$\bar{V}_R = 127 \angle 0° \text{ kV}$$

$$\bar{I}_R = 246 \angle -36.87°$$

$$\bar{V}_S = A\bar{V}_R + B\bar{I}_R$$

$$= (0.9 \angle 0.6° \times 127 \angle 0°) + (153.2 \angle 84.6° \times 246 \angle -36.87° \times 10^{-3})$$

$$= 114.3 \angle 0.6° + 37.7 \angle 47.73°$$

$= (114.29 + j1.2) + (25.36 + j\,27.9)$

$= 139.65 + j29.1 = 142.65 \angle 11.77°$ kV

∴ Sending end line voltage, $\overline{V}_{SL} = \sqrt{3} \times 142.65 \angle 11.77° =$ **247 ∠11.77°. (Ans.)**

(*ii*) **Sending end current, $\overline{I}_S$:**

$$\overline{I}_S = C\overline{V}_R + D\overline{I}_R$$
$$= 0.0012 \angle 90° \times 127 \angle 0° \times 10^3 + 0.9 \angle 0.6° \times 246 \angle -36.87°$$
$$= 152.4 \angle 90° + 221.4 \angle -36.27°$$
$$= (0 + j152.4) + (178.5 - j131)$$
$$= (178.5 + j21.4) = \mathbf{179.8 \angle 6.84°}. \quad \textbf{(Ans.)}$$

(*iii*) **Sending end power factor, cos ϕ_S :**

$\cos \phi_S = \cos (11.77° - 6.84°) = \cos 4.93° =$ **0.996 (lag). (Ans.)**

(*iv*) **Percentage regulation :**

$$\overline{V}_S = A\overline{V}_R + B\overline{I}_R$$

At no-load, $\overline{I}_S = 0$, ∴ $\overline{V}_S = A\overline{V}_{RO}$ (where V_{RO} is the voltage at receiving end at no-load),

or, $\overline{V}_{RO} = \dfrac{\overline{V}_S}{A}$ or $V_{RO} = \dfrac{V_S}{A}$ (in magnitude)

$$\therefore \quad \text{Percentage regulation} = \frac{V_{RO} - V_R}{V_R} \times 100$$
$$= \frac{(V_S/A) - V_R}{V_R} \times 100 = \frac{(142.65/0.9) - 127}{127} \times 100 = \mathbf{24.8\%}. \quad \textbf{(Ans.)}$$

(*v*) **Charging current, $\overline{I}_C$:**

In nominal-T method, the charging current is given as :

$$I_C = \overline{I}_S - \overline{I}_R = 179.8 \angle 6.84° - 246 \angle -36.87°$$
$$= (178.5 + j21.4) - (196.8 - j147.6)$$
$$= -18.3 + j169 = \mathbf{170 \angle 83.8°}. \quad \textbf{(Ans.)}$$

Example 14.19. *A 110 kV, 50 Hz, 3-phase overhead transmission line delivers a load of 40 MW at 0.8 p.f. lagging at the receiving end and has the following constants :*

$A = D = 0.98 \angle 3°$, $B = 100 \angle 75°$ Ω ; $C = 5 \times 10^{-4} \angle 80°$ ℧.

Determine the following :

(i) Sending end voltage, *(ii) Sending end current,*

(iii) Sending end power, and *(iv) Transmission efficiency.* **(N.U.)**

Solution. Given : $\overline{V}_R = \dfrac{110}{\sqrt{3}} \angle 0° = 63.5 \angle 0°$ kV ; $\cos \phi_R = 0.8$, *i.e.*, $\phi_R = \cos^{-1}(0.8) = 36.87°$;

$A = D = 0.98 \angle 3°$; $B = 110 \angle 75°$ Ω, $C = 5 \times 1^{-4} \angle 80°$ ℧ ; Power delivered = 40 MW.

(*i*) **Sending end voltage :**

Receiving end current, $I_R = \dfrac{40 \times 10^3}{\sqrt{3} \times 110 \times 0.8} = 262.43$ A

Taking receiving end phase voltage as the reference phasor, we have

$$\overline{V}_R = 63.5 \angle 0° \text{ kV},$$
$$\overline{I}_R = 262.43 \angle -36.87°$$

$$\overline{V}_S = A\overline{V}_R + B\overline{I}_R$$
$$= (0.98 \angle 3° \times 63.5 \angle 0°) + (110 \angle 75° \times 262.43 \angle -36.87° \times 10^{-3}) \text{ kV}$$
$$= (62.23 \angle 3° + 28.87 \angle 38.13°) \text{ kV}$$
$$= (62.14 + j3.26) + (22.71 + j17.83)$$
$$= 84.85 + j21.09 = 87.43 \angle 14°$$

∴ Sending end line voltage $\overline{V}_{SL} = \sqrt{3} \times 87.43 \angle 14° =$ **151.43 ∠14°. (Ans.)**

(*ii*) **Sending end current, $\overline{I}_S$:**

$$\overline{I}_S = C\overline{V}_R + D\overline{I}_R$$
$$= (5 \times 10^{-4} \angle 80° \times 63.5 \angle 0° \times 10^3) + (0.98 \angle 3° \times 262.43 \angle -36.87°)$$
$$= 31.75 \angle 80° + 257.18 \angle -33.87°$$
$$= (5.51 + j31.27) + (213.54 - j143.33)$$
$$= 219.05 - j112.06 = \mathbf{246.05 \angle -27.09°}. \quad \textbf{(Ans.)}$$

(*iii*) **Sending end power :**

Sending end power factor, $\cos \phi = \cos [14 - (-27.09)] = \cos 41.09° = 0.754$ (lag)

∴ Sending end power $= \sqrt{3}\, V_{SL}\, I_S \cos \phi_S$

$$= \sqrt{3} \times 151.43 \times 246.05 \times 0.754 \times 10^{-3} \text{ MW} = \mathbf{48.66 \text{ MW}}. \quad \textbf{(Ans.)}$$

(*iv*) **Transmission efficiency, η_T :**

$$\eta_T = \frac{\text{Power delivered}}{\text{Power sent out}} = \frac{40}{48.66} = 0.822 \quad \text{or} \quad \mathbf{82.2\%}. \quad \textbf{(Ans.)}$$

14.7. LONG TRANSMISSION LINES

14.7.1. General Aspects

It is a known fact that the constants of a transmission line are uniformly distributed over the whole length of the line. So far electrically short transmission lines less than 160 km in length have been considered wherein the parameters are assumed to be lumped. Such an assumption when applied to long transmission lines causes (l > 150 km) serious errors in the line performance calculations. For such transmission lines rigorous treatment is required. Fig. 14.21 shows the schematic representation of the equivalent circuit of a 3-phase long transmission line. The entire line length is divided into n sections, each section having line constants $\frac{1}{n}$th of those for the entire length.

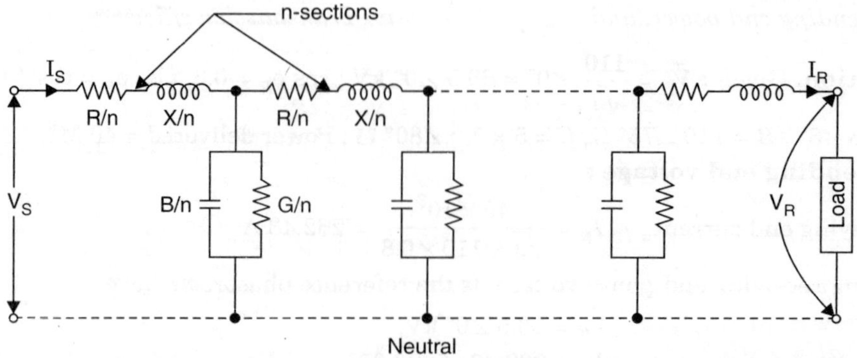

Fig. 14.21. Equivalent circuit of a 3-phase long transmission line.

The following points are worth noting :

- The line constants namely resistance, inductive reactance, capacitive susceptance and conductance are *uniformly distributed over the entire line length.*
- The resistance (R) and inductive reactance (X) are the *series elements.*
- The capacitive susceptance (B) and *leakage* conductance (G) are *shunt elements.* G takes into account the energy losses occurring through leakage over insulators or due to corona effect between conductors (Admittance, $Y = \sqrt{G^2 + B^2}$).
- The leakage current through the shunt admittance is maximum at the sending end of the line and decreases continuously as we approach the receiving end of the line, at which point its value becomes zero.

14.7.2. Analysis of Long Transmission Line–Rigorous Method

Fig. 14.22 shows one phase and the neutral return (of zero impedance) of a transmission line. Let dx be the an elemental section of the line at a distance x from the receiving end.

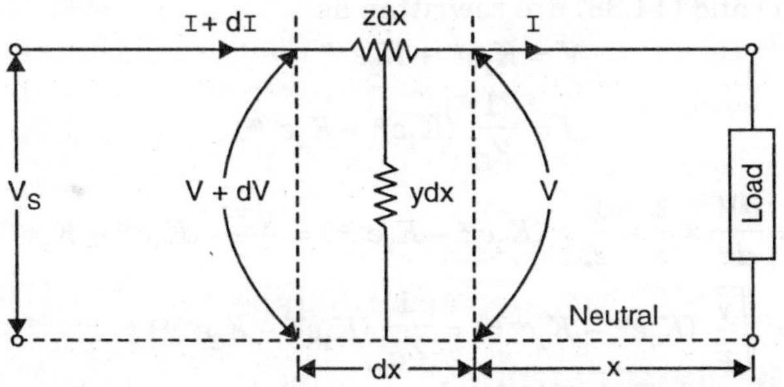

Fig. 14.22. Schematic diagram of a long line.

Let, z = Series impedance of the line per unit length (= $r + jx$, where r and x are respectively resistance and reactance of the line per unit length)

y = Shunt admittance of the line per unit length,

V = Voltage per phase at the end of elements towards receiving end,

$V + dV$ = Voltage per phase at the end of element towards sending end,

$I + dI$ = Current entering the element dx,

I = Current leaving the element dx,

V_S, V_R = Voltages per phase at the sending and receving ends respectively, and

I_S, I_R = Currents per phase at the sending and receiving end respectively,

Then for the small element dx,

zdx = Series impedance, and

ydx = Shunt admittance.

Obviously, the rise in voltage over the element length in direction of increasing x,

$$dV = I\,zdx$$

or,
$$\frac{dV}{dx} = Iz \qquad ...(14.34)$$

Similarly, the difference of current entering the element and that of leaving the element [*i.e.*, $(I + dI) - (I) = dI$],

$$dI = V\,ydx$$

or, $$\frac{dI}{dx} = Vy \quad ...(14.35)$$

Differentiating eqn. (14.34) w.r.t. x, we get

$$\frac{d^2V}{dx^2} = z\frac{dI}{dx} = zyV \qquad \left(\because \text{ From eqn. (14.35), } \frac{dI}{dx} = Vy\right)$$

or, $$\frac{d^2V}{dx^2} - zyV = 0 \quad ...(14.36)$$

The solution of this equation is

$$V = K_1 \exp(\sqrt{yz} \,.\, x) + K_2 \exp(-\sqrt{yz} \,.\, x) \quad ...(14.37)$$

Now, from eqns. (14.34) and (14.37)

let, $$Z_C = \sqrt{\frac{z}{y}} \quad \text{and} \quad \gamma = \sqrt{yz} = \alpha + i\beta \quad ...(14.38)$$

where, Z_C is known as **characteristic impedance** and γ the ***propagation constant*** :

The eqns. (14.37) and (14.38) are rewritten as

$$V = K_1e^{\gamma x} + K_2e^{-\gamma x} \quad ...(14.39)$$

$$I = \frac{1}{Z_C}(K_1e^{\gamma x} - K_2e^{-\gamma x}) \quad ...(14.40)$$

$$\left[\because \quad I = \frac{dV}{dx} \times \frac{1}{z} = \frac{1}{z}\gamma\,(K_1e^{\gamma x} - K_2e^{-\gamma x}) = \frac{\sqrt{yz}}{z}(K_1e^{\gamma x} - K_2e^{-\gamma x}) \right.$$
$$\left. = \sqrt{\frac{y}{z}}(K_1e^{\gamma x} - K_2e^{-\gamma x}) = \frac{1}{Z_C}(K_1e^{\gamma x} - K_2e^{-\gamma x}) \right]$$

Two constants K_1 and K_2 are to be determined, hence two boundary conditions should be known.

At the receiving end where $x = 0$: $V = V_R$ and $I = I_R$

Substituting these values in eqns. (14.39) and (14.40), we have

$$V_R = K_1 + K_2$$

$$I_R = \frac{1}{Z_C}(K_1 - K_2)$$

or, $$K_1 = \frac{V_R + I_RZ_C}{2} \quad \text{and} \quad K_2 = \frac{V_R - I_RZ_C}{2}$$

Substituting the values of K_1 and K_2 in eqns. (14.39) and (14.40), we get

$$V = \frac{V_R + I_RZ_C}{2}e^{\gamma x} + \frac{V_R - I_RZ_C}{2}e^{-\gamma x} \quad ...(14.41)$$

and, $$I = \frac{1}{Z_C}\left[\frac{V_R + I_RZ_C}{2}e^{\gamma x} - \frac{V_R - I_RZ_C}{2}e^{-\gamma x}\right] \quad ...(14.42)$$

Expanding and rearranging eqns. (14.40) and (14.41), we get

$$V = V_R\left(\frac{e^{\gamma x} + e^{-\gamma x}}{2}\right) + I_RZ_C\left(\frac{e^{\gamma x} - e^{-\gamma x}}{2}\right)$$

or, $$V = V_R \cosh \gamma x + I_RZ_C \sinh \gamma x \quad ....(14.43)$$

and,
$$I = \frac{1}{Z_C}\left[V_R\left(\frac{e^{\gamma x} - e^{-\gamma x}}{2}\right) + I_R Z_C\left(\frac{e^{\gamma x} + e^{-\gamma x}}{2}\right)\right]$$

$$= \frac{1}{Z_C}\,[V_R \sinh \gamma x + I_R Z_C \cosh \gamma x]$$

or,
$$I = \frac{V_R}{Z_C} \sinh \gamma x + I_R \cosh \gamma x \quad ...(14.44)$$

Rewriting these equations for $x = l$, where $V = V_S$ and $I = I_S$, we get

$$V_S = V_R \cosh \gamma l + I_R Z_C \sinh \gamma l \quad ...(14.45)$$

$$I_S = \frac{V_R}{Z_C} \sinh \gamma l + I_R \cosh \gamma l \quad ...(14.46)$$

These two equations relate the sending end voltage and current with the receiving end quantities.

Comparing eqns. (14.45) and (14.46) with the general voltage and current equations of the line, we have

$$\left.\begin{aligned} A &= \cos \gamma l \\ B &= Z_C \sinh \gamma l \\ C &= \frac{1}{Z_C} \sinh \gamma l \\ D &= \cosh \gamma l \end{aligned}\right\} \quad ...(14.47)$$

From this it is clear that, $A = D = \cosh \gamma l$

and,
$$AD - BC = \cosh^2 \gamma l - Z_C \sinh \gamma l \times \frac{1}{Z_C} \sinh \gamma l = 1.$$

14.7.3. Evaluation of A, B, C, D constants

ABCD constants can be *evaluated by the following methods* :

1. Convergont series (complex angle) method.
2. Convergent series (Real angle) method.
3. By use of hyperbolic charts.

1. **Convergent series (Complex angle) method.** We know that both, total series impedance $Z(= zl)$ and shunt admittance $Y(= yl)$ of a line are in general complex quantities. Thus expansions of cosh γl are sinh γl may be easily obtained by using Maclaurin's series as follows :

$$\cosh x = 1 + \frac{x^2}{2!} + \frac{x^4}{4!} + \frac{x^6}{6!} + ...$$

and,
$$\sinh x = x + \frac{x^3}{3!} + \frac{x^5}{5!} + \frac{x^7}{7!} + ...$$

where x may be real or complex.

In actual practice, for sufficient accuracy, only a few terms of series (say upto 7th power) need be considered.

As $\gamma l = \sqrt{YZ}$, $\cosh \gamma l = \cosh \sqrt{YZ} = 1 + \frac{YZ}{2} + \frac{Y^2Z^2}{24} + \frac{Y^3Z^3}{720} + ...$

and,
$$\sinh \gamma l = \sinh \sqrt{YZ} = \sqrt{YZ}\left[1 + \frac{YZ}{6} + \frac{Y^2Z^2}{120} + \frac{Y^3Z^3}{5040} + ...\right]$$

Accordingly, $A = D = \cosh \gamma l = 1 + \frac{YZ}{2} + \frac{Y^2Z^2}{24} + \frac{Y^3Z^3}{720} + \ldots$

$$B = Z_C \sinh \gamma l = \sqrt{\frac{Z}{Y}} \sinh \gamma l = \sqrt{\frac{Z}{Y}} \times \sqrt{YZ} \left[1 + \frac{YZ}{6} + \frac{Y^2Z^2}{120} + \frac{Y^3Z^3}{5040} + \ldots\right]$$

$$= Z\left[1 + \frac{YZ}{6} + \frac{Y^2Z^2}{120} + \frac{Y^3Z^3}{5040} + \ldots\right]$$

and,
$$C = \frac{1}{Z_C} \sinh \gamma l = \sqrt{\frac{Y}{Z}} \times \sqrt{YZ}\left[1 + \frac{YZ}{6} + \frac{Y^2Z^2}{120} + \frac{Y^3Z^3}{5040} + \ldots\right]$$

$$= Y\left[1 + \frac{YZ}{6} + \frac{Y^2Z^2}{120} + \frac{Y^3Z^3}{5040} + \ldots\right] \quad \ldots(14.48)$$

2. **Convergent series (Real angle) method.** It has been observed that $A\ B\ C\ D$ constants are hyperbolic functions of γl where $\gamma l = \sqrt{YZ}$ is a complex quantity so that γ may by written in the form

$$\gamma = \alpha + j\beta \quad \ldots(14.49)$$

where both α and β are real. Thus

$$\left.\begin{aligned} \cosh \gamma l &= \cosh(\alpha + j\beta)l = \cosh \alpha l \cosh j\beta l + \sinh \alpha l \sinh j\beta l \\ &= \cosh \alpha l \cos \beta l + j \sinh \alpha l \sin \beta l \\ \text{and,} \quad \sinh \gamma l &= \sinh(\alpha + j\beta)l = \sinh \alpha l \cosh j\beta l + \cosh \alpha l \sinh j\beta l \\ &= \sinh \alpha l \cos \beta l + j \cosh \alpha l \sin \beta l \end{aligned}\right\} \quad \ldots(14.50)$$

3. **By use of hyperbolic charts.** A convenient set of charts for evaluating the hyperbolic functions of complex variables, usually required in transmission line calculations, has been prepared by Woodruff. However, such charts are usually not available.

Generally method-1 is less laborious to use than method-2.

14.7.4. Surge Impedance and Surge Impedance Loading of Transmission Lines.

Surge Impedance (Z_O). It is defined as the *square root of the ratio of line impedance (Z) and shunt admittance (Y).*

i.e.
$$Z_O = \sqrt{\frac{Z}{Y}} \quad \ldots(14.51)$$

where, $Z = R + jX$ and $Y = G + jB$

(*Surge impedance is also sometimes called* ***"Natural impedance"***).

It may be seen that the characteristic impedance was also defined in the same way. Actually characteristic impedance is more correct nomenclature.

Surge impedance is the characteristic impedance of a loss-free line

In case of a line having negligible resistance (such will be the case when conductors are of a large cross-section) and no shunt leakage (*i.e.* $G = O$),

$$Z_O = \sqrt{\frac{L}{C}}, \text{ which is a pure resistance} \quad \ldots(14.52)$$

Its value varies as given below :

Overhead transmission lines .... 400 to 600 Ω

Underground cables .... 40 to 60 Ω

The *phase value of Z_C* for transmission lines is usually between **0° and – 15°.**

A line terminated in its characteristic impedance is called a **flat line** or *an* **infinite line.** The latter term arises from the fact that a line of infinite length cannot have a reflected wave.

The surge impedance can be measured by measuring the line impedance at the sending end when the line at the receiving end is open-circuited and short-circuited respectively as discussed below :

When the line is *open circuited* at the receiving end, $I_R = 0$, so that

$$\bar{V}_S = A\bar{V}_R \quad \text{and} \quad \bar{I}_S = C\bar{V}_R$$

$$\therefore \quad Z_{OC} = \frac{\bar{V}_S}{\bar{I}_S} = \frac{A\bar{V}_R}{C\bar{V}_R} = \frac{A}{C} \quad ...(i)$$

When the line is *short circuited* at the receiving end,

$V_R = 0$, so that

$$\bar{V}_S = B\bar{I}_R \quad \text{and} \quad \bar{I}_S = D\bar{I}_R$$

$$\therefore \quad Z_{SC} = \frac{\bar{V}_S}{\bar{I}_S} = \frac{B\bar{I}_R}{D\bar{I}_R} = \frac{B}{D} \quad ...(ii)$$

Multiplying (*i*) and (*ii*), we have

$$Z_{OC} \times Z_{SC} = \frac{A}{C} \times \frac{B}{D} = \frac{A}{C} \times \frac{B}{A} = \frac{B}{C} \quad ...(iii)$$

Substituting $B = \sqrt{\frac{Z}{Y}} \sinh \sqrt{YZ}$, and $C = \sqrt{\frac{Y}{Z}} \sinh \sqrt{YZ}$ in (*iii*), we get

$$Z_{OC} \times Z_{SC} = \frac{B}{C} = \frac{\sqrt{Z/Y} \sinh \sqrt{YZ}}{\sqrt{Y/Z} \sinh \sqrt{YZ}} = \frac{Z}{Y} = Z_C^{\,2}$$

$$\therefore \text{ Surge impdance, } \quad Z_0 = \sqrt{Z_{OC} \times Z_{SC}} \quad ...(14.53)$$

Surge Impedance Loading :

It is defined as the load, at unity power factor, that can be delivered by the line having no resistance.

$$\therefore \quad \text{Power transmitted, } P_R = \frac{V_{RL}^2}{Z_O} \text{ MW} \quad ...(14.54)$$

where, V_{RL} = Receiving end line voltage in kV,

Z_O = Surge impedance in ohms.

P_R is knowN as the "*surge impedance loading*" or "*natural power*" of the line.

The eqn. (14.53) gives a limit to the maximum power that can be delivered (this may not necessarily be the maximum loading on the line) and is useful in the design of transmission lines.

Its use can be made for comparison of loads that can be carried on the transmission lines at different voltages.

The power transmitted through a long transmission line can be *increased* as follows :

(*i*) *Increasing* V_{RL}. Presently the trend is to use higher and higher voltages for transmission, therefore, this is the most commonly used method to increase the power limit of heavily loaded long transmission line. But there is a limit beyond which it is neither economical nor practicable to increase the value of V_{RL}.

(*ii*) *Decreasing* Z_O. The value of Z_O cannot be varied much since the spacing between the conductors, which depends upon the line voltage employed, cannot be reduced to a greater extent.

In order to decrease Z_O either L is decreased using series capacitor or C is increased using shunt capacitors,

— The use of *series capacitors* improves the system stability limit. These capacitors are also helpful in reducing line drops and hence voltage variations.

— By the use of shunt capacitors although Z_O is reduced but due to increase in phase shift stability conditions become worse specially when synchronous machines are used as loads. Thus this method is not employed where stability limits are present.

Example 14.20. *A 3-phase, 50 Hz, 500 km long transmission line has the following line constants per phase per km uniformly distributed :*

$$r = 0.44\ \Omega\ ;\ x = 0.9\ \Omega\ ;\ g = 8 \times 10^{-9}\ \mho \text{ and } b = 5.06 \times 10^{-6}\ \mho$$

Determine the line constants :

(i) By using convergent series of complex angles ;

(ii) By using convergent series of real angles.

Solution. Resistance per phase per km, $r = 0.44\ \Omega$

Reactance per phase per km, $x = 0.9\ \Omega$

Impedance per phase per km, $z = r + jx = (0.44 + j0.9)\ \Omega$

Conductance per phase per km, $g = 8 \times 10^{-9}\ \mho$

Susceptance per phase per km, $b = 5.06 \times 10^{-6}\ \mho$

Admittance per phase per km, $y = (0.008 + j5.06) \times 10^{-6}\ \mho$

Length of the line, $l = 500$ km

Total series impedance,
$$Z = (0.44 + j0.9) \times 500 = (220 + j450)\ \Omega$$
$$= 500.9 \angle 63.95°\ \Omega$$

Total shunt admittance,
$$Y = (0.008 + j5.06) \times 10^{-6} \times 500$$
$$= (0.004 + j2.53) \times 10^{-3}\ \mho$$
$$= 2.53 \times 10^{-3} \angle 89.91°\ \mho$$

Hence,
$$YZ = 2.53 \times 10^{-3} \angle 89.91° \times 500.9 \angle 63.95°$$
$$= 1.2673 \angle 153.86° = (-1.1376 + j0.5583)$$
$$Y^2Z^2 = (1.2673)^2 \angle 307.72° = 1.606 \angle 307.72° = (0.9826 - j1.27)$$
$$Y^3Z^3 = (1.2673)^3 \angle 461.58° = 2.0353 \angle 461.58° = (-0.408 + j1.9938)$$

(*i*) **Using convergent series of complex angles :**

$$A = D = 1 + \frac{YZ}{2} + \frac{Y^2Z^2}{24} + \frac{Y^3Z^3}{720} \qquad \text{(neglecting other terms)}$$
$$= 1 + \frac{-1.1376 + j0.5583}{2} + \frac{0.9826 + j1.27}{24} + \frac{-0.408 + j1.9938}{720}$$
$$= (1 - 0.5688 + 0.0409 - 0.0005) + j(0.2792 - 0.0529 + 0.0028)$$
$$= 0.4716 + j0.2291\ \Omega = 0.524 \angle 25.9°$$

$$B = Z\left[1 + \frac{YZ}{6} + \frac{Y^2Z^2}{120} + \frac{Y^3Z^3}{5040}\right] \Omega, \text{ neglecting other terms,}$$
$$= (220 + j450)\left[1 + \frac{-1.1376 + j0.5583}{6} + \frac{0.9826 - j1.27}{120} + \frac{-0.408 + j1.9938}{5040}\right] \Omega$$
$$= (220 + j450)\ [(1 - 0.1896 + 0.0081 - 0.00008) + j(0.09305 - 0.01058 + 0.0004)]\ \Omega$$
$$= (220 + j450)(0.81842 + j0.08287)\ \Omega$$
$$= 500.9 \angle 63.95 \times 0.8226 \angle 5.78°\ \Omega$$
$$= 412.04 \angle 69.73° = (142.75 + j386.52)\ \Omega$$

$$C = Y\left[1 + \frac{YZ}{6} + \frac{Y^2Z^2}{120} + \frac{Y^3Z^3}{5040}\right] \mho, \text{ neglecting other terms}$$

$= 2.53 \times 10^{-3} \angle 89.91° \times 0.8226 \angle 5.78°$

$= 2.08 \times 10^{-3} \angle 95.69° = (-0.0002 + j0.00207)$ ℧

(ii) **Using convergent series of real angles :**

We have, $Z = 500.9 \angle 63.95°\ \Omega$

and, $Y = 2.53 \times 10^{-3} \angle 89.91°$ ℧

$\therefore \quad YZ = 2.53 \times 10^{-3} \angle 89.91° \times 500.9 \angle 63.95° = 1.2673 \angle 153.86°$

and, $\sqrt{YZ} = 1.12574 \angle 76.93°$

$= 0.25458 + j1.09658$

Since $\gamma l = (\alpha + j\beta)l = 0.25458 + j1.09658$

$\therefore \quad \alpha l = 0.25458$ radian or $14.586°$

$\beta l = 1.09658$ radian or $62.83°$

$\cosh \gamma l = \cosh \alpha l \cos \beta l + j \sinh \alpha l \sin \beta l$

$= \cosh 0.25458 \times \cosh 62.83° + j \sinh 0.25458 \times \sin 62.83°$

$= 1.03258 \times 0.4566 + j0.25734 \times 0.8896$

$= 0.4715 + j0.2289 = 0.524 \angle 25.9°$

$\sinh \gamma l = \sinh \alpha l \cos \beta l + j \cosh \alpha l \sin \beta l$

$= \sinh 0.25458 \times \cos 62.83° + j \cosh 0.25458 \sin 62.83°$

$= 0.1175 + j0.9186 = 0.9261 \angle 82.71°$

Auxiliary line constants,

$A = D = \cosh \gamma l = 0.524 \angle 25.9°$

$$B = Z_C \sinh \gamma l = \sqrt{\frac{Z}{Y}} \times 0.9261 \angle 82.71°$$

$$= \sqrt{\frac{500.9 \angle 63.95°}{2.53 \times 10^{-3} \angle 89.91°}} \times 0.9261 \angle 82.71° = 412.07 \angle 69.73°\ ℧$$

$$C = \frac{1}{Z_C} \sinh \gamma l = \sqrt{\frac{Y}{Z}} \sinh \gamma l$$

$$= \sqrt{\frac{2.53 \times 10^{-3} \angle 89.91°}{500.9 \angle 63.95°}} \times 0.9261 \angle 82.71° = 2.081 \times 10^{-3} \angle 95.69°$$

It may be noted that results obtained by both the methods are almost same.

Example 14.21. *The following data refer to a 3-phase overhead transmission line :*

Voltage between lines = 220 kV

Total series impedance per phase = 200 ∠80° Ω

Total shut admittance per phase = 0.0013 ∠90° ℧

Load delivered = 100 MW at 0.8 p.f. lagging

Using rigorous method, determine the following :

(i) Sending end voltage ; (ii) Sending end current. **(Delhi University)**

Solution. *Given :* $V_R = \frac{220}{\sqrt{3}} = 127$ kV ; $Z = 200 \angle 80°\ \Omega$;

$Y = 0.0013 \angle 90°$ ℧ ; Load delivered = 100 MW, 0.8 p.f. lagging.

Receiving end current ; $I_R = \frac{100 \times 10^3}{\sqrt{3} \times 220 \times 0.8} = 328$ A

Power factor angle, $\phi = \cos^{-1}(0.8) = 36.87°$ lagging.

Taking receiving end phase voltage as the reference vector ; we have

$$\bar{V}_R = 120 \angle 0° \text{ kV} \quad \text{and} \quad \bar{I}_R = 328 \angle -36.87°$$

Now,
$$YZ = 0.0013 \angle 90° \times 200 \angle 80° = 0.26 \angle 170° = -0.256 + j0.04515$$
$$Y^2Z^2 = 0.26^2 \angle 340° = 0.0676 \angle 340° = 0.06352 - j0.02312$$
$$Y^3Z^3 = 0.26^3 \angle 510° = 0.017576 \angle 510° = -0.01522 + j0.008788$$

Auxiliary line constants :

$$A = D = 1 + \frac{YZ}{2} + \frac{Y^2Z^2}{24} + \frac{Y^3Z^3}{720} + ...$$
$$= 1 + \frac{-0.256 + j0.04515}{2} + \frac{0.06352 - j0.02312}{24} + \frac{-0.01522 + j0.008788}{720} + ...$$
$$= (1 - 0.128 + 0.00265 - 0.00002) + j(0.02257 - 0.00096 + 0.00001)$$
$$= (0.8746 + j0.02162) = 0.8749 \angle 1.42°$$

$$B = Z\left[1 + \frac{YZ}{6} + \frac{Y^2Z^2}{120} + \frac{Y^3Z^3}{5040} + ...\right] \Omega$$
$$= 200 \angle 80° \left[1 + \frac{-0.256 + j0.04515}{6} + \frac{0.06352 - j0.02312}{120} + \frac{-0.01522 + j0.008788}{5040} + ...\right] \Omega$$
$$= 220 \angle 80° \, [(1 - 0.04266 + 0.00053 - 0.000003) + j(0.007525 - 0.00019 + j0.000002)$$
$$= 200 \angle 80° \, (0.95787 + j0.007337) = 200 \angle 80° \times 0.9579 \angle 0.439°$$
$$= 191.58 \angle 80.439° \, \Omega$$

$$C = Y\left[1 + \frac{YZ}{6} + \frac{Y^2Z^2}{120} + \frac{Y^3Z^3}{5040} + ...\right]$$
$$0.0013 \angle 90° \times 0.9579 \angle 0.439° = 0.001245 \angle 90.439° \, \mho$$

(*i*) **Sending end voltage :**

Sending end phase voltage,

$$\bar{V}_S = A\bar{V}_R + B\bar{I}_R$$
$$= 0.8749 \angle 1.42° \times 127 \angle 0° + (191.58 \angle 80.439° \times 328 \angle -36.87° \times 10^{-3}) \text{ kV}$$
$$111.11 \angle 1.42° + 62.84 \angle 43.57° = 163.24 \angle 16.39°$$

∴ Sending end line voltage, $\bar{V}_{SL} = \sqrt{3} \times 163.24 \angle 16.39° = \mathbf{282.74 \angle 16.39° \text{ kV}}$. **(Ans.)**

(*ii*) **Sending end current :**

$$\bar{I}_S = C\bar{V}_R + D\bar{I}_R = 0.001245 \angle 90.439° \times 127 \angle 0° \times 10^3 + 0.8749 \angle 1.42 \times 328 \angle -36.87°$$
$$= 158.11 \angle 90.439° + 286.97 \angle -35.45° = \mathbf{232.7 \angle -2.05° \text{ A}}. \quad \textbf{(Ans.)}$$

14.7.5. Transmission Line with Series Impedance at Receiving End

Fig. 14.23 shows a transmission line with a series impedance at the sending end (Normally series impedance consists of resistance and inductance)

Let *A, B, C* and *D* = Auxiliary constants,

Z_{se} = Value of the series impedance of the line at the receiving end,

V_R = Phase voltage at the load end, and

I_R = Load current.

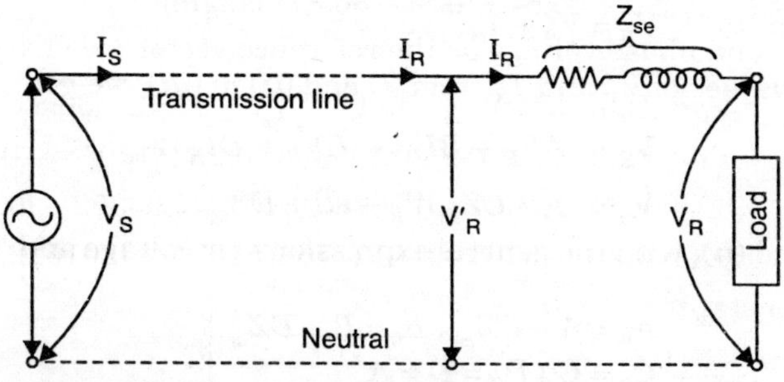

Fig. 14.23. Transmission line with series impedance at the receiving end.

Then, phase voltage at the end of transmission line,

$$\overline{V}_R' = \overline{V}_R + \overline{I}_R \overline{Z}_{se} \qquad ...(i)$$

and,
$$\overline{V}_s = A\overline{V}_R' + B\overline{I}_R = A(\overline{V}_R + \overline{I}_R\overline{Z}_{se}) + B\overline{I}_R$$
$$= A\overline{V}_R + (A\overline{Z}_{se} + B)\overline{I}_R \qquad ...(ii)$$

and,
$$\overline{I}_S = C\overline{V}_R' + D\overline{I}_R$$

Substituting the values of $\overline{V}_R'$ from (*i*), we get

$$\overline{I}_S = C(\overline{V}_R + \overline{I}_R\overline{Z}_{se}) + D\overline{I}_R = C\overline{V}_R + (C\overline{Z}_{se} + D)\overline{I}_R \qquad ...(iii)$$

Comparing with the general voltage and current equations of the transmission line, we have

$$\left.\begin{aligned} A_0 = A \,;\, B_0 = A\overline{Z}_{se} + B \\ C_0 = C \,;\, D_0 = C\overline{Z}_{se} + D \end{aligned}\right\} \qquad ...(14.55)$$

14.7.6. Transmission Line with Series Impedance at the Sending End

Fig. 14.24 shows a transmission line with a series impedance Z_{se} at the sending end. I_S is the sending end current ; this current also passes through Z_{se}. If *A*, *B*, *C* and *D* are the auxiliary constants of the transmission line, V_R and V_S are the phase voltages at the load and supply end respectively, I_R is the load current and V_S' is the phase voltage at the starting end of the line, then

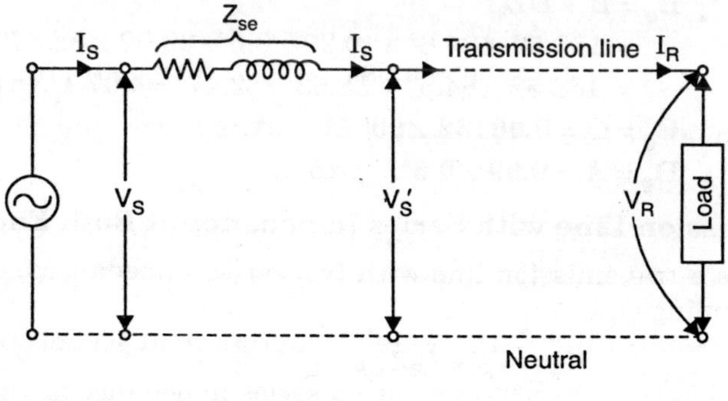

Fig. 14.24. Transmission line with series impedance at the sending end.

$$V_S' = A\overline{V}_R + B\overline{I}_R \qquad ...(i)$$

and,
$$\overline{I}_S = C\overline{V}_R + D\overline{I}_R \qquad ...(ii)$$

Also, $V_S = V_S' + \bar{I}_S \bar{Z}_{se}$ ...(iii)

Substituting the values of V_S' and $\bar{I}_S$ from (*i*) and (*ii*) in (*iii*), we get

$$\bar{V}_S = (A\bar{V}_R + B\bar{I}_R) + (C\bar{V}_R + D\bar{I}_R)\bar{Z}_{se}$$

or $$\bar{V}_S = (A + CZ_{se})\bar{V}_R + (B + DZ_{se})\bar{I}_R \quad ...(iv)$$

Comparing (*iv*) and (*ii*), with the general expressions for voltage and current of the transmission line, we have

$$\left.\begin{array}{l} A_0 = A + CZ_{se}\ ;\ B_0 = B + DZ_{se} \\ C_0 = C\ ;\ D_0 = D = A \end{array}\right\} \quad ...(14.56)$$

Example 14.22. *A 220 kV, 50 Hz, 3-phase transmission line is connected to station bus-bars at the sending end through a 100 MV A, 11/220 kV transformer having impedance of (1 + j8)%. The auxiliary constants of the transmission line are :*

$$A = D = 0.99 \angle 0.6°\ ;\ B = 168.52 \angle 84.6°\ ;\ C = 0.00132 \angle 90°$$

Calculate the A_0, B_0, C_0, D_0 constants of the equivalent circuit of the line and the transformer connected at the supply end.

Solution. *Given :* Transformer impedance, $\bar{Z}_{se} = (1 + j8)\%$;

Auxiliary constants : $A = D = 0.99 \angle 0.6°$; $B = 168.52 \angle 84.6°\ \Omega$; $C = 0.00132 \angle 90°$ ℧ ;

Transformer rating = 100 MVA, 11/220 kV

A_0, B_0, C_0, D_0 constants :

Impedance of the transformer (in ohms) referred to *h.v.* side,

$$\bar{Z}_{se} = \left(\frac{1 + j8}{100}\right) \times \frac{(220 \times 1000)^2}{100 \times 10^6} = 4.84 + j38.72\ \Omega/\text{phase}$$

$$= 39.02 \angle 82.87°\ \Omega/\text{phase}$$

A_0, B_0, C_0, and D_0 constants of the equivalent circuit of the line and transformer connected at the supply end (Fig. 14.24) are given below :

$$\mathbf{A_0} = A + C\bar{Z}_{se}$$
$$= 0.99 \angle 0.6° + (0.00132 \angle 90°)(39.02 \angle 82.87°)$$
$$= 0.99 \angle 0.6° + 0.0515 \angle 172.87° = \mathbf{0.939 \angle 1.02°}. \quad \textbf{(Ans.)}$$

$$\mathbf{B_0} = B + D\bar{Z}_{se}$$
$$= 168.52 \angle 84.6° + (0.99 \angle 0.6°)(39.02 \angle 82.87°)$$
$$= 168.52 \angle 84.6° + 38.63 \angle 83.47° = \mathbf{207.1 \angle 84.39° \Omega}. \quad \textbf{(Ans.)}$$

$$\mathbf{C_0} = C = \mathbf{0.00132 \angle 90°}\ \text{℧}. \quad \textbf{(Ans.)}$$

$$\mathbf{D_0} = A = \mathbf{0.99 \angle 0.6°}. \quad \textbf{(Ans.)}$$

14.7.7. Transmission Line with Series Impedance at Both Ends

Fig. 14.25 shows a transmission line with two series impedances, one at the receiving end and the other at the sending end.

Now, we have $V_R' = \bar{V}_R + \bar{I}_R \bar{Z}_{seR}$ ...(i)

$$V_S' = AV_R' + BI_R = A(\bar{V}_R + \bar{I}_R\bar{Z}_{seR}) + BI_R$$
$$= A\bar{V}_R + (A\bar{Z}_{seR} + B)\bar{I}_R \quad ...(ii)$$

$$\bar{I}_S = CV_R' + D\bar{I}_R = C(\bar{V}_R + \bar{I}_R\bar{Z}_{seR}) + D\bar{I}_R$$
$$= C\bar{V}_R + (C\bar{Z}_{seR} + D)\bar{I}_R \quad ...(iii)$$

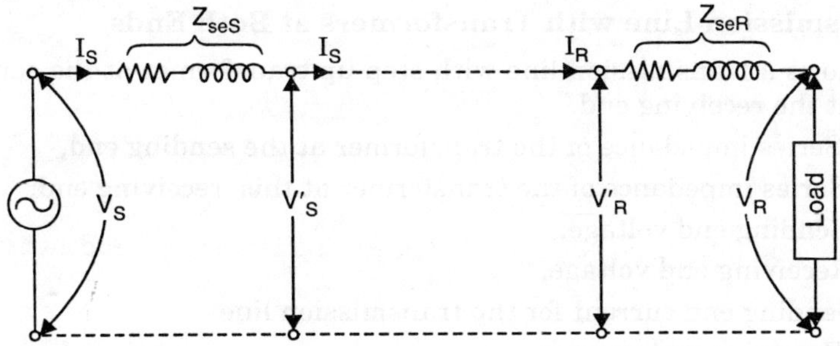

Fig. 14.25. Transmission line with series impedances at both ends.

$$V_S = V'_S + \bar{I}_S Z_{seS}$$
$$= A\bar{V}_R + (A\bar{Z}_{seR} + B)\bar{I}_R + [C\bar{V}_R + C\,(CZ_{seR} + D)\bar{I}_R]\,Z_{seS}$$
$$= (A + C\bar{Z}_{seS})\bar{V}_R + [(A\bar{Z}_{seR} + B) + (C\bar{Z}_{seR} + D)\,\bar{Z}_{seS}]\,\bar{I}_R \quad ...(iv)$$

Comparing (*iv*) and (*iii*) with the general expressions for voltage and current of transmission line, we have

$$\left.\begin{aligned} A_0 &= A + C\bar{Z}_{seS}\ ; \\ B_0 &= (A\bar{Z}_{seR} + B) + (C\bar{Z}_{seR} + D)\,\bar{Z}_{seS} \\ &= (A\bar{Z}_{seR} + B) + (CZ_{seR} + A)\,Z_{seS} \\ &= A\,(\bar{Z}_{seR} + \bar{Z}_{seS}) + B + C\bar{Z}_{seR}\,.\,\bar{Z}_{seS} \\ C_0 &= C \\ D_0 &= C\bar{Z}_{seR} + D. \end{aligned}\right\} \quad ...(14.57) \quad (\because\ A = D)$$

Example 14.23. *Two similar transformers each of impedance 12 ∠80° Ω are interposed at the two ends of a long transmission line having auxiliary constants, A = D = 0.98 ∠2° ; B = 28 ∠69° Ω. Determine the values of A_0, B_0, C_0 and D_0 to represent the line with the transformers.*

Solution. *Given :* Impedance of each transformer, $\bar{Z}_{se} = (\bar{Z}_{seS} = \bar{Z}_{seR})$ = 12 ∠80° Ω

Auxiliary constants : A = D = 0.98 ∠2° ; B = 28 ∠69° Ω

Values of A_0, B_0, C_0 and D_0 :

We know that, $A.D - B.C = 1$

$$\therefore \quad C = \frac{A\,.\,D - 1}{B} = \frac{(0.98\angle 2°)(0.98\angle 2°) - 1}{28\angle 69°} = \frac{0.9604\angle 4° - 1}{28\angle 69°}$$
$$= \frac{0.95806 + j0.067 - 1}{28\angle 69°} = \frac{0.079\angle 122°}{28\angle 69°} = 0.00282\angle 53°\ ℧$$

A_0, B_0, C_0 and D_0 constants of the equivalent circuit of the line and similar transformers connected at both ends (Fig. 14.25) are given below :

$$\mathbf{A_0} = A + C\bar{Z}_{se} = 0.98\angle 2° + (0.00282\angle 53°)(12\angle 80°)$$
$$= 0.98\angle 2° + 0.03384\angle 133° = \mathbf{0.958\angle 3.53°}.\quad \textbf{(Ans.)}$$

$$\mathbf{B_0} = 2A\bar{Z}_{se} + B + C\bar{Z}_{se}^{\ 2} \quad (\because\ Z_{seS} = Z_{seR})$$
$$= 2(0.98\angle 2°)(12\angle 80°) + 28\angle 69° + (0.00282\angle 53°)(12\angle 80°)^2$$
$$= 23.52\angle 82° + 28\angle 69° + 0.40608\angle 213° = \mathbf{50.89\angle 75.24°\ \Omega}.\quad \textbf{(Ans.)}$$

$$\mathbf{C_0} = C = \mathbf{0.00282\angle 53°\ ℧}.\quad \textbf{(Ans.)}$$

$$\mathbf{D_0} = C\bar{Z}_{se} + D = A + C\bar{Z}_{se} = \mathbf{0.958\angle 3.53°}.\quad \textbf{(Ans.)} \quad (\because\ A = D)$$

14.7.8. Transmission Line with Transformers at Both Ends

Fig. 14.26 shows a transmission line with step-up transformer at the sending end and step down transformer at the receiving end.

Let, Z_{TS} = Series impedance of the transformer at the sending end,

Z_{TR} = Series impedance of the transformer at this receiving end,

V_S' = Sending end voltage,

V_R' = Receiving end voltage,

I_S' = Sending end current for the transmission line

I_R' = Receiving end current,

I_{TS} = Magnetising current taken by the transformer at the sending end,

I_{TR} = Magnetising current taken by the transformer at the receiving end,

Y_{TS} = Shunt admittance of the transformer at the sending end, and

Y_{TR} = Shunt admittance of the transformer at the receiving end.

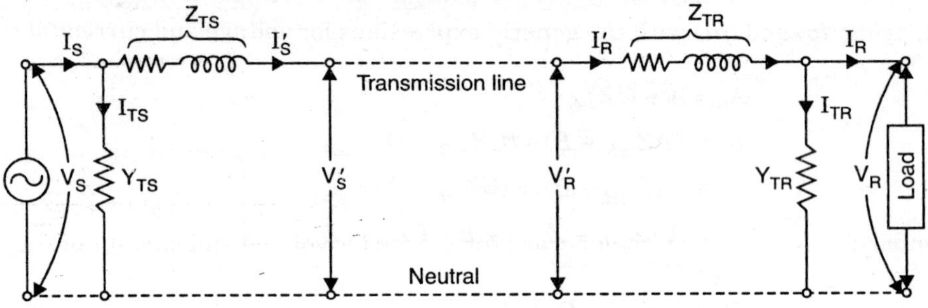

Fig. 14.26. Transmission line with transformers at both ends.

[The shunt admittance of a transformer is usually so high that in most of the cases it may be taken as infinite without much loss of accuracy in which case transformers are just equivalent to series impedance and the equivalent circuit is similar that shown in Fig. 14.25.]

Now, we have :

$$\overline{V}_S' = A\overline{V}_R' + B\overline{I}_R' \quad ...(i)$$

Also, $$\overline{V}_R' = \overline{V}_R + \overline{I}_R' \,.\, Z_{TR} = \overline{V}_R + (\overline{I}_R + \overline{I}_{TR})\overline{Z}_{TR} \qquad (\because \ \overline{I}_R' = \overline{I}_R + \overline{I}_{TR})$$

and, $$\overline{I}_{TR} = \overline{Y}_{TR}\overline{V}_R$$

$$\overline{I}_R' = \overline{I}_R + \overline{Y}_{TR}\overline{V}_R \quad ...(ii)$$

$\therefore$ $$\overline{V}_R' = \overline{V}_R + (\overline{I}_R + \overline{Y}_{TR}\overline{V}_R)\overline{Z}_{TR} \quad ...(iii)$$

Substituting the value of $\overline{V}_R'$ from (*iii*) and value of $\overline{I}_R'$ from (*ii*) in (*i*), we get

$$\overline{V}_S' = A\,[\overline{V}_R + (\overline{I}_R + \overline{Y}_{TR}\overline{V}_R)\overline{Z}_{TR}] + B\,(\overline{I}_R + \overline{Y}_{TR}\overline{V}_R)$$

Now, $$\overline{I}_S' = C\overline{V}_R' + D\overline{I}_R' = C\{\overline{V}_R + (\overline{I}_R + \overline{Y}_{TR}\overline{V}_R)\overline{Z}_{TR}\} + D\,(\overline{I}_R + \overline{Y}_{TR}\overline{V}_R)$$

Also, $$\overline{V}_S = \overline{V}_S' + \overline{I}_S'\,\overline{Z}_{TS} = [A\,\{\overline{V}_R + (\overline{I}_R + \overline{Y}_{TR}\overline{V}_R)\overline{Z}_{TR}\} + B\,(\overline{I}_R + \overline{Y}_{TR}\overline{V}_R)]$$
$$+ [C\overline{Z}_{TS}\,\{\overline{V}_R + (\overline{I}_R + \overline{Y}_{TR}\overline{V}_R)\overline{Z}_{TR}\} + D\overline{Z}_{TS}\,(\overline{I}_R + \overline{Y}_{TR}\overline{V}_R)]$$

$$= [A(1+\overline{Y}_{TR}\overline{Z}_{TR}+\overline{Y}_{TR}\overline{Z}_{TS})+B\overline{Y}_{TR}+C\overline{Z}_{TS}(1+\overline{Y}_{TR}\overline{Z}_{TR})]\overline{V}_R$$

$$+ [A\overline{Z}_{TR}+B+C\overline{Z}_{TS}\overline{Z}_{TR}+A\overline{Z}_{TS}]\overline{I}_R \quad ...(iv)$$

$$(\because \quad A = D)$$

and

$$I_S = I_S' + I_{TS}$$

$$= C\{\overline{V}_R+(\overline{I}_R+\overline{Y}_{TR}\overline{V}_R)\overline{Z}_{TR}\}+D(\overline{I}_R+\overline{Y}_{TR}\overline{V}_R)+\overline{Y}_{TS}\overline{V}_S$$

$$= C\{\overline{V}_R+(\overline{I}_R+\overline{Y}_{TR}\overline{V}_R)\overline{Z}_{TR}\}+A(\overline{I}_R+\overline{Y}_{TR}\overline{V}_R)$$

$$+ [A(1+\overline{Y}_{TR}\overline{Z}_{TR}+\overline{Y}_{TR}\overline{Z}_{TS})+B\overline{Y}_{TR}+C\overline{Z}_{TS}(1+\overline{Y}_{TR}\overline{Z}_{TR})\}\overline{V}_R$$

$$+ \{A\overline{Z}_{TR}+B+C\overline{Z}_{TS}\overline{Z}_{TR}+A\overline{Z}_{TS}\}\overline{I}_R]\overline{Y}_{TS}$$

$$= [C+C\overline{Y}_{TR}\overline{Z}_{TR}+A\overline{Y}_{TR}+A\overline{Y}_{TS}+A\overline{Y}_{TR}\overline{Z}_{TR}\overline{Y}_{TS}+A\overline{Y}_{TR}\overline{Z}_{TS}\overline{Y}_{TS}$$

$$+ B\overline{Y}_{TR}\overline{Y}_{TS}+C\overline{Z}_{TS}\overline{Y}_{TS}+C\overline{Z}_{TS}\overline{Y}_{TR}\overline{Z}_{TR}\overline{Y}_{TS}]\overline{V}_R$$

$$+ [C\overline{Z}_{TR}+A+A\overline{Z}_{TR}\overline{Y}_{TS}+B\overline{Y}_{TS}+C\overline{Z}_{TS}\overline{Z}_{TR}\overline{Y}_{TS}+A\overline{Z}_{TS}\overline{Y}_{TS}]\overline{I}_R$$

$$= C[(1+\overline{Y}_{TR}\overline{Z}_{TR}+\overline{Z}_{TS}\overline{Y}_{TS}+\overline{Z}_{TR}\overline{Z}_{TS}\overline{Y}_{TR}\overline{Y}_{TS})+B\overline{Y}_{TR}\overline{Y}_{TS}$$

$$+ A(\overline{Y}_{TR}+\overline{Y}_{TS}+\overline{Y}_{TR}\overline{Y}_{TS}\overline{Z}_{TR}+\overline{Y}_{TR}\overline{Y}_{TS}+\overline{Z}_{TS})\overline{V}_R$$

$$+ [A(1+\overline{Y}_{TS}\overline{Z}_{TR}+\overline{Z}_{TS}\overline{Y}_{TS})+B\overline{Y}_{TS}+C\overline{Z}_{TR}(1+\overline{Z}_{TS}\overline{Y}_{TS})]\overline{I}_R \quad ...(v)$$

Comparing (*iv*) and (*v*) with general expressions for voltage and current of the transmission line, we have

$$\left.\begin{aligned}
A_0 &= A(1+\overline{Y}_{TR}\overline{Z}_{TR}+\overline{Y}_{TR}\overline{Z}_{TS})+B\overline{Y}_{TR}+C\overline{Z}_{TS}(1+\overline{Y}_{TR}\overline{Z}_{TR}) && ...(vi)\\
B_0 &= A(\overline{Z}_{TR}+\overline{Z}_{TS})+B+C\overline{Z}_{TS}\overline{Z}_{TR} && ...(vii)\\
C_0 &= A(\overline{Y}_{TR}+\overline{Y}_{TS}+\overline{Y}_{TR}\overline{Y}_{TS}\overline{Z}_{TR}+\overline{Y}_{TR}\overline{Y}_{TS}\overline{Z}_{TS})+B\overline{Y}_{TR}\overline{Y}_{TS}\\
&\quad + C(1+\overline{Y}_{TR}\overline{Z}_{TR}+\overline{Z}_{TS}\overline{Y}_{TS}+\overline{Z}_{TR}\overline{Z}_{TS}\overline{Y}_{TR}\overline{Y}_{TS}) && ...(viii)\\
D_0 &= A(1+\overline{Y}_{TS}\overline{Z}_{TR}+\overline{Y}_{TS}\overline{Z}_{TS})+B\overline{Y}_{TS}+C\overline{Z}_{TR}(1+\overline{Z}_{TS}\overline{Y}_{TS}) && ...(ix)
\end{aligned}\right\} \quad ...(14.58)$$

In case of identical transformers impedances Z_{TR} and Z_{TS} will be equal and also shunt admittances $\overline{Y}_{TR}$ and $\overline{Y}_{TS}$ will be equal.

Let $\overline{Z}_{TR} = \overline{Z}_{TS} = \overline{Z}_T$ (say) ;

$\overline{Y}_{TR} = \overline{Y}_{TS} = \overline{Y}_T$ (say), then substituting these values in (*vi*), (*vii*), (*viii*) and (*ix*), we get

$$\left.\begin{aligned}
A_0 &= A(1+2\overline{Y}_T\overline{Z}_T)+B\overline{Y}_T+C\overline{Z}_T(1+\overline{Y}_T\overline{Z}_T) && ...(x)\\
B_0 &= 2A\overline{Z}_T+B+C\overline{Z}_T^2 && ...(xi)\\
C_0 &= 2A\overline{Y}_T(1+\overline{Y}_T\overline{Z}_T)+B\overline{Y}_T^2+C(1+\overline{Y}_T\overline{Z}_T)^2 && ...(xii)\\
D_0 &= A(1+2\overline{Y}_T\overline{Z}_T)+B\overline{Y}_T+C\overline{Z}_T(1+\overline{Y}_T\overline{Z}_T) && ...(xiii)
\end{aligned}\right\} \quad ...[14.58\,(a)]$$

Example 14.24. *Two similar transformers, each of series impedance 100∠70° Ω and admittance 0.0002 ∠– 75° ℧ are connected at the two ends of a long transmission line have auxiliary constants, A = D = 0.92 ∠5.3° ; B = 65.3∠81° Ω.*

Determine A_0, B_0, C_0 and D_0 constants to represents the line with transformers.

Solution. *Given :* $\bar{Z}_T$ ($= \bar{Z}_{TR} = \bar{Z}_{TS}$) = 100∠70° Ω ; $\bar{Y}_T$ ($= \bar{Y}_{TR} = \bar{Y}_{TS}$) = 0.0002∠– 75° ℧

$$A = D = 0.92\angle 5.3° \; ; \; B = 65.3\angle 81° \, \Omega$$

A_0, B_0, C_0 and D_0 :

We know that, $A \,.\, D - B \,.\, C = 1$

$$\therefore \quad C = \frac{A\,.\,D - 1}{B} = \frac{(0.92\angle 5.3°)(0.92\angle 5.3°) - 1}{65.3\angle 81°} = \frac{0.8464\angle 10.6° - 1}{65.3\angle 81°}$$

$$= \frac{0.8319 + j0.1557 - 1}{65.3\angle 81°} = \frac{0.229\angle 137.2°}{65.3\angle 81°} = 0.0035\angle 56.2° \; ℧.$$

A_0, B_0, C_0 and D_0 constants of the equivalent circuit of the transmission line and similar transformers connected at both ends (Fig. 14.26) are given below [Refer to eqn. 14.58(*a*)] :

$$\mathbf{A_0} = \mathbf{D_0} = A\left(1 + 2\bar{Y}_T\bar{Z}_T\right) + B\bar{Y}_T + C\bar{Z}_T(1 + \bar{Y}_T\,\bar{Z}_T)$$

$$= 0.92\angle 5.3° \,(1 + 2 \times 0.0002\angle -75° \times 100\angle 70°) + 65.3\angle 81° \times 0.0002\angle -75°$$
$$+ 0.0035\angle 56.2° \times 100\angle 70° \,(1 + 0.0002\angle -75° \times 100\angle 70°)$$

$$0.92\angle 5.3° \,(1 + 0.04\angle -5°) + 0.01306\angle 6° + 0.35\angle 12.62° \,(1 + 0.02\angle -5°)$$

$$0.92\angle 5.3° \,(1 + 0.03985 - j0.0035) + 0.01306\angle 6° + 0.35\angle 126.2°(1 + 0.02 - j0.0017)$$

$$= (0.92\angle 5.3°)\,(1.0398\angle -0.19°) + 0.01306\angle 6° + (0.35\angle 126.2°)(1.02\angle -0.095)$$

$$= 0.9567\angle 5.1° + 0.01306\angle 6° + 0.357\angle 126.1° = \mathbf{0.8434\angle 26.39°}. \quad \textbf{(Ans.)}$$

$$\mathbf{B_0} = 2A\bar{Z}_T + B + C\bar{Z}_T^{\,2}$$

$$= 2 \times (0.92\angle 5.3°)\,(100\angle 70°) + 65.3\angle 81° + 0.0035\angle 56.2°(100\angle 70°)^2$$

$$= 184\angle 75.3° + 65.3\angle 81° + 35\angle 196.2° = \mathbf{233.87\angle 84.28°} \; \Omega. \quad \textbf{(Ans.)}$$

$$\mathbf{C_0} = 2A\bar{Y}_T\,(1 + \bar{Y}\,\bar{Z}_T) + B\bar{Y}_T^{\,2} + C(1 + \bar{Y}_T\,\bar{Z}_T)^2$$

$$= 2 \times 0.92\angle 5.3° \times 0.0002\angle -75° \,(1 + 0.0002\angle -75° \times 100\angle 70°)$$
$$+ 65.3\angle 81° \times (0.0002\angle -75°)^2$$
$$+ 0.0035\angle 56.2° \,(1 + 0.0002\angle -75° \times 100\angle 70°)^2$$

$$= 0.000368\angle -69.7°(1 + 0.0002\angle -75° \times 100\angle 70°) + 65.3\angle 81° \times (0.0002\angle -75°)^2$$
$$+ 0.0035\angle 56.2°(1 + 0.0002\angle -75° \times 100\angle 70°)^2$$

$$= 0.000368\angle -69.7° \,(1 + 0.02\angle -5°) + 0.0000026\angle -69°$$
$$+ 0.0035\angle 56.2° \,(1 + 0.02\angle -5°)^2$$

$$= 0.000368\angle -69.7° \,(1 + 0.01992 - j0.00174) + \text{zero}$$
$$+ 0.0035\angle 56.2°(1 + 0.01992 - j0.00174)^2$$

$$= 0.000368\angle -69.7° \,(1.01992 - j0.00174) + 0.0035\angle 56.2°(1.01992 - j0.00174)^2$$

$$= 0.000368\angle -69.7° \times 1.02\angle -0.1° + 0.0035\angle 56.2° \times 1.0404\angle -0.2°$$

$$= 0.000375\angle -69.8° + 0.0036414\angle 56° = \mathbf{0.003436\angle 50.9°} \; \Omega. \quad \textbf{(Ans.)}$$

14.7.9. Equivalent T-Network of a Long Transmission Line

The equivalent T-network of a long transmission is determined to represent the line accurately by assuming suitable values of lumped constants. Let the transmission line having auxiliary constants *A, B, C* and *D* be represented by T-network shown in Fig. 14.27.

Now we have,
$$\bar{I}_S = \bar{I}_R + \left(\bar{V}_R + \frac{1}{2}\bar{I}_R\bar{Z}\right)\bar{Y} = \bar{Y}\bar{V}_R + \left(1 + \frac{1}{2}\bar{Y}\,\bar{Z}\right)\bar{I}_R \quad ...(i)$$

and,
$$\bar{V}_S = \bar{V}_R + \frac{1}{2}\bar{I}_R\bar{Z} + \frac{1}{2}\bar{I}_S\bar{Z}$$

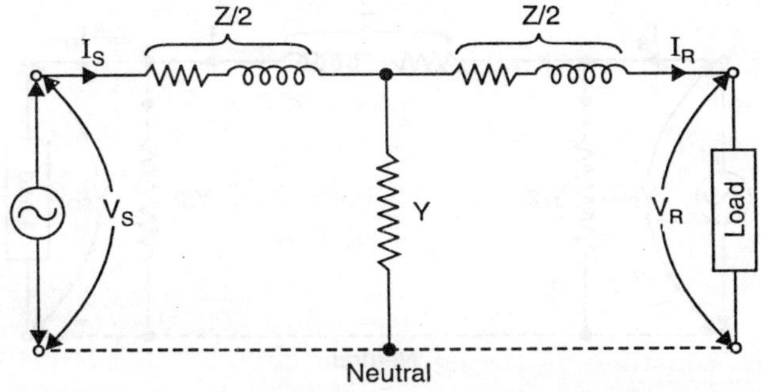

Fig. 14.27. Equivalent T-network of a long transmission line.

$$= \overline{V}_R + \frac{1}{2}\overline{I}_R\overline{Z} + \frac{1}{2}\overline{Z}\left\{\overline{Y}\,\overline{V}_R + \left(1+\frac{1}{2}\overline{Y}\,\overline{Z}\right)\overline{I}_R\right\}$$

$$= \left(1+\frac{1}{2}\overline{Y}\,\overline{Z}\right)\overline{V}_R + \left(\frac{1}{2}\overline{Z} + \frac{1}{2}\overline{Z} + \frac{1}{4}\overline{Y}\,\overline{Z}^2\right)\overline{I}_R$$

$$= \left(1+\frac{1}{2}\overline{Y}\,\overline{Z}\right)\overline{V}_R + \left(\overline{Z} + \frac{1}{4}\overline{Y}\,\overline{Z}^2\right)\overline{I}_R \qquad ...(ii)$$

Comparing (*ii*) and (*i*) with general expressions for voltage and current of transmission line, we have

$$\left.\begin{aligned} A = D &= 1+\frac{1}{2}\overline{Y}\,\overline{Z} && ...(iii)\\ B &= \overline{Z} + \frac{1}{4}\overline{Y}\,\overline{Z}^2 && ...(iv)\\ C &= \overline{Y} && ...(v) \end{aligned}\right\} \qquad ...(14.59)$$

From (*iii*) and (*iv*), we get

$$\left.\begin{aligned} \overline{Y} &= C && ...(vi)\\ \text{and,}\quad \overline{Z} &= \frac{2(A-1)}{C} && ...(vii) \end{aligned}\right\} \qquad ...[14.59(a)]$$

14.7.10. Equivalent π-Network of a Long Transmission Line

Sometimes the equivalent π-network of a long transmission line is required to be determined to represent the line accurately by assuming suitable values of lumped constants. Let the transmission line having auxiliary constants *A*, *B*, *C* and *D* be represented by π-network shown in the Fig. 14.28.

Now, we have

$$\overline{V}_S = \overline{V}_R + \overline{I}\,\overline{Z} = \overline{V}_R + \left(\overline{I}_R + \frac{1}{2}\overline{Y}\,\overline{V}_R\right)\overline{Z} = \left(1+\frac{1}{2}\overline{Y}\,\overline{Z}\right)\overline{V}_R + \overline{I}_R\overline{Z} \qquad ...(i)$$

and
$$\overline{I}_S = \overline{I}_R + \frac{1}{2}\overline{Y}\,\overline{V}_R + \frac{1}{2}\overline{Y}\,\overline{V}_S = \overline{I}_R + \frac{1}{2}\overline{Y}\overline{V}_R + \frac{1}{2}\overline{Y}\left\{\left(1+\frac{1}{2}\overline{Y}\,\overline{Z}\right)\overline{V}_R + \overline{I}_R\overline{Z}\right\}$$

$$= \left(\frac{1}{2}\overline{Y} + \frac{1}{2}\overline{Y} + \frac{1}{4}\overline{Y}^2\overline{Z}\right)\overline{V}_R + \left(1+\frac{1}{2}\overline{Y}\,\overline{Z}\right)\overline{I}_R$$

$$= \left(\overline{Y} + \frac{1}{4}\overline{Y}^2\overline{Z}\right)\overline{V}_R + \left(1+\frac{1}{2}\overline{Y}\,\overline{Z}\right)\overline{I}_R \qquad ...(ii)$$

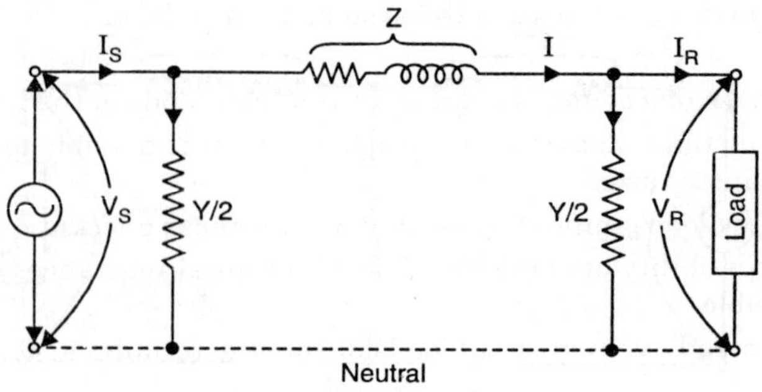

Fig. 14.28. Equivalent π-network of a long transmission line.

Comparing (*i*) and (*ii*) with the general expression for voltage and current of the transmission line, we have

$$\left.\begin{aligned} A = D &= 1 + \frac{1}{2}\overline{Y}\,\overline{Z} && \text{...(iii)} \\ B &= \overline{Z} && \text{...(iv)} \\ C &= \overline{Y} + \frac{1}{4}\overline{Y}^2\overline{Z} && \text{...(v)} \end{aligned}\right\} \quad \text{...(14.60)}$$

From (*iii*) and (*iv*), we get

$$\left.\begin{aligned} \overline{Z} &= B && \text{...(vi)} \\ \text{and} \quad \overline{Y} &= \frac{2(A-1)}{B} && \text{...(vii)} \end{aligned}\right\} \quad \text{...[14.60 (a)]}$$

Example 14.25. *Determine equivalent T-network and π-network for a long transmission line having auxiliary constants, A = D = 0.945∠1.02° ; B = 82.3∠73.03° Ω ; C = 0.001376∠90.4° ℧*

Solution. *Given :* Auxiliary constants $A = D = 0.945\angle 1.02°$; $B = 82.3\angle 73.03°\ \Omega$; $C = 0.001376\angle 90.4°$ ℧.

Equivalent T-network : Refer to Fig. 14.27.

Series impedance,
$$\overline{Z} = \frac{2(A-1)}{C} = \frac{2(0.945\angle 1.02° - 1)}{0.001376\angle 90.4°} = \frac{2(0.9448 + j0.0168 - 1)}{0.001376\angle 90.4°}$$
$$= \frac{-0.1104 + j0.0336}{0.001376\angle 90.4°} = \frac{0.1154\angle 163°}{0.001376\angle 90.4°} = \mathbf{83.87\angle 72.6°\ \Omega}. \quad \textbf{(Ans.)}$$

Shunt admittance, $\overline{Y} = C = \mathbf{0.001376\angle 90.4°}$ **℧. (Ans.)**

Equivalent π-network : Refer to Fig 14.26.

Series impedance, $\overline{Z} = B = \mathbf{82.3\angle 73.03°\ \Omega}$. **(Ans.)**

Shunt admittance,
$$\overline{Y} = \frac{2(A-1)}{B} = \frac{2(0.945\angle 1.02° - 1)}{82.3\angle 73.03°}$$
$$= \frac{0.1154\angle 163°}{82.3\angle 73.03°} = \mathbf{0.001402\angle 89.97°}\ \textbf{℧. (Ans.)}$$

14.7.11. Constant Voltage Transmission

In case of "constant voltage transmission" a *constant voltage drop is always maintained along the line by installing specially designed synchronous motors at the receiving end.* As and when the load changes, the synchronous motors change the power factor of the system, thus maintaining the voltage drop along the line constant.

The *advantages* and *disadvantages* of the system are as follows :

Advantages :

1. For all the loads, *same voltage is maintained* at both ends of the line.
2. In case of long distance heavy power transmission, it is possible to *carry increased power* for a given conductor size.
3. At times of heavy loads improvement in power factor is effected.
4. Owing to the possibility of employing higher terminal reactances a better protection for the line is possible.
5. Availability of steady voltage at all loads at the line terminals.

Disadvantages :

1. There is increased risk of interruption of supply, in case of synchronous phase modifiers falling out of synchronism.
2. Increase of short-circuit current of the system.
3. Lower reserve of lines in case of line trouble.

14.7.12. Power Circle Diagram

The relations between sending end and receiving end voltages and currents for a transmission line, which can be treated as a four-terminal network, are given by the following relations (Refer to Article 14.6) :

$$\bar{V}_S = A\bar{V}_R + B\bar{I}_R \qquad ...(i)$$

$$\bar{I}_S = C\bar{V}_R + D\bar{I}_R \qquad ...(ii)$$

where, A, B, C, D = Auxiliary or generalised constants of the transmission.

$\bar{V}_S$ = Sending end voltage,

$\bar{V}_R$ = Receiving end voltage,

$\bar{I}_S$ = Sending end current, and

$\bar{I}_R$ = Receiving end current.

Also the auxiliary constants A, B, C, D being complex numbers are denoted as :

$\bar{A} = A\angle\alpha$; $\bar{B} = B\angle\beta$; $\bar{D} = D\angle\delta$ (The constant C is not required for the construction of the diagram).

The receiving end voltage and current in terms of standing end voltage and current can be determined as follows :

From eqn. (*ii*), we have

$$\bar{I}_R = \frac{\bar{I}_S - C\bar{V}_R}{D} \qquad ...(iii)$$

Substituting this value of $\bar{I}_R$ in eqn. (*i*), we get

$$\bar{V}_S = A\bar{V}_R + B\left(\frac{\bar{I}_S - C\bar{V}_R}{D}\right)$$

or, $$D\bar{V}_S = AD\bar{V}_R + B\bar{I}_S - BC\bar{V}_R \quad \text{or} \quad \bar{V}_R(AD - BC) = D\bar{V}_S - B\bar{I}_S$$

or, $$\bar{V}_R = \frac{D\bar{V}_S - B\bar{I}_S}{AD - BC} = D\bar{V}_S - B\bar{I}_S \qquad ...(iv)$$

($\because$ $AD - BC = 1$, for a four-terminal network)

Substituting the value of $\bar{V}_R$ in eqn. (*iii*), we have

$$\bar{I}_R = \frac{\bar{I}_S - C(D\bar{V}_S - B\bar{I}_S)}{D} = -C\bar{V}_S + \frac{\bar{I}_S}{D} + \frac{BC}{D}\bar{I}_S = -C\bar{V}_S + \frac{\bar{I}_S}{D}(1+BC)$$

$$= -C\bar{V}_S + \frac{\bar{I}_S}{D} \times AD \qquad (\because\ AD - BC = 1, \text{ or } AD = 1 + BC)$$

or, $$\bar{I}_R = -C\bar{V}_S + A\bar{I}_S \qquad ...(v)$$

Normally V_S and I_S are determined when V_R and I_R are known. The interesting feature of this circuit when operating on fixed frequency is that, they behave according to a circular function.

Philip described the circle diagram for a *short line*, while Dwight extended it for a long transmission line by taking into account the effect of capacitance.

In order to draw the circle diagram for long transmission line the *data required* is *obtained* as follows :

The receiving end current I_R is split into two components, I_P the in-*phase or active component* and I_Q the *quadrature* or *reactive* component respectively, so that

$$\bar{I}_R = I_P \pm jI_Q$$

'+ *ve*' and "– *ve*" signs for '*leading*' and "*lagging*" components respectively.

Substituting, $\bar{I}_R = I_P - jI_Q$ in eqn. (*i*), we get

$$\bar{V}_S = A\bar{V}_R + B(I_P - jI_Q)$$

Also, $$\bar{A} = A\angle\alpha = A\cos\alpha + jA\sin\alpha$$

$$\bar{B} = B\angle\beta = B\cos\beta + jB\sin\beta$$

Substituting these values of $\bar{A}$ and $\bar{B}$ in the equation for $\bar{V}_S$, we get

$$\bar{V}_S = \bar{V}_R(A\cos\alpha + jA\sin\alpha) + (I_P - jI_Q)(B\cos\beta + jB\sin\beta)$$

$$= (\bar{V}_R A\cos\alpha + I_p B\cos\beta + I_Q B\sin\beta) + j(\bar{V}_R A\sin\alpha + I_P B\sin\beta - I_Q B\cos\beta)$$

or $$V_S^2 = (V_R A\cos\alpha + I_p B\cos\beta + I_Q B\sin\beta)^2 + (V_R A\sin\alpha + I_P B\sin\beta - I_Q B\cos\beta)^2$$

The above equation is the equation of the circle and can be reduced to following form :

$$\left[I_p + \frac{V_R(A\cos\alpha \,.\, B\cos\beta + A\sin\alpha \,.\, B\sin\beta}{B^2}\right]^2 + \left[I_Q + \frac{V_R(A\cos\alpha \,.\, B\sin\beta - A\sin\alpha \,.\, B\cos\beta}{B^2}\right]^2 = \left(\frac{V_S}{B}\right)^2$$

or, $$\left[I_p + \frac{V_R A}{B}\cos(\beta - \alpha)\right]^2 + \left[I_Q + \frac{V_R A}{B}\sin(\beta - \alpha)\right]^2 = \left(\frac{V_S}{B}\right)^2$$

Multiplying both the sides by V_R^2, we have

$$\left[I_p V_R + \frac{V_R^2 A}{B}\cos(\beta - \alpha)\right]^2 + \left[I_Q V_R + \frac{V_R^2 A}{B}\sin(\beta - \alpha)\right]^2 = \left(\frac{V_S V_R}{B}\right)^2 \qquad ...(vi)$$

According to eqn. (*vi*), the following are the *co-ordinates of the centre of the circle* and *radius of the circle* :

$$\left.\begin{aligned}\text{Horizontal } (X) &= -\frac{A}{B}\, V_R^2 \cos(\beta - \alpha)\\ \text{Vertical } (Y) &= -\frac{A}{B}\, V_R^2 \sin(\beta - \alpha)\end{aligned}\right\}\ \textit{Coordinates of the centre of the circle}$$

$$\textit{Radius} = \frac{V_S V_R}{B}.$$

Construction of power circle diagram :

(*a*) ***Receiving end power circle diagram.*** The procedures of constructing the receiving end power circle diagram is as follows : Refer to Fig. 14.29.

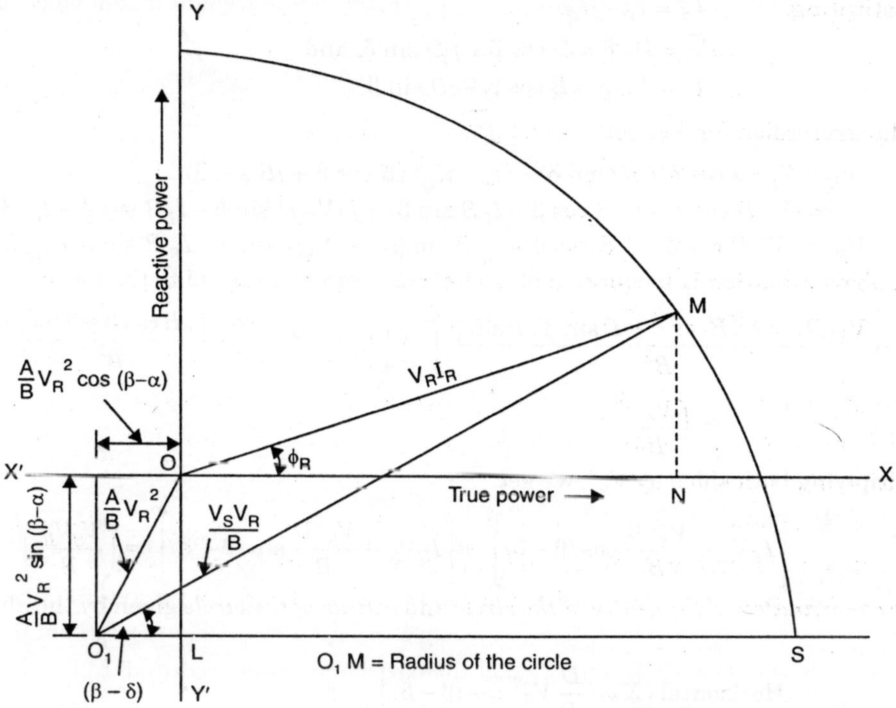

Fig. 14.29. Receiving end power circle diagram.

- Plot the centre of the circle O_1 on a suitable scale.
- From O_1 draw an arc of circle with the calculated radius $\left(\frac{V_R V_S}{B}\right)$.
- From the origin O draw the load line OM inclined at angle ϕ_2 (load power factor angle) with the horizontal. Let it cut the circle at M, then the receiving end *true power* and *reactive power* will be represented by OM and MN respectively.

— If voltage V_S and V_R are taken phase voltages then the power indicated on X-axis and Y-axis will be in W (watts) and VAR (vars) per phase respectively.

— If voltage V_S and V_R and taken line voltages then the power indicated on X-axis and Y-axis and will be W (watts) and VAR (vars) for all the three phases respectively.

— If voltages V_S and V_R are taken as line voltages in kV the power indicated will be MW and MVAR and for all the three phases.

The circle drawn with receiving end true and reactive power as the horizontal and vertical coordinates is called the "*Receiving end power circle diagram*".

- In order to determine the '*maximum power*' draw a horizontal line from the centre of the circle intersecting vertical axis at the point L and the circle at the point S. Then the distance LS *represents the maximum power for the receiving end.*

(*b*) ***Sending end power circle diagram.*** The circle drawn with sending end true and reactive power as the horizontal and vertical coordinates is known as the *sending end power circle diagram.*

The data required for drawing the sending end power circle diagram is determined [as discussed earlier in (*a*)] as follows :

Substituting,
$$\bar{I}_S = I_P - jI_Q,$$
$$\bar{D} = D\angle\delta = D\cos\delta + jD\sin\delta, \text{ and}$$
$$\bar{B} = B\angle\beta = B\cos\beta + jB\sin\beta$$

in the expression for $\bar{V}_R$, get

$$\bar{V}_R = V_S(D\cos\delta + jD\sin\delta) - (I_P - jI_Q)(B\cos\beta + jB\sin\beta)$$
$$= (V_S D\cos\delta - I_P B\cos\beta - I_Q B\sin\beta) + j(V_S D\sin\delta - I_P B\sin\beta + I_Q B\cos\beta)$$

or,
$$V_R^2 = (V_S D\cos\delta - I_P B\cos\beta - I_Q B\sin\beta)^2 + (V_S D\sin\delta - I_P B\sin\beta + I_Q B\cos\beta)^2$$

The above equation is a equation of a circle and can be reduced to the form

$$\left[I_P - \frac{V_S(D\cos\delta\,.\,B\cos\delta + D\sin\delta\,.\,B\sin\beta}{B^2}\right]^2 + \left[I_Q - \frac{V_S(D\cos\delta\,.\,B\cos\beta + D\sin\delta\,.\,B\cos\beta}{B^2}\right]$$
$$= \left(\frac{V_R}{B}\right)^2$$

Multiplying both sides by V_S^2, we get

$$\left[I_P V_S - \frac{V_S^2 D}{B}\cos(\beta - \delta)\right]^2 + \left[I_Q V_S - \frac{V_S^2 D}{B}\sin(\beta - \delta)\right]^2 = \left(\frac{V_S V_R}{B}\right)^2$$

The *co-ordinates of the centre of the circle* and *radius of the circle* given by the above equation are :

$$\left.\begin{aligned}\text{Horizontal } (X) &= \frac{D}{B}V_S^2\cos(\beta - \delta)\\ \text{Vertical } (Y) &= \frac{D}{B}V_S^2\sin(\beta - \delta)\end{aligned}\right\} \quad \textit{Co-ordinates of the centre of the circle}$$

$$\textit{Radius} = \frac{V_S V_R}{B}$$

The sending end power circle diagram is shown in Fig. 14.30, $L'S'$ represents the *maximum power for sending end.*

(*c*) ***Universal power circle diagram.*** In receiving end power circle diagram, if V_R is kept constant, the centre of the circles is fixed and if different sending end voltages circles are drawn, these will be concentric ; however if V_R is not constant, these circle will not be concentric. Similarly, if V_S is not constant, the centre of the sending ending power circle diagram will vary. This *limitation is overcome by drawing the 'universal power circle diagram'.*

- In "*universal power circle diagram*" each distance on the original diagram is *divided* by $\frac{V^2}{B}$ where V is the "*reference or base voltage*" and B is the *generalised constant.* This gives *dimensionless units.*

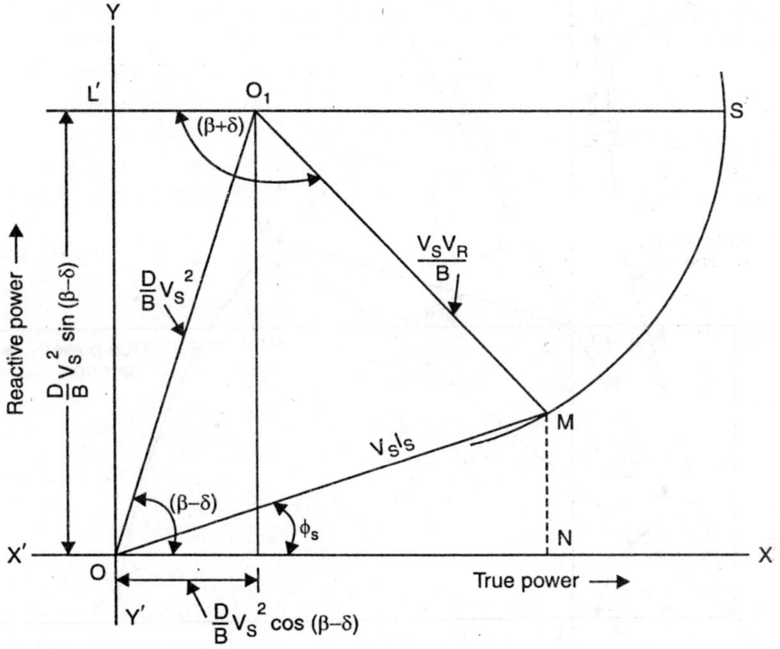

Fig. 14.30. Sending end power circle diagram.

(*i*) ***Receiving end power circle diagram :***

$$\left.\begin{aligned}\text{Horizontal} &= -A\left(\frac{V_R}{V}\right)^2 \cos(\beta-\alpha) \text{ per unit}\\ \text{Vertical} &= -A\left(\frac{V_R}{V}\right)^2 \sin(\beta-\alpha) \text{ per unit}\end{aligned}\right\} \textit{Coordinates of the centre}$$

$$\textit{Radius} = \frac{V_S V_R}{V^2} \text{ per unit.}$$

Receiving end power circle diagram shown in Fig. 14.29 is redrawn as shown in Fig. 14.31 ; the following *observations* may be made :

- The centre of the circle remains at the point O_1 but for different values of receiving end voltage V_R (say V_R') distance O_1O_R varies as the square of the receiving end voltage. In order to have concentric circles for different values of V_SV_R point O_1 is retained as the centre of the circles but measurements of true and reactive power are made from O_R' instead of O_R where $O_RO_R' = O_1O_1'$. Useful portions of the circles will be in the first quadrant with respect to point O, as shown in Fig. 14.32.
- By drawing load line O_RM from O_R at an angle ϕ_R (load power factor angle) we get O_RN as the *true power* (measured from O_R) along the horizontal axis per unit and MN as the reactive power per unit measured vertically above the point N.
- For a new value of V_R above the origin O_R will shift along the line of origins to O_R (as mentioned earlier) but the centre of the circle is still at point O_1. For each value of V_R, power will be measured from different origin.

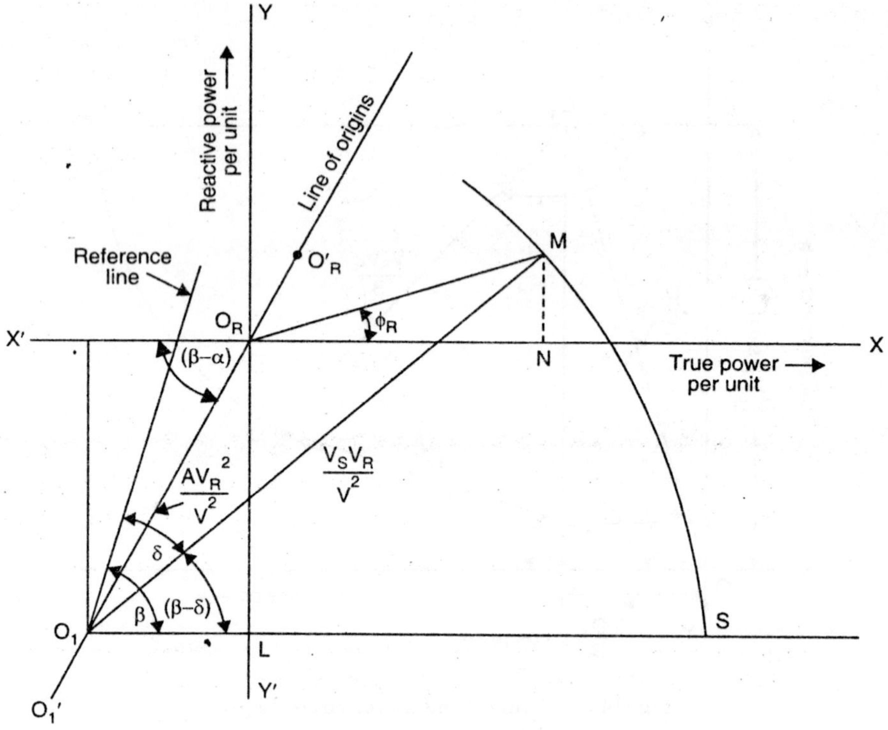

Fig. 14.31. Universal power circle diagram for receiving end.

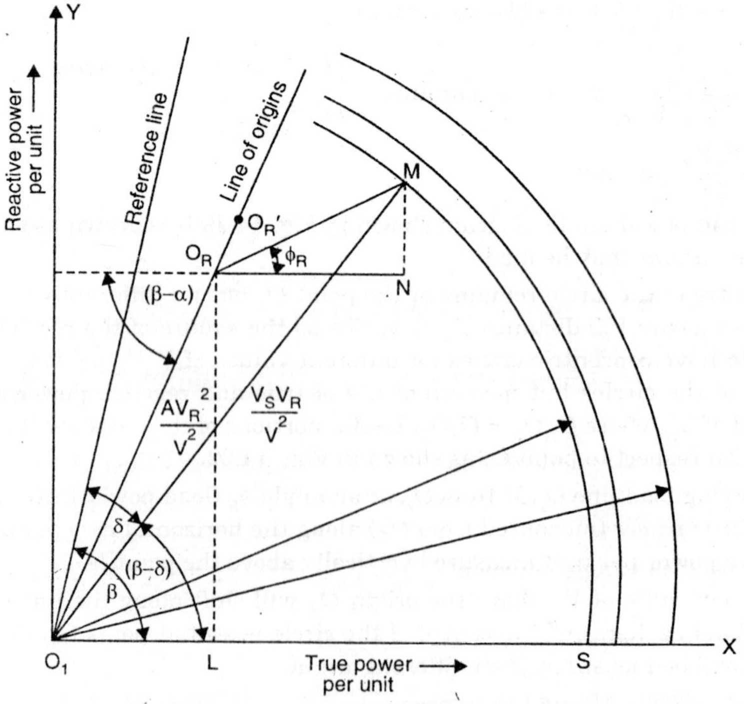

Fig. 14.32. Modified universal circle diagram for receiving end.

(*ii*) ***Sending end power circle diagram :***

Refer to Fig. 14.32.

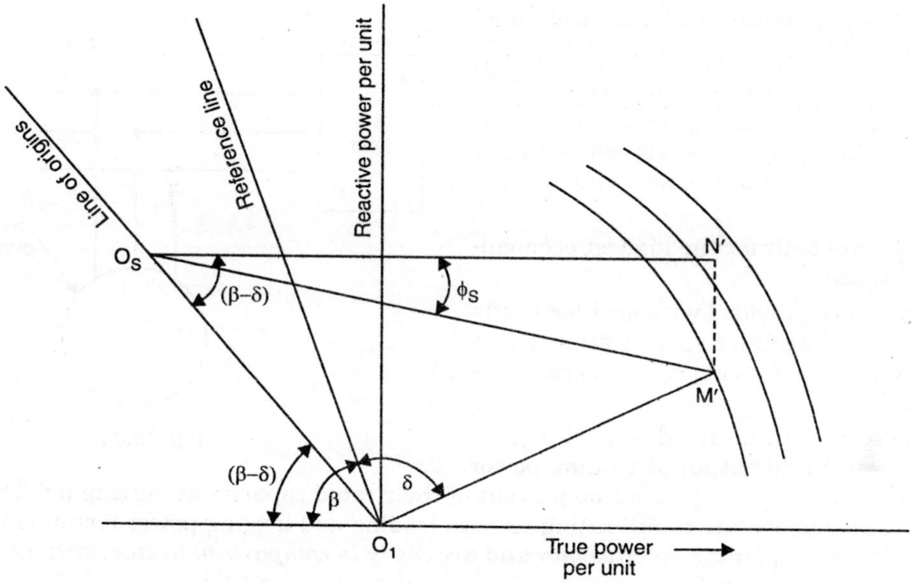

Fig. 14.33. Universal power circle diagram for sending end.

- O_1 is the centre of circle. The point O_S (the origin for measurement of power) now lies in the *second quadrant* ; its coordinates are given by :

$$\left.\begin{aligned}\text{Horizontal}: & -D\left(\frac{V_S}{V}\right)^2 \cos(\beta-\delta) \text{ per unit}\\ \text{Vertical}: & \quad D\left(\frac{V_S}{V}\right)^2 \sin(\beta-\delta) \text{ per unit}\end{aligned}\right\}\textit{Coordinates of the centre}$$

$$\textit{Radius} = \frac{V_S V_R}{V^2} \text{ per unit.}$$

- Line drawn through O_1 and O_S is called *'line of origins'*.
- From O_S draw a load line inclined at power factor angle ϕ_S to the horizontal cutting the arc of the circle at M'. Then,

$Q_S N'$ = True power measured from the point Q_S along the horizontal axis in per unit, and

$N'M'$ = Reactive power measured downward from point M'.

By coinciding the centre points of Figs. 14.31 and 14.32 we get the universal power diagram.

With the use of universal power circle diagram we can get the performance of the whole line readily.

Torque angle. It is the angle between V_S and V_R and is denoted by δ. This angle changes with the change in load.

— In the receiving end universal power circle diagram a reference line is drawn from point O_1 inclined at an angle β to the horizontal, the torque angle δ will be angle measured between the line O_1M and the reference line (Fig. 14.32).

— Similarly, in the sending end universal power circle diagram if a reference line is drawn from the point O_1 inclined at angle β to the horizontal then the torque angle δ will be the angle measured between the line O_1M' and the reference line (Fig. 14.33).

14.7.13. Synchronous Phase Modifiers

Synchronous phase modifiers (Fig 14.34) may be considered as idle-running synchronous motors employed for voltage regulation. These differ from the ordinary synchronous motor in the following ways (and possess the following *characteristics*) :

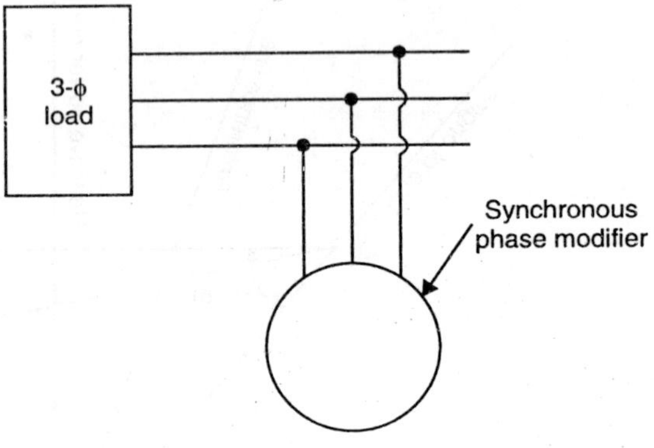

Fig. 14.34

- They are built for the highest economical speeds.
- These are provided with smaller shafts and bearings and special attention is given for achieving high overall efficiency.
- These machines are designed to give their full-load output at leading power factor, and can carry about 50 percent of their rated capacity at lagging p.f. They can be designed to operate at full rating on both leading and lagging power factors but they are *larger in size, have poor efficiency* and are *costly in comparison to standard machines.*

Examples 14.26. *The resistance and reactance per phase of a 3-phase transmission line are 6 Ω and 18 Ω respectively. If the sanding end voltage is 140 kV, determine the load which the line will be supplying at 132 kV and 0.8 power factor lagging.*

If the power factor is improved to 0.95 lagging by connecting a synchronous phase modifier in parallel with the line, determine the leading MVAR supplied by the modifier. The receiving end voltage and load remain unaltered.

Solution. Resistance of line per phase, $R = 6\ \Omega$

Reactance of line per phase, $X = 18\ \Omega$

Receiving end line voltage, $V_{RL} = 132$ kV

Sending end line voltage, $V_{SL} = 140$ kV

Power factor of the load, $\cos \phi_R = 0.8$

Power supplied at 0.8 p.f. lagging :

For *short transmission line,* $A = 1$

$\alpha = 0$

$$B = Z = \sqrt{6^2 + 18^2} = 18.97\ \Omega$$

$$\beta = \tan^{-1} \frac{X}{R} = \tan^{-1}\left(\frac{18}{6}\right) = 71.6°.$$

For receiving end power circle diagram,

$$\textit{Radius of circle} = \frac{V_{SL}V_{RL}}{B} = \frac{140 \times 132}{18.97} = 974.2 \text{ MVA.}$$

Co-ordinates of the centre of the power circle are :

$$\text{Horizontal coordinate} = -\frac{A \times V_{RL}^2}{B} \cos(\beta - \alpha) = -\frac{1 \times 132^2}{18.97} \cos(71.6 - 0) = -290 \text{ MW}$$

$$\text{Vertical coordinate} = -\frac{A \times V_{RL}^2}{B} \sin(\beta - \alpha)$$

$$= -\frac{1 \times 132^2}{18.97} \sin(71.6 - 0) = -871.5 \text{ MVAR}.$$

Using *scale* : 1 cm = 100 MW (Horizontally),
1 cm = 100 MVAR (Vertically)

$$\therefore \quad \text{Radius of the circle} = \frac{974.2}{100} = 9.74 \text{ cm}$$

$$\text{Horizontal coordinate} = \frac{-290}{100} = -2.9 \text{ cm}$$

$$\text{Vertical coordinate} = \frac{-871.5}{100} = -8.71 \text{ cm}.$$

Draw the circle diagram as shown in Fig. 14.35. (Figs. not to the scale)

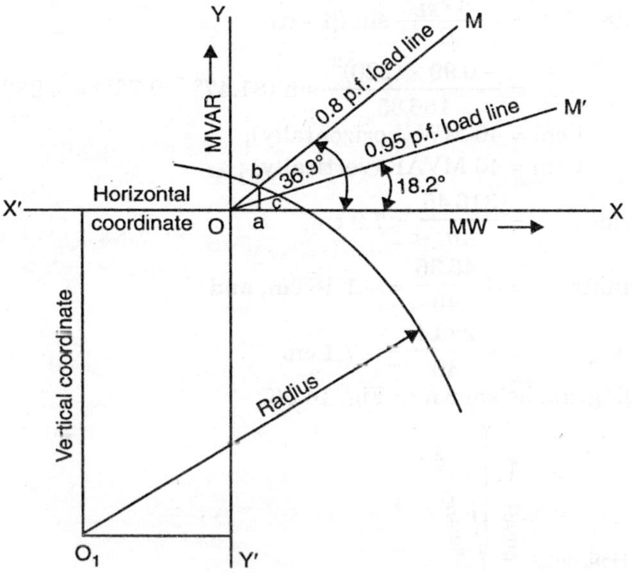

Fig. 14.35

- From origin *O* draw line *OM* inclined at an angle $\cos^{-1}(0.8) = 36.9°$ with the horizontal cutting the circle at the point *b*. From point *b* draw a perpendicular *ba* on the horizontal axis. Then, power supplied at 0.8 p.f. lagging = *Oa* = 0.62 cm = **62 MW. (Ans.)**
- Now draw another line *OM′* inclined at an angle $\cos^{-1}(0.95) = 18.2°$ with the horizontal cutting the vertical line at *c*. Then,

Total leading MVAR supplied by the synchronous modifier

= *bc* = 0.25 × 100 = **25 MVAR. (Ans.)**

Example 14.27. *The sending end and receiving end voltage of a 3-phase line are 240 kV and 220 kV line-to-line respectively. Its generalised constants of one phase are :*

$$A = D = 0.99 + j0.0132 \; ; \; B = (24.75 + j165) \; \Omega \; ; \; C = (-0.000044 + 0.0011) \; \mho.$$

Draw the circle diagram and determine the active and reactive power received when the angle between the sending end and receiving end voltage phasors is 30°. **(Gorakhpur University)**

Solution. *Given* : V_{RL} = 220 kV ; V_{SL} = 240 kV ; $A = D = 0.99 + j\,0.0132$;
$B = (24.75 + j\,165)\ \Omega$; $C = (-\,0.000044 + 0.0011)$ ℧ ; Torque angle, $\delta = 30°$.

Active and reactive power :

Line constants : $A = D = 0.99 + j0.0132 = 0.99 \angle 0.76°$

$B = 24.75 + j165 = 166.85 \angle 81.47°$

∴ $A = 0.99$, $\alpha = 0.76°$; $B = 166.85$; $\beta = 81.47°$.

For receiving end power circle diagram :

Radius of circle $= \dfrac{V_{SL} V_{RL}}{B} = \dfrac{220 \times 240}{166.85} = 316.45$ MVA

Coordinates of the centre of the power circle are :

Horizontal coordinate $= \dfrac{-\,A V_{RL}^2}{B} \cos(\beta - \alpha)$

$= \dfrac{-\,0.99 \times (220)^2}{166.85} \cos(81.47° - 0.76°) = -\,46.36$ MW

Vertical coordinate $= \dfrac{-\,A V_{RL}^2}{B} \sin(\beta - \alpha)$

$= \dfrac{-\,0.99 \times (220)^2}{166.85} \sin(81.47° - 0.76°) = -\,283.4$ MVAR

Using scale : 1 cm = 40 MW (horizontally),
1 cm = 40 MVAR (vertically) ;

Radius of the circle $= \dfrac{316.45}{40} \simeq 7.9$ cm,

Horizontal coordinate $= -\,\dfrac{46.36}{40} = -\,1.16$ cm, and

Vertical coordinate $= -\,\dfrac{283.4}{40} = -\,7.1$ cm.

Draw the circle diagram as shown in Fig. 14.36.

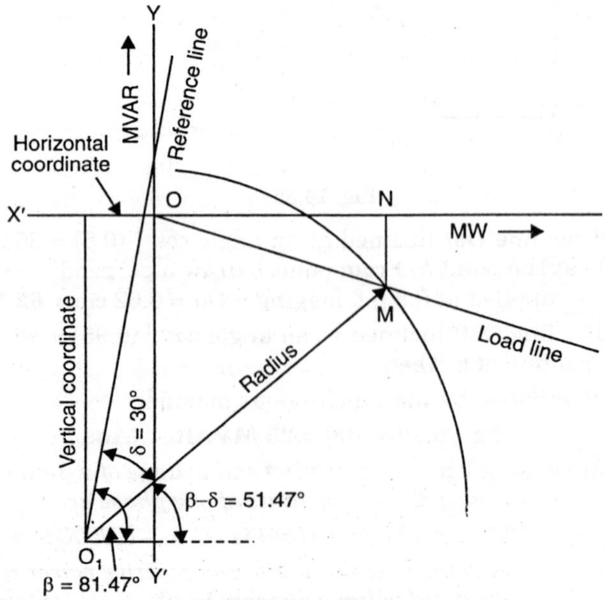

Fig. 14.36

- Draw reference line making an angle $\beta = 81.47°$ with the horizontal and another line O_1M making an angle $\delta = 30°$ with the reference line or $81.47° - 30° = 51.47°$ with the horizontal axis cutting the arc of the circle at point M.
- Draw perpendicular from point M on the X-axis cutting the X-axis at the point N. Then,

Active power received = $ON = 3.7 \times 40 =$ **148 MW. (Ans.)**

Reactive power supplied = $MN = 0.9 \times 40 =$ **36 MVAR (leading). (Ans.)**

Example 14.28. *The per phase resistance and inductive reactance of a 3-phase transmission line are 5 Ω and 12 Ω respectively. The line voltage at the receiving end is 33 kV. If the load at the receiving end is 20 MVA at 0.8 p.f. lagging, determine the voltage at the sending end.*

If a synchronous phase-modifier, which has the same rating at zero load at the receiving end as for the full load of 16 MW, is used to maintain the voltage at the sending end constant, determine the rating of the modifier and the power factor of the full load output. **(AMIE)**

Solution. *Given :* $R_{ph} = 5\ \Omega$; $X_{ph} = 12\ \Omega$; $\cos\phi_R = 0.8$ (lagging) ; Load = 20 MVA.

Sending end voltage :

Impedance of line per phase, $\bar{Z}_{ph} = 5 + j12 = 13 \angle 67.38°$

Load power factor angle, $\phi_R = \cos^{-1}(0.8) = 36.87°$

Load current, $I_R = \dfrac{20 \times 10^6}{\sqrt{3} \times 33000} = 350$ A

Sending end phase voltage, $\bar{V}_S = \bar{V}_R + \bar{I}_R \bar{Z}$

$$= \left(\frac{33000}{\sqrt{3}} + j0\right) + 350 \angle -36.87° \times 13 \angle 67.38°$$

$$= (19052.6 + j0) + 4550 \angle 30.51°$$

$$= 19052.6 \angle 0° + 4550 \angle 30.51°$$

$$= 23088 \angle 5.74° \text{ V}$$

∴ Sending end line voltage, $\mathbf{V_{SL}} = \sqrt{3} \times 23088 = 40{,}000$ V or **40 kV. (Ans.)**

The rating of the modifier and p.f. of the full-load output :

For short transmission line,

$$A = 1, \alpha = 0 ; B = Z = 13\ \Omega, \ \beta = 67.38°$$

For receiving end power circle diagram,

Radius of the circle $= \dfrac{V_{SL}V_{RL}}{B} = \dfrac{40 \times 33}{13} = 101.54$ MVA

Coordinates of the centre of the power circle are :

Horizontal coordinate $= \dfrac{-AV_{RL}^2}{B} \cos(\beta - \alpha)$

$$= \frac{-1 \times (33)^2}{13} \cos(67.38° - 0°) = -32.22 \text{ MW.}$$

Vertical coordinate $= \dfrac{-A \times V_{RL}^2}{B} \sin(\beta - \alpha)$

$$= \frac{-1 \times 33^2}{13} \times \sin(67.38° - 0°) = -77.32 \text{ MVAR.}$$

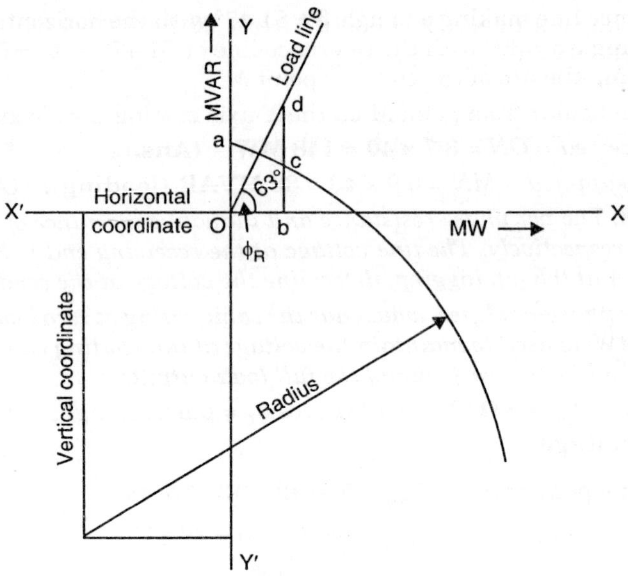

Fig. 14.37

Using scale : 1 cm = 15 MW horizontally
1 cm = 15 MVAR vertically

Radius of circle $= \frac{101.54}{15} = 6.77$ cm

Horizontal coordinate $= -\frac{32.22}{15} = -2.15$ cm

Vertical coordinate $= -\frac{77.32}{15} = -5.15$ cm.

Draw the circle diagram as shown in Fig. 14.37.

- At no-load the *synchronous modifier will be taking lagging MVAR* given by,

$$oa = 1.3 \text{ cm} = 1.3 \times 15 = \mathbf{19.5\ MVAR.}\quad \textbf{(Ans.)}$$

- Take *ob* = 16 MW, the full load supplied. At point *b* draw a perpendicular *bcd* cutting the arc of the circle at the point *c*. The MVAR supplied is given by the line *bc*. The synchronous modifier is supplying the same MVAR at full load of 16 MW as at no-load *i.e.* 19.5 MVAR. Take *cd* = 19.5 MVAR. Joint point *d* to *O* to get the load line.

The receiving end **power factor at full load** = cos ϕ_R = **cos 63° = 0.254. (Ans.)**

Example 14.29. *A 600 km long transmission line has the following parameters :*

$$R = 0.04\ \Omega/km\ ;\ X = 0.2\ \Omega/km\ ;\ Y = 2.575 \times 10^{-6}\ \angle 90°\ \mho/km.$$

(i) Draw the receiving end and sending end power circle diagrams for the line.

(ii) From the diagram find the sending end voltage, current and power factor angle when the line is delivering a load of 192 MW at 0.8 p.f. lagging and 275 kV.

Assume π-configuration.

Solution. *Given :* $R_{ph} = 0.04 \times 600 = 24\ \Omega\ ;\ X_{ph} = 0.2 \times 600 = 120\ \mho\ ;$

$$\overline{Y}_{ph} = 2.575\ \angle 90° \times 10^{-6} \times 600 = 0.001545\ \mho$$

Now, Impedance per phase, $\overline{Z} = 24 + j120 = 122.38\ \angle 78.69°$

Admittance of line per phase, $\overline{Y} = 600 \times 2.575 \times 10^{-6} \angle 90° = 1.545 \times 10^{-3}$ ℧

Representing the line by π-configuration, we have,

Line constants :

$$A = D = 1 + \frac{1}{2}\overline{Y}\,\overline{Z} = 1 + \frac{1}{2} \times (1.545 \times 10^{-3} \angle 90°) \times (122.38 \angle 78.69°)$$

$$= 1 + 0.094539 \angle 168.69° = 1 - 0.0927 + j0.01854$$

$$= 0.9073 + j0.01854 = 0.9075 \angle 1.17°$$

$$B = \overline{Z} = 122.38 \angle 78.69°$$

Receiving end line voltage $V_{RL} = 275$ kV

Load power factor, $\cos \phi_R = 0.8$ (lagging)

∴ $\phi_R = \cos^{-1}(0.8) = 36.87°$

Load current, $I_R = \dfrac{192 \times 10^6}{\sqrt{3} \times 275000 \times 0.8} = 504$ A.

For the given line, $A = 0.9075$

$\alpha = 1.17°$

$B = 122.38\ \Omega$

$\beta = 78.69°$

For *receiving end power circle diagram :*

Horizontal coordinate $= -\dfrac{A}{B} V_{RL}{}^2 \cos(\beta - \alpha)$

$= -\dfrac{0.9075}{122.38} \times (275)^2 \times \cos(78.69° - 1.17°)$

$= -121$ MW

Vertical coordinate $= -\dfrac{A}{B} R_{RL}{}^2 \sin(\beta - \alpha)$

$= -\dfrac{0.9075}{122.38} \times (275)^2 \sin(78.69° - 1.17°)$

$= -548$ MVAR.

Using scale : 1 cm = 80 MW (Horizontally)

1 cm = 80 MVAR (Vertically).

Locate point O_1 having coordinates – 1.51 cm and – 6.85 cm and draw a load line OM at an angle 36.87° inclined to the horizontal and to represent $\dfrac{\sqrt{3}\, V_{RL} I_R}{1000}$ MVAR *i.e.*, $\dfrac{\sqrt{3} \times 275 \times 504}{1000 \times 80}$ cm = 3.0 cm.

Draw circle diagram with O_1 (– 1.51 cm, – 6.85 cm) as centre and OM as radius. From circle diagram (Fig. 14.38) $O_1M = 9.5$ cm ...(By measurement)

∴ $O_1M = \dfrac{V_{SL} V_{RL}}{B} = 80 \times 9.5 = 760$ MVA

$V_{SL} = \dfrac{760 \times B}{V_{RL}} = \dfrac{760 \times 122.38}{275} =$ **338 kV. (Ans.)**

Now, $\beta - \delta = 64°$ ∴ $\delta = \beta - 64° = 78.69° - 64° = 14.69°$

and $\beta + \delta = 78.69 + 14.69° = 93.38°$.

For *sending end power circle diagram* (Fig. 14.39) :

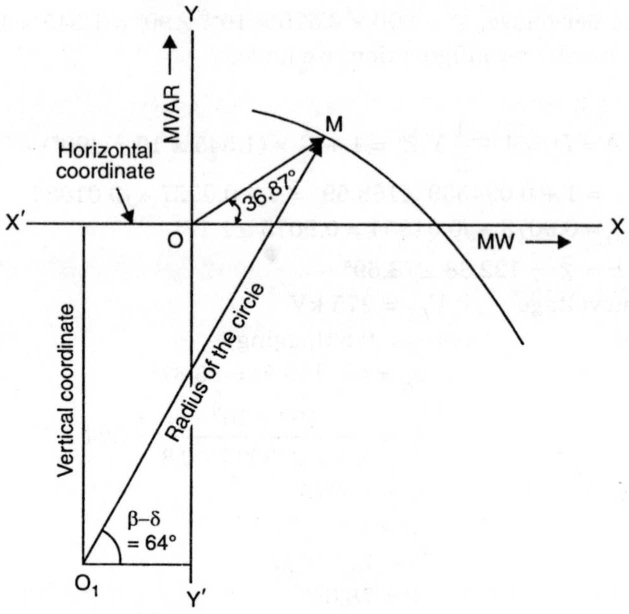

Fig. 14.38

Horizontal coordinate = $\frac{D}{B} V_{SL}^2 \cos(\beta - \alpha) = \frac{0.9075}{122.38} \times 338^2 \times \cos(78.69° - 1.17°) = 183$ MW

Vertical coordinate = $\frac{D}{B} V_{SL}^2 \sin(\beta - \alpha)$

$$= \frac{0.9075}{122.38} \times 338^2 \times \sin(78.69° - 1.17°) = 827 \text{ MVAR}$$

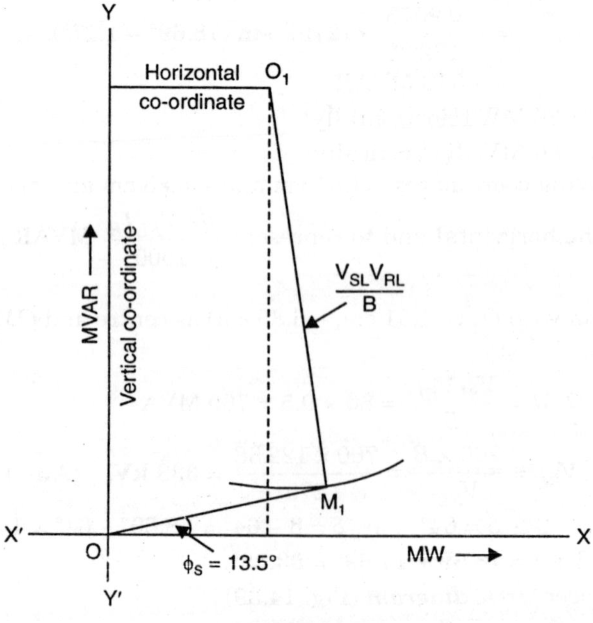

Fig. 14.39

Radius of circle $= \frac{V_{SL}V_{RL}}{B} = 9.5$ cm.

Draw circle diagram with O_1 (2.29 cm, 10.35 cm) as centre and 9.5 cm as radius. Draw O_1M inclined at an angle $\beta + \delta$ *i.e.*, 78.69° + 14.69° = 93.38° to the horizontal cutting the arc of sending end power circle at M_1. Join OM_1.

Sending end power factor angle ϕ_s = **13.5°** (By measurement). **(Ans.)**

∴ Sending end power factor = cos 13.5° = 0.972

$$OM_1 = 3.8 \text{ cm} = 3.75 \times 80 = 300 \text{ } MVA \text{ (By measurement)}$$

or
$$300 = \frac{\sqrt{3}V_{SL}I_S}{1000} \text{ MVA}$$

∴
$$I_S = \frac{300 \times 1000}{\sqrt{3} \times 338} = \mathbf{512.4\ A.} \quad \textbf{(Ans.)}$$

Example 14.30. *A 3-phase transmission line has the following constants :*

$$A = D = 0.945 \angle 1° 01'; B = 82.3 \angle 73° 02' \ \Omega; C = 0.001376 \angle 20° 23' \mho.$$

Construct a universal power circle diagram for the line and find the sending end voltage, current and power factor when the line is supplying a load of 15 MW at 0.8 p.f. lagging, the receiving end voltage being 66 kV. Choose base voltage as 60 kV. **(Agra University)**

Solution. *Given* : $A = D = 0.945 \angle 1° 01°$; $B = 82.3 \angle 73° 02' \ \Omega$; $C = 0.001376 \angle 90° 23' \mho$; V_{RL} = 66 kV ; Load = 15 MW ; cos ϕ_R = 0.8 lagging *i.e.*, $\phi_R = \cos^{-1}(0.8) = 36.87°$

Base voltage, V = 60 kV

Now, for the given line :

$$A = 0.945 \ ; \alpha = \angle 1° 01' \ ; B = 82.3 \ \Omega \ ; \beta \ 73° 02'$$

$$\text{Base MVA} = \frac{(\text{Base voltage})^2}{B} = \frac{60^2}{82.3} = 43.74 \text{ MVA}$$

$$\text{Per unit power} = \frac{15}{43.74} = 0.343$$

In the universal circle diagram the centre of the circle is kept at the *origin* and the power is increased from a point O_R, the coordinates of which are :

Horizontal coordinate
$$= A\left(\frac{V_{RL}}{V}\right)^2 \cos(\beta - \alpha) \text{ per unit (p.u.)}$$
$$= 0.945\left(\frac{66}{60}\right)^2 \cos(73° 02' - 1° 01') = 0.353 \text{ p.u.}$$

Vertical coordinate
$$= A\left(\frac{V_{RL}}{V}\right)^2 \sin(\beta - \alpha) \text{ p.u.}$$
$$= 0.945\left(\frac{66}{60}\right)^2 \sin(73° 02' - 1° 01') = 1.087 \text{ p.u.}$$

Using scale : 1 cm = 0.15 p.u. (Horizontally)
1 cm = 0.15 p.u. (Vertically)

- Locate point O_R having coordinates $\frac{0.353}{0.15}$ and $\frac{1.087}{0.15}$ *i.e.*, (2.35 cm, 7.25 cm) and from point O_R draw load line at an angle $\cos^{-1}$ 0.8 *i.e.*, 36.87° inclined to the horizontal and an horizontal line O_RN of magnitude 0.343 (*i.e.*, 2.29 cm).

- Draw a line NM perpendicular from point N cutting the load line at the point M. O_1M is the radius of the universal power circle (Fig. 14.40).

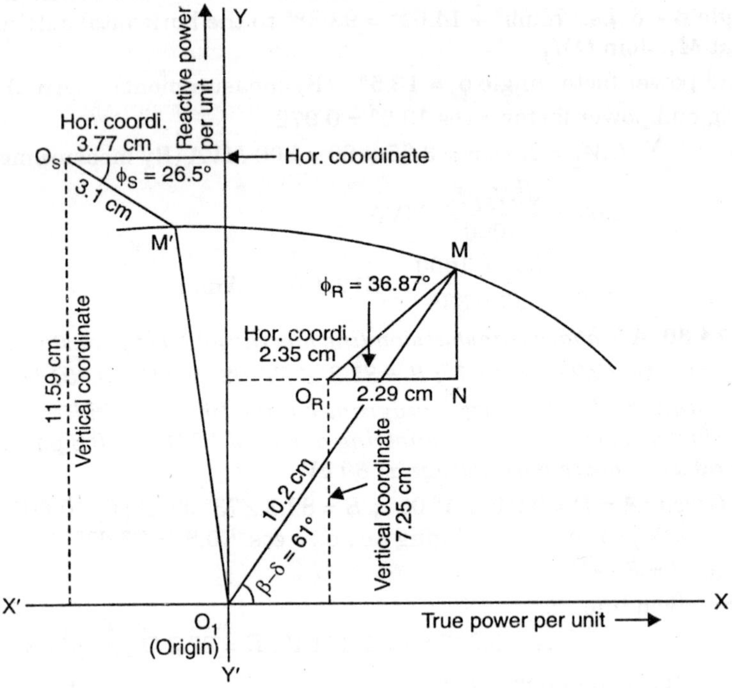

Fig. 14.40

From circle diagram O_1M = 10.2 cm (By measurement) = 10.2 × 0.15 = 1.53 p.u.

$$\therefore \quad O_1M = \frac{V_{RL}V_{SL}}{V^2} \text{ p.u.}$$

i.e., $$V_{SL} = \frac{O_1M \times V^2}{V_{RL}} = \frac{1.53 \times 60^2}{66} = \mathbf{83.45\ kV.} \quad \textbf{(Ans.)}$$

$$\beta - \delta = 61° \qquad \text{...(By measurement)}$$

$$\therefore \quad \delta = \beta - 61° = 73° \ 02' - 61 = 12° \ 02'$$

$$\therefore \quad \beta + \delta = 73° \ 02' + 12° \ 02' = 85° \ 04'.$$

The *coordinates of point* O_S (the points of sending end power measurement) w.r.t. origin (the centre of the circle) are :

Horizontal coordinate $$= -D \frac{V_{SL}^2}{V^2} \cos(\beta - \alpha)$$

$$= -0.945 \left(\frac{83.45}{60}\right)^2 \cos(73° \ 02' - 1° \ 01') = -0.565 \text{ p.u.}$$

Vertical coordinate $$= D \frac{V_{SL}^2}{V^2} \sin(\beta - \alpha)$$

$$= 0.945 \left(\frac{83.45}{60}\right)^2 \times \sin(73° \ 02' - 1° \ 01') = 1.738 \text{ p.u.}$$

- Locate point O_s having co-ordinates $\frac{-0.565}{0.15}$ and $\frac{1.738}{0.15}$ *i.e.*, (– 3.77 cm, 11.59 cm).
- From point O_1 (the centre of the universal circle) draw line O_1M' making an angle $\beta + \delta$ *i.e.*, 85° 04′ with the horizontal so as to cut the universal circle at M'. Join $O_S M'$. Sending end power factor angle, ϕ_S is the angle make by the line O_sM' with the horizontal and the length of line Q_sM' represents sending end MVA *i.e.*, $\frac{\sqrt{3}\, V_{SL} I_S}{1000}$, and

$Q_sM' = 3.1$ cm (By measurement)

$$= 3.1 \times 0.15 \times 43.74 \text{ (Base MVA)} = 20.34 \text{ MVA}$$

∴ *Sending and current* ; $I_s = \frac{20.34 \times 1000}{\sqrt{3} \times 83.45} = \mathbf{140.72\ A.}$ **(Ans.)**

Sending end power factor angle, $\phi_S = 26.5°$ (By measurement)

∴ *Sending end power factor,* $\mathbf{\cos \phi} = \cos 26.5° = \mathbf{0.895.}$ **(Ans.)**

Example 14.31. *A 3-phase, 50 Hz transmission line has the following constants :*

$$A = D = 0.997 \angle 0° \, 6' ; B = 33.5 \angle 68° \, 20' \, \Omega ; C = 0.218 \times 10^{-3} \angle 90° \, 02' \, ℧$$

Construct the universal power circle diagram for the line and compute the following :

(i) If the load supplied at the receiving end is 64 MW at 0.8 p.f. lagging and 132 kV, find the sending end voltage and power factor.

(ii) If the sending end voltage is maintained at 140 kV by a synchronous phase modifier at the receiving end, find the rating of the phase modifier when supplying load as in (i).

(Agra University)

Solution. *Given* : $f = 50$ Hz ; $A = D = 0.997 \angle 0° \, 6'$; $B = 33.5 \angle 68° \, 20' \, \Omega$; $C = 0.218 \times 10^{-3} \angle 90° \, 02' \, ℧$; Load at the receiving end = 64 MW at $\phi = \cos^{-1} (0.8) = 36.87°$ (lagging), $V_{RL} = 132$ kV ; $V_{SL} = 140$ kV.

(*i*) **Sending end voltage and power factor, V_{SL}; $\cos \phi_S$:**

Load current ; $I_R = \frac{64 \times 10^6}{\sqrt{3} \times 132000 \times 0.8} = 350$ A

For the given line :

$$A = 0.997 \; ; \alpha = 0° \, 6' \; ; B = 33.5 \; ; \beta = 68° \, 20°$$

Let the base voltage V be 100 kV.

Then Base MVA = $\frac{V^2}{B} = \frac{(100)^2}{33.5} = 298.5$ MVA

Per unit power = $\frac{64}{298.5} = 0.2144$

In the universal circle diagram the centre of the circle is kept at the origin and the power is measured from a point O_R, the co-ordinates of which are :

Horizontal co-ordinate $= A \left(\frac{V_{RL}}{V} \right)^2 \cos (\beta - \alpha)$ p.u.

$$= 0.997 \left(\frac{132}{100} \right)^2 \cos (68° \, 20' - 0° \, 6') = 0.651 \text{ p.u.}$$

Vertical co-ordinate $= A \left(\frac{V_{RL}}{V} \right)^2 \sin (\beta - \alpha)$

$$= 0.997 \left(\frac{132}{100} \right)^2 \sin (68° \, 20' - 0° \, 6') = 1.61 \text{ p.u.}$$

Using scale : 1 cm = 0.2 p.u. true power (Horizontally)

1 cm = 0.2 p.u. reactive power (Vertically)

- Locate point O_R having coordinates $\left(\frac{0.651}{0.2} \text{ and } \frac{1.61}{0.2}\right)$ *i.e.*, (3.26 cm, 8.05 cm) and from point O_R draw load line at an angle $\cos^{-1}(0.8)$ *i.e.*, 36.87° inclined to the horizontal and an horizontal line $O_R N$ of magnitude 0.2144 *i.e.*, (1.072 cm).
- Draw a line NM perpendicular from the point N cutting the load line at the point M, O_1M is the radius of the universal power circle (Fig. 14.41)

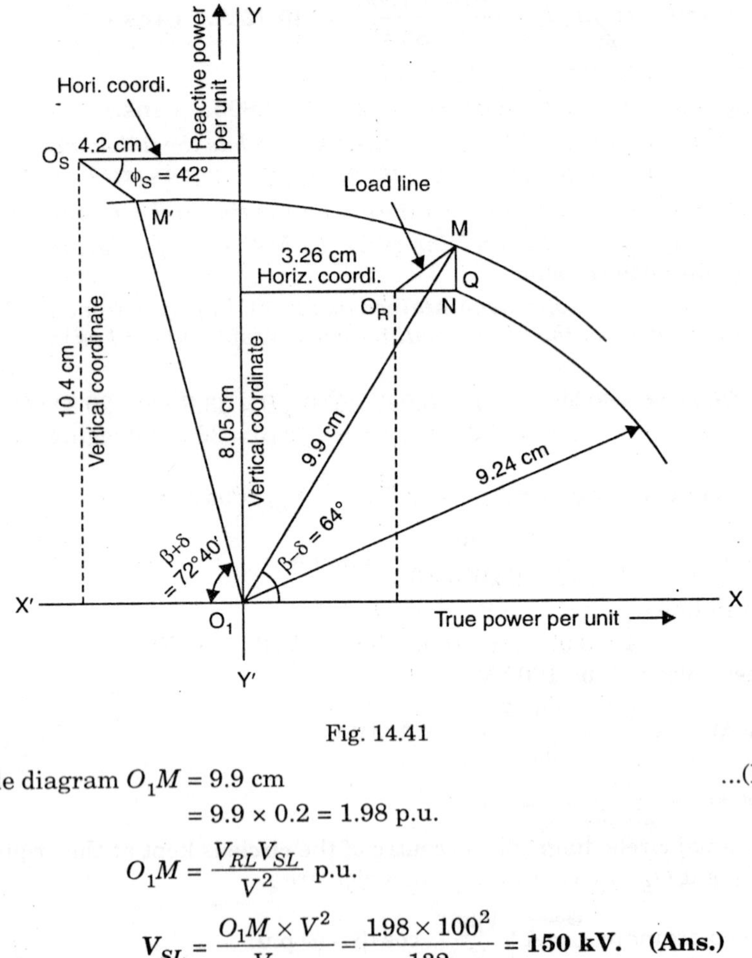

Fig. 14.41

From circle diagram $O_1M = 9.9$ cm ...(By measurement)

$$= 9.9 \times 0.2 = 1.98 \text{ p.u.}$$

$$\therefore \quad O_1M = \frac{V_{RL}V_{SL}}{V^2} \text{ p.u.}$$

$$\therefore \quad V_{SL} = \frac{O_1M \times V^2}{V_{RL}} = \frac{1.98 \times 100^2}{132} = \textbf{150 kV.} \quad \textbf{(Ans.)}$$

$$\beta - \delta = 64° \quad \text{...(By measurement)}$$

$$\therefore \quad \delta = \beta - 64° = 68°\,20' - 64° = 4°\,20'$$

and, $$\beta + \delta = 68°\,20' + 4°\,20' = 72°\,40'$$

The coordinates of point O_S (the point of sending end power measurement) with respect of the origin (the centre of universal power circle) are :

$$\text{Horizontal coordinate} \quad = -D\frac{V_{SL}^{\ 2}}{V^2}\cos(\beta - \alpha)$$

$$= -0.997 \times \frac{150^2}{100^2} \cos (68° 20' - 0° 6') = -0.84 \text{ p.u.}$$

Vertical coordinate $= D \dfrac{V_{SL}^2}{V_{RL}^2} \sin (\beta - \alpha)$

$$= 0.997 \frac{150^2}{100^2} \sin (68° 20' - 0° 6') = 2.08 \text{ p.u.}$$

- Locate point O_S having coordinates $\left(-\frac{0.84}{0.2} \text{ and } \frac{2.08}{0.2}\right)$ *i.e.*, (– 4.2 cm, 10.4 cm).
- From point O_1 (the centre of universal circle) draw line O_1M' making an angle $(\beta + \delta)$ *i.e.*, 72° 40′ with the horizontal so as to cut the universal circle at M'.
- Join O_SM'.

Sending and power factor angle ϕ_S is the angle made by the line O_SM' with the horizontal,

$\phi_S = 42°$ ...(By measurement)

Sending end power factor, **cos** ϕ_S = cos 42° = **0.743. (Ans.)**

(*ii*) **Rating of the phase modifier :**

When sending end voltage is 140 kV the radius of the universal power circle

$$= \frac{V_{RL} V_{SL}}{V^2} = \frac{132 \times 140}{(100)^2} = 1.848 \text{ p.u.}$$

- Draw another arc with point O_1 as centre and $\dfrac{1.848}{0.2}$ = 9.24 cm radius (Refer to Fig.14.41)
- Drop perpendicular MQ so as to cut the circle at point Q.

MVAR supplied by the synchronous phase modifier

$= MQ \times 0.2 \times$ Base MVA

$= 0.8 \times 0.2 \times 298.5 =$ **47.76 MVR (leading). (Ans.)**

14.8. FERRANTI EFFECT

When a long line is operating under no-load or light-load condition, the receiving end voltage is greater than the sending end voltage. This phenomenon is known as **Ferranti effect.** This effect is negligible in case of short lines but increases rapidly with the increase in length of line.

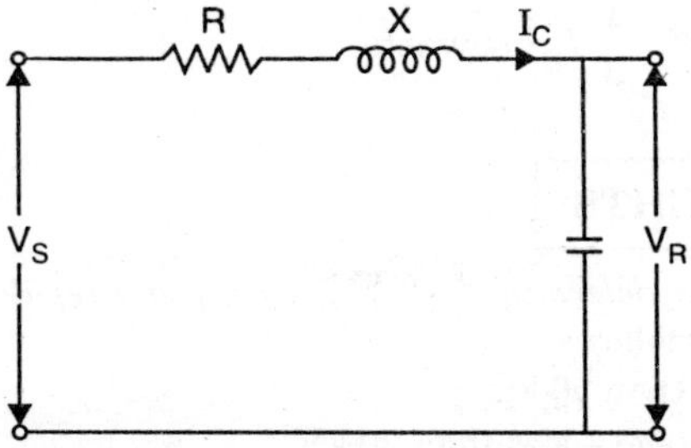

(a) Equivalent circuit of the line (lumped) under no-load condition

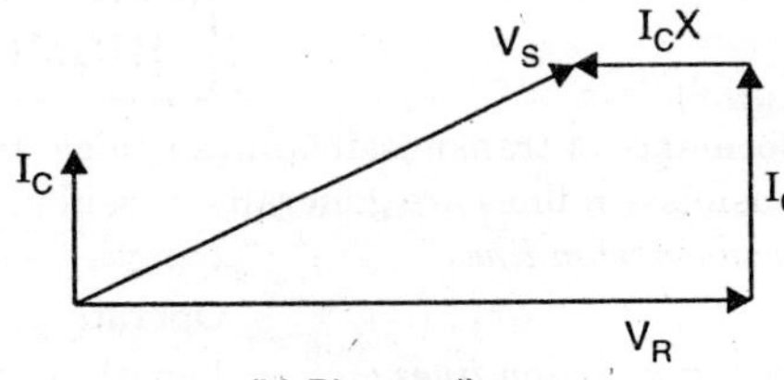

(b) Phasor diagram

Fig. 14.42. Ferranti effect.

A simple explanation of Ferranti effect can be given by approximating the distributed parameters of the line by lumped impedance. Fig. 14.42 (*a*) and (*b*) shows the equivalent circuit of the line (lumped) under no-load condition and the corresponding phasor diagram. The charging current produces drop in the reactance of the line which is in phase opposition to the receiving end voltage and hence the sending and voltage becomes smaller than the receiving end voltage.

14.9. CHARGING CURRENT AND LOSSES IN AN OPEN-CIRCUITED LINE

In the operation of a transmission line the voltage at the receiving end is maintained constant irrespective of the magnitude of the load of the system. The charging current of an open circuited line is the amount of current flowing in the line at the sending end with normal voltage and no load at the receiving end. The charging current will vary from zero, at the receiving end to its maximum value at the sending end.

Consider a single-phase line of length l metres and having charging current of i amperes per unit length, *i.e.*, $i = \frac{I_C}{l}$ (or $I_C = il$), I_C being the total maximum charging current.

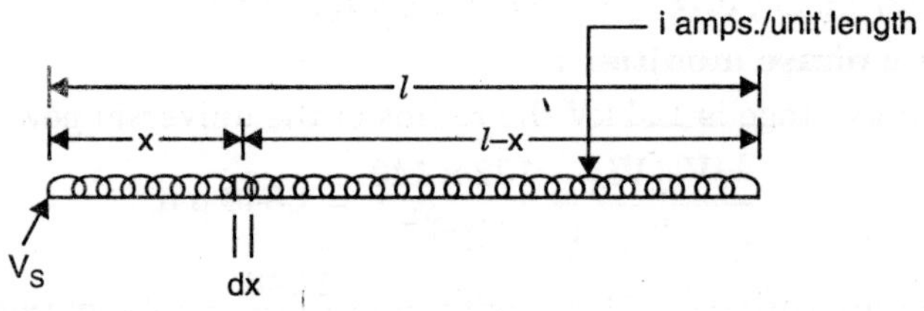

Fig. 14.43

Charging current at distance x (See Fig. 14.43) from the sending end

$$= i(l - x) = \frac{I_C}{l}(l - x)$$

The average value of the square of this quantity

$$= \frac{1}{l}\int_0^l \frac{I_C^2 (l-x)^2}{l^2}\,dx = \frac{I_C^2}{l^3}\int_0^l (l-x)^2\,dx = \frac{I_C^2}{l^3}\left[\frac{(l-x)^3}{(3)(-1)}\right]_0^l = \frac{I_C^2}{3}$$

R.M.S. value of charging current $= \frac{I_C}{\sqrt{3}}$ amperes

Power loss in each line conductor $= \left(\frac{I_C}{\sqrt{3}}\right)^2 R = \frac{1}{3} I_C^2 R$ watts.

HIGHLIGHTS

1. By performance of transmission lines is meant the *determination of efficiency and regulation of lines.*
2. The transmission lines are generally classified as follows :

Short transmission lines :	Length – less than 50 km
	Operating voltage – less than 20 kV.
Medium transmission lines :	Length – between 50 km and 160 km
	Voltage – between 21 and 100 kV (Line)
Long transmission line :	Length – more than 160 km
	Voltage – above 100 kV (Line)

3. The *"Regulation of a line"* is defined as the *change in receiving end voltage, expressed in percent of full load voltage, from no-load to full-load, keeping the sending end voltage and frequency constant.*
4. The *"Efficiency of a transmission line"* is defined as the *ratio of power delivered at the receiving end to the power sent from the sending end.*
5. *ABCD* constants can be evaluated by the following methods :
   (*i*) Convergent series (Complex angle) method.
   (*ii*) Convergent series (Real angle) method.
   (*iii*) By use of hyperbolic charts.
6. *"Surge impedance (Z_0)"* is defined as the square root of the ratio of line impedance (*Z*) and shunt admittance (*Y*) *i.e.*, $Z_0 = \sqrt{\dfrac{Z}{Y}}$.

   "Surge impedance loading" is defined as thè load, at unity p.f., that can be delivered by the line having no resistance.
7. In case of "constant voltage transmission" a constant voltage drop is always maintained along the line by installing specially designed synchronous motors at the receiving end.
8. "Synchronous phase modifiers" may be considered an idle-running synchronous motors employed for voltage regulation.
9. When a long line is operating under no-load or light load condition, the receiving end voltage is greater than the sending end voltage. This phenomenon is known as *"Ferranti effect"*.

THEORETICAL QUESTIONS

1. What do you mean by "Performance of transmission lines" ?
2. How are transmission lines classified ?
3. Define "Regulation" of a transmission line.
4. How is "Efficiency" of a transmission line defined ?
5. Define regulation, efficiency, losses and power factor at each end of the line and explain how these characteristics are affected by the constants of the line.
6. Give the classification of transmission lines. Explain the influence of power factor on the performance of a transmission line.
7. How are transmission lines classified ? Obtain the relation between the sending end and the receiving end voltages and currents of a medium line using nominal-π representation. Draw the equivalent circuit and the vector diagram.
8. Explain the physical significance of the generalised *ABCD* constants of a transmission line ? State the units of these constants. Determine these constants for a medium transmission line with nominal-*T* configuration. Draw neatly corresponding vector diagram.
9. Find the values of *ABCD* constants in the following approximate methods in terms of *Z* and *Y* :
   (*i*) Simple series impedance method ; (*ii*) Nominal-π method ;
   (*iii*) Nominal-*T* method.
10. Derive equations which represent the performance of a long transmission line with its electrical parameters uniformly distributed along its length.
11. Starting from first principles drive an expression for the sending end voltage and current of a long transmission line in terms of the line parameters and receiving and voltage and current.
12. Derive the expressions for voltage and current distribution over a long line. Explain the significance of characteristic impedance loading in connection with the long lines. Deduce the above voltage and current relations in the hyperbolic form and obtain the element values of an equivalent to represent the long lines.
13. Draw the equivalent circuit of a long transmission line. Derive from fundamentals the following relationship between sending and receiving end voltages and currents :

$$\overline{V_S} = A\overline{V_R} + B\overline{I_R}$$
$$\overline{I_S} = C\overline{V_R} + \overline{I_R}$$

14. What are the 'power circle diagrams' ? How are they useful ? Show how a receiving end power circle diagram may be drawn for a transmission line ?

15. Prove that the transmission line performance can be obtained by a power circle diagram. Derive relations to develop receiving end power circle diagram of a transmission line. Explain the construction and its applications.
16. State the meaning of the term 'power circle diagram' of transmission line. What information can be supplied by this as regards performance of line ?
17. Explain in detail how the receiving end power circle diagram can be drawn. Obtain the condition for maximum power.
18. Explain the function of a synchronous phase modifier placed at the receiving end of the transmission line.
19. What is Ferranti effect ? Explain it with the help of phasor diagram.
20. Explain the Ferranti effect and derive an expression for power loss is an open circuited line.
21. Write short notes on the following :
    (*i*) Surge impedance of a transmission line.
    (*ii*) Universal power circle diagram.
    (*iii*) Synchronous phase modifier.

UNSOLVED EXAMPLES

1. A single-phase overhead line is transmitting 1100 kW at 33 kV at 0.8 power factor lagging. The total resistance and loop reactance of the line are 10 Ω and 15 Ω respectively. Determine the following :
   (*i*) The sending end voltage, (*ii*) The sending end power factor, and
   (*iii*) Transmission efficiency. **[Ans.** (*i*) 33709 V ; (*ii*) 0.7956 lagging ; (*iii*) 98.44%]
2. A 3-phase, 50 Hz, 16 km long overhead line supplies 1000 kW at 11 kV, 0.8 power factor lagging. The line resistance and inductance are 0.03 Ω and 0.7 mH per phase per km. Determine :
   (*i*) The line sending end voltage, (*ii*) Percentage regulation, and
   (*iii*) Transmission efficiency of the line. **[Ans.** (*i*) 11284 V ; (*ii*) 2.58% ; (*iii*) 99.38%]
3. Calculate the distance over which a load of 15000 kW at 0.85 power factor lagging can be delivered by a 3-phase transmission line having conductor of steel cored aluminium each of resistance 0.905 Ω/km. The voltage at the receiving end is to be 132 kV and the loss in transmission is to be 7.5 per cent of the load. **[Ans.** 69.55 km]
4. A 3-phase load of 2000 kVA, 0.8 power factor lagging is supplied at 6.6 kV, 50 Hz by means of a 33 kV transmission line 20 km long and 33/6.6 kV step-down transformer. The resistance and reactance of each conductor are 0.4 Ω and 0.5 Ω per km respectively. The resistance and reactance of transformer primary are 7.5 Ω and 13.2 Ω, while those of secondary are 0.35 Ω and 0.65 Ω respectively. Determine the following when 6.6 kV is maintained at the receiving end :
   (*i*) Sending end line voltage, (*ii*) Sending end power factor, and
   (*iii*) Transmission efficiency. **[Ans.** (*i*) 35.6 kV ; (*ii*) 0.7826 lag ; (*iii*) 94.72%]
5. A medium single-phase transmission line 100 km long has the following constants : Resistance/km = 0.25 Ω ; Reactance/km = 0.8 Ω ; susceptance/km = 14×10^{-6} ℧ ; Receiving end line voltage = 66 kV. Assuming that the total capacitance of the line is localised at the receiving end alone, determine :
   (*i*) Sending end current, (*ii*) Sending end voltage,
   (*iii*) Regulation, and (*iv*) Supply power factor.
   The line is delivering 15 MW at 0.8 power factor lagging.
   [Ans. (*i*) 240 A ; (*ii*) 79583 V ; (*iii*) 20.58% ; (*iv*) 0.86 lag]
6. A 3-phase, 50 Hz transmission line, 100 km long delivers 20 MW at 0.9 p.f. lagging and at 110 kV. The resistance and reactance of the line per phase per km are 0.2 Ω and 0.4 Ω respectively, while capacitive admittance is 2.5×10^{-6} mho/km determine :
   (*i*) Sending end current, (*ii*) Sending end line voltage, and
   (*iii*) Transmission efficiency.
   Use the nominal-T method. **[Ans.** (*i*) 110.1 A ; (*ii*) 116.72 kV ; (*iii*) 96.28%]

7. A 3-phase, 50 Hz, 100 km long transmission line has the following line constants :
Resistance/phase/km = 0.1 Ω ; Reactance/phase/km = 0.5 Ω ; Susceptance/phase/km = 10×10^{-6} mho.
If the line supplies load of 20 MW at 0.9 p.f. lagging at 66 kV at the receiving end, determine the following :

(*i*) Sending end power factor,
(*ii*) Percentage regulation, and
(*iii*) Transmission efficiency.

Use nominal-π method **[Ans.** (*i*) 0.905 lag ; (*ii*) 15.27% ; (*iii*) 94%]

8. A 3-phase, 50 Hz, overhead transmission line, 100 km long has 110 kV between lines at the receiving end and has the following constants :
Resistance/km/phase = 0.153 Ω ; Inductance/km/phase = 1.21/mH ; Capacitance/km/phase = 0.00958 μF.
The line supplies a load of 20 MW at 0.9 p.f. lagging. Calculate, using nominal-π representation, the following :

(*i*) Sending end line voltage, (*ii*) Sending end current,
(*iii*) Sending end power factor, (*iv*) Percentage regulation, and
(*v*) Transmission efficiency of the line. **[Ans.** (*i*) 115.645 kV, (*ii*) 109 A ; (*iii*) 0.9423 (lag.) (*iv*) 5.13% ; (*v*) 97.21%]

9. A balanced 3-phase load of 30 MW is supplied at 132 kV, 50 Hz and 0.85 p.f. lagging by means of a transmission line. The series impedance of a single conductor is (20 + *j*52) ohms and the total phase-neutral admittance is 315×10^{-6} mho. Using nominal-*T* method determines.

(*i*) *A, B, C, D* constants, (*ii*) Sending ending line voltage, and
(*iii*) Regulation of the line. **[Ans.** (*i*) $A = D = 0.992 \angle 0.18°$, $C = 0.000315 \angle 90°$, (*ii*) 143 kV, (*iii*) 9.26%]

10. A 132 kV, 50 Hz, 3-phase transmission lines delivers a load of 50 MW at 0.8 p.f. lagging at the receiving end and has the following constants :
$A = D = 0.95 \angle 1.4°$; $B = 96 \angle 78°$; $C = 0.0015 \angle 90°$. Determine the following :

(*i*) Percentage regulation of the line, and
(*ii*) Charging current ; Use nominal-*T* method. **[Ans.** (*i*) 30% ; (*ii*) 128.2 ∠93.1°A]

11. The constants for a 132 kV line are as under :
$A = D = 0.98 \angle 3°$; $B = 110 \angle 75°$ Ω ; $C = 0.0034 \angle 80°$ ℧.
The load at the receiving end is 40 MVA at 132 kV and power factor 0.8 lagging. Draw the circle diagram and find out the following :

(*i*) Sending end voltage.
(*ii*) Leading MVAR for this load at the receiving end, if the sending end voltage is 440 kV.

(Panjab University)
[Ans. (*i*) 157.8 kV ; (*ii*) 23 MVAR]

12. A 3-phase overhead line has resistance and reactance per phase of 25 Ω and 90 Ω respectively. The supply voltage is 145 kV while the load-end voltage is maintained at 132 kV for all loads by an automatically controlled synchronous phase modifier. If the kVAR of the modifier has the same value for zero load as for a load of 50 MW, find the rating of the modifier and the power-factor of the load. **(U.P.S.C)**
[Ans. 19 MVAR ; 0.95 lag]

13. A single-circuit, 50 Hz transmission line is 500 km long and has the following *ABCD* constants :
$A = D = 0.895 \angle 1.5°$; $B = 182.5 \angle 78.5°$; $C = 0.00111 \angle 90.5°$ ℧.
Construct a universal power circle diagram and from the diagram determine :

(*i*) The sending end voltage for a load of 80 MVA at 215 kV and 0.8 p.f. lagging at the receiving end.
(*ii*) The voltage regulation for the loading as in (*i*), and
(*iii*) The reactive power supplied by the line and by a synchronous phase modifier in parallel with the load, when the loading is 80 MVA at 215 kV and 0.9 p.f. lagging. Given that the sending end voltage is 236 kV. Also determine the power factor at the receiving end of the line. **(Agra University)**

[Ans. (*i*) 263 kV ; (*ii*) 22.33 % ; (*iii*) 24.45 MVAR/line) ; 14.3 MVAR lead ; 0.956 lagging]

15

Interference of Power Lines with Neighbouring Communication Circuits

15.1. Introduction. 15.2. Electromagnetic effect. 15.3. Electrostatic effect. 15.4. Transposition. Highlights—Theoretical Questions—Unsolved Examples.

15.1. INTRODUCTION

Usually, communication lines are run along the power lines ; in some cases these lines run on the same support and the same route. Under such conditions the power lines are likely to interfere with the communication lines. The interference is mainly due to *electromagnetic* and *electrostatic* fields of sufficient magnitudes, as given below :

- The "***Electromagnetic effect***" *produces currents due to induction resulting its superposition on speech current of communication line causing distortion.*
- The "***electrostatic effect***" *raises the potential of the communication circuit as a whole and thereby posing a threat of damage to the communication equipments.*

The above mentioned effects *depend on the distance between the power and communication circuits* and length of the route *over which these are in parallel.* Thus, it is essential to ensure that the distortion and rise of potential are within limits for proper communication purpose.

15.2. ELECTROMAGNETIC EFFECT

Consider three power line conductors A, B and C of a 3-phase single-circuit line and two telephone conductors P and Q (running below the power line conductors) running on the same supports as shown in Fig. 15.1.

Let, r = Radius of each power conductor,

d_{AP} = Distance between conductors A and P, and

d_{AQ} = Distance between conductors A and Q.

Firstly, let us consider the loop formed by conductors A and P. Since for this loop only A is contributing to the e.m.f. induced in P, we have *inductance* of this loop (neglecting internal flux),

$$L_{AP} = 0.2 \ln \left(\frac{d_{AP}}{r}\right) \text{ mH/km} \quad \text{...(i)}$$

Similarly, $$L_{AQ} = 0.2 \ln \left(\frac{d_{AQ}}{r}\right) \text{ mH/km} \quad \text{...(ii)}$$

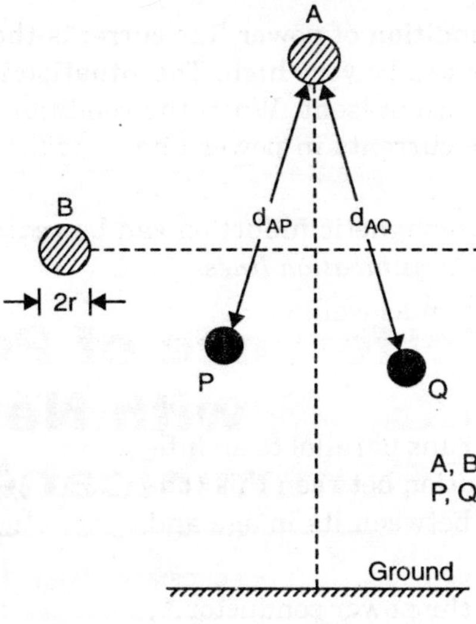

Fig. 15.1. Power and communication conductors on the same support.

∴ Mutual inductance between the conductor A and loop PQ,

$$M_A = L_{AQ} - L_{AP} = 0.2 \ln\left(\frac{d_{AQ}}{r}\right) - 0.2 \ln\left(\frac{d_{AP}}{r}\right) \text{ mH/km}$$

or,
$$M_A = 0.2 \ln\left(\frac{d_{AQ}}{d_{AP}}\right) \text{ mH/km} \quad ...(15.1)$$

Similarly the mutual inductance between conductor B and the loop PQ, M_B and the mutual inductance between conductor C and loop PQ, M_C are given as

$$M_B = 0.2 \ln\left(\frac{d_{BQ}}{d_{BP}}\right) \text{ mH/km} \quad ...[15.1\ (a)]$$

$$M_C = 0.2 \ln\left(\frac{d_{CQ}}{d_{CP}}\right) \text{ mH/km} \quad ...[15.1\ (b)]$$

The mutual inductances M_A, M_B and M_C are the results of fluxes which have a phase displacement of 120°, therefore, the net effect of the magnetic field will be

$$M = M_A + M_B + M_C \quad ...(15.2)$$

where, M is the net mutual inductance which is the *'phasor sum' of the three inductances.*

The voltage induced in the loop PQ is given by,

$$E_m = 2\pi f M I \text{ volts/km}$$

where, I = Current flowing through the power line, and

f = Supply frequency.

In view of the above discussion, the following points are *worth noting* :

(*i*) With the increase in distance between power and communication lines the mutual inductances M_A, M_B and M_C progressively become equal in magnitude so that M diminishes (consequently the voltage induced in telephone wires).

(*ii*) As the voltage induced in telephone wire is directly proportional to frequency, therefore, in case the third harmonic is present a voltage equal to 3 times the voltage due to fundamental frequency will be induced in telephone circuit. Also the higher frequencies may come within the audible range producing a distortion effect.

(*iii*) Under the heavily unbalanced condition of power line currents the flux linkage and induced voltage from the power line can be very high. The situation is further complicated in case third harmonic currents are present. When the condition is balanced the total induced voltage, due to harmonic currents in power line is additive and is hence three times that due to single phase.

— The voltage induced due to electromagnetic induction can be *reduced by increasing the distance between the power and communication lines.*

— *Transposition* improves the situation as well.

15.3. ELECTROSTATIC EFFECT

When a conductor such as (Fig. 15.2) runs parallel to an infinite plane (*e.g.*, earth), the potential distribution between this conductor and plane is exactly similar to that between its image and the plane.

Let, q = Charge per unit length of the power conductor A,

h_A = Height of the conductor A from the ground,

r = radius of conductor A, and

V_A = Potential of conductor A w.r.t. earth.

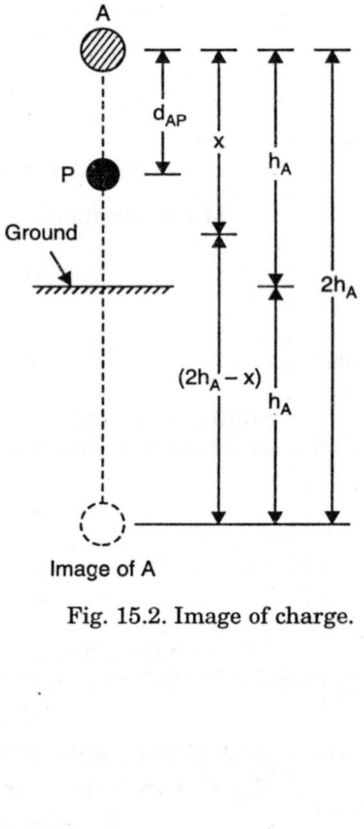

Fig. 15.2. Image of charge.

Then, $$V_A = \frac{1}{2\pi\varepsilon_0}\int_r^{h_A}\left(\frac{q}{x}+\frac{q}{2h_A-x}\right)dx$$

$$= \frac{q}{2\pi\varepsilon_0}\ln\left(\frac{2h_A-r}{r}\right)\text{ volts} \quad ...(i)$$

Similarly potential of conductor P due to conductor A,

$$V_{PA} = \frac{q}{2\pi\varepsilon_0}\int_{d_{AP}}^{h_A}\left(\frac{q}{x}+\frac{q}{2h_A-x}\right)dx$$

or, $$V_{PA} = \frac{q}{2\pi\varepsilon_0}\ln\left(\frac{2h_A-d_{AP}}{d_{AP}}\right) \quad ...(ii)$$

Substituting $q = \dfrac{2\pi\varepsilon_0 V_A}{\ln\left(\dfrac{2h_A-r}{r}\right)}$ from (*i*) in (*ii*), we get

$$V_{PA} = \frac{2\pi\varepsilon_0 V_A}{\ln\left(\dfrac{2h_A-r}{r}\right)}\cdot\frac{1}{2\pi\varepsilon_0}\ln\left(\frac{2h_A-d_{AP}}{d_{AP}}\right)$$

or, $$V_{PA} = V_A\frac{\ln[(2h_A-d_{AP})/d_{AP}]}{\ln[(2h_A-r)/r]}\text{ volts} \quad ...(15.3)$$

Similarly, we can determine the potential of telephone conductor P due to power conductors B and C and finally write the resultant potential of telephone conductor P due to power conductors A, B and C, w.r.t. earth,

$$V_P = V_{PA} + V_{PB} + V_{PC} \quad \text{(phasor addition)} \quad ...(15.4)$$

The potential of telephone conductor Q due to conductor A, B and C can be determined by repeating the above procedure.

Example 15.1. *The spacing between the conductors of a single-phase 50 Hz power line, supported on a horizontal cross arm, is 3.5 m. The spacing between the conductors of a telephone line, supported on a horizontal cross arm 2 m below the power line, is 0.5 m. If the mean radius of the conductors of the power line is 2.5 mm and current flowing through the line is 125 A, calculate the following :*

(*i*) *The mutual inductance between the circuits.*

(*ii*) *The voltage per km induced in the telephone line.*

Sol. *Given :* $d_{AB} = 3.5$ m ; $d_{PQ} = 0.5$ m ; $r = 2.5$ mm $= 0.0025$ m ; $I = 125$ A.

(*i*) **The mutual inductance between the circuits, M :**

$$d_{AP} = \sqrt{2^2 + 1.5^2} = 2.5 \text{ m}$$

$$d_{AQ} = \sqrt{2^2 + 2^2} = 2.83 \text{ m}$$

Due to I_A, $$\lambda_{PQ} = 2 \times 10^{-7} \times I_A \times \ln\left(\frac{2.83}{2.5}\right)$$

Due to I_B, $$\lambda_{PQ} = 2 \times 10^{-7} \times I_A \times \ln\left(\frac{2.5}{2.83}\right)$$

Since I_A and I_B are 180° out of phase,

$$\lambda_{PQ} = 4 \times 10^{-7}\, I_A \times \ln\left(\frac{2.83}{2.5}\right)$$

$\therefore$ $$M = \frac{\lambda_{PQ}}{I_A} = 4 \times 10^{-7} \times \ln\left(\frac{2.83}{2.5}\right)$$

$$= 4.96 \times 10^{-8} \text{ H/km}$$

i.e., $$M = \mathbf{4.96 \times 10^{-5}\ H/km.} \quad \textbf{(Ans.)}$$

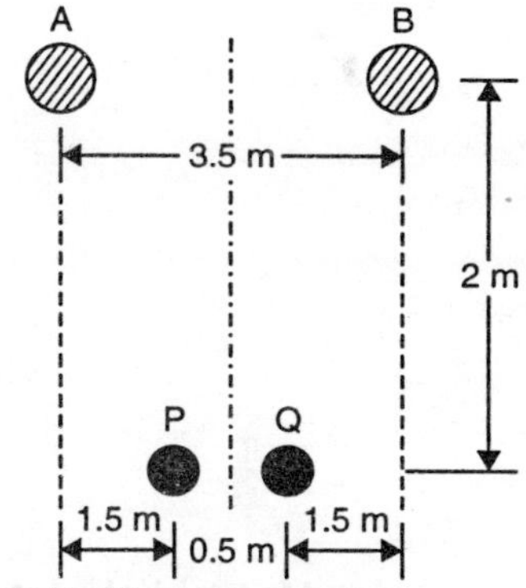

Fig. 15.3

(*ii*) **The voltage per km induced in the telephone line, V_{PQ} :**

$$V_{PQ} = j\omega MI = 2\pi fMI = 2\pi \times 50 \times 4.96 \times 10^{-5} \times 125 = \mathbf{1.948\ V/km.} \quad \textbf{(Ans.)}$$

Example 15.2. *Fig. 15.4 shows the conductor arrangement of a 3-phase 50 Hz 100 km long 132 kV transmission line, delivering a total load of 30 MW at 0.9 p.f. lagging. Each power line conductor is 6 mm in radius and conductors are symmetrically placed in 3.5 metre delta formation. Determine :*

(*i*) *The induced voltage of fundamental frequency in telephone circuit due to electromagnetic effect.*

(*ii*) *The potential of telephone conductor P above earth due to electrostatic effect.*

Sol. *Given :* $l = 100$ km ; supply voltage (line) = 132 kV ; Total load = 30 MW, 0.9 p.f. lagging; Radius of power line conductor = 6 mm ; distance between power line conductors = 3.5 m

(*i*) **The induced voltage at fundamental frequency in telephone circuit due to electromagnetic effect :**

Refer to Fig. 15.4.

$$AO = \sqrt{AB^2 - BO^2} = \sqrt{3.5^2 - 1.75^2} = 3.03 \text{ m}$$

$$d_{AP} = 3.03 + 6 = 9.03 \text{ m}$$

$$d_{AQ} = 3.03 + 6.8 = 9.83 \text{ m}$$

$$d_{BP} = \sqrt{6^2 + 1.75^2} = 6.25 \text{ m}$$

$$d_{BQ} = \sqrt{6.8^2 + 1.75^2} = 7.02 \text{ m}$$

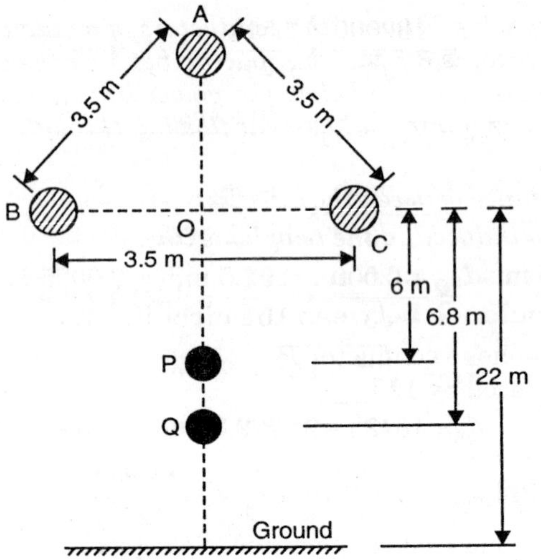

Fig. 15.4

∴ Mutual inductance between conductor A and the loop PQ,

$$M_A = 0.2 \ln \left(\frac{d_{AQ}}{d_{AP}}\right) \text{ mH/km}$$

$$= 0.2 \ln \left(\frac{9.83}{9.03}\right) = 0.017 \text{ mH/km}$$

$$M_B = M_C = 0.2 \ln \left(\frac{d_{BQ}}{d_{BP}}\right) = 0.2 \ln \left(\frac{7.02}{6.25}\right) = 0.0232 \text{ mH/km}$$

Since M_A, M_B and M_C are mutually displaced by 120°,

$$M = 0.0232 - 0.017 = 0.0062 \text{ mH/km}$$

Current flowing through the power conductors,

$$I = \frac{30 \times 10^6}{\sqrt{3} \times 132000 \times 0.9} = 145.8 \text{ A}$$

∴ Induced voltage at fundamental frequency in the telephone circuit due to electromagnetic effect,

$$V_m = 2\pi f M I = 2\pi \times 50 \times 0.0062 \times 10^{-3} \times 145.8 = 0.284 \text{ V/km}$$

Hence, for the whole route of 100 km the induced voltage (at fundamental frequency in the telephone circuit due to electromagnetic effect)

$$= 0.284 \times 100 = \mathbf{28.4 \text{ V}}. \quad \textbf{(Ans.)}$$

(*ii*) **The potential of telephone conductor P above earth due to electrostatic effect :**

We know that, $$V_{PA} = V_A \times \frac{\ln [(2h_A - d_{AP})/d_{AP}]}{\ln [(2h_A - r)/r]}$$ ...[Eqn. 15.3)]

$$= \frac{132000}{\sqrt{3}} \times \frac{\ln [\{2 \times (22 + 3.03) - 9.03\} / 9.03]}{\ln [\{2 \times (22 + 3.03) - 0.006\} / 0.006]}$$

$$= \frac{132000}{\sqrt{3}} \times \frac{1.514}{9.029} = 12779 \text{ volts}$$

and,
$$V_{PB} = V_{PC} = V_B \times \frac{\ln [(2h_A - d_{BP})/d_{BP}]}{\ln [(2h_A - r)/r]}$$

$$= \frac{132000}{\sqrt{3}} \times \frac{\ln [\{2 \times (22 + 3.03) - 6.25\}/6.25]}{\ln [\{2 \times (22 + 3.03) - 0.006\}/0.006]}$$

$$= \frac{132000}{\sqrt{3}} \times \frac{1.947}{9.029} = 16434 \text{ volts}$$

The potential of telephone conductor P w.r.t. earth, V_P being the phasor sum of above three components mutually displaced by 120°,

$$V_P = 16434 - 12779 = \mathbf{3655 \text{ volts.}} \quad \textbf{(Ans.)}$$

15.4. TRANSPOSITION

In order to reduce the induced voltages due to electromagnetic and electrostatic induction the conductors of both the power line and communication line are regularly transposed as shown in Fig. 15.5. Transition *balances the capacitances of the lines*, the result being that electrostatically induced voltages balance out in the length of a complete set of transpositions, called *barrel*. Further due to cancellation of the fluxes due to the, positive and negative phase sequences currents along the barrel, the *electromagnetically induced e.m.f.s. get diminished*. This, however does not apply to zero sequence currents and that is why the *telephone line is also transposed* along with the transposition of power line as shown in Fig. 15.5. The induced voltages on account of electromagnetic and electrostatic effects, under normal working conditions, can be kept down to a low value by a *proper coordination of power and communication lines.* It may observed from Fig. 15.5 that each power line conductor occupies all the three possible positions for $\frac{1}{3}$rd length in one barrel. Fig. 15.6 shows the arrangement of conductors at a transposition tower for location corresponding to A.

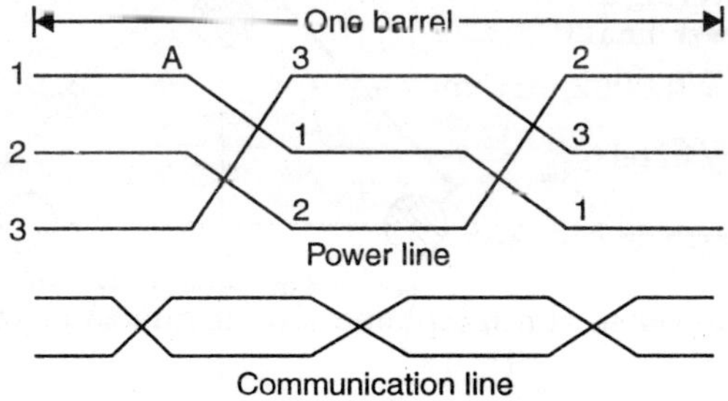

Fig. 15.5. Power line and communication line transposed.

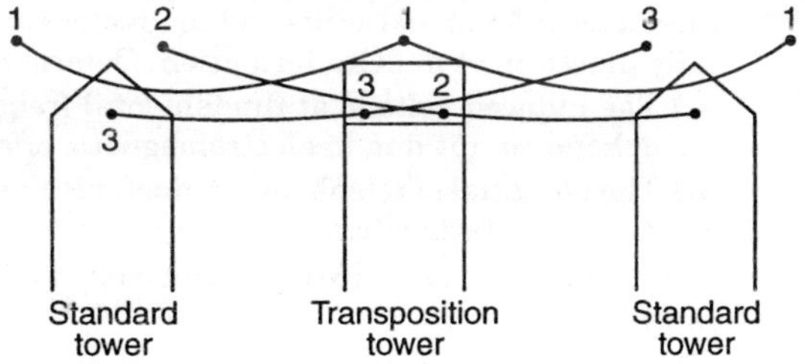

Fig. 15.6. Transposition.

Although the transposition of conductors is helpful under normal working conditions, yet it does not offer a complete solution which arise from such faults as a short circuit to earth. Under such a condition a large e.m.f. is induced on account of electromagnetic induction (since the faults produce zero sequence currents which flow along the wires in parallel and through the earth return). Even though, due to perfect transposition, no voltage between the two wires of the circuit, may exist yet a considerable voltage will be induced between each conductor and earth. This requires that communication line insulation must be so arranged that it can withstand such a voltage. In extreme cases of electrostatic charging it may become necessary to isolate telephone apparatus from the telephone line completely by means of *highly insulated transformers.* The dissipation of induced voltage may be possible by such devices as *earthed drainage coil and lightning arrestors.*

HIGHLIGHTS

1. When the power lines and communication lines run parallel to each in close proximity,
   (*i*) The *electromagnetic effect* produces currents due to induction resulting its superposition on speech current of communication line causing distortion.
   (*ii*) The *electrostatic effect* raises the potential of the communication circuit as a whole and thereby posing a threat of damage to the communication equipments.

THEORETICAL QUESTIONS

1. Discuss in brief how the power circuit interferes with the normal operation of the nearly communication circuit. What steps are taken to avoid this interference ?
2. Derive expressions for voltages induced due to (*i*) electromagnetic and (*ii*) electrostatic effects of interference between power and telephone lines and show how there results can be used for calculating electromagnetically and electrostatically induced e.m.fs on telephone conductors when the power line is 3-phase and there are two telephone conductors.
3. Write short note on "Inductive interference between power and communication lines."

UNSOLVED EXAMPLES

1. The spacing between the conductors of a single-phase 50 Hz power line, supported on horizontal cross-arm is 4 m. The spacing between the conductors of a telephone line, supported on a horizontal cross arm 2 m below the power line, is 0.6 m. If the mean radius of the conductors of power line is 2 mm and current flowing through the power line is 150 A, calculate :
   (*i*) The mutual inductance between the circuits.
   (*ii*) The voltage per km induced in the telephone line. **[Ans.** (*i*) 6×10^{-5} H/km ; (*ii*) 2.82 V/km]
2. Fig. 15.7 shows the conductor arrangement of a 3-phase, 50 Hz, 160 km long 132-kV transmission line, delivering a total load of 25 MW at 0.8 p.f. lagging. Each power line conductor is 5 mm in radius and conductors are symmetrically placed in 4 m delta formation. Determine
   (*i*) The induced voltage at fundamental frequency in telephone circuit due to electromagnetic effect.
   (*ii*) The potential of telephone conductor *P* above earth due to electrostatic effect.
   [Ans. (*i*) 20.8 volts ; (*ii*) 2700 volts]

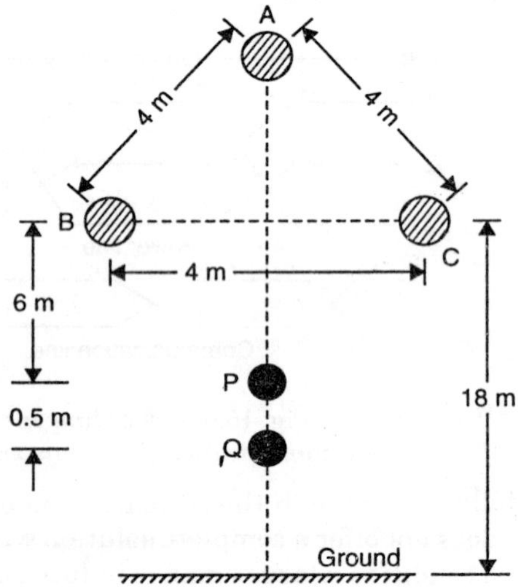

Fig. 15.7.

16

Underground Cables

16.1. Introduction. 16.2. General construction of cable. 16.3. Insulating materials for cables. 16.4. Classification of cables—Low voltage (or L.T.) cables—High voltage (H.T.) cables upto 11 kV—Super tension/screened cables—Extra high tension (E.H.T.) cables. 16.5. Laying of underground cables. 16.6. Insulation resistance of a single-core cable. 16.7. Capacitance of a single-core cable. 16.8. The most economical conductor size in a cable. 16.9. Grading of cables. 16.10. Capacitance of 3-core cables. 16.11. Heating of cables. 16.12. Thermal resistance of cables. 16.13. Permissible current loading. 16.14. Types of cable faults and their location—Types of cable faults—Location of cable fault. 16.15. Causes of failure of underground cables 16.16. Operating problems with cables. 16.17. Direct current (D.C.) cables. 16.18. Cryogenic cables. 16.19. Comparison between overhead lines and underground cables. Highlights—Theoretical Questions—Unsolved Examples.

16.1. INTRODUCTION

- The use of overhead lines, owing to safety regulations, becomes impracticable in big cities and densely populated areas. In such phases insulated conductors are usually laid underground, and are called *cables*. Underground transmission and distribution is *more expensive* than the overhead lines.
- A **cable** *is basically an insulated conductor, externally protected against mechanical injury, moisture entry and chemical reaction*. The conductor is usually aluminium or annealed copper while the insulation is mostly PVC or other chemical compounds.
- Cables developed should be *not only economically attractive but physically they should be able to carry large chunk of power*. The main constraint is the temperature rise of the insulating materials used, the limit of conductor temperature being 90°C. For *flexible cables*, there exists an upper limit to overall diameter, and hence to the size of the conductor. The combination of paper and oil is still the most effective dielectric for high voltage cables.
- The type of cable used will depend upon the voltage and service requirements. Recent improvements in design and manufacture have lead to the development of cables which can withstand voltages associated with modern transmission systems.

16.2. GENERAL CONSTRUCTION OF CABLE

The underground cable employed for transmission of power at high voltage consists of the following :

(*i*) Cores or conductors	(*ii*) Met-llic sheath
(*iii*) Bedding	(*iv*) Armouring, serving.

(*i*) **Cores or conductors.** One central core or number of cores (two, three or four) of tinned stranded copper conductors (use of aluminium conductor is also made) *insulated* from each other by *paper or vanished cambric or vulcanised bitumen or impregnated paper*.

(*ii*) **Metallic sheath.** A "*metallic sheath*" *of lead or alloy or aluminium* is provided around the insulation to *protect it against ingress of moisture*.

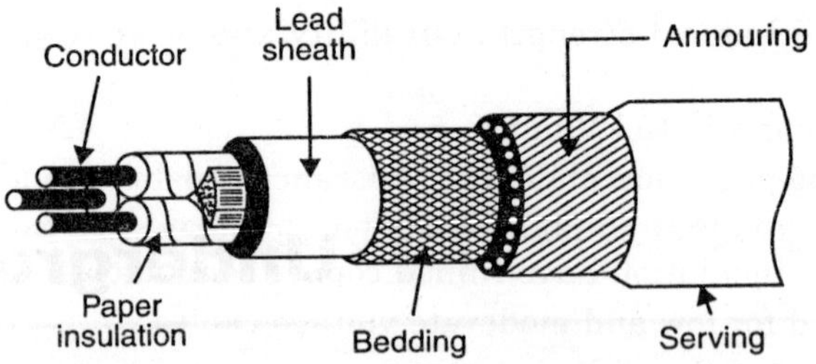

Fig. 16.1. Construction of a cable.

(*iii*) **Bedding.** For protection of sheath against corrosion and from mechanical injury from the armouring a layer of "*bedding*" consisting of paper tape compounded with a fibrous material is provided over the metallic sheath. Also sometimes jute strands or hessian tape is also used for bedding.

(*iv*) **Armouring, serving.** Over the layer of bedding "*armouring*" consisting of one or two layers of galvanised steel wire is provided to save the cable from mechanical injury and over the armouring a layer of fibrous material similar to that of bedding known as "*serving*" is provided in order *to protect the armouring*.

16.3. INSULATING MATERIALS FOR CABLES

The insulating materials should possess the following *properties* :

1. High resistivity.
2. High dielectric strength.
3. Low thermal coefficient.
4. Low water absorption.
5. Low permittivity.
6. Non-flammable.
7. Chemical stability.
8. High mechanical strength.
9. High viscosity at impregnation temperature.
10. Capability to withstand high rupturing voltages.
11. High tensile strength and plasticity.

The various insulating materials used in manufacture of cables are :

1. Rubber
2. Vulcanised India Rubber (V.I.R)
3. Impregnated paper
4. Varnished cambric (or empire tape)
5. Polyvinyl chloride
6. Gutta-percha.

1. Rubber. *Natural rubber* is obtained from the milky sap of tropical trees. *Synthetic rubber* is produced from alcohol or oil products.

- Its relative permittivity is between 2 and 3.
- Its dielectric strength is 30 kV/mm.
- Though it possesses high insulating qualities but it absorbs moisture readily.
- It softens when heated to a temperature of 60 or 70°C.
- It swells under the action of mineral oils and ages when exposed to light.

Uses. It finds little use as an insulating material (due to some major drawbacks mentioned above).

2. Vulcanised India Rubber (V.I.R.). It is prepared by mixing pure rubber with mineral matter such as zinc oxide, red lead etc., and 3 to 5% of sulphur. The compound so formed is rolled into thin sheets and cut into strips. The rubber compound is then applied to the conductor and is heated to a temperature of about 150°C. The whole process is called *vulcanisation* and the product obtained is known as "vulcanised India rubber."

- It has greater mechanical strength, durability and wear resistant property than pure rubber.
- Its dielectric strength is 15 kV/mm.
- It has low melting point, low chemical resistance capability and short span of life.
- Its main drawback is that sulphur reacts very quickly with copper and for this reason, cable using V.I.R. insulation have tinned copper conductor.

Uses. Generally used for low and moderate voltage cables.

3. Impregnated paper. Depending upon the operating voltage, a suitable layer of the paper is lapped on the conductor. It is then dried by combined application of heat and vacuum. This is carried out in a hermetically sealed steam heated chamber. The temperature is 120-130°C before vacuum is created. After the cable is dried an insulating compound having the same temperature as that of the chamber is forced into the chamber. All the pores of the paper cable are completely filled with this compound. After impregnation the cable is allowed to cool under the compound so that the void formation due to compound shrinkage is minimised. After this metal sheath is applied.

- It has almost superseded the rubber insulation because its advantages include *low cost, low capacitance high dielectric strength and higher insulation resistance.*
- Its only *disadvantage* is that paper is hygroscopic and even if it is impregnated with suitable compound, it absorbs moisture and thus lowers the insulation resistance of the cable. Owing to this reason, paper insulated cables are always provided with some protective covering and are never left unsealed.

Uses. Since the paper insulated cables have a tendency to absorb moisture, they are used where the cable route has a few joints.

4. Varnished cambric (or empire tape). This is a cotton cloth impregnated and heated with varnish. The cambric is lapped on to the conductor in the form of a tape and its surfaces are coated with petroleum jelly compound to allow for the sliding of one turn over another as the cable is bent.

- Its dielectric strength is about 4 kV/mm.
- Its permittivity is 2.5 to 3.8
- Such cables do not require sealing.

5. Polyvinyl chloride (P.V.C.). It is a synthetic material and is obtained as *white odourless, tasteless, chemically inert, non-inflammable and insulate powder.* It is chemically combined with a plastic compound and the gel is used over the conductor to form the insulation cover.

- It has dielectric strength of 17 kV/mm.
- It has dielectric constant of 5.
- Maximum continuous temperature rating is 75°C.

Uses. P.V.C. insulated cables are usually employed for low and medium domestic and industrial lights and power installations.

6. Gutta-percha. It is similar to rubber but it becomes soft at about 65°C.

- It is non-hygroscopic.
- It cannot withstand even medium voltages.

Uses. It is mostly used for submarine cables for telephone and telegraph purposes.

16.4. CLASSIFICATION OF CABLES

Cables for underground service may be *classified* as follows :

(*i*) *According to voltage for which they are manufactured.*

(*ii*) *According to the type of insulating material used in their manufacture.*

Usually the former method of classification is generally preferred, and as such, the cables, according to voltage, are classified as follows :

1. *Low voltage (or L.T.) cables.* ...... for operating voltage upto 1 kV.
2. *High voltage (or H.T.) cables.* ...... for operating voltage upto 11 kV.
3. *Supertension (S.T.) cables.* ...... for operating voltage upto 33 kV.

(*a*) *H*-type cables (*b*) H.S.L. type cables

(*c*) H.S.L. types cables.

4. Extra high tension (E.H.T.) cables for operating voltage upto 66 kV.
5. Extra super voltage cables for operating voltage upto 132 kV.

In order to meet the increased voltage demand the extra high tension and extra super voltage power cables useful for 132 kV and above have been developed. *In such cables, the voids have been eliminated by increasing the pressure of the compound* and that is why such cables are also called as **pressure cables.**

Pressure cables are of the following two types :

A. **Oil filled cables :**

...... Single-core oil filled cables used upto 132 kV.

...... Three-core oil filled cables used upto 66 kV.

B. **Gas pressure cables :**

(*i*) External pressure cables

(*ii*) Internal pressure cables

— High pressure gas filled cables

— Gas cushion cables

— impregnated pressure cables.

16.4.1. Low Voltage (or L.T.) Cables

- These are used for voltage below 1 kV.
- The insulation may consist of *paper impregnated with oil.* Resin is sometimes used to increase the viscosity and thus to prevent drainage. Where *varnished cambric* is used for insulation a coating of petroleum jelly is provided between layers ; this allows for the bending of the cable without causing any damage. *Varnished bitumen* in which refined bitumen is treated with sulphur and vegetable oils is also used as insulators.
- *Lead covered cables* are extensively used in house wiring, service mains etc. The insulation is by rubber.

Fig. 16.2 shows the constructional details of a single-core low tension (L.T.) cable.

— It consists of one circular core of tinned stranded copper (or aluminium) insulated by layers of impregnated paper.

— The insulation is surrounded by a *lead sheath* which prevents the entry of moisture into the inner parts.

— In order to protect the sheath from the corrosion, an overall *serving* of compounded fibrous material (fuel etc.) is provided.

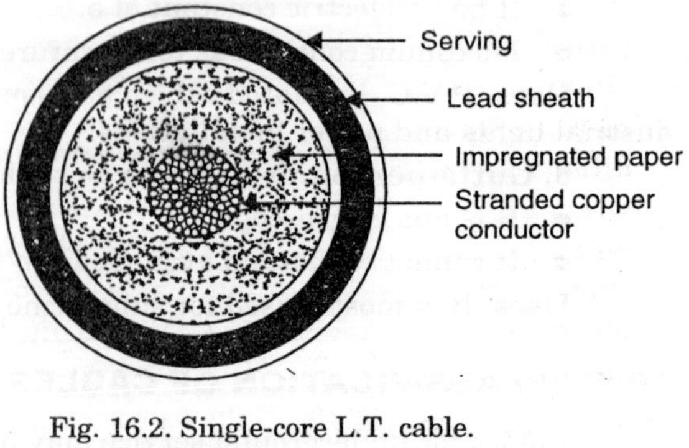

Fig. 16.2. Single-core L.T. cable.

— These types of cables are usually *not* armoured in order to avoid excessive sheath losses.

The main *advantages* of single-core L.T. cables are :

(*i*) Simple in construction.

(*ii*) Availability of large copper section.

16.4.2. High Voltage (H.T.) Cables upto 11 kV

- These are multicore cables and of belt type.
- The belt type cable (See Fig. 16.3) consists of either circular shaped or oval or sector shaped (oval or sector shaped core improves the copper space factor) cores of stranded copper or aluminium conductors wrapped around by impregnated paper. Another layer of impregnated paper tape, called *paper belt* is wound round the grouped insulated cores. The gap between the insulated cores is filled with fibrous insulating material (jute etc.) so as to give circular cross-section to the cable. The cores are generally *stranded*. The belt is covered with *lead sheath* to protect the cable against ingress of moisture and mechanical injury. The lead sheath is covered with one or more layers of armouring with an outer serving (not shown in the figure).

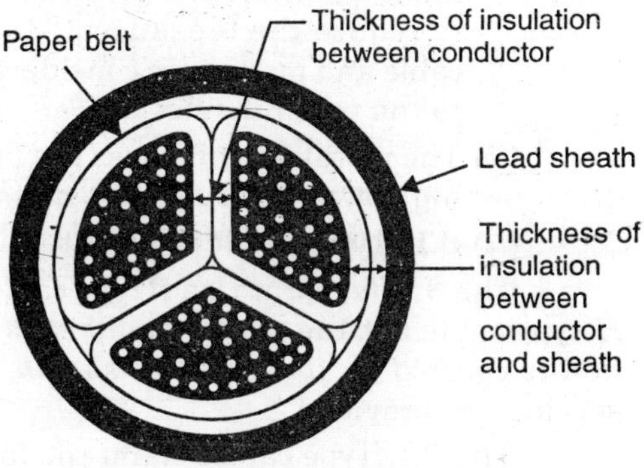

Fig. 16.3. Three-core belted cable (upto lead sheath).

16.4.3. Super Tension/Screened Cables

The major disadvantage of belted type construction is that the electric stress is not purely radial. The existence of tangential stresses forces a leakage current (not the charging current) to flow along the layers of papers and the loss of power sets up local heating. It is to be noted that the *resistance and dielectric strength of laminated paper is much less along the layers as compared to that across the layer.* The *local heating of dielectric may result in breakdown of the material.* In order to overcome this difficulty, *screened cables* are used where leakage currents are conducted to earth through metallic screens.

(*i*) **H-type cables.** Refer to Fig. 16.4.

In this type of cable (invented by M. Hochstadter), no belt-insulation is used, but each of the conductor is insulated with paper perforated to facilitate the process of impregnation. The insulation on each core is covered with a *metallic screen* which usually consists of a perforted aluminium foil. The cores are laid in such a way that metallic screens make contact with one another. An additional conducting belt (copper woven fabric tape) is wrapped round three cores. The cable has no insulating belt but lead sheath, bedding and serving follow as usual (Fig. 16.4). It is easy to see that each core screen is in electrical contact with the conducting belt and lead sheath.

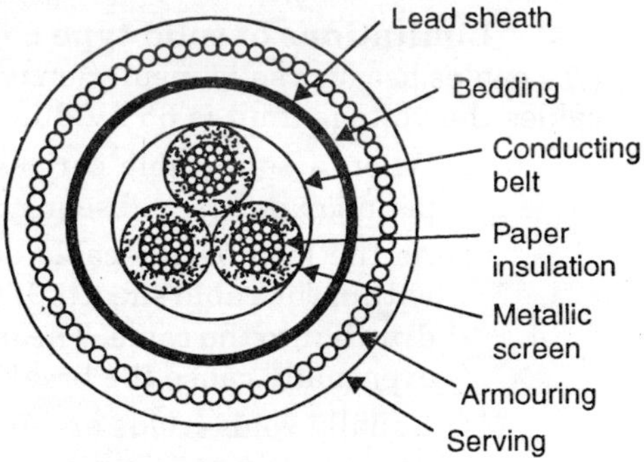

Fig. 16.4. H-type cable.

In this cable as all the four screens (3 core screens and one conducting belt) and lead sheath are at earth potential, therefore electrical stresses are purely radial and consequently dielectric losses are reduced.

The *H*-type cables claim the following ***advantages :***

(*i*) The perforations in the metallic screens assist in the complete impregnation of the cable with the compound and thus *possibility of air pockets or voids* in the dielectric *is eliminated* (The voids if present tend to reduce the breakdown strength of the cable and may cause considerable damage to the paper insulation).

(*ii*) The metal foil increases the heat dissipating power and there are *no sheath losses.*

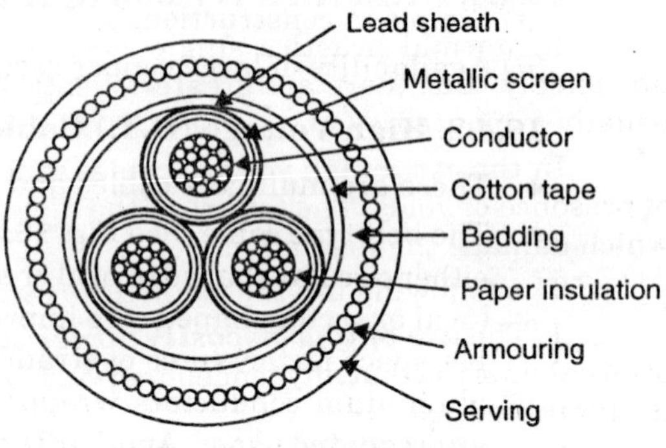

Fig. 16.5. S.L. type cable.

(*ii*) **S.L. type cables.** Refer to Fig. 16.5.

The S.L. (separate lead) type cable is basically *H*-type cable but the *screen round each core insulation is covered by its own lead sheath.* There is no overall lead sheath but only armouring and serving are provided.

The S.L. type cables claim the following ***advantages*** over *H*-type cables :

- The separate sheaths minimise the possibility of core-to-core breakdown.
- The bending of cables becomes easy due to the elimination of overall lead sheath.

However, the "*disadvantage*" of this type of cable is that because of thinner lead sheaths its *manufacturing is difficult*.

(*iii*) **H.S.L. type cables.** This type of a cable is a combination of H-type and S.L. type cables. In such a cable *each conductor is insulated, sheathed with metallized paper* and *is then provided with lead sheath. The* three cores are then laid up and provided with filler, braided, armoured and finally served.

Advantages of screened type cables over belted types :

The screened type cable claim the following advantages over belted types :

- Reduced core to core faults.
- In all sections of the dielectric electric stresses are uniformly radial.
- No possibility of formation of voids within the electric field.
- The current carrying capacity of the cables is increased.

Limitations of solid type cables. The cables considered until now are also called as *solid type* cables because solid insulation is used and no gas or oil circulates in the cable sheath. In these cables the voltage limit is 66 kV due to the following *reasons* :

(*i*) When a solid cable carries load, its conductor temperature increases and cable compound expands. Subsequently the lead sheath stretches and may get damaged.

(*ii*) As the load on the cable decreases, the conductor cools and partial vacuum is formed within the cable sheath. If the pin holes are present in the lead sheath moist air may be drawn into the cable. The moisture reduces the dielectric strength of insulation and may eventually cause the breakdown of the cable.

(*iii*) Usually voids (voids are unintentional spaces in the insulation of cable filled with air or gas, usually at low pressure) are always present in the insulation of a cable (Modern techniques of manufacturing have resulted in void free cables). Under operating conditions, the voids are formed as a result of the differential expansion and contraction of sheath and impregnated compound. The voids nearest to the conductor are first to breakdown, the chemical and thermal effects of ionisation causing permanent damage to the paper insulation(The breakdown strength of voids is considerably less than that of the insulation).

16.4.4. Extra High Tension (E.H.T.) Cables

In normal practice single-core cables are used as three-core cables are too large and unwieldely. But there are greater sheath losses in single-core cables and therefore they are not usually armoured.

In the extra high voltage cables there is a danger of breakdown of the dielectric on account of presence of voids. The failure of the dielectric results in ionization and allied chemical reactions which damage the insulation.

The following are the *methods for elimination of void formation in the cables :*

1. The ***use of low viscosity mineral oil*** for the impregnation of the dielectric and inclusion of oil channels so that any tendency of void formation, due to cyclic heating and cooling of impregnant, is eliminated.

This method is *used in oil-filled cables.* Oil ducts are provided within the cable itself and they communicate with oil tanks provided at suitable locations along the cable route so as to accommodate any changes in the oil volume during heating and cooling process.

2. ***The use of inert gas at high pressure within*** *the metal sheath and indirect contact with the dielectric.*

The pressure cables are placed in metal pipes which are gas-tight and filled with nitrogen at pressures of 12 to 15 atmospheres. *The pressure of gas does not permit formation of voids on account of compression.* As compared to the normal cables of other designs a *pressure cable can be worked at a double the voltage and nearly one and a half times the current.* However, the manufacturing *cost is much higher.*

16.4.4.1. Oil-filled cables

In this type of cables, a *hollow conductor of soft drawn stranded copper, fed by oil reservoirs placed at intervals along the lines is used.* The oil is maintained under pressure by these reservoirs. As the cable heats on load, oil is driven from the cable into the reservoir and vice versa, hence the formation of voids is prevented. *Impregnated paper is used for insulation, and a lead sheath and jute covering are employed to give water proofing.* Fig. 16.6 shows the cross-section of a 220 kV oil filled cable.

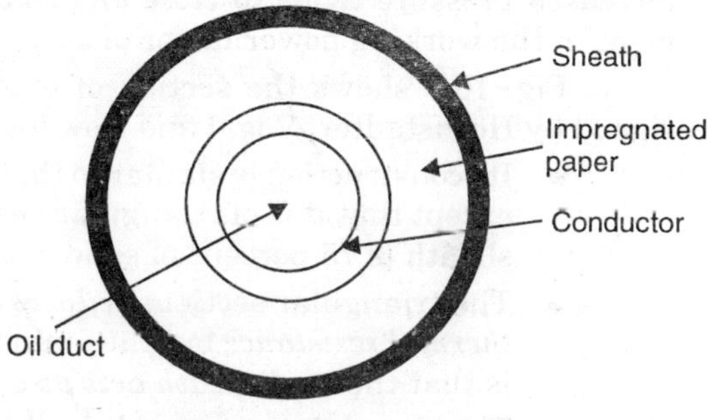

Fig. 16.6. Oil-filled cable.

- Oil-filled cables require relatively *smaller amount of insulation* as compared to the solid type for the same operating voltage and, therefore, are used for higher voltages (66 kV-400 kV).
- The oil-filled cable has a working stress of 15 kV/mm at 500 kV.

Advantages :

The oil-filled cables claim the following *advantages :*

(*i*) Impregnation more perfect.

(*ii*) Impregnation possible after sheathing.

(*iii*) Smaller overall size and smaller weight (owing to reduction in thickness of dielectric required).

(*iv*) Increased temperature range in service possible.

(*v*) No ionisation, oxidation and formation of voids.

(*vi*) Fault detection easy (as leaking of oil can be noticed immediately).

(*vii*) Higher current rating (smaller thermal resistance due to decrease in dielectric thickness).

(*viii*) Higher maximum permissible stresses.

Disadvantages :

(*i*) Higher cost.

(*ii*) Laying of cables and maintenance complicated.

16.4.4.2. Gas pressure cables

To counter the disadvantages of oil-filled cables in terms of expansion and contraction of oil during loading cycles, gas filled cables are used. One *advantage* of the gas-filled cable is its *better heat dissipation by natural convection in the gas.* More revolutionary are the *cables incorporating cryogenic coolants.* An alternative to the superconducting cable is the resistive operation of an Al or Cu conductor at cryogenic temperatures. Such a cable is called ***cryoresistive***.

Types of gas pressure cables :

Gas pressure cables are of the following *two types* :

(*i*) External pressure cables.

(*ii*) Gas-filled cables.

(*i*) ***External pressure cables.*** In such a cable the pressure is applied externally and raised to such an extent that *no ionisation can take place.* At the same time the radial compression due to the increased pressure tends to *close any voids.* There is an improvement in the working power factor of such a cable as well.

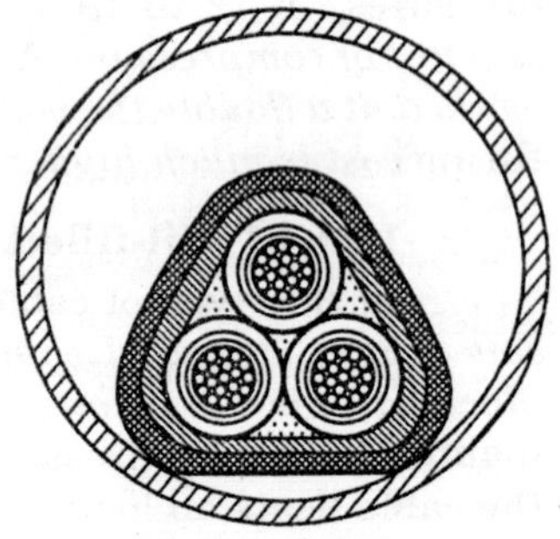

Fig. 16.7. External pressure cable.

Fig. 16.7 shows the section of external pressure cable designed by Hochstadter, Vogal and Bowden.

- Its construction is similar to that of an ordinary solid type except that it is of triangular shape and thickness of lead sheath is 75 percent of solid cable.
- The triangular sections *reduces the weight* and *gives low thermal resistance* but the main reason for triangle shape is that the *lead sheath acts as a pressure membrane.*
- The sheath is protected by a thin *metal tape.*
- The cable is laid in a gas-tight steel pipe, which is filled with dry nitrogen gas at 12 to 15 atmospheres. The gas pressure *produces radial compression and closes the voids* that may have formed between the layers of paper insulation.

Such cables can *carry more load current and operate at higher voltages.* Moreover, *maintenance cost is small and nitrogen gas helps in quenching any flame.* However, it has the *disadvantage* that its *overall cost is high.*

(*ii*) ***Internal pressure cables.*** Internal pressure cables are of the following *three* types :

(*a*) High pressure gas-filled cables.

(*b*) Gas cushion cables.

(*c*) Impregnated pressure cables.

- In *high pressure gas-filled cables* spaces are provided in the dielectric itself for the gas which is inert gas like nitrogen at pressure of about 6 atmospheres for extra H.T. cables and 12 atmospheres for super voltage cables. Pressure is retained by means of a lead sheath.
- In *gas-cushion cables*, screened space is provided all along the length of cable, in between the lead sheath and the dielectric. The screened space is sub-divided by means of barriers

and thus inert gas is stored at various points along the run of a cable. This facilitates the joining of cable without losing gas from the entire cable, because while cutting the cable, loss of gas will be local only.

The major feature of this cable is that *no arrangement is required for the transmission of pressure to the cables from outside, so that the cable is a complete unit with its own armouring, requiring no external pipe protection whatsoever.*

- *Impregnated pressure cable* is similar to solid type excepts that such a cable consists of mass-impregnated paper dielectric and this is maintained at a pressure of 14 atmospheres by means of nitrogen. In order to cater for the large hoop and longitudinal stresses set up, special reinforcement consisting of longitudinal and circumferential metallic tapes is provided.

The *internal pressure cables* claim the following "*advantages*" :

(*i*) The power factor of the cable dielectric is amply improved.

(*ii*) External accessories are eliminated.

(*iii*) By adopting suitable designs the cable can be used for vertical run without any fear of drainage.

16.5. LAYING OF UNDERGROUND CABLES

The laying of underground cables may be done by the following *methods* :

1. Direct laying.
2. Draw-in system or Duct/pipe laying.
3. Solid system.

1. Direct laying. This method of laying underground cables (Fig. 16.8) includes the following *procedure.*

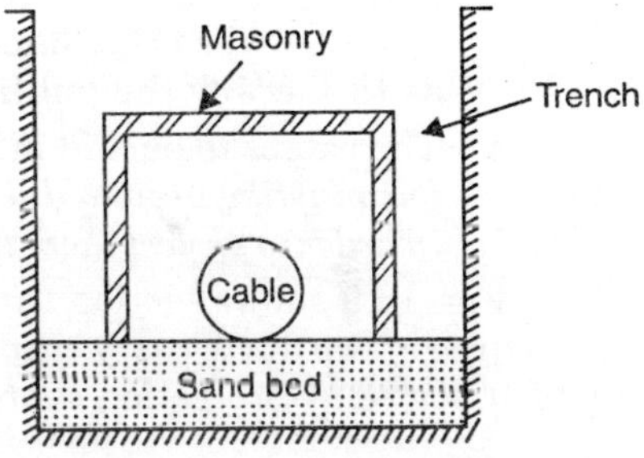

Fig. 16.8. Direct laying of cables.

- A trench of about 1.5 m deep and 45 cm wide is dug.
- The trench is covered with a layer of fine sand, about 10 cm thick, and the cable is laid over this sand bed. The sand does not allow the entry of moisture from the ground and thus protects the cable from decay.
- After the cable has been laid in the trench, it is covered with another layer of sand of about 10 cm thickness.
- The trench is then covered with bricks and other materials in order to protect the cable from mechanical injury.
- When more than one cable is to be laid in the same trench, a horizontal or vertical inter-axial spacing of at least 30 cm is provided in order to reduce the effect of mutual heating and also to ensure that a fault occuring on one cable does not damage the adjacent cable.
- Cables to be laid in this way *must have serving of bituminised paper and hessian tape so as to provide protection against corrosion and electrolysis.*

This method of cable laying is used in open areas where excavation can be done conveniently and at low cost.

Advantages :

1. Simple and cheap method.
2. Clean and safe method.
3. Provides best conditions for dissipating the heat generated in the cables.

Disadvantages :

1. High maintenance cost.
2. Location of fault is difficult.
3. Load extension necessitates completely new excavation costing as much as the original work.
4. The alteration in the cable network cannot be made easily.
5. The cable sheath may sometimes undergo chemical changes and get damaged due to impurities present in the soil.
6. This method cannot be used in congested areas where excavation is expensive and inconvenient.

2. Draw-in system. This method of cable laying is *suitable for congested areas where excavation is expensive and inconvenient.*

It is *most suitable for short length cable routes* such as in workshops railway bridge crossing, road crossing where frequent digging is costlier or impossible.

- In this method a line of conduits, ducts or tubes made of either iron, glazed stoneware, clay or cement concrete are laid in ground with manholes at suitable positions along the cable route. The cables are then pulled into position.
- Separate pipes and ducts (or multiple way ducts) are provided for each cable laid in the same duct.
- Care must be taken that where the duct line changes direction ; depths, dips and offsets be made with very long radius so that a large cable may be pulled easily between the manholes. The distance between the manholes should not be too long so as to simplify the pulling in of the cables.
- The cables to be laid in this way need *not be armoured but must be provided with serving of hessian and jute in order to protect them when being pulled into the ducts.*

Fig. 16.9 shows section through 4-way underground duct line. Three of the ducts carry transmission cables and the fourth duct carries relay protection connection, pilot wires.

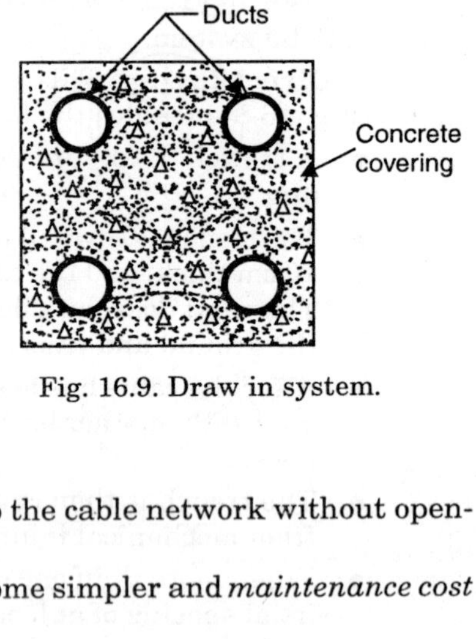

Fig. 16.9. Draw in system.

Advantages :

1. *Less chances of occurrence of fault* (owing to strong mechanical protection provided by the system).
2. It is *easy to carry out repairs, alterations or additions* to the cable network without opening the ground.
3. Since the cables are not armoured, therefore, joints become simpler and *maintenance cost is reduced considerably.*

Disadvantages :

1. *Very high initial cost.*
2. Owing to the close grouping of cables and unfavourable conditions for dissipation of heat the *current carrying capacity of the cables is reduced.*

3. Solid system :

- In this method the cable is laid in open pipes or troughs dug out in earth along the cable route. The troughing is of cast iron, stoneware, asphalt or treated wood. After laying the cable in position, the troughing is filled with a bituminous or asphaltic compound and covered over.

- Cables laid in this manner are usually plain lead covered because troughing affords good mechanical protection.

This method of laying underground cables is rarely used these days because of the following *disadvantages* :

(*i*) Costlier than direct laid system.

(*ii*) Requires skilled labour and favourable weather conditions.

(*iii*) Reduced current carrying capacity owing to poor heat dissipation facilities.

16.6. INSULATION RESISTANCE OF A SINGLE-CORE CABLE

In a cable useful current flows along the axis of the core but there is always present some *leakage* of current. This leakage is *radial i.e.* at right angles to the flow of the useful current. *The resistance offered to the radial leakage of current is called "Insulation resistance" of the cable.* If the length of the cable is greater, the leakage area is also greater meaning thereby that more current will leak. In other words insulation resistance is decreased. Hence the *insulation resistance is inversely proportional to the length of the cable.*

Fig. 16.10 shows a single-core cable of inner and outer radii r_1 and r_2 respectively and length l metres. Consider annular ring of radius 'r' and radial thickness 'dr'. If the volume resistivity of the material is ρ, then resistance of this narrow ring will be,

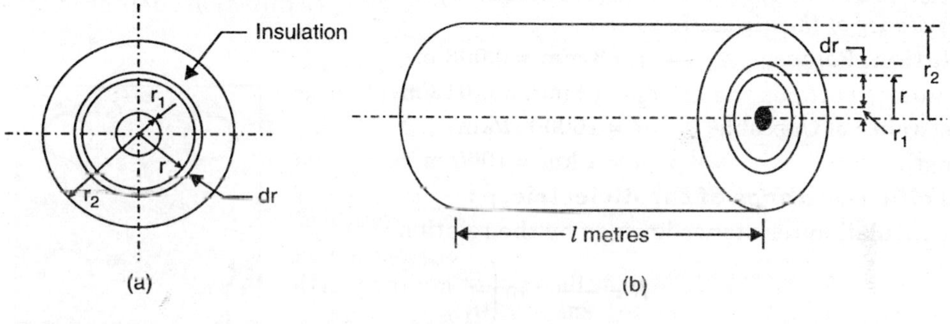

Fig. 16.10. Cross-section of a cable.

$$dR = \frac{\rho dr}{2\pi r l}$$

and *insulation resistance* (R) of the cable is

$$\int dR = \int_{r_1}^{r_2} \frac{\rho dr}{2\pi r l} \quad \text{or} \quad R = \frac{\rho}{2\pi l} \Big| \ln r \Big|_{r_1}^{r_2}$$

i.e.,
$$R = \frac{\rho}{2\pi l} \ln\left(\frac{r_2}{r_1}\right) \Omega \qquad \text{...(16.1)}$$

Measurement of insulation resistance of cables :

The insulation resistance of cable can be measured by the following methods :

1. *Galvanometer method.* In this method a very sensitive galvanometer of high resistance (of 1000 Ω or more) is connected in series with the resistance to be measured, and to a battery supply. The deflection of the galvanometer gives the value of the insulation resistance.

- This method is only sufficient to *indicate whether the insulation* is *faulty or otherwise and cannot be regarded as a precise method.*

2. *By meggar.* A *meggar* is usually used for the finding insulation resistance of underground cables. It carries two terminals *E* and *L*; *L* is connected to the core of the cable and *E* is connected to earth terminal. Now if the handle is rotated at uniform speed, then insultation resistance of the cable between core and earth can be directly read from the deflection of the pointer moving on the calibrated dial.

Example 16.1. *Determine the insulation resistance of a single-core cable of inside diameter 0.03 metre, outside diameter 0.075 m and length 2 km ; the resistivity of the insulating material being 6 × 10^12 ohm metre.*

Solution. Radius, $r_1 = \frac{0.03}{2} = 0.015$ m

Radius, $r_2 = \frac{0.075}{2} = 0.0375$ m

Length, $l = 2$ km $= 2000$ m

Specific resistance, $\rho = 6 \times 10^{12}$ Ωm

Now, *insulation resistance*, $R = \frac{\rho}{2\pi l} \ln\left(\frac{r_2}{r_1}\right) = \frac{6 \times 10^{12}}{2\pi \times 2000} \ln\left(\frac{0.0375}{0.015}\right)$

$= \mathbf{4.37 \times 10^8\ \Omega}$. **(Ans.)**

Example 16.2. *The conductor of single-core cable has a diameter of 6 mm, the diameter over the insulation is 24 mm. If the insulation resistance of the cable is 16000 ohms per km, calculate the specific resistance of the dielectric used.*

Solution. Radius, $r_1 = 3$ mm $= 0.003$ m

Radius, $r_2 = 12$ mm $= 0.012$ m

Resistance of the cable, $R = 16000$ Ω/km

Length, $l = 1$ km $= 1000$ m

Specific resistance of the dielectric, ρ :

The insulation resistance is given by the relation :

$$R = \frac{\rho}{2\pi l} \times \ln\left(\frac{r_2}{r_1}\right)$$

or,

$$16000 = \frac{\rho}{2\pi \times 1000} \ln\left(\frac{0.012}{0.003}\right)$$

$$\rho = \frac{16000 \times 2\pi \times 1000}{\ln\left(\frac{0.012}{0.003}\right)} \ \Omega\text{m} = 72.517 \text{ mega ohm metre.}$$

i.e. Specific resistance = **72.517 mega ohm metre. (Ans.)**

Example 16.3. *A 6.5 km long cable has a conductor diameter of 15 mm and internal sheath diameter of 30 mm. Find the conductor resistance and insulation resistance of each piece if the cable is cut into two pieces of equal length. Specific resistance of conductor material is 0.017 micro-ohm m. Specific resistance of insulation material is 6 megaohm m.*

Solution. Length of the cable $= \frac{6.5}{2}$ km $= 3.25$ km $= 3250$ m

Conductor radius, $r_1 = \frac{15}{2} = 7.5$ mm $= 0.0075$ m

Internal sheath radius, $r_2 = \frac{30}{2} = 15$ mm $= 0.015$ m

Specific resistance of conductor, $\rho_c = 0.017$ micro-ohm m

Specific resistance of insulation material, $\rho_i = 6 \times 10^6$ ohm m.

Resistance of the conductor,

$$R_c = \frac{\rho_c l}{a} = \frac{0.017 \times 10^{-6} \times 3250}{\pi \times (0.0075)^2} = \mathbf{0.3126\ \Omega.} \quad \textbf{(Ans.)}$$

Insulation resistance of each piece,

$$R_i = \frac{\rho_i}{2\pi l} \ln\left(\frac{r_2}{r_1}\right) = \frac{6 \times 10^6}{2\pi \times 3250} \ln\left(\frac{0.015}{0.0075}\right) = \mathbf{203.66\ \Omega.} \quad \textbf{(Ans.)}$$

16.7. CAPACITANCE OF A SINGLE-CORE CABLE

The cable can be considered to be co-axial cylinders of internal radius r and external radius R. In actual cable r represents the radius of the core and R represents inner radius of the lead sheath which is at earth potential.

Fig. 16.11 shows a single-core cable or cylindrical capacitor consisting of two co-axial cylinders of radii r and R metres.

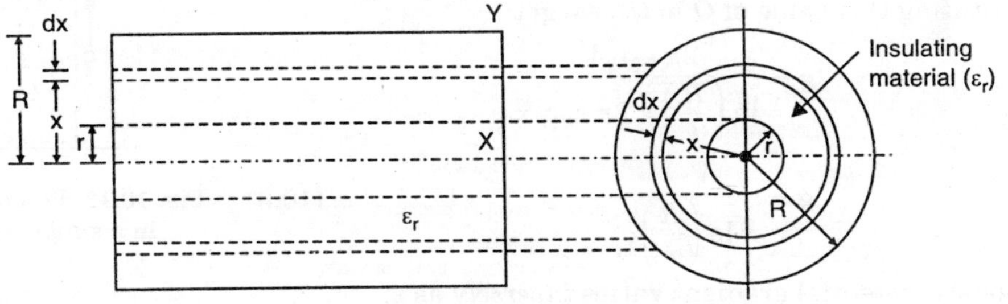

Fig. 16.11. Cylindrical capacitor.

Lot, ε_r = Relative permittivity of insulating material in between the core and load sheath,

V = Potential difference between the cylinders, and

Q = Charge in coulombs *per metre* axial length of the cylinders.

Then, the number of radial lines of electric flux between X and Y per metre length of the cylinder will be Q, and the area on which the flux will be received at a distance x will be $(2\pi x \times 1)\text{m}^2$.

Now, electric flux density at a distance x metres from the axis,

$$D = \frac{Q}{2\pi x \times 1} \quad \text{or} \quad \frac{Q}{2\pi x}\ \text{C/m}^2$$

The value of electric intensity,

$$E = \frac{D}{\varepsilon_0 \varepsilon_r} \quad \text{or} \quad E = \frac{Q}{2\pi\, x \varepsilon_0 \varepsilon_r} \qquad \text{...}(i)$$

Now,

$$dV = E.dx$$

$$V = \int_r^R E.dx = \int_r^R \frac{Q}{2\pi\, x\varepsilon_0\varepsilon_r} . dx$$

$$= \frac{Q}{2\pi\, \varepsilon_0\varepsilon_r} \int_r^R \frac{dx}{x} = \frac{Q}{2\pi\, \varepsilon_0\varepsilon_r} \Big| \ln x \Big|_r^R = \frac{Q}{2\pi\, \varepsilon_0\varepsilon_r} \ln\left(\frac{R}{r}\right) \qquad \text{...}(ii)$$

Now, capacitance, $$C = \frac{Q}{V} = \frac{Q}{\frac{Q}{2\pi\,\varepsilon_0\varepsilon_r}\ln\left(\frac{R}{r}\right)}\ \text{F/m}$$

or, $$C = \frac{2\pi\,\varepsilon_0\varepsilon_r}{\ln\left(\frac{R}{r}\right)}\ \text{F/m} \quad ...(16.2)$$

Dielectric stress/Potential gradient. The potential gradient at a radial distance of x metres,

$$g = \frac{Q}{2\pi\,\varepsilon_0\varepsilon_r\,x}\ \text{V/m} \qquad (\because\ g = E) \quad ...(i)$$

Also, $$V = \frac{Q}{2\pi\,\varepsilon_0\varepsilon_r}\ln\left(\frac{R}{r}\right)$$

or, $$Q = \frac{2\pi\,\varepsilon_0\varepsilon_r V}{\ln\left(\frac{R}{r}\right)}$$

Substituting this value of Q in (i), we get

$$g = \frac{2\pi\,\varepsilon_0\varepsilon_r V}{\ln\left(\frac{R}{r}\right)\times 2\pi\,\varepsilon_0\varepsilon_r x}$$

or, $$\mathbf{g} = \frac{V}{x\ln\left(\frac{R}{r}\right)} \quad ...(16.3)$$

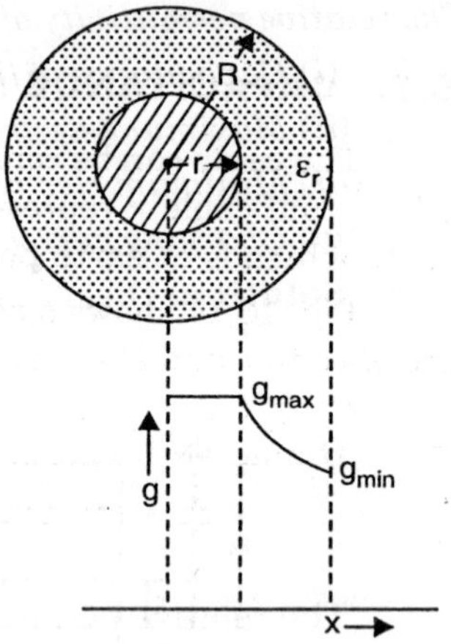

Fig. 16.12. Potential gradient in a single-core cable.

Obviously, potential gradient varies *inversely* as x.

Minimum value of $x = r$, hence maximum value of potential gradient (g_{max}),

$$\mathbf{g_{max}} = \frac{V}{r\ln\left(\frac{R}{r}\right)} \quad ...(16.4)$$

Similarly, $$\mathbf{g_{min}} = \frac{V}{R\ln\left(\frac{R}{r}\right)} \quad ...(16.5)$$

Example 16.4. *A submarine cable is 4000 km long and has a conductor 5 mm in diameter with a gutta-percha covering 5 mm thick. Calculate the total capacitance of the cable, taking the relative permittivity of gutta-percha as 4.*

(Madras University)

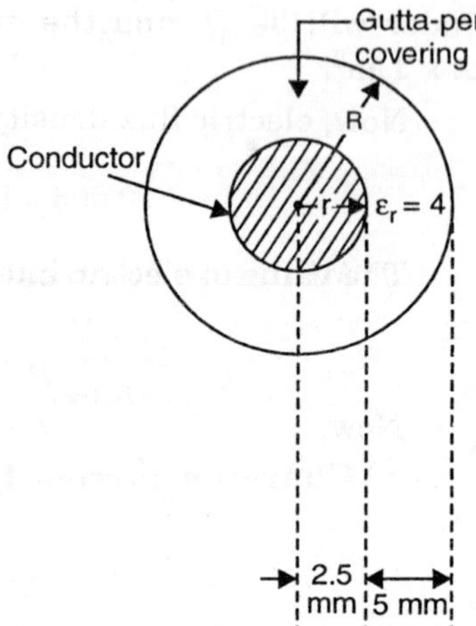

Fig. 16.13

Solution. Dia. of the conductor = 5 mm

$\therefore$ $r = 2.5\ \text{mm} = 2.5 \times 10^{-3}\ \text{m} = 0.0025\ \text{m}$

Thickness of gutta-percha, $t = 5$ mm

$\therefore$ $R = r + t = 2.5 + 5 = 7.5\ \text{mm} = 0.0075\ \text{m}$

Total capacitance :

Using the relation, $$C = \frac{2\pi\varepsilon_0\varepsilon_r}{\ln\frac{R}{r}} \quad ...[\text{Eqn. } (16.2)]$$

$$= \frac{2\pi \times 8.854 \times 10^{-12} \times 4}{\ln (0.0075/0.0025)} \text{ F/m} = 202.5 \times 10^{-12} \text{ F/m}$$

∴ Total capacitance for 4000 km

$$= 202.5 \times 10^{-12} \times (4000 \times 10^3)$$
$$= 810 \times 10^{-6} = \mathbf{810\ \mu F.} \quad \textbf{(Ans.)}$$

Example 16.5. *A single-core cable used on 33 kV, 50 Hz has conductor diameter 10 mm and inner diameter of sheath 25 mm. The relative permittivity of insulating material used is 3.5. Find :*

(i) Capacitance of the cable per km,

(ii) Maximum electrostatic stress in the cable,

(iii) Minimum electrostatic stress in the cable, and

(iv) Charging current per km. **(Panjab University)**

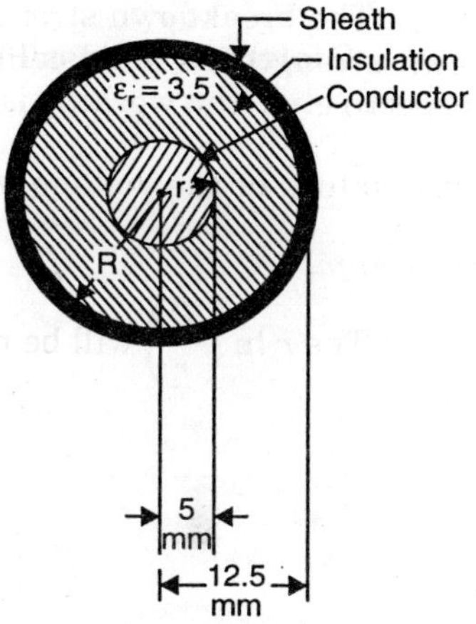

Fig. 16.14

Solution. Dia. of the conductor = 10 mm

$$\therefore \quad r = \frac{10}{2} = 5 \text{ mm} = 0.005 \text{ m}$$

Diameter (inner) of sheath = 25 mm

$$\therefore \quad R = \frac{25}{2} = 12.5 \text{ mm} = 0.0125 \text{ m}$$

The relative permittivity of insulating material, $\varepsilon_r = 3.5$

Applied voltage, $V = 33 \text{ kV} = 33000 \text{ V}$.

(i) **Capacitance of the cable per km :**

Using the relation,

$$C = \frac{2\pi\,\varepsilon_0\varepsilon_r}{\ln \frac{R}{r}} \text{ F/m} = \frac{2\pi\varepsilon_0\varepsilon_r}{\ln \frac{R}{r}} \times 1000 \text{ F/m} = \frac{2\pi \times 8.854 \times 10^{-12} \times 3.5}{\ln \frac{0.0125}{0.005}} \times 1000$$

$$= 212.5 \times 10^{-9} \text{ F} \quad \text{or} \quad \mathbf{0.2125\ \mu F.} \quad \textbf{(Ans.)}$$

(ii) **Maximum electrostatic stress, E_{max} :**

$$E_{max}\ (= g_{max}) = \frac{V}{r \ln \frac{R}{r}} \quad \text{[where } g_{max} = \text{maximum potential gradient]}$$

$$= \frac{33000}{0.005 \times \ln \left(\frac{0.0125}{0.005}\right)} = \mathbf{72 \times 10^5\ V/m.} \quad \textbf{(Ans.)}$$

(iii) **Minimum electrostatic stress, E_{min} :**

$$E_{min}(= g_{min}) = \frac{V}{R \ln \frac{R}{r}} = \frac{33000}{0.0125 \ln \left(\frac{0.0125}{0.005}\right)} = \mathbf{28.8 \times 10^5\ V/m.} \quad \textbf{(Ans.)}$$

(iv) **Charging current I_C :**

$$I_C = \frac{V}{X_C} = \frac{V}{1/2\pi fC} = 2\pi fCV = 2\pi \times 50 \times 0.2125 \times 10^{-6} \times 33000 = \mathbf{2.2\ A.} \quad \textbf{(Ans.)}$$

16.8. THE MOST ECONOMICAL CONDUCTOR SIZE IN A CABLE

We know that the expression for maximum electrostatic stress/potential gradient is,

$$g_{max} = \frac{V}{r \ln\left(\frac{R}{r}\right)} \quad \text{...(Eqn. 16.4)}$$

The breakdown strength of a cable depends upon the maximum stress it can bear (Breakdown strength for solid cables and oil-filled cables is of the order of 4-5 kV/mm and 9 kV/mm respectively). The working is safe if the maximum stress is less than the breakdown strength. For fixed values of the voltage V and overall diameter D, the g_{max} will *minimum* when $r \ln\left(\frac{R}{r}\right)$ is *maximum.*

The $r \ln\left(\frac{R}{r}\right)$ will be maximum if

$$\frac{d}{dr}\left[r \ln\left(\frac{R}{r}\right)\right] = 0$$

$$r \times \left(\frac{r}{R}\right) \times \left(-\frac{R}{r^2}\right) + \ln\left(\frac{R}{r}\right) = 0$$

$$\ln\left(\frac{R}{r}\right) - 1 = 0 \quad \text{or} \quad \ln\left(\frac{R}{r}\right) = 1 \quad \text{or} \quad \frac{R}{r} = e^1 = 2.718$$

∴ Most economical conductor radius is,

$$r = \frac{R}{2.718} \quad \text{or} \quad d = \frac{D}{2.718} \quad \text{...(16.6)}$$

and,

$$g_{max} = \frac{V}{r} = \frac{2V}{d} \quad \text{...(16.7)} \left[\text{Putting } \ln\left(\frac{R}{2}\right) = 1 \text{ in eqn. (16.4)}\right]$$

The diameter, $d = \frac{2V}{g_{max}}$ (from eqn. 16.7), for low and medium voltage cables is somewhat less than that determined from the consideration of *safe current density*, therefore, the *main criterion for determining conductor diameter for such cables is the current carrying capacity.* On the other hand for high voltage cables the value of 'd' determined from the above expression would, in general, give a conductor of cross-section *much larger than* required from the consideration of current carrying capacity. Thus in order to have the desired overall diameter of conductor without increasing the cross-sectional area, the *following methods may be employed* :

(*i*) Using aluminium core instead of copper (core) because for the same current, diameter of aluminium will be more than that of copper.

(*ii*) By stranding copper wires round a hemp centre.

(*iii*) By stranding copper wires over a lead tube.

Example 16.6. *Find the most economical size of a single-core cable working on a 132 kV, 3-phase system, if dielectric stress of 5 kV/mm can be allowed.*

Solution. *Given :* $V_L = 132$ kV ; $g_{max} = 5$ kV/mm

$$V_{ph} = \frac{V_L}{\sqrt{3}} = \frac{132}{\sqrt{3}} = 76.21 \text{ kV}$$

Peak value of phase voltage = $\sqrt{2} \times 76.21 = 107.78$ kV

∴ Most economical conductor diameter,

$$d = \frac{2V}{g_{max}} = \frac{2 \times 107.78}{5} = \textbf{43.1 mm.} \quad \textbf{(Ans.)}$$

Internal diameter of the sheath, $D = 2.718\ d$

$$= 2.718 \times 43.1 = \textbf{117.1 mm.} \quad \textbf{(Ans.)}$$

Example 16.7. *Show that for a given voltage V and maximum stress E_{max} in the dielectric of a single-core cable, the sheath diameter 'D' is a minimum when D : d = e where 'd' is conductor diameter and 'e' is the base of natural logarithms. Determine 'D' and 'd' for V = 10 kV and E_{max} = 23 kV/cm.* **(Panjab University)**

Solution. Maximum stress, $g_{max} = E_{max} = 23$ kV/cm $= 23 \times 10^5$ V/m (Assume r.m.s.)

Applied voltage, $V = 10$ kV $= 10000$ V

Now, $g_{max} = \frac{2V}{d}$ V/m ...[Eqn. (16.7)]

∴ Diameter of conductor, $d = \frac{2V}{g_{max}} = \frac{2 \times 10000}{23 \times 10^5} = 0.0087$ m $=$ **0.87 cm. (Ans.)**

Internal diameter of sheath, $D = 2.718 \times d = 2.718 \times 0.87 =$ **2.365 cm. (Ans.)**

16.9. GRADING OF CABLES

The *process of achieving uniformity in dielectric stress in dielectric of cables is known as* **grading of cables.**

In a cable with a single homogeneous layer of insulation, much of dielectric is being operated at a very much less than the maximum allowable stress. The grading of cables means the subdivision of a cable in such a way that (g_{max} ~ g_{min}) is minimised, with the result that less dielectric is required and overall diameter ($2R$) is *reduced*. It may be mentioned here that there is *little application of methods of grading because of constructional reasons.* However, we shall discuss here briefly to illustrate important basic principles.

The following are the *two methods of grading* :

1. Capacitance grading.
2. Intersheath grading.

1. Capacitance grading. In capacitance grading, the uniformity in dielectric stress is achieved by using various layers of different dielectrics in such a manner that the permittivity ε_r of any layer is inversely proportional to its radius or distance from the centre,

i.e., $\varepsilon_r \propto \frac{1}{x}$ or $\varepsilon_r x = \text{constant}$

or, $g \propto \frac{Q}{2\pi \varepsilon_0}$ which is constant.

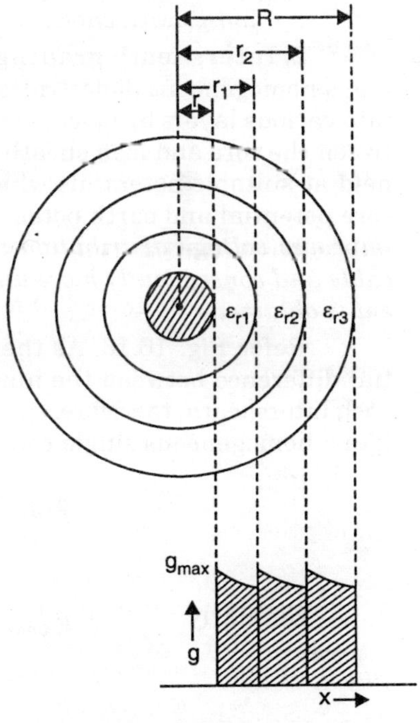

Fig. 16.15. Capacitance grading and voltage distribution.

Thus, if such a condition is satisfied the value of dielectric stress at any point is constant and is independent of distance from the centre and the grading will be ideal one. However, it is not possible to use infinite number of dielectrics for a single cable and in practice, two or three dielectrics are used in the decreasing order of permittivity ; the dielectric of highest permittivity being used near the core.

Let there be three layers of dielectrics of outer radii r_1, r_2 and R and relative permittivities ε_{r1}, ε_{r2} and ε_{r3}. If the permittivities are such that $\varepsilon_{r1} > \varepsilon_{r2} > \varepsilon_{r3}$ and if the maximum stress to which each dielectric is subjected is constant then

$$\varepsilon_{r1}\, r = \varepsilon_{r2} \,.\, \varepsilon r_1 = \varepsilon_{r3} \,.\, r_2$$

Potential difference across the inner layer,

$$V_1 = \int_r^{r_1} g dx = \int_r^{r_1} \frac{Q}{2\pi\, \varepsilon_0 \varepsilon_{r1}\, x}\, dx$$

$$= \frac{Q}{2\pi\, \varepsilon_0\varepsilon_{r1}} \ln\left(\frac{r_1}{r}\right) = g_{max}\, r \ln\left(\frac{r_1}{r}\right) \qquad \left[\because \frac{Q}{2\pi\, \varepsilon_0\varepsilon_{r1}} = {}^*g_{max} \,.\, r\right]$$

Similarly, potential difference across second layer (V_2) and third layer (V_3) is given by

$$V_2 = g_{max}\, r_1 \ln\left(\frac{r_2}{r_1}\right) \quad \text{and} \quad V_3 = g_{max}\, r_2 \ln\left(\frac{R}{r_2}\right)$$

Total potential difference between core and earthed sheath,

$$V = V_1 + V_2 + V_3$$

$$= g_{max}\left[r \ln\left(\frac{r_1}{r}\right) + r_1 \ln\left(\frac{r_2}{r_1}\right) + r_2 \ln\left(\frac{R}{r_2}\right)\right].$$

- The *advantage* of grading is that if the overall diameter is same for a non-graded and graded cable, then permissible safe potential of the cable is increased *i.e.* for the same permissible safe potential, the *size of the graded cable will be smaller than that of the non-graded cable.*

2. Intersheath grading. In intersheath grading, a homogeneous dielectric is used, but it is divided into various layers by placing metallic intersheath between the core and lead sheath. The intersheaths are held at suitable potentials which are in between the core potential and earth potential. This arrangement *improves voltage distribution in the dielectric of the cable and consequently more uniform potential gradient is obtained.*

Refer Fig. 16.16. As there is a definite potential difference between the inner and outer layers of each intersheath, therefore, each sheath can be treated like a homogeneous single-core cable and we have

$$g_{1max} = \frac{V_1}{r \ln\left(\frac{r_1}{r}\right)}$$

Similarly,

$$g_{2max} = \frac{V_2}{r_1 \ln\left(\frac{r_2}{r_1}\right)}$$

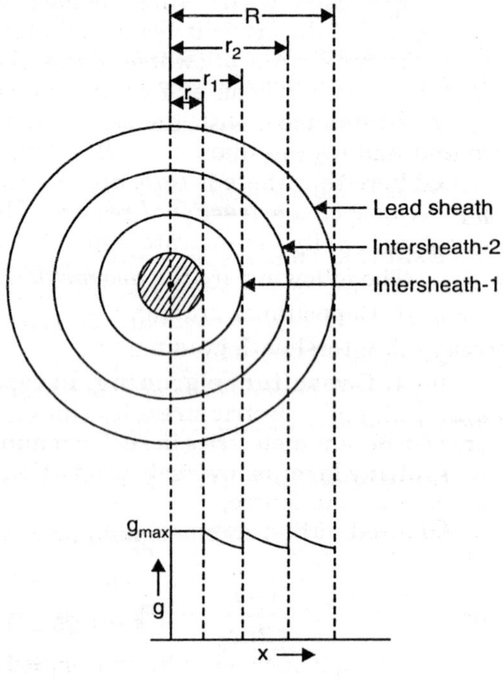

Fig. 16.16. Intersheath grading.

$${}^*g_{max} = \frac{Q}{2\pi\, \varepsilon_0\varepsilon_{r1}\, r}, \quad \therefore \quad g_{max} r = \frac{Q}{2\pi\, \varepsilon_0\varepsilon_{r1}}.$$

and,
$$g_{3\,max} = \frac{V_3}{r_2 \ln\left(\dfrac{R}{r_2}\right)}$$

But for a homogeneous dielectric we have

$$g_{1\,max} = g_{2\,max} = g_{3\,max}$$

$$\therefore \quad \frac{V_1}{r \ln\left(\dfrac{r_1}{r}\right)} = \frac{V_2}{r_1 \ln\left(\dfrac{r_2}{r_1}\right)} = \frac{V_3}{r_2 \ln\left(\dfrac{R}{r_2}\right)} = g_{max} \quad \text{(say)}$$

Since the cable is, in effect, a series of capacitors in series, therefore all these potential differences are in phase and we have

$$V = V_1 + V_2 + V_3$$

The following points in respect of grading are *worth noting*.

- Grading is only useful for very high voltage cables for which the ratio $\frac{R}{r}$ is large (for $R/r = 2$ and $R/r = 5$ the grading will increase the permissible safe working voltage by 30% and 100% respectively).

With "*capacitance grading*" there is the practical difficulty of *obtaining materials having the required values of permittivities.* A second difficulty is the *possibility of change of permittivity with time.* Such a change could completely *alter the distribution of dielectric stress and might even lead to breakdown at the normal working voltage.*

With the "*intersheath method*" there is a possibility of *damage to intersheath* during laying of the cable since the intersheath has to be *thin.* The intersheath also has to carry a *charging current* which in a very long cable may be sufficient to *cause overheating.*

In view of the above reasons the present day tendency is to *avoid grading in favour of oil-filled and gas-filled cables.*

Note. While calculating the safe potential, the *peak values of voltages are to be taken and not the r.m.s. values.*

Example 16.8. *A single-core lead sheathed cable is graded by using three dielectrics of relative permittivities of 4, 3 and 2 respectively. The conductor diameter is 2.5 cm and overall diameter is 10 cm. If the three dielectrics are worked at the same maximum stress of 40 kV/cm, find the safe working voltage of the cable.*

What will be the value of safe working voltage for an ungraded cable, assuming the same conductor and overall diameter and maximum dielectric stress ?

Solution. *Given :* $\varepsilon_{r_1} = 4$; $\varepsilon_{r_2} = 3$; $\varepsilon_{r_3} = 2$; $r = \frac{2.5}{2} = 1.25$ cm ; $R = \frac{10}{2} = 5$ cm ; $g_{max} = 40$ kV/cm.

Graded cable. As the maximum stress in the three dielectrics is the same,

$$\varepsilon_{r_1} \cdot r = \varepsilon_{r_2} \cdot r_1 = \varepsilon_{r_3} \cdot r_2$$

$$4 \times 1.25 = 3 \times r_1 = 2 \times r_2$$

$$\therefore \quad r_1 = 1.67 \text{ cm asnd } r_2 = 2.5 \text{ cm}$$

Permissible peak voltage for the cable

$$= g_{max}\left[r \ln\left(\frac{r_1}{r}\right) + r_1 \ln\left(\frac{r_2}{r_1}\right) + r_2 \ln\left(\frac{R}{r_2}\right)\right]$$

$$= 40\left[1.25 \ln\left(\frac{1.67}{1.25}\right) + 1.67 \ln\left(\frac{2.5}{1.67}\right) + 2.5 \ln\left(\frac{5}{2.5}\right)\right]$$

$$= 40(0.362 + 0.674 + 1.733) = 110.76 \text{ kV}$$

∴ Safe working voltage (r.m.s.) for cable $= \frac{110.76}{\sqrt{2}}$ **= 78.3 kV. (Ans.)**

Ungraded cable

Permissible voltage for the cable

$$= g_{max}\, r \ln\left(\frac{R}{r}\right) = 40 \times 1.25 \ln\left(\frac{5}{1.25}\right) = 69.3 \text{ kV}$$

∴ Safe working voltage (r.m.s.) for cable $= \frac{69.3}{\sqrt{2}}$ **= 49 kV. (Ans.)**

From above results we find that for the same conductor diameter and the same overall dimension the graded cable can be operated at a voltage 29.3 (78.3 – 49) kV higher than the homogeneous ungraded cable–an increase of 59.8%.

Example 16.9. *The inner conductor of a concentric cable has a diameter of 3 cm, the diameter over the insulation being 9 cm. The cable is insulated with two materials having permittivities of 5 and 4 respectively with corresponding safe working stresses of 30 kV/cm and 20 kV/cm. Determine :*

(i) The radial thickness of each insulation.

(ii) The safe working voltage of the cable.

Solution. *Given :* $r = \frac{3}{2} = 1.5$ cm ; $R = \frac{9}{2} = 4.5$ cm ; $\varepsilon_{r_1} = 5$; $\varepsilon_{r_2} = 4$; $g_{1\,max} = 30$ kV/cm ; $g_{2\,max} = 20$ kVcm.

(*i*) **The radial thickness of each insulation :**

Now, $g_{1max} \propto \frac{1}{\varepsilon_{r_1} . r}$; $g_{2\,max} \propto \frac{1}{\varepsilon_{r_2} . r_1}$

∴ $\frac{g_{1max}}{g_{2max}} = \frac{\varepsilon_{r_2} . r_1}{\varepsilon_{r_1} . r}$ or $r_1 = \frac{g_{1max}}{g_{2max}} \times \frac{\varepsilon_{r_1} . r}{\varepsilon_{r_2}} = \frac{30}{20} \times \frac{5 \times 1.5}{4} = 2.81$ cm

∴ Radial thickness of inner dielectric

$= r_1 - r = 2.81 - 1.5 =$ **1.31 cm. (Ans.)**

Radial thickness of outer dielectric

$= R - r_1 = 4.5 - 2.81 =$ **1.69 cm. (Ans.)**

(*ii*) **The safe working voltage of the cable :**

The permissible peak voltage of the cable

$$= g_{1max}\, r \ln\left(\frac{r_1}{r}\right) + g_{2\,max}\, r_1 \ln\left(\frac{R}{r_1}\right)$$

$$= 30 \times 1.5 \ln\left(\frac{2.81}{1.5}\right) + 20 \times 2.81 \ln\left(\frac{4.5}{2.81}\right)$$

$$= 28.25 + 26.46 = 54.71 \text{ kV}$$

Safe working voltage (r.m.s.) $= \frac{54.71}{\sqrt{2}}$ **= 38.68 kV. (Ans.)**

Example 16.10. *A 66 kV (3- phase system) single-core lead sheathed cable has a conductor of 2 cm diameter and two layers of different materials each 1 cm thick. The relative permittivities are 5 (inner) and 3 (outer). Calculate the maximum stress in the two dielectrics.* **(Panjab University)**

Sol. *Given :* Supply voltage (*r.m.s.*) = 66 kV ; $r = \frac{2}{2} = 1$ cm ; $t = 1$ cm ; $r_1 = r + t = 1 + 1 = 2$ cm ; $R = r_1 + t = 2 + 1 = 3$ cm ; $\varepsilon_{r_1} = 5$; $\varepsilon_{r_2} = 3$.

Potential gradients at the surface and other points :

Fig. 16.17 shows the composite dielectric of a capacitance graded cable. The potential difference V between conductor and earthed sheath is given by,

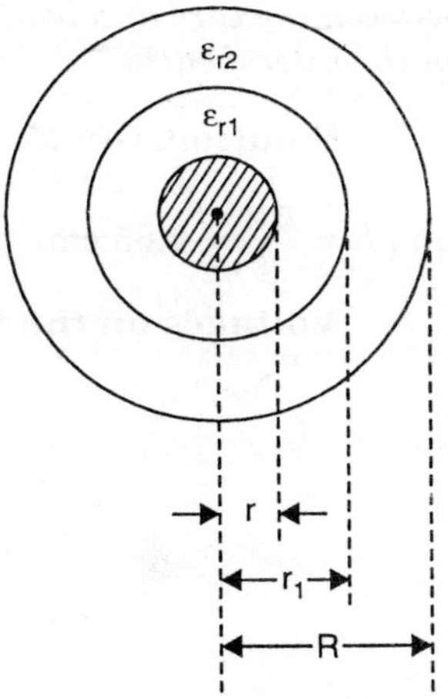

Fig. 16.17

$$V = \int_r^{r_1} g_1 dx + \int_{r_1}^{R} g_2 dx$$

$$= \int_r^{r_1} \frac{Q}{2\pi\varepsilon_0 \varepsilon_{r_1} x} dx + \int_{r_1}^{R} \frac{Q}{2\pi\varepsilon_0 \varepsilon_{r_2} x} dx$$

$$= \frac{Q}{2\pi\varepsilon_0}\left[\frac{1}{\varepsilon_{r_1}} \ln\left(\frac{r_1}{r}\right) + \frac{1}{\varepsilon_{r_2}} \ln\left(\frac{R}{r_1}\right)\right] \quad ...(i)$$

Also, $$g_{1max} = \frac{Q}{2\pi\varepsilon_0 \varepsilon_{r_1} r} \quad ...(ii)$$

Putting the value of $Q = g_{1\,max}\, 2\pi\varepsilon_0 \varepsilon_{r_1} r$ from (*ii*) in (*i*), we get

$$V = \frac{g_{1max}\, 2\pi\varepsilon_0 \varepsilon_{r_1} r}{2\pi\varepsilon_0}\left[\frac{1}{\varepsilon_{r_1}} \ln\left(\frac{r_1}{r}\right) + \frac{1}{\varepsilon_{r_2}} \ln\left(\frac{R}{r_1}\right)\right]$$

$$= g_{1max}\, \varepsilon_{r_1} r \left[\frac{1}{\varepsilon_{r_1}} \ln\left(\frac{r_1}{r}\right) + \frac{1}{\varepsilon_{r_2}} \ln\left(\frac{R}{r_1}\right)\right]$$

or, $$g_{1max} = \frac{V}{r\left[\ln\left(\frac{r_1}{r}\right) + \frac{\varepsilon_{r_1}}{\varepsilon_{r_2}} \ln\left(\frac{R}{r_1}\right)\right]}$$

Substituting the values, we get

$$\mathbf{g_{1max}} = \frac{\frac{66}{\sqrt{3}} \times \sqrt{2}}{1\left[\ln\left(\frac{2}{1}\right) + \frac{5}{3} \ln\left(\frac{3}{2}\right)\right]} = \mathbf{39.36\ kV/cm.} \quad \textbf{(Ans.)}$$

Similarly, it can be proved that ;

$$^*g_{2max} = \frac{V}{r_1\left[\frac{\varepsilon_{r_2}}{\varepsilon_{r_1}} \ln\left(\frac{r_1}{r}\right) + \ln\left(\frac{R}{r_1}\right)\right]} \quad ...(iii)$$

$^*g_{2max} = \frac{Q}{2\pi\varepsilon_0 \varepsilon_{r2} r_1}$. Putting the value of $Q = g_{2max} \times 2\pi\varepsilon_0 \varepsilon_{r_2} r_1$ in (*i*), we get the expression (*iii*).

or $$g_{2max} = \frac{\frac{66}{\sqrt{3}} \times \sqrt{2}}{2\left[\frac{3}{5}\ln\left(\frac{2}{1}\right) + \ln\left(\frac{3}{2}\right)\right]} = \textbf{32.8 kV/cm.} \quad \textbf{(Ans.)}$$

Example 16.11. *A 66 kV, 3-phase system uses a single-core cable of conductor diameter 2 cm and lead sheath of diameter 5.3 cm. Two intersheaths of diameters 3.1 cm and 4.2 cm are introduced between the core and lead sheath. If the maximum stress in the layers is the same, find the voltages on the intersheaths.*

Solution. *Given :* $V = \sqrt{2} \times \frac{66}{\sqrt{3}} = 53.9$ kV ; $r = \frac{2}{2} = 1$ cm ; $r_1 = \frac{3.1}{2} = 1.55$ cm ; $r_2 = \frac{4.2}{2} = 2.1$ cm ; $R = \frac{5.3}{2} = 2.65$ cm.

Voltages on the intersheaths

$$g_{1max} = \frac{V_1}{r\ln\left(\frac{r_1}{r}\right)} = \frac{V_1}{1 \times \ln\left(\frac{1.55}{1}\right)} = 2.28V_1$$

$$g_{2max} = \frac{V_2}{r_1\ln\left(\frac{r_2}{r_1}\right)} = \frac{V_2}{1.55\ln\left(\frac{2.1}{1.55}\right)} = 2.12V_2$$

$$g_{3max} = \frac{V_3}{r_2\ln\left(\frac{R}{r_2}\right)} = \frac{V_3}{2.1\ln\left(\frac{2.65}{2.1}\right)} = 2.05\ V_3$$

As the maximum stress in this layers is same (*given*),

∴ $$g_{1max} = g_{2max} = g_{3max}, \quad \text{or,} \quad 2.28V_1 = 2.12V_2 = 2.05V_3$$

or, $$V_2 = \frac{2.28V_1}{2.12} = 1.075V_1$$

and, $$V_3 = \frac{2.28V_1}{2.05} = 1.112V_1$$

Also, $$V_1 + V_2 + V_3 = V$$

∴ $$V_1 + 1.075V_1 + 1.112V_1 = 53.9 \quad \text{or,} \quad V_1 = 16.91 \text{ kV}$$

and, $$V_2 = 1.075 \times 16.91 = 18.18V$$

∴ Voltage on the *intersheath-1* (near to the core)

$$= V - V_1 = 53.9 - 16.91 = \textbf{36.99 kV.} \quad \textbf{(Ans.)}$$

Voltage on the *intersheath-2*

$$= V - V_1 - V_2 = 53.9 - 16.91 - 18.18 = \textbf{18.81 kV.} \quad \textbf{(Ans.)}$$

Example 16.12. *A single-core 66 kV cable working on 3-phase system has a conductor diameter of 2 cm and a sheath of inside diameter 5.3 cm. If two intersheaths are introduced in such a way that the stress varies between the same maximum and minimum in the three layers, determine :*

(i) Positions of intersheaths,

(ii) Voltage on the intersheaths, and

(iii) Maximum and minimum stresses. **(Roorkee University)**

Solution. *Given :* $r = \frac{2}{2} = 1$ cm ; $R = \frac{5.3}{2} = 2.65$ cm ; $V = \sqrt{2} \times \frac{66}{\sqrt{3}} = 53.9$ kV (max.)

(*i*) **Positions of intersheaths.** Let the radii of the intersheaths are r_1 and r_2 cm respectively. Let V_1, V_2 and V_3 respectively be the voltage between conductor and intersheath-1, between intersheath-1 and intersheath-2 and between inter sheath-2 and outer lead sheath.

Now, $$g_{1max} = \frac{V_1}{r \ln\left(\frac{r_1}{r}\right)}; \quad g_{2max} = \frac{V_2}{r_1 \ln\left(\frac{r_2}{r_1}\right)}; \quad g_{3\,max} = \frac{V_3}{r_2 \ln\left(\frac{R}{r_2}\right)}$$

As the maximum stress in three layers is the same (given),

$\therefore$ $$g_{1max} = g_{2max} = g_{3max}$$

$$\frac{V_1}{r \ln\left(\frac{r_1}{r}\right)} = \frac{V_2}{r_1 \ln\left(\frac{r_2}{r_1}\right)} = \frac{V_3}{r_2 \ln\left(\frac{R}{r_2}\right)} \quad ...(i)$$

In order that stress may vary between the same maximum and minimum in the three layers, we have

$$\frac{r_1}{r} = \frac{r_2}{r_1} = \frac{R}{r_2} \quad ...(ii)$$

$\therefore$ $$\frac{V_1}{r} = \frac{V_2}{r_1} = \frac{V_3}{r_2} \quad ...^*(iii)$$

From eqn. (*ii*), we get

$$r_1^2 = r\,r_2 = r_2 \qquad (\because \; r = 1 \text{ cm})$$

and, $$r_1 r_2 = Rr = 2.65 \times 1 = 2.65 \text{ cm}$$

or, $$r_1^3 = 2.65 \text{ cm} \quad \text{or} \quad \mathbf{r_1 = 1.38 \text{ cm.}} \quad \textbf{(Ans.)}$$

and, $$\mathbf{r_2} = r_1^2 = (1.38)^2 = \mathbf{1.9 \text{ cm.}} \quad \textbf{(Ans.)}$$

(*ii*) **Voltage on intersheaths**

$$V = V_1 + V_2 + V_3$$

$$53.9 = V_1 + \frac{r_1}{r} V_1 + \frac{r_2}{r} V_1 \quad ...[\text{From eqn. } (iii)]$$

$$= V_1\left(1 + \frac{1.38}{1} + \frac{1.9}{1}\right) = 4.28 V_1$$

$\therefore$ $$V_1 = \frac{53.9}{4.28} = 12.6 \text{ kV}$$

$$V_2 = \frac{r_1}{r} \times V_1 = \frac{1.38}{1} \times 12.6 = 17.39 \text{ kV}$$

$\therefore$ *Voltage on Intersheath-1* $= V - V_1 = 53.9 - 12.6 =$ **41.3 kV max. (Ans.)**

Voltage on Intersheath-2 $= V - V_1 - V_2 = 53.9 - 12.6 - 17.39 =$ **23.91 kV max. (Ans.)**

(*iii*) **Maximum and minimum stresses :**

Maximum stress $$= \frac{V_1}{r \ln\left(\frac{r_1}{r}\right)} = \frac{12.6}{1 \times \ln\left(\frac{1.38}{1}\right)} = \mathbf{39.12 \text{ kV/cm.}} \quad \textbf{(Ans.)}$$

Minimum stress $$= \frac{V_1}{r_1 \ln\left(\frac{r_1}{r}\right)} = \frac{12.6}{1.38 \times \ln\left(\frac{1.38}{1}\right)} = \mathbf{28.35 \text{ kV/cm.}} \quad \textbf{(Ans.)}$$

---

*This eqn. is obtained if the values of eqn. (*ii*) are substituted in eqn. (*i*).

Example 16.13. *An 85-kV, single-core metal-sheathed cable is to be graded by means of a metallic intersheath such that overall diameter of the cable is a minimum. The insulating material can be worked at 55 kV per cm. Find the diameter of the intersheath and the voltage at which it must be maintained.*

Compare the conductor and outside diameters of this cable with those of an ungraded cable employing the same insulating material and working under the same conditions. **(N.U)**

Solution. Let, r = Radius of the conductor,

r_1 = Inside radius of the intersheath,

R = Outside radius of the intersheath,

V_1 = Voltage at which intersheath is maintained, and

V = Voltage at which the conductor is maintained.

Considering the conductor and intersheath radii, we have

$$r_1 = 2.718r \quad \text{...}(i) \quad \text{[Eqn. (16.6)]}$$

For the conductor and intersheath maximum stress occurs at a point distant r from the centre and its value is

$$g_{max} = \frac{V - V_1}{r \ln\left(\frac{r_1}{r}\right)} \quad \text{...}(ii)$$

Similarly the maximum stress between the intersheath and metal sheath occurs at the intersheath surface, its value being,

$$g_{max} = \frac{V_1}{r \ln\left(\frac{R}{r_1}\right)} \quad \text{...}(iii)$$

and *both these values are equal.*

$\therefore$ From eqns. (i) and (ii),

$$g_{max} = \frac{V - V_1}{r \ln\left(\frac{r_1}{r}\right)} = \frac{2.718\,(V - V_1)}{r_1 \ln\left(\frac{r_1}{r}\right)}$$

or,

$$r_1 = \frac{2.718\,(V - V_1)}{g_{max} \ln\left(\frac{r_1}{r}\right)} = \frac{2.718\,(V - V_1)}{g_{max} \ln\,(2.718)} = \frac{2.718\,(V - V_1)}{g_{max}} \quad \text{...}(iv)$$

$$\left[\because \frac{r_1}{r} = 2.718 \text{ from } (i) \text{ and } \ln\,(2.718) = 1\right]$$

Substituting the value of r_1 from (iv) in (iii), we have

$$g_{max} = \frac{V_1}{r_1 \ln\left(\frac{R}{r_1}\right)} = \frac{V_1}{\frac{2.718\,(V - V_1)}{g_{max}} \times \ln\left[\frac{R}{\{2.718\,(V - V_1)/g_{max}\}}\right]}$$

or,

$$g_{max} = \frac{V_1 \times g_{max}}{2.718\,(V - V_1) \times \ln\left[\frac{R\,g_{max}}{2.718\,(V - V_1)}\right]} \quad \text{...}(v)$$

or,

$$\frac{V_1}{2.718\,(V - V_1)} = \ln\left[\frac{R\,g_{max}}{2.718\,(V - V_1)}\right]$$

or,

$$Rg_{max} = 2.718(V - V_1)\, e^{\{V_1/2.718\,(V - V_1)\}} \quad \text{...}(vi)$$

In eqn. (*vi*) g_{max} and V are constants. For minimum value of R (the overall radius) $\frac{\partial R}{\partial V_1}$ should be zero. This gives

$$\frac{\partial R}{\partial V_1} = \frac{2.718}{g_{max}} \frac{\partial}{\partial V_1}\left[(V - V_1)\, e^{\{V_1/2.718\,(V - V_1)\}}\right]$$

$$= \frac{2.718}{g_{max}}\left[(V - V_1) \times e^{\{V_1/2.718\,(V - V_1)\}} \times \frac{2.718\,(V - V_1) + 2.718\, V_1}{\{2.718\,(V - V_1)\}^2} + \left\{e^{\{V_1/2.718\,(V - V_1)\}}\right\} \times (-1)\right] = 0$$

Simplifying we get, $$\frac{V}{2.718\,(V - V_1)} = 1$$

or, $$V = 2.718(V - V_1) = 2.718V - 2.718V_1$$

or, $$V_1 = \frac{1.718}{2.718} = 0.632V \qquad ...(vii)$$

From eqn. (*ii*), $$g_{max} = \frac{V - V_1}{r \ln\left(\frac{r_1}{r}\right)} = \frac{V - V_1}{r \ln (2.718)} = \frac{V - V_1}{r}$$

$\therefore$ $$r = \frac{V - V_1}{g_{max}} = \frac{V - 0.632\,V}{g_{max}} = \frac{0.368}{g_{max}}$$

From eqn. (*iv*), $$r_1 = \frac{2.718\,(V - V_1)}{g_{max}} = \frac{2.718\,(V - 0.632V)}{g_{max}} = \frac{V}{g_{max}}$$

From eqn. (*vi*), $$R = \frac{2.718\,(V - V_1)}{g_{max}}\, e^{\{V_1/2.718\,(V - V_1)\}}$$

$$= \frac{2.718 \times 0.368V}{g_{max}}\, e^{V_1/V} = \frac{V}{g_{max}}\, e^{0.632} = 1.88\,\frac{V}{g_{max}}$$

Further when the cable is ungraded,

$$R = er = 2.718r$$

and, $$g_{max} = \frac{V}{r \ln\left(\frac{R}{r}\right)} = \frac{V}{r \ln (2.718)} = \frac{V}{r}$$

Finally substituting the given values, we have

Intersheath potential, $\mathbf{V_1} = 0.632V = 0.632 \times 85 =$ **53.72 kV** to **neutral. (Ans.)**

Core radius, $r = 0.736\,\frac{V}{g_{max}} = 0.736 \times \frac{85}{55} = 1.14$ cm

Intersheath radius, $\mathbf{r_1} = \frac{V}{g_{max}} = \frac{85}{55} =$ **1.545 cm. (Ans.)**

Overall radius, $\mathbf{R} = 1.88 \times \frac{85}{55} =$ **2.905 cm. (Ans.)**

Ungraded cable :

Core radius, $\mathbf{r} = \frac{V}{g_{max}} = \frac{85}{55} =$ **1.545 cm. (Ans.)**

Overall radius, $\mathbf{R} = 2.718r = 2.718 \times 1.545 =$ **4.2 cm. (Ans.)**

16.10. CAPACITANCE OF 3-CORE CABLES

The capacitance of cable is of greater importance than that of an overhead line of the same length due to the *narrow spacing between the conductors themselves and between the conductors and earthed sheath* and due to *their separation by a dielectric medium of higher permittivity as compared to air.*

Since potential difference exists between pairs of conductors and between each conductor and the sheath, electrostatic fields are set up in the cable and they give rise to core-core capacitances C_c and core-sheath capacitances C_s as shown in Fig.16.18 (*i*).

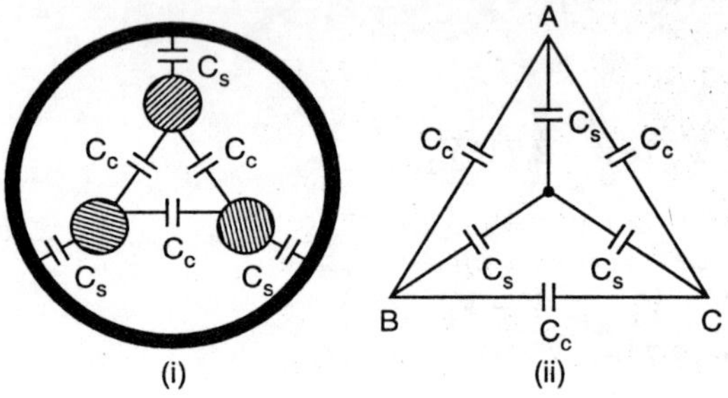

Fig. 16.18

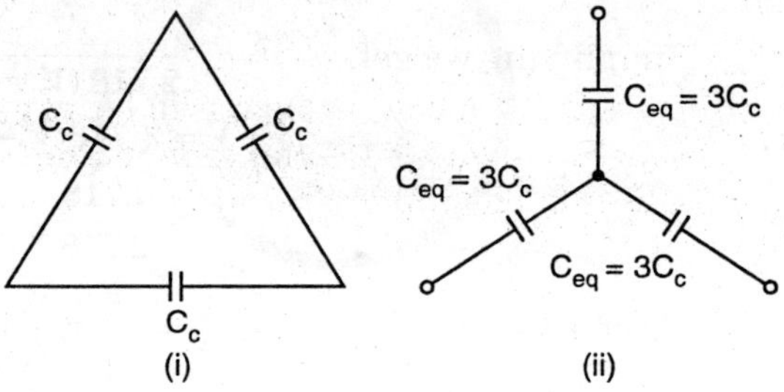

Fig. 16.19

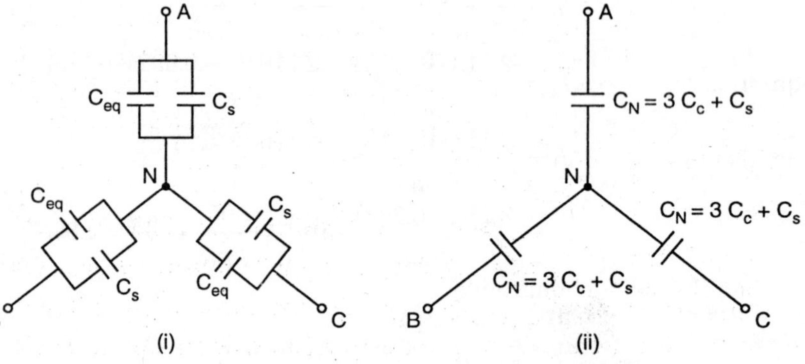

Fig. 16.20

The three C_c are delta connected whereas three C_s are the star connected, the sheath forming the star point, as shown in Fig. 16.18 (*ii*). The lay of a belted cable makes it reasonable to assume equality of each C_c and each C_s.

The delta-connected capacitances (Fig. 16.19 (*i*)] can be converted into equivalent star-connected capacitances, as shown in Fig. 16.19 (*ii*). It can be easily *shown that $C_{eq} = 3C_c$.

The star point may be assumed to be at zero potential and if the sheath is also at zero potential, the capacitance of each core to neutral [Fig. 16.20 (*i*) and (*ii*)],

$$C_N = C_{eq} + C_s = 3C_c + C_s$$

If V_{ph} is the phase voltage, then charging current I_C is given by,

$$I_C = \frac{V_{ph}}{\text{Capacitive reactance per phase}} = \frac{V_{ph}}{\dfrac{1}{2\pi f C_N}}$$

or

$$I_C = 2\pi f V_{ph} C_N = 2\pi f V_{ph}\,(3C_c + C_s) \qquad \text{...(16.8)}$$

Measurement of C_c and C_s. The values of C_c and C_s may be found by actual measurement as follows :

(*i*) The three cores are bunched together (*i.e.* commoned) and the capacitance is measured between the bunched cores and the sheath [Fig. 16.21 (*i*)]. The bunching eliminates all the three capacitances C_c leaving the three capacitances C_s in parallel. Thus, if C_1 is the measured capacitance, this test yields

$$C_1 = 3C_s \quad \text{or} \quad C_S = \frac{C_1}{3} \qquad ...(i)$$

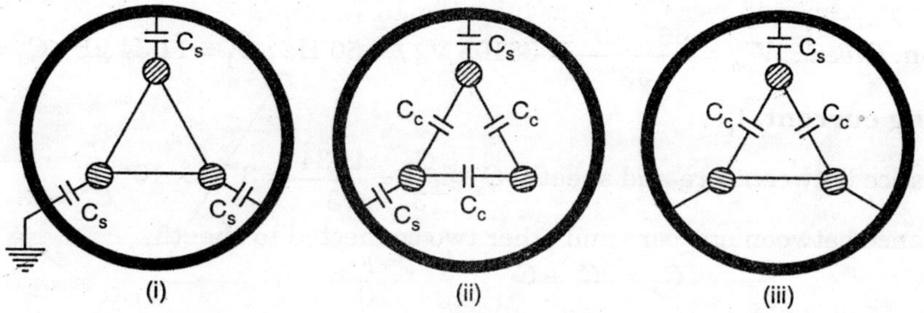

Fig. 16.21. Determination of C_c and C_s.

(*ii*) Capacitance between the two cores or lines is measured with the third core being either insulated or connected to the sheath [Fig. 16.21 (*ii*)]. This eliminates one of the capacitors C_s and measured value C_2 gives $\frac{C_N}{2}$ directly,

i.e., $$C_2 = C_c + \frac{C_c}{2} + \frac{C_s}{2}, \quad \text{or,} \quad C_2 = \frac{1}{2}(3C_c + C_s) = \frac{1}{2}C_N \qquad ...(ii)$$

(*iii*) The capacitance (C_3) is measured between the two cores bunched together and the third core [Fig. 16.21 (*iii*)]. This gives,

$$C_3 = 2C_c + C_s \qquad ...(iii)$$

From the above expressions the values of C_c and C_s can be determined.

It is very difficult to determine the capacitance C_N of the cable from the geometry of the figure. However the empirical formula given below may be used for determining the value of C_N.

$$C_N = \frac{0.03\,\varepsilon_r}{\ln\left[1 + \frac{T+t}{d}\left(3.84 - 1.7\frac{t}{T} + 0.52\frac{t^2}{T^2}\right)\right]}\ \mu\text{F/km} \qquad ...(16.9)$$

where, d = Diameter of conductor,

t = Thickness of belt insulation, and

T = Thickness of conductor insulation.

Example 16.14. *A 3-core, 3-phase belted cable tested for capacity between a pair of cores on single-phase, with the third core earthed, gave a capacity of 0.3 μF per km. Calculate the charging current for 20 km length of this cable when connected to 22 kV, 3-phase, 50 Hz supply.* **(AMIE)**

Solution. Capacitance between a pair of cores on single phase with the third core earthed = 0.3 μF/km

Capacitance between a pair of cores for 20 km length,

$$C_2 = \frac{1}{2}C_N = 20 \times 0.3 = 6 \times 10^{-6}\ \text{F}$$

or, Capacitance of each core to neutral, $C_N = 2 \times 6 \times 10^{-6} = 12 \times 10^{-6}$ F

Phase voltage, $$V_{ph} = \frac{22000}{\sqrt{3}} = 12701.7\ \text{V}$$

Supply frequency, $f = 50$ Hz

∴ Charging current per phase,

$$I_C = 2\pi f V_{ph} C_N = 2\pi \times 50 \times 12701.7 \times 12 \times 10^{-6} = \mathbf{47.88\ A.}\quad \textbf{(Ans.)}$$

Example 16.15. *The capacitances of a 3-phase belted cable are 11.34 μF between the three cores bunched together and the lead sheath and 6.66 μF between one core and other two connected to sheath. Find the charging current drawn by the cable when connected to 66 kV, 50 Hz supply.*

Solution. *Given :* $V_{ph} = \frac{66 \times 10^3}{\sqrt{3}} = 38105$ V ; $f = 50$ Hz ; $C_1 = 11.34$ μF ; $C_3 = 6.66$ μF.

Charging current, I_C :

Capacitance between core and sheath, $C_s = \frac{C_1}{3} = \frac{11.34}{3} = 3.78 \times 10^{-6}$ F

Capacitance between one core and other two connected to sheath,

$$C_3 = 2C_c + C_s$$
$$= 6.66 \times 10^{-6}\ \text{F}$$

or,
$$C_C = \frac{C_3 - C_s}{2} = \frac{6.66 \times 10^{-6} - 3.78 \times 10^{-6}}{2} = 1.44 \times 10^{-6}\ \text{F}$$

Capacitance of each core to neutral,

$$C_N = 3C_c + C_s = 3 \times 1.44 \times 10^{-6} + 3.78 \times 10^{-6} = 8.1 \times 10^{-6}\ \text{F}$$

∴ Charging current, $I_C = 2\pi f V_{ph} C_N = 2\pi \times 50 \times 38105 \times 8.1 \times 10^{-6} = \mathbf{96.96\ A.}$ **(Ans.)**

Example 16.16. *Calculate the kVA taken by a 15 km long, 3-phase, 3-core cable, if the capacitance measured between any two cores is 0.2 μF/km when it is connected to 10 kV, 50 Hz bus-bar.*

Solution. Capacitance between any two cores for 15 km length,

$$C_2 = \frac{1}{2} C_N = 0.2 \times 15 = 3\ \mu\text{F} = 3 \times 10^{-6}\ \text{F}$$

or,
$$C_N = 2 \times 3 \times 10^{-6} = 6 \times 10^{-6}\ \text{F}$$

Phase voltage, $V_{ph} = \frac{10000}{\sqrt{3}} = 5773.5$ V

Supply frequency, $f = 50$ Hz

∴ kVA taken by the cable
$$= \frac{3V_{ph}I_C}{1000} = \frac{3\,V_{ph} \times 2\pi f\, V_{ph}\, C_N}{1000}$$
$$= \frac{3 \times 5773.5 \times 2\pi \times 50 \times 5773.5 \times 6 \times 10^{-6}}{1000}$$
$$= \mathbf{188.5\ kVA.}\quad \textbf{(Ans.)}$$

16.11. HEATING OF CABLES

The rise of temperature of a body depends upon the rate of generation and dissipation of heat by the body. If the rate of generation is greater than the rate of dissipation, the temperature goes on increasing and vice versa.

Generation of heat in the cables. The sources of heat generation within the cables are :

1. Copper losses (I^2R) in the conductors.
2. Dielectric losses in the cable insulation.
3. Sheath losses.

1. *Copper losses in the conductors.* In order to find out the copper losses the value of the resistance of the cable is calculated as follows :

(*i*) Resistance of conductor at an operating temperature of 65°C (assumed) is determined from the resistance given in standard tables (say at ambient temperature) from the following relation ;

$$R_{65} = R_{20}\,[1 + \alpha_{20}(65 - 20)]$$

where, α_{20} is the temperature coefficient of the conductor at 20°C.

(*ii*) As the effective area of section of the cable is smaller than the actual physical section, the effective resistance of the cable is larger. *A factor of 1.02 is multiplied to get the resistance.*

(*iii*) Since the length of the outermost strand is greater than the central strand, the effect of stranding on the resistance is obtained by *multiplying the resistance as calculated according to the length of the central by a factor of 1.02.*

After calculating thus the resistance of the cable the copper losses (core loss) are calculated as I^2R where I is the current carried by the cable.

2. *Dielectric losses :*

- The *insulation resistance* for a single core cable is given by,

$$R = \frac{\rho}{2\pi l}\ln\left(\frac{r_2}{r_1}\right)\,\Omega \qquad \text{...[Eqn. (16.10)]}$$

Here, ρ is the specific resistance and l is the length of the cable. Assuming ρ to remain constant, the impressed voltage V will send a current (in phase with V) through the insulation.

Capacitive current drawn from the supply, $I_C = \dfrac{V}{1/\omega C} = \omega CV$ where, C is the cable capacitance.

The charging current (I_C) leads V by 90°.

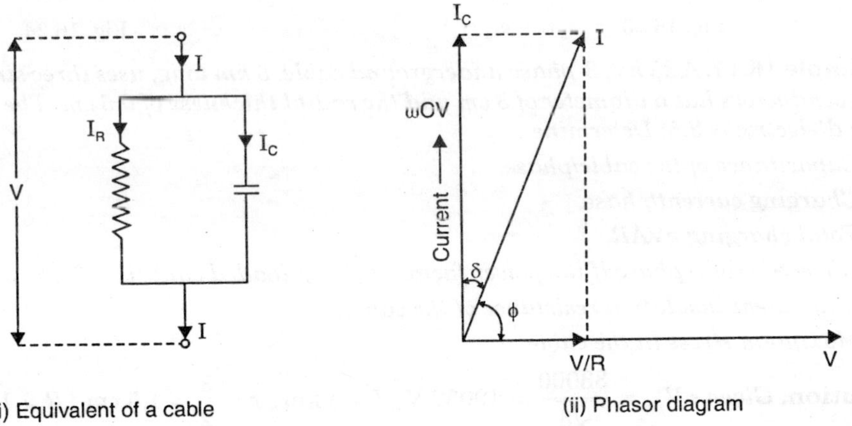

(i) Equivalent of a cable (ii) Phasor diagram

Fig. 16.22

The equivalent circuit with its phasor diagram is shown in Fig. 16.22. The resultant current I leads V by ϕ (≈ 90°) where cos ϕ gives the power factor of the cable.

But, $$\cos\phi = \cos\left(\frac{\pi}{2} - \delta\right) = \sin\delta = \tan\delta \qquad (\because\ \delta \text{ is very small})$$

The "*cable power factor*" thus can be expressed as,

$$\cos\phi = \sin\delta = \tan\delta = \frac{V/R}{\omega CV} = \frac{1}{\omega CR} \qquad \text{...(16.11)}$$

- *Dielectric losses are caused by dielectric absorption or polarisation.* It is small for voltages upto 33 kV, but for higher voltages has an increasingly important effect on current rating. *The loss in dielectric is due to loss in the equivalent leakage resistance.*

$$P = \frac{V^2}{R} \quad \text{...}(i)$$

Again, from phasor diagram, $\frac{V/R}{\omega CV} = \tan \delta$, or, $\frac{V}{R} = \omega CV \tan \delta$ ...(*ii*)

From (*i*) and (*ii*), we have

$$P = \omega CV^2 \tan \delta, \quad \text{or,} \quad P = 2\pi f CV^2 \tan \delta \quad \text{...(16.12)}$$

This shows that dielectric losses are *proportional to the square of the voltage.*

- Fig. 16.23 shows he variation of power factor with the temperature of a cable operating at normal voltage.
- Fig. 16.24 shows the variation of dielectric power factor as a function of electric stress.

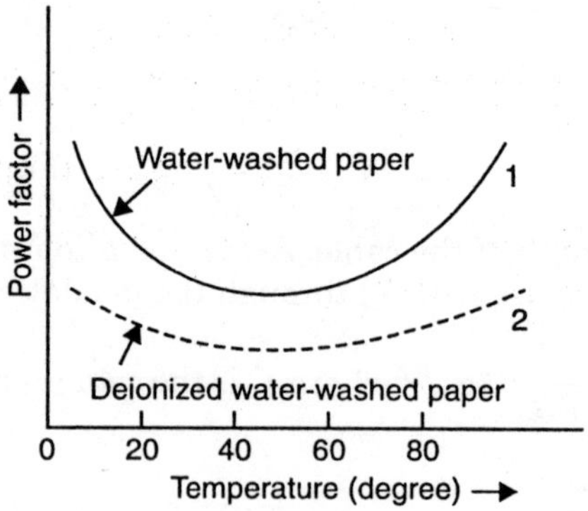

Fig. 16.23

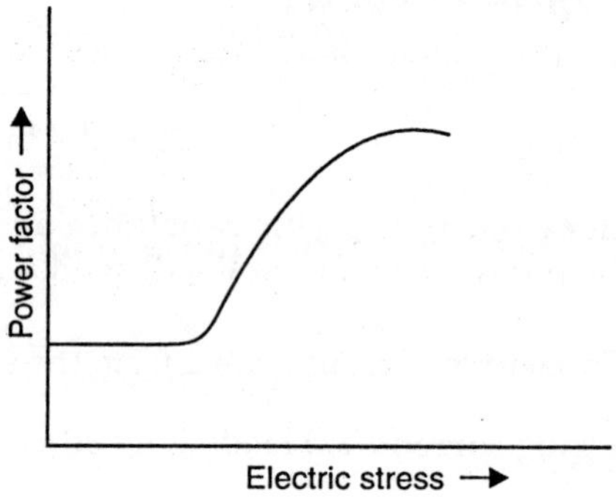

Fig. 16.24

Example 16.17. *A 33 kV, 3-phase underground cable, 5 km long, uses three single-core cables. Each of the conductors has a diameter of 3 cm and the radial thickness of 0.5 cm. The relative permittivity of the dielectric is 2.5. Determine :*

(*i*) *Capacitance of the cable/phase.*

(*ii*) *Charging current/phase.*

(*iii*) *Total charging kVAR.*

(*iv*) *Dielectric loss / phase if the power factor of the unloaded cable is 0.02.*

(*v*) *Equivalent insulation resistance of the cable*

(*vi*) *Maximum stress in the cable.*

Solution. *Given :* $V_{ph} = \frac{33000}{\sqrt{3}} = 19052$ V ; $l = 5$ km ; $r = \frac{3}{2} = 1.5$ cm ; $R = 1.5 + 0.5 = 2$ cm,

$\varepsilon_r = 2.5$; $\cos \phi = 0.02$

(*i*) **Capacitance of the cable/phase :**

$$C = \frac{2\pi\varepsilon_0\varepsilon_r}{\ln\left(\frac{R}{r}\right)} \text{ F/m} \quad \text{...[Eqn. (16.2)]}$$

$$= \frac{2\pi \times 8.854 \times 10^{-12} \times 2.5}{\ln\left(\frac{2}{1.5}\right)} = 483.44 \times 10^{-12} \text{ F/m}$$

For 5 km length, $C = 483.44 \times 10^{-12} \times 5 \times 1000 = \mathbf{2417.2 \times 10^{-9}}$ **F/phase. (Ans.)**

(*ii*) **Charging current, I_C:**

$$\therefore \quad I_C = \frac{V_{ph}}{X_C} = \frac{V_{ph}}{1/2\pi fC} = V_{ph} \times 2\pi fC$$

$$= 19052 \times 2\pi \times 50 \times 2417.2 \times 10^{-9} = \mathbf{14.47\ A/ph.} \quad \textbf{(Ans.)}$$

(*iii*) **Total charging kVAR:**

Total charging kVAR = $3\, V_{ph}\, I_C = 3 \times 19052 \times 14.47 \times 10^{-3}$ kVAR = **827 kVAR. (Ans.)**

(*iv*) **Dielectric loss phase :**

$$\cos \phi = 0.02 \quad (Given)$$

$$\therefore \quad \phi = \cos^{-1}(0.02) = 88.85°$$

i.e. loss angle $\delta = 90° - \phi = 90° - 88.85° = 1.15°$

Dielectric loss per phase $\quad P = 2\pi fCV^2 \tan \delta$ ...(Eqn. 16.9)

$$= 2\pi \times 50 \times 2417.2 \times 10^{-9} \times (19052)^2 \times \tan 1.15°$$

$$= 5.53 \times 10^3\ \text{W} = \mathbf{5.53\ kW.} \quad \textbf{(Ans.)}$$

(*v*) **Equivalent insulation resistance of the cable :**

Equivalent insulation resistance of the cable

$$= \frac{V^2}{\text{Dielectric loss}} = \frac{(19052)^2}{5.53 \times 10^3} = \mathbf{65.64\ k\Omega.} \quad \textbf{(Ans.)}$$

(*vi*) **Maximum stress in the cable, E_{max}**

$$E_{max} = I_{max} = \frac{V}{r \ln\left(\frac{R}{r}\right)} \quad \text{...(Eqn. 16.4)}$$

$$= \frac{19052}{1.5 \ln\left(\frac{2}{1.5}\right)} = 44150\ V \text{ or } \mathbf{44.15\ kV.} \quad \textbf{(Ans.)}$$

3. *Sheath losses.* When power is transmitted by single-core cables, the electrical characteristics of the circuit are influenced by the presence of sheath around each conductor. The effect is negligible in 3-core cables but is of significant importance in single-core cables. The electromagnetic fields produced by the currents flowing in the conductors induce e.m.f. in the sheaths and under certain conditions heavy currents are set up therein. The actual current flowing along the sheath, amongst other things, depends on the following factors :

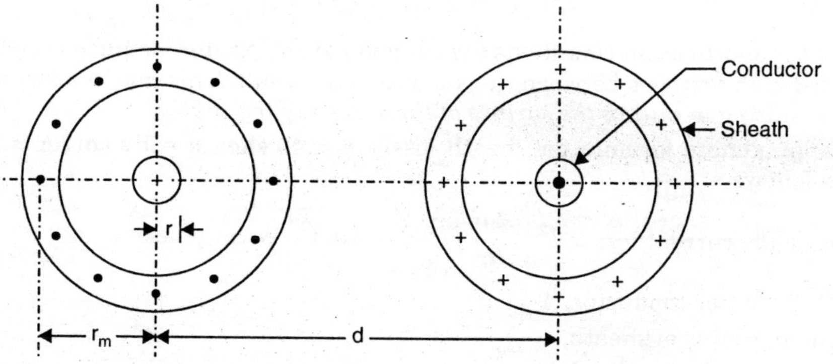

Fig. 16.25. Single-phase underground line-sheath losses.

(*i*) The magnitude and frequency of the current in the conductor,

(*ii*) The arrangement and spacing of the cables,

(*iii*) The sheath resistance, and

(*iv*) The conditions whether the sheath is *bounded* (two different cables having sheath electrically connected at both ends) or *unbounded* (carbon having one end or no end electrically short-circuited).

Since the sheath currents are proportional to the cable core currents, the sheath losses are proportional to the conductor losses. If λ is the ratio of sheath loss to the conductor loss, the equivalent A.C. resistance of the cable will be,

$$R_{eq} = R(1 + \lambda)$$

where, R = Resistance of the cable core.

The sheath currents, according to Cramp and Calder Wood, can be of the following two types :

(*i*) *Sheath eddy currents* (*ii*) *Sheath circuit currents.*

(*i*) ***Sheath eddy currents.*** These currents flow *entirely in the sheath of the same cable.* They will flow through the sheath when the sheaths of the two cables are not connected at both ends or when they are connected only at one end because the currents *do not find a closed path through the sheaths of the two cables.*

Let us first define an expression for the voltage induced in the sheath of individual cable in a two cable system (Fig. 16.25). The voltage induced in the sheath (V_s) due to current I in the core of cable will be,

$$V_s = I\omega M \ (= IX_m)$$

Where M is the mutual inductance between the cable core (conductor) and the sheath (Mutual inductance M between the sheath and core by definition is the flux linkages that link both the core and the sheath due to current I in the core of the cable, per ampere of the current carried by the core).

$$M = 2 \times 10^{-7} \ln\left(\frac{d}{r_m}\right) \text{ H/m} \qquad \text{...(16.13)}$$

$$\therefore \quad V_s = I\omega M = 2 \times 10^{-7} I\omega \ln\left(\frac{d}{r_m}\right) \text{ V/m} \qquad \text{...(16.14)}$$

If the sheaths are bonded at one end, the voltage between the two sheaths at the far end will be

$$2V_s = 2I\omega M = 4 \times 10^{-7} I\omega \ln\left(\frac{d}{r_m}\right) \qquad \text{...(16.15)}$$

Under the short-circuit conditions the currents are of greater magnitudes and in case sheaths are not bonded high voltages between the sheaths may result. This may lead to sparking between the sheaths resulting in *pitting the surface of the sheath of the cables.*

The approximate formula for sheath losses due to sheath eddy currents, as suggested by Arnold, is as follows :

$$\text{Sheath eddy current loss} = I^2\left[\frac{3\omega^2}{R_s}\left(\frac{r_m}{d}\right)^2 \times 10^{-18}\right] \text{ W/cm/phase} \qquad \text{...(16.16)}$$

where, I = Current per conductor, A

r_m = Mean radius of sheath,

d = Spacing between conductors, and

R_s = Sheath resistance.

The sheath loss in open-circuited sheath is about 2% of the total loss and can be neglected.

(*ii*) ***Sheath circuit currents.*** The sheath circuit currents *flow from the sheath of one cable to the sheath of another cable.* These currents flow when the sheaths of both the cables are *electrically connected at both the ends (known as* bonding of the cables).

In a single-phase transmission line, each cable can be considered as an air-cored transformer with loose magnetic coupling and ratio as 1 : 1. The cable core and sheath act as primary and secondary of the transformer respectively Fig. 16.26 shows the equivalent circuit, where

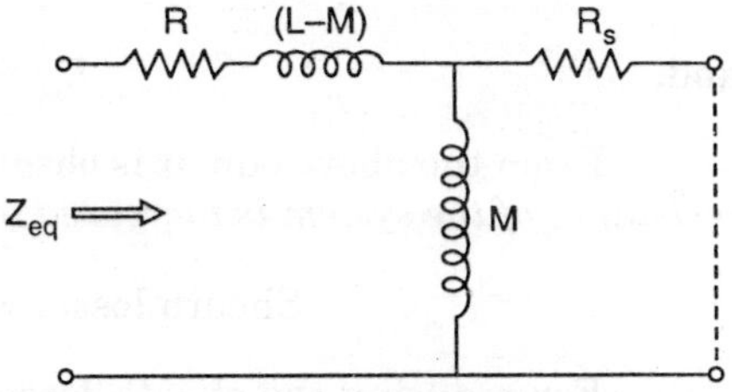

Fig. 16.26. Equivalent of a cable–Air-core transformer.

R = Resistance of the cable core,

$(L - M)$ = Leakage inductance of the core,

M = Mutual inductance between the core and the sheath, and

R_s = Sheath resistance.

The expressions for L and M (see Fig. 16.25) are given as :

$$L = 2 \times 10^{-7} \ln\left(\frac{d}{r}\right) \text{ H/m}$$

$$M = 2 \times 10^{-7} \ln\left(\frac{d}{r_m}\right) \text{ H/m}$$

$$(L - M) = 2 \times 10^{-7} \times \left[\ln\left(\frac{d}{r}\right) - \ln\left(\frac{d}{r_m}\right)\right] = 2 \times 10^{-7} \ln\left(\frac{r_m}{r}\right) \quad \text{...(16.17)}$$

The leakage inductance on the secondary side is not shown since the leakage flux (the flux that does not link core) due to current in the sheath is zero. The secondary side is shorted through the dotted line to represent bonding of the cable sheaths.

The equivalent impedance, as seen through the primary (Fig. 16.26),

$$Z_{eq} = R + j\omega(L - M) + \frac{R_s j\omega M}{R_s + j\omega M}$$

$$= R + j\omega(L - M) + \frac{j R_s \omega M (R_s - j\omega M)}{R_s^2 + \omega^2 M^2}$$

$$= R + j\omega L - j\omega M + \frac{j R_s^2 \omega M}{R_s^2 + \omega^2 M^2} + \frac{\omega^2 M^2 R_s}{R_s^2 + \omega^2 M^2}$$

$$= R + \frac{\omega^2 M^2 R_s}{R_s^2 + \omega^2 M^2} + j\omega L + j\omega M \left(\frac{R_s^2}{R_s^2 + \omega^2 M^2} - 1\right)$$

$$= R + \frac{\omega^2 M^2 R_s}{R_s^2 + \omega^2 M^2} + j\omega L + j\omega M \left(\frac{R_s^2 - R_s^2 - \omega^2 M^2}{R_s^2 + \omega^2 M^2}\right)$$

$$= R + \frac{\omega^2 M^2 R_s}{R_s^2 + \omega^2 M^2} + j\omega L + j\omega M \left(\frac{-\omega^2 M^2}{R_s^2 + \omega^2 M^2}\right)$$

or, $$Z_{eq} = R + \frac{\omega^2 M^2 R_s}{R_s^2 + \omega^2 M^2} + j\omega\left(L - M\frac{\omega^2 M^2}{R_s^2 + \omega^2 M^2}\right) \quad \text{...(16.18)}$$

$\therefore$ $$R_{eq} = R + \frac{\omega^2 M^2 R_s}{R_s^2 + \omega^2 M^2} \quad \text{...(16.19)}$$

and, $$L_{eq} = L - M \frac{\omega^2 M^2}{R_s^2 + \omega^2 M^2} \qquad ...(16.20)$$

From the above eqn. it is observed that by bonding the cables at both the ends the *equivalent resistance of the system is increased whereas inductance is decreased.*

$\therefore$ $$\text{Sheath losses} = I^2 \times \frac{\omega^2 M^2 R_s}{R_s^2 + \omega^2 M^2}$$

For reducing the sheath losses (and thereby to increase the current carrying capacity of a cable) the sheaths of the three single-core cables in a 3-phase system are *cross-bonded.*

16.12. THERMAL RESISTANCE OF CABLES

- The *thermal resistance* between the points in a medium (*e.g.,* insulation) is equal to temperature difference between these points divided by the heat flowing between them in a unit time.

i.e., $$\text{Thermal resistance, } S = \frac{\text{Temperature difference}}{\text{Heat flowing in a unit time}}$$

The SI unit of thermal resistance is °C/W. This is also called *thermal ohm*. Thus 'thermal ohm' may be defined as the resistance which requires a temperature difference of 1°C to cause a heat flow of one watt.

If k is the thermal resistivity of the dielectric of cable having care radius r and sheath radius (inside) r_1 then thermal resistance of the single-core cable.

$$S = \frac{k}{2\pi} \ln \left(\frac{r_1}{r}\right) \text{ thermal ohms per metre length of the cable.} \qquad ...(16.21)$$

The usual values of k are as follows :

Solid type cables below 2200 V ...... 7.5 thermal ohms/m

Oil filled cables ...... 4.5 thermal ohms/m

Impregnated pressure cables ...... 5.5 thermal ohms/m.

- The empirical formula for the thermal resistance of a three-core belted-type cable is given by,

$$S = \frac{k}{6\pi}\left(0.88 + \frac{0.2t}{T}\right) \ln \left[1 + \left(\frac{T+t}{r}\right)\left(4.15 - \frac{1.1t}{T}\right)\right] \qquad ...(16.22)$$

where, r = Radius of the core,

t = Thickness of the belt insulation,

T = Thickness of conductor insulation, and

k = Thermal resistivity.

- The thermal resistance of the ground in which the cable is burried is given by,

$$G = \frac{g}{2\pi} \ln \left(\frac{2k}{R}\right) \text{ thermal ohms/m} \qquad ...(16.23)$$

where, g = Thermal resistivity of the ground,

h = Depth of cable axis below ground, and

R = Radius over the lead sheath.

In practice it is found that the value of g as measured in the laboratory is too large for use in the above equation and it must be applied by a correction factor of $\frac{2}{3}$ in order to correspond with the results of experiments on burried cables. Therefore the amended formula is

$$G = \frac{g}{3\pi} \ln \left(\frac{2h}{R}\right) \qquad ...(16.24)$$

16.13. PERMISSIBLE CURRENT LOADING

Consider a cable laid in soil.

Let, I = Permissible current per conductor,

n = Number of conductors,

R = Electrical resistance per metre length of conductor at the working temperature,

S = Total thermal resistance (*i.e.*, sum of thermal resistances of dielectric and soil) per metre length, and

t = Temperature difference (rise) between the conductor and the soil.

Neglecting the dielectric and sheath losses, we have,

Power dissipated = nI^2R

Also, Power dissipated = $\frac{t}{S}$

$$\therefore \quad nI^2R = \frac{t}{S}, \quad \text{or,} \quad I = \sqrt{\frac{t}{nRS}} \quad ...(16.25)$$

It may be mentioned here that when cables are laid in proximity to each other, the permissible current is reduced further on account of mutual heating.

- Cables should not operated *too hot* for the following *reasons* :

(*i*) Voids may be created in the insulation due to unequal expansion ; this will lead to ionization.

(*ii*) Owing to rapid increase of dielectric losses with temperature, the thermal instability may rise.

(*iii*) The sheath may burst due to expansion of oil.

(*iv*) The oil may lose its viscosity and may start draining off from higher levels.

Example 16.18. *The following data relate to a single-core cable:*

Core diameter ...... 30 mm

Dielectric thickness ...... 40 mm

Thermal resistivity of dielectric ...... 5 thermal ohm-metres

Thermal resistance between the sheath and the ground surface ...... 0.45 thermal ohms per metre length of the cable

Maximum temperature difference between the conductor and the ground surface ...... 55°C

Electrical resistance of the cable ...... 110 μΩ per metre length

Neglecting dielectric and sheath losses, estimate the maximum permissible current.

Solution. *Given* : $r = \frac{30}{2} = 15$ mm or 0.015 m ; $r_1 = 15 + 40 = 55$ mm or 0.055 m ; $k = 5$ thermal ohm-metres ; $t = 55°C$; $R = 110$ μΩ/m length.

Maximum permissible current, I :

Thermal resistance of the dielectric of the cable,

$$S_1 = \frac{k}{2\pi} \ln\left(\frac{r_1}{r}\right) \text{ thermal-ohms per metre length of cable}$$

$$= \frac{5}{2\pi} \ln\left(\frac{0.055}{0.015}\right) = 1.034 \text{ thermal ohms per metre length}$$

Thermal resistance between the sheath and ground surface,

$$S_2 = 0.45 \text{ thermal ohms per metre length}$$

$\therefore$ Total thermal resistance, $S = S_1 + S_2$

$$= 1.034 + 0.45 = 1.484 \text{ thermal ohms per metre length}$$

Now, $$\mathbf{I} = \sqrt{\frac{t}{nRS}} \qquad \text{...[Eqn. (16.25)]}$$

$$= \sqrt{\frac{55}{1 \times 110 \times 10^{-6} \times 1.484}} = \mathbf{580.45\ A.} \quad \textbf{(Ans.)}$$

16.14. TYPES OF CABLE FAULTS AND THEIR LOCATION

16.14.1. Types of Cable Faults

The following types of faults may occur in underground cables :

1. Open-circuit fault :

- An open circuit fault is said to have occurred when there is a *break in the conductor of a cable.*
- The open-circuit fault can be checked by a *megger*. To do so, the three conductors of the 3-core cable at the far end are shorted and earthed. Then resistance between each conductor and earth is measured by a megger. The megger will *indicate infinite resistance if the conductor is broken* (in case the conductor is not broken it will indicate zero resistance)

2. Short-circuit fault :

- When two conductors of a multi-core cable *come in electrical contact with each other due to insulation failure*, it is called a short-circuit fault.
- Again, the fault can be checked by a megger. The two terminals of the megger are connected to any two conductors ; if the megger gives zero reading it indicates short-circuit fault between the conductors. The same procedure is repeated for other conductors taking two at a time.

3. Earth fault :

- An earth fault (or ground fault) is said to have occurred when the *conductor of a cable comes in contact with earth.*
- To identify this fault, one terminal of the megger is connected to the conductor and the other terminal connected to earth. *If the megger indicates zero reading, it means the conductor is earthed.* The same procedure is repeated for other conductors of the cable.

16.14.2. Location of Cable Fault

Out of several methods available for locating the faults in underground cables, the *two commonly used methods are* :

(*i*) Murray loop test.

(*ii*) Varley loop test.

- Both these tests use the principle of Whetstone bridge for locating the faults.
- These tests can be employed to locate the earth fault or short-circuit fault in underground cables *provided that a sound cable runs along the faulty cable.*

16.14.2.1. Murray loop test

This is the most common and accurate method of locating earth or short-circuit faults in the underground cables.

(*i*) ***Earth fault.*** Fig. 16.27 shows the circuit diagram for locating the earth fault by Murray loop test. *AB* is the sound cable and *CD* is the faulty cable. *F* is the point at which earth fault occurs. The far end *D* of the faulty cable is joined to the far end *B* of the sound cable through a low resistance link. Two *variable* resistances *P* and *Q* are joined to ends *A* and *C* respectively and serve as the ratio arms of the Whetstone bridge.

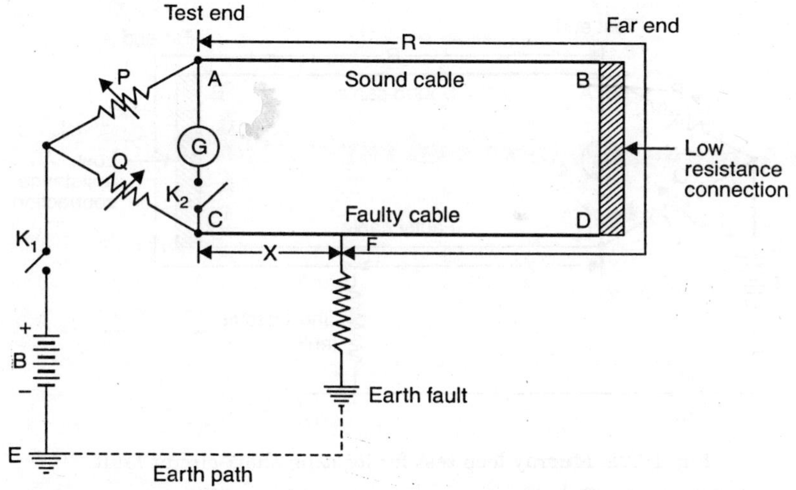

Fig. 16.27. Murray loop test for locating earth fault.

Let, R = Resistance of the conductor loop upto the fault from the test end, and

X = Resistance of the other length of loop.

P, Q, R and X constitute the four arms of the Wheatstone bridge. The resistances P and Q are varied till the galvanometer (G) *indicates zero deflection.*

When the bridge is balanced, we have

$$\frac{P}{Q} = \frac{R}{X}, \quad \text{or,} \quad \frac{P}{Q} + 1 = \frac{R}{X} + 1, \quad \text{or,} \quad \frac{P+Q}{Q} = \frac{R+X}{X}$$

If r is the resistance of each cable, then

$$R + X = 2r$$

$$\therefore \quad \frac{P+Q}{Q} = \frac{2r}{X}, \quad \text{or,} \quad X = \frac{Q}{P+Q} \times 2r$$

If l is the length of each cable in metres, then resistance per metre length of the cable $= \frac{r}{l}$

∴ Distance of fault point (F) from the test end,

$$d = \frac{X}{r/l} = \frac{Q}{P+Q} \times 2r \times \frac{l}{r} = \frac{Q}{P+Q} \times 2l$$

or,

$$d = \frac{Q}{P+Q} \times \text{loop length metres}$$

It may be noted that resistance of the fault is in the battery circuit and not in bridge circuit. Therefore, fault resistance does not affect the balancing of the bridge. However, in case the fault resistance is high, the sensitivity of the bridge is reduced.

(*ii*) ***Short-circuit fault.*** Fig. 16.28 shows the circuit diagram for locating the short-circuit fault by Murray loop test. Again, *P, Q, R* and *X* constitute the arms of the bridge. The fault resistance is in battery circuit and not in the bridge circuit. The bridge is balanced by varying the resistances *P* and *Q*. When the bridge is balanced, we have :

$$\frac{P}{Q} = \frac{R}{X}, \qquad \text{or,} \qquad \frac{P+Q}{Q} = \frac{R+X}{X} = \frac{2r}{X}$$

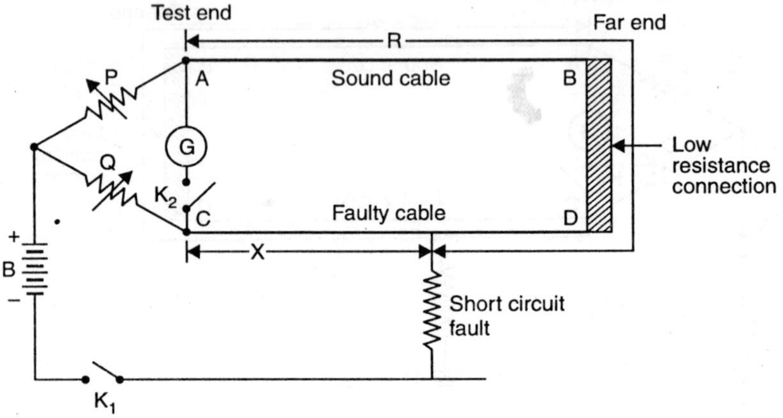

Fig. 16.28. Murray loop test for locating short-circuit fault.

$$\therefore \qquad X = \frac{Q}{P+Q} \times 2r, \qquad \text{or,} \qquad X = \frac{Q}{P+Q} \times (\text{loop length}) \text{ metres}$$

Thus the position of the fault is located.

16.14.2.2. Varley loop test

This test which is also used to locate earth fault or short-circuit fault in the underground cables uses Wheatstone bridge principle.

It *differs* from Murray loop test in that here the *ratio arms P and Q have fixed resistances.* Balance is obtained by adjusting the *variable resistance S* connected to the test end of the faulty cable. The circuit diagrams for locating earth fault and short-circuit fault by Varley loop test are shown in Fig. 16.29 and 16.30 respectively.

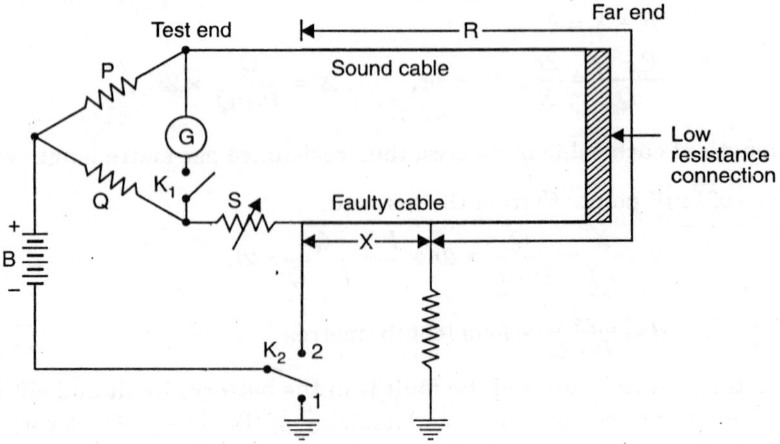

Fig. 16.29. Varley loop test for locating earth fault.

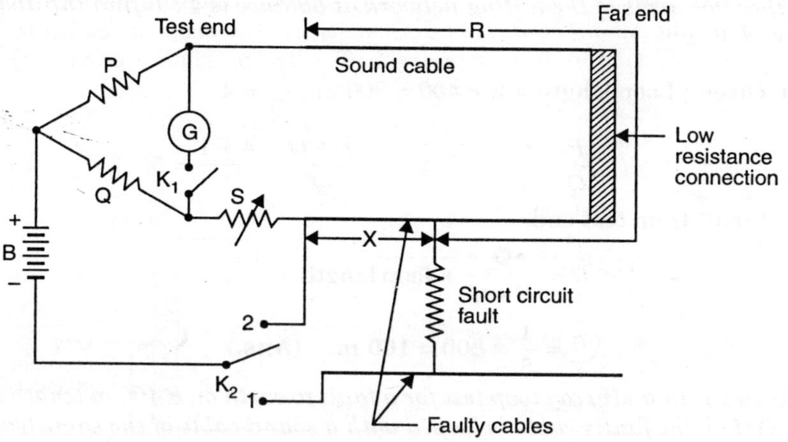

Fig. 16.30. Varley loop test for locating short-circuit fault.

For locating earth fault or short-circuit fault the following *procedure* is adopted :

- The key K_2 is first thrown to position 1. The variable resistance S is varied till the bridge is balanced for resistance value of S_1. Then,

$$\frac{P}{Q} = \frac{R}{X + S_1}, \quad \text{or,} \quad \frac{P+Q}{Q} = \frac{R + X + S_1}{X + S_1}$$

or,
$$X = \frac{Q\,(R + X) - PS_1}{P + Q} \qquad ...(i)$$

- The key is now thrown to position 2 and bridge is balanced with the new value of resistance S_2. Then

$$\frac{P}{Q} = \frac{R + X}{S_2}, \quad \text{or,} \quad Q\,(R + X) = PS_2 \qquad ...(ii)$$

Since the values of P, Q, S_1 and S_2 are known, the value of X can be determined.

Loop resistance $= R + X = \frac{P}{Q} S_2$

If r is the resistance of the cable per metre length, then,

Distance of fault from the test end,

$$d = \frac{X}{r} \text{ metres.}$$

Example 16.19. *Murray loop test was performed on a faulty cable 360 m long. At balance, the resistance connected to the faulty core was set at 10 Ω and the resistance of the resistor connected to the sound core was 30 Ω. Calculate the distance of the fault point from the test end.*

Solution. *Given* : $Q = 10\ \Omega$; $P = 30\ \Omega$; Loop length $= 2 \times 360 = 720$ m

Distance of the fault point from the test end,

$$d = \frac{Q}{P+Q} \times \text{Loop length} = \frac{10}{30+10} \times 720 = \mathbf{180\ m.} \quad \textbf{(Ans.)}$$

Example 16.20. *In a Murray loop test for earth fault on 400 m of cable having a resistance 1.5 Ω/km, the faulty cable is looped with a sound cable of the same length and area of cross-section. If* *the ratio of the other two arms of the testing network at balance is 4 : 1, find the distance of the fault from the testing end of cable.*

Solution. *Given* : Loop length = 2 × 400 = 800 m ; $\frac{P}{Q} = 4$

$$\frac{P}{Q} = \frac{4}{1} \quad \text{or} \quad \frac{P+Q}{Q} = \frac{4+1}{1} = 5$$

Distance of fault from test end,

$$D = \frac{Q}{P+Q} \times \text{loop length}$$

$$= \frac{1}{5} \times 800 = \mathbf{160\ m.} \quad \textbf{(Ans.)}$$

Example 16.21. *In a Murray loop test for a fault to earth on a 400 m length of cable having a resistance of 1.5 Ω/km, the faulty cable is looped with a sound cable of the same length but having a resistance of 3.5 Ω/km. The resistances of the other two arms of the testing network at balance are in the ratio 3 : 1. Calculate the distance of the fault from the testing end of the cable.*

Soluton. $\frac{P}{Q} = \frac{3}{1}$, or, $\frac{P+Q}{Q} = \frac{3+1}{1} = 4$

Loop resistance = $\frac{1.5}{1000} \times 400 + \frac{3.5}{1000} \times 400 = 2\ \Omega$

Resistance of the faulty cable from the test end upto fault point,

$$X = \frac{Q}{P+Q} \times \text{loop resistance} = \frac{1}{4} \times 2 = 0.5\ \Omega$$

Distance of fault point from the testing end,

$$d = \frac{X}{1.5/1000} = \frac{0.5}{15/1000} = \mathbf{333.33\ m.} \quad \textbf{(Ans.)}$$

Example 16.22. *Varley loop test is conducted to locate an earth fault on a 25 km long cable. The resistance per km of single conductor is 15 Ω. The loop is completed with a similar sound conductor. At balance, the variable resistance connected to the faulty conductor is 150 Ω. The fixed resistors have equal values. Calculate the distance of the fault from the test end.*

Solution. Resistance of faulty cable from test end to fault point,

$$X = \frac{Q(R+X) - PS}{P+Q}$$

Here, $P = Q\ ;\ S = 150\ \Omega\ ;\ R + X = 25(15 + 15) = 750\ \Omega$

$$\therefore \quad X = \frac{Q(750) - Q \times 150}{Q+Q} = 300\ \Omega$$

The resistance per km = 15 Ω

∴ Distance of fault from the test end,

$$d = \frac{X}{15} = \frac{300}{15} = \mathbf{20\ km.} \quad \textbf{(Ans.)}$$

16.15. CAUSES OF FAILURE OF UNDERGROUND CABLES

The following are the (probable) *causes of failure of underground cables* :

1. Mechanical puncturing of the lead sheathing of a cable (during excavation and building operations).

2. Failure at cable sealing box (due to poor workmanship of the cable jointer when the end was sealed).

3. Failure due to damage caused to cable insulation on account of improper handling at the time of laying.

4. Failure/damage due to overheating or vibration fatigue.

5. Damage caused due to leakage of the impregnating oil where a cable is led vertically up posts.

6. Damage caused by some chemically active soils due to severe pitting and corrosion.

16.16. OPERATING PROBLEMS WITH CABLES

The following *problems are associated with the operation of cables* :

(*i*) The cables for their operation require high charging current and reactive power. This reactive power is fed to the system and is capacitive. It may lead to rise of sending end voltage of the cable at lightly loaded conditions, due to Ferranti effect.

(*ii*) Generation of capacitive reactive power may hamper the stability considerations of large generators. Further, switching of cable capacitive current may cause rise in over voltage.

(*iii*) Cable gets heated up due to the flow of charging current and consequently its load current capability is reduced.

16.17. DIRECT CURRENT (D.C.) CABLES

In a D.C. cable the losses are *less* than those with A.C. because of the following *reasons :*

(*i*) There is no skin effect.

(*ii*) The dielectric loss is small, since it is only due to leakage current.

(*iii*) The sheath losses are small as they are only due to leakage and ripple currents so that cross-bonding (similar to transposition in an overhead line case) equipment is not required.

Further, there is no continuous charging current so that shunt compensation (reactors at the supply and load end of the cable) is not required.

In view of the above factors, the current carrying capacity is increased for D.C. by about 15 percent so that more power can be transmitted.

- D.C. cables are now available upto ± 600 kV.

16.18. CRYOGENIC CABLES

- Owing to the remote feasibility of superconductivity in electric power transmission, a possible alternative to superconductive cables is the *resistive operations of a conductor (aluminium) at very low temperature*. This is called **"cryogenic cable"**.
- The cryogenic cable *utilizes the decrease in resistivity with decreasing temperature*. A coolant is used to keep the temperature within the cable very low. *Liquid helium or liquid nitrogen are commonly used as coolants.*
- *The cost of a cryogenic cable is very high.*

16.19. COMPARISON BETWEEN OVERHEAD LINES AND UNDERGROUND CABLES

The comparison between the overhead lines and underground cables is given below :

S.No.	Aspects	Overhead lines	Underground cables
1.	*Conductor cost*	*Less* expensive, comparatively. The size of the conductor for the same power transmission is smaller because of better heat dissipation in overhead lines.	*More* expensive.
2.	*Insulation cost*	*Less* insulation cost. The overhead lines use bare conductors supported on steel towers, insulated from the towers through the porcelain insulators. There is sufficient spacing between the conductors depending upon the operating voltage ; the air between the conductors provides insulation.	*More* insulation cost. The insulation in underground cables is provided by various wrappings of high grade paper tapes. A metal sheath is applied over the insulation to prevent moisture from entering the insulation. Oil or inert gas is introduced to fill the voids. Thus for high voltage underground transmission, the *insulation problem is quite complicated and expensive.*
3.	*Erection cost*	Much less comparatively	Erection cost of underground high voltage cables quite high.
4.	*Safety*	Less	More
5.	*Interference with amenities*	More	Less
	Uses	• Quite suitable for a long distance transmission.	• The *large charging current* on very high voltage cables *limits* the use of cables for long length transmission. • Cable installations are suitable for *submarine crossings, railway crossing, inside of power station and substation and in densely populated areas.*

HIGHLIGHTS

1. A *cable* is basically an insulated conductor, externally protected against mechanical injury, moisture entry and chemical reaction.
2. The underground cable employed for transmission of power at high voltage consists of (*i*) Cores or conductors, (*ii*) Metallic sheath, (*iii*) Bedding, and (*iv*) Armouring, serving.
3. Cables, according to voltage, are classified as :
   (*i*) Low voltage (or L.T.) cables-upto 1 kV ; (*ii*) High voltage (or H.T.) cables-upto 11 kV ; (*iii*) Super tension (S.T.) cables-upto 33 kV ; (*iv*) Extra high tension (E.H.T.)-upto 66 kV ; (*v*) Extra supervoltage cables–upto 132 kV.
4. The laying of underground cables may be done by the following methods : (*i*) Direct laying ; (*ii*) Draw-in system or Duct/pipe laying ; (*iii*) Solid system.

5. Insulation resistance of a cable, $R = \frac{\rho}{2\pi l} \ln\left(\frac{r_2}{r_1}\right)$

where r_1 and r_2 are the inner and outer radii of the single-core cable and ρ is the volume resistivity of the material.

6. Capacitance of a single-core is given by, $C = \frac{2\pi \varepsilon_0 \varepsilon_r}{\ln\left(\frac{R}{r}\right)}$ F/m

where r and R represent the radius of the core and inner radius of the lead sheath respectively ; ε_r is the relative permittivity of insulating material in between the core and lead sheath.

7. Dielectric stress/Potential gradient

$$g_{max} (= E_{max}) = \frac{V}{r \ln\left(\frac{R}{r}\right)} ; g_{min} (= E_{min}) = \frac{V}{R \ln\left(\frac{R}{r}\right)}.$$

8. The most economical conductor size : $r = \frac{R}{2.718} ; g_{max} = \frac{V}{r}$.

9. Grading of cables : (*i*) Capacitance grading ; (*ii*) Intersheath grading.

10. Dielectric loss : $P = 2\pi f C V^2 \tan \delta$.

11. Thermal resistance, $S = \frac{\text{Temperature difference}}{\text{Heat flowing in a unit time}}$.

Thermal resistance of the single-core is given by,

$$S = \frac{k}{2\pi} \ln\left(\frac{r_1}{r}\right) \text{ thermal ohms per metre length of the cable.}$$

12. Permissible current loading, $I = \sqrt{\frac{t}{nRS}}$ (neglecting dielectric and sheath losses)

where, n = no. of conductors, R = electrical resistance per metre length of conductor at the working temperature, and t = temperature difference (rise) between the conductor and the soil.

13. The commonly used methods for locating the faults in underground cables are :

(*i*) Murray loop test (*ii*) Varley loop test.

THEORETICAL QUESTIONS

1. What is a cable ?
2. Draw the cross-section of an insulated cable and explain the significance of the various layers.
3. Discuss the qualities of the dielectrics used for cable insulation must have. Also name three dielectrics with their properties used in the manufacture of underground cables.
4. State the classification of cables according to voltage and discuss their general construction.
5. Explain briefly the construction and applications of belted type and *H*-type cables.
6. Draw the cross-section of a 3-core belted high voltage cable and describe its various parts.
7. State the properties which the insulating materials for cables should possess.
8. Explain briefly "supertension/screened cable".
9. State the limitations of solid type cables.
10. Describe briefly "Extra high tension" cables.
11. Explain oil-filled cables, stating their advantages and disadvantages as well.
12. Write short-note on "Gas pressure cables".
13. Name the various methods of laying underground cables. Where 'draw-in-system' of cable laying is used and why ?
14. Explain any one method of laying underground cables. Give suitable sketch.

15. Deduce an expression for insulation resistance of a single-core cable in terms of specific resistance of dielectric, its core and sheath diameters.

16. Find expression for the capacitance of a cable per km length.

17. Find expression for electrostatic stress in a single-core cable. Where does maximum stress occur and where is it minimum and why ?

18. Show that for a given voltage V and maximum stress E_{max} in the dielectric of a single-core cable, the sheath diameter D is a minimum when $D : e = e$ where d is the conductor diameter and e is the base of natural logarithms.

19. Show that the most economical size of conductor in a single-core cable is obtained when radius of cable sheath (R) equals e where e is the base of natural logarithm and r is the radius of conductor.

20. What is meant by grading of cables ? Explain why and how the grading of cables is done ?

21. Describe the reasons for grading of cables.

22. Explain briefly the following methods of grading of cables :
(*i*) Capacitance grading.
(*ii*) Intersheath grading.

23. Explain the meaning of the term 'dielectric loss' and 'power factor angle' in cables with the help of phasor diagram. How does dielectric loss vary with the change in voltage, frequency of supply and capacitance of cable ?

24. What are the sources of heat generation within cables ? Explain them briefly.

25. Write a short note on "Thermal resistance of cables".

26. State the types of faults which may occurs in underground cables.

27. Explain briefly any one of following methods of locating cable faults
(*i*) Murray loop test.
(*ii*) Varley loop test.

28. List the causes of failure of underground cables.

29. Write a short note on D.C. cables.

30. What do you mean by 'Cryogenic cables' ?

31. Given the comparison between overhead lines and underground cables.

UNSOLVED EXAMPLES

1. Determine the insulation resistance of a single-core cable of length 2 km and having conductor radius 12.5 mm, insulation thickness 10 mm and specific resistance of insulation of 5×10^{12} Ωm. [**Ans.** 233.88 MΩ]

2. Calculate the resistivity of the dielectric in a cable, the core diameter of which is 1.5 cm and sheath diameter is 5 cm. For a length of 3000 metres of cable, the insulation resistance is 1820 MΩ. [**Ans.** 28.6×10^{12} Ωm]

3. A single-core cable for use on 11 kV, 50 Hz system has conductor area of 0.645 cm^2 and the internal diameter of sheath is 2.18 cm. The permittivity of the dielectric used in the cable is 3.5. Find :
(*i*) The maximum and minimum electrostatic stresses in the cable,
(*ii*) The capacitance of the above cable per km length, and
(*iii*) The charging current. [**Ans.** (*i*) 27.65 kV/cm (r.m.s.), 11.5 kV/cm (r.m.s.) ; (*ii*) 0.22 μF ; (*iii*) 0.76 A]

4. Find the economic size of a single-core cable working on 220 kV, 3-phase system. The maximum permissible stress in the dielectric is not to exceed 250 kV/cm. [**Ans.** d = 1.43 cm, D = 3.88 cm]

5. Find the economical size of a single-core cable working on a 132 kV, 3-phase system, if a dielectric stress of 60 kV/cm can be allowed. [**Ans.** d = 3.6 cm, D = 9.78 cm]

6. A single-core lead sheathed cable is graded by using three dielectrics of relative permittivities 5,4 and 3 respectively. The conductor diameter is 2 cm and overall diameter is 8 cm. If the three dielectrics are worked at the same maximum stress of 40 kV/cm, find the safe working voltage of the cable.
What will be the value of safe working voltage for ungraded cable, assuming the same conductor and overall diameter and the maximum dielectric stress ? [**Ans.** 57.84 kV ; 39.2 kV]

7. A single-core lead sheathed cable has a conductor diameter of 3 cm ; the diameter of the cable being 8.5 cm. The cable is graded by using two dielectrics of relative permittivities 5 and 3 respectively with corresponding safe working stresses of 38 kV/cm and 26 kV/cm. Determine :
   (*i*) The radial thickness of each insulation.
   (*ii*) The safe working voltage of the cable. [**Ans.** (*i*) 2.15 cm, 0.6 cm ; (*ii*) 46.1 kV]
8. A single-core 66 kV cable has a conductor diameter of 2 cm and a sheath of inside diameter 5.3 cm. The cable has an inner layer of 1 cm thick of rubber of dielectric constant 4.5 and rest impregnated paper of dielectric constant 3.6. Find the maximum stresses in the rubber and in the paper. [**Ans.** 63 kV/cm, 39.5 kV/cm]
9. The capacitance per km of a 3-phase belted cable is 0.3 μF between the two cores with the third core connected to the lead sheath. Calculate the charging current taken by 5 km of this cable when connected to a 3-phase, 50 Hz, 11 kV supply. [**Ans.** 5.98 A]
10. The capacitances of a 3-phase cable are 6.3 μF between the three cores bunched together and earthed sheath and 3.7 μF between one core and the other two connected to sheath. Determine the charging current drawn by the cable when connected a 3-phase, 66 kV, 50 Hz supply. [**Ans.** 54 A]
11. The capacitance per km of a 3-phase belted cable is 0.18 μF between two cores with the third core connected to sheath. Calculate the kVA taken by 20 km long cable when connected to 3-phase, 50 Hz, 3300 V supply. [**Ans.** 24.57 kVA]
12. A 33 kV 3-phase underground cable, 4 km long, uses three single-core cables. Each of the conductors has a diameter of 2.5 cm and radial thickness of insulation is 0.5 cm. The relative permittivity of the dielectric is 3.0. Determine :
   (*i*) Capacitance of the cable per phase.
   (*ii*) Charging current per phase.
   (*iii*) Total charging kVAR.
   (*iv*) Dielectric loss per phase if the power factor of the unloaded cable is 0.02.
   (*v*) Maximum stress in the cable.
   [**Ans.** (*i*) 1984×10^{-9} F/phase ; (*ii*) 11.87 A/phase ; (*iii*) 678.48 kVAR ; (*iv*) 4.52 kW ; (*v*) 45.3 kV/cm]
13. In a test for a fault to earth on a 500 m length of cable having a resistance of 1 Ω/km, the faulty cable is looped with a sound cable of the same length but having a resistance of 2.25 Ω/km. The resistance of the other two arms of the testing network at balance are in the ratio 2.75 : 1. Calculate the distance of the fault from the testing end of the cable. [**Ans.** 433 m]
14. Varley loop test is conducted to locate an earth fault on a 20 km long cable. The resistance per km of the single conductor is 20 Ω. The loop is completed with a similar healthy conductor. At balance, the variable resistance connected to the faulty conductor is 200 Ω. The fixed resistors have equal values. Calculate the distance of the fault from the test end. [**Ans.** 15 km]

17

Power System Stability

17.1. Introduction. 17.2. Classification of stability conditions. 17.3. Steady state stability—Maximum steady state power—power angle diagram—General aspects of steady state stability. 17.4. Transient stability—General aspects—the swing equation—equal area criterion—critical clearing angle—transient stability in multi-machine system—transient stability improvement. Highlights—Theoretical Questions—Unsolved Examples.

17.1. INTRODUCTION

With interconnected systems continually growing in size and extending over vast geographical regions, it is becoming increasingly more difficult to *maintain synchronism* between various parts of a power system.

"Instability" of *a power system means a condition denoting loss of synchronism or falling out of step.*

"Stability" of *a power system is its ability to return to normal or stable operation after having been subjected to some form of disturbance.* Stability considerations are considered as an essential part of the power system planning for a long time.

The stability can be interpreted as '*maintenance of synchronism*'.

Sometimes the term '*power limit*' is also interpreted as '*stability*' because to have the maximum utility of the system it should be capable of supplying maximum power without causing instability.

According to *A.I.E.E.* the terms 'stability' and 'stability limit' are defined as follows :

Stability. *It is that attribute of the power system or part of the system which enables it to develop restoring forces between the elements there of equal to or greater than the disturbing forces so as to restore a state of equilibrium between the elements.*

Stability limit. *It is the maximum power flow possible through some particular point in the system when the entire system or part of the system to which the stability limit refers is operating with stability.*

The following are the *basic features of the dynamics of a power system* :

(*i*) The typical behaviour exhibited by the synchronous tie is that as power transfer is gradually increased a maximum limit is reached beyond which the system cannot remain in synchronium (*i.e.* it falls out of step).

(*ii*) The system is basically a spring-inertia oscillatory system with inertia on the mechanical side and spring action provided by the synchronous tie wherein power transfer is *proportional to sin*δ *or* δ, δ being the relative internal angle of machines.

(*iii*) The equation determining system dynamics is *non linear* (since power transfer is proportional to sin δ) *for disturbances causing large variation in angle* δ. Stability phenomenon peculiar to non-linear systems as distinguished from linear systems is therefore exhibited by power systems (stable upto a certain magnitude of disturbance and unstable for larger disturbances).

17.2. CLASSIFICATION OF STABILITY CONDITIONS

For the purpose of analysis the stability conditions may be classified as follows :

1. Steady state stability.
2. Transient stability.
3. Dynamic stability.

1. Steady state stability :

The ***"steady state stability"*** *is the stability of the system under conditions of gradual or relatively slow change in load.* The load is assumed to be applied at a rate which is slow when compared either with the natural frequency of oscillations of the major parts of the system or with the rate of change of field flux in the rotating machine in response to the change in loading.

- The study of steady state stability is basically concerned with the determination of the upper limit of machine loadings before losing synchronism, provided the loading is gradually increased.
- In case of interconnected systems, synchronism between the ends may be lost once the magnitude of power flow exceeds the steady state stability limit.

2. Transient stability :

"Transient stability" *refers to the maximum flow of power possible through a point without losing the stability with sudden and large changes in the network conditions such as brought about by faults, by sudden large increment of loads.*

Transient stability of a power transmission system is its *inherent ability to recover* normal operation following sudden and/or severe disturbance (*e.g.* the fault).

Transient stability is *characterised by* the highest magnitude of power flow just prior to the transient disturbance for which the system can remain in synchronism once the transient fault is withdrawn or cleared.

- The *transient stability limit* is almost always *lower than the steady state limit*, but unlike the latter, it may exhibit different values depending on the nature, location and magnitude of disturbance.

3. Dynamic stability :

"Dynamic stability" is *the ability of a power system to remain in synchronism after the 'initial swing' (transient stability period) until the system has settled down to the new steady-state equilibrium condition.*

The system is said to be dynamically stable if the oscillations do not acquire more than certain amplitude and die out quickly (*i.e.*, the system is well damped). In a dynamically unstable system, the oscillation amplitude is large and these persist for a long time (*i.e.*, the system is underdamped). This kind of instability behaviour constitutes a serious threat to system security and creates very difficult operating conditions.

Dynamic instability is *more probable than steady state instability.*

Dynamic stability can be significantly improved through the use of power system stabilizers.

- *Computer simulation* is the only effective means of studying dynamic stability problems. The same simulation programmes are, of course, applicable to transient stability studies as well.

Note. In addition to power flow stability, voltage stability is also a major criterion for successful A.C. power transmission. *'Voltage instability'* is characterised by a state of power system when the load end voltage drops abnormally causing withdrawal of high currents from the source even though the power flow starts ceasing and the power angle between the two bus voltages way exceed their rated value for stable power transfer.

17.3. STEADY STATE STABILITY

17.3.1. Maximum Steady State Power

For a transmission line, we have :

$$\bar{V}_S = \bar{A}\,\bar{V}_R + \bar{B}\,\bar{I}_R \qquad ...(17.1)$$

$$\bar{I}_S = \bar{C}\,\bar{V}_R + \bar{D}\,\bar{I}_R \qquad ...(17.2)$$

$$\bar{V}_R = \bar{D}\,\bar{V}_S - \bar{B}\,\bar{I}_S \qquad ...(17.3)$$

$$\bar{I}_R = -\bar{C}\,\bar{V}_S + \bar{A}\,\bar{I}_S \qquad ...(17.4)$$

where, A, B, C, D = Generalised line constants,

$\bar{V}_S, \bar{I}_S$ = Sending end voltage and current respectively, and

$\bar{V}_R, \bar{I}_R$ = Receiving end voltage and current respectively.

Let,

$$\bar{A} = A\angle\alpha\ ;\ \bar{B} = B\angle\beta\ ;\ \bar{D} = D\angle\Delta$$

$$\bar{V}_S = V_S\angle\delta\ ;\ \bar{V}_R = V_R\angle 0°$$

At the receiving end, the *complex power* is given as,

$$\bar{S}_R = P_R + jQ_R = \bar{V}_R \bar{I}_{RC} \qquad ...(17.5)$$

where, P_R = Receiving end power,

Q_R = Receiving end reactive power, and

$\bar{I}_{RC}$ = Conjugate of $\bar{I}_R$.

The receiving end current, from eqn. (17.1),

$$\bar{I}_R = \frac{\bar{V}_S - \bar{A}\,\bar{V}_R}{\bar{B}} = \frac{V_S\angle\delta - A\angle\alpha\, V_R\angle 0°}{B\angle\beta}$$

or,

$$\bar{I}_R = \frac{V_S}{B}\angle(\delta-\beta) - \frac{A}{B}V_R\angle(\alpha-\beta) \qquad ...(17.6)$$

$\therefore$

$$\bar{I}_{RC} = \frac{V_S}{B}\angle(\beta-\delta) - \frac{A}{B}V_R\angle(\beta-\alpha) \qquad ...(17.7)$$

Putting the values $\bar{I}_{RC}$ from eqn. (17.7) and $\bar{V}_R = V_R\angle 0°$ in eqn. (17.5), we get

$$\bar{S}_R = P_R + jQ_R = V_R\angle 0°\left[\frac{V_S}{B}\angle(\beta-\delta) - \frac{A}{B}V_R\angle(\beta-\alpha)\right]$$

or,

$$\bar{S}_R = \frac{V_S V_R}{B}\angle(\beta-\delta) - \frac{A}{B}V_R^2\angle(\beta-\alpha)$$

On separation of real and reactive components, we have

$$P_R = \frac{V_S V_R}{B}\cos(\beta-\delta) - \frac{AV_R^2}{B}\cos(\beta-\alpha) \qquad ...(17.8)$$

and,

$$Q_R = \frac{V_S V_R}{B}\sin(\beta-\delta) - \frac{AV_R^2}{B}\sin(\beta-\alpha) \qquad ...(17.9)$$

The power received will be maximum (for fixed values of V_S and V_R) when,

$$\cos(\beta-\delta) = 1 \quad \text{or} \quad \text{when } \beta-\delta = 0 \quad \text{or} \quad \beta = \delta$$

$\therefore$

$$(P_R)_{max} = \frac{V_S V_R}{B} - \frac{AV_R^2}{B}\cos(\beta-\alpha) \qquad ...(17.10)$$

...(From eqn. 17.8)

The sending end current, from eqn. (17.3),

$$\bar{I}_S = \frac{\bar{D}\bar{V}_S - \bar{V}_R}{\bar{B}} = \frac{D\angle\Delta\, V_S \angle\delta - V_R\angle 0°}{B\angle\beta}$$

or,

$$\bar{I}_S = \frac{D}{B} V_S \angle(\Delta + \delta - \beta) - \frac{V_R}{B}\angle -\beta \qquad ...(17.11)$$

Conjugate of $\bar{I}_S$,

$$\bar{I}_{SC} = \frac{D}{B} V_S \angle(\beta - \Delta - \delta) - \frac{V_R}{B}\angle\beta \qquad ...(17.12)$$

At the sending end, the complex power is given as,

$$\bar{S}_S = P_S + jQ_S = \bar{V}_S \bar{I}_{SC}$$

where, P_S = Sending end power,
Q_S = Sending end reactive power, and

or,

$$\bar{S}_S = V_S \angle\delta\left[\frac{D}{B} V_S \angle(\beta - \Delta - \delta) - \frac{V_R}{B}\angle\beta\right]$$

or,

$$\bar{S}_S = \frac{D}{B} V_S^2 \angle(\beta - \Delta) - \frac{V_R V_S}{B}\angle(\beta + \delta) \qquad ...(17.13)$$

On separation of real and reactive components, we get

$$P_S = \frac{D}{B} V_S^2 \cos\angle(\beta - \Delta) - \frac{V_R V_S}{B}\cos\angle(\beta + \delta) \qquad ...(17.14)$$

and,

$$Q_S = \frac{D}{B} V_S^2 \sin\angle(\beta - \Delta) - \frac{V_R V_S}{B}\sin\angle(\beta + \delta) \qquad ...(17.15)$$

The power sent out will be maximum (for fixed values of V_S and V_R) when,

$$\cos(\beta + \delta) = -1 \quad \text{or} \quad \beta + \delta = 180°$$

i.e., Torque angle, $\delta = 180 - \beta$

$$(P_S)_{max} = \frac{D}{B} V_S^2 \cos(\beta - \Delta) + \frac{V_R V_S}{B} \qquad ...(\text{From eqn. 17.14}) \quad ...(17.16)$$

Let us *consider a transmission line having no losses and negligible shunt admittance* :

Here, $\bar{A} = \bar{D} = 1\angle 0°$; $\bar{B} = X\angle 90°$; $\bar{C} = 0$

In this case, the equations for P_R, $(P_R)_{max}$, P_S and $(P_S)_{max}$ become

$$P_R = \frac{V_S V_R}{X}\sin\delta \qquad ...(17.17)$$

$$(P_R)_{max} = \frac{V_S V_R}{X} \qquad ...(17.18)$$

$$P_S = \frac{V_S V_R}{X}\sin\delta \qquad ...(17.19)$$

$$(P_S)_{max} = \frac{V_S V_R}{X} \qquad ...(17.20)$$

From the above equations we find :

(*i*) $P_R = P_S$ and $(P_R)_{max} = (P_S)_{max}$, since the *line is lossless.*

(*ii*) For fixed values of V_S, V_R and X (line reactance),
P_S α sin of the load angle δ (phase shift between V_S and V_R).

17.3.2. Power-Angle Diagram

As evident from eqns. (17.17) and (17.18), for fixed values of V_S, V_R and X, the P_R and P_S vary as a sin function. As such the *power-angle diagram* is a sine curve as shown in Fig. 17.1. The maximum power occurs when $\delta = 90°$.

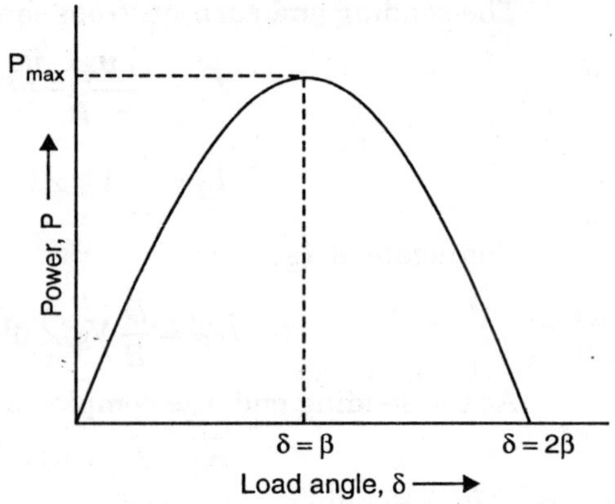

Fig. 17.1. Power angle diagram.

Let us consider a short transmission line having resistance and reactance only and no shunt admittance.

The generalised constants for this line are :

$$\bar{A} = 1\angle 0° \; ; \quad \bar{B} = Z\angle\beta$$

$$\left(\text{where } \beta = \tan^{-1}\frac{X}{R}\right)$$

The power received,

$$P_R = \frac{V_S V_R}{Z}\cos(\beta - \delta) - \frac{V_R^2}{Z}\cos\beta \qquad \text{...From eqn. (17.8)}$$

$$= \frac{V_R}{Z^2}[V_S\, Z\cos(\beta-\delta) - V_R\, Z\cos\beta]$$

$$= \frac{V_R}{Z^2}[V_S\, Z\cos(\beta-\delta) - V_R\, R] \qquad \text{...(17.21)}$$

V_S, V_R and Z being of constant magnitude, the maximum power will occur when $\delta = \beta$,

i.e.
$$(P_R)_{max} = \frac{V_R^2}{Z^2}\left(\frac{V_S}{V_R}Z - R\right) \qquad \text{...(17.22)}$$

$$(P_R)_{max} = \frac{V_R^2}{\sqrt{R^2+X^2}}\left[\frac{V_S}{V_R}\sqrt{R^2+X^2} - R\right]$$

When the line reactance X is varied, keeping the line resistance R constant, the condition for $(P_R)_{max}$ to be maximum is given as,

$$\frac{d}{dx}\{(P_R)_{\max}\} = \frac{d}{dx}\left[\frac{V_R^2}{R^2+X^2}\left(\frac{V_S}{V_R}\sqrt{R^2+X^2} - R\right)\right] = 0$$

$$= \frac{d}{dx}\left[\frac{V_R V_S}{\sqrt{R^2+X^2}} - \frac{V_R^2}{R^2+X^2}\right]$$

$$= V_R V_S \times \left[-\frac{1}{2}(R^2+X^2)^{-3/2}\times 2X\right] - V_R^2\, R\times\left[-1\times(R^2+X^2)^{-2}\times 2X\right]$$

$$= -\frac{V_R V_S X}{(R^2+X^2)^{3/2}} + \frac{2V_R^2 RX}{(R^2+X^2)^2} = 0$$

$$= -V_R V_S\times\sqrt{R^2+X^2} + 2V_R^2 RX = 0$$

or,
$$\frac{V_S}{V_R}\sqrt{R^2+X^2} = 2R \qquad \text{or} \qquad R^2+X^2 = 4R^2\left(\frac{V_R}{V_S}\right)^2$$

or, $$X^2 = 4R^2\left(\frac{V_R}{V_S}\right)^2 - R^2 = R^2\left[4\left(\frac{V_R}{V_S}\right)^2 - 1\right]$$

or, $$X = R\left[4\left(\frac{V_R}{V_S}\right)^2 - 1\right]^{1/2} \qquad \text{...(17.23)}$$

when $V_S = V_R$ (*i.e.* sending and receiving end voltages are equal), we have

$X = R\,(4 - 1)^{1/2} = \sqrt{3}R$

i.e. $$X = \sqrt{3}R \qquad \text{...(17.24)}$$

Conclusions :

- Maximum power will be transferred (from the sending end to receiving end) when $X = \sqrt{3}R$
- Power transfer is possible only when reactance is present.
- The compensation for all the reactance by series capacitance is not necessarily desirable.
- When equal voltages are to be maintained at both the ends, phase shift is essential for transfer of power.

17.3.3. General Aspects of Steady State Stability

- The most fundamental requirements of proper operation of a power system is the operation of the power grid at steady state. Slight insight into the operational pattern suggests that the *steady state operation of the power network is principally governed by the system voltage at the two ends, the transfer reactance of the line and power angle between the two buses.* However, operational aspects vary depending on the nature of the power transmitting network and hence to build up a clear understanding, some analytical tools need to be used in order to classify the steady state operational aspects of power grids existing generally in the service of the utilities.
- The study of steady state stability is basically concerned with the determination of the upper limit of machine loadings before losing synchronism, provided the loading is increased gradually. *Steady state stability limit refers to the maximum flow of power possible through a particular point without causing the loss of stability when the power is increased gradually.*
- The *assessment of steady state stability* is of significant importance in the following cases/ areas :
  - (*i*) Planning and designing or electrical power systems ;
  - (*ii*) Developing special automatic control devices ;
  - (*iii*) Putting into operation new elements of the system or modifying its operating conditions, such as interconnection of power systems and commissioning of new power plants, transformer substitutions, transmission lines etc. ;
  - (*iv*) Power system analysis and synthesis.
- Modern power systems have several interconnected generating stations, each with several generators and many loads. The machines located at any one point in a system normally act in unison. It is, therefore, *common practice in stability studies to consider all the machines at one point as one large machine.* Also machines which are not separated by lines of high reactance are *lumped together and considered as one equivalent machine.* Thus a multimachine system can often be reduced to an equivalent few-machine system. If *synchronism is lost, the machines of each group stay together although they go out of step*

with other groups. Qualitative behaviour of machines in an actual system is usually that of a two-machine system. Because of its *simplicity*, the *two-machine system is extremely useful in describing the general concepts of power system stability and the influence of various factors on stability.*

Consider a system consisting of a generator (G), transmission line and a synchronous motor (M) in form of load as shown in Fig. 17.2.

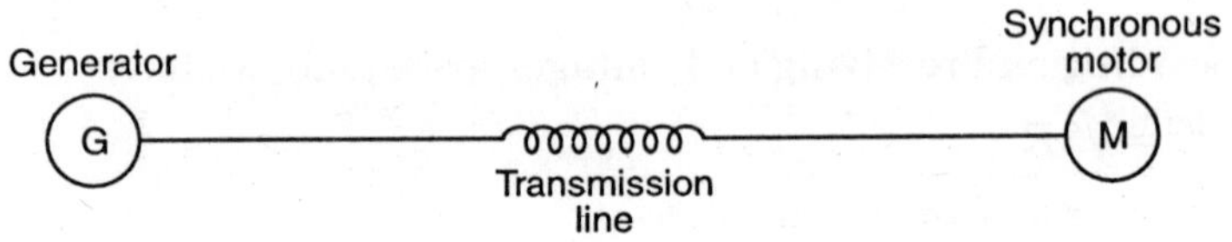

Fig. 17.2

Let, P_G = Power developed by the generator G, and

P_M = Power developed by the synchronous motor M.

P_G, P_M $(P_G)_{max}$, $(P_M)_{max}$ are given by the following expansions [Refer to eqns. (17.14), (17.8), (17.16), (17.10)] :

$$P_G = \frac{D}{B} V_G^2 \cos (\beta - \Delta) - \frac{V_G V_M}{B} \cos (\beta + \delta) \quad ...(i)$$

$$P_M = \frac{V_G V_M}{B} \cos (\beta - \delta) - \frac{A}{B} V_M^2 \cos (\beta - \alpha) \quad ...(ii)$$

$$(P_G)_{max} = \frac{D}{B} V_G^2 \cos (\beta - \Delta) + \frac{V_G V_M}{B} \quad ...(iii)$$

$$(P_M)_{max} = \frac{V_G V_M}{B} - \frac{A}{B} V_m^2 \cos (\beta - \alpha) \quad ...(iv)$$

where, A, B, D = Generalised constants of a two-terminal pair network.

Now consider a *synchronous motor connected to infinite bus* and *running at a constant speed.*

— Power input = Power output + Losses.

— When a small increment of shaft load is added to the motor, the output power of the motor increases as the speed does not change momentarily whereas the input to the motor remains unaltered. Hence there is a net torque on the motor which tends to retard it and the speed falls temporarily. As a consequence of reduction in motor speed, the phase angle between the internal voltage of the motor and the system voltage increases until the electrical power input again equals power output plus losses. Since during this interval (transit) the electrical power input to the motor is less, the *excess power required is supplied by the stored energy in the rotating system.* As such the motor oscillates around the new point of equilibrium and finally may *either come to rest or may lose synchronism* (Synchronism may be lost when the load is too suddenly applied or load is too large).

The maximum power that can be developed by a motor [$(P_M)_{max}$] can be obtained only when $\delta = \beta$; thus the load on the motor may be increased till this condition is attained. If the loading on the motor is continued beyond this limit the angle δ will increase further and the motor may lose synchronism. The excess power required over power developed is available from the stored energy of the rotating system, resulting in decrease of speed. As the value of δ becomes larger and larger, the developed power goes on decreasing till the motor comes to standstill. Although $\delta = \beta$ is not the condition for maximum power developed power by the generator but it needs to be considered since motor will lose synchronism earlier.

For *any value of δ the difference in power developed between generator and motor is equal to 'line losses'*. When the line resistance and shunt admittance are negligible, then the power transferred between generator and the motor,

$$P = P_M = P_G = \frac{V_G V_M}{X} \sin \delta \quad ...(v)$$

when $\delta = 90°$, the power transferred will be maximum,

i.e.
$$P_{max} = \frac{V_G V_M}{X} \quad ...(vi)$$

From the above eqn. it is evident that the *maximum power that can be transferred between a generator/alternator* and a motor is *directly proportional to the products of internal e.m.fs of the two machines and inversely proportional to reactance of the line.*

Methods of improving steady state stability :

In practice the following methods are used to improve steady state stability :

1. Higher excitation voltages.

2. Reducing the impedance between the stations (By adding machines or lines in parallel or by using machines of lower inherent impedance).

3. The optimum conditions of $X = \sqrt{3}R$ for maximum power transfer is approached by the use of series capacitors for overhead lines and series reactors for underground cables.

4. Quick response excitation system.

Example 17.1. *Deduce an expression for the maximum steady state power which can be transmitted over a line (neglecting capacitance of the line) if the voltage at each end is kept constant.*

Show that if the reactance X of the line be varied the resistance R remaining constant the maximum steady state power that could be transmitted over the line would be greatest when $X = \sqrt{3}R$.

Sol. Refer Fig. 17.3.

Let, V_S = Sending end voltage,

V_R = Receiving end voltage,

$I_S = I_R = I$ = Sending end and receiving end currents (same, since capacitance is neglected)

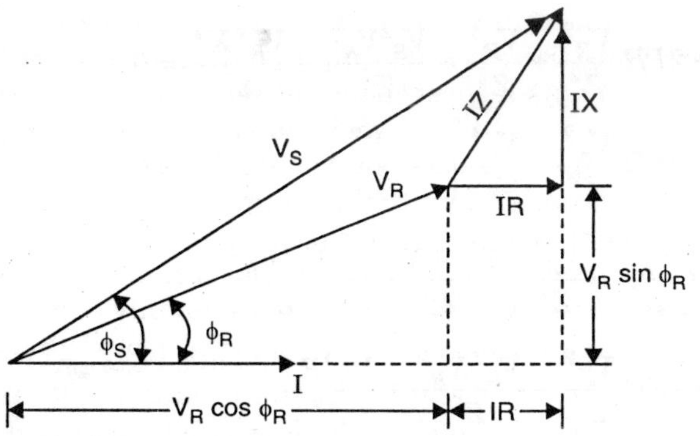

Fig. 17.3. Phasor diagram.

IR, IX = Component voltage drops due to the impedance Z of the line, when current I flows through it. From the phasor diagram, we have

$$V_S \cos \phi_S = V_R \cos \phi_R + IR \quad ...(i)$$

$$V_S \sin \phi_S = V_R \sin \phi_R + IX \quad ...(ii)$$

Squaring (*i*) and (*ii*) and adding, we get

$$V_S^2 = V_R^2 + 2V_R I(R \cos \phi_R + X \sin \phi_R) + I^2(R^2 + X^2)$$

Denoting $V_R I \cos \phi_R = P$ = Power received, and

$V_R I \sin \phi_R = Q$ = Reactive volt-amperes, we have

$$V_S^2 = V_R^2 + 2PR + 2QX + \frac{(P^2 + Q^2)(R^2 + X^2)}{V_R^2}$$

or,

$$-V_S^2 + V_R^2 + 2PR + 2QX + \frac{(P^2 + Q^2)(R^2 + X^2)}{V_R^2} = 0 \quad ...(iii)$$

In the above expression V_S, V_R, R and X are constants, P, Q being the only variables.

Let us now find the maximum value of P when Q is varied. Differentiating the eqn. (*iii*) w.r.t. Q and equating $\frac{dP}{dQ}$ to zero for maximum value of P we get,

$$2R\frac{dP}{dQ} + 2X + \left(\frac{R^2 + X^2}{V_R^2}\right)\left(2P\frac{dP}{dQ} + 2Q\right) = 0$$

so that,

$$2X + 2Q\left(\frac{R^2 + X^2}{V_R^2}\right) = 0 \quad \text{or} \quad Q = -\frac{V_R^2 X}{R^2 + X^2}$$

Substituting this value of Q in eqn. (*iii*), we get

$$-V_S^2 + V_R^2 + 2PR - \frac{2V_R^2 X^2}{R^2 + X^2} + \frac{R^2 + X^2}{V_R^2}\left[P^2 + \frac{V_R^4 X^2}{(R^2 + X^2)^2}\right] = 0$$

This is a quadratic equation in P. Multiplying throughout by $\frac{V_R^2}{R^2 + X^2}$ and putting $R^2 + X^2 = Z^2$, we get

$$P^2 + 2PR\frac{V_R^2}{Z^2} + (V_R^2 - V_S^2)\frac{V_R^2}{Z^2} - \frac{V_R^4 X^2}{Z^4} = 0$$

$$P^2 + 2PR\frac{V_R^2}{Z^2} + \frac{V_R^4}{Z^2} - \frac{V_S^2 V_R^2}{Z^2} - \frac{V_R^4 X^2}{Z^4} = 0$$

$$P^2 + 2PR\frac{V_R^2}{Z^2} + \frac{V_R^4}{Z^2}\left(1 - \frac{X^2}{Z^2}\right) - \frac{V_S^2 V_R^2}{Z^2} = 0$$

$$P^2 + 2PR\frac{V_R^2}{Z^2} + \frac{V_R^4}{Z^2}\left(\frac{Z^2 - X^2}{Z^2}\right) - \frac{V_S^2 V_R^2}{Z^2} = 0$$

$$P^2 + 2PR\frac{V_R^2}{Z^2} - \frac{V_S^2 X_R^2}{Z^2} + \frac{V_R^4 R^2}{Z^4} \quad (\because \; Z^2 - X^2 = R^2)$$

$$\therefore \quad P = \frac{-2R\frac{V_R^2}{Z^2} \pm \sqrt{4R^2\frac{V_R^4}{Z^4} + 4\left(\frac{V_S^2 V_R^2}{Z^2} - \frac{V_R^4 R^2}{Z^4}\right)}}{2}$$

$$= -\frac{RV_R^2}{Z^2} \pm \sqrt{R^2\frac{V_R^4}{Z^4} + \frac{V_S^2 V_R^2}{Z^2} - \frac{V_R^4 R^2}{Z^4}}$$

$$= -\frac{RV_R^2}{Z^2} \pm \frac{V_S V_R}{Z} = \frac{V_R^2}{Z^2}\left(-R \pm \frac{V_S}{V_R} Z\right)$$

So that, $$P_{max} = \frac{\mathbf{V_R}^2}{\mathbf{Z}^2}\left(\frac{\mathbf{V_S}}{\mathbf{V_R}}\mathbf{Z} - \mathbf{R}\right) \quad \textbf{(Ans.)}$$

For the second part we are given that $V_S = V_R$ so that expression of P_{max} in this case becomes,

$$P_{max} = \frac{V_R^2}{Z^2}(Z - R)$$

We want to keep R fixed varying X such that P_{max} is greatest. Evidently $\frac{d}{dX}(P_{max})$ should be zero for this condition so that,

$$\frac{d}{dX}(P_{max}) = \frac{d}{dx}\left[\frac{V_R^2}{Z^2}(Z - R)\right] = 0$$

$$\frac{d}{dX}\left[\frac{V_R^2}{R^2 + X^2}(\sqrt{R^2 + X^2} - R)\right] = 0$$

or, $$\frac{d}{dX}\left[\frac{V_R^2}{(R^2 + X^2)^{1/2}} - \frac{RV_R^2}{R^2 + X^2}\right] = 0$$

$$-\frac{V_R^2 . 2X}{2(R^2 + X^2)^{3/2}} + \frac{RV_R^2 \times 2X}{(R^2 + X^2)^2} = 0$$

or, $$\frac{R}{\sqrt{R^2 + X^2}} = \frac{1}{2} \quad \text{or} \quad 2R = \sqrt{R^2 + X^2} \quad \text{or} \quad 4R^2 = R^2 + X^2$$

or, $$X = \sqrt{3}\, R \quad \textbf{Proved.}$$

17.4. TRANSIENT STABILITY

17.4.1. General Aspects

When a sudden disturbance occurs on a power system rotor speeds, rotor angular differences and power transfer undergo fast changes whose magnitudes are dependent upon the severity of disturbance. For a large disturbance, changes in angular differences may be so large as to cause the machines to fall out of step. This type of instability is known as *transient stability* and is a fast phenomenon usually occurring within 1s for a generator close to the cause of disturbance. There is a large range of disturbances which may occur on a power system, but a fault on a heavily loaded line which requires opening the line to clear the fault is usually of greatest concern. The tripping of a loaded generator or the abrupt dropping of a large load may cause instability.

The effect of short circuit (faults), the most severe type of disturbance to which a power system is subjected, must be determined in nearly all stability studies.

- The transient stability studies are usually carried out over a relatively short-time that will be equal to the time of one swing, the time period normally being one second or less. The analysis is carried out to find out whether the system loses stability during the first swing or not. If the power system does not lose stability, it is assumed that subsequent swings will diminish and the power system will remain stable (as usually happens). However, in some subsequent swing the power system may become unstable.
- In order to know whether the power system which have rotating synchronous machines is stable or not ; it is necessary to drive *swing equation*.

17.4.2. The Swing Equation

Let, θ = Angular displacement, rad.,

$\omega\left(=\frac{d\theta}{dt}\right)$ = Angular velocity, rad./s,

$\alpha\left(=\frac{d\omega}{dt}=\frac{d^2\theta}{dt^2}\right)$ = Angular acceleration, rad./s^2

T = Torque, N.m,

$P(= T.\omega)$ = Power developed, W(watts),

I = Moment of inertia, kg-m^2 or J-s^2/rad.^{2}, and

$M = (= I\omega)$ = Angular momentum.

Generator :

Input ...... Mechanical torque (or shaft torque), T_M
Output ...... Electromagnetic torque, T_E
} Both the torques reckoned as *positive*

Motor :

Input ...... Electromagnetic torque, T_E
Output ...... Shaft torque, T_M
} Both the torques reckoned as *negative*

During discussion, the torques due to rotational losses shall be considered negligible.

The relative position of the rotor axis and stator magnetic field axis is fixed, under normal condition. The angle between the two is known as the load (or torque) angle δ, which *depends upon the machine loading.* More the machine loading, larger is the load angle δ. When the load is added to or removed from the shaft of the synchronous machine the rotor will decelerate or accelerate accordingly w.r.t. the synchronously rotating stator field and a relative motion starts. Under these conditions the rotor is said to be swinging w.r.t. the stator field.

The equation giving the relative motion of the rotor (load angle δ) with respect to stator field as a function of time is called the **swing equation.**

The difference between the input and output torques ($T_M \sim T_E$) will cause acceleration or retardation of the rotor whether the input torque is greater than output torque or otherwise.

For a *generator,*

$$T_{AG} = T_M - T_E \quad \text{... Holds true for a motor also} \qquad ...(17.25)$$

$$P_{AG} = P_M - P_E \qquad ...(17.26)$$

Also,

$$P_{AG} = T_{AG}\,\omega = I\alpha\omega = M\alpha \qquad ...(17.27)$$

where, T_{AG} = Net torque causing acceleration of the rotor (will be positive when $T_M > T_E$),

P_{AG} = Accelerating power,

ω = Angular velocity in mechanical radians/sec., and

α = Acceleration in mechanical radians/sec.^{2}

In eqn. (17.27) ω should be used in mechanical radians/sec. Only and not electrical radians/sec. I is the moment of inertia in kg-m^2.

From eqn. (17.27), $M = I\omega$

where M is in joule-sec./mechanical radian.

As we are interested in studying the rotor motion in terms of electrical degrees or electrical radians we make use of the following *relations* :

No. of electrical radians or degrees

= No. of mechanical radians or degrees × number of pair of poles

If M is to be expressed in *joule-sec./electrical radian* when ω is in mechanical radians/sec., then

$M = I\omega$/No. of pair of poles, and if M is to be expressed in *joule-sec./electrical degree*, then

$M = I\omega$/(No. of pair of poles × 57.32) $\left(\because \text{ 1 radians} = \dfrac{180}{\pi} = 57.3°\right)$

Here M (angular momentum) is expressed in terms of *MJ-s* per Electrical degree if P_{AG} is expressed in *MW* and α is in electrical degrees per second squared.

Also, $\theta = \omega_s t + \delta$ ...(17.28)

where, θ = Angular displacement w.r.t. time, t,

ω_s = Angular velocity of the reference axis *rotating synchronously*, and

δ = Angular displacement from the synchronously rotating reference axis (Fig. 17.4)

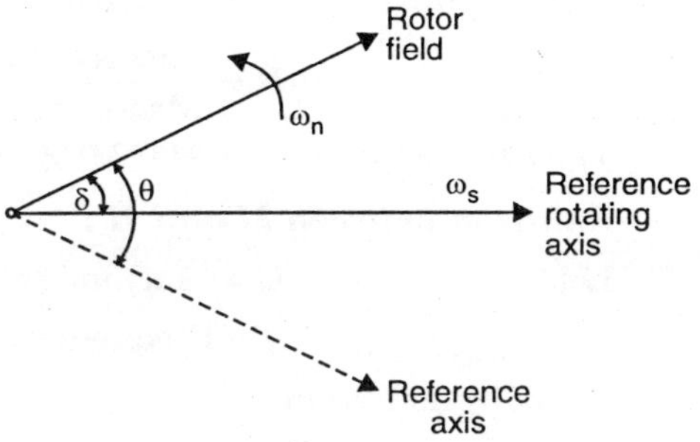

Fig. 17.4. Angular position of rotor with respect to reference axis.

Differentiating eqn. (17.28) w.r.t. time t, we get

$$\frac{d\theta}{dt} = \omega_s + \frac{d\delta}{dt} \quad ...(17.29)$$

and,
$$\frac{d^2\theta}{dt^2} = \frac{d^2\delta}{dt^2} \quad ...(17.30)$$

Again, from eqn. (17.26),

$$M\alpha = P_{AG}$$

Substituting the value $P_{AG} = P_M - P_E$ from eqn. (17.26) and $\alpha = \dfrac{d^2\delta}{dt^2}$ from eqn. (17.30) in the above equation, we get

$$M.\frac{d^2\delta}{dt^2} = P_{AG} = P_M - P_E \quad ...(17.31)$$

This equation is called the ***Swing equation.***

Further,
$$P_E = \frac{V_G V_M}{X} \sin\delta = P_{max} \sin\delta$$

$$\therefore \quad M\frac{d^2\delta}{dt^2} = P_M - P_{max}\sin\delta \quad ...(17.32)$$

— In case of power system comprising several machines, several swing equations need to be solved. The solution is complicated and normally the point-by-point method is used. Even for a single machine system connected to infinite bus with resistance neglected, the formal solution of swing equation is possible only when $P_M = 0$ and by the use of elliptic integrals.

— The solution of eqn. (17.32) gives a graph of load angle δ verses time t, called the '*swing curve*'. This curve is used to determine the stability of the system. The *indefinite increase in* δ *indicates instability*, but if its value reaches maximum and starts decreasing it indicates that system will not lose stability and eventually attain the equilibrium.

- From eqn. (17.27) $M = I\omega$ which therefore is not constant but varies somewhat during the swings due to variation in ω. Normally, in practice, the change in ω from the normal system angular velocity is not much during swing except of course when the machine falls out of step and, therefore, very little error is involved by the assumption that M is constant and is equal to $I\omega_n$ where ω_n is the normal angular velocity of the machine. This value of M is known as *'inertia constant of the machine'*, and is normally used in calculations for stability studies (The inertia constant is truely constant because it is angular momentum at synchronous speed).

H is *another inertia constant* which is quite useful in stability studies. It is *defined as the ratio of K.E. at rated speed to the rated apparent power of the machine.*

i.e.,

$$H = \frac{\text{Stored energy } (MJ)}{\text{Machine rating } [MVA]}$$

Thus H is expressed in MJ/MVA or $MW\text{-}s/MVA$

Relation between M and H :

Let, G = Machine rating, MVA, and

f = Frequency of the system.

Then, by definition,

$$G \times H = \text{Stored energy } (MJ)$$

or,

$$GH = \frac{1}{2} M\omega = \frac{1}{2} M \times 2\pi f = M\pi f \qquad ...(17.33)$$

or,

$$M = \frac{GH}{\pi f} \; MJ\text{-s/rad}$$

or,

$$M = \frac{GH}{180 f} \; MJ\text{-s/elect. degree} \qquad ...(17.34)$$

It may noted that :

— M depends upon the size of the machine as well as on its type ;

— H does not vary widely with size and has a characteristic value or set of values for each class of machines. In this respect H is similar to cent per cent reactance of machines. Whenever value of H is not known a characteristic value may be used.

Some typical inertia constants (H) are given below :

S.No.	*Type of machine*		*Inertia constant H, MJ/MVA*
1.	Synchronous condenser		
	Large		1.25
	Small		1.00
2.	Synchronous motor with load varying from 1.0 to 5.0 and higher for heavy flywheels		2.0
3.	Water wheel generator		
	Slow speed <	200 r.p.m.	2-3
	High speed >	200 r.p.m.	2-4
4.	Turbine generator		
	Condensing	1800 r.p.m.	9-6
		3000 r.p.m.	7-4
	Non-condensing	300 r.pm.	4-3

Example 17.1. *A 200 MVA, 2 pole, 50 Hz alternator has a moment of inertia of* 50×10^3 $kg\text{-}m^2$.

(*i*) *What is energy stored in the rotor at the rated speed ?*

(*ii*) *Find the value of H and determine the corresponding angular momentum.*

Sol. Machine rating, $G = 200$ MVA ; $p = 2$; $f = 50$ Hz ; $I = 50 \times 10^3$ kg-m^2

(*i*) **Energy stored, K. E. :**

$$K.E = \frac{1}{2} I\omega^2$$

where, $\omega = \frac{2\pi f}{p/2} = \frac{2\pi \times 50}{2/2} = 100\ \pi$ rad/s

$$\therefore \quad \text{K.E.} = \frac{1}{2} \times 50 \times 10^3 \times (100\ \pi)^2 = 2.467 \times 10^9 \text{ J}$$

or **2.467×10^3 MJ. (Ans.)**

(*ii*) **Value of H and angular momentum :**

$$\text{K.E.} = G \times H$$

$$\therefore \quad H = \frac{\text{K.E.}}{G} = \frac{2.467 \times 10^3}{200} = \textbf{12.335 MJ/MVA.} \quad \textbf{(Ans.)}$$

Angular momentum, $M = \frac{GH}{180 f}$

$$= \frac{200 \times 12.335}{180 \times 50} = \textbf{0.247 MJ-s/Elect deg.} \quad \textbf{(Ans.)}$$

Example 17.2. *A 4-pole turbogenerator rated at 20 MVA, 13.2 kV, 50 Hz has an inertia constant of H = 8.5 kW-s/kVA. Determine :*

(*i*) *The K.E. stored in the rotor at synchronous speed.*

(*ii*) *The acceleration if the input less the rotational losses is 17300 kW and the electric power developed is 14200 kW.*

(*iii*) *The change in torque angle in that period and the r.p.m. at the end of 10 cycles, the acceleration computed for the generator being constant for a period of 10 cycles.*

Assume that the generator is synchronised and a large system has no accelerating torque before the 10 cycle period begins.

Sol. *Given* : $p = 4$; Rating of Machine, $G = 20$ MVA ; $H = 8.5$ kW-s/kVA or 8.5 MJ/MVA ; Mechanical power input = 17300 kW ; Electric power developed = 14200 kW

(*i*) **K.E. stored in the rotor :**

We know that, $H = \frac{\text{Stored energy } (MJ)}{\text{Machine rating } (MVA)}$

∴ K.E. stored in the rotor = H × Machine rating = 8.5 × 20 = **170 MJ. (Ans.)**

(*ii*) **The acceleration α :**

The accelerating power, P_{AG} = Mechanical power input – electrical power output

= 17300 – 14200 = 3100 kW or 3.1 MW

Now the acceleration is, $\alpha = \frac{d^2\delta}{dt^2} = \frac{P_{AG}}{M} = \frac{3.1}{M}$

where, $M = \frac{GH}{180 f} = \frac{20 \times 8.5}{180 \times 50} = 0.0189$ MJ-s/elect deg.

$$\therefore \quad \alpha = \frac{d^2\delta}{dt^2} = \frac{3.1}{0.0189} = \textbf{164.02 elect. deg/s}^2. \quad \textbf{(Ans.)}$$

(*iii*) **The change in torque angle (δ) and r.p.m. at the end of 10 cycles :**

$$10 \text{ cycles} = \frac{10}{50} = 0.2 \text{ s}$$

$$\therefore \quad \textit{Change in } \delta = \frac{1}{2} \times (164.02) \times (0.2)^2 = \textbf{3.28 elect. deg.} \quad \textbf{(Ans.)}$$

Again,

$$\alpha = 164.02 \text{ elect. deg./s}^2 \quad \text{(calculate above)}$$

$$= \frac{164.02}{\text{pair of poles}} \text{ mech. deg./s}^2$$

$$= \frac{164.02}{2} = 82.01 \text{ mech. deg./s}^2$$

$$= \frac{82.01 \times 60}{360} = 13.67 \text{ r.p.m./s}$$

For 50 Hz system, 10 cycles correspond to $\frac{10}{50}$ *i.e.* $t^2 = 0.2$ second

$$\text{Synchronous speed} = \frac{120\,f}{p} = \frac{120 \times 50}{4} = 1500 \text{ r.p.m.}$$

∴ Speed at the end of 10 cycles

$$= 1500 + 13.67 \times 0.2 = \textbf{1502.7 r.p.m.} \quad \textbf{(Ans.)}$$

Example 17.3. *The 20 MVA generator of example 17.2 is supplying rated power at 0.8 p.f. (lag) when a fault reduces the electric power output by 40 percent. Find the accelerating torque at the time the fault occurs.*

Neglect the losses and assume the generator input to be constant.

Sol. Output power before fault = 20 × 0.8 = 16 MW

$$\text{Output after fault} = \frac{40}{100} \times 16 = 6.4 \text{ MW}$$

$$\therefore \quad \text{Accelerating power,} \quad P_{AG} = 16 - 6.4 = 9.6 \text{ MW}$$

$$\therefore \quad \text{Accelerating torque,} \quad T_A = \frac{P_{AG}}{\omega}$$

$$\text{where,} \quad \omega = \frac{2\pi f}{\text{No. of pair of poles}} = \frac{2\pi \times 50}{4/2} = 50\,\pi \text{ rad/s}$$

$$\therefore \quad T_A = \frac{9.6 \times 10^6}{50\,\pi} \text{ N.m} = \textbf{61115.5 N.m.} \quad \textbf{(Ans.)}$$

17.4.3. Equal Area Criterion

Several times, in a multi-machine system one of the machines may stay in step on the first swing and yet go out of step in the subsequent swing. For a two-machine system under the assumption of constant input, no damping and constant voltage behind transient reactance, the machines either fall out of step in the first swing or never. Under this condition the two machines are said to be running at standstill with respect to each other. There is a *graphical method* of determining whether the *two machines* are running at standstill with respect to each other or not. This method is known as "***Equal Area Criterion***" *for stability*. This method is *not applicable to multi-machine system directly*.

The use of this method *eliminates partially or wholly the calculations of swing curves which thus saves a considerable amount of work.*

The *'equal area criterion'* is derived using the swing equation for a machine connected to an infinite bus. The swing equation is given as

$$M \frac{d^2\delta}{dt^2} = P_M - P_E$$

Multiplying both sides by $2\frac{d\delta}{dt}$, and integrating with respect to time, we get

$$\int 2M \frac{d^2\delta}{dt^2} \cdot \frac{d\delta}{dt}\, dt = \int 2(P_M - P_E)\frac{d\delta}{dt} \cdot dt$$

or,
$$M\left(\frac{d\delta}{dt}\right)^2 = 2\int_{\delta_0}^{\delta} (P_M - P_E)\, d\delta$$

or,
$$\frac{d\delta}{dt} = \sqrt{\frac{2}{M}\int_{\delta_0}^{\delta} (P_M - P_E)\, d\delta} \quad \text{...(17.35)}$$

where, δ_0 is the torque angle at which the machine is operating while running at synchronous speed under normal conditions.

Under the above conditions the torque angle was not changing it before the disturbance $\frac{d\delta}{dt} = 0$. Also if the system has transient stability the machine will again operate at synchronous speed after the disturbance *i.e.* $\frac{d\delta}{dt} = 0$. Hence *condition for transient condition is given by the equation,*

$$\int_{\delta_0}^{\delta} \frac{2}{M}(P_M - P_E)\,.\,d\delta = 0 \quad \text{or} \quad \int_{\delta_0}^{\delta} (P_M - P_E)\, d\delta = 0$$

or,
$$\int_{\delta_0}^{\delta} P_A \,.\, d\delta = 0 \quad \text{...(17.36)}$$

(where P_A is the accelerating power).

This means that area under the curve P_A should be *zero*, which is possible only when P_A has *both accelerating and decelerating powers i.e.* for a part the curve $P_M > P_E$ and for the other $P_E > P_M$ (Fig. 17.5). For a generator action $P_M > P_E$ for positive area A_1 and $P_E > P_M$ for negative area A_2 for stable operation. Hence the name equal area criterion. The *area* A_1 represent the K.E. *stored by the rotor* during acceleration and the area A_2 represents the K.E. *given up by the rotor to the system* and when it is all given up, the machine has returned to its original speed.

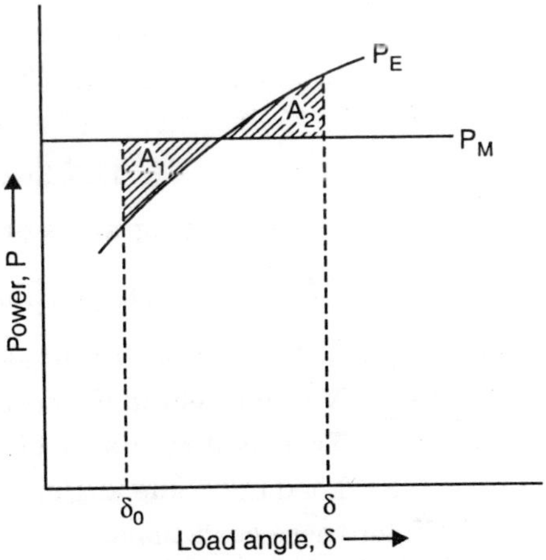

Fig. 17.5. Equal area criterion.

It may be noted that the K.E. involved in the above explanation is fictitious as it is being calculated corresponding to the relative speed rather than the actual speed.

The following *points are worth noting* :

(*i*) When the speed of the rotor is synchronous one, there is no change in the torque angle.

(*ii*) In a motor if $P_M > P_E$ (*i.e.* the mechanical output is more than the electrical input) the angle increases and the speed decreases.

(*iii*) If the speed is more than the synchronous speed the angle decreases.

Applications of equal area criterion *include the following :*

1. *Effects of increase in load.*
2. *Effects of switching operations.*
3. *Effects of faults with subsequent circuit isolation.*

1. Effects of increase in load. From the point of view of transient disturbance and system stability the following are the *two* possibilities of increasing the load :

(*i*) The total load *exceeds* the steady state limit of the system.

(*ii*) The rate of increase in load is so high as to set up *oscillations which cross the critical point.*

Let us consider a synchronous motor connected to an infinite bus-bar and operating at synchronous speed with mechanical output P_0. Let δ_0 (determined from the power-angle diagram shown in Fig. 17.6) be the load angle corresponding to P_0.

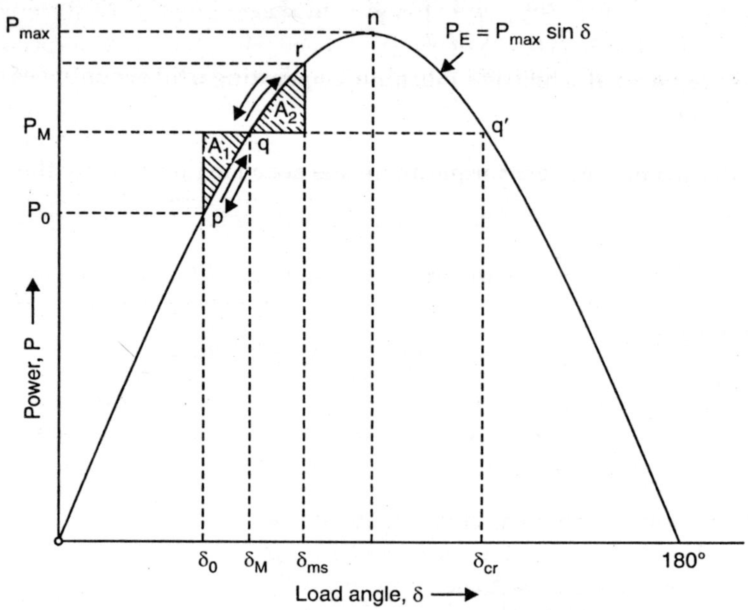

Fig. 17.6. Sudden change of load–equal area criterion.

The power-angle diagram is drawn by using the relation,

$P_E = \frac{V_G . V_M}{X} \sin \delta$, neglecting the system (motor, transmission line, transformers etc.)

where, V_G = Voltage of the infinite bus-bars,

V_M = Voltage behind the transient reactance of the synchronous motor,

X = Transient reactance of the system, and

δ = Load or torque angle.

Under these conditions,

$$P_0 = P_E = \frac{V_G . V_M}{X} \sin \delta$$

where, P_0 = Mechanical power output, and

P_E = Electrical input.

Let this load (on the motor) be suddenly increased to P_M. Momentarily there is no change in δ corresponding to electrical input to the motor $P_E < P_M$. Therefore the motor *decelerates supplying extra power required.* Consequently, the *load angle δ increases and the electrical input power P_E starts increasing. Thus electrical input keeps on increasing till it becomes equal to output power.* This operating point is given by the point q on the diagram and the load angle corresponding to P_M is δ_M. At point q input P_E is equal to P_M and, therefore, *decelerating force is zero but due to inertia of the rotor, the load angle goes on increasing.* The speed of the motor beyond point q starts *increasing. The speed is minimum at the operating point q.*

When the speed of the motor becomes equal to synchronous speed beyond point q, say at point r, the *load angle stops increasing*. But between operating points r and q, $P_E > P_M$; therefore the motor accelerates and the load angle starts *decreasing*. The speed goes on increasing till it reaches the point q where this time speed is maximum and is *more than synchronous speed*. At point q the acceleration ceases and the rotor moves towards a point p with a *finite deceleration*. In this way the *cycle keeps on repeating itself between points p and r till the oscillations are damped out.*

Thus, the speed at p and r is the synchronous speed whereas at q, the speed is below synchronous speed when the rotor oscillates from p to q and is above synchronous speed when it oscillates between r and q.

Applying 'equal area criterion of stability' and neglecting losses, the area of the sectioned part above the curved portion pq must be equal to the sectioned part below the curved portion qr. The first area A_1 represents the *rotational energy lost by thè rotor* which is due to the retarding torque given by,

$$T_R = T_M - T_E$$

The latter area A_2 (under the portion qr) represents the *energy gained by the rotor*. This is due to accelerating torque given by,

$$T_A = T_E - T_M$$

Thc two areas A_1 and A_2 are equal *since the speeds at points p and r are synchronous.*

Let δ_{ms} be the load angle corresponding to the point r, known as the '*point of maximum swing*' of the rotor. This angle should be such that the system stability is maintained.

Now, $$A_1 = \int_{\delta_0}^{\delta_M} (P_M \quad P_E)\, d\delta$$

and, $$A_2 = \int_{\delta_M}^{\delta_{ms}} (P_E - P_M)\, d\delta$$

Hence $A_1 - A_2 = \int_{\delta_0}^{\delta_{ms}} (P_M - P_E)\, d\delta = 0$ if the transient state stability condition given by the expression $\int_{\delta_0}^{\delta} (P_M - P_E)\, d\delta = 0$ is to be satisfied.

Also, at point r, $\dfrac{d\delta}{dt} = 0$.

It follows that point r should be so located that $A_1 = A_2$.

- The system will have transient stability if the point r lies anywhere on the power angle diagram above the line $r\,r'$. The point r' on the curve will give load angle δ_{cr} beyond which $P_E < P_M$ and the *unstable operation will set up.* At this instant the synchronism will be lost and the load angle δ will continue to increase till the motor comes to rest. *The value of the load angle beyond which the operation becomes unstable is called the "critical angle of load", δ_{cr}.*

In order to *maintain transient stability it is necessary that rotor swing must not be above the critical load angle* (δ_{cr}).

- It has been observed from power angle curve shown in Fig. 17.7 that if the load on the motor shaft is increased suddenly from P_0 to P_M, the system is stable. For this system let us now *find out the maximum value of the P_M such that the system is critically stable* (*i.e.*, for any attempt to increase P_M beyond this value the system becomes unstable).

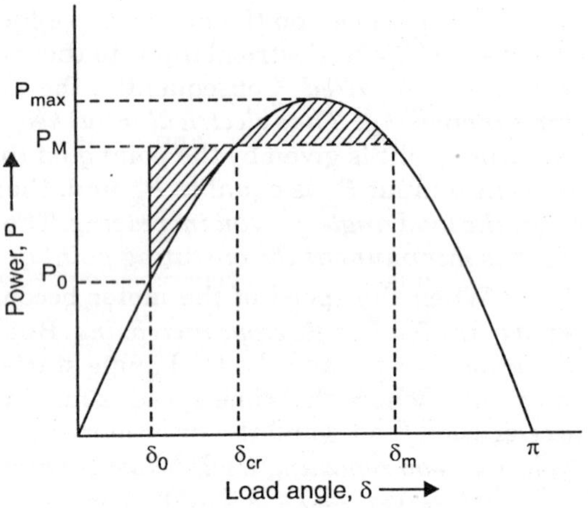

Fig. 17.7. Transient stability limit.

Refer to Fig. 17.7

$$P_M = P_{max} \sin \delta_{cr} = P_{max} \sin \delta_m$$

$$\therefore \quad \delta_m = (\pi - \delta_{cr})$$

where, δ_{cr} is known as critical torque angle.

The two shaded areas will be equal when,

$$P_M(\delta_m - \delta_0) = \int_{\delta_0}^{\delta_m} P_{max} \sin \delta \, d\delta \quad ...(17.37)$$

or,
$$P_{max} \sin \delta_m (\delta_m - \delta_0) = \int_{\delta_0}^{\delta_m} P_{max} \sin \delta \, d\delta \quad (\because \; P_M = P_{max} \sin \delta_m)$$

or,
$$P_{max} \sin \delta_m(\delta_m - \delta_0) = P_{max} (\cos \delta_0 - \cos \delta_m) \quad ...(17.38)$$

From the above equation the only unknown δ_m can be calculated and hence P_M can be determined.

2. Effects of switching operations. Fig. 17.8 shows a generator conveying power to infinite bus-bars through two transmission lines.

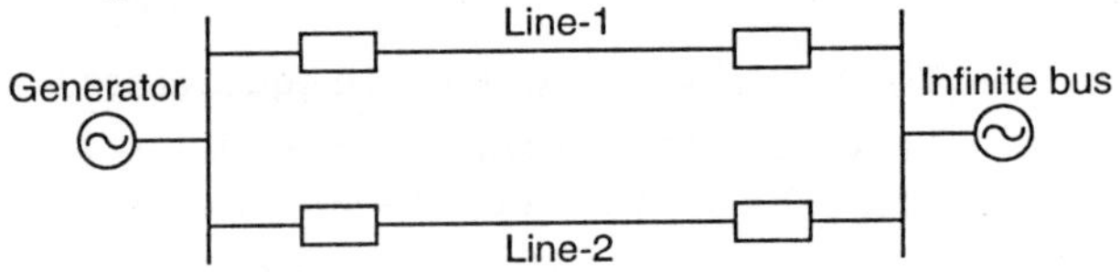

Fig. 17.8

— Curve A represents the power-angle diagram for the two transmission lines operating in parallel.

Curve B represents the power-angle diagram after isolating one of the lines.

— Let P_M be the mechanical power input corresponding to point p on the curve A, the load angle being δ_0. When one of the lines is opened the equivalent impedance between the bus-bars is increased and hence curve B will be lower than curve A. With the opening of the line, the generator output is reduced to say P_1, the operating point q on the curve. Since the input remains constant which is higher than the output, the rotor accelerates and hence the torque angle increases and the operating point moves along the curve B towards r from q. As it reaches r the accelerating power

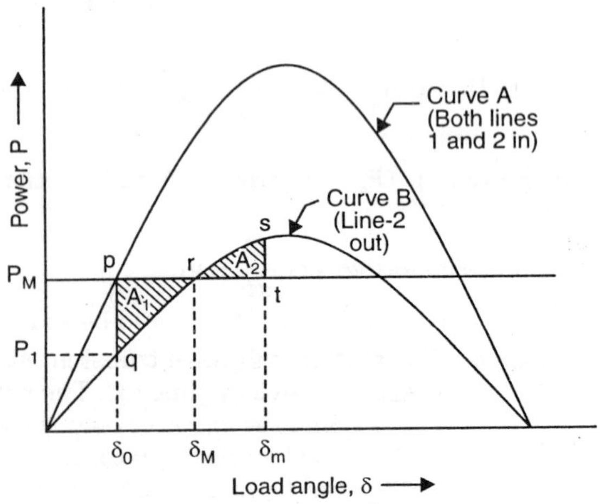

Fig. 17.9. Effects of switching operations.

becomes zero and the speed of the generator is more than the infinite bus-bar and speed continues to increase. At the point *s*, the output is more than input to the generator and hence the rotor decelerates and speed goes down relative to the infinite bus-bar till it reaches point *r* where the speed is minimum. This torque angle continues to decrease till it reaches point *q* where again the speed is equal to the speed of infinite bus-bar. The cycle *repeats itself* if damping is not present. It is found that in practice because of the damping present the rotor operates at point *r* on the curve *B* and the torque angle is δ_M.

— In order to determine the transient stability limit for the above case the input line P_M should be so raised that the areas A_1 and A_2 are equal. This is illustrated in Fig. 17.10. The value of P_M corresponding to this equation is known as the *transient stability limit*. The *transient stability in this case depends upon steady state limit of the remaining system and the difference in initial and final load angles.*

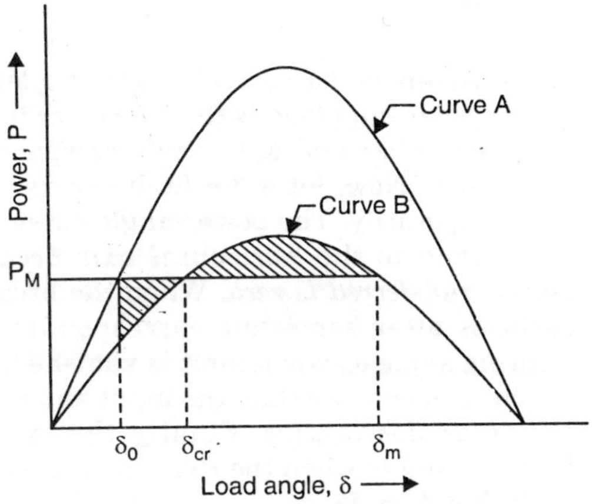

Fig. 17.10. Transient stability limit for the system in Fig. 17.8.

3. Effects of faults with subsequent circuit isolation :

- Refer to Fig. 17.11.

— Curve A represents the power-angle diagram corresponding to *healthy condition of the system* shown in Fig. 17.8.

Curve *B* corresponds to *fault on one of the two lines and fault allowed to exist for sometime.*

Curve *C* corresponds to the situation *when the faulted line is removed.*

— Under normal conditions, let P_M be the power supplied by the generator to the infinite bus-bars and δ_0 be the corresponding load angle.

When the fault occurs, the generator output reduces and becomes as at *p′*. Consequently the rotor accelerates and rotor moves along the curve *B* to point *q′*. Now the faulted line is removed and the operating point becomes *q* on the curve *C* where the output exceeds the input and the rotor decelerates till the speed becomes equal to the speed of the infinite bus and the load angle ceases to increase at point *r*.

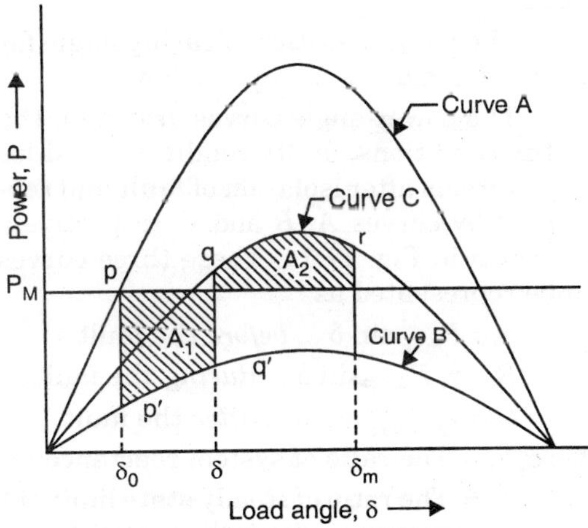

Fig. 17.11. Effects of faults with subsequent circuit isolation.

It is thus evident that the *transient stability limit not only depends upon the type of disturbance but also it depends upon the clearing time of the breaker.* Faster the breaker operation, smaller will be the area A_1 and hence, larger will be the transient limit.

- Let us now consider another case in which a generator is connected to the infinite bus-bars through a single line as shown in Fig. 17.12.

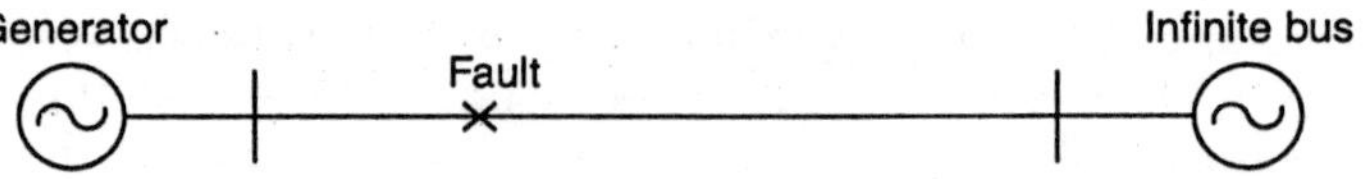

Fig. 17.12

Again, under normal conditions, let P_M be the power supplied by the generator to the infinite bus-bar and δ_0 be the corresponding load angle. Now, let a 3-ϕ fault occurs on a line temporarily. The power-angle curve will correspond to the *horizontal axis because power transferred is zero*. When the breaker recloses after sometime corresponding to clearing angle δ_{cl} when fault is vanished, the output will be more than the input and hence the rotor *decelerates*. Finally the system becomes stable when the clearing angle δ_{cl} is such that $A_1 = A_2$.

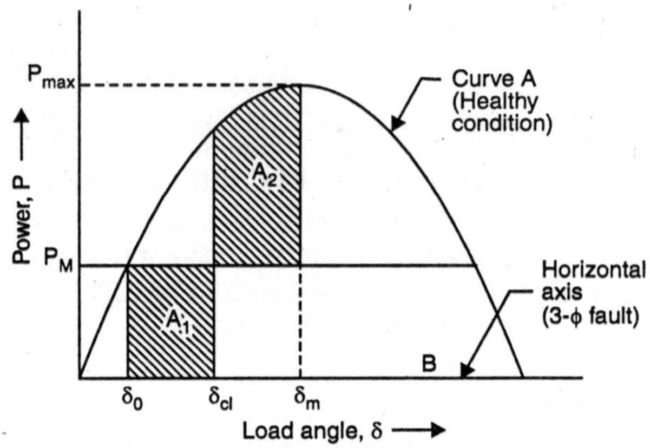

Fig. 17.13

17.4.4. Critical Clearing Angle

We find that for any given load there is a critical clearing angle and if actual clearing angle exceeds critical clearing angle the system becomes unstable, otherwise it is stable.

The value of critical clearing angle for a given load may be found as follows :

The power-angle curves drawn for the healthy conditions, faulty conditions and for the conditions after isolation of fault and represented by curves A, B and C respectively are shown in Fig. 17.14. These three curves can be represented as :

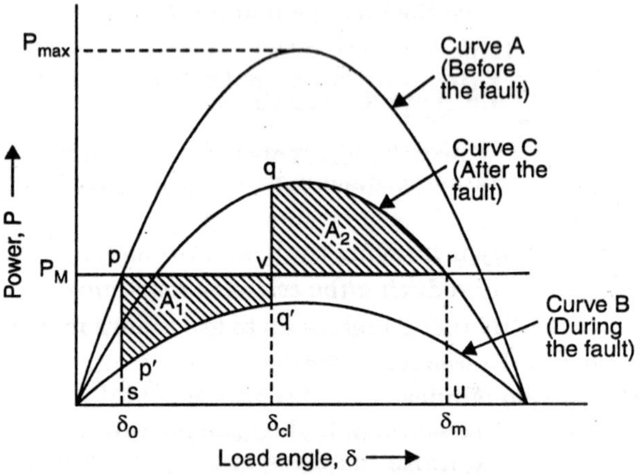

Fig. 17.14. Critical clearing angle.

$A = P_{max} \sin \delta$ ... *before* the fault

$B = \gamma_1 P_{max} \sin \delta$ ... *during* the fault

$C = \gamma_2 P_{max} \sin \delta$ ... *after* the fault

where, γ_1 = The ratio of system reactance in healthy conditions to that during the fault, and

γ_2 = The ratio of steady state limit of the system after the isolation of the fault and that of the system under initial conditions.

For transient stability $A_1 = A_2$ or equivalently the area under the curve *prus* (a rectangle) should be equal to the area under the curve *sp′q′qru i.e.*

$$P_M(\delta_m - \delta_0) = \int_{\delta_0}^{\delta_{cl}} \gamma_1 P_{max} \sin \delta \, d\delta + \int_{\delta_{cl}}^{\delta_m} \gamma_2 P_{max} \sin \delta \, d\delta$$

$$= \gamma_1 P_{max} (\cos \delta_0 - \cos \delta_{cl}) + \gamma_2 P_{max} (\cos \delta_{cl} - \cos \delta_m)$$

Now substituting $P_M = P_{max} \sin \delta_0$, we have

$$(\delta_m - \delta_0) P_{max} \sin \delta_0 = \gamma_1 P_{max}(\cos \gamma_0 - \cos \delta_{cl}) + \gamma_2 P_{max} (\cos \delta_{cl} - \cos \delta_m)$$

or, $$(\delta_m - \delta_0) \sin \delta_0 = \cos \delta_{cl}(\gamma_2 - \gamma_1) + \gamma_1 \cos \delta_0 - \gamma_2 \cos \delta_m$$

$$\therefore \qquad \cos \delta_{cl} = \frac{(\delta_m - \delta_0) \sin \delta_0 - \gamma_1 \cos \delta_0 + \gamma_2 \cos \delta_m}{\gamma_2 - \gamma_1} \qquad ...(17.39)$$

Now from the curves,

$$P_M = P_{max} \sin \delta_0 = \gamma_2 P_{max} \sin \delta_m = \gamma_2 P_{max} \sin (\pi - \delta_m)$$

or,
$$\sin \delta_0 = \gamma_2 \sin (\pi - \delta_m)$$

or,
$$\delta_m = \pi - \sin^{-1}\left(\frac{\sin \delta_0}{\gamma_2}\right) \qquad ...(17.40)$$

Thus if γ_1, γ_2 and δ_0 are known, the critical clearing angle δ_{cl} can be obtained.

17.4.5. Transient Stability in a Multi-machine System

In order to determine the multi-machine stability the following *steps* can be taken :

1. Determine the voltage behind transient reactance for all the generators from the pre fault load data ; this voltage is kept constant. The angle with this voltage is the initial rotor angle.
2. Record the prime mover output P_M (= P_E in steady state).
3. Augment the load flow network by the generator transient reactances. Assume the network buses to be behind the transient reactances.
4. Determine $(Y)_{Bus}$ during fault, and post fault clearing condition.
5. Obtain the generator outputs (using $P - \delta$ equation) during the fault condition.
6. Solve the swing equation. Obtain δ vs t which predicts stability.
7. Repeat steps 5 and 6 for post fault clearing mode.

17.4.6. Transient Stability Improvement

The transient stability can be improved by the following methods :

1. By using machines of higher inertia or by connecting the synchronous motors to heavy flywheels.
2. By increasing the system voltage.
3. By reducing the transfer reactance.
4. By using high neutral grounding impedance.
5. By reducing the severity of faults which can be achieved by using lightning arrestors for protection of lines.
6. By the use of high speed circuit breakers.

- To improve transient stability the governors attached to the turbines driving alternators should be high speed one so as to adjust the generator input quickly as per demand of the load.
- Voltage regulators employed on the lines should be quick acting ones.

Example 17.4. *A 3-phase line has a reactance of 11 Ω per phase. The voltage at each end is maintained at 132 kV (line-to-line). Determine :*

(i) The maximum steady state power that can be transmitted by the line.

(ii) The limits of angular oscillations for transient stability, when the above line develops a sudden jerk at $\frac{2}{5}$th of the steady state limit.

Sol. *Given* : $X = 11\ \Omega$, $V_{SL} = V_{RL} = V_L = 132$ kV

(*i*) **Maximum steady state power :**

Maximum steady state power that can be transmitted by the line for all the three phases,

$$P_{max} = \frac{V_{SL} V_{RL}}{X} MW \qquad [V_{SL}, V_{RL} \text{ are line values in kV}]$$

or,
$$P_{max} = \frac{132 \times 132}{11} MW$$
$$= \mathbf{1584\ MW.} \quad \textbf{(Ans.)}$$

(*ii*) **Limits of angular oscillations :**

- Draw the power angle diagram is drawn, as shown in Fig. 17.15.
- Mark point *A* *i.e.* $\frac{2}{5} P_{max}$ on the graph and read the load angle, which is 24° (elect.). As soon as the jerk occurs on the system the rotor of the alternator at the sending end will tend to advance. If it advances to point *B* the transient stability will be maintained. *B* corresponds to 156°.

Thus limit of angular oscillations for transient stability is **24° – 156° (elect.). (Ans.)**

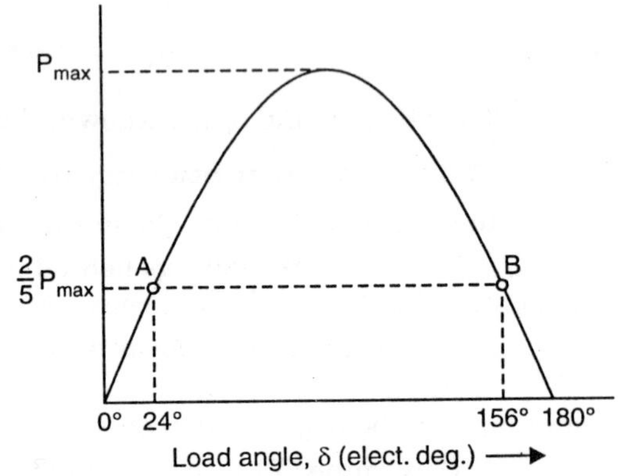

Fig. 17.15

Example 17.5. *A synchronous motor is receiving 25 percent of the power that is capable of receiving from an infinite bus. If the load on the motor is doubled, determine the maximum value of load angle δ during the swinging of the rotor around its new equilibrium position.*

Sol. Initial operating load, $P_0 = 0.25\ P_{max}$

Also, $$P_0 = P_{max} \sin \delta_0,$$

where, δ_0 is the operating load angle corresponding to this load.

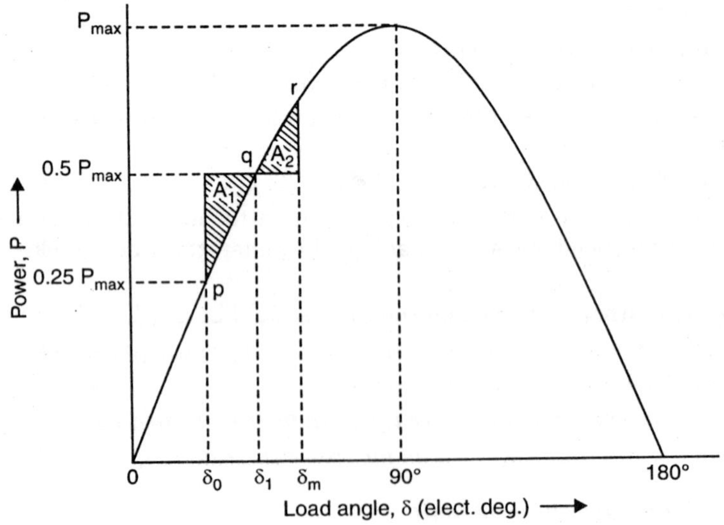

Fig. 17.16

or,
$$\sin \delta_0 = \frac{P_0}{P_{max}} = 0.25$$

$$\delta_0 = \sin^{-1}(0.25) = 14.48° = 0.2527 \text{ rad.}$$

Increased load on the bus,

$$P_1 = 2 \times 0.25\, P_{max} = 0.5\, P_{max}$$

Operating load angle corresponding to this new load,

$$\delta_1 = \sin^{-1}\left(\frac{P_1}{P_{max}}\right) = \sin^{-1}(0.5) = 30°$$

Let the maximum value of load angle during the swinging of the rotor be δ_m.

For transient stability,

Area A_1 = Area A_2

or, $$\int_{\delta_0}^{\delta_1} (0.5\, P_{max} - P_{max} \sin\delta)\, d\delta = \int_{\delta_1}^{\delta_m} (P_{max} \sin\delta - 0.5\, P_{max})\, d\delta$$

After simplification, we get

$$0.5(\delta_m - \delta_0) + \cos\delta_m - \cos\delta_0 = 0$$

or $$\cos\delta_m = \cos\delta_0 - 0.5\,(\delta_m - \delta_0)$$

$$= \cos 14.48° - 0.5(\delta_m - 0.2527)$$

$$= 0.9682 - 0.5\,\delta_m + 0.1263$$

or, $$\cos\delta_m = 1.0945 - 0.5\,\delta_m$$

For solution of this equation we make guess for δ_m such that δ_m should be greater than 30° (as δ_1 = 30°). After some trials, we get $\delta_m \simeq 0.785$ rad (elect.) or **45°. (Ans.)**

Example 17.6. *The steady state limit of a power system is 150 MW. A generator with constant excitation is supplying 75 MW to the system. Estimate the maximum permissible sudden increase in generator output without causing instability.* **(AMIE, Power systems)**

Sol. The maximum power that be transferred, P_{max} = 150 MW

Initial operating load, P_0 = 75 MW

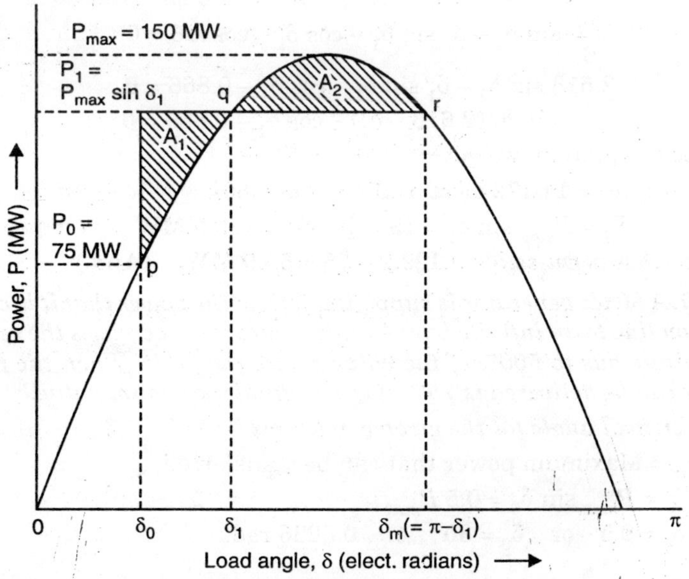

Fig. 17.17

Operating load angle corresponding to this load,

$$\delta_0 = \sin^{-1}\left(\frac{P_0}{P_{max}}\right) = \sin^{-1}\left(\frac{75}{150}\right) = 30° \quad \text{or} \quad 0.5236 \text{ rad.}$$

Let, P_1 = Maximum permissible suddenly increased load on the generator, MW, and

δ_1 = Operating load angle corresponding to load P_1.

Then, $P_1 = P_{max} \sin \delta_1 = 150 \sin \delta_1$

When the load is increased the rotor of the generator will advance beyond its new equilibrium point q and oscillate between the original point p and this extreme point r, and will finally attain new equilibrium point q due to the damping present.

The load angle δ_m corresponding to extreme point r, for transient stability, must be less or equal to $(\pi - \delta_1)$.

Now, Area A_1 = Area A_2 ... For transient stability

$$\int_{\delta_0}^{\delta_1} (P_{max} \sin \delta_1 - P_{max} \sin \delta)\, d\delta = \int_{\delta_1}^{\delta_m} (P_{max} \sin \delta - P_{max} \sin \delta_1)\, d\delta$$

or, $$\int_{\pi/6}^{\delta_1} (150 \sin \delta_1 - 150 \sin \delta)\, d\delta = \int_{\delta_1}^{(\pi-\delta_1)} (150 \sin \delta - 150 \sin \delta_1)\, d\delta$$

or, $$\int_{\pi/6}^{\delta_1} (\sin \delta_1 - \sin \delta)\, d\delta = -\int_{\delta_1}^{(\pi-\delta_1)} (\sin \delta_1 - \sin \delta)\, d\delta$$

or, $$\int_{\pi/6}^{\delta_1} (\sin \delta_1 - \sin \delta)\, d\delta + \int_{\delta_1}^{(\pi-\delta_1)} (\sin \delta_1 - \sin \delta)\, d\delta = 0$$

or, $$\int_{\pi/6}^{(\pi-\delta_1)} (\sin \delta_1 - \sin \delta)\, d\delta = 0 \quad \text{or} \quad \Big[\delta \sin \delta_1 + \cos \delta\Big]_{\pi/6}^{(\pi-\delta_1)} = 0$$

or, $$(\pi - \delta_1) \sin \delta_1 - \pi/6 \sin \delta_1 + \cos (\pi - \delta_1) - \cos \pi/6 = 0$$

or, $$\pi \sin \delta_1 - \delta_1 \sin \delta_1 - \frac{\pi}{6} \sin \delta_1 - \cos \delta_1 - \cos \pi/6 = 0$$

or, $$\frac{5\pi}{6} \sin \delta_1 - \delta_1 \sin \delta_1 - \cos \delta_1 - \cos \pi/6 = 0$$

$$2.618 \sin \delta_1 - \delta_1 \sin \delta_1 - \cos \delta_1 - 0.866 = 0$$

or, $$\sin \delta_1 (2.618 - \delta_1) - \cos \delta_1 - 0.866 = 0$$

Solving the above equation, we get :

$$\delta_1 = 1.0472 \text{ elect. radians} \quad \text{or} \quad 60°$$

∴ $P_1 = P_{max} \sin \delta_1 = 150 \sin 60° = 129.9$ MW

∴ *Sudden increase permissible* = 129.9 – 75 = **54.9 MW.** **(Ans.)**

Example 17.7. *A 50 Hz generator is supplying 50% of the power that it is capable of delivering through a transmission line to an infinite bus. A fault occurs that increases the reactance between the generator and the infinite bus to 500% of the value before the fault. When the fault is isolated, the maximum power that can be delivered is 75% of the original maximum value.*

Determine the critical angle for the given conditions.

Sol. Let, P_{max} = Maximum power that can be transferred.

Then, $P_M = P_{max} \sin \delta_0 = 0.5\, P_{max}$

∴ $\sin \delta_0 = 0.5$ or $\delta_0 = 30°$ or 0.5236 rad.

Since during the fault the reactance is 500% of the value before the fault,

$$\therefore \quad \gamma_1 = \frac{\text{Reactance before the fault}}{\text{Reactance during the fault}} = \frac{100}{500} = 0.2$$

and,
$$\gamma_2 = \frac{\text{Maximum power that can be transferred after isolation of the fault}}{\text{Maximum power that can be transferred initially}} = 0.75 \quad ...Given$$

Now,
$$\delta_m = \pi - \sin^{-1}\left(\frac{\sin \delta_0}{\gamma_2}\right) \quad ...[\text{Eqn. (17.40)}]$$

$$= \pi - \sin^{-1}\left(\frac{0.5}{0.75}\right) = 2.412 \text{ rad. or } 138.2°$$

The expression for the critical clearing angle, δ_{cl} is given by,

$$\delta_{cl} = \cos^{-1}\left[\frac{(\delta_m - \delta_0)\sin \delta_0 - \gamma_1 \cos \delta_0 + \gamma_2 \cos \delta_m}{\gamma_2 - \gamma_1}\right] \quad ...[\text{Eqn. (17.39)}]$$

$$= \cos^{-1}\left[\frac{(2.412 - 0.5236) \times 0.5 - 0.2 \cos 30° + 0.75 \cos 138.2°}{0.75 - 0.2}\right]$$

$$= \cos^{-1}\left[\frac{0.9442 - 0.1732 - 0.559}{0.55}\right] = \cos^{-1}(0.3854) = 67.32°$$

i.e. δ_{cl} **= 67.32°. (Ans.)**

HIGHLIGHTS

1. *Stability* of a power system is its ability to return to normal or stable operation after having been subjected to some form of disturbance.
2. *Stability limit* is the maximum power flow possible through some particular point in the system when the entire system or part of the system to which the stability limit refers is operating with stability.
3. *Steady state stability* is the stability of the system under conditions of gradual or relatively slow change in load.
4. Maximum power will be transferred from the sending end to the receiving end when $X = \sqrt{3}R$. Power transfer is possible only when reactance is present.
5. *Transient stability* refers to the maximum flow of power possible through a point without losing the stability with sudden and large changes in the network condition such as brought about by faults, by sudden large increment of loads.
6. *Dynamic stability* is the ability of a power system to remain in synchronism after the initial swing (transient stability period) until the system has settled down to the new steady-state equilibrium condition.
7. The equation giving the relative motion of the rotor (load angle δ) with respect to the stator field as a function of time is called the "*swing equation*". It is given by :

$$M \cdot \frac{d^2\delta}{dt^2} = P_{AG} = P_M - P_E = P_M - P_{max} \sin \delta.$$

8. The use of 'Equal area criterion' method eliminates partially or wholly with calculations of swing curves, which thus saves a considerable amount of work.
9. The 'critical clearing angle' is given by the relation :

$$\delta_{cl} = \cos^{-1}\left[\frac{(\delta_m - \delta_0)\sin \delta_0 - \gamma_1 \cos \delta_0 + \gamma_2 \cos \delta_m}{\gamma_2 - \gamma_1}\right].$$

THEORETICAL QUESTIONS

1. What do you mean by 'instability of a power system' ?
2. What do you understand by steady state stability limit and transient state stability limit of a power system ?
3. In case of a two-machine system obtain an expression for the steady state stability limit. Show that the maximum steady state power transmitted over the line will be greatest when $X = \sqrt{3}\,R$.
4. Derive the power-angle equation, $P = \frac{E_G E_M}{X} \sin \delta$ with usual notations.
5. Derive the swing equation explaining symbol of each term used.
6. Explain with the help of power-angle diagrams, the difference between steady state and transient stability of a power system.
7. Discuss the methods of improving the transient state stability of a power system.
8. Distinguish between the steady state and transient stability of a power system. Explain clearly the application of equal criterion for studying the transient stability of a system.
9. What is meant by 'power-angle' diagram ? Show how this may be used to determine the stability of a simple system under a transient fault condition.
10. Distinguish between the 'transient' and 'dynamic stability' of a power system. Discuss how the transient stability of a power system could be improved.
11. What is critical fault clearing angle and its effect upon the stability ? Obtain an expression for the same.
12. Write short notes on the following :
    (*i*) Power angle diagram.
    (*ii*) Power angle diagram and system stability.
    (*iii*) Transient stability and methods of its improvement.

UNSOLVED EXAMPLES

1. A 4-pole turbo-generator rated at 20 MVA, 13.2 kV, 50 Hz has an inertia constant of H = 9 MJ/MVA. Determine :
   (*i*) The K.E. stored in the rotor at synchronous speed.
   (*ii*) The acceleration if the input less the rotational losses is 17300 kW and the electric power developed is 14200 kW.
   (*iii*) The change in torque angle in that period and the r.p.m. at the end of 10 cycles, the acceleration computed for generator being constant for a period of 10 cycles.

   Assume that the generator is synchronised and a large system has no accelerating torque before the 10 cycles period begins.

   [**Ans.** (*i*) 180 MJ ; (*ii*) 168.75 elect. deg./s^2 ; (*iii*) 7.59 elect. deg., 1504.2 r.p.m.]
2. A 3-phase transmission line has a reactance of 12 Ω per phase. The voltage at each end is maintained at 132 kV (line-to-line). Determine the following :
   (*i*) The maximum steady state power that can be transmitted by the line.
   (*ii*) The limit of angular oscillations for transient stability, when the above line develops a sudden jerk at $\frac{2}{5}$ th of the steady state limit. [**Ans.** (*i*) 1452 MW, (*ii*) 24° – 156°]
3. A synchronous motor is receiving 35 percent of the power that it is capable of receiving from an infinite bus. If the load on the motor is doubled, determine the maximum value of load angle δ during the swinging of the rotor around its new equilibrium position. [**Ans.** 72°]
4. The steady state limit of a power system is 100 MW. A generator with constant excitation is supplying 50 MW to the system. Estimate the maximum permissible sudden increase in generator output without causing instability. [**Ans.** 36.6 MW]

5. A 50 Hz generator is delivering 40% of the power that it is capable of delivering through a transmission line to an infinite bus. A fault occurs that increases the reactance between the generator and the infinite bus to 600% of the value before the fault. When the fault is isolated, the maximum power that can be delivered is 80% of the original maximum.
Determine the critical angle for the given conditions. [**Ans.** 86.6° elect.]

6. A 50 Hz turbo-generator is delivering 50% of the power that it is capable of delivering through a transmission line to an infinite bus. A fault occurs that increases the reactance between the generator and the infinite bus to 400% of the value before the fault. When the fault is isolated the maximum power that can be delivered is 75% of the original maximum value. Determine the critical clearing angle for the given conditions. [**Ans.** 70.3° elect.]

7. A 50-Hz 4-pole turbo-alternator of rating 40 MVA, 11 kV has an inertia constant of 15 kW-s/kVA. Determine :

(*i*) The K.E. stored in the rotor at synchronous speed.

(*ii*) The acceleration and accelerating torque if the shaft input less the rotational losses is 20 MW and the electrical power developed is 15 MW. [**Ans.** (*i*) 600 MJ ; (*ii*) 37.5 deg/s^2 ; 31831 N-m]

18

Control of Reactive Power, Voltage and Load Frequency in Transmission Systems

18.1. Generation and absorption of reactive power. 18.2. Effect of reactive power on voltage and voltage regulation. 18.3. Relations between δQ, δP and V at a node. 18.4. Reactive compensation in power system. 18.5. Types of compensators. 18.6. Methods of voltage control in transmission system. 18.7. Relations between voltage and reactive power, frequency and active power. 18.8. Control of load frequency. Highlights—Theoretical Questions.

18.1. GENERATION AND ABSORPTION OF REACTIVE POWER

The study of generation and absorption of reactive power in the power system is essential since the reactive power is very precious in keeping the voltage of the power system stable. *Whereas frequency is the indicator of active power balance, voltage is the sole indicator of reactive power balance.*

The components responsible for the generation and absorption of reactive power in the power system are :

1. Alternators.
2. Overhead lines and transformers.

1. Alternators :

- An *overexcited alternator generates reactive power while an underexcited machine absorbs it.* Normally the alternators are run in overexcited mode such that they can act as main source of reactive power supply to the power system.
- The reactive power supply capability is determined by the short circuit ratio.

2. Overhead lines and transformers :

- *A fully loaded* ***overhead line*** absorbs reactive power, given by I^2X, where I and X are the line current and line reactance (Ω) per phase respectively. The distributed shunt capacitances throughout the line, on light loads or at no load, become predominant and consequently the line supplies charging VAR *i.e.* generates reactive power (given by $2\pi fCV^2$, where V is the line voltage, f the system frequency and C the line-to-earth capacitances).
  — ***Cables*** *generate reactive power* due to the various cable capacitances.
- ***Transformers*** *always absorb reactive power*, given by :

$$\frac{I^2VX_t}{I_{rated}} \; VAR\text{/phase}$$

where, I = Phase current,

I_{rated} = Rated phase current,

V = Phase voltage, and

X_t = The p.u. reactance of the transformer.

18.2. EFFECT OF REACTIVE POWER ON VOLTAGE AND VOLTAGE REGULATION

Fig. 18.1 shows the phasor diagram of a simple transmission network.

Let, E = Sending end voltage,

V = Receiving end voltage, and

δ = Power angle.

Then,

$$E^2 = (V + \Delta V)^2 + (\delta V)^2$$

$$E^2 = (V + IR \cos \phi + IX \sin \phi)^2 + (IX \cos \phi - IR \sin \phi)^2$$

Further, let $P = VI \cos \phi$ *i.e.,* $I \cos \phi = \dfrac{P}{V}$, and

$Q = VI \sin \phi$ *i.e.,* $I \sin \phi = \dfrac{Q}{V}$,

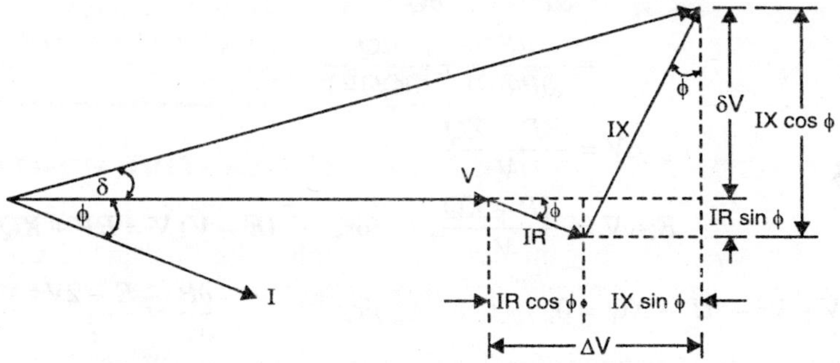

Fig. 18.1. Phasor diagram of a simple transmission network.

Then,

$$E^2 = \left(V + \frac{RP}{V} + \frac{XQ}{V}\right)^2 + \left(\frac{XP}{V} - \frac{RQ}{V}\right)^2$$

$$E^2 = \left(V + \frac{RP + XQ}{V}\right)^2 + \left(\frac{XP - RQ}{V}\right)^2 \quad ...(18.1)$$

Thus,

$$\Delta V = \frac{RP + XQ}{V} \quad ...(18.2)$$

and,

$$\delta V = \frac{XP - RQ}{V} \quad ...(18.3)$$

Assuming $\delta V << \Delta V$, we have

$$E^2 = \left(V + \frac{RP + XQ}{V}\right)^2$$

and,

$$E - V = \Delta V = \frac{RP + XQ}{V} \quad ...(18.4)$$

Assuming line resistance to be zero, we get

$$E - V = \frac{XQ}{V} \quad ...(18.5)$$

- From the above equation we find that *voltage depends on reactive power* (Q).
- The power angle δ is governed by real power (P) and is obtained from,

$$\delta = \sin^{-1}\left(\frac{\delta V}{E}\right) \quad ...(18.6)$$

- When R is assumed to be zero, E is specified and is found from the relation,

$$V = \sqrt{\left(E - \frac{QX}{E}\right)^2 + \left(\frac{PX}{E}\right)^2} \quad ...(18.7)$$

18.3. RELATIONS BETWEEN δQ, δP AND V AT A NODE

The relations between incremental reactive power (δQ), active power (δP) and voltage at a node are discussed below :

The phase voltage at any node is governed by the variations of P and Q (load power) provided there are no external means to control the bus voltage.

i.e. $$V = f(P, Q)$$

$$\therefore \quad dV = \frac{\partial V}{dP}.dP + \frac{\partial V}{\partial Q}.dQ$$

$$= \frac{dP}{(\partial P/\partial V)} + \frac{dQ}{(\partial Q/\partial V)} \quad ...(18.8)$$

Also, $$\Delta V = \frac{RP + XQ}{V} \quad ...[\text{Eqn. } (18.4)]$$

or, $$E - V = \frac{RP + XQ}{V}, \quad \text{or,} \quad (E - V)\, V = RP + XQ$$

or, $$EV - V^2 - RP - XQ = 0, \quad \text{or,} \quad \frac{\partial P}{\partial V} = \frac{E - 2V}{R} \quad ...(18.9)$$

(assuming E to be constant)

Similarly, $$\frac{\partial Q}{\partial V} = \frac{E - 2V}{X} \quad ...(18.10)$$

Putting the values of $\frac{\delta P}{\delta V}$ and $\frac{\delta Q}{\delta V}$ from eqns. (18.9) and (18.10) in (18.8), we get

$$dV = \frac{dP}{\left(\frac{E - 2V}{R}\right)} + \frac{dQ}{\left(\frac{E - 2V}{X}\right)} = \frac{dP.R + dQ.X}{E - 2V} \quad ...(18.11)$$

If the voltage V is constant, $dV = 0$

Then, $$dP.R + dQ.X = 0 \quad \text{or} \quad dQ = \left(-\frac{R}{X}\right) dP \quad ...(18.12)$$

From the relation $\frac{\partial Q}{\partial V} = \frac{E - 2V}{X}$, it may be seen that the smaller the reactance associated with the node, the large will be the value of $\frac{\partial Q}{\partial V}$ for a given voltage drop. This shows that if *the nodal reactance is low the voltage drop will be small. The nodal reactance can be made low by using parallel transmission lines.*

At any short circuit in the node in practical extra high voltage ($EH\ V$) lines, $R << X$, $I_{SC} = \frac{E}{X}$ amp.

When the system is at no load,

$$\frac{\partial Q}{\partial V} = -\frac{E}{X} \quad \text{... from eqn. (18.10) with } E = V \quad ...(18.13)$$

Thus, the magnitude of I_{SC} (short circuit current) becomes numerically equal to the magnitude of $\left(\frac{\partial Q}{\partial V}\right)$.

— With the help of these derivations, we can find relations between active power, reactive power and voltage. All these derivations highlight the role of nodal reactance on the profile of the reactive power voltage.

— The above discussion also reveals that the factor $\frac{\partial Q}{\partial V}$ is a measure of determining I_{SC} at a node.

18.4. REACTIVE COMPENSATION IN POWER SYSTEM

It has been seen in the previous article that for maintaining terminal voltage at the load bus adequate reactive reserve is needed. The devices which supply this reactive power at the receiving end bus in transmission system or load bus in distributing system are known as *"reactive power compensating devices"*. Besides voltage control, these devices exercise the following *effects* :

1. Decrease in system copper loss owing to reduced reactive current.
2. Decrease in supply side reactive burden.
3. Reduction in kVA loading of alternators.
4. Decrease in investment per kW of load supplied.
5. Improvement in system power factor (which helps in achieving better economy in power distribution.)

18.5. TYPES OF COMPENSATORS

The various types of *compensators are* :

1. Shunt capacitors.
2. Synchronous compensator.
3. Series capacitors.

1. Shunt capacitors :

- The small banks of static capacitor usually connected in star/delta are employed in substations for supplying reactive power (Q) to the power system. This arrangement *improves the power factor of the load.* The supply of reactive power to the bus maintains its voltage profile.
- In transmission systems, the static capacitor banks are often connected to the tertiary delta winding of the substation transformer.

The reactive power supplied by the capacitors,

$$Q_C = 2\pi f C V^2 \text{ MVAR/phase,}$$

where f, C and V are frequency, capacitance and voltage per phase respectively.

— Shunt connected inductors are frequently used at the load end of the transmission line to control the over voltage of the receiving end at light load condition (Ferrants effect). The reactive power *absorbed* by these inductors (in star or in delta connection) is given by,

$$Q_L = -V^2/(2\pi f L) \text{ MVAR/phase (–ve sign indicates reactive power absorption).}$$

2. Synchronous compensator :

Synchronous compensator (or synchronous phase modifier) is basically a *synchronous motor running at no load.* Depending on the value of excitation, it can absorb or generate reactive power.

Advantages :

(*i*) Improvement in system stability.

(*ii*) Supply of heavy amount of reactive power during short period.

(*iii*) Reduction of the effect of sudden change in load.

(*iv*) Voltage drips are arrested.

(*v*) Control is continuous.

Disadvantanges :

(*i*) High cost of installation.

(*ii*) Contribution of more short circuit current in the system.

(*iii*) More loss.

(*iv*) Higher maintenance.

3. Series capacitors :

- A series capacitor is a powerful device which *amply improves the power transfer capability as well as the voltage profile of the line as well as the receiving end bus.* Since these capacitors are connected in series with the line conductors, therefore, they reduce the net inductive reactance between the two ends of the line. It improves the voltage stable state and supplies reactive power to the receiving end such that voltage profile is maintained. Power transfer also increases.
- Switched series capacitors can be employed to improve transient stability.

The use of series capacitors is still limited except for EHV systems owing to the following *reasons* :

(*i*) During line fault severe overvoltage occurs.

(*ii*) When the load reactive power requirement is small it is not economical in use.

(*iii*) When the degree of compensation is high, series capacitors compensation may load to mal-operation of distance relays in the transmission line.

18.6. METHODS OF VOLTAGE CONTROL IN TRANSMISSION SYSTEM

The following are the *methods of voltage control in a transmission system* :

1. By transformer tap changing.
2. Booster (or regulating) transformer.
3. Static VAR system.

1. By transformer tap changing :

- Transformer tap changing is the basic and easiest way of voltage control of transmission, sub-transmission and distribution system. The transformer does not generate any reactive power (rather it consumes) and only transfers the reactive power from one side to another side by *changing the in-phase component of the system voltage.*
- In high voltage or extra high voltage lines on-load tap changing (OLTC) transformers are used while ordinary off-load tap changers prevail in distribution circuit.
- *Use of tap changing transformers to improve system voltage profile is limited to smaller voltage variation only.*

2. Booster (or regulating) transformer :

- A single-phase booster consists of two parts ; an exciting transformer connected across the supply along with a series transformer with the supply. The output voltage of the series transformer can be added to the input voltage in phase or in reverse phase by changing the position of a switch. The *output voltage of the regulating transformer can be varied by changing the taps of the exciting transformer.*
- These transformers are oftenly used when it is *inconvenient to have tappings in the main transformers.*
- A booster transformer is costlier not versatile in use (though it can be used in some distribution feeders).

3. Static VAR system :

- Whereas '*shunt capacitor compensation*' is required *to increase the system voltage* when the *transmission line is heavily loaded, the 'reactors'* are needed to reduce *the overvoltage of the line at light load condition.* These two tasks can be performed together by a Static VAR System (SVS) *utilising the static VAR compensator.*
- A '*static VAR compensator (SVC)*' is basically a *parallel combination* of controlled reactor and fixed static capacitor (See Fig. 18.2). The reactor control is done by an anti-parallel thyristor switch assembly. The firing angle of thyristors governs the voltage across the inductor thus controlling the reactor component. Thus the reactive power absorption by the inductor can be controlled.

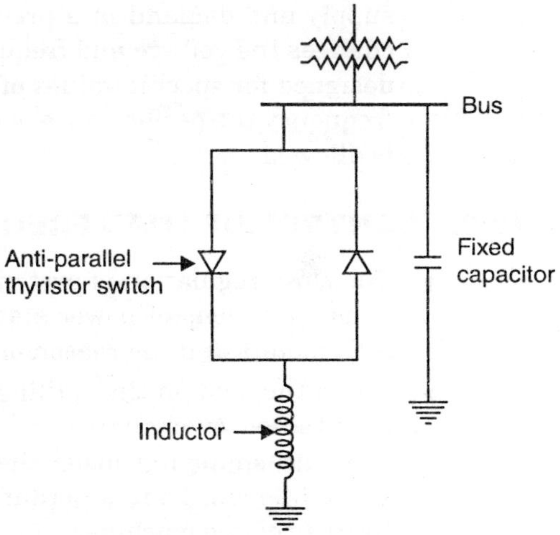

Fig. 18.2. Static VAR compensator.

- Static VAR compensator is much more advantageous than the synchronous compensator. It also *improves system stability, voltage stability and reduces power oscillations.*
- This equipment finds wide application in *transmission, sub-transmission, and even in distribution bus voltage control.*

18.7. RELATIONS BETWEEN VOLTAGE AND REACTIVE POWER, FREQUENCY AND ACTIVE POWER

The regulation of a power transmission line is given by,

$$\Delta V = \frac{PR + QX}{V} \quad ...(18.14)$$

Normally R is negligible in H.V. or E.H.V. lines and hence can be neglected.

$$\therefore \quad \Delta V = \frac{QX}{V}$$

i.e.,

$$\Delta V = f_1(Q)$$

Further, the active power in the transmission line is expressed as,

$$P = \frac{EV}{X} \sin \delta \quad ...(18.15)$$

where, δ in the power angle between sending end and receiving end

$$\therefore \qquad \sin \delta = \frac{PX}{EV}$$

i.e.,
$$\delta = f_2(P) \qquad ...(18.16)$$

Also,
$$\frac{d\delta}{dt} = \omega = 2\pi f = f_2'(P) \qquad ...(18.17)$$

— Thus, the *system voltage is a function of reactive power while the system frequency is a function of real power.*

— In actual power system it is not easy to keep balance between active and reactive power supply and demand in a precise manner. When the active and reactive power demand changes the voltage and frequency of bus also change. Also, the consumer appliances are designed for specific values of voltage and frequency and for change in supply voltage or frequency the performance of the apparatus changes. Some permissible limit of variation is allowed.

18.8. CONTROL OF LOAD FREQUENCY

- The close regulation of system frequency is essential in respect of the need for synchronous operations of power stations, the relation between system frequency and speed of A.C. motors and the reason of governing effect of frequency on electric clocks.
- When the load on the turbine increases, its speed decreases. The nominal speed can be maintained by changing the setting of the governor. If the system consists of a single machine supplying loads then speed and frequency change (normal frequency variation – 3 to 5%) will be in accordance with the governor characteristic. But when the system consists of two machines running in parallel, the system becomes more complex. The load will then be shared by the machines according to their characteristics.

The *frequency control may be done by automatic or by manual regulators.* Since manual regulation is not feasible in a large interconnected system, *load frequency control equipment is installed for each generator. Automatic voltage regulators are also installed to control the voltage of the system.*

- In a power system the load frequency controller and the voltage regulator are modelled and analysed independently. However, the load frequency controller response is slow while the response of voltage regulator is fast. The *frequency regulator should be sensitive to fast and small changes otherwise the system will be prone to hunting of the alternator.*
- The speed governing regulation in a *coherent system* is best achieved when *dynamic frequency response is taken into account.* In a proportional control system, the speed changer setting depends on steady state frequency change. This control will be unable to tolerate large dynamic frequency changes with frequency dependent load. Thus if the speed changer setting is adjusted automatically by monitoring the frequency change then the governor setting will be able to cope with the dynamic variation of frequency. For this purpose, a signal from frequency variation sensor is made through an integrator to the speed changer. The system now becomes a proportional integral controller giving zero steady state error.

Note. (*i*) In an interconnected system, the change in power for a given change in frequency is called *stiffness*. The smaller the change in frequency for a given load change, the stiffer the system.

(*ii*) The *real and reactive loads are proportional to the square of voltage.*

(*iii*) The performance of impedance type of loads depends upon voltage and frequency.

(*iv*) For a change in voltage change in real power is *twice*.

HIGHLIGHTS

1. Whereas frequency is the indicator of active power balance, voltage is the sole indicator of reactive power balance.
2. *Alternators* and *overhead lines and transformers* are the components which are responsible for the generation and absorption of reactive power.
3. The various types of compensators are : (*i*) Shunt capacitors ; (*ii*) Synchronous compensator ; (*iii*) Series capacitors.
4. The methods of voltage control in a transmission system are :
   (*i*) Transformer tap changing ; (*ii*) Booster or regulating transformer ;
   (*iii*) Static VAR system.

THEORETICAL QUESTIONS

1. Name the components of power system which generate or absorb power.
2. Define the relationship between the incremental changes of P, Q and V in a load node.
3. What is reactive compensation ? What are its advantages ?
4. Enumerate various types of compensators and explain any one of them.
5. What is a booster transformer ? Where is it used ?
6. Explain briefly the various methods of voltage control in a transmission system.
7. Explain briefly 'load frequency control'.

19

Power Distribution Systems

19.1. **Distribution systems**—General aspects—Classification of distribution systems—Requirements of a distribution system—Overhead versus underground systems. 19.2. **D.C. distribution**—Types of distributors—Types of loading—D.C. distributor fed at one end—D.C. distributor fed at both ends—Ring distributor—Three-wire D.C. system—Introduction—Advantages and disadvantages of 3-wire distribution—Methods to supply a 3-wire D.C. system—Comparison of 3-wire and 2-wire D.C. distribution. 19.3. **A.C. distribution.** Introduction—Three-phase four-wire distribution system—Voltage tolerances—Service connections—Service mains—Electric supply to our houses and factories—A.C. distribution with concentrated loads—Three-phase unbalanced loads—unbalanced star-connected loads—unbalanced delta-connected load. Highlights—Theoretical Questions—Unsolved Examples.

19.1. DISTRIBUTION SYSTEMS—GENERAL ASPECTS

- Oftenly it is impossible to draw a line between the transmission and distribution systems of a large power network.

That part of the network of a power system which distributes power for local use is known as **distribution system.**

- A distribution system consists of the following : Refer to Fig. 19.1.

(*i*) **Feeders :**

"Feeder" is a conductor which connects the substation (or localised generating station) to the area where power is to be distributed.

— Current through it *remains same*, since no tappings are taken from it.

— The current carrying capacity is the main consideration in the design of a feeder.

(*ii*) **Distributor :**

"Distributor" is a conductor from which tappings are taken for supply to the consumers.

— The current throughout a distributor is *not constant*, because tappings are taken at various places along its length.

— While designing a distributor, voltage drop along its length is the main consideration since the statutory limit of voltage variation is ± 6% of rated value at consumers' terminals.

(*iii*) **Service mains.** Generally service mains is a *small cable which connects the distributor to the consumers' terminals.*

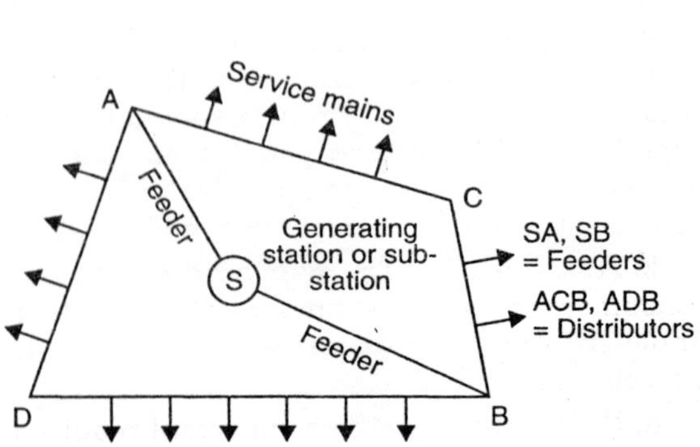

Fig. 19.1. Typical distribution system.

19.1.1. Classification of Distribution Systems

Distribution systems may be *classified* as follows :

1. Nature of current :

(*i*) Direct current (*D.C. distribution*)

(*a*) Two wire (*b*) Three wire.

(*ii*) Alternating current (*A.C. distribution*)

2. Method of connections :

(*i*) Series

(*a*) Open loop (*b*) Parallel loop

(*c*) Combination of open and parallel loops.

(*ii*) Multiple

(*a*) Three system (*b*) Feeder and main

(*c*) Network (*d*) Loop system

(*e*) Ring system.

3. Number of phases :

(*i*) Single

(*a*) Two-wire (*b*) Three-wire

(*ii*) Two

(*a*) Three-wire (*b*) Four-wire

(*c*) Five-wire

(*iii*) Three

(*a*) Threc wire (*b*) Four-wire

4. Mounting :

(*i*) Overhead (*ii*) Underground.

5. Voltage : 115/230, 550, 1100, 2200, 6600, 11000, 12000, 13200, 32000 V.

19.1.2. Requirements of a Distribution System

The following are the *requirements of a good distribution system* :

1. Proper voltage.
2. Availability of power on demand.
3. Reliability.

1. Proper voltage. One of the primary requirements of a distribution system is that voltage variations at consumers' terminals should be as low as possible. While low voltage causes revenue loss, inefficient lighting and possible burning out of motors, the high voltage causes permanent burning out of the lamps and may also cause failure of other appliances. This necessitates that the voltage variations must remain within permissible limits of ± 6% of the rated value.

2. Availability of power on demand. A good distribution system must make available to the consumers the required amount of power efficiently. This necessitates that operating staff must continuously study load patterns to predict in advance those major load changes that follow the known schedules.

3. Reliability. Another important requirement of a good distribution system is its *reliability*. Although cent percent reliability is not possible, yet the reliability can be improved to a considerable extent by :

— Interconnected system ;

— Reliable automatic control system ;

— Providing additional reserve facilities.

19.1.3. Overhead Versus Underground System

The distribution system can be *overhead* or *underground*.

— The *overhead lines* are generally mounted on wooden, concrete or steel poles which are arranged to carry distribution transformers in addition to the conductors.

— The *underground* system uses conduits, cables and manholes under the surface of streets and side walks.

The comparison (for making a choice between those two systems) between the overhead lines and underground system is given in the table below :

S.No.	Aspects	Overhead system	Underground system
1.	*Initial cost*	Low	High (may be 5 to 10 times the overhead system)
2.	*Public safety*	Less	More
3.	*Faults*	More chances of occurrence of faults	Very rare
4.	Fault location and repairs	Easy	difficult
5.	Flexibility	Much more flexible than underground system	Less flexible
6.	*Current carrying capacity and voltage drop*	High	Low comparatively
7.	*Appearance*	Not as good as underground system	Better comparatively (since all the distribution lines are invisible)
8.	*Maintenance cost*	High comparatively	Low
9.	*Interference with communication circuits*	Electromagnetic interference with the telephone lines	No interference with the telephone lines
10.	*Useful life*	25 years approx.	More than 50 years

- In general, overhead system is adopted for distribution and use of underground system is made only where overhead construction is impracticable or prohibited by local laws.

19.2. D.C. DISTRIBUTION

These days all production of power is an A.C. power and nearly all D.C. power is obtained from large A.C. power systems by using converting machinery like *synchronous or rotary converters, solid-state converters and motor-generator sets etc.*

Following are the *reasons for producing power in the form of alternating current rather than direct current.*

(*i*) In practice, it is possible to construct large high-speed A.C. generators of capacities upto 500 MW. Such generators are economical both in matter of cost per kWh of electric energy produced as well in operation. D.C. generators, however, cannot be built of ratings higher than 5 MW because of *commutation trouble.* Moreover, since they must operate at low speeds, it necessitates large and heavy machines.

(*ii*) *A.C. voltage can be efficiently and conveniently raised or lowered for economic transmission and distribution of electric power respectively.* On the other hand, D.C. power has to be generated at comparatively low voltages by units of relatively low power ratings. As yet there is no economical method of raising the D.C. voltage for transmission and lowering it for distribution.

For certain applications, *D.C. supply is absolutely necessary* ; *e.g.*, D.C. supply is required for the *operation of variable speed machinery such as D.C. motors, electro-chemical work and electric traction.*

In a D.C. system the electrical energy may be fed and distributed either by two-wire or by three-wire system.

Two-wire D.C. distribution systems. Fig. 19.2 shows a two-wire D.C. distribution system from generating station or substation to the consumer's terminals.

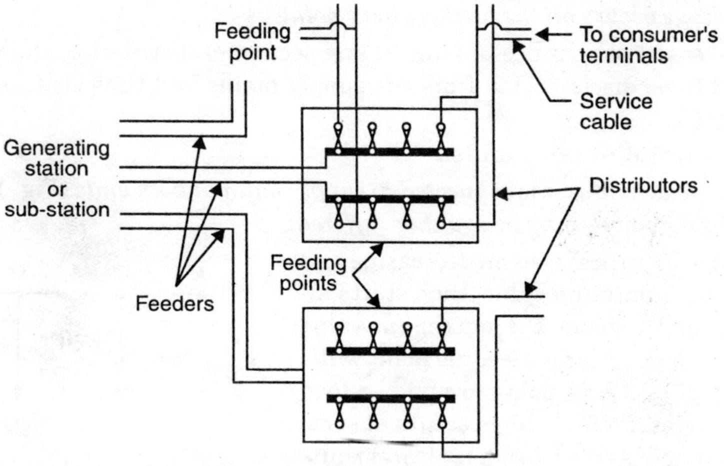

Fig. 19.2. Two-wire D.C. system.

- This system consists of two wires, one outgoing (known as *positive wire*) and another returning one (*negative wire*).
- Untapped feeders run into bus-bars in suitable feeding points in the distribution area.
- Distribution cables are connected to the bus-bars through fuses or links.
- Each separate consumer is fed from the distributor by a service cable tapped on to the distributor at the nearest convenient point.
- The electrical appliances and motors etc. are connected in parallel between the two wires.

Three-wire D.C. distribution systems :

- This system consists of *two outers* and an *earthed middle wire* known as *neutral wire.*
- The lamps or consumer apparatus are connected between the neutral and one of the outers.
- The voltage between the outers being twice that of consumer's terminals *increases the transmission efficiency and reduces the copper cost.*

19.2.1. Types of Distributors

The distributors are of the following *types* :

1. Distributors fed at one end.
2. Distributors fed at both ends.
3. Distributors fed at the centre.
4. Ring distributor.

1. Distributors fed at one end :

- In this type of feeding, *the distributor is connected to supply mains at one end and loads are tapped at different points along the length of the distributor.*
- The current in the various sections of the distributor away from feeding point goes on decreasing, as shown in Fig. 19.3 (which is self explanatory).

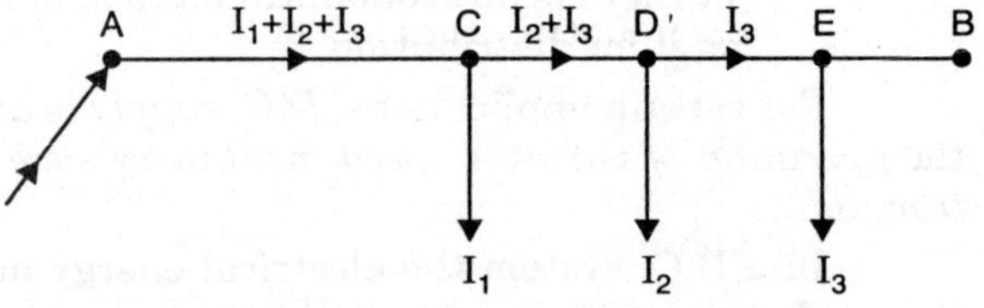

Fig. 19.3. Distributor fed at one end.

- The voltage across the loads away from the feeding point goes on decreasing. The *minimum voltage occurs on the farthest load point.*
- In the event of occurrence of fault in any section of distributor, the whole distributor is required to be disconnected from the supply mains and thus continuity of the supply is interrupted.

2. Distributor fed at both ends :

- In this case distributor is *connected to supply main at both ends* (Fig. 19.4). The voltage at both feeding points may be *equal or different.*
- The load voltage first goes on decreasing and reaches the minimum value, then starts increasing and reaches the maximum value when we reach the other feeding point while going from one load point to another load point. The minimum voltage occurs at some load point and is never fixed. It always shifts with the variation of load or different sections of the distribution.

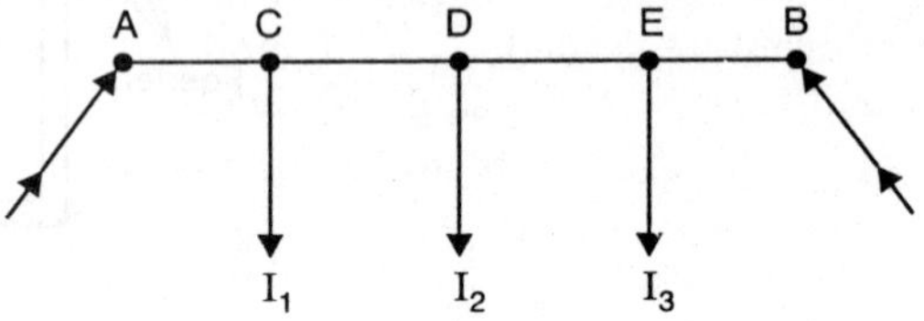

Fig. 19.4. Distributor fed at both ends.

Advantages :

(*i*) If fault occurs in any one feeder (feeding the distributor), the continuity of the supply is maintained from the other feeding point.

(*ii*) If due to some fault if any section of the distributor is isolated, the continuity of the supply is maintained to the remaining sections.

(*iii*) It is economical in cost since cross-section required for doubly fed distributors is much less as compared to singly fed distributor.

3. Distributor fed at the centre :

- In this type of feeding, the *centre of the distributor is connected to the supply mains* as shown in Fig. 19.5.
- This type of distributor is equivalent to two singly fed distributors, each distributor being of one half of its length and having common feeding point.

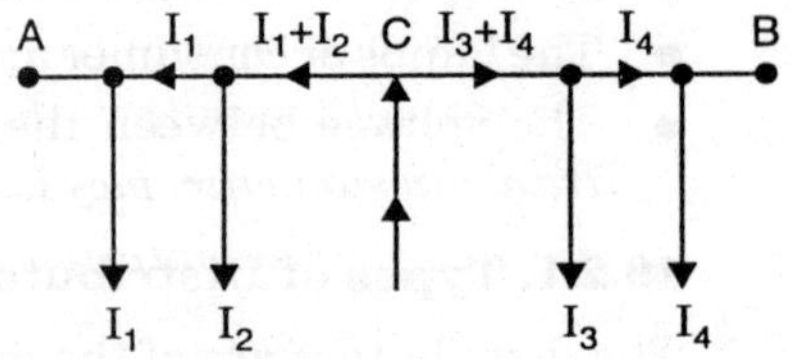

Fig. 19.5. Distributor fed at the centre.

4. Ring distributor. Refer to Fig. 19.6.

- A ring distributor is a *distributor which is arranged to form a closed circuit and which is fed at one or more than one points.*

- For the purpose of calculating voltage distribution, it can be looked upon as consisting of a series of open distributors fed at both the ends.
- If the ring distributor is fed on one point then, for the purpose of calculation, it is equivalent to a straight distributor fed at both ends with equal voltages.
- By using a ring distributor fed properly, great economy in copper can be affected.

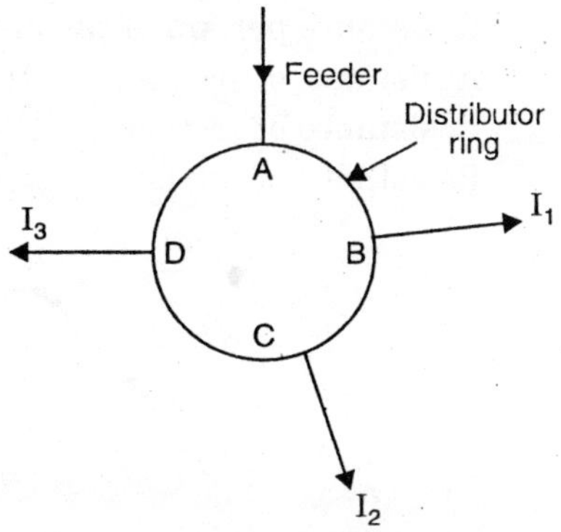

Fig. 19.6. Ring distributor/mains.

19.2.2. Types of Loading

The various types of loading are :

1. Concentrated loading.
2. Uniform loading.
3. Uniform loading in addition to concentrated loading.

19.2.3. D.C. Distributor Fed at One End

(A) Concentrated loading. Consider a 2-wire distributor *AB* fed at one end *A* having concentrated loads I_1, I_2, I_3 and I_4 tapped off at points *C*, *D*, *E* and *F* respectively.

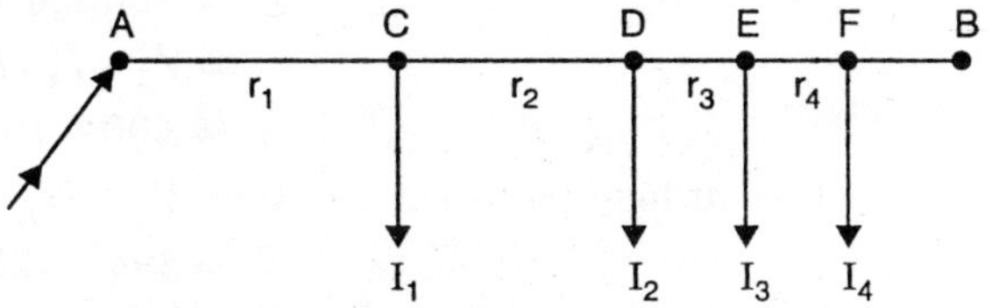

Fig. 19.7. Distributed fed at one end with concentrated loading.

Let r_1, r_2, r_3 and r_4 be the resistances of both wires (go and return) of the sections *AC*, *CD*, *DE* and *EF* of the distributor respectively.

Total current fed from point

$$A = I_1 + I_2 + I_3 + I_4$$

The currents flowing through different sections are :

Sections	*Current flowing*
AC	$I_1 + I_2 + I_3 + I_4$
CD	$I_2 + I_3 + I_4$
DE	$I_3 + I_4$
EF	I_4

Voltage drop in section $AC = (I_1 + I_2 + I_3 + I_4)r_1$

Voltage drop in section $CD = (I_2 + I_3 + I_4)r_2$

Voltage drop in section $DE = (I_3 + I_4)r_3$

Voltage drop in section $EF = I_4 r_4$

∴ Total voltage drop in the distributor

$$= (I_1 + I_2 + I_3 + I_4)r_1 + (I_2 + I_3 + I_4)r_2 + (I_3 + I_4)r_3 + I_4 r_4$$

It may be seen that *minimum potential* will occur at point *F* which is *farthest* from the feeding point.

Example 19.1. *A 2-wire D.C. distributor cable AB is 2.2 km long and supplies loads of 25 A, 50 A, 75 A at 0.4 km, 1 km and 1.6 km from the point A. Each conductor has a resistance of 0.05 Ω/km. Calculate the potential difference at each point if potential difference of 400 volts is maintained at point A.*

Sol. Fig. 19.8 shows the single line diagram of the distributor with tapped currents.

Resistance per km of the distributor = 2 × 0.05 = 0.1 Ω/km

Resistance of section $AC = 0.4 \times 0.1 = 0.04\ \Omega$

Resistance of section $CD = 0.6 \times 0.1 = 0.06\ \Omega$

Resistance of section $DE = 0.6 \times 0.1 = 0.06\ \Omega$

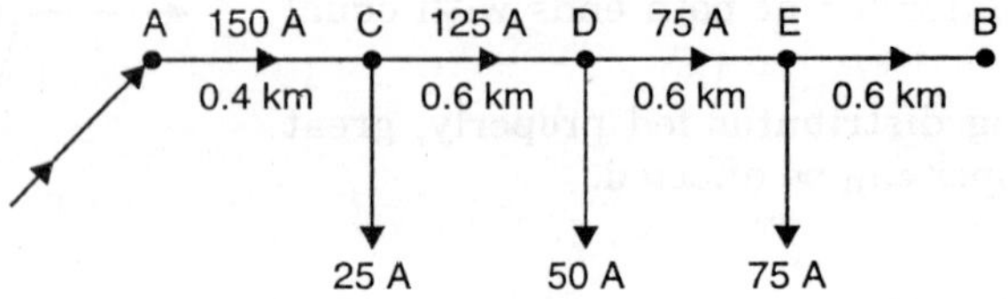

Fig. 19.8

Current in section $AC = 25 + 50 + 75 = 150$ A

Current in section $CD = 50 + 75 = 125$ A

Current in section $DE = 75$ A

P.D. (potential difference) at load point C,

$$V_C = \text{Voltage at } A - \text{voltage drop in } AC$$
$$= V_A - I_{AC} R_{AC}$$
$$= 400 - 150 \times 0.04 = \mathbf{394\ V.}\quad \textbf{(Ans.)}$$

P.D. at load point D, $V_D = V_C - I_{CD} R_{CD}$

$$= 394 - 125 \times 0.06 = \mathbf{386.5\ V.}\quad \textbf{(Ans.)}$$

P.D. at load point E, $V_E = V_D - I_{DE} R_{DE}$

$$= 386.5 - 75 \times 0.06 = \mathbf{382\ V.}\quad \textbf{(Ans.)}$$

Example 19.2. *The load distribution on a 2-wire D.C. distributor is shown in Fig. 19.9. The cross-sectional area of each conductor is 27 mm². The end A is supplied at 240 V. If the resistivity of wire is 1.78 μΩ cm calculate :*

(*i*) *The current in each section of the conductor.*

(*ii*) *The two-core resistance of each section.*

(*iii*) *The voltage at each tapping point.*

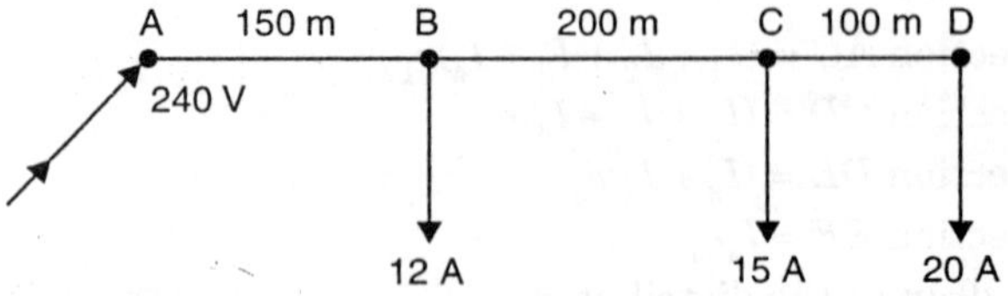

Fig. 19.9

Sol. *Given* : $a = 27\ \text{mm}^2 = 0.27\ \text{cm}^2$; Voltage supplied at end $A = 240$ V ; $\rho = 1.78\ \mu\Omega$ cm.

(*i*) **The current in each section of the conductor :**

Section AB, $I_{AB} = 10 + 15 + 20 =$ **45 A. (Ans.)**

Section BC, $I_{BC} = 15 + 20 =$ **35 A. (Ans.)**

Section CD, $I_{CD} =$ **20 A. (Ans.)**

(*ii*) **The two-core resistance of each section :**

Single core resistance of the section of 100 m length

$$= \frac{\rho l}{a} = \frac{1.78 \times 10^{-6} \times (100 \times 100)}{0.27} = 0.066\ \Omega$$

The resistances of the various sections are :

$$R_{AB} = 150 \times \frac{0.066}{100} \times 2 = \mathbf{0.198\ \Omega.}\quad \textbf{(Ans.)}$$

$$R_{BC} = 200 \times \frac{0.066}{100} \times 2 = \mathbf{0.264\ \Omega.}\quad \textbf{(Ans.)}$$

$$R_{CD} = 100 \times \frac{0.066}{100} \times 2 = \mathbf{0.132\ \Omega.}\quad \textbf{(Ans.)}$$

(*iii*) **The voltage at each tapping point :**

$$V_B = V_A - I_{AB}\,R_{AB} = 240 - 45 \times 0.198 = \mathbf{231.09\ V.}\quad \textbf{(Ans.)}$$

$$V_C = V_B - I_{BC}\,R_{BC} = 231.09 - 35 \times 0.264 = \mathbf{221.85\ V.}\quad \textbf{(Ans.)}$$

$$V_D = V_C - I_{CD}\,R_{CD} = 221.85 - 20 \times 0.132 = \mathbf{219.21\ V.}\quad \textbf{(Ans.)}$$

Example 19.3. *Find the cross-sectional area of the distributor shown in Fig. 19.10. The distances are given in metres. Take $\rho = 1.78 \times 10^{-8}$ Ω-m. The maximum drop is not to exceed 10 V. The conductor is fed from point A.* **(Panjab University)**

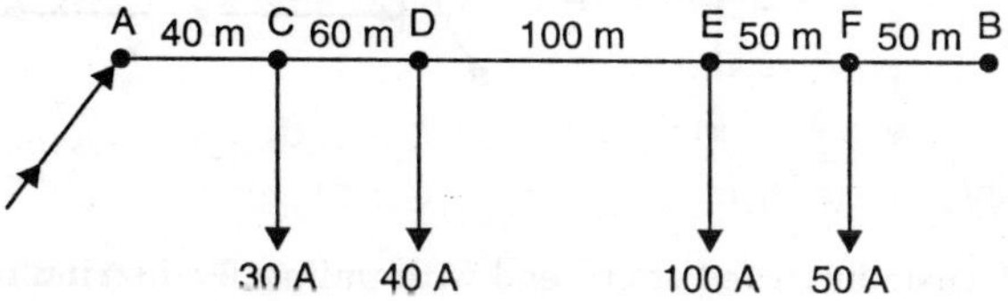

Fig. 19.10

Sol. Let, r = Resistance of 100 m length of the distributor (both conductors), Ω,

Then, resistances of various sections of the distributor are :

$$R_{AC} = 40 \times \frac{r}{100} = 0.4\ r\Omega$$

$$R_{CD} = 60 \times \frac{r}{100} = 0.6\ r\Omega$$

$$R_{DE} = 100 \times \frac{r}{100} = r\Omega\Omega$$

$$R_{EF} = 50 \times \frac{r}{100} = 0.5\ r\Omega$$

The current distribution in the various sections of the distributor shown in Fig. 19.11 is as follows :

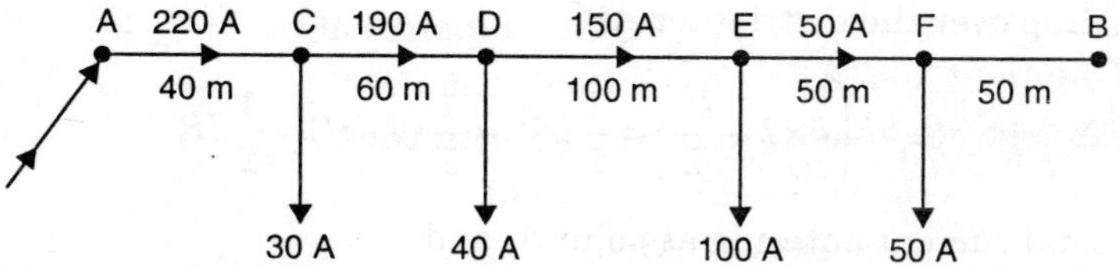

Fig. 19.11. The current distribution in the various sections of the distributor.

$$I_{AC} = 220 \text{ A} ; I_{CD} = 190 \text{ A} ; I_{DE} = 150 \text{ A} ; I_{EF} = 50 \text{ A}.$$

Total voltage drop over the distributor

$$= I_{AC} R_{AC} + I_{CD} R_{CD} + I_{DE} R_{DE} + I_{EF} R_{EF}$$
$$= 220 \times 0.4\, r + 190 \times 0.6\, r + 150 \times r + 50 \times 0.5\, r = 377\, r$$

As the maximum permissible drop in the distributor is 10 V,

$$\therefore \quad 377\, r = 10 \quad \text{or} \quad r = \frac{10}{377} = 0.0265\ \Omega$$

Area of cross-section of the distributor,

$$a = \frac{\rho l}{r} = \frac{1.78 \times 10^{-8} \times (2 \times 100)}{0.0265}$$

$$= 1.343 \times 10^{-4} \text{ m}^2 \quad \left(\because \text{ Resistance of each conductor of length } l = \frac{r}{2}\right)$$

$$= \mathbf{1.343\ cm^2}. \quad \textbf{(Ans.)}$$

(B) Uniform loading. Consider a distributor AB fed at one end A, with uniformly distributed loading, as shown in Fig. 19.12 (a).

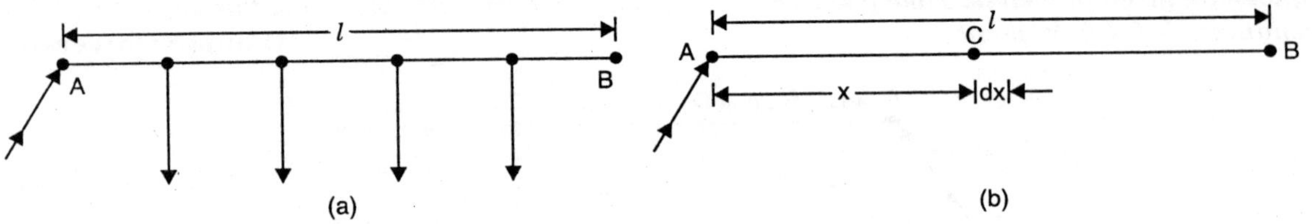

Fig. 19.12. Distributor fed at one end with uniformly distributed loading.

Let, l = Length of the distributor AB,

r = Resistance per metre length, and

i = Amperes per unit meter fed from point A at V_A volts.

Further, consider a point C, x units from the feeding point A, as shown in Fig. 19.12 (b).

Then, current at point $C = (il - ix) = i(l - x)$ amperes.

Now, consider a section of very small length dx near point C. Its resistance will be $r.dx$ and the voltage drop over the length dx,

$$dv = i(l - x)r.dx = ir(l - x)dx$$

Total voltage drop upto point C is given by,

$$v = \int_0^x ir\,(l - x)dx = ir\left(lx - \frac{x^2}{2}\right) \quad \text{...equation of a parabola} \quad \text{...(19.1)}$$

The voltage drop upto point B can be determined by putting $x = l$ in the above expression.

$\therefore$ Voltage drop over the distributor AB

$$= ir\left(l \times l - \frac{l^2}{2}\right) = \frac{1}{2} irl^2 = \frac{1}{2}(il)(rl) = \frac{1}{2} IR$$

where, $il = I$, the total current entering at point A, and

$ir = R$, the total resistance of the distributor.

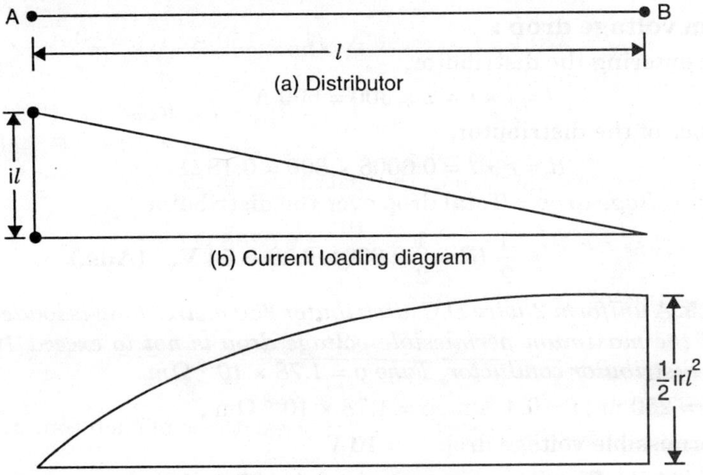

Fig. 19.13

From the expression of voltage drop at point B it is clear that the total voltage drop of an uniform loaded distributor is *equal to that caused by the whole of the load assumed concentrated at the middle point.*

The current loading and voltage drop diagrams are shown in Fig. 19.13 (*b*) and (*c*) respectively.

Power loss :

Power loss in length dx = (Current in length $dx)^2$ × resistance of length dx

$$= [i(l - x)]^2 \times r.dx$$

Total power loss P in the whole distributor,

$$P = \int_0^l [i(l-x)^2]\, r.dx = \int_0^l i^2 (l^2 + x^2 - 2lx)\, r.dx$$

$$= i^2 r \int_0^l (l^2 + x^2 - 2lx)\, dx = i^2 r \left[l^2 x + \frac{x^3}{3} - \frac{2lx^2}{2} \right]_0^l$$

or,
$$P = i^2 r \left[l^3 + \frac{l^3}{3} - l^3 \right] = \frac{1}{3} i^2 r l^3 \quad \text{...(19.2)}$$

Example 19.4. *A 2-wire D.C. distributor 300 metres long is uniformly loaded with 2A/metre. Resistance of single wire is 0.3 Ω/km. If the distributor is fed at one end, calculate :*

(*i*) *Voltage drop upto a distance of 225 m from the feeding point.*

(*ii*) *Maximum voltage drop.*

Sol. *Given* : $l = 300$ m ; $i = 2$ A/m ; $r = \dfrac{2 \times 0.3}{1000} = 0.0006$ Ω/m

(*i*) **Voltage drop upto a distance of 225 m from the feeding point :**

Voltage drop upto a distance x from the feeding point

$$= ir \left(lx - \frac{x^2}{2} \right) \quad \text{...[Eqn. (19.1)]}$$

Here, $x = 225$ m (*Given*)

∴ *Desired voltage drop* $= 2 \times 0.0006 \left(300 \times 225 - \dfrac{225^2}{2} \right) =$ **50.62 V. (Ans.)**

(*ii*) **Maximum voltage drop :**

Total current entering the distributor,

$$I = i \times l = 2 \times 300 = 600 \text{ A}$$

Total resistance of the distributor,

$$R = r \times l = 0.0006 \times 300 = 0.18\ \Omega$$

∴ *Maximum voltage drop* = Total drop over the distributor

$$= \frac{1}{2} IR = \frac{1}{2} \times 600 \times 0.18 = \mathbf{54\ V.}\quad \textbf{(Ans.)}$$

Example 19.5. *A uniform 2-wire D.C. distributor 250 metres long is loaded with 0.4 A / m and is fed at one end. If the maximum permissible voltage drop is not to exceed 10 V, find the cross-sectional area of the distributor conductor. Take ρ = 1.78 × 10^{-8} Ωm.*

Sol. *Given :* l = 250 m ; i = 0.4 A/m, ρ = 1.78 × 10^{-8} Ωm ;

Maximum permissible voltage drop = 10 V

Current entering the distributor = rl = 0.4 × 250 = 100 A

Let r ohm be the resistance per metre length of the distributor (both wires).

$$\text{Maximum voltage drop} = \frac{1}{2} IR$$

$$10 = \frac{1}{2} Irl \qquad (\because\ R = r.l)$$

or,

$$r = \frac{2 \times 10}{I \times l} = \frac{2 \times 10}{100 \times 250} = 0.0008\ \Omega\text{/m length}$$

∴ Area of cross-section of the distributor conductor,

$$a = \frac{\rho l}{(r/2)} = \frac{1.78 \times 120^{-8} \times 1}{(0.008/2)} = 4.45 \times 10^{-5}\ \text{m}^2 = \mathbf{0.445\ cm^2.}\quad \textbf{(Ans.)}$$

...(l is taken as 1 metre since r ohm is the resistance of 1 m length of the distributor)

Example 19.6. *Calculate the voltage at a distance of 250 m of a 350 long distributor uniformly loaded at the rate of 0.8 A / m. The distributor is fed at one end at 250 V. The resistance of the distributor (go and return) per metre is 0.00016 Ω.*

Also find the power loss in the distributor.

Sol. *Given :* x = 250 m ; l = 350 m ; i = 0.8 A/m ; Supply voltage = 250 V ; r = 0.00016 Ω (go and return)

Voltage drop at a distance x from the supply end

$$= ir\left(lx - \frac{x^2}{2}\right) = 0.8 \times 0.00016\left(350 \times 250 - \frac{250^2}{2}\right) = 7.2 \text{ V}$$

∴ *Voltage at a distance of 250 m from supply end*

$$= 250 - 7.2 = \mathbf{242.8\ V.}\quad \textbf{(Ans.)}$$

Power loss in the distributor,

$$P = \frac{1}{3} i^2 r l^3 \qquad \text{...[Eqn. (19.2)]}$$

$$= \frac{1}{3} \times 0.8^2 \times 0.00016 \times (350)^3 = 1463 \text{ W} = \mathbf{1.463\ kW.}\quad \textbf{(Ans.)}$$

19.2.4. D.C. Distributor Fed at Both Ends

A. Concentrated loading

(*a*) ***Distributor fed at two ends with "equal voltages":***

Consider a distributor *AB* fed at both ends with equal voltages, as shown in Fig. 19.14.

Let, V = Voltage at the end *A* and end *B*.

I_1, I_2, I_3, I_4, I_5 = Concentrated loads at points *C, D, E, F* and *G* respectively.

Regarding this type of distributor the *following points are worth noting.*

- As we move away from feeding point *A*, load voltages goes on decreasing, reaches the minimum value at some load point say *E*, and then starts increasing and becomes *V* volts as we reach the other feeding point *B*.

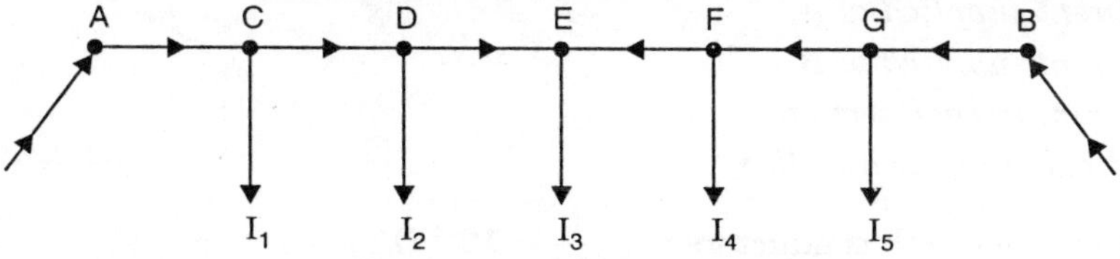

Fig. 19.14. Distributor fed at both ends.

- All the currents tapped off between the feeding point *A* and the point of minimum potential difference and supplied from feeding point *A* whereas the current tapped off between feeding point *B* and point of minimum potential are supplied from feeding point *B*.
- The current tapped at the point of minimum potential difference is supplied partly by feeding point *A* and partly by feeding point *B*.
- The sum of currents fed from feeding points *A* and *B* is equal to the total current tapped off from the distributor,

i.e., $$I_A + I_B = I_1 + I_2 + I_3 + I_4 + I_5.$$

(*b*) ***Distributor fed at two ends with unequal voltages.*** Consider a distributor *AB* fed with unequal voltages, say end *A* at voltage of V_A volts and end *B* at voltage of V_B volts. In order to determine the point of *minimum potential*, let us assume that current of I_A amperes is supplied from the feeding point *A*. Then currents through various sections will be :

Section	*Current flowing*
AC	I_A
CD	$(I_A - I_1)$
DE	$(I_A - I_1 - I_2)$
EF	$(I_A - I_1 - I_2 - I_3)$
FG	$(I_A - I_1 - I_2 - I_3 - I_4)$
GB	$(I_A - I_1 - I_2 - I_3 - I_4 - I_5)$

Now, voltage drop from feeding point *A* upto feeding point *B*

$$= I_A R_{AC} + (I_A - I_1)R_{CD} + (I_A - I_1 - I_2)R_{DE} + (I_A - I_1 - I_2 - I_3)R_{EF} + (I_A - I_1 - I_2 - I_3 - I_4)R_{FG} + (I_A - I_1 - I_2 - I_3 - I_4 - I_5)R_{GB}$$

i.e.
$$V_A - V_B = I_A R_{AC} + (I_A - I_1)R_{CD} + (I_A - I_1 - I_2)R_{DE} + (I_A - I_1 - I_2 - I_3)R_{EF} + (I_A - I_1 - I_2 - I_3 - I_4)R_{FG} + (I_A - I_1 - I_2 - I_3 - I_4 - I_5)R_{GB} \quad ...(19.3)$$

From the above expression the value of I_A can be determined, the other quantities being known. *In case of both the voltages being same $V_A - V_B$ be put equal to zero.*

The load point at which currents are coming from both sides of the distributor will be the point of minimum potential.

Example 19.7. *The resistance of two conductors of a 2-wire D.C. distributor shown in Fig. 19.15 is 0.1 Ω per 1000 mm for both conductors. Both A and B are maintained at 200 V.*

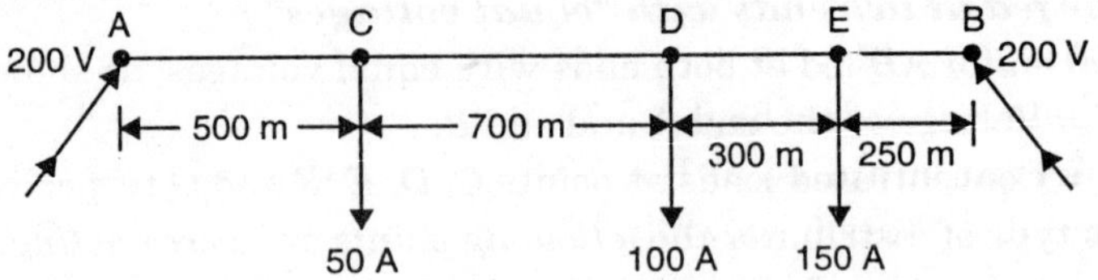

Fig. 19.15

Calculate the following :

(i) The current supplied at A.

(ii) The current supplied at B.

(iii) The current in each section.

(iv) The voltages at C, D and E. **(Gwalior University)**

Sol. Resistance for both conductors $= \dfrac{0.1}{1000} = 10^{-4}\ \Omega/\text{m}$

(i) **Current supplied at A :**

Refer to Fig. 19.16. Let current supplied from feeding point A be I_A amperes, then currents in various sections are :

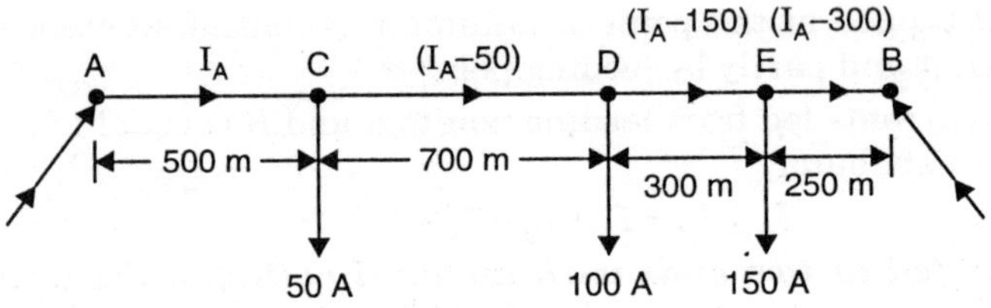

Fig. 19.16

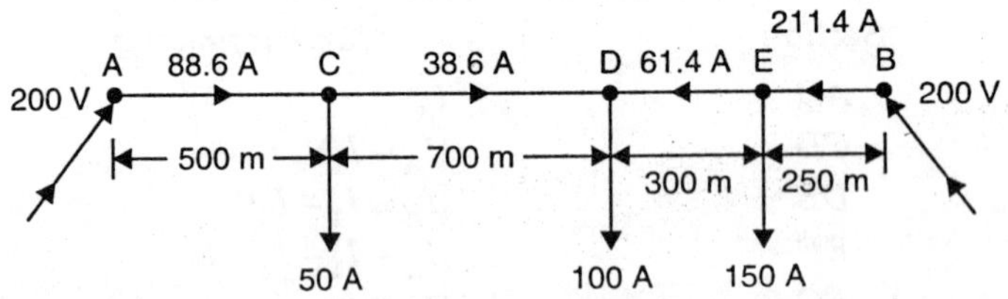

Fig. 19.17

Current in section AC, $I_{AC} = I_A$

Current in section CD, $I_{CD} = I_A - 50$

Current in section DE, $I_{DE} = I_A - 50 - 100 = I_A - 150$

Current in section EF, $I_{EB} = I_A - 50 - 100 - 150 = I_A - 300$

Since voltage at the other feeding end B

= Voltage at feeding end A

– voltage drop in sections AC, CD, DE and EB respectively,

$$\therefore \quad V_B = V_A - (I_{AC} R_{AC} + I_{CD} R_{CD} + I_{DE} R_{DE} + I_{EB} R_{EB})$$

$$200 = 200 - [I_A \times 500 \times 10^{-4} + (I_A - 50) \times 700 \times 10^{-4} + (I_A - 150) \times 300 \times 10^{-4} + (I_A - 300) \times 250 \times 10^{-4}]$$
$$= 200 - 10^{-4}\,[500\, I_A + 700\, I_A - 35000 + 300\, I_A - 45000 + 250\, I_A - 75000]$$
$$200 = 200 - 10^{-4}\,(1750\, I_A - 155000)$$

$\therefore$ *Current supplied at A,* I_A = **88.6 A. (Ans.)**

(*ii*), (*iii*) **The current supplied at B** and **the current in each section :**

From the current distribution shown in Fig. 19.17, we have :

Current in section AC = **88.6 A. (Ans.)**

Current in section CD = 88.6 – 50 = **38.6 A. (Ans.)**

Current in section DB = 38.6 – 100 = – **61.4 A. (Ans.)**

Current in section EB = – 61.4 – 150 = – **211.4 A. (Ans.)**

Hence current entering the cable at point B = **211.4 A.** (*From B to E*) **(Ans.)**

(–ve sign indicates the current flow from B towards A)

(*iv*) **The voltages at C, D and E :**

$$V_C = V_A - I_{AC}\, R_{AC} = 200 - 88.6 \times 500 \times 10^{-4} = \mathbf{195.57\ V.}\quad \textbf{(Ans.)}$$
$$V_D = V_C - I_{CD}\, R_{CD} = 195.57 - 38.6 \times 700 \times 10^{-4} = \mathbf{192.87\ V.}\quad \textbf{(Ans.)}$$
$$V_E = V_D - I_{DE}\, R_{DE} = 192.87 - [-61.4 \times 300 \times 10^{-4}] = \mathbf{194.71\ V.}\quad \textbf{(Ans.)}$$

Example 19.8. *A two-conductor street main AB, 500 metres in length is fed from both ends at 250 volts. Loads of 50 A, 60 A, 40 A and 30 A are tapped at distance of 100 m, 250 m, 350 m and 400 m from A respectively. If the cross-section of the conductors be 1 cm² and specific resistance of the material of the conductors is 1.7 μΩ-cm, determine the minimum consumer voltage.* **(B.T.E. Delhi)**

Sol. *Given :* l = 500 m ; $V_A = V_B$ = 250 V ; a = 1 cm² ; ρ = 1.7 μΩ-cm.

Fig. 19.18 shows the distributor with its tapped currents.

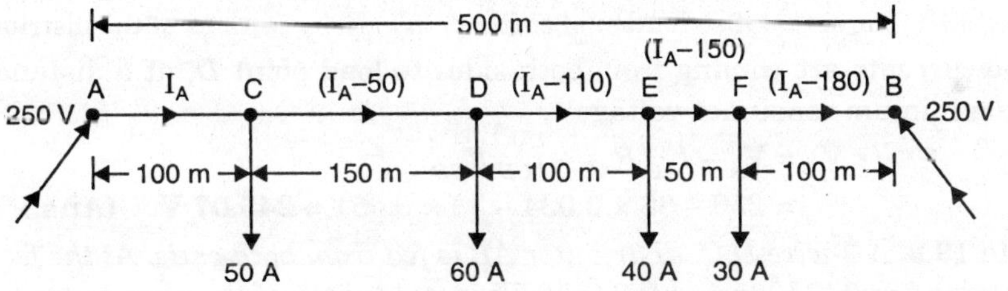

Fig. 19.18

Resistance per 100 metres of street main *both conductors,*

$$r = \frac{\rho l}{a} = \frac{1.7 \times 10^{-6} \times (2 \times 100 \times 100)}{1} = 0.034\ \Omega$$

$$\therefore\quad R_{AC} = 100 \times \frac{0.034}{100} = 0.034\ \Omega$$
$$R_{CD} = 150 \times \frac{0.034}{100} = 0.051\ \Omega$$
$$R_{DE} = 100 \times \frac{0.034}{100} = 0.034\ \Omega$$
$$R_{EF} = 50 \times \frac{0.034}{100} = 0.017\ \Omega$$
$$R_{FB} = 100 \times \frac{0.034}{100} = 0.034\ \Omega$$

Let the current supplied from feeding point A be I_A amperes, then currents in various sections will be as shown in Fig. 19.18.

Now, $V_B = V_A$ – voltage drops in sections AC, CD, DE, EF and FB

or,
$$250 = 250 - (I_{AC} R_{AC} + I_{CD} R_{CD} + I_{DE} R_{DE} + I_{EF} R_{EF} + I_{FB} R_{FB})$$
$$= 250 - [0.034\, I_A + 0.051(I_A - 50) + 0.034(I_A - 110) + 0.017(I_A - 150) + 0.034(I_A - 180)]$$
$$= 250 - [0.034\, I_A + 0.051\, I_A - 2.55 + 0.034\, I_A - 3.74 + 0.017\, I_A - 2.55 + 0.034\, I_A - 6.12]$$
$$= 250 - (0.17\, I_A - 14.96)$$

$\therefore$ $I_A = 88$ A

$\therefore$ Current in section $AC = I_A = 88$ A

Current in section CD, $I_A - 50 = 88 - 50 = 38$ A

Current in section DE, $I_A - 110 = 88 - 110 = -22$ A

Current in section EF, $I_A - 180 = 88 - 180 = -62$ A

Current in section FB, $I_A - 180 = 88 - 100 = -92$ A

(–ve sign indicates that current flow is from B to A)

The actual current distribution in the distributor is shown in Fig. 19.19.

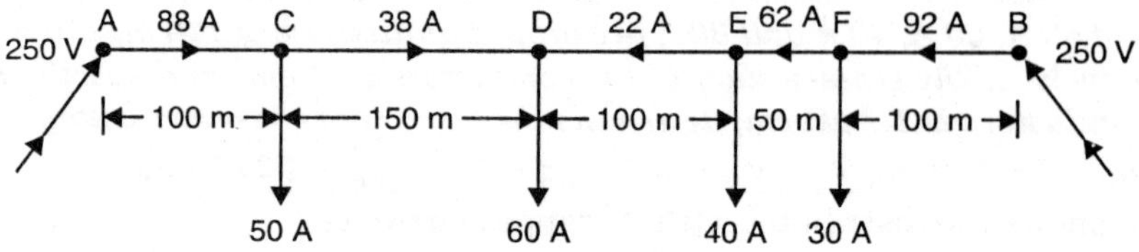

Fig. 19.19. The actual distribution of currents in various sections of the distributor.

Since the currents are coming from both sides to load point D, at a distance of 250 metres from end A, the minimum consumer voltage,

$$V_D = V_A - I_{AC} R_{AC} - I_{CD} R_{CD}$$

or
$$= 250 - 88 \times 0.034 - 38 \times 0.051 = \mathbf{245.07\ V.} \quad \textbf{(Ans.)}$$

Example 19.9. *A 2-wire D.C. distributor AB is fed from both ends. At the feeding point A the voltage is maintained at 240 V and at B 245 V. The total length of the distributor is 200 metres and loads are tapped off as under :*

25 A at 50 metres from A ; 50 A at 75 metres from A ; 30 A at 100 metres from A ; 40 A at 150 metres from A.

If the resistance per km of one conductor is 0.3 Ω, calculate :

(*i*) *The currents in the various sections of the distributor.*

(*ii*) *The minimum voltage and the point at which it occurs.*

(*iii*) *The power dissipated in the distributor.*

Sol. *Given :* $V_A = 240$ V ; $V_B = 245$ V ; $l = 200$ m ; r per conductor = 0.3/km length.

(*i*) **The current in the various sections of the distributor :**

Fig. 19.20 shows the distributor with its tapped currents. Let I_A be current supplied from feeding point A. Then the currents flowing through different sections are as shown in Fig. 19.20.

Resistance of 1000 m length of distributor (both wires)

$$= 2 \times 0.3 = 0.6\ \Omega$$

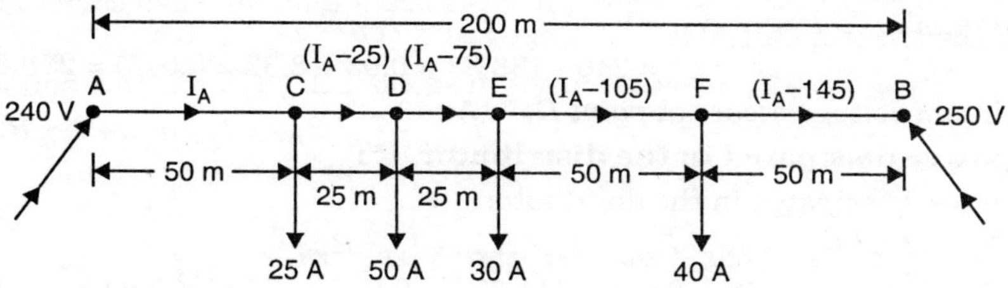

Fig. 19.20.

$\therefore$ Resistance of section AC, $R_{AC} = 50 \times \dfrac{0.6}{1000} = 0.03\ \Omega$

Resistance of section CD, $R_{CD} = 25 \times \dfrac{0.6}{1000} = 0.015\ \Omega$

Resistance of section DE, $R_{DE} = 25 \times \dfrac{0.6}{1000} = 0.015\ \Omega$

Resistance of section EF, $R_{EF} = 50 \times \dfrac{0.6}{1000} = 0.03\ \Omega$

Resistance of section FB, $F_{FB} = 50 \times \dfrac{0.6}{1000} = 0.03\ \Omega$

$$V_B = V_A - \text{voltage drops in the various sections}$$

$$245 = 240 - [I_{AC} R_{AC} + I_{CD} R_{CD} + I_{DE} R_{DE} + I_{EF} R_{EF} + I_{FB} R_{FB}]$$

or,

$$245 = 240 - [0.03\, I_A + 0.015(I_A - 25) + 0.015(I_A - 75) + 0.03(I_A - 105) + 0.03(I_A - 145)]$$

$$= 240 - (0.03\, I_A + 0.015\, I_A - 0.375 + 0.015\, I_A - 1.125 + 0.03\, I_A - 3.15 + 0.03\, I_A - 4.35)$$

$$= 240 - (0.12\, I_A - 9), \quad \text{or,} \quad 5 = -0.12\, I_A + 9$$

$\therefore$ $I_A = 33.33\ A$

Thus, the currents in the various sections of the distributor are :

$I_{AC} = I_A =$ **33.33 A. (Ans.)**

$I_{CD} = I_A - 25 = 33.33 - 25 =$ **8.33 A. (Ans.)**

$I_{DE} = I_A - 75 = 33.33 - 75 =$ **– 41.67 A. (Ans.)**

$I_{EF} = I_A - 105 = 33.33 - 105 =$ **– 71.67 A. (Ans.)**

$I_{FB} = I_A - 145 = 33.33 - 145 =$ **– 111.67 A. (Ans.)**

(–ve sign indicates current flow from end B to end A)

(*ii*) **The minimum voltage and the point at which it occurs :**

The actual distribution of currents in the various sections of the distributor is shown in Fig. 19.21.

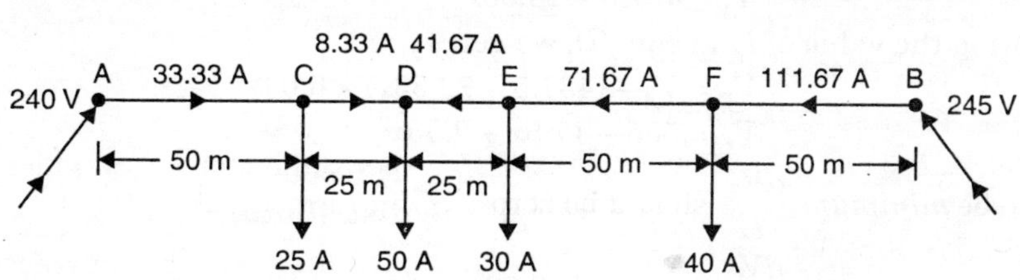

Fig. 19.21. The actual distribution of currents in the various sections of the distributor.

Since the currents are coming from both the sides to load point D, therefore, the point D *is the point of minimal potential*.

∴ *The minimum voltage,* $V_D = V_A - [I_{AC} R_{AC} + I_{CD} R_{CD}]$

$= 240 - (33.33 \times 0.03 + 8.33 \times 0.015) =$ **238.87 V. (Ans.)**

This minimum voltage occurs at *point D*. **(Ans.)**

(*iii*) **The power dissipated in the distributor, P :**

The power loss/dissipated in the distributor,

$$P = I_{AC}^2 R_{AC} + I_{CD}^2 R_{CD} + I_{DE}^2 R_{DE} + I_{EF}^2 R_{EF} + I_{FB}^2 R_{FB}$$

$$= (33.33)^2 \times 0.03 + (8.33)^2 \times 0.015 + (41.67)^2 \times 0.015 + (71.67)^2 \times 0.03 + (111.67)^2 \times 0.03$$

= **588.6 W. (Ans.)**

Example 19.10. *An electric train taking a constant current of 500 A moves on a section of line between two sub-stations 6 kW apart and maintained at 585 V and 600 V respectively. The track resistance is 0.04 Ω/ km both go and return. Find the point of minimum potential along the track and currents supplied by two substations at that instant.*

Sol. The single line diagram is shown in Fig. 19.22 where substation *A* is at 585 V and substation *B* at 600 V. Let the point of minimum potential occur when the train is at a point *C*, distant *x* km from the feed point *A*. Also let current fed from feeding point *A* be I_A so that current drawn from feed point *B* equals $(500 - I_A)$.

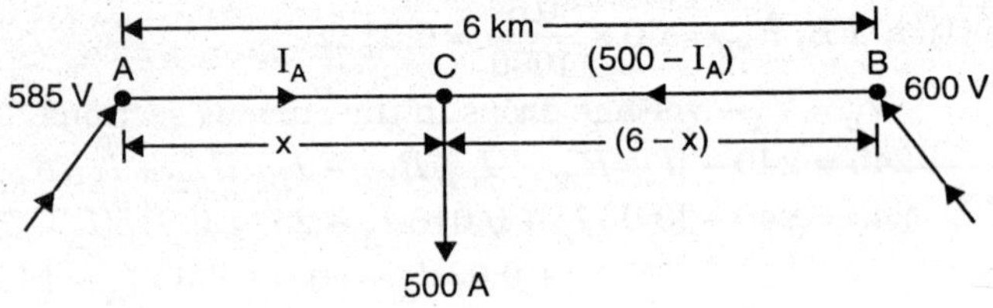

Fig. 19.22

Resistance of track (go and return) = 0.04 Ω/km

∴ Track resistance for section *AC*, $R_{AC} = 0.04\,x\ \Omega$

Track resistance for section *CB*, $R_{CB} = 0.04\,(6 - x)$

Potential at *C*, $V_C = V_A - I_A R_{AC}$ ...(*i*)

Also, Potential at *C*, $V_C = V_B - (500 - I_A)R_{CB}$ ...(*ii*)

From eqn. (*i*) and (*ii*), we get

$$V_A - I_A R_{AC} = V_B - (500 - I_A)R_{CB}$$

or, $$585 - 0.04xI_A = 600 - (500 - I_A) \times 0.04(6 - x)$$

or, $$585 - 0.04xI_A = 600 - 0.04\,(3000 - 500x - 6I_A + xI_A$$

or, $$585 - 0.04xI_A = 600 - 120 + 20x + 0.24I_A - 0.04xI_A$$

or, $$0.24I_A = 105 - 20x$$

or, $$I_A = 437.5 - 83.33x$$

Substituting the value of I_A in eqn. (*i*), we get

$$V_C = V_A - (437.5 - 83.33x) \times 0.04x$$

or, $$V_C = 585 - 17.5x + 3.33x^2$$

For V_C to be *minimum* $\frac{dV_C}{dx}$ should be zero.

∴ $$\frac{dV_C}{dx} = -17.5 + 6.66x = 0$$

or, $$\mathbf{x} = \frac{17.5}{6.66} = \mathbf{2.63\ km.}\ \textbf{(Ans.)}$$

Current supplied by the sub-station A,

$$I_A = 437.5 - 83.33 \times 2.63 = \mathbf{218.34\ A.}\quad \textbf{(Ans.)}$$

Current supplied by the *sub-station B*

$$= 500 - I_A = 500 - 218.34 = \mathbf{281.66\ A.}\quad \textbf{(Ans.)}$$

Example 19.11. *A distributor cable AB is fed at its ends A and B. Loads of 12 A, 24 A, 72 A and 48 A are taken from the cable at points C, D, E and F. The resistances of sections AC, CD, DE, EF and FB of the cable are 8, 6, 4, 10 and 5 milli ohm respectively (for go and return conductors together). The p.d. at point A is 240, the p.d. at the load F is also to be 240 V. Calculate :*

(i) The current supplied by each feeder.

(ii) The voltage at the feeding point B.

(iii) The p.ds. at the loads C, D and E. **(Utkal University)**

Sol. The single line diagram of the distributor is shown in Fig. 19.23.

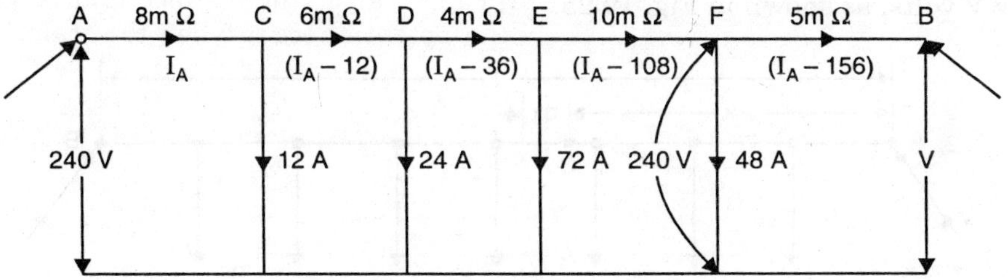

Fig. 19.23

(*i*) **The current supplied by each feeder :**

Let the current fed at the feeding point *A* be I_A. The current distribution in various sections becomes as shown in Fig. 19.23.

Voltage drop on both sides of the distributor over the section *AF* is

$$= [8I_A + 6(I_A - 12) + 4(I_A - 36) + 10(I_A - 108)] \times 10^{-3} = (28I_A - 1296) \times 10^{-3}\ \text{V}$$

Since points *A* and *F* are at the same potential, the p.d. between *A* and *F* is zero.

$$\therefore \quad (28I_A - 1296) \times 10^{-3} = 0 \qquad I_A = 46.29\ \text{A}$$

Thus, currents in various sections are :

Current in section *AC*, $I_A = 46.29$ A

Current in section *CD*, $I_A - 12 = 46.29 - 12 = 34.29$ A

Current in section *DE* = $I_A - 36 = 46.29 - 36 = 10.29$ A

Current in section *EF* = $I_A - 108 = 46.29 - 108 = -61.71$ A

Current in section *FB* = $I_A - 156 = 46.29 - 156 = -109.71$ A

(–ve sign indicates that current flow is from end *B* towards end *A*)

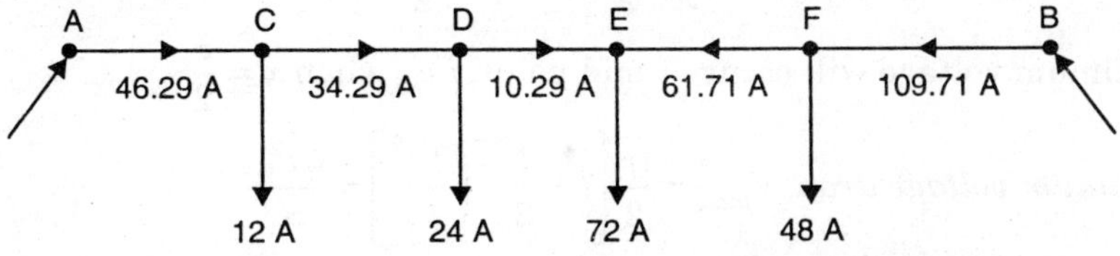

Fig. 19.24

Thus, *current supplied by feeder at point A is* **46.29 A** and *that supplied at point B is* **109.71 A. (Ans.)**

The actual current distribution is as shown in Fig. 19.24.

(*ii*) **The voltage at the feeding point B :**

Voltage at feeding point B = 240 – Voltage drop over section FB

$= 240 - (-109.71 \times 5 \times 10^{-3}) =$ **240.55 V. (Ans.)**

(*iii*) **The p.ds. at the loads C, D and E**

Voltage at the point C = 240 – Voltage drop over section AC

$= 240 - 46.29 \times 8 \times 10^{-3} =$ **239.63 V. (Ans.)**

Voltage at the point D = 239.63 – Voltage drop over section CD

$= 239.63 - 34.29 \times 6 \times 10^{-3} =$ **239.42 V. (Ans.)**

Voltage at the point $E = 239.42 - 10.29 \times 4 \times 10^{-3} =$ **239.38 V. (Ans.)**

(B-I). ***Distributors fed at both ends at equal voltages–Uniform loading*** **:**

Consider a distributor AB of length l metres resistance r ohms per metre run of both the conductors and loaded with i amperes per metre run, fed from feeding points A and B at equal voltages, say V volts, as shown in Fig. 19.25.

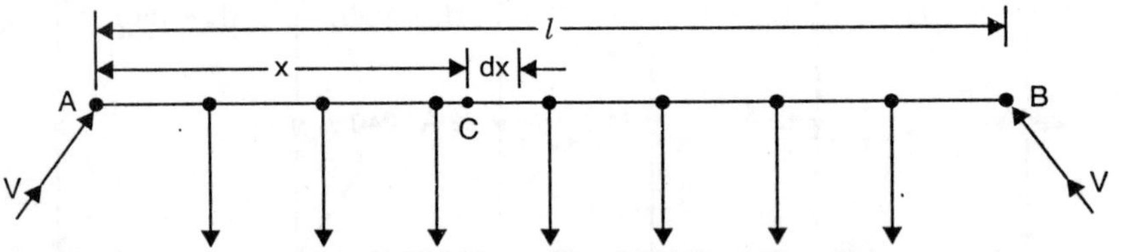

Fig. 19.25. Distributor fed at both ends at equal voltages with uniform distributed loading.

Further, consider a point C at a distance x metres from the feeding point A.

Now, total current flowing through conductor $AB = il$,

and, current supplied from each end $= \dfrac{il}{2}$ ($\because$ Two end voltages are equal *i.e.*, $V_A = V_B = V$)

$\therefore$ The current at $C = \dfrac{il}{2} - ix = i\left(\dfrac{l}{2} - x\right)$

Consider a section of very small length dx near point C. Its resistance is rdx and the voltage drop over length dx is given by,

$$dv = i\left(\frac{l}{2} - x\right) rdx = ir\left(\frac{l}{2} - x\right) dx$$

$\therefore$ Voltage drop in section $AC = \displaystyle\int dv = \int_0^x ir\left(\frac{l}{2} - x\right) dx$

$$= \int_0^x \frac{ir}{2}(l - 2x) = \frac{ir}{2}(lx - x^2) \quad ...(19.4)$$

The maximum voltage will occur at mid point, *i.e.*, when $x = \dfrac{l}{2}$

$$\therefore \quad \textit{Maximum voltage drop, } V_{max} = \frac{ir}{2}\left[l \times \frac{l}{2} - \left(\frac{l}{2}\right)^2\right] = \frac{irl^2}{8} \quad ...(19.5)$$

$$= \frac{il \times rl}{8} = \frac{IR}{8} \quad ...[19.5\ (a)]$$

where, $il = I$, total current fed to the distributor, and

$rl = R$, total loop resistance of the distributor.

$$\textit{Minimum voltage (at mid point) } V_{min} = V - \frac{IR}{8} \qquad ...(19.6)$$

Power loss in the distributor :

The power loss in length dx

$$= (\text{Current in } dx)^2 \times \text{resistance of } dr$$

$$= \left[i\left(\frac{l}{2} - x\right)\right]^2 \times r\,dx$$

Total power loss in the distributor,

$$P = \int_0^l \left[i\left(\frac{l}{2} - x\right)\right]^2 r\,dx = i^2 r \int_0^l \left(\frac{l^2}{4} - lx + x^2\right) dx$$

$$= i^2 r \left[\frac{l^2 x}{4} - \frac{l x^2}{2} + \frac{x^3}{3}\right]_0^l = i^2 r \left[\frac{l^3}{4} - \frac{l^3}{2} + \frac{l^3}{3}\right]$$

or

$$P = \frac{i^2 r l^3}{12} \qquad ...(19.7)$$

Example 19.12. *A 2-wire D.C. distributor cable 800 metres long is loaded with 0.4 A/m. Resistance of each conductor is 0.04 Ω/km. Calculate the maximum voltage drop if the distributor is fed from both ends with equal voltages of 220 V. What is the minimum voltage and where does it occur ?*

Sol. *Given :* $l = 800$ m ; $i = 0.4$ A/m ; $r = 2 \times 0.04 = 0.08$ Ω/km (go and return) ;

$V_A = V_B = V = 220$ volts

Total current supplied by the distributor,

$$I = il = 0.4 \times 800 = 320 \text{ A}$$

$$\text{Total resistance of the distributor, } R = rl = \frac{0.08}{1000} \times 800 = 0.064\ \Omega$$

$$\therefore \textit{ Maximum voltage drop} = \frac{IR}{8} = \frac{320 \times 0.064}{8} = \mathbf{2.56\ V.} \quad \textbf{(Ans.)}$$

Minimum voltage of the distributor,

$V_{min} = 220 - 2.56 =$ **217.44 V.** **(Ans.)**

V_{min} will *occur at the* **mid-point. (Ans.)**

(B-II). *Distributor fed at both ends with different voltages—Uniform loading :*

Consider a distributor AB of length l metres having resistance r ohms per metre run (go and return) and loaded with a uniform loading of i amperes per metre run as shown in Fig. 19.25. Let the distributor be fed from feeding points A and B at voltages V_A and V_B respectively.

Suppose that the point of *minimum* potential C is situated at a distance x metres from the feed point A, so there will be no current at point C and the current fed from feeding point A will be the current in section AC *i.e.* ix

$$\therefore \text{ Voltage drop in section } AC = \frac{irx^2}{2} \text{ volts}$$

As the distance of C from feeding B is $(l - x)$, therefore, current fed from B is $i(l - x)$.

$$\therefore \text{ Voltage drop in section } BC = \frac{ir\,(l - x)^2}{2} \text{ volts}$$

Voltage at point C, $V_C = V_A$ – voltage drop in section AC

or, $$V_C = V_A - \frac{irx^2}{2} \quad ...(i)$$

Also, voltage at point C, $V_C = V_B$ – voltage drop in section BC

or, $$V_C = V_B - \frac{ir\,(l-x)^2}{2} \quad ...(ii)$$

From (*i*) and (*ii*), we have

$$V_A - \frac{irx^2}{2} = V_B - \frac{ir(l-x)^2}{2}$$

$$V_A - V_B = \frac{irx^2}{2} - \frac{ir\,(l-x)^2}{2} = \frac{ir}{2}\,[x^2 - (l-x)^2] = \frac{ir}{2}\,[x^2 - (l^2 + x^2 - 2lx)]$$

$$= \frac{ir}{2}\,(2lx - l^2) = \frac{irl}{2}\,(2x - l) = irlx - \frac{irl^2}{2}$$

or, $$irlx = (V_A - V_B) + \frac{irl^2}{2}$$

or, $$x = \frac{V_A - V_B}{irl} + \frac{l}{2} \quad ...(19.8)$$

Since all the quantities on the right hand side of the equation are known, therefore, the point on the distributor where minimum potential occurs can be calculated.

Example 19.13. *A 800 m long, 2-wire D.C. distributor fed from both ends, is loaded uniformly at the rate of 1.2 A/m run. If the resistance of the distributor is 0.1 Ω/km (go and return) and feed points are maintained at 245 V and 240 V respectively, calculate :*

(*i*) *The minimum voltage and its point of occurrence, and*

(*ii*) *The currents supplied from feeding points A and B.* **(B.T.E. UP)**

Sol. *Given :* $l = 800$ m ; $i = 1.2$ *A/m run* ; $r = \frac{0.1}{1000} = 10^{-4}$ Ω/m, (go and return) ;

$V_A = 245$ V ; $V_B = 240$ V.

(*i*) **The minimum voltage and its point of occurrence :**

Distance of point of minimum potential difference for feeding point A

$$x = \frac{V_A - V_B}{irl} + \frac{l}{2} \quad ...[\text{Eqn. (19.8)}]$$

$$= \frac{(245 - 240)}{1.2 \times 0.0001 \times 800} + \frac{800}{2} = \mathbf{452.08\ m.} \quad \textbf{(Ans.)}$$

Minimum voltage $$= V_A - \frac{irx^2}{2}$$

$$= 245 - \frac{1.2 \times 0.0001 \times (452.08)^2}{2} = \mathbf{232.74\ V.} \quad \textbf{(Ans.)}$$

(*ii*) **The currents supplied from feeding points A and B :**

Current supplied from feeding point $A = ix = 1.2 \times 452.08 =$ **542.5 A. (Ans.)**

Current supplied from feed point $B = i(l - x) = 1.2(800 - 452.08) =$ **417.5 A. (Ans.)**

Example 19.14. *A 2-wire D.C. distributor, 800 metres long, is fed from both ends. It is uniformly loaded at the rate of 1.1 A/m run. Calculate the voltage at the feeding points A and B if the minimum potential of 230 V occurs at point C at a distance of 500 metres from the end A. Resistance of the distributor is 0.1 Ω/km (go and return).*

Sol. *Given* : $l = 800$ m ; $i = 1.1$ A/m run ; $x = 500$ m ;

$$r = 0.1 \ \Omega/\text{km (go and return)} = \frac{0.1}{1000} = 10^{-4} \ \Omega/\text{m (go and return)} ;$$

$$V_C = 230 \text{ V}$$

The single line diagram of the distributor is shown in Fig. 19.26.

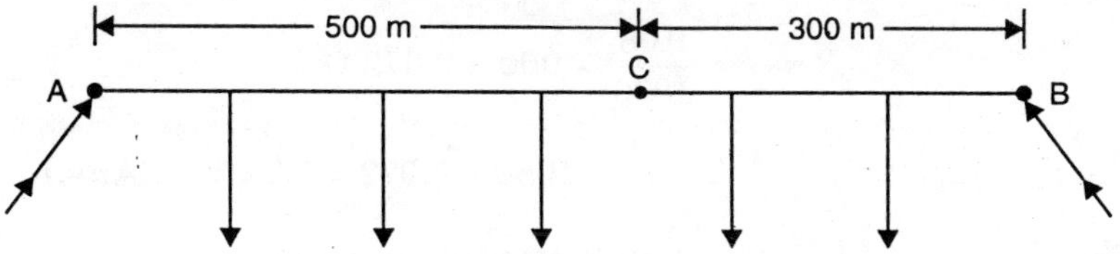

Fig. 19.26

$$\text{Voltage drop in section } AC = \frac{irx^2}{2} = \frac{1.1 \times 0.0001 \times 500^2}{2} = 13.75 \text{ V}$$

$\therefore$ Voltage feeding point A, $V_A = V_C$ + voltage drop in section AC

$= 230 + 13.75 =$ **243.75 V. (Ans.)**

$$\text{Voltage drop in section } BC = \frac{ir\,(l - x)^2}{2}$$

$$= \frac{1.1 \times 0.0001\,(800 - 500)^2}{2} = 4.95 \text{ V}$$

$\therefore$ Voltage at feeding point B, $V_B = 230 + 4.95 =$ **234.95 V. (Ans.)**

Example 19.15. (*a*) *A uniformly loaded distributor is fed at the centre. Show that maximum voltage drop* $= \dfrac{IR}{8}$, *where I is the total current fed to the distributor and R is the total resistance of the distributor.*

(*b*) *A 2-wire D.C. distributor 900 metres long is fed at centre and is loaded uniformly at the rate of 1.2 A/m. If the resistance of each conductor is 0.04 Ω/km, find the maximum voltage drop in the distributor.*

Sol. (*a*) Fig. 19.27 shows distributor AB fed at centre C and uniformly loaded with i amperes/metre. Let l metres be the length of the distributor and r ohms be the resistance per metre run. Obviously, maximum voltage drop will occur at either end.

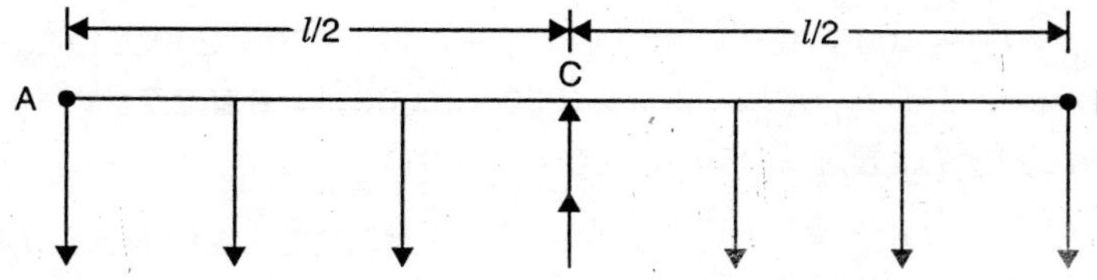

Fig. 19.27

$\therefore$ Maximum voltage drop = Voltage drop in half distributor

$$= \frac{1}{2}\left(\frac{il}{2}\right)\left(\frac{rl}{2}\right) = \frac{1}{8}(il)(rl) = \frac{1}{8} IR$$

where, $il = I$, the total current fed to the distributor, and

$rl = R$, the total resistance of the distributor.

(*b*) *Given :* $l = 900$ m ; $i = 1.2$ A/m ; $r = 2 \times 0.04 = 0.08$ Ω/km

Total current fed to the distributor,

$$I = il = 1.2 \times 900 = 1080 \text{ A}$$

Total resistance of the distributor,

$$R = rl = \frac{0.08}{1000} \times 900 = 0.072\ \Omega$$

$\therefore$ *Maximum voltage drop* $= \frac{1}{8} IR = \frac{1}{8} \times 1080 \times 0.072 =$ **9.72 V. (Ans.)**

Uniform loading in addition to concentrated loading :

In certain problems a distributor has both concentrated and uniform loadings. Under such situation, the total drop over any section of the distributor is *equal to the sum of drops due to concentrated and uniform loading in that section.* A few problems, by way of illustration, will now be solved.

Example 19.16. *A 2-wire D.C. distributor AB, 900 metres long is fed at 410 V. The loads are tapped off as under :*

50 A at 200 m from A ; 100 A at 500 m from A ; 150 A at 800 m from A.

The distributor is also loaded uniformly at the rate of 0.5 A / m. If the resistance of the distributor per metre (go and return) is 0.0001 Ω, calculate voltages at points B and D.

Sol. *Given :* $l = 900$ m ; $i = 0.5$ A/m ; $r = 0.0001$ Ω/m run (go and return).

Fig. 19.28 shows the single line diagram of the distributor.

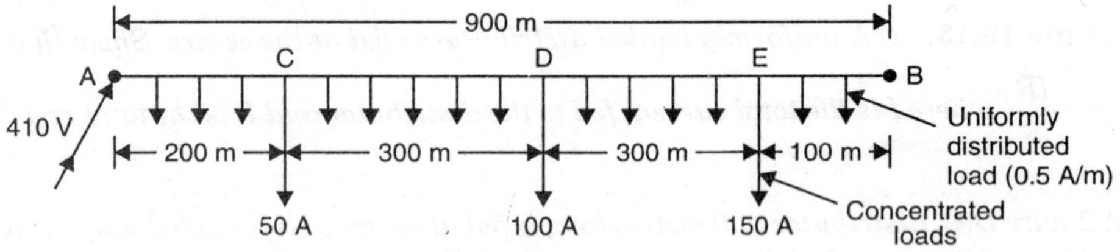

Fig. 19.28

This problem can be solved by calculating, for various sections, the voltage drops due to concentrated loading and drops due to uniform loading separately and then adding them.

Drop due to concentrated load :

The currents in the various sections are :

$$I_{AC} = 50 + 100 + 150 = 300 \text{ A} ; I_{CD} = 100 + 150 = 250 \text{ A} ; I_{DE} = 150 \text{ A}$$

$\therefore$ Drop in section $AC = I_{AC} R_{AC} = 300 \times (200 \times 0.0001) = 6$ V

Drop in section $CD = I_{CD} R_{CD} = 250 \times (300 \times 0.0001) = 7.5$ V

Drop in section $DE = I_{DE} R_{DE} = 150 \times (300 \times 0.0001) = 4.5$ V

Total drop over *AB* (due to concentrated load)

$$= 6 + 7.5 + 4.5 = 18 \text{ V}$$

Drop due to uniform loading :

$$\text{Drop over } AB = \frac{irl^2}{2} = \frac{0.5 \times 0.0001 \times (900)^2}{2} = 20.25 \text{ V}$$

$$\text{Drop over } AD = ir\left(lx - \frac{x^2}{2}\right)$$

Here, $x = 500$ m and $l = 900$ m

$$\therefore \quad \text{Drop over } AD = 0.5 \times 0.0001\left(900 \times 500 - \frac{500^2}{2}\right) = 16.25 \text{ V}$$

$\therefore$ *Voltage at point B* = V_A – Voltage drop over *AB* due to concentrated and uniform loading

= 410 – (18 + 20.25) = **371.75 V.** **(Ans.)**

Voltage at point D = V_A – Voltage drop over *AD* due to concentrated and uniform loading

= 410 – [(6 + 7.5) + 16.25)] = **380.25 V.** **(Ans.)**

Example 19.17. *A 300 m long, 2-wire D.C. distributor AB is fed from both ends and supplies a uniformly distributed load of 0.15 A/m length, together with the following concentrated loads : 50 A at C, 60 A at D and 40 A at E, distances AC, CD and DE being 75 m, 100 m and 50 m respectively. If the supply voltages at A and B are 210 V and 205 V respectively, and the resistance of each conductor is 0.00015 Ω/m, calculate :*

(i) The currents supplied at each end, and

(ii) The point of minimum potential. **(Agra University)**

Sol. *Given :* $l = 300$ m ; $i = 0.15$ A/m ; $V_A = 210$ V ;

$V_B = 205$ V ; $r = 2 \times 0.00015 = 0.0003$ Ω/m

Fig. 19.29 shows the loaded distributor.

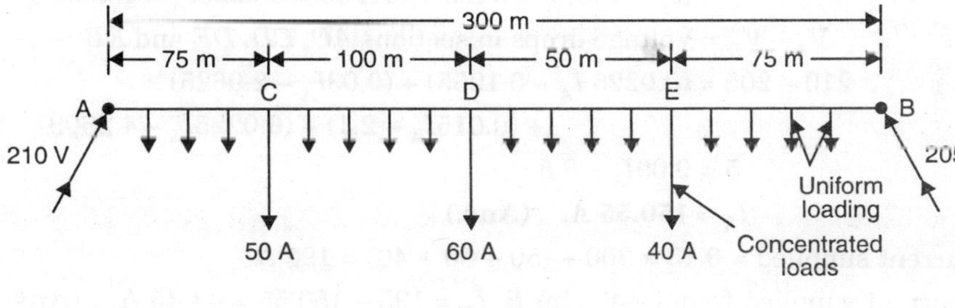

Fig. 19.29

(i) **The currents supplied at each end ; I_A, I_B :**

Let, I_A = Current supplied from end *A*.

Now, Current just before load point *C*

$= I_A - 0.15 \times 75 = I_A - 11.25$

and, Current just after load point *C*

$= I_A - 11.25 - 50 = I_A - 61.25$

Similarly, Current just before load point *D*

$= I_A - 61.25 - 0.15 \times 100 = I_A - 76.25$

Current just after load point *D*

$= I_A - 76.25 - 60 = I_A - 136.25$

Current just before load point E

$$= I_A - 136.25 - 0.15 \times 50 = I_A - 143.75$$

Current just after load point E

$$= I_A - 143.75 - 40 = I_A - 183.75$$

Current just before feeding point B

$$= I_A - 183.75 - 0.15 \times 75 = I_A - 195$$

Voltage drop in section AC

= Voltage drop due to concentrated load

+ voltage drop due to uniform loading of the section

$$= (I_A - 11.25) \times (0.0003 \times 75) + \frac{1}{2}(0.15 \times 75) \times (0.0003 \times 75)$$

$$= (I_A - 11.25) \times 0.0225 + 0.1266 = 0.0225 I_A - 0.1265$$

Voltage drop in section $CD = (I_A - 76.25) \times (0.0003 \times 100) + \frac{1}{2}(0.15 \times 100)(0.0003 \times 100)$

$$= (I_A - 76.25) \times 0.03 + 0.225 = 0.03 I_A - 2.0625$$

Voltage drop in section $DE = (I_A - 143.75) \times (0.0003 \times 50) + \frac{1}{2}(0.15 \times 50)(0.0003 \times 50)$

$$= (I_A - 143.75) \times 0.015 + 0.05625 = 0.015 I_A - 2.1$$

Voltage drop in section EB

$$= (I_A - 195) \times (0.0003 \times 75) + \frac{1}{2}(0.15 \times 75)(0.0003 \times 75)$$

$$= (I_A - 195) \times 0.0225 + 0.1266 = 0.0225 I_A - 4.2609$$

Now, $V_A - V_B$ = Voltage drops in sections AC, CD, DE and EB

$\therefore$ $210 - 205 = (0.0225\, I_A - 0.1265) + (0.03 I_A - 2.0625) + (0.015 I_A - 2.1) + (0.0225 I_A - 4.2609)$

or $5 = 0.09 I_A - 8.55$

$\therefore$ I_A = **150.55 A. (Ans.)**

Total current supplied = 0.15 × 300 + (50 + 60 + 40) = 195 A

Hence current supplied from feed point B, I_B = 195 – 150.55 = **44.45 A. (Ans.)**

(*ii*) **The point of minimum potential :**

At the point of *minimum potential current will be zero.* Since current fed from point A is 150.55 A, therefore current will be zero at load point E because current just before load point E is 150.55 – 143.75 *i.e.*, 6.8 A flowing from D to E and current just after load point E is 150.55 – 183.75 *i.e.*, – 33.2 A or 33.2 A flowing from B to E.

Hence *E is a point of minimum potential.* **(Ans.)**

Example 19.18. *A 2-core distributor cable 400 metres long supplies a uniformly distributed lighting load of 1 A/m run. There are concentrated loads of 120, 72, 48 and 120 A at 40 m, 120 m, 200 m and 320 m respectively from end A. The cable has a resistance of 0.3 Ω/km run (go and return). Determine the position of the lowest-run lamp and its voltage when the cable is fed at 240 V from both ends A and B.*

Sol. *Given :* $i = 1$ A/m run ; $r = \frac{0.3}{1000} = 0.0003$ Ω/m run (go and return)

$V_A = V_B = 240$ V

Fig. 19.32 shows the loaded distributor.

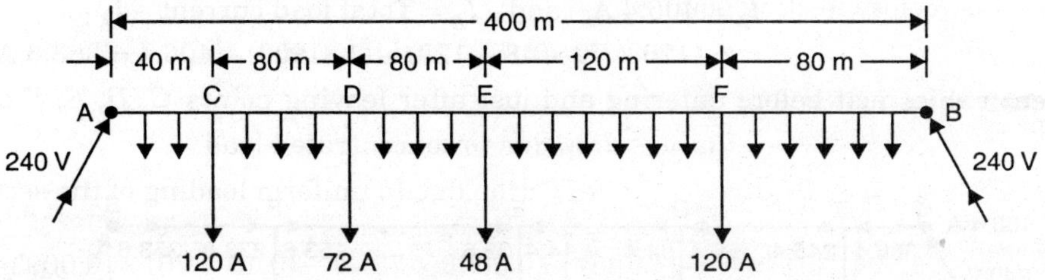

Fig. 19.30

Let, I_A = Current supplied from end A.

Now, Current just before load point $C = I_A - 1 \times 40 = I_A - 40$

Current just after load point $C = I_A - 40 - 120 = I_A - 160$

Similarly, Current just before load point $D = I_A - 160 - 1 \times 80 = I_A - 240$

Current just after load point $D = I_A - 240 - 72 = I_A - 312$

Current just before load point $E = I_A - 312 - 1 \times 80 = I_A - 392$

Current just after load point $E = I_A - 392 - 48 = I_A - 440$

Current just before load point $F = I_A - 440 - 1 \times 120 = I_A - 560$

Current just after load point $F = I_A - 560 - 120 = I_A - 680$

Current just before feeding point $B = I_A - 680 - 1 \times 80 = I_A - 760$

Voltage drop in section AC = Voltage drop due to concentrated load + voltage drop due to uniform loading of the section

$$= (I_A - 40) \times (0.0003 \times 40) + \frac{1}{2} \times (1 \times 40) \times (0.0003 \times 40)$$

$$= (I_A - 40) \times 0.012 + 0.24 = 0.012 I_A - 0.24$$

Voltage drop in section $CD = (I_A - 240) \times (0.0003 \times 80) + \frac{1}{2}(1 \times 80) \times (0.0003 \times 80)$

$$= (I_A - 240) \times 0.024 + 0.96 = 0.024 I_A - 4.8$$

Voltage drops in section $DE = (I_A - 392) \times (0.0003 \times 80) + \frac{1}{2}(1 \times 80) \times (0.0003 \times 80)$

$$= (I_A - 392) \times 0.024 + 0.96 = 0.024 I_A - 8.448$$

Voltage drop in section $EF = (I_A - 560) \times (0.0003 \times 120) + \frac{1}{2}(1 \times 120) \times (0.0003 \times 120)$

$$= (I_A - 560) \times 0.036 + 2.16 = 0.036 I_A - 18$$

Voltage drop in section $FB = (I_A - 760) \times (0.0003 \times 80) + \frac{1}{2}(1 \times 80) \times (0.0003 \times 80)$

$$= (I_A - 760) \times 0.024 + 0.96 = 0.024 I_A - 17.28$$

Now, $V_A - V_B$ = Total voltage drop

or, $240 - 240 = (0.012I_A - 0.24) + (0.024I_A - 4.8) + (0.024I_A - 8.448) + (0.036I_A - 18) + (0.024I_A - 17.28)$

$0 = 0.12I_A - 48.768$

∴ $I_A = 406.4$ A and I_B = Total load current $- I_A$

$= (120 + 72 + 48 + 120) + (1 \times 400) - 406.4 = 353.6$ A

Current values just before entering and just after leaving points *C, D, E, F* are shown in Fig. 19.31.

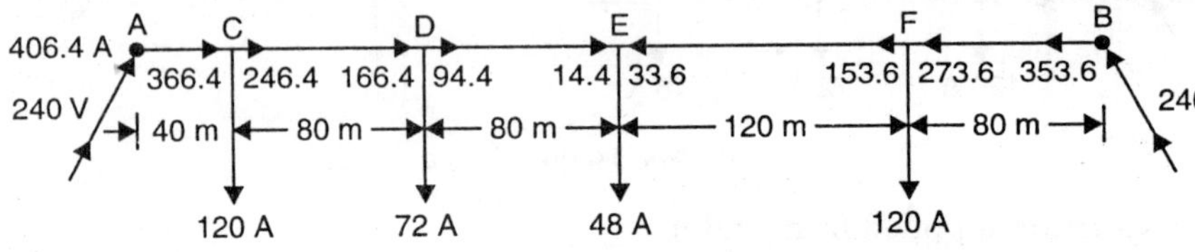

Fig. 19.31

Voltage drop over length *AE*

= Voltage drop in section *AC* + voltage drop in section *CD* + voltage drop in section *DE*

$= (0.012I_A - 0.24) + (0.024I_A - 4.8) + (0.024I_A - 8.448)$

$= 0.06I_A - 13.488 = 0.06 \times 406.4 - 13.488 = 10.9$ V

∴ $V_E = V_A - 10.9 = 240 - 10.9 =$ **229.1 V. (Ans.)**

Example 19.19. (*a*) *A double-circuit three-phase three-wire line is converted into a triple-circuit two-pole two-wire D.C. line with same maximum voltage to earth as that of the A.C. line. Calculate the percentage increase in the capacity of the line if transmission loss remains unchanged. Assume power factor of the A.C. load to be unity.*

(*b*) *A two-wire D.C. distributor AB, 500 m long, fed at both ends at equal potential of 250 volt, is loaded with a uniformly distributed load of 0.5 A/m along with concentrated loads of 100 A, 200 A, 300 A, and 400 A at distances of respectively 100 m, 200 m, 300 m and 400 m from end A. Calculate the distance from A where the potential of the line would be the minimum. Assuming the resistivity of the conductor material to be* 1.7×10^{-8} Ω *m, calculate the minimum cross-section of the conductors so that the maximum drop may not exceed 20 V.*

(*c*) *Explain why the voltage drop is of primary concern in the design of the distributors.*

(AMIE Summer, 2001)

Sol. (*a*) As the existing double-circuit 3-phase 3-wire line containing total 6 conductors is converted into 2-pole 2-wire D.C. line, it is *assumed that the line resistance will remain the same.*

Case 1. ***D.C. 2-wire system.*** Fig. 19.32 shows a 2-wire D.C. system, for which I_1 is line current, *R* is the resistance of conductor.

Power transmitted, $P_1 = VI_1$

∴ Line current, $I_1 = \dfrac{P_1}{V}$

Total I^2R loss for both the conductors $= 2I_1^2R_1 = 2\left(\dfrac{P_1}{V}\right)^2 R$

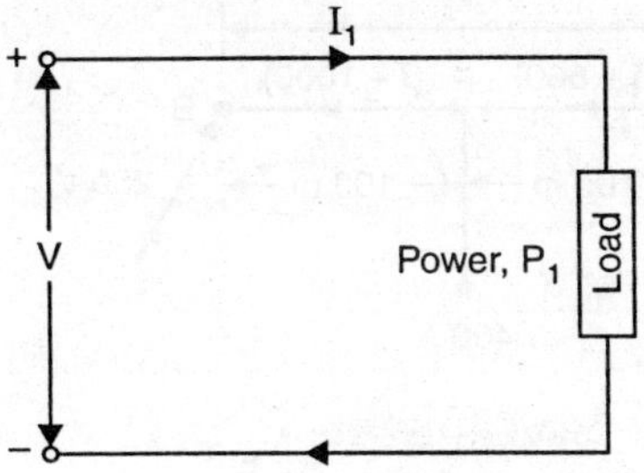

Fig. 19.32. D.C. 2-wire system.

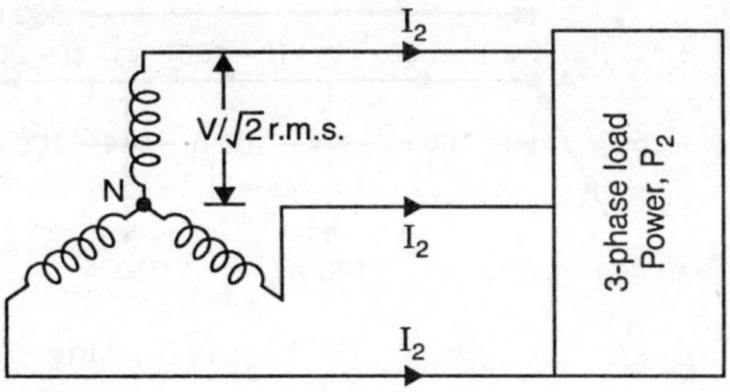

Fig. 19.33. 3-phase, 3-wire, star-connected system.

∴ Total line losses for all the six conductors

$$= 3 \times 2\left(\frac{P_1}{V}\right)^2 R = 6\left(\frac{P_1}{V}\right)^2 R \qquad ...(i)$$

Case 2. *3-phase, 3-wire, star-connected system.* This system is shown in Fig. 19.33.

Maximum voltage per phase = V

∴ R.H.S. value of voltage per phase = $\frac{V}{\sqrt{2}}$

If I_2 is the r.m.s. phase current, R the resistance of each conductor and cos ϕ, the power factor of the load, then

$$P_2 = 3V_{ph}\, I_{ph} \cos \phi = 3 \times \frac{V}{\sqrt{2}} I_2 \times 1 \qquad (\because \cos \phi = 1 \text{ ...Given})$$

$$\therefore \quad I_2 = \frac{\sqrt{2}\, P_2}{3V}$$

Line loss $= 3I_2^2R = 3 \times \left(\frac{\sqrt{2}\, P_2}{3V}\right)^2 R$

Total lines losses for 2 such 3-ϕ circuits

$$= 2 \times 3 \times \left(\frac{\sqrt{2}\, P_2}{3V}\right)^2 R \qquad ...(ii)$$

Since the transmission loss remains unchanged, from (*i*) and (*ii*), we have

$$6\left(\frac{P_1}{V}\right)^2 R = 6\left(\frac{\sqrt{2}\, P_2}{3V}\right)^2 R$$

$$\frac{P_1}{V} = \frac{\sqrt{2}\, P_2}{3V} \quad \text{or} \quad \frac{P_2}{P_1} = \frac{3}{\sqrt{2}} = 2.1213$$

∴ *Percentage increase in the capacity of 3-phase double-circuit line* is by **112.3%. (Ans.)**

(*b*) Fig. 19.34 shows the distributor with concentrated loads only (*i.e.* without uniformly distributed load). It may be noted that the location of point of minimum potential is not affected by the uniformly distributed load of 0.5 A/m run. Thus, by not considering this load, we take I as the current input at end A as shown in Fig. 19.34. The different currents in various sections are also shown in the figure.

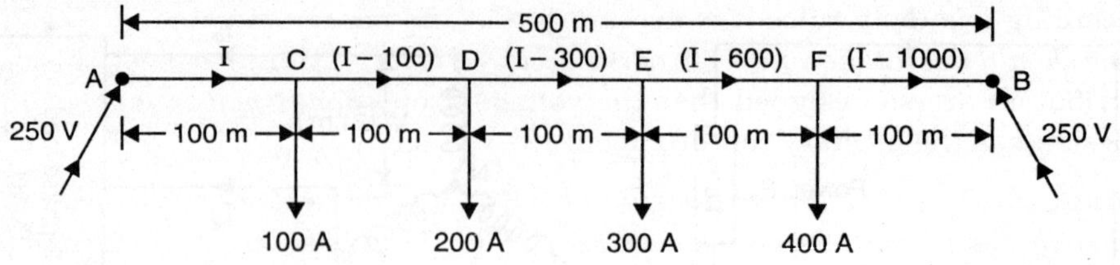

Fig. 19.34

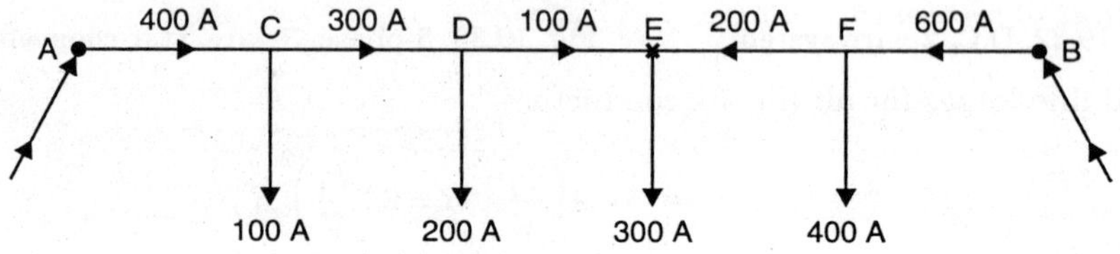

Fig. 19.35

Since the points A and B are fed at equal voltages, total drop over the distributor is *zero.*

The distributor resistance per metre length is, $r = \frac{\rho \times 1}{a} = \frac{\rho}{a}$ Ω/m

$$\therefore \quad r \times 100\,[I + (I - 100) + (I - 300) + (I - 600) + (I - 1000)] = 0$$

or, $$100\,(5I - 2000) = 0$$

$$\therefore \quad I_A = 400 \text{ A}$$

Actual current distribution is now shown in Fig. 19.35 ; from where it is seen that the point of minimum potential is E. The uniform load of 0.5 A/m upto point E will be supplied from A and rest from B.

Uniform load from A to $E = 0.5 \times 300 = 150$ A

Hence, $I_A = 400 + 150 = 550$ A

Similarly, $I_B = 600 + 0.5 \times 200 = 700$ A

Drop due to concentrated loads $= 100(400 \times r + 300 \times r + 100 \times r) = 80000\,r$

Drop due to uniform load can be found by imagining that the distributor is cut into two at point E so that AE can be looked upon as a distributor fed at one end and loaded uniformly. In that case, E becomes the end of the distributor.

Then, Drop upto E due to uniform load $= \frac{irl^2}{2}$

$$= \frac{0.5 \times r \times 300^2}{2} = 22500\,r$$

Total drop upto $E = 80000\,r + 22500\,r = 102500\,r$

Also total drop $= 20$ V *(Given)*

$$\therefore \quad 102500r = 20, \quad \text{or,} \quad 102500 \times \frac{\rho}{a} = 20$$

or, $$\mathbf{a} = \frac{102500\,\rho}{20} = \frac{102500 \times 1.7 \times 10^{-8}}{20} = 8.7125 \times 10^{-5} \text{ m}^2$$

$$= \mathbf{0.871\ cm^2}. \quad \textbf{(Ans.)}$$

(*c*) According to supply rules it is necessary that the minimum value of voltage at the consumer location *should not fall beyond – 5% of the declared voltage.* Hence to meet this it is necessary that the distributors are so designed that the voltage drop *should not be more than 5%.* Hence distributor design has the primary concern about the voltage drop.

Example 19.19. *A D.C. 2-wire distributor AB is 500 metres long and is fed at both ends at 250 V. The distributor is loaded as shown in Fig. 19.36. The resistance of the distributor (go and return) is 0.001 Ω/m. Calculate :*

(*i*) *The point of minimum voltage.*

(*ii*) *The value of minimum voltage.*

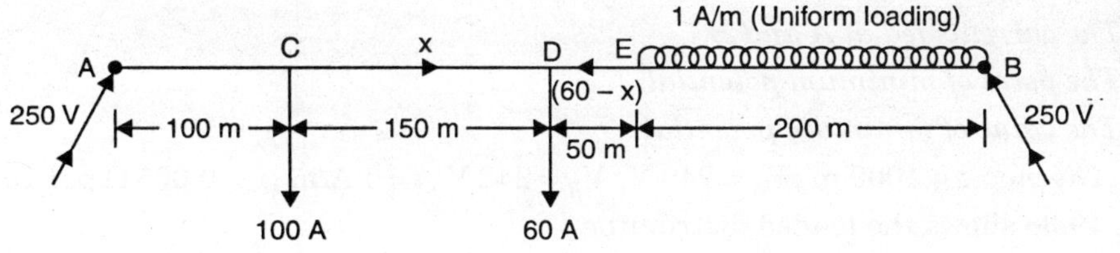

Fig. 19.36

Sol. $l = 500$ m ; $V_A = V_B = 250$ V ; $r = 0.001$ Ω/m

(*i*) **The point of minimum voltage :**

Let *D* be the point of minimum potential and let *x* be current flowing in section *CD*. The current supplied by end *B* will be $(60 - x)$.

Now, Voltage drop in length $AD = I_{AC} R_{AC} + I_{CD} R_{CD}$

$$= (100 + x) \times (100r) + x \times (150r)$$

$$= r[(100 + x) \times 100 + 150x] = r[10000 + 250x]$$

Voltage drop in length $BD = I_{DB} R_{DB} + \frac{1}{2} r \times (200)^2$

$$= (60 - x) \times (250r) + 20000r$$

$$= r[(60 - x) \times 250 + 20000]$$

Since feeding points *A* and *B* are the same potential,

$$\therefore \quad r[10000 + 250x] = r[(60 - x) \times 250 + 20000]$$

or, $$10000 + 250x = 15000 - 250x + 20000$$

$$\therefore \quad x = 50 \text{ A}$$

The actual distribution of the currents in various sections of the distributor is shown in Fig. 19.37. Since the currents supplied by feeding points *A* and *B* meet at *D*, therefore, **point D is the point of minimum voltage. (Ans.)**

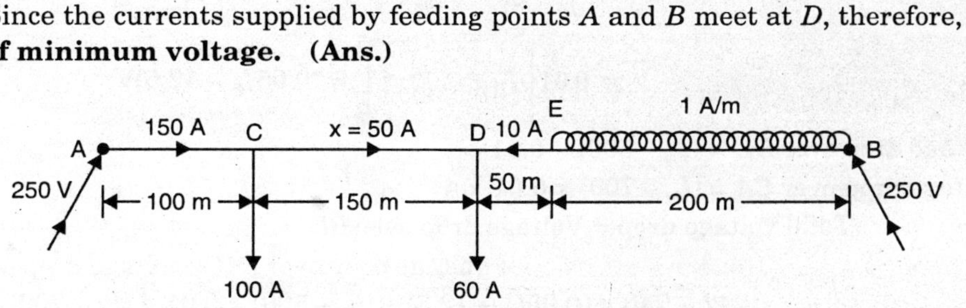

Fig. 19.37

(*ii*) **The value of minimum voltage, V_D :**

$$V_D = V_A - I_{AC} R_{AC} - I_{CD} R_{CD}$$
$$= 250 - 150 \times (0.001 \times 100) - 50 \times (0.001 \times 150) = \mathbf{227.5\ V.} \quad \textbf{(Ans.)}$$

[Check potential of *D* from end *B* :

$$V_D = 250 - 10 \times (250 \times 0.001) - \frac{1}{2} \times 0.001 \times 200^2 = 250 - 22.5 = 227.5 \text{ V}]$$

Example 19.20. *A 2-wire D.C. distributor AB, 1000 metres long, is supplied from both ends, 240 V at A and 242 V at B. There is a concentrated load of 200 A at a distance of 400 metres from A and a uniformly distributed load of 1 A/m between the mid-point and end B. Assuming cable resistance as 0.005 Ω per 100 metre each core, determine :*

(*i*) *The currents fed at A and B.*

(*ii*) *The point of minimum potential.*

(*iii*) *The value of minimum potential.*

Sol. *Given :* $l = 1000$ m ; $V_A = 240$ V ; $V_B = 242$ V ; $i = 1$ A/m ; $r = 0.005\ \Omega$ per 100 m each core.

Fig. 19.38 shows the loaded distributor.

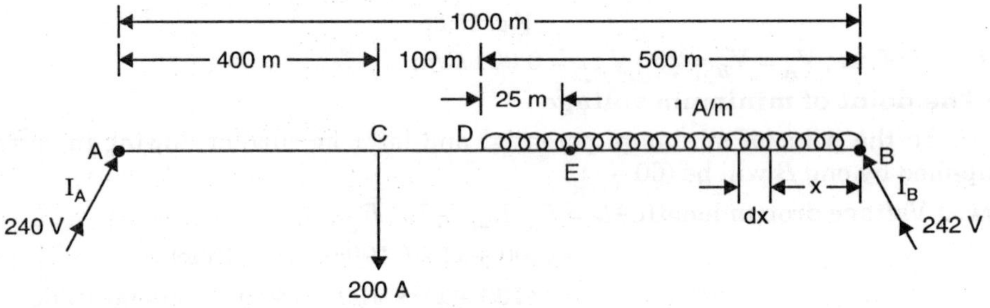

Fig. 19.38

The resistance per 100 metres of both cores = 0.005 × 2 = 0.01 Ω

Let us take 100 m as the unit of length. Let current fed at end *B* be I_B as shown in Fig. 19.38.

(*i*) **The currents fed at A and B :**

Consider any element *dx* anywhere within distance *BD i.e.* distance over which uniformly distributed load is applied. Suppose it is at a distance of *x* units (of 100 metres each). The current through *dx* is $I = I_B - 100 \times 1.0 \times x = I_B - 100x$

∴ Voltage drop over $dx = (I_B - 100x) \times 0.01 \times dx$

$$\text{Voltage drop over distance } BD = \int_0^5 (I_B - 100x) \times 0.01\, dx - 0.01 \int_0^5 (I_B - 100x)\, dx$$

$$= 0.01 \left| I_B x - 50x^2 \right|_0^5 = (0.05 I_B - 12.5) V$$

Voltage drop over $DC = (I_B - 500) \times 0.01$

Voltage drop over $CA = (I_B - 700) \times 0.01 \times 4$

Now, Total voltage drop = Voltage drop over *BD*

+ voltage drop over *DC* + voltage drop over *CA*

$$242 - 240 = (0.05 I_B - 12.5) + (I_B - 500) \times 0.01 + (I_B - 700) \times 0.04$$

$$2 = 0.05I_B - 12.5 + 0.01I_B - 5 + 0.04I_B - 28$$

or, $$2 = 0.1I_B - 45.5$$

$\therefore$ I_B = **475 A. (Ans.)**

Now, total current = 1 × 500 + 200 = 700 A

$\therefore$ I_A = 700 – 475 = **225 A. (Ans.)**

(*ii*) **The point of minimum potential :**

It is obvious that (225 – 200) = 25 A is fed into the distributed load at *D*. Hence, point of minimum potential *E* will be $\frac{25}{10}$ = **25 m from D.** Its distance from *B* = 500 – 25 = 475 m or 4.75 units of length. **(Ans.)**

(*iii*) **The value of minimum potential, V_E :**

Voltage drop upto this point from end *B*

$$= \int_0^{4.75} (475 - 100x) \times 0.01 \times dx = \int_0^{4.75} (4.75 - x)\, dx$$

$$= \left| 4.75x - \frac{x^2}{2} \right|_0^{4.75} = \left(4.75 \times 4.75 - \frac{4.75^2}{2} \right) = 11.28 \text{ V}$$

$\therefore$ V_E = 242 – 11.28 = **230.72 V. (Ans.)**

19.2.5. Ring Distributor

A distributor which is arranged to form a closed circuit and which is fed at one or more than one points is called a **"ring distributor"**.

- For the purpose of calculating voltage distribution, it can be looked upon as consisting of a series of open distributors fed at both ends.
- If the ring distributor is fed at one point then, for the purpose of calculation, it is equivalent to a straight distributor fed at both ends with equal voltages.
- By using a ring distributor *fed properly (proper choice in number of feeding points) great economy in copper can be effected.*

Example 19.21. *A 300 metres ring distributor has loads as shown in Fig. 19.39. The resistance of each conductor is 0.2 Ω/km and the loads tapped off at points B, C and D are as shown. If the distributor is fed at A, find voltages at B, C and D.*

Sol. *Given :* l = 300 m ; r = 2 × 0.2 = 0.4 Ω/km = 0.0004 Ω/m (go and return)

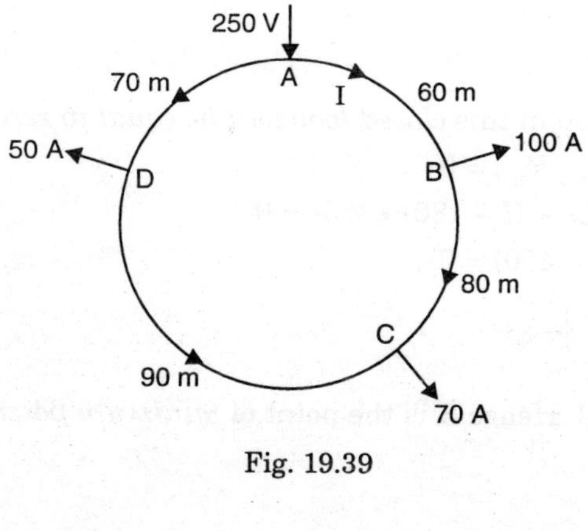

Fig. 19.39

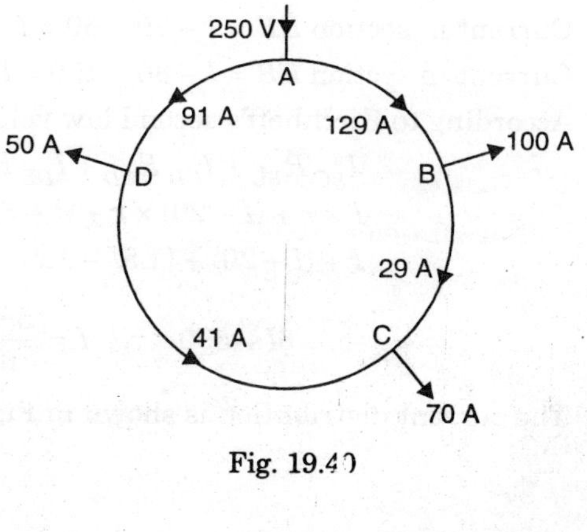

Fig. 19.40

Let us assume a current of I in section AB (Fig. 19.40) and find the total drop which should be *equated to zero.*

i.e., $$I_{AB} R_{AB} + I_{BC} R_{BC} + I_{CD} R_{CD} + I_{DA} R_{DA}$$

$\therefore$ $$I \times 60r + (I - 100) \times 80r + (I - 170)90r + (I - 220)70r = 0$$

or, $$60I + 80I - 8000 + 90I - 15300 + 70I - 15400 = 0$$

$\therefore$ $$I = 129 \text{ A}$$

The current distribution is shown in Fig. 19.40 ; it may be seen that C is the point of minimum potential.

Voltage at $B = V_A - I_{AB} R_{AB} = 250 - 129 \times (0.0004 \times 60) =$ **246.90 V. (Ans.)**

Voltage at $C = V_B - I_{BC} R_{BC} = 246.90 - 29 \times (0.0004 \times 80) =$ **245.97 V. (Ans.)**

Voltage at $D = V_C - I_{CD} R_D = 245.97 - [-41 \times (0.0004 \times 90)] =$ **247.45 V. (Ans.)**

Example 19.22. *A ring distributor is supplied through a feeder AB and is loaded as shown in Fig. 19.41. Calculate the cross-section of the AB and that of the ring main for minimum volume of copper. Assume that maximum voltage drop from A to the point of minimum potential is 15 V. Take ρ = 1.73 μΩ per cm cube.* **(Agra University)**

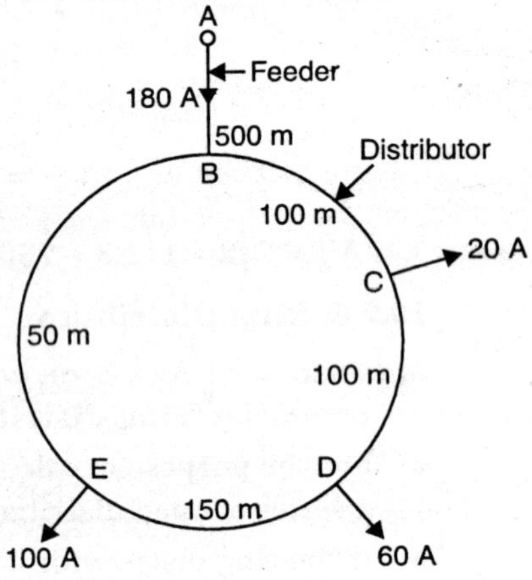

Fig. 19.41

Sol. *Given :* Maximum voltage drop from A to the point of minimum potential = 15 V ; ρ = 1.73 μΩ per cm cube.

Let, r = Resistance per 100 m of ring distributor length (both conductors).

Then, $$R_{BC} = 100 \times \frac{r}{100} = r \; \Omega$$

$$R_{CD} = 100 \times \frac{r}{100} = r \; \Omega$$

$$R_{DE} = 150 \times \frac{r}{100} = 1.5 \, r \; \Omega$$

$$R_{EB} = 250 \times \frac{r}{100} = 2.5 \, r \; \Omega$$

Let I the current flowing in section BC, then

Current in section $CD = I - 20$

Current in section $DE = I - 20 - 60 = I - 80$

Current in section $EB = I - 80 - 100 = I - 180$

According to Kirchhoff's second law voltage drop in this *closed loop will be equal to zero.*

$\therefore$ $$I_{BC} R_{BC} + I_{CD} R_{CD} + I_{DE} R_{DE} + I_{EB} R_{EB} = 0$$

$$I \times r + (I - 20) \times r + (I - 80) \times 1.5r + (I - 180) \times 2.5r = 0$$

or, $$I + (I - 20) + (1.5I - 120) + (2.5I - 450) = 0$$

or, $$6I - 590 \quad \therefore \quad I = \frac{590}{6} = 98.33 \text{ A}$$

The current distribution is shown in Fig. 19.42. Hence E is the point of *minimum potential.*

Let, a_F = Cross-sectional area of the feeder, and
a_D = Cross-sectional area of the distributor.

Resistance of feeder AB (go and return),

$$R_{AB} = \frac{1.73 \times 10^{-6} \times (500 \times 2 \times 100)}{a_F} = \frac{0.173}{a_F}\ \Omega$$

Resistance of distributor section BE (go and return),

$$R_{BE} = \frac{1.73 \times 10^{-6} \times (250 \times 2 \times 100)}{a_D} = \frac{0.0865}{a_D}\ \Omega$$

Maximum voltage drop = Voltage drop in feeder + voltage drop in distributor

$= v_F + v_D = 15$ V *(Given)*

and, $$v_F = I_{AB} \times R_{AB} = 180 \times \frac{0.173}{a_F} = \frac{31.14}{a_F} \quad ...(i)$$

and, $$v_D = I_{BE} \times R_{BE} = 81.67 \times \frac{0.0865}{a_D} = \frac{7.064}{a_D} \quad ...(ii)$$

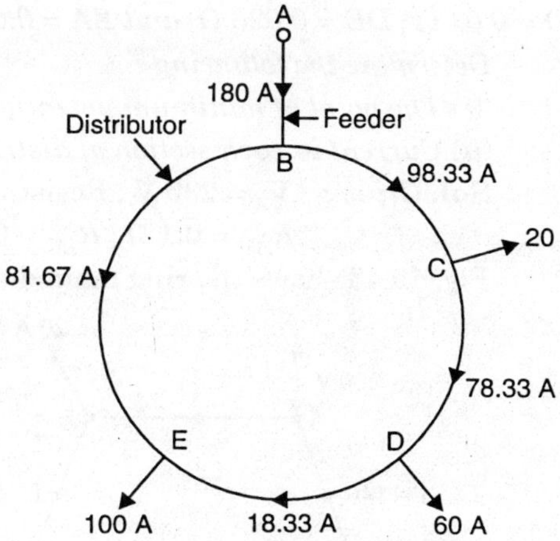

Fig. 19.42

Volume of copper used in feeder

$$= (500 \times 2 \times 100) \times a_F$$

$$= (500 \times 2 \times 100) \times \frac{31.14}{v_F}$$

$$= \frac{3114000}{v_F}\ \text{cm}^3 = \frac{3114000}{15 - v_D} \quad ...(iii)$$

Volume of copper used in distributor

$$= (600 \times 2 \times 100) \times a_D = (600 \times 2 \times 100) \times \frac{7.064}{v_D} = \frac{847680}{v_D}\ \text{cm}^3 \quad ...(iv)$$

Total volume of copper required,

$$\text{Volume} = \frac{3114000}{15 - v_D} + \frac{847680}{v_D}$$

For volume of copper to be *minimum*, $\frac{d\,(\text{Vol.})}{dV_D}$ should be zero.

i.e., $$\frac{3114000}{(15 - v_D)^2} - \frac{847680}{{v_D}^2} = 0$$

or, $$3114000{v_D}^2 - 847680\,(15 - v_D)^2 = 0$$

or, $${v_D}^2 - 0.272(225 + {v_D}^2 - 30v_D) = 0$$

or, $${v_D}^2 - 61.2 - 0.272{v_D}^2 + 8.16v_D = 0$$

or, $$0.728{v_D}^2 + 8.16v_D - 61.2 = 0$$

or, $${v_D}^2 + 11.21v_D - 84.06 = 0$$

$\therefore$ $$v_D = \frac{-11.21 \pm \sqrt{(11.21)^2 + 4 \times 84.06}}{2} = \frac{-11.21 \pm 21.49}{2} = 5.14\ \text{V}$$

(Neglecting –ve sign giving high value)

and, $$v_F = 15 - v_D = 15 - 5.14 = 9.86\ \text{V}$$

Substituting the values of v_D and v_F in (*i*) and (*ii*), we get

$$\mathbf{a_F} = \frac{31.14}{v_F} = \frac{31.14}{9.86} = \mathbf{3.158\ cm^2}. \quad \textbf{(Ans.)}$$

$$\mathbf{a_D} = \frac{7.064}{v_D} = \frac{7.064}{5.14} = \mathbf{1.374\ cm^2}. \quad \textbf{(Ans.)}$$

Example 19.23. *A 2-wire D.C. distributor ABCDEA in the form of a ring main is fed at point A at 230 V and is loaded as under :*

20 A at B ; 40 A at C ; 60 A at D and 20 A at E.

The resistances of various sections (go and return) are : AB = 0.1 Ω ; BC = 0.05 Ω ; CD = 0.01 Ω ; DE = 0.025 Ω and EA = 0.075 Ω.

Determine the following :

(i) The point of minimum potential.

(ii) Current in each section of distributor.

Sol. *Given :* V_A = 230 V ; Resistances of sections (go and return) :

R_{AB} = 0.1 Ω ; R_{BC} = 0.05 Ω ; R_{CD} = 0.01 Ω ; R_{DE} = 0.025 Ω ; R_{EA} = 0.075 Ω.

Fig. 19.43 shows the ring main distributor.

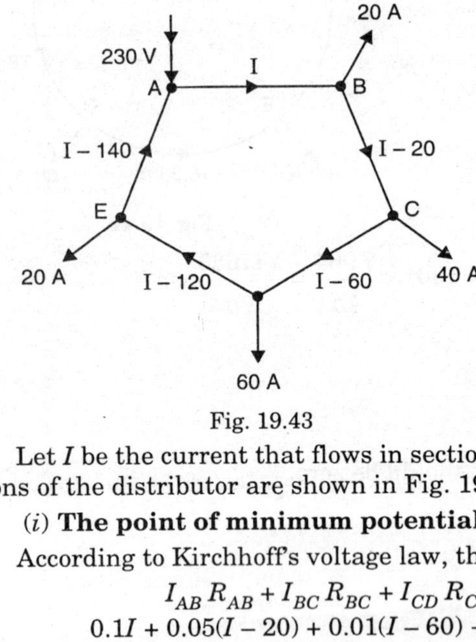

Fig. 19.43

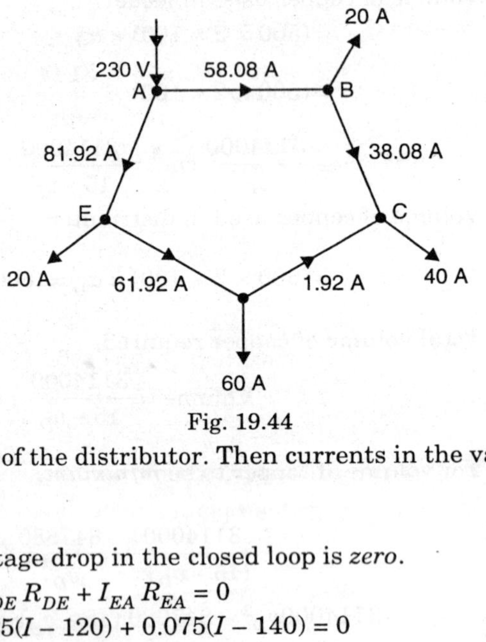

Fig. 19.44

Let I be the current that flows in section AB of the distributor. Then currents in the various sections of the distributor are shown in Fig. 19.43.

(*i*) **The point of minimum potential :**

According to Kirchhoff's voltage law, the voltage drop in the closed loop is *zero*.

i.e., $$I_{AB} R_{AB} + I_{BC} R_{BC} + I_{CD} R_{CD} + I_{DE} R_{DE} + I_{EA} R_{EA} = 0$$

or, $$0.1I + 0.05(I - 20) + 0.01(I - 60) + 0.025(I - 120) + 0.075(I - 140) = 0$$

or, $$0.1I + 0.05I - 1 + 0.01I - 0.6 + 0.025I - 3 + 0.075I - 10.5 = 0$$

or, $$0.26I - 15.1 \quad \therefore \quad I = 58.08 \text{ A}$$

The actual distribution of currents is as shown in Fig. 19.44 from where it is clear that **C is the point of minimum potential. (Ans.)**

(*ii*) **Currents in each section of the distributor :**

Current in section AB = I = **58.08 A** (*A* to *B*) **(Ans.)**

Current in section BC = I – 20 = 58.08 – 20 = **38.08 A** (*B* to *C*) **(Ans.)**

Current in section CD = I – 60 = 58.08 – 60 = – **1.92 A** (*i.e.*, *D* to *C*). **(Ans.)**

Current in section DE = I – 120 = 58.08 – 120 = – **61.92 A** (*i.e.*, *E* to *D*) **(Ans.)**

Current in section EA = I – 140 = 58.08 – 140 = – **81.92 A** (*i.e.*, *A* to *E*) **(Ans.)**

Example 19.24. *A D.C. ring main ABCDE shown in Fig. 19.45 is fed at a point A from a 230 V supply. The resistances (go and return) of the various sections are : AB = 0.1 Ω ; BC = 0.05 Ω ; CD = 0.01 Ω ; DE = 0.025 Ω and EA = 0.075 Ω. The main supplies loads of 10 A at B ; 20 A at C ; 30 A at D and 10 A at E. Find the magnitude and direction of the current flowing in each section and the voltage at each load point.*

If the points C and E are further linked together by a conductor (interconnector) of 0.05 Ω resistance and the output currents from the mains remain unchanged, find the new distribution of the current and voltage in the network. **(Banglore University)**

Sol. *Given :* $V_A = 230$ V ; $R_{AB} = 0.1\ \Omega$; $R_{BC} = 0.05\ \Omega$; $R_{CD} = 0.01\ \Omega$; $R_{DE} = 0.025\ \Omega$; $R_{EA} = 0.075\ \Omega$.

- Refer to Fig 19.45. Let us assume a current of I amperes in section AB and put the total drop round the ring equal to *zero*.

$$\therefore \quad I_{AB} R_{AB} + I_{BC} R_{BC} + I_{CD} R_{CD} + I_{DE} R_{DE} + I_{EA} R_{EA} = 0$$

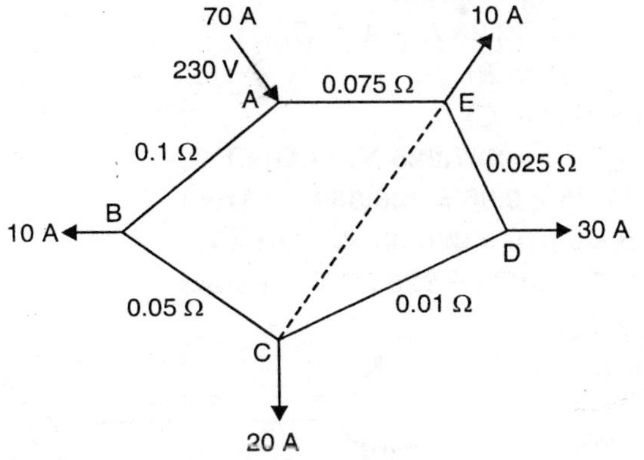

Fig. 19.45

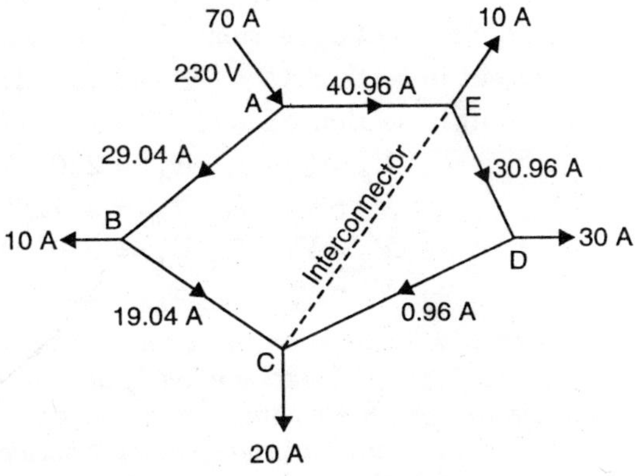

Fig. 19.46

or, $$0.1I + 0.05(I - 10) + 0.01(I - 30) + 0.025(I - 60) + 0.075(I - 70) = 0$$

or, $$0.1I + 0.05I - 0.5 + 0.01I - 0.3 + 0.025I - 1.5 + 0.075I - 5.25 = 0$$

or, $$0.26I - 7.55 = 0 \quad \therefore \quad I = 29.04 \text{ A}$$

The current distribution now becomes as shown in Fig. 19.46

Voltage at each load point is calculated as follows :

$$V_B = V_A - I_{AB} R_{AB} = 230 - 29.04 \times 0.1 = \mathbf{227.1\ V}$$

$$V_C = V_B - I_{BC} R_{BC} = 227.1 - 19.04 \times 0.05 = \mathbf{226.15\ V}$$

$$V_D = V_C - I_{CD} R_{CD} = 226.15 - (-0.96 \times 0.01) = \mathbf{226.16\ V}$$

$$V_E = V_D - I_{DE} R_{DE} = 226.16 - (-30.96 \times 0.025) = \mathbf{226.93\ V}$$

- Fig. 19.46 shows the interconnector between points C and E. *The function of the* **interconnector** *is to reduce the drop of voltage in various sections.*

In order to find the current in the interconnector, first potential difference (p.d.) across its ends is calculated. Then we calculate the resistance, viewed from points E and C of the network composed of the distribution lines only, ignoring the load resistances. Then current through the interconnector is given by,

$$I = \frac{\text{Potential difference between } E \text{ and } C}{\text{Resistance of distribution network + Resistance of interconnector}}$$

P.D. between points E and C = 226.93 – 226.15 = 0.78 V. In order to determine the resistance viewed from C and E, the network is drawn as shown in Fig. 19.47.

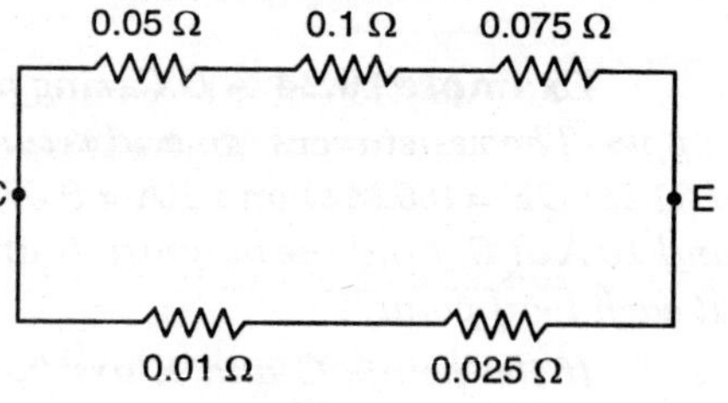

Fig. 19.47

The combined resistance

$$= \frac{(0.05 + 0.1 + 0.075) \times (0.01 + 0.025)}{(0.05 + 0.1 + 0.075) + (0.01 + 0.025)}$$

(Two branches are in parallel)

$$= \frac{0.225 \times 0.035}{0.225 + 0.035} = 0.03\ \Omega$$

$\therefore$ Current in the interconnector $= \frac{0.78}{0.03 + 0.05} = 9.75$ A (E to C)

Let us calculate the currents in other sections again. Further let current I_1, flows in section ED. Then, since the voltage round the closed mesh EDC is zero,

$$\therefore \quad -0.025I_1 - 0.01(I_1 - 30) + 0.05 \times 9.75 = 0$$

or,
$$-0.025I_1 - 0.01I_1 + 0.3 + 0.4875 = 0\ ; \quad \therefore \quad I_1 = 22.5 \text{ A}$$

Thus, current in section AE = 10 + 22.5 + 9.75 = 42.25 A (A to E)

Current in section AB = 70 – 42.25 = 27.75 A (A to B)

Current in section BC = 27.75 – 10 = 17.75 A (B to C)

$V_B = V_A - I_{AB} R_{AB}$ = 230 – 27.75 × 0.1 = **227.225 V. (Ans.)**

$V_C = V_B - I_{BC} R_{BC}$ = 227.225 – 17.75 × 0.05 = **226.337. (Ans.)**

$V_E = V_A - I_{AE} R_{AE}$ = 230 – 42.25 × 0.075 = **226.83 V. (Ans.)**

$V_D = V_E - I_{ED} R_{ED}$ = 226.83 – 22.5 × 0.025 = **226.27 V. (Ans.)**

Example 19.25. *For the same voltage drop between the supply point A and the load point D in the network shown in Fig. 19.48, compare the weights of copper required with and without interconnector between A and D. Conductors have the same cross-section in each scheme.* **(AMIE)**

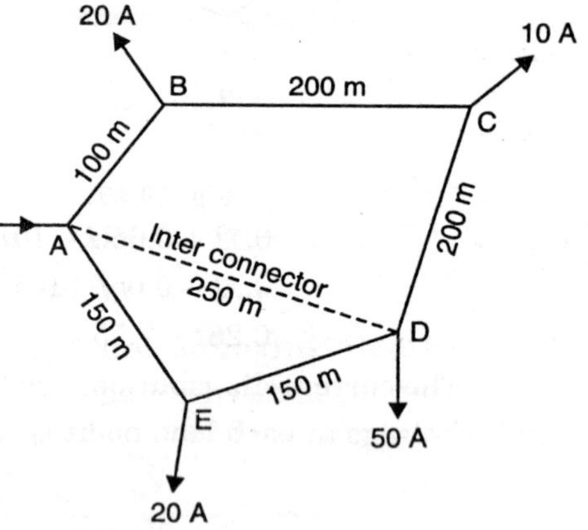

Fig. 19.48

Sol. *Without interconnector*: Refer to Fig. 19.48. Let I be the current in section AB.

Then, Current in section $BC = I - 20$

Current in section $CD = I - 20 - 10 = I - 30$

Current in section $DE = I - 30 - 50 = I - 80$

Current in section $EA = I - 80 - 20 = I - 100$

Applying Kirchhoff's second law round the closed loop $ABCDEA$, we get

$$\frac{2\rho}{a_1}[100I + 200(I - 20) + 200(I - 30) + 150(I - 80) + 150(I - 100)] = 0$$

or,
$$100I + 200I - 4000 + 200I - 6000 + 150I - 12000 + 150I - 15000 = 0$$

or,
$$800I - 37000 = 0 \quad \therefore \quad I = 46.25 \text{ A}$$

Voltage drop From A to D

$$= \frac{2\rho}{a_1}(46.25 \times 100 + 26.25 \times 200 + 16.25 \times 200) = 26250 \times \frac{\rho}{a_1} \qquad ...(i)$$

where a_1 is the cross-section of various sections.

- Addition of interconnector will change the current distribution. So let us now determine the current in the interconnector. If all the conductor (including interconnector) have a cross-section a, then,

Resistance of interconnector $AD = \frac{\rho}{a} \times 250 \times 2 = \frac{500\rho}{a}$

Resistance of network viewed from points A and D

$$= \frac{2\rho}{a}\left[\frac{(100+200+200)(150+150)}{(100+200+200)+(150+150)}\right] = \frac{2\rho}{a} \times \frac{500 \times 300}{800} = \frac{375\rho}{a}$$

According to Thevenin's theorem current flowing in interconnector AD

$$= \frac{\text{Potential different between terminals A and D}}{\text{Resistance of distribution network + Resistance interconnector}}$$

$$= \frac{26250 \times \rho/a}{375\frac{\rho}{a} + 250 \times 2 \times \frac{\rho}{a}} = \frac{26250 \times \rho/a}{875 \times \rho/a} = 30 \text{ A}$$

With interconnector : Refer to Fig. 19.48.

In the second case *i.e.*, with interconnector let the conductor cross-sections of all sections (including interconnector) be a_2. Then voltage drop between A and D

$$= 30 \times \frac{\rho}{a_2} \times 250 \times 2 = 15000\frac{\rho}{a_2} \quad \ldots(ii)$$

For same voltage drop between terminals A and D, equating (i) and (ii), we get

$$26250 \times \frac{\rho}{a_1} = 15000\frac{\rho}{a_2} \quad \text{or} \quad \frac{a_2}{a_1} = \frac{15000}{26250} = 0.5714$$

i.e., the conductor cross-section with interconnector is 0.5714 times the corresponding cross-section without interconnector.

However, total lengths of conductor with and without interconnector are 1050 m and 800 m respectively so that,

$$\frac{\text{Volume (or weight) of copper with interconnector}}{\text{Volume (or weight) of copper without interconnector}} = 0.5714 \times \frac{1050}{800} = \mathbf{0.75.} \quad \textbf{(Ans.)}$$

19.2.6. Three-wire D.C. System

19.2.6.1. Introduction

The major disadvantage of D.C. for general power purposes is that its voltage cannot be readily changed, except by the use of rotating machinery, which in most cases is too expensive. This

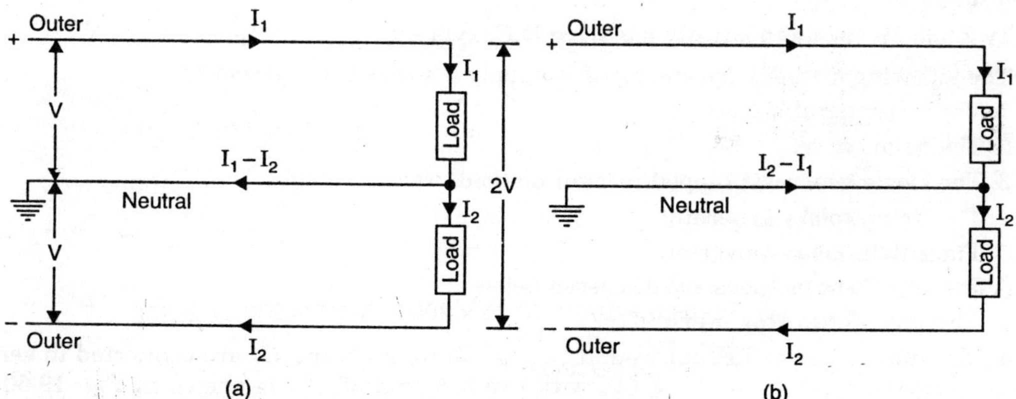

Fig. 19.49

difficulty is largely overcome by using 3-wire D.C. system which consists of *two outers and an earthed middle wire known as neutral wire*, as shown in Fig. 19.49.

- In this system consumer apparatus or lamps are connected between neutral and one of the outers. Motor loads requiring high voltage are connected between the outers. *Due to the availability of two voltages, 3-wire system is preferred over 2-wire system.*
- When the load on both sides is equal, no current flows through the neutral and such a load is known as *balanced load*. But if the load on the two sides is different then difference of load currents on two sides known as *out of balance* current flows in the neutral.
- Since only out of balance current flows through the neutral wire, so generally the area of cross-section of neutral is taken half as compared to either of the outers.
- If the currents flowing through +ve and –ve outers are I_1 and I_2 respectively then current in the middle wire will be difference of currents I_1 and I_2 flowing in the outers. When the load on +ve side is greater than that on –ve side, current equal to $(I_1 - I_2)$ will flow in the neutral wire from load end to supply end, as shown in Fig. 19.49 (*a*). When the load on –ve side is greater than that on +ve side then current equal to $(I_2 - I_1)$ will flow in the neutral wire from supply end to load end, as shown in Fig. 9.49 (*b*).
- A 3-wire D.C. system is usually supplied through a single generator in combination with auxiliary devices or combination of generators. (The 3-wire A.C. system is supplied through the centre-tapped secondary of one phase of a polyphase transformer).

19.2.6.2. Advantages and disadvantages of 3-wire distribution

Following are the advantages and disadvantages of 3-wire distribution :

Advantages :

1. It *provides dual voltages* to an occupancy. It provides the added advantage of the use of the higher voltage (a lower current) for high-power appliances (*e.g.*, ranger and heaters) and the lower voltage for lighting and other smaller loads.

2. Despite the additional line conductor, a *saving in copper results for the same power transmitted.*

3. It encourages the use of electrical energy and new appliances by *providing voltages of greater versatility.*

Disadvantages :

1. In 3-wire system, third wire is required.
2. A balancer is required and as such the cost is increased.
3. The safety is partially reduced.

19.2.6.3. Methods to supply a 3-wire D.C. system

The following methods are employed to supply a 3-wire D.C. system :

1. The two-generator method
2. The balancer set
3. The single-generator tapped-resistor method.
4. The Dobrowolsky generator.
5. The synchronous converter.

A few important methods are discussed below :

1. The two-generator method :

- In this method two shunt wound D.C. generators G_1 and G_2 are connected in series by connecting +ve terminal of G_2 with –ve terminal of G_1, as shown in Fig. 19.50. Both generators may be driven by the same prime mover.

- In this type of combination each generator supplies the load on its own side.
- The major *disadvantage* of this method is that *two separate generators are required.*

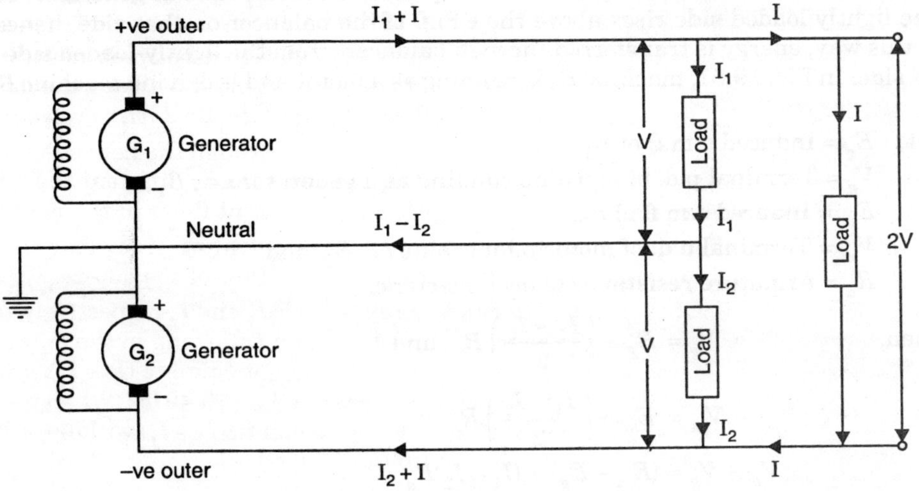

Fig. 19.50. The two-generator method.

2. The balancer set :

A balancer set is used to maintain p.ds (potential differences) on the two sides of neutral equal to each other.

The commonest form of balancer consists of two identical shunt-wound machines which are not only coupled mechanically but have their *armatures, and field circuits joined in series across the outers.* The neutral is connected to the junction of the armatures as shown in Fig. 19.51.

When the system is unloaded or when the loads on the two sides are balanced, then (*i*) *both machines run as unloaded motors,* and (*ii*) since there speeds and field currents are equal, their *back e.m.fs are the same.*

When the two sides are unbalanced *i.e.*, when the load supplied by +ve outer is different from that supplied by –ve outer, then out of balance current $(I_1 - I_2)$ flows through the mid-wire to the balancers where it divides into two halves each equal to $(I_1 - I_2)/2$ as shown in Fig. 19.51.

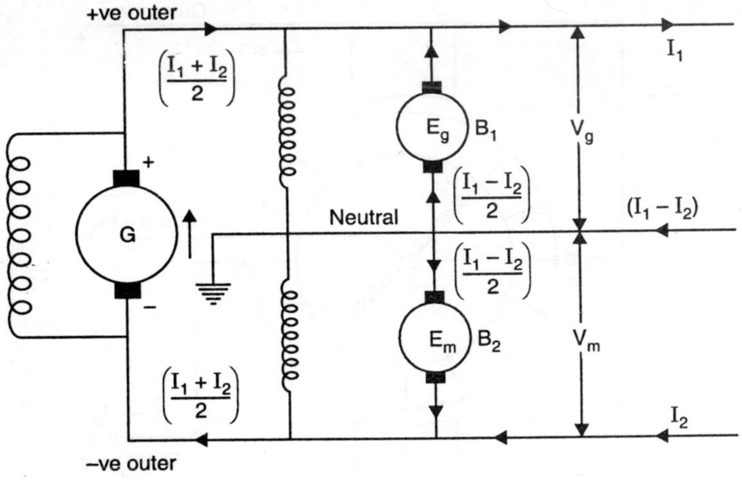

Fig. 19.51

When $I_1 > I_2$, then +ve side is move heavily loaded than –ve side, hence the p.d. on this side tends to fall below the e.m.f. of the balancer on this side so that B_1 *runs as generator*. However, the p.d. on the lightly loaded side rises above the e.m.f. of the balancer on that side, hence B_2 *runs as motor*. In this way, energy is transferred through balancers from the lightly-loaded side to the heavily-loaded side. In Fig. 19.51, machine B_2 is running as a motor and is driving machine B_1 as generator.

Let, E_g = Induced e.m.f. of B_1,

V_g = Terminal p.d. of machine running as a generator *i.e.*, B_1,

E_m = Induced e.m.f. of B_2,

V_m = Terminal p.d. of motoring machine *i.e.* B_2, and

R_a = Armature resistance of each machine.

Then,
$$V_g = E_g - \left(\frac{I_1 - I_2}{2}\right) R_a, \text{ and}$$

$$V_m = E_m + \left(\frac{I_1 - I_2}{2}\right) R_a$$

$$\therefore \quad V_m - V_g = (E_m - E_g) + (I_1 - I_2)R_a \qquad ...(i)$$

Since the speed and excitation of the two machines are equal,

$$\therefore \quad E_g = E_m, \quad \text{or,} \quad V_m - V_g = (I_1 - I_2)R_a \qquad ...(ii)$$

Thus, we find that the difference of voltages between the two sides of system is proportional to (*i*) out of balance current $(I_1 - I_2)$ and (*ii*) armature resistance of the balancer.

Owing to this reason, R_a *is kept very small* and effort is made to arrange the loads on the two sides such that *out-of-balance current is as small as possible*.

The value $(V_m - V_g)$ can be still further *reduced by cross-connecting the field windings of the machines*. A better way for improving the performance of the balancer is to *use a series field with each machine*. The series field shall be so connected that the *machine acting as a generator is cumulatively compounded and that as a motor differentially compounded*.

3. Dobrowolsky generator :

Fig 19.52 shows the connections of Dobrowolsky generator. A 2-pole generator is assumed, and the armature winding is shown only schematically. A coil of wire *PQ*

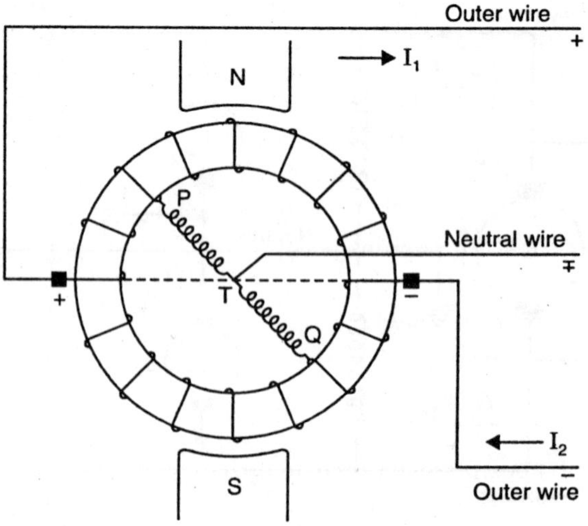

Fig. 19.52. Dobrowolsky generator.

wound on an iron core and *thus having a high reactance* is connected into the armature winding between points P and Q which lie a pole pitch (180 electrical degree) apart. The voltage between points P and Q is alternating. The current in the coil PQ has the frequency $\frac{pN}{120}$, and is very small due to the high reactance of the coil. The mid-point T of the coil is connected to a *slip-ring and the brushes which are connected to a third wire or neutral wire.*

Lamps and small motors are connected between the outer wires and the neutral wire, while the larger motors are connected between the outer wires.

When both halves of the system are loaded equally, *i.e.*, $I_1 = I_2$, no current flows through the neutral wire. When both halves of the system are loaded unequally, the difference $I_1 - I_2$ flows through the neutral wire. This current divides into two parts in the coil PQ, which go through the armature to the positive as well as to the negative brush.

19.2.6.4. Boosters

- A booster may be defined as *a dynamo electric machine, the armature of which is connected in series with a circuit, its generated e.m.f. being added to or subtracted from that of the circuit, depending upon the polarity of its excitation.*
- Boosters may be driven by any form of prime mover but are generally direct-connected to a motor taking current from constant-potential means.
- The primary use of a booster is *to raise the voltage of a generator or of a section of the bus-bars of a central station by an amount sufficient to compensate the ohmic drop in a feeder supplying a distant load*, in case the load is of such character as to require the same voltage as receiving devices at or near the source of supply. Since the line drop is directly proportional to the current, the voltage of the booster should also be proportional to the current ; or the booster should have an external characteristic consisting of a straight line through the origin. It is impossible exactly to realise this from of characteristic without auxiliary devices, but it may be approximated sufficiently closely for practical purposes by designing the booster as a series-wound generator with flux densities well within the point of magnetic saturation.
- *The hysteresis effect is especially objectionable in boosters, and should be reduced to a minimum.*
- If excitation is of such character that the main flux is subject to wide variations, the *magnetic circuit must be laminated throughout*, in order that eddy currents set up by a change in the flux may not be of sufficient magnitude to retard the change of flux and so make the machine sluggish in its action.

The boosters may be *classified* as follows :

(*i*) Series boosters (*ii*) Shunt boosters

(*iii*) None-reversible or constant current booster

(*iv*) Reversible booster.

1. Series booster (*for traction work*) :

The voltage drop (V) in a conductor is given by

$$V = I\,\frac{\rho l}{A}$$

where I, ρ, l and A are current, specific resistance, length and sectional area of the conductor respectively. It follows then for long feeders, as used in low-tension D.C. traction systems, the sectional area of the conductor becomes very heavy if the voltage drop is to be kept within narrow limits, and the conductor has to be worked at a current density much below its normal rating A great economy in the cost of the feeders, and at the same time a much better voltage regulation at the receiving end of the feeder, can be obtained by connecting a small series generator in the feeder circuit as shown in Fig. 19.53. This machine is commonly called a ***'positive series booster'***. It is driven at constant speed by a shunt motor and is designated so that when full-load current flows in the feeder, *the booster generates, and adds to the main voltage, an amount exactly equal to the volt-drop in the positive feeder. The potential of the distant end of the feeder is thus kept the same as that of the positive bus bar at the generating station.* If the booster is designed in such a way as to work entirely on the initial straight-line portion of the magnetisation curve, the voltage generated will be proportional to the line current, so that the volt-drop in the feeder will automatically be compensated at all values of the current.

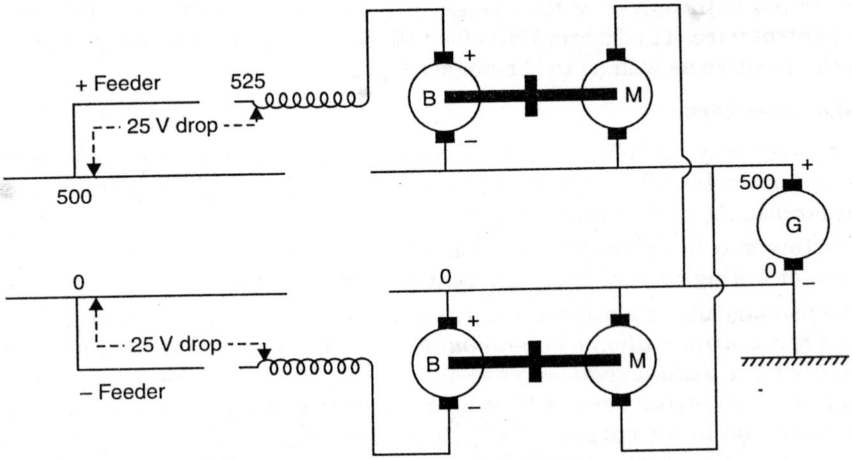

Fig. 19.53. Series booster for traction work.

The series boosters are not only used in circuit with the positive feeders, but similar machines connected in circuit with the negative or track feeders, so that the distant ends of these feeders can be maintained at approximately earth potential. The *negative booster* is thus connected in series with the negative feeder as shown in Fig. 19.53.

Note. *It is imperative only to connect boosters in circuit with long, heavily-loaded feeders.* The voltage of the main part of the system can be maintained sufficiently closely by using over-compounded generators. The voltage of these latter machines is determined by the total load on the whole system and is applied indiscriminately to all the feeders, whether long or short, heavily or lightly loaded. The action of the series booster is, on the contrary, entirely controlled by the load on, and affects solely the feeder to which it is connected.

2. Shunt booster. In case of systems of constant potential having small capacity where the load changes gradually but has a considerable range, a *storage batter may be used in parallel with the bus-bars*. At times of *light load* the *battery* may be connected to carry the entire load and at times of *peak load* it may be connected in parallel with the generator. At other times the battery takes charging current from the generator. In a system of this kind a *shunt booster* is used to force charging current into the battery, the connections being shown in Fig. 19.54. *The booster's field winding is connected across the main bus-bar, never across its own armature.* The booster armature is in series with battery during the charging period and is called upon to supply a relatively small e.m.f., and hence the connection of the field winding to the bus-bars.

The voltage of the booster is manually controlled by means of the field rheostat, adjustment being made when the readings of the ammeters indicate its necessity.

By using a reversing rheostat or by inserting a reversing switch in the field circuit, the booster e.m.f. may be *added* to that of the battery, the battery being assisted to discharge if the voltage is low or if the demand for current is unusually heavy.

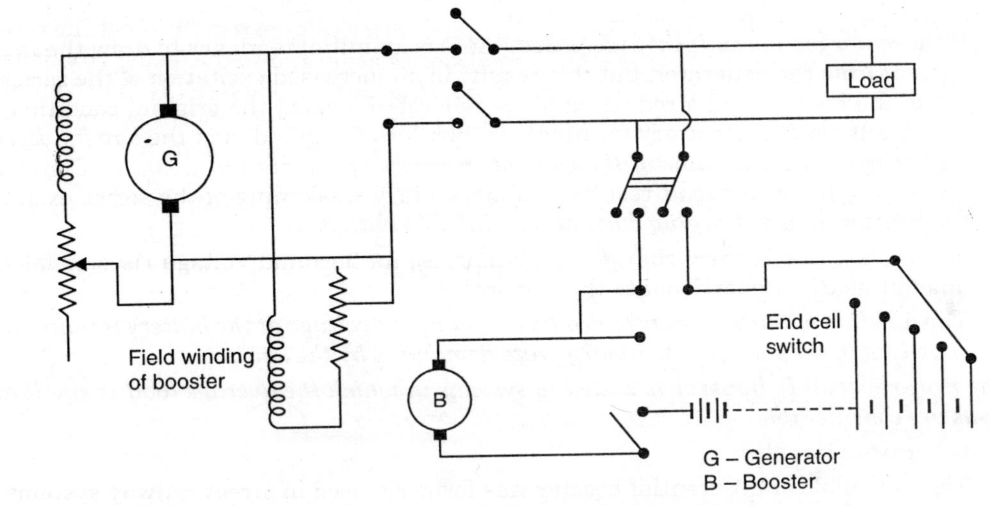

Fig. 19.54. Connections of shunt booster.

The cells shown at the right-hand end of the battery (Fig. 19.54) are the *end cells* which are cut in and out of the circuit by means of an end cell switch. These cells are used to *adjust the battery voltage* to the line requirements to compensate for the changes in voltage due to varying conditions of charge and discharge. The number of end cells may be reduced if the booster field is provided with a reversing switch (for in that case the booster can be made to oppose that of the battery to a sufficient extent to bring the terminal voltage to the proper value).

3. Non-reversible (or Constant current) booster. In case of isolated plants (as in hotels and office buildings) supplying a *lamp load* and *a fluctuating motor load* it is necessary to maintain a constant voltage, and it is permissible or even desirable to allow the voltage of the power circuit to fall when there is a heavy rush of current, as on starting an elevator. Fig. 19.55 shows a type of installation occasionally used in such a case.

The shunt field winding of the booster is connected across the constant potential lighting bus, and its *magnetizing effect is opposed by that of the series winding*, as indicated by the arrows. The excitation due to shunt field winding of the booster is normally greater of the two, and the resultant differential excitation produces a booster voltage that acts in the same direction as the generator and is normally 10 to 15 volts.

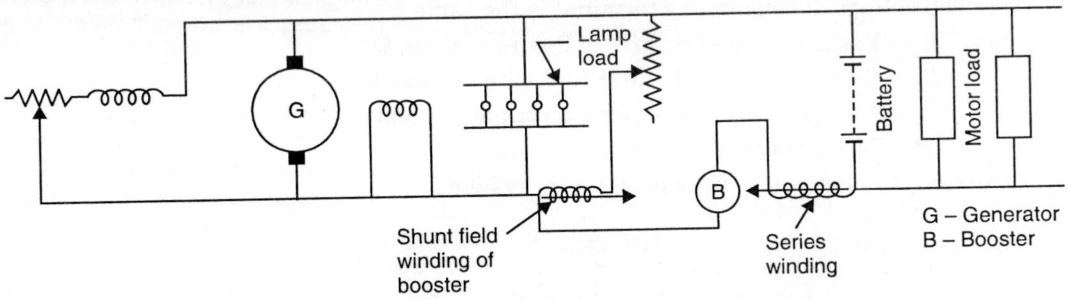

Fig. 19.55. Non-reversible or constant current booster.

- When *load is normal* the adjustments are such that the battery neither charges nor discharges, or the sum of the voltages of generators and booster equals the open-circuit voltage of the battery. The entire lighting and power load is then carried by the main generator.

- If the *motor load is suddenly increased*, there is an initial tendency to draw the increased current from the generator, but this results in an increased excitation of the series winding of the booster and a reduction of its generated e.m.fs., the original condition of balanced voltage at the battery terminals is therefore disturbed, and the *battery discharges and relieves the generator of the current in excess of the normal amount.* Conversely, a *decrease of the motor load* results in a momentary weakening of the series excitation of the booster, and a *charging current flow into the battery.*
- When a storage battery charges or discharges, its terminal voltage rises or falls by an amount nearly proportional to the current.
- The *function of the booster is to produce a change of voltage at the battery terminals, corresponding to the charge or discharge rate demanded by the load.*

The ***non-reversible booster*** *is suited to systems in which the average load is small and the fluctuations are considerable.*

4. Reversible booster

- The reversible or differential booster was formerly used in street railway systems which has a large average load. Its *purpose was to permit the battery to carry the fluctuations above and below the normal load.*
- At *normal load*, the battery current was adjusted to zero value by regulation of shunt field, which opposed the series winding. Under these conditions an *increase in load* caused series winding to overpower shunt field and the resultant booster e.m.f., acting in the same direction as the battery, caused discharge to take place. Similarly a *discharge of load* caused shunt field to overpower series winding, thus reversing the polarity of the booster and causing a charging current to flow into the battery.

19.2.6.5. Comparison of 3-wire and 2-wire D.C. distribution

For making comparison between 3-wire and 2-wire D.C. distribution the following *assumptions* will be made :

(*i*) The amount of power transmitted in both the cases is same.
(*ii*) The distance of transmission is same.
(*iii*) The transmission efficiency (and hence losses) is the same.
(*iv*) The 3-wire system is balanced.
(*v*) In the 3-wire system, the mid-wire is of half the cross-section of each outer.
(*vi*) The voltage at consumer's terminal is the same.

Let, R_2 = Resistance of each wire of 2-wire system, Ω,
R_3 = Resistance of each outer in 3-wire system, Ω,
V = Voltage at the consumer's terminals V, and
P = Power transmitted, W.

Current through outers in case of 3-wire system,

$$I_3 = \frac{P}{2V}$$

Total loss in two outers $= 2I_3^2R_3 = 2\left(\frac{P}{2V}\right)^2 R_3$

Current in 2-wire system, $I_2 = \frac{P}{V}$

Total loss $= 2I_2^2R_2 = 2\left(\frac{P}{V}\right)^2 R_2$

Since the transmission efficiencies are the same, it mean the losses are also the same.

$$\therefore \quad 2\left(\frac{P}{2V}\right)^2 R_3 = 2\left(\frac{P}{V}\right)^2 R_2 \quad \text{or} \quad \frac{R_3}{R_2} = \frac{4}{1} \quad \text{or} \quad R_3 = 4R_2$$

Therefore, the area of cross-section of outers in 3-wire system case will be one-fourth of each conductor in 2-wire case.

Let, a = Area of cross-section of each conductor in 2-wire case.

Then $\frac{a}{4}$ = Area of cross-section of each outer in 3-wire case, and

$\frac{a}{8}$ = Area of cross-section of neutral wire [assumption (*vi*)]

If l is the length of the line, then

$$\text{Volume of Cu for 3-wire system} = l\left(\frac{a}{4} + \frac{a}{4} + \frac{a}{8}\right) = \frac{5}{8}\,al$$

$$\text{Volume of Cu for 2-wire system} = l(a + a) = 2al$$

$$\therefore \quad \frac{\text{Volume of Cu for 3-wire system}}{\text{Volume of Cu for 2-wire system}} = \frac{(5/8)al}{2al} = \frac{5}{16}$$

Hence in 3-wire system to transmit the same amount of power with same efficiency over same distance with same consumer voltage we require $\frac{5}{16}$th or 0.3125 times copper as required in 2-wire system.

19.2.6.6. Ground detectors

- These are the devices that are employed to detect/indicate the ground fault for undergrounded D.C. systems. When a fault occurs on such a system, it must be cleared immediately, otherwise a second ground fault appears which will leads to a short circuit.
- For detection of ground faults lamps are generally used and they are connected as shown in Fig. 19.56. Each lamp should have a voltage rating equal to the line voltage. The two lamps in series (subjected to half their rated value) will glow dimly.

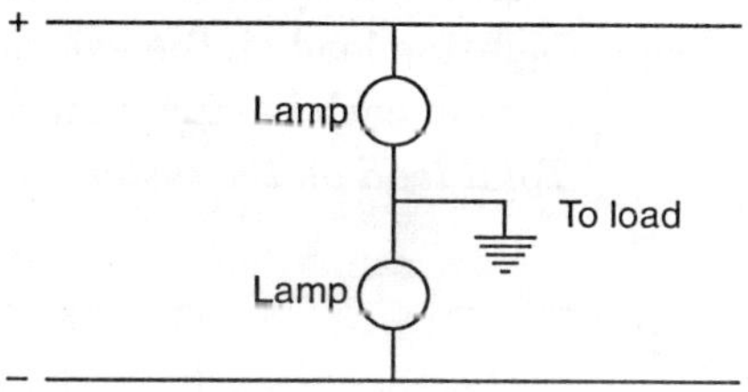

Fig. 19.56

If a ground fault occurs on either wires the lamp connected to grounded wire will not glow while the other lamp will glow brightly.

Example 19.26. *A load supplied on 3-wire D.C. system takes a current of 60 A on the +ve side and 50 A on the negative side. The resistance of each wire is 0.1 Ω and the cross-section of middle wire is one-half that of outer. If the system is supplied at 500/250 V, find the voltages at the load end between each outer and middle wire.*

Sol. Fig. 19.57 shows the current loading.

Current in the neutral wire = 60 – 50 = 10 A

Resistance of neutral wire = 2 × 0.1 = 0.2 Ω

(Since the cross-sectional area of neutral wire is one-half that of outer)

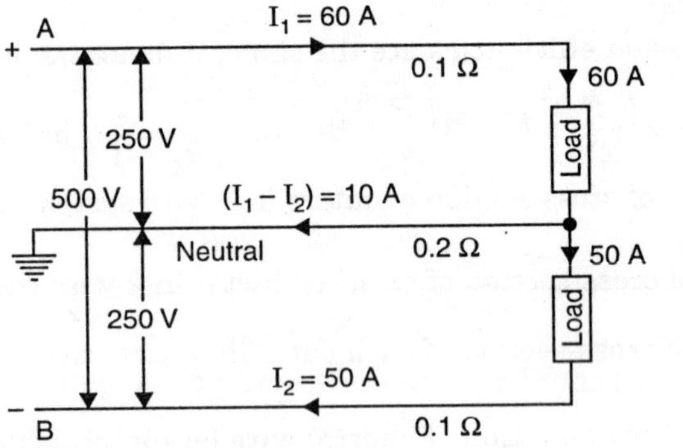

Fig. 19.57

Voltage at the load end on the +ve side

$$= 250 - I_1 \times 0.1 - (I_1 - I_2) \times 0.2$$

$$= 250 - 60 \times 0.1 - 10 \times 0.2 = \mathbf{242\ V.} \quad \textbf{(Ans.)}$$

Voltage at the load end on the –ve side

$$= 250 + (I_1 - I_2) \times 0.2 - I_2 \times 0.1$$

$$= 250 + 10 \times 0.2 - 50 \times 0.1 = \mathbf{247\ V.} \quad \textbf{(Ans.)}$$

Example 19.27. *A D.C. 3-wire system with 500 V between outers has lighting load of 100 kW on the positive and 50 kW on the negative side. If at this loading, the balancer machines have each a loss of 2.5 kW, calculate total load on the system and the kW loading of each balancer machine.*

(Banglore University)

Sol. *Given :* Voltage between outers = 500 V

Lighting load on the +ve side = 100 kW

Lighting load on the –ve side = 50 kW

Loss in each balancer machine = 2.5 kW

Total load on the system = 100 + 50 + (2 × 2.5) = **155 kW. (Ans.)**

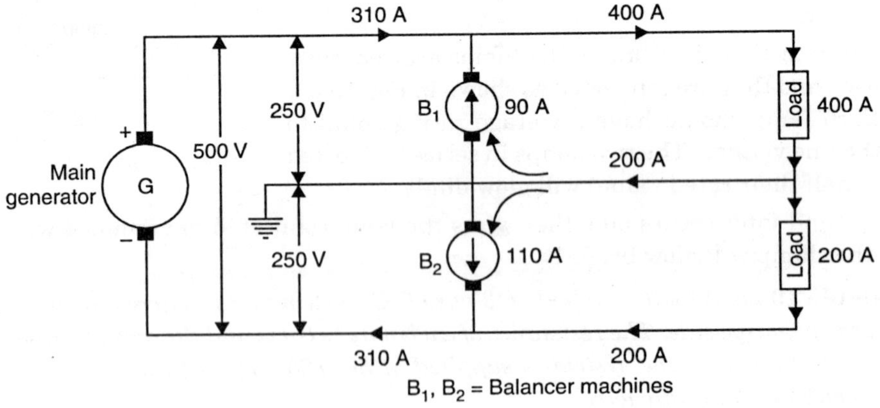

Fig. 19.58

Refer to Fig. 19.58.

$$\text{Output current of main generator} = \frac{155 \times 1000}{500} = 310 \text{ A}$$

Load current on +ve side $= \dfrac{100 \times 1000}{250} = 400$ A

Load current on –ve side $= \dfrac{50 \times 1000}{250} = 200$ A

Out of balance current $= 400 - 200 = 200$ A

Since +ve side is more heavily loaded, B_1 is working as a generator and B_2 as motor.

∴ Current of $B_1 = 400 - 310 = 90$ A

Current of $B_2 = 300 - 90 = 110$ A

∴ *Loading of* $B_1 = \dfrac{250 \times 90}{1000}$ = **22.5 kW. (Ans.)**

Loading of $B_2 = \dfrac{250 \times 90}{1000}$ = **27.5 kW. (Ans.)**

Example 19.28. *In a 500/250 V D.C. 3-wire system, there is a current of 2000 A on the +ve side, 1600 A on the negative side and a load of 300 kW across the outers. The loss in each balancer set is 8 kW. Calculate :*

(i) Total load on the main generator, and

(ii) Current in armature of each of the balancer sets.

Sol. *Given :* Current on +ve side = 2000 A

Current on –ve side = 1600 A

Load across the outers = 300 kW

Loss in each balancer set = 8 kW

(*i*) **Total load on the main generator :**

Total load on the main generator

$$= \frac{2000 \times 250}{1000} + \frac{1600 \times 250}{1000} + 300 + (2 \times 8) = \textbf{1216 kW.} \quad \textbf{(Ans.)}$$

(*ii*) **Current in armature of each of the balancer sets :**

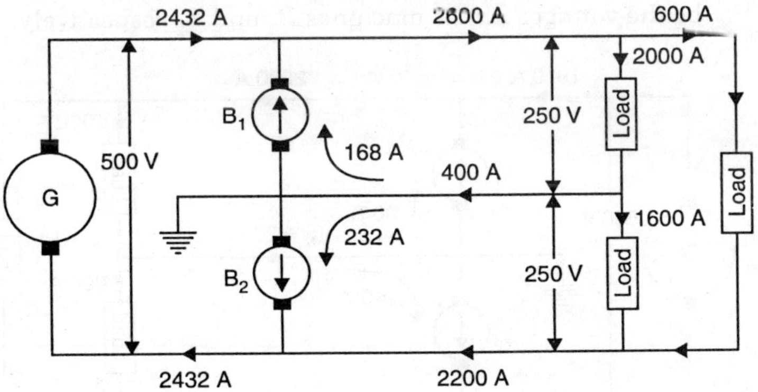

Fig. 19.59

Current in the main generator $= \dfrac{1216 \times 1000}{500} = 2432$ A

Current in the load of 300 kW across the outer $= \dfrac{300 \times 1000}{500} = 600$ A

Out-of-balance current = 2000 – 1600 = 400 A

Current through B_1 = (2000 + 600) – 2432 = **168 A. (Ans.)**

Current through B_2 = 400 – 168 = **232 A. (Ans.)**

Example 19.29. *A D.C. 3-wire system with 500 V across outers supplies 1000 A on +ve side, 800 A on –ve side and 1000 A across outers. The rotary balancer has an armature resistance of 0.1 Ω and takes 10 A on no load. Calculate :*

(i) Current loading of each balancer unit,

(ii) Voltage across each balancer, and

(iii) Load on the main generator. **(Panjab University)**

Sol. *Given :* Voltage across outers = 500 V

Current supplied on +ve side = 1000 A

Current supplied on –ve side = 800 A

Current supplied across outers = 1000 A

Armature resistance, R_a = 0.1 Ω

No-load current of the balancer = 10 A

Total current on +ve outer = 1000 + 1000 = 2000 A

Total current on –ve outer = 800 + 1000 = 1800 A

Current in the middle wire, I_N = 2000 – 1800 = 200 A.

Since the load on +ve side is more, therefore, machine B_1 will run as a *generator* and machine B_2 will run as a *motor.*

(i) **Current loading of each balance unit :**

Refer to Fig. 19.60. Let the current loading of B_1 and B_2 be I_{B1} and I_{B2} respectively.

Since current in the neutral, $I_N = I_{B1} + I_{B2}$

$$\therefore \qquad I_{B1} = I_N - I_{B2} = 200 - I_{B2}$$

Let V_{B1} and V_{B2} be the voltages across machines B_1 and B_2 respectively.

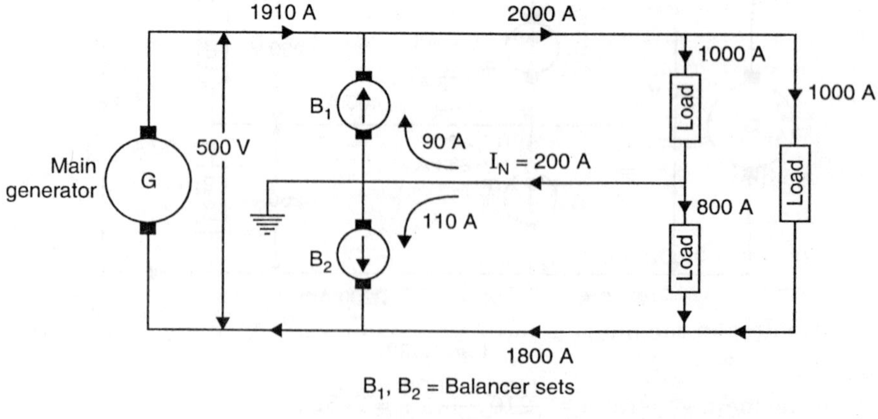

Fig. 19.60

Since machine B_2 is driving machine B_1,

$\therefore$ Output of motor = Output of generator + Losses in the set

or, $$V_{B2}I_{B2} = V_{B1}(200 - I_{B2}) + 500 \times 10 + I_{B2}^2 \times 0.1 + (200 - I_{B2})^2 \times 0.1 \quad ...(i)$$

Since field current and speeds of both the machines are same, therefore, their back e.m.fs. will be same. Let it be E_b.

Back e.m.f. on no load, $E_b = 250 - 10 \times 0.1 = 249$ V

Terminal voltage across machine B_1,

$$V_{B1} = E_b - I_{B1} R_a$$
$$= 249 - (200 - I_{B2}) \times 0.1$$

or, $$V_{B1} = 229 + 0.1 I_{B2}$$

Terminal voltage across machine B_2,

$$V_{B2} = E_b + I_{B2} R_a \quad \text{or} \quad V_{B2} = 249 + 0.1I_{B2}$$

Substituting for V_{B1} and V_{B2} in (i), we get

$$(249 + 0.1I_{B2})I_{B2} = (229 + 0.1I_{B2})(200 - I_{B2}) + 5000 + 0.1I_{B2}^2 + 0.1(40000 - 400I_{B2} + I_{B2}^2)$$

or, $$249I_{B2} + 0.1I_{B2}^2 = (45800 - 229I_{B2} + 20I_{B2} - 0.1I_{B2}^2) + 5000 + 0.1I_{B2}^2 + (4000 - 40I_{B2} + 0.1I_{B2}^2)$$
$$= 54800 - 249I_{B2} + 0.1I_{B2}^2 \quad \text{or} \quad 498I_{B2} = 54800$$

$\therefore$ I_{B2} = **110 A. (Ans.)**

and, I_{B1} = 200 – 110 = **90 A. (Ans.)**

(ii) **Voltage across each balancer :**

Voltage across balancer B_1, $V_{B1} = 229 + 0.1I_{B2} = 229 + 0.1 \times 110$ = **240 V. (Ans.)**

Voltage across balancer B_2, $V_{B2} = 249 + 0.1I_{B2} = 249 + 0.1 \times 110$ = **260 V. (Ans.)**

(iii) **Load on the main generator :**

Load on the main generator, I_a = 2000 – 90 = **1910 A. (Ans.)**

Example 19.30. *A 2-wire D.C. system has the voltage at the supply end maintained at 500 V. The line is 3 km long. If the full load current is 140 A, what must be booster voltage and output in order the far end voltage may also be 500 V. Take the resistance of the cable at the working temperature as 0.5 Ω/km.* **(Calcutta University)**

Sol. Voltage at the supply end = 500 V

Length of the line = 3 km

Full load current = 140 A

Resistance of the cable at working temperature = 0.5 Ω/km

$\therefore$ Total resistance of the line = 0.5 × 3 = 1.5 Ω

Full-load drop in the line = 140 × 1.5 = 210 V (*i.e.*, 210/140 = 1.5 volt per ampere of line current)

$$\text{Booster output} = \frac{210 \times 140}{1000} = \textbf{29.4 kW.} \quad \textbf{(Ans.)}$$

19.3. A.C. DISTRIBUTION

19.3.1. Introduction

- Now-a-days power generation is three-phase, 50-Hz, usually at 11 kV. The generated voltage is stepped upto 132, 220 or 400 kV by means of *step-up transformers.* Then by means of three-wire transmission lines electric power is carried to different places where it is received in sub-stations situated near cities. Here the voltage is stepped down to 66 or 33 kV and further carried through three-wire transmission lines to various sub-stations in the cities where the voltage is further stepped down to 11, 6.6 or 3.3 kV. These voltages are further stepped down to 415 V to make available power to small consumers through 415/240 V, three-phase, four-wire distributors.
- Small consumers receive power at 415/240 V where they can use both three-phase and single-phase supply. Very small domestic consumers are often provided with single-phase supply at 240 V. Distribution of power at 415/240 V is called *secondary or low voltage distribution system.* Electric power to large consumers is, however, often supplied direct at 11 kV, 6.6 kV or 3.3 kV. Such consumers install their own sub-station to step-down voltage to 415/240 volts.

19.3.1.1. Three-phase four-wire distribution system

- Electric power in India is supplied to the consumers by the State Electricity Boards. The following are the different systems by which power is distributed to the consumer :

1. Single-phase A.C. supply using a 2-wire system ;
2. Three-phase A.C. supply using a 3-wire system ;
3. Supply of three-phase and neutral using a 4-wire system.

D.C. supply may also be available in certain areas, on either a two-wire system at 220 V, or a three-wire system with 440 V between the two outer conductors and 220 V between the outer conductors and the centre wire. However, D.C. supply is no longer commonly used in India and, therefore, we rarely come across D.C. installations.

- The standard voltages at which the supply authorities deliver power to the consumer are generally as follows :

  Single phase : 240 V, 50 Hz, 2 wire

  Three phase : 415 V, 50 Hz, 4 wire.

  In India and in many European countries the frequency of the supply is 50 Hz. In the USA the power supply frequency is 60 Hz.
- Standardisation of supply voltage and frequency is extremely important to both electricity supply authorities as well as the manufacturers of electrical equipment. In the 3-phase. 4-wire power distribution system, power is supplied from the sub-station through 4-wires. Three of these wires are called *live or phase or line wires*. The fourth wire is usually at *zero voltage and is called neutral wire. The neutral wire is earthed at the sub-station.*
- The voltage between any two of the line wires in a 3-phase system is $\sqrt{3}$ times the voltage between any phase wire and the neutral.
- Electric loads of the consumers are connected in such a way that all the three phases are equally loaded. If the individual phases of the 3-phase systems are equally loaded the current through the neutral wire will be zero. In case of 3-phase motors, ovens etc. the

three phases are equally loaded by their very design. In such cases the neutral wire may be omitted. But in cases where a large number of single phase loads have to be supplied, loads on the three phases are balanced by connecting various single phase consumers or groups of consumers to different phases of the 3-phase supply.

- Large consumers, particularly those with heavy motor loads, are provided with a 3-phase, 4-wire supply. Motors and heavy heating loads are connected to 415 V, 3-phase supply and are called *three phase loads.* The load is balanced over the three phases by equally distributing the lighting load and small power loads over the three phases. Consumers with load requirement more than 250 kVA are provided with supply at high voltage with a substation installed in the consumer's premises where voltage is stopped down to 415/ 240 V. Fig. 19.61 shows typical 3-phase 4-wire distribution from a sub-station to the consumer.

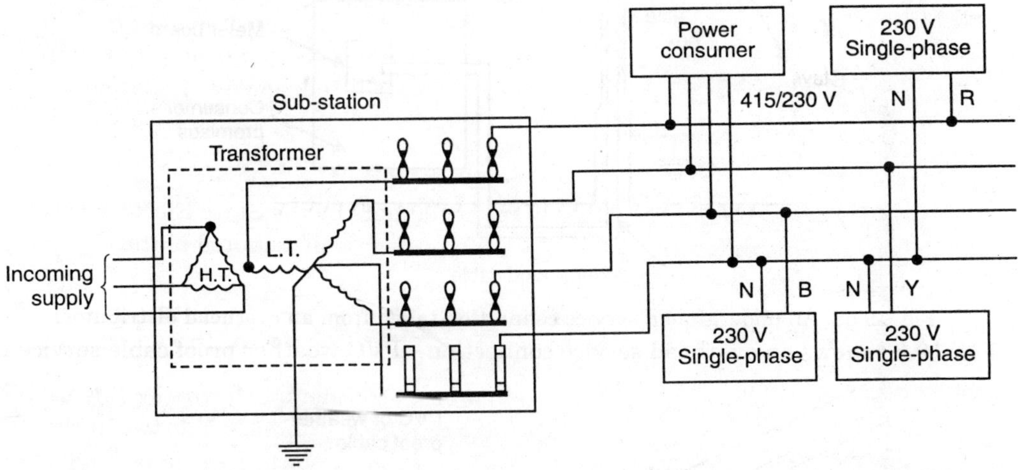

Fig. 19.61. Three phase 4-wire distribution.

19.3.1.2. Voltage tolerances

- Current carried by the electrical power distribution network varies at different times of the day. This leads to varying voltage drops in the supply cables. It is thus impracticable to give each consumer exact nominal supply voltage *i.e.*, 415/240 V at his terminals. Supply authorities are, therefore, permitted certain tolerances. Under the Indian Electricity Rules, the *voltage fluctuation may not vary by more than 5% above or below the declared nominal voltage and the frequency must be within ± 1% of the declared frequency of 50 Hz.* Thus the terminal voltage must be within the range.

  228 V to 252 V for a nominal voltage of 240 V ;

  394.25 V to 435.75 V for a nominal voltage of 415 V.

19.3.1.3. Service connections

- The supplier's distribution system brings power to the consumer through overhead lines or by means of underground cables to a place just outside the consumer's premises.
- The *line bringing electric power from supplier's low voltage distributor upto the energy meter installed at the consumer's premises is called the* **"service connection".**
- Service connection may be by means of *underground cables* or by means of *overhead conductors or cables.*

Fig. 19.62 shows an *underground service connection* taken from an overhead distributor.

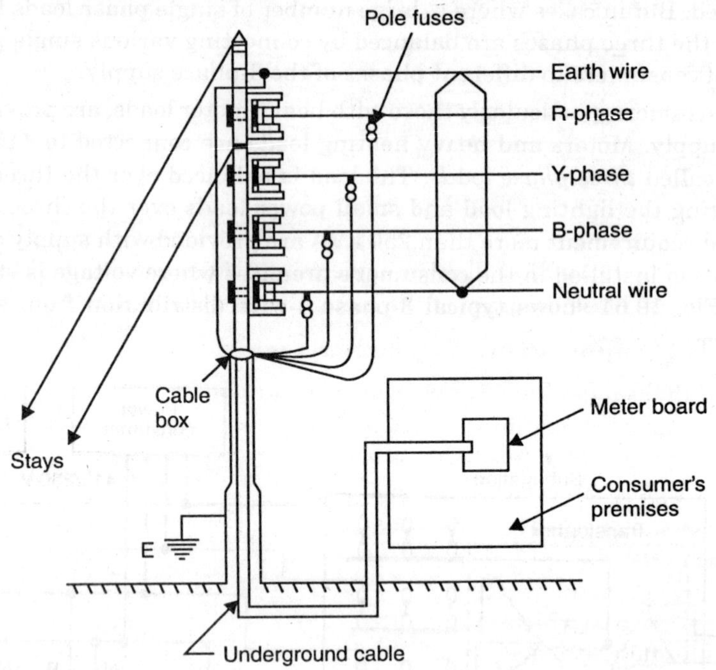

Fig. 19.62. An underground service connection taken from an overhead distributor.

Fig. 19.63 shows an overhead service connection—PVC weather proof cable service line.

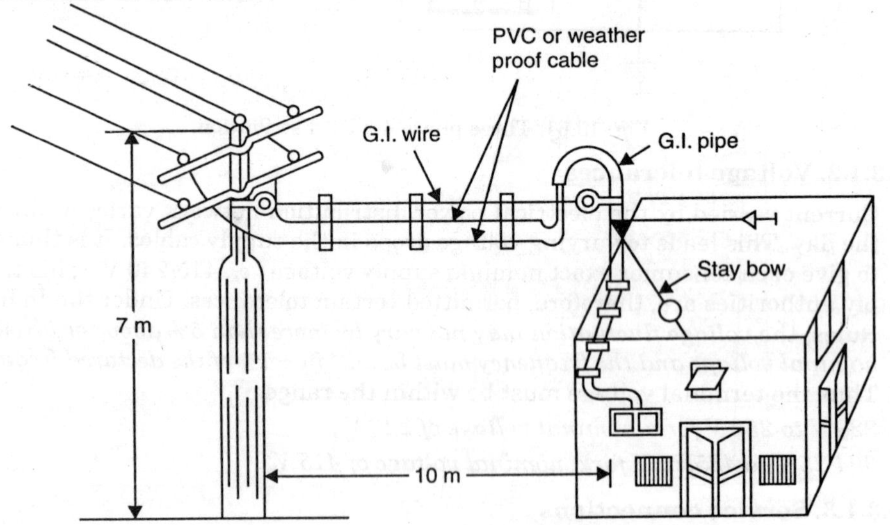

Fig. 19.63. PVC weather proof cable service-line.

19.3.1.4. Service mains

Having brought the supplier's service line into the consumer's premises it is now to be connected with the consumer's internal wiring. The supply authority have to charge the consumer for the electrical energy consumed. For this purpose the supplier's service lines will be connected to the input terminal of the energy meter to be provided by the supply authority.

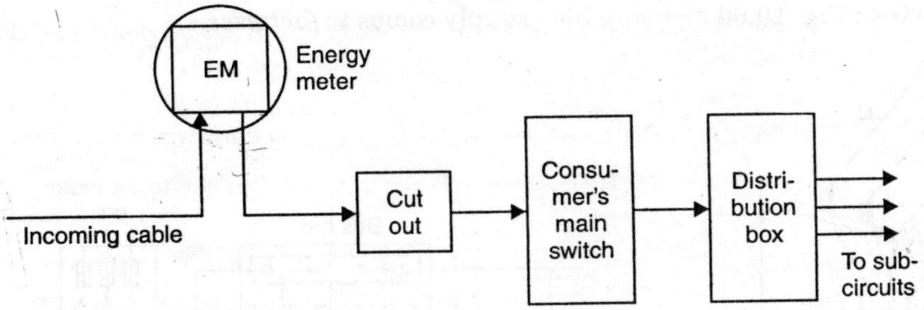

Fig. 19.64. Block diagram of the meter distribution board.

Fig. 19.64 show the block diagram of the meter distribution board.

19.3.1.5. Electric supply to our houses and factories

Houses : Fig. 19.65 shows how supply comes to our *houses.*

- From a pole of overhead lines through service line or from sub-main in the street (L.T. Pillar) through underground cable to the house, electric supply enters our house.
- The supply comes direct in energy meter fixed by supply authority.
- After energy meter the live and neutral wires are controlled in fuse and link supplied by supply authority.

Then comes owner's main switch (I.C.D.P.). After passing through I.C.D.P. wires are again controlled in distribution box which may be three, four, five or more ways. Separate phase and neutral, from distribution board, goes in different rooms for every eight to ten points. Now-a-days instead of fuse grips, miniature circuit breakers are used and instead of I.C.D.P. *automatic isolators are used.*

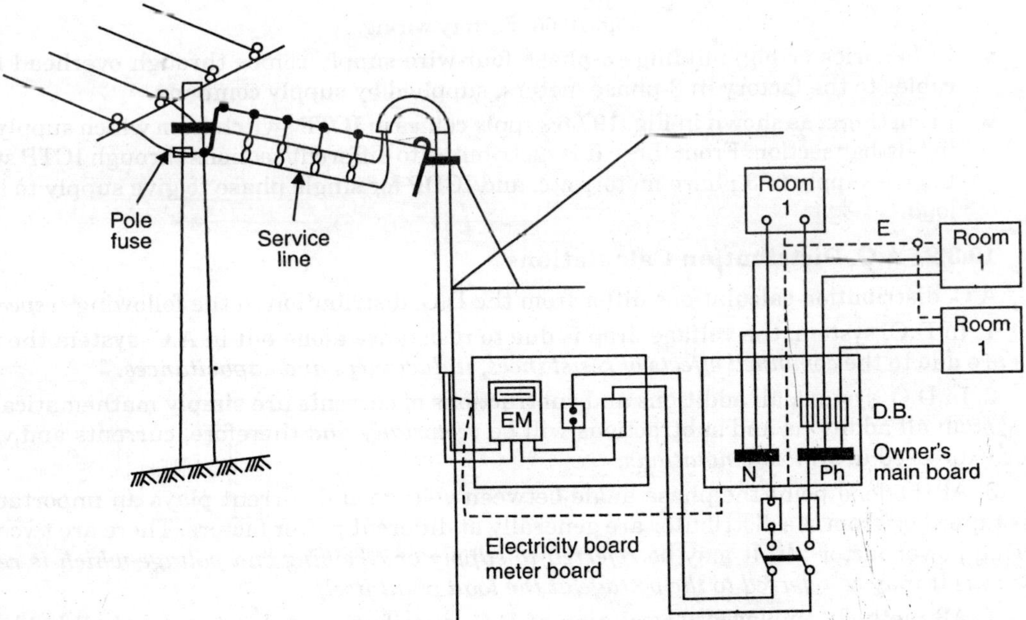

Fig. 19.65. Electric supply to houses.

Factories : Fig. 19.66 showing how supply comes to *factories.*

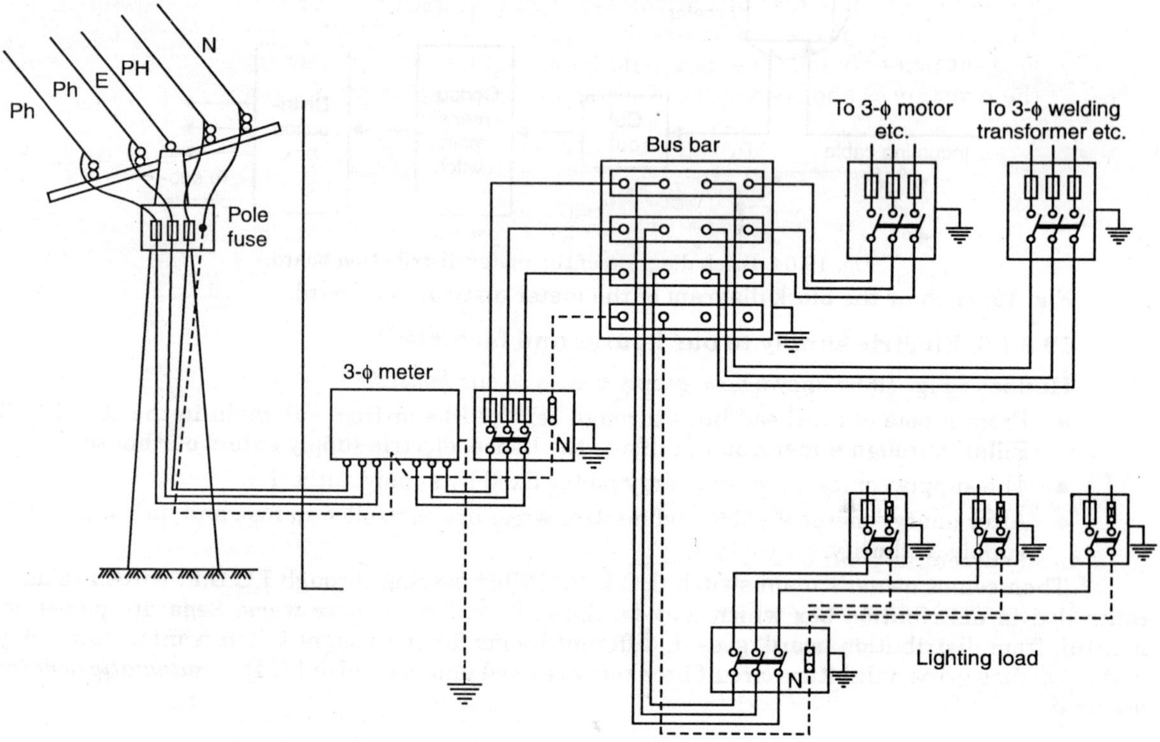

Fig. 19.66. Factory wiring.

- In factories or big buildings 3-phase four-wire supply comes through overhead lines or cables to the factory in 3-phase meters, supplied by supply company.
- From there, as shown in Fig. 19.66 supply comes in ICTP switch from which supply enters in bus-bar section. From there it is distributed to different sections through ICTP switches to give supply to 3 phase motors etc. and ICDP for single phase to give supply to lighting load.

19.3.2. A.C. Distribution Calculations

A.C. distribution calculations differ from the D.C. distribution in the following *respects* :

1. In D.C. system, the voltage drop is due to resistance alone but in A.C. system the voltage drops are due to the *combined effects of resistances, inductances and capacitances.*

2. In D.C. system, all additions and subtractions of currents are simply mathematical but in A.C. system all additions and subtractions will be *vectorially* and therefore, currents and voltages will be expressed in *symbolic notations.*

3. At the load point the phase angle between voltage and current plays an important part. Loads tapped off from the distributor are generally at different power factors. There are two ways of referring power factor : (*i*) It may be *referred to supply or receiving end voltage which is reference vector* ; (*ii*) It may be *referred to the voltage at the load point itself.*

4. All methods employed for solution of D.C. distributors and networks shall hold good in solution of A.C. distributors and networks, the only difference is that all impedances, currents and voltages shall be expressed vectorially (*i.e.*, symbolically) with current or voltage as reference vector.

19.3.3. A.C. Distributors with Concentrated Loads

(*i*) **Power factors referred to receiving end voltage.** Consider an A.C. distributor with concentrated loads I_1 and I_2 at points B and C respectively as shown in the Fig. 19.67. Taking voltage at the receiving end (V_C) as reference vector, let the power factors of the loads at points B and C be $\cos \phi_1$ and $\cos \phi_2$ lagging respectively. Let R_1, X_1 and R_2, X_2 be the resistances and reactances of sections AB and BC of the distributor.

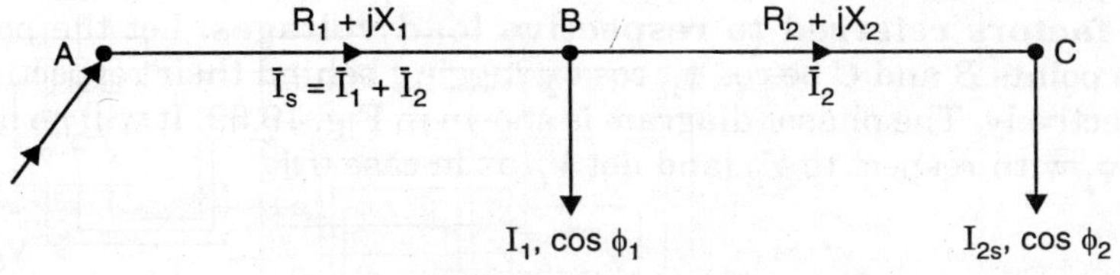

Fig. 19.67. A.C. distributor.

Impedance of section AB, $\overline{Z}_{AB} = \overline{Z}_1 = R_1 + jX_1$

Impedance of section BC, $\overline{Z}_{BC} = \overline{Z}_2 = R_2 + jX_2$

Load current at point B, $\overline{I}_1 = I_1 (\cos \phi_1 - j \sin \phi_1)$

Load current at point C, $\overline{I}_2 = I_2 (\cos \phi_2 - j \sin \phi_2)$

Current in section $AB = \overline{I}_1 + \overline{I}_2$

$$= I_1 (\cos \phi_1 - j \sin \phi_1) + I_2 (\cos \phi_2 - j \sin \phi_2)$$

Voltage drop in section $BC = \overline{I}_2 \overline{Z}_{BC} = I_2 (\cos \phi_2 - j \sin \phi_2)(R_2 + jX_2)$

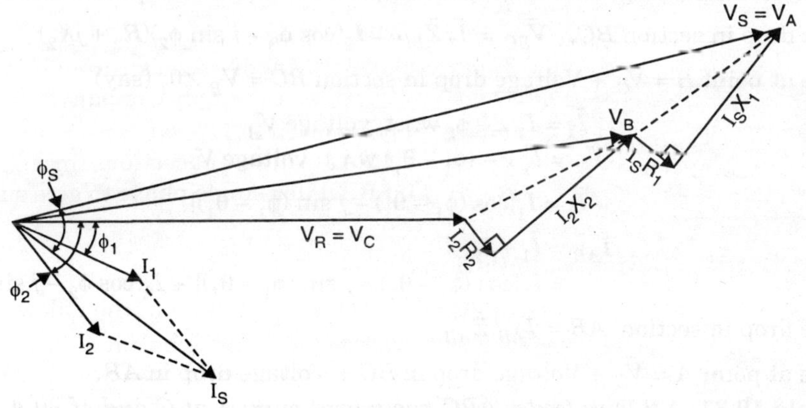

Fig. 19.68. Phasor diagram (of Fig. 19.67) for p.f. with respect to receiving end.

Voltage drop in section $AB = (\overline{I}_1 + \overline{I}_2) \overline{Z}_{AB}$

$$= [I_1(\cos \phi_1 - j \sin \phi_1) + I_2(\cos \phi_2 - j \sin \phi_2)] (R_1 + jX_1)$$

Sending end current, $\overline{I}_S = \overline{I}_1 + \overline{I}_2$

$$= [I_1(\cos \phi_1 - j \sin \phi_1) + I_2(\cos \phi_2 - j \sin \phi_2)]$$

Sending end voltage, $\overline{V}_S = \overline{V}_C$ + Drop in BC + Drop in AB

$$= \overline{V}_C + \overline{I}_2\overline{Z}_{BC} + (\overline{I}_1 + \overline{I}_2)\, \overline{Z}_{AB}$$

$$= \overline{V}_C + I_2(\cos\phi_2 - j\sin\phi_2)(R_2 + jX_2) + [I_1(\cos\phi_1 - j\sin\phi_1) + I_2(\cos\phi_2 - j\sin\phi_2)]\,(R_1 + jX_1)$$

Sending end power factor = ϕ_s

Fig. 19.68 shows the phasor diagram, which is self explanatory.

(*ii*) **Power factors referred to respective load voltages.** Let the power factors of the loads at points B and C be cos ϕ_1, cos ϕ_2, lagging behind their respective voltages V_B and V_C respectively. The phasor diagram is shown in Fig. 19.69. It will be noted that phasor I_1 lags by ϕ_1 with respect to V_B [and not V_C as in case (*i*)]

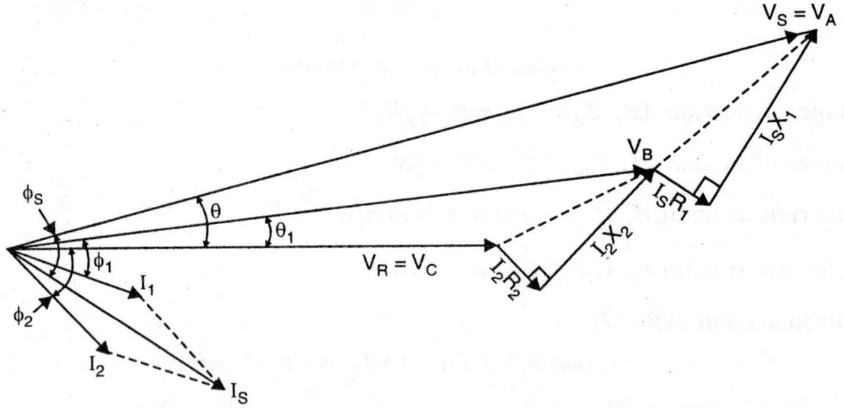

Fig. 19.69. Phasor diagram (of Fig. 19.67) for p.f. with respect to respective load voltages.

Voltage drop in section BC, $\overline{V}_{BC} = \overline{I}_2\, \overline{Z}_{BC} = I_2(\cos\phi_2 - j\sin\phi_2)(R_2 + jX_2)$

Voltage at point B = $\overline{V}_C$ + Voltage drop in section BC = $V_B \angle\theta_1$ (say)

$$\overline{I}_1 = I_1 \angle -\phi_1 \text{ w.r.t. voltage } V_B$$

$$= I_1 \angle -(\phi_1 - \theta_1) \text{ w.r.t. voltage } V_C$$

$$\overline{I}_1 = I_1[\cos(\phi_1 - \theta_1) - j\sin(\phi_1 - \theta_1)]$$

Now, $\overline{I}_{AB} = \overline{I}_1 + \overline{I}_2$

$$= I_1[\cos(\phi_1 - \theta_1) - j\sin(\phi_1 - \theta_1)] + I_2(\cos\phi_2 - j\sin\phi_2)$$

Voltage drop in section $AB = \overline{I}_{AB}\, \overline{Z}_{AB}$

Voltage at point $A = \overline{V}_C$ + Voltage drop in BC + Voltage drop in AB.

Example 19.31. *A 2-wire feeder ABC has a load of 60 A at C and of 30 A at B both at p.f. 0.8 lagging. The impedance AB is (0.8 + j0.16) Ω and that of BC is (0.16 + j0.24) Ω. If the voltage at the far end C is to be maintained at 400 V, determine the voltage at A and voltage at B.*

(Kolkata University)

Sol. Fig. 19.70 shows the loaded feeder.

Impedance of section AB, $\overline{Z}_{AB}$ = (0.08 + j0.16) Ω

Impedance of section BC, $\overline{Z}_{AB}$ = (0.16 + j0.24) Ω

Taking voltage at the far end C, V_C as reference vector and assuming power factors of currents at load points B and C are given referred to the receiving end voltage at C, V_C,

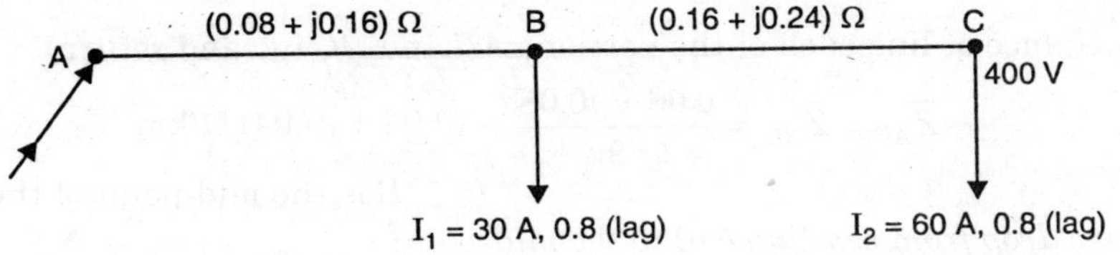

Fig. 19.70

$\bar{V}_C = 400 + j0$

Current through section BC, $\quad I_{BC} = 60(0.8 - j0.6) = (48 - j36)$ A

Current through section AB, $\quad I_{AB} = \bar{I}_1 + \bar{I}_2$

$= 60(0.8 - j0.6) + 30(0.8 - j0.6) = (72 - j54)$ A

Voltage drop in section $BC = \bar{I}_{BC}\,\bar{Z}_{BC}$

$= (48 - j36)(0.16 + j0.24) = (16.32 + j5.76)$ V

Voltage drop in section $AB = \bar{I}_{AB}\,\bar{Z}_{AB}$

$= (72 - j54)(0.08 + j0.16) = (14.4 + j7.2)$ V

Voltage at B, $\quad \bar{V}_B = \bar{V}_C$ + Voltage drop in section BC

$= (400 - j0) + (16.32 + j5.76) = (416.32 + j5.76)$

$= \mathbf{416.36 \angle 0.793° \ V.}$ **(Ans.)**

Voltage at supply end A, $V_A = \bar{V}_B$ + Voltage drop in section AB

$= (416.32 + j5.76) + (14.4 + j7.2)$

$= (430.72 + j12.96)$ V

$= \mathbf{430.9 \angle 1.72° \ V.}$ **(Ans.)**

Example 19.32. *A 2-wire A.C. feeder 1 km long supplies a load of 100 A at 0.8 p.f. lag 200 volts at its far end and a load of 60 A at 0.9 p.f. lag at its mid-point. The resistance and reactance per km (go and return) are 0.06 Ω and 0.08 Ω respectively. Calculate voltage drop along the distributor from sending end to mid-point and from mid-point to far end.*

Sol. Fig. 19.71 shows the feeder AC of 1 km length with B as its middle point and A as its sending end.

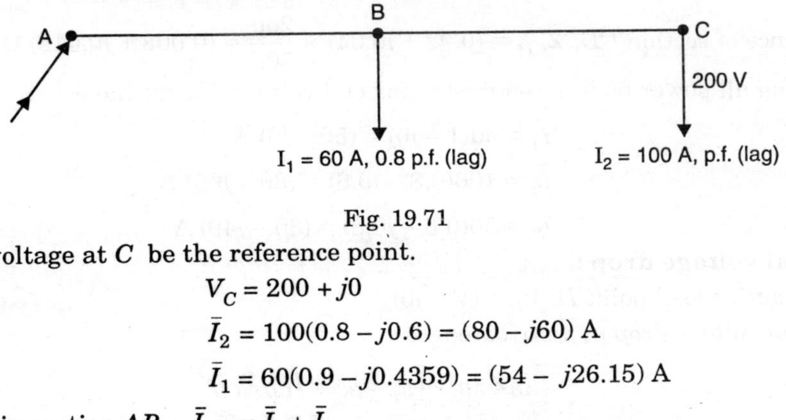

Fig. 19.71

Let the voltage at C be the reference point.

Then, $\quad V_C = 200 + j0$

$\bar{I}_2 = 100(0.8 - j0.6) = (80 - j60)$ A

$\bar{I}_1 = 60(0.9 - j0.4359) = (54 - j26.15)$ A

Current in section AB, $\quad \bar{I}_{AB} = \bar{I}_1 + \bar{I}_2$

$= (54 - j26.15) + (80 - j60) = (134 - j86.15)$ A

Load impedance of line each of the portions AB and BC (go and return)

i.e. $$\overline{Z}_{AB} = \overline{Z}_{BC} = \frac{0.06 + j0.08}{2} = (0.03 + j0.04)\ \Omega/\text{km}$$

($\because$ B is the mid-point of the feeder ... *Given*)

$\therefore$ *Voltage drop from sending end to the mid-point*

$$= \overline{I}_{AB}\,\overline{Z}_{AB} = (134 - j86.15)(0.03 + j0.04) = (7.466 + j2.776)$$
$$= \mathbf{7.96 \angle 20.39°\ V.}\quad \textbf{(Ans.)}$$

Voltage drop from mid-point to far end

$$= \overline{I}_{BC}\,\overline{Z}_{BC} = \overline{I}_2\,\overline{Z}_{BC} = (80 - j60)(0.03 + j0.04)$$
$$= (4.8 + j1.4) = \mathbf{5\angle 16.26°\ V.}\quad \textbf{(Ans.)}$$

Example 19.33. *A 500 m long single-phase A.C. distributor has a total impedance of (0.02 + j0.04) Ω and is fed from one end at 240 V. It is loaded as follows : 50 A at unity power factor, 200 m from feeding point ; 100 A at 0.8 p.f. lagging, 300 m from feeding point ; 50 A at 0.06 p.f. lagging at the far end.*

Calculate : (i) Total voltage drop.

(ii) Voltage at the far end.

Solution. Fig. 19.72 shows the loaded distributor.

Total impedance of the distributor = (0.02 + j0.04) Ω for 500 m length

$\therefore$ Impedance of section AB, $\overline{Z}_{AB} = (0.02 + j0.04) \times \dfrac{200}{500} = (0.008 + j0.016)\ \Omega$

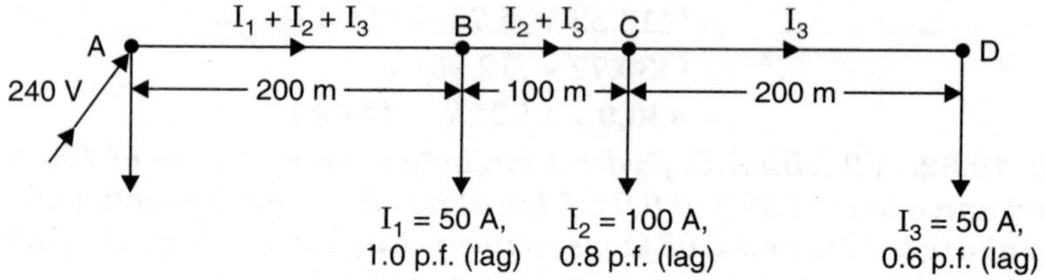

Fig. 19.72

Impedance of section BC, $\overline{Z}_{BC} = (0.02 + j0.04) \times \dfrac{100}{500} = (0.004 + j0.008)\ \Omega$

Impedance of section CD, $\overline{Z}_{CD} = (0.02 + j0.04) \times \dfrac{200}{500} = (0.008 + j0.016)\ \Omega$

Assuming all power factors referred to far end voltage V_D, we have:

$$\overline{I}_1 = 50(1 - j0) = (50 - j0)\ \text{A}$$
$$\overline{I}_2 = 100(0.8 - j0.6) = (80 - j60)\ \text{A}$$
$$\overline{I}_3 = 50(0.6 - j0.8) = (30 - j40)\ \text{A}$$

(*i*) **Total voltage drop :**

Let voltage at load point D, $V_D = (V + j0)$

$\therefore$ *Total voltage drop in distributor*

$$= \overline{I}_{AB}\,\overline{Z}_{AB} + \overline{I}_{BC}\,\overline{Z}_{BC} + \overline{I}_{CD}\,\overline{Z}_{CD}$$
$$= (\overline{I}_1 + \overline{I}_2 + \overline{I}_3)\overline{Z}_{AB} + (\overline{I}_2 + \overline{I}_3)\overline{Z}_{BC} + \overline{I}_3\,\overline{Z}_{CD}$$
$$= [\{50 - j0) + (80 - j60) + (30 - j40)\}(0.008 + j0.016)] + [\{(80 - j60)$$
$$+ (30 - j40)\}(0.004 + j0.008)] + (30 - j40)(0.008 + j0.016)$$

$$= (160 - j100)(0.008 + j0.016) + (110 - j100)(0.004 + j0.008) + (30 - j40)(0.008 + j0.016)$$

$$= (2.88 + j1.76) + (1.24 + j\,0.48) + (0.88 + j0.16)$$

$$= (5 + j2.4) = \mathbf{5.546 \angle 25.64° \ V.} \quad \textbf{(Ans.)}$$

(*ii*) **Voltage at the far end, V_D :**

Voltage at the sending end A, V_A = Voltage at the far end + Voltage drop in distributor

or, $$240 = [(V_D + 5) + j2.4]$$

$$= \sqrt{(V_D + 5)^2 + (2.4)^2}$$

or, $$(240)^2 = (V_D + 5)^2 + (2.4)^2$$

or, $$(V_D + 5)^2 = (240)^2 - (2.4)^2 = 57594.24$$

or, $$V_D = \mathbf{234.98 \ V.} \quad \textbf{(Ans.)}$$

Example 19.34. *A single-phase distributor, one km long has resistance and reactance 0.4 Ω and 0.6 Ω (go and return) respectively. At the far end, the voltage V_B = 240 V and the current is 100 A at a p.f. of 0.8 lag. At the mid-point B of the distributor current of 100 A is tapped at a power factor of 0.6 lag with reference to the voltage V_B at the mid-point. Calculate the supply voltage V_S for the distributor and the phase angle between supply end and receiving end.*

Sol. As shown in Fig. 19.73, let *ABC* be the distributor with *B* as the mid-point.

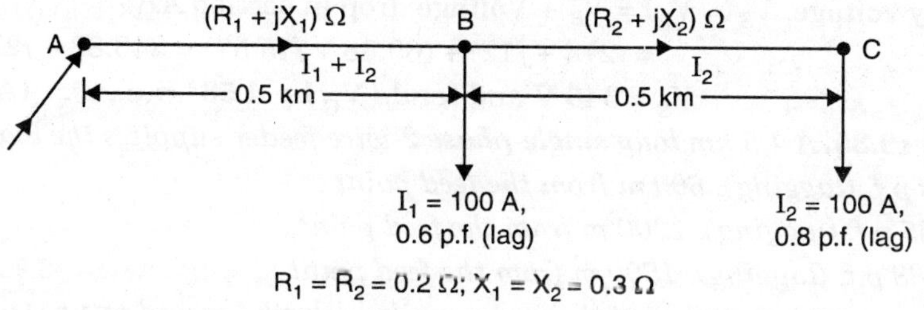

Fig. 19.73

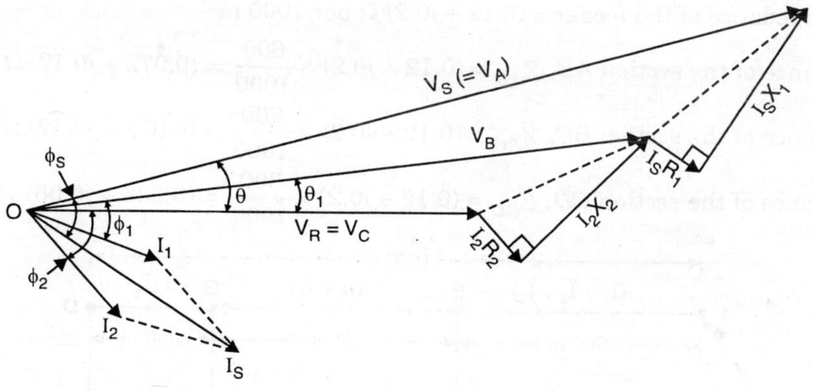

Fig. 19.74. Phasor diagram

Total impedance of the distributor = $(0.4 + j0.6)$ Ω

Let the voltage at C be taken as the reference voltage.

$$V_C = (240 + j0)$$
$$\bar{I}_2 = 100(0.8 - j0.6) = (80 - j60) \text{ A}$$

Voltage drop in section $BC = \bar{I}_2 \bar{Z}_{BC}$
$$= (80 - j60)(0.2 + j0.3) = (34 + j12) \text{ V}$$
$$\bar{V}_B = \bar{V}_C + \text{Voltage drop in section } BC$$
$$= (240 + j0) + (34 + j12) = (274 + j12)$$

The phase difference between V_B and V_C, $\theta_1 = \tan^{-1}\left(\frac{12}{274}\right) = 2.5°$

The load current I_B (*i.e.*, I_1) has lagging power factor of 0.6 w.r.t. V_B. It *lags* V_B by an angle,
$$\phi_1 = \cos^{-1}(0.6) = 53.13°$$

Hence, I_B(*i.e.*, I_1) *lags* behind V_C by an angle (53.13° – 2.5°) = 50.63° as shown in the phasor diagram.

$$\bar{I}_1 = 100 (\cos 50.63° - j \sin 50.63°) = (63.43 - j77.3) \text{ A}$$
$$\bar{I}_S = \bar{I}_{AB} = \bar{I}_1 + \bar{I}_2 = (63.43 - j77.3) + (80 - j60)$$
$$= (143.43 - j137.3) \text{ A}$$

Voltage drop in section $AB = \bar{I}_S \bar{Z}_{AB} = (143.43 - j137.3)(0.2 + j0.3) = (69.88 + j15.57)$

∴ Supply voltage, $\bar{V}_S (= V_A) = \bar{V}_B$ + Voltage drop in section AB
$$= (274 + j12) + (69.88 + j15.57) = 343.88 + j27.57 = 345\angle 4.58°$$

Hence, V_S = **345 V and leads V_C by 4.58° (i.e., θ). (Ans.)**

Example 19.35. *A 1.5 km long single-phase 2-wire feeder supplies the loads as under :*
60A at 0.8 p.f. (lagging), 600 m from the feed point ;
40A at 0.85 p.f. (lagging), 1200 m from the feed point ;
50 A at 0.88 p.f. (lagging), 1500 m from the feed point ;
The resistance and reactance of the feeder per km length (go and return) are 0.12 Ω and 0.2 Ω respectively. If the voltage at the far end is to be maintained at 220 V, calculate the voltage of the sending end, and its phase angle with respect to the receiving end voltage. **(Panjab University)**

Sol. Impedance of the feeder = (0.12 + *j*0.2) Ω per 1000 m

Impedance of the section AB, $\bar{Z}_{AB} = (0.12 + j0.2) \times \frac{600}{1000} = (0.072 + j0.12)\ \Omega$

Impedance of the section BC, $\bar{Z}_{BC} = (0.12 + j0.2) \times \frac{600}{1000} = (0.072 + j0.12)\ \Omega$

Impedance of the section CD, $\bar{Z}_{CD} = (0.12 + j0.2) \times \frac{300}{1000} = (0.036 + j0.06)\ \Omega$

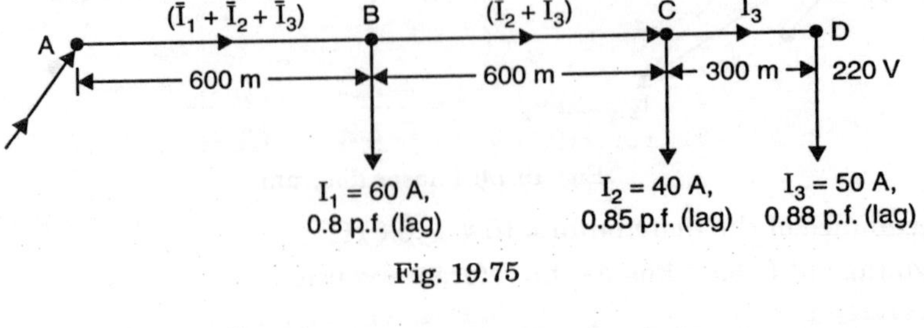

Fig. 19.75

Taking voltage at far end D, V_D as the reference vector,

$$V_D = (220 + j0) \text{ V}$$

Current in section CD, $\bar{I}_3 = 50(0.88 - j0.475) = (44 - j23.75)$

Voltage at load point C, $\bar{V}_C = \bar{V}_D + \bar{I}_3 \bar{Z}_{CD}$

$$= (220 + j0) + (44 - j23.75)(0.036 + j0.06)$$

$$= (220 + j0) + (3.009 + j1.785) = (223.009 + j1.785) \text{ V}$$

Phase angle between V_C and V_D,

$$\theta_1 = \tan^{-1}\left(\frac{1.785}{223.009}\right) = 0.46°$$

The load current I_2 lags behind V_C by an angle

$$= \cos^{-1} 0.85 = 31.79°$$

The load current I_2 *lags* behind voltage at far end V_D by an angle (31.79° – 0.46°) = 31.33°

Load current at point C, $\bar{I}_2 = 40(\cos 31.33° - j \sin 31.33°)$

$$= 40(0.854 - j0.52) = (34.16 - j20.8) \text{ A}$$

Load current in section BC, $\bar{I}_{BC} = \bar{I}_2 + \bar{I}_3$

$$= (34.16 - j20.8) + (44 - j23.75) = (78.16 - j44.55) \text{ A}$$

Voltage at the load point B,

$$\bar{V}_B = \bar{V}_C + (\bar{I}_3 + \bar{I}_2)\,\bar{Z}_{BC}$$

$$= (223.009 + j1.785) + (78.16 - j44.55)(0.072 + j0.12)$$

$$= (223.009 + j1.785) + (10.973 + j6.172) = (233.982 + j7.957) \text{ V}$$

Phase angle between V_B and V_D, $\theta_2 = \tan^{-1}\left(\frac{7.957}{233.982}\right) = 1.95°$

The load current I_1 *lags* behind the voltage V_B by an angle

$$= \cos^{-1}(0.8) = 36.87°$$

The load current I_1 lags behing the voltage at the far end V_D by an angle

$$= 36.87° - 1.95° = 34.92°$$

Load current at point B, $\bar{I}_1 = 60(\cos 34.92° - j \sin 34.92°)$

$$= 60(0.8199 - j0.5724) = (49.194 - j34.344)$$

Load current in section AB,

$$I_{AB} = \bar{I}_1 + \bar{I}_2 + \bar{I}_3$$

$$= (49.194 - j34.344) + (34.16 - j20.8) + (44 - j23.75)$$

$$= (127.354 - j78.894)$$

Voltage at sending end point A,

$$\bar{V}_A = \bar{V}_B + \left(\bar{I}_1 + \bar{I}_2 + \bar{I}_3\right)\bar{Z}_{AB}$$

$$= (233.982 + j7.957) + (127.354 - j78.894)(0.072 + j0.12)$$

$$= (233.982 + j7.957) + (18.637 + j9.602)$$

$$= (252.62 + j17.56) = 253.23 \angle 3.98°$$

i.e., *Voltage at the sending end* = **253.23 V. Ans.**

Phase angle of sending end voltage (V_A) *w.r.t.*

the receiving end voltage (V_D), θ = **3.98°. (Ans.)**

Example 19.36. *A single-phase ring distributor ABC is fed at A. The loads at B and C are 20 A at 0.8 p.f. lagging and 15 A at 0.6 p.f. lagging respectively both expressed with reference to voltage at A. The total impedances of the sections AB, BC and CA are (1 + j1), (1 + j2) and (1 + j3) ohms respectively. Find the total current fed at A and the current in each section.*

(Madras University)

Sol. The ring distributor is shown in Fig. 19.76.

Thevenin's theorem will be used to solve the problem.

Imagine feeder *BC* to be removed (Fig. 19.77).

Current in section *AB* = 20 (0.8 – *j*0.6) = 16 – *j*12

Current in section *AC* = 15(0.6 – *j*0.8) = 9 – *j*12

Voltage drop in section *AB* = (16 – *j*12)(1 + *j*1) = (28 + *j*4) V

Voltage drop in section *AC* = (9 – *j*12)(1 + *j*3) = (45 + *j*15) V

Obviously *C* is at a lower potential as compared to point *B*.

Potential difference between *B* and *C*

= (45 + *j*15) – (28 + *j*4) = (17 + *j*11) V = Thevenin's equivalent circuit e.m.f. (E_0)

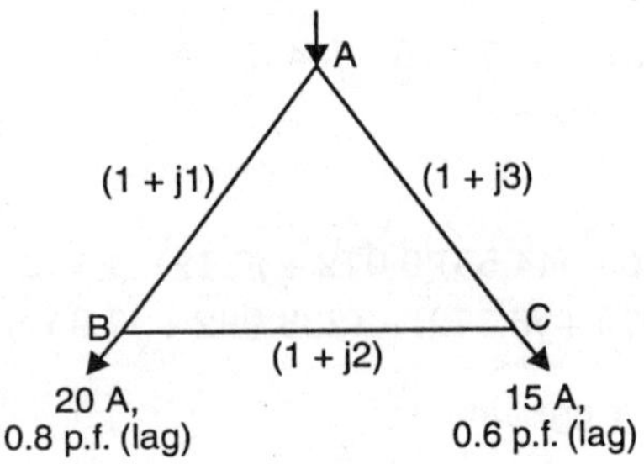

Fig. 19.76

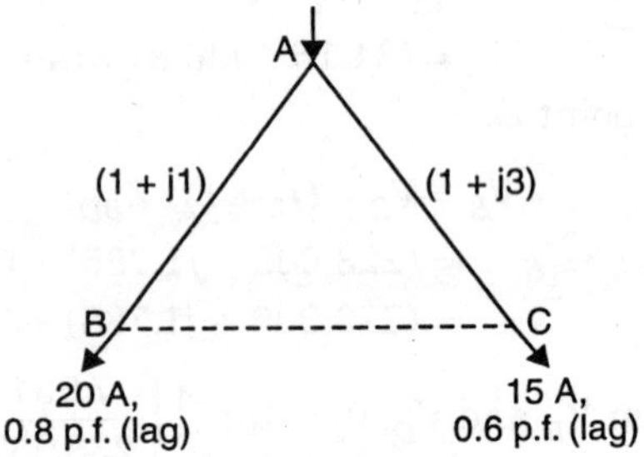

Fig. 19.77

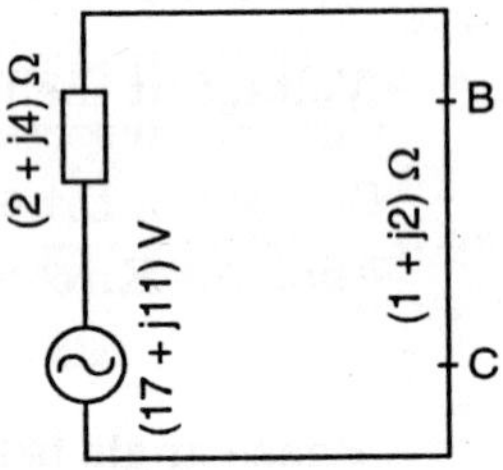

Fig. 19.78

Impedance of network as looked into from points *B* and *C*, Z_0 = (1 + *j*1) + (1 + *j*3) = (2 + *j*4) Ω

The equivalent Thevenin's source is shown in Fig. 19.78 with feeder *BC* connected across it.

$$\text{Current in } BC = \frac{E_0}{Z_0 + \text{Impedance of } BC}$$

$$= \frac{(17 + j11)}{(2 + j4) + (1 + j2)} = \frac{17 + j11}{3 + j6}$$

$$= (2.6 - j1.53) = \mathbf{3.02 \angle -30.48° \ A.} \quad \textbf{(Ans.)}$$

Current in AB = (16 – *j*12) + (2.6 – *j*1.53) = (18.6 – *j*13.53) = **23 ∠– 36.03° A.** **(Ans.)**

Current in AC = (9 – *j*12) – (2.6 – *j*1.53) = (6.4 – *j*10.47) = **12.27∠– 58.56° A.** **(Ans.)**

Total current feed at A = (16 – *j*12) + (9 – *j*12)

= (25 – *j*24) = **34.65 ∠– 43.83° A.** **(Ans.)**

Example 19.37. *A 3-phase ring main ABCD, fed from end A at 11 kV supplies balanced loads of 50 A at 0.8 p.f. lagging at B, 120 A at unity p.f. at C and 70 A at 0.866 p.f. lagging at D, the load currents being referred to the voltage at point A.*

The impedance per phase of the various line sections are :

Section AB = (1 + j0.6) Ω ; Section BC = (1.2 + j0.9) Ω

Section CD = (0.8 + j0.5) Ω ; Section DA = (3 + j2) Ω

Determine : (i) Currents in various sections, and
(ii) Station bus-bar voltages at B, C and D. **(Bombay University)**

Sol. One phase of the ring main is shown in Fig. 19.79.

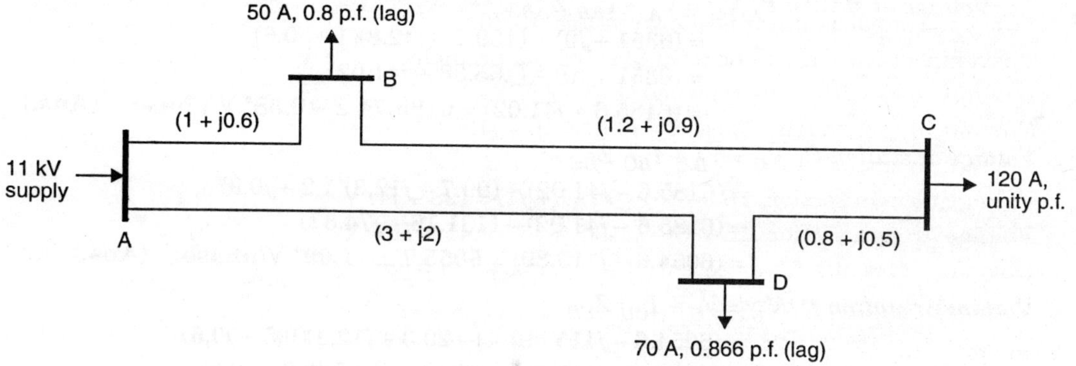

Fig. 19.79

Let the current in section $AB = (x + jy)$

Current in section $BC = (x + jy) - 50(0.8 - j0.6) = (x - 40) + j(y + 30)$

Current in section $CD = [(x - 40) + j(y + 30)] - (120 + j0) = (x - 160) + j(y + 30)$

Current in $DA = [(x - 160) + j(y + 30)] - 70(0.866 - j0.5) = (x - 220.6) + j(y + 65)$

Applying Kirchhoff's voltage law to the closed loop *ABCDA*, we have

$$(x + jy)(1 + j0.6) + [(x - 40) + j(y + 30)](1.2 + j0.9)$$
$$+ [(x - 160) + j(y + 30)]\,(0.8 + j\,0.5) + [(x - 220.6) + j(y + 65)]\,(3 + j2) = 0$$

$$(x + j0.6x + jy - 0.6y) + [1.2(x - 40) + j(1.2y + 36) + j(0.9x - 36) - 0.9(y + 30)]$$
$$+ [0.8(x - 160) + j(0.8y + 24) + j(0.5x - 80) - 0.5(y + 30)]$$
$$+ [3(x - 220.6) + j(3y + 195) + j(2x - 441.2) - 2(y + 65)] = 0$$

or, $$(x - 0.6y + j0.6x + jy) + (1.2x - 48 + j1.2y + j36 + j0.9x - j36 - 0.9y - 27)$$
$$+ (0.8x - 128 + j0.8y + j24 + j0.5x - j80 - 0.5y - 15)$$
$$+ (3x - 661.8 + j3y + j195 + j2x - j441.2 - 2y - 130) = 0$$

or, $$(6x - 4y + 1009.8) + j(4x - 6y - 302.2) = 0$$

Since the real (or active) and imaginary (or reactive) parts have to be separately zero,

$\therefore$ $$6x - 4y - 1009.8 = 0, \quad \text{and}$$
$$4x + 6y - 302.2 = 0$$

Solving for x and y, we get

$$x = 139.7 \text{ A and } y = -42.8 \text{ A}$$

***(i)* The currents in various sections :**

Current in section AB $= (x + jy) =$ **(139.7 – j42.8) A. (Ans.)**

Current in section BC $= (x - 40) + j(y + 30)$

$= (139.7 - 40) + j(-42.8 + 30) =$ **(99.7 – j12.8) A. (Ans.)**

Current in section CD $= (x - 160) + j(y + 30)$

$= (139.7 - 160) + j(-42.8 + 30) =$ **(– 20.3 – j12.8) A. (Ans.)**

Current in section DA $= (x - 220.6) + j(y + 65)$

$= (139.7 - 220.6) + j(-42.8 + 65) =$ **(– 80.9 + j22.2) A. (Ans.)**

(*ii*) **Station bus-bar voltages at B, C and D :**

Voltage at supply and *A*, $V_A = \dfrac{11000}{\sqrt{3}} = 6351$ V/phase

$\therefore$ *Voltage at station B*, $\overline{V}_B = \overline{V}_A - \overline{I}_{AB}\,\overline{Z}_{AB}$

$= (6351 + j0) - (139.7 - j42.8)(1 + j0.6)$

$= (6351 + j0) - (165.38 + j41.02)$

$= (6185.6 - j41.02) =$ **6185.74 ∠ – 0.38° V/phase. (Ans.)**

Voltage at station C, $\overline{V}_C = \overline{V}_B - \overline{I}_{BC}\,\overline{Z}_{BC}$

$= (6185.6 - j41.02) - (99.7 - j12.8)(1.2 + j0.9)$

$= (6185.6 - j41.02) - (131.16 + j74.37)$

$= (6054.6 - j115.39) =$ **6055.7 ∠– 1.09° V/phase. (Ans.)**

Voltage at station D, $\overline{V}_D = \overline{V}_C - \overline{I}_{CD}\,\overline{Z}_{CD}$

$= (6054.6 - j115.39) - (-20.3 - j12.8)(0.8 + j0.5)$

$= (6054.6 - j115.39) + (20.3 + j12.8)(0.8 + j0.5)$

$= (6504.6 - j115.39) + (9.84 + j20.39)$

$= (6064.44 - j95) =$ **6065.18 ∠– 0.897° V/phase. (Ans.)**

19.3.4. Three-phase Unbalanced Loads

19.3.4.1. Unbalanced star-connected load

A. *Three-phase four-wire system :* Refer to Fig. 19.80.

- In this type of system a total of *four conductors* are used, three for three phases and one for neutral.
- This system is *very flexible* and is *used for distribution of electric power to domestic, commercial, industrial and other consumers.*
- The line voltage is $\sqrt{3}$ times the phase voltage. The domestic and other single-phase loads needing higher voltages are connected between lines. *The currents in three lines will not be equal and the phasor sum of line currents would return through neutral. The current in the neutral must be equal and opposite to the sum of line currents.*

In 3-phase, 4-wire star-connected load circuits the star points of load and the generator are tied together through *neutral wire of zero impedance*, therefore they are at the same potential ; voltage across each impedance is same and equal to phase voltage irrespective of the fact that the circuit is balanced or unbalanced.

The three-phase currents or line currents can be determined by dividing the phase voltage by impedance of the concerned phase,

i.e.,
$$\overline{I}_R = \frac{\overline{V}_R}{\overline{Z}_R}\ ;\ \overline{I}_Y = \frac{\overline{V}_Y}{\overline{Z}_Y}\ ;\ \overline{I}_Z = \frac{\overline{V}_B}{\overline{Z}_B}$$

Since the load is unbalanced, the line currents will be different in magnitudes and displaced from one another by unequal angles. The current in the neutral wire will be the phasor sum of the three line currents, *i.e.*,

Current in neutral wire, $\overline{I}_N = -(\overline{I}_R + \overline{I}_Y + \overline{I}_B)$

In the balanced load circuits the line currents are equal and have the same phase angles with their respective phase voltages, so their vector sum is zero and current in neutral wire is zero. In balanced star-connected load circuits, neutral wire carries no current and, therefore, it is usually *omitted*.

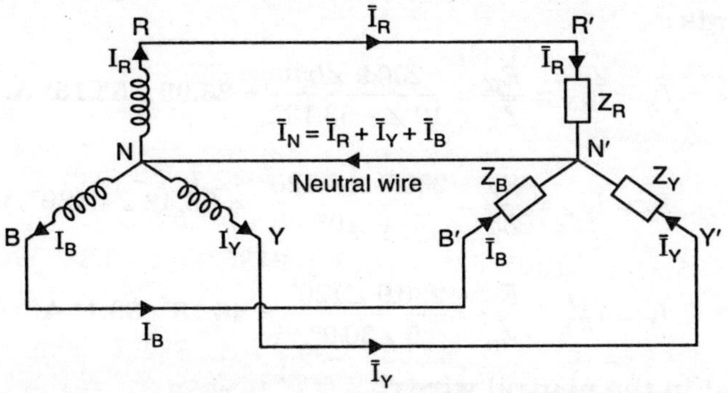

Fig. 19.80

Example 19.38. *A 400 V, 3-phase 4 wire system has the loads 6 – j8, 6 + j0 and 4 + j3 Ω connected in star. Find :*

(*i*) *The line currents,* (*ii*) *Neutral current, and*

(*iii*) *Total power.*

Sol. The connections are shown in Fig. 19.81.

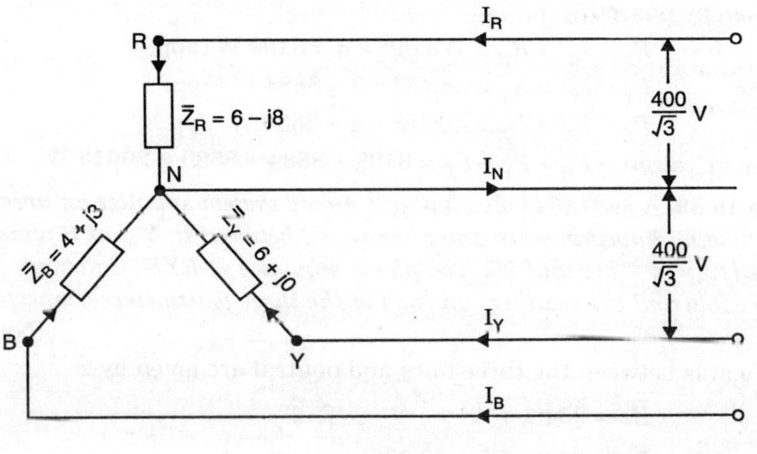

Fig. 19.81

The potential differences across the three phases are :

$$\bar{E}_{RN} = \frac{400}{\sqrt{3}} \angle 0° = 230.9 \angle 0° \text{ V}$$

$$\bar{E}_{YN} = 230.9 \angle -120° \text{ V}$$

$$\bar{E}_{BN} = 230.9 \angle 120° \text{ V}$$

Impedances :

$$\bar{Z}_R = 6 - j8 = 10 \angle -63.13°$$

$$\bar{Z}_Y = 6 + j0 = 6 \angle 0°$$

$$\bar{Z}_B = 4 + j3 = 5 \angle 36.9°$$

(*i*) **Line currents :**

$$\bar{I}_R = \bar{I}_{RN} = \frac{\bar{E}_R}{\bar{Z}_R} = \frac{230.9 \angle 0°}{10 \angle -53.13°} = \mathbf{23.09 \angle 53.13° \ A.} \quad \textbf{(Ans.)}$$

$$\bar{I}_Y = \bar{I}_{YN} = \frac{\bar{E}_Y}{\bar{Z}_Y} = \frac{230.9 \angle -120°}{6 \angle 0°} = \mathbf{38.48 \angle -120° \ A.} \quad \textbf{(Ans.)}$$

$$\bar{I}_B = \bar{I}_{BN} = \frac{\bar{E}_B}{\bar{Z}_B} = \frac{230.9 \angle 120°}{5 \angle 36.9°} = \mathbf{46.18 \angle 83.1° \ A.} \quad \textbf{(Ans.)}$$

(*ii*) **The current in the neutral wire :**

$$\bar{Z}_B = -(\bar{I}_R + \bar{I}_Y + \bar{I}_B)$$
$$= -[23.09 \angle 53.13° + 38.48 \angle -120° + 46.18 \angle 83.1°]$$
$$= -[23.09(0.6 + j0.8) + 36.48(-0.5 - j0.866) + 46.18(0.12 + j0.99)]$$
$$= -[13.85 + j18.47 - 18.24 - j31.59 + 5.54 + j45.72]$$
$$= -(1.15 + j32.6) = -1.15 - j32.6 = \mathbf{32.62 \angle -92° \ A.} \quad \textbf{(Ans.)}$$

(*iii*) **Total power :**

Power taken by phase *R*,

$$P_R = I_R^2 \times R_R = 123.09^2 \times 6 = 3199 \text{ W (app.)}$$

Similarly, $$P_Y = I_Y^2 \times R_Y = 38.48^2 \times 6 = 8884 \text{ W (app.)}$$

and $$P_B = I_B^2 \times R_B = 46.18^2 \times 4 = 8530 \text{ W (app.)}$$

Total power taken $= P_R + P_Y + P_B = 3199 + 8884 + 8530 =$ **20613 W. (Ans.)**

Example 19.39. *A 440/254 volts, 3 phase, 4-core system supplies an unbalanced load represented by the following impedances in ohms connected between R, Y and B phases respectively and the neutral 16 + j12 ; 14 – j21 and 25. The phase sequence is RYB. Calculate the current in each conductor of the cable and the readings on each of the three wattmeters connected in each line and neutral.* **(Nagpur University)**

Sol. The e.m.fs between the three lines and neutral are given by :

$$\bar{E}_R = 254(1 + j0)$$
$$\bar{E}_Y = 254(-0.5 - J0.866)$$
$$\bar{E}_B = 254(-0.5 + j0.866) \text{ taking } E_R \text{ as the reference vector}$$

$$\bar{I}_R = \frac{\bar{E}_R}{\bar{Z}_R} = \frac{254 + j0}{16 + j12} = \frac{(254 + j0) \times (16 - j12)}{(16 + j12) \times (16 - j12)} = \frac{4064 - j3048}{400}$$
$$= (10.16 - j7.62) \text{ A}$$

$$\bar{I}_Y = \frac{\bar{E}_Y}{\bar{Z}_Y} = \frac{254(-0.5 - j0.866)}{(14 - j21)} = \frac{-127 - j220}{14 - j21} = \frac{(-127 - j220) \times (14 + j21)}{(14 - j21)(14 + j21)}$$
$$= \frac{-1778 - j2667 - j3080 + 4620}{637} = (4.46 - j9.02) \text{ A}$$

$$\bar{I}_B = \frac{\bar{E}_B}{\bar{Z}_B} = \frac{254(-0.5 + j0.866)}{25} = \frac{-127 + j220}{25} = (-5.08 + j8.8) \text{ A}$$

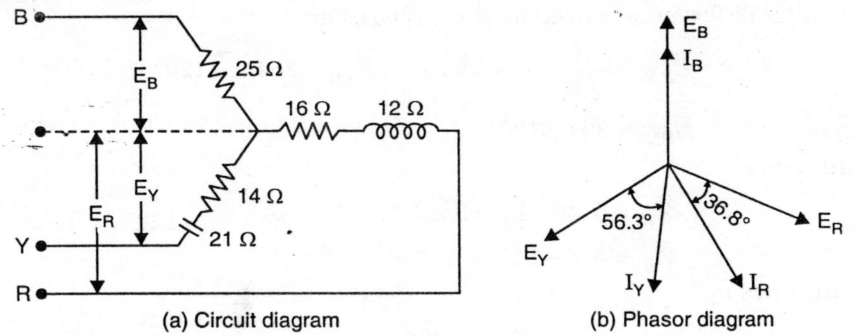

Fig. 19.82

Now current in neutral wire,

$$\bar{I}_N = -(\bar{I}_R + \bar{I}_Y + \bar{I}_B)$$
$$= -[(10.16 - j7.62) + (4.46 - j9.02) + (-5.08 + j8.8)] = -9.54 + j7.84$$

Evaluating these currents, we get

$$\bar{I}_R = \sqrt{10.16^2 + (-7.62)^2} = \mathbf{12.7\ A.}\quad \textbf{(Ans.)}$$
$$\bar{I}_Y = \sqrt{4.46^2 + (-9.02)^2} = \mathbf{10.06\ A.}\quad \textbf{(Ans.)}$$
$$\bar{I}_B = \sqrt{(-5.08)^2 + (8.8)^2} = \mathbf{10.16\ A.}\quad \textbf{(Ans.)}$$
$$\bar{I}_N = \sqrt{(-9.54)^2 + (7.84)^2} = \mathbf{12.35\ A.}\quad \textbf{(Ans.)}$$

Power in R-phase $= I_R^2 \times R_R = (12.7)^2 \times 16 =$ **2580.6 W. (Ans.)**

Power in Y-phase $= I_Y^2 \times R_Y = (10.06)^2 \times 14 =$ **1416.8 W. (Ans.)**

Power in B-phase $= I_B^2 \times R_B = (10.16)^2 \times 25 =$ **2580.6 W. (Ans.)**

Example 19.40. *A 3-phase, 4 wire system having a 440 V line-to-line has the following loads connected between the respective lines and neutral :*

$$Z_R = 5\angle 0°\ \Omega,\ Z_Y = 5\angle 37°\ \Omega,\ Z_B = 5\angle -53°\ \Omega.$$

Calculate the current in neutral wire and power taken by each load when phase sequence is :

(i) RYB *(ii) RBY.*

Sol. (*i*) **Phase sequence RYB** (Figs. 19.83 and 19.84) :

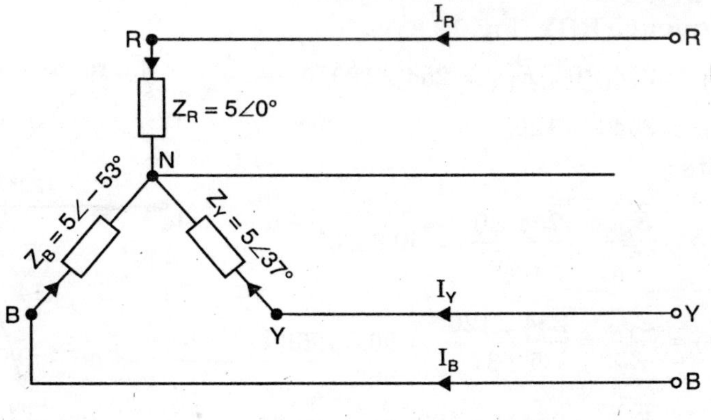

Fig. 19.83

The potential differences across the three phases are :

$$\bar{E}_{RN} = \frac{440}{\sqrt{3}} \angle 0° = 254 \angle 0°, E_{YN} = 254 \angle -120°$$

$$\bar{E}_{BN} = 254 \angle 120°$$

Impedances :

$$\bar{Z}_R = 5 \angle 0°, Z_Y = 5 \angle 37°$$

$$\bar{Z}_B = 5 \angle -53°$$

Line currents :

$$\bar{I}_R = \bar{I}_{RN} = \frac{\bar{E}_{RN}}{\bar{Z}_R} = \frac{254 \angle 0°}{5 \angle 0°} = 50.8 \angle 0°$$

$$\bar{I}_Y = \bar{I}_{YN} = \frac{\bar{E}_{YN}}{\bar{Z}_Y} = \frac{254 \angle -120°}{5 \angle 37°} = 50.8 \angle -157°$$

$$\bar{I}_B = \bar{I}_{BN} = \frac{\bar{E}_{BN}}{\bar{Z}_B} = \frac{254 \angle 120°}{5 \angle -53°} = 50.8 \angle 173°$$

Current in the neutral :

$$I_N = -(\bar{I}_R + \bar{I}_Y + \bar{I}_B)$$
$$= -(50.8 \angle 0° + 50.8 \angle -157° + 50.8 \angle 173°)$$
$$= -[50.8 + 50.8(-0.92 - j0.39) + 50.8(-0.99 + j0.122)]$$
$$= -[50.8 - 46.74 - j19.81 - 50.3 + j6.2)]$$
$$= -(-46.24 - j13.61) = 46.24 + j13.61$$
$$= \mathbf{48.2 \angle 16.4°.} \quad \textbf{(Ans.)}$$

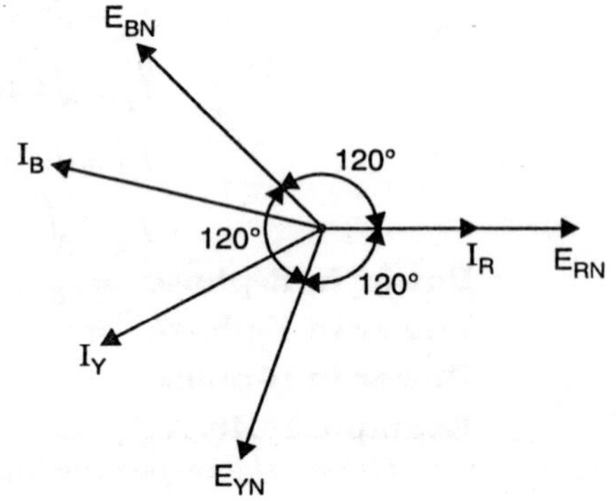

Fig. 19.84

Now $R_R = 5\Omega$, $R_Y = 5 \cos 37° = 4\ \Omega$.

$R_B = 5 \cos(-53°) = 3\ \Omega$

Power taken by phase R,

$P_R = 50.8^2 \times 5 =$ **12903 W. (Ans.)**

Similarly, $P_Y = 50.8^2 \times 4 =$ **10322 W. (Ans.)**

and, $P_B = 50.8^2 \times 3 =$ **7742 W. (Ans.)**

(*ii*) **Phase sequence RBY** (Fig. 19.85) :

$$\bar{E}_{RN} = 254 \angle 0°, \bar{E}_{YN} = 254 \angle 120°$$

$$\bar{E}_{BN} = 254 \angle -120°$$

Line currents :

$$\bar{I}_R = \bar{I}_{RN} = \frac{\bar{E}_{RN}}{\bar{Z}_R} = \frac{254 \angle 0°}{5 \angle 0°} = 50.8 \angle 0°$$

$$\bar{I}_Y = \bar{I}_{YN} = \frac{\bar{E}_{YN}}{\bar{Z}_Y} = \frac{254 \angle 120°}{5 \angle 37°} = 50.8 \angle 83°$$

$$\bar{I}_B = \bar{I}_{BN} = \frac{\bar{E}_{BN}}{\bar{Z}_B} = \frac{254 \angle -120°}{5 \angle -53°} = 50.8 \angle + 67°$$

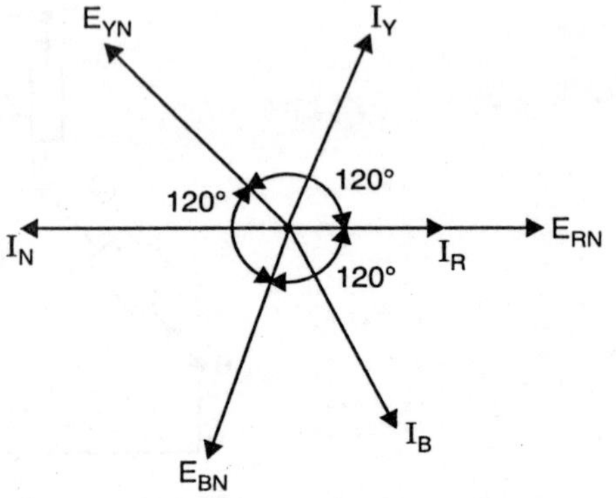

Fig. 19.85

Current in the neutral :

$$\mathbf{I_N} = -(\bar{I}_R + \bar{I}_Y + \bar{I}_B)$$
$$= -(50.8 \angle 0° + 50.8 \angle 83° + 50.8 \angle -67°)$$
$$= -[50.8 + 50.8(0.122 + j0.992) + 50.8(0.39 - j0.92)]$$
$$= -(50.8 + 6.197 + j50.4 + 19.81 - j46.74)$$
$$= -(76.81 + j3.66) = -76.81 - j3.66$$
$$= \mathbf{76.89 \angle -177.3° \ A.} \quad \textbf{(Ans.)}$$

Power would remain the same because magnitude of branch currents is unaltered.

Example 19.41. *In a 3-phase, 4-wire system, the line voltage is 400 V and non-inductive loads of 5, 4 and 2.5 kW are connected between the three lines conductors and the neutral, as shown in Fig. 19.86. Calculate :*

(*i*) *The current in lines.*

(*ii*) *The current in the neutral conductor.*

Sol. Refer to Fig. 19.86.

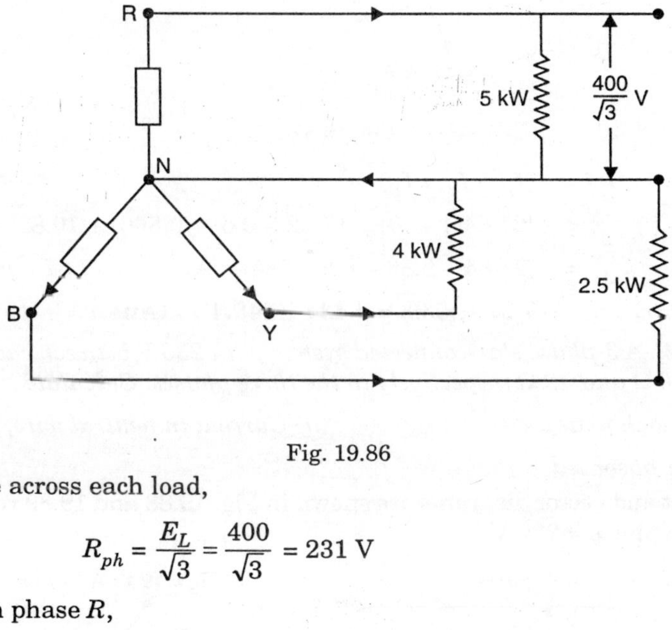

Fig. 19.86

Phase voltage across each load,

$$R_{ph} = \frac{E_L}{\sqrt{3}} = \frac{400}{\sqrt{3}} = 231 \text{ V}$$

(*i*) Current in phase *R*,

$$I_R = \text{(line current)} = \frac{\text{kW} \times 1000}{E_{ph} \cos \phi} = \frac{56 \times 1000}{231 \times 1} = \mathbf{21.64 \ A.} \quad \textbf{(Ans.)}$$

Current in phase *Y*, $I_Y = \dfrac{4 \times 1000}{231 \times 1} = \mathbf{17.32 \ A.}$ **(Ans.)**

Current in phase *B*, $I_B = \dfrac{2.5 \times 1000}{231 \times 1} = \mathbf{10.82 \ A.}$ **(Ans.)**

These currents are mutually 120° out of phase as the circuit is purely resistive. Taking I_R as the reference vector, line currents may be given as :

$$\bar{I}_R = 21.64 \angle 0° = 21.64(1 + j0)$$
$$\bar{I}_Y = 17.32 \angle -120° = 17.32(-0.5 - j0.866)$$
$$\bar{I}_B = 10.82 \angle 120° = 10.82(-0.5 + j0.866)$$

(*ii*) **Current in the neutral wire** (Refer to Fig. 19.87) :

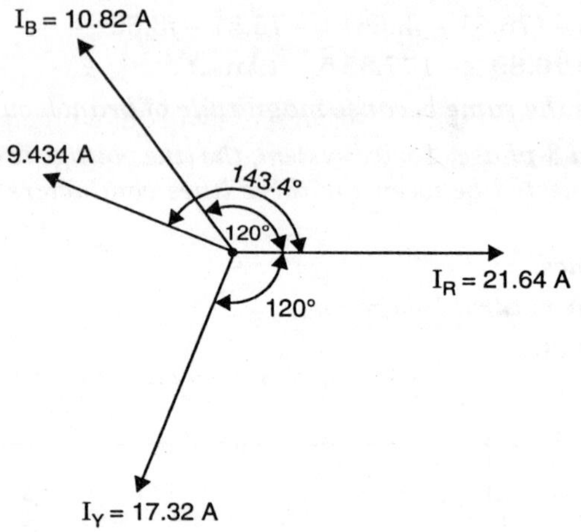

Fig. 19.87

$$\bar{I}_N = -(\bar{I}_R + \bar{I}_Y + \bar{I}_B)$$
$$= -[21.64(1 + j0) + 17.32(-0.5 - j0.866) + 10.82(-0.5 + j0.866)]$$
$$= -[21.64 - 8.66 - j15 - 541 + j9.37] = -(7.57 - j5.63)$$
$$= -7.57 + j5.63 = \mathbf{9.434 \angle 143.4°}. \quad \textbf{(Ans.)}$$

Example 19.42. *A 3-phase star-connected system with 230 V between each phase and neutral has resistances 8 Ω, 10 Ω and 12 Ω respectively in the three phases. Calculate :*

(*i*) *Current in each phase,* (*ii*) *Current in neutral wire, and*

(*iii*) *Total power absorbed.*

Sol. The circuit and vector diagrams are shown in Fig. 19.88 and 19.89 respectively. Potential difference across each phase = 230 V.

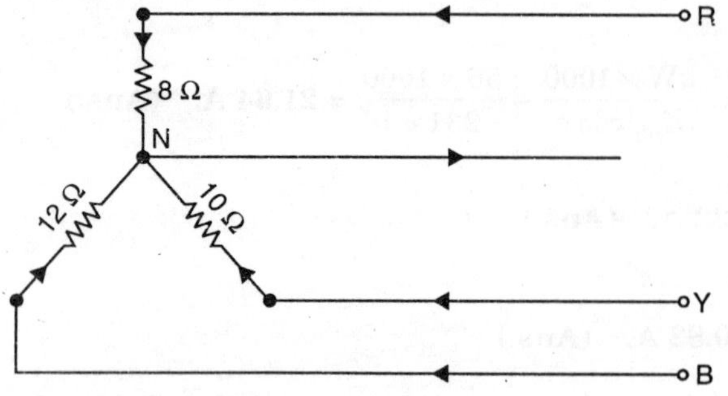

Fig. 19.88

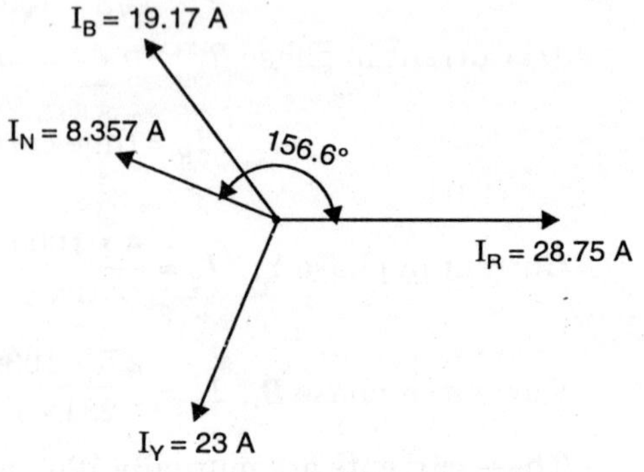

Fig. 19.89

(*i*) **Phase currents :**

Current in 8 Ω resistance $= \frac{230}{8} =$ **28.75 A. (Ans.)**

Current in 10 Ω resistance $= \frac{230}{10} =$ **23 A. (Ans.)**

Current in 12 Ω resistance $= \frac{230}{12} =$ **19.17 A. (Ans.)**

Since the circuit is purely resistive, so currents will be in phase with their respective voltages and will be mutually 120° apart. Taking I_R as the reference vector (Fig. 19.89) the line currents are given as :

$$\bar{I}_R = 28.75 \angle 0° = 28.75(1 + j0)$$

$$\bar{I}_Y = 23 \angle -120° = 23(-0.5 - j0.866)$$

$$\bar{I}_B = 19.17 \angle 120° = 19.17(-0.5 + j0.866).$$

(*ii*) **Current in neutral wire :**

$$\bar{I}_N = -(\bar{I}_R + \bar{I}_Y + \bar{I}_B) = -[28.75(1 + j0) + 23(-0.5 - j0.866) + 19.17(-0.5 + j0.866)$$

$$= -[28.75 - 11.5 - j19.92 - 9.58 + j16.6]$$

$$= -[7.67 - j3.32] = -7.67 + 3.32 = \mathbf{8.357 \angle 156.6° \ A.} \quad \textbf{(Ans.)}$$

(*iii*) **Total power absorbed :**

$$= I_R^2 \times 8 + I_Y^2 \times 10 + I_B^2 \times 12 = 28.75^2 \times 8 + 23^2 \times 10 + 19.17^2 \times 12$$

$$= 6612.5 + 5290 + 4409.8 = \mathbf{16312.3 \ W.} \quad \textbf{(Ans.)}$$

Example 19.43. *A 440 V (line-to-line), 3-phase, 4-core system supplies an unbalanced load represented by the following impedances in ohms connected between R, Y and B phases respectively and the neutral 8 + j6, 7 – j10.5 and 12.5 Calculate :*

(i) Current in each conductor of the cable (ii) Current in the neutral wire, and

(iii) Readings of the three wattmeters connected in each line and neutral.

The phase sequence is RYB.

Sol. The circuit is shown in Fig. 19.90.

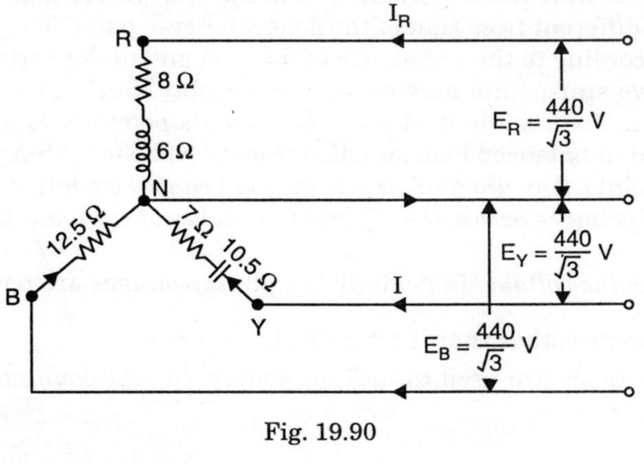

Fig. 19.90

The potential differences across the three phases are :

$$E_{RN} = \frac{440}{\sqrt{3}} = 254 \angle 0°$$
$$E_{YN} = 254 \angle -120°$$
$$E_{BN} = 254 \angle 120°.$$

Impedances :

$$Z_R = 8 + j6 = 10 \angle 36.9°$$
$$Z_Y = 7 - j10.5 = 12.62 \angle -56.3°$$
$$Z_B = 12.5 = 12.5 \angle 0°.$$

(*i*) **Phase currents :**

$$\bar{I}_R = \bar{I}_{RN} = \frac{\bar{E}_{RN}}{\bar{Z}_R} = \frac{254 \angle 0°}{10 \angle 36.9°}$$
$$= \mathbf{25.4 \angle -36.9°\ A.} \quad \textbf{(Ans.)}$$

$$\bar{I}_Y = \bar{I}_{YN} = \frac{\bar{E}_{YN}}{\bar{Z}_Y} = \frac{254 \angle -120°}{12.62 \angle -56.3°}$$
$$= \mathbf{20.12 \angle -63.7°\ A.} \quad \textbf{(Ans.)}$$

$$\bar{I}_B = \bar{I}_{BN} = \frac{\bar{E}_{BN}}{\bar{Z}_B} = \frac{254 \angle 120°}{12.5 \angle 0°}$$
$$= \mathbf{20.32 \angle 120°\ A.} \quad \textbf{(Ans.)}$$

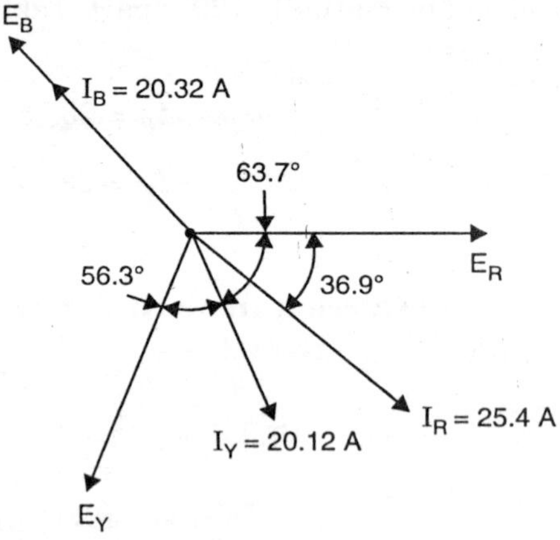

Fig. 19.91

(*ii*) **Current in the neutral wire :**

$$\bar{I}_N = -(\bar{I}_R + \bar{I}_Y + \bar{I}_B)$$
$$= -[25.4 \angle -36.9° + 20.12 \angle -63.7° + 20.32 \angle 120°]$$
$$= -[25.4(+0.8 - j0.6) + 20.12(0.44 - j0.896) + 20.32(-0.5 + j0.866)]$$
$$= -(20.32 - j15.24 + 8.85 - j18.03 - 10.16 + j17.6)$$
$$= -(19.01 - j15.67) = -19.01 + j15.67 = \mathbf{24.64 \angle 39.5°\ A.} \quad \textbf{(Ans.)}$$

(*iii*) **Readings of wattmeters :**

Power in phase $R = I_R^2 \times R_R = 25.4^2 \times 8 =$ **5161.3 W. (Ans.)**

Power in phase $Y = I_Y^2 R_Y = 20.12^2 \times 7 =$ **2833.7 W. (Ans.)**

Power in phase $B = I_B^2 R_B = 20.32^2 \times 12.5 =$ **5161.3 W. (Ans.)**

B. *Three-phase three-wire system.* In case a star-connected load is unbalanced and if there is *no neutral wire*, then its start point is isolated from the star-point of the generator. The potential of the load star-point is different from that of the generator star-point. The potential of the former is subject to variations according to the unbalance of the load and under certain conditions of loading, the potentials of the two star-points may differ considerably. Such an isolated load star-point or neutral point is called a '*floating*' neutral point *because its potential is always changing and not fixed.* All star-connected unbalanced load supplied from polyphase systems without a neutral wire, have floating neutral points. *Any unbalancing of the load causes variations not only of the potential of star-point but also of voltages across the different branches of the load.* Hence in that case, *phase voltage is not* $\frac{1}{\sqrt{3}}$ *of the line voltage.* In general, the phase voltages are not only unequal in magnitude but also subtend angles other than 120° with one another.

The following methods are used to tackle *unbalanced star-connected loads having isolated neutral points.*

1. By Kirchhoff's laws.
2. By star-delta conversion.
3. By applying Milliman's Theorem.
4. By using Maxwell's Mesh or Loop Current Method.

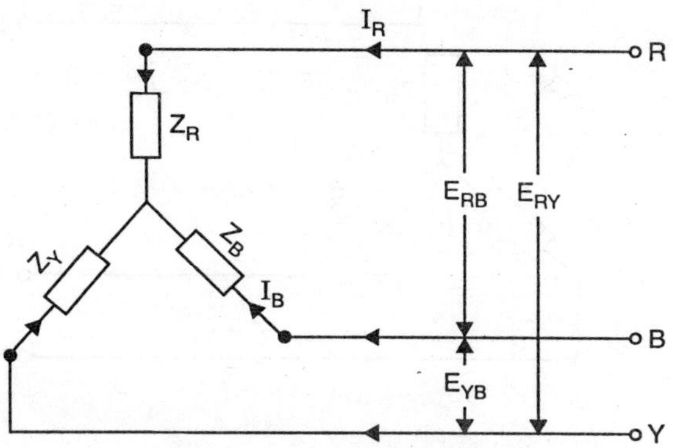

Fig. 19.92

1. Kirchhoff's laws. By applying Kirchhoff's laws, we get (Refer to Fig. 19.92)

$$E_{RY} = I_R Z_R - I_Y Z_Y \quad ...(i)$$

$$E_{YB} = I_Y Z_Y - I_B Z_B \quad ...(ii)$$

$$E_{BR} = I_B Z_B - I_R Z_R \quad ...(iii)$$

and, $$I_R + I_Y + I_B = 0 \quad ...(iv)$$

Adding (i) and (iii), we get

$$E_{RY} + E_{BR} = I_R Z_R - I_Y Z_Y + I_B Z_B - I_R Z_R$$

$$= I_B Z_B - I_Y Z_Y$$

or, $$I_B = \frac{E_{RY} + E_{BR} + I_Y Z_Y}{Z_B} \quad ...(v)$$

Substituting $I_Y + I_B = -I_R$ from (iv) in (iii), we get

$$E_{BR} = I_B Z_B + (I_Y + I_B) Z_R = I_B (Z_B + Z_R) + I_Y Z_R \quad ...(vi)$$

Substituting the value of I_B from (iv) in (vi), we get

$$E_{BR} = \left(\frac{E_{RY} + E_{BR} + I_Y Z_Y}{Z_B} \right) (Z_B + Z_R) + I_Y Z_R$$

or, $$E_{BR} Z_B = E_{RY} Z_B + E_{BR} Z_B + I_Y Z_Y Z_B + E_{RY} Z_R + E_{BR} Z_R + I_Y Z_Y Z_R + I_Y Z_R Z_B$$

or, $$I_Y (Z_R Z_Y + Z_Y Z_B + Z_B Z_R) = -Z_R (E_{RY} + E_{BR}) - E_{RY} Z_B$$

or, $$I_Y (Z_R Z_Y + Z_Y Z_B + Z_B Z_R) = Z_R E_{YB} - Z_B E_{RY} \quad (\because \; E_{RY} + E_{YB} + E_{BR} = 0)$$

or, $$I_Y = \frac{E_{YB} Z_R}{Z_R Z_Y + Z_Y Z_B + Z_B Z_R} - \frac{E_{RY} Z_B}{Z_R Z_Y + Z_Y Z_B + Z_B Z_R}$$

i.e., $$I_Y = \frac{E_{YB}}{Z_Y + Z_B + \dfrac{Z_Y Z_B}{Z_R}} - \frac{E_{RY}}{Z_R + Z_Y + \dfrac{Z_R Z_Y}{Z_B}}$$

Similarly, $$I_B = \frac{E_{BR}}{Z_B + Z_R + \dfrac{Z_B Z_R}{Z_Y}} - \frac{E_{YB}}{Z_Y + Z_B + \dfrac{Z_Y Z_B}{Z_R}}$$

and, $$I_R = \frac{E_{RY}}{Z_R + Z_Y + \dfrac{Z_R Z_Y}{Z_B}} - \frac{E_{BR}}{Z_B + Z_R + \dfrac{Z_B Z_R}{Z_Y}}$$

2. Star-delta conversion. In case of unbalanced star-connected load it is somewhat more convenient to convert it into equivalent delta-connected load and solve it.

If Z_R, Z_Y and Z_B be the impedances of the given star-connected load and Z_{RY}, Z_{YB}, Z_{BR} be the impedances of the equivalent delta-connected load (Fig. 19.93), then

$$Z_{RY} = Z_R + Z_Y + \frac{Z_R Z_Y}{Z_B} = \frac{Z_R Z_B + Z_{YB} + Z_{RY}}{Z_B}$$

$$Z_{YB} = Z_Y + Z_B + \frac{Z_Y Z_B}{Z_R} = \frac{Z_Y Z_R + Z_B Z_R + Z_Y Z_B}{Z_R}$$

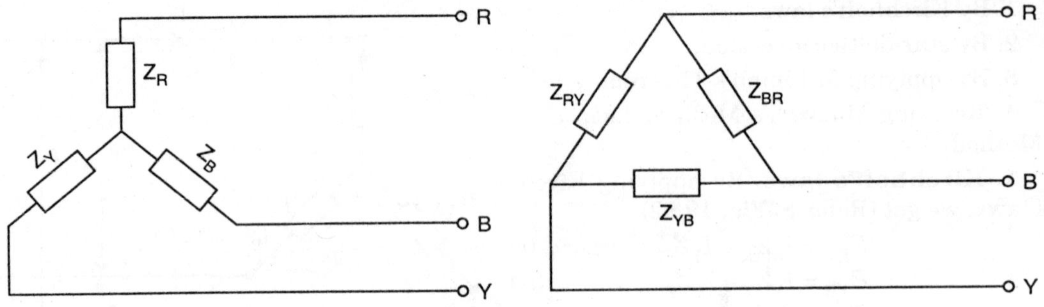

Fig. 19.93

$$Z_{BR} = Z_B + Z_R + \frac{Z_B Z_R}{Z_Y} = \frac{Z_B Z_Y + Z_R Z_Y + Z_B Z_R}{Z_Y}.$$

Now the phase currents and line currents can be determined (Refer to worked examples).

3. Milliman's theorem. Refer to Fig. 19.94. A number of linear bilateral admittances, Y_1, Y_2, Y_3, ..., Y_n are connected to a common point (or mode) O'. The voltages of the free ends of these admittances with respect to another common point O are V_{1O}, V_{2O}, V_{3O}, ..., V_{nO}. Then according to this theorem, the voltage of O' is given by

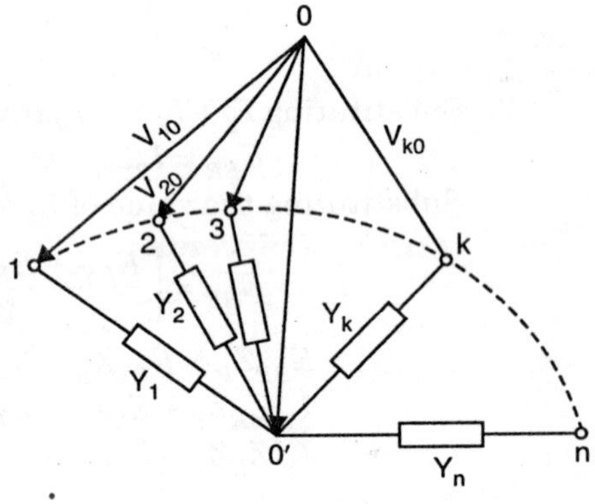

Fig. 19.94

$$V_{O'O} = \frac{V_{1O}Y_1 + V_{2O}Y_2 + Y_{3O}Y_3 + \ldots\ldots V_{nO}Y_n}{Y_1 + Y_2 + Y_2 + \ldots\ldots + Y_n}$$

$$V_{O'O} = \frac{\sum_{k=1}^{n} V_{kO}Y_k}{\sum_{k=1}^{n} Y_k}.$$

19.3.4.2. Unbalanced delta-connected load

When the load impedance in the three phases are not equal in magnitude or phase or both, the load is said to be *unbalanced*. If three unequal loads are connected to form a delta and connected across a 3-phase supply the currents in the three loads will not be equal in magnitude and/or phase. The three-phase currents and the line currents will also be unbalanced.

Example 19.44. *A 440 V, 50 Hz, 3-phase supply has delta-connected load having 50 Ω between R and Y, 159 mH between Y and B and 15.9 μF between B and R. Find :*

(i) The line currents for the sequence RYB.

(ii) The value of star-connected balanced resistor for the same power.

Sol. Phase voltage (E_{ph}) = Line voltage (E_L) = 440 V

The potential difference across three-phases are given by

$$E_{RY} = 440(1 + j0) = 440 \angle 0° \text{ V}$$

$$E_{YB} = 440(-0.5 - j0.866) \text{ or } 440 \angle -120° \text{ V}$$

$$E_{BR} = 440(-0.5 + j0.866) \text{ or } 440 \angle 120° \text{ V}$$

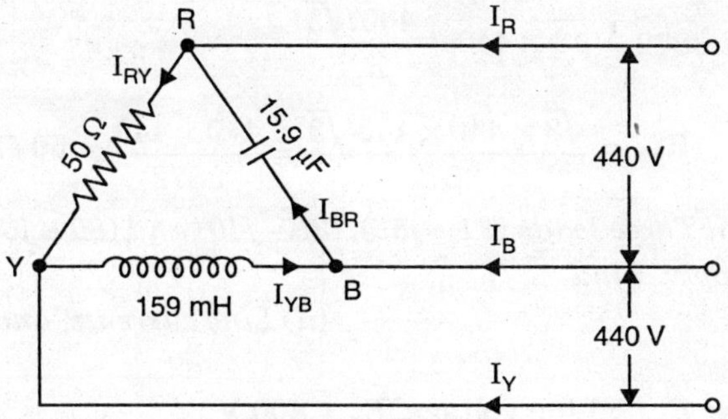

Fig. 19.95

Impedance, $\bar{Z}_{RY} = (50 + j0) = 50 \angle 0° \ \Omega$

Impedance, $\bar{Z}_{YB} = (0 + j2\pi fL) = (0 + j2\pi \times 50 \times 159 \times 10^{-3})$

$= (0 + j50)$ or $50 \angle 90° \ \Omega$

Impedance, $\bar{Z}_{BR} = \left(0 - \dfrac{j}{2\pi fC}\right)$

$$= \left(0 - \frac{j}{2\pi \times 50 \times 15.9 \times 10^{-6}}\right) = (0 - j200) = 200 \angle -90°$$

(*i*) **Phase currents :**

Phase current $\bar{I}_{RY} = \dfrac{\bar{E}_{RY}}{\bar{Z}_{RY}} = \dfrac{440 \angle 0°}{50 \angle 0°} = 8.8(1 + j0)$ A

Phase current $\bar{I}_{YB} = \dfrac{\bar{E}_{YB}}{\bar{Z}_{YB}} = \dfrac{440 \angle -120°}{50 \angle 90°} = 8.8 \angle -210° = 8.8\,(-0.866 + j0.5)$

Phase current $\bar{I}_{BR} = \dfrac{\bar{E}_{BR}}{\bar{Z}_{BR}} = \dfrac{440 \angle 120°}{200 \angle -90°} = 2.2 \angle 210° = 2.2(-0.866 - j0.5)$ A.

Line currents :

Line current, $\bar{I}_R = \bar{I}_{RY} - \bar{I}_{BR} = 8.8(1 + j0) - 2.2(-0.866 - j0.5)$

$= 8.8 + j0 + 1.905 + j1.1 = 10.7 + j1.1 =$ **10.75 ∠5.8° A. (Ans.)**

Line current, $\bar{I}_Y = \bar{I}_{YB} - \bar{I}_{RY} = 8.8(-0.866 + j0.5) - 8.8(1 + j0)$

$= -7.62 + j4.4 - 8.8 - j0$

$= -16.42 + j4.4 =$ **16.99 ∠164.9° A. (Ans.)**

Line current, $\bar{I}_B = \bar{I}_{BR} - \bar{I}_{YB} = 2.2(-0.866 - j0.5) - 8.8(-0.866 + j0.5)$

$= -1.905 - j1.1 + 7.62 - j4.4 = 5.715 - j5.5$

= **7.93 ∠– 43.9° A. (Ans.)**

(*ii*) **Power supplied :**

$$P = I_{RY}^2 \times R = 8.8^2 \times 50 = 3872 \text{ W}$$

Let the resistance per phase be R_{ph}, which when connected in star across 440 V, 3-phase supply takes power of 3872 W.

Line current, $I_L = I_{ph} = \dfrac{E_L/\sqrt{3}}{R_{ph}} = \dfrac{440/\sqrt{3}}{R_{ph}}$.

Power supplied, $P = \sqrt{3}\, E_L I_L$

$$\therefore \quad 3872 = \sqrt{3} \times 440 \times \frac{400/\sqrt{3}}{R_{ph}}$$

$$\therefore \quad \mathbf{R_{ph}} = \frac{\sqrt{3} \times 440 \times 440/\sqrt{3}}{3872} = \frac{440 \times 440}{3872} = \mathbf{50\ \Omega.} \quad \textbf{(Ans.)}$$

☞**Example 19.45.** *Three loads (31 + j59), (30 – j40) and (80 + j60) Ω are connected in delta across 400 V, 3-phase supply. Find :*

(i) Phase currents, *(ii) Line currents, and*

(iii) Power consumed.

Sol. Phase voltage, E_{ph} = Line voltage, E_L = 400 V

The potential differences across three-phases are given by :

$$\overline{E}_{RY} = 400 \angle 0°$$

$$\overline{E}_{YB} = 400 \angle -120°$$

$$\overline{E}_{BR} = 400 \angle 120°$$

The phase impedances are :

$$\overline{Z}_{RY} = 31 + j59 = 66.6 \angle 62.3°$$

$$\overline{Z}_{YB} = 30 - j40 = 50 \angle -53.1°$$

$$\overline{Z}_{BR} = 80 + j60 = 100 \angle 36.9°$$

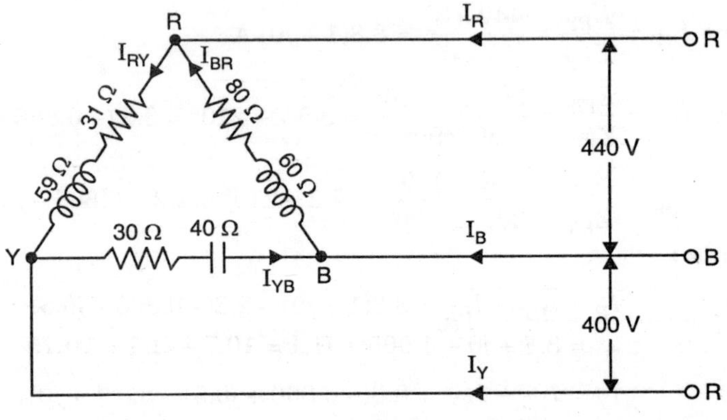

Fig. 19.96

(*i*) **Phase currents :**

$$\mathbf{\overline{I}_{RY}} = \frac{\overline{E}_{RY}}{\overline{Z}_{RY}} = \frac{400 \angle 0°}{66.6 \angle 62.3°} = \mathbf{6 \angle -62.3°\ A.} \quad \textbf{(Ans.)}$$

$$\mathbf{\overline{I}_{YB}} = \frac{\overline{E}_{YB}}{\overline{Z}_{YB}} = \frac{400 \angle -120°}{50 \angle -53.1°} = \mathbf{8 \angle -66.9°\ A.} \quad \textbf{(Ans.)}$$

$$\mathbf{\overline{I}_{BR}} = \frac{\overline{E}_{BR}}{\overline{Z}_{BR}} = \frac{400 \angle 120°}{100 \angle 36.9°} = \mathbf{4 \angle 83.1\ A.} \quad \textbf{(Ans.)}$$

(*ii*) **Line currents :**

$$\mathbf{\overline{I}_R} = \overline{I}_{RY} - \overline{I}_{BR} = 6 \angle -62.3° - 4 \angle 83.1°$$

$= 6(0.465 - j0.885) - 4(0.12 + j0.993)$
$= 2.79 - j5.31 - 0.48 - j3.97 = 2.31 - j9.28 = \mathbf{9.56 \angle -76° \text{ A.}}$ **(Ans.)**

$\bar{\mathbf{I}}_{\mathbf{Y}} = \bar{I}_{YB} - \bar{I}_{RY} = 8 \angle -66.9° - 6 \angle -62.3°$
$= 8(0.392 - j0.92) - 6(0.465 - j0.885)$
$= (3.136 - j7.36) - (2.79 - j5.31)$
$= 0.346 - j2.05 = \mathbf{2.07 \angle -80.4° \text{ A.}}$ **(Ans.)**

$\bar{\mathbf{I}}_{\mathbf{B}} = \bar{I}_{BR} - \bar{I}_{YB} = 4 \angle 83.1° - 8 \angle -66.9°$
$= 4(0.12 + j0.993) - 8(0.392 - j0.92)$
$= 0.48 + j3.97 - 3.136 + j7.36$
$= -2.656 + j11.33 = \mathbf{11.637 \angle 103.2° \text{ A.}}$ **(Ans.)**

(*iii*) **Power consumed :**

Total power is the sum of power in difference phases.

$\therefore \quad P = E_{RY} \times I_{RY} + E_{YB} \times I_{YB} + E_{BR} \times I_{BR}$
$= 400 \angle 0° \times 6 \angle -62.3° + 400 \angle -120° \times 8 \angle -66.9° + 400 \angle 120° \times 4 \angle 83.1°$
$= 400 \times 6 \cos(-62.3°) + 400 \times 8 \cos(53.1°) + 400 \times 4 \cos(-36.9°)$
$= 1115.6 + 1921.3 + 1279.5 = 4316.4 \text{ W}$

[Or $\quad P = I^2_{RY} R_{RY} + I^2_{YB} \times R_{YB} + I^2_{BR} \times R_{BR}$
$= 6^2 \times 31 + 8^2 \times 30 + 4^2 \times 80 = 1116 + 1920 + 1280 = 4316 \text{ W}$]

Hence, **total power = 4316. W. (Ans.)**

HIGHLIGHTS

1. That part of the network of a power system which distributes power for local use is known as *distribution system.*
2. The following are the requirements of a good distribution system :
   (*i*) Proper voltage ; (*ii*) Availability of power on demand ;
   (*iii*) Reliability.
3. The various types of distributors are :
   (*i*) Distributors fed at one end (*ii*) Distributors fed at both ends
   (*iii*) Distributors fed at the centre (*iv*) Ring distributor.
4. A distributor which is arranged to form a closed circuit and which is fed at one or more than one points is called a *ring distributor.*
5. The various methods used to supply a 3-wire D.C. system are :
   (*i*) The two-generator method (*ii*) The balancer set
   (*iii*) The single-generator tapped-resistor (*iv*) The Dabrowolsky generator method
   (*v*) The synchronous converter.
6. In a 3-wire D.C. system to transmit the same amount of power with the same efficiency over same distance with same consumer voltage we require $\frac{5}{16}$ th or 0.3125 times copper as required in 2-wire D.C. system.
7. The line bringing electric power from supplier's low voltage distributor upto the energy meter installed at the consumer's premises is called the *service connection.*
8. In A.C. system the voltage drops are due to the combined effects of resistances, inductances and capacitances.

THEORETICAL QUESTIONS

1. Explain briefly a "distribution system" ?
2. State the main types of distribution systems used and compare their applications.
3. How are distribution systems classified ?
4. What are the requirements of a distribution system ?
5. Give the comparison between overhead system and underground system.
6. Explain 2-wire D.C. system of distributing electric power.
7. Explain clearly the advantages of feeding a distributor at two points over feeding at one point.
8. Prove that the voltage drop diagram for a uniformly loaded distributor fed at one end is parabola.
9. State the advantages and disadvantages of a 3-wire D.C. distribution system over a 2-wire D.C. distribution system.
10. Compare the 3-wire and 2-wire systems of D.C. distribution in respects of weights of copper conductors, all other conditions being the same.
11. Explain clearly the function of a balancer in a 3-wire distribution system.
12. What is balancer ? State its functions with suitable example.
13. How balancers are used to maintain equal voltages on two sides of neutral in a 3-wire D.C. distribution system ?
14. Enumerate the methods to supply a 3-wire D.C. system.
15. What is a booster ?
16. How are boosters classified ?
17. Explain briefly series and shunt boosters.
18. Write short notes on the following :
    (*i*) Ring main distributor.
    (*ii*) Current distribution in 3-wire D.C. system.
    (*iii*) Balancers.
19. How does A.C. distribution differ from D.C. distribution ?
20. What is the importance of load power factors in A.C. distribution.
21. How are A.C. distributors solved ?
22. Explain briefly 3-phase, 4-wire system of distributing electrical power.
23. Write short notes on the following :
    (*i*) Single-phase 2-wire A.C. distribution system.
    (*ii*) Three-phase, four-wire A.C. distribution system.
    (*iii*) Methods of solving A.C. distributors.

UNSOLVED EXAMPLES

1. A 2-wire distributor cable *AB* is 2 km long and supplies loads of 50 A, 100 A and 150 A situated at 0.4 km, 1 km and 1.6 km from the point *A*. Each conductor has a resistance of 0.05 Ω/km. Calculate the potential difference at each point if a potential difference of 400 V is maintained at point *A*.
   [**Ans.** 388 V, 373 V, 364 V]
2. A 2-wire D.C. distributor *AB* 500 metres long is fed from point *A* and is loaded as under :

Distance from feeding point A (metres)	100	300	400	500
Load (amperes)	20	40	40	50

   If the specific resistance of copper is 1.7×10^{-8} Ω-m, what must be the cross-section of each wire in order that the voltage drop in the distributor shall not exceed 10 volts ? [**Ans.** 1.734 cm²]
3. A 2-wire D.C. distributor 500 m long is loaded uniformly at the rate of 0.4 A/m. If the voltage drop in the distributor is not to exceed 5 V, calculate the area of cross-section of each conductor required when the distributor is fed at one end. Take resistivity of a conductor material as 1.7×10^{-8} Ωm.
   [**Ans.** 3.4 cm²]

4. Calculate the voltage at a distance of 200 m of a 300 m long distributor uniformly loaded at the rate of 0.75 A/m. The distributor is fed at one end at 250 V. The resistance of the distributor (go and return) per metre is 0.00018 Ω.
Also find the power loss in the distributor. [**Ans.** 244.6 V ; 911.25 W]

5. A distributor of 500 metres length fed at both ends at 230 V is loaded with loads of 100, 60, 40 and 100 A at distances of 50, 150, 250 and 400 metres respectively. If the maximum voltage drop of 5.5 V is to be allowed, find the cross-sectional area of each conductor and point of minimum voltage. Specific resistance of conductor material may be taken as 1.73×10^{-6} ohms per cm cube.
[**Ans.** 1.06 cm^2 ; Minimum voltage point is *E*]

6. A 2-wire distributor *AB*, 400 metres long is fed at both ends at 200 V and is loaded as follows :
120 A at a distance of 100 metres from *A* ; 150 A at a distance of 250 metres from *A*, and 100 *A* at a distance of 325 metres from *A*. The resistance of a single conductor is 0.03 Ω per 100 metres length. Calculate :
(*i*) The current distribution in the distributor, and
(*ii*) The voltage across each load.
[**Ans.** (*i*) 165 A ; 45 A ; 105 A ; 205 A ; (*ii*) 190.1 V ; 186.05 V ; 190.775 V]

7. A D.C. distributor *AB* is fed at both ends. At feeding point *A*, the voltage is maintained at 235 V and at *B* 236 V. The total length of the distributor is 200 metres and loads are tapped off as under :
20 A at 50 m from *A* ; 40 A at 75 m from *A* ; 25 A at 100 m from *A* ; 30 A at 150 m from *A*.
If the resistance per km of one conductor is 0.4 Ω, calculate the minimum voltage and the point at which it occurs. [**Ans.** 232.17 V ; 75 m from point *A*]

8. An electric train taking a constant current of 600 A moves on a section of line between two substations 8 km apart and maintained at 575 V and 590 V respectively. The track resistance is 0.04 Ω/km both go and return. Find the point of minimum potential along the track and currents supplied by two substation at that instant. [**Ans.** 3.67 km ; 278 A ; 322 A]

9. A 2-wire D.C. distributor cable 1000 metres long is loaded with 0.5 A/m. Resistance of each conductor is 0.05 Ω/km. Calculate the maximum voltage drop if the distributor is fed from both ends with equal voltages of 220 V. What is minimum voltage and where does it occurs ?
[**Ans.** 213.75 V, occurs at mid-point]

10. A 2-wire distributor *AB* 500 metres long is fed from both ends and is loaded uniformly at the rate of 1.0 A/m. At feeding point *A*, the voltage is maintained at 255 V and at *B* 250 V. If the resistance of each conductor is 0.1 Ω/km, determine the minimum voltage and the point where it occurs, and the currents supplied from feeding points *A* and *B*.
[**Ans.** 246 V at 300 m from feeding point A ; 300 A, 200 A]

11. A 800 metres 2-wire distributor *AB* fed from both ends is uniformly loaded at the rate of 1.25 A/m run. Calculate the voltage at the feeding points *A* and *B* if the minimum potential of 220 V occurs at point *C* at a distance of 450 metres from the end *A*. Resistance of each conductor is 0.05 Ω/km.
[**Ans.** 232.65 V ; 227.65 V]

12. A 2-wire D.C. distributor 1000 metres long is fed at the centre and is loaded uniformly at the rate of 1.25 A/metre. If the resistance of each conductor is 0.05 Ω/km, find the maximum voltage drop in the distributor. [**Ans.** 15.62 V]

13. A 2-wire D.C. distributor *AB* 900 metres long is fed at 400 V and loads of 50 A, 100 A and 150 A are tapped off from *C*, *D* and *E* which are at distances of 200 m, 500 m and 800 m from *A* respectively. The distributor is also loaded uniformly at the rate of 0.5 A/m. If the resistance of distributor per metre (go and return) is 0.0001 Ω, calculate the voltages at points *B* and *D*. [**Ans.** 361.75 V ; 370.25 V]

14. A 2-wire D.C. distributor 500 metres long has a uniformly distributed land of 1 A/m. There are concentrated loads of 200 A, 150 A, 50 A and 100 A at 100 m, 200 m, 300 m and 400 m respectively from the end *A*. The distributor resistance is 0.1 Ω/km (go and return). If the distributor is fed at both ends at 250 V determine :
(*i*) The value of minimum potential, and
(*ii*) The position of minimum potential on the distributor. [**Ans.** (*i*) 243.2 V ; (*ii*) 200 m from A]

15. A D.C. 2-wire distributor *AB* is 450 m long and is fed at both ends at 250 V. It is loaded as follows : 20 A at 60 m from *A*, 40 A at 100 m from *A* and a uniform loading of 1.5 A/m from 200 to 245 m from *A*. The resistance of each conductor is 0.05 Ω/km. Find the point of minimum potential and its potential.
[**Ans.** 261.7 m from end *A* ; 247.35 V]

16. A 2-wire distributor *AB*, 300 m long, is fed at both ends *A* and *B*. The distributor supplies a uniformly distributed load of 0.25 A per metre together with concentrated loads of 40 A at *C* and 60 A at *D*, *AC* and *BD* being 120 m each. *A* and *B* are maintained at 300 V, the loop resistance of the distributor is 0.1 Ω/100 m. Determine the currents fed at *A* and *B* and also the potentials of points *C* and *D*.

[**Ans.** 85.5 A, 89.5 A, 291.54 V, 291.06 V]

17. A 2-wire D.C. ring distributor is 300 m long and is fed at 240 V at point *A*. At point *B*, 150 m from *A*, a load of 120 A is taken and at *C*, 100 m in the opposite direction, a load of 80 A is taken. If the resistance per 100 m of single conductor is 0.03 Ω find the currents in each section of distributor and voltages at points *B* and *C*.

[**Ans.** 86.67 A (*A* to *B*), 33.33 A (*C* to *B*), 113.33 (*A* to *C*) ; V_B = 232.2 V, V_C = 233.2 V]

18. A 2-wire D.C. distributor *ABCDEA* in the from of a ring main is fed at point *A* at 220 V and is loaded as under :

10 A at *B*, 20 A at *C*, 30 A at *D*, and 10 A at *E*. The resistances of various sections (go and return) are : *AB* = 0.1 Ω ; *BC* = 0.05 Ω ; *CD* = 0.01 Ω ; *DE* = 0.025 Ω and *EA* = 0.075 Ω. Determine : (*i*) The point of minimum potential and (*ii*) Current in each section of distributor.

[**Ans.** (*i*) *C* is the point of minimum potential ; (*ii*) 29.04 A, 19.04 A, – 0.96 A, – 30.96 A, – 40.96 A]

19. A distributor cable in the form of a ring main *ABCDEA*, supplies loads of 20 A, 60 A, 30 A and 40 A taken at the points *B*, *C*, *D* and *E* respectively, the feeding point being at *A*. The resistances of the section are *AB* = 0.1 Ω, *BC* = 0.15 Ω, *CD* = 0.1 Ω, *DE* = 0.05 Ω and *EA* = 0.1 Ω. The points *E* and *C* are connected by a cable having a resistance of 0.3 Ω. Calculate the current in each section of the network.

[**Ans.** 60 A (*A* to *B*) ; 40 A (*B* to *C*) ; 10 A (*E* to *C*) ; 10 A (*D* to *C*) ; 40 A (*E* to *D*) ; 90 A (*A* to *E*)]

20. In the D.C. network shown in Fig. 19.97, *A* is the feeding point and is maintained at 250 V. The resistances of the various branches (go and return) are indicated in the figure. Determine the current in each branch. [**Ans.** *AB* = 14 A ; *BC* = 2A, *DC* = 5 A, *AD* = 13 A]

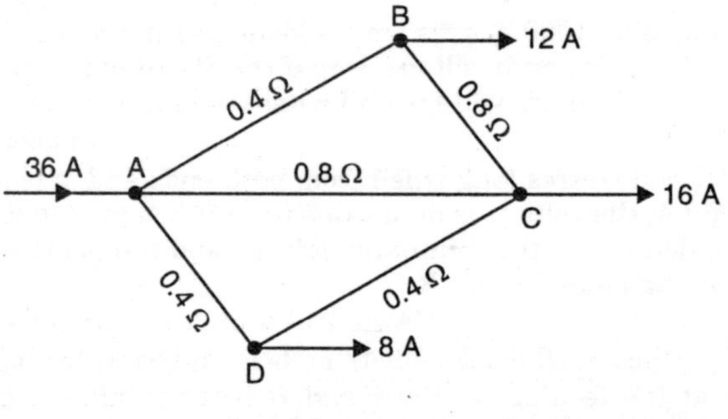

Fig. 19.97

21. A load supplied on 3-wire D.C. system takes a current of 50 A on the positive side and 40 A on the negative side. The resistance of each outer wire is 0.1 Ω and the cross-section of middle wire is one half of that of outer. If the system is supplied at 500/250 V, find the voltage at the load end between each outer and middle wire. [**Ans.** 243 V, 248 V]

22. A load on 3-wire D.C. system with 500 V between outers consists of a lighting current of 1500 A on positive and 1300 A on negative side while motors connected across outers absorb 500 kW. Assuming each balancer machine having loss of 3.75 kW, calculate load on main generator and each balancer machine. [**Ans.** 1207.5 kW ; 21.25 kW , 28.75 kW]

23. A 2-wire feeder *ABC* has a load of 120 A at *C* and of 60 A at *B* both at p.f. 0.8 lagging. The impedance *AB* is (0.04 + *j*0.08) Ω and that of *BC* is (0.08 + *j*0.12)Ω. If the voltage at the far end *C* is to be maintained at 400 V, determine the voltage at *A* and voltage at *B*.

[**Ans.** 430.9 ∠1.72° V ; 416.36 ∠0.793° V]

24. A single-phase distributor 2 km long supplies a load of 120 A at 0.8 p.f. lagging at its far end and a load of 80 A at 0.9 p.f. lagging at its mid-point. Both power factors are referred to the voltage at the far end. The resistance and reactance per km (go and return) are 0.05 Ω and 0.1 Ω respectively. If the voltage at the far end is maintained at 230 V, calculate :

(*i*) The voltage at the sending end, and

(*ii*) The phase angle between voltages at the two ends. [**Ans.** (*i*) 261.67 V ; (*ii*) 3.82°]

25. The loading on a distributor is shown in Fig. 19.98. The distributor is a two-core cable for which resistance and reactance are 0.25 Ω and 0.125 Ω per km of cable run respectively. What should be the voltage at the point *A* to maintain 400 V at the point *D*. **[Ans.** 406.54 V]

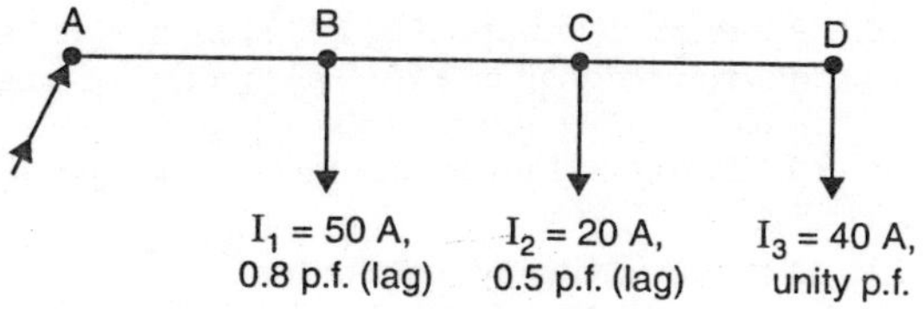

Fig. 19.98

26. A single-phase, ring distributor *PQR* is fed at *P*. The loads at *Q* and *R* are 50 A at p.f. 0.8 lagging and 80 A at p.f. 0.6 lagging respectively, both expressed relative to the voltage at *P*. The total impedance of the three sections *PQ*, *QR* and *RP* are $(1.5 + j1.5)\Omega$, $(2 + j3)\Omega$ and $(2 + j1)\Omega$ respectively. Determine the current in each section with respect to the voltage at *P*.

[Ans. $I_{PQ} = 56 \angle 49.6°$ A ; $I_{PR} = 72.8 \angle 44.8°$ A]

27. A 3-phase ring main *ABCD*, fed from end *A* at 11 kV supplies balanced loads of 30 A at 0.9 p.f. lagging at *B*, 50 A at 0.707 p.f. lagging at *C* and 40 A at 0.8 p.f. lagging at *D*, the load currents being referred to the voltage at point *A*.

The impedances of the various sections are : Section *AB* = 3 + *j*4 ; Section *BC* = 1 + *j*1 ; Section *CD* = 2 + *j*3 ; Section *DA* = 1 + *j*2. Determine :

(*i*) Currents in various sections, and

(*ii*) Station bus-bar voltages at *B*, *C* and *D*.

[Ans. (*i*) (40.55 – *j*26.45) A ; (13.55 – *j*13.35) A ; (– 21.8 + *j*22) A ; (– 53.8 + *j*46) A ; (*ii*) (6130.6 – *j*82.85) V/phase; (6103.7 – *j*83.05) V/phase ; (6213.3 – *j*61.65)V/phase]

28. A 440/254 V, 3-phase, 4-core system supplies an unbalanced load represented by the following impedances in ohms between *R*, *Y* and *B* phases respectively and the neutral 16 + *j*12, 14 – *j*21 and 25 Ω. The phase sequence is *RYB*. Calculate :

(*i*) The current in each conductor of the cable, and

(*ii*) Readings of three wattmeters connected in each line and neutral.

[Ans. 12.7 A, 10.06 A, 10.16 A ; 2580 W, 1413 W ; 2580 W]

29. Calculate the voltage across each of the following star-connected impedances when they are connected to a 400 V, 3-phase system, $Z_R = j20\ \Omega$, $Z_Y = j1\ \Omega$, $Z_B = -j1\ \Omega$.

Also find the potential of the star point above ground.

The phase sequence is *RYB*. **[Ans.** (4000 + *j*6928) V ; (3600 + *j*6928) V ; (3800 + *j*7274.4) V ; – 8000 V]

30. An unbalanced star-connected load comprising of the impedances $Z_R = (10 + j0)$, $Z_Y = (3 + j4)$, $Z_B = (0 - j10)$ is connected across a 3-phase 200 V circuit with balanced voltages. Find :

(*i*) Three line currents, and (*ii*) Voltage across each impedance.

The phase sequence is *RYB*. **[Ans.** (– 20.2 + *j*4.54) ; (7.24 + *j*5.48) ; (12.78 – *j*10.02 ; (– 200.2 + *j*45.4) ; (– 0.2 + *j*45.4) ; (– 100.2 – *j*127.8)]

31. An unbalanced star-connected load is supplied from a symmetrical 3-phase, 440 V, 3-wire system. The branch impedances of the load are $Z_1 = 5 \angle 30°\ \Omega$, $Z_2 = 10 \angle 45°\ \Omega$ and $Z_3 = 10 \angle 60°\ \Omega$. Assuming the positive phase sequence of *RYB* determine the line currents and voltage across each impedance. Draw the phasor diagram.

[Ans. 35.75 ∠– 71.3° A ; 32.77 ∠156.1° A ; 27.67 ∠48° A ; 178.7 ∠– 41.3° V, 327.7 ∠201.1° V, 276.7 ∠108° V]

32. An unbalanced star-connected load has branch impedances of $Z_1 = 10 \angle 30°\ \Omega$, $Z_2 = 30 \angle -45°\ \Omega$, $Z_3 = 20 \angle 60°\ \Omega$ and is connected across a balanced 3-phase, 3-wire supply of 200 V. Find the line currents and the voltage across each impedance using star-delta conversion method.

[Ans. 5.39 ∠8° 32′ A, 15.79 ∠–148° 6′ A, 10.73 ∠44° 16′ A, 53.9 ∠38° 32′ V, 157.9 ∠–193.6°, 214.6 ∠104° 16′]

33. A 400 V, 50 Hz, 3-phase supply has delta-connected load having 100 Ω between *R* and *Y*, 318 mH between *Y* and *B* and 31.8 μF between *B* and *R*. Find the line currents for the phase sequence *RYB*. Calculate the value of star-connected balanced resistor for the same power.

[Ans. 7.73 ∠15° A ; 7.73 ∠165° A ; 4 ∠270° A ; 100 Ω]

20

Extra High Voltage Transmission

20.1. Introduction. 20.2. Extra high voltage A.C. transmission—Advantages of E.H.V. transmission—Limitations/Problems involved in E.H.V. transmission. 20.3. High voltage direct current (H.V.D.C.) transmission—Classification of H.V.D.C. transmission systems—Advantages, limitations and applications of H.V.D.C. transmission systems. Highlights—Theoretical Questions.

20.1. INTRODUCTION

The problem of ever increasing demand of electric power throughout the world is presenting unusual problems to the power engineers. Although excellent hydro-power sites are available in some countries but they are at a far distance from the load centres. The thermal power plants also need to be located near coal fields in order to avoid pollution hazards and from economic consideration, which are also at large distances from the load centres. The nuclear power stations also need to be located far away from large cities and thickly populated areas. All these problems involve transmission of large amounts of over long distances. This can be done most economically only by using Extra High Voltages (E.H.V.).

The present trend is to transmit large amounts of power over medium and long lines at very high voltages (above 300 kV) and then r.m.s. values of these voltages are termed as "*Extra High Voltage*". The voltages whose values exceed 750 kV are termed as "*Ultra High Voltages*" (U.H.V.). Generally almost all aspects of characteristics and operations of E.H.V. lines are applicable to U.H.V. lines.

- Extra High Voltage (E.H.V.) A.C. transmission may be considered to have come of age in 1952 when the first 380-400 kV line was put into service in Sweden. Since then, industrialised countries all over the world have adopted this and higher voltage levels.

In our country the first 400 kV transmission line from Obra to Sultanpur and 400 kV substation at Sultanpur (U.P.) was commissioned in December, 1977.

The single serious problem encountered with E.H.V. voltage level is the *over voltages during switching* operations, commonly called "*switching-surge over voltages*". Very soon it was found that a long air gap was weakest for positive polarity switching-surges. The co-ordination of insulation must now be based on *switching impulse levels (SIL) and not on lightning impulse levels only.*

- The high D.C. voltages find applications in industries, research, medical sciences etc. H.V.D.C. transmission over both overhead lines and underground cables is becoming more and more popular.
  - The first commercially used High Voltage D.C. (H.V.D.C.) link of 20 MW, 100 kV capacity in the world, was built in 1954 between the mainland of Sweden and the island of Gotland. Since then the technique of power transmission by H.V.D.C. has been continuously developed. In 1970 *Thyristor valves* replaced the valves based on the mercury-arc technique.

— To date the biggest H.V.D.C. transmission is ITAIPU in Brazil (two bipoles, 6300 MW and ± 300 kV).

— In India the first H.V.D.C. line is Rihand-Delhi (± 500 kV, 800 MW).

— The highest voltage transmission reached is ± 600 kV.

20.2. EXTRA HIGH VOLTAGE A.C. TRANSMISSION

Use of E.H.V. A.C. transmission is increasing day by day for interconnections of two or more power systems to achieve sharing of installed reserves ; construction of large power plants, and for development of integrated systems and grids.

20.2.1. Advantages of E.H.V. Transmission

The following are the *advantages* of E.H.V. A.C. transmission for transmitting bulk power over long distances :

1. With the increase in transmission voltage, the *transmission efficiency increases* for a given amount of power to be transmitted over a given distance (line losses are reduced since these are inversely proportional to the transmission voltage).

2. *Voltage regulation is improved* because of reduction in line losses.

3. *The volume of conductor material decreases*, being *inversely* proportional to the square of transmission voltage.

4. The transmission capacity of the line *increases* tremendously, since the transmission capacity is proportional to the square of the operating voltages.

Although the costs of the tower, installation and terminal equipments increase but in general use these costs are proportional to the voltages rather than square of the transmission voltage. Consequently, *overall capital cost of transmission decreases as the voltage increases.*

5. With the increase in voltage level, the *installation cost* of the transmission line per km *decreases.*

6. Since Surge Impedance Loading (SIL) is proportional to the square of the voltage $\left(P_{SIL} = \frac{V^2}{Z_0}, \text{ where } Z_0 \text{ is the surge impedance of line}\right)$, therefore, with increase of voltage level, *SIL* itself increases which indicates that *power transfer increases.*

7. The interconnections of the power systems on a large scale is possible with EHV transmission only.

8. Flexibility for future system growth.

9. Reduction in rights of way.

20.2.2. Limitations/Problems involved in E.H.V. Transmission

Following are some *limitations/problems involved in E.H.V. transmission* :

1. Corona loss and radio interference.

2. Heavy supporting structures and erection difficulties.

3. Insulation requirements.

4. Suitability considerations.

5. Current carrying capacity.

6. Ferranti effect.

7. Environmental and biological aspects.

1. Corona loss and radio interference :

- Since transmission line voltage level is the governing factor in corona loss, it is more acute in E.H.V. This loss further intensifies in bad weather conditions. Corona is not only a source of loss but is also source of interference with radio and television.
- To reduce the corona losses and radio interference large size diameter conductors have been employed. The cost of manufacture of such conductors however is high and then handling is expensive as well as difficult.
- The corona loss and radio interference can be *reduced* to reasonable value in E.H.V. transmission only by *using the bundle conductors* (A bundle conductor is a conductor made up of two or more conductors called the *sub-conductors,* per phase in close proximity compared with the spacing between phases).

2. Heavy supporting structures and erection difficulties :

- E.H.V. lines carry large mechanical loadings on towers due to bundle conductors, large air and ground clearances, considerable dynamic forces due to broken conductors etc. The towers with fabricated steel members are usually employed in E.H.V. transmission. Since the cost of steel towers varies from 30 to 50% of the total cost of line for voltages upto 500 kV, therefore, better and cheaper designs must be evolved to effect economy.
- For the erection of E.H.V. lines, problems of transportation and erection arise as the supporting structures are to be transported over long distances, moreover a high standard workmanship is required.

3. Insulation requirements :

- The required line insulation level depends upon the magnitude of likely voltage surges which are caused due to internal causes (switching operations) or due to external causes (atmospheric disturbances like lightning).
- The protection of E.H.V. transmission lines against lightning etc. in most cases is obtained by the use of ground wire and rapid auto-reclosing circuit breakers.
- *Switching surges*, especially those due to arc restriking in circuit breakers are dangerous as they may cause over-voltage of 2-3 times the normal operating voltage. It is possible to control and minimise switching over-voltage with a proper and effective design of relay-breaker systems.

4. Stability considerations :

- It can be proved that the theoretical length of a transmission line in km, that can be operated in its natural load and load angle of 30° $\left(\frac{V_S V_R}{X}\right.$, sin δ, assuming $\delta = 30°$ and $\left. X = 0.3475\ \Omega/\text{km}\right)$ without loss of stability is only 1.45 times surge impedance (which comes to be about 500 km). Thus, the transmission distance in E.H.V. A.C. systems is limited.
- This limitation is overcome by employing special equipment such as series capacitors or shunt reactors.

5. Current carrying capacity :

- The use of cable is limited in high voltage systems since the cost of cable is very high in comparison with overhead line.
- Whereas the loading of overhead line conductors does not depend upon the thermal considerations, the cable, however, is never operated at load beyond its natural load as in

case of underground cables thermal current limit and charging current are of much significance.

6. Ferranti's effect :

It is a known fact that when a line is loaded with capacitive load its receiving end voltage exceeds the sending and voltage (Ferranti effect). Furthermore, whenever the load on the generator is thrown off suddenly, there is a rise in the sending end voltage. Care needs to be taken from these aspects in case of E.H.V. A.C. transmission system.

7. Environmental and biological aspects :

- The recent researches in this field show that E.H.V. and U.H.V. lines generate electrostatic and electromagnetic fields. These fields can induce current and voltage in animals, human beings and birds. However, fortunately these effects are minimum and within tolerate limits.
- In E.H.V. range, transmission lines also produce audible noise in line conductors, however, these noises do not exceed the satisfactory noise level upto 500 kV.

Note. ***Loadability*** *of a transmission line is defined as the optimum power capability of a transmission line under specific operating criteria (like prescribed voltage drop etc.). The power transfer decreases with increase in length.*

20.3. HIGH VOLTAGE DIRECT CURRENT (H.V.D.C.) TRANSMISSION

- Since long back D.C. used to be employed for generation, transmission and distribution of electric power. The first 110 V. D.C. Central Electric Station was installed by Edison in New York in 1882. However, later, for several reasons such as facility of transformation of voltage from one value to another, better performance of A.C. motors and superiority of A.C. generators in comparison to D.C. generators, the power was generated, transmitted and utilised in the form of A.C.
- Presently, the developments in the design of mercury arc rectifier and Silicon Controlled Rectifier (SCR) have made H.V.D.C. transmission all the more feasible and attractive, though the generation and distribution still continue to be carried out by A.C.

20.3.1. Classification of H.V.D.C. Transmission Systems

The most efficient method of generating high D.C. voltage is through the *process of rectification employing voltage multiplier circuits. Electrostatic generators* have also been used for generating higher D.C. voltages.

H.V.D.C. transmission systems are *classified* as follows :

1. Two-pole, one-wire system (See Fig. 20.1)
2. Two-pole, two-wire system (See Fig. 20.2)
3. Three-pole, two-wire system (See Fig. 20.3)
4. Three-pole, three-wire system (See Fig. 20.4)

Fig. 20.1. Two-pole, one-wire system.

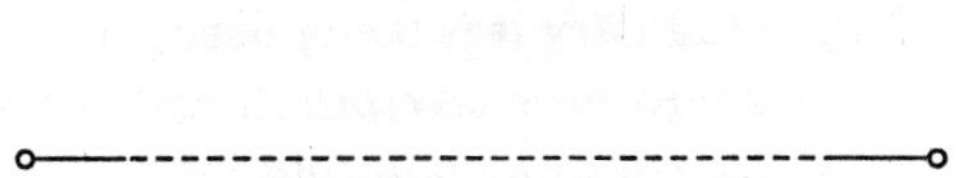

Fig. 20.2. Two-pole, two-wire system.

Fig. 20.3. Three-pole two-wire system.

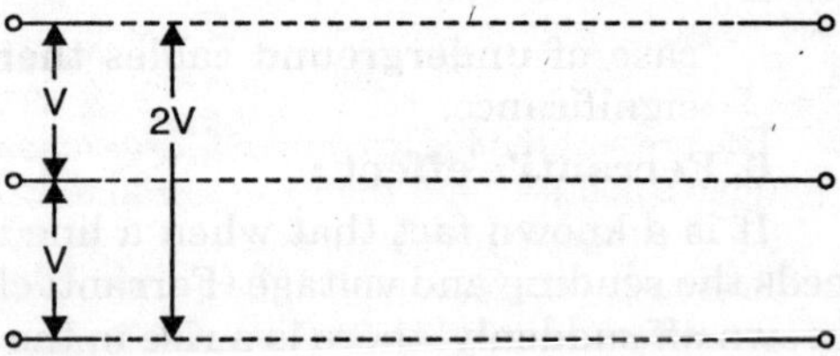

Fig. 20.4. Three-pole, three-wire system.

20.3.2. Advantages, Limitations and Applications of H.V.D.C. Transmission systems

Advantages :

Following are the *advantages, limitations* and *applications* of H.V.D.C. transmission systems :

1. *Cheaper in cost.*
2. *Requires less insulation* comparatively (since the potential stress on the insulation in case of D.C. system is $\frac{1}{\sqrt{2}}$ times of that in A.C. system).
3. The designs of the supporting structures are *simpler and cheaper.*
4. *Sheath losses negligible* (since in case of D.C. only leakage current flows in the cable sheath.) whereas in the case of A.C. charging, circulating and eddy currents flow through the cable sheath).
5. *Lesser dielectric power loss.*
6. *Higher current carrying capacity.*
7. *Higher* natural dielectric strength and *longer life* of cable insulation.
8. *No stability limit*, and as such the transmission lines may be of any length (whereas in A.C. systems the theoretical length of a transmission line which can be operated without loss of stability is only 1.43 times its natural impedance, mostly it is about 500 km).
9. *Less corona loss and radio interference, comparatively.*
10. Use as a non-synchronous link between two A.C. systems.

Limitations :

1. Costly terminal equipment.
2. Non-availability of reliable D.C. circuit breakers.
3. Distortion in wave form, presence of ripples and harmonics.
4. Non-transmission of reactive power over the D.C. link.

Applications :

The following are a few important applications of H.V.D.C.

1. H.V.D.C. is used for *testing H.V.A.C. cables of long lengths.*

Even though D.C. tests on A.C. cables is convenient and economical, these suffer from the fact that stress distribution within the insulating material is different from the normal operating condition.

2. In industry it is being used for :

— *electrostatic precipitation* of ashing in thermal power plants ;

— *Electro-static painting* ;

— *Cement industry* ;

— *Communication systems* ; etc.

3. H.V.D.C. is also being used extensively in physics for *particle acceleration and in medical equipments (X-rays).*

HIGHLIGHTS

1. The present tend is to transmit large amounts of power over medium and long lines at very high voltages (above 300 kV) and then r.m.s. values of these voltage are termed as "*Extra High Voltages*" (E.H.V.). The voltages whose values exceed 750 kV are termed as "*Ultra High Voltages*" (U.H.V.).
2. Presently, the developments in the design of mercury arc rectifier and Silicon Controlled Rectifiers (SCR) have made H.V.D.C. transmission all the more feasible and attractive, though the generation and distribution still continue to be carried out by A.C.
3. H.V.D.C. transmission systems are classified as follows :
   (*i*) Two-pole, one-wire system ;
   (*ii*) Two-pole, two-wire system ;
   (*iii*) Three-pole, two-wire system ;
   (*iv*) Three-pole, three-wire system.

THEORETICAL QUESTIONS

1. Why is E.H.V. transmission preferred ?
2. What are the problems involved in E.H.V. A.C. transmission.
3. Give the advantages, drawbacks and limitations of E.H.V. A.C. transmission.
4. What is stability limitation in E.H.V. A.C. long line transmission ?
5. Give the advantages and disadvantages of H.V.D.C. transmission.
6. What are the applications of H.V.D.C. transmission ?
7. Give reasons for the following :
   (*i*) Extra high voltages are used for transmission of electric power.
   (*ii*) Extra high voltage D.C. system is used for transmission of electric power.
   (*iii*) Extra high voltage A.C. system is used for transmission of electric power.
   (*iv*) Bundled conductors are used for transmission voltages above 200 kV.

PART–III : SWITCHGEAR AND PROTECTION

Chapters :

21. Symmetrical Components and Fault Calculations
22. Switches, Fuses and Circuit Breakers
23. Protective Relays
24. Protection of Alternators and Transformers
25. Protection of Bus-bars and Lines/Feeders
26. Protection Against Overvoltages and Insulation Co-ordination
27. Neutral Earthing (Grounding)
28. Sub-stations

21

Symmetrical Components and Fault Calculations

21.1. Introduction. 21.2. Types of faults. 21.3. **Symmetrical fault calculations**—Percentage reactance—Reactances of synchronous machines—Reactor control of short-circuit currents—Steps for symmetrical fault calculations. 21.4 **Unsymmetrical fault calculations**—Unsymmetrical faults—*Symmetrical components*—Operator λ—Properties of operator λ—symmetrical components in terms of currents and voltages—Zero sequence components of current and voltage—Unbalanced star load supplied from unbalanced three-phase three-wire system, Unbalanced star load supplied from balanced three-phase, three-wire system—Average 3-phase power in terms of symmetrical components—*Sequence impedances*—Sequence impedances of power system elements—Measurement of sequence impedances of rotating machines—*Analysis of unsymmetrical faults*—Single line-to-ground fault—Line-to-line fault—Double line-to-ground fault—*Sequence networks.* Highlights—Theoretical Questions—Unsolved Examples.

21.1. INTRODUCTION

- In a power system the *faults* occur because of insulation failure in plant which may be caused by a system over-voltage such as a switching surge or a lightning stroke, or may be due to broken insulators or conductors, and various other causes on the transmission line.
- The most common and dangerous fault that occurs in a power system is the short circuit or shunt fault. Such a fault involves power conductor or conductors-to-ground or short-circuit between conductors and causes a flow of heavy current, called the *short-circuit current*, through the power system and equipment. It results in considerable damage to the equipment and interruption of supply to the consumers.
- In *symmetrical faults* the fault currents in the three lines (in a 3-ϕ circuit) are equal in magnitude and displaced 120° electrical from one another. Although symmetrical faults are the most severe and impose heavy duty on the circuit breakers, yet the analysis of such faults can be made with a fair degree of ease. It is because the balanced nature of fault permits to consider only one phase in calculations; the condition in the other two phases being similar.
- The majority of faults which occur in a power system are *unsymmetrical faults,* the most common being a short-circuit from one line to ground. When such a fault occurs, it gives rise to unsymmetrical currents *i.e.,* the magnitudes of fault currents in the three lines are different having unequal phase displacement. The calculation procedure known as method of *symmetrical components* is used to determine the currents and voltages in this type of fault.
- The *sources of fault power* originate in all the generating and other connected plants which under normal conditions take power from the system.
  - — A drop in frequency or voltage under short-circuit conditions is common and in this event, *synchronous machines* will feed back into the system for a short period. Simi-

larly, *large induction motors* where considerable flywheel effect is available act as generators in tne event of reduced frequency. Where machines such as these are connected, and they are of size as to have an effect particularly where they are connected to a point close to that for which short circuit values are being calculated, they should be included in the calculation.

21.2. TYPES OF FAULTS

In a power system the following types of *faults* generally occur :

1. **Over current.** It may occur due to :
   — short-circuit, or
   — leakage due to corona effect, or
   — overload on the system.
2. **Under voltage.** It occurs due to :
   — short circuits (due to more voltage drop in lines) and machines, or
   — failure of alternator's field.
3. **Unbalance.** It occurs due to :
   — grounding of one or two phases, or
   — short-circuit of two phases, or
   — breaking of one of the conductors.
4. **Reversed power.** It occurs only in interconnected systems.
5. **Surges.** Whenever lightning takes place or severe fault occurs in a neighbouring circuits, some short-lived waves of very high voltage and current are set up in the lines. Such a fault is known as *surge.* The faults which invariably occur in a 3-phase power systems are enumerated below :

(*i*) Single phase to earth. (*ii*) Phase-to-phase. (*iii*) Two phases to earth. (*iv*) Phase-to-phase and at the same time from third phase to earth.	Give rise to *unsymmetrical* fault currents
(*v*) All the three phases to earth. (*vi*) All the three phases short-circuited.	Give rise to *symmetrical* short-circuit currents.

- Such faults cause heavy currents, called *short-circuit currents,* to flow in the system. The determination of the values of such currents enables us to *make proper selection of circuit breakers, protective relays* and helps us to ensure that the associated apparatus *e.g.,* bus-bars, connections, current transformers will *withstand the forces* which arise due to the fault currents during the period prior to the interrupting device clearing the fault.

21.3. SYMMETRICAL FAULT CALCULATIONS

The fault on a power system which gives rise to symmetrical currents (i.e., equal fault currents in the lines with 120° displacement) is called a **symmetrical fault.**

In a 3-phase system the symmetrical fault occurs when all the three conductors of the 3-phase line are brought together simultaneously into a short circuit conditions, as shown in Fig. 21.1. While making calculations, because of balanced nature of fault only one phase need be considered since the conditions in the other two phases will also be similar.

- The symmetrical fault rarely occurs in practice, however, it is the most severe and imposes more heavy duty on the circuit breaker.
- The value of fault current which can flow in any system under short circuit is *limited only by the impedances in that system.* Therefore, it is necessary in any calculation to have knowledge of these impedances.
  — In several situations, the impedances limiting the fault current are largely reactive, such as transformers, reactors and generators. Cables and lines are mostly resistive, but where the total reactance in calculation exceeds three times the resistance, the latter is usually, *neglected.* The error introduced by this assumption will *not exceed five per cent.*

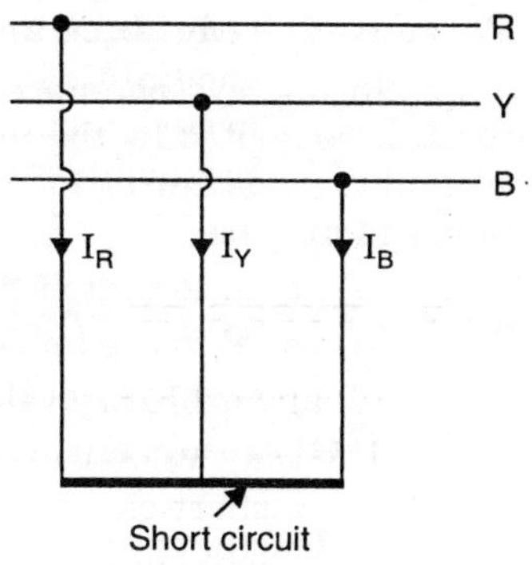

Fig. 21.1

21.3.1. Percentage Reactance

Oftenly the resistance is very small in comparison to reactance and therefore in most cases reactance is considered for determination of fault currents.

The reactance of transformers, alternators, reactors etc. is usually expressed as a percentage.

Percentage reactance is defined as follows :

"The ratio of reactive voltage drop at full load to normal voltage when multiplied by 100, gives the percentage reactance".

Mathematically, if

I = Full load current, A,
V = Normal voltage, V, and
X = Reactive, Ω,

Then, Percentage reactance, $\% \ X = \dfrac{IX}{V} \times 100$

or,
$$X = \frac{(\% X)V}{100\, I} \quad \text{...(21.1)}$$

$$X = \frac{(\% X)V \times V}{100\, VI} = \frac{(\% X)\left(\dfrac{V}{1000}\right)\left(\dfrac{V}{1000}\right) \times 1000}{100\left(\dfrac{V}{1000}\right) \times I}$$

or,
$$X = \frac{(\% X)(\text{kV}) \times 10}{\text{kVA}} \quad \text{...(21.2)}$$

$$\therefore \quad \% X = \frac{X(\text{kVA})}{10\,(\text{kV})^2} \quad \text{...(21.3)}$$

where, X is the reactance in ohms.

If X is the only reactance in the circuit, the short-circuit is given by,

$$I_{sc} = \frac{V}{X} = I \times \left(\frac{100}{\% X}\right) \quad \text{...(From eqn. 21.1)} \quad \text{...(21.4)}$$

Percentage reactance and base kVA :

In any system since different plants have different kVA ratings and since the percentage reactances quoted by the manufacturers are for the respective kVA ratings, it is necessary to convert all the reactances to a common kVA rating base. The value of this base kVA is quite unimportant and may be :

(*i*) equal to that of the largest plant ;

(*ii*) equal to the total plant capacity ;

(*iii*) any arbitrary value.

The conversion can be affected by using the expression :

$$\% \text{ Reactance at base kVA} = \frac{\text{Base kVA}}{\text{Rated kVA}} \times \% \text{ Reactance at rated kVA} \quad ...(21.5)$$

Short-circuit kVA :

Although the potential at the point of fault is zero, it is normal practice to express the short-circuit current in terms of short-circuit kVA based on the normal system voltage at the point of fault. *The product of normal system voltage and short-circuit current at the point of fault expressed in kVA is known as* ***short-circuit kVA.***

Let, V = Normal phase voltage, volts,

I = Full-load current, at base kVA, amperes, and

$\% X$ = Percentage reactance of the system on base kVA upto the fault point.

Then, short-circuit kVA for 3-phase circuit

$$= \frac{3VI_{sc}}{1000} = \frac{3VI}{1000} \times \frac{100}{\% X} \qquad \left[\because I_{sc} = I\left(\frac{100}{\% X}\right) \quad ...\text{Eqn. (21.4)}\right]$$

$$= \text{Base } kVA \times \frac{100}{\% X}$$

i.e.,

$$\text{Short-circuit } kVA = \text{Base } kVA \times \frac{100}{\% X} \quad ...(21.6)$$

Reactance and base voltage :

In certain cases it may be more convenient to work in ohmic values of various reactance. The method would be simple if all the reactances relate to the same voltage but if step-up or step-down transformers or other equipments which operate at different voltages are also included, *all the ohmic values will have to be reduced to a common voltage base.* For this conversion, the following expression may be used :

$$X_2 = \left(\frac{V_2}{V_1}\right)^2 X_1 \quad ...(21.7)$$

where, X_1 and X_2 are the reactances in ohms at system voltages V_1 and V_2 respectively.

- It may be mentioned here that the *advantage* of using percentage reactance instead of ohmic resistance in short-circuit calculations is that the *percentage reactance values remain unchanged as they are referred through transformers,* unlike ohmic reactances which become multiplied or divided by the square of transformation ratio. This makes the procedure *simple and permits quick calculations.*

Per unit quantities :

The ***"per unity quantity"*** *is defined as the ratio of the quantity to its base value.* The per unit method of calculation has an advantage over the percent method because the product of two quantities expressed in per unit directly comes out in per unit whereas in percent method it is to be divided by 100 to obtain the result in percent.

Per unit (p.u.) impedance of an element is

$$= \frac{\text{Actual impedance in ohms}}{\text{Base impedance in ohms}}$$

If we take base voltage in kV and base volt-amperes in kVA, then

$$\text{Base impedance} = \frac{(\text{kV})^2}{\text{kVA}} \times 100 \qquad ...(21.8)$$

The per unit (p.u.) impedances of a transformer is the same regardless whether it is determined from h.v. or l.v. side.

21.3.2. Reactances of Synchronous Machines

The current flowing in a power network due to occurence of a fault is determined by the internal e.m.fs of the machines in the network, their impedances and impedances in the network, between the machines and the fault. Owing to the effect of the armature current on the flux, the current flowing in a synchronous machine immediately after occurence of fault, that flowing a few cycles later and the steady state value of the fault current differ considerably.

Fig. 21.2 shows the short-circuit current as function of time for a synchronous generator. During sub-transient period, the current decrement is very rapid and this period lasts for few cycles only. The transient period covers a relatively longer time during which the current decrement is moderate.

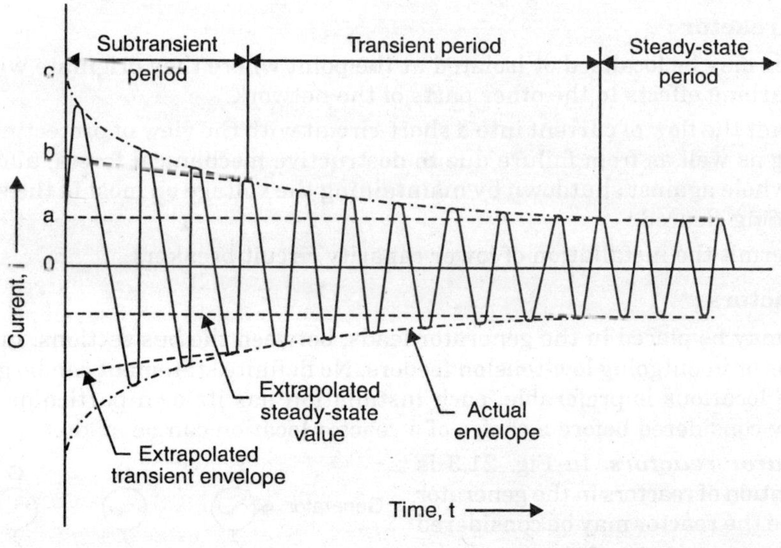

Fig. 21.2. Oscillogram of symmetrical short-circuit current.

The three types of reactances of synchronous machines are :

1. **Synchronous reactance, X_d.** In Fig. 21.2 *Oa* represents the value of steady state current, hence,

$$X_d = \frac{\text{No-load voltage}}{Oa}$$

2. **Transient reactance, X_d'.** If the envelope of the current wave is extended back to zero time, neglecting first few cycles, the value of the current is given by *Ob* and the transient reactance,

$$X_d = \frac{\text{No-load voltage}}{Ob}$$

3. **Subtransient reactance, X_d''.** The value of current at zero time is represented by *Oc* and,

$$X_d'' = \frac{\text{No-load voltage}}{Oc}$$

When a synchronous generator is short-circuited on all the three phases, the initial short-circuit current is limited by the subtransient reactance and in a very short time this current decreases to the transient value limited by the transient reactance and then it settles down to the steady state short circuit value limited by the synchronous reactance.

21.3.3. Reactor Control of Short-circuit Currents

In order to limit the short-circuit currents to a value which the circuit breakers can handle, additional reactances known as *reactors* are connected in series with the system at suitable points,. It may be added that due to very small resistance of reactors, there is a very little change in the efficiency of the system.

Reactor is *a coil which has high inductive reactance as compared to its resistance and is used to limit the short-circuit current during fault conditions.* To perform this function it is essential that magnetic saturation at high current does not reduce the coil reactance. If an iron-cored inductor is expected to maintain constant reactance for currents two to three times its normal value it will turn out to be very costly and heavy. Therefore, air-coiled coils having constant inductance are generally used for current limiting reactors.

Functions of a reactor :

1. Troubles may be localised or isolated at the point where they originate without communicating their disturbing effects to the other parts of the network.

2. They limit the flow of current into a short-circuit with the view of protecting the equipment from overheating as well as from failure due to destructive mechanical forces, and also protecting the system as a whole against shutdown by maintaining the voltage on most of the system while the short circuit is being cleared.

3. They permit the installation of lower capacity circuit breakers.

Location of reactors :

Reactors may be placed in the generator leads, between the bus sections, in the low-tension transformer leads or in outgoing low-tension feeders. No definite statement can be given as to which one of the above locations is preferable, each installation has its own particular demands which must be carefully considered before a choice of a reactor location can be made.

1. ***Generator reactors.*** In Fig. 21.3 is shown the application of reactors in the generator leads. In this case the reactor may be considered as a part of the transient reactance of the generator, hence its effects is to protect the generator in case of any short-circuit beyond the reactors. In case of slow speed alternators, as for example, in some of the hydroelectric units, it is possible to incorporate as much reactance as necessary in the generator itself, hence no reactors are needed. For this reason *reactors are*

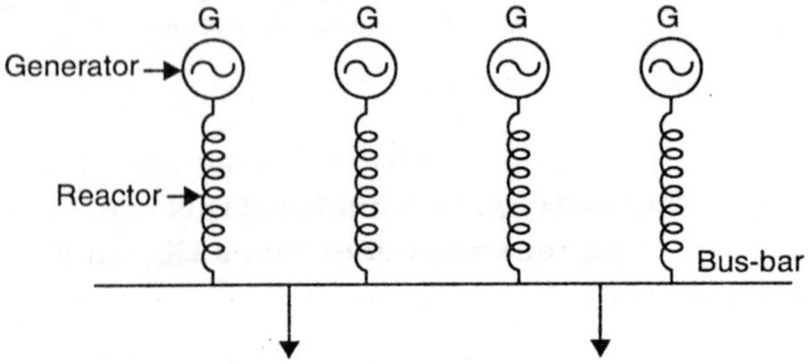

Fig. 21.3. Generator reactors.

rarely used in the generator leads of hydraulic plants. In the case of high-speed alternators it may often be desirable to use reactors in the generator circuit.

Disadvantages :

(*i*) There is constant voltage drop and power loss in the reactors even during normal operation.

(*ii*) If a bus-bar or feeder fault occurs close to the bus-bar, the voltage at the bus-bar will be reduced to a low value, thereby causing the generators to fall out of step.

(*iii*) If a fault occurs on any feeders, the continuity of supply to other is likely to be affected.

2. ***Transformer reactors.*** In a few cases reactors have been installed in the low-tension side of power transformers as shown in Fig. 21.4, *but as a general rule transformers can be designed with enough inherent reactance so that reactors are seldom necessary in the transformer circuits.*

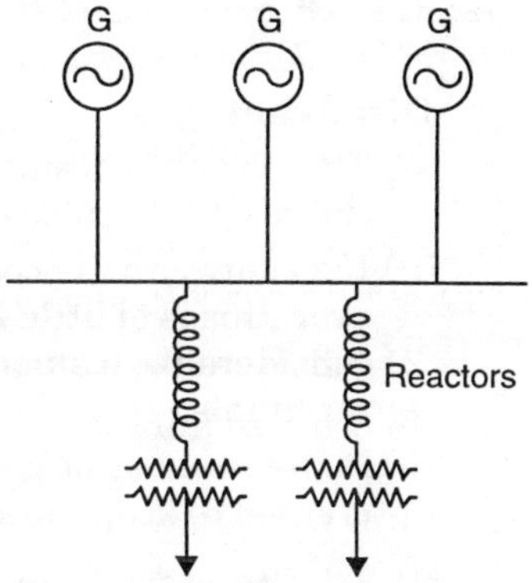

Fig. 21.4. Transformer reactors.

3. ***Bus-bar reactors.*** The constant voltage and constant power loss in reactors may be avoided by inserting reactors in the bus-bars. There are two methods for this purposes namely ; *Ring system* and *Tie-bar system.*

(*i*) *Ring system.* In this system sections are made of generators and feeders and these sections are connected to each other at the common bus-bar as shown in Fig. 21.5 (*i*). Generally, one feeder is fed from one generator only. Under normal operating conditions small amount of power flows though the reactors, therefore, voltage drop and power loss in the reactors is low. In the event of occurence of fault on any one feeder, only one generator feeds the fault while the current from other generators is limited because of presence of the bus-bar reactors.

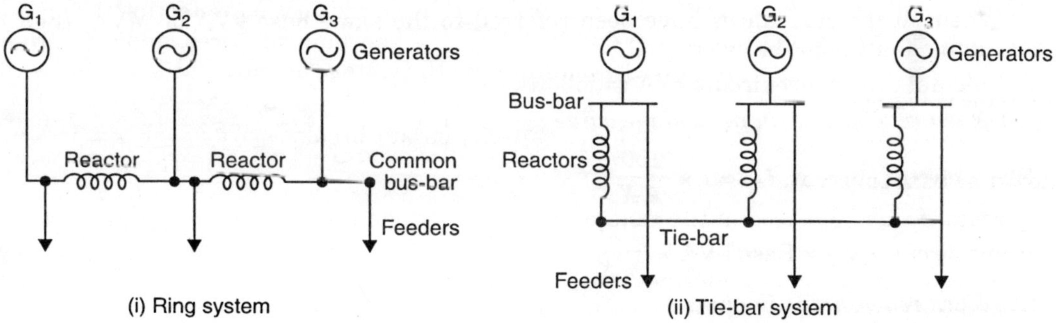

Fig. 21.5. Bus-bar reactors.

(*ii*) *Tie-bar system.* Fig. 21.5 (*ii*) shows the tie-bar system. Here, the generators are connected to the common bus-bar (tie-bar) through the reactors but the feeders are fed from the generator side of the reactors. The operation of this system is similar to ring system. The additional *advantage* of this system is that if the number of sections is increased, the fault current will not exceed a certain value, which is fixed by the size of individual reactors. Thus the switchgear designed to operate successfully on this limiting current will continue operating for any number of extension of sections and requires no modification.

However, this system has the *disadvantage* that it requires an additional bus-bar *i.e.,* the tie-bar.

4. ***Feeder reactors.*** Most of the disturbances and short circuits occur in the low-tension distribution feeders either feeding from a power plant or distributing substation, hence it is not surprising that a large number of reactors are used for such circuits. In the case of a short circuit in a particular feeder, the *reactor prevents the communication of the trouble to the remainder of the system* (Fig. 21.6).

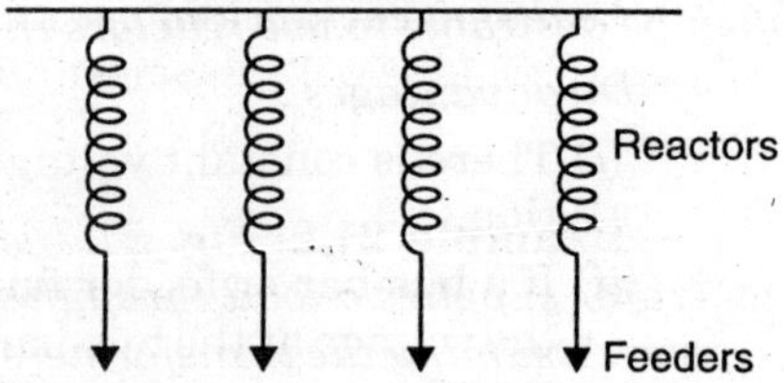

Fig. 21.6. Feeder reactors.

Disadvantages

(*i*) There is a constant power loss and voltage drop in the reactor even during normal operation.

(*ii*) If a short-circuit occurs at the bus-bars, no protection is provided to the generators. However, this is of little importance because such faults are rare and modern generators have considerable leakage reactance to enable them to withstand short-circuit across their terminals.

(*iii*) If the number of generators is increased, the size of feeder reactors will have to be increased to keep the short-circuit currents within the ratings of the feeder circuit breakers.

21.3.4. Steps for Symmetrical Fault Calculations

The various steps involved in symmetrical fault calculations are :

1. Draw a *single line diagram* of the network and indicate on it (*i*) the capacity, voltage, reactance, resistance of all synchronous plants ; (*ii*) the capacity, type, reactance, resistance and length of all cables and onverhead lines with spacings where necessary ; (*iii*) the capacity and reactance of transformers ; (*iv*) the values of any artificial reactance (reactors) introduced in the system.

2. Choose a common kVA and convert all reactances into its common base level and express them as percentage.

3. Corresponding to the single line diagram of the network, draw the reactance diagram showing one phase of the system and the neutral.

4. When all the reactances have been referred to the same base kVA or MVA, then total reactance upto a fault point be calculated.

5. Calculate the short-circuit kVA as follows :

(*i*) *When reactance is in percentage value :*

$$\text{Short-circuit current, } I_{SC} = I \times \frac{100}{\% X}$$

$$\text{Short-circuit kVA} = \text{Base kVA} \times \frac{100}{\% X} \quad \text{...(21.9)}$$

(*ii*) *When reactance is in ohms :*

$$\text{Short-circuit current (r.m.s.)} = \frac{V}{\sqrt{3} \times \text{reactance}} \text{ amperes}$$

$$\text{Short-circuit kVA} = \frac{\text{Short-circuit current} \times V \times \sqrt{3}}{10^3} \quad \text{...(21.10)}$$

Example 21.1. *A single-phase transformer is rated as 3 kVA, 11/0.6 kV. If its leakage reactance is 1.44 Ω when referred to low voltage side, then determine its leakage reactance in p.u.*

Solution. Base voltage = 0.6 kV ; Base kVA = 3 kVA

$$\text{Base impedance on l.v. side} = \frac{(\text{kV})^2 \times 100}{\text{kVA}} = \frac{(0.6)^2 \times 1000}{3} = 120\ \Omega$$

Leakage reactance in per unit $= \dfrac{\text{Actual impedance}}{\text{Base impedance}}$

$$= \frac{1.44}{120} = \mathbf{0.012\ p.u.} \quad \textbf{(Ans.)}$$

Example 21.2. *Fig. 21.7 shows one line diagram for a single-phase power system where a generator supplies a load through a step-up transformer, a transmission line and a step-down transformer. Assuming the transformers to be ideal, calculate the per unit current.* **(Panjab University)**

Solution. Rating of the generator G = 500 V, 10 kVA

Transformation ratio of step-up transformer, $K = 2$

Transformation ratio of step-down transformer,

$$K = \frac{1}{10}$$

Line impedance $= 1 + j2$

Load supplied $= 2 + j4$

Generator

$$V_{\text{base (gen.)}} = 500 \text{ V} = 1 \text{ p.u.}$$

$$kVA_{\text{base (gen.)}} = 10 \text{ kVA} = 1 \text{ p.u.}$$

$$\therefore \quad I_{\text{base (gen.)}} = \frac{10 \times 1000}{500} = 20 \text{ A} = 1 \text{ p.u.}$$

and, $$Z_{\text{base (gen.)}} = \frac{500}{20} = 25\ \Omega = 1 \text{ p.u.}$$

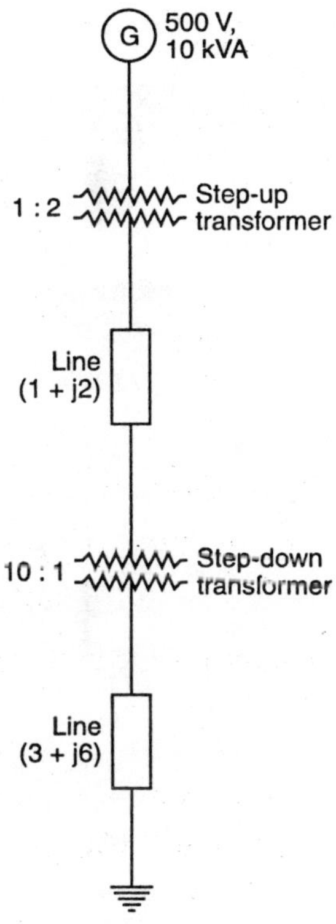

Fig. 21.7

Transmission line :

$$V_{\text{base (line)}} = K \times 500 = 2 \times 500 = 1000 \text{ V} = 1 \text{ p.u.}$$

$$kVA_{\text{base (line)}} = 10 \text{ kVA} = 1 \text{ p.u.}$$

$$I_{\text{base (line)}} = \frac{kVA_{\text{base (line)}}}{V_{\text{base (line)}}} = \frac{10 \times 1000}{1000} = 10 \text{ A} = 1 \text{ p.u.}$$

$$Z_{\text{base (line)}} = \frac{V_{\text{base (line)}}}{I_{\text{base (line)}}} = \frac{1000}{10} = 100\ \Omega = 1 \text{ p.u.}$$

$$Z_{\text{line}} = \frac{1 + j2}{100} = (0.01 + j0.02) \text{ p.u.}$$

Load :

$$V_{\text{base (load)}} = K \times V_{\text{base (line)}} = \frac{1}{10} \times 1000 = 100 \text{ V} = 1 \text{ p.u.}$$

$$kVA_{\text{base (load)}} = 10 \text{ kVA} = 1 \text{ p.u.}$$

$$I_{\text{base (load)}} = \frac{kVA_{\text{base (load)}}}{V_{\text{base (load)}}} = \frac{10 \times 1000}{100} = 100 \text{ A} = 1 \text{ p.u.}$$

$$Z_{\text{base (load)}} = \frac{V_{\text{base (load)}}}{I_{\text{base (load)}}} = \frac{100}{100} = 1\ \Omega = 1 \text{ p.u.}$$

$$Z_{\text{load}} = \frac{3 + j6}{1} = (3 + j6) \text{ p.u.}$$

As the transformers are assumed to be ideal, hence, the impedances of these are neglected.

Then, $$Z_{total} = Z_{line} + Z_{load}$$
$$= (0.01 + j0.02) + (3 + j6) = (3.01 + j6.02) \text{ p.u.}$$
$$= 6.73 \angle 63.43°$$

Per unit current, $$\mathbf{I} = \frac{1\angle 0°}{Z_{total}} = \frac{1\angle 0°}{6.73\angle 63.43°} = \mathbf{0.1486 \angle{-63.43°} \text{ p.u.}} \quad \textbf{(Ans.)}$$

Example 21.3. *A 3-phase transmission line operating at 33 kV and having a resistance and reactance of 6 ohms and 24 ohms respectively is connected to the generating station bus-bar through a 6000 kVA step-up transformer which has a reactance of 6%. Connected to the bus-bars are two synchronous generators, one 12000 kVA having 10% reactance, and another 6000 kVA having 7.5% reactance. Calculate the kVA at short-circuit fault between phases occuring at the high voltage terminals of the transformers and at load end of transmission line.*

Solution. The single line diagram of the network is shown in Fig. 21.8 (*i*).

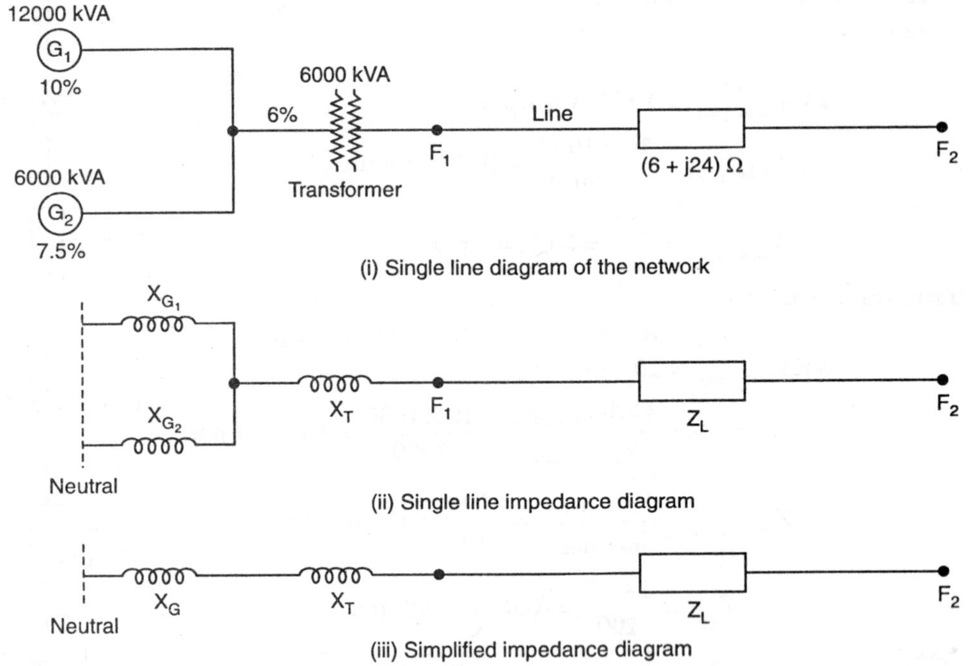

Fig. 21.8

Choosing 12000 kVA as base *kVA*, we have :

Reactance of generator G_1, $X_{G_1} = 10\%$

Reactance of generator G_2, $X_{G_2} = 7.5 \times \dfrac{12000}{6000} = 15\%$

Reactance of transformer, $X_T = 6 \times \dfrac{12000}{6000} = 12\%$

The line impedance in ohms can be converted to percentage impedance (using eqn. 21.3) as follows :

$$\% X = \frac{X\,(kVA)}{10\,(kV)^2} = \frac{24 \times 12000}{10 \times (33)^2} = 26.45\%$$

$$\% R = \frac{6 \times 12000}{10 \times (33)^2} = 6.61\%$$

The single line impedance diagram corresponding to Fig. 21.8 (*i*) can now be drawn as in Fig. 21.8 (*ii*) which when further simplified reduces to 21.8 (*iii*).

For a 3-phase fault at high voltage terminals (F_1) of the transformer, the impedance drop upto the fault

$$= \frac{15 \times 10}{15 + 10} + 12 = 18\%$$

∴ *Short-circuit kVA fed into the fault at* $F_1 = 12000 \times \frac{100}{18}$ = **66667 kVA. (Ans.)**

For a fault at the end (F_2) of the transmission line, the % resistance upto F_2

= 6.61% and % reactance upto $F_2 = 18 + 26.45 = 44.45\%$

∴ Total impedance upto $F_2 = \sqrt{(6.61)^2 + (44.45)^2} = 44.94\%$

∴ Short-circuit kVA fed into fault at $F_2 = \frac{100}{44.94} \times 12000$ = **26702 kVA. (Ans.)**

Example 21.4. *Two generator G_1 and G_2 rated 10 MVA, 11 kV and 5 MVA, 11 kV respectively are connected to a 15 MVA, 11Δ/66 Y kV step-up transformer with 10 percent reactance. If both the generators have a subtransient reactance of 20 percent calculate the subtransient current in eash generator when a 3-phase fault occurs on h.v. side of the transformer assuming that there is no circulating current in the machines.*

Solution. Rating of generator G_1 = 10 MVA, 11 kV

Rating of generator G_2 = 5 MVA, 11 kV

Rating of the step-up transformer = 15 MVA, 11Δ/66 Y kV ;
10% reactance

Subtransient reactance of each generator = 20%

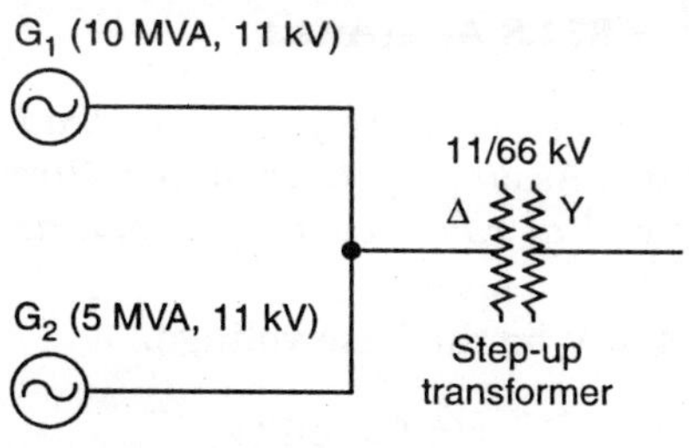

Fig. 21.9. Single line diagram.

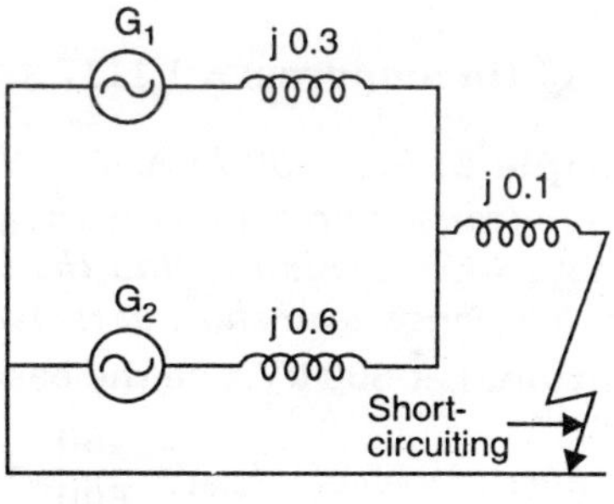

Fig. 21.10. Impedance diagram.

Subtransient current in each generator :

Fig. 21.9 shows the single line diagram of the network.

Let us have a base of 15 MVA, 66 kV on h.v. side then base voltage for l.v. side will be 11 kV.

Subtransient reactance of generator G_1,

$$X''_{d_1} = \frac{\text{Base MVA}}{\text{Rated MVA}} \times \%\text{ reactance at rated MVA}$$

$$= \frac{15}{10} \times 20 = 30\% \text{ or } 0.3 \text{ p.u.}$$

Subtransient reactance of generator G_2,

$$X''_{d_2} = \frac{15}{5} \times 20 = 60\% \text{ or } 0.6 \text{ p.u.}$$

Transformer reactance,

$$X_T = \frac{15}{15} \times 10 = 10\% \text{ or } 0.1 \text{ p.u.}$$

Impedance diagram is shown in Fig. 21.10. The internal voltages of the two machines may be considered to be in parallel and since there is no circulating current between the machines so these internal voltages must be identical in magnitude as well as in phase.

The equivalent parallel subtransient reactance

$$= \frac{0.3 \times 0.6}{j0.3 + j0.6} = 0.2 \text{ p.u.}$$

Subtransient short-circuit current,

$$I'' = \frac{E_g}{j0.2 + j0.1} = \frac{1}{j0.3} = -j3.333$$

The voltage drop on Δ side of the transformer,

$$(\Delta V)_{\Delta \text{ side}} = -j3.333 \times j0.1 = 0.3333 \text{ p.u.}$$

Subtransient current in *generator* G_1,

$$\bar{I}_1'' = \frac{E_g - (\Delta V)_{\Delta \text{side}}}{X''_{d_1}} = \frac{1 - 0.3333}{j0.3} = -j2.2223 \text{ p.u.}$$

$$\bar{\mathbf{I}}_1'' \textbf{ (in amperes)} = 2.2223 \times \frac{(15 \times 1000)}{11\sqrt{3}} = \mathbf{1749.6 \text{ A.}} \quad \textbf{(Ans.)}$$

Subtransient current in *generator* G_2,

$$\bar{\mathbf{I}}_2'' = \frac{E_g - (\Delta V)_{\Delta \text{side}}}{X''_{d_2}} = \frac{1 - 0.3333}{j0.6} = -j1.11117 \text{ p.u.}$$

$$\bar{\mathbf{I}}_2'' \textbf{ (in amperes)} = 1.1117 \times \frac{(15 \times 1000)}{11\sqrt{3}} = \mathbf{874.8 \text{ A.}} \quad \textbf{(Ans.)}$$

Example 21.5. *A 600 kVA, 450 V alternator supplies a purely resistive load of 480 kW at 450 V. The sub-transient reactance of the generator is 8%. Assuming the load to be directly connected across the generator terminals, find the initial symmetrical r.m.s. current in p.u. at the generator terminal for a 3-phase dead short at its terminals.*

Solution. Let 600 kVA be the base kVA and 450 V be the base voltage.

$$\therefore \quad I_{\text{load}} = \frac{480}{600} = 0.8 \text{ p.u.} \angle 0° \qquad \text{(p.f. = unity)}$$

Let the terminal voltage (450 V) = $1\angle 0°$ p.u. (reference)

$$\therefore \quad E'' = V + jIX''_d$$

$$= (1 + j0) + 0.8(j0.08) \qquad (X''_d = 8\% = 0.08 \text{ p.u.})$$

$$= (1 + j0.064) \text{ p.u.}$$

Hence,

$$\bar{\mathbf{I}}'' = \frac{E''}{X''_d} = \frac{1 + j0.064}{j0.08} = \mathbf{(0.8 - j12.5) \text{ p.u.}} \quad \textbf{(Ans.)}$$

Example 21.6. *A generator is connected to a sychronous motor through a transformer. Reduced to the same base, the p.u. sub-transient reactances of the generator and motor are 0.18 and 0.42 respectively, the leakage reactance of the transformer being 0.12 p.u. Three phase fault occurs at the terminals of the motor when the terminal voltage of the generator is 0.9 p.u. and the output current of the generator is 1.0 p.u. at 0.8 p.f. load. Find the sub-transient current in p.u. in the fault in (i) Generator, and (ii) Motor.*

Solution. Sub-transient reactance of the generator, $X''_{dg} = 0.18$ p.u.

Sub-transient reactance of the motor, $X''_{dm} = 0.42$ p.u.

Leakage reactance of the transformer, $X_T = 0.12$ p.u.

Terminal voltage of the generator, $V_g = 0.9$ p.u.

Output current $I = 1.0$ p.u. at 8 p.f. (lag)

For a generator,

$$E''_g = V + jIX''_{dg}$$
$$= 0.9 + j1(0.8 + j0.6)(0.18)$$
$$= 0.9 + j(0.144 + j0.108) = 0.9 + j0.144 - 0.208$$
$$= (0.792 + j0.144) \text{ p.u.}$$

and

$$E''_m = V - jIX''_{dm}$$
$$= 0.9 - j1(0.8 + j0.6)(0.42)$$
$$= 0.9 - j(0.336 + j0.252) = 0.9 - j0.336 + 0.252$$
$$= (1.152 - j0.336) \text{ p.u.}$$

Sub-transient current in the fault, $\mathbf{I''_g}$, $\mathbf{I''_m}$:

(*i*) **Generator :**

$$I''_g = \frac{E''_g}{X''_{dg} + X_T} = \frac{0.792 + j0.144}{(j0.18 + j0.12)}$$
$$= \frac{0.792 + j0.144}{j0.3} = (0.48 - j2.64) \text{ p.u.}$$
$$= \mathbf{2.683 \angle -79.69° \text{ p.u.}} \quad \textbf{(Ans.)}$$

(*ii*) **Motor :**

$$I''_m = \frac{E''_m}{X''_{dm}} = \frac{(1.152 - j0.336)}{j0.42}$$
$$= -0.8 - j2.743 = 2.857 \angle -106.26°$$

For a fault at the motor terminal, the fault current will be fed by the generator as well as motor.

The total fault current (sub-transient),

$$I''_f = I''_g + I''_m$$
$$= (0.48 - j2.64) + (-0.8 - j2.743)$$
$$= (-0.32 - j5.383) = \mathbf{5.39 \angle -93.4° \text{ p.u.}} \quad \textbf{(Ans.)}$$

Example 21.7. *A 625 kVA, 2.4 kV generator with sub-transient reactance of 8% is connected to a bus through a circuit breaker (A) shown in Fig. 21.11. Three synchronous motors, each rated 250 H.P., 2.4 kV, u.p.f., 88.7% efficiency, are connected to the bus. The sub-transient reactance of each of the motors being 20%, find the sub-transient fault current in p.u. in the fault for a 3-phase dead short at the bus at F. Disregarding pre-fault currents determine the fault current rating of the generator breaker and each motor breaker (B).* **(Calcutta University)**

Solution. Rating of the generator = 625 kVA, 2.4 kV

Sub-transient reactance of the generator = 8%

Rating of each synchronous motors = 250 H.P., 2.4 kV, 4 p.f., 88.7% efficiency.

Sub-transient of each motor = 20%

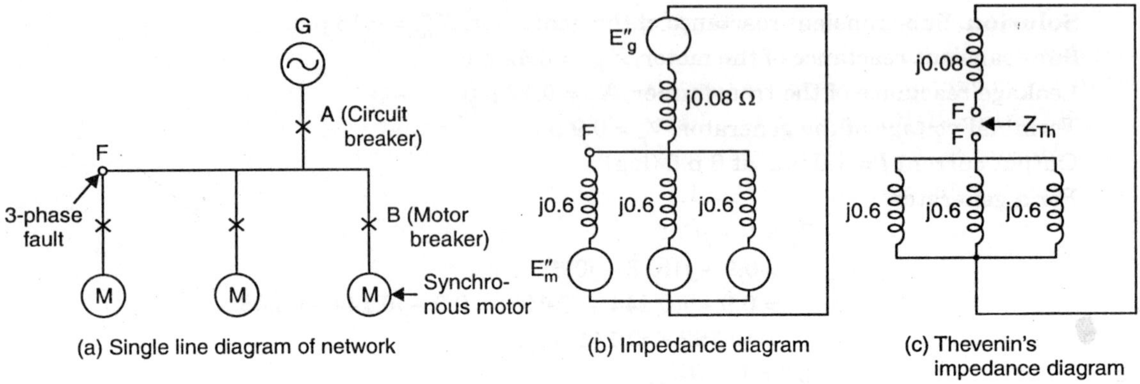

Fig. 21.11

The fault current rating of the generator and each motor breaker :

Input of each motor $= \dfrac{250 \times 0.7355}{0.887} = 207.3$ kVA (∵ 1 H.P. = 0.7355 kW)

Sub-transient reactance of motor,

$$X''_m = j0.2 \times \frac{625}{207.3} = 0.6 \text{ p.u.} \qquad \text{...assuming 625 kVA as base rating}$$

$$\text{Base current} = \frac{625}{\sqrt{3} \times 2.4} = 150.35 \text{ A}$$

The impedance diagram is shown in Fig. 21.11(*b*) . The Thevenin's impedance diagram is shown in Fig. 21.11 (*c*).

Here,
$$Z_{Th} = \frac{j0.08 \times j\dfrac{0.6}{3}}{j0.08 + j\dfrac{0.6}{3}} = \frac{-0.016}{j0.28} = j0.057 \text{ p.u.}$$

Assuming 2.4 kV as 1 p.u., the initial symmetrical current at *F*,

$$I'' = \frac{1}{j0.057} = -j17.54 \text{ p.u.}$$

In the system, this sub-transient fault current (of 17.54 p.u.) will be fed by one generator and three synchronous motors.

The generator contribution will be

$$I''_g = -j17.54 \left[\frac{j\dfrac{0.6}{3}}{j\dfrac{0.6}{3} + j0.08} \right] = -\,j12.53 \text{ p.u.}$$

The total contribution of motors will be,

$$3I''_m = -j17.54 - (-j12.53) = -j5.01 \text{ p.u.}$$

∴ Contribution of each motor

$$= -j\frac{5.01}{3} = -j1.67 \text{ p.u.}$$

Thus the *generator fault current rating in the sub-transient period is* **12.53 p.u.** *and that of each motor is* **1.67 p.u.** **(Ans.)**

Example 21.8. *A 3-phase 15 MVA, 10 kV alternator has internal reactance of 6% and negligible resistance. Find the external reactance per phase to be connected in series with the alternator so that steady current on short-circuit does not exceed 6 times the full load current.*

Solution. Rating of the alternator = 15 MVA, 10 kV

Internal reactance of the alternator = 6%

Short-circuit current = 6 × Full-load current

$$\text{Full-load current} = \frac{15 \times 10^6}{\sqrt{3} \times (10 \times 1000)} = 866 \text{ A}$$

$$\text{Voltage/phase} = \frac{10 \times 1000}{\sqrt{3}} = 5773.5 \text{ V}$$

Total percentage reactance required

$$= \frac{\text{Full-load current}}{\text{Short-circuit current}} \times 100$$

$$= \frac{1}{6} \times 100 = 16.67\%$$

∴ External percentage reactance required

$$= 16.67 - 6 = 10.67\%$$

Let X Ω be the phase external reactance required.

Then, % Reactance $= \frac{IX}{V} \times 100$

$$= 10.67 - \frac{866X}{5773.5} \times 100$$

∴ $$\mathbf{X} = \frac{10.67 \times 5773.5}{865 \times 100} = \mathbf{0.711\ \Omega.}\quad \textbf{(Ans.)}$$

Example 21.9. *A 3-phase transmission line operating at 10 kV and having a resistance of 1 Ω and reactance of 4 Ω is connected to the generating station bus-bars through a 5 MVA step-up transformer having a reactance of 6%. The bus-bars are supplied by a 10 MVA alternator having 12% reactance. Calculate the short-circuit kVA fed to symmetrical fault between phases if it occurs at the*

(*i*) *high voltage terminals of the transformer ;*

(*ii*) *load end of transmission line.*

Solution. Fig. 21.12 shows the single line diagram of the network.

Let 10000 kVA be the base kVA.

% Reactance of alternator on base kVA,

$$\% X_A = \frac{10000}{10 \times 10^3} \times 12 = 12\%$$

% Reactance of transformer on base kVA,

$$\% X_T = \frac{10000}{5 \times 10^3} \times 6 = 12\%$$

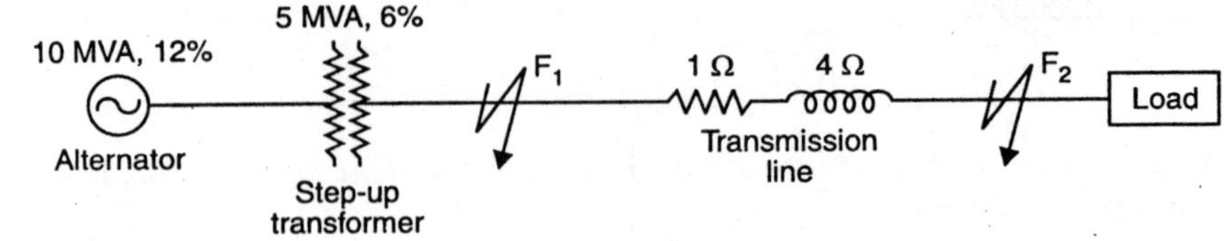

(a) Single line diagram of the network

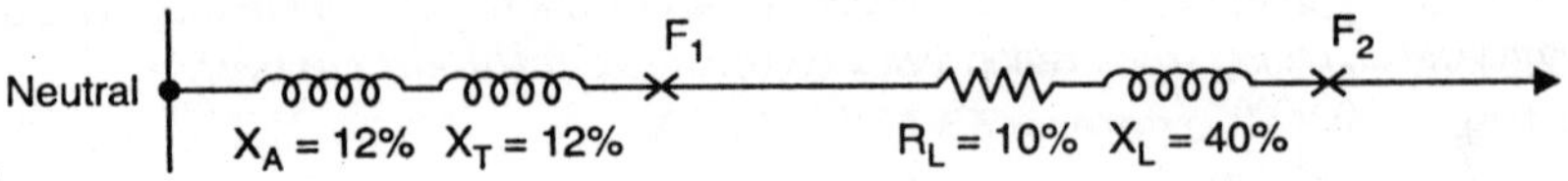

(b) Impedance diagram of the network

Fig. 21.12

The line impedance is given in ohms. It can be converted into percentage impedance (by using eqn. 21.3) as follows :

% Resistance of transmission line,

$$\% R_L = \frac{R\,(\text{kVA})}{10\,(\text{kV})^2} = \frac{1 \times 10000}{10 \times (10)^2} = 10\%$$

$$\% X_L = \frac{X\,(\text{kVA})}{10\,(\text{kV})^2} = \frac{4 \times 10000}{10 \times 10^2} = 40\%$$

The impedance diagram of the network is shown in Fig. 21.12 (*b*).

(*i*) *Fault at the high voltage terminals of the transformer (Point F_1) :*

Total % reactance from alternator neutral upto fault point F_1

$$= \% X_A + \% X_T = 12 + 12 = 24\%$$

$\therefore$ *Short-circuit kVA* $= 10000 \times \frac{100}{24} =$ **41667 kVA. (Ans.)**

(*ii*) *Fault at the load end of transmission line (Point F_2) :*

Total % reactance $= \% X_A + \% X_T + \% X_L$

$$= 12 + 12 + 40 = 64\%$$

% Resistance = 10%

$\therefore$ % Impedance from alternator neutral upto fault point F_2

$$= \sqrt{(64)^2 + (10)^2} = 64.78\%$$

$\therefore$ *Short circuit kVA* $= 10000 + \frac{100}{64.78} =$ **15437 kVA. (Ans.)**

Example 21.10. *The plant capacity of a 3-phase generating station consists of two 12000 kVA generator of reactance 10% each and one 6000 kVA generator of reactance 15%. The generators are connected to the station bus-bars from which load is taken through three 6000 kVA step-up transformers each having a reactance of 5%. Determine the maximum fault MVA which the circuit breakers on (i) low voltage side and (ii) high voltage side may have to deal with.*

Solution. Fig. 21.13 shows the single line diagram of the network.

Let 12000 kVA be the base kVA.

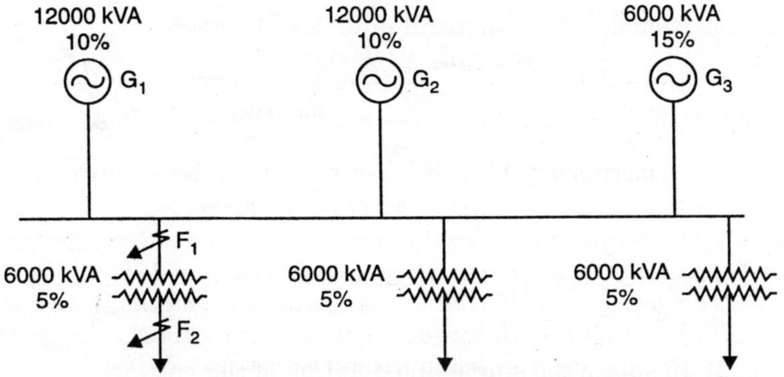

Fig. 21.13. Single line diagram of the network.

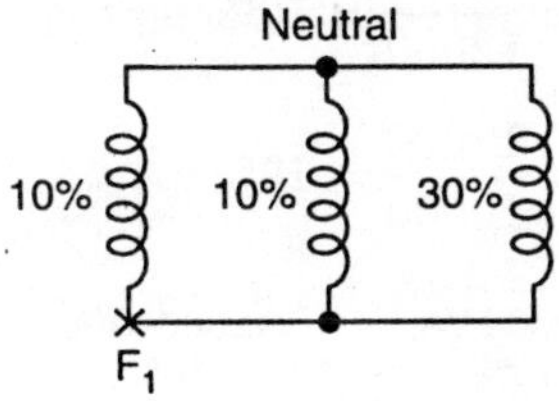

Fig. 21.14

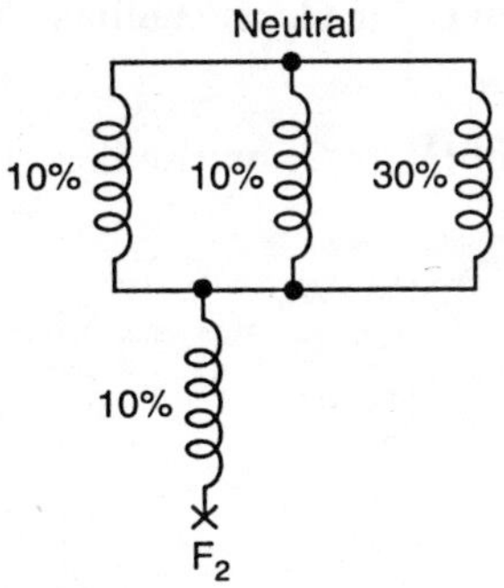

Fig. 21.15

The % reactances of generators G_1, G_2 and G_3 and that of each transformer on the selected base kVA are :

$$\% X_{G_1} = 10 \times \frac{12000}{12000} = 10\%$$

$$\% X_{G_2} = 10 \times \frac{12000}{12000} = 10\%$$

$$\% X_{G_3} \quad 15 \times \frac{12000}{6000} = 30\%$$

$$\% X_T = 5 \times \frac{12000}{6000} = 10\%.$$

(i) When the fault occurs on the low voltage side of the transformer (Point F_1) :

The reactance diagram at the selected base kVA is shown in the Fig. 21.14.

Total % reactance from generator neutral upto fault point F_1

$$= \% X_{G_1} \,||\, \% X_{G_2} \,||\, \% X_{G_3}$$

$$= 10\% \,||\, 10\% \,||\, 30\% = 4.286\%$$

$$\therefore \quad \textit{Fault MVA} = 12000 \times \frac{100}{4.286} \times 10^{-3} \text{ MVA} = \textbf{280 MVA.} \quad \textbf{(Ans.)}$$

(ii) When the fault occurs on the high voltage side of the transformer (Point F_2) :

The reactance diagram is shown in Fig. 21.15.

Total % reactance from generator neutral upto fault point F_2

$$= 4.286 + 10 = 14.286\%$$

$$\therefore \quad \textit{Fault MVA} = 12000 \times \frac{100}{14.286} \times 10^{-3} \text{ MVA} = \mathbf{84\ MVA.} \quad \textbf{(Ans.)}$$

It may be noted that circuit breakers of lower rating will be required on the h.v. side of the transformer.

Example 21.11. *The section bars A and B are linked by a bus-bar reactor rated at 6000 kVA with 10% reactance. On bus-bar A, there are two generators each of 12000 kVA with 10% reactance and on B two generators each of 10000 kVA with 10% reactance. Find the steady MVA fed into a dead short circuit between all phases on B with bus-bar reactor in the circuit.*

Solution. Fig. 21.16 shows the single line diagram of the network.

Let 12000 kVA be the base kVA

$$\% \text{ Reactance of } G_1 \text{ or } G_2 \text{ on the base kVA} = 10 \times \frac{12000}{12000} = 10\%$$

$$\% \text{ Reactance of } G_3 \text{ or } G_4 \text{ on the base kVA} = 10 \times \frac{12000}{10000} = 12\%$$

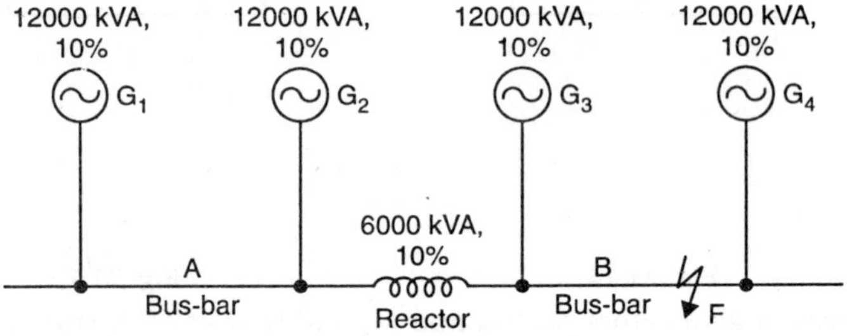

Fig. 21.16. Single line diagram of the network.

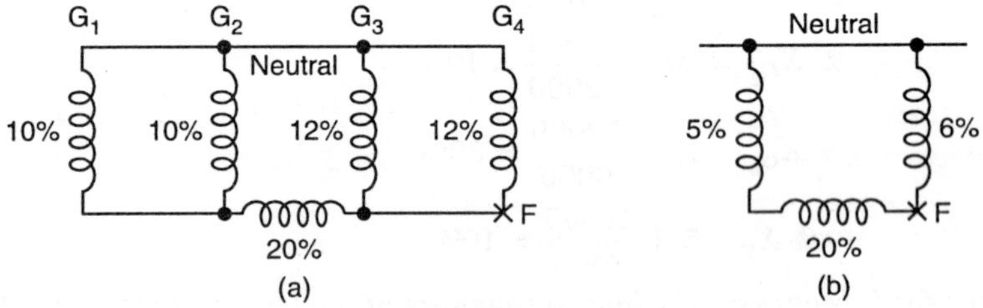

Fig. 21.17. Reactance diagram.

% Reactance of bus-bar reactor on the base kVA

$$= 10 \times \frac{12000}{6000} = 20\%$$

When fault occurs on section B (Point F in Fig. 21.16), the reactance diagram at the selected base kVA will be as shown in Fig. 21.17 (*a*). This series-parallel circuit is further reduced to Fig. 21.17 (*b*).

Total % reactance from generator neutral upto fault point F

$$= (5\% + 20\%) \ || \ 6\% = \frac{25 \times 6}{25 + 6} = 4.839\%$$

$$\therefore \quad \textit{Fault MVA} = 12000 \times \frac{100}{4.839} \times 10^{-3} = \mathbf{248\ MVA.} \quad \textbf{(Ans.)}$$

Examle 21.12. *A 30 MVA generator with 15% reactance is connected to a bus-bar. A 25 MVA transformer with 10% reactance is also connected through a 10% bus-bar reactor to the same bus-bar. Both these reactances are based on 25 MVA.*

If a feeder taken out from the bus-bars through a circuit breaker develops a line-to-ground fault, what should be the appropriate rating of a circuit breaker ? **(AMIE)**

Solution. Fig. 21.18 shows the single line diagram of the network.

Let 30 MVA be base MVA.

% Reactance of generator G on the base MVA,

$$X_G = 15 \times \frac{30}{30} = 15\%$$

% Reactance of the transformer on the base MVA,

$$X_T = 10 \times \frac{30}{25} = 12\%$$

% Reactance of the reactor on the base MVA

$$= 10 \times \frac{30}{25} = 12\%$$

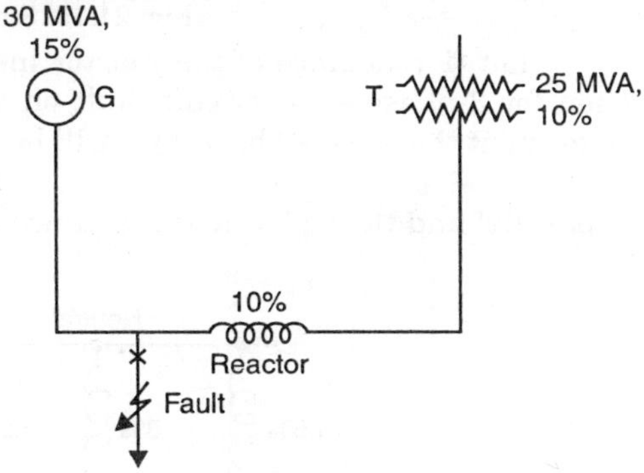

Fig. 21.18. Single line diagram of the network.

Equipment % reactance (12% + 12%) || 15%],

$$X_{eq} = \frac{24 \times 15}{(24 + 15)} = 9.23\%$$

$$\therefore \quad \text{Short-circuit MVA} = 30 \times \frac{100}{9.23} = 325 \text{ MVA}$$

Hence, *Rating of circuit breaker* = **325 MVA. (Ans.)**

Example 21.13. *A small generating station has two alternators of 3000 kVA and 4500 kVA and percentage reactances 7% and 8%. The circuit breakers are used at 150 MVA. It is intended to extend the system by a supply from the grid via a transformer of 7500 kVA rating and 7.5% reactance. Find the reactance necessary to protect the switchgear. Assume bus voltage = 33 kV.* **(AMIE)**

Solution. Fig. 21.19 shows the single line diagram of the network.

Let 4500 kVA be the base kVA

% Reactance of generator G_1 on the base kVA $= 7 \times \frac{4500}{3000} = 10.5\%$

% Reactance of generator G_2 on the base kVA $= 8 \times \frac{4500}{4500} = 8\%$

% Reactance of transformer T on the bse kVA $= \frac{4500}{7500} \times 7.5 = 4.5\%.$

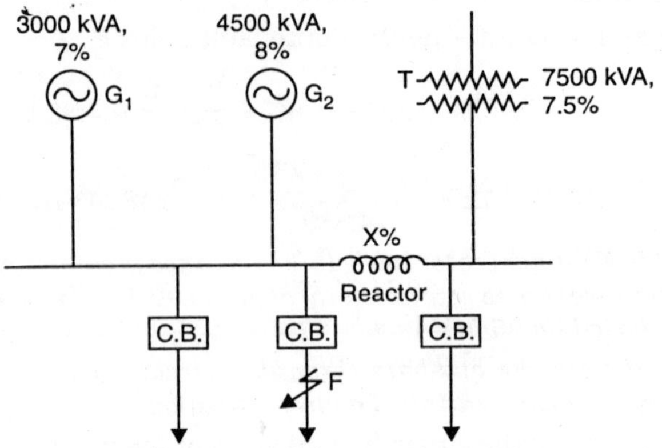

Fig. 21.19. Single line diagram of the network.

Let % reactance of the reactor inserted to limit the short-circuit MVA to 150 MVA be X%. Then for 3-phase short-circuit fault on an outgoing feeder (Point F in Fig. 21.19), the reactance diagram at the selected base kVA will be as shown in Fig. 21.20 (*b*). Reactances of 10.5% and 8% are in parallel and the equivalent reactance $\left(\frac{10.5 \times 8}{10.5 + 8}\right) = 4.54\%$.

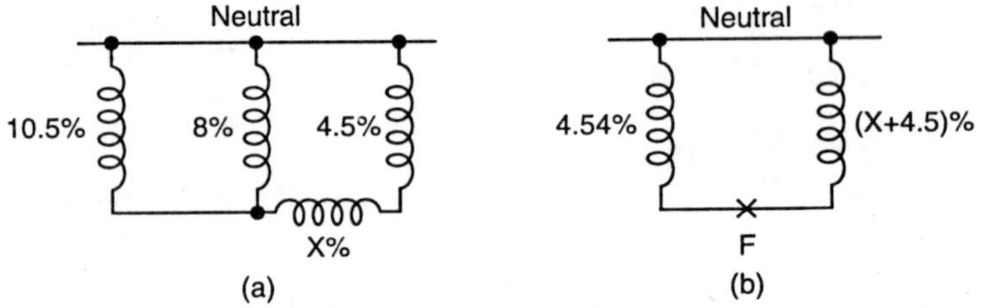

Fig. 21.20. Reactance diagram.

Total % reactance from generator neutral to fault point F

$$= 4.54\% \;||\; (X + 4.5)\%$$

$$= \frac{4.54\,(X + 4.5)}{4.54 + (X + 4.5)} = \frac{4.54\,(X + 4.5)}{9.04 + X}\ \%$$

$$\text{Short-circuit kVA} = 4500 \times 100 \times \frac{(9.04 + X)}{4.54\,(X + 4.5)}$$

But short-circuit kVA should not exceed 150×10^3 kVA, the rupturing capacity of the breaker.

$$\therefore \quad 150 \times 10^3 = \frac{4500 \times 100 \times (9.04 + X)}{4.54\,(X + 4.5)}$$

or, $$4.54(X + 4.5) = \frac{4500 \times 100}{150 \times 10^3}\,(9.04 + X) = 3(9.04 + X)$$

or, $$4.54X + 20.43 = 27.12 + 3X$$

or, $$X = \frac{(27.12 - 20.43)}{(4.54 - 3)} = 4.344\%$$

Also, $$\%X = \frac{X\,(kVA)}{10\,(kV)^2}, \text{ where } X \text{ is in ohms.}$$

$$\therefore \qquad 4.344 = \frac{X(4500)}{10(3.3)^2}$$

or
$$\mathbf{X} = \frac{4.344 \times 10(3.3)^2}{4500} = \mathbf{0.105\ \Omega.} \quad \textbf{(Ans.)}$$

Example 21.14. *The estimated short-circuit MVA at the bus-bars of a generating station A is 1200 MVA and of another station B is 1000 MVA. The generated voltage at each station is 33 kV. If these station are interconnected through a line having a reactance of 1.2 Ω and negligible resistance, calculate the possible short-circuit MVA at both stations.*

Solution. Fig. 21.21(*a*) shows the single line diagram of the system.

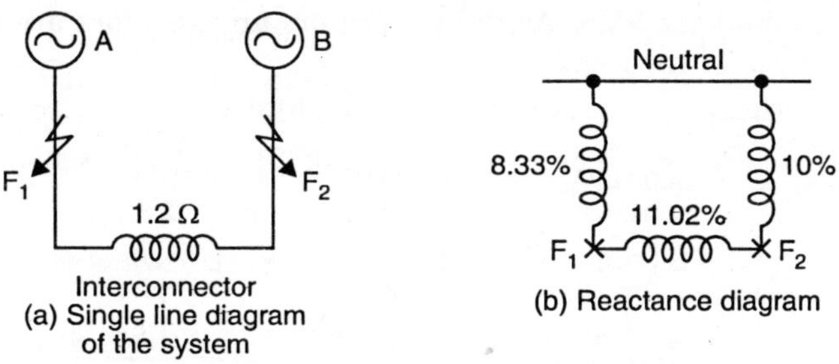

Fig. 21.21

Let the base MVA be 100.

% Reactance of station *A* on the base MVA,

$$\% X_A = \frac{\text{Base MVA}}{\text{Short-circuit MVA}} \times 100$$

$$= \frac{100}{1200} \times 100 = 8.33\%$$

% Reactance of station *B* on the base MVA,

$$\% X_B = \frac{100}{1000} \times 100 = 10\%$$

% Reactance of interconnector on base MVA,

$$\% X_I = \frac{X\,(kVA)}{10\,(kV)^2} = \frac{1.2\,(100 \times 10^3)}{10 \times (33)^2} = 11.02\%.$$

Fault on station A. When a fault occurs on generating station *A* (Point F_1), the total percentage reactance from generator neutral upto the fault point F_1 is the parallel combination of 8.33% and (10% + 11.02% = 21.02%)

i.e. Total % reactance upto fault point F_1

$$= \frac{8.33 \times 21.02}{8.33 + 21.02} = 5.966\%$$

$$\therefore \quad \text{Short-circuit MVA} = \text{Base MVA} \times \frac{100}{\%\ \text{Reactance}}$$

$$= 100 \times \frac{100}{5.966} = \mathbf{1676\ MVA.} \quad \textbf{(Ans.)}$$

Fault on station B. When a fault occurs on power station *B* (Point F_2), the total percentage reactance from generator neutral upto fault point F_2 is the parallel combination of (8.33% + 11.02% = 19.35%) and 10%, *i.e.,* Total % reactance upto fault point F_2

$$= \frac{19.35 \times 10}{19.35 + 10} = 6.593\%$$

$\therefore$ *Short-circuit MVA* $= 100 \times \frac{100}{6.593} =$ **1517 MVA. (Ans.)**

Example 21.15. *A generating station has five section bus-bars connected with a lie bar through 8% reactor rated at 3000 kVA. Each generator is of 3000 kVA with 12% reactance and is connected to each section bus-bar. Find the total steady input to a dead short-circuit between the lines on one of the the section bus-bars (i) with reactors and (ii) without reactors.*

Solution. Fig. 21.22 shows the line diagram of the network.

Let 3000 kVA be the base kVA. As the generators and reactors are rated at this kVA, their percentage reactances remain the same.

Let the fault occur on section 5, as shown in the figure.

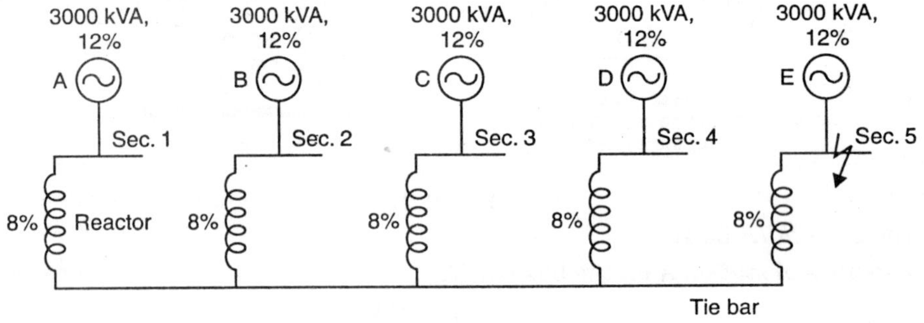

Fig. 21.22

(*i*) **With reactors.** Refer to Fig. 21.23 (*a*) The reactances of generators *A, B, C, D* and their associated reactors are in parallel so that their combined reactance is $\frac{1}{4}$ (12% + 8%) *i.e.* 5%. This is in series with reactance *E*.

$\therefore$ Total reactance upto fault due to generators *A, B, C, D* and reactors *A, B, C, D* and *E*.

$$= 5 + 8 = 13\%$$

This reactance Fig. 21.23 (*b*) is in parallel with that of generator *E* (12%).

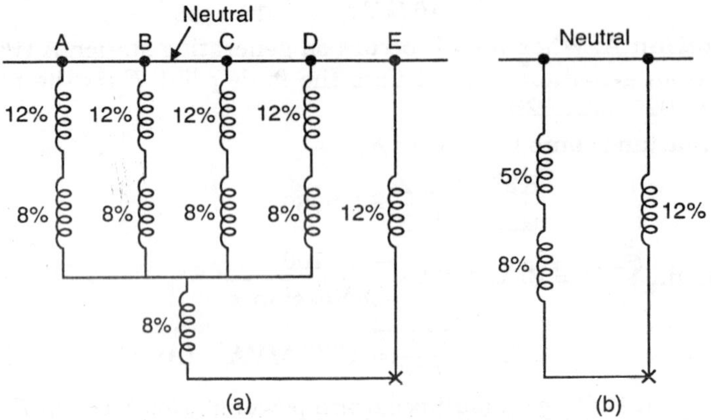

Fig 21.23. Reactance diagram.

∴ Total reactance upto fault $= \dfrac{13 \times 12}{13 + 12} = 6.24\%$

∴ *Short-circuit input* $= 3000 \times \dfrac{100}{6.24} \times 10^{-3}$ MVA = **48.07 MVA. (Ans.)**

(*ii*) **Without reactors.** When there is no reactor, total reactance upto fault $= \dfrac{12}{5} = 2.4\%$.

∴ *Short-circuit input* $= 3000 \times \dfrac{100}{2.4} \times 10^{-3}$ MVA = **125 MVA. (Ans.)**

Example 21.16. *Show that a generating plant having N section bus-bars each rated at Q kVA with x% reactance, connected on the tie-bar system through bus-bar reactances of 6% has a total short-circuit kVA on one section of*

$$\left[\frac{Q}{x} + \frac{Q(N-1)}{bN + x}\right] \times 100$$

If the section rating is 50000 kVA ; x = 20% and b = 10% find the short-circuit kVA with (i) 3 sections (ii) 9 sections and (iii) show that with a very large number of sections, the short-circuit kVA would not exceed 750000 kVA.

Solution. Fig. 21.24 shows the single line diagram of a generating station having N bus-bar sections connected on the tie-bar system. Let us assme that a symmetrical 3-phase short-circuit occurs on sec.-1 as indicated by point *F* in Fig. 21.24.

The reactance diagram is shown in Fig. 21.25 (*a*). For a fault on sec.-1 (point *F*), the remaining $(N - 1)$ sections are in parallel and their equivalent percentage reactance is $\left(\dfrac{x+b}{N-1}\right)\%$. The circuit of Fig. 21.25 (*a*) is, therefore, reduced to the one shown in Fig. 21.25 (*b*). Refering to Fig. 21.25 (*b*), the total percentage reactance between the generator neutral and fault point *F* is given by :

% Reactance from generator neutral upto fault point *F*

$$= x\% \;||\; \left[b + \frac{x+b}{N-1}\right]\%$$

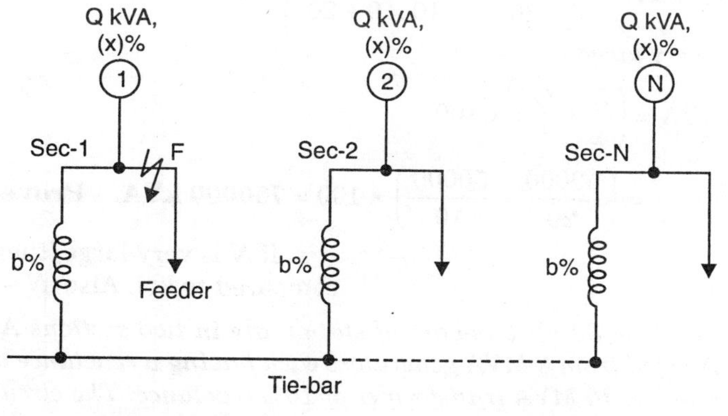

Fig. 21.24. Single line diagram of the network.

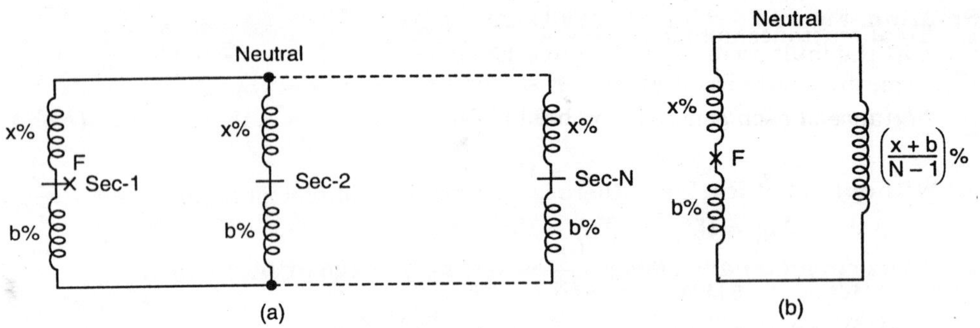

Fig. 21.25. Reactance diagram.

$$= \frac{x\left[b + \frac{x+b}{N-1}\right]}{x + \left(b + \frac{x+b}{N-1}\right)} \% = \frac{x\,(bN - b + x + b)}{xN - x + (bN - b + x + b)} \% = \frac{x(x+bN)}{N(x+b)} \%$$

Suppose the base kVA is Q.

$\therefore$ Total short-circuit kVA $= Q \times 100 \times \dfrac{N(x+b)}{x(x+bN)}$

$$= Q \times 100 \times \left[\frac{(x+bN) + x(N-1)}{x(x+bN)}\right]$$

$$= Q \times 100 \times \left[\frac{1}{x} + \frac{N-1}{x+bN}\right]$$

$$= \left[\frac{Q}{x} + \frac{Q(N-1)}{bN+x}\right] \times 100. \quad \textbf{Proved.}$$

(*i*) With *3 sections*, $N = 3$

$\therefore$ Short-circuit kVA $= \left[\dfrac{50000}{20} + \dfrac{50000\,(3-1)}{10 \times 3 + 20}\right] \times 100 =$ **450000 kVA. (Ans.)**

(*ii*) With *9 sections*, $N = 9$

$\therefore$ Short-circuit kVA $= \left[\dfrac{50000}{20} + \dfrac{50000(9-1)}{10 \times 9 + 20}\right] \times 100 =$ **613636 kVA. (Ans.)**

(*iii*) When N is *very large*,

Short-circuit kVA $= \left(\dfrac{Q}{x} + \dfrac{Q}{b}\right) \times 100$

$$= \left(\frac{50000}{20} + \frac{50000}{10}\right) \times 100 = \textbf{750000 kVA.} \quad \textbf{Proved.}$$

($\because$ If N is very large, then x can be neglected as compared to kN. Also $(N-1)$ may be taken as N)

Example 21.17. *The 33 kV bus-bars of station are in two sections A and B separated by a reactor. The section A is fed from 8 MVA generators each having a reactance of 15%. The section B is fed from the grid through a 40 MVA transformer of 10% reactance. The circuit breakers have a rupturing capacity of 400 MVA. Find the reactance of the reactor to prevent the circuit breakers from being overloaded if a symmetrical short-circuit occurs on an outgoing feeder connected to the generator G_1. Take base MVA as 40 MVA.*

Solution. Fig. 21.26 shows the single line diagram of the network.

Support the fault occurs at point F on the outgoing feeder connected to section A. As per the given condition, the short-circuit MVA at F should not exceed 400 MVA.

% Reactance of each generator on base MVA

$$= 15 \times \frac{40}{8} = 75\%$$

$$\therefore \qquad \% X_{G_1} = 15\% \; ; \; \% X_{G_2} = 15\% \; ; \; X_{G_3} = 15\% \; ; \; X_{G_4} = 15\%$$

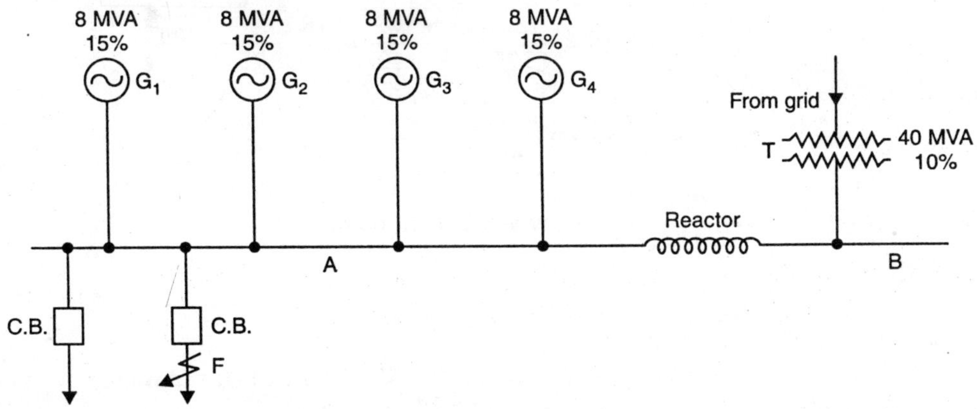

Fig. 21.26. Single line diagram of the network.

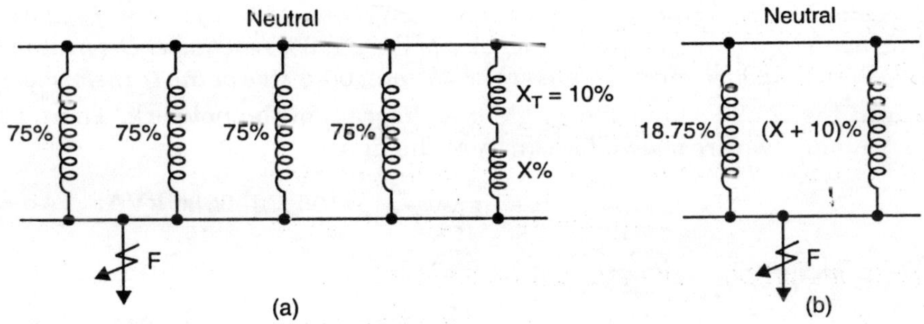

Fig. 21.27. Reactance diagram.

% Reactance the transformer on base 40 MVA,

$$\% X_T = 10 \times \frac{40}{40} = 10\%$$

Let us assume that the required reactance of the reactor is $X\%$ on 40 MVA base.

When the fault occurs at point F, the reactance diagram at the selected base MVA will be as shown in Fig. 21.27 (*a*). Evidently the reactances of four generators are in parallel and their equivalent reactance $= \frac{75}{4} = 18.75\%$. The reactance diagram reduces to that shown in Fig. 21.27 (*b*).

Total percentage reactance from generator neutral to fault point F

$$= \frac{18.75\,(X + 10)}{18.75 + (X + 10)}$$

Now, Fault MVA $= \text{Base MVA} \times \dfrac{100}{\text{Required \% reactance}}$

$\therefore \quad 400 = 40 \times \dfrac{100}{\text{Required \% reatance}}$

(Since fault MVA at F is not to exceed 400 MVA ...*Given*)

$\therefore$ Required % reactance $= \dfrac{40 \times 100}{400} = 10\%$

This means that total % reactance from generator neutral to fault point F should be 10%,

i.e., $$10 = \frac{18.75\,(X + 10)}{18.75 + (X + 10)} = \frac{18.75X + 187.5}{28.75 + X}$$

or, $$10(28.75 + X) = 18.75X + 187.5$$

or, $$287.5 + 10X = 18.75X + 187.5$$

$\therefore$ $$X = 11.43\%$$

Now, % reactance $= \dfrac{X\,(\text{Base kVA})}{10\,(\text{kV})^2}$, where X is in ohms

$$11.43 = \frac{X\,(40 \times 10^3)}{10 \times (33)^2}$$

$\therefore$ $$X = \frac{11.43 \times 10 \times (33)^2}{40 \times 10^3} = \mathbf{3.112\ \Omega.} \quad \textbf{(Ans.)}$$

Example 21.18. *The bus-bars of a power station are in two sections A and B separated by a reactor. Connected in section A are two 12 MVA generators of 15% reactance each and to B 8 MVA generator of reactance 10%. The reactor is rated at 8 MVA and 15% reactance. Feeders are connected to bus-bar A through transformers, each rated at 4 MVA and 5% reactance. Determine the maximum short-circuit MVA with which circuit breakers on the outgoing side of the transformers have to deal.*

Solution. Fig. 21.28 shows the single line diagram of the network. There may be several transformers but only two are shown for clarify of the figure.

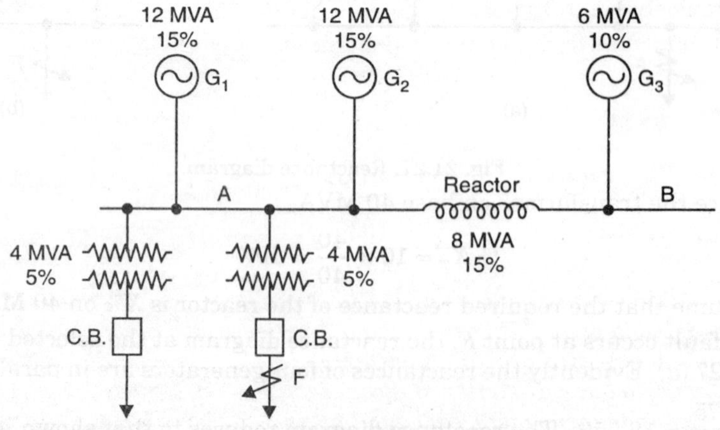

Fig. 21.28. Single line diagram of the network.

Let us assume that the fault occurs at point F on the outgoing side of one transformer as shown in Fig. 21.28. The Circuit Breaker (C.B.) will have to handle short-circuit MVA at F.

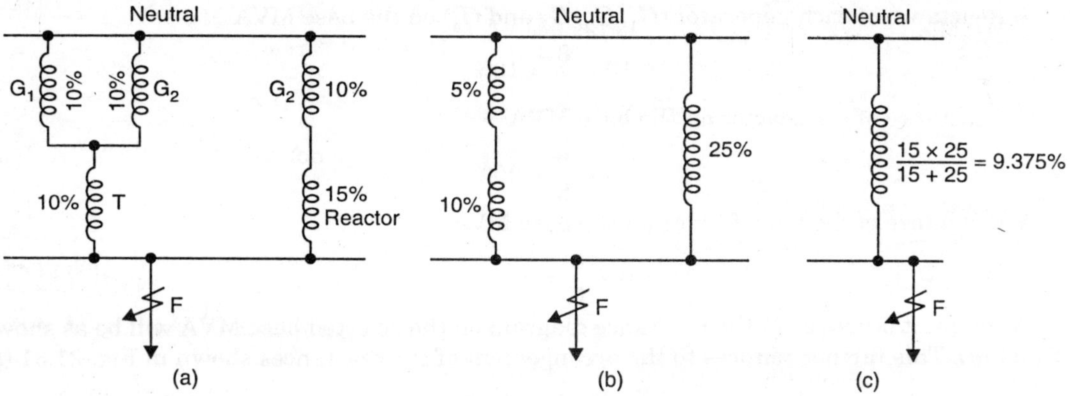

Fig. 21.29. Reactance diagram.

Let 8 MVA be the base MVA. The percentage reactances of various elements on the selected base MVA will be as under :

$$\% X_{G_1} = 15 \times \frac{8}{12} = 10\%$$

$$\% X_{G_2} = 15 \times \frac{8}{12} = 10\%$$

$$\% X_{G_3} = 10 \times \frac{8}{8} = 10\%$$

$$\% X_T = 5 \times \frac{8}{4} = 10\%$$

When fault occurs at point F, the reactance diagram at the selected base MVA will be as shown in Fig. 21.29 (a). The reactances of generators G_1 and G_2 are in parallel and their equivalent reactance $= \frac{10}{2} = 5\%$. The reactance diagram then reduces to that shown in Fig. 21.29 (b). This further reduces to one shown in Fig. 21.29 (c).

∴ Total % reactance from generator neutral upto fault point F

$$= 15\% \;||\; 25\% = \frac{15 \times 25}{15 + 25} = 9.375\%$$

∴ $$\textit{Fault MVA} = \text{Base MVA} \times \frac{100}{\% \text{ Reactance}}$$

$$= 8 \times \frac{100}{9.375} = \mathbf{85.33\ MVA.} \quad \textbf{(Ans.)}$$

Example 21.19. *An 11 kV generating station has four identical 3-phase synchronous generators G_1, G_2, G_3 and G_4, each of 8 MVA capacity and 10% reactance. There are two sections of bus-bar, A and B linked by a reactor rated at 8 MVA with 20% reactance. G_1 and G_2 are connected to bus-bar section A and G_3 and G_4 to bus-bar section B. From each section, load is taken through a number of 5 MVA, 11 kV /6 kV step-up transformers, each having a reactance of 5%. Calculate the current fed into fault if a short-circuit occurs on all phases near the high-voltage terminals of the transformers at the bus-bar section B.*

Solution. Fig. 21.30 shows the single line diagram of the network.

Let 8 MVA be the base MVA.

% Reactance of each generator (G_1, G_2, G_3 and G_4) on the base MVA

$$= 10 \times \frac{8}{8} = 10\%$$

% Reactance of the reactor on the base MVA

$$= 20 \times \frac{8}{8} = 20\%$$

% Reactance of the transformer on the base MVA

$$= 5 \times \frac{8}{5} = 8\%$$

When fault occurs at F, the reactance diagram on the selected base MVA will be as shown in Fig. 21.31 (a). This further reduces to the arrangement of the reactances shown in Fig. 21.31 (b).

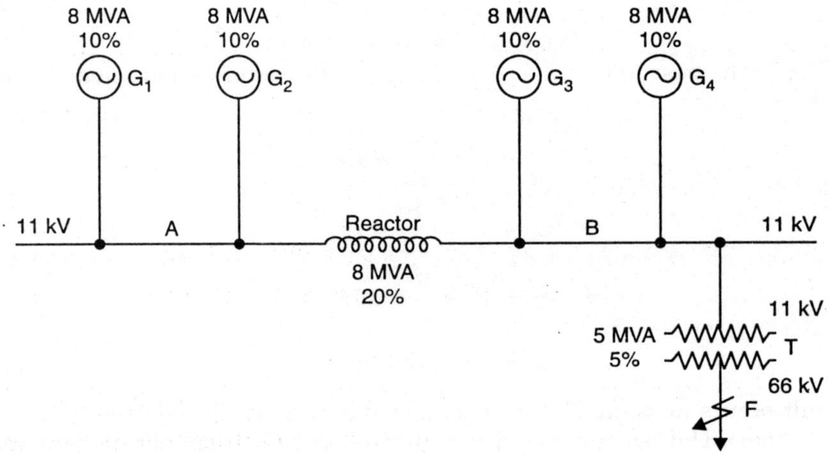

Fig. 21.30. Single line diagram of the network.

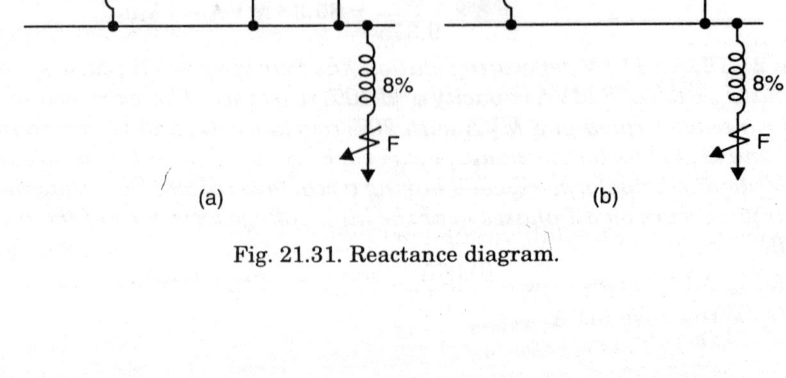

Fig. 21.31. Reactance diagram.

Total % reactance from generator neutral upto fault point F

$$= \frac{25 \times 5}{25 + 5} + 8 = 12.167\%$$

Fault MVA $$= 8 \times \frac{100}{12.167} = 65.75 \text{ MVA}$$

Short-circuit current, $$I_{SC} = \frac{65.75 \times 10^6}{\sqrt{3} \times 66000} = \mathbf{575.2\ A.}\quad \textbf{(Ans.)}$$

Example 21.20. *The main bus-bars in a generating station are divided into three sections, each section being connected to tie-bar by a similar reactor. One 20 MVA, 3-phase, 50 Hz, 11 kV generator with a short-circuit reactance of 15% is connected to each section bus-bar. When a short-circuit takes place between the phases of one of the section bus-bars, the voltage of the remaining sections falls to 55% of the normal value. Calculate the reactance of each reactor.*

(Panjab University)

Solution. Fig. 21.32 shows the single line diagram of the network.

Let G_1, G_2 and G_3 be the three generators, each of 20 MVA and of 15% (0.15 p.u.) reactance. A, B and C are the three reactors of reactance X per unit based on 20 MVA.

Let us assume that the fault occurs in section 3.

Fig. 21.33 shows the reactance diagram. G_1 and A in series are in parallel with G_2 and B in series.

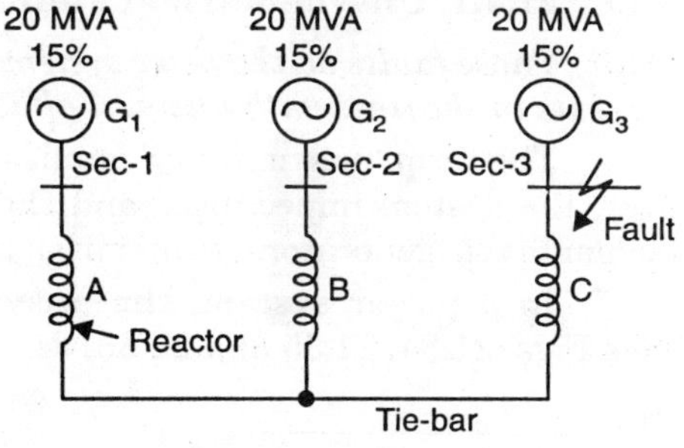

Fig. 21.32. Tie-bar.

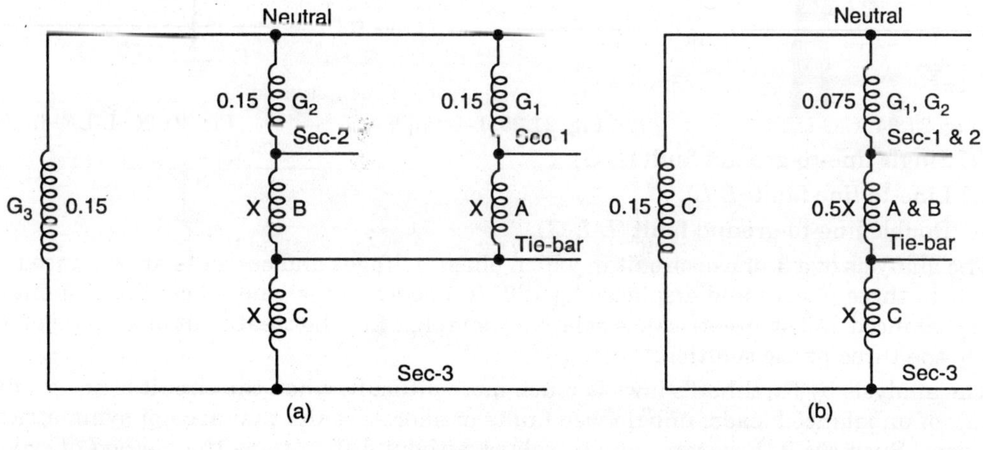

Fig. 21.33

∴ Reactance of G_1, G_2, A and B between the tie-bar and neutral

$$= \frac{0.15 + X}{2} = (0.075 + 0.5X) \text{ p.u.}$$

Reactance from neutral to fault via tie-bar

$$= (0.075 + 0.5X + X) \text{ p.u.} = (1.5X + 0.075) \text{ p.u.}$$

The voltage on sections 1 and 2 drops to 55% *i.e.* 0.55 time of the normal value, 0.45 time of normal value is dropped in the reactances of G_1 and G_2 *i.e.* the reactances of G_1 and G_2 which is 0.075 p.u. is 0.45 time of the total reactance from neutral to fault

i.e. $0.075 = 0.45\ (1.5X + 0.075)$

$\therefore$ $X = 0.0611$ p.u.

Hence reactance of each reactor

$$= \frac{\text{Per unit reactance} \times 1000 \times (\text{kV})^2}{\text{Base kVA}}$$

$$= \frac{0.0611 \times 1000 \times (11)^2}{20 \times 10^3} = \mathbf{0.37\ \Omega.} \quad \textbf{(Ans.)}$$

21.4. UNSYMMETRICAL FAULT CALCULATIONS

21.4.1. Unsymmetrical Faults

Those faults on the powr system which give rise to unsymmetrical currents (i.e., unequal fault currents in the lines with unequal phase displacements) are known as **unsymmetrical faults.**

The term 'unsymmetry' applies only to the fault itself and the resulting line currents. However, the system impedances and the source voltages are always symmetrical through its main elements *viz.* generators, transmission lines, synchronous reactors etc.

In a power system, the unsymmetrical faults may occur in the following *three ways* (See Figs. 21.34, 21.35 and 21.36) :

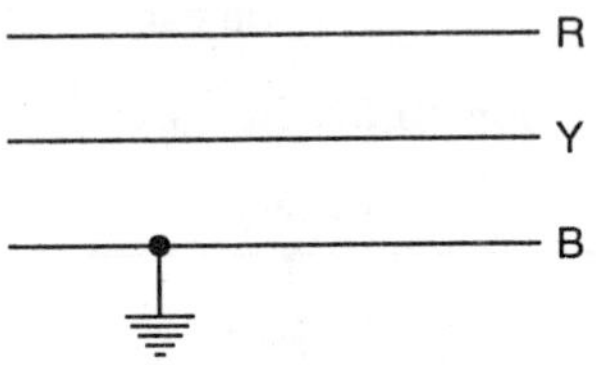

Fig. 21.34. L-G fault.

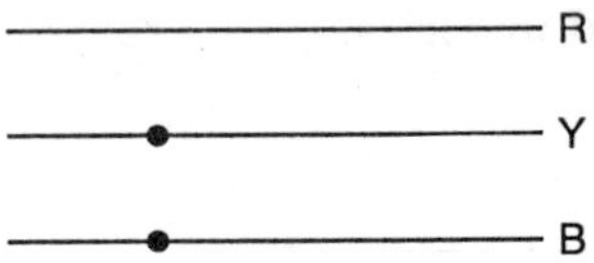

Fig. 21.35. L-L fault.

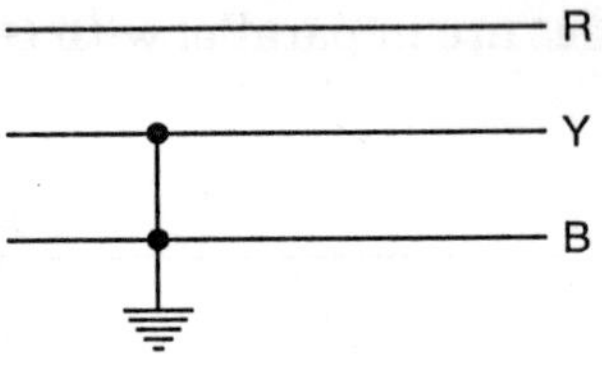

Fig. 21.36. L-L-G fault.

(*i*) Single line-to-ground fault (*L-G*).

(*ii*) Line-to-line fault (*L-L*).

(*iii*) Double line-to-ground fault (*L-L-G*).

The analysis of a 3-phase circuit in which phase voltages and currents are balanced (of equal magnitude in three phases and displaced by 120° from each other) and all circuit elements in each phase are balanced and symmetrical, is relatively simple since the treatment of a single phase leads directly to the three phase solution.

The analysis by Kirchhoff's laws is much more difficult, when the circuit is not symmetrical, as a result of unbalanced loads, unbalanced faults or short-circuits that are not symmetrical in the three phases. Such cases, however, may be solved without difficulty by the *method of symmetrical components.*

21.4.2. Symmetrical Components

The "*method of symmetrical components*" for the solution of unbalanced 3-phase A.C. circuits was first proposed by Dr. C.L. Fortescue in 1918.

According to "*Fortescue theorem*", *any unbalanced three phase system of currents, voltages or other sinusoidal quantities can be resolved into three balanced systems of vectors which are called* **"symmetrical components".** The calculation of currents and voltages in these symmetrical systems is a simple process, and the superposition or vector addition of these currents is easily carried out to obtain the actual current in the original unsymmetrical system.

In the Fig. 21.37 is shown a set of three unbalanced current vectors I_R, I_Y, I_B having phase sequence $R \to Y \to B$. These can be regarded as made up of the following *"symmetrical components"* :

(*i*) A set of three vectors, $\bar{I}_{R1}$, $\bar{I}_{Y1}$, $\bar{I}_{B1}$ equal in magnitude, displaced from each other by *120°* in phase and having the same phase sequence ($R \to Y \to B$) as the original vectors ; forming a 3-phase balanced system of **positive sequence** (Fig. 21.38).

(*ii*) A set of three vectors ($\bar{I}_{R2}$, $\bar{I}_{X2}$, $\bar{I}_{B2}$) equal in magnitude, displaced from each other by 120° in phase and having the phase sequence opposite ($R \to B \to Y$) to that of the original vectors ; forming a 3-phase balanced system of **negative sequence** (Fig. 21.39).

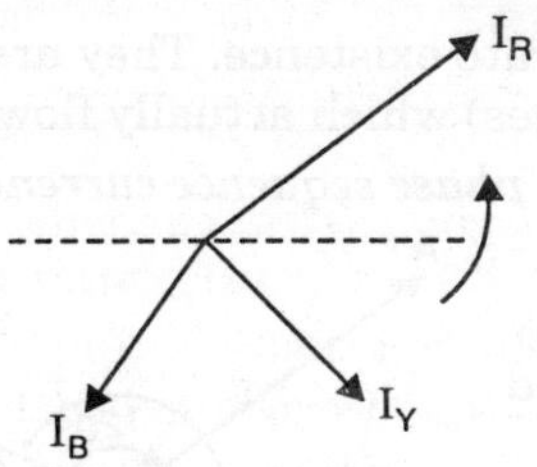

Fig. 21.37. Unbalanced vectors.

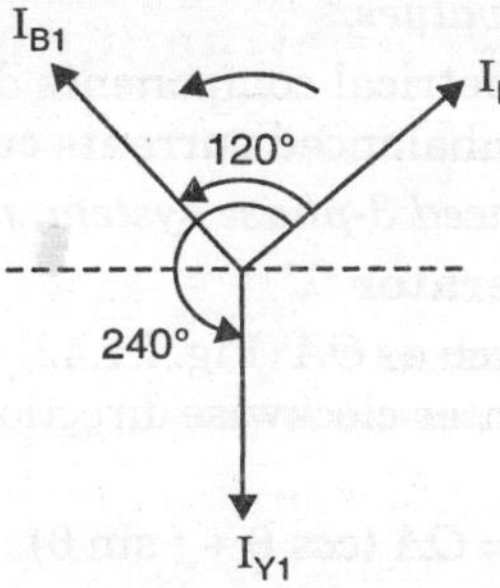

Fig. 21.38. Positive sequence components of the vectors.

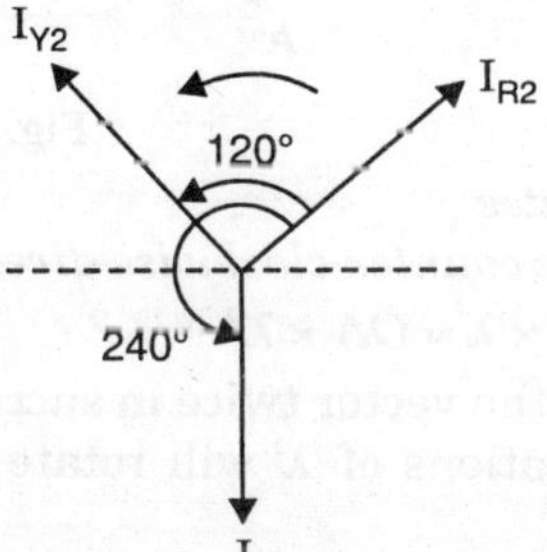

Fig. 21.39. Negative sequence components of the vectors.

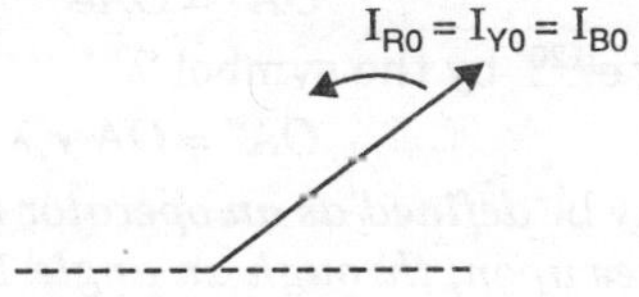

Fig. 21.40. Zero sequence components of the vectors.

(*iii*) A set of three vectors, ($\bar{I}_{R0}$, $\bar{I}_{Y0}$, $\bar{I}_{B0}$) equal in magnitude with *zero* phase displacement from each other (Fig. 21.40). These three co-phasal vectors form a uniphase system and are known as **zero sequence** components.

The current in any phase is equal to the vector sum of positive, negative and zero phase sequence currents in that phase (star-connected system being considered) as shown in Fig. 21.41.

$$\left.\begin{aligned}\bar{I}_R &= \bar{I}_{R1} + \bar{I}_{R2} + \bar{I}_{R0}\\ \bar{I}_Y &= \bar{I}_{Y1} + \bar{I}_{Y2} + \bar{I}_{Y0}\\ \bar{I}_B &= \bar{I}_{B1} + \bar{I}_{B2} + \bar{I}_{B0}\end{aligned}\right\} \qquad ...(21.11)$$

I_{R1}, I_{R2}, I_{R0} ...... are known as the symmetrical components or more precisely the *positive sequence component, negative sequence component* and the *zero sequence component* respectively of I_R ....

The *following points are worth noting :*

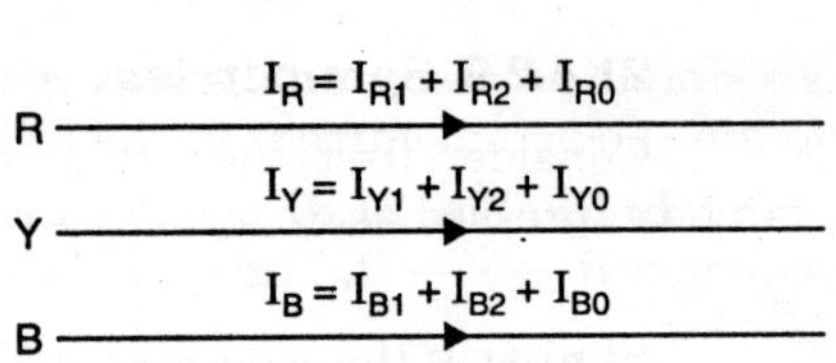

Fig. 21.41

(*i*) The positive sequence currents ($\bar{I}_{R1}, \bar{I}_{Y1}, \bar{I}_{B1}$), negative phase sequence currents ($\bar{I}_{R2}, \bar{I}_{Y2}, \bar{I}_{B2}$) and zero phase sequence currents ($\bar{I}_{R0}, \bar{I}_{Y0}, \bar{I}_{B0}$) separately form balanced system of currents. Hence, they are called symmetrical components of the unbalanced system.

(*ii*) The theory of symmetrical components applies equally to 3-phase *currents and voltages, both phase and line values.*

(*iii*) The symmetrical components do not have separate existence. They are only mathematical components of unbalanced currents currents (or voltages) which actually flow in the system.

(*iv*) In *a balanced 3-phase system, negative and zero phase sequence currents are zero.*

21.4.2.1. Operator 'λ'

Any vector such as *OA* (Fig. 21.42) when displaced by an angle θ in the counter-clockwise direction can be expressed as,

$$OA' = OA\ (\cos\theta + j\sin\theta)$$
$$= (OA)\ e^{j\theta} \text{ (OA is the reference vector)}$$

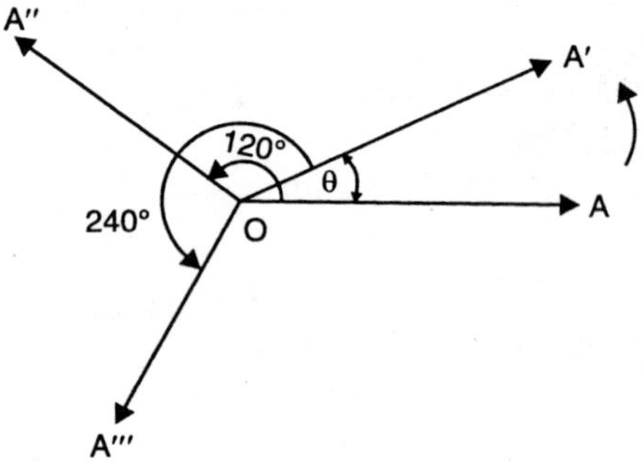

Fig. 21.42

If the angle of displacement is 120° so that vector *OA* occupies the position *OA″*, then

$$OA'' = OAe^{j120°}$$

Designating $e^{j120°}$ by the symbol 'λ'

$$OA'' = OA \times \lambda$$

Thus "λ" may be *defined as an operator which rotates the vector, it operates upon, through an angle 120°, in the counter-clockwise direction.*

Applying this condition, $OA''' = OA'' \times \lambda = OA \times \lambda \times \lambda = OA \times \lambda^2$.

It may be seen that if the operator λ is applied to the vector twice in succession, the vector is rotated through 240°. Similarly three successive applications of 'λ' will rotate the vector through 360°.

Properties of Operator **λ :**

$$\lambda = e^{j120°} = (\cos 120° + j\sin 120°) = -0.5 + j0.866$$
$$\lambda^2 = e^{j240°} = (\cos 240° + j\sin 240°) = -0.5 - j.866$$
$$\lambda^3 = e^{j360°} = (\cos 360° + j\sin 360°) = 1$$
$$\lambda^4 = \lambda^3 \times \lambda = \lambda = -0.5 + j0.866$$
$$\lambda^5 = \lambda^3 \times \lambda^2 = \lambda^2 = -0.5 - j0.866$$
$$1 + \lambda + \lambda^2 = 1 + (-0.5 + j0.866) + (-0.5 - j0.866) = (0 + j0)$$
$$\lambda - \lambda^2 = (-0.5 + j0.866) - (-0.5 - j0.866)$$
$$= j\sqrt{3} = \sqrt{3}\angle 90° = 0 + j1.732$$
$$\lambda^2 - \lambda = -j\sqrt{3} = \sqrt{3}\angle -90° = 0 - j1.732$$
$$\lambda + \lambda^2 = 1\angle 180° = -1 - j0$$
$$1 + \lambda = 1\angle 120° = 0.5 + j0.866 = -\lambda^2$$
$$1 - \lambda = 1 - (-0.5 + j0.866) = 1.5 - j0.866 = \sqrt{3}\angle -30°.$$

21.4.2.2. Symmetrical components in terms of currents and voltages

Consider, first, the symmetrical components of the *positive phase sequence* (Fig. 21.38). These can be expressed as :

$$I_{R1} = I_{R1}\,; \qquad I_{Y1} = \lambda^2 I_{R1}\,; \qquad I_{B1} = \lambda I_{R1}$$

Similarly the symmetrical components of the *negative phase sequence* (Fig. 21.39) can be expressed as :

$$I_{R2} = I_{R2}\,; \qquad I_{Y2} = \lambda I_{R2}\,; \qquad I_{B2} = \lambda^2 I_{R2}$$

And the *zero phase sequence* components (Fig. 21.40) can be expressed as

$$I_{R0} = I_{R0}\,; \qquad I_{Y0} = I_{R0}\,; \qquad I_{B0} = I_{R0}$$

Eqn. (21.11) can, therefore, be written as :

$$\left.\begin{aligned} \bar{I}_R &= \bar{I}_{R1} + \bar{I}_{R2} + \bar{I}_{R0} && ...(i)\\ \bar{I}_Y &= \lambda^2 \bar{I}_{R1} + \lambda \bar{I}_{R2} + \bar{I}_{R0} && ...(ii)\\ \bar{I}_B &= \lambda \bar{I}_{R1} + \lambda^2 \bar{I}_{R2} + \bar{I}_{R0} && ...(iii) \end{aligned}\right\} \qquad ...(21.12)$$

In eqn. (21.12) the *first term* on the right-hand side constitute the *positive sequence system,* the *second term* constitute the *negative sequence system,* and the *third term* constitute the *zero sequence system.*

In Matrix form eqn. (21.12) can be written as :

$$\begin{bmatrix} \bar{I}_R \\ \bar{I}_Y \\ \bar{I}_B \end{bmatrix} = \begin{bmatrix} 1 & 1 & 1 \\ 1 & \lambda^2 & \lambda \\ 1 & \lambda & \lambda^2 \end{bmatrix} \begin{bmatrix} \bar{I}_{R0} \\ \bar{I}_{R1} \\ \bar{I}_{R2} \end{bmatrix}$$

$$\begin{bmatrix} \bar{I}_R \\ \bar{I}_Y \\ \bar{I}_B \end{bmatrix} = [Z] \begin{bmatrix} \bar{I}_{R0} \\ \bar{I}_{R1} \\ \bar{I}_{R2} \end{bmatrix}$$

where
$$[Z] = \begin{bmatrix} 1 & 1 & 1 \\ 1 & \lambda^2 & \lambda \\ 1 & \lambda & \lambda^2 \end{bmatrix}$$

$$\therefore \qquad \begin{bmatrix} \bar{I}_{R0} \\ \bar{I}_{R1} \\ \bar{I}_{R2} \end{bmatrix} = [Z]^{-1} \begin{bmatrix} \bar{I}_R \\ \bar{I}_Y \\ \bar{I}_B \end{bmatrix}$$

where
$$[Z]^{-1} = \frac{1}{3}\begin{bmatrix} 1 & 1 & 1 \\ 1 & \lambda & \lambda^2 \\ 1 & \lambda^2 & \lambda \end{bmatrix}$$

Determination of sequence components :

Let us consider eqn. (21.12)

Adding (*i*), (*ii*) and (*iii*), we have

$$\bar{I}_R + \bar{I}_Y + \bar{I}_B = (1 + \lambda^2 + \lambda)\bar{I}_{R1} + (1 + \lambda + \lambda^2)I_{R2} + 3I_{R0} = 3I_{R0}$$

$$\therefore \qquad I_{R0} = \frac{1}{3}(\bar{I}_R + \bar{I}_Y + \bar{I}_B) \qquad (\because\ 1 + \lambda + \lambda^2 = 0)$$

Multiplying (*ii*) by λ and (*iii*) by λ^2 and adding these eqns. to (*i*), we get

$$\bar{I}_R + \lambda\bar{I}_Y + \lambda^2\bar{I}_B = \bar{I}_{R1}(1+\lambda^3+\lambda^3)$$

$$+\ \bar{I}_{R2}(1+\lambda^2+\lambda^4) + I_{R0}(1+\lambda+\lambda^2) = 3\bar{I}_{R1}$$

$$\therefore \quad \bar{I}_{R1} = \frac{1}{3}(\bar{I}_R + \lambda\bar{I}_Y + \lambda^2\bar{I}_B)$$

Again, multiplying (*ii*) by λ^2 and (*iii*) by λ and then adding these eqns. to (*i*), we get

$$\bar{I}_R + \lambda^2\bar{I}_Y + \lambda\bar{I}_B = \bar{I}_{R1}(1+\lambda^4+\lambda^2)$$

$$+\ \bar{I}_{R2}(1+\lambda^3+\lambda^3) + I_{R0}(1+\lambda^2+\lambda) = 3\bar{I}_{R2}$$

$$\therefore \quad \bar{I}_{R2} = \frac{1}{3}(\bar{I}_R + \lambda^2\bar{I}_Y + \lambda\bar{I}_B)$$

Thus for any unbalanced system of voltages,

$$\bar{I}_{R1} = \frac{1}{3}(\bar{I}_R + \lambda\bar{I}_Y + \lambda^2\bar{I}_B) \quad ...(21.13)$$

$$\bar{I}_{R2} = \frac{1}{3}(\bar{I}_R + \lambda^2\bar{I}_Y + \lambda\bar{I}_B) \quad ...(21.14)$$

$$\bar{I}_{R0} = \frac{1}{3}(\bar{I}_R + \bar{I}_Y + \bar{I}_B) \quad ...(21.15)$$

Similarly, for any unbalanced system of voltages,

$$E_{R1} = \frac{1}{3}(\bar{E}_R + \lambda\bar{E}_Y + \lambda^2\bar{E}_B) \quad ...(21.16)$$

$$E_{R2} = \frac{1}{3}(\bar{E}_R + \lambda^2\bar{E}_Y + \lambda\bar{E}_B) \quad ...(21.17)$$

$$E_{R0} = \frac{1}{3}(\bar{E}_R + \bar{E}_Y + \bar{E}_B) \quad ...(21.18)$$

Absence of zero sequence components :

- From eqn. (21.15) it follows that there will no zero-sequence component of current when $\bar{I}_R + \bar{I}_Y + \bar{I}_B = 0$. From eqn. (21.16) it follows that there will be no zero-sequence component of voltage when $\bar{E}_R + \bar{E}_Y + \bar{E}_B = 0$.
  - Since the *phasor sum* of the three line currents in 3-phase 3-wire system is zero, there will be no zero-sequence component in the line currents of a 3-phase 3-wire system.
  - Similarly, since the *phasor sum* of the line-to-line voltage in a 3-phase system is always zero, there will be no zero-sequence component in the line voltages of a 3-phase system.
- *Zero-sequence components may, however, be present in the phase voltages of a star circuit or phase currents of a delta circuit.*
  - In a *star-circuit* the zero-sequence components of the phase voltages are all in phase. The line voltage being the vector difference of two phase voltages, the zero sequence components cancel out when the line voltage is determined.
  - In a *delta-circuit* the zero-sequence components of the phase currents are all in phase and consequently form a local circulating current inside the mesh. There can, therefore, be no zero-sequence components in the line currents.

The following summarised *facts* about sequence currents are worth noting :

(*i*) A balanced 3-phase system consists of positive sequence components only ; the *negative and zero-sequence components being zero.*

(*ii*) In a 3-phase system, the *presence of negative or zero sequence currents introduces unsymmetry* and is indicative of an abnormal condition of the circuit in which these components are found.

(*iii*) In an unbalanced 3-phase system the vector sum of positive and negative sequence of currents is zero. The resultant solely consists of three zero sequence currents *i.e.,*

Vector sum of all sequence currents in 3-phase unbalanced system = $\bar{I}_{R0} + \bar{I}_{Y0} + \bar{I}_{B0}$

(*iv*) In a 3-phase 4-wire unbalanced system,

$$\text{Zero sequence current} = \frac{1}{3}\ \text{(Current in neutral wire)}$$

— In the absence of path through the neutral of a 3-phase system, the neutral current is zero and the line currents contain no zero-sequence components.

— A delta-connected load provides no path to the neutral and the *line currents flowing to delta-connected load can contain no zero-sequence components.*

(*v*) In a 3-phase unbalanced system, the magnitude of negative sequence components *cannot exceed* that of the positive sequence components. If the negative sequence components were the greater, the phase sequence of the resultant system would be *reversed.*

(*vi*) The current of a *single phase load* drawn from a 3-phase system *comprises equal positive, negative and zero sequence components.*

21.4.2.3. Zero sequence components of current and voltage

The following are some typical 3-phase connections with reference to zero sequence components of current and voltage.

1. Four-wire star connection
2. Three-wire star connection
3. Three-wire delta connection.

1. *Four-wire star connection.* Because of the presence of fourth wire, the *zero sequence current may flow. The neutral wire carries only the zero sequence which is the sum of the zero sequence currents in the three lines.*

2. *Three-wire star connection :*

- In this type of connection since there is no fourth or return wire, *zero-sequence components cannot flow.*
- The absence of zero-sequence currents may be explained by considering that the impedance offered to these currents is infinite and this impedance is situated between the star points of the generator and the load. If the two star points were joined by a neutral, only zero sequence currents will flow through it so that only zero sequence voltage can exist between the load and generator star points. Obviously, no zero sequence component of voltage appears across the load.

3. *Three-wire delta connection :*

- The zero sequence components of currents, due to the absence of fourth wire, cannot be fed into the delta-connected load. However, though line currents have to sum upto zero (whereas phase currents need not do so) it is possible to have a zero sequence current circulating in the delta-connected load.

- Similarly, individual phase voltages will generally possess zero sequence components though components are absent in the line-to-line voltage. (However, under balanced conditions, the phase voltages will possess no sequence components).

21.4.2.4. Unbalanced star load supplied from unbalanced three-phase three-wire system

In this type of arrangement, line voltages and load currents will consist of only positive and negative sequence components, but no zero sequence component. But the load voltages will consist of positive, negative and zero sequence components.

Let, $\overline{V}_{RY}, \overline{V}_{YB}, \overline{V}_{BR}$ = Line voltages,

$\overline{I}_R, \overline{I}_Y, \overline{I}_B$ = Line (and load) currents,

$\overline{V}_{RN}, \overline{V}_{YN}, \overline{V}_{BN}$ = Phase voltages, and

$\overline{Z}_R, \overline{Z}_Y, \overline{Z}_B$ = Load impedances (their values being the same for currents of any sequence)

Then,
$$\overline{V}_{RN} = \overline{I}_R\overline{Z}_R \ ; \ \overline{V}_{YN} = \overline{I}_Y\overline{Z}_Y \ ; \ \overline{V}_{BN} = \overline{I}_B\overline{Z}_B$$

If $\overline{V}_{RN1}$, $\overline{V}_{RN2}$ and $\overline{V}_0$ are the symmetrical components of V_{RN}, then we have

$$\overline{V}_0 = \frac{1}{3}(\overline{V}_{RN} + \overline{V}_{YN} + \overline{V}_{BN}) = \frac{1}{3}(\overline{I}_R\overline{Z}_R + \overline{I}_Y\overline{Z}_Y + \overline{I}_B\overline{Z}_B)$$
$$= \frac{1}{3}[\overline{Z}_R(\overline{I}_{R1} + \overline{I}_{R2}) + \overline{Z}_Y(\overline{I}_{Y1} + \overline{I}_{Y2}) + \overline{Z}_B(\overline{I}_{B1} + \overline{I}_{B2})]$$
$$= \frac{1}{3}[\overline{Z}_R(\overline{I}_{R1} + \overline{I}_{R2}) + \overline{Z}_Y(\lambda^2\overline{I}_{R1} + \lambda\overline{I}_{R2}) + \overline{Z}_B(\lambda\overline{I}_{R1} + \lambda^2\overline{I}_{R2})]$$
$$= \overline{I}_{R1} \times \frac{1}{3}(\overline{Z}_R + \lambda^2\overline{Z}_Y + \lambda\overline{Z}_B) + \overline{I}_{R2} \times \frac{1}{3}(\overline{Z}_R + \lambda\overline{Z}_Y + \lambda^2\overline{Z}_B) \quad ...(21.19)$$

$$\overline{V}_{RN1} = \frac{1}{3}(V_{RN} + \lambda V_{YN} + \lambda^2 V_{BN})$$
$$= \frac{1}{3}(\overline{I}_R\overline{Z}_R + \lambda\overline{I}_Y\overline{Z}_Y + \lambda^2\overline{I}_B\overline{Z}_B)$$
$$= \frac{1}{3}[\overline{Z}_R(\overline{I}_{R1} + \overline{I}_{R2}) + \lambda\overline{Z}_Y(\overline{I}_{Y1} + \overline{I}_{Y2}) + \lambda^2\overline{Z}_B(\overline{I}_{B1} + \overline{I}_{B2})]$$
$$= \frac{1}{3}[\overline{Z}_R(\overline{I}_{R1} + \overline{I}_{R2}) + \lambda\overline{Z}_Y(\lambda^2\overline{I}_{R1} + \lambda\overline{I}_{R2}) + \lambda^2\overline{Z}_B(\lambda\overline{I}_{R1} + \lambda^2\overline{I}_R)]$$
$$= \frac{1}{3}[\overline{I}_{R1}(\overline{Z}_{R1} + \lambda^3\overline{Z}_Y + \lambda^3\overline{Z}_B) + \overline{I}_{R2}(\overline{Z}_R + \lambda^2\overline{Z}_Y + \lambda^4\overline{Z}_B)]$$
$$= \overline{I}_{R1} \times \frac{1}{3}(\overline{Z}_{R1} + \overline{Z}_Y + \overline{Z}_B) + \overline{I}_{R2} \times \frac{1}{3}(\overline{Z}_R + \lambda^2\overline{Z}_Y + \lambda\overline{Z}_B)$$

$[\because \lambda^4 = \lambda^3 \times \lambda = 1 \times \lambda = \lambda]$

$$= \overline{I}_{R1}\overline{Z}_0 + \overline{I}_{R2}\overline{Z}_{R2} \quad ...(21.20)$$

$$\overline{V}_{RN2} = \overline{I}_{R2}\overline{Z}_0 + \overline{I}_{R1}\overline{Z}_{R1} \quad ...(21.21)$$

Similarly,
$$\overline{V}_{YN1} = \overline{I}_{Y1}\overline{Z}_0 + \overline{I}_{Y2}\overline{Z}_{Y2} = \lambda^2(\overline{I}_{R1}\overline{Z}_0 + \overline{I}_{R2}\overline{Z}_{R2}) = \lambda^2\overline{V}_{RN1} \quad ...(21.22)$$
$$\overline{V}_{BN1} = \overline{I}_{B1}\overline{Z}_0 + \overline{I}_{B2}\overline{Z}_{B2} = \lambda(\overline{I}_{R1}\overline{Z}_0 + \overline{I}_{R2}\overline{Z}_{R2}) = \lambda\overline{V}_{RN1} \quad ...(21.23)$$

$$\bar{V}_{YN2} = \bar{I}_{Y2}\bar{Z}_0 + \bar{I}_{Y1}\bar{Z}_{Y1} = \lambda(\bar{I}_{R2}\bar{Z}_0 + \bar{I}_{R1}\bar{Z}_{R1}) = \lambda\bar{V}_{RN2} \quad ...(21.24)$$

$$\bar{V}_{BN2} = \bar{I}_{B2}\bar{Z}_0 + \bar{I}_{B1}\bar{Z}_{B1} = \lambda^2(\bar{I}_{R2}\bar{Z}_0 + \bar{I}_{R1}\bar{Z}_{R1}) = \lambda^2\bar{V}_{RN2} \quad ...(21.25)$$

Now, V_{RN1} and V_{RN2} may be determined from the relation between the line and phase voltages as given below :

$$\begin{aligned}
\bar{V}_{RY} &= \bar{V}_{RN} + \bar{V}_{NY} = \bar{V}_{RN} - \bar{V}_{YN} = \bar{V}_0 + \bar{V}_{RN1} + \bar{V}_{RN2} - (\bar{V}_0 + \bar{V}_{YN1} + \bar{V}_{YN2})\\
&= \bar{V}_0 + \bar{V}_{RN1} + \bar{V}_{RN2} - (\bar{V}_0 + \lambda^2\bar{V}_{RN1} + \lambda\bar{V}_{RN2})\\
&= \bar{V}_{RN1}(1-\lambda^2) + \bar{V}_{RN2}(1-\lambda)\\
&= \bar{V}_{RN1}\,[1-(-0.5-j0.866)] + \bar{V}_{RN2}\,[1-(-0.5+j0.866)]\\
&= \bar{V}_{RN1}\,(1.5+j0.866) + \bar{V}_{RN2}(1.5-j0.866)\\
&= \frac{1}{2}\sqrt{3}\,[\bar{V}_{RN1}\,(\sqrt{3}+j1) + \bar{V}_{RN2}\,(\sqrt{3}-j1)] \quad ...(21.26)
\end{aligned}$$

$$\begin{aligned}
\bar{V}_{YB} &= \bar{V}_{YN} + \bar{V}_{NB} = \bar{V}_{YN} - \bar{V}_{BN} = V_0 + \bar{V}_{YN1} + \bar{V}_{YN2} - (\bar{V}_0 + \bar{V}_{BN1} + \bar{V}_{BN2})\\
&= \bar{V}_0 + \lambda^2\bar{V}_{RN1} + \lambda\bar{V}_{RN2} - (\bar{V}_0 + \lambda\bar{V}_{RN1} + \lambda^2\bar{V}_{RN2})\\
&= V_{RN1}\,(\lambda^2-\lambda) + V_{RN2}\,(\lambda-\lambda^2) = j\sqrt{3}\,\bar{V}_{RN1} + j\sqrt{3}\,V_{RN2} \quad ...(21.27)
\end{aligned}$$

or, $$\bar{V}_{RN2} = \frac{1}{j\sqrt{3}}(\bar{V}_{YB} + j\sqrt{3}\,\bar{V}_{RN1})$$

or, $$\bar{V}_{RN2} = \bar{V}_{RN1} - j\bar{V}_{YB}\,/\,\sqrt{3}$$

Substituting this value of $\bar{V}_{RN2}$ in eqn. (21.26), we have

$$\begin{aligned}
\bar{V}_{RY} &= \frac{1}{2}\sqrt{3}\left[\{\bar{V}_{RN1}\,(\sqrt{3}+j1)\} + \left\{\bar{V}_{RN1} - \frac{j\bar{V}_{YB}}{\sqrt{3}}\right\}(\sqrt{3}-j1)\right]\\
&= \frac{\sqrt{3}}{2}\bar{V}_{RN1}\,(\sqrt{3}+j1) + \left(\bar{V}_{RN1} - \frac{j\bar{V}_{YB}}{\sqrt{3}}\right)\left(\frac{\sqrt{3}(\sqrt{3}-j1)}{2}\right)\\
&= \frac{\sqrt{3}}{2}V_{RN1}\,(\sqrt{3}+j1) + \bar{V}_{RN1}\left(\frac{3-j\sqrt{3}}{2}\right) - \bar{V}_{YB}\left(\frac{1+j\sqrt{3}}{2}\right)
\end{aligned}$$

or, $$\bar{V}_{RY} + \bar{V}_{YB}\left(\frac{1+j\sqrt{3}}{2}\right) = \bar{V}_{RN1}\left[\frac{3}{2} + j\frac{\sqrt{3}}{2} + \frac{3}{2} - j\frac{\sqrt{3}}{2}\right] = 3\bar{V}_{RN1}$$

$\therefore$ $$\bar{V}_{RN1} = \frac{1}{3}\left[\bar{V}_{RY} + \frac{1}{2}\bar{V}_{YB}\,(1-j\sqrt{3})\right] \quad ...(21.28)$$

$$\bar{V}_{RN2} = \frac{1}{3}\left[\bar{V}_{RY} + \frac{1}{2}\bar{V}_{YB}\,(1-j\sqrt{3})\right] \quad ...(21.29)$$

Having known $\bar{V}_{RN1}$ and $\bar{V}_{RN2}$, the currents $\bar{I}_{R1}$ and $\bar{I}_{R2}$ can be determined from the following equations :

$$\bar{I}_{R1} = \frac{\bar{V}_{RN1}\bar{Z}_0 - \bar{V}_{RN2}\bar{Z}_{R2}}{\bar{Z}_0^2 - \bar{Z}_{R1}\bar{Z}_{R2}} \quad ...(21.30)$$

$$\bar{I}_{R2} = \frac{\bar{V}_{RN2}\bar{Z}_0 - \bar{V}_{RN1}\bar{Z}_{R1}}{\bar{Z}_0^2 - \bar{Z}_{R1}\bar{Z}_{R2}} \quad ...(21.31)$$

Alternatively, when $\bar{I}_{R1}$ becomes known, $\bar{I}_{R2}$ may be found from the relation,

$$\bar{I}_{R2} = \frac{\bar{V}_{RN2} - \bar{I}_{R1}\bar{Z}_{R1}}{\bar{Z}_0} \quad ...(21.32)$$

The symmetrical components of the currents in other phases can be calculated from $\bar{I}_{R1}$ and $\bar{I}_{R2}$ by using the relations given in Art. 21.4.2.2.

The *phase voltages* may be calculated by any of the following *two methods* :

(*i*) Directly by calculating the products $I_R Z_R$, $I_Y Z_Y$ and $I_B Z_B$.

(*ii*) By first calculating the zero sequence component $\bar{V}_0$ of the phase or load voltages and then adding this to the approximate positive and negative sequence components, for example : $\bar{V}_{RN} = \bar{V}_0 + \bar{V}_{RN1} + \bar{V}_{RN2}$.

21.4.2.5. Unbalanced star load supplied from balanced three-phase three-wire system

In this case (a special case of the general case considered in Art. 21.4.2.5 above), the *symmetrical components of the load voltages consist only of positive and zero sequence components.* This fact may be verified by substituting the value of $\bar{V}_{YB}$ in eqn. (21.29).

Now in a balanced or symmetrical system of positive-phase sequence,

$$\bar{V}_{YB} = \bar{V}_{RY}\left(-\frac{1}{2} - j\frac{\sqrt{3}}{2}\right)$$

Substituting this value in eqn. (21.28), we have

$$\bar{V}_{RN2} = \frac{1}{3}\left[\bar{V}_{RY} + \frac{1}{2}\bar{V}_{YB}\,(1 - j\sqrt{3})\right]$$

$$= \frac{1}{3}\left[\bar{V}_{RY} + \frac{1}{2}\bar{V}_{RY}\left(-\frac{1}{2} - j\frac{\sqrt{3}}{2}\right)(1 - j\sqrt{3})\right] = \frac{1}{3}(\bar{V}_{RY} - \bar{V}_{RY}) = 0$$

Hence, substituting this value of $\bar{V}_{RN2}$ in eqns. (21.28) and (21.30), we get

$$\bar{V}_{RN1} = j\bar{V}_{YB}\sqrt{3} = \frac{1}{2}\bar{V}_{RY} - j\frac{1}{2}\bar{V}_{RY} / \sqrt{3} \quad ...(21.33)$$

$$\bar{I}_{R1} = \bar{V}_{RN1}Z_0 / (Z_0{}^2 - \bar{Z}_{R1}\bar{Z}_{R2}) = \frac{\bar{Z}_R + \bar{Z}_Y + \bar{Z}_B}{\bar{Z}_R\bar{Z}_Y + \bar{Z}_Y\bar{Z}_B + \bar{Z}_B\bar{Z}_R} \quad ...(21.34)$$

$$\bar{I}_{R2} = -\bar{I}_{R1} \,.\, \bar{Z}_{R1} / \bar{Z}_0 = -\bar{I}_{R1}\frac{\bar{Z}_R + \lambda\bar{Z}_Y + \lambda^2\bar{Z}_B}{\bar{Z}_R + \bar{Z}_Y + \bar{Z}_B} \quad ...(21.35)$$

Example 21.21. *In a 3-phase, 4-wire system, the currents in R, Y and B lines under abnormal conditions of loading are as under :*

$$\bar{I}_R = 200 \angle 30° \text{ A} ; \; \bar{I}_Y = 100 \angle 300° \text{ A} ; \; \bar{I}_B = 60 \angle 180° \text{ A}$$

Calculate the zero, positive and negative sequence currents in the R line and return current in the neutral wire.

Solution. Let $\bar{I}_{R0}$, $\bar{I}_{R1}$, $\bar{I}_{R2}$ be the zero, positive and negative sequence currents respectively of the line current in red line.

Then, $\mathbf{\bar{I}_{R0}} = \frac{1}{3}(\bar{I}_R + \bar{I}_Y + \bar{I}_B)$ ...[Eqn. (21.15)]

$= \frac{1}{3}[200 \angle 30° + 100 \angle 300° + 60 \angle 180°]$

$= \frac{1}{3}[(173.2 + j100) + (50 - j86.6) + (-60 + j0)]$

$= \frac{1}{3}(163.2 + j13.4) = 54.4 + j4.47 =$ **54.58 ∠4.7° A. (Ans.)**

$\mathbf{\bar{I}_{R1}} = \frac{1}{3}(\bar{I}_R + \lambda\bar{I}_Y + \lambda^2\bar{I}_B)$ ...[Eqn. (21.22)]

$= \frac{1}{3}(200 \angle 30° + 1 \angle 120° \times 100 \angle 300° + 1 \angle 240° \times 60 \angle 180°)$

$= \frac{1}{3}[(173.2 + j100) + (50 + j86.6) + (30 + j51.96)]$

$= \frac{1}{3}(253.2 + j238.56) = 84.4 + j79.52 =$ **115.96 ∠43.3° A. (Ans.)**

$\mathbf{\bar{I}_{R2}} = \frac{1}{3}(\bar{I}_R + \lambda^2\bar{I}_Y + \lambda\bar{I}_B)$ ...[Eqn. (21.14)]

$= \frac{1}{3}[200 \angle 30° + 1 \angle 240° \times 100 \angle 300° + 1 \angle 120° \times 60 \angle 180°]$

$= \frac{1}{3}[(173.2 + j100) + (-100 + j0) + (30 - j51.96)]$

$= \frac{1}{3}(103.2 + j48.04) = 34.4 + j16.01 = 37.94 \angle 24.96°$

Current in central wire $= \bar{I}_R + \bar{I}_Y + \bar{I}_B = 163.2 + j13.4 =$ **163.75 ∠4.69° A. (Ans.)**

Example 21.22. *The line currents in a 3-ϕ supply to an unbalanced load are represented by $\bar{I}_R = (6 + j12)$ A, $\bar{I}_Y = (8 - j1)$ A and $\bar{I}_B = (-2 - j3)$ A. The phase sequence is RYB. Calculate the positive, negative and zero sequence currents.*

Solution. $\bar{I}_R = 6 + j12 = 13.42 \angle 63.43°$ A

$\bar{I}_Y = 8 - j1 \text{ A} = 8.06 \angle -7.125°$ A

$\bar{I}_B = -2 - j3 \text{ A} = 3.6 \angle -123.69°$ A

Positive sequence components are :

$\bar{I}_{R1} = \frac{1}{3}(\bar{I}_R + \lambda\bar{I}_Y + \lambda^2\bar{I}_B)$

$= \frac{1}{3}(13.42 \angle 63.43° + 1 \angle 120° \times 8.06 \angle -7.125°$

$+ 1 \angle 240° \times 3.6 \angle -123.69°)$

$= \frac{1}{3}[(6.0 + j12) + (-3.133 + j7.426) + (-1.6 + j3.23)]$

$= \frac{1}{3}(1.267 + j22.656) = 0.42 + j7.55 =$ **7.56 ∠86.8° A. (Ans.)**

$$\bar{I}_{Y1} = 7.56 \angle(86.8° - 120°) = \mathbf{7.56 \angle -33.2° \text{ A.}} \quad \textbf{(Ans.)}$$

$$\bar{I}_{B1} = 7.56 \angle(86.8° + 120°) = \mathbf{7.56 \angle 206.8° \text{ A.}} \quad \textbf{(Ans.)}$$

Negative sequence components are :

$$\bar{I}_{R2} = \frac{1}{3}(\bar{I}_R + \lambda^2 \bar{I}_Y + \lambda \bar{I}_B)$$

$$= \frac{1}{3}(13.42 \angle 63.43° + 1 \angle 240° \times 8.06 \angle -7.125° + 1 \angle 120° \times 3.6 \angle -123.69°)$$

$$= \frac{1}{3}[(6.0 + j12) + (-4.86 - j6.43) + (3.59 - j0.23)]$$

$$= \frac{1}{3}(4.73 + j5.34) = 1.577 + j1.78 = \mathbf{2.38 \angle 48.5° \text{ A.}} \quad \textbf{(Ans.)}$$

$$\bar{I}_{Y2} = 2.38 \angle(48.5° + 120°) = \mathbf{2.38 \angle 168.5° \text{ A.}} \quad \textbf{(Ans.)}$$

$$\bar{I}_{B2} = 2.38 \angle(48.5° + 240°) = \mathbf{2.38 \angle 288.5° \text{ A.}} \quad \textbf{(Ans.)}$$

Zero sequence components are :

$$I_{R0} = I_{Y0} = I_{B0} = \frac{1}{3}(\bar{I}_R + \bar{I}_Y + \bar{I}_B)$$

$$= \frac{1}{3}[(6 + j12) + (8 - j1) + (-2 - j3)]$$

$$= \frac{1}{3}(12 + j8) = 4 + j2.67] = \mathbf{4.81 \angle 33.72° \text{ A.}} \quad \textbf{(Ans.)}$$

As a check it may be found that,

$$\bar{I}_R = \bar{I}_{R1} + \bar{I}_{R2} + \bar{I}_{R0} \; ; \; \bar{I}_Y = \bar{I}_{Y1} + \bar{I}_{Y2} + \bar{I}_{Y0} \; ; \; \bar{I}_B = \bar{I}_{B1} + \bar{I}_{B2} + \bar{I}_{B0}.$$

Example 21.23. *The sequence voltages in the red phase are as under :*

$$\bar{E}_{R1} = (400 - j200) \text{ V} ; \; \bar{E}_{R2} = -200 \text{ V} ; \; \bar{E}_{R0} = 200 \text{ V}$$

Find the phase voltages $\bar{E}_R, \bar{E}_Y, \bar{E}_B$.

Solution. In polar form, we have

$$\bar{E}_{R1} = 447.2 \angle -26.56° \; ; \; \bar{E}_{R2} = 200 \angle 180° \text{ V} \; ; \; \bar{E}_{R0} = 200 \angle 0°$$

$$\bar{E}_R = \bar{E}_{R1} + \bar{E}_{R2} + \bar{E}_{R0}$$

$$= (400 - j200) + (-200) + 200 = 400 - j200 = 447.2 \angle -26.56° \text{ V}$$

$$\bar{E}_Y = \lambda^2 \bar{E}_{R1} + \lambda \bar{E}_{R2} + E_{R0}$$

$$= 1 \angle 240° \times 447.2 \angle -26.56° + 1 \angle 120° \times 200 \angle 180° + 200 \angle 0°$$

$$= (-373.17 - j246.43) + (100 - j173.2) + (200 + j0)$$

$$= -73.17 - j419.63 = \mathbf{426 \angle -99.9° \text{ V.}} \quad \textbf{(Ans.)}$$

$$\bar{E}_B = \lambda E_{R1} + \lambda^2 E_{R2} + E_{R0}$$

$$= 1 \angle 120° \times 447.2 \angle -26.56° + 1 \angle 240° \times 200 \angle 180° + 200 \angle 0°$$

$$= (-26.83 + j446.4) + (100 + j173.2) + (200 + j0)$$

$$= 273.17 + j619.6 = \mathbf{677.14 \angle 66.2° \text{ V.}} \quad \textbf{(Ans.)}$$

Example 21.24. *A 3-phase, 4-wire system supplies loads which are unequally distributed in the three phases. An analysis of the current flowing in R, Y and B lines show that in R-line, positive phase sequence component is 100 ∠0° A and the negative phase sequence component is 50 ∠60° A. The total observed current flowing back to the supply in the neutral conductor is 150 ∠300° A. Calculate the currents in three lines.*

Solution. Positive phase sequence current in R-line,

$$\bar{I}_{R1} = 100 \angle 0° A \quad \text{...(Given)}$$

Negative phase sequence current in R-line,

$$\bar{I}_{R2} = 50 \angle 60° A \quad \text{....(Given)}$$

Zero phase sequence current in R-line,

$$\bar{I}_{R0} = \frac{1}{3} \times \text{Current in neutral wire}$$

$$= \frac{1}{3} \times 150 \angle 300° = 50 \angle 300° A$$

Current in the R-line,

$$\bar{\mathbf{I}}_{\mathbf{R}} = \bar{I}_{R1} + \bar{I}_{R2} + \bar{I}_{R0} \quad \text{...(Eqn. 21.12)}$$
$$= 100 \angle 0° + 50 \angle 60° + 50 \angle 300°$$
$$= (100 + j0) + (25 + j43.3) + (25 - j43.3) = \mathbf{150 \angle 0° A.} \quad \textbf{(Ans.)}$$

Current in Y-line,
$$\bar{\mathbf{I}}_{\mathbf{Y}} = \lambda^2 \bar{I}_{R1} + \lambda \bar{I}_{R2} + \bar{I}_{R0}$$
$$= 1 \angle 240° \times 100 \angle 0° + 1 \angle 120° \times 50 \angle 60° + 50 \angle 300°$$
$$= (-50 - j86.6) + (-50 + j0) + (25 - j43.3)$$
$$= -75 - j129.9 = \mathbf{150 \angle -120° A.} \quad \textbf{(Ans.)}$$

Current in B-line,
$$\bar{\mathbf{I}}_{\mathbf{B}} = \lambda \bar{I}_{R1} + \lambda^2 \bar{I}_{R2} + \bar{I}_{R0}$$
$$= 1 \angle 120° \times 100 \angle 0° + 1 \angle 240° \times 50 \angle 60° + 50 \angle 300°$$
$$= (-50 + j86.6) + (25 - j43.3) + (25 - j43.3) = \mathbf{0 A.} \quad \textbf{(Ans.)}$$

Example 21.25. *Find out the positive, negative and zero phase sequence components of the following set of three unbalanced voltage vectors :*

$$V_A = 10 \angle 30° ; V_B = 30 \angle -60° ; V_C = 15 \angle 145°$$

Indicate on an approximate diagram how the original vectors and their different sequence components are located. **(Jadavpur University)**

Solution. (*i*) **Positive sequence vectors :**

Unbalanced vectors $V_A = 10 \angle 30°$; $V_B = 30 \angle -60°$; $V_C = 15 \angle 145°$ are shown in Fig. 21.43 (a)

$$\bar{V}_1 = \frac{1}{3}(\bar{V}_A + \lambda \bar{V}_B + \lambda^2 \bar{V}_C) \quad \text{...[Eqn. (21.16)]}$$

$$= \frac{1}{3}(10 \angle 30° + 1 \angle 120° \times 30 \angle -60° + 1 \angle 240° \times 15 \angle 145°)$$

$$= \frac{1}{3}(10 \angle 30° + 30 \angle 60° + 15 \angle 385°)$$

$$= \frac{1}{3}(37.25 + j37.32) = 12.42 + j12.44 = 17.6 \angle 45°$$

$\therefore$ $\bar{V}_{A1} = 17.6 \angle 45°$; $\bar{V}_{B1} = \lambda^2 V_{A1} = 1 \angle 240° \times 17.6 \angle 45° = 17.6 \angle 285°$ or

or $17.6 \angle -75°$;

$$V_{C1} = \lambda V_{A1} = 1 \angle 120° \times 17.6 \angle 45° = 17.6 \angle 165° \quad \text{...[Eqn. (21.12)]}$$

These are shown in Fig. 21.43 (*b*)

(*ii*) **Negative sequence vectors :**

$$\bar{V}_2 = \frac{1}{3}(\bar{V}_A + \lambda^2 \bar{V}_B + \lambda \bar{V}_C)$$

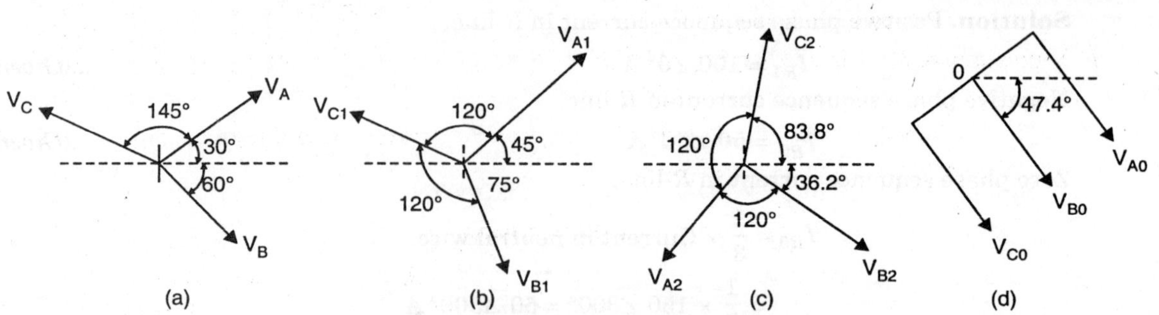

Fig. 21.43

$$= \frac{1}{3}(10 \angle 30° + 1 \angle 240° \times 30 \angle -60° + \angle 120° \times 15 \angle 145°)$$

$$= \frac{1}{3}(10 \angle 30° + 30 \angle 180° + 15 \angle 265°) = 8.24 \angle -156.3°$$

$$V_{A2} = 8.24 \angle -156.2° \; ; V_{B2} = \lambda V_{A2} = 1 \angle 120° \times 8.24 \angle -156.2°$$
$$= 8.24 \angle -36.2° \; ;$$
$$V_{C2} = \lambda^2 V_{A2} = 1 \angle 240° \times 8.24 \angle -156.2° = 8.24 \angle 83.8°$$

These vectors are shown in Fig. 21.43 (*c*)

(*iii*) **Zero sequence vectors :**

$$\overline{V_0} = \frac{1}{3}(\overline{V_A} + \overline{V_B} + \overline{V_C})$$

$$= \frac{1}{3}(10 \angle 30° + 30 \angle -60° + 15 \angle 145°) = 5.6 \angle -47.4°$$

i.e., $V_{A0} = V_{B0} = V_{C0} = 5.6 \angle -47.4°$

These vectors are shown in Fig. 21.43 (*d*).

Example 21.26. *Prove that in a 3-phase system if V_1, V_2 and V_3 are the three balanced voltages whose phasor sum is zero, the positive and negative sequence components can be expressed as*

$$V_{1P} = \left\{\frac{1}{\sqrt{3}}(V_1 + V_2 \angle 60°)\right\} \angle 30° \; ; V_{1N} = \left\{\frac{1}{\sqrt{3}}(V_1 + V_2 \angle -60°)\right\} \angle -30°$$

Phase sequence is 1-2-3.

A system of 3-phase current is given as $I_1 = 10 \angle 180°$, $I_2 = 14.14 \angle -45°$ and $I_3 = 10 \angle 90°$. Determine phasor expression for the sequence components of these currents. Phase sequence is 1-2-3.

(Bombay University)

Solution. *Positive sequence component,* $\overline{V_{1P}} = \frac{1}{3}(\overline{V_1} + \lambda \overline{V_2} + \lambda^2 \overline{V_3})$ ...[Eqn. (21.15)]

Now, $\overline{V_1} + \overline{V_2} + \overline{V_3} = 0 \quad \therefore \quad \overline{V_3} = -(\overline{V_1} + \overline{V_2})$

$$\therefore \quad \overline{V_{1P}} = \frac{1}{3}[\overline{V_1} + \lambda \overline{V_2} - \lambda^2(\overline{V_1} + \overline{V_2})] = \frac{1}{3}[\overline{V_1}(1 - \lambda^2) + \overline{V_2}(\lambda - \lambda^2)]$$

Also, $1 - \lambda^2 = [1 - (-0.5 - j0.866)] = 1.5 + j0.866 = \frac{3}{2} + j\frac{\sqrt{3}}{2}$

Refer to Art 2.4.2.1.

and, $\lambda - \lambda^2 = j\sqrt{3}$

$$\therefore \quad \mathbf{V_{1P}} = \frac{1}{3}\left[\bar{V}_1\left(\frac{3}{2} + j\frac{\sqrt{3}}{2}\right) + j\sqrt{3}\,\bar{V}_2\right] = \frac{1}{\sqrt{3}}\left[\bar{V}_1\left(\frac{\sqrt{3}}{2} + j\frac{1}{2}\right) + j\bar{V}_2\right]$$

$$= \frac{1}{\sqrt{3}}[\bar{V}_1 \angle 30° + \bar{V}_2 \angle 90°] = \left[\frac{1}{\sqrt{3}}(\bar{V}_1 + V_2 \angle 60°\right] \angle 30°. \quad \text{...}\mathbf{Proved}$$

Similarly, the *negative sequence* component is given by,

$$\bar{V}_{1N} = \frac{1}{3}(\bar{V}_1 + \lambda^2\bar{V}_2 + \lambda\bar{V}_3)$$

$$= \frac{1}{3}[\bar{V}_1 + \lambda^2\bar{V}_2 - \lambda(\bar{V}_1 + \bar{V}_2)] = \frac{1}{3}[\bar{V}_1(1-\lambda) + \bar{V}_2(\lambda^2 - \lambda)]$$

Now $\lambda^2 - \lambda = -j\sqrt{3}$

$$\therefore \quad \mathbf{\bar{V}_{1N}} = \frac{1}{3}\left[\bar{V}_1\left(\frac{3}{2} - j\frac{\sqrt{3}}{2}\right) - j\sqrt{3}\,\bar{V}_2\right] = \frac{1}{\sqrt{3}}\left[\bar{V}_1\left(\frac{\sqrt{3}}{2} - j\frac{1}{2}\right) - j\bar{V}_2\right]$$

$$= \frac{1}{3}[\bar{V}_1 \angle -30° + \bar{V}_2 \angle -90°] = \left\{\frac{1}{\sqrt{3}}(\bar{V}_1 + V_2 \angle -60°\right\} \angle -30°$$

...**Proved.**

Now, $\bar{I}_1 = 10 \angle 180°\ ;\ \bar{I}_2 = 14.14 \angle -45°\ ;\ \bar{I}_3 = 10 \angle 90°$

$$\therefore \quad I_{1P} = \frac{1}{3}(\bar{I}_1 + \lambda\bar{I}_2 + \lambda^2\bar{I}_3)$$

$$= \frac{1}{3}(10 \angle 180° + 1 \angle 120° \times 14.14 \angle -45° + 1 \angle 240° \times 10 \angle 90°)$$

$$= \frac{1}{3}(10 \angle 180° + 14.14 \angle 75° + 10 \angle 330°) = \mathbf{2.988 \angle 75°\ A.} \quad \textbf{(Ans.)}$$

$$I_{1N} = \frac{1}{3}(\bar{I}_1 + \lambda^2\bar{I}_2 + \lambda\bar{I}_3)$$

$$= \frac{1}{3}(10 \angle 180° + 1 \angle 240° \times 14.14 \angle -45° + \angle 120° \times 10 \angle 90°)$$

$$= \frac{1}{3}(10 \angle 180° + 14.14 \angle 195° + 10 \angle 210°) = \mathbf{11.15 \angle -165°\ A.} \quad \textbf{(Ans.)}$$

$$I_{10} = \frac{1}{3}(\bar{I}_1 + \bar{I}_2 + \bar{I}_3)$$

$$= \frac{1}{3}(10 \angle 180° + 14.14 \angle -45° + 10 \angle 90°) = \mathbf{0.} \quad \textbf{(Ans.)}$$

Example 21.27. *Three resistors of 10 Ω, 20 Ω and 40 Ω are connected in delta across the three phases of a balanced 200 volts supply. What are the sequence components in the resistors and in supply lines.*

Solution. Fig. 21.44 (*a*) shows the arrangement of the resistors. Let the voltage across 10 Ω, 20 Ω and 40 Ω by $\bar{E}_R, \bar{E}_Y$ and $\bar{E}_B$ respectively and the corresponding currents in the resistors be $\bar{I}_R, \bar{I}_Y$ and $\bar{I}_B$. These voltage can be represented by the phasor diagram shown in Fig. 21.44 (*b*).

$$\bar{E}_R = -200 \angle 0°\ V\ ;\ \bar{E}_Y = 200 \angle 60°\ V\ ;\ \bar{E}_B = 200 \angle -60°\ V$$

Current in 10 Ω resistor, $\bar{I}_R = \dfrac{\bar{E}_R}{10} = \dfrac{-200 \angle 0°}{10} = -20 \angle 0°\ A$

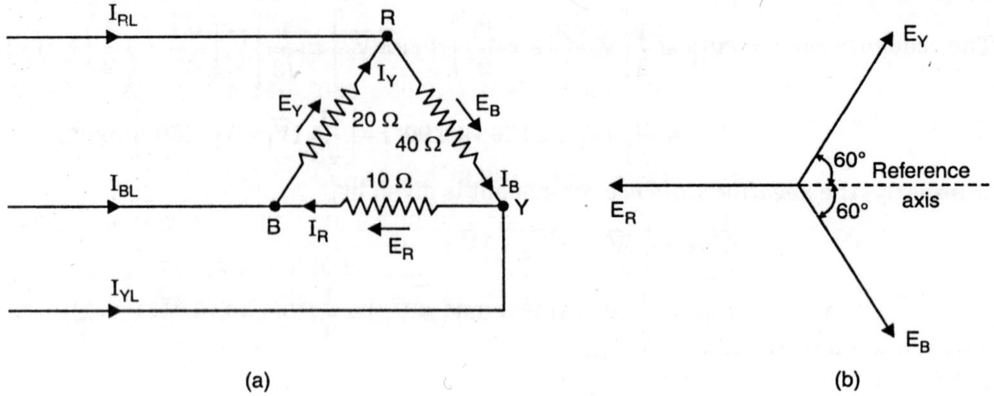

Fig. 21.44

Current in 20 Ω resistor, $\bar{I}_Y = \dfrac{\bar{E}_Y}{20} = \dfrac{200 \angle 60°}{20} = 10 \angle 60°$ A

Current in 40 Ω resistor, $\bar{I}_B = \dfrac{\bar{E}_B}{40} = \dfrac{200 \angle -60°}{40} = 5 \angle -60°$ A

Sequence current in resistors :

Positive sequence component of $\bar{I}_R$,

$$\mathbf{\bar{I}_{R1}} = \frac{1}{3}(\bar{I}_R + \lambda \bar{I}_Y + \lambda^2 \bar{I}_B)$$

$$= \frac{1}{3}(-20 \angle 0° + 1 \angle 120° \times 10 \angle 60° + \angle 240° \times 5 \angle -60°)$$

$$= \frac{1}{3}(-20 \angle 0° + 10 \angle 180° + 5 \angle 180°) = \frac{1}{3}(-20 \angle 0° + 15 \angle 180°)$$

$$= \frac{1}{3}(35 \angle 180°) = \mathbf{11.66 \angle 180° \text{ A.}} \quad \textbf{(Ans.)}$$

Negative sequence components of $\bar{I}_R$,

$$\mathbf{\bar{I}_{R2}} = \frac{1}{3}(\bar{I}_R + \lambda^2 \bar{I}_Y + \lambda \bar{I}_B)$$

$$= \frac{1}{3}(-20 \angle 0° + 1 \angle 240° \times 10 \angle 60° + 1 \angle 120° \times 5 \angle -60°)$$

$$= \frac{1}{3}(-20 \angle 0° + 10 \angle 300° + 5 \angle 60°)$$

$$= \frac{1}{3}(13.23 \angle -160.9°) = \mathbf{4.41 \angle -160.9° \text{ A.}} \quad \textbf{(Ans.)}$$

Zero sequence component of $\bar{I}_R$,

$$\mathbf{\bar{I}_{R0}} = \frac{1}{3}(\bar{I}_R + \bar{I}_Y + \bar{I}_B)$$

$$= \frac{1}{3}(-20 \angle 0° + 10 \angle 60° + 5 \angle -60°)$$

$$= \frac{1}{3}(13.23 \angle 160.9°) = \mathbf{4.41 \angle 160.9° \text{ A.}} \quad \textbf{(Ans.)}$$

The sequence components of $\bar{I}_Y$ and $\bar{I}_B$ can be found as under :

$$\bar{\mathbf{I}}_{\mathbf{Y1}} = \lambda^2 \bar{I}_{R1} = 1\angle 240° \times 11.66\angle 180° = \mathbf{11.66\angle 60°\ A.}\quad \textbf{(Ans.)}$$

$$\bar{\mathbf{I}}_{\mathbf{Y2}} = \lambda \bar{I}_{R2} = 1\angle 120° \times 4.41\angle -160.9° = \mathbf{4.41\angle -40.9°\ A.}\quad \textbf{(Ans.)}$$

$$\bar{\mathbf{I}}_{\mathbf{Y0}} = \bar{I}_{R0} = \mathbf{4.41\angle 160.9°\ A.}\quad \textbf{(Ans.)}$$

$$\bar{\mathbf{I}}_{\mathbf{B1}} = \lambda \bar{I}_{R1} = 1\angle 120° \times 11.66\angle 180° = \mathbf{11.66\angle -60°\ A.}\quad \textbf{(Ans.)}$$

$$\bar{\mathbf{I}}_{\mathbf{B2}} = \lambda^2 \bar{I}_{R2} = 1\angle 240° \times 4.41\angle -160.9° = \mathbf{4.41\angle 79.1°\ A.}\quad \textbf{(Ans.)}$$

$$\bar{\mathbf{I}}_{\mathbf{B0}} = \bar{I}_{R0} = \mathbf{4.41\angle 160.0°\ A.}\quad \textbf{(Ans.)}$$

Sequence current in supply lines :

Line current, $\bar{\mathbf{I}}_{\mathbf{RL}} = \bar{I}_B - \bar{I}_Y = 5\angle -60° - 10\angle 60° = \mathbf{13.23\angle -100.9°\ A.}$ **(Ans.)**

Line current, $\bar{\mathbf{I}}_{\mathbf{YL}} = \bar{I}_R - \bar{I}_B = -20\angle 0° - 5\angle -60° = \mathbf{22.91\angle 169.1°\ A.}$ **(Ans.)**

Line current, $\bar{\mathbf{I}}_{\mathbf{BL}} = \bar{I}_Y - \bar{I}_R = 10\angle 60° - (-20\angle 0°) = \mathbf{26.45\angle 19.1°.}$ **(Ans.)**

Positive sequence component of $\bar{I}_{RL}$,

$$\bar{\mathbf{I}}_{\mathbf{RL1}} = \frac{1}{3}(\bar{I}_{RL} + \lambda\bar{I}_{YL} + \lambda^2\bar{I}_{BL})$$

$$= \frac{1}{3}(13.23\angle -100.9° + 1\angle 120° \times 22.91\angle 169.1° + 1\angle 240° \times 26.45\angle 19.1°)$$

$$= \frac{1}{3}(13.23\angle -100.9° + 22.91\angle 289.1° + 26.45\angle 259.1°)$$

$$= \frac{1}{3}(60.6\angle -90°) = \mathbf{20.2\angle -90°\ A.}\quad \textbf{(Ans.)}$$

Negative sequence component of $\bar{I}_{RL}$,

$$\bar{\mathbf{I}}_{\mathbf{RL2}} = \frac{1}{3}(\bar{I}_{RL} + \lambda^2\bar{I}_{YL} + \lambda^2\bar{I}_{BL})$$

$$= \frac{1}{3}(13.23\angle -100.9° + 1\angle 240° \times 22.91\angle 169.1° + 1\angle 120° \times 26.45\angle 19.1°)$$

$$= \frac{1}{3}(13.23\angle -100.9° + 22.91\angle 409.1° + 26.45\angle 139.1°)$$

$$= \frac{1}{3}(22.9\angle 109.1°) = \mathbf{7.63\angle 109.1°\ A.}\quad \textbf{(Ans.)}$$

Zero sequence component of $\bar{I}_{RL}$,

$$\bar{\mathbf{I}}_{\mathbf{RL0}} = \frac{1}{3}(\bar{I}_{RL} + \bar{I}_{YL} + \bar{I}_{BL})$$

$$= \frac{1}{3}(13.23\angle -100.9° + 22.91\angle 169.1° + 26.45\angle 19.1°) = 0$$

This shows that is delta formation, the zero sequence currents are present in phases but they disappear in line currents. As line current is the difference of two phase currents, therefore, the zero sequence components cancel out.

Example 21.28. *A balanced star-connected load takes 90 A from a balanced 3-phase, 4-wire supply. If the fuses in two of the supply lines are removed, find the symmetrical components of the line currents before and after the fuses are removed.*

Solution. The circuit is shown in Fig. 21.45.

Before fuses are removed :

$$\bar{I}_R = 90 \angle 0° \,;\, \bar{I}_Y = 90 \angle -120° \,;\, \bar{I}_B = 90 \angle 120°$$

$$\bar{I}_{R1} = \frac{1}{3}(\bar{I}_R + \lambda \bar{I}_Y + \lambda^2 \bar{I}_B)$$

$$= \frac{1}{3}(90 \angle 0° + 1 \angle 120° \times 90 \angle -120° + 1 \angle 240° \times 90 \angle 120°)$$

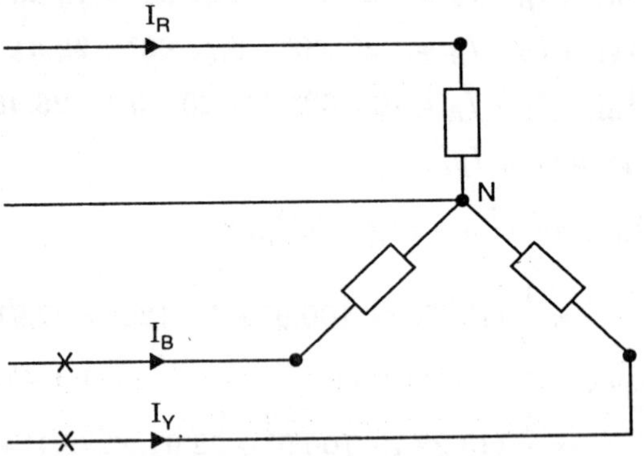

Fig. 21.45

$$= \frac{1}{3}(90 \angle 0° + 90 \angle 0° + 90 \angle 0°) = \mathbf{90° \ A.} \quad \textbf{(Ans.)}$$

$$\bar{I}_{R2} = \frac{1}{3}(\bar{I}_R + \lambda^2 \bar{I}_Y + \lambda \bar{I}_B)$$

$$= \frac{1}{3}(90 \angle 0° + 1 \angle 240° \times 90 \angle -120° + 1 \angle 120° \times 90 \angle 120°))$$

$$= \frac{1}{3}(90 \angle 0° + 90 \angle 120° + 90° \angle 240°) = 0$$

$$= \frac{1}{3}(90 + (-45 + j77.94) + (-45 - j77.94) = \mathbf{0.} \quad \textbf{(Ans.)}$$

$$\bar{I}_{R0} = \frac{1}{3}(\bar{I}_R + \bar{I}_Y + \bar{I}_B) = \frac{1}{3}(90 \angle 0° + 90 \angle -120° + 90 \angle 120°) = \mathbf{0.} \quad \textbf{(Ans.)}$$

After fuses are removed :

$$\bar{I}_R = 90 \angle 0° \,;\, \bar{I}_Y = \bar{I}_B = 0.$$

$$\bar{I}_{R1} = \frac{1}{3}(90 \angle 0° + 0 + 0) = \mathbf{30 \angle 0° \ A.} \quad \textbf{(Ans.)}$$

$$\bar{I}_{R2} = \mathbf{30 \angle 0° \ A.} \quad \textbf{(Ans.)}$$

$$\bar{I}_{R0} = \frac{1}{3}(\bar{I}_R + \bar{I}_Y + \bar{I}_B) = \frac{1}{3}(90 \angle 0° + 0 + 0) = \mathbf{30 \angle 0°.} \quad \textbf{(Ans.)}$$

Example 21.29. *A 3-phase star-connected load as shown in Fig. 21.46 is connected to a 3-phase supply system having a line voltage of 440 V. Calculate the line current I_R.*

Solution. This is a case of unbalanced 3-phase star-connected load supplied from a balanced phase supply. Since the phase sequence is *RYB*,

$$\bar{V}_{RY} = 440\angle 0^\circ \text{ V} ; \bar{V}_{YB} = \lambda^2 \bar{V}_{RY} = 440\angle 240^\circ \text{ V}$$

Let $\bar{V}_R, \bar{V}_Y$ and $\bar{V}_B$ be the voltages across impedances in R, Y and B phase respectively ; $\bar{I}_R, \bar{I}_Y$ and $\bar{I}_B$ be the resulting currents.

$\therefore$ $$\bar{V}_R - \bar{V}_Y = \bar{V}_{RY} = 440 + j0$$

and $$\bar{V}_Y - \bar{V}_B = \bar{V}_{YB} = 440\,(-0.5 - j0.866)$$

Since $$\bar{I}_R + \bar{I}_Y + \bar{I}_B = 0$$

$\therefore$ $$\bar{I}_{RO} = \bar{I}_{YO} = \bar{I}_{BO} = 0$$

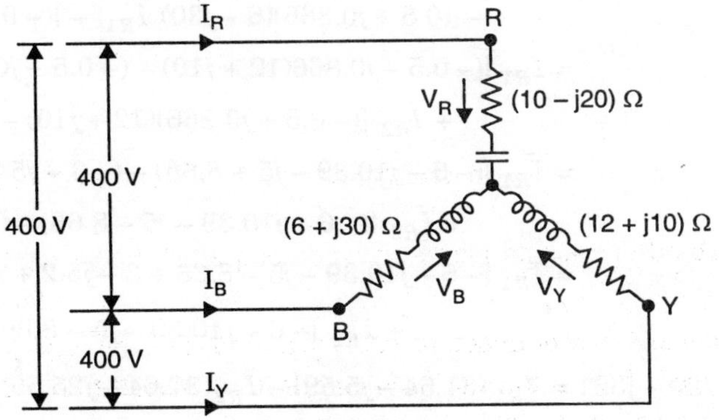

Fig. 21.46

$$\bar{V}_R = \bar{I}_R(10 - j20)$$
$$= (\bar{I}_{R0} + \bar{I}_{R1} + \bar{I}_{R2})(10 - j20) = (\bar{I}_{R1} + \bar{I}_{R2})(10 - j20) \quad ...(i)$$
$$\bar{V}_Y = \bar{I}_Y(12 + j10)$$
$$= (\bar{I}_{Y0} + \bar{I}_{Y1} + \bar{I}_{Y2})(12 + j10) = (\bar{I}_{Y1} + \bar{I}_{Y2})(12 + j20)$$
$$= (\lambda^2 \bar{I}_{R1} + \lambda \bar{I}_{R2})(12 + j10) \quad ...(ii)$$
$$\bar{V}_B = \bar{I}_B(6 + j30) = (\bar{I}_{B0} + \bar{I}_{B1} + \bar{I}_{B2})(6 + j30) = (\bar{I}_{B1} + \bar{I}_{B2})(6 + j30)$$
$$= (\lambda \bar{I}_{R1} + \lambda^2 \bar{I}_{R2})(6 + j30) \quad ...(iii)$$

Subtracting (*ii*) from (*i*), we get

$$\bar{V}_R - \bar{V}_Y = (\bar{I}_{R1} + \bar{I}_{R2})(10 - j20) - (\lambda^2 \bar{I}_{R1} + \lambda \bar{I}_{R2})(12 + j10)$$

or, $$(440 + j0) = (\bar{I}_{R1} + \bar{I}_{R2})(10 - j20) - [(-0.5 - j0.866)\,\bar{I}_{R1} + (-0.5 + j0.866)\,\bar{I}_{R2}]\,(12 + j10)$$

$$= \bar{I}_{R1}(10 - j20) + \bar{I}_{R2}(10 - j20) - \bar{I}_{R1}(-0.5 - j0.866)(12 + j10) - \bar{I}_{R2}(-0.5 + j0.866)(12 + j10)$$

$$= \bar{I}_{R1}\,[(10 - j20) - (-0.5 - j0.866)(12 + j10)] + \bar{I}_{R2}\,[(10 - j20) - (-0.5 + j0.866)(12 + j10)]$$

$$= \bar{I}_{R1}\,[10 - j20 - (-6 - j10.39 - j5 + 8.66)] + \bar{I}_{R2}\,[(10 - j20) - (-6 + j10.39 - j5 - 8.66)]$$

$$= \bar{I}_{R1}\,[10 - j20 + 6 + j10.39 + j5 - 8.66] + \bar{I}_{R2}\,[10 - j20 + 6 - j10.39 + j5 + 8.66]$$

or,
$$440 + j0 = \bar{I}_{R1}\,(7.32 - j2.3) + \bar{I}_{R2}\,(24.66 - j25.39)$$

Subtracting (*iii*) from (*ii*), we get

$$\bar{V}_Y - \bar{V}_B = (\lambda^2 \bar{I}_{R1} + \lambda \bar{I}_{R2})(12 + j10) - (\lambda \bar{I}_{R1} + \lambda^2 I_{R2})(6 + j30)$$

$$440\,(-0.5 - j0.866) = [(-0.5 - j0.866)\,\bar{I}_{R1} + (-0.5 + j0.866)\,\bar{I}_{R2}](12 + j10) - [(-0.5 + j0.866)\,\bar{I}_{R1} + (-0.5 - j0.866)\,\bar{I}_{R2}]\,(6 + j30)$$

$$-220 - j381 = [\bar{I}_{R1}\,(-0.5 - j0.866)(12 + j10)] + [\bar{I}_{R2}(-0.5 + j0.866)(12 + j10)] - [(0.5 + j0.866)(6 + j30)\,\bar{I}_{R1}] - [(-0.5 - j0.866)(6 + j30)]\,\bar{I}_{R2}$$

$$= \bar{I}_{R1}\,[(-0.5 - j0.866(12 + j10) - (-0.5 + j0.866)(6 + j30)] + \bar{I}_{R2}\,[(-0.5 + j0.866)(12 + j10) - (-0.5 - j0.866)(6 + j30)]$$

$$= \bar{I}_{R1}\,[(-6 - j10.39 - j5 + 8.66) - (-3 + j5.2 - j15 - 25.98)] + \bar{I}_{R2}\,[(-6 + j10.39 - j5 - 8.66) - (-3 - j5.2 - j15 + 25.98)]$$

$$= \bar{I}_{R1}\,[-6 - j10.39 - j5 - 8.66 + 3 - j5.2 + j15 - 25.98] + \bar{I}_{R2}\,[-6 + j10.39 - j5 - 8.66 + 3 + j5.2 + j15 + 25.98]$$

or,
$$-220 - j381 = \bar{I}_{R1}\,(31.64 - j5.59) - \bar{I}_{R2}(37.64 - j25.59)$$

Solving (*iv*), and (*v*), we get

$$\bar{I}_{R1} = (8.865 - j4.265)\,A$$

$$\bar{I}_{R2} = (6.33 - j9.4)\,A$$

$\therefore$
$$\mathbf{I_R} = \bar{I}_{R1} + \bar{I}_{R2} = (8.865 - j4.265) + (6.33 - j9.4)$$
$$= 15.195 - j13.665 = \mathbf{20.43\,\angle{-41.96°}\ A.}\quad \textbf{(Ans.)}$$

Example 21.30. *A star-connected load consists of three equal resistors of 1 Ω resistance. The load is assumed to be connected to unsymmetrical 3-phase supply, the line voltages are 200 V, 346 V, 400 V. Find the magnitude of current in any phase by the method of symmetrical components. Prove the formulae used.* **(AMIE)**

Solution. Refer to Fig. 21.47 (*a*). The line voltages form a closed right-angled triangle ($200^2 + 346^2 = 400^2$) with an angle $= \tan^{-1}\left(\frac{346}{200}\right) = 60°$.

Hence, if $\bar{V}_{RY} = 200\,\angle 0°$; then $\bar{V}_{YB} = 346\,\angle{-90°} = -j346$

and $\bar{V}_{BR} = 400\,\angle 120°$

Now,
$$\bar{V}_{RN1} = \frac{1}{3}\left[\bar{V}_{RY} + \frac{1}{2}\,\bar{V}_{YB}\,(1 + j\sqrt{3})\right] \qquad \text{...(Eqn. (21.28)]}$$

$$= \frac{1}{3}\left[200 + \left(\frac{1}{2} + j\,\frac{\sqrt{3}}{2}\right)(-j346)\right] = \frac{1}{3}\,(200 - j173 + 299.64) = 166.5 - j57.7$$

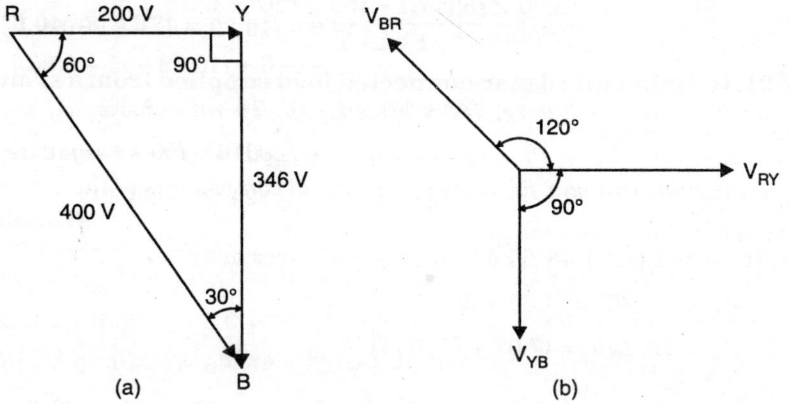

Fig. 21.47

$$\overline{V}_{RN2} = \frac{1}{3}\left[\overline{V}_{RY} + \frac{1}{2}\overline{V}_{YB}(1 + j\sqrt{3})\right] = \frac{1}{3}\left[\overline{V}_{RY} + \overline{V}_{YB}\left(\frac{1}{2} - j\frac{\sqrt{3}}{2}\right)\right]$$

$$= \frac{1}{3}\left[200 + \left(\frac{1}{2} - j\frac{\sqrt{3}}{2}\right)(-j346)\right] = \frac{1}{3}(200 - j173 - 299.64) = -33.2 - j57.7$$

Now, $\overline{Z}_A + \overline{Z}_B + \overline{Z}_C = 1 + 1 + 1 = 3$

$$\overline{Z}_{R1} = \overline{Z}_R + \lambda\overline{Z}_Y + \lambda^2\overline{Z}_B = 0 \qquad (\because\ \overline{Z}_A = \overline{Z}_B = \overline{Z}_C \text{ and } 1 + \lambda + \lambda^2 = 0)$$

Similarly, $\overline{Z}_{R2} = \overline{Z}_R + \lambda^2\overline{Z}_Y + \lambda\overline{Z}_B = 0$

Now, $\overline{V}_{RN1} = \overline{I}_{R1}\overline{Z}_0 + \overline{I}_{R2}\overline{Z}_{R2}$ ...[Eqn (21.20)]

$= \overline{I}_{R1}\overline{Z}_0$ $(\because\ \overline{Z}_{R2} = 0)$

$$\therefore \quad \overline{I}_{R1} = \frac{\overline{V}_{RN1}}{\overline{Z}_0} = \frac{\overline{V}_{RN1}}{\frac{1}{3}(\overline{Z}_R + \overline{Z}_Y + \overline{Z}_B)} = \frac{166.5 - j57.7}{\frac{1}{3}(1 + 1 + 1)} = (166.5 - j57.7)\,A$$

$$\left[\because\ \overline{Z}_0 = \frac{1}{3}(\overline{Z}_R + \overline{Z}_Y + \overline{Z}_B)\right]$$

$\overline{V}_{RN2} + \overline{I}_{R2}\overline{Z}_0 + \overline{I}_{R1}\overline{Z}_{R1}$ ...[Eqn. (21.21)]

$= \overline{I}_{R2}\overline{Z}_0$ $[\because\ \overline{Z}_{R1} = 0]$

$$\overline{I}_{R2} = \frac{\overline{V}_{RN2}}{Z_0} = \frac{-33.2 - j57.7}{\frac{1}{3}(1 + 1 + 1)} = (-33.2 - j57.7)\,A$$

$$\therefore \quad \mathbf{I_R} = \overline{I}_{R1} + \overline{I}_{R2} = (166.5 - j57.7) + (-33.2 - j57.7)$$

$$= 133.3 - j115.4 = \mathbf{176.3 \angle -40.9\ A.}\ \ \textbf{(Ans.)}$$

- This question can also be solved by using *Millman's theorem as follows :*

Using Millman's theorem and taking the line terminal R as reference point, the voltage between R and the neutral point N is

$$V_{NR} = \frac{Y_{YR}Y_Y + V_{BR}Y_B}{Y_R + Y_Y + Y_B}$$

$$= \frac{200 \angle 180° \times 1 + 400 \angle 120° \times 1}{1+1+1} = \mathbf{176.3 \angle -40.9° \ A.} \quad \textbf{(Ans.)}$$

Example 21.31. Unbalanced star-connected load supplied from a symmetrical 3-wire system). *A symmetrical 3-phase, 3-wire, 440-V system supplies an unbalanced Y-connected load of impedance $\overline{Z}_R = 5 \angle 30°\ \Omega$; $\overline{Z}_Y = 10 \angle 45°\ \Omega$, $\overline{Z}_B = 10 \angle 60°\ \Omega$. Phase sequence is $R \to Y \to B$. Calculate in the rectangular complex form the symmetrical components of the currents in the R-line.*

(Bombay University)

Solution. Refer to Fig. 21.48. The branch impedances are :

$$\overline{Z}_R = 5 \angle 30° = (4.33 + j2.5)\ \Omega$$

$$\overline{Z}_Y = 10 \angle 45° = (7.07 + j7.07)\ \Omega$$

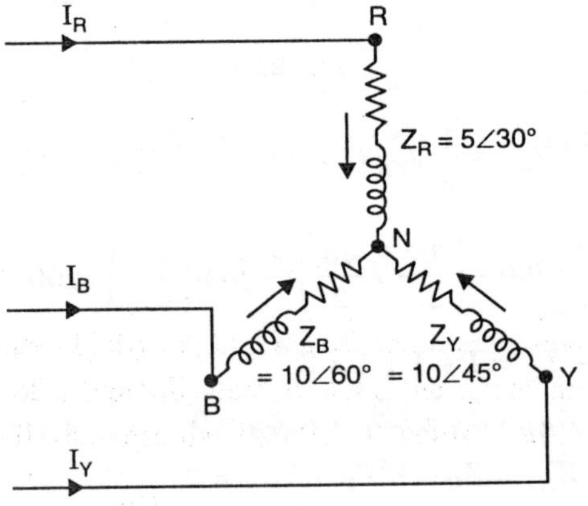

Fig. 21.48

$$\overline{Z}_B = 10 \angle 60° = (5 + j8.66)\ \Omega$$

$$\lambda \overline{Z}_Y = 1 \angle 120° \times 10 \angle 45° = 10 \angle 165° = (-9.66 + j2.59)\ \Omega$$

$$\lambda^2 \overline{Z}_Y = 1 \angle 240° \times 10 \angle 45° = 10 \angle 285° = (2.59 - j9.66)$$

$$\lambda \overline{Z}_B = 1 \angle 120° \times 10 \angle 60° = 10 \angle 180° = (-10 + j0)$$

$$\lambda^2 \overline{Z}_B = 1 \angle 240° \times 10 \angle 60° = 10 \angle 300° = (5 - j8.66)$$

The symmetrical component impedances required for calculation purposes are as follows :

$$\overline{Z}_0 = \frac{1}{3}(\overline{Z}_R + \overline{Z}_Y + \overline{Z}_B) = \frac{1}{3}\ [(4.33 + j2.5) + (7.07 + j7.07) + (5 + j8.66)] = (5.47 + j6.08)\ \Omega$$

$$\overline{Z}_{R1} = \frac{1}{3}(\overline{Z}_R + \lambda \overline{Z}_Y + \lambda^2 \overline{Z}_B) = \frac{1}{3}\ [(4.33 + j2.5) + (-9.66 + j2.59) + (5 - j8.66)] = (-0.11 - j1.19)\ \Omega$$

$$\overline{Z}_{R2} = \frac{1}{3}(\overline{Z}_R + \lambda^2 \overline{Z}_Y + \lambda \overline{Z}_B)$$

$$= \frac{1}{3}\ [(4.33 + j2.5) + 2.59 - j9.66) + (-10 + j0)] = (-1.03 - j2.39)$$

Now, $\bar{Z}_0^2\,\bar{Z}_{R1}\,\bar{Z}_{R2} = \frac{1}{3}(\bar{Z}_R\bar{Z}_Y + \bar{Z}_Y\bar{Z}_B + \bar{Z}_B\bar{Z}_R)$

$$= \frac{1}{3}\,[5\angle 30° \times 10\angle 45° + 10\angle 45° \times 10\angle 60° + 10\angle 60° \times 5\angle 30°]$$

$$= \frac{1}{3}\,[50\angle 75° + 100\angle 105° + 50\angle 90°] = \frac{1}{3} \times 195.3\angle 93.8°$$

$$= \frac{1}{3}\,(-12.94 + j194.87) = -4.31 + j65$$

Let $\bar{V}_{RY}$ be taken as the reference vector, so that

$$\bar{V}_{RY} = (440 + j0)$$

Then, $\bar{V}_{RN1} = \frac{1}{2}\bar{V}_{RY} - j\frac{1}{2}\bar{V}_{RY}/\sqrt{3}$ ....[Eqn. (21.33)]

$$= \frac{1}{2}(440 + j0) - j\,\frac{1}{2}\,(440 + j0)/\sqrt{3} = (220 - j127)$$

Now, *Positive sequence component,*

$$\bar{\mathbf{I}}_{\mathbf{R1}} = \frac{\bar{V}_{RN1}\,.\,\bar{Z}_0}{\bar{Z}_0^2 - \bar{Z}_{R1}\bar{Z}_{R2}}$$ ...[Eqn. (21.34)]

$$= \frac{(220 - j127) \times (5.47 + j6.08)}{(5.47 + j6.08)^2 - (-0.11 - j1.19)(-1.03 - j2.39)}$$

$$= \frac{254\angle -30° \times 8.18\angle 48°}{(8.178\angle 48°)^2 - 1.195\angle 84.7° \times 2.6\angle 66.7°}$$

$$= \frac{2077.7\angle 18°}{66.88\angle 96° - 3.11\angle 151.4°} = \frac{2077.7\angle 18°}{65.16\angle 93.75°}$$

$$= \mathbf{31.88\angle -75.75°\ A.}\quad \textbf{(Ans.)}$$

Negative sequence component ; $\bar{\mathbf{I}}_{\mathbf{R2}} = -\frac{\bar{I}_{R1}\bar{Z}_{R1}}{\bar{Z}_0}$

$$= -\frac{31.88\angle -75.75° \times (-0.11 - j1.19)}{(5.47 + j6.08)}$$

$$= \frac{31.88\angle -75.75° \times 1.195\angle 84.72°}{8.2\angle 48°} = \mathbf{4.64\angle -39°.}\quad \textbf{(Ans.)}$$

- Symmetrical components of other currents are :

$$\bar{I}_{Y1} = \lambda^2\bar{I}_{R1} \text{ and } \bar{I}_{Y2} = \lambda\bar{I}_{R2}\ ;$$
$$\bar{I}_{B1} = \lambda\bar{I}_{R1} \text{ and } \bar{I}_{B2} = \lambda^2\bar{I}_{R2}.$$

Example 21.32. *The three equal impedances of (4 + j3) Ω are connected in star across a 3-phase, 3-wire supply. The phase voltages are* $V_R = (220 + j0)$, $V_Y = (-j220)$ *and* $V_B = (-100 + j220)$ *V. If there is no connection between the load neutral and supply neutral, calculate the symmetrical components of R-phase current and three line currents.*

Solution. *Given :* $\bar{Z}_R = \bar{Z}_Y = \bar{Z}_R = (4 + j3)\ \Omega\ ;\ \bar{V}_R = (220 + j0)$ V ;
$\bar{V}_Y = -j220$ V ; $\bar{V}_B = (-100 + j220)$ V.

Since there is no fourth wire, there is *no zero-component current.* Moreover $\bar{I}_R + \bar{I}_Y + \bar{I}_B = 0$.
The symmetrical components of the R-phase voltages are :

$$V_0 = \frac{1}{3}(\bar{V}_R + \bar{V}_Y + \bar{V}_B)$$

$$= \frac{1}{3}[(220 + j0) + (-j220) + (-100 + j200)] = 40 \text{ V}$$

(However, this is not required in the problem)

$$\bar{V}_{R1} = \frac{1}{3}[\bar{V}_R + \lambda\bar{V}_Y + \lambda^2\bar{V}_B]$$

$$= \frac{1}{3}[(220 + j0) + (1\angle 120° \times -j220° + 1\angle 240° (-100 + j\,220)]$$

$$= \frac{1}{3}[220\angle 0° + 1\angle 120° \times 220\angle -90° + 1\angle 240° \times 241.7\angle 114.4°]$$

$$= \frac{1}{3}[220\angle 0° + 220\angle 30° + 241.7\angle 354.4°]$$

$$= \frac{1}{3} \times 656.78\angle 7.56° = 218.9\angle 7.56° \text{ V}$$

$$\bar{V}_{R2} = \frac{1}{3}[\bar{V}_R + \lambda^2\bar{V}_Y + \lambda\bar{V}_B]$$

$$= \frac{1}{3}[220\angle 0° + 1\angle 240° \times 220\angle -90° + 1\angle 120° \times 241.7\angle 114.4°]$$

$$= \frac{1}{3}(220\angle 0° + 220\angle 150° + 241.7\angle 234.4°)$$

$$= \frac{1}{3} \times 140.9\angle -142° = 46.96\angle -142° \text{ V}$$

The component current in phase R are :

$$\bar{\mathbf{I}}_{\mathbf{R1}} = \frac{\bar{V}_{R1}}{(4 + j3)} = \frac{218.9\angle 7.56°}{5\angle 36.87°} = \mathbf{43.78\angle -29.3°A.} \quad \textbf{(Ans.)}$$

$$\bar{\mathbf{I}}_{\mathbf{R2}} = \frac{\bar{V}_{R2}}{4 + j3} = \frac{46.96\angle -142°}{5\angle 36.87°} = \mathbf{9.39\angle -178.87°\ A.} \quad \textbf{(Ans.)}$$

$$\therefore \quad \bar{\mathbf{I}}_{\mathbf{R}} = \bar{I}_{R1} + \bar{I}_{R2} = 43.78\angle -29.3° + 9.39\angle -178.87° = \mathbf{36\angle -36.9°.} \quad \textbf{(Ans.)}$$

$$\bar{\mathbf{I}}_{\mathbf{Y}} = \lambda^2\,\bar{I}_{R1} + \lambda\bar{I}_{R2} = 1\angle 240° \times 43.78\angle -29.3° + 1\angle 120° \times 9.39\angle -178.87°$$

$$= 43.78\angle 21.07° + 9.39\angle -58.87° = \mathbf{44.7\angle -137°\ A.} \quad \textbf{(Ans.)}$$

$$\bar{\mathbf{I}}_{\mathbf{B}} = \lambda\bar{I}_{R1} + \lambda^2\bar{I}_{R2} = 1\angle 120° \times 43.78\angle -29.3° + 1\angle 240° \times 9.39\angle -178.87°$$

$$= 43.78\angle 90.7° + 9.39\angle 61.13° = \mathbf{52.15\angle 85.6°\ A.} \quad \textbf{(Ans.)}$$

Example 21.33. *The line to line voltages of a 3-phase system measure 200, 300 and 400 volts. Find the magnitude of the positive and negative sequence components of the delta voltages and the star voltages.*

Solution. Let V_a, V_b, V_c = Star voltages, and
V_A, V_B, V_C = Delta voltages.

Then,

$$\overline{V}_A = -\overline{V}_b + \overline{V}_c = \overline{V}_c - \overline{V}_b$$
$$\overline{V}_B = -\overline{V}_c + \overline{V}_a = \overline{V}_a - \overline{V}_c$$
$$\overline{V}_C = -\overline{V}_a + \overline{V}_b = \overline{V}_b - \overline{V}_a$$

Zero sequence components of line voltages (or delta voltages)

$$\overline{V}_{A0} = \frac{1}{3}(\overline{V}_A + \overline{V}_B + \overline{V}_C) = 0$$

The delta voltages can therefore, never contain a zero sequence component. The star voltages on the other hand may contain a zero sequence component.

Positive sequence component of line voltages (delta voltages),

$$\overline{V}_A = \frac{1}{3}[\overline{V}_A + \lambda\overline{V}_B + \lambda^2\overline{V}_C] = \frac{1}{3}[(\overline{V}_c - \overline{V}_b) + \lambda(\overline{V}_a - \overline{V}_c) + \lambda^2(\overline{V}_b - \overline{V}_a)]$$
$$= \frac{1}{3}[(\overline{V}_c + \lambda\overline{V}_a + \lambda^2\overline{V}_b) - (\overline{V}_b + \lambda\overline{V}_c + \lambda^2\overline{V}_a)]$$
$$= \frac{1}{3}(\lambda^3\overline{V}_c + \lambda\overline{V}_a + \lambda^2\overline{V}_b) - \frac{1}{3}(\lambda^3\overline{V}_b + \lambda^4\overline{V}_c + \lambda^2\overline{V}_a)$$
$$= \frac{\lambda}{3}(\overline{V}_a + \lambda\overline{V}_b + \lambda^2\overline{V}_c) - \frac{\lambda^2}{3}(\overline{V}_a + \lambda\overline{V}_b + \lambda^2\overline{V}_c)$$
$$= (\lambda - \lambda^2)(\overline{V}_{a1}) = j\sqrt{3}\overline{V}_{a1}$$

$$\therefore \quad \overline{V}_{A1} = j\sqrt{3}\overline{V}_{a1}$$

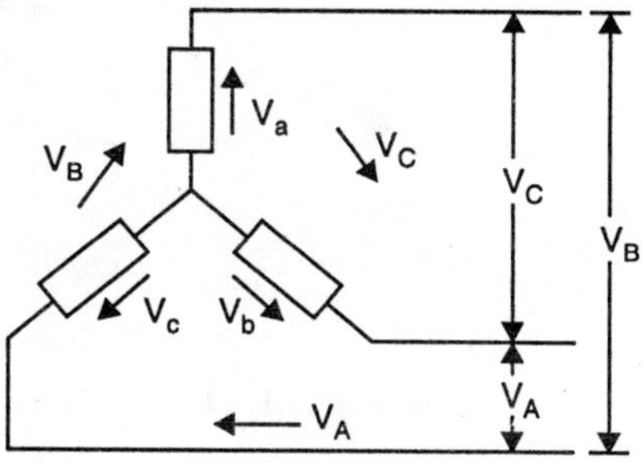

Fig. 21.49

Negative sequence component of line voltages (delta voltages),

$$\overline{V}_{A2} = \frac{1}{3}[\overline{V}_A + \lambda^2\overline{V}_B + \lambda\overline{V}_C] = \frac{1}{3}[(\overline{V}_c - \overline{V}_b) + \lambda^2(\overline{V}_a - \overline{V}_c) + \lambda(\overline{V}_b - \overline{V}_a)]$$
$$= \frac{1}{3}[(\overline{V}_c + \lambda^2\overline{V}_a + \lambda\overline{V}_b) - (\overline{V}_b + \lambda^2\overline{V}_c + \lambda\overline{V}_a)]$$
$$= \frac{\lambda^2}{3}(\overline{V}_a + \lambda^2\overline{V}_b + \lambda\overline{V}_c) - \frac{\lambda}{3}(\overline{V}_a + \lambda^2\overline{V}_b + \lambda\overline{V}_c)$$
$$= (\lambda^2 - \lambda)\,\overline{V}_{a2} = -j\sqrt{3}\,\overline{V}_{a2}$$

$$\therefore \quad \overline{V}_{A2} = -j\sqrt{3}\,\overline{V}_{a2}$$

Thus it may be seen that star delta sequence voltages are related as follows :

Positive sequence voltages, $\overline{V}_{a1} = \dfrac{-j}{\sqrt{3}}\overline{V}_{A1}$

Negative sequence voltages, $\overline{V}_{a2} = \dfrac{j}{\sqrt{3}}\overline{V}_{A2}$

Where $\overline{V}_{a1}, \overline{V}_{a2}$ are the star voltage and $\overline{V}_{A1}, \overline{V}_{A2}$ are the delta voltages.

Now, the solution of the problem is as follows :

Since the vector sum of the line voltages is always zero, therefore, the line voltages can be conveniently represented by the three sides of a triangle as shown in Fig. 21.50.

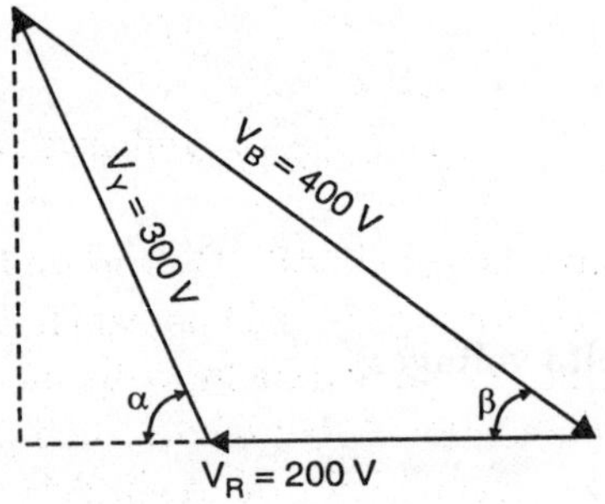

Fig. 21.50

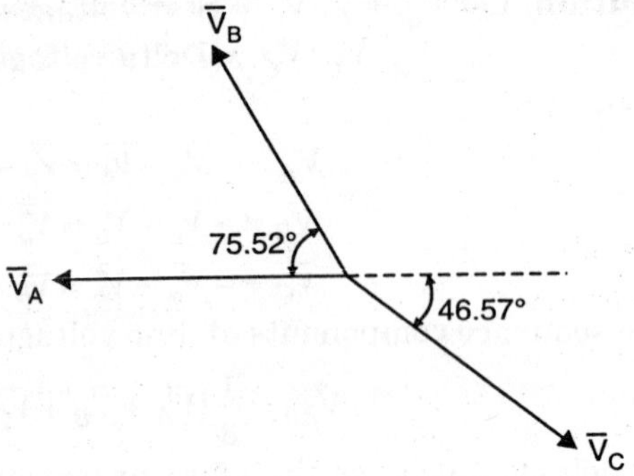

Fig. 21.51

Now,
$$(400)^2 = (200 + 300 \cos \alpha)^2 + (300 \sin \alpha)^2$$
$$(2)^2 = (1 + 1.5 \cos \alpha)^2 + (1.5 \sin \alpha)^2 \quad \text{...Dividing both sides by } (200)^2$$
$$4 = 1 + 2.25 \cos^2 \alpha + 3 \cos \alpha + 2.25 \sin^2 \alpha$$
$$= 1 + 2.25 (\cos^2 \alpha + \sin^2 \alpha) + 3 \cos \alpha$$
$$= 1 + 2.25 + 3 \cos \alpha$$

$\therefore$
$$\cos \alpha = 0.25 \quad \text{or} \quad \alpha = \cos^{-1} (0.25) = 75.52°$$

Also,
$$\cos \beta = \frac{200 + 300 \cos \alpha}{400} = \frac{200 + 300 \cos 75.52°}{400} = 0.6875$$

$\therefore$
$$\beta = \cos^{-1}(0.6875) = 46.57°$$

From Fig. 21.51,

$$\overline{V}_A = 200 \angle 180° \ V \ ; \ \overline{V}_B = 300 \angle 104.48° \ V \ ; \ \overline{V}_C = 400 \angle -46.57° \ V$$

$$\overline{V}_{R0} = \frac{1}{3}\left(\overline{V}_A + \overline{V}_B + \overline{V}_C\right)$$
$$= \frac{1}{3}(200\angle 180° + 300\angle 104.48° + 400\angle -46.57°) = 0$$

$$\overline{\mathbf{V}}_{\mathbf{A1}} = \frac{1}{3}\left(\overline{V}_A + \lambda\overline{V}_B + \lambda^2\overline{V}_C\right)$$
$$= \frac{1}{3} \ (200 \angle 180° + 1 \angle 120° \times 300 \angle 104.48°$$
$$+ 1 \angle 240° \times 400 \angle -46.57°)$$
$$= \frac{1}{3} \ (200 \angle 180° + 300 \angle 224.48° + 400 \angle 193.43°)$$
$$= \frac{1}{3} \ (858.4 \angle -159°) = \mathbf{286.1 \angle -159° \ A.} \quad \textbf{(Ans.)}$$

$$\overline{V}_{a1} = -\frac{j}{\sqrt{3}} \ \overline{V}_{A1} = -\frac{j}{\sqrt{3}} \times 286.1 \angle -159° = -\frac{j}{\sqrt{3}} \ (-267.1 - j102.53)$$
$$= (j154.2 - 59.2) = \mathbf{165.17\angle 111° V.} \quad \textbf{(Ans.)}$$

$$\left(\text{Check } \frac{296.1}{\sqrt{3}} = 165.17 \text{ V}\right)$$

$$\overline{V}_{A2} = \frac{1}{3}\left(\overline{V}_A + \lambda^2\overline{V}_B + \lambda\overline{V}_C\right)$$

$$= \frac{1}{3}\ (200\ \angle 180° + 1\ \angle 240° \times 300\ \angle 104.48° + 1\ \angle 120° \times 400\ \angle -46.57°)$$

$$= \frac{1}{3}\ (200\ \angle 180° + 300\ \angle 344.48° + 400\ \angle 73.43°)$$

$$= \frac{1}{3}\ (364.89\ \angle 56.17°) = \mathbf{121.63\ \angle 56.17°\ V.}\quad \textbf{(Ans.)}$$

$$\overline{V}_{a2} = \frac{j}{\sqrt{3}}\ \overline{V}_{A2} = \frac{j}{\sqrt{3}} \times 121.63\ \angle 56.17° = \frac{j}{\sqrt{3}}(67.71 + j101)$$

$$= (j39.1 - 58.31) = \mathbf{70.2\ \angle 146.16°\ V.}\quad \textbf{(Ans.)}$$

Example 21.34. Balanced 3-phase star-load supplied from 3-phase unbalanced supply. *Three identical resistors are star-connected and rated for 2300 V, 250 kVA as a 3-phase unit. The central point is not available. The resistor unit is connected to an unsymmetrical 3 phase system whose line voltages are $\overline{V}_A$ = 2760 ∠0° V, $\overline{V}_B$ = 2300 ∠– 138.6° V, $\overline{V}_C$ = 1840 ∠124.2° V. Find the current in each line by method of symmetrical components.*

Solution. Let R be the phase resistance in ohms.

Then, $\dfrac{(2300/\sqrt{3})^2}{R} = 250 \times 10^3 \times \dfrac{1}{3}$ or $R = \dfrac{3 \times (2300/\sqrt{3})^2}{250 \times 10^3} = 21.16\ \Omega$

Taking V_A as the reference vector,

$$\overline{V}_A = 2760\ \angle 0°\ \text{V}\ ;\ \overline{V}_B = 2300\ \angle -138.6°\ \text{V}\ ;\ \overline{V}_C = 1840\ \angle 124.2°\ \text{V}$$

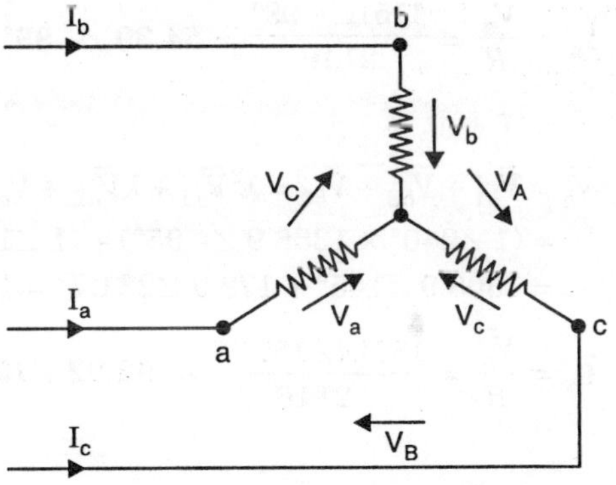

Fig. 21.52

Now, $$\overline{V}_{A1} = \frac{1}{3}(\overline{V}_A + \lambda\overline{V}_B + \lambda^2\overline{V}_C)$$

$$= \frac{1}{3}\ (2760\ \angle 0° + 1\ \angle 120° \times 2300\ \angle -138.6° + 1\ \angle 240° \times 1840\ \angle 124.2°)$$

$$= \frac{1}{3}(2760\ \angle 0° + 2300\ \angle -18.6° + 1840\ \angle 364.2°)$$

$$= \frac{1}{3}\ (6\,801.34\ \angle -5.05°) = 2267\ \angle -5.05°\ \text{V}$$

$$\overline{V}_{A2} = \frac{1}{3}(\overline{V}_A + \lambda^2\overline{V}_B + \overline{V}_C)$$

$$= \frac{1}{3}(2760\angle 0° + 1\angle 240° \times 2300\angle -138.6° + 1\angle 120° \times 1840\angle 124.2°)$$

$$= \frac{1}{3}(2760\angle 0° + 2300\angle 101.4° + 1840\angle 244.2°)$$

$$= \frac{1}{3}(1619\angle 21.68°) = 539.7\angle 21.68°$$

$\overline{V}_{A0} = 0$, since there is no zero sequence component present in the line voltages of 3-phase system.

Since the neutral point is *not* available, therefore, *no zero sequence component is present in the line voltages of 3-phase system.* Thus, the *phase voltage contains only positive and negative sequence components.*

To find $\overline{\mathbf{I}}_a$:

$$\overline{V}_{a1} = -\frac{1}{\sqrt{3}}\overline{V}_{A1} = \frac{-j}{\sqrt{3}} \times 2267\angle -5.05 = \frac{-j}{\sqrt{3}}(2258.2 - j199.55)$$

$$= (-j1303.8 - 115.18) = 1308.9\angle -95° \text{ V}$$

$$\overline{V}_{a2} = \frac{j}{\sqrt{3}}\overline{V}_{A2} = \frac{j}{\sqrt{3}} \times 539.7\angle 21.68° = \frac{j}{3}(501.5 + j199.4)$$

$$= (j167.17 - 66.47) = 179.9\angle 111.7° \text{ V}$$

$$\overline{V}_a = \overline{V}_{a1} + \overline{V}_{a2} + \overline{V}_{a0}$$

$$= 1308.9\angle -95° + 179.9\angle 111.7° + 0 = 1151\angle -99° \text{ V}$$

$$\overline{\mathbf{I}}_a = \frac{\overline{V}_a}{R} = \frac{1151\angle -99°}{21.16} = \mathbf{54.39\angle -99°\ A.}\quad \textbf{(Ans.)}$$

To find $\overline{\mathbf{I}}_b$:

$$\overline{V}_b = \overline{V}_{b1} + \overline{V}_{b2} + \overline{V}_{b0} = \lambda^2\overline{V}_{a1} + \lambda\overline{V}_{a2} + \overline{V}_{a0}$$

$$= (1\angle 240° \times 1308.9\angle -95°) + (1\angle 120° \times 179.9\angle 111.7°) + 0$$

$$= 1308.9\angle 145° + 179.9\angle 231.7° = 13314.\angle 152.75° \text{V}$$

$$\therefore \quad \overline{\mathbf{I}}_b = \frac{\overline{V}_b}{R} = \frac{1331.4\angle 152.75°}{21.16} = \mathbf{62.92\angle 152.75°\ A.}\quad \textbf{(Ans.)}$$

To find $\overline{\mathbf{I}}_c$:

$$\overline{V}_c = \overline{V}_{c1} + \overline{V}_{c2} + \overline{V}_{c0} = \lambda\overline{V}_{a1} + \lambda^2\overline{V}_{a2} + \overline{V}_{a0}$$

$$= (1\angle 120° \times 1308.9\angle -95°) + 1\angle 240° \times 179.9\angle 111.7°) + 0$$

$$= 1308.9\angle 25° + 179.9\angle 351.7° = 1462.6\angle 21.13° \text{ V}$$

$$\overline{\mathbf{I}}_c = \frac{\overline{V}_c}{21.16} = \frac{1462.6\angle 21.13°}{21.16} = \mathbf{69.12\angle 21.13°\ A.}\quad \textbf{(Ans.)}$$

Example 21.35. *A 3-phase line supplies a delta-connected load. The line current in 1 is 150 amperes. Taking the current in line-1 as reference, find the symmetrical components of line currents if phase-3 is only switched-off. Assume line current in 2 to be 150 ∠180° amperes.*

Solution. Refer to Fig. 21.53.

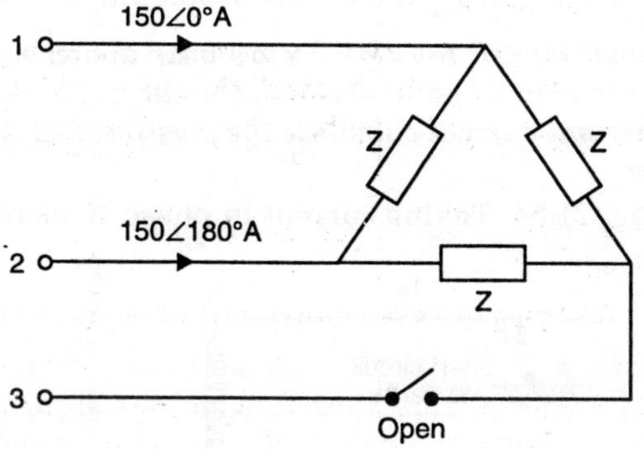

Fig. 21.53

$$\bar{I}_1 = 150 \angle 0° \; ; \; \bar{I}_2 = 150\angle 180° \; ; \; \bar{I}_3 = 0$$

For phase-1 :

$$\bar{\mathbf{I}}_{10} = \frac{1}{3}(\bar{I}_1 + \bar{I}_2 + \bar{I}_3)$$

$$= \frac{1}{3}(150 \angle 0° + 150 \angle 180° + 0) = \mathbf{0.} \quad \textbf{(Ans.)}$$

$$\bar{\mathbf{I}}_{11} = \frac{1}{3}(\bar{I}_1 + \lambda\bar{I}_2 + \lambda^2\bar{I}_3)$$

$$= \frac{1}{3}(150 \angle 0° + 1 \angle 120° \times 150 \angle 180° + 0)$$

$$= \frac{1}{3}(150 \angle 0° + 150 \angle 300°)$$

$$= \frac{1}{3}(25.98 \angle -30°) = \mathbf{86.6 \angle -30° \ A.} \quad \textbf{(Ans.)}$$

$$\bar{\mathbf{I}}_{12} = \frac{1}{3}(\bar{I}_1 + \lambda^2\bar{I}_2 + \lambda\bar{I}_3)$$

$$= \frac{1}{3}(150 \angle 0° + 1 \angle 240° \times 150 \angle 180° + 0)$$

$$= \frac{1}{3}(150 \angle 0° + 150 \angle 420°)$$

$$= \frac{1}{3}(259.8 \angle 30°) = \mathbf{86.6 \angle 30° \ A.} \quad \textbf{(Ans.)}$$

For phase-2 :

$$\mathbf{I}_{21} = \lambda^2\bar{I}_{11} = 1 \angle 240° \times 86.6 \angle -30° = \mathbf{86.6 \angle 210° \ A.} \quad \textbf{(Ans.)}$$

$$\bar{I}_{22} = \lambda\bar{I}_{12} = 1 \angle 120° \times 86.6 \angle +30° = \mathbf{86.6 \angle 150° \ A.} \quad \textbf{(Ans.)}$$

For phase-3 :

$$\bar{I}_{31} = \lambda\bar{I}_{11} = 1 \angle 120° \times 86.6 \angle -30° = \mathbf{86.6 \angle 90° \ A.} \quad \textbf{(Ans.)}$$

$$\bar{I}_{32} = \lambda^2\bar{I}_{12} = 1 \angle 240° \times 86.6 \angle 30° = \mathbf{86.6 \angle 270° \ A.} \quad \textbf{(Ans.)}$$

It may be noted that even if $\bar{I}_3 = 0$, $\bar{I}_{31}$ and $\bar{I}_{32}$ have definite values, but a little consideration reveals that the vector sum of $\bar{I}_{31}$ and $\bar{I}_{32}$ ($86.6 \angle 90° + 86.6 \angle 270° = 0$) is zero, showing $\bar{I}_3 = 0$.

Example 21.36. *A short circuit to earth occurs near the terminals of phase R of a 3-phase alternator, start connected with neutral point earthed, the current to earth being 1200 A. If the alternator is not supplying any normal current calculate the positive, negative and zero sequence components of current of all phases.*

Solution. Refer to Fig. 21.54. Taking current in phase 'R' as reference,

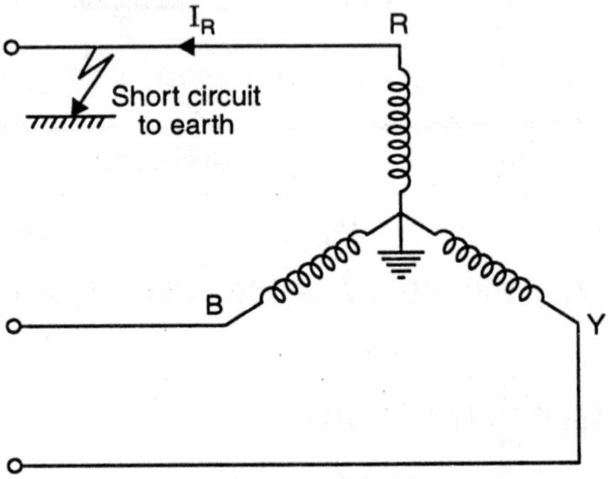

Fig. 21.54

$$\bar{I}_R = 1200 \angle 0° \; ; \; \bar{I}_Y = 0 \; ; \; \bar{I}_B = 0$$

Zero sequence component, $I_0 = I_{R0} = I_{Y0} = I_{B0}$

$$= \frac{1}{3}(\bar{I}_R + \bar{I}_Y + \bar{I}_B) = \frac{1}{3}(1200 \angle 0°) = \mathbf{400 \angle 0° A.} \quad \textbf{(Ans.)}$$

Positive sequence components are :

$$\bar{I}_{R1} = \frac{1}{3}(\bar{I}_R + \lambda \bar{I}_Y + \lambda^2 \bar{I}_B)$$

$$= \frac{1}{3}(1200 \angle 0° + 1 \angle 120° \times 0 + 1 \angle 240° \times 0) = \mathbf{400 \angle 0° \; A.} \quad \textbf{(Ans.)}$$

$$\bar{I}_{Y1} = \lambda^2 \bar{I}_{R1} = 1 \angle 240° \times 400 \angle 0° = \mathbf{400 \angle 240° \; A.} \quad \textbf{(Ans.)}$$

$$\bar{I}_{B1} = \lambda \bar{I}_{R1} = 1 \angle 120° \times 400 \angle 0° = \mathbf{400 \angle 120° \; A.} \quad \textbf{(Ans.)}$$

Negative sequence components are :

$$\bar{I}_{R2} = \frac{1}{3}(\bar{I}_R + \lambda^2 \bar{I}_Y + \lambda \bar{I}_B)$$

$$= \frac{1}{3}(1200 \angle 0° + 1 \angle 240° \times 0 + 1 \angle 120° \times 0) = \mathbf{400 \angle 0° \; A.} \quad \textbf{(Ans.)}$$

$$\bar{I}_{Y2} = \lambda \bar{I}_{R2} = 1 \angle 120° \times 400 \angle 0° = \mathbf{400 \angle 120° \; A.} \quad \textbf{(Ans.)}$$

$$\bar{I}_{B2} = \lambda^2 \bar{I}_{R2} = 1 \angle 240° \times 400 \angle 0° = \mathbf{400 \angle 240° \; A.} \quad \textbf{(Ans.)}$$

Example 21.37. *An unbalanced star-connected 3-phase supply having negligible internal impedance is connected to three identical voltmeters in star connection. Each voltmeter has a resistance of 25000 ohms and negligible reactance. The supply voltages of the three-phase from line-to-neutral are respectively $V_R = 100 \angle 0°$, $V_Y = 200 \angle 270°$, $V_B = 100 \angle 120°$.*

Using the method of symmetrical components, determine :

(i) The reading of the voltmeter connected to the Y-line.

(ii) The current through the voltmeter.

Solution. *Given :* $V_{RN} = 100 \angle 0°$; $V_{YN} = 200 \angle 270°$;

$Y_{BN} = 100 \angle 120°$; Resistance of each voltmeter, $R_v = 25000\ \Omega$.

(*i*) **The reading of the voltmeter connected to the Y-line,** $\bar{V}_Y$ **:**

$$\bar{V}_{R0} = \frac{1}{3}(\bar{V}_{RN} + \bar{V}_{YN} + \bar{V}_{BN})$$

$$= \frac{1}{3}(100 \angle 0° + 200 \angle 270° + 100 \angle 120°)$$

$$= \frac{1}{3}(123.93 \angle -66.2°) = 41.31 \angle -66.2° \text{ V}$$

$$\bar{V}_{R1} = \frac{1}{3}(\bar{V}_{RN} + \lambda\bar{V}_{YN} + \lambda^2\bar{V}_{BN})$$

$$= \frac{1}{3}(100 \angle 0° + 1 \angle 120° \times 200 \angle 270° + 1 \angle 240° \times 100 \angle 120°)$$

$$= \frac{1}{3}(386.37 \angle 15°) = 128.8 \angle 15° \text{ V}$$

$$\bar{V}_{R2} = \frac{1}{3}(V_{RN} + \lambda^2 V_{YN} + \lambda V_{BN})$$

$$= \frac{1}{3}(100 \angle 0° + 1 \angle 240° \times 200 \angle 270° + 1 \angle 120° \times 100 \angle 120°)$$

$$= \frac{1}{3}(123.93 \angle 173.8°) = 41.3 \angle 173.8° \text{ V.}$$

so that, $$\bar{I}_{R1} = \frac{\bar{V}_{R1}}{R_v} = \frac{\bar{V}_{R1}}{25000} ; \bar{I}_{R2} = \frac{\bar{V}_{R2}}{25000} \quad \text{and} \quad \bar{I}_R = \bar{I}_{R1} + \bar{I}_{R2}$$

The current flowing in voltmeter connected in Y-phases,

$$\bar{I}_Y = \bar{I}_{Y1} + \bar{I}_{Y2} = \lambda^2\bar{I}_{R1} + \lambda\bar{I}_{R2} = \lambda^2 \frac{\bar{V}_{R1}}{25000} + \lambda \frac{\bar{V}_{R2}}{25000}$$

or, $$\bar{I}_Y = \frac{\bar{V}_{Y1}}{25000} + \frac{\bar{V}_{Y2}}{25000}$$

or, $$25000\,\bar{I}_Y = \bar{V}_{Y1} + \bar{V}_{Y2}$$

or, $$\bar{V}_Y = \bar{V}_{Y1} + \bar{V}_{Y2} \qquad (\text{where } \bar{V}_Y = 25000\ \bar{I}_Y)$$

$$= (\lambda^2\bar{V}_{R1} + \lambda\bar{V}_{R2}) = (1 \angle 240° \times 128.8 \angle 15° + 1 \angle 120° \times 41.3 \angle 173.8°) = \mathbf{163 \angle -95.86° \text{ V.}}$$

Hence *reading of the voltmeter connected to the Y-line* = **163 V. (Ans.)**

(*ii*) **The current through the voltmeter :**

$$I_Y = \frac{163}{25000} = \mathbf{6.52 \text{ mA.}} \quad \textbf{(Ans.)}$$

21.4.2.6. Average 3-phase power in terms of symmetrical components

The power consumed in a 3-phase circuit can be computed directly from the symmetrical components provided the current and voltage are known. The power expression can be easily derived using matrix manipulations an follows :

Let S be the total complex power flowing into a 3-phase circuit through its phases R, Y, B while P and Q are the real and reactive powers in three phases. E_R, E_Y and E_B indicate phase voltages at terminals and I_R, I_Y and I_B the currents flowing into the circuit.

Now,
$$S = P + jQ = V_R I_R^* + V_Y I_Y^* + V_B I_B^*$$

$$= [V_R \quad V_Y \quad V_C] \begin{bmatrix} I_R \\ I_Y \\ I_B \end{bmatrix}^* = \begin{bmatrix} V_R \\ V_Y \\ V_B \end{bmatrix}^T \begin{bmatrix} I_R \\ I_Y \\ I_R \end{bmatrix}^*$$

Also,
$$\begin{bmatrix} V_R \\ V_Y \\ V_B \end{bmatrix} = \begin{bmatrix} 1 & 1 & 1 \\ 1 & \lambda^2 & \lambda \\ 1 & \lambda & \lambda^2 \end{bmatrix} \begin{bmatrix} V_{R0} \\ V_{R1} \\ V_{R2} \end{bmatrix} = AV$$

and,
$$\begin{bmatrix} V_R \\ V_Y \\ V_B \end{bmatrix} = (AV)^T = V^T A^T$$

$\therefore$
$$S = P + jQ = \begin{bmatrix} V_R \\ V_Y \\ V_B \end{bmatrix} \begin{bmatrix} I_R \\ I_Y \\ I_B \end{bmatrix}^* = (V_{R0} \quad V_{R1} \quad V_{R2}) \begin{bmatrix} 1 & 1 & 1 \\ 1 & \lambda^2 & \lambda \\ 1 & \lambda & \lambda^2 \end{bmatrix} \begin{bmatrix} I_R \\ I_Y \\ I_R \end{bmatrix}^*$$

Now substituting for the phase currents the corresponding symmetrical components, noting that λ and λ^2 are conjugate,

$$\begin{bmatrix} I_R \\ I_Y \\ I_B \end{bmatrix}^* = \begin{bmatrix} 1 & 1 & 1 \\ 1 & \lambda^2 & \lambda \\ 1 & \lambda & \lambda^2 \end{bmatrix}^* \begin{bmatrix} I_{R0} \\ I_{R1} \\ I_{R2} \end{bmatrix}^* = \begin{bmatrix} 1 & 1 & 1 \\ 1 & \lambda & \lambda^2 \\ 1 & \lambda^2 & \lambda \end{bmatrix} \begin{bmatrix} I_{R0} \\ I_{R1} \\ I_{R2} \end{bmatrix}^*$$

$\therefore$
$$S = P + jQ = [V_{R0} \quad V_{R1} \quad V_{R2}] \begin{bmatrix} 1 & 1 & 1 \\ 1 & \lambda^2 & \lambda \\ 1 & \lambda & \lambda^2 \end{bmatrix} \begin{bmatrix} 1 & 1 & 1 \\ 1 & \lambda & \lambda^2 \\ 1 & \lambda^2 & \lambda \end{bmatrix} \begin{bmatrix} I_{R0} \\ I_{R1} \\ I_{R2} \end{bmatrix}^*$$

$$= [V_{R0} \quad V_{R1} \quad V_{R2}] \begin{bmatrix} 3 & 0 & 0 \\ 0 & 3 & 0 \\ 0 & 0 & 3 \end{bmatrix} \begin{bmatrix} I_{R0} \\ I_{R1} \\ I_{R2} \end{bmatrix}^*$$

$$= 3[V_{R0} \quad V_{R1} \quad V_{R2}] \begin{bmatrix} I_{R0} \\ I_{R1} \\ I_{R2} \end{bmatrix}^*$$

$$= 3[V_{R0} I_{R0}^* + V_{R1} I_{R1}^* + V_{R2} I_{R2}^*]$$

i.e., sequence power is one-third of the power in terms of phase quantities.

$\therefore$
$$P = 3[\,|V_{R0}|\,|I_{R0}|\cos\phi_0 + |V_{R1}|\,|I_{R1}|\cos\phi_1 + |V_{R2}|\,I_{R2}\cos\phi_2]$$

21.4.3. Sequence Impedances

The sequence impedances of an equipment or a component of power system are :

(*i*) Positive sequence impedance (Z_1)

(*ii*) Negative sequence impedance (Z_2)

(*iii*) Zero sequence impedance (Z_0).

The impedance offered by a circuit when positive sequence current alone is flowing is called the *positive sequence impedance.* Similarly with only negative sequence current, the impedance offered is called the *negative sequence impedance* and with only zero sequence current, the impedance offered is called *zero sequence impedance.*

The following points are *worth noting :*

- Each piece of equipment or circuit, in a *3-phase balanced system* offers *only one impedance*—the one offered to positive or normal sequence current. This is expected because of the absence of negative and zero sequence currents in the 3-phase balanced system.
- Each piece of equipment or circuit, in a *3-phase unbalanced system,* will have *three values of impedances viz.,* positive sequence impedance, negative sequence impedance and zero sequence impedance.
- The positive and negative sequence impedance of *linear, symmetrical and static circuits* (*e.g.,* transmission lines, cables, transformers and static loads) are *equal and are the same* as those used in the analysis of balanced conditions. This is due to the fact that impedance of such circuits is independent of the phase order, provided the applied voltages are balanced.

  — The *positive and negative sequence impedances of rotating machines* (*e.g.,* synchronous and induction motors) *are normally different.*
- The zero sequence impedance depends upon the path taken by the zero sequence current. As this path is generally different from the path taken by the positive and negative sequence currents, therefore, zero sequence impedance is usually different from positive and negative sequence impedances.

21.4.3.1. Sequence impedances of power system elements

To determine the fault currents in a 3-phase unbalanced system, the concept of impedances of various elements of power system (*e.g.,* generators, transformers, transmission lines etc.) to positive, negative and zero sequence currents is of considerable importance. A brief explanation of synchronous generators, transformers and transmission lines is given below :

(*i*) ***Synchronous generators :***

- The *positive sequence* of a synchronous generator is *equal* to the synchronous impedance of the machine.
- The *negative sequence* is *much less than the positive sequence impedance.*
- The *zero sequence impedance is a variable item* and if its value is *not given,* it may be *assumed equal to the positive sequence impedance.*
- Any impedance Z_e in the earth connection of a star-connected system has the effect to introduce an impedance of $3Z_e$ per phase. It is owing to the fact that the three equal zero-sequence currents, being in phase, do not sum to zero at the star point, but they flow back along the neutral earth connection.

(*ii*) ***Transformers :***

- In case of transformers the *positive and negative sequence impedances are equal* since they (transformers) have the same impedance with reversed phase rotation.
- The *zero sequence* depends upon *earth connection.* If there is a *through circuit for earth current, zero sequence impedance will be positive sequence impedance otherwise it will be infinite.*

(*iii*) ***Transmission lines :***

- In a transmission line the positive sequence and negative sequence are same and each is equal to normal impedance of the line (this is due to the fact that the phase rotation of the currents does not make any difference in the line constants).
- The *zero sequence impedance* is usually *much greater than the positive or negative sequence impedances.*

21.4.3.2. Measurement of sequence impedances of rotating machines

(*i*) ***Measurement of positive sequence impedance :***

- The positive sequence impedance depends upon the working of the machine, *i.e.*, whether it is working under subtransient, transient or steady state condition.
- The *impedance under steady state condition is known as "synchronous impedance".* It is measured by open circuit—short-circuit test.

The *synchronous impedance in p.u. is defined as the ratio of field current at rated armature current on sustained symmetrical short circuit to field current at normal open circuit voltage on the air gap line (i.e., the extended straight line part of the magnetisation curve).*

Method of test for synchronous impedance :

The following procedure is adopted to test synchronous impedance :

- To perform the open-circuit test run the machine at synchronous speed in proper direction with the help of a prime mover (Fig. 21.55). Keep the switch in off-position.

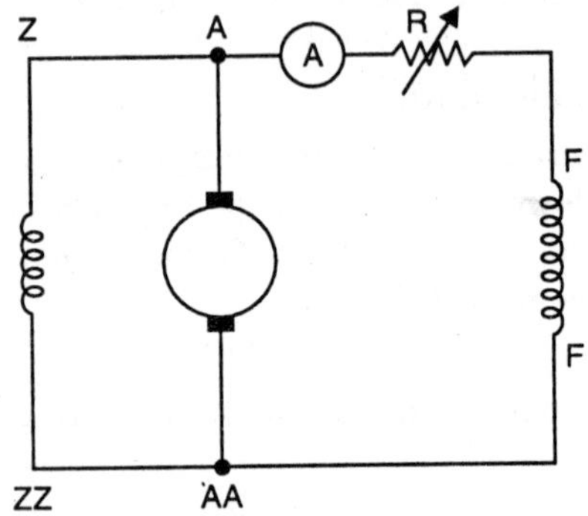

Fig. 21.55. Connection diagram for open-circuit test.

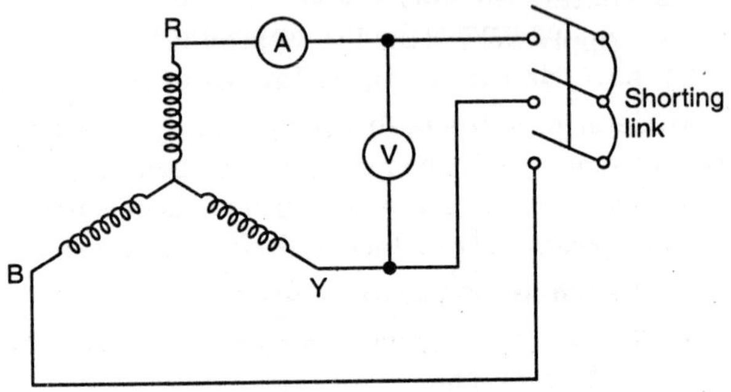

Fig. 21.56. Connection diagram for short-circuit test.

Note down the readings of voltmeter for various field currents.

- Now, to perform the short-circuit test reduce the excitation to minimum by putting the total resistance in the field circuit. Since this short-circuit test is under saturation condition of the machine, it will be a linear characteristic passing through the origin. Note down the reading (Single reading is enough).
- Plot the two characteristics (*O.C* and *S.C*) and according to the definition of the synchronous impedance calculate the value from the graph.

(*ii*) ***Method of test for sub-transient reactance :***

- With the rotor at rest and short-circuited on itself through an ammeter (Fig. 21.57) apply voltage across any two terminals except the neutral.

- Rotate the rotor by hand and it will be observed that for a fixed voltage applied, the *field current varies with the position of the rotor.*
  — When the rotor is in the position of *maximum induced field current* (the *direct axis* position of rotor), one half the voltage required to circulate the rated current is *equal to the direct axis sub-transient reactance* X_d'' *in p.u. value.*

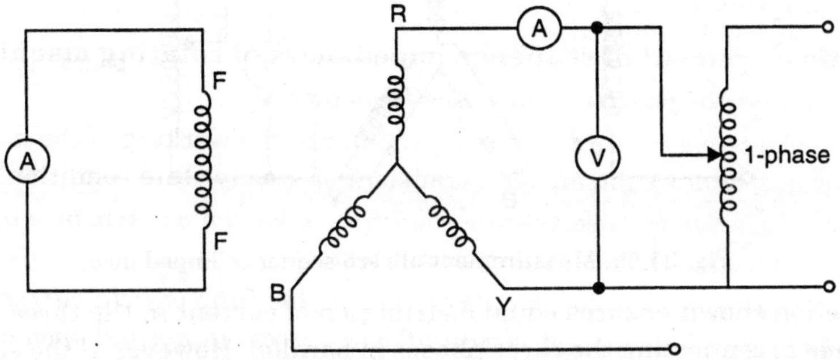

Fig. 21.57. Measurement of sub-transient reactance of an alternator.

  — When the rotor is in the position of *minimum induced field current* the *quadrature axis sub-transient reactance* X_d'' *is obtained.*

(*iii*) ***Measurement of negative sequence reactance :***

- Rotate the machine at the rated speed.
- Apply the reduced voltage so that rated current (approximately) circulates.

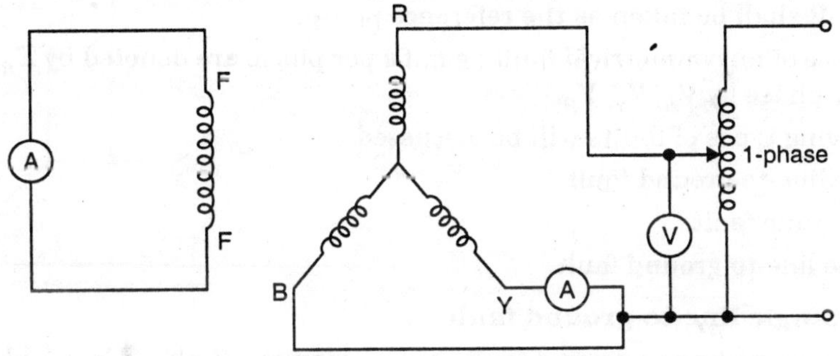

Fig. 21.58. Measurement of negative sequence impedance.

- Note down the reading of voltmeter and ammeter.

The negative impedance (reactance) is given by : $Z_2 = \dfrac{V}{\sqrt{3}I}$ where V and I are voltmeter and ammeter readings respectively.

It may be noted here that since negative sequence current flows in this case, there is a possibility of hunting which will result in oscillation of the pointer of the ammeter and as such mean reading may be taken.

(*iv*) ***Measurement of zero-sequence impedance :***

For measuring zero sequence impedance the machine *must be star-connected* (for otherwise the term zero sequence impedance has no significance as no zero-sequence currents can flow).

Fig. 21.59 shows the connections for the measurement of zero-sequence impedance. The machine is at standstill and a reduced voltage is applied. The zero sequence impedance, $Z_0 = \frac{V}{3I}$.

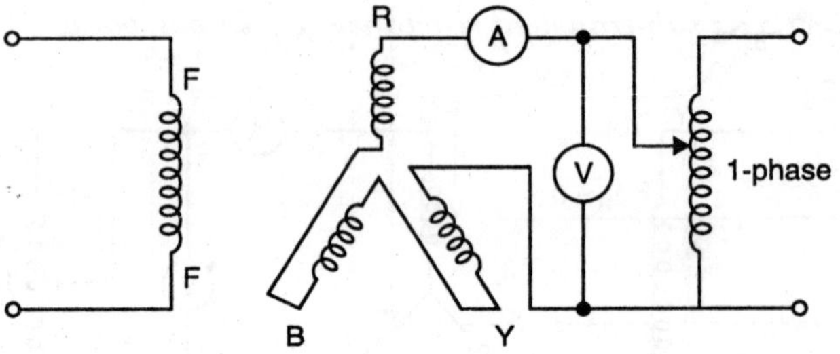

Fig. 21.59. Measurement of zero sequence impedance.

The connection shown ensures equal distribution of current in the three phases and for this reason is preferable to connecting the three phases in parallel. However, if the six terminals are not available the three phases are connected in parallel and the experiment is conducted in the same way.

21.4.4. Analysis of Unsymmetrical Faults

The following *assumptions* will be made in the analysis of unsymmetrical faults :

(*i*) The generated e.m.f. system is of positive sequence only.

(*ii*) The impedance of the fault in zero.

(*iii*) No current flows in the network other than due to fault *i.e.*, load currents are neglected.

(*iv*) Phase-R shall be taken as the reference phase.

In each case of unsymmetrical fault ; e.m.f.s per phase are denoted by E_R, E_Y and E_B and the terminal p.d. per phase by V_R, V_Y, Y_B.

The following types of faults will be discussed :

(*i*) Single line-to-ground fault.

(*ii*) Line-to-line fault.

(*iii*) Double line-to-ground fault.

21.4.4.1. Single line-to-ground fault

Consider a single line-to-ground fault occurring on the *R*-phase in a 3-phase system with an earthed neutral, as shown in Fig. 21.60.

Here, p.d. between *N* and *R*, $\bar{V}_R = 0$ and $\bar{I}_Y = \bar{I}_B = 0$

The sequence currents in *R*-phase in terms of line currents will be :

$$\bar{I}_{R0} = \frac{1}{3}(\bar{I}_R + \bar{I}_Y + \bar{I}_B) = \frac{1}{3}\bar{I}_R$$

$$\bar{I}_{R1} = \frac{1}{3}(\bar{I}_R + \lambda\bar{I}_Y + \lambda^2\bar{I}_B) = \frac{1}{3}\bar{I}_R$$

$$\bar{I}_{R2} = \frac{1}{3}(\bar{I}_R + \lambda^2\bar{I}_Y + \lambda\bar{I}_B) = \frac{1}{3}\bar{I}_R$$

$$\therefore \quad \bar{I}_{R0} = \bar{I}_{R1} = \bar{I}_{R2} = \frac{1}{3}\bar{I}_R \qquad (\text{or} \quad \bar{I}_R = 3\bar{I}_{R0})$$

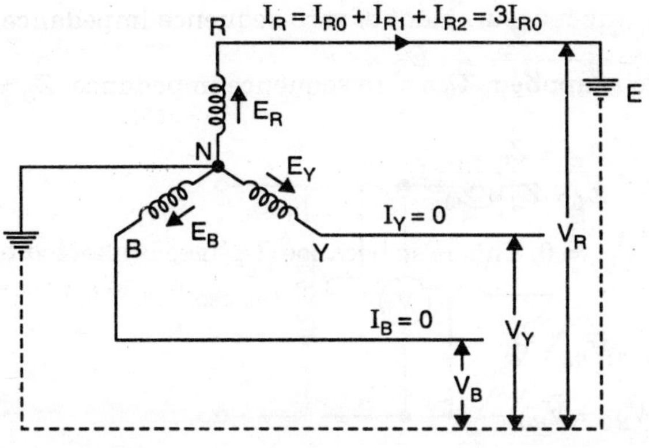

Fig. 21.60

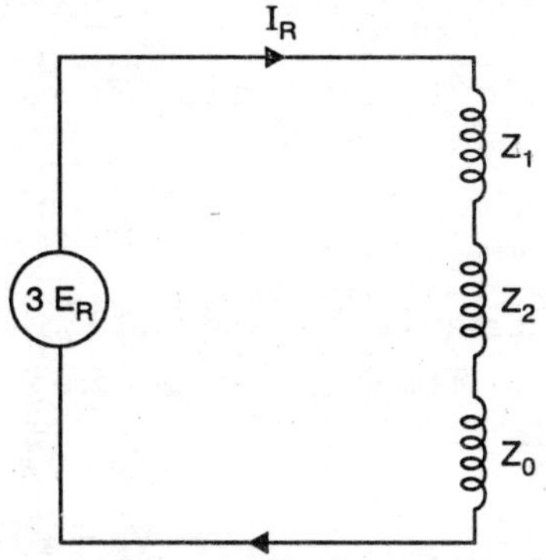

Fig. 21.61

Fault current $\overline{I}_R$:

Let $\overline{Z}_1, \overline{Z}_2, \overline{Z}_0$ = Positive, negative and zero sequence impedances respectively.

Considering closed loop NREN, we have :

$$\overline{E}_R = \overline{I}_{R1}\overline{Z}_1 + \overline{I}_{R2}\overline{Z}_2 + \overline{I}_{R0}\overline{Z}_0 + \overline{V}_R$$

or, $$\overline{E}_R = \overline{I}_{R0}(\overline{Z}_1 + \overline{Z}_1 + \overline{Z}_0) \qquad (\because \ \overline{V}_R = 0 \, ; \overline{I}_{R1} = \overline{I}_{R2} = \overline{I}_{R0})$$

$$\therefore \qquad \overline{I}_{R0} = \frac{\overline{E}_R}{\overline{Z}_1 + \overline{Z}_2 + \overline{Z}_0}$$

$\therefore$ Fault current, $$\overline{I}_R = 3\overline{I}_{R0} = \frac{3\overline{E}_R}{\overline{Z}_1 + \overline{Z}_2 + \overline{Z}_0} \qquad ...(21.36)$$

Fig. 21.61 shows equivalent circuit.

The above eqn. has been derived by assuming fault impedance to be zero. However, if the fault impedance is Z_e, the expression for the fault current becomes

$$\overline{I}_R = \frac{3\overline{E}_R}{\overline{Z}_1 + \overline{Z}_2 + \overline{Z}_0 + \overline{Z}_e} \qquad ...(21.37)$$

It may be added that if the neutral is not grounded the zero sequence impedance will be infinite and the fault current is zero (It is so because now no path exists for flow of fault current).

Phase voltages at fault :

As the generated e.m.f. system is of positive sequence only, the sequence components of e.m.f. in R-phase are :

$$\overline{E}_1 = \overline{E}_R \, ; \overline{E}_2 = 0 \, ; \overline{E}_0 = 0$$

The sequence voltages at the fault for R-phase are :

$$\overline{V}_{R1} = \overline{E}_R - \overline{I}_{R1}\overline{Z}_1 = \overline{E}_R - \frac{\overline{E}_R\overline{Z}_1}{\overline{Z}_1 + \overline{Z}_2 + \overline{Z}_0}$$

$$\therefore \qquad \overline{V}_{R1} = \frac{\overline{Z}_2 + \overline{Z}_0}{\overline{Z}_1 + \overline{Z}_2 + \overline{Z}_0} \overline{E}_R \qquad ...(21.38)$$

$$\bar{V}_{R2} = 0 - \bar{I}_{R2}\bar{Z}_2 = \frac{-\bar{Z}_2}{\bar{Z}_1 + \bar{Z}_2 + \bar{Z}_0}\,\bar{E}_R \quad \text{...(21.39)}$$

$$\bar{V}_{R0} = 0 - \bar{I}_{R0}\bar{Z}_0 = \frac{-\bar{Z}_0}{\bar{Z}_1 + \bar{Z}_2 + \bar{Z}_0}\,\bar{E}_R \quad \text{...(21.40)}$$

It may be observed that $\bar{V}_{R1} + \bar{V}_{R2} + \bar{V}_{R0} = 0$. This is so because R-phase is shorted to ground.

∴ The *phase voltages* are :

$$\bar{V}_R = \bar{V}_{R1} + \bar{V}_{R2} + \bar{V}_{R0} = 0 \quad \text{...(21.41)}$$

$$\bar{V}_Y = \lambda^2\bar{V}_{R1} + \lambda\bar{V}_{R2} + \bar{V}_{R0} \quad \text{...(21.42)}$$

$$\bar{V}_B = \lambda\bar{V}_{R1} + \lambda^2\bar{V}_{R2} + \bar{V}_{R0} \quad \text{...(21.43)}$$

21.4.4.2. Line-to-line fault

Fig. 21.62 shows a line-to-line fault between the B (Blue) and Y (Yellow) lines.

Here, $\bar{V}_Y = \bar{V}_B \,;\, \bar{I}_R = 0 \,;\, \bar{I}_Y + I_B = 0$

Taking P-phase as the reference, we have

$$\bar{I}_0 = \frac{1}{3}(\bar{I}_R + \bar{I}_Y + \bar{I}_B) = 0$$

Now, $\bar{V}_Y = \bar{V}_B$

Expressing in terms of sequence components of red (R), line, we get

$$\lambda^2\bar{V}_{R1} + \lambda\bar{V}_{R2} + \bar{V}_{R0} = \lambda\bar{V}_{R1} + \lambda^2\bar{V}_{R2} + \bar{V}_{R0}$$

or, $$\bar{V}_{R1}(\lambda^2 - \lambda) = \bar{V}_{R2}(\lambda^2 - \lambda)$$

∴ $$\bar{V}_{R1} = \bar{V}_{R2} \quad \text{...(21.44)}$$

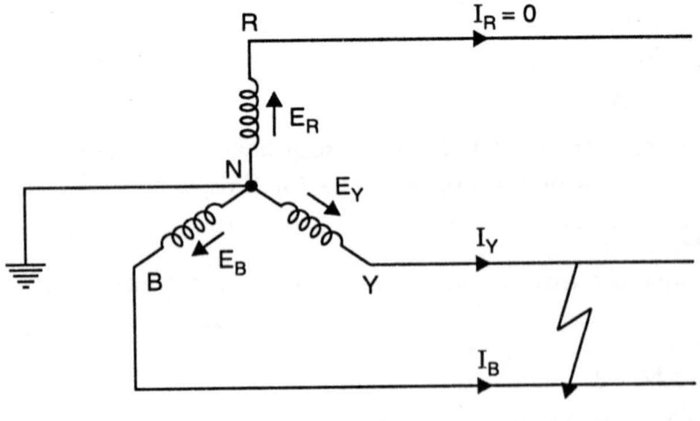

Fig. 21.62

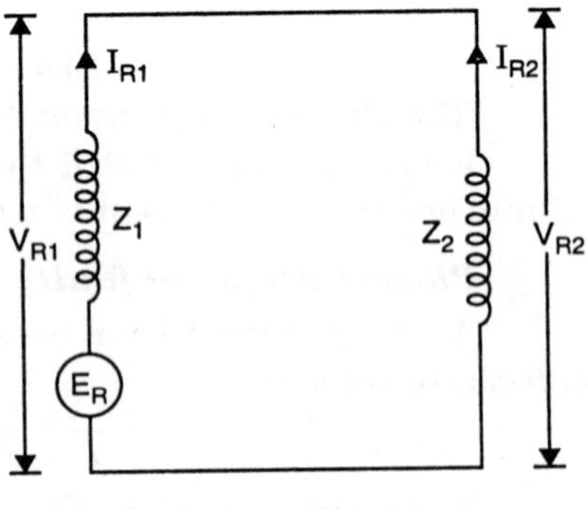

Fig. 21.63

Also, $\bar{I}_Y + \bar{I}_B = 0$

or, $$(\lambda^2\bar{I}_{RI} + \lambda\bar{I}_{R2} + \bar{I}_{R0}) + (\lambda\bar{I}_{R1} + \lambda^2\bar{I}_{R2} + \bar{I}_{R0}) = 0$$

or, $$(\lambda^2 + \lambda)(\bar{I}_{R1} + \bar{I}_{R2}) + 2\bar{I}_{R0} = 0$$

or, $$\bar{I}_{R1} + \bar{I}_{R2} = 0 \quad \text{...(21.45)} \quad [\because \bar{I}_{R0} = 0]$$

Fault current $\bar{I}_Y$ $(= -\bar{I}_B)$ **:**

The connection diagram for the sequence impedances is shown in Fig. 21.62.

Here, $$\bar{I}_{R1} = -\bar{I}_{R2} = \frac{\bar{E}_R}{\bar{Z}_1 + \bar{Z}_2} \qquad \text{...[21.45 (a)]}$$

Fault current, $\bar{I}_Y (= -\bar{I}_B) = \lambda^2 \bar{I}_{R1} + \lambda \bar{I}_{R2} + \bar{I}_{R0}$

$$= \lambda^2 \times \frac{\bar{E}_R}{\bar{Z}_1 + \bar{Z}_2} + \lambda \times \left(\frac{-\bar{E}_R}{\bar{Z}_1 + \bar{Z}_2} \right) + 0$$

$$= (\lambda^2 - \lambda) \frac{\bar{E}_R}{\bar{Z}_1 + \bar{Z}_2}$$

or, $$\bar{I}_Y (= -\bar{I}_B) = \frac{-j\sqrt{3}\, \bar{E}_R}{\bar{Z}_1 + \bar{Z}_2} \qquad \text{...(21.46)}$$

Phase voltages :

As the generated e.m.f. system is of positive phase sequence only, the sequence components of e.m.f. in R-phase are :

$$\bar{E}_0 = 0 \,;\, \bar{E}_2 = 0 \quad \text{and} \quad \bar{E}_1 = \bar{E}_R$$

The sequence voltages at the fault for B-phase are :

$$\bar{V}_{R1} = \bar{E}_R - \bar{I}_{R1}\bar{Z}_1 = \bar{E}_R - \left(\frac{\bar{E}_R}{\bar{Z}_1 + \bar{Z}_2} \right) \bar{Z}_1$$

$\therefore$ $$\bar{V}_{R1} = \frac{\bar{Z}_2}{\bar{Z}_1 + \bar{Z}_2} \bar{E}_R$$

$$\bar{V}_{R2} = 0 - \bar{I}_{R2}\bar{Z}_2 = \frac{\bar{Z}_2}{\bar{Z}_1 + \bar{Z}_2} \bar{E}_R$$

$$\bar{V}_{R0} = 0 - \bar{I}_{R0}\bar{Z}_0 = 0$$

The phase voltages at fault are :

$$\bar{V}_R = \bar{V}_{R1} + \bar{V}_{R2} + \bar{V}_{R0}$$

$$= \frac{\bar{Z}_2}{\bar{Z}_1 + \bar{Z}_2} \bar{E}_R + \frac{\bar{Z}_2}{\bar{Z}_1 + \bar{Z}_2} \bar{E}_R + 0$$

$\therefore$ $$\bar{V}_R = \frac{2\bar{Z}_2}{\bar{Z}_1 + \bar{Z}_2} \bar{E}_R \qquad \text{...(21.47)}$$

$$\bar{V}_Y = \lambda^2 \bar{V}_{R1} + \lambda \bar{V}_{R2} + \bar{V}_{R0}$$

$$= \lambda^2 \left(\frac{\bar{Z}_2}{\bar{Z}_1 + \bar{Z}_2} \bar{E}_R \right) + \lambda \left(\frac{\bar{Z}_2}{\bar{Z}_1 + \bar{Z}_2} \bar{E}_R \right) + 0$$

$$= (\lambda^2 + \lambda) \left(\frac{\bar{Z}_2}{\bar{Z}_1 + \bar{Z}_2} \bar{E}_R \right)$$

$$\therefore \qquad \overline{V}_Y = -\frac{\overline{Z}_2}{\overline{Z}_1 + \overline{Z}_2}\overline{E}_R \qquad (\because \quad \lambda^2 + \lambda = -1) \quad ...(21.48)$$

$$\overline{V}_B = \lambda \overline{V}_{R1} + \lambda^2 \overline{V}_{R2} + \overline{V}_{R0}$$

$$= \lambda \left(\frac{\overline{Z}_2}{\overline{Z}_1 + \overline{Z}_2} \overline{E}_R \right) + \lambda^2 \left(\frac{\overline{Z}_2}{\overline{Z}_1 + \overline{Z}_2} \overline{E}_R \right)$$

$$= (\lambda^2 + \lambda) \left(\frac{\overline{Z}_2}{\overline{Z}_1 + \overline{Z}_2} \overline{E}_R \right)$$

$$\therefore \qquad \overline{V}_B = -\frac{\overline{Z}_2}{\overline{Z}_1 + \overline{Z}_2}\overline{E}_R \qquad ...(21.49)$$

21.4.4.3. Double line-to-ground fault

Double line-to-ground fault involving *Y-B* lines and earth is shown in Fig. 21.64.

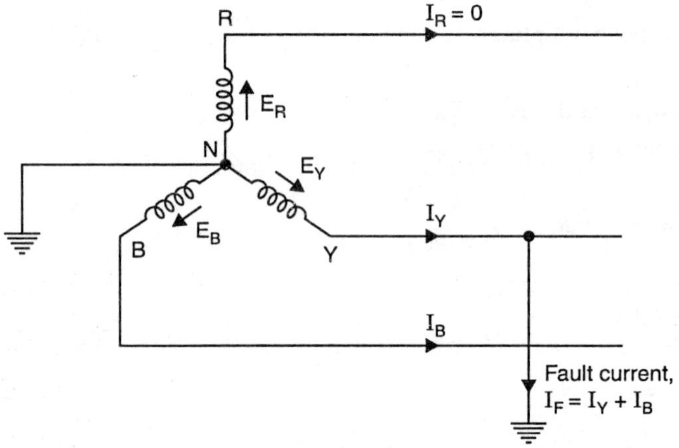

Fig. 21.64

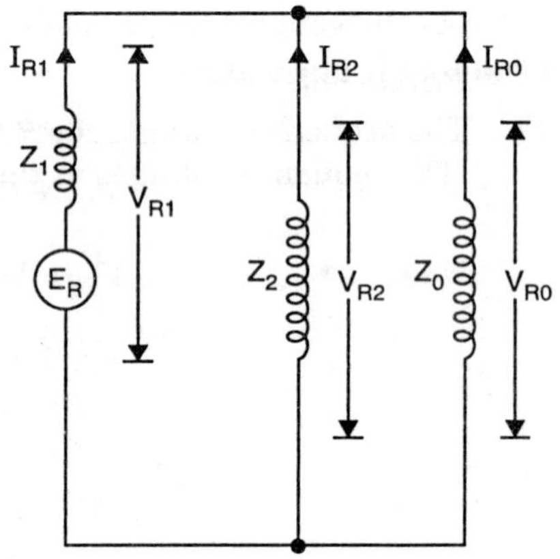

Fig. 21.65

Here, $\overline{I}_R = 0 \,;\, \overline{V}_Y = \overline{V}_B = 0$

Since, $\overline{V}_Y = \overline{V}_B$, therefore,

$$\overline{V}_{R1} = \overline{V}_{R2} = \overline{V}_{R0} = \frac{1}{3}\overline{V}_R \qquad ...(21.50)$$

Also, $\overline{I}_R = \overline{I}_{R1} + \overline{I}_{R2} + \overline{I}_{R0} = 0$ ...(*Given*) ...(21.51)

Fault current, I_F ***(=*** $\overline{I}_Y + \overline{I}_B$***) :***

The connection diagram for the sequence impedances is shown in Fig. 21.65.

Here,
$$\overline{I}_{R1} = \frac{\overline{E}_R}{\overline{Z}_1 + \dfrac{\overline{Z}_2 \overline{Z}_0}{\overline{Z}_2 + \overline{Z}_0}} \qquad ...(21.52)$$

$$\overline{I}_{R2} = -\overline{I}_{R1}\left(\frac{\overline{Z}_0}{\overline{Z}_2 + \overline{Z}_0}\right) \qquad ...(21.53)$$

$$\bar{I}_{R0} = -\bar{I}_{R1}\left(\frac{\bar{Z}_2}{\bar{Z}_2 + \bar{Z}_0}\right) \quad ...(21.54)$$

Fault current, $I_F = \bar{I}_Y + \bar{I}_B = 3\bar{I}_{R0}$ ...[21.54 (a)]

$$\left[\begin{array}{l} \because \ \bar{I}_{R0} = \frac{1}{3}(\bar{I}_R + \bar{I}_Y + \bar{I}_B) = \frac{1}{3}(0 + I_F) \\ \text{or} \ \ I_F = 3\bar{I}_{R0} \end{array}\right]$$

$$= 3\left(-\bar{I}_{R1}\frac{\bar{Z}_2}{\bar{Z}_2 + \bar{Z}_0}\right)$$

$$= -\frac{3\bar{Z}_2}{\bar{Z}_2 + \bar{Z}_0} \times \frac{\bar{E}_R}{\left(\bar{Z}_1 + \frac{\bar{Z}_2\bar{Z}_0}{\bar{Z}_2 + \bar{Z}_0}\right)}$$

$$\therefore \quad I_F = -\frac{3\bar{Z}_2\bar{E}_R}{\bar{Z}_1\bar{Z}_2 + \bar{Z}_1\bar{Z}_0 + \bar{Z}_2\bar{Z}_0} \quad ...(21.55)$$

Phase voltages :

The sequence voltages for R-phase are :

$$\bar{V}_{R1} = \bar{E}_R - \bar{I}_{R1}\bar{Z}_1 \ ; \ \bar{V}_{R2} = 0 - \bar{I}_{R2}\bar{Z}_2 \ ; \ \bar{V}_{R0} = 0 - \bar{I}_{R0}\bar{Z}_0$$

Now, $\bar{V}_{R1} = \bar{V}_{R2} = \bar{V}_{R0} = \frac{1}{3}\bar{V}_R$

$$\therefore \quad \bar{V}_R = \bar{V}_{R1} + \bar{V}_{R2} + \bar{V}_{R0} = 3\bar{V}_{R2} \quad ...(21.56)$$

$$\bar{V}_Y = \lambda^2\bar{V}_{R1} + \lambda\bar{V}_{R2} + \bar{V}_{R0} = (\lambda^2 + \lambda + 1)\,\bar{V}_{R2} \quad (\because \ \bar{V}_{R1} = \bar{V}_{R2} = \bar{V}_{R0})$$

$$= 0 \times \bar{V}_{R2} = 0 \quad ...(21.57)$$

$$\bar{V}_B = \lambda\bar{V}_{R1} + \lambda^2\bar{V}_{R2} + \bar{V}_{R0} = (\lambda + \lambda^2 + 1)\,\bar{V}_{R2} = 0 \quad ...(21.58)$$

Example 21.38. *A 3-phase, 8 MVA, 11 kV generator with solidly earthed neutral point supplies a feeder. The relevant impedances of the generator and feeder in ohms are as under :*

	Impedances		
	Positive sequence ($\bar{Z}_1$)	*Negative sequence ($\bar{Z}_2$)*	*Zero sequence ($\bar{Z}_0$)*
Generator	*j2.4*	*j1.8*	*j0.8*
Feeder	*j2.0*	*j2.0*	*j6.0*

If a fault from one phase to earth occurs on the far end of the feeder, calculate :

(i) The magnitude of fault current.

(ii) Line-to-neutral voltage at the generator terminal.

Solution. The circuit diagram is shown in Fig. 21.66.

Let us assume that the fault occurs on R-phase. Taking R-phase as the reference,

Phases e.m.f. of R-phase $= \frac{11 \times 10^3}{\sqrt{3}} = 6350$ V (*i*) **The magnitude of fault current, I_R :**

The total impedance to any sequence current is the sum of generator and feeder impedances.

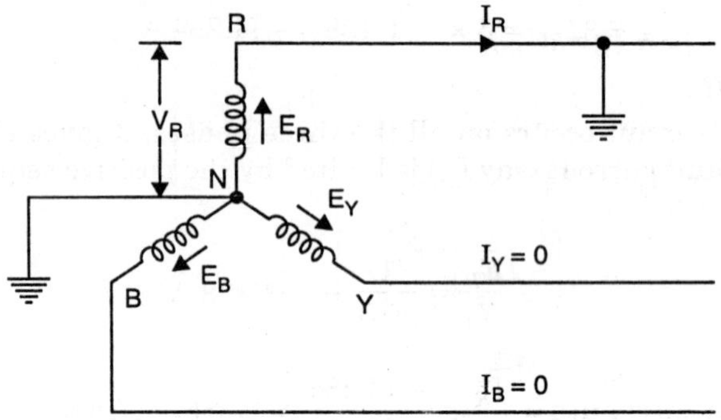

Fig. 21.66

∴ Total $\bar{Z}_1 = j2.4 + j2.0 = j4.4\ \Omega$

$\bar{Z}_2 = j1.8 + j2.0 = j3.8\ \Omega$

$\bar{Z}_0 = j0.8 + j6.0 = j6.8\ \Omega$

In case of a line-to-ground fault, we have,

$$\bar{I}_{R1} = \bar{I}_{R2} = \bar{I}_{R0} = \frac{\bar{E}_R}{\bar{Z}_1 + \bar{Z}_2 + \bar{Z}_0} \qquad \text{...[Eqn. (21.36)]}$$

$$= \frac{6350}{j4.4 + j3.8 + j6.8} = \frac{6350}{j15} = -j423.33 \text{ A}$$

∴ *Fault current,* $\bar{I}_R = 3\bar{I}_{R0} = 3 \times -j423.33 = -$**j1270 A. (Ans.)**

(*ii*) **Line-to-neutral voltage at the generator terminal :**

Line-to-neutral voltage of *R*-phase,

$$\bar{V}_R = \bar{E}_R - \bar{I}_{R1}\bar{Z}_1 - \bar{I}_{R2}\bar{Z}_2 - \bar{I}_{R0}\bar{Z}_0$$
$$= \bar{E}_R - I_{R0}(\bar{Z}_1 + \bar{Z}_2 + \bar{Z}_0)$$
$$= 6350 - (-j423.33)(j2.4 + j1.8 + j0.8)$$
$$= 6350 - (-j423.33) \times j5 = \textbf{4233.35 V.} \quad \textbf{(Ans.)}$$

Example 21.39. *A 3-phase, 11 kV, 12 MVA alternator has positive, negative and zero sequence reactances as 0.3 p.u, 0.3 p.u. and 0.1 p.u. respectively. If the generator has no load, find the ratio of fault currents for line-to-ground fault to that when all the 3-phases are dead short-circuited.*

Solution. *Given :* Rating of the alternator = 11 kV, 12 MVA ;

$$X_1 = 0.3 \text{ p.u.} ; X_2 = 0.3 \text{ p.u.} ; X_0 = 0.1 \text{ p.u.}$$

Taking *R*-phase as the reference, let its phase e.m.f. be $\bar{E}_R = 1$ p.u.

Line-to-ground fault :

Assume that the fault occurs on the *R*-phase. Then,

$$\bar{I}_{R1} = \bar{I}_{R2} = \bar{I}_{R0} = \frac{\bar{E}_R}{\bar{X}_1 + \bar{X}_2 + \bar{X}_0}$$

$$\therefore \quad \bar{I}_{R0} = \frac{1}{j0.3 + j0.3 + j0.1} = \frac{1}{j0.7} = -j1.428 \text{ A}$$

$\therefore$ Fault current, $\bar{I}_R = 3\bar{I}_{R0} = 3 \times -j1.428 = -j4.284$ A.

Three-phase fault :

When a dead short-circuit occurs on all the three phases, it gives rise to symmetrical fault currents. Therefore, the fault current (say I_{sc}) is limited by the positive sequence reactance (*i.e.*, X_1) only.

$\therefore$ Fault current,
$$I_{sc} = \frac{\bar{E}_R}{\bar{X}_1} = \frac{1}{j0.3} = -j3.333 \text{ A}$$

Ratio of two fault currents $= \dfrac{\bar{I}_R}{I_{sc}} = \dfrac{-j4.284}{-j3.333} =$ **1.285. (Ans.)**

Example 21.40. *A 3-phase, 11 kV, 20 MVA generator with positive, negative and zero sequence reactances as 0.4 p.u., 0.4 p.u and 0.1 p.u. respectively is grounded through a reactance of 0.6 Ω. Calculate the fault current for a single line-to-ground fault.*

Solution. *Given :* Rating of the generator = 11 kV, 20 MVA ;

$X_1 = 0.4$ p.u. ; $X_2 = 0.4$ p.u. ; $X_0 = 0.1$ p.u., $X_N = 0.6\ \Omega$.

The circuit diagram is shown in Fig. 21.67. Assume that the fault occurs on R-phase. Taking R-phase as the reference, let its phase e.m.f. be $\bar{E}_R = 1$ p.u.

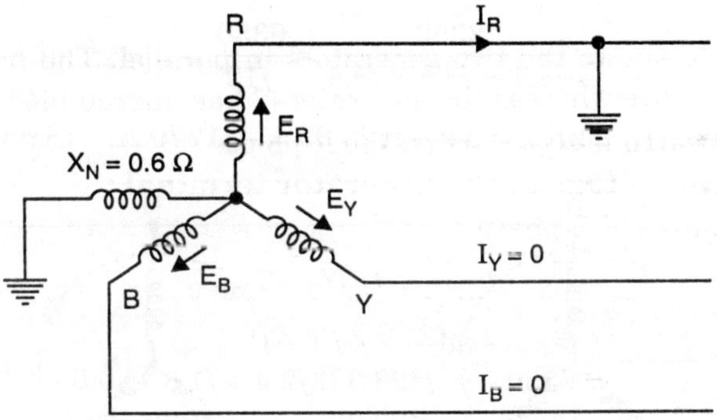

Fig. 21.67

Let as first convert the reactance X_N into p.u. values as follows :

p.u. value of
$$X_N = X_N \text{ (in ohms)} \times \frac{\text{kVA rating}}{(\text{kV})^2 \times 1000}$$
$$= 0.6 \times \frac{20 \times 1000}{(11)^2 \times 1000} = 0.1 \text{ p.u.}$$

For a line-to-ground fault, we have

$$\bar{I}_{R1} = \bar{I}_{R2} = \bar{I}_{R0} = \frac{\bar{E}_R}{\bar{X}_1 + \bar{X}_2 + (\bar{X}_0 + 3X_N)}$$
$$= \frac{1}{j0.4 + j0.4 + j(0.1 + 3 \times 0.1)} = \frac{1}{j1.2} = -j0.833$$

Fault current, $\bar{I}_R = 3\bar{I}_{R0} = 3 \times (-j0.833) = -j2.5$ p.u.

∴ Fault current in amperes = Rated current × p.u. value

$$= \frac{20 \times 10^6}{\sqrt{3} \times 11 \times 10^3} \times 2.5 = \mathbf{2624.3\ A} \quad \textbf{(Ans.)}$$

Example 21.41. *The per unit values of positive, negative and zero sequence reactances of a network at fault are 0.16, 0.14 and 0.1. Determine the fault current if the fault is double line-to-ground.*

Solution. *Given :* $\bar{X}_1 = j0.16$ p.u. ; $\bar{X}_2 = j0.14$ p.u. ; $\bar{X}_0 = j0.1$ p.u.

For a double line-to-ground fault, we have

Fault current, $$I_F = \bar{I}_Y + \bar{I}_B = -\frac{3\bar{X}_2\bar{E}_R}{\bar{X}_1\bar{X}_2 + \bar{X}_1\bar{X}_0 + \bar{X}_2\bar{X}_0} \quad \text{...[Eqn. (21.55)]}$$

$$= -\frac{3 \times j0.14 \times 1}{j \times 0.16 \times j0.14 + j0.16 \times j0.1 + j0.14 \times j0.1}$$

$$= \frac{-j0.42}{-(0.0224 + 0.016 + 0.014)} = \frac{j0.42}{0.0524} = \mathbf{j8\ p.u.} \quad \textbf{(Ans.)}$$

Example 21.42. *A 25 MVA, 11 kV, 3-phase, 50 Hz generator has its neutral earthed through a 4% reactor. It is in parallel with another identical generator having isolated neutral. Each generator has a positive sequence reactance of 16%, negative sequence reactance of 8% and zero sequence reactance of 12%. If a line-to-ground short circuit occurs in the common bus-bar, determine the fault current.*

Solution. Fig. 21.68 shows the two generators in parallel. The neutral of generator–1 is earthed through a 5% reactance whereas the generator–2 has ungrounded neutral.

Let as assume that earth fault occurs on the R-phase

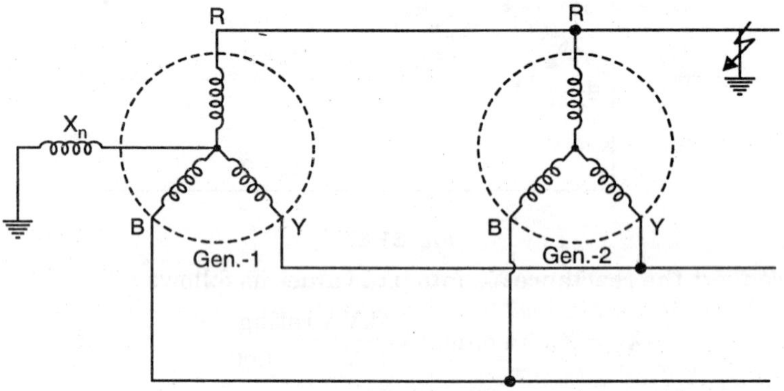

Fig. 21.68. Two generators connected in parallel.

Taking R-phase as the reference, its phase e.m.f. $\bar{E}_R = \frac{11 \times 10^3}{\sqrt{3}} = 6351$ V. Fig. 21.69 shows the equivalent circuit diagram. It may be noted that the positive sequence reactances (16%) of two generators are in parallel and so are their negative reactances (8%). The zero sequence reactance of generator–2 is zero because its neutral is ungrounded. However, the zero sequence reactance of generator = 12% + 3 × 4% = 24%.

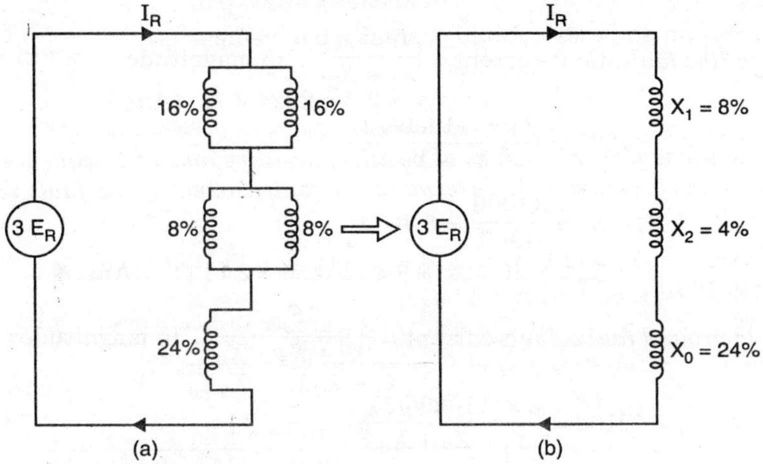

Fig. 21.69. Connection diagram for reactances.

The percentage reactances in Fig. 21.69 (*b*) can be converted into ohmic values as under :

$$X_1 = \% \text{ reactance} \times \frac{(\text{Voltage in kV})^2 \times 10}{\text{Base kVA}}$$

$$= 8 \times \frac{(11)^2 \times 10}{25 \times 10^3} = 0.3872\ \Omega$$

$$X_2 = 4 \times \frac{(11)^2 \times 10}{25 \times 10^3} = 0.1936\ \Omega$$

$$X_0 = 24 \times \frac{(11)^2 \times 10}{25 \times 10^3} = 1.1616\ \Omega$$

$\therefore$ Fault current, $\bar{I}_R = \dfrac{3\bar{E}_R}{\bar{X}_1 + \bar{X}_2 + \bar{X}_0}$

$$= \frac{3 \times 6351}{j0.3872 + j0.1936 + j1.1616} = \frac{19053}{j1.7424} = -\mathbf{j10935\ A.} \quad \textbf{(Ans.)}$$

Example 21.43. *A 25 MVA, 11 kV, 3-phase alternator was subjected to the following of faults : 3-phase fault = 1000 A ; Line-to-line fault = 1300 A ; Line-to-ground fault = 2100 A. The generator neutral is solidly grounded. Ignoring resistances calculate the values of three sequence reactances of the alternator.*

Solution. Let X_1, X_2 and X_0 be the positive, negative and zero sequence reactances respectively of the alternator.

For a *3-phase fault*, fault current = $\dfrac{E_{ph}}{X_1}$ ... in magnitude

or $$1000 = \frac{11000/\sqrt{3}}{X_1}$$

$\therefore$ $$\mathbf{X_1} = \frac{11000/\sqrt{3}}{1000} = \mathbf{6.35\ \Omega.} \quad \textbf{(Ans.)}$$

For a *line-to-line fault*, fault current = $\dfrac{\sqrt{3}\, E_{ph}}{X_1 + X_2}$ ... in magnitude

or, $$1300 = \frac{\sqrt{3} \times (11000/\sqrt{3})}{X_1 + X_2}$$

or, $$X_1 + X_2 = \frac{11000}{1300} = 8.46\ \Omega$$

∴ $$\mathbf{X_2} = 8.46 - X_1 = 8.46 - 6.35 = \mathbf{2.11\ \Omega.}\quad \textbf{(Ans.)}$$

For a *line-to-ground fault*, fault current = $\dfrac{3E_{ph}}{X_1 + X_2 + X_0}$ ... in magnitude

or, $$2100 = \frac{3 \times (11000/\sqrt{3})}{X_1 + X_2 + X_0}$$

or, $$X_1 + X_2 + X_0 = \frac{3 \times (11000/\sqrt{3})}{2100} = 9.073$$

∴ $$\mathbf{X_0} = 9.073 - X_1 - X_2 = 9.073 - 6.35 - 2.11 = \mathbf{0.613\ \Omega.}\quad \textbf{(Ans.)}$$

Example 21.44. *A dead earth fault occurs on one conductor of a 3-conductor cable supplied by a 12 MVA 3-phase alternator with earthed neutral. The alternator has positive, negative and zero phase sequence impedances of (0.25 + j2.35), (0.1 + j0.3) and (0 + j0.215) ohm per phase. The corresponding line-to-neutral values for the cable upto the fault position are (0.18 + j0.125), (0.18 + j0.125) and (1.45 + j0.475) ohms. If the generator is excited to give 6600 V between lines on open circuit, calculate :*

(i) The fault current.

(ii) Sequence components of the current.

(iii) Voltages of the sound lines to earth at the fault.

Solution. Refer to Fig. 21.70. $\overline{Z}_{g1} = (0.25 + j2.35)\ \Omega$; $\overline{Z}_{g2} = (0.1 + j0.3)\ \Omega$; $\overline{Z}_{g0} = (0 + j0.215)$;

$$\overline{Z}_{l1} = (0.18 + j0.125)\ \Omega\ ;\ \overline{Z}_{l2} = (0.18 + j0.125)\ \Omega\ ;\ \overline{Z}_{l0} = (1.45 + j0.475)$$

(*i*) **The fault current :**

Now, $$\overline{Z}_1 = \overline{Z}_{g1} + \overline{Z}_{l1} = (0.25 + j2.35) + (0.18 + j0.125) = (0.43 + j2.475)\ \Omega$$

$$\overline{Z}_2 = \overline{Z}_{g2} + \overline{Z}_{l2} = (0.1 + j0.3) + (0.18 + j0.125) = (0.28 + j0.425)\ \Omega$$

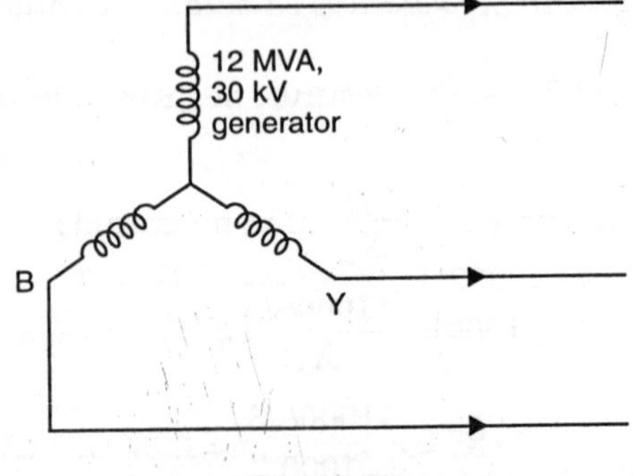

Fig. 21.70

$$\bar{Z}_0 = \bar{Z}_{g0} + \bar{Z}_{l0} = (0 + j0.215) + (1.45 + j0.475) = (1.45 + j0.69)\ \Omega$$

$$\bar{I}_R = \bar{I}_{R1} + \bar{I}_{R2} + \bar{I}_{R0} = \frac{3\bar{E}_R}{\bar{Z}_1 + \bar{Z}_2 + \bar{Z}_0}$$

or, *Fault current,* $\bar{\mathbf{I}}_{\mathbf{R}} = \dfrac{3 \times (6600/\sqrt{3})}{(0.43 + j2.475) + (0.28 + j0.425) + (1.45 + j0.69)}$

$$= \frac{11431.5}{(2.16 + j3.59)} = \frac{11431.5}{4.19 \angle 59°} = \mathbf{2728 \angle{-59°}\ A.} \quad \textbf{(Ans.)}$$

(*ii*) **Sequence components of the current :**

$$\bar{I}_{R1} = \bar{I}_{R2} = \bar{I}_{R0} = \frac{\bar{I}_R}{3} = \frac{2728 \angle -59°}{3} = \mathbf{909.3 \angle{-59°}\ A.} \quad \textbf{(Ans.)}$$

Note, $\bar{I}_Y = \bar{I}_{Y1} + \bar{I}_{Y2} + \bar{I}_{Y0} = \lambda^2\bar{I}_{R1} + \lambda\bar{I}_{R2} + \bar{I}_{R0} = \bar{I}_{R1}(\lambda^2 + \lambda + 1) = 0$

$(\because\ \lambda^2 + \lambda + 1 = 0)$

$$\bar{I}_B = \bar{I}_{B1} + \bar{I}_{B2} + \bar{I}_{B0} = \lambda\bar{I}_{R1} + \lambda^2\bar{I}_{R2} + \bar{I}_{R0} = \bar{I}_{R1}(\lambda + \lambda^2 + 1) = 0$$

(*iii*) **Voltages of the sound lines to earth at the fault :**

$$\bar{V}_Y = \bar{V}_{Y1} + \bar{V}_{Y2} + \bar{V}_{Y0} = \lambda^2\bar{V}_{R1} + \lambda\bar{V}_{R2} + \bar{V}_{R0}$$

$$= \lambda^2(E_R + \bar{I}_{R1}\bar{Z}_1) + \lambda(0 - \bar{I}_{R2}\bar{Z}_2) + (0 - \bar{I}_{R0}\bar{Z}_0)$$

$$= \lambda^2(\bar{E}_R - \bar{I}_{R0}\bar{Z}_1) + \lambda(0 - \bar{I}_{R0}\bar{Z}_2) + (0 - \bar{I}_{R0}\bar{Z}_0)$$

$$= \lambda^2\bar{E}_R - \bar{I}_{R0}(\lambda^2\bar{Z}_1 + \lambda\bar{Z}_2 + \bar{Z}_0)$$

$$= \lambda^2\bar{E}_R - \frac{\bar{E}_R}{\bar{Z}_1 + \bar{Z}_2 + \bar{Z}_0}(\lambda^2\bar{Z}_1 + \lambda\bar{Z}_2 + \bar{Z}_0)$$

$$= \frac{\lambda^2\bar{E}_R(\bar{Z}_1 + \bar{Z}_2 + \bar{Z}_0) - \bar{E}_R(\lambda^2\bar{Z}_1 + \lambda\bar{Z}_2 + \bar{Z}_0)}{\bar{Z}_1 + \bar{Z}_2 + \bar{Z}_0}$$

$$= \frac{\bar{E}_R}{\bar{Z}_1 + \bar{Z}_2 + \bar{Z}_0}[\lambda^2(\bar{Z}_1 + \bar{Z}_2 + \bar{Z}_0) - (\lambda^2\bar{Z}_1 + \lambda\bar{Z}_2 + \bar{Z}_0)]$$

$$= \frac{\bar{E}_R}{\bar{Z}_1 + \bar{Z}_2 + \bar{Z}_0}[\lambda^2\bar{Z}_1 + \lambda^2\bar{Z}_2 + \lambda^2\bar{Z}_0 - \lambda^2\bar{Z}_1 - \lambda\bar{Z}_2 - \bar{Z}_0]$$

$$= \frac{\bar{E}_R}{\bar{Z}_1 + \bar{Z}_2 + \bar{Z}_0}[(\lambda^2 - \lambda)\bar{Z}_2 + (\lambda^2 - 1)\bar{Z}_0]$$

$$= \frac{6600/\sqrt{3}}{4.19 \angle 59°}[(0 - j1.732)(0.28 + j0.425) + (-0.5 - j0.866 - 1)(1.45 + j0.69)]$$

$(\because\ \bar{Z}_1 + \bar{Z}_2 + \bar{Z}_0 = 4.19 \angle 59°$ ...Calculated earlier)

$$= 909.4 \angle{-59°}\ [(1.732 \angle{-90°} \times 0.509 \angle 56.6°) + (1.732 \angle{-150°} \times 1.606 \angle 25.45°)]$$

$$= 909.4 \angle{-59°}\ [0.882 \angle{-33.4°} + 2.78 \angle{-124.55°}]$$

$$= 909.4 \angle{-59°} \times 2.9 \angle{-106.8°} = 2637.3 \angle{-165.8°}$$

i.e., $\bar{\mathbf{V}}_{\mathbf{Y}} = \mathbf{2637.3 \angle{-165.8°}\ V.}$ **(Ans.)**

Similarly, $\overline{V}_B = \overline{V}_{B1} + \overline{V}_{B2} + \overline{V}_{B0} = \lambda\overline{V}_{R1} + \lambda^2\overline{V}_{R2} + \overline{V}_{R0}$

$$= \lambda(\overline{E}_R - \overline{I}_{R1}\overline{Z}_1) + \lambda^2(0 - \overline{I}_{R2}\overline{Z}_2) + (0 - \overline{I}_{R0}\overline{Z}_0)$$

$$= \lambda(\overline{E}_R - \overline{I}_{R0}\overline{Z}_1) + \lambda^2(0 - \overline{I}_{R0}\overline{Z}_2) + (-\overline{I}_{R0}\overline{Z}_0)$$

$$= \lambda\overline{E}_R - \overline{I}_{R0}(\lambda^2\overline{Z}_2 + \lambda\overline{Z}_1 + \overline{Z}_0)$$

$$= \lambda\overline{E}_R - \frac{\overline{E}_R}{\overline{Z}_1 + \overline{Z}_2 + \overline{Z}_0}(\lambda^2\overline{Z}_2 + \lambda\overline{Z}_1 + \overline{Z}_0)$$

$$= \frac{\lambda\overline{E}_R(\overline{Z}_1 + \overline{Z}_2 + \overline{Z}_0) - \overline{E}_R(\lambda^2\overline{Z}_2 + \lambda\overline{Z}_1 + \overline{Z}_0)}{\overline{Z}_1 + \overline{Z}_2 + \overline{Z}_0}$$

$$= \frac{\overline{E}_R}{\overline{Z}_1 + \overline{Z}_2 + \overline{Z}_0}[(\lambda - \lambda^2)\overline{Z}_2 + (\lambda - 1)\overline{Z}_0]$$

$$= \frac{6600/\sqrt{3}}{4.19\angle 59°}[(0 + j1.732)(0.28 + j0.425) + (-1.5 + j0.1866)(1.45 + j0.69)]$$

$$= 909.4\angle -59° [1.732\angle 90° \times 0.509\angle 56.6° + 1.732\angle 150° \times 1.606\angle 25.45°]$$

$$= 909.4\angle -59° [0.882\angle 146.6° + 2.78\angle 175.45°]$$

$$= 909.4\angle -59° \times 3.58\angle 168.6° = \mathbf{3255.6\angle 109.6°\ V.}\quad \textbf{(Ans.)}$$

Example 21.45. *Two 11 kV, 40 MVA, 3-phase alternators are connected in parallel and supply a sub-station feeder having an impedance of 0.32 + j0.56 ohms to positive and negative sequence currents and 0.56 + j2.4 ohms to zero sequence currents. The reactances of the alternators to positive, negative and zero sequence currents are 0.48 ohm, 0.32 ohm and 0.16 ohm respectively. Both machines have their neutrals earthed through reactances of 0.16 ohm. If an earth fault occurs simultaneously on the blue and yellow phases at the sub-station, calculate :*

(i) The fault currents in each line, and

(ii) The potential above earth attained by the alternator neutrals.

Solution. The connection diagram of alternators is shown in Fig. 21.71

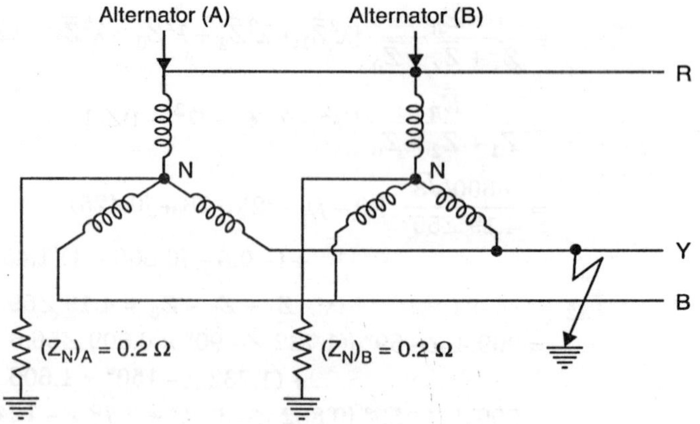

Fig. 21.71

Given : $\bar{Z}_{l1} = (0.32 + j0.56)\ \Omega$; $\bar{Z}_{l2} = (0.32 + j0.56)\ \Omega$; $\bar{Z}_{l0} = (0.56 + j2.4)\ \Omega$;

$(\bar{Z}_{g1})_A = (\bar{Z}_{g1})_B = 0.48\ \Omega$; $(\bar{Z}_{g2})_A = (\bar{Z}_{g2})_B = 0.32\ \Omega$; $(\bar{Z}_{g0})_A = (\bar{Z}_{g0})_B = 0.16\ \Omega$;

$(\bar{Z}_N)_A = (\bar{Z}_N)_B = 0.16\ \Omega$

The alternators are in *parallel*, as such their impedances can be treated in parallel and equivalent circuit found out, *i.e.*,

$$\bar{Z}_{g1} = \frac{(\bar{Z}_{g1})_A \times (\bar{Z}_{g1})_B}{(\bar{Z}_{g1})_A + (\bar{Z}_{g1})_B} = \frac{0.48 \times 0.48}{0.48 + 0.48} = 0.24\ \Omega$$

Similarly, $\bar{Z}_{g2} = 0.16\ \Omega$; $\bar{Z}_{g0} = 0.08\ \Omega$; $\bar{Z}_N = 0.08\ \Omega$

where, $\bar{Z}_{g1}, \bar{Z}_{g2}, \bar{Z}_{g0}, \bar{Z}_N$ are the *equivalent* positive sequence, negative sequence, zero sequence and the neutral impedances respectively.

Hence,

$$\bar{Z}_1 = \bar{Z}_{g1} + \bar{Z}_{l1} = j0.24 + (0.32 + j0.56) = (0.32 + j0.8)\ \Omega$$

$$\bar{Z}_2 = \bar{Z}_{g2} + \bar{Z}_{l2} = j0.16 + (0.32 + j0.56) = (0.32 + j0.72)\ \Omega$$

$$\bar{Z}_0 = \bar{Z}_{g0} + \bar{Z}_{l0} + 3Z_N = j0.08 + (0.56 + j2.4) + 3 \times j0.08 = (0.56 + j2.72)\ \Omega$$

(*i*) **Fault current in each line ; $\bar{I}_R, \bar{I}_Y, \bar{I}_B$:**

For double line-to-ground fault on blue (*B*) and yellow (*Y*) phases, we have

$$\bar{I}_{R1} = \frac{\bar{E}_R}{\bar{Z}_1 + \dfrac{\bar{Z}_2 \bar{Z}_0}{\bar{Z}_2 + \bar{Z}_0}} \qquad \text{...[Eqn. (21.52)]}$$

$$\bar{I}_{R2} = -\bar{I}_{R1}\left(\frac{\bar{Z}_0}{\bar{Z}_2 + \bar{Z}_0}\right) \qquad \text{...[Eqn. (21.53)]}$$

$$\bar{I}_{R0} = -\bar{I}_{R1}\left(\frac{\bar{Z}_2}{\bar{Z}_2 + \bar{Z}_0}\right) \qquad \text{...[Eqn. (21.54)]}$$

Now,

$$\frac{\bar{Z}_2 \bar{Z}_0}{\bar{Z}_2 + \bar{Z}_0} = \frac{(0.32 + j0.72)(0.56 + j2.72)}{(0.32 + j0.72) + (0.56 + j2.72)} = \frac{0.788\ \angle 66.04° \times 2.78\ \angle 78.37°}{0.788\ \angle 66.04° + 2.78\ \angle 78.37°}$$

$$= \frac{2.19\ \angle 144.4°}{3.55\ \angle 75.7°} = 0.617\ \angle 68.7°$$

$$\bar{I}_{R1} = \frac{(11 \times 10^3)/\sqrt{3}}{(0.32 + j0.8) + 0.617\ \angle 68.7°} = \frac{6351}{0.862\ \angle 68.2° + 0.617\ \angle 68.7°}$$

$$= \frac{6351}{1.48\ \angle 68.4°} = 4291.2\ \angle -68.4°$$

$$\bar{I}_{R2} = -4291.2\ \angle -68.4° \times \left(\frac{0.56 + j2.72}{3.55\ \angle 75.7}\right)$$

$$= -4291.2\ \angle -68.4° \times \left(\frac{2.78\ \angle 78.37°}{3.55\ \angle 75.7°}\right) = -3360.4\ \angle -65.7°$$

$$\bar{I}_{R0} = -4291.1 \angle -68.4° \times \left(\frac{0.32 + j0.72}{3.55 \angle 75.7°}\right)$$

$$= -4291.2 \angle -68.4° \times \left(\frac{0.788 \angle 66.04°}{3.55 \angle 75.7°}\right) = -952.5 \angle -78.1°$$

Now, $\bar{\mathbf{I}}_{\mathbf{R}} = \bar{I}_{R1} + \bar{I}_{R2} + \bar{I}_{R0} = \mathbf{0}$. **(Ans.)**

(This is evident from the condition of fault.)

$$\bar{\mathbf{I}}_{\mathbf{Y}} = \bar{I}_{Y1} + \bar{I}_{Y2} + \bar{I}_{Y0}$$
$$= \lambda^2 \bar{I}_{R1} + \lambda \bar{I}_{R2} + \bar{I}_{R0}$$
$$= 1 \angle 240° \times 4291.2 \angle -68.4° + 1 \angle 120° \times (-3360.4 \angle -65.7°) + (-952.5 \angle -78.1°)$$
$$= 4291.2 \angle 171.6° - 3360.4 \angle 54.3° - 952.5 \angle -78.1°$$
$$= \mathbf{6508.5 \angle -169.6° \ A.} \quad \textbf{(Ans.)}$$

$$\bar{\mathbf{I}}_{\mathbf{B}} = \bar{I}_{B1} + \bar{I}_{B2} + \bar{I}_{B0}$$
$$= \lambda \bar{I}_{R1} + \lambda^2 \bar{I}_{R2} + \bar{I}_{R0}$$
$$= 1 \angle 120° \times 4291.2 \angle -68.4° + 1 \angle 240° \times (-3360.4 \angle -65.7°) + (-952.5 \angle -78.1°)$$
$$= 4291.2 \angle 51.6° - 3360.4 \angle 174.3° - 952.5 \angle -78.1°$$
$$= \mathbf{7034 \angle 34.3° \ A.} \quad \textbf{(Ans.)}$$

(*ii*) **The potential above earth attained by the earth neutrals :**

Current through the earth reactance

$$= 3\bar{I}_{R0} = 3 \times (-952.5 \angle -78.1°) = 2857.5 \angle -78.1°$$

∴ Magnitude of potential above earth attained by the earth neutrals

$$= (2857.5 \angle -78.1°) \times 0.08 = \mathbf{228.6 \angle -78.1° \ V.} \quad \textbf{(Ans.)}$$

Example 21.46. *Three 6.6 kV, 12 MVA, 3-phase alternator are connected to a common set of bus-bars. Each has a rectance to positive sequence current of 15%. The reactances to negative and zero sequence currents are 75% and 30% of the positive sequence value. If an earth fault occurs on one bus-bar, determine the fault currents in the following cases :*

(*i*) *When all the alternator neutrals are solidly earthed,*

(*ii*) *When only one of the alternators neutral is solidly earthed and others are isolated, and*

(*iii*) *When one of the alternators neutral is earthed through a resistance of 0.25 ohm and the others are isolated.*

Solution. (*i*) ***When all the alternator neutrals are solidly earthed :***

Refer to Fig. 21.72. Full load current of each alternator

$$= \frac{12 \times 10^6}{\sqrt{3} \times (6.6 \times 1000)} = 1050 \text{ A}$$

Let the value of positive sequence reactance of one alternator be X.

Then, $\frac{X \times 1050}{6600/\sqrt{3}} = \frac{15}{100}$ or $X = \frac{15 \times (6600/\sqrt{3})}{1050 \times 100} = 0.544\ \Omega$

∴ $Z_{g1} = j0.544\ \Omega$; $Z_{g2} = 0.75 \times 0.544 = j0.408$; $\bar{Z}_{g0} = 0.3 \times 0.544 = j0.163\ \Omega$

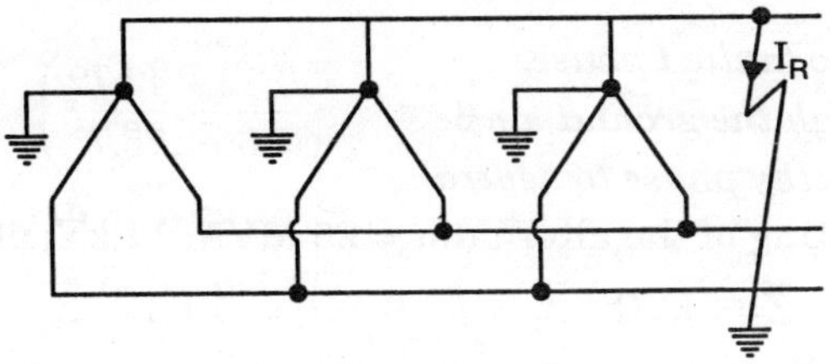

Fig. 21.72

When the three alternators are running in parallel, the resultant reactances will be $\frac{1}{3}$ rd, *i.e.*,

$$\overline{Z}_1 = j0.181\ \Omega\ ;\ \overline{Z}_2 = j0.136\ \Omega\ ;\ \overline{Z}_0 = j0.054\ \Omega$$

$\therefore$ Fault current, $$\overline{\mathbf{I}}_\mathbf{R} = \frac{3\overline{E}_R}{\overline{Z}_1 + \overline{Z}_2 + \overline{Z}_0}$$

$$= \frac{3 \times (6600/\sqrt{3})}{(j0.181 + j0.136 + j0.054)} = -\mathbf{j30813\ A.}\quad \textbf{(Ans.)}$$

(*ii*) ***When only one of the alternators neutral is solidly earthed and others are isolated :***

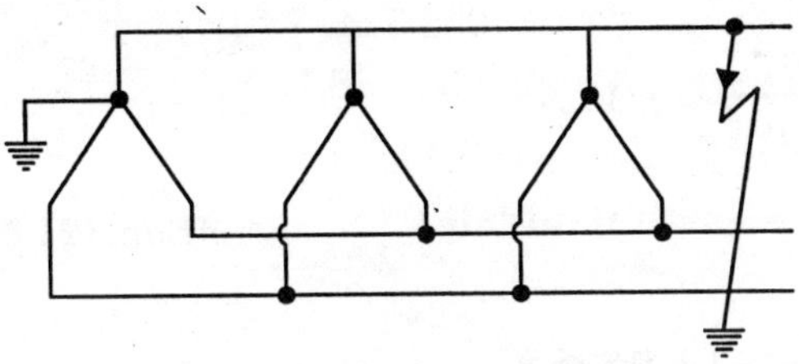

Fig. 21.73

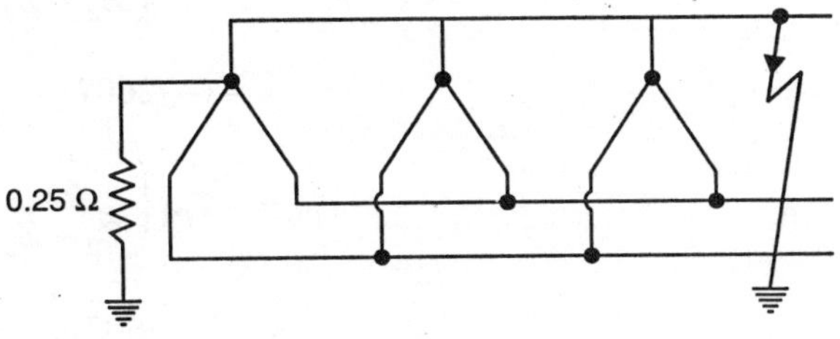

Fig. 21.74

Refer to Fig. 21.73. In this case, $\overline{Z}_0 = \overline{Z}_{g0} = j0.163\ \Omega$

$\therefore$ Fault current, $$\overline{I}_R = \frac{3 \times 6600/\sqrt{3}}{(j0.181 + j0.138 + j0.163)} = -\mathbf{j23717\ A.}\quad \textbf{(Ans.)}$$

(*iii*) ***When one of the alternator neutrals is earthed through a resistance of 0.25 Ω and others are isolated :***

Refer to Fig. 25.74. In this case,

$$\overline{Z}_0 = (3Z_N + \overline{Z}_{g0}) = (3 \times 0.25 + j0.163) = (0.75 + j0.163)\ \Omega$$

$\therefore$ Fault current = $$\frac{3 \times (6600/\sqrt{3})}{j0.181 + j0.138 + (0.75 + j0.163)}$$

$$= \frac{11431.5}{(0.75 + j0.482)} = \frac{11431.5}{0.89\ \angle 32.7^\circ} = \mathbf{12844\ \angle -32.7^\circ\ A.}\quad \textbf{(Ans.)}$$

Example 21.47. *A 25 MVA, 3-phase, 11 kV, 50 Hz alternator having the neutral solidly grounded, is operating at no-load. It has reactance to positive sequence currents of 2 ohms. The reactances to negative and zero sequence currents are 80% and 30% percent of the positive sequence value respectively. For a double line-to-ground fault, determine :*

(i) *The currents in the faulted phase,*

(ii) *The current through the ground, and*

(iii) *The voltage of healthy phase to neutral.*

Solution. *Given :* Rating of the alternator = 25 MVA, 11 kV, 50 Hz ;

$$\bar{Z}_1 = j2\ \Omega\ ;\ \bar{Z}_2 = j(0.8 \times 2) = j1.6\ \Omega\ ;\ \bar{Z}_0 = j(0.3 \times 2) = j0.6\ \Omega.$$

Now,
$$\frac{\bar{Z}_2 \bar{Z}_0}{\bar{Z}_2 + \bar{Z}_0} = \frac{j1.6 \times j0.6}{j1.6 + j0.6} = -\frac{0.96}{j2.2} = j0.436\ \Omega$$

Double line-to-ground fault :

(i) **The currents in the faulted phase :**

Assuming phases Y and B to be under fault,

$$\bar{I}_{R1} = \frac{\bar{E}_R}{\bar{Z}_1 + \dfrac{\bar{Z}_2 \bar{Z}_0}{\bar{Z}_2 + \bar{Z}_0}} \quad \text{...[Eqn. (21.52)]}$$

$$= \frac{11000/\sqrt{3}}{(j2 + j0.436)} = -j2607\ \text{A}$$

Also,
$$\bar{I}_{R2} = -\bar{I}_{R1}\left(\frac{\bar{Z}_0}{\bar{Z}_2 + \bar{Z}_0}\right) \quad \text{...[Eqn. (21.53)]}$$

$$= -(-j2607) \times \frac{j0.6}{j1.6 + j0.6} = j711\ \text{A}$$

and,
$$\bar{I}_{R0} = -\bar{I}_{R1}\left(\frac{\bar{Z}_2}{\bar{Z}_2 + \bar{Z}_0}\right) \quad \text{...[Eqn. (21.54)]}$$

$$= -(-j2607)\left(\frac{j1.6}{j1.6 + j0.6}\right) = j1896\ \text{A}$$

$\therefore$
$$\mathbf{I_R} = \bar{I}_{R1} + \bar{I}_{R2} + \bar{I}_{R0} = -j2607 + j711 + j1896 = \mathbf{0.} \quad \textbf{(Ans.)}$$

$$\bar{I}_Y = \bar{I}_{Y1} + \bar{I}_{Y2} + \bar{I}_{Y0} = \lambda^2 \bar{I}_{R1} + \lambda \bar{I}_{R2} + \bar{I}_{R0}$$
$$= 1\angle 240° \times (-j2607) + 1\angle 120° \times j711 + j1896$$
$$= 1\angle 240° \times 2607\angle -90° + 1\angle 120° \times 711\angle 90° + 1896\angle 90°$$
$$= 2607\angle 150° + 711\angle 210° + 1896\angle 90° = \mathbf{4042.9\angle 135.3°\ A.} \quad \textbf{(Ans.)}$$

$$\bar{I}_B = \bar{I}_{B1} + \bar{I}_{B2} + \bar{I}_{B0}$$
$$= \lambda \bar{I}_{R1} + \lambda^2 \bar{I}_{R2} + \bar{I}_{R0}$$
$$= 1\angle 120° \times (-j2607) + 1\angle 240° \times j711 + j1896$$
$$= 1\angle 120° \times 2607\angle -90° + 1\angle 240° \times 711\angle 90° + 1896\angle 90°$$
$$= 2607\angle 30° + 711\angle 330° + 1896\angle 90° = \mathbf{4042.9\angle 44.7°\ A.} \quad \textbf{(Ans.)}$$

(ii) **The current through the ground :**

As ground serves as the return path for zero sequence currents, therefore,

$$\bar{I}_{RN} = 3\bar{I}_{R0} = 3 \times j1896 = 3 \times 1896\angle 90° = \mathbf{5688\angle 90°\ A.} \quad \textbf{(Ans.)}$$

(iii) **The voltage of healthy phase to neutral :**

The voltage of healthy phase to neutral,

$$\bar{V}_R = \bar{V}_{R1} + \bar{V}_{R2} + \bar{V}_{R0}$$

Also, $\bar{V}_{R1} = \bar{V}_{R2} = \bar{V}_{R0}$

$\therefore$ $\bar{V}_R = 3\bar{V}_{R2}$

But, $\bar{V}_{R2} = \bar{I}_{R1} \times \dfrac{\bar{Z}_2\bar{Z}_0}{\bar{Z}_2 + \bar{Z}_0} = -j2607 \times j0.436 = 1136.6 \text{ V}$

$\therefore$ $\mathbf{V_R = 3 \times 1136.6 = 3410\ V.}$ **(Ans.)**

$(\bar{V}_Y = \bar{V}_B = 0)$

21.4.5. Sequence Networks

The analysis of an unsymmetrical fault can be done by symmetrical components of the unbalanced current. Current of any one sequence may be considered to flow in an independent network having that particular sequence impedance.

The single phase equivalent circuit of one particular sequence voltages, currents and impedances is called the **"sequence network".**

Based on positive, negative and zero sequence currents there will be following three sequence networks for a given power system :

1. Positive sequence network ; 2. Negative sequence network ;
3. Zero sequence network.

1. **Positive sequence network.** It shows all the paths for the flow of positive sequence currents in the system. It is represented by one-line diagram and is composed of impedances offered to the positive sequence currents.

While drawing positive sequence network the *following points need be considered* :

(*i*) In the system, each generator is represented by the generated voltage in series with appropriate reactance and resistance.

(*ii*) Since the *current limiting impedances between generator's neutral and ground* pass no positive sequence current, therefore, these are *not included* in the positive sequence network.

(*iii*) For the sake of simplicity all resistances and magnetising currents for each transformer are neglected.

(*iv*) Shunt capacitances and resistances for transmission lines are generally neglected.

2. **Negative sequence network.** This type of network for a given power system shows all the paths for the flow of negative sequence currents in the system. It is also represented by one-line diagram and consists of impedances offered to the negative sequence currents.

This network can be easily obtained from the positive sequence network with the following *modifications.*

(*i*) Omit the e.m.f.s of 3-phase generators and motors in the positive sequence network since these devices have only positive sequence networks.

(*ii*) If necessary, change the impedances that represent rotating machinery in the positive sequence network, since the negative sequence impedance of rotating machinery is generally different from that of positive sequence impedance.

(*iii*) *Current limiting impedances* between generator' neutral and ground pass no negative sequence current and hence are *not included* in the negative sequence network.

(*iv*) The negative sequence impedance for static devices such as transmission lines and transformers have the same value as the corresponding positive sequence impedances.

3. **Zero sequence network.** For a given power system, a zero sequence network shows all the paths for the flow of zero sequence currents. This network depends upon the nature of connections of the 3-phase windings of the components in the system.

It may be noted that zero sequence currents will flow only if there is a return path (the path from neutral to ground or to another neutral point in the circuit). These currents cannot exist in a system with no return path for zero sequence currents.

Reference bus for sequence networks :

While drawing the sequence network, it is necessary to specify the reference potential w.r.t. to which all sequence voltage drops are to be taken. Following points are worth noting :

- *For positive or negative sequence networks, the neutral of the generator is taken as the reference bus.*
- *For zero sequence network, the reference bus is the ground at the generator.*

Example 21.48. *Fig. 21.75 shows an unloaded generator grounded through a reactance Z_N. If a single line-to-ground (L-G) fault occurs, draw :*

(i) Positive sequence network,

(ii) Negative sequence network, and

(iii) Zero sequence network.

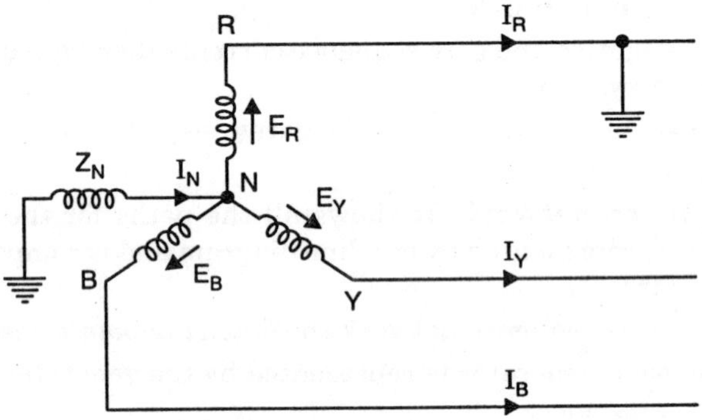

Fig. 21.75

Solution. Fig. 21.75 shows the unloaded generator with single line-to-ground fault.

(*i*) **Positive sequence network.** As the generator is designed to supply 3-phase balanced voltages, the generated voltages are of positive sequence only. Thus, the positive sequence network is composed of phase e.m.fs. in series with sequence impedances of the generator. Fig. 21.76 (*a*) shows the positive sequence current paths whereas Fig. 21.76 (*b*) shows the *single-phase positive sequence network.*

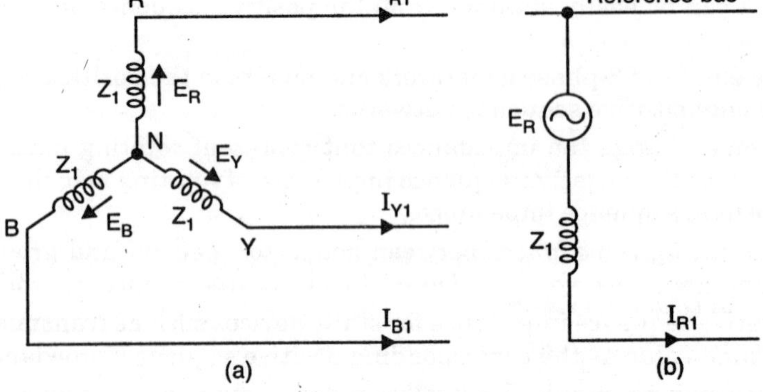

Fig. 21.76

(ii) **Negative sequence network.** This type of network *contains no e.m.f.* but includes the impedances of the generator to negative sequence currents. Thus negative sequence network is easily obtained by omitting e.m.fs. in the positive sequence network. The negative sequence current paths and the single-phase negative sequence network are shown in Fig. 21.77 (*a*) and (*b*) respectively.

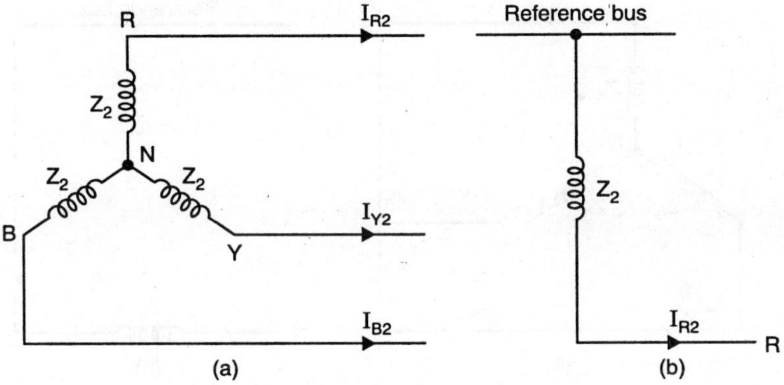

Fig. 21.77

(iii) **Zero sequence network.** The zero sequence currents flow through the phases as well as through the reactance Z_N as shown in Fig. 21.78 (*a*). The current flowing in impedance Z_N is $3I_{R0}$, since $I_{R0} = I_{Y0} = I_{B0}$.

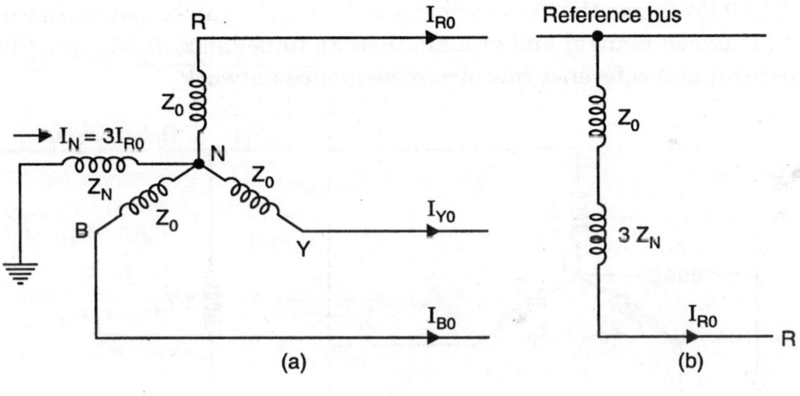

Fig. 21.78

$\therefore$ Voltage drop of zero sequence current from R-phase to ground

$$= -3\bar{I}_{R0}\bar{Z}_N - \bar{I}_{R0}\bar{Z}_0 = -\bar{I}_{R0}(3\bar{Z}_N + \bar{Z}_0)$$

Thus, the per phase impedance to zero sequence current is $(3Z_N + Z_0)$.

The zero sequence network is shown in Fig. 21.78 (*b*).

Example 21.49. *Draw the zero sequence network for the following cases :*

(i) Star-connected load with no earth connection,

(ii) Star-connected load with Z_N from neutral to ground, and

(iii) Delta-connected load.

Solution. (*i*) **Star-connected load with no earth connection :**

- Fig. 21.79 (*a*) shows the star-connected load with no earth connection. In this case, *neutral current is zero and zero sequence current does not exist.*
- Fig. 21.79 (*b*) shows the zero sequence network.

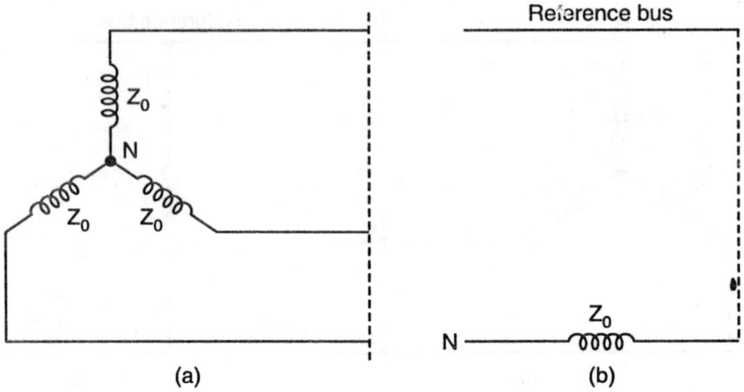

Fig. 21.79

(*ii*) **Star-connected load with Z_N from the neutral to ground :**

- Fig. 21.80 (*a*) shows a star-connected load with an impedance Z_N between neutral and ground.
- Fig. 21.80 (*b*) shows the zero sequence network. It may be noted that if impedance Z_N is placed between neutral and ground, then an impedance of $3Z_N$ must be placed between the neutral and reference bus of zero sequence network.

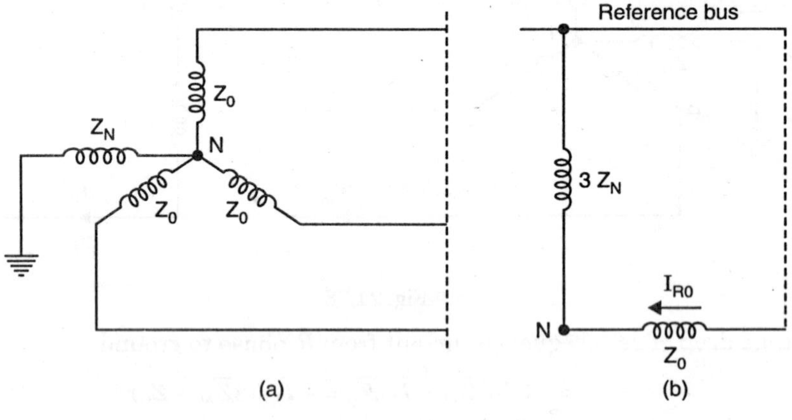

Fig. 21.80

(*iii*) **Delta-connected load.** As a delta-connected load does not provide return path, therefore, *zero sequence currents cannot exist in the phase windings.* In other words, a delta-connected load provides infinite impedance to zero sequence line currents Fig. 21.81 (*b*) show the zero sequence network-for a delta-connected circuit.

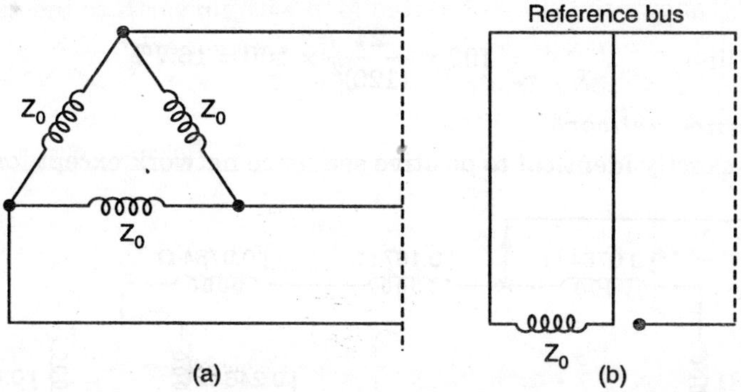

Fig. 21.81

Example 21.50. *A 24 MVA, 13.8 kV, 3-phase alternator has a sub-transient reactance of 15% and negative and zero reactances of 15% and 5% respectively. The alternator supplies two motors over a transmission line having transformers at both ends as shown on the one-line diagram. The motors have rated inputs of 16 MVA and 8 MVA both 12.5 kV with 20% sub-transient reactance and negative and zero sequence reactances are 20% and 5% respectively. Current limiting reactors of 2.5 ohms each are in the neutral of the alternator and the larger motor. The 3-phase transformers are both rated 28 MVA, 13.2Δ – 115Y kV with leakage reactance of 10%. Series reactance of the line is 100 ohms. The zero reactance of line is 250 ohms. Assuming V_f = 120 kV, calculate the faults currents, when the following faults take place at point P :*

(i) Line-to-ground (L-G)

(ii) Line-to-line (L-L) fault, and

(iii) Double line-to ground (L-L-G) fault.

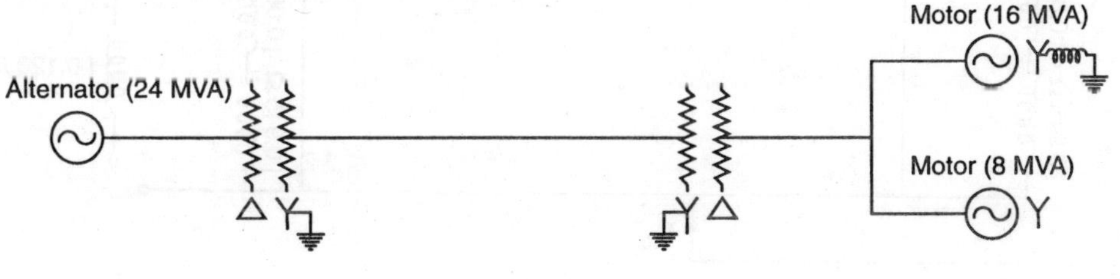

Fig. 21.82

Solution. Assume base of 24 MVA and base voltage of 13.8 kV in generator circuit.

Positive sequence network :

The base voltage on the *line side of the transformer* $= 13.8 \times \frac{115}{13.2} = 120$ kV

∴ The base voltage on the motor side of the transformer $= 120 \times \frac{13.2}{115} = 13.8$ kV

% Reactance of transformer $= 10 \times \left(\frac{13.2}{13.8}\right)^2 \times \frac{124}{28} = 7.84\%$

% Reactance of motor $= 20 \times \left(\frac{12.5}{13.8}\right)^2 \times \frac{24}{16} = 24.6\%$

% Reactance of line $= 100 \times \dfrac{24}{(120)^2} \times 100 = 16.7\%$

Negative sequence network :

The network is exactly identical to positive sequence network except for the sources.

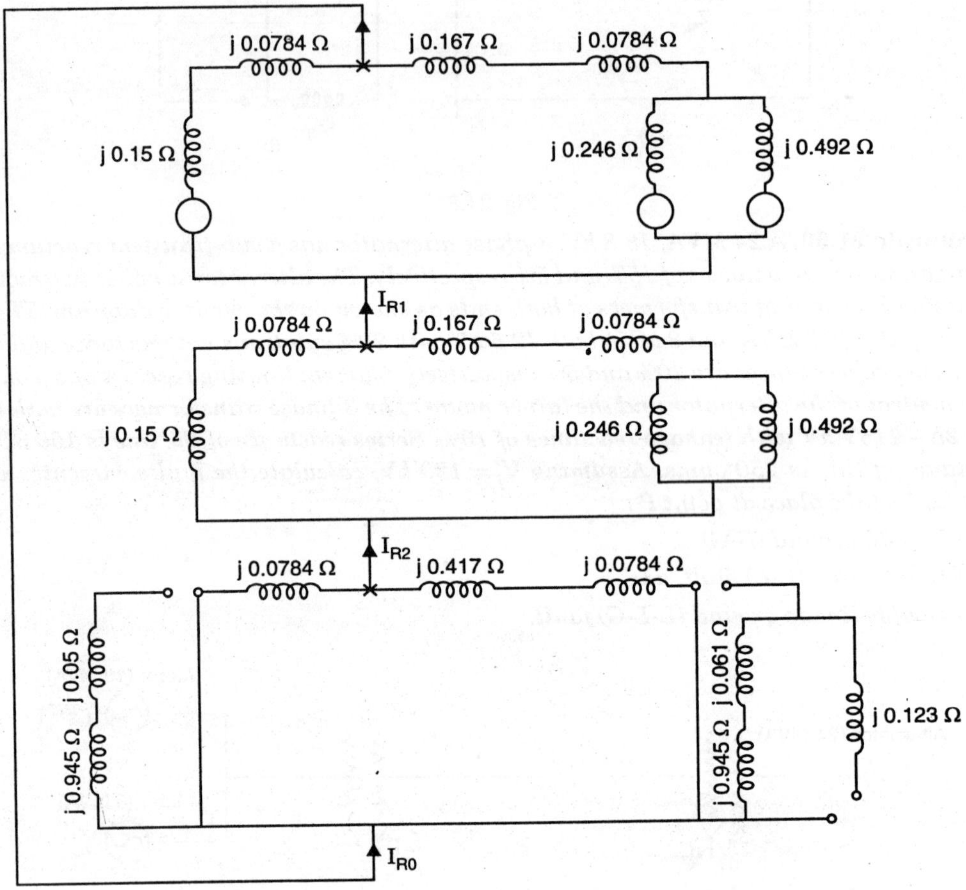

Fig. 21.83. Positive, negative and zero sequence networks.

Zero sequence network :

The neutral reactance $= 2.5 \times 3 \times \dfrac{24}{(13.8)^2} \times 100 = 94.5\%$

The zero sequence reactance of line $= 250 \times \dfrac{24}{(120)^2} \times 100 = 41.7\%$

The three sequence networks are shown in Fig. 21.83.

Positive sequence impedance between *P* and *ZPB* (when sources are short circuited),

$$\bar{Z}_1 = (j0.15 + j0.0784) \parallel [(j0.167 + j0.0784) + (j0.246 \parallel j0.492)]$$

$$= j0.2284 \parallel [j0.2454 + j0.164] = j0.2284 \parallel j0.4094 = j0.147\ \Omega$$

Similarly, Negative sequence impedance, $\bar{Z}_2 = j0.147\ \Omega$

Zero sequence impedance, $\bar{Z}_0 = j0.0784 \parallel (j0.417 + j0.0784)$

$= j0.078 \parallel (j0.4954 = j0.0677\ \Omega$

(*i*) **Line-to-ground (L-G) fault current :**

$$\bar{I}_{R0}\ (= \bar{I}_{R1} = \bar{I}_{R2}) = \frac{\bar{E}_R}{\bar{Z}_1 + \bar{Z}_2 + \bar{Z}_0} + \frac{1 + j0}{j0.147 + j0.147 + j0.0677} = \frac{1 + j0}{j0.3617}$$

$$= -j2.765 \text{ p.u.}$$

Fault current, $\bar{I}_R = 3\bar{I}_{R0} = 3 \times -j2.765 = -j8.295$

Base current $= \dfrac{24 \times 1000}{\sqrt{3} \times 13.8} = 1004$ A

or, on the line side,

Base current $= \dfrac{24 \times 1000}{\sqrt{3} \times 120} = 115.47$ A

∴ *Fault current* $= 115.47 \times 8.295 =$ **957.8 A. (Ans.)**

(*ii*) **Line-to-line (L-L-G) fault current :**

Here *only positive and negative sequence networks are required.*

$$\therefore \quad \bar{I}_{R1}\ (= -\bar{I}_{R2}) = \frac{\bar{E}_R}{\bar{Z}_1 + \bar{Z}_2} \qquad \text{...[21.45 (a)]}$$

$$= \frac{1 + j0}{j0.147 + j0.147} = \frac{1 + j0}{j0.294} = -j3.4$$

$$\therefore \quad \bar{I}_{R1} = -\bar{I}_{R2} = -j3.4$$

Fault current, $\bar{I}_Y = -\bar{I}_B = \lambda^2 \bar{I}_{R1} + \lambda \bar{I}_{R2}$, as $I_{R0} = 0$

or, $\bar{I}_Y = (-0.5 - j0.866)(-j3.4) + (-0.5 + j0.866)(j3.4)$

$= (j1.7 - 2.944) + (-j1.7 - 2.944) = 5.89$ p.u.

∴ *Fault current* $= 5.89 \times 115.47 =$ **680 A. (Ans.)**

(*iii*) **Double line-to-ground (L-L-G) fault current :**

Fault current $= \bar{I}_Y + \bar{I}_B = 3\bar{I}_{R0}$

$$= -\frac{3\bar{Z}_2\bar{E}_R}{\bar{Z}_1\bar{Z}_2 + \bar{Z}_1\bar{Z}_0 + \bar{Z}_2\bar{Z}_0} \qquad \text{...[Eqn. (21.55)]}$$

$$= -\frac{3 \times j0.147 \times (1 + j0)}{(j0.147 \times j0.147) + (j0.147 \times j0.0677) + (j0.147 \times j0.0677)}$$

$$= -\frac{j0.441}{(-0.0216) + (-0.009952) + (-0.009952)} = \frac{j0.441}{0.0415} = j10.63$$

$= 10.63 \times 115.47 =$ **1227.4 A. (Ans.)**

Example 21.51. *Fig. 21.84 shows a network in which data about the system is as follows :*

G = 2.5 MVA, reactance = 10%, solidly earthed, T_1 *= 2.5 MVA, 6.66/11 kV, reactance 5%, solidly earthed on 11 kV side ; Cable, jX = 0.4 Ω at 11 kV ;* T_2 *= 2.5 MVA, 11/6.6 kV, reactance 5%, solidly earthed on 6.6 kV, side.*

If a fault occurs at F in the network, find the fault current for (i) line-to-line fault, and (ii) line-to-ground fault.

Take negative sequence impedance of the generator equal to 70% of its positive sequence impedance.

Solution. Let 2500 kVA be base kVA and 6.6 kV be the base kV.

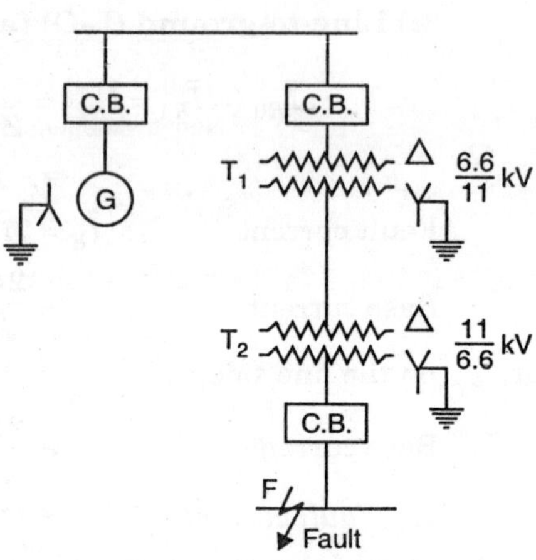

G = Generator ;
T_1, T_2 = Transformers ;
C.B. = Circuit breakers

Fig. 21.84

Then, 10% reactance at 6.6 kV

$$= \frac{\% \text{ Reactance} \times (\text{kV})^2 \times 10}{(\text{kVA})}$$

$$= \frac{10 \times (6.6)^2 \times 10}{2500} = 1.74\ \Omega$$

5% reactance at 6.6 kV = 0.87 Ω

0.5 Ω at 11 kV when referred to 6.6 kV

$$= \left(\frac{6.6}{11}\right)^2 \times 0.4 = 0.144\ \Omega$$

Positive sequence impedance of generator,

$$\overline{Z}_{g1} = j1.74\ \Omega$$

Assuming negative sequence impedance of the generator equal to 70% of the positive sequence impedance,

$$\overline{Z}_{g2} = 0.7 \times j1.74 = j1.218\ \Omega$$

Positive sequence impedance of transformer T_1, $(\overline{Z}_{T1}) = j0.87\ \Omega$

Negative sequence impedance of transformer T_1, $(Z_{T1})_2 = j0.87\ \Omega$

Positive sequence impedance of cable, $\overline{Z}_{C1} = j0.144\ \Omega$

Negative sequence impedance of cable, $\overline{Z}_{C2} = j0.144\ \Omega$

Positive sequence impedance of T_2, $(\overline{Z}_{T2})_1 = j0.87\ \Omega$

Negative sequence impedance of T_2, $(\overline{Z}_{T2})_2 = j0.87\ \Omega$.

Positive, negative and zero sequence networks along with their Thevenin's equivalent are shown in Figs. 21.85, 21.86 and 21.87 respectively.

$$\overline{Z}_1 = j1.74 + j0.87 + j0.144 + j0.87 = j3.624\ \Omega$$

$$\overline{Z}_2 = j1.218 + j0.87 + j0.144 + j0.87 = j3.102$$

$$\overline{Z}_0 = j0.87\ \Omega$$

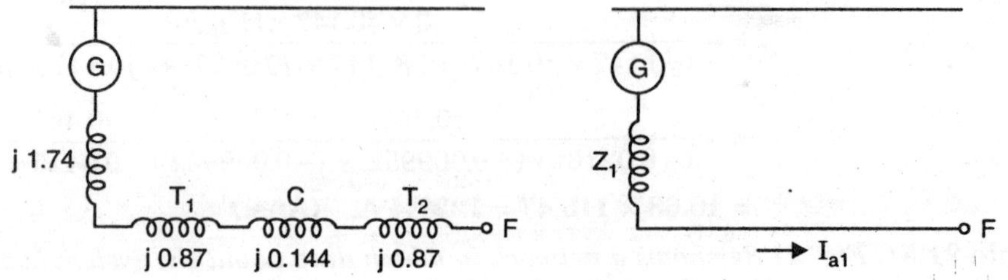

Positive sequence network Thevenin's equivalent

Fig. 21.85

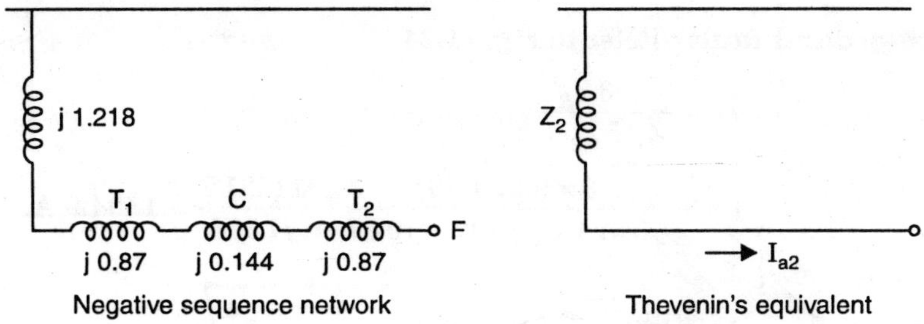

Fig. 21.86

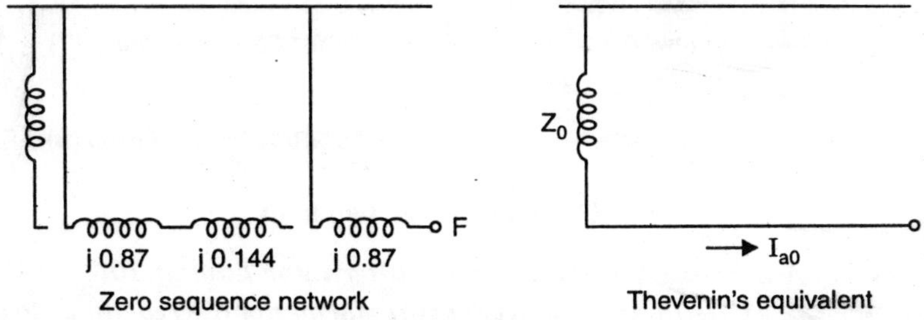

Fig. 21.87

(*i*) **Line-to-line fault** = Refer Fig. 21.83

$$\bar{I}_{R1} = -\bar{I}_{R2} = \frac{\bar{E}_R}{\bar{Z}_1 + \bar{Z}_2} \quad \text{...[Eqn. 21.45 (a)]}$$

$$= \frac{6600/\sqrt{3}}{j3.624 + j3.102} = \frac{3810}{j6.726} = -j566.4 \text{ A}$$

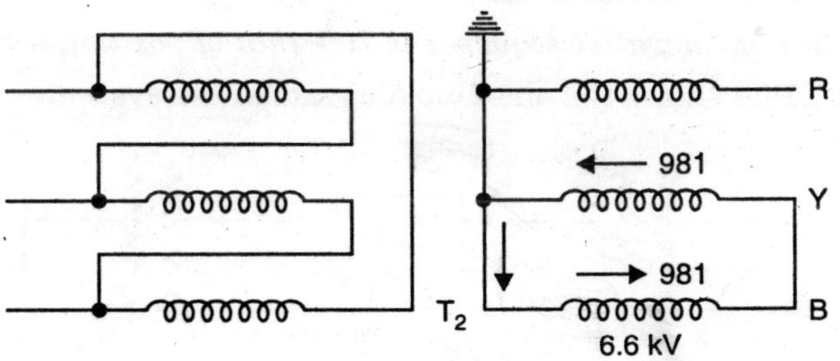

Fig. 21.88. Line-to-line fault.

Now, $\bar{I}_{R0} = 0$

$\mathbf{\bar{I}_B = 0.}$ **(Ans.)**

$\bar{I}_Y = \bar{I}_{Y1} + \bar{I}_{Y2} + \bar{I}_{Y0} = \lambda^2 \bar{I}_{R1} + \lambda \bar{I}_{R2} + \bar{I}_{R0} = (\lambda^2 - \lambda)\bar{I}_{R1}$

$= (0 - j1.732) \times (-j566.4) =$ **981 A. (Ans.)**

$\bar{I}_B = -\bar{I}_Y =$ **981 A. (Ans.)**

(*ii*) **Line-to-ground fault :** Refer to Fig. 21.84.

$$\bar{I}_R = \frac{3\bar{E}_R}{\bar{Z}_1 + \bar{Z}_2 + \bar{Z}_0} \quad \text{...[Eqn. (21.36)]}$$

$$= \frac{3 \times (6600/\sqrt{3})}{j3.624 + j3.102 + j0.87} = \frac{11431.5}{j7.596} = \mathbf{1504.9\ A.} \quad \textbf{(Ans.)}$$

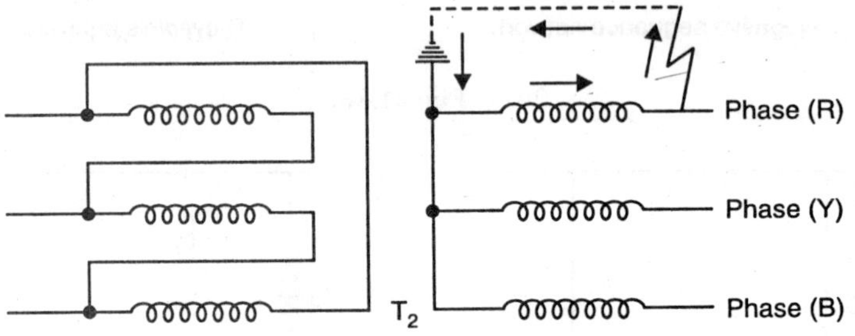

Fig. 21.89. Line-to-ground fault.

Example 21.52. *A 3-phase, 24 MVA, 6.6 kV alternator having 10% reactance is connected through a 24 MVA, 6.6/33 kV delta-star connected transformer of 6% reactance to a 33 kV transmission line having a negligible resistance and a reactance of 5 ohms. At the receiving end of the line there is a 24 MVA, 33/6.6 kV delta-star connected transformer of 5% reactance stepping down the voltage to 6.6 kV. Both transformers have their neutral solidly grounded.*

Draw the one-line diagram and the positive, negative and zero sequence networks of this system and determine the fault currents for the following types of fault at the receiving station l.v. busbars :

(i) All the three phases are short circuited.

(ii) A single line grounded.

(iii) A short circuit between two lines.

For generator, assume negative sequence is 70% that of positive sequence.

Solution. Fig. 21.90 shows the one–line diagram of the system.

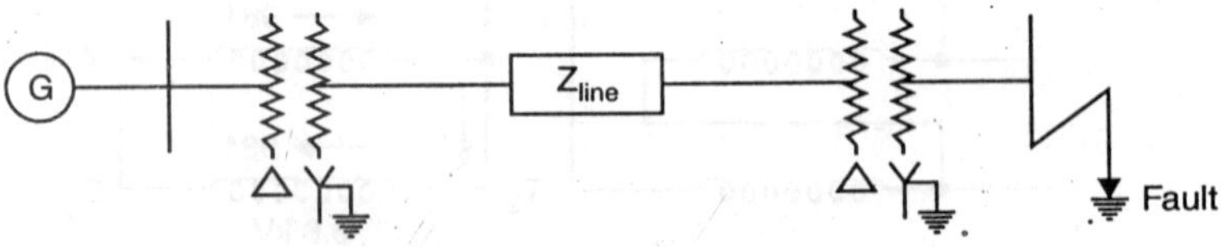

Fig. 21.90. One-line diagram of the system.

Let us choose 24 MVA (*i.e.*, 24000 kV), 6.6 kV, as the base.

Then base impedance $= \frac{(\text{kV})^2}{\text{MVA}} = \frac{(6.6)^2}{24} = 1.815\ \Omega$

10% (0.1 p.u.) impedance gives 0.1815 Ω and 6% (0.06 p.u.) impedance gives 0.1089 Ω

For transmission line, the impedance of 4 Ω when referred to 6.6 V becomes

$$j\left(\frac{6.6}{33}\right)^2 \times 5 = j0.2\ \Omega.$$

Thus, we have the following impedances :

For *Alternator :* $\overline{Z}_{g1} = j0.1815\ \Omega$; $\overline{Z}_{g2} = 0.7 \times 0.1815 = 0.127\ \Omega$;

$\overline{Z}_{g0}$ (Not necessary to know)

For *transformers :* $\overline{Z}_{T1} = 0.1089\ \Omega = \overline{Z}_{T2} = \overline{Z}_{T0}$

For *transmission line* (*L*) = $\overline{Z}_{L1} = j0.2\ \Omega = \overline{Z}_{L2}$; $\overline{Z}_{L0}$ (Not necessary to know)

Positive, negative and zero sequence networks are shown in Figs. 21.91, 21.92 and 21.93.

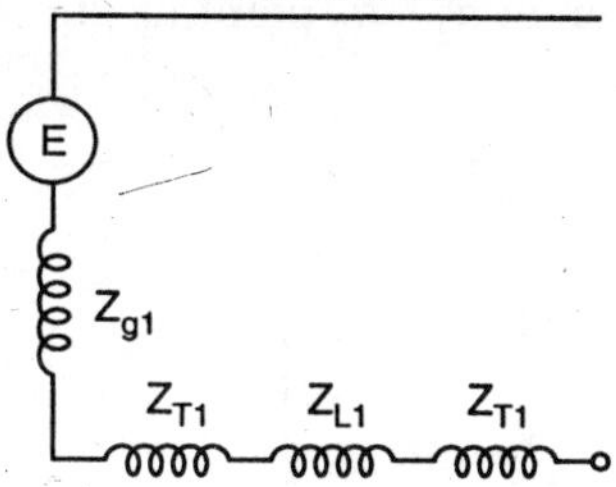

Fig. 21.91

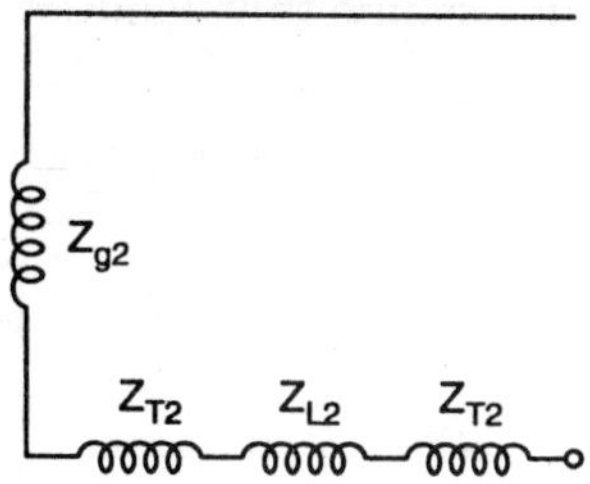

Fig. 21.92

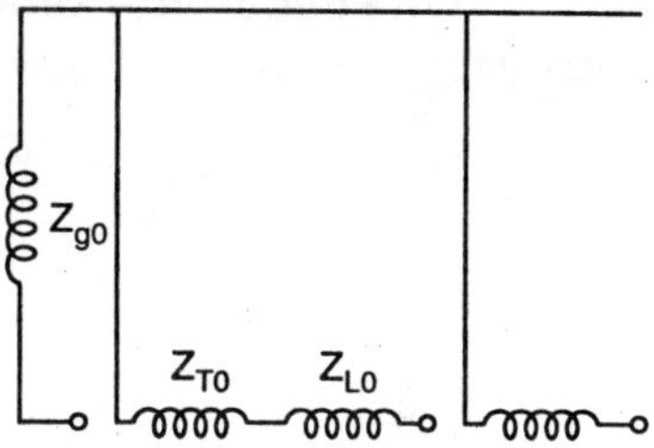

Fig. 21.93

Now,
$$\overline{Z}_1 = \overline{Z}_{g1} + \overline{Z}_{T1} + \overline{Z}_{L1} + \overline{Z}_{T1} = \overline{Z}_{g1} + 2\overline{Z}_{T1} + \overline{Z}_{L1}$$
$$= j0.1815 + 2 \times j0.1089 + j0.2 = j0.5993\ \Omega$$
$$\overline{Z}_2 = \overline{Z}_{g2} + \overline{Z}_{T2} + \overline{Z}_{L2} + \overline{Z}_{T2} = \overline{Z}_{g2} + 2\overline{Z}_{T2} + \overline{Z}_{L2}$$
$$= j0.127 + 2 \times j0.1089 + j0.2 = j0.5448\ \Omega$$
$$\overline{Z}_0 - \overline{Z}_{T0} - j0.1089\ \Omega$$

Fault currents :

(i) All the three phases are short circuited :

When all the three phases are short circuited it is symmetrical fault so that *only the positive sequence network need be considered.*

$$\therefore \quad \text{Short-circuit current, } \overline{I}_{SC} = \frac{6600/\sqrt{3}}{\overline{Z}_1} - \frac{6600/\sqrt{3}}{j0.5993} = -j6358.3 = \mathbf{6358.3 \angle -90^\circ\ A.} \quad \textbf{(Ans.)}$$

(ii) A single line grounded :

For a single line-to-ground fault, assuming '*R*' phase to be ground,

$$\overline{I}_R = \frac{3\overline{E}_R}{\overline{Z}_1 + \overline{Z}_2 + \overline{Z}_0} = \frac{3 \times 600/\sqrt{3}}{j0.5993 + j0.5448 + j0.1089}$$
$$= \frac{11431.5}{j1.253} = -j9123.3 = \mathbf{9123.3 \angle -90^\circ A.} \quad \textbf{(Ans.)}$$

(iii) A short circuit between two lines :

For a short circuit between two lines (say '*Y*' and '*B*'.) we have,

$$\overline{I}_Y = (-\overline{I}_B) = \frac{-\sqrt{3}\ \overline{E}_R}{\overline{Z}_1 + \overline{Z}_2} \qquad \text{...[Eqn. (21.46)]}$$

$$= \frac{-\sqrt{3}\times(6600/\sqrt{3})}{j0.5993 + j0.5448} = -\frac{6600}{j1.144} = j5769.2$$

$$= \mathbf{5769.2 \angle 90°A.} \quad \textbf{(Ans.)}$$

21.4.6. Open Conductor Faults

An open conductor fault is taken to be a fault in series with the line. Let us consider the following types of open conductor faults :

(*i*) One conductor open (*ii*) Two conductors open.

(*i*) **One conductor open.** Let us assume that conductor is open at $a_1 - a_2$ point as shown in Fig. 21.94.

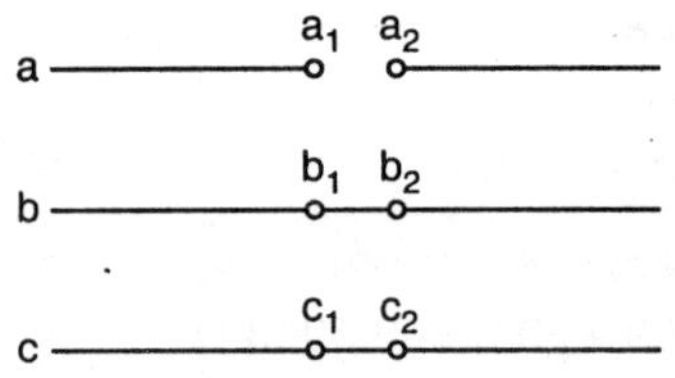

Fig. 21.94. One conductor open fault.

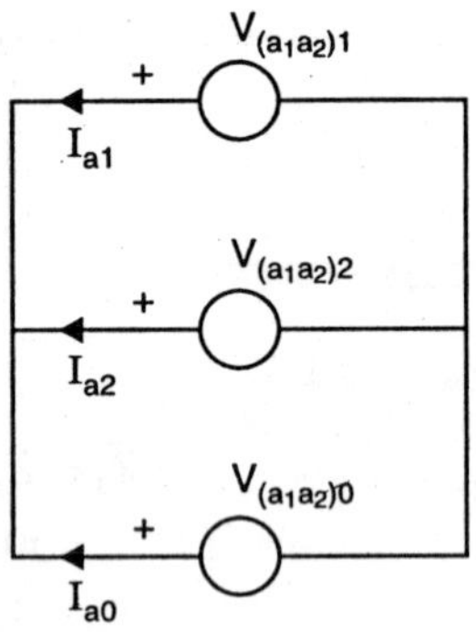

Fig. 21.95. Sequence network of one conductor open fault.

The current and voltages are given as :

$$\left.\begin{aligned} \bar{I}_a &= 0 \\ \bar{V}_{b1b2} = \bar{V}_{c1c2} &= 0 \end{aligned}\right\} \qquad ...(21.59)$$

i.e., $I_{a1} + I_{a2} + I_{a0} = 0$

and, $\bar{V}_{(a1a2)1} = \bar{V}_{(a1a2)2} = \bar{V}_{(a1a2)0} = \frac{1}{3}\bar{V}_{a1a2}$

The equations show a parallel connection of the sequence network (Fig. 21.94)

(*ii*) **Two conductors open.** Fig. 21.96 shows the diagram of a two conductor open fault. The current and voltage are given by,

$$\left.\begin{aligned} \bar{V}_{a1a2} &= 0 \\ \bar{I}_b = \bar{I}_c &= 0 \end{aligned}\right\} \qquad ...(21.60)$$

In terms of symmetrical components,

$$\bar{V}_{(a1a2)1} + \bar{V}_{(a1a2)2} + \bar{V}_{(a1a2)0} = 0$$

$$\bar{I}_{a1} = \bar{I}_{a2} = \bar{I}_{a0} = \frac{1}{3}\bar{I}_a$$

These equations suggest a series connection of sequence networks (Fig. 21.97).

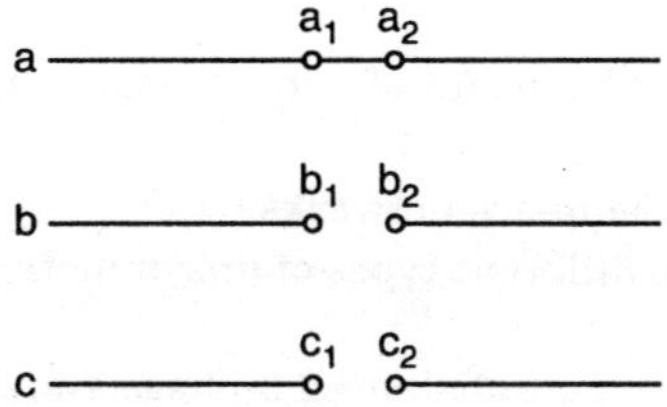

Fig. 21.96. Two conductors open fault.

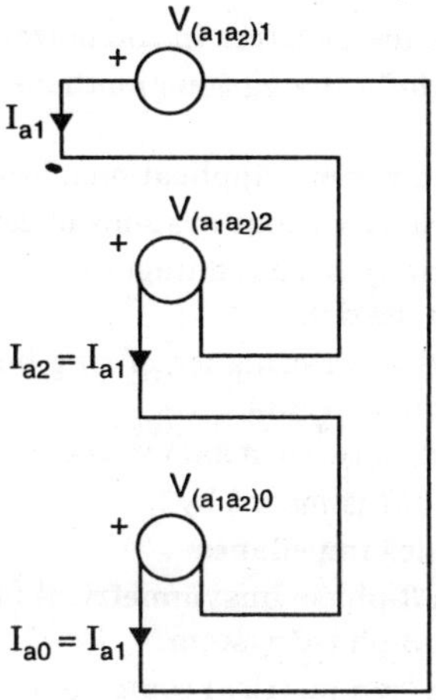

Fig. 21.97. Sequence connection of two conductors open fault.

HIGHLIGHTS

1. In *symmetrical faults* the fault currents in the three lines (in a 3-phase circuit) are equal in magnitude and displaced 120° electrical from one another.
2. The following types of faults generally occur in a power system :
   (*i*) Overcurrent ; (*ii*) Under voltage ; (*iii*) Unbalance ; (*iv*) Reversed power ; (*v*) Surges.
3. *Percentage reactance* is defined as : "The ratio of reactive voltage drop at full load to normal voltage when multiplied by 100, gives the "percentage reactance".
4. The product of normal system voltage and short-circuit current at the point of fault expressed in kVA is known as *"short-circuit kVA"*.
5. When a synchronous generator is short-circuited on all the three phase, the initial short-circuit current is limited by the sub-transient reactance and in a very short line this current decreases to the transient value limited by the transient reactance and then it settles down the steady state short-circuit value limited by the synchronous reactance.
6. *"Reactor"* is a coil which has high inductive reactance as compared to its resistance and is used to limit the short-circuit current during fault conditions.

THEORETICAL QUESTIONS

1. What do you understand by a short-circuit ? Discuss the possible causes of short-circuit in a power system.
2. Explain briefly the harmful effects of short-circuit fault on the power system.
3. What is meant by percentage reactance ? Why do we prefer to express the reactances of various elements in percentage values for short-circuit calculations.
4. What is meant by 'per unit' quantity ? Why the per unit method is considered superior to percent method for short-circuit calculations.
5. What is the importance of base kVA in short-circuit calculations ?
6. Draw the diagram showing the short-circuit current as a function of time for a synchronous generator and explain the following terms :
   (*i*) Synchronous reactance (*ii*) Transient reactance
   (*iii*) Sub-transient reactance.

7. Why do we use reactors in the power system ? Discuss their advantages.
8. Explain briefly the various methods of connecting short-circuit current limiting reactors in the power system.
9. Discuss the various applications, locations and design features of current limiting reactors.
10. Discuss briefly various systems of connecting current limiting reactors in a power system.
11. Explain briefly the advantages to be gained by insertion of reactances in the various locations of a large generating station.
12. Compare the performance of ring bus-bar reactors and tie-bar reactors in power systems for limiting short-circuit currents.
13. Distinguish between a balanced and unbalanced system by means of symmetrical components of vectors.
14. Explain what is meant by :
    (*i*) Sequence impedance (*ii*) Sequence networks.
15. What is a 3-phase unsymmetrical fault ? Discuss the different types of unsymmetrical faults that can occur on a 3-phase system.
16. Discus the 'symmetrical component method' to analyse an unbalanced 3-phase system.
17. What is operator 'λ' ? Show that :
    (*i*) $\lambda^2 = -0.5 - j8.66$ (*ii*) $\lambda^3 = 1$
    (*iii*) $1 + \lambda + \lambda^2 = 0$ (*iv*) $\lambda - \lambda^2 = j\sqrt{3}$.
18. What do you understand by positive, negative and zero sequence impedances ? Discuss them with reference to synchronous generators, transformers and transmission lines.
19. What is meant by (*i*) sequence impedance, and (*ii*) sequence networks.
20. Derive an expression for fault current for single line-to-ground fault by symmetrical components method.
21. Derive an expression for fault current for double line-to-ground fault by symmetrical components method.
22. What do you understand by sequence networks ? What is their importance in unsymmetrical fault calculations.
23. Write short-notes on the following :
    (*i*) Common faults in power systems. (*ii*) Short-circuit and short-circuit currents.
    (*iii*) Percent and percent systems. (*iv*) Symmetrical components.
    (*v*) Unsymmetrical faults on unloaded generators.
    (*vi*) Use of reactors in power station. (*vii*) Negative sequence network.

UNSOLVED EXAMPLES

1. A single-phase transformer is rated as 2.5 kVA, 11/0.4 kV. If its leakage reactance is 0.96 Ω when referred to l.v. side, then determine its leakage reactance in p.u. [**Ans.** 0.015 p.u.]
2. Two generators G_1 and G_2 are rated 10 MVA, 11 kV and 5 MVA, 11 kV respectively. These generators are connected to a 15 MVA, 11Δ/66*Y* kV step-up transformer with a reactance of 10%. It both the generators have a sub-transient reactance of 25%, calculate the sub-transient current in each generator when a 3-phase fault occurs on the h.v. side of the transformer assuming that there is no circulating current in the machines. [**Ans.** 1500 A ; 750 A]
3. A 3-phase transmission line operating at 33 kV and having a resistance and reactance of 5 ohms and 20 ohms respectively is connected to the generating station bus-bar through a 5000 kVA step-up transformer which has a reactance of 6%. Connected to the bus-bars are two synchronous generators, one 10000 kVA having 10% reactance, and another 5000 kVA having 7.5% reactance. Calculate the kVA at a short-circuit fault between phases occurring at the high voltage terminals of the transformers and at the load end of transmission line. [**Ans.** 55560 kVA ; 27260 kVA]
4. A 625 kVA, 480 V alternator supplies a purely resistive load of 500 kW at 480 V. The sub-transient reactance of the generator is 8%. Assuming the load to be directly connected across the generator terminals, find the initial symmetrical r.m.s. current in p.u. at the generator terminals for a 3-phase dead short at its terminals. [**Ans.** $(0.8 - j12.5)$ p.u.]

5. A generator is connected to a synchronous motor through a transformer. Reduced to the same base, the p.u. sub-transient reactances of the generator and motor are 0.15 and 0.35 respectively, the leakage reactance of the transformer being 0.10 p.u. Three phase fault occurs at the terminals of the motor when the terminal voltage of the generator is 0.9 p.u. and the output current of the generator is 1.0 p.u. at 0.8 p.f. load. Find the sub-transient current in p.u. in the fault in (*i*) generator (*ii*) motor.
[**Ans.** 3.25 ∠–81.598° p.u. ; 6.61 ∠– 94.77°]
6. A 3-phase, 20 MVA, 10 kV alternator has internal reactance of 5% and negligible resistance. Find the external reactance per phase to be connected in series with the alternator so that steady current on short-circuit does not exceed 8 times the full-load current. [**Ans.** 0.375 Ω]
7. A 3-phase transmission line operating at 10 kV and having a resistance of 1 Ω and reactance of 4 Ω is connected to the generating station bus-bars through 5 MVA step-up transformer having a reactance of 5%. The bus-bars are supplied by a 10 MVA alternator having 10% reactance. Calculate the short-circuit kVA fed to symmetrical fault between phases if it occurs at the (*i*) high voltage terminals of the transformer and (*ii*) load end of transmission line. [**Ans.** (*i*) 50000 kVA ; (*ii*) 16440 kVA]
8. The plant capacity of a 3-phase generating station consists of two 10000 kVA generators of reactance 12% each and one 5000 kVA generator of reactance 18%. The generators are connected to the station bus-bars from which load is taken through three 5000 kVA step-up transformers each having a reactance of 5%. Determine the maximum fault MVA which the circuit breakers on low voltage side and high voltage side may have to deal with. [**Ans.** 194.5 MVA ; 66]
9. A 30 MVA generator with 20% reactance is connected to a bus-bar. A 25 MVA transformer with 15% reactance is also connected through a 10% bus-bar reactor to the same bus-bar. Both there reactances are based on 25 MVA rating.
If a feeder taken out from the bus-bars through a circuit breaker develops a line-to-ground fault ; what should be the approximate rating of a circuit breaker ? [**Ans.** 250 MVA]
10. The estimated short-circuit MVA at the bus-bars of a generating station *A* is 1500 MVA and of another station *B* is 1200 MVA. The generated voltage at each station is 33 kV. It these stations are interconnected through a line having a reactance of 1 Ω and negligible resistance, calculate the possible short-circuit MVA at both stations. [**Ans.** 2070 MVA, 1831 MVA]
11. A generating station has five section bus-bars connected with a tie-bar through 7.5% reactors rated at 3000 kVA. Each generator is of 3000 kVA with 10% reactance and is connected to each section bus bar. Find the total steady input to dead short-circuit between the lines on one of the section bus-bars (*i*) with reactors and (*ii*) without reactors. [**Ans.** (*i*) 55.3 MVA, (*ii*) 150 MVA]
12. The 33 kV bus-bars of a station are in two sections *A* and *B* separated by a reactor. The section *A* is fed from four 10 MVA generators each having a reactance of 20%. The section *B* is fed from the grid through a 50 MVA transformer of 10% reactance. The circuit breakers have a rupturing capacity of 500 MVA. Find the reactance of the reactor to percent the circuit breakers from being overloaded if a symmetrical short-circuit occurs on outgoing feeder connected to generator G_1. Take base MVA as 50 MVA.
[**Ans.** 1.452 Ω]
13. The bus-bars of a power station are in two sections *A* and *B* separated by a reactor. Connected in section *A* are two 15 MVA generators of 15 MVA generators of 12% reactance each and to *B* one 8 MVA generator of 10% reactance. The reactance is rated at 10 MVA and 15% reactance. Feeders are connected to bus-bar *A* through transformers, each rated at 5 MVA and 4% reactance. Determine the maximum short-circuit MVA with which circuit breakers on the outgoing side of the transformers have to deal.
[**Ans.** 119.8 MVA]
14. The 11 kV generating station has four identical 3-phase synchronous generators G_1, G_2, G_3 and G_4, each of 10 MVA capacity and 12% reactance. There are two sections of bus-bar, *A* and *B* linked by a reactor rated at 10 MVA with 24% reactance. G_1 and G_2 are connected to bus-bar section *A* and G_3 and G_4 to bus-bar section *B*. From each station, load is taken through a number of 6 MVA, 11 kV/66 kV step-up transformers, each having a reactance of 3%. Calculate the current fed into fault if a short-circuit occurs on all phases near the high-voltage terminals of one of the transformers at the bus-bar section *B*.

15. The line currents in a 3-phase supply to an unbalanced load are represented by $\bar{I}_R = (12 + j24)$ A, $\bar{I}_Y = (16 - j12)\,A$ and $\bar{I}_B = (-4 - j6)\,A$. The phase sequence is *RYB*. Calculate the positive, negative and zero sequence currents.

[**Ans.** $\bar{I}_{R1} = 15.13 \angle 86.82°$ A, $\bar{I}_{Y1} = 15.13 \angle -33.18°\,A$; $\bar{I}_{B1} = 15.13 \angle 206.82°$
$\bar{I}_{R2} = 4.76 \angle 48.4°$ A, $\bar{I}_{Y2} = 4.76 \angle 168.4°$ A ;
$\bar{I}_{B2} = 4.76 \angle 288.4°$ A ; $I_{R0} = I_{Y0} = I_{B0} = 9.515 \angle 33.69°$ A]

16. A 3-phase, 11 kVA, 10 MVA alternator has sequence reactances of $X_1 = 0.15$ p.u., $X_2 = 0.15$ p.u., $X_0 = 0.05$ p.u. If the generator is on no load, find the ratio of fault currents for line-to-ground fault to that when all the 3-phases are dead short-circuited. [**Ans.** 1.284]

17. A 20 MVA, 11 kV, 3-phase, 50 Hz generator has its neutral earthed through a 5% reactor. It is in parallel with another identical generator having isolated neutral. Each generator has positive sequence reactance of 20%, negative sequence reactance of 10% and zero sequence reactance of 15%. If a line-to-ground short-circuit occurs in the common bus-bar, determine the fault currents. [**Ans.** $-j6998\,A$]

18. A 50 MVA, 11 kV, 3-phase alternator was subjected to the following faults : 3-phase fault = 2000 A ; line-to-line fault = 2600 A ; line-to-ground fault = 4200 A.
The generator neutral is solidly grounded. Ignoring resistances, calculate the values of the three sequence reactances of the alternator. [**Ans.** $X_1 = 3.175\ \Omega$; $X_2 = 1.056\ \Omega$; $X_0 = 0.305\ \Omega$]

22

Switches, Fuses And Circuit Breakers

22.1. **Switchgear—general aspects**—Definition of switchgear—Functions of switchgear—Essential features of switchgear—Principles of layout of switchgear—Types of switchgear—Short-circuits in electrical installations and limiting methods. 22.2. **Switches.** 22.3. **Fuses**—Definition of fuse—Advantages and disadvantages of a fuse—Desirable characteristics of fuse element—Fuse element materials—Important terms relating to fuses—Types of fuses—Fuse selection—Difference between sub-circuit fuses, main circuit fuses, meter board fuses and pole fuses in an electrical installation—Difference between fuse and circuit breaker. 22.4. **Circuit breakers**—Introduction—Functions of a circuit breaker—Principle of circuit breaker—Arc phenomenon—Methods of arc extinction—Arc, restriking and recovery voltages—Current chopping—Resistance switching—Classification of circuit breakers—Oil circuit breakers—Water type circuit breaker—Air blast circuit breakers—Sulphur hexafluoride (SF_6) circuit breakers—Vacuum circuit breakers—Low voltage circuit breakers—Ratings of circuit breakers—Testing of circuit breakers—Autoreclosing—Maintenance of circuit breakers—Switchgear components—Bus-bar arrangements. 22.5. Control room—power station. Highlights—Theoretical Questions—Unsolved Examples.

22.1. SWITCHGEAR—GENERAL ASPECTS

22.1.1. Definition of Switchgear

Switchgear is *a term which covers wide range of equipment as regards switching and interrupting the currents in the power system during normal and abnormal conditions.*

Or

The apparatus used for switching, controlling and protecting the electrical circuits and equipment is known as **switchgear.**

"*Switchgear*" in general *consists of the following :*

(*i*) Switches
(*ii*) Fuses
(*iii*) Circuit breakers
(*iv*) Isolators
(*v*) Relays
(*vi*) Control panels
(*vii*) Metering panels
(*viii*) Lightning arrestors
(*ix*) Current transformers
(*x*) Potential transformers etc.

22.1.2. Functions of Switchgear

The *functions* performed by a switchgear are listed below :

1. To localise the effects of faults by operation of protective equipment and so automatically disconnect faulty point from the system.

2. To break efficiently short circuits without giving rise to dangerous conditions.

3. To facilitate redistribution of loads, inspection and maintenance on the system.

22.1.3. Essential Features of Switchgear

The essential features of switch gear are :

1. *Complete reliability.*
2. *Absolutely certain discrimination* (The switch gear must be able to discriminate between faulty and healthy sections).
3. *Quick operation.*
4. *Provision for manual control.*
5. *Provision for instruments.*

22.1.4. Principles of Layout of Switchgear

The following *principles* should be followed while carrying out layout of switch gear :

1. The layout should be such that any section may be isolated without unduly affecting the service.
2. There should be an easy and safe access for general routine inspection and for maintenance.
3. The individual units should be so designed that the risks of failure are reduced to minimum.
4. Fire protection arrangement must be made adequately.
5. To keep the breaking duty within the capacity of the circuit breakers reactors should be used where necessary.

22.1.5. Types of Switchgear

The various types of switchgear are enumerated and discussed below :

(*i*) Cellular (*ii*) Cubicle
(*iii*) Truck (*iv*) Metal clad
(*v*) Outdoor switchgear.

1. Cellular. The components of this type of switchgear are enclosed in *cells* made of brick, concrete or moulded stone. These components can be easily inspected and modifications can be made without any difficulty. The cellular type gear has the following *disadvnatages* :

(*i*) The erection of a cellular type gear is a lengthy process.
(*ii*) It occupies large space.
(*iii*) The interlocking systems may be complicated.

2. Cubicle. The whole of the equipment, in this case, is enclosed in steel plate cubicles with complete or partial sub-dividing barriers. It entails the following *advantages* :

(*i*) It requires relatively small floor area.
(*ii*) It presents a neat and simple layout.

3. Truck. In this type of switchgear a movable truck carries the circuit breaker moving isolator contacts, and the potential and current transformers. The front panel of the truck fits into the stationary steel cubicle which houses the busbars, cable box, and fixed isolator contacts. When inspection is to be carried out, the truck is withdrawn from the cubicle.

4. Metal clad. Here all conductors and insulators are *enclosed by an earthed metal case.* A metal clad switchgear is of two types :

(*i*) Horizontal draw out (*ii*) Vertical drop down.

A metal clad switchgear has the following *advantages* :

(*i*) As all the live parts are enclosed in metal it ensures safety to operators and reduces maintenance charges.

(*ii*) Maintenance work can be easily carried out (since working parts are easily and safely accessible).

(*iii*) To prevent operating mistakes simple and efficient interlocks can be installed.

(*iv*) Erection on the site can be easily carried out.

(*v*) Since it occupies less space, therefore, building cost is reduced.

5. Outdoor switchgear :

- Although it requires large ground space but there is a saving in building cost.
- Maintenance work is at the mercy of the weather.
- The damage from lightning is likely to be more since the equipment is exposed.
- Frequent insulation cleaning.
- As compared to an indoor switchgear its maintenance cost is more.

22.1.6. Short-circuits in Electrical Installations and Limiting Methods

A ***'short circuit'*** *is any solid contact established between the different phases of an electrical installation or any bridging of them by a very low resistance.* When a *short circuit* takes place a short circuit current starts flowing which is usually many times excess of normal rated current in the currents carrying parts, circuit apparatus and even the alternators.

Cause of short circuits :

Primarily a short circuit is *caused due to failure or breakdown of the insulation beteween phases which allow the conductors to come into electrical contact with each other.* In electrical equipment and installations the insulation failures take place due to the following :

(*i*) Defects in material

(*ii*) Poor workmanship

(*iii*) Natural ageing

(*iv*) Mechanical injury and various causes.

Effects of short circuits :

1. During a short circuit a large amount of current (exceeding the normal load current) flows causing *overheating* which can prove very harmful to the electrical equipments. To avoid overheating of the current conducting parts it is generally *necessary to considerably increase their cross section areas.*

2. Supply to the consumers is interrupted.

3. The generator e.m.f. and the voltage in the stator winding drop due to considerable large short circut current flow through the circuit. As a consequence of the drop in the generator e.m.f., and also as a result of considerable voltage drop in the stator winding due to the flow of large short-circuit current, *every short circuit is found to cause a serious drop in voltage at the busses of the generating stations and in power circuit.*

4. Due to short circuit when there is significant decrease in voltage the motors may stall ; to prevent the stalling of motors it is necessary to switch off the faulty section of the circuit as quickly as possible.

Limiting of short circuit currents :

In installations of large capacity, short circuit currents can attain such high values that unless these are limited by some means, the selection of electrical equipment capacity of withstanding them is very difficult. This can be achieved by *interposing auxiliary inductive reactances called* ***reactors*** in each phase of the given installation.

Functions of a reactor :

1. Troubles may be localized or islated at the point where they originate without communicating their disturbing effects to the other parts of the network.

2. They limit the flow of current into a short circuit with the view of protecting the equipment from overheating as well as from failure due to destructive mechanical forces, and also protecting

the system as a whole against shutdown by maintaining the voltage on most of the system while the short circuit is being cleared.

3. They permit the installation of lower capacity circuit breakers.

Location of reactors :

Reactors may be placed in the generator leads, between bus sections, in the low-tension transformer leads, or in out going low-tension feeders. No definite statement can be given as to which one of the above locations is preferable, each installation has its own particular demands which must be carefully considered before a choice of reactor location can be made.

22.2. SWITCHES

A **switch** is *a device which is used to open or close an electrical circuit in a convenient way*. It can be used under full-load or no-load conditions *but it cannot interrupt the fault currents*. When the contacts of a switch are opened, an arc is produced in the air between the contacts. This is particularly true for circuits of high voltage and large current capacity.

The most important switches fall in one of the following *classes* :

1. Knife switches
2. Disconnecting switches
3. Air-break switches
4. Control switches
5. Auxiliary switches
6. Oil switches
7. Magnet impulse switches.

1. Knife switches :

- These switches are used to open and close circuit of *low voltage and current capacity.*
- They are extensively used in lighting and small power circuits. In order to protect such low capacity circuits against overloads and short circuits, it is necessary that proper fuses be connected in series with the switch blades.

2. Disconnecting switches. For power circuits it is necessary to use some type of circuit breaker to open or close the circuit. In order to isolate the circuit breakers it is generally considered good practice to connect knife switches in series with the circuit breakers. Such knife switches are known as *disconnecting switches.*

- They *should never be opened until the circuit breaker in the same circuit has been opened, and should always be closed before the circuit breaker is closed.*
- They are *not designed to break currents* and should therefore, *never, be opened while current is flowing in the line.*

3. Air-break switches. By "*air-break*" switches are meant switches designed to *open circuits under load.* They are generally used outdoors for circuits of medium capacity, such as lines supplying an industrial load from a main transmission line or feeder. In order to take off the arc that occurs on opening such a switch, special arcing horns are provided, so that the arc may rise and be ruptured.

- Air-break switches are built for about 135000 volts maximum but their use is in general confined to lower voltage.
- *These switches are not designed to operate under abnormal conditions, such as short circuits.*

4. Control switches. Under this class are included all the switches that are used to *control the operation of other equipment.*

- As a general rule, they are designed for operating voltages less than 250 volts and very small current capacities.

5. Auxiliary switches. Under this class are included all switches or contractors that are *actuated by some other control switch or device.* In the control of power equipment, such as circuit breakers, it is impracticable to handle the operating current through the control switches on the switch-board, hence an auxiliary operating switch is placed near the circuit breaker, this switch being electrically operated by a control switch that may be located at any desired distance from the circuit breaker. Thus, the control circuits may be designed for very small currents, while the operating circuits as a general rule must handle much large currents and very often at higher voltages.

6. Oil switches. For certain applications at high voltages and large current capacities it is desirable to immerse the switch contacts under oil. The effect of the oil is to cool and quench the arc that tends to form when the circuit is broken.

7. Magnet impulse switches. In this type of switch the arc is extinguished by blowing it magnetically into arc chutes where it is lengthened, cooled, and interrupted. The magnetic effect is produced by the circuit current which is passed through suitable coils, setting up a strong magnetic field across the space between the switch contacts as they are opened. The basic principle involved in lengthening the arc is that of a simple motor, in which force is directly proportional to the product of magnetic flux and current.

22.3. FUSES

22.3.1. Definition of Fuse

Fuse *is a device used in circuit for protecting electrical equipments against overloads and/or short circuits.*

"Fuse element" or **"Fuse wire"** *is that part of the fuse device which melts when an excessive current flows in the circuit and thus isolates the faulty device from the supply circuit.*

- The time for blowing out of fuse depends upon the magnitude of the excessive current. The greater the currents, the smaller is the time taken by the fuse to blow out. In other words, a fuse has inverse time-current characteristics. Such a characteristic permits its use for over current protection.
- Fuses are used on low and high-voltage circuits of moderate to high capacity where frequent operation is not expected *e.g.* for the protection of distribution transformers, small and medium size motors, lighting circuits, branch circuits of distribution lines, and industrial plants and commercial buildings upto 600 A, at 250 V ; 500 A at 600 V, and in high voltages, upto 100 A at 160 kV.

22.3.2. Advantages and Disadvantages of a Fuse

Advantages :

1. A fuse is the cheapest from of protection.
2. It requires no maintenance.
3. It interrupts enormous short circuit currents without noise and also without flame, gas and smoke.
4. Its operation is inherently completely automatic (unlike a circuit breaker which requires an elaborate equipment for automatic action).
5. The minimum time of operation can be made much smaller than that with circuit breakers.
6. The inverse time current characteristic of a fuse enables it to be used for overload protection.
7. It affords current limiting effect under short-circuit conditions due to cut-off.

Disadvantages :

1. Time is lost in replacing fuse after operation.
2. On heavy short-circuits discrimination between fuses in series cannot be obtained unless there is considerable difference in the relative sizes of the fuses concerned.
3. The current-time characteristic of a fuse cannot always be co-related with that of the protected apparatus.

22.3.3. Desirable Characteristics of Fuse Element

The fuse element should possess the following *desirable characteristics* :

1. Low melting point.
2. Low ohmic losses.
3. High conductivity.
4. Free from deterioration due to oxidation.
5. Low cost.

Since all the characteristics are not available in a single material, therefore, compromise is made in the selection of fuse material.

22.3.4. Fuse Element Materials

Lead, tin, copper, zinc and *silver* are the most commonly used materials for fuse elements.

- For *small currents* upto 10 A, tin or an alloy of lead and tin (lead 37%, tin 63%) is used for making the fuse element.
- For *large currents*, copper or silver is employed.
  — It is usual practice to tin the copper to protect it from oxidation.
- Zinc, in strip form only, is good if a fuse with a considerable time-lag is required *i.e.*, one which does not melt very quickly with a small overload.

The present trend is to use "*silver*" despite its high cost due to the following *reasons* :

(*i*) It does not deteriorate when used in dry air.

(*ii*) It is comparatively free from oxidation.

(*iii*) Due to its very small coefficient of expansion no critical fatigue occurs and as such the fuse element can carry the rated current continuously for a long time.

(*iv*) Silver vaporises at a temperature much lower than the one at which its vapour will readily ionise. Therefore, when an arc is formed through the vaporised portion of the element, the arc path has high resistance ; consequently, the short-circuit current is quickly reduced.

(*v*) Due to its low specific heat silver fusible elements can be raised from normal operating temperature to vaporisation very much quicker than other fusible elements. Also, as the temperature of the silver reaches the melting point its resistance increases abruptly and thus the transition from melting to vaporisation becomes nearly instantaneous. As a result, operation becomes very much faster at higher currents.

(*vi*) Due to heavy conductivity of silver, the mass of silver metal required, for a given rating of fuse element, is smaller than that of other materials. This minimises the problem of clearing the mass of vaporised material set free on fusion and thus permits fast operating speed.

22.3.5. Important Terms relating to Fuses

The important terms relating to fuses are discussed below :

1. Current rating of fuse element. *It is the current which the fuse element can carry without overheating or melting.*

- It depends upon the permissible temperature rise of the contacts of the fuse holder, fuse and upon the deterioration of fuse caused by oxidation.

2. **Fusing current.** *It is the minimum value of current at the which the fuse element or fuse wire melts.*

- Its value is *more* than the current rating of fuse element.
- For a *round wire* the approximate value of fusing current is given by.

$$I = kd^{3/2} \text{...... called the ordinary "fuse law"} \quad ...(22.1)$$

where, k = Constant (called fuse constant), depends on the metal of wire, and

d = Diameter of the wire.

Preece has given the value of the constant k, as under :

Material	*Value of 'k' for 'd' in mm*
Copper	80
Aluminium	59
Tin	12.8
Lead	10.8

The fusing current depends upon the following *factors* :

(*i*) Material of the fuse element.

(*ii*) Length—the shorter the fuse the greater the current because all the heat is easily conducted away.

(*iii*) Diameter.

(*iv*) Shape of cross-section of the fuse element.

(*v*) Size and location of terminals.

(*vi*) Previous history.

(*vii*) The type of enclosure employed.

(*viii*) Type of surface—the fusing current for a stranded fuse will be less than the product of the fusing current for one strand and the number of strands.

3. **Fusing factor.** *It is the ratio of minimum fusing current to the current rating of the fuse element.* It is *always greater than one.*

i.e.,
$$\text{Fusing current} = \frac{\text{Minimum fusing current}}{\text{Current rating of fuse element}}$$

- The smaller the fusing factor, the greater is the difficulty in avoiding deterioration due to overheating and oxidation at rated carrying current.
- Fusing factor of a fuse can be determined from the time-current characteristic of the fuse.

4. **Prospective current.** It *is defined as the r.m.s value of current which would flow in a circuit immediately following the fuse when a short-circuit occurs assuming that the fuse has been replaced by a link of negligible resistance.*

Fig. 22.1 shows how A.C. current is cut off by a fuse. The fault current would normally have a very large first loop, but it actually generates sufficient energy to melt the fuseable element well before the peak of this first loop is reached. The *r.m.s. value of the first loop of fault current is known as "prospective current".*

5. **Cut-off current.** *It is the maximum value of fault current actually reached before the fuse melts.*

6. **Pre-arcing time.** *This is the time between the commencement of the current which tends to cause a break in a fuse element and the instant that the arc is initiated.*

7. **Arcing time.** *This is the time between the end of the pre-arcing time and the instant when the circuit is broken and the current becomes permanently zero.*

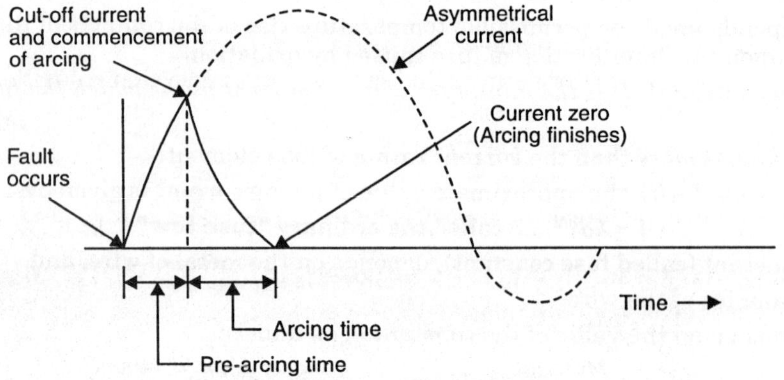

Fig. 22.1. Cut-off A.C. current by fuse.

8. **Total operating time.** This is the *sum of the pre-arcing and arcing time.*

9. **Breaking capacity.** Breaking capacity of a fuse is *the rating corresponding to the r.m.s. value of the A.C. component of the maximum prospective current and the system voltages.*

22.3.6. Types of Fuses

In general, fuses may be *classified* as follows :

A. *Low voltages fuses*

1. Rewireable or kit-kat fuse.
2. High rupturing capacity (H.R.C.) cartridge fuse.

B. *High voltage fuses*

1. Cartridge type.
2. Liquid type.
3. Metal clad fuses.

- In houses, blocks of flats, personnel amenity rooms and public buildings, *i.e.* whenever the electrical circuits are not constantly supervised by trained personnel the fuse elements must satisfy the condition :

Continuous current rating of fuse element = 0.8 × permissible continuous current capacity of the conductors forming the circuit to be protected.

22.3.6.1. Low voltage fuses

1. ***Rewirable or kit-kat fuse.*** Refer to Fig. 21.2

- Rewirable fuse consists of a "*fuse base*" and a "*fuse carrier*". The *base* is of porcelain and carries the fixed contacts to which the incoming and outgoing phase wires are connected. The *fuse carrier* is also of porcelain and holds the fuse elements (tinned copper wire) between its terminals. The fuse carrier can be inserted or taken out of the base when desired.
- Different diameters of fuse wire are used for different current ratings. The wire is normally stretched between two terminals of a fuse carrier which is inserted into *fuse base.*
- These fuses are standardized for rated currents upto 200 A.

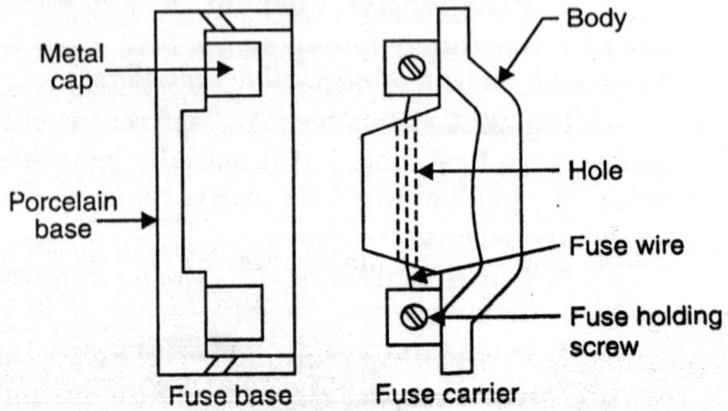

Fig. 22.2. Rewirable fuse.

Advantages :

(*i*) These fuses are *cheap and simple in construction*.

(*ii*) The detachable fuse carrier permits the replacement of fuse element without any danger of coming in contact with live parts.

Disadvantages :

(*i*) There is a possibility of renewal by the fuse wire of wrong size.

(*ii*) As the fuse has a low-breaking capacity, hence it cannot be used in circuits of high fault level.

(*iii*) Through continuous heating up of the element it is subjected to deterioration due to oxidation. Consequently, after a relatively short period fuse operates at a lower current than originally rated (*i.e.* current rating of the fuse is decreased.)

(*iv*) Accurate calibration of the fuse wire is not possible, as a longer fuse operates earlier than one of shorter length.

(*v*) The protective capacity of this type of fuse is uncertain as it is affected by the ambient conditions.

2. ***High rupturing capacity (H.R.C) cartridge fuse.*** The primary objection of low and uncertain breaking capacity of semi-enclosed rewireable fuses is overcome in H.R.C cartridge fuse.

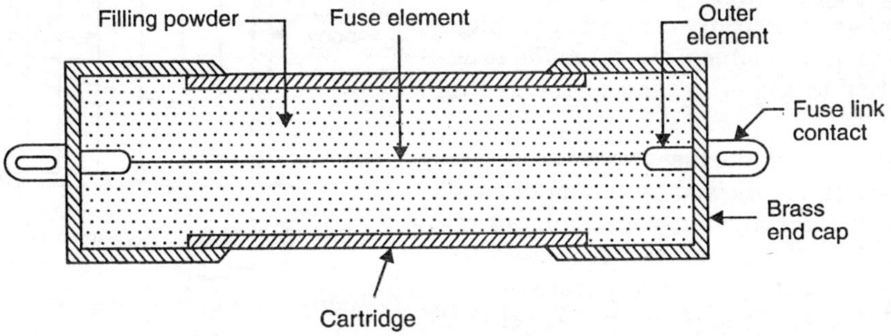

Fig. 22.3 H.R.C. cartridge fuse.

In its simplest form (Fig. 22.3) it consists of a heat resisting ceramic body heaving metal (brass) end caps to which are welded fusible silver (or bimetallic) current carrying elements. The space within the body, surrounding the element, is completely packed with a powder, usually quartz, which acts as an arc extinguishing agent.

The fuse element, under normal load conditions, is at a temperature below its melting point. On occurence of a fault short circuit current flows through the fuse element and the fuse element melts before the fault current reaches its first peak. The heat produced in the process vaporises the melted fuse element. The chemical reaction between the silver vapour and filling powder results in the formation of a high resistance substance which helps in quenching the arc.

Advantages :

The main advantages of H.R.C. cartridge fuse are :

(*i*) Provides consistence performance.

(*ii*) Requires no maintenance.

(*iii*) Capable of clearing high as well as low fault currents.

(*iv*) Provides reliable discrimination.

(*v*) Does not deteriorate with age.

(*vi*) High speed of operation.

(*vii*) Cheaper than other circuit interrupting devices of equal breaking capacity.

(*viii*) Enclosed fuse wire, therefore no chances of its arc doing any damage exist.

(*ix*) Low temperature rise at full-load.

Disadvantages :

(*i*) Need to be replaced after each operation.

(*ii*) This type of fuse causes overheating of the adjacent contacts

(*iii*) Interlocking is not possible.

(*iv*) Little difficult to ascertain which of the fuses has blown.

H.R.C. fuse with tripping device :

Fig. 22.4 shows a H.R.C. cartridge fuse incorporating tripping device.

- The body of the fuse is of ceramic material with a metallic cap rigidly fixed as each end. These are connected by a number of silver fuse elements. At one end is a plunger which *under fault conditions hits the tripping mechanism of the circuit breaker and causes it to operate.* The plunger is electrically connected through a fusible link, chemical charge and a tungsten wire to the other end of the cap as shown.

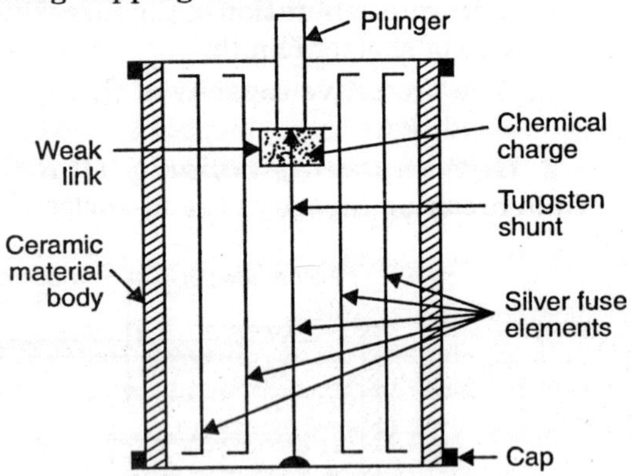

Fig. 22.4. H.R.C. fuse with tripping device.

When a fault occurs, the fuse elements are first to be blown off which is followed by fusing of weak link in series with the tungsten wire, causing the chemical charge to be detonated. Subsequently the plunger is pushed out with a force to operate the circuit breaker. The travel of the plunger is so set that it is not ejected from the fuse body under fault conditions.

Advantages :

(*i*) In case of a single-phase fault on a 3-phase system, the plunger operates the tripping mechanism of circuit breaker to open all the three phases and thus prevents "*single phasing*".

(*ii*) The fuse-tripped circuit breaker is generally capable of dealing with fairly small fault currents itself. This avoids the necessity for replacing the fuse except after highest corrents for which it is intended.

(*iii*) A relatively inexpensive circuit breaker can be used (since the effects of full short circuit need not be considered in the choice of circuit breaker).

- Low voltage H.R.C. fuse may be built with a breaking capacity of 16000 A to 30,000 A at 440 V. They are *extensively used on low-voltage distribution system against overload and short-circuit conditions.*

22.3.6.2. High voltage fuses

The main problem that starts at high voltage is that of corona. Therefore, high voltage fuses of special designs have been developed ; these are discussed below :

1. ***Cartridge type :***

- In this type of fuse, the element is wound in the shape of *helix* so as to *avoid corona effect.*
- In some designs, *two fuse elements in parallel* are employed-one of *low resistance* (silver wire) and the other of *high resistance* (tungsten wire). When a fault occurs, the low-

resistance element is blown out and the high resistance element reduces the short circuit current and finally breaks the circuit.

- High voltage cartridge fuses are used upto 33 kV with breaking capacity of about 8700 A at that voltage. Rating of the order of 200 A at 6.6 kV and 11 kV and 50 A at 33 kV are also available.

2. ***Liquid type :***

- In this type of fuses, the fuse elements are surrounded by liquid (carbon tetrachloride) and thus corona is avoided
- It has the widest range of application to *h.v.* systems. They may be used for voltage transformer protection or for circuit upto about 400 A rated current on systems upto 132 kV or higher and may have breaking capacities of 6100 A (350 MVA, 3 phase) at 33 kV.

3. ***Metal clad fuses :***

- Metal clad oil-immersed fuses have developed with the object of providing a substitute for the oil circuit breaker.
- Such fuses can be used for very high voltage circuits and operate most satisfactorily under short-circuit conditions approaching their rated capacity.

22.3.7. Fuse Selection

The selection of fuse in respect of steady and fluctuating loads is discussed below :

1. **Steady load :**

- When fuse is required to give *both overload and short-circuit protection* it should have a steady current rating *not greatly in excess of* the full-load current of the circuit, in order to give close protection.
- When fuse is required to give *only short circuit operation,* the fuse of much *greater rating than the normal load current can be selected.* Such a fuse will run cooler, have a longer life and will have a good margin in hand for some temporary overload without risk of deterioration

2. **Fluctuating load :**

(*i*) *Motor loads :*

- For the protection of motors, the cables and starter itself from damage due to short circuit, *H.R.C. fuse* may be used.
- Protection against *overheads* should be provided by separate devices like *overhead trips.*

(*ii*) *Switching of transformers, fluorescent lighting :*

- The fuse selected should have such a rating as will withstand the transient current surge and switching in.
- Generally such a fuse is selected which has a *normal rating 25 to 50% above the normal full load current of the apparatus to be protected.*

(*iii*) *Switching capacitor circuit.* For capacitor circuit similar selection as above is recommended but here the tendency should always be towards the higher percentage.

22.3.8. Difference between Sub-circuit Fuses, Main Circuit Fuses, Meter Board Fuses and Pole Fuses in an Electrical Installation

The difference between sub-circuit fuse, main circuit fuses, meter board fuses and pole fuses in an electrical installations is given below :

(*i*) A **sub-circuit fuse** is inserted after every 10 or less electrical points on a phase wire feeding those points. The fuse is housed in a sub-distribution board.

(*ii*) A **main circuit fuse** is of higher rating inserted in a circuit feeding the sub-distribution board *i.e.,* it should be capable to withstand the load of several sub-circuits.

(*iii*) **Meter board fuse** is of much higher rating to withstand the load of electrical points installed in the premises where meter is installed. It is a single fuse placed on phase wire for the protection of energy meter from line surges. The fuse is housed in a sealed metal box to avoid theft of electricity. The fuse unit is also of comparatively larger size.

(*iv*) The **pole fuse** is installed on pole itself at a reasonable height to provide protection to several energy meters installed in various buildings where service connections terminate in the case of domestic connections. It may be a common fuse to cover several domestic connections. The pole fuses are also installed on 3-phase, 4-wire distribution line to cut-off supply of individual phase.

22.3.9. Difference between Fuse and Circuit Breaker

The differences between a fuse and a circuit breaker are given below :

S. No.	Aspects	Fuses	Circuit breaker
1.	*Function*	Performs both detection and interruption functions.	Performs interruption function only. The detection of the fault made by relay system.
2.	*Operation*	Inherently completely automatic.	Requires elaborate equipment (relays etc.) for automatic action.
3.	*Breaking capacity*	Small	Very large
4.	*Operating time*	Very small (0.002 s or so)	Comparatively large (0.1 to 0.2s)
5.	*Replacement*	Needs replacement after every operation.	No replacement after operation.

Comparison between kit-kat fuse and miniature circuit breaker. The current at which a fuse wire melts varies considerably with the design of fuse holder. For *rewirable fuse holders*, the fuse wire should not be permitted to carry more than half the fusing current, because the fuse holder may heat up and there may be frequent breakdown. *The fuses are successful when currents are small. Fuses are cheapest source of protection with very little initial cost as compared to circuit breakers.*

Re-wirable kit kat fuse unit	Miniature Circuit Breaker (MCB)
1. The fuse wire melts on excessive current flow.	1. It operates to cut-off the circuit even on small overload current flow.
2. The fuse wire available at the time of rewiring may not be to standard rating.	2. These are available in standard rating and will have to be kept ready on demand.
3. The fuse wire unit has to be located amongst other alive units.	3. The switching knob provides direct indication of faulty circuit.
4. It melts on 50% to 100% (or even more) excessive overloads.	4. It operates on just 5% to 15% excessive overloads.
5. It is hazardous to re-wire a fuse unit at night and a person may require hand tool kit.	5. No hand tools required, just switch on the unit. Even ladies and children can operate to revive the circuit.
6. It is cheapest source of protection so far as initial installation cost is concerned.	6. The initial cost is more but it is bearable keeping trouble free and safe protection in view.

22.4. CIRCUIT BREAKERS

22.4.1. Introduction

There is probably no other part of a power system that is more important than the *equipment that controls the system*. The arrangement of switches alongwith fuses is suitable for only relatively small capacities, but for large capacities it is necessary to employ more dependable means of control, such as is obtained by the use of *circuit breakers*.

Circuit breakers are *mechanical devices designed to close or open contact members, thus closing or opening an electrical circuit under normal or abnormal conditions.*

- The *"interrupting or rupturing capacity"* of a circuit breaker is the *maximum value of current which can be interrupted by it without any damage.* The circuit breakers are also rated in MVA which is the product of interrupting current, rated voltage and 10^{-6}.

22.4.2. Function of a Circuit Breaker

The function of a circuit breaker is to break a circuit when various abnormal conditions arise and create a danger for the electrical equipment in an installation. The heaviest duty circuit breaker is required to interrupt a short-circuit current which may reach a value of several tens of thousands and even more (> 10^5 amperes) in a large capacity system. At the same time, in order to quickly eliminate the source of the fault, a circuit breaker must open the circuit with least possible delay.

- A circuit incorporates manual (or remote control) as well as automatic control for switching operations.
- *Automatic circuit breakers,* which are usually employed for the protection of electrical circuits, are equipped with a *trip coil connected to a relay or other means*, designed to open the breaker automatically under abnormal conditions, such as over-current. An automatic circuit breaker performs the following *duties* : (*i*) It carries the full load current continuously without overheating or damage ; (*ii*) It opens and closes the circuit on no load, ; (*iii*) It makes and breaks the normal operating current ; (*iv*) It makes and breaks the short-circuit currents of magnitude upto which it is designed for.

22.4.3. Principle of Circuit Breaker

Circuit breakers consist, "essentially, of current carrying contacts called *electrodes*. These are normally engaged but, under predetermined conditions, separate to interrupt the circuit.

The breaking of circuit means *rapid conversion of predetermined section of circuit from a conductor to insulator*. When the current carrying contacts are separated, an arc (which contains an isonised gas) is produced between them. This arc provides for the gradual change over from current carrying to voltage isolating states of the contacts. Therefore, it plays an important part in circuit interruption process. The arc has to be carefully controlled because a good deal of energy of the form of heat is produced in it. The produced *arc may be extinguished by high resistance or low resistance methods*.

The basic construction of any circuit breaker requires the separation of contacts in an *'insulating fluid'* which performs two *functions* :

(*i*) It *extinguishes the arc* drawn between the contacts when the circuit breaker opens ; (*ii*) It *provides insulation* between the contacts and from each contact to earth.

Several insulating fluids are used for arc extinction and the fluid chosen *depends upon the rating and type of circuit breaker*. The insulating fluids commonly used for circuit breakers are :

1. Air at atmospheric pressure.
2. Compressed air.

3. Oil which produces hydrogen for arc extinction.
4. Ultra high vacuum.
5. Sulphur hexa-fluoride (SF_6).

The gases which have been considered for circuit breakers are :

Simple gases : Air, oxygen, hydrogen, nitrogen, carbondioxide ;

Electronegative gases : Sulphur hexafluoride, arcton.

The *fluids used in circuit breakers should have the following properties* :

(*i*) High dielectric strength.
(*ii*) Thermal and chemical stability.
(*iii*) Non-inflammability.
(*iv*) High thermal conductivity (this assists cooling of current carrying conductors immersed in the gas and also assists the arc-extinction process).
(*v*) Arc extinguishing ability.
(*vi*) Commercial availability at moderate cost.

- '*Air*' is the cheapest and *most widely used for circuit breaking.*
- '*Hydrogen*' has better arc extinguishing property but it has lower dielectric strength as compared to air. Also, if hydrogen is contaminated with air, it forms an explosive mixture.
- '*Nitrogen*' has similar properties as air.
- '*Carbon-dioxide*' has almost the same dielectric strength as air but is a better arc extinguishing medium at moderate currents.
- '*Oxygen*' is a good extinguishing medium but is chemically active.
- '*Sulphur hexafluoride* (SF_4)' has *outstanding arc-quenching properties and good dielectric strength.*
  — Of all these gases SF_6 and air are used in commercial gas blast circuit breakers.

22.4.4. Arc Phenomenon

On occurence of a short-circuit, a heavy current flows through the contacts of the circuit breaker before they are opened by the protective system. At the instant when the contacts begin to separate, the contact area decreases rapidly and large fault current causes increased current density and hence rise in temperature. The heat produced in the medium (usually oil or air) between the contacts is sufficient to ionise the air or vaporise and ionise the oil. The ionised air or vapour act as conductor and an arc is struck between the contacts. The potential difference between the contacts is quite small and is just sufficient to maintain the arc. The arc provides a low resistance path and consequently the current in the circuit remains uninterrupted so long as the arc persists.

Initiation and maintenance of arc. The arc consists of a column of ionised gas *i.e.*, a gas in which molecules have lost one or more of their electrons resulting in *positive ions and electrons*. The electrons which have a negative charge are attracted towards the positive contact (*i.e.* a anode), and being light move towards it very rapidly. The positive ions, however, are attracted towards the negative contact (*i.e.* cathode), but, as they comprise almost the entire weight of atom, they move towards it relatively slowly. The *electron movement thus constitutes the current flow.*

- In order to ***initiate an arc***, *electrons must be emitted from the cathode as soon as the contacts begin to separate.* It is a thought that the initiating electrons are produced by the following processes :

(*i*) ***Field emission.*** When the voltage gradient at the surface is sufficiently high (10^6 V/cm) electrons can be dragged from the surface ; it is conceivable that such gradient can occur at the first moment of parting when the voltage is appreciable but the distance infinitesimally small.

(*ii*) ***Thermal emission.*** When the contacts part, current tends to concentrate at few points leading to very high local current densities (10^6 A/cm^2) and correspondingly high temperatures which would result in emission of electrons.

The electrons so liberated from cathode make many collisions with the atoms and molecules of the gases and vapours existing between the contacts during their journey towards the anode. These collisions cause ionization of atoms and the molecules and thus liberating more electrons. Ionization is facilitated by : (*i*) An *increase of field strength* ; (*ii*) An *increase of mean free path* (it is the distance through which the electron moves with freedom) ; (*iii*) *High temperature.*

The process, namely field emission, thermal emission and ionization, may start either one after the other or almost simultaneously and enable the arc to the started and maintained and finally if the arc current is high, the are may attain a temperature high enough for thermal ionization to become the main source of electrical conductivity.

22.4.5. Methods of Arc Extinction

During the arcing period, the current flowing between the contacts depends upon the *arc resistance*. The greater the arc resistance, the smaller the current that flows between the contacts. The arc resistance depends upon the following factors :

(*i*) *Degree of ionization.* The arc resistance increases with the decrease in the number of ionized particles between the contacts.

(*ii*) *Arc length.* The arc resistance increases with the arc length *i.e.*, separation of contacts.

(*iii*) *Arc cross-section.* The arc resistance increases with decrease in cross-section of the arc.

The following are the *two methods of extinguishing the arc* :

1. High resistance method.
2. Low resistance or current zero method.

1. **High resistance method.** In this method the *arc resistance is increased in time to such a high value that it forces the current to reach zero without possibility of arc being restruck thereafter.* The rate at which the resistance is increased or the current is decreased is not abnormal so as to cause harmful induced voltages in the system.

The resistance of an arc can be *increased* by :

(*i*) *Lengthening the arc.* The arc length can be increased by increasing the gap length between the contacts but it is not practicable to draw the arc out to such a length that the voltage available becomes inadequate to maintain the arc.

(*ii*) *Constraining the arc.* If the arc can be constrained into a very narrow channel the voltage necessary to maintain it is increased.

(*iii*) *Cooling the arc.* The voltage required to maintain ionization increases with a decrease of temperature so that cooling effectively increases the resistance.

(*iv*) *Splitting the arc.* The arc resistance can be increased by splitting the arc into a number of smaller arcs in series. Each one of these arcs experiences the effect of lengthening and cooling. The arc may be split by introducing some conducting plates between the contacts.

Because of the resistive nature of the arc discharge, most of the energy in the system will be received by the circuit breaker. Therefore, while designing the circuit breaker, provision of mechanical strength to *withstand such sudden release of large quantities of energy must be made.*

- *This method is employed in D.C. circuit breakers and low and medium power industrial type air circuit breakers.*

2. **Low resistance or current zero method.** This method is *used only in A.C. circuit interruption because there is natural zero of current present in such system* (In case of 50 Hz supply there are 100 zeroes per second). In this method the arc resistance is kept low until the current is zero where the arc extinguishes naturally and is prevented from restriking after it has gone out at current zero.

This method is employed in *all modern high power A.C. circuit breakers.*

The phenomenon of extinction of arc is explained by the following *two theories* :

(*i*) Energy balance theory.

(*ii*) Voltage race theory.

(*i*) ***Energy balance theory (Cassie's theory).*** According to this theory if *the rate of heat dissipation between the contacts is greater than the rate at which heat is generated, the arc will be extinguished.* Hence by increasing the rate of heat dissipation such as by cooling, lengthening and splitting the arc, the arc can be extinguished.

(*ii*) ***Voltage race theory.*** The arc, as is said earlier, is due to the ionization of the gap between the contacts. In an A.C. system, current drops to zero after every half-cycle. At every current zero, the arc extinguishes for a moment. Now the medium between the contacts contains ions and electrons so that it has small dielectric strength and can be easily broken down by the rising contact voltage known as *restriking voltage.* If such a breakdown does occur, the arc will persist for another half cycle. If immediately after current zero, the dielectric strength of the medium between the contacts is *built up more rapidly than the voltage across the contacts,* the *arc fails to restrike and the current will be interrupted.*

The rapid increase of dielectric strength of the medium near current zero can be achieved by : (*i*) Causing the ionized particles in the space between contacts to *recombine into neutral molecules* ; (*ii*) *Sweeping the ionized particles away and replacing them* by *unionized particles.*

The deionization of the medium can be achieved by :

(*i*) *Lengthening of the gap.* The dielectric strength is proportional to the length of gap so that lengthening by rapid opening of the contacts is an obvious procedure. The permissible arc length is limited, however, by other considerations, *e.g.*, arc energy and possibility of transient voltages due to current chopping.

(*ii*) *High pressure.* If the pressure is increased, the density of particles constituting the discharge also increases. The increased density of particles causes higher rate of deionization and thus the dielectric strength of the medium between the contacts is increased.

(*iii*) *Cooling.* Natural combination of ions and electrons takes place more rapidly if they are allowed to cool. Cooling by conduction to adjacent parts, *e.g.,* baffles or by the use of gas such as hydrogen that has a high diffusion and heat absorption rate is, therefore, effective.

(*iv*) *Blast effect.* If the ionized particles between the contacts are swept away and replaced by un-ionized particles, the dielectric strength of the medium can be increased considerably. This may be achieved by forcing the cool un-ionized gas or liquid into the contact space.

These methods are used in *oil and air-blast breakers.*

Extinction of D.C. arc. The extinction of D.C. arc is more difficult than that of an A.C. arc since in an A.C. circuit the current wave passes through zero point twice during each cycle whereas *in D.C. circuit the full current has to be broken.*

When a direct current passing through a highly inductive circuit is broken, the arc tends to persist (this tendency is more marked if the voltage is high). Once an arc is formed between two points, the air molecules in the path become extremely hot and get *ionized* (*i.e.,* the normal insulation property of air is destroyed and the hot air molecules become conductors of electricity). Therefore, the arc is maintained even if the contacts are further drawn apart, resulting in a flash over ultimately. The *heat generated by the arc is intense.* To overcome this difficulty, general methods have been devised :

- When *small currents are involved,* the *contacts are broken quickly before the air can get ionized.*

- When heavier currents are to be broken, the *arc is blown off by creating a magnetic field heaving the property of deflecting the arc.* The magnetic field is created by '***blow out coils***' through which the current to be broken flows. The *arc path is deflected until it is blown upwards in special arc chutes.*

22.4.6. Arc, Restriking and Recovery Voltages

Arc voltage. The *voltage across the contacts during the arcing period is known as the **"arc voltage"***.

- It is relatively low with heavy current arcs of short length.
- At current zero, the arc voltage rises rapidly to peak value (since a short circuit current is almost 90° lagging) and this peak voltage tends to maintain the current flow in the form of arc.

Restriking voltage. *The transient voltage appearing across the contacts during arc period is called the **"restriking voltage"***. If restriking voltage rises more rapidly than the dielectric strength of the medium between the contacts, the arc will persist for another half-cycle.

The expression for the *restriking voltage* is given by (for a lossless system) :

$$v = V_{max}\left(1 - \cos\frac{t}{\sqrt{LC}}\right) \qquad ...(22.2)$$

where, v = Restriking voltage at any instant t,

V = Voltage at the instant of interruption, and

L, C = Series inductance and shunt capacitance upto the fault point.

Maximum restriking voltage would occur when

$$\cos\frac{t}{\sqrt{LC}} = -1 \quad \text{or} \quad \cos\frac{t}{\sqrt{LC}} = \cos\pi = \cos\pi$$

or

$$t = \pi\sqrt{LC} \qquad ...(22.2a)$$

It can be seen from the eqn. (22.2) that lower the value of the inductance and capacitance the higher will be the natural frequency of oscillation and more sever will be the effect of restriking voltage. Therefore, a *fault near the source is more severe from the view-point of arc interruption as compared to a fault far from the source.*

Further, it is seen that v is a function of V which is turn depends upon the power factor of the system. When the system is highly lagging, this voltage will correspond to the peak system voltage. The variation of this voltage is shown in Fig. 22.5.

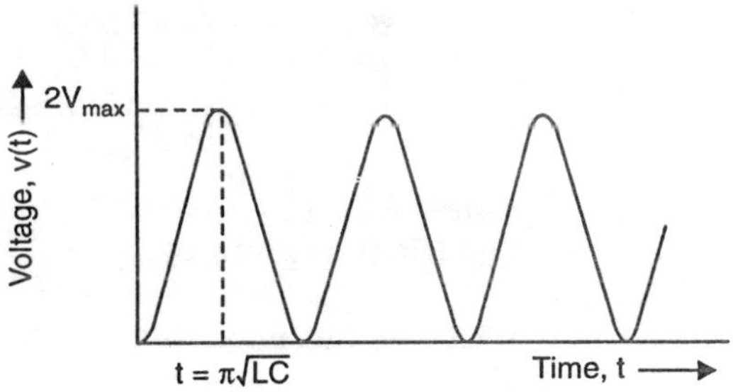

Fig. 22.5. Restriking voltage across breaker contacts.

The above figure gives the restriking voltage across the breaker contacts only when the arc voltage (being a small percentage of the system voltage) and the losses are *neglected*. But taking losses into account the restriking voltage will be *damped out depending upon the resistance of the system* as shown in Fig. 22.6.

The Fig. 22.6 illustrates the restriking voltage assuming that the current lags behind the voltage by 90°. However, in actual practice it is not always true. The fault current can lag by any degree of asymmetry depending upon the time in the cycle at which fault occurs. In case when the angle of symmetry between current and voltage is ϕ, then restriking voltage will be obtained by substituting V_{max} by $V_{max} \sin\phi$ which is certainly by low as compared to when $\phi = 90°$.

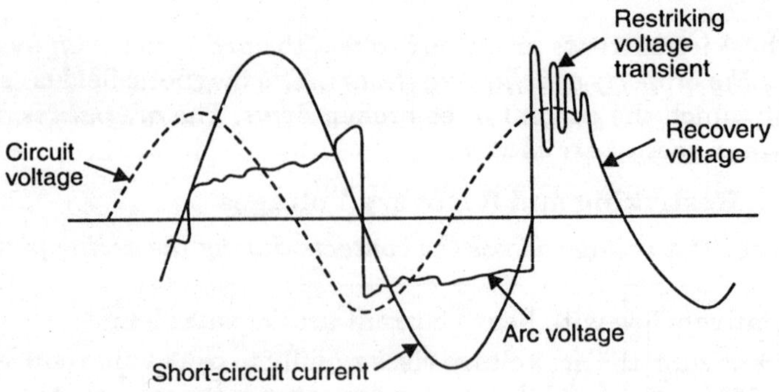

Fig. 22.6. Restriking voltage wave shape.

Recovery voltage. *It is normal frequency (50 Hz) r.m.s. voltage that appears across the contacts of the circuit breaker after final arc extinction.* It is approximately equal to the system voltage.

The voltage at the instant of arc extinction is known as the "active recovery voltage (V_{ar})." It is given by the expression :

$$V_{ar} = kV_{max} \sin \phi \quad ...(22.3)$$

where, $k = 1$ ...... for 3-phase *ground fault.*

and $k = 1.5$ ...... for 3-phase *isolated fault.*

- The active recovery voltage depends upon :
  (*i*) Power factor ; (*ii*) Armature reaction ; (*iii*) Circuit conditions.

Rate of rise of restriking voltage (RRRV) :

$$\text{The average RRRV} = \frac{\text{Peak value of restriking voltage}}{\text{Time taken to reach the peak value}}$$

$$= \frac{2V_{max}}{\pi\sqrt{LC}} \quad ...(22.4)$$

Again, $$v = V_{max}\left(1 - \cos\frac{t}{\sqrt{LC}}\right)$$

The RRRV is given by, $$\frac{dv}{dt} = \frac{V_{max}}{\sqrt{LC}} \sin\frac{t}{\sqrt{LC}}$$

RRRV will be maximum when

$$\sin\frac{t}{\sqrt{LC}} = 1$$

or, $$\frac{t}{\sqrt{LC}} = \frac{\pi}{2} \quad \text{or} \quad t = \frac{\pi}{2}\sqrt{LC}$$

$$\therefore \quad (\text{RRRV})_{max} = \frac{V_{max}}{\sqrt{LC}} \quad ...(22.5)$$

Thus if RRRV is smaller than the rate at which the dielectric strength between the contacts is developed, the arc will be quenched otherwise it will further restrike in next half cycle. This theory is called the *recovery rate theory* advocated by Dr. J. Slepian.

22.4.7. Current Chopping

When a low inductive current (*e.g.*, current to a shunt reactor or magnetising current of a transformer) is *interrupted by a circuit breaker, current falls to zero before the natural current zero.* It *is termed as* **current chopping**. It so happens since the circuit breaker exerts the same force of arc quenching as it were with heavy short circuit current.

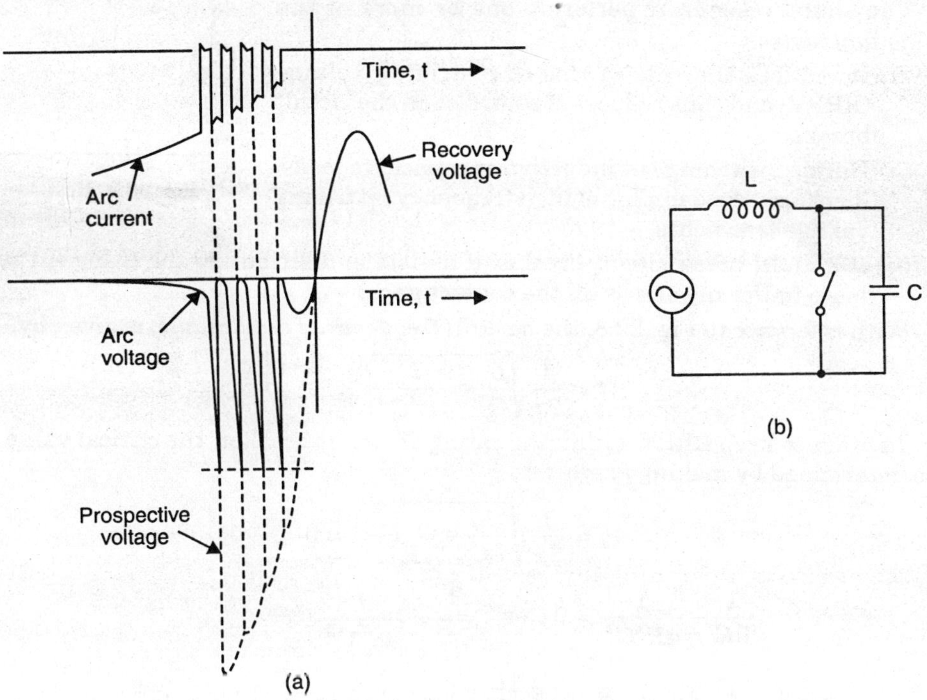

Fig. 22.7. Current chopping.

The effect of a practically instantaneous collapse of the arc current, even of only a few amperes, is potentially very serious from point of view of over-voltages which may result in the system. As shown in Fig. 22.7(*a*) arc current gradually reduce with low arc voltage but at a certain time it *suddenly becomes zero due to large de-ionizing force.* The current in the arc was flowing from the source through the inductance and the circuit breaker contacts. The energy contained in the electromagnetic field cannot become zero instantaneously, and as such, it changes into some other form of energy. The only possibility is the conversion from electromagnetic to electrostatic form of energy *i.e., current is diverted from the arc to the capacitor* (Fig. 22.7 (*b*)). If i_a is the arc current, L and C are the inductance of the circuit and capacitance between breaker contacts respectively, then, energy stored in L, $\frac{1}{2} L i_a^2 = \frac{1}{2} C v^2$, energy transformed to C where v is the voltage developed across the breaker contacts,

or,
$$v = i_a \sqrt{\frac{L}{C}} \text{ volts.} \qquad \ldots(22.6)$$

The *voltage developed across the breaker contacts due to transfer of energy is called* ***Prospective voltage***. This voltage may be extremely high as compared to the normal system. The deionizing force is still in action and current will again be chopped. Successive chops may occur (as shown in the figure) until a final chop brings the current to zero prematurely with no further restrike since the gap is now in an advance of deionization.

22.4.8. Resistance Switching

It has been observed that during chopping a very high voltage appears across the circuit breaker contacts which may cause damage to it. It can be prevented by connecting a resistance across the breaker contacts as illustrated in Fig. 22.8.

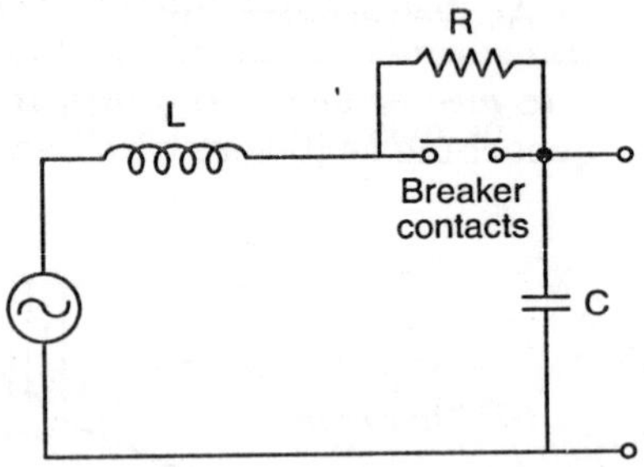

Fig. 22.8. Resistance switching.

The shunt resistor R performs one or more of the following *functions* :

(*i*) It reduces the rate of rise of restriking voltage (RRRV) and thus reduces the burden on the circuit breaker.

(*ii*) During switching out inductive or capacitive loads, it ensures the damping of high frequency restriking voltage transients.

(*iii*) It a multi-break circuit breaker it assists in distributing the transient recovery voltage more uniformly across all the contact gaps.

With reference to Fig. 22.8, the *natural frequency of oscillations*, is given by,

$$f = \frac{1}{2\pi}\sqrt{\frac{L}{C} - \frac{1}{4R^2C^2}}$$

In order to keep RRRV within the rating of circuit breaker, the critical value of resistance R can be determined by making f zero, *i.e.*,

$$f = \frac{1}{2\pi}\sqrt{\frac{1}{LC} - \frac{1}{4R^2C^2}} = 0$$

or, $$\frac{1}{LC} - \frac{1}{4R^2C^2} = 0 \quad \text{or} \quad \frac{1}{LC} = \frac{1}{4R^2C^2}$$

$$\therefore \quad R = 0.5\sqrt{\frac{L}{C}} \quad \quad ...(22.7)$$

Example 22.1. *In a short-circuit on 132 kV 3-phase system, the breaker gave the following results :*

Power factor of the fault = 0.45

Recovery voltage = 0.9 time full line voltage

The breaking current is symmetrical and the restriking transient had a natural frequency of 15 kHz.

Determine the rate of rise of restriking voltage (RRRV) in the following types of faults :

(*i*) *Grounded fault.* (*ii*) *Ungrounded fault.*

Solution. *Given* : Power factor of the fault, $\cos \phi = 0.45$;

Recovery voltage = 0.9 × Full line voltage ; $f_n = 15$ kHz

Peak value of phase voltage (line-to-neutral voltage),

$$V = \frac{132}{\sqrt{3}} \times \sqrt{2} = 107.8 \text{ kV}$$

Recovery voltage, $V_{max} = 0.9\ V = 0.9 \times 107.8 = 97$ kV

Time taken to reach the peak value, $t = \frac{1}{2f_n} = \frac{1}{2 \times 15 \times 10^3} = 33.33\ \mu s$

(*i*) ***Grounded fault :***

Active recovery voltage = $kV_{max} \sin \phi$

$= 1 \times 97 \times \sin (\cos^{-1} 0.5) = 84$ kV (∵ For grounded fault $k = 1$)

The maximum (peak value) restriking voltage = 2 × 84 = 168 kV

$$\text{RRRV} = \frac{\text{Peak value of restriking voltage}}{\text{Time taken to reach the peak value}}$$

$$= \frac{168}{33.33} = \mathbf{5.04\ kV/\mu s.} \quad \textbf{(Ans.)}$$

(*ii*) ***Ungrounded fault :***

Active recovery voltage $= kV_m \sin \phi$

$= 1.5 \times 97 \sin (\cos^{-1} 0.5) = 126$ kV

(∵ For ungrounded fault, $k = 1.5$)

$$\therefore \quad \text{RRRV} = \frac{2 \times 126}{33.33} = \mathbf{7.56\ kV/\mu s.} \quad \textbf{(Ans.)}$$

Example 22.2. *The following data relate to a 50 Hz generator :*

E.m.f. to neutral = 7.5 kV (r.m.s.) ;

Reactance of generator and connected system = 5 Ω

Distributed capacitance to neutral = 0.02 μF ; Resistance negligible

Determine the following :

(*i*) *Maximum voltage across the contacts of circuit breaker when it breaks a short-circuit current at zero current.*

(*ii*) *Frequency of the transient oscillation.*

(*iii*) *Average rate of rise of voltage upto the first peak of the oscillation.*

Solution. *Given* : E. m.f. to neutral = 7.5 kV (r.m.s.) ; $X_L = 5\ \Omega$, $C = 0.02\ \mu F$.

(*i*) **Maximum voltage across contacts :**

Active recovery voltage (phase to neutral),

$$v = \sqrt{2} \times 7.5 = 10.6 \text{ kV}$$

∴ Maximum re-striking voltage (phase-to-neutral) = 10.6 × 2 = 21.2 kV

(*ii*) **Frequency of the transient oscillation f_n :**

$$X_L = 2\pi f L = 5 \quad \therefore \quad L = \frac{5}{2\pi f} = \frac{5}{2\pi \times 50} = 0.0159 \text{ H}$$

Now,

$$f_n = \frac{1}{2\pi\sqrt{LC}} = \frac{1}{2\pi\sqrt{0.0159 \times 0.02 \times 10^{-6}}} = 8.92 \text{ kHz}$$

(*iii*) **Average rate of rise of voltage :**

$$v = V\left(1 - \cos \frac{t}{\sqrt{LC}}\right) \qquad \text{...[Eqn. (22.2)]}$$

Maximum restriking voltage would occur when,

$$\cos \frac{t}{\sqrt{LC}} = -1 \quad \text{or} \quad \cos \frac{t}{\sqrt{LC}} = \cos \pi$$

$$\therefore \quad t = \pi\sqrt{LC} = \frac{\pi}{2\pi f_n} = \frac{1}{2 f_n} = \frac{1}{2 \times 8.92 \times 10^3} = 56.05\ \mu s$$

∴ *Average rate of rise of restriking voltage*

$$= \frac{\text{Maximum restriking voltage}}{t} = \frac{21.2}{56.05} = \mathbf{0.378\ kV/\mu s.} \quad \textbf{(Ans.)}$$

Example 22.3. *In a 132 kV, 3-phase, 50 Hz power system the line-to-ground capacitance is 0.02 μF and the inductance is 4.5 H.*

Determine : (i) Voltage appearing across the breaker pole when a magnetising current of 5 A (instantaneous value) is interrupted.

(ii) Resistance to be connected across the contacts to eliminate the restriking voltage.

Solution. *Given :* $C = 0.02\ \mu F = 0.02 \times 10^{-6}$ F ; $L = 4.5$ H ; Magnetising current to be interrupted, $i = 5$A.

(*i*) **Voltage appearing across the breaker pole :**

The voltage appearing across breaker contacts

= Voltage across the capacitor ;

$$v = i\sqrt{\frac{L}{C}} = 5 \times \sqrt{\frac{4.5}{0.02 \times 10^{-6}}} = 75000 \text{ V} \quad \text{or} \quad \mathbf{75\ kV.} \quad \textbf{(Ans.)}$$

(*ii*) **Resistance to be connected across contacts, *R* :**

$$R = 0.5\sqrt{\frac{L}{C}} \qquad \text{...(Eqn. (22.7))}$$

$$= 0.5\sqrt{\frac{4.5}{0.02 \times 10^{-6}}} = 7500\ \Omega \quad \text{or} \quad \mathbf{7.5\ k\Omega.} \quad \textbf{(Ans.)}$$

22.4.9. Classification of Circuit Breakers

Circuit breakers may be *classified* as follows :

1. **High voltage circuit breakers :**

(*a*) *Oil circuit breakers :*

(*i*) Bulk oil circuit breakers (using a large quantity of oil)

(*ii*) Low oil contact circuit breakers (which operate with a minimum amount of oil)

(*b*) *Oilless circuit breakers :*

(*i*) Water circuit breakers (in which water is used for arc extinction)

(*ii*) Hard gas circuit breakers.

(*iii*) Air blast circuit breakers (in which compressed air extinguishes the arc)

(*iv*) Sulphur hexafluoride circuit breaker (in which SF_6 is used as an arc extinguisher)

(*v*) Vacuum circuit breakers (in which vacuum is used for arc extinction).

2. **Low voltage circuit breakers :**

- In *bulk oil circuit breakers* the transformer oil, with which they are filled, is used to extinguish the arc during opening of the contacts and to insulate the current-conducting parts from one another and from the earthed tank.
- In the *low oil circuit breakers* the oil serves only for arc extinction ; their current-conducting parts are insulated by air and porcelain, or organic insulating materials.
- In *hard gas circuit breakers* the arc is extinguished by gases generated when the wall surfaces of the arc extinguishing chambers, lined with insulating materials capable of producing large amounts of gases, are decomposed by the arc.

Characteristics of high voltage rating circuit breakers :

The *high voltage rating circuit breakers* should possess the following *characteristics* :

1. They should have high reliability-electrically and mechanically.

2. They should be capable of interrupting inductive and capacitive circuits and fault currents of all values within their rating.

3. The opening *i.e.*, the time interval between receipt of tripping impulse and contact separation should be the minimum mechanically possible.

22.4.10. Oil Circuit Breakers (O.C.B.)

These are the most common and oldest type of circuit breakers. The rating range of these circuit breakers lies between 25 MVA at 2.5 kV and 5000 MVA at 250 kV.

In these breakers, separating contacts are made to separate within insulating oil, which has better insulating properties than air. The bubbles of the gas formed prevents restriking of the arc after the current reaches zero point of the cycle.

The ***advantages*** *of using oil as a quenching medium* are :

1. It has high dielectric strength.
2. It absorbs arc energy while decomposing.
3. Good cooling property of gas formed (as a result of decomposition of oil).
4. Surrounding oil in close proximity to the arc presents a large cooling surface.
5. It acts as an insulator between live part and earth.
6. The ability of cool oil of high dielectric strength to flow into the arc space after the current is zero.

Disadvantages :

1. It is easily inflammable.
2. It can cause explosion by mixing with air.
3. It requires maintenance and periodic replacement.

- The movement of the oil into the arc space after the current zero, which is an essential part of the action of an oil circuit breaker, may be effected by *the pressure*,

(*i*) due to the natural head of oil above the contacts,

(*ii*) generated by the action of arc current itself, and

(*iii*) set up by external means.

The oil circuit breakers are *correspondingly called* :

(*i*) Plain break oil circuit breakers.

(*ii*) Self-generated pressure oil circuit breakers (also called *self-blast* oil circuit breakers or arc control oil circuit breakers)

(*iii*) Externally generated pressure oil circuit breakers (also called *forced-blast* oil circuit breakers or impulse oil circuit breakers).

22.4.10.1. Plain break oil circuit breaker

Fig. 22.9. shows the arrangement of a *double break plain oil circuit breaker. It is called a double break breaker since it provides two breaks as shown in series. It consists of fixed and moving contacts enclosed in a strong weather-tight earthed tank containing oil upto a certain level and an air cushion above the oil level. The air cushion provides sufficient room to allow for the reception of the arc gases without the generation of unsafe pressure in the dome of the circuit breaker. It also absorbs the mechanical shock of the upward oil movement.

* This type of construction increases the effective speed of arc lengthening and permits to divide the transient re-striking voltage over two breaks so that only half appears across each.

A gas outlet from the tank is essential and some form of vent is fitted in the tank cover. The position of the vent is carefully chosen so that the partially ionized gases which come out of the vent do not harm the personnel and do not cause flash over to the neighbouring equipments under normal operating conditions, the fixed and moving contacts remain closed and the breaker carries the normal circuit current. On occurence of the fault, moving contacts are pulled down by the protective system and an arc is struck which vaporises the oil mainly into hydrogen gas. The gas thus liberated surrounds the arc and its explosive growth around it displaces the oil violently, As the moving contacts separate, the length of the arc increases, the rate of gas formation decreases due to fall in temperature. The arc is extinguished when the distance between the fixed and moving contacts reaches a certain critical value. The arc is extinguished when the arc current passes through zero.

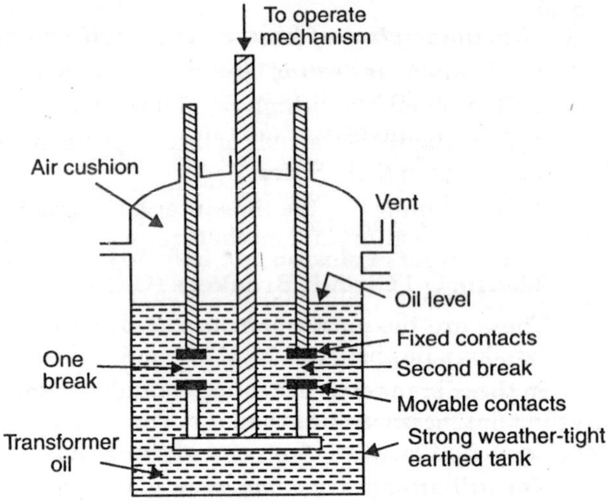

Fig. 22.9. Plain oil circuit breaker.

- The ***major drawback*** of this breaker is that it *suffers from the defect of permitting rather long and inconsistent arcing times.* In view of this limitation, such circuit breakers are *suitable only for low current and low voltage operations not exceeding 150 MVA at 11 kV.* Also, since such breakers are not suitable for high speed interruption, therefore these cannot be employed for auto-reclosing.

22.4.10.2. Self-generated pressure oil circuit breakers

In this type of circuit breaker the pressure generated by the arc is utilised in speeding up the movement of oil into the contact space immediately after the current zero. This is done by surrounding the contact by a pressure chamber or pot. The *breaking capacity of such a circuit breaker is, therefore, increased while the arcing time reduced.*

The design of chamber or pot should be such that the pressure developed is sufficient to quench the arc even at low values of current but not so much as to break the pot on heavy current. This has lead to manufacture of a variety of pots, some of which are described below :

1. ***Plain explosion type .*** Refer Fig. 22.10.

In this type of explosion pot the fixed and moving contacts are enclosed in a strong shell of insulating material. The moving contact comprises a stud which moves vertically out of fixed finger contacts (not shown here) and the relatively close-fitting throat at the lower end of the pot.

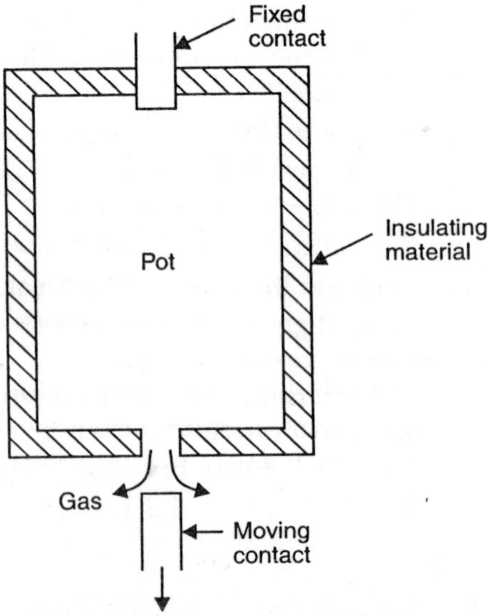

Fig. 22.10. Plain explosion pot.

On occurrence of a fault, the contacts separate and arc is produced which breaks the oil into gas at very high pressure in the pot. This high pressure forces the oil and gas through and around the arc to extinguish it. If the arc is not extinguished while the moving contact is still within the pot, it occurs immediately after the moving contact leaves the spot due to the axial high velocity blast of gas which is released through the throat.

- The main *drawback* of this type of pot is that it *cannot* be used either for very low or for very high currents. With low currents the arcing time is long while with heavy currents the pot is liable to burst due to pressure. Thus the pressure set up in the pot limits the breaking capacity.

2. ***Cross-jet explosion pot.*** Refer to Fig. 22.11.

The cross-jet explosion pot is so named since it *performs cross or transverse extinction of the arc.*

It consists of channels made up of metal on one side in order to *act as arc splitters*, whenever the length of arc increases.

On occurence of a fault the contacts separate and an arc is initially struck in the top of the pot. The gas generated by the arc exerts pressure on the oil in the back passage. When the moving contact uncovers the arc splitter ducts, fresh oil is forced *across* the path. The arc is, therefore, driven sideways into the "arc splitters" which increase the arc length, causing arc extinction.

- This type of oil circuit breaker is *suitable for interrupting heavy current at high voltage (66 kV).*

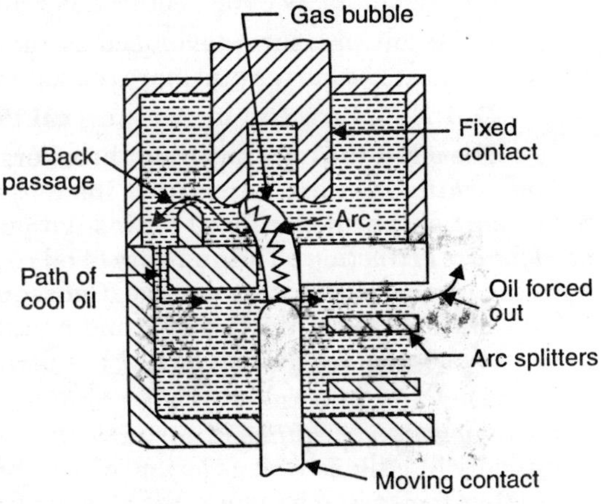

Fig. 22.11. Cross-jet explosion pot.

3. ***Self compensated explosion pot.*** Refer to Fig. 22.12.

This type of pot is a *combination* of the cross-jet explosion pot and plain explosion pot and thus operates quite satisfactorily both at heavy as well as low currents.

It consists of two chambers, the upper being a cross-jet explosion pot while the lower is a plain explosion pot.

When the *short-circuit current is heavy*, the rate of generation of gas is very high and the device operates as a cross-jet explosion pot, arc extinction occurring when the first or second lateral orifice is uncovered.

When the *short-circuit current is low*, the rate of generation of gas is low so that the tip of the moving contact has time to reach the bottom chamber while the pressure builds up therein. During this time very little pressure leakage occurs through the orifices since the movement of the oil towards these openings is obstructed by the right angle bends and also by the presence of arc. When the tip of the moving contact emerges from the throat, the arc is extinguished by the plain explosion pot action.

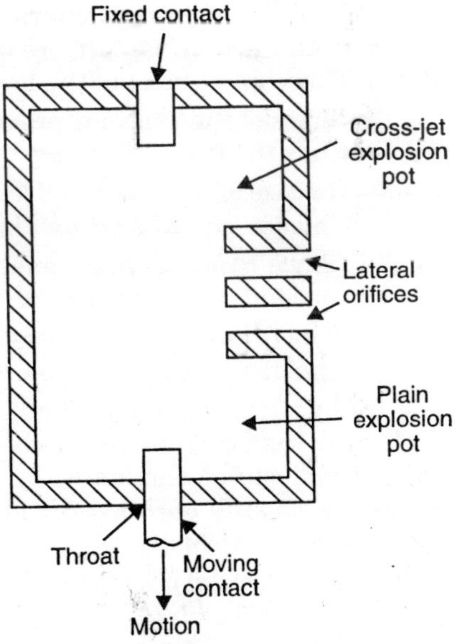

Fig. 22.12. Self compensated explosion pot.

22.4.10.3. Externally generated pressure type circuit breakers

These are also called *force blast oil* or *impulse oil* circuit breakers . In these breakers the oil pressure instead of being created by the *arc is created mechanically by the piston-cylinder arrangement*. The movement of the piston is mechanically coupled to the moving contact, which automatically creates the oil pressure and leads to high speed interruption. The possibility of current chopping can be avoided by employing resistance switching.

Advantages :

(*i*) The quantity of oil required is reduced considerably (say one-quarter),

(*ii*) As oil pressure developed is independent of the fault current to be interrupted, the performance at low currents is more consistent than with self-generated pressure oil circuit breakers.

22.4.10.4. Low (on minimum) oil circuit breakers

The oil in the bulk oil circuit breakers (discussed earlier) serves two purpose. (*i*) It *acts as an arc quenching medium* ; and (*ii*) It insulates the live parts from earth. It has been observed that major part of the oil is utilised for insulation purposes whereas a *small percentage of oil is actually used for arc extinction* . The quantity of oil required for bulk oil circuit breakers reaches a high figure with the increase in system voltage. Consequently, besides increases in the expenses, tank size and weight of the breaker the fire risk and maintenance problems are also increased. These factors led to the development of low oil circuit breaker.

A low circuit breaker *employs solid materials for insulation purposes and uses a small quantity of oil which is just sufficient for extinction of the arc.* As regards quenching arc, the oil behaves identically in bulk as well as low oil circuit breaker. The arc extinction (in a low circuit breaker) can be further facilitated by using suitable control devices.

It can be self-blast type or external blast type or a combination of the two. In case of *self-blast type* the gas pressure developed depends upon the current to be interrupted. The higher the current to be interrupted the larger is the gas pressure developed and hence more effective is the beaker for arc quenching. But this puts a limit on the design of the arc chamber for mechanical stresses. With the use of better insulating materials for arcing chambers such as glass fibre, reinforced synthetic resins etc., the low oil circuit breakers are able to meet easily the increased fault levels of the systems. These days, most of the low oil circuit breakers are of the self-blast type.

In terms of the ventings provided, two different designs of arcing chambers provided are : (*i*) Axial venting ; (*ii*) Radial venting.

— In case of '*axial venting*' the gases produced sweep the arc in *longitudinal direction* whereas in case '*radial venting*' they sweep the arc in *transverse direction*.

— Since *axial venting* generates high gas presures and has high dielectric strength it is used mainly for the *interruption of low currents at low voltages.*

— The *radial venting* is used for interruption of *relatively heavy currents at low voltages* as the gas pressures developed are low and also the dielectric strength is low.

Many a times a *combination* of both is used so that the arc chamber is equally efficient at low as well as at high currents. Such chambers, however, suffer from the disadvantages of longer arcing periods which can be eliminated by providing oil injection devices in addition to the self-blast. *The contacts are usually operated by pull rods or rotating insulators actuated in turn by solenoid or pneumatic mechanisms.*

- This type of circuit breaker is *available upto 8000 MVA* at *245 kV with a total break time of 3 to 5 cycles.*

Advantages. A low oil circuit breaker claims the following *advantages* over a bulk oil circuit breakers :

(*i*) Requires lesser quantity of oil. (*ii*) Requires smaller space.

(*iii*) Reduced risk of fire. (*iv*) Reduced maintenance problems.

Disadvantages :

(*i*) Owing to smaller quantity of oil, the *degree of carbonisation is increased.*

(*ii*) Due to high degree of decarbonisation, the *dielectric strength of the oil deteriorates rapidly.*

(*ii*) It is difficult to remove gases from contact space in time.

22.4.10.5. Maintenance of oil circuit breakers

Generally the maintainenance work required on oil circuit breakers is *replacement of contacts and oil.* After a circuit has interrupted fault currents a few times or load currents several times, its contacts may get burnt by arcing and oil may lose some of its dielectric strength due to carbonisation. This may lead to reduction in rupturing capacity of the breaker. Thus, it is imperative to inspect the circuit break at regular intervals (say 3 to 6 months).

The following points showed be checked during inspection of the circuit breaker : (*i*) Check the oil level ; (*ii*) Check the closing and tripping mechanism ; (*iii*) Check the current carrying parts and arcing contacts. If burning is severe, the contacts should be replaced ; (*iv*) Check the dielectric strength of the oil. If the oil is badly discoloured, it should be changed or reconditioned. The oil in good condition should withstand 30 kV for one minute in a standard oil testing cup with 4 mm gap between electrodes.

22.4.11. Water Type Circuit Breaker

Fig. 22.13 shows a water type circuit breaker, in which an insulating cover divides the chamber into two parts :

(*i*) The lower part-*Pressure chamber*

(*ii*) The upper part-*Expansion chamber*

The contacts are kept in waler. When a fault occurs, the moving contact is pulled up and the water of the pressure chamber is changed into steam due to heat of the arc. The steam so produced reaches a pressure of about 12 bar, lifts the cover and expands suddenly into the expansion chamber. Consequently, there is a rapid fall in pressured and temperature of steam in the pressure chamber, leading to interruption of the arc.

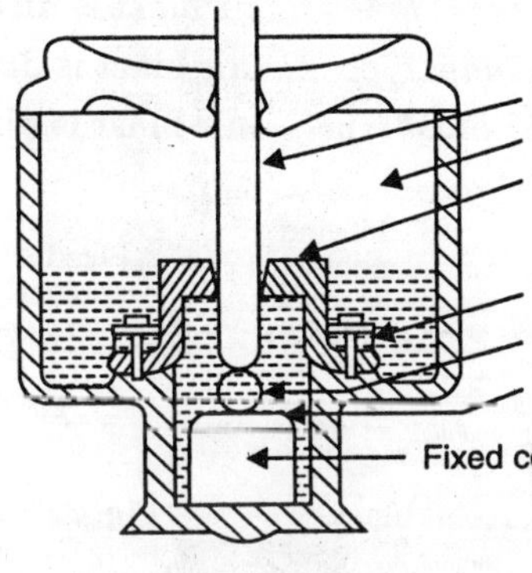

Fig. 22.13. Water type circuit breakers.

In case of fault not being severe and not sufficient to produce a steam of pressure high enough to raise the cover, expansion and cooling take place as the moving contact rod is withdrawn from the throat.

- These circuit breakers are *much smaller* than oil circuit breakers when they are used for *interrupting small currents* and operate within one or two half cycles.

22.4.12. Air Blast Circuit Breakers

These breakers use a high pressure air blast (at a pressure of 20 bar) as *an arc quenching medium.*

They find wide applications in systems operating at 132 kV and above (upto 400 kV) with breaking capacity upto 7500 MVA (during short-circuit fault) and above, although such breakers have also been designed to cover the wide range of 6.6 to 132 kV.

The advantage and disadvantages of these breakers over oil circuit breakers as under :

Advantages :

(*i*) No risk of fire hazard and explosion.

(*ii*) Short and consistent arc duration.

(*iii*) The burning of contacts is less due to less arc energy (since the arc duration is short and consistent).
(*iv*) Less maintenance required.
(*v*) High speed operation.
(*vi*) Facility of high speed reclosure.
(*vii*) Very suitable for conditions where frequent operation is required (since the cooling medium is replaced constantly).

Disadvantages :

(*i*) A compressor is constantly required to remain in operation.
(*ii*) The air leakage at the pipeline fittings creates problems.
(*iii*) Sensitivity to restriking voltage.
(*iv*) Current chopping.

Types of air blast circuit breakers :

Air blast circuit breakers, according to type of flow of blast of compressed air, are classified as follows :

(*i*) *Axial-blast type* ..... air blast is directed along the arc path (Fig. 22.14).
(*ii*) *Cross-blast type* ...... air blast is directed at right angles to the arc path (Fig. 22.15).
(*iii*) *Radial-blast type*......air blast is directed radially (Fig. 22.16).

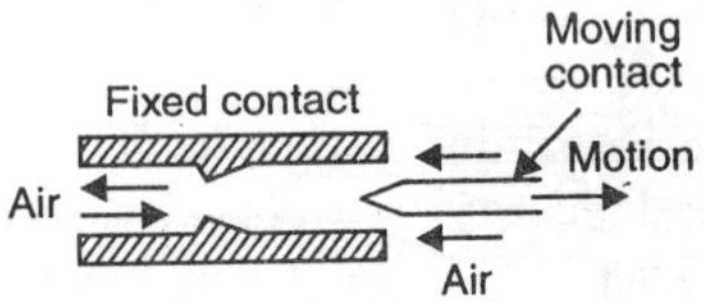

Fig. 22.14. Axial blast.

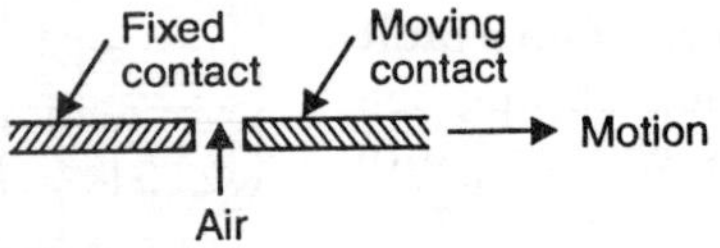

Fig. 22.15. Cross blast.

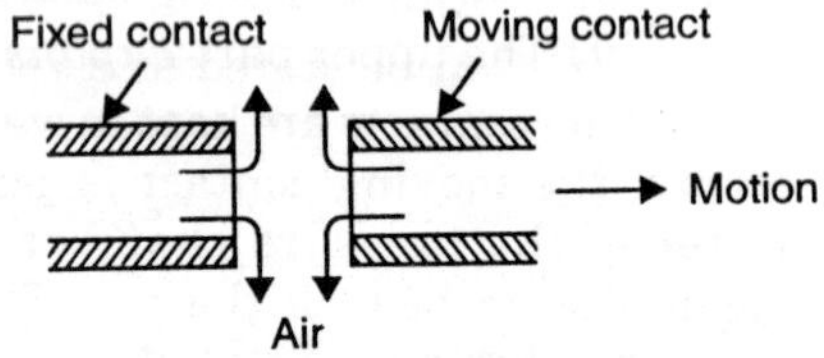

Fig. 22.16. Radial blast.

Two principle types of air blast circuit breakers namely axial blast and cross blast type breaker arc described as follows :

22.4.12.1. Axial blast air circuit breaker

The most common type of air blast circuit breaker employing axial flow of blast air for extinction of air is shown in Fig. 22.17. The moving contact is connected to a piston and the shaft of the contact is guided by guide springs. The circuit breaker is closed by opening lower air valve.

When the fault occurs, the relay operates and opens the upper air valve. The compressed air (at a pressure of about 15 bar) enters the arcing chamber and piston chamber and accelerates the moving contact. On the separation of contacts an arc is struck, the cold air rushes round the contacts axially and blows out the arc.

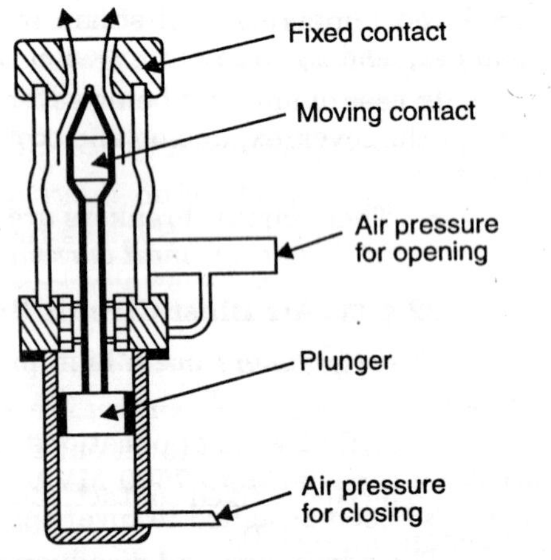

Fig. 22.17. Axial blast air circuit breaker.

22.4.12.2. Cross blast air circuit breaker

In this type of circuit breaker, an air-blast is directed at right angles to the arc. The cross blast lengthens and forces the arc into a suitable chute for arc extinction.

The essential parts of a typical cross blast breaker is shown in Fig. 22.18.

When the moving contact is withdrawn, an arc is struck between the fixed and moving contacts. The high pressure cross blast forces the arc into a chute consisting of arc splitters and baffles. The splitters serve to increase the arc length and baffles provide improved cooling. Consequently, the arc is blown out and flow of current is interrupted.

Since blast pressure is same for all currents, the inefficiency at low currents is eliminated. The final gap for interruption is quite adequate to give normal insulation clearance so that there is no necessity of a series isolating switch.

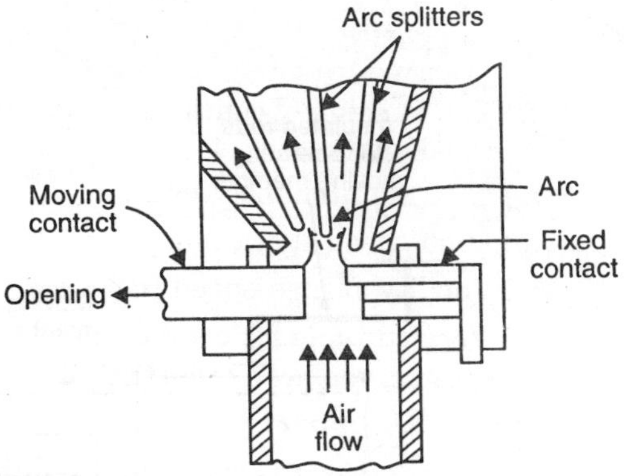

Fig. 22.18. Cross blast air circuit breaker.

22.4.13. Sulphur Hexafluoride (SF_6) Circuit Breakers

Owing to the following properties, SF_6 proves superior to other medium such as oil or air for use in circuit breakers :

(*i*) Very high dielectric strength (about 24 times that of air and comparable to oil)

(*ii*) Non-reactive (being an inert gas) to the other components of circuit breaker.

(*iii*) About 100 times more effective than air is extinction of arc.

(*iv*) Owing to its high density, its heat transfer property is about 1.6 times that of air.

(*v*) Can be stored at a relatively smaller pressure than that of air (since thermal time constant of SF_6 is low).

SF_6 is an electro-negative gas and has a strong tendency to absorb free electrons. The contacts of the breakers are opened in a high pressure flow of SF_6 gas and an arc is stuck between them. The conducting free electrons in the arc are rapidly captured by the gas to form relatively immobile negative irons. This loss of conducting electrons in the arc quickly builds up enough insulation strength to blow out the arc.

SF_6 breakers can withstand severe rate of rise of restriking voltage (RRRV) and thus are *most suitable for short line faults without switching resistors and can interrupt capacitive currents without restriking*.

Construction. A typical SF_6 circuit breaker is shown in Fig. 22.19.

— It consists of an *interruption chamber* which houses fixed and moving contacts and is connected to SF_6 gas reservoir. When the breaker contacts open, the valve mechanism permits a high pressure SF_6 gas from the reservoir to flow towards the interruption chamber.

— The *fixed contact* is a hollow cylinder current carrying contact fitted with an arc horn.

— The *moving contact* is also a hollow cylinder with rectangular holes in the sides to permit the SF_6 gas to let out through these holes after flowing along and across the arc.

The tips of fixed contact, moving contact and arcing horn are *coated with copper-tungsten* (arc resistant material).

SF_6 gas being costly, is reconditioned and reclaimed by suitable auxiliary system after each operation of the circuit breaker.

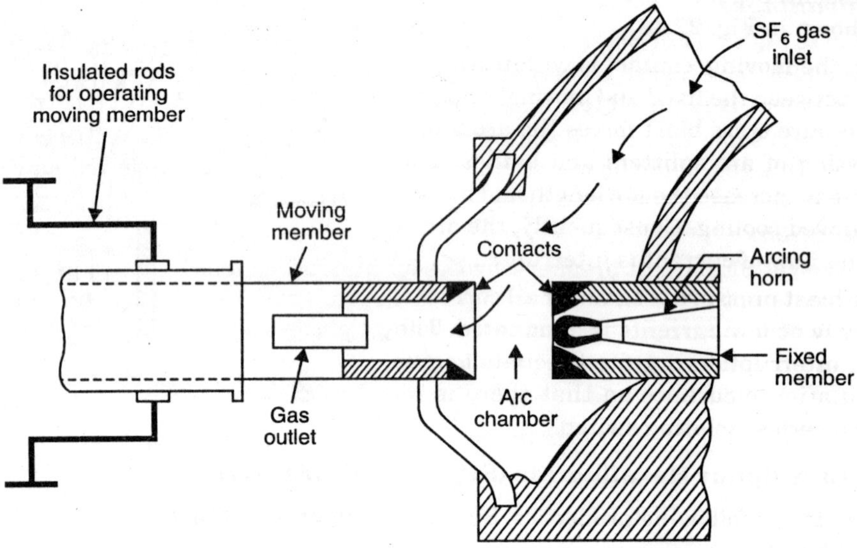

Fig. 22.19. Sulphur hexafluoride (SF_6) circuit breaker.

Working. When the circuit breaker is in *closed position*, the contacts remain surrounded by SF_6 gas at a pressure of about 2.8 bar. As the circuit breaker *operates*, the moving contact is pulled apart and arc is struck between the contacts. Immediately the valve opens and permits SF_6 gas at 14 bar pressure to flow from the reservoir to the arc interruption chamber. The high pressure flow of SF_6 rapidly absorbs the free electrons in the arc path to form immobile negative ions which are ineffective as charge carriers. Consequently, the medium between the contacts quickly builds high dielectric strength and causes the arc to extinguish. After the arc is extinguished, the valve is closed by the action of a set of springs.

Advantages of SF_6 breakers. The following arc the advantages of SF_6 breakers over the conventional breakers :

(*i*) Very short arcing time ; this reduces contact erosion.

(*ii*) No risk of fire (since SF_6 gas in non-inflammable).

(*iii*) No reduction of dielectric strength (since no CO_2 is formed).

(*iv*) Silent operation.

(*v*) Performance of the circuit breaker is not due to atmospheric conditions.

(*vi*) Breaker is compact in size and electrical clearances are drastically reduced.

(*vii*) The current chopping tendency is minimised by using SF_6 gas at low pressure and low velocity.

(*viii*) The closed gas enclosure keeps the interior dry so that moisture problem is almost eliminated.

(*ix*) Since the dielectric strength of SF_6 gas is 2 to 3 times that of air, such breakers can interrupt much larger currents.

(*x*) Being totally enclosed, these breakers are particularly suitable where explosion hazard exists (*e.g.*, coal mines).

Disadvantages :

(*i*) Due to high cost of SF_6 gas these breakers are costly.

(*ii*) Since SF_6 gas has to be reconditioned after every operation of the breaker, additional equipment is required for this purpose.

(*iii*) SF_6 is to some extent suffocating (but non-poisonous). In case of leakage in the breaker tank, this gas, being heavier than air settles in the surroundings and may lead to suffocation of the operating personnel.

Applications :

- A typical SF_6 circuit breaker consists of interrupter units, each capable of dealing with currents upto 60000 A and voltages in the range of 50 to 80 kV. A number of units are connected in series according to the voltage of the system.
- These breakers have been developed for voltages 115 kV to 230 kV, power ratings of 10 MVA to 20 MVA and interrupting time less than 3 cycles.

22.4.14. Vacuum Circuit Breakers (VCB)

In a vacuum (*vacuum* means the pressure below atmospheric pressure which is 760 mm of Hg) circuit breaker *vacuum* of the order of 10^{-5} to 10^{-7} torr (1 torr = 1 mm of Hg) is *used as the quenching medium.* In such a low pressure mean free path of the electrons is of the order of few metres and thus when the electrodes are separated by few mm an electron crosses the gap without any collision. It is due to this reason that the dielectric strength of the vacuum is 1000 times more than that of any other medium.

The arc which is formed in vacuum is of different type from those formed in other mediums, as it is formed by neutral atoms, ions and electrons emitted from the electrodes themselves. The surface of the cathode which is normally not smooth has many micro-projections. When contact separate the current will concentrated in these micro-projects as they are the last points of contact. Owing to their small *X*-sectional area the projections will suffer explosive evaporation by resistive heating. This evaporation of cathode leads to the formation of arc.

Construction. Fig. 22.20 shows the parts of a vacuum circuit breaker.

— It consists of an outer envelope which is made of glass and is joined with metallic end caps. The glass envelope facilitates the examining of the breaker from outside after operation. When it becomes milky white (from its original finish of silver mirror) it indicates that it is losing vacuum.

— After the envelope there is a '*sputter shield*' made of stainless steel to prevent the metal vapour reaching the envelope.

— Inside the shield the breaker has two points one '*fixed*' and the other '*moving contact*' which moves through a distance of 5 to 10 mm depending upon the operating voltage. The '*metallic bellows*' made of stainless steel are used to move the lower contact.

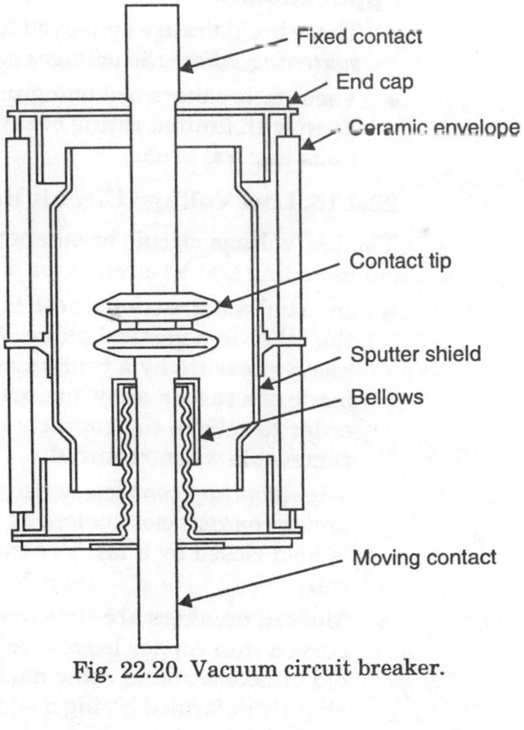

Fig. 22.20. Vacuum circuit breaker.

— The periphery of the end-cap is sealed to the envelope and the fixed contact stem is an integral part of one end-cap. One end of the fixed as well as moving contact is brought out of the chamber for external connections.

— The lower end of the breaker is fixed to a spring-operated or solenoid operated mechanism so that the metallic bellows inside the chamber are moved downward and upwards during opening and closing operation respectively.

Working. As the breaker operates, the moving contact separates from the fixed contact and an arc is struck between the contacts. The *production of arc is due to the ionisation of metal ions and depends very much upon the material of contacts.* The arc is quickly extinguished because the metallic vapour, electrons and ions produced during the arc are diffused in a short-time and seized by the surfaces of moving and fixed members and shields. Since vacuum has very fast rate of recovery of dielectric strength, the arc extinction in a vacuum breaker occurs with a short contact separation, say 6.25 mm.

Advantages. Vacuum circuit breakers possess the following *advantages* :

(*i*) Compact, reliable and have longer life.

(*ii*) No fire hazards.

(*iii*) Less maintenance required.

(*iv*) Quiet in operation.

(*v*) No generation of gas during and after operation.

(*vi*) Can break any heavy fault current perfectly just before the contacts reach the definite open position.

(*vii*) Low arc energy.

(*viii*) Require smaller power for control mechanism (due to low inertia).

(*ix*) Can withstand lightning, surges successfully.

Applications :

- These breakers are employed for reactor *switching, transformer, switching, capacitor-bank switching where the voltages are high and the current to be interrupted is low.*
- Vacuum breakers are being used for outdoor applications ranging from 22 kV to 66 kV. Even with limited rating of say 60 to 100 MVA, they are suitable for a majority of applications in rural areas.

22.4.15. Low Voltage Circuit Breakers

The low voltage circuit breakers are designed for use on D.C. circuits and low voltage A.C. (upto and including 600 V) circuits for *protection of general lighting, power and motor circuits.*

- In usual construction, these breakers make use of two fixed terminals mounted one above the other in a vertical plane. As the breaker is closed, these terminals are bridged under heavy pressure by a bridging member operated by a system of linkages. Arcing contacts, made of a carbon alloy, arc resistance silver tungsten or copper tungsten alloy are used in order to relieve the contacts from damage due to arcing. The arcing contacts are easily removable when required.

  The *secondary contacts*, when used, are usually of copper or silver alloy. The *auxiliary and arcing contacts* close before and open after the main contacts during operation. The breaker is held closed by a latch or the equivalent which may be tripped electrically or mechanically.
- Modern breakers are "*trip free*". Brush type bridging member, which consist of a group of curved thin copper leaves backed by a bronze spring, is used for closing main contacts in old breakers and in some modern breakers. As the breaker is pressed closed, the brush is slightly deformed giving a sliding or wiping action on the stationary blocks to ensure good

contact. Many modern breakers make use of solid bridging member with spring mounted self aligning contacts. The contact surfaces are of silver, as with this material oxidation does not cause excessive resistance and over-heating. The contacts are often of the line type.

- Breakers on A.C. and D.C. circuits of 250 V and above are generally furnished with *barriers between poles* ; *special arc chutes, quenches, or deionising chambers are also used throughout the available lines of modern air circuit breakers.*
- These breakers are usually provided with *over-current tripping mechanism* which may be of the instantaneous or time delay type or a combination of both. These may also be provided with other tripping ranges and arrangements such as low voltage trip, shunt trip connected to over voltage, reverse current, or over current relays.

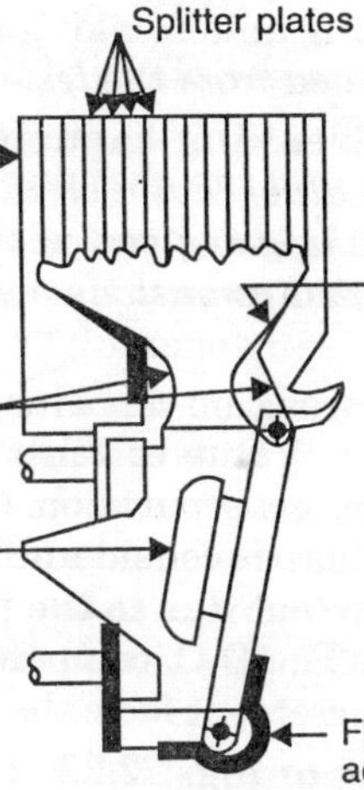

Fig. 22.21. Low voltage circuit breaker-elementary arrangement.

- In such breakers, hand operation is achieved by means of a *lever*, even in large size breakers an electric operation by means of a solenoid or motor mechanism for either 125 or 250 V direct current is obtainable on all except smallest size of breakers.

Fig. 22.21 shows an elementary arrangement of a low voltage circuit breakers.

Ratings of these breakers are as under :

Current............. 6000 A A.C., and 12000 A D.C.

Voltage............... 250 to 600 V A.C. and 250-750 D.C.

Special breakers are available upto 3 kV for D.C. services.

22.4.16. Ratings of Circuit Breakers

The rating of a circuit breaker depends upon the duties it performs. A circuit breaker is required to perform the following duties (besides normal working) under short-circuit/fault conditions :

(*i*) It must be capable of breaking the circuit and isolating the faulty section.

(*ii*) It must be capable of making circuit in the greatest symmetrical peak in current wave.

(*iii*) It must be capable of carrying fault currents safely for a short time while another circuit (in series) is clearing fault corresponding to above mentioned duties, the circuit breakers have following *three* ratings,

(*i*) Breaking capacity ;

(*ii*) Making capacity ;

(*iii*) Short-time capacity ;

In addition to the above ratings, a circuit breaker should be specified in terms of :

— Number of poles ;

— Service voltage ;

— Normal current ;

— Frequency ;

— Operating duty.

Making and breaking capacities are expressed as the making and breaking currents, which are determined from the *test oscillograms.*

(*i*) **Breaking capacity.** *It is the current (r.m.s value) that a circuit breaker is capable of breaking at given recovery voltage and under specified conditions (e.g., power factor, RRRV).*

The breaking capacity is always stated as the r.m.s. value of fault current at the instant of contact separation. On the occurence of a fault ; there is considerable asymmetry in the fault current due to the presence of D.C. component. The D.C. component dies rapidly (a typical decrement factor being 0.8 per cycle)

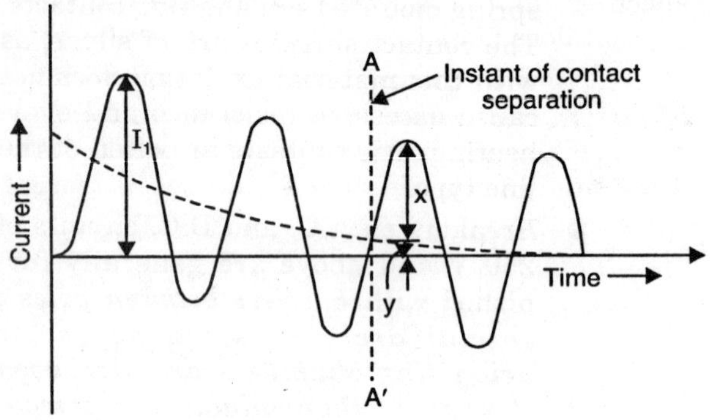

Fig. 22.22. Determination of current at the instant of contact separation.

Refer to Fig. 22.22. The contacts are separated at AA'. At this instant, the fault current has,

x = Maximum value of A.C. component (short-circuit wave), and

y = D.C. component (short-circuit wave).

Now, symmetrical breaking current = r.m.s. value of A.C. component

$$= \frac{x}{\sqrt{2}}$$

Asymmetrical breaking current = r.m.s. value of combined sums of A.C. and D.C. components

$$= \sqrt{\left(\frac{x}{\sqrt{2}}\right)^2 + y^2}$$

Generally the breaking capacity is expressed in terms of MVA and is equal to the *product of rated breaking current and rated service voltage.*

Breaking capacity for a 3-phase circuit breaker

$$= \sqrt{3}\, V \times I \times 10^{-6} \text{ MVA}$$

where, V = Rated service line voltage in volts, and

I = Rated breaking current (symmetrical or asymmetrical) in ampers.

The Indian (or British) practice is to take breaking current equal to the r.m.s. value of symmetrical breaking current whereas American practice is to use asymmetrical value of braking current in place of symmetrical value.

(*ii*) **Making capacity.** It is possible that the circuit breaker is closed under short-circuit conditions. The making capacity of a circuit breaker depends upon its ability to withstand the effects of electromagnetic forces. These force are proportional to the square of the peak value of the making current. Thus the making capacity is stated in terms of a peak value of current instead of r.m.s. value.

The peak value of current (including the D.C. component) during first cycle of current wave after the closure of circuit breaker is known as **making capacity.**

Making capacity = 2.55 × symmetrical breaking capacity

(The total multiplication factor 2.55 = $\sqrt{2}$ × 1.8, where $\sqrt{2}$ is used to convert r.m.s. to peak value and 1.8 is used to include the *"doubling effect" of maximum asymmetry*) (*iii*) **Short-time rating.** The short time rating of a circuit breaker depends upon its *ability to withstand the electromagnetic force effects and the temperature rise.* The oil circuit breakers have a specified limit of 3 seconds when the ratio of symmetrical breaking current to the rated normal current does not exceed 40. However if this ratio exceeds 40, then the specified limit is 1 second.

Normal current rating. *It is the r.m.s. value of current which the circuit breaker is capable of carrying continuously at its rated frequency under specified conditions.* The important condition for normal working of an oil circuit breaker is that the temperature of oil should not be more than 40° C and that of contacts should not be more than 35° C.

Example 22.1. *A circuit breaker is rated as 1500 A, 1500 MVA, 33 kV, 3-seconds, 3-phase oil circuit breaker. Determine :*

(*i*) *Rated normal current.* (*ii*) *Breaking capacity.*
(*iii*) *Rated symmetrical breaking current.* (*iv*) *Rated making current.*
(*v*) *Short-time rating.* (*vi*) *Rated service voltage.*

Solution. (*i*) Rated normal current = **1500 A. (Ans.)**

(*ii*) Breaking capacity = **1500 MVA. (Ans.)**

(*iii*) Rated symmetrical braking current $= \dfrac{1500 \times 10^6}{\sqrt{3} \times 33 \times 10^3}$ = **26243 A (r.m.s.). (Ans.)**

(*iv*) Rated making current = 2.55 × 26243 = **66920 A (peak). (Ans.)**

(*v*) Short time rating = **26243 A for 3 seconds. (Ans.)**

(*vi*) Rated service voltage = **33 kV (r.m.s. line value). (Ans.)**

Example 22.2. *Explain what do you understand by* (*i*) *rated symmetrical breaking current,* (*ii*) *making capacity,* (*iii*) *short-time rating. Obtain their values for a 3-phase oil circuit breaker having the following data :*

Operating voltage = 11 kV ; Rupturing capacity = 500 MVA ; Normal current = 500 A ; Time = 3 seconds. **(Panjab University)**

Solution. (*i*) Symmetrical breaking current

$$= \frac{\text{Rupturing capacity in MVA} \times 10^6}{\sqrt{3} \times \text{operating voltage}} = \frac{500 \times 10^6}{\sqrt{3} \times 11 \times 10^3}$$

= **26243 A (r.m.s.). (Ans.)**

(*ii*) Rated making current = 2.55 × 26243 = **66920 A. (Ans.)**

(*iii*) Short time rating = **26243 A for 3 seconds. (Ans.)**

22.4.17. Testing of Circuit Breakers

The circuit breakers are subjected to the following *tests* :

I. Type tests :

These tests are performed on a single specimen picked up at random from the finished products so as the determine whether the units are manufactured as per design or not. These tests are classified as follows :

1. ***Short circuit tests*** **:**

(*i*) Making capacity test.

(*ii*) Breaking capacity test.

(*iii*) Short time current test.

(*iv*) Operating duty test.

These tests may be conducted either in field or in laboratory. In the field, the power required for testing is directly taken from a large power system whereas in laboratory, specially designed generators are available for conducting short-circuit tests on circuit breakers.

2. ***Dielectric tests :***

(*i*) *Power frequency test* :

(*a*) One minute dry withstand test.

(*b*) One minute wet withstand test.

- In power frequency withstand test, the test voltage is applied between : (*i*) *each phase and earth*, (*iii*) *each set of terminals* for one minute when the breaker is open and between each two of the three phases when the breaker is closed also for one minute. Power frequency tests are conducted under normal atmospheric condition and are termed as *dry withstand test*.
- Wet withstand test is performed by spraying the external insulation by water for two minutes while the rated service voltage is applied. The test overvoltage is maintained for one minute. This test, however, is not required for indoor breakers.

There *should not be any puncture or flashover in the power frequency tests.*

(*ii*) *Impulse withstand voltage test :*

- This test is *not mandatory for indoor breakers* since the indoor breakers are not usually electrically exposed to impulse surges produced by lightning.
- In case of *outdoor breakers* the effect of impulse voltage is very serious and hence the breaker is tested to prove its immunity to impulse voltage conditions. The impulse withstand test voltage should be standard (1/50 μs) wave. Depending on grounding system, where the breaker is to be installed, the peak value of the impulse wave is selected. The magnitude of the selected peak of the impulse voltage is high for non-effectively grounded system.

The test voltages are applied between : (*i*) each phase terminal and earth of the breaker with the breaker closed and other two phases grounded ; (*ii*) all the poles of the breaker on one side while the poles of the breaker on other side are earthed, the breaker being open.

The breaker should not puncture or flashover for at least ten applications of such a voltage.

3. ***Thermal tests :***

These tests are conducted to check the thermal behaviour of the breakers.

- In this test the rated current through all three phase of the switchgear is passed continuously for a period long enough to achieve steady state conditions. Temperature readings are obtained by means of thermocouples whose hot junctions are placed in appropriate positions. The temperature rise, above an ambient, should not rise above the recommended value.
- In this test, the contact drops or the contact resistances are also measured as these contact surfaces are responsible for production of heat and subsequent temperature rise.

4. ***Mechanical tests :***

A circuit breaker must open and close at the correct speed and perform such operation without mechanical failure. The breaker is thus tasted for mechanical endurance test as per relevant standards.

II. **Routine tests :**

After the type test are conducted and a particular design is found to be satisfactory the product becomes phototype and a large number of circuit breakers of similar design are manufactured.

These tests are classified as :

(*i*) *Operation test.* This test is performed on the breaker by stimulating its tripping by artificially closing the contacts of the relays.

(*ii*) *Millivolt drop test.* This test is conducted to determine the voltage drops within the current path of the breaker mechanism.

(*iii*) *Power frequency voltage tests :*

(*a*) at manufacturer's premises ;

(*b*) after erection on site.

22.4.18. Autoreclosing

The *faults which exist on the system, depending upon the time, are classified* as under :

(*i*) Transient fault ; (*ii*) Semi-permanent fault ;

(*iii*) Permanent fault.

It has been found that about 80% of the faults are transient faults, 12% semi-permanent and 8% are permanent faults.

Transient fault exists only for a short time. It can be removed faster still if the line is disconnected from the system momentarily so that arc blows out. After the arc is deionized, the line can be reclosed to restore normal service.

Semi-permanent fault may occur due to twig falling on the power conductor or bird spanning the power conductors. The reclosing could be restored with some delay so that the cause of the fault could be dispensed with during a time delay trip and the line could be reclosed to restore normal service.

In case of a *permanent-fault* reclosing does not help as it has to be attended and removed, and the line is to be taken out till the fault is cleared. Therefore, if the fault is not cleared after first reclosure, a double or triple shot reclosing is desired. If the fault still persists, the line is taken out of service.

Autoreclosing may be single or three phase type. Mostly *single-phase autoreclosing breakers are preferred as most of the transmission faults are single-phase to ground faults.* Autoreclosing in single phase also improves stability as the power remains transmitted through the two healthy phases when one phase is interrupted.

— Mostly, single shot or double shot auto reclosure breakers are available. The breakers may be autoreclosing type or delayed autoreclosing type. In case of autoreclosing breakers synchronous check is essential.

22.4.19. Maintenance of Circuit Breakers

Depending upon the usage, every circuit breaker is required to be inspected at regular internals of 3 or 6 months, during which the following checks should be made :

1. Check closing, tripping and interlock mechanism.
2. Check indicating devices and lamps.
3. Check all current carrying parts and attend the arcing contacts, if necessary.
4. Check the dielectric strength, condition and level of oil. In case the dielectric strength is low or the oil is badly discoloured , replace it.
5. Check the insulation for possible damage. Clean the surface and remove deposits of carbon.
6. Before closing the tank ensure that :
   — no tools have been left behind ;
   — tank linings and barriers are in position and secure ;
   — tank gasket is in good condition.

22.4.20. Switchgear Components

Some important components common to most of the circuit breakers are :

1. Bushings.
2. Circuit breaker contacts.
3. Instrument transformers.
4. Bus-bars and conductors.

1. **Bushings :**

- The primary function of a bushing (made of some insulating material *e.g.,* porcelain, steatite) is to *prevent electrical breakdown between the enclosed conductor and the surrounding earthed metal work.*
- Although there are several types of bushings (*i.e.* condenser type, oil filled type etc.) they perform the same function of insulating the conductor from earthed tank.
- The bushing can fail in the following two ways :

(*i*) The breakdown may be caused by *puncture i.e.* dielectric failure of the insulating material of the bushing.

(*ii*) The breakdown may occur in the form of a *flash-over* between the exposed conductor at either end of the bushing and the earthed metal.

The bushings are so designed that the flash-over takes place before they get punctured.

2. **Circuit breaker contacts.** The function of the circuit breaker contacts is to carry normal as well as short-circuit current. In carrying normal current it is desirable that the temperature should not rise above the specified limits and that there should be low voltage drop at the point of contact. In carrying breaking and making short-circuit currents, the main effects to be dealt with are melting and vaporisation by the heat of arc and those due to electromagnetic forces. Thus for the satisfactory operation of circuit breaker, the design of contacts is of significant importance.

The various types of circuit breaker contacts are :

(*i*) *Tulip type contacts.* In this type of contacts the moving contact moves inside the fixed contacts. When the contacts separate, the arc is generally established between the tips of fixed contacts and tip of the moving contact. The advantage of this arrangement is that arcing is confined to the regions which are not in contact in the fully engaged position.

(*ii*) *Finger and wedge contacts.* These types of contacts are largely used for low-voltage circuit breakers due to the general unsuitability for use with control devices.

(*iii*) *Butt contacts.* These contacts (formed by the spring and moving contact) are suitable for use in single-break oil circuit breakers and air blast circuit breakers (where relatively small "loop" forces are available to assist in opening). This type of contact is also especially suitable for higher short-circuit rating.

3. **Instrument transformers :**

It is not practicable to connect instruments and meters directly to the lines in high voltage circuits. Instead instrument transformers are used. The following are the *two basic advantages inherent* in this method :

(*i*) Standard rated instruments may be used.

(*ii*) Operating personnel coming in contact with the instruments are not subject to high voltage and current of the lines, and so there is less danger to them. Even with a low-voltage system, instrument transformers are used for measuring large currents, so that heavy leads to the instrument panel and to the ammeter and other current terminals are avoided.

The principle of the instrument transformer is fundamentally the same as that of the power transformer. The instrument transformers are classified as follows :

1. Potential transformers
2. Current transformers.

Potential transformers (P.T.) :

— A potential transformer is a *step down transformer* used along with a low range voltmeter for measuring a high voltage. The primary is connected across the high voltage supply and the secondary to the voltmeter or potential coil of the wattmeter. Since the voltmeter (or potential coil) impedance is very high, the secondary current is very small and the potential transformer behaves as an ordinary two winding transformer operating on no-load. Fig. 22.23 shows a potential transformer used to measure the voltage of a circuit. It may be noted that the secondary is grounded. This is done so that if the insulation breaks down, the high voltage does not endanger personnel who may be reading the meters.

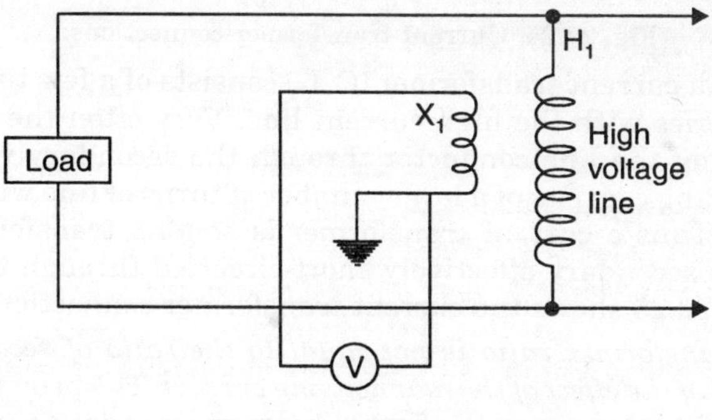

Fig. 22.23. Potential transformer connections.

— These transformers are *made with high quantity iron core operating at very low flux densities so that the magnetising current may be very small*. Careful design ensures minimum variation of voltage ratio with load and minimum phase shift between input and output voltages. Potential transformer secondaries are commonly designed for an output of 110 V.

Current transformers (C.T.) :

— Just as a shunt extends the range of a D.C. ammeter, so does the current transformer perform the same function in A.C. circuits. Thus a *high magnitude alternating current can be easily measured by a combination of a current transformer and a low range ammeter.*

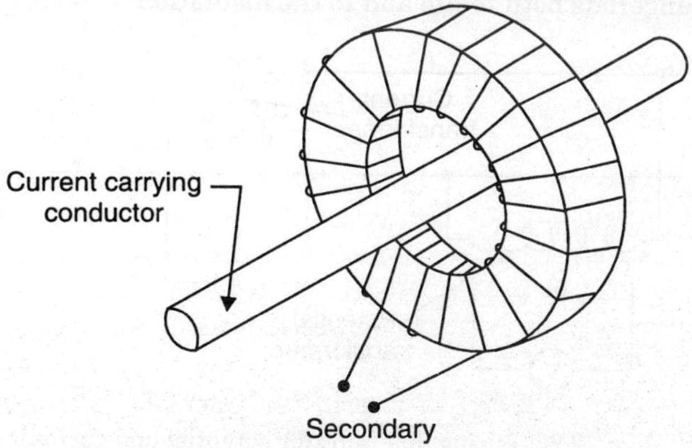

Fig. 22.24. Line conductor acting as primary.

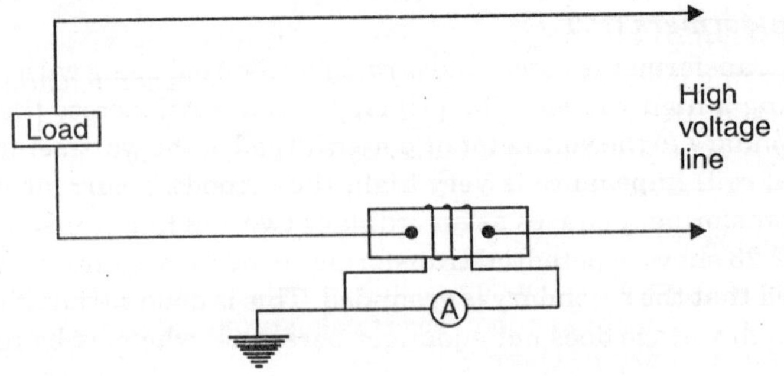

Fig. 22.25. Current transformer connections.

— The primary of a current transformer (C.T.) consists of a few turns of thick cross-section connected in *series* with the high current line. Very often the primary is just one turn formed by taking the line conductor through the secondary winding (Fig. 22.24). The secondary winding consists of a large number of turns of fine wire designed for either 5 A or 1 A rating. Thus a current transformer is *step-up* transformer. The current transformer has the secondary effectively short-circuited through the low impedance of the ammeter. Fig. 22.25 shows the current transformer connections.

— The *current transformer ratio is not equal to the ratio of secondary to primary turns, mainly because of the effect of the magnetising current.* The primary current can be thought of as the sum of two currents, the first to balance secondary current so that primary and secondary m.m.fs. may balance and the second being the no-load current I_0. The component I_0 besides being responsible for a slight error in the current ratio, is also responsible for a phase angle error. The transformer *must be carefully designed to minimise the ratio and phase angle error.*

— It may be noted that current transformer *must never be operated on open-circuit* for the following two reasons :

(*i*) There will be no secondary m. m. f. and since the primary current (and m.m.f.) is fixed, the core flux will increase enormously. This will cause large eddy current and hysteresis losses and the *resulting high temperature may damage the insulation or even the core.*

(*ii*) A very high voltage will be induced in the multi-turn secondary and this high voltage may be dangerous both to life and to the insulation.

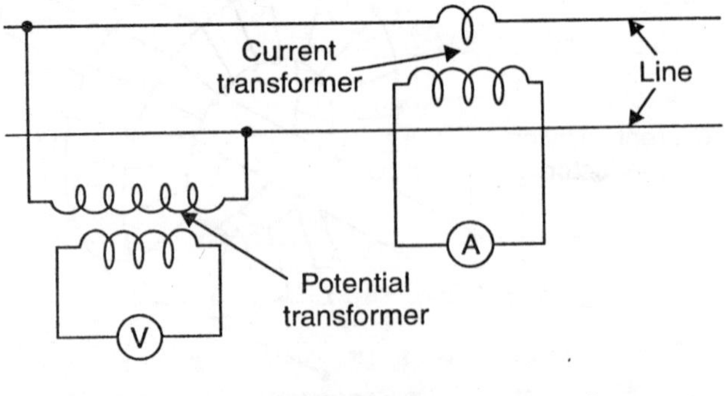

Fig. 22.26. Wiring diagrams for potential and current instrument transformers.

Fig. 22.26 shows the wiring diagrams for potential and current instrument transformers.

4. **Bus-bars and conductors :**

The current carrying members in a circuit breaker consists of fixed and moving contacts and the conductors connecting these to the points external to the breaker. In case of outdoor type switchgear, these connections are connected directly to the overhead lines. If the switchgear is of indoor type, the incoming conductors to the circuit breaker are connected to the bus bars.

22.4.21. Bus-bar Arrangements

When several generators or feeders operating the same voltage have to be directly connected electrically, bus-bars are used as the common electrical equipment. *Bus-bars are copper rods* or *thin walled tubes* and operate at constant voltage.

The aim of any particular arrangement of bus-bars is to *achieve adequate operating flexibility, sufficient reliability and minimum cost.* The cost can be minimised by reducing the number of circuits breakers to a minimum but by doing so the cost and complication of the protective gear are increased.

Important bus-bar arrangements are enumerated and discussed below :

1. Single bus-bar system.
2. Double bus-bar system with sectionalisation.
3. Ring bus-bar system.

1. **Single bus-bar system.** Refer to Fig. 22.27. In case of a power plant which has a number of generators and a single busbar arrangement, the busbar is sectionalised by circuit breakers. The main *advantage* of this system is *that fault on one part of the busbar or system does not completely shutdown the whole station.* However, the use of such a large number of circuit breakers is out of date and presently there is a tendency to use fewer.

G = generator, CB = circuit breaker, S = Isolator or switch, LVB = Low voltage bus bar, HVB = high voltage bus bar, T = step up transformer.

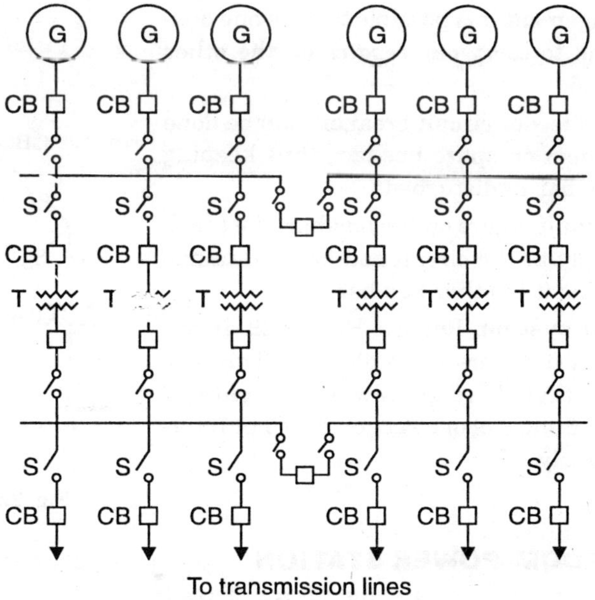

Fig. 22.27. Single bus-bar system.

2. **Double bus-bar system.** In the Fig. 22.28 is illustrated a doubled busbar system. In this system both low voltage and high bus-bars are duplicated, any one of the busbar sections can be used as desired. There is a provision of a *busbar coupling switch* for transferring operation from one busbar to another.

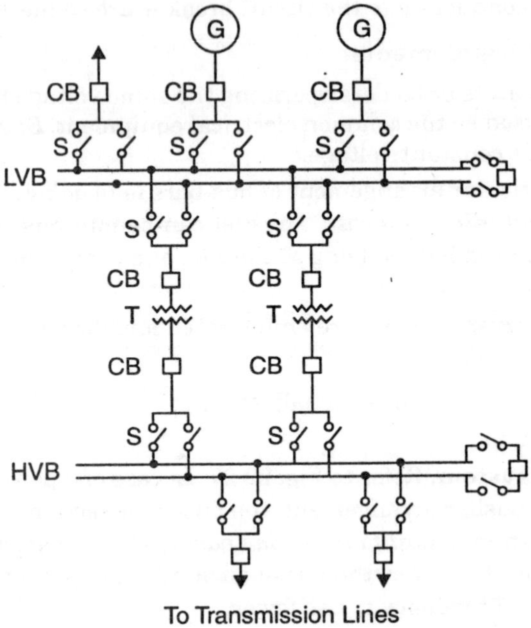

Fig. 22.28. Double bus-bar system.

Advantages **:**

(*i*) In this arrangement it is possible to have one bus-bar *"live"* and to carry out *repairs* on the other when required.

(*ii*) The testing of feeder circuit breakers can be done by putting them on spare bus-bar, thus keeping the main bus-bar undisturbed.

(*iii*) Whenever a fault occurs on the bus-bar, the continuity of supply to the circuit can be maintained by transferring it to other bus-bar.

3. **Ring bus-bar system.** Refer to Fig. 22.29. In the arrangement two circuit breakers serve one line. This system has the *advantage* that there are always two parallel paths to the circuit and failure of one section does not interrupt the service completely.

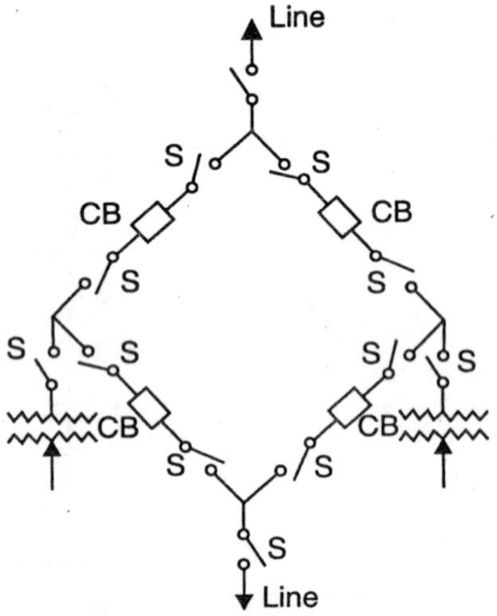

Fig. 22.29. Ring bus-bar system.

22.5. CONTROL ROOM–POWER STATION

In a power station the control room (or the operating room) acts as the *nerve centre*. The following *controls* are located in a control room.

(*i*) Circuit breakers
(*ii*) Load and voltage adjustment
(*iii*) Transformer tap changing
(*iv*) Emergency tripping of the turbines etc.
(*v*) The instruments for indicating the load, voltage, frequency, power factor, winding temperatures and water levels in the case of hydrostations and so on.
(*vi*) Synchronising equipment
(*vii*) Voltage regulators
(*viii*) Relays
(*ix*) Integrating meters and other appliances
(*x*) A mimic diagram and suitable indicating equipment to show the open or closed position of circuit breakers, isolators etc.

The control room location in relation to other sections of the station, important and suitable position should be obtained. It should be located in the following manner :

(*i*) Should be located *near the switch house* so that lengths of the multicore cables are shortened.
(*ii*) Should be located *away from noise sources.*
(*iii*) From the control room there should be an access to the turbine house.
(*iv*) The location of the control room should be such that is should not be affected if any fire errupts in the switch house.

Fig. 22.30 shows the location of control room.

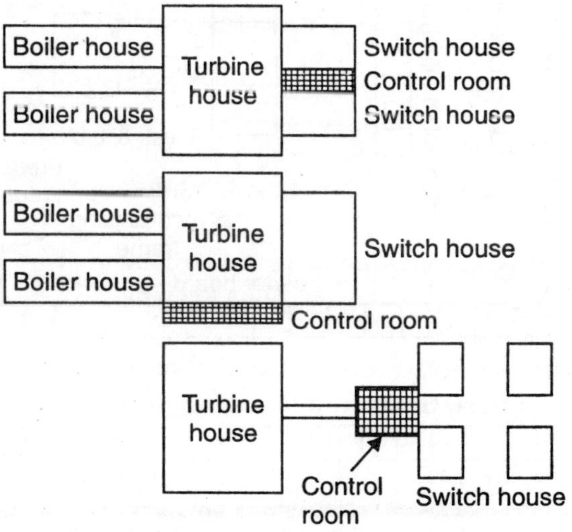

Fig. 22.30. Location of control room.

The control room should be well arranged as follows :

(*i*) Control room should be clean and comfortable.
(*ii*) Should be ventilated and well lighted.
(*iii*) Should be free form draughts.
(*iv*) There should be no glare.
(*v*) The colour schemes should be soothing to eyes.

(*vi*) The instruments should have clear scales properly calibrated.

(*vii*) All the apparatus and circuits should be labelled so that they are clearly visible.

Control boards. These are of the following three forms :

(*i*) *Linear* (*ii*) *Horse shoe shaped*

(*iii*) *Semi-circular*.

The *horse shoe shaped* and *semi-circular control boards* occupy more floor space but they enable all the instruments to be seen from one point.

In some cases, the control boards are of *desk type* in which the control and indicating equipment are located on the front, and the relays and integrating instruments are placed behind.

Fig. 22.31 shows the layout of a control room.

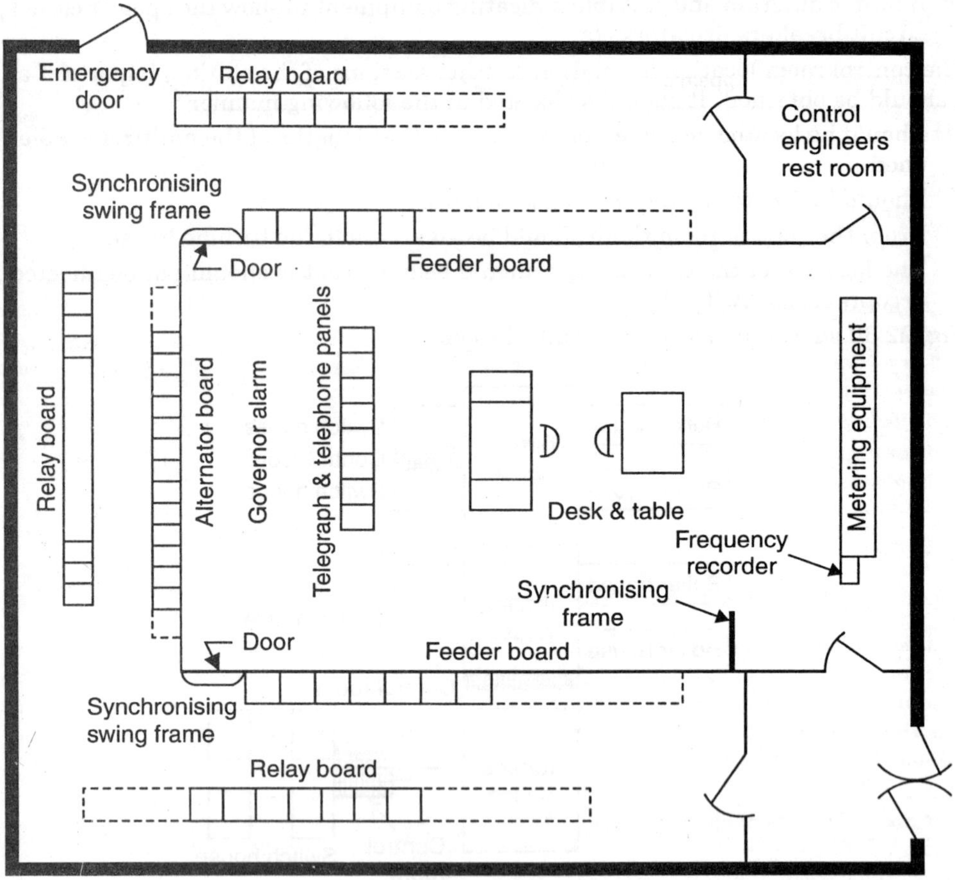

Fig. 22.31. Layout of a control room.

HIGHLIGHTS

1. The apparatus used for switching, controlling and protecting the electrical circuits and equipment is known as a '*switchgear*'.
2. A *short-circuit* is any solid contact established between the different phases of an electrical installation or any bridging of them by a very low resistance.

3. A *switch* is a device which is used to open or close an electrical circuit in a convenient way.
4. *Fuse* is a device used in circuit for protecting electrical equipments against overloads and/or short circuits.
5. For a *round wire* the approximate value of fusing current is given by : $I = k(d)^{3/2}$ ...... called ordinary *"Fuse law"*
where, k = Constant, depends on the metal of wire, and
d = Diameter of the wire.
6. *Low voltage H.R.C. fuse* may be built with a breaking capacity of 16000 A to 30000 A at 440 V. They are extensively used on low-voltage distribution system against overload and short-circuit conditions
High voltage cartridge fuses are used upto 33 kV with breaking capacity of about 8700 A at that voltage. Rating of the order of 200 A at 6.6 kV and 11 kV and 50 A at 33 kV are also available.
7. *Circuit breakers* are mechanical devices designed to close or open contact members, thus closing or opening an electrical circuit under normal or abnormal conditions.
8. Arc in a circuit breaker can be extinguished by the following two methods :
(*i*) High resistance method (*ii*) Low resistance or current zero method.
9. The transient voltage appearing across the contacts during arc period is called the *"Restriking voltage"*.
10. *"Recovery voltage"* is the normal frequency (50 Hz) r.m.s. voltage that appears across the contacts of the circuit breaker after final arc extinction.
11. The *rate of rise of restriking voltage (RRRV)* is the ratio of the peak value of restriking voltage to time taken to reach the peak value.
12. When a low inductive current is interrupted by a circuit breaker, the current falls to zero before the natural current zero. It is termed as *current chopping*.
13. The voltage developed across the breaker contacts due to transfer of energy is called *prospective voltage.*
14. Circuit may be classified as
A. *High voltage circuit breakers* : (*i*) Oil circuit breakers ; (*ii*) oilless circuit breakers—water circuit breakers—hard gas circuit beakers—air blast circuit breakers—sulphur hexafluoride circuit breakers —vacuum circuit breakers.
B. *Low voltage circuit breakers.*
15. Important bus-bar arrangements are
(*i*) Single bus-bar system (*ii*) Double bus-bar system
(*iii*) Ring bus-bar system.

THEORETICAL QUESTIONS

1. What do you understand by 'switchgear' ? Enumerate various types of switchgears.
2. State the functions of switchgear.
3. What are the essential features of switchgear ?
4. State the principles of layout of switchgear.
5. Define a 'short-circuit'.
6. What are the causes of a short-circuit ?
7. What are the effects of short circuits ?
8. What is a 'switch' ?
9. Discuss the various types of switches.
10. Define 'fuse' and 'fuse element'.
11. State the advantages and disadvantages of a fuse.
12. What are the desirable characteristics of fuse element ?
13. Describe briefly the fuse element materials.
14. Define the following terms :
(*i*) Current rating of fuse element (*ii*) Fusing current
(*iii*) Fusing factor (*iv*) Prospective current
(*v*) Cut-off current (*vi*) Arcing time.

15. What are the limitations of a fuse ?
16. Why do we prefer silver as a fuse element ?
17. Explain briefly the following fuses :
    (*i*) Rewirable or kit-kat fuse (*ii*) H.R.C. cartridge fuse.
18. State the difference between sub-circuit fuse, main circuit fuses, meter board fuses and pole fuses in an electrical installation.
19. Give the differences between a fuse and a circuit breaker.
20. What do you understand by "switchgear" ? Enumerate various types of switchgears.
21. What is meant by a circuit breaker ? Discuss the phenomenon of arc formation in a circuit breaker.
22. Explain various duties to be performed by a circuit breaker. Describe various methods of arc extinction.
23. Explain how arc is initiated and sustained in a circuit breaker when the circuit breaker contacts break.
24. Discuss different methods of interrupting the arc current in circuit breakers. Explain two main theories of current zero interruption.
25. Explain what do you understand by
    (*i*) Recovery voltage (*ii*) Restriking voltage, and (*iii*) RRRV.
26. Distinguish clearly between the recovery voltage and restriking voltage and explain the significance of RRRV in the operation of a circuit breaker.
27. Derive the relation for rate of rise of restriking voltage (RRRV).
28. What is the purpose of circuit breakers ? What is current chopping ?
29. What is current chopping and explain how can the effect of current chopping be minimised ?
30. Describe the working of a plain break oil circuit breaker.
31. Explain with the help of a neat sketch the function of a plain explosion pot. What are its advantages over plain circuit breaker. What are its limitations ?
32. Explain the construction, principle of operation and applications of a minimum oil circuit breaker.
33. What is a minimum oil circuit breaker ? What are its main advantages and disadvantages ? Draw a diagram illustrating the construction of this circuit breaker.
34. Explain with diagrams the working principle of the arc extinction device in a low oil content circuit breaker.
35. Compare the relative merits and demerits of oil circuit breaker with air blast circuit breaker.
36. Give the construction and principle of working of (*a*) air circuit breaker (*b*) air blast circuit breaker. Explain the principle of arc interruption in these circuit breakers.
37. Explain the working of an air blast circuit breaker with the help of neat diagram.
38. Describe the operating principle of an air blast circuit breaker.
39. Describe with the help of neat diagram the function of air blast circuit breaker.
40. Describe the construction, principle of operation and application of air blast circuit breakers.
41. Explain the reasons behind formation of an arc across the contacts of a circuit breaker while breaking a large short-circuit current.
    How is this arc controlled in a vacuum circuit breaker ?
42. Explain the principle of a vacuum breaker ?
43. Describe the construction, principle of operation and applications of SF_6 circuit breakers. Explain clearly the current chopping process in SF_6 circuit breakers.
44. Describe the various types of ratings of circuit breakers.
45. Explain the following terms in reference to circuit breakers :
    1. Symmetrical breaking current. 2. Asymmetrical breaking current.
    3. Making current.
46. Draw a sketch of outdoor isolator switch and describe its working.
47. What is an isolating switch ? Given its functions. How is it rated ?
48. Discuss the sequence of opening and closing of isolator switch with circuit breaker. In addition to isolator what other switches are used in power system, describe briefly.
49. What is a bus-bar ? Briefly explain the common arrangement of bus-bars.
50. Discuss the different types of bus-bar systems used in generating stations. Bring out their relative advantages and disadvantages.

51. Write short notes on the following :

(*i*) Arc formation and extinction in circuit breaker.

(*ii*) Oil circuit breakers.

(*iii*) Minimum oil circuit breaker.

(*iv*) Air circuit breaker.

(*v*) Airblast circuit breakers.

(*vi*) Vacuum circuit breakers.

(*vii*) SF_6 circuit breakers.

(*viii*) Rating of circuit breakers.

(*ix*) Making current and breaking current.

(*x*) Different type of isolators.

(*xi*) Oil circuit reclosers.

(*xii*) Rating and maintenance of circuit breaker.

(*xiii*) Bus-bar arrangements.

52. Differentiate between

(*i*) Switch and switchgear.

(*ii*) Switch and circuit breaker.

(*iii*) Air isolator and circuit breaker.

(*iv*) Restriking and recovery voltages.

(*v*) Bulk oil and low oil content circuit breakers.

(*vi*) High resistance and low resistance methods of extinguishing arc.

(*vii*) Making capacity and breaking capacity of a circuit breaker.

(*viii*) Oil circuit breakers and oil circuit reclosers.

UNSOLVED EXAMPLES

1. In a short-circuit test on 220 kV, 3-phase system the breaker gave the results as : Power factor of the fault = 0.5 ; Recovery voltage = 0.9 time the full line voltage.
   The breaking current is symmetrical and restriking transient has a natural frequency of 10 kHz. Determine the rate of rise of restriking voltage for (*i*) grounded fault and (*ii*) ungrounded fault.
   [**Ans.** 5.6 kV/μs ; 8.4 kV/μs]
2. In a system of 132 kV, the line-to-ground capacitance is 0.01 μF and the inductance is 5 henries. Determine the voltage appearing across the pole of a circuit breaker if a magnetising current of 5 A (instantaneous value) is interrupted. Determine also the value of resistance to be used across the contacts to eliminate restriking voltage. [**Ans.** 111.8 kV ; 11.18 kΩ]
3. A circuit breaker is rated at 1500 amperes, 2000 MVA, 33 kV, 3-sec, 3-phase oil circuit breaker. Determine the rated normal current, breaking current, making current and short time rating (current).
   [**Ans.** 1500 A ; 34.99 kA ; 89.22 kA 34.99 kA for 3-seconds]
4. A circuit breaker is rated at 1000 amperes, 1200 MVA, 33 kV, 3-sec, 3-phase oil circuit breaker. Determine : (*i*) Rated normal current, (*ii*) Breaking capacity ; (*iii*) Rated symmetrical breaking current ; (*iv*) Rated making capacity ; (*v*) Short-time rating ; and (*vi*) Rated service voltage.
   [**Ans.** (*i*) 1000 A ; (*ii*) 1200 MVA ; (*iii*) 21000 A (r.m.s.) ; (*iv*) 53600 A ; (*v*) 21000 A for 3 secs ; 33 kV (r.m.s. line value)]

23
Protective Relays

23.1. Protective system—General aspects. 23.2. Protective relays—Introduction. 23.3. Classification of relays. 23.4. Electromagnetic attraction relays—Attracted armature type relay—Solenoid type relay—Balanced beam type relay. 23.5. Thermal relay. 23.6. Buchholz relay. 23.7. Electromagnetic induction relays. 23.8. Induction type non-directional overcurrent relay. 23.9. Induction type directional power relay. 23.10. Induction type directional overcurrent relay. 23.11. Differential relays—current differential relay—voltage balance differential relay. 23.12. Translay relay. 23.13. Distance or impedance relays—Definite–distance type impedance relay—time–distance impedance relay. 23.14. Reactance type distance relay. 23.15. Mho type distance relay. 23.16. Universal torque equation. 23.17. Static relays—Introduction—Merits and demerits of static relays—Applications of static relays—Electronic circuits commonly used in static relays—Analog circuits used in static relay—Different static relays. Highlights—Theoretical Questions.

23.1. PROTECTIVE SYSTEM—GENERAL ASPECTS

In an electrical power system, as and when a fault occurs or any parameter becomes abnormal, it becomes necessary to isolate the faulty section instantaneously from the rest of the system. The isolation should be automatic, such that the supply must get disconnected from the faulty section leaving healthy remainder in service. This can be achieved by using the *protective devices* like relays or fuses in the circuit to isolate the circuit from damage to occur due to short circuit, *overhead, earth fault etc.*

The use of fuse for isolation or breaking the circuit for protection of the system is limited to low voltage circuits *i.e.*, upto 11 kV supply system for breaking the circuit in the air ; for higher voltages say from *11 kV onwards*, the circuit is protected by providing the *protective relays*.

Main features of good protective system/gear. The main features of a good protective system/gear are :

(*i*) ***Sensitivity.*** *It is the ability of the relay system to operate with low value of actuating quantity.* The protective system should be so sensitive that it operates even at low values of fault current.

(*ii*) ***Selectivity.*** *It is the ability of the protective system to select correctly that part of the system in trouble and disconnect the faulty part without disturbing the rest of the system.* The protective system should select correctly the faulty part of the power system and disconnect the same without disturbing the rest of the system. Thus it is absolutely necessary for a protective system to have the *property of discrimination.*

(*iii*) ***Reliability.*** The reliability is one of most important factors in the system. The component installed as a part of the protective system should *operate definitely under pre-determined condition.*

(*iv*) ***Simplicity.*** The relaying system should be simple so that it can be easily maintained. The simpler the protection system, the greater will be its reliability.

(v) ***Quickness.*** The protect system should respond as quickly as possible in order to improve the quality of service, increase safety of life and equipment and increase stability of operation.

(vi) ***Non-interference with future extension.*** The installation of a protective system should be carried out in such a way that future extension is possible without any interference with the original one.

(vii) ***Economy.*** While selecting a protective gear, due consideration should be given to its cost. However, when utmost important and sensitive apparatus (like alternator, transformer etc.) is to be protected, the cost factor or economic considerations are ignored.

Types of protection :

The ***protection*** is of the following *two types :*

(i) ***Main or primary protection.*** This type of protection is the *first line of defence* and ensures quick acting and selective clearing of faults within the boundary of the circuit section or element it produces. Main protection as a rule, is provided for each section of an electrical installation.

(ii) ***Back-up or secondary protection.*** The back-up protection is quite important for the proper functioning of a good system of electrical protection since cent-percent reliability of protective system as well as of the associated CTs, PTs and circuit breakers cannot be guaranteed. It is the *second line of defence* which functions to isolate a faulty section of the system when the *main protections fails to function properly*. Back-up protection is only provided where the main protection of the adjacent circuit is unable to back-up the main protection of the given circuit.

23.2. PROTECTIVE RELAYS—INTRODUCTION

The **protective relay** *may be defined as an electrical device connected between the main circuit and the circuit breaker that detects the fault and initiate the operation of the circuit breaker to isolate the defective section from the rest of the system.*

Whenever fault occurs on the power system, the relay detects that fault and closes the trip coil circuit. This results in opening of circuit breaker which disconnects the faulty circuit. Thus the relay ensures the safety of circuit equipment being damaged.

A typical relay circuit is shown in Fig. 23.1. On occurrence of a short-circuit at point F on the transmission line, there is an enormous increase in the value of current flowing in the line. This causes a heavy flow of current through the relay coil (R.C.), making the relay to operate by closing its contacts. This in turn closes the trip circuit of the breaker, consequently the circuit breaker opens and the faulty section is isolated from the rest of the system.

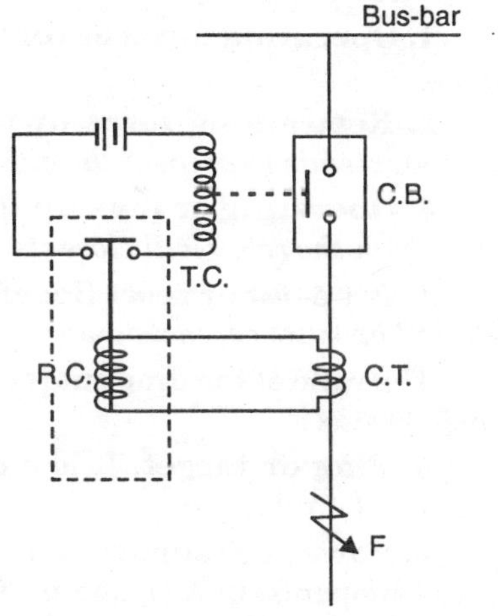

T.C. = Trip coil;
R.C. = Relay coil;
C.B. = Circuit breaker;
C.T. = Current transformer

Fig. 23.1. Typical relay circuit.

Essential fundamental elements of the relays. Refer to Fig. 23.2. All the relays have the following three essential fundamental elements :

1. *Sensing element.* It is also called *measuring element.* It is the element *which responds to the change in the actuating quantity, the current in a protected system, in case of over-current relay.*

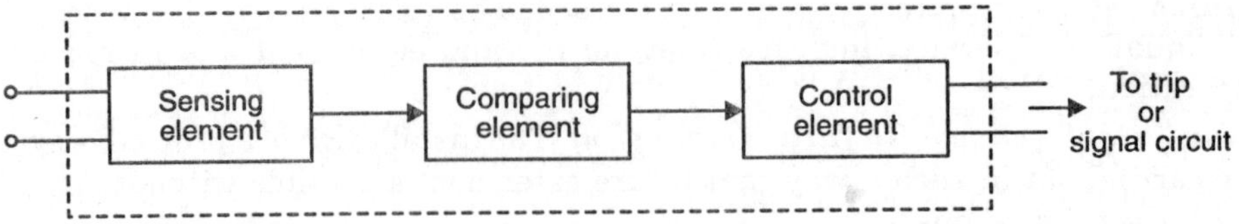

Fig. 23.2. Essential fundamental elements of a relay.

2. *Comparing element.* This element *compares the action of the actuating quantity on the relay with a pre-designed relay setting.* The relay picks up only if the actuating (operating) quantity is more than the relay setting.

3. *Control element.* This *element accomplishes a sudden change in control quantity* such as closing of the operative current circuit.

- **Operation of relay** *means either closing or opening of its contacts.* Most of the relays are restrained by string so that they are at a predetermined position when correctly de-energised. At this position if the status of the contact is *closed* it is said ***'NC'*** (normally closed) contact and if it is *open* it is said ***'NO'*** (normally open) contact.
  - — If a relay operates to open NC contact or close a NO contact it is said that the relay '*picks up*' and the *smallest value* of this actuating quantity is called the *pick up value.*
  - — If a relay operates to return back to its original position it is said that the relay resets and the *largest value* of the actuating quantity at which this occurs is called *reset value. Reset value is less than pick-up value.*

Terms Connected with Relays :

Some commonly used terms relating to relays are discussed below :

1. Operating force or torque. It is torque (or force) which *tends to close* the contacts of the relay.

2. Restraining force or torque. It is the torque (or force) which opposes the operating torque and *tends to prevent the closure of relay contacts.*

3. Operating or pick up (level). It is defined as the *threshold value of current, voltage etc. above which the relay will close its contacts.*

4. Drop out or reset (level). It is defined as the *value of current, voltage etc. below which the relay will open its contacts and return to normal position.*

The ratio of the drop out (or reset) value to the operating (or pick-up) value is called *drop out (or reset) ratio.*

5. Flag or target. It is a device (usually spring or gravity operated) which indicates the operation of a relay.

6. Power consumption (burden). *It is power absorbed by circuits* of the relay expressed in VA (volt-amperes) in A.C. and in W (watts) in D.C. circuits at the rated current or voltage.

7. Operating time. It is defined as the *time which elapses from the moment when the actuating quantity attains a value equal to the pick-up value until the relay operates its contacts.*

8. Reset time. The time which *elapses between the instant when the actuating quantity becomes less than the reset value to the instant when the relay contact returns to its normal position.*

9. Seal-in-coil. It is the *coil which does not permit the relay contacts to open when the current is flowing through them.*

10. Maximum torque angle. It is defined as the *design angle of the relay that will yield maximum torque.* It is also known as *characteristic angle of relay.*

11. Over-reach. A relay is said to over-reach when it *operates at a current which is lower than its setting.*

'Under-reach' is just reverse of 'over-reach'.

12. Pick-up current. If the *minimum current in relay coil at which the relay starts to operate.*

13. Current setting. Oftenly it is desirable to adjust the pick-up current to any required value. This known as *current setting* and is usually *achieved by the use of tappings on the relay operating coil.* The values assigned to each tap are expressed in terms of percentage full-load rating of C.T. with which the relay is associated and represents the value *above* which the disc commences to rotate and finally closes the trip circuit.

∴ Pick-up current = Rated secondary current of C.T. × Current setting.

14. Plug-setting multiplier (P.S.M.). It is the ratio of fault current in relay coil to the pick-up current *i.e.*,

$$\text{P.S.M.} = \frac{\text{Fault current in relay coil}}{\text{Pick-up current}} = \frac{\text{Fault current in relay coil}}{\text{Rated secondary current of C.T.} \times \text{Current setting}}.$$

15. Time-setting multiplier. Generally, a relay is provided with control to adjust the time of operation. This adjustment is known as *time-setting multiplier.* The time-setting dial is calibrated from 0 to 1 in steps of 0.05 second.

- ***Primary relays.*** These are the relays which are *connected directly in the circuit to be protected.*
- ***Secondary relays.*** These are the relays which are connected in the circuit to be protected *through current and potential transformers.*
- ***Auxiliary relays.*** These are the relays which operate in response to the opening or closing of its operating circuit *to assist another relay* in the performance of its function. This relay may be instantaneous or may have a time delay.
- ***Instantaneous relay.*** An intantaneous relay is one in which *no intentional time delay is provided.* In this case, the relay contacts are closed immediately after current in the relay coil exceeds the minimum calibrated value.
  - — This *relay is effective only where the impedance between the relay and source is small* compared to the protected section impedance.
  - — Its operating time is sometimes expressed in cycles based on the power-system frequency *e.g.*, in a 50-cycle system one-cycle would be $\frac{1}{50}$ second.
- ***Inverse-time relay.*** In this type of relay the operating time is *approximately inversely proportional to the magnitude of the actuating quantity.* The inverse-time delay can be achieved by associating mechanical accessories (*e.g.*, Drag magnet, oil dash pot or time limit fuse).
  - — At values of current less than pick-up, the relay never operates.
  - — At higher values, the time of operation of relay decreases steadily with the increase of current.
- ***Definite time lag relay.*** In this type of relay, there is a *definite time elapse between the instant of pick up and the closing of relay contacts.* This particular time setting is independent of the amount of current through the relay coil ; being the same for all values of current in excess of the pick-up value.
  - — It may be mentioned here that all inverse-time relays are also provided with definite minimum time feature in order that the relay may never became instantaneous in its action for very long overloads.

Calculation of relay operating time. For calculating the actual relay operating time, the following parameters must be known :

(*i*) Time/P.S.M. curve. (*ii*) Current setting.
(*iii*) Time setting. (*iv*) Fault current.
(*v*) Current transformer ratio.

The following is the *procedure to calculate the actual relay operating time*

- By the use of current transformer ratio, convert the fault current into the relay coil current.
- Express the relay current as a multiplier of current setting *i.e.* calculate P.S.M.
- Using Time/P.S.M. curve of the relay, read-off the time of operation for the calculated P.S.M.
- Determine the actual time of operation by multiplying the above time of the relay by time-setting multiplier in use.

23.3. CLASSIFICATION OF RELAYS

There are many kinds of relays used in the power system. The relays can be designed and constructed to operate in response to one or more electrical quantities such as voltage, current, phase angle etc.

The types of relays may be *classified* as under :

1. According to their construction and principle of operation :

(*i*) *Electromagnetic attraction type relays :*

(*a*) 'Attracted armature type' relays... Operation depends on the movement of an armature under the influence of attractive force due to magnetic field set up by current flowing through the relay winding.

(*b*) 'Solenoid type' relays... Operation depends on the movement of an iron plunger core along the solenoid axis.

(*ii*) *Electromagnetic induction or simply induction relays...* Operation depends on the movement of a metallic disc or cylinder free to rotate by the interaction of induced eddy currents and the alternating magnetic field producing them.

(*iii*) *Electro-dynamic relays...* In such a relay, moving member consists of a coil free to rotate in the air gap of a permanent magnet.

(*iv*) *Moving coil type relays...* In such a relay moving member consists of a coil free to rotate in the air gap of a permanent magnet.

(*v*) *Thermal relays...* In this type of relay movement depends upon the action of heat produced by the current flowing through the element of the relay.

(*vi*) *Physio-electric relays...* Buchholz's relay is an example of this type of relay.

(*vii*) *Static relays...* These relays employ thermionic valves, transistors or magnetic amplifiers to obtain the operating characteristics.

2. According to their applications :

(*i*) *Under-voltage, under-current, and under-power relays...* Operation occurs when the voltage, current or power falls below a specified value *(mostly instantaneous or induction relays).*

(*ii*) *Over-voltage, over-current and over-power relays...* Operation takes place when the voltage, current or power rises above a specified value *(mostly instantaneous or induction relays).*

(*iii*) *Directional or reverse current relays...* Operation occurs when the applied current assumes a specific phase displacement with respect to the applied voltage and the relay is compensated for fall in voltage (*induction current relays*).

(iv) *Directional or reverse power relays...* Operation takes place when the applied current and voltage assume a specific phase displacement with respect to the applied voltage and the relay is compensated for fall in voltage (*induction current relay*).

(v) *Differential relays...* Operation occurs at some specific phase or magnitude difference between two or more electrical quantities.

(vi) *Distance relays...* Operation depends upon the ratio of voltage to current.

3. According to their time of operation or timing characteristics :

(i) *Instantaneous relays...* Operation occurs after a negligibly small interval of time from the incidence of the current or other quantity which causes operation.

(ii) *Definite time-lag relays...* Operation is quite independent of the magnitude of the current or other quantity which causes operation.

(iii) *Inverse time-lag relays...* The time of operation is approximately inversely proportional to the magnitude of the current or other quantity causing operation.

(iv) *Inverse-definite minimum time lag (I.D.M.T.) relays...* The time of operation is approximately inversely proportional to the smaller values of current or other quantities causing operation and tends to a definite minimum time as the value increases without limit.

- The *time lag in 'induction type relays'* may be achieved by using a permanent magnet which is so arranged that the relay rotor cuts the flux between the poles of this magnet. This type of magnet is called "*Drag magnet*".

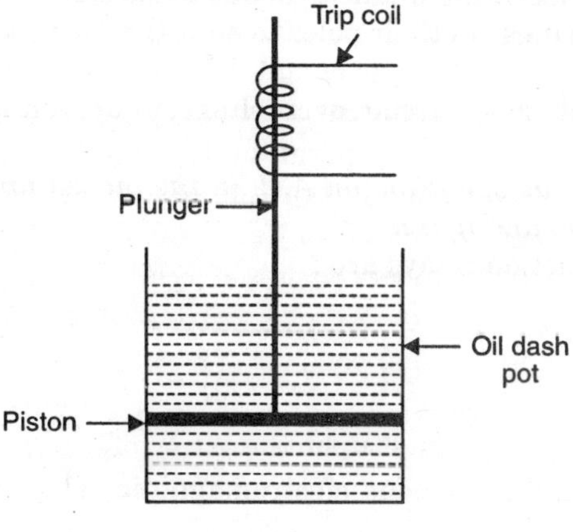

Fig. 23.3

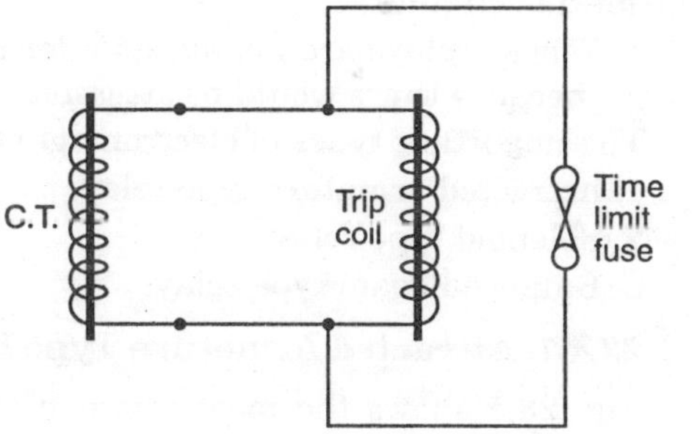

Fig. 23.4

In other relays it may be achieved in the following *two ways* :

(i) When a series connected overload trip coil is used, an oil dash pot can be attached in which a piston is connected to the lower end of the solenoid plunger of the relay (Fig. 23.3). On the occurrence of a fault, when the plunger is pulled, the piston, immersed in the oil, retards plunger motion and thus provides the necessary drag or delay in the operation of the relay.

(ii) If the trip coil is *transformer operated, time delay is achieved by connecting a time-limit fuse across the trip coil terminals* (Fig. 23.4). This fuse should have inverse time-current characteristic and will normally carry the C.T. (Current Transformer) secondary current to by-pass the trip coil. When sufficient current flows to "blow" this fuse, the whole current is then transferred to the trip coil which operates to trip the circuit breaker.

23.4. ELECTROMAGNETIC ATTRACTION RELAYS

These relays are also termed as "*magnitude*" *relays* as they compare the magnitude of the quantity with the set value.

Operating principle. In such relays the electromagnetic *force exerted on the moving element is proportional to the square of the flux in the air gap.* The total actuating force, neglecting saturation, may be expressed as

$$F = kI^2 - k' \qquad ...(23.1)$$

where, F = Net force,

k = A constant of current coil converting the current to force,

k' = The restraining force including friction, and

I = *r.m.s.* value of the operating current.

When the relay is on the verge of pick-up, the net force is zero,

i.e., $$kI^2 - k' = 0$$

or, $$I = \sqrt{\frac{k'}{k}} = \text{constant} \qquad ...(23.2)$$

Characteristics, applications and limitations :

(*i*) These relays are *non-directional.*

(*ii*) Simpler in construction.

(*iii*) Inherently very fast in operation (hence they are also called *instantaneous relays*)

- These relays are normally used for D.C. operating quantities, but these can also be used for A.C. operation by providing 'shading rings' on their poles to split the air flux into two out-of-phase components.

These are normally employed as undercurrent, short-circuit, overvoltage, undervoltage relays in distribution circuits.

- These relays are *not suitable for continuous operation on A.C. in the picked up position* because there would be *excessive vibration and noise.*

The important types of electromagnetic attraction relays are :

1. Attracted armature type relay.
2. Solenoid type relay.
3. Balanced beam type relay.

23.4.1. Attracted Armature Type Relay

Fig. 23.5 shows the most common type of instantaneous electromagnetic relay (attracted armature type).

The relay consists of a laminated electromagnet and a pivoted laminated armature. The armature is usually balanced by a counter weight and carries a pair of spring contact fingers at its free end.

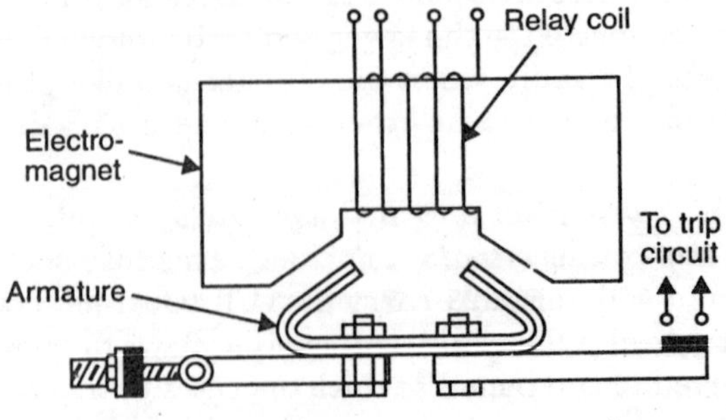

Fig. 23.5. Attracted armature type relay.

Under normal operating conditions, the current through the relay coil is such that the counterweight holds the armature in the balanced position as shown in the figure. However, when a fault takes place, the current through the relay coil increases considerably and the relay armature is attracted upwards. Subsequently, the stationary contacts attached to the relay frame are bridged and trip circuit is completed.

Usually a numbers of tappings on the relay coil are provided so that the number of turns in use and the setting value at which the relay operates can be varied.

The minimum current at which the relay armature is attracted to close the trip circuit is called *pick-up current.*

23.4.2. Solenoid Type Relay

Fig. 23.6 shows the schematic arrangement of a solenoid type relay. It consists of a solenoid and movable iron plunger arranged as shown.

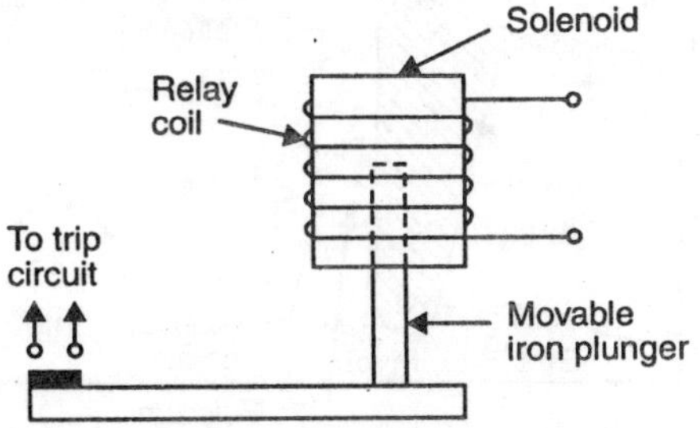

Fig. 23.6. Solenoid type relay.

Under normal working conditions, the current through coil is such that it holds the plunger by gravity or spring in the position shown. However, when a fault occurs, the current through the relay coil exceeds the pick-up value, causing the plunger to be attracted to the solenoid. The upward movement of the plunger closes the trip circuit, consequently circuit breaker opens and the faulty circuit is disconnected.

23.4.3. Balanced Beam Type Relay

The schematic arrangement of a balanced beam type relay is shown in Fig. 23.7. It consists of an armature fastened to a balance beam.

Under normal working conditions, the current through the coil is such that the beam is held in the horizontal position by the spring.

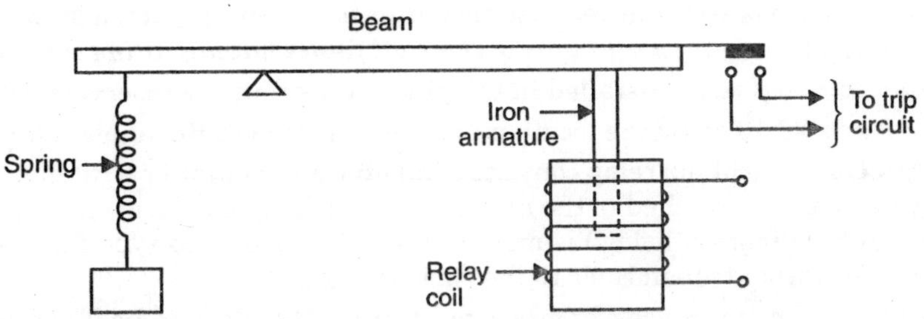

Fig. 23.7. Balanced beam type relay.

However, when a fault takes place, the current through the relay coil becomes more than the pick-up value and the beam is attracted to close the circuit. Consequently the circuit breaker opens and the faulty circuit is isolated.

23.5. THERMAL RELAY

The Fig. 23.8 shows a schematic arrangement of a thermal relay. Working principle of this relay is based upon *heating effect of the electric current.*

The thermal relay consists of a *bimetallic strip* which is heated by a *heating element.* The heating element gets supply from a current transformer (C.T.). An insulated *contact arm* carrying a *moving contact* is pivoted and is held by the spring.

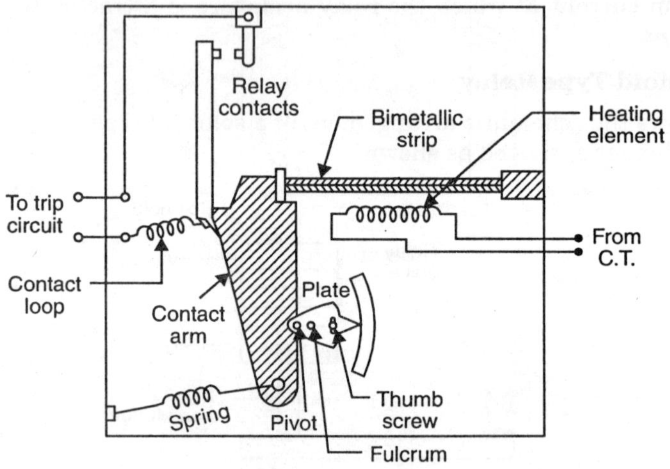

Fig. 23.8. Thermal relay.

The outer contact of trip circuit has a *fixed contact*. The spring tension can he varied by changing the position of the contact arm with the help of sector plate.

Under normal conditions, the current flowing through the heating element is proportional to the normal full-load current of the circuit. The heat produced by the heating element under this condition is not sufficient to bend the bimetallic strip. However, when fault occurs, current flowing the heating element increases which produces *heat sufficient to bend the bimetallic strip*. This releases the contact arm and because of the spring tension the relay contacts are closed which closes the trip coil circuit. As a consequence the circuit breaker opens and the faulty circuit is isolated.

The thermal relays are now obsolete as magnetic relays are available.

23.6. BUCHHOLZ RELAY

It is another form of thermal relay named after its inventor Buchholz *for use in protection of oil immersed transformer against all kinds of internal faults having rating more than 750 kVA.*

It is a *gas actuated relay* installed in the pipe connecting the reservoir to the main tank.

Figs. 23.9 and 23.10 show the location and construction of Buchholz relay respectively.

Construction. Buchholz relay consists of an oil tight container with mercury switches. One of the mercury switches is attached to the upper float which closes alarm circuit. The other mercury switch (as seen in the diagram below) is mounted on a lower hinged type flap located in the direct path of the flow of oil from transformer to the conservator.

Working. On the occurence of a *minor fault*, current leaks and heat is produced. Due to this heat, some of the oil evaporates in the transformer tank (about 70% hydrogen). The gas being light tries to go into the conservator tank via relay. In the process, some gas and oil vapours are collected in the top chamber while passing to the conservator and the oil level in the tank falls. The mercury type switch attached to the float is tilted, thereby closing the alarm circuit and the bell rings in the control room as a warning to the attendant that a serious fault is developing inside the transformer.

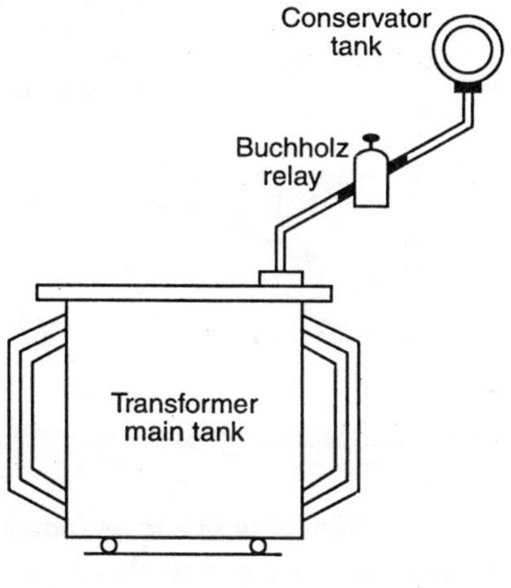

Fig. 23.9. Location of Buchholz Relay.

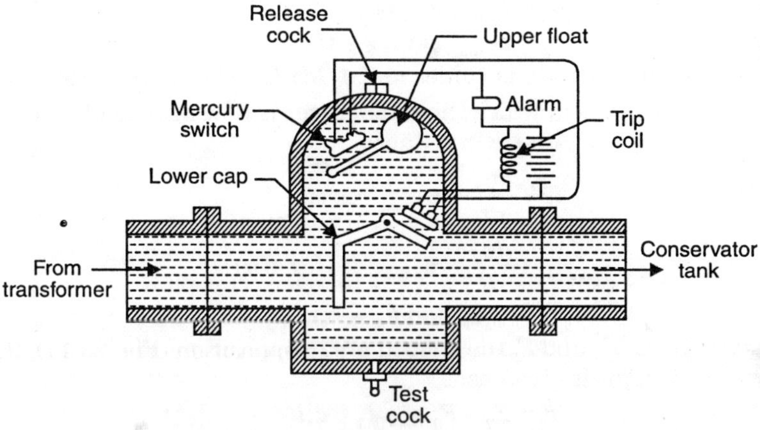

Fig. 23.10. Construction of Buchholz Relay.

When any severe fault takes place, large volume of gas is generated in the main tank. The oil gases rush towards the conservator via Buchholz relay, this result in tilting of mercury switch fitted on hinged type gap. The trip coil contacts of the circuit breaker are closed. The pressure in the chamber can be released and the gas emitted to allow the chamber to refil with oil. A test cock is installed at the bottom of the chamber to allow air to be pumped into the chamber for test purposes.

23.7. ELECTROMAGNETIC INDUCTION RELAYS

These relays are widely used for protective relaying purposes *involving A.C. quantities* (not D.C. quantities).

The electromagnetic induction relays work on the *principle of split-phase induction motor*. The actuating force is developed on a movable element, that may be a disc or other form of rotor of non-magnetic current-conducting material (such as aluminium) by the *interaction of electromagnetic fluxes with eddy currents that are induced in the rotor by these fluxes.*

Fig. 23.11 illustrates how force is developed in a disc (or section of rotor).

The split phase arrangement (or two separate sources) creates two A.C. fluxes ϕ_1 and ϕ_2. These fluxes differing in phase by an angle α induce e.m.fs. in the disc/rotor and cause the circulation of eddy currents i_1 and i_2 respectively. These currents lag behind their respective fluxes by 90°.

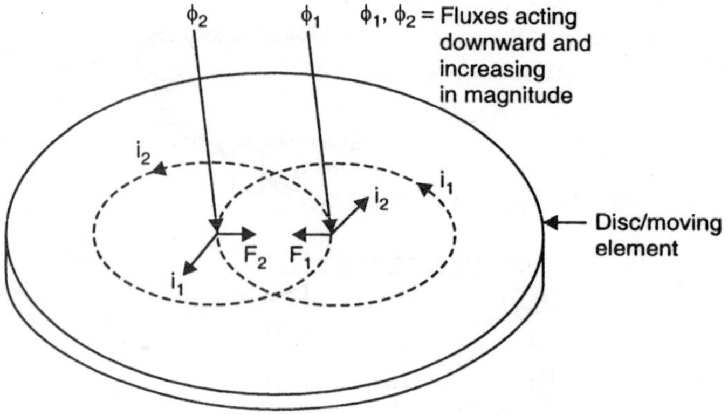

Fig. 23.11. Production of torque in an induction relay.

The current i_1 (produced by flux ϕ_1) reacts with flux ϕ_2 and develops force F_2 and current i_2 (produced by flux ϕ_2), reacts with flux ϕ_1 and develops force F_1.

Now,
$$\phi_1 = \phi_{1\,max} \sin \omega t$$
$$\phi_2 = \phi_{2\,max} \sin (\omega t + \alpha),$$

where ϕ_1 and ϕ_2 are the instantaneous values and ϕ_2 *leads* ϕ_1 *by an angle* 'α'.

Assuming that the paths in which the rotor currents flow have negligible self-inductance, the rotor currents will be in phase with their voltages.

Now,
$$i_1 \propto \frac{d\phi_1}{dt} \propto \phi_{1max} \cos \omega t$$

and,
$$i_2 \propto \frac{d\phi_2}{dt} \propto \phi_{2\,max} \cos (\omega t + \alpha)$$

Also,
$$F_1 \propto \phi_1 i_2, \text{ and } F_2 \propto \phi_2 i$$

Since the two forces (F_1 and F_2) developed are in opposition (Fig. 23.11), therefore net force F acting on the moving element is given as

$$F = F_2 - F_1 \propto \phi_2 i_1 - \phi_1 i_2$$
$$\propto \phi_{2\,max} \sin (\omega t + \alpha)\, \phi_{1\,max} \cos \omega t - \phi_{1\,max} \sin \omega t\, \phi_{2\,max} \cos (\omega t + \alpha)$$
$$\propto \phi_{1\,max}\, \phi_{2\,max} [\sin (\omega t + \alpha) \cos \omega t - \cos (\omega t + \alpha) \sin \omega t]$$
$$\propto \phi_{1\,max}\, \phi_{2\,max} [\sin (\omega t + \alpha - \omega t)]$$
$$\propto \phi_{1\,max}\, \phi_{2\,max} \sin \alpha$$

Substituting the r.m.s. values of the fluxes for the maximum values

$$F \propto \phi_1\, \phi_2 \sin \alpha \qquad \text{...(23.3)}$$

where ϕ_1 and ϕ_2 are the r.m.s. values of the fluxes.

- Thus the net force or torque on the disc/moving element is the *same at every instant* so that *average net force or torque* is $\phi_1\, \phi_2 \sin \alpha$.
- The direction of force and hence *direction motion of the moving element depends on which flux is leading the other*.

Structures used in induction relays :

In order to obtain the phase difference in the fluxes and hence the operating torque in induction relays, the following three types of structures are commonly used :

(*i*) Shaded pole (disc type) structure.

(*ii*) Watt-hour meter or double winding structure (Disc type) structure.

(*iii*) Induction cup structure.

(*i*) ***Shaded pole structure.*** The shaded pole structure is illustrated in Fig. 23.12. It consists of a pivoted aluminium disc free to rotate in the air-gap of an electromagnet. A portion of the pole face of each pole is surrounded by a copper band known as "*shading ring*". The alternator flux ϕ_s in the shaded portion of the poles will, owing to the reaction of the current induced in the ring, lags behind the flux ϕ_{us} in the unshaded portion by an angle α. These two A.C. fluxes differing in phase will produce the necessary torque to rotate the disc.

Now, $$T \propto \phi_s \phi_{us} \sin \alpha$$

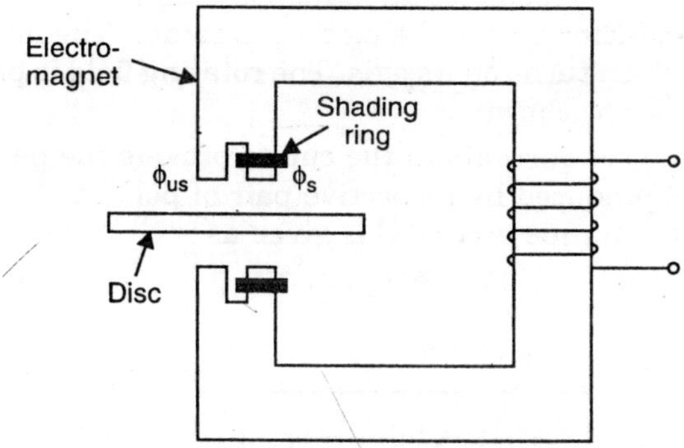

Fig. 23.12. Shaded pole structure.

Assuming ϕ_s and ϕ_{us} to be each proportional to the current in the coil,

Torque, $$T \propto I^2 \sin \alpha$$

Thus the torque is proportional to the square of the current, sin α being constant depending upon the design.

∴ $$T = kI^2 \quad ...(23.4)$$

(*ii*) ***Watt-hour meter or double winding structure.*** Fig. 23.13 shows the arrangement of watt-hour or double winding structure. It consists of two electromagnets.

The upper magnet carries two windings ; the primary and secondary . The primary winding carries the relay current I_1, while the secondary winding is connected to the lower magnet. The primary current (I_1) induces e.m.f. in the secondary and so circulates a current I_2 in it. The flux ϕ_2 induced in the lower magnet (by the current in the secondary winding of the upper magnet) will lag behind ϕ_1 by an angle α.

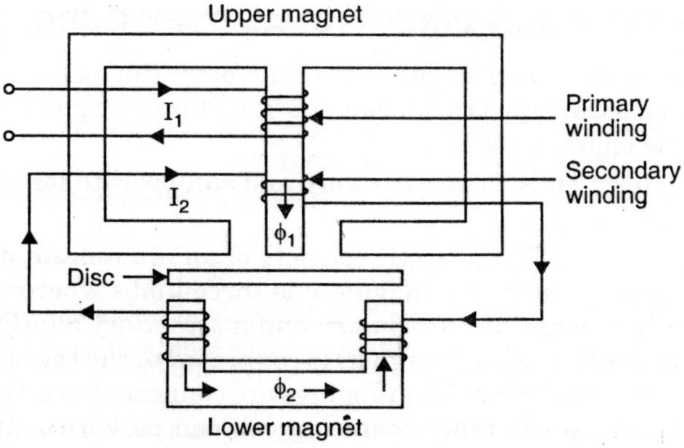

Fig. 23.13. Watt-hour meter or double winding structure.

Assuming that the whole of flux ϕ_1 enters the *disc* from the upper magnet and the whole of flux ϕ_2 enters the disc from the lower magnet, the torque on the disc is given by,

$$T \propto \phi_1 \phi_2 \sin \alpha$$

- An important feature of this type of relay is that its operation can be controlled by opening or closing the secondary winding circuit. If this circuit is opened, no torque will be produced and thus the relay can be made inoperative.
- *Majority of the induction type relays are of double winding structure.*

(*iii*) ***Induction cup structure.*** The sehematic arrangement of induction cup structure is shown in Fig. 23.14. It closely resembles an induction motor, except that rotor iron is stationary, only the rotor conductor portion being free to rotate. The moving iron is a hollow metal cylinder or cup, which turns on its axis. The rotating field is produced by two pairs of coils wound on four poles as shown.

The rotating field induces currents in the cup to provide the necessary driving torque. If ϕ_1 and ϕ_2 represent the fluxes produced by respective pair of poles and α is the phase difference between the two fluxes then the torque produced is given as :

$$T \propto \phi_1 \phi_2 \sin \alpha$$

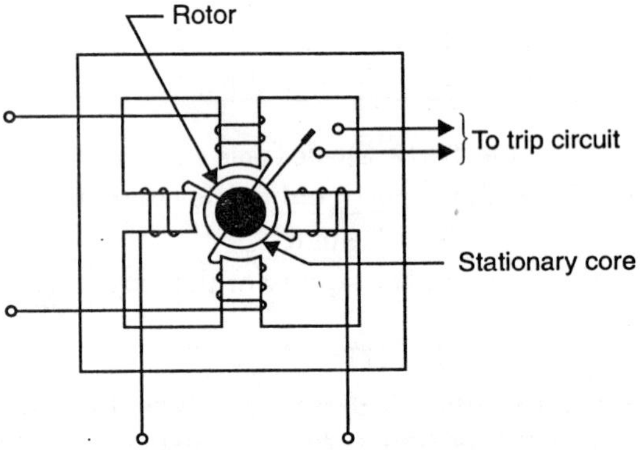

Fig. 23.14. Induction cup structure.

- This type of structure is more efficient torque producer than either the shaded pole or the watt-hour structures.

This type of relay has *very high speed and may have an operating time less than 0.1 second.*

23.8. INDUCTION TYPE NON-DIRECTIONAL OVER-CURRENT RELAY

This relay operates on *induction principle.* The actuating source is a current in the circuit supplied to the relay from a Current Transformer (C.T.). This relay gives *inverse time operation* with a definite minimum time characteristic.

These relays are used on *A.C. circuits only* and can operate for *fault current flow in either direction.*

Construction. Refer to Fig. 23.15. It consists of an aluminium disc which is free to rotate between two magnets. The upper electromagnet has three limbs whereas the lower electromagnet has two limbs. The upper magnet has a *primary* and a *secondary winding.* The '*primary winding* (wound on the central limb of the upper magnet) is connected to the secondary of a C.T. in the line to be protected and is tapped at intervals. The tappings are connected to a plug-setting bridge by which the number of active turns on the relay operating coil can be varied, thereby giving the *desired* *current setting*. The secondary winding is a closed one and wound on the central limb of the upper magnet and the lower magnet and this winding is energised by the primary winding. The controlling torque is provided by a spiral spring.

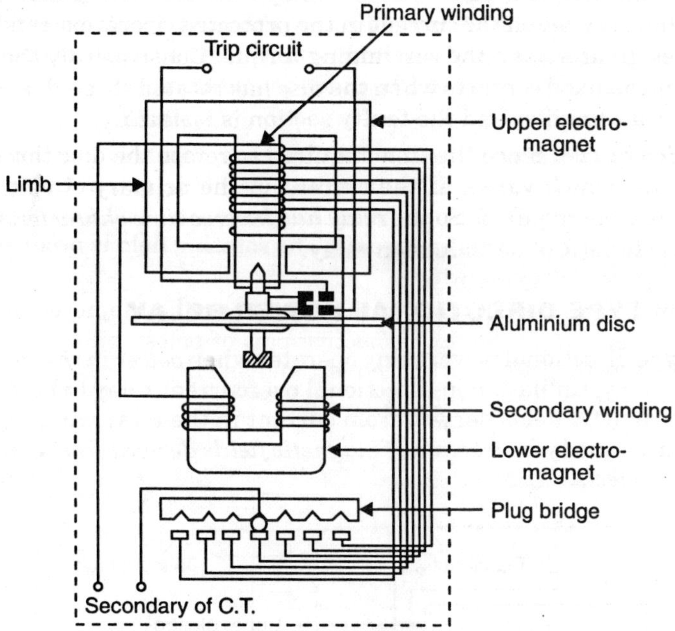

Fig. 23.15. Induction type non-directional over-current relay.

The spindle of the disc carries a moving contact which bridges two fixed contacts (connected to trip circuit) when the disc rotates through a preset angle. This angle can be adjusted to any value between 0° and 360°. By adjusting this angle, the travel of the moving contact can be adjusted and hence the relay can be given any desired time setting.

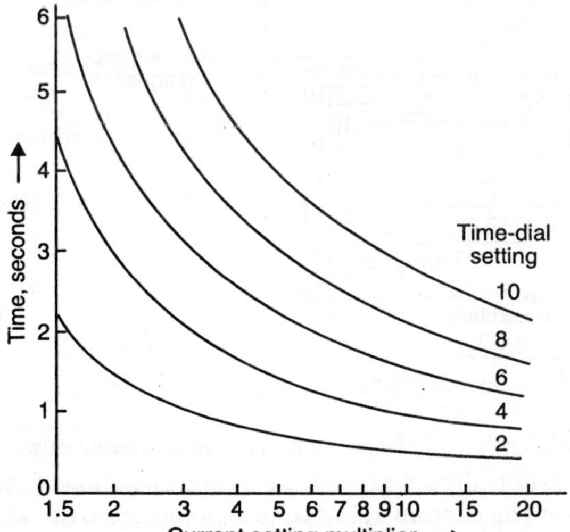

Fig. 23.16. Current-time characteristic of an induction relay.

Operation. The fluxes produced by the primary and secondary windings are separated in phase and space and a rotational torque is set up on the aluminium disc. This torque is opposed by the restraining torque provided by the spring. Under normal operating conditions, restraining torque is greater than the driving torque produced by the relay coil current and hence the aluminium disc remains stationary. However, when the current in the protected circuit exceeds the preset value, the driving torque becomes greater than the restraining torque. Consequently the disc rotates and the moving contact bridges the fixed contacts when the disc has rotated through a preset angle. The trip circuit operates the circuit breaker and the faulty section is isolated.

Characteristics curve. Since the time required to rotate the disc through the preset angle depends upon the torque which varies as the current in the primary circuit, therefore, more the torque lesser will be the time required. So the *relay has inverse time characteristic.* Fig. 23.16 shows the current-time characteristic of an induction relay.

23.9. INDUCTION TYPE DIRECTIONAL POWER RELAY

An induction type directional power relay operates when *power in the circuit flows in a specific direction.* This type of relay, unlike a non-directional overcurrent relay (where the driving torque is due to interaction of magnetic fields derived from current in the relay winding), is so designed that it obtains its operating torque *by interaction of magnetic fields derived from both voltage and current source of the circuit it protects.*

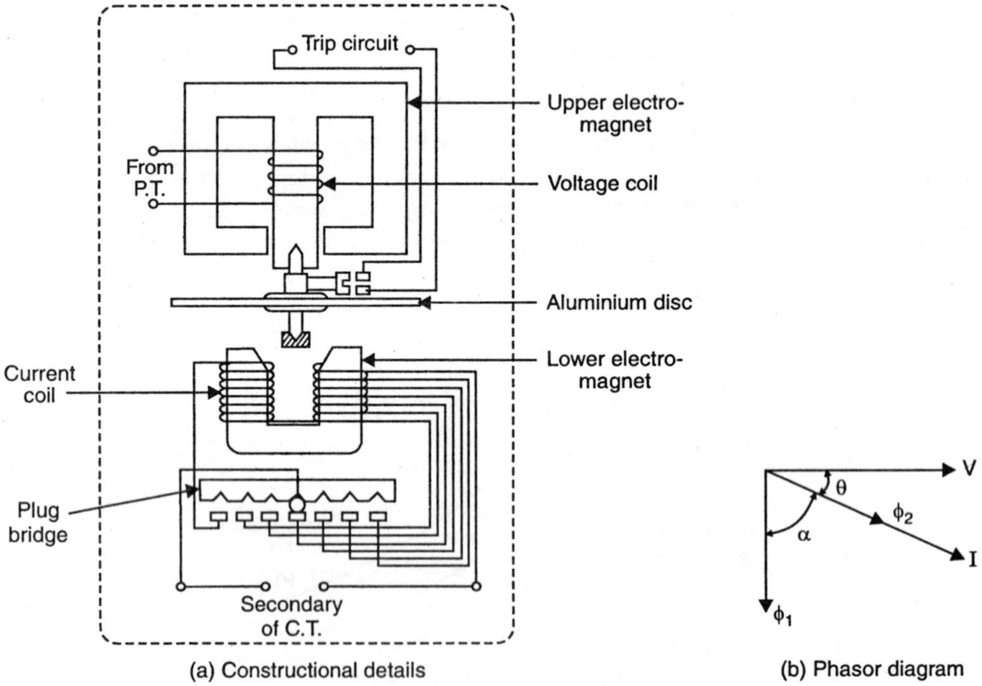

Fig. 23.17. Induction type directional power relay.

This type of relay is essentially a wattmeter and the direction of the torque set up in the relay depends upon the direction of the current relative to the voltage with which it is associated.

Construction. Fig. 23.17 (*a*) shows the constructional details of a typical induction type directional power relay. It consists of an aluminium disc which is free to rotate between the poles of two electromagnets. The upper electromagnet carries a winding, (called *'potential coil'*) on the central limb which is connected through a P.T. (Potential Transformer) to the circuit voltage source. The lower electromagnet has a separate winding (called *'current coil'*) connected to the secondary of C.T. (Current Transformer) in the line to be protected. The current winding is provided with tappings which are connected to the plug bridge ; this permits to have any desired current setting. The restraining torque is provided by a spiral spring.

The spindle of the disc carries a moving contact which bridges two fixed contacts when the disc has rotated through a preset angle. By adjusting this angle, the travel of the moving disc can be adjusted and hence any desired time-setting can be given to the relay.

Working. Refer to Fig. 23.17 (*b*). The flux ϕ_1 due to current in the potential coil will *lag* behind the applied voltage V by nearly 90°. The flux ϕ_2 due to current coil will be nearly *in phase* with the operating current I. The interaction of fluxes ϕ_1 and ϕ_2 with the eddy currents induced in the disc produces a driving torque which is given by,

$$T \propto \phi_1 \phi_2 \sin \alpha$$

Also, $\phi_1 \propto V, \phi_2 \propto I \quad \text{and} \quad \alpha = 90° - \theta$

$$\therefore \quad T \propto VI \sin (90° - \theta)$$
$$\propto VI \cos \theta$$
$$\propto \text{power in the circuit}$$

Obviously the direction of driving torque on the disc depends upon the direction of power flow in the circuit to which the relay is associated.

If the power in the circuit flows in the normal direction, the driving torque and the restraining torque (due to spring) help each other to turn away the moving contact from the fixed contacts, and consequently the relay does not operate. However, when the *current in the circuit reverses the direction of driving torque on the disc also reverses.* When the reversed driving torque is large enough, the disc rotates in the reverse direction and the moving contact closes the trip circuit. Subsequently the circuit breaker operates and the faulty section is isolated.

23.10. INDUCTION TYPE DIRECTIONAL OVERCURRENT RELAY

Under shortcircuit conditions, the directional power relay (discussed earlier) is unsuitable for use as a directional protective relay, since when short-circuit occurs, the system voltage falls to a low value and the torque developed may be insufficient to operate the relay. This difficulty is overcome in the directional over current relay which is designed to be almost independent of voltage and power factor of the system.

Construction. Fig. 23.18 shows the constructional details of an induction type directional overcurrent relay. It consists of two relay elements (mounted on a common base) namely *'directional element'* and *'non-directional element'*.

Directional element is essentially a directional power relay which operates when power flows in a specific direction. In order to operate the overcurrent element, the directional element must operate first (*i.e.,* contacts 1 and 2 should close).

The directional element is made as sensitive as possible to ensure positive operation-even 2% of the power in the reverse direction operates it.

Non-directional element is an overcurrent element similar in all respects to a non-directional overcurrent relay (described in Art. 23.8). The spindle of the disc of this element carries a moving contact which closes the fixed/trip circuit contacts *after the operation of directional element.*

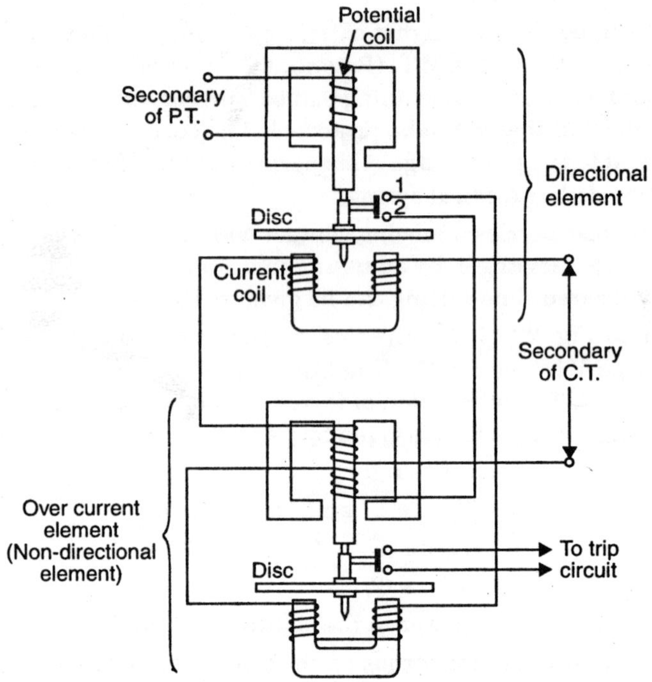

Fig. 23.18. Induction type directional overcurrent relay.

A plug-setting bridge (not shown in the figure) is also provided in the relay for current setting. The tappings are provided on the upper magnet of overcurrent relay and are connected to the bridge.

Working. When the power flows in its normal direction the directional power relay (upper element) does not operate, thereby, keeping the overcurrent element (lower element) un-energised. But as soon as there is a reversal of current or power (due to short-circuit) the disc of the upper element starts rotating and bridges the fixed contacts 1 and 2. This completes the circuit for the overcurrent element. Due to overcurrent a torque is set up on the disc (lower element) and the action closes the trip contacts, thereby enabling the circuit breaker to operate.

The relay operates only when the following *conditions are satisfied :*

(*i*) The direction of current is in reverse direction.

(*ii*) The current in reverse direction exceeds the preset value.

(*iii*) The excessive current (greater than the preset value) persists for duration longer than its time setting (overcurrent element).

Operating characteristic of a directional relay :

The operating characteristic of a directional relay is shown in Fig. 23.19.

Voltage V and current I are the two independent sources. On the polar co-ordinate, the operating characteristic of the directional relay is drawn. Here voltage V is taken the polarizing quantity and its magnitude is assumed constant.

The torque equation of the relay will be equal to

$$T = K_1 VI \cos(\theta - \alpha) - K_2$$

where, K_1 = Force conversion constant, and

K_2 = Restraining force.

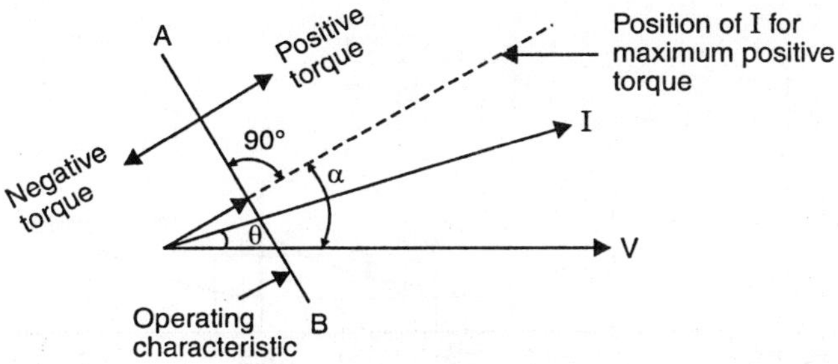

Fig. 23.19

Now at the balance point, when the relay is on the verge of operating, the net torque will be zero *i.e.*,

$$VI \cos (\theta - \alpha) = \frac{K_2}{K_1}$$

On the polar co-ordinate this expression will indicate a straight line '*AB*' as shown in Fig. 23.19.

Time grading in the overcurrent protection system. The main objective of the protective relay is *to remove the fault with smallest possible disturbance of the whole system.* That is why operating time of the operation of the relay near the fault should be quicker than that of relay near the source. Therefore, some time grading of the overcurrent protection system is made.

The *different types of time grading* in the overcurrent system are as follows :

1. Definite time current characteristics.
2. Inverse time current characteristic.
3. Inverse definite time characteristics.

1. ***Definite time current characteristics.*** Fig 23.20 shows the definite time characteristics of an electrical system. Here the time settings are *increased* toward the source.

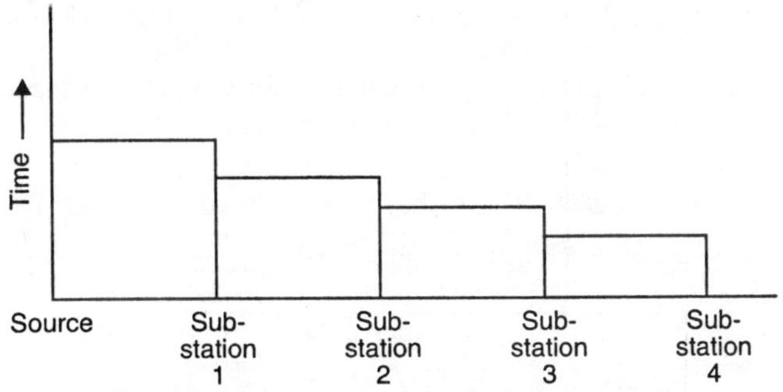

Fig. 23.20. Definite time current characteristics.

2. ***Inverse time current characteristics. Fig.*** 23.21 shows the inverse time current characteristics.

In the definite time characteristics, in an electrical system of many sections in series the tripping time for a fault near the power source will be very much high and that is not desirable. That is why inverse time characteristics as shown in Fig. 23.21 are developed. The *basic principle of this type of characteristics is that a relay whose time is inversely proportional to the current will trip faster for a fault at the end of the section nearer the power source since there is appreciable difference between the current for a fault at the far end of the section and the current for a fault at the near end. Mathematically,*

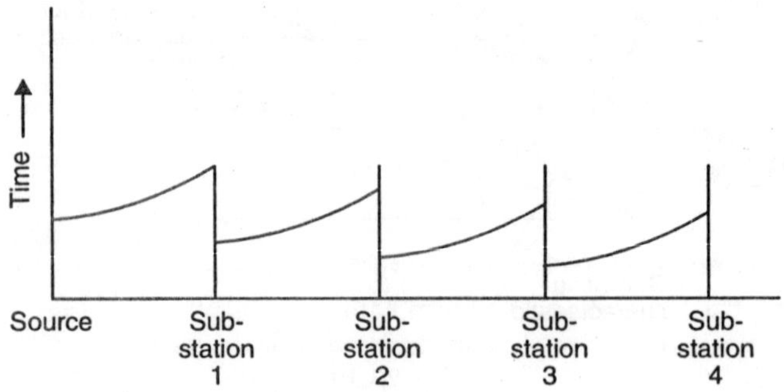

Fig. 23.21. Inverse time current characteristics.

The current for fault at the for end of the section $= \dfrac{E}{Z_S + Z_L}$

The current for a fault at the near end $= \dfrac{E}{Z_S}$

where, E = Power source e.m.f.,

Z_S = Impedance between relay and power source e.m.f., and

Z_L = Impedance of the protected section.

3. *Inverse definite time characteristics.* The inverse time definite time characteristic is called the *inverse curve* with definite time lag. If the source impedance is found very high, then it is quite obvious that the tripping time may be appreciably reduced. That is why, a compromise curve, *i.e.*, the inverse definite minimum time characteristic has been developed.

Fig. 23.22 shows the inverse definite minimum time characteristic where some *definite minimum time is kept for the operation of the relay.*

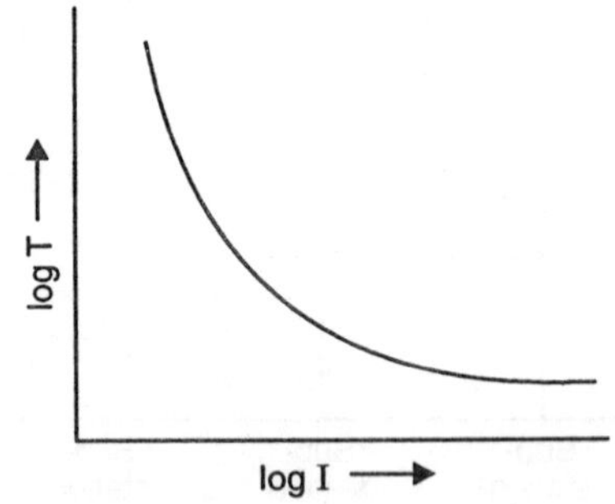

Fig. 23.22. Inverse definite time characteristic.

- **Calculation of the actual time of relay operation.** The actual time of relay operation can be calculated in the following manner :

Suppose, I_f is the fault current, the time of operation of the relay is to be calculated for this fault current.

Let, the current setting is $S\%$

Time multiplier setting $= f$

Ratio of current transformer $= I : i$

Now the relay fault current $= I_f \times \dfrac{i}{I} = i_x$ (say)

Now current multiplier of the relay $= \dfrac{i_x \times 100}{i \times S}$

Since, the current setting of the relay for 100% is i so the current setting for $S\%$ is

$$\frac{i_x}{100} \times S = x \text{ (say)}.$$

The current time characteristic for overcurrent relay is provided where current multiplier and time of operation of the relay are represented by abscissa and ordinate respectively.

From that curve, the time of operation of relay can be obtained (say, y). Since the time multiplier setting is f, the actual time of relay operation will be $y \times f$.

23.11. DIFFERENTIAL RELAYS

Differential relay is *that type of relay which operates when the phasor difference of two or more similar electrical quantities exceeds a predetermined value.*

Atmost any type of relay, when connected in a certain way can be made to operate as a differential relay.

The two fundamental systems of differential or balanced protection are :

(*i*) *Current balance protection.*

(*ii*) *Voltage balance protection.*

- Most differential relays are of the "current differential" type.

23.11.1. Current Differential Relay

Fig. 23.3 shows the arrangement of an overcurrent relay connected to work on the *current differential principle.* In this arrangement a pair of current transformers (C.Ts.) are fitted on either ends of the element to be protected and secondary windings of C.Ts. are connected in series so that they carry induced currents in the same direction.

Under normal conditions, where there is no fault or there is external fault, the currents in two C.Ts. secondaries are *equal* and relay operating coil, therefore, does not carry any current.

Whenever there is an *internal fault*, currents in the two secondaries of C.Ts. (fitted on either end) are *different*, the relay operating coil gets energised by the *current equal to their difference* and the trip circuit is completed to operate the circuit breaker.

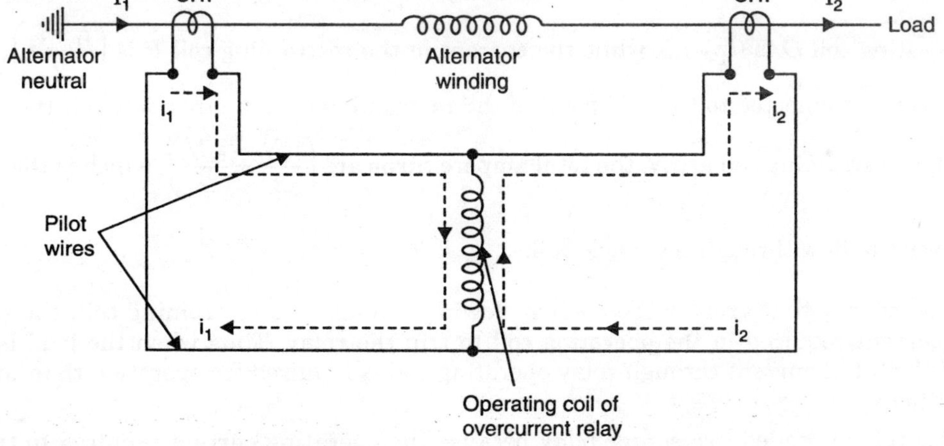

Fig. 23.23. Principle of current differential relay.

- This system is employed for the *protection of feeders, alternators and transformers.*
  — The C.Ts. of equal ratio are employed when used either at the two ends of an alternator winding or at the two ends of a feeder with no tapping.
  — When this system is to be used for protection of transformers, correction must be made for different currents determined approximately by the transformer turn-ratio.

Such a system used for a single-phase alternator is shown in Fig. 23.23.

Biased beam relay. The most extensively used form of differential relay is the *percentage-differential* or *biased beam relay* (Fig. 23.24). This relay is designed to respond to the differential current in terms of its fractional relation to the current flowing through the protected section. It is essentially an overcurrent balanced beam type relay with an additional *restraining coil.* The restraining coil produces a bias force in the opposite direction to the operating force.

Under normal and through conditions, the bias force due to restraining coil is greater than the operating force. Therefore, the relay remains inoperative.

When an internal fault occurs, the operating force *exceeds* the bias force and consequently the trip contacts are closed to open the circuit breaker.

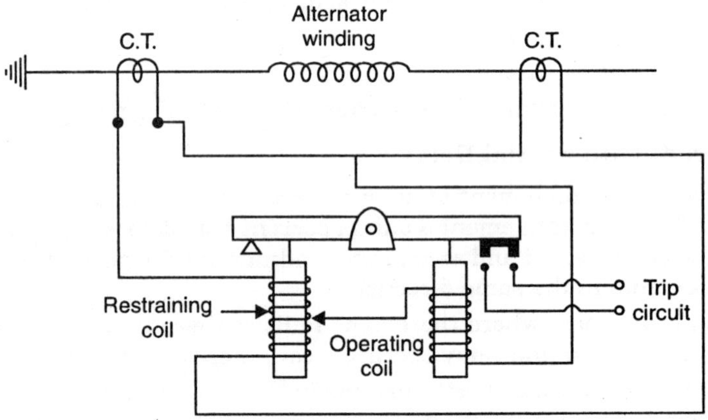

Fig. 23.24. Beam type percentage differential relay.

The equivalent circuit of a biased beam relay is shown in Fig. 23.25. The differential current in the operating coil O is $(i_1 - i_2)$, while the current in the restraining coil R is $\left(\frac{i_1 + i_2}{2}\right)$, since the operating coil is connected to the mid point of the restraining coil, in other words if the number of turns on the restraining coil are N, the total ampere turns are $\left(\frac{i_1 N}{2} + \frac{i_2 N}{2}\right)$, which is the same as if $\left(\frac{i_1 + i_2}{2}\right)$ were to flow through the whole coil.

It is evident that greater the current flowing through the restraining coil, the higher the value of current required in the operating coil to trip the relay. Thus when the load is heavy, a greater differential current through relay operating coil is required for operation than under light load conditions.

This relay is called *percentage relay* because the operating current required to trip can be expressed as a percentage of load current.

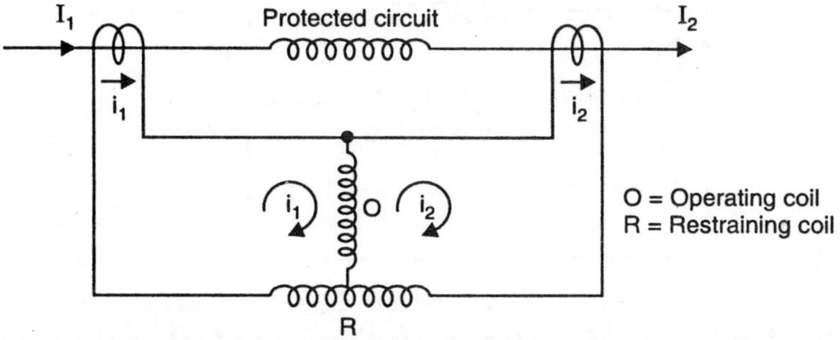

Fig. 23.25. Equivalent circuit of a percentage differential relay.

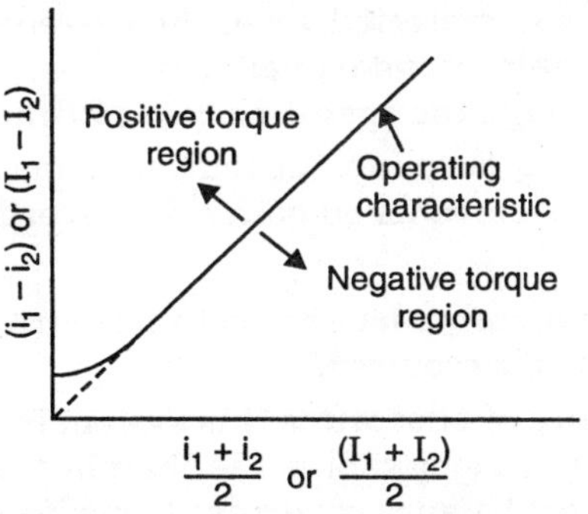

Fig. 23.26. Characteristic of percentage differential relay.

Since this relay has a rising operating characteristic *i.c.*, the pick up value increases as the magnitude of through current increases, the relay is restrained or biased against operating inaccurately. Due to this the relay is also called a biased relay.

- The operating characteristic of this type of relay is shown in Fig. 23.26. It shows that except at low currents, the ratio of the differential operating current to the average restraining current is a fixed percentage.
- The ***advantage*** of this relay is that it is *less likely to operate inaccurately than a differentially connected overcurrent relay when a short circuit occurs external to the protected zone.*
  — The current transformers of types normally employed do not transform their primary currents *so accurately* under transient conditions as for a short time after a short circuit occurs. This is particularly true when the short-circuit current is offset. Under such conditions, suppose identical C.Ts. may not have identical secondary currents. This difference will be enough to operate the relay for the larger magnitude of short-circuit current. But percentage differential relay will not operate because of its rising pick up characteristic ; as the magnitude of current increases the relay is restrained against operating improperly.

23.11.2. Voltage Balance Differential Relay

Fig. 23.27 shows a voltage balance differential relay. This is an alternative arrangement of obtaining a differential protective gear. In this scheme of protection, two similar current transformers are connected at either end of the element (*e.g.*, alternator winding) to be protected by means of pilot curves. The secondaries of current transformer are connected in series with a relay in such a way that under normal conditions, their induced e.m.fs. are in opposition.

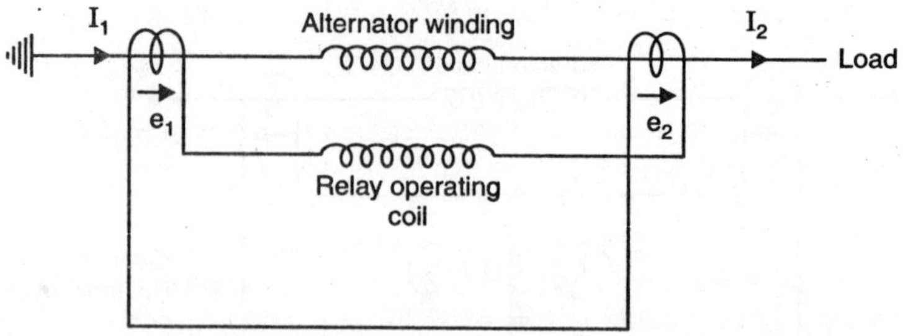

Fig. 23.27. Voltage balance differential relay.

Under *normal/healthy conditions* (*i.e.*, when there is no fault in the system) equal currents ($I_1 = I_2$) flow in both the primary windings. Therefore, the secondary voltages of the two transformers are balanced against each other and *no current will flow through the relay operating coil.*

Whenever *fault occurs* in the protected zone, the currents in the two primaries will *differ* from one another (*i.e.* $I_1 \neq I_2$) and their secondary voltages will *no longer be in balance*. Consequently, a circulating current will flow through the operating coil, causing the trip circuit to close.

In order to provide for capacity currents, the relay used may be overcurrent type which should operate only when the difference of currents on both sides exceeds certain value.

Disadvantages :

(*i*) In order to achieve the accurate balance between current transformer pairs, a multi-gap transformer construction is required.

(*ii*) Owing to the capacitance of pilot wires this system is suitable for protection of cables of relatively short lengths. On long cables, the charging current may be sufficient to operate the relay even if a perfect balance of current transformers is attained.

The above disadvantages have been overcome in Translay (modified) balanced voltage system.

- This method may be employed *for protection of feeders, alternators and transformers*.

23.12. TRANSLAY RELAY

Construction. The schematic arrangement of a translay relay (a differential relay) is shown in Fig. 23.28. It is similar to overcurrent relay except that, (*i*) secondary is not closed on itself, and also (*ii*) the central limb is provided with copper loop or ring.

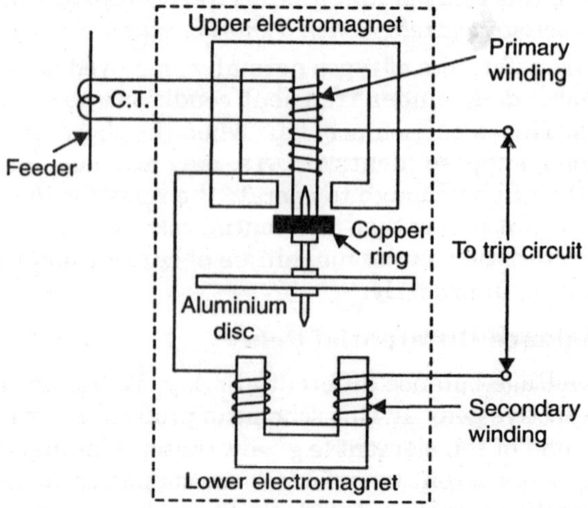

Fig. 23.28. Schematic arrangement of translay relay.

The secondary winding is connected to *two pilot wires to which is again connected the secondary winding of the similar relay at the other end of the feeder.*

Working. Under healthy conditions, when the current at the two ends of the feeder is the same, the primaries of the two relays carry the same currents inducing equal voltages in the secondaries, which are so connected that their voltages are in opposition. Hence no current flows in the two secondary circuits under the condition and so no torque results.

- Whenever the current leaving the feeder differs from the current entering the feeder, unequal voltages are induced in the secondaries. Consequently, a circulating current flows in the secondary circuit causing torque to be exerted on the disc of each relay. Since the direction of secondary current will be opposite in the two relays, therefore, torque in one relay will tend to close the trip circuit while, in other relay, the torque will hold the movement in the operated condition. It may be noted that the resulting operating torque depends upon the position and nature of the fault in the protected zone and *at least one element of either relay will operate under any fault condition.*

The closed copper rings not only neutralise the effects of pilot capacitive currents but are also effective in preventing the relays from operating when a through fault current flows in the feeder. The ring is adjusted in position until the torque due to current induced in the copper ring by the relay primary winding is restraining. The restraining torque is sufficient to compensate for the lack of balance of current transformers and reaches a maximum under through fault conditions.

- A *"Translay relay" is extensively used in Translay protection applied to* **feeders.**

23.13. DISTANCE OR IMPEDANCE RELAYS

The relays, in which the *operation is governed by the ratio of applied voltage to current in the protected circuit* are called **Distance or impedance relays** (*"Impedance"* is an electrical measure of distance along transmission line). A distance or impedance relay is essentially an ohmmeter and operates whenever the impedance of the protected zone falls below a predetermined value.

In an impedance relay, the torque produced by a current element is *opposed* by the torque produced by a voltage magnet. The relay operates when the ratio *V/I* is less than a predetermined value.

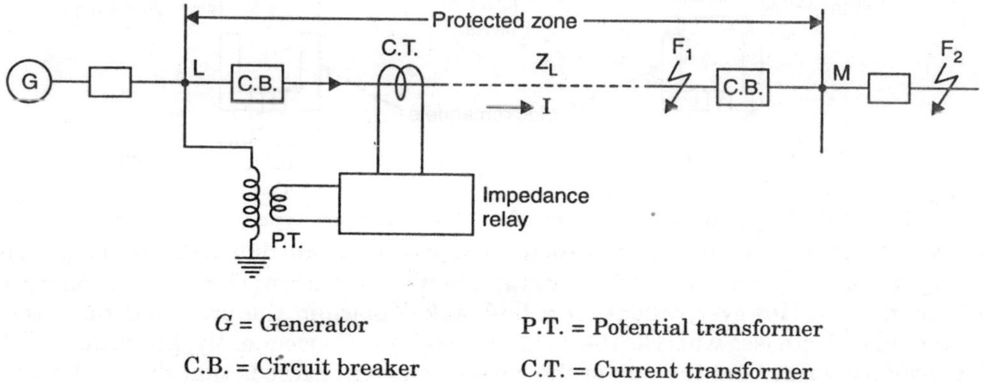

G = Generator P.T. = Potential transformer

C.B. = Circuit breaker C.T. = Current transformer

F_1, F_2 = Faults (location)

Fig. 23.29. Basic principle of operation of an impedance relay.

The basic principle of operation of an impedance relay is illustrated in Fig. 23.29. The voltage element of the relay is excited through a P.T. from the line to be protected. The current element of the relay is excited from a C.T. in series with the line. The portion LM of the line is he protected zone.

- Under normal operating conditions, the impedance of the protected zone is Z_L. The relay is so designed that it closes its contacts whenever impedance of the protected zone falls below the predetermined value (Z_L in this case). On the occurence of fault at point F_1 in the protected zone, the impedance $Z\left(=\frac{V}{I}, V \text{ and } I \text{ being substation voltage and the fault current respectively}\right)$ between the point where the relay is installed and the point of fault being less than Z_L, the relay will operate. However, the relay will not operate if the fault occurs beyond the protected zone (say point F_2), since Z is greater than Z_L at this point.

Types of distance relays. For the protection of power supply the following two types of relays are in use :

1. *Definite-distance relay.*
2. *Time-distance relay.*

The distance relays are produced by modifying either of two types of basic relays : The balance beam or the induction disc.

23.13.1. Definite-distance Type Impedance Relay

Construction. The schematic arrangement of a definite-distance type impedance relay is shown in Fig. 23.30. It consists of a pivoted beam and two electromagnets energised respectively by a voltage transformer (P.T.) and current transformer (C.T.) in the protected circuit. The armatures of the two electromagnets are mechanically coupled to the beam on the opposite sides of the fulcrum. The beam is provided with a bridging piece for the trip contacts. The design of the relay is such that the *torques produced by the two electromagnets are in opposite direction.*

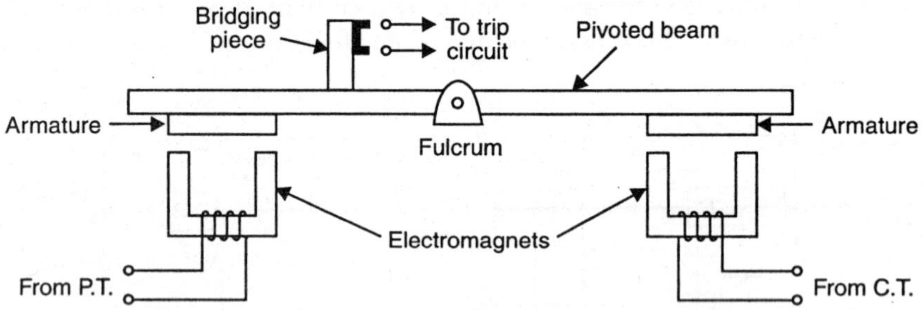

Fig. 23.30. Schematic arrangement of a definite-distance type impedance relay.

Working. When the operating conditions are normal the pull due to voltage magnet is *greater* (because current I corresponds to normal load) than that of the current element. Therefore, the relay contacts remain *open*. However, whenever a fault takes place in the protected zone, the applied voltage to the relay decreases whereas the current increases. Consequently, the ratio V/I falls below the pre-determined value, and the pull of the current element exceeds that due to the voltage element, thereby causing the beam to tilt in a direction to complete the trip circuit.

Since the pull of the voltage element is proportional to V^2 and that of current element to I^2, therefore, relay will operate when

$$k_1V^2 < k_2I^2$$

or, $$\frac{V^2}{I^2} < \frac{k_2}{k_1} \quad \text{or} \quad \frac{V}{I} = \sqrt{\frac{k_2}{k_1}}$$

or, Impedance, $Z < \sqrt{\dfrac{k_2}{k_1}}$

where, k_1, k_2 = Constants, their values depend upon the ampere-turns of the two electromagnets.

The setting value of the relay can be changed by providing tappings on the coils.

Operating characteristics. Fig. 23.31 shows the operating characteristic of the impedance type distance relay. Here the *effect of the control spring is neglected.* Now, from the mathematical analysis, it is obvious that the impedance Z will be a *constant* and therefore the operating characteristic will be a *circle*. The value of Z which is less than the radius of the circle will produce positive torque and the value of Z which is greater than the radius of the circle will produce negative torque.

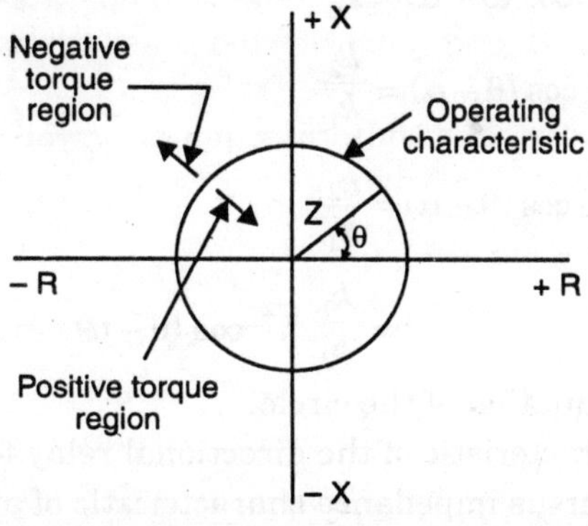

Fig. 23.31. Operating characteristic of the impedance type distance relay.

Behaviour of operating characteristic of an impedance type distance relay with the application of directional relay with the distance relay:

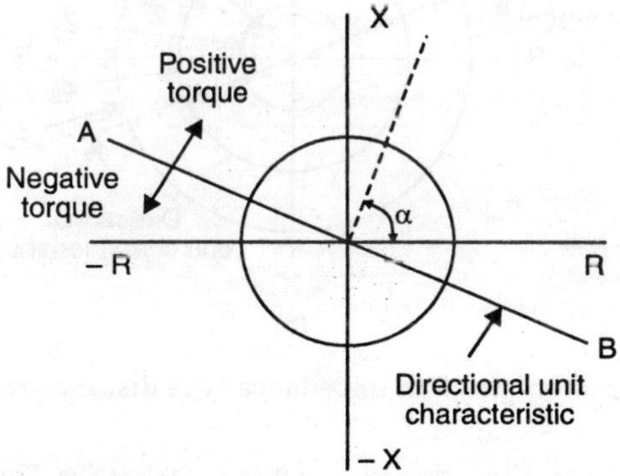

Fig 23.32.

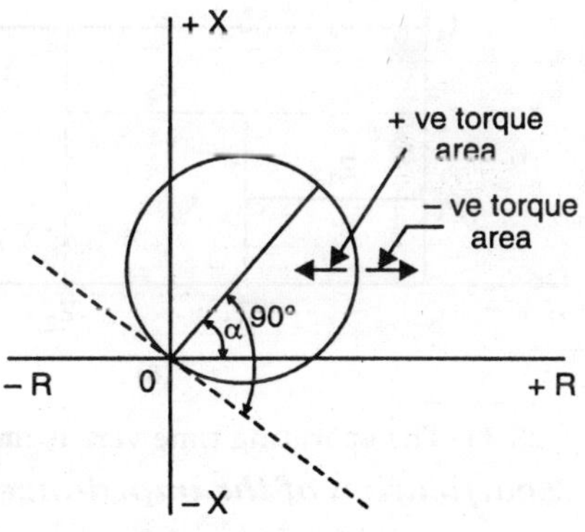

Fig. 23.33.

The operating characteristic of an impedance type distance relay behaves with the application of a directional unit with the distance relay in the manner shown in Fig. 23.32.

If the control spring effect is neglected, the torque of the directional unit can be expressed as follows :

$$T = k_1 \, VI \cos (\theta - \alpha)$$

Since k_1, V and I are not equal to zero, at the balance point

$$\cos (\theta - \alpha) = 0$$

or, $$\theta - \alpha = \pm \frac{\pi}{2}$$

This proves that the head of any radius vector z at 90° from the angle of maximum torque must lie on the operating characteristic.

Therefore, it is quite obvious that the directional unit characteristic of the impedance type distance relay will be a straight line as shown in Fig. 22.33.

On the other hand if the control spring effect is considered, the torque of the directional unit can be expressed as follows :

$$T = k_1 \, VI \cos (\theta - \alpha) - k_2$$

At the balance point $k_1 VI \cos (\theta - \alpha) = k_2$

or, $$VI \cos (\theta - \alpha) = \frac{k_2}{k_1}$$

or, $$\frac{V^2}{z} \cos (\theta - \alpha) = \frac{k_2}{k_1} \qquad \left(\because \; I = \frac{V}{z}\right)$$

or, $$z = \frac{k_1}{k_2} V^2 \cos (\theta - \alpha).$$

This is nothing but the equation of the circle.

Fig. 23.33 shows the characteristic of the directional relay for one value of voltage.

- The operating time versus impedance characteristic of an impedance type distance relay is shown in Fig. 23.34 (*a*) and (*b*).

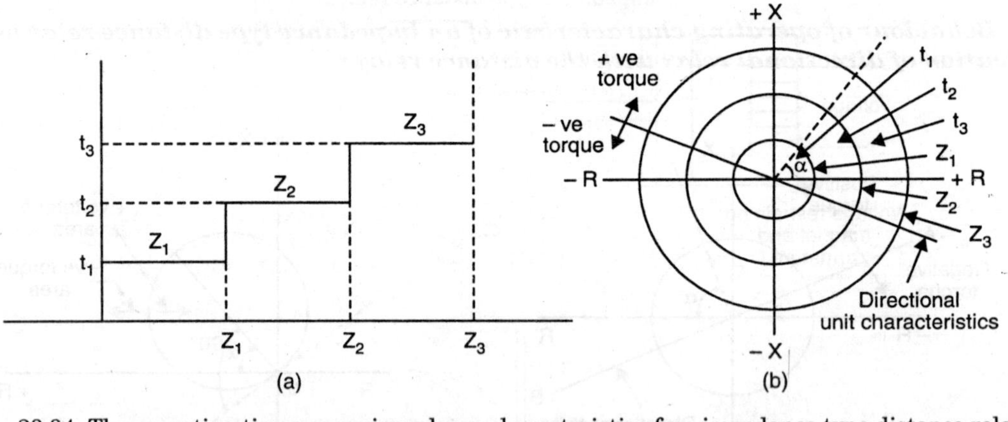

Fig. 23.34. The operating time versus impedance characteristic of an impedance type distance relay.

Modification of the impedance relay :

- The *need of the modification* of the impedance type distance relay is due to the fact that the saturation of the circuit elements may cause slight variation in the biasing and therefore it is not practicable to try to make the circles passing through the origin. That is why some modification is made in the impedance type distance relay for shifting its operating characteristic as shown in the Fig. 23.35.
- The impedance type distance relay is *modified by introducing an additional voltage proportional to the current with the voltage supply.* This can be expressed mathematically as follows :

$$T = k_1 I^2 - k_2 (V + CI)^2$$

where T, V, I and C are torque, voltage, current and a constant respectively. This is nothing but an equation of circle having a centre which is *offset from the region*.

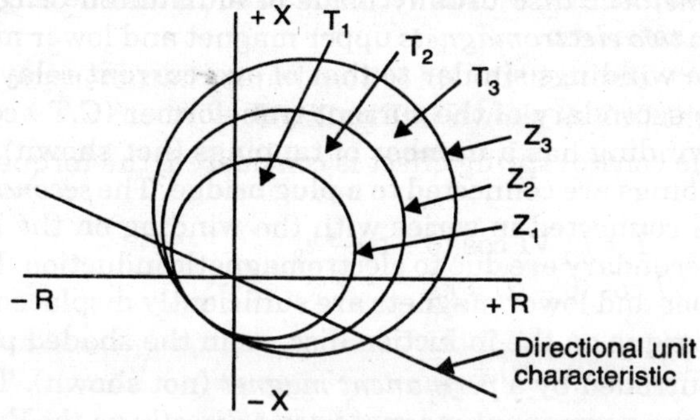

Fig. 23.35. Operating characteristic of a modified impedance-type distance relay.

23.13.2. Time-distance Impedance Relay

This type of relay automatically adjusts its operating time according to the distance of the relay from the fault point *i.e., operating time,* $T \propto \frac{V}{I} \propto Z \propto$ *Distance*

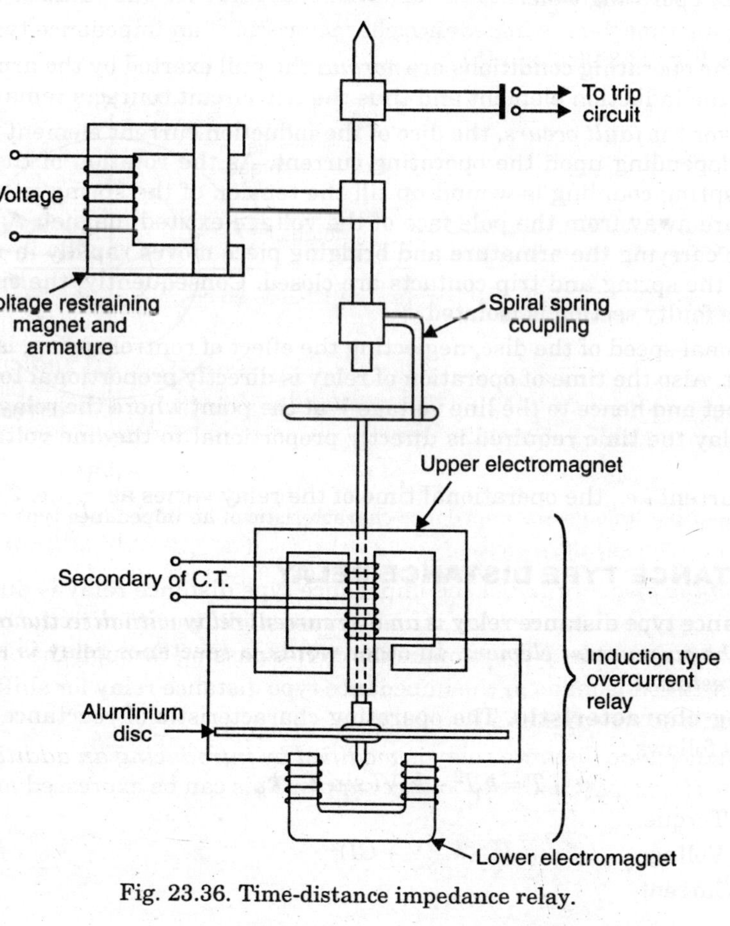

Fig. 23.36. Time-distance impedance relay.

Construction. Fig. 23.36 shows the schematic arrangement of a typical induction type time-distance impedance relay.

- It consists of a *metallic disc* usually made of aluminium or copper which is capable of rotating between *two electromagnets* upper magnet and lower magnet. The upper magnet has two separate windings similar to that of overcurrent relay. The *primary winding* is connected to the secondary of the current transformer (C.T.) connected to the line to be protected. The winding has a number of tappings (not shown) so as to vary the current settings, the tappings are connected to a plug bridge. The *secondary winding* on the upper electromagnet is connected in series with the winding on the lower electromagnet. The currents in the secondary are due to electromagnetic induction. By this arrangement leakage fluxes of upper and lower magnets are sufficiently displaced in space and phase to set up a rotational torque on the induction disc, as in the shaded pole induction disc motor.
- The torque is controlled by a *permanent magnet* (not shown). The controlling or braking torque cause by the permanent magnet *varies directly as the driving torque.*
- The spindle which carries the induction disc is shown connected by means of a spiral coupling to the second spindle which carries the bridging piece or the moving contacts of the relay trip contacts. The bridge is normally held in the open position by an *armature held against the pole face of an electromagnet energised by the voltage of the line.* In actual practice the spindle carrying the induction disc is attached to the inner end of the spiral spring through a geared counter shaft in order to obtain the required characteristic for the relay operating time.

Working :

- When the operating conditions are *normal* the pull exerted by the armature is more than that of the induction element and thus the trip circuit contacts remain open.
- Whenever the *fault occurs*, the disc of the induction current element starts to rotate at a speed depending upon the operating current. As the rotation of the disc proceeds, the spiral spring coupling is wound up till the tension of the spring is sufficient to pull the armature away from the pole face of the voltage-excited magnet. As soon it occurs, the spindle carrying the armature and bridging piece moves rapidly in response to the tension of the spring and trip contacts are closed. Consequently, the circuit breaker opens and the faulty section is isolated.

The rotational speed of the disc, neglecting the effect of control spring, is proportional to the operating current. Also the time of operation of relay is directly proportional to the pull of the voltage-excited magnet and hence to the line voltage V at the point where the relay is connected. Thus, in this type of relay the time required is directly proportional to the line voltage V and inversely proportional to current *i.e.*, the operational time of the relay varies as $\frac{V}{I}$ *i.e.* Z-distance.

23.14. REACTANCE TYPE DISTANCE RELAY

The reactance type distance relay is *an overcurrent relay with directional element that either aids or opposes the overcurrent element.* In other words, a reactance relay is an overcurrent relay with *directional restraint.*

Operating characteristic. The operating characteristic of reactance type distance relay can be derived as follows :

$$T = k_1 I^2 - k_2 VI \sin \theta - k_3 \qquad (\theta \text{ is + when } I \text{ lags})$$

where T = Torque,

V = Voltage,

I = Current,

k_1, k_2 = Constants, and

k_3 = Spring constant.

k_1I^2 represents the characteristic of overcurrent relay, and $k_2VI \sin \theta$ represents the characteristic of directional element.

At balance point,

$$k_1I^2 = k_2 VI \sin \theta + k_3$$

Dividing both sides by I^2, we get

$$k_1 = k_2 \frac{V}{I} \sin \theta + \frac{k_3}{I^2}$$

or, $$\frac{V}{I} \sin \theta = Z \sin \theta = \frac{k_1}{k_2} - \frac{k_3}{k_2 I^2}$$

Neglecting control spring effect,

$$X = \frac{k_1}{k_2} = \text{Constant}$$

This is an equation of a straight line and the operating characteristic will be as shown in Fig. 23.37.

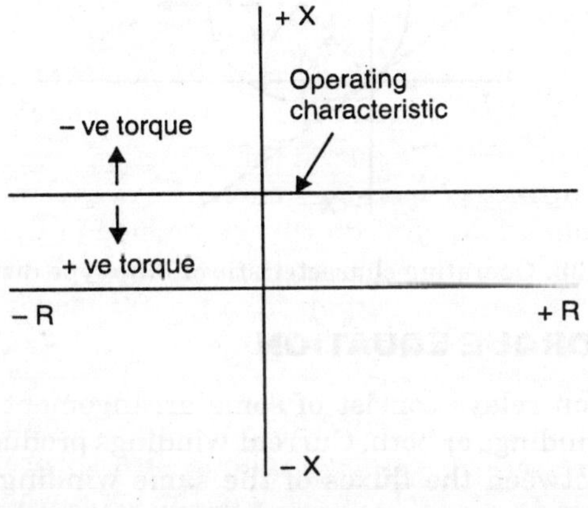

Fig. 23.37. Operating characteristic of a reactance relay.

- In this type of relay the resistance component of the impedance has no effect on its operation and the relay responds solely to the reactance component.
- The *structures* which are generally used for reactance type distance relay are *induction cup and double induction loop structures.*

23.15. MHO TYPE DISTANCE RELAY

The significance of the name of mho relay is that the mho characteristic when plotted in the admittance instead of impedance axis indicates a *straight line.* The Mho relay superedes impedance and reactance relays, due to the fact that it is *more stable on power swings* than reactance or impedance relays.

This relay has a *voltage restraining element that opposes a directional element.* The operation of this relay is the same as that of the impedance type distance relay but with a difference that no separate directional unit is required because *mho units are inherently directional.*

The torque equation is given as :

$$T = k_1 VI \cos(\theta - \alpha) - k_2 V^2 - k_3 \qquad (\theta \text{ and } \alpha \text{ being +ve, where } I \text{ lags } V)$$

At balance point,

$$k_2 V^2 = k_1 VI \cos(\theta - \alpha) - k_3$$

Dividing both sides by $k_2 VI$, we get

$$\frac{V}{I} = Z = \frac{k_1}{k_2}\cos(\theta - \alpha) - \frac{k_3}{k_2 VI}$$

If the control spring effect is neglected,

$$Z = \frac{k_1}{k_2}\cos(\theta - \alpha)$$

The operating characteristic of such a relay, which is circular one, is shown in Fig. 23.38.

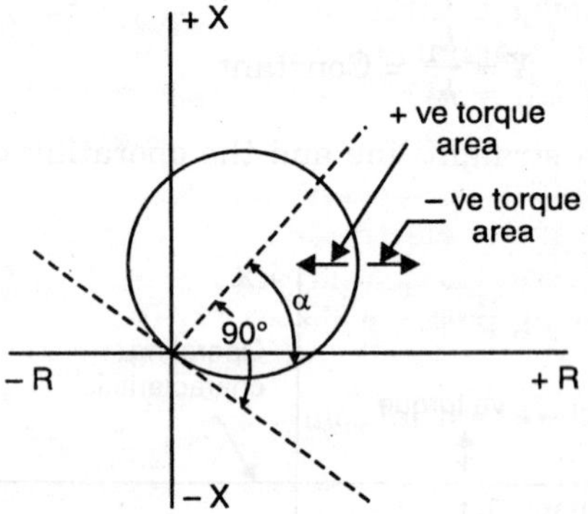

Fig. 23.38. Operating characteristic of Mho-type distance relay.

23.16. UNIVERSAL TORQUE EQUATION

Most of the protection relays consist of some arrangement of electromagnets which have current windings, voltage windings or both. Current windings produce magnetic fluxes and torque is developed by interaction between the fluxes of the same winding or between fluxes of both the windings.

Torque developed by current winding = $k_1 I^2$

Torque developed by voltage winding = $k_2 V^2$

If both the current and voltage windings are employed, torque developed by the interaction between the fluxes

$$= k_3 VI \cos(\theta - \alpha)$$

where, θ = Angle between V and I, and

α = Value of torque angle (known as *relay maximum torque angle*) ; it is the design constant of the relay.

When all the elements are present, the torque will be produced by all the three cases, so that the total torque produced in general case is

$$T = k_1 I^2 + k_2 V^2 + k_3 VI \cos(\theta - \alpha) + k$$

where, k_1, k_2, k_3 = Tap settings or constants of I and V, and

k = Mechanical constraint due to a spring or gravity.

By assigning plus or minus sign to some of the terms and letting others be zero and sometimes adding some terms having a combination of voltage and current, the operating characteristics of all types of relays can be obtained.

For example, for *'over current relay'* $k_2 = 0, k_3 = 0$ and the spring torque will be $= k$. Similarly for 'directional relay', $k_1 = 0, k_2 = 0$.

23.17. STATIC RELAYS

23.17.1. Introduction

A static relay is the *relay which incorporates solid state components like transistors, diodes etc., for the measurement* or *comparison of electrical quantities.* The static network is so designed that it gives an output signal in the tripping direction whenever a threshold condition is reached. The output signal in turn operates a tripping device which may be electromagnetic or electronic.

Static relays are capable of performing the same functions with the use of electronic circuit control which an electromagnetic relay performs with the use of moving parts or elements.

Fig. 23.39 illustrates the essential components of a static relay.

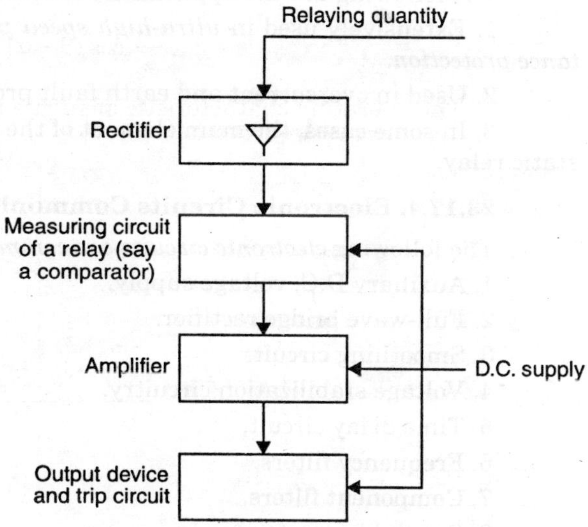

Fig. 23.39. Block diagram of a static relay.

- The relaying quantity (*i.e.,* the output of a C.T. or P.T. or a transducer) is first *rectified* and then *compared* with the set values.
- The output is actuated when the dynamic input (*i.e.,* relaying quantity) reaches the threshold value. This output is *amplified* and given to the output device which is usually an electromagnetic one to energise the trip coil (when the relay operates).

23.17.2. Merits and Demerits of Static Relays

Merits. The static relays claim the following *merits over electromagnetic relays* :

1. More accurate and fast in operation.
2. The moving parts and the contacts are largely eliminated. The only moving element in static relay is final tripping contact.
3. *VA* rating of C.Ts. and P.Ts. is comparatively lower (since static relays require a very little *VA* for their operation).
4. Low power consumption.
5. Resetting time and overshoots can be reduced.
6. Very compact.
7. Superior characteristic and accuracy.
8. Several functions can be accommodated in a static relay.
9. Simplified testing and servicing is possible.

Demerits. The following are the *demerits of static relays* :

1. Unpredictable reliability.
2. Construction is not very robust.
3. Easily affected by surrounding interference.
4. Auxiliary D.C. supply is required, constituting a constant loss of power.
5. Susceptible to voltage transients.

23.17.3. Applications of Static Relays

The following are the applications of static relays :

1. Extensively used in *ultra-high speed protection schemes of EHV-A.C. lines utilising distance protection.*
2. Used in overcurrent and earth fault protection schemes.
3. In some cases, the main element of the differential relay in equipment protection is also a static relay.

23.17.4. Electronic Circuits Commonly used in Static Relays

The following *electronic circuits are commonly used in static relays* :

1. Auxiliary D.C. voltage supply.
2. Full-wave bridge rectifier.
3. Smoothing circuit.
4. Voltage stabilization circuitry.
5. Time delay circuit.
6. Frequency filters.
7. Component filters.
8. Saturable reactors.
9. Output device.
10. Phase determination using transistors.
11. Amplitude comparator.
12. Phase comparator.
13. Level detectors.

23.17.5. Analog Circuits Used in Static Relay

The following analog circuits are used in static relay :

1. Operational amplifier-as inverter.
2. Non-inverting amplifier.
3. Voltage follower.
4. Added circuitry.
5. Subtraction circuitry.
6. Integrator circuitry.
7. Differentiator circuitry.
8. Detector.

23.17.6. Different Static Relays

The static relays encompasses the following relays :

Transducer relays. These are magnetic amplifier relays which consists of a control and operating winding. The control winding is energised with D.C. and the operating winding with A.C.

These relays are *mechanically very simple and quite reliable*. Since the relay rectifies and smoothens a signal, a delay is introduced because of the time constant of the smoothing circuit and the relays are *slow*, therefore, are discarded for protection applications.

Rectifier bridge relays. This type of relay consists of two rectifier bridges and a moving coil or polarized moving iron relay.

Transistor relays. These are *most widely used static relays*. The fact that a transistor can be used both as an amplifying device and as a switching device, makes this component suitable for achieving any functional characteristic. The transistor circuits cannot only perform the essential functions of a relay such as comparison of inputs, summation and integrating them but they also provide necessary flexibility to suit the various relay requirements.

The static relays can also be categorised as follows :

1. *Static overcurrent relays.*
2. *Static time overcurrent relays.*
3. *Static instantaneous overcurrent relay.*
4. *Directional static overcurrent relay.*
5. *Static differential relay.*

- Mostly used in protection of generators and transformers against any type of internal fault similar to that of electromagnetic type differential relay.

— Highly sensitive for internal faults.
— Low power consumption.
— Low *VA* burden.
— Inrush current proof characteristics.
— Mostly, a permanent magnet moving coil relay is used as a tripping device.

6. *Static distance relay.* These relays are used extensively for protection of medium and long transmission lines, parallel feeders and unit back up protection as well as interconnected and *T* connected lines.

7. ***Microprocessor based digital relays :***

- These relays can process and display the signals very efficiently, accurately and in a fastest possible manner. Due to their programmable characteristics, microprocessor based relays can be applied extensively in the *protection of electrical systems*. Moreover, one microprocessor unit may be able to perform relaying function of several systems.
- The microprocessors should be properly shielded from external influences and the system earthing must be very good from which they receive control voltage.

HIGHLIGHTS

1. The *protective relay* may be defined as an electrical device connected between the main circuit and the circuit breaker that detects the fault and initate the operation of the circuit breaker to isolate the defective section from the rest of the system.
2. The three essential fundamental elements of all the relays are :
   (*i*) Sensing element (*ii*) Comparing element
   (*iii*) Control element.
3. The following structures are used in induction relays :
   (*i*) Shaded pole (disc type) structure.
   (*ii*) Watt-hour meter or double winding structure (disc type) structure.
   (*iii*) Induction cup structure.

4. A *differential relay* is that type of relay which operates when the phasor difference of two or more similar electrical quantities exceeds a predetermined value.
5. The relays in which the operation is governed by the ratio of applied voltage to current in the protected circuit are called *"Distance or impedance relays."*
6. The *static relay* is the relay which incorporates solid state components like transistors, diodes etc., for the measurement or comparison of electrical quantities.

THEORETICAL QUESTIONS

1. What are the main features of a good protective system ? What is meant by a relay ?
2. What is a protective relay ? Explain its functions in an electrical system.
3. What are the fundamental requirements of a relay ?
4. Explain the following terms connected with relays :
   Operating force or torque ; Operating time ; Maximum torque angle ; Reset time ; Pick-up current ; Current setting.
5. Explain briefly the following :
   (*i*) Main a primary protection. (*ii*) Back up or secondary protection.
6. State the procedure for calculating the actual relay operating time.
7. Give the classification of relays. Describe and define the following terms :
   (*i*) Pick-up (*ii*) Reset or Drop off (*iii*) Drop off/Pick-up ratio.
8. What is a relay ? How are relays classified ?
9. Describe the essential features of a protective relay with reference to reliability, selectivity, speed of operation and discrimination.
10. Explain the working principle of an overcurrent relay.
11. With the help of neat sketch describe the working of a thermal type relay.
12. Describe with the help of neat diagram the functioning of Buchholz relay.
13. Explain with the help of a neat sketch the working of an electromagnetic relay.
14. Explain with the help of neat diagrams the construction and working of :
    (*i*) Induction type directional power relay.
    (*ii*) Non-directional induction type overcurrent relay.
15. Explain with a neat sketch the operation of an induction type overcurrent relay. What are the functions of current and time multiplier settings associated with such a relay ?
16. Discuss the time-current characteristics of an induction type overcurrent relay.
17. Discuss briefly the structures used in induction relays.
18. What do you mean by "Time grading in the overcurrent protection system" ?
19. What is a 'differential relay' ?
20. Explain briefly the following :
    (*i*) Current differential relay. (*ii*) Voltage balance differential relay.
21. Explain with a neat diagram construction and working of Translay relay.
22. What is meant by percent bias ? How is this achieved in a practice in a differential relay ? Under what circumstances is a percentage differential relay preferred over the differential one.
23. What is an impedance relay ? Explain its principle of operation. Show its characteristics on *R-X* diagram. What is the advantage of this relay for transmission line protection.
24. Derive operating characteristics of (*i*) an impedance and (*ii*) a reactance relay. Explain how you provide direction features to these relays.
25. Draw schematic diagrams for (*i*) impedance relay (*ii*) reactance relay and (*iii*) mho relay. Discuss operating characteristics of these relays.
26. How does a distance relay derive its name from its function ? Draw a neat sketch of a time-distance relay and explain its working.
27. What is a 'static relay' ?
28. What are the merits and demerits of static relays over electromagnetic relays ?
29. Enumerate different static relays.
30. State the applications of static relays.

24

Protection of Alternators And Transformers

24.1. **Protection of Alternators**—Types of faults in alternators—Protection against stator faults by percentage differential relays (Merz-Price protection system)—Balanced earth–fault protection—Stator inter-turn protection—Unbalanced loading of the alternator—Overload (overcurrent) protection—Failure of prime mover—Overvoltage protection—Restricted earth fault protection of alternator—Standby earth fault protection—Rotor fault protection. 24.2. **Protection of transformers**—Buchholz protection—core-balance leakage protection—Combined leakage and overhead protection—Differential protection for transformers—Interlocked current protection in the transformer protection scheme. Highlights—Theoretical Questions—Unsolved Examples.

24.1. PROTECTION OF ALTERNATORS/SYNCHRONOUS GENERATORS

24.1.1. Types of Faults in Alternators

The following faults may occur in the stator and rotor of an alternator :

Stator faults :

1. Failure of prime mover (turbine) resulting in operation of the alternator as synchronous motor.
2. Overvoltage at the terminals of the alternator.
3. Overloading.
4. Unbalanced loading and subsequent heating of alternator.
5. Insulation failure of stator winding, either between phase-to-phase or between phase to ground.
6. Inter-turn faults in any of the phases.
7. Bearing oil pressure failure.
8. Current leakage in the body of the alternator.
9. Ventilation failure.

Rotor faults :

1. Excitation failure.
2. Rotor earth fault.
3. Overcurrent in rotor circuit due to abnormally high excitation.

- The ***reasons of overcurrent*** in the alternator are :

  (*i*) External faults ;

  (*ii*) Thermal overloading ;

  (*iii*) Unbalanced loading.

- The *effects of overcurrent* in the alternator are :
  - (*i*) At the time of external faults, large short-circuit currents may be developed and severe mechanical stresses may act upon the stator windings. If there is already any mechanical defect in the winding, that will be more aggravated. If the fault exists, dangerous temperature level may be attained.
  - (*ii*) Continuous overloading may increase the temperature of the winding to such a value that the winding insulation gets damaged and life of the generator will be reduced.
  - (*iii*) Dangerous heating of the cylindrical rotor of turbo-generators may result due to continuous unbalanced loads being equal to or more than 10 percent of the rated current.
- The ***causes of overvoltage*** *on alternator* are :
  - (*i*) Atmospheric surge voltage ;
  - (*ii*) Switching surge ;
  - (*iii*) Arcing grounds.
  - — For protecting 'surge voltages', *lightning arrestors* and *surge capacitors* are generally used.
  - — For avoiding 'switching surge', *lightning arrestor* is generally used between generator circuit breaker and generator. Furthermore, modern circuit breaker is also used to avoid this.
  - — The 'arcing grounds' can be reduced with the help of *resistance earthing of the alternator neutral*.
- ***Causes of overspeeding*** *of the alternator* are :
  - (*i*) Fault in the turbine governor or its associated equipment.
  - (*ii*) Tripping of the main generator circuit breaker while full electrical power is being supplied to the network.
  - — In general, the normal action of governor is sufficient for the protection of overspeeding of the governor. Besides, to supervise the normal working of the governor, some additional protective device is used.
- The reasons of ***excessive bearing temperature*** *of alternator* are :
  - (*i*) Mechanical fault ;
  - (*ii*) Impurities in the lubricating oil ;
  - (*iii*) Defect in the oil circulation system.
  - — The excessive bearing temperature can be *detected* in the alternator with the help of *temperature monitoring device* embedded in the bearing.
- The ***bearing current*** develops in the alternator due to the following fact :

For some magnetic dissimilarities in the armature field some voltage is induced on the shaft of the generator. Now this e.m.f. will be impressed across the thin oil films of the bearings when the bearing pedestals at each side of the alternator are earthed. If there occurs a breakdown of the oil film insulation in the two bearings, a heavy bearing current will flow due to the very small resistance of the shaft and the external circuit.

- The ***motoring action of alternator*** occurs when the *driving torque of the prime-mover is reduced below the total losses of the turbo-generator unit.* At that time active power will be taken by the alternator from the network to maintain synchronism and thus the alternator will act as motor.

Vibration detector is generally mounted on the alternator bearing pedestal for detecting vibration which may be produced due to unbalanced loads or certain types of mechanical faults.

- The *unbalance on alternator is usually due to leakage* of one conductor to earth.

24.1.2. Protection Against Stator Faults by Percentage Differential Relays (Merz-Price Protection System)

"Merz-Price protection system" is the most common form of protection used for stator winding faults such as *phase to phase* or *phase to ground*. It is the standard practice to use *'differential protection'* for alternators of rating 1 MVA or higher and *percentage differential protection* for 10 MVA and higher.

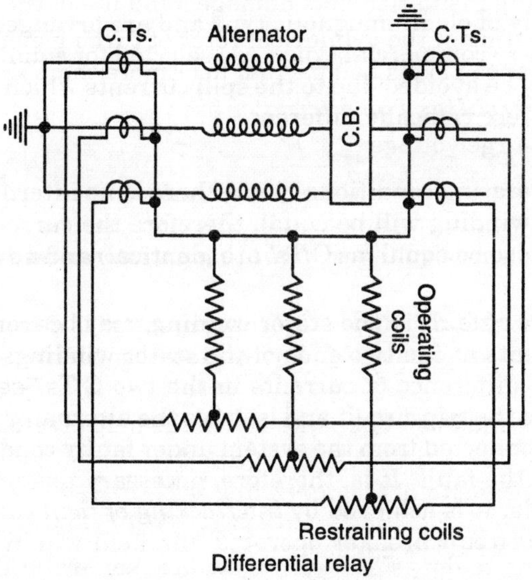

Fig. 24.1. Differential protection for star-connected alternators.

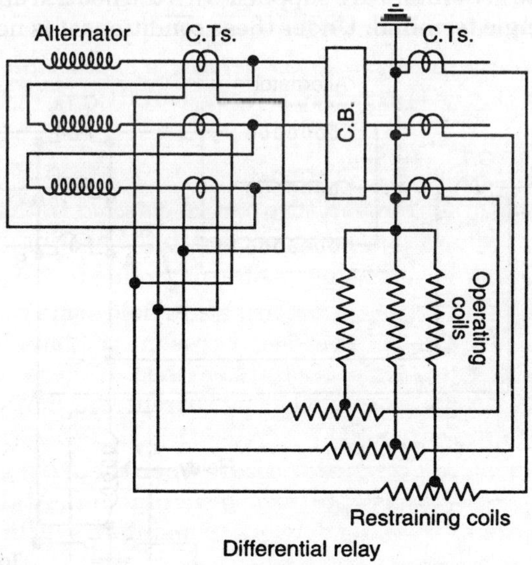

Fig. 24.2. Differential protection for delta-connected alternators.

Figs. 24.1 and 24.2 show the arrangement of C.Ts. and percentage differential relays for star-connected and delta-connected machines respectively.

Schematic arrangement. There are two sets of current transformers (C.Ts.), each set is mounted on either end of the stator phase windings. The secondaries of these C.Ts. are connected in star and their ends are connected together through pilot wires. The relays are connected across equipotential points of three pilot wires which would naturally be in the centre of these wires, thus the relays are located midway as illustrated in Fig. 24.1.

The relays are usually of electromagnetic type and are arranged for instantaneous tripping. The C.Ts. while being ordinary commercial types, are selected for equality of characteristics in order that unwanted tripping may be avoided due to the spill currents which would occur if the characteristics of any opposing pair were radically different.

Working :

- Under normal operating conditions (when there is no internal fault) the currents at both the ends of each winding will be equal, therefore the currents in the two corresponding secondaries will also be equal (as C.Ts. are identical) and no current will flow through the relays.
- On the occurrence of fault in the stator winding, may be from phase to phase or phase to ground, the currents at the two ends of the stator windings become different so the current equal to the difference of currents in the two C.Ts.' secondaries flows through the relay which closes the trip circuit and isolates the alternator from the system. Though the alternator is disconnected from the system under faulty conditions but it still continues to feed current into the fault. It is, therefore, necessary to suppress the induced voltage as quickly as possible. It is achieved by *interlocking of field current with the circuit breaker* (C.B.). When the circuit breaker operates, the field winding of the alternator is short-circuited.

24.1.3. Balanced Earth-fault Protection

Sometimes, small size generators are supplied with the neutral ends of their 3-phase windings internally connected to a single terminal. Under these conditions it is not possible to use Merz-Price

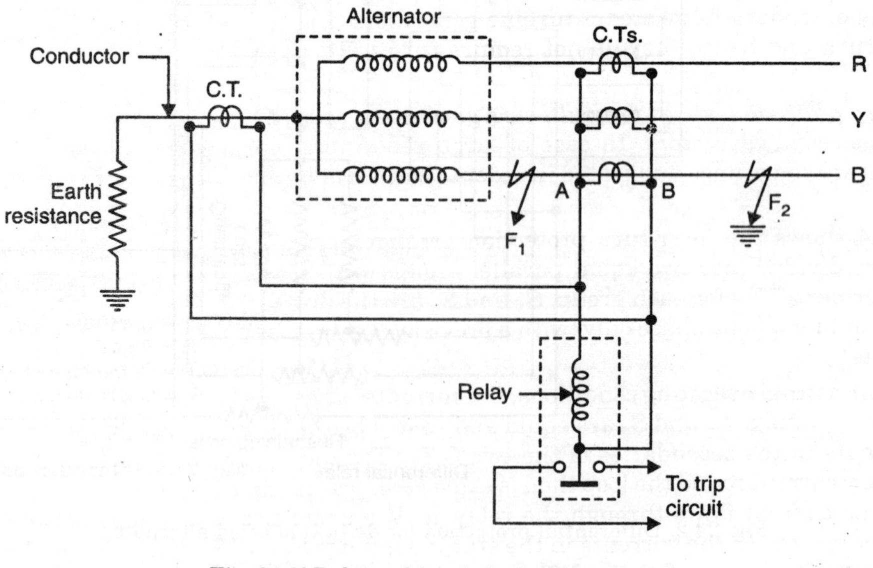

Fig. 24.3. Balanced earth fault protection.

circulating current principle (described in article 24.1.2) and protection is provided against earth faults only by the use of balanced earth fault protection system. This system provides no protection against phase-to-phase faults, unless and until they develop into earth-faults, as most of them will.

Schematic arrangement. Fig. 24.3 shows the schematic arrangement of balanced earth fault protection for a 3-phase alternator. It consists of the line current transformers (C.Ts.), one mounted in each phase, having their secondaries connected in parallel with that of a single current transformer (C.T.) in the conductor joining the star point of the alternator to earth. A relay is connected across the transformer secondaries. The protection against earth faults is limited to the region between the neutral and the line current transformers.

Working :

- Under normal operating conditions (when the windings are sound), the secondary currents in three current transformers (C.Ts.) sum up to zero at the points *AB* while the secondary current in the single transformer (C.T.) is also zero ; thus no current flows through the relay.
- When an earth fault occurs at F_1 or within the protected zone (*i.e.* left of C.Ts.), the fault current passes through the primary of C.T. and the corresponding secondary current passes through the relay. Subsequently the relay closes its contacts to disconnect the alternator from the system.

However, if an earth fault occurs at F_2 (external to the protected zone) the current passes through the primary winding of the appropriate current transformer (C.Ts.) and also through C.T. Thus the two secondary currents are balanced and no current passes through the relay, and it does not operate.

24.1.4. Stator Inter-turn Protection

Merz-Price protection system does not provide protection against short-circuit between turns on the same phase, since this condition does not cause unbalance between currents in C.Ts. pairs. However, it is considered unnecessary to have such protection since short-circuits between turns on the same phase invariably develop into earth fault.

The coils of modern large steam-turbine generators usually have one turn and hence they do not require turn-to-turn protection.

Inter-turn protection is used for multi-turn generators, such as large hydro-electric generators. In case of large generators, stator windings are sometimes duplicated in order to carry heavy current.

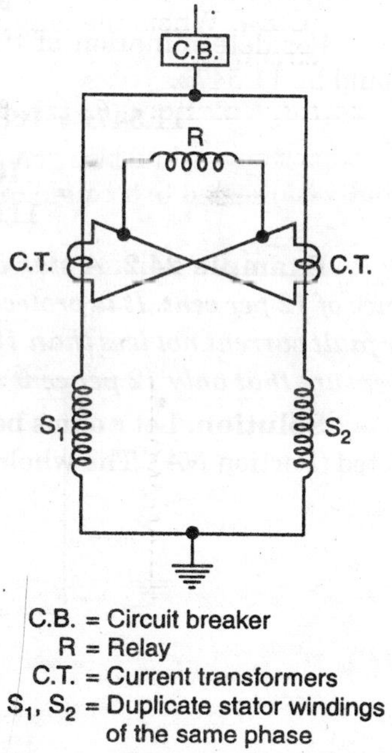

Fig. 24.4. Inter-turn protection.

Fig. 24.4 shows the inter-turn protection arrangement. The circuits are divided into two equal parallel groups with a current transformer (C.T.) for each group. S_1 and S_2 are the duplicate stator windings of one phase only (with a provision against inter-turn faults).

- Under normal operating conditions, the currents in the stator windings S_1 and S_2 are equal and so will be the currents in the secondaries of the two C.Ts. The secondary current round the loop then is same at all points and no current flows through the relay *R*. If a short-circuit develops between adjacent turns, say on S_1, the currents in the stator windings S_1 and S_2 will no longer be equal and a current proportional to the difference will be diverted into the relay operating coil which will close the trip circuit and isolate the alternator from the system.

Example 24.1. *A 3-phase, 15 MVA, 11 kV, star-connected alternator is protected by the current balancing system of protection. If the ratio of the C.Ts. is 1200/5, the minimum operating current of the relay is 0.7 A and the neutral point earthing resistance is 5.5 Ω, calculate the percentage of each phase of stator winding which is unprotected against earth faults when the machine is operating at normal voltage.* **(Panjab University)**

Solution. *Given* : Rating of alternator = 15 MVA, 11 kV, 3-phase, star-connected ; Ratio of C.Ts. = 1200/5 ; Minimum operating current of the relay = 0.7 A ; Neutral point earthing resistance = 5.5 Ω.

% age of each phase of stator winding unprotected :

Let x% of the winding be unprotected. The e.m.f. induced in this part of winding

$$= \frac{11 \times 10^3}{\sqrt{3}} \times \frac{x}{100} = \frac{110}{\sqrt{3}} x \text{ volts}$$

The fault current, which this unprotected portion of the winding will cause

$$= \frac{\text{E.m.f. induced in the faulty portion of the winding}}{\text{Earth resistance}}$$

$$= \frac{110x}{\sqrt{3} \times 5.5} = 11.547x \text{ amps.}$$

The minimum current which will operate the relay during faulty conditions

$$= \frac{1200}{5} \times 0.7 = 168 \text{ A}$$

For determination of the unprotected portion of the winding, the minimum fault current should be $11.547x$.

$$\therefore \quad 11.547x = 168$$

or $$x = \frac{168}{11.547} = 14.55\%. \quad \textbf{(Ans.)}$$

Example 24.2. *A star-connected 3-phase, 12 MVA, 6.6 kV alternator has a per phase reactance of 12 per cent. It is protected by Merz-Price circulating current principle which is set to operate for fault current not less than 180 A. Calculate the value of earthing resistance to be provided in order to ensure that only 12 per cent of the alternator winding remains unprotected.*

Solution. Let r ohms be the earthing resistance required to leave 12% of the winding unprotected (Portion *NA*). The whole arrangement (simplified) is shown in Fig. 24.5.

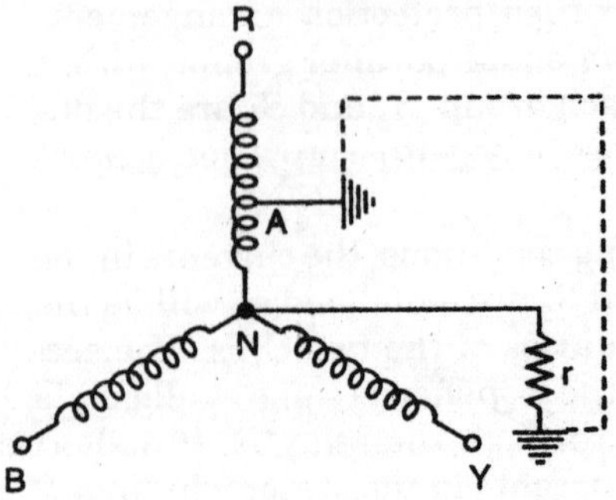

Fig. 24.5

Voltage per phase, $V_{ph} = \dfrac{6.6 \times 10^3}{\sqrt{3}} = 3810$ V

Full-load current, $I = \dfrac{12 \times 10^6}{\sqrt{3} \times 6.6 \times 10^3} = 1049.7$ A

Let the reactance per phase be x ohms

$$\therefore \quad 12 = \frac{\sqrt{3} \times x \times 1049.7}{6.6 \times 10^3} \times 100$$

or, $x = 0.4356\ \Omega$

Reactance of 12% winding $= x \times 0.12 = 0.4356 \times 0.12 = 0.0523\ \Omega$

E.m.f. induced in 12% winding $= V_{ph} \times 0.12 = 3810 \times 0.12 = 457.2$ V

Impedance offered to fault by 12% winding,

$$Z_f = \sqrt{(0.0523)^2 + r^2}$$

Earth-fault current due to 12% winding

$$= \frac{457.2}{Z_f} = \frac{457.2}{\sqrt{(0.0523)^2 + r^2}}$$

When this fault current becomes 180 A, the relay will trip.

$$\therefore \quad 180 = \frac{457.2}{\sqrt{(0.0523)^2 + r^2}}$$

or, $(0.0523)^2 + r^2 = \left(\dfrac{457.2}{180}\right)^2 = 6.45$

$\therefore \quad r = [6.45 - (0.0523)^2]^{1/2} = \mathbf{2.54\ \Omega}$. **(Ans.)**

Example 24.3. *A 13.8 kV, 150 MVA, star-connected alternator has a synchronous reactance of 1.68 per unit per phase and a negligible resistance. It is protected by a Merz-Price balanced current system which operates when out-of-balance current exceeds 10% of the full-load current. If the neutral point is earthed through a resistance of 2.5 Ω, determine what proportion of winding is protected against earth fault.* **(Panjab University)**

Solution. Rating of the alternator = 13.8 kV, 150 MVA, star-connected

Synchronous reactance = 1.68 per unit per phase

Earthing resistance = 2.5 Ω

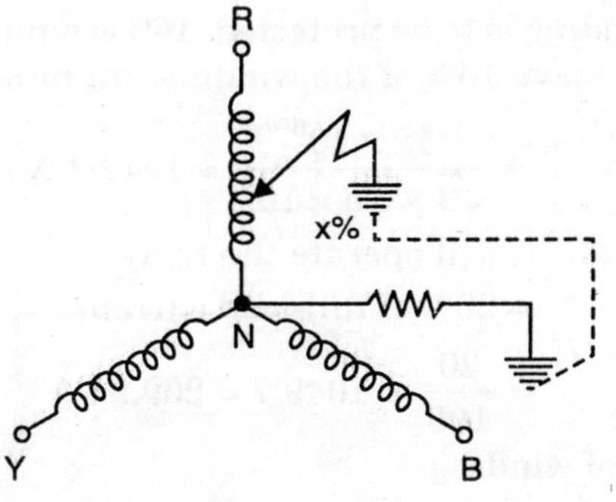

Fig. 24.6

Let x% of the winding be unprotected. The arrangement is shown in Fig. 24.6.

Voltage per phase, $$V_{ph} = \frac{13.8 \times 10^3}{\sqrt{3}} = 7967 \text{ V}$$

Full-load current, $$I = \frac{150 \times 10^6}{\sqrt{3} \times 13.8 \times 10^3} = 6276 \text{ A}$$

Reactance per phase, $$X = \frac{\text{Reactance per unit} \times V_L}{\sqrt{3} \times I}$$

$$= \frac{168 \times 13.8 \times 10^3}{\sqrt{3} \times 6276} = 2.133\ \Omega$$

E.m.f. induced in the unprotected winding

$$= 7967 \times \frac{x}{100} = 79.67x \text{ volts}$$

Earth fault current caused by unprotected winding

$$= I \times \frac{10}{100} = 6276 \times \frac{10}{1000} = 627.6 \text{ A}$$

Earth impedance $$= \frac{\text{E.m.f. induced}}{\text{Earth fault current}} = \frac{79.67x}{627.6}\ \Omega = 0.1269x\ \Omega$$

Reactance of unprotected winding $$= \frac{x \times X}{100}$$

$$= \frac{x}{100} \times 2.133 = 0.02133x\ \Omega$$

Now, $(\text{Earthing impedance})^2 = (\text{Earthing resistance})^2 + (\text{Earthing reactance})^2$

or, $$(0.1269x)^2 = (2.5)^2 + (0.02133x)^2$$

or, $$(0.1269x)^2 - (0.02133x)^2 = 6.25$$

or, $$0.01565x^2 = 6.25$$

or, $$x = \sqrt{\frac{6.25}{0.01565}} = 20\%$$

∴ **Protected winding** = 100% – 20% = **80%. (Ans.)**

Example 24.4. *A star-connected 3-phase, 12 MVA, 6.6 kV alternator is protected by circulating current protection, the star being earthed via a resistance r. Calculate the value of earthing resistor if 84% of the stator winding is protected against earth faults. Assume an earth fault setting of 20 percent. Neglect the impedance of the alternator winding.*

Solution. Since 84% winding is to be protected, 16% would be unprotected. Let r ohms be the earthing resistance required to leave 16% of the winding unprotected.

Full-load current $$= \frac{12 \times 10^6}{\sqrt{3} \times 6.6 \times 10^3} = 1049.7 \text{ A}$$

Minimum fault current which will operate the relay

= 20% of full-load current

$$= \frac{20}{100} \times 1049.7 = 209.94 \text{ A}$$

Voltage induced in 16% of winding

$$= \frac{16}{100} \times \frac{6.6 \times 10^3}{\sqrt{3}} = 609.7 \text{ V}$$

$$\text{Earth fault current which 16\% winding will cause} = \frac{609.7}{r}$$

This current must be equal to 209.94 A

$$\therefore \qquad 209.94 = \frac{609.7}{r}$$

$$\therefore \qquad r = \frac{609.7}{209.94} = \mathbf{2.904\ \Omega.} \quad \textbf{(Ans.)}$$

Example 24.5. *Fig. 24.7 shows an alternator winding protected by a percentage differential relay. The relay has a 0.16 amp. minimum pick-up and a 10% slope. While the alternator is carrying load, a high resistance ground fault occurs near the grounded neutral end of the alternator winding. The currents flowing at each end of the alternator winding are shown in the figure. If the C.Ts. have 400/5 ampere ratio and no inaccuracies, determine whether the relay will trip the alternator circuit breaker or not under this fault condition.*

Solution. The minimum pick-up of the relay = 0.16 A

The slope line of relay = 10%

C.Ts. ampere ratio = 400/5

$$i_1 = 320 \times \frac{5}{400} = 4.0\ \text{A}$$

$$i_2 = 280 \times \frac{5}{400} = 3.5\ \text{A}$$

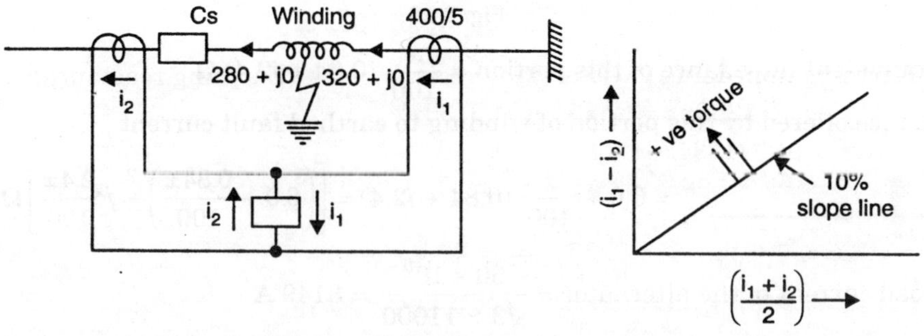

Fig. 24.7

$$i_1 - i_2 = 4.0 - 3.5 = 0.5\ \text{A}$$

$$\frac{i_1 + i_2}{2} = \frac{4.0 + 3.5}{2} = 3.75\ \text{A}$$

$$\therefore \qquad \frac{(i_1 - i_2)}{\left(\frac{i_1 + i_2}{2}\right)} = \frac{0.5}{3.75} = 0.1333 \quad \text{or} \quad 13.33\%$$

Thus, this point would lie above 10% slope characteristic *i.e.* it would lie on the *+ve torque regime* and hence the **relay would operate. (Ans.)**

Example 24.6. *A 3-phase 60 MVA, 11 kV star-connected alternator has a synchronous reactance of 2.4 ohms per phase and resistance of 0.84 ohm per phase. It is being protected by a Merz-Price balanced current system.*

Determine what portion of the winding remains unprotected if the neutral of the alternator is earthed through a resistance of 6 ohms. Assume that relay operates when the out-of-balance current exceeds 20% of the load current.

Solution. Rating of the alternator = 60 MVA, 11 kV, 3-phase

Synchronous reactance per phase = 2.4 Ω

Resistance per phase = 0.84 Ω

Earthing resistance = 6 Ω

Let x% of the winding be unprotected (Fig. 24.8)

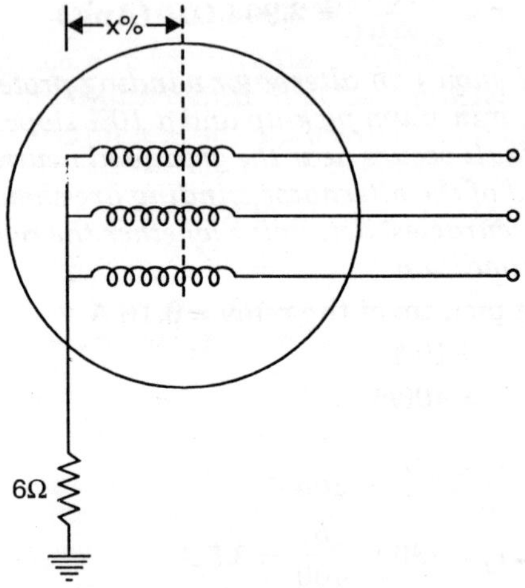

Fig. 24.8

The equivalent impedance of this portion $= \dfrac{x}{100}\,(0.84 + j2.4)\ \Omega$

Impedance offered by this portion of winding to earthed fault current

$$= 6.0 + \frac{x}{100}\,(0.84 + j2.4) = \left[\left(6.0 + \frac{0.84x}{100}\right) + j\,\frac{2.4x}{100}\right]\Omega$$

Full-load current of the alternator $= \dfrac{60 \times 10^6}{\sqrt{3} \times 11000} = 3149$ A

Out of balance current which can cause relay operation

$$= 3149 \times \frac{20}{100} = 629.8 \text{ A} \qquad \text{...}(i)$$

Again, e.m.f. induced in x% of the winding

$$= \frac{11000}{\sqrt{3}} \times \frac{x}{100} = 63.5x \text{ volts}$$

The earth fault current caused by this induced e.m.f.

$$= \frac{63.5x}{\text{Earth fault resistance}} = \frac{63.5x}{\left[\left(6.0 + \dfrac{0.84x}{100}\right) + j\,\dfrac{2.4x}{100}\right]} \qquad \text{...}(ii)$$

Equating eqns. (i) and (ii), we get

$$\frac{63.5x}{\left[\left(6.0 + \dfrac{0.84x}{100}\right) + j\,\dfrac{2.4x}{100}\right]} = 629.8$$

On neglecting the alternator reactance, we have

$$\frac{63.5x}{6.0 + \frac{0.84x}{100}} = 629.8$$

or, $$63.5x = \left(6.0 + \frac{0.84x}{100}\right) \times 629.8 = 3778.8 + 5.29x$$

$\therefore$ **x = 64.9%. (Ans.)**

Example 24.7. *A 11 kV, 3-phase turbo-alternator has a maximum rating of 120 MW at 0.8 p.f. and its reactance is 0.12 p.u. It is equipped with differential current protection scheme. It is set to operate at fault current not less 600 A. Determine the magnitude of the neutral earthing resistance that leaves the 12% of the winding unprotected.*

Solution. Turbo-alternator voltage = 11 kV
Maximum rating = 120 MW at 0.8 p.f.
Alternator reactance = 0.12 p.u.
Minimum operational fault current, i = 600 A
Percentage of unprotected winding = 12%

Magnitude of the neutral earthing resistance, r :

Rated current of the alternator, $I = \dfrac{120 \times 10^6}{\sqrt{3} \times 11000 \times 0.8} = 7873$ A

Relay setting, $s = \dfrac{i}{I} = \dfrac{600}{7873} = 0.0762$

Portion of the winding unprotected $= \dfrac{sI}{I_n}$

$\therefore$ $0.12 = \dfrac{0.0762 \times 7873}{I_N}$ (where, I_N is the current through neutral)

or $I_N = \dfrac{0.0762 \times 7873}{0.12} = 4999.36$ A

Hence, earthing resistance, $r = \dfrac{V}{I_N} = \dfrac{11000}{\sqrt{3} \times 4999.36} = \mathbf{1.27\ \Omega.}$ **(Ans.)**

24.1.5. Unbalanced loading of the Alternator

Unbalanced loading of the stator winding *develops negative phase sequence components of currents in itself.* This causes *additional heating of the stator winding*, which is not recommended for the sake of life of alternator. Negative sequence current protection scheme is mostly applied in order to sense degree of unbalancing in stator current.

24.1.6. Overload (or Overcurrent) Protection

For alternators, overcurrent protection is not considered necessary, since modern alternators are capable of withstanding their complete short-circuit current for sufficient time without serious overheating and damage. On such faults these are disconnected manually.

24.1.7. Failure of Prime-mover

When the input to the prime-mover fails, the alternator runs as a synchronous motor and draws some current from the supply system. This motoring condition is known as "***inverted running***".

- If the steam supply fails to a *turbo-alternator set*, the alternator will continue running as a synchronous motor and draw only a small fraction of full-load power from the supply system. It will remain in synchronism and operate as an alternator when steam supply is restored, so protection against the failure of prime mover is not required.
- In case of *hydro-generator sets*, protection against inverted running is achieved by providing mechanical devices on the water-wheel. When the flow of water drops to an insufficient rate to maintain the electrical output, the alternator is disconnected from the system. Thus, in this case also electrical protection is not necessary.
- *Diesel-engine driven alternators* should be protected against inverted running to guard against motoring in the event of a mechanical seizure. This is achieved by *applying reverse power relays to the alternators* which isolate the latter during their motoring action.

24.1.8. Overvoltage Protection

- The field excitation system of modern alternators is so designed that overvoltage conditions at normal running speeds cannot occur. However, overvoltage in an alternator occurs when the speed of the prime mover increases due to sudden loss of the alternator load.
- If there is an overvoltage in the alternator terminals and if it persists beyond a specific time, it is not safe for the generator to run under this condition. An *overload relay* (time delayed attracted armature type) picks up if the voltage of the generator terminal exceeds 110 – 130%. The generator main circuit breaker and the generator exciter breaker should be tripped to save the system from this overvoltage problem.

24.1.9. Restricted Earth Fault Protection of Alternator

Normally, an alternator is saved from an internal earth fault by a differential protection (circulating current). However, to safeguard the costly machine with an additional scheme, for large alternators, restricted earth fault protection scheme is applied. This scheme is so called, since it provides protection of only the stator winding against earth fault in the stator, and does not operate in external earth fault.

The schematic arrangement of the scheme is shown in Fig. 24.9, which is self explanatory.

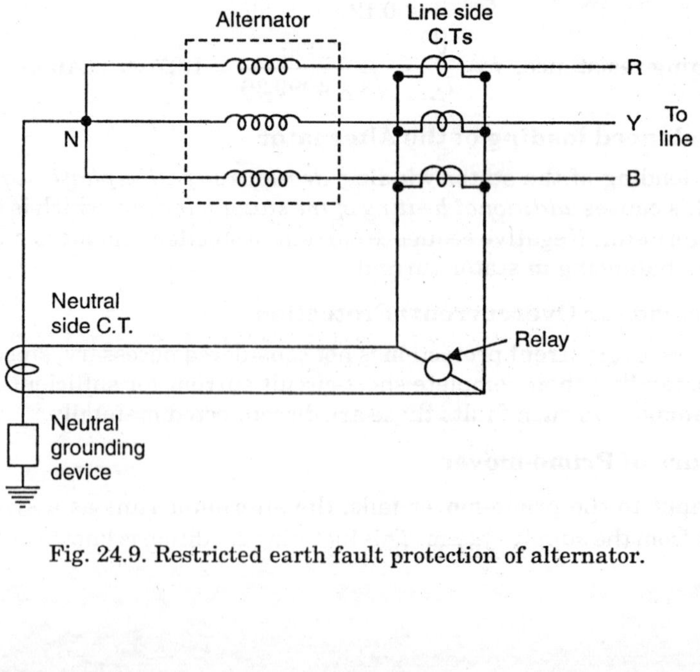

Fig. 24.9. Restricted earth fault protection of alternator.

24.1.10. Stand-by Earth Fault Protection

For providing an effective earth fault protection scheme even for 7% – 10% of the winding of the stator that remains unprotected utilizing restricted earth fault protection, stand-by earth fault relay is the provided at the neutral side of the alternator, as shown in Fig. 24.10.

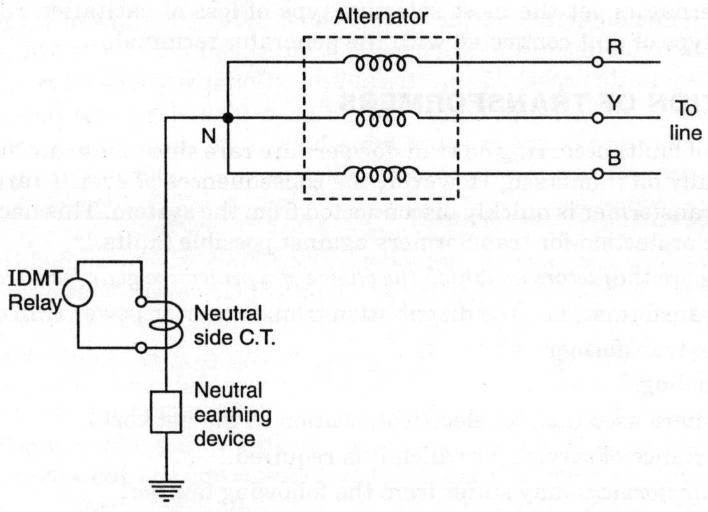

Fig. 24.10. Stand-by earth fault protection scheme.

24.1.11. Rotor Fault Protection

(*i*) **Rotor earth fault protection.** Usually the rotor circuit of an alternator is unearthed and D.C. voltage is imposed on it. The equivalent circuit is shown in Fig. 24.11. The single earth fault does not cause any circulating current to flow through the rotor circuit and hence does not cause any damage to it.

In order to protect the rotor from the second ground fault, either A.C. or D.C. voltages may be impressed between the field circuit and ground through an *overvoltage relay* (Fig. 24.11). A ground contact anywhere in the field circuit, will pick up the relay. An alarm may be operated after a single ground fault in the rotor circuit takes place. However, in cases of large alternators, when the first ground fault occurs, usually the alternator's main and field breakers are tripped immediately.

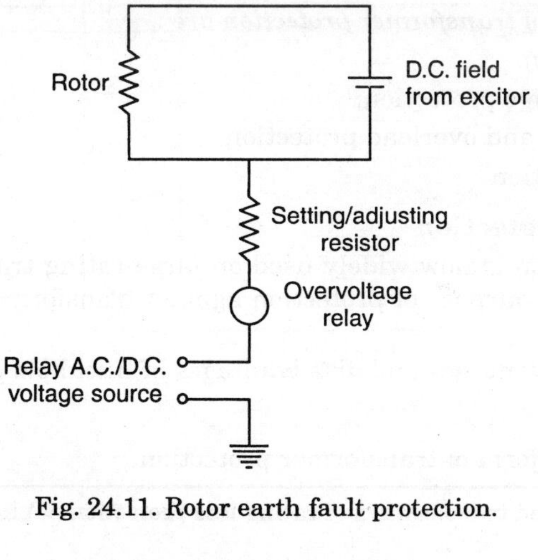

Fig. 24.11. Rotor earth fault protection.

(*ii*) **Loss of excitation relay.** In the event of an alternator losing excitation, it operates as an induction generator, running above the synchronous speed. This operational mode is usually not recommended since it requires large reactive powers to be supplied to the alternator operating as an induction generator. Although undercurrent relays connected in series with the field circuit have been used quite extensively for low and medium MVA rated alternators yet the most selective type of loss of excitation relay is a directional distance type of unit connected with the generator terminals.

24.2. PROTECTION OF TRANSFORMERS

The chances of faults occurring on transformers are rare since these are static devices, totally enclosed and generally oil immersed. However, the consequences of even a rare fault may be very serious unless the transformer is quickly disconnected from the system. This necessitates to provide adequate automatic protection for transformers against possible faults.

The following are the *factors on which the choice of a protective gear for a transformer depends* :

(*i*) Type of transformer, *i.e.*, the distribution transformer or power transformer.

(*ii*) Size of the transformer.

(*iii*) Type of cooling.

(*iv*) System where used (*i.e.*, its electrical location in the network).

(*v*) The importance of service for which it is required.

The *power transformers* may suffer from the following faults :

(*i*) *Open circuits.*

(*ii*) *Overheating.*

(*iii*) *Winding short-circuits* (*internal faults*) *e.g.*, earth faults, phase-to-phase faults and inter-turn faults.

For ***distribution transformers*** employed in rural area, the normal practice is to use the fuses for its protection against external faults but for urban distribution work, where discrimination is absolutely necessary, fuses will not serve the purpose.

For ***power transformers***, the protection is to beprovided usually against dangerous overloads and excessive temperature rise. Dangerous overloads may be due to external or internal faults. External faults, however, are cleared by the relay system outside the transformer within the shortest possible time in order to avoid any danger to the transformer due to these faults. Hence the *protection for internal faults is to be* provided in such transformers.

The various *systems of transformer protection are* :

1. Buchholz protection.
2. Core-balance leakage protection.
3. Combined leakage and overload protection.
4. Differential protection.

24.2.1. Buchholz Protection

This type of protection is now widely used on large rating transformers (having ratings in *excess of 750 kVA) for the purpose of protection against transformer internal faults. For details refer Article 23.6.

Following are the advantages and disadvantages of Buchholz protection :

Advantages :

(*i*) It is the simplest form of transformer protection.

* Its use for oil immersed transformers of rating less than 750 kVA is generally *uneconomical*.

(*ii*) It detects the incipient faults at a stage much earlier than is possible with other forms of protection.

Disadvantages :

(*i*) It can only be used with oil immersed transformers equipped with conservator tanks.

(*ii*) The device can detect only faults below oil level in the transformer, thus separate protection is needed for connecting cables.

24.2.2. Core-balance Leakage Protection

This system is employed to *provide protection against earth faults on high voltage winding.*

It consists of three primary conductors surrounded by the magnet of a current transformer (C.T.). This has a single secondary winding which is connected to the relay.

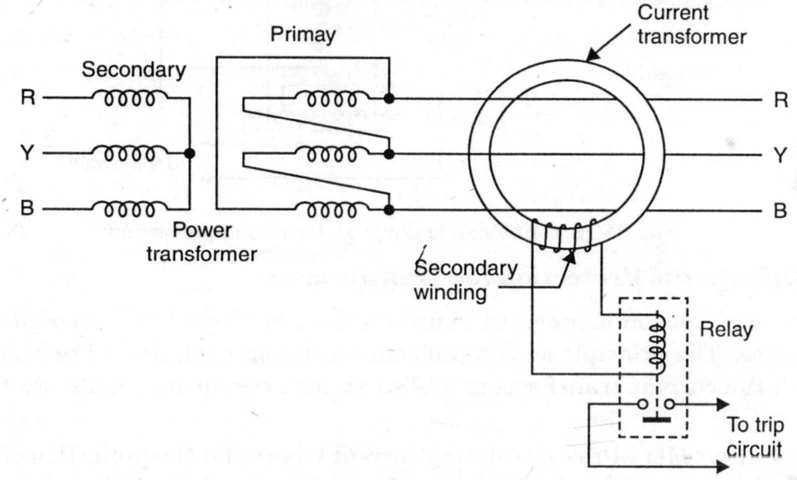

Fig. 24.12. Core-balance leakage protection.

Under *normal operating conditions* (*i.e.*, when there is no fault) the instantaneous sum of the currents in the three phases is always zero, and there is no resultant flux in the core of the C.T. no matter how much the load is out of balance. Consequently, no current flows through the relay and it remains inoperative.

However, when the earth fault occurs, the sum of the three currents is no longer zero and a current is induced in the secondary of the C.T. causing the relay to operate and isolate the transformer from the bus-bars.

24.2.3. Combined Leakage and Overload Protection

The drawback of core-balance leakage protection is that it cannot provide protection against overloads ; if a fault or leakage occurs between phases, the core-balance relay will not operate. This shortcoming is overcome by incorporating three separate current transformers (C.Ts.).

In this system of protection two overload relays and one earth leakage relay are connected as shown in Fig. 24.13. The *two overload relays are sufficient to protect all the three phases, while the leakage relay is energized by the resultant current from all the three C.Ts. in case of leakage fault.* Hence protection against faults and short-circuits either to earth or between phases is achieved.

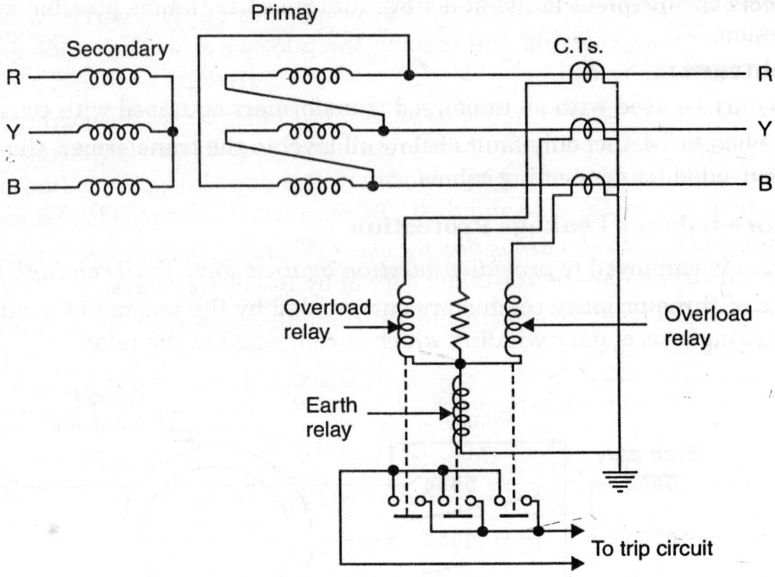

Fig. 24.13. Combined leakage and overload protection.

24.2.4. Differential Protection for Transformers

This type of protection is provided to protect the transformer in case of *phase-to-phase and phase-to-earth fault.* The principle of this protection is based upon *Merz-Price circulating-current scheme* in which the current transformers (C.Ts.) at each output terminals are *balanced against each other.*

Fig. 24.14 shows Merz-Price circulating current scheme for the protection of a 3-phase *delta / delta power* transformer (against phase-to-ground and phase-to-phase faults). Current transformers (C.Ts.) on the two sides of the transformer are connected in star. This compensates for the phase difference between the power transformer primary and secondary. The C.Ts. on the two sides are connected by pilot wires and for each pair of C.Ts. one relay is used.

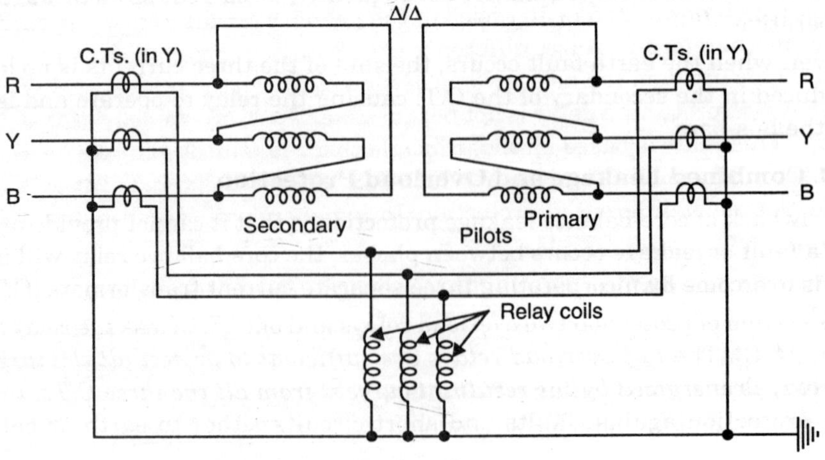

Fig. 24.14. Merz-Price circulating-current scheme.

Under *normal operating conditions*, the secondaries of C.Ts. carry *identical currents*. As a consequence, the currents entering and leaving the pilot wires at both ends are the same and no current flows through the relays.

However, when a *ground or phase-to-phase fault takes place*, the currents in the secondaries of C.Ts. will no longer be the same and differential current flowing through the relay circuit will clear the breaker on both sides of the transformer. The protected zone is limited to the region between C.Ts. on the high-voltage side and the C.Ts. on the low-voltage side of the power transformer.

This scheme *also provides protection for short-circuits between turns on the same phase winding*. However, such circuits are better taken care of by Buchholz relays.

- Protection C.Ts. must be connected in the following order for the different power transformer connections :

S.No.	*Power transformers*		*Protection transformers*	
	Connection of primary	*Connection of secondary*	*Connected on primary side in*	*Connected on secondary side in*
1.	Star with neutral earthed	Delta	Delta	Star
2.	Delta	Star with neutral earthed	Star	Delta
3.	Star	Star with neutral earthed	Delta	Delta
4.	Delta	Delta	Star	Star

- The following *factors affect the operation of differential relay protection :*

1. *Tap changing.* During tap changing voltage ratio differs from original one and the operation of the differential relay is affected. This problem is overcome by using *biasing coil* (*i.e.* percentage differential relay protection).

2. *Magnetising inrush current.* When the transformer is energised it takes current for its magnetisation which flows for 5 to 10 cycles in the primary circuit. Since this current flows only in primary winding, therefore, it causes difference in C.T. output and makes the relay to operate which is undesirable. This problem can be overcome by slowing down the relaying selecting T.M.S. But for E.H.V. transformers protection delay is not permissible hence only remedy is to *use second harmonic restraint relay* (This relay is based on the principle that magnetizing inrush current consists about 60 percent of second harmonic component which can be filtered out and sent to restraining coil and hence malfunctioning of differential protection during energisation of transformer is prevented).

3. *Magnetising characterics.*

Example 24.8. *A 3-phase 220/11000 V star-delta transformer is protected by Merz-Price system. The C.Ts. on the low-voltage side have a ratio of 500/5. Find the ratio of C.Ts. on the high voltage side.*

(Panjab University)

Solution. Transformer rating = 220/11000 V, 3-phase

Ratio of C.Ts. on low-voltage side = 500/5.

Ratio of C.Ts. on the high-voltage side :

For star-delta transformers, C.Ts. will be connected in delta for star-side power transformer (*i.e.*, on 220 V side) and in star for delta side of power transformer (*i.e.*, on 11000 V), as shown in Fig. 24.15. Assume that line current on 220 V side is 500 A.

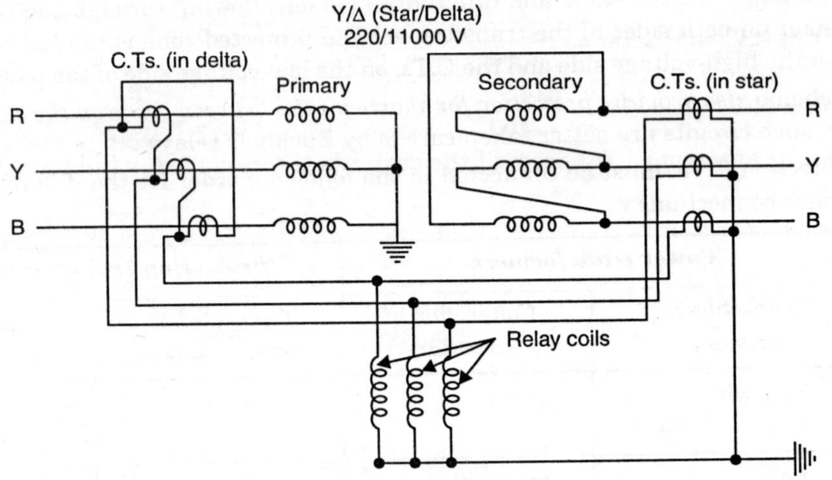

Fig. 24.15

∴ Phase current of delta-connected C.Ts. on 220 V side = 5 A

Line current of delta-connected C.Ts. on 220 V side = $\sqrt{3} \times 5 = 5\sqrt{3}$ A. This current of $5\sqrt{3}$ A will flow through the pilot wires. Obviously this will be the current which flows through the secondary of C.Ts. on the 11000 V side.

∴ Phase current of star-connected C.Ts. on 11000 V side = $5\sqrt{3}$ A

If *I* is the *line current* on 11000 V side, then,

Primary apparent power = Secondary apparent power

or, $$\sqrt{3} \times 220 \times 500 = \sqrt{3} \times 11000 \times I$$

or, $$I = \frac{\sqrt{3} \times 220 \times 500}{\sqrt{3} \times 11000} = 10 \text{ A}$$

∴ Turn-ratio of C.Ts. on 11000 V side

$$= 10 : 5\sqrt{3} = \mathbf{1.155 : 1.} \quad \textbf{(Ans.)}$$

Example 24.9. *A 120 MVA, delta/star connected, 11/220 kV transformer is to be protected by percentage differential scheme. C.Ts. used are of 5000/5 and 400/1 respectively. Draw the sketch of complete scheme.*

Solution. Rating of transformer = 120 MVA ; delta/star connected ; 11/220 kV

Ratio of C.Ts. on L.V. side = 5000/5

Ratio of C.Ts. on H.V. side = 400/1

Rated current for star (220 kV) side

$$= \frac{120 \times 10^6}{\sqrt{3} \times 220 \times 10^3} = 314.9 \text{ A}$$

Rated current for delta (11 kV) side

$$= \text{Transformation ratic} \times \text{star-side current}$$

$$= \frac{220}{11} \times 314.9 = 6298 \text{ A}$$

Current in L.V. side C.Ts. $= \dfrac{6298 \times 5}{5000} = 6.298$ A

Current in H.V. side C.Ts. $= \dfrac{314.9 \times 1}{400} = 0.787$ A

Relays are rated upto 1 A hence all the currents exceeding 1 A will be reduced further by using auxiliary C.Ts.

For normal operation of transformer current output of C.Ts. should be balanced (in phase as well as in magnitude) so that relay remains inoperative.

Take 1 A current for the rated current in transformer.

- Auxiliary C.Ts. used on L.V. side should be of ratio 6.298/1.
- On H.V. side we compensate phase difference because of star/delta connection of transformer through using auxiliary C.Ts. in reverse manner *i.e.*, delta/star.

If C.Ts. used are of ratio $x/1$ then we will get $\sqrt{3}$A output for x amperes in primary winding because of delta-connected secondary winding of C.Ts. Hence to get 1 A output for 0.787 A input to C.Ts. ratio required is $\sqrt{3} \times 0.787 = 1.363$.

The arrangement of the complete scheme is shown in Fig. 24.16.

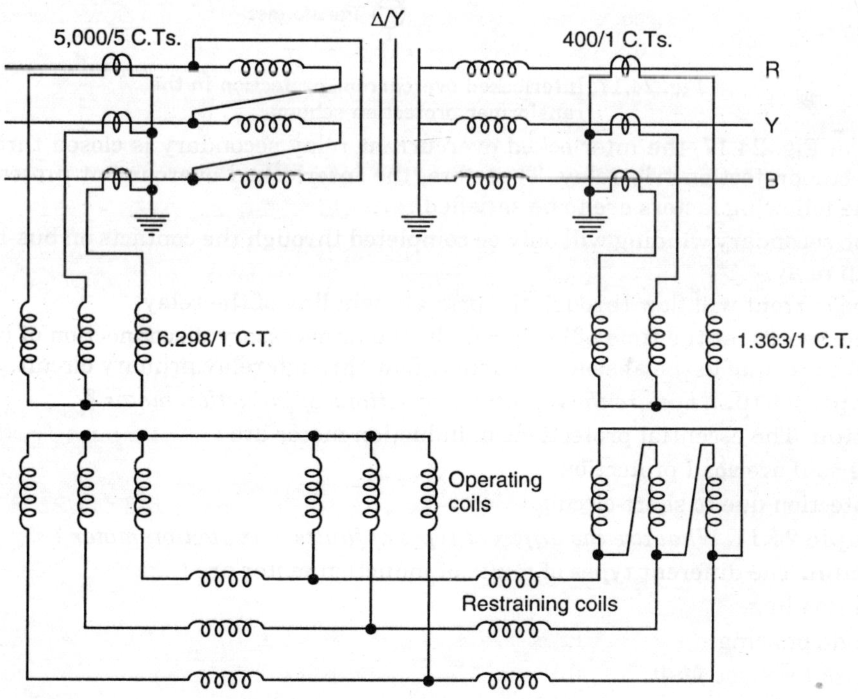

Fig. 24.16

24.2.5. Interlocked Overcurrent Protection in the Transformer Protection Scheme

The interlocked overcurrent protection in a transformer protection scheme is nothing but *a protection for discrimination between a bus-bar fault and a fault between current transformer and circuit breaker.* When protective current transformer is utilised on one side of the circuit breaker from the point of view of economy, some arrangement should be made so that if fault occurs between current transformer and breaker (as shown in Fig. 24.17), both sides of the transformers will be tripped. One arrangement may be made such that the protective device of bus-bar will trip both sides of the transformer. In that case for only bus fault the whole healthy system is to be disconnected. That is why interconnected overcurrent protection is employed for discrimination between bus-bar fault and fault between current transformer and circuit breaker.

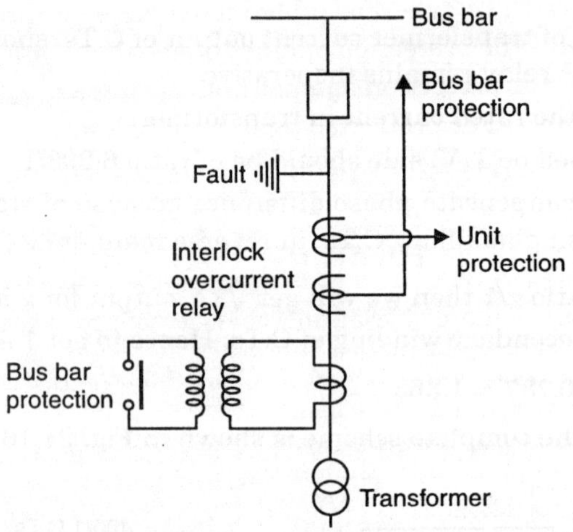

Fig. 24.17. Interlocked overcurrent protection in the transformer protection scheme.

Here, in Fig. 24.17, the interlocked overcurrent relay secondary is closed through the contacts on bus-bar protection trip relay. Therefore, the interlocked overcurrent protection will trip only when the following factors are to be satisfied :

(*i*) The secondary winding will only be completed through the contacts on bus-bar protection trip relay.

(*ii*) The current will flow through the primary winding of the relay.

In other words, at the time of busbar fault, the unnecessary disconnection of healthy equipment will not occur due to the absence of current flow through relay primary circuit.

Example 24.10. *What are the essential protections of induction motor ?*

Solution. The essential protections of induction motor are :

1. Thermal overload protection.
2. Protection due to short-circuit.

Example 24.11. *What are the different types of faults of induction motor ?*

Solution. The different types of faults of induction motor are :

1. Overloading.
2. Single phasing.
3. Phase-to-phase fault.
4. Rotor fault.
5. Stator earth fault.

6. Bearing failure.
7. Unbalanced supply voltage.
8. Undervoltage supply.
9. Stalling.

Example 24.12. *What are the different types of relays used for induction motor ?*

Solution. The different types of relays used for induction motor are as follows :

1. *Thermal overload relays*—For protection against overload, stalling and single phasing.
2. *Inverse overcurrent relays*—For protection against overloads.
3. *High set instantaneous overcurrent relay*—For protection against phase faults and earth faults.
4. *Instantaneous overcurrent relays*—For protection against stalling and rotor fault.
5. *Differential protection*—For protection against phase faults and earth faults.
6. *Undervoltage relays*—For protection against undervoltage.
7. *Negative phase sequence*—For protection against unbalanced voltage supply.
8. *Phase reversal relay*—For protection against reverse phase sequence.

HIGHLIGHTS

1. *Types of faults in alternators.* Failure of prime mover ; Overvoltage at the terminals of the alternator ; Overloading ; Unbalanced loading ; Insulation failure of stator winding ; Inter-turn faults ; Bearing oil pressure failure ; Current leakage in the body of the alternator ; Ventilation failure ; Excitation failure ; Rotor earth fault ; Overcurrent in the rotor circuit.
2. *The various systems of transformer protection are :*
   (*i*) Buchholz protection.
   (*ii*) Core-balance leakage protection.
   (*iii*) Combined leakage and overload protection.
   (*iv*) Differential protection.

THEORETICAL QUESTIONS

1. Discuss the different faults that can occur in a 3-phase alternator.
2. What are the abnormal conditions in a large alternator against which protection is necessary ? Discuss them briefly.
3. Discuss protection schemes for a large generator against the following faults :
   (*i*) Overloading (*ii*) Unbalanced loading.
4. Explain with a neat diagram the application of Merz-Price circulating current principle for the protection of alternator.
5. How is an alternator protected from turn-to-turn fault on the same phase winding ?
6. Enumerate the various types of protections used in an alternator. Describe one method for the protection of stator of an alternator against faults.
7. Explain the principle of operation of differential scheme. How do you apply it in the protection of alternators ?
8. Explain with the help of a line diagram the connections and functioning of differential relay for generator protection.
   What are the other protection schemes generally used for the protection of generators ?
9. Explain biased differential protection for all alternators.
10. What are the common types of generator faults ? How is the generator stator protected against an in-turn fault ?
11. Enumerate main features of a good protective system. Explain the working of Merz-Price protection scheme for a 3-phase alternator. Draw a neat sketch in support of your answer.

12. Discuss the different transformer faults. What are the various protection schemes available for transformers ?
13. What are the various protections usually recommended for power transformers ? In what respect 'core-balance leakage protection' is inferior to 'combined leakage and overload protection' for power transformers.
14. Describe the construction and working of Buchholz relay.
15. Describe the Merz-Price circulating-current system for protection of transformers.
16. Describe with the help of neat diagram the Merz-Price protection of *Y/Y* transformer.
17. Discuss the faults in transformer and describe with sketch Merz-Price scheme of transformer protection.
18. Explain the principle of Merz-Price system of protection used for power transformers. What are the limitations of this scheme and how are they overcome ?
19. Explain the basic principle of the differential scheme of protection and state its advantages. Sketch the connection diagram of a scheme of differential protection for a 3-phase delta/star transformer and explain its operation.
20. Explain interlocked overcurrent protection in the transformer protection scheme.
21. Write short notes on the following :
    (*i*) Protection of alternators
    (*ii*) Overload protection of alternator
    (*iii*) Overspeed protection of alternators
    (*iv*) Inter-turn fault protection of alternators
    (*v*) Buchholz protection
    (*vi*) Merz-Price protection for power transformers
    (*vii*) Restricted earth fault protection for power transformers
    (*viii*) Loss of excitation protection of alternators.

UNSOLVED EXAMPLES

1. A star-connected, 3-phase, 10 MVA, 6.6 kV alternator is protected by Merz-Price circulating-current principle using 1000/5 amperes current transformers. The star point of the alternator is earthed through a resistance of 7.5 Ω. If the minimum operating current for the relay is 0.5 A, calculate the percentage of each phase of the stator winding which is unprotected against earth-faults when the machine is operating at normal voltage. **[Ans.** 19.69%]
2. A star-connected, 3-phase, 10 MVA, 6.6 kV alternator has a per phase reactance of 10%. It is protected by Merz-Price circulating-current principle which is set to operate for fault currents not less than 175 A. Calculate the value of earthing resistance to be provided in order to ensure that only 10% of the alternator winding remains unprotected. **[Ans.** 2.17 Ω]
3. A 13.8 kV, 125 MVA, star-connected alternator has a synchronous reactance of 1.4 per unit per phase and a negligible resistance. It is protected by a Merz-Price balanced current system which operates when out-of-balance current exceeds 10% of the full-load current. If the neutral point is earthed through a resistance of 2Ω, determine what proportion of winding is protected against earth fault. **[Ans.** 86.74%]
4. A star-connected, 3-phase, 10 MVA, 6.6 kV alternator is protected by circulating current protection, the star point being earthed *via* a resistance '*r*'. Calculate the value of earthing resistor if 85% of the stator winding is protected against earth faults. Assume an earth fault setting of 20%. Neglect the impedance of the alternator winding. **[Ans.** 3.27 Ω]
5. A 11 kV, 3-phase turbo-alternator has a maximum rating of 100 MW at 0.8 p.f. and its reactance is 0.1 p.u. It is equipped with differential current protection scheme. It is set to operate at fault current not less than 500 amps. Determine the magnitude of the neutral earthing resistance that leaves the 10 per cent of the windings unprotected. **[Ans.** 1.27 Ω]
6. A 3-phase, 66/11 kV star-delta connected transformer is protected by Merz-Price system. The C.Ts. on low-voltage side have a ratio of 420/5 A. Find the ratio of C.Ts. on the high-voltage side. **[Ans.** 24.25 : 1]

25

Protection of Bus-bars And Lines/Feeders

25.1. Introduction. 25.2. **Bus-bar protection**—Differential protection—Fault bus protection. 25.3. **Protection of lines/feeders**—Requirements of line protection—Protection schemes for line/feeder protection—Time-graded overcurrent protection—Differential pilot-wire protection—Distance protection. Highlights—Theoretical Questions.

25.1. INTRODUCTION

In an electric power system, bus-bars and lines are important elements which need to be protected against the possible faults occurring on them. The methods employed for the protection of alternators and transformers can also be employed, with slight modifications, for the bus-bars and lines. The modifications are required to cope with the protection problems arising out of greater length of lines and a large number of circuits connected to a bus-bar. Although differential protection can be used, but it becomes too expensive for longer lines due to the greater length of pilot wires required. However, less expensive methods are available which provide effective protection to the bus-bars and lines.

25.2. BUS-BAR PROTECTION

It is very important to protect the bus-bars since they form a vital part of the supply system and if a fault develops in this part considerable damage and disruption of supply will occur. For the purpose of protection, the bus-bar area includes the bus-bars the circuit breakers and isolating switches. When the fault occurs within the bus-bar area of any particular section of a bus-bar all circuit equipments connected to that section must be tripped out to give complete isolation.

Although bus faults are rare due to very high standard of construction of bus-bars, yet the possibility of damage and service interruption from even a rare bus fault is so great that more attention is now given to the bus-bar protection.

- In a power system the *bus fault occurs due to the following reasons :*

(*i*) Supporting insulation failure.

(*ii*) At time of overvoltage, flashover across support insulator.

(*iii*) Flashover due to heavily polluted insulator.

(*iv*) Failure of equipment connected.

(*v*) Mechanical damage.

- At *the time of bus protection the following factors should be considered :*

(*i*) The protection should be stable enough so that no operation can occur at external fault.

(*ii*) The protection should be made at high speed.

(*iii*) There should not be any mal-operation due to current transformer saturation or power swings.

(*iv*) Trip circuit of each circuit breaker should be separately controlled.

Schemes of bus-bar protection :

The various schemes of bus-bar protection are :

1. Differential protection (Bus protective by differential relays)
2. Fault bus protection.
3. Bus protection by overcurrent relays.
4. Static protection.
5. Overcurrent protection or distance protection as back up protection of bus-bar.

25.2.1. Differential Protection

This system of protection of the bus-bars is extensively used in modern power station or substations.

In this system the currents entering and leaving the bus are totalised. Under normal operating conditions, the sum of these currents is zero. However, on occurence of the fault, the fault current upsets the balance and produces a differential current which operates a relay.

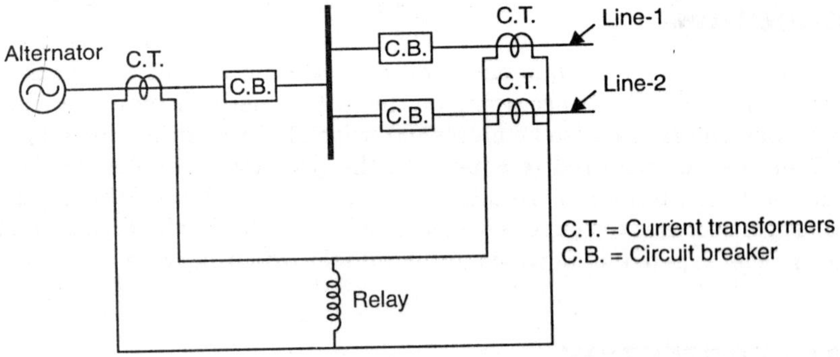

Fig. 25.1. Differential protection–Station bus-bar.

Fig. 25.1 shows the schematic arrangement of differential (circulating current) protection to a station bus-bar. The secondaries of C.Ts. in the alternator load in line-1 and line-2 are connected in parallel. The protective relay is connected across this parallel connection. Regardless of the capacities of the various circuits, *all C.Ts. must be of the same ratio* in the scheme.

Under *normal load conditions* (or external fault conditions), the sum of the currents entering the bus is *equal* to those leaving it and no current flows through the relay and it remains inoperative.

However, when fault occurs within the protected zone, the currents entering the bus will no longer be equal to those leaving it. The differential current (difference of entering and outgoing currents) will flow through the relay and cause opening of generator circuit breaker and each of the line circuit breakers.

- Circulating current protection to bus-bar can be *adopted to discriminate operation on duplicate bus-bar*. In this case *auxiliary switches* are mounted on the bus-bar selector switches of each equipment and so connected that the current transformer residual circuit of the equipments connected to the same set of bus-bars are parallel to the bus-bar protective relay controlling that particular zone.

25.2.2. Fault Bus Protection

The design of such a station in which the faults that develop are mostly earth faults is feasible by *providing earthed metal barrier* (known as *fault bus*) surrounding each conductor throughout its length in the bus structure. In such an arrangement every fault that might occur *must involve a connection between a conductor and an earthed metal. The faults can be detected and located by directing the flow of earth-fault current. Such a scheme of protection is called* **fault bus protection.**

The schematic arrangement of fault bus protection is shown in Fig. 25.2. The metal supporting structure or fault bus is earthed through a current transformer (C.T.). Across the secondary of this C.T. is connected a relay.

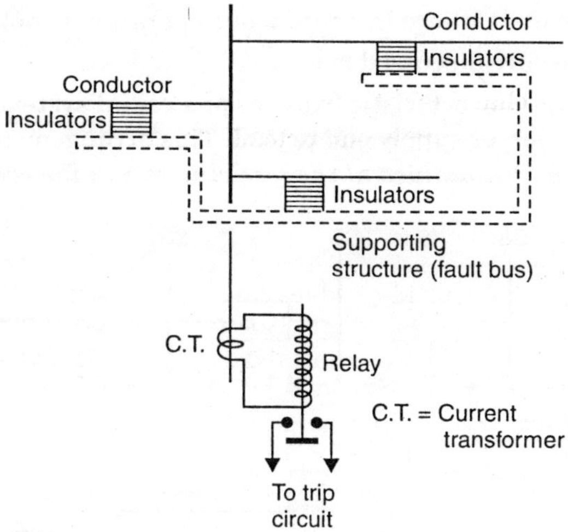

Fig. 25.2. Fault bus protection.

Under normal operating conditions (*i.e.*, when there is no fault), no current flows from fault bus to ground and the relay does not operate.

However, when a fault involving a connection between a conductor and earthed supporting structure occurs, fault current flows to ground through the fault bus. Consequently the relay operates and trips all circuit breakers connecting equipment to the bus.

25.3. PROTECTION OF LINES/FEEDERS

The word **'Feeder'** *here means the connecting link between two circuits.* The feeder could be in the form of transmission line, short, medium or long, or this could be a distribution circuit. The transmission lines are much more prone to faults, owing to their long lengths and exposure to atmospheric conditions. As such several protective schemes have been developed (which have no application to the comparatively simple cases of alternators and transformers).

25.3.1. Requirements of Line Protection

The following are the *requirements of line protection* :

1. On the occurrence of a short-circuit, the circuits breaker closest to the fault should open, all other circuit breakers remaining in closed position.

2. If the nearest breaker to the fault fails to open, back-up protection should be provided by the adjacent circuit breakers.

3. Line circuit breakers should not trip due to loss of synchronism or hunting of generators.

4. In order to preserve system stability (without unnecessary tripping of circuits) the relay operating time should be just as short as possible.

25.3.2. Protection Schemes for Line/Feeder Protection

The *most common methods of line/feeder protection are* :

1. Time-graded overcurrent protection.
2. Differential protection.
3. Distance protection.

25.3.2.1. Time-graded overcurrent protection

The conditions required for time-graded overcurrent protection are :

(*i*) This protection is to be used where time lag is permitted.

(*ii*) This protection is used where instantaneous operation is not necessary.

A few important cases are discussed below.

1. Radial feeder. The characteristic features of a radial system is that power can flow only in one direction, from generator, or supply end to load. This arrangement has the disadvantage that *continuity of supply cannot be maintained at the receiving end in the event of occurrence of fault.*

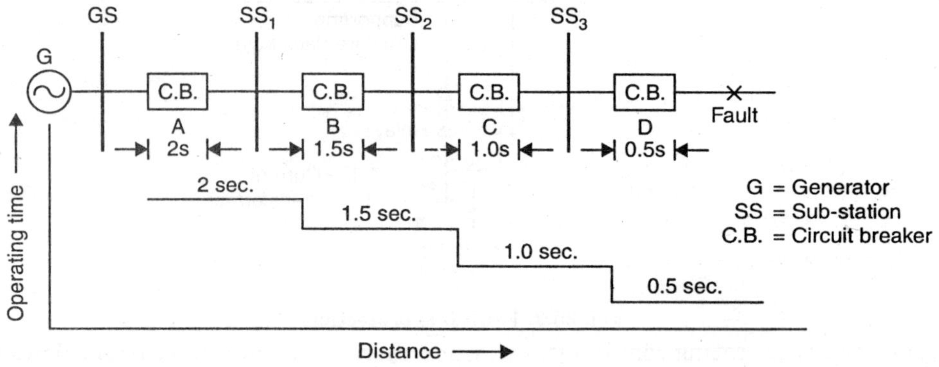

Fig. 25.3. Time-graded protection for radial feeders (by definite time relays).

In a radial system when number of feeders are connected in series (Fig. 25.3) it is desired that the smallest possible part of the system should be off. This can be conveniently achieved by time-graded protection.

Fig. 25.3 shows the overcurrent protection of a radial feeder by ***definite time relays.*** In this protection system the time setting of overcurrent relays is adjusted in such a way that *farther* the relay from the generating station the *lesser the time* of operation. The time of operation of each relay is fixed and is independent of the operating current. Thus the relay at *D* has an operating time of 0.5 second while for other relays, time delay (usually varies from second 0.25 to 0.5 second) is successively increased by 0.5 second.

When fault occurs on the far side SS_3, it will be cleared in 0.5 second by the relay and circuit breaker at *D* because all other relays have higher operating time. In this way only section of the feeder/line beyond SS_3 will be isolated. If the relay at *D* fails to trip, the relay at *C* will operate after a time delay of 0.5 second *i.e.* after 1 second from the occurence of fault.

The *demerit* of this system is that if there are a number of feeders in series, the *tripping time* for faults near the supply end becomes *high* (2 seconds in this case). However, in most cases, it is necessary to limit the maximum tripping time to 2 seconds. This disadvantage can be overcome to a reasonable extent by using inverse-time relays.

Fig. 25.4 shows overcurrent protection of a radial feeder using ***inverse-time relays*** in which *operating time is inversely proportional to the operating current.*

The relay-time settings are adjusted so that the first circuit breaker to trip in the event of a fault, is the one nearest the fault. The selectivity between substations is obtained by the time interval '*S*'.

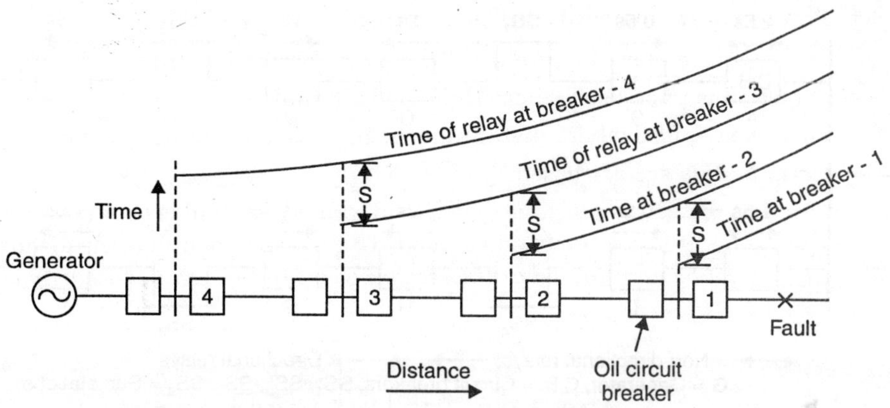

Fig. 25.4. Overcurrent protection of a radial feeder using inverse-time relays.

2. Parallel feeders. Fig. 25.5 shows the two feeders (fed from one end) connected in parallel to increase the transmission capacity. It is required that in case of a fault on any one of the feeders, that feeder should be isolated from the supply and the load must receive power through the other healthy feeder.

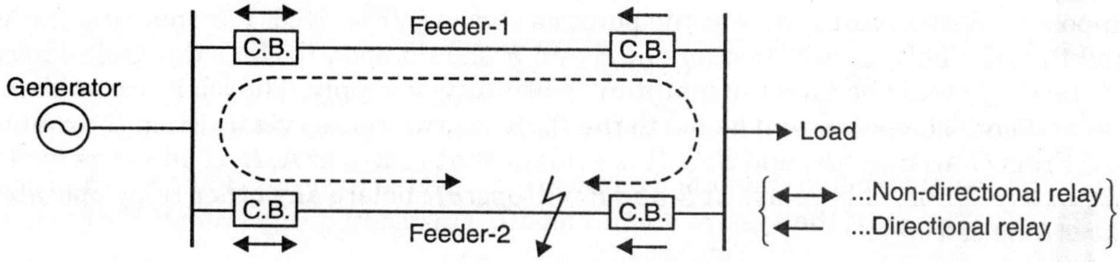

Fig. 25.5. Protection of parallel feeders.

When a fault occurs on any one feeder, say feeder-2, the current will be fed to the fault as shown in Fig. 25.5 by the arrows. It is to be seen that as far as the relays near the source are concerned, the direction of the current is same as the normal direction of current, whereas the direction of the current in the relay near the load end of the faulty feeder is reversed. Therefore, for proper coordination the *relays near the source-end are 'non-directional' relays whereas relays near the load-end are 'directional' relays.* The direction of the current for which the directional relays will operate is indicated by the corresponding arrow heads.

In this case as soon as fault takes place in feeder-2, the directional relay in feeder-1 corresponds to load current and after some time the non-directional relay in feeder-2 will operate, thereby isolating feeder-2 from the source.

3. Ring main system. In this protection system, various power stations or substations are interconnected by *alternate routes*, thus *forming a closed ring*. In the event of damage to any section of the ring, that section may be disconnected for repairs, and the power will be supplied from both ends of the ring, thereby keeping the supply intact.

A typical ring main system consisting of one generator (G) and four substations (SS_1, SS_2, SS_3, SS_4) is shown in Fig. 25.6. Since the power can flow in both directions in this arrangement, therefore, it is necessary to grade in both directions round the ring and also to use directional relays. In order to isolate only faulty section of the ring under faulty conditions, the types of relays and their time settings should be as under :

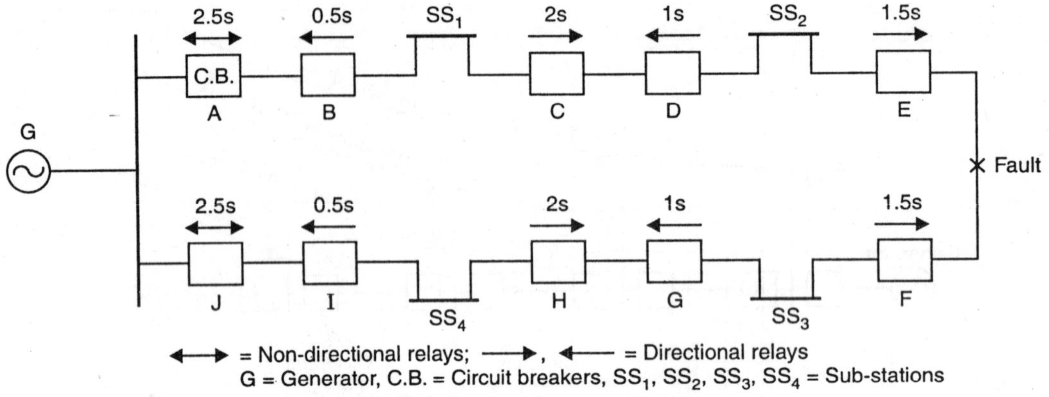

Fig. 25.6. Ring main protection system.

(*i*) The two lines leaving the generating station should be equipped with non-directional overcurrent relays (relays at *A* and *J* in this) ; (*ii*) At each substation, reverse power or directional relays should be placed in both incoming and outgoing lines (relays at *B, C, D, E, F, G, H* and *I* in this case) ; (*iii*) There should be relative proper time-setting of the relays.

Suppose a short-circuit occurs at the point as shown in Fig. 25.6. For ensuring the selectivity it is desired that the only circuit breakers at *E* and *F* should open to clear the fault whereas other sections of the ring should be intact to maintain continuity of supply. The job is accomplished by the above arrangement. The power will be fed to the fault via two routes viz (*i*) From *G* around SS_1 and SS_2 and (*ii*) From *G* around SS_4 and SS_3. It is evident that relays at *A, B, C* and *D* as well as *J, I, H* and *G* *will not trip*. Thus, only relays at *E* and *F* will *operate* before any other relay operates *because of their lower time-setting*.

25.3.2.2. Differential pilot-wire protection

This protection scheme is based on the principle that under normal conditions, the current entering one end of a line is equal to that leaving the other end. On the occurence of fault between the two ends, this condition no longer holds and the difference of incoming and outgoing current is arranged to flow through a relay which operates the circuit breaker and the faulty line is isolated. Out of the several differential protection systems the following two systems will be discussed :

1. Merz-Price system
2. Translay system.

1. Merz-Price system. Probably the best known of the differential systems is Merz-price system, which, as applied to feeder/line protection, *utilizes the principle of "voltage balance"*.

Fig. 25.7 shows Merz-Price protection system for a 3-phase system. Here, each conductor has its own pair of current transformers (C.Ts.) and relays. The secondaries of C.Ts. are connected in series by means of pilot wires.

Under *normal operating conditions* (*i.e.*, when there is no fault on the feeder), equal current flows at the two different ends, so induced voltages in the secondaries of C.Ts. are equal. As the secondaries are connected in opposition their secondary e.m.fs. are equalised resulting into no flow of circulating current in the relays.

However, when *fault occurs*, currents differ at two ends and consequently induced e.m.fs. in the secondaries of C.Ts. also differ. A circulating current then flows through the pilot wires and relays and the feeder is isolated.

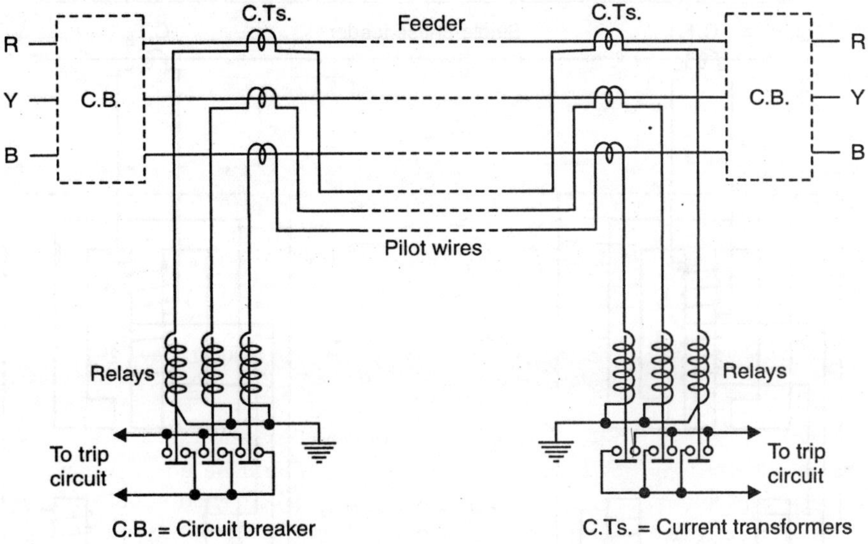

Fig. 25.7. Merz-Price protection system.

Advantages :

(*i*) This system can be employed for protection of both ring mains as well as feeders.

(*ii*) This system is independent of operating voltage and fault power factor.

(*iii*) This system provides instantaneous protection for ground faults, due to which the possibility of these faults involving other phases is reduced.

(*iv*) This system provides instantaneous relaying which reduces the amount of damage to overhead conductors resulting from arcing faults.

Disadvantages :

(*i*) A break in the pilot-wire makes the system inoperative.

(*ii*) It is essential to have accurate matching of current transformers.

(*iii*) In long transmission lines, charging current due to pilot-wire capacitance effects may be sufficient to cause relay operation even under normal working conditions.

(*iv*) Owing the constructional difficulties in matching the C.Ts., this system cannot be used for line voltages beyond 33 kV.

2. Translay system. The name *"Translay"* relates to the fact that the relay embodies a transformer feature.

- This system is similar to voltage balance system except that here *balance or opposition is between the voltages induced in the secondary windings wound on the relay magnets and not between the secondary voltages of the line current transformers.*

This system can be used for protection of single or 3-phase feeders, transformer feeders, feeders with a tee-off and parallel feeders against both earth and phase faults.

Fig. 25.8 shows a simple form of Translay protection for a ***single-phase feeder.*** Under healthy conditions, current transformers $C.T_1$ and $C.T_2$ carry equal currents and the coils 1 and 1′, induce equal e.m.fs. in the windings 2 and 2′ respectively. These latter windings are in opposition via the pilot wires with the operating windings 3 and 3′, in series with them. Thus no forward torque is exerted on the disc. When a fault occurs, the current through one C.T. is greater than that through the other. A small current circulates through the operating windings and pilot-wires, and when it reaches the set value it causes the relay to close the tripping circuit and to isolate the feeder.

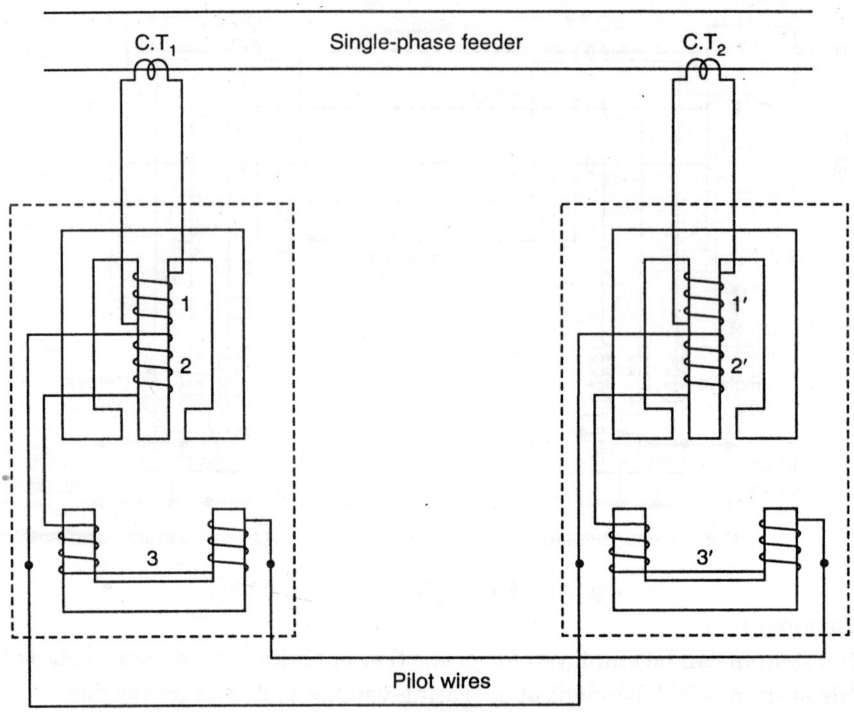

Fig. 25.8. Translay protection to a single-phase feeder.

The translay relay employed is quite similar in construction to an overcurrent induction type relay.

The Translay system employed for a ***3-phase circuit*** has a single element relay at each end of the feeder which protects against faults between phases and faults to earth. Each relay has two electromagnetic elements. The upper element carries a winding (A or A') which is energised as a summation transformer from the secondaries of the line C.Ts. connected in the phases of the line to be protected. The upper element also carries a secondary winding (4 or 4′) which is connected in series with the operating winding (5 or 5′) on the lower magnet. The secondary windings 4, 4′ and operating windings 5, 5′ are connected in series in such a way that voltages induced in them oppose each other (The discs and tripping circuits are not shown in the figure for the sake of clarity in the diagram).

Under *normal conditions* (*i.e.*, when the feeder is healthy) the currents at the two ends are equal so that the secondary currents in both sets of C.Ts. are equal. Consequently, the currents flowing in the relay primary winding A and A' will be equal and they will induce equal voltages in the secondary windings 4 and 4′. Since these windings are connected in opposition, no current flows in them and in the operating windings 5 and 5′.

On the *occurence of a fault* on the protected line, the line current at one end carries a greater current than that at the other end. As a result, the voltages induced in the secondary windings 4 and 4′ will be different and the current will flow through the operating coils 5 and 5′ and the pilot circuit. Under these conditions, both upper and lower elements of each relay are energised and a forward torque acts on the each relay disc. As a consequence, the relays will operate and open the circuit breakers at both ends of the line.

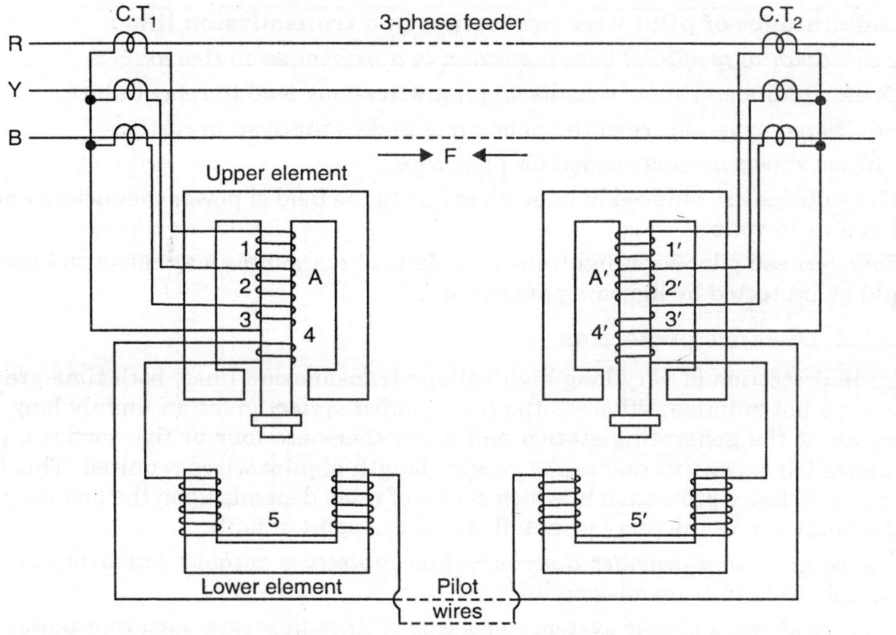

Fig. 25.9. Translay protection to a 3-phase feeder.

- Assume that a fault '*F*' occurs between phases *R* and *Y* and is fed from both sides as shown in Fig. 25.9. This will energise only section 1 of primary winding *A* and *A′* and induce voltages in the secondary windings 4 and 4′. As these windings are now *additives* (since the fault is being fed from both sides), therefore, current will circulate through the operating coils 5 and 5′ and the pilot circuit. This will cause the relay contacts to close and open the circuit breakers at both ends.
  - — When a fault occurs between phases *Y* and *B* section 2 of the primary windings *A* and *A′* gets energised.
  - — When a fault takes place between phase *R* and *B* sections 1 and 2 of primary winding are energised.
- Now assume that fault takes place on phase *R*. This will energise sections 1, 2 and 3 of the primary windings *A* and *A′*. Again if fault is fed from both ends, the voltages induced in the secondary windings 4 and 4′ are additive and cause a current to flow through the operating coils 5 and 5′. The relays, therefore, open the circuit breakers at both ends of the line.
  - — On the occurence of the fault on phase *Y*, sections 2 and 3 of primary winding *A* and *A′* will be energised and cause the relay to operate
  - — When an earth fault occurs on phase *B*, only section 3 of primary winding *A* and *A′* will be energised.

This system claims the following ***advantages*** :

1. The current transformers of normal design (*i.e.*, with air gap) can be used.

2. The system is economical as only two pilot wires are required for protection of a 3-phase line.

3. The capacitive currents do not affect the operation of the relays.

4. The pilot resistance does not affect the operation as major part of the energy required to operate the relay is obtained from current transformation.

Disadvantages of pilot wire protection in a transmission line :

The disadvantages of pilot wire protection in a transmission line are :

1. Open circuits and short circuits on pilot wires may lead to relay failure.
2. Additional expenditure of the pilot wires makes the system costly.
3. Always supervision is needed for pilot wire.
4. The voltages are induced in pilot wires due to the field of power conductors and those are to be limited to 5 to 15 volts.
5. For overhead pilot wires lightning and high voltage surges may cause problem. That is why those should be protected by lightning arresters.

25.3.2.3. Distance protection

For the protection of very long high voltage transmission lines, both time-graded and pilot wire system are not suitable. Whereas the *time-graded system* gives an unduly long time delay in fault clearance at the generating station end when there are four or five sections, the *pilot-wire system* becomes too expensive due to the greater length of pilot wires required. This has led to the development of distance protection in which action of relay depends upon the distance or impedance between the point where the relay is installed and the point of fault.

Distance protection provides discrimination protection without employing pilot wires. This system is widely used in transmission lines.

Fig. 25.10 shows a simple system consisting of lines in series, such that power can flow only from left to right. If a short-circuit occurs at F between sub-stations SS_2 and SS_3, the fault loop impedances at Power-Station (PS) and substation SS_1 and SS_2 and $Z_1 + Z_2 + Z$, $Z_2 + Z$, and Z respectively. The relays at A, B, C etc. are set to operate with impedances less than Z_1, Z_2 and Z_3 respectively, so that under the condition shown only relay C will operate. Similarly if the fault occurs between sub-stations SS_1 and SS_2, only relay B will operate.

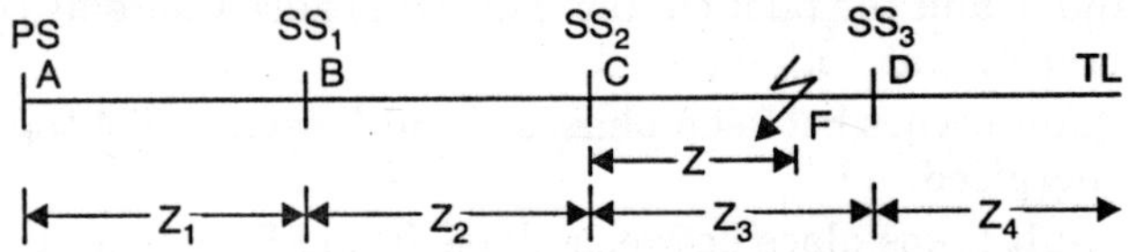

PS = Power station, SS_1, SS_2, SS_3 = Sub-stations ;
TL = Transmission line, Z_1, Z_2, Z_3 = Impedances

Fig. 25.10. Distance protection of line.

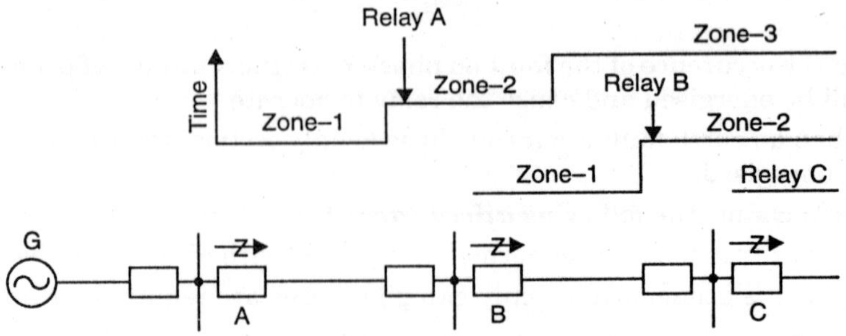

Fig. 25.11. 'Three-zone' distance protection scheme.

In actual practice, it is not possible to obtain instantaneous protection for complete length of the line due to inaccuracies in the relay elements and instrument transformers. Thus the relay *A* (Fig. 25.10) would not be very reliable in distinguishing between a fault at 99 per cent of the distance *AB* and the one at 101 percent of distance *AB*. This problem is overcome by using '***three-zone***' *distance protection* shown in Fig. 25.11. In this system of protection, three distance elements are used at each terminal.

— The *'Zone–1' element* covers first 90 percent of the line and is arranged to trip simultaneously for faults in this portion.

— The *'Zone–2' element* trips for faults in the remaining 10 per cent of the line and for faults in the next line section, but a time delay in introduced to prevent the line from being tripped if the fault is in the next section.

— The *'Zone–3' element* provides back-up protection in the event a fault in the next section is not cleared by its circuit breaker.

- Distance protection schemes are non-unit type of protection systems *offering considerable economical and technical advantages.* These can be applied in a simple form and is of the high speed class. These can provide primary and back-up facilities in a single scheme.
- The distance relay is termed a distance one because it protects the certain length of line of transmission line by comparing the measured impedance value of the line with the set value at the relay.
- The *factors to be considered for choosing a distance protection scheme* are :
  (*i*) The number of stations which are in series.
  (*ii*) The distance between points where relays are existing.
  (*iii*) The configuration of the system.
  (*iv*) The other protections which will exist in the line.
  (*v*) Whether there is necessity of any directional feature.
  (*vi*) Whether high speed auto reclosure is provided.
  (*vii*) Consideration of stability of the system.
- The *major items of distance protection scheme are :*
  (*i*) Starting element (*ii*) Measuring element
  (*iii*) Zone timer (*iv*) Tripping relay.

The "*starting element*" of distance relay for protection of transmission line is a *device for detecting the fault and switching the measuring element.*

The different types of starting elements are :

(*a*) *Overcurrent starters*—Generally utilised in most of the cases.

(*b*) *Undervoltage starters*—Generally utilised for resistance earthed system.

(*c*) *Impedance starters*—Generally utilised to detect accurately the faulty phases in case of phase fault. This type of starter is preferred for extra high voltage system.

The *functions of starting element are* :

(*i*) When a single measuring element is used, the function of the starting element is to switch the measuring element to correct input quantity.

(*ii*) When single phase auto-reclosure is used, the function of the starting element is to choose correct phase for providing tripping instructions.

(*iii*) The function of the starting element is also to change distance steps or reverse the direction of measuring element after a certain time lag.

(*iv*) The starting element provides non-directional back up to measuring elements.

(*v*) The starting element also makes the carrier equipment able to have a possible instruction.

- The *effect of power swing* on the distance protection scheme of transmission line can be summed up as follows :

(*i*) The starting element of the distance protection scheme may be operated by the power swing.

(*ii*) Specially, the overcurrent starting elements respond readily due to the large flow of equalising current in transmission lines.

(*iii*) Even due to the change of ratio of voltage and current, the minimum impedance starting element also operates.

(*iv*) The undervoltage starting element may also operate at heavy power swing due to the voltage drops at certain points of the system.

Choice between relays of impedance, reactance or mho types :

The distance relays are used for both phase faults and ground fault protection and preferred to overcurrent relays, latter being slow and not so selective. *Distance relays are high speed relays and are also independent of changes in magnitude of the short-circuit current and are not much affected by changes in generation capacity.*

— The "***reactance relay***" is used where the arc resistance is likely to be comparable with the impedance of the protected section. The *reactance relay ignores the arc resistance and hence is indispensable for short lines and for protection against ground faults.*

— The "***Mho type relay***" is best suited for phase fault relaying for *longer lines* and particularly where severe synchronising swings may occur. These relays have the advantage over other two other relays of an *inherent directional characteristics.*

— The ***impedance relay*** is better suited for phase fault relaying for lines of *moderate length* than for either very short or very long lines. Arc resistance affects an impedance relay more than a reactance relay but less than a mho relay. Synchronising power swings affect an impedance relay less than reactance relay but more than a Mho relay.

Carrier distance protection :

In case of long lines carrier-pilot relaying is *cheaper and more reliable* than that required for use with pilot relays. The main difference between this scheme and the voltage balance (Translay) pilot systems employing pilots is that, in the former, *only the phase angles of the currents* at the two ends of the feeder are compared instead of the actual currents as in the latter case and this *phase angle decides whether the fault is internal or external.* Improvements in carrier-current techniques have made it possible to use carrier waves, transmitted over the actual feeders instead of pilots. It is worth knowing that when a voltage of positive polarity is impressed on the control circuit of the transmitter, it generates a high frequency output voltage. This voltage is impressed between one-phase conductor of the transmission line and the earth.

The carrier distance protections which are generally used in the transmission lines are :

1. Carrier transfer.
2. Carrier blocking.
3. Carrier acceleration.

1. *Carrier transfer* :

The '*carrier transfer*' is the transfer of carrier signal from one of the transmission line section to the other end for simultaneous tripping of the line section.

The necessity of carrier transfer in the transmission line is due to the following *fact* :

From Fig. 25.12 it is quite obvious that if fault occurs in the middle of the section, the distance relays R_1 and R_2 will trip with time t_1, but if the fault occurs near any end of the line section, the relay at local end will trip at time t_1 and the relay at remote end will trip at time t_2. Thus, variation of time of openings of circuit breakers at both ends will occur. This is not at all desirable, *carrier transfer helps to avoid this* and that is why the carrier transfer is necessary in the transmission line.

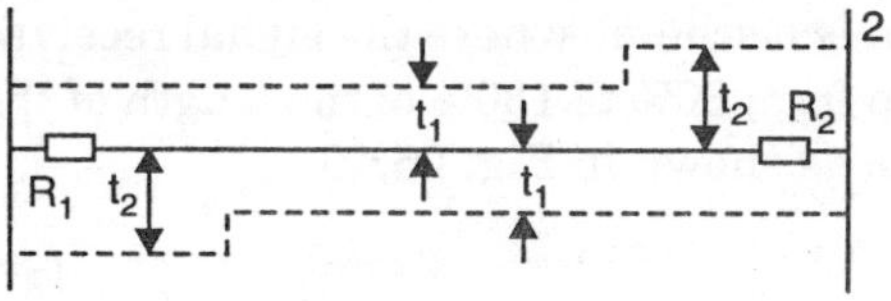

Fig. 25.12

The "*scheme of transfer relaying*" (Fig. 25.13) is as follows :

Here, the transmission signal is sent by the transmitter from one end of the transmission line to another end receiving initiation from the relay R_1 of Zone–1. Moreover, the relay R_1 also completes the trip circuit at the local end (as shown in Fig. 25.14) by the control circuit (CS).

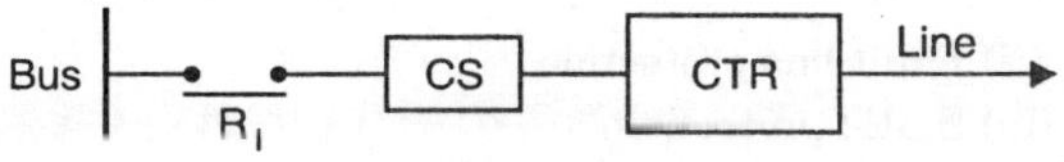

R_1 = Relay of the zone-1;
CS = Control circuit;
CTR = Carrier transmitter

Fig. 25.13

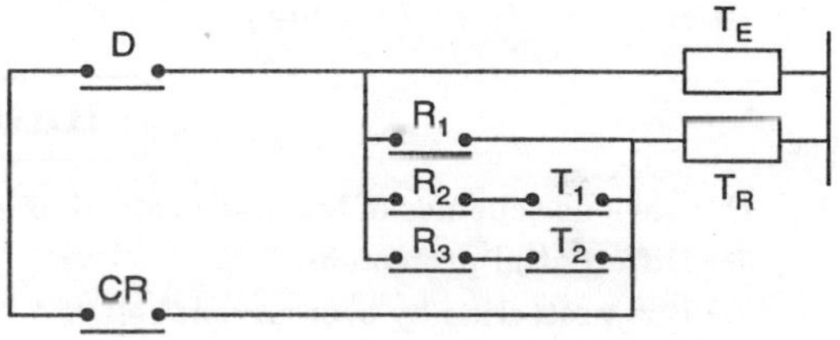

D = Faulty detector; CR = Carrier receiver;
R_1, R_2, R_3 = Relays; T_1, T_2 = Timer elements;
T_E = Timer element; T_R = Tripping relay

Fig. 25.14

2. *Carrier blocking* :

The main function of carrier blocking scheme is *to prevent the tripping of the circuit breakers for faults on the next section.* It does not act as primary protecting device, but it only operates in conjunction with the distance relay at the time of the fault in the line.

Fig. 25.15 shows the operation of the carrier blocking scheme. Suppose the fault has occurred at the point F on the zone '2-3'. Now the directional relay at Y_1 will actuate and send signal to the section-1 for blocking the trip of the circuit breakers at X_2 and Y_1. On the other hand if fault occurs in the zone '1-2', the signal is not sent by Y_1 to block tripping of the circuit breakers at X_2 and Y_1.

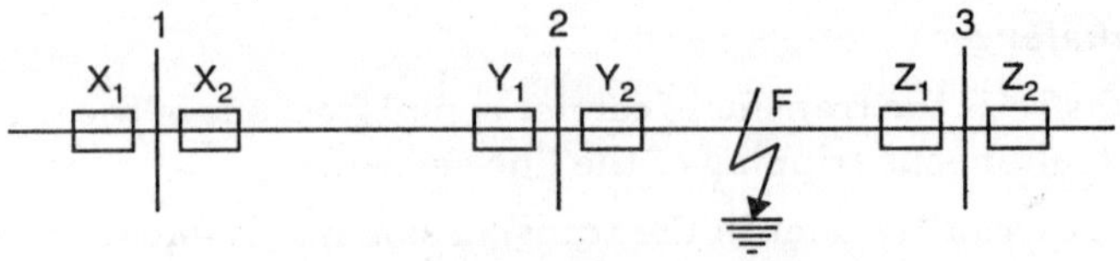

Fig. 25.15. Carrier blocking scheme.

The following are the "*advantages*" of carrier blocking scheme :

(*i*) The carrier signal is transferred over healthy line.

(*ii*) The carrier signal is sent only at fault condition.

3. *Carrier acceleration :*

The carrier acceleration is a scheme where the signal received from a relay at opposite end is utilised to extend the first step from 80% to 130% of the length of the protected line by shunting the timer element of the next zone as shown in Fig. 25.16.

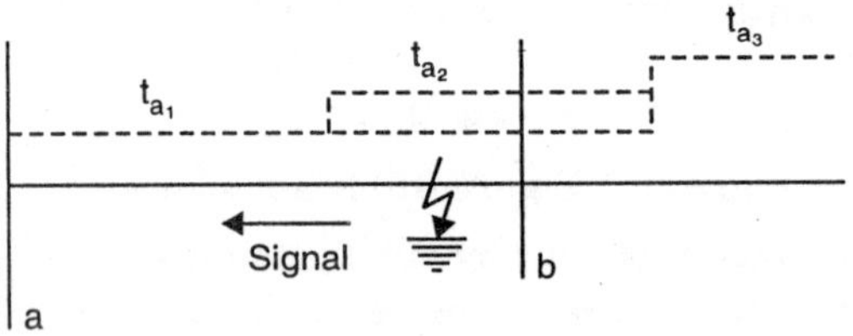

Fig. 25.16

HIGHLIGHTS

1. The various schemes of bus-bar protection are :
   (*i*) Differential protection. (*ii*) Fault bus protection.
   (*iii*) Bus protection by overcurrent relays. (*iv*) Static protection.
   (*v*) Overcurrent protection or distance protection as back up protection of bus-bar.
2. The most common methods of line/feeder protection are :
   (*i*) Time-graded overcurrent protection. (*ii*) Differential protection.
   (*iii*) Distance protection.

THEORETICAL QUESTIONS

1. State the importance of bus-bar protection.
2. State the reasons due to which bus fault occurs in a power system.
3. Name the various schemes of bus-bar protection.
4. Explain the following systems of bus-bar protection :
   (*i*) Differential protection (*ii*) Fault bus protection.
5. What are the requirements of protection of lines ?
6. Describe the time-graded overcurrent protection for
   (*i*) Radial feeders ; (*ii*) Parallel feeders ; (*iii*) Ring main system.
7. Describe the differential pilot-wire method of protection of feeders.

8. Explain translay protection scheme for feeders.
9. Describe distance protection scheme for the protection of the feeders.
10. Explain briefly major items of distance protection schemes.
11. Explain briefly the effect of power swing on the distance protection scheme.
12. Describe briefly the following carrier distance protection schemes :
    (*i*) Carrier transfer ; (*ii*) Carrier blocking ; (*iii*) Carrier acceleration.
13. What is meant by time-graded system ?
14. Explain why distance protection is superior to other types of protection for an overhead line.
15. What are the advantages of distance protection over types of protection of feeders ? Describe any type of impedance relay and show how these relays provide discriminatory protection.
16. Describe feeder protection using carrier current relaying.
17. What is Merz-Price voltage balance protection system ? How does it protect feeders ? Explain it with a neat diagram.
18. Describe a neat sketch of the Translay system of feeder protection and describe critically its working principle.
19. Explain a scheme of protection for a ring main.
20. Write short notes on the following :
    (*i*) Fault bus protection.
    (*ii*) Merz-Price system for protection of feeders.
    (*iii*) Translay scheme.

26

Protection Against Overvoltages And Insulation Co-ordination

26.1. Introduction. 26.2. Causes of overvoltages. 26.3. Internal causes of overvoltages. 26.4. External causes of overvoltages—Lightning—Process of lightning discharge—Types of lightning strokes—Protection against overvoltages and lightning—Types of lightning arresters—Surge absorber—Rating of lightning arrester—Tests on lightning arresters. 26.5. Insulation coordination. Highlights—Theoretical Questions.

26.1. INTRODUCTION

Protection against overvoltages. At several instances the elements of a power system (*viz.* generators, transformers, transmission lines etc.) during operation may be subjected to overvoltages (*i.e.,* voltages above the normal values). These overvoltage may occur due to the grounding of a conductor, opening of a circuit breaker, lightning etc. The overvoltages, in general, are not of significant magnitude but may still be important because of their effect on the performance of circuit interrupting equipment and protective devices. Quite often these overvoltage may be of such magnitude as to cause insulation breakdown of the power system equipment. Therefore, it is of paramount important to devise ways and means to limit the magnitude of these overvoltages to check their detrimental effects on the power system elements.

Insulation coordination. It means the correlation of the insulation of the various equipments in a power system to the insulation of the protective devices used for protection of those equipments against overvoltages. The various equipments like generators, transformers, circuit breakers etc., in a power system, have different breakdown voltages and hence the volt-time characteristics. Therefore, in order, that all these equipments should be properly protected it is essential that the *insulation of the various protective devices must be properly coordinated.* The problem of coordinating the insulation of the protective equipment involves not only guarding the equipment insulation but also it is desired that the protecting equipment should not be damaged.

26.2. CAUSES OF OVERVOLTAGES

Overvoltages may be caused by conditions external or internal to the system ; these are broadly categorised as follows :

I. Internal causes :

1. Switching surges.
2. Insulation failure.
3. Arcing ground.
4. Resonance.

II. External causes :

These are mainly due to atmospheric conditions *i.e.* due to *'lightning.'*

The surges produced by *internal causes* are not of a large magnitude. It has been observed that internal causes hardly increase the voltage of system twice the normal value. Normally such surges (caused by internal causes) can be taken care of with the provision of proper insulation to the equipment in the power system. On the other hand the surges produced by lightning (external cause) are very severe and may increase the system voltage to several times of the normal value. Thus, if the equipment is not protected against the lightning surges, these surges may damage it considerably. Hence the protective devices incorporated in a power system check mainly the surges caused due to lightning.

26.3. INTERNAL CAUSES OF OVERVOLTAGES

These are primarily due to oscillations set up by sudden changes of circuit conditions or by resonance. The normal system insulation is suitably designed to withstand such surges.

1. Switching surges. The switching surges are the overvoltages produced on the power system due to switching operations. By way of illustration, the following cases will be discussed :

(*i*) Case of an open line.

(*ii*) Case of a loaded line.

(*iii*) Current chopping.

(*i*) ***Case of an open line.*** When an open ended line is connected to a source of voltage, travelling waves are set up which rapidly charge the line. These waves, on reaching the open end of the line, are totally reflected without change of sign, thereby producing *voltage–doubling at that end.* These reflected waves travel back to the supply end, giving rise to further reflections. Due to losses, the wave gradually attenuate and ultimately a normal voltage is attained by the line. Thus, the maximum potential attained by the line at the time of switching cannot exceed twice the maximum supply voltage V_{max} or $2\sqrt{2}$ $V_{r.m.s.}$ (phase). However, in actual practice, the potential attained by the line is somewhat less than this theoretical value owing to dissipation of energy.

Similarly, at the instant of switching off the unloaded line, travelling waves are set up and the line momentarily attains the voltage of the magnitude not exceeding twice the supply voltage.

(*ii*) ***Case of a loaded line.*** When a line carrying line is suddenly opened, a transient voltage(e) of value given below is set up,

$$e = iZ_0$$

where, i = Instantaneous value of the current, and

Z_0 = The natural or surge impedance of the line.

Thus if i = 200 A and Z_0 = 1000 Ω, the transient rise of voltage possibly cannot exceed 200 × 1000 = 200 kV (irrespective of the fact how rapidly the current is interrupted). If V_{max} is the peak value in kV, the maximum voltage to which the line may be subjected is (V_{max} + 200 kV) kV.

It is evident from the above discussion that rise of transient voltage due to sudden interruption of load is not a function of the line voltage, and therefore, *low voltage transmission systems are liable to have overvoltages of the same magnitude as high voltage systems.* Thus a high factor of safety is justified in the case of low voltage system.

Under the worst conditions (*e.g.* interruption of short-circuit current of the system at its peak value) voltage rise of several hundred kV might take place. However, fortunately, oil circuit breakers have the capability to eliminate all danger on account of short-circuits by opening the circuit when the current wave attains zero value.

2. Insulation failure. In a power system, the insulation failure may take place in the following ways :

(*i*) Between the conductors of an overhead line.

(*ii*) Between the cores of an insulated cable.

(*iii*) Between one conductor, or one core, and earth.

The insulation failure between the line and earth is very frequent. In case of a breakdown of insulation to earth, the potential at fault suddenly falls from maximum to zero and consequently a negative voltage wave of very steep front in the form of surge travels from the fault in both directions.

Suppose a line at potential V is earthed at point F (Fig. 26.1). The earthing of the line causes two equal voltages of $-V$ along FM and FL containing currents $-V/Z_0$ and $+V/Z_0$ respectively. Both these currents pass through F to earth so that current to earth is $\frac{2V}{Z_0}$.

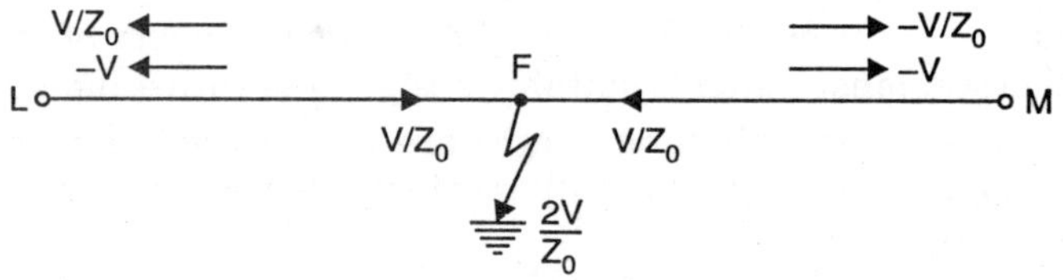

Fig. 26.1. Surge due to insulation failure.

3. Arcing ground. *The phenomenon of intermittent arc taking place in line-ground fault of a 3-phase system with consequent production of transients is known as* ***arcing ground.***

The transients produced due to arcing ground are cumulative and may cause serious damage to the equipment in the power system by causing insulation breakdown.

Arcing ground can be prevented by *earthing the neutral*.

4. Resonance. In the usual transmission lines, the resonance cannot occur at the fundamental supply frequency since the capacitance is usually very small. However, if the generator e.m.f. wave is distorted, trouble may errupt due to resonance of one of the higher harmonics. Resonance causes high voltages in the electrical system.

26.4. EXTERNAL CAUSES OF OVERVOLTAGES—LIGHTNING

Lightning *is an electric discharge between cloud and earth between clouds or between the charge centres of the same cloud.* It is basically a *huge spark*. A large number of discharges occur between or within clouds than to earth, and enough of them terminate on the earth causing serious hazards.

Several theories have be propounded to explain how the clouds acquire charge. The ***most accepted theory*** is as follows :

- During the uprush of warm moist air from earth, the friction between the air and the tiny particles of water causes the building up of the charges.
- When drops of water are formed, the *larger drops become positively charged and the smaller drops negatively charged.*
- When the drops of water accumulate, they form clouds, and hence the cloud may possess either a positive or a negative charge, depending upon the charge of drops of water they contain. *The charge on a cloud may become so great that it may discharge to another cloud or to earth and this discharge is called as* ***lightning.***
- The thunder which accompanies lightning is due to the fact that lightning suddenly heats up air, thereby causing it to expand. The surrounding air pushes the expanded air back and forth causing the wave motion of air, recognised as ***thunder***.

26.4.1. Process of Lightning Discharge

The process / mechanism of lightning discharge is as follows :

When a charged cloud passes over the earth, it induces equal and opposite charges on the earth below. Fig. 26.2 shows a negatively charged cloud inducing a +ve charge on the earth below it. With the increase in charge of the cloud, the potential between cloud and earth increases and hence the gradient in the air also increases.

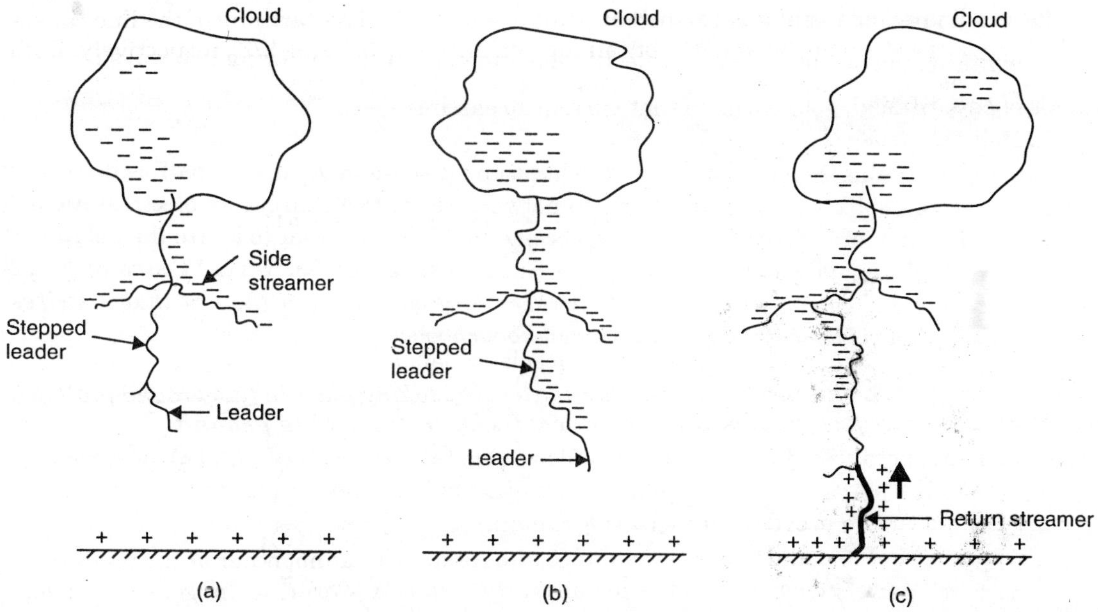

Fig. 26.2. Process of lightning discharge.

- *Refer to Fig. 26.2 (a).* Immediately after the air near the clouds breaks down, a streamer called a *leader / pilot streamer* starts from the cloud, and carrying charge with it moves towards the earth. The leader streamer with continue moving toward earth so long as the cloud, from which it originates feeds enough charge to it to maintain gradient at the tip of leader streamer above the strength of air. In case the gradient is not maintained, the leader streamer stops and the charge is dissipated without completion of the lightning stroke. As such the leader streamer will not reach the earth [Fig. 26.2 (*a*)].
- *Refer to Fig. 26.2 (b).* The leader streamer is continuing its journey towards earth until it makes contact with earth or some object on the earth. As the leader streamer moves towards earth, it is accompanied by luminescence which travel in jumps giving rise to stepped leaders (their velocity exceeds one-sixth of that of light and distance travelled is about 50 m in one step). The stepped leaders have adequate luminosity and give rise to first visual phenomenon of discharge.
- *Refer to Fig. 26.2 (c).* As soon as the leader streamer reaches near the earth, a *return streamer* shoots up from the earth to the cloud, following the same path as the main channel of the downward leader. This phenomenon causes a *sudden spark*, called *'lightning'*. Owing to neutralisation of much of the negative charge on the cloud, any further discharge from the cloud may have to originate from some other portion of it.

The following points about lightning discharge are worth noting :

(*i*) The currents during lightning discharge may be in the range of 10,000 to 90,000 A.

(*ii*) 87 percent of all the strokes come from negatively charged clouds and only 13 percent from positive ones.

(*iii*) A lightning discharge which usually appears to the eye as a single flash is actually made up of a number of separate strokes that travel down the path. The interval between them varies from 0.0005 to 0.5 second. Each separate stroke starts as a downward leader from the cloud.

26.4.2. Types of Lightning Strokes

A lightning may strike/affect the electrical equipment in the following two ways :

1. Direct stroke.
2. Indirect stroke

1. Direct stroke. In a direct stroke, the lightning discharge or current path is directly from the cloud to the tall objects or nearby clouds. In between these two charges an electrostatic field is set up and when intensity of electrostatic field becomes sufficiently great to ionize the neighbouring air, the air breaks down and discharge takes place. Such a discharge is known as ***'A' stroke*** (Fig. 26.3) is *characterised by the comparatively long time taken to produce it and the fact that it strikes the highest and most sharply pointed object in the neighbourhood.*

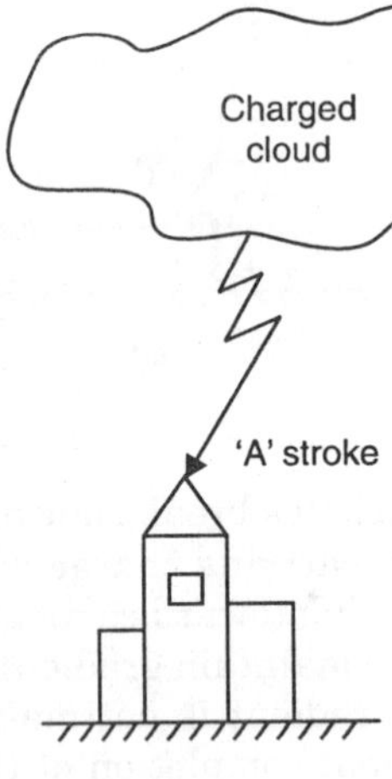

Fig. 26.3

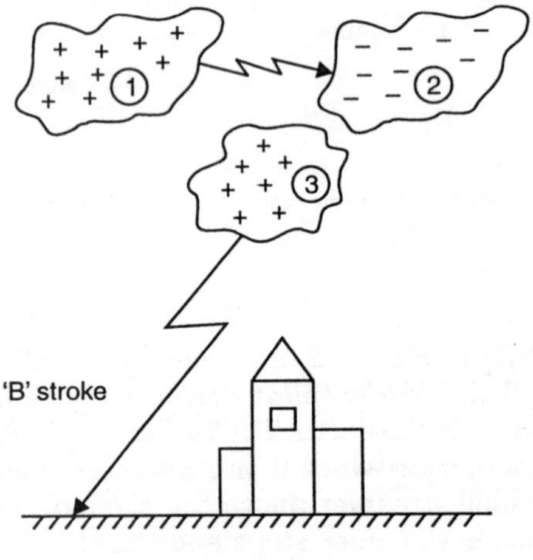

Fig. 26.4

The ***stroke 'B'*** is much more a suddent stroke and is produced in the manner shown in Fig. 26.4. Here three clouds 1, 2 and 3 are involved. Clouds 1 and 3 are positively charged and cloud 2 is negatively charged. The potential of cloud 3 is reduced due to the presence of charged cloud 2. When flashover occurs from cloud 1 to cloud 2, both these clouds are discharged rapidly and cloud 3 assumes a much higher potential and flashes to earth very rapidly. This is the 'B' stroke and is characterised by its rapidity and the fact that it ignores tall objects and reaches earth in a random manner. This stroke is *most dangerous one.*

A direct stroke may cause a potential of *many million volts.* If the overhead line is struck at *a long distance* from the station or sub-station the overvoltage and associated current will flow along the line in both directions shattering insulators and even wrecking poles until its energy is spent up in paths to earth so formed. If the overhead line is struck *immediately adjacent* to station then damage to plant is almost certain. Fortunately direct strokes do not occur frequently.

- For all purposes these are only *travelling waves through the lines which are considered for protection.*

2. Indirect stroke. Indirect strokes may result from electromagnetically induced currents due to a lightning discharge in the immediate vicinity of the line or result from electrostatically induced charges on the conductors due to the presence of charged clouds.

Fig. 26.5 shows how a conductor may be indirectly struck when charged electrostatically by a neighbouring charged cloud. A negatively charged cloud is over a conductor which is high '*h*' above the ground. The charges induced on the conductor are + Q and – Q having a magnitude $Q = CV$ where C is the capacitance between conductor and earth, and $V = Eh$, E being the gradient of the field.

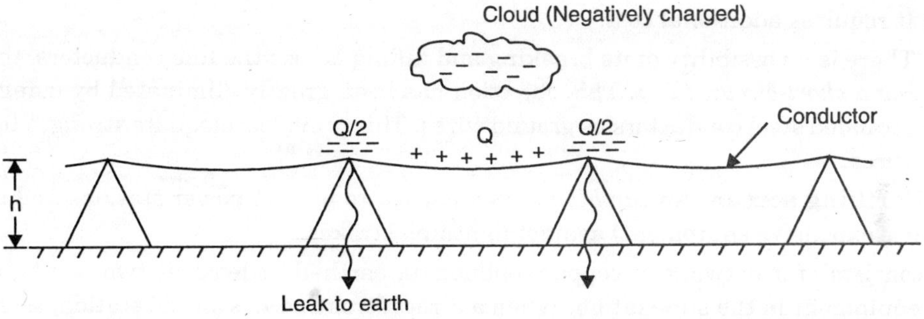

Fig. 26.5. Indirect lightning stroke.

The –ve charge leaks slowly to earth via the insulators. When the cloud discharges to earth or another cloud, or passes away, the bound +ve charge on the line is released in the form of *travelling wave in both directions.*

The **harmful effects of lightning** *may be summed as follows.*

1. The travelling wave produced due to lightning surges will shatter the insulators and may even damage the poles.

2. When the travelling waves produced as a result of lightning hit the windings of transformer or generator, they cause a considerable damage to the equipment.

3. The arc initiated in any part of the power system by the lightning stroke will set up very disturbing oscillations in the line. This may result in damage to other equipment connected to the line.

26.4.3. Protection Against Overvoltages and Lightning

The devices employed for protection of electrical equipment from surges or overvoltages are known as *overvoltage protection devices* ; these are enumerated and described below.

1. Ground wire.
2. Earthing screen.
3. Surge diverters or lightning arresters.

1. Ground wire. The protection of transmission line from direct strokes may be done by running a conductor known as "*ground wire*" over the towers or poles and earthed at regular intervals, preferably at every pole or tower. The wire can be run either above or below the line but *running of earth wire above the line is better as it provides more effective shield.*

- The ground wire *not only handles a direct stroke* but also *provides a certain amount of electrostatic screening* as it causes the capacitance of ground to increase and so the voltages induced in the conductors, owing to the discharge of a neighbouring cloud, to decrease.
- Another protective action of the ground wire is its *damping effect on any wave or other disturbance travelling along the line* by its action as a short-circuited secondary.
- The selection of ground wire should be *based on mechanical rather than electrical considerations.* These should have a *high strength and be non-corrosive.* Ground resistance, insulation and clearances between the ground wire and the lines are important in the design and construction of lines with ground wires.

Demerits :

(*i*) It requires additional cost.

(*ii*) There is a possibility of its breaking and falling across the line conductors, thereby *causing a short-circuit fault*. This objection has been greatly eliminated by using galvanised stranded steel conductors as ground wires. This provides adequate strength to the ground wires.

2. Earthing screen. An *earthing screen* is used to protect power stations and sub-stations (which house expensive equipment) against lightning strokes.

It consists of a network of copper conductors, earthed at least at two points, over all the electrical equipment in the sub-station. When a direct stroke occurs on the station, screen provides a low resistance path by which lightning surges are conducted to ground. In this way, station equipment is protected against damage.

The ***limitation*** of *this method is that it does not provide protection against the travelling waves which may reach the equipment in the station.*

3. Surge diverters or lightning arresters. The ground wire and earth screen can protect electrical equipment against lightning strokes but *cannot* provide protection against the high voltage waves reaching the terminal equipment. The protection against such surges can be provided by *surge diverter* or *lightning arrester.*

A ***lightning/surge diverter*** *is a protective device which conducts the high voltage surges on the power system to the ground.*

A surge diverter acts similar to a *safety valve.* When a high voltage wave reaches the surge diverter, it sparks over and provides a conducting path of relatively low impedance between the line and earth so the resulting current flows to the earth. As soon as the voltage attains normal value, it stops the flow of current.

Fig. 26.6 shows a *basic form of a surge diverter*. It consists of spark gap in series with a non-linear resistor (resistance decreases as the voltage or current increases and vice-versa). One end of the diverter is connected to the terminal of the equipment to be protected and the other end is effectively grounded. The gap length is so set that normal line voltage is insufficient to cause an arc across the gap but a dangerously high voltage will breakdown the air insulation and form an arc.

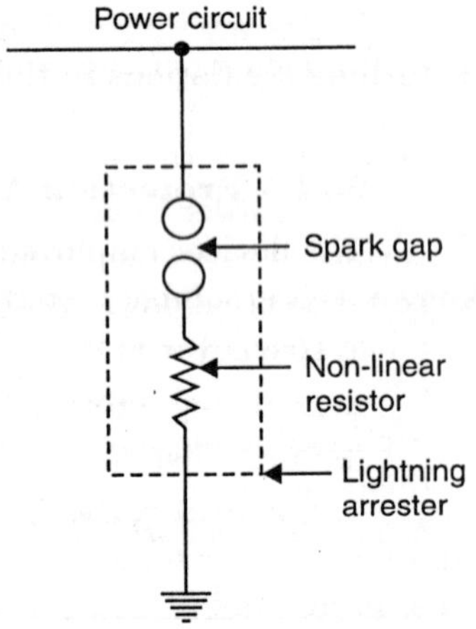

Fig. 26.6. Basic form of a single diverter.

Under normal operating conditions the surge diverter conducts no current to the earth (or the gap is non-conducting). However, when overvoltage occurs, the air insulation across the gap breaks down and an arc is formed, providing a low resistance path for the surge to the ground. Thus, the excess charge on the line due to the surge is safely conducted through the diverter to the ground.

Characteristics of an ideal surge diverter :

An ideal surge diverter should possess the following *characteristics* :

1. It must not take any current at any normal system voltage, that is, its power frequency breakdown (or spark over) voltage must be above any normal or abnormal fundamental frequency voltage that may occur.

2. Any abnormal transient voltage above the breakdown value must cause it to breakdown as quickly as possible so that it may provide a conducting path to earth.

3. After the breakdown has taken place, it must be capable of carrying the resulting discharge current without damage to itself and without the voltage across it exceeding the breakdown voltage.

4. The power frequency current following the breakdown must be interrupted as soon as the transient voltage has fallen below the breakdown value.

26.4.4. Types of Lightning Arresters

There are several types of surge arresters which are used to protect the power system. The choice of a *lightning arrester depends upon the following *factors* :

1. Line voltage
2. Lightning frequency
3. Cost of the arrester
4. Weather conditions
5. Reliability etc.

Various *types of lightning arresters* are :

1. Rod gap arrester
2. Horn gap arrester
3. Electrotype arrester
4. Oxide film arrester
5. Thyrite arrester
6. Expulsion type lightning arrester
7. Valve type lightning arrester.

1. Rod gap arrester. Fig. 26.7 shows a rod gap across a transformer bushing insulator. It is a very simple type of diverter and consists of two rods with ends facing each other, one connected to the line and the second connected to the earth. These are generally connected across the bushings of various equipment.

In order to avoid ascending across the insulator surface of very steep fronted waves, the rod gap should be set to breakdown at about 20 percent *below* the impulse sparkover voltage of the insulation at the point where it is installed. Further, in order to prevent the arc from being blown on the insulator, the distance S (distance between the gap and the insulator) should *not be less than one-third of the rod gap length.*

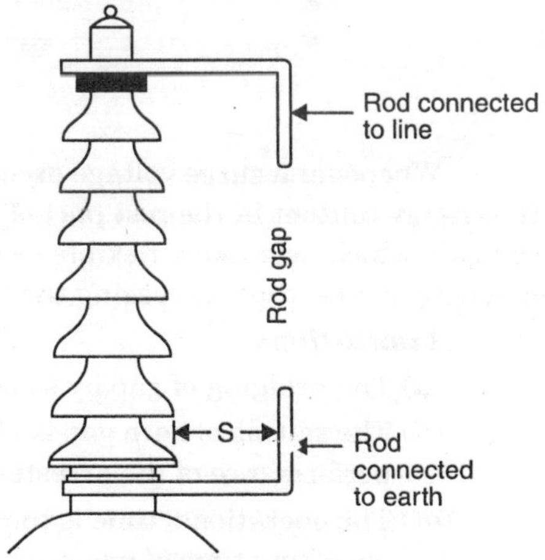

Fig. 26.7. Rod gap on bushing insulator.

---

* Lightning arrester is a misnomer. The device is actually a lightning diverter and a follow-current arrester.

Under healthy operating conditions, the rod gap remains non-conducting. However, when a high voltage surge on the line occurs, the gap sparks over and surge current is conducted to earth. Thus, the excess charge on the line due to the surge is safely conducted to earth.

Limitations. The rod gap (very *cheap and simple device*) has the following *limitations* :

1. It can afford protection against waves with a relatively small sloping front only.
2. It cannot seal off power frequency follow current.
3. The rod may melt or get damaged due to excessive heat produced by the arc.
4. The performance characteristic of the rod gap arrester is adversely affected by climatic conditions (*e.g.*, humidity, pressure, temperature etc.)
5. The gap is also influenced considerably by the polarity of surge.

In view of the above limitations, the rod gap arrester is only used as a 'back-up' protection in case of main arresters.

2. Horn gap arrester. It is one of the earliest types of surge diverters to be developed, and is still used to a certain extent on *low-voltage lines* on account of its great simplicity.

It consists of two horn-shaped pieces of metal separated by a small air-gap. One end of this is connected to the line and the other to the earth as shown in Fig. 26.8 with or without a series *resistance. The choke connected between the equipment to be protected and the horn gap serves the two purposes : (*i*) The steepness of the wave incident on the equipment to be protected is reduced ; (*ii*) It reflects the voltage surge back on to the horn.

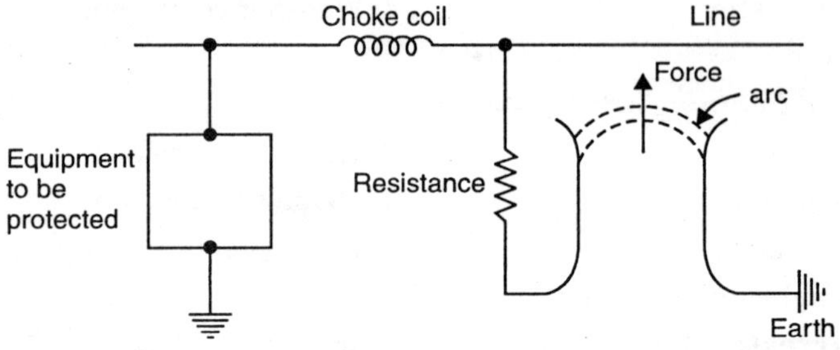

Fig. 26.8. Horn gap arrester.

Whenever a surge voltage exceeds the breakdown value of the gap discharge takes place and the energy content in the rest part of the wave is bypassed to the ground. An arc is set up between the gap, which acts like a flexible conductor and rises upwards under the influence of the electro-magnetic forces, thus increasing the length of the arc which eventually blows out.

Limitations

(*i*) The bridging of gap by some external agency (*e.g.*, birds) can render the device useless.

(*ii*) The setting of horn gap is likely to change due to corrosion and pitting, owing to which the performance of the arrester may be adversely affected.

(*iii*) The operational time is comparatively long, say about 3 seconds (in view of the very short operating time of modern protective gear for feeders).

- Like rod gap arrester it is also employed as an *auxiliary protection.*

---

* The horn gap is unable to rupture currents beyond about 10 A and, therefore, in order to limit this current a non-inductive resistance in inserted in series with the horn gap.

3. Electrolyte arrester. An electrotype arrester operates on the fact that a thin film of aluminium hydroxide deposited on the aluminium plates immersed in electrolyte *acts as a high resistance to low voltage but a low resistance to voltage above a critical value.* Voltages more than 400 V (critical breakdown voltage) causes a puncture and flow of current to earth. When the voltage regains its normal value of 400 V, the arrester again offers high resistance in the path and leakage stops. The total critical value of voltage of such an arrester can be increased by arranging a number of films one above the other.

Since these arresters are very delicate, require daily supervision, and the film is required to be reformed whenever destroyed, therefore, these have become obsolete now-a-days.

4. Oxide film arrester. Its operating principle is based on the fact that *certain chemicals have the property to change rapidly from a good conductor to almost perfect insulator when slightly heated.*

It consists of 2.4 mm diameter pellets of *lead per oxide* with a thin porous coating of *litharge* arranged in a column and enclosed in a tube of diameter of about 6 cm and of height of 5 cm per kV of rating. Out of the two leads of the arrester upper is connected to the line, while the lower is connected to earth. The tube contains a series spark-gap. A single tube system is available for voltages upto 25 kV when the neutral is solidly earthed and 18 kV when neutral is isolated or earthed through an inductive coil. For use on higher voltages several units in series are employed

On the occurrence of overvoltage, an arc passes through the series spark gap and an additional voltage is applied to the pellet column and a discharge takes place. After the discharge, the resistance of the pellet column increases till only very small current can flow through it. This small current is finally interrupted by the series spark gap.

The *main advantage* of this type of arrester is that it does not require daily charging, and it may thus be installed at points on transmission systems where daily attendance is difficult or expensive to provide.

5. Thyrite arrester. *It is the most commonly used arrester for the protection against high dangerous voltages.*

It operates on the fact that *thyrite*, a dense inorganic compound of ceramic nature *has high resistance decreasing rapidly from high value to low value for currents of low value to those of high value. The current decreases 12.6 times on doubling the voltage.*

It consists of discs of 15 cm diameter and 19 mm thickness. Both the sides are metal sprayed so as to give electrical contact between consecutive discs. These discs are assembled inside the glazed porcelain container. It is used in conjunction with porcelain container.

On the occurrence of lightning stroke, a voltage is raised and breakdown of gap takes place, the resistance falls to a very low value and wave is discharged to earth. After the surge has passed the thyrite again comes back to its original position (there being no chemical change occurring simultaneously).

Advantages :

(*i*) The thyrite arrester discharges several thousands amperes without the slightest tendency to flash-over on the edges.

(*ii*) There is absolutely no time lag in its performance.

6. Expulsion type lightning arrester. This type of arrester is also called "***protector tube***" and is commonly used on system operating at voltages upto 33 kV. It is an improvement over the rod gap in that *it seals the flow of power frequency follow current.* Its application is normally limited to circuits in which the possible range of power current magnitude following a sparkover is not more than three-to-one, since an excessive current may cause bursting of the tube while a small current may not set up sufficient pressure to extinguish the arc.

Construction. It consists of the following : Refer to Fig. 26.9.

(*i*) A *series gap* external to the tube which is good enough to withstand normal system voltage, thereby there is no possibility of corona or leakage current across the tube ;

(*ii*) A *fibre tube* which has a fibre lining on the inner side which is a highly gas evolving material ;

(*iii*) A *spark gap* (internal gap) in the tube ; and

(*iv*) An *open vent* at the lower end for the gases to be expelled.

It is desired that the breakdown voltage of a tube must be lower than that of the insulation for which it is used.

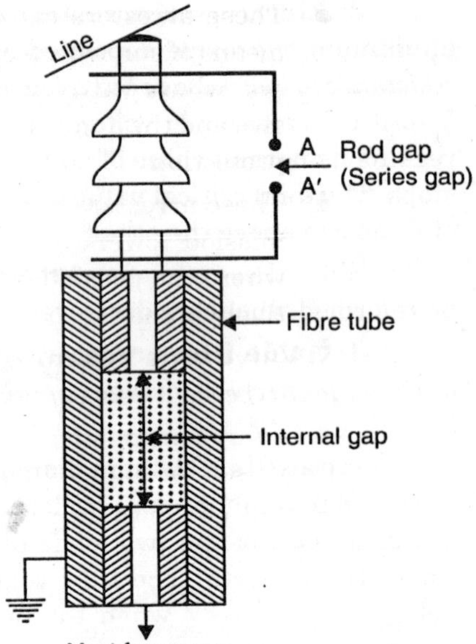

Fig. 26.9. Expulsion type lightning arrester.

Working. When a surge voltage is incident on the expulsion tube the series gap is spanned and an arc is formed between the electrodes within the tube. The heat of the arc vaporises some of the organic material of the tube wall causing a high pressure *gas to build up in the tube. The resulting neutral gas creates lot of turbulence within the tube and is expelled out from the open bottom vent of the tube and it extinguishes the arc at the first current zero. At this instant the rate of build up of insulation strength is greater than the RRRV. Very high currents have been interrupted using these tubes.

Advantages :

(*i*) Not very expensive.

(*ii*) Can be easily installed.

(*iii*) Better than rod gap arresters as they block the flow of power frequency follow currents.

Disadvantages :

(*i*) Totally unsuitable to protection of expensive equipment on account of their poor volt-time characteristic.

(*ii*) Unsuitable for mounting within equipment enclosure or in close proximity to other electrical apparatus.

(*iii*) This type of arrester can perform only limited number of operations as during each operation some of the fibre material is used up.

Applications :

(*i*) *Transmission line type* expulsion arresters are used for the protection of transmission line insulators.

(*ii*) *Distribution type* are used largely for protection of distribution transformers and similar apparatus on distribution circuits.

*The gas evolved is an un-ionized mixture of water vapour and decomposition products of the fibre.

- These arresters can be considered as economical means of surge protection for small rural transformers, where valve-type arresters may prove expensive and the application of air-gaps yield inadequate protection.
- These arresters can also be used on special transmission towers of extra height on river crossings where the possibility of lightning strokes are relatively high.

7. Valve type lightning arrester. It is also known as *non-linear type lightning arrester.*

It is very cheap, effective and robust and is, therefore extensively used for high voltage systems.

Fig. 26.10 shows the schematic arrangement of a valve type lightning arrester.

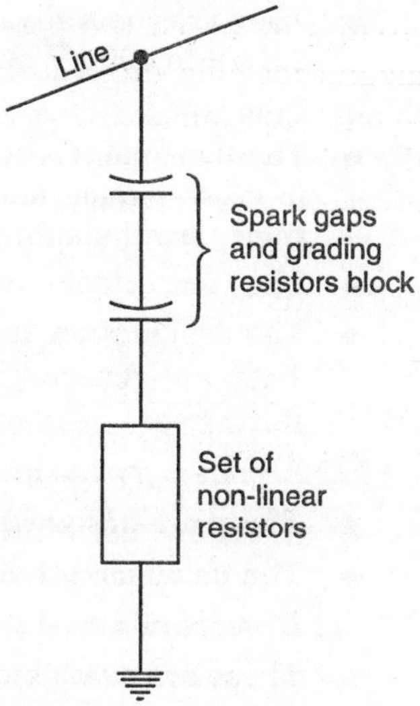

Fig. 26.10. Schematic arrangement of a valve type lightning arrester.

Construction. It consists of a stack of spark gaps and resistor discs, the latter having the unique characteristics and *their ohmic value decreases rapidly as the magnitude of current flowing through them increases.* Both assemblies, *i.e.*, spark gaps and set of resistors, are accommodated in series within a completely tight porcelain housing ensuring reliable protection against atmospheric moisture, condensation and humidity.

Working. Under normal operating conditions, the normal system voltage is insufficient to cause the breakdown of air gap assembly. However, when overvoltage occurs, the breakdown of the series spark gap takes place and the surge current is conducted to earth via the non-linear resistors. As the *magnitude of surge current is very large, the non-linear elements will offer a low resistance to the passage of surge.* Consequently the surge will rapidly move to earth instead of being sent back over the line. When the surge is over, the non-linear resistors assume high resistance and flow of the current ceases.

Advantages :

(*i*) These arresters provide very effective protection against surges (especially for transformers and cables).

(*ii*) The impulse ratio (ratio of breakdown voltage under surge conditions to breakdown voltage under low frequency conditions) is practically *unity*.

(*iii*) Operate very rapidly, taking less than a second.

Drawbacks. Although these arresters possess very superior characteristics, they suffer from the following *drawbacks* :

(*i*) The performance characteristic is adversely affected by the ingress of moisture into the enclosure.

(*ii*) These arresters may fail to respond quick enough to check the surge voltage of very steep wavefront from finding access into the station equipment.

The *various types of valve type arresters and their applications are* :

(*i*) *Station type :*

- These are the most efficient and most expensive type. The word 'station' indicates their general application in large station.

- They are generally used for the protection of important power equipment in the circuits of 2.2 kV to 400 kV and higher.

(*ii*) *Line type :*

- These are constructed in a way similar to that of the station type arresters but are smaller in cross-section and weight and less in cost. They permit higher surge voltages across their terminals than do the station types and have lower surge current capacity.
- They are seldom used on system voltage above 66 kV.
- The designation 'line type' is confusing as these are not used for the protection of lines. They are generally used *for protecting large transformers and immediate sub-stations.* Ratings are available upto 5000 A.

(*iii*) *Arresters for the protection rotating machines :*

- These are designed particularly for the protection of motors and generators.
- The usual circuit voltages are 2.2 to 22 kV.

(*iv*) *Distribution type arresters :*

- These are available for voltages varying between 2.2 to 15 kV.
- These are mostly pole mounted and employed for protecting distribution circuits/transformers.

(*v*) *Secondary arresters :*

- These are available for protection of low voltage apparatus (120 to 750 V).

26.4.5. Surge Absorber

The travelling waves set up on the transmission line by the surges may reach the terminal equipment and cause damage to it. The damage caused depends not only the amplitude of the incoming wave but also on the steepness of its wavefront. *The device, which reduces the steepness of the wavefront of a particular surge and thus minimises the danger due to overvoltage is known as surge* ***absorber*** *or* ***modifier.***

Whereas a surge diverter diverts the surge to earth, the surge absorber absorbs the surge energy.

A few case of surge absorption are discussed below :

1. Fig. 26.11 shows a condenser, connected between the line and earth, acting as surge absorber to protect the equipment, say the transformer winding. Since the reactance of a condenser is inversely proportional to frequency, it will be high at low frequency and low at high frequency. As the surges are of high frequency, the *capacitor acts as a short circuit and passes them directly to the earth. However, in case of power frequency the reactance of the capacitor is very high and practically no current flows to the ground.

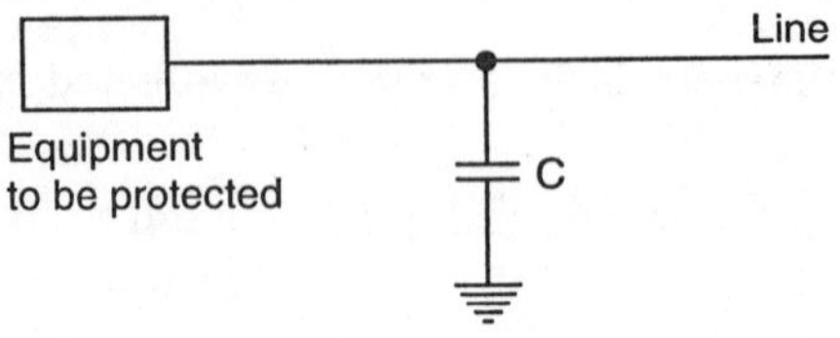

Fig. 26.11

*A pure capacitor cannot dissipate the energy in the wave front of travelling wave or in a high frequency discharge ; it merely reflects the wave energy away from the equipment to be protected and the energy is dissipated in the resistance of the line conductors and earthing resistances.

2. Fig. 26.12 shows a surge absorber which consists of a parallel combination of choke and resistance connected in series with the line. The choke offers high reactance to surge frequencies and as such the surges are forced to flow through the resistance R where they are dissipated.

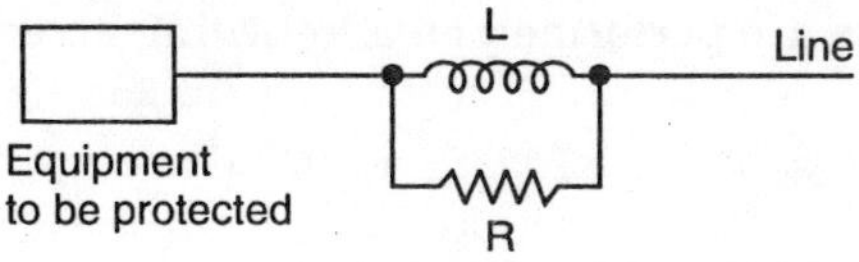

Fig. 26.12

3. Fig. 26.13 shows a *Ferranti surge absorber* which consists of an air-cored inductor connected in series with the line. The inductor is surrounded by but insulated from an earthed metallic sheet called *dissipator*. Such an arrangement is equivalent to a transformer with short-circuited secondary. The surge energy is used up in the form of heat in the dissipator due to transformer action.

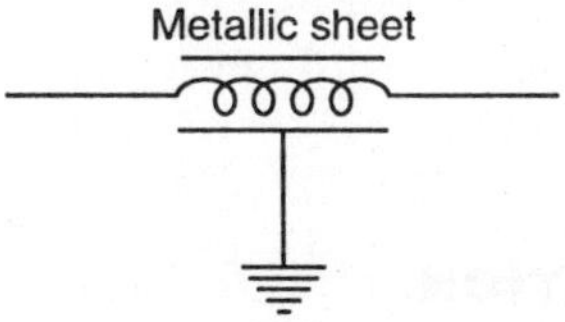

Fig. 26.13

This type of absorber is mainly used for the *protection of transformers.*

26.4.6. Rating of Lightning Arrester

Normally, it is expected that a lightning arrester will discharge currents of very large magnitude (thousands of amperes), but since the time is very short in terms of micro-seconds, the energy that is dissipated through the lightning arrester is small compared with what it would have been if a few amperes of power frequency current had been flown for a few cycles. Hence, the *main considerations in selecting the rating of a lightning arrester is the line-to-ground dynamic voltage to which the arrester may be subjected for any condition of system operation.*

An arrester of 105 percent is used on a system where the line to ground voltage may reach line-to-line value during line-to-ground fault condition (An allowance of 5 per cent is normally assumed, to take into account the light operating condition under no-load at the far end of the line due to Ferranti effect and the sudden loss of load on water wheel generators).

The overvoltages on a system depend upon the neutral grounding condition which is determined by the parameters of the system. The overvoltages can actually be obtained with the help of pre-calculated curves.

The voltage rating of an arrester

= Highest line-to-line r.m.s. value × coefficient of earthing

= Line-to-line r.m.s. voltage × 1.1 × coefficient of earthing

The coefficient of earthing for *effectively earthed system* = 0.8.

For *non-effectively* earthed systems and isolated or unearthed neutral systems it is usual practice to apply arresters rated at 100% of system highest voltage.

In case of systems grounded through Peterson coil, the overvoltages may be 100 percent, if it is properly tuned and, therefore, it is customary to apply an arrester of 105 percent for such systems.

Depending upon the neutral grounding condition, the voltage rating of the arrester, ranges from 75 percent to 105 percent.

26.4.7. Tests on Lightning Arresters

The following types of tests are performed on a lightning arrester.

1. $\frac{1}{50}$ impulse spark over test.
2. Wave-front impulse spark over test.
3. Impulse current withstand test.
4. Switching impulse voltage test.
5. Peak discharge residual voltage at rated current of the unit.
6. Discharge capability test.
7. Power duty-cycle test.
8. Transmission line discharge test.
9. Pressure relief test.

26.5. INSULATION CO-ORDINATION

In a power station or sub-station the insulation of electrical equipment is subjected from time to time, to momentary overvoltage. The protective devices such as lightning arresters, are used to prevent flashover or breakdown of equipment insulation from overvoltages of sufficient magnitude.

Insulation co-ordination *is the selection of suitable values for the insulation levels of the various components in any electrical system and their arrangement in a rational manner.*

The "***insulation level of an apparatus***" *is defined as that combination of voltage values (both power frequency and impulse) which characterise its insulation with regard to its capability of withstanding the dielectric stresses.*

A proper insulation co-ordination seeks to ensure the following :

(*i*) The efficient discharge of overvoltage due to internal and external causes.

(*ii*) The insulation provided shall withstand all normal working stresses and majority of abnormal ones ;

(*iii*) The positions at which breakdown occurs shall be those where breakdown may cause no or comparatively little damage ;

(*iv*) The breakdown shall occur due to external flashover.

In a power system, the insulation problem involves : (*i*) Determination of line insulation, (*ii*) Selection of Basic Impulse Insulation Level (BIL) and insulation levels of other equipment, and (*iii*) Selection of lightning arrester.

Fig. 26.14 illustrates the *basic concept of insulation co-ordination.* Curve A is the volt-time curve of the protective device and curve B is the volt-time curve of the equipment to be protected. Thus, any insulation having a withstand voltage strength in excess of insulation strength of curve B is protected by the protective device of curve A.

Proper insulation co-ordination would ensure that volt-time curve of the equipment will lie above the volt-time curve of the protective device. Thus, any rise is voltage beyond the protective level will cause the protective device, like lightning arrester, to operate and the voltage will be prevented to exceed the impulse strength of the equipment to be protected.

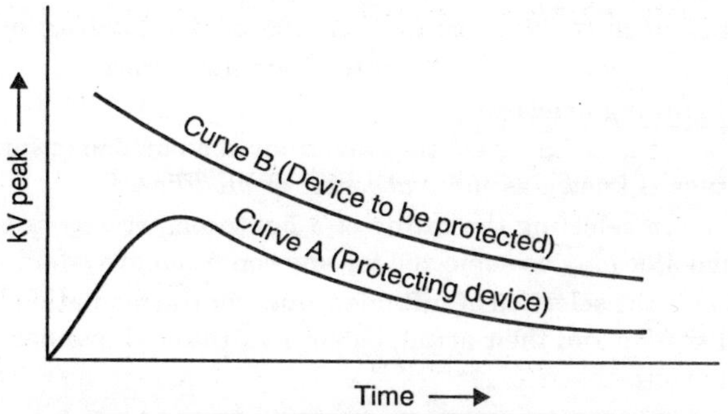

Fig. 26.14. Basic concept of insulation co-ordination.

The basic impulse level (BIL) should be selected, which can be protected with a suitable protective device like lightning arrester. For all practical purposes impulse testing, in order to establish insulation level, has been standardised.

Impulse wave shape gives the time in micro-seconds (μs) for the impulse to reach crest followed by the time in μs for wave to reach half of the crest magnitude.

Universally it has been adopted that basic impulse insulation levels (BILS) expressed in crest voltage (E in Fig. 26.15) with a standard wave not longer than 1.5 × 50 μs are reference levels. In Fig. 26.15 O_1A_1 = 1.5 μs and O_1A_2 = 50 μs.

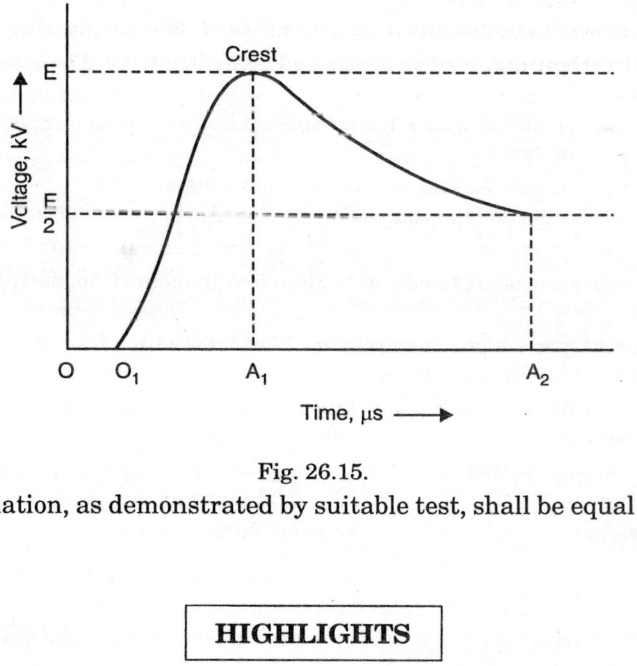

Fig. 26.15.

Apparatus insulation, as demonstrated by suitable test, shall be equal to or greater than the BIL.

HIGHLIGHTS

1. *Causes of overvoltage* :
   I. Internal causes :
   (*i*) Switching surges (*ii*) Insulation failure
   (*iii*) Arcing ground (*iv*) Resonance.
   II. External cause–Lightning

2. *Overvoltage protection devices :*
   (*i*) Ground wire (*ii*) Earthing screen
   (*iii*) Surge diverter or lightning arresters.
3. The device, which reduces the steepness of the wavefront of a particular surge and thus minimises the danger due to overvoltage is known as *surge absorber* or *modifier.*
4. The main considerations in selecting the rating of a lightning arrester is the line-to-ground dynamic voltage to which the arrester may be subjected for any condition of system operation.
5. *Insulation co-ordination* is the selection of suitable values for the insulation levels of the various components in any electrical system and their arrangement in a rational manner.

THEORETICAL QUESTIONS

1. What are the causes of overvoltages in an electrical system ?
2. Discuss the various internal and external causes of overvoltages in a power system.
3. Explain the phenomenon of lightning stroke.
4. What are the various methods of overvoltage protection of overhead transmission lines ?
5. Explain the use of ground wire.
6. What is lightning ? Describe the mechanism of lightning discharge.
7. Describe the various types of lightning strokes.
8. How do earthing screen and ground wires provide protection against direct lightning strokes ?
9. How do we use lightning arresters on transmission lines ? How many types of lightning arresters are generally used ? Name the best type.
10. Describe various types of lightning arresters used for protection against surge voltages.
11. Briefly explain the working and relative merits and demerits of (*i*) rod gap (*ii*) expulsion arrester, and (*iii*) valve arresters.
12. What is horn gap arrester ? Explain how it works. What is the purpose of inserting a resistance between horn gap arrester and the line ?
13. Describe the construction and working of a thyrite lightning arrester.
14. (*a*) Explain how a substation and the equipments in the sub-station are protected from lightning strokes.
    (*b*) What are surge absorbers and how do they differ from surge diverters ?
    (*c*) Explain how surge absorbers provide protection to equipments from destructive effects of travelling waves.
15. What are the various types of lightning arresters ? Explain, with a neat sketch, the working of thyrite lightning arrester and give its application.
16. Why is insulation co-ordination needed in a large power system ? What is meant by Basic Impulse Level (BIL) of an equipment ?
17. Write short notes on the following :
    (*i*) Switching surges (*ii*) Lightning arresters
    (*iii*) Rod-gap arrester (*iv*) Surge absorbers.

27

Neutral Earthing (Grounding)

27.1. Introduction. 27.2. Objects of earthing. 27.3. Advantages of earthing. 27.4. Necessity of neutral earthing. 27.5. Methods of neutral earthing. 27.6. Earthed transformer. 27.7. Specifications required for earthing as per I.S.I. 27.8. Sizes of earth wire and earth plate for domestic and motor installations. 27.9. Measurement of earth resistance by earth faster. 27.10. Indian electricity rules. Highlights—Theoretical Questions—Unsolved Examples.

27.1. INTRODUCTION

- *The process of connecting the neutral point of a supply system on the non-current carrying parts of electrical apparatus to the general mass of earth in such a manner that at all times an immediate discharge of electrical energy takes place without danger is called* **earthing.**

Or

Earthing *means connecting earth-terminals to electrodes installed solidly in the mass of earth.*

Or

A *wire coming from the ground 2.5 to 3 metres deep from an electrode (plate or so) is called* **earthing.**

- The earth's potential is always taken *as zero* for all practical purposes. The electrical appliances or machines when connected with earth attain zero potential and are said to be **earthed.**
- *Good earthing* is that earthing which gives *very low resistance* to the flow of heavy current (short-circuit current) of a circuit.
- *Double earth* is used to give minimum resistance to the flow of whole current of the apparatus in case short-circuit to leakage or any other such fault occurs. Second reason is, if one earth is out of order, second will do the work.
- The earth resistance for copper wire is 1 Ω and for G.I. wire it should not be more than 3 Ω.
- *The earth resistance should be kept as low as possible.*

27.2. OBJECTS OF EARTHING

Earthing is carried out to achieve the following *objectives* :

1. To save human life from danger or shock or death by blowing fuse of any apparatus which becomes leaky.

2. To protect all machines fed from overhead lines from lightning.

3. To protect large buildings from atmospheric lightning.

4. To maintain the line voltage constant (since neutral of every alternator, transformer is earthed).

27.3. ADVANTAGES OF EARTHING

Earthing of neutrals of all industrial power systems is always preferable. Earthing is necessary as it offers many ***advantages*** given below :

1. Persistent arcing grounds is eliminated.
2. Overvoltage due to restriking is minimized.
3. The ground faults can be located and isolated fastly.
4. Steady state voltage stress to earth is reduced.
5. Sensitive protective apparatus can be used.
6. The maintenance expenditure is reduced.
7. Better safety is ensured.
8. Service reliability is improved.
9. Earthing provides improved lightning protection.

The earthing of systems should be done at the neutral of the supply transformers and generators. If the supply transformers and generators are delta connected, separate earthing transformers may be used.

27.4. NECESSITY OF NEUTRAL EARTHING

These days majority of 3-phase systems operate with an earthed neutral, *earthing being achieved either directly or through an impedance.* However, in early days, quite a number of systems used to operate with an *insulated neutral*. The main argument which goes in favour of insulated systems is the possibility of maintaining continuity of supply in case one phase gets earthed until it is convenient to disconnect and repair the faulty line. This argument is specifically true for an overhead transmission system when the failure of one line to earth is not likely to develop into a fault between two or more lines.

Fig. 27.1 shows a simple 3-phase system with insulated neutral. The line conductors have capacitance between one another and to earth, the former being delta connected, while latter are star-connected, as shown in Fig. 27.1 (*a*). The effect of line capacitances on the grounding characteristic of the system is little and, therefore, can be neglected.

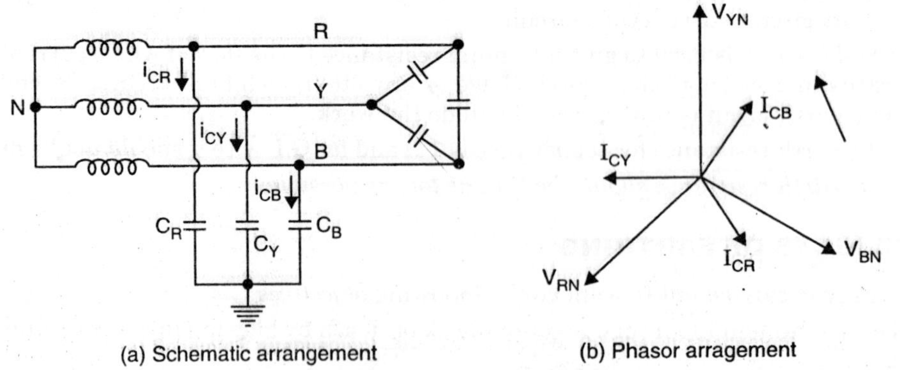

Fig. 27.1. Insulated neutral system.

Let us first consider a 3-phase line (perfectly transposed) having *same capacitances to ground.* In this case, the charging currents for each line to earth capacitor lead the phase voltage by 90° and are equal. The magnitude of these currents is given by $\frac{V_{ph}}{X_c}$, where V_{ph} is the phase voltage and X_C

is the reactance due to the capacitance of the line to ground. The charging currents I_{CR}, I_{CY} and I_{CB} are balanced and their *result is zero, and no current flows to the earth.* Now consider a phase of earth fault in the line B say at point F. Under these circumstances, the faulty line takes up the earth potential while the potentials of the remaining two healthy phases R and Y rise from phase values to line values. The capacitance currents become unbalanced and fault current I_F flows through the faulty line, into the fault and return to the system via earth and through the earth capacitances C_R and C_Y. Thus fault current I_F has two components I_{BR} and I_{BY} which flow through capacitances C_R and C_Y respectively under the potential differences of V_{BR} and V_{BY} respectively. These currents lead their respective voltages by 90° and their vector sum is equal to fault current I_F as shown in Fig. 27.2 (*b*).

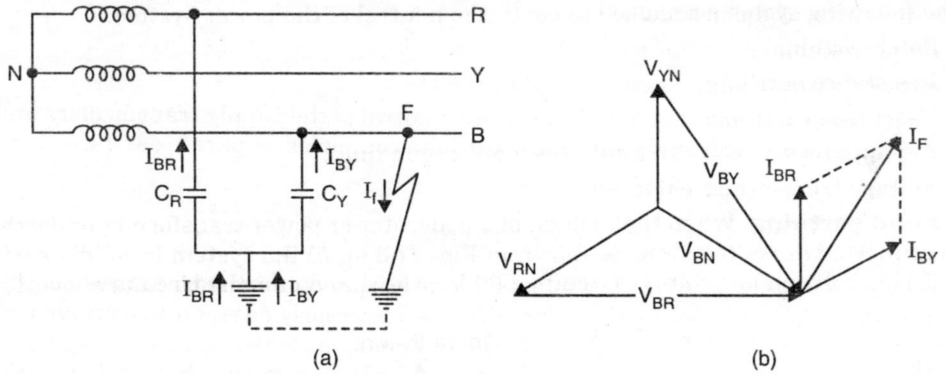

Fig. 27.2. A ground fault in one of the phases of the ungrounded system.

Since phase voltages are equal, say equal to V_{ph} and capacitances from line to ground are also equal, say equal to X_C, we have :

$$I_{BR} = \frac{V_{BR}}{X_C} = \frac{\sqrt{3}\, V_{ph}}{X_C}$$

Similarly,
$$I_{RY} = \frac{V_{BY}}{X_C} = \frac{\sqrt{3}\, V_{ph}}{X_C},$$

and thus,
$$I_F = \sqrt{3}\, I_{BR} = \sqrt{3} \times \frac{\sqrt{3}\, V_{ph}}{X_C} = \frac{3V_{ph}}{X_C} \quad ...(27.1)$$

From the above discussion, the following *conclusions can be drawn :*

1. No zero sequence current.

2. Little interference with communication lines (since the fault current is small).

3. In case of one phase becoming earthed the voltages of the remaining two healthy phases to earth rise from their normal phase to neutral.

4. The capacitive current in the faulty phase is 3 times its normal value.

5. The capacitive current in the two healthy phases increased to $\sqrt{3}$ times their normal values.

6. The fault current in the overhead lines may be so small as to render automatic isolation by protective means difficult, if not impossible.

7. A capacitive fault current I_F flows into the earth. Such a current if exceeds 4 A, is sufficient to maintain an arc in the ionized path of the fault, even though the medium causing the fault has cleared itself. The persistency of the arc due to flow of capacitive current gives rise to a condition known as "*arcing ground*" in which cycle charging and discharging of the system capacity through

the fault results in high frequency oscillations being superimposed on the whole system and a build up of very high voltages c an occur. This results in phase voltage to rise to 5 to 6 times of normal voltage. The build up of high voltage may cause breakdown of the insulation.

The systems are *not* operated with ungrounded neutrals because of the following *reasons* :

(*i*) The insulation of such a system in likely to be overstressed by overvoltage.

(*ii*) Such a system cannot be adequately protected for fault to earth.

(*iii*) Insulation overstress may give rise to insulation failure on other parts of the system which may lead to heavy phase-to-phase fault conditions.

27.5. METHODS OF NEUTRAL EARTHING

The following systems are used to earth the neutral of the power system.

1. Solid earthing.
2. Resistance earthing.
3. Reactance earthing.
4. Arc-suppression coil earthing (Resonant grounding).
5. Voltage transformer earthing.

1. Solid earthing. When the neutral of a generator or power transformer or earthing transformer is connected *direct* to earth, as shown in Fig. 27.3 (*a*, *b*) the system is solidly earthed. Solid earthing is generally on low voltage circuits, 600 V or less, and on most circuits above 15 kV.

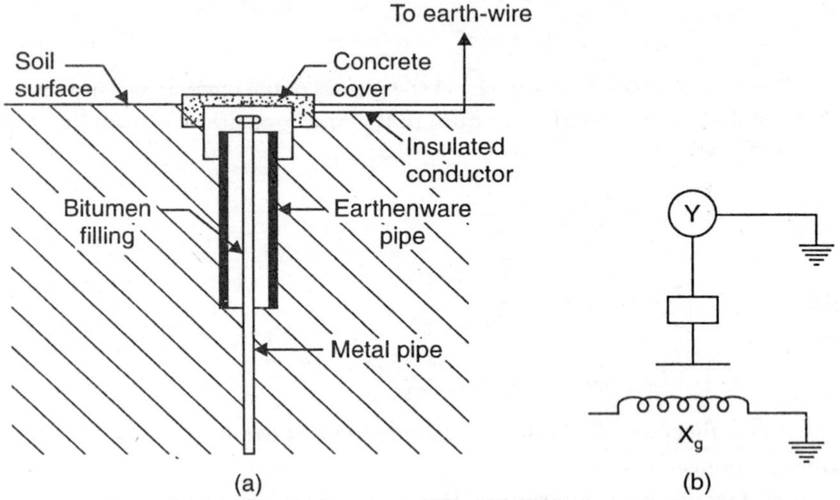

Fig. 27.3. Solid earthing.

When a neutral point is connected to earth it is ensured that the neutral point will be at *earth potential under all operating conditions* and consequently the voltage of any conductor to earth under fault conditions will not exceed the normal phase voltage of the system.

In solid earthing, a direct metallic connection is made from the system neutral to one or more earth electrodes consisting of *plates*, *rods* or *pipes* buried in or driven into the ground, generally in a bed of *coke* below the permanent water level. Where permanent moist ground cannot be ensured a supply of water should be made available so that the periodic watering can be done. The value of contact resistance or resistance of the earthing system should as small as possible and this should be checked from time to time. Now-a-days the term 'solidly earthed' has been replaced by the term "*Effectively earthed.*"

A transformer neutral may be 'solidly earthed' in that there may be no impedance between the neutral and earth. However, the transformer capacity thus solidly earthed may be too small in comparison with the sizes of the system to be effective in stabilizing the voltage from phases to ground when the fault takes place.

In case of solidly grounded system, the following points are worth noting :

1. The flow of heavy fault current will completely multiply the effect of capacitive current at the fault and so *no arcing ground phenomena or overvoltage can occur.*

2. The flow of heavy current permits the use of discriminative protective gear.

3. Increased earth fault current results in greater influence on neighbouring communication circuits.

4. Since the fault has to be cleared by circuit breakers, heavy earth fault currents may result in burning of contacts.

5. When a fault to earth occurs on any one phase of the system, the voltage to earth of the faulty phase becomes zero, but the healthy phases, in general, remain at their normal phase value. As such *lightning arresters* rated for phase voltage can be employed. Also, the equipment should be insulated for phase voltage. Thus a *saving in cost of equipment results.*

- The use of 'solid earthing' is *limited only to systems where the normal circuit impedance is sufficient to prevent very high value of fault current being reached.* This is necessary in order to avoid excessive damage at the fault location.

2. Resistance earthing. When it becomes necessary to limit the earth fault current, a current limiting device is introduced in the neutral and earth. One method of introducing a current limiting device is *resistance earthing.* The resistors used in this case may be metallic carbon powder or liquid type. The magnitude of the resistance to be used should be such that it should limit the earth fault current to a value which will cause minimum damage at the point of fault.

Fig. 27.4 shows the neutral of a generator connected to the earth through a resistor.

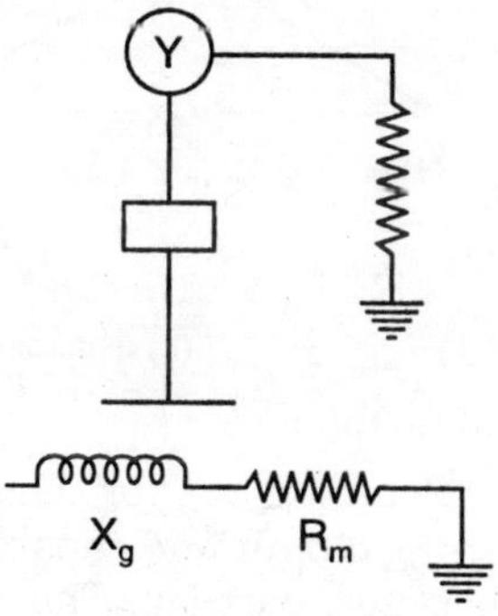

Fig. 27.4. Resistance earthing.

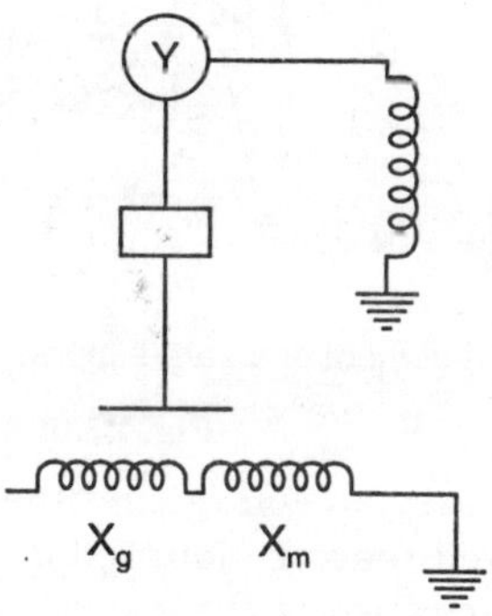

Fig. 27.5. Reactance earthing.

Advantages :

1. The ground faults are readily relayed.
2. The hazards of arcing grounds are minimised.
3. The mechanical stresses in the circuit carrying fault currents are reduced.
4. Least inductive influence on neighbouring circuit.
5. Electric shock hazards to the persons, caused by stay-earth fault currents in the earth return path, are reduced.
6. Line voltage drop caused by the occurrence of earth fault is reduced.

Disadvantages :

(*i*) This system is costlier than solidly grounded system.

(*ii*) The system neutral will almost invariably be displaced in case of a ground fault, thereby necessitating the use of lightning arrester at increased cost and sacrifice in protective performance.

3. Reactance earthing. In this case a *reactor* is connected between the neutral of the machine and earth as shown in Fig. 27.5. Sometimes a low reactance is connected in series with the neutral of the machine to limit the earth fault current. This current should not be greater than the 3-phase fault current of the generator. The earth-fault-current of this system should not be less than 25% of the 3-phase fault current in order to minimise the transient voltages.

A reactance earthed system *ensures satisfactory relaying, partial grading of the apparatus insulation, reduced interference to neighbouring communication as compared with that in solidly earthed system.*

The main *drawback* of this system is *very high transient overvoltages*, due to which this system is not in common use.

4. Arc-suppression coil earthing (Resonant grounding). In this method of earthing the danger of *arcing grounds is eliminated* and the system is approximated to the isolated neutral system, in which one or two healthy phases continue supply power and *complete shut down is avoided.*

It works on the *principle* that if an inductance of appropriate value is connected in parallel with the capacitance, the fault current can be reduced considerably or even it can be neutralised.

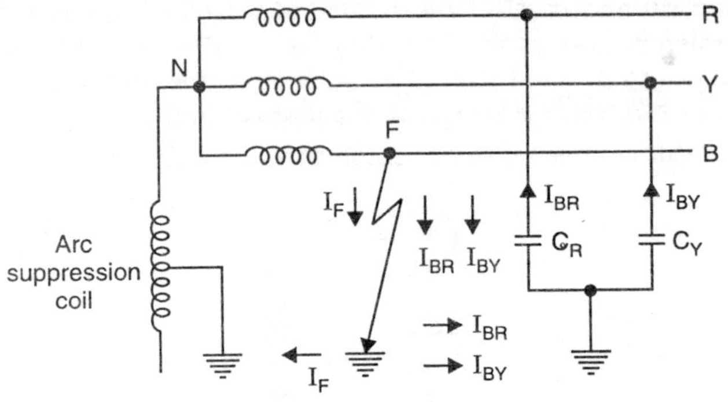

(a) Schematic arragement of an arc-suppression coil earthing

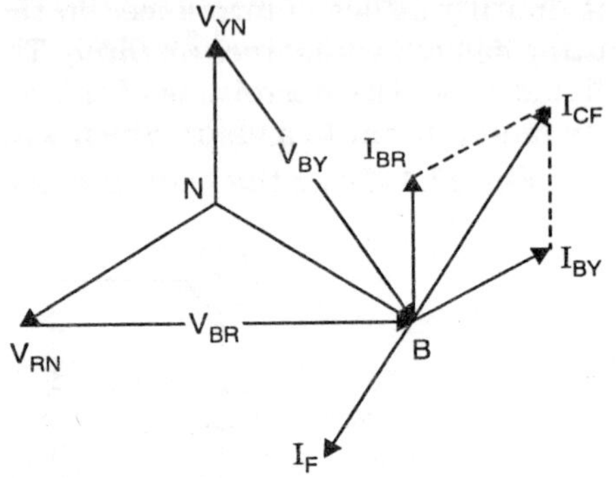

(b) Phasor diagram

Fig. 27.6. Arc suppression coil earthing.

An arc-suppression coil (sometimes referred to as "Peterson coil" or "Ground fault neutralizer") is an iron cored reactor mounted in the neutral earthing circuit and capable of being tuned to *resonate* with capacitance of the system when one line becomes earthed. The function of the arc-suppression coil is to make arcing earth faults self-extinguishing and in the case of sustained faults to reduce the earth current to low value so that system can be kept in operation with one line earthed.

This method of neutral earthing is usually confined to medium voltage overhead transmission lines which are connected to generating source through intervening power transformers. The reason for this limitation is application is due to the fact that the *higher level insulation is required on apparatus associated with arc-suppression coil earthed systems.*

Fig. 27.6 (*a*) and (*b*) show schematic arrangement of an arc-suppression coil earthing, and phasor diagram respectively ; B-phase is earthed by a single line to earth fault on an arc-suppression coil neutral earthed system. The resultant capacity current is 3 times the normal line to neutral charging current of one phase as derived below :

$$I_{CF} = \sqrt{3}\, I_{BR} = \sqrt{3}\left(\frac{V}{X_C}\right)$$

$$= \sqrt{3}\left(\frac{\sqrt{3}\, V_{ph}}{X_C}\right) = 3 \times \frac{V_{ph}}{X_C} = 3 \times \text{line-to-neutral charging current}$$

The inductance of the coil can be determined as follows :

$$I_{CF} = 3 \times \frac{V_{ph}}{X_C}$$

Also, $I_F = \dfrac{V_{ph}}{X_L}$, where X_L is the reactance of the coil

At resonance, $I_F = I_{CF}$

i.e., $$\frac{V_{ph}}{X_L} = \frac{3\, V_{ph}}{X_C}$$

or, $$X_L = \frac{X_C}{3} \quad \text{or} \quad \omega L = \frac{1}{3\,\omega C}$$

or, $$L = \frac{1}{3\omega^2 C} \qquad \text{...(27.2)}$$

Some difficulty arises when due to varying operational conditions the capacitance of the network varies from time to time. It can be overcome, however, by using a *"tapped coil"*, the approximate tapping being used for each change in network conditions (Fig. 27.7).

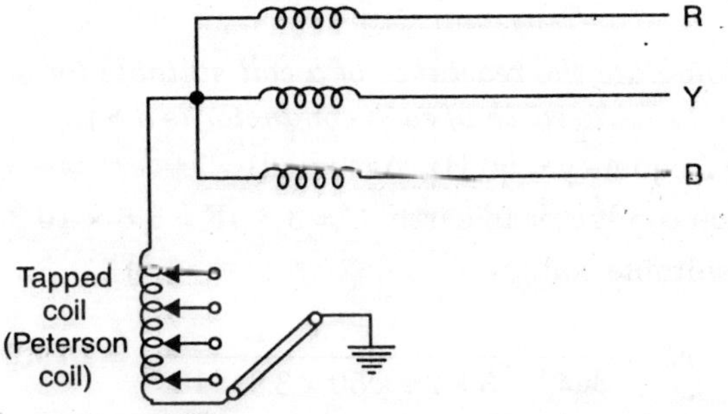

Fig. 27.7

The current rating of coil is given by :

$$I_F = \frac{3V_{ph}}{X_C} \qquad \text{...(27.3)}$$

5. Voltage transformer earthing. Fig. 27.8 shows the schematic arrangement of voltage transformer earthing. Here, primary of a voltage transformer is connected between the neutral and the ground. A low resistor and a relay combination is connected across the secondary of the voltage transformer.

In earth fault on the primary side will produce a voltage across the relay that will cause the protective device to operate.

- The voltage transformer acts as a reflection point for travelling waves passing through the machine winding and in order to prevent high voltage built-up it *is necessary to connect a surge diverter between the machine neutral point and the earth.*

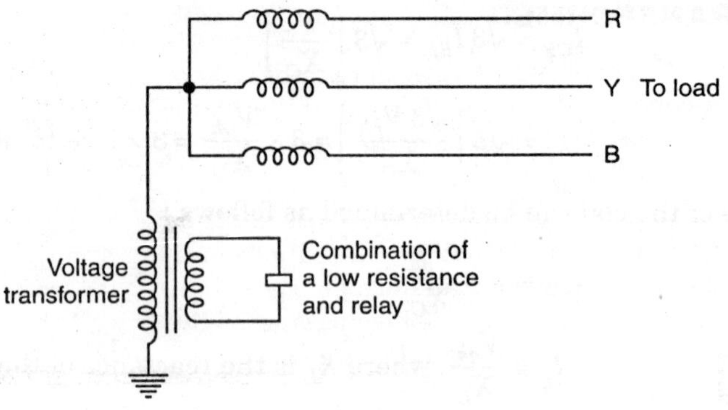

Fig. 27.8. Voltage transformer earthing.

Advantages :

(*i*) Arcing grounds are eliminated.

(*ii*) The neutral is earthed through a voltage transformer and thus its operational characteristics are similar to an isolated neutral system.

(*iii*) Since voltage transformer provides extremely high reactance to the earth path, therefore, the transient overvoltages due to switching and arcing ground are *reduced*.

Disadvantage :

The major drawback of the system is that earthed neutral *acts as a reflection point for travelling waves* through the machine winding and it should be prevented with a surge diverter (as earlier stated).

Example 27.1. *Calculate the reactance of a coil suitable for a 33 kV, 3-phase transmission system of which the capacitance to earth of each conductor is 3.8 μF.* **(Panjab University)**

Solution. Supply frequency = 50 Hz (Assumed)

Capacitance of each conductor to earth, $C = 3.8\ \mu F = 3.8 \times 10^{-6}$ F

Reactance of the suitable coil,

$$\omega L = \frac{1}{3\omega C} = \frac{1}{3 \times 2\pi \times 50 \times 3.8 \times 10^{-6}} = \mathbf{279.2\ \Omega.} \quad \textbf{(Ans.)}$$

Example 27.2. *A 230 kV, three-phase, 50 Hz 250 km transmission line has a capacitance to earth of 0.025 μF/km per phase. Calculate the inductance and kVA rating of the Peterson coil used for earthing the above system.* **(AMIE)**

Solution. Supply frequency, $f = 50$ Hz,

Capacitance of each line to earth = $(0.025 \times 10^{-6}) \times 250 = 6.25 \times 10^{-6}$ F

Inductance of Peterson coil,

$$L = \frac{1}{3\omega^2 C} = \frac{1}{3 \times (2\pi \times 50)^2 \times 6.25 \times 10^{-6}} = \mathbf{0.54\ H.} \quad \textbf{(Ans.)}$$

$$\text{Current through the Peterson coil, } I_L = \frac{V_{ph}}{\omega L} = \frac{(230 \times 1000)/\sqrt{3}}{2\pi \times 50 \times 0.54} = 782.7 \text{ A}$$

$$\therefore \quad \text{Rating of the Peterson coil} = \frac{230 \times 1000}{\sqrt{3}} \times 782.7 \times \frac{1}{1000} \text{ kVA} = \mathbf{103935\ kVA.} \quad \textbf{(Ans.)}$$

27.6. EARTHED TRANSFORMER

When the access to the existing neutral is not possible or when the system is delta-connected and availability of neutral is essential *earthed transformers are used to create an artificial neutral.*

Fig. 27.9 shows the basic connections of an earthed transformer. It is a core type transformer having three limbs built up in the same fashion as that of power transformer. Each limb accommodates two equally spaced windings. The currents in two halves of the winding of each limb are in *opposite directions.* Consequently there will *not be undesirable harmonic prevailing in the circuit and so the stresses on the insulation of the transformer are considerably reduced.*

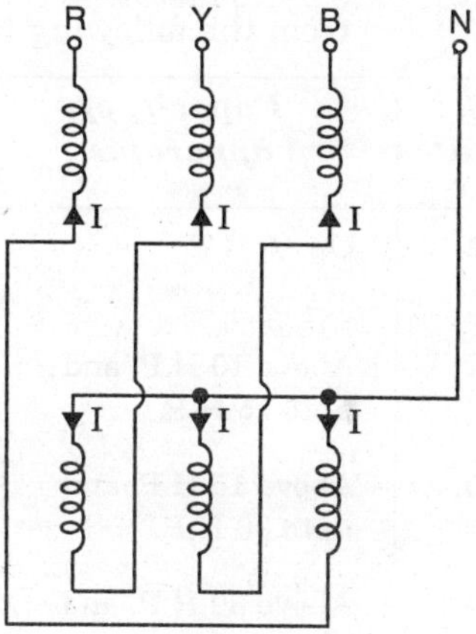

Fig. 27.9

Since the impedance of earthing transformer is quite low, the fault current is quite high. In order to limit the magnitude of the fault current, the resistance or resistances are inserted in the neutral circuit or in the windings of earthing transformer.

Under healthy conditions (*i.e.*, when there is no fault), only iron losses will be occurring continuously but at the *time of fault copper losses will also occur in addition*, because of heavy fault current, in the earthing transformer. The duration of occurrence of copper losses is quite short, being the duration of fault, which is usually of 30 to 60 seconds duration.

27.7. SPECIFICATIONS REQUIRED FOR EARTHING AS PER I.S.I.

Following are *recommended specifications as per I.S.I.* for providing good earthing :

1. The earthing electrode should be situated at a place atleast $1\frac{1}{2}$ metres away from the building (outside) whose installation system is being earthed.

2. The earth wire should be of same material as that of earth electrode used.

3. The minimum sectional area of earth lead wire should not be less than 0.02 sq. inch (No. 8 S.W.G.) and not more than 0.1 sq. inch.

4. The size of earth conductor as a general rule should not be less than half of the section of the line conductor.

5. The earth wire should be taken through G.I. pipe of 12 mm dia. for at least 32 cm length above and below ground surface to the earth electrode to safeguard against mechanical wear and tear.

6. Loose earth and coal salt mixture should be filled around the earth electrode for effective earthing.

7. The earth wire connected to the earth electrode *shall not be* necessarily run along the whole wiring system. All the earth wire run along the sub-circuits should be terminated and hooked firmly at the main board and from where the main earth wires should be run to the earth electrode. The loop earth wires should be of 14 S.W.G. copper wire.

8. All the joints in the earth wire should be firmly done with nuts and bolts of the same material as of earth wire.

27.8. SIZES OF EARTH WIRE AND EARTH PLATE FOR DOMESTIC AND MOTOR INSTALLATIONS

- For *household wiring or installation*, a 14 SWG hand drawn bare copper conductor is employed as earth wire.

- For *power installations*, the size of the earth wire is decided on the basis of rating of motors installed. The size of the earth wire, for different ratings of motors, can be selected from the following table :

S. No.	*Capacity of apparatus*	*Size of earth wire in SWG*		*Size of earth electrode*	
		Copper	*G.I.*	*Copper*	*G.I.*
1.	Up 10 H.P.	No 8	No 8	60 cm × 60 cm × 3 mm	60 cm × 60 cm × 6 mm
2.	Above 10 H.P. and upto 15 H.P.	No 8	No 6	–do–	–do–
3.	Above 15 H.P. and upto 30 H.P.	No 6	No 2	–do–	90 cm × 90 cm × 6 mm
4.	Above 30 H.P. and upto 50 H.P.	No 4	—	90 cm × 99 cm × 6 mm	—
5.	Above 50 H.P. and upto 100 H.P.	No 2 or strip 13 mm × 2.5 mm	—	–do–	—
6.	Above 100 H.P.	Strip 25 mm × 2.5 mm	—	–do–	—

27.9. MEASUREMENT OF EARTH RESISTANCE BY EARTH TESTER

The earth resistance can be measured with the help of *Earth tester* or a *Megger earth tester.*

The **Megger earth tester** (Fig. 27.10) is essentially a *direct reading ohmmeter and a hand driven generator which supplies the testing current.*

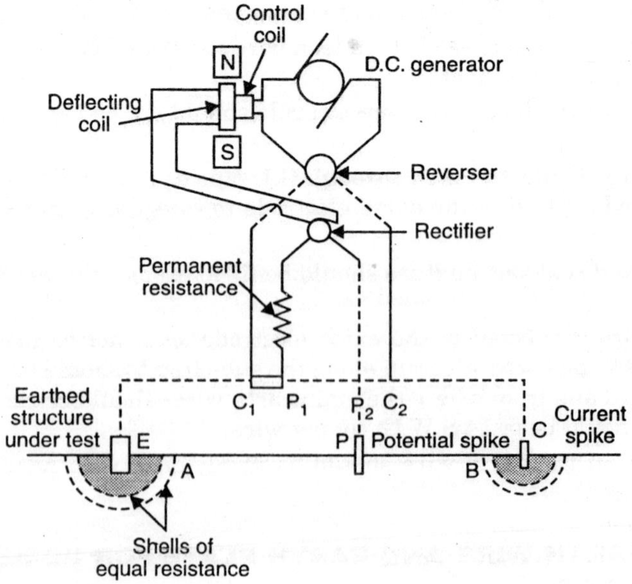

Fig. 27.10. Megger tester circuit containing direct current shown by full lines ; circuit containing A.C. current shown by dotted lines.

Construction :

- The ohmmeter consists of two coils (current coil and potential coil) mounted at a fixed angle to each other on a common axle. The current coil carries current proportional to the current flowing in the test circuit, while the potential coil carries current proportional to the potential across the resistance under test. Thus the potential coil acts as a voltmeter while current coil acts as an ammeter. Since the deflection of the needle is proportional to the ratio of the current in the two coils, it gives *resistance directly*.
- The hand operated generator produces the direct current, but to eliminate the effect of electrolytic *e.m.f.*, it is necessary to pass alternating current through the soil, so to change the D.C. into an alternating supply a rotary current reverser is mounted on the same shaft of the generator. The alternating current in the soil will produce an alternating drop in the soil but the potential to be applied across the moving coil must be direct because the ohmmeter in a moving coil instrument working on D.C. alone, so for changing the alternating drop into the direct drop a synchronous rotary rectifier is also attached as shown in Fig. 27.10. Although the instrument is immune to stray currents, yet while taking readings sometimes it will be observed that the needle vibrates ; this happens only at the instant when the stray alternating current has the same frequency as the frequency generated. For such occasions either *increase the speed of the generator a little or lower its speed.*

Working :

- For *measurement of earth resistance* two spikes acting as current and potential electrodes are driven into the ground at a suitable distance (C_2 at a distance of about 25 m from earth electrode under test). P_1 and C_1 terminals of the meter are shorted and are connected to the earth electrode under test.
- First spike P_2 is fixed between the spike C_2 and the earth electrode and the resistance is determined ; then this spike taken 3 to 4 m near to the earth electrode and then for the same distance near to electrode C_2 and the two readings are again noted. The average of these three readings is the required earth resistance.

27.10. INDIAN ELECTRICITY RULES

A few important/relevant Indian Electricity Rules are mentioned below :

- **Rule 33. Earthed terminal or consumer's premises :**

1. The supplier shall provide and maintain on the consumer's premises for the consumer's use a suitable earthed terminal in an accessible position at or near the point of commencement of supply as defined under rule 58.

Provided that in the case of medium, high or extra high voltage installation, the consumer shall, in addition to the afore-mentioned earthing arrangement, provide his own earthing system with an independent electrode,

Provided further that the supplier may not provide any earthed terminal in the case of installations already connected to his system on or before the date to be specified by the State Government in this behalf if he is satisfied that the consumer's earthing arrangement is efficient.

2. The consumer shall take all reasonable precautions to prevent mechanical damage to the earthed terminal and its lead belonging to the supplier.

3. The supplier may recover from the consumer the cost of installation of such earthed terminal on the basis laid down in the sub-rule (2) of rule 82.

- **Rule 46. Periodical inspection and testing of consumer's installation :**

1. (*a*) Where an installation is already connected to supply system of the supplier, every such installation shall be periodically inspected and tested at intervals not exceeding

five years either by the Inspector or by the supplier as may be directed by the State Government in this behalf or in the case of installations in mines, oil-fields and railways by the Central Government.

(*b*) Where the supplier is directed by the Central or the State Government, as the case may be, to inspect and test the installation, he shall report on the condition of the installation to the consumer concerned in a form approved by the Inspector and shall submit a copy of such report to the Inspector.

2. (*a*) The fees for such inspection and test shall be determined by the Central or the State Government, as the case may be, in the case of each class of consumers and shall be payable by the consumer in advance.

(*b*) In the event of the failure of any consumer to pay the fees on or before the date specified in the fee-notice, supply to the installation of such consumer shall be liable to be disconnected under the direction of Inspector. Such disconnection, however, shall not be made by the supplier without giving to the consumer even clear day's notice in writing of his intention so to do.

3. Notwithstanding the provisions of this rule, the consumer shall at all times be solely responsible for the maintenance of his installation in such condition as to be free from danger.

- **Rule 47. Testing of consumer's installation :**

1. Upon receipt of an application for a new or additional supply of energy and before connecting the supply or reconnecting the same after a period of six months, the supplier shall inspect and test the applicant's installation.

The supplier shall maintain a record of test results obtained at each supply point to a consumer, in a form to be approved by the inspector.

2. If as a result of such inspection and test, the supplier is satisfied that the installation is likely to constitute danger, he shall serve on the applicant a notice in writing requiring him to make such modifications as are necessary to render the installation safe. The supplier may refuse to connect or reconnect the supply until the required modifications have been completed and he has been notified by the applicant.

- **Rule 61. Connection with earth :**

1. The following provisions shall apply to the connection with earth of systems at low voltage in cases where the voltage normally exceeds 125 volts and of systems at medium voltage.

(*a*) The neutral conductor of a three-phase four-wire system and the middle conductor of a two-phase three-wire system shall be earthed by not less than two separate and distinct connections with earth both at the generating station and at the sub-station. It may also be earthed at one or more points along the distribution system or service line in addition to any connection with earth which may be at the consumer's premises.

(*b*) In case of a system comprising electric supply lines having concentric cables, the external conductor of such cable shall be earthed by two separate and distinct connections with earth.

(*c*) The connection with earth may include a link by means of which the connection may be temporarily interrupted for the purpose of testing or for locating a fault.

(*d*) (*i*) In a direct current three-wire system the middle conductor shall be earthed at the generating station only, and the current from the middle conductor to earth shall be continuously recorded by means of a recording ammeter, and if at any time the current exceeds one-thousandth part of the maximum supply current, immediate steps shall be taken to improve the insulation of the system.

(*ii*) Where the middle conductor is earthed by means of a service-breaker with a resistance connected in parallel the resistance shall not exceed 10 ohms and on the opening of the circuit-breaker immediate steps shall be taken to improve the insulation of the system and the circuit-breaker shall be reclosed as soon as possible.

(*iii*) The resistance shall be used only as a protection for the ammeter in the case of earths on the system and until such earths are removed, immediate steps shall be taken to locate and remove the earth

(*e*) In the case of an alternating system, there shall not be inserted in the connection, with earth any impedance (other than that required solely for the operation of switch-gear or instruments), cut-out or circuit-breaker, and the result of any test made to ascertain whether current (if any) passing through the connection with earth is normal, shall be duly recorded by the supplier.

(*f*) No person shall make connection with earth by the aid of, nor shall he kept it contact with, any water main not belonging to him except with the consent of the owner thereof and of the Inspector.

(*g*) Alternating current systems which are connected with earth as aforesaid may be electrically interconnected.

Provided that each connection with earth is bounded to the metal sheathing and metallic armouring (if any) of the electric supply lines concerned.

2. The frame of every generator, stationary motor, and so far as is practicable, portable motor, and the metallic parts (not intended as conductors) of all transformers and any other apparatus used for regulating or controlling energy and all medium voltage energy consuming apparatus shall be earthed by the owner by two separate and distinct connections with earth.

3. All metal casings or metallic coverings containing or protecting any electric supply line or apparatus shall be connected with earth and shall be so joined and connected across all junction boxes and other openings as to make good mechanical and electrical connection throughout their whole length :

Provided that where the supply is at low voltage, the sub-rule shall not apply to the isolated wall tubes or to brackets, switches, fans regulator covers or other fittings (other than portable hand lamps and portable and transportable apparatus) unless provided with earth terminal.

This sub-rule shall come into force immediately in the case of new installations and in case of existing installation ; the provision of this sub-rule shall be complied with before the expiry of a period of two years from the commencement of those rules.

4. All earthing systems shall, before electric supply lines or apparatus are energised, be tested for electrical resistance to ensure efficient earthing.

5. All earthing systems belonging to the supplier shall, in addition, be tested for resistance on dry day during the dry season not less than once every two years.

6. A record of every earth test made and the result there of shall be kept by the supplier for a period of not less than two years after the day of testing and shall be available to the Inspector when required.

- **Rule 67. Connection with earth :**

1. The following provisions shall apply to the connection with earth of three-phase systems for use at high or extra-high voltages :

In the case of star connected systems with earthed neutrals or delta connected systems with earthed artificial neutral point :

(*a*) the neutral point shall be earthed by not less than two separate and distinct connections with earth each having its own electrode at the generating station and at the sub-station and may be earthed at any other point, provided that no interference of any description is caused by such earthing ;

(*b*) in the event of an appreciable harmonic current flowing in the neutral connections so as to cause interference with communication circuits, the generator or transformer neutral shall be earthed through a suitable impedance.

2. Single-phase or extra-high voltage systems shall be earthed in a manner approved by the Inspector.

3. In the case of a system comprising electric supply lines having concentric cables, the external conductor shall be the one to be connected with earth.

4. Where a supplier purposes to connect with earth an existing system for use at high or extra-high voltage which has not either been so connected with earth, he shall give not less than fourteen days notice in writing together with particulars to the telegraph authority of the proposed connection with earth.

5. Where the earthing lead and earth connection are used only in connecting earthing guards created under high or extra high voltage overhead lines where they cross a telecommunication line or a railway line, and where such lines are equipped with earth leakage relays of a type and setting approved by the Inspector, the resistance shall not exceed 25 ohms.

6. In so far as the provisions of rule 61 are consistent with the provisions of this rule, all connections with earth shall also comply with the provisions of that rule.

HIGHLIGHTS

1. The process of connecting the neutral point of a supply system on the non-current carrying parts of electrical apparatus to the general mass of the earth in such a manner that at all times an immediate discharge of electric energy takes place without danger is called *earthing*.
2. *Methods of neutral earthing :*
   (*i*) Solid earthing
   (*ii*) Resistance earthing
   (*iii*) Reactance earthling
   (*iv*) Arc-suppression coil earthing (Resonant grounding)
   (*v*) Voltage transformer earthing.
3. When the access to the existing neutral is not possible or when the system is delta-connected and availability of neutral is essential *earthed transformers are used to create an artificial neutral.*

THEORETICAL QUESTIONS

1. Define the term "Earthing".
2. What are the functions of grounding in power system ?
3. State the advantages of neutral grounding of an electrical system.
4. What are the objects of earthing.
5. What is the necessity of neutral earthing ? Explain.

6. What is earthing of neutral ? Why is it done ? Briefly discuss different types of earthing the neutral.
7. Discuss and compare various methods of neutral grounding.
8. Explain the reasons leading to the general practice of earthing the neutral point of a power system and discuss he relative merits and demerits of earthing it (*i*) solidly and (*ii*) through a resistance.
9. Explain different types of earthing the neutral point of a power system. Derive an expression for the reactance of the Peterson coil in term of the capacitance of the protected line.
10. Explain the meaning of the term "Resonant grounding". How is this achieved by the use of Peterson coil ? What are its advantages ?
11. Explain the functions of Persons coil and how it helps in reducing "arcing grounds".
12. Write short notes on :
    (*i*) Earthing of neutral.
    (*ii*) Peterson coil.
    (*iii*) Earthed transformer.
    (*iv*) Neutral earthing arrangement in power stations.

UNSOLVED EXAMPLES

1. Calculate the reactance of a coil suitable for a 33 kV, 3-phase transmission system of which the capacitance to earth of each conductor is 4.5 μF. **[Ans.** 235.8 Ω)
2. A 230 kV, 3-phase, 50 Hz, 200 km transmission line has a capacitance to earth of 0.01 μF/km per phase. Calculate the inductance and kVA rating of the Peterson coil used for earthing the above system. **[Ans.** 1.7 H ; 33012 kV]
3. Derive an expression for the reactance of the Peterson coil in terms of the capacitance of the protected line. Calculate the reactance of a coil suitable for a 33 kV, 3-phase transmission system of which the capacitance to earth of each conductor is 5.0 μF. **[Ans.** 212.2 Ω]

28

Sub-Stations

28.1. Introduction. 28.2. Functions of sub-stations. 28.3. Classification of sub-stations. 28.4. Indoor sub-stations. 28.5. Outdoor sub-stations. 28.6. Transformer sub-stations. 28.7. Symbols for circuit elements/equipment in sub-stations. 28.8. Equipment for sub-stations and switchgear installations. 28.9. Key diagram of 11 kV/400 V indoor sub-station. 28.10. Earthing. Highlights—Theoretical Questions.

28.1. INTRODUCTION

The assembly of apparatus used to change some characteristic (*e.g. voltage, A.C. to D.C., frequency, p.f., etc.*) *of electric supply is called a* **sub-station.**

The sub-stations *serve as sources of energy supply* for local areas of distribution in which these are located.

28.2. FUNCTIONS OF SUB-STATIONS

The main functions of sub-stations are :

1. To *receive energy transmitted at high voltage* from the generating stations.
2. To *reduce the voltage* to a value appropriate for local distribution.
3. To provide facilities for switching.

Sub-stations have some following *additional functions :*

They provide points where safety devices may be installed to disconnect equipment or circuit in the event of fault. A sub-station is a convenient place for installing synchronous condensers at the end of transmission line for the purpose of improving power factor and make measurements to check the operations of various parts of the power system. Street-lighting equipment as well as switch control for street lights can be installed in a sub-station.

Some sub-stations are simply *switching stations* where different connections between various transmission lines are made, others are *converting sub-stations* which convert either A.C. to D.C. or vice versa or convert frequency from higher to lower or versa.

28.3. CLASSIFICATION OF SUB-STATIONS

The sub-stations may be *classified* as follows :

I. According to service :

1. Transformer sub-stations

(*i*) Transmission and primary sub-stations.

(*ii*) Sub-transmission and secondary sub-stations.

(*iii*) Distribution sub-stations.

2. Industrial sub-stations.
3. Switching sub-stations.
4. Power factor correction or synchronous sub-stations.
5. Frequency charger sub-stations.
6. Converting sub-stations.

II. According to design :

1. Indoor type sub-stations.
2. Outdoor sub-stations.
   (*i*) Pole-mounted sub-stations.
   (*ii*) Foundation-mounted sub-stations.

28.4. INDOOR SUB-STATIONS

According to construction indoor distribution transformer sub-stations and high voltage switch boards are further *subdivided* as follows :

1. Sub-stations of integrally built type. In such sub-stations the apparatus is installed on site. The all structures are constructed of concrete or brick.

2. Sub-stations of the composite built-up type. Here the assemblies and parts are factory or workshop fabricated, but are assembled on site within a sub-station switchgear room. The components of substations take form of metal cabinets or enclosures, each of which contains the equipment of one main connection cell. Within such cabinets or enclosures an oil circuit breaker, a load interrupter switch, one or more voltage transformers may be mounted.

3. Unit type factory fabricated sub-stations and metal clay switch boards. These are built in electrical engineering workshops and are shipped to the site of installation fully pre-assembled. After installation of sub-stations switch board only connection to the incoming and outgoing power circuits are required to be made.

28.5. OUTDOOR SUB-STATIONS

Outdoor sub-stations are of the following *two types* :

1. Pole-mounted sub-stations.
2. Foundation-mounted sub-stations.

1. Pole-mounted sub-stations :

- These sub-stations are erected for mounting *distribution transformers of capacity upto 300 kVA.*
- Such sub-stations are *cheapest*, *simple* and *smallest of sub-stations.*
- All the equipment is of outdoor type and mounted on the supporting structure of high tension distribution line. Triple Pole Mechanically Operated (T.P.M.O.) switch is used for switching "on" and "off" of high tension transmission line. Lightning arresters are installed over the high tension line to protect the high tension line and to protect the transformer from the surges.
- Sub-station is *earthed at two or more places.*
- Generally transformers upto 125 kVA are mounted on double pole structure and for transformers of capacity above 125 kVA but not exceeding 300 kVA, 4-pole structure with suitable platform is used.

- The pole-mounted sub-station is *erected in very thickly populated location.* The *maintenance cost* of such sub-stations is *low* and by using a large number of sub-stations in a town it is possible to lay the distributors, at a lower cost.

2. Foundation-mounted sub-stations :

- Such sub-stations are *built entirely in the open* and all the equipment is assembled into one unit usually *enclosed by a fence* from the point of view of safety.
- Sub-stations for primary and secondary transmission and for secondary distribution (above 300 kVA) are foundation mounted outdoor type. Since equipment required for such sub-stations is heavy, therefore, site selected for these sub-stations must have a good access for heavy transport.
- The switchgear consists of circuit breakers of suitable type on both the sides but with the increased reliability of the modern transformers, the practice is to dispense with the circuit breaker on the incoming side from the economic consideration. The isolating switches thus serve the purpose.

While ***selecting the sites*** for such sub-stations, the following factors should be given due *considerations :*

1. Nearness to the load centres of distribution areas.
2. Availability of land.
3. Cost of land.
4. Local zoning laws.
5. Future load growth.

Advantages and disadvantages of "Outdoor sub-stations" :

Advantages. The outdoor sub-stations have the following *main advantages over indoor sub-stations* :

(*i*) Fault location is easier since all the equipment is within view.

(*ii*) The time required to erect such substations is less.

(*iii*) The extension of the installation is easier.

(*iv*) The smaller amount of building material is required.

(*v*) Switchgear installation cost is low.

(*vi*) There is practically no danger of a fault which appears at one point being carried over to another point in the installation because the apparatus of the adjoining connections can be spaced liberally, without any appreciable increase in costs.

Disadvantages :

(*i*) More space required.

(*ii*) The various switching operations with the insulators, as well as supervision and maintenance of the apparatus are to be performed in the open air during all kinds of weather.

(*iii*) The influence of rapid fluctuation in ambient temperature and dust and dirt deposits upon the outdoor sub-station equipment makes it necessary to install apparatus *specially designed for outdoor service* and, therefore, *more costly*.

- Not-withstanding the disadvantages, outdoor stations are *very widely used* in the power system.

Comparison between "outdoor" and "indoor" sub-stations :

The comparison between outdoor and indoor sub-stations, in tabular form, is given below :

S. No.	*Aspects*	*Outdoor sub-stations*	*Indoor sub-stations*
1.	*Capital cost*	Low	High
2.	*Space required*	More	Less
3.	*Operation*	Difficult	Easier
4.	*Fault location*	Easier, since the equipment is in full view	Difficult, since the equipment is enclosed
5.	Possibility of fault escalation	Less, since greater chances clearances can be provided	More
6.	Time required for erection	Less	More
7.	Future extension	Easy	Difficult

28.6. TRANSFORMER SUB-STATIONS

In a power system, majority of the sub-stations are concerned with changing of voltage level of electric supply. These are known as transformer sub-stations because transformer is the main component employed to change the voltage level.

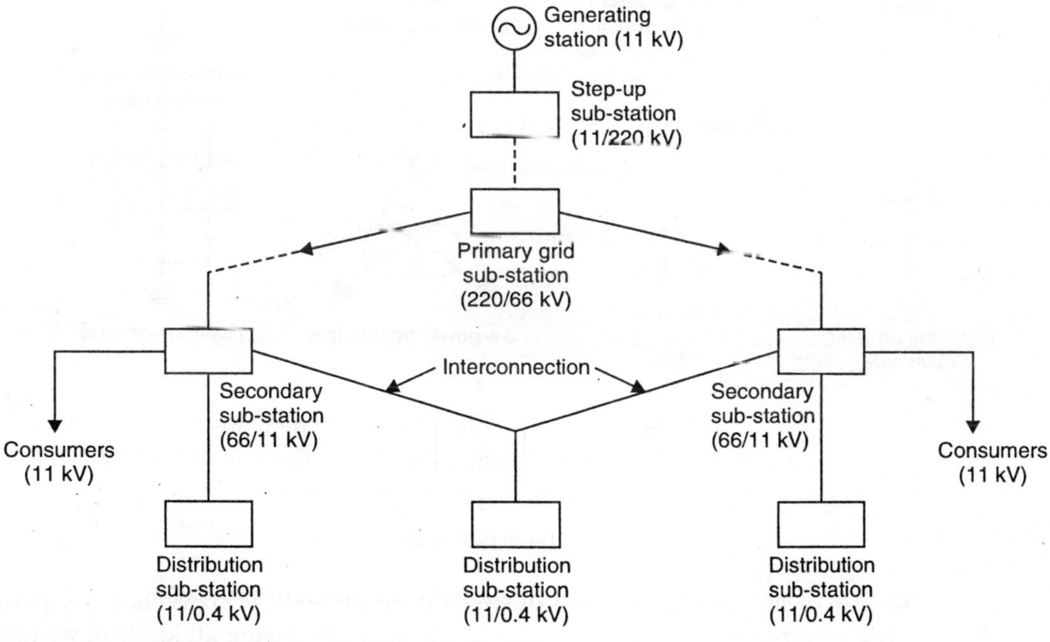

Fig. 28.1. Block diagram of a typical electric supply system indicating position of various types of transformer substations.

On the basis of purpose served, transformer sub-stations may be *classified* as follows :

1. ***Step-up sub-station***—Generally located in the power house and of *outdoor type*.
2. ***Primary grid sub-station***—Generally of *outdoor type*.
3. ***Secondary sub-station***—Generally of *outdoor type*.
4. ***Distribution sub-station***—Majority of these sub-stations are of *pole-mounted type*.

Fig. 28.1 shows the block diagram of a typical electric supply system indicating the position of the various types of sub-stations, which is self explanatory.

28.7. SYMBOLS FOR CIRCUIT ELEMENTS/EQUIPMENT IN SUB-STATIONS

Symbols of important circuit elements/equipment in sub-station are given below :

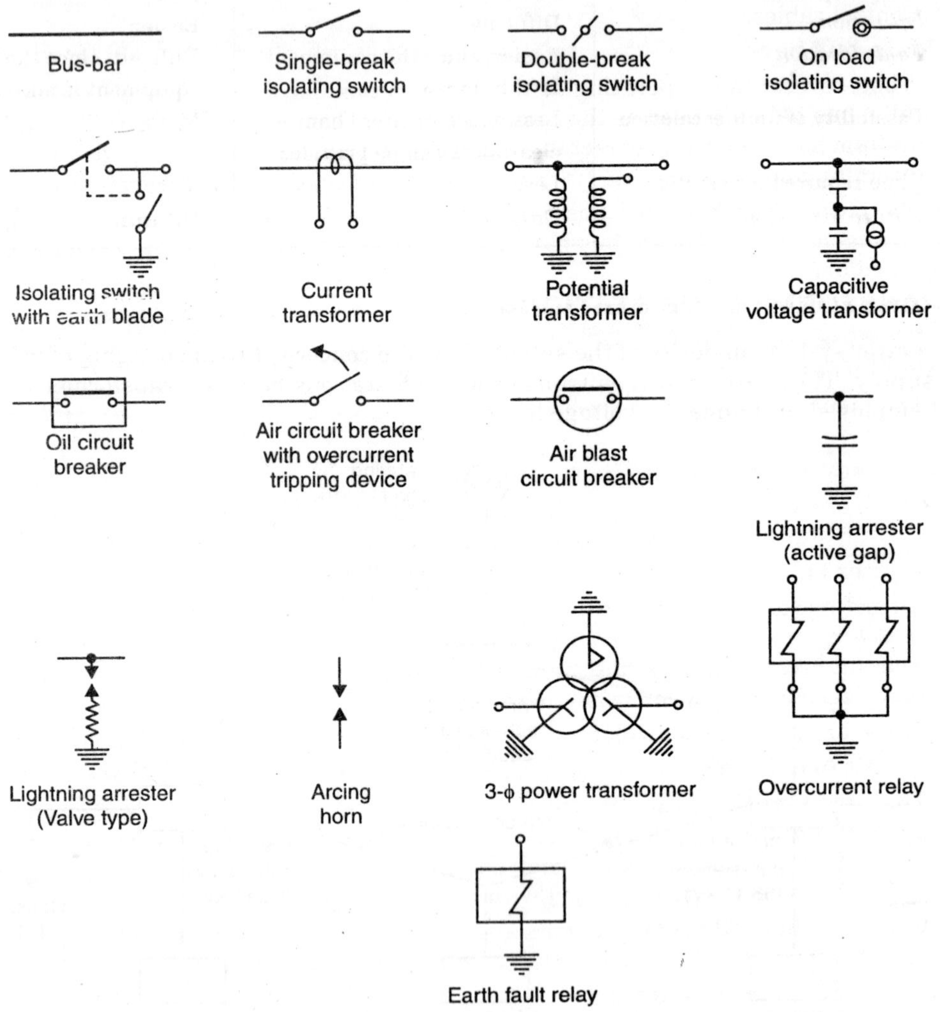

Fig. 28.2. Symbols of important circuit elements/equipment in sub-station.

28.8. EQUIPMENT FOR SUB-STATIONS AND SWITCHGEAR INSTALLATIONS

The main equipment required for sub-stations and switchgear insulations is given below :

1. Main bus-bars
2. Insulators
3. Isolators
4. Circuit breakers

5. Load-interrupter switches
6. Fuses
7. Power transformers
8. Current and potential transformers
9. Indicating and metering instruments
10. Protective relays
11. Carrier-current equipment
12. Control cables.

1. Main bus-bars :

- When several lines operating at the same voltage have to be directly connected electrically, bus-bars are used as the common electrical component. Bus-bars are copper or aluminium bars, generally of rectangular cross-section, and operate at constant voltage. The incoming and outgoing lines in a sub-station are connected to the bus-bars.
- The most commonly used bus-bar arrangement in sub-stations are : (*i*) Single bus-bar arrangement ; (*ii*) Single bus-bar system with sectionalisation ; (*iii*) Double bus-bar arrangement.
- The switches used for connecting feeder or equipment to one bus-bar (or bus in short) or the other are called "*selector*" or "*transfer*" switches.

2. Insulators :

- The insulators *support the conductors* (*or bus-bars*) *and confine the current to the conductors.*
- The most commonly used material for the manufacture of insulators is *porcelain.*
- There are several types of insulators (*e.g.*, pin type, suspension type etc.) and their use in the sub-station will depend upon the service requirement.

3. Isolators :

- An isolator or disconnecting switch is *used to open some given part of a power circuit after switching of the load by means of a circuit breaker.* In some cases isolators may be used as a circuit breaking device but their use for this purpose is strictly limited by definite conditions, such as the power rating of the given circuit.
- There are two types of isolators : (*i*) Single-pole isolators ; (*ii*) Three-pole isolators.

4. Circuit breakers :

- The circuit breaker is an equipment *which can open* or *close a circuit under normal as well as fault conditions.* It is so designed that it can be operated manually (or by remote control) under normal conditions and automatically under fault conditions. For the latter operation, a relay circuit is used with a circuit breaker.
- Generally, bulk oil circuit breakers are used for voltages upto 66 kV while for high voltages (greater than 66 kV), low oil circuit breakers are used.

5. Load-interrupter switches :

- The switches are designed and used to close and open high voltage circuits under normal working conditions (at normal load). The arc extinguishing device of the load interrupter is made in the form of a split, moulded plastic chute fitted with organic glass inserts. This chute surrounds the moving knife of the arc extinguishing system. The stationary acting contact is located in the lower part of the chute.

When the switch is opened, the working contacts between the arc are drawn separate. Acted upon by the high temperature of arc, the walls of the organic material inserts generate gases (mainly hydrogen), which create a longitudinal blast serving to extinguish the arc.

- For closing and opening the load interrupter switches lever-arm *manually operating mechanisms* are employed.

6. Fuse. Fuses have already been discussed in the earlier part of this book.

7. Power transformers :

- These transformers are used for stepping-up the voltage for transmission at generating stations and for stepping-down voltage for further distribution at main step-down transformer sub-stations. Usually naturally cooled, oil immersed, known as 'ON' type, two winding, 3–ϕ transformers, are used upto the rating of 10 MVA. The transformers of rating higher than 10 MVA, are usually air blast cooled. For very high rating, the forced oil, water cooling and air blast cooling may be used.
- For regulating the voltage the transformers used are provided with on load tap changer.
- The transformers are generally installed upon lengths of rails fixed on concrete slabs having foundation 1 m to 1.5 m deep.

8. Current and potential transformers (Instrument transformers) :

Current transformer (C.T.) :

- It is essentially a step-up transformer which steps down the current to a known ratio. The primary of this transformer consists of one or more turns of thick wire connected in series with the line. The secondary consists of a large number of turns of fine wire and provides for the measuring instruments and relays a current which is a constant fraction of the current in the line.
- The current transformers are *rated* for rated voltage of the installation, the rated currents of the primary and secondary windings and the accuracy class (the accuracy class indicates the limit of the error in the percentage of the rated turn-ratio of the given current transformers).

Potential transformer (P.T.) :

- It is essentially a step-down transformer and steps down the voltage to a known ratio. The primary of this transformer consists of a large number of turns of fine wire connected across the line. The secondary winding consists of a few turns and provides for measuring instruments and relays a voltage which is a known fraction of the line voltage.
- These instruments make the ordinary low voltage instruments suitable for measurement of high voltage and isolate them for high voltage.
- The potential transformers are rated for primary and secondary rated voltage, accuracy class, number of phases, system of cooling.

9. Indicating and metering instruments. A large number of indicating and metering instruments (*e.g.*, ammeters, voltmeters, energy meters etc.) are installed in a sub-station to maintain watch over the circuit quantities. The instrument transformers are invariably used with them for satisfactory operation.

10. Protective relays. These have already been discussed earlier.

11. Carrier-current equipment. Such equipment is installed in the sub-stations for communication, relaying, telemetering or for supervisory control. This equipment is suitably mounted in a room known as *carrier room* and connected to the high voltage power circuit.

12. Control cables. For affecting automatic controls, the control cables and conduit system are required. The control system generally operates at 110 V or 220 V and the cables employed for this purpose are multi-core cables having 10 or 37 or 61 conductors according to requirement. In order to lay these cables generally ducts are run from control room basement to centrally located junction box from where the conduits are run to the required points.

28.9. KEY DIAGRAM OF 11 kV/400 V INDOOR SUB-STATION

The key diagram of 11 kV/400 V indoor station is shown in Fig. 28.3, which is self explanatory.

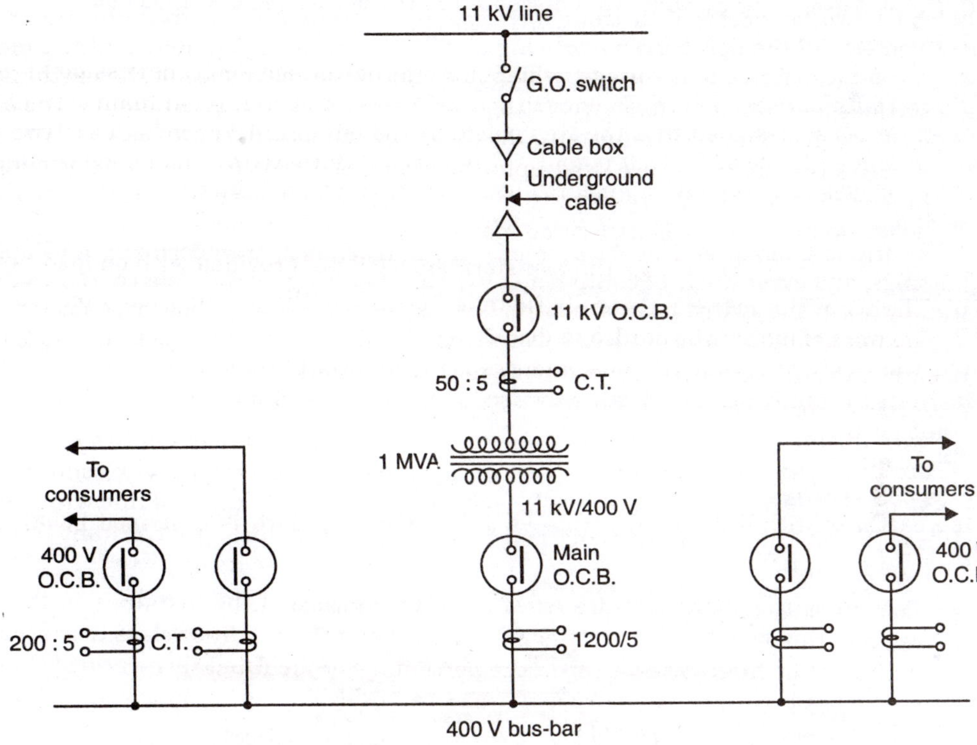

Fig. 28.3. Key diagram of 11 kV/400 V indoor station.

28.10. EARTHING

"**Earthing**" or "**grounding**" *is the connecting of an electrical equipment or apparatus to the earth with the help of a connecting wire of negligible resistance.*

If a metallic part of an electric appliance in an electric installation comes in direct contact with a base or live wire, the metal being a good conductor of electricity is charged and static charge on it will accumulate. Now if any person comes in contact with this charged metal part, he will get a severe shock. But if the metallic parts of the equipment/apparatus are earthed, the charge will be given to earth immediately as the metallic part comes in direct contact with a bare or live wire or breakdown occurs. As discharge takes place to the earth, the impedance of the path of the current being low, a heavy current flows to the earth and the protective device (usually a fuse) operates (the instant the current exceeds the limiting value) and isolates the equipment/apparatus from the supply. Thus it is very essential to provide earthing to ensure safety to the users as well as to the operating and maintenance personnel.

The earthing can be categorised in the following two parts way :

1. System earthing.

2. Safety earthing.

1. System earthing. It is required to provide low fault impedance to the ground fault currents for proper operation of the protective relays and for meeting the system requirements by effectively earthed system.

2. Safety earthing. It is required to provide protection to the operating staff working in the yard and sub-station from any injury during fault conditions by keeping the voltage gradient within safe limits.

The above two categories of earthing have common earthmat from which flat iron risers are taken out to connect all the non-current carrying metal parts of the equipment. At the time of fault the earthmat conductors rise to a voltage, which is equal to the resistance of the earthmat multiplied by ground fault current. This difference in potential results in voltage gradients. The earthmat of the sub-station is so designed that the *step potential* (potential difference between two feet of a human at the time of fault) and *touch* potential (potential difference to which body is subjected to when a person comes in contact with an earth body) are within safe limits.

The *following components in electrical installations must be earthed :*

(*i*) The frames, tanks and enclosures of electric machines, transformers, and apparatus, lightning fittings, and other items of equipment ; (*ii*) The operating mechanisms of the switch gear ; (*iii*) The framework of the switch boards, control boards, individual panel boards, cubicles ; (*iv*) The structural steel work of indoor and outdoor substations, metal cable jointing boxes, the metal sheaths of the cable, the rigid metal conduit runs, and similar metal work.

Methods of earthing. The commonly used methods of earthing are :

1. Pipe earthing
2. Plate earthing.

1. *Pipe earthing :*

Pipe earthing with full specifications as per Indian Standard is described below. Refer to Fig. 28.4.

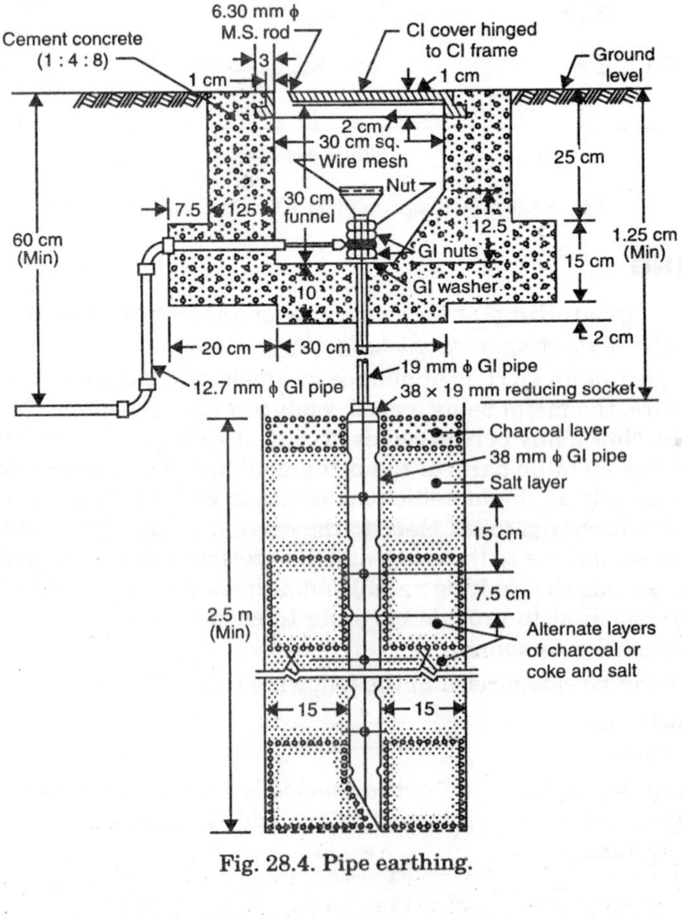

Fig. 28.4. Pipe earthing.

- In this system of earthing a G.I. pipe of 38 mm dia. and 2 metres length is embedded vertically in ground to work as earth electrode but the depth depends upon the soil conditions ; there is no hard and fast rule for this.
- The earth wires are fastened to the top section of the pipe with nut bolts.
- The pit area around G.I. pipes is filled with *salt and coal mixture* for *improving the soil condition and efficiency of the earthing system.*

The contact surface area of G.I. pipe with soil is more in comparison to the *plate* because of its circular section and hence can *take up heavy leakage current for the same electrode size.* The earth wire connection with the G.I. pipe being above the ground level can be checked for carrying out continuity test as and when desired (while in plate earthing it is difficult).

2. *Plate earthing :*

- In plate earthing, the looping earth wire is bolted effectively with the earth plate made up of copper size 60 cm × 60 cm × 3.18 mm (2′ × 2′ × 1/4″) and embedded 3 metres in ground. Copper plates are found to be most effective earth electrodes and are not affected by the soil moisture, *i.e.*, these do not get rusted. But on account of its high material cost, galvanized iron plates are preferred and usually used for normal work.
- In case the soil resistivity is high, the plate should be placed vertically in ground at a higher depth. A line diagram showing the plate earthing system is shown in Fig. 28.5.

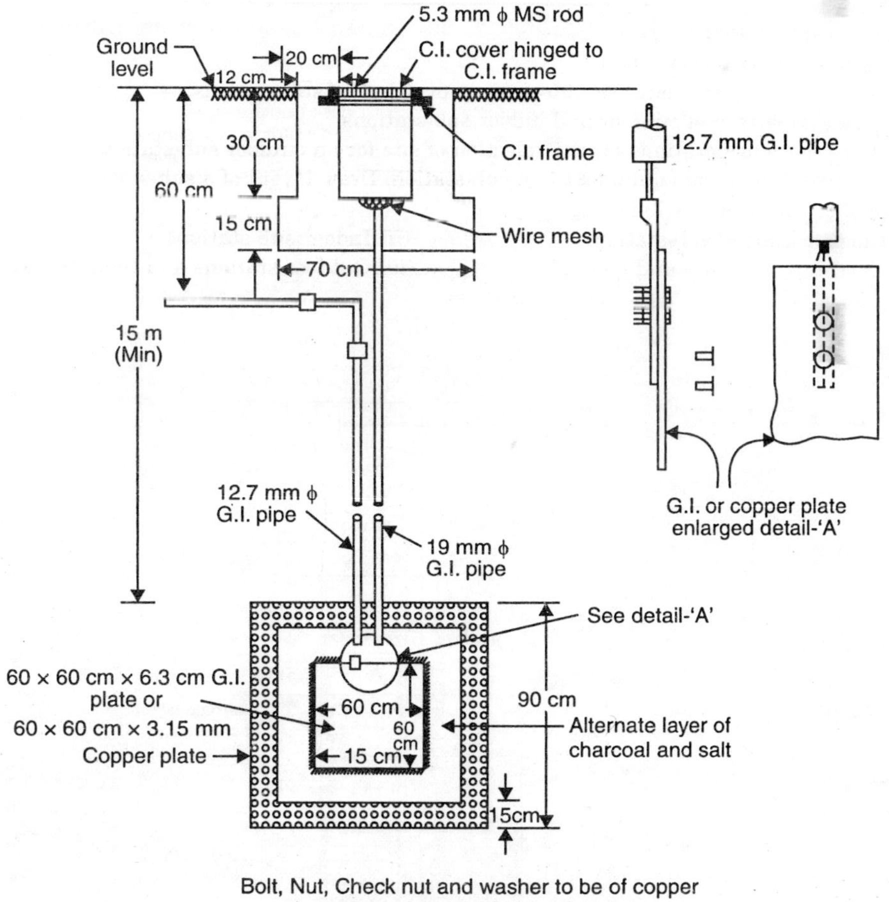

Bolt, Nut, Check nut and washer to be of copper for copper plate and G.I. for G.I. plate

Fig. 28.5. Plate earthing.

The plate is kept with its face vertical and is so arranged that it is embedded in an alternate layer of coke and salt for a minimum thickness of about 15 cm. The coke salt decreases the earth resistance. It should be remembered that the nuts and bolts must be of copper for copper plate and should be of G.I. for G.I. plates. Usually, the earth wire is drawn through a G.I. pipe fitted with a tunnel on the top through which salty water is poured in the pit of earth plate from time to time in summer sea on when the moisture of the soil will decrease to a larger extent which will increase the earth resistance.

HIGHLIGHTS

1. The assembly of apparatus used to change some characteristics (*e.g.*, voltage, A.C. to D.C., frequency, p.f. etc.) of electric supply is called a *sub-station*.
2. The main equipment required for sub-stations and switchgear installation is as under :
   (*i*) Main bus-bars ; (*ii*) Insulators ; (*iii*) Isolators ; (*iv*) Circuit breakers ; (*v*) Load-interrupter switches ; (*vi*) Fuses ; (*vii*) Power transformers ; (*vii*) Current and potential transformers ; (*ix*) Indicating and metering instruments ; (*x*) Protective relays ; (*xi*) Carrier-current equipment.

THEORETICAL QUESTIONS

1. What is a sub-station ?
2. How are sub-stations classified ?
3. State the advantages of outdoor sub-stations over indoor sub-stations.
4. Give the comparison of outdoor and indoor sub-stations.
5. What are the considerations for the selection of site for an outdoor sub-station ?
6. Make a list of the main equipment in a sub-station. Draw layout of a sub-station.
7. Write short notes on the following :
   (*i*) Classification of sub-stations.
   (*ii*) Indoor sub-stations.
   (*iii*) Earthing–necessity and methods.
   (*iv*) Outdoor stations and their necessity.

29

Universities' Questions (Latest) with "Answers and Solutions"

I. SHORT ANSWER QUESTIONS

GENERATION OF POWER

Q. 1. *Enumerate principal types of power plants.*

Ans. The principal types of power plants are:

1. Steam plant using coal, oil or nuclear fission.
2. Internal combustion engines plants.
3. Gas turbine plants.
4. Hydro-electric plants.

Q. 2. *What is the basic principle on which a thermal power station works?*

Ans. A ***thermal power station*** works on the *basic principle that heat liberated by burning fuel is converted into mechanical work by means of a suitable working fluid. The mechanical work is converted into electric energy with the use of generator.*

A working fluid goes through a repetitive cyclic change and this cyclic change involving heat and work is known as the *thermodynamic cycle.* Thus a *thermodynamic cycle is a series of operations, involving a heat source, a heat receiver, a machine or utilizer between the source and the receiver and a working substance.*

In a *steam power station*, heat is released by burning fuel, this heat is taken up by water which works as the working fuel. Water is converted into steam as it receives heat in a boiler. The steam then expands in a turbine producing mechanical work which is then converted into electrical energy through a generator. The exhaust steam from the turbine is then condensed in a condenser and condensate thereafter pumped to the boiler where it again receives heat and the cycle is repeated.

Q. 3. *How are power plant-cycles classified?*

Ans. Thermal power plants, in general, may work on vapour-cycles or gas power cycles.

Vapour power cycles—Classification:

1. Rankine cycle
2. Reheat cycle
3. Regenerative cycle
4. Binary vapour cycle.

Gas power cycles—Classification:

1. Otto cycle
2. Diesel cycle
3. Dual combustion cycle
4. Gas turbine cycles
   (*i*) Open cycle;
   (*ii*) Closed cycle.

Q. 4. *What is the function of a steam power plant?*

Ans. A ***steam power plant*** *converts the chemical energy of the fossil fuels (coal, oil, gas) into mechanical / electrical energy.* This is achieved by raising the steam in the boilers, expanding it through the turbines and coupling the turbines to the generators which convert mechanical energy to electrical energy as shown in Fig. 29.1.

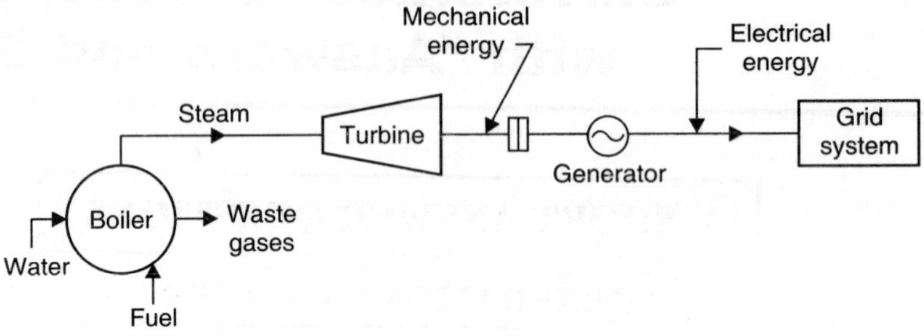

Fig. 29.1. Production of electric energy by a steam power plant.

The following two purposes can be served by a steam power plant:

(*i*) To produce electric power;

(*ii*) To produce steam for industrial purposes besides producing electric power. The steam may be used for varying purposes in the industries such as textiles, food manufacture, paper mills, sugar mills and refineries etc.

Q. 5. *List the points which should be taken into consideration while selecting the site for a steam power station.*

Ans. The following points should be taken into consideration while *selecting the site* for a steam power station:

1. Availability of raw material;
2. Nature of land;
3. Cost of land;
4. Availability of water.

Q. 6. *What are the factors on which the choice of steam conditions depend?*

Ans. The choice of steam conditions depends upon the following factors:

1. Price of coal.
2. Capital cost of the plant.
3. Time available for erection.
4. Thermal efficiency obtainable.
5. The station 'load factor'.

The present trend is towards adoption of high pressures and high temperatures.

Q. 7. *What are the requirements of a good handling plant?*

Ans. The requirements of a good handling plant are:

1. It should need minimum maintenance.
2. It should be reliable.
3. It should be simple and sound.
4. It should require a minimum of operatives.
5. It should be able to deliver requisite quantity of coal at the destination during peak periods.
6. There should be minimum wear in running the equipment due to abrasive action of coal particles.

Q. 8. *What are the various types of 'overfeed stokers'?*

Ans. *The "overfeed stokers" are used for large capacity boiler installation where the coal is burnt with pulverisation.*

These stokers are mainly of two types:

1. Travelling grate stoker
   (*a*) Chain grate type (*b*) Bar grate type.
2. Spreader stoker.

Q. 9. *What is pulverised fuel firing?*

Ans. In ***pulverised fuel firing system*** the *coal is reduced to a fine powder with the help of grinding mill and then projected into the combustion chamber with the help of hot air current.* The amount of air required (known as *secondary air*) to complete the combustion is supplied separately to the combustion chamber. The resulting turbulence in the combustion chamber helps for uniform mixing of fuel and air and through combustion. *The amount of air which is used to carry the coal and to dry it before entering into the combustion chamber is known as 'Primary air'* and the amount of air which is supplied separately for completing the combustion is known as *'Secondary air'*.

The *efficiency* of the pulverised fuel firing system mostly *depends upon the size of the powder.* The fineness of the coal should be such as 70% of it would pass through a 200 mesh sieve and 90% through 50 mesh sieve.

Q. 10. *What is a pulveriser?*

Ans. ***Pulveriser.*** Coal is pulverised in order to increase its surface exposure, thus promoting rapid combustion without using large quantities of excess air. A pulveriser is the most important part of a pulverised coal system. Pulverisers (sometimes called *mills*) are classified as follows:

1. *Attrition mills:*
   (*i*) Bowl mills
   (*ii*) Ball and race mills.
2. *Impact mills:*
   (*i*) Ball mills
   (*ii*) Hammer mills.

Pulverisers are driven by electric motors with the feeders either actuated by the main drive or by a small d.c. motor, depending upon the control used.

Q. 11. *What are 'Ball and race mill' and 'Bowl mill'?*

Ans. ***Ball and race mill.*** This a low speed unit in which grinding pressure is maintained by adjustable springs. The coal passes between the two rotating elements again and again until it has been pulverised to the desired degree of fineness. Mill, feeder and fan require upto 14 kWh per tonne of coal pulverised.

Bowl mill. The bowl mill grinds the coal between a whirling bowl and rolls mounted on pivoted axes. Coal fed into the centre is thrown by centrifugal force against the sides of the bowl where it is pulverised between the sides of the bowl and the grinding ring. The fine and intermediate sizes are picked up from the top by an air current and carried into the separator above for classification. It consumes about 5 kWh of electricity per tonne of coal.

Q. 12. *What are burners?*

Ans. Primary air that carries the powdered coal from the mill to the furnace is only about 20% of the total air needed for combustion. Before the coal enters the furnace, it must be mixed with additional air, known as *secondary air,* in burners mounted in the furnace wall. *In addition to the prime function of mixing, burners must also maintain stable ignition of fuel-air mix and control the flame shape and travel in the furnace.* Ignition depends on the rate of flame propagation. To prevent flash back into the burner, the coal air mixture must move away from the burner at a rate equal to flame-front travel. *Too much secondary air can cool the mixture and prevent its heating to ignition temperature.*

The *requirements of a burner* can be summarised as follows:

(*i*) The coal and air should be so handled that there is stability of ignition.

(*ii*) The combustion is complete.

(*iii*) In the flame the heat is uniformly developed avoiding any superheat spots.

(*iv*) Adequate protection against overheating, internal fires and excessive abrasive wear.

Q. 13. *How are pulverised fuel burners classified?*

Ans. Pulverised fuel burners may be *classified* as follows:

1. Longflame burners
2. Turbulent burners
3. Tangential burners
4. Cyclone burners.

Q. 14. *Give the classification of 'Dust collectors'.*

Ans. The dust collectors may be *classified* as follows:

1. *Mechanical dust collectors:*

(*i*) Wet type (Scrubbers)

(*a*) Spray type (*b*) Packed type

(*c*) Impingement type

(*ii*) Dry type

(*a*) Gravitational separators (*b*) Cyclone separators.

2. *Electrical dust collectors:*

(*i*) Rod type (*ii*) Plate type.

Q. 15. *What are the uses of 'Ash' and 'Dust'?*

Ans. The *uses of ash and dust* are listed below:

1. Ash is widely used in the production of cement.
2. Ash is used in the production of concrete. 20 percent fly-ash and 30 percent bottom ash are presently used constructively in U.S.A.
3. Because of their better alkali values, they are used for treating acidic soils. It has been found that if ash is used in limited quantity in soil, it increases the yield of corn, turnip etc.
4. From the ash, the metals such as Al, Fe, Si and titanium can be recovered.

Q. 16. *What is a draught?*

Ans. *The small pressure difference which causes a flow of gas to take place is termed as a* ***draught****.* The function of the draught, in case of a boiler, is to force air to the fire and to carry away the gaseous products of combustion. In a boiler furnace proper combustion takes place only when sufficient quantity of air is supplied to the burning fuel.

The draught may be classified as:

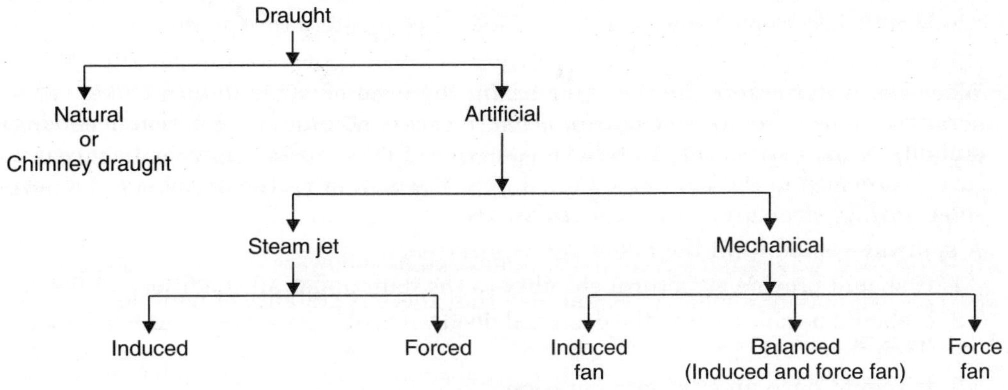

Fig. 29.2

Q. 17. *What are the major fields of application of gas turbines?*

Ans. The following are the major fields of application of gas turbines:

1. Aviation
2. Power generation
3. Oil and gas industry
4. Marine propulsion.

The efficiency of a gas turbine is not the criteria for the choice of this plant. A gas turbine is used in *aviation* and marine fields because *it is self contained, light weight not requiring cooling water and generally fit into the overall shape of the structure.* It is selected for *'power generation'* because of its *simplicity, lack of cooling water, needs quick installation and quick starting.* It is used in *oil and gas industry* because of *cheaper supply* of *fuel and low installation cost.*

Q. 18. *What are the limitations of gas turbines?*

Ans. The gas turbines have the following ***limitations:***

1. They are not self starting.
2. Low efficiencies at part loads.
3. Non-reversibility.
4. Higher rotor speeds.
5. Low overall plant efficiency.

Q. 19. *List the points which should be given due considerations while selecting the site for a gas turbine plant.*

Ans. While selecting the site for a gas turbine plant the following *points* should be given due consideration:

1. The plant should be located near the load centre to avoid transmission costs and losses.
2. The site should be away from business centres due to noisy operations.
3. Cheap and good quality fuel should be easily available.
4. Availability of labour.
5. Availability of means of transportation.
6. The land should be available at a cheap price.
7. The bearing capacity of the land should be high.

Q. 20. *Which factors should be considered while selecting the site for an hydroelectric plant?*

Ans. The following factors should be considered while selecting the site for an hydroelectric plant.

1. Availability of water
2. Water storage
3. Water head
4. Accessibility of the site
5. Distance from load centre
6. Type of the land of site.

Q. 21. *What is a spillway?*

Ans. When the water enters the reservoir basin, the level of water in basin rises. This rise is arranged to be of temporary nature because excess accumulation of water endangers the stability of dam structure. To relieve reservoir of this excess water contribution, a structure is provided in the body of a dam or near the dam or on the periphery of a basin. This *safeguarding structure is called a* ***spillway***.

A spillway should fulfil the following *requirements*:

1. It should provide structural stability to the dam under all conditions of floods.
2. It should be able to pass the designed flood without raising reservoir level above H.F.L. (high flood level).
3. It should have an efficient operation.
4. It should have an economical section.

Q. 22. *What is a surge tank?*

Ans. *A* ***surge tank*** *is a small reservoir or tank in which the water level rises or falls to reduce the pressure swings so that they are not transmitted in full to a closed circuit.* In general a surge tank serves the following *purposes*:

To reduce the distance between the free water surface and turbine thereby reducing the water-hammer effect (the *water hammer* is defined as the change in pressure rapidly above or below normal pressure caused by sudden changes in the rate of water flow through the pipe according to the demand of prime mover) on penstock and also protect upstream tunnel from high pressure rises.

To serve as a *supply tank* to the turbine when the water in the pipe is accelerating during increased load conditions and as a *storage tank* when the water is decelerating during reduced load conditions.

Q. 23. *What are the functions of a 'draft tube'?*

Ans. The draft tube serves the following *two purposes*:

1. It allows the turbine to be set above tail-water level, without loss of head, to facilitate inspection and maintenance.
2. It regains, by diffuser action, the major portion of the kinetic energy delivered to it from the runner.

The following two types of draft tubes are commonly used:

(*i*) The straight conical or concentric tube; (*ii*) The elbow type.

Q. 24. *What is 'Run-of-river plant with pondage'?*

Ans. ***Pondage*** usually refers to the collection of water behind a dam at the plant and increases the stream capacity for a short period, say a week. *Storage* means collection of water in up stream reservoirs and this increases the capacity of the stream over an extended period of several months. Storage plants may work satisfactorily as base load and peak load plants.

This type of plant, as compared to that without pondage, is *more reliable* and its generating capacity is *less dependent on the flow rates of water available.*

Q. 25. *What do you mean by a storage type plant?*

Ans. A storage type plant is one with a reservoir of sufficiently large size to permit carry-over storage from the wet reason to the dry reason, and thus to supply firm flow substantially more than the minimum natural flow. This plant can be used as base load plant as well as peak load plant as water is available with control as required. The majority of hydro-electric plants are of this type.

Q. 26. *Define the term 'Hydrology'.*

Ans. ***Hydrology*** may be defined as the *science which deals with the depletion and replenishment of water resources.* It deals with the surface water as well as the ground water. It is also concerned with the *transportation of water* from one place to another, and from one form to another. It helps us in determining the occurrence and availability of water.

Q. 27. *What is 'Radioactivity'?*

Ans. Radioactivity was originally discovered by Becquerel in 1896. This phenomenon is *confined almost entirely to the heaviest element* from 83 to 106 in the periodic table.

The phenomenon of spontaneous emission of powerful radiations exhibited by heavy elements is called ***radioactivity***. Radioactivity is essentially a *nuclear phenomenon* and is a *drastic process* because the element changes its kind. It is *spontaneous* and an *irreversible self-disintegrating activity* because the element breaks itself up for good. Those elements which exhibit this activity are called *radioactive elements.* Examples are: *Uranium, polonium, radium, radon, ionium, thorium, actinium and mesothorium.*

The radioactive radiations emitted by the radioactive elements are found to consist of the following:

(*i*) Alpha (α) rays or α-particles

(*ii*) β-rays or β-particles

(*iii*) γ-rays or photons.

The radioactivity may be *natural* or *artificial.*

Q. 28. *What are the general properties of radioactive radiations?*

Ans. The general *properties* of radioactive radiations are:

1. These radiations are *highly penetrating, they affect photographic plates, ionise gases, cause scintillations on fluorescent screen, develop heat and produce chemical changes.*
2. As radiations are given out, *new elements are formed* in an irreversible process—the new elements themselves being usually radioactive.
3. The emission of radiations is *spontaneous* and is not affected by external agents.
4. The emission is *not instantaneous* but is prolonged *i.e.,* it is extended over a period of time otherwise it would not have been discovered at all.
5. Exercise for radioactivity, there is nothing abnormal about the radioactivity elements as regards their physical and chemical properties.

Q. 29. *What are fertile materials?*

Ans. It has been found that some materials are not fissionable by themselves but they *can be converted to the fissionable materials,* these are known as ***fertile materials***.

Pu^{239} and U^{233} are not found in nature but U^{238} and Th^{232} can produce them by nuclear reactions.

Q. 30. *What is the main purpose of design and operation of the plant?*

Ans. *The main purpose of design and operation of the plant is to bring the cost of energy produced to minimum.* Among many factors, the efficiency of the plant is one of the factors that determines the energy cost. In majority of cases, unfortunately, the most thermally efficient plant is not economic one.

Q. 31. *How can power generation cost be reduced?*

Ans. The cost of power generation can be *reduced by:*

1. Using a plant of simple design that *does not need highly skilled personnel.*
2. Selecting equipment of *longer life and proper capacities.*

3. Carrying out *proper maintenance* of power plant equipment to avoid plant breakdowns.
4. Running the power stations at *high load factors.*
5. Increasing the efficiency of the power plant.
6. Keeping proper supervision, which ensures *less breakdowns* and extended plant life.

Q. 32. *What are the drawbacks of gas turbine power plants as compared to steam power plants?*

Ans. The drawbacks are:

(*i*) The operating cost of a gas turbine power plant is *much higher* than that of a steam power plant.

(*ii*) Gas turbines can be employed *only in small sizes* (about 50 MW or so).

(*iii*) Like steam turbine, the gas turbine is *not readily reversible.*

(*iv*) In case of a gas turbine power plant, the compressor has to be started before starting of turbine and the former requires power from some external source.

Q. 33. *What are 'direct energy conversion devices'?*

Ans. The devices which convert naturally available energy into electricity, without an intermediate conversion, into mechanical energy (energy source may be thermal, solar or chemical) are called ***Direct energy conversion (DEC) devices.***

Q. 34. *What is a 'thermoelectric generator'?*

Ans. Thermoelectric generator uses '*seebeck effect*' to produce electrical energy directly from the available heat input. Its *thermal efficiency is very low,* of the order of 1 to 3%. In any heat engine the efficiency of thermoelectric generator *depends upon the temperatures of hot and cold junctions.*

Q. 35. *What are the advantages and disadvantages of thermoelectric power generator?*

Ans. *"Advantages":* (*i*) Highly reliable; (*ii*) Free from noise due to the absence of parts; (*iii*) Compact and durable; (*iv*) Minimum maintenance; (*v*) Portable; (*vi*) Uses low grade thermal energy.

"Disadvantages": (*i*) Low output; (*ii*) Low efficiency; (*iii*) High cost.

Q. 36. *What is a thermoinic generator/converter?*

Ans. *Thermionic generator* converts heat energy directly into electrical energy by *utilizing thermionic emission effect.* In this device, *electrons act as the working fluid* in place of a vapour (or gas) and electrons are emitted from the surface of the heated metal.

Q. 37. *What is MHD generator?*

Ans. *MHD generator* is a device which converts heat energy of a *conducting fluid* directly into electrical energy without a conventional electric generator.

Q. 38. *What are 'Renewable energy sources'?*

Ans. The ***renewable energy sources*** are:

(*i*) Solar photo-voltaic cell; (*ii*) Wind;

(*iii*) Hydrogen-fuel cell.

Q. 39. *What are commercial (conventional) and non-commercial energy sources?*

Ans. These energy sources are:

- *Coal, oil, gas, uranium and hydro* are commonly known as ***commercial or conventional energy sources.*** These represent about 92% of the total energy used in the world.
- *Firewood, animal dung and agricultural waste,* etc. are called as ***non-commercial energy sources.***

Q. 40. *Name the various non-conventional energy sources.*

Ans. The various ***non-conventional energy sources*** are:

(*i*) Solar energy; (*ii*) Wind energy; (*iii*) Energy from biomass and bio-gas; (*iv*) Ocean thermal energy conversion (OTEC); (*v*) Tidal energy; (*vi*) Geothermal energy; (*vii*) Hydrogen energy; (*viii*) Fuel cells; (*ix*) MHD generator; (*x*) Thermionic converter; (*xi*) Thermo-electric power.

Q. 41. *What is 'wind energy'?*

Ans. • ***Wind*** *is air set in motion by small amount of insolation reaching the upper atmosphere of earth.*

— Nature generates about 1.67×10^5 kWh of wind energy annually over land area of earth and 10 times this figure over the entire globe.

— Wind contains *kinetic energy* which can easily be *converted to electrical energy*.

• The ***wind energy***, which is an indirect source of energy, can be used to run a wind will which in turn drives a generator to *produce electricity*.

Q. 42. *What are the main applications of wind plants?*

Ans. The main applications of wind plants are:

(*i*) Electrical generation; (*ii*) Pumping;
(*iii*) Drainage; (*iv*) Grinding grains;
(*v*) Saw milling.

Q. 43. *What is 'tidal energy'?*

Ans. The periodic rise and fall of the water level of sea which are carried by the action of sun and moon on water of the earth is called the ***tide.***

• ***Tidal energy*** can furnish a significant portion of all such energies which are renewable in nature. The large scale up and down movement of sea water represents an unlimited source of energy. If some part of this vast energy can be converted into electrical energy, it would be an important source of hydropower.

• The *main feature of the tidal cycle is the difference in water surface elevations at the high tide and at the low tide.* If this differential head could be utilized in operating a hydraulic turbine, the tidal energy could be converted into electrical energy by means of an attached generator.

Q. 44. *What is 'wave energy'?*

Ans. ***Wave energy*** *comes from the interaction between the winds and the surfaces of oceans.* The energy available varies with the size and frequency of waves. It is estimated that about 10 kW of power is available for every metre width of the wave front.

Wave energy when active is very concentrated, therefore, wave energy conversion into useful energy can be carried out at high power densities. A large variety of devices (*e.g., hydraulic accumulator wave machine; high-level reservoir machine; Dolphin-type wave-power machine; Dam-Atoll wave machine)* have been developed for harvesting of energy but these are complicated and fragile in face of gigantic power of ocean storms.

Q. 45. *What is an 'Ocean Thermal Energy Conversion (OTEC) plant?*

Ans. The oceans cover about 70% of the global surface and are particularly extensive in the tropical zones. Therefore, most of the sun's radiations are absorbed by sea water. Thus warm water on the ocean's surface flows from the tropics towards poles. Cold water circulates at the ocean bottom from the poles to the tropics. Hence, in the tropical regions the water temperature is around **5°C** at a depth of 1000 m, whereas at the surface, it remains almost constant at **25°C** (for the first few metres because of mixing; subsequently it decreases and asymptotically approaches the value at the lower level).

Thus, we can employ a *carnot-type process to generate power between these two steady temperatures.* Such plants are called ***Ocean Thermal Energy Conversion (OTEC)*** plants.

Q. 46. *What is 'geothermal energy'?*

Ans. As we travel down earth's surface radially, there exists a temperature gradient of 0.03°C per metre. Thus a 30°C increase in temperature can be obtained per kilometre depth from the earth's crust. *There are many local hot spots just below the surface where the temperatures are much higher than expected.* Ground water, when comes into contact with these hot spots, either dry or wet steam is formed. By drilling holes to these locations, hot water and steam can be tapped and these can be used for power generation or space heating.

Geothermal energy is primarily energy from the earth's own interior, it is classified as *renewable* because the earth's interior is and will continue in the process of cooling for the indefinite future. Hence, geothermal energy from the earth's interior is almost inexhaustible as solar or wind energy, so long as its sources are actively sought and economically tapped. Geothermal energy can be used for *heat* and *power generation*. Geothermal energy is present over the entire extent of earth's surface except that it is nearer to the surface in the volcanic areas. Heat transfer from the earth's interior is by *three primary means:*

(*i*) Direct heat conduction;

(*ii*) Rapid injection of ballistic magma along natural rifts penetrating deep into earth's mantle; and

(*iii*) Bubble like magma that buoys upwards towards the surface.

Q. 47. *List the general categories of geothermal sources.*

Ans. The following five general categories of geothermal sources have been identified:

1. Hydrothermal convective systems

(*i*) Vapour-dominated or dry steam fields.

(*ii*) Liquid-dominated system or wet steam fields.

(*iii*) Hot-water fields.

2. Geopressure resources.

3. Petro-thermal or hot dry rocks (HDR).

4. Magma resources.

5. Volcanoes.

The hydro-thermal convective systems are best resources for geothermal energy exploitation at present. Hot dry rock is also being considered.

Q. 48. *What do you mean by 'Biomass' and 'Biomass energy'?*

Ans. ***Biomass*** *is an organic matter from plants, animals and micro-organism grown on land and water and their derivatives.* The energy obtained from biomass is called ***biomass energy.***

Biomass is considered as a *renewable source of energy* because the organic matter is generated every day. Coal, petroleum oil and natural gas do not come in the category of 'biomass', because they are produced from dead, burried biomass under pressure and temperature during millions of year. Biomass can also be considered a form of solar energy as the latter is used indirectly to grow these plants by *'photosynthesis'*.

Q. 49. *What are 'Direct energy conversion systems'?*

Ans. The energy conversion devices that have been in use for a long time are those that accept energy as heat and produce mechanical work, which is transformed into electric power distribution at large. *Direct energy conversion devices convert naturally available energy into electricity without an intermediate conversion into mechanical energy.* (The energy source may be thermal, solar or chemical). Until now, their use has been confined to small scale, special purpose applications, since the voltage output available with them is rather small and no inexpensive device that is reliable like a turbine or alternator has been built.

Examples:

(*i*) Thermoelectric conversion system
(*ii*) Thermionic conversion system
(*iii*) Photovoltaic power system
(*iv*) Magnetohydrodynamic system
(*v*) Electrostatic mechanical generators
(*vi*) Electro Gas-Dynamic Generators (EGD)
(*vii*) Fuel cells
(*viii*) Nuclear batteries.

Q. 50. *What is a 'solar or photo voltaic cell'?*

Ans. Solar energy can be directly converted to electric energy by means of *photovoltaic effect*, which is defined as the generation of an electromotive force as a result of the absorption of ionizing radiation. Energy conversion devices which are employed to convert sunlight into electricity by the use of photovoltaic effect are ***solar cells***.

Q. 51. *What is a 'Magnetohydrodynamics (MHD) generator?*

Ans. ***MHD generator*** is a device which converts the heat energy of a fuel directly into electrical energy without a conventional generator.

Q. 52. *What is a 'fuel cell'?*

Ans. A ***fuel cell*** is an electrochemical device in which the chemical energy of a conventional fuel is converted directly and efficiently into low voltage.

ECONOMICS OF POWER GENERATION

Q. 53. *What is the main purpose of design and operation of the plant?*

Ans. The main purpose is to *bring the cost of energy to minimum.*

Q. 54. *What is 'diversity factor'?*

Ans. It is the ratio of the maximum power demands of the subdivisions of the system, or part of a system, to the minimum demand of the whole system, or part of the system under consideration, measured at the point of supply.

$$\text{Mathematically, Diversity factor} = \frac{\text{Sum of individual maximum demands}}{\text{Maximum demand of entire group}}$$

It is always *greater than unity.*

Q. 55. *What is 'dump power'?*

Ans. This term is used in hydroplants and it shows the power in *excess* of the load requirements and is made available by *surplus water.*

Q. 56. *What is the significance of load curves?*

Ans.
- ***Load curves*** give full information about the incoming load and help to decide *the installed capacity* of the power station and to decide the *economical sizes* of various generating units.
- These curves also help to *estimate* the generating cost and to decide the *operating schedule* of the power station *i.e.*, the *sequence* in which different units should be run.

Q. 57. *What is the significance of load and diversity factors?*

Ans. *'Higher'* the values of load and diversity factors, *'lower'* will be the overall cost per unit generated.

Q. 58. *Define the terms 'base load' and 'peak load'.*

Ans. The unvarying load which occurs *almost the whole day* on the power station is called ***base load***.

The various peak demands of the load *over and above the base load* is called ***peak load.***

Q. 59. *What are the advantages of 'combined plants'?*

Ans. The following advantages accrue from combined plants:

(*i*) Flexibility of operation; (*ii*) Security of supply;

(*iii*) Improved utilization of hydro-power; (*iv*) Spare plants.

Q. 60. *What are tariffs'?*

Ans. ***Tariffs*** or ***energy rates*** are the different methods of charging the consumers for consumption of electricity.

The tariff chosen should recover the fixed cost, operating cost and profit, etc. incurred in generating the electrical energy.

Q. 61. *What is 'depreciation reserve'?*

Ans. ***Depreciation charge*** represents the amount which is set aside from income every year and the amount so accumulated is called the ***depreciation reserve***.

Q. 62. *What is the difference between chronological load curve and load duration curve?*

Ans. The ***chronological load curve*** is the curve obtained by plotting the daily variations in load on the power station, plotted on a graph from time-to-time hourly or half hourly basis, taking load on Y-axis and time on X-axis.

The ***load duration curve*** indicates the variation of load, but with the load arrangement in *descending order* of magnitude *i.e.*, the greatest load on the left, lesser loads towards the right hand and the least load at the extreme right.

Q. 63. *What information can be supplied by load curves?*

Ans. (*i*) Variation of load during different hours of day; (*ii*) Total number of units generated; (*iii*) Maximum demand on the power station; (*iv*) Average load; (*v*) Load factor.

Q. 64. *How does diversity factor influence the cost of generation?*

Ans. The "*diversity factor*" is the ratio of sum of individual maximum demands of all the consumers supplied by it to the maximum demand of the power station. It is *always greater than unity*.

Higher the diversity factor better it is with a given number of consumers, the *higher* the diversity factor of their loads, the *smaller* will be capacity of the power plant required and consequently the fixed charges due to capital investment will be much *reduced*.

Q. 65. *What are the different types of loads on a power system?*

Ans. The main types of loads are: Domestic; Commercial; Agricultural; Industrial; Municipal, etc.

Q. 66. *Can a power factor be included in a tariff?*

Ans. *Yes*, power factor is included in power factor tariffs such as kVA maximum demand tariff, kWh and kVARh tariff and sliding scale or average power factor tariff.

Q. 67. *What type of tariff is employed for domestic consumers?*

Ans. Block rate tariff.

Q. 68. *What type of tariff is usually applied to bulk consumers?*

Ans. Three-part-tariff.

Q. 69. *Which type of tariff encourages the consumers to keep the load factor and power factor high?*

Ans. kVA maximum demand tariff.

Q. 70. *What is the necessity of power factor (p.f.) improvement?*

Ans. The low power factor *causes the increase* in the ratings of generators and transformers, cross-sectional area of the bus-bars and contact surface of the switchgear, the size of the feeders and distributors, energy losses, voltage drops in generators, transformers, transmission lines and distributors.

TRANSMISSION AND DISTRIBUTION

Q. 71. *What do you mean by 'transmission and distribution of electric power'?*

Ans. By transmission and distribution of electric power is meant its *conveyance* from the central station, where it is generated, to the places where it is demanded by the consumers (like pumping stations, residential and commercial buildings, mills, factories, etc.)

Q. 72. *What are transmission voltages in India? What is grid?*

Ans. The transmission voltages in India are: 400 kV, 220 kV and 132 kV.

— The high voltage transmission lines transmit electrical power from the generating stations to main receiving end sub-stations. At these stations the voltage is stepped down to a lower value of 66 kV or 33 kV.

— The secondary transmission system forms the link between the main receiving end sub-stations and secondary sub-stations. At the secondary sub-stations the voltage is stepped down to 33 kV or 11 kV and the power is fed into the primary distribution system.

— The 33 kV or 11 kV distribution lines (usually known as *feeders*) emanate from the secondary sub stations and terminate in distribution sub-stations. The distribution sub-stations consist of step-down transformers and are located at convenient places in the area in which the power is to be supplied. Sometimes these distribution sub-stations consist of pole mounted transformers located on the road side. These transformers step down the voltage to 400 V.

— The 400 V distribution lines (usually known as *distributors*) are laid along the roads and service connections to consumers are tapped off from the distributors.

— All *transmission and distribution systems are 3-phase systems.*

— The transmission lines and feeders are 3-phase 3 wire circuits.

— The distributors are 3-phase 4 wire circuits because a neutral wire is necessary to supply the single phase loads of domestic and commercial consumers.

● *The transmission network is commonly known as* ***grid.***

Q. 73. *Show a typical power supply network.*

Ans. Fig. 29.3 shows a typical power supply network

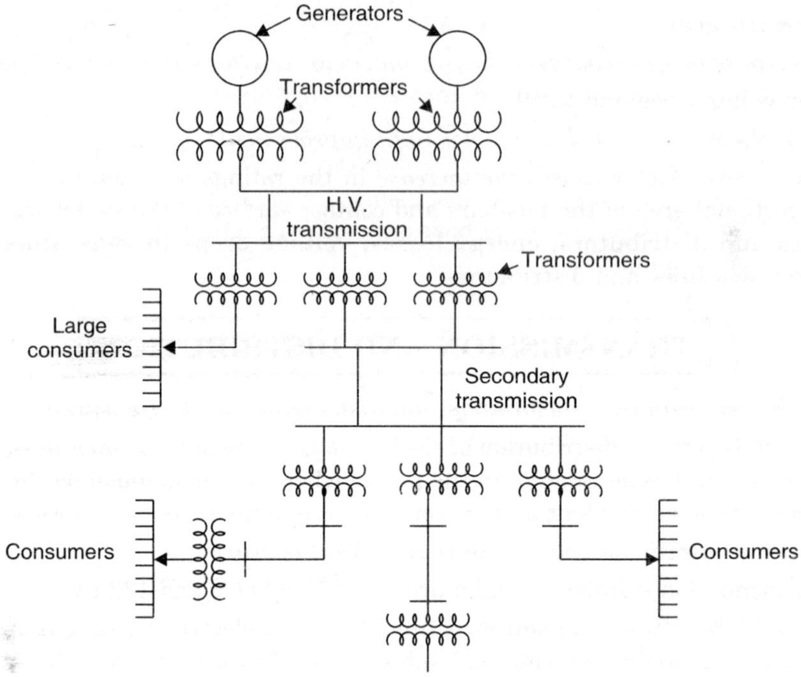

Fig. 29.3. A typical power supply network.

Q. 74. *What is the voltage that has been selected for HVDC transmission?*

Ans. HVDC transmission voltage: ± 500 kV.

Q. 75. *What are the methods by which electric power may be transmitted?*

Ans. Electric power may be transmitted by two methods:

1. *Overhead system.*
2. *Underground system.* Its main advantage is that it is less prone to natural hazards like rain, wind, lightning and that it does not interfere with other amenities. However it is *more costly* as compared to an overhead system.

Because of cost consideration the transmission systems in India, are generally overhead. For distribution also the *use of underground cables is limited to densely populated areas.*

Q. 76. *What is Kelvin's law?*

Ans. The statement of Kelvin's law is as follows:

"If variable part of annual cost on account of interest and depreciation on the capital outlay is equal to the annual cost of electrical energy wasted in the conductors, the total annual cost will be minimum and the corresponding size of the conductor will be most economical."

Q. 77. *List the characteristics which the conductor materials used for transmission and distribution of electric power must possess?*

Ans. The conductor materials used for transmission and distribution of electrical power must have the following characteristics:

1. High conductivity
2. High tensile strength
3. Low cost
4. Low specific gravity
5. Should not be brittle.

Q. 78. *Which conductor materials are commonly used?*

Ans. The commonly used conductor materials are:

(*i*) Copper (*ii*) Aluminium

(*iii*) Steel and steel cored aluminium (A.C.S.R.)

(*iv*) Galvanised steel conductors.

- All conductors used for overhead lines are preferably *stranded* in order *to increase the flexibility.*

Q. 79. *Name the various components of an overhead line.*

Ans. Supports/powers; cross-arms and clamps; Insulators; conductors, stays; Lightning arresters, Earth wire; V-guards; Guard wires; Phase plates; Bird guards, Danger plates; Anticlimbing devices; Vibration damper, etc.

Q. 80. *Enumerate various types of line supports.*

Ans. Wooden poles; R.C.C. or P.C.C. poles; Rail poles; steel structures or towers.

Q. 81. *What for ACSR stands?*

Ans. ACSR stands for 'Aluminium conductor steel reinforced'.

Q. 82. *Name insulator materials usually used for overhead line.*

Ans. (*i*) Porcelain; (*ii*) Glass; (*iii*) Stealite.

Q. 83. *Enumerate various types of insulators employed in overhead lines.*

Ans. (*i*) Pin type insulators (*ii*) Suspension type insulators

(*iii*) Strain insulators (*iv*) Shackle insulators

(*v*) Egg or stay insulators.

Q. 84. *What is 'string efficiency'?*

Ans. ***String efficiency*** is the ratio of spark over voltage for n insulators and n times the spark over voltage of one insulator (where n is the total number of insulators in a string).

The *string efficiency* is also given by the ratio of voltage across the string and n times the voltage across single unit adjacent to line.

Q. 85. *What is 'Corona'?*

Ans. When an alternating potential difference is applied across two conductors whose spacing is large as compared to their diameter, there is no apparent change in the condition of the atmospheric air surrounding the wires if voltage is low. However, when the potential difference is increased, then a point is reached when a faint luminous glow of bluish colour appears along the lengths of conductors and at the same time a hissing sound is heard. This bluish discharge is known as ***corona.*** *Corona is always accompanied by the production of ozone* which is readily detected because of its characteristic odour.

- Corona occurs when the electrostatic stress in the air around the conductors exceeds 30 kVA (maximum)/cm or 21.1 kV (r.m.s.)/cm.

Q. 86. *What is 'skin effect'?*

Ans. ***Skin effect*** is that effect which does not allow equal distribution of current over the cross-section of the conductor in alternating current system. Since the total flux linking with the inner portion of the conductor is more than the outside portion of the conductor, the inductance of the inner portion is more than the outer portion of the conductor. Hence current flows with greater density at the outer surface of the conductor. Now this increase in current density in the outer layers results in a greater I^2R loss for a given current and the *effective resistance increases*.

- At low frequency the skin effect is very small, but at high frequency it is very large.

Q. 87. *What is 'proximity effect'?*

Ans. ***Proximity effect*** is an effect of non-uniform current distribution in a conductor due to the effect of linkage of flux of *another current carrying conductor placed nearby.* More flux will link with the *nearer half portion of the conductor and thus an increase in the resistance of the conductor results.*

Q. 88. *What are 'stranded conductors'?*

Ans. ***Stranded conductors*** *are the conductors made of thin wires of small cross-section and bunched together.* They are flexible, not rigid and can be coiled very easily. They eliminate to a large extent the risk of breaking through insulation.

Q. 89. *Name the tests by which overhead line insulators are tested.*

Ans. The insulators are tested by the following *tests*:

1. Mechanical tests
2. Electrical insulation tests
3. Environmental tests
4. Temporary cycle tests
5. Corona and radio interference tests.

Q. 90. *What are the causes of failure of insulators?*

Ans. (*i*) Mechanical stresses
(*ii*) Porosity of material
(*iii*) Cracking of insulator
(*iv*) Improper vitrification
(*v*) Short circuits
(*vi*) Flash-over
(*vii*) Dust deposition.

Q. 91. *What are the methods of improving string efficiency?*

Ans. String efficiency can be improved by the following methods:

(*i*) Long cross-arm method;
(*ii*) Capacitance grading method;
(*iii*) Static shielding.

Q. 92. *What are 'stringing charts'?*

Ans. The curves of tension and sag versus temperature are called ***stringing chart.***

These charts are quite useful while erecting the transmission line conductor for adjusting the sag and tension properly.

Q. 93. *What is 'Sag'?*

Ans. ***Sag*** is the difference in level between the points of line supports and the lowest point on the conductor.

Q. 94. *Name the factors which affect the sag in a transmission line.*

Ans. ***Factors:*** (*i*) Weight of conductor; (*ii*) Span length (distance between the poles); (*iii*) Working tensile strength; (*iv*) Ice coating; (*v*) Wind pressure; (*vi*) Temperature.

Q. 95. *What are the causes of failure of insulators?*

Ans. Causes of failure are: (*i*) Cracking of porcelain; (*ii*) Porosity; (*iii*) Puncture; (*iv*) Mechanical stresses; (*v*) Flashover, etc.

Q. 96. *How does an insulator fail?*

Ans. The electric failure of an insulator may occur either by puncture or flashover. In case of puncture the arc passes through the insulator body while flashover is caused by an arc discharge between the line conductor and earth through air surrounding the insulator.

Q. 97. *What functions are performed by a 'grading ring'?*

Ans. The ***grading ring*** performs the following *functions*:

(*i*) It *equalises* the voltage distribution across the *insulator units.*

(*ii*) When used in conjunction with arcing horn fixed at the top end of the string, serves the purpose of arcing shield and *protects* the insulator string from flashover.

Q. 98. *Enumerate the factors which affect 'corona'.*

Ans. ***Factors:*** (*i*) Number of ions, size and charge per ion, mean free path depending on atmospheric conditions; (*ii*) Line voltage; (*iii*) Ratio of conductor spacing to conductor radius; (*iv*) Contour of the conductor surface; (*v*) State of conductor surface.

Q. 99. *What are the adverse effects of corona?*

Ans. ***Adverse effects:*** (*i*) Power loss but not very important except under abnormal weather conditions; (*ii*) Corrosion due to production of ozone gas; (*iii*) Interference with neighbouring communication circuits.

Q. 100. *Enumerate factors which affect 'corona loss'.*

Ans. (*i*) System frequency; (*ii*) System voltage; (*iii*) Air conductivity; (*iv*) Air density; (*v*) Conductor diameter; (*vi*) Conductor surface conditions; (*vii*) Atmospheric conditions; (*viii*) Load current.

Q. 101. *How can the corona effect be minimised in a transmission line?*

Ans. Corona effect can be minimised by using *large diameter conductors,* which may be accomplished by using hollow conductors with hemp core, ACSR conductors or bundled conductors.

This effect can also be minimised by *increasing the spacing* between the conductors.

Q. 102. *What are the factors on which 'skin effect' depends?*

Ans. "Skin effect" depends upon:

(*i*) Type of material, (*ii*) Frequency of current, (*iii*) Diameter of conductor, and (*iv*) Shape of conductor.

This effect *increases* with increase of cross-section, permeability and supply frequency.

Q. 103. *What is GMR of a conductor?*

Ans. GMR of a conductor is the product $re^{-1/4}$ and is equal to 0.7788 lines the radius of the conductor.

Q. 104. *What are the various types of vibrations in overhead line conductors?*

Ans. ***Types of vibrations:*** (*i*) Simple swinging; (*ii*) Aeroline vibration; (*iii*) Galloping or dancing of conductors (vibration).

Q. 105. *What do you mean by 'bundling of conductors'?*

Ans. The use of *more than one conductor per phase* is called the ***bundling of conductors.***

Q. 106. *What is 'transposition of line conductors'.*

Ans. ***Transposition*** *of line conductors* means changing the positions of three phases on the line supports *twice* over the total length of the line.

The line conductors, in practice, are so transposed that each of the three possible arrangements of the conductors exists for $\frac{1}{3}$ rd of the total length of the line.

Q. 107. *How are transmission lines classified?*

Ans. The transmission lines are classified according to distance of transmission as follows:

1. ***Short transmission line:*** Length less than 60 km and operating voltage less than 20 kV.
2. ***Medium transmission line:*** Length between 60 km and 150 km and line voltage between 20 kV and 100 kV.
3. ***Long transmission line.*** Length above 150 km and line voltage above 100 kV.

Q. 108. *Define the terms 'Regulation' and 'Efficiency' of a transmission line.*

Ans. ***Regulation.*** Regulation is defined as *the change in voltage at the receiving end when the full load is thrown off, the sending end voltage remaining constant.*

Mathematically percentage regulation $= \frac{V_S - V_R}{V_S} \times 100$

where, V_S = Voltage at sending end, and

V_R = Voltage at receiving end.

Efficiency. Efficiency of a transmission line is defined as the *ratio of power received to the power sent out.*

Mathematically, percent $\eta_T = \frac{V_R I_R \cos \phi_R}{V_S I_S \cos \phi_S} \times 100$

where V_R, I_R and $\cos \phi_R$ are the receiving end voltage, current and power factor while V_S, I_S and $\cos \phi_S$ are the sending end voltage, current and power factor.

Q. 109. *What are the factors which govern the performance of a transmission line?*

Ans. (*i*) Series resistance R and inductance L; (*ii*) Shunt capacitance C and conductance G, the effect of conductance is, however, negligible.

Q. 110. *What is 'Ferranti effect'?*

Ans. ***Ferranti effect*** is the phenomenon of rise in voltage at the receiving end of the open circuited or lightly loaded line.

Q. 111. *Why in medium and long transmission lines, regulation is greater than voltage?*

Ans. Due to 'Ferranti effect'.

Q. 112. *What are the units of generalised constants of a transmission line?*

Ans. Generalised constants A and D are dimensionless, whereas the units of generalised constants B and D are ohms and siemens (mho) respectively.

Q. 113. *What is 'surge impedance loading' of a line?*

Ans. ***Surge impedance loading*** is the unity power factor load that can be delivered by a line of negligible resistance.

Q. 114. *What is mean by 'receiving-end' and 'sending-end' power circle diagrams?*

Ans.
- ***Receiving-end power circle diagram*** is the circle diagram drawn with receiving-end true and reactive power components as the horizontal and vertical ordinates.
- ***Sending-end power circle diagram*** is the circle diagram drawn with sending-end true and reactive power components as the horizontal and vertical ordinates.

Q. 115. *How are cables classified?*

Ans. Cables are *classified* as follows:

1. Low voltage (L.T.) cables ... operating voltage upto 1 kV
2. High voltage (H.T.) cables ... operating voltage upto 11 kV
3. Super tension (S.T.) cables ... operating voltage upto 33 kV
4. Extra high tension (E.H.T.) cables ... operating voltage upto 66 kV
5. Extra super voltage power cables ... operating voltage beyond 132 kV

Q. 116. *Why is a metallic sheath provided in an underground cable?*

Ans. The metallic sheath is provided around the insulation to protect it against ingress of moisture, gas or other damaging liquids (acid or alkalies) from the soil and atmosphere.

Q. 117. *Copper conductors are not used in power cables, why?*

Ans. Because of scarcity of copper.

Q. 118. *Why 'armouring' is usually not provided in single core cables?*

Ans. To avoid excessive loss in the armour.

Q. 119. *What are the factors on which insulation resistance of a cable depends?*

Ans. (*i*) Core radius, (*ii*) Insulation thickness, (*iii*) Length of cable, and (*iv*) Resistivity of insulating material.

Q. 120. *On what factors does the capacitance of a single core depend?*

Ans. (*i*) Length, (*ii*) Relative permittivity of its dielectrics, and (*iii*) Ratio of core diameter to sheath inner diameter.

Q. 121. *Name the methods by which insulation resistance of a cable can be measured.*

Ans. (*i*) Galvanometer method; (*ii*) By Meggar

Q. 122. *What is 'Surge impedance'?*

Ans. ***Surge impedance*** (or natural impedance) is defined as the square root of the ratio of the line impedance (Z) and shunt admittance (Y).

i.e.,

$$Z_0 = \sqrt{\frac{Z}{Y}}$$

where, $Z = R + jX$ and $Y = G + jB$

In case of a line having negligible resistance (such will be the case when conductors are of a large cross-section) and no shunt leakage (*i.e.,* $G = 0$),

$$Z_0 = \sqrt{\frac{L}{C}}\text{, which is a pure resistance.}$$

Its values are given below:

Overhead transmission lines ... 400 to 600 Ω

Underground cables ... 40 to 60 Ω

Q. 123. *Why loss angle of a cable should be very small?*

Ans. As the dielectric losses occurring in a cable are very small, the value of loss angle δ is very small.

Q. 124. *On what factors does the primary distribution voltage depend?*

Ans. (*i*) The amount of power to be supplied, and (*ii*) Distances of the substations required to be supplied

Q. 125. *As per IE rules, what is the declared consumer voltage?*

Ans. According IER, the declared consumer voltage is 415/240 V (*i.e.,* 415 between phases and 240 V between phase and neutral) with a possible variation of ± 6%.

Q. 126. *Define the terms: (i) Feeder; (ii) Distributor.*

Ans. (*i*) ***Feeder*** is a line which supplies a point of distribution network without being tapped at any intermediate point.

(*ii*) ***Distribution*** is a line from which tappings are taken along its length for providing supply to the consumer.

Q. 127. *Why is D.C. 2-wire system is never employed for transmission purpose?*

Ans. Because the *efficiency is low / poor*.

Q. 128. *Why is radial system of distribution not generally used?*

Ans. Because the *service reliability is the poorest*.

Q. 129. *What is a 'Booster'?*

Ans. ***Booster*** is a low-voltage, high-current series wound D.C. generator inserted into a D.C. circuit to add or inject a certain voltage proportional to circuit current.

Q. 130. *What principle do Murray loop and Varley loop tests employ?*

Ans. Principle of Wheatstone bridge.

Q. 131. *What are the transmission voltages that are used in India?*

Ans. 33, 66, 110, 132, 220 and 400 kV. 765/800 kV is also used in India.

Q. 132. *What are the advantages of adopting extra high voltage/ultra high voltage for transmission of A.C. electric power?*

Ans. ***Advantages:*** (*i*) Reduced line losses; (*ii*) High transmission efficiency; (*iii*) Improved voltage regulation; (*iv*) Reduced conductor material equipment; (*v*) Flexibility for future system growth; (*vi*) Increase in transmission capacity of the line.

Q. 133. *What are the problems associated with an EHV transmission?*

Ans. ***Problems:*** (*i*) Corona loss and radio interference, (*ii*) Requirements of heavy supporting structures and erection difficulties, and (*iii*) Insulation requirements.

- *The conductor size of an EHV-line is decided on the basis of 'corona'.*

Q. 134. *In EHV transmission systems, what are the 'sources of audible noise generation'?*

Ans. ***Sources:*** (*i*) Corona; (*ii*) Humming of transformers; (*iii*) Cooling systems; (*iv*) Mechanical and electrical auxiliaries.

Q. 135. *What type of conductors are universally employed for overhead transmission lines?*

Ans. ACSR (Aluminium conductor steel reinforced) conductors.

Q. 136. *Why is voltage regulation better in case of D.C. transmission?*

Ans. Due to absence of inductance in D.C. systems.

SWITCHGEAR AND PROTECTION

Q. 137. *What is 'Switchgear'?*

Ans. ***Switchgear*** is a term where covers wide range of equipment as regards switching and interrupting the currents in power system during normal and abnormal conditions. Controlling, protecting, regulating and measuring instruments are also covered by switchgear. Switchgear in general consists of *switches, fuses, circuit breakers, isolators, relays, control panel, metering panels, lightning arrestors, current transformers, potential transformers, etc.*

Q. 138. *What types of faults generally occur in a power system?*

Ans. ***Faults:*** (*i*) Over-current; (*ii*) Under-voltage; (*iii*) Unbalance; (*iv*) Reversed power; (*v*) Surges.

Q. 139. *What are 'Circuit-breakers'?*

Ans. ***Circuit breakers*** *are mechanical devices designed to close or open contact members, thus closing or opening an electrical circuit under normal or abnormal conditions.*

A circuit breaker has *fixed* and *moving contacts.* These contacts remain in contact whenever the circuit breaker is in normal operation. Whenever fault occurs the contacts separate out *resulting in arc between the electrodes. The production of arc not only delays the current interruption process but it also generates enormous heat which may cause damage to the system or the breaker itself.* Thus the *main problem is to extinguish the arc* whenever it occurs is circuit breaker. The temperature of arc may be as high as 6000°C.

Q. 140. *What are the commonly used methods of arc extinction?*

Ans. Commonly used methods of *arc extinction* are:

1. *High resistance interruption:*

By increasing the effective resistance with time, current can be reduced to a value insufficient to maintain it. The arc resistance can be increased by

(*i*) Arc lengthening;

(*ii*) Arc cooling;

(*iii*) Arc splitting;

(*iv*) Arc constraining.

2. *Low resistance interruption:*

Here, the arc resistance is kept low, in order to keep the arc energy to a minimum. This is achieved by:

(*i*) Cooling;

(*ii*) Gap lengthening;

(*iii*) Blast effect.

Q. 141. *What is 'recovery voltage'?*

Ans. The normal frequency r.m.s voltage that appears across the breaker contacts after final arc extinction has occurred, is called the ***recovery voltage.***

Q. 142. *What is 'restriking voltage'?*

Ans. The transient voltage that appears across the contacts at the instant of arc extinction is called the ***restriking voltage.*** The rate of rise of restriking voltage depends on:

(*i*) *Active recovery voltage* (the instantaneous value of the recovery voltage at the instant of arc extinction).

(*ii*) *Natural frequency of oscillations.*

Q. 143. *What are the various types of circuit breakers?*

Ans. (*i*) Low voltage air circuit breakers;

(*ii*) Oil circuit breakers;

(*iii*) Water type circuit breakers;

(*iv*) Air blast circuit breakers.

Q. 144. *What are the various ratings of circuit breakers?*

Ans. (*i*) Breaking capacity; (*ii*) Making capacity; (*iii*) Short-time capacity.

Q. 145. *Show by a diagram the clearing and closing time of a circuit-breaker.*

Ans. Fig. 29.4 illustrates the clearing and closing time of a circuit breaker.

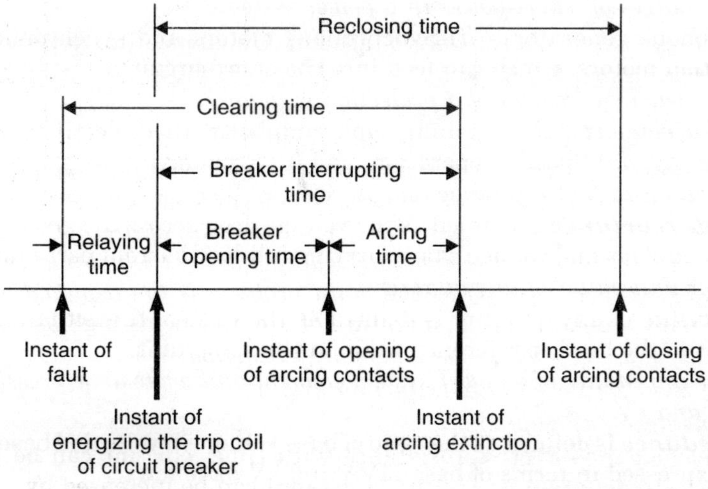

Fig. 29.4. Clearing and closing time of a circuit breaker.

Q. 146. *What are 'reactors'?*

Ans. The purpose of ***reactors*** incorporated in circuit breakers is *to limit the short-circuit current flowing to a safe value thus providing protection of instrument.* They consist of *large coils of high self-inductance and very low resistance.*

The reactors are of two types namely:

1. Open type
2. Oil immersed type .... used for voltages above 33 kV.

Reactors in a power system can be located

(*i*) in series with generators; (*ii*) in series with feeders;

(*iii*) in bus-bars (in ring system or in tie bar system).

Q. 147. *Name the various bus-bar arrangements.*

Ans. Typical bus-bar arrangements are:

1. *Single bus-bar system* used in D.C. and small A.C. power stations.
2. *Single bus-bar system with sectionalisation.*
3. *Ring bus-bar system.*
4. *Duplicate bus-bar system.*

Q. 148. *What is the difference between a short-circuit and an overload?*

Ans. When a ***short-circuit*** occurs, the voltage at the fault point is *reduced to zero* and current of abnormally high magnitude flows through the network to the fault point.

An ***overload*** means that loads greater than the designed value have been imposed on the system. In case of an overhead, the voltage at the overload point may become low, but *cannot be zero.* The overload conditions may extend to some distance beyond the overload point in the remainder of the system. In case of an overload currents are high but substantially lower than in case of a short-circuit.

Q. 149. *Why the probability of the failure or occurrence of abnormal conditions is more on the power line?*

Ans. Due to the reasoning that the power lines are widely branched, have greater length, operate under variable weather conditions and are subject to the action of atmospheric disturbances of electrical nature.

Q. 150. *Name the fault which occurs more frequently in overhead lines.*

Ans. Single-phase to ground (*L-G*).

Q. 151. *Name the 'sources of short-power in a power system'.*

Ans. (*i*) Synchronous generators, (*ii*) Synchronous motors and synchronous condensers, and (*iii*) Induction motors, which can feed into the short-circuit.

Q. 152. *Enlist the basic requirement of a protection system.*

Ans. ***Basic requirements:*** (*i*) Reliability and simplicity, (*ii*) Selectivity and discrimination, (*iii*) Sensitivity, (*iv*) Speed of operation, and (*v*) Economy.

Q. 153. *What do you mean by 'percentage reactance'?*

Ans. ***Percentage reactance*** is defined as the voltage drop across the given reactance expressed as percentage of normal voltage when carrying full-load current pertaining to normal rating.

Q. 154. *What is 'per-unit value' of any quantity?*

Ans. ***Per-unit value*** of any quantity is defined as the ratio of the actual value of the quantity in any unit to the base or reference value in the same unit.

Q. 155. *Define base impedance of a power system element and write an expression for it in terms of base MVA and kV.*

Ans. ***Base impedance*** is defined as the ratio of base voltage in volts and base current in amperes and it is expressed in terms of base MVA and kV as follows:

$$\text{Base impedance (Ohms)} = \frac{(\text{Base kV})^2}{\text{Base MVA}}.$$

Q. 156. *What are 'Base quantities'?*

Ans. For common representation, the quantities like voltage, current, power or impedance, chosen as common or reference quantities, are called the ***base quantities.***

Q. 157. *What is the relation between percentage impedance and pu impedance?*

Ans. Percentage impedance = pu impedance × 100.

Q. 158. *What is meant by fault level?*

Ans. When a fault occurs at a point in a power system, the corresponding MVA is referred to as the ***fault level*** at that point.

Q. 159. *What are current limiting reactors?*

Ans. ***Current limiting reactors*** are inductive coils having *large inducting reactances* in comparison to their resistances and are commonly installed on *large power systems to limit* the magnitude of short-circuit currents and *reduce* the resulting voltage disturbances on the rest of the systems.

Q. 160. *What are symmetrical components?*

Ans. According to the theory of symmetrical components, any unbalanced system of phasors (voltages, currents or impedances, etc.) can be resolved into balanced system of phasors equal in number to the number of phases. These *balanced systems of phasors are called symmetrical components of the original unbalanced system.*

Q. 161. *What is negative-sequence impedance?*

Ans. The impedance of the network offered to flow of negative-sequence currents is called the ***negative-sequence impedance*** of the network.

Q. 162. *What are sequence networks?*

Ans. The single-phase equivalent circuit composed of impedances to current of any one sequence only is called the ***sequence network*** for that particular sequence. Hence corresponding to positive-, negative- and zero-sequence currents we have positive-, negative- and zero-sequence networks.

Q. 163. *What are the various types of faults which occur in a power system?*

Ans. The most common and dangerous faults that occur in a power system are:

- The short-circuit or shunt faults such as *L-G, L-L, L-L-G, L-L-L* and *L-L-L-G* faults.
- The other types of faults that occur in a power system are series type such as one conductor open fault and two conductors open fault.

Q. 164. *What do you mean by 'Unsymmetrical faults'?*

Ans. ***Unsymmetrical faults*** are the faults that give rise to unsymmetrical currents *i.e.,* currents differing in magnitude and phases in the three phases of the power system.

These faults may also be defined as the faults involving one or two phases (not involving three phases) such as *L-G, L-L-L-G* faults.

Q. 165. *For a fault at given location rank, what are the various faults in the order of severity?*

Ans. The faults in *increasing* order of severity are:

L-G, L-L, L-L-G, and symmetrical three-phase faults (*L-L-L or L-L-L-G*).

Q. 166. *Which are the faults that do not have zero-sequence currents flowing?*

Ans. Line-to-line faults and symmetrical faults (*L-L-L* or *L-L-L-G* faults), *no* zero-sequence current flows in these faults.

Q. 167. *In which fault, all the three sequence currents are present and are equal to each other?*

Ans. *L-G* fault.

Q. 168. *Which of the faults would cause a greater electromagnetic interference between the power line and a nearby communication line?*

Ans. Single phase-to-ground (*L-G*) fault, due to heavy unbalancing of line currents, causes greater electromagnetic interference between the power line and a nearby communication line.

Q. 169. *What is a 'Fuse'?*

Ans. ***Fuse** is a wire of short length or thin strip of material having low melting point and inserted in an electric circuit as protective device to the flow of an excessive current through the circuit.* The time for blowing out of fuse depends upon the magnitude of the excessive current.

Q. 170. *What are the essential characteristics of the material to be used for fuse wire?*

Ans. ***Characteristics:*** (*i*) Low melting point; (*ii*) Low ohmic losses; (*iii*) High conductivity; (*iv*) Free from deterioration.

Q. 171. *What is 'Fuse law'?*

Ans. For a round wire the approximate value of fusing current is given by:

$I = Kd^{3/2}$ ...called the ordinary **Fuse law**

where, K = constant (depending on the metal of wire), and d = diameter of the wire.

Q. 172. *What is 'Fusing factor'?*

Ans. The ratio of minimum fusing current and current rating of fuse element is known as ***fusing factor***. The factor is always *greater than unity.*

Q. 173. *What material is mostly used for an ordinary fuse wire?*

Ans. Copper or lead-tin alloy.

Q. 174. *What are the factors on which current carrying capacity of a fuse depend?*

Ans. ***Factors:*** (*i*) The permissible temperature rise of the contacts of the fuse holder, (*ii*) Size of fuse wire, (*iii*) Fuse material, and (*iv*) The surrounding of the base.

Q. 175. *What are the factors on which fusing current depends?*

Ans. ***Factors:*** (*i*) Diameter of wire; (*ii*) Length of wire; (*iii*) Fuse material; (*iv*) *X*-sectional area (round or rectangular); (*v*) Size and location of terminals; (*vi*) Type of enclosure used; (*vii*) Type of surface, etc.

Q. 176. *What is mean by 'Discrimination'?*

Ans. ***Discrimination*** means the breaking of the faulty section of the line to an extent restricted to minimum; healthy sections of the circuit remaining in operation.

Q. 177. *What is meant by 'Current chopping'?*

Ans. ***Current chopping*** is a phenomenon of current interruption before the natural current zero is reached.

- 'Current chopping' is considered *serious drawback* because it *sets up high voltage transient across the breaker contacts.*

Q. 178. *What is meant by 'kilometric fault'?*

Ans. ***Kilometric fault*** is the fault occurring between a distance of a few km to few km from the circuit breaker.

Q. 179. *What is meant by 'making capacity' of a circuit breaker?*

Ans. The ***making capacity*** of a circuit breaker, when closed on a short-circuit, is the peak value of the maximum current wave (including D.C. component) in the first cycle of the current after the circuit is closed by the circuit breaker.

Rated making capacity = 2.55 × symmetrical rating breaking capacity

Q. 180. *Why a decaying D.C. offset current wave appears when an A.C. circuit breaker is suddenly switched to an inductive circuit at the instant when voltage wave is passing through zero?*

Ans. In an inductive circuit, current lags behind the voltage by a large angle *i.e.*, when voltage is passing through zero, current is around maximum value. Also current in an inductive circuit cannot change instantaneously. So when A.C. circuit breaker is suddenly switched on to an inductive circuit, the current may be around maximum value which cannot change instantaneously, to compensate this D.C. off current appears which dies out as time passes and a steady-state condition is reached.

Q. 181. *What is a 'protective relay'?*

Ans. ***Protective relay*** is an electrical device designed to initiate isolation of a part of an electrical installation, or to operate an *alarm signal*, in the event of an *abnormal condition* or *fault.*

Q. 182. *What are the functions of protective relays?*

Ans. The main functions of protective relays are to *detect* the presence of faults, their locations and *initiate the action* for quick removal from service of the faulty section of the system. Relays operate to *trip* the power circuit breakers to disconnect the faulty section.

Q. 183. *Name the essential elements of a protective relay.*

Ans. ***Essential elements:*** (*i*) Sensing element; (*ii*) Comparison element; (*iii*) Control element.

Q. 184. *What is 'relay time'?*

Ans. ***Relay time*** is the time interval between the occurrence of fault and closure of relay contacts.

Q. 185. *What are shortcomings of electromagnetic relays?*

Ans. The vibration of relay armature at double the supply frequency, due to pulsating nature of electromagnetic force, causes the relay to *hum* and *produce noise* and also is a source of *damage* to the relay contacts. This *leads to sparking* and *unreliable operation* of the relay operative circuit contacts due to make and break of the circuit.

Q. 186. *What is 'Buchholz relay'? Which equipment is protected by it?*

Ans.
- ***Buchholz relay*** is a gas actuated relay.
- It is used for the *protection of power transformers of ratings exceeding 500 kVA.*

Q. 187. *What are the essential features of directional relays?*

Ans. ***Essential features:*** (*i*) High speed operations; (*ii*) High sensitivity; (*iii*) Short-time thermal rating; (*iv*) Ability to operate with low values of voltage; (*v*) Burden must not be excessive; (*vi*) There should be no voltage and current creep.

Q. 188. *Where are direction relays used?*

Ans. ***Direction relays*** are used when graded-time overload protection is applied to ring mains and interconnected networks.

Q. 189. *In what way a distance relay is superior to over-current protection for the protection of transmission lines?*

Ans. Distance relay is superior to over-current protections for the protection of transmission lines due to the following *reasons*:

(*i*) Faster operation; (*ii*) Simpler coordination; (*iii*) Simpler application; (*iv*) Permanent setting without need for readjustments; (*v*) Less effect of amount of generation and fault levels; (*vi*) Fault current magnitude; (*vii*) Permits the high line loading.

Q. 190. *Why a 'Mho relay' is normally used for the protection of a long distance transmission line?*

Ans. ***Mho relays*** are best suited for phase fault relaying for long lines and particularly where a severe synchronising power surge may occur. They are most sensitive of all the distance relays. The Mho relay is *'reliable'* because it *combines both the directional and distance measuring functions in one unit.*

Q. 191. *What is 'differential protection'?*

Ans. The ***differential protection*** is that which responds to the phase difference between the two or more electrical quantities.

Q. 192. *What type of protective device is used for the protection of alternator against overheating of its stator?*

Ans. Resistance temperature detector.

Q. 193. *What types of protective devices are used for protection of an alternator against overheating of its rotor?*

Ans. D.C. relay and resistance measuring circuit.

Q. 194. *What is 'Field suppression'?*

Ans. The opening of the field circuit breaker and discharge of stored energy in the field winding through a resistor is called the ***field suppression.***

Q. 195. *What is 'through fault'?*

Ans. A ***through fault*** is one which is beyond the protected zone of the transformer but fed through it.

Q. 196. *What are the shortcomings of differential protection scheme as applied to power transformers?*

Ans. ***Shortcomings:*** (*i*) Unmatched characteristic of CTs, (*ii*) Change in ratio due to tap-changing, and (*iii*) Magnetising in-rush currents.

Q. 197. *What is mean by 'time-graded protection'?*

Ans. ***Time-graded protection*** is a scheme of over current protection, in which discrimination is incorporated *i.e.*, the time setting of relays is *so graded* that in the event of fault, the *smallest possible part of the system is disconnected.*

Q. 198. *What type of relay is suitable for radial feeders?*

Ans. Induction type inverse definite minimum time (IDMT) relays are most suitable for radial feeders.

Q. 199. *Why Merz-Price protection is not suitable beyond 33 kV?*

Ans. Difficulties are experienced in balancing the secondaries of the two CTs and that is why Merz-Price protection system *cannot be* used beyond 33 kV.

Q. 200. *What is meant by 'voltage surge'?*

Ans. By ***voltage surge*** is meant any sudden excessive rise in voltage that may be dangerous to the electrical equipment of an installation.

Q. 201. *What is 'Lightning'?*

Ans. ***Lightning*** is huge spark, which is due to the electrical discharge taking place between the clouds, within the same cloud and between the clouds and the ground.

Q. 202. *What is 'earthing screen'? What type of protection is provided by it?*

Ans. ***Earthing screen*** consists of a network of copper conductors, earthed at least at two points, over all electrical equipment in the substation. An electrostatic shield against external field is presented by such a screen and so protects the system.

Earthing screen protects the system from direct lightning strokes, reduces the voltages induced electrostatically or electromagnetically and *does not* provide any protection against voltage waves which may still reach the terminal equipments.

- Substations, interconnectors and power houses are protected from direct strokes by earthing screen.

Q. 203. *What is an 'earth wire'?*

Ans. ***Earth wire*** is a form of lightning protection using a conductor or conductors, well grounded at regular intervals preferably at each support, and attached from support to support above the transmission line.

Q. 204. *What is 'lightning arrestor' or 'surge detector'?*

Ans. ***Lightning arrestor or surge detector*** is a device that is connected between line and earth, *i.e.*, in parallel with the equipment under protection at the substation. It *limits* the duration and amplitude of the follow current.

Q. 205. *What is meant by 'surge absorber'?*

Ans. ***Surge absorber*** is a device that reduces the steepness of the wavefront of a particular surge and thus *minimises the danger due to over-voltage.*

Q. 206. *What is the difference between 'lightning conductor' and 'lightning arrestor'?*

Ans. ***Lightning conductor*** is installed at the *top* of all structures and *connected to general mass of earth* through a system of connecting conductors and the earth electrodes so that the structure under protection is not subjected to the dangerously high potential of the cloud discharge.

Lightning arrestor is a device that is *connected in parallel with the equipment,* under protection from lightning strokes, *at the substation.*

Q. 207. *What is the application of 'Bewley's lattice diagram'?*

Ans. By using ***Bewley's lattice diagram*** one can know at a glance the position and direction of motion of every incident, reflected and transmitted wave on the system at every instant of time.

Q. 208. *What is the average resistance of a human body? What is the current that is considered safe for human body?*

Ans.
- Average resistance: 1000 Ω
- Safe current: 9 mA

Q. 209. *What is 'harmonic suppressor'?*

Ans. ***Harmonic suppressor*** is an iron-cored reactor and is used in the neutral ground connections of a generator so that third harmonic currents are reduced to safe value.

Q. 210. *What are the different methods of controlling the voltages in A.C. supply systems?*

Ans. Methods of controlling the voltage in A.C. supply systems are:

(*i*) Excitation control at generating stations, and by using; (*ii*) Tap-changing transformers; (*iii*) Induction regulators; (*iv*) Booster transformers; (*v*) Series compensation; (*vi*) Shunt compensation; (*vii*) Synchronous phase modifiers.

Q. 211. *Why the excitation control is not possible beyond certain limits?*

Ans. Excitation below a certain level may cause instability of the system and excitation above a certain limit may cause overheating of the rotor. So excitation control is *not* possible *beyond* certain limit.

Q. 212. *What is a 'booster transformer'?*

Ans. ***Booster transformer*** is one which is often *used at the end of a power line* in order to *raise* the voltage to the desired value.

Q. 213. *State the relative merits of indoor and outdoor substations.*

Ans. ***Merits of indoor substations:***

(*i*) Less space requirement; (*ii*) Less maintenance and control cable length; (*iii*) Protection from lightning; (*iv*) Flexibility in installation; (*v*) No dust and dirt.

Merits of outdoor substations:

(*i*) No building requirement; (*ii*) Short erection time; (*iii*) Easy fault finding due to visibility of equipment; (*iv*) Easy repair work; (*v*) Easier installation extension; (*vi*) Availability of sufficient space between equipments.

Q. 214. *What is 'flexible bus'?*

Ans. Flexible bus consists of flexible ACSR or all aluminium alloy stranded conductors supported by strain insulators from each end.

Q. 215. *Define steady state stability.*

Ans. ***Steady state stability*** may be defined as the capability of an electric power system to *maintain synchronism* between machines within the system and external tie lines following a small slow disturbance.

Q. 216. *Distinguish between 'steady state stability' and 'transient stability' of a power system.*

Ans. ***Steady state stability*** refers to the conditions of stability in response to *small and gradual changes of load,* while ***transient stability*** refers to the conditions of *stability* in response to *large sudden disturbances* (*e.g.,* application of faults, clearing of faults, sudden load changes, and inadvertent dripping of lines and generators).

Q. 217. *Define dynamic stability of a power system.*

Ans. ***Dynamic stability*** is the ability of a power system to remain is synchronism after the initial swing (transient stability period) until the system has settled down to the new steady state equilibrium condition.

Q. 218. *What is 'power angle curve'?*

Ans. ***Power angle curve*** is a curve of power P versus power angle δ. It is a sine curve.

Q. 219. *Define infinit bus in a power system.*

Ans. ***Infinit bus*** has constant voltage, constant electrical angle of the voltage, constant frequency, infinite fault level and, therefore, the quantities of bus do not get affected by connecting or disconnecting individual machines on transmission lines.

Q. 220. *Write swing equation for a single machine system.*

Ans. For a single machine system, the *'swing equation'* can be written as:

$$M\frac{d^2\delta}{dt} = P_S - P_E = P_S - P_{max} \sin \delta$$

where, M = Angular momentum; P_S = Shaft power (mechanical power),

P_E = Electromagnetic power; δ = Load angle between E and V, and

P_{max} = The maximum steady state power.

Q. 221. *What is the purposes served by power flow analysis?*

Ans. ***Power flow analysis*** provides a systematic mathematical approach for determination of various bus voltages, their phase angles, active and reactive power flows through different branches, generators and loads under steady state conditions.

Q. 222. *Name three categories of buses.*

Ans. ***Categories:*** (*i*) Generation or voltage-controlled or P–V buses; (*ii*) Load or P–Q buses; (*iii*) Slack or swing reference buses.

Q. 223. *When is one of the buses taken as slack bus in load flow studies?*

Ans. One of the buses is designated as a ***slack bus*** due to the fact that the system power losses are *not known initially* and, therefore, the net power flow into the system *cannot* be fixed in advance. At this bus, the magnitude and phase angle of the voltage are specified. The *phase angle of the voltage* is *usually set equally to 'zero'.* The active and reactive power at this bus are to be determined through the solution of equations.

Q. 224. *How is the swing bus selected in a load flow study?*

Ans. One of the generation bus is selected as slack or swing bus and the generators at this bus *supply the difference* between the specified real power injected into the system at other buses and the total output of the system plus system power loss.

Q. 225. *Give any two operating constraints imposed in load flow studies.*

Ans. Two ***operating constraints*** may be *'voltage'* and *'reactive power'.*

Q. 226. *Name any two iterative methods used for solution of load flow problems.*

Ans. ***Iterative methods:*** (*i*) Gauss-Seidel method, and (*ii*) Newton-Raphson method.

Q. 227. *What is meant by 'acceleration factor' in load flow solution and its best value?*

Ans. ***Acceleration factor*** is used in Gauss-Seidel method of load flow solution to *increase the rate of convergence.*

Acceleration factor = 1.6 is considered to be a good value for power flow studies.

II. CONVENTIONAL QUESTIONS WITH "SOLUTIONS"

Q. 29.1. (*a*) *What factors should be taken into consideration while selecting the site for steam power station:*

(b) ***Operation of different power plants.*** *In a system the load duration curve is a straight line, the maximum loads being 125 MW and 20 MW respectively. The costs of the base load and peak load points which supply the load, are given below.*

Base load plant: ₹250/kW-year + 6 p/kWh

Peak load plant: ₹65/kW-year + 12p/kWh

Determine, for the minimum overall cost of the following:

(i) The load shared by peak load point.

(ii) The annual load factors for both stations.

Solution. (*a*) The following factors should be taken into consideration while *selecting the site* for a steam power station:

1. Availability of raw material,
2. Nature of land,
3. Cost of land,
4. Availability of water,
5. Transport facilities,
6. Ash disposal facilities,
7. Availability of labour,
8. Size of plant,
9. Load centre,
10. Public problems, and
11. Future extensions.

(*b*) (*i*) **Load shared by peak load plant:**

Let,

C_{base} = Operating cost of base load plant,

C_{peak} = Operation cost of peak load plant;

P_{base} = Load (peak) on the base load plant, and

P_{peak} = Load (peak) on the peak load plant.

The operating costs of base load and peak load plants may be expressed as follows:

$$C_{base} = A_1 \text{ kW} + B_1 \text{ kWh}$$

$$C_{peak} = A_2 \text{ kW} + B_2 \text{ kWh}$$

From the given data: $A_1 = 250,\ B_1 = 0.06;$

$A_2 = 65,\ B_2 = 0.12.$

The time, h hours, for which base load to be operated for *minimum overall cost* is given by:

$$h = \frac{A_1 - A_2}{B_2 - B_1} = \frac{250 - 65}{0.12 - 0.06} = 3083 \text{ hours}$$

Now, $P_{base} + P_{peak} = 125$

$\therefore \quad P_{peak} = 125 - P_{base}$

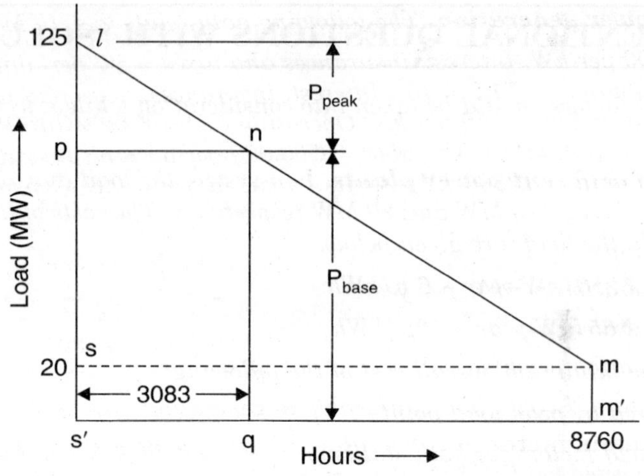

Fig. 29.5

Now, from Δs lms and lnp, we have:

$$\frac{125 - P_{\text{base}}}{125 - 20} = \frac{3083}{8760}$$

or,

$$P_{base} = 125 - (125 - 20) \times \frac{3083}{8760} = \mathbf{88\ MW} \quad \textbf{(Ans.)}$$

$$P_{peak} = 125 - 88 = \mathbf{37\ MW} \quad \textbf{(Ans.)}$$

(*ii*) **Annual load factors (L.F.) for both the stations:**

$$\text{L.F. (base load plant)} = \frac{\text{Average load}}{\text{Base load}} \times \frac{8760}{8760} = \frac{\text{Area '}pnmm's'p\text{'}}{P_{\text{base}} \times 8760}$$

$$= \frac{\text{Area '}pnmsp\text{' + area '}smm's's\text{'}}{P_{\text{base}} \times 8760}$$

$$= \frac{\frac{1}{2}(3083 + 8760) \times (88 - 20) + 20 \times 8760}{88 \times 8760} \simeq \mathbf{0.75 \text{ or } 75\%}$$

i.e., Load factor for base load plant = **75% (Ans.)**

$$\text{L.F. (Peak load plant)} = \frac{\text{Area '}lnpl\text{'}}{P_{\text{peak}} \times 8760} = \frac{\frac{1}{2} \times 3083 \times 37}{37 \times 8760} = 0.176 \text{ or } 17.6\%$$

i.e. Load factor for peak load plant = **17.6% (Ans.)**

Q. 29.2. (*a*) *Define the following terms:*

(*i*) *Demand factor;* (*ii*) *Load factor;* (*iii*) *Utilisation factor.*

(*b*) ***Economics of power generation.*** *The following data relate to a 15 MW power station:*

Cost of plant = ₹ 2400 per kW, Interest, insurances and taxes = 5% per annum; Depreciation = 5%; Cost of primary distribution = ₹ 1.2 × 10^6; Interest, insurances, taxes and depreciation = 5%; Cost of coal including transporation = ₹ 10 per kN; Operating cost = ₹ 1.2 × 10^6; Plant maintenance cost: (a) Fixed = ₹ 40,000 per annum, (b) Variable = ₹ 60,000 per annum; Installed plant capacity = 15000 kW; Maximum demand = 13500 kW; Annual load factor = 0.6; Consumption of coal = 390000 kN

Determine the following:

(i) Cost of power generation per kW per year.

(ii) Cost per kWh generated.

(iii) Total cost of generation per kWk.

Transmission or primary distribution chargeable to generation.

Solution. (*a*) (*i*) **Demand factor.** *It is the ratio of maximum demand of the system, or a part of the system, to the total connected load of the system, or of the part of the system, under consideration.*

$$\text{Mathematically, Demand factor} = \frac{\text{Maximum demand}}{\text{Connected load}}$$

(*ii*) **Load factor.** *It is the ratio of the average power to the maximum demand.*

$$\text{Mathematically, Load factor} = \frac{\text{Average load}}{\text{Maximum demand}}.$$

(*iii*) **Utilisation factor.** *It is the ratio of the maximum power demands of the substations of the system, or part of a system, to the maximum demand of the whole system, or part of system, under consideration, measured at the point of supply.*

$$\text{Mathematically, Utilisation factor} = \frac{\text{Sum of individual maximum demands}}{\text{Maximum demand of entire group}}$$

(***b***) Installed capacity of plant = 15 MW or 15000 kW

Total cost of plant = 15000 × 2400 = ₹ 36 × 10^6

Annual interest, insurances and taxes = ₹ 0.05 × 36 × 10^6 = ₹ 1.8 × 10^6

Annual depreciation = ₹ 0.05 × 36 × 10^6 = ₹ 1.8 × 10^6.

Annual interest, insurance taxes and depreciation on primary distribution

= ₹ 0.05 × 1.2 × 10^6 = ₹ 60,000 = 0.06 × 10^6

Annual plant maintenance (fixed) = ₹ 40,000 or 0.04 × 10^6

Total fixed cost (per anuum) = ₹ (1.8 × 10^6 + 1.8 × 10^6 + 0.06 × 10^6 + 0.04 × 10^6)

= ₹ 3.7 × 10^6

Annual operating cost = ₹ 1.2 × 10^6

Annual plant maintenance cost (variable) = ₹ 60,000 = 0.06 × 10^6

Annual cost of coal = ₹ 10 × 390000 = 3.9 × 10^6

Total annual running cost (per annum) = ₹ (1.2 × 10^6 + 0.06 × 10^6 + 3.9 × 10^6)

= ₹ 5.16 × 10^6

Maximum demand = 13500 kW

Annual load factor = 0.6

Average load $= 13500 \times 0.6 = 8100$ kW

Annual energy generated $= 8100 \times 365 \times 24 = 70.956 \times 10^6$ kWh

(*i*) **Cost of power generation per kW per year:**

Annual cost per kW of maximum demand

$$= \frac{\text{Fixed cost per annum}}{\text{Maximum demand}}$$

$$= \frac{3.7 \times 10^6}{13500} \simeq ₹\ \mathbf{274} \quad \textbf{(Ans.)}$$

(*ii*) **Cost per kWh generated:**

$$\textit{Annual cost/kWh} = \frac{\text{Annual running cost}}{\text{Annual energy generated}}$$

$$= \frac{5.16 \times 10^6}{70.956 \times 10^6} \simeq ₹\ \mathbf{0.0737} \quad \textbf{or} \quad \mathbf{7.3\ p} \quad \textbf{(Ans.)}$$

(*iii*) **Total cost per kWh:**

$$\textit{Total cost per kWh} = \frac{\text{Total annual cost}}{\text{Annual energy generated}}$$

$$= \frac{3.7 \times 10^6 + 5.16 \times 10^6}{70.956 \times 10^6} = ₹\ \mathbf{0.125} \ \textbf{or} \ \mathbf{12.5\ p.} \quad \textbf{(Ans.)}$$

Q. 29.3. (*a*) *State the advantages and limitations of interconnected system of power stations.*

(*b*) ***Interconnected systems.*** *Two identical 3-phase alternators work in parallel and supply a total load of 1500 kW at 11 kV at a power factor of 0.9. Each machine supplies half the total power. The synchronous reactance of each is 45 Ω/phase and resistance is 2.4 Ω/phase. The field excitation of the first machine is adjusted so that armature current is 50 A lagging. Determine the following:*

(i) The armature current of the machine-2.

(ii) The power factor at which each machine is working.

(iii) Generated voltage of machine-1.

Solution. (*a*) Some of the important *advantages* and *limitations demerits of Interconnected system* are listed below.

Advantages:

1. Reserved capacity required is reduced.
2. Reliability of the supply is increased.
3. Reduction in total installed capacity.
4. The load factor, diversity factor and efficiency are improved.
5. The interconnected system makes it possible to use the older and less efficient plants to carry peak loads of short durations,
6. Economical operation of stations is ensured.
7. Peak loads of combined system can be carried at *much lower cost* than that what is possible with small individual system.
8. Low capital and maintenance costs.

Limitations/Demerits:

1. The interconnection of stations involves *extra cost.*
2. The *problems of load and frequency control are more difficult* in a large interconnected system with many power stations scattered over a wide area in comparison with a system having one or two generating stations.

(*b*) Total load supplied = 1500 kW (at 11000 V and 0.9 p.f.)

Synchronous reactance of each machine/phase = 45 Ω

Resistance per phase = 2.4 Ω

$$\text{Load current at 0.9 p.f.} = \frac{1500 \times 1000}{\sqrt{3} \times 11000 \times 0.9} = 87.48 \text{ A}$$

$$\cos \phi = 0.9, \sin \phi = 0.436$$

∴ Working component = 87.48 × cos ϕ = 87.48 × 0.9 = 78.7 A

and, Wattless component = 87.48 × sin ϕ = 87.48 × 0.436 = 38.1 A

Each alternator supplies half of each component when conditions are identical as shown in Fig. 29.6.

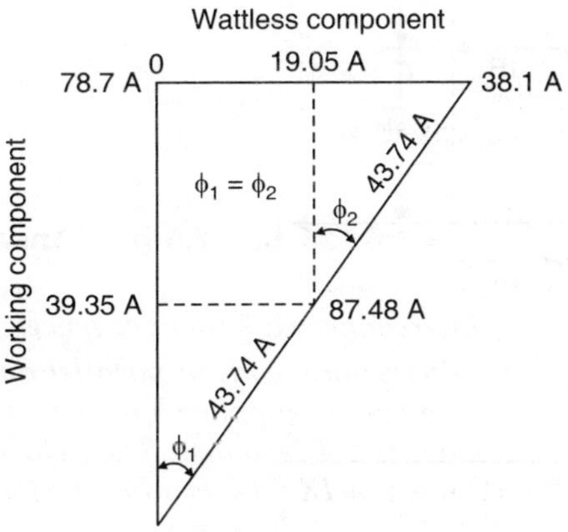

Fig. 29.6

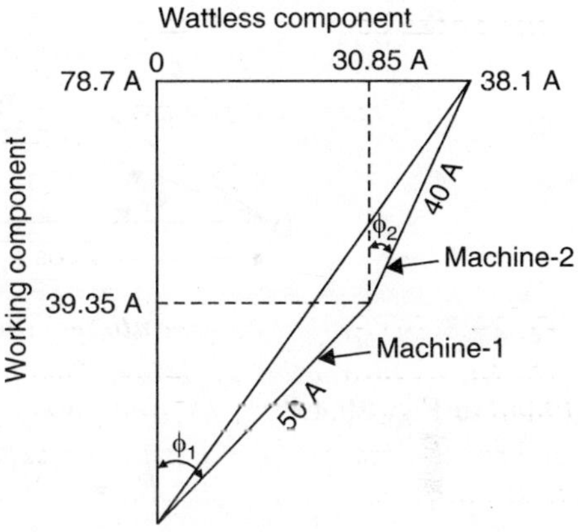

Fig. 29.7

$$\text{Total current supplied by each machine} = \frac{87.48}{2} = 43.74 \text{ A}$$

Since the steam supply of machine-1 is not changed hence the working components of both machines would remain the same at 39.35 A. But the reactive/wattless component will be redivided due to change in excitation.

The total current taken by machine-1 is changed from 43.74 A to 50 A.

∴ Wattless component for machine-1

$$= \sqrt{50^2 - 39.35^2} = 30.85 \text{ A}$$

∴ Wattless current delivered by machine-2

$$= 38.1 - 30.85 = 7.25 \text{ A}$$

New current diagram is shown in Fig. 29.7.

(*i*) **New total current for machine-2 (I_2):**

$$I_2 = \sqrt{(39.35)^2 + (7.25)^2} = \mathbf{40\ A} \quad \textbf{(Ans.)}$$

(*ii*) **Power factor at which each machine works, cos ϕ_1, cos ϕ_2:**

$$\cos \phi_1 = \frac{39.35}{50} = \mathbf{0.787} \quad \textbf{(Ans.)}$$

$$\cos \phi_2 = \frac{39.35}{40} = \mathbf{0.984} \quad \textbf{(Ans.)}$$

(*iii*) **Generated voltage of machine-1:**

$$\text{Terminal voltage/phase} = \frac{11000}{\sqrt{3}} = 6350 \text{ V}$$

$$\cos \phi_1 = 0.787, \ \sin \phi_1 = 0.617$$

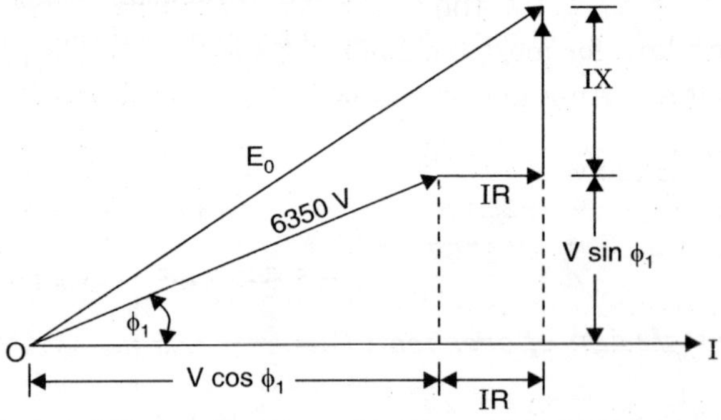

Fig. 29.8

Refer to Fig. 29.4.

$$E_0 = \sqrt{(V \cos \phi_1 + IR)^2 + (V \sin \phi_1 + IX)^2}$$

$$= \sqrt{(6350 \times 0.787 + 50 \times 2.4)^2 + (6350 \times 0.617 + 50 \times 45)^2}$$

$$= \sqrt{(5117.45)^2 + (6167.95)^2} = 8014.5 \text{ V}$$

$\therefore$ *Line voltage* $= \sqrt{3} \times 8014.5 =$ **13881.5 V (Ans.)**

Q. 29.4. *Electric power systems.* *(a) Explain Kelvin's law for determination of conductor size. Give its practical limitations.*

(b) A two-conductor cable 1.2 km long, is required to supply a constant load of 175 A throughout the year. The cost of cable is ₹ (110a + 55) per metre, where 'a' is the area of cross-section of the conductor in cm². The cost of energy is 30 p per kWh and interest and depreciation charges amount to 10%. Specific resistivity of copper is 1.84 μΩ cm. Find the most economical cross-section of the cable.

Solution. (*a*) Refer to Art. 11.8

(*b*) *Given:* l = 1.2 km; I = 175 A; cost of cable per metre length = ₹ (110a + 55); cost of energy = 20 p/kWh; Interest and depreciation charges = 10%; ρ = 1.84 μΩ cm.

Most economical X-section of the cable, a (cm²):

Resistance of each conductor, $R = \frac{\rho l}{a}$

$$= \frac{1.84 \times 10^{-6} \times 1.2 \times (1000 \times 100)}{a} = \frac{0.221}{a} \Omega$$

$$\text{Energy cost per annum} = \frac{2I^2 R \times 8760}{1000} = \frac{2 \times (175)^2 \times 0.221 \times 8760}{a \times 1000} = \frac{118577}{a} \text{ kWh}$$

Annual cost due to energy lost = ₹ $\frac{30}{100} \times \frac{118577}{a}$ = ₹ $\frac{35573}{a}$

Capital cost (variable) of cable = ₹ $110a \times (1.2 \times 1000) = 132000a$

Annual charges on account of interest and depreciation on variable cost of cable

= ₹ $\frac{10}{100} \times 132000a = 13200a$

According to Kelvin's law, for most economical *X*-section of the conductor,

Variable annual charge = Annual cost of energy lost

i.e., $$13200a = \frac{35573}{a}$$

∴ $$a = \left(\frac{35573}{13200}\right)^{1/2} = \mathbf{1.642\ cm^2} \quad \textbf{(Ans.)}$$

Q. 29.5. *Mechanical design of overhead lines.* *(a) List the basic tests that are carried out on insulators.*

(b) For a string insulator with four discs, the capacitance of disc is 12 times the capacitance between pin and earth. Calculate the voltage across each disc when used on a 62 kV line. Also calculate the string efficiency.

Solution. (*a*) Basic tests carried out on insulators:

1. *Design tests:*
   (*i*) Dry flash over tests.
   (*ii*) Wet flash over tests.
2. *Performance tests:*
   (*i*) Puncture test.
   (*ii*) Mechanical test.
   (*iii*) Porosity test.
3. *Routine tests:*
   (*i*) High voltage test.
   (*ii*) Proof load test.
   (*iii*) Corrosion test.

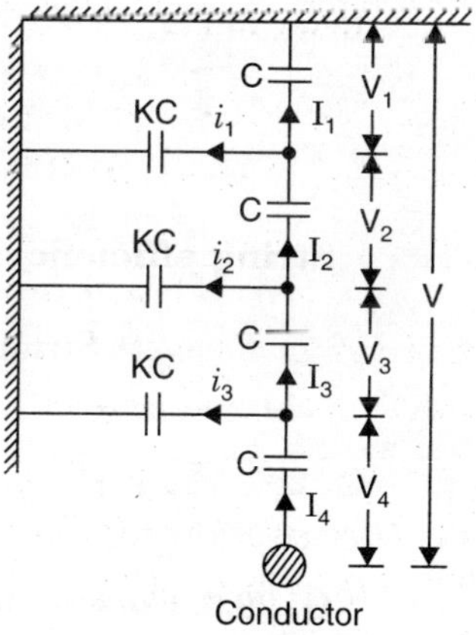

Fig. 29.9

(*b*) Refer to Fig. 29.9.

Let *C* be the self capacitance of each disc and *KC* be the capacitance between each pin and earth; $K = \frac{1}{12}$ (given)

Voltage across each disc:

In terms of voltages shown in Fig. 29.9, we know that:

$$V_2 = (1 + K)\,V_1 = \left(1 + \frac{1}{12}\right)V_1 = \frac{13}{12}V_1$$

$$V_3 = (1 + 3K + K^2)\,V_1 = \left[1 + 3\times\frac{1}{12} + \left(\frac{1}{12}\right)^2\right]V_1 = \frac{181}{144}V_1$$

$$V_4 = (1 + 6K + 5K^2 + K^3)\,V_1$$

$$= \left[1 + 6\times\frac{1}{12} + 5\times\left(\frac{1}{12}\right)^2 + \left(\frac{1}{12}\right)^3\right]V_1 = \frac{2653}{1728}V_1$$

$$V = V_1 + V_2 + V_3 + V_4$$

$$= V_1 + \frac{13}{12}V_1 + \frac{181}{144}V_1 + \frac{2653}{1728}V_1$$

$$= V_1\left(1 + \frac{13}{12} + \frac{181}{144} + \frac{2653}{1728}\right)$$

or,

$$V = V_1\left(\frac{1728 + 1872 + 2172 + 2653}{1728}\right) = \frac{8425}{1728}V_1$$

$$V_1 = V \times \frac{1728}{8425} = \frac{66}{\sqrt{3}} \times \frac{1728}{8425} = \mathbf{7.816\ kV} \quad \textbf{(Ans.)}$$

$$V_2 = \frac{13}{12}V_1 = \frac{13}{12}\times 7.816 = \mathbf{8.467\ kV} \quad \textbf{(Ans.)}$$

$$V_3 = \frac{181}{144}V_1 = \frac{181}{144}\times 7.816 = \mathbf{9.824\ kV} \quad \textbf{(Ans.)}$$

$$V_4 = \frac{2653}{1728}V_1 = \frac{2653}{1728}\times 7.816 = \mathbf{11.999\ kV} \quad \textbf{(Ans.)}$$

String efficiency:

$$\%\text{ String efficiency} = \frac{\text{Voltage across string}}{\text{No. of units (discs)} \times \text{voltage across the unit nearest to the conductor}}\times 100$$

$$= \frac{(66/\sqrt{3})}{4\times V_4}\times 100 = \frac{(66/\sqrt{3})}{4\times 11.999}\times 100 = \mathbf{79.39\%} \quad \textbf{(Ans.)}$$

Q. 29.6. *Corona.* (*a*) *What is corona? Explain briefly the theory of corona formation.*

(*b*) *A 3-phase, 220 kV, 50 Hz transmission line consists of 10 mm diameter conductor spaced 2.2 m apart in the form of an equilateral triangle. If the temperature is 36°C and atmospheric pressure is 76 cm calculate the corona loss per km of line. Assume the irregularity factor as 0.82.*

Solution. (*a*) *The phenomenon of appearance of violet luminous glow and production of hissing noise and ozone gas in an overhead transmission line is known as* **corona.**

Theory of corona formation. The phenomenon of corona is the *ionization of air surrounding the power conductor.* Owing to radioactivity and cosmic rays, free electrons are normally present in free surface. When the potential difference between the conductors is increased the gradient around the surface of the conductor increases. Assuming the spacing between the conductors to be large as compared to the diameter of the conductors, the free electrons will move with certain velocity depending upon the strength of the field. These electrons will have collision with molecules of air, and in case the speed is large, will dislodge the electrons from these molecules and thereby increase the number of electrons. This process being cumulative, will *lead to an avalanche of electrons.* Eventually there will be *ionization of air surrounding the conductor.*

If the spacing of the conductors is *'less' than 15 times the diameter of conductor, the flash-over will take place between the conductors before corona phenomenon occurs;* in practice the spacing is *much large* as compared to diameter, so the flash-over is *almost impossible* to occur.

(b) *Given:* $f = 50$ Hz; $r = \frac{10}{2} = 5$ mm $= 0.5$ cm; $D = 2.2$ m $= 220$ cm; $t = 36°$C;

Atmospheric pressure, $b = 76$ cm; Irregularity factor, $m_0 = 0.83$.

Corona loss per km of line, P:

Corona loss is given by:

$$P = (241 \times 10^{-5})\left(\frac{f+25}{\delta}\right)\sqrt{\frac{r}{D}}\,(V_{ph} - V_c)^2 \text{ kV/km/ph}$$

Now, $$\delta = \frac{3.92b}{273+t} = \frac{3.92 \times 76}{273+36} = 0.964$$

Assuming $g_0 = 21.21$ kV/cm (r.m.s), we have critical disruptive voltage,

$$V_c = g_0\, \delta m_0 r \ln\left(\frac{D}{r}\right) \text{ kV/phase}$$

$$= 21.21 \times 0.964 \times 0.83 \times 0.5 \ln\left(\frac{220}{0.5}\right) = 51.6 \text{ kV/phase}$$

$$\text{Supply voltage per phase} = \frac{220}{\sqrt{3}} = 127 \text{ kV}$$

Substituting the values, we get:

$$P = 241 \times 10^{-5}\left(\frac{50+25}{0.964}\right)\sqrt{\frac{0.5}{220}}\,(127 - 51.6)^2$$

$$= 50.81 \text{ kW/km/phase}$$

Total corona loss per km for three phases

$$= 3 \times 50.81$$

$$= \mathbf{152.43 \text{ kW}} \quad \textbf{(Ans.)}$$

Q. 29.7. ***Performance of transmission lines.*** *(a) Define 'Regulation' and 'Efficiency' of a transmission line.*

(b) A 3-phase load of 2200 kVA, 0.8 p.f. lagging is supplied at 6.6 kV, 50 Hz by means of a 33 kV transmission line 30 km long and 33/6.6 kV step-down transformer. The resistance and reactance of each conductor are 0.38 Ω and 0.45 Ω per km respectively. The resistance and reactance of transformer

primary are 6.5 Ω and 10.6 Ω, while those of secondary are 0.34 Ω and 0.65 Ω respectively. Determine the following when 6.6 kV is maintained at the receiving end:

(i) Sending end line voltage.

(ii) Sending end power factor.

(iii) Transmission efficiency.

Solution. (*a*) Regulation: The ***"regulation of a line"*** is defined as the *change in receiving end voltage, expressed in percent of full load voltage, from no-load to full-load, keeping the sending end voltage and frequency constant.*

$$\text{Mathematically, \% Regulation} = \frac{V_{R0} - V_R}{V_R} \times 100$$

where, V_{R0} = Receiving end voltage under *no-load* condition.
(= *sending end voltage in a short line*),
V_R = Receiving end voltage under *full-load* condition.

It may be noted here that V_{R0} and V_R are the magnitudes of voltages.

- Knowledge of regulation helps in maintaining the voltage at the load terminals *within presented limits* (± 5 percent of declared voltage) by using suitable *voltage control equipment*.

Efficiency:

"Efficiency of a transmission line" *is defined as the ratio of power delivered at the receiving end to the power sent from the sending end.*

$$\text{Mathematically, \% Transmission efficiency, } \eta_T = \frac{V_R I_R \cos \phi_R}{V_S I_S \cos \phi_S} \times 100$$

where, V_R, I_R, $\cos \phi_R$ are the receiving end voltage, current and power factor while V_S, I_S, $\cos \phi_S$ are the corresponding values at the sending end.

(***b***) *Given:* P(kVA = 2200; $\cos \phi_R = 0.8$ lagging; Transformation ratio, $K = \frac{6.6}{33} = \frac{1}{5}$

Resistance of each conductor = 0.38 × 30 = 11.4 Ω

Reactance of each conductor = 0.45 × 30 = 13.5 Ω

Let us transfer the impedance of transformer secondary to H.V. side *i.e.*, 33 kV side. Equivalent resistance of transformer referred to 33 kV side

$$= \text{Primary resistance} + 0.34 \times \frac{1}{K^2}$$

$$= 6.5 + 0.34 \times 25 = 15\ \Omega$$

Equivalent reactance of transformer referred to 33 kV side

$$= \text{Primary reactance} + 0.65 \times \frac{1}{K^2}$$

$$= 10.6 + 0.65 \times 25 = 26.85\ \Omega$$

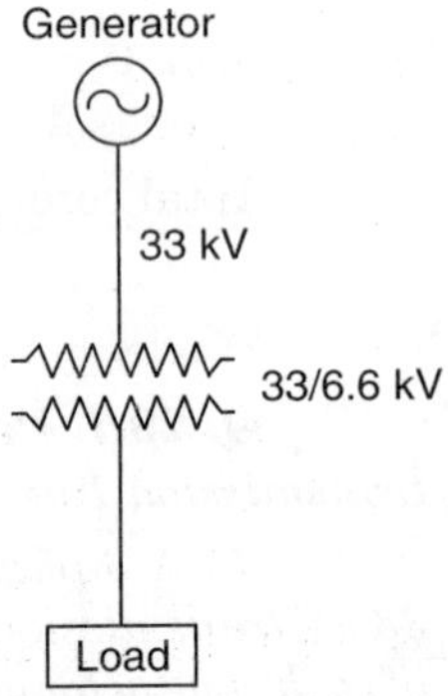

Fig. 29.10

∴ Total resistance of the line and transformer,

$$R = 11.4 + 15 = 26.4\ \Omega$$

Total reactance of the line and transformer,

$$X = 13.5 + 26.85 = 40.35\ \Omega$$

(*i*) **Sending end voltage V_{SL}:**

Receiving end voltage/phase, $V_R = \dfrac{33 \times 1000}{\sqrt{3}} = 19052$ V

Line current, $I = \dfrac{2200 \times 1000}{\sqrt{3} \times 33000} = 38.49$ A

Also, sending end voltage/phase,

$$V_S = V_R + IR \cos \phi_R + IX \sin \phi_R \text{ (app.)}$$
$$= 19052 + 38.49 \times 26.4 \times 0.8 + 38.49 \times 40.35 \times 0.6$$
$$= 20797 \text{ V or } 20.797 \text{ kV}$$

$\therefore$ $V_{SL} = \sqrt{3} \times 20.797 =$ **36.02 kV (Ans.)**

(*ii*) **Sending end power factor, $\cos \phi_s$:**

$$\cos \phi_s = \frac{V_R \cos \phi_R + IR}{V_S} = \frac{19052 \times 0.8 + 38.49 \times 26.4}{20797}$$

= **0.782 lag. (Ans.)**

(*iii*) **Transmission efficiency, η_T:**

Line losses $= 3I^2R = 3 \times (38.49)^2 \times 26.4 = 117333$ W or 117.333 kW

$$\therefore \quad \eta_T = \frac{\text{Power delivered}}{\text{Power delivered + line losses}}$$

$$= \frac{2200 \times 0.8}{(2200 \times 0.8) + 117.333} = 0.9375 \text{ or } \mathbf{93.75\%} \text{ (Ans.)}$$

Q. 29.8. (*a*) *What is Ferrantic effect? Explain it with the help of phasor diagram.*

(*b*) ***Underground cables.*** *A single-core cable used on 33 kV, 50 Hz has conductor diameter 12 mm and inner sheath diameter 30 mm. The relative permittivity of insulating material used is 3.5. Find:*

(*i*) *Capacitance of the cable per km,*

(*ii*) *Maximum electrostatic stress in the cable,*

(*iii*) *Minimum electrostatic stress in the cable, and*

(*iv*) *Charging curve per km.*

Solution. (*a*) Ferranti effect. *When a long line is operating under no-load or light-load condition, the receiving end voltage is greater than the sending end voltage. This phenomenon is known as* **Ferranti effect.** This effect is negligible in case of short lines but increases rapidly with the increase in length of line.

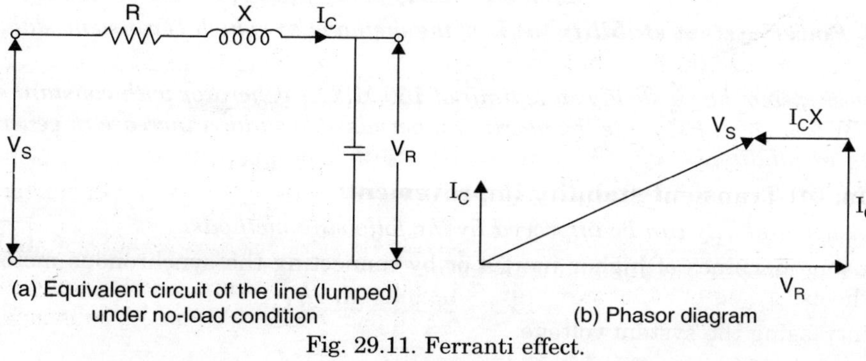

(a) Equivalent circuit of the line (lumped) under no-load condition

(b) Phasor diagram

Fig. 29.11. Ferranti effect.

A simple explanation of Ferranti effect can be given by approximating the distributed parameters of the line by lumped impedance. Fig. 29.11 (*a*) and (*b*) shows the equivalent circuit of the line (lumped) under no-load condition and the corresponding phasor diagram. The charging current produces drop in the reactance of the line which is in phase opposition to the receiving end voltage and hence the sending and voltage becomes smaller than the receiving end voltage.

(*b*) *Given:* Radius of conductor, $r = \frac{d}{2} = \frac{12}{2} = 6$ mm = 0.006 m; Inner radius of the sheath, $R = \frac{30}{2} = 15$ mm = 0.015 m; Relative permittivity of insulating material, $\varepsilon_r = 3.5$. Applied voltage, $V = 33$ kV = 30,000 V. Refer to Fig. 29.11 (*a*) and (*b*).

(*i*) **Capacitance of the cable per km:**

Using the relation:

$$C = \frac{2\pi\varepsilon_0\varepsilon_r}{\ln(R/r)} \text{ F/m}$$

$$= \frac{2\pi\varepsilon_0\varepsilon_r}{\ln(R/r)} \times 1000 \text{ F/m}$$

$$= \frac{2\pi \times 8.854 \times 10^{-12} \times 3.5}{\ln(0.015/0.006)} \times 1000$$

$$= 212.5 \times 10^{-9} \text{ or } \mathbf{0.2125\ \mu F} \quad \textbf{(Ans.)}$$

(*ii*) **Maximum electrostatic stress, E_{max}:**

$$E_{max} (= g_{max}) = \frac{V}{r\ln(R/r)}$$

(where g_{max} = maximum potential gradient)

$$= \frac{33000}{0.006 \times \ln(0.015/0.006)} = \mathbf{60 \times 10^5\ V/m} \quad \textbf{(Ans.)}$$

(*iii*) **Minimum electrostatic stress, E_{min}:**

$$E_{min} (= g_{min}) = \frac{V}{R\ln(R/r)} = \frac{33000}{0.015\ln(0.015/0.006)} = \mathbf{24 \times 10^5\ V/m} \quad \textbf{(Ans.)}$$

(*iv*) **Charging current, I_C:**

$$I_C = \frac{V}{X_C} = \frac{V}{(1/2\pi fC)} = 2\pi fCV$$

$$= 2\pi \times 50 \times 212.5 \times 10^{-9} \times 33000 = \mathbf{2.2\ A} \quad \textbf{(Ans.)}$$

Q. 29.9. ***Power system stability*** *(a) List the methods by which 'Transient stability' can be improved.*

(b) A power system has a steady state limit of 160 MW. A generator with constant excitation is supplying 80 MW to system. Estimate the maximum permissible sudden increase in generator output without causing instability.

Solution. (*a*) Transient stability improvement:

The transient stability can be improved by the following methods:

1. By using machines of higher inertia or by connecting the synchronous motors to heavy flywheels.
2. By increasing the system voltage.

3. By reducing the transfer reactance.
4. By using high neutral grounding impedance.
5. By reducing the severity of faults which can be achieved by using lightning arrestors for protection of lines.
6. By the use of high speed circuit breakers.
   - To improve transient stability the governors attached to the turbines driving alternators should be high speed one so as to adjust the generator input quickly as per demand of the load.
   - Voltage regulators employed on the lines should be quick acting ones.

(***b***) The maximum power that can be transferred, P_{max} = 160 MW; Initial operating load, P_0 = 80 MW

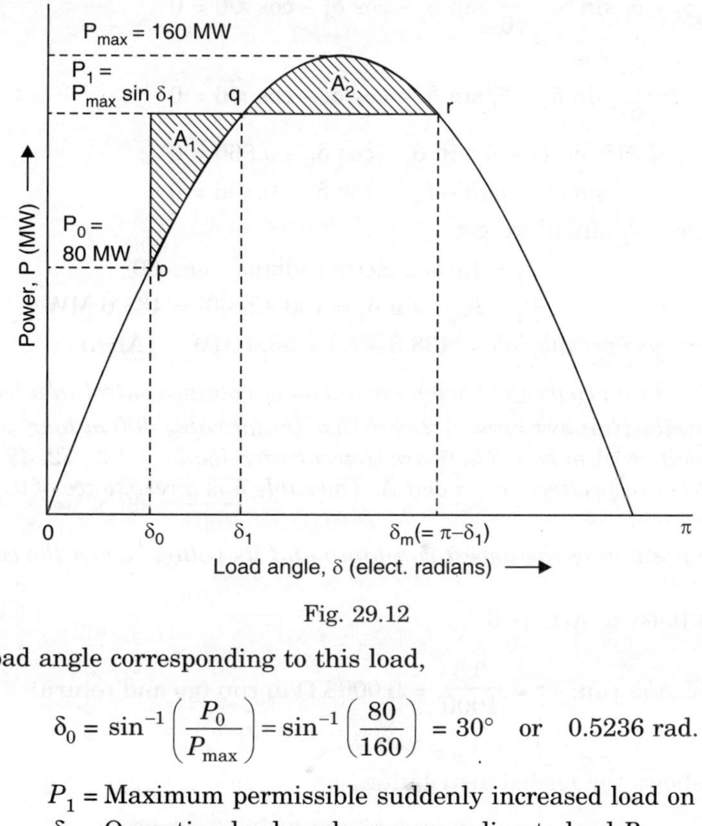

Fig. 29.12

Operating load angle corresponding to this load,

$$\delta_0 = \sin^{-1}\left(\frac{P_0}{P_{max}}\right) = \sin^{-1}\left(\frac{80}{160}\right) = 30° \quad \text{or} \quad 0.5236 \text{ rad.}$$

Let, P_1 = Maximum permissible suddenly increased load on the generator, MW,

and, δ_1 = Operating load angle corresponding to load P_1.

Then, $P_1 = P_{max} \sin \delta_1 = 160 \sin \delta_1$

When the load is increased the rotor of the generator will advance beyond its new equilibrium point *q* and oscillate between the original point *p* and this extreme point *r*, and will finally attain new equilibrium point *q* due to the damping present.

The load angle δ_m corresponding to extreme point *r*, for transient stability, must be less or equal to $(\pi - \delta_1)$.

Now, Area A_1 = Area A_2 ... For transient stability

$$\int_{\delta_0}^{\delta_1} (P_{max} \sin \delta_1 - P_{max} \sin \delta)\, d\delta = \int_{\delta_1}^{\delta_m} (P_{max} \sin \delta - P_{max} \sin \delta_1)\, d\delta$$

or, $$\int_{\pi/6}^{\delta_1} (160 \sin \delta_1 - 160 \sin \delta)\, d\delta = \int_{\delta_1}^{(\pi-\delta_1)} (160 \sin \delta - 160 \sin \delta_1)\, d\delta$$

or, $$\int_{\pi/6}^{\delta_1} (\sin \delta_1 - \sin \delta)\, d\delta = -\int_{\delta_1}^{(\pi-\delta_1)} (\sin \delta_1 - \sin \delta)\, d\delta$$

or, $$\int_{\pi/6}^{\delta_1} (\sin \delta_1 - \sin \delta)\, d\delta + \int_{\delta_1}^{(\pi-\delta_1)} (\sin \delta_1 - \sin \delta) d\delta = 0$$

or, $$\int_{\pi/6}^{(\pi-\delta_1)} (\sin \delta_1 - \sin \delta) d\delta = 0 \quad \text{or} \quad [\delta \sin \delta_1 + \cos \delta]_{\pi/6}^{(\pi-\delta_1)} = 0$$

or, $$(\pi - \delta_1) \sin \delta_1 - \pi/6 \sin \delta_1 + \cos (\pi - \delta_1) - \cos \pi/6 = 0$$

or, $$\pi \sin \delta_1 - \delta_1 \sin \delta_1 - \frac{\pi}{6} \sin \delta_1 - \cos \delta_1 - \cos \pi/6 = 0$$

or, $$\frac{5\pi}{6} \sin \delta_1 - \delta_1 \sin \delta_1 - \cos \delta_1 - \cos \pi/6 = 0$$

$$2.618 \sin \delta_1 - \delta_1 \sin \delta_1 - \cos \delta_1 - 0.866 = 0$$

or, $$\sin \delta_1 (2.618 - \delta_1) - \cos \delta_1 - 0.866 = 0$$

Solving the above equation, we get:

$$\delta_1 = 1.0472 \text{ elect. radians} \quad \text{or} \quad 60°$$

$\therefore$ $$P_1 = P_{max} \sin \delta_1 = 180 \sin 60° = 138.6 \text{ MW}$$

$\therefore$ *Sudden increase permissible* = 138.6 – 80 = **58.6 MW. (Ans.)**

Q. 29.10. (*a*) *Explain briefly the various methods of voltage control in a transmission system.*

(*b*) ***Power distribution systems.*** *A 2-core distributor cable 400 m long supplies a uniformly distributed lighting load of 1A/m run. There are concentrated loads of 120, 72, 48 and 120 A at 40 m, 120 m, 200 m and 320 m respectively from end A. The cable has a resistance of 0.3 Ω/km run (go and return).*

Determine the position of the lowest-run lamp and its voltage when the cable is fed at 240 V from both ends A and B.

Solution. (*a*) Refer to Art. 18.6

(*b*) *Given:* i = 1 A/m run; $r = \frac{0.3}{1000} = 0.0003$ Ω/m run (go and return)

$$V_A = V_B = 240 \text{ V}$$

Figure 29.13 shows the loaded distributor.

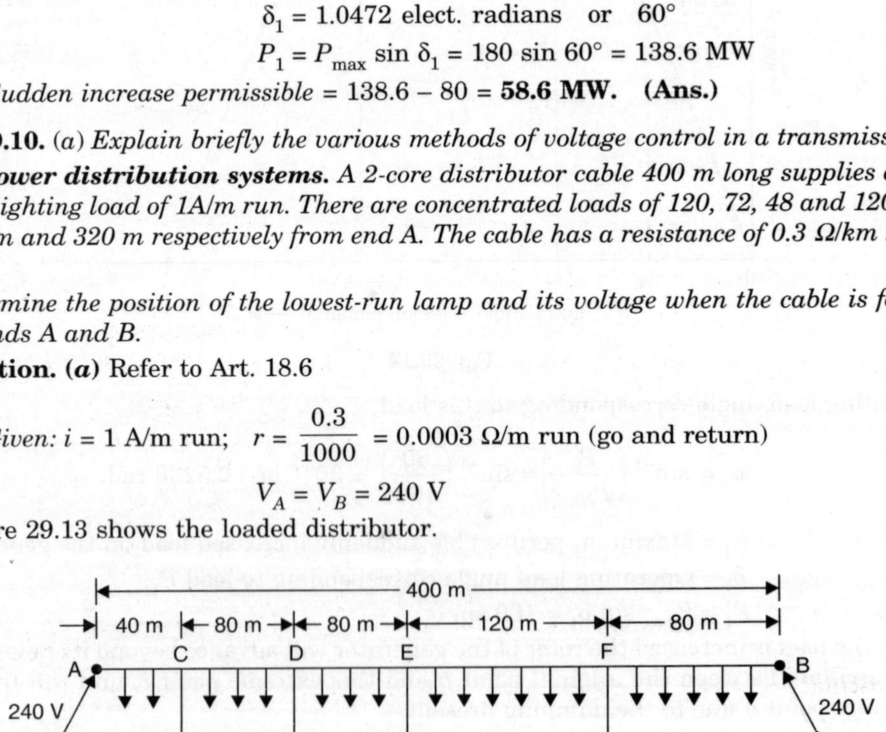

Fig. 29.13.

Let, I_A = Current supplied from end A.

Now, Current just before load point $C = I_A - 1 \times 40 = I_A - 40$

Current just after load point $C = I_A - 40 - 120 = I_A - 160$

Similarly, Current just before load point $D = I_A - 160 - 1 \times 80 = I_A - 240$

Current just after load point $D = I_A - 240 - 72 = I_A - 312$

Current just before load point $E = I_A - 312 - 1 \times 80 = I_A - 392$

Current just after load point $E = I_A - 392 - 48 = I_A - 440$

Current just before load point $F = I_A - 440 - 1 \times 120 = I_A - 560$

Current just after load point $F = I_A - 560 - 120 = I_A - 680$

Current just before feeding point $B = I_A - 680 - 1 \times 80 = I_A - 760$

Voltage drop in section AC = Voltage drop due to concentrated load + voltage drop due to uniform loading of the section

$$= (I_A - 40) \times (0.0003 \times 40) + \frac{1}{2} \times (1 \times 40) \times (0.0003 \times 40)$$

$$= (I_A - 40) \times 0.012 + 0.24 = 0.012I_A - 0.24$$

$$\text{Voltage drop in section } CD = (I_A - 240) \times (0.0003 \times 80) + \frac{1}{2} (1 \times 80) \times (0.0003 \times 80)$$

$$= (I_A - 240) \times 0.024 + 0.96 = 0.024I_A - 4.8$$

$$\text{Voltage drops in section } DE = (I_A - 392) \times (0.0003 \times 80) + \frac{1}{2} (1 \times 80) \times (0.0003 \times 80)$$

$$= (I_A - 392) \times 0.024 + 0.96 = 0.024I_A - 8.448$$

$$\text{Voltage drop in section } EF = (I_A - 560) \times (0.0003 \times 120) + \frac{1}{2} (1 \times 120) \times (0.0003 \times 120)$$

$$= (I_A - 560) \times 0.036 + 2.16 = 0.036I_A - 18$$

$$\text{Voltage drop in section } FB = (I_A - 760) \times (0.0003 \times 80) + \frac{1}{2} (1 \times 80) \times (0.0003 \times 80)$$

$$= (I_A - 760) \times 0.024 + 0.96 = 0.024I_A - 17.28$$

Now, $V_A - V_B$ = Total voltage drop

or, $$240 - 240 = (0.012I_A - 0.24) + (0.024I_A - 4.8) + (0.024I_A - 8.448) + (0.036I_A - 18) + (0.024I_A - 17.28)$$

$$0 = 0.12I_A - 48.768$$

$\therefore$ $I_A = 406.4$ A and I_B = Total load current $- I_A$

$$= (120 + 72 + 48 + 120) + (1 \times 400) - 406.4 = 353.6 \text{ A}$$

Current values just before entering and just after leaving points C, D, E, F are shown in Fig. 29.14.

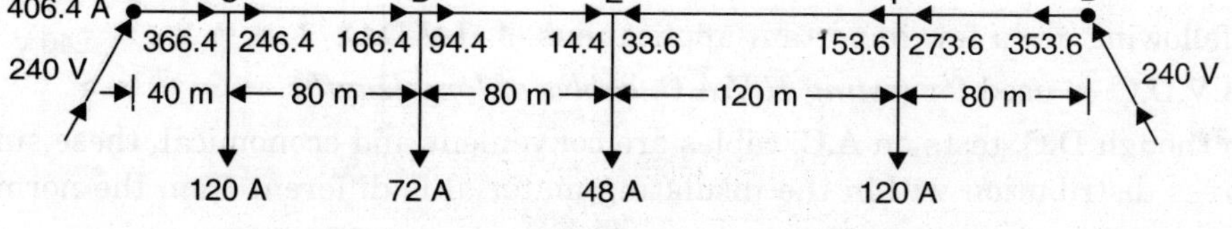

Fig. 29.14

Voltage drop over length AE = Voltage drop in section AC + voltage drop in section CD + voltage drop in section DE

$$= (0.012I_A - 0.24) + (0.024I_A - 4.8) + (0.024I_A - 8.448)$$

$$= 0.06I_A - 13.488 = 0.06 \times 406.4 - 13.488 = 10.9 \text{ V}$$

$$\therefore \quad V_E = V_A - 10.9 = 240 - 10.9 = \mathbf{229.1\ V.} \quad \textbf{(Ans.)}$$

Example 29.11. (*a*) *List the advantages, disadvantages and applications of H.V.D.C. transmission.*

(*b*) ***Power distribution systems.*** *A 600 m long single-phase A.C. distributor has a total impedance of (0.02 + 0.04) and is fed from one end at 240 V. It is loaded as follows: 50 A at unity p.f., 240 m from feeding point; 100 A at 0.8 p.f. lagging, 360 m; 50 A at 0.6 p.f. lagging at the far end. Calculate: (i) Total voltage drop; (ii) Voltage at the far end.*

Solution. (*a*) Advantages, Limitations and Applications of H.V.D.C. Transmission systems Advantages:

Following are the *advantages, limitations* and *applications* of H.V.D.C. transmission systems:

1. *Cheaper in cost.*
2. *Requires less insulation* comparatively (since the potential stress on the insulation in case of D.C. system is $\frac{1}{\sqrt{2}}$ times of that in A.C. system).
3. The designs of the supporting structures are *simpler and cheaper.*
4. *Sheath losses negligible* (since in cases of D.C. only leakage current flows in the cable sheath.) whereas in the case of A.C. charging, circulating and eddy currents flow through the cable sheath).
5. *Lesser dielectric power loss.*
6. *Higher current carrying capacity.*
7. *Higher* natural dielectric strength and *longer life* of cable insulation.
8. *No stability limit,* and as such the transmission lines may be of any length (whereas in A.C. systems the theoretical length of a transmission line which can be operated without loss of stability is only 1.43 times its natural impedance, mostly it is about 500 km).
9. *Less corona loss and radio interference, comparatively.*
10. Use as a non-synchronous link between two A.C. systems.

Limitations:

1. Costly terminal equipment.
2. Non-availability of reliable D.C. circuit breakers.
3. Distortion in wave form, presence of ripples and harmonics.
4. Non-transmission of reactive power over the D.C. link.

Applications:

The following are a few important applications of H.V.D.C.

1. H.V.D.C. is used for *testing H.V.A.C. cables of long lengths.*

Even though D.C. tests on A.C. cables are convenient and economical, these suffer from the fact that stress distribution within the insulating material is different from the normal operating condition.

2. In industry it is being used for:
   — *Electrostatic precipitation* of ashing in thermal power plants;
   — *Electrostatic painting;*
   — *Cement industry;*
   — *Communication systems;* etc.
3. H.V.D.C. is also being used extensively in physics for *particle acceleration and in medical equipments (X-rays).*

(b) Fig. 29.15 shows the loaded distributor.

Total impedance of the distributor = (0.02 + *j*0.04) Ω for 600 m length Impedance of section

$$AB, \bar{Z}_{AB} = (0.02 + 0.04) \times \frac{240}{600} = (0.008 + j0.016)\ \Omega$$

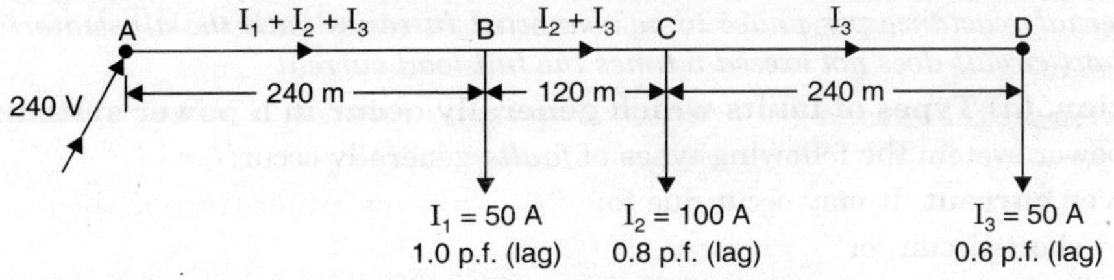

Fig. 29.15

$$\text{Impedance of section } BC, \bar{Z}_{BC} = (0.02 + j0.04) \times \frac{120}{600} = (0.004 + j0.008)\ \Omega$$

$$\text{Impedance of section } CD, \bar{Z}_{CD} = (0.02 + j0.04) \times \frac{240}{600} = (0.008 + j0.016)\ \Omega$$

Assuming all power factors referred to far end voltage V_D, we have:

$$\bar{I}_1 = 50\,(1 - j0) = (50 - j0)\text{ A}$$

$$\bar{I}_2 = 100\,(0.8 - j0.6) = (80 - j60)\text{ A}$$

$$\bar{I}_3 = 50\,(0.6 - j0.8) = (30 - j40)\text{ A}$$

(i) Total voltage drop:

Let voltage at load point *D*, $V_D = (V + j0)$

∴ **Total voltage drop in distributor**

$$= \bar{I}_{AB}\bar{Z}_{AB} + \bar{I}_{BC}\bar{Z}_{BC} + \bar{I}_{CD}\bar{Z}_{CD}$$

$$= (\bar{I}_1 + \bar{I}_2 + \bar{I}_3)\bar{Z}_{AB} + (\bar{I}_2 + \bar{I}_3)\bar{Z}_{BC} + \bar{I}_3\bar{Z}_{CD}$$

$$= [\{(50 - j0) + (80 - j60) + (30 - j40)\}\,(0.008 + j\,0.016)] + [\{(80 - j60) + (30 - j40)\}\,(0.004 + j0.008)] + [(30 - j40)\,(0.008 + j\,0.016)]$$

$$= [(160 - j100)\,(0.008 + j0.016)] + [(110 - j100)\,(0.004 + j0.008)] + [(30 - j40)\,(0.008 + j0.016)]$$

$$= (2.88 + j1.76) + (1.24 + j0.48) + (0.88 + j0.16)$$

$$= (5 + j2.4) = \mathbf{5.546\angle 25.64^\circ}\ \textbf{(Ans.)}$$

(*ii*) **Voltage at the for end V_D:**

Voltage at the sending end A, V_A

= Voltage at the far end + voltage drop in distributor

or, $$240 = [(V_D + 5) + j2.4]$$

$$= \sqrt{(V_D + 5)^2 + (2.4)^2}$$

or, $$(240)^2 = (V_D + 5)^2 + (2.4)^2$$

or, $$(V_D + 5)^2 = (240)^2 - (2.4)^2 = 57594.24$$

or, $$V_D = \mathbf{234.98\ V} \quad \textbf{(Ans.)}$$

Q. 29.12. *Symmetrical components.* (*a*) *List the types of faults which generally occur in a power system.*

(*b*) *A 3-phase 15 MVA, 10 kV alternator has internal reactance of 7% and negligible resistance. Find the external reactance per phase to be connected in series with the alternator so that steady current on short-circuit does not exceed 5 times the full load current.*

Solution. (*a*) Types of faults which generally occur in a power system:

In a power system the following types of *faults* generally occur:

1. **Over current.** It may occur due to:
   — short-circuit, or
   — leakage due to corona effect, or
   — overload on the system.
2. **Under voltage.** It occurs due to:
   — short circuits (due to more voltage drop in lines) and machines, or
   — failure of alternator's field.
3. **Unbalance.** It occurs due to:
   — grounding of one or two phases, or
   — short-circuit of two phases, or
   — breaking of one of the conductors.
4. **Reversed power.** It occurs only in interconnected systems.
5. **Surges.** Whenever lightning takes place or severe fault occurs in a neighbouring circuits, some short-lived waves of very high voltage and current are set up in the lines. Such a fault is known as *surge*. The faults which invariably occur in a 3-phase power systems are enumerated below:

(*i*) Single phase to earth. (*ii*) Phase-to-phase. (*iii*) Two phases to earth. (*iv*) Phase-to-phase and at the same time from third phase to earth.	Give rise to *unsymmetrical* fault currents
(*v*) All the three phases to earth. (*vi*) All the three phases short-circuited.	Give rise to *symmetrical* short-circuit currents.

- Such faults cause heavy currents, called *short-circuit currents,* to flow in the system. The determination of the values of such currents enables us to *make proper selection of circuit breakers, protective relays* and helps us to ensure that the associated apparatus *e.g.,* busbars, connections, current transformers will *withstand the forces* which arise due to the fault currents during the period prior to the interrupting device clearing the fault.

(*b*) *Given:* Rating of the alternator = 10 MVA, 10 kV alternator, internal reactance of the alternator = T%; Short-circuit current = 5 × full load current.

External reactance to be connected:

$$\text{Full-load current} = \frac{10\times10^6}{\sqrt{3}\times(10\times1000)} = 577.35 \text{ A}$$

$$\text{Voltage/phase} = \frac{10\times1000}{\sqrt{3}} = 5773.5 \text{ V}$$

Total percentage reactance required

$$= \frac{\text{Full-load current}}{\text{Short-circuit current}}\times100$$

$$= \frac{1}{5}\times100 = 20\%$$

∴ External percentage reactance required

$$= 20 - 7 = 13\%$$

Let $X\Omega$ be the phase external reactance required.

Then, $$\% \text{ Reactance} = \frac{IX}{V}\times100$$

or, $$13 = \frac{577.35X}{5773.5}\times100$$

or, $$X = \frac{13\times5773.5}{577.35\times100} = \mathbf{1.3\ \Omega} \quad \textbf{(Ans.)}$$

Q. 29.13. *Circuit breakers* *(a) What is a circuit breaker? State its function.*

(b) In a short-circuit on 132 kV 3-phase system, the breaker gave the following results: Power factor of the fault = 0.48; Recovery voltage = 0.85 times full line voltage; The breaking current is symmetrical and the restriking transient had a natural frequency of 15 Hz.

Determine the rate of rise of restriking voltage (RRRV) in the following types of faults:

(i) Grounded fault *(ii) Ungrounded fault.*

Solution. (*a*) Refer to Art. 22.4

(*b*) *Given:* Power factor of the fault, cos φ = 0.48; Recovery voltage = 0.85 × Full line voltage; f_n = 15 Hz

Peak value of phase voltage (line-to-neutral voltage),

$$V = \frac{132}{\sqrt{3}}\times\sqrt{2} = 107.8 \text{ kV}$$

Recovery voltage, $$V_{max} = 0.85 \times 107.8 = 91.63 \text{ kV}$$

Time taken to reach the peak value, $$t = \frac{1}{2f_n} = \frac{1}{2\times15\times10^3} = 33.33\ \mu\text{s}$$

(*i*) **Grounded fault:**

$$\text{Active recovery voltage} = kV_{max} \sin\phi$$

$$= 1 \times 91.63 \times \sin(\cos^{-1} 0.48) = 80.38 \text{ kV}$$

(∵ For ground fault, k = 1)

The maximum (peak value) restriking voltage = 2 × 80.38 = **160.76 kV**

$$\textbf{\textit{RRRV}} = \frac{\text{Peak value of restriking voltage}}{\text{Time taken to reach the peak value}}$$

$$= \frac{160.76}{33.33} = \textbf{4.82 kV/}\boldsymbol{\mu}\textbf{s} \quad \textbf{(Ans.)}$$

(*ii*) **Ungrounded fault:**

Active recovery voltage = $kV_m \sin \phi$

$= 1.5 \times 91.63 \times \sin(\cos^{-1} 0.48)$

($\because$ For ungrounded fault $k = 1.5$)

$= 120.57$ kV

$$\textbf{\textit{RRRV}} = \frac{2 \times 120.57}{33.33} = \textbf{7.23 kV/}\boldsymbol{\mu}\textbf{s} \quad \textbf{(Ans.)}$$

Q. 29.14. *Interconnected systems.* (*a*) *What is load despatching? Explain briefly.*

(*b*) *Two 3-phase generating stations A and B are linked through a 33kV interconnected having a resistance of 0.8 Ω and a reactance of 4.5 Ω per phase. At station-A the load on the generators is 75 MW at a p.f of 0.8 lagging and the local load taken by the consumers connected to the A bus-bars is 50MW at a p.f of 0.707 lagging. Calculate kW received from station-A by station-B, its p.f and phase difference between the voltage of A and B.*

Solution. (*a*) Load despatching. Due to electric system expansion and interconnection, load despatching (or centralised control of system operation) has become a necessity. In large power stations, central load despatcher is essential. The *functions* of a load despatcher include:

(*i*) Loading of generating stations.

(*ii*) Loading of units within the generating stations.

(*iii*) Control of system voltage and receive power flow

(*iv*) Control of system frequency and active power flow.

(*v*) Loading of various interconnectors and transfer of power from one station to another and

(*vi*) Maintaining continuity of service.

Very large systems are sometimes divided into load districts each having its own load despatcher, often with a central load despatching agency for general supervision over the districts.

Centralised operation supervises *reliability of service, quality of service, and economy of production.*

(***b***) *Given:* Interconnector voltage = 33 kV

Interconnector importance = $0.8 + j4.5$

Load on the generator at station-A = 75 MW at 0.8 p.f. lagging

Load connected to A bus-bars = 50 MW at 0.707 lagging

Refer to Fig. 29.16.

Let the phase voltage at A $\left(= \dfrac{33000}{\sqrt{3}} \simeq 19050\right)$ be taken as a reference phasor, so that

$$V_A = 19050\,(1 + j0) \text{ volts}$$

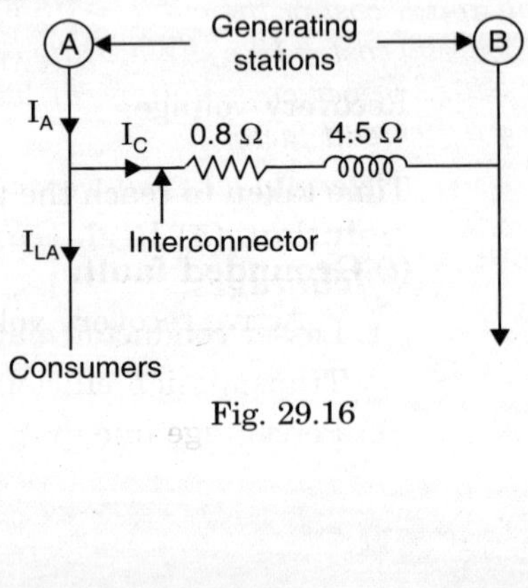

Fig. 29.16

Load current on generators at stations A,

$$I_A = \frac{75 \times 10^6}{\sqrt{3} \times (33 \times 10^3) \times 0.8} = 1640.2\ A$$

$$= 1640.2\ (0.8 - j0.6) = (1312.2 - j984.1) A$$

Current due to local load at A,

$$I_{LA} = \frac{50 \times 10^6}{\sqrt{3} \times (33 \times 10^3) \times 0.707} = 1237.3\ A$$

$$= 1237.3\ (0.707 - j0.707) = 875.1 - j875.1$$

Current transferred through the interconnector,

$$I_C = I_A - I_{LA}$$
$$= (1312.2 - j984.1) - (875.1 - j875.1)$$
$$= 437.1 - j109 = 450.5 \angle -14^\circ$$

Voltage drop in the interconnector

$$= (437.1 - j109)\ (0.8 + j4.5)$$
$$= 349.68 + j1966.95 - j87.2 + 490.5 = (840.18 + j1879.75) V$$

Voltage at generating station-B,

$$V_B = 19050\ (1 + j0) - (840.18 + j1879.75)$$
$$= 18209.82 - j1879.75 \simeq 18307 \angle -5.89^\circ$$

Phase difference between V_B and I_C

$$= -14^\circ - (-5.89^\circ) = -8.11^\circ$$

i.e., 8.11° with I_C *lagging* V_B

Power factor of the current received by station B

$$= \cos 8.11^\circ = \mathbf{0.99\ (lagging)} \quad \textbf{(Ans.)}$$

Power received by station B

$$= 3V_B I_C \cos 8.11^\circ = 3 \times 18307 \times 450.5 \times 0.99 \times 10^{-6}\ \text{MW}$$
$$= \mathbf{24.49\ MW} \quad \textbf{(Ans.)}$$

Phase angle between voltage of station-A and station-B

$$= \mathbf{5.89^\circ\ with\ V_A\ lagging} \quad \textbf{(Ans.)}$$

Q. 29.15. *Electric power systems.* *(a) List the advantages and limitations of high transmission voltage.*

(b) The following data refer to a 3-phase cable supplying power to an industry at 11 kV:

Maximum demand = 1200 kW; Total annual energy consumption = 6 × 10^6 kWh at 0.82 p.f.; Capital cost of the cable = ₹(85000 a + 25000) per km, "a" being X-additional area in cm^2/per core; Energy cost = 10 p/kWh; Rate of interest and depreciation = 10%; Specific resistance of copper = 1.72 × 10^{-6} Ω per cm^3; Voltage at which power is supplied = 11 kV; Loss load factor = 0.22 load factor + 0.72 (load factor)^2.

Determine the most economical section of the conductor.

Solution. (*a*) High transmission voltage:

Advantages:

1. Lesser conductor material required.
2. Transmission efficiency is increased.
3. Percentage line drop decreases, this leads to better voltage regulation.

Limitations:

1. The cost of insulating the conductor increases.
2. The cost of transformers, switchgear and other terminal apparatus increases.

The limit to the use of higher transmission voltage (which can be economically used in a particular case) is *reached when the saving* in cost of conductor material due to higher voltage is offset by the increased cost of insulation, transformer, switchgear etc. Thus, the choice of proper transmission voltage is essentially a question of economics.

(*b*) Let, l = Length of the cable, km

Max. economical section of the conduction, '*a*'.

$$\text{Annual load factor} = \frac{\text{Annual energy consumption}}{\text{Maximum demand} \times 8760} = \frac{6 \times 10^6}{1200 \times 8760} = 0.57$$

$$\text{Loss load factor} = 0.22 \times 0.57 + 0.72\,(0.57)^2 = 0.359$$

current corresponding to maximum demand,

$$I = \frac{1200 \times 1000}{\sqrt{3} \times 11000 \times 0.82} = 76.8 \text{ A}$$

Capital cost (variable) of the cable = 85000 al

Annual cost on account of interest and depreciation on capital cost (variable) of the cable

$$= ₹\,\frac{10}{100} \times 85000\,al = 8500\,al$$

$$\text{Energy lost per annum} = \frac{3I^2R \times 8760}{1000} \times \text{loss load factor}$$

$$= \frac{3 \times (76.8)^2 \times (\rho l/a) \times 8760}{1000} \times 0.359$$

$$= \frac{3 \times (76.8)^2 \times (1.72 \times 10^{-6} \times l \times 1000 \times 100) \times 8760}{a \times 1000} \times 0.359$$

$$= \frac{9571.3\,l}{a}$$

$$\text{Annual cost due to energy lost} = ₹\,\frac{10}{100} \times \frac{9571.3\,l}{a} = ₹\,\frac{957.13\,l}{a}$$

For most economical section of the conductor,

Variable annual charge = Annual cost of energy lost

i.e.,
$$8500\,al = \frac{957.13\,l}{a}$$

$$\therefore \quad a = \left(\frac{957.13}{8500}\right)^{1/2} \simeq \mathbf{0.336\ cm^2} \quad \textbf{(Ans.)}$$

Q. 29.16. *Mechanical design of overhead lines* *(a) What is a sag? List the factors which affect the sag.*

(b) The following data refer to a transmission line.

Length of span = 220 m; Effective diameter of the conductor = 1.75 cm; Weight of the conductor = 0.85 kg / m run; Ultimate strength = 8000 kg; Radial thickless of ice coating on the conductor = 1.1.cm; Wind pressure = 35 kg / m² of projected area; F.O.S = 2; Density of ice = 915 kg / m³.

Calculate the sag.

Solution. (*a*) Sag in overhead lines: In overhead transmission lines, the conductors are supported at the towers (or poles for low voltage lines). The conductors are pulled and stringing effected. When the conductor is supported in this fashion it will sag or dip under its own weight and it takes the shape of catenary. *The distance between the adjacent supporting tower is called the **"span"**.*

The difference in level between the points of supports and the lowest point is known as ***sag***.

In an overhead line, the following factors *affect the sag:*

(*i*) *Weight of the conductor:* Heavier the conductor, greater will be the sag.

(*ii*) *Span length:* Sag is *directly proportional* to the square of the span length, other condition remaining unchanged.

(*iii*) *Working tensile strength:* Sag is *inversely proportional* to the *working tensile strength*, other conditions remaining the same.

(*iv*) *Temperature:* Sag *increases* with increase in temperature.

(*b*) *Given:* l = 220 m; d = 1.8 cm; w_C = 0.85 kg/m run; Ultimate strength = 8000 kg; Radial thickness of ice coating on the conductor, t = 1.1 cm; p = 35 kg/m²; F.O.S. = 2. Density of ice = 915 kg/m³.

Slant and vertical slag:

Working tension,
$$T = \frac{\text{Ultimate strength}}{\text{F.O.S.}} = \frac{8000}{2} = 4000 \text{ kg}$$

Weight of ice per metre length of conductor,

$$w_2 = \text{Volume of conductor per metre length} \times \text{density of ice}$$

$$= \frac{\pi}{4}[(d + 2t)^2 - d^2] \times 1 \times 915 \qquad \text{[where } d \text{ and } t \text{ are in cm]}$$

$$= \frac{\pi}{4}[(1.8 + 2 \times 1.1)^2 - (1.8)^2] \times 10^{-4} \times 1 \times 915$$

$$= 0.917 \text{ kg/m run}$$

Horizontal unit force per metre length

$$w_w = \text{Wind pressure} \times \text{projected area of the ice covered conductor}$$

$$= 35 \times [(d + 2t) \times 10^{-2}] \times 10 \qquad \text{[where } d \text{ and } t \text{ are in cm]}$$

$$= 35 \times [(1.8 + 2 \times 1.1) \times 10^{-2}] \times 1 = 1.4 \text{ kg/m}$$

Resultant force per metre conductor length

$$w_r = \sqrt{(w_c + w_i)^2 + (w_w)^2}$$

$$= \sqrt{(0.85 + 0.917)^2 + (1.4)^2} = 2.254 \text{ kg/m}$$

Slant sag, S_{slant}:

Using the relation, $$S_{slant} = \frac{w_r l^2}{8T}$$

i.e., $$S = \frac{2.254 \times (220)^2}{8 \times 4000} = \mathbf{3.41\ m} \quad \textbf{(Ans.)}$$

Vertical sag, $S_{vertical}$:

$$S_{vertical} = S_{slant} \cos \theta$$

$$= 3.41 \times \frac{(w_c + w_i)}{w_r}$$

$$= 3.41 \times \frac{(0.85 + 0.917)}{2.254} = \mathbf{2.67\ m} \quad \textbf{(Ans.)}$$

Q. 29.17. *Constant of overhead lines.* *(a) How is the size of the conductor and spacing between conductor affect the line performance, given reasons.*

(b) A 3-phase overhead line is designed with an equilateral spacing of 3.8 m with a conductor diameter of 1.2 cm. If the line is constructed with horizontal spacing with suitably transposed conductors, find spacing between adjacent conductors which would give the same value of inductance as in the equilateral arrangement.

Solution. (*a*) For conductor of radius r and having distance of separation D in between them, the values of inductance and capacitance between two parallel conductors are given by:

$$L \propto \ln\left(\frac{D}{r}\right), \text{ and } C \propto \frac{1}{\ln\left(\frac{D}{r}\right)}$$

- Hence, with *increased value of D i.e.,* large value of distance of separation, *inductance increases* and *capacitance decreases.*
- *With large value of r i.e.,* with larger size of conductor, *inductance decreases and capacitance increases.*

Thus with changes in values of inductance and capacitance the performance of line will be accordingly affected.

(*b*) *Given:* Spacing, $D = 3.8$ m; $r = \frac{1.2}{2} = 0.6$ cm $= 0.006$ m

For equilateral triangular arrangement, the inductance of each conductor is given by:

$$L = 10^{-7}\left[0.5 + 2\ln\left(\frac{D}{r}\right)\right] \text{ H/m}$$

$$= 10^{-7}\left[0.5 + 2\ln\left(\frac{3.8}{0.006}\right)\right] = 13.4 \times 10^{-7} \text{ H/m}$$

If the line is constructed with horizontal spacing with suitably transposed conductor, then inductance of each conductor is given by:

$$L = 10^{-7}\left[\frac{1}{2} + 2\ln\left(\frac{\sqrt[3]{D_{AB} \times D_{BC} \times D_{CA}}}{r}\right)\right] \text{ H/m}$$

Let distance be D meters, then:

$$D_{AB} = D_{BC} = D \text{ and } D_{CA} = 2D$$

$$\therefore \quad L = 10^{-7}\left[\frac{1}{2} + 2\ln\left(\frac{\sqrt[3]{D \times D \times 2D}}{r}\right)\right] \text{H/m} = 10^{-7}\left[\frac{1}{2} + 2\ln\frac{\sqrt[3]{2}\, D}{r}\right]$$

$$= 10^{-7}\left[\frac{1}{2} + 2\ln\left(\frac{1.26\, D}{r}\right)\right] \text{H/m}$$

For same value of inductance in both the arrangements, we have:

$$10^{-7}\left[\frac{1}{2} + 2\ln\left(\frac{1.26\, D}{r}\right)\right] = 13.4 \times 10^{-7}$$

or,
$$\ln\left(\frac{1.26D}{r}\right) = \left(\frac{13.4 - 0.5}{2}\right) = 6.45$$

or,
$$\frac{1.26D}{r} = e^{6.45} = 632.7$$

or,
$$D = \frac{632.7 \times r}{1.26} = \frac{632.7 \times 0.006}{1.26} = \mathbf{3.013\ m} \quad \textbf{(Ans.)}$$

Q. 29.18. *Performance of transmission lines.* (*a*) *What do you mean by performance of transmission lines? Explain briefly.*

(*b*) *A 3-phase, 50 Hz overhead transmission line 120 km long has the following constants.*

Resistance/km= 0.28 Ω; Reactance/km = 0.9 Ω; Susceptance/km = 5.8 × 10^{-6} ℧.

The line voltage at the receiving end is 132 kV. The transmission line is delivering 50 MVA at 0.82 p.f. lagging at the receiving end. Using nominal T-method, calculate:

(*i*) *Sending end current,* (*ii*) *Sending end voltage,*

(*iii*) *Sending end p.f. and,* (*iv*) *Transmission efficiency.*

Solution. (*a*) ● *By performance of transmission lines* is meant the *determination of efficiency and regulation of lines.*

- The voltage drop, power losses occurring in the line and transmission efficiency are largely influenced by the transmission line constants *i.e.*, R (resistance), L (inductance) and C (capacitance). The line *resistance* causes voltage drop (IR) and power loss (I^2R) in the line, the line *inductance* causes voltage drop in quadrature with the current flowing in the conductor ($2\pi fL \times I$) whereas the *capacitance* produces charging current ($2\pi fCV$), in quadrature with the voltage. The value of charging current is maximum at the sending end of the line and decreases continuously and its value becomes zero at the receiving end. It may be added that in case of overhead lines inductive reactance effects are relatively more important due to wide spacing of the conductors while in the case of underground cables capacitive reactance effects are more predominant in comparison to inductive reactance effects.

(*b*) Refer to Fig. 29.17

Total resistance/phase, $R = 0.28 \times 120 = 33.6\ \Omega$

Total reactance/phase $X = 0.9 \times 120 = 108\ \Omega$

Total susceptance/phase, $Y = 5.8 \times 10^{-6} \times 120 = 6.96 \times 10^{-4}$ ℧

Receiving end line voltage, $V_{RL} = 132 \text{ kV} = 132000 \text{ V}$

Receiving end phase voltage, $V_R = \dfrac{132000}{\sqrt{3}} = 76210$ V

Load current, $I_R = \dfrac{50 \times 10^6}{\sqrt{3} \times 132000} = 218.7$ A

$\cos \phi_R = 0.82$, $\sin \phi_R = \sqrt{1 - \cos \phi_R^2} = \sqrt{1 - (0.82)^2} = 0.572$

Total voltage at receiving end as the reference phasor, we have:

$$\bar{V}_R = V_R\,(1 + j0) = 76210\,(1 + j0) \text{ V}$$

Load current, $\bar{I}_R = I_R\,(\cos \phi_R - j \sin \phi_R)$

$$= 218.7\,(0.82 - j0.572) = (179.33 - j125.1) \text{ A}$$

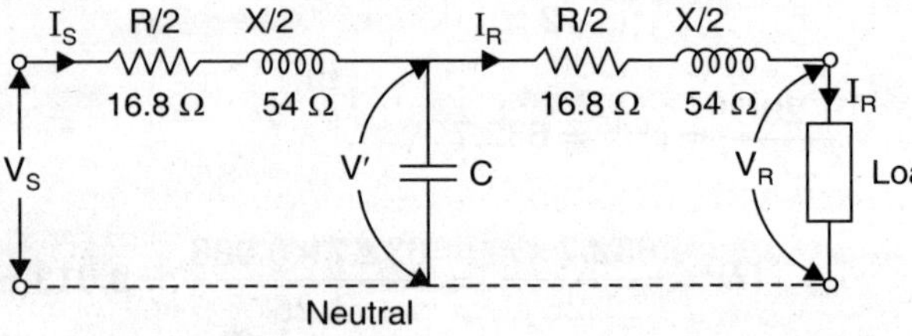

Fig. 29.17 Nominal-*T* method (circuit diagram)

Line independence, $\bar{Z} = R + jX = (33.6 + j108)\ \Omega$

$$\frac{\bar{Z}}{2} = (16.8 + j54)\ \Omega$$

Phase voltage across the capacitor, C, $\bar{V}' = \bar{V}_R + \left(\bar{I}_R \times \dfrac{\bar{Z}}{2}\right)$

$$= 76210 + (179.33 - j125.1)\,(16.8 + j54)$$
$$= 76210 + (3012.7 + j9683.8 - j2101.7 + 6755.4)$$
$$= 85978.1 + j7582.1$$

Changing current; $\bar{I}_C = jY\bar{V}' = j6.96 \times 10^{-4}\,(85978.1 + j7582.1)$

$$= j59.84 - 5.28$$

(*i*) **Sending end current, $\mathbf{I_S}$** :

$$\bar{I}_S = \bar{I}_R + \bar{I}_C = (179.33 - j125.1) + (j59.84 - 5.28)$$
$$= 174.05 - j65.26 = \mathbf{185.9 \angle -20.55° \text{ A (Ans.)}}$$

(*ii*) **Sending end line voltage, $\bar{\mathbf{V}}_{\mathbf{SL}}$** :

$$\bar{V}_S = \bar{V}' + \left(\bar{I}_S \times \frac{\bar{Z}}{2}\right)$$
$$= (85978.1 + j7582.1) + (174.05 - j65.26)\,(16.8 + j54)$$

$$= (85978.1 + j7582.1) + (2924.04 + j9398.7 - j1096.4 + 3524.04)$$

$$= 92426.2 + j15884.4 = \mathbf{93781 \angle 9.75°V}$$

(*iii*) **Sending end *p.f.*, cos ϕ_s:**

$$\cos \phi_s = \cos [9.75° - (-20.55°)] = \mathbf{0.863} \quad \textbf{(Ans.)}$$

(*iv*) **Transmission efficiency, η_T:**

$$\text{Line losses} = 3\left(I_s^2 \times \frac{R}{2} + I_R^2 \times \frac{R}{2}\right)$$

$$= 3\,[(185.9)^2 \times 16.8 + (218.7)^2 \times 16.8] \times 10^{-3} \text{ kW} = 4152.4 \text{ kW}$$

$$\eta_T = \frac{\text{Power delivered}}{\text{Power delivered + line losses}}$$

$$= \frac{(50 \times 10^3) \times 0.82}{[(50 \times 10^3) \times 0.82 + 4152.4]} = 0.908 \quad \text{or} \quad \mathbf{90.8\%} \quad \textbf{(Ans.)}$$

Q. 29.19. *Power distribution systems.* *(a) Give the comparison between overhead system and underground system.*

(b) A D.C. 2-wire distributor AB is 500 metres long and is fed at both ends at 250 V. The distributor is loaded as shown in Fig. 29.18. The resistance of the distributor (go and return) is 0.001 Ω/m. Calculate:

(i) The point of minimum voltage, and (ii) the value of minimum voltage.

Solution. (*a*) Overhead versus underground system: The comparison (for making a choice between those two systems) between the overhead system and underground system is given in the table below:

S.No.	Aspects	Overhead system	Underground system
1.	*Initial cost*	Low	High (may be 5 to 10 times the overhead system)
2.	*Public safety*	Less	More
3.	*Faults*	More chances of occurrence of faults	Very rare
4.	Fault location and repairs	Easy	Difficult
5.	Flexibility	Much more flexible than underground system	Less flexible
6.	*Current carrying capacity and voltage drop*	High	Low comparatively
7.	*Appearance*	Not as good as underground system	Better comparatively (since all the distribution lines are invisible)
8.	*Maintenance cost*	High comparatively	Low
9.	*Interference with communication circuits*	Electromagnetic interference with the telephone lines	No interference with the telephone lines
10.	*Useful life*	25 years approx.	More than 50 years

- In general, overhead system is adopted for distribution and use of underground system is made only where overhead construction is impracticable or prohibited by local laws.

(b) $l = 500$ m; $V_A = V_B = 250$ V; $r = 0.001\ \Omega$/m

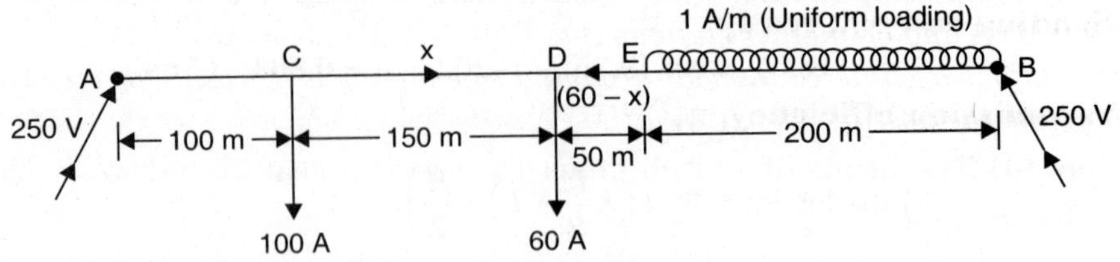

Fig. 29.18

(*i*) **The point of minimum voltage:**

Let D be the point of minimum potential and let x be current flowing in section CD. The current supplied by end B will be $(60 - x)$.

Now, Voltage drop in length $AD = I_{AC} R_{AC} + I_{CD} R_{CD}$

$$= (100 + x) \times (100r) + x \times (150r)$$

$$= r[(100 + x) \times 100 + 150x] = r[10000 + 250x]$$

Voltage drop in length $BD = I_{DB} R_{DB} + \frac{1}{2} r \times (200)^2$

$$= (60 - x) \times (250r) + 20000r$$

$$= r\,[(60 - x) \times 250 + 20000]$$

Since feeding points A and B are at the same potential,

$$\therefore \quad r\,[10000 + 250x] = r\,[(60 - x) \times 250 + 20000]$$

or,
$$10000 + 250x = 15000 - 250x + 20000$$

$$\therefore \quad x = 50 \text{ A}$$

The actual distribution of the currents in various sections of the distributor is shown in Fig. 29.19. Since the currents supplied by feeding points A and B meet at D, therefore, **point D is the point of minimum voltage. (Ans.)**

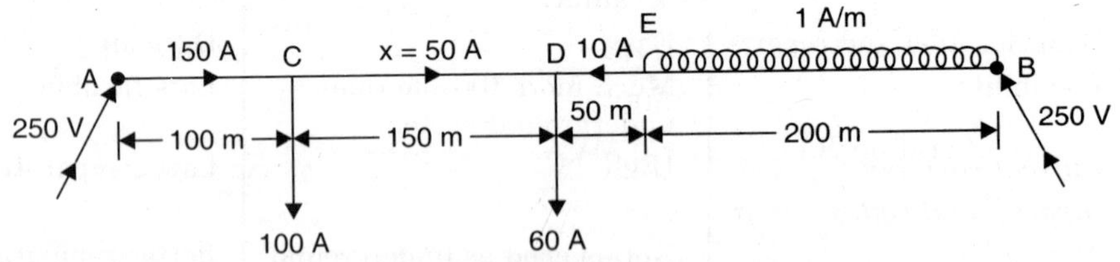

Fig. 29.19

(*ii*) **The value of minimum voltage, V_D:**

$$V_D = V_A - I_{AC} R_{AC} - I_{CD} R_{CD}$$

$$= 250 - 150 \times (0.001 \times 100) - 50 \times (0.001 \times 150)$$

$$= \mathbf{227.5\ V.}\quad \textbf{(Ans.)}$$

[Check potential of D from end B:

$$V_D = 250 - 10 \times (250 \times 0.001) - \frac{1}{2} \times 0.001 \times 200^2 = 250 - 22.5 = 227.5 \text{ V}$$]

Q. 29.20. (*a*) *What are the essentials of switch gear?*

(*b*) ***Symmetrical components.*** *The estimated short-circuit MVA at the bus-bars of a generating station A is 1500 MVA and of another stations B is 1200 MA. The generated voltage at each station is 33 kV. If these stations are interconnected through a line having a reactance of 1.4 Ω and negligible resistance, calculate the possible short-circuit MVA at both stations.*

Solution. (*a*) Essentials of switch gear: The essential features of switch gear are:

1. *Complete reliability.*

2. *Absolutely certain discrimination* (The switch gear must be able to discriminate between faulty and healthy sections).

3. *Quick operation.*

4. *Provision for manual control.*

5. *Provision for instruments.*

(***b***) Refer to Fig. 29.20.

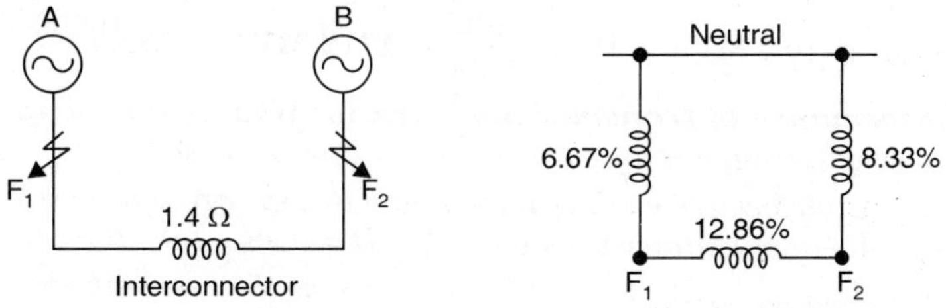

(a) Single line diagram of the system

(b) Reactance diagram

Fig. 29.20

Let the base MVA be 100.

% reactance of station *A* on the base MVA,

$$\% X_A = \frac{\text{Base MVA}}{\text{Short-circuit MVA}} \times 100$$

$$= \frac{100}{1500} \times 100 = 6.67\%$$

% reactance of station *B* on the base MVA,

$$\% X_B = \frac{100}{1200} \times 100 = 8.33\%$$

% reactance of interconnector on base MVA,

$$\% X_I = \frac{X(\text{kVA})}{10(\text{kV})^2}$$

$$= \frac{1.4 \times (100 \times 10^3)}{10 \times (33)^2} = 12.86\%$$

Fault on station A. When a fault occurs on generating station *A* (point F_1), the total percentage reactance from generator neutral upto the fault point F_1 is in the parallel combination of 6.67% and (8.33% + 12.86% = 21.19%)

i.e., Total % reactance upto fault point F_1

$$= \frac{6.67 \times 21.19}{6.67 + 21.19} = 5.073\%$$

$$\therefore \quad \text{Short-circuit MVA} = \text{Base MVA} \times \frac{100}{\%\ \text{Reactance}}$$

$$= 100 \times \frac{100}{5.073} = \mathbf{1971\ MVA} \quad \textbf{(Ans.)}$$

Fault on station B. What a fault occurs on power station B (Point F_2), the total percentage reactance from generator neutral upto fault point F_2 is parallel combination of (6.67% + 12.86% = 19.53%) and 8.33%

i.e., Total % reactance upto fault point F_2

$$= \frac{8.33 \times 19.53}{8.33 + 19.53} = 5.84\%$$

$$\therefore \quad \text{Short-circuit MVA} = 100 \times \frac{100}{5.84} = \mathbf{1712\ MVA} \quad \textbf{(Ans.)}$$

Q. 29.21. *Performance of transmission lines.* *(a) What is surge impedance loading? How can the power transmitted through a long transmission line increased.*

(b) Two similar transformers each of impedance $10\angle 85°$ *are interposed at the two ends of a long transmission line having auxiliary constants,* $A = D = 0.96\ \angle 1.9°$; $B = 26\ \angle 66°\ \Omega$.

Determine the values of A_0, B_0, C_0 *and* D_0 *to represent the line with the transformers.*

Solution. (*a*) Surge impedance loading: *It is defined as the load, at unity power factor, that can be delivered by the line having no resistance.*

$$\therefore \quad \text{Power transmitted,} \quad P_R = \frac{V_{RL}^2}{Z_O}\ \text{MW}$$

where, V_{RL} = Receiving end line voltage in kV,

Z_O = Surge impedance in ohms.

P_R is known as the "*surge impedance loading*" or "*natural power*" of the line.

The eqn.: Surge impedance, $Z_O = \sqrt{Z_{OC} \times Z_{SC}}$ gives a limit to the maximum power that can be delivered (this may not necessarily be the maximum loading on the line) and is useful in the design of transmission lines.

Its use can be made for comparison of loads that can be carried on the transmission lines at different voltages.

The power transmitted through a long transmission line can be *increased* as follows:

(*i*) *Increasing* V_{RL}. Presently the trend is to use higher and higher voltages for transmission, therefore, this is the most commonly used method to increase the power limit of heavily loaded long transmission line. But there is a limit beyond which it is neither economical nor practicable to increase the value of V_{RL}.

(*ii*) *Decreasing* Z_O. The value of Z_O cannot be varied much since the spacing between the conductors, which depends upon the line voltage employed, cannot be reduced to a greater extent.

In order to decrease Z_O either L is decreased using series capacitor or C is increased using shunt capacitors,

— The use of *series capacitors* improves the system stability limit. These capacitors are also helpful in reducing line drops and hence voltage variations.

— By the use of shunt capacitors although Z_O is reduced but due to increase in phase shift stability conditions become worse specially when synchronous machines are used as loads. Thus this method is not employed where stability limits are present.

(*b*) *Given:* Impedance of each transformer,

$$\bar{Z}_{se} = (\bar{Z}_{seS} = \bar{Z}_{seR}) = 10 \angle 85°$$

(where $\bar{Z}_{seS}$ and $\bar{Z}_{seR}$ are series impedances at sending end and receiving end respectively)

Auxiliary constants: $A = D = 0.96 \angle 1.9°$; $B = 26\angle 66°\ \Omega$

Values of A_0, B_0, C_0 and D_0:

We know that, $A.D - B.C = 1$

$$\therefore \quad C = \frac{A.D - 1}{B} = \frac{(0.96\angle 1.9°)(0.96\angle 1.9°) - 1}{26\angle 66°}$$

$$= \frac{0.9216\angle 3.8° - 1}{26\angle 66°} = \frac{0.9196 + j0.061 - 1}{26\angle 66°}$$

$$= \frac{-0.0804 + j0.061}{26\angle 66°} = \frac{0.1009\angle 142.8°}{26\angle 66°} = 0.00388 \angle 76.8°\ ℧$$

A_0, B_0, C_0 and D_0 constants of equivalent circuit of the line and similar transformers connected at both ends (Fig. 29.21) are given below

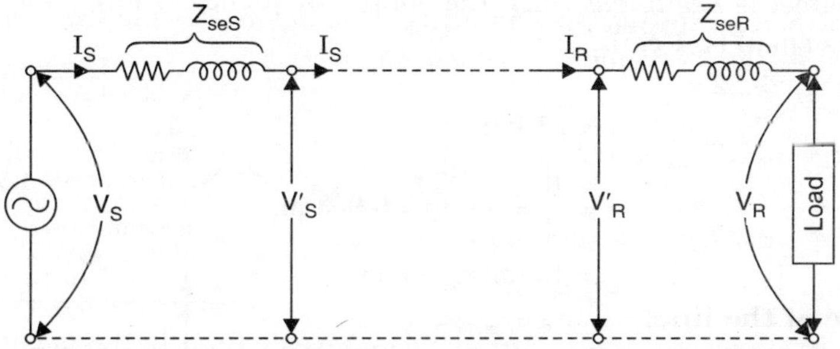

Fig. 29.21. Transmission line with series impedances at both ends.

$$A_0 = A + C\bar{Z}_{se} = 0.96 \angle 1.9° + (0.00388 \angle 76.8°)(10 \angle 85°)$$

$$= 0.96 \angle 1.9° + 0.0388 \angle 161.8°$$

$$= (0.9595 + j0.0318) + (-0.0368 + j0.0121)$$

$$= 0.9227 + j0.0439 = \mathbf{0.9237 \angle 2.72°} \quad \textbf{(Ans.)}$$

$$B_0 = 2A\bar{Z}_{se} + B + C\bar{Z}_{se}^2 = 2(0.96\angle 1.9°)(10\angle 85°) + 26\angle 66° + 0.00388\angle 76.8°(10\angle 85°)^2$$

$$= 19.2 \angle 86.9° + 26 \angle 66° + 0.388 \angle 246.8°)$$

$$= (1.038 + j19.17) + (10.575 + j23.75) + (-0.153 - j0.357)$$

$$= 11.46 + j42.56 = \mathbf{44.07 \angle 74.93°\ \Omega} \quad \textbf{(Ans.)}$$

$$C_0 = C = \mathbf{0.00388 \angle 76.8°\ ℧} \quad \textbf{(Ans.)}$$

$$D_0 = C\bar{Z}_{se} + D = A + C\bar{Z}_{se} = \mathbf{0.9237 \angle 2.72} \quad \textbf{(Ans.)} \quad (\because A = D)$$

Q. 29.22. Constants of overhead lines. *(a) Explain briefly the 'skin effect' in a transmission line. What is its effect on the resistance of the line?*

(b) Calculate capacitance of a single phase transmission line 25 km long consisting of two parallel wires each 6 mm in diameter and 1.5 m apart. The height of the each conductor above ground is 6 m.

Solution. (*a*) Skin effect: Direct currents distribute themselves uniformly over the cross-section of the conductor and therefore use the centre of the conductor just as effectively as they use the periphery. *Alternating currents, however, owing to inductance effects within the conductor, crowd towards the outside the conductor. This behaviour is known as '**skin effect**'. It raises the apparent resistance of the conductor,* because power loss varies as the current and higher current density of current flowing near the skin of the conductor produces more power loss than the lower density of current flowing through the centre of the conductor.

The resistivity of the conducting material is unchanged by skin effect; only the *resistance of the conductor increases as the frequency of the current and the diameter of the conductor increases. The increase in resistance should not be neglected in very large conductors at commercial power frequencies and in most conductors at carrier and radio frequencies.* Since skin effect reduces the amount of current carried by the centre portion of the conductor, *hollow conductors* are sometimes employed to use them more effectively. *Aluminium cable steel reinforced (ACSR)* is used even more than hollow conductors. This cable has steel core which strengthens the cable, and an outer covering of aluminium which, because of skin effect and high resistivity of the steel, carries most of the current. These conductors also have the advantage of a *large surface area available for cooling.*

- The skin effect is *negligible* when the supply frequency is low (< 50 Hz) and conductor diameter is small (< 1 cm).

(*b*) Refer to Fig. 29.22.

Given: $l = 25$ km;

$$r = \frac{6}{2} = 3 \text{ mm} = 0.003 \text{ m};$$

$D = 1.5$ m; $h = 6$ m

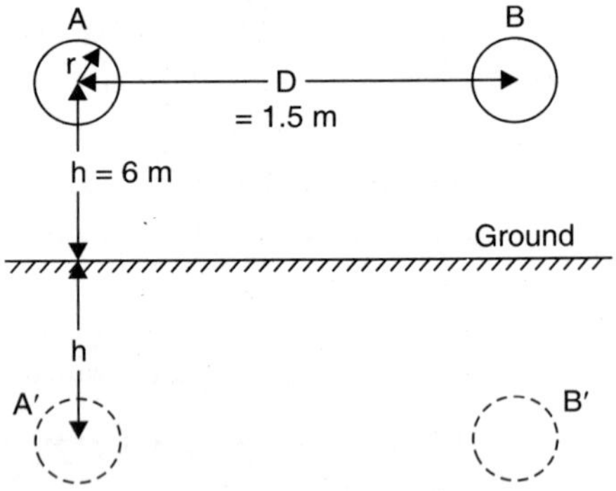

Fig. 29.22

Capacitance of the line:

The capacitance of the line taking into consideration the presence of ground, is given as:

$$C = \frac{\pi\varepsilon_0}{\ln\left[\dfrac{D}{r\sqrt{1+(D^2/4h^2)}}\right]} \text{ F/m (line-to-line)}$$

$$= \frac{\pi \times 8.854 \times 10^{-12}}{\ln\left[\dfrac{1.5}{0.003\sqrt{1+(1.5^2/4\times 6^2)}}\right]}$$

$$= \frac{\pi \times 8.854 \times 10^{-12}}{6.207} = 4.48 \times 10^{-12} \text{ F/m}$$

Capacitance for 25 km = $4.48 \times 10^{-12} \times (25 \times 1000)$ = **0.112 μF (Ans.)**

Q. 29.23. *Performance of transmission lines.* *(a) How are transmission lines classified?*

(b) A 3-phase, 50 Hz, 100 km long overhead line has the following constants:

Resistance/phase/km = 0.12 Ω; Inductance/phase/km 1.1 mH;

Capacitance/phase/km = 0.0085 μF.

The line supplies a load of 25 MW at 0.85 p.f. lagging at a line voltage of 132 kV at the receiving end. Using nominal-π method, determine:

(i) Sending end line voltage, *(ii) Sending end current, and*

(iii) Sending end power factor.

Solution. (*a*) Classification of transmission lines:

The transmission lines are generally *classified* as follows:

1. Short transmission lines.
2. Medium transmission line.
3. Long transmission line.

1. **Short transmission lines:**

- Transmission lines having *length less than 50 km* and operating voltage *less that 20 kV are generally* considered as "*short transmission lines*".
- Due to smaller distance and lower line voltage the capacitance effects are extremely small and can be neglected. Hence the performance of such lines depend upon resistance and inductance only. Further, although resistance and inductance are distributed along the line, yet in short lines these constants are *assumed to be lumped at one phase.*

2. **Medium transmission lines:**

- The transmission lines having *length between 50 km and 160 km* and *line voltage between 21 kV and 100 kV* are considered as "*medium transmission lines*".
- Due to sufficient length and voltage of the line the capacitance effects need to be considered. Although capacitance is distributed uniformly over the entire length of the line, yet it is *assumed to be concentrated at one or more points.*

3. **Long transmission lines:**

- The transmission lines having *length more than 160 km* and line voltage above 100 kV are considered as "*long transmission lines*".
- Here the *line constants are considered uniformly distributed* (not lumped) *over the entire length of the line and rigorous methods are employed for solution.*

(***b***) Fig. 29.23, shows the circuit diagram of the transmission line.

Total resistance/phase, $R = 0.12 \times 100 = 12\ \Omega$

Total inductance/phase, $L = (1.1 \times 10^{-3}) \times 100 = 0.11$ H

Inductive reactance/phase, $X = 2\pi fL = 2\pi \times 50 \times 0.11 = 34.56\ \Omega$

Total capacitance phase, $C = (0.0085 \times 10^{-6}) \times 100 = 0.85 \times 10^{-6}$ F

$$\text{Capacitive reactance/phase, } X_C = \frac{1}{2\pi fC} = \frac{1}{2\pi \times 50 \times 0.85 \times 10^{-6}} = 3744.8\ \Omega$$

$$\text{Total susceptance/phase, } \overline{Y} = j\frac{1}{Z} = j\frac{1}{X_C} = j\left(\frac{1}{3744.8}\right) = j2.67 \times 10^{-4}\ \mho$$

Receiving end current, $I_R = \dfrac{25 \times 10^6}{\sqrt{3} \times (132 \times 1000) \times 0.85} = 128.6$ A

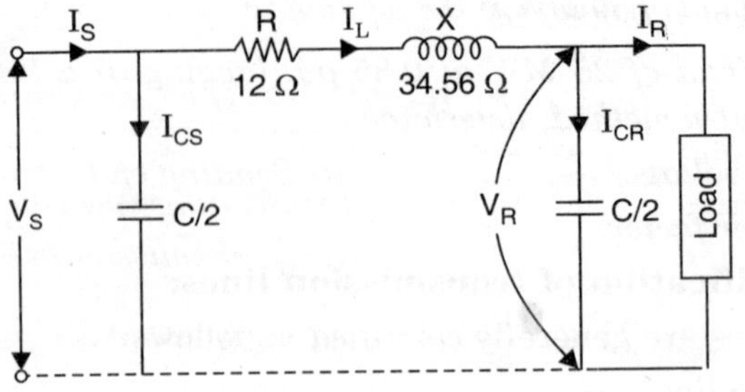

Fig. 29.23. Nominal-π method (Circuit diagram)

Phase voltage at the receiving end,

$$V_R = \frac{132 \times 1000}{\sqrt{3}} = 76210 \text{ V (or 76.21 kV)}$$

$$\cos \phi_R = 0.85, \sin \phi_R = \sqrt{1 - (\cos \phi_R)^2} = 0.527$$

Taking phase voltage at receiving end as reference vector, we have:

$$\bar{V}_R = V_R (1 + j0) = 76210 (1 + j0) = (76210 + j0)$$

$$\bar{I}_R = I_R (\cos \phi_R - j \sin \phi_R) = 128.6 (0.85 - j0.527)$$

$$= (109.31 - j67.77) \text{ A}$$

$$\bar{I}_{CR} = \frac{1}{2} \bar{Y} \bar{V}_R = \frac{1}{2} (j2.67 \times 10^{-4})(76210 + j0) = j10.17 \text{ A}$$

Line current, $\bar{I}_L = \bar{I}_R + \bar{I}_{CR} = (109.31 - j\,67.77) + j10.17$

$$= (109.31 - j57.6) \text{ A}$$

Impedance drop in line $= \bar{I}_L \bar{Z} = (109.31 - j57.6)(12 + j34.56)$

$$= 1311.72 + j\,3777.75 - j691.2 + 1990.66$$

$$= (3302.38 + j3086.55) \text{ V}$$

(*i*) **Sending end line voltage, $\bar{V}_{SL}$:**

Sending end phase voltage,

$$\bar{V}_S = \bar{V}_R + \bar{I}_L \bar{Z} = 76210 + (3302.38 + j3086.55)$$

$$= (79512.38 + j3086.55) \text{ V} = 79.57 \angle 2.22° \text{ kV}$$

Sending end line voltage, $\bar{V}_{SL} = \sqrt{3}(79.57 \angle 2.22) =$ **137.82 ∠2.22° kV (Ans.)**

(*ii*) **Sending end current, $\bar{I}_S$:**

$$\bar{I}_{CS} = \frac{1}{2} \bar{Y} \bar{V}_S = \frac{1}{2} j(2.67 \times 10^{-4})(79512.38 + j3086.55) = (j10.61 - 0.41) \text{A}$$

$$\bar{I}_S = \bar{I}_L + \bar{I}_{CS} = (109.31 - j57.6) + (j10.61 - 0.41)$$

$= (108.9 - j46.99) =$ **118.6 ∠–23.34°A (Ans.)**

(*iii*) **Sending end power factor, cos ϕ_s:**

$$\cos [2.22 - (-23.34°)] = \mathbf{0.9021} \quad \textbf{(Ans.)}$$

Q. 29.24. ***Performance of transmission lines.*** *A 110 kV, 50 Hz, 3-phase overhead transmission line delivers a load of 50 MW at 0.8 p.f. lagging at the receiving end and has the following constants:*

$$A = D = 0.92 \angle 2.8°,\ B = 95 \angle 78°\ \Omega;\ C = 4.5 \times 10^{-4} \angle 75°\ ℧.$$

Determine the following:

(*i*) *Sending end voltage,* (*ii*) *Sending end current,*

(*iii*) *Sending end p.f., and* (*iv*) *Transmission efficiency.*

Solution. *Given:* $\bar{V}_R = \frac{110}{\sqrt{3}} \angle 0° = 63.5 \angle 0°$ kV; $\cos \phi_R = 0.8$ *i.e.,*

$\phi_R = \cos^{-1}(0.8) = 36.87°$; $A = D = 0.92 \angle 2.8°$; $B = 95 \angle 78°\ \Omega$;

$C = 4.5 \times 10^{-4} \angle 75°\ ℧$; Power delivered = 50 MW.

(*i*) **Sending end voltage:**

Receiving end current $I_R = \frac{50 \times 10^3}{\sqrt{3} \times 110 \times 0.8} = 328.04$ A

Taking receiving end phase voltage as the reference phasor, we have:

$$\bar{V}_R = 63.5 \angle 0° \text{ V}$$

$$\bar{I}_R = 328.04 \angle -36.87°$$

$$\begin{aligned}\bar{V}_S &= A\bar{V}_R + B\bar{I}_R \\ &= (0.92 \angle 2.8° \times 63.5 \angle 0°) + (95 \angle 78° \times 328.04 \angle -36.87° \times 10^{-3}] \\ &= (58.42 \angle 2.8° + 31.16 \angle 41.13°) \\ &= (58.35 + j2.85) + 23.47 + j20.49) \\ &= 81.82 + j23.34 = 85.08 \angle 15.92°\end{aligned}$$

$\therefore$ Sending end voltage $\bar{V}_{SL} = \sqrt{3} \times 85.08 \angle 15.92° = \mathbf{147.36 \angle 15.92°}$ **(Ans.)**

(*ii*) **Sending end current, I_S:**

$$\begin{aligned}\bar{I}_S &= C\bar{V}_R + D\bar{I}_R \\ &= (4.5 \times 10^{-4} \angle 75° \times 63.5 \angle 0° \times 10^3) \\ &\quad + (0.92 \angle 2.8° \times 328.04 \times \angle -36.87°) \\ &= 28.57 \angle 75° + 301.79 \angle -34.07° \\ &= (7.39 + j27.59) + (249.99 + j169.06) \\ &= 257.38 - j141.47 = \mathbf{293 \angle -28.79°} \quad \textbf{(Ans.)}\end{aligned}$$

(*iii*) **Sending end power:**

Sending end power factor, $\cos \phi = \cos [15.92° - (-28.79°)] = 0.71$ (lag)

$\therefore$ Sending end power $= \sqrt{3}\, V_{SL} I_S \cos \phi_s$

$= \sqrt{3} \times 147.36 \times 293 \times 0.71 \times 10^{-3} =$ **53.1 MW** **(Ans.)**

(*iv*) **Transmission efficiency, η_T:**

$$\eta_T = \frac{\text{Power delivered}}{\text{Power sent out}} = \frac{50}{53.1} = 0.9416 \text{ or } \mathbf{94.16\%} \quad \textbf{(Ans.)}$$

"COMPETITIVE EXAMINATIONS' OBJECTIVE TYPE QUESTIONS" WITH

"ANSWERS"

(Including Questions from UPSC, GATE, AMIE, and Other Engineering Examinations)

OBJECTIVE TYPE QUESTIONS' BANK

Choose the Correct Answer :

1. A hydroelectric power station is commonly found in
   (*a*) desert areas (*b*) hilly areas
   (*c*) swamps (*d*) grasslands.
2. Bio-gas plants are suitable for
   (*a*) metallurgical industries (*b*) commercial complexes
   (*c*) rural areas (*d*) coal mines.
3. Condensers in thermal power plants for condensing
   (*a*) steam to water (*b*) water to ice
   (*c*) hydrogen gas to liquid hydrogen (*d*) carbon dioxide to dry ice.
4. Capacitance of a transmission line
   (*a*) increases (*b*) decreases
   (*c*) remains same (*d*) with increases in its length.
5. The angle of A, constant of the transmission line normally lies between
   (*a*) 90°—70° (*b*) 70°—40°
   (*c*) 40°—10° (*d*) 10°—0°.
6. For a long distance e.h.v. transmission line the receiving-end voltage under unloaded condition is
   (*a*) much lower than (*b*) lower than
   (*c*) equal to than (*d*) higher than.
   the sending-end voltage.
7. By increasing the transmission voltage to double of its original value the same power can be despatched keeping the line loss
   (*a*) equal to original value (*b*) half the original value
   (*c*) double the original value (*d*) one-fourth of original value.
8. ACSR conductors have
   (*a*) all conductors made of aluminium (*b*) outer conductors made of aluminium
   (*c*) inner conductors made of aluminium (*d*) no conductors made of aluminium.
9. Shunt capacitors in a sub-station
   (*a*) consume lagging var (*b*) deliver lagging var
   (*c*) consume active power (*d*) deliver active power.
10. Wavy structure of pin insulator increases its
    (*a*) mechanical strength (*b*) puncture strength
    (*c*) flashover strength (*d*) thermal strength.
11. Increase in frequency of transmission line causes
    (*a*) no change in line reactance (*b*) increase in line resistance
    (*c*) decrease in line resistance (*d*) decrease in line series reactance.

12. Diversity factor in a power system is
(*a*) always less than unity (*b*) normally less than unity
(*c*) always more than unity (*d*) normally more than unity.

13. Graphite is used in nuclear power plant as a
(*a*) fuel (*b*) coolant
(*c*) moderator (*d*) electrode.

14. Constant power locus of a transmission line at a particular sending-end and receiving-end voltage is
(*a*) a straight line (*b*) a circle
(*c*) a parabola (*d*) an ellipse.

15. The allowable limit of variation of frequency in a power system is
(*a*) ± 10% (*b*) ± 1%
(*c*) ± 5% (*d*) ± 0.5%.

16. The null point of a uniformly loaded distributor fed at equal voltage at both end lies at
(*a*) mid-point (*b*) either end
(*c*) two-third distance from one end (*d*) one-fourth distance from one end.

17. Transposition of transmission line is done to
(*a*) reduce line loss (*b*) reduce skin effect
(*c*) balance line voltage drop (*d*) reduce corona.

18. Power transmission by cable is generally adopted for line lengths
(*a*) above 100 km (*b*) upto 200 km
(*c*) less than 50 km (*d*) less than 10 km.

19. Power despatch through a line can be increased by
(*a*) installing series capacitor (*b*) installing shunt capacitor
(*c*) installing series reactor (*d*) installing shunt reactor.

20. A bus coupler circuit breaker is utilised in a sub-station for
(*a*) joining the transmission line with station bus-bar
(*b*) joining main and transfer bus in a sub-station
(*c*) joining the generator with transformers
(*d*) joining the neutral of generator with earth.

21. Plant capacity factor is
(*a*) actual energy produced/maximum possible energy that could have been produced (based on installed capacity)
(*b*) actual energy produced/(plant capacity hours for which the plant has been in operation)
(*c*) energy generated in a given period (maximum demand × hours of operation in a given period)
(*d*) none of the above.

22. Spinning reserve is
(*a*) the reserve generating capacity which is available for service but not in operation.
(*b*) the reserve generating capacity which is connected to the bus and is ready to take load.
(*c*) the reserve generating capacity which is in operation but is not in service.
(*d*) none of the above.

23. Unit of permeability is
(*a*) pure number (*b*) farad
(*c*) henry (*d*) henry/metre
(*e*) farad/metre.

24. Reactive power is
(*a*) rate of energy transfer
(*b*) the product of r.m.s. volt and quadrature component of r.m.s. current
(*c*) the product of r.m.s. volt and r.m.s. current
(*d*) the product of r.m.s. volt and in phase component of r.m.s. current
(*e*) none of these.

25. The advantage of neutral earthing is
(*a*) safety of personnel
(*b*) reduction of earthfault current
(*c*) elimination of arcing ground
(*d*) none of these.

26. Sulphur hexafluoride cable is insulated by
(*a*) impregenated paper
(*b*) polyvinyl chloride
(*c*) high pressure oil
(*d*) compressed gas.

27. When the load on a transmission line is equal to the surge impedance loading
(*a*) the receiving end voltage is less than the sending end voltage
(*b*) the sending end voltage is less than the receiving end voltage
(*c*) the receiving end voltage is equal to the sending end voltage
(*d*) none of these.

28. In a long transmission line under no-load condition
(*a*) the receiving end voltage is less than the sending end voltage
(*b*) the sending end voltage is less than the receiving end voltage
(*c*) the sending end voltage is equal to the receiving end voltage
(*d*) none of these.

29. The permissible variation of frequency in power system is
(*a*) ± 1%
(*b*) ± 3%
(*c*) ± 5%
(*d*) ± 10%.
(*e*) none of the above.

30. The *ABCD* constants of a three-phase transposed transmission line with linear and passive elements
(*a*) are always equal
(*b*) never equal
(*c*) *A* and *D* are equal
(*d*) *B* and *C* are equal.

31. 66 kV is suitable for transmission of power over
(*a*) 30 km
(*b*) 60 km
(*c*) 120 km
(*d*) 200 km
(*e*) None of the above.

32. When an alternator is delivering a balanced load at unity power factor, the phase angle between line voltage and line current is
(*a*) 90°
(*b*) 60°
(*c*) 30°
(*d*) 0°
(*e*) none of the above.

33. For stable operation of interconnected system, the passive element that can be used as the interconnecting element is
(*a*) reactor
(*b*) resistor
(*c*) capacitor
(*d*) resistor and capacitor.

34. If the insulation resistance of a cable of length 10 km is 1 mega ohm its insulation resistance for 50 km length will be
(*a*) 1 mega ohm (*b*) 5 mega ohms
(*c*) 0.2 Maga ohm (*d*) none of the above.

35. If a synchronous machine is underexcited it takes lagging vars from the system when it is operated as a
(*a*) synchronous motor (*b*) synchronous generator
(*c*) synchronous motor as well as generator (*d*) none of the above.

36. Corona is accompanied by
(*a*) violet visible-discharge in darkness (*b*) hissing sound
(*c*) vibration (*d*) power-loss
(*e*) radio interference (*f*) ozone
(*g*) all of the above.

37. With bundled conductors
(*a*) the corona inception voltage increases
(*b*) the corona inception voltage decreases
(*c*) the corona inception voltage remains unaffected.

38. Sheaths are used in cables to
(*a*) provide proper insulation (*b*) provide mechanical strength
(*c*) prevent ingress of moisture (*d*) none of the above.

39. An *RLC* series circuit at resonance frequency remains
(*a*) predominantly inductive (*b*) predominantly capacitative
(*c*) purely resistive (*d*) none of the above.

40. A synchronous phase modifier as compared to synchronous motor of same rating has
(*a*) larger shaft diameter and higher speed
(*b*) smaller shaft diameter and higher speed
(*c*) larger shaft diameter and smaller speed
(*d*) smaller shaft diameter and smaller speed.

41. Diversity factor × Maximum demand is
(*a*) average demand (*b*) sum of consumers' maximum demand
(*c*) installed capacity (*d*) generated power.

42. A generating station which has a high investment cost and low operating cost is usually operated as a
(*a*) peak load station (*b*) base load station.
(*c*) medium load station (*d*) none of the above.

43. To meet the reactive power requirements at load centres usually
(*a*) shunt capacitors are used (*b*) series capacitors are used
(*c*) tap changing transformers are used (*d*) shunt reactors are used.

44. Characteristic impedance of an overhead transmission line is usually in the range of
(*a*) 100 to 200 ohms (*b*) 200 to 300 ohms
(*c*) 0 to 100 ohms (*d*) 400 to 500 ohms.

45. For synchronous generator connected to a power station, increasing the excitation will
(*a*) decrease reactive power generation (*b*) increase reactive power generation
(*c*) decrease active power generation (*d*) increase active power generation.

46. Increasing the frequency of transmission line will
(*a*) increase shunt reactance (*b*) decrease line resistance
(*c*) increase line resistance (*d*) decrease series reactance.

47. The power factor of an open ended cable can be improved by
(*a*) increasing the capacitance (*b*) decreasing the capacitance
(*c*) increasing the conductor resistance (*d*) increasing the insulation resistance.

48. For the same voltage drop, increasing the voltage of a distributor by n-times
(*a*) reduces the cross-section of the conductor by n-times
(*b*) increases the cross-section of the conductor by n-times
(*c*) reduces the cross-section of the conductor by n^2-times
(*d*) increases the cross-section of the conductor by n^2-times.

49. For a medium length transmission line, A is
(*a*) equal to B (*b*) equal to C
(*c*) equal to D (*d*) not equal to any of the above.

50. If L, C and Y are the inductance, capacitance and shunt admittance of a line for unit length, then for length l
(*a*) the shunt admittance is $Y \cdot l$ (*b*) the inductance is L/l
(*c*) the capacitance is $C \cdot l$ (*d*) the shunt admittance is Y/l.

51. To reduce corona effect usually
(*a*) the distance between the conductors is reduced
(*b*) the conductor diameter is reduced
(*c*) hundled conductors are used
(*d*) stranded conductors are used.

52. The dielectric strength of air under normal conditions is around
(*a*) 30 kV per cm (*b*) 100 kV ker cm
(*c*) 150 kV per cm (*d*) 200 kV per cm.

53. Bundled conductors are mainly used
(*a*) to increase the shunt capacitance (*b*) to decrease the shunt capacitance
(*c*) to increase the series reactance (*d*) to decrease the series reactance.

54. According to Indian Electricity Act, the permissible voltage drop at the consumer terminals of a distribution feeder is
(*a*) 10% (*b*) 12%
(*c*) 3% (*d*) 6%.

55. To increase the transmission capability of a high voltage long line
(*a*) the resistance can be increased (*b*) the resistance can be decreased
(*c*) the series reactance can be reduced (*d*) the shunt admittance can be reduced.

56. For a short line if the receiving end voltage is equal to sending voltage under loaded conditions
(*a*) the sending power factor is unity (*b*) the receiving end power factor is unity
(*c*) the sending end power factor is leading (*d*) the receiving end power factor is leading.

57. Series capacitors can be used in distribution lines
(*a*) to provide reactive power compensation
(*b*) to reduce the receiving end voltage under light load conditions
(*c*) to reduce the voltage drop
(*d*) to reduce line losses.

58. Under frequency operation of power systems is undesirable as it causes
(*a*) damage to hydro units
(*b*) damage to thermal units
(*c*) increased line charging requirements
(*d*) reduces the line reactance.

59. Back-to-back HVDC is used to
(*a*) increase the transmission capability
(*b*) decrease line losses
(*c*) provide stable interconnection
(*d*) reduce voltage drop.

60. Most of the steam turbo alternators are wound for
(*a*) two poles
(*b*) ten to twenty poles
(*c*) twenty to thirty poles
(*d*) six poles.

61. If the frequency of a transmission system is changed from 50 Hz to 100 Hz, the string efficiency
(*a*) will increase
(*b*) will decrease
(*c*) remains unchanged
(*d*) may increase or decrease depending on the line parameters.

62. The surge impedance of 50 miles long under-ground cable is 50 ohms. For a length of 25 miles, the impedance will be
(*a*) 26 ohms
(*b*) 50 ohms
(*c*) 100 ohms
(*d*) 12.5 ohms
(*e*) none of these.

63. In order to have lower cost of electrical energy generation it is required to have
(*a*) low load factor and diversity factor
(*b*) low load factor but high diversity factor
(*c*) high load factor but low diversity factor
(*d*) high load factor and high diversity factor.

64. Shunt compensation in a EHV line is resorted to
(*a*) improve the stability
(*b*) reduce the fault level
(*c*) improve the voltage profile
(*d*) as a substitute to synchronous phase modifier.

65. The size of conductor on modern EHV lines is obtained based on
(*a*) voltage drop
(*b*) current density
(*c*) corona
(*d*) skin effect
(*e*) none of the above.

66. The presence of earth in case of overhead lines
(*a*) increases the capacitance
(*b*) increases the inductance
(*c*) decreases the capacitance
(*d*) decreases the inductance.

67. Phase modifier is normally installed in the case of
(*a*) short transmission lines
(*b*) medium length lines
(*c*) long length lines
(*d*) all length of lines.

68. Increase in temperature in overhead transmission lines causes
(*a*) increase in stress and length
(*b*) decrease in stress and length
(*c*) decrease in stress but increase in length
(*d*) none of the above.

69. High voltage long lines are transposed because then
(*a*) corona losses can be minimised
(*b*) computation of inductance becomes easier
(*c*) voltage drops in the lines can be minimised
(*d*) phase voltage imbalances can be minimised.

70. If the separation between the three phases of a transmission system is increased then
(*a*) the inductance will increase and capacitance will remain unchanged
(*b*) both the inductance will increase and the capacitance will decrease

(*c*) the inductance will increase and the capacitance will decrease
(*d*) the inductance will decrease and capacitance will increase.

71. Bundled conductors are used to
(*a*) reduce inductance of the line
(*b*) reduce both inductance and capacitance
(*c*) reduce corona loss
(*d*) reduce corona loss and the line inductance.

72. The function of the economiser is to
(*a*) heat up the incoming water with exhaust steam
(*b*) heat up the pulverised fuel by exhaust gases
(*c*) heat up the incoming air by exhaust gases
(*d*) heat up the incoming water by exhaust gases
(*e*) none of the above.

73. The purpose of the 'Boiler Feed Pump' is to
(*a*) pump hot air into the boiler
(*b*) pump pulverised coal into the boiler
(*c*) pump water from the boiler
(*d*) pump water into the boiler
(*e*) none of the above.

74. The purpose of Moderator in a Nuclear Power Plant is to
(*a*) control the flow of water inlet
(*b*) control the amount of nuclear fuel into the reactor
(*c*) control the nuclear fusion or fusion rate by slowing down the neutrons
(*d*) control the steam flow to the turbine
(*e*) none of the above.

75. In a load duration curve for an integrated power system the uppermost crust represents the energy contributed by
(*a*) base power stations
(*b*) major thermal stations
(*c*) peaking hydro or gas turbine stations
(*d*) non-conventional power stations
(*e*) none of the above.

76. In a transmission system, the weight of copper used is proportional to
(*a*) E^2
(*b*) E
(*c*) $1/E^2$
(*d*) $1/E$
(*e*) none of the above.

77. In a cable of conductor diameter d and overall diameter with dielectric material D, the maximum dielectric stress
(*a*) occurs at the conductor surface and is proportional to d
(*b*) occurs at the conductor surface and is proportional to $1/d$
(*c*) occurs at the middle of the dielectric and is proportional to $1/D$
(*d*) occurs at the outer surface of the dielectric and is proportional to D
(*e*) none of the above.

78. In an interconnected power system the increase in the field current in any generating unit causes
(*a*) increase in active power flow from the unit to the rest of the system
(*b*) increase in reactive power flow from the unit to the rest
(*c*) decrease in active power and increase in reactive power flow from the unit
(*d*) increase in both active and reactive power flow from the unit.

79. The effect of bonding the cable is
(*a*) to increase the effective resistance and inductance
(*b*) to increase the effective resistance but reduce inductance
(*c*) to reduce the effective resistance and inductance
(*d*) to reduce the effective resistance but increase the inductance.

80. The insulation resistance of a cable of length 10 km is 1 $M\Omega$; its resistance for 50 km length will be
(*a*) 1 $M\Omega$ (*b*) 5 $M\Omega$
(*c*) 0.2 $M\Omega$ (*d*) none of the above.

81. The surge impedance of a 50 miles long under-ground cable is 50 ohms. For a 25 miles length it will be
(*a*) 25 ohms (*b*) 50 ohms
(*c*) 100 ohms (*d*) none of these.

82. Shunt compensation in an EHU line is resorted to
(*a*) improve the stability (*b*) reduce fault level
(*c*) improve the voltage (*d*) none of these.

83. The impulse ratio of a rod gap is
(*a*) unity (*b*) between 1.2 and 1.5
(*c*) between 2 and 2.2 (*d*) none of these.

84. The design of insulation of modern EHU line is based on
(*a*) the lightning voltage (*b*) corona
(*c*) switching voltage (*d*) radio interference.

85. Phase modifier is normally installed in the case of
(*a*) short transmission line (*b*) medium length of lines
(*c*) long lines (*d*) for any length of lines.

86. Ferranti effect on long overhead line is experienced when it is
(*a*) lightly loaded (*b*) on full load at unity p.f.
(*c*) on full load at 0.8 p.f. load (*d*) on any load.

87. Stringing chart is useful for
(*a*) finding the sag in the conductor (*b*) in the design of tower
(*c*) in the design of insulator string (*d*) finding the distance between towers.

88. The presence of earth in case of overhead lines
(*a*) increases the capacitance
(*b*) increase the inductances
(*c*) decreases the capacitance and increases the inductance
(*d*) does not effect any of the line constants.

89. Effect of increase in temperature in overhead line is to
(*a*) increase the stress and the length (*b*) decrease the stress but increase in length
(*c*) decrease the stress and length (*d*) any of these.

90. The voltage of a bus can be controlled by controlling the
(*a*) phase angle (*b*) reactive power of the bus
(*c*) active power of the bus (*d*) phase angle and the reactive power.

91. The leakage resistance of a 50 km long cable is 1 MΩ. For a 100 km long cable it will be
(*a*) 1 MΩ (*b*) 2 MΩ
(*c*) 0.66 MΩ (*d*) none of the above.

92. The connected load of a consumer is 2 kW and his maximum demand is 1.5 kW. The load factor of the consumer is
(*a*) 0.75 (*b*) 0.375
(*c*) 1.33 (*d*) none of the above.

93. Sheath are used in cables to
(*a*) provide proper insulation (*b*) provide mechanical strength
(*c*) prevent ingress of moisture (*d*) none of the above.

94. An *R—L—C* circuit remains predominantly inductive
(*a*) at resonance frequency (*b*) below resonance frequency
(*c*) above resonance frequency.

95. As per Indian electricity rules the maximum allowable variation between declared and actual voltage at consumer premises should be
(*a*) ± 6% (*b*) ± 8%
(*c*) ± 5%.

96. The inductance of a line is minimum when
(*a*) GMD is high (*b*) GMR is high
(*c*) both GMD and GMR are high (*d*) GMD is low but GMR is high.

97. The insulating material most commonly used for power cables is
(*a*) PVC (*b*) paper
(*c*) rubber.

98. For the stable operation of interconnected system, the passive element that can be used at interconnecting element is
(*a*) reactor (*b*) resistor
(*c*) capacitor (*d*) resistor and capacitor.

99. If two synchronous generators are connected, loss of synchronism will result in
(*a*) stalling of generators (*b*) wild fluctuations is current
(*c*) wild fluctuations is current and voltage (*d*) none of the above.

100. The capacitance between any two conductors of a 3-core cable with sheath earthed is 3 μF. The capacitance per phase will be
(*a*) 1.5 μF (*b*) 6 μF
(*c*) 1 μF (*d*) none of the above.

101. Which of the following plants has the maximum capital cost ?
(*a*) Steam plant (*b*) Hydro-plants
(*c*) Diesel plant (*d*) Nuclear plants.

102. The largest size of Hydroelectric unit in INDIA is
(*a*) 350 MW (*b*) 500 MW
(*c*) 165 MW (*d*) 90 MW.

103. Large size steam plants and nuclear plants are suitable for
(*a*) base loads (*b*) Intermediate loads
(*c*) peak loads (*d*) both base and peak loads.

104. One kg of natural Uranium gives energy equivalent to about
(*a*) 100 kg coal (*b*) 1000 kg of coal
(*c*) 5000 kg of coal (*d*) 10000 kg of coal.

105. The spacing between sub-conductor of a bundle is approximately
(*a*) 20 cm (*b*) 40 cm
(*c*) 80 cm (*d*) 1.20 cm.

106. Under frequency operation of power system is undesirable as it causes
(*a*) damage to hydro units (*b*) damage to thermal units
(*c*) increased line charging requirement (*d*) reduction in line reactance.

107. The regulation of a line at full-load 0.8 of lagging is 12%. The regulation at full-load 0.8 pf leading can be
(*a*) 24% (*b*) 18%
(*c*) 12% (*d*) 4%.

108. In a three-core cable the capacitance between two conductors (with sheath earthed) is 3 μF. The capacitance per phase is
(*a*) 1.5 μF (*b*) 3 μF
(*c*) 6 μF (*d*) none of these.

109. The surge impedance loading for a single circuit 220 kV line is around
(*a*) 90 MW (*b*) 120 MW
(*c*) 220 MW (*d*) 400 MW.

110. To obtain the minimum value of stress in cable R/r ratio should be
(*a*) 2.13 (*b*) 3.14
(*c*) 2.718 (*d*) 1.96.

111. As the load factor of a generating plant increase, the generation cost per kWh energy generated
(*a*) decreases (*b*) increases
(*c*) remains the same.

112. As the size of a thermal generating unit increases, the capital cost per kW of installed capacity
(*a*) increases (*b*) decreases
(*c*) remains the same (*d*) may increase or decrease.

113. For a short line with R/X ratio 1.0, the regulation will be zero when the load power factor is
(*a*) unity (*b*) 0.707 lagging
(*c*) 0.707 leading (*d*) 0 leading.

114. For a good voltage profile under no-load condition, a long line needs
(*a*) shunt capacitors at receiving end (*b*) shunt reactors at receiving end
(*c*) shunt resistance at receiving end.

115. For a 400 kV line the spacing between phase conductors is approx.
(*a*) 8 m (*b*) 11 m
(*c*) 14 m (*d*) 17 m.

116. The string efficiency of a high voltage line is around
(*a*) 100% (*b*) 80%
(*c*) 40% (*d*) 10%.

117. Frequency of resonant vibrators lies in the range of
(*a*) 10-20 Hz (*b*) 100-200 Hz
(*c*) 50-100 Hz (*d*) 200-300 Hz.

118. The insulation resistance of a 2 km long cable is 200 MΩ. For a length of 20 km, the insulation resistance will be

(*a*) 20 M Ω (*b*) 400 M Ω

(*c*) 2000 M Ω (*d*) 1000 MΩ.

119. At NTP the breakdown strength of air is

(*a*) 30 kV-rms/cm (*b*) 30 kV-peak/cm

(*c*) 40 kV-peak/cm (*d*) 60 kV-peak/cm.

120. The velocity of propagation of electromagnetic waves on overhead lines is

(*a*) 3×10^8 m/s (*b*) 3×10^8 km/s

(*c*) 3×10^{10} m/s (*d*) 3×10^8 km/hr.

121. The addition of a pumped storage plant to an existing system

(*a*) improves the load factor without any additional generation capacity

(*b*) improves the load factor with additional generation capacity

(*c*) increases the total generation capacity with increase in the unit cost of generation

(*d*) reduces the unit cost of generation with increase in the total generation capacity.

122. Hydroplants are more efficient than thermal plants

(*a*) Yes, because (*b*) No, because.

123. Bundle conductors in transmission lines

(*a*) reduce corona inception voltage (*b*) increase corona inception voltage.

124. The presence of earth in case of overhead lines

(*a*) increases the line capacitance (*b*) decreases the line capacitance

(*c*) has no effect on the line capacitance.

125. At higher voltages, belted cables are more suitable than screened cables

(*a*) Yes, because (*b*) No, because.

126. Grading of cables

(*a*) reduces insulation cost and increases current rating

(*b*) reduces insulation cost but decreases current rating

(*c*) increases both

(*d*) None of the above.

127. Effect of temperature rise in overhead lines is to

(*a*) increase the sag and decrease the tension

(*b*) decrease the sag and increase the tension

(*c*) increases both

(*d*) decreases both.

128. The maximum tension in a section of overhead line conductor between two supports of unequal height occurs at

(*a*) the higher support (*b*) the lower support

(*c*) the midpoint of the conductor (*d*) None of the above.

129. The leakage resistance of a 100 km long insulated cable will be

(*a*) twice that of a 50 km long cable (*b*) half of that of a 50 km long cable

(*c*) equal to that of a 50 km long cable (*d*) four times that of a 50 km long cable.

130. The value of the '*A*' parameter of a transmission line

(*a*) increases with the increase in length of the line

(*b*) decreases with the increase in line length

(*c*) is independent of line length.

131. Surge impedance values for cables are
(*a*) higher than those of overhead lines (*b*) lower than those of overhead lines
(*c*) of the same order as those of overhead lines.

132. Multiple conductor lines have
(*a*) lower surge impedance than single conductor lines
(*b*) higher surge impedance than single conductor lines
(*c*) higher surge impedance than single conductor lines for short lengths and lower surge impedance for long lines.

133. The inductance of a line
(*a*) increases with increase in line spacing (*b*) decreases with the increase in line spacing
(*c*) does not depend on the line spacing.

134. The receiving-end voltage of a transmission line will be greater than the sending-end voltage if the load is
(*a*) greater than SIL (surge impedance loading)
(*b*) less than SIL
(*c*) equal to SIL.

135. Sheaths are used in cables to
(*a*) provide proper insulation (*b*) protect the cable from moisture
(*c*) provide mechanical strength (*d*) None of the above.

136. Skin effect in a conductor becomes more pronounced
(*a*) at higher frequency (*b*) at lower frequency
(*c*) at d.c.

137. Spinning reserve is required for
(*a*) economic operation of power system (*b*) reliable operation of power system
(*c*) None of the above.

138. The insulation design of EHV line is based on
(*a*) switching overvoltage (*b*) lightning overvoltage
(*c*) power frequency overvoltage (*d*) corona inception voltage.

139. In an HVDC link, power flows
(*a*) from the rectifier end to the inverter end
(*b*) from the inverter end to the rectifier end.

140. In HVDC system a converter acts as a rectifier when the firing angle α has a value
(*a*) $0 < \alpha < 120°$ (*b*) $0 < \alpha < 90°$
(*c*) $90° < \alpha < 180°$.

141. Corona is
(*a*) partial breakdown of air (*b*) complete breakdown of air
(*c*) sparking between lines.

142. Surge impedance of a transformer is in the range of
(*a*) 80 to 100 ohms (*b*) 400-500 ohms
(*c*) 1000-2000 ohms.

143. Bundled conductors
(*a*) reduce surface electric stress of conductor (*b*) increase the line reactance
(*c*) decrease the line capacitance (*d*) none of these.

144. Back-to-back *hvdc* is used to
(*a*) provide stable interconnection (*b*) bulk power transmission
(*c*) reduce voltage drop (*d*) any of these.

145. We have
(*a*) several *hvdc* systems in India (*b*) no *hvdc* systems in India
(*c*) two *hvdc* systems in India (*d*) any of these.

146. Series capacitors
(*a*) increase the power transmission capability of the line
(*b*) provide voltage control
(*c*) increase the shunt admittance of the line.

147. No load compensation of a high voltage line involves
(*a*) shunt capacitors (*b*) shunt reactors
(*c*) series capacitors.

148. Full load compensation of a line requires
(*a*) shunt capacitors (*b*) series capacitors
(*c*) transformers (*d*) shunt reactors.

149. Extra-high voltage transmission means
(*a*) voltage less than 400 kV (*b*) voltage in the range of 400 kV-750 kV
(*c*) voltage greater than 750 kV.

150. A d.c. line carries
(*a*) less power than an equivalent a.c. line (*b*) more power than the a.c. line
(*c*) same power as an a.c. line.

151. Steam turbo alternators have
(*a*) 2 poles (*b*) 4 poles
(*c*) 6 poles.

152. The capital cost on a *hvdc* converter is
(*a*) less than a.c. substation (*b*) more than a.c. substation
(*c*) same as that of a.c. substation.

153. Improving p.f.
(*a*) reduces current for a given power (*b*) increases losses in line
(*c*) increases the cost of station equipment.

154. 1-ph a.c. power
(*a*) is constant
(*b*) pulsates around an average value at double the supply frequency
(*c*) pulsates around the average value at the supply frequency.

155. Large loads in industry use
(*a*) single-phase induction motors (*b*) 3-phase induction motors
(*c*) d.c. motors.

156. Transmission of power by a.c. cable is impossible beyond
(*a*) 40-60 km (*b*) 400 km
(*c*) 200 km.

157. The use of a guard ring
(*a*) equalises the voltage division between insulator discs
(*b*) is unnecessary complication
(*c*) decreases string-efficiency.

158. Wind loading in coastal regions is in the range of

(*a*) 40-50 kg/m^2 (*b*) 150 kg/m^2

(*c*) 96 kg/m^2.

159. Inductance of a line is of the order of

(*a*) 1 mH/km (*b*) 1 H/km

(*c*) 10 mH/kM.

160. HVDC transmission is

(*a*) economically and technically superior to a.c. transmission

(*b*) not a good alternative to a.c. transmission

(*c*) not economically viable.

161. The capacity factor of a plant is equal to

(*a*) maximum load/average load (*b*) average load/maximum load

(*c*) average load/plant capacity (*d*) maximum load/plant capacity.

162. A mass curve can be plotted from

(*a*) load duration curve (*b*) chronological load curve

(*c*) energy load curve

(*d*) both load duration curve and chronological load curve.

163. The capital cost of a plant depends upon

(*a*) total installed capacity only (*b*) total number of units only

(*c*) total installed capacity as well as number of units

(*d*) neither the installed capacity nor the number of units.

164. With the increase of load factor of a generating plant, the generation cost of per unit of energy

(*a*) increases (*b*) remains the same

(*c*) decreases (*d*) may increase or decrease.

165. Two areas *A* and *B* have equal connected loads. However load diversity in area *A* is more than in area *B*. Then

(*a*) maximum demands of the two areas would be equal

(*b*) maximum demand of *A* would be more than that of *B*

(*c*) maximum demand of *B* would be more than that of *A*

(*d*) maximum demand of *A* may be more or less than that of *B*.

166. Which of the following generating plants has the minimum operating cost ?

(*a*) Steam plant (*b*) Hydro-electric plant

(*c*) Nuclear plant (*d*) Diesel plant.

167. The modern trend in electric power generation is

(*a*) to have a large number of small size thermal plants located at different places

(*b*) to have large size thermal plants near load centre

(*c*) to have large size thermal plants located near coalfields

(*d*) None of the above.

168. Large size steam plants and nuclear plants are suitable for

(*a*) peak loads (*b*) intermediate loads

(*c*) base loads (*d*) both base and peak loads.

169. The advantage of hydro plants is

(*a*) low operating cost (*b*) they can be started and loaded very quickly

(*c*) they can be used as base load and peak load plants as well

(*d*) All the above.

170. An alternator is operating on an infinite bus bar. If the excitation of the alternator is increased

(*a*) real power output of the alternator will be increased

(*b*) power factor of the alternator will decrease

(*c*) power factor of the alternator will increase

(*d*) both the real power and power factor of the alternator will decrease.

171. The conductor of an EHV line is selected on the basis of

(*a*) current carrying capacity (*b*) corona and *RI* performance

(*c*) line voltage (*d*) None of these.

172. The maximum permissible value of fair weather corona loss for an HV line is

(*a*) 0.6 kW/3-phase km (*b*) 1.2 kW/3-phase km

(*c*) 0.3 kW/3-phase km (*d*) 2.4 kW/3-phase km.

173. The inductance of a line is minimum when

(*a*) GMD is high (*b*) GMR is high

(*c*) both GMD and GMR are high (*d*) GMD is low and GMR is high.

174. The transmission capacity of a line at 50 Hz frequency as compared to that at 60 Hz frequency is

(*a*) lower (*b*) higher

(*c*) the same.

175. The SIL of a single circuit 220 kV line is around

(*a*) 120 MW (*b*) 90 MW

(*c*) 220 MW (*d*) 400 MW.

176. The reactive power transfer over a line mainly depends on

(*a*) power angle δ (*b*) $|V_s| - |V_r|$

(*c*) V_s (*d*) V_r.

177. For a 400 kV line, the spacing between phase conductors is around

(*a*) 8 m (*b*) 11 km

(*c*) 14 m (*d*) 17 m.

178. The insulation resistance of a 2 km long cable is 150 mega-ohms. For a length of 20 km, the insulation resistance will be

(*a*) 1500 mΩ (*b*) 150 mΩ

(*c*) 15 mΩ (*d*) 3000 mΩ.

179. To obtain the maximum value of stress in cables, the ratio (*R*/*r*) should be

(*a*) 2.13 (*b*) 2.718

(*c*) 1.96 (*d*) 1.5.

180. A 300 km long overhead line has a surge impedance of 400 Ω. For 600 km length, the surge impedance will be

(*a*) 100 Ω (*b*) 200 Ω

(*c*) 400 Ω (*d*) 800 Ω.

181. Diversity factor is the ratio of

(*a*) sum of maximum demand of consumers/system maximum demand

(*b*) maximum demand of consumers/average demand

(*c*) demand of all consumers/average demand

(*d*) None of the above.

182. Surge impedance of an overhead transmission line is in the range of
(*a*) 50-100 ohms
(*b*) 300-500 ohms
(*c*) 1000-1500 ohms.

183. A generating station has a maximum demand of 20 MW and a connected load of 40 MW. Total units generated are 4.36×10^7 per annum. The demand factor is
(*a*) 2.0
(*b*) 2/3
(*c*) 1/2
(*d*) 1/3.

184. We have
(*a*) two h.v.d.c. system in India
(*b*) no h.v.d.c. system in India
(*c*) several h.v.d.c. systems in India
(*d*) None of the above.

185. A synchronous condenser is usually
(*a*) d.c. generator
(*b*) induction motor
(*c*) overexcited synchronous motor
(*d*) underexcited synchronous motor.

186. The advantage of static capacitor for improvement of power factor is
(*a*) low loss
(*b*) little maintenance cost
(*c*) case in installation
(*d*) All of the above.

187. Generators for base load plants are usually designed for maximum efficiency around
(*a*) 20% overload
(*b*) full load
(*c*) 75% full load.

188. The area under the load curve divided by 24 hours gives
(*a*) average load
(*b*) peak load
(*c*) least load.

189. Back-to-back connection is used in h.v.d.c. systems to
(*a*) provide a synchronous tie
(*b*) provide bulk power transmission
(*c*) reduce voltage drop.

190. Which support for overhead transmission line has the least life ?
(*a*) Wooden poles
(*b*) Fabricated steel structure
(*c*) RCC poles
(*d*) Steel poles.

191. As compared to a.c. transmission, d.c. transmission is free from
(*a*) inductance
(*b*) capacitance
(*c*) phase displacement
(*d*) All of the above.

192. The inductance of a single-phase two-wire line is given by (D is the distance between conductors and r is the radius of conductor)
(*a*) $0.4 \log_2 (D/r)$ mH/km
(*b*) $0.55 \log_e (D/r)$ mH/km
(*c*) $0.4 \log_e (r/D)$ mH/km
(*d*) $0.55 \log_e (r/D)$ mH/km.

193. Due to skin effect
(*a*) the portion of the conductor near the surface carries more current and core of conductor carries less current
(*b*) portion of the conductor near the surface carries less current and core of conductor carries more current
(*c*) None of the above.

194. String efficiency can be improved by
(*a*) using long cross-arm
(*b*) grading the insulator
(*c*) using a guard ring
(*d*) Any of the above.

195. The transfer of power between two stations is maximum when the phase displacement between the voltage of two stations is
(*a*) zero (*b*) 90°
(*c*) 120° (*d*) 180°.

196. A 100 MW alternator is connected to an infinite bus and its excitation is increased, then the terminal voltage of the alternator will
(*a*) rise (*b*) fall
(*c*) remain unaltered.

197. For bulk power transmission over long distance HVDC transmission is
(*a*) economically and technically superior to a.c. transmission
(*b*) not a good alternative to a.c. transmission
(*c*) not economically viable.

198. Series capacitors on transmission lines are of the little use when
(*a*) the load VAR requirement is small
(*b*) the load VAR requirement is large
(*c*) the load VAR requirement is fluctuating
(*d*) Series capacitors are never used on transmission lines.

199. Transmission of power by a.c. cables is impossible beyond
(*a*) 34-45 km (*b*) 500 km
(*c*) 300 km.

200. Inductance of a transmission line is of the order of
(*a*) 1 H/km (*b*) 1 mH/km
(*c*) 15 mH/km.

201. Standard domestic a.c. supply voltage in India is
(*a*) 220 volt (*b*) 230 volt
(*c*) 240 volt.

202. The total instantaneous power in a balanced three phase line
(*a*) varies at power frequency (*b*) varies at twice the power frequency
(*c*) is constant, independent of time.

203. A large diversity factor of the load in a power system
(*a*) reduces the installation cost (*b*) increases the installation cost
(*c*) does not affect the installation cost.

204. In a nuclear reactor thermal energy is obtained from
(*a*) fission of radioactive materials (*b*) fusion of radioactive materials
(*c*) burning of the fuel rods in oxygen.

205. Inductance of an overhead line, in comparison to that of a cable of same capacity is
(*a*) larger (*b*) smaller
(*c*) of the same order.

206. Proximity of a line to the earth surface
(*a*) does not affect its capacitance to neutral
(*b*) increases the capacitance to neutral
(*c*) decreases the capacitance to neutral.

207. For a given conductor the value of GMR is
(*a*) larger for capacitance calculation (*b*) large for inductance calculation
(*c*) same for both capacitance and inductance calculation.

208. The skin effect in conductor results in
(*a*) increase in its dc resistance (*b*) decrease in its ac resistance
(*c*) increase in its ac resistance.

209. Post type insulators are generally used in lines operating
(*a*) above 100 kV (*b*) below 33 kV
(*c*) at any voltage level, hv or ehv.

210. In suspension type insulator the potential drop is
(*a*) maximum across the lowest disc (*b*) maximum across the topmost disc
(*c*) uniformly distributed over the discs.

211. The sag of a transmission line conductor in summer is
(*a*) less than that in winter (*b*) more than that in winter
(*c*) same as in winter.

212. The generalised constant *A* for a medium long line has magnitude
(*a*) close to but less than 1 (*b*) close to but greater than 1
(*c*) nearly equal to the series impedance of the line.

213. Flow of active power through a line in a power system may be controlled by
(*a*) Tap changing transformer (*b*) Phase shifting transformer
(*c*) Synchronous phase modifier.

214. Transmission of power between India and Sri Lanka may be effected through submarine
(*a*) ac or dc link (*b*) dc link only
(*c*) ac link only.

215. The lead sheath in a cable is used to
(*a*) provide mechanical protection (*b*) control the field distribution
(*c*) prevent entry of moisture.

216. For economy measure the generators at a power plant operates at
(*a*) equal loads (*b*) loads proportional to the ratings
(*c*) equal incremental cost.

217. A synchronous phase modifier supplies
(*a*) both active and reactive powers (*b*) both lagging and leading reactive power
(*c*) inductive reactive power only.

218. Corona loss in a dc line is
(*a*) less than that in ac line operating at same rms voltage
(*b*) more than that in ac line at same rms voltage
(*c*) same as in ac line at equal rms voltage.

219. The function of moderator in a nuclear reactor is
(*a*) to absorb the excess neutrons (*b*) to increase the energy of the neutrons
(*c*) to slow down the neutrons.

220. The pumped storage scheme is employed to supply
(*a*) during the peak hours (*b*) during the of-peak hours
(*c*) the system base load.

221. Which of the following is a vector quantity ?
(*a*) Electrical potential (*b*) Electric field intensity
(*c*) Electric charge (*d*) None of the above.

222. Which power plant cannot have single unit of 100 MW capacity ?
(*a*) Steam power plant (*b*) Nuclear power plant
(*c*) Hydroelectric power plant (*d*) Diesel engine power plant.

223. A 120 MW generator is usually
(*a*) air cooled (*b*) hydrogen cooled
(*c*) oxygen cooled (*d*) nitrogen cooled.

224. Which of the following plants will take least time in starting from cold conditions to full load operation ?
(*a*) Nuclear power plant (*b*) Steam power plant
(*c*) Hydroelectric plant (*d*) Gas turbine plant.

225. During load shedding
(*a*) System voltage is reduced (*b*) System frequency is reduced
(*c*) Some loads are switched off (*d*) System power factor is changed.

226. Sparking occurs when a load is switched off because the circuit has high
(*a*) inductance (*b*) resistance
(*c*) capacitance (*d*) magnetism.

227. Domestic supply of 220 volt. 50 Hz. signifies that peak value of voltage wave from is
(*a*) 220 volt (*b*) 155 volt
(*c*) 185 volt (*d*) 310 volt.

228. Which of the following is the protective device against lightning overvoltages ?
(*a*) Rod gaps (*b*) Surge absorbers
(*c*) Horn gaps (*d*) All of the above.

229. The effect of corona is
(*a*) increased energy loss (*b*) increased reactance
(*c*) increased inductance (*d*) all of the above.

230. Which of the following is usually not the general voltages ?
(*a*) 6.6 kV (*b*) 9.9 kV
(*c*) 11 kV (*d*) 13.2 kV.

231. Highest voltage for transmitting electrical power in India is
(*a*) 735 kV (*b*) 700 kV
(*c*) 550 kV (*d*) 400 kV.

232. Which of the following voltage regulation is considered to be the best ?
(*a*) 2% (*b*) 30%
(*c*) 70% (*d*) 98%.

233. If 3 MW power is to be transmitted over a distance of 30 km, the desirable transmission voltage will be
(*a*) 11 kV (*b*) 33 kV
(*c*) 66 kV (*d*) 3.3 kV.

234. Sheaths are used in cables to
(*a*) provide proper insulation (*b*) provide mechanical strength
(*c*) prevent ingress of moisture (*d*) none of the above.

235. The insulation of the modern EHV lines is obtained based on
(*a*) The lighting voltage (*b*) The switching voltage
(*c*) Corona (*d*) *RI*.

236. Which distribution system is more reliable ?

(*a*) Ring main system (*b*) Tree system

(*c*) Radial system (*d*) All are equally reliable.

237. Which type of insulators are used on 132 kV transmission lines ?

(*a*) Pin type (*b*) Disc type

(*c*) Shackle type (*d*) Pin and Shackle type.

238. For 11 kV line, the inductance per km per phase will be of the order of

(*a*) 1 H (*b*) 0.1 H

(*c*) 1 mH (*d*) 0.1 mH.

239. Which domestic utility item has highest power rating ?

(*a*) Refrigerator (*b*) Ceiling fan

(*c*) Mixi (*d*) Electric iron.

240. Current rating is not necessary in case of

(*a*) isolators (*b*) circuit breakers

(*c*) load break switches (*d*) circuit breakers and load break switches.

241. Load factor of the curve of the figure is

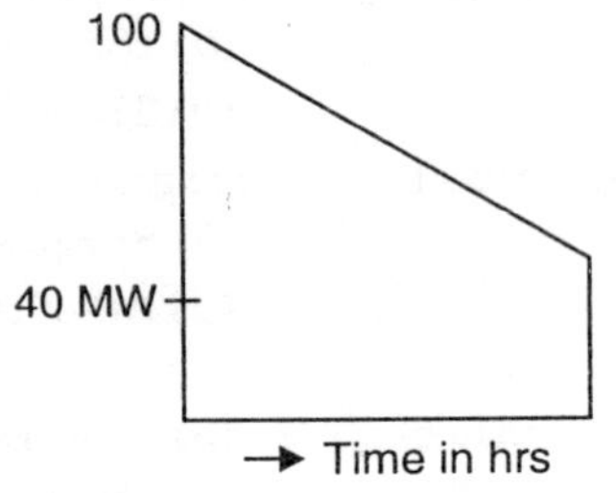

(*a*) 80% (*b*) 70%

(*c*) 60% (*d*) 50%.

242. Which of the following is *correct* ?

(*a*) Load factor = Capacitor factor × utilisation factor

(*b*) Utilisation factor = Capacity factor × load factor

(*c*) Capacitor factor = Load factor/utilisation factor

(*d*) Capacity factor = Load factor × utilisation factor.

243. A mass curve can be plotted from

(*a*) load duration curve

(*b*) chronological load curve

(*c*) energy load curve

(*d*) both load duration curve and chronological load curve.

244. The capital cost of a power plant depends on

(*a*) total installed capacity only

(*b*) total number of units only

(*c*) total installed capacity and number of units as well

(*d*) neither the installed capacity nor number of units.

245. As the load factor of a power plant increase, the cost per kWh of energy generated

(*a*) increases (*b*) decreases

(*c*) may increase or decrease (*d*) remains the same.

246. The power which must be available even under emergency conditions is known as
(*a*) spinning reserve (*b*) cold reserve
(*c*) firm reserve (*d*) hot reserve.

247. An industrial installation has a power factor of 0.8 lagging. It would be economical to improve p.f. to :
(*a*) unity (*b*) about 0.8 leading
(*c*) about 0.95 lagging (*d*) about 0.9 leading.

248. Hydrograph is similar to
(*a*) load duration curve (*b*) mass curve
(*c*) energy load curve (*d*) chronological load curve.

249. One kilogram of natural Uranium gives energy equivalent to about
(*a*) 100 kg of coal (*b*) 1000 kg of coal
(*c*) 5000 kg of coal (*d*) 10000 kg of coal.

250. A moderator material should have
(*a*) small atomic mass (*b*) large atomic mass
(*c*) none of the above.

251. In h.v. transmission, the spacing between sub-conductors of a bundle is approximately
(*a*) 20 cm (*b*) 40 cm
(*c*) 80 cm (*d*) 3.5 m.

252. The approximate cost ratio of a 220 kV underground cable transmission and 220 kV O/H transmission is
(*a*) 50 (*b*) 25
(*c*) 13 (*d*) 5.

253. For a standard conductor, the ratio of GMR to actual radius is
(*a*) equal to 1 (*b*) more than 1
(*c*) 0.7788 (*d*) less than 0.7788.

254. The inductance of a transmission line is minimum when
(*a*) GMD is high (*b*) GMR is high
(*c*) Both GMD and GMR are high (*d*) GMD is low and GMR is high.

255. The p.u. impedance of a line to a 50 MVA, 132 kV base is 0-4, the p.u. impedance to a 100 MVA, 132 kV base will be
(*a*) 0.2 (*b*) 0.8
(*c*) 0.4 (*d*) 1.6.

256. For a 400 kV line, the spacing between phase conductors is around
(*a*) 20 m (*b*) 11 m
(*c*) 14 m (*d*) 8 m.

257. The Corona loss in a 50 Hz system is 0.25 kW/ph/km. At a frequency of 60 Hz, the Corona loss would be
(*a*) 0.21 kW/ph/km (*b*) 0-30 kW/ph/km
(*c*) 0.36 kW/ph/km.

258. In a 3 core cable, the capacitance between 2 conductors (with sheath earthed) is 3μF. The capacitance/phase is
(*a*) 1.5 μF (*b*) 3 μF
(*c*) 6 μF (*d*) 12 μF.

259. For a 400 kV line, the number of discs in an insulator string is around

(*a*) 37 (*b*) 31
(*c*) 25 (*d*) 16.

260. In a cable the sheath radius is R and conductor radius is r. As r changes from $0.5\,R$ to $0.25\,R$ the maximum voltage gradient in the dielectric

(*a*) decreases by about 6% (*b*) increases by about 6%
(*c*) increases by about 15% (*d*) decreases by about 15%.

261. The power dissipated in a pure capacitor is

(*a*) zero (*b*) minimum
(*c*) maximum (*d*) depends on the size and voltage.

262. Overall thermal efficiency of a steam power station is in the range

(*a*) 18-24% (*b*) 30-40%
(*c*) 44-62% (*d*) 68-79%.

263. In pumped storage plant

(*a*) power is produced by means of pumps
(*b*) water is stored by pumping to high pressures
(*c*) downstream water is pumped upstream during off load periods
(*d*) water is recirculated through turbine.

264. In a steam power plant water is used for cooling purposes in

(*a*) boiler (*b*) economiser
(*c*) condenser (*d*) superheaters.

265. In which of the following power plants the availability of power is least reliable ?

(*a*) Solar power plant (*b*) Wind energy
(*c*) Tidal power plant (*d*) Geothermal power plant.

266. A load curve is a plot of

(*a*) load versus generation capacity (*b*) load versus current
(*c*) load versus time (*d*) load versus cost of power.

267. In a system, if the base load is the same as the maximum demand, the load factor will be

(*a*) 1.0 (*b*) zero
(*c*) infinity (*d*) 1 per cent.

268. A 120 MW generator is usually

(*a*) air cooled (*b*) hydrogen cooled
(*c*) oxygen cooled (*d*) nitrogen cooled.

269. A nuclear power plant is invariably used as a

(*a*) base load plant (*b*) peak load plant
(*c*) stand by plant (*d*) spinning reserve plant.

270. Load shedding is possible through

(*a*) voltage reduction (*b*) frequency reduction
(*c*) switching off the loads (*d*) any of the above.

271. The fact that a conductor carries more current on the surface as compared to core, is known as

(*a*) skin effect (*b*) corona
(*c*) permeability (*d*) unsymmetrical fault.

272. Pin insulators are normally used up to voltages of about
(*a*) 100 kV (*b*) 66 kV
(*c*) 33 kV (*d*) 25 kV.

273. Which distribution system is more reliable ?
(*a*) ring main system (*b*) tree system
(*c*) radial system (*d*) all are equally reliable.

274. A bus bar is rated by
(*a*) current only (*b*) current and voltage only
(*c*) current, voltage and frequency (*d*) none of the above.

275. Bundled conductors are used in transmission lines to
(*a*) reduce the line capacitance to ground (*b*) reduce the weight of the lines
(*c*) reduce regulation (*d*) reduce corona losses.

276. The effect of electric shock on human body depends on
(*a*) voltage (*b*) current
(*c*) duration of contact (*d*) all of the above.

277. Which of the following equipment is not installed in a substation ?
(*a*) exciters (*b*) shunt reactors
(*c*) voltage transformers (*d*) series capacitors.

278. Shunt capacitance is usually neglected in the analysis of
(*a*) short transmission lines (*b*) medium transmission lines
(*c*) long transmission lines (*d*) medium and long transmission lines.

279. The chances of corona are maximum during
(*a*) summer heat (*b*) winter
(*c*) dry weather (*d*) humid weather.

280. The power transmitted will be maximum when
(*a*) sending end voltage is more (*b*) receiving end voltage is more
(*c*) reactance is high (*d*) corona losses are least.

281. Load factor for the load duration curve shown in figure is
(*a*) 100% (*b*) 70%
(*c*) 60% (*d*) 50%.

282. The long-term load forecast is needed for

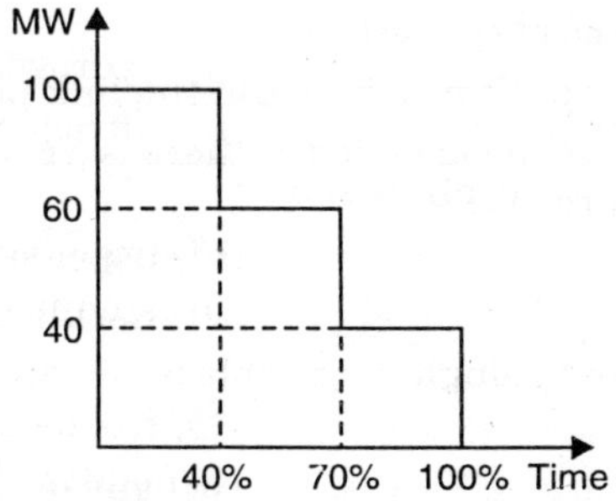

(*a*) load-frequency control (*b*) economic load despatch
(*c*) plant operation (*d*) planning the addition in generation capacity.

283. The inductance of a single-phase two-conductor power transmission line per km becomes double when the
(*a*) radius of the conductor is doubled
(*b*) distance between the conductors is doubled
(*c*) distance between the conductors is increased four fold
(*d*) distance between the conductors is increased as square of the original distance.

284. In a high head hydroelectric power station a surge tank is provided to
(*a*) reduce the length of the penstock
(*b*) augment water at the forebay
(*c*) control pressure variations in the penstock due to sudden load changes
(*d*) control the water flow through the turbine.

285. Transmission lines are transposed to
(*a*) reduce transmission loss
(*b*) reduce skin effect
(*c*) reduce conductor vibrations
(*d*) reduce interference with neighbouring communication lines.

286. For a transmission line with resistance $R\Omega$, reactance $X\Omega$, and negligible capacitance, the parameter A is
(*a*) 0 (*b*) 1
(*c*) $R + jX$ (*d*) $(R^2 + X^2)^{1/2}$.

287. Bundled conductors in EHV transmission system provide
(*a*) reduced capacitance (*b*) increased capacitance
(*c*) increased voltage gradient (*d*) increased inductance.

288. The charging current in a transmission line increases due to corona effect because corona increases
(*a*) line current (*b*) power loss in the line
(*c*) effective line voltage (*d*) effective conductor diameter.

289. Control rods used in nuclear reactor are made of
(*a*) boron (*b*) lead
(*c*) beryllium (*d*) zirconium.

290. The non-uniform distribution of voltage gradient across the units in a string of suspension type insulators is due to
(*a*) unequal distance between the cross-arm and the units
(*b*) unequal self capacitance of the units
(*c*) the existence of stray capacitance between the link pins of the units and the tower body

291. Whenever the conductors are dead ended or there is a change in the direction of transmission line, the insulators used are of the
(*a*) pin type (*b*) suspension type
(*c*) strain type (*d*) shackle type.

292. The insulation resistance of a single-core cable is 150 MΩ/km. The insulation resistance for 3 km length is
(*a*) 16.67 MΩ (*b*) 50 MΩ
(*c*) 150 MΩ (*d*) 450 MΩ.

293. A moderator material in a nuclear reactor should
(*a*) have small atomic mass (*b*) have large atomic mass
(*c*) have large neutron absorption section (*d*) be chemically active.

294. Mini hydroelectric power plant generally use
(*a*) Pelton turbine (*b*) Francis turbine
(*c*) kaplan turbine (*d*) bulb turbine.

295. The voltage regulation of a long EHV transmission line at no load is
(*a*) negative (*b*) positive
(*c*) zero (*d*) unity.

296. Corona loss increases with
(*a*) increase in supply frequency and conductor size
(*b*) increase in supply frequency and reduction in conductor size
(*c*) decrease in supply frequency and reduction in conductor size
(*d*) decrease in supply frequency and increase in conductor size.

297. The electrostatic stress in a single-core belted cable is
(*a*) minimum at the conductor surface and maximum at the sheath
(*b*) maximum at the conductor surface and minimum at the sheath
(*c*) same throughout the insulation thickness
(*d*) maximum both at the conductor surface and at the sheath and minimum at the armouring layer.

298. The alternate layers of an ACSR conductor are spiralled in opposite direction to
(*a*) prevent unwinding (*b*) reduce inductance
(*c*) increase capacitance (*d*) make it mechanically stronger.

299. The efficiency of transmission increases as
(*a*) system voltage and load power factor both decrease
(*b*) system voltage and load power factor both increase
(*c*) system voltage decreases but load power factor increases
(*d*) system voltage increases but load power factor decreases.

300. A stringing chart is useful
(*a*) in the location of transmission towers
(*b*) in the determination of number of units in a string of insulators
(*c*) in finding the sag at different temperatures
(*d*) for checking uplift force on towers.

301. A diesel power plant is best suited as
(*a*) base load plant (*b*) stand-by plant
(*c*) peak load plant (*d*) general purpose plant.

302. A gas turbine power plant usually suits for
(*a*) peak load operation (*b*) base load operation
(*c*) casual run (*d*) none of the above.

303. Which of the following plant is expected to have the longest expected life ?
(*a*) hydroelectric (*b*) steam
(*c*) diesel (*d*) nuclear.

304. Hydrogen is used for cooling of large size generators, because
(*a*) it has high thermal conductivity (*b*) it is light
(*c*) it offers reduced fire risk (*d*) of all above reasons.

305. Major share of power is produced in India through
(*a*) thermal power plants (*b*) diesel power plants
(*c*) nuclear power plants (*d*) hydroelectric power plants.

306. Efficiency is secondary consideration in case of
(*a*) peak load plants (*b*) base load plants
(*c*) both peak load and base load plants (*d*) none of the above.

307. For the same plant size, initial cost of which plant is the highest ?
(*a*) steam power plant (*b*) diesel power plant
(*c*) nuclear power plant (*d*) gas turbine plant.

308. The fact that a conductor carries more current on the surface as compared to core, is known as
(*a*) skin effect (*b*) corona
(*c*) permeability (*d*) unsymmetrical fault.

309. Presence of ozone as a result of corona is harmful because
(*a*) it gives bad odour (*b*) it corrodes the material
(*c*) it transfers energy to the ground (*d*) reduces power factor.

310. In a transmission line having negligible resistance, the surge impedance is
(*a*) $\sqrt{L+C}$ (*b*) $\sqrt{C/L}$
(*c*) $\sqrt{\frac{1}{CL}}$ (*d*) $\sqrt{L/C}$.

311. For 11 kV line, the capacitance per km per phase will be of the order of
(*a*) 1 Farad (*b*) 0.10 Farad
(*c*) 0.01 Farad (*d*) 0.01 μF.

312. Which distribution system is more reliable ?
(*a*) ring main system (*b*) tree system
(*c*) radial system (*d*) all are equally reliable.

313. The function of steel wire in an ACSR conductor is
(*a*) to take care of surges (*b*) to prevent corona
(*c*) to reduce inductance (*d*) to provide additional mechanical strength.

314. For a 66 kV line having span of 200 metres between towers, the approximate sag will be :
(*a*) 0.02 m (*b*) 0.2 m
(*c*) 2 m (*d*) 20 m.

315. In a sub-station the equipment used to limit short circuit current level is
(*a*) series reactor (*b*) coupling capacitor
(*c*) lightning switch (*d*) isolator.

316. A 30 km transmission line carrying power at 33 kV is known as
(*a*) short transmission line (*b*) long transmission line
(*c*) high power line (*d*) ultra high voltage line.

317. The permissible voltage variation in a distribution system is
(*a*) ± 0.1% (*b*) ± 1%
(*c*) ± 10% (*d*) ± 20%.

318. 750 kV is termed as
(*a*) medium high voltage (*b*) high voltage
(*c*) extra high voltage (*d*) ultra high voltage.

319. The sag of a transmission line is least affected by
(*a*) self weight of conductors (*b*) temperature of surrounding air
(*c*) current through conductor (*d*) ice deposited on conductor.

320. ACSR conductor having 7 steel stands surrounded by 25 aluminium conductors will be specified as

(*a*) 7/25 (*b*) 7/32

(*c*) 25/7 (*d*) 25/32.

321. The following sequence currents were recorded in a power system under a fault condition

$$I_{\text{positive}} = j1.653 \text{ p.u.}$$
$$I_{\text{negative}} = -j0.5 \text{ p.u.}$$
$$I_{\text{zero}} = -j1.153 \text{ p.u.}$$

The fault is

(*a*) line to ground (*b*) three-phase

(*c*) line to line to ground (*d*) line to line.

322. Steady-state stability of a power system is improved by

(*a*) reducing fault clearing time

(*b*) using double circuit line instead of single circuit line

(*c*) single pole switching

(*d*) decreasing generator inertia.

323. Which one of the following statements is true ?

(*a*) Steady-state stability limit is greater than transient stability limit.

(*b*) Steady-state stability limit is equal to transient stability limit.

(*c*) Steady-state stability limit is less than the transient stability limit.

(*d*) No generalisation can be made regarding the equality or otherwise of the steady-state stability limit and transient stability limit.

324. Tho minimum oil circuit breaker has less volume of oil because

(*a*) there is insulation between contacts

(*b*) the oil between the breaker contacts has greater strength

(*c*) solid insulation is provided for insulating the contacts from earth

(*d*) none of the above.

325. When two transformers of different kVA ratings are connected in parallel they share the load in proportion to their respective kVA rating only when their

(*a*) kVA ratings are identical (*b*) Efficiencies are equal

(*c*) pu impedances are equal (*d*) equivalent impedances are equal.

326. When bundle conductors are used in place of single conductors, the effective inductance and capacitance will respectively

(*a*) increase and decrease (*b*) decrease and increase

(*c*) decrease and remain unaffected (*d*) remain unaffected and increase.

327. The voltages of a generator and an infinite bus are given as 0.92 $\angle 10°$ and 1.0 $\angle 0°$ respectively. The active power will flow from

(*a*) generator to infinite bus (*b*) infinite bus to generator

(*c*) the data is insufficient to judge (*d*) none of the above.

328. If the inertia constant H of a machine of 200 MVA is 2 p.u. its value corresponding to 400 MVA will be

(*a*) 4 p.u. (*b*) 2 p.u.

(*c*) 1.0 p.u. (*d*) 0.5 p.u.

329. The operation of a nuclear reactor is controlled by controlling the multiplication-factor (K), defined as

$$K = \frac{\text{Number of neutrons of any one generation}}{\text{Number of neutrons of immediately preceding generation}}$$

The power-level of the reactor can be increased by
(*a*) raising the value of K above 1 and, keeping it at that raised value
(*b*) raising the value of K above 1, but later bringing it back to $K = 1$
(*c*) lowering the value of K below 1 and, keeping it at that lowered value
(*d*) lowering the value of K below 1, but later bringing it back to $K = 1$.

330. The following data pertain to two alternators working in parallel and supplying a total load of 80 MW
Machine 1 : 40 MVA with 5% speed regulation
Machine 2 : 60 MVA with 5% speed regulation
The load sharing between machines 1 and 2 will be

(*a*) $\frac{P_1}{48 \text{ MW}}, \frac{P_2}{32 \text{ MW}}$ (*b*) 40 MW, 40 MW
(*c*) 30 MW, 50 MW (*d*) 32 MW, 48 MW.

331. Whenever the conductors are dead-ended or there is a change in the direction of transmission line, the insulators used are of the
(*a*) pin type (*b*) suspension type
(*c*) strain type (*d*) shackle type.

332. Equal area criterion gives the information regarding
(*a*) stability region (*b*) absolute stability
(*c*) relative stability (*d*) swing curves.

333. A three-phase transmission line of negligible resistance and capacitance has an inductive reactance of 100 ohms per phase. When the sending-end and receiving-end voltages are maintained at 110 kV, the maximum power that can be transmitted will be
(*a*) 121 MW (*b*) $121\sqrt{3}$ MW
(*c*) 363 MW (*d*) 363 kW.

334. One current transformer (CT) is mounted over a 3-phase 3-core cable with its sheath and armour removed from the portion covered by the CT. An ammeter placed in the CT secondary would measure
(*a*) the positive sequence current (*b*) the negative sequence current
(*c*) the zero sequence current (*d*) three times the zero sequence.

335. The charging current in a transmission line increases due to corona effect because corona increases
(*a*) line current (*b*) effective line voltage
(*c*) power loss in lines (*d*) the effective conductor diameter.

336. For a transmission line with negligible losses, the lagging reactive power (VAR) delivered at the receiving-end, for a given receiving-end voltage is directly proportional to the
(*a*) square of the line voltage drop (*b*) line voltage drop
(*c*) line inductance reactance (*d*) line capacitive reactance.

337. A power system is subjected to a fault which makes the zero sequence component of current equal to zero. The nature of fault is
(*a*) double line to ground fault (*b*) double line fault
(*c*) line-to-ground fault (*d*) three-phase to ground fault.

338. Load frequency control is achieved by properly matching the individual machine's
(*a*) reactive powers (*b*) generated voltages
(*c*) turbine inputs (*d*) turbine and generator ratings.

339. The surge impedance of a 100 km long underground cable is 50 ohms. The surge impedance of a 40 km long similar cable would be
(*a*) 20 ohms (*b*) 50 ohms
(*c*) 80 ohms (*d*) 125 ohms.

340. Bundled conductors in EHV transmission system provide
(*a*) reduced capacitance (*b*) increased capacitance
(*c*) increased inductance (*d*) increased voltage gradient.

341. A string insulator has 4 units. The voltage across the bottom-most unit is 33.33% of the total voltage. Its string efficiency is
(*a*) 25% (*b*) 33.33%
(*c*) 66.67% (*d*) 75%.

342. Earth wire on EHV overhead transmission line is provided to protect the line against
(*a*) lightning surge (*b*) switching surge
(*c*) excessive fault voltages (*d*) corona effect.

343. "Expanded ACSR" are conductor composed of
(*a*) larger diameter individual strands for a given cross-section of the aluminium strands
(*b*) larger diameter of the central steel strands for a given overall diameter of the conductor
(*c*) larger diameter of the aluminium strand only for a given overall diameter of the conductor
(*d*) a filler between the inner steel and the outer aluminium strands to increase the overall diameter of the conductor.

344. If, in a short transmission line, resistance and inductive reactance are found to be equal and regulation appears to be zero, then the load will
(*a*) have unity power factor (*b*) have zero power factor
(*c*) be 0.707 leading (*d*) be 0.707 lagging.

345. In the solution of load-flow equation, Newton-Raphon (NR) method is superior to the Gauss-Seidal (GS) method, because the
(*a*) time taken to perform one iteration in the NR method is less when compared to the time taken in the GS method
(*b*) number of iteration required in the NR method is more when compared to that in the GS method
(*c*) number of iterations required is not independent of the size of the system in the NR method
(*d*) convergence characteristic of the NR method are not affected by the selection of slack bus.

346. The inertia constant of two groups of machines which do not swing together are M_1 and M_2 such that $M_1 > M_2$. It is proposed to add some inertia to one of the two groups of machines for improving the transient stability of the system. It should be added to :
(*a*) M_1 (*b*) M_2
(*c*) It does not matter whether to add to M_1 or M_2
(*d*) any of the above.

347. A hydel power plant of run-off-river type should be provided with a pondage so that the
(*a*) firm-capacity of the plant is increased
(*b*) operating head is controlled

(*c*) pressure inside the turbine casing remains constant

(*d*) kinetic energy of the running water is fully utilised.

348. Three insulating materials with breakdown strengths of 250 kV/cm, 200 kV/cm, 150 kV/cm and permittivities of 2.5, 3.0 and 3.5 are used in a single core cable. If the factor of safety for the materials is 5, the location of the materials with respect to the core of the cable will be

(*a*) 2.5, 3.0, 3.5 (*b*) 3.0, 2.5, 3.5

(*c*) 3.5, 3.0, 2.5 (*d*) 3.5, 2.5, 3.0.

349. Galloping in transmission line conductors arises generally due to

(*a*) asymmetrical layers of ice formation

(*b*) vortex phenomenon in light winds

(*c*) heavy weight of the line conductors

(*d*) adoption of horizontal conductor configurations.

350. In a high voltage dc transmission scheme, reactive power is needed both for the rectifier at the sending-end and, for the inverter at the receiving-end. During the operation of such a dc link the rectifier receives

(*a*) lagging reactive power and the inverter supplies leading reactive power

(*b*) leading reactive power and the inverter supplies lagging reactive power

(*c*) lagging reactive power and the inverter supplies lagging reactive power

(*d*) leading reactive power and the inverter supplies leading reactive power.

351. Consider the following statements

To provide reliable protection for a distribution transformer against overvoltages using lightning arresters, it is essential that the

(*a*) lead resistance is high

(*b*) distance between the transformer and the arrester is small

(*c*) transformer and the arrester have a common inter-connecting ground

(*d*) spark overvoltage of the arrester is greater than the residual voltage.

352. A line trap in carrier current relaying tuned to carrier frequency presents

(*a*) high impedance to carrier frequency but low impedance to power frequency

(*b*) low impedance to both carrier and power frequency

(*c*) high impedance to both carrier and power frequency

(*d*) low impedance to carrier frequency but high impedance to power frequency.

353. The use of fast acting relays and circuit breakers for clearing a sudden short-circuit on a transmission link between a generator and receiving-end bus improves the transient stability of the machine because

(*a*) short-circuit current becomes zero

(*b*) post-fault transfer impedance attains a value higher than that during the fault

(*c*) ordinate of the post-fault power-angle characteristic is higher than that of during-fault characteristic

(*d*) voltage behind the transient reactance increases to a higher value.

354. For a *Y*-delta transformer with *Y*-side grounded, the zero sequence current

(*a*) has no path to ground (*b*) exists in the lines on the delta side

(*c*) exists in the lines on the *Y* side (*d*) exists in the lines on both *Y* and delta sides.

355. Lightning arresters are used in power systems to protect electrical equipments against

(*a*) direct strokes of lightning

(*b*) power frequency overvoltages

(*c*) overvoltages due to indirect lightning stroke
(*d*) overcurrents due to lightning stroke.

356. The effect of bonding the cable is
(*a*) to increase the effective resistance and inductance
(*b*) to increase the effective resistance but reduce inductance
(*c*) to decrease the effective resistance and inductance
(*d*) to decrease the effective resistance but increase the inductance.

357. In EHV transmission lines, efficiency of transmission can be increased by decreasing the corona loss. This is achieved by :
(*a*) Increasing the distance between the line conductors
(*b*) Using bundled conductors
(*c*) Using thick conductors
(*d*) Using thin conductors.

358. A voltmeter gives 120 oscillations per minute when connected to the rotor. The stator frequency is 50 Hz. The slip of the motor is
(*a*) 2% (*b*) 4%
(*c*) 5% (*d*) 2.5%.

359. The phase comparators in case of static relays and electro-mechanical relays normally are :
(*a*) Sine and cosine comparators respectively
(*b*) Cosine and sine comparators respectively
(*c*) Both are cosine comparators
(*d*) Both are sine comparators.

360. A synchronous alternator is connected to an infinite bus bar and is initially operating at lead p.f. If the steam input is increased
(*a*) the p.f. increases (*b*) the current delivered increases
(*c*) the frequency increases (*d*) (*a*) and (*c*).

361. A balanced 3-phase, 3-wire supply feeds balanced star connected resistors. If one of the resistors is disconnected, then the percentage reduction in load will be
(*a*) $33\frac{1}{3}$ (*b*) 50
(*c*) $66\frac{2}{3}$ (*d*) 75.

362. The most suitable C.B. for short line fault without switching resistor is
(*a*) Air blast C.B. (*b*) M.O.C.B.
(*c*) SF_6 Breaker (*d*) None of the above.

363. A transmission line of 80 km length is operating at 400 Hz, it can be classified as :
(*a*) Short length line (*b*) Medium length line
(*c*) Long length line (*d*) All of the above.

364. The most efficient torque-producing actuating structure for induction-type relays is
(*a*) shaded-pole structure (*b*) watt-hour-meter structure
(*c*) induction-cup structure (*d*) single-induction-loop structure.

365. The capacitance of a 3-core cable between any two conductors with sheath earthed is 2 μF. The capacitance per phase will be
(*a*) 1 μF (*b*) 4 μF
(*c*) 0.667 μF (*d*) 1.414 μF.

366. Control rods used in nuclear reactors are made of
(*a*) zirconium (*b*) boron
(*c*) beryllium (*d*) lead.

367. For a given power delivered, if the working voltage of a distributor line is increased to *n* times, the cross-sectional area *A* of the distributor line, would be reduced to
(*a*) $\frac{1}{n}A$ (*b*) $\frac{1}{n^2}A$
(*c*) $\frac{1}{2n^2}A$ (*d*) $\frac{1}{2n}A$.

368. A 100 kVA transformer has maximum efficiency of 98% when operating at half full load. Its full load losses are
(*a*) Cu = 2.04 kW and Iron loss = 0.51 kW (*b*) Cu = 4.08 kW and Iron loss = 1.02 kW
(*c*) Cu = 3.0 kW and Iron loss = 0.75 kW (*d*) None of the above.

369. The charging reactance of 50 km length of line is 1500 Ω. The charging reactance for 100 km length of line will be
(*a*) 1500 Ω (*b*) 3000 Ω
(*c*) 750 Ω (*d*) 600 Ω.

370. The line trap unit employed in carrier current relaying
(*a*) offers high impedance to 50 Hz power frequency signals
(*b*) offers high impedance to carrier frequency signals
(*c*) offers low impedance to carrier frequency signals
(*d*) offers low impedance to carrier frequency signals.

371. A generator is connected to a synchronous motor. From stability point of view it is preferable to have
(*a*) generator neutral reactance grounded and motor neutral resistance grounded
(*b*) generator and motor neutrals resistance grounded
(*c*) generator and motor neutrals reactance grounded
(*d*) generator neutral resistance and motor neutral reactance grounded.

372. An alternator of 300 kW is driven by a prime mover of speed regulation 4% and another alternator of 200 kW driven by a prime mover of speed regulations 3%, the total load they can take is
(*a*) 500 kW (*b*) 567 kW
(*c*) 425 kW (*d*) 257 kW.

373. The most appropriate operating speeds in rpm of generators used in Thermal, Nuclear and Hydro-power plants would respectively be
(*a*) 3000, 300 and 1500 (*b*) 3000, 3000 and 300
(*c*) 1500, 1500 and 3000 (*d*) 1000, 900 and 750.

374. For complete protection of a 3-phase line
(*a*) three-phase and three-earth fault relays are required
(*b*) three-phase and two-earth fault relays are required
(*c*) two-phase and two-earth fault relays are required
(*d*) two-phase and one-earth fault relays are required.

375. Induction generator works between the slip
(*a*) $1 < s < 2$ (*b*) $0.1 < s < 1.0$
(*c*) $s < 0.0$ (*d*) none of the above.

376. Tick out the correct statement :

(*a*) The negative and zero sequence voltages are maximum at the fault point and decrease towards the neutral

(*b*) The negative and zero sequence voltages are minimum at the fault point and increase towards the neutral

(*c*) The negative sequence is maximum and zero sequence minimum at the fault point and decrease and increase respectively towards the neutral

(*d*) None of the above.

377. If the frequency of input voltage of a transformer is increased keeping the magnitude of voltage uncharged, then

(*a*) both hysteresis loss and eddy current loss in the core will increase

(*b*) hysteresis loss will increase but eddy current loss will decrease

(*c*) hysteresis loss will decrease but eddy current loss will increase

(*d*) hysteresis loss will decrease but eddy current loss will remain unchanged.

378. The magnetising-inrush-current in a transformer is rich in

(*a*) 3rd harmonics (*b*) 5th harmonics

(*c*) 7th harmonics (*d*) 2nd harmonics.

379. Tick out the correct one :

(*a*) Higher the SCR of a machine the heavier is its rotor

(*b*) Higher the SCR of a machine the lighter is its rotor

(*c*) The SCR and size of the rotor are not at all related

(*d*) None of the above.

380. For the stable operation of interconnected system, the passive element that can be used as interconnecting element is :

(*a*) Reactor (*b*) Resistor

(*c*) Capacitor (*d*) Resistor and capacitor.

381. If the normal system frequency is 50 Hz and if it is operating at 53 Hz, the equipment on the system, most adversely affected is :

(*a*) Power transformer (*b*) Alternator

(*c*) Turbine (*d*) All the above are equally affected.

382. For a two-bus system if the change in load at bus 2 is 5 MW the corresponding change in generation at bus 1 is 8 MW, the penalty factor of bus 1 is

(*a*) 0.6 (*b*) 1.67

(*c*) 0.625 (*d*) None of the above.

383. If transformer frequency is changed from 50 HZ to 60 HZ, the ratio of eddy current loss 50 HZ to 60 HZ at constant voltage is

(*a*) 5/6 (*b*) 25/36

(*c*) 6/5 (*d*) 1.0.

384. Corona loss is less when the shape of the conductor is :

(*a*) Circular (*b*) Flat

(*c*) Oval (*d*) Independent of shape.

385. The sending end voltage of a feeder with reactance 0.2 p.u. is 1.2 p.u. If the reactive power supplied at the receiving end of the feeder is 0.3 p.u., the approximate drop of volts in the feeder is :

(*a*) 0.2 p.u. (*b*) 0.06 p.u.

(*c*) 0.05 p.u. (*d*) 0.072 p.u.

386. In a 3-phase extra-high voltage cable, a metallic screen around each core-insulation is provided to

(*a*) facilitate heat dissipation
(*b*) give mechanical strength
(*c*) obtain radial electric stress
(*d*) obtain longitudinal electric stress.

387. If a synchronous machine is underexcited it takes lagging vars from the system when it is operated as a

(*a*) synchronous motor
(*b*) synchronous generator
(*c*) synchronous motor as well as generator
(*d*) none of the above.

388. A voltage of 1000 kV is applied to an overhead line with its receiving-end open. If the surge impedance of the line is 500 ohms, then the total surge power in the line is

(*a*) 2000 MW
(*b*) 500 MW
(*c*) 2 MW
(*d*) 0.5 MW.

389. Transmission lines are transposed to

(*a*) reduce copper loss
(*b*) reduce skin effect
(*c*) prevent interference with neighbouring telephone lines
(*d*) prevent short-circuit between any two lines.

390. Hollow conductors are used in transmission lines to

(*a*) reduce weight of copper
(*b*) improve stability
(*c*) reduce corona
(*d*) increase power transmission capacity.

391. Series compensation on EHV lines is resorted to

(*a*) improve the stability
(*b*) reduce the fault level
(*c*) improve the voltage profile
(*d*) as a substitute for synchronous phase modifier.

392. The effect of increasing gating angle in thyristorised controlled reactor is

(*a*) to decrease power loss in thyristor controller
(*b*) to decrease power loss in the reactor
(*c*) to make current wave form less sinusoidal
(*d*) all of the above.

393. If the penalty factor of a plant is unity, its incremental transmission loss is

(*a*) 1.0
(*b*) – 1.0
(*c*) Zero
(*d*) None of the above.

394. A synchronous machine has higher capacity for

(*a*) leading p.f.
(*b*) lagging p.f.
(*c*) it does not depend upon the p.f. of the machine
(*d*) it depends upon the p.f. of the load.

395. The inertia constants of two groups of machines which do not swing together are M_1 and M_2. The equivalent inertia constant of the system is :

(*a*) $M_1 + M_2$
(*b*) $M_1 - M_2$ if $M_1 > M_2$
(*c*) $\dfrac{M_1M_2}{M_1 + M_2}$
(*d*) $\sqrt{M_1M_2}$.

396. The effect of series capacitance compensation is
(*a*) to decrease the virtual surge impedance
(*b*) to decrease the effective length of the line
(*c*) to increase virtual surge impedance loading
(*d*) all of the above.

397. If a line is 100% series compensated it may result into series resonance at
(*a*) power frequency (50 Hz) (*b*) 100 Hz
(*c*) 150 Hz (*d*) none of the above.

398. A synchronous motor connected to an infinite bus takes power at a lag p.f. If its excitation is increased
(*a*) the terminal voltage increases (*b*) the load angle increases
(*c*) the p.f. of motor increases (*d*) (*c*) and (*b*).

399. The phase angle error of a current transformer can be reduced by connecting an element across its secondary which is predominantly
(*a*) resistive (*b*) inductive
(*c*) capacitive (*d*) independent of the nature of element.

400. The reflection coefficient of a short-circuited line for voltage is
(*a*) – 1 (*b*) + 1
(*c*) 0.5 (*d*) zero.

401. The normal practice to specify the making current of a circuit breaker is in terms of
(*a*) r.m.s. value (*b*) Peak value
(*c*) Average value (*d*) Both r.m.s. and peak value.

402. The inertia constant H of a synchronous condenser is
(*a*) greater than that of hydro generator (*b*) greater than that of turbo alternator
(*c*) equal to that of turbo alternator (*d*) none of the above.

403. The most economic load on an underground cable is
(*a*) Greater than its surge loading (*b*) Less than the surge loading
(*c*) Equal to the surge loading (*d*) None of the above is necessary.

404. The rotation of disc of an induction disc relay under the poles is
(*a*) from unshaded pole to shaded pole
(*b*) from shaded pole to unshaded pole
(*c*) it depends upon the magnitude of current
(*d*) it depends upon the C.T. secondary connection.

405. A single-phase transmission line of impedance j 0.8 ohm supplies a resistive load of 500 A at 300 V. The sending-end power factor is
(*a*) unity (*b*) 0.8 lagging
(*c*) 0.8 leading (*d*) 0.6 lagging.

406. The rate of rise of resisting voltage depend upon :
(*a*) The type of circuit breaker
(*b*) The inductance of the system only
(*c*) The capacitance of the system only
(*d*) The inductance and capacitance of the system.

407. In case of a 3-phase short circuit in a system, the power fed into the system is
(*a*) mostly reactive (*b*) mostly active
(*c*) active and reactive both equal (*d*) reactive only.

408. If the effect of earth is taken into account, then the capacitance of line to ground

(*a*) decreases (*b*) increases

(*c*) remains unaltered (*d*) becomes infinite.

409. In the optimum generator scheduling of different power plants, the minimum fuel cost is obtained when

(*a*) only the incremental fuel cost of each plant is the same

(*b*) the penalty factor of each plant is the same

(*c*) the ratio of the incremental fuel cost to the penalty factor of each plant is the same

(*d*) the incremental fuel cost of each plant multiplied by its penalty factor is the same.

410. To limit current chopping in vacuum circuit breakers, the contact material used has

(*a*) high vapour pressure and low conductivity properties

(*b*) high vapour pressure and high conductivity properties

(*c*) low vapour pressure and high conductivity properties

(*d*) low vapour pressure and low conductivity properties.

411. Where voltages are high and current to be interrupted is low the breaker preferred is

(*a*) air blast C.B. (*b*) oil C.B.

(*c*) vacuum C.B. (*d*) any one of the above.

412. A 60 Hz 320 km lossless line has sending end voltage 1.0 p.u. the receiving end voltage on no load is

(*a*) 1.1 p.u. (*b*) 1.088 p.u.

(*c*) 1.116 p.u. (*d*) none of the above.

413. With the help of a reactive compensator it is possible to have

(*a*) constant voltage operation only (*b*) unity p.f. operation only

(*c*) both constant voltage and unity p.f. (*d*) either constant voltage or unity p.f.

414. The bus-bars of each of the two alternators of 15% reactance each, are interconnected through tie-bar reactors of 15% each. The equivalent impedance to fault current for a 3-phase fault in any alternator bus-bar will be

(*a*) 75% (*b*) 10%

(*c*) 11.25% (*d*) 15%.

415. An alternator has a phase sequence of *RYB* for its phase voltages. In case the field current is reversed, the phase sequence will become

(*a*) *RBY* (*b*) *RYB*

(*c*) *YRB* (*d*) None of the above.

416. A transmission line of 210 km length has certain values of parameters *A*, *B*, *C*, *D*. If the length is made 100 km then the parameter

(*a*) *A* increase *B* decreases (*b*) *A* decreases *B* also decreases

(*c*) *A* and *B* both increase (*d*) *A* decreases *B* increases.

417. The voltages across the various discs of a string of suspension insulators having identical discs is different due to

(*a*) surface leakage currents (*b*) series capacitance

(*c*) shunt capacitances to ground (*d*) series and shunt capacitances.

418. Tick out the correct statement :

(*a*) The higher the initial load, the larger the critical clearing angle.

(*b*) The higher the initial load, the lower the critical clearing angle.

(*c*) The initial load has nothing to do with the critical clearing angle.
(*d*) The higher the operating time of the circuit breaker, the large will be the critical clearing angle.

419. Arcing on transmission lines is prevented by connecting a suitable
(*a*) circuit breaker (*b*) protective relay
(*c*) inductor in the neutral (*d*) capacitor in the neutral.

420. A Buchholz relay is used for
(*a*) protection of a transformer against all internal faults
(*b*) protection of a transformer against all external faults
(*c*) protection of a transformer against both internal and external faults
(*d*) protection of induction motors.

421. For protection of rotating machines against lightning surges a combination of
(*a*) lightning conductor and capacitor (*b*) lightning conductor and lightning arrester
(*c*) lightning arrester and capacitor (*d*) lightning arrester alone is used.

422. For a synchronous phase modifier the load angle is
(*a*) 0° (*b*) 25°
(*c*) 30° (*d*) None of the above.

423. If one phase of supply goes off in the case of 3-phase induction motor, the motor
(*a*) comes to a stop
(*b*) draws double the initial current and continues to run
(*c*) draws $1/\sqrt{3}$ times tho initial current and continues to run
(*d*) none of the above.

424. A salient pole synchronous motor is running at no load, its field current is switched off. The motor will
(*a*) come to a stop
(*b*) continue to run at synchronous speed
(*c*) continue to run at a speed slightly less than synchronous speed
(*d*) none of these.

425. The in-rush current of a transformer at no load is maximum if the supply voltage is switched on
(*a*) at zero voltage value (*b*) at peak voltage value
(*c*) at $V/2$ value (*d*) at $\sqrt{3/2}\ V$ value.

426. Tick out the correct one :
(*a*) The insulators and lightning arresters should have high impulse ratio.
(*b*) The insulators and lightning arresters should have low impulse ratio.
(*c*) The insulator should have high impulse ratio and lightning arrester low.
(*d*) The lightning arrester should have high impulse ratio but insulator low.

427. For successful parallel operation of two single-phase transformers, the most essential condition is that their
(*a*) percentage impedances are equal (*b*) polarities are properly connected
(*c*) turns ratios are exactly equal (*d*) kVA ratings are equal.

428. A 2 kVA transformer has iron loss of 150 Watts and full-load copper loss of 250 Watts. The maximum efficiency of the transformer would occur when the total loss is

(*a*) 500 W (*b*) 400 W
(*c*) 300 W (*d*) 275 W.

429. The insulation resistance of a cable of length 10 km is 1 MΩ, its resistance for 50 km length will be
(*a*) 1 MΩ (*b*) 5 MΩ
(*c*) 0.2 MΩ (*d*) None of the above.

430. The size of conductor on modern EHV lines is obtained based on
(*a*) voltage drop (*b*) current density
(*c*) corona (*d*) (*a*) and (*b*).

431. In a power station, the cost of generation of power reduces most effectively when
(*a*) diversity factor alone increases (*b*) both diversity factor and load factor increase
(*c*) load factor along increases (*d*) both diversity factor and load factor decrease.

432. A thyrite type lightning arrester
(*a*) blocks the surge voltage appearing in a line
(*b*) absorbs the surge voltage appearing in a line
(*c*) offers a low resistance path to the surge appearing in line
(*d*) returns the surge back to the source.

433. If the fault current is 2 kA, the relay setting is 50% and the C.T. ratio is 400/5, then the plug setting multiplier of a relay will be
(*a*) 5 (*b*) 7
(*c*) 8 (*d*) 10.

434. Compared with a solid conductor of the same radius, corona appears on a stranded conductor at a lower voltage, because stranding
(*a*) assists ionisation
(*b*) makes the current flow spirally about the axis of the conductor
(*c*) produces oblique sections to a plane perpendicular to the axis of the conductor
(*d*) produces surfaces of smaller radius.

435. A 100 V/10 V, 50 VA transformer is converted to 100 V/110 V auto transformer, the rating of the auto transformer is
(*a*) 550 VA (*b*) 500 VA
(*c*) 110 VA (*d*) 100 VA.

436. A synchronous generator connected to an infinite bus delivers power at a lag p.f. If its excitation is increased
(*a*) the terminal voltage increases (*b*) voltage angle δ increases
(*c*) Current delivered increases (*d*) (*b*) and (*c*).

437. A 100 km transmission line is designed for a nominal voltage of 132 kV and consists of one conductor per phase. The line reactance is 0.726 ohm/km. The static transmission capacity of the line, in megawatts, would be
(*a*) 132 (*b*) 240
(*c*) 416 (*d*) 720.

438. An RLC series circuit remains predominantly inductive
(*a*) at resonance frequency (*b*) below resonance frequency
(*c*) above resonance frequency (*d*) at the lower half power frequency.

439. The stability of arc in vacuum depends upon
(*a*) the contact material only (*b*) the contact material and its vapour pressure
(*c*) the circuit parameters only (*d*) the combination of (*b*) and (*c*).

440. If a combination of HRC fuse and circuit breaker is used, the C.B. operates for
(*a*) Low overload currents (*b*) Short circuit current
(*c*) Under all abnormal current (*d*) The combination is never used in practice.

441. The typical value of SCR for modern alternators is
(*a*) 1.5 (*b*) 1.2
(*c*) 1.0 (*d*) 0.5.

442. The main consideration for higher and higher operating voltage of transmission is to
(*a*) increase efficiency of transmission (*b*) reduce power losses
(*c*) increase power transfer capability (*d*) (*a*) and (*b*).

443. SF_6 gas has excellent heat transfer properties because of its
(*a*) higher molecular weight (*b*) low gaseous viscosity
(*c*) higher dielectric strength (*d*) a combination of (*a*) and (*b*)
(*e*) a combination of (*b*) and (*c*).

444. The insulation of modern EHV lines is designed based on
(*a*) the lightning voltage (*b*) corona
(*c*) radio interference (*d*) switching voltage.

445. Effect of increase in temperature in overhead transmission lines is to
(*a*) increase the stress and length
(*b*) decrease the stress and length
(*c*) decrease the stress but increase the length
(*d*) none of the above.

446. For a long uncompensated line the limit to the line loading is governed by
(*a*) thermal limit (*b*) voltage drop
(*c*) stability limit (*d*) corona loss.

447. For any fixed degree of series compensation additional capacitive shunt compensation
(*a*) increases the effective length of line
(*b*) increases virtual surge impedance of line
(*c*) decreases virtual surge impedance loading of the line
(*d*) (*b*) and (*c*).

448. Standard impulse testing of a power transformer requires
(*a*) two applications of chopped wave followed by one application of a full wave
(*b*) one application of chopped wave followed by one application of a full wave
(*c*) one application of chopped wave followed by two applications of a full wave
(*d*) none of the above.

449. The order of the lightning discharge current is
(*a*) 10,000 amp. (*b*) 100 amp.
(*c*) 1 amp. (*d*) 1 micro ampere.

450. A short length of cable between the dead end tower and the power transformer
(*a*) always reduces the steepness of the incident wave
(*b*) reduces the steepness of the wave under certain conditions only
(*c*) it does not change the steepness
(*d*) none of the above.

451. If δ is the loss angle of the cable, its power factor is
(*a*) sin δ
(*b*) cos δ
(*c*) power factor is independent of δ
(*d*) power factor depends upon (δ) but it is not as per (*a*) and (*b*).

452. A reactance relay is
(*a*) voltage restrained directional relay
(*b*) directional restrained over-current relay
(*c*) voltage restrained over-current relay
(*d*) none of the above.

453. The impedance of a circuit is given by $z = 3 + j4$, its conductance is given by
(*a*) 1/3
(*b*) 3/5
(*c*) 3/25
(*d*) 4/5.

454. A large-size alternator is protected against overloads by providing
(*a*) overcurrent relays
(*b*) temperature sensitive relays
(*c*) thermal relay
(*d*) a combination of (*a*) and (*b*)
(*e*) a combination of (*b*) and (*c*).

455. An overhead line with surge impedance 400 ohms is terminated through a resistance *R*. A surge travelling over the line does not suffer any reflection at the junction if the value of *R* is
(*a*) 20 ohms
(*b*) 200 ohms
(*c*) 800 ohms
(*d*) None of the above.

456. A Mho relay is a
(*a*) voltage restrained directional relay
(*b*) voltage controlled over current relay
(*c*) directional restrained over current relay
(*d*) directional restrained over voltage relay.

457. For measuring positive, negative and zero sequence voltages in a system, the reference is taken as :
(*a*) Neutral of the system only
(*b*) Ground only
(*c*) For zero sequence neutral and for positive and negative the ground
(*d*) None of the above.

458. The current chopping tendency is minimised by using the SF_6 gas at relatively
(*a*) high pressure and low velocity
(*b*) high pressure and high velocity
(*c*) low pressure and low velocity
(*d*) low pressure and high velocity.

459. For stability and economic reasons we operate the transmission line with power angle in the range
(*a*) 10° to 25°
(*b*) 30° to 45°
(*c*) 60° to 75°
(*d*) 65° to 80°.

460. The leakage resistance of a 50 km long cable is 1 MΩ. For a 100 km long cable it will be
(*a*) 1 MΩ
(*b*) 2 MΩ
(*c*) 0.66 MΩ
(*d*) None of the above.

461. A synchronous phase modifier as compared to a synchronous motor used for mechanical loads has
(*a*) larger shaft and higher speed
(*b*) smaller shaft and higher speed
(*c*) larger shaft and smaller speed
(*d*) smaller shaft and smaller speed.

462. Shunt compensation in an EHV line is used to
(*a*) improve stability
(*b*) reduce fault level
(*c*) improve the voltage profile
(*d*) substitute for synchronous phase modifier.

463. A lightning arrestor connected between the line and earth in a power system
(*a*) protects the terminal equipment against travelling surges
(*b*) protects the transmission line against lightning stroke
(*c*) suppresses high frequency oscillations in the line
(*d*) reflects back the travelling wave approaching it.

464. The flow-duration curve at a given head of a hydro-electric plant is used to determine the
(*a*) total power available at the site
(*b*) total units of energy available
(*c*) load-factor at the plant
(*d*) diversity-factor for the plant.

465. In order to eliminate sheath losses, a successful method is
(*a*) to transpose the cable along with cross bonding.
(*b*) transpose the cables only.
(*c*) cross bonding the cables is enough.
(*d*) none of the above is effective.

466. Gas turbines can be brought to the bus bar from cold in about
(*a*) 2 minutes
(*b*) 30 minutes
(*c*) 1 Hour
(*d*) 2 Hours.

467. The non-uniform distribution of voltage across the units in a string of suspension type insulators is due to
(*a*) unequal self-capacitance of the units
(*b*) non-uniform distance of separation of the units from the tower body
(*c*) the existence of stray capacitance between the metallic junctions of the units and the tower body
(*d*) non-uniform distance between the cross-arm and the units.

468. High voltage d.c. testing for HV machines is resorted because
(*a*) certain conclusion regarding the continuous ageing of an insulation can be drawn
(*b*) the stress distribution is a representation of the service condition
(*c*) standardisation on the magnitude of voltage to be applied is available
(*d*) the stresses do not damage the coil end insulation.

469. The voltage of a particular bus can be controlled by controlling the :
(*a*) phase angle
(*b*) reactive power of the bus
(*c*) active power of the bus
(*d*) phase angle and reactive power.

470. The capacitor switching is easily done with :
(*a*) Air blast circuit breaker
(*b*) Oil C.B.
(*c*) Vacuum C.B.
(*d*) Any one of the above.

471. The motor which can be used on both a.c. and d.c. is :
(*a*) Reluctance motor
(*b*) Induction motor
(*c*) d.c. series motor
(*d*) None of the above.

472. If a voltage-controlled bus is treated as load bus, then which one of the following limits would be violated ?
(*a*) Voltage
(*b*) Active power
(*c*) Reactive power
(*d*) Phase angle.

473. Stringing chart is useful for
(*a*) finding the sag in the conductor
(*b*) in the design of tower
(*c*) in the design of insulator string
(*d*) finding the distance between the tower.

474. The insulation of the modern EHV lines is designed based on :
(*a*) The lighting voltage.
(*b*) The switching voltage.
(*c*) Corona.
(*d*) *RI*.

475. The design of insulation for systems above 400 kV, is based upon
(*a*) lightning overvoltage
(*b*) switching surges
(*c*) system voltage level
(*d*) system load level.

476. A synchronous generator is connected to an infinite bus bar and is initially operating at a lag p.f. If the steam input to the alternator is increased
(*a*) the p.f. of the alternator improves
(*b*) reactive power decreases
(*c*) the frequency increase
(*d*) (*a*) and (*b*).

477. The coefficient of reflection of voltage for a short circuited line is
(*a*) 1.0
(*b*) – 1.0
(*c*) 0
(*d*) 2.0.

478. A power system network with a capacity of 100 MVA has a source impedance of 10% at a point. The fault level at that point is
(*a*) 10 MVA
(*b*) 30 MVA
(*c*) 3000 MVA
(*d*) 1000 MVA.

479. The coefficient of reflection for current for an open ended line is
(*a*) 1.0
(*b*) 0.5
(*c*) – 1.0
(*d*) Zero.

480. A large ac generator, supplying power to an infinite bus, has a sudden short-circuit occurring at its terminals. Assuming the prime mover input and the voltage behind the transient reactance to remain constant immediately after the fault, the acceleration of the generator rotor is
(*a*) inversely proportional to the moment of inertia of the machine
(*b*) inversely proportional to the square of the voltage
(*c*) directly proportional to the square of the short-circuit current
(*d*) directly proportional to the short-circuit power.

481. A shunt fault is characterized by :
(*a*) Increase in current, frequency and p.f.
(*b*) Increase in current reduction in frequency and p.f.
(*c*) Increase in current and frequency but reduction in p.f.
(*d*) None of the above.

482. For protection of parallel feeders fed from one end the relays required are :
(*a*) Non-directional relays at the source end and directional relays at the load end.
(*b*) Non-directional relays at both the ends.
(*c*) Directional relays at the source end and non-directional at the load end.
(*d*) Directional relays at both the ends.

483. Pure inductive circuit takes power (reactive) from the a.c. line when
(*a*) both applied voltage and current rise
(*b*) both applied voltage and current decrease
(*c*) applied voltage decreases but current increases
(*d*) (*a*) and (*b*).

484. The operation of the relay which is most affected due to arc resistance is
(*a*) mho relay (*b*) reactance relay
(*c*) impedance relay (*d*) all are equally affected.

485. For the same rupturing capacity, the actual current to be interrupted by an HRC fuse is
(*a*) much less than any CB (*b*) much more than any CB
(*c*) equal to the CB (*d*) none of the above is necessary.

486. Resistance switching is normally resorted in case of :
(*a*) Bulk oil circuit breakers (*b*) Minimum oil circuit breakers
(*c*) Air blast circuit breakers (*d*) All types of breakers.

487. If two synchronous generators are connected, loss of synchronism will result in
(*a*) stalling of generators (*b*) wild fluctuations in current
(*c*) wild fluctuations in current and voltage (*d*) none of the above.

488. An alternator is delivering a balanced load at unity p.f. The phase angle between line voltage and line current is :
(*a*) 90° (*b*) 60°
(*c*) 30° (*d*) 0°.

489. With 100% series compensation of lines
(*a*) the circuit is series resonant at power frequency
(*b*) low transient voltage
(*c*) high transient current
(*d*) (*a*) and (*c*).

490. Carrier current protection scheme is normally used for :
(*a*) HV transmission lines only (*b*) HV cables only
(*c*) HV transmission and cables.

491. Corona loss increases with
(*a*) increase in supply frequency and conductor size.
(*b*) increase in supply frequency but reduction in conductor size.
(*c*) decrease in supply frequency and conductor size.
(*d*) decrease in supply frequency but increase in conductor size.

492. A voltage-controlled bus is treated as a load bus in subsequent iteration when its
(*a*) voltage limit is violated (*b*) active power limit is violated
(*c*) reactive power limit violated (*d*) phase angle limit is violated.

493. Sheaths are used in cables to
(*a*) provide proper insulation (*b*) provide mechanical strength
(*c*) prevent ingress of moisture (*d*) none of the above.

494. For effective use of a counterpoise wire
(*a*) its leakage resistance should be greater than the surge impedance
(*b*) its leakage resistance should be less than the surge impedance
(*c*) its leakage resistance should be equal to the surge impedance
(*d*) the two resistances may have any relation.

495. With 100% inductive shunt compensation the voltage profile is flat for
(*a*) 100% loading of line (*b*) 50% loading of line
(*c*) zero loading of line (*d*) none of the above.

496. For economic operation, the generator with highest positive incremental transmission loss will operate at

(*a*) the lowest positive incremental cost of production

(*b*) the lowest negative incremental cost of production

(*c*) the highest positive incremental cost of production

(*d*) none of the above.

497. The connected load of a consumer is 2 kW and his maximum demand is 1.5 kW. The load factor of the consumer is

(*a*) 0.75 (*b*) 0.375

(*c*) 1.33 (*d*) None of the above.

498. The incremental transmission loss of a plant is

(*a*) positive always (*b*) negative always

(*c*) can be positive or negative (*d*) all of these.

499. Efficiency of a power transformer under no load condition is approximately

(*a*) 75% (*b*) 50%

(*c*) 25% (*d*) none of the above.

500. The cost of generation is theoretically minimum if

(*a*) the system constraints are considered (*b*) the operational constraints are considered

(*c*) (*a*) and (*b*) (*d*) the constraints are not considered.

501. For the same voltage boost, the reactive power capacity is more for a :

(*a*) shunt capacitor (*b*) series capacitor

(*c*) it is same for both series and shunt (*d*) any of these.

502. The relay which is most sensitive to power swings (maloperation) is

(*a*) mho relay (*b*) reactance relay

(*c*) impedance relay (*d*) all are equally affected.

503. A lossless line terminated with its surge impedance has

(*a*) flat voltage profile

(*b*) transmission line angle is greater than actual length of the line

(*c*) transmission line angle is less than the actual length

(*d*) (*a*) and (*b*).

504. If the phase angle of the voltage coil of a directional relay is 50° the maximum torque angle of the relay is

(*a*) 130° (*b*) 100°

(*c*) 25° (*d*) None of the above.

505. A synchronous machine has its field winding on the stator and armature winding on the rotor. Under steady running condition, the air gap field

(*a*) rotates at synchronous speed w.r.t. stator

(*b*) rotates at synchronous speed in the direction of rotation of the rotor

(*c*) remains stationary w.r.t. stator

(*d*) remains stationary w.r.t. rotor.

506. The unit protection scheme provides

(*a*) primary protection (*b*) back up protection

(*c*) simultaneous protection (*d*) remote protection.

507. An overhead line with series compensation is protected using :
(*a*) Impedance relay. (*b*) Reactance relay.
(*c*) Mho relay. (*d*) None of the above.

508. Phase modifier is normally installed in the case of :
(*a*) Short transmission lines. (*b*) Medium length lines.
(*c*) Long length lines. (*d*) For all length lines.

509. Ferranti effect on long overhead lines is experienced when it is
(*a*) lightly loaded. (*b*) on full load at unity p.f.
(*c*) on full load at 0.8 p.f. lag. (*d*) in all these cases.

510. Three phase transformers which can not be connected in parallel are
(*a*) YY with $\Delta\Delta$ (*b*) $Y\Delta$ with ΔY
(*c*) YY with Y (*d*) (*b*) and (*c*).

511. Standard voltage ratings of the electromagnetic relay are
(*a*) 110 V, 63.5 (L-N) (*b*) 220 V 110 V
(*c*) 230 V 154 V (*d*) 400 V 231 V.

512. If a transformer is switched on to a voltage more than the rated voltage, for a certain loading condition (lag p.f.)
(*a*) its p.f. will improve (*b*) its p.f. will deteriorate
(*c*) no effect on p.f. (*d*) it may or may not improve.

513. The presence of earth in case of overhead lines
(*a*) increases the capacitance (*b*) increases the inductance
(*c*) decreases the capacitance (*d*) decreases the inductance.

514. The load flow solution is always assured in case of ;
(*a*) Newton-Raphson method. (*b*) Gauss method.
(*c*) Gauss-Seidel method. (*d*) None of these methods guarantees.

515. Auto reclosing is used in case of :
(*a*) Lightning arrester. (*b*) Air blast C.B.
(*c*) Bulk oil C.B. (*d*) Minimum oil C.B.

516. A distance relay is said to be inherently directional if its characteristic on *R-X* diagram
(*a*) is a straight line of-set from the origin
(*b*) is a circle that passes through the origin
(*c*) is a circle that encloses the origin
(*d*) always a separate directional relay is required.

517. Corona loss can be reduced by the use of hollow conductor, because
(*a*) the current density is reduced
(*b*) the eddy-current in the conductor is eliminated
(*c*) for a given cross-section, the radius of the conductor is increased
(*d*) of better ventilation in the conductor.

518. For certain geometry and operating voltage of the uncompensated transmission line the ratio of power transfer capability to the surge impedance loading with increase in length
(*a*) increases (*b*) remains unchanged
(*c*) decreases (*d*) no fixed criterion.

519. In a synchronous machine, in case the axis of field flux is in line with the armature flux, the machine is working

(a) as synchronous motor
(b) as synchronous generator
(c) as floating machine
(d) the m/c will not work.

520. For reducing tower footing resistance it is better to use :
(a) chemical and ground rods only.
(b) chemical and counterpoise only.
(c) ground rod and counterpoise only.
(d) chemical, ground rods and counterpoise.

521. Mho relay is normally used for the protection of :
(a) Long transmission lines.
(b) Medium length lines.
(c) Short length lines.
(d) No length criterion.

522. Load flow study is carried out for :
(a) Fault calculations.
(b) Stability studies.
(c) System planning.
(d) Load frequency control.

523. The number of pilot wires required for protection 3-phase transmission lines using Translay system of protection is
(a) 6
(b) 4
(c) 3
(d) 2.

524. The shape of the disc of an induction disc relay is :
(a) Circular.
(b) Spiral.
(c) Elliptic.
(d) none of the above.

525. Four identical alternators each rated for 20 MVA, 11 kV having a subtransient reactance of 16% are working in parallel. The short-circuit level at the bus-bars is
(a) 500 MVA
(b) 400 MVA
(c) 125 MVA
(d) 80 MVA.

526. If the normal frequency is 50 Hz, the setting of the under-frequency relay for the most important load could be
(a) 49 Hz
(b) 48.5 Hz
(c) 47.5 Hz
(d) there is no grading for under frequency relays.

527. When a line-to-ground fault occurs, the current in a faulted phase is 100 A. The zero sequence current in this case will be
(a) zero
(b) 33.3 A
(c) 66.6 A
(d) 100 A.

528. Reactance relay is normally preferred for protection against :
(a) Earth faults only.
(b) Phase faults only.
(c) None of the above.

529. Impulse ratios of insulators and lightning arresters should be
(a) both low
(b) high and low respectively
(c) low and high respectively
(d) both high.

530. A 66-kV system has string insulator having five discs and the earth to disc capacitance ratio of 0.10. The string efficiency will be
(a) 89%
(b) 75%
(c) 67%
(d) 55%.

531. A 100 kVA transformer has 4% impedance and 50 kVA transformer has 3% impedance. When they are operated in parallel, which transformer will reach full load first :
(a) 3%
(b) 4%
(c) The data is insufficient to judge
(d) any of these.

532. Zero sequence currents can flow from a line into a transformer bank if the windings are in
(*a*) grounded star/delta (*b*) delta/star
(*c*) star/grounded star (*d*) delta/delta.

533. The $X : R$ ratio for 220 kV line as compared to 400 kV line is
(*a*) greater (*b*) smaller
(*c*) equal (*d*) it could be anything.

534. While using air-blast circuit breaker, current chopping is a phenomenon often observed when
(*a*) a long overhead line is switched off (*b*) a bank of capacitors is switched off
(*c*) a transformer on no-load is switched off (*d*) a heavy load is switched off.

535. Match List I (Type of relay) with List II (For the protection of) and select the correct answer using the codes given below in the Lists

List I	*List II*
A. Buchholtz relay	1. Feeder
B. Translay relay	2. Transformer
C. Negative sequence relay	3. Generator
D. Directional over current relay	4. Long Overhead line

Codes :

	A	B	C	D
(*a*)	1	2	3	4
(*b*)	2	1	3	4
(*c*)	2	1	4	3
(*d*)	1	4	2	3

536. Match List I with List II and select the correct answer using the codes given below the lists :

List I (*Equipment*)	*List II* (*Function*)
A. Circuit Breaker	1. Voltage control
B. Lightning Arrester	2. Power Control
C. Governor	3. Overvoltage protection
D. Exciter	4. Overcurrent protection

Codes :

	A	B	C	D
(*a*)	1	2	3	4
(*b*)	4	1	2	3
(*c*)	2	3	4	1
(*d*)	4	3	2	1

537. Match List I (Devices) with List II (Application) and select the correct answer using the codes given below the lists :

List I	*List II*
A. Microprocessor	1. Monitoring surge voltage
B. Breather	2. Digital protection
C. Magnetic links	3. Monitoring surge current
D. Klydonograph	4. Power transformer

Codes :

	A	B	C	D
(*a*)	4	2	1	3
(*b*)	4	2	3	1
(*c*)	2	4	1	3
(*d*)	2	4	3	1

538. Match List I with List II and select the correct answer using the codes given below the lists :

List I	*List II*
A. Thyrite arrester	1. Tower location
B. Sag template	2. Cross bonding
C. Cable sheaths	3. Restriking voltage
D. Circuit breaker	4. Non-linear resistor

Codes :

	A	B	C	D
(*a*)	4	1	3	2
(*b*)	4	1	2	3
(*c*)	1	4	3	2
(*d*)	4	3	1	2

539. Match List I (Parameter) with List II (Effect) and select the correct answer using the codes given below the lists :

List I	*List II*
A. Percent power lost in transmission	1. Decreases with system voltage
B. For a given current density the conductor size	2. Reduces with line length
C. Power handling capacity of a line at a given voltage	3. Remains independent of line length
D. Surge impedance of a transmission line	4. Increases with line length

Codes :

	A	B	C	D
(*a*)	1	2	4	3
(*b*)	3	4	2	1

(*c*)	A	B	C	D
	3	2	4	1
(*d*)	A	B	C	D
	1	4	2	3

540. Match List 1 with List 2 and select the correct answer using the codes given below the lists

List 1

A. Mho relay

B. Plain impedance relay

C. Directional relay

D. Angle impedance relay

List 2

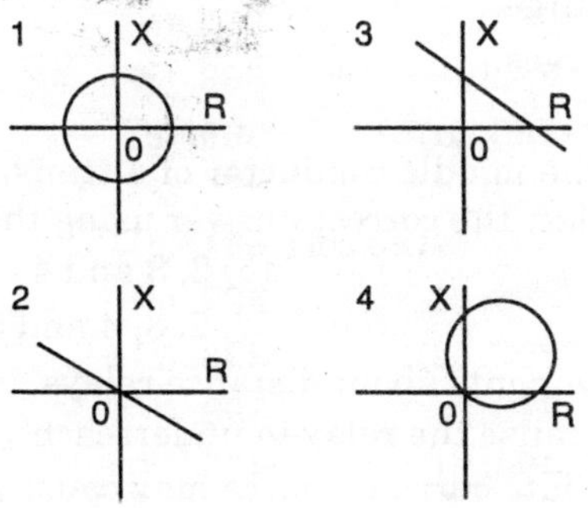

Codes :

(*a*)	A	B	C	D
	4	3	2	1
(*b*)	A	B	C	D
	4	1	2	3
(*c*)	A	B	C	D
	3	2	1	4
(*d*)	A	B	C	D
	3	2	4	1

541. Consider the following statements

The transient stability of the power system under unbalanced fault conditions can be effectively improved by

1. Excitation control.
2. phase-shifting transformer.
3. single-pole switching of circuit breakers.
4. increasing the turbine input.

Of these statements

(*a*) 1 and 2 are correct (*b*) 2 and 3 are correct

(*c*) 3 and 4 are correct (*d*) 1 and 3 are correct.

542. Consider the following statements
In the case of suspension-type insulators, the string efficiency can be improved by
1. using a longer cross arm.
2. using a guard ring.
3. grading the insulator discs.
4. reducing the cross-arm length.

Of these statements
(*a*) 1, 2 and 3 are correct (*b*) 2, 3 and 4 are correct
(*c*) 2 and 4 are correct (*d*) 1 and 3 are correct.

543. Which of the following statements regarding corona are true ?
1. It causes radio interference.
2. It attenuates lightning surge.
3. It amplifies switching surges.
4. It causes power loss.
5. It is more prevalent in the middle conductor of a transmission line employing a flat conductor configuration. Select the correct answer using the codes given below :

(*a*) 1, 3 and 5 (*b*) 2, 3 and 4
(*c*) 1, 2, 4 and 5 (*d*) 2, 3, 4 and 5.

544. Consider the following statements about distance relays
1. The effect of an arc may cause the relay to underreach
2. The effect of an intermediate current source may cause relay to over-reach
3. The effect of an intermediate load may cause the relay to under-reach.
4. The presence of dc off-set in the fault current may cause the relay to over-reach.

Of these statements
(*a*) 2, and 3 are correct (*b*) 1 and 4 are correct
(*c*) 1, 2 and 3 are correct (*d*) 2, 3 and 4 are correct.

545. In a power system, each bus or node is associated with four quantities, namely
1. real power.
2. reactive power.
3. bus-voltage magnitude and
4. phase-angle of the bus voltage.

For load-flow solution, among these four, the number of quantities to be specified is
(*a*) any one (*b*) any two
(*c*) any three (*d*) all the four.

ANSWERS

1. (*b*)	**2.** (*c*)	**3.** (*b*)	**4.** (*a*)	**5.** (*d*)	**6.** (*d*)
7. (*d*)	**8.** (*b*)	**9.** (*b*)	**10.** (*c*)	**11.** (*b*)	**12.** (*c*)
13. (*c*)	**14.** (*b*)	**15.** (*b*)	**16.** (*a*)	**17.** (*c*)	**18.** (*d*)
19. (*a*)	**20.** (*b*)	**21.** (*a*)	**22.** (*b*)	**23.** (*d*)	**24.** (*b*)
25. (*c*)	**26.** (*c*)	**27.** (*c*)	**28.** (*b*)	**29.** (*a*)	**30.** (*c*)
31. (*b*)	**32.** (*c*)	**33.** (*a*)	**34.** (*c*)	**35.** (*d*)	**36.** (*g*)
37. (*a*)	**38.** (*b*)	**39.** (*c*)	**40.** (*d*)	**41.** (*b*)	**42.** (*b*)

43. (*a*)	**44.** (*d*)	**45.** (*d*)	**46.** (*c*)	**47.** (*d*)	**48.** (*b*)
49. (*c*)	**50.** (*a*)	**51.** (*c*)	**52.** (*a*)	**53.** (*d*)	**54.** (*d*)
55. (*c*)	**56.** (*d*)	**57.** (*a*)	**58.** (*d*)	**59.** (*c*)	**60.** (*a*)
61. (*c*)	**62.** (*b*)	**63.** (*c*)	**64.** (*a*)	**65.** (*d*)	**66.** (*a*)
67. (*c*)	**68.** (*c*)	**69.** (*d*)	**70.** (*c*)	**71.** (*d*)	**72.** (*d*)
73. (*d*)	**74.** (*c*)	**75.** (*c*)	**76.** (*c*)	**77.** (*b*)	**78.** (*d*)
79. (*b*)	**80.** (*c*)	**81.** (*b*)	**82.** (*a*)	**83.** (*d*)	**84.** (*c*)
85. (*d*)	**86.** (*a*)	**87.** (*a*)	**88.** (*a*)	**89.** (*b*)	**90.** (*b*)
91. (*d*)	**92.** (*a*)	**93.** (*c*)	**94.** (*c*)	**95.** (*a*)	**96.** (*d*)
97. (*c*)	**98.** (*a*)	**99.** (*a*)	**100.** (*c*)	**101.** (*b*)	**102.** (*a*)
103. (*a*)	**104.** (*a*)	**105.** (*a*)	**106.** (*b*)	**107.** (*d*)	**108.** (*c*)
109. (*b*)	**110.** (*c*)	**111.** (*a*)	**112.** (*b*)	**113.** (*c*)	**114.** (*b*)
115. (*a*)	**116.** (*b*)	**117.** (*c*)	**118.** (*a*)	**119.** (*a*)	**120.** (*a*)

121. (*b*) **122.** (*a*) because hydraulic turbine efficiency is high

123. (*a*) **124.** (*a*)

125. (*b*) at high voltages a breakdown may occur due to the non-radial dielectric stress

126. (*a*)	**127.** (*a*)	**128.** (*a*)	**129.** (*b*)	**130.** (*a*)	**131.** (*b*)
132. (*a*)	**133.** (*a*)	**134.** (*b*)	**135.** (*c*)	**136.** (*a*)	**137.** (*b*)
138. (*a*)	**139.** (*a*)	**140.** (*b*)	**141.** (*b*)	**142.** (*c*)	**143.** (*a*)
144. (*a*)	**145.** (*b*)	**146.** (*a*)	**147.** (*b*)	**148.** (*a*)	**149.** (*b*)
150. (*b*)	**151.** (*a*)	**152.** (*b*)	**153.** (*a*)	**154.** (*b*)	**155.** (*b*)
156. (*a*)	**157.** (*a*)	**158.** (*b*)	**159.** (*a*)	**160.** (*a*)	**161.** (*c*)
162. (*b*)	**163.** (*a*)	**164.** (*c*)	**165.** (*b*)	**166.** (*b*)	**167.** (*c*)
168. (*c*)	**169.** (*b*)	**170.** (*a*)	**171.** (*b*)	**172.** (*d*)	**173.** (*d*)
174. (*b*)	**175.** (*a*)	**176.** (*b*)	**177.** (*a*)	**178.** (*c*)	**179.** (*b*)
180. (*c*)	**181.** (*a*)	**182.** (*b*)	**183.** (*c*)	**184.** (*a*)	**185.** (*c*)
186. (*a*)	**187.** (*b*)	**188.** (*c*)	**189.** (*b*)	**190.** (*a*)	**191.** (*d*)
192. (*a*)	**193.** (*a*)	**194.** (*d*)	**195.** (*b*)	**196.** (*c*)	**197.** (*a*)
198. (*a*)	**199.** (*a*)	**200.** (*b*)	**201.** (*b*)	**202.** (*c*)	**203.** (*a*)
204. (*b*)	**205.** (*a*)	**206.** (*b*)	**207.** (*c*)	**208.** (*c*)	**209.** (*c*)
210. (*a*)	**211.** (*b*)	**212.** (*a*)	**213.** (*a*)	**214.** (*a*)	**215.** (*b*)
216. (*c*)	**217.** (*b*)	**218.** (*a*)	**219.** (*c*)	**220.** (*a*)	**221.** (*b*)
222. (*d*)	**223.** (*b*)	**224.** (*c*)	**225.** (*c*)	**226.** (*a*)	**227.** (*d*)
228. (*d*)	**229.** (*a*)	**230.** (*b*)	**231.** (*d*)	**232.** (*a*)	**233.** (*c*)
234. (*c*)	**235.** (*b*)	**236.** (*a*)	**237.** (*b*)	**238.** (*c*)	**239.** (*d*)
240. (*a*)	**241.** (*b*)	**242.** (*c*)	**243.** (*d*)	**244.** (*a*)	**245.** (*b*)
246. (*a*)	**247.** (*a*)	**248.** (*b*)	**249.** (*d*)	**250.** (*a*)	**251.** (*a*)
252. (*b*)	**253.** (*a*)	**254.** (*c*)	**255.** (*c*)	**256.** (*c*)	**257.** (*b*)
258. (*c*)	**259.** (*a*)	**260.** (*a*)	**261.** (*a*)	**262.** (*a*)	**263.** (*c*)
264. (*c*)	**265.** (*c*)	**266.** (*c*)	**267.** (*a*)	**268.** (*b*)	**269.** (*a*)
270. (*c*)	**271.** (*a*)	**272.** (*b*)	**273.** (*a*)	**274.** (*a*)	**275.** (*d*)
276. (*b*)	**277.** (*a*)	**278.** (*a*)	**279.** (*d*)	**280.** (*a*)	**281.** (*b*)
282. (*d*)	**283.** (*c*)	**284.** (*c*)	**285.** (*d*)	**286.** (*b*)	**287.** (*b*)

288. (*b*)	**289.** (*a*)	**290.** (*a*)	**291.** (*c*)	**292.** (*b*)	**293.** (*a*)
294. (*c*)	**295.** (*a*)	**296.** (*a*)	**297.** (*b*)	**298.** (*b*)	**299.** (*b*)
300. (*c*)	**301.** (*b*)	**302.** (*a*)	**303.** (*d*)	**304.** (*d*)	**305.** (*a*)
306. (*a*)	**307.** (*b*)	**308.** (*a*)	**309.** (*a*)	**310.** (*d*)	**311.** (*d*)
312. (*a*)	**313.** (*d*)	**314.** (*c*)	**315.** (*a*)	**316.** (*a*)	**317.** (*c*)
318. (*c*)	**319.** (*c*)	**320.** (*c*)	**321.** (*c*)	**322.** (*b*)	**323.** (*a*)
324. (*c*)	**325.** (*c*)	**326.** (*b*)	**327.** (*a*)	**328.** (*c*)	**329.** (*b*)
330. (*d*)	**331.** (*c*)	**332.** (*b*)	**333.** (*a*)	**334.** (*d*)	**335.** (*d*)
336. (*d*)	**337.** (*b*)	**338.** (*c*)	**339.** (*b*)	**340.** (*b*)	**341.** (*d*)
342. (*a*)	**343.** (*d*)	**344.** (*c*)	**345.** (*d*)	**346.** (*b*)	**347.** (*a*)
348. (*a*)	**349.** (*a*)	**350.** (*a*)	**351.** (*c*)	**352.** (*a*)	**353.** (*c*)
354. (*c*)	**355.** (*a*)	**356.** (*b*)	**357.** (*b*)	**358.** (*b*)	**359.** (*b*)
360. (*b*)	**361.** (*b*)	**362.** (*c*)	**363.** (*c*)	**364.** (*c*)	**365.** (*b*)
366. (*b*)	**367.** (*b*)	**368.** (*a*)	**369.** (*c*)	**370.** (*b*)	**371.** (*d*)
372. (*c*)	**373.** (*b*)	**374.** (*d*)	**375.** (*c*)	**376.** (*a*)	**377.** (*d*)
378. (*d*)	**379.** (*a*)	**380.** (*a*)	**381.** (*c*)	**382.** (*d*)	**383.** (*d*)
384. (*a*)	**385.** (*c*)	**386.** (*c*)	**387.** (*c*)	**388.** (*a*)	**389.** (*c*)
390. (*c*)	**391.** (*a*)	**392.** (*d*)	**393.** (*c*)	**394.** (*a*)	**395.** (*c*)
396. (*d*)	**397.** (*a*)	**398.** (*c*)	**399.** (*b*)	**400.** (*a*)	**401.** (*b*)
402. (*d*)	**403.** (*b*)	**404.** (*a*)	**405.** (*d*)	**406.** (*d*)	**407.** (*a*)
408. (*b*)	**409.** (*d*)	**410.** (*a*)	**411.** (*c*)	**412.** (*b*)	**413.** (*d*)
414. (*c*)	**415.** (*b*)	**416.** (*a*)	**417.** (*c*)	**418.** (*b*)	**419.** (*c*)
420. (*a*)	**421.** (*c*)	**422.** (*a*)	**423.** (*d*)	**424.** (*b*)	**425.** (*a*)
426. (*c*)	**427.** (*b*)	**428.** (*c*)	**429.** (*c*)	**430.** (*c*)	**431.** (*b*)
432. (*c*)	**433.** (*d*)	**434.** (*c*)	**435.** (*a*)	**436.** (*c*)	**437.** (*b*)
438. (*c*)	**439.** (*d*)	**440.** (*a*)	**441.** (*d*)	**442.** (*c*)	**443.** (*d*)
444. (*d*)	**445.** (*c*)	**446.** (*c*)	**447.** (*a*)	**448.** (*a*)	**449.** (*a*)
450. (*a*)	**451.** (*a*)	**452.** (*b*)	**453.** (*c*)	**454.** (*b*)	**455.** (*d*)
456. (*a*)	**457.** (*d*)	**458.** (*c*)	**459.** (*b*)	**460.** (*d*)	**461.** (*b*)
462. (*c*)	**463.** (*a*)	**464.** (*a*)	**465.** (*a*)	**466.** (*a*)	**467.** (*c*)
468. (*a*)	**469.** (*b*)	**470.** (*c*)	**471.** (*c*)	**472.** (*c*)	**473.** (*a*)
474. (*b*)	**475.** (*b*)	**476.** (*d*)	**477.** (*b*)	**478.** (*d*)	**479.** (*c*)
480. (*a*)	**481.** (*b*)	**482.** (*a*)	**483.** (*c*)	**484.** (*c*)	**485.** (*a*)
486. (*c*)	**487.** (*c*)	**488.** (*c*)	**489.** (*d*)	**490.** (*a*)	**491.** (*b*)
492. (*c*)	**493.** (*b*)	**494.** (*b*)	**495.** (*c*)	**496.** (*a*)	**497.** (*d*)
498. (*c*)	**499.** (*d*)	**500.** (*d*)	**501.** (*a*)	**502.** (*b*)	**503.** (*a*)
504. (*d*)	**505.** (*c*)	**506.** (*a*)	**507.** (*d*)	**508.** (*c*)	**509.** (*a*)
510. (*c*)	**511.** (*a*)	**512.** (*b*)	**513.** (*a*)	**514.** (*d*)	**515.** (*b*)
516. (*b*)	**517.** (*c*)	**518.** (*c*)	**519.** (*c*)	**520.** (*c*)	**521.** (*a*)
522. (*c*)	**523.** (*d*)	**524.** (*b*)	**525.** (*a*)	**526.** (*c*)	**527.** (*b*)
528. (*a*)	**529.** (*b*)	**530.** (*c*)	**531.** (*a*)	**532.** (*a*)	**533.** (*b*)
534. (*c*)	**535.** (*b*)	**536.** (*d*)	**537.** (*d*)	**538.** (*b*)	**539.** (*d*)
540. (*b*)	**541.** (*c*)	**542.** (*a*)	**543.** (*c*)	**544.** (*b*)	**545.** (*b*).

Index

A

A.C. distribution system, 728
  calculations, 732
  introduction, 728
  — electric supply to our houses and factories, 731
  — service connections, 729
  — service mains, 730
  — three-phase four-wire, 728
  — voltage tolerances, 729
  three-phase unbalanced loads, 742
  with concentrated loads, 733
Ash handling, 42

B

Boilers, 45
Bus-bar arrangements, 905

C

Circuit breakers, 877
  air blast, 891
  arc, restriking and recovery voltages, 881
  arc phenomenon, 878
  — methods of arc extinction, 879
  autoreclosing, 901
  bus-bar arrangements, 905
  classification, 886
  current chopping, 883
  function, 877
  low voltage, 896
  maintenance, 901
  oil, 887
  ratings, 897
  resistance switching, 884
  sulphur hexafluoride, 893
  switchgear components, 902
  testing, 899
  vacuum, 895
  water type. 891
Constants of overhead transmission lines, 470
Capacitance of transmission lines, 496
  inductance of a transmission line, 471
  resistance of a transmission line, 494
  skin effect and proximity effect, 495
Control of load frequency, 676
Corona, 458
  advantages and disadvantages, 459
  factors affecting, 459
  important terms, 460
  loss and its control, 460
  theory, 459

D

D.C. distribution system, 680
  — fed at both ends, 689
  — fed at one end, 683
  — ring distributor, 709
  — three-wire D.C. system, 715
  — types of distributors, 681
  — types of loading, 683
Diesel engine power plant, 70
  applications, 71
  essential components, 73
  layout, 78
  operation, 77
  site selection, 71
Direct energy conversion, 223

E

Earthed transformer, 1009
Economics of power generation, 258
  cost analysis, 263
  economic load sharing, 276
  economics in plant selection, 270

location of power plant, 262
principles of power plant design, 261
tariff for electrical energy, 307
Electro gas-dynamic generators (EGD), 234
Electromagnetic effect, 590
Electrostatic effect, 592
Electrostatic mechanical generators, 234
Electric power systems, 383
— choice of transmission voltage, 408
— electric supply system, 383
— comparison of volume of conductor material, 390
— conductor size and Kelvin's Law, 409
Energy, 3
Extra high voltage transmission, 760
H.V.D.C. transmission, 763
— advantages, limitations, 764
— classification, 763
A.C. transmission, 761
— advantages, 761
— limitations, 761

F

Feed water treatment, 59
Ferranti effect, 585
Fuels, 4
calorific value, 9
gaseous, 8
liquid, 7
solid, 5
Fuel cells, 235
Fluidised bed combustion, 40
Fuses, 869
advantages and disadvantages, 869
characteristics of fuse element, 870
definition, 869
fuse element materials, 870
selection, 875
types, 872
— high voltage, 874
— low voltage, 872

G

Gas turbine power plants, 82
advantages and disadvantages, 82
applications, 82
classification, 85
components, 93
layout, 92
operation, 91
site selection, 83
various arrangements, 97
Generalised circuit constants, 541
Geothermal energy, 13
Geothermal power plants, 219

H

Hydro-electric power plant, 104
advantages and disadvantages, 105
applications, 105
auxiliaries, 130
average life of components, 131
classification, 108
— base load, 111
— high head, 109
— mini and microhydel, 113
— medium head, 110
— peak load, 111
— pumped storage, 112
— run-of-river plants with pondage, 111
— run-of-river plants without pondage, 111
— storage type, 111
combined hydro and steam power plants, 133
controls, 131
cost, 131
cost of hydro-power, 138
hydraulic turbines, 113
hydrology, 139
— flow duration curve, 145
— hydrograph, 141
— hydrologic cycle, 139
— mass curve, 145
— measurement of run-off, 140
layout, 130
preventive maintenance, 137
safety measures, 137
underground, 135

I

Inductance of a transmission line, 471
Insulation co-ordination, 998
Interconnected systems, 340
  interconnected stations, 369
  — load despatching, 378
  — load sharing, 369
  — power limit of interconnectors, 371
  — voltage, frequency and load control, 376
  parallel operation of alternators, 341

K

Kelvin's law, 409

L

Lightning, 986
  arresters, 991
  — tests, 998
Load despatching, 378

M

Magnetohydrodynamics (MHD) system, 231
Mechanical design of overhead lines, 418
  corona, 458
  main component of overhead lines, 418
  — conductors, 418
  — line supports, 423
  — overhead line insulators, 426
  sag in overhead lines, 446
  stringing chart, 457
  vibration and vibration dampers, 457

N

Neutral earthing, 1001
  advantages, 1002
  methods, 1004
  necessity, 1002
  objects, 1001
Nuclear batteries, 237
Nuclear energy, 10
Nuclear power plant, 158
  advantages and disadvantages, 193
  general aspects, 158
  — atomic mass unit, 160
  — atomic structure, 158
  — binding energy, 163
  — comparison of fission and fusion processes, 172
  — fertile materials, 168
  — fission of nuclear fuel, 169
  — isotopes, 160
  — nuclear cross-sections, 167
  — nuclear fusion, 172
  — nuclear radiation, 161
  — nuclear reactions, 166
  — radioactivity, 161
  — radioactive decay, 164
  main components, 182
  site selection, 193
Nuclear power systems, 173
Nuclear reactors, 175
  — boiling water reactor, 184
  — breeder reactor, 189
  — CANDU reactor, 184
  — gas-cooled reactor, 186
  — liquid metal cooled reactors, 188
  — pressurised water rector, 182

O

Overvoltages, 984
  internal causes, 985
  external causes, 986

P

Photovoltaic power system, 228
Power, 4
Power distribution systems, 678
  classification, 679
  requirements, 679
Power system stability, 642
  steady state stability, 644
  — maximum steady state power, 644
  — power angle diagram, 646
  transient stability, 651
  — critical clearing angle, 662
  — equal area criterion, 656
  — swing equation 652

Power systems' structure, 14
Pulverised fuel firing, 34
Protection of alternators, 947
Protection of bus-bars, 969
— differential protection, 970
— Fault bus protection, 970
Protection of lines/feeders, 971
distance, 978
differential pilot-wire, 974
time-graded overcurrent, 972
Protective relays, 913
— Buchholz, 920
— differential, 931
— distance or impedance, 935
— electromagnetic attraction, 918
— electromagnetic induction, 921
— induction type directional overcurrent, 927
— induction type directional power, 926
— induction type non-directional overcurrent, 924
— mho type distance, 941
— reactance type distance, 940
— static, 943
— translay, 934
Protection of transformers, 960
— Buchholz protection, 960
— combined leakage and overload protection, 961
— core-balance leakage protection, 961
— differential protection, 962
— interlocked overcurrent protection, 966

R

Resistance of a transmission line, 494
Rigorous method, 551
Ring distributor, 709

S

Sag in overhead lines, 446
Skin effect, 495
Solar energy, 11
Solar power plants, 209
Sources of energy, 4
Steam power plant 16
ash handling, 42
boilers, 45
burners, 39
capacity, 21
classification, 17
dust collection, 43
feed water treatment, 59
fluidised bed combustion, 40
fuel handling, 23
pulverised fuel firing, 34
selection of site, 19
steam turbines, 49
stoker firing, 28
Stringing chart, 457
Swing equation, 652
Switches, 868
Switchgear, 865
definition, 865
essential features, 866
functions, 865
principles of layout, 866
types, 866
Switchgear components, 902
Surge absorber, 996
Sub-stations, 1016
classification, 1016
functions, 1016
indoor, 1017
outdoor, 1017
transformer, 1019
Symmetrical components and fault calculations, 769
analysis of unsymmetrical faults, 832
— double line-to-ground fault, 836
— line-to-line fault, 834
— single line-to-ground fault, 832
Sequence impedances, 828
Sequence networks, 849
Symmetrical fault calculations, 770
types of faults, 770
unsymmetrical fault calculations, 798
— symmetrical components, 798
— unsymmetrical faults, 798

T

Thermionic conversion system, 226
Thermoelectric conversion system, 223
Thermo-electric power, 14
Tidal power, 12
Tidal power plants, 206
Transient stability, 651
— critical clearing angle, 662
— equal area criterion, 656
— swing equation, 652
Transmission lines, 518
capacitance, 496
classification, 518
Ferranti effect, 585
generalised circuit constants, 541
inductance, 471
long, 550
— analysis, 551
— constant voltage transmission, 566
— equivalent T-network, 564
— equivalent π-network, 565
— evaluation of, A, B, C, D constants, 553
— power circle diagram, 567
— synchronous phase modifiers, 574
medium, 528
— end condenser method, 528
— nominal-T method, 531
— nominal-π method, 536
regulation and efficiency, 519
short, 519
— single-phase lines, 519
— three-phase lines, 521
Transposition, 595

U

Underground cables, 597
capacitance, 609, 622
causes of failure, 636
classification, 599
— cryogenic cables, 637
— extra high tension, 603
— high voltage, 601
— low voltage, 600
— supertension, 601
direct current, 637
faults and their location, 632
— Murray loop test, 632
— Varley loop test, 634
grading, 613
heating, 624
insulating materials, 598
insulation resistance, 607
laying, 605
most economical conductor size, 612
permissible current loading, 631
thermal resistance, 630
Universal torque equation, 942

W

Wind power, 10
Wind power plants, 199